THE OXFORD THESAURUS

American Edition

THE OXFORD THESAURUS

American Edition

Laurence Urdang

OXFORD UNIVERSITY PRESS
New York 1992

Oxford University Press

Oxford New York Toronto
Delhi Bombay Calcutta Madras Karachi
Kuala Lumpur Singapore Hong Kong Tokyo
Nairobi Dar es Salaam Cape Town
Melbourne Auckland

and associated companies in
Berlin Ibadan

Published by Oxford University Press, Inc.,
200 Madison Avenue, New York, NY 10016

Oxford is a registered trademark of Oxford University Press

Library of Congress Cataloging-in-Publication Data
Urdang, Laurence.
The Oxford Thesaurus: American Edition /
Laurence Urdang.
p. cm.
Includes index.
ISBN 0-19-507354-1
1. English language—United States—Synonyms and antonyms.
2. Americanisms—Dictionaries. I. Title.
PE2832.U7 1992
423'.1—dc20 91-3938 CIP

For Irena, Nicole, and Alexa

2 4 6 8 9 7 5 3 1

Printed in the United States of America
on acid-free paper

Foreword

The biggest problem with a book of this length and complexity is that one's competence and uniformity of judgment improve in the course of its preparation, gaining adequacy in the task only as the deadline for delivery of the manuscript draws nigh. Of course, given all the time in the world one would never complete any project, for each revision would yield not only greater experience but, theoretically, the temptation to achieve perfection, clearly unattainable.

The idea of publishing the *Oxford Thesaurus* was first broached to me in 1986. The book was originally to have been some 300,000 words in length; for various reasons, it was subsequently agreed to increase its size, and the book you are holding contains approximately 650,000 words (not counting the Index). The ultimate responsibility for the *Thesaurus* rests with me, not only for errors of omission and commission but for any benefits that I hope might be derived from its use.

January 1992
Old Lyme, Connecticut

Laurence Urdang

Introduction

In its narrowest sense, a synonym is a word or phrase that is perfectly substitutable in a context for another word or phrase. People who study language professionally agree that there is no such thing as an ideal synonym, for it is virtually impossible to find two words or phrases that are identical in denotation (meaning), connotation, frequency, familiarity, and appropriateness. Indeed, linguists have long noted the *economy of language*, which suggests that no language permits a perfect fit, in all respects, between any two words or phrases. Many examples of overlapping can be cited; the more obvious ones in English are those that reflect a duplication arising from Germanic and Romance sources, like *motherly* and *maternal, farming* and *agriculture, teach* and *instruct.* In such pairs the native English form is often the one with an earthier, warmer connotation. In some instances, where a new coinage or a loanword has been adopted inadvertently duplicating an existing term, creating "true" synonyms, the two will quickly diverge, not necessarily in meaning but in usage, application, connotation, level, or all of these. For example, scientists some years ago expressed dissatisfaction with the term *tidal wave*, for the phenomenon was not caused by tides but, usually, by submarine seismic activity. The word *tsunami* was borrowed from Japanese in an attempt to describe the phenomenon more accurately, but it was later pointed out that *tsunami* means 'tidal wave' in Japanese. Today, the terms exist side by side in English, the older expression still in common use, the newer more frequent in the scientific and technical literature.

Any synonym book must be seen as a compromise that relies on the sensitivity of its users to the idiomatic nuances of the language. In its best applications, it serves to remind users of words, similar in meaning, that might not spring readily to mind and to offer lists of words and phrases that are alternatives to and compromises for those that might otherwise be overused and therefore redundant, repetitious, and boring. The *Oxford Thesaurus* goes a step further by offering example sentences to illustrate the uses of the headwords and their alternatives in natural, idiomatic contexts. In addition, by marking those synonyms that appear as main entries in the book and by providing a detailed Index of all words and expressions, the *Thesaurus* offers an effective, useful networking feature that unifies the language covered into an integrated whole.

1. Selection of headwords

Two criteria have been employed: first, headwords have been selected because of their frequency in the language, on the assumption that synonyms are more likely to be sought for the words that are most used; second, some headwords of lower frequency have been included because it would be otherwise impossible to find a suitable place to group together what are perceived as useful sets of synonyms with their attendant illustrative sentences. Obvious listings have been omitted on the grounds that users of the *Thesaurus* can easily find synonyms for, say, *abdication* by making nouns out of the verbs listed under *abdicate.* This deliberate attempt to avoid duplication is mitigated in the case of very common words. For the convenience of the user, both *shy* and *bashful* are main entries, as are *method, manner,* and *mode,* which, though much the same in some respects, differ in detail and application. In contrast to other books, however, *mitigate* is a main entry but not *mitigation, mistake* and *mistaken* are main entries but not *mistakenly,* etc. Where it is determined that such derivations are neither automatic nor semantically obvious, separate listings have been provided. In all instances, it is wise to consult the Index first to find a word.

2. Index

The Index lists, in alphabetical order, each occurrence of every word and phrase that appears among the synonyms in the main text and as main entries, the latter being easily identified by a degree sign. Many Index entries have several references to main entries where the sought-after word or phrase can be found; these references serve as gist-words that pinpoint the particular sense of the word being sought, enabling the user to go directly to the most appropriate main entry. For a fuller description of the principles and layout of the Index, see "Guide to Use of the Index" on page 599.

3. Illustrative sentences

On the principle that a word is known by the company it keeps, one or more sentences showing the main entry word in context are provided for each sense discrimination. These have been carefully selected to demonstrate the use of the main entry in a context likely to be encountered in ordinary written or spoken English. (See also **8. Substitutability,** below.)

4. Synonym lists

Each main entry is followed by one or more sense groupings, each illustrated by one or more sentences. An effort has been made to group the synonyms semantically as well as syntactically and idiomatically: that is, each synonym listed within a given set should prove to be more or less substitutable for the main entry in the illustrative sentence.

In some instances, idiomatic congruity may, unavoidably, beeome strained; where it is felt to be stretched too far—though still properly listed among its accompanying synonyms—a semicolon has been inserted to separate subgroups of synonyms, and, in many cases, additional illustrative sentences have been provided. Such subgroupings have been confined largely to distinctions between literal and figurative uses, between transitive and intransitive verbs, and between synonyms that differ in more subtle aspectual characteristics of meaning or syntax. (See also **8. Substitutability,** below.)

Not all senses of all words are covered for either or both of the following reasons: the sense, though it exists, is relatively rare in ordinary discourse and writing; there are no reasonable synonyms for it. Thus, this sense of *mercy,*

> an affecting or moving of the mind in any way; a mental state brought about by any influence; an emotion or feeling: *Mercy is an affection of the mind.*

is not covered for the first reason, as it is a literary and somewhat archaic usage. The same can be said for the sense,

> a bodily state due to any influence

and for other senses listed in the largest dictionaries but rarely encountered except in literary contexts. Even in such contexts it would be unusual to need a synonym for this word and others like it.

5. Cross references

There are very few cross references between main listings in the *Thesaurus,* for the Index serves as the chief finding mechanism, and it was felt desirable to avoid unnecessary duplications of listings. Where such cross references do occur, they are simple and straightforward:

> **superior** *adj.* . . . **3** See **supercilious,** above. —*n.* **4** See **supervisor,** below.

A number of cross references occur within entries, between variant forms of an expression. At the entry for *take,* for example, as one can say either *take* or *take it* in the sense of 'understand' etc., the option is shown in the following way:

> **take** *v.* . . . **19** °understand, °gather, °interpret, °perceive, apprehend, °deduce, conclude, °infer, °judge, deem, °assume, °suppose, °imagine, °see: *I take it from your expression that you've had bad news.*
>
> **33** ***take it.*** **(a)** withstand *or* tolerate *or* survive punishment *or* abuse, °survive: *The Marines are extremely tough and can take it.* **(b)** See **19,** above.

In a few entries, the form "See also" is used.

A unique feature of the *Thesaurus* is the cross reference system between synonyms and main entries: every word listed as a synonym that also appears as a main entry is preceded by a degree mark (°). Users seeking additional synonyms or a different aspect or nuance of meaning of a synonym so marked are at once alerted to the fact that it has its own main entry and can look it up directly, without recourse to the Index.

> **locale** *n.* °site, °location, °spot, °place °setting, venue, °neighborhood, °situation, locality: *We moved the locale of the film to Spain.*

When a degree mark precedes a term beginning with a word or words in parentheses, the cross reference is to the word following the parentheses. Thus, in the following example, the degree marks refer the reader to the entries for *play, show, piece, production,* and *scenario.*

> **dress** *n.* **1** °play, stage play, photoplay, screenplay, °(stage) show, °(theatrical) piece, °(stage) production; °scenario: *He plays only in dramas, never in musicals.*

Some words that are main entries are not marked when they appear in a phrase, for reference to them would not yield useful synonyms. In the following entry, for example, marking the words *take, pleasure, put, edge,* and *enjoyment* would be misleading, hence they are unmarked:

> **overshadow** *v.* . . . **2** °spoil, °blight, °ruin, °mar, take (all) the pleasure from, put a damper on, take the edge off, °impair, take the enjoyment out of: *The news from the east overshadowed everyone's spirits at the party.*

6. Labels

A. All words and phrases that are recognized as belonging to a particular variety of English, whether geographical or stylistic, are labeled. It might at first seem that a large number of colloquial, slang, and taboo words have been included, but that is less to provide such alternatives to the main entries than it is to ensure their presence in the Index so that users who need alternatives to them can find them readily. The labels used are those commonly encountered in ordinary dictionaries:

Colloq Colloquial; informal; used in everyday conversation and writing, especially in the popular press and in dramatic dialogue; sometimes avoided where more formal language is felt to be appropriate, as in business correspondence, scholarly works, technical reports, documents, etc.

Slang Belonging to the most informal register and characteristic chiefly of spoken English; often originating in the cult language of a particular socio-cultural group. Not sufficiently elevated to be used in most writing (aside from dialogue), but often found in the popular press and frequently heard on popular radio and television programs.

Taboo Not used in polite society, usually because of the risk of offending religious, sexual, or cultural sensibilities; occasionally encountered on late-night television and radio (chiefly in dialogue from a script but not usually tolerated on call-in shows where it is edited out by time-delay devices); often occurring in graffiti and in dialogue in novels, plays, and films.

Archaic Describing an obsolete word or phrase (like *coxcomb, lounge lizard*) that is used deliberately to invoke the feeling of a bygone time.

Old-fashioned Used of a synonym (like *comfit*) that is no longer current but might occasionally be encountered among older speakers and in older writing.

Technical Used of a somewhat specialized word that is not commonly encountered in ordinary, everyday English, like *defalcator,* which appears as a synonym under *swindler.*

Literary Describes a word, like *euchre* 'cheat,' that is not usually met with in everyday language, even of the formal genre, but may be found in poetry and other literary works.

Brit, US, Australian, Canadian, New Zealand Marks a word or phrase that occurs typically in the designated variety.

The meanings of other labels are self-evident.

B. All labels can occur in combination. Usage labels always take precedence over dialect labels. For example,

> **pushover** *n.* **1** sure thing, *Colloq* piece of cake, child's play, snap, picnic, walkover, *US* breeze, *Slang* cinch, *Brit* doddle, *US* lead-pipe cinch: *The Latin exam was a pushover.*

Here, "sure thing" is standard universal English. All words and phrases following *Colloq* up to the *Slang* label are colloquial: "piece of cake, . . . walkover" are universal colloquial English, "breeze" is US colloquial. All synonyms following the *Slang* label are slang: "cinch" is universal English slang, "doddle" is confined to British slang, and "lead-pipe cinch" is confined to American slang.

> **talented** *adj.* . . . *Colloq* ace, crack, topnotch, *Brit* wizard, whizzo, *US* crackerjack. . . .

In this entry, all synonyms shown are colloquial, "ace, crack, topnotch" being universal English, "wizard, whizzo" British, and "crackerjack" US.

It must be emphasized that such labels are to some extent impressionistic and are based in the *Thesaurus* on a consensus of several sources: that is, there is no implication that "breeze" is never used in the sense of "pushover" except in the US, nor should such an inference be made.

C. Comments regarding what might be viewed as "correct" in contrast to "incorrect" usage are generally avoided. For example, the nonstandard use of *between* in contexts referring to more than two of anything or of *among* in contexts involving fewer than three goes unmarked. However, if the usage question is confined to what can easily be represented in a "lexical" environment, then suitable treatment is accorded it; thus 'infer' is labeled *Nonstandard* under *imply* but, because the substitution of *imply* for *infer* is not attested, there is no reciprocal listing under *infer.* To take another example, "different to," in the typically British usage *His house is different to mine,* is rarely encountered in American English; in American English, purists condemn "different than," as in *His house is different than mine,* which is increasingly heard in British English; purists on both sides of the Atlantic prefer "different from." Such matters are best left to usage books and to usage notes in dictionaries and are not treated in the *Thesaurus.*

D. Main entry words and subentries are not labeled, only the synonyms. Thus, under *push* appears the idiomatic expression, (*push up*) *daisies,* which is not labeled:

> **push** *v.* . . . **12** ***push (up) daisies.*** be dead (and buried), be six feet under: *Her husband's pushing up daisies.*

The idiom is not labeled because it is assumed that the user has looked it up to find a substitute for it, hence needs no information about it other than a listing of its alternatives (which are labeled, when appropriate) and an illustrative example. Should users need more detailed information about a given word or phrase, they should resort to a dictionary, the proper repository for such information. A rare exception to the foregoing occurs where a headword has one meaning in American English and quite a different meaning in another regional variety. Thus, in the following entry, the labels apply to the different senses of the main entry word, not to the synonyms following (since "tunnel," for example, has the same meaning in British and American English):

> **subway** *n.* **1** *In US:* underground (railway), tube: *She takes the subway to work.* **2** *In Britain:* tunnel, underpass: *Use the subway to cross the road in safety.*

E. Synonyms bearing any kind of label appear at the end of the set in which they are listed, except in the case described immediately above.

7. Spelling and other variants

The spellings shown throughout are those preferred by most modern American writers. American variant spellings are shown; if they are variants of the main entry word, they appear as the first word in the set(s) of synonyms following:

mousy *adj.* **1** mousey, . . .
movable *adj.* moveable, . . .

Such variants are also shown when they appear within an entry:

movable *adj.* . . . transferable *or* transferrable, . . .

Common British spelling variants (*humour, traveller, unravelled*) are not shown, but less common ones are listed for convenience. Where both forms are variants in American spelling, they are described by *"or US also"*:

. . . accouterments *or US also* accoutrements, . . .
. . . phony *or US also* phoney, . . .

This should be understood to mean "the normal American spelling is *accouterments* (or *phony*)" but that alternative spellings occur occasionally.

8. Substitutability

A. The purpose of a synonym book is to provide the user with a collection of words that are as close as possible in meaning to a designated word. The *Oxford Thesaurus* tries to go a step further by providing examples that not only illustrate the main entry word in a natural contextual environment but also allow the user to substitute as many of the synonyms as possible into the framework of the context. For example:

porous *adj.* spongy, spongelike, permeable, pervious, penetrable: *The rainwater runs through the porous rock and collects in the pools below.*

It is possible to substitute for *porous* in the sample sentence any of the words given as synonyms without any adjustment of the grammar or phrasing of the example. That is not to suggest that the synonyms are identical: "permeable" and "pervious" are far less frequent in English than "spongy, spongelike." Some might argue that "penetrable" is not synonymous with the other listed words; but it is the function of this book to provide synonyms for the main entries, not for the other synonyms listed. No claim is made—nor could it be made—that synonyms are identical, either to one another or to another word, merely that they fall well within the criteria of what, for practical purposes, is viewed as synonymy in the language.

It is certainly true that substituting for *porous* any of the five listed synonyms will yield five standard English sentences.

B. Some judgment is required of the user in determining the syntax and idiomaticity with which a given word or expression can be substituted in an illustrative context: words are rarely as readily interchangeable in a context as might be components in a chemical or mathematical formula. Moreover, while such formulae are reflective of science, language offers its users the virtually infinite variety available only in art, with each individual speaker of any language being presented with the opportunity to become an artist.

In the following example, all terms can be substituted for *adjoining* in the first illustrative sentence; to create idiomatic parallels to the second sentence, the parenthetical prepositions must be used:

adjoining *adj.* neighboring, contiguous (to), adjacent (to), abutting, bordering, next (to): *We have bought the adjoining land and will build our new house there. The land adjoining the supermarket is not for sale.*

Interpreting this, the following are all idiomatic: *adjoining land, neighboring land, contiguous land, adjacent land, abutting land,* and *bordering land.* But if the context requires the adjective to come after *land* (with a following noun), then the parenthetical words must be added to yield constructions that are idiomatic, like *land adjoining the supermarket, land neighboring the supermarket, land contiguous to the supermarket, land adjacent to the supermarket, land abutting the supermarket, land bordering the supermarket,* and *land next to the supermarket.*

As this is intended as a synonym book and not a work on English collocations, the matter of idiomaticity cannot be taken much further.

C. There are other reasons why direct substitutability is not always possible within a single semantic concept. The following extract demonstrates this:

> **possess** *v.* . . . **3** dominate, control, govern, consume, take control of, preoccupy, obsess; charm, captivate, enchant, cast a spell over, bewitch, enthrall: *What possessed her to think that I could help? He behaves as if possessed by the devil.*

Here, two aspects of the same sense have been divided by a semicolon, with the synonyms preceding the semicolon illustrated by the first contextual example and those following it by the second. While it may be argued that in this instance the synonyms following the semicolon, with their illustrative sentence, might better have been listed in a separately numbered set, the close semantic association of the two groups would thereby have been lost.

D. Sometimes, where the subsense is familiar enough not to require its own example yet needs to be set off from the other synonyms because of a subtle or aspectual semantic distinction, a semicolon is inserted among the synonyms and only one example is provided:

> **practice** *n.* . . . **2** exercise, discipline, drill, practicing, repetition, rehearsal, training, preparation, work-out, warm-up; application, study: *She needs more practice on the beginner's slope before going down the main piste.*

The idiomatic usage of this sense of "study" and "application" is sufficiently familiar not to require a separate example.

On the other hand, a second example is needed for the next sense of *practice:*

> . . . **3** pursuit, exercise, work, profession, career, vocation, conduct; business: *He genuinely enjoys the practice of law. I heard of a veterinary practice for sale in Yorkshire.*

It would be difficult—perhaps impossible—to defend such fine distinctions in every instance: indeed, as a comparison of the different lengths of the entries in any dictionary will quickly reveal, language does not provide the same levels of sense discrimination for all words. The metaphorical focus and diversity of a language provide for polysemy in some semantico-cultural spheres but not in others. The classic observation often cited to demonstrate this linkage is that of the Inuit language that has a large number of distinguishing words for types of snow or of the African language that has an extensive vocabulary to describe the kinship among its speakers. On the grounds that the lexicon of a language is molded by speakers who, quite naturally, use it to talk (and write) about things that are important to them, one might be tempted to draw conclusions about the voracity of English speakers by reflecting that the entry for *take* has about twice as many definitions in most dictionaries as that for *give.*

E. Often, the semicolon may be used to separate transitive uses of a verb from intransitive:

> **preach** *v.* . . . **2** moralize, sermonize, advise, counsel, admonish; reprimand, lecture, harangue, pontificate; urge, inculcate, advocate: *Practice what you preach. Mother used to preach to us about being charitable. Father preached restraint in all things.*

It can be seen that the three subsets of synonyms fit, *mutatis mutandis,* with the three illustrative sentence examples.

Because of the behavior of verbs in English, different synonyms may be required depending on what the object of the verb is and, often, whether the object is a word or phrase or a clause:

> **predict** *v.* foretell, prophesy, forecast, foresee, augur, prognosticate, forewarn, presage, vaticinate; portend, foreshadow, foretoken, forebode; intimate, hint, suggest: *My mother predicted that there would be moments like this. The term for predicting coming events from the entrails of animals is "haruspicy."*

In the following set, it will be seen that "sanction, approve" do not fit idiomatically into the second illustrative sentence:

> **allow** *v.* . . . **3** permit, let; sanction, approve: *Leaving the grounds after dinner was not allowed. The headmaster allowed us to practice football after classes. The teacher would not allow talking in class.*

The preceding also illustrates how important it is to have a good grasp of English idiom, for the use of "let" in the last example would be acceptable English only if the syntax were converted to *The teacher would not let us talk in class,* for "let talking" is not English.

F. Wherever possible, the proper prepositional or adverbial particle normally accompanying a verb in a certain sense has been supplied, though it must be emphasized that the one offered is the most frequently used and not, necessarily, the only one acceptable in standard usage. Particles used with some words may vary considerably, owing not only to dialect variation but also to whether the verb is used actively or passively as well as to which nuance of meaning, sometimes far too subtle to be dealt with adequately in a book of this kind, is to be expressed. The following entry illustrates the full treatment that can be accorded to words that occur in a wide variety of grammatical environments:

> **persevere** *v.* Often, ***persevere in*** *or* ***with*** *or* ***at.*** persist, resolve, decide, endure, continue, carry on *or* through, keep at *or* on *or* up, be steadfast *or* staunch *or* constant, keep going, stand fast *or* firm, see through, be *or* remain determined *or* resolved *or* resolute *or* stalwart *or* purposeful *or* uncompromising, be tenacious *or* persistent *or* constant *or* pertinacious *or* assiduous *or* sedulous, be tireless *or* untiring *or* indefatigable, show determination *or* pluck *or* grit, be plucky, be patient *or* diligent *or* stubborn *or* inflexible *or* adamant *or* obstinate *or* obdurate, show *or* exhibit *or* demonstrate patience *or* diligence *or* stubbornness *or* inflexibility *or* obstinacy *or* obduracy, remain dogged, pursue doggedly, be intran-

> sigent *or* intractable, cling to, stick to, support, stop at nothing, sustain, *Colloq* stick with, stick (it) out: *We must persevere to win. I shall persevere in my loyalty.*

G. In some adjective senses, a split might occur between attributive and predicative uses, though in most such cases, where the syntax is open, only one, usually common illustration is given. For example, *alone* is used only predicatively or postpositively, not attributively; that is, one cannot say **An alone woman . . .* In this particular case, the normal attributive form would be *lone,* but *lone* is not listed as a synonym for *alone* because they are not mutually substitutable. It is acknowledged that the detailed description of the special syntactic ways in which certain words (like *alone, agog, galore*) behave lies outside the province of this book.

Although similar cautions must be observed and adjustments made throughout, it is hoped that the illustrative sentences will providc a substantial basis for the user to identify idiomatic contexts and to discriminate senses that are not always carefully distinguished in dictionaries.

A

abandon *v.* **1** give up *or* over, °yield, °surrender, °leave, °cede, let go, °deliver (up), °turn over, °relinquish: *I can see no reason why we should abandon the house to thieves and vandals.* **2** °depart from, °leave, °desert, °quit, go away from: *The order was given to abandon ship.* **3** °desert, °forsake, °jilt, walk out on: *He even abandoned his fiancée.* **4** °give up, °renounce; °discontinue, °forgo, °drop, desist, abstain from: *She abandoned cigarettes and whiskey after the doctor's warning.* —*n.* **5** recklessness, intemperance, wantonness, lack of restraint, unrestraint: *He behaved with wild abandon after he received the inheritance.*

abandoned *adj.* **1** left alone, °forlorn, forsaken, °deserted, neglected; rejected, shunned, cast off *or* aside, jilted, dropped, °outcast: *An abandoned infant was found on the church steps. Totally alone, she felt abandoned by her friends.* **2** °bad, °immoral, amoral, °wicked, °sinful, °evil, °corrupt, unprincipled, unrestrained, °uninhibited, °reprobate; °loose, °wanton, debauched, °wild, °dissolute, dissipated, °profligate; depraved, °lewd, °lascivious, flagitious: *His abandoned behavior soon landed him in jail.*

abbreviate *v.* **1** °shorten, compress, °contract, truncate, °trim, °reduce, °curtail: *We abbreviated some of the longer words to save space.* **2** °shorten, °cut, condense, °abridge, °abstract, °digest, epitomize, summarize, *US* synopsize: *The school presented an abbreviated version of* A Midsummer Night's Dream.

abbreviated *adj.* skimpy, °brief, revealing: *The dancers' abbreviated costumes shocked some members of the audience.*

abbreviation *n.* initialism; acronym; shortening, contraction: *USA is one kind of abbreviation, or initialism; NATO, which is pronounced as a word, is another, usually called an acronym.*

abdicate *v.* °give up, °renounce, disclaim, °waive, disown, °surrender, °yield, °relinquish, °abandon, °resign, °quit: *He abdicated all responsibility for care of the children. She abdicated the throne to marry a commoner.*

abduct *v.* °kidnap, °carry off, make away *or* off with, °seize, *Slang US* °snatch, °grab: *The child that was abducted is safe.*

abet *v.* **1** °encourage, °urge, instigate, °incite, °provoke, egg on, °prod, goad; °aid, °help, °assist: *The jury found that his wife had abetted him in the murder.* **2** countenance, °approve (of), °support, °endorse, °second, °sanction, condone; °further, °advance, °promote, °uphold: *By failing to inform on the terrorists, the neighbors abetted the bombing.*

abeyance *n.* ***in abeyance.*** °pending, abeyant, °reserved, °in reserve, shelved, pushed *or* shoved *or* shunted aside, postponed, °put off, suspended, *US* tabled; temporarily inactive, °dormant; latent; *Colloq* in a holding pattern, on the back burner; *Slang* on hold, in the deep freeze, on the shelf, on ice, hanging fire: *Legal proceedings were held in abeyance pending talks aimed at reaching an out-of-court settlement.*

abhor *v.* °hate, °loathe, °detest, abominate, execrate; regard *or* view with horror *or* dread *or* fright *or* repugnance *or* loathing *or* disgust, shudder at, recoil *or* shrink from; be *or* stand aghast at: *He abhorred any violation of human rights.*

abhorrent *adj.* °hateful, detestable, abhorred, °abominable, °contemptible, odious, °loathsome, horrid, heinous, execrable, °repugnant; °repulsive, °repellent, °revolting, °offensive, °disgusting, horrifying, °obnoxious: *The idea of war was totally abhorrent to her.*

abide *v.* **1** °stand, °endure, °suffer, °submit to, °bear, °put up with, °accept, °tolerate, °brook: *How can you abide the company of such a fool?* **2** °live, °stay, reside, °dwell, °sojourn: *His green card entitles him to enter and abide in the US.* **3** °remain, °stay, °continue, °tarry; °linger, °rest: *He'll abide in my care till he can walk again.* **4** ***abide by.*** °consent to, °agree to, comply with, °observe, °acknowledge, °obey, °follow, °submit to, conform to, keep to, remain true to, stand firm by, adhere to, hold to: *You must abide by the rules of the club if you become a member.*

abiding *adj.* °lasting, °permanent, °constant, °steadfast, °everlasting, unending, °eternal, °enduring, °indestructible; unchanging, °fast, hard and fast, °fixed, °firm, immutable, °changeless: *Her abiding love is a solace to him.*

ability *n.* **1** adeptness, °aptitude, °facility, °faculty, °capacity, °power, °knack, °proficiency, *Colloq* know-how: *I have perceived your ability to manipulate situations to your own advantage.* **2** °talent, °skill, cleverness, °capacity, °wit, °gift, °genius, °capability: *He has such extraordinary ability it is difficult to see why he doesn't accomplish more.* **3** ***abilities.*** °faculty, faculties, talent(s), gift(s), skill(s): *Her abilities have made her one of the finest cellists of our time.*

ablaze *adj.* **1** aflame, afire, °burning, °on fire, alight, blazing: *By the time the firemen arrived, the roof was ablaze.* **2** lit up, alight, brilliantly *or* brightly lit, sparkling, gleaming, aglow, °bright, °brilliant, °luminous, illuminated, °radiant: *The ballroom was ablaze with the light from thousands of candles.*

able *adj.* **1** °capable, °qualified, °competent, °proficient: *I feel quite able to take care of myself, thank you. He is an able tennis player.* **2** °talented, °clever, skilled, °masterful, masterly; °adept, °skillful, °gifted, °superior, °expert, °accomplished: *There is no doubt that Wellington was a very able general.*

abnormal *adj.* **1** °deviant, deviating, °irregular, °unusual, unconventional, aberrant, *Psych jargon* °exceptional: *The wing of a bat is an abnormal structure. Facilities for abnormal children are available.* **2** °peculiar, °unusual, °odd, °strange, °queer, freakish, °unnatural, °extraordinary, °weird, °eccentric, °bizarre, anomalous, aberrant, °perverse, °deviant, °irregular, *Colloq* off-beat, *Slang* oddball, °kinky, °weirdo: *The format of that TV show requires the contestants to do some very abnormal things.*

abnormality *n.* **1** irregularity, unconformity, unusualness, °singularity, °eccentricity, unconventionality, uncommonness, deviation, aberration, idiosyncrasy: *The desire in a man to wear women's clothing is viewed as an abnormality.* **2** distortion, anomaly, malformation, deformity: *The child was born with an abnormality of the right foot.*

abode *n.* °residence, °dwelling, dwelling place, °house, °home, °domicile, habitation, °quarters, °lodging, °accommodations, *Military* billet; *Colloq Brit* digs, diggings: *He was described as being of no fixed abode.*

abolish *v.* °eliminate, °end, put an end to, °terminate, °destroy, annihilate, annul, °void, make void, °demolish, do away with, nullify, °repeal, °cancel, °obliterate, liquidate, °destroy, °stamp out, °quash, °extinguish, °erase, °delete, expunge; eradicate, extirpate, deracinate, °uproot: *The best way to abolish folly is to spread wisdom. Prohibition in the US was abolished in 1933.*

abolition *n.* elimination, °end, °termination, annulment, nullification, repudiation, °cancellation; °destruction, annihilation: *1837 marks the abolition of the slave trade in the British Empire.*

abominable *adj.* **1** °offensive, °repugnant, °repulsive, °vile, °monstrous, °loathsome, odious, execrable, detestable, °despicable, °base, °disgusting, °nauseous, nauseating, °foul, °abhorrent, horrid, °deplorable: *He was accused of crimes too abominable to detail in open court.* **2** °terrible, unpleasant, °disagreeable; °awful, °distasteful, in bad taste, °horrible, °frightful, *Colloq Brit* °beastly: *No one wants to go out in this abominable weather. The decor in this hotel is simply abominable.*

aboriginal *n.* °native, indigene, autochthon; *Colloq Australian* Abo, *Offensive Australian* aborigine, *Slang Australian contemptuous* boong: *Many aboriginals are not assimilated to modern life.*

abound *v.* **1** °prevail, °thrive, °flourish: *Disease abounds among the undernourished peoples of Africa.* **2** ***abound in.*** be crowded *or* packed *or* jammed with, be abundant *or* rich in, proliferate (in *or* with): *The ship abounds in conveniences.* **3** ***abound with,*** teem *or* swarm *or* throng with, be filled *or* infested with, overflow with: *The ship abounds with rats.*

about *adv.* **1** °round, °around, close *or* near by, on every side: *Gather about, for I have something to tell you.* **2** °approximately, °around, °nearly, °roughly, more or less, °almost, close to *or* upon; give or take: *In 1685 London had been, for about half a century, the most populous capital in Europe. Light travels at about 186,000 miles a second.* **3** to and fro, up and down, back and forth, here and there, hither and yon, °far and wide, hither and thither: *He wandered about aimlessly for several days.* **4** here and there, °far and wide, hither and yon, hither and thither, °helter-skelter: *My papers were scattered about as if a tornado had struck.* **5** °around, °prevalent, in the air: *There is a lot of flu about this year.* **6** °approximately, °nearly, °close to, not far from, °almost, just about, °around: *It is about time you telephoned your mother.* —*prep.* **7** °around, °surrounding, encircling: *There is a railing about the monument.* **8** °round, °around, all round, °everywhere, in all directions, all over: *Please look about the room for my hat.* **9** °near, °nearby, adjacent to, °beside, alongside, °close by, nigh: *There were a lot of trees about the garden.* **10** with, °at hand, *Colloq* on: *I am sorry, but I haven't my checkbook about me.* **11** °touching, °concerning, connected with, involving, in *or* with reference to, in *or* with regard to, °regarding, in the matter of, with respect to, respecting, °relative to, relating to, apropos, *Formal* anent: *He wrote a book about the Spanish Armada.*

about-face *n.* °reversal, °reverse, °turnabout, turnaround, U-turn, volte-face, *Brit* about-turn: *There has been a complete about-face in the policy concerning immigration.*

above *adv.* **1** °overhead, on high, °aloft, in the sky *or* heavens: *Far above, the clouds scudded swiftly by.* **2** upstairs: *They lived on the ground floor and the landlady lived above.* —*prep.* **3** on, on (the) top of, upon, °over, atop: *The plume of smoke remained fixed above the volcano. He hasn't got a roof above his head for the night.* **4** °over, more than, °exceeding, in excess of, beyond, greater than, °surpassing: *The operations are controlled by gears, of which there are above fifty in number.* **5** insusceptible to, unaffected by, out of reach of, not susceptible *or* vulnerable *or* exposed to, superior to: *The judge is above bribery or other influence.* **6** ***above all.*** before *or* beyond everything, first of all, °chiefly, °primarily, in the first place, °mainly, essentially, °at bottom: *Above all, serve God and country before you serve yourself.*

aboveboard *adv.* **1** °openly, candidly, °freely, publicly, frankly, straightforwardly, plainly, for all to see, out in the open, in the open: *Donald has always been completely aboveboard with everyone.* —*adj.* **2** °open, °candid, °frank, °straight, °direct, °honorable, straightforward, °forthright, guileless, undeceiving, °artless, °ingenuous, undeceptive, undeceitful, °straight from the shoulder; °honest, °genuine: *The company's dealings have always been aboveboard.*

abridge *v.* °shorten, °reduce, condense, °cut, °abbreviate, °cut back, °trim, °curtail, °pare down, °contract, compress, °digest, summarize, epitomize, °abstract, *US* synopsize: *We abridged the original edition of 1,000 pages to 480 pages.*

abridgment *n.* **1** shortening, reduction, °abbreviation, condensation, contraction, truncation, trimming: *The abridgment took ten years.* **2** curtailment: *We protested against the abridgment of our right to picket.* **3** °digest, condensation, °epitome, compendium, concise edition *or* version, cut edition *or* version; °synopsis, °abstract, °summary, °précis, °outline, °résumé: *The one-volume abridgment of the dictionary is easier to use.*

abroad *adv.* **1** °overseas, in foreign lands *or* parts: *We were abroad on assignment for a few years.* **2** broadly, °widely, °at large, near and far, °far and wide, °everywhere, extensively, publicly: *Don't spread rumors abroad.* **3** °outside, out-of-doors, away, out and about: *There are few people abroad this early in the morning.*

abrupt *adj.* **1** °sudden, °hasty, °quick, °precipitate, °snappy; unexpected, unannounced, unplanned, °unforeseen, unanticipated: *The general's abrupt departure has been linked with the disappearance of a great deal of money.* **2** °precipitous, °steep, °sheer, °sudden: *From the ridge there is an abrupt drop of 3000 feet into the valley.* **3** °curt, °short, °brusque, °blunt, °bluff, °gruff, uncivil, °rude, °discourteous, °impolite, unceremonious, °snappish: *My bank manager gave me an abrupt reply when I asked for an additional loan.*

absence *n.* **1** nonattendance, nonpresence, nonappearance, truancy: *This is Bart's third absence from class in a week. Who runs the office in your absence?* **2** °lack, °want, deficiency, nonexistence; insufficiency, scantiness, paucity, °scarcity, °dearth: *In the absence of new evidence, the matter must remain undecided.*

absent *adj.* **1** away, °out, °off, °elsewhere, not present, missing, gone: *Twenty people attended, but Frobisher was conspicuously absent.* **2** missing, lacking, °wanting, °deficient: *All warmth is absent from her singing.* —*v.* **3** ***absent (oneself) from.*** keep *or* stay away from; withdraw *or* retire from: *He absented himself from the court during his father's trial for murder. Absent thee from felicity awhile.*

absent-minded *adj.* °preoccupied, °inattentive, unattentive, °absorbed, unmindful, °absent, °off, °withdrawn, unheeding, °heedless, unheedful, °inadvertent; distracted, abstracted, daydreaming, in a brown study, in the clouds, °unaware, °oblivious, in a trance, distrait(e), mooning, (far) away (somewhere), stargazing, woolgathering: *The absent-minded professor delivered his lecture to an empty lecture hall.*

absolute *adj.* 1 °perfect, °complete, °total, finished, °thorough, through and through, consummate, °flawless, °faultless, unadulterated, °pure, unmixed, unalloyed, °undiluted; °rank: *Alan behaved like an absolute gentleman.* 2 °complete, °outright, °downright, °genuine, °real, °pure, °out-and-out, °transparent, °unmitigated, °categorical, °unqualified, unconditional, utter, °veritable, unconditioned: *Peace is an absolute requirement for prosperity.* 3 unrestricted, unrestrained, unconstrained, °unlimited, °unmitigated, °arbitrary, °despotic, °dictatorial, °totalitarian, °supreme, almighty, °arbitrary, autocratic, °tyrannical: *The days of absolute monarchy are numbered.* 4 °positive, °certain, °sure, unambiguous, °unquestionable, °authoritative, verifiable, uncompromised: *Few intelligent people would claim absolute knowledge of anything.*

absolutely *adv.* 1 unqualifiedly, unconditionally, unreservedly, unexceptionally, unequivocally, unquestionably, °positively, °definitely, °really, genuinely, decidedly, °surely, °truly, certainly, categorically: *She is absolutely the best dancer I have ever seen. I absolutely refuse to go.* 2 °totally, °utterly, °completely, °entirely, fully, °quite, °altogether, °wholly: *It is absolutely necessary that you undergo surgery.* —*interj.* 3 certainly, assuredly, °positively, °definitely, °of course, °naturally, indubitably, yes, to be sure: *"Are you sure you want to go?" "Absolutely!"*

absorbed *adj.* engrossed, °lost, wrapped up, occupied, °engaged, immersed, buried, °preoccupied, concentrating, °rapt: *He was absorbed in his reading.*

absorbing *adj.* engrossing, °engaging, °riveting, captivating, fascinating, °spellbinding, gripping: *Maria was watching an absorbing thriller on television.*

abstract *adj.* 1 °theoretical, unapplied, notional, ideational, conceptual, metaphysical, unpractical, °intellectual: *It is difficult to capture abstract ideas on paper.* 2 nonrepresentational, °symbolic, nonrealistic: *Museums began buying abstract art in the 1930s.* —*n.* 3 °summary, °epitome, °synopsis, °essence, °digest, condensation, °survey, conspectus, °extract; °outline, °précis, °résumé: *By reading the abstracts, you can determine which articles merit reading in full.* —*v.* 4 epitomize, °abbreviate, °digest, summarize, condense, °shorten, °abridge, °cut, °cut down, *US* synopsize: *The service abstracts articles that appear in scientific journals.*

absurd *adj.* 1 °ridiculous, °silly, °nonsensical, °senseless, °outlandish, °preposterous, °farcical, °mad, °stupid, °foolish, idiotic, imbecilic *or* imbecile, moronic, °childish; laughable, °ludicrous, risible, °inane, *Colloq* °crazy, nutty, nuts, *Chiefly Brit* °daft: *The notion that the moon is made of green cheese is absurd.* 2 asinine, °senseless, illogical, irrational, unreasoned, °unreasonable, °incongruous, °paradoxical, °unsound, °meaningless: *Today, most people view it absurd to believe that the earth is flat.*

absurdity *n.* 1 °folly, silliness, ridiculousness, foolishness, ludicrousness, °nonsense, senselessness, meaninglessness, illogicality, irrationality, unreasonableness, incongruity, °stupidity, *Colloq* craziness, nuttiness, *Chiefly Brit* daftness: *Many comics rely on absurdity rather than cleverness for humor.* 2 °paradox, self-contradiction, °error, °fallacy: *No one can abide the man's pretentiousness and other absurdities.*

abundance *n.* overflow, °superfluity, overabundance, superabundance, °excess, °surplus, oversupply, °glut, °satiety, oversufficiency; °plenty, plenteousness, plentifulness, plenitude, copiousness, °profusion, *Formal* nimiety: *The days when there was an abundance of fresh drinking water have come to an end.*

abundant *adj.* 1 °plentiful, overflowing, °ample, copious, oversufficient, superabundant, plenteous, °profuse, °inexhaustible, °replete, °bountiful, bounteous: *The abundant rainfall fills the reservoirs every day.* 2 abounding (in), °full (of), °rich (in), °luxuriant, °lavish: *We know a stream that is abundant in trout. The abundant vegetation of the rain forest is an ecological wonder.*

abuse *v.* 1 °misuse, misemploy, °pervert, misapply, °exploit: *The officer abused his authority in ordering the forced march at midnight.* 2 maltreat, ill-use, °injure, °wrong, °hurt, °mistreat, °manhandle, °ill-treat; °damage: *I cannot stand by and watch that drunk abuse his wife and family.* 3 malign, revile, censure, °upbraid, assail, objurgate, °lambaste, °berate, °rebuke, °scold, reproach, °disparage, traduce, defame, °insult, swear at, °curse (at), calumniate, °slander, °libel, decry, deprecate, °vilify, rail against: *In the report the director was abused in the most virulent terms.* —*n.* 4 °misuse, misusage, misemployment, °perversion, misapplication, misappropriation, *Rhetoric* catachresis: *Beware of imitating his abuse of the language.* 5 addiction, dependence: *They are being treated for drug abuse at the local clinic.* 6 maltreatment, ill-treatment, ill-use, °fault: *It seemed perfectly natural that he should defend abuses by which he profited.* 7 °self-abuse, self-pollution, masturbation, °violation, defilement; corruption: *The schoolmasters consistently lectured the boys against any abuse of themselves.* 8 revilement, reviling, execration, vituperation, malediction, imprecation, °tongue-lashing, calumny, calumniation, vilification, obloquy, scurrility, invective, maligning, upbraiding, berating, objurgation, scolding; billingsgate: *The two parties, after exchanging a good deal of abuse, came to blows.*

abused *adj.* 1 misused: *Permission to use the office copying machine has become an abused privilege.* 2 maltreated, ill-treated, mistreated, °hurt: *It was explained that he had been an abused child.*

abusive *adj.* 1 insulting, °scurrilous, °vituperative, calumnious, °offensive, °slanderous, libelous, defamatory, censorious, opprobrious, disparaging, deprecatory, depreciatory, °derogatory, °derisory, derisive, reviling, vilifying, °vituperative, °reproachful; °profane; °rude, °filthy, °dirty, °foul, °vulgar, °obscene, smutty, °vile, thersitical: *The Crown refuses to tolerate abusive satire directed at the king. If I hear another word of abusive language out of you, I'll wash out your mouth with soap!* 2 °perverted, misapplied, °improper, °wrong, °incorrect; exploitive, exploitative, exploitatory; °brutal, °cruel, °injurious, °hurtful, °harmful, °destructive: *Despite the abusive treatment of wives, married women commanded much respect.* 3 °corrupt, °venal, °dishonest, °crooked: *The politicians exercised abusive power over the townspeople.*

abysmal *adj.* 1 °awful, appalling, °dreadful, °terrible, °profound: *The government of Nero presented a spectacle of abysmal degradation.* 2 abyssal, °bottomless, °profound, unfathomable, unfathomed: *The abysmal depths have been plumbed in the diving bell.*

abyss *n.* °deep, abysm, bottomless gulf, yawning chasm, gaping void, unfathomable cavity, impenetrable depth(s): *The path led straight down into the abyss. In the scandal the Congressman was plunged into the abyss of disgrace.*

academic *adj.* 1 scholastic, collegiate; °scholarly, °learned, °lettered, erudite: *Comstock's academic background qualifies him for the professorship. The university began publishing academic journals in the nineteenth century.* 2 °theoretical, °hypothetical, conjectural, °speculative, °abstract; ivory-tower, °visionary, °idealistic; °impractical, °unrealistic, unpractical: *The car* *doesn't run, so the question of miles per gallon is purely academic.*

accent *n.* **1** °emphasis, °stress, °force, °prominence, accentuation; °intensity, inflection; °cadence, °beat: *The accent is on the second syllable in "reward."* **2** diacritic, diacritical mark, °mark, accent mark: *There is an acute accent on the "e" in "cliché."* **3** °pronunciation, articulation, °intonation, speech pattern, inflection: *Even after forty years in the country, he still speaks English with an Italian accent.* —*v.* **4** accentuate, °emphasize, °stress, give prominence to, °mark, underline, underscore, °distinguish, highlight, set off *or* apart: *In her speech, the psychologist accented the "id" in "idiot." Why must he always accent the negative aspect of everything?*

accept *v.* **1** °receive, °take, °allow, °permit: *Sorry, but we cannot accept any more applications.* **2** accede (to), °agree (to), assent (to), °consent (to), °acknowledge, °admit, °allow, °recognize: *We accept your request for a hearing.* **3** °assume, °undertake, take on *or* up, agree to bear: *I'll accept the responsibility for replying.* **4** reconcile oneself to, °suffer, °undergo, °experience, °stand, °withstand, °stomach, °endure, °bear, °resign oneself to, °brook, °allow, °tolerate, °take: *I think I have accepted enough criticism for one day.*

acceptable *adj.* **1** °satisfactory, °adequate, °tolerable, all right, °sufficient, admissible, °passable, *Colloq* O.K., okay: *The bread and meat were acceptable, but the beer was awful.* **2** °agreeable, °pleasing, °welcome, °satisfying, °delightful, °pleasant, °pleasing: *Most people find her compliments quite acceptable.*

accessible *adj.* approachable, °open, °available, attainable, obtainable, reachable, °ready, °at hand, *Colloq* get-at-able: *The president is always accessible to those seeking help. The mechanism is accessible if the cover is removed.*

accessory *n.* **1** °extra, °addition, adjunct, °attachment, component, °frill, *Slang* bells and whistles, doodah, *US and Canadian* doodad: *My food processor has enough accessories to equip a symphony orchestra.* **2** accessary, °accomplice, helper, °assistant, confederate, °colleague, abettor, °aide, collaborator, conspirator, coconspirator, fellow criminal, associate *or* partner in crime: *Although he did not rob the bank, he drove the getaway car, which legally makes him an accessory before the fact. A seller of stolen goods is an accessory after the fact.* —*adj.* **3** °extra, °subordinate, °auxiliary, additional, ancillary, supplemental, °supplementary, °secondary, adventitious, *Formal* adscititious: *For no apparent reason, the salamander grew an accessory limb near its hind leg.*

accident *n.* **1** °mishap, °misfortune, mischance, misadventure, °blunder, °mistake; °casualty, °disaster, °catastrophe, °calamity: *A high percentage of the road accidents were caused by drunken drivers.* **2** °chance, °fortune, °luck, fortuity, °fluke; serendipity: *I came across the gold ring by accident, when cleaning out a disused cupboard.* **3** °nonessential, accessory *or* accessary, °extra, °addition: *Melancholy is an almost inseparable accident of old age.*

accidental *adj.* °chance, fortuitous, °lucky, unlucky, serendipitous; undesigned, °unpremeditated, uncalculated, unintended, unintentional, unwitting, °inadvertent; unexpected, unplanned, °unforeseen, unanticipated, adventitious; °casual, °random: *Our meeting was entirely accidental.*

accommodate *v.* **1** °fit, °suit, °adapt, °adjust, °modify; customize: *I shall do my best to accommodate the equipment to your needs.* **2** harmonize, make consistent, °reconcile, °adapt: *It is uncertain whether his version of the incident can be accommodated to ours.* **3** °equip, °supply, °provide, °furnish: *The bank will be unable to accommodate you with a loan, Mr. Smeedle.* **4** °put up, °house, °lodge, °shelter, °quarter, *Military* billet: *The innkeeper is unable to accommodate us for the night.* **5** °suit, °oblige, convenience, °serve: *I was willing to accommodate you by selling your old car.*

accommodating *adj.* **1** °obliging, cooperative, °helpful, °hospitable; °considerate, conciliatory, easy to deal with, pliant, °yielding, compliant, °polite, °friendly, complaisant, °kind, °kindly: *The lady at the complaints desk in the store was most accommodating.* **2** °pliable, °accessible, corruptible, subornable, get-at-able; bribable: *If you want to get off scot-free, we'll have to find an accommodating judge.*

accommodation *n.* **1** °adaptation, °adjustment, modification, °change, °alteration, conformation, conformity: *Her skillful accommodation to her boss's demands kept peace in the office.* **2** °settlement, °treaty, compromise: *Negotiations were now opened for an accommodation between the belligerents.* **3** convenience, °favor: *Would you take the mail to the post office as an accommodation to me?* **4** Usually, ***accommodations.*** lodging(s), room(s), °quarters, °shelter, °housing; °facility, premises, *Brit* digs: *We were able to arrange for accommodations at the hotel. Have you seen our new office accommodations?* **5** °loan, (financial) assistance *or* aid; °grant, grant-in-aid: *The man was able to obtain an accommodation from his brother-in-law.*

accompany *v.* **1** convoy, °escort, chaperon *or* chaperone, °go along with; °attend; usher, °squire: *Allow me to accompany you to your taxi.* **2** go (along) with, come with, be associated with, belong with, go together with, be linked with: *Accompanying the roast was a bottle of claret.*

accomplice *n.* accessory *or* accessary, partner in crime, confederate, °ally, °associate, °colleague, °fellow, °henchman, collaborator, conspirator, coconspirator, abettor, °assistant, fellow criminal, *Colloq US* °cohort: *The police arrested the safecracker and three accomplices within hours of the robbery.*

accomplish *v.* °fulfill, °perform, °achieve, °carry out, °execute, °carry off, do, °complete, °carry through, °finish, °effect, bring to an end, conclude, °wind up, °end; attain, °reach, °gain; *Colloq* °bring off, °knock off, °polish off, *Slang* °pull off, *US* °swing, °hack, °cut: *I don't know how she accomplished it, but she sailed around the world single-handed. Has he accomplished his goal yet?*

accomplished *adj.* consummate, °perfect, °expert, °adept, °skillful, °proficient, °practiced, °gifted, °talented, skilled, °professional: *Did you know that she is also an accomplished flautist?*

accomplishment *n.* **1** °fulfillment, °consummation, °completion, °realization, attainment, °achievement, conclusion, culmination, °realization: *Winning the gold medal was the accomplishment of years of training.* **2** coup, °feat, °exploit, °triumph, tour de force: *Among her many accomplishments was climbing Mount Everest.* **3** °skill, skillfulness, °talent, °gift, °ability: *Playing the violin is another of his accomplishments.*

accord *v.* **1** °agree, harmonize, concur, be at one, °correspond, °agree, be in harmony, be consistent, °go (together), °coincide, °conform: *His principles and practices do not accord with one another.* —*n.* **2** °agreement, unanimity, concord, °reconciliation, °harmony, mutual understanding, conformity, accordance, °rapport, concert: *The countries are in accord on a beneficial trade balance.* **3** °agreement, °treaty, °pact, °contract: *The accords will be signed at the summit meeting in May.* **4** °agreement, °harmony, congruence; correspondence: *The colors of the curtains are in perfect accord with those of the carpet.*

accordingly *adv.* **1** °hence, °therefore, °consequently, °thus, in consequence (where)of, (and) so: *Smoking was forbidden; accordingly, we put out our cigars.* **2** suitably, in conformity, in compliance; conformably, °appropriately, compliantly: *Dinner jackets were required, and the men dressed accordingly.*

according to *adv.phr.* **1** on the authority of, consistent with, in conformity *or* agreement with, as said *or* believed *or* maintained etc. by: *We are going to play this game according to Hoyle. According to his lawyer, he should never have been acquitted.* **2** conformable to, consistent with, in conformity with, commensurate with: *The queen greeted them in order, according to rank.*

account *v.* **1** ***account for.*** °explain, give a reason for, give *or* render a reckoning for, °answer for, °justify, reckon for: *The treasurer has been able to account for every penny of expense. His desire to conceal his background accounts for his secrecy.* —*n.* **2** °calculation, accounting, °reckoning, computation, °(financial) statement; enumeration: *The account shows that the company has ample funds in reserve. Williams hasn't submitted his expense account for the trip.* **3** °interest, °profit, °advantage, °benefit, °favor; °sake: *Nigel turned his convalescence to good account by writing a bestseller. Don't read the book on my account.* **4** °explanation, °statement, °description, °report, °recital, °narrative, °history, °chronicle: *The defendant gave a credible account of his whereabouts at the time of the crime.* **5** °consideration, °use, °worth, °importance, consequence, °note, value, °merit; °standing, °significance, °estimation, °esteem: *The committee decided that length of service is of some account in determining retirement pensions.* **6** °story, °narration, °narrative, °report, °tale, °relation, °description: *Alice's account of the rabbit wearing a waistcoat is unbelievable.* **7** ***take into account*** *or* ***take account of.*** °notice, take note of, °consider, take into consideration, allow for: *In passing sentence, the judge took into account the child's poverty and that it was Christmas time.*

accountability *n.* answerability, °responsibility, °liability, culpability, accountableness: *In a democracy, there can be no reducing of the accountability of the government to the citizens.*

accountable *adj.* answerable, °responsible, °liable, °obliged, obligated: *I am accountable to no man, but the greatest man in England is accountable to me.*

accumulate *v.* °collect, °gather, °amass, °mass, pile *or* heap up, aggregate, cumulate; °assemble, °store, °stock, °hoard, stockpile, put *or* lay away: *Overnight, the snow accumulated in six-foot drifts about the house. Ill fares the land, to hast'ning ills a prey, / Where wealth accumulates, and men decay.*

accumulation *n.* **1** collecting, amassing, °gathering, aggregation, heaping *or* piling up: *One effect of the strike was the accumulation of rubbish in the streets.* **2** °growth, °increase, buildup: *The accumulation of wealth has never proved a valid purpose in life.* **3** °heap, °pile, °mass, °collection, °hoard, °store, stockpile, °stock, aggregation; assemblage: *Our gardener made sure that there was an ample accumulation of compost.*

accuracy *n.* exactness, correctness, *Loosely* °precision, preciseness: *The translation from the Greek has been accomplished with great accuracy. Rifling the inside of the barrel of a firearm increases its accuracy.*

accurate *adj.* **1** °exact, °correct, error-free, °precise: *She gave an accurate description of the events. There is a nice distinction between "accurate" and "precise."* **2** °careful, °meticulous, °nice, with an eye to *or* for detail, °scrupulous, °conscientious: *Marvin is a very accurate speed typist.* **3** unerring, on target, *Colloq* on the mark, *Brit* spot on (target): *This rifle is accurate if you allow for the wind.*

accusation *n.* °charge, °allegation, indictment, °charge, citation, arraignment, °complaint; °imputation, incrimination, denunciation, impeachment: *The politician denied the accusation of having accepted a bribe.*

accuse *v.* **1** ***accuse (of or with).*** °blame, censure, hold responsible (for), °charge (with), °denounce (for), point the finger (at), cite, call to account: *She accused the Knave of Hearts of lying.* **2** ***accuse (of or with).*** °charge, °indict, °impeach, arraign, °incriminate; °attribute, °impute: *The prisoner is accused of carrying an unlicensed firearm.*

accustom *v.* °familiarize, °acquaint, habituate, °train, °season; acclimatize *or* acclimate: *You must wear red goggles to accustom your eyes to the dark. She soon accustomed herself to the new surroundings.*

accustomed *adj.* **1** °customary, °habitual, °usual, °traditional, °normal, °regular, °set, °routine, °ordinary, °familiar, wonted, °common, habituated: *The old man took his accustomed place near the fire.* **2** °used: *I've grown accustomed to her face.*

ache *v.* **1** °pain, °hurt, °smart, throb, °pound; °sting: *My jaw has been aching since that tooth was extracted.* **2** °yearn, °long, °hunger, °hanker, pine; crave: *A hostage for a year, he was aching to see his wife and children.* —*n.* **3** °pain, °pang, throbbing, pounding, smarting, soreness: *I have had this ache in my back, Doctor, and I can't stand up straight.* **4** °pang, °pain; °distress; °longing: *There's been an ache in my heart, my darling, ever since you went away.*

achieve *v.* **1** °accomplish, °carry out, °execute, °succeed in, °complete, °fulfill, bring off *or* about; °realize, °effect: *When the fund reaches its goal, we shall have achieved our purpose.* **2** °accomplish, attain, °reach, °gain, °get, °acquire, °win, °obtain: *She achieved her ends by cheating and conniving.*

achievement *n.* **1** attainment, °accomplishment, °acquisition, acquirement: *As he was still in his thirties, the achievement of great fame still lay ahead for him.* **2** °accomplishment, attainment, °feat, °deed, °exploit, °victory: *The winning of the Nobel prize was her greatest achievement.* **3** °fulfillment, °realization, °accomplishment, attainment, °completion: *What virtue lies more in achievement than in the desire for it?*

acknowledge *v.* **1** °admit, °confess, °allow, °concede, own, °recognize, °accept, accede, acquiesce; own up to: *We acknowledge that we might have been mistaken. She finally acknowledged my presence by looking up.* **2** °answer, reply to, °respond to, react to: *She couldn't possibly acknowledge personally every letter she receives.*

acknowledgment *n.* **1** acknowledging, confessing, admitting, owning, °admission, confession, avowal, affirmation: *His acknowledgment of his involvement in the crime saved the police a great deal of time.* **2** °approval, acceptance, °recognition, °allowance: *By acknowledgment of the parliament, the king was the commander of the army and navy.* **3** °reply, °response, °answer, °recognition: *Our acknowledgment will be in tomorrow's post.*

acme *n.* °peak, apex, °top, °summit, °pinnacle, °zenith; °climax, culmination: *Roger has reached the acme of perfection as a diamond-cutter.*

acquaint *n.* ***acquaint with.*** familiarize with, inform of *or* about, make aware of, apprise of, advise of: *The management requires employees to acquaint themselves with the safety rules.*

acquaintance *n.* **1** °familiarity, °knowledge, acquaintanceship, °understanding, awareness; °experience: *His acquaintance with the works of Coleridge is sparse at best.* **2** °associate, °fellow, °colleague: *She's not a friend of mine, only an acquaintance.*

acquainted *adj.* **1** known to each other *or* one another, familiar with each other *or* one another, on speaking terms: *I have known Flamsteed for years, but his wife and I are not acquainted.* **2** ***acquainted with.*** °familiar with, known to, aware of, informed of, knowledgeable of, conversant with: *I have studied trigonometry, but I am not acquainted with calculus.*

acquire *v.* °get, °obtain, °gain, °win, °earn, °procure, °secure, come by *or* into; °receive, come into possession of; °buy, °purchase: *He acquired great wealth by marrying rich, old, dying widows.*

acquisition *n.* **1** obtaining, getting, acquiring, acquirement, °gain, procurement: *The acquisition of property entails many obligations.* **2** possession(s), °property, °purchase; °object: *This first edition is a recent acquisition.*

act *n.* **1** °deed, °action, °undertaking, °operation, °step, °move; °feat, °exploit; °accomplishment, °achievement: *The first act of the new commission was to ban smoking in public places.* **2** °performance, °show, °bit, skit, °stand, °routine, °turn, sketch, *Colloq* °thing, *Slang US* shtick: *Stand-up comedians do their acts in nightclubs.* **3** °performance, °pretense, °posture, °stance, feigning, °front, °fake, °dissimulation, °show, °deception, °hoax, °affectation: *She didn't mean what she said—that was just an act.* **4** °bill, °law, °decree, edict, statute, °order, ordinance, °command, mandate, °resolution, °measure, enactment: *The city council has passed an act banning overnight parking in the streets.* —*v.* **5** °behave (oneself), °carry on, deport oneself, comport oneself, °conduct oneself: *I don't know how she'll act when we're in public.* **6** °perform, °play, do: *She is acting on Broadway.* **7** °portray, °represent, impersonate, act out, °personify, take *or* play the part *or* role of, personate: *Reginald acts the fool whenever he has had too much to drink.* **8** feign, °pretend, °counterfeit, °fake, dissemble, °make believe, °sham, simulate, °dissimulate, °posture: *You may think him sincere, but I know he is just acting.* **9** °take effect, °work, °operate, °function, °perform: *This drug will act only if taken with meals.*

action *n.* **1** °activity, °performance, °movement, °motion, °energy, liveliness, vim, °vigor, °spirit, °vitality; °enterprise, °initiative: *Being a man of action, he hates just sitting and reading.* **2** °influence, °effect, °power, °force, °strength: *The action of the moon's gravitational pull causes tides on earth.* **3** °deed, °act, °undertaking, °exertion, °exercise: *The very action of breathing caused me pain.* **4** °remedy, °proceeding, °process: *If they don't stop beating their dog we shall take action against them.* **5** fighting, °combat: *We saw action in the Far East.* **6** °fight, °battle, °engagement, °encounter, °clash, °fray, sortie, °skirmish, affray: *How many men were lost in last night's action?* **7** °effect, effectiveness, °activity, °function, °performance, functioning, °reaction: *What is the action of steroids on the lymph system?* **8** ***actions.*** °behavior, °conduct, deportment, demeanor, ways, °manner, °manners: *She must be held responsible for her actions.*

activate *v.* °move, actuate, set in motion, get started, °energize, get *or* set going, °start, °initiate, switch *or* turn on, trigger; °motivate, °rouse, °arouse, °prompt, °stimulate, °stir, °mobilize, °animate, impel, galvanize, *Colloq US* light a fire under: *The sensor in the pavement activates the traffic signal. Her enthusiasm activated him to go into business for himself.*

active *adj.* **1** °strenuous, °vigorous, °full, °dynamic, °physical; °energetic, °lively, °busy, °brisk, bustling, occupied, °on the move, *Colloq* on the go, °running: *She is healthier for having led a very active life. He always seems to be active.* **2** acting, °effective, °efficacious, °effectual, working, functioning, °operative, °potent, °influential; °powerful: *The active ingredient in her medicine is an antihistamine.* **3** °energetic, °lively, hyperactive, °animated, spry, °nimble, °quick, °agile, °sprightly: *There is no keeping up with an active child.*

activity *n.* **1** °action, °movement, °motion, °vigor, vim, °energy, liveliness, bustle: *Last week there wasn't much activity in the stock market.* **2** °pursuit, °occupation, °vocation, °work, °function, °operation, °job, °labor, °endeavor, °enterprise, °project, °undertaking, °venture, °interest: *What sort of business activity are you engaged in?*

actual *adj.* **1** existing, existent, °real, °genuine, °factual, °true, °authentic, verified, verifiable, true-to-life, °manifest, realized, °realistic, *Colloq* °solid: *The actual cost of the project turned out to be double the estimate.* **2** °present, °current, existent, °real, °genuine, °physical, °tangible: *No telescope has detected any actual volcanic eruption on the moon.*

actually *adv.* °really, °in reality, °in fact, in actuality, in point of fact, °in truth, °absolutely, as a matter of fact, °indeed, °truly, °literally: *The interest rates actually charged by banks may vary from those quoted publicly.*

acute *adj.* **1** °sharp, °pointed, °narrow: *The two roads meet at an acute angle.* **2** °severe, °intense, °critical, °crucial, °dangerous, °grave, °serious, °severe: *This is the ward for patients with acute illnesses.* **3** °sharp, °cutting, °intense, °severe, °violent, °penetrating, °exquisite, °excruciating, °fierce, shooting, stabbing, °piercing, °sudden: *The onset of the disease is marked by acute pains in the abdomen.* **4** °keen, °sharp, °sensitive: *The bloodhound is known for its acute sense of smell.* **5** °keen, sharp-witted, °shrewd, °clever, °ingenious, °astute, °sharp, canny, °incisive, discerning, °perceptive, perspicacious, °intelligent, °penetrating, insightful, percipient, °wise, °sensitive, °discriminating; °alert, °aware, on the qui vive: *Such a circumstance could not be lost upon so acute an observer.*

adapt *v.* **1** °suit, °fit, make suitable, °qualify: *The structure of the outer ear is adapted to collect and concentrate the vibrations.* **2** °alter, °modify, °change, remodel, °tailor, reshape, °shape, °fashion; °adjust, °accommodate, °accustom, acclimatize *or* acclimate, habituate: *He adapted the play from an old French comedy. The whale adapts itself to great changes in pressure when it dives thousands of feet.*

adaptable *adj.* °flexible, °pliable, pliant, compliant, accommodative, °tractable, malleable, ductile, °versatile; alterable, °changeable: *Men, generally, are not as adaptable as women.*

adaptation *n.* **1** °fitting, suiting, modifying, adjusting, conversion: *In 1831 electricity was ripe for adaptation to practical purposes.* **2** modification, °change, °adjustment, °accommodation, reworking, customization, °alteration: *She was responsible for the adaptation of her short story to a television play.*

add *v.* **1** °join, °unite, °combine, annex: *The expression 5 + 3 denotes that 3 is to be added to 5.* **2** °total, °sum, °sum up, °combine, count up, °reckon, *US* tote (up), *Brit* °tot (up): *The computer can add all those figures in a few seconds.* **3** °continue, °go on: *"And I won't take no for an answer," she added.* **4** ***add to.*** °increase, °enlarge, °amplify, augment, °supplement: *His articles have added greatly to his reputation as a financial analyst.*

addict *n.* **1** °(habitual) user, *Slang* junkie, dopefiend, doper, °head, pothead, acid-head, pill popper, tripper,

Chiefly US hophead: *His contributions helped set up the halfway houses for addicts.* **2** °devotee, aficionado, °fan, °admirer, °follower, adherent, °supporter, °enthusiast, *Colloq* buff, °hound, °fiend, groupie, *Slang* °freak, °bug, nut, teeny-bopper: *She became a rock 'n' roll addict in the '60s.*

addition *n.* **1** adding, joining, putting together, uniting, combining: *The addition of this paragraph is uncalled for.* **2** totaling, adding up, summing up, summation, counting up, °reckoning, *US* toting up, *Brit* totting up: *You have made an error in addition.* **3** addendum, appendix, appendage, °supplement, °increment, augmentation, °extension: *This addition contributes nothing to the manuscript.* **4** °extension, ell, *US* annex, wing, *Brit* annexe: *We used our lottery winnings to pay for an addition to the house.* —*prep.* **5** ***in addition to.*** as well as, °besides, beyond, over and above: *In addition to books, the shop sold greeting cards.* —*adv.phr.* **6** ***in addition.*** °moreover, furthermore, additionally, °besides, withal, °to boot, in *or* into the bargain, too, also, as well: *We were compelled to take a ten-mile run every Saturday in addition to exercising every morning.*

address *n.* **1** °speech, °talk, discourse, °oration, °lecture; °sermon: *The Prime Minister's address to the nation was broadcast last night.* **2** °location, °whereabouts: *She couldn't write to me because she didn't have my address.* —*v.* **3** speak *or* talk to; deliver *or* give a speech to; °lecture: *After the coup, the general addressed the crowd in the square.* **4** °greet, °hail, accost, °approach: *She was addressing strangers in the street to ask their views on women's rights.* **5** ***address oneself to.*** devote *or* direct *or* apply oneself to: *After the holidays, I again addressed myself to studying for examinations.*

adept *adj.* **1** °versed, °proficient, skilled, well-skilled, °expert, °accomplished, °skillful, adroit, dexterous *or* dextrous, °able, °masterful, masterly, °polished: *She is an adept pianist, and her husband is adept at carpentry.* —*n.* **2** °expert, °master, °specialist, °authority, *Colloq* °dab hand, old hand: *He is an adept at anything that he can do with his hands.*

adequate *adj.* **1** °sufficient, °enough, °ample; °satisfactory, °fitting, °equal, °suitable: *Is there language adequate to describe my feelings?* **2** °passable, °fair, fair to middling, middling, °average, °tolerable, °(barely) acceptable, °(barely) satisfactory, all right, °competent, not (at all) bad, °so-so, *Colloq* OK *or* okay, up to snuff, not that *or* too bad, no great shakes: *The music was good, the band only adequate.* **3** °equal, °suitable, suited, °fitted, up, °proper, °qualified, °competent, good enough: *Johnson was unsure that he was adequate to the task at hand.*

adjoining *adj.* °neighboring, contiguous (to), adjacent (to), abutting, bordering; next (to): *We have bought the adjoining house. The land adjoining the supermarket is for sale.*

adjust *v.* **1** set right, °arrange, °settle, harmonize, °reconcile, °resolve, set *or* put to rights; arbitrate, mediate; redress, °rectify, °correct, °patch up: *Four were named on each side to adjust their differences.* **2** °change, °alter, °modify, °regulate, °set: *After he adjusted the pendulum, the clock kept good time.* **3** °adapt (to), accommodate (oneself) (to), accustom (oneself) (to); get used (to), acclimatize *or* acclimate (to), reconcile (oneself) (to): *If she travels a distance east or west, it takes her a few days to adjust to the local time. Army life was very different, but I was able to adjust quickly.* **4** put in order, °arrange, rearrange, close *or* fasten *or* zip *or* button (up): *She adjusted the children's coats and tied their shoes.*

adjustment *n.* **1** adjusting, altering, °alteration, °setting, regulating, °regulation, setting *or* putting right *or* aright *or* to rights, correcting, °correction, calibrating, calibration; tuning: *The adjustment of the clocks is my responsibility.* **2** °arrangement, °balance, coordination, °order, alignment, °harmony, harmonization: *The inspector requires everything to be in perfect adjustment.*

administer *v.* **1** administrate, °manage, °control, °run, °direct, °conduct, superintend, °supervise, °oversee: *The president said that she had administered the department well during her year as its head.* **2** °execute, °carry on, °carry out; °apply, °implement, °prosecute: *It is the responsibility of the police to administer the law, not to make it.* **3** °dispense, °supply, °furnish, °give (out), °provide (with), mete out, °distribute, °deliver, °deal, °hand out: *Doctors sometimes administer drugs that have side effects.*

administration *n.* **1** °management, °direction, °conduct, supervision, °oversight, superintendence, °regulation, °charge: *Lord Hampden was given administration of her affairs till she came of age.* **2** °authority, °management, *US* °government: *The current administration is in favor of a better health program.* **3** dispensation, administering, supplying, furnishing, °provision, °delivery, °distribution, °application: *The judge is charged with the administration of justice.*

admirable *adj.* wonderful, awe-inspiring, °excellent, °estimable, °splendid, °marvelous, °superior, °first-rate, first-class, °of the first water, °great, °fine, *Colloq* top-drawer, rip-snorting, A-1 *or* A-one, *Brit* smashing, °magic: *His performance in Harper's new play is admirable.*

admiration *n.* °wonder, awe; °delight, °pleasure; °esteem, °regard, °appreciation, °respect: *She is lost in admiration of her mother's latest painting. Randolph was presented with a gold medal as a token of his colleagues' admiration.*

admire *v.* **1** wonder *or* marvel (at), °delight in: *Typically, he most admires people who are wealthy.* **2** °esteem, regard *or* respect highly, °look up to, °revere, °idolize, °venerate, °worship: *The queen is one of the most admired people in the country.*

admirer *n.* **1** °devotee, aficionado, °fan, °supporter, °enthusiast, adherent, °follower, *Slang* groupie: *Rock stars always seem to be accompanied by a retinue of admirers.* **2** beau, °suitor; °lover, °sweetheart, °darling: *Scarlett was always surrounded by many admirers.*

admission *n.* **1** access, °admittance, entrée, °entry: *The special card gives me admission to the rare book room of the library.* **2** °reception, acceptance, °appointment, °institution, induction, °installation, investiture: *The committee has approved the formal admission of women into the club.* **3** acknowledging, °acknowledgment, allowing, °allowance, admitting, °admittance, conceding, concession: *The court refuses to consider the admission of testimony taken under duress.* **4** °acknowledgment, confession, concession, °profession, °declaration, disclosure, affirmation, concession, divulgence *or* divulgement, °revelation: *The police were able to extract an admission of guilt from the suspect.* **5** ticket, (entry *or* entrance) fee, °tariff: *Admission is free for senior citizens.*

admit *v.* **1** °let in, allow to enter, take *or* allow in; °accept, °receive: *I opened the window to admit some air. The harbor is too small to admit even one more ship.* **2** °allow, °permit, °grant, °brook, °tolerate: *The governor will admit no delay in the execution of the sentence, and I shall be hanged at dawn.* **3** °accept, °concede, acquiesce, °allow, °grant, °accept, °recognize, take cognizance of: *Descartes' principle admitted nothing but what his own consciousness obliged him to admit.* **4** °confess, own, °concede, divulge, °reveal, °acknowledge, °declare: *She readily admitted to having incited the riot.*

admittance *n.* leave *or* permission to enter, °entry, entering, °entrance, access, entrée: *Admittance to the club is restricted to members.*

adolescent *n.* **1** °teenager, °youth, °juvenile, °minor, °stripling, youngster, *US* teen, *Colloq* kid; *Slang* teeny-bopper: *A group of adolescents volunteered to work at the home for the elderly.* —*adj.* **2** teenaged, °young, youthful, maturing, pubescent; °immature, °puerile, °juvenile: *Adolescent growth is often dramatic, a gain of two inches in height being not unusual.*

adopt *v.* **1** °take (in), °accept, take *or* accept as one's own: *Carpathia and her husband adopted two children.* **2** °take, take up *or* on *or* over, °embrace, espouse; arrogate, °appropriate: *All Garmisch's ideas are adopted from others—he's never had one of his own.*

adorable *adj.* °lovable, °beloved, loved, °darling, °sweet, °dear; °delightful, appealing, °attractive, charming, captivating, °fetching: *To look at him now, it is hard to imagine what an adorable child he once was.*

adore *v.* **1** °esteem, °honor, °respect, °admire; °idolize, dote on: *An entire generation adored the Beatles.* **2** °worship, °venerate, °reverence, °revere, °exalt; °hallow: *O! Come let us adore him—Christ, the Lord!* **3** °love, be in love with, °cherish, °fancy, °revere, adulate, *Colloq* have a crush on, carry the *or* a torch for: *Katie just adores the captain of the football team at school.*

adult *adj.* **1** °mature, grown up, full-grown, matured, of age: *Now that you are adult, you come into a large inheritance.* —*n.* **2** grown-up: *Tiger cubs are cute, but the adults are very dangerous.*

adulterate *v.* °falsify, °corrupt, °alloy, °debase, °water (down), °weaken, °dilute, bastardize, °contaminate, °pollute, °taint, *Colloq* °doctor; *Slang US* °cut: *They are apt to adulterate the bread, sometimes with alum, sometimes with chalk.*

advance *v.* **1** move *or* put *or* push *or* go forward; °approach: *Man has advanced the frontier of physical science. The battalion advanced toward the fort with guns blazing.* **2** °further, °promote, °forward, °help, °aid, °abet, °assist, °benefit, °improve; °contribute to: *The terrorists' dynamiting of the school has done nothing to advance their cause.* **3** go *or* move forward, °move (onward), °go on, °proceed, °get ahead: *As people advance in life, they acquire what is better than admiration—judgment.* **4** °hasten, accelerate, °speed: *We have advanced the date of our departure from December to October.* **5** move up, °promote: *In less than a year, Mrs. Leland has been advanced from supervisor to manager of the production department.* **6** prepay, °lend: *Could you advance me some money till payday?* —*n.* **7** °progress, °development, °progress, forward movement; °improvement, betterment; °headway: *Who has done more for the advance of knowledge?* **8** °rise, °increase, °appreciation: *Any advance in prices at this time would reduce our sales.* **9** prepayment, °deposit; °loan: *I cannot understand why George is always asking for an advance on his allowance.* **10** ***in advance.*** **(a)** beforehand, °ahead (of time), °before: *You will have to make reservations well in advance.* **(b)** °before, °in front (of), °ahead (of), beyond: *The colonel rode in advance of the cavalry.*

advantage *n.* **1** °superiority, °upper hand, dominance, °edge, head start; °sway; *Colloq US and New Zealand* °drop: *After a year, the advantage was with the Royalists. His height gives him an advantage at basketball.* **2** °gain, °profit, °benefit, °interest; °asset, betterment, °improvement, advancement; °use, °usefulness, utility, °help, °service: *I have information that will be of advantage to her.* **3** ***to advantage.*** °better, °(more) favorably, advantageously: *The dress sets off her figure to advantage.*

advantageous *adj.* °profitable, °worthwhile, °gainful, °opportune, °beneficial, °favorable, °useful, valuable: *The minister signed an advantageous treaty of commerce with Russia.*

adventure *n.* **1** °exploit, escapade, °danger, °peril; °affair, °undertaking, °feat, °deed; °experience, °incident, °event, °occurrence, °happening, °episode: *We shared many wartime adventures.* **2** °speculation, °hazard, °chance, °risk, °venture, °enterprise: *I lost a fortune in some of his financial adventures.* —*v.* **3** °venture, °hazard, °risk, imperil, °endanger, °jeopardize, °threaten: *Would you adventure your pension money in such a scheme?* **4** °dare, wager, °bet, °gamble, °stake, try one's luck, *Brit* punt: *She adventured a whole week's salary on the lottery.*

adventurer *n.* **1** adventuress, soldier of fortune, swashbuckler, °hero, °heroine, °daredevil; °mercenary: *Errol Flynn often played the role of the adventurer.* **2** adventuress, °cheat, °swindler, charlatan, trickster, °rogue, °scoundrel, knave; cad, bounder, °philanderer, fortune hunter, opportunist: *That adventuress is just after Nelson's money.*

adventurous *adj.* °daring, °rash, °brash, °reckless, devil-may-care, °bold, °foolhardy, °hazardous, °risky, °daredevil, °venturesome, adventuresome, temerarious, °audacious, °bold, °intrepid, °brave, °courageous: *She was adventurous enough to sail round the world single-handed.*

adversary *n.* **1** foe, °enemy, °opponent, °antagonist, °competitor, °rival: *Before beginning to fight, each adversary sized up the other.* —*adj.* **2** °opposed, °hostile, antagonistic, competitive: *Why does she always take the adversary position in every argument?*

advertisement *n.* **1** °notice, handbill, blurb, broadside, °bill, °circular, °brochure, °poster, placard, *affiche*, classified, commercial, °spot (announcement), *US* car card, billboard, *Colloq* ad, °plug, *Brit* advert: *The company has placed advertisements in all major media.* **2** advertising, °promotion; publicity; °propaganda, ballyhoo, hoopla, *Colloq* hype, beating the drum, *US* puffery: *Advertisement on TV may be very effective, but it is very expensive.*

advice *n.* **1** °counsel, °guidance, °recommendation, °suggestion, °opinion, °view; °warning, admonition, *Technical* paraenesis: *His lawyer's advice is to say nothing.* **2** °information, °news, °intelligence, °notice, notification; communication: *Advice has reached us that someone has cornered the market in pork bellies.*

advisable *adj.* recommendable, °expedient, °prudent, °practical, °sensible, °sound, °seemly, °judicious, °wise, °intelligent, °smart, °proper, °politic: *It would be advisable for you to keep out of sight for a few days.*

advise *v.* **1** °counsel, °guide, °recommend, °suggest, commend; °caution, admonish, °warn; °urge, °encourage: *I advised him to be careful driving at night in that area.* **2** °tell, °announce (to), °inform, apprise, °register, °make known (to), °intimate (to), °notify: *We advised her of our disapproval. The police have advised the defendants of their rights.*

adviser *n.* °counselor, mentor, °guide, cicerone, °counsel, °consultant, confidant(e): *The chairman always consults his advisers before making a decision.*

advisory *adj.* **1** consultive, consultative, counseling, hortatory, monitory, admonitory, *Technical* paraenetic(al): *Our firm has been engaged in an advisory capacity on the unification of the utility companies.* —*n.* **2** °bulletin, °notice, °warning, admonition, °prediction:

The Weather Bureau has issued a storm advisory for the weekend.

advocate *v.* **1** °support, °champion, °back, °endorse, °uphold, °recommend, stand behind, °second, °favor, speak *or* plead *or* argue for *or* in favor of: *Don't you advocate the policies of the Party?* —*n.* **2** °supporter, °champion, °backer, upholder, °second, exponent, °proponent, °patron, defender, apologist: *She is an enthusiastic advocate of free speech.* **3** °lawyer, °counsel; intercessor; *US* attorney, counselor-at-law, *Brit* barrister, solicitor: *The advocate for the opposition is not in court.*

aesthete *n.* connoisseur, art lover, lover of beauty, aesthetician *or* esthetician, *US* tastemaker: *It was the aesthetes who set the standard for the art purchased by the museum.*

aesthetic *adj.* **1** artistic, °tasteful, °beautiful; in good, °excellent, etc. taste: *Daphne always does such aesthetic flower arrangements.* **2** °sensitive, artistic, °refined, °discriminating, °cultivated: *These paintings might be realistic, but they are an aesthetic disaster.*

affair *n.* **1** °matter, °topic, °issue; °business, °concern, °interest, °undertaking, °activity: *These are affairs of state and require the approval of a minister.* **2** °concern, °business, *US slang* beeswax: *Who wiped the fingerprints off the weapon is none of your affair.* **3** °event, °business, °occurrence, °happening, °proceeding, °incident, °operation: *Last night's farewell party was truly a dull affair.* **4** Also, ***affaire.*** °love affair, amour, °romance, °intrigue, °fling, °liaison, °relationship, *affaire d'amour, affaire de coeur. Lady Constance is having an affair with the gamekeeper.*

affect[1] *v.* **1** °attack, act upon, °lay hold of, °strike: *Arthritis has affected his hands and he can no longer play the piano.* **2** °move, °stir, °impress, °touch, °strike; °perturb, °upset, °trouble, °agitate: *The sportsman was not affected by all the taunts and jeers.* **3** °influence, °sway, °change, °transform, °modify, °alter: *Her sudden fame has affected her view of herself.*

affect[2] *v.* **1** °assume, °adopt, °put on, °pretend (to), feign, °sham, °fake, °counterfeit: *Charles affects a knowledge of high finance.* **2** °choose, °select; °use, °wear, °adopt: *When he was trying to impress people, he affected a British accent.*

affectation *n.* **1** affectedness, pretentiousness, artificiality, insincerity, posturing: *She behaves with so much affectation that I never can be sure of her real feelings.* **2** °pretense, simulation, false display, °show, °front, °pose, °pretension, façade; °act, °airs: *Some people's charitable concern for others is mere affectation. Using a long cigarette-holder is one of her many affectations.*

affected *adj.* **1** °unnatural, °artificial, °specious, °stilted, °stiff, °studied, °awkward, nonnatural, contrived, °mannered: *But does not Dryden find Shakespeare's style stiff and affected?* **2** °pretended, simulated, °hollow, °assumed, feigned, °fake, faked, °false, °counterfeit, °insincere, °spurious, °sham, °bogus, *Colloq* phoney *or US also* phony: *The heir's affected grief concealed his secret exultation.* **3** °pretentious, °pompous, high-sounding, °mincing, niminy-piminy, *Colloq* la-di-da: *Smallpiece's affected airs were enough to make his classmates detest him.* **4** attacked, seized, afflicted, °stricken, gripped, touched; °diseased, laid hold of: *Her affected lungs never quite recovered.* **5** afflicted, moved, touched, stirred, distressed, troubled, °upset, °hurt; influenced, swayed, impressed, struck, played *or* worked *or* acted upon: *Many affected theatergoers enjoyed her performances.*

affection *n.* goodwill, °(high) regard, °liking, fondness, °attachment, loving attachment, tenderness, °warmth, °love: *The affection she felt toward her stepchildren was returned many times over.*

affectionate *adj.* °fond, loving, °tender, caring, °devoted, doting, °warm: *She gave her mother an affectionate embrace and boarded the train.*

affiliated *adj.* associated; °attached, connected, combined, °united, joined: *For our members' convenience, the club is now affiliated with one that serves meals.*

affinity *n.* **1** °relationship, °kinship, closeness, °alliance, connection *or Brit* connexion; °sympathy, °rapport: *He felt an affinity with other redheaded people.* **2** friendliness, fondness, °liking, °leaning, °bent, °inclination, °taste, °partiality, attractiveness, °attraction: *I have an affinity for the sea.*

afflict *v.* °affect, °bother, °distress, °oppress, °trouble, °torment: *Last winter's intense cold afflicted everyone, but those in the north especially.*

affliction *n.* **1** °hardship, °misery, °misfortune, °distress, °ordeal, °trial, tribulation, adversity, °suffering, °woe, °pain, °grief, °distress, °torment, wretchedness: *Moses saw the affliction of his people in Egypt.* **2** °curse, °disease, °calamity, °catastrophe, °disaster, °plague, °scourge, tribulation, °trouble: *He often observed that greed was the affliction of the middle class.*

afford *v.* **1** have the means, be able *or* rich enough, °manage, bear the expense, °pay, °provide: *We cannot afford to send the children to better schools.* **2** °give, °spare, °give up, °contribute, °donate; °sacrifice: *The loss of a single day's work was more than I could afford.* **3** °yield, °give, °supply, °produce, °provide, °furnish, °grant, °offer; give forth: *May kind heaven afford him everlasting rest. The poems afford no explanation.*

afoul *adv.* ***afoul of.*** entangled with, in trouble with, in conflict with, at odds with: *Farnsworth fell afoul of the new tax regulations.*

afraid *adj.* **1** °fearful, frightened, °scared, intimidated, apprehensive, lily-livered, white-livered, terrified, °panic-stricken, °fainthearted, weak-kneed, °timid, timorous, °nervous, °anxious, jittery, °on edge, edgy, °jumpy; °cowardly, pusillanimous, craven, *Colloq* yellow: *Don't be afraid, the dog won't bite you.* **2** °sorry, °unhappy, °regretful, °apologetic, rueful: *I'm afraid I cannot help you find a cheap apartment in New York.*

age *n.* **1** lifetime, duration, length of existence; lifespan: *The age of a stag is judged chiefly by the furniture on its head. The days of our age are three score years and ten. She is sixteen years of age.* **2** °maturity, °discretion; °majority, adulthood, seniority: *When he comes of age he will inherit millions.* **3** °period, °stage, °time: *Among those people, both boys and girls undergo the rites of passage at the age of puberty. He is a man of middle age.* **4** long time, aeon *or, esp. US,* eon; years: *I haven't seen you for an age! The noise went on for ages.* **5** °era, epoch, °period, °time: *The eighteenth century was known as the Augustan Age in England.* —*v.* **6** grow old(er), °mature, °ripen: *O, Matilda, I age too fast for my years! You must first age the whiskey in the barrel, then bottle it.*

aged *adj.* °old, °elderly, superannuated, °ancient, age-old, °gray, °venerable: *The three aged women crouched in their chairs, each with her memories.*

agency *n.* °means, °medium, instrumentality; intervention, intercession, °action, intermediation; °operation, °mechanism, °force, °power, °activity, working(s), °energy: *Pollen is carried from flower to flower by the agency of certain insects.*

agent *n.* **1** °representative, °intermediary, °go-between, °proxy, emissary, °delegate, spokesman, spokeswoman, spokesperson, °deputy, °substitute, surrogate,

°advocate, emissary, legate, °envoy, °factor: *My agent in Tokyo will look after the matter for you.* 2 °factor, °agency, °cause, °means, °force, °instrument, °power, °vehicle, °ingredient, catalyst: *The active agent in this cleaner is ammonia.*

aggravate *v.* 1 °worsen, °intensify, exacerbate, °heighten, °magnify, °increase; °inflame: *They introduce new problems and aggravate the old ones.* 2 °exasperate, °frustrate; °anger, incense, °infuriate; °provoke, °irritate, nettle, rile, vex, °annoy, °harass, hector, °bother; embitter, °rankle, *Colloq* peeve, needle, get on one's nerves; *Slang Brit* give (someone) aggro: *Threats only serve to aggravate people.*

aggression *n.* 1 aggressiveness, °hostility, belligerence, combativeness, *Slang Brit* aggro: *The mere crossing of the river is an act of aggression.* 2 °attack, °assault, onslaught, °invasion, encroachment: *The conflict had become a war of aggression.*

aggressive *adj.* 1 combative, °warlike, °martial, °belligerent, bellicose, °pugnacious, °quarrelsome, disputatious, litigious; °hostile, unfriendly: *The Germanic tribes were known to the Romans to be the most aggressive.* 2 °forward, °assertive, °forceful, °bold, *Colloq* °pushy: *Owing largely to his aggressive nature, Dennis might make a good salesman.*

aggressor *n.* °assailant, attacker, instigator, initiator, provoker; °belligerent: *You will find that the Nazis were the aggressors in Poland in 1939.*

agile *adj.* 1 °nimble, °quick, °brisk, °swift, °active, °lively, lithe, limber, spry, °sprightly: *Sofia is an agile dancer.* 2 °keen, °sharp, °alert, dexterous *or* dextrous, °resourceful, °acute: *With his agile mind Richard is able to solve the problems in no time at all.*

agitate *v.* 1 °excite, °arouse, °rouse, °move, °perturb, °stir up, disquiet, °fluster, °ruffle, °rattle, disconcert, °discomfit, unsettle, °upset, °rock, °unnerve, °shake (up), *Colloq* discombobulate: *Rachel was agitated to learn of the bank's threat to foreclose on the mortgage.* 2 °push, °press, °campaign; °promote: *The miners have been agitating for better safety measures.* 3 °stir (up), churn, °disturb, °shake, roil: *The calm lake was agitated by the motorboats.*

agitated *adj.* moved, stirred (up), shaken (up), rattled, °disturbed, °upset, °nervous, perturbed, jittery, °jumpy, uneasy, °ill at ease, fidgety, disquieted, discomfited, ruffled, flustered, °unsettled, unnerved, wrought-up, discomposed, °disconcerted, aroused, roused, °excited, *Colloq* discombobulated: *The sheriff was in a very agitated state about the mob forming outside the jail.*

agitation *n.* 1 shaking, °disturbance, churning, °stirring, turbulence: *The agitation made the solution become cloudy.* 2 °excitement, arousal, rabble-rousing, °provocation, stirring-up, °incitement, °ferment, stimulation, overstimulation, commotion: *The organized agitation of the crowds continued for weeks after the coup.*

agitator *n.* activist, °rabble-rouser, incendiary, *agent provocateur,* insurrectionist, °troublemaker, demagogue, firebrand: *The opposition party hires professional agitators to incite the people to riot.*

agog *adj.* °eager, avid, °keen, °enthusiastic, °expectant, °impatient, °breathless: *The children were all agog waiting for Santa Claus to come.*

agonizing *adj.* °painful, distressful, distressing, °harrowing, torturous, racking, °excruciating, tortured, tormented: *We went through an agonizing reappraisal of our policy on immigration.*

agony *n.* °anguish, °trouble, °distress, °suffering, °misery, wretchedness, °pain, pangs, °woe, °torment, throes, torture, °affliction: *For two days his parents experienced the agony of not knowing whether he was dead or alive.*

agree *v.* 1 concur, °conform, come *or* go together, °coincide, °correspond, harmonize, °reconcile; °accord, °tally, *Colloq* jibe: *At last my checkbook agrees with my bank statement!* 2 Often, ***agree on*** *or* ***upon*** *or* ***to.*** °consent to, °favor, acquiesce in *or* to, °approve of, accede to, settle (on *or* upon), assent to: *The terms of the contract have been agreed.* 3 °concede, °grant, °consent, °admit, °approve, °allow, °accept, concur; accede (to), acquiesce (in *or* to), assent (to), see eye to eye: *The committee agreed that she should be given time to comply with the request. I objected and they agreed with me.* 4 ***agree with.*** °suit: *The climate in England agrees with me, strange to say.*

agreeable *adj.* 1 °pleasing, °pleasant, enjoyable, °pleasurable, °favorable, °delightful, °satisfying, °satisfactory, °good, °nice, °acceptable; to one's liking *or* taste: *He found the Caribbean an agreeable place for a vacation.* 2 in favor, approving, °willing, consenting, acquiescent, complying, compliant, in agreement *or* accord, concurring, amenable, °sympathetic, well-disposed; °accommodating, accommodative: *If Anne's agreeable, we can leave tomorrow.*

agreement *n.* 1 °understanding, covenant, °treaty, °pact, °accord, °compact, °settlement, concordat; °contract, °bargain, *Colloq* °deal: *They drew up a ten-year agreement to be signed at the summit in Geneva.* 2 concord, °harmony, compatibility, °unity, concurrence, unanimity: *Agreement in error is far worse than division for the sake of truth.*

ahead *adv.* 1 at the *or* in front, °in advance, in the lead *or* vanguard, up ahead, °before, to the fore: *The general rode ahead.* 2 °winning: *At half time, our team was ahead by two points.* 3 °onward, °forward, on: *Please move ahead if you can.*

aid *v.* 1 °help, °support, °assist, °facilitate, °back, °abet, °uphold, °promote; succor, °relieve, °subsidize: *The invasion was aided by Richard's subjects. These charities aid the poor all year round, not just at Christmas.* —*n.* 2 °help, °support, °assistance, °backing, °relief, °benefit, °service, succor, °comfort: *He was convicted of giving aid to the enemy in time of war. Can you tell me what your scheme is supposed to be in aid of?* 3 funding, °subsidy, subvention; grant money, °grant, grant-in-aid, °scholarship: *He could never have earned his degree without aid from the endowment.*

aide *n.* aide-de-camp, °assistant, helper, coadjutor; good *or* strong right arm, right hand, right-hand man; °colleague, °partner, °ally, °comrade, comrade-in-arms, *US* °cohort, *Colloq* man Friday, girl Friday, *US* gal *or US* guy Friday: *The general's aides are always at his side.*

ail *v.* 1 °trouble, °afflict, °affect, °bother, °distress, °upset, °worry, make ill *or* sick, °pain, °hurt: *I cannot imagine what ails him, and the doctor can find nothing wrong.* 2 °suffer, be *or* feel ill *or* sick *or* poorly *or* unwell *or* indisposed, *US* be sick: *Granny has been ailing lately.*

ailment *n.* °illness, sickness, °affliction, °disease, °disorder, indisposition, malady; °disability, °infirmity; malaise, queasiness: *Pecksniff's ailment has been diagnosed as influenza.*

aim *v.* 1 °direct, °point, °focus, °train, °level: *The guns of the fort are aimed at the narrow pass.* 2 ***aim at.*** focus on, have designs on, °aspire to, plan for *or* on, set one's sights on, °seek, strive for, try for, °wish, °want: *Edward aimed at absolute dominion over that kingdom.* 3 °seek, °intend, °plan: *I aim to mosey down to Chauncy's for a game of pool.* —*n.* 4 °direction, pointing, °focus, focusing *or* focussing, sighting: *His aim is so bad he can't hit the side of a barn with a shotgun.* **5** °purpose, °goal, °ambition, °desire, °aspiration, °object, °end, °objective, °target, °intent, °intention, °plan: *It was never her aim in life to be rich. The aim of the book is set forth in the Foreword.*

aimless *adj.* **1** °purposeless, °pointless, °frivolous: *After receiving the inheritance she led an aimless life of ease and luxury.* **2** undirected, °erratic, °chance, °haphazard, °random, vagrant, wayward; °wanton: *We were concerned about her aimless wandering in the streets at night.*

air *n.* **1** °atmosphere, ambiance *or* ambience, aura, °climate, °feeling, °sense, °mood, °quality: *This restaurant has a delightful air about it.* **2** °breeze, zephyr, °current, °draft; °breath, °puff, °wind: *Light airs sprang up from the south.* **3** °manner, °style, °appearance, °aura, °feeling, °bearing, °quality, flavor: *Louis has a lugubrious air about him.* **4** °melody, °tune, °song, music: *She was humming airs from some Italian opera.* **5** ***airs.*** °pretension, °pretense, °show, affectedness; haughtiness, hauteur, °arrogance, °superiority, superciliousness: *He puts on such airs since he got his knighthood.* —*v.* **6** ventilate, °freshen, °refresh, aerate: *The chambermaid is airing the room, so you can't go in now.* **7** °show off, °parade, °display, °exhibit; °publish, °broadcast, °circulate, °publicize, make public *or* known, °reveal, °expose, °disclose, divulge, °tell, °express, °declare: *Carbuckle is again airing his ideas on modern art.*

akin *adj.* ***akin (to).*** °related (to), allied *or* connected *or* affiliated (to *or* with), associated (with), germane (to), °like, °alike, °similar (to): *The two ideas are akin. Desultoriness is akin to indolence. The decision to ban the film smacks of something akin to censorship.*

alarm *n.* **1** °warning, °alert, danger *or* distress signal; tocsin, bell, gong, °siren, whistle, horn: *At the approach of the storm, the lookouts gave the alarm. The alarm is set to wake me at four o'clock.* **2** °fear, °fright, apprehension, °dismay, trepidation, °terror, °dread, °anxiety, °excitement, °panic, consternation, °distress, nervousness, uneasiness, °discomfort: *He viewed with alarm the arrest of his next-door neighbors.* —*v.* **3** °frighten, °scare, °daunt, °startle, °terrify, °panic; °unnerve, °dismay, °disturb, °upset: *Don't be alarmed—such delays are quite normal. She was alarmed at the news of the car crash.*

alcohol *n.* °spirits, °liquor, °the bottle, the cup that cheers, demon rum, John Barleycorn, *Colloq* °booze, hard stuff, °juice, °moonshine, firewater, *Slang* rotgut, *US and Canadian* hooch: *Alcohol and driving do not mix.*

alcoholic *adj.* **1** °intoxicating, inebriating: *His doctor has forbidden him any alcoholic beverage.* —*n.* **2** drunkard, °drunk, dipsomaniac, sot, toper, drinker, winebibber, serious *or* problem drinker, tippler, *Colloq* barfly, °soak, *Slang* boozer, alchy *or* alkie *or* alky, dipso, °stew, rummy, *US and Canadian* °lush, boozehound, wino: *The community runs a center for rehabilitating alcholics.*

alert *adj.* **1** °awake, wide-awake, watchful, °vigilant, °attentive, heedful, °wary, °cautious, on the qui vive, °aware, on guard, on the lookout, °observant, *Colloq* on the ball, on one's toes: *The sentinels must remain alert throughout the night. Kenneth is alert to the perils of smoking cigarettes.* **2** °active, °nimble, °lively, °agile, °active, °quick, spry, °sprightly, °vivacious: *He is an alert and joyous old soul.* —*n.* **3** °lookout: *She is always on the alert for new ways of saving money.* **4** °alarm, °warning, °signal, °siren: *Sound the air-raid alert!* —*v.* **5** °warn, °caution, °advise, °alarm, forewarn, °signal, °notify: *We must alert people to the fact that the fugitive is a vicious killer.*

alibi *n.* **1** °excuse, °explanation: *Your alibi places you very close to the scene of the crime.* —*v.* **2** °excuse, °explain: *Caught red-handed, she couldn't alibi her way out of it.*

alien *adj.* **1** °foreign, °strange, °exotic, °outlandish, °unfamiliar: *The customs of the country were alien to me.* —*n.* **2** °foreigner, °stranger, outlander, °outsider, nonnative, °immigrant, °newcomer: *Aliens are required to register during January.*

alienate *v.* **1** Usually, ***alienate from.*** disabuse (of *or* from), wean away (from), °detach (from), °distance (from): *Gradually people were alienated from belief in the Olympian divinities.* **2** °distance, estrange: *Snyder alienated his friends by his weird behavior.*

alike *adj.* **1** °similar, °akin, resembling *or* like one another, akin to *or* similar to one another, showing *or* exhibiting a resemblance: *They began to think all religions were alike.* —*adv.* **2** in like manner, in the same manner *or* way, similarly, equally, uniformly, identically: *She believes that all people should be treated alike.*

alive *adj.* **1** living, °live, breathing, among the living, in the land of the living: *Branfield is not only alive, but he's a very lively octogenarian.* **2** ***alive to.*** sensitive *or* alert to, aware *or* conscious of, aware *or* cognizant of: *She is alive to every slight nuance in the poem.* **3** °alert, °active, °lively, °vivacious, °quick, °spirited, °animated, °brisk, spry, °sprightly, °vigorous, °energetic: *Look alive, my lads, and hoist away!* **4** astir, teeming, swarming, thronging, crowded, °packed, buzzing, crawling, jumping, bustling, humming, *Colloq* °lousy: *In a few minutes the water around the corpse was alive with deadly piranha.*

allegation *n.* °charge, °accusation, °complaint; °assertion, avowal, asseveration, °claim, °declaration, °statement, deposition: *I resent the allegation that I don't bath often enough.*

allege *v.* °declare, aver, °state, assert, °charge, affirm, avow, asseverate, depose, °say: *The guard alleged that he had caught the boy climbing in a basement window.*

alleged *adj.* described, designated; claimed, avowed, stated; purported, °so-called, suspected, °supposed, °assumed, presumed; °hypothetical, conjectural: *The press reported that the alleged assailant had confessed. He is awaiting trial for his alleged involvement in the bombing.*

alliance *n.* **1** °union, confederation, °combination, °federation, °pact, °league, °association, coalition, affiliation, °connection, °bond; °unity, °affinity: *The alliance between the two empires has been faithfully maintained.* **2** °marriage, °affinity: *The alliance between the two families was welded by the children born of it.*

allot *v.* °distribute, apportion, allocate, earmark, °assign, parcel *or* dole out, °deal (out), °divide, °share (out), °dispense: *The millionaire allotted an equal share of his fortune to each of his children.*

allotment *n.* °share, apportionment, °ration, °portion, °quota, °allowance, °measure: *Each prisoner was given a daily allotment of four ounces of black bread and a cup of water.*

allow *v.* **1** °acknowledge, °admit, °grant, °concede, own: *He allowed that he had not been completely truthful in his earlier testimony.* **2** °agree to, °concede, cede to, °admit, °permit, °authorize, °entertain, °consent to: *The judge said that he would allow a plea of "guilty with an explanation."* **3** °permit, °let, °suffer: *Please allow the children to select their own friends.* **4** °tolerate, °stand (for), °brook, °sanction, countenance, °permit, °consider, °put up with: *The headmaster refuses to al-*

low such goings-on at his school. **5** °give, let (someone) have, °appropriate, °grant, budget, earmark, °assign, allocate, °assign, °approve: *It was arranged that the emperor would allow him a hundred florins for expenses.* **6** make allowance *or* concession for, set apart *or* aside, °put aside, take into account *or* consideration; °add; °deduct: *You must allow at least an extra hour for the traffic during rush hour. The shipper allows ten pounds for the weight of the container.*

allowance *n.* **1** °permission, toleration, °tolerance, sufferance, °admission, concession, °sanction; allowing, permitting, tolerating, °suffering, sanctioning, brooking, countenancing: *There were many causes of difference between them, the chief being the allowance of slavery in the south.* **2** °payment, recompense, °remuneration, reimbursement, °remittance: *Allowance will be made for all reasonable expenses.* **3** °stipend, °dole, pin *or* pocket money, °quota, °ration; °pension, annuity, allocation: *Bill gets a liberal weekly allowance for expenses.* **4** °deduction, °discount, reduction, °rebate; °credit; tret; tare: *You must make allowance for the weight of the crate.* **5** excuse(s), concession, °consideration: *Allowance must be made for his poor eyesight.*

alloy *n.* **1** °mixture, °mix, °combination, °compound, composite, °blend, °amalgam, admixture; aggregate: *Brass is an alloy of copper and zinc.* —*v.* **2** °contaminate, °pollute, °adulterate, °debase, °diminish, °impair, °vitiate: *Their external prosperity was not alloyed by troubles from within.* **3** °change, °modify, °temper, °alter, °moderate, allay: *Gentle persons might by their true patience alloy the hardness of the common crowd.*

ally *n.* **1** °comrade, confederate, collaborator, coadjutor; accessory *or* accessary, °accomplice; °associate, °partner, °friend: *I had hoped to have you as an ally in my proposal for reorganization. The Allies finally defeated the Nazi war machine in 1945.* —*v.* **2** °league, °combine, °unite, °join (up), °team (up), °side, °band together, °associate, affiliate, °collaborate, confederate: *In their attempt at a takeover of our company, the raiders have allied themselves with two banks. We shall ally the Romans to us and conquer the territory.*

almost *adv.* °nearly, °about, °approximately, °practically, °virtually, well-nigh, bordering on, on the brink of, verging on, °on the verge of, little short of; not quite, all but; °barely, °scarcely, °hardly; *Colloq* damn near: *We are almost ready to go. You almost broke the window!*

aloft *adv.* °above, °overhead, (up) in the air, in flight, up (above); on high; heavenward, skyward: *The plane was overloaded, but we finally made it aloft. Five women held the banner aloft.*

alone *adj.* **1** °unaccompanied, unescorted, °solitary, by oneself, *tout(e) seule,* °solo, unattended, unassisted; °abandoned, °desolate, °deserted: *I am alone in the world. Leave me alone.* **2** unequaled, °unparalleled, °unique, °singular, unexcelled, unsurpassed, without equal, °peerless, °matchless: *As a poet, Don stands alone.* —*adv.* **3** solitarily, by oneself, °solo: *I'll walk alone.* **4** °only, solely, exclusively, °simply, °just, °merely: *You alone can help me.*

aloof *adv.* **1** °apart, away, at a distance, °separate; at arm's length: *We invited Martha to join us but she preferred to remain aloof.* —*adj.* **2** °private, °reticent, °reserved, °withdrawn, °haughty, °supercilious, °standoffish, °formal, unsociable, °unsocial; °distant, °remote: *Deirdre is quite an aloof sort of person—not what you would call a "mixer."* **3** °standoffish, °distant, °remote, °cool, °chilly, unresponsive, unfriendly, antisocial, °unapproachable; °unsympathetic, apathetic, °indifferent, undemonstrative: *Roger keeps himself aloof from the needs of those less fortunate than he.*

alter *v.* °change, °revise, °modify, °vary, °transform; °adjust, °adapt, °convert, remodel: *After the attack, we altered our opinion of the rebels. The dress fits better since being altered.*

alteration *n.* °change, modification, °revision, °transformation; °adjustment, °adaptation, conversion, remodeling: *We found places where alterations had been made in the original document. My new suit needs alteration to fit properly.*

alternate *v.* **1** °rotate, °exchange, °change, interchange, °take turns, go *or* take, etc. in turn, *US* change off, interexchange: *To help out, we could alternate our days off.* **2** °succeed, be in succession *or* rotation: *This wallpaper has alternating stripes of pink, gray, and maroon.* —*adj.* **3** in rotation, °successive; every other, every second: *The embankment revealed alternate layers of clay and gravel. The nurse visited our district on alternate days.* **4** °alternative, °second, other: *The alternate selection contains only milk chocolate.* —*n.* **5** °variant, °alternative, °(second) choice, *US and Canadian* °substitute, °deputy, °stand-in, backup, °understudy; pinch hitter, *Baseball* designated hitter: *I prefer the alternate to the featured model. My alternate takes over if I am ill.*

alternation *n.* rotation, °succession; °exchange, interchange: *In a temperate climate there is the advantage of the alternation of the seasons.*

alternative *adj.* **1** °alternate, °variant, (an)other, °different, additional; °substitute, surrogate: *Alternative models are available.* —*n.* **2** °alternate, °variant, °choice, °option, °selection; °possibility; °substitute, surrogate: *The alternative was to remain at home and do nothing. You leave me no alternative. "Esthetic" is an American spelling alternative.*

altogether *adv.* °entirely, °utterly, °completely, °wholly, °totally, fully, in all respects, °absolutely, °perfectly, °quite; all in all, in all: *I don't altogether agree with you. Altogether, you may be right.*

altruism *n.* selflessness, self-sacrifice, unselfishness, °philanthropy, generosity, °charity, charitableness, humanitarianism, humaneness, °benevolence, °humanity, public-spiritedness: *Carver does things for others out of altruism, expecting nothing in return.*

always *adv.* **1** at all times, again and again, on all occasions, every *or* each time, each and every time, without exception, unexceptionally; °often, many times, °usually: *He that indulges hope will always be disappointed. I have always made coffee this way and see no reason for changing.* **2** °forever, continually, °ever, perpetually; unceasingly, unendingly, eternally, evermore, ever after, everlastingly, till the end of time, in perpetuity: *You are a fool, Filbert, and you will always be a fool.* **3** °in any case, as a last resort: *You could always refuse to pay.*

amalgam *n.* °mixture, °blend, °combination, °alloy, °mix, composite, admixture, °amalgamation; °compound: *The population was an amalgam of original settlers and new immigrants.*

amalgamate *v.* °blend, °combine, °unite, °mix, °join, consolidate, °compound, °integrate, °merge: *The four sentences of the original are amalgamated into two.*

amalgamation *n.* °blend, fusion, °combination, °mixture, mingling, admixture, composite, °compound, blending, joining, consolidating, consolidation, compounding, commingling, fusing, coalescing, coalescence, °union, uniting, unification, integration, °merger, °association, °composition: *The directors voted for an amalgamation of the two companies that would benefit both.*

amass *v.* °accumulate, °mass, pile *or* heap *or* rack up, °collect, °gather (together), °assemble, aggregate, cumulate, stock *or* store up, °hoard, °set aside: *How many points have you amassed? Owing to the bountiful harvest, the farmers amassed huge amounts of grain.*

amateur *n.* **1** layman, nonprofessional, lay person, tyro; dabbler, °dilettante, bungler; *Colloq US* bush leaguer: *When it comes to repairing cars, I'm a mere amateur.* —*adj.* **2** °lay, nonprofessional, untrained; °unpaid; °dilettante, amateurish, °unprofessional, unskilled, inexpert, unskillful, °clumsy, °mediocre, °inferior, °crude, bungling, second-rate; *Colloq US* bush league: *The amateur theater group's performance received excellent reviews. These paintings are strictly amateur and totally without merit.*

amaze *v.* °astound, °astonish, °surprise, awe, °stun, °stagger, °take aback, °floor, dumfound *or* dumbfound, confound, °nonplus, stupefy, *Colloq* flabbergast, °dazzle: *Annie Oakley amazed audiences with her fancy shooting. I was amazed that she still cared for me.*

amazement *n.* °astonishment, °surprise, awe, °wonder, stupefaction: *We gazed in amazement as the table rose by itself into the air.*

amazing *adj.* astonishing, astounding, surprising, wonderful, °remarkable, °extraordinary, °marvelous, °fabulous, °stunning, °dazzling, staggering, °awesome: *The Cossacks put on an amazing display of horsemanship.*

ambassador *n.* °envoy, °delegate, legate, emissary, °minister, plenipotentiary, diplomat; °agent, °deputy, °representative, (papal) nuncio, °messenger: *The ambassador must present his credentials to the queen.*

ambiguity *n.* **1** equivocalness, equivocacy, amphibology *or* amphiboly; vagueness, indistinctness, uncertainty, indefiniteness, imprecision, inconclusiveness: *Ambiguity of language must be avoided in writing laws.* **2** equivocation, double-talk, doublespeak, equivoque; °pun, double entendre, amphibologism: *The minister's speech was filled with ambiguities.*

ambiguous *adj.* **1** °equivocal, amphibological, amphibolic *or* amphibolous; misleading: *If one says "Taylor saw Tyler drunk," which one was drunk is ambiguous.* **2** °doubtful, dubious, °questionable, °obscure, °indistinct, unclear, °indefinite, indeterminate, °uncertain, undefined, °inconclusive, °uncertain, °vague, °misty, foggy, unclear; °cryptic, delphic, enigmatic(al), oracular, °mysterious, °puzzling; confusable *or* confusible: *The soothsayer's prophecies were sufficiently ambiguous to allow for several conflicting interpretations.* **3** °unreliable, undependable: *How can the doctor decide on a correct diagnosis when the symptoms are ambiguous?*

ambition *n.* **1** °hunger, °thirst, craving, °appetite, arrivisme: *Herbert's relentless ambition may yet be his undoing.* **2** °drive, °enterprise, °energy, °initiative, °vigor, °enthusiasm, zeal, avidity, *Colloq* get-up-and-go: *The company is seeking young men of ambition. You'll never get anywhere, since you are totally lacking in ambition.* **3** °goal, °object, °aim, °aspiration, °hope, °desire, °dream, °objective, °wish, °purpose: *It is Ophelia's ambition to marry someone with a title.*

ambitious *adj.* **1** aspiring, °hopeful; °enthusiastic: *My son, I am just as ambitious for you as you are for yourself.* **2** °energetic, °enterprising, °vigorous, zealous, °enthusiastic, °eager: *We prefer ambitious young people to those who are seeking a sinecure.* **3** °greedy, °avaricious, overzealous, overambitious, *Colloq* °pushy, yuppie: *Fred may be a trifle too ambitious, expecting to be department head after only one year.*

ambush *n.* **1** °trap, ambuscade *or Archaic* ambuscado: *The company set up an ambush near the crossroads.* —*v.* **2** lie in wait, °trap, °waylay, ensnare, entrap, °lurk, ambuscade, °intercept, *Colloq* lay in wait, *US* bushwhack: *The guerrillas were ready to ambush the soldiers.*

amend *v.* **1** °reform, change for the better, °improve, °better, ameliorate: *The prisoner believes he could amend his ways if given the chance.* **2** °correct, emend, emendate, °rectify, set to rights, °repair, °fix, °revise: *Take whatever time you need to amend the text.*

amendment *n.* **1** °correction, emendation, reformation, °change, °alteration, rectification, °repair, °reform, °improvement, amelioration, betterment, enhancement: *The committee approved the amendment of the constitution by the addition of the suggested paragraphs.* **2** °attachment, °addition, addendum; clause, paragraph; °alteration: *A two-thirds majority in the Congress is needed to pass the amendment.*

amends *n.* ***make amends.*** °compensate, °pay, °repay, make reparation *or* restitution, recompense, redress, °remedy, °requite: *How can the bus driver ever make amends for the loss of a beloved kitten?*

amiable *adj.* °friendly, well-disposed, °kindly, °kind, °amicable, °agreeable, congenial, °genial, °warm, winsome, °winning, affable, °agreeable, °pleasant, °obliging, °tractable, approachable, °benign, °good-natured, good-hearted, kindhearted; °affectionate: *Melissa is well named for her sweet and amiable disposition.*

amicable *adj.* °friendly, °amiable, congenial, °harmonious, °brotherly, kindhearted; °warm, °courteous, °cordial, °polite, °civil, °pleasant; °peaceful, °peaceable: *Our countries have always enjoyed the most amicable relations.*

amid *prep.* mid, in *or* into the middle *or* midst *or* center of, amongst, °among, surrounded by, in the thick of, *Literary* amidst: *She is sitting in her cottage, / Amid the flowers of May. Without further ado, she plunged amid the waves.*

amiss *adj.* **1** °wrong, °at fault, awry, °out of order, °faulty, °defective, °improper, °untoward; astray, °erroneous, fallacious, °confused, °incorrect, °off: *Something is amiss with the ignition. If I am amiss in my thinking, let me know.* —*adv.* **2** °wrong, awry, °badly, °poorly, imperfectly; inopportunely, unfavorably, unpropitiously: *Everything possible has already gone amiss with the rocket launch.* **3** wrongly, improperly, °badly; incorrectly, inappropriately: *A word of advice might not come amiss here.* **4** ***take*** **or *think* (*it*) *amiss.*** °mistake, °misinterpret, °misunderstand, °take offense (at): *I trust that you will not take amiss what I intended as constructive criticism.*

among *prep.* **1** amongst, °amid, amidst, mid, in the midst *or* middle *or* center of, surrounded by: *Please take a seat among the people over there. We lay down among the flowers.* **2** amongst, to each *or* all (of): *The examination booklets were passed out among the students.*

amount *v.* **1** ***amount to.*** **(a)** add up to, °total, aggregate, come (up) to: *Waiter, what does my bill amount to, please?* **(b)** °become, develop into: *That son of his will never amount to much.* —*n.* **2** °quantity, °volume, °mass, °expanse, °bulk, °supply, °lot; °number; °magnitude: *What amount of water is needed to fill the container? She eats a huge amount of chocolates every day.* **3** °(sum) total, aggregate, °extent, °entirety: *What is the amount of the invoice without the tax?*

ample *adj.* **1** °broad, °wide, °spacious, °extensive, °expansive, °great: *Her ample bosom heaved with sobs.* **2** wide-ranging, °extensive, °broad: *In one ample swoop they snatched up all the land.* **3** °abundant, °extensive,

°fruitful: *The event proved a very ample subject for history.* **4** °abundant, °full, °complete, °plentiful, copious, °generous, °substantial; °sufficient, °adequate, °enough: *He had stored ample provision of food for the winter.* **5** °liberal, unsparing, unstinted, unstinting, °generous, °substantial, °large, °lavish: *Barking's contributions have always been ample, especially at Christmastime.* **6** copious, °full, °broad, °detailed, °extensive, extended, °thorough: *The subject deserves more ample treatment.*

amplify *v.* **1** broaden, °widen, °extend, °increase, °expand (on), °enlarge (on), expatiate on, °detail; °add to, augment, °supplement: *Let no man comfort him / But amplify his grief with bitter words.* **2** °exaggerate, °overstate, °magnify, °stretch: *The reports have amplified the number of horsemen slain in the encounter.* **3** °enlarge (on), °elaborate (on), °stretch, °lengthen, °detail, °embellish, embroider: *He amplifies every point in microscopic detail.*

amply *adv.* **1** °widely, broadly, extensively, greatly, expansively: *This fabric stretches amply enough to fit over the couch.* **2** to a great extent, °largely, fully, abundantly: *My confidence in her was amply recompensed by her success.* **3** abundantly, fully, copiously: *The prophecy was amply fulfilled.* **4** fully, °well, liberally, unstintingly, generously, °richly, °substantially, lavishly; sufficiently: *He has been amply paid for his work.*

amulet *n.* °charm, °talisman, good-luck piece; °fetish: *Whenever the man rubbed the silver amulet, his number would win.*

amuse *v.* **1** °divert, °entertain, please, °beguile, °interest, °occupy: *Perhaps the crossword puzzle will amuse her while she is waiting.* **2** make laugh, °delight, °cheer, *Colloq* °tickle: *That form of rowdy slapstick doesn't amuse me.*

amusement *n.* **1** °entertainment, °diversion, °recreation, °pleasure, °relaxation, °distraction, °enjoyment, °fun, °sport, °joke, °lark, beguilement: *No, I do not regard Coney Island as the amusement capital of the world. Lampwick used to pull the wings off flies for amusement.* **2** °entertainment, °diversion, divertissement, °recreation, °distraction, °pastime; °game, °sport: *During the festival there are concerts, plays, and other amusements.*

anachronism *n.* misdate, misdating, misapplication; antedate, antedating, prochronism; postdate, postdating, parachronism: *The poster showing Cleopatra smoking a cigarette is an anachronism—a prochronism, to be specific.*

analysis *n.* **1** °examination, °investigation, °study, °scrutiny, enquiry *or* inquiry, dissection, assay, °breakdown, °division: *The analysis showed the presence of arsenic in her soup.* **2** °interpretation, °opinion, °judgment, °criticism, critique; °review: *She disagrees with our analysis of the poem.*

analyze *v.* **1** take apart *or* to pieces, °separate, dissect, °break down, anatomize: *If we analyze these figures for England and Wales, we find no pattern. The scientists are analyzing the internal constitution of a glacier.* **2** °examine, °investigate, °study, °scrutinize, °interpret; assess, °evaluate, critique, °criticize, °review; °estimate, assay, °test: *We must first explicitly define and analyze the nature of the sample we found.*

ancestor *n.* forebear, forefather; °forerunner, °precursor, antecedent, *Formal* °progenitor, primogenitor: *His ancestors were transported to Australia in a prison ship, and he's proud of it. The eohippus, only a foot high, was the ancestor of the horse.*

anchor *n.* **1** mooring: *The ship rode at anchor in the harbor.* **2** °stability, °security, °mainstay, °support, stabilizer, holdfast, sheet anchor: *Marie is an anchor to windward for George, who tends to be a bit irresponsible.* —*v.* **3** °attach, affix, °secure, °moor, °fix, °fasten; °pin, rivet, °glue: *You must anchor the foundation before adding the walls. She remained anchored to the spot, refusing to move.*

ancient *adj.* **1** °old, °bygone, °past, °former, earlier, *Literary* olden: *In ancient times there were very few books.* **2** °old, °antique, antediluvian, °primitive, °prehistoric, primeval, primordial, Noachian, *Literary* Ogygian: *In those ancient days man had only just come down from the trees.* **3** °old, °old-fashioned, archaic, °timeworn, °aged, aging, °obsolescent, °antiquated, °elderly, °venerable, °gray, hoary, superannuated, °obsolete, fossil, fossilized: *We were accosted by an ancient crone at the mouth of the cave.*

anger *n.* **1** °rage, wrath, ire, °fury, pique, spleen, choler; °antagonism, irritation, vexation, °indignation, °displeasure, °annoyance, irritability, °resentment, °outrage: *Her anger got the better of her, so she simply punched him.* —*v.* **2** °enrage, °infuriate, °madden, pique, incense, raise one's hackles, make one's blood boil, rile, °gall; °annoy, °irritate, vex, nettle, °displease, °exasperate, °provoke: *Father was so angered by the insult that he refused to pay.*

angle[1] *n.* **1** °slant, °oblique, corner, °edge, intersection; °bend, cusp, °point, apex, °projection: *The two walls meet at an angle.* **2** °slant, °point of view, °aspect, °viewpoint, °standpoint, °approach, °position, °side, °perspective: *The managing editor told me he's looking for a new angle on the kidnapping story.*

angle[2] *v.* ***angle for.*** fish for; °look for, °seek, be after, try for, °hunt for: *On our vacation we went angling for perch. Fran is angling for compliments on her new dress.*

angry *adj.* **1** enraged, °furious, irate, °resentful, ireful, wrathful, piqued, incensed, infuriated, fuming; irritated, °irritable, annoyed, vexed, irascible, provoked, °indignant, exasperated, splenetic, *Literary* wroth, *Colloq* livid, hot under the collar, on the warpath, (all) steamed up, up in arms, °mad: *Father was angry with me for letting the cat out.* **2** °inflamed, irritated, °sore, smarting: *He has an angry lesion where the fetters rubbed against his ankles.*

anguish *n.* **1** °suffering, °pain, °agony, °torment, torture, °misery: *She endured the anguish of toothache rather than go to the dentist.* **2** °suffering, °grief, °distress, °woe, °anxiety: *He underwent terrible anguish in the waiting room till the surgeon arrived.* —*v.* **3** °disturb, °upset, °distress, °afflict, °trouble; °torment, torture: *The anguished cries of prisoners could be heard.*

animal *n.* **1** °creature, being, mammal, °organism: *Scientists are unlikely to employ the popular division of all things into animal, vegetable, or mineral.* **2** °beast, °brute, °savage, °monster: *Think of the poor girl married to that animal!* —*adj.* **3** zoological, zooid, animalistic: *The sponge is a member of the animal kingdom.* **4** °physical, fleshly, °sensual, °gross, °coarse, °unrefined, uncultured, uncultivated, °rude, °carnal, °crude, bestial, beastlike, subhuman: *His animal appetites occasionally got the better of him.*

animate *v.* **1** °activate, °enliven, invigorate, °stimulate, inspirit, °excite, °stir, °vitalize, °spark, vivify, revitalize, breathe life into, innervate: *A little enthusiasm would have animated their dull relationship.* **2** °inspire, inspirit, °stimulate, actuate, °move, °motivate, °incite, °rouse, °arouse, °excite, °fire (up), °encourage, °energize, °vitalize, spur (on *or* onward(s)): *He spent the few minutes before the battle in animating his soldiers.* —*adj.* **3** °lively, °spirited, °vivacious, °animated, °quick: *A courser more animate of eye, / Of form more faultless never had been seen.* **4** °alive, °moving, breathing, *Archaic* °quick: *Although they move, plants are not considered to be animate.*

animated *adj.* **1** °lively, °quick, °spirited, °active, °vivacious, °energetic, °vigorous, °excited, °ebullient, °enthusiastic, °dynamic, vibrant, °ardent, enlivened, °passionate, °impassioned, °fervent: *In the corner, Trembly was engaged in an animated conversation with Minerva.* **2** °mechanical, automated, °lifelike, °moving: *Each Christmas, the shop has an animated window display.*

animation *n.* **1** °spirit, spiritedness, °vitality, °dash, élan, °zest, °fervor, °verve, liveliness, °fire, °ardor, ardency, exhilaration, °intensity, °energy, °pep, °dynamism, °enthusiasm, °excitement, °vigor, vivacity: *Johnson was in high spirits, talking with great animation.* **2** enlivenment, liveliness, energizing, invigoration, enlivening, innervation: *The scout leader was credited with the animation of the youths in his care.*

animosity *n.* °hostility, °antagonism, antipathy, °ill will, malevolence, enmity, hatred, animus, °loathing, detestation, °contempt; bad blood, malice, °bitterness, acrimony, °resentment, °rancor: *The animosity he felt for his brother soon disappeared.*

announce *v.* **1** °proclaim, °make public, °make known, set *or* put forth, °put out, °publish, advertise, °publicize, promulgate, °broadcast, herald; °circulate; °tell, °reveal, °disclose, divulge, °declare, °propound: *The appointment of a new prime minister has been announced in Graustark.* **2** °intimate, °suggest, hint at, °signal: *The sight of a gamboge top hat announced Niblett's presence in the club.* **3** °declare, °tell, °state, aver, assert, asseverate; °notify; °confirm: *The president announced that he was resigning because of the scandal.* **4** foretell, betoken, augur, portend, presage, °harbinger, herald, °signal; °precede: *The sighting of the first crocus announces spring.*

announcement *n.* **1** °declaration, °pronouncement, °proclamation, °statement: *Ladies and gentlemen, I wish to make an announcement.* **2** notification, °notice, °word: *We received an announcement of the wedding but no invitation.* **3** commercial, °advertisement, advert, ad, °spot: *The window was filled with announcements of houses for sale.* **4** °report, °bulletin, communiqué, disclosure: *An announcement has just been received from the firefighters at the scene.*

announcer *n.* presenter, master of ceremonies, MC, emcee, *Literary or Brit hist* Master of the Revels; °reporter, anchorman, anchorwoman, anchorperson, °anchor; newsreader, newscaster, sportscaster, weatherman, weathergirl: *The announcer didn't get my name right.*

annoy *v.* **1** °irritate, °bother, °irk, vex, nettle, get on (someone's) nerves, °exasperate, °provoke, incense, rile, °madden, *Colloq* °get at: *The anonymous telephone calls were beginning to annoy us.* **2** °pester, °harass, harry, badger, °nag, °plague, °molest, bedevil, *Colloq* °bug, needle, hassle, *Slang* get up someone's nose: *Stop annoying me with your persistent requests for money.*

annoyance *n.* **1** irritation, °bother, vexation, exasperation, pique, aggravation, *Colloq* botheration: *Must I put up with the annoyance of that constant bickering?* **2** °nuisance, °pest, irritant, °bore, *Colloq* °pain, pain in the neck *or US* ass *or Brit* arse: *He's such an annoyance, I wish he'd leave.*

answer *n.* **1** °reply, °response; rejoinder, °retort, riposte, *Colloq* comeback: *The boy's answer is unprintable.* **2** *Law* °defense, counter-statement, °plea, °explanation; *Technical* °declaration, °plea, replication, rejoinder, surrejoinder, rebutter *or* rebuttal, surrebutter *or* surrebuttal: *Her answer to the charge was "Not Guilty."* **3** °solution, °explanation: *Ten points were taken off because I had the wrong answer to question three.* —*v.* **4** °reply, °respond; °retort, rejoin, riposte: *When I ask you a question, I expect you to answer.* **5** °satisfy, °fulfill, suffice for, °meet, °suit, °serve, °fit, °fill, conform to, correlate with: *The bequest answered my needs for the moment.* **6** ***answer back.*** talk back (to): *How dare you answer back to your father!* **7** ***answer for.*** **(a)** be accountable *or* responsible *or* answerable for, be to blame for; take *or* undertake responsibility for; °sponsor, °support, °guarantee: *I answer alone to Allah for my motives. So shall my righteousness answer for me.* **(b)** make amends for, atone for, suffer the consequences of: *Caesar was ambitious and he answered for it with his life.* **(c)** take *or* accept the blame for: *Ferdinand shouldn't have to answer for his brother's shortcomings.*

antagonism *n.* **1** °opposition, °animosity, enmity, °rancor, °hostility, antipathy: *It is difficult to understand your antagonism toward classical music.* **2** °conflict, °rivalry, °discord, °dissension, °friction, °strife; contention: *Giving jobs only to personal friends has engendered antagonism.*

antagonist *n.* °adversary, °opponent, °enemy, foe; contender, °competitor, °competition, °opposition: *The antagonists squared off against one another.*

anticipate *v.* **1** °forestall, °intercept, °preclude, obviate, °prevent; nullify: *She anticipated her opponent's maneuver by moving the queen's bishop one square.* **2** foretell, °forecast, °predict, °prophesy, foretaste, °foresee: *He anticipated that flying would be a future mode of locomotion.* **3** °expect, °look forward to, prepare for; count *or* reckon on: *We eagerly anticipated the arrival of Uncle Robert.*

anticipation *n.* **1** °expectation, expectancy; °hope: *In anticipation of the arrival of Santa Claus, we hung up our stockings.* **2** foreknowledge, precognition; °intuition, presentiment, °feeling; °foreboding, apprehension: *His anticipation of the solar eclipse by a week established him as the foremost scientist of his day.*

antidote *n.* antitoxin, antiserum, antivenin; counteractant, counterirritant; °cure, °remedy, °specific; medication, °medicine, °drug, medicament, *Technical* alexipharmic: *The old prospector says that the best antidote against snakebite is whiskey.*

antiquated *adj.* °old, °old-fashioned, outmoded, °passé, out-of-date, dated, archaic, °obsolescent, °antique, °obsolete, °quaint, °ancient, antediluvian, mediaeval *or* medieval, °primitive; °extinct; *Colloq* old hat: *Antiquated laws list penalties for practicing witchcraft.*

antique *adj.* **1** °old, °old-fashioned; °antiquated, outmoded, °passé, out-of-date, °obsolete: *She wore the antique clothing she had found in the trunk.* —*n.* **2** collectable *or* collectible, collector's item, bibelot, objet d'art, objet de vertu, object *or* article of vertu, heirloom, curio, °rarity: *His hobby is collecting antiques.*

anxiety *n.* **1** °solicitude, °concern, uneasiness, disquiet, nervousness, °worry, °dread, angst, apprehension, °foreboding: *Philip began to feel genuine anxiety over Tanya's safety.* **2** °appetite, °hunger, °thirst, °desire, °eagerness, °longing, °ache, °concern: *It is every person's anxiety to obtain for himself the inestimable pearl of genuine knowledge.*

anxious *adj.* **1** troubled, uneasy, disquieted, °uncertain, apprehensive; °solicitous, °concerned, °worried, distressed, °disturbed, °nervous, °tense, °fretful, °on edge, °restless, edgy, perturbed, °upset; °wary, °cautious, °careful, watchful: *She has been terribly anxious about the diagnosis. We were anxious for her safety.* **2** °desirous, °eager, °keen, °enthusiastic, °ardent, °agog, avid, yearning, °longing, aching, °impatient: *I was anxious to visit the Pitti Palace once again.*

apart *adv.* **1** aside, to one side, by oneself, at a distance, °separate, °separately: *He stood apart when the awards were given out.* **2** °separately, distinctly, °individually, °singly, °alone, independently: *The provisions of the bill should be seen together as a whole, not viewed apart.* **3** to *or* into pieces, asunder: *At the touch of the button, the building blew apart.* **4** ***apart from.*** °except for, excepting, separately from, aside from, °besides, but for, not including, excluding, not counting: *Apart from the immediate family, no one knows of your indiscretions.*

aperture *n.* °opening, °space, °gap, cleft, °chink, °crevice, °crack, fissure, °hole, chasm: *As much water ran through as the aperture could accommodate.*

apologetic *adj.* °regretful, °sorry, contrite, °remorseful, °penitent, rueful, °repentant, conscience-stricken: *The lad was most apologetic for having run his car into my wheelbarrow.*

apologize *v.* **1** beg *or* ask pardon, express regret(s), feel sorry *or* regretful *or* remorse(ful): *You needn't apologize for sneezing.* **2** make *or* give excuses *or* explanation(s), °defend, °justify, °vindicate, espouse: *You don't have to apologize for her.*

appall *v.* °dismay, °shock, °discomfit, °unnerve, °intimidate, °terrify, °frighten, °scare, °horrify, °alarm, °startle, °daunt: *The council was appalled to discover that the police superintendent was accepting bribes.*

apparatus *n.* °equipment, requisites, °tool, °instrument, utensil, °device, °implement, °machine, machinery, °gear, °paraphernalia, °tackle, °outfit; appliance, *Colloq* °contraption, gadgetry, °gadget: *The apparatus needed for the experiment is here.*

apparel *n.* clothing, attire, °clothes, °dress, raiment, °garments, *Colloq* °gear, °rags, glad rags, duds, *Slang US* threads: *The police found various items of apparel strewn about the apartment.*

apparent *adj.* **1** °evident, °plain, °clear, °obvious, °patent, unmistakable; °conspicuous, °marked, °manifest, °visible, °discernible: *It was apparent to all of us that she would become a successful opera singer.* **2** appearing, °seeming, °illusory, ostensible, °superficial, °outward: *In an apparent show of strength, he ordered his forces to attack the capital.*

apparently *adv.* **1** °evidently, plainly, °clearly, °obviously, patently, °manifestly: *There is apparently no cure in sight for the disease.* **2** °seemingly, °ostensibly, superficially, °outwardly: *In stop-action photography, the bullet apparently hangs in midair.*

appeal *v.* **1** entreat, supplicate, °solicit, °plead, °petition, °apply, °sue; °beseech, °beg, implore, °pray: *She appealed to the king to release her son from the dungeon.* **2** °attract, be attractive to, allure, please; invite, °tempt, °beguile, °fascinate, °interest: *He seems to appeal to older women.* —*n.* **3** °application, °suit; entreaty, °call, °request, °supplication, solicitation, °petition, °plea; °prayer: *Her appeal to the court has been dismissed. I don't know if God heard our appeal.* **4** °attraction, °lure, allurement, °charm, °fascination: *It is not hard to see why his type would have some appeal.*

appear *v.* **1** come forth, become visible *or* manifest, put in an appearance, °materialize, °surface, °emerge, °rise, °arise, °come up, enter (into) the picture, show oneself, °turn up, °arrive, °come, *Colloq* crop *or Colloq* show up; *Slang* °show: *Suddenly, a vision appeared before me. His wife appeared after an absence of ten years.* **2** °perform, °act, °play, take the role *or* part of: *She has appeared as Roxanne in dozens of productions of* Cyrano de Bergerac. **3** °occur, °happen, °come up, be included, °figure, °arrive: *That four-letter word does not appear in written form till the twentieth century.* **4** °seem, be clear *or* evident *or* plain *or* manifest; °look: *It appears that the money was taken while the manager was at lunch.* **5** be published, °come out, become available: *The next issue will appear in March.*

appearance *n.* **1** °arrival, advent; °presence; °publication: *I was awaiting the appearance of the book in the shops.* **2** °aspect, look(s), °form; mien, °air, demeanor; °bearing, °manner: *The doorman would not let him in because of his shabby appearance.* **3** °display, °show: *Their fine horses with their rich trappings made a splendid appearance.* **4** °semblance, °show, °hint, °suggestion; °illusion: *She disdained the slightest appearance of coveting recognition for her contribution.*

appetite *n.* **1** °desire, °inclination, proclivity, °tendency, °disposition, °bent, °preference, °liking, predilection, °zest, fondness, °love, zeal; °enthusiasm; °taste, °relish; *Formal* appetency, appetence: *I have never lost my appetite for chocolate. They tried to suppress their bodily appetites, such as hunger and lust.* **2** craving, °hunger, °thirst, °desire, keenness, hankering, yearning, °longing, °passion, °demand, *Formal* edacity: *She developed an insatiable appetite for reading.*

applaud *v.* **1** °approve, express approval, °clap, °cheer, give (someone) a hand, *Colloq* °root (for): *The audience applauded when the villain was caught.* **2** express approval of, °praise, °laud, °hail, commend: *Susan's parents applaud her decision to go to college.*

applause *n.* clapping, acclamation, acclaim, éclat; cheering, cheers; °approval, commendation, approbation, °praise, °kudos, plaudit(s): *At the curtain there was applause from the audience.*

applicable *adj.* °fit, °suitable, suited, °appropriate, °proper, apropos, °fitting, °befitting, °pertinent, apt, germane, °right, °seemly, °relevant, apposite: *Are the laws of the mainland applicable to the islands?*

application *n.* **1** °use, °employment, utilization, °practice, °operation: *The committee wants to see a sterner application of the law with respect to mail-order offers.* **2** relevancy, °relevance, °reference, pertinence, germaneness, appositeness; °bearing: *The application of the regulation to present circumstances is somewhat vague.* **3** °attention; diligence, industriousness, °effort, °perseverance, °persistence, assiduity, °devotion, °dedication, commitment, attentiveness, *Colloq* stick-to-itiveness; °industry: *Her application to her studies leaves little time for recreation. Without application, you will never develop much skill at the piano.* **4** °request, solicitation; °appeal, °petition, °claim: *Grover sent out six job applications. The board will consider your application.*

apply *v.* **1** °fasten, °fix, affix, °stick, °cement, °glue: *The signs were applied to the window with a special substance.* **2** °administer, rub in *or* on, embrocate: *The doctor said to apply this ointment before retiring.* **3** °appropriate, °assign, °allot, °credit; °use, utilize, °employ, put to use: *The money raised for food was illegally applied to paying the administrators.* **4** °bear, have bearing; be relevant, °refer, °pertain, appertain, °relate, °suit: *I am not sure that the law applies to this situation.* **5** °devote, °dedicate, °commit, °focus, °concentrate, pay attention, °address; do, °attend, °tend, *Colloq* buckle down (to): *He stubbornly applies himself to the task at hand.* **6** °seek, go after; °register, °bid, try out, °put in; audition, °interview, make application: *Are you qualified to apply for a job as a nanny?* **7** °petition, °solicit; °appeal, °request: *Geraldine applied to the court for compensation.*

appoint *v.* **1** °fix, °set, °settle, °determine, ordain, °authorize, °establish, °destine, °arrange, °assign, °allot, °prescribe, °decree: *The time appointed for the execu-*

tion has been delayed. **2** °name, °designate, °nominate, °elect; °assign, °delegate, commission, deputize; °select, °choose: *I was delighted to have been appointed as chairman.* **3** °equip, fit out, °furnish, °decorate: *They live comfortably in a well-appointed home in the suburbs.*

appointment *n.* **1** °meeting, °date, rendezvous, °engagement; assignation, tryst: *You are again late for your appointment.* **2** nomination, °election; °assignment, designation; °selection, °choice: *We fully approve of his appointment as dogcatcher.* **3** °job, °position, °post, °situation, °office, °place, °assignment, *Colloq* berth, °slot: *He got the appointment as manager.*

appreciate *v.* **1** value, find worthwhile *or* valuable; °esteem, °cherish, °enjoy, °admire, rate *or* regard highly, °prize, °treasure, °respect: *I appreciate all you have done for me. Bosley's contribution is not really appreciated.* **2** increase *or* rise *or* gain in value *or* worth: *The property in this area has been appreciating at a rate of about 10 percent a year.* **3** °understand, °comprehend, °recognize, °perceive, °know, be aware *or* cognizant *or* conscious of: *Do you appreciate the implications of the new tax law?*

appreciation *n.* **1** °gratitude, thankfulness, gratefulness, °thanks; °acknowledgment; °obligation: *She is trying to think of an appropriate way to express her appreciation for all he has done.* **2** °increase, °rise, °advance, °growth, enhancement, °gain; aggrandizement: *The appreciation in the value of the shares made me very wealthy on paper—till the stock-market crash.* **3** °understanding, comprehension, °perception, °recognition, °knowledge, awareness; °realization, °enjoyment; °admiration: *It's fortunate that Richard's appreciation of the finer things in life is supported by his income.*

apprentice *n.* **1** °novice, tyro, °learner, starter, beginner, °greenhorn, *Colloq US* rookie: *Lever served as an apprentice in the soap factory.* —*v.* **2** indenture, °contract, °bind: *Cartwright was apprenticed to a carpenter before becoming a journeyman cabinetmaker.*

approach *v.* **1** °near, °advance, draw *or* come near *or* nearer *or* close *or* closer, *Formal* come nigh: *Claude approached the table. As night approached, the sky darkened. With approaching manhood, you must take on more responsibilities.* **2** °approximate, nearly equal, come close to, °compare with: *The total is beginning to approach your estimate.* **3** make advances *or* overtures to, °proposition, propose to, °sound out, make (a) proposal to, °solicit, *Colloq* °chat up: *Thewless makes mincemeat of any man who tries to approach his daughter.* —*n.* **4** ***approaches.*** advances, overtures, proposals, propositions: *Michelle had no intention of discouraging Pierre's approaches.* **5** access, °passage, °way, °path, °course; °entry: *The approach to the house was overgrown with brambles.* **6** °advance, °movement: *Our approach to the gates was being watched very carefully.* **7** °method, °procedure, modus operandi, °way, °technique, °style, °manner, °attitude, *Slang US* MO (= 'modus operandi'): *Our approach in dealing with the problem is different.*

appropriate *adj.* **1** °suitable, apt, °fitting, °fit, °proper, °right, °meet, °becoming, °befitting, °seemly, suited, apropos, °correct, germane, °pertinent, °happy, felicitous: *Will a dinner jacket be appropriate attire? She has written a poem appropriate to the occasion.* —*v.* **2** °take, °take over, °seize, expropriate, arrogate, annex, impound; commandeer; °steal, °pilfer, filch, usurp, make away *or* off with, *Colloq* °pinch, °lift, *Brit* °nick, *US* °boost: *The police appropriated the paintings. Somebody has appropriated my dictionary without permission.* **3** set aside *or* apart, °devote, °assign, earmark, °allot, apportion: *Most of the money has been appropriated for back taxes.*

appropriately *adv.* fittingly, suitably, °properly, correctly, aptly, rightly, becomingly, meetly: *She came down appropriately dressed for dinner.*

approval *n.* °sanction, approbation, °blessing, °consent, °agreement, concurrence; °endorsement, acceptance, imprimatur, affirmation, éclat, confirmation, mandate, authorization; °license, °leave, °permission, rubber stamp, *Colloq* OK *or* okay, °go-ahead, green light: *I don't think that the plan will meet with the committee's approval. We gave our approval to proceed.*

approve *v.* **1** Often, ***approve of.*** °allow, countenance, condone, °permit, °sanction, °authorize, °endorse, put one's imprimatur on, °agree (to), °accept, assent (to), °go along with, *Colloq* OK *or* O.K. *or* okay, give the green light *or* go-ahead *or* one's blessing (to), rubber-stamp: *The headmistress would never approve your leaving the building during classes.* **2** °confirm, affirm, °support, °ratify, °uphold, °subscribe to, °second, give the stamp of approval to; °favor, commend, °recommend: *Sheridan's appointment to the commission has been approved unanimously.* **3** ***approve of.*** °sanction, consider fair *or* good *or* right, °accept, °favor, °respect, be partial to, °like, have regard for, have a preference for, °tolerate, reconcile oneself to: *I always had the feeling that her father didn't quite approve of me.*

approximate *adj.* **1** °rough, °inexact, °loose, °imprecise, estimated, *Colloq* guesstimated, ballpark: *The figures are only approximate, not exact.* —*v.* **2** °near, °approach, come close to, °verge on: *Your estimates approximate those of the budget committee.* **3** °resemble, °approach, look *or* seem like; simulate: *The laboratory tests on rats approximate the way the virus behaves in humans.*

approximately *adv.* **1** approaching; °nearly, °almost, °close to, °about, °around, give or take, °roughly, °generally: *I haven't seen Sally for approximately three weeks. There are approximately fifty people in the audience.*

aptitude *n.* **1** °fitness, suitability, appropriateness, °relevance, applicability, suitableness, aptness: *One need only look at an albatross in the air to appreciate its aptitude for flight.* **2** °tendency, propensity, °disposition, predilection, °bent, proclivity; °talent, °gift, °ability, °capability, °facility, °faculty, °flair: *Griselda displays a natural aptitude for the violin.* **3** °intelligence, quick-wittedness, °intellect; °capacity, aptness: *The aptitude of that new student sets her apart from the others in the class.*

arbitrary *adj.* **1** °capricious, varying, °erratic, °uncertain, °inconsistent, °doubtful, unpredictable, °whimsical, irrational, °chance, °random, °inconsistent, °subjective, unreasoned, irrational, *Colloq* chancy, iffy: *The choices are entirely arbitrary, totally at the whim of the council and not based on research or knowledge.* **2** °absolute, °tyrannical, °despotic, °authoritarian, magisterial, °summary, °peremptory, autocratic, °dogmatic, imperious, uncompromising, °inconsiderate, highhanded, °dictatorial, *Rare* thetic(al): *The conduct of the archbishop appears to have been arbitrary and harsh.*

arch *adj.* **1** °chief, °principal, °prime, °primary, °preeminent, °foremost, °first, greatest, consummate, °major: *Moriarty was Holmes's archenemy.* **2** °clever, cunning, °crafty, roguish, °tricky, °shrewd, °artful, °sly, °designing: *Brendan loves to play his arch pranks on unsuspecting friends.* **3** waggish, saucy, °mischievous, prankish: *I could tell from the boy's arch expression that he had thrown the snowball.*

ardent *adj.* °eager, °intense, zealous, °keen, °fervent, fervid, °passionate, avid, °fierce, °impassioned, °hot,

°warm; °enthusiastic: *Diane was carrying on ardent love affairs with at least three men.*

ardor *n.* °eagerness, °desire, zeal, fervency, burning desire, keenness, °fervor, °passion, °heat, °warmth; °enthusiasm: *Bernadette has supported the cause with great ardor.*

arduous *adj.* **1** °laborious, °difficult, °hard, °tough, °strenuous, onerous, °burdensome, backbreaking, °painful, *Formal* operose; tiring, °exhausting, wearisome, fatiguing, taxing, grueling, °trying, °formidable: *The Sherpas were well equipped for the arduous climb. What an arduous task it is to read the proofs of a dictionary!* **2** °energetic, °strenuous, °vigorous: *Montrose made arduous efforts to reconstruct his army.*

area *n.* **1** °space, °room: *Is there enough floor area here for the carpet?* **2** °extent, °limit, compass, °size, square-footage, acreage: *The area of my greenhouse is thirty by fifteen feet.* **3** °space, °field, °region, °tract, °territory, °district, °zone, °stretch; °section, °quarter, °precinct, arrondissement, °neighborhood, locality, bailiwick, *US* °block: *An area was set aside for a garden. There has been a lot of crime in that area lately.* **4** °scope, °range, °extent, °breadth, compass, °section: *His studies cover only one area of Scottish history.* **5** court, courtyard, °enclosure, °close, yard; °square, °ground, arena, °field, parade ground, °parade: *The soldiers drill in the area behind the barracks.*

argue *v.* **1** °dispute, °debate, °disagree, °bicker, wrangle, °quarrel, squabble, °spar, °fight, remonstrate, altercate, *Colloq chiefly Brit* °row, °scrap: *The couple next door are continually arguing with each other at the tops of their lungs.* **2** °discuss, °reason, °debate, wrangle: *He would argue by the hour, but never for arguing's sake.* **3** make a case, °talk, °plead, °debate, contend: *I cannot tell whether she's arguing for or against the proposition.* **4** °prove, evince, °indicate, °denote, °demonstrate, °show, °establish, °suggest, °signify, betoken: *The increase in street crime argues that the police are not visible enough.* **5** °say, assert, °hold, °maintain, °reason, °claim, contend: *The defendant argues that he has never met the witness.* **6** ***argue into** or **out of.*** persuade *or* dissuade, talk out of *or* into, °prevail upon; °convince: *I argued him out of sailing to Bermuda alone. She succeeded in arguing me into going to the tea dance.*

argument *n.* **1** °debate, °dispute, °disagreement, °quarrel, °controversy, polemic, wrangle, squabble, °tiff, spat, altercation; °conflict, °fight, °fracas, affray, °fray, donnybrook, °feud, *Colloq* °row, falling-out, °scrap, barney: *The argument was about who had invented the wheel. The argument spilled out into the street.* **2** °point, °position, °(line of) reasoning, °logic, °plea, °claim, pleading, °assertion, contention, °case; °defense: *His argument has merit. Arthur's argument falls apart when he brings in the phlogiston theory.*

argumentative *adj.* °quarrelsome, disputatious, °belligerent, combative, contentious, litigious, disagreeable, °testy: *Evelyn is irritable and argumentative.*

arise *v.* **1** °rise, °get up, °stand up, get to one's feet; °wake up, get out of bed, °awake: *We arose when Lady Baffleigh entered the room. I have arisen before dawn all my life.* **2** °rise, °go up, °come up, ascend, °climb; °mount: *The full moon arose in the eastern sky.* **3** °come up, be brought up, be mentioned, *Colloq* crop up: *The subject never would have arisen if the waiter hadn't spilled the wine on me.* **4** °spring up, °begin, °start (up), °originate, °come up, *Colloq* crop up: *A very unpleasant situation has arisen regarding the missing funds.*

aroma *n.* **1** °smell, °odor, °fragrance, °scent, °perfume, °savor, °bouquet; redolence: *Don't you just love to be awakened by the aroma of fresh coffee?* **2** °smell, °odor, °character, °aura, °atmosphere, °flavor, °hint, °suggestion: *There is an aroma of dishonesty about them that I can't quite identify.*

aromatic *adj.* °fragrant, °spicy, perfumed, °savory, °pungent: *A most agreeable scent came from a bowl of aromatic herbs.*

around *adv.* **1** °about, °approximately, °nearly, °almost, °roughly; circa: *There were around a dozen of us in the place.* **2** °about, °everywhere, in every direction, on all sides, all over, °throughout: *By this time the savages were all around and we couldn't move.* **3** 'round, °round, °about, in a circle *or* ring, on all sides: *Gather around and I'll tell you a story. Pass the hat around for Harry's farewell party.* **4** 'round, °round, °about, for everyone *or* all: *I don't think we have enough food to go around.* **5** 'round, °round, °about, all about, °everywhere, here and there, hither and thither, hither and yon, °far and wide: *The tinker traveled around selling his wares and repairing pots.* **6** in perimeter *or* periphery *or* circumference: *The indoor track is exactly one fifth of a mile around.* **7** 'round, °round, from beginning to end, from start to finish, °through: *He now lives in France the year round.* **8** 'round, °round, °about, in the neighborhood *or* vicinity *or* (general) area: *There isn't a soul for miles around.* **9** 'round, in a circle *or* circuit, in *or* by a circular *or* circuitous route *or* path, circuitously, in a roundabout way: *The music goes round and round and it comes out here! This road goes round to the other side of the lake.* —*prep.* **10** 'round, °round, °about, °surrounding, encompassing, enveloping, encircling, on all sides of, in all directions from, enclosing; orbiting: *The fields around the castle were cultivated by tenant farmers. The new bypass around town ought to reduce local traffic.* **11** °about, 'round, °approximately, °roughly; circa: *He was born around the turn of the century.* **12** 'round, °round, here and there in, °about, °throughout, all over, everywhere in: *Dad used to smoke and kept dozens of ashtrays around the house.* **13** 'round, °about, °nearby, in the neighborhood *or* vicinity *or* (general) area of: *Whenever I want to see you alone, there are always people around you. Just look around you if you don't believe me.*

arouse *v.* **1** °awaken, °raise (up), °wake up, waken, °rouse, °revive, °stir (up): *I was aroused by the noise and reached for my pistol.* **2** °excite, °stir up, °stimulate, °awaken, °summon up, °spark, *Colloq* °turn on: *My suspicions were aroused because she was carrying my umbrella.* **3** °provoke, °encourage, °quicken, °foster, °call forth, °stir up, °kindle, °foment: *The song aroused feelings of patriotism among the recruits.*

arrange *v.* **1** °order, °dispose, array, °organize, °sort (out), systematize, °group, °set up, °rank, °line up, align, °form, °position: *The teachers arranged the children according to height. The flowers were arranged in a vase so as to conceal the listening device.* **2** °settle, °plan, °set (up), °organize, orchestrate, °manipulate, choreograph; predetermine, °decide, °prepare, °determine, prearrange, °devise, °bring about, contrive; fix it: *Everything has been arranged—you won't have to lift a finger. For a small fee I can arrange for you to win the first prize.* **3** orchestrate, °score, °adapt: *Flemburgh has arranged music for some of the best-known modern composers.*

arrangement *n.* **1** °order, °disposition, grouping, °organization, array, °display, °structure, structuring, ordering, alignment, line-up, *Colloq* °setup: *Don't you care for the arrangement of the furniture, Ma'am?* **2** °structure, °combination, construction, contrivance, °affair, °setup: *An arrangement of bricks served as a hearth.* **3** °settlement, °agreement, °terms, °plan, °contract, covenant, °compact: *The arrangement called for Bosworth to get 10 percent of the gross.* **4** orchestration,

°score, instrumentation, °adaptation, °interpretation, °version: *I prefer Fats Waller's arrangement of "Sugar Blues."* **5 *arrangements.*** °preparations, plans; °groundwork, planning: *Arrangements have been made for the limousine to pick you up at five.*

arrest *v.* **1** °stop, °halt, °check, °stall, °forestall, detain, °delay, °hinder, °restrain, °obstruct, °prevent, °block, °interrupt: *The progress of the train has been arrested.* **2** °catch, °capture, °seize, apprehend, °take, °take in, take into custody, detain, *Colloq* °nab, °pinch, collar, bust, °run in, *Brit* °nick: *Foxworthy was arrested crossing the border.* **3** °slow, °retard, °stop: *I'm afraid that we have here a case of arrested mental development.* —*n.* **4** °seizure, °capture, apprehension, °detention; °restraint, *Colloq* bust, *US* collar: *The police have made six arrests.* **5** °stop, stoppage, °check, cessation: *The doctor said it was a case of cardiac arrest.* **6 *under arrest.*** in custody, under legal restraint, in the hands of the law, imprisoned, arrested: *You are under arrest for the murder of Lord Foppington, and anything you say may be used in evidence against you.*

arresting *adj.* °striking, °shocking, °remarkable, °impressive, electrifying, °stunning, °extraordinary, surprising, °dazzling: *It is indeed an experience to be in the presence of such an arresting beauty.*

arrival *n.* **1** coming, advent, °appearance: *We have been awaiting your arrival for weeks.* **2** °newcomer; °immigrant; °traveler, °passenger; °tourist; *Australian* °migrant, new chum: *The arrivals on flight 422 were questioned about a bearded passenger on the plane.*

arrive *v.* **1** °come, make one's appearance, °appear, °turn up, *Colloq* °show up; *Slang* °hit (town), blow in: *She arrived only two minutes before the plane was to take off.* **2** °succeed, °prosper, °get ahead (in the world), reach the top, *Colloq* °make it, °make the grade, get somewhere, get there: *Yuppies believe that once they own a fur coat and a Mercedes, they've arrived.* **3 *arrive at.*** come *or* get to, °reach; attain: *I think that Crumley has arrived at the stage in his career where he merits a promotion.*

arrogance *n.* self-assertion, °impertinence, insolence, °presumption, °nerve, °effrontery, °gall, presumptuousness, self-importance, °conceit, egotism, hauteur, haughtiness, loftiness, °pride, hubris, pompousness, pomposity, °pretension, pretentiousness, °bluster, °snobbery, snobbishness, *Colloq* snottiness, *Slang Brit* °side: *He has the arrogance to assume that I wish to see him again.*

arrogant *adj.* **1** °presumptuous, assuming, self-assertive, °conceited, °egotistical, °pompous, °superior, °brazen, bumptious, cavalier: *It would be most arrogant of me to take for myself the glory that belongs to the whole team.* **2** °haughty, °overbearing, imperious, high-handed, overweening, °disdainful, °contemptuous, °scornful, °snobbish, °supercilious, °lofty, swaggering, *Brit* toffee-nosed; *Colloq* uppity, on one's high horse, high and mighty, snotty: *Since her husband was made a company director, she's become unbearably arrogant.*

art *n.* **1** °skill, skillfulness, °ingenuity, °aptitude, °talent, artistry, craftsmanship; °knowledge, °expertise; °craft, °technique, adroitness, °dexterity, *Colloq* know-how: *Little art is required to plant turnips.* **2** artistry; °taste, tastefulness: *High art differs from low art in possessing an excess of beauty in addition to its truth.* **3** °craft, °technique, °business, °profession, °skill: *The fishermen can't employ their art with much success in so troubled a sea.* **4** °knack, °aptitude, °faculty, °technique, mastery; °dexterity, adroitness: *Conversation may be esteemed a gift, not an art. You have acquired the art of insulting people without their realizing it.* **5** °trickery, craftiness, cunning, wiliness, slyness, guile, °deceit, duplicity, artfulness, cleverness, astuteness: *You have to admire the art with which she wraps him round her little finger.* **6 *arts.*** wiles, schemes, stratagems, artifices, subterfuges, tricks; maneuvers, : *She was expert in the arts which ladies sometimes condescend to employ for captivation.*

artful *adj.* **1** °scheming, °wily, °sly, cunning, °foxy, °tricky, °crafty, °deceitful, underhand *or* underhanded, double-dealing, guileful, °disingenuous: *That artful fellow tried to sell me a three-dollar bill.* **2** °ingenious, °clever, °astute, °shrewd, °dexterous: *She has practiced her artful deceptions so long that nobody believes anything she says.*

artifice *n.* **1** °skill, cunning, °trickery, °craft, craftiness, artfulness, guile, duplicity, °deception, °chicanery, underhandedness, shrewdness, slyness, wiliness, trickiness: *He used artifice to get control of the firm.* **2** °stratagem, °device, °maneuver, °trick, contrivance, °wile, °ruse, °subterfuge, °expedient, contrivance, *Colloq* °dodge: *They were deluded by artifices to cheat them out of their money.*

artificial *adj.* **1** °unnatural, °synthetic, man-made, manufactured, simulated, °imitation, °plastic: *The odd man was a collector of artificial teeth.* **2** made-up, concocted, °bogus, °fake, °sham, °false, °counterfeit, *Colloq* phoney *or US also* phony: *The figures used in the sample survey are entirely artificial.* **3** °affected, °unnatural, °forced, °pretended, high-sounding, feigned, °assumed, contrived, °factitious; meretricious, °insincere, °sham, faked, *Colloq* phoney *or US also* phony: *I tell you that Alan's concern for you is entirely artificial.*

artless *adj.* **1** °innocent, °sincere, guileless, °ingenuous, °true, °natural, °open, unartificial, °genuine, °simple, °direct, °candid, °frank, °honest, straightforward, °aboveboard, uncomplicated, undevious, undeceptive, *Colloq* °up front, °on the level, on the up and up: *Imitation is a kind of artless flattery.* **2** unpretentious, unassuming, °unaffected, °natural, °simple, naive *or* naïve, °unsophisticated, °plain, °ordinary, °humble: *The remarks were those of an artless young man who meant nothing sinister.* **3** unskilled, untalented, unskillful, unpracticed, °inexperienced, inexpert, °primitive, unproficient, °incompetent, °inept, °clumsy, °crude, °awkward, bungling: *Jogging shoes must be the most artless footwear ever made.*

ashamed *adj.* °embarrassed, abashed, humiliated, chagrined, mortified, blushing, °shamefaced, °sheepish, red-faced: *I was ashamed to admit that it was I who had written the nasty letter.*

ask *v.* **1** °question, interrogate, °query, °quiz; inquire *or* enquire (of): *Let's ask the policeman for information. Just ask directions of any passerby. I merely asked if you were going my way.* **2** °demand, °require, °expect, °request: *Doing his laundry is a lot to ask.* **3** °beg, °apply (to), °appeal (to), °seek (from), °solicit (from), °petition, °plead (to), °beg, °beseech, °pray, entreat, implore: *In the streets, thousands of beggars ask passersby for alms.* **4** invite, °bid, °summon: *Nellie asked me to dinner.* **5 *ask after*** *or* ***about.*** inquire *or* enquire after *or* about: *My sister asked after you—wanted to know how you were getting along.* **6 *ask for.*** **(a)** invite, °attract, °encourage, °provoke: *You're asking for trouble if you walk alone through that neighborhood after dark.* **(b)** °request, °seek: *We asked for more time to finish the project.*

aspect *n.* **1** °viewpoint, °point of view, °position, °standpoint, °side: *Looked at from a different aspect, the problem seems quite easy to solve.* **2** complexion, °light, °angle, °interpretation, mien, °face: *His conviction for robbery put a different aspect on hiring him as a security guard.* **3** °exposure, °prospect, °outlook, °orientation: *The western aspect of the room made it sunny in the afternoons.* **4** °side, °feature, °attribute, °characteristic, °quality, °detail, °angle, facet, °manifestation, °element, °circumstance: *There are many aspects of Buddhism that you do not understand.*

aspersion *n.* °slander, °libel, false insinuation, calumny, °imputation, °allegation, detraction, °slur, obloquy, defamation, disparagement: *He resented my casting aspersions on the legitimacy of his birth.*

aspiration *n.* °desire, °longing, yearning, craving, hankering, °wish, °dream, °hope; °ambition, °aim, °goal, °objective, °purpose, °intention, °plan, °scheme, °plot: *It was his lifelong aspiration to marry someone with money.*

aspire *v.* ***aspire to.*** °desire, °hope, °long, °wish, °aim, °yearn; dream of: *I'd never aspire to anything higher. He still aspired to being a full professor.*

assailant *n.* attacker, assaulter, mugger: *My assailant threatened me with a knife.*

assault *n.* **1** °attack, onslaught, °onset, °charge, °offensive, blitzkrieg, blitz, °strike, °raid, incursion, sortie; °aggression, °invasion: *At dawn we launched the assault on the fort.* **2** beating, battering, °holdup, mugging; °rape, °violation, molestation; *Law* battery: *The defendant is accused of assault.* —*v.* **3** °attack, assail, set *or* fall upon, °pounce upon, °storm, °beset, °charge, °rush, °lay into: *The elderly couple were assaulted near their home.* **4** °rape, °violate, °molest: *Three women were assaulted in that neighborhood last night.* **5** °beat (up), °batter, °bruise, °harm, °hit, °strike, °punch, smite: *She complained that her husband continually assaulted the children.*

assemble *v.* **1** convene, °gather, call *or* bring *or* get together, convoke, °summon, °muster, marshal, °rally, levy, °round up, °collect, congregate, forgather *or* foregather; °meet: *The forces were assembled along the waterfront. A small crowd assembled at the airport.* **2** °accumulate, °gather, °amass, °collect, bring *or* group *or* lump together, °compile, °unite, join *or* draw together: *The paintings were assembled from many sources.* **3** °construct, put together, °erect, °set up, fit *or* join *or* piece together, °connect, °fabricate, °manufacture, °make: *The sculpture was assembled from so-called* objets trouvés, *or "found" objects.*

assembly *n.* **1** °gathering, °group, °meeting, assemblage, °body, °circle, °company, congregation, °flock, °crowd, °throng, multitude, °host; horde: *A huge assembly of well-wishers greeted the candidate.* **2** convocation, °council, °convention, congress, °association, conclave; °diet, synod: *The assembly voted to reelect the incumbent officers.* **3** construction, putting together, erection, °connection, setting up, °setup, fitting *or* joining *or* piecing together, °fabrication; °manufacture, making: *Assembly of the bicycle can be completed in an hour.*

assertion *n.* **1** °statement, °declaration, affirmation, contention, asseveration, averment, avowal, °pronouncement; *Law* affidavit, deposition: *He made the assertion that he had never seen the defendant before.* **2** insistence, °proclamation, °representation, affirmation, confirmation: *The kings exercised their jurisdiction in the assertion of their regal power.*

assertive *adj.* declaratory, affirmative, asseverative; °definite, °certain, °sure, °positive, °firm, °emphatic, °bold, °aggressive, °confident, °insistent; °dogmatic, doctrinaire, °domineering, °opinionated, °peremptory, *Colloq* °bossy, °pushy: *Harold won't obey unless his mother takes an assertive tone with him.*

asset *n.* **1** Also, ***assets.*** °property, °resources, °possessions, holdings, °effects, °capital, °means, valuables, °money, °wealth: *We have to pay a tax on the company's assets. Her only liquid asset was some shares in the Suez Canal Company.* **2** °talent, °strength, °advantage, °resource, °benefit: *His main asset is that he speaks fluent Japanese.*

assign *v.* **1** °allot, allocate, apportion, consign, °appropriate, °distribute, °give (out), °grant: *A water ration was assigned to each person.* **2** °fix, set (apart *or* aside), °settle (on), °determine, °appoint, °authorize, °designate, ordain, °prescribe, °specify: *Have they really assigned Thursday as the day of worship? Please sit in the seats assigned to you.* **3** °appoint, °designate, °order; °name, °delegate, °nominate, °attach; °choose, °select; *Brit* °second: *The men have been assigned to their posts. I assigned David to look after the champagne.* **4** °attribute, ascribe, accredit, °put down; °refer: *To which century did the curator assign this vase?*

assignment *n.* **1** °allotment, allocation, apportionment, giving (out), °distribution: *Assignment of the political plums is up to the governor.* **2** °task, °obligation, °responsibility, chore, °duty, °position, °post, °charge, °job, °mission, commission; °lesson, homework: *Every agent is expected to carry out his assignment. The school assignment for tomorrow is an essay on Alexander Pope.* **3** °appointment, designation, naming, nomination: *The assignment of Evelyn to the post was a stroke of genius.* **4** designation, °specification, ascription: *In ancient medicine, the assignment of the functions of the organs was often wrong.*

assist *v.* **1** °aid, °help, °second, °support: *He could walk only if assisted by the nurse.* **2** °further, °promote, °abet, °support, °benefit, °facilitate: *The rumors will not assist his election.* **3** °help, succor, °serve, work for *or* with; °relieve: *She has always assisted the poor.*

assistance *n.* °help, °aid, °support, succor, °backing, °reinforcement, °relief, °benefit: *Can you get up without my assistance? The scholarship fund offered financial assistance.*

assistant *n.* **1** helper, helpmate *or* helpmeet, °aid, °aide; aide-de-camp, °second: *These systems make use of rhymes as assistants to the memory.* **2** °deputy, °subordinate, °subsidiary, °auxiliary; underling: *He is now the assistant to the sales manager.*

associate *v.* **1** ***associate (with).*** **(a)** ally with, °link, join *or* unite (with), combine *or* confederate (with), °connect (with), conjoin (with): *In the 1930s Abe was associated with Dutch and Louis in Murder, Incorporated. I always associate them with fast cars and hard drinking.* **(b)** °see, be seen with, socialize *or* fraternize (with), mix *or* mingle (with), go (out) with, consort with, have to do with, *Colloq* hang out with, *Brit* pal with *or* about, pal up (with), *US* pal around (with): *Mother told me not to associate with boys who use that kind of language.* —*n.* **2** °colleague, °partner; °fellow, fellow worker: *I'd like you to meet my associate, Ian Owen.* **3** confederate, °ally, collaborator; °accomplice, accessory *or* accessary: *He and his associates have been sent to prison for conspiracy.* **4** °comrade, °companion, °friend, °mate, buddy; confidant(e): *We have been close associates for many years.* —*adj.* **5** °subsidiary, °secondary: *She is an associate professor at an American university.* **6** allied, affiliate, °affiliated, associated; accessory *or* accessary: *Publication is under the direction of an associate company.*

association *n.* **1** °society, °organization, confederation, confederacy, °federation, °league, °union, °alliance, guild, coalition, °group; °syndicate, °combine, consortium, cooperative: *The society became known as the American Association for the Advancement of Science.* **2** °connection, °link, affiliation, °relationship, °bond, °tie, linkage, linking, pairing, joining, conjunction, bonding: *The association between princes and frogs is probably lost on anyone so literal-minded.* **3** °fellowship, intimacy, °friendship, camaraderie, comradeship, °relationship: *There has been a longstanding association between Peter and Wendy.*

assortment *n.* **1** °group, °class, °category, °batch, °set, °lot, classification, grouping: *Which assortment contains only plain chocolates?* **2** °collection, °potpourri, °mixture, mélange, array, agglomeration, conglomeration, °medley, farrago, °variety, °miscellany, °jumble, salmagundi, gallimaufry, °mishmash, *Colloq* mixed bag: *That hat looks like something from the assortment at a rummage sale. A bizarre assortment of people attended the meeting.*

assume *v.* **1** °accept, °adopt, °take, °use, °employ; arrogate, °appropriate, take over *or* up, °undertake: *Who will assume the leadership of the party?* **2** °take on (oneself), take upon (oneself), put *or* try on, don, °adopt; °acquire: *Whenever she delivered the information, she assumed the disguise of an old man. That trivial dispute has assumed gargantuan proportions.* **3** °presume, °suppose, °believe, °fancy, °expect, °think, °presuppose, °take, take for granted, °surmise, *Chiefly US* °guess: *When I saw the knife in his hand, I assumed the chef was going to slice a lemon. The president assumed he had congressional support for the veto.* **4** pretend to, feign, °sham, °counterfeit, simulate, °sham, °affect, °fake: *Though she cared deeply, she assumed a devil-may-care attitude.*

assumed *adj.* **1** appropriated, °taken, usurped, expropriated, preempted, usurped, seized: *He functions in his assumed capacity as a judge.* **2** °pretended, °put on, °sham, °false, feigned, °affected, °counterfeit, simulated, °spurious, °bogus, °fake; pseudonymous, made-up, *Colloq* phoney *or US also* phony: *She morosely stared at the floor in assumed contrition. He wrote poetry under an assumed name.* **3** °taken, taken for granted, presumed, °supposed, accepted, expected, presupposed; °hypothetical, °theoretical, suppositional: *The payment depends materially on the assumed rate of interest.*

assurance *n.* **1** °promise, °pledge, guarantee *or* guaranty, °warranty, commitment, °bond, surety; °word, word of honor, °oath, °vow: *You have the bank's assurance that the money will be on deposit.* **2** °certainty, °confidence, °trust, °faith, reassurance, surety, assuredness, certitude; °security: *There is no assurance that Herr Kleister will get the job done.* **3** audacity, °impudence, °presumption, boldness, brazenness, °nerve, °effrontery, insolence, *Colloq* °brass, °gall, cheek, chutzpah: *With an air of assurance they quote authors they have never read.* **4** °self-confidence, self-reliance, °confidence, steadiness, intrepidity, self-possession, °poise, aplomb, coolness, °control, °self-control, °resolve, *Colloq* °gumption, °guts, gutsiness: *He has the assurance of one born to command.*

assure *v.* **1** °secure, stabilize, °settle, °establish, °confirm, °certify, °warrant, °guarantee, °ensure, be confident of, make *or* be sure *or* certain: *Force, fear, and the multitude of his guard do less to assure the estate of a prince than the good will of his subjects.* **2** °encourage, inspirit, °reassure, hearten: *Your humanity assures us and gives us strength.* **3** °convince, °persuade, °reassure, make (someone) certain; °ensure: *What can I do to assure you of my love?* **4** assert, °state, asseverate, °promise: *I assure you that we shall do everything possible to find your dog.*

astonish *v.* °amaze, °surprise, °shock, °astound, °stun, °stagger, dumbfound *or* dumfound, bowl over, °floor, stupefy, °daze, *Colloq* flabbergast: *She astonished the audience with her gymnastic skill. I was astonished when told my wife had given birth to quintuplets.*

astonishment *n.* °amazement, °surprise, °shock, stupefaction, °wonder, wonderment: *I'll never forget that look of astonishment on his face when he learned he had won.*

astound *v.* °surprise, °shock, °astonish, °stun, °stagger, dumbfound *or* dumfound, bowl over, °floor, stupefy, °bewilder, °overwhelm, *Colloq* flabbergast: *We were astounded to learn that he had survived all those years on a desert island. The Great MacTavish performs astounding feats of magic and levitation!*

astute *adj.* **1** °shrewd, °subtle, °clever, °ingenious, adroit, °wily, cunning, °calculating, canny, °crafty, °artful, °arch, °sly, °foxy, guileful, underhand, underhanded; *Rare* astucious: *He had a reputation for being an astute businessman.* **2** °sharp, °keen, °perceptive, °observant, °alert, °quick, °quick-witted, °sage, sagacious, °wise, °intelligent, insightful, perspicacious, discerning, °knowledgeable: *That was a very astute comment, Smedley.*

atmosphere *n.* **1** °air, heaven(s), °sky, aerosphere: *It is gravity that prevents the earth's atmosphere from drifting off into outer space.* **2** °air, ambiance *or* ambience, °environment, °climate, °mood, °feeling, °feel, °spirit, °tone: *There's such a friendly atmosphere in the bar of the Golden Cockerel. The atmosphere became very chilly when she told him she was marrying someone else.*

atone *v.* expiate, °make amends, °pay, °repay, °answer, °compensate; redress, °remedy, propitiate, °redeem: *Nothing can atone for her betrayal of him to the enemy. He has atoned for his sins many times over.*

atonement *n.* °amends, propitiation, reparation, repayment, compensation, °payment, °restitution, recompense, expiation, °penance, °satisfaction: *In atonement for his behavior, David sent flowers and an apology.*

atrocious *adj.* **1** °cruel, °wicked, iniquitous, °villainous, °fiendish, execrable, appalling, °abominable, °monstrous, °inhuman, °savage, barbaric, °brutal, barbarous, heinous, °dreadful, °flagrant, flagitious, °gruesome, °grisly, °ruthless, °ghastly, unspeakable, horrifying, °horrible, °awful, °infamous, °infernal, °satanic, hellish: *They will never forget the atrocious crimes that took place during the war.* **2** °awful, °terrible, °bad, °rotten, horrid, appalling, °frightful, horrendous, *Colloq* °lousy: *That's the most atrocious book it has ever been my misfortune to review. Priscilla has atrocious taste.*

atrocity *n.* **1** °enormity, wickedness, flagitiousness, iniquity, °infamy, cruelty, heinousness, °horror, °evil, inhumanity, °barbarity, savagery: *The atrocity of the "Final Solution" was a well-kept secret during the war.* **2** °evil, °outrage, °crime, villainy, °offense: *She could not listen when the prosecutor read a list of the atrocities perpetrated at the camp.*

attach *v.* **1** °fasten, °join, °connect, °secure, °fix, affix; tack *or* hook *or* tie *or* stick on, °pin, rivet, °cement, °glue, °bond, solder, °weld, braze; °unite; *Nautical* °bend: *The tag is still attached to your dress. Attach this to the wall. How do I attach the sail to the spar?* **2** °connect, °associate, °assign, affiliate, °enlist, °join, °add, subjoin, *Brit* °second: *Her brother-in-law has been attached to my regiment.* **3** endear, °attract: *I won't say that we were in love, but I was very closely attached to her.* **4** °fix, affix, °pin, °apply, ascribe, °assign, °attribute, °put, °place: *Why do you attach so much importance to what Dora says?* **5** adhere, °cleave, °stick: *Many legends have attached themselves to Charlemagne.* **6** °seize, °lay hold of, °confiscate, °appropriate: *If he cannot meet the mortgage payments the bank will attach his house.*

attached *adj.* **1** connected, joined, *Brit* seconded: *She has been attached to the State Department for many years.* **2** °united, fastened, °fixed: *That knob attached to the outside of the door might come off.* **3** Often, ***attached to.*** °devoted (to), °partial (to), °fond (of), °devoted (to): *I feel closely attached to her. I became attached to the painting and did not wish to sell it.* **4** °spoken for, married, unavailable, °engaged, betrothed: *I would have asked Suzanne out, but I gather she's attached.*

attachment *n.* **1** °fastening; °connection, °tie, °link, °bond: *The attachment of this fitting is too flimsy. William cannot understand how an attachment could have been formed between his wife and his brother.* **2** attaching, °fastening, linking, joining, affixing, fixing, °connection: *The mode of attachment to the wall is not immediately apparent.* **3** °affection, °regard, fidelity, faithfulness, °devotion, °regard, °liking, fondness, °affinity, friendliness, °loyalty, °admiration, tenderness, °partiality, °friendship, °love: *We still feel a deep attachment, despite the divorce.* **4** adjunct, °addition, accessory *or* accessary, °device, appliance, °extra, accoutrement *or US also* accouterment, appendage, °part; °ornament, °decoration; *Colloq* °gadget: *With this attachment, the film is advanced automatically. Attachments are available at extra cost.*

attack *v.* **1** assail, °assault, fall *or* set *or* pounce upon; °charge, °rush, °raid, °strike (at), °storm; °engage (in battle), °fight; *Colloq* °mug, °jump: *They were attacked on their way home by a gang of boys. Helicopter gunships were sent out to attack the bunker.* **2** °criticize, censure, °berate; °abuse, revile, inveigh against, °denounce, °condemn, malign, denigrate, decry, °disparage, deprecate, °vilify: *His article attacked the minister for his views on housing.* **3** °begin, °start; °approach, °undertake: *We attacked the meal with gusto.* **4** °affect, °seize; infect: *Rheumatism attacks young and old alike.* **5** °waste, °devour, °destroy, °eat; °erode, corrode, °decompose, °dissolve: *Termites have attacked the beams of the house. Watch how the acid attacks the areas on the plate that have not been protected.* —*n.* **6** °assault, °onset, °offensive, onslaught, incursion, °raid, °strike, °inroad, °invasion: *The enemy responded to our attack with a smokescreen. After capturing the pawn, Karpov launched an attack on the queen.* **7** °criticism, censure; °abuse, denunciation, revilement, denigration, decrial, disparagement, deprecation, vilification: *The quarterly's attack is totally uncalled for.* **8** °seizure, °spell, °spasm, °paroxysm; °fit, °bout: *Preston has had another attack of gout. How do you stop an attack of hiccups?* **9** °destruction, wasting; °erosion, corrosion: *Noting the attack on the planks by shipworm, the surveyor declared the vessel unseaworthy. Aluminum will not withstand the attack of the salt air in this area.*

attempt *v.* **1** °try, °essay, °undertake, °take on, °venture; °endeavor, °strive, *Colloq* have *or* take a crack at, try on, have a go *or* shot at: *It is too stormy to attempt the crossing tonight. Is she going to attempt to dive off the cliff tomorrow?* —*n.* **2** °endeavor, °try, °essay; °effort, °undertaking, °bid, *Colloq* °crack, °go, °shot: *The weather cleared sufficiently for another attempt at the summit. He made a feeble attempt to wave.* **3** °attack, °assault: *An abortive attempt was made on the life of the vice president tonight.*

attend *v.* **1** be present (at), go to, be at, °appear (at), put in an appearance (at), °turn up (at), °haunt, °frequent; °sit in (on), *US and Canadian* audit: *Are you attending the concert? She attends Miss Ribble's elocution class.* **2** °turn to, pay attention to, °serve, tend to, °take care of, °deal with, °handle, °heed, °fulfill: *I shall attend to your request as soon as possible, madam.* **3** Also, ***attend to.*** watch over, wait on *or* upon, °care for, °take care of, °minister to, occupy oneself with, °look after, look out for, devote oneself to: *Mrs. Atterbury attends the patients in the cancer ward on weekdays. The clergyman has his own flock to attend to.* **4** °escort, °accompany, °conduct, convoy, °squire, usher, °wait upon, °follow; chaperon *or* chaperone: *The actress arrived, attended by her entourage of toadies.* **5** be associated with, °accompany, result in *or* from, °give rise to: *A departure in the midst of the battle would be attended by great peril, Sir.*

attendance *n.* **1** °presence, °appearance, being: *Your attendance at chapel is required.* **2** audience, °crowd, °assembly, assemblage, °gathering, °turnout, °gate, °house: *The attendance at the fete was greater than we expected.* **3** ***in attendance.*** waiting upon, attending, serving: *The king always has at least four people in attendance.*

attendant *adj.* **1** waiting upon, accompanying, following; resultant, resulting, °related, consequent, concomitant, depending, accessory *or* accessary: *The circumstances attendant on your acceptance of the post are immaterial.* —*n.* **2** °escort, °servant, °menial, helper, usher *or* usherette, chaperon *or* chaperone; °aide, °subordinate, underling, °assistant; °follower, *Denigrating* lackey, °flunky, °slave; *Colloq US* °cohort: *He dismissed his attendants and entered the church alone.*

attention *n.* **1** °heed, °regard, °notice; concentration: *Please give your attention to the teacher. Pay attention! Don't let your attention wander.* **2** publicity, °notice, °distinction, acclaim, °prominence, *réclame,* °notoriety; limelight: *She seems to have been getting a lot of attention lately.*

attentive *adj.* **1** heedful, °observant, °awake, °alert, °intent, watchful, concentrating, assiduous; °mindful, °considerate: *James is very attentive in class. You really must be more attentive to the needs of others.* **2** °polite, °courteous, courtly, °gallant, °gracious, °accommodating, °considerate, °thoughtful, °solicitous, °civil, °respectful, deferential: *Bret is always very attentive to the ladies.*

attest *v.* °bear witness (to), °bear out, °swear (to), °vow, °testify, °certify, °vouchsafe, °declare, assert, asseverate, aver, affirm, °confirm, °verify, °substantiate, vouch for, *Law* depose, depose and say, depone: *I attest to the fact that they left the restaurant together. The merits of chateau-bottled Bordeaux are attested by most epicures.*

attitude *n.* **1** °posture, °position, °disposition, °stance, °bearing, °carriage, °aspect, demeanor: *The attitude of the figures in the sculpture was one of supplication.* **2** °posture, °position, °disposition, °opinion, °feeling, °view, °point of view, °viewpoint, °approach, °leaning, °thought, °inclination, °bent, °tendency, °orientation: *What is your attitude toward the situation in South Africa?*

attract *v.* °draw, invite; °entice, °lure, allure, appeal to, °charm, °captivate, °fascinate, *Colloq* °pull: *Our attention was attracted by a slight noise in the cupboard. Melissa attracts men the way flowers attract bees.*

attraction *n.* **1** °draw, °appeal; °magnetism; gravitation, *Colloq* °pull: *David confided to Joan that he felt a strong attraction to her. There is an attraction between the north and south poles of these magnets.* **2** °draw, °lure, °enticement, attractant, °inducement; °show, °entertainment, °presentation, °performance, *Colloq* °come-on, crowd-puller, crowd-pleaser: *The presence of the movie stars has been a powerful attraction. The producer has planned to repeat the attraction every evening.*

attractive *adj.* attracting, °drawing, pulling, captivating, °taking, °fetching, appealing, luring, °inviting, enticing, °seductive, °inviting, °engaging, charming,

°interesting, °pleasing, °winning, alluring, good-looking, °pretty, °handsome: *The person I'd like to meet needn't be beautiful or stunning—attractive will do nicely.*

attribute *n.* **1** °quality, °character, °characteristic, °property, °feature, °trait, °virtue: *It is surprising how soon historical personages become invested with romantic attributes.* —*v.* **2** ascribe, °impute, °assign, put down to, trace to, °charge, °credit: *The shriveled arm of Richard the Third was attributed to witchcraft. To what do you attribute your interest in birds?*

attribution *n.* °assignment, ascription, °credit: *The curator disagreed with the expert's attribution of the painting to Canaletto.*

audacious *adj.* **1** °daring, °bold, °confident, °intrepid, °brave, °courageous, °adventurous, °venturesome, °reckless, °rash, °foolhardy, °daredevil, devil-may-care, °fearless, doughty, mettlesome: *The troop launched an audacious daylight attack. Cranshaw made an audacious bid for the chairmanship.* **2** °presumptuous, °shameless, °bold, °impudent, °pert, saucy, °defiant, °impertinent, °insolent, °brazen, °unabashed, °rude, °disrespectful, °cheeky, °forward: *Charlotte was so audacious as to assume that she could win.*

aura *n.* °air, °atmosphere, °feeling, ambiance *or* ambience, °spirit, °character, °quality, °odor, °aroma, emanation: *There is an aura of elegance about the woman that impresses everyone she meets.*

auspices *n.* aegis, sponsorship, °authority, °protection, °support, °backing, supervision, °guidance, °patronage, °sanction, °approval, °control, °influence: *The competition is under the auspices of the astronomical society.*

authentic *adj.* °genuine, °real, °actual, °bona fide, °factual, °accurate, °true, °legitimate, °authoritative, °reliable, °veritable, °trustworthy, °faithful, °undisputed: *This is an authentic Chippendale chair.*

authenticate *v.* °verify, validate, °certify, °substantiate, °endorse, vouch for, °confirm, corroborate: *You will have to go to the consul to have your passport authenticated.*

author *n.* °creator, originator, inventor, °father, °founder, framer, initiator, maker, prime mover, architect, °designer; °writer, novelist, littérateur: *Adolfo was the author of the plot to kill the governor.*

authoritarian *adj.* °dictatorial, imperious, °totalitarian, autocratic, °arbitrary, °absolute, °dogmatic, °domineering, °strict, °severe, unyielding, °tyrannical, °despotic, *Colloq* °bossy: *Why do people ever elect an authoritarian government?*

authoritative *adj.* **1** °official, valid, °authentic, documented, certified, validated, °legitimate, sanctioned; conclusive: *The second edition is usually considered the authoritative one.* **2** °scholarly, °learned, °authentic, valid, °sound, °veritable, verifiable, °accurate, °factual, °faithful, dependable, °reliable, °trustworthy, °true, °truthful: *There is no more authoritative source than Professor Fitzhugh on early Egyptian history.*

authority *n.* **1** °power, °jurisdiction, °dominion, °right, °control, °prerogative, authorization; hegemony: *Who gave you the authority to tell me what to do? By the authority vested in me, I now pronounce you man and wife.* **2** °word, °testimony, °evidence, *Colloq* °say-so: *Do not accept anything solely on the authority of the* Herald. **3** °expert, °specialist, °scholar, °sage, °judge, arbiter: *Gardner is an authority on Scottish history.* **4** ***authorities.*** °government, °establishment, officials, officialdom, °powers that be, °police: *The authorities lowered the speed limit.*

authorize *v.* empower, commission; °sanction, °approve, countenance, °permit, give leave, °allow, °license, °entitle, consent *or* subscribe to, °endorse, *Colloq* OK *or* okay, give the green light *or* go-ahead to: *Who authorized you to speak for all of us?*

automatic *adj.* self-acting, self-governing, self-regulating, °mechanical, °robot, automated: *Most modern cars are equipped with an automatic choke.* **1** °mechanical, °involuntary, °unconscious, instinctive *or* instinctual, °natural, °spontaneous, °impulsive, conditioned, reflex, robotlike, *Slang* knee-jerk: *Flinching is an automatic reaction to a threatening gesture.* **2** °unavoidable, °inevitable, inescapable, ineluctable: *It is automatic for the tax auditor to suspect people of hiding something.*

auxiliary *adj.* **1** °helping, assisting, °supportive, aiding, abetting; °helpful, accessory *or* accessary, °supplementary: *In a well-balanced mind, imagination and understanding are auxiliary to each other.* **2** °subordinate, additional, °subsidiary, °secondary, ancillary, °extra, °reserve; °accessory: *Larger sailing vessels have an auxiliary motor in case the wind fails.* —*n.* **3** °help, °assistance, °aid, °support, °accessory: *Knowing another language is a useful auxiliary in the study of your own.* **4** helper, °assistant, °aide, alter ego, °supporter, *Colloq* man Friday, girl Friday: *Let me introduce Pat, my auxiliary, who will help you if I am not available.*

available *adj.* at *or* to hand, at one's disposal, °accessible, °handy, °present, °ready, (readily) obtainable, °convenient, °nearby, °close by, within reach, °on tap, at one's fingertips *or* elbow: *Running water is available. If you need me for anything, I am available.*

avant-garde *adj.* innovative, advanced, °progressive, °experimental, °original, °new, ground-breaking, pioneering, precedent-setting; °revolutionary, °extreme, extremist, *Colloq* far-out, °way-out: *We disapprove of your avant-garde notions of teaching. Some modern art, avant-garde not very long ago, seems quite conventional today.*

avarice *n.* °greed, acquisitiveness, cupidity, craving, covetousness, °desire, greediness, °rapacity, selfishness; stinginess, meanness, miserliness, parsimony, tightfistedness, closefistedness, niggardliness, penuriousness: *The classic tale of avarice is that of King Midas, whose touch turned everything to gold.*

avaricious *adj.* °greedy, acquisitive, °grasping, covetous, °mercenary, °selfish; pennypinching, stingy, °miserly, °mean, parsimonious, tightfisted, closefisted, niggardly, °penurious, °tight: *She fell into the clutches of an avaricious lawyer.*

average *n.* **1** °mean, °norm, °usual, °standard: *The Bell Inn is certainly far above average in accommodation, food quality, and service.* **2** ***on average.*** °in the main, °generally, normally, °usually, °ordinarily, typically, customarily, °as a rule, °for the most part: *On average, I go abroad twice a year on business.* —*adj.* **3** °normal, °common, °usual, °customary, °general, °typical, °ordinary, °regular: *On an average day, the museum has about two thousand visitors.* **4** °mediocre, middling, °run-of-the-mill, commonplace, °common, °ordinary, °undistinguished, unexceptional, *Colloq* °so-so: *Boris is only an average violinist, but he's a virtuoso on the harmonica.*

averse *adj.* °disinclined, unwilling, °reluctant, °resistant, °loath, °opposed, anti, antipathetic, ill-disposed, °indisposed: *I can understand her being averse to going out with the likes of you.*

aversion *n.* **1** °dislike, abhorrence, repugnance, antipathy, °antagonism, °animosity, °hostility, °loathing, hatred, odium, °horror; disinclination, unwillingness, °reluctance, °dislike, °distaste: *Does Anne have an aversion to people who smoke? Your aversion to the theater* *might be explained as agoraphobia.* **2** °dislike, hatred, °hate, °loathing: *Turnips are a particular aversion of mine.*

avoid *v.* °shun, keep (away) from, keep off, leave alone, keep *or* steer clear of, °refrain from, °dodge, circumvent, °sidestep, °elude, °escape, °evade: *The doctor suggested that I avoid chocolate. Why does Bennie avoid looking me straight in the eye?*

awake *v.* **1** °wake (up), °awaken, °get up, rouse *or* bestir oneself: *When I awoke, she was standing over me with a pistol.* **2** °awaken, °animate, °arouse, °rouse, °stimulate, °revive, °incite, °excite, °activate, °alert, °stir up, °fan, °kindle, ignite, °fire: *Marches awake my sense of patriotism.* **3** ***awake to.*** awaken to, wake up to, °realize, °understand, become aware *or* conscious of: *I finally awoke to the fact that my tax return was overdue.* —*adj.* **4** up, aroused, roused, wide-awake, up and about, °alert, on the alert, on the qui vive, watchful, on guard, °attentive, °conscious; heedful, °alive: *I am always awake a few minutes before the alarm goes off.*

awaken *v.* See **awake, 1, 2,** above.

award *v.* **1** °grant, °give, °confer, °bestow, present, °accord, °furnish, endow with; °assign, apportion: *Her dog was awarded the blue ribbon in the club show.* —*n.* **2** °prize, °trophy, °reward: *The award for the tidiest boats has been won by the Bristol Yacht Club.* **3** °grant, bestowal, °presentation, °endowment, awarding: *Before the award of the prizes, we listened to speeches.*

aware *adj.* **1** °informed, apprised, °knowledgeable, °knowing, posted, in the know, °enlightened, *au fait, au courant,* cognizant, *Slang* °hip, hep, °wise: *She is well aware of the consequences.* **2** °sensitive, °sensible, °conscious: *I became aware that someone was watching us.*

awesome *adj.* awe-inspiring, °awful, °imposing, °amazing, wonderful, breathtaking, °marvelous, wondrous, °moving, °stirring, affecting, °overwhelming, °formidable, daunting, °dreadful, °fearsome, °fearful, °frightening, horrifying, °terrifying, °terrible; °unbelievable, °incredible; alarming, °shocking, °stunning, stupefying, astounding, astonishing: *The eruption of Vesuvius in* AD *67 must have been a truly awesome spectacle.*

awful *adj.* **1** °bad, °terrible, °inferior, °base, °abominable, °rotten, °horrible, horrid; °tasteless, °unsightly, °ugly, °hideous, °grotesque, *Slang* °lousy, *Brit* naff: *That is an awful piece of sculpture. I feel awful this morning.* **2** °frightful, °shocking, execrable, unpleasant, °grotesque, °nasty, °ghastly, °gruesome, horrendous, horrifying, horrific, °horrible, unspeakable: *That was an awful thing to do.*

awfully *adv.* °very (much), °badly, °terribly, °extremely, greatly, remarkably, in the worst way, dreadfully, extraordinarily, °exceedingly, excessively, °really, °fearfully, inordinately; incomparably: *I get awfully tired running in the marathon. I am awfully sorry I stepped on your toe. Doreen is an awfully good horsewoman.*

awkward *adj.* **1** °clumsy, ungainly, °left-handed, ham-handed, ham-fisted, blundering, bungling, maladroit, uncoordinated, undexterous, inexpert, gauche, unhandy, °inept, oafish, unskilled, unskillful, *Colloq* °all thumbs, butterfingered, *Brit* cack-handed: *In his awkward attempt at putting the watch back together, Sam left out a few parts.* **2** °ungraceful, ungainly, inelegant, °wooden, gawky: *The ballerina made an awkward, flat-footed pirouette and stumbled off stage.* **3** °embarrassed, °shamefaced, uncomfortable, °ill at ease, uneasy, °out of place, discomfited, °confused: *He felt awkward being the only boy in the class. Terry made an awkward excuse and left the room.* **4** °dangerous, °hazardous, °risky, °precarious, °perilous: *Be careful, there's an awkward step here.* **5** °difficult, °touchy, °sensitive, °embarrassing, °delicate, unpleasant, uncomfortable, °ticklish, °tricky, °trying, °troublesome, *Colloq* °sticky: *He's got himself into a very awkward situation indeed.*

B

babble *v.* **1** °prattle, twaddle, °jabber, gibber, °chatter, °blab, blabber, °gurgle, burble, gabble, *Colloq* °blab, blabber, °gab, yak, natter, witter, *Brit* rabbit: *The silly fellow kept babbling away, but no one was listening. Madelaine, too young to talk, babbles to herself.* **2** divulge, °tell, °disclose, °repeat, °reveal, °tattle, °gossip, °blurt (out), *Colloq* °blab: *Don't tell Nick about the affair—he'll babble it all over town.* —*n.* **3** °gibberish, °nonsense, twaddle, °prattle, chatter(ing), gibber, °jabber, gibber-jabber, °drivel, °rubbish, *bavardage;* °murmur, hubbub: *Ella's conversation about the financial market is just so much babble.*

baby *n.* **1** infant, neonate, newborn, babe, babe in arms, °child, toddler, °tot: *The baby is just beginning to teethe.* —*v.* **2** cosset, °coddle, °pamper, mollycoddle, °indulge, °spoil, °pet: *He turned out that way because he was babied till he was ten. I know you like to be babied when you're ill.*

back *v.* **1** invest in, wager *or* bet on: *She backed a 35-to-2 long shot in the Derby, and she won.* **2** Also, ***back up.*** **(a)** °support, °uphold, stand behind, °promote, °encourage, °help, °uphold, °second, °side with, °endorse, °aid, °abet, °assist; °sponsor, °subsidize, °underwrite, subvene, °finance, *Slang US and Canadian* bankroll: *If you want to start your own business, your mother and I will back you.* **(b)** °reverse, go *or* move in reverse, go *or* move backward(s): *He backed into the driveway.* **3** ***back down (from) or off (from) or away (from) or out (of) or up.*** °withdraw (from), °retreat (from), °abandon, °retire (from), backtrack (from), shy away (from), °recoil (from), °turn tail (from): *When Percy stood up to him, the bully backed down. The investment sounded risky, so I backed off. Philippa backed out of singing the leading role. Back up and give me room!* —*n.* **4** backside, °rear, *Technical* dorsum: *She stood with her back towards me.* **5** ***at the back of*** or ***at someone's back.*** behind, following, pursuing, chasing, *US* in back of: *Here come the hounds at the back of the fox. You were at my back in the line a minute ago.* **6** ***behind the back of*** or ***behind someone's back.*** surreptitiously, °secretly, clandestinely, privately, furtively, sneakily, slyly; treacherously, traitorously, perfidiously, deceitfully, insidiously: *Granville is always telling tales about you behind your back.* **7** ***break the back of.*** **(a)** °overcome, °master: *Now that he's broken the back of that problem he can get on with his work.* **(b)** *US* °crush, °ruin, bankrupt, °destroy, °defeat, vanquish, *Colloq* °break: *The government has tried on many occasions to break the back of the Mafia operation.* **8** ***on (someone's) back.*** *US* weighing (down) on *or* upon (someone), burdening (someone), lodged with (someone), resting with (someone): *The responsibility for the decision is on your back.* **9** ***turn one's back on*** or ***upon.*** °abandon, °forsake, °ignore, °disregard, °repudiate, °reject, °cast off, disown, °deny: *He turned his back on her when she needed him most.* **10** ***with one's back to*** or ***against the wall.*** hard pressed, struggling (against odds), without hope, with little *or* no hope, °helpless, in dire straits, in (serious) trouble: *After the stock-market crash, some brokers*

found themselves with their backs to the wall. —*adj.* **11** °rear; °service, servants': *Both back tires are flat. Please use the back staircase from now on.* **12** *US and Australian and New Zealand* °outlying, °remote, °isolated, °distant; °undeveloped, °primitive, °raw, °rough, °uncivilized: *We raised three boys in the back country, and all of them became doctors.* **13** in arrears, °overdue, past due, °late; behindhand: *The IRS has advised me that I owe thousands in back taxes.* —*adv.* **14** to *or* toward(s) the rear, rearward(s), backward(s); away: *We beat back the enemy in severe hand-to-hand fighting. I accepted his offer at once, lest he should draw back. Get back from the edge!* **15** in return *or* repayment *or* requital *or* retaliation; again: *I'll pay you back when I have the money. She gave him back as good as he had given.* **16** ago, in time(s) past: *Two generations back, his was the finest house in the town.* **17** behind, behindhand, in arrears, °overdue: *We are a week back in the rent.* **18** ***go back on.*** °renege, °fail; °deny, disavow, °break, °repudiate: *He has gone back on his promise to send the payment on the first of every month.*

backbone *n.* **1** °spine, spinal column: *He's much better since the surgery on his backbone.* **2** °mainstay, chief *or* main support, °buttress, °pillar: *Drusilla, who has been the backbone of the society, has moved to Melbourne.* **3** resoluteness, sturdiness, firmness, °determination, °strength (of character), mettle, purposefulness, °resolution, °courage, °fortitude, °resolve, °will, willpower, °strength, °stability, °stamina, staying power, °grit: *Has she the backbone to run the company alone?*

backer *n.* **1** °supporter, °advocate, promoter, °sponsor, °patron: *Mrs. Henderson has always been an enthusiastic backer of adult education.* **2** investor, benefactor *or* benefactress, °supporter, underwriter, *Colloq* angel: *The play's backers have made huge profits.* **3** °bettor, *Brit* punter: *His backers are offering odds of 10 to 1.*

background *n.* **1** °history, °experience, °qualifications, credentials, grounding, training; °breeding, °upbringing, °family; curriculum vitae, *Colloq* CV: *His background suits him admirably for the post of ambassador to Nauru.* **2** °distance, offing, °horizon, °obscurity: *I like the way the coastline disappears into the background toward the edge of the painting.* **3** ***in the background.*** °inconspicuous, °unnoticed, °unobtrusive, °behind the scenes, out of the limelight *or* spotlight, unseen, out of the public eye, backstage: *Abernathy prefers to remain in the background, letting his dealer bid at the auctions.*

backing *n.* **1** °support, °help, °aid, °assistance, succor; °approval, °endorsement, °patronage, °approval, sponsorship: *At least you can rely on the backing of the Republicans in your district.* **2** investment, °money, °funds, funding, °subsidy, °grant; sponsorship: *How can you launch the company without backing?*

backlash *n.* °reaction, °repercussion, °recoil, counteraction, °rebound, °kickback, backfire; °boomerang: *There was a strong backlash in the USA against giving minorities preferred instead of equal job opportunities.*

backward *adj.* **1** °bashful, °shy, °reticent, diffident, °retiring, °coy, °timid, unwilling, °loath, chary, °reluctant, °averse: *She took him to be a bit backward when he didn't respond to her smile.* **2** °slow, dimwitted, °dull, °stupid, slow-witted, °dumb, °feebleminded, *Colloq Brit* gormless, °dim: *Some of the more backward students will need help.* **3** °slow, °late, behindhand, retarded: *Millie seemed a bit backward in learning to walk.* **4** rearward; to the rear, behind; to the past: *She gave him a backward glance. He went on through life with never a backward look.* **5** retrograde, retrogressive, °reverse, regressive: *The ancients were unable to account for the apparently backward motion of the planets.* —*adv.* **6** backwards, rearward(s), in reverse, regressively, retrogressively; withershins *or* widdershins, *Brit* anticlockwise, *US* counterclockwise: *Walk backward to the door with your hands up. The general, who refused to acknowledge defeat, explained that his troops were "advancing backward." Do the clocks run backward in Australia, Daddy?* **7** backwards, in reverse; back to front: *She can even ride sitting backward on a galloping horse. In back slang, the words are pronounced backward, as* yob *for "boy" or* loof *for "fool." I think you're wearing your pullover backward.*

bad *adj.* **1** °poor, °wretched, °inferior, °defective, °awful, °worthless, °miserable, egregious, execrable, substandard, °unsatisfactory, °disappointing, °inadequate, nonstandard, *Colloq* °lousy, °rotten, crummy, *Slang Brit* grotty, naff: *Sometimes they would send him a letter, but he was a bad correspondent. We went to see a rather bad play the other night.* **2** °corrupt, polluted, vitiated, debased, °base, °vile, °foul, °rotten, miasmic, noxious, mephitic, °unhealthy, °poisonous, °injurious, °dangerous, °harmful, °hurtful, pernicious, deleterious, °ruinous: *The fugitive cannot long breathe the bad air of the sewer.* **3** °evil, °ill, °immoral, °wicked, °vicious, °vile, °sinful, depraved, °awful, °villainous, °corrupt, amoral, °criminal, °wrong, unspeakable: *Do you really think there could have been a scintilla of good in a man as bad as Hitler?* **4** unpleasant, °offensive, °disagreeable, °inclement, °severe, °awful, unfavorable, adverse, °inclement, unpleasant, *Colloq* °lousy, °rotten: *Surely you're not going sailing in this bad weather?!* **5** unfavorable, unlucky, unpropitious, °unfortunate, °inauspicious, troubled, °grim, distressing, discouraging, unpleasant: *Agreeing to do that job may yet turn out to have been a bad decision.* **6** °off, tainted, spoilt *or* spoiled, °moldy, °stale, °rotten, decayed, putrefied, °putrid, contaminated: *The fridge isn't working and the food has gone bad. She ate a bad egg and felt ill.* **7** irascible, ill-tempered, grouchy, °irritable, °nasty, °peevish, °cross, crotchety, crabby, °cranky, curmudgeonly: *Don't go near the boss—he's been in a bad mood all day.* **8** °sorry, °regretful, °apologetic, contrite, rueful, °sad, conscience-stricken, °remorseful, °upset: *She felt bad about having invited me.* **9** °sad, depressed, °unhappy, °dejected, °downhearted, disconsolate, °melancholy; °inconsolable: *I feel bad about the loss of your cat.* **10** °naughty, ill-behaved, misbehaving, °disobedient, °unruly, °wild; °mischievous: *Ronnie isn't a bad boy, he's just bored.* **11** distressing, °severe, °grave, °serious, °terrible, °awful, °painful: *He is laid up with a bad case of the mumps.*

badly *adv.* **1** °poorly, defectively, insufficiently, inadequately, unsatisfactorily, carelessly, ineptly, shoddily, inadequately, deficiently: *We lived in a badly furnished apartment on the other side of the tracks.* **2** unfortunately, unluckily, unsuccessfully, unfavorably, °poorly: *These are an improvement on the former rules, which worked badly.* **3** incorrectly, faultily, defectively, °poorly, improperly, inaccurately, erroneously, unacceptably; ineptly, inartistically, amateurishly, °awfully: *He speaks English badly. He sings badly.* **4** immorally, wickedly, viciously, mischievously, naughtily, shamefully, improperly, villainously: *The school has had its share of badly behaved pupils.* **5** °dangerously, °severely, gravely, critically, grievously, °seriously: *Her father was badly wounded in the war.* **6** unkindly, cruelly, harshly, °severely, wretchedly, dreadfully, improperly, atrociously, horribly, unspeakably: *The prisoners were treated so badly that few survived.* **7** unfavorably, damagingly, critically: *Even her friends spoke badly of her.* **8** very much, greatly, °seriously: *From hearing him* *sing, I'd say that Peter is badly in need of voice lessons.* **9** distressfully, emotionally, °hard: *He took the news badly.*

bag *n.* **1** °sack, shopping bag, reticule, string bag, *Chiefly Brit* carrier bag, *Scots and dialectal* °poke, °pocket: *They have helpers at the market who carry your bags to your car.* **2** baggage, °luggage, valise, satchel, °grip, °suitcase, overnight bag, carry-on luggage *or* bag, Gladstone bag, carpetbag, portmanteau, toilet kit *or* case, sponge bag; brief case, attaché case, dispatch *or* despatch case: *Boarding in London I flew to Buenos Aires while my bag went to Seoul.* **3** °purse, handbag, evening bag, °wallet, *Scots Highland dress* sporran: *She reached into her bag and felt the gun that Brian had given her.* **4** crone, °hag, °beast, ogress, gorgon, nightmare, °witch, harridan, *Archaic* beldam, *Slang* old bat, dog, °monster, *US* two-bagger: *Derek has been romancing some old bag for her money.* **5** °occupation, °hobby, avocation, °business, °vocation, °department, °concern, °affair, *Colloq* °lookout, °worry, *Slang* °thing: *Peter's bag at the moment is learning to play the violin.* —*v.* **6** °catch, °trap, ensnare, °snare, entrap, °capture, °land; °kill, °shoot: *We bagged six pheasants and two partridges this morning.*

balance *v.* **1** °weigh, °estimate, °ponder, °consider, °deliberate, assess, °compare, °evaluate: *One must balance the advantages against the disadvantages.* **2** °steady, °poise; °equalize, stabilize, °level, °match, even out *or* up: *The seesaw will balance better if both of you get on the other end.* **3** °compensate (for), °make up for, counterbalance, °offset, °match, °equal; counterpoise: *The column of mercury in the barometer balances the atmospheric pressure on the surface of the bowl. The total of expenses seems to balance the total of income.* —*n.* **4** scale(s), steelyard: *According to the balance, the package weighs two pounds.* **5** °control, °command, °authority, °weight, °preponderance: *Britain held the balance of power during those decades.* **6** equilibrium, °stability, steadiness, °footing; equiponderance; °equality, °harmony: *The acrobat almost lost his balance on the high wire. The judges tried to maintain some balance between the beauty and talent of the contestants.* **7** °remainder, °residue, °rest; °excess, °surplus, °difference: *You take these and I'll follow with the balance. My bank balance is down to zero.*

ban *v.* **1** °prohibit, °forbid, °outlaw, proscribe, interdict, °bar, disallow, debar: *They have banned smoking in all public places.* —*n.* **2** °prohibition, °taboo, proscription, interdiction, interdict; °embargo, °boycott: *They have put a ban on the sale of alcoholic beverages. The ban against importing firearms is strictly enforced.*

banal *adj.* trite, hackneyed, stereotyped, clichéd, stereotypical, commonplace, old hat, °stock, °common, °everyday, °ordinary, °pedestrian, °humdrum, °tired, unoriginal, unimaginative, platitudinous; trivial, °petty, jejune, *Slang* corny: *The book was blasted as banal and boring. The plot of boy-meets-girl, though banal, still brings in the audiences.*

band[1] *n.* **1** °strip, ribbon, °belt, bandeau, fillet, °tie; °stripe, °line, °border: *He wears a cloth band round his head to keep the sweat out of his eyes. There is a decorative band at the top of each page.* —*v.* **2** °line, °stripe, °border: *The column is banded at intervals with bas-reliefs depicting scenes from the emperor's life.* **3** °tie, °keep, °bind: *Only those papers that are banded together should be sent to my lawyer.*

band[2] *n.* **1** °company, troop, °platoon, °corps, °group, °body, °gang, horde, °party, °pack, °bunch: *We were set upon by a band of robbers in the forest.* **2** °group, °ensemble, °combination, orchestra, *Colloq* combo: *A jazz band plays at the civic center every Tuesday evening.* —*v.* **3** ***band together.*** °unite, confederate, gather *or* join *or* league together, °team up, affiliate, °merge, federate: *We must band together if we expect to accomplish anything.*

banish *v.* **1** °exile, expatriate, deport, extradite, °transport, °eject, oust, °expel, rusticate, send to Siberia, drive out *or* away, °dismiss, excommunicate, °outlaw, °ostracize: *After ten years in prison, the thief was released and banished from the kingdom.* **2** °drive, drive out *or* away, °expel, °cast out, °dismiss, °reject: *He tried to banish suspicion from his mind.*

banner *n.* **1** °standard, °flag, °pennant, ensign, burgee, gonfalon, pennon, °streamer, banderole; °symbol: *The flag of the United States is called the star-spangled banner. He is seeking election under the banner of the Tories.* —*adj.* **2** °leading, °foremost, °momentous, °memorable, °notable, °important, °noteworthy: *The firm had a banner year, with profits up 25 percent.*

banquet *n.* **1** °feast, sumptuous repast *or* meal, ceremonial dinner, lavish dinner: *After the banquet, the guest of honor arose to make a few remarks.* —*v.* **2** °feast, °indulge, wine and dine, °regale, °carouse: *The winners of the trophy banqueted night after night on champagne and caviar.*

banter *n.* °raillery, badinage, persiflage, pleasantry, jesting, joking, °repartee; , chaffing, teasing, °chaff; *Colloq* kidding, ribbing: *Despite the good-natured banter between them, Farmer knew that Fellowes really detested him.*

bar *n.* **1** °rod, °shaft, °pole, °stick, °stake: *A heavy iron bar is used to tamp the dynamite into place in the hole.* **2** °strip, °stripe, °band, °belt; °streak, °line: *The company trademark is a narrow red bar around the barrel of every ballpoint pen.* **3** °barrier, °obstacle, °obstruction, barricade, °hindrance, °block, °deterrent, °impediment; °ban, °embargo: *A steel bar was across the entrance. Her pride proved a bar to her success. There is a bar against importing spirits.* **4** sandbar, °shallow, shoal, bank, sandbank: *Because the keel is too deep, the sloop will be unable to cross the bar till high tide.* **5** °tribunal, court, courtroom, law court, bench: *The former mayor was found guilty of corruption at the bar of public opinion.* **6** barroom, saloon, public house, °café, °lounge, cocktail lounge, tavern, taproom, canteen, *Brit* °local, wine bar; *Colloq* °pub; *Slang* boozer, gin mill: *I was at the bar on my third beer when she walked in.* **7** °counter: *We had a quick lunch at the sandwich bar.* —*v.* **8** °fasten, °close up, °secure, °shut up; °lock, °lock up, padlock: *We tried to get in through the window, but they had barred it.* **9** °block, °obstruct, °stop, °stay, °hinder, °keep (out), °shut out, °exclude, °prevent, °forbid, °prohibit, °set aside; °forestall, °impede, °hamper, °retard, balk, barricade; °ban, °embargo: *After her behavior, she was barred from the club for a year. A huge man in an ill-fitting dinner jacket barred my way. The new regulations bar the import of firearms.* —*prep.* **10** °except (for), excepting, excluding, °barring, °outside (of), save for, aside from, but: *It's all over now bar the shouting.*

barbarian *n.* **1** °savage, °brute: *The barbarians wore animal skins.* **2** °boor, lowbrow, lout, oaf, °clod, churl, °philistine, ignoramus, °yahoo; hooligan, vandal, ruffian, °tough, *Brit slang* yob, yobbo, skinhead: *Those barbarians ought to be denied admittance to the games.* —*adj.* **3** °uncivilized, uncultivated, uncultured, °philistine, °savage; barbarous, barbaric, °coarse, °vulgar, uncouth, °rude; °boorish, loutish, oafish, °crude, °rough, insensitive, churlish, uncivil: *Geologists believed that barbarian man existed at an enormously remote period. Technically, that may be true, but one sees a great deal of barbarian behavior every day.*

barbarity *n.* cruelty, inhumanity, ruthlessness, savagery, brutishness, barbarousness, heartlessness, viciousness, coldbloodedness, bloodthirstiness: *The barbarity of this mass murderer cannot be overstated.*

bare *adj.* **1** unclothed, °naked, °nude, stark-naked, unclad, exposed, uncovered, undressed, hatless, unshod, discalced, *Brit* starkers; *Colloq* in the altogether, in one's birthday suit, in the buff; *Slang US* bareass: *He stood completely bare in the middle of the room.* **2** unconcealed, °undisguised, °open, revealed, °literal, bald, °manifest, °out-and-out, °overt, uncovered, straightforward, °direct, °unvarnished, unembellished, °cold, °hard, °plain, °unadorned, °basic, °simple: *The bare facts indicate that she was the culprit.* **3** unfurnished, undecorated, °vacant, stripped, °empty: *The tenants not only disappeared without paying the rent but took the furniture, leaving the apartment totally bare.* **4** denuded, stripped, leafless, defoliated, shorn, °barren; bared: *After the storm, the trees were bare of foliage. The hurricane began blowing in earnest, and the little ketch was driving forward under bare poles.* **5** °plain, °mere, °simple, °minimal, °essential, °absolute, °basic; °meager, scant, °scanty: *For years we scraped by with only the bare necessities of life.* —*v.* **6** °expose, °lay bare, uncover, °reveal, °open; undress, °unveil: *The torrential rain had washed away the soil, baring the clay and rock beneath. He tore off his shirt, baring his hairy chest.* **7** °disclose, °reveal, °lay bare, uncover, divulge, °unfold, °tell, °expose, unmask, °bring to light: *Because of the way he had treated her, she decided to bare his secrets to the police. Benson bares his soul in his book.* **8** °strip, °divest, denude; defoliate: *The autumn winds bared all the trees in the arboretum.*

barefaced *adj.* **1** unconcealed, °open, °undisguised, °blatant, °manifest, °unmitigated, °outright, °downright, °out-and-out, °sheer, unalloyed, °undiluted: *His proposal is a barefaced attempt by the military to gain control.* **2** °audacious, °impudent, °shameless, °insolent, °impertinent, °immodest, °bold, arrant, °unabashed, °forward, °brazen, °brassy, saucy, °pert, unblushing, *Colloq* °cheeky: *You, sir, are a barefaced liar, and I'll see you in court to answer a charge of libel!*

barely *adv.* °scarcely, °only, °just, not quite, °hardly, only just, no more than: *I barely had my coat off when she said she'd forgotten to shop for dinner.*

bargain *n.* **1** °agreement, °contract, °understanding, °arrangement, covenant, °pact, °compact, °settlement, °transaction, °deal: *We made a bargain—I would provide the materials and he would do the work.* **2** good deal, *Colloq* giveaway, *US* °steal: *If you paid only $100 for this painting, you got a real bargain.* —*v.* **3** °negotiate, °trade, °haggle, barter, °dicker, chaffer: *We bargained far into the night, and I finally traded my car for his motorcycle.* **4** ***bargain for.*** °expect, °count on, °anticipate, °foresee, °take into account, allow for, be prepared for: *Even though the storm had been predicted, it was windier than we had bargained for.*

barren *adj.* **1** °sterile, childless, °infertile: *We won't have any calf from this barren cow.* **2** unproductive, °sterile, °bare, °infertile; °fruitless, °dry, unfruitful, °unprofitable, °poor: *The land was exceedingly stony and barren. The period was the most barren in the history of English literature.*

barrier *n.* **1** °bar, °fence, railing, °wall; ditch, ha-ha: *A barrier was erected at each end of the street. I put up a barrier to keep the chickens from straying into the neighbors' yard.* **2** °obstacle, °bar, °obstruction, °block, °impediment, °hindrance: *Neither race, nor creed, nor color shall be a barrier to success.* **3** °boundary, boundary line, °limit, °frontier: *No mountain barrier lay between France and Flanders.*

barring *prep.* excluding, °exclusive of, °bar, omitting, leaving out, excepting, °except (for), save for, aside from, °besides, but: *Barring another stock market crash, your money is safe. Nobody else, barring the author, knew the truth.*

base[1] *n.* **1** °bottom, foot, °support, °stand, °pedestal: *The base of the statue cracked and the whole thing fell down. Have you been able to find a teak base for the new lamp?* **2** °groundwork, °background, fundamental principle, °principle, °foundation, underpinning; infrastructure, °basis: *Henry's charter was at once welcomed as a base for the needed reforms.* **3** °root, °theme, °radical, °stem, °core: *In the word* interdigitation *the base is* -digit-. **4** °home, °station, °camp, starting point, point of departure, °post, °center: *Using the Sherpa village as a base of operations, we set up smaller camps as we began to climb the mountain.* —*v.* **5** °establish, °found, °secure, °build, °ground, °anchor, °fix, hinge, °form; °derive, °draw: *All our hopes are based on his making the sale.* **6** °establish, headquarter, °post, °station, °position, °place: *The company is based offshore.*

base[2] *adj.* **1** °low, undignified, °cowardly, °selfish, °mean, °despicable, °contemptible, °filthy, °evil: *He must have had some base motive in revealing to her what Martha had said.* **2** degraded, °degrading, °menial, °inferior, °mean, °unworthy, °lowly, °low, °groveling, °servile, slavish, subservient, °downtrodden, abject, °miserable, °wretched, °sordid, undignified, ignoble, °dishonorable, °disreputable, °vile, °scurrilous, °wicked, *Colloq* infra dig: *Foolish sinners will submit to the basest servitude, and be attendants of swine.* **3** °mean, °cheap, °sorry, °common, °poor, °shabby, °shoddy: *He cast off his base attire, revealing a splendid suit of armor, like burnished gold.* **4** °sordid, °offensive, °lewd, °lascivious, °obscene, °profane, °rude, °raw, ribald, °unseemly, °vulgar, °coarse, °rude, °dirty, °indecent, °evil-minded, °filthy, °pornographic: *The entertainment in that theater caters to the basest appetites.* **5** °poor, °shoddy, °cheap, °fake, pinchbeck, °inferior, °counterfeit, °fraudulent, debased, forged, °spurious, °worthless, °bad: *Her jewels look valuable but are, in fact, made of base materials.* **6** °wicked, °evil, °wretched, °corrupt, °shameful, currish, °loathsome, °scurvy, °insufferable, °villainous: *"Base infidel know / You have trod on the toe / Of Abdul, the Bulbul, Ameer!"*

bashful *adj.* **1** °shy, °retiring, °embarrassed, °meek, abashed, °shamefaced, °sheepish, °timid, diffident, self-effacing, unconfident; °ill at ease, uneasy, uncomfortable, °nervous, °self-conscious, °awkward, °confused, *Colloq* in a tizzy, *US and Canadian* discombobulated: *Mr. Holly is so bashful around women that he blushes merely talking to them.* **2** °modest, °coy, unassuming, unostentatious, demure, °reserved, restrained, *Rare* verecund: *When we first met, she was a bashful young girl of fifteen who had no notion of her own beauty.*

basic *adj.* °fundamental, °essential, °key, °elementary, underlying, °prime, °primary, °root; °principal, °central, °focal, °vital: *He enrolled for a basic course in Sanskrit. We must reconsider the basic facts.*

basis *n.* **1** °foundation, °base, °bottom, °heart, °footing, °principle, underpinning; infrastructure: *At the basis of elementary education are the three R's. Does modern Western society still rest on the basis of the family?* **2** °essence, main ingredient *or* constituent, point of departure: *The basis of the discussion is that the hospitals are understaffed.*

batch *n.* **1** °quantity, °lot; °amount, °volume: *Mother baked a huge batch of bread.* **2** °set, °group, °number, °quantity, °assortment, °bunch, °pack, °collection: *Please sort this batch of cards into alphabetical order.*

batter *v.* **1** °beat, °hit, °strike, clout, °belabor, °pound, pummel *or* pommel, °pelt, bash, smite, thrash, *Colloq* wallop, clobber: *He was battered till he was black and blue.* **2** °bombard, °attack, °assault: *Battering by the cannons finally breached the wall of the fort.* **3** maltreat, °mistreat, °ill-treat, °abuse; maul, °bruise, °harm, °mangle, disfigure: *The police report an increase in complaints about battered wives and children.*

battle *n.* **1** °fight, °conflict, °combat, °action, °encounter, °clash, °engagement, °struggle, donnybrook, °fray, *Law* affray; °brawl, °fracas, melee *or* mêlée; °contest; duel, hand-to-hand encounter: *You won the battle, but you lost the war. And now, the battle between the world champion and the challenger!* **2** °argument, °dispute, altercation, °quarrel, °war; °contest, °competition; °struggle, °fight, °crusade, °campaign: *The battle spilled out of the restaurant and into the street. We are not yet winning the battle against AIDS.* —*v.* **3** Usually, ***battle against.*** °fight, contend *or* struggle *or* fight with *or* strive against, °combat: *We must battle against ignorance at every opportunity.*

bauble *n.* °gewgaw, trinket, °ornament, °trifle, °toy, bagatelle, knickknack, °plaything, kickshaw: *Wear your diamonds to the ball, my dear, not those cheap baubles.*

bawdy *adj.* °lewd, °obscene, °taboo, °vulgar, °dirty, smutty, °filthy, °coarse, °earthy, °gross, °raw, scatological, °rude, °lascivious, salacious, indelicate, °indecent, indecorous, °broad, °crude, ribald, °risqué, °suggestive, Rabelaisian, °uninhibited, unrestrained, °lusty, *Literary* lubricious *or Literary* lubricous: *Afterward, each of us had to recite a bawdy limerick.*

bawl *v.* **1** °shout, °bellow, vociferate, °roar, °yell, trumpet, °thunder, *Colloq* holler: *Fishwives bawled out their wares continuously, creating a deafening din.* **2** °cry, wail, °weep, °keen, squall, blubber, whimper; yelp, *Colloq* yammer: *Stop that bawling or I'll really give you something to cry about!* **3** ***bawl out.*** °scold, °reprimand, °upbraid: *My father bawled me out because I stayed out past midnight.*

beach *n.* **1** shore, lakeshore, bank, seashore, seaside, strand, °coast, °margin, *Formal* littoral: *The children wanted to go to the beach and build sandcastles.* —*v.* **2** °ground, run aground, strand; °careen: *Despite the heavy surf, we finally beached the boat safely.*

beacon *n.* °signal, °sign, °fire, °light, bonfire, °flare, signal fire, Very light, °rocket; lighthouse, pharos: *Beacons blazed at the tops of the hills to spread the news of the victory. The drunkard's nose shone like a beacon.*

beam *n.* **1** °timber, scantling, girder, rafter; °bar, °brace, °plank, °board, stud, trestle: *Are you sure that these beams will support the weight of the upper stories?* **2** °ray, °gleam; °shaft; pencil: *I could just make out his face in the beam of the electric torch.* —*v.* **3** °radiate, °shine; smile radiantly: *The door opened and the firelight beamed forth onto the snowdrifts. "I'm so happy to meet you at last," she beamed.*

beamy *adj.* °broad, °wide, broad in the beam; °big, °heavy, °chubby, chunky, °fat, °obese: *She's quite a beamy boat, with accommodation for eight below. A beamy gentleman sat on my homburg, squashing it flat.*

bear *v.* **1** °carry, °transport, convey, °move, °take, *Colloq* tote: *She was borne round the stadium on the shoulders of her teammates.* **2** °carry, °support, °sustain, °shoulder, °hold up, °uphold; °suffer, °undergo, °experience, °endure: *Looking after her invalid mother while working is a heavy burden to bear.* **3** °merit, be worthy of, °warrant; °provoke, invite: *Gordon's suggestion bears looking into.* **4** °stand, °abide, °tolerate, °brook, °survive, °endure, °stand up to; reconcile oneself to, admit of, *Colloq* °put up with: *How can you bear such boring people? His actions will not bear examination. I cannot bear to see you unhappy.* **5** °have, °carry, °show, °exhibit, °display, °sustain: *The getaway car bore German license plates. The knight bore the scars of many battles. She bears her grandmother's name.* **6** °produce, °yield, °develop, °breed, °generate, engender; give birth to, °spawn, °bring forth: *Are you certain that this apple tree will bear fruit this year? She bore thirteen children and still had time to write books.* **7** °entertain, °harbor, °wish: *He bore her no ill will, despite her accusations.* **8** ***bear on*** *or* ***upon.*** relate *or* have relevance *or* be relevant to *or* pertain to, touch on *or* upon, °affect, °concern, have a bearing on *or* upon, °influence: *I don't quite see how your illness bears on which school James attends.* **9** ***bear out.*** °confirm, °support, corroborate, °substantiate, °uphold, °back up: *The evidence bears out what I said.* **10** ***bear up.*** **(a)** °survive, °hold out, °stand up, °hold up, °withstand: *Can Alex bear up under the strain of keeping two jobs?* **(b)** °support, °cheer, °encourage: *What hope have you to bear you up?* **11** ***bear with.*** °put up with, be patient with, make allowance(s) for: *If you'll bear with me, you won't be disappointed.*

bearable *adj.* °tolerable, °supportable, endurable, °acceptable, °manageable: *The heat last summer was made bearable only by frequent dips into the swimming pool.*

bearing *n.* **1** °carriage, deportment, °manner, °behavior, °conduct, °aspect, demeanor, °posture, °stance, °air, °attitude, mien, °presence: *Drumforth's noble bearing makes him noticeable, even in a crowd.* **2** sustaining, supporting, °endurance, °enduring: *Thomas Jefferson considered the government of England totally without morality and insolent beyond bearing.* **3** °aspect; °relation, °reference, °relationship, correlation, pertinence, °relevance, °connection, relevancy, applicability, °application, germaneness, °significance: *The legal bearing of the case will become obvious in court. It is unclear exactly what bearing your remarks have on the situation.* **4** Often, ***bearings.*** °direction, °orientation, °(relative) position: *The bearing of the lighthouse is now 180 degrees. Which way is north? I have lost my bearings entirely.*

beast *n.* **1** °animal, °creature, being: *He loves all the beasts of the field, of the sea, and of the air.* **2** °brute, °savage, °animal, °monster: *That beast has been known to strike his wife in public.*

beastly *adj.* **1** °uncivilized, uncultivated, uncivil, °rude, °crude, °boorish, °unrefined, °coarse; °cruel, °inhuman, °savage, barbaric, barbarous, bestial, °brutal: *Priscilla treats Cyril in a beastly way.* **2** °abominable, intolerable, °offensive, unpleasant, °awful, °terrible, °ghastly, horrid, °disagreeable, °horrible, °hateful, execrable; °foul, °vile, °nasty, °rotten, °dirty, °filthy: *If this beastly weather keeps up, the plane won't take off.*

beat *v.* **1** °strike, °pound, bash, smite, °batter, pummel *or* pommel, °belabor, °pelt, clout, thrash, give (someone) a thrashing *or* beating, drub, °manhandle, thump, whack, cane, °scourge, °whip, bludgeon, °club, cudgel, fustigate; °whip, °flog, °lash, *Colloq* clobber, wallop, give (someone) a once-over: *At first he refused to tell them, but then they beat it out of him.* **2** °defeat, °best, worst, win (out) over, vanquish, trounce, °rout, °outdo, °subdue, °overcome, °overwhelm, °preempt; °surpass, °conquer, °crush, °master, *US* beat out: *Can they beat the Dodgers for the pennant? He first beat the Danes, then the Russians.* **3** throb, °pulsate, palpitate, °pound, thump: *I could feel my heart beating against my ribs.* **4** *Nautical* °tack: *Closehauled, the sloop was beating to windward against the howling gale.* **5** hammer, °forge, °shape, °form, °fashion, °make, °mold: *They shall beat their swords into plowshares.* **6** °mix, °whip, °stir, °blend: *Beat two eggs, then add the flour* *and sugar.* 7 tread, °wear, °trample: *Build a better mousetrap and the world will beat a path to your door.* 8 ***beat it.*** °depart, °leave, abscond, run off *or* away, *Slang US* take it on the lam, lam out of here, *US* hit the road: *You'd better beat it before the cops come.* 9 ***beat off.*** drive off *or* away, °rout: *We beat off our attackers, who fled into the forest.* —*n.* 10 °stroke, °blow: *The signal was to be three beats of a tin cup on the pipes.* 11 °rhythm, °tempo, °measure; °pulse, throb, °stress, pulsation: *In boogie-woogie the beat is eight to the bar.* 12 °course, °round, °tour, °route, °circuit, °run, °path; °area, bailiwick: *In the old days, it was the cop on the beat who prevented a lot of crime. As a reporter, my beat is the financial news.* —*adj.* 13 dead beat, °exhausted, °spent, drained, worn-out, °weary, bone-tired, °fatigued, fagged: *I was really beat after winning the marathon.*

beautiful *adj.* 1 °attractive, charming, °comely, °lovely, good-looking, °fair, °pretty, alluring, appealing, °handsome, °radiant, °gorgeous, *Formal* pulchritudinous, *Scots* °bonny; *Colloq* smashing: *She's not only intelligent, she's beautiful. She entered on the arm of a beautiful youth.* 2 °excellent, °first-rate, unequaled, °skillful, °admirable, °magnificent, well done; °superb, spectacular, °splendid, °marvelous, wonderful, °incomparable, °superior, °elegant, °exquisite, °pleasant, °pleasing, °delightful, *Colloq* smashing: *The garage did a beautiful job in tuning the engine. Armand's arranged a beautiful wedding reception for us.*

beautifully *adv.* 1 attractively, chicly, fashionably, delightfully, charmingly, splendidly, magnificently, *Colloq* smashingly: *The princess was beautifully dressed in a rose satin ball-gown appliquéed with pearls.* 2 admirably, superbly, excellently, wonderfully, marvelously, splendidly, spectacularly, magnificently, *Colloq* smashingly: *Emil did beautifully in the decathlon, even though he came in second.*

beautify *v.* adorn, °embellish, °decorate, °ornament, titivate, °elaborate, garnish, deck (out), bedeck: *The old façade was removed and the building beautified by refacing it with white marble.*

beauty *n.* 1 loveliness, attractiveness, handsomeness, pulchritude: *The beauty of the actress took my breath away.* 2 belle, *Colloq* looker, °knockout, °dream, dreamboat, stunner: *She was one of the great beauties of her day.* 3 °attraction, °strength, °advantage, °asset: *The beauty of the plan is that they won't suspect an old fogy in a wheelchair of being a bank robber.*

beckon *v.* °signal, °gesture, °motion; °summon, °bid, °call: *The manager beckoned to me and I went over to see what he wanted.*

become *v.* 1 turn *or* change *or* transform into: *The princess kissed the prince, who immediately became a frog.* 2 grow *or* develop *or* evolve into; mature *or* ripen into: *At last the acorn has become an oak tree.* 3 °enhance, °suit, °fit, befit, be proper *or* appropriate for, °behoove: *Moonlight becomes you, / It goes with your hair.* 4 °grace, adorn: *Wallingford was a man who became the dignity of his function as a doorman.* 5 ***become of.*** come of, happen to: *What will become of you if you don't go to school?*

becoming *adj.* enhancing, beautifying, °seemly; °attractive, °comely, °fetching, °chic, °stylish, °fashionable, °tasteful; °appropriate, °fitting, °fit, °meet, °befitting, °proper, °suitable: *Your new hairdo is most becoming, Frances.*

bedlam *n.* °pandemonium, °uproar, hubbub, commotion, °confusion, °tumult, turmoil, °furor, °chaos; madhouse: *The announcement created instant bedlam in the Oval Office.*

bedraggled *adj.* soiled, °dirty, °muddy, muddied, °untidy, stained, disheveled, scruffy, messy; °wet, °sloppy, soaking *or* sopping *or* wringing wet, soaked, drenched, *Colloq Brit* gungy, *US* grungy: *We took the two bedraggled waifs in out of the pouring rain.*

befitting *adj.* °fitting, °becoming, °due, suitable *or* suited (to), °appropriate (to), apropos, °proper (to), °seemly (for): *He really ought to behave in a manner befitting his position as chairman. This must be done with a befitting sense of awe.*

before *adv.* 1 °previously, earlier, already, beforehand; °formerly, in the past; °once: *I have told you before, don't count your chickens.* 2 °ahead, °in advance, °in front, in the forefront, °first, in the vanguard, *Colloq* °up front: *He let his wife walk before, as he knew the road was mined.* 3 °ahead, in the future, to come: *Before us lie the prospects of surrendering or dying.* —*prep.* 4 ahead of, in advance of, in front of, forward of: *The king indicated that the page should go before him.* 5 in front of; in the presence of: *The entire valley was spread out before me.* 6 °preceding, previous *or* anterior to, °prior to; on the eve of: *Before my departure I have to kiss Annie goodbye.* 7 in preference to, rather than, sooner than, more willingly than: *They said they would die before yielding.* —*conj.* 8 previous to *or* preceding the time when: *This was a good book before the copy editors got their hands on it.*

beg *v.* 1 entreat, °beseech, °plead (with), crave, implore, importune, °wheedle, °cajole, supplicate (with), °pray; °ask for, °request: *She begged me to stay.* 2 °solicit, sponge, *Colloq* cadge, °scrounge, *US* panhandle: *When he was an alcoholic, he used to beg drinks off everyone.*

beggar *n.* 1 mendicant, °supplicant, suppliant, almsman, sponger, °tramp, vagrant, °pauper, *Colloq* cadger, scrounger, *US* panhandler: *We were approached by beggars on every street corner.* —*v.* 2 impoverish; °want, °challenge, °defy, baffle: *The misery of those people beggars description.*

begin *v.* 1 start (out *or* off *or* in *or* on), °initiate, enter on *or* upon, set out *or* about, set out on *or* upon, *Somewhat formal* °commence: *We began the journey full of enthusiasm.* 2 °start (off), °inaugurate, °originate, °open, °launch, °create, °establish, °found, °set up; °go into: *We began the company five years ago.* 3 °arise, °start, °originate, *Somewhat formal* °commence: *The greatness of the Prussian monarchy begins with Frederick II. The paragraph begins in the middle of the page.*

beginning *n.* 1 °start, commencement, °outset, °onset, inception, °dawn, dawning, °birth, genesis, °origin, °creation, day one; origination, °source, wellspring: *There are several competing theories about the beginning of life on earth. The beginning of the idea can be traced to Galileo.* 2 °opening, °start, inception, commencement: *I have plenty of energy at the beginning of the day. The book is good at the beginning, but then it gets boring.*

begrudge *v.* 1 °resent, °envy, °grudge: *She doesn't begrudge him his success.* 2 give (be)grudgingly *or* unwillingly *or* reluctantly, °deny, °refuse: *He begrudges her the slightest consideration.*

beguile *v.* 1 delude, °deceive, °cheat, °swindle, °dupe, °fool, °mislead, °hoodwink, bamboozle, °take in: *She was easily beguiled by his solicitude.* 2 °defraud (of), °deprive (of), cheat (out of *or* into), °swindle (out of): *Let no man beguile you of your reward.* 3 °charm, °divert, °amuse, °distract, °fascinate, engross, °engage, allure: *I always meet the most beguiling people at Daphne's parties.*

behalf *n.* ***on*** *or* ***US in behalf of*** *or* ***on*** *or* ***US in one's behalf.*** °for, as a representative of, in place of, °instead of, in the name of, on the part of; in the interest of, for the benefit *or* advantage of: *The lawyer is acting on behalf of the heirs.*

behave *v.* °act, °react, °function, °operate, °perform, °work, conduct *or* deport *or* comport *or* bear (oneself); act obediently, act properly, be good: *The boy behaved with great insolence. I wish the children would behave themselves.*

behavior *n.* °conduct, demeanor, deportment, °bearing, °manners, comportment; action(s): *His behavior in the presence of the royal couple was abominable.*

behead *v.* decapitate, guillotine, *Archaic* decollate: *Criminals and enemies of the state were formerly beheaded.*

behold *v.* °see, °look at, °regard, set *or* lay eyes on, descry, °notice, °note, espy, °perceive, discern, °remark, °view: *As we emerged from the gorge, we beheld the mountain looming above us.*

beholden *adj.* °obliged, obligated, °indebted, °grateful, °in debt, under (an) obligation: *She said that she was beholden to him for everything he had done.*

behoove *v.* be required of, be incumbent on, be proper of, be fitting of *or* for, befit; be advisable for, be worthwhile for, be expeditious for *or* of, be advantageous to *or* for, be useful to *or* for, be beneficial to *or* for: *It behooves you to be respectful to the governor of the prison.*

belabor *v.* thrash, °beat, pummel *or* pommel, buffet, °pelt, °lambaste: *We tried to stop the drover from belaboring the poor horse with a whip.*

belated *adj.* °late; behind time, behindhand, °out of date; delayed, detained: *I forgot your birthday, so here's a belated gift.*

belief *n.* **1** °trust, dependence, °reliance, °confidence, °faith, °security, °assurance: *He retains his belief in the divine right of kings.* **2** acceptance, credence; assent: *His statements are unworthy of belief.* **3** °tenet, °view, °idea, °sentiment, °conviction, °doctrine, dogma, principle(s), axiom, °maxim, °creed, °opinion, °persuasion: *The belief that there is no God is as definite a creed as the belief in one God or in many gods.* **4** °intuition, °judgment: *It is her belief that nuclear energy will eventually prove economical.*

believe *v.* **1** °accept, put faith *or* credence in *or* into, find credible, find creditable; °allow, °think, °hold, °maintain, °feel; °take it, °suppose, °assume: *He still believes that the moon is made of green cheese.* **2** ***believe in.*** trust to *or* in, rely upon *or* on, have faith *or* confidence in, put one's trust in, be convinced of, °swear by, °credit: *Do you believe everything you read in the papers? The chairman believes in your ability to carry out the plan.* **3** ***make believe.*** °pretend, °suppose, °imagine, °fancy, conjecture, °assume: *I used to make believe I was a great detective.*

belittle *v.* °diminish, °minimize, °disparage, °slight, decry, °detract from, °depreciate, °trivialize, deprecate, °degrade, denigrate, °downgrade, deemphasize, °discredit, °criticize, derogate; °reduce, °mitigate, lessen, undervalue, °underestimate, underrate, °minimize, *Colloq* °play down, pooh-pooh: *He belittles the efforts of others but accomplishes nothing himself.*

belligerent *adj.* **1** warring; °warlike, °militant, warmongering, hawkish, jingoistic, bellicose, °martial: *The belligerent nations have agreed to discuss an accord.* **2** °quarrelsome, °pugnacious, contentious, disputatious, °truculent, °aggressive, °hostile, combative, antagonistic, bellicose: *I cannot see why you have to take such a belligerent attitude toward the chairman.* —*n.* **3** warring party, °antagonist, °contestant; warmonger, hawk, jingoist, °militant: *Our country has refused to sell arms to the belligerents in the conflict.*

bellow *v.* **1** °roar; °yell, °shout, °blare, trumpet, °howl, *Colloq* holler: *Father was bellowing that he couldn't find his pipe. The public-address system bellowed out my name.* —*n.* **2** °roar; °yell, °shout, *Colloq* holler: *The bull gave a bellow and charged.*

belong *v.* **1** be a member (of), be affiliated *or* associated *or* connected (with), be attached *or* bound (to), be a part (of): *Didn't he belong to the Democratic party? He didn't want to belong while his wife was a member.* **2** have a (proper) place (in), be proper (to): *Do you ever get the feeling that you don't belong here?* **3** ***belong to.*** be owned by, be the property *or* possession of: *That coat belongs to me.*

belonging *n.* °association, °connection, °alliance, °relationship, °affinity, °relation: *The church gives me a strong sense of belonging.*

belongings *n.* °(personal) property, °effects, °possessions, °goods, °things, chattels: *He returned home to find all his belongings in the street.*

beloved *adj.* **1** loved, cherished, adored, °dear, dearest, °darling, °precious, treasured; admired, worshiped, revered, esteemed, idolized, respected, esteemed; valued, prized: *He denied nothing to his beloved children. She was their beloved queen.* —*n.* **2** °sweetheart, °darling, dearest, °love; °lover, °paramour, inamorata *or* inamorato, *Colloq* °flame: *He wrote poems to his beloved.*

below *adv.* **1** lower down, farther down, further down: *Please see the explanation given below. The department head could no longer resist the pressures from below.* **2** °beneath, underneath, °under; downstairs, *Nautical* below-decks, *Brit* below-stairs: *Can you hear someone walking about below? Put the captain in irons below.* **3** on earth, here, in this world, under the sun: *Man wants but little here below.* —*prep.* **4** °under, underneath, °beneath: *Below the sea live creatures we have never even seen. Barely discernible below his nose was a tiny moustache. Sign your name below "Yours truly."* **5** less *or* lower *or* cheaper than: *The sale price is below cost.* **6** deeper *or* farther down than: *The current is strongest about six feet below the surface.* **7** °under, °beneath, underneath: *Her bright eyes peered at him from below the wide hat.* **8** lower *or* less than, °under: *The temperature was twenty degrees below zero.* **9** inferior *or* subordinate to, lower than: *He gives orders to the servants below him.* **10** inferior *or* secondary to, °under, °beneath, lower than: *In exports, the USA and UK are below Japan.* **11** °beneath, °unworthy of, unbefitting, not worth: *Mugging old ladies is below contempt.*

belt *n.* **1** sash; *Literary* girdle, cestus, cincture, °zone: *At her belt she wore a dagger in a golden scabbard.* **2** °zone, °band, °strip, °circuit, °perimeter; °area, °swath, °tract, °region, °district: *The planners ensured that each city would be surrounded by a green belt.* —*v.* **3** °strike, °hit, °punch; °beat, thrash: *When he insulted her, I simply belted him.* **4** ***belt out.*** sing *or* perform stridently *or* loudly; put over *or* across: *Sophie Tucker was there, belting out "One of These Days."*

bemoan *v.* °lament, mourn *or* grieve *or* weep *or* moan for: *She bitterly bemoaned the loss of her sole companion, her canary.*

bemuse *v.* **1** °confuse, °muddle, °mix up, addle, befuddle, °perplex, °bewilder, °puzzle, *Colloq US and Canadian* discombobulate: *The actors were thoroughly bemused by the sudden appearance of a horse on stage.* **2** stupefy, benumb, °numb, °paralyze: *I found him, completely bemused, with the empty bottle beside him.*

bend *n.* **1** curve, °turn, turning, corner; °bow, °angle, crook, °hook, curvature, flexure: *Go left at the bend in the road. If you put a bend in a wire hanger, you can fish out the obstruction.* —*v.* **2** °arch, °bow, curve, crook: *Soak the branch in water and it will bend easily. Stop bending my arm—it hurts!* **3** °bow; curtsy *or* curtsey; °kowtow, salaam; kneel, genuflect: *The cannibal bent down before a pile of skulls.* **4** °incline, °channel, °focus, °direct, °steer, °set; °fix: *He bent his attention on more important matters. She bent her steps towards the cemetery.* **5** °submit, °bow, °yield, °give way, be pliant *or* subservient *or* tractable: *The cabinet bends to the will of the prime minister.* **6** °incline, °turn, °deflect: *As you can see, the ray is bent by the lens.*

bender *n. Colloq* °drunk, °spree, °bout, °revel, carousal, °carouse, bacchanal; *Slang* binge, °jag, *US* toot: *He goes off on a bender whenever his wife leaves him.*

beneath *adv.* **1** low *or* lower down, °below, °under, underneath: *Please sign beneath if you agree the terms.* **2** °below, underneath, °under; °underground: *The flowers are above the ground, the roots beneath.* —*prep.* **3** °under, underneath, °below: *Beneath that gruff exterior beats a heart of gold.* **4** °below, °unworthy of, unbefitting, undeserving of, not (even) meriting, lower than: *Your behavior is beneath criticism.*

benefactor *n.* °patron, °supporter, °sponsor, °donor, °philanthropist; °backer, investor, °supporter, *Colloq* angel: *Our benefactor has made a donation that will enable the mission to carry on its work.*

beneficial *adj.* **1** °advantageous, °serviceable, °useful, °profitable, °helpful, °supportive, °favorable, °constructive, °good: *No measures could have been more beneficial to the kingdom.* **2** healthful, °healthy, salutary, salubrious; °efficacious, °effective: *A certain amount of sunshine is quite beneficial.*

benefit *n.* **1** °advantage, °profit, °good, °sake, °gain, °aid, °help, °service: *It would be to their benefit to call off the strike.* **2** Often, ***benefits.*** perquisite(s), emolument(s), allowance(s), extra(s), fringe benefit(s), *Colloq* perk(s): *We offer one of the best schemes in the industry for employee benefits.* —*v.* **3** °improve, °aid, °help, °better, °promote, °further, °advance, °forward: *Enrolling in a management course could benefit your chances for advancement.* **4** °profit, °gain: *No one has ever personally benefited a penny from these contributions.*

benevolence *n.* **1** °charity, °kindness, kindliness, °humanity, humanitarianism, beneficence, charitableness, goodness, °altruism, goodwill, unselfishness, °philanthropy, generosity, magnanimity: *The poor rely on our benevolence.* **2** °gift, °grant, contribution, °donation, beneficence: *The victims of the famine were recipients of the benevolence of the British people.*

benevolent *adj.* °charitable, well-disposed, °gracious, °good, °kind, °kindly, humane, °humanitarian, well-wishing, °thoughtful, °considerate, °sympathetic, caring, kindhearted, warmhearted, compassionate, °benign, benignant; °liberal, °generous, magnanimous, openhanded; °beneficial, °helpful, salutary: *That hypocrite has cast himself in the role of a benevolent despot.*

benighted *adj.* unenlightened, naive *or* naïve, °uninformed, °ignorant: *That poor, benighted fool believes that the doctors can cure him.*

benign *adj.* **1** °kindly, °gracious, °good, °kind, kindhearted, °benevolent, benignant, °warm, warmhearted, °cordial, °genial, congenial, °tender, tenderhearted, compassionate, °sympathetic, °softhearted: *It was Grandad's benign goodwill that kept us together in those hard times.* **2** °bland, °gentle, °mild, °warm: *A benign smile lit the headmaster's face as he announced the awards for scholastic achievement.* **3** °kind, °favorable, °fortunate; salutary, salubrious, °mild, congenial, °propitious: *She recovered rapidly in that most benign climate.* **4** nonfatal, nonmalignant, nonvirulent, curable, °harmless: *Fortunately, the biopsy showed that the tumor is benign.*

bent *adj.* **1** curved, deflected, bowed, °crooked, distorted, twisted, warped: *As the twig is bent, the tree's inclined. He complained to the waiter because his fork was bent.* **2** °strange, °weird, °peculiar, twisted, °deviant, warped, °wry, awry, °corrupt, corrupted; °perverted, °perverse, °abnormal: *You'd be bent, too, if you'd been in prison for fifteen years.* **3** °dishonest, °crooked, °illegal: *That share deal sounds a bit bent to me—I'll pass.* **4** °determined, °intent, °set, resolved, °resolute, °decided, °set: *Garvey is bent on running in the marathon, despite his sprained ankle.* —*n.* **5** °turn, °inclination, °direction, °disposition, predisposition, °tendency, °bias, °leaning, proclivity, propensity, °partiality, °prejudice; °ability, °aptitude, °talent, °gift: *She wished to follow the bent of her own taste. He has a natural bent for the violin.*

bequeath *v.* °leave, °make over, °will, °pass on, hand down *or* on, °transmit, *Law* °devise: *Aunt Margaret has bequeathed her collection of music boxes to the museum.*

bequest *n.* legacy, °inheritance: *A huge bequest was received by the hospital.*

berate *v.* °scold, chide, °rate, °upbraid, revile, °abuse, rail at, excoriate, °castigate, objurgate; °harangue: *In the square an ancient virago was berating a butcher.*

bereave *v.* °deprive; °strip, °rob, °dispossess: *The accident bereaved him of his child.*

berserk *adj.* amok, °mad, °violent, °wild, crazed, frenzied, °maniacal: *He went berserk, destroying tables and chairs.*

beseech *v.* supplicate, entreat, implore, °plead (with), °beg, importune, obsecrate: *The prisoners beseeched the king to have mercy on them.*

beset *v.* encompass, °surround, °besiege; assail, °attack, °harass, harry, hector, °bother, °afflict, °trouble: *She was beset by all the problems involved in having a job and a family.*

beside *prep.* **1** alongside, °near, next to, with, °close to, hard by, °nearby, °by: *Along came a spider, / And sat down beside her.* **2** away from, wide of, °apart from, unconnected with, °off: *The fact that I owe you money is entirely beside the point.* **3 *beside oneself.*** out of one's mind *or* wits, at the end of one's tether, °overwrought, °agitated, °upset, °crazy, °mad: *She was beside herself with grief when she heard the news.*

besides *adv.* **1** °in addition, additionally, also; °further, furthermore, °moreover, as well, too; °to boot, on top of everything else, into the bargain: *Maria is our choice for the post and, besides, she's the only qualified person available. On their anniversary he gave her a diamond ring and a sapphire brooch besides.* —*prep.* **2** over and above, above and beyond, °in addition to, additionally to, as well as; aside from, °barring, excepting, °except for, excluding, °exclusive of, not counting or including, beyond, °apart from, other than: *St. Paul became acquainted with many Christians besides his converts.*

besiege *v.* **1** lay siege to, beleaguer: *For ten years Troy was besieged by the Greeks.* **2** blockade, °block, block off *or* up, hem in, °cut off; °surround, crowd round: *The strikers have besieged the plant gates, not allowing anyone in or out.* **3** importune, °sue, °petition, assail, *Brit* pressurize *or US* pressure, °press, °overwhelm, inundate: *The prosecutor's office has been besieged by requests for leniency in your case.*

best *adj.* **1** °superlative, unexcelled, finest, °preeminent, °first, °superb, unsurpassed, °superior, °excellent,

°paramount, °first-rate, *Colloq* A-1 *or* A-one: *Henry VIII was the best rider, the best lance, and the best archer in England.* **2** kindest, most beneficent, nicest: *Which of your brothers is the best to you?* **3** °foremost, choicest, °preeminent, most suitable, most appropriate, most qualified, most talented, most desirable, most outstanding: *We want the best person to fill the job.* **4** largest, most, greatest: *She had traveled the best part of the way by ship.* **5** richest, wealthiest; first-class, upper-crust, °upper-class: *He associates only with those he considers to be the best people.* —*n.* **6** finest; °first: *The best is yet to come.* **7** °finery, best clothes, *Colloq* best bib and tucker: *He was all decked out in his Sunday best.* **8** greatest *or* maximum effort: *His did his best but it wasn't good enough.* —*adv.* **9** most excellently, to the fullest extent, in the most suitable way, most adroitly, most skillfully, most superbly, most artistically: *He prayeth best who loveth best / All things both great and small.* **10** with greatest satisfaction, most successfully: *He who laughs last laughs best.* —*v.* **11** win (out) over, °conquer, °beat, °surpass, °overpower, get the better of, °subdue, °defeat, worst, vanquish, trounce, °rout, °crush, °master, °outdo, °overwhelm, °overcome, °outwit: *He was bested in three falls out of four.*

bestow *v.* °confer; °give, °award, °present, °donate, °grant: *The country has bestowed its highest honors on her.*

bet *n.* **1** wager, °stake, °gamble, °risk, °venture, °speculation, *Brit* punt, *Colloq Brit* °flutter: *He could not afford more than a small bet.* —*v.* **2** wager, °stake, °gamble, °risk, °hazard, °play, lay (a bet *or* stake *or* wager), °put, °chance, °speculate, °venture, *Brit* punt: *Every week he bet a small amount on the lottery.*

betray *v.* **1** be *or* prove false *or* disloyal to, °sell out, break faith with, °let down, °fail, °inform on, *Colloq* sell down the river, *Slang Brit* °shop: *He betrayed her to the enemy.* **2** °reveal, °disclose, divulge, °impart, °tell; °expose, °lay bare: *She betrayed our hideout to the police. He betrayed an unsuspected streak of cowardice.* **3** °lead astray, °mislead, misguide, °deceive, °dupe, °fool, °hoodwink: *He has been betrayed by his own arrogance.*

betrayal *n.* **1** treachery, treason, disloyalty, °perfidy, traitorousness, faithlessness, bad faith, breach of faith, *Slang* °sell out: *His delivery of the country into the hands of an invader was an outright act of betrayal.* **2** °revelation, divulging, disclosure, divulgence: *Men should not be led into betrayals of their secret opinions.*

better *adj.* **1** °superior: *You're a better man than I am, Gunga Din. I don't think you'll find a better wife, Matthew. I know of no better invention than the wheel.* **2** more; greater, larger, bigger: *I waited for her the better part of two hours.* **3** wiser, safer, °well-advised, more intelligent: *It would be better to wait till tomorrow to tell her.* **4** healthier, haler, heartier, less ill *or US* sick, improved; cured, recovered: *You will feel better after you have eaten something.* —*adv.* **5** preferably, °best; more wisely, more advisedly, more safely: *We had better go before the trouble starts.* **6** ***better off.*** **(a)** improved, happier, °well-advised: *You'd be better off attending a technical college.* **(b)** wealthier, richer: *She is better off than any of us.* **7** ***think better of.*** reconsider, think twice, change one's mind: *He was going to fight but thought better of it.* —*n.* **8** °advantage, mastery, °superiority, °control: *Don't let the obstacle course get the better of you.* **9** ***betters.*** superiors: *That young imp should learn how to address his betters!* —*v.* **10** °improve, ameliorate, °advance, °raise, elevate: *It was almost impossible in those days for laborers to better their condition.* **11** °surpass, °excel, °outdo, °outstrip, °beat, °improve: *She bettered her record for the 100-meter race by half a second.*

bettor *n.* gambler, speculator, wagerer, gamester, °player, *Brit* °better, punter, *Colloq* crapshooter, °sport: *The bettors were gathered round the craps table.*

bewail *v.* °lament, °mourn, °bemoan, moan *or* mourn over, shed tears *or* whimper over, weep *or* cry *or* keen over, beat one's breast over: *Instead of bewailing his condition, why doesn't he do something about it?*

beware *v.* take heed, be careful, be wary, be cautious, be on one's guard, exercise caution, °mind, watch out, °look out, take care: *There are shoals nearby, so beware. Beware the ides of March!*

bewilder *v.* °confuse, confound, °perplex, °puzzle, °mystify, befuddle, baffle, °bemuse: *I was bewildered by differential calculus.*

bewitch *v.* °enchant, °entrance, spellbind, °charm, °fascinate, °beguile, cast a spell over, °captivate, °enrapture: *She easily bewitches men with her sultry good looks and her husky, low voice.*

bias *n.* **1** °prejudice, °partiality; °inclination, °leaning, °bent, °disposition, propensity, °tendency, predilection, predisposition, proclivity: *She shows a marked bias in favor of the Irish.* **2** °angle, °slant, diagonal: *Cut the fabric on the bias.* **3** °influence, impulse, °weight: *If he is under any bias, it is on the side of fairness.* —*v.* **4** °influence, affect unduly *or* unfairly, °sway, °incline, °prejudice, °color, °taint, predispose: *Artists are seldom good critics of art because they have been biased.*

biased *adj.* biassed, °prejudiced, °partial; warped, distorted, °jaundiced: *I don't want to listen to your biased opinions about women's rights.*

bicker *v.* °dispute, °quarrel, wrangle, °argue, squabble, °tiff, *Colloq* spat: *The couple next door are always bickering about trifling matters.*

bid *v.* **1** °offer, make an offer (for), °tender, proffer: *We bid $500 for the painting.* **2** *Archaic or literary* °ask, °pray, °press, entreat, °beg, °request, °suggest, invite: *She bade me leave her to mourn alone.* **3** *Formal* °command, °demand, °order, °tell, enjoin, °dictate: *Please bid him enter. Custom bade him blow his horn.*

bidding *n.* **1** °invitation, summons: *We attended the ceremony at the bidding of the President.* **2** °command, °order, °dictate, °direction, °instruction, °demand: *The letter went out to all members at the bidding of the finance committee.*

big *adj.* **1** °large, °great, °grand; °huge, °enormous, °immense, °gigantic, °giant, tremendous, °colossal, Brobdingnagian, °jumbo, *Colloq Brit* socking *or* whacking big *or* great, *US* humongous: *We live in a big house in the country.* **2** °ample, °hefty, °huge, °bulky, °fat, °obese; °large, °hulking, beefy, °burly, °brawny, strapping, gargantuan, elephantine, °enormous, °gigantic, °immense, °monstrous: *The sumo wrestler was one of the biggest men I'd ever seen.* **3** °tall, grown, °mature, grown up, °large: *Cecil is certainly big for his age.* **4** °important, °significant, °outstanding, °weighty, consequential, °major, °grave, °momentous, °notable, °noteworthy, °telling: *Changing careers can be one of the biggest decisions of your life.* **5** °important, °prominent, °illustrious, °noteworthy, °notable, °renowned, °eminent, °distinguished, esteemed: *Johnson is a big man in our town.* **6** °generous, magnanimous, °charitable, °unselfish, giving: *It was very big of her to take on the support of the orphanage.* **7** *Colloq* °popular, °famous, °well-known, °successful: *That singing group is very big with the teenagers these days.* **8** °large, °capital, upper-case, majuscule: *Chicago is always spelled with a big 'C.'* —*adv.* **9** *Colloq* pompously, boastfully, conceitedly, arrogantly, pretentiously: *Alfred talks big at home, but he's very modest when he's with his friends.* **10** successfully, °well, outstandingly, effectively: *I think* *your speech went over big with those opposed to the new budget.*

bigoted *adj.* °prejudiced, °intolerant, °biased, °jaundiced, °one-sided, °partial: *Just one bigoted judge makes the entire judicial system appear to be unfair.*

bigotry *n.* °prejudice, °intolerance, °bias, °partiality: *No government that practices bigotry can survive long.*

bigwig *n.* **1** °boss, kingpin, °king, °queen, nabob, VIP (= 'very important person'), *Colloq* big shot, big gun, big cheese, big wheel, hotshot, °chief, brass hat, *US* (chief) honcho, Mr. Big: *The bigwig sits here, at the head of the table.* **2** ***bigwigs.*** °brass, brass hats: *Don't let the bigwigs find out what we've done.*

bilious *adj.* ill-tempered, bad-tempered, ill-natured, °peevish, °testy, °cross, °petulant, tetchy, choleric, dyspeptic, °angry, wrathful: *The director was absolutely bilious when he heard we had lost the account.*

bill[1] *n.* **1** invoice, °account; °tally, °reckoning, tabulation, *US* °(restaurant) check, *Colloq US* °tab: *Have you paid the telephone bill?* **2** *US and Canadian* °note, banknote, paper money, *Colloq* folding money: *The robbers took only small bills, which they could spend easily.* —*v.* **3** invoice, °charge: *I don't have my checkbook with me—Could you please bill me?*

bill[2] *n.* beak, neb, nib, pecker; jaws: *That bird has a fish in its bill.*

bind *v.* **1** °tie, °fasten, °secure, make fast, °tie up: *The thieves bound him hand and foot.* **2** constrain; °hold, °oblige, °obligate: *The contract we signed is equally binding on both parties. The union is bound by an agreement that expires in a month.* **3** gird, °encircle, wreathe, °wrap, °cover, °swathe, bandage: *They were binding his wounded head.* **4** °cement, °stick, cause to adhere; °attach, °connect: *Ordinary glue will bind these pieces together.* —*n.* **5** *Colloq US* °dilemma, °predicament, tight spot, °(difficult) situation, *Colloq* pickle, °fix, °jam: *I'm in a real bind because I've invited two girls to the party.* **6** *Brit* °annoyance, irritant, °bother, °bore, °trial, °ordeal, irritation, vexation, *Colloq* pain (in the neck *or Colloq* arse): *It was a bit of a bind having to wait three hours for the bus.*

birth *n.* **1** childbirth, °delivery, *Technical* parturition, *Old-fashioned* confinement: *Nicole is expecting the birth of her second child to take place in late March.* **2** °origin, °creation, °emergence, °beginning, °start, origination: *I believe we may be present at the birth of a powerful idea.* **3** nativity, °origin, °extraction; °parentage, °line, °lineage, ancestry, descent, °family, blood: *She is Scottish by birth. Hortense is of noble birth.*

bisexual *adj.* **1** hermaphrodite *or* hermaphroditic(al), androgynous: *Many of these microscopic animals are bisexual and self-fertilizing.* **2** *Colloq* AC/DC, swinging both ways, *Facetious* ambisextrous: *He was known to have had bisexual relationships.* —*n.* **3** androgyne, hermaphrodite: *Statistics showed that more women than previously believed were bisexual.*

bit *n.* **1** °morsel, °piece, °scrap, °fragment, °shred, °particle, °grain, °crumb: *We didn't have a bit of food in the house.* **2** °jot, tittle, whit, scintilla, °trace, °touch, °hint, °suggestion, °suspicion, °particle, iota, °speck, atom: *There's not the slightest bit of evidence to link her with the crime.* **3** °moment, °minute, °second, °flash, *Colloq* two shakes (of a lamb's tail): *I'll be with you in a little bit.* **4** °piece, °share, °equity, °segment, °portion, °part, fraction: *He owns a little bit of the business.*

bitch *n.* **1** °shrew, °nag, termagant, virago, harpy, °fury, spitfire, °scold: *She pestered him so much that he refused to have anything to do with the bitch. That greedy bitch has the house, and now she's suing me for half my income.* **2** whore, °prostitute, bawd, harlot, call girl, trollop, strumpet, trull, °drab, °tart, floozy, streetwalker, *Colloq* bimbo, pro, *US* hooker, °tramp, hustler: *He roamed the street every night, ending up with some bitch he found in a bar.* —*v.* **3** °complain, °object, °protest, grumble, *Colloq* °gripe: *Oh, stop your bitching, Heinrich, and keep digging!* **4** °bungle, °botch, °ruin, °spoil: *They bitched the job by using too little paint.*

bite *v.* **1** °nip; °chew, °gnaw: *That dog of yours bit a piece out of my ankle.* **2** °sting: *She was bitten by a mosquito.* —*n.* **3** °mouthful, °morsel, °scrap, °bit, °piece, °taste; °snack, *Slang* nosh: *The survivors hadn't had a bite of food for three days. Stop over for a bite on Sunday evening.* **4** °sting: *These mosquito bites itch horribly.*

biting *adj.* °severe, °harsh, °cutting, °piercing, °penetrating, °keen, °sharp, °bitter; °cold, °wintry, °freezing: *The biting wind went right through his thin coat.*

bitter *adj.* **1** °harsh, acerbic, acrid, °sharp, °caustic, mordant: *Everyone agrees to call vinegar sour, honey sweet, and aloes bitter.* **2** unappetizing, °distasteful, °unsavory, unpleasant, hard (to swallow *or* take), irritating, °obnoxious, °disagreeable, °nasty, °painful, °unwelcome, °unpalatable: *The additional tax demand was a bitter pill.* **3** °miserable, °grievous, dispiriting, distressing, °cruel, distressful: *Dismissal after all those years was a bitter experience.* **4** °resentful, °embittered, °rancorous; °hateful: *Andrew felt bitter at not being selected as chairman.* **5** stinging, °cutting, °biting, °harsh, °reproachful, °vicious, acrimonious, °virulent; °cruel, °unkind, unpleasant, °nasty: *His bitter denunciation of other candidates lost him the campaign.* **6** °sharp, °keen, °cutting, °severe, °biting, °cold, °wintry, °freezing: *A bitter gale lashed at the rigging.*

bitterness *n.* **1** harshness, acerbity, acrimony, acrimoniousness, °bitterness, spleen, *Literary* gall and wormwood: *The bitterness of his opponent's attack was totally uncalled for.* **2** °animosity, hatred, °resentment; °hostility, °antagonism: *She felt bitterness in her heart over the way she had been treated.*

bizarre *adj.* **1** °eccentric, °unusual, unconventional, °extravagant, °whimsical, °strange, °odd, °curious, °peculiar, °queer, °offbeat, °fantastic, °weird, °incongruous, °deviant, °erratic, *Slang* °kinky: *The police thought his behavior sufficiently bizarre to invite him to the station house for a chat.* **2** °grotesque, °irregular, °nonconformist, nonconforming, °outlandish, °outré, °quaint, °fantastic, unconventional: *His house is a bizarre mixture of baroque and modern design.*

blab *v.* °broadcast, °tattle, °babble, °betray, °reveal, °disclose, divulge, °expose: *Don't tell Frieda—she'll blab your secrets all over town.*

blabbermouth *n.* telltale, babbler, chatterer, °gossip, *Colloq* °blab, tattletale, bigmouth: *Oscar is such a blabbermouth that you can't tell him anything you don't want everyone to know.*

black *adj.* **1** jet, jet-black, coal-black, inky, sooty, swart, °swarthy, raven, ebony, °dusky, *Literary* ebon: *Her hair was as black as coal.* **2** Negro, negroid, colored, dark-skinned: *Most of the black races live near or south of the equator.* **3** °dark, °pitch-black, jet-black, coal-black, Stygian; starless, moonless: *He bundled his coat round himself and walked into the black night.* **4** °dark, °somber, °dusky, °gloomy, °menacing, glowering, louring *or* lowering, °threatening, °funereal: *The sky became black with storm clouds.* **5** °malignant, baleful, baneful, °deadly, deathly, °sinister, °dismal, °hateful, °disastrous: *It was a black day when he came into my life.* **6** °bad, °foul, iniquitous, °wicked, °evil, diabolical, °infernal, hellish, °atrocious, °awful, malicious, °abominable, °outrageous, °vicious, °villainous, flagitious, °vile, °disgraceful, °unscrupulous, °unconscionable, unprincipled, blackguardly, knavish, °perfidious, insidious, nefarious, dastardly, treacherous, unspeakable, °disgraceful, °shameful, °scurvy, °criminal, felonious: *You have told the blackest lies about me.* **7** °angry, wrathful, °furious, frowning, bad-tempered, sulky, °resentful, clouded, °threatening, glowering: *She gave him a black look and he withered in abject fear.*

blacken *v.* **1** darken, smudge, begrime: *The chimney sweep's face was blackened with soot.* **2** °slander, °libel, asperse, cast aspersions on, traduce, °smear, °sully, °soil, besmirch, °taint, °tarnish, defame, revile, malign, °vilify, °discredit, denigrate: *His article has blackened my reputation.*

blackleg *n.* scab, strikebreaker: *Despite the strike, the plant is being operated by blackleg labor.*

blackmail *n.* **1** extortion, °ransom, °tribute, *US* °graft: *Even if you pay the blackmail, that is no guarantee that he won't demand more later.* —*v.* **2** extort money from; °force, coerce, compel, °make: *They had discovered his indiscretions and were blackmailing him. He blackmailed her into signing the alimony settlement.*

blade *n.* **1** °knife, cutting edge: *Hold it by the handle, not by the blade.* **2** *Literary* sword, rapier, sabre, °dagger, poniard, stiletto, cutlass, bayonet, °knife, penknife, jackknife: *She plunged the blade in up to the hilt.* **3** leaf, °leaflet, frond, °shoot: *The blades of grass were depressed where he had stepped.* **4** *Somewhat old-fashioned* °playboy, ladies' man, man about town, fop, °dandy: *He was quite a gay blade in his youth.*

blame *v.* **1** find fault with, censure, °criticize, °fault; °accuse, °charge, °indict, °condemn, °point to, point (the finger) at, °rebuke, °reprimand, recriminate, reproach, °scold, reprehend, °reprove: *Don't blame me if you can't get to school on time.* **2** hold responsible, fix (the) responsibility upon or on, put or place or lay (the) blame on, lay at someone's door, °denounce, °incriminate: *Why blame Casey for these graffiti?* —*n.* **3** censure, °criticism, °reproof, °rebuke, °recrimination, °disapproval, disapprobation, reproach, objurgation, condemnation, reprehension: *Latimer put the blame for the incident on me.* **4** culpability, °responsibility; °guilt, *Slang* °rap: *Why should you take the blame for something that Donald did?*

blameless *adj.* °faultless, guiltless, °innocent, °irreproachable, unimpeachable: *Hugh has led a blameless life.*

bland *adj.* **1** °gentle, °soothing, °smooth, °mild, °suave, urbane, °cool, unruffled, °calm, composed, unemotional, °nonchalant, insouciant: *His reaction to the news of the invasion was bland indifference.* **2** insipid, °boring, °dull, uninteresting, ennuyant, °tasteless, *Colloq US* plain vanilla; *Slang US* blah: *The play is a bland mixture of clichés embedded in a tired plot.*

blank *adj.* **1** °empty, °plain, °bare: *I stared at the blank paper, unable to write even my name.* **2** unornamented, °unadorned, undecorated, °void: *Whenever Irena sees a blank wall she feels compelled to hang a painting on it.* **3** °vacant, °empty: *The actor's mind went completely blank—he had forgotten his lines.* **4** °passive, °impassive, expressionless, emotionless, vacuous, °mindless, unexpressive: *He gave us a blank look when asked about the missing rare stamps.* **5** °disconcerted, discomfited, nonplussed, °confused, °helpless, resourceless, perplexed, dazed, bewildered: *The two old men looked at each other with blank and horror-stricken faces.* **6** unrelieved, °stark, °sheer, utter, °pure, unmixed, °absolute, °unqualified: *Fernet faced the blank prospect of solitary confinement.* —*n.* **7** °space; °line, °box: *Please fill in the blanks on the form.* **8** °nothing, °zero, °nil; °void, °emptiness: *I asked her when the baby was coming, but I drew a blank.*

blare *v.* **1** °blast, °bellow, trumpet, °ring, °boom, °thunder, °roar, bray; °resound, °echo, reverberate, resonate: *Everyone in the neighborhood can hear your hi-fi blaring.* —*n.* **2** °blast, °bellow, °ring, °roar, °boom, °noise, °sound, clamor: *The games began with a blare of trumpets.*

blasé *adj.* **1** bored, °jaded, °weary, unimpressed, ennuyé: *Her blasé attitude does little to endear her at job interviews.* **2** °indifferent, °cool, °superior, °supercilious, °sophisticated, °unmoved, °nonchalant, emotionless, °phlegmatic, apathetic, pococurante, °carefree, lighthearted, insouciant, devil-may-care: *His blasé behavior hides his basic feelings of insecurity.*

blaspheme *v.* **1** °curse, °swear, imprecate, execrate, °profane, °damn: *They were denounced for blaspheming against God.* **2** °abuse, malign, calumniate, defame, °disparage, revile, °put down, decry, deprecate, °depreciate, °belittle: *The ungrateful wretches blaspheme the charitable soul who would help them.*

blasphemous *adj.* °profane, °impious, °irreverent, °disrespectful, °sacrilegious, irreligious, °sinful, °wicked, °evil, iniquitous: *He was excommunicated for his blasphemous writings.*

blast *n.* **1** °blow, °gust, °wind, °gale: *The door opened and a blast of icy air made us shiver.* **2** °blare, °sound, °noise, °racket, °din, °bellow, °roar; °boom: *At the trumpet blast thousands of Goths descended screaming on the camp.* **3** °explosion, °burst, °eruption, °discharge; detonation: *A blast of dynamite leveled all the houses in the vicinity.* **4** ***(at*** *or* ***in) full blast.*** fully, at full tilt, at the maximum, °completely, °thoroughly, °entirely, maximally, *Slang* with no holds barred, *US* to the max: *The factory was going full blast before the strike.* —*v.* **5** °blow up, °explode, dynamite, °demolish, °destroy, °ruin, °waste, °lay waste, °shatter, °devastate: *The pillbox was blasted out of existence by our guns.* **6** defame, °discredit, °denounce, °criticize, °attack; °ruin, °destroy: *The candidate has been blasted by the press.* **7** °curse, °damn: *The minister continued to blast the proposal till the legislature dropped it.*

blatant *adj.* **1** °obvious, °flagrant, palpable, °obtrusive, arrant, °shameless, unashamed, °brazen, °overt, °glaring: *Those hooligans have shown a blatant disregard for the law.* **2** °noisy, clamorous, °loud, bellowing, °strident, vociferous, °rowdy, °boisterous, °obstreperous, °uproarious: *The blatant radical faction insists on making itself heard.*

blaze *n.* **1** °flame, °fire, °holocaust, inferno, conflagration: *The fuel barrels exploded, feeding the blaze.* **2** °outburst, °eruption, flare-up: *Her speech fanned the lower house into a blaze of resentment.* **3** °light, brightness, °brilliance, brilliancy, °glow: *The blaze of thousands of neon signs lights the Great White Way.* —*v.* **4** °burn, °flare up, °flame: *In a few minutes the logs were blazing merrily.* **5** ***blaze away (at).*** °fire, °shoot, open fire, °blast; °bombard, °shell: *The enemy appeared, and we just blazed away at them.*

bleach *v.* **1** whiten, °lighten, °fade, blanch, blench, *Technical* etiolate: *My jeans are all bleached by the sun.* —*n.* **2** whitener, chlorine: *Add a little bleach to the laundry.*

bleak *adj.* **1** cheerless, °dreary, depressing, °dismal, °gloomy, °somber, °melancholy, °sad, °unhappy, °mournful: *The year 1940 was one of the bleakest periods in British history.* **2** °cold, °chilly, °raw, °bitter: *The days were getting shorter, and the bleak winter was setting in.* **3** °barren, °bare, exposed, windswept, °desolate: *How depressing the bleak landscape of the Russian steppes can be in winter!*

blemish *v.* **1** °deface, °mar, °scar, °impair, disfigure: *The company said they would do nothing to blemish the* *natural beauty of the landscape.* **2** °tarnish, °stain, °sully, °spoil, °mar, °flaw, °harm, °damage, °scar, °injure, °bruise, besmirch: *She has blemished my reputation by spreading those stories about me.* —*n.* **3** disfigurement, °scar, °mark, °impairment, °stain, °smear, °blot; °defect, °flaw, °error, °fault, °imperfection, °error, erratum: *Her complexion was entirely without blemish.*

blend *v.* **1** °mix, °mingle, °combine, meld, commingle, intermingle: *The Latakia is blended with the Virginia to produce a fine smoking tobacco.* **2** °shade, °grade, gradate, °graduate, °merge, coalesce, °fuse, °unite: *Note how the pink sky and the tinted clouds blend in this Turner painting.* —*n.* **3** °mixture, °mix, °combination, mingling, meld, commingling, intermingling: *His humor is a fine blend of the sardonic with slapstick.*

bless *v.* **1** consecrate, °hallow, °sanctify; °extol, °glorify, °praise, °revere, °adore: *God bless this ship and all who sail in her. Bless the Lord.* **2** °give, make happy or fortunate, endow, °favor, °furnish, °provide, °supply, °grace: *She was blessed with one of the most beautiful soprano voices we ever heard.*

blessing *n.* **1** benediction, °prayer, consecration: *Each spring the vicar officiated at the blessing of the fleet.* **2** °boon, °favor, °advantage, good fortune, °godsend, °luck, °profit, °gain, °help, °asset, °gift, °bounty: *Hot, sunny days are a blessing for wine growers.*

blight *n.* **1** °affliction, °disease, °plague, infestation, °pestilence, °scourge: *The crops were visited with a blight that lasted seven years.* **2** °misfortune, °curse, °trouble, °woe, °calamity: *Genius may suffer an untimely blight.* —*v.* **3** °afflict, °infest, °plague, °scourge; wither, °mar, °taint, °blast: *Central Africa has been blighted by famine after famine.*

blind *adj.* **1** sightless, eyeless, unsighted, purblind, stone-blind: *He has been blind from birth.* **2** imperceptive, °slow, insensitive, °thick, °dense, °obtuse, °stupid, weak-minded, dull-witted, slow-witted, dimwitted, *Colloq Brit* gormless: *How blind some parents are! There's another case of the blind leading the blind.* **3** °indiscriminate, undiscriminating, °heedless, °reckless, °rash, °impetuous, °inconsiderate, unreasoning, °mindless, °senseless, °thoughtless, °unthinking, irrational, delusional: *He did her bidding with the blind obedience of a dog.* **4** ***blind to.*** unaware or unconscious of, impervious or insensible to, unaffected or untouched or unmoved by: *The critics were blind to her merits as a novelist till many years had passed.* —*v.* **5** °deceive, blindfold, blinker; bamboozle, °hoodwink, °fool: *Wolsey could not blind himself to the true condition of the church. How jealousy blinds people!* **6** °conceal, °hide, °eclipse, °overshadow; °dazzle, blindfold: *The bright lights of the city blinded our view of the airport runway. Her beauty blinded him to her greed.* —*n.* **7** °shade, curtain, °screen, °cover, shutter(s), awning: *The sun is too bright—please draw the blind.* **8** °pretense, °pretext, °front, °cover, smoke screen, °stratagem, °subterfuge, °ruse, °trick, °deception, *Colloq* °dodge; *Slang* scam: *The plumbing service is merely a blind for getting into houses to rob them.*

blindly *adv.* recklessly, heedlessly, deludedly, indiscriminately, rashly, impetuously, irrationally, thoughtlessly, mindlessly, senselessly, unthinkingly: *Despite his parents' warnings, he went blindly on, till one day he was arrested.*

blink *v.* **1** wink, °flicker, *Technical* nictitate: *She blinked in the strong light.* **2** °twinkle, °flicker, °gleam, glimmer, °shimmer, °flash, °sparkle, scintillate, coruscate: *A billion stars blinked in the wintry sky.* **3** °flinch, wince, °shrink, quail, blench, °recoil, °start, °move: *She didn't even blink when she got the injection.* **4** ***blink at.*** wink at, °ignore, °overlook, °disregard: *That inspector has been known to blink at health violations.* —*n.* **5** wink, °flicker: *He switched the cards in the blink of an eye.* **6** ***on the blink.*** *Colloq* °out of order, °broken, in disrepair, not working *or* operating, not operational, *Slang US* out of whack, on the fritz: *Our new fridge is on the blink again.*

bliss *n.* °happiness, blitheness, gladness, °joy, blessedness, °delight, felicity, °glee, °enjoyment, °pleasure, joyousness, °cheer, exhilaration, °gaiety, blissfulness, °rapture, °ecstasy: *The bliss of our honeymoon has remained with us throughout our marriage.*

blithe *adj.* **1** blissful, °happy, °cheerful, joyous, °merry, lighthearted, well-pleased, °delighted, °gay, °joyful, °elated, jubilant: *His spirit was blithe and his heart unquenchable.* **2** happy-go-lucky, insouciant, °heedless, °carefree, unconcerned, °blasé, °casual, °detached, °indifferent, uncaring, °careless: *She goes through life with a blithe disregard for the feelings of others.*

bloated *adj.* °swollen, distended, fully, puffy; puffed-up, °overgrown, °inflated, °pompous: *He felt bloated after the enormous banquet. That bloated, conceited, self-important petty official had the gall to refuse me a visa.*

blob *n.* °gob, gobbet, globule, °drop, droplet, °bit, gout, °lump, °dab, *Colloq* glob, *Chiefly US and Canadian* smidgen *or Chiefly US and Canadian* smidgin: *There's a blob of jelly on your tie.*

block *n.* **1** °piece, chunk, hunk, °lump, °slab; °stump; °brick, cube: *The figure is carved out of a solid block of stone.* **2** °bar, °obstacle, °obstruction, °hindrance, °stumbling block, °deterrent, °impediment, °barrier: *Her arrival should be no block to your leaving.* —*v.* **3** °obstruct, °close off, barricade; °bar, °shut off; °hinder, °hamper, balk, °impede, °prevent: *Entry to the playing field was blocked by the police.* **4** ***block out.*** **(a)** °rough out, °design, °outline, sketch, °lay out, °plan: *The colonel blocked out a strategy for our escape.* **(b)** °mask, °screen, °blank (out), °erase, °eliminate, °exclude, °blot out, °deny: *She has blocked out that part of her life from her memory.* **5** ***block (up).*** °stuff (up), congest, °clog, *Colloq Brit* bung up: *My nose is all blocked up because of this awful cold.*

bloodshed *n.* °slaughter, °carnage, butchery, °killing, °murder, bloodletting; °violence; genocide: *Let's settle this peaceably and avoid bloodshed.*

bloodsucker *n.* leech, extortionist, extortioner, blackmailer; °parasite, barnacle, *Colloq* sponge, freeloader, °scrounge, scrounger; *Slang US* moocher: *He's nothing but a bloodsucker, always demanding more and more money.*

bloodthirsty *adj.* °murderous, °homicidal, °savage, feral, °cruel, °ruthless, pitiless, °vicious, °brutal, °sadistic, °ferocious, °fierce, *Formal* °sanguinary, *Literary* °fell: *Bloodthirsty pirates had slaughtered the whole crew.*

blot *n.* **1** °stain, °spot, °mark, smudge, blotch, °blemish, disfigurement, °smear, smirch, °scar, *Colloq* splodge *or US also* splotch: *That ink blot covers the date of the document. There are a few blots on his record from his time in the army.* —*v.* **2** °stain, °spot, °spatter, smudge, °mark, °blur: *You have blotted these pages where you wrote with a fountain pen.* **3** ***blot one's copybook.*** °err, destroy *or* ruin *or* mar *or* spoil one's reputation, commit an indiscretion, °transgress, °sin: *She's certainly blotted her copybook by having an affair with that subaltern.* **4** ***blot out.*** **(a)** °obscure, °conceal, °cover (up), °hide, °eclipse, °dim: *The clouds blotted out the sun for a few minutes.* **(b)** °obliterate, °destroy, °erase, °demolish, efface, annihilate, °delete, rub *or* wipe out: *The psychiatrist told me that I had subconsciously blotted out all memory of the gory event.*

blow[1] *v.* **1** °breathe, °puff, °exhale; °expel: *If the yellow crystals turn green when you blow into the tube, it means that you've had too much to drink. Blow some air into the balloon.* **2** °waft, °puff, whistle, whine, °blast: *An icy wind blew through the cracks in the windows.* **3** °bungle, °botch, make a mess of, muff, mismanage, *Colloq* °screw up, °mess up, °fluff, °bugger up, *Taboo* fuck up: *It was my last chance to win and I blew it.* **4** *Colloq* °spend, °lavish, squander, °waste, throw out or away: *She blew hundreds on that dress and now she won't wear it.* **5** short-circuit, burn out: *All the fuses blew when I turned on the electric heater.* **6** ***blow hot and cold.*** vacillate, °hesitate, dither, *Colloq* °shilly-shally: *The sales manager has been blowing hot and cold over my proposal for a month now.* **7** ***blow out.*** **(a)** °extinguish: *Blow out all the candles in one breath. The match blew out in the wind.* **(b)** °explode, °burst: *My new tire blew out on the way over here.* **(c)** short-circuit, burn out: *The lights blew out during the storm.* **8** ***blow up.*** **(a)** become furious or angry or enraged, °flare up, lose one's temper, *Slang* blow one's top *or US also* stack, flip one's lid *or Brit* top: *She really blew up when I said I was going to the pub.* **(b)** °explode, °burst, °shatter, *Colloq* bust; detonate, dynamite, °destroy, °blast: *The bridge blew up with a roar. Demolition experts will blow up the dam.* **(c)** °enlarge, °inflate, embroider, °magnify, °expand, °exaggerate, °overstate: *The tabloid press has blown up the story out of all proportion.* **(d)** °enlarge, °magnify, °amplify, °expand, °increase: *Can you blow up just this corner of the photograph?* **(e)** °inflate; distend, °swell: *Before the party, we blew up some balloons.* —*n.* **9** °gale, °storm, °tempest, °whirlwind, tornado, cyclone, °hurricane, typhoon, northeaster, nor'easter: *We can expect a big blow tonight—winds of gale force, they say.*

blow[2] *n.* **1** °stroke, °punch, clout, whack, °hit, °knock, thump, thwack, *Colloq* wallop: *He was felled by a blow to the chin in the fourth round.* **2** °shock, °surprise, °bombshell, °jolt, bolt from the blue, °revelation: *It came as a blow to learn that she was an embezzler.*

blue *adj.* **1** depressed, low-spirited, dispirited, °sad, °dismal, down, °down in the mouth, °gloomy, °unhappy, °glum, downcast, crestfallen, chapfallen, °dejected, °melancholy, °despondent, °downhearted, morose: *I've been feeling blue since Kathleen left me.* **2** °obscene, °vulgar, °indecent, titillating, °pornographic, °dirty, °filthy, °lewd, smutty, °risqué, °bawdy, °sexy, X, X-rated, 18, *US* XXX; indelicate, °suggestive, °off-color, °erotic, °coarse, °offensive, °improper: *There's a place nearby that shows blue movies.*

bluff[1] *v.* **1** °deceive, °hoodwink, °dupe, °mislead, delude, °trick, cozen, °fool, *Colloq* bamboozle: *I bluffed him into believing that I held four aces.* **2** °pretend, feign, °bluster, °fool, *Colloq* kid; *Slang* bullshit: *She became frightened, unaware that I was only bluffing.* —*n.* **3** °bombast, °bravado, boasting, bragging, °bluster, °show, puffery; °deception, °blind; *Literary* rodomontade, gasconade; *Colloq* °hot air: *Roger's ranting is all bluff—he's really very timid.*

bluff[2] *adj.* **1** blustering, °gruff, °rough, °abrupt, °blunt, °curt, °short, °crude: *Fred's bluff manner puts many people off.* **2** °frank, °open, °hearty, straightforward, °plain, plain-spoken, °outspoken, affable, approachable, °good-natured, °friendly: *That comment is typical of his bluff honesty.* —*n.* **3** °cliff, escarpment, °precipice, scarp, headland, promontory, palisades: *Along the lower Hudson River, tall bluffs form a natural wall along the western bank.*

blunder *v.* **1** °stumble, °flounder: *I had somehow blundered into a meeting of the local crime syndicate. She blundered upon the truth when she saw them together.* —*n.* **2** °mistake, °error, gaffe, faux pas, °slip, slip-up, °howler, *Colloq* boo-boo, screw-up, °fluff, boner, *US* goof, goof-up: *Telling the tax auditor that I'd bought a Porsche was a stupid blunder.*

blunt *adj.* **1** °dull, °worn: *This knife is too blunt to cut butter.* **2** °abrupt, °curt, rough-spoken, plain-spoken, °short, °direct, °candid, °frank, unceremonious, undiplomatic, °inconsiderate, °thoughtless, °brusque, °outspoken, °bluff, °brash, indelicate, °rude, uncivil, °ungracious, °discourteous, °impolite; straightforward, °straight, uncomplicated, uncompromising: *Ralph may be blunt, but at least you know exactly where you stand.* —*v.* **3** °dull, take the edge off: *You've blunted the scissors cutting that cardboard.* **4** °soften, °mitigate, mollify, soothe; efface, °dim, °obscure, °blur, °weaken: *Mother's love is an absorbing delight, blunting all other sensibilities.*

blur *n.* **1** indistinctness, dimness, haziness, cloudiness, fogginess: *We were unable to pick out the star from the blur of the galaxy.* **2** °fog, haze, *Brit* fuzz: *Without my spectacles, everything is a blur.* —*v.* **3** °dim, °befog, °obscure, bedim; efface: *My vision was momentarily blurred, and I didn't see the oncoming car.* **4** °obscure, °hide, °conceal, °veil, °mask; °weaken: *The gentleman from Idaho has blurred the distinction between the unemployed and the unemployable.*

blurt *v.* Usually, ***blurt out.*** burst out with, utter; °reveal, °disclose, °give away, divulge, *Colloq* °blab: *She blurted out the name of her accomplice.*

blush *v.* be *or* act ashamed, redden, °flush, °color: *He blushed when asked if he still loved Belinda.*

bluster *v.* **1** °storm, °rage, °harangue: *It won't do any good to bluster on about the postal service.* **2** °swagger, °strut, °talk big, °boast, °brag, blow one's own horn *or* trumpet, °show off, crow: *He's always blustering about his conquests.* —*n.* **3** swaggering, storming, raging, °raving, haranguing, °tumult; °hot air, puffery, °bravado, grandiloquence, *Literary* rodomontade: *He's all bluster and will do nothing despite his threats.*

board *n.* **1** °plank, scantling, °timber: *We nailed the last board in place and the house was finished.* **2** °table, gaming *or* game table *or* surface: *I have the chessmen, have you brought the board?* **3** °food, meals, °provisions: *I am moving to the country, where room and board are cheaper.* **4** °council, °committee, directors, directorship, °management, °cabinet, panel, trustees, advisers *or* advisors: *That issue will be discussed at the meeting of the board.* **5** ***on board.*** aboard, on: *Women aren't allowed on board my boat wearing high heels.* —*v.* **6** go aboard, ship aboard; °enter, °embark on: *We all boarded the ship but it didn't leave for an hour.* **7** °feed; °eat, take meals; °accommodate, °lodge, °house, billet, °quarter; °lodge, °stay, °live, °room; *Colloq* °put up: *Mrs. O'Brien boards three gentlemen at her house, and I board there now.*

boast *n.* **1** °brag, bragging: *They did not make good their boasts of being the fastest in the competition.* —*v.* **2** °brag, vaunt, crow, °showoff, *Colloq* blow one's own trumpet, *US* blow one's (own) horn *or* trumpet, toot one's (own) horn *or* trumpet; *Slang* lay it on thick, °talk big: *He boasted that he was the best poker player in the casino.*

boastful *adj.* °ostentatious, showoffish, bragging, vainglorious, °egotistical, °vain, °conceited: *She's boastful about her wealth, but it was left to her by her mother.*

boat *n.* [See also **ship, 1.**] °vessel, °craft, skiff, small craft, motorboat, speedboat, knockabout, runabout, yacht, motor yacht, sailing yacht, *Brit* rowing boat, sailing boat, *US* rowboat, sailboat, *Colloq* °ship: *I bought a thirty-foot boat at this year's show. They went off on a slow boat to China.*

bode *v.* portend, °promise, augur, betoken, forebode, presage; °foreshadow: *The weather bodes well for the picnic.*

body *n.* **1** °corpse, °cadaver, °remains, carcass, *Slang* °stiff: *A body has been dragged up from the lake.* **2** °trunk, torso: *They found the body, but the arms, legs, and head were missing.* **3** main part *or* portion, °hull, fuselage: *The body of the plane remained intact, though the wings and superstructure broke away.* **4** °substance, essentials, main part, °essence, °heart, °center, °core: *The body of the book is all right, but the index needs work.* **5** °majority, °bulk, main part *or* portion, mass(es): *Under Henry VIII the main body of the people were prosperous.* **6** °association, °league, °band, °corps, confederation, °fraternity, °society; °committee, °council; °group, assemblage, °assembly, congress, °company: *It is not within the power of this body to do more than vote on the proposal.* **7** richness, °substance, firmness, consistency, solidity, thickness, density, fullness, viscosity, fullness: *This wine has excellent body. Add a little cornstarch to give the sauce more body.*

bog *n.* **1** °swamp, fen, °marsh, quagmire: *The peat bogs have yielded interesting fossils.* **2** ***bog down.*** °impede, °slow, °hamper, °encumber, °stymie, °stick, °handicap, °clog, °check, °set back, °hold back: *Traffic is bogged down because they are repaving the road.*

bogus *adj.* °counterfeit, °spurious, °fake, °false, °fraudulent, °sham, °imitation, °fictitious, *Colloq* phoney *or US also* phony: *The police reported that a gang was trying to pass bogus money to unsuspecting shopkeepers in the area.*

Bohemian *adj.* °nonconformist, unconforming, unconventional, °unorthodox, °casual, free-and-easy: *In the '60s she became a flower-person and adopted a Bohemian way of life.*

boil[1] *v.* **1** °bubble, °seethe; °simmer, °stew, steam: *A pot of soup boiled on the kitchen stove.* **2** °seethe, °fume, sizzle, °smolder, °chafe, fulminate, °ferment, sputter, splutter, °bluster: *When she learned what he had been saying about her, she boiled with furious indignation.*

boil[2] *n.* abscess, carbuncle, pustule, *Technical* furuncle: *The doctor said the boil had to be lanced at once.*

boisterous *adj.* °rowdy, clamorous, °rough, °noisy, °lively, °exuberant, °unruly, °wild, °undisciplined, °tempestuous, °stormy, turbulent, *Colloq* rambunctious: *That kind of boisterous behavior is not condoned in the library.*

bold *adj.* **1** °courageous, °brave, plucky, °confident, stouthearted, lionhearted, °daring, °enterprising, °audacious, °fearless, unafraid, °intrepid, °resolute, °dauntless, undaunted, valiant, °stout, valorous, °stalwart, °adventurous, °venturesome; °reckless, °foolhardy, incautious, °daredevil, °rash: *It would take a bold man to enter the ring with the champion.* **2** °audacious, °presumptuous, °forward, °immodest, °brazen, °impudent, temerarious, °impertinent, °shameless: *It was very bold of you to speak your mind to the foreman.* **3** °pronounced, °outstanding, °striking, °vigorous, °clear, °strong, °vivid, °distinct, °conspicuous: *He wrote down their demands in a good, bold hand.*

bolster *v.* °support, °prop (up), °brace, shore up, °buttress, °uphold, °back (up), °reinforce, °aid, °help, °assist, °further, °advance: *The miners cited a lack of safety measures to bolster their arguments.*

bolt *n.* **1** arrow, dart, °projectile, °missile, *Historical* °quarrel: *He had only three bolts remaining for the crossbow.* **2** °pin, °bar, °rod, °catch; latch: *We hoped that the bolt would prevent their opening the door.* **3** machine screw: *Bolts can be tightened or removed, unlike rivets.* **4** °roll, °length: *We sell only full bolts of fabric.* **5** lightning flash, thunderbolt, *Formal* fulguration: *One bolt traveled down the television aerial and blew out the set.* **6** ***bolt from*** *or* ***out of the blue.*** °surprise, °shock, °bombshell, °bomb, °blow, °revelation, eye-opener, *Colloq* shocker: *The news of her resignation came like a bolt from the blue.* **7** ***shoot one's bolt.*** exhaust *or* use up one's resources, *Slang* burn out, *US* poop out: *He was fast early in the marathon, but he'd shot his bolt long before the finish line.* —*v.* **8** °spring, dart, shoot off, °take flight, °run (away or off), °rush (off or away), °break away, °flee, decamp, abscond, °escape, °fly, °dash (off or away), *Colloq* skedaddle, scram, *Brit* scarper, do a bunk, do a moonlight flit, *US* take a (runout) powder: *When they saw the policeman the two suspicious figures bolted. The couple in room 315 bolted without paying their bill.* **9** °gulp (down), swallow whole: *When the bell rang, she bolted her breakfast and ran out of the back door.* **10** °fasten, °lock, latch, °secure: *Make sure you bolt your door at night and not let anyone in.* **11** °fix, °attach, °fasten, °connect, make fast to: *The motor must be securely bolted to the workbench.* —*adv.* **12** ***bolt upright.*** °erect, °straight, rigidly, stiffly: *When her name was called, Penny sat bolt upright in her chair.*

bomb *n.* **1** °bombshell, °shell, °explosive: *One of the bombs blew up the school.* —*v.* **2** °bombard, °shell, °batter, °blow up: *Last night the air force bombed the rail center.*

bombard *v.* **1** °batter, °bomb, °shell: *The artillery continued to bombard the enemy with everything in its arsenal.* **2** assail, °attack, °assault, °set upon; °besiege: *The president was bombarded with requests to amend the law.*

bombast *n.* pretentious language, flatulence, °bluster, °show, grandiloquence, magniloquence, °hot air, °bravado, °boast, boasting, *Literary* gasconade, rodomontade; *Colloq* puffery: *The speaker continued to bore the audience with his pompous bombast.*

bombastic *adj.* high-flown, °extravagant, °pompous, °grandiose, grandiloquent, magniloquent, °inflated, fustian, flatulent, turgid, *Literary* euphuistic: *The Honorable Gentleman from Buncombe is again delivering himself of a bombastic speech.*

bombshell *n.* °surprise, °shock, eye-opener, °bomb, °blow, °revelation, bolt from *or* out of the blue, *Colloq* shocker: *Then came the bombshell—she and Tony had been married the week before.*

bona fide *adj.* °genuine, °authentic, attested, °real, °veritable, °legitimate, °true, valid; in good faith, °sincere, °honest: *The expert affirmed that it was a bona fide ruby. I believe her sorrow is bona fide.*

bond *n.* **1** tie(s), °shackles, chains, fetters, °manacles, °handcuffs, °trammels, thongs, cord(s), rope(s); restraint(s), constraint(s), check(s), control(s), rein(s): *The council is hampered by the bonds of the old regulations.* **2** covenant, °pact, °contract, °agreement, °engagement, °compact, °treaty: *To unite the party a bond of confederacy was formed.* **3** °connection, °link, linkage, °union, °tie, °relationship: *The bond between us was a shared love of the theater. With more glue, the bond between the veneer and the board will hold.* —*v.* **4** °cement, °bind, hold together, °stick, cohere: *They use mortar to bond the bricks to one another.*

bondage *n.* °slavery, °servitude, °subjection, subjugation, enslavement, serfdom, thralldom; vassalage, villeinage: *Those poor souls were kept in bondage for most of their lives, forced to row in the galleys till they died.*

bonny *adj.* °beautiful, °comely, °attractive, °pretty, °lovely: *She's grown into quite a bonny lass.*

bonus *n.* °reward, °largess, handout, °perquisite, °extra, °honorarium, °tip, gratuity, °remuneration, compensation, *Colloq* °perk: *Employees often receive a Christmas bonus.*

book *n.* **1** °volume, tome, °work, °publication; hardcover, softcover, paperback: *Our personal library contains more than five thousand books.* **2** libretto, °words, °lyrics: *Richard Rodgers wrote the music and Oscar Hammerstein the book for several hit shows.* **3** rules, laws, regulations: *Why does he always insist that we have to go by the book?* —*v.* **4** °engage, °reserve; earmark, ticket; °order, °register, °enroll, °list, °enlist, log, °record, °post: *Please phone the restaurant and book a table for four for seven-thirty.*

bookkeeper *n.* clerk; accountant, *Brit* chartered accountant, CA, cashier; *US* CPA, certified public accountant: *The office was haunted by the melancholy ghosts of departed bookkeepers.*

bookworm *n.* bibliophile, booklover, inveterate *or* ardent reader, *Formal* bibliophage: *Fiona is such a bookworm, she hardly does anything but read.*

boom *v.* **1** °sound, °resound, resonate, °blast, rumble, °thunder, °roar, bang, °explode: *We heard the cannons booming in the distance.* **2** °prosper, °thrive, °flourish, °progress, °grow, °increase, burgeon *or* bourgeon: *Business is booming in every sector.* —*n.* **3** °blast, rumble, °explosion: *There was a resounding boom and the car went up in flames.* **4** °prosperity, profitability; °growth, °increase, burgeoning *or* bourgeoning: *In this business it is either boom or bust.*

boomerang *v.* °rebound, °recoil, backfire, °miscarry, redound: *His plan boomeranged.*

boon *n.* °gift, °favor, °award, °reward, gratuity, °present; °blessing, °benefit, °advantage: *The mobile library service is a great boon to the elderly.*

boor *n.* **1** °rustic, °peasant, yokel, (country) bumpkin, °provincial, backwoodsman, *US* hayseed, hillbilly, Juke, Kallikak, *Slang* hick: *The boor is blind to the beauties of nature.* **2** °barbarian, °yahoo, oaf, °clod, clodhopper, °philistine, °clown, grobian; hoyden, *Colloq* lummox, *Slang* galoot, °slob, *US* goop, slobbovian: *The guests behaved like boors, throwing their food at each other.*

boorish *adj.* °rustic, °barbarian, °rude, °crude, °ill-mannered, uncultured, °coarse, clownish, uncouth, loutish, oafish, gawky, °vulgar, ill-bred: *How can you even think of inviting such a boorish fellow?*

boost *n.* **1** °lift, shove *or* push up *or* upward(s), *Colloq* leg up; °rise, °raise: *If you give me a boost, I can reach the window sill.* **2** °encouragement, °help, °aid, °assistance, °support: *With a boost from your constituency, Trevor should win the vote.* **3** °increase, °rise, *US* °raise, hike: *The boost in the price of milk could be inflationary.* —*v.* **4** °lift, shove *or* push up *or* upward(s), °raise: *He boosted her over the fence. The second stage is intended to boost the rocket beyond the atmosphere.* **5** °encourage, °promote, °help, °aid, °support, °assist, °improve: *A talk before the game will boost the players' morale.* **6** °increase, °raise: *Her salary was boosted twice in one year.*

boot *n.* **1** *to boot.* °in addition, into the bargain, °in addition, °besides, °moreover, as well, also, too, additionally: *He's stingy and cruel—and ugly to boot.* **2** shoe, riding boot, bootee: *I need a new pair of boots.* —*v.* **3** °eject, °expel, shove, °propel, °push, *Colloq* °kick: *The landlord booted three rowdies out of the pub.* **4** *Literary* °profit, avail, °help, be in aid of: *What boots it to complain?*

booth *n.* **1** °stall, °stand: *Our company has taken three booths at the book fair.* **2** °compartment, cubicle, °box, *Brit* kiosk: *The threatening calls were all made from a public phone booth.*

bootless *adj.* °pointless, unavailing, °vain, °purposeless, °useless, °futile, °worthless, unproductive, °ineffective, °inefficacious, °fruitless, °unprofitable, profitless, unremunerative, unrewarding, °wasteful, time-wasting, Sisyphean: *Trying to persuade him of the importance of an education is a bootless task.*

booty *n.* °plunder, °gain, spoil(s), contraband, takings, °loot, *Slang* swag, boodle, °(hot) goods, °take: *The pirates fought over the booty seized from the Spanish galleon.*

booze *n.* **1** °drink, °(hard) liquor, spirit(s), whisk(e)y, °alcohol, *US* demon rum, John Barleycorn, mountain dew, white lightning, white mule; *Slang* rotgut, °poison, firewater, mother's ruin, *US and Canadian* °sauce, °juice, hooch, red-eye: *I ordered plenty of booze for the party.* —*v.* **2** °drink, tipple; *Humorous* bibulate; *Slang* hit the bottle, *US* hit the sauce: *After boozing all last night he has a terrible hangover.*

border *n.* **1** °edge, °margin, hem, binding, trimming, °trim, edging, °periphery, purfle, purfling: *The border of the tablecloth is beautifully embroidered with flowers.* **2** Usually, ***borders.*** limit(s), bound(s), confines: *Sometimes Tony exceeds the borders of good taste.* **3** °boundary, °frontier: *You won't be able to cross the border without a passport.* **4** °frame, frieze, molding; dado, wainscot *or* wainscoting *or* wainscotting: *The border of the fresco is in Greek fretwork design.* **5** borderline, °edge, °verge, °brink: *She is just on the border of becoming a born-again Christian.* **6** bed, flowerbed, herbaceous border: *Hollyhocks are growing in the border.* —*v.* **7** °edge, °trim, °bind, °fringe, purfle: *The hem of the skirt is bordered with lace.* **8** °resemble (closely), °approach (closely), verge upon *or* on: *Isabel's attempts at playing the tuba border on the ludicrous.* **9** lie alongside, adjoin, abut (on *or* upon), verge upon *or* on, °touch, be adjacent to: *The territory of the Gauls bordered the western lands of the Germans.*

bore[1] *n.* **1** °hole, drill-hole, bore-hole: *A bore of six inches was carried to a depth of 2086 feet.* —*v.* **2** °pierce, °perforate, °drill, °penetrate, °puncture, °tap, °punch, °stab, °prick; °sink, °tunnel, °dig (out), °gouge (out); hollow out: *Bore a hole through this sheet metal. The oil rig is boring through solid rock.*

bore[2] *n.* **1** °annoyance, °nuisance: *Reggie is such a bore—always talking about himself.* —*v.* **2** °weary, °wear out, °tire, °exhaust, °jade: *The program so bored me that I fell asleep.*

boredom *n.* dullness, dreariness, ennui, °tedium, monotony: *I have to look forward to the boredom of an evening of chamber music.*

boring *adj.* °dull, °monotonous, °tedious, °humdrum, °tiresome, °dreary, °flat, °dead, uninteresting, unexciting, ennuyant, °stale, °tired, °dry, dry as dust, arid; tiring, wearying, wearisome, °exhausting, soporific; °repetitious, °wordy, prolix, unending, long-drawn-out: *Felicity's boring old stories put me to sleep.*

borrow *v.* °take, °appropriate, °draw, °adopt, °refer to, °obtain, *Colloq* sponge, cadge, touch (someone) for, *US* °bum; *Slang* mooch: *Has Fenwick borrowed my lawn mower again?*

bosom *n.* **1** °breast, °chest, bust; *Slang* boobs, knockers, tits, titties, °pair, jugs, bazoom(s), *Brit* Bristols: *The sex goddess's lack of talent was more than compensated for by her ample bosom.* **2** °midst, °interior, °heart, °core, °center: *She was welcomed into the bosom of the family as if she had been their own child.* **3** °soul, °heart, heart of hearts, °bowels, blood, *Colloq* °gut: *I know he loves me, I can feel it in my bosom.* —*adj.* **4** °close,

°intimate, °dear, °beloved, cherished, °boon, °special, °confidential: *We were once bosom companions.*

bosomy *adj.* big-busted, busty, well-endowed: *This tabloid always has a bosomy model on page 3.*

boss *n.* **1** °chief, °supervisor, °head, administrator, °manager, °foreman, °superintendent, °overseer; °employer, °director, °proprietor, °owner, *Brit* managing director, *US* president, *Dialect* himself, *Colloq* supremo, *Brit* governor, gov, gaffer, *US* °super, °leader, kingpin, big cheese, the man, *Slang* honcho, head *or* chief honcho, Mr. Big, prexy *or* prex: *If you have to leave early, check with the boss. The company boss was interviewed on television last night.* —*v.* **2** °supervise, °head, °manage, °run, °oversee, °overlook, °direct, °control, superintend, °command, take charge, be in charge: *Clive has been here only a year and he's already bossing a department.* **3** domineer, push *or* shove around *or* about, °dominate, order about, °lord it over: *That slave driver had better stop bossing me about or I'll quit!*

bossy *adj.* °overbearing, °domineering, °dictatorial, °tyrannical, °despotic, imperious, lordly: *Her boyfriend is awfully bossy—and they're not even married yet.*

botch *v.* °bungle, mismanage, °spoil, *Colloq* screw *or* louse up, °blow, °mess up, °muck up, make a mess *or* hash *or* muddle of; *Slang Brit* bollocks *or* ballocks *or US* bollix up: *Give Gordon an assignment and he's sure to botch it.*

bother *v.* **1** °annoy, °pester, °worry, °irritate, °trouble, hector, °harass, °hound, dog, °nag, °plague, needle, *Colloq* hassle; *Slang US* °nudge: *I wish they'd stop bothering me about paying the telephone bill.* **2** °trouble (about), °fuss (at), make a fuss (about), concern oneself (with), °burden: *Too few people are interested in bothering about the welfare of others.* **3** °confuse, °bewilder, °perplex, °perturb, °upset, disconcert, °discomfit: *She was increasingly bothered by her inability to understand the local language.* —*n.* **4** °trouble, °inconvenience: *Those pets must be a lot of bother.* **5** °worry, °annoyance, vexation, °nuisance, irritation, °trouble, °effort, °disturbance, °upset, *Slang* hassle: *Painting the lattice was more bother than it's worth.* **6** dither, °flutter, *Colloq* tizzy, °pet, °stew, °lather, °sweat: *She seems to have worked herself into quite a bother about something quite insignificant.* **7** °pest, irritant, °nag, °nuisance, *Colloq* °pain, pain in the neck *or Brit taboo* arse *or US taboo* ass; *Slang US* °nudge: *Mother is such a bother, always asking if I wear my galoshes when it rains.* **8** °disturbance, to-do, ado, commotion, °fuss, °trouble, °disorder, °stir, hubbub: *We ran into a bit of bother at the pub last night.*

bottle *n.* **1** flask, container; °fiasco, decanter: *The milkman left two bottles of milk.* **2** °courage, °nerve, manliness, manfulness, °grit, °backbone, °gumption, mettle, °pluck, Dutch courage, *Slang* °guts; *Colloq* °spunk, starch, *US* moxie: *He was going to tell her off but lost his bottle at the last minute.* **3** ***the bottle.*** °alcohol, alcoholic drink, spirit(s), °liquor, °booze, °sauce: *He's back on the bottle after only two weeks of being on the wagon.* —*v.* **4** ***bottle up.*** **(a)** °contain, °restrain, °hold back, °control, °suppress, °repress, hold *or* keep in check, °stifle: *All the emotions, bottled up for so long, burst upon him at once, and he wept pitiably.* **(b)** °trap, °cut off, hem in, °box in: *With the help of the posse, we can bottle up the gang in the canyon.*

bottom *n.* **1** °seat, °buttocks, °rear, behind, rear end, derrière, rump, °posterior, hindquarters, breech, fundament, gluteus maximus, *Colloq* backside, °butt, buns, prat, *Brit* °bum, *US* can, °duff, keister *or* keester, hinie; *Taboo Brit* arse, *US* ass; *Slang US* tokus *or* tochis *or* tuchis, tushie *or* tushy *or* tush: *He just sits there on his bottom, never doing a bit of work.* **2** °base, foot, °foundation, °groundwork, substructure, °footing, underpinning, fundament: *A ditch was dug along the bottom of the wall.* **3** °basis, °foundation, °source, °origin, °cause, °heart, °nub: *We have to get to the bottom of the problem.* **4** °depths, Davy Jones's locker; bed: *The ship sank to the bottom of the sea.* **5** ***at bottom.*** basically, fundamentally, in the final *or* last analysis, °really, °in reality, °truly, °in truth, essentially: *Despite her behavior at the party, at bottom she is very reserved.* **6** ***Bottoms up!*** Prosit!, To your (very good) health!, Cheers!, Here's to—!, Skoal!: *I propose a toast to the best man—Bottoms up!*

bottomless *adj.* unfathomed, unfathomable, abyssal, °abysmal, °inexhaustible, °unlimited, °immeasurable, unplumbable: *The bottomless ignorance of the man is incredible.*

bounce *n.* **1** °bound, °leap, °hop, °recoil, ricochet, °rebound: *The ball took a bad bounce and the infielder missed it.* **2** °vitality, °energy, °verve, °zest, vivacity, liveliness, °animation, °dynamism, °life, *Colloq* °pep, zip, °go, get-up-and-go: *Betty has so much bounce, that she's a bit tiring to have around.* —*v.* **3** °bound, °rebound, °hop; °recoil, ricochet: *The ball bounced over the wall and into the river.*

bound[1] *n.* **1** Usually, ***bounds.*** °boundary, boundary line, limit(s), °extent, border(s), confines: *Please try to keep the dogs within the bounds of the estate. Cartwright's plan is beyond the bounds of common sense.* —*v.* **2** °limit, °restrict, confine, delimit, °define, circumscribe: *The river bounds the property on the east.*

bound[2] *n.* **1** °leap, °jump, vault, °spring; °bounce, °hop: *With a great bound, the leopard was upon me. The ball took some peculiar bounds and disappeared into the bushes.* **2** ***by leaps and bounds.*** See **leap, 7,** below. —*v.* **3** °leap, °jump, °hop, °spring, vault, gambol, °caper, romp, °frolic, °bounce, *Colloq* galumph: *The wolfhound came bounding toward me across the meadow.*

bound[3] *adj.* **1** tied, °fast, °fixed, fastened, confined, secured: *We were bound hand and foot and left in the cave.* **2** °obliged, obligated, required, constrained, °forced, compelled: *In the circumstances, Philippa was bound to do as she was told.* **3** °determined, resolved: *Otto is bound to go to the party if Antonia is going.* **4** °likely, °certain, °sure, °destined, predestined, °fated, °doomed: *He is bound to die if he keeps on playing Russian roulette.* **5** °destined, scheduled, booked; headed, directed: *We were bound for Cardiff.*

boundary *n.* border(s), limit(s), frontier(s); bound(s), confines, °perimeter: *If you cross that boundary, you will be safely in Switzerland.*

boundless *adj.* °limitless, unbounded, °unlimited; illimitable, °vast, °endless, unending, °infinite, °immense, °enormous, °immeasurable, incalculable, measureless, unrestricted, unchecked, °inexhaustible, unstoppable, unbridled, °uncontrolled, *Literary* vasty: *How can you keep up with a teenager's boundless energy?*

bountiful *adj.* **1** °generous, beneficent, munificent, °liberal, unsparing, unstinting, °charitable, eleemosynary, magnanimous, *Literary* bounteous: *We are grateful to Mr. Carnegie, our most bountiful patron, for endowing this library.* **2** °ample, °abundant, plenteous, °plentiful, copious, °rich, *Literary* bounteous: *Till the stock exchange debacle, these shares paid bountiful dividends.*

bounty *n.* **1** generosity, liberality, munificence, charitableness, °philanthropy, °charity, unselfishness, beneficence, goodness: *The poor were formerly dependent solely on the bounty of the local gentry.* **2** °gift, °present, °largess, °grant, °subsidy, °endowment, subvention: *People should be given work and not live off the bounty* *of the state.* **3** °reward, °award, °premium, °bonus, gratuity: *In America they paid a bounty of fifty dollars for every dead wolf.*

bouquet *n.* **1** nosegay, posy, °bunch, °arrangement, °spray: *I sent her a bouquet of spring flowers for her birthday.* **2** °aroma, °scent, °odor, °fragrance, °perfume: *This '83 burgundy certainly has a fine bouquet.* **3** compliment(s), °praise, commendation: *Mrs. Campbell has received many bouquets for her performances.*

bourgeois *adj.* **1** middle-class, °conventional, °philistine, capitalistic, propertied; °materialistic, °greedy, money-grubbing, money-hungry: *The yuppies constitute the modern bourgeois element in society.* **2** working-class, proletarian, °plebeian: *In his bourgeois mind, he had only his labor to offer.*

bout *n.* **1** °turn, °round, °time, °occasion, °spell, °period, °session: *He's just got over a bout of pneumonia.* **2** °chance, °spree, °stint, °opportunity, innings: *We had long planned this bout of shopping.* **3** °contest, °match, boxing match, prizefight, °meet, set-to, °struggle, °encounter, °engagement; duel: *A bout has been arranged between the heavyweight champion of the world and the challenger from Puerto Rico.*

bow *n.* **1** °nod; curtsy *or* curtsey, salaam, °kowtow, genuflection, °prostration, °obeisance: *We all bowed respectfully before the emperor.* —*v.* **2** °defer, °yield, °submit, °give in, °bend, bow down, °capitulate: *I bow to your greater knowledge of the subject.* **3** °bend, °incline, °lower: *The servants bowed their heads when the master entered.* **4** °weigh down, °crush, °overload, bend down, °burden: *Michael was bowed down by the responsibilities of his new family.* **5** °nod, curtsy *or* curtsey, salaam, °kowtow, genuflect, °prostrate oneself, make obeisance: *The natives bowed as the king passed by.*

bowels *n.* °interior, °insides, °depths; °heart, °center, °core, intestines, viscera, vitals, belly, °gut, *Colloq* innards, °guts: *We descended the shaft into the very bowels of the earth. She hates me, I can feel it in my bowels.*

bowl[1] *v.* °move, trundle, °wheel, °roll, °spin: *We saw him in his car, bowling along at about forty.*

bowl[2] *n.* dish; basin, °pan: *She brought me a bowl of cereal.*

box[1] *n.* **1** °case, °receptacle, crate, carton, container, °casket, coffer, caddy, °chest: *She keeps her valuables in a small tortoise-shell box on the dressing table.* —*v.* **2** crate, encase, °package: *The candles are boxed in dozens.* **3** ***box in*** *or* ***up.*** °trap, confine, °bottle up, hem in, °enclose, °surround; °pin down: *They have the horses boxed in and are now driving them into the corral.*

box[2] *v.* **1** °fight, engage in fisticuffs, °spar, °battle: *When he was in the army, he boxed for his regiment.* **2** °strike, buffet, °punch, °hit, *Colloq* slug, sock, whack, thwack, clout, °belt, thump, °lambaste, whomp: *Every time she heard him swear, she'd box his ears.* —*n.* **3** °blow, buffet, °punch, °hit, °strike, *Colloq* slug, sock, whack, thwack, clout, °belt, thump, whomp: *How would you like a box on the ear, you jerk!*

boy *n.* **1** °lad, °youth, young man, °stripling, youngster, schoolboy, °fellow, urchin, brat, *Colloq* kid, °guy, small fry, little shaver: *There were two girls and five boys in my family.* **2** °servant, houseboy, °attendant; lackey, °slave, *Archaic* knave, varlet, °rogue, °wretch, caitiff: *Here! Boy! Bring me another gin and tonic.* **3** ***old boy.*** (in Britain) (public) schoolmate; °friend, °chum, °pal, *Archaic* old bean, old egg, old crumpet, dear boy; crony: *I say there, old boy, care for a rubber of bridge? Carruthers wouldn't be where he is now if it weren't for the old-boy network.*

boycott *v.* **1** blacklist, °embargo; °avoid, °refuse, °shun, °reject, eschew, pass over *or* by: *They are boycotting Fern's Dairy because it won't hire women. The US government is still boycotting cigars from Havana.* —*n.* **2** °embargo, blacklist, blacklisting, °ban: *A boycott of their products soon forced them to change their policies.*

boyish *adj.* **1** °young, youthful, °juvenile, °adolescent: *She liked his boyish good looks.* **2** °childish, °puerile, °juvenile, °immature: *Don't be too hard on them—it was just a boyish prank.*

brace *n.* **1** °bracket, stiffener, °reinforcement, reinforcer, °support, °buttress, °prop, °stay, °strut, truss: *Two steel braces have been installed to steady the columns.* **2** °drill: *"With his brace and auger in her side he bored holes three."* **3** °clasp, °clamp, °buckle, fastener, °clip, holdfast, °catch, coupler, coupling: *Another brace will be needed here to strengthen the handle.* **4** °pair; °couple, °span, °team (of two): *A brace of dueling pistols was sold at auction last week for twenty thousand dollars. Her carriage was drawn by a brace of palominos.* —*v.* **5** °steady, °reinforce, °support, °strengthen, prop *or* shore up: *Iron bars are used to brace the arches.* **6** ***brace oneself.*** steady *or* gird *or* prepare oneself; hold *or* hang on: *I braced myself against the likelihood that she would refuse.*

bracing *adj.* °invigorating, °tonic, °stimulating, °refreshing, °exhilarating, fortifying, restorative: *I cannot live in the tropics and need the more bracing climate of the north.*

bracket *n.* **1** °support, corbel, °console: *The mantelpiece rests on a pair of stone brackets.* **2** shelf: *Her collection of glass paperweights was arrayed on a bracket in the sitting room.* **3** °category, °class, °set, °group, grouping, classification, °division, °level; °order, °grade, °rank: *He comes from an altogether different bracket of society.* —*v.* **4** classify, °rank, °group; °unite, °combine, °join, °link: *I wish you wouldn't bracket her with me—our politics are as different as day and night.*

brag *v.* °boast, crow, trumpet, vaunt, °strut, °swagger, °show off, *Colloq* °talk big, blow *or* toot one's own horn *or* trumpet, go on about: *He's always bragging about what he did in the war.*

braggart *n.* boaster, bragger, braggadocio, windbag, peacock, °showoff, Scaramouch *or* Scaramouche, *Slang* bigmouth, loudmouth, gasbag: *That braggart Wilford talks about himself incessantly.*

braid *n.* **1** plait: *Katrina wore her blond hair in a tightly coiled braid on top of her head.* **2** trimming, embroidery, soutache, °lace, fillet, °band, ribbon: *The edges are decorated with narrow braid containing gold thread.* —*v.* **3** plait, intertwine, interlace, °weave, °twist: *At school we learned how to braid leather laces into a belt.*

brain *n.* **1** brains, °intelligence, °intellect, °understanding, °sense, °thought, °imagination, °capacity, perspicacity, perceptiveness, °perception, percipience; °wisdom, sagacity, °wit, discernment, acumen; °knowledge, cognition: *Although she's not yet ten, she has the brain to become a great physicist.* **2** °genius, °mastermind, °intellectual; °leader, planner: *Most people regard Einstein as the big brain of the twentieth century. Igor was clearly the brains of the operation.*

brake *n.* **1** °curb, °check, °restraint, °restriction, constraint, °control, °rein: *The central bank applied a brake to the upward trend of the dollar by buying Deutschmarks.* —*v.* **2** °slow, slow up *or* down, put on *or* apply the brakes, reduce speed, decelerate, slacken, °hold up: *He braked before the bad curve. She braked the car going down the steep hill.*

branch *n.* **1** °offshoot, arm; limb, bough, °stem, °shoot, °twig, sprig: *The branches of this tree need trim-*

ming. **2** °department, °section, subsection, °division, subdivision, °office, °part, °ramification; affiliate, °subsidiary; spinoff: *Which branch of medicine interests you most? The company maintains branches in London and Melbourne.* —*v.* **3** ramify, °divide, subdivide, °diverge; °diversify: *This road branches off in three directions. The company will branch out into electronics this year.*

brand *n.* **1** °kind, °make, °type, °sort, °variety; brand name, °manufacturer, maker, trade name, trademark, °label, °mark, marque, *Chiefly US and Canadian* name brand: *Which brand of toothpaste do you prefer? Our advertising agency is conducting a survey of brand loyalty.* —*v.* **2** °mark, °stamp, °identify, °tag, °label, trademark: *We sell only branded merchandise in our shops.* **3** °label, °characterize; °stigmatize, °discredit, °disgrace: *Because of his actions at the front, Tremble has been branded as a coward.*

brand-new *adj.* °new, °unused, °fresh, firsthand, °mint, virgin: *We bought a brand-new car last week.*

brash *adj.* **1** °hasty, °rash, °impetuous, °precipitate, °impulsive, headlong, °reckless: *He may be a brash young man, but I think he's going places.* **2** °impudent, °rude, °impertinent, °disrespectful, °insolent, °forward, °audacious, °brassy, °brazen, °bold, °tactless, undiplomatic, °presumptuous, *Colloq* °cheeky, °fresh: *Her brash behavior has already landed her in trouble with the headmistress.*

brass *n.* °effrontery, °gall, °nerve, temerity, °impudence, insolence, rudeness, *Colloq* cheek, °nerve: *He had the brass to turn down a knighthood.*

brassy *adj.* **1** °impudent, °forward, °insolent, saucy, °brash, °rude, °brazen, °shameless; °coarse, °flashy, florid, °flamboyant; *Colloq* °cheeky, °fresh: *Our landlady was a big brassy blonde.* **2** °harsh, °strident, °tinny, °grating, dissonant, °shrill, °loud: *She has just the right kind of brassy voice for belting out songs like "There's No Business Like Show Business."*

bravado *n.* boldness, °bluster, boasting, braggadocio, °swagger, °front, self-assurance, *Literary* rodomontade, gasconade; °arrogance, pretentiousness, *Colloq* °machismo, *Slang Brit* °side: *With an attempt at bravado, the union leader refused to meet the management representatives.*

brave *adj.* **1** °fearless, °intrepid, °bold, °courageous, °daring, °gallant, °stout, stouthearted, valiant, valorous, °stalwart, plucky, °staunch, undaunted, °dauntless, unafraid, unfearing, °indomitable, °heroic, *Colloq* °macho; *Slang* gutsy: *Despite her misgivings about her proposal, she put on a brave face in the boardroom. He was brave to face the enemy alone.* **2** °fine, °handsome, °grand, °splendid, °showy, colorful, spectacular, °smart: *The colonel made a brave appearance in full Highland regalia.* —*v.* **3** °challenge, °defy, °dare; °brazen (out), °face, confront, °encounter, °meet: *We had to brave the elements in the open boat. I had to brave my father at breakfast.*

bravery *n.* °daring, °courage, valor, heroism, °fortitude, fearlessness, intrepidity, intrepidness, °pluck, °determination, staunchness, firmness, resoluteness, °resolution, indomitability, stalwartness, *Colloq* °machismo: *One has to admire the bravery of a woman who supported suffrage in the early 1900s.*

brawl *n.* **1** °fight, melee *or* mêlée, °battle, battle royal, donnybrook, °fray, wrangle, °dispute, °disorder, brannigan, °fracas, °row, °quarrel, squabble, *Colloq* punch-up, free-for-all, °scrap, ruckus: *The police had to be brought in to break up the brawl.* —*v.* **2** °fight, wrangle; °row, °quarrel, squabble, *Colloq* °scrap: *The two brothers always seem to be brawling.*

brawn *n.* muscle(s), °strength, robustness, brawniness, °might, °power, *Colloq* huskiness: *Can you imagine what brawn it takes to lift those weights?*

brawny *adj.* °muscular, °strong, °tough, °robust, °mighty, °powerful, °burly, strapping, beefy, °hefty, °bulky, *Colloq* °husky: *That brawny fellow tossing the shotput is my brother.*

brazen *adj.* °brassy, °shameless, °barefaced, °brash, °outspoken, °forward, °immodest, unashamed, °audacious, °candid, °open, °unabashed, brazenfaced; °rude, °impudent, °impertinent, °insolent, saucy, *Colloq* °cheeky, °fresh, *US* sassy: *That's the last time I'll let that brazen hussy near my husband!*

breach *n.* **1** °break, °violation, infraction, disobedience, nonobservance, °infringement, contravention: *Their failure to comply with paragraph 3 is a clear breach of our contract.* **2** °break, °rift, °gulf, °split, breakup, °separation, °rupture, severance, °schism, °split, alienation, estrangement: *There seems to be no way to heal the breach between them.* **3** °gap, fissure, °crack, °split, °hole, °opening; chasm: *Their cannon opened a breach in the castle wall.* —*v.* **4** °rupture; °break through, invade: *The sea has breached the dike. Someone breached the security measures set up for the missile design.*

breadth *n.* **1** °width, wideness, broadness, °beam, °span, °spread, thickness: *The breadth of the cloth is fifty-four inches.* **2** °extent, °magnitude, °degree, °amount, °scope, °expanse, °range, °area, °depth, °detail: *I like the breadth of coverage of the six o'clock news.* **3** liberality, largeness, catholicity, latitude: *Great breadth of vision was exhibited in the conference papers.*

break *v.* **1** break apart *or* up *or* asunder, °fracture, °rupture, break into bits, °come apart, °shatter, °shiver, °crack, °crash, °splinter, °fragment, °split, °burst, °explode, *Colloq* bust: *The ball flew over the fence and broke my neighbor's window. I broke a blood vessel in my arm.* **2** °reveal, °announce, °disclose, divulge, °tell, °make public: *Break the news to him gently.* **3** °relax, ease up, °improve, ameliorate, change for the better: *When will this spell of wet weather break?* **4** °demolish, smash, °destroy, °crush, °ruin, °defeat, °foil, °frustrate: *The power of the dictator was finally broken.* **5** °ruin, bankrupt: *He's the man that broke the bank at Monte Carlo.* **6** °weary, °exhaust, °wear out, °weaken, debilitate: *Twenty years in the chain gang had broken him completely.* **7** °crush, °overcome; cow, °cripple, °demoralize, °weaken, °undermine, °discourage: *The divorce has broken her spirit.* **8** °break in, °tame, °discipline, °train, °condition: *I used to break horses for a living.* **9** °violate, °transgress, °disobey, contravene, °defy, °infringe, fail to observe, °ignore, °disregard, °flout: *If you break the law, you'll regret it. They broke the contract.* **10** °break off, °discontinue, °interrupt, °sever, °cut off; °give up, °suspend, °disrupt: *We broke relations with Spain after the incident. It is very difficult to break a habit of a lifetime. The narrative breaks at this point, to be taken up later.* **11** °break up, °divide, °disperse, °scatter: *The rain is over and the clouds are breaking.* **12** break loose *or* away *or* forth, separate from, °break out (of), °escape (from), °depart (from): *The ship broke from its moorings during the storm.* **13** break forth, burst forth; emerge *or* come out suddenly: *The storm broke in all its fury. After a little while, the sun broke through.* **14** demote, *Colloq* bust: *He was broken from sergeant to private.* **15** ***break away.*** °leave, °depart, °separate (oneself): *A small group broke away from the established church to worship as they saw fit.* **16** ***break down.*** **(a)** °demolish, °destroy: *All right, men, let's break down that wall.* **(b)** °decompose, °break up; °analyze: *The carbon dioxide molecules and water are* *broken down by photosynthesis.* **(c)** °collapse, °give way, °disintegrate, be crushed, be prostrated: *His health has broken down completely.* **17** ***break ground.*** °initiate, °begin, °commence, °found, °set up, °establish, °inaugurate, be innovative, innovate, *Colloq* °break the ice, take the plunge, start the ball rolling: *Laser printers have broken new ground in the area of computer printout.* **18** ***break in.*** **(a)** °interrupt, interpose, interject, burst in, °intrude, °intervene, °interfere, °disturb: *If the results of the election become known, we shall break in to keep you informed.* **(b)** °train, °educate, °prepare; °accustom, °condition, habituate, °wear: *We'll break you in for a week or two on the new machine. Wear the new boots for a few days to break them in.* See also **8,** above. **(c)** °rob, burgle, burglarize, break and enter: *Someone broke in and stole my typewriter.* **19** ***break off.*** **(a)** °discontinue, °stop, °cease, °end: *Sally broke off in mid sentence. After the Fashoda Incident, Britain broke off relations with France.* **(b)** °disengage; °sever, °detach: *A branch broke off from a tree and crashed down, narrowly missing me.* See also **10,** above. **20** ***break out.*** **(a)** °escape; °emerge, °appear: *She broke out of prison in 1985 and hasn't been seen since.* **(b)** °erupt, come out in, break out in *or* into: *He breaks out in a rash from eating strawberries. A war could break out any minute.* **21** ***break the ice.*** See **17,** above. **22** ***break through.*** °penetrate, force *or* get through: *Wit, like beauty, can break through the most unpromising disguise.* **23** ***break up.*** See also **11, 16(b),** above. **(a)** °disband, °disperse; °disintegrate: *Heraclius succeeded in breaking up the Persian power.* **(b)** °fracture, °fragment, comminute: *In the spring, the ice on the river breaks up.* **(c)** See **24(a),** below. **24** ***break with.*** **(a)** °break up (with), separate from, °leave, °depart from: *The leader broke with the party and established a new organization. Sally has broken up with Michael.* **(b)** °renounce, °repudiate, disavow: *They have broken entirely with the traditions we valued so highly.* —*n.* **25** °fracture, °split, °separation, °rupture, °breach, °rift, °schism: *There was a break in a gas pipe. Disagreement over the fishing grounds has resulted in a break in relations.* **26** °gap, °opening, °hole; °crack, °slit: *You can escape through a break in the wall near the bridge.* **27** °interruption, discontinuity, discontinuation, hesitation, °suspension, hiatus, °gap, lacuna, unevenness, irregularity: *There was a five-minute break in transmission from the ship.* **28** °rest, °respite, rest period, coffee break, tea break, °intermission, °interlude, °lull, °pause, playtime, *US* °recess, *Colloq* breather: *We take a break at ten o'clock.* **29** °chance, stroke of luck, °opportunity, °opening: *All he needs is a break to get started.*

breakdown *n.* **1** °collapse, °downfall, °failure, foundering; °destruction, °ruin: *There was a breakdown of our computer system. The arbitrators blamed a breakdown of communication between union and management.* **2** °(mental) collapse, nervous breakdown, *Colloq* crackup: *She had a bad breakdown after her daughter was killed.* **3** °analysis, °rundown, detailing, °review; decomposition, itemization, classification, dissection, distillation, fractionation: *I want a breakdown of these figures by noon. The chemical breakdown of the substance indicated the presence of arsenic.*

breakneck *adj.* °reckless, °dangerous, °daredevil; °excessive, °careless, headlong, °rash, *Colloq* hell-for-leather: *The car came round the corner at breakneck speed on two wheels.*

breast *n.* **1** °chest, °bosom, bust; teat, *Technical* mamma, *Slang* boob, knocker, tit, titty: *He clasped the child to his breast. On some beaches in Europe, women bare their breasts when sunbathing.* **2** °soul, °core, °heart, heart of hearts: *I feel in my breast it is the right thing to do.*

breath *n.* **1** °gust, zephyr, °breeze, °puff, whiff, °stirring, °stir: *There wasn't a breath of air in the tent.* **2** °hint, °suggestion, °indication, °touch, °murmur, °whisper, soupçon: *She never allowed the breath of scandal to affect her behavior.* **3** ***take one's breath away.*** °astound, °astonish, °surprise, °amaze, °dazzle, °startle, °shock, °stagger: *The sheer beauty of the waterfall takes your breath away.*

breathe *v.* **1** °live, °exist: *Breathes there a man with soul so dead, / Who never to himself has said, / "This is my own, my native land"?* **2** inhale and exhale, respire, suspire: *He was breathing regularly.* **3** °exhale, °expel, °puff, °blow: *The banner depicts a fierce dragon breathing fire.* **4** °whisper, °murmur, °hint (at), °suggest, °tell, °speak, °say: *She told me not to breathe a word of it to anybody.*

breathless *adj.* **1** panting, out of breath, winded, gasping, °exhausted, °spent, worn-out, tired out, *Colloq Brit* puffed: *We were breathless after carrying the piano up two flights of stairs.* **2** surprised, amazed, astonished, astounded, awe-struck, staggered: *The news of Penny's having given birth to twins left me breathless.* **3** °eager, °agog, °feverish, in suspense: *We were all breathless with anticipation when the emcee said, "May I have the envelope, please."*

breed *n.* **1** °kind, °sort, °type, °variety, species; °race, °lineage, °stock, °family, °strain: *What breed of dog won at kennel club show this year?* —*v.* **2** °produce, °generate, °bring forth, °create, engender, °hatch, beget, °give rise to, °develop, °cause: *The cheese is so old it's breeding maggots. Familiarity breeds contempt.* **3** °raise, °rear, °cultivate, °propagate: *Charollais cattle are widely bred in Europe today.* **4** °arise, °originate, °appear; °develop, °grow, °increase, multiply: *The sergeant allowed discontent and jealousy to breed within his platoon.*

breeding *n.* **1** rearing, bringing-up, raising, cultivation, °development, propagation: *The breeding of sheepdogs has been Tom's hobby for years.* **2** °(good) upbringing, °(good) manners, °civility, politeness, politesse, gentility, °(good) behavior: *You can tell from the way she treats people that she has breeding.*

breeze *n.* **1** °breath, °puff, zephyr, °wind, °draft, °gust, *Nautical* cat's-paw: *A breeze sprang up from the north, and the little boat moved forward.* **2** easy *or* simple job *or* task, °nothing, *Colloq* °snap, *Slang* cinch, *US* lead-pipe cinch: *It ought to be a breeze to find someone at that salary.*

breezy *adj.* **1** airy, °fresh, °windy, drafty, °brisk, gusty: *The afternoon was breezy and warm, ideal for walking.* **2** °casual, °carefree, lighthearted, °cheerful, cheery, airy, °lively, °spirited, blithesome, °buoyant: *The chairman's breezy opening of the annual meeting made everyone feel comfortable.*

brevity *n.* shortness, briefness, conciseness, concision, terseness, succinctness, pithiness, compactness, laconicism *or* laconism, °economy: *Brevity is the soul of wit.*

brew *v.* **1** °ferment, cook, °boil; infuse: *Our beer is brewed using the best hops.* **2** concoct, °devise, °plan, *Colloq* cook up; contrive, °prepare, °bring about, °cause, °produce, °hatch: *They are brewing up a plot to unseat the financial director.* **3** °start, °go on, °hatch, °begin, °form; °stew, °simmer, *Colloq* cook: *A storm is brewing.* —*n.* **4** beer, ale, °stout; tea; beverage, °drink; concoction, °mixture: *She served me some strange brew in which I could detect nutmeg.*

bribe *n.* **1** °graft, °inducement, *Colloq* °kickback, *Chiefly US* payola, *US* plugola: *Some judges were offered bribes for reducing the sentences of convicted felons.* —*v.* **2** pay *or* buy off, °buy; °corrupt, suborn,

Colloq °fix; *Slang* °oil, grease (someone's) palm, *Brit* nobble: *The guards were bribed to look the other way during the prison break.*

bric-a-brac *n.* bric-à-brac, curiosities, knick knacks, collectables *or* collectibles, trinkets, gewgaws, gimcracks; bibelots, curios, objets d'art, objets de vertu: *On Saturday she went to an antiques fair and bought still more bric-a-brac to clutter up the house.*

brick *n.* **1** °block, cube, chunk, hunk, °slab; stone: *Don't leave that brick of ice cream out or it will melt. A university is not just bricks and mortar.* **2** °pal, °comrade, °friend, *Colloq* °chum, *US and Canadian* buddy: *You're a real brick to watch the children for me till I get back.*

bridal *adj.* °nuptial, °wedding; conjugal, connubial, °marriage: *The bridal gown was white, with lace appliqués. Because of the accident, the bridal bed was never occupied.*

bridge *n.* **1** °span: *We could build a bridge over the river here.* **2** °link, °connection, °tie, °bond: *She regarded teaching as a bridge between her studies and a post in school administration.* —*v.* **3** °span, °cross (over), go *or* pass over, °traverse: *The viaduct bridges the swamp.* **4** °connect, °link, °unite, °join, °tie: *The gap between being poor and being rich is not easily bridged.*

bridle *n.* **1** °restraint, °curb, °check, °control: *Man has need of a bridle on his passions.* —*v.* **2** °curb, °check, °restrain, °hold in, °control: *You must learn to bridle your temper.* **3** °bristle, draw oneself up, be *or* become indignant, take offense *or* umbrage *or* affront (at), be affronted *or* offended (by): *She bridled at the suggestion that she was responsible for Keith's departure.*

brief *adj.* **1** °short, °momentary, °little, °fleeting; °short-lived, transitory, °transient, evanescent, °passing, °temporary, ephemeral, °fugitive: *The lights went back on after a brief interval. His glory was brief.* **2** °short, °concise, °succinct, °to the point; condensed, shortened, °cut, curtailed, °abbreviated, compressed, abridged, °thumbnail, compendious: *The chairman made a few brief remarks. Here is a brief description of what happened.* **3** °curt, °abrupt, °terse, °short, °blunt, °brusque: *You mustn't be so brief with little children.* —*n.* **4** °summary, °outline, °digest, °précis, °résumé, compendium, °abstract, condensation, °abridgment, °synopsis, °extract: *This is merely a brief; the full document will follow.* **5** ***in brief.*** °briefly, concisely, in sum, in summary, to sum up, succinctly, °in a word: *He is a cutthroat, too—in brief, the greatest scoundrel living.* —*v.* **6** °advise, °inform, °fill in, °coach, °instruct, °enlighten; °explain, run through *or* down: *Howard will brief you on the details.*

briefly *adv.* **1** concisely, tersely, succinctly, °in a word, °in short; bluntly, curtly, in a nutshell, in a few words, to sum up: *Briefly, Greenwick, we don't want you drinking here any longer.* **2** momentarily, for a few moments *or* seconds *or* minutes, fleetingly, hurriedly, °hastily, °quickly: *I stopped briefly at the post office on my way home.*

bright *adj.* **1** °light, shining, gleaming, °radiant, °brilliant, resplendent, glittering, flashing, *Formal* refulgent, effulgent, fulgent, fulgid, fulgorous; alight, aglow, beaming, °dazzling, °glowing, °luminous, lambent, °incandescent, ablaze with: *We arrived on a bright, sunny day. The water was bright with phosphorescence.* **2** °clear, cloudless, °fair, unclouded: *It certainly is a bright night—you can see every star.* **3** °shiny, °polished, °lustrous, °glossy, sparkling: *I want that brass so bright I can see my face in it.* **4** °hopeful, °optimistic, °favorable, °propitious, auspicious, °promising, °rosy: *Glenys's prospects for a new job are not very bright.* **5** °brilliant, °vivid, °intense, fluorescent, *US trademark* Day-Glo: *But you said you wanted the room bright orange, Madam.* **6** °intelligent, °clever, °quick-witted, °witty, °brilliant, keen-minded, sharp-witted, °gifted, °astute, °ingenious, °alert, °smart; °precocious; *Colloq* brainy, on the ball: *No one can deny that Ashley is a bright young woman.* **7** °illustrious, °glorious, °splendid, °magnificent, °distinguished, °outstanding: *Today has been one of the brightest days in the history of Britain.* **8** °cheerful, °gay, °happy, °exuberant, °lively, °animated, °vivacious, °spirited: *It is a pleasure to see so many bright faces in the audience.*

brighten *v.* **1** °illuminate, °enliven, °lighten, cheer up, liven up, *Colloq* °perk up: *Replacing those heavy draperies with thinner curtains ought to brighten the room.* **2** °shine, °polish, burnish: *The silver could use a bit of brightening up.*

brilliance *n.* **1** brightness, °radiance, °luster, °splendor, magnificence, °sparkle, °dazzle, °glitter, effulgence, °light: *The brilliance of the opening night rivaled that of Hollywood.* **2** °intelligence, °wit, °intellect, keenness, sharpness, acuteness, °genius, °talent, sagacity; precocity: *Her brilliance shows in her books.*

brilliant *adj.* **1** °bright, shining, °lustrous, °radiant, resplendent, °dazzling, °luminous; °incandescent, glittering, sparkling, °scintillating, coruscating, °twinkling, *Formal* effulgent: *At the show I saw the most brilliant display of diamonds.* **2** °splendid, °magnificent, °superb, °beautiful, °distinguished, °striking, °glorious, °remarkable, °exceptional, °outstanding: *The audience rose for a standing ovation after the brilliant last movement of the concerto.* **3** °illustrious, °famous, °noted, °celebrated, °eminent, °prominent, °renowned, °accomplished: *Potter is one of the country's most brilliant chemists.* **4** °intelligent, °clever, °gifted, °bright, °talented, °smart, °expert, °masterful, °accomplished, °ingenious, °imaginative, °creative; °quick-witted, sharp-witted, keen-witted, °enlightened; °resourceful, discerning, °able, °competent: *Goddard's brilliant mind understood principles of practical rocket flight.*

brim *n.* **1** °edge, °margin, lip, °rim; °brink: *I filled the cup to the brim.* —*v.* **2** be full *or* filled, overflow: *His cup was brimming with steaming mulled wine. They were brimming over with confidence before the game.*

bring *v.* **1** °carry, °bear, °fetch, °get, °take; °deliver: *Don't forget to bring some wine home for dinner tonight.* **2** °lead, °conduct, convey; °escort, invite, °accompany: *The road brought me to your house. You may bring whomever you like to the party.* **3** °draw, °attract, °lure, allure: *What brings you to London?* **4** °carry, °bear, convey; °report: *She brought word of the uprising.* **5** °bring on, °bring about, °occasion, °give rise to, be the source *or* cause of, °create, °cause, engender, °produce; °contribute to: *The thought of his mother brought tears to his eyes.* **6** °institute, °advance; invoke: *She is bringing charges against him for slander.* **7** ***bring about.*** °occasion, °cause, °bring on, °accomplish, °effect, °achieve, °produce: *The government has brought about changes in the health service.* **8** ***bring around.*** **(a)** °revive, resuscitate, bring to; °restore: *The smelling salts brought her around when she fainted.* **(b)** °persuade, °win over, °convince, °influence: *Can he be brought around to our way of thinking?* **9** ***bring down.*** **(a)** °overthrow, depose, oust, unseat, dethrone, °overturn, °topple: *A military faction has brought down the government.* **(b)** °reduce, lessen, °diminish, cut (back *or* down): *The chancellor promised to bring down taxes in the next budget.* **10** ***bring forth.*** **(a)** °bear, give birth to, °produce; °yield: *The kangaroo brings forth young less than an inch in size.* **(b)** °set forth, bring out *or* in *or* up, °introduce, present, °produce, °put out, °submit, °offer, °advance: *Hanson has brought forth a new sales plan.*

11 ***bring in.*** **(a)** °earn, °yield, °produce, °realize, °fetch, °return, sell for: *Advertising brings in more revenue than subscriptions.* **(b)** See **15,** below. **12** ***bring off.*** °succeed (in), °carry out, °achieve, °accomplish, do, carry out *or* off, °perform, °succeed, °pull off; *Colloq* °put over: *Do you really think she'll be able to bring off her masquerade?* **13** ***bring on.*** **(a)** °produce, °put on, °introduce, °bring in: *If the audience gets restless, bring on the clowns.* **(b)** °induce, °produce, °occasion, °bring about: *Eating parsnips may bring on a rash.* **14** ***bring out.*** **(a)** °display, °feature, focus on, °illuminate, °set off, make noticeable *or* conspicuous, °emphasize, °develop: *The color of the dress brings out the blue of your eyes.* **(b)** °publish, °issue, °release, make known *or* public, °produce; °put on, °stage: *They've brought out a new edition of Dickens's works.* **15** ***bring up.*** **(a)** °rear, °raise, °care for, °look after, nurture, °breed; °educate, °teach, °train, °tutor: *She has brought up six children on her own.* **(b)** °introduce, °broach, °bring in, °raise, °pen (up), °set forth, °mention, °touch on, °talk about, °discuss; reintroduce, °recall: *Why bring up irrelevant matters like her age?* **(c)** °raise, elevate: *They have brought up only three survivors from the mine.* **(d)** °vomit, °throw up, °regurgitate, disgorge: *A raw egg in milk will make him bring up the poison.*

brink *n.* **1** °edge, °brim, °rim, °margin, lip, °border: *He lost his footing and almost went over the brink into the gorge.* **2** °verge, °point: *He was on the brink of telling them everything, but the torture suddenly stopped.*

brisk *adj.* **1** °active, °lively, °busy, °vigorous: *The poachers are doing a brisk trade in rhinoceros horn.* **2** °quick, °animated, °sprightly, spry, °energetic, °spirited: *Patrick was a brisk lad, fresh from Oxford.* **3** °strong, °steady, °fresh, °refreshing, °bracing, °invigorating, °stimulating, °crisp, °biting, °bracing, °keen, nippy, °chill, °chilly, °cool, °cold: *A brisk breeze had started up from the north, chilling us through.* **4** °energetic, vibrant, °invigorating, °stimulating: *After a brisk massage, Mariette felt completely revitalized.*

bristle *n.* **1** °hair, whisker, barb, °prickle, °thorn, quill, *Technical* seta: *Shaving brushes are often made from badger bristles.* —*v.* **2** °prickle, °rise, °stand up, *Formal* horripilate: *He could feel the hair on the back of his neck bristle.* **3** °seethe, become angry *or* infuriated *or* furious *or* maddened, °boil, °flare up, see red, °bridle: *He bristled with enraged frustration.* **4** °teem, °crawl, be thick, °swarm, be alive: *The sea urchin was bristling with sharp spines.*

brittle *adj.* **1** °fragile, frangible, breakable; friable: *My fingernails become brittle in the cold and break easily.* **2** °frail, °weak, °delicate, °sensitive, °fragile, °insecure: *She might seem strong, but she has a very brittle nature and is easily upset.*

broach *v.* °introduce, °raise, °open (up), °suggest, °mention, hint at, touch on *or* upon, bring up *or* in, °talk about, °advance: *I didn't dare broach the subject of money.*

broad *adj.* **1** °wide, °expansive, °large, °extensive; °spread out, °ample, °spacious: *The broad highway stretched out for miles before them. Cattle graze in the broad pastures.* **2** °bright, °plain, °open, °full; unshaded: *He had the nerve to kiss me in broad daylight, in front of everyone!* **3** °plain, °clear, °obvious, °emphatic, °explicit, °pronounced, °direct, unconcealed, °undisguised, unsubtle, °evident: *His wink gave a broad hint of what he really had in mind.* **4** °main, °general, generalized, °rough, unspecific, nonspecific, °approximate, °sweeping: *Without the details, here is a broad outline of what happened.* **5** plainspoken, °outspoken, °forthright, °direct, unreserved, °frank, °candid, unrestrained: *When he reached the witness box, he repeated the accusation in broad terms.* **6** °inclusive, °general, widely applicable, °extensive, wide-ranging, °comprehensive, wholesale; °vague, °imprecise, °indefinite, unfocused, nonspecific, unspecified: *We have broad support for these policies. She formulated a broad rule to fit all imaginable cases.* **7** °liberal, °tolerant, °catholic, ecumenical, latitudinarian: *The term "Broad Church" is said to have been coined by A. H. Clough. The judge feels that he must give the broadest possible interpretation of the law.* **8** °dirty, °blue, °coarse, °rude, °indecent, °vulgar, °improper, indelicate, °off-color, °loose, °gross, °obscene, °lewd, °lascivious, °filthy, °pornographic; inelegant, °unrefined, unladylike, ungentlemanly, titillating: *Penrose was in the corner telling some of his broad jokes.* —*n.* **9** °woman, °girl, *Slang* dame, cookie *or* cooky, skirt, bimbo, bird, chick, °number, doll, °piece (of baggage): *We picked up a couple of broads at the dance hall last night.*

broadcast *v.* **1** °air, °transmit, relay; °radio; televise, telecast: *The program will be broadcast tonight.* **2** °announce, advertise, °publish, °proclaim; disseminate: *It may be a bad idea to broadcast your plans in advance.* **3** °sow, °scatter, °seed: *The farmer broadcasts this seed instead of planting it.* —*n.* **4** °program, °show; °transmission, telecast: *I heard the broadcast on my car radio.*

brochure *n.* °pamphlet, booklet; catalog; folder, °leaflet; °tract: *The brochure advertising the company's products will be ready tomorrow.*

broil *v.* grill, barbecue: *I think hamburgers taste better broiled than fried.*

broke *adj.* penniless, °indigent, °down-and-out, poverty-stricken, °penurious, °impoverished, °insolvent, °destitute, °poor, °needy, bankrupt, ruined, *Colloq* on one's beam ends, °on one's uppers, strapped, flat *or* dead *or* stony broke, °hard up, °short, up against it, *US* °flat, on the skids; *Slang Brit* skint: *I was broke after paying the rent—didn't even have food money.*

broken *adj.* **1** fragmented, shattered, shivered, splintered, ruptured, cracked, °split, smashed, pulverized, disintegrated, destroyed, demolished: *A broken Ming vase cannot be worth much.* **2** fractured: *With a broken leg, she certainly won't be competing in the slalom.* **3** enfeebled, weakened, crushed, defeated, beaten, ruined; dispirited, °dejected, discouraged, demoralized, defeated, °subdued, debilitated, *Colloq* licked: *Rosa's running away with a sailor left Hugh a broken man.* **4** tamed, trained, disciplined, °obedient, docile, domesticated, °subdued; conditioned: *Of what use is a horse that isn't broken?* **5** violated, transgressed, disobeyed, contravened, defied, flouted, disregarded, ignored, infringed: *The rules of this club are broken too often.* **6** interrupted, °disturbed, discontinuous, °disjointed, °disconnected, fragmented, °fragmentary, °intermittent, °erratic, °sporadic: *I had a terrible night of broken sleep.* **7** Also, ***broken down.*** out of order *or* commission, not working *or* functioning, in disrepair, *Slang* °on the blink, out of kilter, kaput, *US* on the fritz, out of whack: *My watch is broken. Why waste money repairing that broken-down car of yours?*

brokenhearted *adj.* °heartbroken, depressed, °downhearted, °dejected, devastated, crushed, overwhelmed, heartsick, downcast, °upset; °forlorn, °sorrowful, disconsolate, °inconsolable, grief-stricken, °miserable, °wretched, °melancholy, heavy-hearted, °sad, °doleful, dolorous, woeful, °woebegone, °gloomy, morose, °glum, cheerless, *Colloq* down: *She was brokenhearted when her puppy was lost.*

broker *n.* stockbroker; °agent, °dealer, middleman, °intermediary, °go-between, *Brit* stockjobber: *I have phoned my broker to tell him to sell all my shares.*

brooch *n.* °clasp, °pin; °fastening: *She was wearing the cameo brooch I had given to her mother.*

brood *n.* **1** °young, °offspring, °progeny; children, °family: *A mallard was tending her brood among the rushes.* —*v.* **2** °incubate, °hatch, °set, °sit, °cover: *The old hen was brooding three eggs.* **3** Also, ***brood on** or **over.*** ponder (on *or* over), meditate (on *or* over), °contemplate, ruminate (on *or* over), muse (on *or* over): *He just sits there brooding over the subject of his next novel.* **4** mope, °sulk, °pout, pine, eat one's heart out, °fret, °worry, agonize, °despair: *Don't just brood over the problem, do something about solving it!*

brook[1] *n.* °stream, rivulet, °run, runnel, rill, *US, Canadian, Australian, and New Zealand* creek; *No. England dialectal* beck, gill *or* ghyll, *Northern, Midwestern, and Western US* crick; *Scots* °burn: *The river is fed by numerous brooks from every part of the country.*

brook[2] *v.* °endure, °tolerate, °stand, °abide, °put up with, °suffer, °allow: *She runs the business in her own way and brooks no interference from anyone.*

broth *n.* °stock, bouillon, consommé; soup; decoction: *We tossed food scraps into the large, simmering pot to make a broth.*

brothel *n.* bordello, whorehouse, house of ill fame *or* ill repute, bawdyhouse, bagnio; seraglio, harem, *Obsolete* °stew, *Colloq US* sporting house, *Slang Brit* knocking-shop, *US* cathouse: *On their first night in Paris, they visited a brothel.*

brother *n.* sibling; °relation, °relative, °kin, kinsman; °fellow, fellow man, fellow clansman, fellow citizen, fellow countryman, fellow creature; °associate, °colleague, confrère, °companion, *Colloq* °pal, °chum, *Brit and Australian* °mate, *US* buddy: *Some day, perhaps all men will regard each other as brothers.*

brotherhood *n.* **1** brotherliness, °fellowship, °companionship, °alliance, °friendship, comradeship, camaraderie, °kinship: *We should all live together in harmony and brotherhood.* **2** °fraternity, guild, °society, °association, °order, °league, °union, °organization, °club, community, °circle, °set, °clique: *The tribes fused into a united and enthusiastic brotherhood.*

brotherly *adj.* °fraternal, °kind, °affectionate, °cordial, °friendly, °amicable, °amiable, °neighborly, °loyal, °devoted: *The boys grew up together and maintained a brotherly relationship throughout their lives.*

browbeat *v.* °bully, °intimidate, °threaten, badger, °dominate, cow, °frighten, °discourage, °tyrannize, hector, °harass, keep after, °nag, *Colloq* hassle: *The foreman constantly browbeat anyone who wasn't one of his drinking cronies.*

browse *v.* look over *or* through, °skim (through), °scan, thumb *or* flip *or* flick through: *I was browsing through some recent acquisitions at the secondhand bookshop.*

bruise *n.* **1** °injury, °hurt, contusion, °bump, °welt, °scrape, abrasion, °scratch, °wound, black-and-blue mark, blotch, °blemish, °mark, °spot, discoloration, °damage, *Technical* ecchymosis: *I got this bruise from walking into the corner of the table. The price is lower if the fruit has a few bruises.* —*v.* **2** °injure, contuse, °hurt, °scrape, °harm; °wound, °damage: *I bruised my knee when I fell down. Being arrested can bruise one's self-esteem.*

bruiser *n.* prizefighter, boxer, fighter; °tough, ruffian, bodyguard, °thug, °hoodlum, bouncer, *Colloq* hooligan, tough guy, toughie, *Brit* °minder, *US* roughneck, hood, gorilla, plug-ugly, torpedo, enforcer: *The heavyweight contender is really a big bruiser. Mr. Big strode in with two of his bruisers.*

brunt *n.* °(full) force, °burden, onus, °weight, °impact; °shock, °stress, °violence, onslaught: *As the head of the department was on holiday, I had to take the full brunt of the customers' complaints.*

brush[1] *n.* **1** brushwood, shrubs, undergrowth, branches, °scrub, °brush, bracken, brambles, underbrush, underwood: *It took us three days to clear the brush from around the house.* **2** °thicket, °brake, copse, grove, boscage: *The fox disappeared into the brush, which was too dense for the dogs to follow.*

brush[2] *n.* **1** hairbrush, toothbrush, clothesbrush, shoebrush, nailbrush, paintbrush; broom, dustbroom, besom, *US* whiskbroom: *This brush is too coarse and may damage your teeth.* **2** See **brush-off.** **3** °encounter, °engagement, °skirmish, *Colloq Brit* spot of bother: *Millworth has had some brushes with the law.* —*v.* **4** °scrub, °clean; °groom, curry; °sweep, °whisk, °gather: *Brush your teeth twice a day. Brush down the mare before saddling her. Brush the crumbs off the table.* **5** graze, °touch: *He deliberately tried to brush against her in the corridor.* **6** ***brush aside** or **away.*** °disregard, °dismiss, °put aside, shrug off: *Brushing aside the members' objections, he tried to force the committee's acceptance of the new rules.* **7** ***brush off.*** °dismiss, °ignore, °rebuff, send off *or* away *or* packing: *He proposed to her, but she brushed him off.* **8** ***brush up (on).*** °review, restudy, °go over, °refresh, °study, *Archaic* con: *You should brush up on your geometry before taking trigonometry.*

brushoff *n.* °dismissal, °rebuff, °rejection, snub, *Colloq* cold shoulder, putdown, °slap in the face, the (old) heave-ho; *Slang US and Canadian* walking papers: *Tanya has given Theo the brushoff—said she never wants to see him again.*

brusque *adj.* °blunt, °rude, °overbearing, °impolite, uncivil, °discourteous, °ungracious, °ill-mannered, unmannerly; churlish, °gruff, °abrupt, °short, °curt, °sharp, °terse, °brash, °bluff: *By her brusque attitude, I could tell that the interviewer had already decided against me.*

brutal *adj.* **1** °inhuman, °savage, °cruel, pitiless, °harsh, °severe, barbaric, barbarous, °beastly, bestial, °sadistic, °murderous; inhumane, °heartless, hardhearted, °unkind, °fierce, stonyhearted, insensitive, unfeeling, °coldblooded, °unsympathetic, °remorseless, °ruthless, °ferocious, °atrocious, Draconian *or* Draconic, *Literary* °fell: *Few survived the brutal treatment in the concentration camps.* **2** °rude, °ill-mannered, °coarse, °unrefined, °boorish, ill-bred, °rustic, crass, uncouth, uncultured, uncultivated, °rough, °crude: *His brutal behavior made him unfit to represent the government.*

brute *adj.* **1** brutish, °dull, unfeeling, °senseless, °blind, unintelligent, °unthinking, °thoughtless, °mindless, unreasoning, irrational, °instinctive, °physical, °material; insensate, °unconscious: *He was able to lift the safe without help, by sheer brute strength.* —*n.* **2** °animal, °beast, °savage: *The human race to be placed as the middle link between angels and brutes. Patrick was an absolute brute to her.*

bubble *n.* **1** blister, air pocket, globule, droplet: *This painted surface is full of air bubbles.* **2** ***bubbles.*** °froth, °foam, suds, °lather, spume; effervescence, carbonation, °fizz: *The beer is full of bubbles.* —*v.* **3** °foam, °froth, °boil, °seethe, °fizz: *A pot of soup was bubbling on the fire.*

bubbly *adj.* **1** °effervescent, foamy, frothy, fizzy, sparkling: *I could see that the surface was all bubbly.* **2** °effervescent, °merry, °ebullient, bouncy, °animated, °vivacious, °cheerful, cheery, °lively, °excited: *Janet is known for her bubbly personality.* —*n.* **3** champagne, sparkling wine, sparkling burgundy, asti spumante, *Colloq Brit* champers: *Let's open a bottle of bubbly and celebrate.*

bucket *n.* pail, scuttle: *Keep this bucket of coal near the hearth.*

buckle *n.* **1** °clasp, fastener, °clip, °fastening, °hook, °catch: *The buckle broke on my belt and my trousers fell down.* —*v.* **2** °collapse, °cave in, °crumple, °bend, °warp, °distort, °twist, °bulge: *The support gave way and the entire wall buckled.*

bug *n.* **1** insect, beetle, larva, grub, caterpillar, butterfly, mosquito, °fly, spider, *Colloq Brit* creepy-crawly, *US* no-see-um: *There's a bug on your collar.* **2** °microbe, °germ, virus; °disease, °affliction, °illness, sickness, °ailment, °disorder, malady, infection; °condition, °complaint, °infirmity, indisposition: *She caught some kind of bug and won't be in for a few days.* **3** °obsession, °craze, °fad, °mania, °rage: *Almost everyone in those days succumbed to the hula hoop bug.* **4** °enthusiast, faddist, °fan, °fanatic; hobbyist: *She's turned into a fruit machine bug.* **5** listening device; microphone, transmitter, electronic eavesdropper, °tap: *They planted a bug in the ambassador's telephone.* **6** °fault, °error, °mistake, °failing, °shortcoming, *Colloq* hang-up, glitch: *There's a bug in the program that's preventing the list from being sorted. They can't market the device till they've ironed out all the bugs.* —*v.* **7** °annoy, °irritate, °pester, °irk, °harass, °bother: *I wish Mom'd stop bugging me about my homework.* **8** °tap, °spy on: *They bugged her phone and recorded all her conversations.*

bugger *n.* **1** buggerer, sodomite. **2** °chap, °fellow, °man; °boy, °lad, °child, °tot; *Slang chiefly Brit* geezer, *US* °jerk; *Colloq* °guy, *Brit* bloke, °fool, idiot: *He's a cute little bugger, isn't he? Who's that silly-looking bugger with Christina?* —*v.* **3** Also, ***bugger up.*** °ruin, °destroy, °botch, °bungle, °wreck; make a mess of, *Colloq* mess *or* screw up, *Brit* bollocks *or* ballocks up, balls up, make a balls-up of, cock up, *US* ball up, bollix up; *Taboo* fuck up: *He's buggered the recording, so we'll have to start again at the beginning. Why does she bugger up everything I try to do?* **4** ***bugger about*** *or* ***around.*** **(a)** °fool about, waste time, °dawdle, *Colloq US* lallygag *or Colloq US* lollygag; *Taboo* fuck about *or* around: *He buggers about the house all the time instead of looking for a job. Don't bugger about with my hi-fi.* **(b)** cause complications for, create difficulties for: *She pretends to be helping me, but she's just buggering me about.* **5** ***bugger off.*** °go away, °depart, °leave, clear off *or* out, *Colloq* make tracks, skedaddle, °beat it, *Slang* piss off; *Taboo* fuck off: *Oh, bugger off and leave me alone!*

build *v.* **1** °construct, °erect, °raise, °set up, °assemble: *I hope to build my own house in another year or so.* **2** °found, °establish, °base: *The theory is built on the principle that light travels at 186,000 miles per second.* **3** °develop: *She built the company in about five years.* **4** Also, ***build up.*** °intensify; °increase, °develop, °enlarge, °strengthen: *The distant hum of voices gradually built to a mighty roar.* —*n.* **5** °physique, °figure, °body, °shape, *Slang* bod: *He has a good build from working out at the gym.*

building *n.* edifice, °structure, construction, erection: *The building where I work is air-conditioned.*

bulge *n.* **1** °lump, °hump, protuberance, °bump, °swelling, °projection: *This wallet is making a bulge in my jacket.* —*v.* **2** °protrude, °stick out, °swell (out): *His stomach bulges out over his belt.*

bulk *n.* **1** °volume, °magnitude, °mass, enlargement, largeness, °size: *The sausage-makers add bread just for bulk.* **2** °majority: *The bulk of the people voted for the proposal.*

bulky *adj.* °large, °voluminous, °unwieldy, °awkward, ungainly, cumbersome, *Brit* chunky: *The package, though quite bulky, didn't cost much to post.*

bulletin *n.* °message, °notice, communication, °announcement, communiqué, dispatch *or* despatch, °report, °account, °flash, news item *or* flash: *And now, here's a bulletin from the center court at Wimbledon.*

bully *n.* **1** persecutor, intimidator, °tyrant: *That bully Roderick is always beating up the younger boys.* —*v.* **2** °persecute, °intimidate, °tyrannize, °torment, °browbeat, °daunt, awe, cow, terrorize; hector, °harass, °push around: *Roderick even bullied his best friend into parting with his allowance.* —*adj.* **3** *Old-fashioned* °jolly, °worthy, °admirable: *Ah, there you are, my bully boy!* —*interj.* **4** Usually, **Bully for (someone)!** Bravo!, Great!, Fantastic!, Fabulous!, Marvelous!, Spectacular!; So what?, What of it?; *US* Peachy!, Dandy!, Neat-oh!; *Old-fashioned* Fantabulous!: *"David's won the snooker competition again." "Bully for him!"*

bulwark *n.* **1** °defense, °safeguard, redoubt, bastion, buffer, °barrier, °rampart, fortification: *A strong defense is the best bulwark against aggression from outside.* —*v.* **2** °defend, °protect, °shelter: *Marnie's indifference to others bulwarks her against any feelings of contrition.*

bum *n.* **1** °buttocks, °posterior, hindquarters, fundament, behind, rump, °bottom, behind, derrière, rear end, backside, °seat, °rear, *Colloq Brit* arse, *US* fanny, can, hinie, tush, tushy *or* tushie, tokus *or* tochis *or* tuchis, keister *or* keester, ass: *Why don't you get off your fat bum and go out and get a job?!* **2** °tramp, panhandler, °beggar, vagrant, °loafer, °drifter, °vagabond, hobo, °derelict, gypsy; *Brit* caird, °tinker, °traveler; *US* (shopping) bag lady: *Along the Bowery the doorways and pavements are strewn with bums.* —*adj.* **3** °improper, unjustified, °false, °fraudulent, trumped-up, °untrue, fabricated, made-up, °bogus: *That conviction for auto theft was a bum rap, but he still served eighteen months.* **4** °bad, °awful, unfair, °dishonest, °poor, °rotten, *Slang* °lousy, crummy: *I still think you got a bum deal on that toaster.* —*v.* **5** °borrow, °beg, sponge, *Colloq* °scrounge, cadge, °touch, put the touch on, *US* mooch, °hit, °hit up: *Can I bum a cigarette from you?*

bump *n.* **1** °blow, °collision, °thud, °hit, °knock, buffet, clunk, whack: *That bump on the head seems to have affected Beazley.* **2** °lump, protuberance, °welt, °swelling, tumescence, °knob, °bulge: *How did you get that bump on your forehead?* —*v.* **3** °knock (against), °strike, °hit, °collide (with), °run into, °ram; smash, °crash, *Colloq* wallop: *We met in front of the supermarket when my car bumped into hers as I was parking.* **4** ***bump into.*** °meet, °encounter, run into *or* across, °come across, stumble over: *I bumped into Philippa at the hairdresser's.* **5** ***bump off.*** °murder, °kill, °put away, assassinate, do away with, °execute, liquidate, dispatch *or* despatch, *Slang* °take for a ride, °destroy, °eliminate, °rub out, °wipe out, do in, *US* °waste, ice: *They bumped off Wimpy, boss; he was pulled out of the river wearing concrete overshoes.*

bumpy *adj.* °lumpy, °rough, uneven, °irregular, knobby, knobbly, °pitted; potholed, bouncy, jarring, jerky, rutted: *The skin on his forehead is a bit bumpy. This is the bumpiest road in the town.*

bunch *n.* **1** °bundle, °cluster, °batch, °clump; °bouquet, nosegay, posy, °spray: *That's a nice-looking bunch of grapes. Mr. Herbert arrived with a bunch of flowers for me.* **2** °crowd, °knot, °collection, °group, °lot, °gathering, °cluster, °clutch, °batch, °assortment, °mass: *A bunch of people stood outside the courtroom, awaiting news of the verdict.* —*v.* **3** °sort, °class, classify, °categorize, assort, group together, °bracket: *It would be a mistake to bunch all different kinds of liberals into the* *same category.* **4** ***bunch up.*** °gather; smock; °collect, °crowd, °group, °cluster: *The fabric is all bunched up at the bottom. Don't let the people bunch up in front of the exits.*

bundle *n.* **1** °bunch, °collection, °package, °parcel, °packet, °pack; bale, sheaf; *Archaic* fardel: *I have to leave this bundle at the laundry today. Bring this bundle of hay to the horse.* —*v.* **2** °gather (together), °tie up (together), °collect, °pack, °package: *He bundled up all his belongings.* **3** ***bundle off*** *or* ***out.*** dispatch *or* despatch, °pack off, hustle *or* hurry off *or* away, send away *or* off; decamp, scurry off *or* away, *Colloq Brit* do a moonlight flit: *We bundled Aunt Mary off home as soon as the storm subsided. That couple have bundled out of room 429.*

bungle *v.* °spoil, °botch, mismanage, °stumble, bumble, *Golf* foozle, *Colloq* foul *or* screw *or* louse up, °blow, mess *or* muck up, make a mess *or* hash *or* muddle of, muff, *Slang Brit* °bugger, *US* snafu, *Taboo* fuck up: *Thewless has bungled the job again, and we must replace him.*

buoy *n.* **1** (navigational *or* channel) mark *or* marker, °float; nun (buoy), can (buoy), bell (buoy), gong (buoy), °siren, °signal, mooring buoy, spar buoy, lollipop: *Returning to port, always leave the red buoys to starboard.* —*v.* **2** Often, ***buoy up.*** °lift, °raise, elevate, °support, hearten, °sustain, keep up: *We sang songs to buoy up our spirits while the rescuers dug their way toward us.*

buoyant *adj.* **1** afloat, floating, floatable: *Waterlogged wood is not usually buoyant.* **2** °light, resilient, °lively, °vivacious, °bright, °cheerful, °carefree, °blithe, °animated, °jaunty, bouncy, °ebullient, lighthearted, *Colloq* peppy: *One had to admire his buoyant optimism, even under adverse conditions.*

burden *n.* **1** °load, °weight, gravamen; °strain, °pressure, °trouble, onus, millstone, °cross, albatross: *The old man put down his burden. The burden of the evidence is against them. Her death was a terrible burden for him to bear.* —*v.* **2** °load, °weigh down, saddle with, °encumber; °tax, °oppress: *The mules are heavily burdened with a month's supply of food. Don't burden me with your problems.*

burdensome *adj.* onerous, cumbersome, °oppressive, °weighty, °troublesome, wearisome, bothersome, distressing, worrying, worrisome, vexatious, °irksome: *A tax on food is burdensome for those on a low income.*

bureau *n.* **1** *Brit* (writing) desk, *US* chest of drawers, °chest, dresser, chifferobe, chiffonier: *Simon has a beautiful antique bureau in his office. One of my cufflinks rolled under the bureau.* **2** °office, °agency, °department, °division, °section, subdivision, subsection, desk: *I sent the form to the bureau a month ago, but I still don't have my visa.*

bureaucracy *n.* officialdom, officialism, °government, red tape, °administration, °authorities: *The bureaucracy survives because the officials rely on graft for their income.*

burglar *n.* housebreaker, °thief, °robber; sneak thief, cat burglar, *US* second-story man: *The burglars, remarkably, didn't take the most valuable paintings.*

burial *n.* interment, °funeral, entombment, obsequies, sepulture: *His six ex-wives attended the burial.*

burlesque *n.* **1** °caricature, °lampoon, spoof, °parody, °satire, °mockery, travesty, *Colloq* °takeoff; °(grotesque) imitation, vulgarization, °exaggeration: *In the mid-nineteenth century, burlesques drove pantomimes off the stage.* **2** *US* striptease, strip show, nudie *or* girlie show: *The old comedians insist that burlesque acts were an art form, but the audience went just for the girls.* —*v.* **3** °satirize, °take off, °lampoon, spoof, °parody, °caricature, travesty: *Cervantes burlesqued the old romances in* Don Quixote. —*adj.* **4** °satirical, derisive, mock-heroic, mock-pathetic: *She sang a burlesque opera based on* Hamlet, *called "Omelette."*

burly *adj.* °stout, °sturdy, corpulent, °large, °big, °hefty, °stocky, thickset, °brawny, chunky, °heavy, beefy, °muscular, °strong, strapping, °rugged, °tough, *Colloq* °husky: *Two rather burly gentlemen were called in to help me out of the place.*

burn *v.* **1** °blaze, °flame, °flare, °smolder: *A fire was burning on the hearth.* **2** ignite, set on fire, °fire, °light, °kindle, incinerate, *Slang* torch: *He burnt the incriminating papers in the fireplace.* **3** °desire, °yearn, °wish, °long, °itch: *He wrote "Darling, I am burning to be with you tonight."* **4** °waste, throw *or* fritter away, squander: *Don't worry about Norman, he has money to burn.* **5** overcook, °blacken, char, °singe: *If you're not careful, you'll burn the toast again.*

burning *adj.* **1** °flaming, blazing, °fiery; °ablaze, aflame, afire, °on fire: *When we arrived, the entire building was burning.* **2** vehement, °ardent, °excited, °passionate, °fervent, fervid, °intense, °fiery, °enthusiastic: *She had a burning desire to join that illustrious company.* **3** raging, °violent, parching: *His burning fever had finally subsided a little.* **4** °hot, blazing, °scorching, seething, °withering: *She was married on a burning hot day in July.*

burrow *n.* **1** °excavation, °hole, warren, °tunnel: *The rabbit retreated to its burrow under the hedge.* —*v.* **2** °dig, delve, °tunnel, °bore; °excavate: *The larvae burrow into the wood where the birds can hear them moving about.*

burst *v.* °break (asunder), °rupture, °shatter, °explode, °blow up; °puncture; *Slang* bust: *If it keeps raining the dam will burst and the valley will be flooded.*

bury *n.* **1** inter, inhume, lay to rest: *Yesterday they buried the last person who really knew the truth about the scandal.* **2** °abandon, °forget, consign to oblivion, eradicate, extirpate: *The residents buried their differences and united to repel the town planners.* **3** °submerge (oneself), °exile (oneself), °plunge, become engrossed *or* absorbed: *She buried herself in her book.* **4** °conceal, °obscure, °hide, °cover up: *The real story was by now completely buried beneath the mass of legend.* **5** °overwhelm, °overcome, inundate: *I'm so buried by work I can't even take a day off, let alone a week.*

business *n.* **1** °duty, °function, °occupation, °calling, °vocation, °trade, °profession, °work, °province, °area, °subject, °topic, °concern, °affair, °responsibility, °role, °charge, °obligation: *Her business is supplying models for fashion shows. Mind your own business and don't be such a nosy parker.* **2** °matter, °job, °task, °subject, °question, °problem, °issue, °point, °affair: *Gentlemen, let us call the meeting to order and attend to the business at hand.* **3** dealing, °transaction; °trade, °commerce, °traffic: *We've never done any business with that company.* **4** °concern, °establishment, °organization, °company, °firm, °house, °enterprise; corporation, partnership, proprietorship: *Rodney wants to sell the business and retire to Spain.*

busy *adj.* **1** occupied, °engaged, employed, °involved: *I can't talk to you now, I'm busy.* **2** working, °industrious, °active, °diligent; bustling, °hectic, °lively, hustling, °energetic: *Are you very busy at the office these days? The diamond district is certainly a busy place.* **3** °ornate, °elaborate, °detailed, °complicated, complex, (over)decorated, °intricate, Baroque, Rococo: *Some of the late Victorian architecture is far too busy for my taste.* —*v.* **4** °occupy, °involve, °employ, °divert, absorb, engross: *She has busied herself with charity work to get her mind off the tragedy.*

busybody *n.* °pry, snoop(er), peep(er), °gossip, meddler, Paul Pry, *Colloq* Nosy Parker, *Slang US* buttinsky: *If he so much as sees us talking together, that busybody will probably cook up some sex scandal.*

butcher *n.* **1** °murderer, slaughterer, °killer, ripper, °cutthroat, executioner, annihilator: *That coldblooded butcher dismembered his victims after strangling them.* **2** destroyer, bungler, muddler: *Look what that butcher of a tailor has done to my suit!* —*v.* **3** °slaughter, °massacre, °murder, cut *or* hack *or* hew to pieces, dismember, disembowel, °exterminate, annihilate, °kill, liquidate: *The entire crew was butchered by the islanders.* **4** °botch, °bungle, °foul up, *Colloq* °mess up, make a mess *or* hash of; *Slang* louse up, °screw up, *Brit* bollocks *or* ballocks up, *US* bollix up; *Taboo* fuck up: *He butchered the restoration of my antique cabinet.*

butt[1] *n.* °target, °end, °object, °prey, °victim, °dupe; gull, *Colloq* pigeon, °sucker; *Brit* Aunt Sally, *Slang US and Canadian* patsy: *He was always the butt of their jokes.*

butt[2] *v.* **1** abut, °join, °meet: *This wall butts up against my garage.* **2** ***butt in** or **into**.* °interfere, °intrude, °interrupt, *Colloq US* kibitz; °meddle: *Please let me finish a sentence without butting in. Don't butt into my affairs.*

buttocks *n.* °bottom, behind, derrière, °seat, °rear, rear end, backside, °posterior, hindquarters, fundament, *Colloq Brit* °bum, arse, *US* hinie, can, tush *or* tushy *or* tushie, tokus *or* tochis *or* tuchis, keister *or* keester, °butt, °tail, prat, ass; *Slang* cheeks, °duff: *A person with large buttocks should not wear tight shorts.*

buttonhole *v.* **1** corner, detain, accost, importune, °waylay: *A reporter buttonholed one of the senators for details of the new tax bill.* —*n.* **2** corsage, *US* boutonniere *or US* boutonnière: *He wore a rose for a buttonhole.*

buttress *v.* °sustain, °support, °strengthen, °prop (up), °brace, °reinforce, shore up: *Huge beams were needed to buttress the walls after the bombing.*

buxom *adj.* **1** °hearty, °healthy, °vigorous, °lusty, °attractive, °comely, °plump, *Colloq* °hefty: *Sylvia was a buxom serving-wench at the Bugle Horn.* **2** busty, °bosomy, chesty, well-endowed, big-busted: *The centerfolds in this magazine usually show quite buxom women.*

buy *v.* **1** °purchase; °acquire, °obtain, °get, °procure, °gain, °come by, °secure: *Where did you buy that hat?* **2** °accept, °allow, °take, °believe, °swallow, °go for: *Do you buy his story about the tooth fairy leaving him ten thousand dollars?* **3** °bribe, suborn, °pay off, buy off, °corrupt: *That customs man wouldn't have let the package through unless he'd been bought.* —*n.* **4** °purchase, °acquisition: *We made a bad buy at the last auction.* **5** Also, ***good buy***. °bargain, *Colloq US and Canadian* °steal: *If you paid only two thousand dollars, that was a real buy.*

buyer *n.* °customer, consumer, °client, purchaser: *Is it likely that you will find many buyers of Coptic dictionaries in Boston?*

buzz *n.* **1** °hum, °murmur, drone, buzzing: *I lay listening to the buzz of the bees.* **2** °stir, °ferment, °talk, °undercurrent: *Her voice could be heard above the buzz of conversation.* **3** phone call, °ring: *I think I'll give him a buzz to see if our appointment is still on.* **4** °thrill, feeling of excitement, °sensation, stimulation, °kick, *Colloq* °high: *I got quite a pleasant buzz from that drink.* —*v.* **5** °hum, °murmur, drone: *The flies were buzzing around the dead squirrel.* **6** fly down on, zoom onto: *The pilot was grounded for a month for buzzing the airfield.* **7** °telephone, °ring (up), °call (up), phone; °summon, °signal, buzz *or* ring for: *She said she'd buzz me if she needed anything.*

by *prep.* **1** °near, °beside, next to, °close to, alongside: *I park my car by my house.* **2** via, °by way of, °through; °past: *I go home by High Wycombe.* **3** °by means of, on: *I often travel by train.* **4** °before, not later than, sooner than: *I have to leave by Monday.* **5** during, at: *We travel only by night.* —*adv.* **6** Often, ***close by***. °near, °nearby, °at hand, °close, °about, °around, *Literary* nigh: *When she is close by I get a tingling sensation.* **7** °past, °nearby: *When he walks by I nearly die.* **8** away, aside: *We put by a little for a rainy day.*

bygone *adj.* °past, °former, olden; of old, of yore: *In bygone times, the fashion was for high-button shoes.*

bypass *v.* **1** °avoid, °evade, circumvent, °sidestep, skirt, go *or* get round, °detour; °ignore, *Slang* give the go-by: *I shall bypass many problems if I take that route.* —*n.* **2** °detour, alternative (way *or* route): *Take the bypass and avoid the town traffic.*

bystander *n.* °spectator, °onlooker, °observer, °witness, nonparticipant, passerby, °eyewitness: *He has always claimed he was an innocent bystander, but I'm not so sure.*

byword *n.* °proverb, proverbial saying, °parable, °maxim, adage, °motto, °slogan, apophthegm *or* apothegm, aphorism, catchword, catch phrase: *My byword has always been "Honesty is the best policy."*

C

cab *n.* °taxi, taxicab, *Obsolete* (horse-drawn) hackney, hansom (cab); *Old-fashioned US* °hack: *A cab picked me up and dropped me at the hotel.*

cabal *n.* **1** °intrigue, °plot, °conspiracy, °scheme: *The cabal against Washington found supporters exclusively in the north.* **2** junta *or* junto, °clique, °set, coterie, °faction, °band, °league; °unit, °party, caucus, °club; °ring, °gang: *A cabal of artists was formed.* —*v.* **3** °intrigue, °plot, conspire, connive, machinate: *The barons began to sow dissension and to cabal against his succession.*

cabaret *n.* **1** nightclub, , nightspot: *The ever-popular entertainer Mimi opened at the Golden Palm cabaret last night.* **2** floor show, °show, °entertainment, °amusement: *The dinner was poor, but the cabaret was marvellous.*

cabin *n.* **1** °hut, °shack, °cottage, °cot, shanty; bungalow, °lodge, chalet; *Scots* bothy: *The old trapper lives in a cabin in the forest. You are welcome to come skiing with us and stay in our cabin.* **2** stateroom, °compartment, berth: *Our cabin was uncomfortable, but we weren't in it for very long.*

cabinet *n.* **1** cupboard, °bureau, chifferobe, commode, chiffonier, °chest (of drawers), chest-on-chest, tallboy, *US* highboy, lowboy: *The aspirin is in the medicine cabinet. Our china cabinet is, unfortunately, not a genuine Chippendale.* **2** °council, °ministry, °committee, advisers *or* advisors, senate: *At the age of thirty, John became the youngest member of the cabinet.*

cable *n.* **1** wire, °line, °rope, hawser, °chain, mooring, strand, °guy: *The cable broke and we were set adrift.* **2** °telegram, wire, cablegram, radiogram, *US* Mailgram: *Send a cable to Jones about the meeting.* —*v.* **3** telegraph, wire; °radio: *Cable Jones to come at once.*

cache *n.* **1** hiding place, °hole, vault, repository: *There was a small cache concealed by the paneling in the library.* **2** °store, °hoard, °supply, °reserve, nest egg, stockpile, *Colloq US and Canadian* stash: *The wise hunter keeps a cache of supplies buried along his route.* —*v.* **3** °hide, °store, °conceal, squirrel away, °secrete, °bury, *Colloq* stash (away): *I cached the money in a biscuit tin.*

cachet *n.* **1** °stamp, °feature, distinguishing mark, °identification: *The cachet of good taste is simplicity of design.* **2** °distinction, °prominence, °importance, °prestige, °dignity: *Her new job doesn't pay much, but it has a certain cachet.*

cadaver *n.* °corpse, °(dead) body, °remains, *Slang* °stiff: *The medical students and doctors once paid to have cadavers exhumed for anatomical study.*

cadence *n.* °measure, °beat, °rhythm, °tempo, °accent, °pulse, meter, lilt, °swing: *The snaredrum marked the cadence for the marching band.*

café *n.* coffee house, coffee bar, coffee shop, bistro, snack bar, brasserie; tearoom, lunchroom, restaurant, eating house, canteen; cafeteria, *US* diner, *Colloq* eatery; *Slang Brit* caff, *US* greasy spoon: *We stopped at a café for refreshment.*

cage *n.* **1** crate, °enclosure, °pen, °pound, coop, hutch: *He keeps rooks in a cage.* —*v.* **2** Also, ***cage up*** *or* ***in.*** confine, °enclose, °pen, impound, °shut up, *or* in, coop (up), °imprison; °restrict, °restrain, hem in: *They keep the kitten caged like a wild animal. I don't like to stay caged up in my office all day.*

cajole *v.* °wheedle, °coax, °beguile, °jolly (along), cozy along, °seduce, inveigle, °persuade, *Colloq* soft-soap, butter (up), °stroke, sweet-talk: *Robert's wife always has to cajole him to go and visit her mother.*

cajolery *n.* wheedling, coaxing, blandishment, beguilement, jollying, °persuasion, seduction, inveigling, inveiglement, *Colloq* soft soap, buttering-up, sweet talk: *She uses cajolery rather than threats to get what she wants.*

cake *n.* **1** pastry, bun, *Brit* gateau: *Right now, I'd like to have a glass of milk and a piece of chocolate cake.* **2** °piece, chunk, °bar, °block, cube, °lump, °loaf, °slab: *Barbara gave me a cake of fancy perfumed soap for my birthday.* —*v.* **3** °harden, °solidify, °thicken, congeal, °dry, °coagulate, encrust, consolidate: *You can see where the paint has caked.*

calamitous *adj.* distressful, dire, °tragic, °disastrous, °destructive, °awful, °devastating, °fatal, °deadly, pernicious, cataclysmic, catastrophic, °ruinous, °dreadful, °terrible: *They seemed unaware of the calamitous consequences of what they were doing to the environment.*

calamity *n.* **1** °disaster, °destruction, °ruin, °catastrophe, cataclysm, devastation, °tragedy, misadventure, mischance, °mishap: *Calamity has befallen the town, which was entirely destroyed by a landslide.* **2** °distress, °affliction, °trouble, °hardship, °misery, °tragedy, °misfortune, adversity, °reverse, °ruin, ruination, °desolation, wretchedness: *So full is the world of calamity that every source of pleasure is polluted.*

calculate *v.* °compute, °reckon, add up, assess, °evaluate, °count, °figure (out), °estimate, °gauge, °determine, ascertain, °work out: *Bradley was able to calculate the velocity of light. They calculated where the sun would come up at the equinox and built their temple accordingly.*

calculated *adj.* **1** arranged, designed, planned, °prepared, adjusted, adapted, °fit, °fitted, intended, suited: *The coach was calculated to carry six regular passengers.* **2** °deliberate, °purposeful, °intentional, °premeditated, planned: *The so-called accident was really a calculated attempt to kill me.*

calculating *adj.* °shrewd, conniving, °crafty, °sly, °scheming, °designing, Machiavellian, manipulative, canny, contriving: *She is a calculating woman, who knows what she wants and maneuvers people to help her get it.*

calculation *n.* **1** computation, °reckoning, counting, °estimation, figuring, determining: *We needn't number them one by one, for the total can be arrived at by calculation.* **2** °answer, °product, °result, °figure, °count, °estimate, °amount: *This calculation is wrong, so please do it again.* **3** °estimate, °forecast, °expectation, °prediction, deliberation; circumspection, cautiousness, wariness, °caution, °prudence, °forethought, °discretion: *His attack was not the inspiration of courage but the result of calculation.*

calculator *n.* computer, adding machine; abacus: *According to my calculator, the answer should be 7.1592.*

calendar *n.* **1** appointment book, °schedule, slate, *Brit* °diary, *US* datebook, *US law* docket: *I have next week's lunch date in my calendar.* **2** almanac, °chronology, °chronicle, annal(s): *The ecclesiastical calendar lists today as St. David's Day.*

caliber *n.* **1** diameter, °size, °bore, °gauge: *You need a .38 caliber bullet to fit a .38 caliber pistol.* **2** °merit, °ability, °talent, °capability, competence, °capacity, °quality, °strength, stature: *They should be playing against a team of their own caliber.* **3** °degree, °measure, °stamp, °quality: *I doubt that you will find anyone of equal caliber to Julia in artistic sensibility.*

calibrate *v.* °adjust, °graduate; °standardize: *This balance has been dropped on the floor, and you'll have to calibrate it again.*

call *v.* **1** °shout, °cry (out), °hail, °yell, °roar, °bellow, call out, *Colloq* holler: *I heard someone calling my name.* **2** °name, °designate, °denote, denominate, °term, °style, °nickname, °label, °title, °entitle, °tag, °identify, dub, °christen, baptize: *My name is Angus, but they call me Scotty. A person from Glasgow is called a Glaswegian.* **3** °call up, °telephone, phone, °ring (up), dial, *Colloq* °buzz: *As it's her birthday, I shall call my mother in Australia. Don't call us, we'll call you.* **4** °summon, invite, °assemble, convoke, convene, °bid, °gather, °collect, °muster, °rally: *From the minaret, the muezzin was calling the faithful to prayer. Many are called but few are chosen.* **5** °visit, °attend; call in; °call on: *My great aunt Frederica came to call last Sunday. May I call on your daughter, Mr. Brummel?* **6** °awake, °awaken, °wake up, °rouse, *Colloq Brit* °knock up: *Please call me at six.* **7** ***call down.*** **(a)** appeal to, invoke, °petition, °request, entreat, supplicate: *He called down the wrath of God on the Philistines.* **(b)** °reprimand, °chastise, °castigate, °upbraid, °scold, °reprove, °rebuke: *He was called down for having left the house after curfew.* **8** ***call for.*** **(a)** °demand, °request, °ask for, °order, °require, °claim: *The people in room 429 have called for more hot water. The problem calls for your urgent attention.* **(b)** °pick up, °fetch, come for, °get, °accompany, *Colloq* °collect: *I'll call for you at seven o'clock.* **9** ***call forth.*** °summon, invoke, draw on *or* upon, °evoke; °elicit, °inspire: *Susan called forth her courage and faced her accusers. He failed to call forth much enthusiasm in his listeners.* **10** ***call off.*** °cancel; °discontinue; °postpone: *The picnic has been called off because of rain.* **11** ***call on*** *or* ***upon.*** **(a)** request of, entreat, °ask, °address; apostrophize: *The teacher called on me today to recite Hamlet's soliloquy.* **(b)** supplicate, apostrophize, appeal to: *He called on Aeolus, god of the winds, for a fair breeze to carry his ship home.* **(c)** °visit:

The vicar called on us when we first moved in. **12 *call up.*** **(a)** °summon, °enlist, °recruit, conscript, *US* °draft: *Father was called up as soon as war was declared.* **(b)** °call, °telephone, phone, °ring (up): *Call me up sometime.* —*n.* **13** °shout, °cry, °yell, °whoop, *Colloq* holler: *I'll be out in the garden, so give me a call if you want me.* **14** summons, °invitation, °bidding, °notice, notification, °order, °request, °demand, °command; telephone call, phone call, *Brit* °ring; *Colloq* tinkle: *She received a call to report at once for duty.* **15** °reason, justification, °cause, °need, °occasion, °right, °excuse; °requirement: *You have no call to be nasty, regardless of what he said.* **16 *on call.*** °ready, on duty, standing by, on standby, awaiting orders: *You are to remain on call from midnight till eight o'clock.* **17 *within call.*** within earshot *or* hearing *or* (easy) reach: *Please stay within call in case I need you.*

calling *n.* °vocation, °occupation, °profession, °business, °trade, °employment, °work, °line, °job, métier, °pursuit, °career, °area, °province, °(area of) expertise, °specialty, *Colloq* °racket: *He found his calling as a veterinary surgeon very satisfying.*

callous *adj.* hardened, °thick-skinned, unfeeling, uncaring, °insensible, insensitive, °hard, hardhearted, °tough, hard-bitten, °cold, °coldhearted, °heartless, °indifferent, °unsympathetic, apathetic, *Colloq* hard-boiled, hard-nosed: *It was callous of Gerry to go off to the pool hall right after the funeral.*

callow *adj.* °inexperienced, °immature, °juvenile, naive *or* naïve *or* naïf, °green, guileless, °unsophisticated, °innocent, °raw, °unfledged, °untried, *Colloq* (still) wet behind the ears: *It was a mistake to let a callow youth take out the boat alone.*

calm *n.* **1** °quiet, stillness, tranquillity, °serenity, °hush, °peace, peacefulness: *A storm raged outside, but in the harbor was a breathless calm.* **2** calmness, composure, placidity, placidness, °peace, °repose, °sangfroid, coolness, °self-control, equanimity, self-possession: *The calm exhibited by the passengers during the hijacking was admirable.* —*adj.* **3** °quiet, °still, °tranquil, °serene, °peaceful, balmy, halcyon, °mild, undisturbed, unagitated, placid, pacific; motionless, °smooth, °even; windless: *The sea is never calm in the same way as a mountain lake.* **4** composed, °cool, coolheaded, self-controlled, °impassive, °dispassionate, °unmoved, unruffled, °serene, °tranquil, °sedate, °staid, °stoical, *Colloq* together: *She remained calm while the others panicked.* —*v.* **5** Also, ***calm down.*** °quiet, quieten, °still, soothe, °hush, °lull, pacify; mollify, appease, placate, become *or* make quiet *or* pacified *or* less agitated, *Colloq* cool off *or Colloq* down: *The arbitrator did his best to calm the two litigants by suggesting a compromise. After everyone had calmed down, the speaker continued.*

camouflage *n.* **1** °disguise, concealment, coverup, °cover, °guise, °cloak, °mask, °screen, °blind, °(false) front, °show, façade, °pretense, °trickery, °deception; protective coloring *or* coloration, *Technical* apatetic *or Technical* aposematic *or Technical* cryptic coloring *or Technical* coloration: *Camouflage prevented the enemy from seeing our tanks.* —*v.* **2** °disguise, °cloak, °mask, °cover (up), °hide, °conceal, °screen, °veil; °misrepresent, °falsify: *We camouflaged our movements by fastening twigs and leaves to our helmets.*

camp[1] *n.* **1** camping ground, campground, bivouac, encampment, campsite; °settlement; camping site, *Brit* caravan site: *The name* Chester *derives from Latin* castrum, *meaning 'camp,' for the British city was originally the site of a Roman camp. Is there a camp where we can stay overnight?* **2** °faction, °set, coterie, °clique, °group, °party, °body: *The politicians are divided into opposing camps on the issue.* —*v.* **3** encamp, pitch camp, tent: *Our family likes to go camping in the mountains during the summer.* **4** °lodge, bivouac, °settle: *The platoon camped by the river.* **5 *camp out.*** *Slang* °crash: *Mind if I camp out in your pad tonight?*

camp[2] *adj.* **1** °outré, °outrageous, exaggerated, °artless, °affected, inartistic, °extravagant, °artificial, dadaistic, °theatrical, °mannered, °flamboyant, °showy, °ostentatious, °effeminate, *Colloq* campy: *Some of the kitsch produced in the 1930s was the epitome of camp. Please don't invite Wembley—his behavior is a bit too camp for my taste.* —*v.* **2** °exaggerate, °show off, °strut, °flaunt, °flounce, °prance, °posture, *Colloq* ham: *Clarence just loves to camp it up when he gets away from the squares at his office.*

campaign *n.* **1** operation(s), maneuver(s), °crusade, °action; °drive, °offensive, °push, °effort; °struggle: *Napoleon's Russian campaign ended in disaster. Our next sales campaign will be aimed at teenagers.* **2** °competition, °contest, °rivalry, °race: *Presidential campaigns last for more than a year.* —*v.* **3** °run, °electioneer, °compete, *Brit* °stand; *US and Canadian* °stump; *Colloq* throw *or Colloq* toss one's hat in the ring: *Next week the Labour candidate will campaign in Yorkshire.*

cancel *v.* **1** °void, annul, invalidate, nullify, °quash; °revoke, rescind, °redeem, °repeal, °abolish, °retract, °withdraw, °recall, °repudiate, abrogate, countermand, °deny: *The bonds have been canceled and are worthless. She canceled the incorrect check.* **2** °delete, °obliterate, cross *or* strike *or* blot out, dele, °rub out, °erase, expunge, efface, eradicate, °quash, deracinate; °eliminate, do away with: *I was forced to cancel the chapter of my book that dealt with CIA activities.* **3** Sometimes, ***cancel out.*** °neutralize, nullify, counterbalance, countervail, °compensate (for), °make up for, °offset, °counteract: *His later kindnesses cancel his previous injustices.*

cancellation *n.* **1** cancelation, canceling, annulment, nullification, rescinding, voiding, rescission, revocation, °abolition, abandonment, withdrawal, abrogation; °repeal: *Didn't you receive the cancellation of my hotel reservation?* **2** cancelation, invalidation, revocation, °abolition, discontinuance, °termination, °suppression: *If you fail to pay the premium, the policy is subject to cancellation. Where is the cancellation of the order to evacuate?* **3** cancelation, elimination, °abolition; stoppage, cessation: *Owing to the storm, some trains are subject to cancellation.*

candid *adj.* **1** °frank, °open, °plain, °sincere, °ingenuous, °straight, straightforward, °truthful, °forthright, °direct, unequivocal, plain-spoken, plain-speaking, °outspoken, °honest, °artless, °blunt, guileless, openhearted, °aboveboard, undeceitful, undeceiving, undeliberative, uncalculating, uncalculated, °unpremeditated, uncontrived, *Colloq* upfront: *Henry offered a very candid account of his feelings. Let us be candid and speak our minds.* **2** °just, °impartial, °objective, °fair, °equitable, unbiased, °unprejudiced, evenhanded; unbigoted: *The speaker expressed a candid view of all of the proposals.* **3** unposed, °informal, impromptu: *Here is a candid photo of the two of us in Rome.*

candidate *n.* aspirant, seeker, officeseeker, °runner, °nominee; applicant, entrant; °prospect, °possibility: *There are quite a few candidates for the post.*

candor *n.* **1** openness, frankness, ingenuousness, °simplicity, naiveté *or* naïveté *or* naivety, outspokenness, unreservedness, forthrightness, °honesty, °sincerity, directness, straightforwardness, unequivocalness: *I admire her candor, but the truth sometimes hurts.* **2** impartiality, fairness, °justice, °objectivity, open-mindedness: *In criticism candor is as rare as bigotry is frequent.*

candy *n.* sweet(s), bonbon(s), sweetmeat(s), confectionery: *Eating candy can be bad for your teeth.*

cannibal *n.* anthropophagite, man-eater: *The cannibals had captured us and we thought we were about to be eaten.*

cant *n.* **1** °hypocrisy, insincerity, °sham, °pretense, humbug, sanctimony, sanctimoniousness, lip service, affectedness, °pretension: *He wasn't really enthusiastic—all that talk was just cant.* **2** °jargon, °shop, shoptalk, argot, °vernacular, slang, °dialect, patois, creole, pidgin, gobbledegook *or* gobbledygook, *Colloq* °lingo: *The beggars use a cant not understood by those outside their fraternity.*

cantankerous *adj.* ill-natured, °quarrelsome, °perverse, °cross, choleric, cross-grained, crabby, curmudgeonly, crusty, grumpy, °surly, irascible, °snappish, bad-tempered, ill-tempered, bearish, °bilious, °peevish, °testy, °irritable, °touchy, °disagreeable, tetchy, °contrary, *Colloq* crotchety, grouchy, *US* °cranky: *Simon used to be so friendly, but he's become a cantankerous old codger.*

canvass *v.* **1** °solicit, °electioneer, °campaign, °poll, *US and Canadian* °stump: *The candidates will be canvassing in farming areas next week.* **2** °survey, °poll, °study, °analyze, °examine, °investigate, °interview, °question: *The statisticians are not satisfied that enough women were canvassed to provide an accurate sample.* —*n.* **3** solicitation, °campaign: *The politicians' canvass of rural areas was unsuccessful.* **4** °survey, °study, °investigation, °poll, °examination, °tally: *A canvass of editors shows they have a conservative view of the language.*

canyon *n.* °gorge, °ravine, gully *or* gulley, °pass, defile, *Brit dialectal* gill *or* ghyll, *US and Canadian* coulee, gulch; *US* °gap, arroyo: *The canyon created by the river is more than a thousand feet deep.*

cap *n.* **1** hat, head covering: *In those days one could tell where a man came from and his trade by his cap.* **2** lid, °top, °cover: *Screw the cap on tight.* **3** ***cap in hand.*** humbly, meekly, servilely, submissively, subserviently, docilely, respectfully: *He went, cap in hand, for permission to take a short leave.* —*v.* **4** °surpass, °outdo, °outstrip, °better, °beat, °exceed, °top, °excel: *Betty capped her earlier triumphs by winning the semifinals.* **5** °cover, °protect: *As it's begun to rain, you'd best cap the camera lens.*

capability *n.* °ability, °power, °potential, °capacity, °means, °faculty, wherewithal; °talent, °proficiency, °aptitude, adeptness, °skill, competence: *Deirdre has the capability to be first in her form.*

capable *adj.* **1** °able, °competent, °efficient, °proficient, °qualified, °talented, °gifted, skilled, °skillful, °accomplished, apt, °adept, °clever, °effective, °effectual; °expert, masterly, °masterful: *Halliwell is quite capable of speaking for himself. He is a capable violinist, but scarcely a virtuoso.* **2** ***capable of.*** disposed to, inclined to, predisposed to: *Though violent, he is not capable of murder.*

capacity *n.* **1** °volume, °content, °size, dimensions; °room, °space: *What is the capacity of this bottle in quarts? The car is of sufficient capacity to hold only four adults.* **2** °potential, °ability, °capability, competence, °intelligence, °wit, brain(s), °talent, °aptitude, acumen, °understanding, °sense, °judgment, perspicacity, perceptiveness, °perception, mother wit, °intellect, °genius, °skill, °gift, °faculty, °power, °potential, *Colloq chiefly US* right stuff, the goods: *They don't yet have the capacity to absorb advanced theory.* **3** °position, °condition, °character, °place, °post, °role, °job, °office, °duty, °responsibility, °province, °sphere, °function; *Law* competency, °qualification: *She has every right to sign checks in her capacity as director.*

cape[1] *n.* headland, promontory, peninsula, neck, °point, *Archaic* ness: *We sailed round the cape and made for the harbor.*

cape[2] *n.* °mantle, shawl, °stole, °cloak: *His black cape reached to the floor.*

caper *n.* **1** °skip, °leap, °spring, °frolic, °hop, gambol, °frisk, curvet, gambado: *He can dance, though he does not cut capers.* **2** escapade, °stunt, °mischief, °prank, high jinks, *US* °crime, burglary, °robbery, *Colloq* shenanigan, dido, °lark, *Slang US and Canadian* °job: *The capers we used to get up to after lights-out in the dormitory!* —*v.* **3** °skip, °hop, °frolic, °leap, °jump, °frisk, romp, gambol, °prance, °cavort, curvet: *She capered about like a lamb in a meadow.*

capital *n.* **1** °head, °top, °crown, °cap: *The column was surmounted by a finely carved capital.* **2** °seat (of government): *Winnipeg is the capital of Manitoba.* **3** °money, °assets, °funds, finance(s), °cash, wherewithal; °wealth, °means, °property, °resources, °savings, °principal: *My capital is invested in land at the moment.* **4** majuscule, °upper case, large letter, °initial, *Colloq* °cap: *The chapter titles should be set in capitals.* —*adj.* **5** °chief, °main, °major, °important, °cardinal, °central, °principal, °prime, °primary, °paramount, °preeminent, °foremost, °leading: *Our capital responsibility is to ensure the passengers' safety.* **6** first-class, °first-rate, °excellent, °superior, °matchless, °peerless, °choice, °select, °outstanding, °fine, °superb, °splendid, °marvelous, °extraordinary, *Colloq* smashing, °great, °super, *Brit* brill, *Old-fashioned* topping, top-hole, °ripping, ripsnorting: *Ordering calamari was a capital idea.*

capitulate *v.* **1** °surrender, °yield, °give up, °submit, °succumb: *Want of provisions quickly obliged the fortress to capitulate.* **2** acquiesce, °concede, °relent, °give in, °yield: *He begged so piteously that the king finally capitulated and allowed him to live.*

capricious *adj.* °whimsical, °erratic, °flighty, °fickle, mercurial, unsteady, °variable, °unstable, wayward, unpredictable, undependable, °changeable, °impulsive, crotchety, quirky, °unreliable, °inconstant, °fanciful, °wanton: *His decisions are capricious and not based on sound judgement. The weather in March is capricious—as Mark Twain said, if you don't like it, just wait five minutes.*

capsize *v.* °upset, °overturn, turn turtle *or* upside down, °tip (over), keel over, invert: *When the wind capsized the boat, we lost all our gear overboard.*

captivate *v.* enthral *or US* enthrall, °enslave, °fascinate, °hypnotize, °entrance, °beguile, °charm, enamor, °enchant, °bewitch, °enrapture, °dazzle, infatuate, °attract, allure, °seduce, °win: *Her beauty captivated filmgoers everywhere.*

captive *n.* **1** °prisoner, °convict, °hostage, detainee, internee; °slave, bondman *or* bondsman, bondservant: *The captives were kept in a wretched hole.* —*adj.* **2** imprisoned, incarcerated, confined, caged, locked up, under lock and key: *Captive animals lose their free spirit.*

captivity *n.* confinement, °imprisonment, internment, °detention, °custody, incarceration, °restraint; °bondage, °slavery, thralldom, enslavement, °servitude; *Archaic* durance: *Some wild creatures do not survive in captivity. Entire populations of conquered territories were taken into captivity in ancient times.*

capture *n.* **1** °seizure, °taking, °catching, °arrest, apprehension, *Slang* °pinch, collar: *We celebrated the capture of the Spanish galleon. The state has offered a reward for the capture of the bank robbers.* —*v.* **2** °seize, °take, °catch, lay *or* take hold of, °grab, apprehend,

°arrest, *Slang* °pinch, collar, °nab, *Brit* °nick: *Fogarty captured the thief on the roof.*

car *n.* **1** °(motor) vehicle, motorcar, automobile, passenger car, *Old-fashioned or slang* motor; *Chiefly US* auto; *Colloq* jalopy, °heap, °pile, crate, °machine, buggy, °transport; *Slang* wheels: *Borrow a car and drive down for the weekend.* **2** °(railway) carriage: *The body was found in a sleeping car of the Orient Express.*

card *n.* **1** playing card, *Slang* pasteboard: *The winning card was the ten of diamonds.* **2** calling card, visiting card, *carte de visite,* business card: *Visitors used to leave their cards on the silver tray at the front door.* **3** greeting card, Christmas card, birthday card, anniversary card, condolence card, Easter card, New Year card: *I sent Jacquelyn a card for her birthday last year.* **4** postcard, *US* postal card: *Drop me a card when you get there, just so I'll know you're all right.* **5** index card, file card: *The names and addresses of our members, formerly on cards, are now stored in the computer.* **6** membership card; press card; union card: *I showed my card at the door and they let me in without any problem.* **7** dance card: *She told me that her dance card was full and was likely to be for the next ten years.* **8** credit card; bank card: *You may pay by card or check. They won't accept your check without a card.* **9** identity *or* identification card, ID (card): *The police asked to see my card.* **10** °joker, prankster, practical joker, °wag, humorist, °comedian, funnyman: *That Oscar—he's quite a card, isn't he?* **11** ***in*** *or, esp Brit* ***on the cards.*** °destined, °fated, slated, in the offing; °likely, °probable, °possible, °liable: *I doubt that a change of government is in the cards for some time to come.* **12** ***play one's cards right, well, badly, etc.*** °act, °behave, take action; °plan, use strategy: *If Thomas plays his cards right, Phyllis might agree to marry him.* **13** ***put*** *or* ***lay one's cards on the table*** *or* ***show one's cards.*** act openly, reveal all, be forthright, be direct, be open, be honest, be unsecretive, *Colloq* °come clean: *I'm going to put my cards on the table, Rodney, and let you know all my plans.*

cardinal *adj.* °important, °chief, °key, °special, °main, °central, °principal, °prime, °primary, °essential, °necessary, °fundamental; °supreme, °paramount, highest, °first, °foremost, °leading, °preeminent: *The cardinal virtues are justice, prudence, temperance, and fortitude, to which some writers add faith, hope, and charity.*

care *n.* **1** °anxiety, °worry, °trouble, °anguish, disquiet, °distress, °grief, °sorrow, dolor, °sadness, °suffering, °misery, °woe, tribulation: *His haggard look reflected a life of care.* **2** °concern, °regard, °vigilance, mindfulness, °heed, °solicitude; heedfulness, °attention, °pains, carefulness, meticulousness, punctiliousness; °caution, circumspection: *The essence of public spiritedness is care for the common good. He looks after his moustache and beard with great care. Open with care.* **3** °responsibility, °charge, °protection, guardianship, °custody, keeping, °safekeeping; °control, °direction, supervision: *The child has been released into our care.* **4** ***take care of.*** °look after, °attend to, be responsible for, take charge of, take responsibility for; °tend, °nurse: *You should take care of your money. Does she have enough experience to take care of someone who is ill?* —*v.* **5** be concerned, trouble oneself, feel interest, °worry, °fret, °trouble, *Brit* °mind: *Do you care whether Arnold gets the job he wants? I don't care who you are, you can't come in here!* **6** ***care for.*** **(a)** °look after, °tend, °attend (to), watch over, °protect, °take care of, °provide for; °nurse: *He cared for his ailing parents for about twenty years.* **(b)** °like, °fancy, be attracted to, be fond of, °love, be keen on, be enamored of: *Jennifer told me last night that she cares for me, not David.*

careen *v.* heel over, keel over; *US loosely* °career, °sway, °tip, °pitch, veer, °swerve, °lurch: *We hauled out the boat, careened her, and proceeded to caulk her seams.*

career *n.* **1** °employment, °occupation, °calling, °vocation, °pursuit, °(life's) work, °job, °business, livelihood; °profession, °trade, °craft, métier: *She has made a career out of helping others. He is undecided whether to choose brain surgery or watchmaking as a career.* —*v.* **2** °speed, °race, °rush, °dash, °fly, °tear, °hurtle, °bolt, °shoot, *Colloq* zoom: *A bicycle came careering around the corner and knocked him down.*

carefree *adj.* °nonchalant, °easy, easygoing, insouciant, lighthearted, °blithe, happy-go-lucky, °breezy, airy; °blasé, °indifferent, unconcerned, unworried, trouble-free, worry-free, contented, °happy: *Till he graduated from university, he had lived an entirely carefree life.*

careful *adj.* **1** °cautious, °wary, circumspect, chary, °prudent, watchful, °aware, °alert, °vigilant: *These days one cannot be too careful about walking in the city at night.* **2** °meticulous, °painstaking, °attentive, punctilious, (well-)organized, °systematic, °precise, °fastidious, °thorough, °scrupulous, °conscientious, °particular, °finicky, finical, °fussy: *The police conducted a careful search for weapons.*

careless *adj.* **1** unconcerned, untroubled, unworried, °casual, °indifferent, °heedless, °thoughtless, °inconsiderate, uncaring, devil-may-care, °irresponsible, °cursory, °lackadaisical, °perfunctory: *No one could approve of the careless way he treats his family.* **2** °inattentive, negligent, °thoughtless, °absent-minded, neglectful, °remiss; unobservant, °unthinking, °imprudent, unmindful, incautious, °unwary, °reckless, slapdash, °rash: *Many of the errors come from being careless.* **3** °inaccurate, °imprecise, °inexact, °incorrect, °wrong, error-ridden, °erroneous, *Colloq* °sloppy: *You won't get a good mark for such a careless paper.* **4** unstudied, °ingenuous, °artless, °casual, °nonchalant: *I dislike his careless way of dressing, but it does show some style.*

caress *n.* **1** °pat, °stroke, fondling, blandishment; °cuddle, °embrace, °hug; nuzzle, °kiss: *Cranborne submitted willingly to her caresses.* —*v.* **2** °touch, °pat, °pet, °fondle, °stroke; °cuddle, °embrace, °hug; nuzzle, °kiss: *The fawn approached warily and Isabella caressed it.*

cargo *n.* shipment, consignment, shipload, truckload, wagonload, °load, trainload, *US* carload; °freight, °goods, °merchandise: *The cargo of rifles was delivered to the warehouse. The ship was lost with all its cargo.*

caricature *n.* **1** cartoon, °parody, °burlesque, °lampoon, °satire, pasquinade, *Colloq* °takeoff, spoof, *Brit* sendup: *The cartoon in the newspaper showed a caricature of the president.* —*v.* **2** °parody, °satirize, °lampoon, °burlesque, °ridicule, °mock, °distort, *Colloq* °take off, *Brit* °send up: *Hogarth caricatured Churchill in the form of a bear.*

carnage *n.* °slaughter, butchery, °massacre, blood bath, °holocaust, °killing, *Shoah, Churban* or *Hurban*: *The carnage of the Nazi concentration camps was unbelievable.*

carnal *adj.* fleshly, °sensual, °animal, bodily, °lustful, °voluptuous, libidinous, °lecherous, concupiscent, °sexual, °erotic, °lascivious, licentious, °lewd, °prurient: *She is intent on satisfying her carnal desires.*

carouse *v.* **1** °make merry, °revel, *Colloq* °party, pub-crawl, make whoopee, go on a bender *or* tear *or* binge *or* toot, °paint the town red, binge, °booze: *After the cup final, we all caroused till the wee hours.* —*n.* **2** °revel, °spree, °fling, wassail, carousal, °drunk, bacchanal, *Colloq* binge, °bender, °booze, boozer, *Brit* knees-up, *US*

°tear, toot: *They went on a carouse on New Year's Eve, then needed a day to sleep it off.*

carp *v.* °find fault, °criticize, °cavil, °complain, °nag, °pick at, °pick on, °bully, bullyrag *or* ballyrag, *Colloq* °knock, pick holes (in), °gripe, *Brit* whinge: *She said that she would leave him if he kept on carping at her about her cooking.*

carriage *n.* **1** °(railway) coach, *US* °car: *We moved our belongings into the carriage ahead.* **2** °bearing, mien, °air, °manner, deportment, °conduct, demeanor, °attitude, °posture, °stance, °presence, °behavior, comportment: *His upright carriage immediately identified him as a military man.* **3** °freight, freightage, transportation, cartage, shipping; postage: *How much is the carriage for that crate of balloons?*

carrier *n.* **1** bearer, °porter; transporter, drayman, shipper, hauler *or Brit* haulier; carter: *The company we use as a carrier is expensive.* **2** transmitter, *Immunology* vector, *US* typhoid Mary: *She couldn't have caught the disease directly, only through some carrier.*

carry *v.* **1** °transport, convey, °bear, °lug, °drag, °cart, °move, *Colloq tote, Slang US* schlep: *He shouldn't carry such heavy packages at his age.* **2** °conduct, convey, °lead, °take, °transport, °transfer, °transmit: *This cable carries the power to the town.* **3** °drive, impel, °conduct, convey, °take, °move: *He traveled aimlessly, wherever the wind carried his ship.* **4** °support, °maintain, °finance: *I had a wife and four children and was unable to carry my brother's family as well.* **5** °bear, °hold up, °uphold, °maintain: *Despite her troubles, she carried her head high.* **6** °win, °take, °sweep, °capture, °gain, °secure: *Gimpleigh carried the election easily.* **7** °stock, °sell, °offer; °display: *We don't carry purple shoes in this shop, madam.* **8** °broadcast, disseminate, °offer, °release; °communicate, °present, °read, °report, °announce; °give: *The news is carried on this station every night at nine.* **9** ***carry away.*** °transport, °excite, °enrapture, °delight: *He was quite carried away by her attentions.* **10** ***carry off.*** **(a)** °win, °gain, °capture, °secure: *Thripps has carried off the first prize once more with his cherry tarts.* **(b)** abscond with, °kidnap, °take, purloin, *Colloq Brit* °pinch, °nick: *I fear that your chickens, sir, have been carried off by a fox.* **(c)** °accomplish, °perform, °effect, do, °succeed, handle *or* manage successfully, °bring off, °carry out: *We carried off the raid without loss of a single man.* **(d)** °kill, be *or* cause the death of, cause to die: *He was carried off by yellow fever in his eightieth year.* **11** ***carry on.*** **(a)** °continue, °proceed, °go on, °persist, keep on *or* at, °persevere: *Don't stop—just carry on with what you were doing.* **(b)** °manage, °conduct, °operate: *Despite the fire, we are carrying on our business as usual.* **(c)** °misbehave, *Colloq* act up, °fool around, *Brit* °play up: *The children are carrying on so, I can't get any work done.* **12** ***carry out*** or ***through.*** °perform, °effect, °implement, °complete, °execute, °accomplish, °continue, conclude: *Henry is carrying out his father's wishes according to the will.*

cart *n.* **1** handcart, pushcart, trolley, barrow, wagon *or Brit also* waggon: *You'll need a cart to carry all these things to the car.* —*v.* **2** °carry, convey, °move, °lug, °drag, tote, °transport, °bring, °haul, *Colloq US* schlep: *Why do you cart that heavy bag everywhere you go?*

carte blanche *n.* °license, °permission, °sanction, free rein, °authority, °discretion: *She was given carte blanche to spend the money any way she wished.*

carve *v.* **1** hew, °cut, sculpt, °sculpture, °shape, °chisel, °model, °fashion, °engrave, incise, °grave, °whittle, °chip: *The bust is carved out of solid marble.* **2** Often, ***carve up*** *or* ***out.*** °divide (up), °cut (up), subdivide, apportion, °parcel out, °allot, °partition: *The gang-leaders carved up the territory, and the killings stopped for a while.*

case[1] *n.* **1** °instance, °example, °event, °occurrence; °happening, °occasion, °circumstance, °state, °situation: *In a recent case a farmer was attacked by a man-eating tiger. Holmes is investigating a case of a missing necklace.* **2** °action, °suit, lawsuit, °dispute; °cause: *I lost my case.* **3** °patient, °invalid, °victim: *Four new cases were admitted to the hospital yesterday.* **4** °specimen, °instance, °example: *That Hadley is an odd case, isn't he?* **5** ***in any case.*** °in any event, come what may, °at all events, anyhow, anyway: *In any case, your decision won't affect me.* **6** ***in case.*** **(a)** lest, for fear that: *He was worried in case his wife found out where he had been.* **(b)** if, in the event that, if it happens *or* proves *or* turns out that, if it should happen *or* prove *or* turn out that: *In case you were thinking of leaving, remember that we have your car keys.* **7** ***in case of.*** in the event of; for fear of: *In case of fire, you must use the staircase. We insured the house in case of fire.* **8** ***the case.*** the fact, the actuality, the truth, the reality, what really happened *or* took place: *She said he was drunk, but that's not the case.*

case[2] *n.* **1** °box, container, carton, crate; °chest, holder, °receptacle; °trunk, °suitcase, °casket: *Please order two cases of paper for the copying machine. The cosmetics came in a fitted case lined in velvet.* **2** covering, °cover, °protection, casing, envelope, °wrapper: *The engraving was on the inside of the watch case. The book came in a case of fine calfskin.* —*v.* **3** encase, °box, crate, °pack, °package, containerize: *The computer arrived, completely cased in rigid foam.*

cash *n.* **1** °money, currency, bills, °notes, banknotes, °change, hard cash *or* money, specie, coin of the realm, legal tender, *Slang* moolah, dough, bread, °loot, spondulix *or* spondulicks, *Brit* lolly, °ready, °readies, *US* °scratch, gelt, mazuma: *The shop accepts only cash, no charge cards.* —*v.* **2** Also, ***cash in.*** °change, °sell, liquidate, °exchange; °realize: *She cashed some bonds to pay off the loan.*

casket *n.* **1** °chest, °box, container, °case, coffer, °receptacle: *She keeps her jewels in a leather casket on the dressing table.* **2** °coffin; sarcophagus: *After the funeral, the pallbearers carried the casket from the church.*

cast *n.* **1** °throw, °toss, °pitch, °shy, °lob, °thrust, chuck: *In his next cast, the quoit landed right around the stick.* **2** dramatis personae, actors and actresses, players, performers, troupe, °company: *We invited the cast to a party after the show.* **3** °form, °shape, °mold; °formation, formulation, °arrangement: *She can appreciate the turn of the phrase, the happy cast and flow of the sentence.* **4** °model, casting, °mold; °stamp, °type: *The Ming vase was copied from a cast. There are not many men of the cast of Crocker.* **5** °twist, °turn, irregularity, °warp; squint: *The mare had a cast in her gallop. The pirate had a cast in his left eye.* **6** °turn, °inclination, °bent, °hint, °touch; tinge, °tint, coloring: *He has a melancholy cast of mind.* —*v.* **7** °throw, °toss, °pitch, °fling, °sling, °hurl, °dash, °send, *Colloq* chuck, °shy: *She tore off the gold necklace and cast it into the lake.* **8** °assign, °delegate, °appoint, °designate, °name, °nominate, °choose, °pick, °select: *He has cast me as the villain in his little drama.* **9** °form, °mold, °found: *The king's death mask, cast in plaster, was on the floor of the tomb.* **10** ***cast about for.*** °search for, °look for, °seek: *He was casting about for an excuse to avoid going to the Fordyces' for dinner.* **11** ***cast aside.*** °reject, °discard, cast *or* throw away *or* out, °get rid of: *The expensive toys had been cast aside and the children were playing with the boxes and wrappings.* See also **14,** below. **12** ***cast away.*** °maroon, shipwreck: *Jim O'Shea was cast away* *upon an Indian isle.* See also **11,** above. **13** ***cast off.*** °throw off, °shed, doff: *One's upbringing cannot be cast off like an old overcoat.* **14** ***cast out.*** °expel, drive out, °throw out, °evict, °eject, oust, °exile, °remove: *She was cast out of the house by her mother, who had married a biker.* See also **11,** above.

castaway *n.* °reject, castoff, °outcast, pariah, °exile: *She always looked after the moral well-being of the castaways of society.*

caste *n.* °(social) class, °rank, °order, °level, °stratum, °standing, °position, °station, °status, °estate: *The women of her caste practiced suttee.*

castigate *v.* °chastise, °punish, °correct, °penalize, °discipline, °rebuke, °reprimand, read the riot act (to), keelhaul, °chasten, °criticize, *Colloq* °tell off, °dress down, *Chiefly Brit* tick off, *Brit* carpet, haul over the coals, *US and Canadian* °chew out, rake over the coals, put *or* call on *or* on to the carpet: *The pranksters who sprayed paint on the school windows will be severely castigated.*

castle *n.* **1** fortress, °stronghold, citadel: *The king moved to Windsor Castle during the winter.* **2** mansion, °palace, manor house, °hall, chateau: *Mr. Mooney lives alone in his castle and has nothing to do with his neighbors.*

casual *adj.* **1** °accidental, °chance, °random, fortuitous, unexpected, °unforeseen, °unpremeditated, unplanned, unforeseeable, unpredictable, serendipitous: *Are these differences merely casual or have they a meaning and a purpose? We interviewed casual passersby.* **2** °uncertain, unsure, °haphazard, °occasional, °random, °irregular, unsystematic, °sporadic, °erratic: *The budget includes provision for both certain and casual revenues.* **3** °indifferent, °nonchalant, °offhand, insouciant, apathetic, °cool, unconcerned, uninterested, pococurante, °dispassionate, °blasé, °relaxed, °lackadaisical: *He may seem casual, but he is genuinely concerned about the patients.* **4** °informal; °lounge: *Dinner dress is not required—you may come in casual clothes.* **5** °offhand, happy-go-lucky, °natural, °easy, easygoing, devil-may-care, unconcerned, °relaxed, dégagé, unconstrained: *Bert is quite casual about losing money at roulette.*

casualty *n.* **1** °disaster, °catastrophe, °calamity, °accident, mischance, misadventure, °mishap: *The company insures against casualties at sea.* **2 (a)** °victim, °fatality, *Colloq* statistic: *I'm afraid that Jeffers was a casualty of last year's personnel cut. She was only one of thousands of casualties of the earthquake.* **(b)** Usually, ***casualties.*** *Chiefly military* wounded, injured, missing, missing in action, °dead, fatalities, *US* MIA(s), body count: *The casualties were mounting.*

catastrophe *n.* **1** °disaster, °calamity, cataclysm: *The eruption of Vesuvius was one of the major catastrophes in recorded history.* **2** °misfortune, bad luck, °shock, °blow, °tragedy, °disaster; °mishap, mischance, misadventure, °accident, °fiasco, °failure: *Thresher's failure to attend our party was a catastrophe. The cook reported a catastrophe with the layer cake.*

catch *v.* **1** °capture, °seize, apprehend, take *or* get (hold of), °grab, °grip, °grasp, take captive, °hold, °arrest, take prisoner, *Colloq* °nab, °pinch, collar, *Brit* °nick: *The police caught him when he returned to the scene of the crime.* **2** °trap, ensnare, entrap, °snare, °net, °bag, °hook, °round up, corral: *We caught three trout this morning. I caught all the horses that had broken through the fence.* **3** °take, get on *or* on to, °board: *You can catch the London train at Aylesbury.* **4** °surprise, °discover, °find: *They fired him after catching him with his hand in the till.* **5** be seized *or* taken hold of by *or* with, °come down with, be afflicted by *or* with, °contract, °get, °suffer from: *You'll catch a cold if you don't wear a hat.* **6** °strike, °hit, °deliver, °fetch, °box: *She caught him a great blow on the ear and he went down.* **7** °tangle, become entangled *or* stuck *or* trapped *or* hooked: *His foot caught in the stirrup when he fell, and he was dragged along.* **8** °restrain, °stop, °check, °curb: *She caught herself before telling the police where the thief was hiding.* **9** °intercept, °grab, °seize, °snatch: *He caught the ball before it touched the ground.* **10** °understand, °see, °comprehend, °grasp, apprehend, °fathom, °perceive, discern, °follow, °take in, °gather, *Colloq* °figure out, °get, °catch on (to), get the drift (of), *Brit* °twig: *I didn't quite catch what you said—please repeat it.* **11** °captivate, °charm, °bewitch, °enchant, °fascinate, °seduce, °attract, °entice, allure: *She knows how to use her charms to catch a man.* **12** °attract, °draw: *A very slight movement caught my eye.* **13** ***catch on.*** **(a)** °understand, °grasp, °see (through), °comprehend, °get (it), *Brit* °twig: *I didn't catch on to what she planned to do it till it was too late. The joke's on you and you still don't catch on, do you?* **(b)** take hold, °succeed, become popular *or* fashionable: *Do you think the hula hoop will catch on again?* **14** ***catch up.*** **(a)** absorb, °involve, enthrall, °immerse: *He was all caught up in the plot of the new novel.* **(b)** °reach, °overtake, °overhaul: *I finally caught up with her as she neared the house.* —*n.* **15** °capture, °take, °bag, °prize, °trophy: *The catch of the day was a twenty-pound pike.* **16** °acquisition; °conquest: *She was considered quite a catch.* **17** °clasp, °hook, °pin, °clip, °fastening, fastener: *The catch on my necklace opened and pearls spilled all over the floor.* **18** °trick, °disadvantage, °hitch, °snag, °fly in the ointment, catch-22, °trap, °problem, °drawback, *Colloq US* hooker: *The first book is free, but the catch is that you have to buy four more at the regular price.*

catching *adj.* **1** contagious, °infectious, transmissible, transmittable, communicable: *The doctor said that what I have is not catching.* **2** °attractive, captivating, fascinating, °enchanting, bewitching, entrancing, °winning, enticing, alluring, °fetching: *The strange object in the shop window was most catching to the eye of a passerby.*

categorical *adj.* °direct, °explicit, °express, unconditional, °firm, °positive, unreserved, unrestricted, °absolute, °outright, °downright, unequivocal, unambiguous, °specific; °emphatic, °unqualified, °authoritative, °dogmatic, *Technical* apodeictic *or Technical* apodictic: *His denial was clear and categorical.*

categorize *v.* classify, °class, °sort, °organize, assort, °rank, °order, °section, departmentalize, °group, °arrange: *Should plankton be categorized under zoology or botany?*

category *n.* °class, classification, °type, °sort, °kind, °variety, °group, grouping, listing, °rank, ranking, °list, °grade, °department, °division, °section, sector, °area, °sphere; °head, heading: *Into which category would you put the partially disabled?*

cater *v.* **1** °provision, victual; purvey, °provide: *We cater exclusively for housebound gourmets.* **2** ***cater for or to.*** °indulge, °humor, °serve, dance attendance on, °pamper, °baby, °coddle, °minister to, °spoil, mollycoddle, cosset, °pander to: *She caters for him night and day. Our music group caters for all levels of ability.*

catholic *adj.* °universal, °general, (all-)inclusive, °broad, °wide, °comprehensive, widespread, all-embracing, eclectic, °liberal: *Her musical tastes are catholic—they range from Bach to Berry (Chuck, that is).*

cattle *n.* livestock, °stock, beef; cows, bulls, bullocks, steers, bovines, oxen: *He spent twenty years as a cowboy, herding cattle in Texas.*

cause *n.* **1** °origin, °occasion, °source, °root, genesis, °agent, prime mover, wellspring: *Your behavior is the cause of my grief.* **2** originator, °creator, °producer, °agent, °agency: *Indecision was the cause of most of our problems.* **3** ground *or* grounds, justification, °reason, °basis, °call, °motive: *You have no cause to be dissatisfied.* **4** °case, °matter, °issue, °concern; °movement, °undertaking; °ideal, °belief: *We appealed the miners' cause to the high court.* —*v.* **5** °make, °induce: *What causes hot air to rise?* **6** °effect, bring on *or* about, °give rise to, °result in, °produce, °create, °precipitate, °occasion, lead to, °induce, °generate, °provoke, °promote; engender; °motivate, compel: *Overeating causes indigestion.*

caustic *adj.* **1** °burning, corrosive, °destructive, mordant, astringent: *Sulfuric acid is the caustic agent that eats away the metal.* **2** °sarcastic, °biting, acrimonious, °sharp, °bitter, °biting, °sardonic, °cutting, °trenchant, °critical, °scathing, acid, °harsh, °pungent, °virulent: *His caustic remarks do not earn him many friends.*

caution *n.* **1** °warning, admonition, admonishment, caveat, monition, °advice, °counsel, °injunction: *A word of caution before you enter that den, Daniel.* **2** wariness, °prudence, °care, °vigilance, °forethought, °heed, watchfulness, alertness, circumspection, °discretion: *Please exercise caution in opening this box, Pandora.* —*v.* **3** °warn, admonish, forewarn, °tip (off); °advise, °counsel: *Some employees had to be cautioned about arriving on time.*

cautious *adj.* °wary, heedful, °careful, °prudent, circumspect, watchful, °vigilant, °alert, circumspect, °discreet, °guarded: *They are very cautious about letting their children go out alone after dark.*

cave *n.* **1** cavern, grotto, °hollow, °hole, °cavity, den: *In the cave were prehistoric wall paintings.* —*v.* **2** ***cave in.*** **(a)** °collapse, °break down, °give way, °subside, fall in *or* inwards: *The earthquake caused the walls of the house to cave in.* **(b)** °yield, °submit, °give way; °surrender; *Colloq* °buckle, knuckle under: *After eight hours of questioning, he caved in and told them everything.*

cavil *n.* **1** °quibble, °complaint: *There is a minor cavil at the wording of the statutes.* —*v.* **2** °carp, °quibble, split hairs, °complain, °find fault, censure, °criticize, °dispute, °object, demur, *Colloq* nitpick: *They caviled at a mere misspelling.*

cavity *n.* °pit, °hole, °hollow, °opening, crater, °gap; °space: *The limestone is marked with a pattern of cavities. Vowel sounds resonate in the oral cavity.*

cavort *v.* curvet, °prance, °caper, °frisk, °bound, gambol, romp, °skip, °leap, °jump, °dance: *Stop cavorting about and settle down.*

cease *v.* **1** °stop, °end, °finish, °leave off, °terminate, °halt, °discontinue, desist (from), °break off (from), °refrain (from): *Hostilities ceased at midnight. Will that noise never cease?* —*n.* **2** ***without cease.*** ceaselessly, endlessly, unendingly, incessantly, interminably, continuously, continually, constantly, ad infinitum, infinitely, perpetually, °forever, eternally, everlastingly, °nonstop, unremittingly: *Sisyphus was condemned to roll his burden uphill without cease.*

cede *v.* °yield, °give way, °give up, °grant, °give, °surrender, deliver up, turn *or* make *or* hand over, convey, °transfer, °relinquish, °abandon, °renounce, °abdicate: *Let private concerns always cede to the common good. The territory was ceded to our government in 1792.*

celebrant *n.* officiant, °official; °priest: *The celebrant at High Mass was the archbishop.*

celebrate *v.* **1** °hold, °perform, solemnize, ritualize, °observe, °keep, °honor, officiate at; °sanctify, °hallow, consecrate, °dedicate: *The archbishop himself celebrated holy communion.* **2** rejoice (in *or* at), °memorialize; have a party, °revel, °make merry, wassail, *Colloq* °party, °paint the town red, whoop it up: *The entire town celebrated the opening of the bridge with a huge party.* **3** °extol, °praise, °exalt, °glorify, °laud, °eulogize, °honor; lionize: *His poems celebrated mountains for their majesty. She was widely celebrated for her achievements.* **4** °publicize, advertise, °broadcast: *The stones themselves would find a Voice / To celebrate his Praise.*

celebrated *adj.* °famous, °renowned, °well-known, famed, °prominent, °noted, °eminent, °noteworthy, °distinguished, °illustrious, acclaimed: *Their son became a celebrated chemist.*

celebration *n.* **1** °observance, °observation, °performance, solemnization, hallowing, sanctification, memorialization, commemoration: *The celebration of the Eucharist was delayed because the vicar had been called to a sickbed.* **2** praising, extolling, honoring: *The ceremony will be held in celebration of their achievements in space exploration.* **3** °party, °fête, °gala, °festivities, °frolic, °revelry, merrymaking: *The New Year's celebration is planned at Jill's house this year.*

celebrity *n.* **1** °renown, °fame, repute, °reputation, °prominence, eminence, °distinction, °prestige, famousness, °popularity, notability, stardom; °notoriety: *Dr. Johnson was not enriched in proportion to his celebrity.* **2** °notable, °dignitary, °star, luminary, toast of the town, °personage, °name, °personality, superstar: *The hotel lobby was packed with celebrities from show business.*

celestial *adj.* **1** °heavenly, °divine, °spiritual, °godly, paradisiac(al) *or* paradisaic(al), °sublime, empyrean, elysian, ethereal, °immortal, °supernatural: *They worshiped Jupiter and other celestial beings.* **2** astronomical, astral: *We studied celestial navigation and how to use a sextant.*

celibacy *n.* **1** bachelorhood, spinsterhood, singleness: *People who say they enjoy celibacy may simply have not met the right partner.* **2** °chastity, virginity, continence, (self-)restraint, abstinence, °purity: *They have to take a vow of celibacy before entering the priesthood.*

celibate *adj.* **1** °unmarried, °single, unwed: *Both brother and sister remained celibate all their lives.* **2** abstinent, abstemious, continent, ascetic; virgin(al), °pure, °chaste, unsullied, undefiled, °virtuous, °immaculate: *He led the celibate life of a monk for twenty years.* —*n.* **3** bachelor, spinster: *She belonged to an order of female celibates.*

cell *n.* °chamber, °room, apartment, cubicle; °stall: *As a friar, he lived in a small, plain cell for most of his life.*

cellar *n.* basement, vault: *She led me down to the cellar to show me her wine collection.*

cement *n.* **1** mortar, °bond, °glue, gum, paste, solder; adhesive: *You'll need a special kind of cement to stick metal to glass.* —*v.* **2** °stick, °glue, paste, solder, °weld, braze, °bond; °join, °bind, °combine, °unite; cohere, °hold, °cling, adhere: *First cement the tiles to the wall. The ashes and cinders cement readily into a compact mass.*

central *adj.* **1** °middle, medial, median; inner, °inside: *The central divider between the roads will be planted with bushes.* **2** °main, °principal, °important, °chief, °key, °leading, °dominant, °prime, °primary, °preeminent, °cardinal, °significant, °essential: *Odysseus is the central figure of the poem.*

center *n.* **1** °middle, °core, °heart; °nucleus, focal point, °hub, °pivot, nave; midpoint: *He stood in the center of the road. We journeyed to the center of the* *earth. The tower is at the center of the market square. The amount to be paid was at the center of the controversy. Mark the center of the line.* —*v.* **2** °focus, °converge, °meet, °concentrate, °cluster: *The business of the meeting centered on the nomination of a chairperson. All my hopes were centered on getting the job as supervisor.*

ceremonial *adj.* **1** °ritual, celebratory, commemorative: *We followed the ceremonial procession.* **2** °formal, °solemn, °stately, °dignified; °ceremonious, august: *The ceremonial robes of his office were white. The ceremonial rites of passage involve many participants.* —*n.* **3** °rite, °ritual, °formality, °ceremony, °service, °observance: *These are the ceremonials prescribed in the Anglican service.*

ceremonious *adj.* **1** °ceremonial, °formal, °dignified, °solemn, *Colloq* °stuffy, °stiff, starchy: *There are many ceremonious procedures involved in a coronation.* **2** courtly, °courteous, °polite, °civil, °correct, °proper, °conventional, punctilious, °careful: *He entered the room and made a ceremonious bow.*

ceremony *n.* **1** °rite, °observance, °solemnity, °service, °ceremonial, °ritual, °formality, °function; obsequies: *I had to attend my grandmother's funeral ceremony and was absent from school.* **2** motions, formalities *or* formality, conventions *or* convention, niceties, proprieties, °form, °protocol; lip service, appearances, proformas, °etiquette, °decorum: *Going through the ceremony is all they want—they don't care what you believe. Please sign even if only for ceremony's sake.*

certain *adj.* **1** °determined, °set, °fixed, °predetermined, °decided, settled, °firm, °stable, °invariable, established, °standard, °constant, unchanging, °steady, unfluctuating, nonfluctuating, °traditional: *He agreed to pay a certain yearly rent. She met him there every day at a certain time.* **2** °sure, unerring, °definite, dependable, °trustworthy, unfailing, °infallible, °reliable, assured, guaranteed: *How do you know that the dividend is certain?* **3** °sure, °inevitable, inescapable, °destined, predestined, ineluctable, inexorable, °unavoidable, °definite, °firm; unchanging, °changeless, °infallible, °permanent, *Colloq* on the cards, a sure thing, *US* °in the cards: *It is not always certain that justice will triumph. Nothing is certain but death and taxes.* **4** indubitable, °indisputable, °undisputed, undoubted, °sure, °doubtless, unequivocal, incontestable, undeniable, °incontrovertible, °absolute, irrefutable, °unquestionable, unquestioned, unarguable, valid: *It is certain only that we exist, according to Descartes.* **5** °confident; assured, °sure, °positive, °definite: *I am certain that she did not steal the money.* **6** °specific, °particular, °definite; unnamed, unspecified, nonspecified, nonspecific: *He gave us certain information which we now have reason to doubt.*

certainty *n.* **1** °fact, actuality, °reality, °truth, *Colloq* sure thing: *I would not advise you to neglect a certainty for something doubtful.* **2** °assurance, self-assurance, definiteness, °confidence, °conviction, °faith, authoritativeness, positiveness, certitude: *The certainty with which he played the card showed he expected it to be a winner.* **3** ***for a certainty.*** assuredly, °definitely, certainly, °surely, °positively; °undoubtedly, indubitably, without (a) doubt, undeniably, unquestionably, °absolutely, *Colloq* for sure: *I know for a certainty that I cannot fly.*

certify *v.* **1** °confirm, °attest (to), °verify, °testify (to), affirm, aver, asseverate, corroborate, °substantiate, °endorse, °guarantee, °warrant; °swear (to), °bear witness (to), °vouchsafe, °vouch (for): *I will certify the accuracy of the report. She certified that she was the owner of the car.* **2** °declare, classify, °establish, °confirm: *The magistrate certified the man insane.*

chafe *v.* **1** °rub, °warm (up), °heat (up): *I took the chill off my hands by chafing them a bit.* **2** °rub, abrade, °fret, °gall, °irritate, make sore: *The skin is very tender where it was chafed.* **3** °fume, °rage, °seethe; °ruffle, vex, °fret, °irritate: *I chafed in impotent rage and exasperation at the ridiculous regulations. To chafe and vex me is a part of her nature.* —*n.* **4** °sore, abrasion, °bruise, soreness, irritation: *The saddle caused a chafe on the inside of my thigh.*

chaff *n.* **1** °banter, °raillery, °ridicule, badinage, joking, teasing, twitting, *Colloq* kidding, ragging, *Chiefly US and Canadian* joshing: *After the speech, he had to put up with chaff from the good-natured audience.* —*v.* **2** °banter, °tease, °twit, rail at, *Colloq* kid, °rag, *Chiefly US and Canadian* josh: *When he was in the navy, his family chaffed him for having a girl in every port.*

chain *n.* **1** °string, series, °combination; °sequence, °succession, °train, °course, °set, concatenation: *He owns a chain of bookshops. Interruption of the food chain can cause serious ecological consequences. A curious chain of circumstances led me to a small hotel at Torquay in February.* **2** °restraint, °check, °trammel, °control, confinement, fetter, °bond, °manacle, °shackle, gyve: *In sunny Spain are streams unbound by winter's chains. The family finally threw off the chains of poverty.* —*v.* **3** °shackle, °secure, °fasten, °bind, gyve; confine, fetter, °restrain, confine, °restrict, °tie, °limit: *Prometheus was chained to a rock as punishment for having brought fire to man. Marguerite felt chained after twenty years of marriage.*

chair *n.* **1** °seat, armchair, stool, bench, easy chair, rocking chair: *He offered me a chair so I sat down.* **2** throne, bench, °position, cathedra, °authority; professorship, directorship: *Sue has been offered a chair on the board.* **3** chairperson, chairman, chairwoman, presiding officer, °leader, °moderator: *The chair ruled on the matter after due consideration.* —*v.* **4** °preside, °lead, °govern, °moderate, °run, °direct, °manage, °oversee: *Katherine will chair the meetings during the absence of the president.*

challenge *v.* **1** °question, °dispute, °defy, object to, °take exception to, °contest, °doubt, call into doubt, call into *or* to question, impugn: *I challenge the validity of your accusation.* **2** invite, °dare, °summon, call out, °provoke: *The duke was challenged to a duel.* **3** °brave, °dare, confront, °defy, °contest: *We could challenge criticism with an easy confidence.* —*n.* **4** °question, °dispute, °doubt: *His opinions are open to challenge.* **5** °invitation, °dare, summons, °provocation, confrontation, defiance; °ultimatum: *An older opponent might not have issued such a challenge.* **6** °problem, °demand, stimulation, °trial, °test: *Are you sure that Mr. Willow will be able to meet the challenge of the new position?*

chamber *n.* **1** °assembly, °body, legislature, judicature, °house, congress, judiciary, senate, °diet; consortium: *She is entitled to sit in the Upper Chamber.* **2** meeting hall, reception room, assembly room: *The council chamber was packed with people.* **3** °compartment, °niche, °nook, °cavity: *We hid the gold in a small chamber in the cave.* **4** °room, apartment; bedroom, bedchamber: *On the second floor are the magnificent royal chambers.*

champion *n.* **1** °victor, °winner, conqueror, titleholder, prizewinner, titleist: *She is the women's singles champion for the fourth year in a row.* **2** defender, °guardian, °protector, °hero, °supporter, °backer, °protagonist, °advocate: *He acquired a reputation as a champion of the underdog.* **3** fighter, combatant, °hero, warrior, campaigner, °veteran: *A stouter champion never handled a sword. The boar is the absolute champion among beasts.* —*v.* **4** °defend, °protect, °guard; °support, °back, °stand up for, fight for, °maintain, °sustain, °uphold; espouse, °forward, °promote, °advocate: *He has always championed the cause of the underdog.*

chance *n.* **1** °fortune, °luck, °fate: *We met, as chance would have it, at the supermarket. Life is but a game of chance for those who cannot control their destiny.* **2** °opportunity, °time, °turn; °occasion: *You have had your chance to return the money, now it is too late.* **3** Also, ***chances.*** °likelihood, °probability, °prospect, °odds, °certainty, predictability; conceivability, °possibility: *Chances are that he'll be late. The chance of winning the lottery is pretty remote.* **4** Also, ***chances.*** °risk, °speculation, °gamble: *You are taking a chance going out there without a weapon. I'll take my chances.* **5** ***by chance.*** **(a)** accidentally, unintentionally, inadvertently: *By chance the witness saw him talking to the suspect.* **(b)** °perhaps, °maybe, °possibly, conceivably: *Have you by chance a match?* —*adj.* **6** °casual, °incidental, °accidental, unintentional, °inadvertent; unplanned, °unpremeditated, unexpected, °unforeseen; unlooked-for: *The affair began with a chance meeting at a bar.* —*v.* **7** °happen; °occur, °come to pass, °take place, °come about; befall, betide: *We chanced to see him jogging in the park. It chanced that a passerby called the police.* **8** °risk, °hazard; imperil, °endanger, °jeopardize, °stake, °bet, wager: *Few would chance severe penalties or jail by lying on a tax return. Don't chance everything you've worked for!*

change *n.* **1** °substitution, replacement, °exchange, interchange, °switch: *You have five minutes for a change of costume. This sunny weather is certainly a change for the better.* **2** °variation, °difference, °switch, °variety, °novelty: *We prefer to live where there is a change of the seasons, not in the tropics.* **3** °variation, °alteration, changeover, °mutation, °shift, modulation, modification, °transformation, metamorphosis, °revolution: *I can't believe the change that has come over Betty since the divorce.* **4** coin(s), coppers, °silver; °(hard) cash: *I need some change for the coffee machine.* —*v.* **5** °exchange, interchange, °switch, °trade; °replace (with), °substitute, *Colloq* swap *or Colloq* swop: *I won't be a minute, I just want to change my shoes. I'd like to change this shirt for a larger size.* **6** °modify, °alter, °modulate; mutate, °transform, metamorphose: *Antoinette has changed since her marriage. I never thought Foley would be able to make her change her mind.* **7** °fluctuate, °shift, °vary; vacillate: *The temperature often changes very rapidly here.* **8** ***change to*** *or* ***into.*** °turn into, °become, °transform, mutate, transmute, °convert, metamorphose: *The alchemists tried to change base metal into gold. Every winter changes into spring—sooner or later.*

changeable *adj.* **1** °variable, mutable, °protean, °inconstant, °unstable, °unsettled, shifting, °uncertain, °irregular, uneven, unpredictable, labile, °capricious, °erratic, °fickle, °unreliable, undependable, mercurial, °volatile: *The weather has been changeable for the past week.* **2** alterable, modifiable, transformable, convertible: *Their meeting places were changeable and known only to them.*

changeless *adj.* **1** unchanging, unvaried, °eternal, °permanent, °fixed, °stable; unchangeable, immutable, unalterable, °inevitable, °uniform: *We gazed out over the changeless Martian landscape. The fundamental truths of the Gospel are changeless.* **2** °abiding, °permanent, °constant, °perpetual, °everlasting, °steadfast, unvarying, unchanging: *Nothing could alter my changeless love for you.*

channel *n.* **1** watercourse, canal, waterway, ditch, aqueduct, sluice, trench, trough, gutter, moat; riverbed, streambed: *The engineers dug a channel to drain the swamp.* **2** °strait, °narrows, neck: *The English Channel connects the North Sea with the Atlantic Ocean.* **3** °furrow, °groove, flute: *The channels cut into this column are not straight.* **4** °course, °means, °way, °approach, avenue, °medium, °path, artery, conduit: *We have to open a new channel of communication with the terrorists.* —*v.* **5** °direct, convey, °pass, °guide, °lead, °conduct: *Their grievances are being channeled through the information officer.*

chant *n.* **1** °song, psalm, hymn, canticle, plainsong, plainchant, mantra, paean, dirge, monody, descant, carol; singsong: *The war chant of the natives, echoing over the water, struck fear into their enemies.* —*v.* **2** °sing, intone, descant, carol: *The choir chanted the verses of a lugubrious threnody.*

chaos *n.* formlessness, °disorder, °confusion; °pandemonium, °bedlam, turmoil, °tumult; entropy: *The universe arose out of chaos. If you want to see chaos, look in any teenager's bedroom. There was complete chaos when the bank locked its doors and forbade withdrawals.*

chaotic *adj.* **1** formless, °shapeless, °incoherent, disordered, °disorderly, disorganized, unorganized, unsystematic, unsystematized, unmethodical, °haphazard, °irregular, °helter-skelter, °confused, °topsy-turvy, jumbled, higgledy-piggledy, *Brit* shambolic: *The present solar system condensed from a chaotic mass of nebulous matter. The rules may seem chaotic at first sight.* **2** °tumultuous, °noisy, clamorous, °uproarious, °wild, °riotous, frenzied, °hectic, turbulent, unstuck: *The press conference became chaotic when the president announced his resignation.*

chap *n.* °fellow, °lad, °man, °boy, *Colloq* °guy, geezer, °customer, gink, *US* buddy, *Brit* bloke, *Australian* cove, *Old-fashioned US* gazabo *or* gazebo; *Old-fashioned Brit* (old) egg, (old) bean, (old) crumpet, °(old) boy; *Slang US* bozo: *I went with some of the chaps from the club.*

character *n.* **1** °brand, °stamp, °mark, °symbol, monogram, insigne, badge, °emblem, °sign, °seal, °label; °letter, °number, °figure, °type, °sort, °arbitrary, °peculiar, rune, hieroglyphic *or* hieroglyph: *There was once a custom of branding the character of a horse in the forehead of a bondslave. We shall need Cyrillic characters to print the Russian text.* **2** °characteristic, °quality, °distinction, °trait, °feature, °mark; °sort, °kind, °type, °nature, °description, °attribute; idiosyncrasy, °peculiarity: *He now tried to give the war the character of a crusade. It is the character of some people to be curious.* **3** °morality, °honesty, °integrity, respectability, °rectitude, °honor, °courage, goodness: *Everyone agrees that she is a person of outstanding character.* **4** °person, °personage, °personality, °individual: *Cobbett had more sagacity and foresight than any other public character of his time.* **5** °role, °part, °personality, characterization, dramatis persona: *He played the character of Caesar.* **6** °eccentric, °card, *Colloq* oddball, nut, nutter, loony, bat, °weirdo, nutcase, screwball, crackpot, fruitcake, *Australian or old-fashioned Brit* cove: *The character in the next apartment takes his canary out for a walk every night.* **7** °role, °position, °status, °capacity: *He assumes the character of a Dutch uncle when he speaks to me.* **8** ***in character.*** °fitting, °proper, °suitable, in keeping, °typical, °normal, expected, °characteristic: *It is completely in character for Witlow to criticize everything he encounters.* **9** ***out of character.*** untypical, atypical, uncharacteristic, °abnormal, unexpected, unfitting: *It would be out of character for Janet to refuse help to someone in need.*

characteristic *adj.* **1** °typical, °representative; °emblematic, °symbolic, °distinctive, idiosyncratic, °symptomatic: *How characteristic it is of him to refuse to go to the dance! These subjects are characteristic of the Hudson River School of painting.* —*n.* **2** °mark, °trait,

°attribute, °feature, °quality, °property, °peculiarity, idiosyncrasy, °character, earmark: *It is a characteristic of bees to swarm.*

characterize *v.* delineate, °describe, °portray, depict, °represent, °define, °brand, °label, °mark, °note, °identify: *She has consistently characterized him as a buffoon. "Virago" is the term that would best characterize Felicity.*

charade *n.* travesty, °absurdity, °mockery, farce, °parody: *He has made a charade of what could have been a serious relationship.*

charge *n.* **1** °load, °burden, °weight, onus, °impediment; °care, °concern, °obligation: *Thank God I shall never be a charge on my children.* **2** °price, °fee, °cost: *What is the charge for admission?* **3** °debt, debit, °expense, assessment, °liability: *Any charge against the estate will be paid. I settled all the charges billed by Charles's school.* **4** °care, °custody, °protection, guardianship, wardship, supervision, °jurisdiction, °control, °responsibility, °safekeeping: *We left the children in the charge of Nanny and Nanny in charge of the children.* **5** °order, mandate, °injunction, °precept, °command, °dictate, °direction, °instruction, °demand, exhortation: *The judge's charge to the jury was to ignore the evidence given by the caretaker.* **6** °accusation, °imputation, indictment, °allegation: *The charge is murder.* **7** °attack, °onset, °action, °assault, sally, °raid, foray, sortie: *At the signal the cavalry charge was under way.* —*v.* **8** °fill, imbue, °load, °instill, pervade, °permeate, °saturate, °suffuse: *The air was highly charged with a stench from the kitchen.* **9** °burden, °entrust, commission, °assign; °afflict, °tax: *He was charged with the supervision of all the military schools.* **10** °command, °order, °bid, enjoin, exhort, °urge, °require, °instruct, °direct: *I charge you not to speak of this matter to anyone.* **11** °blame, censure, °accuse; °indict, cite, °name; °allege, assert: *She charged him with being a hypocrite. He has been charged with assault. It is charged that she was present at the commission of the crime.* **12** °bill, invoice, assess, debit: *Please charge my account for the price of the dinner. Do not charge me for merchandise not shipped.* **13** °ask, °demand, °claim, °require, °expect: *How much do they charge for asparagus at the supermarket?* **14** °attack, °assault, °storm, assail, do battle (with): *Four thousand horsemen stood ready to charge the enemy.*

charitable *adj.* **1** °generous, °liberal, °bountiful, munificent, °unselfish, openhanded, bighearted, magnanimous, °philanthropic, public-spirited, unsparing, eleemosynary: *Despite her income, Irena has always been most charitable when it comes to worthwhile causes.* **2** well-disposed, °kindly, °kind, beneficent, °benevolent, well-wishing, °lenient, °tolerant, °forgiving, °indulgent, °understanding, compassionate, humane, °sympathetic, °considerate, well-meaning, °good: *Few men are charitable who remember not that they have sinned. They took a charitable view of the matter and decided not to complain to the authorities.*

charity *n.* **1** generosity, almsgiving, munificence, liberality, openhandedness, magnanimity, beneficence, °philanthropy, unselfishness, °humanity, humanitarianism, goodwill: *Your charity toward this hospital has been unequaled by any other donor, Sir Keith.* **2** leniency, bigheartedness, largeheartedness, °benevolence, magnanimity, °indulgence, considerateness, °consideration, compassion, °understanding, °sympathy, kindheartedness: *I beseech all that find fault or error in this book to correct or amend it out of charity.* **3** alms, °donation, contribution, largess *or* largesse, *Colloq Brit* °dole, *US* °welfare, °relief: *I want the opportunity to work, not the government's charity.*

charm *n.* **1** °amulet, °talisman, °fetish, rabbit's foot, good-luck piece: *She wears a charm to ward off evil spirits.* **2** attractiveness, °appeal, °fascination, allure, °magnetism, desirability, °elegance, urbanity, °sophistication, sophisticatedness, suavity, °grace, °refinement, cultivatedness, cultivation, °culture, °polish; °magic, °enchantment, °spell, °sorcery: *To Diderot we go not for charm of style but for a store of fertile ideas. To get ahead Cooksley relies more on his charm than on his ability.* **3** ***charms.*** °beauty, attractiveness, pulchritude, prettiness, handsomeness, °appeal, allure, °magnetism, °pull, °draw: *For all her charm, I wouldn't trust her an inch.* **4** ***like a charm.*** successfully, °perfectly, miraculously, marvelously, extraordinarily, especially well: *After the repairs, the fridge works like a charm. His appeal to their egos worked liked a charm.* —*v.* **5** °influence, °control, °subdue, °bind, put a spell on, °bewitch, °enchant, °seduce, °hypnotize, mesmerize, enthral *or US* enthrall, °captivate, °delight, °fascinate, *Literary* °enrapture: *He charmed them with some tale and they gave him their money.* **6** °overcome, °subdue, °calm, soothe, allay, assuage, °hypnotize, mesmerize: *Music is said to have qualities capable of charming savages.*

charmed *adj.* **1** bewitched, spellbound, °enchanted, magical: *Apollonius considered the use of charmed rings essential to quackery.* **2** fortified, protected: *He must lead a charmed life to have survived all those battles.* **3** pleased, °delighted, °enchanted, °happy: *I am charmed to meet you at last, Madam President.*

charmer *n.* enchanter, enchantress, °sorcerer, sorceress, °magician; vamp, °siren, Circe, Cleopatra, Lorelei, °temptress, °seductress; °seducer, Romeo, Valentino, Don Juan, Lothario, Casanova, lady-killer, ladies' man; °flatterer; smooth talker, *Colloq* °(big-time) operator, con artist *or* man, *Old-fashioned* smoothie, wolf: *Charlotte has run off with some charmer from Chester.*

chart *n.* **1** sea-chart, map: *According to the chart, we are fifty miles west of Diamond Head.* **2** map, °table, tabulation, graph, diagram; blueprint: *A weather chart appears on page 23. Here is a chart of the highest-yielding stocks. She drew up a genealogical chart of the descendants of Queen Victoria.* —*v.* **3** °plot, °plan, map (out), °design: *He charted a course of action for the company. Have you charted the shortest route between Gibraltar and Cyprus?*

charter *n.* **1** °document, °contract, °compact, °agreement, covenant: *This year we again commemorate the signing of the United Nations charter.* **2** °permit, °permission, °license, °authority, franchise, °right, °privilege, concession: *He was given an exclusive charter to export furs in 1679.* **3** °lease, °contract: *We have the yacht under charter for the summer.* —*v.* **4** °license, °authorize, °document, commission, °approve, °certify, franchise, °qualify; °recognize: *He is a chartered accountant, she a chartered surveyor.* **5** °let, °lease, °rent, °hire, °engage, °contract: *I chartered the sloop for the weekends only.*

chase *n.* **1** hunting, °hunt, °pursuit: *Three bloodhounds entered the chase, and the escaped prisoner was quickly caught.* **2** °run after, °follow, °pursue, °track, go (out) after; court, woo: *The police were chasing a man down the street. Stop chasing women and settle down.* **3** ***chase away, off, out, etc.*** °rout, °put to flight, °hound; drive away, °off, °out, etc.: *I chased the cat away from the birdcage.*

chaste *adj.* **1** °pure, virginal, virgin, °celibate, abstinent, continent, °virtuous, undefiled, stainless, unstained, unsullied, unblemished, °decent, °clean, °good, °wholesome, °moral: *Only a knight who was wholly chaste would find the Holy Grail.* **2** °subdued, °severe, restrained, °unadorned, austere, unembellished, °simple, °pure, undecorated, °clean: *In some respects, modern architecture emulates the chaste style of the ancient Egyptians.*

chasten *v.* **1** °discipline, °correct, °chastise, °punish, °castigate: *He used every means to chasten the unruly and disobedient.* **2** °moderate, °temper, °subdue, °curb, °restrain, °repress, °tame, °suppress: *I am not as sanguine as I was—time and experience have chastened me.*

chastise *v.* °punish, °beat, thrash, °belabor, °spank, °whip, °flog, °scourge, birch, cane; °discipline, °chasten, °correct, censure, °berate, °scold: *Pupils are not being chastised as in the old days.*

chastity *n.* °purity, continence, virginity, maidenhood, maidenhead, °virtue, °celibacy, abstinence, abstention, abstemiousness, °restraint, self-restraint, forbearance: *As a nun, she had taken a vow of chastity.*

chat *n.* **1** °conversation, colloquy, °talk, small talk, °gossip, °palaver, chitchat, °tête-à-tête, heart-to-heart, *Colloq* °gab, *Chiefly Brit* chinwag, confab, *Brit* witter, natter, *US and Canadian* °rap, gabfest, bull session: *We'd get together for a chat every now and then.* —*v.* **2** °converse, °gossip, °talk, chitchat, *Colloq* °gab, chew the fat *or* the rag, jaw, *Brit* witter, natter; *US and Canadian slang* °rap, bullshit: *We were just chatting when I smelled something burning.* **3** ***chat up.*** flirt *or* dally with, °persuade, °induce, °prevail upon, °tempt, °lure, °entice, inveigle, °seduce, °proposition: *Renfrew chats up every girl he meets.*

chatter *v.* **1** °prattle, gabble, °jabber, prate, °patter, gibber, cackle, gibber-jabber, *Brit* chaffer, *Colloq* °gab, jaw, *Brit* natter, witter, rabbit on *or* away, °waffle: *He just kept chattering on about nothing.* **2** clatter, °rattle: *It was so cold my teeth were chattering.* —*n.* **3** °prattle, prate, °patter, °gossip, cackle, jabbering, chattering: *I don't want to hear any more chatter in the library.*

cheap *adj.* **1** °inexpensive, low-priced, bargain-priced, low-cost, sale-priced, cut-price, °reasonable; °economy, budget(-priced): *He was chewing on a cheap cigar. Everything used to be a lot cheaper when I was younger.* **2** °economical; reduced: *Eggs are cheaper by the dozen.* **3** °shoddy, °base, °shabby, °tawdry, °sleazy, tatty, °seedy; °inferior, low-grade, °poor, second-rate, trashy, °worthless, *Brit* twopenny *or* tuppenny; *Colloq* °tacky, *Brit* tinpot, *Slang US* two-bit, °lousy, chintzy: *Those cheap pictures ruin the look of the place.* **4** stingy, °miserly, °penurious, niggardly, penny-pinching, cheeseparing, °frugal, °tight, tightfisted, Scroogelike, skinflinty: *That cheap brother of yours wouldn't give even a penny to a beggar.* —*adv.* **5** inexpensively, cheaply, °for a song, *Brit* for twopence *or* tuppence: *You can buy those cheap from any street vendor.* **6** cheaply, °easily, reasonably, °for a song, *Brit* for twopence *or* tuppence: *She has sold cheap that which she holds most dear.* —*n.* **7** ***on the cheap.*** inexpensively, reasonably, cheaply, at *or* below cost, °for a song, *Brit* for twopence *or* tuppence; *Slang* for peanuts: *We buy these watches on the cheap and sell them to tourists.*

cheat *n.* **1** °swindler, deceiver, °impostor, °fraud, faker, °fake, °swindler, trickster, confidence man, con man, °operator, charlatan, mountebank, °rogue, shark, *Colloq* phoney *or US* phony, snake-oil artist: *It is amazing how many cheats are out there waiting to take advantage of you.* —*v.* **2** °swindle, °deceive, bilk, °trick, con, °take, °fleece, °defraud, euchre, °hoax, °hoodwink, *Colloq* con, °take in, rook, flimflam, finagle, diddle, °fiddle, move the goal posts, bamboozle, °take for a ride; *Slang* °rip off: *His own solicitor cheated him out of his inheritance. If you paid ten dollars for that painting, you were cheated.*

check *v.* **1** °stop, °arrest, °stay, °halt, °obstruct, °block, °limit; °retard, °slow, °brake, °curb, °hinder, °hamper, °impede, °thwart: *What can we do to check the spread of the disease?* **2** °restrain, °control, °repress, °stay, °inhibit, °contain, °curb, °restrict: *The animal population is checked only by availability of food.* **3** °authenticate, °verify, °confirm, °substantiate, validate, corroborate, °check into, °check out, °check up on: *Please check his story to make sure he's not lying.* **4** enquire about *or* after *or* into, °check into, check (up) on, °examine, °investigate, °inspect, make sure of, °verify, °monitor, °test, °study, °scrutinize: *You'd best check the temperature in the kiln.* **5** °correspond, °coincide, °agree, jibe, °tally, °conform, °check out, °fit, °mesh; °compare: *His alibi doesn't check with the statements of the witnesses.* **6** ***check in.*** °arrive, °report: *We check in for work at eight o'clock.* **7** ***check in*** *or* ***into.*** °register, sign in *or* on, °enroll, log in: *We checked into the hotel.* **8** ***check into.*** °investigate, °check out, °check up on, °verify, °check: *The detective checked into the backgrounds of all applicants.* **9** ***check off.*** tick (off), °mark, °check: *Check off the names in red.* **10** ***check out.*** **(a)** °depart, °leave; °go: *He checked out of the hotel and took a taxi to the airport.* **(b)** °investigate, °research, °explore, enquire into, look into *or* at *or* over, °scrutinize, °examine, °inspect, °probe, °survey, °check up on, °check, °check into, check over: *You had best check out her references before hiring her.* **(c)** °pass, pass muster *or* scrutiny, meet approval, be verified, °check: *According to our records, his story checks out.* **(d)** *Slang* cash in one's checks *or* chips, kick the bucket, croak: *Sam checked out last week—heart attack, I think.* **11** ***check over*** *or* ***out.*** °review, °verify, °authenticate, °check: *Please check over my figures before I submit them to the accountant.* **12** ***check up*** **(*on*).** **(a)** °investigate, do research, °probe, °explore, °check: *I don't know her name, but I'll check up and let you know.* **(b)** °determine, °discover, °find out, °look into, °check: *I want you to check up on where they eat lunch.* —*n.* **13** °stop, stopping, °cease, surcease, hesitation, cessation, stoppage, °interruption, °break, °pause, balk *or* baulk, discontinuity, discontinuation, discontinuance, °suspension: *The visitors continued to arrive without check, far into the night.* **14** °restraint, °repression, °inhibition, limitation, °curb, °restriction, °control, constraint, °hindrance, °obstruction, °impediment, damper: *He keeps a good check on the foreman. This tax will serve as a check against free trade.* **15** °control, °test, inspection, °examination, °scrutiny, verification, substantiation, authentication, confirmation, validation, corroboration: *We do a thorough check on the quality of every product.* **16** tick, °mark, °dash, X: *Place a check in the box alongside your choice.* **17** °token, °receipt, counterfoil, °stub; voucher, chit, certificate: *Don't lose your baggage check.* **18** °chip, °counter: *Let's cash in our checks and go home.* **19** °tab, charge(s), *Chiefly Brit* °bill: *Please add 15 percent to the check for a tip.*

checkered *adj.* **1** checkerboard, checked; °patchwork; plaid, tartan: *You cannot use a checkered tablecloth for a formal dinner.* **2** °variegated, diversified, alternating, °variable, good and bad, varying, fluctuating, up and down; °uncertain: *Waybright had a rather checkered career in the army.*

cheeky *adj.* °impudent, °impertinent, °insolent, °audacious, °disrespectful, °rude, uncivil, °forward, °brazen, °pert, saucy: *That cheeky little brat told me to get lost!*

cheer *n.* **1** °disposition, °frame of mind, °spirit: *Be of good cheer, My birthday is here!* **2** cheerfulness, gladness, °mirth, °joy, °gaiety, blitheness, °happiness, buoyancy, lightheartedness, merrymaking: *There wasn't much cheer at the pub when we learned of what had* *befallen poor Grover.* **3** °comfort, °solace, °encouragement, consolation: *She brought in a little breath of cheer from the outside world.* **4** °shout, °cry, hurrah, rah, huzzah, hurray *or* hooray: *Three cheers for Penelope!* —*v.* **5** °comfort, °console, °solace, °encourage, inspirit, °warm, *Colloq* buck up: *Your friendly note cheered me considerably.* **6** °gladden, °enliven, cheer up, hearten, °buoy up, °brighten, elate, °brighten, uplift, °lift up: *Let thy heart cheer thee in the days of thy youth.* **7** °applaud, °shout, hurrah, °clap, °yell; *Colloq Brit, Australian, New Zealand* barrack for: *The crowd cheered for five minutes when Mr. Flews stood to speak.*

cheerful *adj.* **1** joyous, °glad, gladsome, blithesome, °blithe, °happy, cheery, of good cheer, °joyful, °jolly, °exuberant, jubilant, °gleeful, °gay, lighthearted, °merry: *I am pleased to see that Agatha is so cheerful. Why do you cry at weddings, which are supposed to be such cheerful occasions?* **2** cheering, gladdening, animating, °bright, enlivening, cheery, °gay, °buoyant, °invigorating: *She has redecorated the bedroom in more cheerful colors.*

cherish *v.* **1** °treasure, hold *or* keep dear, °prize: *I know she cherishes every moment you were together, right till the end.* **2** °foster, °tend, °cultivate, °preserve, °sustain, nurture, °nourish, °nurse, cosset: *For its smell, sweet alyssum is cherished in gardens. We cherish little Ned and probably spoil him a bit too much.*

chest *n.* **1** °box, coffer, °trunk, strongbox, caddy, °casket, °case: *It took four men to carry the chest outside, where we could open it. Martha kept her jewels in a small chest on the dresser.* **2** °breast; thorax: *The wrestler was pounding his chest shouting "I am the greatest!"*

chew *v.* **1** masticate, °grind, °munch; °bite, °gnaw: *Make sure to chew each mouthful thoroughly. The puppy chewed up my slipper.* **2** ***chew the fat*** *or* ***rag.*** °gossip, °palaver, °chat, °converse, °talk, *Slang US and Canadian* bullshit: *We sat round the fire and chewed the fat all evening.* **3** ***chew out.*** °scold, °rebuke, °reprimand: *The sergeant chewed out the recruit because his boots were dirty.* **4** ***chew over.*** think about *or* on *or* over, °consider, °review, °ponder, ruminate on, meditate on *or* over: *I'll chew over your proposal and let you know.*

chic *adj.* **1** °stylish, °fashionable, à la mode, modish, °smart, °tasteful, °elegant; °sophisticated; *Colloq* °trendy: *Susanna was always a chic dresser.* —*n.* **2** °style, °fashion, good taste, tastefulness, °elegance, stylishness, modishness: *There is an air of chic about him that repels many men but attracts many women.*

chicanery *n.* °trickery, sophistry, °deception, quibbling, sharp practice, cheating, deviousness, duplicity, pettifoggery, double-dealing, °artifice, skulduggery: *They lost the case because of the chicanery of their lawyers.*

chief *n.* **1** °head, °leader, °principal, °superior, °supervisor, °superintendent, °manager, °overseer, captain, °master, ringleader, chieftain, *Dialect* himself, *Colloq* boss, bossman, *Brit* governor, gov, supremo, *US* °man, kingpin, (head *or* chief) honcho, number one, numero uno, head man, big White Chief, big Chief, Great White Father, big Daddy, °super; *Slang* big cheese, *Brit* gaffer, *Chiefly US* Mr. Big: *You'd best ask the chief for permission to fly to Rome.* —*adj.* **2** °head; °leading, ranking, °superior, °supreme, °foremost, °premier, °first, greatest, °outstanding: *Flambeau has been appointed chief justice. Terence is your chief competition for the singles trophy.* **3** °principal, most important, °essential, °key, °paramount, °(first and) foremost, °primary, °prime, °main: *The chief reason I came was to see you. Here is a list of the chief crimes committed in the area last year.*

chiefly *adv.* °mainly, °in particular, °especially, °particularly, °above all, most of all, °preeminently, °principally, °primarily, mostly, predominantly, °largely, by and large, °on the whole, °in the main, °generally, in general, °usually, °as a rule: *Inflation affected chiefly the price of food.* The Anatomy of Melancholy *consists chiefly of quotations.*

child *n.* **1** °offspring, °descendant, son *or* daughter, little one, youngster, *Formal* °progeny, °issue, *Colloq* kid, nipper, *Slang Brit* sprog: *How many children do you have?* **2** fetus, newborn, neonate, infant, °baby, babe, toddler, boy *or* girl, lad *or* lass, °stripling, youngster, °youth, °juvenile, °adolescent, °teenager, young man *or* woman, young gentleman *or* lady, *Chiefly Scots* laddie *or Chiefly Scots* lassie: *No children were born in the village for five years. These miscreants are mere children, who should not be punished as adults.*

childhood *n.* °infancy, babyhood, boyhood *or* girlhood, °youth, °puberty, minority, adolescence, teens: *During her childhood the family moved to Kent. She spent most of her childhood dreaming about traveling to the moon.*

childish *adj.* °childlike, °juvenile, °puerile, °infantile, babyish; °immature, °inexperienced, naive *or* naïve *or* naïf, °undeveloped, underdeveloped, retarded; °silly, *US* sophomoric: *A bit childish to be organizing an international hopscotch competition, isn't it, Wimpole?*

childlike *adj.* youthful, °young, °innocent, trustful, °ingenuous, °unsophisticated, naive *or* naïve, °trusting, credulous, °open, undissembling, unassuming, guileless, °artless: *There is a childlike simplicity to some primitive paintings.*

chill *n.* **1** coldness, °cold, coolness, sharpness, °nip: *We put on our jackets to ward off the chill of the evening.* **2** °cold, flu, influenza, (la *or* the) grippe, ague, *Technical* coryza, *Colloq* (the) sniffles, sneezles and wheezles: *Take off those wet clothes before you catch a chill.* **3** coolness, iciness, frigidity, aloofness; unfriendliness, °hostility: *Mrs. Marlow felt the chill in the stare of her husband's ex-wife.* —*adj.* **4** °cold, °cool, numbing, chilling, °chilly, °raw, °penetrating, °icy, °frigid, °wintry, frosty, arctic, °polar, glacial: *A chill easterly wind made me shiver.* **5** shivering, chilled (through), °numb, numbed, numbing, benumbed: *She kissed me with a lip more chill than stone.* **6** °cold, °coldblooded, °aloof, °indifferent, insensitive, unemotional, °unsympathetic; °chilly: *The prison commandant viewed the cadavers with chill detachment.* —*v.* **7** °cool, °freeze, °refrigerate, ice: *The fruit tastes better if it has been chilled.* **8** °dampen, dispirit, °depress, deject, dishearten, °distress: *The news of Mother's illness chilled us all. The atmosphere in his house chills me so that I rarely visit him.*

chilly *adj.* **1** °cool, coldish, °cold, °frigid, nippy, frosty, °icy, °crisp, °chill: *The weather has been quite chilly for May.* **2** °chill, °unenthusiastic, unresponsive, unreceptive, frosty, unwelcoming, °crisp, °cool, °cold, unfriendly, °hostile; °distant, °aloof: *The suggestion that the charity fair be held in her garden met with a chilly response from the Grinnells.*

chime *n.* **1** bell, set of bells, carillon, °ring, °peal: *Our church has a full chime of eight bells.* **2** ringing, °peal, chiming, tolling, tintinnabulation, clanging, ding-dong, °striking; tinkle, °jingle, °jangle: *We could hear the chimes of Big Ben from our hotel room.* —*v.* **3** °ring, °peal, °toll, °sound, tintinnabulate, clang, °strike: *The clock chimed on the hour.* **4** °mark, °denote, °indicate, °announce: *The carillon chimed the hour at noon.* **5** ***chime in.*** **(a)** join in, °blend, harmonize: *When singing this round, chime in at the third bar.* **(b)** °interrupt, intercede, °intrude, °interfere, °break in, *Colloq* °chip in; *Slang* °butt in: *I was about to speak when he chimed in with some silly remark.*

chink *n.* fissure, °rift, °crack, °crevice, °gap, °opening, cleft, °slit, °aperture, °cranny: *The chain caught in a chink in the rock and saved me from certain death.*

chip *n.* **1** °fragment, °piece, shard *or* sherd, °splinter, °flake, °sliver, °scrap, °morsel, °bit: *A chip of slate rattled down off the roof.* **2** °counter, marker, °token; °plaque, *US* °check: *He put his chip on number 14.* —*v.* **3** °chisel, °whittle, hew: *He chipped away at the stone till it fitted perfectly into the hole.* **4** ***chip in.*** **(a)** °contribute; °participate: *All the neighbors chipped in to pay for the street decorations.* **(b)** °interrupt, °break in, °intrude, °interfere, intercede, interpose, *Colloq* °chime in: *Clive chipped in with his usual silly comment.*

chirp *v.* **1** tweet, °peep, °twitter, chirrup, warble, trill, cheep, chitter, chirr, °pipe: *I was awakened by the birds, chirping away in the forest.* —*n.* **2** tweet, °peep, °twitter, chirrup, warble, trill, cheep, chitter, chirr: *The canary gave two chirps and jumped onto its perch.*

chisel *v.* **1** °carve, °cut, °pare, °groove, °shape, °engrave, °grave: *He was chiseling the figure of an eagle out of the board.* **2** °cheat, °defraud, °swindle, bilk, °trick, °fool, °dupe, gull, *Colloq* bamboozle: *The gamblers chiseled him out of a week's wages.*

chivalrous *adj.* courtly, °gracious, °courteous, °polite, °gallant, °noble, knightly, gentlemanly, °considerate, °kind, °charitable, magnanimous: *It was quite chivalrous of you to drive me home.*

chivalry *n.* knight-errantry; °honor, °bravery, °courage, °courtesy, politeness, courtliness, gallantry, °nobility, virtuousness, righteousness, justness, fairness, impartiality, equitableness: *All the noble sentiments blended together constitute chivalry.*

choice *n.* **1** °selection, °election, °preference, choosing, °pick, acceptance: *I don't care for his choice of language.* **2** °option, realm of possibilities; °alternative, °voice, °determination: *She was given no choice in selecting her husband.* **3** °pick, elite *or* élite, °flower, °best, °select, cream, *crème de la crème*: *The king's guard is made up from the choice of the kingdom.* —*adj.* **4** °select, °exquisite, °special, °superior, °prime, high-quality, °excellent, °preeminent, °best, °prize, °first-rate, °exceptional, preferred, °desirable, °ideal, °rare, *Colloq Brit* plummy: *She has the choicest wines in her cellar.* **5** selected, °select, handpicked, well-chosen, °fit, °appropriate, °fitting: *The eulogy was disposed of in a few choice words.*

choke *v.* **1** suffocate, asphyxiate, °smother, °stifle, strangle, throttle, garrotte *or* garrote *or* garotte, burke: *He choked his elderly victims, then stole their money.* **2** °stop, °fill (up), °block (up), °obstruct, congest, °clog, dam (up), constrict: *The channel is completely choked with weeds.* **3** Also, ***choke off.*** °smother, °suppress, °stifle, °prohibit, °frustrate, °deny, obviate, °cut off, °stop, put a stop to; dissuade, °discourage: *His policies choked off any chance for innovation.* **4** ***choke back*** *or* ***down.*** °suppress, °repress, °stifle, °restrain: *He choked back the tears when he saw the gravestone.*

choose *v.* °select, °elect, °pick (out), °determine, °judge; °decide, °prefer, opt, settle upon *or* on: *She had the right to choose the course that seemed the best to her. Given the options, I chose to stay.*

choosy *adj.* °selective, °discriminating, discerning, °fastidious, finicky *or* finical, °particular, °fussy, °demanding, °exacting, °difficult, hard to please, *Colloq* picky: *If you weren't so choosy, you wouldn't have to pay so much.*

chop *v.* **1** Also, ***chop away*** *or* ***down*** *or* ***off.*** °cut, °hack, hew, °lop, crop, °cleave, °sever: *Chop away that underbrush. I tried to chop off the end. Don't chop down that tree!* **2** Also, ***chop up.*** mince; dice, cube; °hash: *Chop up the parsley very fine before adding it to the sauce.* —*n.* **3** °cut, °blow, °stroke: *With a quick chop of the axe, the branch was severed.*

christen *v.* **1** baptize, anoint: *They are going to christen the baby next week.* **2** °name, °call, dub: *The child was christened Madelaine. The highest peak in Wales is christened Snowdon.*

chronic *adj.* **1** longlasting, longstanding, °lingering, inveterate, °persistent, continuing, °lasting, long-lived: *The doctor said that the condition, for which there is no cure, is chronic.* **2** inveterate, °persistent, dyed-in-the-wool, confirmed, °habitual, hardened: *Abby is a chronic liar.*

chronicle *n.* **1** °record, °history, °diary, °chronology, °account, °narrative, °description, °report, °register, annal(s), archive: *She has written a chronicle of the events leading up to the War of Jenkins' Ear.* —*v.* **2** °record, °register, °list, °enter, archive, °document, °describe; °tell, °recount, °narrate, °report, °relate, retail: *In the* Iliad *Homer chronicled the legends of the Trojan War.*

chronology *n.* °account, °record, °calendar, almanac, °journal, log; °sequence: *Describe the chronology of events preceding your discovery of the body.*

chubby *adj.* podgy *or US* pudgy; stumpy, stubby, chunky, tubby, °plump, °dumpy, thickset, heavyset, °heavy, °ample, overweight: *She might have been a bit chubby when a teenager, but she later became a professional fashion model.*

chuckle *v.* **1** °laugh, chortle, crow, snigger, °giggle, °titter: *Robin always chuckled when the subject of embezzlement came up.* —*n.* **2** chuckling, °laugh, chortle, crowing, °giggle; °laughter, snigger, sniggering: *He found it hard to resist a satisfied chuckle when he saw the children's faces light up.*

chum *n.* **1** °friend, °comrade, °companion; confidant(e), °familiar; °fellow, °colleague, *Colloq* °pal, sidekick, *Chiefly Brit, Australian, New Zealand* mate, *Chiefly US and Canadian* buddy: *I invited a chum of mine for the weekend.* —*v.* **2** Often, ***chum around.*** °associate, *Colloq* °pal (around): *Yes, we used to chum around together in the army.* **3** ***chum up with.*** ally (oneself) with, be friendly with, °go with, °associate with, *Colloq* pal (up *or* about *or* around) with, *US* team up with: *Lionel chummed up with Ashley to go swimming.*

chummy *adj.* °friendly, °sociable, °intimate, °close, °thick, *Colloq* pally, *US* palsy-walsy, buddy-buddy: *You and Freddy were once very chummy, weren't you?*

chute *n.* **1** °waterfall, °rapid: *The canoe skimmed down the chute with lightning speed.* **2** °slide, °shaft, °channel, °ramp, runway, trough, °incline: *The parcels come down this chute and you have to sort them by zip code.*

circle *n.* **1** disc *or chiefly US* disk, °ring, hoop, °loop, °band, °wheel, annulus, ringlet; cordon: *Using a compass, he carefully drew a circle. We formed a circle around the speaker.* **2** °set, coterie, °clique, °class, °division, °group, °crowd; °society, °fellowship, °fraternity, °company: *John and I don't move in the same circles.* —*v.* **3** °encircle, circumambulate, go round *or* around, °tour; circumnavigate: *For exercise, I circle the lake in the park every morning.* **4** °encircle, °surround, gird, °enclose, circumscribe: *Twenty small diamonds circle each star sapphire.*

circuit *n.* **1** compass, circumference, °perimeter, °periphery, °girth, °border, °boundary, °edge, °limit, ambit, °margin, °outline, confine(s), °bound, °pale: *The* *circuit of the area amounts to seventy-two miles.* **2** °round, °tour, ambit, °circle, °orbit, °course, °lap: *The rider completed the circuit of the ranch, mending the fence as he went.*

circular *adj.* **1** °round, disc-shaped *or chiefly US* disk-shaped, disklike *or* disclike, discoid; ring-shaped, ringlike, hooplike, hoop-shaped, annular: *Notice the circular pattern of growth of this ivy.* **2** °roundabout, °indirect, circuitous, °tortuous, twisting, twisted, anfractuous; periphrastic, circumlocutory; °devious: *We had to take a circular route because the road was closed. Why can't she say what she means instead of being so circular?* **3** illogical, °inconsistent, °redundant, fallacious, irrational, *Formal* sophistic *or* sophistical: *To say that you exist because you think and that you think because you exist is an example of circular reasoning.*

circulate *v.* **1** move *or* go about *or* round *or* around, °orbit, °flow, °course, °run, °circle: *The blood circulates from the heart through the arteries and veins and back to the heart.* **2** °spread, °distribute, disseminate, °issue, °publish, °air, °announce, °proclaim, °make known, noise abroad, bruit about, °report, °broadcast, °reveal, divulge, advertise, °publicize, promulgate, °put about, bring *or* put out, pass out *or* round *or* around: *He has been circulating the story that his ex-wife cheated on her income tax.* **3** °spread, go around *or* round, be bruited about, °come out: *A rumor has been circulating about your behavior at the office party.*

circulation *n.* **1** °circuit, °course, °orbit, °flow, flowing, °motion: *It was Harvey who discovered the circulation of the blood.* **2** °spread, spreading, dissemination, °transmission, °passage, °distribution, diffusion, °publication, °advertisement, °announcement, issuance, issuing, °pronouncement, °proclamation, promulgation, °broadcast, broadcasting: *The state has again forbidden the free circulation of information.*

circumstance *n.* **1** Often, ***circumstances.*** °situation, condition(s), °state (of affairs); °status, °station, °resources, °income, °finances: *In the circumstances, all leaves are canceled. Each person will be helped according to individual circumstance.* **2** °event, °incident, °episode, °occurrence, °affair, °happening, °occasion: *Any unforeseen circumstance could set off a shooting war.*

circumstantial *adj.* **1** °indirect, °presumptive, evidential *or* evidentiary, interpretive, deduced, presumed, °presumptive, presumable, implicative, implied, inferred, inferential: *Some circumstantial evidence is very strong, as when you find a gun in the bedside table.* **2** °accidental, °incidental, hearsay, °indirect, unimportant, adventitious, °provisional, °secondary, unessential, °nonessential, fortuitous, °chance, °extraneous: *Such circumstantial trivia have no bearing on the case.* **3** °detailed, °particular, °precise, °explicit, °specific: *We cannot believe that Cavendish invented so circumstantial a narrative.*

citizen *n.* **1** voter; °native; householder, °resident, °inhabitant, °denizen, dweller, freeman; *Brit* patrial, ratepayer; *US* taxpayer: *All citizens are entitled to certain rights.* **2** city dweller, town dweller, townsman, townswoman, villager, burgess, oppidan: *She considers herself a citizen of Oxford.*

city *n.* °metropolis, °municipality, borough, burgh; conurbation, megalopolis; *Brit* urban district; °see, diocese, bishopric; *New Zealand* urban area; *Colloq* °town, *US* burg: *We gave up our apartment in the city and moved to the country.*

civil *adj.* **1** civilian, nonmilitary, °lay, laic, laical, °secular: *One must distinguish between civil law and canon law.* **2** °domestic, °internal; °public: *The economic conditions have led to civil strife.* **3** °polite, °courteous, °respectful, well-mannered, °proper, civilized, °cordial, °formal, courtly, urbane, °polished, °refined: *They are civil enough, but I always have the feeling they really despise tourists.*

civility *n.* °courtesy, politeness, °respect, comity, urbanity, amiability, °consideration, courteousness, cordiality, °propriety, °tact, °diplomacy, politesse, °protocol: *Despite his rude behavior at her dinner party, she treats him with great civility.*

civilization *n.* **1** °culture, °refinement, cultivation, enlightenment, °edification, °sophistication, °polish: *The Romans brought civilization to many peoples who had been quite barbarous.* **2** °culture, mores, custom(s): *He has studied Egyptian civilization all his life.*

civilize *v.* **1** °enlighten, °refine, °polish, edify, °educate, acculturate: *Civilized people do not behave in such a boorish way.* **2** °tame, domesticate; broaden, elevate, acculturate: *The claim that they civilized the Aborigines means only that they forced them to conform to the White man's notion of civilization.*

claim *n.* **1** °demand, °assertion, °request, °requisition, °petition, °application; °requirement: *As far as the land is concerned, his claim has been denied.* **2** right(s), °call, °title: *What possible claim could the Miss Dashwoods have on his generosity?* —*v.* **3** °demand, °seek, ask *or* call (for), °exact, insist (on *or* upon), °require, °command, be entitled to: *She has every right to claim a share in the estate.* **4** °declare, assert, °allege, °state, put *or* set forth, affirm, contend, °maintain: *These measurements lack the degree of accuracy claimed for them. She claims that she was the first person to ring the police.*

clammy *adj.* **1** °moist, °sticky, gummy, °pasty, viscous, °slimy: *In the swamp the police found a clammy pistol that they believe to be the weapon.* **2** °moist, °damp, °humid, °close, °muggy, °wet, °misty, foggy: *It was the kind of clammy summer's day when your shirt sticks to your back.*

clamp *n.* **1** °clasp, °vice, °brace, °clip, fastener: *Use a clamp to hold the pieces together till the glue dries.* —*v.* **2** °fasten (together), °clip (together), °bracket, make fast, °clasp: *You should clamp the planks together and plane the edges of both.*

clan *n.* **1** °tribe, °family, °dynasty, °line, °house: *Among American Indians a man is forbidden to marry in his own clan.* **2** °fraternity, °brotherhood, °party, °set, °clique, coterie, °circle, °crowd, °group, °fellowship, °society, °faction, °family, °tribe; °band, °ring, °gang: *He regards social scientists as a clan quite separate from other scientists.*

clap *v.* **1** °applaud; °cheer, acclaim: *Everyone clapped when the boxer climbed into the ring.* **2** °slap, °strike, °pat: *He clapped me on the shoulder in the friendliest way.* **3** °put, °place, °slap, °fling, °toss, °cast, *Colloq* °stick: *He had no sooner set foot in the town when he was clapped in jail.* **4** °impose, °lay, °apply: *The magistrate clapped a severe fine on me for speeding.* —*n.* **5** °crack, °slap, °report, °crash, bang, °snap: *There was a loud clap of thunder, making the house shake.*

clapper *n.* °tongue: *We attached a rope to the clapper of the bell.*

clarify *v.* **1** elucidate, make clear, °simplify, make plain, °clear up, °explain, shed *or* throw light on *or* upon, °illuminate, explicate: *She offered to clarify any points about which we had questions.* **2** °clear, °purify, °clean: *The trout should be lightly basted with clarified butter.*

clarity *n.* **1** clearness, transparency, limpidity, pellucidity: *The clarity of the sea in the tropics is owing to a lack of plankton.* **2** lucidity, °definition, definiteness, distinctness; comprehensibility, understandability, intelligibility, unambiguousness: *One cannot argue with the clarity of her explanation of Hegelianism.*

clash *n.* **1** °crash, clang, clank, clangor: *The concerto ends with a clash of cymbals.* **2** °collision, smash, °(hostile) encounter, °conflict, °engagement, °fight, °battle, °disagreement, °difference, °argument, °dispute, altercation, °quarrel, squabble: *He has survived many a clash between the president and the Congress.* —*v.* **3** °conflict, °fight, °battle, °disagree, °differ, °argue, °dispute, °quarrel, squabble, °feud, wrangle, cross swords: *My brother and I always clash on the question of who should pay Father's hospital bills.* **4** °conflict, disharmonize, °jar, be at odds *or* out of keeping: *The pink of the blouse and the fuchsia of the skirt clash badly.*

clasp *n.* **1** fastener, °fastening, °hook, °catch, °clip, °pin, °brooch: *The ends were fastened together with a diamond clasp.* **2** °embrace, °hug, °hold, °grasp, °grip: *He held her tight in his clasp.* —*v.* **3** °fasten, °secure, °close, °hold, °hook, °clip, °pin, °clamp: *Her robe was clasped by an emerald pin.* **4** °hold, °embrace, take hold of, °hug, °enclose, °envelop: *His arms clasped her tightly.* **5** °grab, °grasp, °seize, °clutch, °grip: *The beggar clasped my hand, his eyes seeking mine in piteous supplication.*

class *n.* **1** °rank, °grade, °level, °order, °stratum; °caste, °lineage, °birth, °pedigree, °stock, °extraction, descent: *He was born in the 1940s into a family of the middle class. Aunt Maggie has little to do with anyone outside her own class.* **2** °category, °division, classification, °group, °genre, °league, °realm, °domain; °kind, °sort, °type: *Madam Kropeczka is in a class by herself.* **3** °excellence, °merit, °refinement, °elegance, °prestige, °importance, °taste, discernment, °distinction, °bearing, °presence, °savoir-faire, °savoir-vivre, °breeding: *He may be a good drinking companion, but he has no class whatsoever.* **4** year, °form, *US* °grade: *We were in the same class at school.* —*v.* **5** classify, °group, °arrange, assort, °type, °categorize, °rank, °grade, °rate, °order: *They are classed as self-employed for these purposes.*

classic *adj.* **1** °standard, °leading, °outstanding, prototypical, °definitive, °model, °ideal, archetypal, paradigmatic: *Eppington's military career is a classic example of what family connections can achieve.* **2** °legendary, °immortal, °enduring, °deathless, ageless, °timeless, undying, °venerable, °time-honored; °outstanding, °first-rate, °superior, °excellent, °noteworthy, °notable, °exemplary: *By the time she was ten, she had read most of the classic works of English literature. He collects classic cars.* —*n.* **3** °paragon, °epitome, outstanding example, exemplar, °model, paradigm, °prototype: *When it comes to comedians, Ronnie is a classic.* **4** °masterpiece, masterwork: *The Rolls-Royce Silver Ghost is regarded as a classic by collectors.*

classical *adj.* **1** °standard, °model, °exemplary, °traditional, established, °influential, °authoritative, °serious, °weighty: *Classical authors are those who are regarded as being of good credit and authority in the schools. Montaigne is the earliest classical writer in the French language.* **2** Greek, Latin, Roman: *The styles of the nineteenth century harked back to Classical designs.*

claw *n.* **1** talon, °nail: *She had been badly scratched by the cat's claws.* —*v.* **2** °scratch, °tear, °scrape, °rake, °slash: *She clawed at his face to break his grip on her throat.* **3** °grapple, °grab, °catch, °scrape, scrabble: *He tried to climb up the embankment, clawing at the steep wall.*

clean *adj.* **1** °pure, undefiled, unsullied, unmixed, unadulterated, uncontaminated, unpolluted, uninfected, °unspoiled, °sanitary, disinfected; antiseptic, decontaminated, purified, °sterile: *The laboratory reports that our well water is absolutely clean. You must use a clean bandage.* **2** unsoiled, untainted, unstained; unsullied; cleansed, cleanly, (freshly) laundered *or* washed, scrubbed; °spotless, °immaculate: *She puts on clean underwear every day in case she's involved in an accident.* **3** clean-cut, °neat, °simple, °definite, uncomplicated, °smooth, °even, °straight, °trim, °tidy: *The edges of the fracture are clean and will mend quickly.* **4** °innocent, °blameless, °inoffensive, °respectable; °decent, °chaste, °pure, °honorable, °good, undefiled, °virtuous, °moral: *Pinky is completely clean—he was out of town at the time of the robbery.* **5** nonradioactive: *They say they've produced a clean bomb, but I don't believe it.* **6** °unarmed, weaponless: *Frisk that suspect and make sure he's clean.* —*adv.* **7** °completely, °entirely, °thoroughly, fully, °totally, °wholly, °altogether, °quite, °utterly, °absolutely: *Gunderson's clean out of his mind if he believes that. With one blow he cut the orange clean through.* **8** ***come clean.*** °confess, °acknowledge, make a clean breast, °admit, make a revelation, °reveal, *Colloq* own up, °spill the beans; *US dialect* fess up; *Slang* °sing: *They said I could expect a lot of trouble if I didn't come clean.* —*v.* **9** °cleanse, °wash, lave, °(take a) shower, sponge, mop, °scrub, °scour, °sweep, dust, vacuum, °polish, °launder, dry-clean, *Brit* hoover; °tidy, °neaten, do up, straighten up *or* out, unclutter; *Brit* bath; *US and Canadian* bathe: *When we cleaned the urn, we could read the inscription. I have told Richard a thousand times to clean his room.* **10** ***clean out.*** **(a)** °exhaust, deplete: *The gambler cleaned her out of every penny she had in the world.* **(b)** °empty, leave bare, °clear out, °evacuate: *We must clean out the larder before it can be painted.* **11** ***clean up.*** **(a)** °clean, °cleanse, °wash, °(take a) shower, *Brit* bath; *US* take a bath, bathe, wash up: *Clean up, please, dinner is almost ready. It took us a week to clean up the stables.* **(b)** °purge, °purify, °disinfect, depollute, decontaminate, °clear, sanitize: *The council has led the way toward cleaning up the wetlands of chemical waste.*

cleanse *v.* **1** °clean, absterge, deterge, °wash, °scour, °scrub: *You need a scrubbing brush to cleanse the tub.* **2** °purify, depurate; °purge, °wash away, expiate: *Each prayer repeated has a certain value in cleansing away sin.*

clear *adj.* **1** unclouded, cloudless, °sunny, °fair, sunlit, °fine: *On a clear day, you can see the lighthouse several miles away.* **2** °transparent, limpid, crystalline; translucent, uncloudy, unclouded, pellucid: *The water was clear enough to see the bottom.* **3** °bright, °lustrous, shining, °shiny, sparkling, *Formal* nitid: *We painted the bathroom a lovely clear blue.* **4** °bright, °fresh, unblemished, unscarred: *She has a lovely clear complexion.* **5** °distinct, °sharp, well-defined, °definite; °legible, °readable; °acute, °vivid: *Her notes are written in a clear hand.* **6** understandable, °intelligible, perspicuous, lucid, comprehensible, apprehensible, °discernible, °plain, °obvious, unambiguous, unequivocal, °explicit, °definite, unmistakable, °indisputable, °undisputed, °unquestionable, °incontrovertible: *He has made himself very clear on that point.* **7** °distinct, unclouded, unconfused, °explicit, °plain, °definite, clear-cut, palpable: *I have a clear recollection of her words.* **8** °evident, °plain, °obvious, °patent, °manifest, °apparent: *It has become clear that someone was serious about killing her.* **9** °perceptive, °acute, °sensitive, perspicacious, discerning, °keen: *It was only his clear vision of the situation that saved us all.* **10** °certain, °sure, convinced, °confident, °positive, °determined, °definite, assured: *I am not clear that the subject was a good one.* **11** °pure, unwavering, well-defined, °distinct, clarion, bell-like: *We heard father's clear voice calling from* *below.* **12** °pure, guileless, °unsophisticated, °innocent, °blameless, °faultless; not guilty: *Fame is the spur that the clear spirit doth raise. I still cannot look her in the eye with a clear conscience.* **13** unencumbered, °free, °net: *In our first year we made a clear profit of 25 percent.* **14** °unlimited, °unqualified, unquestioned, °unquestionable, °absolute, °complete, °entire; °pure, °sheer, °perfect: *You must allow three clear days for the ascent.* **15** disengaged, disentangled, unentangled, °free, freed, °rid, °quit, °loose, unencumbered, released: *When the line is clear of any obstruction, hoist sail and let's be off.* **16** °open, unencumbered, °free, unblocked, unobstructed, °unimpeded, °direct: *There is a clear view of the park from here.* —*adv.* **17** brightly, effulgently, radiantly, luminously, lambently: *The stars shone clear in the night sky.* **18** distinctly, °clearly, starkly, perceptibly, discernibly, understandably, prominently: *When you see the reef, sing out loud and clear.* **19** °completely, °utterly, °entirely, cleanly, °clean, °wholly, °totally: *The thief got clear away in the confusion.* —*v.* **20** °clarify, °cleanse, °clean, °purify: *The chemical soon cleared the water of all sediment.* **21** exonerate, absolve, acquit; °excuse, °forgive: *He has been cleared of all charges and released.* **22** Also, ***clear up.*** become fair *or* cloudless *or* sunny: *I hope the weather clears in time for the game.* **23** °open (up), °free; unblock, unclog, unstop; disencumber, dislodge: *We were able to clear a path through the jungle. He cleared his throat and began to speak.* **24** °empty: *Clear the land of trees before farming it.* **25** Also, ***clear away*** *or* ***out.*** °remove, °eliminate, °take; cut away *or* down: *Clear those branches from the paths.* **26** disburden, unburden, °purge, °free, °rid: *He has cleared his conscience of any responsibility in the matter.* **27** leap *or* jump over, vault: *She cleared the fence easily.* **28** Also, ***clear up.*** °settle, °discharge, °pay, °square, °defray, °satisfy: *The company has cleared all its debts.* **29** ***clear off*** *or* ***out.*** °leave, °depart, decamp, go *or* run off, get out, °withdraw, *Slang* °beat it, scram, *Taboo Brit* sod off, *Chiefly Australian* shoot through, *US and Canadian* take a (runout) powder: *I told them to clear off and stop bothering me.* **30** ***clear up.*** **(a)** °eliminate, °remove, °settle; °clarify: *I wish to clear up any misunderstanding between us.* **(b)** °tidy (up), °neaten (up), put *or* set in order: *I'll clear up after dinner.* **(c)** °explain, elucidate, explicate, °clarify, make plain *or* clear, disambiguate: *Please clear up the difference between* imply *and* infer. See also **22, 28,** above. —*n.* **31** ***in the clear.*** °innocent, not guilty; exonerated, forgiven, absolved; unburdened, disburdened, unencumbered, °free: *The other chap confessed, leaving me in the clear.*

clearance *n.* **1** °space, °gap, °hole, °interval, °separation, °room, °margin, °leeway, °allowance: *You must allow a clearance of two millimeters.* **2** °approval, °endorsement, authorization, °consent; °license, °leave, °permission: *He cannot get security clearance with his prison record.*

clearly *adv.* **1** distinctly; starkly, plainly: *With spectacles, I can see everything more clearly.* **2** °evidently, °apparently, °manifestly, °obviously, certainly, °definitely, °positively, unequivocally, unquestionably, incontestably, without doubt, °undoubtedly, indubitably, demonstrably, °absolutely, °utterly: *His statement is clearly untrue.* **3** audibly, distinctly, understandably: *I wish she would speak more clearly.*

cleave *v.* °split, °divide, °cut, cut *or* chop *or* hew in two *or* asunder, bisect, halve, °separate, °slit, rive: *With a mighty blow the log was cleaved cleanly in two.*

clergyman *n.* **1** clergywoman, ecclesiastic, churchman, churchwoman, cleric, reverend, °divine, man of the cloth, holy man, °priest, °minister, chaplain, °father, rabbi, °pastor, parson, rector, vicar, dean, canon, presbyter, prebendary *or* prebend, deacon, sexton, sacristan, guru, ayatollah, imam: *In his black frock coat he looked like a clergyman.* **2** °monk, friar, °brother, monastic, °religious: *A clergyman was responsible for his religious education.* **3** °preacher, gospeler, evangelist, revivalist, °missionary, sermonizer: *She was moved by the clergyman's sermon.*

clerical *adj.* **1** ecclesiastical, churchly, °pastoral, sacerdotal, °priestly, hieratic, rabbinical, ministerial, monastic, apostolic, prelatic, papal, pontifical, episcopal, canonical: *He was wearing his clerical vestments.* **2** white-collar, °office, °professional, secretarial, stenographic, accounting, bookkeeping: *He runs the shop and his wife has the clerical responsibilities.*

clever *adj.* **1** skilled, °talented, °skillful, adroit, °dexterous, °gifted, °agile, °quick-witted, °intelligent, °perceptive, discerning, °sharp, sharp-witted, °adept, °able, *Colloq* brainy: *Be good, sweet maid, and let who will be clever.* **2** °shrewd, cunning, guileful, canny, °artful, °crafty, °sly, °wily, °foxy: *Kerleigh thinks he is very clever getting insider information and then buying up the shares.* **3** °intelligent, °wise, °sage, sagacious; °ingenious, °original, °resourceful, Daedalian, inventive, °creative, °smart, °imaginative: *It was very clever of her to memorize the boxing record book for the competition.* **4** deft, adroit, nimble-fingered, °dexterous, °handy, °skillful: *The old woman is clever with her hands.*

cliché *n.* stereotype, bromide, trite saying, old saw *or* maxim, °truism, platitude, commonplace, banality, *Colloq* chestnut: *The writing is so full of clichés that it is hard to discern an original thought.*

client *n.* °customer, °patron, shopper; °patient: *She opened a law office and already has four clients.*

clientele *n.* clients, patrons, customers; °custom, °business, °trade, °patronage, following: *What sort of clientele do you expect to attract?*

cliff *n.* °precipice, °bluff, escarpment, scarp, °crag, rock face, cuesta, scar *or Scots* scaur: *The commandos are trained to scale a hundred-foot cliff.*

climate *n.* **1** °weather, *Literary* clime: *We are retiring to the Maldives because we like a sunny climate.* **2** °atmosphere, ambiance *or* ambience, °air; °feeling, °mood, °aura, °milieu, °feel: *In the present climate of opinion, we'd best delay introducing the bill.*

climax *n.* **1** culmination, °height, °acme, apex, °summit, °zenith, apogee, °peak, high point, °maximum, supreme moment: *The war reached its climax at the battle of Arbela.* **2** turning point, °crisis, crossroads: *The climax of the play occurs in the third act.* **3** orgasm: *She told her psychiatrist that she had never reached a climax with her husband.* —*v.* **4** culminate, °peak, °crest, come to a head: *The week's events climaxed with the presentation of the gold medal.*

climb *v.* **1** Also, ***climb up.*** °mount, ascend, °go up, °scale, °shin (up), clamber up, *US* shinny (up): *In one of the games we had to climb a greased pole. Two Japanese teams have climbed Mount Everest.* **2** °creep, °trail, °twine; °grow: *The ivy has climbed all over the garden wall.* **3** °rise, °arise, ascend, °go up, °mount; °advance: *Watch the smoke climb into the sky.* **4** ***climb along.*** °creep, °edge, clamber, °crawl, inch: *The cat burglar climbed along the ledge till he reached the window.* **5** ***climb down.*** **(a)** °descend, °go down: *We shall need a rope to climb down from here.* **(b)** Usually, ***climb down from.*** °retreat (from), °withdraw (from), back away (from), °give up, °abandon, °renounce: *He has climbed down from his earlier position regarding women in the priesthood.* —*n.* **6** °grade, °incline, °route, °pitch; ascent; descent: *It was a steep climb to Camp Four.*

clinch *v.* **1** °secure, °settle, °confirm, °determine, conclude, °dispose of, °complete, °wind up, °finalize, *Colloq* sew up: *He clinched the argument by resigning.* —*n.* **2** close quarters, °hug, °clasp, °embrace; °cuddle: *The boxers went into a clinch to regain their breath.*

clincher *n.* finishing touch, °payoff, punch line, coup de grâce, final *or* crowning blow: *Agreement on the date of completion proved to be the clincher, and we started work at once.*

cling *v.* **1** °stick, adhere, °attach, °fasten, °fix: *The detectives found that one of the victim's hairs had clung to the suspect's lapel.* **2** °favor, be *or* remain devoted *or* attached to, °embrace, hang onto, °retain, °keep, °cherish: *He still clung to his old-fashioned notions of honor.* **3** ***cling together*** *or* ***to one another.*** °embrace, °hug, cleave to one another, clasp one another, clutch one another, hold (fast) to one another, grasp one another: *The children clung together in the darkness.*

clip[1] *v.* **1** °clasp, °fasten, °fix, °attach, °hold, °clinch; °staple: *Please clip these papers together.* —*n.* **2** °clasp, fastener: *That clip isn't strong enough to hold all these papers.*

clip[2] *v.* **1** °trim (off), °lop (off), °cut (off), crop, bob, °snip: *The barber clipped my hair short. Roger has clipped two seconds off the record.* **2** °shorten, °reduce, °abbreviate, °diminish, °cut (short): *The film was clipped by fifteen minutes for television. In her rapid-fire way of speaking, she clipped each word.* **3** °strike, °hit, °punch, smack, °box, cuff, whack, *Colloq* wallop, clout; *Slang* sock: *He was clipped on the jaw and knocked out.* **4** °cheat, °swindle, bilk, overcharge, *Slang* rook: *Using the old Spanish handkerchief scam, they clipped him out of a week's wages. She was clipped for a 10 percent "service fee."* —*n.* **5** °segment, °interval, °section, °part, °portion, °extract, °cutting, °excerpt, °cutting, °bit, snippet, °scrap, °fragment: *We were shown a film clip of the cheese-making process.* **6** °blow, cuff, °punch, °hit, °strike, smack, whack, °box, *Colloq* wallop, clout; *Slang* sock: *She gave him a clip at the side of the head and he went down.* **7** °pace, °rate, °speed: *I was riding along at a good clip when my horse shied and I was thrown off.*

clique *n.* °set, coterie, °crowd, °circle, °group: *The usual clique from the sales department were crowded around the bar.*

cloak *n.* **1** °mantle, °cape, °robe, °wrap, poncho; °coat, overcoat: *She pulled her cloak around her in the chill night air.* **2** °mantle, concealment, °cover, °screen, °shroud, °veil: *He stole away under the cloak of darkness.* —*v.* **3** °conceal, °hide, °mask, °screen, °veil, °shroud, °cover up; °disguise: *All of the spies' activities were cloaked in secrecy.*

clod *n.* **1** °lump, °mass, °gob, °wad, hunk, chunk; piece of sod *or* turf; *Colloq* glob: *A clod of earth was stuck between the spikes of my golf shoes.* **2** idiot, °fool, °dolt, blockhead, simpleton, dunce, °dope, oaf, lout, ass, °boor, °clown, ninny, ninnyhammer, bumpkin, clodhopper, *Slang* vegetable, *US and Canadian* °jerk: *He felt like a clod standing there with no idea which way to go.*

clog *v.* °hamper, °encumber, °impede; °obstruct, °choke (up), °block, congest, °jam: *The road was clogged with returning vacationers. A piece of orange peel had clogged the sink drain.*

close *v.* **1** °shut, °close up, °seal; °close off, °lock, padlock, °secure, °fasten: *I closed my eyes. Please close the door behind you.* **2** make inaccessible, °shut, *Chiefly US* place off limits: *The Bodleian Library will be closed for a week.* **3** conclude, °end, °finish, °complete, bring to a close *or* end, °terminate, °climax, *Colloq* °wind up: *A brilliant flourish closes the first movement of the symphony.* **4** conclude, °sign, °seal, °make, °settle, °clinch, °agree, °arrange, °work out, °establish: *Union and management closed a deal, and the strike was called off.* **5** Also, ***close down.*** °discontinue, °terminate, °stop, °suspend, °shut down, go out of business, cease operation(s), °close (up), *Colloq* °wind up, shut up shop, put up the shutters: *Competition from the supermarket forced the greengrocer to close down.* **6** Also, ***close off.*** °seal, make inaccessible, °shut (off), °obstruct, obturate: *This wing of the museum has been closed off temporarily.* **7** ***close one's eyes to.*** °ignore, °overlook, °disregard: *You have always closed your eyes to any of his faults.* **8** ***close up.*** **(a)** °close, °shut (up), °lock up, °close (down): *It's time to close up for the night.* **(b)** °close, come *or* draw *or* bring together, °unite, °join; °connect: *We closed up ranks and stood at attention.* —*adj.* **9** °near; adjacent, proximate, proximal: *He claims to have had a close encounter with an extraterrestrial. There certainly is a close resemblance between Kathy and her daughter.* **10** closed, °shut (up), °fixed, °fast, °secure, °tight: *The hostages spent a month in close confinement.* **11** °dense, °compact, °tight, °cramped, compressed, °tiny, minuscule, °minute: *I could hardly read the close writing on the matchbox.* **12** °stuffy, °musty, °stale, fusty, confining, °oppressive, airless, unventilated, confined, stifling, suffocating: *They locked me in a room that was so close I could hardly breathe.* **13** nearly equal *or* even, close-matched, neck and neck, °tight: *It was a close race, but Flanagan won by a hair.* **14** °careful, assiduous, °precise, °detailed, concentrated, °strict, rigorous, °minute, searching, °attentive, °alert, °intent, °intense, °thorough, °painstaking: *Close analysis has revealed that the handwriting is yours, Mrs. Gridley.* **15** °attached, °intimate, °devoted, °familiar, inseparable, close-knit, °solid, °confidential; °fast; *Colloq* °thick, thick as thieves, pally, *US and Canadian* palsy-walsy, buddy-buddy: *They are a very close family. She and her father are very close.* **16** °private, °privy, °secret, °guarded, closely guarded, °confidential: *What was to have been a close secret has been leaked to the press.* **17** °secretive, °reticent, °taciturn, °reserved, close-mouthed, °tight-lipped, °silent: *She is very close about the whereabouts of her husband.* **18** stingy, °mean, °miserly, niggardly, tightfisted, closefisted, parsimonious, °penurious, penny-pinching, cheeseparing, Scroogelike, skinflinty, *Colloq* °near, *Brit* mingy: *He's so close he charges his own mother rent.* **19** °secluded, concealed, shut up *or* away, °hidden: *She advised the fugitives to lie close till nightfall.* —*adv.* **20** °near, in the neighborhood (of), not far (from), adjacent (to); alongside; °at hand, °nearby, °close by: *The murder took place close to my house. I'm frightened, so please stay close by.* **21** ***close to*** *or* ***on*** *or* ***onto.*** °nearly, °almost, °about, °practically, °approximately, nigh onto, approaching: *For close to two thousand years the site lay untouched.* —*n.* **22** °end, °termination, conclusion, °finish, °completion, cessation; culmination: *At the close of trading, the market had gained two percent.*

cloth *n.* **1** °fabric, °material, textile, *Chiefly Brit* °stuff: *The curtains are made of cloth, not plastic.* **2** ***the cloth.*** the clergy, the (religious) ministry, the priesthood: *He is a man of the cloth.*

clothe *v.* **1** °dress, attire, garb, °apparel, °outfit, fit out *or* up, accoutre *or US also* accouter, *Brit* kit out *or* up, *Colloq* tog up *or* out: *He earns barely enough to clothe and feed his family.* **2** endow, °invest, caparison, endue: *They tried to clothe their transactions with the raiment of honesty.*

clothes *n.* pl. clothing, °apparel, attire, °wear, °dress, °garments, raiment, °wardrobe, °outfit, °ensemble, vestment(s), *Old-fashioned* duds, *Colloq* togs, °gear, get-up, *Slang* glad rags, *Brit* clobber; *Slang US* (set of) threads: *Put on some old clothes and make yourself comfortable.*

clown *n.* **1** jester, °fool, °zany, °comic, °comedian, funnyman: *Of all the performers at the circus, I like the clowns best.* **2** buffoon, °boor, °rustic, °yahoo, oaf, lout, °clod, °dolt, bumpkin, clodhopper, °provincial, °peasant, yokel, *Colloq* lummox; *Slang chiefly US* °jerk; *Old-fashioned* galoot *or Old-fashioned* galloot; *Slang chiefly US and Canadian* hick: *That's the kind of language we expect to hear only from the most ignorant clowns.* —*v.* **3** Often, ***clown around*** *or* ***about.*** °fool (around), play the fool, horse around *or* about, °caper, cut a caper *or* capers, engage in high jinks *or* hijinks, *US* °cut up, cut didos: *Stop clowning around with that hose and help water the garden.*

club *n.* **1** cudgel, bat, bludgeon, mace, billy, truncheon, baton, °staff, °stick; cosh, *Chiefly US and Canadian* blackjack: *That blow from the club required six stitches.* **2** °association, °society, °organization, °fraternity, sorority, °fellowship, °brotherhood, sisterhood, °federation, °union, guild, °lodge, °alliance, °league, °order, consortium, °company: *Our sailing club holds an annual race.* **3** clubhouse: *The society's club is near Pall Mall.* **4** nightclub, °cabaret, *Colloq* nightspot: *After the theater, we stopped at Oscar's club for a nightcap and to see the new acts.* —*v.* **5** °beat, cudgel, bludgeon, bat, °belabor; °lambaste, baste, thrash, trounce: *The guerrillas caught the traitor and clubbed him to death.* **6** Often, ***club together.*** band *or* join *or* league (together), °team (up), join forces, °combine, °ally, °associate, confederate, °cooperate: *We clubbed together to purchase the antique clock.*

clue *n.* **1** °hint, °suspicion, °trace, intimation, °suggestion, °inkling, °indication, °pointer, °lead, °tip, tip-off, °evidence, °information, °advice; °key, °answer, °indicator: *There is no clue pointing to anyone in particular. Any clue to the solution of the mystery disappeared in the fire.* —*v.* **2** ***clue someone in*** *or Brit also* ***up.*** °hint, °suggest, °imply, °intimate, °inform, °advise, °indicate: *She clued us in as to who might have sent the letter.*

clump *n.* **1** °lump, °mass, °clod, chunk, hunk, °wad, °gob, *Colloq* glob: *As the soup cooled, clumps of fat formed on its surface.* **2** °bunch, °cluster; °thicket, copse; wood, *Chiefly Brit* spinney: *The deer disappeared into a clump of trees. That clump will provide good cover for us.* —*v.* **3** °lump, °mass, °heap, °collect, °gather, °bunch, °pile: *She wore her hair clumped on top of her head.*

clumsy *adj.* °awkward, ungainly, °unwieldy, °ungraceful, gawky, maladroit, unhandy, unskillful, °inept, bungling, bumbling, cloddish, oxlike, bovine, uncoordinated, lubberly, oafish; gauche, *Colloq* butterfingered, ham-fisted, ham-handed, cack-handed: *He made a clumsy attempt to put the key in the lock. She gave a clumsy excuse for being in the bank after hours.*

cluster *n.* **1** °collection, °bunch, °clutch, tuft, °bundle: *Notice that cluster of flowers near the top of the plant.* **2** °collection, °bunch, °group, °knot, °body, °band, °company, °gathering, °crowd, °assembly, congregation, °throng, °flock, assemblage, °swarm: *A cluster of well-wishers stood talking with the minister.* —*v.* **3** °collect, °bunch, °group, °band, °gather, °crowd, congregate, °throng, °assemble, °accumulate, °mass, aggregate: *A number of people clustered round the new sculpture. Why do these flowers cluster at this tree?*

clutch *v.* **1** °seize, °snatch, °grab, °grasp, take *or* lay hold of; °hold; *US* °snag: *She clutched feebly to the rope before losing her grip and plunging into the abyss below. He clutched the child to his bosom.* —*n.* **2** ***clutches.*** **(a)** grasp at, °hold; °embrace: *We watched as the gazelle deftly eluded the cheetah's clutches.* **(b)** °influence, °control, °power, °domination, dominance, °possession: *Once again she has managed to avoid the clutches of the evil Dr. Fu Manchu.*

clutter *n.* **1** °mess, °litter, °jumble; °mishmash, olla podrida, °confusion, °hash, gallimaufry, hotchpotch *or US also* hodgepodge, °muddle, farrago, °medley: *I must insist that you clear up the clutter in your room at once. That philosophy is a clutter of competing ideas.* **2** °confusion, °tangle, °chaos, disarray: *The old part of town is a clutter of narrow, crooked, dark, and dirty lanes.* —*v.* **3** Often, ***clutter up.*** °mess up, °litter, °strew, make a shambles of: *Please don't clutter up my desk with newspaper clippings.*

coach *n.* **1** °carriage, bus, omnibus, motorcoach: *The sightseeing coach broke down near Exeter.* **2** °tutor, trainer, °instructor, °teacher, mentor, *Brit* crammer: *Her voice coach says she's not yet ready for grand opera.* —*v.* **3** °tutor, °train, °instruct, °guide, °direct, °drill, °prepare, °prompt, °school, °exercise, *Brit* °cram: *Her lawyers have coached her in what to say in court.*

coagulate *v.* congeal, gel, °jell, jellify, clot, curdle, °set: *The white of the egg had coagulated, but the yolk was still runny.*

coarse *adj.* **1** °rough, uneven, °scratchy, °prickly, bristly; °crude, rough-hewn, unfinished, °unrefined: *He has a three-day coarse growth of beard. The surface on the furniture is still too coarse.* **2** °rude, °boorish, loutish, °crude, °ill-mannered, unpolished, °rough, uncouth, °impolite, uncivil, °unrefined: *Nigel's behavior is coarse, and he spits when he talks.* **3** °rude, °indecent, °improper, indelicate, °obscene, °lewd, °vulgar, °gross, smutty, °dirty, °filthy, °foul, °offensive, °lascivious, ribald, °bawdy; foulmouthed: *Don't use that kind of coarse language around me.* **4** °inferior, low-quality, second-rate, °shoddy, °tawdry, trashy; kitschy: *That shop stocks only the coarsest merchandise.*

coast *n.* **1** seaside, seashore, shore, °seacoast, strand, °beach, littoral, coastline, seaboard: *Be careful of rocks if you sail near the coast.* —*v.* **2** °glide, °skim, °slide, °sail: *The children coasted down the hill on the toboggan.*

coat *n.* **1** overcoat, greatcoat; jacket, anorak, parka, *Brit* cagoule *or* kagoul *or* kagoule, *Colloq Brit* cag: *Put on your coat, it's cold out.* **2** coating, layer, covering, overlay; °film: *Two coats of paint ought to be enough. There was a coat of dust on everything.* —*v.* **3** °cover, °paint, °spread: *We coated the floor with three layers of varnish.*

coax *v.* °persuade, °urge, °wheedle, °cajole, °beguile, °charm, inveigle, °jolly, °manipulate: *She coaxed me into spending a weekend with her at Blackpool.*

cocky *adj.* °overconfident, °arrogant, °haughty, °conceited, °self-important, °egotistical, °proud, °vain, prideful, cocksure, saucy, °cheeky, °brash: *Since winning the beauty contest, Claire has been entirely too cocky.*

coddle *v.* °pamper, baby, cosset, mollycoddle, °indulge, °humor, °spoil, *Brit* cocker: *Give him a cold bath and don't coddle him so much.*

code *n.* **1** law(s), regulation(s), rule(s), jurisprudence, jus canonicum (= 'canon law'), jus civile (= 'civil law'), jus divinum (= 'divine law'), jus gentium (= 'universal law'), jus naturale (= 'natural law'), corpus juris (= 'body of law'), pandect, lex, non scripta (= 'common law, unwritten law'), lex scripta (= law'): *In the present code there is no statute that forbids keeping a pet gnu.* **2** cipher *or* cypher, cryptogram: *Our agents send all their messages in code.* **3** °system, practice(s), convention(s), standard(s), criterion (criteria), principle(s), rule(s), maxim(s), custom(s), pattern(s), °structure, tradition(s), °organization, °protocol, orthodoxy: *Our code of behavior is completely foreign to the islanders.* —*v.* **4** encode, encipher *or* encypher, encrypt: *It took an hour to code the information.*

coffin *n.* °casket, °pall, (pine) °box; sarcophagus: *The coffin was slowly lowered into the grave.*

cog *n.* **1** tooth, geartooth, sprocket, ratchet: *Stripping the gears means breaking the cogs off them.* **2** underling, °pawn, °subordinate, nonentity, °zero, cipher *or* cypher, °nothing, °nobody, small fry: *He's only a small cog in the organization--we're after the big wheel himself.*

cognizance *n.* °knowledge, awareness, °knowledge, °perception, °notice, consciousness, mindfulness: *They ran the gambling den with the full cognizance of the police.*

coherent *adj.* **1** °consistent, °orderly, organized, well-organized, °logical, °rational, °reasonable, well-ordered: *The representative from our district set forth a coherent argument against an excise tax.* **2** understandable, °sensible, comprehensible, °intelligible, articulate, lucid, °clear: *He was so frightened and hysterical that he was unable to tell a coherent story.*

cohort *n.* **1** troop, °squad, squadron, °platoon, brigade, °unit, cadre, wing, legion, °detachment, contingent: *Ten select Roman cohorts were sent against the Mitanni.* **2** °company, °band, °group, °faction, °set, °body, °corps: *She was a member of a small cohort of suffragettes.* **3** °companion, confederate, °accomplice, °associate, °fellow, °comrade, °friend, confrère: *Grimsby arrived with a few of his cohorts.*

coil *v.* **1** °wind, °twist, °coil, °snake, °wrap, enwrap, °spiral, *Nautical* fake *or Nautical* flake (down): *Coil the rope round the bitts.* —*n.* **2** winding(s), circle(s), °loop, whorl, °spiral, helix, °twist: *His foot caught in the coil of rope and he was carried overboard.*

coin *n.* **1** specie, °money, currency; °change, °cash, °silver: *That is a real half-crown, genuine coin of the realm. I have only a few coins in my pocket.* —*v.* **2** °mint, °stamp: *The US government has stopped coining silver dollars.* **3** °invent, °create, °conceive, °originate, °start, °make up, °fabricate, °frame, concoct, think *or* dream up: *Joyce coined the word "quark."* **4** ***coin money.*** earn *or* make money, become wealthy, enrich oneself, *Colloq* °rake it in: *Those rock stars really coin money from their record sales.*

coincide *v.* fall *or* come *or* go together, °line up, co-occur, °correspond, synchronize, °match, °tally, °agree, °(be in) accord, °equal, jibe: *This year, Easter and Passover coincide. The southern boundary coincides with the Thames.*

coincidence *n.* **1** co-occurrence, simultaneity, correspondence, concurrence, consistency, contemporaneity, synchronism, synchrony, coextension, coevality, coinstantaneity: *The coincidence of twelve by the clock with noon by the sundial is exact only four times in the year.* **2** congruence, °matching, jibing, °agreement, concord, °accord, °harmony, accordance, conformity, congruity, consonance, concomitance: *Brewster noted the coincidence between the religious views of Newton with those of Locke.* **3** chance occurrence, °fluke, °chance, °accident, °luck, fortuity, fortuitousness, *US and Canadian* happenstance: *By sheer coincidence, I met my next-door neighbor from Croydon first in Bangkok and then again in Perth.*

coincidental *adj.* °chance, lucky *or* unlucky, fortuitous, °accidental, unexpected, unpredicted, unpredictable, °unforeseen: *The earthquake, the flood, and your purchase of the house were purely coincidental, you say.*

cold *adj.* **1** °chill, °chilly, frosty, °icy, °keen, nippy, °freezing, °frigid, ice-cold, stone-cold, °bitter (cold), °raw, °biting, biting-cold, numbing, gelid; °wintry, hibernal, brumal; arctic, glacial, °polar, hyperborean *or* hyperboreal, Siberian: *It was so cold that the canal had completely frozen over.* **2** °chilly, chilled; unheated, heatless: *The room is cold; turn on the heat.* **3** °indifferent, apathetic, °chilly, chilling, °cool, °icy, °dispassionate, °unsympathetic, °aloof, unresponsive, spiritless, °frigid, unfriendly, uncordial, °lukewarm, °frigid; °coldblooded, insensitive, uncaring, unemotional, undemonstrative, °reserved, °unmoved, spiritless, °callous, °remote, °distant, stand-offish, °unapproachable, stonyhearted, emotionless, unfeeling, °coldhearted: *My ideas received rather a cold reception. Because she had offended him, he was quite cold to her.* **4** depressing, cheerless, chilling, °gloomy, dispiriting, deadening, disheartening, °bleak, °dismal, discouraging: *The sweat stood out on his brow in cold apprehension.* **5** unmoving, °stale, trite, stereotyped; °dead: *The coldest word was once a glowing new metaphor.* **6** °weak, °faint, °stale, °old, °dead: *The trail of the tiger had grown cold.* **7** °unprepared, unready: *She hadn't studied and went into the exam cold.* **8** Often, ***getting cold.*** °far, °distant, °remote, off the track: *As I searched for the weapon, I felt I was getting cold the further I went from the night table.* —*n.* **9** coldness, frigidity, iciness: *Last winter, the cold killed off many of our shrubs.* **10** head *or* chest *or* common cold, influenza, ague, (the *or* la) grippe, *Technical* coryza, gravedo, *Colloq* sniffles, the flu, °bug, sneezles and wheezles: *I caught a cold waiting for you in the rain.* —*adv.* **11** °completely, °thoroughly, °entirely, °absolutely, unhesitatingly, °promptly, °immediately, unreservedly, abruptly: *His application to join the police was turned down cold.*

coldblooded *adj.* **1** *Technical* poikilothermic *or Technical* poikilothermal: *Reptiles are coldblooded.* **2** unexcited, unemotional, °cool, unimpassioned, unfeeling, °callous, °thick-skinned, insensitive, °heartless, uncaring, °stony, °steely, stonyhearted, °coldhearted, imperturbable, °unmoved, °indifferent, unresponsive, °unsympathetic, apathetic, °dispassionate: *Carefully planned for weeks, the murder was an act by a coldblooded killer.* **3** °cruel, °brutal, °savage, °inhuman, barbarous, °vicious, barbaric, °merciless, pitiless, °ruthless: *They were victims of the coldblooded policies of the Nazi extermination camps.*

coldhearted *adj.* insensitive, °unsympathetic, apathetic, °indifferent, unfeeling, uncaring, °callous, °thick-skinned, °cold, °cool, °frigid, hardhearted, °heartless, °unkind, °thoughtless, unthoughtful, uncharitable, °ruthless, pitiless, °unmerciful, °cruel, °merciless, °mean: *Putting that kitten out on a snowy night is the most coldhearted thing you ever did.*

collaborate *v.* °cooperate, °join (forces), work together, °team up: *We collaborated in writing both the lyrics and music.*

collapse *v.* **1** fall (down *or* in *or* apart), °crumple, °cave in, deflate, °crumble, tumble down, °break down, °go: *When he opened the valve, the balloon collapsed. Hundreds of buildings collapsed in the earthquake.* **2** °fail, °(come to an) end, °fall through, °peter out, °disintegrate, °dissolve, °fall flat, °founder, come to naught *or* nought, break up *or* down; °decline, °diminish; °disappear, °evaporate, go up in smoke, go bankrupt, °go under, *Brit* °go to the wall, *Colloq* °fizzle out: *After the imprisonment of their leader, the entire movement collapsed. Owing to the recession, many businesses collapsed.* **3** °pass out, °faint, °drop, *Colloq* keel over; *Old-fashioned or literary* swoon: *He collapsed onstage and they took him to his dressing room.* **4** °break down (mentally), have a (nervous) breakdown, °go to pieces, come *or* fall apart, *Colloq* crack up, *US also* °crack: *Between overwork and lack of sleep she finally collapsed and is now in a sanatorium.* —*n.* **5** cave-in, °breakdown: *The collapse of the house was attributed to termites.* **6** °failure, °downfall, °ruin; disappearance, disintegration, °dissolution, bankruptcy: *Will he be able to survive the collapse of his financial empire?* **7** °(mental) breakdown, °prostration, *Colloq* crackup: *He suffered a mental collapse when his family was killed in a car crash.*

colleague *n.* teammate, fellow worker, coworker; °associate, °comrade, °ally, confrère, °mate, consociate, *Chiefly Brit and Australian* °mate, *US* buddy: *I have asked some of my colleagues from the office to join us for dinner.*

collect *v.* **1** °gather (together), get *or* bring *or* come *or* together, °amass, °accumulate, °assemble, °compile, °pile up, heap up, rack up; convene, congregate, °converge, °rally, °meet: *You have not yet collected enough examples of so-called victimless crimes. A crowd had collected outside the mayor's home.* **2** °summon (up), °draw (up), °muster, °gather (up), °concentrate: *She collected all her courage to ask for an increase in salary.*

collected *adj.* °calm, °serene, controlled, °cool, °sedate, composed, °nonchalant, °poised, unruffled, unperturbed, at ease, °comfortable, °tranquil, unexcited; imperturbable; °confident: *Considering what she's just gone through, Tanya seems quite collected.*

collection *n.* **1** collecting, °gathering, solicitation, garnering, gleaning, °accumulation, amassment, aggregation, *Colloq Brit* whip-round: *The collection of donations in this neighborhood is going well.* **2** °accumulation, °hoard, °store, assemblage, omnium-gatherum; anthology, chrestomathy: *Would you like to come up to see my collection of etchings? They have published some very interesting collections.*

collector *n.* gatherer, accumulator; connoisseur, art lover: *The rent collector is coming tomorrow. We are collectors of paintings by unknown artists.*

collide *v.* **1** °crash, strike *or* dash together: *The cars collided at the bridge.* **2** ***collide with.*** crash into, smash into, °run into, °bump into, smack into: *The car collided with the bus at the crossing.*

collision *n.* smashup, smash, °crash, °wreck, °pileup, *Colloq Brit* prang; *US* crackup: *There has been a major collision on the Tring road.*

colossal *adj.* **1** °huge, °vast, °enormous, °gigantic, °giant, mammoth, °massive, gargantuan, cyclopean, Brobdingnagian, °immense, °monumental, titanic, herculean, elephantine, °jumbo: *Mystery surrounds the exact methods used in moving the colossal stones used in the pyramids.* **2** spectacular, stupendous, wonderful, awe-inspiring, staggering, °extraordinary, °incredible, °overwhelming, °unbelievable: *The old Hollywood extravaganzas were described by press agents as "colossal." I think you have made a colossal mistake in failing to hire Cynthia.*

color *n.* **1** °hue, °tint, tincture, °shade, tone, cast, tinge, pigmentation; pigment, dye: *The colors of the curtains don't match the wall.* **2** ***colors.*** **(a)** °flag, ensign, °standard, °pennant, °banner, burgee: *The sloop hoisted the British colors.* **(b)** °device, badge, °emblem, insigne *or pl.* insignia, symbol(s), °identification; °identity, °appearance, °face; loyalties: *The investigators found he'd been operating under false colors. She has shown her true colors at last.* —*v.* **3** °tint, dye, °stain, °paint, crayon, tincture, tinge; pigment: *They will stand out better if you color these sections red.* **4** °influence, °affect, °distort, °falsify, °taint, °warp, °twist, °slant, °pervert, °bias: *Jealousy colors his opinion of his supervisor.* **5** °blush, redden, °flush: *After their affair, she visibly colored whenever they met.* **6** °falsify, °distort, °misrepresent, °disguise, °mask, °conceal: *He feigns confusion when he wishes to color his true feelings.*

colorless *adj.* **1** °pale, pallid, blanched, °white; °wan, ashen, sallow, waxen, °sickly, °washed-out: *You could see from his colorless complexion that he had not been outside for months.* **2** °dull, °drab, uninteresting, vacuous, °vapid, °lifeless, °boring, °tedious, spiritless, °dry, dry as dust, °dreary, characterless, insipid, °bland, namby-pamby, °lackluster, uninspiring, uninspired: *She has led a colorless life. Few people have so colorless a personality as he.*

combat *n.* **1** °fight, °encounter, °engagement, duel, °battle, °conflict, °war, warfare; °skirmish: *The difference would be decided by a combat between thirty men from each clan.* **2** °struggle, °contest, °strife, °controversy, °dispute, °quarrel, °disagreement, altercation, °vendetta, °feud: *The combat between the Hatfields and the McCoys was finally settled.* **3** °opposition, °difference, confrontation: *The combat between good and evil can never end.* **4** °action, fighting, °battle, °war: *He was in combat on three occasions. Have you seen any combat?* —*v.* **5** °fight, °(do) battle, °war, °clash, contend, duel, joust, °wrestle, come to blows, °spar, °grapple (with): *The soldiers are combating hand to hand in the trenches.* **6** °fight, struggle *or* strive against, °contest, °oppose, °defy, enter the lists against, °withstand: *He is remembered for having combated prejudice all his life.*

combination *n.* **1** °union, conjunction, °mixture, °mix, grouping, °set, array: *They always serve the same combination of foods.* **2** °association, °alliance, coalition, °union, °federation, confederation, °combine, syndication, °syndicate, consortium, °trust, bloc, cartel, °party, °society, °organization, °league, °cabal, °conspiracy, °clique; claque: *When they get together they form an unbeatable combination.* **3** °mixture, °amalgam, °compound, compounding, °mix, °alloy, conglomerate, conglomeration, aggregate, aggregation, °amalgamation, °blend, emulsion, °suspension, colloid, °solution, °composition, *Technical* parasynthesis, parathesis; mosaic, °patchwork: *From a combination of ingredients the witch made a slimy love potion. A combination of every color of the rainbow covered the walls.*

combine *v.* **1** °unite, °unify, °join, °connect, °relate, °link, conjoin, °band, °ally, °associate, °integrate, °merge, °pool: *Combine forces, and we'll win.* **2** °blend, °mix, °amalgamate, °mingle, consolidate, °compound, °incorporate, put together: *Combine the water, butter, and salt in a saucepan.* **3** °blend, °fuse, synthesize, °bind, °bond, °compound, °unite, coalesce, come together, commingle, °mingle: *When heated, the silver combines with the chlorine.*

come *v.* **1** °approach, °advance, °(draw) near, °move, *Archaic and literary* draw nigh: *The car came toward us. She has come to me for comforting words. Winter is coming.* **2** °arrive, °appear, make *or* put in an appearance, *Colloq* blow in, °report (in), turn *or* show up, °check in, sign in, clock on *or* in, °roll in: *Winter has come. When Cora comes, we'll ask her.* **3** °enter: *Come into the light, where I can see you.* **4** ***come about.*** **(a)** °occur, °happen, °take place, °come up; befall, *Loosely* °transpire: *I cannot imagine how this state of affairs came about.* **(b)** *Nautical* °tack, °go about: *After the marker, come about and hoist the spinnaker.* **5** ***come across.*** **(a)** °find, °discover, °encounter, meet (up *or* up with), run across *or* into, happen *or* chance upon *or* on, hit *or* light on *or* upon, stumble upon *or* on, *Colloq* °bump into: *I came across some information about Charles.* **(b)** °pay (up), °settle; °yield, °give up, °submit: *Frank owes me money but refuses to come across.* **(c)** be communicated *or* understandable, °penetrate, °sink in: *I am not sure that my points came across.* **6** ***come along.***

°fare, do, °progress, move along: *How is William coming along at his new school?* 7 ***come apart.*** °disintegrate, °crumble, fall *or* fly to pieces, °separate, break (apart *or* up *or* down): *The carburetor came apart in my hands.* 8 ***come at.*** °attack, °assault, °charge, °rush (at), fly at, descend upon *or* on, *Colloq* go *or Colloq* make for: *She came at me waving her umbrella.* 9 ***come by.*** **(a)** °acquire, °obtain, °get, °procure, °secure, °find, take *or* get possession of, get *or* lay hold of, get *or* lay *or* put (one's) hands *or US also* fingers on; be given: *The police wanted to know how she came by such a valuable ring.* **(b)** °win, °earn, attain; be awarded: *I came by that trophy fair and square.* 10 ***come clean.*** See **clean, 8,** above. 11 ***come down on*** *or* ***upon.*** pounce on *or* upon, °rebuke, °criticize, revile, °reprimand, bear down on, °blame: *Mother really came down on us when she discovered who had taken the pie.* 12 ***come down with.*** succumb to, °contract, °catch, be stricken *or* afflicted with, °acquire: *He's come down with pneumonia.* 13 ***come in.*** **(a)** °win, °succeed; *Colloq* °finish (in the money): *My horse came in.* **(b)** be, °prove, turn out *or* prove to be: *Knowing someone on the council can come in handy.* **(c)** °finish, end up, °arrive: *Donald came in first in the backstroke.* **(d)** °enter: *Don't come in, I'm dressing.* 14 ***come off.*** °occur, °happen, °come to pass, °take place, *Loosely* °transpire: *I doubt that the performance will ever come off.* **(a)** °emerge, result as: *We came off the winners in Saturday's game.* 15 ***come out.*** **(a)** be revealed, become public *or* common knowledge, become known, get about *or* around, get *or* leak out, °emerge: *The story has come out that he tried to bribe the inspector.* **(b)** be published *or* issued *or* produced *or* distributed, be shown, be in print, °première: *The new edition of the dictionary has just come out.* **(c)** °end, conclude, °turn out, °terminate, °finish: *How did the chess match come out?* 16 ***come over.*** **(a)** °go over, °communicate, °come across, be communicated, °succeed, be received: *How did my speech come over?* **(b)** °affect, °influence, °possess: *I can't imagine what's come over Louis.* **(c)** °visit, drop *or* stop by *or* in: *Quentin and his wife came over for dinner last night.* 17 ***come through.*** **(a)** °recover (from), °recuperate (from), get well *or* better: *He came through his operation with flying colors.* **(b)** conclude *or* end (up) *or* finish *or* wind up successfully *or* satisfactorily, °succeed, °arrive, not fail *or* disappoint: *I knew he'd come through.* 18 ***come to.*** **(a)** °amount to, add up to, °total, aggregate: *My bill came to more than I had with me.* **(b)** regain *or* recover consciousness, awake(n), °revive, °wake up, come (a)round: *When I came to, I was on the floor with a terrific headache.* **(c)** °regard, °concern, °relate to, be a question of, °involve, be relevant to, be involved: *When it comes to real ale, Mario is the expert.* 19 ***come up.*** **(a)** °arise, °surface, present itself, be brought up, be broached, °come about, °turn up, °rise, *Colloq* crop up: *The question of religion never came up.* **(b)** °grow, °thrive, °appear: *None of my tulips came up this year.* **(c)** °rise, °arise: *The moon came up just as the sun was setting.*

comedian *n.* comedienne, humorist, °comic, °wit, °wag, jokesmith; °clown, buffoon, funnyman, funster, jester, °fool, °zany, merry-andrew: *The new comedian at the variety show is very funny.*

comely *adj.* good-looking, °pretty, °bonny, °lovely, °fair, °beautiful, °handsome, °attractive, appealing, °wholesome, winsome, °buxom: *Your young friend is comely enough to be a model.*

come-on *n.* °lure, °attraction, °enticement, °inducement, °temptation, bait; loss leader: *The free glassware is a come-on to buy a tankful of gas.*

comfort *v.* 1 °console, °solace, soothe, assuage, °reassure, °relieve, hearten, °cheer, °gladden: *It might comfort you to know that Roderick has recovered completely. He comforted her when the pain became unbearable.* —*n.* 2 consolation, °solace, °relief, °cheer: *I derived some comfort from knowing that my attacker had been caught.* 3 °ease, °luxury, °security, °abundance, °plenty, opulence: *Cordelia lived out her days in comfort after winning the lottery.*

comfortable *adj.* 1 at ease, °easy, °tranquil, °serene, °relaxed, contented, untroubled, undisturbed: *After the surgery, the nurse did everything she could to make me comfortable.* 2 °well-off, °carefree, insouciant, contented, satisfied; self-satisfied, complacent, °smug: *They don't have a lot of money, but they're comfortable enough.* 3 °likable, °easy, congenial, °amiable, °cordial, °warm, °pleasant, °agreeable, enjoyable, relaxing: *Daphne is a very comfortable sort of person to be with.* 4 °suitable, °acceptable, °adequate, °satisfactory, °reasonable: *Adjust the volume of the hi-fi to a more comfortable level, if you don't mind.*

comic *adj.* 1 °funny, droll, comical, °humorous, °hilarious, sidesplitting, mirthful, jocose, jocular, °witty, waggish, °clever, facetious, amusing: *Barry's comic routines have made him a popular performer for years.* —*n.* 2 See **comedian.**

command *v.* 1 °order, °direct, °bid, enjoin, °charge, °request, °require, °demand, °instruct; °say, °prescribe, °decree: *The court has the power to command the execution of the sentence. What the Queen commands must be done.* 2 °control, °dominate, have *or* maintain *or* wield authority *or* control *or* sway *or* influence over, hold sway over; °lead, °rule, °govern, have under one's thumb, call the tune; °head (up): *Whoever commands the sea commands the town. He commanded a battalion during the war.* 3 °master, draw upon *or* on, °control, °summon: *That work required all the skill the sculptor could command.* 4 °attract, °earn; °exact, compel, °demand: *Gunga Din's bravery commanded the respect of the entire regiment.* 5 °dominate, °control, °overlook, °look down on; °have, °enjoy, °possess: *The tower commands a view of the entire valley.* —*n.* 6 °order, °direction, behest, mandate, °charge, °bidding, °instruction: *Your wish is my command.* 7 °control, °authority, °power, °sovereignty, °dominion, °regulation, °direction, °management, °government, °oversight, °leadership, °charge, °sway, stewardship, °jurisdiction: *The unit is under a colonel's command.* 8 mastery, °control, (thorough) grasp *or* knowledge: *He has a command of three languages.*

commemorate *v.* °memorialize, °remember, °celebrate, °observe, °dedicate, consecrate, solemnize, °sanctify, °hallow, °reverence, °revere, °honor, °venerate, pay tribute *or* homage to, °salute; °immortalize: *We are here to commemorate deeds of valor and the men who performed them.*

commence *v.* 1 °begin, enter upon, °start, °initiate, °launch, embark on *or* upon: *Tomorrow morning, we commence the ascent of Mont Blanc.* 2 °begin, °start, °open: *The ceremonies are about to commence.* 3 °begin, °start, °initiate, °launch, °inaugurate, °establish: *We commenced operations at this plant last year.*

comment *n.* 1 °remark, °reference, animadversion, °note, annotation, °criticism, °exposition, °explanation, °expansion, elucidation, clarification, footnote: *The author's comments on his sources appear in the appendix.* 2 commentary, °opinion, °remark, °view, °observation, °reaction: *The judge's comments are not for publication.* —*v.* 3 °remark, °observe, opine, °say: *He commented that he knew nothing about the minister's private life.* 4 ***comment on*** *or* ***about.*** °discuss, °talk about, remark on; °reveal, °expose: *She refused to comment on what had happened the previous night.*

commerce *n.* °trade, °business, mercantilism, marketing, merchandising, °traffic, trafficking: *All commerce consists in the exchange of commodities of equal value. My husband is in commerce.*

commit *v.* **1** °entrust, consign, °transfer, °assign, °delegate, °hand over, °deliver, °give; °allot, °pledge, allocate: *They committed the goods to traders with strong distribution facilities.* **2** °sentence, °send (away), confine, °shut up, intern, °put away, °imprison, incarcerate: *The judge committed her to prison. You can be committed for such behavior.* **3** °perpetrate, do, °perform, °carry out: *They committed murder for money.* **4** ***commit oneself.*** °pledge, °promise, covenant, °agree, °assure, °swear, give one's word, °vow, °vouchsafe, °engage, °undertake, °guarantee, bind oneself: *He committed himself to buy the company after seeing the books.*

committee *n.* °council, °board, °cabinet, panel, °body, commission: *They have set up a committee to oversee park planning.*

common *adj.* **1** °ordinary, °everyday, commonplace, °prosaic, °usual, °familiar, °customary, °prevalent, °frequent, °run-of-the-mill, °general, °normal, °standard, °conventional, °regular, °routine, °stock, °average, °proverbial; °plain, °simple, garden-variety, common-or-garden, workaday, °undistinguished, unexceptional: *Intermarriage is a common occurrence among the members of the sect. We planted a common variety of carrot.* **2** °mutual, °reciprocal, °joint, shared: *Our common heritage must be protected.* **3** low-class, °ordinary, °plain, °simple, °plebeian, °bourgeois, proletarian, °run-of-the-mill, °vulgar, °unrefined: *Kings avoid dealing with the common people.* **4** °inferior, low-grade, °mean, °cheap, °base: *He was smoking a cigar of the commonest type.* **5** °public, °general, community, communal, collective, nonprivate, °universal; well-known: *The contents of the library are the common property of everyone. Their romance is common knowledge in the village.* **6** trite, °stale, hackneyed, worn-out, °banal, °tired, overused, stereotyped, clichéd, stereotypical: *The term "yuppie" has become too common to have much impact any longer.*

communicate *v.* **1** °make known, °impart, °confer, °transmit, °transfer, hand on *or* down, °share, pass on *or* along, send on, °spread; °tell, divulge, °disclose, °reveal, °announce, °transmit, promulgate, proffer, °tender, °offer, convey, °deliver, present, °give, °yield, °supply: *Moral and physical qualities are communicated by descent. I communicated to her the information about the missiles.* **2** Also, ***communicate with.*** be in communication (with), °converse (with), °talk (with), °chat (with); °correspond (with); °associate (with), be in contact *or* touch (with), °reach: *Donald and I haven't communicated in years. Instead of communicating with him by telephone, she did so via personal notices in the newspaper.* **3** get *or* put across, make understandable; get through to, °reach, be of one mind, be in tune, °relate, be in *or* en rapport, make oneself understood, *Slang* be *or Slang* vibrate on the same frequency *or Slang* wavelength: *He cannot communicate his ideas to his students. We might talk, but are we communicating?*

compact *adj.* **1** °packed, compacted, closely knit, condensed, concentrated, consolidated, compressed; °dense, °solid, °firm, °thick: *The sesame seeds are mixed with honey and pressed into a compact block.* **2** °tight, °small, °snug, °little: *Strap these overcoats into one compact parcel.* **3** condensed, °terse, laconic, °close, pithy, °succinct, °concise, °brief, compendious, laconic, °epigrammatic, aphoristic: *The best of Voltaire's tragedies are compact and full of feeling.*

companion *n.* **1** °fellow, °associate, °comrade, °colleague, confrère, *Colloq chiefly Brit and Australian* °mate, *US and Canadian* buddy: *Companions I have enough, friends few.* **2** °vade mecum, °manual, handbook, °guide, reference book, enchiridion: *They publish a pocket companion listing the month's events.* **3** °escort, chaperon(e), °attendant, (in Spain, Portugal) duenna: *Aunt Dinah is too old to be alone, so we have engaged a companion for her.*

companionship *n.* °fellowship, camaraderie, comradeship, °company, °society, amity, °friendship, °fraternity: *I often enjoy the companionship of an older person.*

company *n.* **1** °companionship, °society, °fellowship; °attendance, °presence; associates, friends, companions, comrades: *It was a stormy night, and I was only too glad to have his company. A man is known by the company he keeps.* **2** assemblage, °party, °band, °group, °retinue, entourage, °suite, °train, coterie, °ensemble, troop, followers, following, °flock; °circle, °assembly, °gathering, °convention, °body, °crowd, °throng, *Theater* troupe, °cast, players, actors, performers: *The king arrived with his company at the gate of the city. The speaker addressed the assembled company. The company leaves today for a month on the road.* **3** guest(s); visitor(s), caller(s): *Are you having company for dinner tonight?* **4** °firm, °business, °house, °concern, °institution, °establishment, °enterprise; proprietorship, partnership, corporation, *Brit* public limited company, plc, *Australian and New Zealand and South African* private limited company, Pty: *The company was founded in 1867.*

compare *v.* **1** °liken, °associate, make (an) analogy (with), °refer, analogize: *How can you compare your collection to mine?* **2** ***compare with.*** °resemble, be *or* look like, be on a par with, be in a class *or* the same class with, °correspond, °match, °parallel, °approach, °approximate, bear *or* merit comparison (with); °rival, compete with *or* against, be a match for: *Your paintings compare well with Picasso's. The shirt washed in their detergent cannot compare with the one washed in ours.* **3** °contrast, measure against, °weigh, juxtapose, set side by side, °relate, correlate: *Compare the motion of the eastern and western halves of the glacier.*

comparison *n.* **1** contrasting, °contrast, juxtaposing, juxtaposition, balancing, °balance, weighing: *His comparison of the value of the books is prejudiced.* **2** °match, similarity, °resemblance, °likeness, comparability, °relation, °relationship, commensurability, °kinship, point of agreement *or* correspondence: *There is no comparison between a racing car and a family car.*

compartment *n.* °division, °section, °partition, °part, °space, °chamber, bay, alcove, °cell, pigeonhole, locker, cubbyhole, °niche, cubicle, °slot: *Each specimen is in its own separate compartment.*

compensate *v.* **1** recompense, °make up (for), make restitution *or* reparation, °offset, °make good, indemnify, °repay, °reimburse, redress, °requite; expiate, °atone, °make amends (for): *The company compensated us for the loss of the car. Arriving at school early today, Gerard, does not compensate for having been late yesterday.* **2** °balance, counterpoise, counterbalance, °equalize, °neutralize, °even (up), °square, °offset: *Deduct six ounces to compensate for the weight of the container.* **3** °pay, remunerate, °reward, °repay, recompense: *Two dollars does not compensate me adequately for an hour's work.*

compensatory *adj.* compensative, remunerative, restitutive *or* restitutory, expiatory, reparative *or* reparatory, piacular: *His huge donation to the church was a compensatory offering for his past sins.*

compete *v.* contend, °vie, °struggle, °strive, °conflict, joust, °fence; °fight, °battle, °clash, °collide: *They are* *competing to see who will become chairman of the company. These two designs compete with one another for the viewer's attention.*

competent *adj.* **1** °adequate, °suitable, °sufficient, °satisfactory, °acceptable, all right, *Colloq* OK *or Colloq* okay: *Colquhon will make a competent bureau chief.* **2** °qualified, °fit, °capable, °proficient, °able, °prepared: *Do you really think that fellow Johnson competent to write a dictionary?*

competition *n.* **1** °rivalry, contention, striving, °struggle: *The competition for newspaper circulation becomes keener every day.* **2** °contest, °match, °meet, °game, °tournament, °event; championship: *We entered the competition as underdogs.* **3** See **competitor.**

competitor *n.* °rival, °opponent, °competition, °opposition, °adversary; °antagonist, °contestant, contender: *Our competitor has just announced a new chocolate-flavored toothpaste.*

compile *v.* °collect, put together, °gather, °accumulate, °assemble, °amass, collate, °organize, °order, systematize; anthologize, °compose: *He has compiled a large butterfly collection. Every year she compiles a volume of the best stories.*

complain *v.* grumble, °moan, °groan, wail, grouse, °carp (at), whimper, °cry, °lament, °bemoan, *Colloq* °gripe, °squawk, grouch, *Brit* whinge, *Slang* °bitch, beef, *US* °kick: *What are you complaining about now?*

complaint *n.* grumble, °grievance, *Colloq* °gripe, °squawk, *Slang* beef, *US* °kick: *I have no complaints about my treatment while in prison—but I was only a visitor.*

complement *v.* **1** °completion, °perfection, confirmation, finishing touch, °consummation: *The grand tour was once considered the necessary complement of English education.* **2** °crew, °team, °company, °band, °outfit; °quota, °allowance, quorum: *The regiment's full complement was attained by selecting from among the recruits.* —*v.* **3** °complete, °perfect, round out *or* off, °set off, top off; °flesh out: *The setting was complemented by a huge floral arrangement. His argument was complemented by evidence from rare documents.* **4** °supplement, °enhance, °add to: *These are facts that complement but do not contradict her story.*

complete *adj.* **1** °entire, °whole, °intact, uncut, unbroken, °undivided, °unabridged, °full, undiminished, unabated, unreduced: *They played the complete symphony. I have read the complete version of* A Study of History, *by Arnold Toynbee.* **2** finished, ended, concluded, °over, done, °accomplished, terminated; settled, executed, performed: *The company's figures are not yet complete. When will your building plan be complete?* **3** °entire, °total, °thorough, °absolute, utter, °unqualified, unmixed, unalloyed, °pure, °unmitigated, °rank: *I attribute the disaster to a complete failure of communication and Fern's complete stupidity.* **4** °perfect, consummate, °exemplary, °ideal, °model, °superior, °superlative, °superb, °faultless, °flawless: *Her dissertation is a work of complete scholarship.* —*v.* **5** conclude, °finish, °end, bring to an end, °accomplish, °achieve, do, *Colloq* °wrap up; °finalize: *After completing a circuit of the globe, they rested. Have you completed the prospectus?* **6** °round out, °round off, °perfect; °crown, culminate: *The unit was completed by the addition of five platoons. A golden cupola completed the top of the dome.*

completely *adv.* **1** °entirely, fully, °quite, °wholly, °totally, °altogether, in toto, °thoroughly, °perfectly, °exactly, °precisely, down to the ground, from start to finish, from beginning to end, from A to Z, from the word "go," in full; °lock, °stock, and barrel; °hook, line and sinker; heart and soul; °root and branch; en masse: *The currency does not completely represent the wealth of the country.* **2** unqualifiedly, unconditionally, °thoroughly, °utterly, °totally, °absolutely, °quite, °altogether, unreservedly: *He's completely mad if he thinks that thing will fly.* **3** °clearly, °expressly, explicitly, unambiguously, °entirely, fully, °totally, °wholly, °altogether, unequivocally, °truly, categorically, flatly: *I am completely in agreement with your policy.*

completion *n.* **1** conclusion, °end, °close, °termination, °fulfillment, °consummation, culmination, °realization, °accomplishment, °finish: *The completion of the building phase is scheduled for next July.* **2** finishing, finalization, windup, finishing-off, completing: *The completion of the house is scheduled for next October.*

complexity *n.* **1** °complication, convolution: *We are finding it difficult to understand the complexities of the agreement.* **2** intricacy, involvement, complicatedness; inscrutability: *The complexity of her theory makes it difficult to interpret.*

complicate *v.* **1** °mix up, °entangle, °snarl, °tangle, confound, °muddle, °confuse: *You only complicate matters by bringing up the question of religion in a discussion of money.* **2** make complicated *or* complex, make involved *or* intricate, make a shambles *or* mess *or* muddle of, °mess up, *Colloq* °screw up: *The phenomena of tides and currents greatly complicate coastwise navigation.*

complicated *adj.* °involved, °intricate, complex, °compound, °elaborate; °ornate, Byzantine, Daedalian, tangled, knotty, °confused, °labyrinthine: *In birds the eye is a more complicated organ than in our own species. His plan is too complicated to understand.*

complication *n.* **1** °complexity, involvement, intricacy, convolution: *The complication of the diagram makes it almost indecipherable.* **2** °difficulty, °problem, °predicament, °dilemma, °obstacle, °obstruction, °snag, °drawback: *Asking for more money might create complications.*

compliment *n.* **1** °praise, °homage, commendation, °honor, °tribute, °flattery, °bouquet, °favor: *The greatest compliment given to my work has been its success.* **2** Usually, ***compliments.*** °respects, °regards, good *or* best wishes, felicitations, salutations, °greetings: *I stopped by to pay my compliments to your mother.* —*v.* **3** °honor, °praise, pay homage *or* tribute to, commend, °laud, °congratulate, felicitate; °flatter: *She came backstage to compliment me on my performance.*

complimentary *adj.* **1** °laudatory, commendatory, encomiastic, panegyrical, eulogistic, congratulatory, °flattering: *The duke was most complimentary about my sculpture.* **2** °free, gratis, °on the house: *The shoehorn is complimentary when you buy a pair of shoes.*

comply *v.* °agree, °obey, °conform, °consent, acquiesce, concur, °submit, °yield, accede; °accord: *They require your signature and I hope you'll comply.*

compose *v.* **1** constitute, °form, °make (up), be a constituent *or* ingredient *or* component *or* element of, be a part of: *The countless particles of water rise to compose a cloud.* **2** °write, °create, °imagine, °think up, °originate, °frame, °formulate, °make (up), °author, °devise; contrive; set to music, °arrange: *He composed the poem while traveling on the Flying Scotsman. That symphony was composed in 1873.* **3** ***be composed of.*** consist of *or* in, comprise, be formed *or* made (up) of, be constituted of: *The new government is composed of princes and bishops.* **4** ***compose oneself.*** °calm (down), quiet *or Chiefly Brit* quieten (down), pacify, control oneself, get control of *or* over oneself: *They stopped crying and composed themselves.*

composition *n.* **1** °theme, °essay, article, °paper, °story: *The teacher required each of us to write a thousand-word composition every week.* **2** °combination, °makeup, °structure, °form, °assembly, °setup, °organization, layout, °arrangement, configuration, shaping; °balance, °harmony, °proportion, °placement, placing, construction: *Notice the composition of the various elements in this painting.* **3** °combination, aggregate, °mixture, °compound, compounding, °mix, formulation, °formula, composite, °amalgam, °alloy, mélange, °medley: *The medication was a composition of several odd ingredients.* **4** °creation, origination, formulation, fashioning: *The composition of the opera was begun in 1837.* **5** °makeup, constitution: *What goes into the composition of brass?*

compound *v.* **1** put together, °combine, °mix, concoct, °compose, °make (up), °formulate, °blend: *They compound curry from different spices.* **2** °blend, °merge, coalesce, °combine, °unite, fuse *or US also* fuze, come *or* go together: *Sometimes two words compound to form one, as in "ingrown," "outgrow," and "uptake."* **3** °aggravate, °intensify, exacerbate, °heighten, augment, °add to, °worsen, °increase; °enhance, multiply: *Demanding your money back now will only compound the problem.* —*adj.* **4** °intricate, complex, °involved, °complicated; composite, multiple, multiform, multifaceted, *Technical* parasynthetic, parathetic: *The compound eye of the fly / Lets it see far better than I. A compound sentence is composed of two or more clauses joined by one or more coordinating conjunctions, express or understood.* —*n.* **5** composite, °blend, °synthesis, °combination, consolidation, *Technical* parasynthesis, parathesis; °mixture, °amalgam, °alloy, merging, °merger, °mix: *Table salt is a compound of the metallic element sodium and the gaseous element chlorine. "Slithy" is a compound of "slimy" and "writhe."*

comprehend *v.* °understand, °see, °grasp, °conceive, °take in, apprehend, °realize, °fathom, °perceive, discern, absorb, assimilate, °appreciate: *Do you comprehend how serious the matter has become?*

comprehensive *adj.* °inclusive, encompassing, °thorough, °extensive, °full, °exhaustive, °complete, °sweeping, °wide, °broad, encyclopedic *or* encyclopedic: *We hope to give comprehensive coverage to all aspects of the subject in our new book.*

compulsive *adj.* compelling, °obsessive, coercive, °urgent, °forceful, °overwhelming, constrained: *She is such a compulsive workaholic that she double-checks everything done by her staff.*

compunction *n.* **1** °remorse, contrition, °regret, uneasiness of mind, pang *or* pricking of conscience, self-reproach: *He has no compunction about hurting your feelings.* **2** hesitation, °reluctance, °reserve, disinclination, °qualm, °misgiving, unwillingness, °fear: *She has no compunction about speaking her mind.*

compute *v.* °calculate, °reckon, °figure (out), °work out, °determine, ascertain, °estimate: *My accountant computed my income tax this year.*

comrade *n.* °colleague, °associate, °friend, °companion, °chum, crony, confrère, *Colloq* °pal, °chum, *Chiefly Brit and Australian* °mate, *Australian* cobber, *US* buddy: *None of my comrades from the old regiment attended the reunion this year.*

conceal *v.* **1** °hide, °secrete, °bury, °cover, °disguise, °camouflage: *Packets of a white powdery substance were concealed inside each doll.* **2** keep secret *or* hidden, keep quiet about, °disguise, not reveal; dissemble: *He concealed his true identity even from his wife.*

concede *v.* **1** °admit, °allow, °grant, °acknowledge, °confess, own (up *or* to *or* up to), °accept: *I conceded that I had no business in the bank after closing.* **2** °grant, °yield, °surrender, °cede, °give up, °submit, °resign, °relinquish, °abandon, °waive: *In chess, upon the loss of a queen, many players will concede. She has conceded any right to the estate of her uncle.*

conceit *n.* **1** °vanity, °pride, egotism, °self-esteem, self-admiration, self-love, narcissism, vainglory, *amour-propre;* °arrogance: *His conceit is matched only by his incompetence.* **2** °fancy, whim, caprice: *Some have a conceit their drink tastes better / In an outlandish cup than their own.* **3** elaborate figure (of speech), °affectation, strained *or* far-fetched metaphor: *A conceit would be calling the waves "nodding hearse-plumes."*

conceited *adj.* °vain, °egotistical, self-centered, egocentric, self-admiring, narcissistic, prideful, °proud, °arrogant, self-involved, °self-important, self-satisfied, °smug, complacent, vainglorious, °snobbish, *Colloq* stuck-up; *Slang* snotty: *That conceited ass really thinks the world of himself.*

conceive *v.* **1** °have, °bear, beget, sire, °father, give birth to; become pregnant (with): *After the twins, they conceived three boys.* **2** °formulate, °devise, °plan, contrive, °create, °plot, °hatch, °develop, evolve, °fabricate, think *or* make up, °form, °frame, °design: *He conceived a scheme for bilking that poor woman out of her life savings.* **3** °think (up), °imagine, °speculate (on), °perceive, °see, °understand, °realize, °comprehend, °envision, °envisage, conjure up, dream up, hypothesize, postulate, °posit, °suggest, °suppose: *I cannot conceive of any reason why she shouldn't be allowed to go on the picnic. I conceived an umbrella might be of some use today.*

concentrate *v.* **1** °focus, °direct, °center, centralize, °converge, consolidate: *The council concentrated its efforts on refurbishing the schools.* **2** condense, °reduce, distill, °intensify, °refine, °strengthen: *The sap of the sugar maple is concentrated by boiling.* **3** °gather, °collect, congregate, draw *or* bring together, °crowd, °cluster, °group: *Much of the population is concentrated around the large cities.* **4** °think, focus one's thoughts *or* attention, apply oneself: *I cannot concentrate with the radio on.*

conception *n.* **1** °birth, °beginning, genesis, inception, commencement, °emergence, °start, inauguration, °initiation, °launch, launching, °origin, origination, °formation, formulation, introduction: *We were excited to be in at the conception of the scheme.* **2** °idea, °notion, °inkling, °clue, concept; *idée reçu;* °understanding, °knowledge, °appreciation, comprehension: *He has no conception of what is involved in maintaining a yacht.* **3** °design, °plan, °scheme, °proposal, °outline: *Transporting a building from another site instead of constructing one was a bold conception.*

concern *v.* **1** refer *or* relate to, have relation *or* reference to, be about, pertain *or* appertain to, be pertinent or relevant to, °regard, apply to, be connected *or* involved with, °bear on, be germane to, °involve, apply to, °touch (on): *The matter concerns your inheritance, Cosgrove.* **2** °affect, have (a) bearing *or* (an) influence on, °involve, °touch; °interest, be of importance *or* interest to: *This war concerns us all.* **3** °worry, °trouble, °disturb, °bother, °perturb, unsettle, °upset, °distress: *He doesn't let anything concern him.* —*n.* **4** °business, °affair, °problem; °responsibility, °duty, °charge, °task, involvement, *Colloq* °thing; *Slang* °bag, shtick: *What she does is no concern of yours. The safety of the passengers is his concern.* **5** °interest, °regard, °consideration, °care, °thought, awareness, °attention: *You should show more concern for those less fortunate than you.* **6** °anxiety, °worry, °solicitude, apprehension, °distress, apprehensiveness, uneasiness, malaise, disquiet, disquietude: *It's only a cold, and no cause for concern.* **7** °business,

°firm, °company, °house, °establishment, °enterprise, °organization: *The business is a going concern.* **8** °matter, °affair, °issue: *The preservation of wildlife is an international concern.*

concerned *adj.* **1** °involved, °responsible, °interested, °active; caring, °solicitous: *The best governments are run by a concerned citizenry.* **2** troubled, vexed, °anxious, °worried, distressed, uneasy, perturbed, bothered, °upset, °disturbed: *They were not at all concerned about the state of my health.*

concerning *prep.* °about, °regarding, relative *or* relating to, referring to, with *or* in reference to, as regards, in *or* with regard to, with an eye to, with respect to, respecting, apropos (of), as to *or* for, in the matter of, on the subject of, re, *Formal* anent: Concerning *your recent application, please phone this office.*

concise *adj.* °brief, °terse, laconic, °compact, °direct, °succinct, °epigrammatic, cogent, pithy, compendious, °summary, °epigrammatic, °trenchant, compressed, condensed, °short; shortened, abridged, curtailed, °abbreviated: *He gave a concise summary of the finding. This is a concise edition of the dictionary.*

concrete *adj.* °real, °actual, °literal, °realistic, °authentic, valid, °genuine, °bona fide, °reliable; °specific, °particular, °definite, °definitive, clear-cut, °material, °physical, °tangible, °substantial: *Have you any concrete evidence for the existence of UFOs?*

condemn *v.* **1** censure, °blame, °criticize, remonstrate with *or* against, °denounce, °disparage, reproach, °rebuke, °reprove, °scold, °reprimand, °upbraid: *The council was condemned for failing to provide adequate health care.* **2** °convict, find guilty; °sentence, °doom: *The judge condemned them to twenty years in prison.* **3** Usually, ***condemned.*** °doomed, damned, °destined, °fated, ordained, foreordained; consigned: *He has been condemned to wander forever.*

condescend *v.* °stoop, °deign, lower *or* humble *or* demean oneself, come down off one's high horse: *She wouldn't condescend to talk to the stableboy directly.*

condescending *adj.* patronizing, belittling, °disdainful, °contemptuous, °pompous, °overbearing, high-handed, imperious, °snobbish, °haughty, *Colloq* snooty, *Brit* toffee-nosed, *Slang* snotty: *He thinks he's better than everyone, and I can't stand his condescending manner.*

condition *n.* **1** °state; circumstance(s), °shape: *In what condition is the house? My bank account is in poor condition.* **2** °stipulation, °proviso, °demand, °requirement, °term, °qualification, contingency, requisite, °prerequisite: *The terrorists have announced their conditions for releasing the hostages.* **3** working order, °fitness, °shape, °form, fettle; °health: *He's in good condition, but his car isn't.* **4** ***conditions.*** °circumstances; °quarters; °environment: *The ship's crew live in very crowded conditions.* —*v.* **5** °ready, get *or* make ready, °prepare, °equip, °outfit, fit (out *or* up), °adapt, °modify: *The mechanics are conditioning the plane for high-altitude flights.* **6** °train, °educate, °teach; brainwash; °influence, °mold, °persuade: *The children were conditioned to avoid talking to strangers.* **7** °accustom, inure, °adapt, acclimate, acclimatize; mithridatize: *At this training base, we condition the commandos to all kinds of hardships.*

conduct *n.* **1** °behavior, action(s), demeanor, °manners, deportment, comportment, °attitude: *Such conduct will not be allowed in this school.* **2** °guidance, °direction, °management, supervision, °leadership, °administration, °government, °running, handling, °control, °command, °regulation, °operation: *Had the conduct of the war been left up to him, we should have lost.* —*v.* **3** °guide, °direct, °supervise, °manage, °carry on, °run, °control, °administer, °regulate, °operate: *They conduct a remarkably successful business.* **4** °lead, °guide, °escort, show (in *or* out), usher: *We were conducted through the gallery by the curator herself.* **5** °channel, °carry, °transmit, convey; °direct: *Electrical power is conducted by the cable.* **6** ***conduct oneself.*** °behave, °act, demean, deport, comport, acquit: *For a six-year-old, he conducted himself very well.*

confer *v.* **1** °converse, °consult, °deliberate, °talk (over), °discuss, take counsel: *I shall have to confer with my colleagues on that matter.* **2** When transitive, ***confer on.*** °give, °grant, °present, °award; °bestow (on): *The prizes will be conferred after the dinner. He was bewildered by the honors conferred on him.*

conference *n.* °meeting, °convention, symposium, congress, seminar, forum, colloquium; °discussion, °talk, colloquy, *US* bull session: *In 1988, the conference was held in Budapest.*

confess *v.* °disclose, °acknowledge, °admit, own (up *or* to *or* up to), °declare, avow, make a clean breast (of); °reveal, divulge, °confirm, °concede, affirm, aver, °testify; disbosom oneself, *Colloq* °come clean: *She confessed her part in the swindle. Confronted with the evidence, he confessed.*

confidence *n.* **1** °trust, °reliance, °faith; °belief: *Your parents have a great deal of confidence in you.* **2** °assurance, °self-confidence, self-assurance, self-reliance, °poise, aplomb, coolness; °conviction, certitude, boldness, °courage, °nerve: *We admire the confidence she shows in her daring plan.* **3** ***in confidence.*** °in secrecy, in privacy, privately, confidentially, intimately, *Colloq* on the Q.T.: *I am telling you this in confidence.*

confident *adj.* **1** °secure, °sure, °certain, assured, °positive, convinced: *I feel confident that we shall get the contract.* **2** °self-confident, self-assured, °self-possessed, reliant, self-reliant, °dauntless, °bold, °cool, cocksure, °fearless, °courageous, *Colloq* °cocky: *He strode into the room with a confident air.*

confidential *adj.* °private, °secret, °intimate; classified; *Colloq* hush-hush: *These confidential papers must never be out of your possession.*

confirm *v.* **1** °ratify, °sanction, °authorize, °endorse, °support, °sustain, °approve, °uphold, °back up, validate, °verify, °recognize; °authenticate, accredit: *By-laws shall not take effect unless confirmed by local governments.* **2** °establish, °settle, affirm, °ensure, °clinch, °substantiate, °guarantee, °bind, °seal: *The king thereby confirmed his control over the islands.* **3** °strengthen, °encourage, °fortify, °reinforce, corroborate, °substantiate, °buttress, °prove: *Later events confirmed his opinion.*

confiscate *v.* °appropriate, °seize, impound, sequester, sequestrate, expropriate, °take (away), commandeer: *The police confiscated my car as evidence.*

conflict *n.* **1** °fight, °battle, °combat, °engagement, °struggle, °war, °fray, °fracas, affray, °brawl, donnybrook: *Gurkha troops entered the conflict.* **2** °dispute, °argument, °controversy, wrangle, contention, °disagreement, altercation, °feud, °quarrel, °row; squabble, °tiff, *Colloq* spat: *The counselor was unable to resolve the conflict between the sisters regarding the will.* **3** °clash, °antagonism, °difference, °opposition, °disagreement, °variance, °discord: *There is a basic conflict between the interests of labor and of management.* —*v.* **4** °clash, °disagree, °differ, be incompatible *or* at odds *or* at variance, be in opposition (to): *The plans conflict on only one point.*

conform *v.* **1** °comply (with), °follow, °observe, °obey, °respect, °abide by, adapt *or* adjust (to): *We agree to conform to the rules of the club.* **2** °accord (with), °agree (with), concur (with), °coincide (with), °correspond (with), harmonize (with), °square (with), °match, °tally (with), °fit (in with), be consistent (with), be in accord *or* in accordance (with), ring true: *Their behavior did not conform with what is expected in such circles. The two plans do not conform.*

confuse *v.* **1** disconcert, °perplex, °puzzle, °bewilder, °mystify, baffle, °bemuse, befuddle, °discomfit, confound, °fluster, °flummox, °upset, disorient, °embarrass, abash, °shame, °dismay, *Colloq* °rattle, °throw, *Chiefly US* discombobulate, *US and Canadian* buffalo: *She was completely confused by his offer to help.* **2** °disorder, confound, disorganize, throw into disarray, °muddle, °mix up, °snarl (up), ensnarl, °tangle (up), °entangle, °botch, *Colloq* °mess up, make a mess of, °screw up, *Brit* make a balls-up of, *US* ball up: *He has done more to confuse the situation than to clear it up.* **3** °mix up, confound, °muddle, °jumble, °snarl (up), ensnarl; °blur: *The identities of the children were confused at birth.*

confused *adj.* **1** mixed up, jumbled, disordered, disorganized, °disorderly, muddled, muddle-headed, snarled (up), messy, baffling, confusing, mystifying, °puzzling, °perplexing; °contradictory, °ambiguous, misleading, °inconsistent, mixed up, botched (up), *Colloq* higgledy-piggledy: *The accountants have provided a confused set of figures.* **2** bewildered, perplexed, puzzled, baffled, (be)fuddled, mystified, disoriented, discomposed, at sea, flummoxed, dazed, muddled, bemused, mixed up, nonplussed, °disconcerted, abashed, °put off, °put out, °disturbed, flustered, °ill at ease, °upset, at sixes and sevens, at a loss, *Rare* metagrobolized, *Colloq* screwed up, muzzy, out of it, not with it, *Chiefly US* discombobulated, fouled up; *Slang* (all) balled up, *Brit* (all) bollocksed *or* ballocksed (up), *US* (all) bollixed (up), *US and Canadian* snafu: *I have never seen anyone so confused about a simple question of astrophysics.* **3** jumbled, mixed up, muddled, °disorderly, confusing, messy, disorganized, °topsy-turvy; °miscellaneous, motley, *Brit* shambolic: *A confused array of merchandise is on offer at the sale.*

confusion *n.* **1** °disorder, °mix-up, °mess, °jumble, °muddle, disarray, disarrangement, °chaos, °shambles: *The files are in complete confusion.* **2** °tumult, commotion, °disorder, turmoil, °pandemonium, °bedlam, °chaos: *Untold confusion resulted from sounding the alarm.* **3** °mix-up, confounding; °ambiguity, ambiguousness, °misunderstanding, contradiction, inconsistency: *One must contend with the confusion between the name of a thing and the thing itself.* **4** mixing, combining, mixing-up, intermingling: *The removal firm is responsible for the confusion of your books with mine.* **5** °assortment, °mixture, °potpourri, gallimaufry, hotchpotch *or US and Canadian also* hodgepodge: *A confusion of products lines the shelves.* **6** °embarrassment, discomfiture, mortification, abashment, shamefacedness, chagrin: *He felt terrible confusion when confronted with the evidence.*

congested *adj.* (over)crowded, blocked (up), jammed, crammed, plugged, stopped *or* stuffed (up), choked: *The police are trying to deal with traffic at congested intersections.*

congratulate *v.* felicitate, °compliment: *Her friends congratulated her on winning the award.*

congratulations *interj.* Felicitations!, Best wishes!, Well done!, Many happy returns!, *Colloq* Nice going!, Good show!: *Heartiest congratulations! You've come in first!*

connect *v.* **1** join *or* link *or* tie (together), °unite: *An old road connects the two towns.* **2** °associate, affiliate, °link, °relate, °league, °tie (in): *The police connected him with the break-in. Dr. Basing is connected with the Stoke Clinic.* **3** °fasten, °bind, °unite, °tie, °link, °join, °attach, °couple, put together, °secure, °fit, °fix, affix, °stick, °anchor, °lock; rivet, °weld, braze, solder, °screw, °nail, stitch, °sew, °pin, °hook, °staple, °tack, °glue, °cement, °fuse, °seal, °buckle, strap, °bolt, °lash, °chain, °moor: *Connect the parts to the frame.*

connection *n.* **1** uniting, joining, linking, connecting, coupling; °union, °bond, °joint, °link: *The US constitution forbids a connection between church and state. The connection between the fittings has broken.* **2** °link, °tie, (inter)relation(ship), interplay, °bearing, °reference, °relevance, appropriateness, correlation, °tie-in; coherence, consistency, °association: *Your answer had no connection with the question.* **3** Often, ***connections.*** °contact, °ally, °acquaintance, °friend (at court); °influence, *Colloq* °pull; *US slang* °drag: *With their connections, they can get away with anything.* **4** ***connections.*** relatives, °relations, °family, °kin, kith and kin: *They have connections in Australia.*

conquer *v.* **1** °overcome, vanquish, °beat, °defeat, °subdue, °crush, °subjugate: *The Moors conquered most of Spain.* **2** °capture, °seize, °win, °gain, °acquire, °obtain; °occupy, annex, °overrun: *They conquered the territory by force of arms.* **3** °overcome, triumph *or* prevail over, °beat, surmount, °master, win out (over): *He has finally conquered the habit of biting his fingernails.*

conquest *n.* **1** vanquishment, subjugation, °defeat, °domination, °subjection: *Hernando Cortés is famous for his conquest of Mexico.* **2** °victory, °triumph, mastery, °win: *He is credited with the conquest of a number of diseases by means of this drug.*

conscience *n.* °morality, °morals, °judgment, fairness, sense of right and wrong, ethics, °honor, standards, °principles, scruples: *In such matters, your conscience must be your guide.*

conscientious *adj.* **1** °scrupulous, °principled, °fair, °moral, °ethical, °strict, °righteous, right-minded, upstanding, °upright, °honorable, °just, °responsible, high-minded; incorruptible: *Fetherby is a conscientious arbitrator.* **2** °cautious, °careful, °scrupulous, °exacting, °meticulous, punctilious, °painstaking, °diligent, °particular, rigorous, °thorough: *A conscientious effort was made to restore the painting to its original condition.* **3** °prudent, °discreet, °politic, °careful, circumspect, heedful, °attentive, °serious: *Fred is conscientious about keeping secrets.*

conscious *adj.* **1** °aware, °awake, °alert: *I was conscious of an eerie presence. The victim of the attack is now conscious.* **2** °deliberate, °intentional, purposive, °purposeful, °willful, °studied: *Lydia has been making a conscious effort to be friendlier to me.*

consent *v.* **1** °agree, °comply, concur, accede, acquiesce, °concede, °yield, °submit, °cede, °conform, °give in: *He asked for payment in advance and I consented.* **2** ***consent to.*** °permit, °allow, °agree to, give in to, °approve, °authorize: *Richard's parents consented to his going on the outing.* —*n.* **3** °approval, assent, °permission, °sanction, authorization, imprimatur, seal of approval, *Colloq* °OK, okay, °go-ahead: *Have Richard's parents given their consent?* **4** °agreement, acceptance, acquiescence, compliance, °approval, concurrence: *Taxes cannot be raised without the consent of Congress.*

consequently *adv.* so, °therefore, as a result *or* consequence, °accordingly, ergo, °hence, °thus: *He was found guilty and, consequently, sentenced to death.*

conservation *n.* °preservation, °protection, °safekeeping, °maintenance, °upkeep, °management, safeguarding; husbandry, °economy: *The conservation of natural resources is foremost.*

conservative *adj.* **1** °reactionary, °right, right-wing, rightist, Tory: *In his conservative view, no change is ever for the better.* **2** °cautious, °careful, °prudent, °moderate, °temperate, middle-of-the-road, °sober, °stable; unprogressive, °orthodox, °traditional, conformist, °hidebound, °conventional, °standard, fundamentalist, trueblue, dyed-in-the-wool: *The conservative approach would be to study the problem before making a change. The conservative members voted against electing certain members to the club.* —*n.* **3** °reactionary, rightist, right-winger, Tory, fundamentalist; °moderate, middle-of-the-roader: *He's a conservative and favors a classical education.*

conserve *v.* **1** °keep, °preserve, hold on to, °save, °spare, °reserve: *Conserve your energy for later, when we near the top.* **2** °preserve, °maintain, keep up, °take care of: *These buildings should be conserved for later generations.*

consider *v.* **1** think about *or* over, take into *or* under consideration, deliberate (over *or* about), contemplate (on *or* over), °weigh, °ponder, mull over, cogitate on, meditate (on *or* upon *or* over), reflect (on *or* upon), ruminate (on *or* over), °chew over, °study, °examine: *The council will consider your proposal.* **2** °heed, °mark, take into account *or* consideration, °reckon with, °bear in mind, °note, °observe, make allowance for; °esteem, °respect, have regard for: *Consider your mother's feelings in the matter.* **3** °regard, look upon; °judge, deem, take to be, °think, °believe, °gauge, °rate, °estimate, °reckon: *Consider yourself under arrest. I don't consider Speckerly the best person for the job.*

considerable *adj.* **1** sizeable *or* sizable, °substantial, °large, °big, °great; appreciable, °respectable, °noticeable, largish, biggish, °goodly, °decent, °fair, *Colloq* °tidy: *A considerable crowd was gathered outside.* **2** °important, °worthy, of consequence, of distinction, °distinguished, °illustrious, °noteworthy, °notable, °remarkable, °estimable, °influential, °respectable: *Some of the most considerable citizens were banished.*

considerate *adj.* °thoughtful, °kind, °kindly, kindhearted, goodhearted, °helpful, °friendly, °neighborly, °gracious, °obliging, °accommodating, °charitable, °generous, °unselfish; °sympathetic, compassionate, °sensitive; °attentive; °solicitous: *It was very considerate of you to offer your car.*

consideration *n.* **1** °regard, °concern, attentiveness, °solicitude, thoughtfulness, compassion, °kindness, kindliness, kindheartedness, considerateness, °respect, caring, °care: *Out of consideration for your father, you should complete your studies.* **2** °reward, compensation, °remuneration, °fee, °payment, recompense, emolument, °tip, gratuity, *pourboire,* baksheesh *or* backsheesh; °honorarium: *The boy will look after your luggage for a small consideration, madam.* **3** °thought, deliberation, °reflection, contemplation, rumination, cogitation, °study, °examination: *After some consideration, we have decided that we will finance the project.*

considering *prep.* °in view of, in (the) light of, bearing in mind, making allowance for, taking into consideration *or* account, looking at, all in all, all things *or* everything considered, inasmuch as, insomuch as: *Considering your background, I doubt that you are qualified.*

consistent *adj.* **1** agreeing, in agreement, in harmony, in keeping, °harmonious, in concordance, conforming, in conformance, accordant, compatible, in accord *or* accordance, consonant: *Her story is not consistent with the facts.* **2** dependable, °regular, °predictable, undeviating, °steady, °steadfast, unchanging, °uniform, unswerving, °constant: *His behavior, even under pressure, has been quite consistent.*

consistently *adv.* **1** steadily, constantly, regularly, uniformly, °daily, day by day: *Her piano technique is improving consistently.* **2** dependably, unswervingly, staunchly, devotedly, °firmly, resolutely, faithfully, uniformly, unfailingly: *The courts have consistently upheld our claim to the treasure.*

console *v.* °comfort, soothe, °calm, assuage, °solace, °cheer (up): *Ivan made an effort to console the grieving widow.*

conspicuous *adj.* **1** °obvious, °clear, °evident, °plain, palpable, °perceptible, °patent, °prominent, °apparent, clear-cut, °unquestionable, incontestable, °incontrovertible: *The sultan played a conspicuous role in the kidnapping of the envoy.* **2** °obvious, unmistakable, °prominent, °outstanding, °noticeable, °impressive, °vivid, °obtrusive; °striking, °showy, °garish, °gaudy, °loud, °tawdry, °blatant, °lurid, °vulgar, °flashy, °ostentatious: *The silhouette of the castle was conspicuous against the sky. Fingal was again conspicuous, this time in a green wig and bowler.* **3** °notable, °noteworthy, °exceptional, °outstanding, °eminent, °unusual, °marked, °extraordinary, °remarkable, °distinguished, °impressive, °awesome, awe-inspiring, °glorious: *The medal is awarded for conspicuous bravery.*

conspiracy *n.* °plot, °scheme, °stratagem, °intrigue, collusion, °cabal, connivance, °foul play, dirty work: *He suspected them of a conspiracy to defraud their clients.*

constable *n.* policeman, policewoman, °(police) officer, *US* patrolman, *Colloq* cop, copper, *Brit* bobby; *Slang* flatfoot, fuzz: *Constables don't carry pistols in England.*

constant *adj.* **1** °resolute, °immovable, °steadfast, °firm, dependable, unshakeable *or* unshakable, °determined, unswerving, undeviating, persevering, unwearying, unwearied, °untiring, indefatigable, °tireless, unflagging, unwavering, unfailing, unfaltering, °persistent; °loyal, °true, tried and true, °devoted, °staunch, trusty, °faithful: *He was her constant companion during her troubles.* **2** incessant, unceasing, ceaseless, °perpetual, °persistent, uninterrupted, °steady, °regular, °invariable, unremitting, unvarying, °relentless, unrelenting, °continuous, °continual; unending, °endless, never-ending, °nonstop, °perennial, °eternal, °everlasting, *Literary* sempiternal: *The constant pain almost made me cry out. Their constant bickering is getting on my nerves.* **3** unchanging, unchanged, °invariable, unvarying, °fixed, °uniform, unalterable, immutable, °changeless, °persistent: *The numbers might change, but the ratio is constant.*

construct *v.* **1** °build, °erect, °make, put together, °frame, °set up, °put up, °assemble: *We constructed a summerhouse in the garden.* **2** °fabricate, °devise, °create, °forge, °invent, °formulate, °compose, °shape, °set up, °fashion: *He has constructed a complex argument to support his theory.*

constructive *adj.* **1** °helpful, °useful, °practicable, °advantageous, °practical, °productive, °beneficial, °positive: *She provided much constructive advice on how to design the factory.* **2** °virtual, inferential, °implicit, inferred, derived, deduced: *As it turned out, the shareholders were the constructive victims of the fraud.*

consult *v.* **1** Often, ***consult with.*** °confer (with), °discuss (with), °deliberate (with), °talk over (with), inquire *or* enquire of, seek advice from, °ask (of), °question, interrogate, take counsel (with *or* of): *I shall have to consult a doctor about that medicine. You should consult with your lawyer.* **2** °refer to, °look up, seek information from: *If in doubt, consult the dictionary.*

consultant *n.* **1** °physician, °doctor, °specialist, °expert: *You ought to get the opinion of another consultant.*

2 adviser *or* advisor, °expert, counsellor *or US* counselor: *Our financial consultant tells us how to handle the company funds.*

consume *v.* **1** °devour, °eat (up), °gulp (down), °swallow, °drink (up), °put away, gobble (up); °digest: *When those teenagers come home, they consume everything in sight.* **2** °use up, °exhaust, deplete, °drain, °expend, °diminish, °reduce: *The new car has consumed all our savings.* **3** °waste, °occupy, squander, °fritter away, °dissipate, absorb, °lose, °throw away, °lavish, *Slang* °blow: *Too much of your time has already been consumed by that problem.* **4** °destroy, °ruin, °(lay) waste, °demolish, °wreck, °gut, °raze, *Slang US and Canadian* °total: *Fire consumed the entire house.* **5** °overcome, °overwhelm, °devastate, °destroy, annihilate, °ravage, °(lay) waste, °wear out, °ruin, eat up, °devour, do in; preoccupy, °obsess: *He is consumed by jealousy.*

consummation *n.* **1** °completion, °accomplishment, °fulfillment, °finish, °end, °realization, attainment, °achievement, °success; completing, accomplishing, fulfilling, finishing, ending, realizing, attaining, achieving: *Owning a Rolls-Royce was the consummation of her dreams.* **2** °acme, °perfection, °peak, culmination, finishing touch, conclusion, grand finale, °climax: *The Nobel prize was the consummation of an arduous life of research.*

contact *n.* **1** °junction, conjunction, °connection: *The wires make contact in this terminal.* **2** °acquaintance, °friend, °connection, *Colloq US* in: *I have a contact on the board of directors.* **3** °touch, communication, °association: *Are you still in contact with Gunilla?* —*v.* **4** get in touch with, °communicate with, °reach, get hold of; phone, °ring (up), °telephone, speak to *or* with, write to, correspond with: *Try to contact Mr. Bottley at his home.*

contain *v.* **1** °hold, have in it; °bear, °carry: *The capsule contained a deadly poison.* **2** °hold, have the capacity for, °accommodate, °admit, °carry; °seat: *This bottle contains no more than a quart. The theater can contain two hundred.* **3** °restrain, °restrict, confine, °repress, °control, hold back *or* in, °curb, °bridle, keep under control, °suppress, °check, °stifle: *He could hardly contain himself when he learned he had passed the examination.*

contaminate *v.* defile, °sully, °pollute, °corrupt, °rot, °stain, °soil, °taint, infect, °poison, °foul, °spoil, befoul; °debase, °adulterate, °vitiate: *Just one rotten apple can contaminate an entire barrel.*

contemplate *v.* **1** look *or* gaze at *or* on *or* upon, °behold, °view, °survey, °observe, °regard, °eye; °scan, °scrutinize, °inspect: *I contemplated the scene of the Grand Canal from my hotel room.* **2** ruminate (over), ponder (on *or* over), °deliberate (over), muse (on *or* over), meditate *or* reflect (on), think (about *or* over), mull over, cogitate (over), turn over in one's mind, brood on *or* over, chew on *or* over, °consider, °study, °examine: *She was contemplating the events of the past night. Give me a moment to contemplate.* **3** °plan, °intend, think of *or* about, °consider, entertain the idea *or* notion of: *After we broke up, I contemplated moving to Tierra del Fuego.*

contemporary *adj.* **1** of the time, contemporaneous, coeval, coexistent, concurrent, concomitant, °parallel, synchronous, synchronic, °coincidental, coetaneous: *We examined some of the documents contemporary with his reign.* **2** °modern, °current, present-day, °new, up-to-date, °stylish, °fashionable, modish, à la mode, °latest, in; °novel, newfangled, *Colloq* °trendy: *She always keeps up with contemporary fads in dress and makeup. Much of the contemporary metal-and-glass architecture is boring.*

contempt *n.* °loathing, abhorrence, hatred, odium, °hate; °scorn, disdain, contumely, °disgust: *She has nothing but contempt for cowards.*

contemptible *adj.* °despicable, °loathsome, detestable, °scurvy, °low, °mean, °base, °inferior, currish, °wretched, °vile, abject, ignominious, °unworthy, °shabby, °shameful: *It was contemptible of you to tattle to the teacher.*

contemptuous *adj.* °scornful, °disdainful, sneering, derisive, insulting, contumelious, °insolent: *The maestro was contemptuous of my piano-playing.*

content[1] *n.* **1** °capacity, °volume, °size, °measure: *The content of the barrel is exactly fifty-five gallons.* **2** Usually, ***contents.*** ingredients, components, constituents; °load: *The bottle broke and its contents spilled on the floor.* **3** °substance, subject matter; °significance, purport, °import, °essence, °text, °theme, °topic, °thesis: *The book is amusing but its content is quite trivial.*

content[2] *n.* **1** °pleasure, °satisfaction, °gratification, °happiness, contentment, contentedness, felicity, °delight: *He kept on singing to his heart's content.* **2** °ease, °comfort, tranquillity, °serenity, °peace, peacefulness, contentedness: *I have a feeling of such content merely being with you.* —*adj.* **3** pleased, satisfied, °happy, °delighted, contented, gratified, °glad, °cheerful; °comfortable, fulfilled: *I was quite content to be home once more.* —*v.* **4** °satisfy, please, °gratify, soothe, °cheer, °gladden, °delight: *It contented him to be near her.*

contest *n.* **1** °competition, °match, °tournament, championship, tourney, °meet, °game, °rivalry, °trial: *The contest was won by a woman from Shropshire.* **2** °strife, °controversy, °dispute, contention, °debate, altercation, °argument, velitation; °conflict, °struggle, °fight, °battle, °combat, °war: *The contest is between those for and those against capital punishment.* —*v.* **3** contend, °argue, °dispute, °debate; °challenge, °(call into) question, °oppose, °counter, confute, object to, refute: *Cranston has decided to contest his father's will.*

contestant *n.* contender, °competitor, °opponent, °rival, °adversary, entrant, °player, °participant: *The contestant from Chearsley has won.*

context *n.* °structure, framework, °environment, °situation, circumstance(s); ambiance *or* ambience, °surround, surroundings, °frame (of reference), °setting, °background: *It is often hard to understand something taken out of its context.*

continual *adj.* °constant, incessant, °perpetual, °nonstop, °persistent, uninterrupted, °regular, °steady, unbroken, unceasing, ceaseless, °constant, °eternal, unremitting, interminable, °endless, unending; *Loosely* °continuous: *She has this continual ringing in her ears.*

continue *v.* **1** °carry on, °proceed (with), keep up *or* on *or* at, °go on (with), °pursue, °persist (in), °persevere (in): *Please continue whatever it was you were doing.* **2** °endure, °last, °go on, °persist, be prolonged, °remain: *How long will the curfew continue?* **3** °maintain, °keep (on), °prolong, °perpetuate, °carry on (with), persist in *or* with, °sustain, °extend: *Mother continued her career throughout my childhood.* **4** °resume, °pick up, °take up, °carry on (with): *Allow me to continue my story and don't interrupt again.* **5** °proceed, °go (on), °extend: *The road continues for about a mile, ending at the sea.*

continuous *adj.* **1** connected, unbroken, uninterrupted: *The wall is continuous except for one gate.* **2** incessant, °persistent, °perpetual, °nonstop, unceasing, ceaseless, °constant, unremitting, interminable, °endless, unending; *Loosely* °continual: *A continuous stream of refugees passed through the camp.*

contract *n.* **1** °agreement, °understanding, °deal, °bargain, °arrangement, °pact, commitment, °obligation, °compact: *I have here a valid contract to supply horseshoes to the cavalry.* —*v.* **2** °engage, °agree, °promise, covenant, °undertake: *Our company contracted to maintain the roads in this area.* **3** °catch, °acquire, °get, °come down with, °develop, become infected with, *Brit* go down with: *Eunice has contracted the Asian flu.* **4** °diminish, °shrink, draw together, °roll (oneself), °narrow, °squeeze, constrict, compress, condense, °decrease, °reduce: *When disturbed, the animal contracts itself into a ball.* **5** °wrinkle, °knit, crease, corrugate, °pucker: *His brow contracted into a frown.*

contradict *v.* **1** °deny, gainsay, °dispute, controvert, argue against; °oppose: *When I say the sun is shining, don't contradict me.* **2** contravene, belie, refute, disallow, °forbid, disaffirm, °counter, abrogate, nullify, annul, °reverse, °counteract: *The evidence yields nothing that contradicts my argument.*

contradictory *adj.* °inconsistent, °paradoxical, °incongruous, conflicting, °incompatible, discrepant; °ambiguous, ambivalent: *The witnesses' descriptions of the robbers are contradictory.*

contraption *n.* contrivance, °device, °gadget, °mechanism, °apparatus, *Colloq* widget, thingumabob *or* thingamabob, thingumajig *or* thingamajig, thingummy, whatsit, doodah, thingy, *US* gizmo *or* gismo, Rube Goldberg (invention), whatchamacallit, *Colloq Brit* gubbins: *Those Wright brothers have built a contraption that they think will fly.*

contrary *adj.* **1** °opposite, °opposing, °opposed, °different, °contradictory, conflicting, antagonistic: *Set aside enough time to hear the contrary side of the argument.* **2** antagonistic, °perverse, contrarious, °hostile, unfriendly, inimical, cross-grained, refractory, contumacious, °self-willed, °argumentative, unaccommodating, antipathetic, *Literary* froward: *He can disagree, but why must he be so contrary?* **3** adverse, unfavorable, °inauspicious, unlucky, °unfortunate, unpropitious, °untoward, °inopportune, °bad, °foul: *We ran into contrary winds and were delayed.* —*n.* **4** °opposite, °reverse: *Her present position is the direct contrary of that she took yesterday.* —*adv.* **5** perversely, oppositely, contrariwise, contrarily, in opposition to: *The rat in the maze acted contrary to the expected pattern.*

contrast *v.* **1** juxtapose, °oppose, °compare, °distinguish, °differentiate, °discriminate, set *or* place against; °set off: *Contrast life in the eighteenth century with life today.* **2** °conflict, differ *or* diverge *or* deviate (from): *The two styles contrast sharply. Australian speech contrasts with that of Canada.* —*n.* **3** °comparison; °difference, °distinction, °disparity, °dissimilarity: *The author emphasizes the contrasts between the two economic policies.*

contribute *v.* **1** °give, °furnish, °donate, °bestow, °grant, °present, °provide, °supply: *We contributed three paintings by Longchamp to the museum.* **2** ***contribute to.*** °add to, °promote, °advance, °help, °aid, °support, °forward, have a hand in, play a part *or* role in: *They believe that poor parental supervision contributes to juvenile delinquency.*

control *v.* **1** °command, °dominate, °direct, °steer, °pilot, hold sway over, °rule, exercise power *or* authority over, °govern, °manage, °lead, °conduct, be in control (of), call the tune, °guide, °oversee, °supervise: *Does she really control the future of the company?* **2** °check, hold back *or* in check, °curb, °repress, °contain: *Try to control yourself.* **3** °suppress, °put down, °master, °subdue, °restrain, °curb, °manage: *They were totally unable to control the unruly teenagers.* —*n.* **4** °command, °direction, °power, °authority, °leadership, °management, °guidance, supervision, °oversight, °charge; °sway, °rule, °jurisdiction: *Turn control of the mission over to Mrs. Beale. The court is under the control of the state.* **5** °restraint, °check, °curb, mastery, °command, dominance, °domination: *You must get better control over your emotions.* **6** °knob, button, dial, °handle, lever, °switch; °device, °mechanism: *This control opens the bomb bay doors.*

controversial *adj.* **1** °debatable, °disputable, °questionable, °moot, °doubtful, °unsettled: *Who will run the department is a controversial matter.* **2** polemical, dialectic, litigious, °factious: *She has studied the controversial writings of the nineteenth-century feminists.* **3** disputatious, °argumentative, contentious; °provocative: *Race relations have remained a controversial issue for centuries.*

controversy *n.* **1** °dispute, °debate, contention, °argument, argumentation, disputation, wrangling, confrontation, questioning, °disagreement: *The pirates enjoyed the control of the seas without controversy.* **2** °argument, °dispute, °disagreement, °quarrel; squabble, °tiff, *Colloq* spat: *The controversy still rages over the theory of natural selection.*

convalesce *v.* °recover, °improve, get better, °recuperate: *The doctor said I needed only a week to convalesce after the surgery.*

convenient *n.* **1** °suitable, commodious, °useful, °helpful, °handy, °serviceable, °expedient, °opportune, °advantageous: *The bus is quite convenient for getting to and from the airport.* **2** °handy, °nearby, within (easy) reach, at one's fingertips, close at hand, °available, °accessible; °(at the) ready: *There's a convenient post office round the corner.*

convention *n.* **1** °assembly, °meeting, °gathering, congregation, congress, °conference, symposium, °council, conclave, °diet, synod, seminar: *The annual convention of cat fanciers will take place in June.* **2** °rule, °practice, °custom, °tradition, °usage, °formality, conventionalism: *According to convention, this year's vice president becomes president next year.*

conventional *adj.* °customary, °habitual, °usual, °normal, °regular, °standard, °orthodox, °traditional, established, °ordinary, °everyday, °common, commonplace, °accustomed, received, agreed; °reactionary, °old-fashioned, °stodgy, °stuffy, old hat: *Conventional methods of teaching mathematics are being criticized.*

converge *v.* come *or* go together, °meet, °join, °unite, °merge, °coincide; °blend: *The roads converge in the valley.*

conversation *n.* °discussion, °talk, °chat, °dialogue, colloquy, °parley; chitchat, °gossip, discourse, °palaver, *Colloq chiefly Brit* chinwag: *The conversation about the situation in the Middle East ended abruptly. I want action, not conversation.*

conversationalist *n.* deipnosophist: *It was a pleasure to have dinner with an intelligent conversationalist for a change.*

converse *v.* °discuss, °talk, °speak, °chat, °parley, discourse, °gossip, °chatter: *The men were conversing about her over dinner.*

convert *v.* **1** °change, °modify, °alter, °transform, transmute, mutate, transfigure, transmogrify, remodel, remake, metamorphose: *We converted our rowing boat into a sailing dinghy.* **2** proselytize, °switch, °change (over): *To avoid the horrors of the Inquisition, many Spanish Jews converted to Catholicism.* —*n.* **3** proselyte; neophyte, catechumen, °disciple: *Converts are often the most passionate believers.*

convict *v.* **1** find *or* prove guilty, *Slang* °nail: *She was convicted of theft.* —*n.* **2** °prisoner, °captive, *Slang* con, jailbird *or Brit also* gaolbird, °lag: *The rioting convicts burnt down two prison buildings.*

conviction *n.* **1** proof of guilt: *After his conviction, he was sentenced to life imprisonment.* **2** °belief, °opinion, °view, °persuasion, °position: *It is her conviction that the painting is by Titian.* **3** °certainty, sureness, positiveness, °confidence, °assurance, certitude: *He doesn't have the courage to back up his convictions.*

convince *v.* °win over, °talk into, °persuade, bring (a)round, °sway: *I convinced them to be more sympathetic to older people.*

cool *adj.* **1** °chilly, °chill, chilling, cooling, unheated; chilled, °cold, °refreshing, °fresh: *It is cool when you're not close to the fire. I'd prefer some cool lemonade.* **2** °calm, °serene, °collected, °levelheaded, °quiet, unexcited, unemotional, undisturbed, unexcitable, unruffled, unflappable, coolheaded, °relaxed, controlled, under control, °self-possessed, self-controlled, unperturbed, °phlegmatic, composed, imperturbable: *He remains cool even in a crisis.* **3** °dispassionate, °cold, °coldblooded, emotionless, °deliberate, °coldhearted, °calculated, °willful, °premeditated, °purposeful, purposive: *It was clearly the cool act of a professional criminal.* **4** uninvolved, °distant, °remote, °aloof, °detached, removed, uninterested, unconcerned, °unsympathetic, apathetic, °cold, °coldhearted, °coldblooded: *How can you be so cool where human lives are concerned?* **5** °lukewarm, °distant, uncordial, unfriendly, unsociable, °unapproachable, °standoffish, °forbidding, unwelcoming, °cold, °frigid: *After the affair, she was distinctly cool toward him.* **6** °bold, °audacious, °brazen, °overconfident, °presumptuous, °shameless, °unabashed, °impertinent, °impudent, °insolent: *I cannot account for the cool way he insulted his host.* —*n.* **7** coolness, °chill, chilliness, *Colloq* coolth: *I shall have a sherry to ward off the cool of the evening.* **8** aplomb, °poise, sedateness, °control, °self-control, composure, °sang-froid: *He really lost his cool when she told him he was a lousy driver.* —*v.* **9** °chill, °refrigerate, ice: *Cool the pudding before serving.* **10** °diminish, °reduce, lessen, abate, °moderate: *Her interest quickly cooled when she discovered he was married.*

cooperate *v.* **1** °collaborate, work together, °join, °unite, interact, °team up, join forces, act jointly *or* in concert: *If we cooperate, the work will be done in half the time.* **2** °participate, °contribute, lend a hand, °help, °assist: *You must learn to cooperate and not just sit there.*

cooperation *n.* **1** collaboration, teamwork, interaction, synergism *or* synergy: *Only through cooperation will we be able to achieve success.* **2** °support, °help, °aid, °assistance, °patronage, °backing, advocacy, °favor, helping hand, °friendship, °blessing, sponsorship, °auspices, backup: *We needed the cooperation of people like you to mount the exhibition.*

coordinate *v.* **1** °organize, classify, °order, °arrange; systemize, systematize, codify, °categorize, °group, °match (up), °dispose, °rate, °rank, °grade: *Coordinate the information before preparing the report.* **2** harmonize, correlate, °unify, °mesh, synchronize, °integrate, *Colloq* pull together: *We must coordinate our efforts for the best results.* —*adj.* **3** °equivalent, °parallel, °correspondent, complementary, correlative, °equal, °reciprocal, coordinating, coordinative, *Technical* paratactic: *The two systems are coordinate and operate in parallel.*

cope *v.* **1** °manage, get along *or* by, °make do, °survive, subsist, °come through: *Even with seven children to care for, she copes very well.* **2** ***cope with.*** be a match for, °withstand, contend with *or* against, °handle, °deal with, °dispose of: *Our small force cannot cope with those bands of guerrillas.*

copy *n.* **1** °reproduction, °replica, °facsimile, °likeness, °imitation, °double, °twin, duplication, °duplicate, °transcript, replication, carbon (copy), photocopy, °print: *This is a copy of the lost manuscript.* **2** °example, °sample, °specimen: *How many copies of the book have been sold?* **3** °text, °writing: *The copy is ready; we are waiting for the illustrations.* —*v.* **4** °reproduce, °duplicate, replicate, °transcribe, *Colloq* °knock off: *Don't copy others' work—they might be wrong.* **5** °imitate, °mimic, impersonate, emulate, ape, °parrot, °echo: *Ted copies the rock stars in every possible detail of their dress and behavior.*

cord *n.* °string, °line, °twine; °rope: *Tie the cord around the parcel twice.*

cordial *adj.* °friendly, °warm, affable, °amiable, °kindly, °genial, °gracious, welcoming, °pleasant, °good-natured, °nice; °courteous, °polite: *After a cordial greeting at the door, the guests were served champagne.*

core *n.* **1** °center, °heart, °middle, °nucleus, inside(s): *Remove the core of the apple first.* **2** °essence, marrow, °heart, °pith, °gist, °quintessence, sum and substance: *Her refusal to consider any alternative is the core of the problem.* —*v.* **3** °pit, °seed: *The pie might have tasted better if you'd cored the apples first.*

corps *n.* body of men *or* women, troop, cadre, °unit, °detachment, °cohort, °division, battalion, brigade, °platoon, °squad, column, squadron: *We delivered supplies to the medical corps.*

corpse *n.* °body, °remains, °cadaver, *Slang* °stiff; (of an animal) carcass: *The corpses were buried in a mass grave.*

correct *v.* **1** °right, set *or* put right, °amend, redress, °rectify, °remedy, °repair, °fix, right wrongs; °cure: *A good mechanic can correct the faults in this engine.* **2** °scold, admonish, °rebuke, °reprimand, °berate, chide, °reprove; censure, °blame: *You mustn't correct people for their bad manners.* **3** °punish, °chastise, °chasten, °discipline, °castigate: *The boys were corrected for swearing at the teacher.* **4** °reverse, °offset, °counteract, counterbalance, °neutralize, nullify, °make up for, annul, °cancel; °adjust, °change, °modify: *Adding this fertilizer will correct the acid content of the soil.* **5** °mark, °grade: *The exam papers haven't yet been corrected.* —*adj.* **6** °proper, °decorous, °decent, °appropriate, °suitable, °fit, °right, °meet, °fitting, °befitting, apt, *de rigueur, comme il faut, Old-fashioned Brit* tickety-boo: *I have found her behavior correct at all times.* **7** °conventional, established, °set, °standard, °normal, °orthodox, approved, °in order, *de rigueur, comme il faut,* °usual, °natural, °customary, °traditional, done, °right, *Old-fashioned Brit* tickety-boo: *Sending flowers to the funeral parlor would be the correct thing to do.* **8** °accurate, °right, °precise, °exact, °factual, valid, °true, °proper, °fitting, apt, °suitable, °appropriate; °faultless, °perfect, unimpeachable: *Joanna gave the correct answer.*

correction *n.* **1** °improvement, emendation, rectification, redress, °remedy, reparation, °amendment; corrigendum: *With these corrections, the work is vastly better.* **2** °punishment, castigation, chastisement: *He resented her continual correction of him for trivial things.*

correspond *v.* **1** °agree, °conform, °tally, °comply, °accord, harmonize, be congruous, °match, °coincide: *The results of the surveys correspond.* **2** °write, °communicate, be in touch *or* contact: *We have been corresponding for years.*

correspondent *n.* newspaperman, newspaperwoman, pressman, presswoman, °journalist, °reporter, stringer, newsman, newsperson: *Here is a report from our correspondent in Sydney.*

corridor *n.* °hall, hallway, °passage, passageway: *We met in the corridor outside my room.*

corrupt *adj.* **1** °dishonest, untrustworthy, °dishonorable, underhand(ed), °venal, *Colloq* °crooked: *He got off by bribing a corrupt judge.* **2** debased, depraved, °perverted, subverted, °evil, °wicked, °degenerate, degraded: *The inhabitants practiced a corrupt form of Christianity.* —*v.* **3** °debase, °pervert, °subvert, °degrade, deprave, °warp: *A funds manager could easily be corrupted by all that money.* **4** °adulterate, °contaminate, °pollute, °taint, defile, infect, °spoil, °poison: *Drainage from the site has corrupted the purity of the water.* **5** °bribe, suborn, °buy (off): *He thought he knew a juror who might be corrupted.*

cost *n.* **1** °price, °outlay, °payment, °charge, °expense, °expenditure, °rate, °tariff: *How can you afford the cost of this watch? If the cost increases, the selling price must go up.* —*v.* **2** sell for, °get, °fetch, °bring in, *Colloq* set (someone) back: *This would cost twice as much in London.*

costume *n.* °dress, clothing, attire, °clothes, garb, °apparel, raiment, °garments, °outfit, vestment, livery, °uniform, °kit, *Colloq* °gear, togs, get-up; *Slang* °rags, *US* threads: *What kind of costume is that you're wearing?*

cozy *adj.* **1** °comfortable, °snug, °warm, °restful, °secure, relaxing, °easy, *US* °cozy, *Colloq* comfy: *They bought a cozy little rose-covered cottage in the Cotswolds.* **2** °convenient, °expedient, self-serving, underhand(ed): *He has a cozy arrangement with the planning board.*

cot *n.* bed, crib; cradle, bunk: *Baby is asleep in his cot.*

cottage *n.* °hut, °shack, °cabin, bungalow, shanty, *Literary* °cot; *US and Canadian* °lodge, chalet: *She's going to stay at our cottage for a week.*

couch *n.* **1** sofa, settee, °settle, divan, love seat, chaise (longue); daybed; °tête-à-tête, vis-à-vis, siamoise; *US* davenport: *Come sit by me on the couch.* —*v.* **2** embed, °frame, °style, °express, °phrase: *Her warning was couched in friendly words.*

council *n.* **1** °assembly, °meeting, conclave, °conference, synod, consistory, °convention, congress, congregation, °gathering, convocation, *US* caucus: *The vote of the council is to ban nuclear arms.* **2** °board, °ministry, directors, °cabinet, panel, °committee, °body, directorate, directory, caucus: *She was elected to the council last year.*

counsel *n.* **1** °advice, °judgment, °direction, °opinion, °guidance, °instruction, °recommendation, exhortation, *Technical* paraenesis: *Your counsel has always been wise in the past.* **2** consultation, °discussion, deliberation, °consideration: *We took counsel with the cabinet on the matter.* **3** adviser *or* advisor, °guide, °counselor; °lawyer, *Brit* barrister; *US* attorney: *My counsel suggests we settle out of court.* —*v.* **4** °advise, recommend to, suggest to, °instruct; °guide: *I have counseled her to pursue the matter.*

counselor *n.* adviser *or* advisor, °counsel, °lawyer, *Brit* counsellor-at-law, barrister, *US* counselor-at-law, attorney: *We have retained Vestley and Stock as our counselors.*

count *v.* **1** count up *or* off, °enumerate, °number, °calculate, add up, °total, °reckon, °compute, °tally, °figure up, quantify, *Colloq* °figure out: *Maddie counted the number of pencils in the box.* **2** °include, °consider, °regard, deem, °judge, look on *or* upon: *You can count me among those who favor the idea.* **3** ***count on*** *or* ***upon.*** rely on *or* upon, depend on *or* upon, be sure of, °trust, bank on, be confident of, *Chiefly Brit* (*dialectal in US*) reckon on *or* upon, *Chiefly US* figure on *or* upon: *I knew I could count on Moira to do the right thing.*

counter *n.* **1** °token, disk *or* disc, °chip, °piece, marker: *She placed three counters on the number 14.* **2** °table, °bar: *We do not serve beer at this counter.*

counteract *v.* counterbalance, °neutralize, °correct, annul, nullify, °cancel, °oppose, °mitigate: *The coffee counteracted the effect of the sleeping pill.*

counterfeit *adj.* **1** forged, °fake, °fraudulent, °imitation, °bogus, °spurious, *Colloq* phoney *or US also* phony: *The bank refuses to accept counterfeit money.* **2** make-believe, °sham, °pretended, °pretend, feigned, °insincere, °fake, faked, °false, °artificial, meretricious, pseudo, °factitious, °synthetic, °unreal, simulated: *You were warned about his counterfeit sincerity.* —*n.* **3** °fake, °imitation, °forgery, °reproduction, *Colloq* phoney *or US also* phony: *This is the original deed, that one is a counterfeit.* —*v.* **4** °forge, °copy, °reproduce, °falsify, °imitate; *Slang* hang paper: *He made a living counterfeiting passports.* **5** feign, °pretend, simulate, °put on, °fake: *The suspects have shown signs of wealth that are difficult to counterfeit.*

counterfeiter *n.* *Slang* paperhanger: *The counterfeiter, who forged only five-dollar bills, was arrested today.*

country *n.* **1** °nation, °state, °power; °territory, °realm: *How many countries belong to the British Commonwealth?* **2** °(native) land, homeland, °fatherland, motherland, mother country: *I would gladly fight for my country.* **3** countryside, rural area *or* surroundings, °provinces, hinterlands; mountains, woods, wilderness, outback, *Colloq* sticks, *US* boondocks, boonies: *We are spending our vacation in the country.*

couple *n.* **1** °pair, duo, twosome; °brace, °span, yoke, °team: *They certainly make a nice couple.* **2** ***a couple of.*** a few, °several, a handful (of), one or two, three or four: *I'll be with you in a couple of minutes.* —*v.* **3** °join, °link, yoke, °combine, °unite, match up, °connect: *The two carriages are easily coupled together.*

courage *n.* °bravery, valor, boldness, intrepidity, gallantry, dauntlessness, °daring, fearlessness, heroism, °nerve, *Colloq* °grit, °guts, °pluck, °spunk, *US* moxie, sand, *Slang Brit* °bottle: *She had the courage to face the two of them alone.*

courageous *adj.* °brave, valiant, valorous, °bold, °intrepid, °gallant, °dauntless, °daring, °fearless, °heroic, *Colloq* plucky: *The soldiers were very courageous and fought against tremendous odds.*

course *n.* **1** °path, °way, °orbit, °route, °run, °track, ambit, °line, °circuit, °passage: *We continued on our course. The sun pursued its fiery course across the heavens.* **2** °movement, °progress, °headway, °advance, °progression; °speed: *The driver slackens his course at the curves.* **3** °procedure, °process, °performance, °routine, °conduct, °order, °practice, dispatch *or* despatch, °execution: *In the course of her duties, she handles a great deal of money.* **4** °direction, °tack: *If we stay on this course we'll run aground.* **5** °class, °lecture, seminar, °program: *You should sign up for a course in English grammar.* **6** ***of course.*** °naturally, °surely, certainly, °positively, °obviously, °definitely, assuredly, °by all means; °undoubtedly, indubitably, without (a) doubt, no doubt, *Colloq US* °sure: *Of course I shall go to the theater with you!*

courteous *adj.* °polite, well-mannered, well-behaved, gentlemanly, °ladylike, °well-bred, °polished, urbane, civilized, °respectful, °civil, courtly, °proper, °decorous, °tactful, °considerate, °diplomatic: *He might have been rude to you, but he was always quite courteous to me.*

courtesy *n.* politeness, °elegance, courtliness, politesse, courteousness, °respect, respectfulness, good manners, °formality, °civility, °ceremony; red-carpet treatment: *I much appreciated the courtesy with which they treated me.*

cover *v.* **1** °protect, °shelter, °shield, °screen; °guard, °defend, °command: *The guns cover the approaches to the town.* **2** Also, ***cover up*** or ***over.*** °conceal, °hide, °bury, °mask, °shroud, °obscure; dissemble; °enclose, °envelop: *I was unable to cover my embarrassment. Her face was covered by the hood of the cloak.* **3** overlie, spread over, overspread, lie on, layer, °coat, blanket: *Oil covers the surface of the lake.* **4** °wrap, swaddle: *Mother covered us with warm blankets.* **5** °dress, °clothe, garb, attire, °robe, sheathe: *She was covered in silk from neck to ankle.* **6** extend *or* stretch over, °occupy, engulf, inundate, °submerge: *A lake has covered the original site of Abu Simbel.* **7** °include, °comprehend, °provide for, comprise, extend over, °contain, °embody, °incorporate, °account for, °take into account, °take in, °deal with: *This report covers our activities over the past year.* **8** °act, take responsibility *or* charge, stand *or* sit in, °substitute, °take over, run things, °double: *Go and get some coffee—I'll cover for you.* **9** °traverse, °complete, pass *or* travel over, °travel, °cross: *With frequent stops, we could not cover more than fifty miles a day.* **10** compensate for, °defray, be enough *or* sufficient for, °counter, °offset, counterbalance, °make up for, insure *or* protect against: *The policy covers losses of up to a million.* —*n.* **11** lid, °top, °cap, covering: *I can't find the cover for this pot.* **12** binding, boards, °wrapper, dust jacket, jacket: *You can't tell a book by its cover.* **13** Often, ***covers.*** blanket, quilt, eiderdown, duvet, bedclothes, bedding, °(bed) linen; coverlet, counterpane; *US* comforter: *I crept into bed and pulled the covers over my head.* **14** °shelter, °protection, concealment, hiding place, hide-out, °retreat, °refuge; °hide, *US and Canadian* °blind; *Colloq Brit* hidey-hole: *We tried to find some sort of cover till the sun went down.* **15** °cloak, °screen, °disguise, concealment, °pretense, °front, °camouflage, smoke screen, coverup, °mask, covering: *His bluster and bullying were only a cover for his cowardice.*

coward *n.* poltroon, craven, dastard, sissy *or* cissy, °baby, mouse, °milksop; Scaramouch *or* Scaramouche; *Colloq* chicken, *Slang* yellow-belly; *US and Canadian* milquetoast: *He's such a coward that he's afraid of his own shadow.*

cowardice *n.* cowardliness, chicken-heartedness, faintheartedness, timidity, timorousness, pusillanimity: *Owing to the cowardice of the lieutenant, the troop surrendered without a shot being fired.*

cowardly *adj.* °timid, °fearful, frightened, °afraid, °scared, °fainthearted, timorous, chicken-hearted, chicken-livered, lily-livered, white-livered, craven, namby-pamby, dastardly, pusillanimous, vitelline, *Slang* yellow, yellow-bellied: *The cowardly rascals ran from the battle.*

coy *adj.* °shy, °modest, diffident, demure, °timid, °bashful, °self-conscious, °sheepish, timorous, unassuming, unpretentious; °reserved, self-effacing, °retiring, °evasive, °reluctant, °recalcitrant: *She was so coy she would disappear whenever we had guests.*

crack *n.* **1** °break, °fracture, °chink, °crevice, °rift, °gap, °flaw, °split, fissure, °slit, cleft, °split, °check, °rupture, °breach: *The crack in the dam was caused by an earthquake.* **2** °snap, °report, bang, °clap, °shot: *I ducked when I heard the crack of the rifle.* **3** °moment, °instant, °time, °second: *She gets up at the crack of dawn. He'll be here in a crack.* —*v.* **4** °snap: *He cracks the whip and the horses start up.* **5** °break, °fracture, °rupture; °shiver, °shatter, smash: *He fell backward on the pavement and cracked his skull.* **6** fissure, °craze, crackle, *US* alligator: *The sun's heat has caused the paint to crack.*

craft *n.* **1** °skill, °ability, artisanship, handiwork, °ingenuity, skillfulness, °art, °talent, °dexterity, cleverness, mastery, expertness, °expertise, °flair, °genius, *Colloq* know-how: *Considerable craft is required to make that kind of jewelry.* **2** °deceit, guile, cunning, °fraud, °trickery, wiliness, foxiness, artfulness, craftiness, duplicity: *He exhibits that crooked wisdom called craft.* **3** °trade, °occupation, °calling, °vocation, métier; °profession: *He was a member of one of the medieval craft guilds.* **4** °vessel, °ship, °boat; hovercraft; aircraft, airplane, °plane; spaceship, spacecraft, °rocket: *One day there will be at least as many craft in space as there now are in the air.* —*v.* **5** °make, °fashion, °fabricate: *She crafted these figures out of solid wood.*

crafty *adj.* °artful, cunning, °clever, °shrewd, °foxy, canny, °wily, °sly, °scheming, °calculating, °designing, plotting, °tricky, °sneaky, °deceitful, °shifty, °dodgy, guileful, insidious, double-dealing, °two-faced, duplicitous, treacherous: *That crafty beggar has made off with my life's savings.*

crag *n.* °cliff, °bluff, tor, °peak, °rock, escarpment, scarp, °precipice, *US* palisade: *Soaring above us was a huge crag that we still had to climb.*

cram *v.* **1** °pack, °stuff, overstuff, overcrowd, °jam, °fill: *The car was crammed to the top with suitcases.* **2** °study, burn the midnight oil, *Literary* lucubrate, *Colloq* °grind, *Brit* swot: *Bob cannot go out because he's cramming for an exam.*

cramped *adj.* °tight, crowded, incommodious, uncomfortable, °close: *The tiny cabin was too cramped to hold all of us at once.*

crank *n.* **1** °eccentric, °character, °oddity, *Colloq* nut; *Brit slang* nutter, nutcase: *Pay no attention to Thirlew, he's just a crank.* **2** monomaniac, °zealot, °fanatic: *That restaurant is patronized mainly by health-food cranks.*

cranky *adj.* **1** °eccentric, °odd, °weird, °strange, °queer, °peculiar, quirky, °capricious, °whimsical: *That hermit is a cranky old bird.* **2** °testy, grouchy, crabby, °short-tempered, °surly, irascible, °waspish, churlish, °gruff, curmudgeonly, °cantankerous, choleric, °snappish, °petulant, °peevish, contentious, °querulous, °irritable, splenetic, *Colloq* crotchety: *He's always cranky before breakfast.*

cranny *n.* °chink, °crevice, °crack, fissure, °check, °fracture, °break, °furrow, °split, cleft: *Flowers grew from the crannies in the ancient wall.*

crash *v.* **1** °fall, °topple: *The vase crashed to the stone floor.* **2** °force, °drive, °run, smash: *He crashed the car into a wall.* **3** bang, °boom, °explode: *The thunder crashed all around us.* —*n.* **4** °boom, bang, smash, °explosion, °blast: *We heard a great crash as the building collapsed.* **5** °disaster, °collapse, °failure: *The stock-market crash has had a devastating effect.*

crawl *v.* **1** °creep, worm, °wriggle, wiggle, °squirm; °edge: *A spider is crawling on your collar.* **2** inch, °creep, °drag: *For a solid hour the cars just crawled at a snail's pace.* **3** cower, °cringe, grovel, toady, fawn: *Don't worry, he'll come crawling, begging you to take him back.* **4** °teem, °abound, °swarm, be overrun *or* swamped: *The scene of the crime crawled with police.*

craze *n.* °fad, °fashion, °trend, °enthusiasm, °rage, °mania, °thing, °obsession; last word, *dernier cri*: *The craze for printed T-shirts goes on and on.*

crazy *adj.* **1** °mad, °insane, demented, °deranged, °unbalanced, unhinged, lunatic, *non compos mentis,*

°daft, certifiable, °mental, touched (in the head), out of one's mind *or* head, mad as a March hare *or* hatter, maddened, crazed, *Colloq* balmy, cuckoo, cracked, crackers, crackbrained, dotty, daffy, dippy, gaga, goofy, crackpot, loony, off one's rocker, have a screw loose, screwy, batty, bats, bats-in-the-belfry, *Brit* barmy (in the crumpet), potty, bonkers, round the bend *or* twist, off one's chump, doolally, *US* off one's trolley, out of one's gourd, screwball, nuts, nutty (as a fruitcake); *Slang* bananas, *US* out to lunch, meshuga, out to lunch, flaky, flaked-out, (plumb) loco: *His wife thinks he's crazy to want to walk around the world.* **2** °silly, °absurd, °foolish, °nonsensical, °inane, °ridiculous, °preposterous, laughable, risible, °ludicrous, asinine, °stupid, moronic, imbecile *or* imbecilic, idiotic, °feebleminded, °harebrained, *Colloq* crackpot: *Someone came up with a crazy idea of a square tennis ball to slow down the game.* **3** °impractical, °impracticable, unworkable, °unsound, °pointless, °imprudent, °rash, °reckless, ill-considered: *They once thought that Columbus' plan to sail round the world was crazy.* **4** °enthusiastic, °eager, avid, zealous, °keen, °excited: *I'm really crazy about windsurfing.* **5** °infatuated, keen on *or* about, °wild, °mad, *Colloq* dotty, *US* nuts, nutty; *US slang* ape: *Marjorie, I'm absolutely crazy about you.*

create *v.* **1** °make, °produce, °form, bring into being, °originate, °conceive; sire, °father: *The question remains whether God created Man or vice versa.* **2** engender, beget, °spawn, °generate, °invent, °imagine, °think up, °frame, °forge, °fashion, °fabricate, °manufacture, °develop, °design, contrive, °devise, °produce, dream up, °initiate: *Here is where they create many of the most successful advertising programs.*

creation *n.* **1** °beginning, °origin, °birth, °start, inception, genesis, making, °formation: *The creation of the lake began with the damming of the stream.* **2** the world, the universe, the cosmos: *In all creation there'll never be another you.*

creative *adj.* °imaginative, inventive, originative, artistic, °original, °ingenious, °resourceful: *A truly creative artist seldom lacks for inspiration.*

creator *n.* **1** originator, °author, initiator, °founder, °father, inventor, architect, °designer, framer, maker, prime mover: *The creator of this painting must have been a genius.* **2** God, Supreme Being, the Deity: *Some day, you will have to answer to your Creator for your treachery.*

creature *n.* **1** being, °organism, °entity, living thing: *He created every creature that ever lived.* **2** ***creature comforts.*** (physical *or* bodily *or* material *or* mundane *or* superficial *or* nonspiritual) luxuries: *He has the money to enjoy all the creature comforts.*

credit *n.* **1** °belief, °faith, °trust, credence: *I don't give much credit to what they say.* **2** creditation, °acknowledgment, °attribution, ascription: *Credit for inventing the telegraph goes to Guglielmo Marconi.* **3** °trust, °confidence, faithfulness, reliability, trustworthiness, °honesty, °probity, dependability; solvency: *Her credit rating at the bank is excellent.* **4** °honor, commendation, °praise, °tribute, acclaim, °esteem, °recognition, °merit: *The team's victory in the finals has brought credit to the school.* —*v.* **5** °believe, °trust, hold accountable, put *or* place one's faith *or* confidence in, have faith *or* confidence in, °rely on, °accept, depend on *or* upon: *If you credit the Bible, the world was created in six days.* **6** ascribe, °acknowledge, °attribute, °assign; °impute: *The goal was credited to Beauchamp.*

creed *n.* °tenet, dogma, °doctrine, credo, teaching, °principles, °belief, set of beliefs: *She adheres to the creed of the Golden Rule.*

creek *n.* **1** (*in Brit. usage*) inlet, bay, cove, °harbor: *Overnight we moored in a little creek, sheltered from the sea.* **2** °(in US and Canadian usage) stream, streamlet, °brook, rivulet, rill, runnel, °run, °burn: *We used to fish in the creek behind the house.*

creep *v.* **1** °crawl, °slither, inch, °squirm, °wriggle, wiggle: *A tiny lizard was creeping up the wall.* **2** °crawl, °drag: *The hours creep by slowly when you have nothing to do.* **3** °steal, °sneak; °slink, skulk, tiptoe, *Colloq* °pussyfoot: *The thief must have crept in through this window. Someone is creeping about out there in the dark.*

crescent *n.* **1** demi-lune, semi-lune, lune, lunette: *The moon's crescent hung low in the western sky.* —*adj.* **2** crescent-shaped, demi-lune, semi-lune, biconcave, concavo-concave: *For chopping, the chef uses a crescent blade that just fits the wooden bowl.*

crest *n.* **1** °top, °summit, °pinnacle, °peak, °head, °ridge: *The surfers rode in on the crest of a wave.* **2** °seal, °device, °figure, badge, °emblem, insigne, °symbol, °design: *The school crest shows an inkpot and a scroll.* —*v.* **3** °top, °crown, surmount, °cap: *The ancient walls were crested with ivy.* **4** culminate, °reach, °top, *US* top out: *The floodwaters crested at nine feet.*

crevasse *n.* °gorge, chasm, °abyss, °ravine, fissure, °crack, °furrow: *One of the climbers fell into a crevasse in the glacier.*

crevice *n.* °crack, fissure, °chink, cleft, °cranny, °groove, °furrow, °break, °split, °rift: *Water ran down the crevices in the rocks.*

crew *n.* °group, °company, °band, troupe, °party, °gang, °team, °corps, °body: *We shall need a crew of twenty for tomorrow's job.*

crime *n.* °offense, °violation, °misdeed, °wrong; felony, misdemeanor; lawlessness: *The number of crimes of violence is increasing.*

criminal *adj.* **1** °illegal, °unlawful, °illicit, °lawless, °dishonest, *Colloq* °crooked: *Arson is a criminal act. We have to weed out the criminal element.* **2** °wicked, °evil, °bad, °wrong, °corrupt, °vile, °black, °immoral, amoral, °sinful, °villainous, iniquitous, flagitious, depraved; °disgraceful, reprehensible: *The way they treat their children is absolutely criminal.* —*n.* **3** °felon, °convict, lawbreaker, °outlaw, °culprit, °offender, °miscreant, malefactor, wrongdoer, °villain, °scoundrel, knave, blackguard; °gangster, mafioso, desperado, °racketeer; °hoodlum, °thug, hooligan, °tough, ruffian, °terrorist, *Colloq* roughneck, bad guy, black hat, bad hat, baddie *or* baddy, crook; *Slang* hood, *US* mobster: *He was arrested for consorting with known criminals.*

cringe *v.* **1** cower, wince, °flinch, quail, °recoil, blench, °tremble, °quiver, quake *or* shake in one's boots *or* shoes, °shrink: *That dirty little coward cringed even when they called his name.* **2** °defer, °kowtow, grovel, °crawl, fawn, bootlick, *US* apple-polish; *Slang* kiss someone's *Brit* arse *or US and Canadian* ass, *Taboo slang* brown-nose: *Flewitt cringed before the magistrate, his eyes downcast, tugging his forelock.*

cripple *n.* **1** amputee, paralytic: *He has been a cripple since the accident.* —*v.* **2** disable, °lame, °incapacitate, °handicap, °maim; °impair, °damage, °weaken, debilitate, emasculate, °enervate: *She was crippled when a child. The dictator's power was crippled by the revolt.*

crippled *adj.* **1** °disabled, °lame, handicapped, incapacitated; weakened, °weak, debilitated: *He takes care of his crippled mother. The crippled party platform succumbed to attack from the far left.* **2** damaged, immobilized, inoperative: *There is a crippled car blocking the road.*

crisis *n.* **1** turning point, critical time *or* moment: *She has passed the crisis and will be better tomorrow.* **2** °disaster, °emergency, °calamity, °catastrophe, °danger: *The storm has created a crisis and the residents are being evacuated.*

crisp *adj.* **1** °brittle, crunchy, friable, breakable, crumbly, frangible: *Keep the biscuits crisp in this special jar.* **2** curly, crispy, crinkly, frizzy, frizzled: *His hair is brown and crisp, just like his father's.*

critical *adj.* **1** carping, °faultfinding, censorious, disparaging, depreciatory *or* depreciative, depreciating, deprecatory *or* deprecative, deprecating, judgmental: *The article was highly critical of the council.* **2** °crucial, °important, °essential, °basic, °key, decisive, °pivotal, °vital, °momentous: *The meeting at the bank will be critical for us.* **3** °grave, °serious, °dangerous, °uncertain, °perilous, °severe, touch-and-go, °ticklish, °sensitive, °touchy, *Colloq* °parlous: *His illness has reached the critical stage.*

criticism *n.* **1** °judgment, °evaluation, appraisal, °analysis, assessment, °estimation, valuation: *Their criticism was generally favorable.* **2** censure, °disapproval, condemnation, disparagement: *I was very upset by her criticism of my behavior.* **3** critique, °review, commentary: *My sister writes the theater criticisms for the* Journal.

criticize *v.* **1** °judge, °evaluate, value, assess, appraise, °estimate; °discuss, °analyze: *He criticizes books for the quarterly.* **2** censure, °find fault (with), °carp (at), °cavil (at), °condemn, °attack, °denounce, °disapprove (of), animadvert on *or* upon, °put down, impugn, °blast, °lambaste, *Colloq* °pan, °knock, *Brit* slate: *His book was criticized because of its poor scholarship. Why must he constantly criticize, even when there is nothing wrong?*

crooked *adj.* **1** °criminal, °dishonest, °illegal, °unlawful, °illicit, °wrong, °perverse, *Brit slang* °bent: *Selling a stolen painting is crooked.* **2** °bent, bowed, askew, awry, °deformed, distorted, contorted, °lopsided, twisted, °misshapen, °disfigured, warped, °gnarled: *Because of the constant west wind, those trees are all crooked.*

cross *n.* **1** crucifix, rood: *In ancient times, it was common to execute certain criminals by nailing them to a cross.* **2** °hybrid, cross-breed, °mongrel; °blend, °combination: *This is a cross between a plum and a pear.* —*v.* **3** ***cross off*** *or* ***out.*** °strike out, °erase, °cancel, °rub out, °delete, °wipe out: *After that remark, I'm crossing you off my list.* **4** °meet, intersect, °join: *The roads cross further on.* **5** cross over, go across, °pass over, °span, °traverse: *The bridge crosses the river here.* —*adj.* **6** °peevish, irritated, annoyed, piqued, °irritable, °testy, °snappish, irascible, °surly, choleric, splenetic, grouchy, huffish *or* huffy, pettish, °cranky, grumpy, °touchy, °moody, fractious, vexed, curmudgeonly, °petulant, °waspish, °querulous, °cantankerous, crusty, °short-tempered, on a short fuse, hot-headed, *Colloq* crotchety, *Slang Brit* shirty: *He's cross because he has a headache.* **7** annoyed, irritated, °angry, irate, °furious: *I was very cross that you took the car without permission.*

crouch *v.* °bend (down), squat (down), hunker down, °stoop (down): *If you crouch down, you can make it through the window.*

crowd *n.* **1** °throng, multitude, horde, °swarm, °mass, °press, °flood, °mob, °flock, °pack: *A huge crowd descended on the village square.* **2** °company, °set, °circle, °lot, °bunch, °group, coterie, °clique, claque, °faction: *She doesn't associate with our crowd any longer.* —*v.* **3** °throng, °swarm, °herd, °pour, °pile, °press, °cluster, °gather, °get together, °flood, °flock, °assemble, congregate: *People crowded into the stadium.* **4** °push, °press, °drive, shove, °thrust, °force, °load, °pack, °cram, °jam, corral: *The police crowded the hooligans into vans.* **5** compress, °squeeze, °pack, °jam, °cram, °collect; °stuff: *"We were crowded in the cabin, not a soul would dare to speak."*

crown *n.* **1** coronet, diadem, wreath, fillet, circlet, tiara: *The princess wore a golden crown set with jewels.* **2** °sovereignty, °rule, °dominion, °authority, °government, °realm, rulership, °jurisdiction: *They discovered many lands and annexed them to the Crown.* **3** °monarch, ruler, °sovereign, potentate; °king, °queen, emperor, empress, His *or* Her Majesty, His *or* Her Highness: *The resentment of the crown is a serious thing.* —*v.* **4** enthrone, *Colloq US* coronate: *He was crowned on the death of his father.* **5** °cap, °top, surmount, culminate, °climax, consummate, °fulfill, °reward: *All her years of practicing the violin were finally crowned with success.*

crucial *adj.* °critical, decisive, °pivotal, °vital, °momentous, °major, °important, °essential: *It is crucial that you press the right button.*

crude *adj.* **1** °unrefined, °raw, °natural, °original, unprocessed: *Those are the prices of crude oil, not of the gasoline used in cars.* **2** °rough, unpolished, °rudimentary, °immature, °undeveloped, °primitive, °unrefined, unfinished: *At this stage, she has only a crude idea of the design.* **3** °rough, °coarse, °rude, °unrefined, uncouth, crass, °gross, °rustic, uncivil: *Don't you despise his crude manners?* **4** °blunt, °brusque, °unsophisticated, °inconsiderate, °tasteless, indelicate, °offensive, °improper, °vulgar: *How crude of him to ask her how long since her husband had "croaked"!*

cruel *adj.* **1** °merciless, pitiless, hardhearted, °harsh, stonyhearted, °heartless, unsparing, °callous, °beastly, °coldblooded, °ruthless, °unkind, °hard: *It was cruel of you to refuse to help.* **2** °ferocious, °inhuman, barbaric, barbarous, °brutal, °savage, °bloodthirsty, °vicious, °sadistic, °fiendish, diabolical, hellish, °atrocious, Neronian *or* Neronic *or* Neroic: *The Nazis subjected him to the cruelest tortures.*

cruise *v.* **1** °sail, °coast, °travel, °journey, voyage; yacht: *We cruise in the Caribbean during the winter.* —*n.* **2** °sail, voyage, °journey, boat *or* yachting trip: *I took a three-day cruise around the Isle of Wight.*

crumb *n.* °fragment, °morsel, °bite, °scrap, °particle, °shred, snippet, °sliver, °bit, °speck, scintilla, mote, molecule, atom: *There isn't a crumb of food in the house.*

crumble *v.* °disintegrate, °fragment, break apart, °break up, °shiver, come to pieces: *Acid rain has caused the stone façade to crumble. In the face of the attack, his resolve crumbled.*

crumple *v.* °wrinkle, °crush, crease, °rumple, °mangle, crinkle: *Your jacket is all crumpled.*

crunch *v.* **1** °chew, °bite, °crush, °grind, °munch: *If you are caught, crunch the vial between your teeth to release the poison.* —*n.* **2** moment of truth, decision time, °crisis, critical moment, °showdown, crux, °juncture: *You can count on me when it comes to the crunch.*

crusade *n.* **1** °campaign, °expedition, holy war; jihad *or* jehad: *He joined the crusade against the Saracens.* —*v.* **2** °campaign, °war, °battle; take up a cause, °lobby, °fight: *She is crusading for equal rights for women.*

crush *n.* **1** °break, smash, °crunch, °pulverize, °shiver, °splinter, °pound, °grind: *The vandals crushed the statue to bits with hammers.* **2** °crumple, °wrinkle, crease, crinkle, °rumple, °mangle: *This shirt came back crushed from the laundry.* **3** squash, °pulp, mash,

°squeeze, compress, °press: *The machine crushes the oranges and extracts the juice.* **4** °overcome, °defeat, °conquer, vanquish, °beat, thrash; °subdue, °put down, °quash, °quell, °overwhelm, °squelch, °suppress, °repress: *The titleholder has thoroughly crushed the challenger. The junta crushed the uprising without bloodshed.* **5** abash, °embarrass, °shame, °mortify, °depress, °devastate, °humiliate, °disgrace: *She was really crushed when he refused to see her.* —*n.* **6** °press, °pressure, °crowd: *When the fire alarm sounded, I was almost caught in the crush of the people trying to escape.*

cry *v.* **1** °weep, °sob, wail, °keen, °bawl, shed tears: *Paul cried when they took his mother away.* **2** whimper, °snivel, pule, mewl, whine, °moan, °groan, °fret, *Colloq* turn on the waterworks, *Brit* grizzle: *Don't cry over spilt milk.* **3** ***cry out for.*** °demand, °need, °call for, beg for, °plead for: *Wanton murder cries out for vengeance.* —*n.* **4** °scream, °shriek, wail, °howl, yowl: *I heard the mournful cries of those being tortured.* **5** °shout, °whoop, °yell, °howl: *With bloodcurdling cries, the rebels attacked.* **6** °call, °sound, °note: *The noise was the cry of the lesser grebe to its mate.* **7** war cry, battle cry, °slogan, watchword: *"Down with the king!" was the cry used to rally the rabble.* **8** ***a far cry.*** a long way, quite a distance, °remote, °distant, very different; not, not quite: *This report is a far cry from what I had expected.*

crypt *n.* °tomb, vault, mausoleum, °sepulcher, °grave, catacomb; °cellar, basement: *He is buried in the crypt of St. Paul's.*

cryptic *adj.* **1** °secret, °occult, °mystical, °hidden, esoteric, mystic, cabalistic: *The sarcophagus was covered with cryptic symbols.* **2** °obscure, °mysterious, unclear, °nebulous, °vague, inscrutable, °recondite, arcane, enigmatic, °puzzling: *I cannot make head or tail of her cryptic remarks.*

cuddle *v.* **1** snuggle up (to), nestle *or* huddle (against): *She cuddled the baby closer to her.* **2** °caress, °embrace, °fondle, °hug, °pet, bill and coo, make love (to), *Colloq* neck, smooch, *Australian and New Zealand* smoodge *or* smooge, *Slang US* °make out (with), watch the submarine races: *The couple were cuddling in the back seat.* —*n.* **3** °hug, °embrace, °snuggle: *Give us a cuddle, Janie.*

cue *n.* **1** °prompt, °hint, °reminder, °signal, °sign: *Give her the cue to start singing.* —*v.* **2** °signal, °prompt, °remind: *She was cueing me to begin, but I'd lost my voice.*

culprit *n.* **1** accused, °prisoner: *How does the culprit plead?* **2** °offender, °criminal, malefactor, wrongdoer: *They caught the culprit red-handed.*

cultivate *v.* **1** °till, °plow, °farm, °work: *These lands have been cultivated since time immemorial.* **2** °grow, °raise, °tend, °produce: *What crops can be cultivated in this climate?* **3** °develop, °promote, °further, °encourage, °foster, °advance: *She's been cultivating a friendship with her boss.* **4** woo, make advances to, ingratiate oneself with, court, pay court to, curry favor with, *Colloq* °work on, *Slang* suck up to, butter up, *US* shine up to; *Taboo slang* brown-nose: *He is cultivating Trevor only because he wants something from him.*

cultivated *adj.* °sophisticated, cultured, °educated, °refined, °elegant, soigné(e), civilized, °polished, aristocratic, urbane, °suave, cosmopolitan: *She prefers to go out with cultivated older men.*

culture *n.* **1** cultivation, °refinement, °sophistication, urbanity, suavity, °elegance, °(good) breeding, °background, erudition, °education, enlightenment, °learning, °taste, °discrimination, °savoir-faire, °savoir-vivre, discernment: *She is a lady of considerable culture as well as beauty.* **2** °civilization, mores, °customs, lifestyle, way of life, (sense of) values: *In their culture, biting someone is a sign of love and respect.*

curb *n.* **1** °check, °restraint, °control: *You should put a curb on your tongue.* —*v.* **2** °check, °restrain, °bridle, °control, °contain, °repress, °subdue, °suppress: *Try to curb your exuberance.*

cure *n.* **1** course of treatment, °therapy, °remedy, medication, medicament, °medicine, °drug, °prescription; cure-all, nostrum, panacea: *What cure did the doctor recommend?* —*v.* **2** °heal, °mend, restore to health *or* working order, °remedy, °rectify, °correct, °repair, °fix: *What can't be cured must be endured.* **3** smoke, pickle, °dry, °salt, °preserve, corn, marinate: *That cured ox tongue is simply delicious!*

curiosity *n.* **1** inquisitiveness, °interest: *His insatiable curiosity next led him to study astronomy.* **2** snooping, prying, peeping, intrusiveness, meddlesomeness, °interference, *Colloq* nosiness, Nosy Parkerism: *Curiosity killed the cat.* **3** curio, °oddity, °rarity, conversation piece, *objet de virtu* or *vertu, objet d'art,* found object; bric-a-brac *or* bric-à-brac, knickknack, °bauble, trinket, °gewgaw: *The shop sells curiosities, like goblets made from ostrich eggs. In those days, every house had a curiosity cabinet.*

curious *adj.* **1** °inquisitive, inquiring, °interested: *I am curious to know what you were doing in there all night.* **2** snooping, prying, °intrusive, meddlesome, interfering, *Colloq* °nosy: *Our neighbors are entirely too curious about our activities.* **3** °odd, °peculiar, °eccentric, °strange, °outré, °queer, °unusual, °outrageous, °offbeat, °weird, °bizarre, unconventional, freakish, °exotic, °singular, °out of the ordinary, °extraordinary, °erratic, pixilated, °quaint, °outlandish, °grotesque, aberrant, °abnormal, °singular, °irregular, °deviant, °deviate, *Colloq* °kinky, nuts, nutty; *Slang Brit* barmy: *How do you explain Frieda's curious behavior?*

current *adj.* **1** °contemporary, °ongoing, °present, contemporaneous, °simultaneous, coeval: *The current issue of the magazine came out last week.* **2** °prevalent, °prevailing, °common, °popular, accepted, known, widespread, reported, in circulation, going round *or* around, bruited about, widely known, in the air, present-day: *The current theories reject those of a decade ago.* **3** °fashionable, °stylish, à la mode, modish, in vogue, °latest, up-to-date, *Colloq* °trendy: *The current trend is toward shorter skirts.* **4** *US* °up to date, in the know, °informed, advised, in touch, °aware, posted, *au courant, au fait,* on the qui vive: *The* Financial Journal *keeps me current on stock prices.* —*n.* **5** °stream, °flow, °undercurrent: *The canoe was caught in the current and carried away.* **6** °course, °progress, °tendency, °tenor, °drift, °trend, °inclination, mainstream: *The current of public opinion is turning in favor of your policies.*

curse *n.* **1** malediction, imprecation, denunciation, damnation, execration, °oath: *He heaped curses on all those who opposed him.* **2** °evil, bane, °misfortune, °affliction, °torment, °harm, °scourge, cross to bear, hex, *Colloq* °jinx: *The curse of our generation is that so few of us deeply believe anything.* **3** °profanity, °oath, blasphemy, obscenity, bad language, dirty word, swearword, curse word: *A stream of curses issued from the bathroom when Joe cut himself shaving.* —*v.* **4** °damn, execrate, °blast, °denounce, anathematize, excommunicate: *He was cursed by the priests and forbidden ever to enter a temple again.* **5** swear at, blaspheme at: *The muleteer was cursing his team.* **6** °burden, saddle, °weigh down, °handicap: *She was cursed with perfect pitch, which made most singing sound off key.*

cursory *adj.* °superficial, °hasty, °hurried, °passing, °quick, slapdash, °perfunctory, °rapid, °summary: *She gave the note a cursory glance and threw it away.*

curt *adj.* °abrupt, °short, °terse, °brief, laconic, °concise; °blunt, °gruff, °harsh, °brusque, unceremonious, °snappish, crusty, °rude: *His answer was a curt "No," without any explanation.*

curtail *v.* °shorten, °abbreviate, °cut short, °abridge, °diminish, °reduce, °cut, °cut back, °cut down: *Both our working week and our salaries were curtailed.*

cushion *n.* **1** pillow, °bolster, °pad: *The women of the harem sat on cushions on the floor.* —*v.* **2** °soften, absorb, °mitigate, °reduce, buffer, °insulate, mollify, lessen: *They offered me a month's paid vacation to help cushion the blow of being transferred.*

custody *n.* **1** °care, custodianship, °safekeeping, °protection, °charge, guardianship, keeping: *She was granted custody of the children.* **2** °imprisonment, °detention, incarceration, confinement: *The police took three troublemakers into custody.*

custom *n.* **1** °practice, °habit, °usage, °fashion, °way, wont, °tradition, °routine, °convention, °form: *According to custom, warriors paint their bodies.* **2** ***customs.*** °toll, °duty, impost, °tax, excise, levy, °dues, °tariff: *He was assigned to collect customs and port duties.* **3** °patronage, °support, °business, °trade: *The new butcher needs as much custom as possible.* —*adv.* **4** °specially, °especially, °expressly, exclusively, °particularly; to order: *All her clothes are custom made.*

customary *adj.* **1** °usual, °normal, °conventional, °routine, °everyday, °common, commonplace, °ordinary: *Is it customary for them to eat lunch here?* **2** °accustomed, °habitual, °regular, °traditional, wonted: *I took the customary way home.*

customer *n.* **1** °client, °patron, °buyer, purchaser; consumer: *Mrs. Kosciuszko is one of our regular customers.* **2** °chap, °fellow, °character, °person, °guy, *Colloq Brit* bloke: *She would never go out with an ugly customer like that.*

cut *v.* **1** °gash, °slash, °slit; °open: *I cut my finger on the glass.* **2** °slice, °cut off, °carve: *Please cut me a piece of steak.* **3** Often, ***cut up.*** °hurt, °wound, °pain, °upset, °grieve, °distress, aggrieve, °slight, °insult, °offend, affront: *I was really all cut up by her nasty remarks.* **4** °trim, °snip, °lop, °clip, crop, °shorten, shear, °chop off; °mow; °dock: *The barber has cut too much off the sides.* **5** °abbreviate, °shorten, crop, condense, °abridge, °edit, °cut back, °reduce, °cut down; epitomize, °abstract, °digest, summarize, °curtail: *Cut this script to make it fit into the allotted time.* **6** °dilute, °thin, °water (down), °weaken; °degrade, °adulterate: *They cut the rum with water to make grog.* **7** °avoid, fail to attend, eschew: *Cynthia cut classes three days this week.* **8** °lower, °reduce, lessen, °cut back (on), °cut down (on), °slash, °diminish, °decrease, retrench (on), °curtail: *We shall have to cut expenses if the company is to survive.* **9** conclude, °settle, °agree: *Once the deal was cut, there was no going back on it.* **10** °prepare, °draw, °write, °sign: *We'll have accounting cut your check and send it at once.* **11** ***cut back.*** **(a)** See **5**, above. **(b)** See **8**, above. **12** Often, ***cut dead.*** snub, °slight, °spurn, °shun, °ignore, give the cold shoulder (to): *Cornelia had the gall to cut Jason dead at his own party.* **13** ***cut down.*** **(a)** °fell, chop *or* hew down: *Don't cut down that tree!* **(b)** °kill, °cut off, °murder, assassinate: *He was cut down in his prime.* **14** ***cut in.*** °interrupt, °intrude, °interfere, *Colloq* °butt in: *Please don't cut in on our conversation.* **15** ***cut off.*** **(a)** °cleave, °sever, chop *or* lop *or* hack off: *Cut the branch off near the trunk.* **(b)** °intercept, °interrupt, °discontinue, °end, °stop, °terminate, °break off: *There was a click and our phone conversation was cut off.* **(c)** °separate, °sever, °split, estrange: *He's been cut off from the family for years.* **(d)** disinherit, disown, °reject: *She was cut off without a penny.* **16** ***cut out.*** **(a)** °delete, °remove, excise, strike *or* cross out, °edit out, °omit, °cut, °kill, *Technical* dele: *The publisher made me cut out the libelous parts.* **(b)** °extract, excise, °remove, resect: *My appendix was cut out years ago.* **(c)** °stop, °cease, desist (from), °quit: *Whenever it rains the engine cuts out. Cut out the clowning around.* **(d)** °suit, °equip, °fit: *I don't think Cyril's cut out to be a lumberjack.* **(e)** °plan, °prepare, °ready, °organize, °destine: *He certainly has his work cut out for him!* **17** ***cut up.*** **(a)** °chop (up), dice, cube, mince, °cut, °divide (up), °carve (up): *Cut the celery up very small.* **(b)** °misbehave: *The football hooligans began to cut up.* **18** ***cut up rough.*** get angry, lose one's temper, show resentment: *The driver began to cut up rough when I refused to pay.* —*n.* **19** °gash, °slash, °incision, °nick, °wound: *I got a nasty cut from that razor.* **20** °share, °portion, °percentage, °piece, dividend, commission: *Blatchley gets a cut on every car we sell.* **21** reduction, cutback, curtailment, °decrease: *The government's cuts in spending affect us all.* **22** deletion, excision, °omission: *The author refuses to approve the cuts in the script.* **23** affront, °insult, °offense, °slight, snub, °dig, jibe, °slap in the face, cold shoulder: *The unkindest cut was the accusation of cheating.* **24** °engraving, °plate: *This cut is too badly worn for use.* **25** artwork, °picture, °illustration, °plate, °drawing, line cut, line engraving, halftone: *The book has three hundred cuts and only twenty pages of text.* —*adj.* **26** separated, °detached, severed: *I prefer cut flowers to a plant.* **27** abridged, °abbreviated, cut-down, shortened, edited, curtailed: *The magazine published a cut version of the novel.* **28** reduced, diminished, lowered, discounted: *This shop sells everything at cut prices.* **29** ***cut and dried.*** **(a)** clearcut, settled, arranged, °decided, °predetermined, prearranged: *Unfortunately, the solution to this problem is not cut and dried.* **(b)** °stale, unoriginal, trite, hackneyed, °old; °dull, °boring: *His suggestions for improvement are always cut and dried.* **(c)** manufactured, °automatic, unchanging, unchanged: *Percy will again present his old cut-and-dried plan.*

cute *adj.* **1** °pretty, °attractive, °adorable, °dainty, °lovely, °beautiful, *Colloq US* cunning: *There were a lot of cute children at the beauty competition.* **2** °clever, °shrewd, °ingenious, adroit, °crafty, cunning: *It was very cute of him to suggest his brother for the job.*

cutthroat *n.* **1** °murderer, °pirate, °killer, °thug, hatchet man, gunman, assassin, *Slang US* gunsel, torpedo, hit man: *Those streets are frequented by thieves and cutthroats.* —*adj.* **2** °merciless, °ruthless, °unmerciful, unprincipled, °relentless, pitiless, °brutal, °coldblooded, °coldhearted: *Her cutthroat tactics call for dismissing all executives.* **3** °murderous, °homicidal, °lethal, °deadly, barbaric, °fierce, °cruel, barbarous, °savage, °inhuman, °brutal, brutish, °violent, °ferocious, °bloodthirsty, °sanguinary, bloody, feral, °vicious, °truculent: *He was once a member of a gang of cutthroat hoodlums.*

cutting *adj.* **1** °severe, °biting, °chill, °cold, °icy, °frigid, °freezing, °raw, °piercing, °penetrating: *A cutting wind seemed to go right through me.* **2** °sarcastic, °sardonic, °bitter, °scornful, sneering, acid, °scathing, acerb(ic), wounding, °stern, °harsh, °caustic, mordant, acrimonious, °contemptuous; malevolent, malicious, invidious, °vicious, °venomous: *Her cutting remarks completely devastated me.* —*n.* **3** scion, °slip, clipping: *Mrs. Galloway allowed me to take a cutting from her rose bush.*

cycle *n.* **1** °round, rotation, °circle, °course; series, °sequence, °run, °succession, °pattern: *We must learn to break the continuous cycles of war and peace.* —*v.* **2** °recur, °return, °rotate, recycle, °circle: *The water from the fountain is cycled back to the reservoir.*

D

dab *v.* **1** daub, °pat, °tap, tamp, °touch: *Dab a little more paint into the crevices.* —*n.* **2** °touch, °drop, °trace, °bit, mite, °hint, °suggestion, °pinch, °dash, °spot, tinge, *Colloq* dollop, smidgen: *Add just a dab of mustard to the sauce.* **3** daub, °poke, °pat, °tap, °touch: *Wipe it away with a dab of a damp cloth.*

dabble *v.* **1** °dip, °splash, °spatter, sprinkle, bespatter, besprinkle, bedabble: *I sat on a rock, dabbling my toes in the pool.* **2** ***dabble in*** *or* ***with*** *or* ***at.*** °tinker, °trifle (with), putter *or Brit* potter, dally, *Colloq* fool (around *or* about *or* with *or* about with): *She's not seriously interested in music—she just dabbles in it.*

dab hand *n.* phr. past master, °expert, °master, °adept, °authority, wizard, *Colloq* ace: *Oscar is a dab hand at woodcarving.*

daft *adj.* **1** °foolish, °silly, °giddy, °senseless, °absurd, °ridiculous, °stupid, °nonsensical, fatuous, fatuitous, imbecile *or* imbecilic, idiotic, moronic, °obtuse, cretinous, boneheaded, fatheaded, dimwitted, witless, asinine, attocerebral, weak-minded, simple-minded, brainless, °feebleminded, featherbrained, rattlebrained, °harebrained, slow-witted, °half-witted, fatwitted, addlepated, addlebrained, *Brit* gormless, *Colloq* °dumb, dopey *or* dopy, daffy; *Slang* cockeyed, *US* cockamamie *or* cockamamy, running on empty: *He has the daft idea that he will be appointed managing director.* **2** See **crazy, 1,** above. **3** ***daft about.*** °mad about, infatuated with, besotted by *or* with, °sweet on, *Colloq* nuts about, crazy about: *Those two are daft about each other.*

dagger *n.* °knife, poniard, skean, short sword, stiletto, dirk, °blade, kris, bowie knife, bayonet: *It was his dagger that was sticking out of the man's back.*

daily *adj.* **1** °diurnal, circadian, °everyday, quotidian: *The daily papers reported nothing about the fire.* **2** °ordinary, °common, commonplace, °everyday, °routine, °regular: *Her trip to the market has become a daily occurrence.* —*adv.* **3** constantly, °always, habitually, day after day, regularly, every day, continually, continuously: *The trains run daily between here and London.*

dainty *adj.* **1** °delicate, °graceful, °fine, °elegant, °exquisite, °neat: *The value lies in this dainty border painted round the edge of the cup.* **2** °fastidious, °sensitive, °squeamish, finicky *or* finical, overnice, overrefined, °genteel, °mincing: *He seems a bit dainty in his choice of words.* **3** °choice, °delicious, delectable, °tasty, appetizing, palatable, toothsome: *They were given a few dainty morsels to nibble while waiting.* —*n.* **4** °delicacy, sweetmeat, °treat, comfit, tidbit *or Brit* titbit, °morsel: *A plate of dainties was placed beside the bed each night.*

damage *n.* **1** °harm, °injury, °hurt, °impairment, mutilation, °destruction, devastation: *Fortunately, there was little damage from the storm.* **2** °expense, °price, °cost; °bill, invoice, *US* °check: *At the restaurant, the damage came to fifty dollars.* **3** ***damages.*** compensation, reparation, °indemnity: *We won the suit and were awarded damages of ten thousand dollars for defamation of character.* —*v.* **4** °harm, °hurt, °injure; °wound; °mutilate, disfigure, °mar, °deface; °wreck, °ruin, °spoil, °impair: *Although the car was badly damaged, the passengers escaped unscathed. Will this news damage your chances of a promotion?*

damn *v.* **1** °condemn, °criticize, find fault with, °berate, °castigate, °upbraid, °attack, °blast, °reprimand, °reprove, remonstrate, °denounce; °blame: *Some would damn him for saving the murderer from drowning, others would damn him if he didn't.* **2** °doom, °condemn, °sentence: *The Danaids were damned for all eternity to fill jars with water gathered in sieves.* **3** °curse (at), °swear (at), execrate: *I damned the day I first set foot in that house.* —*n.* **4** jot or tittle, brass farthing, *Slang* hoot, two hoots (in hell), *Slang* tinker's damn *or Slang* cuss: *His opinion isn't worth a damn.* **5** ***give a damn.*** °care, °mind, be concerned, °worry, *Slang* give a hoot: *Why should he give a damn if the critics panned his play?*

damnable *adj.* °awful, °terrible, °horrible, horrid, °atrocious, °abominable, °dreadful, °hideous, execrable, accursed, cursed, detestable, °hateful, °abhorrent, °despicable, °loathsome, °wicked, °sinful, °offensive, heinous, pernicious, °infernal, malicious, malevolent, °outrageous, °foul, °rotten, °base, °vile, odious: *He has been telling the most damnable lies about her since they broke up.*

damp *adj.* **1** °clammy, °moist, wettish; °humid, dank, °misty, dewy, °steamy, °muggy: *Wipe off the table with a damp cloth. Nothing dries out in this damp weather.* —*n.* **2** moistness, moisture, dampness, clamminess, humidity: *The mold on the walls is the result of the damp.*

dampen *v.* **1** °damp, moisten, sprinkle, bedew: *Dampen the clothes before ironing them.* **2** °stifle, °deaden, °damp, °check, °chill, °cool, °restrain, °retard, lessen, °diminish, °reduce, °suppress, abate, °moderate, allay, °subdue, °temper, °dull, °discourage: *His constant chattering on about himself dampened her ardor.*

dance *v.* **1** °cavort, gambol, °caper, °skip, °leap, romp, trip the light fantastic (toe), *US* cut a rug, sashay, *Colloq* bop, hoof it: *We danced for joy when we heard the news. Would you care to dance?* —*n.* **2** ball, °social, dancing party, *thé dansant*, *US* tea dance, °promenade, *Colloq* shindig *or* shindy, °hop, bop, *US and Canadian* prom: *I have invited her to the dance on Saturday evening.*

dandy *n.* **1** fop, coxcomb, °(gay) blade, beau, °gallant, lady-killer, ladies' *or* lady's man, °rake, *Colloq* °swell, clotheshorse, *Brit* toff, blood, *US* °dude: *He was a great dandy and spent hours dressing every day.* —*adj.* **2** °fine, °splendid, °first-rate, °great, °marvelous, °neat, spectacular: *Penny's father bought her a dandy new car.*

danger *n.* **1** °peril, °risk, °threat, °hazard, °jeopardy: *The danger of an avalanche is too great to go skiing.* **2** ***in danger of.*** °likely (to be), °liable (to be): *If you drink and drive you are in danger of causing an accident.*

dangerous *adj.* **1** °risky, °perilous, °hazardous, unsafe, °precarious, °rickety, *Colloq* chancy, iffy: *Rock climbing is very dangerous.* **2** °threatening, °menacing, °harmful, treacherous: *He is a dangerous criminal, wanted for murder.*

dangerously *adv.* **1** perilously, hazardously, unsafely, precariously, recklessly: *He's a mountain climber who likes to live dangerously.* **2** ominously, alarmingly: *She is standing dangerously close to the edge.*

dangle *v.* **1** °hang (down), °droop, °depend, °swing, °sway: *The rope dangled from the top of the flagpole.* **2** °flaunt, brandish, °wave, °flourish: *Competitors often dangle big salary increases in front of those who agree to leave our company.* **3** °wait, *Slang* cool one's heels: *They have kept me dangling for weeks for their decision.*

dapper *adj.* °neat, °spruce, °smart, °trim, well-dressed, well-turned-out, °stylish, °fashionable, °elegant, °chic, °dressy; *Colloq* got up *or* dressed to the nines, dressed to kill, swanky *or* swank, ritzy; *Slang* snazzy, °nifty, spiffy, °sharp, °swell, classy, *US and Canadian* spiffy: *Tony looks very dapper in his new Savile Row suit.*

dapple *adj.* **1** spotted, dotted, °mottled, °speckled, °flecked, dappled; brindled; pied, piebald, skewbald, °paint, fleabitten, *US* pinto: *Take the chestnut mare—I'll ride the dapple gray.* —*v.* **2** °spot, °dot, mottle, speckle, bespeckle, stipple: *Dapple paint on the wall with a sponge to get a mottled effect.*

dare *v.* **1** °challenge, °defy, °provoke; throw down the gauntlet: *She dared me to jump, so I jumped.* **2** °risk, °hazard, °gamble, °venture, °face, make bold, be so bold as: *I would never dare to talk to my father that way.* —*n.* **3** °challenge, °provocation, °taunt; °ultimatum: *She took the dare and swam across the lake.*

daredevil *n.* **1** exhibitionist, °showman, stuntman, stuntwoman; °adventurer, soldier of fortune, *Colloq* °showoff: *Wilbert finally got a job as a daredevil in the circus.* —*adj.* **2** °reckless, °rash, death-defying, °impulsive, °daring, °dashing, °impetuous, incautious, °imprudent, °wild, °foolhardy, madcap, devil-may-care; °audacious, °bold, °brave, °fearless, °gallant, °courageous, °intrepid: *Do you consider ski-jumping a sport or an example of daredevil madness?*

daring *n.* **1** °courage, boldness, °bravery, valor, intrepidity, fearlessness, °grit, °pluck, °spirit, mettle, adventurousness, derring-do, *Colloq* °guts, °spunk, °nerve; *Slang Brit* °bottle: *Diving from that cliff into the sea takes a lot of daring.* —*adj.* **2** °bold, °audacious, °courageous, °brave, valorous, °intrepid, °fearless, unafraid, plucky, mettlesome, °adventurous, °venturesome, °hardy; °rash, °reckless, *Colloq* gutsy, *US* nervy: *In the nineteenth century, a few daring explorers penetrated the jungles of Africa.*

dark *adj.* **1** unlit, unlighted, unilluminated, ill-lighted, ill-lit, °sunless; °black, stygian, pitch-dark, inky, jet-black: *We cowered in a recess in the dark cave.* **2** °dim, °murky, tenebrous, °shady, °shadowy: *I could scarcely see ahead of me in the dark forest.* **3** °gloomy, °dismal, °dreary, °dull, °drab, subfuscous, subfusc, °bleak, cheerless, °mournful, °dour, °pessimistic, °somber, °doleful, °joyless, °grim, °sad, °melancholy, °sorrowful: *Why do you always look at the dark side of things?* **4** °evil, °wicked, °vile, °base, °foul, iniquitous, nefarious, blackhearted, °villainous, °sinister, °satanic, °devilish, hellish: *Dark forces control the spirits dwelling in that house.* **5** °murky, °overcast, cloudy, °threatening, °black, °dusky, lowering *or* louring; foggy, °misty; *US* glowering: *Another dark day on the moor and I thought I'd go mad.* **6** °mysterious, °deep, °profound, °incomprehensible, enigmatic, °puzzling, impenetrable, unfathomable, abstruse, °recondite, arcane, °obscure: *She took her dark secret to the grave.* **7** °hidden, concealed, °secret, °occult, mystic(al), °cryptic: *The true reason for his leaving was always kept dark in the family.* **8** brunette; °black, °swarthy, brown; (sun-)tanned, *Old-fashioned* swart: *One is fair with dark hair, the other has dark skin.* **9** °ignorant, unenlightened, °benighted: *Our culture passed through a dark phase before the Renaissance.* —*n.* **10** °night, nighttime, nightfall: *We waited till dark to make good our escape.* **11** darkness, blackness, °gloom, gloominess, murk, murkiness: *At fifty, isn't he a bit old to be afraid of the dark?* **12** °obscurity, °ignorance: *She was always kept in the dark about his true identity.*

darling *n.* **1** °sweetheart, °beloved, °love, °dear, dearest, truelove: *She insists on buying all her darling's clothes.* **2** °pet, °favorite, apple of (someone's) eye, *Brit* blue-eyed boy; *US* fair-haired boy: *Franz might have been the black sheep of the family, but he was always his mother's darling.* —*adj.* **3** °beloved, loved, cherished, adored, °dear, °precious, treasured: *He traveled everywhere with his darling niece.* **4** °pleasing, °fetching, °attractive, °adorable, °enchanting, °lovely, alluring, °engaging, bewitching, charming: *Josephine was wearing a darling frock she'd just bought at the Corner Boutique.*

dash *v.* **1** °crash, smash, °shatter, °break, °shiver, °fragment, °split; °destroy, °ruin, °spoil, °frustrate, °obliterate: *The mirror was dashed to smithereens when it fell. The ship didn't see our raft, and our hopes of rescue were dashed.* **2** °hurl, °toss, °throw, °fling, °cast, °pitch, *Colloq* chuck: *We drank a toast, then dashed our glasses into the fireplace.* **3** °rush, °run, dart, °spring, °bolt, °bound, °race, sprint; °hasten, °fly, °hurry, °speed: *I'll have to dash to catch my train.* **4** ***dash off.*** scribble: *I've just dashed off a note to mother.* —*n.* **5** dart, °bolt, °rush, °run, °spurt, °spring, °bound, sprint: *He made a dash for the door but it was too late.* **6** °flourish, élan, °flair, liveliness, °style, °panache, °spirit, brio, °verve, °zest, °spice; °ardor, °fervor, °vigor, °energy: *She is known for her beauty as well as her dash and courage.* **7** °bit, °pinch, soupçon, °hint, °suggestion, °touch, °trace, tinge, °taste, °drop, °piece, *Colloq* smidgen, *US* tad: *Add a dash of nutmeg at the end.*

dashing *adj.* **1** °spirited, °lively, °impetuous, °energetic, °vigorous, °dynamic, °animated, *Colloq* peppy: *She is now going out with a dashing young fellow from London.* **2** °fashionable, °stylish, °chic, à la mode, modish, °smart, °elegant, °dapper, *Colloq Brit* °swish: *That's a dashing coat, Felicia.* **3** °flamboyant, °showy, °ostentatious, °pretentious: *Josiah Bigglesworth was a bit too dashing for her taste.*

data *n.* °facts, °information, statistics, figures, °details, °matter, observations, material(s); °text; °evidence: *We shall process the data on the computer and print out the results.*

date *n.* **1** °time, year, °season, °period, °day; °age, °era, epoch, °stage, °phase: *These artifacts are from an earlier date than was first supposed.* **2** °appointment, °meeting, °engagement, rendezvous, assignation, tryst; °fixture: *She already has a date for Saturday night.* **3** °escort, °companion, °friend, boyfriend, girlfriend, °girl, °woman, °boy, °man, swain, beau, °lover, *Colloq* °steady: *Bob is Sally's date for the dance.* **4** ***out of date.*** °old-fashioned, °old, °ancient, archaic, °antiquated, dated, °passé, outmoded, °obsolete, °obsolescent, *Colloq* old hat: *This timetable is out of date. Why do you wear those out-of-date clothes?* **5** ***up to date.*** °modern, °latest, °current, °contemporary, à la mode, °fashionable, *Colloq* °trendy: *Her taste in music is quite up to date. Use this up-to-date edition of the encyclopedia.* —*v.* **6** show one's age, make obsolete *or* obsolescent *or* old-fashioned: *That pompadour hairdo really dates her.* **7** °entertain, °escort, °go out (with), °go steady (with): *Does Michael still date Patsy? Those two are still dating.*

daunt *v.* °intimidate, cow, °discourage, dishearten, dispirit, °unnerve, °shake, °upset, disconcert, °discomfit, °put off, awe, °overawe, °appall, °alarm, °threaten, °frighten, °terrify, °scare, terrorize: *He was daunted by the prospect of facing the entire council.*

dauntless *adj.* °fearless, undaunted, unafraid, unflinching, °stalwart, °brave, °courageous, °bold, °audacious, °intrepid, valorous, °daring, °gallant, °heroic, °venturesome, plucky, stouthearted, valiant: *Dauntless, the knight rode into the thick of the fray.*

dawdle *v.* °linger, loiter, °straggle, °delay, °procrastinate, dally, °lounge, laze, °idle, °lag, lie about, waste time, *Colloq* dillydally, °shilly-shally: *We have to catch the next train, so stop dawdling.*

dawn *n.* **1** daybreak, sunrise, break of day, crack of dawn, first light, dawning, cockcrow *or* cockcrowing, *Literary* aurora, dayspring, *US* sunup: *We shall attack the castle at dawn.* **2** dawning, °beginning, commencement, °start, °birth, awakening, inception, genesis, °outset, °onset, °origin, °appearance, °arrival, advent, °emergence, inauguration, °rise, first occurrence: *The dawn of Western civilization has been placed in Anatolia.* —*v.* **3** °gleam, °break, °brighten, °lighten: *The day dawned on the deserted beach.* **4** °begin, °originate, °commence, °arise, °appear, °emerge, °start, °arrive, °develop, °unfold: *The day of the computer had not yet dawned when I was a child.* **5** ***dawn on*** *or* ***upon.*** °occur to, come to mind, become apparent *or* evident to: *It dawned on us that he had been lying all along.*

day *n.* **1** daytime, °daylight, broad daylight, light of day: *Sunrise quickly turned night into day.* **2** °time, hour, °age, °period, °era, epoch, °date, °prime, heyday; lifetime: *Her day will come. In his day, there was no telephone.*

daydream *n.* **1** °reverie, woolgathering, °fantasy, °fancy, °dream, musing, castle in the air *or* in Spain, pipe dream: *The realities of life have cured me of many daydreams.* —*v.* **2** °fantasize, °imagine, °fancy, °envision, °dream: *She still daydreams that a knight in shining armor will come and carry her away.*

daylight *n.* **1** sunlight, °sun, sunshine, °light: *Coming from the cave, we were blinded by the daylight.* **2** °open, broad daylight, light of day, full view, full knowledge, °clarity: *We must bring his treachery out into the daylight.*

daze *v.* **1** °stun, stupefy, °blind, °dazzle, bedazzle, °shock, °stagger, °startle, °astonish, °astound, °amaze, °surprise, °overcome, °overpower, °dumbfound, benumb, °paralyze, *Colloq* bowl over, °floor, flabbergast; *Slang* blow one's mind: *She was dazed to learn her husband was still alive.* **2** befuddle, °confuse, °bemuse, °bewilder, °puzzle, °mystify, baffle, °perplex, °nonplus, °blind: *He was dazed by the difficulty of the examination.* —*n.* **3** °confusion, °flurry, °spin, whirl: *The entire week was a continuous daze of cocktail parties and dinner parties.* **4** ***in a daze.*** stupefied, in a trance, bewildered, °confused, perplexed, disoriented, °dizzy, dazzled, bedazzled, °overcome, overpowered, nonplussed, befuddled, flustered; startled, surprised, shocked, stunned, astonished, astounded, amazed, staggered; bemused, baffled, puzzled, mystified, *Colloq* flabbergasted, bowled over, floored: *Arthur was in a daze to find himself the center of attention.*

dazzle *v.* **1** °impress, °bewitch, °enchant, °charm, °beguile, °intrigue, °captivate, °fascinate, spellbind, °entrance, °hypnotize, mesmerize: *Every man in the room was dazzled by Mrs. d'Arcy's brilliant wit and good looks.* **2** See **daze, 1,** above. —*n.* **3** °brilliance, °splendor, magnificence, °sparkle, °glitter, *Slang* razzle-dazzle, razzmatazz: *Many actors are lured to New York by the dazzle of Broadway.*

dazzling *adj.* °bright, °brilliant, resplendent, blinding, bedazzling, °radiant, °splendid, °magnificent, °glorious, sparkling, °scintillating; °stunning, °overwhelming, °overpowering, stupefying, dizzying; °gorgeous; *Colloq* splendiferous, mind-boggling: *In the chest was a dazzling collection of the finest jewels.*

dead *adj.* **1** deceased, °defunct, °extinct, gone, departed, °late, °lifeless, no more, *Colloq* done for, *Slang* pushing up daisies, *Brit* gone for a burton: *Lycidas is dead, dead ere his prime. His parents are dead, but he has a brother who lives in Cairo.* **2** insensate, °insensible, °numb, paralyzed, benumbed, unfeeling: *After the accident, my left thumb was completely dead.* **3** °insensible, °unconscious, °out, dead to the world, deathlike, deathly: *At the news of her son, she fell in a dead faint.* **4** insensitive, unemotional, unfeeling, emotionless, apathetic, °lukewarm, °cool, °cold, °frigid, unresponsive, °unsympathetic, °indifferent, unconcerned, uninterested; °numb, °wooden, °callous, hardened, impervious, inured, °inert: *He has always been dead to others' problems.* **5** °out, smothered, extinguished: *The fire is dead.* **6** °inanimate, °lifeless, °inert, inorganic: *Dead stones speak volumes to the geologist.* **7** °extinct, °obsolete, perished, °past, outmoded, °disused, expired, °passé: *Latin is a dead language.* **8** °barren, unfruitful, °infertile, unproductive: *That area off the coast is dead as far as fishing goes.* **9** °tired (out), °exhausted, worn-out, °fatigued, °spent, collapsing, in a state of collapse, *Slang* bushed, °beat, *Brit* knackered, *US and Canadian* pooped: *We were completely dead after the hike into town.* **10** °dull, lusterless, °flat, °neutral, °vapid, °empty, °bland, °colorless, °gray, beige, °dun: *The walls of the prison were painted a dead white.* **11** °stagnant, motionless, °still, °standing, °static, °inert, unmoving, °inactive, °quiet, °calm: *There were small pools of dead water covered with a green slime. Without a breath of air stirring, the boat was dead in the water.* **12** °boring, °dull, °tedious, °tiresome, °monotonous, °prosaic, uninteresting, °run-of-the-mill, °ordinary, commonplace, °dry, insipid, °bland, °flat, two-dimensional, °lifeless, °stiff, °rigid, °stony: *The play was bad, the performance dead.* **13** °dull, muffled, deadened, anechoic, unresounding, nonresonant: *One room in the laboratory was built to be dead to all sound.* **14** °complete, °entire, °total, °absolute, °downright, °thorough, through and through, utter, all-out, °out-and-out, °unqualified, unrelieved, unbroken, °categorical, °outright: *My investment in the antigravity pill has so far been a dead loss.* **15** °profound, °deep: *I fell into a dead sleep.* **16** °sudden, °abrupt, °complete, °full: *The train came to a dead stop.* **17** °certain, °sure, unerring, °exact, °precise, °accurate, °crack: *According to the records, Calamity Jane was a dead shot.* —*adv.* **18** °completely, °entirely, °absolutely, °totally, °utterly, categorically, °thoroughly, unconditionally, unqualifiedly: *You are dead right about Pontefract.* **19** °completely, °entirely, °absolutely, °totally; abruptly, °suddenly: *He stopped dead in his tracks and stared at me.* **20** °directly, °exactly, °precisely: *An enormous maelstrom lay dead ahead of our ship.* —*n.* **21** depth(s), °extreme, °midst, °middle: *She used to visit his room in the dead of night.*

deaden *v.* **1** °numb, benumb, °paralyze, anesthetize, desensitize, °dull; °damp: *This injection will deaden your hand and you'll feel no pain.* **2** °weaken, °moderate, soothe, °mitigate, assuage, °reduce, lessen, °diminish, alleviate, °cushion, °soften, mollify, °blunt, °dull: *He took to drink to deaden the shock of losing his only son.*

deadlock *n.* **1** °standstill, °impasse, °stalemate, standoff, °draw, stoppage, *Colloq US* Mexican standoff: *Union and management negotiators have reached a deadlock on the pension issue.* —*v.* **2** bring *or* come to a standstill *or* impasse, °stall, °stop, °halt: *The Congress is likely to deadlock on the question of expanding national health benefits.*

deadly *adj.* **1** °lethal, °fatal; °dangerous, pernicious, °poisonous, noxious, toxic; baleful, °harmful, nocuous: *This drug is deadly if taken in large doses.* **2** °mortal, °implacable, °ruthless, °savage: *They were deadly enemies long after the war was over.* **3** °murderous, °homicidal, °bloodthirsty, °brutal, °vicious, °ferocious, barbarous, barbaric, °savage, °inhuman, °coldblooded, °heartless, °ruthless, pitiless, °merciless: *Two deadly killers have escaped from Dartmoor prison.* **4** deathly, deathlike, °pale, pallid, °ghostly, cadaverous, °ghastly, °wan, °white, livid, ashen: *He turned a deadly hue, as if* *he had seen a ghost.* **5** °boring, °excruciating, °dull, °tiresome, °tedious, °dreary, °humdrum, °lackluster, wearying, wearisome: *It was a deadly play put on by deadly actors.* **6** °exact, °precise, °accurate, °true, unerring, unfailing: *Each arrow hit the bull's-eye with deadly accuracy.*

deaf *adj.* **1** hard of hearing, stone-deaf: *Sean is slightly deaf in his left ear.* **2** unhearing, unheedful, °heedless, °insensible, insensitive, impervious, °indifferent, °oblivious, unresponsive, °unmoved, unconcerned, unyielding: *The judge was deaf to all appeals for clemency.*

deal *v.* **1** °distribute, dole out, °give out, °parcel out, mete out, °allot, apportion, °administer, °dispense: *Deal thirteen cards to each of four players. She dealt out her own brand of justice to criminals.* **2** buy and sell, °handle, °stock, do business, °trade, °traffic: *This shop deals only in the most expensive linens.* **3** °behave, °act, °conduct oneself: *Simon has never dealt openly, so you mustn't trust him.* **4** ***deal with.*** °treat, °handle, °take care of, have to do with, °attend to, °see to, °reckon with, °grapple with, act on; °practice, °administer, °engage in: *I shall deal with the matter tomorrow.* —*n.* **5** °transaction, °arrangement, °negotiation, °agreement, °contract, °bargain, °understanding: *The deal to sell the textbook division is off.* **6** Often, ***great deal.*** (large *or* great) amount, °lot, (large *or* huge) quantity; °extent: *There's been a great deal of crime in that neighborhood.*

dealer *n.* °trader, businessman, businesswoman, °merchant, °tradesman, retailer, shopkeeper, vendor, merchandiser; wholesaler, jobber, distributor, stockist, supplier; °broker, °agent, salesman, *US* storekeeper: *He has been a dealer in precious gems for years.*

dealings *n. pl.* °business, °commerce, °exchange, °trade, °traffic, °transactions, negotiations; °relations, relationships, affairs: *All his business dealings are reviewed by his solicitor.*

dear *adj.* **1** °beloved, loved, adored, °darling, °precious, cherished, prized, valued, treasured, °favored, °favorite, °pet, esteemed, admired, venerated, honored: *He was my nearest and dearest friend.* **2** °expensive, costly, high-priced, highly priced, *Colloq* °pricey: *Tomatoes are much too dear at this time of the year.* —*n.* **3** °darling, °sweetheart, °beloved, °love, truelove, °sweet, honey, °precious, °pet, °favorite, °treasure, °precious, *Colloq* sweetie, sweetiepie, *Slang* °baby: *My dear, I hope we'll be together always.* —*adv.* **4** °dearly; at great cost *or* expense, at a high *or* excessive price: *That little error will cost you dear, my friend.*

dearly *adv.* **1** greatly, very much, °indeed, °sincerely: *I should dearly like to go, but I cannot.* **2** affectionately, °fondly, lovingly, tenderly: *He loves his mother very dearly.* **3** expensively, °dear, at great cost *or* expense, at a high *or* excessive price, punitively: *The victory at Thalamos was dearly bought.*

dearth *n.* °scarcity, °want, °need, °lack, deficiency, sparseness *or* sparsity, scantiness, insufficiency, inadequacy, °shortage, paucity, exiguity, °poverty, exiguousness, °absence: *There is a dearth of major roles for black actors.*

death *n.* **1** demise, decease, °passing, °dying, °end: *The death of the boy is to be avenged.* **2** °end, °termination, cessation, °expiration, expiry: *Nobody mourned the death of the bill in the lower house.* **3** °end, °finish, °termination; extinction, °destruction, extermination, annihilation, eradication, obliteration, eradication, extirpation, liquidation, °ruin, °downfall, °undoing: *The invasion marked the death of the Roman Empire.*

deathless *adj.* °eternal, °everlasting, °immortal, undying, imperishable, °permanent, unending, °timeless, never-ending: *In his opinion, his novel was another example of his deathless prose.*

debase *v.* **1** °lower, °degrade, devalue, °depreciate, °depress, demote, deprecate, °belittle, °diminish, °reduce, °disparage: *Words which convey fine qualities are in time debased.* **2** °adulterate, °contaminate, °taint, °pollute, °corrupt, °mar, °spoil, °impair, °vitiate, abase, defile, bastardize; °poison: *To increase profits, the manufacturer has debased the traditional formula.*

debatable *adj.* °controversial, arguable, °questionable, °doubtful, dubious, problematic *or* problematical, °disputable, open *or* subject to dispute *or* doubt *or* question, in dispute *or* doubt *or* question, °moot, polemic *or* polemical, unsure, °uncertain, °unsettled, undecided: *Whether he is the best person is debatable.*

debate *n.* **1** °discussion, °argument, °dispute, altercation, °controversy, wrangle, contention, polemic; argumentation: *I refuse to take sides in the debate over social services.* **2** deliberation, °consideration, °(careful) thought, °reflection, cogitation, meditation, contemplation: *Payment of reparations to the victims of the disaster is a matter for debate.* —*v.* **3** °argue, wrangle, °dispute, °contest, contend; °discuss, °moot, °question: *We debated only the most important issues.* **4** °deliberate, °consider, °reflect (on), mull over, °ponder (over), °weigh, ruminate (over), meditate (on *or* over), think (over *or* on), think through: *I have often debated in my own mind the question of capital punishment.*

debonair *adj.* **1** °suave, soigné(e), °elegant, urbane, °refined, °dapper, °genteel, °well-bred, °courteous, °civil, mannerly, °gracious, °polite, affable, °obliging, °pleasant, *Colloq* °smooth: *Despite his vicious temper, he was most debonair in company.* **2** °carefree, insouciant, °gay, °nonchalant, lighthearted, °dashing, charming, °cheerful, °buoyant, °jaunty, °sprightly: *Being handsome and debonair, he was much sought after by hostesses.*

debt *n.* **1** °obligation; °due, indebtedness, °liability, °responsibility, °accountability, °encumbrance: *He owes a debt of gratitude to his wife for her moral support. The company takes care of all debts promptly.* **2** ***in debt.*** under obligation, owing, °accountable, °beholden, °indebted, °responsible, answerable for, °liable, encumbered, in arrears, °straitened, in dire straits, in (financial) difficulty *or* difficulties, in the red, *Colloq US and Canadian* in hock: *I shall always be in debt to you for your help. The London branch is in debt for ten million pounds.*

debut *n.* **1** °première, introduction, °initiation, inauguration, launch *or* launching, coming out: *Madame Kropeczka's debut at La Scala was a disaster.* —*v.* **2** °launch, °come out, °enter, °appear: *His plan is to debut with a zither accompaniment.*

decadent *adj.* **1** declining, decaying, deteriorating, debased, degenerating, falling off, °on the wane, °withering, degenerative: *The decadent literature of the period was a reflection of the decline in moral standards.* **2** °corrupt, °dissolute, °immoral, debauched, dissipated, °self-indulgent, °degenerate: *His decadent behavior brought him to the attention of the police.*

decay *v.* **1 (a)** decline, wane, ebb, dwindle, diminish, decrease: *The magnetic field rapidly decays when the power is removed.* **(b)** °decline, °waste away, atrophy, °weaken, wither, °degenerate, °deteriorate, °disintegrate; °crumble: *Her great beauty decayed quickly.* **2** °rot, °decompose, molder, °putrefy, °spoil; °turn, go bad, °go off: *The flesh has decayed and only a skeleton remains.* —*n.* **3** °decline, weakening, °failing, fading, deterioration, decadence, degeneration, wasting, atrophy, dilapidation, disintegration, °collapse; °downfall: *The buildings were in an advanced state of decay.* **4** °rot, rotting, decomposition, °mold, putrefaction, mortification: *The decay has weakened the timbers supporting the bridge.*

deceit *n.* **1** °deception, deceitfulness, °fraud, fraudulence, cheating, °trickery, chicanery *or* chicane, °dissimulation, dishonesty, misrepresentation, double-dealing, duplicity, °hypocrisy, treachery, underhandedness, guile, °craft, slyness, craftiness, cunning, knavery, funny business, *Colloq* °hanky-panky, monkey business: *Inside traders on the stock exchange profit enormously from deceit.* **2** °trick, °subterfuge, °stratagem, ploy, °ruse, maneuver, °artifice, °wile, °hoax, °swindle, °double-cross, misrepresentation, °pretense, °sham, contrivance, °shift, confidence trick, subreption, gloze, *Dialectal Brit and colloq US* flam; *Colloq* flimflam; *Slang* scam, con, con trick, con game: *One of the spy's many deceits was to dress as a policeman. She was sick of all his lies and deceits.*

deceitful *adj.* °dishonest, underhand(ed), untrustworthy, misleading, °crooked, °insincere, °false, °fraudulent, °counterfeit, °disingenuous, °lying, mendacious, untruthful; °wily, °crafty, °sly, cunning, °scheming, guileful, °artful, °sneaky, double-dealing, °two-faced, °hypocritical, duplicitous, *Colloq* phoney *or US also* phony: *It was deceitful of you to pretend you loved her when all you wanted was her money.*

deceive *v.* °mislead, delude, impose on *or* upon, °fool, °hoax, °trick, °cheat, °swindle, °betray, °double-cross, °lead on, lead up *or* down the garden path, °lead astray, pull the wool over (someone's) eyes, inveigle, °cajole, *Archaic* cozen; *Colloq* con, bamboozle, °take in, °take for a ride, two-time, move the goal posts; *Slang US* °take: *He deceived even his friends and family into believing he had been a war hero.*

decent *adj.* **1** °becoming, °suitable, °appropriate, °proper, °seemly, °fitting: *Despite the life she led, the woman should have a decent burial.* **2** °seemly, °decorous, °tasteful, °dignified, mannerly, °nice, °clean, °respectable, °polite, °modest, °presentable, °acceptable: *Hereafter, you will use only decent language when speaking to me!* **3** °adequate, °acceptable, °passable, °fair, °competent, °mediocre, middling, fair to middling, °moderate, °respectable, not bad, °ordinary, °so-so, not outstanding, unimpressive, °average, neither here nor there, all right, °reasonable, °tolerable, °satisfactory, good enough, *Colloq* OK *or Colloq* okay: *Sales in the first quarter were decent but hardly outstanding.* **4** °courteous, °proper, °right, °fair, °honest, °honorable, °friendly, °considerate, °gracious, °nice, °thoughtful, °obliging, °kind, °generous, °accommodating: *You can count on David to do the decent thing.* **5** °chaste, °pure, °virtuous, °modest, °well-bred, °decorous, well-brought-up, °nice, °respectable: *Caroline is a decent girl, but no great brain or beauty.*

deception *n.* **1** duplicity, °deceit, °intrigue, °hypocrisy, °fraud, cheating, °trickery, chicanery *or* chicane, °dissimulation, double-dealing, °subterfuge, sophistry, treachery, knavery, tergiversation: *He practised deception even in his family relationships.* **2** °trick, °ruse, °artifice, °stratagem, °subterfuge, °maneuver, °wile, imposture, °hoax, °sham, °pretense: *He tried every deception in the book to separate her from her money.*

deceptive *adj.* **1** misleading, °false, °illusory, deceiving, °unreliable: *He has the look of an athlete, but appearances can be deceptive.* **2** °fraudulent, °deceitful, °dishonest, untruthful, °fake, °false, °shifty, fallacious, °specious, °spurious, °bogus, °counterfeit, pseudo, sophistical; °tricky, °dodgy, °evasive, °elusive, °slippery, *Colloq* phoney *or US also* phony: *The bank is being deceptive about his credit rating.*

decide *v.* **1** °determine, °settle, °resolve, conclude, take *or* reach *or* come to a decision, reach *or* come to a conclusion, make up one's mind, arbitrate, °judge, adjudicate, referee, °umpire: *She decided that you were right. They decided the case in my favor.* **2** ***decide on*** *or* ***upon.*** fix *or* fasten *or* settle on *or* upon, °choose, °select, °pick (out), °elect, opt (for), °commit oneself (to): *I have decided on a British-made car.*

decided *adj.* **1** °definite, °pronounced, °marked, unmistakable, unambiguous, unequivocal, °certain, °sure, °absolute, °obvious, °clear, °evident, °unquestionable, unquestioned, °indisputable, °undisputed, undeniable, irrefutable, incontestable, °unqualified, unconditional, °incontrovertible, °solid: *The party was a decided success.* **2** °fixed, °firm, °resolute, °determined, adamant(ine), °stony, °unhesitating, decisive, °definite, unfaltering, °assertive, asseverative, unswerving, unwavering: *They are decided in their approval of her plan.*

decipher *v.* **1** decode, decrypt; unravel, unscramble, disentangle, °translate, °work out, °explain, °solve, *Colloq* °figure out: *It was Champollion who deciphered the Rosetta Stone.* **2** °read, °interpret, °make out, *Colloq* °figure out: *I can't decipher Gilbert's handwriting or what he's trying to say.*

decision *n.* **1** °settlement, °determination, °resolution, settling, resolving, arbitration: *The decision is the umpire's responsibility.* **2** °judgment, conclusion, °resolution, verdict, °sentence, ruling, °finding, °decree, °settlement, °outcome: *According to the decision, the victims will receive compensatory damages.* **3** °determination, firmness, decidedness, °resolve, decisiveness, conclusiveness, steadfastness, °purpose, purposefulness: *She asserted her position with decision.*

declaration *n.* **1** °statement, °assertion, attestation, deposition, asseveration, affirmation, avowal, °announcement, °proclamation, °pronouncement, °profession: *Henrietta desperately wanted to believe Henry's declaration of love.* **2** °proclamation, °announcement, °pronouncement, promulgation, pronunciamento, edict, ukase, manifesto, °notice: *The colonists issued a declaration of independence.*

declare *v.* **1** assert, °say, °offer, °submit, affirm, °state, aver, asseverate, avow, avouch, °profess, °protest, °swear, °claim, °proclaim; °confirm, °certify, °ratify: *I solemnly declare that the testimony I am to give is true, so help me God.* **2** °announce, °make known, °pronounce, °decree, °rule, °proclaim, herald, promulgate, °publish, °broadcast, trumpet (forth): *Roberto has declared his intention to make Marianna his wife.*

decline *v.* **1** °refuse, °turn down, °deny, °reject, demur, °forgo, °veto, °avoid, abstain from: *She declined help with the packages. Roger was offered a professorship at the university but he declined.* **2** °diminish, lessen, °decrease, °wane, °flag, °go down, fall *or* taper off, °subside, °ebb, abate, °dwindle, °shrink, °fade, *Colloq* °peter out, run out of steam, *US* run out of gas: *Demand for hula hoops declined.* **3** slope *or* slant (downward(s)), °descend, drop *or* fall off, °dip, °sink: *The meadow declines toward the river.* **4** °deteriorate, °degenerate, °worsen, °fail: *My health has declined over the last year.* **5** go *or* drop down, °settle, °dip, °sink, °set: *The sun was declining as I went home.* —*n.* **6** diminution, °decrease, lessening, °ebb, downturn, falloff, falling off, reduction, abatement, °slump, descent: *The decline in the value of the dollar has made vacationing in America quite attractive.* **7** degeneration, deterioration, °loss, diminution, weakening, debility, °weakness, worsening, °decay, °failing: *We noted a decline in the physical condition of those living near the nuclear plant.* **8** declivity, (downward) slope *or* slant, descent, °downgrade, °incline: *The path led down a steep decline to the pond.*

decompose *v.* **1** °disintegrate, °separate, fall *or* come apart, break up *or* down, take apart, dissect, anatomize, atomize, °resolve, decompound, °analyze: *By absorption the scientist decomposed the green light into yellow and blue.* **2** °rot, °disintegrate, °decay, molder, °putrefy; °spoil, go off *or* bad, turn sour: *The meat will decompose if left outside the refrigerator.*

decorate *v.* **1** °embellish, adorn, °ornament, garnish, embroider, °elaborate, bedeck, deck (out), °trim, °dress (up), spruce *or* smarten up, °beautify, *Literary* caparison, *Colloq Brit* tart up: *We decorated the pub for the Christmas holidays.* **2** *Brit* °paint, wallpaper, redecorate, furbish, °refurbish, °renovate, °fix up, °restore: *All the bedrooms have been decorated.*

decoration *n.* **1** garnish, °trim, trimming, adornment, °embellishment, °ornament, ornamentation, garnishment: *There's a bit too much decoration on the cake.* **2** medal, laurel, °award, badge, °colors, °order, ribbon, °star, garter: *Captain Harder won many decorations in the war.*

decorous *adj.* °becoming, °dignified, °decent, °correct, mannerly, °seemly, °refined, °elegant, °polite, well-behaved, °genteel, demure, °polished, gentlemanly, °ladylike, °seemly: *Your behavior was less than decorous at last night's party.*

decorum *n.* **1** °etiquette, proper behavior, °propriety, good form, mannerliness, politeness, °dignity, gentility, good manners, respectability, courtliness, deportment: *The decorum of the meeting was disturbed by rabble-rousers.* **2** correctness, °propriety, °protocol, punctilio, conformity: *Please observe proper decorum when visiting the embassy.*

decoy *n.* **1** bait, °lure, °trap, °attraction, °enticement, °inducement, stool pigeon: *The hunters set out their decoys and waited for the ducks.* —*v.* **2** °lure, entrap, °entice, °attract, °induce, °seduce, bait, °trick, °tempt, ensnare, inveigle, allure: *He was decoyed into a dark alley and robbed.*

decrease *v.* **1** °diminish, °reduce, °decline, lessen, °lower, abate, °fall off, °shrink, °shrivel (up), °contract, °dwindle, °ebb, °subside, °wane, °taper off, de-escalate, slacken, °let up, ease (off *or* up), °curtail, cut (down *or* back), *Colloq* run out of steam, *US* run out of gas: *Demand for tickets to rock concerts has decreased over the years. The number of applicants for work is decreasing.* —*n.* **2** diminution, reduction, °decline, lessening, lowering, abatement, falling off, shrinking, shriveling, contraction, decrement, dwindling, °ebb, subsidence, tapering off, °wane, de-escalation, slackening, easing (off *or* up), curtailment, °cut, cutback: *There has been no noticeable decrease in the price of houses in the southeast. Have you noticed the decrease in arrests for dangerous driving?*

decree *n.* **1** °order, mandate, directive, ordinance, edict, °law, statute, °regulation, enactment, °act, ruling, dictum, °dictate, °injunction, °sanction, manifesto, °proclamation, promulgation, °determination, °decision, °judgment, rescript, °prescription, pronunciamento, firman, ukase, *Rom Cath Ch* decretal: *The star chamber issued a decree restricting the freedom of the press.* —*v.* **2** °order, °command, °direct, °rule, mandate, ordain, °dictate, °charge, enjoin, °proclaim, °pronounce, °prescribe, °decide, °determine, adjudge, *Scots law* decern: *The council has decreed that no spirits can be sold on Sundays.*

decrepit *adj.* **1** °feeble, enfeebled, °weak, weakened, °frail, °infirm, wasted, worn-out, unfit, debilitated, enervated, °disabled, incapacitated, °crippled, °doddering; out of shape, in bad shape; °aged, °old, °elderly, °ancient, superannuated, senescent, °senile, *Colloq* gaga: *The old man was so decrepit he was unable to lift the cup to his lips.* **2** °dilapidated, deteriorated, crumbling, decayed, decaying, withered, wasted, °antiquated, tumbledown, broken-down, °rickety, °unstable, °shaky, °ramshackle, °derelict, creaking, creaky, °rundown: *The barn was so decrepit we had to tear it down.*

decrepitude *n.* **1** feebleness, °weakness, °infirmity, debilitation, enervation, incapacity, old age, superannuation, senescence, °senility, caducity, dotage: *In his decrepitude he was unable to understand most of what was said to him.* **2** dilapidation, deterioration, °decay, °ruin: *The house is in an advanced state of decrepitude.*

dedicate *v.* **1** °devote, consecrate, give (up *or* over), °yield, °offer, °surrender, °commit, °pledge, °assign: *She dedicated her life to helping the poor.* **2** consecrate, °bless, °sanctify, °hallow: *There stands the temple dedicated to Apollo.* **3** inscribe; °address, °assign: *This book has been dedicated to you.*

dedication *n.* **1** °devotion, °assignment, °pledge, commitment, allegiance, adherence, faithfulness, fidelity, °loyalty, devotedness, wholeheartedness, single-mindedness, fixedness, fealty: *I admire Rudolph's dedication to duty.* **2** inscription, °address; °message: *The dedication reads, "To my mother and father."* **3** consecration, sanctification, hallowing: *The ceremony for the dedication of the youth center will be held tomorrow.*

deduce *v.* conclude, °infer, °understand, °gather, °assume, °presume, °derive, °draw, °work out, °divine, glean, °take it, °suppose, °surmise, °suspect, *Brit slang* suss out: *From your smile, I deduce you thought that was funny.*

deduct *v.* °subtract, take away *or* out *or* off, take from, °remove, °withdraw, *Colloq* °knock off: *Deduct six from ten and you're left with four.*

deduction *n.* **1** subtraction, diminution, °decrease, reduction, withdrawal, °removal, abstraction: *After deductions for expenses, you'll be left with nothing.* **2** conclusion, °inference, °finding, °reasoning, °result: *After considering the evidence, my deduction is that the butler didn't do it.*

deed *n.* **1** °act, °action; °performance: *Their deeds did not match their promises.* **2** °exploit, °feat, °achievement, °accomplishment: *We are here to honor her for her many deeds, both on and off the tennis court.* **3** °title (deed), °document, °instrument, indenture, °contract, °agreement: *The bank holds the title deed till you pay off the mortgage.*

deep *adj.* **1** °extensive, °bottomless, abyssal, unfathomable, °profound; °wide, °broad, yawning, chasmal *or* chasmic: *All our supplies were lost in a deep crevasse in the glacier.* **2** °profound, arcane, °recondite, °difficult, abstruse, °obscure, esoteric, °incomprehensible, beyond *or* past comprehension, impenetrable, unfathomable, inscrutable, °mysterious, mystic(al), °occult, °weighty, °serious, *Colloq* °heavy: *Religious philosophy is too deep a subject to discuss at breakfast.* **3** °wise, °learned, °sage, sagacious, °astute, perspicacious, °profound, discerning, °acute, °intense, °penetrating, °knowledgeable, °knowing: *Margaret is one of the deepest thinkers on the subject.* **4** °rapt, °absorbed, engrossed, occupied, °preoccupied, °intent, °intense, °involved, °engaged, immersed, °lost, *Colloq* into: *Don't disturb him when he's deep in thought.* **5** °devious, cunning, °shrewd, °crafty, canny, °clever, °knowing, °scheming, °artful, °designing: *He thinks there is a deep plot against him.* **6** °profound, °intense, °sincere, °serious, °heartfelt, °earnest, °ardent, °fervent, °poignant, deep-rooted: *I know of your deep concern for her.* **7** °low, °resonant, booming, resounding, sonorous, rumbling: *The deep sound of thunder rolled across the valley.* **8** °rich, °dark, °intense, °strong: *The sky was a deep blue.* —*n.* **9** ***the deep.*** the ocean, the main, the sea, the waters, the high seas, the briny (deep), the wave(s), Davy Jones's locker, Neptune's *or* Poseidon's kingdom *or* domain: *It was midnight on the waters and a storm was on the deep.* —*adv.* **10** °deeply, far down, °profoundly, intensely, earnestly, heavily: *We drank deep of the cooling liquid.*

deepen *v.* **1** °dig out, °burrow, °sink, dredge, °excavate, °scoop (out): *We'll have to deepen the hole to support the flagpole.* **2** °intensify, °increase, °concentrate, °strengthen, °expand, °magnify: *The program reflects a deepening interest in education.*

deeply *adv.* **1** °deep, (far) downward(s) *or* inward(s), way down, deep down: *The glacier was deeply fissured. She swam to the surface and inhaled deeply.* **2** °profoundly, intensely, strongly, powerfully, very much, acutely, keenly, gravely, greatly, to a great extent, °extremely, °thoroughly, °completely, °entirely, °seriously, °severely, irrevocably, unreservedly; passionately, heavily, emotionally: *She is deeply involved with a man from Kent. They were deeply committed to the labor movement.*

deface *v.* °mar, disfigure, °spoil, °ruin, deform, °blemish, °damage, °mutilate, °harm, °impair, °injure, °destroy: *Nothing should be allowed to deface the beauty of these buildings.*

default *n.* **1** °failure, °fault, °defect, °neglect, °negligence, dereliction, °lapse, °oversight, nonperformance, nonfulfillment, inaction: *We won the case because of their default on the contract. The other contestant failed to appear, so Gordon won the match by default.* **2** nonpayment, delinquency: *Default in the rent may result in eviction.* —*v.* **3** °fail, °neglect, °dishonor, °lapse, °fall short, come (up) short: *He has defaulted on a car payment.*

defeat *v.* **1** °overcome, °conquer, vanquish, be victorious over, get the better of, °beat, °subdue, °overwhelm, °overpower, prevail over, °triumph over, bring down, worst, thrash, °rout, °repulse, °overthrow, trounce, °whip, °crush, °destroy, do in, °best: *The champion defeated the contender in a ten-round match.* **2** °thwart, °frustrate, °disappoint, °check, balk, °stop, °terminate, °end, °finish, °foil: *He never let his handicap defeat his hopes of winning.* —*n.* **3** °conquest, °overthrow, beating, °repulse, trouncing, °rout, vanquishment: *The defeat of our team must be charged to lack of training.* **4** frustration, °undoing, °failure, °miscarriage, °setback; Waterloo: *The stock market crash spelled the defeat of the company's plan for a share issue.*

defecate *v.* °void (excrement), move the bowels, excrete, °eliminate, °evacuate (the bowels), have a (bowel) movement *or* bm, open the bowels, relieve oneself, *Babytalk* do number two, *Euphemistic* go to the men's *or* ladies' (room), go to the toilet *or* bathroom *or* lavatory, °excuse (oneself), wash (one's) hands, go to the bathroom, go to the powder room; *Mincing* go to the little boys' *or* girls' room; *Colloq Brit* spend a penny, go to the loo, pass a motion, *Taboo slang* (take a) crap *or* shit: *The first symptoms of bowel disease are problems when defecating.*

defect *n.* **1** °shortcoming, deficiency, °lack, °want, inadequacy, insufficiency, shortfall, °failure, °weakness, °frailty, °weak point, °imperfection, irregularity, °liability: *See the doctor about that hearing defect.* **2** °blemish, °imperfection, °failing, °weakness, °flaw, °fault, °mark, °stain, irregularity, °mistake, °error: *The inspector is supposed to find any defects before the product is shipped.* —*v.* **3** desert, change sides *or* loyalties, turn traitor, °go over; °escape: *Ropovich tried to defect, but the Albanians sent him back.*

defective *adj.* **1** °imperfect, °faulty, °flawed, °deficient, °broken, °out of order, impaired, marred, *Colloq* °on the blink, *US* on the fritz: *The brakes on his car were defective.* **2** retarded, °simple, °feebleminded, (mentally) deficient *or* incompetent, °backward, subnormal, *Brit education* ESN (= 'educationally subnormal'), *US education* °exceptional: *Found to be defective, he could not stand trial.*

defector *n.* °deserter, apostate, °turncoat, °traitor, °renegade, *Colloq* rat: *Some political defectors were suspected of being spies.*

defense *n.* **1** °shelter, °protection, °cover, °guard, °safeguard, °shield: *There is no defense against certain illnesses.* **2** fortification, armor, barricade, °screen, °bulwark, °rampart: *The shore defenses included concrete pillboxes spaced fifty yards apart.* **3** °excuse, apology, °reason, apologia, °explanation; justification, vindication, °argument, °plea, advocacy, °support: *His defense for decreasing welfare payments was inadequate. She spoke in defense of nationalizing industry.*

defenseless *adj.* unprotected, exposed, °vulnerable, °unguarded; °helpless, °weak, °powerless, °impotent: *Would you take advantage of a poor, defenseless creature?*

defend *v.* **1** °protect, watch over, °guard, °safeguard, °keep (safe), °shelter, °shield, °screen, °preserve; fight for: *We must defend our civil rights.* **2** °fortify, arm, °secure; fend *or* ward off, parry: *Can you defend your position from attack?* **3** °plead for, speak *or* stand up for, °stick up for, go to bat for, °support, °uphold, °stand by, °champion, stand with *or* behind *or* beside, argue for *or* in behalf of, hold a brief for, espouse: *The lawyers defended her right to free speech.*

defer[1] *v.* °put off, °postpone, °delay, °shelve, lay *or* put aside, adjourn, *US* °table; *Colloq Brit* kick into touch: *The judge has deferred his decision.*

defer[2] *v.* Often, **defer to.** give in (to), give ground *or* way (to), °yield (to), °submit (to), °bow (to), °capitulate (to), °cede (to), accede (to), acquiesce (to); °comply (with), °agree (to): *I'll defer to your decision in the matter.*

deference *n.* **1** °respect, °regard, politeness, °civility, °courtesy, °consideration, °esteem: *They treated him with deference owing to his age.* **2** °obeisance, °submission, acquiescence, °obedience, compliance: *Considering her accomplishments, she is paid little deference.*

defiant *adj.* challenging, °bold, °brazen, °audacious, °daring; °rebellious, °disobedient, °stubborn, °recalcitrant, °obstinate, refractory, unyielding, °insubordinate, °mutinous, °unruly, °self-willed, °aggressive, headstrong, contumacious, °pugnacious, °hostile, °belligerent, antagonistic, *Slang* gutsy, spunky: *His defiant attitude toward authority often gets him into trouble.*

deficient *adj.* **1** °wanting, lacking, °defective, °incomplete, unfinished, °short, °insufficient, °inadequate, °sketchy, skimpy, °scarce: *Some foods are deficient in vitamins.* **2** °faulty, impaired, °flawed, °imperfect, °incomplete, °defective, °inferior, °unsatisfactory: *Many have a deficient knowledge of their legal rights.*

deficit *n.* °loss, deficiency, shortfall, °shortage, °default: *At the end of the year there was a considerable deficit.*

define *v.* **1** °determine, °establish, °fix, demarcate, mark off *or* out, delimit, °limit, lay *or* set down, circumscribe, °specify, °identify, delineate, °describe: *You must first define the subjects to be covered.* **2** °describe, °explain, °interpret, °spell out, °detail, °clarify, delineate, °expand on, expatiate on *or* upon, delineate; °characterize, °state, °name: *No one dictionary defines all the words of a language. Please define exactly what you want me to do.*

definite *adj.* **1** °specific, °particular, °exact, °pronounced, °explicit, °express, °precise: *She came here with a definite purpose.* **2** °sure, °positive, °certain, assured, °fixed, settled, confirmed: *Then we have a definite appointment for two o'clock?* **3** °clear, °plain, well-defined, unambiguous, unequivocal, °distinct, clear-cut, °obvious: *The plans for revision are definite.*

definitely *adv.* °positively, °absolutely, °surely, to be sure, assuredly, certainly, indubitably, °undoubtedly, categorically, unequivocally, unquestionably, decidedly, °finally, °once and for all; plainly, °clearly, °obviously, patently: *Then you're definitely not going to the dance with Waldo? That was definitely the worst movie of the year.*

definition *n.* **1** delineation, delimitation, demarcation, outlining; acutance, °resolution, distinctness, °clarity, sharpness, °focus, °precision: *The definition at the edge of the photograph is fuzzy.* **2** °description, °explanation, explication, clarification, °statement (of meaning), °sense, °meaning: *How many definitions are there for the word "good"?*

definitive *adj.* **1** decisive, °final, conclusive, °ultimate: *My definitive answer will be given tomorrow.* **2** °thorough, through and through, °exhaustive, °ultimate, consummate, °complete, °authoritative, °reliable: *She has written the definitive work on the axolotl.* **3** clarifying, unambiguous, °categorical, °absolute, °unqualified, °accurate, °exact, °precise: *We expect a definitive statement from the union regarding their claims.*

deflect *v.* avert, turn away *or* aside, °deviate, °change, °swerve, °switch, °divert, °shy, veer, °sidetrack; °fend off: *The trajectory of the bullet is deflected by gravity. Deflect a bit to your left and you will be on course.*

deformed *adj.* **1** °misshapen, malformed, distorted, twisted, °grotesque, °gnarled, °crooked, contorted, awry, warped, °bent: *This tree is deformed because of the constant wind.* **2** °disfigured, °crippled, °lame, °misshapen; °abnormal: *He was born with a deformed foot.* **3** distorted, warped, °bent, °perverted, twisted, °grotesque; °abnormal: *The deformed personalities of her patients are the subject of my book.*

defraud *v.* °cheat, °swindle, °trick, °beguile, cozen, °dupe, delude, °fool, bilk, °fleece, °victimize, °take in, °deceive, humbug, °hoodwink, flimflam, *Colloq* do, diddle, con, slip one over on, put (something) over on, pull a fast one on, fast-talk, °rope in, *US* °take; *Slang* °take for a ride, gyp, °rob, °rip off, rook; *Dialect* flam: *Shareholders are defrauded by insider trading schemes.*

defray *v.* °pay, °settle, °meet, °discharge, liquidate, °clear, °cover, °reimburse, *Colloq* pick up the bill *or* tab *or US* check (for), foot the bill (for): *The company defrays the cost of all traveling expenses.*

defunct *adj.* **1** °dead, deceased, °extinct: *The dinosaurs have been defunct for millions of years.* **2** inoperative, °inapplicable, °unused, unusable, °invalid, expired, °obsolete, °passé, °dead, expired, °nonexistent, outmoded, °out: *Although still on the books, that law is defunct.*

defy *v.* **1** °challenge, °dare, °face, confront, °brave, °stand up to, °flout, brazen out, °thumb one's nose at, *Colloq Brit* cock a snook at: *The defendant defied the prosecutor to prove the allegations.* **2** °frustrate, °thwart, baffle, °resist, °withstand, °repel, °disobey, °repulse: *Her feats of legerdemain defy the imagination. Copeley has invented a device that defies the law of gravity.*

degenerate *adj.* **1** debased, degraded, °corrupt, corrupted, vitiated, °decadent, depraved, °reprobate, °dissolute, ignoble, °base, °low, °inferior, °vile: *He was a degenerate descendant of a once noble lineage. Bletchly sank into the depths of a degenerate existence after Penelope left him.* —*v.* **2** °decline, °deteriorate, °decay, °sink, °worsen; backslide, regress, retrogress, °weaken, go to the dogs, go to rack and ruin, *Colloq* go to pot: *He felt that art had degenerated since the days of Rembrandt.* —*n.* **3** °reprobate, debauchee, °wastrel, °profligate, °rake, rakehell, °roué; °pervert, °deviate: *The detective said that only a degenerate could have committed such a crime.*

degradation *n.* **1** degeneracy, degeneration, deterioration, corruptness, corruption, vitiation, baseness, depravity, turpitude: *History records the moral degradation of a whole society.* **2** disrepute, °discredit, °shame, °humiliation, ignominy, °dishonor, °disgrace, abasement, debasement: *He had to face the degradation of an accusation of child molestation.*

degrade *v.* **1** °downgrade, demote, °break, *Military* cashier, *Ecclesiastic* unfrock, *Law* disbar; depose, unseat; disfranchise *or* disenfranchise; *Military* drum out (of the corps), *esp naval* disrate; *US military* bust: *They degraded him from captain to lieutenant.* **2** °disgrace, °dishonor, °humble, °shame, °discredit, °debase, demean, abase; °humiliate, °mortify, °belittle, deprecate, °depreciate, cheapen, °reduce, °lower: *He has been degraded to mopping the floor.* **3** °dilute, °adulterate, °weaken, °thin, °water (down), °alloy: *Cologne is, essentially, degraded perfume.*

degrading *adj.* demeaning, humiliating, °shameful, shaming, debasing, lowering, discreditable: *Why should you deem selling a degrading occupation?*

degree *n.* **1** °grade, °level, °stage, °class, °caste, °rank, °order, °scale, °standing, °status, °station, °position, °situation, °estate, °condition: *He is entertaining a lady of high degree.* **2** °measure, °magnitude, °extent, °limit, °point; lengths, °step: *Up to a certain degree, I think you're right about the choice of words.* **3** ***by degrees.*** little by little, bit by bit, °step by step, inch by inch, inchmeal, °gradually, slowly, (almost) imperceptibly: *By degrees, her health has improved.* **4** ***to a degree.*** **(a)** °rather, °somewhat, °quite: *She is to a degree a better dancer than he.* **(b)** °substantially, considerably, °highly, decidedly, °exceedingly, to a considerable extent: *She must be stupid to a degree if she believes in levitation.*

deign *v.* °condescend, °stoop, °vouchsafe, °concede; °yield, °agree: *Do you think Lord Worthington would deign to sit at our table?*

deity *n.* °god, goddess, supreme being, °creator, demiurge: *Deities in various religions are represented as men, women, or animals.*

dejected *adj.* downcast, °downhearted, depressed, dispirited, discouraged, °despondent, down, °low, chapfallen, crestfallen, °melancholy, °sad, °unhappy, °gloomy, °glum, °miserable, °blue, low-spirited, in low spirits, °forlorn, °woebegone, disconsolate, °sorrowful, morose, °heartbroken, heavy-hearted, in the doldrums, *Colloq* down in the dumps, °down in the mouth: *She was bound to feel dejected when she couldn't find a job.*

delay *v.* **1** °postpone, put off *or* aside, °defer, temporize, °suspend, °shelve, hold off *or* up (on), put on hold, hold in abeyance, put *or* keep in a holding pattern, pigeonhole, put on ice, put in *or* into the deepfreeze, *Colloq* put on the back burner, *Brit* kick into touch, *US* hold off *or* up (on), °table, put on hold: *We shall delay our decision till next month.* **2** hold up *or* back, detain, °impede, °hinder, °retard, °keep, °bog down, °set back, slow (up *or* down); °stop, °arrest, °halt, °check; °obstruct: *Delivery of the mail has been delayed by the strike. We were delayed by traffic.* **3** loiter, °procrastinate, °hesitate, poke *or* drag (along), °tarry, °wait, °lag (behind), °dawdle, hang back, °stall, °linger, dally, mark time, putter *or Brit* potter; vacillate; *Colloq* dillydally, shilly-shally, °drag one's feet: *Stop delaying and get to work.* —*n.* **4** °postponement, deferral, deferment, °wait, °holdup; °setback: *There will be a ten-day delay in paying the rent.* **5** °lull, °interlude, hiatus, °interruption, °gap, °interval, lacuna, °stop, stoppage, °wait, waiting, °holdup, suspension: *After an hour's delay, service was resumed.* **6** tarrying, loitering, dawdling, *Colloq* dillydallying, shilly-shallying: *There should be no further delay in shipping the order.*

delectation *n.* °delight, °enjoyment, °amusement, °entertainment, °diversion, °pleasure, °satisfaction: *For your delectation, Le Moulin Rouge presents La Goulue!*

delegate *n.* **1** °envoy, °agent, legate, °representative, °ambassador, plenipotentiary, °minister, emissary, commissioner, (papal) nuncio, (papal) internuncio, spokesperson, spokesman, spokeswoman, °go-between: *They kowtowed to the delegate from His Imperial Highness.* —*v.* **2** depute, commission, °appoint, °designate, °assign, °name, °nominate, accredit, °authorize, empower, mandate: *The chairman has delegated the sales manager to represent him at the meeting.* **3** °assign, °give, hand over *or* on, pass over *or* on, depute, °transfer, °entrust, °relegate, *Colloq* pass the buck for, *US* buck: *She has delegated the responsibility to one of the directors.*

delete *v.* °erase, °cancel, rub *or* cross out *or* off, °remove, °blot out, expunge, efface, °eliminate, °obliterate, °wipe out, eradicate, °strike out, cut *or* edit (out), *Publishing* blue-pencil; *Printing* dele: *Delete the old address and insert the new one.*

deliberate *adj.* **1** °intentional, planned, °studied, °willful, intended, °premeditated, °calculated, °conscious, prearranged, °purposeful, °preconceived, considered; °coldblooded: *The insult was deliberate, not a slip.* **2** °slow, °methodical, °careful, °unhurried, paced, °measured, °regular, °even, °steady, °sure, °unhesitating, unfaltering, °confident: *He moved across the room with a deliberate step and tore the medals from the general's tunic.* **3** °careful, °prudent, °cautious, °painstaking, °discreet, considered, °considerate, °thoughtful, well-thought-out, °thorough, °methodical, °systematic, °fastidious, °orderly, punctilious, °dispassionate, °cool, composed, °collected, °calm, °serene, unruffled: *A deliberate speaker, he chose his words with care.* —*v.* **4** °consider, °ponder, think (about *or* over), °weigh, °debate, meditate (on *or* over), reflect (on *or* over), cogitate (on *or* over), °study: *I shall need a few days to deliberate on that question.*

deliberately *adv.* intentionally, °on purpose, purposely, willfully, consciously, wittingly, calculatedly, calculatingly, knowingly, pointedly, resolutely, of one's (own) free will, on one's own, with one's eyes (wide) open: *She did that deliberately, not by accident.*

delicacy *n.* **1** fineness, exquisiteness, gracefulness, °beauty, lightness, daintiness: *Notice the delicacy of the tracery in the rose window.* **2** fragility, fragileness, °frailty, frailness, °weakness, °infirmity, feebleness, tenderness; susceptibility: *Because of the delicacy of his constitution, even a cold might be fatal.* **3** °sensitivity, °difficulty, ticklishness, °finesse, nicety, °sensibility: *The delicacy of the situation demands the utmost diplomacy.* **4** °luxury, sweetmeat, °dainty, tidbit *or Brit* titbit, °savory: *The table was laden with delicacies from all over the world.*

delicate *adj.* **1** °fragile, breakable, °frail, °tender, frangible, °dainty; perishable, °flimsy: *This filament is extremely delicate, so be careful.* **2** °fine, °exquisite, °dainty, °graceful, °elegant, °subtle: *A delicate border of lace sets off the collar.* **3** °feeble, °weak, °sickly, °frail, debilitated, weakened, enfeebled, °unhealthy: *Her condition is too delicate for her to be moved.* **4** °critical, °ticklish, °sensitive, °dangerous, °tricky, °precarious, °touchy, *Slang* °hairy; *Colloq* °sticky: *Rescuing the survivors of the avalanche was a delicate operation.* **5** °dainty, °squeamish, °queasy, °fastidious, °prudish, Victorian, °finicky, finical, °refined, °discriminating, discerning, °sensitive, puristic, °proper, °coy, °modest, demure: *In those days ladies were thought to be too delicate to read books of that sort.* **6** °gradual, °subtle, °nice, °precise, muted, °soft, °faint, °subdued: *The delicate shading at the horizon is characteristic of this artist.*

delicious *adj.* **1** delectable, °luscious, ambrosial, °savory, mouthwatering, toothsome; °choice, flavorful, °tasty, appetizing, palatable, *Colloq* scrumptious; *Slang* °yummy: *Larry's fried chicken is quite delicious.* **2** enjoyable, °delightful, °pleasurable, °pleasing, °pleasant, °choice, °enchanting, fascinating; °agreeable, charming, °engaging; amusing, °entertaining: *I heard the most delicious bit of gossip about the Armbrusters.*

delight *v.* **1** please, °gratify, °satisfy, °gladden, °cheer, °tickle, °amuse, °entertain, °divert, °excite, °thrill, °captivate, °entrance, °fascinate: *We were delighted to hear the Mighty Allen Art Players once again.* **2** ***delight in.*** °enjoy, °appreciate, °like, °relish (in), °savor, °revel in, glory in; °love, °adore; *Colloq* get a kick from *or Colloq* out of; *Slang* °get off on: *She delights in any kind of jazz.* —*n.* **3** °pleasure, °gratification, °joy, °satisfaction, °enjoyment, °delectation; °bliss, °ecstasy, °rapture: *She takes great delight in playing practical jokes on her guests. In his dreams he visited the garden of earthly delights.*

delighted *adj.* pleased, °happy, °charmed, thrilled, °enchanted, *enchanté(e)*: *I am delighted to meet you. "Miss Smith, meet Mr. Brown." "Delighted!"*

delightful *adj.* **1** °pleasing, °agreeable, °pleasurable, enjoyable, °joyful, °pleasant, °lovely, amusing, °entertaining, diverting, °exciting, °thrilling: *We spent a delightful evening together.* **2** °attractive, congenial, °winning, winsome, charming, °engaging, °exciting; captivating, °ravishing, fascinating, °enchanting: *Georgina is one of the most delightful people I have met in a long time.*

delinquent *n.* **1** malefactor, (young *or* youthful) offender, wrongdoer, lawbreaker, °culprit, °criminal, °miscreant; hooligan, ruffian, roughneck: *The police rounded up six juvenile delinquents and charged them with rowdyism.* —*adj.* **2** neglectful, negligent, °derelict, °remiss, °failing, defaulting: *I have been delinquent in my obligations to my mother.* **3** °overdue, past due, in arrears, °late, °unpaid: *All these delinquent accounts should be collected as soon as possible.*

delirious *adj.* **1** °wild, °hysterical, distracted, °incoherent, °rambling, irrational, °raving, ranting, frenzied, °frantic, °disturbed, demented, °deranged, unhinged, °mad, °insane, °crazy, lunatic: *He is still delirious and doesn't know what he's saying.* **2** °wild, °excited, crazed, thrilled, °ecstatic: *She was delirious with joy that Ken was coming home.*

deliver *v.* **1** °carry, °bring, convey, °distribute, give *or* hand out; purvey, take round; °cart, °transport: *Only in a few places in the world do they still deliver milk to the door.* **2** °hand over, °give, °surrender, °cede, °yield, °make over, °relinquish, give up *or* over, °commit, °transfer, °turn over, °resign: *We were forced to deliver our children to the enemy as hostages.* **3** set free, °liberate, enfranchise, °extricate, °release, °save, °rescue; °emancipate, manumit, °redeem; disencumber, disburden, °ransom: *We were delivered from certain death by the arrival of the helicopter. Modern appliances have delivered millions of women from the drudgery of housework.* **4** °give, °present, utter, °read, °broadcast; °proclaim, °announce, °declare, °set forth, °communicate, °make known, °express, °publish, °hand over, °hand out, promulgate, °pronounce, °enunciate: *He has to deliver a speech tonight. The police delivered an ultimatum to the terrorists.* **5** °give, °administer, °inflict, °deal, °direct, °send, °launch, °impart, °throw; °cast, °hurl, °shoot, °discharge, °fire: *He delivered a blow on the chin that knocked me out. The ball was delivered with enormous speed.* **6** °bring forth, °bear, give birth to, °bring into the world: *In the next three years, she delivered three more girls.*

delivery *n.* **1** °distribution, delivering, deliverance, conveyance, transportation, °transport: *The strikers have caused delivery of newspapers to stop.* **2** °liberation, °release, deliverance, emancipation: *His delivery from poverty was still a few years away.* **3** childbirth, parturition; confinement: *The second child is often an easier delivery.* **4** °presentation, °performance; utterance, enunciation, articulation, °pronunciation, °expression, °execution: *He is an accomplished orator, with a spellbinding delivery.*

delusion *n.* **1** °deception, °trick, °stratagem, °artifice, °ruse, °pretense: *It was a snare and a delusion to represent the painting as genuine.* **2** false *or* mistaken impression, °fallacy, °illusion, °mistake, °error, °misconception, misbelief, °hallucination: *He suffers under the delusion that he is a great pianist.*

demand *v.* **1** °require, °order, °bid, °call (for); °insist, °command: *I demand that you retract that remark! She demanded to know where he was going.* **2** °claim, °ask (for), °require, insist on; °exact: *They had paid for tickets and demanded entrance.* **3** °require, °call for, °need, °want, necessitate, °cry out for: *This superb dish demands an excellent claret.* **4** °ask (for), inquire *or* enquire, °request; °requisition: *We demanded help from the police.* —*n.* **5** °request, °bid, behest, °requisition, °order, insistence; °outcry: *Our demand for service went unheeded.* **6** °want, °need, °requirement, °desire; °market (demand), marketability; consumer *or* customer acceptance: *The demand for typewriters is diminishing as more people turn to word processors.* **7** ***in demand.*** wanted, needed, requested, coveted, °popular, sought after, desired, °desirable, *US* on request, *Brit* in request: *Is a better mousetrap really in demand?* **8** ***on demand.*** °on call, on request, on presentation, when requested *or* required; °at once, °immediately, without delay: *These notes are payable on demand.*

demanding *adj.* **1** °difficult, °hard, exigent, °tough, °exacting, °trying, taxing: *Legree is a demanding boss. Diamond cutting is demanding work.* **2** °insistent, clamorous, °urgent, °nagging, °persistent: *Your demanding fans want another encore.*

democratic *adj.* egalitarian, classless; republican, °representative, °popular, self-governing, autonomous: *The colonists voted for a democratic form of government.*

demolish *v.* **1** tear *or* pull down, dismantle, reduce to ruin(s), smash, pull to pieces, °knock down, °raze, °topple, °destroy, °level: *This building will have to be demolished to make room for the new shopping mall.* **2** °destroy, °end, bring to an end, make an end of, put an end to, °devastate, °terminate, annihilate, °overturn, °overthrow, °crush, °defeat, refute, °disprove, °dispose of, °suppress, °squelch, °quash: *With just one phrase he demolished their entire argument.*

demon *n.* **1** °devil, evil spirit, °fiend, cacodemon *or* cacodaemon; °monster, ghoul, °ogre, harpy, vampire: *Medieval demons are generally depicted as having horns, hoofs, and tails.* **2** °fanatic, °fiend, °enthusiast, °addict, *Colloq* °freak: *He's a real speed demon when he gets onto the highway.*

demonstrable *adj.* provable, confirmable, attestable, verifiable; °evident, °self-evident, °obvious, undeniable, °apparent, °manifest, °indisputable, °unquestionable, °positive, °certain, conclusive: *The judge showed a demonstrable bias against my client.*

demonstrate *v.* **1** °show, °prove, make evident, °establish, evince, °evidence, °exhibit, °manifest: *The increase in arrests demonstrates the efficiency of the police.* **2** °display, °explain, °expose, °describe, °present; °illustrate: *The salesman demonstrated the new camera for us.* **3** °picket, °march, °parade, °rally, °protest: *More than five thousand people demonstrated against the fraudulent election.*

demonstration *n.* **1** °proof, °evidence, °testimony, confirmation, verification, substantiation; °manifestation, °exhibition, °display, °illustration, °indication: *I have seen sufficient demonstration of her ineptitude.* **2** °presentation, °display, °show, °explanation, °description, clarification, elucidation, °exposition, *Colloq* demo: *The student gave an excellent demonstration of how a computer works.* **3** picketing, °march, °parade, °protest, °rally, sit-in, *Colloq Brit* demo: *There have been numerous demonstrations against the government's policies.*

demonstrative *adj.* **1** °open, unrestrained, unconstrained, unreserved, °expansive, °effusive, °emotional, °warm, °tender, °affectionate, loving: *Porphyria is quite demonstrative, often causing me to blush.* **2** illustrative, indicative, °representative, probative, evidential; provable, °evident: *Her point was proved by several demonstrative arguments. The hostility of these few is demonstrative of what to expect of the entire group.*

demoralize *v.* **1** dispirit, °daunt, dishearten, °discourage, °defeat; °weaken, °cripple, °enervate, devitalize, °depress, °subdue, °crush: *Wittery's loss of the election thoroughly demoralized his supporters.* **2** °corrupt, °pervert, deprave, °vitiate, °debase, debauch: *The committee consider him a demoralizing influence and insist he should resign.* **3** °bewilder, °discomfit, °unnerve, °shake (up), °confuse, °fluster, disconcert, °perturb, °disturb, °upset, *Colloq* °rattle: *The demonstrators were completely demoralized when arrested for loitering.*

denial *n.* **1** contradiction, negation, repudiation, refutation, disavowal, disclaimer, disaffirmation: *Her denials notwithstanding, she was found guilty.* **2** retraction, recantation, renunciation, withdrawal: *The arbitrary denial of civil rights to some is unconscionable.* **3** °refusal, °rejection, negation; °veto: *The boy's persistent denial of authority went into his record.*

denizen *n.* °inhabitant, dweller, °occupant, frequenter, °resident; °citizen: *The bathyal depths of the seas harbor some strange denizens. William is a denizen of some of the seedier bars on the South Side.*

denomination *n.* **1** °sect, °persuasion, °school, church, °order: *He is a member of the Mormon denomination.* **2** °sort, °kind, °type, °nature, °variety, °unit, °size, value, °grade, °class, genus, species, °order, classification: *The kidnappers demanded the ransom money in used notes of small denomination.* **3** designation, appellation, °name, °identification, °style, °title, °tag, °term; designating, naming, identifying, styling, classifying, titling, entitling, tagging, terming, denominating: *The denomination of people by race, creed, color, or sex is discriminatory.*

denote *v.* **1** °indicate, °specify, °designate, °distinguish, °signify, °mark, °note: *Hypothetical word forms are denoted by an asterisk.* **2** °mean, °name, °symbolize, °represent, betoken: *The word* mother *denotes "female parent," but its connotations are far more extensive.*

denounce *v.* 1 °accuse, °brand, °stigmatize, °charge, °blame, °incriminate, °implicate, complain about: *He has been denounced for the blackguard he is.* 2 °betray, °inform against, °report, °reveal: *He denounced his own son to the authorities.* 3 °criticize, °condemn, decry, denunciate, °attack, assail, censure, impugn, declaim *or* rail (against), °vituperate, revile, °vilify, inveigh against; °ridicule, °(hold up to) shame, pillory, °(heap) scorn (upon), cast a slur on: *The playwright was denounced as a neo-Nazi.*

dense *adj.* 1 °compact, °thick, compressed, condensed, °close, °solid, °heavy, impenetrable: *The fox escaped into a dense thicket.* 2 crowded, °packed, °tight, impenetrable, impassable: *There was a dense crowd blocking the exit.* 3 °stupid, °slow, slow-witted, thickheaded, °dull, thick-witted, °obtuse, °stolid, cloddish, °dim, dimwitted, °foolish, *Colloq* °thick, °dumb: *He may be a gifted artist but he is dense when he has to balance his checkbook.*

deny *v.* 1 °contradict, gainsay, refute, controvert, disaffirm, disclaim, confute, negate, °dispute: *She denies ever having met the defendant.* 2 °reject, °refuse, °withhold, °forbid, °turn down, °decline, disallow; °recall, °revoke, °recant: *He asserts that his right to counsel was denied.* 3 disavow, °repudiate, °renounce, disown, forswear, disclaim: *The witch doctor demanded sacrifices, saying that the angry gods would not be denied.*

depart *v.* 1 °go, go away *or* out *or* from *or* off, °leave, °quit, °retire (from), °retreat (from), °withdraw (from), °exit (from), set out *or* forth *or* off, decamp, get out, abscond, °fly, cut and run, °skip (out), run off *or* away *or* out, take to the road, take one's leave, °check out, °disappear, vanish, °evaporate, *Jocular* toddle off, *Imperative* Begone!, *Colloq* °beat it, scram, shove off, make oneself scarce, *Brit* scarper, *US* hit the road, be out of (someplace), *Slang* °split, *Imperative* get lost, *US* cut (on) out, vamoose, take a (runout) powder, lam (on) out, take it on the lam, *Brit* do a moonlight flit, *Usually imperative* °bugger off, buzz off, *Taboo imperative* fuck off: *Our bags are packed and we depart at noon. If you're going to talk that way, you must depart at once!* 2 Often, ***depart from.*** °deviate (from), °change, °diverge (from), turn (aside *or* away) (from), °differ (from), °vary (from), °break away (from), °leave, °abandon, °stray (from), veer (from): *She refused to depart from established practices.*

department *n.* 1 °division, subdivision, °branch, °office, °bureau, °section, °segment, °unit, °part: *Some departments are in another building.* 2 °responsibility, °concern, °worry, °sphere, bailiwick, °jurisdiction, °domain, °control, area *or* sphere of influence *or* activity: *Wernher Von Braun just sent up the V-2 missiles—where they came down was not his department.*

depend *v.* 1 ***depend (on or upon).*** be contingent *or* dependent *or* conditional on, °turn on, hinge on, pivot on, °hang on, be subject to, rest on, be influenced *or* determined *or* conditioned by: *The plans for our picnic depend on the weather.* 2 ***depend on*** *or* ***upon.*** °trust (in), °rely on, °count on, °reckon on, bank on, be sure of, put one's faith *or* trust in: *I knew we could depend on you, Giles, to do the right thing.*

deplorable *adj.* 1 °lamentable, °regrettable, °sad, woeful, °grievous, °wretched, °miserable, °unfortunate, °awful, distressing, °disturbing, troubling, upsetting, °grave, °serious, °oppressive, °difficult, °desperate, °hopeless, °tragic, °disastrous: *Orphaned at six, he had a deplorable childhood.* 2 °shameful, °disgraceful, °scandalous, °disreputable, °awful, °bad, appalling, °dreadful, °abominable, execrable, °terrible, reprehensible: *What did you think of Annie's deplorable behavior at last week's dance? That's a deplorable painting.*

deposit *v.* 1 °place, °leave, set *or* put *or* lay (down), °drop, *Colloq US* plunk down: *Deposit used cups on this table, please. We deposited twenty thousand dollars on a house.* 2 °entrust, °leave, °lodge, consign, °keep, °place, °put; °store, °save, °set aside, bank, lay *or* put away, *Brit* pay in, *Colloq* stash away: *Each morning she deposits the children at the day nursery and goes to work. He deposits money every week in a pension fund.* —*n.* 3 down payment, part *or* partial payment, advance payment: *A small deposit will hold your purchase till you are ready to pay for it in full.* 4 °precipitate, °sediment, °silt, alluvium, °dregs, lees, °accumulation, deposition: *There is a dark deposit at the bottom of the coffeepot.*

depreciate *v.* 1 devalue, devaluate, °decrease, °diminish, lessen, °reduce, °lower, °depress, cheapen, °mark down: *The abundant harvest has depreciated the price of commodities.* 2 °disparage, °diminish, °deride, decry, underrate, undervalue, °underestimate, °minimize, °belittle, °slight, derogate, deprecate, °discredit, denigrate, °run down, vilipend, *Colloq* °play down, *US* °talk down: *When he depreciates another's work he adds nothing to the value of his own.*

depredation *n.* °plunder, plundering, °pillage, pillaging, despoliation, despoiling, ravaging, sacking, laying waste, devastation, °destruction; ransacking, °robbery, looting; °ravages: *The depredation caused by ten years of war is unimaginable.*

depress *v.* 1 deject, dispirit, °oppress, °sadden, °grieve, cast down, dishearten, °discourage, °dampen, cast a gloom *or* pall over, °burden, °weigh down: *He's very depressed right now because he failed to get a promotion.* 2 °weaken, °dull, debilitate, °enervate, °sap; °depreciate, cheapen, devalue, devaluate; °diminish, °lower, bring down, °reduce: *The news about a new oil field depressed the market today.* 3 °press (down), push (down) (on), °lower: *If the pressure gets too high, just depress this lever.*

depression *n.* 1 °indentation, dent, dimple, °impression, °pit, °hollow, °recess, °cavity, concavity, °dip: *When the box fell, its corner left a small depression in the top of the metal cabinet.* 2 dejection, °despair, °gloom, downheartedness, °sadness, °melancholy, discouragement, despondency, gloominess, glumness, the blues, unhappiness; *Colloq* the dumps: *A general feeling of depression came over us at the doctor's words.* 3 °recession, °slump, °(economic) decline, downturn, *US, Canadian* bust: *The analysts are unable to predict either booms or depressions.*

deprive *v.* °withhold, °deny, °refuse; °withdraw, °remove, °strip, °dispossess, take away, expropriate, °divest; mulct: *They deprived him of the right to have visitors.*

deprived *adj.* °needy, in want, in need, °impoverished, badly off, °destitute, °poor, poverty-stricken, *Euphemistic* underprivileged, disadvantaged: *As a deprived family, they are entitled to certain benefits.*

depth *n.* 1 deepness, °extent, °measure, °profundity, profoundness: *The depth of the cavern was at least three miles.* 2 °profundity, profoundness, abstruseness, °obscurity, reconditeness, °complexity, intricacy: *There is great depth of meaning in many proverbs.* 3 °profundity, °wisdom, sagacity, sageness, °understanding, °perception, astuteness, perspicacity, perspicaciousness, °insight, °intuition, acumen, °penetration: *One would scarcely characterize Mickey Mouse as possessed of great depth.* 4 °intensity, °profundity, °strength; vividness, °brilliance, brilliancy, brightness, richness: *It is hard for me to express the depth of my feeling for you. The depth of color is much better in this picture.* 5 ***depths.*** deep(s), °abyss, abysm, chasm, bowels of the earth, °(bottomless) pit, nethermost reaches *or* regions, nadir: *As we descended into the depths the temperature increased. She is in the depths of despair and needs your moral support.* **6 *in depth.*** °thoroughly, comprehensively, °in detail, °profoundly, °deeply, extensively, intensively, concentratedly, probingly: *The specialists have looked into the problem in depth and have no answer yet.*

deputy *n.* °substitute, replacement, surrogate, °stand-in, °reserve, °proxy; °agent, °operative, °representative, °go-between, °intermediary, spokesperson, spokesman, spokeswoman, °delegate, °ambassador, °minister, emissary, °envoy, legate, (papal) nuncio; *Chiefly US* °alternate: *The head of the department sent a deputy to the conference in his place.*

deranged *adj.* °mad, °insane, demented, lunatic, unhinged, °unbalanced, °berserk, °crazy, crazed, °psychotic, irrational, *non compos mentis,* out of one's mind *or* senses *or* head, not all there, of unsound mind, crackbrained, mad as a hatter *or* March hare, off the rails, *Colloq* touched, dotty, °daft, cracked, bats, cuckoo, balmy, *US* have nobody home (upstairs), out to lunch, off the wall, *Brit* barmy (in the crumpet), potty; *Slang* bonkers, dippy, barmy *or* balmy, batty, screwy, loony, nuts, nutty, wacky, bananas, off one's rocker, off one's trolley, °mental, missing a few marbles, not having all one's marbles, kooky, with a screw loose, *Chiefly US* (plumb) loco, meshuga, *Chiefly Brit* off one's chump: *You are deranged if you think I'm going in there with that lion.*

derelict *adj.* **1** °deserted, °abandoned, forsaken, neglected; ruined, °dilapidated, °run-down, tumbledown: *The commission funds the renovation of derelict buildings in the inner city.* **2** negligent, °remiss, neglectful, °delinquent, dilatory, °careless, °heedless, °lax, °slack, °irresponsible, °slipshod, slovenly, *Colloq* °sloppy: *He was accused of having been derelict in his duty.* —*n.* **3** vagrant, °tramp, °outcast, pariah, °loafer, °wastrel, °good-for-nothing, ne'er-do-well, malingerer, °vagabond, °slacker, °down-and-out, *US and Canadian* hobo, *Colloq US* °bum: *Because of alcohol, he ended up as a derelict.*

deride *v.* °mock, °ridicule, °scoff (at), °jeer (at), °laugh (at), make fun *or* sport (of), °tease, °taunt, °twit, poke fun (at), make a laughingstock (of), °sneer (at), °scorn, °flout, disdain, pooh-pooh, °belittle, °diminish, °disparage, laugh off, *Brit* °rally, *Colloq* °knock, *Brit* take the mickey *or Brit* micky out of: *His classmates had always derided his attempts at getting anywhere with the girls.*

derision *n.* °ridicule, °mockery, °raillery, °laughter, °sarcasm, scoffing, °contempt, °scorn, contumely, °disrespect; °satire, °lampoon, pasquinade, °burlesque, °caricature, travesty: *Her suggestion was greeted with derision.*

derisory *adj.* mocking, ridiculing, °scornful, derisive, °disdainful, °contemptuous, taunting, insulting, contumelious, jeering; °sardonic, °sarcastic, ironic(al), °satirical: *He felt crushed by their derisory laughter.*

derivation *n.* °origin, descent, °extraction, °source, °beginning, °foundation, ancestry, genealogy, etymology, °root: *The derivations of many words are unknown.*

derivative *adj.* **1** derived, borrowed, procured, obtained, acquired; unoriginal, °secondhand, copied, imitative, plagiarized, plagiaristic: *He created nothing of his own—all his compositions were highly derivative.* —*n.* **2** °derivation, °offshoot, °development, spinoff, by-product: *The Romance languages are derivatives from Latin.*

derive *v.* **1** °draw, °extract, °get, °obtain, °acquire, °procure, °receive, °secure, °gain, °elicit, °deduce, educe, °infer, °gather, °collect, °harvest, glean, cull, winnow: *I derive no pleasure from punishing you. I derived from her remark that she didn't like the play.* **2 *derive from.*** arise from *or* out of, originate in *or* with *or* from, emerge from *or* out of, come (forth) from *or* out of, arrive from, issue from, proceed from, develop from, spring from, flow from, emanate from, stem from, be traceable *or* traced to: *I believe that the word* Schadenfreude *derives from German. All our knowledge is derived from experience.*

derogatory *adj.* depreciatory, depreciating, depreciative, disparaging, abasing, debasing, lowering, denigrating, belittling, diminishing, demeaning, detracting, deflating, minimizing, °mitigating; uncomplimentary, °offensive, insulting: *The family took a somewhat derogatory attitude toward commerce. He said something derogatory about my wife, so I punched him.*

descend *v.* **1** come *or* go down, move down, °climb down, °get down: *He descended from the roof when his work was done.* **2** °decline, incline (downward(s)), °slope, °slant, °dip, °drop, °fall, °plunge, °plummet: *Beyond the curve, the road descends suddenly for a mile.* **3** °stoop, °condescend, °sink, lower oneself: *If you start shouting, you're just descending to Basil's level.* **4 *descend on.*** °attack, °assault, invade, pounce on *or* upon, swoop down on *or* upon: *Fighter planes descended on us in droves and destroyed the base completely.*

descendant *n.* °offspring, °progeny, °issue, °heir, °posterity, °family; °child, son, daughter, grandchild, scion; °offshoot: *They claim to be descendants of Tsar Nicholas.*

describe *v.* **1** °tell (of), °recount, °relate, give an account (of), °narrate, °recite, °report, °chronicle; retail: *He described his adventures in Rio.* **2** °detail, °define, °explain, °specify, delineate: *Please describe exactly where you found the body.* **3** °characterize, °portray, °paint, depict, °identify, °label, °style; °represent: *I would describe her as careless rather than uncaring.* **4** °trace, mark out, °outline, °traverse, °draw: *The trail of the comet described a perfect arc in the black sky.*

description *n.* **1** portrayal, characterization, depiction, (thumbnail) sketch, °portrait: *Her description of Mr. Bilker was not flattering.* **2** °account, °narrative, °story, °report, °representation, °statement, °definition; °explanation, commentary; °chronicle, °history, °record, °narration; °memoir: *I want your detailed description of what led up to the argument.* **3** °sort, °kind, °nature, °character, °type, °variety, °brand, °breed, species, °category, genus, ilk, °genre, °class; °stripe, kidney, feather: *Carstairs is a roué of the worst description.*

desecrate *v.* °profane, defile, °blaspheme (against), °dishonor, °degrade, °debase, befoul, °contaminate, °pollute, °corrupt, °violate, °pervert, °vitiate: *Vandals desecrated the temple of Minerva.*

desert[1] *n.* **1** °waste, wilderness, wasteland, dust bowl: *After nine days in the desert my camel died.* —*adj.* **2** °barren, °desolate, °uninhabited, unpeopled, °lonely, °deserted, arid, °bare, °vacant, °empty, °wild, uncultivated: *I was marooned on a desert island.* —*v.* **3** °forsake, °leave, °abandon; °jilt, °throw over; °maroon, strand, leave to twist (slowly) in the wind; *Colloq* run *or* walk out on, leave flat *or* in the lurch, leave high and dry: *His courage deserted him when he saw the child's eyes. He has deserted his wife for a younger woman.* **4** abscond, °quit, °run away (from), °defect, °abandon; *Military slang* go over the hill: *He deserted and will be court-martialed.*

desert[2] *n.* Often, **deserts.** payment, recompense, °requital, compensation, °due, °right; °retribution, °justice, *Slang* comeuppance, what's coming to one: *She'll get her just deserts one of these days.*

deserted *adj.* °abandoned, °desolate, forsaken, neglected, °uninhabited, unpeopled, °vacant, vacated, unfrequented, unvisited, unoccupied, °empty; stranded, rejected, godforsaken, °isolated, °solitary, °lonely, friendless: *The reef where I was marooned was deserted except for a seagull or two. At that hour, the streets are totally deserted.*

deserter *n.* °runaway, °fugitive, escapee, absconder, °defector, °renegade, °outlaw; °traitor, °turncoat, *Colloq* rat: *Deserters are shot when caught.*

deserve *v.* °merit, °earn, be entitled to, be worthy of, °rate, °warrant, °justify: *You ought to be nicer to Bosworth—he doesn't deserve such unkind treatment.*

deserved *adj.* merited, earned, °just, °rightful, °suitable, °fitting, °fit, °appropriate, °proper, °right, °fair, °equitable, °meet, warranted, condign: *Carla was never given her deserved credit for catching the thief.*

deserving *adj.* °meritorious, °worthy, merited, commendable, °laudable, °praiseworthy, creditable, °estimable: *Why not leave your money to a deserving charity?*

design *v.* **1** °plan, °draw up, °think of, conceive of, °contemplate, °devise, °lay out, visualize, °envisage, °envision, sketch (out), °pattern, °set up: *The building was originally designed as a skyscraper.* **2** °plan, sketch (out), delineate, °outline, °draft, work *or* map *or* block out, °lay out, °devise, °invent, contrive, °create, °conceive, °originate, °think up, °develop, °organize, °frame, °shape, °mold, °forge, °make, °construct, °form, °fashion: *Smathers has designed a new sales strategy for the company.* **3** sketch, °draft, °lay out, °draw; °form, °devise: *Did you design that new trademark for Cosmic Hairpins?* **4** °intend, °mean, °plan; °purpose, °destine; °scheme, °plot: *The building was originally designed to be a school. The book was designed for children.* —*n.* **5** °plan, °scheme, °conception, °study, °project, °proposal, °undertaking, °enterprise; blueprint, °pattern, °chart, diagram, layout, map, °drawing, °draft, sketch, °model, °prototype: *The grand design for rebuilding the city was not approved.* **6** °form, °shape, configuration, °pattern, °style, °motif, °format, layout, °makeup, delineation, °arrangement, °organization, °composition, °structure, construction: *I don't much care for her new design of my monogram.* **7** °aim, °purpose, °intention, °objective, °object, °goal, °point, °target, °intent: *My design had been to go at once to London.* **8 *designs.*** °plot, °intrigue, °stratagem, °cabal, °conspiracy, conniving, manipulation, connivance, evil intent *or* intentions: *His designs against me have borne bitter fruit.*

designate *v.* **1** °indicate, °specify, pinpoint, particularize, delineate, °point out, °identify, °state, °set forth, write *or* put down, °name: *You should designate your heirs in your will.* **2** °appoint, °nominate, °name, °identify, denominate, °select, °pick, °choose, °elect, °assign, °appropriate, °delegate, depute: *She has not yet designated her successor.* **3** °mean, °stand for, °symbolize, °denote, °represent: *The Greek letter* pi *designates the ratio of the circumference of a circle to its diameter.* **4** °call, °name, °style, °term, °label, °christen, dub, °nickname, °entitle: *By acclamation he was designated Lord High Executioner.*

designer *n.* **1** °creator, originator, architect, artificer, °author, deviser, inventor; (interior) decorator, artist; draftsman: *Raymond Loewy was a designer of locomotives and fountain pens. Lady Mendl was the best-known interior designer of the 1920s.* **2** intriguer, schemer, conniver, plotter, conspirator: *That cunning designer has wormed his way into the favor of the caliph.*

designing *adj.* °scheming, plotting, conniving, conspiring, intriguing, °calculating, °wily, °tricky, cunning, °sly, underhand(ed), °crafty, °artful, °shrewd, Machiavellian, guileful, °deceitful, double-dealing, °devious, treacherous, *Colloq* °crooked: *The prince has fallen prey to designing courtiers.*

desirable *adj.* **1** sought-after, wanted, coveted, longed-for, looked-for, desired: *Few things are more desirable than security in old age.* **2** °attractive, °pleasant, °pleasing, °agreeable, °winning, winsome, captivating, °seductive, alluring, °fetching: *I saw Penfirth at the theater with a most desirable young lady.* **3** °good, °goodly, °excellent, °choice, °fine, °superior, °superb, *Colloq Brit* plummy: *The company has produced some very desirable automobiles.* **4** °profitable, °worthwhile, °beneficial, °advantageous, valuable, °worthy, °estimable, commendable, °admirable: *Lady Chelm's plan possesses many desirable attributes.*

desire *v.* **1** crave, °want, °fancy, covet, wish for, hope for, long *or* yearn for, pine *or* sigh for, °hanker after, have an eye *or* taste for, hunger *or* thirst for *or* after, die for, have one's heart set on, give one's eyeteeth for, *Colloq* have a yen for, *US slang* have the hots for: *I desire nothing but your happiness. He desired her more than anything else in the world.* **2** °ask for, °request, °order, °demand, °solicit, importune, °summon, °require: *Do you desire anything further, sir?* —*n.* **3** °longing, craving, yearning, hankering, °hunger, °thirst, °appetite; °passion, °lust, libido, lustfulness, concupiscence, lecherousness, lechery, lasciviousness, salaciousness, prurience, *Slang* hot pants, *US* the hots; *Colloq* yen: *He felt desire rising in him like a fever.* **4** °wish, °request, °urge, °requirement, °order, °requisition, °demand, desideratum; °appeal, entreaty, °petition: *He fulfills her every desire.*

desirous *adj.* wishful, desiring, °longing, yearning, °hopeful, hoping: *I was desirous to learn more about his whereabouts.*

desolate *adj.* **1** °solitary, °lonely, °isolated, °deserted, °forlorn, forsaken, friendless, °alone, °abandoned, neglected; °desert, °uninhabited, °empty, unfrequented, °bare, °barren, °bleak, °remote: *He felt desolate after his wife's death. Tristan da Cunha is a group of four desolate islands in the Atlantic.* **2** laid waste, ruined, devastated, ravaged, destroyed: *After the disaster, we wandered through the desolate land.* **3** °dreary, °dismal, °wretched, °joyless, cheerless, comfortless, °miserable, °unhappy, down, disconsolate, °sad, °melancholy, °sorrowful, °forlorn, °mournful, °woebegone, °gloomy, °brokenhearted, heavy-hearted, °inconsolable, °dejected, downcast, °downhearted, dispirited, low-spirited, depressed, °melancholy, spiritless, °despondent, °dismal, distressed, discouraged, °hopeless: *He has brought some happiness into her desolate existence.* —*v.* **4** depopulate: *The country was desolated by the invaders.* **5** °destroy, °devastate, °ruin, °lay waste, despoil, °ravage, °demolish, °obliterate, annihilate, °raze, °gut: *Invaders desolated the countryside.* **6** °dismay, dishearten, °depress, °daunt, dispirit, °sadden, deject, dispirit, °discourage: *He was either buoyed up by renewed hope or desolated by despair.*

desolation *n.* **1** °destruction, °ruin, devastation, °waste, spoliation, despoliation, °sack, °depredation, extirpation, obliteration, ravagement, barrenness, °havoc, °chaos: *We had to shape a new life from the desolation left by the war.* **2** °grief, °sorrow, dreariness, °despair, °gloom, °distress, °melancholy, °sadness, °misery, °woe, °anguish, wretchedness, dolor, dolefulness, unhappiness: *She felt the desolation of loneliness after Ashton's death.*

despair *n.* **1** hopelessness, °desperation, discouragement, disheartenment, despondency, dejection, °depression, °gloom, gloominess, °misery, °melancholy, wretchedness, °distress, miserableness, °anguish; °resignation: *The despair of the prisoners was evident in their ravaged faces.* —*v.* **2** give up *or* lose hope; °surrender, °quit: *We despaired of ever seeing our children again.*

desperate *adj.* **1** °reckless, °foolhardy, °rash, °impetuous, °frantic, frenzied, °panic-stricken: *A desperate situation requires desperate measures.* **2** °careless, °hasty, devil-may-care, °wild, °mad, frenetic, °furious: *They made a last desperate attack on the fort.* **3** °anxious (for), craving, °hungry (for), °thirsty (for), needful (of), °desirous (of), covetous (of), °eager (for), longing *or* yearning (for), wishing (for), hoping (for), aching (for), pining (for): *She is desperate for attention.* **4** °urgent, °pressing, compelling, °serious, °grave, °acute, °critical, °crucial, °great: *There is a desperate need for medicines at the disaster site.* **5** °precarious, °perilous, life-threatening, °hazardous, °dangerous, °tenuous, °hopeless, beyond hope *or* help: *Avalanches are making the climbers' situation even more desperate.* **6** at one's wit's end, °forlorn, despairing, °despondent, °wretched, at the end of one's tether *or* rope, °frantic: *With no one to turn to for help, he was truly desperate.*

desperation *n.* **1** recklessness, impetuosity, rashness, foolhardiness, imprudence, heedlessness: *Penniless and half-starved, he was driven to desperation and stole a loaf of bread.* **2** °despair, °anxiety, °anguish, anxiousness, despondency, °depression, dejection, discouragement, defeatism, pessimism, hopelessness, °distress, °misery, °melancholy, wretchedness, °gloom, °sorrow: *In a final act of desperation, he tried to commit suicide.*

despicable *adj.* °contemptible, below *or* beneath *or* beyond contempt *or* scorn *or* disdain, °mean, detestable, °base, °low, °scurvy, °vile, °sordid, °wretched, °miserable, ignoble, ignominious, °shabby; °shameful, °shameless, reprehensible: *The depths of despicable degradation were plumbed in the concentration camps.*

despise *v.* disdain, °scorn, look down on *or* upon, be contemptuous of, sneer at, °spurn, contemn; °hate, °loathe, °detest, °abhor: *She despised her servants and treated them badly. He despised anyone who did not have a college degree.*

despite *prep.* °in spite of, °notwithstanding, undeterred by, °regardless of, in the face *or* teeth of, in defiance of, without considering, without thought *or* consideration *or* regard for, ignoring: *We went sailing despite the forecast of a storm.*

despondent *adj.* °dejected, °sad, °sorrowful, °unhappy, °melancholy, °blue, depressed, down, downcast, °downhearted, °low, morose, °miserable, disheartened, discouraged, dispirited, low-spirited, °down in the mouth, *Colloq* down in the dumps: *He's been despondent since she went away.*

despot *n.* absolute ruler *or* monarch, °dictator, °tyrant, °oppressor, autocrat: *History has painted Ivan the Terrible as one of the cruelest despots of all time.*

despotic *adj.* °dictatorial, °tyrannical, °oppressive, °authoritarian, imperious, °domineering, °totalitarian, °absolute, autocratic, °arbitrary: *The country was under the despotic rule of a callous tyrant.*

despotism *n.* autocracy, monocracy, autarchy, totalitarianism, absolutism, dictatorship, °tyranny, °oppression, °suppression, °repression: *The stringent new laws were just another instance of the brutal despotism of the regime.*

dessert *n.* °sweet, *Brit* pudding, *Colloq Brit* pud, afters: *For dessert, I had ice cream and she had a fruit tart.*

destination *n.* journey's end, terminus, °stop, stopping place; °goal, °end, °objective, °target: *Our destination is Bristol.*

destine *v.* **1** °fate, predetermine, predestine, ordain, foreordain, preordain; °doom: *The gods destined him for greater things.* **2** °design, °intend, °mean, °devote, °assign, °appoint, °designate, °purpose, °mark, earmark, °set aside: *I beheld the chariot destined to carry me heavenward.*

destined *adj.* **1** meant, intended, designed, °predetermined, foreordained, predestined, °fated; °doomed, written; *US* °in the cards: *His destined end was to be shot while escaping. Oliver was destined to fail at everything he tried. It was destined that the boy would become king.* **2** °certain, °sure, °bound, ineluctable, °unavoidable, °inevitable, inescapable: *Being devoured by monsters is the destined demise of all who dare to enter there.*

destiny *n.* °fate, °doom, °fortune, °lot, kismet, karma: *It is my destiny to be ignored when living and forgotten when dead.*

destitute *adj.* **1** in want, °impoverished, poverty-stricken, °poor, °indigent, °down-and-out, °needy, °on one's uppers, badly off, penniless, °penurious, impecunious, °insolvent, bankrupt, *Colloq* °hard up, °broke, *US* on skid row: *Why distribute food to destitute families only at Christmas?* **2** Usually, ***destitute of.*** bereft of, deficient in, deprived of, devoid of, lacking (in), °wanting (in), in need, needful (of), without: *Bandling is destitute of any moral conscience. The landscape was destitute of any greenery.*

destroy *v.* **1** °demolish, tear *or* pull down, °raze, °wipe out, °ravage, °wreck, smash, °ruin, break up *or* down, annihilate, °crush, eradicate, extirpate, °exterminate, °devastate, commit mayhem, °lay waste, vandalize, *Slang US* °trash: *The invading hordes destroyed everything, leaving desolation in their wake. The storm destroyed fifty houses.* **2** °ruin, do away with, °end, make an end of, bring to an end, bring *or* put an end to, °terminate, °finish, °kill: *Realizing what he had done, he destroyed himself. The trial destroyed his career.* **3** °counteract, °neutralize, nullify, annul, °cancel (out), °reverse; °stop, interfere with: *Caught embezzling, Martin destroyed everything he had worked for. Sunspot activity destroyed radio transmission this week.* **4** °disprove, refute, confute, °deny, °contradict, negate, °overturn, °overthrow, °ruin, °spoil, °undermine, °weaken, enfeeble, devitalize, °exhaust, disable, °cripple: *By pointing out just one flaw, she destroyed his entire argument.*

destruction *n.* **1** demolition, razing, wrecking, °ruin, ruining, ruination, breaking up *or* down, °mayhem, °havoc, annihilation, devastation, tearing *or* knocking down, laying waste, ravagement; rack and ruin, *Colloq* wiping out: *The destruction of the city took place in 1942.* **2** °slaughter, annihilation, °killing, eradication, °murder, extermination, °holocaust, liquidation, °massacre, extinction, genocide, assassination, slaying, putting to death, putting an end to, making an end of, doing away with, putting away, *Colloq* doing in, wiping out; *Slang US* rubbing out, rubout: *They were bent on the destruction of an entire people.* **3** °undoing, °end, °ruin, ruination, °downfall, °termination, breakup, °breakdown, °collapse: *The imprisonment of the bosses spelled the destruction of the entire crime network.*

destructive *adj.* **1** °harmful, °injurious, baneful, pernicious, °dangerous, °hurtful, toxic, °poisonous, °virulent, noxious, °bad, °malignant, baleful, °unwholesome, damaging, °detrimental, deleterious, °devastating; °deadly, °fatal, °lethal, °fell, °killing, internecine: *The spray keeps away insects but is destructive of the plant life.* **2** °negative, adverse, °opposing, °opposed, °contrary, °contradictory, antithetical, conflicting, unfavorable, condemnatory, °derogatory, disparaging, disapproving, °critical: *The playwrights feared and disliked him because of his destructive criticism.*

desultory *adj.* shifting, °devious, unsteady, °irregular, wavering, °inconstant, °fitful, °spasmodic, unmethodical, °disconnected, unsystematic, °disorderly, disordered, unorganized, disorganized, °inconsistent, °random, °haphazard, °chaotic, °erratic, °shifty: *He made no more than a desultory effort to stop smoking. The countries engaged in intermittent, desultory warfare for decades.*

detach *v.* °separate, uncouple, °part, disjoin, °disengage, disunite, °disconnect, disentangle, °free, unfasten, °undo, °cut off, °remove: *Detach the tag without tearing it.*

detached *adj.* **1** °disconnected, °unattached, separate(d), °free, °isolated, disentangled, unfastened, removed, °cut off, divided, disjoined: *A young man was standing there, detached from the crowd. The doctor said she had a detached retina.* **2** °disinterested, °aloof, uninvolved, unemotional, °dispassionate, *dégagé(e)*, °reserved, °impersonal, °impartial, °neutral, °objective, unbiased, °unprejudiced: *She exhibited a slight, though detached concern for the children's welfare.*

detachment *n.* **1** separating, unfastening, disconnecting, detaching, disengaging; °separation, disconnection, disengagement: *Most young birds cannot survive a prolonged period of detachment from their parents.* **2** aloofness, unconcern, °indifference, coolness, inattention, insouciance: *He viewed the carnage of the battle with regal detachment.* **3** See **detail, 3,** below.

detail *n.* **1** °particular, °element, °factor, °point, °fact, °specific, technicality, component, °item, °feature; °aspect, °respect, °count: *He gave us a general idea of the plan but not a single detail.* **2** ***details.*** °particulars, minutiae, niceties, fine points, specifics, technicalities: *Must we go into all the details of his dismissal?* **3** °detachment, °squad, °party, cadre, °duty, °fatigue, °group: *The sergeant appointed a detail to police the area.* **4** ***in detail.*** specifically, °particularly, °thoroughly, °in depth, item by item, point by point, exhaustively, comprehensively, °inside out, °perfectly: *We examined the report in detail.* —*v.* **5** °specify, °spell out, °itemize, delineate, catalogue, °list, °tabulate, °enumerate, particularize, °recount, cite (chapter and verse): *She detailed every little move I was to make.* **6** °assign, °appoint, °charge, °delegate, °name, °specify, °send: *We have been detailed to act as your bodyguard during your visit.*

detailed *adj.* **1** itemized, °exhaustive, °comprehensive, °thorough, °full, °complete, °inclusive, particularized, °precise, °exact, °minute, blow-by-blow, °circumstantial: *He kept a detailed report of everything that happened on D-Day.* **2** °intricate, complex, °complicated, °elaborate, °ornate: *Note the detailed scrollwork on this screen.*

detect *v.* **1** uncover, °find (out), °discover, °locate, °learn of, ascertain, °determine, °dig up, °unearth: *The pathologist detected the presence of prussic acid in the victim's bloodstream.* **2** °perceive, °note, °notice, °identify, °spot, °observe, °sense, °read, °scent, °smell, discern, °feel, °catch, °find: *Did I detect a tone of sarcasm in your reply, young man?*

detective *n.* investigator, private investigator, CID man, policeman, °constable, *Colloq* private eye, °sleuth, sherlock, °snoop, snooper, *US* P.I., dick, Hawkshaw, *Brit* tec; *Slang* cop, copper, *US and Canadian* gumshoe, peeper: *A detective has solved the case of the missing weapon.*

detention *n.* °custody, confinement, °imprisonment, °captivity, internment, incarceration, °restraint, *Literary, archaic* durance: *The culprit was kept in detention for a week.*

deter *v.* dissuade, °discourage, °inhibit, °intimidate, °daunt, frighten off *or* from *or* away, scare off *or* from; °prevent, °stop, °obstruct, °check, °hinder, °impede: *I was deterred from entering by three large dogs. Regular airing of the bedclothes helps deter mold.*

detergent *n.* **1** cleaner, cleanser, soap (powder *or* flakes *or* liquid); surfactant, surface-active agent, detersive: *You put too much detergent into the washing machine and it overflowed.* —*adj.* **2** cleaning, cleansing, washing, purifying, detersive: *The detergent effect is reduced if too much soap is used.*

deteriorate *v.* **1** °worsen, °decline, °degenerate, °degrade, °spoil, °worsen, get worse, °depreciate, °slip, °slide, *Colloq* go to pot, go to the dogs, go downhill: *We have watched their relationship deteriorate over the years.* **2** °decay, °decline, °disintegrate, °fall apart, °decompose, °crumble, °erode: *The building slowly deteriorated and is now uninhabitable.*

determination *n.* **1** resoluteness, °resolution, firmness, °resolve, steadfastness, °tenacity, °perseverance, °fortitude, doggedness, °persistence, constancy, single-mindedness, °will (power), *Colloq* °grit, °guts: *The idea is a good one, if only she has the determination to see it through.* **2** °settlement, °resolution, resolving, °decision, °solution, °judgment, verdict, °outcome, °result, °upshot, conclusion, °end, °termination: *None of us could live in peace till the determination of the border dispute.* **3** fixing, settling, ascertainment, ascertaining, delimitation, °definition: *The determination of our position is critical in setting our course.*

determine *v.* **1** °settle, °decide, °clinch, arbitrate, °judge, adjudge, conclude, °terminate, °end: *The ambiguity must be determined one way or the other.* **2** ascertain, °find out, °discover, conclude, °infer, °draw, °learn, °detect; °verify: *From the evidence, they determined the identity of the intruder.* **3** °decide, °choose, °select, °resolve, make up one's mind, settle on *or* upon, fix on *or* upon: *You alone must determine which candidate you want to vote for.* **4** °affect, °influence, act on, °shape, °condition, °govern, °regulate, °dictate: *There were many factors determining my choice.*

determined *adj.* **1** °decided, °resolute, resolved, °purposeful, dogged, strong-willed, °strong-minded, °single-minded, °tenacious, °intent, °firm, unflinching, unwavering, °fixed, °constant, °persistent, persevering, °steady, unfaltering, °unhesitating, unyielding, °stubborn, °obstinate, adamant: *He was determined not to go. We made a determined effort to locate the wreck.* **2** °fixed, determinate, °definite, °exact, °precise, °distinct, °predetermined, ascertained, identified: *The researchers worked to a previously determined schedule. They agreed to pay a percentage of the determined price.*

deterrent *n.* °hindrance, °impediment, discouragement, disincentive, dissuasion, °check, °hitch, °obstacle, °obstruction, °stumbling block; °catch, °snag, °rub, °fly in the ointment, °bar, °drawback: *Some experts hold that the death penalty is no deterrent to murder. The only deterrent to your plan is that we are likely to be caught.*

detest *v.* °despise, °loathe, °hate, °abhor, execrate, abominate: *They served turnips, which I detest, and sat me next to Ida, whom I also detest.*

detour *n.* **1** °diversion, deviation, circuitous route *or* way, roundabout way, °bypass: *The detour took us five miles out of our way.* —*v.* **2** °deviate, turn (away) from, °divert, °bypass: *I detoured from the main road and took a shortcut.*

detract *v.* ***detract from.*** °diminish, °reduce, take away from, °subtract from, lessen, °depreciate, °disparage: *Once you are in the public eye, your slightest fault detracts from your reputation.*

detriment *n.* °disadvantage, °drawback, °liability; °damage, °harm, °ill, °impairment, °injury, °hurt, °loss: *He has supported some very unpopular causes to his detriment. Seeds survive without detriment where their plants would perish.*

detrimental *adj.* disadvantageous, °harmful, °injurious, °hurtful, damaging, deleterious, °destructive, °prejudicial, adverse, unfavorable, inimical, pernicious: *I know nothing detrimental about either one of them.*

devastate *v.* **1** °lay waste, °ravage, °destroy, °waste, °sack, °raze, °ruin, °desolate, °spoil, °wreck, °demolish, °level, °flatten, °gut, °obliterate: *The island was completely devastated by the tidal wave that followed the typhoon.* **2** disconcert, confound, °discomfit, °take aback, °nonplus, °shatter, °overwhelm, abash, °shock; °humiliate, °mortify, °embarrass, chagrin, *Colloq* °floor, *US* discombobulate: *She was devastated by the news of Bertie's expulsion from college.*

devastating *adj.* **1** °keen, °incisive, mordant, °penetrating, °trenchant, °telling; °sardonic, °sarcastic, °bitter, acid, °caustic, °savage, °satirical, °virulent, vitriolic: *Because of his bland manner, his devastating wit often caught people by surprise.* **2** °ravishing, captivating, °enthralling, °stunning, °overpowering, bewitching, °spellbinding; spectacular: *Kathy was wearing a devastating black silk dress.*

develop *v.* **1** bring out *or* forth, °advance, expand (on *or* upon), broaden, enlarge (on *or* upon), °amplify, evolve, expatiate (on *or* upon), elaborate (on *or* upon), °reveal, lay open, °expose, °unfold, °disclose, °bare, °(cause to) grow, realize the potential (of); °cultivate, °improve, °promote, °exploit, °strengthen: *The plot is fine, but the characters need to be developed more fully. It is the aim of the school to develop the students' natural abilities.* **2** °(make) grow, °mature, °ripen, °age, °expand; °flower, blossom, bloom, °increase: *You can't develop that idea without financial backing. These shrubs will be fully developed next year.* **3** °exhibit, °display, °show, °demonstrate, °manifest: *She has recently developed an interest in cooking.* **4** °emerge, °arise, °appear, °come out, °come to light, evolve, °originate, °begin, °commence, °happen, °occur, °come about; come forth, °result: *A serious fault has developed in the rocket's fuel line. His natural talent for music developed when he joined the school band.*

development *n.* **1** °occurrence, °happening, °event, °incident, °circumstance, °situation, °condition, °phenomenon: *William Nye will report new developments from the scene.* **2** °evolution, °growth, evolvement, maturation, unfolding, maturing, °maturity, °increase, °expansion, enlargement, °increment; °advance, advancement, °progress; °improvement: *She has studied the region's economic development.*

deviant *adj.* **1** deviating, °divergent, °different, °abnormal, °strange, uncommon, °unusual, °odd, °peculiar, °curious, aberrant, °eccentric, idiosyncratic, °deviate, °queer, quirky, °weird, °bizarre, °offbeat, °singular, *Slang* °kinky, freaky, *Chiefly Brit* °bent: *They have been observing his deviant behavior for some time.* **2** °perverse, °perverted, twisted, °unnatural, depraved, °degenerate, licentious, °lascivious, °lewd; °homosexual, °gay, lesbian, tribadistic, *Offensive and disparaging* °queer, homo, butch: *According to the police, more victims of the person branded "the deviant killer" have been found.* —*n.* **3** °homosexual, °gay, lesbian, tribade, sapphist, homophile; *All the following are offensive and denigrating* °pervert, °degenerate, °deviate, *Slang* °queer, homo, °queen, fairy, pansy, nancy (boy), nance, butch, (bull) dyke, *US and Canadian* °fag, faggot, °fruit, aunt, auntie, *Brit* poof, poofter, ginger (beer), *Caribbean* auntie man: *In former days, such people would have been called deviants.*

deviate *v.* **1** turn aside *or* away, °swerve, veer, °wander, °stray, °drift, digress, °diverge; °divert: *He has chosen a path that deviates from the straight and narrow.* —*adj., n.* **2** See **deviant, 1, 3,** above.

device *n.* **1** contrivance, °mechanism, °machine, machinery, °implement, utensil, °apparatus, °instrument, appliance, °tool, °gadget, °gimmick, *Colloq* °contraption, widget, thingamajig, *Brit* gubbins: *She has patented a device for peeling hard-boiled eggs.* **2** °stratagem, °scheme, °trick, °artifice, °ruse, °plot, ploy, gambit, °strategy, °maneuver, °machination; machinery, °apparatus, °mechanism, contrivance, °gimmick, °tool, weapon: *What kinds of devices will they resort to in achieving their ends? That lawyer used every device in the devil's arsenal to separate Cornelia from her inheritance.* **3** °design, °emblem, °figure, °(heraldic) bearing, insigne, cadency mark, mark of cadency, °hallmark, trademark, °symbol, badge, coat of arms, °seal, °crest, colophon, logotype, logo, monogram, °charge, °cognizance, signet; °motto, °slogan, °legend: *The device—a closed eye—is that of Lord Boring.* **4** ***devices.*** °pleasure, °disposition, °will, °inclination, °fancy, °desire, whim: *Left to his own devices, he'll survive very well indeed.*

devil *n.* **1** Satan, Lucifer, Mephistopheles, Beelzebub, Asmodeus, Abaddon, Apollyon, Belial, Lord of the Flies, prince of darkness, spirit of evil, evil spirit, cacodemon *or* cacodaemon, evil one, wicked one, archfiend, Fiend, deuce, *Scots* Clootie; *Colloq* Old Harry, (Old) Nick, *US* (Old) Scratch: *In medieval times the devil was given horns, a tail, and cloven hoofs.* **2** °brute, °fiend, °demon, °beast, °ogre, °monster, °rogue, °scoundrel, °rake, knave, rakehell, °villain, ghoul, hellhound, vampire, °barbarian; °witch, hellcat, °shrew, termagant, vixen, virago, ogress, harpy, °hag, Xanthippe *or* Xantippe, crone: *If you hit me again I'll phone the police, you devil!* **3** °fellow, °person, °chap, °wretch, bloke, °guy, °beggar, °unfortunate, *Colloq* °bugger, *Brit* sod: *The poor devil lost an arm at Gallipoli.* **4** °imp, scamp, °rascal, fox, slyboots, sly dog, rapscallion, confidence man, trickster, *Colloq* °operator, smoothie, smooth *or* slick operator, con man, con artist: *The little devil has stolen our hearts. The devil wormed his way into our confidence and then made off with our money.* **5** ***like the devil.*** °exceedingly, °extremely, excessively, violently, speedily, confoundedly, deucedly: *The car was going like the devil when it hit the tree. She fought like the devil to protect the house.* **6** ***—the devil.*** in heaven's name, the dickens, in the world, on God's green earth, in hell: *What the devil do you think you are doing? Who the devil is she? Where the devil have you put my trousers?*

devilish *adj.* diabolic(al), °satanic, Mephistophelian, °fiendish, demonic, cacodemonic, demoniac(al), °infernal, hellish, °villainous, °sinister, °wicked, °evil, iniquitous, °sinful, flagitious, heinous, malign, malevolent, °malignant, °cruel, maleficent; impish, °mischievous, prankish, °naughty, °crazy, madcap: *He has come up with a devilish plan for stealing the secret formula.*

devilry *n.* **1** deviltry, °mischief, mischievousness, roguery, naughtiness, rascality, roguishness, diablerie, archness, knavery, knavishness: *His latest bit of devilry is hiding father's bedroom slippers.* **2** deviltry, devilishness, wickedness, °evil, fiendishness, diablerie, cruelty, malice, malevolence, viciousness, perversity, iniquity, hellishness, villainy: *That traitor is up to some devilry.*

devious *adj.* 1 °deceitful, underhand(ed), °insincere, °deceptive, misleading, subreptitious, °sneaky, furtive, °surreptitious, °secretive, double-dealing, treacherous, °dishonest, °shifty, °smooth, °slick, °slippery, °scheming, plotting, °designing, °foxy, vulpine, °wily, °sly, °crafty, °tricky, *Colloq* °crooked: *What is the latest product of your devious mind, Moriarty?* 2 °indirect, °roundabout, zigzag, °evasive, circuitous, °crooked, °rambling, °serpentine, °tortuous, sinuous, anfractuous: *That is about the most devious bit of reasoning I have ever heard!*

devise *v.* 1 concoct, °make up, °conceive, °scheme, contrive, dream up, °design, °draft, °frame, °form, °formulate, °plan, °arrange, °work out, °think up, °originate, °invent, °create, *Colloq* cook up: *He thinks he has devised a method for making gold out of lead.* 2 °bequeath, °will, convey, °hand down, °give, °assign, °dispose of, °transfer, °bestow: *I devise to my nephew, Faragut Ferguson, my property in Oklahoma.*

devote *v.* 1 °apply, °appropriate, °assign, °allot, °commit, allocate, set aside *or* apart, put away *or* aside, °dedicate, consecrate: *Each of the chapels was devoted to a separate sect.* 2 °apply, °pledge, °dedicate, °commit, °give up: *She has devoted her life to helping others.*

devoted *adj.* °faithful, °true, dedicated, committed, °devout, °loyal, loving, doting, °staunch, °tender, °steadfast, °constant; °ardent, caring, °fond, °earnest, zealous, °enthusiastic: *My most devoted friend was your brother Amos.*

devotee *n.* °fan, aficionado, adherent, votary, °enthusiast, °addict, *Colloq* buff, °fiend, *US* °hound; *Slang* °bug, nut, °freak, *US* °head, junkie, groupie: *The band was followed about on tour by scores of screaming devotees of rock music.*

devotion *n.* 1 devotedness, devoutness, °reverence; earnestness, religiousness, °piety, religiosity, pietism, godliness, holiness, spirituality, °sanctity; °worship, °prayer, °observance, °ritual: *The sect was noted for its devotion to martyrs and their relics. It is gratifying to see such devotion amongst the younger members of the congregation. They interrupted the holy man at his devotions.* 2 °dedication, consecration, °attachment, °loyalty, devotedness: *His devotion to duty will be remembered by his fellow soldiers.* 3 zeal, °ardor, °fervor, ardency, °intensity, °fanaticism, °eagerness, °enthusiasm, earnestness, °readiness, willingness; °love, °passion, infatuation, fondness, °affection, °attachment, adherence, °loyalty, allegiance: *They would dedicate themselves with slavish devotion to some brutal master.*

devour *v.* 1 wolf (down), °gulp (down), °bolt, °swallow (up), °gorge, gobble (up), gormandize, °cram, °stuff, eat (up) greedily, *Archaic* gluttonize; *Colloq Brit* pig, *US and Canadian* pig out (on): *To celebrate his winning the eating competition, he devoured two more whole apple pies.* 2 °consume, °waste, °destroy, °wipe out, °ravage, annihilate, °demolish, °ruin, wreak havoc upon, °devastate, °obliterate, eradicate: *We would then fall prey to any other nation that would conquer, rob, and devour us. A quarter of Europe was already devoured by the plague.* 3 °relish, °revel in, absorb, be absorbed by; engulf, °consume, drink in, eat up, °swallow up, °take in; °swamp, °overcome, °overwhelm: *He eagerly devoured all of Dickens's novels. The sea devoured its victims silently.*

devout *adj.* 1 °devoted, °pious, °religious, reverent, worshipful, °faithful, dedicated, °staunch, churchgoing; °holy, °godly, °saintly, °pure: *When I last saw him, he had become a devout whirling dervish.* 2 devotional, reverential, °religious, °solemn: *Through devout prayer one might see the kingdom of heaven.* 3 °earnest, °sincere, °genuine, °hearty, °heartfelt, °devoted, °ardent, zealous: *You have my devout best wishes for your happiness.*

dexterity *n.* 1 °touch, nimbleness, adroitness, deftness, °facility, °knack, °skill, °proficiency; sleight of hand: *Much fine rug-weaving is done by little children because of the dexterity of their small fingers.* 2 cleverness, °ingenuity, ingeniousness, °tact, astuteness, keenness, sharpness, shrewdness, cunning, guile, canniness, artfulness: *I admire his dexterity in arguing the case in court. He exercised great dexterity in eluding capture.*

dexterous *adj.* 1 dextrous, deft, lithe, °nimble, °supple, °agile, °quick, °skillful: *He was a dexterous archer.* 2 °clever, °ingenious, °astute, °keen, °sharp, °shrewd, cunning, guileful, canny, °artful, °crafty, °slick: *She was devout in religion, decorous in conduct, and dexterous in business. He was the most dexterous of our political leaders.*

diabolic *adj.* 1 diabolical, °devilish, °satanic, Mephistophelian, demonic, demoniac(al), °fiendish, hellish, °infernal: *His interest in the supernatural included participation in diabolic rituals of the most repulsive kind.* 2 diabolical, °cruel, °wicked, iniquitous, °evil, °fiendish, appalling, °dreadful, °inhuman, °atrocious, execrable, °abominable, °awful, °terrible, °damnable, accursed, horrid, °horrible, °hideous, °monstrous, odious, °vile, °base, °corrupt, °foul, depraved, flagitious, heinous, malicious, malevolent, malign, maleficent, °sinister, °sinful, °impious, °bad: *The prisoners suffered the most diabolic treatment.*

diagnose *v.* °identify, °name, °determine, °recognize, °distinguish, pinpoint, °interpret; °analyze: *The doctor diagnosed the symptoms as those of Lyme arthritis.*

dialect *n.* °speech (pattern), phraseology, °idiom, °accent, °pronunciation, patois, °vernacular; °jargon, °cant, slang, argot, °language, °tongue, creole, pidgin; brogue, burr, *Colloq* °lingo: *Some of the regional dialects are hard to understand.*

dialogue *n.* 1 duologue, °conversation, °discussion, °conference, °talk, °chat, colloquy, communication: *I wrote down that dialogue—it was hilarious!* 2 °parley, °conference, °meeting, °huddle, °tête-à-tête, colloquy, *Colloq US and Canadian* rap session: *A meaningful dialogue between labor and management could easily settle the question.*

diary *n.* appointment book, datebook, °calendar, engagement book; °journal, °chronicle, log, °record, annal(s): *According to my diary, the date we dined was the first.*

dicey *n.* °risky, °tricky, °dangerous, °difficult, °ticklish, unpredictable, °uncertain, unsure, °doubtful, *Colloq* iffy, chancy, °hairy: *Asking for a raise at this time could be pretty dicey, Danny.*

dicker *v.* 1 °bargain, °trade, barter, °deal, °haggle, °negotiate: *If I dicker with him, he may drop his price.* —*n.* 2 °bargain, °deal, °haggle, °negotiation: *We had a bit of a dicker but finally settled on a figure.*

dictate *v.* 1 °say, °prescribe, ordain, °decree, °demand, °command, °lay down (the law), °order, °direct, °pronounce, °impose: *It is our leader who dictates what we may say and do.* —*n.* 2 °decree, °demand, °command, °order, °direction, °instruction, °charge, °pronouncement, edict, fiat, ukase, mandate, caveat, °injunction, °requirement, °bidding, behest: *Each must act in accord with the dictates of his conscience.*

dictator *n.* autocrat, absolute ruler *or* monarch, °despot, overlord, °oppressor, tsar *or* czar, °tyrant, Fuehrer *or* Führer: *Among monarchs, Henry VIII certainly could have been characterized as a dictator.*

dictatorial *adj.* **1** °absolute, °arbitrary, °totalitarian, °authoritarian, autocratic, all-powerful, omnipotent, °unlimited: *The peoples of some countries often confer dictatorial powers on their leaders.* **2** °despotic, °tyrannical, °authoritarian, ironhanded, °domineering, imperious, °overbearing, *Colloq* °bossy: *The dictatorial way she runs the department makes those who work there miserable.*

diction *n.* **1** °language, °wording, (verbal *or* writing) style, °expression, °usage, expressiveness, °terminology, word choice, vocabulary, phraseology, phrasing, °rhetoric: *Please go over my paper and correct the diction.* **2** articulation, °pronunciation, enunciation, °delivery, elocution, °oratory, °presentation, °speech, °intonation, inflection: *That course in public speaking improved Brian's diction enormously.*

dictionary *n.* lexicon, °glossary, wordbook; °thesaurus: *My dictionary gives the pronunciation, etymology, and meanings of hundreds of thousands of words.*

die *v.* **1** lose one's life, lay down one's life, °perish, °expire, decease, suffer death, *Euphemistic* °depart, give up the ghost, be no more, (go to) meet one's maker, breathe one's last, go to the happy hunting grounds, go to one's reward, go to one's final resting place, go west, pay the debt of nature, pay one's debt to nature, pass through the pearly gates, pass away *or* on, join the majority, go the way of all flesh; *Slang* pop off, bite the dust, kick the bucket, croak, *Brit* snuff it, go for a burton, pop one's clogs, *US* turn up one's toes, cash in one's chips *or* checks: *He died of old age, a rare affliction these days.* **2** Often, ***die down*** *or* ***out*** *or* ***away.*** °dwindle, lessen, °diminish, °decrease, °ebb, °decline, °wane, °subside, wither (away), °wilt, °melt (away), °dissolve, °peter out, °fail, °weaken, °deteriorate, °disintegrate, °degenerate, °fade (away), °droop, molder, °sink, vanish, °disappear: *We lost the boat race because the breeze died down. After the third try, her enthusiasm died. The sound of the flute died away among its echoes.* **3** °expire, °end, °stop, °cease: *Your secret will die with me.* **4** Usually, ***die off*** *or* ***out.*** become extinct, °perish: *By about 200 million years ago, all the dinosaurs had died out.* **5** °long, pine, °yearn, crave, °hanker, °want, °desire, °hunger, °ache: *He said he was dying to meet a real movie star.*

diet[1] *n.* **1** °fare, °food, °nourishment, nutriment, °sustenance, °subsistence, victuals, intake, aliment: *A well-balanced diet is very important.* **2** regimen, °regime: *She is on a diet of bread and water.* —*v.* **3** °fast, abstain; °slim; °reduce: *I am dieting to lose weight.*

diet[2] *n.* °council, congress, °parliament, senate, legislature, °house, °chamber, °assembly: *In Japan, the legislature is called the* diet.

differ *v.* **1** °diverge, °deviate, be separate *or* distinct; be dissimilar *or* different, °contrast; °depart: *Even the leaves of the same tree differ from one another. These substances differ in their magnetic properties.* **2** °disagree, °conflict, °contradict, be contradictory, °vary, be at variance, °take issue, part company, °fall out, °quarrel, °argue: *Opinions differ on how to bring up children. She differed with me on many subjects.*

difference *n.* **1** °distinction, °dissimilarity, °discrepancy, unlikeness, °disagreement, inconsistency, °diversity, °variation, imbalance; °inequality, dissimilitude, incongruity, °contrast, contradistinction, contrariety: *Difference of opinion is the one crime which kings never forgive. Being colorblind, he cannot tell the difference between red and green.* **2** Often, ***differences.*** °dispute, °quarrel, °argument, °disagreement, °dissension, °conflict: *We were able to settle our differences amicably.* **3** °change, °alteration, metamorphosis, reformation, °transformation, conversion, °adjustment, modification: *Since her operation, the difference in Philippa is surprising.* **4** idiosyncrasy, °peculiarity, °characteristic, °character, °nature: *There are important differences between socialism and communism.* **5** °rest, °remainder, °leftover, °balance: *After each had taken his share, the difference was twelve, which we divided equally among the four of us.*

different *adj.* **1** °unlike, unalike, °dissimilar, conflicting; °contrary, °discrete, contrastive, contrasting, disparate, °divergent, °diverse, °distinct, °opposite, °separate, distinguishable; another *or* other: *We both enjoy boating but in different ways. When modeling, she assumes a different pose every few seconds.* **2** °unique, °unusual, °peculiar, °odd, °singular, °particular, °distinctive, °personal, °extraordinary, °special, °remarkable, °bizarre, °rare, °weird, °strange, unconventional, °original, °out of the ordinary; °new, °novel, °exceptional, °unheard-of: *And now, for something completely different, we present a juggling act.* **3** assorted, °manifold, multifarious, numerous, °abundant, °sundry, °various, °varied, °divers, °many, °several: *Different kinds of breakfast cereal are now available.*

differentiate *v.* **1** °distinguish, °discriminate, contradistinguish, °separate, °contrast, °oppose, set off *or* apart, tell apart: *They must learn how to differentiate one species from another.* **2** °modify, specialize, °change, °alter, °transform, transmute, °convert, °adapt, °adjust: *All organisms possess the power to differentiate special organs to meet special needs.*

difficult *adj.* **1** °hard, °arduous, °toilsome, °strenuous, °tough, °laborious, °burdensome, onerous, °demanding; *US* like pulling teeth, *Brit* like drawing teeth; *Colloq US* tough sledding, hard going, *Chiefly Brit* hard slogging: *He found it difficult to work the longer hours. The first birth is sometimes difficult.* **2** °puzzling, °perplexing, baffling, enigmatic(al), °profound, abstruse, °obscure, °recondite, complex; °thorny, °intricate, °sensitive, knotty, problematic(al), °ticklish, scabrous: *Some of the questions on the exam were very difficult. The analyst raised a lot of difficult issues which I had to confront.* **3** intractable, °recalcitrant, obstructive, °stubborn, unmanageable, °obstinate, °contrary, unaccommodating, refractory, unyielding, uncompromising; °naughty, ill-behaved; *Colloq Brit* bloody-minded: *Tessa has three difficult teenagers in the house these days.* **4** troubled, troubling, °tough, °burdensome, onerous, °demanding, °trying, °hard, °grim, °dark, unfavorable, straitening: *We have been through some difficult times together.* **5** °fussy, °particular, °demanding, °finicky, finical, °fastidious, °critical, °troublesome, *difficile,* °awkward, *Colloq* nitpicking: *I'll go wherever you like; I don't want to be difficult about it. Sharon can be a very difficult person to be with.*

difficulty *n.* **1** °strain, °hardship, arduousness, laboriousness, formidableness, tribulation, painfulness: *Despite much difficulty she succeeded.* **2** °hardship, °obstacle, °problem, °distress, °pitfall, °dilemma, °predicament, °problem, °snag, °hindrance; Gordian knot: *He has encountered difficulties during his career.* **3** Often, ***difficulties.*** °embarrassment, °plight, °predicament, °mess, strait(s), °trouble, °scrape, *Colloq* hot water, °jam, pickle, °fix; hot potato: *She always seems to be in financial difficulties.*

diffuse *adj.* **1** spread (out *or* about *or* around), scattered, dispersed, widespread; °sparse, °meager, °thin (on the ground): *A few diffuse clouds could be seen on the horizon.* **2** °wordy, verbose, prolix, long-winded, loquacious, °discursive, digressive, °rambling, circumlocutory, °meandering, °roundabout, circuitous, periphrastic, ambagious, diffusive, pleonastic: *The style of the book is very diffuse, being extravagantly uneconomic of expression.* —*v.* **3** °spread, °circulate, °distribute,

°dispense, °disperse; dispel, °scatter, °broadcast, °sow, disseminate; °dissipate: *The color rapidly diffused, turning the liquid crimson. She has successfully diffused her ideas of female equality throughout the community.*

dig *v.* **1** °excavate, °burrow, °gouge, °scoop, hollow out; °tunnel: *He dug a hole in which to set the post.* **2** °nudge, °thrust, °stab, °jab, °plunge, °force, °prod, °poke: *I dug my spurs into my horse and rode off. He kept digging me in the ribs with his finger.* **3** °appreciate, °enjoy, °like, °understand: *They really dig the jazz of the big-band era.* **4** °notice, °note, °look at, °regard: *Hey, man, dig those crazy threads!* **5** ***dig into.*** °probe (into), delve into, go deeply into, °explore, °look into, °research, °study: *We dug into many books of forgotten lore to find the words of the magic spell.* **6** ***dig out*** *or* ***up.*** °unearth, disinter, exhume, °bring up, °find, °obtain, °extract, ferret out, winkle out, °discover, °bring to light, °expose, dredge up, °extricate, come up with, *Australian* fossick: *I dug out an old book on witchcraft. She has dug up some interesting information about your friend Glover.* —*n.* **7** °thrust, °poke, °jab, °stab, °nudge: *She playfully gave him a dig in the ribs.* **8** °insult, insinuation, °gibe, °slur; °taunt, °jeer; *Colloq* °slap (in the face), °wisecrack, °crack, *US* low blow: *Referring to him as a graduate of Leavenworth was a nasty dig.*

digest *v.* **1** assimilate: *She has trouble digesting milk.* **2** °bear, °stand, °endure, °survive, assimilate, °accept, °tolerate, °brook, °swallow, °stomach: *The attack was too much for even him to digest.* **3** °comprehend, assimilate, °understand, °take in, °consider, °grasp, °study, °ponder, meditate (on *or* over), °reflect on, think over, °weigh: *I need a little time to digest the new regulations.* **4** °abbreviate, °cut, condense, °abridge, compress, epitomize, summarize, °reduce, °shorten: *Her assistant had digested the report into four pages by noon.* —*n.* **5** condensation, °abridgment, °abstract, °précis, °résumé, °synopsis, °summary, conspectus, °abbreviation: *I never did read the original novel, only a digest.*

dignified *adj.* °stately, °noble, °majestic, °formal, °solemn, °serious, °sober, °grave, °distinguished, °honorable, distingué, °elegant, august, °sedate, °reserved; °regal, courtly, lordly, °lofty, °exalted, °grand: *Despite the abuse, he maintained a dignified demeanor.*

dignify *v.* °distinguish, ennoble, elevate, °raise, °exalt, °glorify, upraise, °lift, uplift, °enhance, °improve, °better, upgrade: *The critic wrote that he wouldn't deign to dignify the book by calling it a novel.*

dignitary *n.* °personage, °official, °notable, °worthy, magnate, °power, higher-up; °celebrity, lion, luminary, °star, superstar, *Colloq* VIP, °bigwig, big shot, big wheel, big name, big gun, hotshot, hot stuff, big noise, big White Chief, big Chief, big Daddy, *Brit* Lord *or* Lady Muck, high-muck-a-muck, *Slang* big cheese, *Chiefly US* Mr. Big, biggie, fat cat: *Anyone with a lot of money is treated today as a dignitary.*

dignity *n.* **1** °nobility, majesty, °gravity, gravitas, °solemnity, courtliness, °distinction, stateliness, °formality, °grandeur, eminence; hauteur, loftiness: *She entered and walked with dignity to the throne.* **2** °worth, worthiness, °nobility, nobleness, °excellence, °honor, honorableness, respectability, respectableness, °standing, °importance, greatness, °glory, °station, °status, °rank, °level, °position: *The real dignity of a man lies not in what he has but in what he is.* **3** °self-respect, self-regard, *amour-propre,* °self-confidence, °self-esteem, °pride, self-importance: *It was beneath her dignity to speak directly to a footman.*

digression *n.* **1** aside, departure, deviation, °detour, obiter dictum, parenthesis, apostrophe, excursus: *His numerous digressions made it difficult to focus on the main points of the speech.* **2** digressing, deviating, divergence, going off at a tangent, °rambling, °meandering, straying, wandering, deviation: *Digression from the main theme of his speech only diluted his argument.*

dilapidated *adj.* ruined, broken-down, in ruins, gone to rack and ruin, wrecked, destroyed, falling apart, °decrepit, °derelict, battered, tumbledown, °run-down, °ramshackle, crumbling, decayed, decaying, °rickety, °shaky, °shabby, *Brit* raddled: *We shall have to fix up that dilapidated barn if we expect to use it.*

dilemma *n.* °predicament, °quandary, double bind, catch-22, °impasse, °deadlock, °stalemate; °plight, °difficulty, °trouble; °stymie, snooker; *Colloq* °bind, °box, °fix, °jam, °spot, pickle, °squeeze: *He was faced with the dilemma of killing the injured animal or allowing it to die in agony.*

dilettante *n.* dabbler, trifler, °aesthete, °amateur: *You know art like a curator; I am a mere dilettante.*

diligent *adj.* persevering, °persistent, °industrious, assiduous, sedulous, °intent, °steady, °steadfast, focused, concentrated, °earnest, °attentive, °conscientious, hard-working, indefatigable, °tireless, °constant, °painstaking, °careful, °thorough, °scrupulous, °meticulous, punctilious: *Only through diligent application was she able to get through law school.*

dilute *v.* °water (down), thin (down *or* out), °cut, °weaken, °doctor, °adulterate; °mitigate, lessen, °diminish, °decrease: *For the table, wine was often diluted with water. He dilutes his argument by citing irrelevancies.*

dim *adj.* **1** °obscure, obscured, °vague, °faint, °weak, weakened, °pale, °imperceptible, °fuzzy, °indistinct, ill-defined, indiscernible, undefined, °indistinguishable, foggy, clouded, cloudy, °nebulous, blurred, blurry, unclear, °dull, °hazy, °misty, °dark, °shadowy, °murky, tenebrous, °gloomy, °somber, °dusky, crepuscular: *Her beauty made / The bright world dim. We could barely see in the dim light of the cave.* **2** °stupid, °obtuse, doltish, °dull, dull-witted, °foolish, slow-witted, dim-witted, °dense, *Colloq* °thick, °dumb: *Anyone who couldn't understand that must be really very dim.* —*v.* **3** °obscure, °dull, becloud: *His natural feelings of compassion had been dimmed by neglect.* **4** darken, bedim, °shroud, °shade: *Twilight dims the sky above. The stagelights dimmed and the curtain fell.*

diminish *v.* **1** °decrease, °decline, abate, lessen, °reduce, °lower, °shrink, °curtail, °contract, °lop, crop, °dock, °clip, °prune, °cut, truncate, °cut down, °abbreviate, °shorten, °abridge, compress, condense, °pare (down), °scale down, boil down: *As the height increases, the pressure diminishes. The need for police patrols was diminished when we hired security guards.* **2** °belittle, °disparage, °degrade, °downgrade, °discredit, °detract (from), °vitiate, °debase, deprecate, demean, derogate, °depreciate, vilipend, devalue, cheapen, °put down, °dismiss, °humiliate, demean, °reject: *His abuse by the authorities did not diminish him in her eyes.* **3** °wane, °fade, °dwindle, °ebb, die out *or* away, °peter out, soap out, °recede, °subside; slacken, °let up, °wind down, °slow (down), °ease (off), *Colloq* run out of steam: *Soaking in the hot water, I felt the tensions of mind and body gradually diminishing. The campaign finally diminished to a negligible effort.*

diminutive *adj.* °small, °tiny, °little, °miniature, °petite, °minute, minuscule, mini, °compact, °undersized, °pocket, pocket-sized, pygmy, °elfin, Lilliputian, midget, °wee, microscopic; micro, infinitesimal; *US* vest-pocket, vest-pocket-sized, *Colloq* teeny, teeny-weeny *or* teensy-weensy, itty-bitty *or* itsy-bitsy: *When she opened her eyes, she saw a diminutive being standing next to the clock on the table.*

din *n.* **1** °noise, clamor, °uproar, shouting, screaming, yelling, babel, clangor, clatter, commotion, °racket, °row, hullabaloo, hubbub, hurly-burly, °rumpus, °blare, blaring, bray, braying, °bellow, bellowing, °roar, °blast, roaring, °pandemonium, °tumult, *Colloq* hollering: *We couldn't hear the speech above the din of the crowd.* —*v.* **2** °instill, drum, hammer: *The names and dates of the British monarchs were dinned into me in childhood.*

dine *v.* °eat, °banquet, °feast, sup, break bread, breakfast, lunch, have a bite *or* snack, nibble, *Colloq* °feed, *Slang* nosh: *We'll dine at eight o'clock, so don't be late.*

dingy *adj.* °dark, °dull, °gloomy, °dim, °lackluster, faded, discolored, °dusky, °drab, °dreary, °dismal, cheerless, depressing, °gloomy, °shadowy, tenebrous, smoky, sooty, gray-brown, smudgy, grimy, °dirty, soiled: *He was a dingy man, in dingy clothes, who lived in a dingy house.*

dip *v.* **1** °immerse, °plunge, °duck, dunk, douse, bathe, °submerge: *He dipped each dish into the soapy water.* **2** °decline, °go down, °fall, °descend, °sag, °sink, °subside, °slump: *The road dips after the next curve. The price of shares has dipped again.* **3** ***dip in*** or ***into.*** °dabble in, °play at; °skim, °scan: *I haven't had time to read it, but I dipped into it here and there.* —*n.* **4** swim, °plunge; immersion; *Brit* bathe: *We are going for a dip in the pool before dinner.* **5** lowering, °sinking, °depression, °drop, °slump, °decline: *This dip in the price of oil means nothing.*

diplomacy *n.* **1** °tact, tactfulness, adroitness, °discretion: *She was able to get rid of that rude boor with her customary diplomacy.* **2** statecraft, statesmanship, °negotiation; °intrigue, Machiavellianism, °machination, maneuvering: *Cardinal Richelieu is considered the founder of modern diplomacy.*

diplomatic *adj.* °tactful, °discreet, °prudent, °wise, °sensitive, °politic, °courteous, °polite, discerning, °perceptive, perspicacious, °thoughtful: *How diplomatic it was of you to have invited Frances's husband!*

direct *v.* **1** °manage, °handle, °run, °administer, °govern, °regulate, °control, °operate, superintend, °supervise, °command, head up, °rule; *Colloq* °call the shots: *She directs the company with an iron hand.* **2** °guide, °lead, °conduct, °pilot, °steer, show *or* point (the way), be at the helm; °advise, °counsel, °instruct, °mastermind; usher, °escort: *He has directed the company for forty years, through good times and bad. Can you direct me to the post office?* **3** °rule, °command, °order, °require, °bid, °tell, °instruct, °charge, °dictate, enjoin; °appoint, ordain: *He directed that the attack be launched at dawn.* **4** °aim, °focus, °level, °point, °train; °turn: *That bullet was directed at my heart. Direct your attention to the front of the room.* **5** °send, °address, °post, °mail: *Please direct the letter to my home.* —*adj.* **6** °straight, unswerving, shortest, undeviating, °through: *We turned off the direct road to take in the view.* **7** uninterrupted, unreflected, unrefracted, without interference, unobstructed: *She cannot remain in direct sunlight for very long.* **8** unbroken, lineal: *He claims to be a direct descendant of Oliver Cromwell's.* **9** straightforward, °unmitigated, °outright, °matter-of-fact, °categorical, °plain, °clear, unambiguous, unmistakable, °to the point, without *or* with no beating about the bush, °unqualified, unequivocal, °point-blank, °explicit, °express: *I expect a direct answer to my direct question. Have you direct evidence of his guilt? That was a gross insult and a direct lie!* **10** straightforward, °frank, °candid, °outspoken, plain-spoken, °honest, °blunt, °open, °uninhibited, unreserved, °forthright, °honest, °sincere, unequivocal; undiplomatic, °tactless: *She is very direct in commenting about people she dislikes.*

direction *n.* **1** directing, aiming, pointing, guiding, °guidance, conducting, °conduct, instructing, °instruction, managing, °management, administering, °administration, governing, °government, supervising, supervision, operating, °operation, °running, °leadership, directorship, directorate, °control, captaincy, handling, manipulation, °regulation, °rule, °charge: *The Freedom Party's direction of the country has led to many reforms.* **2** Often, ***directions.*** instruction(s), °information; °bearing, °road, °way, °route, avenue, °course: *You can't assemble the computer without the directions. Can you give me directions to the nearest gas station?*

directly *adv.* **1** °straight, in a beeline, unswervingly, undeviatingly, as the crow flies: *This road should take me directly to the beach.* **2** °immediately, °at once, °straightaway, right away, °quickly, °promptly, without delay, speedily, instantly, *Colloq US and Canadian* momentarily: *She called and I went directly.* **3** °soon, later (on), anon, °presently, in a (little) while, °shortly: *The doctor will be here directly.* **4** °exactly, °precisely, °just; °completely, °entirely: *My garage is directly opposite. The playground is directly in the center of the park.* —*conj.* **5** as soon as, when: *The police arrested him directly he tried to leave the building.*

director *n.* **1** °executive, administrator, °official, °principal; chairman, president, vice president; governor; °head, °chief, °boss, °manager, °superintendent, °supervisor, °overseer, °foreman, headman, *Colloq* kingpin, number one, numero uno, Mr. Big, the man; *Slang* top dog, top banana, *Brit* gaffer, *US* big cheese, head *or* chief honcho: *The sale of the company was announced at the meeting of the board of directors.* **2** °guide, °leader; steersman, helmsman, °pilot, °skipper, commander, commandant, captain; cicerone; maestro, concertmaster, conductor; impresario: *We were lucky to have a director who really knew what he was doing.*

dirt *n.* **1** °soil, °mud, °muck, °mire, °grime, slime, °sludge, °ooze, slop; dust, soot; excrement, ordure; °filth, °waste, °refuse, °trash, °garbage, °rubbish, offal, °junk, dross, sweepings; leavings, °scrap, orts; *Slang Brit* gunge, *US* grunge: *This vacuum cleaner is guaranteed to pick up any kind of dirt.* **2** °soil, °earth, loam, °ground, clay: *Hydroponics is the technique of farming without dirt, using only liquid nutrients.* **3** indecency, obscenity, smut, °pornography, foulness, corruption, °filth, vileness: *Customs confiscated much of the dirt before it could enter the country.* **4** °gossip, °scandal, °talk, °rumor, inside information, *Colloq* °lowdown, °(inside) dope, *Slang US* scuttlebutt: *I got the dirt from David about what really happened at the party.*

dirty *adj.* **1** °foul, unclean, befouled, soiled, begrimed, sooty, grimy, °filthy, mucky, besmeared, besmirched, polluted, squalid, sullied, stained, spotted, smudged, slovenly, °unwashed, °bedraggled, slatternly, °untidy, *Slang Brit* gungy, *US* grungy: *If you think his shirt was dirty, you should have seen his body!* **2** smutty, °indecent, °obscene, ribald, °off-color, °prurient, °risque, salacious, °lewd, °lascivious, salacious, °pornographic, °coarse, licentious, °rude, °blue, scabrous: *He learned that from those dirty books.* **3** unfair, °unscrupulous, unsporting, °dishonest, °mean, underhand(ed), unsportsmanlike, °dishonorable, °deceitful, °corrupt, treacherous, °perfidious, °villainous, °disloyal; malicious, malevolent, °rotten, °filthy: *It was a dirty trick of Sue's to tell the teacher.* **4** °bad, °foul, °nasty, °stormy, rainy, °windy, blowy, blowing, squally, °sloppy: *We're in for some dirty weather, Mr. Christian, so you'd best reduce sail.* **5** °bitter, °resentful, °angry, °furious, wrathful, smoldering: *She gave me a dirty look when I said anything about her sister.* **6** °sordid, °base, °mean, °despicable, °contemptible, ignoble, °scurvy, °low,

°lowdown, ignominious, °vile, °nasty, °infamous: *That villain has done his dirty work and now we must all suffer. He's nothing but a dirty coward!* —*v.* **7** °stain, °sully, befoul, °soil, begrime, besmirch, °pollute, °muddy, °smear, defile; °blacken, °tarnish: *She refused to so much as dirty her hands to help us. Are you afraid it will dirty your reputation to be seen with me?*

disability *n.* **1** °handicap, °impairment, °defect, °infirmity, disablement: *James is unable to play tennis owing to his disability.* **2** °inability, incapacity, unfitness, °impotence, powerlessness, helplessness: *The author labors under many disabilities when writing a book.*

disabled *adj.* incapacitated, °crippled, °lame; damaged, ruined, impaired, harmed, nonfunctioning, inoperative, *Slang Brit* scuppered: *Disabled ex-servicemen ought to receive compensation. Parts are unavailable for the disabled machines.*

disadvantage *n.* **1** deprivation, °setback, °drawback, °liability, °handicap, °defect, °flaw, °shortcoming, °weakness, weak spot, °fault: *Being colorblind has not been a disadvantage in his kind of work.* **2** °detriment, °harm, °loss, °injury, °damage; °prejudice, °disservice: *Failure to send in a tax return will be to your distinct disadvantage.*

disagree *v.* **1** °differ, dissent, °diverge: *She disagrees with most of my ideas. I said the painting was by Hockney, but he disagreed.* **2** °conflict, °dispute, °quarrel, °argue, contend, °contest, °bicker, °fight, °fall out, squabble, wrangle, °debate: *Those who agree on major principles often disagree about trifles, and vice versa.*

disagreeable *adj.* **1** unpleasant, unpleasing, °offensive, °distasteful, °repugnant, °obnoxious, °repellent, °repulsive, objectionable, °revolting, odious: *He found the heat and humidity in the tropics most disagreeable.* **2** °offensive, noxious, °unsavory, °unpalatable, nauseating, °nauseous, °nasty, sickening, °disgusting, °revolting, °repellent, °abominable, objectionable: *A disagreeable odor arose from the beggar on the doorstep.* **3** bad-tempered, ill-tempered, disobliging, uncooperative, unfriendly, uncivil, °abrupt, °blunt, °curt, °brusque, °short, uncourtly, °impolite, bad-mannered, °ill-mannered, °discourteous, °rude, °testy, grouchy, splenetic, °cross, ill-humored, °peevish, morose, sulky, °sullen: *Bottley became quite disagreeable, and I did not see him again.*

disagreement *n.* **1** °difference, °discrepancy, °discord, discordance, discordancy, °dissimilarity, disaccord, °diversity, incongruity, nonconformity, incompatibility: *Can you resolve the disagreement between the results of these experiments?* **2** dissent, °opposition, °conflict, contradiction, °difference, °disparity: *The problem arises from a basic disagreement in their principles.* **3** °quarrel, °strife, °argument, °dispute, velitation, altercation, °controversy, contention, °dissension, °debate, °clash, *Colloq US* rhubarb: *Their mother had to settle the disagreement between the brothers.*

disappear *v.* **1** vanish, °evaporate, vaporize, fade (away *or* out), evanesce, *Poetic* evanish: *After granting my wish, the leprechaun disappeared, laughing diabolically.* **2** die (out *or* off), become extinct, °cease (to exist), °perish (without a trace): *The dinosaurs, though enormously successful as a species, suddenly disappeared from the earth.*

disappoint *v.* **1** °let down, °fail, dissatisfy: *Miss Sheila disappointed her public by refusing to sing.* **2** °mislead, °deceive, disenchant, *Colloq* °stand up: *She disappointed me by saying she would be there and then not arriving.* **3** °undo, °frustrate, °foil, °thwart, balk, °defeat: *How can I answer you truthfully without disappointing your expectations?*

disappointed *adj.* **1** frustrated, unsatisfied, °dissatisfied, disillusioned, °disenchanted, discouraged, °downhearted, disheartened, downcast, saddened, °unhappy, °dejected, °discontented, °let down: *There will be a lot of disappointed children at Christmas this year.* **2** foiled, thwarted, balked, defeated, °undone, failed, °let down: *Though she campaigned energetically, Theodora was among the disappointed candidates.*

disappointing *adj.* discouraging, dissatisfying, °unsatisfactory, unsatisfying, °disconcerting; °poor, second-rate, °sorry, °inadequate, °insufficient, °inferior, °pathetic, °sad: *The former champion turned in a disappointing performance yesterday evening.*

disappointment *n.* **1** frustration, nonfulfillment, unfulfillment, unsatisfaction, °dissatisfaction, °setback, °failure, °letdown, °defeat, °blow, °fiasco, °calamity, °disaster, °fizzle, *Brit* damp squib, *Colloq* °washout: *His recent history has been of one disappointment after another.* **2** dejection, °depression, discouragement, disenchantment, °distress, °regret, mortification, chagrin: *I cannot tell you the disappointment your father and I felt when you failed to get into university.*

disapproval *n.* disapprobation, condemnation, censure, °criticism, °reproof, reproach, °objection, °exception, °disfavor, °displeasure, °dissatisfaction: *The council voiced their disapproval of holding a carnival in the village square.*

disapprove *v.* °condemn, °criticize, censure, object to, decry, °denounce, animadvert on *or* upon, put *or* run down, deplore, deprecate, °belittle, °look down on, frown on *or* upon, *Colloq* °knock, look down one's nose at, tut-tut: *I don't care if you disapprove of my marrying Eustace. The monopolies commission has disapproved the merger.*

disarm *v.* **1** unarm; demilitarize, demobilize, °disband, deactivate: *After the war, most—but not all—European countries disarmed.* **2** °win over, put *or* set at ease, mollify, appease, placate, pacify, °reconcile, conciliate, propitiate, °charm: *I was completely disarmed by her friendly disposition. Many people found his naivety disarming.*

disaster *n.* °catastrophe, °calamity, cataclysm, °tragedy, °misfortune, debacle, °accident, °mishap, °blow, act of God, adversity, °trouble, °reverse: *The flooding of the river was as much of a disaster as the earlier drought.*

disastrous *adj.* **1** °calamitous, catastrophic, cataclysmic, °tragic, °destructive, °ruinous, °devastating, appalling, °harrowing, °awful, °terrible, dire, horrendous, °horrible, horrifying, °dreadful, °fatal: *There has been a disastrous earthquake which killed thousands.* **2** °awful, °terrible, unlucky, °unfortunate, °detrimental, °grievous, °harmful: *The postal strike has had disastrous effects on the mail-order business.*

disband *v.* °disperse, disorganize, °scatter, °break up, °dissolve, demobilize, deactivate, °retire: *After the war, the special spy force was disbanded.*

discard *v.* **1** °get rid of, °dispense with, °dispose of, throw away *or* out, toss out *or* away, °abandon, jettison, °scrap, *Colloq* °trash, °dump, *Slang* ditch: *We discarded boxes of old photographs when we moved house.* —*n.* **2** °reject, castoff: *I felt like a discard from the lonely hearts club.*

discernible *adj.* **1** °perceptible, °visible, seeable, perceivable, °apparent, °clear, °observable, °plain, detectable; °conspicuous, °noticeable: *A small sailing boat was discernible on the horizon.* **2** distinguishable, recognizable, identifiable, °distinct: *To me there is a discernible difference between puce and burgundy.*

discharge *v.* **1** °release, °let out, °dismiss, let go, send away; °pardon, exonerate, °liberate, °(set) free, acquit,

°let off, absolve: *She was discharged from hospital yesterday. He was discharged from police custody last week.* **2** °expel, oust, °dismiss, cashier, °eject, °give notice, *Colloq* °sack, give (someone) the sack, °fire, kick out: *He was discharged from his job yesterday.* **3** °shoot, °fire (off); set *or* let off, detonate, °explode: *It is illegal to discharge a firearm or other explosive device in this area.* **4** °emit, send out *or* forth, pour out *or* forth, °gush; disembogue; °ooze, °leak, exude; excrete, °void: *The sore in his leg continued to discharge pus. We can ill afford to discharge those effluents into the sea.* **5** °carry out, °perform, °fulfill, °accomplish, do, °execute: *He faithfully discharges the duties of his office.* **6** °pay, °settle, liquidate, °clear, °honor, °meet, °square (up): *Before going off on holiday, we discharged all our financial obligations.* **7** °unload, off-load, disburden, °empty: *After discharging its cargo, the vessel rode high in the water.* —*n.* **8** °release, °dismissal: *What is the date of his discharge from the clinic?* **9** °expulsion, ouster, °dismissal, °ejection, °notice, *Colloq* the ax, °the sack, the boot, *Slang US and Canadian* one's walking papers, the bounce, the gate: *Her discharge from the firm was rather ignominious.* **10** shooting, firing (off), °report, °shot; salvo, fusillade, °volley; detonation, °explosion, °burst: *The discharge of a pistol could not be heard at that distance. I heard the discharge from the guns of the firing squad in the courtyard below. The discharge of the bomb maimed three children.* **11** emission, °release, °void, voiding, excretion, excreting, emptying, °flow; °ooze, oozing, pus, suppuration, °secretion, seepage: *The discharge of blood from the wound continued.* **12** °performance, °fulfillment, °accomplishment, °execution, °observance, °achievement: *The discharge of my family responsibilities will have to await my return from the front.* **13** °payment, °settlement, liquidation, squaring (up), °clearance: *The bank expects full discharge of all debts before they lend any money.* **14** unloading, disburdening, offloading, emptying: *The customer will pay in full after the discharge of his cargo.*

disciple *n.* **1** °apprentice, °pupil, °student, proselyte, °learner, °scholar: *Pietro Zampollini was a disciple of the great artist Ravelli.* **2** °follower, adherent, °devotee, °admirer, votary; °partisan, °fan, aficionado: *She is a disciple of Louis Armstrong's.*

disciplinarian *n.* taskmaster, martinet, drill sergeant; °tyrant, °despot, °dictator: *The headmaster at Briarcliffe was a stern disciplinarian who beat the students regularly.*

discipline *n.* **1** training, drilling, regimen, °exercise, °practice, °drill, inculcation, indoctrination, °instruction, °schooling: *Strict discipline is good for young people, according to my father.* **2** °punishment, °penalty, chastisement, castigation, °correction: *The discipline meted out to senior students was very harsh.* **3** °order, °routine, °(proper) behavior, °decorum: *The sergeant is there to maintain discipline among the recruits.* **4** °direction, °rule, °regulation, °government, °control, °subjection, °restriction, °check, °curb, °restraint: *There was far too much discipline during my childhood, both at school and at home.* **5** °subject, °course, branch of knowledge, °area, °field, °specialty: *In which discipline is Eulalia's doctorate?* —*v.* **6** °train, °break in, °condition, °drill, °exercise, °instruct, °coach, °teach, °school, °indoctrinate, inculcate; edify, °enlighten, °inform: *The aim of his education is to discipline him to respond to orders.* **7** °check, °curb, °restrain, °bridle, °control, °govern, °direct, °run, °supervise, °manage, °regulate, hold *or* keep in check, *US* ride herd on: *You have to discipline those children or they will always misbehave.* **8** °punish, °chastise, °castigate, °correct, °penalize, °reprove, °criticize, °reprimand, °rebuke: *Discipline that boy or he will just do it again.*

disclose *v.* **1** °reveal, °impart, divulge, °betray, °release, °tell, °blurt out, °blab, °leak, °let slip, °report, °inform, *Colloq* °spill the beans, blow the gaff, *Slang* squeal, snitch, squeak, rat, peach, *US* fink: *To get a shorter sentence, he disclosed all to the police.* **2** °bare, °reveal, °expose, uncover, °show, °unveil: *When the pie was opened, twenty-four blackbirds were disclosed.*

discomfit *v.* **1** °embarrass, abash, disconcert, °disturb, °confuse, make uneasy *or* uncomfortable, discompose, °fluster, °ruffle, confound, °perturb, °upset, °worry, unsettle, °unnerve, *Colloq* °rattle, *US* faze, discombobulate: *Being short, she was discomfited by references to her height.* **2** °frustrate, °foil, °thwart, baffle, °check, °defeat, trump, °outdo, °outwit, °overcome: *Discomfited by her violent reaction, her attacker fled.*

discomfort *n.* **1** uneasiness, °hardship, °difficulty, °trouble, °care, °worry, °distress, vexation: *She hasn't known the discomfort of being the wife of a miner.* **2** °ache, °pain, °twinge, soreness, irritation; °bother, °inconvenience, °nuisance: *Some discomfort persisted in my leg long after the accident.*

disconcerted *adj.* discomposed, discomfited, ruffled, uneasy, put out *or* off, uncomfortable, °queasy, flustered, °agitated, °upset, shaken, °unsettled, perturbed, °confused, bewildered, perplexed, baffled, puzzled, *US* thrown off, *Colloq* rattled, *US* fazed, discombobulated; *Slang* (all) shook (up): *They were really disconcerted by the arrival of the police.*

disconcerting *adj.* °awkward, discomfiting, off-putting, upsetting, unnerving, °unsettling, °disturbing, confusing, confounding, bewildering, °perplexing, baffling, °puzzling: *I found his persistence quite disconcerting.*

disconnect *v.* °separate, disjoin, disunite, uncouple, °detach, unhook, °undo, °disengage, unhitch; cut *or* break off; cut *or* pull apart, °part, °divide, °sever: *They disconnected the engine after pushing the passenger cars onto a siding. Disconnect the power before changing the light bulb.*

disconnected *adj.* **1** unconnected, °separate, °apart, °unattached; °split, separated: *A totally disconnected thought suddenly occurred to me.* **2** °incoherent, irrational, °confused, illogical, garbled, °disjointed, °rambling, mixed-up, unintelligible, uncoordinated, °random: *He lost the debate because his argument was disconnected and lacked cogency.*

discontent *n.* °displeasure, unhappiness, °dissatisfaction, discontentment, °distaste, uneasiness; malaise: *He felt discontent at being barred from the club.*

discontented *adj.* displeased, °dissatisfied, °discontent, annoyed, vexed, °fretful, irritated, °testy, piqued, °petulant, °disgruntled, exasperated, *Colloq* fed up, *Slang* browned off, pissed off, *Brit* cheesed *or Brit* brassed off: *The umpire's decision made many fans quite discontented.*

discontinue *v.* °cease, °break off, °give up, °stop, °terminate, put an end to, °quit, °leave off, °drop; °interrupt, °suspend: *Please discontinue newspaper delivery till further notice.*

discord *n.* °strife, °dissension, °disagreement, °conflict, disharmony, contention, disunity, discordance, °division, incompatibility: *The seeds of discord between the families were sown generations before.*

discordant *adj.* **1** °contrary, disagreeing, °divergent, °opposite, °opposed, adverse, °contradictory, °incompatible, differing, °different, conflicting, °at odds, °incongruous, in conflict, in disagreement, °at variance, °dissimilar: *The testimony of the fossils is discordant* *with the evidence in the legend.* **2** inharmonious, dissonant, jarring, cacophonous, unmelodious, unmusical, °harsh, °strident, jangling, °grating: *He struck some discordant notes on his zither.*

discount *v.* **1** °reduce, °mark down, °deduct, °lower, take *or* knock off: *As I was buying a dozen, he discounted the price by 10 percent.* **2** °diminish, lessen, °minimize, °detract from: *One must discount what she says when she's angry.* **3** °disregard, °omit, °ignore, pass *or* gloss over, °overlook, °brush off, °dismiss: *Those statistics are old and can be discounted.* —*n.* **4** reduction, markdown, °deduction, °rebate, °allowance: *The shop overstocked the item and is offering it at a big discount.*

discourage *v.* **1** dispirit, dishearten, °daunt, unman, °dismay, cow, °intimidate, awe, °overawe, °unnerve: *We were discouraged by the arrival of more enemy troops.* **2** °deter, °put off, dissuade, advise *or* hint against, talk out of, divert from; °oppose, °disapprove (of), °dampen, *Colloq* throw cold water on: *They discouraged me from applying again.* **3** °prevent, °inhibit, °hinder, °stop, °slow, °suppress, obviate: *This paint is supposed to discourage corrosion.*

discourteous *adj.* uncivil, °impolite, °rude, unmannerly, °ill-mannered, bad-mannered, °disrespectful, misbehaved, °boorish, °abrupt, °curt, °brusque, °short, ungentlemanly, unladylike, °insolent, °impertinent, °ungracious: *He had been discourteous and would not be invited again.*

discover *v.* **1** °find (out), °learn, °perceive, °unearth, uncover, °bring to light, turn *or* dig up, smoke *or* search out, root *or* ferret out; °determine, ascertain, °track down, °identify; °locate: *He discovered the ninth moon of Saturn. We discovered why the tire had gone flat.* **2** °see, °spot, catch sight *or* a glimpse of, lay eyes on, °behold, °view, °encounter, °meet (with); °notice, espy, descry, °detect, discern: *He discovered Madagascar lying right on our course.* **3** °originate, °conceive (of), °devise, contrive, °invent, °make up, °design, °pioneer; come *or* chance *or* stumble upon: *Wasn't it Jenner who discovered a vaccine against smallpox?*

discovery *n.* **1** °finding, °recognition, uncovering, determining, ascertaining, unearthing; origination, °invention, °conception, °idea; °development: *Who is credited with the discovery of Christmas Island? That year marks the discovery of a cure for tuberculosis.* **2** °exploration, disclosure, detection, °revelation: *He's off on a voyage of discovery.*

discredit *v.* **1** °detract, °disparage, defame, °dishonor, °disgrace, °degrade, bring into disfavor *or* disrepute, deprecate, demean, °lower, devalue, °depreciate, devaluate, °belittle, °diminish, °reduce; °slur, °slander, °vilify, calumniate, °sully, °smear, °blacken, °taint, °tarnish, besmirch, smirch, °stigmatize, asperse, malign, °libel: *Both of them were thoroughly discredited by the scandal.* **2** disbelieve, °deny, °dispute, °doubt, °question, raise doubts about, °distrust, °mistrust, give no credit *or* credence to: *As he's a known liar, you can discredit whatever he tells you.* **3** °disprove, °reject, refute, invalidate; °mock, °ridicule: *The phlogiston theory is generally discredited by most modern chemists.* —*n.* **4** °dishonor, °degradation, °disfavor, disrepute, ill repute, °disgrace, ignominy, °infamy, odium, °stigma, °shame, °smear, °slur, °scandal, obloquy, opprobrium, °humiliation: *Her performance has brought discredit to all female saxophonists.* **5** °damage, °harm, °reflection, °slur, °aspersion, °slander, defamation, °blot, °brand, °tarnish, °blemish, °taint: *The discredit to her reputation is irreparable.* **6** °doubt, °skepticism, dubiousness, doubtfulness, °qualm, °scruple, °question, incredulity, °suspicion, °distrust, °mistrust: *These answers throw discredit on the validity of the previous testimony.*

discreet *adj.* °careful, °cautious, °prudent, °judicious, °considerate, °guarded, °tactful, °diplomatic, circumspect, °wary, chary, heedful, watchful, circumspect: *She has always been very discreet in her business dealings with me.*

discrepancy *n.* °gap, °disparity, lacuna, °difference, °dissimilarity, deviation, divergence, °disagreement, incongruity, incompatibility, inconsistency, °variance; °conflict, discordance, contrariety: *There is a great discrepancy between what he says and what he means.*

discrete *adj.* °separate, °distinct, °individual, °disconnected, °unattached, discontinuous: *These items must be treated as discrete entities and not taken together.*

discretion *n.* **1** °tact, °diplomacy, °prudence, °care, discernment, sound judgment, circumspection, sagacity, common sense, good sense, °wisdom, °discrimination: *You can rely on my discretion not to reveal the club's secrets.* **2** °choice, °option, °judgment, °preference, °pleasure, °disposition, °volition; °wish, °will, °liking, °inclination: *Buyers may subscribe to insurance coverage at their own discretion.*

discriminate *v.* **1** °distinguish, °separate, °differentiate, discern, draw a distinction, tell the difference: *He cannot discriminate between good art and bad.* **2** °favor, °disfavor, °segregate, show favor *or* prejudice *or* bias for *or* against, be intolerant: *It is illegal here to discriminate against people on the basis of race, creed, or color.*

discriminating *adj.* discerning, °perceptive, °critical, °keen, °fastidious, °selective, °particular, °selective, °fussy, °refined, °cultivated: *From the wine you chose, I see you are a lady of discriminating tastes.*

discrimination *n.* **1** °bigotry, °prejudice, °bias, °intolerance, °favoritism, one-sidedness, unfairness, inequity: *In Nazi Germany discrimination was practiced against everyone except the Nazis.* **2** °taste, °perception, perceptiveness, discernment, °refinement, acumen, °insight, °penetration, keenness, °judgment, °sensitivity; connoisseurship, aestheticism: *He exercises excellent discrimination in his choice of paintings.*

discursive *adj.* wandering, °meandering, digressing, digressive, °rambling, circuitous, °roundabout, °diffuse, long-winded, verbose, °wordy, prolix, °windy: *Frobisher was again boring everyone with his discursive description of life in an igloo.*

discuss *v.* converse about, talk over *or* about, chat about, °deliberate (over), °review, °examine, consult on; °debate, °argue, thrash out: *We discussed the problem but came to no conclusion.*

discussion *n.* °conversation, °talk, °chat, °dialogue, colloquy, °exchange, deliberation, °examination, °scrutiny, °analysis, °review; confabulation, °conference, powwow; °debate, °argument; *Colloq chiefly Brit* chinwag, *US and Canadian* bull session: *The subject of your dismissal came up for discussion yesterday.*

disdainful *adj.* °contemptuous, °scornful, contumelious, derisive, sneering, °superior, °supercilious, °pompous, °proud, prideful, °arrogant, °haughty, °snobbish, lordly, °regal; jeering, mocking, °insolent, insulting, *Colloq* °hoity-toity, high and mighty, stuck-up, highfalutin *or* hifalutin; *Slang* snotty: *She was most disdainful of our efforts to enter the cosmetics market.*

disease *n.* **1** sickness, °affliction, °ailment, malady, °illness, infection, °complaint, °disorder, °condition, °infirmity, °disability, *Archaic* murrain, *Colloq* °bug: *The colonel contracted the disease while in Malaysia.* **2** °blight, cancer, virus, °plague; contagion: *Panic spread through the stock market like an infectious disease.*

diseased *adj.* °unhealthy, unwell, °ill, °sick, ailing, °unsound, °infirm, °out of sorts, abed, infected, contaminated; afflicted, °abnormal: *We must care for the diseased patients before those with broken bones.*

disembark *v.* °land, alight, go *or* put ashore, get *or* step off *or* out, °leave; debark, detrain, deplane: *Tomorrow we disembark at Ouagadougou.*

disembodied *adj.* **1** incorporeal, bodiless; °intangible, °immaterial, insubstantial *or* unsubstantial, impalpable, °unreal; °spiritual, °ghostly, °spectral, °phantom, wraithlike: *She wafted before his eyes, a disembodied spirit.*

disenchanted *adj.* disillusioned, disabused, undeceived, °disappointed; °blasé, °indifferent, °jaundiced, sour(ed), cynical: *I'm afraid she's now thoroughly disenchanted with rock singers.*

disengage *v.* °loose, °loosen, unloose, °detach, unfasten, °release, °disconnect, disjoin, °undo, disunite, °divide, °cleave (from), °separate, uncouple, °part, disinvolve, °extricate, get out (of), °get away (from), cut loose, °throw off, °shake (off), °get rid of, break (with *or* from), break (up) (with); unbuckle, unhitch, unclasp, unlatch, unbolt, unlock, unleash, unfetter, unchain, unlace, unhook, unbind, untie; °(set) free, °liberate, disentangle: *She was holding on to me so tenaciously that I could hardly disengage myself.*

disfavor *n.* **1** °disapproval, °dislike, °displeasure, disapprobation, unhappiness: *Katerina regards your decision with disfavor.* **2** disesteem, °discredit, °dishonor, °disgrace, disrepute: *After last night's events, we are really in disfavor with the management.* —*v.* **3** °disapprove (of), °dislike, discountenance, frown on *or* upon: *We strongly disfavor the merger.*

disfigured *adj.* marred, damaged, scarred, defaced, mutilated, injured, impaired, blemished, disfeatured, °deformed, distorted, spoilt *or* spoiled, ruined: *Plastic surgery has repaired her disfigured face.*

disgrace *n.* **1** ignominy, °shame, °humiliation, °embarrassment, °degradation, debasement, °dishonor, °discredit, °disfavor, disrepute, vitiation, °infamy; disesteem, °contempt, odium, obloquy, opprobrium: *His conduct has brought disgrace on his family.* **2** °blemish, °harm, °aspersion, °blot, °scandal, °slur, °stigma, vilification, smirch, °smear, °stain, °taint, black mark: *The way she has been treated by the company is a disgrace.* —*v.* **3** °shame, °humiliate, °embarrass, °mortify: *He has been disgraced by his son's cowardice.* **4** °degrade, °debase, °dishonor, °discredit, °disfavor, °vitiate, defame, °disparage, °scandalize, °slur, °stain, °taint, °stigmatize, °sully, besmirch, smirch, °tarnish, °smear, asperse, °vilify, °blacken, drag through the mud, reflect (adversely) on: *Once again his actions have disgraced the family name.*

disgraceful *adj.* **1** °shameful, humiliating, °embarrassing, °dishonorable, °disreputable, °infamous, ignominious, °degrading, debasing, degraded, debased, °base, °low, °vile, °corrupt, °bad, °wrong, °sinful, °evil, °low, °mean, °despicable, °contemptible, opprobrious: *He was forced to submit to the most disgraceful punishment.* **2** °shameless, °outrageous, °notorious, °shocking, °scandalous, °improper, °unseemly, °unworthy; °indecent, °rude, °flagrant, °lewd, °lascivious, °delinquent, objectionable: *That drunken carousal last night was a disgraceful performance.*

disgruntled *adj.* displeased, °dissatisfied, irritated, peeved, vexed, °cross, exasperated, annoyed, °unhappy, °disappointed, °discontented, °put out; malcontent, °discontent, °testy, °cranky, °peevish, grouchy, grumpy, °moody, °sullen, sulky, ill-humored, bad-tempered, ill-tempered, *Colloq* fed up, *Slang* browned off, *Brit* cheesed off: *He was disgruntled at the thought of having to go shopping in the pouring rain.*

disguise *v.* **1** °camouflage, °cover up, °conceal, °hide, °mask: *The van was disguised as a hay wagon.* **2** °misrepresent, °falsify, °counterfeit, °fake, °deceive: *They have disguised the true profits to avoid paying taxes.* —*n.* **3** °guise, °identity, coverup, °camouflage, °appearance, °semblance, °form, °outfit, °costume: *She appeared in the disguise of a nightclub dancer.* **4** °pretense, °deception, °dissimulation, façade, °semblance, *Colloq* °front: *Disguise is seldom resorted to by spies these days.*

disgust *v.* **1** °sicken, °offend, °nauseate, °repel, °revolt, °put off, °outrage, °appall, *Slang* gross out: *His patronizing attitude disgusts those who work for him.* —*n.* **2** °revulsion, nausea, sickness, repugnance, fulsomeness, °outrage, °distaste, °aversion: *One look at the food filled me with disgust.* **3** °loathing, °contempt, hatred, abhorrence, odium, animus, °animosity, enmity, °antagonism, antipathy, °dislike: *Some feel disgust at the thought of eating insects.*

disgusted *adj.* °nauseated, sickened, °nauseous, °queasy; offended, outraged, *Colloq* fed up (with), °sick (of), sick and tired (of); *Slang US* grossed out: *Disgusted customers complain about delays in service.*

disgusting *adj.* nauseating, sickening, °offensive, °outrageous, sick-making, fulsome, °repulsive, °revolting, °repugnant, off-putting, °repellent, °obnoxious, °loathsome, °gross, °vile, °foul, °nasty; unappetizing, °unsavory, objectionable, °distasteful: *Mrs. Ekelhaft has the most disgusting table manners.*

dishonest *adj.* untrustworthy, underhand(ed), °dishonorable, °fraudulent, °fake, °counterfeit, deceiving, °deceptive, unfair, double-dealing, thieving, thievish, knavish, cheating, °deceitful, °lying, untruthful, mendacious, treacherous, °perfidious, °corrupt, °unscrupulous, unprincipled; °two-faced, °hypocritical; *Colloq* °crooked, °shady; *Chiefly Brit slang* °bent: *He was so dishonest he stole from his mother's purse.*

dishonor *v.* **1** °insult, °abuse, affront, °outrage, °slight, °offend, °injure: *His slaughter of the prisoners has dishonored our flag.* **2** °disgrace, °degrade, °shame, °debase, °humiliate, °mortify, abase, °vitiate, °humble: *We were all dishonored by our colleague's defection.* **3** defile, °violate, °ravish, °rape, °seduce, deflower, debauch: *The general learned that his wife had been dishonored by one of his adjutants.* —*n.* **4** disesteem, °disrespect, irreverence, °slight, °indignity, ignominy, °disgrace, °shame, disrepute, °discredit, °insult, °offense, affront, loss of face, depreciation, belittlement, disparagement, detraction, derogation, obloquy: *You cannot retreat without dishonor.* **5** °aspersion, defamation, °libel, °slander, °blot, °slur, °smear, smirch, black mark, °blemish, denigration: *His actions have brought us dishonor.*

dishonorable *adj.* **1** °disgraceful, °degrading, inglorious, ignominious, °shameful, shaming, °base, debased: *After the court-martial, he received a dishonorable discharge.* **2** unprincipled, °shameless, °corrupt, °unscrupulous, untrustworthy, treacherous, °traitorous, °perfidious, °dishonest, °hypocritical, °two-faced, duplicitous, °disreputable, discreditable, °base, °despicable; °disloyal, unfaithful, °faithless: *A double agent is considered dishonorable by both governments.* **3** °improper, °unseemly, °unbecoming, °unworthy, °outrageous, objectionable, reprehensible, °flagrant, °bad, °evil, °vile, °low, °mean, °contemptible, below *or* beneath criticism, °foul, heinous, °dirty, °filthy: *Informing on your classmates is the most dishonorable thing you can do.*

disillusion *v.* disabuse, °disappoint, disenchant, break the spell, °enlighten, set straight, disentrance, disenthrall, undeceive: *When I saw her without make-up, I was thoroughly disillusioned.*

disinclined *adj.* °averse, °indisposed, °reluctant, unwilling, °loath, °opposed, unwilling; °hesitant: *I was disinclined to try skydiving.*

disinfect *v.* °clean, °cleanse, °purify, °purge, sanitize, °fumigate, decontaminate, °sterilize: *The bedding will have to be disinfected before it can be used.*

disinfectant *n.* germicide, antiseptic, sterilizer, bactericide, sanitizer, fumigant, decontaminant, decontaminator, purifier, cleaner, cleanser: *Most disinfectants are poisonous.*

disingenuous *adj.* °clever, °artful, °crafty, °sly, on the qui vive, cunning, insidious, °foxy, °wily, °slick, °smooth; °insincere, °false, °dishonest, °tricky, °devious, °deceitful, underhand(ed), guileful, °shifty; double-dealing, °two-faced, duplicitous, °hypocritical, °scheming, plotting, °calculating, °designing, contriving: *It is disingenuous to ask for advice when what you want is assistance.*

disintegrate *v.* break up *or* apart, °shatter, come *or* fall apart, come *or* go *or* fall to pieces, °crumble; °decompose, °rot, °decay, molder: *The fossil disintegrated in my hands.*

disinterested *n.* unbiased, °impartial, °unprejudiced, altruistic, °objective, °fair, °neutral, open-minded, °equitable, °just, °dispassionate, °detached, even-handed, °impersonal, uninvolved: *The judge is supposed to be a disinterested party.*

disjointed *adj.* **1** disjoined, separate(d), °disconnected, unconnected, dismembered, disunited, divided, °split (up): *The disjointed parts of the building were kept in a warehouse.* **2** ununified, °loose, °incoherent, °confused, °aimless, directionless, °rambling, muddled, jumbled, mixed up, °fitful, discontinuous, disorganized, unorganized, °disorderly: *His speech was disjointed—total gibberish.*

dislike *v.* **1** be averse to, °mind, turn from, °disfavor, disesteem, be put *or* turned off by; °hate, °loathe, °scorn, °despise, contemn, °detest, abominate, execrate, *Brit* take a scunner (to): *I no longer dislike spinach.* —*n.* **2** °aversion, °displeasure, °distaste, °disfavor, disesteem, disrelish, disaffection, disinclination; °loathing, hatred, animus, °animosity, antipathy, detestation, °contempt, execration, °ill will; °disgust, repugnance; °hostility, °antagonism, *Brit* scunner: *I took an instant dislike to the fellow. She feels an intense dislike for her father.*

disloyal *adj.* unfaithful, °faithless, °untrue, °false, untrustworthy, recreant; treasonable *or* treasonous, treacherous, °traitorous, unpatriotic, °subversive, °perfidious, °deceitful; °renegade, apostate, °heretical: *It would be disloyal of you not to vote along party lines.*

dismal *adj.* depressing, °gloomy, cheerless, °melancholy, °somber, °dreary, °sad, °bleak, °funereal, lugubrious, °forlorn, morose, °solemn, °dark, °grim, °wretched, °woebegone, woeful, °black, °blue, °joyless, °doleful, dolorous, °unhappy, °miserable, lowering; °pessimistic: *She was alone, alone on the dismal moor. The prospects for the company looked very dismal.*

dismay *v.* **1** °alarm, °frighten, °scare, °terrify, appall *or* appal, °panic, °horrify, °petrify, °intimidate, cow, disconcert, °unnerve: *We were dismayed when the motorcycle gang came to the house.* **2** unsettle, discompose, °upset, °discourage, °take aback, °startle, °shock, °put off, dishearten: *I was dismayed to hear she was still married to Grimsby.* —*n.* **3** consternation, °alarm, °anxiety, °agitation, °terror, °panic, °horror, °shock, °fright, °fear, trepidation, apprehension, °dread, awe: *The thought of the children alone in the boat filled me with dismay.*

dismiss *v.* **1** °discharge, oust, °release, °give notice (to), let go, °lay off, °throw out, toss out, °remove, *Chiefly military* cashier, *Old-fashioned military* drum out, *Brit politics* deselect, *Colloq* °fire, send packing, kick out, send to the showers, *Brit* °sack, give (someone) the sack, °boot (out), °turn off, *US* give (someone) (his/her) walking papers, give (someone) a pink slip, can; *Slang* give (someone) the (old) heave-ho: *Gabney has been dismissed without notice.* **2** °reject, °set aside, °repudiate, °spurn, °discount, °disregard, lay aside, put out of one's mind, think no more of, °write off, °banish, have *or* be done with, °scorn, °discard, °ignore, shrug off; °belittle, °diminish, pooh-pooh: *She dismissed the story as just so much gossip.* **3** °disperse, °release, °disband, send away: *After returning from the mission, the commando unit was dismissed.*

dismissal *n.* **1** °discharge, °expulsion, °notice, *Colloq* firing, °bounce, marching orders, walking papers, *Brit* °sack, sacking, one's cards, *US* pink slip; *Slang* the (old) heave-ho, *Brit* the boot: *Cholmondley got his dismissal yesterday.* **2** °cancellation, adjournment, °discharge, °end, °release; congé: *The judge ordered dismissal of the charge of murder.*

disobedient *adj.* **1** °insubordinate, °unruly, °naughty, °mischievous, °bad, ill-behaved, badly behaved, °obstreperous, unmanageable, refractory, fractious, °ungovernable, uncomplying, unsubmissive, wayward, noncompliant, incompliant, intractable, °defiant; °delinquent, °derelict, disregardful, °remiss, undutiful: *Disobedient children will be kept after school.* **2** °contrary, °perverse, °willful, headstrong, °stubborn, °recalcitrant, obdurate, °obstinate, contumacious, wayward, cross-grained, °opposed, °mutinous, °rebellious, °revolting, anarchic, *Colloq* pigheaded: *We cannot tolerate disobedient recruits.*

disobey *v.* °defy, °break, contravene, °flout, °disregard, °ignore, °resist, °oppose, °violate, °transgress, °overstep, go counter to, °fly in the face of, °infringe, °thumb one's nose at, °snap one's fingers at, *Brit* cock a snook at; °mutiny, °rebel, °revolt, °strike: *You cannot play because you disobeyed the rules. If anyone disobeys, throw him in irons.*

disorder *n.* **1** disarray, °confusion, °chaos, disorderliness, disorganization, untidiness, °mess, °muddle, °jumble, °hash, °mishmash, °tangle, hotchpotch *or US and Canadian also* hodgepodge, derangement, °shambles, °clutter: *After the party, the place was in terrible disorder.* **2** °tumult, °riot, °disturbance, °pandemonium, °upheaval, °ferment, °fuss, °unrest, °uproar, hubbub, hullabaloo, commotion, clamor, turbulence, turmoil, turbulence, °violence, °bedlam, free-for-all, °rumpus, brouhaha, °fracas, affray, °fray, °brawl, donnybrook, scuffle, °fight, melee *or* mêlée, battle royal, °battle, civil disorder, breach of the peace, *Colloq Brit* kerfuffle *or* carfuffle *or* kurfuffle, *Slang Brit* bovver: *The army had to be called out to quell the disorder.* **3** °ailment, °illness, sickness, °affliction, malady, °affection, °complaint, °disease: *The doctors diagnosed it as a liver disorder.* —*v.* **4** °upset, disarrange, °muddle, °confuse, confound, unsettle, disorganize, discompose, °shake up, °disturb, °mix (up), befuddle, °jumble, °scramble, °tangle, °snarl: *You obscure the sense when you disorder the words.*

disorderly *adj.* **1** °confused, °chaotic, scrambled, muddled, disordered, °irregular, °untidy, messy, messed up, disarranged, disorganized, unorganized, jumbled, cluttered, °haphazard, in disarray, °pell-mell, °helter-skelter, *Colloq* °topsy-turvy, higgledy-piggledy:

The books lay about in disorderly array. **2** °unruly, °uncontrolled, °undisciplined, ungoverned, °disobedient, °mutinous, °rebellious, °lawless, °obstreperous, refractory, turbulent, °violent, °tumultuous, unrestrained, °boisterous, °noisy, °rowdy, °wild; unmanageable, °ungovernable, uncontrollable, intractable: *He was charged with being drunk and disorderly.*

disorientated *n.* °confused, bewildered, °lost, adrift, (all) at sea, mixed up, °uncertain, unsure, °insecure, disoriented, *Colloq* out of it, in a fog, *Brit* off (the) beam, *US* off the beam: *I left by another door and was completely disorientated for a moment.*

disparage *v.* **1** °belittle, °diminish, °depreciate, devalue *or* devaluate, cheapen, °talk down, °discredit, °dishonor, decry, demean, °criticize, denigrate, deprecate, derogate, underrate, undervalue, °downgrade, °reduce, °minimize: *She keeps making remarks that disparage her husband.* **2** °run down, °slander, °libel, defame, traduce, malign, backbite, °vilify, °insult, °stab in the back, *US* backstab; *Colloq* poor-mouth; *Slang US and Canadian* bad-mouth: *A loving person never disparages others.*

disparity *n.* °difference, °discrepancy, °gap, °inequality, unevenness, imbalance, °dissimilarity, °contrast, imparity, inconsistency, incongruity: *Our interests differ owing to the disparity in our ages.*

dispassionate *adj.* **1** °cool, °calm, composed, °self-possessed, unemotional, unexcited, unexcitable, unflappable, °levelheaded, °sober, self-controlled, even-tempered, unruffled, °unmoved, °tranquil, °equable, placid, °peaceful, °serene: *You can count on Hedwig for dispassionate treatment of the subject.* **2** °fair, °impartial, °neutral, °disinterested, °detached, °equitable, evenhanded, unbiased, °just, °objective, °unprejudiced, open-minded, °candid, °frank, °open: *Judge Bean is known to be completely dispassionate in his decisions.*

dispatch *v.* **1** send off *or* away *or* out, send on one's way: *We dispatched a messenger with the parcel.* **2** °send, °mail, °post, °transmit, °forward, °ship, °express, °remit, convey, *Chiefly US and Canadian* °freight: *Please dispatch this letter as quickly as possible.* **3** °kill, °murder, slay, °dispose of, put to death, °execute, do away with, do in, assassinate, liquidate, °finish (off), put an end to, °put away (for good), *Slang* °polish off, °bump off, °eliminate, gun down, °silence, °get, °erase, °rub out, °knock off, °bury, *US* ice, °hit, °take for a ride, °waste, °zap: *The gang soon dispatched all their rivals.* **4** °hasten, °hurry, °speed up, accelerate, get done, °accomplish, °get through, conclude, °finish off, °complete, °execute, do: *The task was dispatched in just two days.* —*n.* **5** °haste, °speed, promptness, quickness, °expedition, expeditiousness, celerity, alacrity, swiftness, °hurry, °rapidity: *She concluded the interview with dispatch and sent me away.* **6** communiqué, °report, °bulletin, °story, °news (item), communication, °message, °piece, °document, °instruction, °missive: *Here is a dispatch from our correspondent on Pitcairn Island.* **7** °execution, °killing, °murder, disposal, assassination, dispatching, slaying: *The dispatch of the consul left us without a representative.*

dispensable *adj.* °disposable, °nonessential, unessential, inessential, °unnecessary, unneeded, °expendable, °superfluous, °needless, °useless: *Some regard a dishwasher as a luxury and entirely dispensable.*

dispense *v.* **1** °distribute, °give out, hand *or* pass out, °furnish, °supply, °provide, °give away, °deal (out), dole out, °parcel out, disburse, mete out, °share (out), °issue, apportion, allocate, °allot, °assign, *Colloq* dish out: *The Red Cross dispensed medicines to the stricken villagers.* **2** °administer, °conduct, °direct, °operate, superintend, °supervise, °carry out, °execute, °discharge, °apply, °implement, °enforce: *It is the governor who dispenses justice in these islands.* **3** ***dispense with.*** **(a)** do without, °forgo, °give up, eschew, °relinquish, °refuse, °waive, forswear, abstain (from), °renounce, °reject: *Can we dispense with the jokes and get to work?* **(b)** do away with, °get rid of, °eliminate, do without, °dispose of, °abolish, manage *or* do without, °remove, °cancel, °ignore, render unnecessary *or* superfluous: *Building on solid rock will dispense with the need for a foundation.*

disperse *v.* **1** °spread (out), °scatter, °broadcast, °distribute, °circulate, °diffuse, disseminate: *The practice is now widely dispersed throughout Asia.* **2** °disband, °spread out, °scatter, °dissipate, °break up; °disappear, vanish; dispel, °dismiss, °rout, send off *or* away: *The crowd dispersed quietly.*

displace *v.* **1** °move, °transfer, °shift, relocate, dislocate, misplace, °disturb, disarrange, °disorder, unsettle: *The entire population of the village was displaced when the dam was built.* **2** °expel, unseat, °eject, °evict, °exile, °banish, depose, °remove, oust, °dismiss, °discharge, cashier, *Colloq* °fire, kick *or* throw out, *Brit* °sack: *The voters displaced the corrupt council.* **3** take the place of, °supplant, °replace, °supersede, °succeed: *Television has displaced books in many modern homes.*

display *v.* **1** °show, °exhibit, °air, put *or* set forth, make visible, °expose, evince, °manifest, °demonstrate, °betray, °reveal, °unveil, °disclose; advertise, °publicize: *Her paintings are being displayed at the gallery today.* **2** unfurl, °unfold, spread *or* stretch *or* open out, °present: *The ship suddenly displayed the Jolly Roger.* **3** °show off, °flaunt, °parade, °flourish, vaunt, *Colloq* °flash: *He goes on those quiz programs only to display his knowledge.* —*n.* **4** °show, °exhibition, °exhibit, °presentation, array; °demonstration; °exposition, °manifestation, °revelation: *We visited a display of weapons at the armory. I have seldom seen such a display of ignorance.* **5** °ostentation, °spectacle, °flourish, °show, °parade, °ceremony, °pageantry, °pageant, °splendor, array, panoply, magnificence, °grandeur, °pomp, °splash, éclat, élan, °dash: *The display put on for Queen Victoria's jubilee was truly lavish.*

displease *v.* °offend, °put out, dissatisfy, °upset, °provoke, °exasperate, °worry, °trouble, vex, °annoy, °irritate, pique, °irk, nettle, peeve, °chafe, rile, °ruffle, °anger, °infuriate, °frustrate, get (someone's) goat, *Colloq* miff; *Slang US* °bug: *Having to listen to rock 'n' roll on your damned hi-fi is what displeases me most.*

displeasure *n.* **1** °dissatisfaction, °disapproval, °disfavor, discontentment, °distaste, °dislike, discountenance: *Your parents view your giving up college with displeasure and disappointment.* **2** °annoyance, irritation, vexation, chagrin, °indignation, dudgeon, ire, °anger, exasperation: *Whoever incurred the king's displeasure was at once banished from the land.*

disposable *adj.* **1** discardable, throwaway, nonreturnable, °paper, °plastic, biodegradable: *The new product is packaged in a disposable container.* **2** °available, °liquid, spendable, usable, °expendable, obtainable: *Her disposable assets include valuable government bonds.*

dispose *v.* **1** °place, °arrange, °move, °adjust, °order, array, °organize, °set up, °situate, °group, °distribute, °put: *She is planning how to dispose the furniture in the room.* **2** °incline, °influence, °persuade, °induce, °bend, °tempt, °move, °motivate, °lead, °prompt, °urge: *Her actions disposed me to cut her out of my will.* **3** ***dispose of.*** **(a)** °deal with, °settle, °decide, °determine, conclude, °finish (with): *I hope we can dispose of these matters quickly.* **(b)** throw away *or* out, °discard, °get rid of, jettison, °scrap, *Colloq* °dump, °junk, *US* °trash: *Dispose of the remains of the broken chair.*

(c) °distribute, °give out, deal out, °give (away), °dispense, apportion, °parcel out, °allot, °part with, °transfer, °make over, °bestow, °sell: *My grandfather disposed of his wealth before he died.* **(d)** do away with, °finish off, °put away, °demolish, °destroy, °consume, °devour, °eat, *Slang* °kill (off), °knock off, °polish off: *She could dispose of four hamburgers at one sitting. The boys disposed of Louie because he knew too much.*

disposed *adj.* °likely, °inclined, apt, °liable, °given, tending *or* leaning toward(s), °prone, °subject, of a mind to, minded, °willing, °ready, predisposed: *Great-aunt Agatha was not disposed to associate with rock musicians.*

disposition *n.* **1** °character, °temper, °attitude, temperament, °nature, °personality, °bent, °frame of mind, °humor, °makeup, °spirit: *Alan's son David has a cheerful disposition.* **2** °arrangement, °organization, °placement, disposal, ordering, grouping, °set, placing: *I don't care much for the disposition of the furniture.* **3** °transfer, transference, dispensation, disposal, °assignment, °settlement, °determination, bestowal, parceling out, °distribution: *The disposition of father's assets is not your affair.* **4** °determination, °choice, disposal, °power, °command, °control, °management, °discretion, °decision, °regulation: *Distribution of favors is at the disposition of the crown.*

dispossess *v.* °evict, °expel, oust, °eject, turn *or* drive out, dislodge, *Colloq* kick *or* throw out, *Brit* boot out, *US* °bounce: *The landlord dispossessed them for nonpayment of rent.*

disproportion *n.* °inequality, unevenness, °disparity, imbalance, asymmetry, irregularity, lopsidedness, °dissimilarity, inconsistency, incongruity: *Now that we're older, there isn't such a disproportion in our ages.*

disproportionate *adj.* °unbalanced, out of proportion, asymmetrical, °irregular, °lopsided, °dissimilar, °inconsistent, incommensurate, °incongruous; unfair, unequal, uneven, disparate: *The windows are disproportionate to the size of the house. The contractor was paid a disproportionate amount for his work.*

disprove *v.* refute, confute, invalidate, °contradict, negate, °rebut, °discredit, controvert, °puncture, °demolish, °destroy, *Colloq* shoot *or Colloq* poke full of holes: *Modern science has disproved the phlogiston theory.*

disputable *n.* °debatable, °moot, °doubtful, °uncertain, dubious, °questionable, °uncertain, undecided, °unsettled, °controversial; arguable: *His claim to ownership of the property is disputable.*

dispute *v.* **1** argue with *or* against, °question, °debate, °challenge, impugn, gainsay, °deny, °oppose, °fight (against), object to, °take exception to, disagree with, °contest, confute, quarrel with, °doubt, raise doubts about, dissent (from): *The council disputes his right to build a hotel on that land.* **2** °argue (about), °debate, °discuss, quarrel about, wrangle over, differ (on *or* about): *The question of capital punishment was again being disputed.* —*n.* **3** °argument, °debate, °disagreement, °difference (of opinion), °controversy, polemic, °conflict, °quarrel, wrangle, velitation; °discussion; *Colloq Brit* argy-bargy *or Colloq Brit* argie-bargie *or Colloq Brit* argle-bargle: *There is a dispute about the runner's qualifications to enter the race.* **4** °conflict, °disturbance, °fight, altercation, °row, °disagreement, °brawl, donnybrook, °feud, °rumpus, °fracas; °strife, °discord; °tiff, velitation, *US s*°pat: *Four people have been injured in the dispute.*

disqualify *v.* declare ineligible *or* unqualified, turn down *or* away, °reject, °exclude, °bar, debar, °rule out: *He was disqualified from voting because of his age.*

disregard *v.* **1** °ignore, °overlook, pay little *or* no heed *or* attention to, take little *or* no notice *or* account of, dismiss from one's mind *or* thoughts, turn a blind eye *or* deaf ear to, °brush aside, °pass up, wink *or* blink at, °make light of, let go by, °gloss over, *Rare* pretermit: *I shall disregard those insulting remarks.* **2** snub, °slight, turn up one's nose at, °disparage, °despise, contemn, disdain, °scorn, give the cold shoulder to, cold-shoulder, °cut; underrate, °underestimate, take little *or* no account of, undervalue, °minimize, °dismiss, °sneeze at, *Slang* °brush off, give the go-by: *Visitors often disregard the cultural attractions of Las Vegas.* —*n.* **3** °disrespect, °contempt, °indifference, inattention, nonobservance, °neglect, heedlessness, *Rare* pretermission; disdain, low regard, disesteem: *Some drive with a profound disregard for the law.*

disrepair *n.* °decay, °ruin, °collapse, dilapidation, deterioration, ruination: *The house is in a terrible state of disrepair.*

disreputable *adj.* **1** °low, °base, abject, °contemptuous, unrespectable, disrespectable, untrustworthy, discreditable, °dishonorable, °disgraceful, reprehensible, °shameful, °despicable, ignominious, °bad, °wicked, heinous, °vicious, iniquitous, °vile, opprobrious, °scandalous, louche, °questionable, dubious, *Colloq* °shady: *She keeps disreputable company.* **2** disheveled, °unkempt, slovenly, °untidy, °shabby, disordered, messy, °dirty, °bedraggled, scruffy, °seedy, °threadbare, °tattered, *Brit* down at heel, raddled, *US* down at the heel(s), *Colloq* °sloppy, *Slang Brit* grotty: *That disreputable beggar is your brother?*

disrespect *n.* rudeness, impoliteness, discourtesy, °incivility, unmannerliness, irreverence, °impudence, °impertinence, insolence, indecorum, *Colloq* cheek: *I meant no disrespect keeping my hat on, ma'am.*

disrespectful *adj.* °impolite, °rude, °discourteous, uncivil, unmannerly, °ill-mannered, bad-mannered, °irreverent, °impudent, °insolent, indecorous, °pert, saucy, °forward, *Colloq* °fresh, °cheeky: *Sara is sometimes disrespectful to her elders.*

disrobe *v.* undress, °strip, bare oneself: *She disrobed and put on a swimsuit.*

disrupt *v.* **1** °disorder, °upset, disorganize, °disturb, unsettle, °shake up, disconcert, °agitate: *You've disrupted my plan completely.* **2** °interrupt, break in *or* into, °interfere (with): *They disrupted the meeting by their loud outbursts.*

dissatisfaction *n.* **1** °discontent, discontentment, unhappiness, °displeasure, nonfulfillment, °disappointment, frustration, °discomfort, uneasiness, disquiet, malaise: *I was left with a feeling of dissatisfaction at the end of the play.* **2** °annoyance, irritation, °dismay, °displeasure: *Complaints concerning dissatisfaction with the food plagued the hospital administrators.*

dissatisfied *adj.* °discontented, displeased, °disappointed, unsatisfied, °discontent, °disgruntled, °unhappy, unfulfilled, ungratified, frustrated: *We return the full purchase price to any dissatisfied customer.*

dissension *n.* °disagreement, dissent, °discord, contention, °strife, °conflict, discordance, °friction: *The issue has sown dissension among the members.*

disservice *n.* °harm, °damage, °injury, °wrong, unkindness, °bad turn, °disfavor, °injustice: *It was a disservice to tell my boss about my expense account.*

dissident *n.* **1** dissenter, °nonconformist, protester *or* protestor, heretic, °rebel, apostate, recusant; °revolutionary: *Many dissidents were released and allowed to leave the country.* —*adj.* **2** disagreeing, °nonconformist, nonconforming, dissenting, dissentient, apostate, noncompliant, heterodox, °discordant, conflicting, contentious: *The couple spent ten years in Siberia for promoting their dissident philosophy.*

dissimilar *adj.* °different, °unlike, unalike, °distinct, °separate, contrasting, °diverse, °unrelated, heterogeneous: *The styles are entirely dissimilar.*

dissimilarity *n.* °difference, dissimilitude, unlikeness, °disparity; °discrepancy: *The dissimilarities between Art Deco and Art Nouveau are too numerous to mention.*

dissimulate *v.* °pretend, dissemble, feign, °disguise, °camouflage, °cover up, °conceal, °deceive, °misrepresent, °fake, °counterfeit: *She's dissimulating her real attitude toward the wealthy.*

dissimulation *n.* °deception, misrepresentation, dissembling, °deceit, °deception, °hypocrisy, °sham, °pretense, duplicity, double-dealing: *There can be no dissimulation between honest people.*

dissipate *v.* **1** °scatter, °spread (out), °disperse, be dispelled, °diffuse; disseminate, °sow, °distribute; °break up: *The crowd had dissipated by noon.* **2** spread thin, °evaporate, vanish, °disappear, vaporize, °peter out, °diminish: *By the time we were ready to go, the clouds had dissipated.* **3** squander, °waste, °fritter away, °throw away, burn up, °use up, °exhaust, °run through: *By the time he was twenty, he had dissipated a huge fortune.* **4** °revel, °carouse, °party, sow one's wild oats, burn the candle at both ends, roister, °make merry, debauch, go on a spree: *Before their marriage, they were seen dissipating in the fleshpots of Europe.*

dissipation *n.* **1** squandering, °waste, wastefulness, °profligacy, °abandon, abandonment, self-indulgence, self-gratification, overindulgence, intemperance, hedonism, fast *or* high living, *la dolce vita,* voluptuousness, sensualism, sybaritism, dissoluteness, °dissolution, excess(es), wantonness, debauchery, carousing, °prodigality, recklessness, °extravagance, rakishness: *Owing to my dissipation, I had become an alcoholic vagrant.* **2** disappearance, dispersion, dispersal, diffusion, °scattering, vanishing: *The dissipation of the tear gas was rapid in the strong breeze.* **3** °distraction, °amusement, °diversion, °entertainment: *Reading, once a dissipation, had become an obsession.*

dissociate *v.* °separate, °cut off, °sever, disassociate, disjoin, °disconnect, °abstract, °disengage, °detach, °isolate, °distance, °break off (from), °break up (with), °divorce, °set apart, °segregate: *I have carefully dissociated myself from any political party.*

dissolute *adj.* dissipated, debauched, °abandoned, °corrupt, °degenerate, °rakish, °profligate, °wanton, rakehell, intemperate, °incontinent, °loose, licentious, overindulgent, carousing, °self-indulgent, hedonistic, pleasure-bound, °immoral, amoral, libidinous, unrestrained, depraved: *He has paid dearly for his dissolute life.*

dissolution *n.* **1** disintegration, °separation, breakup, °breakdown, °separation, breaking up, breaking down, °collapse, °undoing: *Much ill will attended the dissolution of our marriage.* **2** °destruction, decomposition, °decay, °ruin, °overthrow, dissolving, disbandment, °dismissal, dispersal, disorganization, discontinuation; adjournment, ending, °end, °termination, conclusion, °finish: *A vote of no confidence led to the dissolution of the parliament.*

dissolve *v.* **1** °melt (away), liquefy, °disperse, °disintegrate, °diffuse, °decompose, °thaw (out), °fuse, deliquesce; °sublime; vanish, °disappear, °fade (away), °diminish, °decline, °peter out: *Dissolve one tablet in water. The sugar dissolved in the tea.* **2** °collapse, break into, °melt into: *She dissolved into tears whenever he shouted at her.* **3** °break up, °disperse, °dismiss, °terminate, °finish, conclude, adjourn, °recess, °disband, °wind up; liquidate: *We took a vote and dissolved the meeting.*

distance *n.* **1** remoteness, °space, °gap, °interval, mileage, footage, °stretch: *What is the distance from here to your house?* **2** aloofness, °detachment, °reserve, coolness, haughtiness, hauteur, stiffness, rigidity: *He maintains a distance between himself and the servants.* —*v.* **3** °separate, °detach, °dissociate, disassociate: *She distanced herself from her students.*

distant *adj.* **1** °far, far-off, °remote, °faraway, long-way-off; removed: *The creature said he had come from a distant star.* **2** away, °off: *The ship is ten miles distant.* **3** °aloof, °detached, °reserved, °cool, °cold, °haughty, °standoffish, °unapproachable, °inaccessible, °withdrawn, °reticent, °ceremonious, °formal, °stiff, °rigid, °frigid, unfriendly: *You find him warm, but I think him very distant.*

distaste *n.* **1** °dislike, °disfavor, antipathy, disrelish, disinclination; °dissatisfaction, °displeasure, discontentment: *You know of my distaste for long engagements.* **2** °aversion, °revulsion, °disgust, nausea, abhorrence, °loathing, repugnance, °horror: *She has a distinct distaste for fried witchetty grubs, which she had eaten when in Australia.*

distasteful *adj.* °disgusting, °revolting, sick-making, nauseating, °nauseous, °repugnant, °repulsive, °loathsome, fulsome, °nasty, °disagreeable, °foul, off-putting, °unpalatable, °obnoxious, objectionable, °offensive, unpleasing, unpleasant, displeasing: *I found their children's table manners quite distasteful.*

distinct *adj.* **1** °clear, °perceptible, °plain, understandable, °vivid, °definite, well-defined, °precise, °exact, unmistakable *or* unmistakeable, °noticeable, recognizable, °obvious, °patent, °marked, °manifest, °evident, °apparent, °explicit, unambiguous, clear-cut, palpable, unequivocal, lucid, °sharp, pellucid, limpid, °transparent: *There is a distinct outline of a figure on the Turin shroud.* **2** °separate, °detached, °discrete, °different, °dissimilar, distinguishable, °distinguished; °individual, sui generis, °unique, °special, °singular; °peculiar, °unusual, uncommon, contrasting: *The government of Puerto Rico is distinct from that of the US. He has been charged with three distinct offenses.*

distinction *n.* **1** differentiation, °discrimination, °difference, °contrast, °separation, °division, dividing line; distinctiveness: *Any distinction between them is difficult to discern.* **2** °honor, °credit, °prominence, eminence, °preeminence, °superiority, uniqueness, greatness, °excellence, °quality, °merit, °worth, value, °prestige, °note, °importance, °significance, consequence, °renown, °fame, repute, °reputation, °celebrity, °glory, °account: *We all know her as a scholar of distinction.*

distinctive *adj.* distinguishing, °characteristic, °unique, °singular, °distinct, °individual, °typical, idiosyncratic, °peculiar: *She has developed a distinctive style of her own.*

distinguish *v.* **1** °differentiate, °discriminate, tell the difference, tell apart, °determine, °judge, °decide, tell who's who *or* what's what: *He is still unable to distinguish between his own twin daughters.* **2** classify, °categorize, °characterize, individualize, °mark, °identify, °define, °designate, °denote, °indicate, °separate, °single out, °set apart; °grade, °group: *The male is distinguished by his brighter coloring.* **3** °sense, °make out, °perceive, discern, °pick out, °recognize, °identify, °detect, °notice; °see, espy, descry; °hear; °smell; °taste; °feel: *I could distinguish two people in the dark.* **4** call attention to, °identify, °mark, °set apart, °separate,

°segregate, °indicate, particularize: *She distinguished herself by her great beauty and her awful soprano voice.*

distinguished *adj.* **1** °celebrated, °famous, °illustrious, °noted, °renowned, °notable, °noteworthy, °pre-eminent, °eminent, °prominent, honored, respected, °honorable: *Gladstone was one of the most distinguished men of his day.* **2** °dignified, °noble, °grand, °stately, *distingué,* °royal, °regal, aristocratic: *What is he doing in this distinguished gathering?*

distort *v.* **1** °twist, °warp, deform, misshape, contort, gnarl, °bend, disfigure, °wrench: *The car was completely distorted in the crash.* **2** °twist, °warp, °slant, tamper with, °color, varnish, torture, °pervert, °misrepresent, °fabricate, °falsify, misstate, °alter, °change, °bend, °garble, °violate: *She distorted the facts if she said it was Bill who had a gun.*

distract *v.* **1** °divert, °deflect, °sidetrack, °turn aside, draw away: *Sorry, I was distracted for a moment—where were we?* **2** °divert, °amuse, °entertain, °gratify, °delight, °occupy, °interest, absorb, engross: *We found the belly dancers quite distracting.* **3** °bewilder, °confuse, confound, °perplex, °puzzle, discompose, befuddle, °mystify, disconcert, °fluster, °rattle, °bemuse, °daze, °disturb, °agitate, °trouble, °bother: *I am distracted with doubts about whether to phone the police.*

distraction *n.* **1** bewilderment, befuddlement, °disorder, °disturbance, °upset, °confusion, °agitation: *The princess loves you to distraction.* **2** °diversion, °entertainment, °amusement: *I was never really interested in him, he was merely a temporary distraction.*

distraught *adj.* distracted, °agitated, troubled, °disturbed, °upset, perturbed, wrought *or* worked up, °excited, °frantic, at one's wit's end, °overwrought, frenetic, °nervous, frenzied, °feverish, °wild, °hysterical, °delirious, irrational, °crazy, °mad, °insane, °berserk, run(ning) amok *or* amuck: *He is distraught with grief.*

distress *n.* **1** °anguish, °anxiety, °affliction, angst, °grief, °misery, °torment, °ache, °pain, °suffering, °agony, torture, °woe, woefulness, wretchedness; unhappiness, °sorrow, °sadness, °depression, heartache, °desolation: *It is impossible to imagine the distress of a bereaved parent.* **2** °calamity, °trouble, adversity, °catastrophe, °tragedy, °misfortune, °difficulty, °hardship, °straits, °trial, °disaster: *Has he no sympathy for the distresses that have beset his people?* —*v.* **3** °bother, °disturb, °perturb, °upset, °trouble, °worry, harrow, harry, vex, °harass, °plague, °oppress, °grieve, °torment, torture, °afflict: *The thought of Miss Camberley as a hostage distressed us all.*

distribute *v.* **1** deal *or* dole out, °parcel out, °give (out), mete out, °dispense, apportion, °allot, °share (out), °partition, °divide up, °assign, °issue, °circulate, °pass out, pass round *or* around, °hand out, °deliver, convey, *Colloq* dish *or Colloq* spoon out: *Emergency rations were distributed to the flood victims.* **2** °disperse, °scatter, °strew, spread (round *or* around *or* about), °diffuse, disseminate: *Mammals are uniformly distributed over the globe.* **3** °sort, classify, °class, °categorize, assort, °arrange, °group, °file, °order: *Distribute the packages according to their size.*

distribution *n.* **1** apportionment, °allotment, allocation, °assignment, parceling out, sharing; deployment: *She supervised the distribution of the prizes.* **2** issuance, °circulation, dissemination, giving (out), dispersal, dispensation; deployment: *The distribution of food parcels is being handled by charities.* **3** °arrangement, °disposition, grouping, classification, °order, ordering, °division, cataloging, codification; deployment: *What is the distribution of scientists among the population?*

district *n.* °territory, °region, °section, sector, °division, °partition, °part, °precinct, locality, °area, °locale, °department, °province, community, °quarter, °neighborhood, °ward: *We need a new hospital in our district.*

distrust *v.* **1** °mistrust, °doubt, °question, be skeptical of, be circumspect *or* cautious about, °suspect, be suspicious *or* wary of, °discredit, disbelieve, *Colloq* smell a rat; *Colloq* be leery of: *I distrusted her motives from the very beginning.* —*n.* **2** °mistrust, °doubt, doubtfulness, uncertainty, misgiving(s), °skepticism, °suspicion, disbelief, incredulity, incredulousness, hesitation, °caution, wariness, °qualm, hesitancy: *His claims were greeted with distrust.*

distrustful *adj.* distrusting, untrusting, mistrustful, doubting, chary, °wary, °cautious, °suspicious, °skeptical, °doubtful, dubious, cynical, disbelieving, °unbelieving, uneasy, °nervous, °hesitant, hesitating, unsure, °uncertain, *Colloq* °leery: *She is distrustful of men who bring her flowers.*

disturb *v.* **1** °interrupt, °disrupt, °intrude (on), °inconvenience, °put out, °interfere (with); °bother, °pester, °annoy, °irritate, °irk, °upset, °plague, hector, harry, °harass, °worry, vex, °provoke, pique, peeve, get on (someone's) nerves, *Colloq* °bug, miff, get under (someone's) skin, get in (someone's) hair, drive nuts *or* crazy *or* bats *or* batty *or* bananas *or* up the wall, hassle: *The sound of dripping water disturbed me. Please do not disturb the animals.* **2** °agitate, stir *or* churn (up), °shake (up), unsettle, roil, °disorder: *The lake's surface was violently disturbed by an enormous creature.* **3** unsettle, °affect, °upset, °damage, °harm, °destroy: *We put the delicate mechanism where it wouldn't be disturbed by curious visitors.* **4** °trouble, disconcert, °discomfit, °perturb, °ruffle, °fluster, °upset, °agitate, °put off, °bother, discommode, °put out, unsettle, °distress; °alarm, *Colloq* °shake (up): *He was greatly disturbed by the death of his father.* **5** °affect, °upset, confound, °confuse, °change, °put off, °ruin, °destroy, °cancel, make ineffectual *or* ineffective, negate: *Any change in temperature will disturb the results of the experiment.*

disturbance *n.* **1** disruption, °disorder, disorganization, disarrangement, disarray; °upheaval, °interruption, °upset, intrusion, °interference: *She won't tolerate any disturbance to her schedule.* **2** commotion, °disorder, °upset, °outburst, °tumult, turmoil, turbulence, °violence, hubbub, hullabaloo, hurly-burly, °uproar, brouhaha, °rumpus, °brawl, melee *or* mêlée, breach of the peace, donnybrook, °fray, affray, °fracas, °trouble, *Colloq* ruckus, *Brit* spot of bother, *Slang Brit* spot of bovver: *There was a disturbance at the pub yesterday.*

disturbed *adj.* **1** °upset, uneasy, uncomfortable, discomfited, troubled, °worried, bothered, °agitated, °anxious, °concerned, apprehensive, °nervous: *He's disturbed that Marie didn't come home last night.* **2** psychoneurotic, °neurotic, °unbalanced, psychopathic, °psychotic, maladjusted, °mad, °insane, °out of one's mind, depressed, *Colloq* °crazy, unable to cope, *Brit* bonkers, *Slang* nuts, screwy, batty, off one's rocker, off the deep end, messed up, screwed-up: *She looks after her sister, who is disturbed.*

disturbing *adj.* upsetting, off-putting, perturbing, troubling, °unsettling, worrying, °disconcerting, disquieting, alarming, distressing: *There is disturbing news from the front.*

disused *adj.* °abandoned, neglected, °unused; discontinued, °obsolete, archaic: *We had to sleep in a disused railway carriage.*

diurnal *adj.* °daily, circadian; day-to-day, °regular, °everyday, quotidian; daytime: *Jet lag is a disturbance of the body's diurnal rhythms. Are these animals nocturnal or diurnal?*

dive *v.* **1** °plunge, nose dive, °sound, °descend, °dip, °submerge, °go under, °sink; °jump, °leap, °duck;

°swoop, °plummet: *The submarine dived at once.* —*n.* 2 °plunge, nose dive: *The plane went into a dive.* 3 °bar, saloon, nightclub, bistro, °club, *Colloq* nightspot, *Slang* °joint, *US* °dump, honky-tonk, juke joint: *He met the woman in a dive on the West Side.*

diverge *v.* 1 °separate, °radiate, °spread (apart), °divide, subdivide, fork, branch (off *or* out), ramify, °split: *The roads diverge farther on.* 2 °deviate, turn aside *or* away, °wander, digress, °stray, °depart, °drift, divagate: *Our policy diverges from that set up by the committee.*

divergent *adj.* differing, °different, °dissimilar, disparate, °variant, °separate, diverging, disagreeing, conflicting, discrepant: *There are divergent theories about the origin of the universe.*

divers *adj.* °various, °several, °sundry; °miscellaneous, multifarious, °manifold, °varied, assorted, °variegated, differing, °different; some, numerous, °many: *We have the divers statements of the witnesses.*

diverse *adj.* °different, °varied, diversified, multiform, °various, assorted, °mixed, °miscellaneous; °distinctive, °distinct, °separate, varying, °discrete, °dissimilar, differing, °divergent, heterogeneous: *Diverse subjects are available for study.*

diversify *v.* °vary, variegate, °change, °mix, °change; °spread, °distribute, °divide, °break up, °separate; branch out: *We must diversify our investments to hedge against losses. Perhaps this is not a good time to diversify into other areas.*

diversion *n.* 1 °digression, deviation, departure, °distraction: *George created a diversion, while we robbed the safe.* 2 °detour, °sidetrack, deviation, °bypass, deviation: *Owing to the construction, we had to take a diversion off the main road.* 3 °amusement, °distraction, °entertainment, °pastime, °recreation, divertissement, °game, °play, °relaxation: *She prefers chess for diversion.*

diversity *n.* 1 °difference, °dissimilarity, dissimilitude, unlikeness, °disparity, deviation, divergence, departure, distinctiveness, diverseness, °variation, °variety, individuality, inconsistency, contrariety, °discrepancy, °contrast: *Flowers are impressive in their diversity.* 2 °variety, °range, °extent, heterogeneity, multiplicity, multifariousness, variegation, multiformity: *Democracy encourages diversity of opinion.*

divert *v.* 1 °switch, rechannel, redirect; °change, °alter, °deflect: *Funds for the new civic center have been diverted to housing. We must divert the course of the river.* 2 turn away, °turn aside, avert, reroute, °deflect; change course, swerve (off *or* away), °shift, °sidetrack, °depart, °deviate: *Cars were diverted to avoid flooded areas. We diverted from our route because of the roadworks.* 3 °entertain, °amuse, °distract, °interest, °beguile, °engage, °occupy, absorb: *We found the standup comedian mildly diverting but not really funny.*

divest *v.* 1 °strip, denude, °rid, get rid, °relieve, disencumber, °deprive, °dispossess; despoil, mulct. *The company has been divested of all its assets.* 2 ***divest oneself of.*** take *or* put off, doff, °remove; °disrobe, unclothe, undress: *She divested herself of her fur coat.*

divide *v.* 1 °separate, °split (up), °break up, °cleave, cut up *or* asunder, °partition, °segregate, subdivide; °disconnect, disjoin, °detach, °sever, sunder, °part: *Argyle divided his mountaineers into three regiments. A divided nation cannot stand. Some would like to see Britain divided from continental Europe.* 2 Sometimes, ***divide up.*** °distribute, °share (out), °measure out, °parcel out, °partition, °dole (out), °deal (out), mete out, allocate, °allot, apportion, °dispense, °give (out): *The remaining food was divided among us.* 3 °separate, °split, cause to disagree, °alienate, disunite, set at odds, sow dissension (among), pit *or* set against one another, disaffect: *Racial issues still divide the people.* 4 °branch (out), ramify, °split, °separate: *The road divides there and passes on each side of that huge rock.* 5 °categorize, classify, °sort, assort, °grade, °group, °(put in) order, °rank, °organize, °arrange: *You have to divide the books into several piles according to size.*

divine *adj.* 1 °godlike, °godly, °holy, deiform, deific, angelic, °seraphic, °saintly; °heavenly, °celestial; °sacred, sanctified, hallowed, consecrated, °religious, °spiritual: *They believe in the divine right of kings. He receives divine inspiration at divine services.* 2 °superhuman, °supernatural, °gifted, °preeminent, °superior, °excellent, °supreme, °exalted, °transcendent, °extraordinary: *Even the divine Homer nods.* 3 °great, °marvelous, °splendid, °superlative, °glorious, °superb, °admirable, wonderful, °awesome, °perfect, °excellent, °beautiful, *Colloq* °super, °great, °terrific, smashing, °fantastic, splendiferous, *Colloq Brit* ace, °magic: *They say that the new musical is simply divine.* —*v.* 4 intuit, °imagine, conjecture, °guess, °assume, °presume, °infer, °suppose, hypothesize, °surmise, °suspect, °understand, °perceive, °speculate, °theorize, °predict, foretell, have foreknowledge of; °determine, °discover: *He had divined that she might be there.* —*n.* 5 holy man, °priest, °clergyman, cleric, ecclesiastic, °minister, °pastor, reverend, churchman, prelate: *At his club, Colquhon enjoys the company of bishops, archbishops, and other divines.*

division *n.* 1 dividing, °split, splitting (up), breaking up, °partition, partitioning, partitionment, °separation, separating, diremption, segmentation, segmenting, compartmentation, sectioning, apportioning, apportionment, °allotment: *In England a division between church and state is not recognized.* 2 °section, °compartment, °segment; °partition, °separation: *Egg crates have 144 divisions.* 3 °branch, °department, sector, °section, °unit, °group, arm; °part, °set, °category, °class, classification: *The textile division of the company lost money last year.* 4 °boundary (line), °border, borderline, °frontier, °margin, °line, dividing line: *Where is the division between good and evil?* 5 °discord, °disagreement, °upset, °conflict, °strife, disunity, disunion: *The issue of equal rights has led to much division within the movement.*

divorce *n.* 1 °separation, °split, split-up, °dissolution, severance, disunion, breakup: *Their divorce after twenty years surprised everyone.* —*v.* 2 °separate, °divide, °split (up), °part, °sever, °detach, °dissociate, disassociate; °dissolve: *A splinter group has divorced itself from the main party. We were divorced last year.*

dizzy *adj.* 1 °giddy, vertiginous, lightheaded, °faint, dazed, tottering, unsteady, reeling, tipsy, *Colloq* woozy: *I felt dizzy after riding on the Boomerang.* 2 °confused, °silly, °giddy, empty-headed, °scatterbrained, muddled, befuddled, °flighty, featherheaded, featherbrained, rattlebrained, °harebrained, °frivolous: *He is dizzy with power.*

dock *n.* 1 wharf, °pier, berth, jetty, quay: *We went to the dock to see them off.* —*v.* 2 °(drop) anchor, berth, °tie up, °moor, °land, °put in: *The ship docks at noon.*

doctor *n.* 1 °physician, medical practitioner, M.D., general practitioner, G.P., *Colloq* medic, medico, doc, sawbones, bones; shrink: *You ought to see a doctor for that cough.* —*v.* 2 °treat, °attend, medicate; °cure, °heal; practice medicine: *She's been doctoring herself for a head cold.* 3 °mend, °repair, °patch (up), °fix: *We doctored the tire as best we could.* 4 °falsify, tamper with, °adulterate, °disguise, °change, °modify, °alter; °cut, °dilute, °water (down); °spike; °drug, °poison: *This sauce has been doctored.*

doctrine *n.* teaching, body of instruction, °precept; °principle, °tenet, dogma, article of faith, canon, °conviction, °creed, °belief, credo, °opinion, °idea, concept, theory, °proposition, °thesis, postulate: *Few believed the doctrine that all men are created equal.*

document *n.* **1** °paper, certificate, °instrument, °report, °chronicle, °record: *All the legal documents are at my lawyer's office.* —*v.* **2** °record, °chronicle, particularize, °detail, °describe; °verify, validate, °certify, °authenticate, corroborate, °substantiate: *Detectives have documented every move you made since the murder.*

doddering *adj.* shaking, quaking, palsied, trembling, trembly, quivering, quavering, reeling, unsteady, °shaky, staggering, °shambling, °decrepit, faltering; °feeble, °weak, °frail, °infirm; °aged, °old, superannuated, °senile, anile: *Once a vigorous sportsman, his illness has reduced him to a doddering octogenarian.*

dodge *v.* **1** dart, °shift, move aside, °sidestep, °duck, bob, °weave, °swerve, veer: *He dodged here and there across the traffic.* **2** °avoid, °elude, °evade, escape from: *He neatly dodged the punches of his opponent.* **3** escape from answering, °sidestep, °duck, °evade, hedge; °quibble, tergiversate, double-talk, *Colloq* °waffle: *She dodged the questions put to her by the interviewer.* —*n.* **4** °trick, °subterfuge, ploy, °scheme, °ruse, °device, °stratagem, °plan, °plot, °machination, chicane, °deception, prevarication, contrivance, °evasion, *Slang* wheeze, °racket: *Crenshaw worked out a new dodge to avoid paying tax.*

dodgy *adj.* °tricky, °dangerous, °perilous, °risky, °difficult, °ticklish, °sensitive, °delicate, °touchy, °dicey; °uncertain, °unreliable; °rickety, *Colloq* chancy, °hairy, *Brit* dicky: *Climbing up the sheer face of that rock could be a bit dodgy. I shouldn't be climbing stairs in my dodgy condition.*

dogmatic *adj.* °arbitrary, °categorical, °dictatorial, imperious, °peremptory, °overbearing, doctrinaire, °authoritarian, °emphatic, °insistent, °assertive, °arrogant, °domineering; obdurate, °stubborn; °opinionated, °positive, °certain, *Rare* thetic(al), *Colloq* °pushy: *Patrick tends to be quite dogmatic when he is sure of his ground.*

dole *n.* **1** °portion, °allotment, °share, °quota, °lot, °allowance, °parcel; compensation, °benefit, °grant, °award, °donation, °gift, largess *or* largesse, alms, gratuity; *Slang* handout: *The prisoners received a daily dole of bread. If you've lost your job, are you eligible for unemployment insurance?* **2** °distribution, apportionment, allocation, dispensation: *The money was given to the disaster victims by dole.* —*v.* **3** °give (out), °deal (out), °distribute, °hand out, mete out, °share (out), °dispense, °allot, allocate, apportion, *Colloq* dish out: *They dole out the reparations on the basis of need.*

doleful *adj.* °sad, °sorrowful, °melancholy, °gloomy, °mournful, cheerless, °joyless, °somber, depressed, disconsolate, °blue, down, distressed, °dejected, °downhearted, °forlorn, °unhappy, lugubrious, dolorous, °wretched, °miserable, °woebegone, °dreary, woeful, *Colloq* °down in the mouth, down in the dumps; distressing, °funereal, depressing, °grievous, °harrowing: *From his doleful expression I thought he would cry any minute. She lives in the most doleful surroundings.*

dolt *n.* °fool, ass, blockhead, dunce, dullard, idiot, nitwit, ignoramus, numskull *or* numbskull, donkey, nincompoop, ninny, ninnyhammer, simpleton, dunderpate, dunderhead, bonehead, simpleton, °twit, fathead, goon, moron, imbecile, *Colloq* °dope, dumbbell, dimwit, chump, °dummy, °half-wit, birdbrain, pinhead, clot, °clod, chucklehead, *Brit* muggins, *US* thimble-wit, °jerk, knucklehead, lunkhead, meathead, lamebrain, dingbat, ding-a-ling, °flake: *The dolt actually tried to buy striped paint!*

domain *n.* **1** °realm, °dominion, °territory, °property, land(s), °province, °kingdom, empire: *At one time his domain included most of Europe.* **2** °province, °realm, °territory, °field, bailiwick, °area, °department, °sphere, °discipline, °specialty, specialization, °concern: *As a dentist, he considered diseases of the throat outside his domain.*

domestic *adj.* **1** °home, °private, °family, familial; residential, household: *Her domestic life is a shambles. This toaster is for domestic use.* **2** °tame, domesticated, house-trained, housebroken: *Tenants are forbidden to keep domestic animals.* **3** °home, °native, °indigenous, °internal, autochthonous: *The domestic market accounts for most of the company's income.* —*n.* **4** °servant, °(hired) help, housekeeper, major-domo, steward: *Her domestics left and she now does the cleaning herself.*

domicile *n.* **1** °dwelling (place), °residence, °abode, °home, habitation, °(living) quarters, °housing, accommodation(s), lodging(s), *Colloq Brit* digs, diggings, *Slang* °pad: *Domiciles in southeast England increased enormously in value.* —*v.* **2** °locate, °quarter, °lodge, °settle, °establish, °situate, domiciliate: *She is domiciled abroad, hence pays no income tax here.*

dominant *adj.* **1** commanding, °authoritative, controlling, governing, ruling, °leading, reigning, °influential, °assertive, °supreme, °superior, ascendant: *He has taken a dominant role in promoting foreign language teaching.* **2** °predominant, °chief, °main, °principal, °primary, °prevailing, °outstanding, °preeminent, °paramount: *A large nose is a dominant characteristic in their family.*

dominate *v.* **1** °command, °control, °govern, °rule, °direct, °lead, °reign (over), exercise command *or* authority *or* control *or* rule over, have the whip *or* upper hand (over), °run (things), be in *or* have under control, rule the roost *or* roast, *Colloq* call the shots *or* the tune, wear the trousers *or US* the pants, be in the driver's seat, rule with an iron hand, have under one's thumb: *She clearly dominates the board of directors.* **2** °overlook, look (out) over, tower over *or* above, rise above, °overshadow; °predominate: *The Eiffel Tower dominates the Parisian skyline.*

domination *n.* **1** °authority, °control, °rule, °power, °command, °influence, °sway, °supremacy, ascendancy, hegemony, the whip *or* upper hand, °preeminence, mastery: *The tsar's domination lasted for more than thirty years.* **2** °oppression, °subjection, °repression, °suppression, subordination, enslavement, enthrallment; dictatorship, °despotism, °tyranny: *The Allies finally brought to an end the Fascist domination of Europe.*

domineering *adj.* °overbearing, imperious, °officious, °arrogant, autocratic, °authoritarian, highhanded, high and mighty, °masterful, °arbitrary, °peremptory, °dictatorial, °despotic, °tyrannical, °oppressive, °strict, °hard, °harsh, °tough, *Colloq* °bossy, °pushy: *A classic character in humorous writing is the domineering spouse.*

dominion *n.* **1** °rule, °authority, °control, dominance, °domination, °grasp, mastery, °grip, °command, °jurisdiction, °power, °sovereignty, °sway, ascendancy, °preeminence, primacy, °supremacy, hegemony: *The magician claimed dominion over the entire universe.* **2** °domain, °realm, °territory, °region, °area, °country, °kingdom: *For six generations the dynasty ruled over its dominions on five continents.*

donate *v.* °give, °provide, °supply, °present, °contribute, subscribe (to *or* for), °pledge, °award, °bestow, °confer, °grant, °vouchsafe, °will, °bequeath: *Lady*

Crayford donated two silver candlesticks to our charity drive.

donation *n.* **1** °gift, contribution, largess *or* largesse, °present, °grant, °award, alms, °offering, °bequest: *Donations have exceeded our expectations.* **2** giving, contribution, bestowal, °allotment, °provision, °offer: *We are seeking the donation of a piano for our theatre group.*

donor *n.* giver, provider, supplier, benefactor *or* benefactress, contributor, °supporter, °backer: *Blood donors receive a suitably inscribed certificate.*

doom *n.* °fate, karma, °destiny, °fortune, °lot, kismet; °downfall, °destruction, °death, °ruin, extinction, annihilation, °death, °end, °termination, terminus: *You have defied the Snake God, and your doom is sealed!*

doomed *adj.* **1** °fated, cursed, °condemned, damned, °destined, ordained, foreordained, predestined: *She was doomed to live forever.* **2** accursed, bedeviled, ill-fated, luckless, star-crossed, bewitched, °condemned: *The doomed ship sank to the bottom of the sea.*

dope *n.* **1** See **dolt.** **2** °narcotic, °drug, opiate, hallucinogen, psychedelic, *Slang* °upper, downer: *He was caught trying to smuggle dope past customs.* **3** °information, °data, °facts, °news, °details, °story, °scoop, *Slang* info, °lowdown, °score, *Brit* gen, *US and Canadian* poop: *The real dope on the minister is sensational!*

dormant *adj.* **1** asleep, sleeping, slumbering, resting, at rest, °quiet, °inactive, °still, °inert, unmoving, motionless, stationary, immobile, quiescent, comatose, °torpid, hibernating, slumberous, somnolent, °sleepy, °lethargic, °dull, sluggish: *The bears are dormant during much of the winter.* **2** latent, °potential, °hidden, concealed, undisclosed, unrevealed, unexpressed: *The theory lay dormant for centuries and has only recently been revived.*

dose *n.* **1** °portion, °quantity, °amount, °measure, dosage: *How big a dose of the medication did the doctor prescribe?* —*v.* **2** °dispense, °administer, °prescribe: *I was dosed with medicine and slept all day.*

dot *n.* **1** °spot, °speck, °point, °jot, °mark, iota, fleck, °dab; decimal point, *Brit* full stop, *US* °period: *Use three dots to denote text omissions.* **2** ***on the dot.*** °exactly, °precisely, punctually, to the minute *or* second, °on time, *Colloq* on the button: *She arrived at noon on the dot.* —*v.* **3** °spot, fleck, speckle, stipple, bespeckle: *The wallpaper is dotted with tiny squares of color.*

dote *v.* Often, **dote on** *or* **upon.** be fond of, be infatuated with, °love, °idolize, hold dear, °adore, °make much of; °coddle, °pamper, °spoil, °indulge: *I think she dotes on her husband at the expense of the children. What we need is a doting grandmother to babysit when we want to go out.*

double *adj.* **1** twofold, paired, coupled, duplicate(d), doubled: *The forms* banned *and* banning *are spelled with a double* n. **2** folded *or* doubled *or* bent over, overlapped, two-ply: *This wound needs a double bandage.* **3** dual, twofold, °ambiguous, double-barreled: *He pronounced it "de-seat," giving* deceit *a double meaning.* **4** twice: *The plant had grown to double its size.* **5** °deceitful, °dishonest, treacherous, °traitorous, °insincere, °hypocritical, double-dealing, °false: *It was Maria who exposed Flint as a double agent.* —*v.* **6** °duplicate, replicate; °copy; °increase, °enlarge; °magnify: *We'll have to double our milk order.* —*n.* **7** °twin, °duplicate, °copy, °replica, °facsimile, clone, °copy, counterpart, doppelgänger, °lookalike, °stand-in, °understudy, *Slang* (dead) ringer, spitting image *or* spit and image: *He could be Clint Eastwood's double.* **8** ***at*** *or* ***on the double.*** °quickly, °on the run, at full speed *or* tilt, briskly, °immediately, °at once, without delay, *Slang* PDQ: *Put down that book and come over here on the double!*

double-cross *v.* °cheat, °defraud, °swindle, °hoodwink, °trick, °betray, °deceive, °mislead, play false with, *Colloq* two-time: *He swore he'd give me the money but he double-crossed me and kept it himself.*

doubt *v.* **1** disbelieve, °discredit, °mistrust, °distrust, have misgivings (about), °question, °suspect: *I doubted his ability to beat the record.* **2** °hesitate, waver, vacillate, °fluctuate, °scruple, be uncertain, entertain doubts, have reservations: *Who ever doubted about her honesty?* —*n.* **3** uncertainty, hesitation, °misgiving, reservation(s), °qualm, °anxiety, °worry, apprehension, disquiet, °fear: *He has harbored doubts about the success of the enterprise.* **4** °distrust, °mistrust, °suspicion, incredulity, °skepticism, dubiousness, dubiety *or* dubiosity, lack of faith *or* conviction, irresolution: *Her doubts about his intentions have evaporated.* **5** ***in doubt.*** See **doubtful,** below.

doubtful *adj.* **1** °in doubt, dubious, °questionable, open to question, °problematic, °debatable, °disputable, °uncertain, unpredictable, indeterminate, °unsettled, °unresolved, conjectural, °indefinite, unclear, °obscure, °vague, anybody's guess, *Colloq* up in the air: *The result is very doubtful.* **2** °skeptical, unconvinced, °distrustful, mistrustful, °suspicious, °uncertain, unsure, °hesitant, hesitating, vacillating, °indecisive: *I am doubtful whether an investigation will yield anything.* **3** dubious, °questionable, °shady, louche, °disreputable, °controversial: *Those are people of doubtful reputation.*

doubtless *adv.* **1** doubtlessly, °undoubtedly, no doubt, indubitably, indisputably, unquestionably, °surely, for sure, certainly, for certain, °naturally, without (a) doubt, beyond *or* without (a *or* the shadow of) (a) doubt, °truly, °positively, °absolutely, *Colloq* absotively, posolutely, *US* make no mistake: *You doubtless remember Jack the Ripper?* **2** °probably, most *or* very likely, in all probability, °supposedly, °presumably: *He will doubtless be refused entry into the country.*

dour *adj.* **1** °sullen, °sour, unfriendly, °cold, °gloomy, morose, °dreary, °grim, cheerless, °dismal, °forbidding: *We went to Spain, away from the dour northern climate.* **2** °hard, °tough, austere, °severe, °hardy, °inflexible, °obstinate, °stubborn, unyielding, uncompromising, °strict, °rigid, obdurate, °stern, °harsh, adamant, *Colloq* hard-nosed: *Her father was a dour Scot who wouldn't let me in the house.*

dowdy *adj.* frowzy, frumpy, °drab, °dull, °seedy, °shabby, °unseemly, °unbecoming; slovenly, °sloppy, messy, °unkempt; °old-fashioned, unfashionable, *Colloq US* °tacky: *Aunt Patricia looked particularly dowdy in her robe and slippers.*

down-and-out *adj.* **1** °indigent, poverty-stricken, °poor, penniless, °destitute, °impoverished, *Colloq* °broke, *US* on the skids, on skid row, on the bum, *Slang Brit* skint: *Those vagrants are down and out and need help, not pity.* —*n.* **2** °derelict, °beggar, °outcast, °tramp, vagrant, °vagabond, *US* °bum: *He was a complete down-and-out till he stopped drinking.*

downfall *n.* °ruin, °undoing, debacle, °collapse, °degradation, °defeat, °overthrow, °breakdown: *Selling the company to the conglomerate spelled its downfall.*

downgrade *v.* **1** demote, dethrone, °humble, °lower, °reduce, °displace, depose, °dispossess, disfranchise *or* disenfranchise, *US military* bust; *Colloq* bring *or Colloq* take down a peg: *He was downgraded from supervisor to foreman.* **2** °belittle, °minimize, °play down, °disparage, decry, denigrate, °run down, *US and Canadian* downplay: *How could she downgrade her own sister?* —*n.* **3** descent, °decline, declension, °(downward) slope, gradient, °grade, °inclination: *Apply the brake as* *you approach the downgrade.* **4** ***on the downgrade.*** °on the wane, waning, declining, falling, slipping, falling off, losing ground, going downhill, *US and Canadian* on the skids: *After the scandal, her popularity was on the downgrade.*

downhearted *adj.* discouraged, depressed, low-spirited, °miserable, °blue, °sad, downcast, °dejected: *Don't be so downhearted, we know you can win the gold medal.*

downpour *n.* rainstorm, deluge, inundation, cloudburst, thundershower, thunderstorm, torrential rain, °torrent; monsoon: *We got caught in that downpour without an umbrella.*

downright *adj.* **1** °direct, straightforward, °plain, °frank, °open, °candid, plain-spoken, °explicit, °blunt, °brash, °bluff, not roundabout *or* circuitous, unambiguous, °out-and-out, °outright, °categorical, °flat, unequivocal, °outspoken, unreserved, °unabashed, unrestrained, unconstrained, °bold: *She speaks with a downright honesty you have to admire.* —*adv.* **2** °completely, °entirely, °totally, °thoroughly, certainly, °surely, (most) assuredly, °definitely, °absolutely, unconditionally, unequivocally; °very, °extremely, unqualifiedly, °perfectly, uncompromisingly, unmitigatedly, °utterly, unquestionably, °profoundly, °undoubtedly, indubitably: *It's downright stupid of you to leave in this weather.*

downtrodden *adj.* subjugated, oppressed, burdened, plagued, afflicted, exploited, overwhelmed, cowed, °overcome, beaten, °abused, mistreated, maltreated, tyrannized, *Colloq* °beat: *This poor, downtrodden wreck of a man had once been on top.*

downward *adj.* **1** declining, sliding, slipping, spiraling, descending, going *or* heading *or* moving down: *This downward trend in the market will soon be reversed.* —*adv.* **2** downwards, down, °below, °lower: *We moved downward, toward the center of the earth.*

doze *v.* **1** Often, ***doze off.*** (take *or* have a) nap, catnap, drowse, °sleep, slumber, *Colloq* snooze, have forty winks, drop *or* nod off, grab some shuteye, *Chiefly Brit* (have *or* take a) zizz, *Brit* kip, *US* catch a few Z's: *I dozed before the fat lady started to sing.* —*n.* **2** °nap, catnap, siesta, °sleep; °rest; *Colloq* snooze, forty winks, shuteye, *Brit* zizz, kip, lie-down: *I'll have a short doze before dinner.*

drab *adj.* °dull, °colorless, °dreary, °dingy, °lackluster, lusterless, °dismal, cheerless, °gray, °somber: *She wore drab clothes and no makeup.*

draft *n.* **1** °plan, sketch, °drawing, °outline, °rough (sketch), blueprint, diagram, °prospectus: *We must have the draft of the new design by morning.* **2** bill of exchange, °check, money order, postal order; letter of credit: *Our customer issued a draft in full payment.* **3** °breeze, °breath (of air), °(light) wind, °current (of air), puff (of air *or* wind): *You'll get a cold sitting in the draft.* **4** °dose, °portion, °measure, °quantity, °drink, °swallow, °sip, °nip, °tot, potation, dram, °gulp, *Colloq* swig, tipple: *The doctor recommended a draft of this tonic before meals.* —*v.* **5** sketch, delineate, °outline, °design, °plan, °frame, °block out, °compose, diagram, °draw (up): *The art department has drafted the layout for the new encyclopedia.*

drag *v.* **1** °pull, °draw, °haul, °tow, °tug, °trail, °lug: *It took the two of us to drag the desk into the other office.* **2** °pull, °distract, °draw; °induce, °persuade, °coax, °wheedle: *She's been unable to drag him away from the TV.* **3** trudge, slog, °crawl, °creep, inch, °shuffle, shamble: *He's looking for a job and just drags along from one employment agency to another.* **4** °trail (behind), °linger, °dawdle, °lag (behind), °straggle, draggle, °putter, loiter, °poke (along), dillydally, *US* lallygag: *She just drags along after us wherever we go.* **5** (be) prolong(ed), (be) extend(ed), °(be)draw(n) out, (be) protract(ed), (be) stretch(ed) out, spin out *or* be spun out: *Why drag out the agony of uncertainty any longer? His speech dragged on for another hour.* **6** ***drag one's feet*** or ***heels.*** °delay, °procrastinate, hang back; °obstruct, °block, °stall: *The committee is dragging its feet on the housing issue.* —*n.* **7** °bore, °nuisance, °annoyance; °pest; *Colloq* °drip, °pain (in the neck), °headache: *That course in botany is a real drag.*

drain *n.* **1** ditch, °channel, trench, culvert, conduit, °pipe, gutter, °outlet, watercourse, sewer, cloaca: *The storm drains have overflowed.* **2** depletion, reduction, sapping, °sap, °exhaustion, °strain, °drag; outgo, outflow, withdrawal, disbursement, °expenditure: *The cost of the new roof was a drain on our resources.* **3** ***down the drain.*** wasted, gone, thrown away, °lost, *Slang* °up the spout: *All that money spent on his education went down the drain.* —*v.* **4** °draw off, °tap, °extract, °remove, take away, °withdraw, pump off *or* out; °empty, °evacuate, drink up *or* down, quaff, °swallow, °finish: *After washing the lettuce, drain off the water. He drained the glass in one gulp.* **5** °consume, °use up, °exhaust, °sap, deplete, bleed, °strain, °tax, °spend; °weaken, debilitate, °impair, °cripple: *The car repairs drained my bank account. After climbing to the top of the cliff, we were completely drained.* **6** seep, °trickle, °ooze, °drip, °leave, go *or* flow from *or* out of, °disappear (from), °ebb: *Let the pus drain from the boil. The blood drained from his face when he saw her.*

drama *n.* **1** °play, stage play, photoplay, screenplay, °(stage) show, °(theatrical) piece, °(stage) production; °scenario: *He plays only in dramas, never in musicals.* **2** dramaturgy, stagecraft, theater art(s), thespian *or* histrionic art(s), acting, °theater, dramatic art: *She studied drama at the Royal Academy.* **3** histrionics, dramatics, theatrics, theatricalism, play-acting: *There's always a drama over who's going to wash the dishes.*

dramatic *adj.* **1** theatric(al), dramaturgic(al), °thespian, histrionic, °stage: *This volume contains Shakespeare's dramatic works only. There will be a festival of dramatic arts at the center next week.* **2** °vivid, °sensational, °startling, breathtaking, °sudden, °striking, °noticeable, °extraordinary, °impressive, °marked, °shocking, °expressive, °graphic, °effective; °complete, °considerable, °radical, °major: *A dramatic change has come over him since meeting her.* **3** °flamboyant, °melodramatic, colorful, °showy, °stirring, spectacular; °theatrical, histrionic, exaggerated, overdone: *His presentation was quite dramatic, well staged and with much arm-waving.*

dramatist *n.* °playwright, dramaturge, screenwriter, scriptwriter, scenarist, tragedian, melodramatist: *The actors failed to carry out the dramatist's intentions.*

dramatize *v.* °exaggerate, overplay, °overstate, °overdo, make a production *or* show (out) of, *Colloq* °lay it on (thick), °pile it on, ham (something *or* it) up: *He always dramatizes everything way out of proportion.*

drape *v.* **1** °hang, festoon, °swathe, deck, array, bedeck, adorn, °ornament, °decorate: *The coffin was draped with the national flag.* —*n.* **2** °drapery, curtain; hanging, tapestry: *The drapes match neither the carpet nor the wallpaper.*

drapery *n.* °drape, curtain; hanging, valance, pelmet, tapestry, arras, portière, lambrequin, °drop: *Which color will you choose for the drapery?*

drastic *adj.* °violent, °severe, °extreme, °strong, °powerful, °potent, puissant, °fierce, °forceful, °vigorous, rigorous, °harsh, °radical, Draconian, °desperate, dire: *I shall have to take drastic measures if this misbehavior continues.*

draw *v.* **1** °pull, °tug, °tow, °drag, °haul, °lug: *The gypsy caravan was drawn by two horses.* **2** pull *or* take out, °extract; unsheathe, unholster: *When he said he'd have to draw out my teeth, I drew my pistol.* **3** °draw off; °pour; drain off *or* out: *She drew two pails of water for the horses.* **4** °attract, °gather, allure, °lure, bring out *or* forth, °elicit, *Colloq* °pull: *Anything will draw a crowd in New York.* **5** depict, sketch, °portray, °outline, delineate, °design, limn, °paint: *The artist was drawing pictures in chalk on the pavement.* **6** °devise, °draw up, °draft, °create, contrive, °frame, °compose, °prepare: *The plans for the new civic center have not yet been drawn.* **7** °inhale, °breathe (in), °inspire; suck in: *She's very ill and may draw her last breath any minute.* **8** °draw out, °withdraw, °take, °receive, °get, °acquire, °obtain, °secure, °procure, °extract, °remove: *I have to draw some money from my bank account for groceries.* **9** °choose, °pick, °select, °take: *It is your turn to draw a card.* **10** ***draw back.*** °retreat, °recoil, °shrink (from), °withdraw: *He drew back quickly when he saw the snake.* **11** ***draw in.*** °arrive, °pull in: *The train drew in to the station.* **12** ***draw off.*** **(a)** °tap, °pour: *The barmaid drew off two large beers from the keg.* **(b)** °withdraw, draw *or* go away, °depart, °leave: *The Indians drew off and waited to see what we would do.* **13** ***draw on.*** **(a)** °employ, °use, make use of, °exploit, have resort *or* recourse to, °resort to, °fall back on, rely *or* depend on: *She drew on her years of experience as a doctor.* **(b)** come close *or* near, °near, draw nigh, °approach, °advance: *With the cold season drawing on, we had to get in the crops.* **14** ***draw out.*** **(a)** °extend, drag out, °prolong, protract, °lengthen, °stretch, °spin out: *Her visit has been drawn out to a week.* **(b)** °elicit, °evoke, induce to talk: *I drew him out on his feelings about social security.* **(c)** See **8**, above. **15** ***draw up.*** **(a)** °halt, °stop, pull up *or* over: *A taxi drew up and I got in.* **(b)** °draft, °compose, °prepare, °put down (in writing), °frame, °compile, put together, °formulate: *We drew up the agreement only yesterday.* **(c)** °arrange, deploy, °position, °order, °rank, marshal: *The troops were drawn up in full battle array.* *—n.* **16** °magnetism, °attraction, °lure, °enticement, *Colloq* °pull, drawing power: *The draw of the rock concert was extraordinary.* **17** °tie, °stalemate, dead heat, °deadlock: *The race ended in a draw for second place.*

drawback *n.* °disadvantage, °hindrance, °stumbling block, °obstacle, °impediment, °hurdle, °obstruction, °snag, °problem, °difficulty, °hitch, °catch, °handicap, °liability, °flaw, °defect, °detriment, *Colloq* °fly in the ointment; *Taboo* nigger in the woodpile: *Lack of education is a serious drawback to getting a good job.*

drawing *n.* °picture, depiction, °representation, sketch, °plan, °outline, °design, °composition, black-and-white, monochrome: *The book is illustrated by some delightful pen-and-ink drawings.*

drawn *adj.* °haggard, °worn out, °tired, °fatigued, °strained, pinched, °tense, °exhausted: *Sidonia looks a bit drawn after her ordeal.*

dread *v.* **1** °fear, be afraid of, apprehend, °anticipate, °flinch, shrink *or* recoil from, cringe *or* quail *or* blench *or* wince at, view with horror *or* alarm: *She dreads any kind of surgery.* *—n.* **2** °fear, °fright, fearfulness, trepidation, apprehension, apprehensiveness, uneasiness, °anticipation, °alarm, nervousness, °qualm, queasiness, °misgiving, °dismay, °worry, °anxiety, consternation, °concern, °distress, perturbation, disquiet, °aversion, °horror, °terror, °panic, *Colloq* cold feet, butterflies (in the stomach), the jitters; *Slang* the heebie-jeebies, the willies, the collywobbles: *I regarded the history exam with dread.* *—adj.* **3** feared, dreaded, °dreadful, °terrifying, °terrible: *Before us, breathing fire, was the dread dragon of the Druids.*

dreadful *adj.* **1** °bad, °awful, °terrible, *Colloq* °rotten, *Slang* °lousy: *That TV soap opera is simply dreadful.* **2** °grievous, dire, °horrible, horrendous, horrifying, horrid, °monstrous, °fearful, feared, °frightful, °dread, °frightening, °shocking, alarming, appalling, °fearsome, °hideous, °ghastly, °atrocious, heinous, °wicked, °evil, iniquitous, °villainous, flagitious, °fiendish, diabolical, °devilish, demonic, malevolent, maleficent, malefic, *Colloq* °scary: *They did the most dreadful things to political prisoners.*

dream *n.* **1** °reverie, °daydream, °delusion, °fantasy, °hallucination, °illusion, °vision, mirage, pipe dream, °(flight of) fancy, °speculation: *When I awoke I realized that my winning the lottery had just been a dream.* *—v.* **2** °imagine, °fancy, conjure up, hallucinate: *I dreamt I dwelt in marble halls.*

dreamer *n.* fantasizer, °visionary, idealist, °romantic, romanticist, idealizer, utopian; daydreamer, escapist, stargazer: *If you think people change, you're a dreamer.*

dreamlike *adj.* °unreal, °fantastic, °unbelievable, phantasmagoric(al), hallucinatory *or* hallucinative *or* hallucinational, surreal, delusionary *or* delusional, illusionary *or* illusional, delusive *or* delusory, illusory *or* illusive, insubstantial *or* unsubstantial, °imaginary, chimeric(al), °fanciful, °fancied, °visionary: *His plans have a dreamlike quality about them that make them impractical.*

dreamy *adj.* **1** °dreamlike, °vague, °indefinite, °indistinct, undefined, °intangible, °misty, °shadowy, °faint: *He has a dreamy recollection of being awakened in the middle of the night.* **2** °absent-minded, °absent, °faraway, abstracted, °pensive, °thoughtful; daydreaming, musing, occupied, in a reverie, in a brown study, in the clouds; *Colloq* off somewhere: *I was in a dreamy mood, my mind wandering through old memories.* **3** relaxing, °soothing, calming, lulling, °gentle, °tranquil, °peaceful, °peaceable, °quiet; °lazy, °sleepy, °drowsy: *It was one of those dreamy, hot midsummer days.*

dreary *adj.* **1** °dismal, °joyless, cheerless, °gloomy, °bleak, drear, °somber, °doleful, depressing, °wretched; °sad, °melancholy, downcast, depressed, °funereal, °glum, °unhappy, °forlorn, °mournful, morose, °blue, °miserable: *One more day on these dreary moors and I shall go mad. Carpathia was again in a dreary mood.* **2** °boring, °lifeless, °colorless, ennuyant, °drab, °dull, arid, °dry, uninteresting, °dead, °monotonous, °prosaic, °tedious, °tiresome, tiring, wearisome, wearying, °humdrum, °ordinary, °vapid, °run-of-the-mill, unstimulating, unexciting: *Do you mean to tell me that that dreary book is a bestseller?*

dregs *n.pl.* **1** °sediment, °grounds, lees, °deposit, °residue, solids, °remains; °precipitate: *Filter the coffee to remove the dregs.* **2** outcasts, pariahs, °rabble, °riffraff, scum, tramps, down-and-outs, losers: *That park is frequented by the dregs of society.*

drench *v.* °soak, °saturate, °wet, °flood, inundate, °immerse, °drown: *She had no coat or umbrella and got completely drenched in the storm.*

dress *v.* **1** °clothe, put on (clothing *or* clothes), attire, °apparel, °outfit, fit out, garb, accoutre *or US also* accouter; array, bedeck, deck out, °rig out, smarten up: *They dressed him to look like a prince.* **2** array, °equip, adorn, °decorate, deck out, °arrange: *He has a job dressing shop windows.* **3** bandage, °treat, medicate, °doctor: *After dressing my wound they gave me a sedative.* **4** ***dress down.*** °reprimand, °scold, °berate, °castigate, °rebuke, °reprove, °upbraid, *Colloq* °tell off, haul (someone) over the coals, *US and Canadian* °chew out, *US* rake (someone) over the coals, tee off on (someone), *Brit* tear (someone) off a strip: *The colonel dressed us* *down and canceled all leaves.* **5** ***dress up.*** **(a)** put on dinner *or* formal clothes, put on one's (Sunday) best (clothes), *Colloq* put on one's best bib and tucker *or* one's glad rags: *On the cruise, we dressed up in our dinner jackets every night.* **(b)** °(put on a) costume, °disguise, °masquerade, °camouflage, put on fancy dress: *The children dressed up as goblins for Hallowe'en.* —*n.* **6** frock, gown, °outfit, °costume, *Colloq* get-up: *Why not wear your new dress to the dance tonight?*

dressmaker *n.* seamstress, °tailor, couturier *or* couturière, modiste: *She's at the dressmaker's having a ball gown fitted.*

dressy *adj.* **1** °formal, dressed-up, °elegant, °fancy, °chic: *A black suit is too dressy to wear tonight—it's not a dressy party.* **2** °elegant, °smart, °stylish, *Colloq* classy, ritzy, *Brit* °swish: *That's a very dressy outfit, I must say!*

drift *v.* **1** °coast, °float, °waft: *A log drifted by on the tide.* **2** °wander, °roam, °meander, °stray, rove, °ramble, *Colloq* mosey: *He seems just to drift through life, without a purpose.* —*n.* **3** °trend, °tendency, °direction, °course, °current, °bias, °inclination, °flow, °sweep, °bent: *The drift of the conversation seemed to be towards politics.* **4** °intention, °meaning, purport, °purpose, °aim, °object, °tenor, °tone, °spirit, °color, °essence, °gist, °significance, °import: *Offended by the drift of her remarks, I excused myself.* **5** °accumulation, °pile, °heap, °mass, bank, °mound, dune: *After the snowstorm, a huge drift blocked the door.*

drifter *n.* vagrant, °tramp, °vagabond, beachcomber, rambler, wanderer, *Colloq* knight of the road, *US* °bum, hobo: *A drifter, he had no place to call home.*

drill *v.* **1** °bore, °penetrate, °pierce, cut a hole: *The thieves drilled into the safe.* **2** °rehearse, °train, practise, °exercise, °teach, °instruct, °school, °tutor, °coach, °indoctrinate; °discipline: *We were thoroughly drilled in the Latin conjugations and declensions.* —*n.* **3** auger, °(brace and) bit, gimlet: *The bit for this drill is no longer sharp.* **4** °practice, training, °repetition, °exercise, °rehearsal; °discipline: *Tomorrow there will be a complete drill of the parts of speech.*

drink *v.* **1** quaff, imbibe, °sip, °gulp, °swallow, °swill, guzzle, toss off, °lap (up), *Colloq* wet one's whistle, swig, knock back, *US* °belt: *She prefers not to drink beer.* **2** tipple, °nip, °indulge, tope, chug-a-lug, °carouse, *Colloq* °booze, bend the elbow, hit the bottle, go on a binge *or* bender, drown one's sorrows, *US and Canadian* go on a toot, *Chiefly Brit* pub-crawl: *He threatened to leave her if she continued to drink.* **3** ***drink to.*** °toast, °salute, °celebrate, °pledge: *Let's drink to friendship!* —*n.* **4** beverage, potation, liquid refreshment, °liquid, potable, °draft: *After the match I was dying for a drink.* **5** °alcohol, °spirits, °liquor, the cup that cheers; stirrup cup; *Colloq* °booze, °the bottle, hard stuff, mother's ruin, eye-opener, nightcap, *US* hooch; *Slang* rotgut, *US* the sauce, red-eye: *After the accident, he took to drink.* **6** °tot, °nip, draft *or chiefly Brit* draught, schooner, pint, bumper, jigger, snifter, °sip, °taste, °glass, °gulp, °swallow, *Scots* (wee) deoch-an-doruis *or* doch-an-doris, (wee) dram, *Brit* sundowner, cordial, highball, shot, after-dinner drink, liquor, brandy, cognac, beer, wine; *Colloq* snort, slug, swig: *Granny likes a drink before retiring.* **7** ***the drink.*** the sea, the ocean, the main, °the deep, *Nautical* Davy Jones's locker, *Colloq* the briny: *The canoe tipped and our picnic went right into the drink!*

drip *v.* **1** dribble, °trickle, °drop; drizzle, sprinkle: *The tap began to drip and kept me up all night.* —*n.* **2** dribble, °trickle, °drop, dripping: *Yes, it was the drip from the tap that kept me awake.* **3** °milksop, °bore, wet blanket, °killjoy, damper, *Colloq Brit* °wet, weed, *Colloq* wimp, *Slang* °pill, °drag, *US and Canadian* milquetoast: *Must you invite that drip George?*

drive *v.* **1** °push, °propel, impel, °urge, °press, °thrust, °move, °motivate, actuate, °prod, °spur, goad, °urge, °force, °make, compel, coerce, constrain, °oblige, pressure *or Brit* pressurize, high-pressure, °induce, °require; °demand: *What drove you to become a traitor?* **2** °operate, °conduct, °maneuver, °manipulate, °handle, °steer, °control; °pilot: *Have you a valid license to drive this car?* **3** °ride, °travel, motor, °go, °move, °proceed, °journey, °tour, *Colloq* tool along: *Luckily, when the tire blew out, we were driving at only 20 mph.* **4** °stab, °plunge, °thrust, °sink, °push, °send, °dig, °ram: *He has driven the dagger deep into the monster's heart.* **5** °herd, drove, °shepherd, ride herd (on): *We used to drive the cattle up the old Chisholm Trail to market in Abilene.* **6** ***drive at.*** °hint (at), °suggest, °imply, °intimate, allude *or* refer to, °intend, °mean, have in mind, °indicate, *Colloq* °get at: *He was so naive he had no idea what she was driving at.* —*n.* **7** °ride, °trip, °outing, °journey, °run, °tour, °excursion, *Colloq* °spin, whirl: *On Sundays we would go for a drive in the country.* **8** °energy, °effort, °impetus, °vigor, vim, °spunk, °enterprise, °industry, °initiative, °ambition, ambitiousness, °determination, °persistence, °urgency, zeal, °enthusiasm, keenness, aggressiveness, *Colloq* get-up-and-go, °pep, zip, °push, °hustle: *She owes her success to her drive as well as her talent.* **9** driveway, °approach, (private) road *or* street, lane, byway, °(scenic) route: *The drive up to the house is lined with trees.* **10** °campaign, °effort, °appeal, °crusade: *The club has had a successful membership drive this year.*

drivel *v.* **1** dribble, drool, slobber, °slaver: *You're driveling all over the front of your shirt!* **2** °babble, prate, °prattle, gibber, °jabber, burble, gabble, °chatter, blether *or US* blather, *Colloq* gibber-jabber, °gab, *US* run off at the mouth, *Brit* rabbit *or* witter *or* natter on: *She keeps driveling on about her operation.* —*n.* **3** °gibberish, °rubbish, °(stuff and) nonsense, twaddle, balderdash, hogwash, *Colloq* eyewash, tripe, °garbage, malarkey, hooey, °hot air, bosh, baloney *or* boloney, *Slang* crap, bull, bilge (water), codswallop, *US* horsefeathers, *Taboo* bullshit, balls, *Brit* (load of old) cobblers: *I've never heard so much drivel from a candidate in my entire life!*

droop *v.* **1** °sag, °hang (down), °wilt, °dangle: *Flags drooped in the windless heat.* **2** languish, °weaken, °flag, °wilt, wither, be limp, °slump, °sag: *Halfway through the marathon she began to droop a bit.*

drop *n.* **1** globule, bead, °drip, droplet, °tear: *A drop of sweat hung from his nose.* **2** °bit, °spot, °particle, °taste, dram, °sip, °nip, °pinch, °dash, °dab, *Colloq* smidgen: *Add a drop or two of milk before kneading the dough.* **3** descent, °fall: *There's a sheer drop of a thousand feet from this ledge into the chasm below.* **4** °decline, °slope, fall-off, drop-off, declivity, °incline: *The drop in the road is about 15 feet in a hundred.* —*v.* **5** °drip, °trickle, dribble: *As the water drops, filling the tube, the float rises.* **6** °fall, °descend, °sink, drop away *or* down *or* off, °dive, °plunge, °plummet, °decline, °collapse: *The barometer dropped 10 millibars in 10 minutes. Near that rock, the road drops to the beach. At the first shot, we dropped to the ground.* **7** °desert, °forsake, °give up, °abandon, °leave, °quit, °throw over, °jilt, °discard, °reject, °repudiate, °renounce, *Colloq* chuck, ditch, °dump; °relinquish, let go, °discontinue, °stop, °cease, °end: *After what he said, she dropped him like a hot potato. I wish you'd drop the subject of my disability.* **8** °release, let go of, °shed, °cast off, °discard, doff: *The deciduous trees drop their leaves.* **9** °omit, °leave out, °exclude, °eliminate: *To avoid confusion with his father, he dropped his middle initial.* **10** °dismiss, let go, °fire,

°discharge, oust, *Colloq chiefly Brit* °sack, give (someone) the sack: *They dropped her after a week's trial.* **11** °decline, °decrease, drop *or* fall off, °diminish, slacken, slack *or* taper off, °subside, lessen: *Demand for swimsuits drops during the winter.* **12** ***drop in (on).*** °visit, °call (on), °pop in (on), °come by, stop in: *Viola dropped in for tea yesterday.* **13** ***drop out.*** °withdraw (from), °leave; rusticate, °depart, decamp, go away *or* off, °take off, °turn off: *She dropped out of school. After winning the award, Crater dropped out and hasn't been seen since.*

drown *v.* **1** °flood, inundate, °swamp, deluge, °drench, °immerse, °submerge, engulf: *The village was completely drowned in the tidal wave.* **2** °overwhelm, °overcome, °overpower, engulf, °swamp, deluge, inundate: *We were almost drowned by the responses to our advertisement.*

drowsy *adj.* °sleepy, heavy-lidded, °groggy, somnolent, dozy, oscitant; °nodding, yawning; °torpid, sluggish, °tired, °weary, °listless, °lethargic, °lazy: *We all felt a bit drowsy after that big dinner.*

drudgery *n.* toil, °labor, moil, travail, °(hack) work, donkey-work, chore, slog, slogging, °slavery, *Colloq* °grind, °sweat, *Brit* skivvying, °fag: *She wanted some relief from the sheer drudgery of housework.*

drug *n.* **1** medication, °medicine, medicament, pharmaceutical, °remedy, °cure, °treatment; cure-all, panacea: *My doctor prescribes too many drugs.* **2** opiate, °narcotic, °stimulant, °tranquilizer, antidepressant, hallucinogen(ic), psychedelic, hypnotic, soporific, °sedative, analgesic, pain-killer, *Slang* °dope, downer, °upper: *Can they control the traffic in drugs?* —*v.* **3** °dose, medicate, °treat: *I was drugged with antihistamines and was unable to drive.* **4** anesthetize, °dope, °deaden, °knock out, °sedate, stupefy, °numb, benumb, °dull, narcotize; °poison, *Slang* slip (someone) a Mickey (Finn): *The victim had been drugged and kidnapped.*

druggist *n.* °pharmacist, apothecary, *Brit* chemist: *Only a druggist is qualified to dispense this medication.*

drunk *adj.* **1** drunken, intoxicated, inebriated, besotted, tipsy, °groggy, sotted, crapulent *or* crapulous, in one's cups, °under the weather, °under the influence, °maudlin, ebriate, ebriose, ebrious, *Colloq* soused, pickled, °high (as a kite), °tight, boozed, boozy, lit (up), half seas over, three *or* four sheets to the wind, °out (cold), under the table, *Brit* squiffy; *Slang* pie-eyed, °loaded, stoned, stewed (to the gills), (well-)oiled, bombed (out of one's mind), crocked, plastered, tanked, sloshed, polluted, stinko, smashed, blotto, pissed, *Taboo slang chiefly US* shit-faced: *He was so drunk he tried to fly.* **2** exhilarated, °excited, °exuberant, invigorated, inspirited, °animated, °ecstatic; flushed, °feverish, °inflamed, aflame, °fervent, fervid, °delirious: *Since he became a director, he's been drunk with power.* —*n.* **3** drunkard, drinker, toper, tippler, sot, °soak, bibber, winebibber; dipsomaniac, °alcoholic, problem drinker; *Colloq* guzzler, swiller, sponge, *Slang* wino, boozer, dipso, °lush, souse, alky, *US* juicer, juice head, rummy: *The drunks who volunteered were registered for treatment.* **4** °carouse, bacchanal, carousal, bacchanalia, °revel, *Slang* °bender, °tear, °jag, bat, *US and Canadian* toot, *Chiefly Brit* pub-crawl: *I went off on a wild drunk the night before my wedding.*

drunkenness *n.* intoxication, insobriety, intemperance, sottishness, bibulousness, inebriety, crapulence, crapulousness, tipsiness, ebriety; dipsomania, alcoholism, ebriosity; *Colloq* boozing, *Slang* hitting the bottle *or US* the sauce: *Only a psychiatrist could help cure his drunkenness.*

dry *adj.* **1** dehydrated, desiccated, arid, sear, parched, waterless, moistureless; °barren, °bare, °fruitless: *With no rain for a month, the dry earth yielded no crops.* **2** °dreary, °boring, °tedious, °tiresome, wearisome, wearying, tiring, °dull, uninteresting, °monotonous, °prosaic, commonplace, °stale, uninspired; °plain, °unadorned, unembellished: *The minister's speech was as dry as could be, a litany of dry statistics.* **3** °witty, droll, °wry, cynical, °biting, °sarcastic, °cutting, °keen, °sly, ironic: *Oscar Wilde was known for his dry witticisms.* —*v.* **4** dehydrate, desiccate, °parch: *As the rainfall subsided, the land dried and changed into a desert.* **5** dry up *or* out, wither, °shrivel, °shrink, °wilt: *The plants dried because they weren't watered.*

duck *v.* **1** bob, °dodge, °dip, °dive, °stoop, °bow, °bend, °crouch: *I ducked to avoid hitting my head on the beam.* **2** °plunge, °submerge, °immerse, dunk: *In the pool, she ducked me when I least expected it.* **3** °avoid, °sidestep, °evade, °dodge, °elude, °shun, °steer clear of, shy away from; °shirk: *He is known for ducking his responsibilities.*

dud *n.* **1** °failure, *Colloq* °flop, lead balloon, lemon, °washout, *Colloq US and Canadian* dog, clinker: *Her second novel certainly proved a dud--it sold only ten copies.* —*adj.* **2** °worthless, valueless, °broken, unusable, °useless, inoperative, nonfunctioning, inoperative, malfunctioning, *Colloq* kaput, bust(ed), *Brit* °duff: *They deliberately supplied us with dud ammunition.*

dude *n.* **1** °dandy, fop, fancy dresser, Beau Brummell, popinjay, boulevardier, man about town, *Archaic* coxcomb, macaroni; *Slang* °swell, *Brit* toff: *He dressed like a real dude—zoot suit and all.* **2** °man, °fellow, °chap, *Colloq* °guy: *Hey, man, who's that dude in the tartan suit?*

due *adj.* **1** °payable, owed, owing, °unpaid, °outstanding, in arrears: *The rent is due tomorrow.* **2** °fitting, °right, °rightful, °correct, °proper, °appropriate, apropos, apposite, °suitable, apt, °meet; °deserved, (well-)earned, merited, °just, justified: *Was she treated with due respect?* **3** °necessary, needed, °adequate, °sufficient, °enough, °satisfactory; °ample, plenty of: *I do not think my case was given due consideration.* **4** expected, scheduled, anticipated: *He was due on the two o'clock plane.* —*adv.* **5** °directly, °exactly, °precisely, °straight: *Go due east to the river, then turn north.*

dues *n.* pl. (membership) fee, charge(s): *If you have not paid your dues, you may not use the club's facilities.*

duff *adj.* °bad, °useless, °worthless, unworkable, inoperable, inoperative, °broken; °fake, °false, °counterfeit, *Colloq* °dud, phony *or Brit also* phoney: *We couldn't get that duff radio to work. They were provided with duff papers for crossing the border.*

duffer *n.* °incompetent, blunderer, bungler, oaf, *Colloq* ox, lummox: *He may be an expert at computers but he's a duffer at golf.*

dull *adj.* **1** °stupid, slow-witted, °dense, °stolid, bovine, cloddish, clodlike, °backward, °obtuse, doltish, crass, °dumb, *Colloq* °thick, °dim, dimwitted, *Brit* dim as a Toc H lamp: *He might be a dull student but he's a brilliant artist.* **2** insensitive, °numb, °insensible, imperceptive *or* impercipient, unresponsive, °indifferent, unfeeling, °unsympathetic, °callous, hardened, °hard, inured, obtundent: *He knew that he could expect only a dull response to his pleading.* **3** °lifeless, °indifferent, unresponsive, sluggish, °slow, °listless, °inactive, °torpid: *The market for luxury cars is a little dull now.* **4** °boring, °tiresome, °tedious, °monotonous, uninspired, uninspiring, unoriginal, uninteresting, °humdrum: *All work and no play make Jack a dull boy.* **5** °dismal, °dreary, depressing, °somber, °gray, °dark, °murky, °gloomy, cloudy, clouded, °overcast, °sunless: *If the day is dull, the photographs will show it.* **6** blunted, °blunt; °obtuse: *I nicked myself with that dull razor.*

7 °hazy, blurry, °opaque, °drab: *Rub the dull film off that silver goblet.* **8** muffled, numbing, deadened, muted, °indistinct: *I've had a dull pain in my arm all day.* —*v.* **9** allay, assuage, °relieve, °mitigate, lessen, °reduce: *Weeping dulls the inner pain.* **10** °dim, °tarnish, °obscure, bedim, °blur, cloud, becloud: *A mist dulled the rich colors of the glen.* **11** stupefy, narcotize, °numb, benumb, desensitize, °deaden, °blunt, obtund: *His war experiences had dulled his feelings towards others.*

duly *adv.* **1** °properly, fittingly, deservedly, °appropriately, suitably, befittingly, rightly, correctly, °accordingly: *Those elected were duly installed in office.* **2** punctually, °on time: *The train duly arrived.*

dumb *adj.* **1** °mute, °speechless, voiceless; °silent, °quiet, °taciturn, °mum, wordless; °inarticulate: *She was struck dumb with astonishment.* **2** °dull, °stupid, *Colloq* °thick: *He's too dumb to understand what you are saying.*

dumbfound *v.* dumfound, °amaze, °shock, °surprise, °startle, °astonish, °astound, °bewilder, °stagger, °stun, °floor, °nonplus, °confuse, confound, *Colloq* flabbergast, bowl over: *Their offer for the house dumbfounded us.*

dumbfounded *adj.* dumfounded, amazed, shocked, surprised, startled, astonished, astounded, bewildered, staggered, floored, nonplussed, overwhelmed, °speechless, stunned, °thunderstruck, dazzled, dazed, dumbstruck, taken aback, °confused, confounded, bemused, perplexed, baffled, °disconcerted, *Colloq* bowled over, flabbergasted, knocked out, thrown (off), *US* thrown for a loss, *Brit* knocked for six, knocked sideways: *She is dumbfounded that he proposed marriage.*

dummy *n.* **1** mannequin, manikin *or* mannikin, °model, °figure: *I saw the coat on a dummy in the shop window.* **2** °sample, °copy, reprint, °reproduction, °likeness, °substitution, °imitation, °sham, mock-up, simulation, *Colloq* phoney *or US also* phony: *Those aren't the real crown jewels, they're just dummies.* **3** °fool, idiot, dunce, blockhead, ninny, ass, °dolt, numskull *or* numbskull, simpleton, *Colloq* dimwit, *US* thimble-wit: *They're such dummies they don't know that you're joking.* **4** *US* pacifier: *Give the baby the dummy to suck.*

dump *v.* **1** °unload, offload, °empty, °drop, °deposit, throw *or* fling down, °tip: *They dumped the topsoil all over the walk.* **2** °get rid of, °throw away, °scrap, °discard, ditch, jettison, °dispose of, °reject, °tip, toss out *or* away, *Colloq* °junk, chuck out *or* away: *We dumped all the food when the fridge broke down.* —*n.* **3** junkyard, rubbish heap, *US* garbage dump, *Brit* °tip: *You'll have to take this old icebox to the town dump.*

dumpy *adj.* °stocky, pudgy, squat, chunky, °chubby, tubby, °stout, °plump, portly, °fat: *No one with a dumpy figure looks good in shorts.*

dun *v.* °press, importune, °solicit, °plague, °nag, °pester, *Slang US* °bug: *The gas company has been dunning me to pay the bill.*

dung *n.* manure, °muck, droppings, cowpats, fertilizer, guano, excrement, faeces *or US chiefly* feces, *US* cow *or* buffalo chips, horse apples, *Slang* cowflop, *Taboo* shit, horseshit, cowshit: *The dung is used as a fertilizer on the fields.*

dungeon *n.* donjon, °keep, °cell, °prison, lockup, oubliette, black hole, °stronghold: *Throw the infidels into the dungeon and give them twenty lashes!*

dupe *n.* **1** °fool, gull, °victim, fair game, *Colloq* chump, *Chiefly US and Canadian* fall guy; *Slang* °sucker, °sap, boob, °pushover, pigeon, °mark, *Brit* °mug, *Chiefly US and Canadian* patsy: *Swindlers often choose tourists as likely dupes.* **2** cat's-paw, °pawn, °tool, °puppet, *Slang* stooge: *I'm not going to be the dupe in your little game!* —*v.* **3** °deceive, °fool, °outwit, °cheat, °trick, °take in, °defraud, humbug, °hoax, °swindle, °hoodwink, bilk, gull, cozen, delude, °mislead, snooker, °victimize, *Colloq* bamboozle, flimflam, put one over on, pull a fast one on; *Slang* con, °rip off, rook, °take, *US and Canadian* snow, do a snow job on: *She was duped into believing she had won the lottery.*

duplicate *adj.* **1** °identical; °twin, °matching: *They sent me duplicate tickets by mistake.* —*n.* **2** (exact *or* carbon) copy, photocopy, machine copy, °double, clone, °(perfect) match, °facsimile, °twin, °reproduction, °replica, replication, look-alike, *Trademark* Xerox (copy), *Slang* (dead) ringer: *This painting looks like a duplicate of the one you bought.* —*v.* **3** °copy, photocopy, clone, °match, replicate, °imitate, °reproduce, °double, *Trademark* Xerox; °repeat, °equal: *Would you please duplicate this letter for me? Can he duplicate his performance in the Commonwealth Games?*

durable *adj.* °enduring, long-lasting, °stable, wear-resistant, heavy-duty, hard-wearing, long-wearing, °lasting, °persistent, °indestructible, °substantial, °sturdy, °tough, °stout, °strong, °firm, °sound, °fixed, °fast, °permanent, dependable, °reliable: *The product is durable, guaranteed to last a lifetime.*

duress *n.* **1** coercion, °threat, °pressure, constraint, compulsion; °force, °power: *The boys wash the dishes only under duress.* **2** confinement, °imprisonment, incarceration, °captivity, °restraint, *Literary* durance: *There were workhouses, prisons, and other forms of duress.*

dusk *n.* °twilight, sundown, nightfall, °evening, sunset, °dark, eventide: *The workers came in from the fields at dusk.*

dusky *adj.* **1** °dark, °black, ebony, sable, jet-black; °swarthy, swart, dark-complected, dark-complexioned: *'Dusky diamonds' is another name for coal. A dusky gentleman offered to see her home safely.* **2** °shadowy, °shady, °dim, °dark, unilluminated, unlit, °murky, subfusc, subfuscous, °gloomy, °obscure: *An ominous figure was lurking in the dusky area under the stairs.*

dutiful *adj.* °obedient, compliant, °willing, °obliging, filial, °faithful, °conscientious, °reliable, °responsible, °diligent, °attentive, punctilious, °respectful, °polite, °considerate, deferential, °submissive, °yielding, acquiescent, malleable, °flexible, pliant, °accommodating, *Formal, archaic* duteous: *A dutiful son, he visits his parents weekly.*

duty *n.* **1** °responsibility, °obligation, °burden, onus, devoir, °office, °work, °task, °assignment, °job, °stint, chore, °occupation, °calling, °function, °role, °part, °bit, °charge: *Every man is expected to do his duty.* **2** °respect, °deference, °loyalty, fealty, fidelity, faithfulness, allegiance: *I think she did it out of a sense of duty to her family.* **3** °tax, excise, °tariff, impost, levy, °customs: *You will have to pay duty on that whisky.*

dwarf *v.* °overshadow, °dominate, °diminish, °minimize: *The new tower dwarfs the older buildings.*

dwell *v.* **1** reside, °abide, °live, °lodge, °stay, °remain, °rest, *Formal* °domicile: *After the father's death, the mother dwelt with her daughter.* **2** ***dwell on*** *or* ***upon.*** harp on, °persist in, °emphasize, °stress, focus on, linger *or* tarry over, °elaborate (on); °labor: *Why must you always dwell on a person's shortcomings?*

dwelling *n.* °abode, habitation, dwelling-place, °house, °domicile, °lodging, °quarters, °home, °residence, homestead: *His dwelling is just a shanty in old shantytown.*

dwindle *v.* °diminish, °decrease, °shrink, lessen, °wane, °fade, °contract, condense, °reduce, °peter out,

°waste away, die out *or* down *or* away, °ebb, °decline, °subside, °taper off, shrivel (up *or* away): *The last days of summer dwindled away. His funds dwindled till today he has nothing.*

dying *adj.* expiring; °sinking, slipping away, °going, °failing, fading (fast), at death's door, on one's deathbed, with one foot in the grave, *in extremis;* °moribund: *The doctor said the dying man was in no pain.*

dynamic *adj.* dynamical, °vigorous, °active, °forceful, °energetic, °potent, °powerful, high-powered, °lively, spry, °vital, °electric, °spirited, zealous, °eager, °emphatic: *We are seeking a dynamic salesman for the Reading office.*

dynamism *n.* °energy, °vigor, °pep, °vitality, liveliness, °spirit, spiritedness, forcefulness, °power, °drive, °initiative, °enterprise, *Colloq* get-up-and-go, zip, °push: *That woman has the dynamism needed to get ahead in this organization.*

dynasty *n.* °line, °family, °heritage, °house: *The Ming dynasty ruled China for more than 300 years.*

E

eager *adj.* avid, zealous, °ardent, °earnest, °keen, °enthusiastic, °hot, °hungry, °fervent, fervid, °passionate, °spirited, inspirited, °energetic, energized, vehement, °animated, °excited, vitalized, stimulated; °desirous, yearning, desiring, craving, °wanting, °longing, itchy, °impatient; °anxious; *Colloq* °dying, *Slang US* hot to trot: *We were particularly eager to spend our holiday in Spain.*

eagerness *n.* **1** avidity, zeal, earnestness, keenness, °enthusiasm, °fervor, °hunger, vehemence, °animation, °vitality, °appetite, °zest, °relish, °spirit, spiritedness, °gusto, °verve, °dash, élan, vim, °vigor, °energy, *Colloq* get-up-and-go, zip, °go: *Such eagerness for learning is rare. If her eagerness could be properly channelled she'd make a great executive.* **2** °desire, °longing, wishing, yearning: *In his eagerness to please everyone he satisfies no one.*

eagle-eyed *adj.* °sharp-eyed, sharp-sighted, keen-eyed, keen-sighted, lynx-eyed, hawk-eyed; °perceptive, perspicacious, discerning, °sharp, watchful, °alert: *It would be impossible to deceive our eagle-eyed supervisor.*

ear *n.* **1** °attention, °heed, °notice, °regard, °consideration: *See if you can get his ear for a moment between meetings.* **2** °sensitivity, °appreciation, °taste, °discrimination: *She has an excellent ear for the right expression.*

early *adv.* **1** beforehand, °ahead (of time), °prematurely: *I arrived too early and had to wait.* **2** anciently, initially, °originally, at *or* near the start *or* beginning: *Plants appeared early in the development of life forms on earth.* **3** betimes, at cockcrow *or* cock's crow, at (the crack *or* break of) dawn, at daybreak: *You're up early this morning!* —*adj.* **4** untimely, °premature; °inopportune, °inappropriate: *The early fruit isn't as sweet.* **5** °initial, °beginning, °original, °first, °pioneer, advanced: *He was one of the earliest writers on the subject.* **6** primeval, °primitive, primordial, °ancient, °old, °prehistoric, antediluvian, °original; °antique, °antiquated: *The Olduvai Gorge has yielded up many early humanoid fossils.*

earn *v.* **1** °merit, °deserve, be worthy of, be entitled to, °win, °warrant, °rate, qualify for, have a claim *or* right to: *Pinckley has earned everyone's respect.* **2** °make, °pocket, °gross, °net, °clear, °realize, °receive, °get, °procure, °collect, °reap, °bring in, take home; °draw, *Colloq US* °pull down: *Carple earns more because he has an education.*

earnest *adj.* **1** °serious, °solemn, °grave, °sober, °intense, °steady, °resolute, resolved, °firm, °determined, assiduous, °sincere, dedicated, committed, °devoted, °thoughtful: *He made an earnest promise to do his best.* **2** zealous, °ardent, °diligent, assiduous, °industrious, hard-working, °devoted, °eager, °conscientious, °keen, °fervent, fervid, °enthusiastic, °passionate: *Burbridge is an earnest pupil.* **3** ***earnest money.*** °deposit, down payment, binder, handsel, °guarantee, °security, °pledge: *The company paid $10,000 earnest money to secure the bid.* —*n.* **4** ***in earnest.*** °serious, °sincere: *She said she would come, but I doubt whether she was in earnest.*

earnings *n.* pl. wages, °salary, °income, compensation, °pay, °stipend, emolument, °proceeds, °return, °revenue, °yield, takings, *Slang* °take: *The interest from tax-free bonds was not included in my earnings.*

earth *n.* **1** °globe, mother earth, planet, °world, blue planet, terra: *The earth turns at about 1,000 miles per hour on its axis.* **2** °soil, °dirt, loam, sod, clay, °turf, °ground, °mold: *Pack the earth firmly around the roots.*

earthly *adj.* **1** °terrestrial, terrene, telluric: *Extraterrestrial beings might not survive in our earthly atmosphere.* **2** °worldly, mundane, °material, °materialistic, °physical, nonspiritual, °sensual, °carnal, fleshly, corporeal, °base, °natural: *He has forsaken earthly pleasures.* **3** °human, °temporal, °secular, °profane, °mortal, °physical, nonspiritual, °material: *His earthly remains were committed to the sea.* **4** conceivable, imaginable, °feasible, °possible: *What earthly reason could she have had for kissing me?*

earthy *adj.* ribald, °bawdy, °unrefined, °coarse, °crude, °shameless, °wanton, °uninhibited, °abandoned, °vulgar, °lusty, °rough, °dirty, °indecent, °obscene: *She found Henry Miller's books a bit too earthy for her tastes.*

ease *n.* **1** °comfort, °repose, well-being, °relaxation, °leisure, °rest, contentment, calmness, tranquillity, °serenity, peacefulness, °peace, peace and quiet: *After 50 years of hard work, she felt entitled to a few years of ease.* **2** easiness, °simplicity, °facility, effortlessness, adeptness: *He passed the other runners with ease.* **3** affluence, °wealth, °prosperity, °luxury, opulence, °abundance, °plenty: *He has always led a life of ease, never having had to work.* **4** naturalness, informality, unaffectedness, ingenuousness, casualness, artlessness, insouciance, nonchalance, aplomb; unconcern: *Note the ease with which she moves among her constituents.* —*v.* **5** °comfort, °relax, °calm, °tranquilize, quieten, °still, pacify, soothe, disburden: *Learning his son was safe certainly eased his mind.* **6** lessen, °diminish, abate, °mitigate, °reduce, °decrease, allay, alleviate, assuage, mollify, appease, palliate, °quiet, °relieve: *Her anxiety was considerably eased by the news of the successful moon launch.* **7** °maneuver, °manipulate, inch, °guide, °steer, °slip: *The helmsman eased the ship into her dock.* **8** °facilitate, °expedite, °simplify, °smooth, °further, °clear, °assist, °aid, °advance, °forward, °help: *Having a wealthy father eased her way in life.*

easily *adv.* **1** smoothly, effortlessly, °readily, °simply, °handily, without a hitch, °hands down, without even trying, comfortably, with no *or* without difficulty, *Colloq* easy as pie: *She easily passed all her exams.* **2** °by far, beyond *or* without (a *or* the shadow of) (a) doubt,

°without question, indisputably, indubitably, °undoubtedly, doubtless(ly), unquestionably, °clearly, far and away, °definitely, definitively, conclusively, certainly, °surely, undeniably, °obviously, patently: *He is easily the best lawyer in the firm.* **3** °probably, most *or* very likely, °well, almost certainly: *We may easily be the first in Hampton to have plaster flamingos on the lawn.*

easy *adj.* **1** °simple, °effortless, °plain, °clear, straightforward, hands-down, uncomplicated, °elementary, °foolproof; easy as pie, easy as ABC, easy as falling off a log, easy as 1, 2, 3, easy as can be: *Feeding goldfish is normally quite an easy job. Copying that painting turned out to be easy.* **2** °carefree, °easy-going, °casual, °lenient, undemanding, °relaxed, °quiet, °serene, °restful, °tranquil, °peaceful, untroubled, undisturbed, unoppressive, °gentle, °mild, °calm, °comfortable, °cozy, °unhurried, leisurely: *He has a pretty easy life now that he's retired.* **3** °light, °lenient, undemanding, °mild, °flexible, °indulgent, °tolerant: *You really should be easy on him after what he's been through.* **4** °tractable, pliant, docile, compliant, °submissive, acquiescent, amenable, °accommodating, °soft, °suggestible, credulous, °trusting, °weak, °easy-going: *He was an easy victim for the confidence tricksters. She has a reputation of being of easy virtue.* **5** unstrained, °gentle, °moderate, °unhurried, leisurely, °even, °steady, undemanding, °comfortable, unexacting: *We kept up an easy pace of about five miles a hour.* **6** affable, °friendly, °amiable, °amicable, °agreeable, °outgoing, °informal, unceremonious, down-to-earth, unreserved, relaxing, °natural, °relaxed, °easy-going: *We found them easy to be with.* —*adv.* **7** effortlessly; calmly, unexcitedly, temperately, peacefully, tranquilly, serenely, nonchalantly, casually: *Take it easy and don't get all worked up.*

easy-going *adj.* °relaxed, °casual, °mellow, °carefree, undemanding, °easy, even-tempered, forbearing, °lenient, °tolerant, °permissive, overtolerant, overpermissive, °lax, °weak, *Colloq* °wishy-washy, laid-back: *If you're too easy-going, the students won't respect you.*

eat *v.* °dine, lunch, breakfast, sup, break bread, °snack, have a bite; °consume, °devour, take (in) nourishment, *Colloq* put *or* pack away, *Slang* nosh, put *or* tie on the nosebag *or US and Canadian* feedbag: *I've already eaten and I'm not hungry, thank you.*

eavesdrop *v.* listen in, °tap, overhear, °snoop; °spy, °pry: *That couple was eavesdropping on our conversation.*

ebb *v.* **1** °recede, flow back, °subside, °go out, °go down; fall back *or* away, °retreat, retrocede, °retire: *The tide ebbed and left the boat stranded.* **2** °decline, °flag, °decay, °wane, °diminish, °decrease, °drop, slacken, °fade (away), °drain (away), °dwindle, °peter out, °waste (away), °deteriorate: *His enthusiasm for exercise is beginning to ebb.* —*n.* **3** low tide, low water, ebb tide, low point: *The rocks appear when the sea is at its ebb.* **4** °decline, °decay, °decrease, diminution, °wane, °drop, slackening (off), dwindling, lessening, deterioration, degeneration: *She was no longer able to contend with the ebb and flow of his temper.*

ebullient *adj.* bubbling, overflowing, °effervescent, °excited, °effusive, exhilarated, °elated, °buoyant, °exuberant, °enthusiastic, zestful: *The crowd was ebullient at the news from Mafeking.*

eccentric *adj.* **1** unconventional, °unusual, uncommon, idiosyncratic, anomalous, °unorthodox, °out of the ordinary, °irregular, atypical, °incongruous, errant, aberrant, °exceptional, °individual, °singular, °unique; °abnormal, °odd, °peculiar, °strange, °curious, °bizarre, °outlandish, °queer, °quaint, quirky, °weird, °offbeat, *Colloq* far-out, °kinky, °cranky: *Yes, I would agree that walking a canary is a bit eccentric.* —*n.* **2** °original, °individualist, °nonconformist, queer fellow, odd fish, *Colloq* °character, °card, °freak, °(nut) case, °crank, oddball, weirdo *or* weirdie, *US* oner: *In my opinion, anyone who isn't a carbon copy of his neighbor is called an eccentric.*

eccentricity *n.* **1** unconventionality, unusualness, uncommonness, irregularity, nonconformity, individuality, individualism, °singularity, uniqueness, strangeness, oddness, bizarreness, distinctiveness, capriciousness, weirdness: *Why should someone be criticized for eccentricity?* **2** idiosyncrasy, °quirk, °peculiarity, °mannerism, crotchet, aberration, anomaly, °oddity, °curiosity, caprice: *Eating crackers in bed is only one of her eccentricities.*

echo *n.* **1** reverberation, °repercussion, °repetition, iteration, reiteration: *The echo of the church bells could be heard throughout the valley.* **2** °imitation, °copy, replica *or* replication, duplication, °reproduction, simulation, °facsimile; °reflection, mirror image, °repetition: *Modern Rome is but a feeble echo of its glorious past.* —*v.* **3** °resound, reverberate, °ring: *The chapel echoed with children's laughter.* **4** °imitate, ape, °parrot, °mimic, °copy, °duplicate, °reproduce, simulate, °repeat, emulate, °mirror, °reflect: *These poems are unoriginal and merely echo others' works.*

eclipse *v.* **1** °conceal, °hide, °blot out, °obscure, °block, °veil, °shroud, °cover, darken: *A black cloud eclipsed the moon.* **2** °overshadow, °obscure, °surpass, °top, outshine: *His career was eclipsed by his wife's brilliant successes.* —*n.* **3** concealment, covering, hiding, blocking, blockage, occultation, obscuring, obscuration, darkening, shading, dimming: *Though good may suffer an eclipse, it can never be extinguished.* **4** °decline, downturn, °slump; °recession: *After the scandal, her career went into a ten-year eclipse.*

economic *adj.* **1** °financial, °fiscal, pecuniary, °monetary, budgetary; commercial, °mercantile, °trade: *The economic indicators for July affected the markets.* **2** °profitable, cost-effective, money-making, remunerative, °productive; °solvent: *Demand for our product has made the company economic.*

economical *adj.* **1** cost-effective, money-saving, °thrifty, unwasteful; °cheap, °inexpensive, °reasonable; °economic: *This car is very economical to run.* **2** °provident, °thrifty, °sparing, economizing, °prudent, °conservative, °frugal; parsimonious, °penurious, stingy, °cheap, °miserly, niggardly, °tight, closefisted, tightfisted, °mean, penny-pinching, scrimping: *Aunt Gertrude was always a very economical housekeeper.*

economize *v.* °save, °cut back, °husband, retrench; tighten one's belt, cut corners *or* costs, scrimp, skimp, °pinch pennies: *Without the extra income, we'll have to economize on something.*

economy *n.* **1** °thrift, husbandry, thriftiness, °conservation, conservatism, °saving, °restraint, °control, frugality: *We'll have to exercise economy to get through the winter.* **2** °brevity, briefness, succinctness, terseness, conciseness, concision, compactness, °restraint, curtness: *He manages to get his ideas across with an admirable economy of language.*

ecstasy *n.* **1** °delight, °joy, °rapture, °bliss, °transport, nympholepsy *or* nympholepsia, °happiness, gladness, elation, °pleasure, °enjoyment, °gratification; heaven on earth: *The prospect of being with her again filled me with ecstasy.* **2** exaltation, °frenzy, °thrill, elation, °paroxysm, °excitement: *The ecstasy of space flight is impossible to describe.*

ecstatic *adj.* exhilarated, thrilled, °exultant, blissful, euphoric, °rapturous, enraptured, nympholeptic, °enchanted, transported, °rhapsodic, °excited, °elated,

°delighted, °joyful, °gleeful, °overjoyed, °happy, °glad, °beside oneself, °delirious, orgasmic, *Colloq* on cloud nine, *Brit* over the moon, in the seventh heaven, cock-a-hoop, *US* in seventh heaven, flying: *Our team had won the Olympic gold medal and we were truly ecstatic.*

eddy *n.* **1** °swirl, whirl, vortex, gurgitation; °whirlpool, maelstrom, Charybdis; dust devil, °whirlwind, °twister, tornado, cyclone, typhoon, °hurricane; waterspout: *These treacherous eddies are caused by the tidal change.* —*v.* **2** °swirl, whirl, °turn, °spin: *The wind eddied round us, driving the dinghy this way and that.*

edge *n.* **1** °brink, °verge, °border, °side, °rim, lip, °brim; °fringe, °margin, °boundary, °bound, °limit, bourn, °perimeter, °periphery: *I was hanging on to the very edge of the cliff. This is the edge of my property.* **2** acuteness, sharpness, keenness: *That hunting knife has quite an edge.* **3** harshness, sharpness, acrimony, pungency, °force, °urgency, effectiveness, incisiveness, causticity, °virulence, vehemence: *There was an edge to her voice when she told me not to bother her.* **4** °advantage, head start, °superiority, °lead, °upper hand: *You're holding the gun, so I guess you have the edge on me.* **5** ***on edge.*** on tenterhooks, °nervous, °touchy, °sensitive, °prickly, itchy, °tense, irascible, crabbed, °irritable, °peevish, apprehensive, with one's heart in his mouth, edgy, °anxious, °ill at ease, °restive, °restless, fidgety, *Colloq* uptight, like a cat on a hot tin roof: *She's on edge waiting for the exam results.* —*v.* **6** inch, °move, °sidle, °crawl, °creep, °steal, worm, °work (one's way): *The burglar was edging along the wall, 30 storeys above the street.*

edible *adj.* eatable, esculent, palatable, good *or* fit (to eat), °wholesome, *Rare* comestible: *After the 'sell by' date, fresh food is not edible.*

edification *n.* enlightenment, °improvement, uplifting, enlightening, °guidance, °education, °information, °tuition, teaching, °schooling, °instruction: *For the edification of your parents, please tell us why you are a day late for dinner.*

edit *v.* **1** redact, copy-edit, rewrite, rephrase, °modify, °alter, °adapt, °change, °revise, °correct, emend, °style, restyle, °polish, °touch up: *His job is to edit the stories into idiomatic English.* **2** Often, ***edit out.*** blue-pencil, °cut (out), °delete, censor, °erase, bleep, blip; bowdlerize, expurgate, °clean up: *They edited out all the dirty words before broadcasting the interview.* **3** °cut, condense, compress, °shorten, crop, °reduce: *The boss said to edit this story to fit on one page.* **4** °prepare, °compile, °assemble, °select, °arrange, °organize, °order, reorganize, reorder: *Gumthorpe edited the collected letters of Penelope Shinlappe.*

edition *n.* °number, °issue, printing, print run; °copy; °version: *The first Monday edition carried the news of the break-in, and a later edition reported the capture of a suspect.*

editor *n.* rewrite man *or* woman, rewriter, copy editor, redactor, reviser, °writer, columnist, °journalist, editorial writer, *Brit* leader writer; editor in chief, managing editor, senior editor; compiler, °collector: *The editor sent the reporter out to check the facts of the story.*

editorial *n.* *Brit* °leader, leading article; op-ed article, think piece, opinion piece, position statement; °essay, article, column: *Did you see the editorial about the Middle East situation in today's paper?*

educate *v.* °teach, °train, °instruct, edify, °tutor, °school, °inform, °enlighten, °indoctrinate, inculcate, °coach, °drill, °prepare, °ready, °rear, °bring up, °cultivate, °develop, °civilize: *It is the responsibility of the state to educate the people.*

educated *adj.* **1** °cultivated, cultured, erudite, well-read, °lettered, °literary, °scholarly, °learned; (well-)informed, °knowledgeable, °enlightened: *They might seem ignorant to an educated person like you. I haven't got the answer, but I could make an educated guess.* **2** °refined, °polished, °cultivated, civilized; discerning, °critical, °sensitive: *She has an educated palate and really knows her wines.*

education *n.* **1** teaching, °schooling, training, °instruction, °tuition, tutelage, °edification, tutoring, cultivation, °upbringing, indoctrination, drilling: *My education was in the hands of monks till I was twelve.* **2** °learning, °lore, °knowledge, °information, erudition: *Education is gained from the experience of others.* **3** °lesson, °course (of study): *The inhabitants gave her an education in how to survive in the wilderness.*

educational *adj.* **1** °academic, scholastic, pedagogical, instructional: *This is one of the best educational centers in the world.* **2** °informative, °instructive, enlightening, edifying, eye-opening, revelatory, educative: *A year spent abroad is always educational.*

eerie *adj.* °frightening, °weird, °strange, uncanny, °ghostly, °spectral, °dreadful, °unearthly, °frightful, *Literary and Scots* eldritch, *Scots* °mysterious, *Colloq* °scary, creepy, spooky: *I had the eeriest feeling I was being watched.*

effect *n.* **1** °result, consequence, °outcome, conclusion, °upshot, aftermath, °impact: *The effects of the storm could be seen everywhere.* **2** effectiveness, efficacy, °force, °power, °capacity, potency, °influence, °impression, °impact, *Colloq* clout, °punch: *The threats have no effect.* **3** °significance, °meaning, signification, purport, °sense, °essence, °drift, °implication, °import, °tenor, °purpose, °intent, °intention, °object, °objective: *She said 'Get lost', or words to that effect.* **4** °impact, °impression, °form, °sensation: *He didn't mean that, he just said it for effect.* **5** ***in effect.*** effectively, °virtually, for (all) practical purposes, °so to speak, more or less; °actually, in (point of) fact, °really, essentially, basically, °at bottom, °in truth, °truly, °to all intents and purposes, at the end of the day, any way you look at it: *Because the tax is included in the retail price, the consumer in effect pays it.* **6** ***take effect.*** become operative *or* operational, come into force, begin *or* start to work *or* function *or* operate: *The regulation doesn't take effect till next week.* —*v.* **7** °bring about, °cause, make happen *or* take place, °effectuate, °achieve, °accomplish, °secure, °obtain, °make, °execute, °carry out, °produce, °create: *The opposition was unable to effect any change in the law.*

effective *adj.* **1** °effectual, °efficacious, °productive; °capable, °useful, °serviceable, °competent, °operative, °able, °functional, °efficient: *The effective life of the battery is six months.* **2** °impressive, °remarkable, °noticeable, °conspicuous, °outstanding, °striking, °powerful, compelling, °moving, °telling, °effectual: *The stage setting was very effective.* **3** °operative, operational, °in operation, functioning, °real, °actual, °essential, °basic, °true: *The monthly interest may seem low, but the effective annual rate is much higher.*

effects *n.* pl. belongings, °(personal) property, °gear, °possessions, °stuff, °things, °paraphernalia, chattels, °goods, *Colloq* °junk, crap, *Brit* clobber, *Taboo slang US* shit: *Gather up all your effects and get out of my house!*

effectual *adj.* **1** °effective, °efficacious, °efficient, °functional, °productive, °useful, °telling, °influential, °powerful, °forcible, °forceful; °capable, °adequate: *What is the most effectual way of stopping a takeover of the company?* **2** °effective, °in force, °legal, °lawful, binding, °sound, valid: *You must register the agreement for it to be effectual.*

effectuate *v.* °bring about, °effect, °carry out, °implement, °accomplish, do, °execute, °realize, °achieve; °cause, make happen: *It is unclear just how she will effectuate her escape.*

effeminate *adj.* unmanly, womanish, womanly, sissyish, °weak, campy; °gay, °homosexual; *Slang US* limp-wristed, faggy, faggoty, *Brit* °bent, poncy: *He's too effeminate to play the pirate lead.*

effervescent *adj.* **1** bubbling, fizzy, carbonated, sparkling, fizzing, gassy; foaming, foamy, frothing, frothy, °bubbly: *I prefer the effervescent water to the still.* **2** bubbling, °bubbly, high-spirited, °vivacious, °ebullient, °lively, °exuberant, °buoyant, °animated, °lively, exhilarated, °excited, °enthusiastic, °irrepressible: *Jeanette's effervescent personality is infectious.*

efficacious *adj.* °effective, °effectual, °productive, °competent, °successful, °efficient, °useful, °serviceable; °capable: *He was not very efficacious in getting the council to change its policy.*

efficiency *n.* **1** effectiveness, efficacy, competence, °capability, adeptness, °proficiency, expertness, °expertise, know-how, °experience, °skill, skillfulness, °dexterity, adroitness: *The efficiency of the staff has been greatly improved.* **2** productivity, effectiveness, efficaciousness: *How does the efficiency of a diesel compare with that of other engines?*

efficient *adj.* unwasteful, °economic, °thrifty; °effective, °efficacious, °effectual, °competent, °productive, °proficient, °operative: *The technique for producing electrical power from tidal action has not yet proved efficient.*

effort *n.* **1** °exertion, striving, °struggle, °strain, °labor, °pains, °energy, toil, °application, °trouble, travail, °work, *Colloq* elbow grease: *He's gone to a lot of effort to please her. Getting the place cleaned up took a lot of effort.* **2** °attempt, °endeavor, °essay, °try, °venture, *Colloq* °stab, °crack: *Her effort to be pleasant was unsuccessful.* **3** °achievement, °accomplishment, °creation, °feat, °deed, attainment, °exploit: *Last year's fundraising fair was a superb effort.*

effortless *adj.* easy (as pie *or* as A, B, C, *or* as 1, 2, 3), °simple, °painless, °smooth, trouble-free, uncomplicated: *You have to admire the effortless way she plays the cello.*

effrontery *n.* °impertinence, °impudence, audacity, °nerve, °presumption, presumptuousness, brazenness, boldness, insolence, temerity, brashness, rashness, °arrogance, °front, °indiscretion, *Archaic* frowardness, *Colloq* °gall, °brass, °nerve, cheek, lip, °mouth, *Slang Brit* °side: *He had the effrontery to call her "Queenie"!*

effusive *adj.* °demonstrative, gushing, (over)enthusiastic, unrestrained, unchecked, unreserved, °expansive, °emotional, °exuberant, °rhapsodic, °ebullient, °lavish, °voluble, °profuse; fulsome: *Her fans greeted her with an effusive outpouring of affection.*

egoistic *adj.* egoistical, self-centered, egocentric, narcissistic, self-seeking, self-absorbed, °selfish, self-serving, °self-indulgent, °self-important, solipsistic: *She is too egoistic to consider anyone but herself.*

egotistical *adj.* egotistic, °conceited, °proud, overweening, bragging, °boastful, boasting, swelled-headed *or* swellheaded *or* swollen-headed, °vain, vainglorious, self-worshiping, self-admiring, vaunting, crowing: *Considering his sparse accomplishments, he has nothing to be egotistical about.*

eject *v.* **1** force *or* drive out, °expel, oust, °remove, °get rid of, °evict, *Colloq* throw *or* kick *or* boot out, send to the showers: *They were ejected for causing a disturbance.* **2** °expel, °emit, throw up *or* out, °spew (forth), °discharge, °spout, disgorge, vomit (up *or* forth), send out *or* forth; °ooze, exude, extravasate: *The volcano ejected boulders the size of houses.* **3** °discharge, °dismiss, cashier, drum out, °lay off, declare *or* make redundant, *Colloq* °fire, °sack, boot out, ax, give the sack *or* boot *or* ax, give (someone) his marching orders *or US also* walking papers, send packing: *He was ejected for stealing paper clips.*

ejection *n.* **1** °expulsion, casting out *or* up, disgorgement, vomiting forth, throwing out *or* up, °discharge, emission, disgorging: *The ejection of lava was preceded by loud rumblings.* **2** °exile, °expulsion, banishment, deportation, ouster, °removal; °eviction, dispossession: *His ejection from the meeting angered his supporters. My ejection by the landlord was illegal.* **3** °dismissal, °discharge, congé, cashiering, layoff, *Colloq* firing, sacking, *Slang* °the sack, the boot, the ax, the (old) heave-ho, *US* the bounce: *Business was bad, and the entire staff was faced with ejection.*

elaborate *adj.* **1** °detailed, °painstaking, °meticulous, punctilious, °comprehensive, °thorough, °complete, °exhaustive, °intricate, °involved, °minute, °precise, °exact: *The escape had been worked out to the last elaborate particular.* **2** °complicated, complex, convoluted, °ornate, °fancy, Byzantine, °laborious, °labored, °extravagant, °showy; ornamented, decorated, baroque, rococo, °busy, °fussy, gingerbread: *We marveled at the elaborate mosaics. His plan is too elaborate and should be simplified.* —*v.* **3** °ornament, °decorate, °complicate, °embellish, garnish, adorn: *Later craftsmen elaborated the earlier Greek motifs.* **4** °enlarge, expand (upon *or* on), expatiate, °develop, °cultivate, °flesh out, °enhance, °refine, °enrich, °improve, °better, ameliorate, emend, °polish: *Later on, I shall elaborate on the recruitment methods we used.*

elaboration *n.* **1** enhancement, °refinement, enrichment, °improvement, amelioration, melioration, betterment; °embellishment, adornment, garnish, garnishment, °decoration, overdecoration, gingerbread, *Slang* bells and whistles: *The customers always welcome elaboration of the equipment.* **2** enlargement, °development, amplification, °expansion: *The argument is subject to elaboration.*

elapse *v.* °pass (by), °go (by), slip by *or* away, °pass away, slide by, glide by, °transpire: *Three weeks have elapsed since I last saw her.*

elastic *adj.* **1** °flexible, stretchable, stretchy, °stretch, bendable, °pliable, springy, °plastic, extensile, extensible, expansible, expandable, contractile, resilient, bouncy, compressible: *Is it elastic enough to stretch round that box?* **2** adjustable, °adaptable, °accommodating, °flexible: *Our schedule is elastic and we can fit in an appointment.*

elasticity *n.* **1** °flexibility, °resilience, rubberiness, plasticity, ductility, springiness, stretchability, stretchiness, °stretch, suppleness, pliancy, *Colloq* °give: *There's enough elasticity in the band to hold the papers.* **2** °flexibility, adjustability, adaptability, °tolerance, suppleness: *My plan has the elasticity needed to accommodate changes.*

elated *adj.* exhilarated, uplifted, °elevated, °gleeful, °joyful, jubilant, joyous, °exultant, °ecstatic, blissful, °happy, °delighted, euphoric, °overjoyed, °excited, thrilled, transported, pleased (as Punch), °on top of the world, on cloud nine, *Colloq* tickled, tickled pink, *Brit* chuffed, over the moon, in the seventh heaven, *US* in seventh heaven: *We were elated to learn that our daughter had won first prize.*

elder *adj.* **1** older, °senior: *My elder brother was born in 1930.* **2** °venerable, respected, °preeminent, °eminent; °experienced, °veteran: *Clemenceau was an elder statesman at the Geneva Convention.* —*n.* **3** °senior,

°superior; patriarch, elder statesman, dean, doyen *or* doyenne: *Everyone paid great respect to the elders of the tribe.*

elderly *adj.* **1** °old, past middle age, oldish, advanced in years *or* age, of advanced age, along in years, °gray, aging *or* ageing, °aged, °venerable; hoary, °ancient, senescent, °decrepit, superannuated; °senile, anile; *Colloq* over the hill, past it, long in the tooth, having one foot in the grave, old-fogyish *or* old-fogeyish: *Today, a person isn't considered elderly till he's past sixty-five.* —*n.* **2** ***the elderly.*** the retired, the old, senior citizens, golden agers, *Chiefly Brit* pensioners, old age pensioners, *Brit* OAP's, *Colloq* old-timers, (old) geezers, (old) fogies *or* fogeys, *Brit* wrinklies: *The elderly constitute an increasingly large body of voters.*

elect *v.* **1** °choose, °select, °pick, °vote (for), °determine, °designate: *We must elect a new chairperson before the next meeting.* —*adj.* **2** chosen, elected, selected, picked out: *The president-elect takes office next month.* **3** °select, °choice, °first-rate, first-class: *An elect few make up the executive committee.*

election *n.* °poll, °vote, referendum, °plebiscite; °selection, °choice, choosing, nomination, designation, °appointment; voting, electing: *We held an election and you lost. The election of a member of the committee is required.*

electioneer *v.* °campaign, °canvass, °support, °back, °promote: *They are electioneering for their candidate.*

electric *adj.* charged, °tense, energized, °stimulating, °exciting, °thrilling, galvanizing, electrifying, °moving, °stirring: *As the jury filed in, the atmosphere in the courtroom was electric.*

electricity *n.* °excitement, °verve, °energy, °tension, tenseness, fervency, °intensity, °ardor; vibrations: *I could feel the electricity between us.*

electrify *v.* **1** °startle, °shock, °stun, °jolt, °stagger, °astound, °jar, °astonish, °amaze: *We were electrified by the news of the disaster.* **2** °excite, galvanize, °animate, °move, °rouse, °stir, °stimulate, °vitalize, °fire, °thrill, °arouse: *Fred was inspired by the electrifying sermon.*

elegance *n.* **1** °refinement, °grace, tastefulness, good taste, gentility, °polish, courtliness, °culture, politeness, politesse, °propriety, °dignity: *Where but in eighteenth-century France could one find such elegance?* **2** °luxury, °grandeur, luxuriousness, sumptuousness, exquisiteness, °splendor, °beauty: *The overwhelming elegance of the palace took our breath away.*

elegant *adj.* **1** °tasteful, °exquisite, °handsome, °beautiful, °comely, °dapper, °smart, well-turned-out; °graceful, °choice, °superior, °fine, °select, °refined, °delicate, discerning, artistic; °dignified, °genteel, °sophisticated, °cultivated, °polished, urbane, Chesterfieldian, °suave, soigné(e), °debonair, courtly, to the manner born, °well-bred, well-born, highborn: *Desmond and Elizabeth are such an elegant couple!* **2** artistic, °stylish, modish, à la mode, °chic, °fashionable, *Colloq* in, with it: *The newly refurbished rooms are very elegant.* **3** °luxurious, °sumptuous, °grand, °opulent, °plush, *Colloq* °posh, swank, °swanky, ritzy, °fancy: *We stayed in a very elegant West End hotel.* **4** apt, °clever, °ingenious, °neat: *Dr. Fingerling devised a truly elegant answer to Knoedel's Theorem.*

element *n.* **1** component, constituent, °ingredient, °essential, °fundamental, °part, °unit, °piece, °segment, °feature, °factor, °detail, °particular: *We must investigate the elements that make up the whole.* **2** °environment, °atmosphere, °situation, °locale, °territory, °sphere, °habitat, °medium, °domain: *Ordway is really in his element at a party.* **3** ***elements.*** **(a)** (adverse *or* unfavorable) weather, climatic conditions: *Stay here tonight—there's no need to brave the elements.* **(b)** °rudiments, basics, fundamentals, foundations, essentials, °principles: *It was she who taught me the elements of flying a helicopter.*

elemental *adj.* °basic, °fundamental, primal, °original, primordial, °primitive: *Elemental religion focused on worship of the sun and fertility.*

elementary *adj.* **1** °simple, °rudimentary, °easy, straightforward, uncomplicated, °clear, understandable, °plain: *The solution to the problem is elementary, but only if one knows how to arrive at it.* **2** °basic, °fundamental, °primary, °introductory, °initial, °beginning; °elemental: *She received her elementary education in France.*

elevated *adj.* **1** raised, upraised, uplifted, lifted (up): *Only three hands were elevated in opposition and the motion was carried.* **2** uplifted, °noble, °lofty, °high, °grand, °exalted, °dignified, °eminent, °preeminent, ennobled, °prominent, °notable, °illustrious, °distinguished, °imposing, °impressive, °sublime: *He has rather elevated notions of morality.* **3** °elated, °cheerful, °happy, exhilarated, °animated, °joyful, °glad: *Her elevated spirits and pleasant countenance are welcome everywhere.*

elevation *n.* **1** altitude, °height: *The elevation of Denver, Colorado, is one mile above sea level.* **2** °swelling, °lump, wen, eminence, °prominence; °hill, °height, °rise: *The doctor noticed a slight elevation in the skin near the eye. Standing on this elevation, you can see for miles.* **3** advancement, °promotion, uplifting, enhancement, °advance: *There is a rumor about her elevation to a peerage.* **4** °grandeur, nobleness, loftiness, exaltation, sublimity, °distinction, °dignity, °refinement, cultivation: *The difficulty of interpretation of his style owes much to its elevation.*

elfin *adj.* **1** elvish, elfish, elflike, impish, puckish, °frolicsome, spritelike, °arch, °playful, °mischievous, °tricky: *I suppose she has a certain elfin charm, but I don't like her.* **2** °small, °wee, °diminutive, °tiny, °little, °dainty, Lilliputian: *Three elfin beings dressed in green stood near my teacup.*

elicit *v.* °draw out, °call forth, °evoke, bring out *or* forth, °bring to light, °extract, wring, wrest, °wrench: *They finally elicited the truth by torture.*

eligible *adj.* **1** °fit, °worthy, °qualified, °proper, °suitable, °appropriate, °fitting: *Is he an eligible candidate for the post?* **2** °single, °unmarried, unwed, °available: *He is one of the few eligible bachelors in the town.*

eliminate *v.* **1** °remove, °exclude, °rule out, °reject, °drop, °leave out, °omit: *She has been eliminated from consideration.* **2** take out *or* away, °omit, °get rid of, °dispose of, °expel, °knock out: *He was eliminated in the first heat of the hundred-meter run.* **3** °erase, eradicate, expunge, °obliterate, °strike (out), cross out *or* off, °cut (out), excise, °delete, °throw out, °edit (out), blue-pencil, °cancel: *The censors have eliminated all references to sex.* **4** °kill, °murder, assassinate, slay, °terminate, °exterminate, °dispose of, liquidate, °finish off, annihilate, °stamp out, °destroy, *Slang* °bump off, °polish off, *US* °rub out, °take for a ride, °bury, ice, °waste: *They used a submachine gun to eliminate the competition.*

elite *n.* **1** élite, °gentry, aristocracy, aristocrats, °elect, upper class, °nobility, privileged class, blue bloods, *crème de la crème, haut monde,* jet set, jet-setters, *US* Four Hundred, F.F.V., First Families of Virginia, *Colloq* °upper crust, beautiful people, *Brit* nobs: *One could scarcely say that Botworth belongs among the elite.* —*adj.* **2** aristocratic, °elect, °upper-class, °privileged, blue-blooded, °noble, °exclusive, °choice, °best, °top: *She socializes with rather an elite group of people.*

elixir *n.* **1** panacea, cure-all, nostrum, wonder drug, miracle drug, sovereign remedy: *Formerly, alchemists sought a universal cure, an elixir.* **2** °essence, °potion, °extract, tincture, compound, °mixture: *Ophidia's Elixir was the name of a patent snake-oil medicine.* **3** °pith, °core, °kernel, heart, °essence, °quintessence, °principle, °extract, °base, °basis, °soul: *The elixir of life is wisdom and its mystic ingredient is not knowledge but understanding.*

eloquent *adj.* **1** °expressive, articulate, silver-tongued, °fluent, well-spoken, °effective, °persuasive, convincing, cogent, °trenchant, °incisive, °graphic, °vivid, °striking, facile, °smooth, °glib, oratorical, °rhetorical: *An eloquent speaker, he rose and gave an eloquent speech.* **2** °suggestive, °meaningful, °pregnant: *His eyebrow was raised in eloquent skepticism.*

elsewhere *adv.* somewhere else, to another place; in another place, °abroad, °absent, away: *Instead of going to church he went elsewhere. She's not in the office, so I assume she's elsewhere.*

elude *v.* **1** °evade, °escape, °avoid, °dodge, slip away from, *Colloq* °duck, give the slip, °shake off: *The suspect has eluded the police for a year.* **2** °evade, °escape; baffle, °puzzle, °confuse, °bewilder, confound; °frustrate, °stump, °thwart: *The point of your argument eludes me.*

elusive *adj.* **1** °evasive, °slippery, °tricky, °shifty: *They sought the elusive Pimpernel everywhere.* **2** °evasive, evanescent, °fleeting, °fugitive, transitory, indefinable, elusory, °intangible, impalpable: *The notion of truth has always been elusive.*

emaciated *adj.* emaciate, atrophied, shriveled, °wizened, shrunken, °haggard, °gaunt, °drawn, pinched, bony, skeletal, cadaverous, withered, wasted, consumptive, phthisic, anorexic *or* anorectic, wasting (away), °scrawny, °skinny, °thin, °lean, °spare, undernourished, underfed, °starved, half-starved: *The inmates of the camp were so emaciated they could scarcely stand.*

emanate *v.* **1** °issue, °come (out), °emerge, °proceed, °flow, °ooze, exude; °radiate: *Black smoke emanated from the mouth of the idol.* **2** °radiate, give off *or* out, send out *or* forth, disseminate, °discharge, °put out, °emit; °exhale, °ooze, exude: *The idol's eyes emanated a blue light.*

emancipate *v.* °release, set free, °liberate, enfranchise, manumit, °loose, °let loose, let go, set free, disenthrall, unfetter, unchain, unshackle; °deliver: *Britain emancipated the slaves almost fifty years before Lincoln's Emancipation Proclamation.*

embargo *n.* **1** °restraint, °block, blockage, °bar, °ban, stoppage, cessation, proscription, °prohibition, interdiction, °check, °restriction, °barrier; °hindrance, °impediment: *The government placed an embargo on the export of oil.* —*v.* **2** °restrain, °block, °bar, °ban, °stop, °cease, proscribe, °prohibit, interdict, °check, °restrict, °hold back, °prevent; °hinder, °impede, °retard, °hold up: *They may embargo oil shipments to drive up the price.*

embark *v.* **1** °board, go aboard; entrain; enplane: *We embarked and the ship sailed.* **2** Often, ***embark on.*** °commence, °begin, °enter (upon), °initiate, °launch, °start, °go into, °set about, take up *or* on, °engage in, °assume, °tackle: *He will embark on the new enterprise next month.*

embarrass *v.* disconcert, °discomfit, chagrin, abash, °shame, °mortify, °humble, °humiliate, discountenance, discompose, °fluster, °upset, °distress, °disgrace, *Colloq* °show up: *He embarrassed his colleagues by his bad manners.*

embarrassed *adj.* **1** °ashamed, °shamefaced, blushing, °disconcerted, discomfited, chagrined, abashed, shamed, mortified, humiliated, discountenanced, discomposed, flustered, distressed, red-faced, uncomfortable, °self-conscious, °sheepish, red in the face; humbled, disgraced: *The master of ceremonies had an extremely embarrassed young lady to deal with.* **2** °in debt, in the red, °straitened, °insolvent, *Colloq* °short, °hard up, °broke, *Brit* skint: *I'm financially embarrassed now but I'll pay you later.*

embarrassing *adj.* °awkward, humiliating, mortifying, shaming, °shameful, uncomfortable, discomfiting, °disconcerting, °touchy, distressing, worrying: *I had a very embarrassing moment when I forgot the client's name.*

embarrassment *n.* **1** bashfulness, awkwardness, clumsiness, discomposure, abashment, uneasiness, °discomfort, self-consciousness, mortification, chagrin: *My embarrassment made me blush.* **2** °difficulty, °mess, °predicament, °dilemma, °problem, °trouble, *Colloq* hot water, pickle, °fix, °scrape, °bind: *Finding us together presented me with a real embarrassment.* **3** °excess, °superfluity, superabundance, overabundance, *embarras de richesse, embarras de choix,* oversupply, °surplus, °profusion: *An embarrassment of options was open to me.*

embellish *v.* **1** °beautify, °improve, titivate *or* tittivate, °dress (up), trick out *or* up, °enhance, °elaborate, °enrich, embroider, gild, furbish (up), garnish, °decorate, adorn, °ornament, deck, bedeck, °trim, °elaborate, caparison, rubricate, varnish; gild refined gold, paint the lily, *Misquote* gild the lily: *The saddles are embellished with silver studs.* **2** °elaborate, °exaggerate, °overdo, embroider, °enhance, °dress up: *His reports are so embellished that you cannot separate fact from fiction.*

embellishment *n.* **1** °decoration, ornamentation, °ornament, °elaboration, adornment, embroidery: *The basic design, which is shoddy, is unimproved by embellishment.* **2** °exaggeration, enhancement, tinsel, garnish, gilding, °frill: *All the embellishments make her story totally unbelievable.*

embers *n.* pl. live coals; cinders, ashes; °remains, remnants: *The dying embers of the fire symbolized her waning love for Darrin.*

embezzle *v.* °misappropriate, peculate, misapply, °misuse, °steal, make off *or* away with, filch, °pilfer, purloin, *Law* defalcate; *Colloq* have one's hand in the till: *The treasurer had embezzled half a million from the company.*

embezzlement *n.* misappropriation, peculation, misapplication, °misuse, misusing, abstraction, °stealing, °theft, thievery, larceny, filching, purloining, pilferage, pilfering, *Law* defalcation: *Wanted for embezzlement, he fled to Brazil.*

embittered *adj.* °bitter, °resentful, °sour, soured, °caustic, acrimonious, acid, envenomed; °angry, choleric, °rancorous: *His conviction for fraud left him an embittered man.*

emblem *n.* badge, insigne, °symbol, °representation, °device, °seal, °crest, °token, °sign; trademark, logotype *or* logo: *The white knight's emblem was a pair of crossed flaming swords.*

emblematic *adj.* emblematical, symbolic(al), °representative, representational: *The white dove is emblematic of peace.*

embodiment *n.* **1** incarnation, °realization, concretization, °manifestation, °expression, personification, materialization, actualization, reification, substantiation: *The golden section is the embodiment of an ancient principle of proportion.* **2** consolidation, °collection, unification, incorporation, °inclusion, integration, °combination, concentration, systematization, °organization, codification, °synthesis, °epitome: *This book is the embodiment of Arnolfo's theories of aesthetics.*

embody *v.* **1** concretize, °realize, °manifest, °express, °personify, °materialize, reify, actualize, externalize, incarnate: *Her ideas of feminism are embodied in her lifestyle.* **2** °exemplify, °typify, °represent, °symbolize, °stand for: *Note how this painting embodies Longchamp's unique notions of form.* **3** consolidate, °collect, °unite, °unify, °incorporate, °include, °integrate, °combine, °concentrate, systematize, °organize, comprise, codify, epitomize, synthesize: *The teachings of the sect are embodied in its scriptures.*

embrace *v.* **1** °hug, °clasp, °grasp, °hold, enfold, °cuddle, cleave together, *Archaic* °clip: *She embraced him warmly when they met again.* **2** °adopt, espouse, take up *or* in, avail oneself of, °use, make use of, °employ, °accept, °receive, °welcome: *They embraced Christian ideals.* **3** °include, comprise, °embody, °incorporate, °comprehend, encompass: *Their tribal territory embraces all areas south of the mountains.* —*n.* **4** °hug, °squeeze, °clutch, *Slang* °clinch: *He gave her a tender embrace and left.*

emerge *v.* **1** °appear, °come out, come forth, °come up, °rise; °arise, °surface, come into view *or* notice, °come to light, be revealed, crop up, °turn out, °develop, become known, become apparent, °transpire, °happen, evolve: *It emerged that she had bought a pistol that morning.* **2** °issue, °emanate, come forth, °proceed: *The salmon emerged at the other end of the fish ladder.*

emergence *n.* °rise, surfacing, °appearance; °development, materialization, °manifestation: *His emergence as a leader was eagerly awaited.*

emergency *n.* °crisis, exigency, °danger, °predicament, °difficulty, °pinch: *In an emergency, put on lifebelts.*

emigrant *n.* emigre *or* émigré, expatriate, displaced person, DP, °refugee, boat person; colonist, °settler: *Many European emigrants sailed to the US.*

emigrate *v.* °migrate, °move, relocate, resettle; °leave, °quit, °depart, °forsake: *Her parents emigrated from Turkey and settled in Scotland.*

eminent *adj.* **1** °distinguished, esteemed, °exalted, respected, revered, honored, °dignified, °notable, °noteworthy, °important, °noted, °outstanding, °prominent, °preeminent, °conspicuous, °superior, °great, °illustrious, °famous, °renowned, °well-known, °celebrated: *McLeod is an eminent meteorologist.* **2** °conspicuous, °outstanding, °marked: *His suggestion made eminent good sense.*

eminently *adv.* °very, °exceedingly, °extremely, exceptionally, remarkably, singularly, °notably, signally: *This man is eminently well suited for his job.*

emit *v.* °discharge, °eject, °expel, °emanate, send out *or* forth, pour out *or* forth, give off *or* out, °issue, °vent, °radiate; °exhale; exude, °ooze: *The factory has been emitting toxic gases into the atmosphere.*

emotion *n.* °feeling, °passion, °sentiment, °sensation: *They say there is a fine line between the emotions of love and hate.*

emotional *adj.* **1** °passionate, °impassioned, °ardent, °enthusiastic, °heated, zealous, °heartfelt, °excited, °fervent, fervid: *She has an emotional reaction to such suggestions.* **2** °tense, °nervous, °excitable, highly strung, high-strung, °temperamental, °volatile, °hotheaded, °demonstrative: *He is a very emotional person who should not be driving a bus.* **3** °sensitive, °warm, °sentimental, °tender, °moving, °poignant, °stirring, emotive, affective, °touching: *Their meeting after fifty years was certainly emotional.* **4** °frantic, °agitated, irrational, °hysterical, °wild, ranting: *She became very emotional when the police took away her son.*

emphasis *n.* °importance, °stress, °significance, °prominence, °attention, °weight, °gravity, °force, °moment, °preeminence, °priority, underscoring, underlining, *Technical* paralipsis: *He places too much emphasis on the social aspect of school.*

emphasize *v.* °stress, accentuate, °accent, underscore, °point up, underline, call *or* draw attention to, °mark, highlight, °play up, °spotlight, °feature: *The new procedures emphasize safety.*

emphatic *adj.* °firm, uncompromising, °determined, °decided, °resolute, dogged; °earnest, °definite, unequivocal, unambiguous, °distinct, °dogmatic, °categorical, °peremptory, °explicit, °incisive, °insistent, affirmative, °positive, °sure, °certain, unmistakable *or* unmistakeable, °specific, °definitive, °direct; °forceful, °vigorous, °energetic, °assertive, °intense; °express, °pronounced, °strong: *She was emphatic about leaving.*

empirical *adj.* empiric, experiential, °practical, observed, pragmatic, °experimental: *He has been there and has empirical knowledge of the system.*

employ *v.* **1** °hire, °engage, °enlist, °recruit, °enroll, °sign (up), °take on, °retain, commission: *I have employed a solicitor to look after my affairs while I am away. The company employed 120 engineers.* **2** °use, make use of, utilize, °apply: *We plan to employ the most modern equipment.* **3** °occupy, °take (up), °engage, °involve: *He is employed with his stamp collection.*

employee *n.* °worker, staff member, wage earner; °hand: *The employees are on strike.*

employer *n.* **1** °proprietor, °owner, °patron, °manager, °director, °chief, °head, *Colloq* °boss, *Brit* gaffer, governor, *Eye dialect* guv'nor, guv: *My employer comes in late every day.* **2** °company, °firm, corporation, °business, °establishment, °organization, *Colloq* °outfit: *She took her employer to court for unfair practices.*

employment *n.* **1** °occupation, °job, °trade, °work, °business, °profession, °vocation, °calling, livelihood, °pursuit, métier, °skill, °craft, *Colloq* °line, *Slang* °racket: *My employment for years has been as a furniture polisher.* **2** °hire, hiring, °engagement, °engaging, employing, taking on, retaining, enlistment, enlisting: *The employment of two hundred people by the new firm will help the town.* **3** °use, utilization, °application, °operation, implementation: *That job involves the employment of specialized equipment.*

emptiness *n.* **1** voidness, hollowness, vacantness, °vacancy, vacuity, blankness, bareness, barrenness, °desolation, desertedness, vacuum, °void: *He was again alone in the vast emptiness of space. After she left, I felt a terrible emptiness.* **2** senselessness, meaninglessness, pointlessness, aimlessness, purposelessness, futility, uselessness, worthlessness, hollowness: *The emptiness of the candidate's words was apparent to all.* **3** vacuity, vacuousness, vacantness, blankness, expressionlessness, emotionlessness: *The emptiness of her facial expression told me that she hadn't understood a word.*

empty *adj.* **1** °void, unfilled, °hollow, °bare, °barren, °vacant, unfurnished, °unadorned, undecorated; emptied, drained, °spent, °exhausted: *That empty space on the wall needs a painting.* **2** °vacant, unoccupied, °uninhabited, untenanted: *He finally found an empty apartment with two bedrooms.* **3** °deserted, °desolate, °uninhabited, °wild, °waste, °bare, °barren; forsaken: *He wandered on the empty moor.* **4** trivial, °insincere, °hypocritical, °hollow, °cheap, °worthless, valueless, °meaningless, °insignificant, °insubstantial, unsatisfying, °idle: *His promises were merely words, empty* *words.* **5** °vacant, °blank, deadpan, expressionless, poker-faced; vacuous, fatuous, °stupid, °foolish, °inane: *He looked at me with that empty expression of his.* **6** °blank, °clean, °new, °unused, °clear: *On an empty page write your name and the date.* **7** ***empty of.*** devoid of, lacking (in), °wanting, in want of, deficient in, °destitute of, without, sans: *Their hearts are empty of compassion.* —*v.* **8** °clear, °remove, take out *or* away, °put out, cast *or* throw out, °eject; °vacate, °evacuate; °dump, °drain, °exhaust, pour out, °void, °discharge, °unload: *Thieves emptied everything out of the house. The police are emptying the building because of the bomb scare. Empty these bottles outside.*

enable *v.* **1** °qualify, °authorize, °entitle, °permit, °allow, °sanction, °approve, empower, °license, commission, °entrust, depute, °delegate, °assign, °charter, franchise: *An act was passed to enable them to build the railway.* **2** capacitate, °facilitate, °expedite, °help, °aid, °assist: *The grant enabled me to continue my studies.* **3** °permit, °allow, °approve, assent to, °go along with, °agree to, give the go-ahead *or* green light, *Colloq* okay *or Colloq* O.K.: *Her press pass enables her to get through police lines.*

enact *v.* **1** °pass, legislate, °ratify; ordain, °decree, °rule, °command, °order, °authorize: *The law was enacted to protect consumers.* **2** °act (out), °represent, °play, °portray, depict, °perform, appear as: *She enacted the role of a modern Lady Macbeth.*

enchant *v.* **1** °bewitch, cast a spell on, ensorcell *or* ensorcel, spellbind, °hypnotize, mesmerize, voodoo, *Brit* °magic, *US* hex, *Colloq* hoodoo: *Circe enchanted Ulysses' men and turned them into swine.* **2** °charm, °fascinate, °beguile, °captivate, enthrall, °enrapture, °attract, allure, °delight, °entrance: *With her sultry looks she has enchanted the most cynical of men.*

enchanted *adj.* pleased, °delighted, °happy, thrilled, *French* enchanté(e): *I was enchanted to meet you at last, Mrs. Calabash.*

enchanting *adj.* beguiling, bewitching, entrancing, °spellbinding, charming, fascinating, captivating, intriguing, °enthralling, alluring, °delightful, hypnotic, °attractive, appealing, winsome, °ravishing, °seductive: *Caesar found Cleopatra utterly enchanting.*

enchantment *n.* **1** witchcraft, °sorcery, °magic, wizardry, thaumaturgy, conjuration *or* conjury; °spell, °charm, °jinx, *US* hex: *When she worked her enchantment on me, I succumbed completely.* **2** °charm, beguilement, allure, °fascination, °rapture, mesmerism, °bliss: *She was completely carried away by the enchantment of the moment.*

encircle *v.* °surround, gird, °circle, °enclose, °ring, encompass, compass, confine, hem *or* hold in; wreathe: *When the horde encircled the castle we thought we were doomed.*

enclose *v.* **1** inclose, confine, °shut in, close *or* hem in, °surround, °pen, °encircle, encompass, °bound, °envelop, hedge in, wall in, immure, fence in *or* off, *US and Canadian* corral: *A high wall encloses the garden.* **2** °insert, °include, °contain; °wrap: *The check was enclosed in the same envelope.*

enclosure *n.* **1** °fold, °pen, cote, °run, sty, yard, farmyard, barnyard, courtyard, quadrangle *or* quad, °square, °compound, *Brit* °close, *US and Canadian* corral: *We had trouble keeping the dogs in the enclosure.* **2** °fence, °wall, °rail, railing, °barrier, hedge, barricade, °boundary: *The buildings serve as an enclosure.*

encounter *v.* **1** °meet, come upon, run into *or* across, °happen upon, chance upon, °hit upon, °light upon, °stumble upon, *Colloq* °bump into: *She encountered him again in the supermarket.* **2** °face, °experience, °meet with, contend with, be faced with, come into contact with, wrestle with: *She encounters such problems every day.* **3** come into conflict with, contend with, assail, cross swords (with), °grapple with, °engage, joust with, do battle with, confront, clash with, °join, °meet: *He encountered the black knight in single combat.* —*n.* **4** °meeting: *It was a chance encounter that brought us together.* **5** confrontation, °brush, °quarrel, °disagreement, °dispute, altercation, °engagement, °action, °battle, °fight, °clash, °conflict, °skirmish, °contest, °competition, duel, contention, °struggle, °war, *Colloq* dust-up, °scrap, °run-in, set-to: *My brother took part in the bloody encounter in the Ardennes.*

encourage *v.* **1** hearten, embolden, °reassure, °buoy (up), °stimulate, °animate, °support, inspirit, °inspire, °cheer (up), urge *or* spur on *or* onward(s), °incite, *Colloq* egg on, °pep up: *She encouraged him in his study of medicine.* **2** °promote, °advance, °aid, °support, °help, °assist, °abet, °foster, °forward, °boost, *Colloq* give a shot in the arm: *The continued success of the team encouraged attendance at the games.*

encouragement *n.* **1** heartening, reassuring, reassurance, buoying up, °stimulating, stimulation, stimulus, animating, °animation, supporting, °support, promoting, °promotion, inspiring, °inspiration, cheering, urging, °spur, spurring, exhorting, exhortation, prodding, urging, innervation, inciting, °incitement: *Without her encouragement, I could never have won.* **2** °boost, stimulus, °help, °aid, °support, *Colloq* pep talk: *The team gets a lot of encouragement from local merchants.*

encroach *v.* Often, **encroach on** *or* **upon.** intrude, trespass, °infringe, invade, make inroads: *When you mention arctophily, you're encroaching on my territory.*

encumber *v.* **1** °burden, °weigh down, load (up *or* down), °overload, overburden, °strain, °oppress, saddle, °tax, overtax: *Encumbered with intolerable taxes, the people revolted. She almost drowned because she was encumbered with her numerous petticoats.* **2** °hamper, °impede, °hinder, °handicap, °inconvenience, °trammel, °retard, °slow down: *Though encumbered by an invalid husband, she managed to get a college degree.*

encumbrance *n.* °weight, °burden, onus, °cross (to bear), albatross, millstone, °handicap, °impediment, °hindrance, °obstacle, °obstruction, °liability, °disadvantage, °drag: *The poor were regarded merely as an encumbrance.*

encyclopedic *adj.* encyclopaedic, °comprehensive, °inclusive, °broad, °extensive, °universal, °thorough, °exhaustive, wide-ranging, °complete: *She has an encyclopedic knowledge of Chinese art.*

end *n.* **1** °extremity, °extreme, °extent, °bound, °boundary, °tip, °limit, terminus: *That fence marks the southern end of the property.* **2** °close, °termination, conclusion, cessation, °expiration, °finish, °completion, finale, ending, windup; denouement *or* dénouement: *At the end of the film, the lights came on.* **3** °aim, °purpose, °intention, °intent, °objective, °object, °goal, °point, °reason, raison d'être, °destination, °motive, motivation, °aspiration: *To what end does she persist so vehemently?* **4** consequence, °result, °outcome, °effect, °upshot: *The end of the affair was that he had to leave town.* **5** °destruction, °death, °ruin, extermination, annihilation, °death, °termination, conclusion: *That would spell the end to all life on earth.* **6** ***at loose ends*** or *Brit* ***at a loose end.*** °unsettled, unoccupied, °unemployed, uncommitted, undecided, °indecisive, ambivalent, vacillating, °purposeless, °aimless, adrift, drifting, betwixt and between, neither here nor there: *My wife has gone off for a holiday and I'm at a loose end.* **7** ***on end.*** **(a)** °upright, °erect, °standing: *He can balance a ruler on end.* **(b)** continuously, uninterruptedly, unceasingly, incessantly, consecutively: *It rained for days on end.* **8** ***the end.*** **(a)** the worst, the last straw, the final blow, *Colloq* °the limit, too much: *Brian's winning first prize is the absolute end.* **(b)** the best, the greatest: *That record by The What is the living end.* —*v.* **9** °terminate, conclude, bring to an end, °stop, °halt, °cease, wind up *or* down, °settle, put an end to, °discontinue, °break off, °cut off, °close, °finish, culminate, end up, *Brit* put paid to; °die (out), °expire, °climax, °peter out, vanish: *We have ended our relationship. The book ends with her returning to him. The year ends on December 31.* **10** °kill, put to death, annihilate, °exterminate, °terminate, °extinguish; °destroy, °ruin: *He ended his life last night with a bullet.* **11** °surpass, °outdo, outclass, outshine, °outstrip, °supersede: *It is a disaster film to end all disaster films.*

endanger *v.* imperil, °threaten, °jeopardize, °risk, put at risk, °hazard, °expose (to danger), put in jeopardy, tempt fate: *She endangered her life while trying to save his.*

endangered *adj.* imperiled, threatened, near extinction: *We must protect the endangered species.*

endearing *adj.* °attractive, °engaging, likable *or* likeable, appealing, winsome, captivating, °winning: *He has a few endearing qualities, I suppose.*

endeavor *v.* **1** °try, °attempt, °strive, make an effort, do one's best, °struggle, °exert oneself, °undertake; °aim, °aspire; *Colloq* take a stab at, have a go *or* crack *or* whack *or* shot at: *For years he's endeavored to see her.* —*n.* **2** °effort, °pains, °attempt, °try, striving, °struggle, °venture, °enterprise, *Colloq* °stab, °crack, whack, °shot: *Her endeavors to be published have come to naught.*

endless *adj.* **1** °limitless, °unlimited, °boundless, unbounded, °infinite, °immeasurable, °eternal, unending: *We found ourselves in the endless reaches of outer space.* **2** ceaseless, uninterrupted, incessant, unceasing, unending, °constant, °perpetual, interminable, unremitting, °nonstop, °continuous, °continual, °everlasting: *I wish you two would stop your endless bickering.*

endorse *v.* **1** indorse, °approve, °sanction, °authorize, °advocate, °support, °back, °subscribe to, °sustain, °confirm, countenance, put one's stamp *or* seal (of approval) on, set one's seal (of approval) to, give (something) one's imprimatur, *Colloq* okay, O.K.: *The council endorsed our application for planning permission.* **2** countersign: *Endorse the check to cash it.*

endorsement *n.* **1** indorsement, °approval, affirmation, °sanction, authorization, confirmation, ratification, °support, °backing, approbation, seal *or* stamp of approval, imprimatur, *Colloq* okay, O.K.: *He is acting with the full endorsement of his union.* **2** countersignature: *His endorsement is on the back of the check.*

endowment *n.* **1** °grant, (financial) °aid, °subsidy, subvention, °allowance, °allotment, contribution, °donation, °gift, °present, °award; °bequest, °inheritance, dowry: *The endowment was enough to support me for a year.* **2** °gift, °presentation, bestowal, °award, awarding, °settlement: *The endowment of the grant was approved by the college.* **3** ***endowments.*** qualities, talents, gifts, °abilities, aptitudes, capabilities, capacities, °qualifications, strengths; attributes, properties, characteristics: *At the party, he met a woman of considerable endowments.*

endurance *n.* **1** °stamina, staying power, °perseverance, °persistence, °resolution, °fortitude, °tenacity, °patience, °tolerance, *Colloq US* stick-to-itiveness: *He showed remarkable endurance in the pentathlon trials.* **2** lasting quality, durability, longevity, lifetime, continuation: *The endurance of the gear is improved by regular oiling.*

endure *v.* **1** °last, °persist, °stay, °remain, °abide, °prevail, °survive, °continue, °hold, °live (on), *Colloq* go the distance: *Her fame as a poet will endure.* **2** °stand, °abide, °tolerate, °face, °survive, °withstand, °bear, °weather, °take (it), °suffer, °stomach, °undergo, °hold out (against), *Colloq* hang in (there), stick *or* sweat (it *or* something) out: *He endured the pressure of his job as long as he could.* **3** °suffer, °undergo, °bear, °face, °stand, °put up with, °stomach, °take: *Consider the tyranny that Europe endured under Hitler.*

enduring *adj.* °lasting, long-lasting, °durable, °abiding, continuing, long-standing, persisting, °persistent, °remaining, °steady, °steadfast; °eternal, °immortal, °permanent: *Their enduring faith carried them through the ordeal.*

enemy *n.* foe, °antagonist, °adversary, °opponent, °rival, °competitor, °contestant, contender; the opposition, the other side: *His political enemies were trying to blackmail him.*

energetic *adj.* °lively, °active, °vigorous, invigorated, °dynamic, °animated, °spirited, °untiring, °tireless, indefatigable, °sprightly, spry, °vital, high-powered, °brisk, vibrant, zesty, zestful, *Colloq* °hot, peppy, full of pep, full of get-up-and-go, zippy, on one's toes, zingy, full of beans: *I feel most energetic at the start of the day.*

energize *v.* °enliven, liven up, °stimulate, °animate, invigorate, °activate, actuate, °move, °motivate, galvanize, °electrify, °inspire, inspirit, °pep up, waken, °rouse, °stir, °arouse, °excite, egg on, °urge: *Max's enthusiasm energized the whole of the research team.*

energy *n.* °vitality, forcefulness, vivacity, liveliness, °vigor, °animation, °spirit, °force, °dynamism, °drive, °verve, °dash, élan, °intensity, °power, °determination, puissance, °strength, °might, *Colloq* °pep, vim and vigor, *US* stick-to-itiveness, get-up-and-go, zip, zing: *At seventy, she just couldn't put as much energy into her performance as she once did.*

enervate *v.* °weaken, °tire, °strain, enfeeble, debilitate, °fatigue, °exhaust, °drain, °sap, °wear out, devitalize, °break, °defeat: *I have always found the tropics quite enervating.*

enforce *v.* **1** insist upon *or* on, °stress, °require, °impose, °support, put into effect, °apply, °administer, °carry out, °inflict, bring to bear, °implement, °prosecute, °discharge; °reinforce; *Colloq* crack *or* clamp down: *The police will enforce the curfew tonight.* **2** °force, compel, pressure *or Brit* pressurize, °press, coerce, lay stress upon *or* on, impose upon *or* on, impress upon *or* on, insist upon *or* on, °demand, °require; °intimidate, °browbeat, °bully, °railroad; *Colloq* °lean on, twist (someone's) arm: *They hired thugs to enforce their claim.*

engage *v.* **1** °employ, °hire, enrol *or US also* enroll, °enlist, °retain, °sign (up), contract with *or* for, indenture; °rent, °book, °reserve, °secure, bespeak: *She was engaged on a part-time basis. We engaged rooms for the night.* **2** °occupy, engross, °busy, absorb, °involve, °tie up, preoccupy, °employ: *This job will engage all my available time till next month.* **3** °pledge, °undertake, °bargain, °agree, covenant, °promise, °guarantee, °contract: *I engaged to complete the work by Tuesday.* **4** °attract, °hold, °capture, °catch, °draw: *The museum has many exhibits that will engage the interest of children.* **5** join (in) combat *or* battle with, °meet, °encounter, °fight, °combat, °attack, °battle, clash with, °grapple with: *We engage the enemy at dawn.* **6** ***engage in.*** °participate (in), °partake in, take part (in), °enter (into), °undertake, °embark on: *She engages in many school activities.*

engaged *adj.* **1** betrothed, affianced, plighted, pledged, promised; °spoken for: *An engaged young* *woman does not go out with other men.* **2** °busy, occupied, tied up, °involved, employed, °absorbed, °preoccupied, wrapped up: *He is otherwise engaged and cannot meet me today.*

engagement *n.* **1** °appointment, °meeting, °date, rendezvous, °arrangement, commitment: *I'm sorry, I have an engagement tonight.* **2** betrothal: *Her parents have announced her engagement.* **3** °agreement, °bargain, °obligation, °promise, °pledge, covenant, °contract: *The company undertook an engagement to provide the steel.* **4** °job, °position, °post, commission, booking; °employment, °work; *Colloq* °spot, *Slang* (of a musician) gig: *He has an engagement with the newspaper.* **5** °fight, °battle, °conflict, °encounter, °combat: *The naval engagement lasted three days and nights.*

engaging *adj.* charming, °pleasant, °attractive, winsome, °winning, appealing, °agreeable, °delightful, °pleasing, likable *or* likeable, °friendly, °open: *With her engaging personality, it is no wonder she has many friends.*

engine *n.* motor, °machine, °mechanism, appliance, °apparatus; locomotive: *The internal combustion engine is used universally in cars.*

engineer *n.* **1** °designer, originator, inventor, contriver, architect, planner, °mastermind: *Who is the engineer behind this health scheme?* **2** (engine) driver, conductor, °operator: *He was the engineer on the afternoon train out of Washington.* **3** mechanic, technician, repairman: *The telephone engineers will be here on Monday to install the new phone system.* —*v.* **4** °devise, °plan, °develop, °originate, contrive, °invent, °mastermind, °construct, °build, °make: *It is said that Daedalus engineered the first man-made wings.* **5** °manipulate, °scheme, °plot, machinate, °intrigue, connive, conspire, °maneuver, °rig, °set up, °organize, °arrange, °put over, *Colloq* finagle, °wangle, °swing: *She tried to engineer the laundering of the money but was caught.*

engrave *v.* **1** °cut, °carve, °chisel, inscribe; °etch: *He engraves wood block designs for a living.* **2** °impress, °stamp, °set, °lodge, °fix, embed, imprint, ingrain: *The horror of that night was forever engraved in his mind.*

engraving *n.* **1** intaglio, cameo, etching, drypoint, woodcut, linocut, wood *or* steel engraving, anaglyph, block *or US also* cut: *She uses a burin to make these engravings.* **2** °print, °impression, etching, drypoint: *The exhibit of Dürer engravings closes tomorrow.*

enhance *v.* °improve, °better, augment, °boost, °raise, elevate, °lift, °heighten, °exalt, °enlarge, °swell, °magnify, °increase, °add to, °amplify, °intensify, °enrich, °embellish, °complement, °reinforce, °strengthen: *His public image was greatly enhanced by his support of charities.*

enigma *n.* °puzzle, conundrum, °mystery, °riddle, poser, °problem: *How he escaped is an enigma to the police.*

enjoy *v.* **1** °delight in, °appreciate, °like, take *or* derive pleasure *or* satisfaction in *or* from, °relish (in), °fancy, °take to, *Slang* °dig, get a kick *or* lift *or* charge out of, get high on, °get off on: *Bernard really enjoys Wagner.* **2** benefit *or* profit from, take advantage of, °use, utilize, make use of, use to advantage, °have, °possess: *He cannot be charged because he enjoys diplomatic immunity.* **3** ***enjoy oneself.*** have a good time, °make merry, *Colloq* have a ball *or Colloq* the time of one's life: *I enjoyed myself at your party.*

enjoyment *n.* **1** °pleasure, °delight, °joy, °gratification, °satisfaction, °relish, °zest, °delectation, °recreation, °entertainment, °diversion, °amusement: *The public has derived much enjoyment from Shaw's plays.* **2** °use, utilization, °exercise, °possession; °benefit, °advantage: *As a member, you are entitled to the enjoyment of all club privileges.*

enlarge *v.* **1** °increase, °expand, °magnify, °amplify, °extend, °swell, dilate, °spread, wax, °widen, broaden, °lengthen, elongate, °stretch, distend; °add to, °supplement, augment; °inflate; *Colloq* °blow up: *The government will enlarge the area devoted to public parks. Enlarge the photographs and details appear.* **2** ***enlarge on or upon.*** °expand on, expatiate on, °amplify, expound; °detail, °elaborate (on): *The speaker was asked to enlarge on her plans for building new hospitals.*

enlighten *v.* °inform, edify, °instruct, °teach, °tutor, °educate, °coach, apprise, make aware, °advise, °counsel: *You must enlighten George on how to behave in public.*

enlightened *adj.* °well-informed, °informed, °educated, °aware, °knowledgeable, literate, °rational, °reasonable, °sensible, common-sense, common-sensical, broad-minded, open-minded, °liberal; °cultivated, civilized, °sophisticated, *Colloq* in the know: *In a democracy, the press serves an enlightened public.*

enlist *v.* **1** °enroll, °register, °join (up), volunteer, sign up *or* on; °engage, °recruit, °induct, °muster, conscript, °impress, °call up, *US* °draft: *He enlisted in the navy at seventeen.* **2** °employ, °hire, °engage, °retain, make available, °secure, °obtain, °get, °procure, °gather, drum up, °mobilize, *Colloq* °organize: *We must enlist all the help we can.*

enliven *v.* **1** invigorate, inspirit, °animate, °pep up, °stimulate, °energize, vivify, °vitalize, °quicken, exhilarate, °arouse, °rouse, °awaken, °wake up, °spark (off), °kindle, enkindle, °fire (up), °inspire: *Enlivened by the coach's talk, we were determined to win.* **2** °brighten, °cheer (up), °buoy (up), °gladden, uplift: *The room was considerably enlivened by the new curtains.*

enormity *n.* outrageousness, °outrage, atrociousness, °atrocity, wickedness, heinousness, flagitiousness, horribleness, °horror, °barbarity, savagery, monstrousness, horridness, °evil, viciousness: *The enormity of the crime shocked us all.*

enormous *adj.* °huge, °immense, °gigantic, elephantine, gargantuan, mammoth, titanic, °colossal, tremendous, °vast, °massive, stupendous, Brobdingnagian, °gross, °monstrous, °prodigious: *An enormous dragon blocked our way out of the cave.*

enormousness *n.* immensity, hugeness: *The enormousness of the hybrid rose was astonishing.*

enough *adj.* **1** °sufficient, °adequate, °ample: *There isn't enough food to go around.* —*n.* **2** sufficiency, adequacy, ample supply, °plenty: *I have enough for myself.* —*adv.* **3** sufficiently, adequately, reasonably, satisfactorily, tolerably, passably: *Your word is good enough for me.*

enquire *v.* **1** °inquire, °ask, °question, °query: *The police enquired whether we had noticed any strange goings-on next door.* **2** See **inquire.**

enrage *v.* °anger, °infuriate, °madden, incense, °provoke, °inflame, make (someone's) blood boil, *Colloq* get (someone's) back *or* Irish *or* hackles *or* dander up, make (someone) see red, wave a red flag before (someone), make (someone's) blood boil, *US* burn (someone) up, *Slang US* tick (someone) off, *Taboo* piss (someone) off, *Brit* put (someone's) monkey up, *US* tee (someone) off: *Lord Thimble was enraged at being made to wait his turn.*

enrapture *v.* °enchant, °entrance, °transport, °thrill, °bewitch, spellbind, °fascinate, °charm, enthrall, °captivate, °beguile, °delight: *The soprano quite enraptured her audience.*

enrich *v.* **1** endow, °enhance, °improve, upgrade, °better, ameliorate, °refine, °add to: *His novels have enriched our literature.* **2** °ornament, adorn, °decorate, °embellish; °beautify, °grace: *Before us rose a lofty dome enriched with precious stones.*

enroll *v.* **1** °enlist, °register, sign up *or* on (for), °join; volunteer; °recruit; *Colloq* join up: *He enrolled at the university.* **2** °record, °register, °chronicle, °put down, °list, °note, inscribed, catalog: *Their names will be enrolled forever in the Book of Heroes.*

ensemble *n.* **1** °outfit, °costume, clothing, °clothes, attire, °apparel, garb, °garments, coordinates, *Colloq* get-up: *The hat, gloves, and shoes completed the ensemble.* **2** °band, °combination, orchestra, °group; chorus, choir; *Colloq* combo: *The chamber music ensemble performs tonight.* **3** assemblage, composite, aggregate, °collection, °set, °whole, °entirety, °totality; agglomeration, conglomeration: *A strange ensemble of objects was on display.*

enslave *v.* °subjugate, yoke, fetter, enchain, °shackle, °trammel, °dominate; °bind, indenture, *Archaic* enthrall: *The enslaved masses rose and overcame their masters.*

ensure *v.* **1** insure, °assure, make sure *or* certain, °confirm, °certify, °guarantee; °secure, °effect: *He will ensure there will be no leak.* **2** insure, °protect, make safe, °safeguard, °guard, °secure: *The captain ensures the safety of the passengers.*

entail *v.* °involve, °require, °call for, necessitate, °demand, °occasion, °give rise to, °impose; lead to, °cause: *What is entailed in my accepting the post?*

entangle *v.* **1** °tangle, ensnarl, °snarl, enmesh, °catch (up), entrap, °snag, °foul, °implicate, °knot (up), °twist; °impede; °involve, embroil: *The seaweed entangled the propeller and the boat couldn't move.* **2** °confuse, °mix (up), °twist, °snarl, ensnarl, ensnare, °hamper, °complicate, confound, °bewilder, °perplex, °embarrass: *She was entangled by unforeseen consequences.*

enter *v.* **1** go *or* come (in *or* into), °pass (into): *Please enter at the left.* **2** °penetrate, °pierce, °stick (into), °stab (into), °puncture; invade, infiltrate: *The nail entered the tire here. The iron entered into his soul.* **3** °insert, inscribe, °write, set *or* write *or* put down, °note, °record, take *or* jot down, °register; log, °document, °minute: *Enter your name on the dotted line.* **4** enter on *or* upon, °begin, °start, °commence, °undertake, set out on, °take up: *I didn't want to enter the race if I couldn't finish.* **5** °enroll, °enlist, sign on *or* up, °join, become a member of: *He has entered the ranks of the unemployed.* **6** present, °offer, proffer, °tender, °submit: *More than one hundred poems were entered in the competition.* **7** °file, °register, °record, °submit: *The defendant entered a plea of "Not guilty."* **8** ***enter into.*** engage *or* participate in, °sign, be (a) party to, cosign, countersign: *We entered into an agreement to buy the house.*

enterprise *n.* **1** °undertaking, °project, °venture, °adventure, °effort, °program, °plan, °scheme: *His latest enterprise involves establishing a chain of pizza shops.* **2** boldness, °daring, °courage, mettle, adventurousness, audacity, °enthusiasm, zeal, °energy, °spirit, °drive, °vigor, °ambition, °initiative, °push, °eagerness, °determination, °resolve, purposefulness, °purpose; aggressiveness; *Colloq* get-up-and-go, zip, °pep, °gumption, °guts, *US* starch: *It takes a lot of enterprise to start your own business.* **3** °business, °operation, °firm, °company, °concern, °establishment: *We began this enterprise on a shoestring.*

enterprising *adj.* °resourceful, °venturesome, °adventurous, °daring, °courageous, °bold, °brave, mettlesome, °audacious, °enthusiastic, °eager, °keen, zealous, °energetic, °spirited, °vigorous, °ambitious, °determined, resolved, °resolute, °earnest, °purposeful, purposive, goal-oriented; °aggressive, hard-working, indefatigable, °tireless, °diligent, assiduous, °industrious, persevering; *Colloq* °pushy, °go-ahead: *She is an enterprising young lawyer with her own practice.*

entertain *v.* **1** °amuse, °divert; °delight, please; °occupy: *While we waited the boy entertained us with his juggling.* **2** °receive, °accommodate, °treat, be host (to), cater (for *or* to); have *or* see people *or* guests *or* visitors *or* company; *Colloq* °host: *We entertain on Tuesdays only.* **3** °contemplate, °consider, °have, °hold, °harbor, °foster, °tolerate, °allow, °maintain, °sustain, °support: *How could they entertain any idea of her marrying Pembroke Krudder?*

entertaining *adj.* amusing, diverting, °delightful, enjoyable, °pleasant, °fun, °pleasing, °pleasurable, °interesting, °engaging; °funny, °comic, °humorous, °witty: *We find Laurel and Hardy films entertaining.*

entertainment *n.* **1** °amusement, °diversion, °distraction, °pastime, °recreation, °sport, °play, °fun, °pleasure, °enjoyment, °relaxation, °relief: *What do you do for entertainment?* **2** °performance, °presentation, °diversion, °amusement, divertissement, °exhibition, °pageant, °spectacle, °show, °production, spectacular, °extravaganza: *They put on a lavish entertainment for the queen.*

enthralling *adj.* captivating, entrancing, °spellbinding, °enchanting, bewitching, beguiling, fascinating, gripping, °absorbing, intriguing, hypnotizing, mesmerizing, °riveting: *There was an enthralling melodrama on TV last night.*

enthusiasm *n.* **1** °eagerness, keenness, earnestness, °fervor, avidity, zeal, °excitement, °passion, °ardor, °interest, °relish, °devotion, devotedness, °gusto, °exuberance, °zest; °fanaticism, °mania, °rage: *No one matches her enthusiasm for grand opera.* **2** °rage, °passion, °craze; °hobby, °interest, °pastime, °diversion, °amusement; *Colloq* °fad: *His current enthusiasm is acid rock.*

enthusiast *n.* °fan, °devotee, aficionado, °lover, °admirer, °zealot, °addict, °fanatic, promoter, °supporter, °champion, °follower, °disciple, adherent, *US* booster, *Colloq* teeny-bopper, °bug, °hound, buff, °fiend, *Slang* nut, °freak, groupie, *US* °head: *They are model railway enthusiasts.*

enthusiastic *adj.* °eager, °keen, °fervent, fervid, °hearty, °ardent, avid, °energetic, °vigorous, °devoted, °earnest, °passionate, °spirited, °exuberant, zealous, fanatic(al), °unqualified, unstinting, °irrepressible: *He's an enthusiastic supporter of the prime minister's policies.*

entice *v.* °lure, allure, °tempt, °attract, °draw, °seduce, °coax, °persuade, °prevail on, °beguile, °cajole, blandish, °wheedle; °decoy, °lead on, inveigle, *Colloq* sweet-talk, soft-soap, *Slang* suck in: *He enticed us to invest in his scheme.*

enticement *n.* **1** °temptation, allurement, beguilement, seduction, °cajolery, wheedling, blandishment, coaxing, °persuasion: *Do you approve of the system of enticement used to get you to buy time-share holidays?* **2** °lure, bait, °decoy, °trap, °inducement, °attraction, °temptation, *Colloq* °come-on, soft soap: *Among the enticements offered was a free weekend in Torquay.*

entire *adj.* **1** °complete, °whole, °total, °full, °undivided, °absolute, °thorough, unreserved, unrestricted, undiminished, unconditional, °express, unexceptional, unmixed, unalloyed: *That problem has commanded my entire attention all week.* **2** °intact, °whole, °sound, unbroken, undamaged, unimpaired, inviolate, without a scratch, °unscathed, in one piece: *These amphorae were dredged up entire from the wreck.* **3** °continuous,

°full, °whole, °complete, uninterrupted: *She has lasted an entire year in her new job.*

entirely *adv.* **1** °completely, °wholly, °altogether, fully, °totally, °utterly, unreservedly, unqualifiedly, unexceptionally, in every respect, in all respects, °thoroughly, to a T, *in toto,* exhaustively, all-out, from head to toe *or* foot, (right) down to the ground, from A to Z, °lock, stock and barrel, °root and branch, without exception *or* reservation: *She was entirely satisfied. This job suits him entirely.* **2** solely, exclusively, °only, unambiguously, unequivocally; °positively, °definitely, °clearly: *It was entirely my fault.*

entirety *n.* **1** completeness, °unity, °totality, wholeness, fullness, °integrity, °perfection: *Any chance of completing his mission in its entirety had vanished.* **2** °whole, sum total, °everything, all: *The entirety of the bequest amounted to a million francs.*

entitle *v.* **1** °allow, °permit, °qualify, make eligible, °authorize, °fit; enfranchise, °license, empower: *This document entitles her to collect half of the estate.* **2** °name, °title, °call, °label, °nickname, dub, °designate, °term; °christen, baptize: *His first novel was entitled* Out of the Depths.

entity *n.* **1** °thing, °object, being, °existence, °quantity, article, °individual, °organism: *Many regard God as an entity.* **2** °essence, real nature, quiddity, °quintessence, *Metaphysics* ens: *Every living creature has a distinct entity.*

entrance[1] *n.* **1** °(right of) entry, access, °admission, °admittance, entrée, introduction: *You need security clearance to gain entrance.* **2** °entry, entryway, access, door, °gate, °passage, °way (in); ingress: *The entrance is locked after midnight.* **3** °arrival, °appearance; coming, °entry, coming *or* going in: *Her entrances were accompanied by cheers. He opposed our entrance into the war.* **4** °beginning, °start, commencement: *Today marks his entrance into his new duties.*

entrance[2] *v.* °enchant, °enrapture, °fascinate, °bewitch, spellbind, °transport, °delight, °charm, °captivate, enthrall, °overpower, mesmerize, °hypnotize: *He was entranced by her beauty.*

entrenched *adj.* °rooted, deep-rooted, embedded, °fixed, (firmly) planted, established, °set, deep-seated, unshakable *or* unshakeable, ineradicable, °ingrained: *We were unable to combat the politicians' entrenched interests.*

entrust *v.* intrust, °trust, °charge, °assign, °delegate, confide: *We entrusted her to keep her own account of what she spent.*

entry *n.* **1** access, °entrance, entrée, °admittance, °admission: *The burglar gained entry through the skylight.* **2** access, °entrance, entryway, door, inlet, °passage, way in: *This entry is open only to staff.* **3** °entrance, °arrival, coming *or* going in: *His entry was met with jeers.* **4** °record, °item, °memorandum, °note, °account, listing; registration; *Colloq* memo: *There is no entry in his diary for April 15.* **5** °competitor, °contestant, °player, entrant, °participant, °candidate; °rival, °adversary, °opponent: *Dennis was a late entry in the marathon.*

entwine *v.* intwine, interlace, °braid, interweave, intertwine, °weave, °knit, plait, °twist, °coil, °twine, °splice; °entangle, °tangle: *Her hair was entwined with wildflowers.*

enumerate *v.* **1** °list, °name, °itemize, °specify, °detail, °spell out, catalogue, tick off, take stock of, cite, °quote, °recite, °recount, °relate, °narrate, *US* °check off: *I'll enumerate the reasons why you may not go.* **2** °count, °calculate, °compute, °reckon, °tally, °add, °number: *The researcher could enumerate only seven different species.*

enunciate *v.* **1** articulate, °pronounce, utter, °voice, °say, °speak, vocalize, °express, °deliver, °present, *Formal* enounce: *She enunciates the foreign names clearly.* **2** °state, °proclaim, °declare, promulgate, °announce, °broadcast, °pronounce, °propound: *The party platform was enunciated in last night's speech.*

envelop *v.* **1** °wrap, °enclose, enfold, enwrap, °cover, engulf, °swathe, °shroud, enshroud, swaddle: *The body was enveloped in a white robe.* **2** °shroud, enshroud, °cover, °conceal, °hide, °screen, °shield, °obscure, °veil, °cloak: *The motive for the murder was enveloped in a mass of misleading clues.*

enviable *adj.* °desirable, wanted, desired, sought-after, covetable, °in demand: *He is in an enviable financial position.*

envious *adj.* °jealous, covetous, °resentful, begrudging, green-eyed, °green (with envy), °desirous: *He is envious of his wife because she has a better job.*

environment *n.* surroundings, environs, °atmosphere, ecosystem, °conditions, °habitat, °circumstances, °medium, °milieu; °territory, °locale, °setting, *mise en scène,* °situation: *Car exhaust fumes are ruining our environment.*

environmentalist *n.* ecologist, conservationist, naturalist, preservationist, nature lover, green *or* Green: *We have environmentalists to thank for clean-air laws.*

envisage *v.* **1** visualize, °contemplate, °imagine, °picture, °conceive (of), °fancy, think *or* dream *or* conjure up, *Chiefly US* °envision: *Envisage a city built entirely of glass.* **2** °foresee, °see, °predict, °forecast, °anticipate: *I envisage a time when all people will be free.*

envision *v.* °envisage, visualize, °imagine, conceive of, °foresee, °anticipate, °predict, °forecast, °prophesy: *I envision great success for you.*

envoy *n.* °delegate, legate, °ambassador, diplomat, °minister, (papal) nuncio, attaché; °representative, emissary, °agent; *Formal* envoy extraordinary, minister plenipotentiary: *The government sent an envoy to discuss trade.*

envy *n.* **1** jealousy, enviousness, °resentment: *She was consumed with envy of anyone with more money than she.* **2** covetousness, °desire, °longing: *Success excites my envy.* —*v.* **3** covet, °begrudge, °resent: *He envies his brother and his new car.*

epicure *n.* °gourmet, connoisseur, °aesthete, °epicurean, °sybarite, hedonist, gastronome, bon viveur, *bon vivant,* Lucullus; gourmand: *An epicure, he refuses to eat any food not prepared by his own chef.*

epicurean *adj.* **1** °sensual, sybaritic, °luxurious, °voluptuous, °carnal, °self-indulgent, pleasure-seeking, pleasure-oriented, hedonistic, °gluttonous, intemperate, overindulgent, crapulent *or* crapulous, swinish, porcine, piggish, °immoderate, orgiastic, libidinous, °wild, unrestrained, unconfined, °dissolute, dissipated, Bacchanalian, saturnalian: *The king's epicurean life contrasted with that of his starving subjects.* **2** Lucullan, °gourmet: *That was truly an epicurean repast.* —*n.* **3** °epicure: *Scheunendrescher is a true epicurean!*

epidemic *adj.* **1** widespread, °universal, °prevalent, °prevailing, °rampant, °general, wide-ranging, pandemic: *The disease has reached epidemic proportions.* —*n.* **2** °plague, °pestilence, °scourge, °rash, °growth, upsurge, outbreak, °spread: *An epidemic of anthrax has affected the cattle.*

epigram *n.* **1** °witticism, bon mot, °quip, mot, turn of phrase, *jeu d'esprit,* Atticism; °pun, double entendre, *jeu de mots,* play on words, equivoque; paronomasia:

His epigram, characterizing Eskimos as "God's frozen people," was widely quoted. 2 °proverb, aphorism, °maxim, °saw, saying, adage, apophthegm *or* apothegm: *"Nothing succeeds like success" is his favorite epigram.*

epigrammatic *adj.* pithy, °terse, laconic, °concise, °succinct, compendious, piquant, °pungent, °trenchant, sententious, °witty, °pointed, °proverbial, aphoristic, apophthegmatic *or* apothegmatic, *Colloq* °snappy, punchy: *Heinrich Kurtz is known for his epigrammatic verses.*

episode *n.* 1 °event, °incident, °occurrence, °happening, °experience, °adventure, °affair, °matter: *Please do not remind me of the episode with the chicken.* 2 chapter, °scene, installment, °part: *Don't miss tonight's episode of your favorite soap opera.*

epitome *n.* 1 °essence, °quintessence, °embodiment, personification, archetype, exemplar, °(typical) example, °model, °prototype: *My secretary is the epitome of laziness.* 2 °summary, °abstract, condensation, °synopsis, °digest, compendium, °abridgment, °abbreviation, conspectus, °résumé, contraction; °outline, °précis, syllabus: *Pingle wants an epitome of the report today.*

equable *adj.* 1 even-tempered, easygoing, °serene, °calm, placid, composed, °cool, imperturbable, °collected, unruffled, °tranquil, °peaceful, °levelheaded, *Colloq* unflappable: *With his equable disposition, Edgar is the man we need.* 2 °uniform, unvarying, unvaried, °consistent, °stable, °steady, °regular, °even, unchanging, °invariable, °constant: *We moved to the Caribbean because of its equable climate.*

equal *adj.* 1 °identical, the same (as), interchangeable, one and the same, coequal, selfsame; °like, °alike, tantamount, °similar (to), °equivalent, commensurate: *This year's sales figures are equal to last year's.* 2 °uniform, °regular, corresponding, °correspondent, congruent, congruous, (evenly) balanced, (evenly) matched, °matching; °equivalent, °even; commensurate, comparable, proportionate, (evenly) proportioned, °harmonious, °symmetrical; *Colloq* fifty-fifty, *Brit* level pegging, *US* even-steven: *Women are entitled to equal employment opportunities. The scores are equal.* 3 ***equal to.*** up to, °capable of, fit(ted) *or* suited *or* suitable for, adequate for, *Archaic or literary* sufficient unto: *Are you sure that Renwick is equal to the responsibility?* —*n.* 4 °peer, °colleague, °fellow, °brother, °mate, counterpart, °equivalent, alter ego, compeer: *Constance is certainly anyone's equal in intelligence.* —*v.* 5 °match, °meet, °even, °correspond (to), °square (with), °tally (with), °tie (with), °parallel, come up to; °rival: *He will never be able to equal the world's record.*

equality *n.* 1 °parity, sameness, °identity, coequality, °uniformity: *The equality of the two bids was very suspicious.* 2 similarity, °likeness, °resemblance, equivalence, correspondence, conformity, congruence, similitude, analogy, comparability, °comparison, °coincidence: *The equality between their performances is surprising.* 3 impartiality, fairness, °justice; egalitarianism: *Surely we all deserve equality of treatment under the law.*

equalize *v.* regularize, °even up, °square, °balance, equate, °match, °standardize, °(make) equal: *Equalize the amounts of liquid in all the containers.*

equip *v.* °furnish, °provide, °supply, °stock, °outfit, fit (out *or* up), rig (out *or* up), accouter *or* accoutre, array, attire, °dress, deck (out), caparison, °clothe, *Chiefly Brit* kit out *or Chiefly Brit* up: *We can equip you with any scuba gear you may require.*

equipment *n.* °gear, °apparatus, furnishings, accouterments *or* accoutrements, appurtenances, °paraphernalia, °kit, materiel *or* matériel, °tackle, °outfit, °trappings, °tack, equipage, *Colloq Brit* clobber: *They spent a fortune on mountain climbing equipment.*

equitable *adj.* °fair, evenhanded, °just, °impartial, °objective, unbiased, °unprejudiced, °square, fair-minded, open-minded, °disinterested, °dispassionate, °neutral, °tolerant, unbigoted, °reasonable, °judicious, °ethical, °principled, °moral, °proper, right-minded, *Colloq* fair and square: *Suspects have the right to equitable treatment.*

equity *n.* fairness, impartiality, evenhandedness, °justice, fair play, °objectivity, disinterest, fair-mindedness, equitableness, open-mindedness, disinterestedness, neutrality, °tolerance, judiciousness, rightmindedness, high-mindedness: *This court recognizes the equity of your claim.*

equivalent *adj.* 1 tantamount, commensurate, °alike, °similar, °close, comparable, corresponding, interchangeable, °equal, °synonymous, of a piece *or* a kind: *Do you believe that women's rights should be equivalent to men's?* —*n.* 2 °match, °equal, °peer, counterpart, °twin: *The garage could not supply the same part but they offered an equivalent.*

equivocal *adj.* 1 °evasive, misleading, °roundabout, hedging, °suspicious, duplicitous, °questionable, °oblique, circumlocutory, ambagious, ambivalent, amphibolic *or* amphibolous, *Colloq* waffling, °wishy-washy: *When asked about their guarantee, they gave an equivocal answer.* 2 °ambiguous, °vague, °hazy, °indefinite, unclear, °indistinct, enigmatic(al), °puzzling, °perplexing, indeterminate, °uncertain, *Colloq* waffling: *Just say Yes or No—none of your equivocal responses.*

equivocate *v.* °evade, °mislead, hedge, °deceive, °quibble, °dodge, weasel out (of), double-talk, °fence, °sidestep, skirt, °avoid, tergiversate, prevaricate, *Colloq* °waffle, beat about *or* around the bush, °pussyfoot: *I wish she'd confirm or deny it and stop equivocating.*

era *n.* °age, °period, time(s), day(s), epoch, °stage; °generation, °cycle, °date: *They lived in an era of peace.*

erase *v.* 1 expunge, rub *or* scratch *or* blot *or* wipe out, °delete, °cancel, efface, °scratch, cross *or* strike out *or* off, °obliterate: *The scribe erased one line and substituted another.* 2 °abolish, °destroy, °obliterate, °remove, °eliminate, °(get) rid of, eradicate, efface: *We erased every trace of evidence that we had been there.*

erect *adj.* 1 °upright, °standing, upstanding, °straight, vertical, °perpendicular, °plumb: *Naughty children are made to remain erect in the corner.* —*v.* 2 °build, °construct, °put up, °raise; °pitch: *I could swear that they erected that building overnight!* 3 °establish, °found, °set up, °form, °institute, °organize, °create: *Their religion was erected on the principles of utilitarianism.*

erode *v.* wear (down *or* away), eat away, °grind down, abrade, gnaw away (at), °consume, corrode, °wash away; °deteriorate, °destroy, deplete, °reduce, °diminish: *Water has eroded the rock. Their continued lying has eroded my confidence in their honesty.*

erosion *n.* °wear (and tear), wearing (down *or* away), wasting away, washing *or* grinding *or* rubbing away, corroding, corrosion, abrading, abrasion, eating *or* gnawing away, chafing, fraying, weathering, attrition: *Erosion by rainwater has washed away the topsoil.*

erotic *adj.* 1 °sensual, °stimulating, °suggestive, titillating, °risqué, °bawdy, ribald, °seductive, °voluptuous, °lustful, *Colloq* °sexy: *Some insist that there is a difference between pornographic and erotic literature.* 2 amatory, °venereal, amorous, Anacreontic: *Many classical poets wrote erotic verse.* 3 erogenous, °naughty, °carnal, arousing, °rousing, aphrodisiac, libidinous, lubricious *or* lubricous, °prurient, °lascivious, °lewd, concupiscent, salacious, °obscene, °pornographic, °dirty, °filthy, °nasty, *Colloq* °blue: *He has a collection of photos that he calls erotic art.*

err *v.* **1** be wrong, be in error, be mistaken, be inaccurate, be incorrect, be in the wrong, °go wrong, go astray, make a mistake, °miscalculate, °(make a) blunder, °bungle, °botch, °fumble, muff, make a mess of, make a faux pas, °mess up, *US* bobble; *Colloq* goof (up), °slip (up), drop a clanger, °foul up, *Brit* drop a brick, °blot one's copybook, *Slang* °screw up, *Brit* boob, *Taboo slang* fuck up: *The referee erred in ruling that the ball was out.* **2** °misbehave, °sin, °transgress, trespass, °lapse, °fall, do wrong: *She has erred many times in her long life.*

errand *n.* **1** °trip, °journey: *She was on an errand of mercy.* **2** °mission, °charge, °assignment, commission, °task, °duty: *Flavius is running some errands for me.*

erratic *adj.* **1** °irregular, unpredictable, °inconsistent, °unreliable, °capricious, °changeable, °variable; wayward, °unstable, aberrant, °flighty: *Without a timetable, these trains run on an erratic schedule.* **2** °peculiar, °abnormal, wayward, °odd, °eccentric, °outlandish, °strange, °unusual, °unorthodox, °extraordinary, °queer, °quaint, °bizarre, °weird, unconventional: *He thinks that his erratic behavior marks him as an individualist.* **3** wandering, °meandering, directionless, planetary, °aimless, °haphazard, °discursive, errant, divagatory: *Having lost our compass, our ship followed an erratic course.*

erroneous *adj.* °wrong, °mistaken,. °incorrect, °inaccurate, °inexact, °imprecise, °amiss, awry, °false, °faulty, misleading, °flawed, botched, bungled, °unsound, °invalid, °untrue, fallacious, °spurious, °counterfeit, *Colloq* off the mark, off course, *Brit* off beam, *US* off the beam: *He gives the erroneous impression of being intelligent. "Seperate" is an erroneous spelling of "separate."*

error *n.* **1** °mistake, inaccuracy, °fault, °flaw, °blunder, °slip, gaffe; °misprint, typographical error, erratum, °solecism; *Brit* °literal, *Colloq* slip-up, goof, clanger, °fluff, boo-boo, °howler, *Brit* bloomer, *Slang* foul-up, boner, *Brit* boob: *I cannot accept a report so full of errors.* **2** °sin, °transgression, trespass, °offense, °indiscretion, wrongdoing, misconduct, iniquity, °evil, wickedness, flagitiousness: *He seems to have seen the error of his ways.* **3** ***in error.*** **(a)** °wrong, °mistaken, °incorrect, °at fault: *She was in error about the date of the conference.* **(b)** mistakenly, incorrectly, by mistake, erroneously: *I caught the earlier train in error.*

erupt *v.* **1** °eject, °discharge, °expel, °emit, burst forth *or* out, °blow up, °explode, spew forth *or* out, °break out, °spout, vomit (up *or* forth), throw up *or* off, spit out *or* up, belch (forth), °gush: *The volcano erupted ash and lava.* **2** °appear, °come out, °break out: *A boil erupted on his chin.*

eruption *n.* **1** outbreak, °outburst, °discharge, °expulsion, emission, bursting forth, °explosion, spouting, vomiting (up *or* forth), belching forth: *The eruption of Vesuvius killed thousands in Pompeii.* **2** outbreak, °rash: *The doctor said the eruption would disappear in a day.*

escape *v.* **1** °get away, break out *or* free, °bolt, °flee, °fly, run away *or* off, elope, decamp, abscond, steal *or* slip off *or* away, °take to one's heels, take French leave, °disappear, vanish, *Brit* levant, *Colloq* °take off, °clear out, cut and run, duck out, make oneself scarce, do a disappearing act, *Brit* do a moonlight flit, *US* vamoose, hightail it, skedaddle, *US and Canadian* °skip (town), fly the coop, °cut out; *Slang* vamoose, *Brit* do a bunk, °bugger off, mizzle off, *US and Canadian* scram, °blow, lam out, take it on the lam, take a (runout) powder, *Chiefly Australian* shoot through: *They escaped when I wasn't looking.* **2** °evade, °elude, °avoid, °dodge: *By hiding, they escaped detection for years.* **3** °drain, °leak, °issue, seep, °discharge, °emanate: *Steam was escaping through a hole.* **4** °elude, °evade, baffle, °stump, °mystify, °puzzle, be forgotten by, be beyond (someone): *How the thing started escapes me for the moment.* —*n.* **5** °flight, °getaway, departure, decampment, °bolt, jailbreak, prisonbreak, *Colloq* °break, breakout: *The escape was planned for midnight.* **6** °distraction, °relief, °diversion, °recreation: *He watches westerns as an escape.* **7** leakage, leaking, seepage, seeping, drainage, draining, °leak, °discharge, °outpouring, outflow, effluence, efflux, effluxion: *The escape of radioactive waste from the nuclear power station was reported.*

escort *n.* **1** °guard, convoy, bodyguard, °protection, °guardian, °protector, chaperon, cortege *or* cortège, °retinue, entourage, safe-conduct, usher, °companion: *The king rode in with his armed escort.* **2** °guide, °attendant, conductor, °leader, cicerone: *The curator acted as our escort through the museum.* **3** °companion; °date, boyfriend, beau: *Donald is Thea's escort to the ball.* —*v.* **4** °accompany, °shepherd, °squire, usher, °conduct, °guide, °attend: *Would you please escort Denise in to dinner?* **5** °guard, convoy, °protect, watch over: *The oil tankers were escorted by destroyers.*

especially *adv.* **1** °particularly, °specially, specifically, exceptionally, conspicuously, singularly, remarkably, extraordinarily, unusually, uncommonly, peculiarly, outstandingly, uniquely, °notably, strikingly, noticeably, markedly, signally: *She was especially good at mathematics.* **2** °chiefly, °mainly, predominantly, °primarily, °principally, °first, firstly, first of all, °above all: *He is especially interested in music.*

essay *n.* **1** article, °composition, °paper, °theme, °piece; °thesis, dissertation, disquisition, °tract: *Her essay is on the mating habits of fleas.* **2** °attempt, °effort, °try, °endeavor, °venture; *Colloq* °shot, °go: *This is his first essay into the financial world.* —*v.* **3** °try, °attempt, °endeavor, °strive, make an effort, °undertake, °venture, °tackle, °test, °go about, *Colloq* take a crack *or* whack *or* stab at, *Slang* have a go at, give (it *or* something) a shot, have a go *or* bash (at): *Let him essay to do better.*

essence *n.* **1** °quintessence, quiddity, °(essential) nature, °substance, °spirit, being, °heart, °core, °pith, °kernel, marrow, °soul, °significance, °(active) principle, crux, cornerstone, foundation stone, *Colloq* bottom line: *The essence of her argument is that animals have the same rights as people.* **2** °extract, °concentrate, distillate, °elixir, tincture: *The essence of the flowers is extracted in perfumery.* **3** ***in essence.*** essentially, basically, fundamentally, °materially, °substantially, °at bottom, in the final analysis, *au fond;* °in effect, °virtually: *In essence, there is little to choose between Trotskyism and Stalinism.* **4** ***of the essence.*** °essential, °critical, °crucial, °vital, °indispensable, requisite, °important: *In this contract, time is of the essence.*

essential *adj.* **1** °indispensable, °necessary, requisite, required, °important, °imperative, °vital, °material, quintessential: *A strong defense is essential to peace.* **2** °fundamental, °basic, °intrinsic, °elemental, °elementary, °principal, °primary, °key, °main, °leading, °chief: *Yeast or baking powder is an essential ingredient of bread.*

establish *v.* **1** °found, °create, °institute, °set up, °start, °begin, °inaugurate, °organize, °form, constitute; °decree, °enact, ordain, °introduce: *The company was established in 1796. That law was established to protect consumers.* **2** °secure, °settle, °fix, entrench, install *or* instal, °seat, ensconce; °lodge, °locate; °station: *Hitler* *became established as dictator in 1933. Are you established in your new house?* **3** °prove, °confirm, °certify, °verify, affirm, °determine, °authenticate, °demonstrate, °show, °substantiate, corroborate, validate, °support, °back (up): *It will be difficult to establish exactly how the crime was committed.*

establishment *n.* **1** °foundation, founding, °formation, °organization, construction, °creation, °origin, origination, °institution, inauguration, setting up: *We favor the establishment of diplomatic relations with our neighbors.* **2** °business, °concern, °firm, °company, °enterprise, °institution, °organization; °office; °shop, °store, °market: *He works for a retail establishment.* **3** ***the Establishment.*** the system, the government, the authorities, the administration, the power structure, the ruling class, the (established) order, the conservatives, the powers that be; the Church: *The press must not be under the control of the Establishment.*

estate *n.* **1** °property, holdings, °domain, demesne, °land, landed estate, manor, mansion: *They live on a large estate in the south of France.* **2** °property, holdings, °assets, °capital, °resources, °wealth, °fortune; °belongings, °possessions, chattels: *The estate was divided among the heirs.* **3** estate of the realm, °class, °caste, °order, °standing, °position, °(social) status, °state, °station, °place, °situation, °stratum, °level, °rank: *By virtue of her high estate, she is entitled to certain privileges.*

esteem *v.* **1** °respect, value, °treasure, °prize, °cherish, hold dear, °appreciate, °admire, °look up to, regard highly, °venerate, °revere, °reverence, °honor, defer to; °like, °love, °adore: *I esteem greatly the friendship we have enjoyed.* **2** °consider, °judge, deem, °view, °regard, °hold, °estimate, °account, °believe, °think, °rate, °rank, °reckon, °evaluate: *He esteemed the plan to be the most harebrained he'd ever heard.* —*n.* **3** °estimation, °(high) regard, °respect, °(high) opinion, °favor, °admiration, °appreciation, °approval, approbation: *He holds her talents in very high esteem. My esteem for your father's accomplishments is undiminished.*

estimable *adj.* esteemed, °respectable, respected, °admirable, admired, valuable, valued, creditable, °worthy, °meritorious, °reputable, honored, °honorable, °laudable, °praiseworthy, commendable, °excellent, °good: *I want you to meet my estimable friend, Esterhazy.*

estimate *v.* **1** °approximate, °gauge, °determine, °judge, °guess; assess, appraise, value, °evaluate, °reckon, °calculate, °work out, *Colloq* guesstimate *or Colloq* guestimate: *The experts estimated the costs of restoration at too high a figure.* **2** °consider, °think, °believe, °guess, conjecture, °judge: *I estimate our chances of success as very low.* —*n.* **3** approximation, °gauge, °guess, conjecture, assessment, appraisal, °evaluation, °reckoning, °calculation, *Colloq* guesstimate *or Colloq* guestimate: *What is your estimate of the company's value?* **4** °estimation, °belief, °opinion, °judgment, °thinking, °feeling, °sentiment, °sense, °(point of) view, °viewpoint: *My estimate of his abilities has nothing to do with it.*

estimation *n.* **1** °opinion, °judgment, °view, °(way of) thinking, °mind: *In my estimation, the scheme will fail.* **2** °esteem, °regard, °respect, °admiration: *Her estimation of his talent is unflagging.* **3** °estimate, approximation, °guess, °gauge: *You must make an estimation of the value for insurance purposes.*

estranged *adj.* alienated, divided, separated, °withdrawn, disaffected, driven apart, dissociated, disassociated: *Carstairs may be getting together again with his estranged wife.*

etch *v.* **1** °engrave, incise, °carve, inscribe, °grave, °cut, °score, °scratch, corrode, eat into: *After painting on the design, use acid to etch the metal plate.* **2** °impress, imprint, °engrave, ingrain: *The scene will be etched into my memory forever.*

eternal *adj.* **1** °everlasting, °timeless, °infinite, °endless, °immortal, °limitless: *They pledged eternal love.* **2** unending, °endless, ceaseless, unceasing, incessant, °perpetual, °constant, °continuous, interminable, uninterrupted, °nonstop, unremitting, °persistent, °relentless; °continual, °recurrent: *I hear my neighbors' eternal arguing.* **3** unchanged, unchanging, immutable, °invariable, unvarying, unalterable, °permanent, °fixed, °constant, °everlasting, °enduring, °lasting, undiminished, unfaltering, unwavering: *Eternal elements of tragedy beset man.*

eternity *n.* endlessness, everlastingness, unendingness, boundlessness, °perpetuity, timelessness, infinity: *Many religions believe in the eternity of the soul.*

ethical *adj.* °moral, °upright, °righteous, °right, °just, °principled, °correct, °honest, °proper, °open, °decent, °fair, °good, °virtuous, straightforward, high-minded, °noble: *It wasn't ethical of him to cheat in the exam.*

etiquette *n.* °code (of behavior), °form, °convention, °ceremony, formalities, °protocol, rules, custom(s), °decorum, °propriety, politesse, politeness, °courtesy, °(good) manners, °civility, seemliness: *Etiquette requires that you address me as "Sir."*

eulogize *v.* °praise, °extol, °laud, °applaud, °compliment, sound *or* sing the praises of, acclaim; °appreciate, °honor; °flatter: *Kirk was eulogized for his contribution to space travel.*

eulogy *n.* °praise, commendation, acclaim, acclamation, °tribute, °compliment, °applause, °homage, plaudits, encomium, accolade, paean, panegyric: *The eulogy listed Wotton's many achievements.*

euphemism *n.* amelioration, mollification, mitigation, cushioning, *Technical* paradiastole: *Euphemism is saying "not too good" when you mean "bad" or "awful."*

evacuate *v.* **1** °empty, °clear (out), °exhaust, °drain, deplete, °purge, °get rid of, °void, °discharge, °vent; °divest, °deprive: *In poisoning cases, they first evacuate the stomach. The term "spirit" has been evacuated of all its proper meaning.* **2** °vacate, °desert, °leave, °depart (from), withdraw *or* retire (from), °go away (from), °quit, °relinquish, °abandon, decamp (from), move *or* pull out (of *or* from): *When the enemy approached, the troops evacuated the area.* **3** relocate, °move: *Thousands were evacuated to a safe distance when the bomb was discovered.*

evade *v.* **1** °avoid, °elude, °dodge, °sidestep, °escape (from); °get away (from), °get out of, °duck, circumvent, °shirk, *Colloq chiefly US and Canadian* weasel out (of): *The prisoner evaded capture. Don't evade your responsibilities.* **2** °quibble, °equivocate, tergiversate, °maneuver, hedge, °shuffle, fudge, °fence, parry, *Colloq* °waffle, *Slang* cop out: *The witness continued to evade the barrister's questions.*

evaluate *v.* **1** value, appraise, assess: *You must be an expert to evaluate netsuke.* **2** °judge, °rank, °rate, °gauge, °estimate, °approximate, °calculate, °reckon, °compute, °figure, quantify, °determine, ascertain: *How can they evaluate your importance to the project?*

evaluation *n.* **1** appraisal, valuation, assessment: *The insurance company refused to accept a higher evaluation on my house.* **2** °estimate, °estimation, approximation, rating, °opinion, ranking, °judgment, °reckoning, figuring, °calculation, computation, °determination: *The committee meets annually to discuss the evaluation of each employee's contribution.*

evaporate *v.* 1 vaporize; boil off *or* out; dehydrate, desiccate; *Technical* °sublime: *Evaporate some of the water to strengthen the soup.* 2 °disappear, °disperse, °dissipate, vanish, evanesce, evanish, dispel; °fade (away), °melt away, °dissolve: *Opposition to his appointment seems to have evaporated.*

evaporation *n.* 1 vaporization, drying (up *or* out), dehydration, desiccation, exsiccation, parching, searing: *Clouds are formed by the evaporation of terrestrial water.* 2 disappearance, dispersion, dispelling, °dissipation, evanescence, dematerialization, °dissolution, fading (away), melting (away): *She rejoiced at the evaporation of all charges against her.*

evasion *n.* 1 °escape, avoidance, shirking, dodging: *We disapprove of his evasion of his civil duties.* 2 °subterfuge, °deception, °deceit, chicane *or* chicanery, °artifice, cunning, °trickery, sophistry, °excuse, dodging, prevarication, °lying, fudging, evasiveness, quibbling, equivocation, double-talk: *All the interviewers' questions were met with evasion.*

evasive *adj.* °devious, °indirect, equivocating, °equivocal, misleading, °oblique, °ambiguous, sophistical, casuistic, °shifty, dissembling, cunning, °tricky, °deceitful, *Colloq* cagey, jesuitical: *Asked if he had visited her, he gave an evasive reply, and she was evasive about how she had spent the evening.*

eve *n.* 1 evening *or* day *or* night before, time *or* period before; vigil: *It was Christmas Eve, so we went out caroling.* 2 °verge, °threshold, °brink: *We met on the eve of my departure for Hungary.*

even *adj.* 1 °smooth, °flat, °plane, °level, °regular, °uniform, °flush, °straight, °true: *Sand the edges till they are even.* 2 Sometimes, ***even with.*** level *or* uniform (with), coextensive (with), °flush (with), parallel (with *or* to): *Is that board even with the others? Make sure that the lines at the bottom of the columns are even.* 3 °steady, °regular, °consistent, °constant, °uniform, unvaried, unvarying, °methodical, unchanging, °set, °equable, °stable, °measured, metrical, rhythmical, °orderly, ordered, °monotonous, unbroken, uninterrupted: *We walked along at an even pace.* 4 even-tempered, °calm, °equable, composed, placid, °serene, °peaceful, °cool, °tranquil, unruffled, imperturbable, undisturbed, °impassive, °steady, °temperate, equanimous, °self-possessed, °sober, °staid, °sedate, sober-sided: *People of even disposition are unexcited, unexcitable, and unexciting.* 5 balanced, °equal, the same, °identical, coequal, °level, °drawn, on a par, tied, neck and neck; °equivalent, *Colloq* fifty-fifty, *Brit* level pegging, *US* even-steven: *At half time the scores were even. I have an even chance of getting the job.* 6 °square, quits, °equal: *If I pay for this beer we'll be even.* 7 °fair (and square), °square, °impartial, °disinterested, °neutral, °just, even-handed, °equitable, straightforward, °on the level, °honest, °upright, unbiased, °unprejudiced: *See that you make an even distribution of the food parcels.* 8 °exact, °precise, °round, rounded off *or* out *or* up *or* down: *The bill came to an even fifty dollars.* 9 ***get even (with).*** °repay, revenge oneself (on), even *or* settle accounts *or* the score (with), °requite, °reciprocate, °retaliate, be revenged: *I'll get even with her for telling mother. What will you do to get even for that insult?* —*adv.* 10 °still, °yet; all the (more), °indeed, (more) than ever: *He is even dumber than I had thought. He is in debt to everyone, even his tailor.* 11 Sometimes, ***even with*** *or* ***though.*** °notwithstanding, °despite, °in spite of, disregarding: *Even with delays, we arrived on time.* 12 ***even so.*** °nevertheless, °nonetheless, °still, °yet, °notwithstanding, °all the same, in spite of that, despite that: *He refused to attend; even so, we sent him an invitation.* —*v.* 13 Usually, ***even up*** *or* ***out.*** °smooth, °flatten, °level, °equalize; align: *This road will be fine when they even out the bumps.* 14 ***even out*** *or* ***up.*** °equalize, °balance (out), °settle; °compensate: *Unfortunately, our profits and our expenses evened out.*

evening *n.* nightfall, eventide, °dusk, sunset, sundown, p.m., *Literary* gloaming: *The Klincks will join us for dinner this evening.*

event *n.* 1 °occurrence, °happening, °incident, °episode, °occasion, °circumstance, °affair, °experience: *An event then took place that changed the course of his life.* 2 °issue, °outcome, consequence, °result, conclusion, °upshot, °end, °effect: *There is no merit in preparing for a disaster after the event.* 3 ***at all events*** *or* ***in any event.*** come what may, °in any case, °at any rate, °regardless, anyhow, anyway: *At all events, we were ready and waiting when the raid started.* 4 ***in the event.*** in the reality *or* actuality, as it *or* things turned out, at the time, when it happened: *In the event, we left as soon as we could.*

eventful *adj.* °busy, °full, °active, °lively, °exciting, °interesting; °important, °significant, °signal, consequential, °notable, °noteworthy, °momentous, °memorable: *Between your wedding and the birth of your son, it certainly has been an eventful week.*

eventual *adj.* 1 °ultimate, °final, °last, concluding, resulting: *The eventual outcome will be the same.* 2 °due, expected, anticipated, °inevitable, °likely, consequent, resulting, resultant, foreordained, preordained, °unavoidable, °destined, predestined, °unavoidable, ineluctable, °probable: *As they cannot afford the mortgage payments, they are faced with the eventual loss of their home.*

eventuality *n.* °circumstance, contingency, °event, °occurrence, °happening, °case; °likelihood, °chance, °possibility, °probability: *We must prepare for the eventuality of war.*

eventually *adv.* °ultimately, °finally, °at last, in the end *or* long run, at the end of the day, °sooner or later, when all is said and done, in the final analysis, in due course, in (the course of) time, after all: *We must all die eventually.*

ever *adv.* 1 at all, (at) any time, at any point *or* period, on any occasion, °in any case, by any chance: *Do you ever visit London?* 2 °always, °forever, °yet, °still, °even, at all times, in all cases, eternally, perpetually, endlessly, everlastingly, constantly, continuously, continually, forever and a day, till the end of time, till the cows come home, till Doomsday; °all the time: *He is ever the one to make us laugh. Literacy is becoming ever more important every day.*

everlasting *adj.* °eternal, °deathless, undying, °immortal, °infinite, °timeless; never-ending, °perpetual, °constant, °continual, °continuous, °permanent, unceasing, incessant, interminable, °endless: *They believed in everlasting punishment after death. I wish that dog would stop its everlasting barking.*

everyday *adj.* 1 °daily, day-to-day, quotidian, °diurnal; circadian: *In our family a big breakfast was an everyday occurrence.* 2 commonplace, °common, °ordinary, °customary, °regular, °habitual, °routine, °usual, °run-of-the-mill, unexceptional, °accustomed, °conventional, °familiar: *Bathing is an everyday activity in many homes.* 3 °prosaic, mundane, °dull, unimaginative, unexciting, °mediocre, °inferior: *These are very everyday paintings of little value.*

everyone *pron.* everybody, all (and sundry), one and all, each and every one *or* person, the whole world, everybody under the sun, every Tom, Dick, and Harry: *Everyone will want to come to my party.*

everything *pron.* all, all things, the aggregate, the (whole *or* entire) lot, the total, the entirety, *Colloq* the whole kit and caboodle, the whole shooting match,

Chiefly US and Canadian the whole shebang: *Everything was destroyed in the earthquake.*

everywhere *adv.* in all places, in *or* to each *or* every place *or* part, in every nook and cranny, high and low, °far and wide, near and far; ubiquitously, °universally, globally; °throughout: *I went everywhere searching for clues. Love is everywhere. Errors occur everywhere in his writings.*

evict *v.* oust, dislodge, °turn out (of house and home), °expel, °eject, °remove, °dispossess, °put out, *Law* disseize, *Colloq* toss *or* throw *or* kick *or* boot out, *Brit* turf out: *The landlord evicted us for nonpayment of rent.*

eviction *n.* ouster, dispossession, dislodgment, °expulsion, °ejection, °removal, *Law* disseizin, *Colloq* the boot: *My eviction from the club was for refusing to wear a tie.*

evidence *n.* **1** °proof, ground(s), fact(s), °data, °basis, °support, verification, attestation, affirmation, confirmation, validation, corroboration, substantiation, documentation, certification: *Do we have enough evidence to convict the suspects?* **2** °testimony, °statement, deposition, affidavit, averment, °assertion: *The prosecution will present its evidence tomorrow.* **3** °indication, °sign, °mark, °token, °manifestation, °demonstration, °hint, °suggestion, °clue, °trace, smoking gun: *There is evidence that there are mice in the house.* —*v.* **4** °demonstrate, °show, °display, °manifest, °signify, °exhibit, °reveal, °denote, °attest, °prove, evince, °testify, °(bear) witness: *The destruction of the forests is evidenced by the open plains.*

evident *adj.* °clear, °obvious, °plain, °apparent, °manifest, °patent, palpable, °conspicuous, clear-cut, °express, unmistakable, °incontrovertible, understandable, comprehensible, recognizable, °perceptible, perceivable, °discernible, °noticeable: *It is evident that you stole the tarts.*

evidently *adv.* **1** °clearly, °obviously, plainly, °manifestly, palpably, °apparently, patently, indubitably, °undoubtedly, doubtless(ly), without a doubt, indisputably, incontestably, incontrovertibly, undeniably, unquestionably, °surely, certainly, to be sure: *He is evidently the culprit.* **2** °apparently, °outwardly, °seemingly, it would seem, so it seems, as far as one can see *or* tell, to all appearances, °ostensibly: *Evidently, there were two people here, not just one.*

evil *adj.* **1** °bad, °awful, °wrong, °immoral, °wicked, °sinful, nefarious, iniquitous, °base, °corrupt, °vile, accursed, °damnable, °villainous, heinous, °infamous, flagitious, °foul, °nasty, °abominable, °atrocious, °horrible, horrid, °ghastly, °grisly, °dreadful, depraved, °vicious, malevolent, maleficent, malefic, black-hearted, °evil-minded: *The evil tyrant killed any who opposed him.* **2** treacherous, °traitorous, °perfidious, insidious, °unscrupulous, unprincipled, °dishonest, °dishonorable, °crooked, °criminal, felonious, knavish, °sinister, underhand(ed), °dirty, °corrupt: *She has an evil plan for killing her boss's wife.* **3** °harmful, °destructive, °hurtful, °injurious, °mischievous, °detrimental, °ruinous, deleterious, °disastrous, catastrophic, pernicious, noxious, °malignant, malign, °virulent, toxic, °poisonous, °deadly, °lethal: *The Nazis promulgated evil policies against the Jews and other minorities.* **4** °unfortunate, unlucky, °ominous, °inauspicious, dire, unpropitious, °calamitous, infelicitous, woeful: *Our business has fallen on evil days.* **5** °bad, °offensive, °disgusting, °repulsive, °awful, °nasty, mephitic, noxious, °foul, pestilential, °putrid, °vile; °disagreeable, unpleasant: *An evil odor permeated the crypt.* —*n.* **6** badness, °sin, °vice, wickedness, iniquity, turpitude, immorality, °profligacy, depravity, degeneracy, corruption, °degradation, devilry *or* deviltry, villainy, nefariousness, viciousness, vileness, heinousness, flagitiousness, baseness, foulness: *The evil that men do lives after them, the good is oft interred with their bones.* **7** °harm, °hurt, °injury, °mischief, °damage, °ruin, °calamity, °misfortune, °catastrophe, °destruction, °disaster, cataclysm; °ill, °misery, °suffering, °pain, °sorrow, °woe, °agony, °anguish: *Evil befell the residents of Pompeii.*

evil-minded *adj.* **1** dirty(-minded), smutty, °obscene, depraved, °lewd, °lascivious, °lecherous, salacious, licentious, °filthy, °nasty; foulmouthed: *Those anonymous phone calls were made by some evil-minded degenerate.* **2** °wicked, °sinful, flagitious, °vicious, °hateful, malicious, °spiteful, malevolent, °evil, °bad: *The evil-minded old witch grabbed Hansel and Gretel.*

evoke *v.* °summon (up), call up *or* forth, °elicit, conjure up, invoke, °recall, reawake(n), (a)wake, wake(n), (a)rouse, °raise: *Seeing her again evoked fond memories.*

evolution *n.* °development, °advance, °growth, °progress, °progression, phylogeny, evolvement, developing, growing, evolving, °formation, maturation, °production: *This book traces the evolution of the airplane. His treatise is on the evolution of insects.*

exact *adj.* **1** °precise, °accurate, °correct, °faithful, °true, °faultless, °identical, °literal, °perfect, consummate: *This is an exact copy of the original. Were those her exact words?* **2** °careful, °meticulous, °strict, rigorous, °accurate, °exacting, °severe, °fastidious, °scrupulous, °thorough, °painstaking, °rigid, punctilious: *He has conducted the most exact experiments.* —*v.* **3** °demand, °extort, °require, °enforce, insist on *or* upon, °extract, °impose, wrest, compel, enjoin, °call for, °requisition, °claim: *The chief exacted tribute before releasing the prisoners.*

exacting *adj.* °demanding, rigorous, °difficult, °rigid, °stern, °hard, °tough, °severe, °harsh, °burdensome, taxing, stringent, °imperative, unsparing, °oppressive, °tyrannical: *I know of no more exacting job than that of air traffic controller.*

exactly *adv.* **1** accurately, °precisely, strictly, °perfectly, correctly, unerringly, faultlessly, faithfully, scrupulously, °literally, °to the letter, word for word, °verbatim, closely; methodically, systematically: *She translated the passage exactly, with no errors.* **2** °definitely, °absolutely, °positively, undeniably, °surely, certainly, unequivocally, °completely, in every respect, in all respects, °particularly, specifically, explicitly, °just, °quite, °expressly, °precisely, accurately, °truly, *Colloq Brit* bang on: *This is exactly the kind of house we want.*

exaggerate *v.* °overstate, °magnify, °inflate, overdraw, embroider, °embellish, °elaborate, °enlarge, °stretch, °romance, overemphasize, overstress, overplay, °overdo, °exalt, hyperbolize, °paint, *Colloq* lay it on thick, °play up, °pile it on: *She exaggerates when claiming to be the best actress in the world.*

exaggeration *n.* overstatement, °magnification, inflation, embroidery, °embellishment, °elaboration, enlargement, °stretch, romanticization, °extravagance, overemphasis, °excess, exaltation, enhancement, hyperbole; empty talk, °bombast, bragging, boasting, boastfulness, magniloquence, *Literary* gasconade, rodomontade, *Colloq* fish story, puffery, *Slang* bull(shit), °hot air: *It was a press agent's exaggeration to call the movie a "colossal epic."*

exalt *v.* **1** elevate, raise *or* lift (up *or* on high), upraise, uplift, upgrade, °boost, °promote, °advance: *The headmaster was exalted to heroic status by the students.* **2** °praise, °honor, °extol, °glorify, °idolize, °dignify, ennoble, °revere, °reverence, °venerate, pay homage *or* tribute to, °celebrate; lionize: *O magnify the Lord with me, and let us exalt His name together.* **3** °stimulate,

°excite, °animate, (a)rouse, °fire, °inspire, °electrify, °awaken, °spur, °stir (up), inspirit: *Certain drugs have the effect of exalting the imagination.*

exalted *adj.* **1** °elevated, °lofty, °high, °eminent, °notable, °noted, °prominent, °famous, famed, °celebrated, °distinguished, °dignified, honored, °prestigious, °glorified, °sublime, °grand: *During his later life he enjoyed an exalted reputation.* **2** °elevated, °noble, °lofty, °superior, uplifting, heightened, high-flown; exaggerated, °pretentious, overblown, °inflated: *His writing is euphuistic, that is, flowery or exalted. She has an exalted notion of her importance.* **3** °elated, °excited, °exultant, °ecstatic, jubilant, °overjoyed, °joyful, °rapturous, transported, blissful, °happy, joyous, in seventh heaven, uplifted, *Colloq* on cloud nine, *Brit* over the moon: *The news of the award made him feel exalted.*

examination *n.* **1** °investigation, °scrutiny, °study, °analysis, inspection, inquiry *or* enquiry, °probe, °search, °exploration, °research, °survey, going over, checkup, check(out), appraisal, assessment: *Examination of the finances revealed a secret Swiss bank account.* **2** testing, °test, °quiz, exam: *There will be a written examination on Friday.* **3** °interrogation, inquisition, inquiry *or* enquiry, catechism, cross-examination, *Colloq* third degree, grill(ing): *Examination of the prisoners is left to intelligence officers.*

examine *v.* **1** °investigate, °scrutinize, °study, °peruse, °scan, °pore over, °analyze, °sift, °inspect, inquire *or* enquire into, go over *or* through *or* into, look over *or* into, °probe, °search, °explore, °research, °survey, °check up on, °check (out), appraise, assess, °weigh, *Brit* °vet, *Slang* °case: *The theory will be examined thoroughly.* **2** °test; interrogate, °quiz, catechize, cross-examine, °question, °sound out, *Colloq* grill, °pump: *The director closely examined all applicants for the position.*

example *n.* **1** °instance, °case, °sample, °specimen, °illustration: *If this is an example of your work, I'm afraid we aren't interested.* **2** °model, °prototype, °standard, archetype, exemplar, °pattern, benchmark, °norm, criterion: *You should set an example to the children.* **3** °warning, admonition, °lesson: *The judge made an example of him by giving him the maximum sentence.* **4** ***for example.*** °for instance, as a case in point, as an illustration *or* example, by way of illustration, to illustrate, e.g., *exempli gratia*: *Consider, for example, the poem by Mangiacavallo.*

exasperate *v.* **1** °anger, °infuriate, °enrage, incense, °madden, rile, drive mad; embitter; °inflame; *Colloq* drive crazy, °drive up the wall: *Dealing with the bureaucracy can exasperate anyone.* **2** °irritate, °irk, °annoy, °bother, °harass, pique, °gall, nettle, °rankle, °provoke, vex, °pester, °torment, °plague; hector, badger, *Colloq* °bug, needle, peeve, °get, get under (someone's) skin, rub the wrong way, °aggravate, *Brit* rub up the wrong way, *Slang* get (someone's) goat, piss (someone) off: *The child's constant questions were beginning to exasperate me. It is the bill collectors who exasperate me the most.*

excavate *v.* **1** dig (out *or* up), hollow *or* gouge (out), °scoop out, burrow *or* cut (out): *They are excavating a great hole for the Channel tunnel.* **2** °unearth, uncover, °expose, °clear, °lay bare, °dig up, disinter, °bring up, exhume: *A large area of Pompeii has been excavated.*

excavation *n.* °cavity, °hole, °pit, crater, °cut, ditch, trench, trough, °burrow, °hollow, °shaft, °tunnel; °mine, °quarry: *The concrete for the foundation will be poured into this excavation.*

exceed *v.* **1** °surpass, °top, °excel, be superior to, go beyond, °beat, °overwhelm, °better, outdistance, °pass, °overtake, °outstrip, outrank, outrun, °outdo, outpace, °transcend, outshine, outreach, °overshadow, °eclipse: *The success of the new product exceeded our expectations.* **2** °overstep, go beyond, overextend: *His behavior exceeded the bounds of decency.*

exceeding *adj.* °great, °huge, °enormous, °extraordinary, °excessive, °exceptional, °surpassing: *The exceeding poverty of the people is heart-rending.*

exceedingly *adv.* °very, °extremely, °especially, exceptionally, considerably, incomparably, immeasurably, extraordinarily, remarkably; excessively, greatly, hugely, enormously: *She plays the violin exceedingly well. Even for a dog, that is an exceedingly ugly dog.*

excel *v.* °surpass, be superior (to), °dominate, °top, °exceed, go beyond, °beat, °outstrip, outrank, °outdo, outpace, outshine, °overshadow, °eclipse; °shine, be preeminent: *Few places excel the Caribbean islands for beauty. He really excels when it comes to swimming.*

excellence *n.* °superiority, °merit, °(high) quality, goodness, fineness, greatness, °prominence, eminence, °preeminence, °distinction, value, °worth, °supremacy: *Those who attain excellence often devote their lives to one pursuit.*

excellent *adj.* °superb, °outstanding, °exceptional, °superior, °matchless, °peerless, unequaled, without equal, °nonpareil, °supreme, °superlative, °sterling, °capital, first-class, °first-rate, °prime, °choice, °select, °distinguished, °noteworthy, °notable, °worthy, the best, tiptop, °admirable, °splendid, °remarkable, °marvelous, °extraordinary, *Colloq* A-1 *or* A-one, °great, smashing, °super, °terrific, °fantastic, *Brit* °magic, *Dialectal* °champion, *Old-fashioned* top-hole, °ripping, tickety-boo, *US* A number 1, °major, *Australian* bonzer, *Slang* °cool, ripsnorting: *Thank you for an excellent dinner. He is an excellent pianist.*

except *prep.* **1** Sometimes ***except for.*** excepting, °save, but, excluding, °exclusive of, °barring, °bar, with the exception of, omitting, not counting, °apart from, but for, other than, °saving: *There was no one there except us. Except for us, no one came.* —*conj.* **2** ***except that.*** except *or* but (for the fact) that, but, save that: *He would have gone except that he has no car.* —*v.* **3** °exclude, °omit, °leave out, °excuse: *As usual, the wealthy were excepted from the tax increase.*

exception *n.* **1** °exclusion, °omission: *With the exception of those who are absent, all members are entitled to a free T-shirt.* **2** debarment, blockage, lockout, shutout: *His exception from membership in the country club was a great offense to his wife.* **3** departure, anomaly, irregularity, special case; °oddity, °freak, °rarity, °peculiarity, °quirk: *The exception proves the rule. Why does Vanni always have to be an exception?* **4** ***take exception (to).*** make *or* raise (an) objection (to *or* against), °object (to), demur (at), °find fault (with), take offense *or* umbrage (at), be offended (at); °(call into) question, °cavil, °quibble, °challenge, °oppose, °disagree (with): *She takes exception to everything I suggest. If you propose to ban smoking, will some people take exception?*

exceptionable *adj.* objectionable, °disputable, °questionable, criticizable, °unacceptable, °unsatisfactory: *We found nothing exceptionable about the service.*

exceptional *adj.* **1** °special, °unusual, especial, °out of the ordinary, °extraordinary, uncommon, °rare, °singular; °strange, °irregular, aberrant, °odd, °peculiar, anomalous: *Those candlesticks were an exceptional value. He insisted that rainstorms in the Sahara were not exceptional.* **2** °gifted, °talented, °superior, °outstanding, above average, °excellent, °prodigious, °extraordinary: *Your child is exceptional, Mrs. Einstein.* **3** handicapped, below average, °deficient, *Brit* ESN (= 'educationally subnormal'): *The boy attends a school for exceptional children.*

excerpt *n.* **1** °extract, °selection, °quotation, citation, °passage, pericope: *The speaker read excerpts from the best writing.* —*v.* **2** °extract, °select, °quote, cite, cull (out), °pick (out), °take: *Parts of this book were excerpted from his earlier writings.*

excess *n.* **1** °surplus, overabundance, overflow, superabundance, overabundance, nimiety, °superfluity, °surfeit, plethora, °glut, redundancy, oversufficiency, supererogation, °leftover; overkill: *The excess of income over expenses constitutes profit. There is an excess of water on the road after a rainstorm.* **2** Often, ***excesses.*** debauchery, °extravagance, immoderation, °prodigality, overindulgence, intemperance, °dissipation, dissoluteness: *Because of their excesses they have been shunned by their friends.* —*adj.* **3** °surplus, °extra, °superfluous, °excessive, °leftover, °residual, °remaining: *After paying the bills, any excess money goes into savings.*

excessive *adj.* **1** °immoderate, °inordinate, °disproportionate, °extravagant, °exorbitant, °superfluous, °excess, undue, °enormous, °extreme, °unreasonable, °unwarranted, unjustifiable, °outrageous, °unconscionable: *This job is making excessive demands on my time.* **2** overdone, fulsome, cloying, nauseating, °disgusting: *I was sickened by her excessive sweetness.*

exchange *v.* **1** °trade, barter, °switch, °change, interchange, °reciprocate, °return, *Colloq* swap *or Colloq* swop: *We exchange gifts at Christmas.* —*n.* **2** °trade, barter, °change, °traffic, °commerce, dealing, °truck, °transfer, interchange, reciprocity, reciprocation, °switch, quid pro quo, tit for tat, *Colloq* swap *or Colloq* swop: *An exchange of prisoners was swiftly completed. What can I do for you in exchange?* **3** altercation, °argument, °quarrel, °disagreement, unpleasantness: *We had a brief exchange, then he struck me.* **4** °market, stock market, stock exchange, securities exchange, the Market, the Board, the Big Board, the Exchange, the Bourse, Wall Street, *US* the Street: *The exchange reported little activity in anticipation of the finance minister's speech.*

excitable *adj.* °volatile, °jumpy, apprehensive, °nervous, °restive, °restless, fidgety, edgy, °touchy, highly strung, high-strung, mercurial, °emotional, °quick-tempered, °testy, hot-blooded, °feverish, °hysterical, *US* on a short string: *The director becomes very excitable before each performance.*

excite *v.* **1** (a)rouse, °spur (on), °stir (up), °move, °animate, °enliven, °activate, °motivate, invigorate, °energize, °stimulate, °cause, °provoke, °prod, °agitate, °incite, °quicken, °urge, (a)wake, (a)waken, °call forth, °summon (up), °elicit, °inspire, inspirit, °rally, galvanize, °electrify, °foment, °fire (up), °inflame, °kindle, ignite, °initiate, instigate, °generate, °occasion, °begin, °start, °bring about, °effect, set in motion, *Colloq* get going, °spark, °wind up, get (someone) (all) steamed up, hop up, *US* kick-start, light a fire under: *The speech excited the crowd's patriotic fervor.* **2** °agitate, °disturb, °perturb, °stir up, discompose, °fluster, °ruffle, °upset, disconcert: *Don't excite the horses.* **3** °thrill, °stir up, titillate, °work up, °arouse, °inflame: *Risqué films are intended to excite viewers.*

excited *adj.* **1** (a)roused, stirred (up), stimulated, °agitated, °disturbed, perturbed, °upset, worked up, wrought up, wound up, keyed up, °overwrought, discomposed, °disconcerted, discomfited, °nervous, edgy, °on edge, uneasy, flustered, ruffled, fidgety, °frantic, frenetic, aflame, °feverish, frenzied, °hysterical, °beside oneself, *Colloq* itchy, (all) hot and bothered, °high, on a high, off the deep end, °out of one's mind: *Thewless was in a very excited state by the time the police arrived. Don't get so excited just because he called you a name.* **2** °ardent, zealous, °impassioned, °passionate, °eager, energized, °energetic, °active, °brisk, °animated, °lively, °spirited, fervid, °fervent, vehement, stimulated, °enthusiastic, galvanized, electrified, intoxicated, *Colloq* turned on: *The excited children scrambled into the boat. She becomes excited listening to rock 'n' roll.*

excitement *n.* **1** restlessness, disquiet, disquietude, °tension, °agitation, °unrest, malaise, °discomfort, jumpiness, nervousness, freneticness, excitation: *He had to give the patient an injection to quell her excitement.* **2** perturbation, °upset, °action, ado, °activity, °ferment, °furor, turmoil, °tumult, to-do, °stir, commotion, hubbub, brouhaha, °fuss, hurly-burly, *Colloq* fireworks: *The robbery caused a lot of excitement at the bank.* **3** °animation, °eagerness, °enthusiasm, exhilaration, ebullience: *She could hardly contain her excitement at winning the lottery.*

exciting *adj.* **1** °stimulating, °intoxicating, heady, °thrilling, °stirring, °moving, inspiring, °rousing, °exhilarating, electrifying, galvanizing, energizing, °invigorating; °overwhelming, °overpowering, astounding, astonishing, °amazing, mind-boggling, *Colloq* far-out, rip-roaring, mind-blowing: *Orbiting in the space shuttle was the most exciting experience of my life.* **2** °seductive, °sensuous, °voluptuous, °ravishing, captivating, charming, °tempting, enticing, alluring, °provocative, titillating, *Colloq* °sexy: *Farfella is truly an exciting young woman.*

exclaim *v.* call *or* cry (out), °proclaim, vociferate, utter, °declare, ejaculate, °shout, °yell, °bawl, °bellow, burst out (with), °blurt out, *Colloq* holler: *The robber exclaimed, "Hands up!"*

exclamation *n.* °outcry, °call, °cry, utterance, ejaculation, °interjection, vociferation, °shout, °yell, °bellow, *Colloq* holler: *She gave an exclamation of surprise at seeing me there.*

exclude *v.* **1** Often, ***exclude from.*** keep out *or* away, lock *or* shut out, °ban, °bar, debar, °prohibit, interdict, °forbid, proscribe, °deny, °refuse, disallow: *Women are excluded from participating in certain church services. Must we exclude Maria, too?* **2** °eliminate, °leave out, °reject, °omit, °except, °preclude, °repudiate, count out: *Exclude them from consideration as candidates.* **3** °eject, °evict, °expel, oust, °get rid of, °remove, °throw out, *Colloq* toss out, *Slang* °bounce: *He was excluded from the club because of his behavior.*

exclusion *n.* **1** lockout, shutout, °ban, °bar, °prohibition, interdiction, forbiddance, °denial, °refusal, disallowance, proscription: *He would not join a club that supported the exclusion of women members.* **2** elimination, °rejection, °omission, repudiation, °exception, preclusion: *We must ensure the exclusion of unqualified candidates.* **3** °ejection, °eviction, °expulsion, ouster, °removal, riddance: *The exclusion of rowdies and hooligans is in the club's best interests.*

exclusive *adj.* **1** °incompatible, inimical; unshared, °unique, °absolute, restricted, °limited: *Exclusive concepts, like "animal" and "vegetable," do not coincide save in their both being "alive." Our newspaper printed the exclusive story of the disaster.* **2** °chic, clannish, °choice, °upper-class, aristocratic, closed, restricted, restrictive, °private, °snobbish, °fashionable, °elegant, °stylish, °select, *Colloq* °trendy, *Slang* classy: *They are seen in the most exclusive nightclubs.* **3** °only, °single, °one, °sole, °singular, °unique: *The exclusive way of getting home was by boat.* **4** Usually, ***exclusive of.*** excluding, excepting, °except for, omitting, ignoring, leaving aside, °apart from, (de)barring, not counting, eliminating: *Exclusive of minors, all citizens are entitled to vote.*

excruciating *adj.* tormenting, torturing, torturous, °agonizing, °painful, racking, °intense, °extreme, °unbearable, unendurable, °severe, °acute, °exquisite, °harrowing, distressful, distressing, °piercing, °insufferable: *That hangnail is causing me excruciating pain.*

excursion *n.* **1** °trip, °tour, °outing, airing, °expedition, voyage, °cruise, °journey, junket, jaunt; °ramble, °stroll, °walk, hike, trek, °drive, °ride, °sail: *We left the city and took an excursion to the mountains.* **2** °detour, deviation, side trip, °diversion, °digression, excursus: *Forgive the excursion, but I'll return to the subject in a minute.*

excuse *v.* **1** °forgive, °pardon, °overlook, absolve, °clear, exonerate, acquit, exculpate, °pass over, °disregard, wink at, °ignore, be blind to, look the other way, pay no attention *or* heed (to), find *or* prove innocent (of): *Please excuse my tardiness.* **2** °release, let go *or* off, °liberate, °free, °relieve, °exempt, absolve; °dismiss, *Colloq* let off the hook: *She was excused from attending class today. You may be excused.* **3** condone, °allow, °permit, °defend, apologize for, °justify, °warrant, °explain, °vindicate, °rationalize, °mitigate, extenuate, palliate: *Poverty does not excuse your stealing.* —*n.* **4** apology, °explanation, °story, °reason, justification, °defense, °plea, vindication, condonation, rationalization, extenuation, mitigation, palliation; °basis, °grounds, °foundation, °cause: *Can you think of any excuse for your behavior?* **5** °forgiveness, °remission, °pardon, °indulgence, °reprieve, clearing, exculpation, absolution, exoneration, acquittal, °disregard, heedlessness, vindication, °clearance, acquittance: *Ignorance of the law is not an excuse.* **6** °evasion, °subterfuge, °pretense, °pretext, °makeshift, °escape, °loophole, way out, *Colloq* °alibi, °stall, *Slang* cop-out: *He always uses his grandmother's illness as an excuse for missing school.*

execute *v.* **1** °accomplish, do, carry out *or* off *or* through, °perform, °discharge, dispatch *or* despatch, bring about *or* off, °implement, °engineer, °cause, *Colloq* °pull off, °put over, *Slang US* °swing, °cut, °hack (out): *She executes her duties satisfactorily.* **2** °complete, °finish, °deliver; °achieve, consummate, °fulfill, °effect, °effectuate; °sign, °seal, validate, countersign: *He executed the portrait in an hour. The contract was executed yesterday.* **3** put to death, send to the gallows *or* gas chamber *or* electric chair; °kill, put to the sword, °butcher; liquidate, assassinate, °murder, °remove, slay, *Slang* °bump off, rub *or* wipe out, snuff (out), °knock off, *US* °waste, ice: *Convicted murderers are no longer executed here. The mob executed the rival gangleaders.*

execution *n.* **1** °accomplishment, °performance, carrying out, doing, °discharge, dispatch *or* despatch, implementation, prosecution, °realization, enactment: *His drinking interferes with the execution of his responsibilities.* **2** °completion, °fulfillment, °consummation, °achievement, attainment, implementation, bringing about, °administration, pursuance: *The execution of my obligations rests with me.* **3** °killing, capital punishment; assassination, °murder, °removal, liquidation, slaying: *The traitor's execution is tomorrow. There has been another gangland style execution.* **4** °skill, °art, mastery, °technique, °style, °manner, °mode, °touch, °approach, °delivery, °rendering, °rendition, °production: *The cellist's execution was superb.*

executive *n.* **1** chairman (of the board), °director, managing director, chief executive, president, °chief (executive officer), CEO, °manager, °head, °leader, °principal, administrator, °official; °supervisor, °foreman, °superintendent, °overseer, °boss, °master, *Colloq* Mr. Big, (chief *or* head) honcho, number one, kingpin, *Slang* top banana, big cheese, numero uno, top dog: *She is an executive in the local power company.* **2** °administration, °management, directorship, directorate, °government, °leadership, supervision: *Five people form the executive of the society.* —*adj.* **3** administrative, managerial, °supervisory, °official, governing, governmental, gubernatorial, regulatory: *He works in the executive department of the government.*

exemplary *adj.* **1** illustrative, °typical, °characteristic, °representative, archetypal; paradigmatic: *The text contains passages exemplary of good writing.* **2** °model, °meritorious, °outstanding, °noteworthy, °admirable, commendable, °praiseworthy, °excellent, °superior: *Private Feller is an exemplary soldier.* **3** cautionary, admonitory, °warning, monitory: *In addition to payment for the damage, we were awarded exemplary damages.*

exemplify *v.* **1** °illustrate, °typify, °represent, epitomize, °instance; °embody, °personify: *This harlequin painting exemplifies Picasso's blue period.* **2** °demonstrate, °display, °show, °exhibit, °model, depict: *Early Mediterranean civilization is exemplified at Knossos.*

exempt *v.* **1** Often, ***exempt from.*** free *or* liberate *or* release from, excuse *or* relieve from, spare from, °let off, absolve, °except, *Colloq* let off the hook: *The doctor exempted Becker from strenuous activity.* —*adj.* **2** exempted, °free, liberated, released, excused, relieved, spared, °let off, excepted, °immune, *Colloq* °off the hook: *It seems that Becker is now exempt from doing any work.*

exemption *n.* °exception, °immunity, °freedom, °release, impunity, dispensation, °exclusion: *This document confirms his exemption from jury service.*

exercise *v.* **1** °employ, °use, °apply, practise, bring to bear, put to use *or* effect; °discharge, °exert, °wield, °execute; utilize, °effect: *Try to exercise better judgment next time. The chairman exercises entirely too much authority.* **2** °work out, limber up, warm up, °train, °drill: *She exercises for an hour every day.* **3** °harass, °annoy, °irritate, vex, harry, °distress, °worry, °concern, °burden, °try, °trouble, °perturb, °disturb, °agitate, make nervous, *Colloq* drive crazy, °drive up the wall: *We have been much exercised over the issue of acid rain.* —*n.* **4** °activity, workout, working out, warm-up, warmingup, callisthenics, aerobics, isometrics, gymnastics; training, °drill, drilling: *Exercise is good for the heart.* **5** °action, °application, °practice, °performance, °discharge, °use, utilization, °employment, °execution, °operation: *We expect the exercise of your best endeavors.*

exert *v.* **1** °exercise, °use, utilize, put to use *or* work *or* effect, °employ, °wield, bring to bear, bring into play, °expend: *Please exert your influence to have my parking ticket fixed.* **2** ***exert oneself.*** °attempt, °try, °endeavor, make an effort, apply oneself, °strive, do one's best, °work, °strain, °struggle, toil, °push, °drive (oneself), go all out, give one's all, *Colloq* knock oneself out, cudgel one's brains, beat one's brains out, do one's damnedest, *Slang* bust a gut: *If Scumble exerted himself a bit more, we might see some results.*

exertion *n.* °action, °effort, striving, °strain, °work, °struggle, toil, °drive, °push, diligence, °industry, assiduity, assiduousness, sedulousness, sedulity, *Colloq US* stick-to-itiveness: *By considerable exertion, we finished on time.*

exhalation *n.* **1** °expiration, exhaling, °breath, respiration, suspiration: *The doctor could detect the wheezing only on exhalation.* **2** °vapor, °breath, °air, °puff, whiff, °exhaust, emission, steam, °mist, gas, °fog, °fume, emanation, effluvium, °evaporation: *The poisonous exhalations from the volcano destroy all vegetation.*

exhale *v.* °breathe (out), °blow, °puff, °huff, °gasp, °evaporate, °pass off, °discharge, °emit, °emanate, °issue (forth), respire, suspire, give forth, blow off, °eject, °expel, exsufflate: *Please exhale into this balloon. The volcano continued to exhale noxious fumes.*

exhaust *v.* **1** °use (up), °expend, °consume, °finish, deplete, °spend, °dissipate, °run through, squander, °waste, °fritter away, *Slang* °blow: *In less than a year, he had exhausted all his funds.* **2** °tire (out), °fatigue, °weary, °wear out, °enervate, °fag, overtire, °sap, °strain, °tax, °weaken, °prostrate, debilitate, disable, *Colloq* frazzle: *Working sixty hours a week would exhaust anyone.* **3** °empty, °drain, °evacuate, °void, clean *or* clear out: *The pump is used to exhaust the chamber of air.* **4** °overdo, °overwork, treat thoroughly, deplete, °drain, °empty: *I think we've exhausted that topic.* **5** °empty (out), drain (off *or* out), °vent, °issue, °escape, °discharge, °run out: *The fumes exhaust through this tube.* —*n.* **6** emanation, effluent, emission, °fumes, gas: *Car exhausts are polluting the atmosphere.*

exhausted *adj.* **1** °(dead) tired, °fatigued, °weary, wearied, worn-out, enervated, debilitated, overtired, °weak, weakened, °prostrate, fagged *or* played *or* burnt out, °spent, all in, out on one's feet, *Colloq* dog-tired, °dead (on one's feet), wiped out, drained, knocked out, all in, done in, frazzled, *Slang* °(dead) beat, *Brit* knackered, *US and Canadian* pooped: *She was exhausted from running the marathon.* **2** °empty, emptied, °bare; depleted, consumed, done, gone, at an end, finished: *The air is now exhausted from the tire. Our supply of paperclips is exhausted.* **3** °spent, worn-out, depleted, °impoverished, °poor, °infertile, °barren: *It is impossible to grow anything in this exhausted soil.*

exhausting *adj.* **1** tiring, fatiguing, wearying, enervating, wearing, debilitating: *Proofreading these figures can be exhausting.* **2** °arduous, °laborious, backbreaking, °strenuous, °hard, grueling, crippling, °difficult, °burdensome, onerous: *Harvesting sugar is exhausting work.*

exhaustion *n.* **1** emptying, drawing out *or* forth, °discharge, draining, evacuation, voiding, depletion, consumption, finish(ing): *The exhaustion of air from the cylinder is accomplished by a powerful pump.* **2** tiredness, °fatigue, enervation, debilitation, weariness, lassitude: *The survivors were suffering from exposure and exhaustion.*

exhaustive *adj.* °complete, °comprehensive, (all-)inclusive, °thorough, all-encompassing, °encyclopedic, °extensive, thoroughgoing, far-reaching, °sweeping, full-scale, in-depth, maximal, °maximum, *Colloq* all-out: *The police conducted an exhaustive investigation of the company's finances.*

exhibit *v.* **1** °show, °display, °present, °offer, °expose; °show off, °parade, brandish, °flaunt; °demonstrate, °reveal, °betray, °manifest, °exemplify, evince, °evidence, °disclose, °express: *Her paintings are widely exhibited. Such behavior exhibits poor judgment.*

exhibition *n.* °exposition, °fair, show(ing), °display, °demonstration, °presentation, °offering, *US* °exhibit, *Colloq* expo, demo: *The craft exhibition will be held at the civic center.*

exhilarating *adj.* **1** °invigorating, °bracing, °stimulating, vivifying, enlivening, rejuvenating, °refreshing, vitalizing, fortifying, restorative, °tonic: *We took an exhilarating walk around the lake.* **2** cheering, uplifting, gladdening, elating, inspiriting, heartening, comforting, reassuring; °happy, °good, °delightful: *We received the exhilarating news of Phoebe's complete recovery.*

exile *n.* **1** expatriation, banishment, °expulsion, deportation, transportation; °separation: *Napoleon was sentenced to exile on Elba.* **2** expatriate, emigre *or* émigré(e), °emigrant, °outcast, deportee, pariah, displaced person, DP; °alien, °foreigner, °outsider: *Some exiles from Nazi Germany settled in Britain.* —*v.* **3** deport, °expel, °alienate, °banish, expatriate, oust, °eject, °displace, °transport, drive *or* run *or* cast out, °outlaw, °exclude, oust, °evict, °bar, °ban; extradite; °maroon: *Many criminals were exiled to Australia.*

exist *v.* **1** be, °continue, °prevail, °endure, °abide; °live, °breathe: *Some believe that the universe exists only in our imagination.* **2** °survive, subsist, eke out a living *or* an existence, stay alive, get by *or* along: *How can the family exist only on his pension?* **3** °occur, °happen, be found, be present, °remain, °persist; °obtain, °prevail: *A two-foot gap exists between the wall and the roof. "Status quo" refers to the circumstances that exist.*

existence *n.* **1** being, °presence, actuality, °fact: *Most cultures believe in the existence of one god or more.* **2** °life, living; continuance, continuation, °persistence, °permanence, duration, °endurance: *The first settlers had to struggle for existence.* **3** °entity, being, °creature; ens, quiddity, °essence: *In Hindu philosophy, there is no limit to the number of existences.*

exit *n.* **1** way out, egress, door, °gate; °outlet, °vent: *Everyone left by the exit when the alarm rang. This is the exit for the smoke.* **2** departure, leave-taking, withdrawal, leaving, °retreat, retirement; °flight, exodus, evacuation, °escape: *Terribly embarrassed, they made their exit. The villain's exit from the stage was marked by catcalls.* —*v.* **3** go (out *or* away), °(take one's) leave, °depart, take *or* make one's departure, °retire, °(beat a) retreat, bid adieu, °withdraw, °run, °(take a) walk, °walk out (on), °quit, °escape, °take to one's heels, show a clean pair of heels, vanish, °disappear, *Colloq* °take off, skedaddle, °kiss goodbye, *US* °cut out; *Slang* °beat it, *US and Canadian* take it on the lam, lam (on) out of *or* from, take a (runout) powder: *He exited from the party as soon as he could.*

exorbitant *adj.* °extraordinary, °excessive, °extravagant, °outrageous, °immoderate, extortionate, °extreme, °unreasonable, °inordinate, °disproportionate, °unconscionable, °preposterous, undue, °unwarranted, unjustifiable, unjustified: *The price of gasoline in Britain is exorbitant.*

exotic *adj.* **1** °foreign, °alien, nonnative, imported: *She raises exotic plants.* **2** °strange, °unfamiliar, °unusual, °bizarre, °odd, °peculiar, °unique, °singular, °extraordinary, °remarkable, °out of the ordinary, °different, °outlandish, °weird, °crazy: *His exotic clothes make him stand out in a crowd.* **3** striptease, belly, go-go, topless, °bottomless, °nude: *Her daughter is an exotic dancer at the new disco.*

expand *v.* **1** °enlarge, °spread (out), °extend, °increase, open (out *or* up), °swell, °inflate, distend; °unfold: *Pump in air to expand the balloon.* **2** °prolong, °lengthen, °stretch, dilate: *An elastic valve expands easily.* **3** °increase, °extend, °amplify, °magnify, broaden, °widen, augment, °heighten, °develop: *The minister sought to expand his influence.* **4** Often, ***expand on.*** °detail, °enlarge on, °embellish, °develop, °amplify, expatiate on *or* upon, °elaborate (on); °flesh out: *The speaker expanded on the topic of health insurance.*

expanse *n.* °stretch, °extent, °area, °space, °range, °sweep, °reach, length and breadth, °spread: *The vast expanses of space are yet to be explored.*

expansion *n.* **1** °increase, augmentation, °development, enlargement, °extension, burgeoning *or* bourgeoning, °flourishing, °growth, °spread: *How will the company finance its expansion in other markets?* **2** dilatation *or* dilation, stretching, distension *or* distention, inflation, °swelling: *The animal moves by the alternate expansion and contraction of the muscle.*

expansive *adj.* **1** expansible *or* expandable, inflatable, dilatable, extensible *or* extendible *or* extendable; extending, expanding, enlarging, spreading, opening *or* stretching (out): *Increased temperature causes the enlargement of expansive materials.* **2** °effusive, °open, °free, °easy, °genial, °amiable, °friendly, °warm, affable, °sociable, °outgoing, communicative, °outspoken, extrovert(ed) *or* extravert(ed), °talkative, loquacious, garrulous, °frank, unreserved: *In his own expansive way McGiver considers everyone his friend.* **3** °broad, °extensive, far-reaching, wide-ranging; °comprehensive, widespread, all-embracing, (all-)inclusive: *We stood on the edge of an expansive desert. The company has expansive interests in Europe.*

expect *v.* **1** °anticipate, look forward *or* ahead to, have *or* keep in view, await, °envisage, watch *or* look for, wait for, °contemplate, °foresee, *US* °envision: *She expects to leave. He was expecting her at noon.* **2** °assume, °presume, °suppose, °imagine, °believe, °think, °trust, °surmise, conjecture; °foresee, *US and Canadian* °guess: *I expect you were hungry when you returned.* **3** °look for, °want, °require, °wish, °need, °demand, reckon on *or* upon, hope for, calculate *or* count on *or* upon: *What did you expect me to do?*

expectant *adj.* °expecting, (a)waiting, °ready, °eager, apprehensive, °anxious, with bated breath, °hopeful, looking, watchful, anticipating: *The understudy waited, expectant, in the wings.*

expectation *n.* **1** °anticipation, °confidence, hopefulness, watchfulness, apprehension, apprehensiveness, expectancy, °suspense: *There was an air of expectation in the room.* **2** °hope, assumption, °presumption, °surmise, °supposition, °belief, conjecture, *US and Canadian* °guess: *Our expectation is that he will come on the next train.* **3** °demand, °requirement, °wish, °desire, °want, insistence, °reliance: *I think your expectations might be too optimistic.* **4** °prospects, °outlook: *I have the greatest expectations from my rich uncle.*

expecting *adj.* °pregnant, gravid, with child, in a family way, *enceinte, Brit* in the family way, *US* in a family way, *Colloq* in the club, *Brit* preggers, *Slang US* with a bun in the oven: *Monica is expecting and will give birth any day.*

expedient *adj.* **1** °suitable, °appropriate, °fitting, °fit, °befitting, °proper, apropos, °right, °correct, °meet, °pertinent, °applicable, °practical, pragmatic, °worthwhile, °politic: *To cross the river, a bridge would prove expedient.* **2** °advantageous, °beneficial, °advisable, °desirable, recommended, °useful, °practical, utilitarian, °prudent, °wise, °propitious, °opportune, °helpful, °effective: *Count on him to do what is expedient and ignore what is honest.* —*n.* **3** °device, °resource, °means, °measure, contrivance, °resort, °recourse: *Life jackets are a useful expedient if the boat sinks.*

expedite *v.* **1** °hasten, °rush, °hurry, speed *or* step up, accelerate; dispatch *or* despatch: *The shipment was expedited and should reach you tomorrow.* **2** °advance, °facilitate, °promote, °forward, °ease, °enable: *The growing process can be expedited by adding this chemical.*

expedition *n.* **1** °exploration, °journey, voyage, field trip, °trip, °tour, °excursion; °enterprise, °undertaking, °mission, °quest: *The tiny ship set out on an expedition up the Amazon to research herbal medicines.* **2** °speed, promptness, celerity, alacrity, dispatch *or* despatch, °haste, °rapidity, swiftness, quickness: *He was ordered to complete his chores with expedition.*

expeditious *adj.* °ready, °quick, °rapid, °swift, °fast, °brisk, °speedy, °fleet, °efficient, °diligent: *We made an expeditious passage around Cape Horn. The problem requires an expeditious solution.*

expel *v.* **1** °eject, dislodge, throw *or* cast out, drive *or* force out, °evict, put *or* push out, °remove, run (someone) off *or* out, °displace, °dispossess, show the door, °suspend, °dismiss, let go, *Colloq* °fire, *Brit* °sack, turf out: *They will expel the tenant at the earliest opportunity.* **2** °banish, deport, °exile, expatriate, °outlaw, °maroon; proscribe, °ban, °bar, debar, °dismiss, °exclude, blackball, drum out, cashier, °discharge, oust, *Brit* °send down: *The leader of the opposition was expelled from the country. The committee voted to expel Horace from the club.*

expend *v.* **1** °pay out, °spend, disburse, °use, °employ, *Slang* lay *or Slang* dish out, fork *or* shell out: *Reduce the money expended on entertainment.* **2** °use up, °consume, °exhaust, deplete, °finish (off), °dissipate, °sap, °drain: *I expended all my energies eking out a bare living.*

expendable *adj.* °dispensable, °disposable, °nonessential, inessential *or* unessential, °unnecessary, replaceable; unimportant, °insignificant: *In terms of military strategy, the commando unit was expendable.*

expenditure *n.* °outlay, outgoings, disbursement, spending, °payment, °expense, °cost; °price, °charge, °fee: *The financial director must closely examine all expenditures. I cannot justify that kind of expenditure for a car.*

expense *n.* **1** Often, ***expenses.*** °payment, costs, °outlay, outgoings, disbursement, °expenditure, spending, out-of-pocket (expenses); °cost, °price, °charge, °fee, °rate: *Keep your expenses to a minimum. Expenses for travel and entertainment will be reimbursed by the company.* **2** °detriment, °sacrifice, °cost, °loss, °impairment, °ruin, °destruction: *They continue to smoke at the expense of their health.*

expensive *adj.* costly, °dear, high-priced, up-market, valuable, °precious, °priceless, °extravagant; overpriced: *Richard bought Elizabeth a few expensive baubles.*

experience *n.* **1** °knowledge, °contact, involvement, °practice, °familiarity, °acquaintance, °exposure; participation, °observation: *Her book on nursing is based on personal experience.* **2** °incident, °event, °happening, °affair, °episode, °occurrence, °circumstance, °adventure, °encounter; °trial, °test, °ordeal: *He has lived through some harrowing experiences.* **3** common sense, °wisdom, sagacity, °knowledge, know-how, savoir-faire, savoir-vivre, °sophistication, °skill, °judgment, *Slang* savvy: *She is a woman of experience.* —*v.* **4** °undergo, live *or* go through, °suffer, °endure, °sustain, °face, °encounter, °meet (with), °feel, °sense, °taste, °sample, be familiar with, °know: *Your problem is that you have never experienced genuine hunger.*

experienced *adj.* **1** °adept, skilled, °skillful, °accomplished, °practiced, °proficient, °knowledgeable, °knowing, °wise, °sage, sagacious, °shrewd, °prepared, (well-)informed, trained, (well-)versed, °expert, °master, masterly, °qualified, °professional, °competent, °efficient, °capable, *au fait; Slang* on the ball, *US* savvy: *She is an experienced surgeon.* **2** °mature, °seasoned, °sophisticated, battle-scarred, °seasoned, °veteran; *Slang* in the know, *US* savvy: *We need an experienced leader.*

experiment *n.* **1** °test, °trial, °investigation, inquiry *or* enquiry, °examination, experimentation, °research, °proof: *Experiment has shown that the ointment cures athlete's foot.* **2** °procedure, °policy: *Totalitarianism seems to have failed as a political experiment.* —*v.* **3** ***experiment on*** *or* ***with.*** °test, °try, °examine, °investigate, °research, °probe: *The time has come to experiment on humans. Artists now experiment with many media, from canvas to computers.*

experimental *adj.* **1** °hypothetical, °theoretical, °tentative, °speculative, conjectural, exploratory: *Man's first attempts at flying were purely experimental.* **2** °empirical, experiential: *She has profound experimental knowledge of what is involved.*

expert *n.* **1** °authority, °professional, °specialist, °scholar, °master, connoisseur, pundit, *Colloq* wizard, whiz, pro, ace, *Brit* °dab hand, boffin, *Slang US* maven *or Slang US* mavin: *We rely on experts to verify the age of works of art.* —*adj.* **2** °skillful, skilled, trained, °knowledgeable, °learned, °experienced, °practiced, °qualified, °adept, °proficient, °accomplished, *au fait,* adroit, °dexterous, °polished, finished, °masterful, masterly, °first-rate, °excellent, °superb, wonderful, °superior, champion(ship), A-1 *or* A-one, °virtuoso, *Colloq* topnotch, *Brit* whizzo, wizard, *US* crackerjack, °crack: *Davis is an expert snooker player.*

expertise *n.* expertness, °skill, °knowledge, knowhow, °judgment, mastery; °dexterity, adroitness, *Slang* savvy: *Does Cappacino have the expertise to be a museum curator?*

expiration *n.* expiry, °finish, °(coming to an) end, °termination, running out, ending, conclusion, concluding, °close, closing, discontinuance, discontinuation: *The expiration of the option is in March.*

expire *v.* **1** °cease, °(come to an) end, °close, °finish, °terminate, °run out, conclude, °discontinue: *Your subscription expires with the next issue.* **2** °die, breathe one's last, decease, °perish, °pass away: *His grandmother expired only last year, at the age of 110.* **3** °exhale, breathe out, °expel: *In the daytime we expire more carbon dioxide than during the night.*

explain *v.* **1** °interpret, °define, explicate, °detail, delineate, make plain, °simplify, °spell out, °resolve, °get across, °clarify, °clear up, elucidate, °illustrate, expound, °describe, °disclose, °unfold, unravel, untangle: *The teacher explained the theory so that even I could understand it.* **2** Also, ***explain away.*** °justify, °account for, °excuse, °rationalize, °legitimate, legitimatize, extenuate, palliate: *You will be required to explain your absence.*

explanation *n.* **1** °interpretation, °definition, explication, delineation, simplification, °resolution, clarification, elucidation, °description, °illustration, °exposition, °account, disclosure; exegesis, commentary, °criticism, °analysis: *That encyclopedia contains explanations of how machines work.* **2** °excuse, rationalization, justification, vindication: *What is your explanation for such outrageous behavior?* **3** °cause, °motive, °reason, °key, signification, °solution: *The explanation for aberrant behavior often lies in hormonal imbalance.*

explanatory *adj.* explanative, elucidative, revelatory, interpretive *or* interpretative, expository, descriptive, °critical, exegetic(al): *I need an explanatory text when reading Chaucer.*

expletive *adj.* **1** °wordy, verbose, prolix, °repetitious, °redundant, tautological, iterative, reiterative, pleonastic; °unnecessary, unneeded, °needless, unessential, °nonessential, °gratuitous, °superfluous: *Her speech is punctuated by expletive words and phrases, like "like, you know," and so forth.* —*n.* **2** °oath, swearword, °curse, curse word, obscenity, epithet, *Colloq* cuss word, dirty word, four-letter word: *Expletives sometimes lend an air of naturalness to written dialogue.* **3** filler, padding, redundancy, °tautology, pleonasm: *Grammarians term the "It" in "It is raining" an expletive.*

explicit *adj.* **1** °specific, °categorical, (crystal-)clear, °definite, well-defined, °distinct, unambiguous, °precise, °exact, unequivocal, °express, stated, °plain, °manifest, unmistakable, °positive, °absolute, °final, °peremptory, °unqualified, unconditional: *The children had explicit instructions to keep away from the canal.* **2** °open, °outspoken, unreserved, unrestrained, °candid, °frank, °direct, °forthright, straightforward, °definite: *She was explicit in her orders.*

explode *v.* **1** °blow up, °burst, °blast, fly apart, °go off, °erupt, fulminate; °set off, detonate: *At the moment we exploded the bomb, the volcano exploded.* **2** °reject, °discredit, refute, °repudiate, °disprove, debunk, belie, give the lie to, *Slang* pick holes in, poke *or* shoot full of holes: *Ptolemy's geocentric theory has been thoroughly exploded.* **3** lose one's temper, °rant, °rave, °rage, °storm, throw a tantrum, *Colloq* get into a tizzy, blow one's top, °fly off the handle, go through *or* hit the roof, hit the ceiling; *Slang* lose one's cool, go up the wall, *US* blow one's stack *or* cool, °flip (one's lid), freak out: *He exploded when he learned his car had been smashed up.*

exploit *n.* **1** °achievement, °deed, °feat, attainment, °accomplishment: *The speaker regaled them with tales of his exploits as an explorer.* —*v.* **2** °use, take advantage of, °manipulate, make capital out of, °profit from, utilize, turn to account, °maneuver, °work: *Far from being your friends, they are exploiting you for their own purposes. They need capital to exploit the country's natural resources.*

exploration *n.* °examination, °investigation, °search, °probe, inquiry *or* enquiry, °study, °research, °analysis, °review, °scrutiny, inspection, °survey, °reconnaissance, °observation; °expedition: *Exploration of the intestine revealed the presence of polyps. He is noted for his exploration of the sources of the Nile.*

explore *v.* **1** °survey, °tour, °travel, °reconnoiter, °traverse: *They are on safari, exploring the Congo.* **2** °investigate, °scrutinize, °examine, inquire *or* enquire into, °inspect, °probe, °search, °research, °look into, °study, °analyze, °review, °observe: *We ought to explore the possibilities of your working for us.*

explosion *n.* **1** °blast, bang, °report, °burst, °boom, °clap, °crack, °eruption, °crash, °outburst, fulmination; detonation: *The explosion shook the entire house.* **2** °outburst, outbreak, °paroxysm, °upheaval, flare-up, °eruption, °burst, °fit, °spasm, °tantrum, *Colloq Brit* paddy *or* paddy-whack, wax: *The unrest among the people is building up to an explosion.* **3** °increase, burgeoning *or* bourgeoning, °expansion, welling-up, mushrooming: *The population explosion has caused a housing shortage.*

explosive *adj.* **1** °volatile, °sensitive, °delicate, °tense, °anxious, °fraught, touch-and-go, °touchy, °inflammable, (highly) charged, °unstable, °uncertain, °unsound, °shaky, °hazardous, chancy, unpredictable, °precarious, °dangerous, °perilous, °critical, °nasty, ugly, *Colloq* °dicey, iffy: *The explosive situation between the strikers and management may lead to bloodshed.* —*n.* **2** dynamite, TNT, gunpowder, gelignite, °plastic, plastique: *The car was blown up by a powerful charge of explosive.*

expose *v.* **1** °(lay) bare, °reveal, uncover, °show, °exhibit, °present, °display, °disclose; divulge, °unveil, unmask, °discover, °air, ventilate, °let out, °leak, °betray, °bring to light, °make known: *He unbuttoned his shirt, exposing his hairy chest. The facts of the case will be exposed tomorrow. Will he expose the names of his accomplices?* **2** °risk, imperil, °endanger, °jeopardize, °hazard: *Do not expose your house to danger of flooding by building it so close to the river.* **3** ***expose to.*** °subject to, introduce to, °acquaint with, bring into contact with: *Today children are not taught, but "exposed to education."*

exposition *n.* **1** °exhibition, show(ing), °presentation, °display, °demonstration, *US* °exhibit, *Colloq* expo: *Have you seen the exposition of medieval farming implements?* **2** °description, °declaration, °statement, °explanation, explication, clarification, °interpretation, exegesis: *The exposition of his argument was remarkably lucid.* **3** °paper, °theme, article, °essay, °thesis, dissertation, treatise, disquisition, °study, critique, commentary: *She received high marks for her exposition on Zoroastrianism.*

exposure *n.* **1** baring, uncovering, laying open, unveiling, disclosure, disclosing, unmasking, revealing, °revelation, exposé, airing, °publication, publishing, communicating, communication, leaking, °leak, divulging: *The exposure of the spy was part of the plan.* **2** °jeopardy, °risk, °hazard, endangerment, vulnerability, imperilment; °danger, °peril: *By diversifying your investments, you reduce your exposure to a loss in just one.* **3** °familiarity, °knowledge, °acquaintance, °experience, °contact, conversancy: *My exposure to Chinese philosophy has been negligible.* **4** °aspect, °view, °outlook, °orientation, frontage; °setting, °location, °direction: *I prefer a breakfast room with an eastern exposure.*

express *v.* **1** articulate, verbalize, °phrase, utter, °voice, °state, °word, °put (into words), set *or* put forth, put *or* get across, °communicate, depict, °portray; °say, °speak, °tell: *She expressed her thoughts on the subject very clearly.* **2** °show, °indicate, °demonstrate, °manifest, °exhibit, evince, °evidence, °reveal, °expose, °disclose, divulge, °make known, °intimate, betoken, °signify, °embody, depict, °designate, °denote, convey: *His tone of voice expresses his resentment.* **3** °symbolize, °represent, °signify, °stand for, °denote, °designate: *The ratio can be expressed in the form of a fraction.* **4** press *or* squeeze *or* wring *or* force out, °expel, °extract: *The oil is expressed from the olives.* —*adj.* **5** °explicit, °clear, °plain, unambiguous, unmistakable, °unqualified, °outright, °definite, °out-and-out, °downright, straightforward, °categorical, °direct, °specific, well-defined, °distinct, °precise, °accurate, °exact, °positive: *We had an express understanding, not just a gentleman's agreement.* **6** °specific, °special, °particular, clearcut; °true: *Our express purpose in coming was to see you.* **7** °quick, °speedy, °swift, °fast, °rapid, °prompt, °immediate; °direct, °nonstop: *The documents must be sent by express delivery.*

expression *n.* **1** verbalization, airing, °representation, °declaration, utterance, °assertion, enunciation, asseveration, °pronouncement, communication, voicing, °announcement: *Any expression of criticism of government is forbidden in many countries.* **2** °representation, °manifestation, °sign, °token, °symbol, °show, °demonstration, °indication, °evidence: *A curled lip is an expression of scorn.* **3** °look, mien, °air, °appearance, °face, °aspect, countenance: *He always seems to have an annoyed expression.* **4** °tone, °note, nuance, °intonation, °accent, °touch, nuance, shading, loudness, softness; expressiveness, °emotion, °feeling, °sensitivity, °passion, °spirit, °depth, °ardor, °intensity, pathos: *There was much expression in her playing of the fugue.* **5** °word, °term, °phrase, °idiom, turn of phrase, locution, saying: *Some find the expression "Have a nice day" to be irritating.* **6** °wording, phrasing, phraseology, °language, °style, °diction, °usage, °speech, °delivery: *Writers are not the only ones who should study effective expression.*

expressive *adj.* **1** indicative, °suggestive, allusive, °eloquent, revealing, °meaningful, °significant, denotative: *Her frown is expressive of her disapproval.* **2** °pointed, pithy, °explicit: *Leave it to him to make some expressive remark about Frieda's hat.* **3** °striking, °vivid, °telling, °pregnant, °loaded, °forceful, °moving, °emotional, °poignant, °provocative, thought-provoking: *His poetry contains much expressive language.*

expressly *adv.* **1** distinctly, °definitely, categorically, explicitly, °absolutely, °positively, °directly, unambiguously, unequivocally, unmistakably, plainly, pointedly, °exactly, °clearly: *He expressly denied the rumor.* **2** purposely, °especially, purposefully, °particularly, specifically, °specially; °on purpose: *The gift was intended expressly for you.*

expulsion *n.* expelling, °ejection, °eviction, repudiation, ouster, °removal, °dismissal, °discharge; *Colloq* (the old) heave-ho, *Brit* °(the) boot, °(the) sack, sacking, *US* °(the) bounce: *His expulsion from the pub was accompanied by cheers from the patrons.*

exquisite *adj.* **1** °delicate, °fine, °elegant, °graceful, °excellent, °choice, well-crafted, well-made, well-executed, °refined, °elaborate: *They have a collection of exquisite miniature portraits.* **2** °ingenious, °detailed, recherché, °rare, °subtle, °deep, abstruse; °far-fetched: *He is expert in the exquisite points of seduction.* **3** °beautiful, °perfect, °lovely, °attractive, °handsome, °comely, good-looking; °smart, °chic, °elegant, °striking: *In her youth, she was an exquisite blonde goddess. He wore an exquisite silk costume.* **4** °acute, °sharp, °keen, °excruciating, °agonizing, °intense; °elaborate: *The most exquisite pain I have known is occasioned by gout. Spies, subjected to exquisite torture, usually revealed their secrets.* **5** °superb, °superior, °peerless, °matchless, °incomparable, unequaled, °rare, °precious, °choice, consummate, °outstanding, °superlative, °excellent, °select, °flawless, °perfect, wonderful, °splendid, °marvelous: *She wore a tiara of exquisite diamonds.*

extemporaneous *adj.* unstudied, °unpremeditated, extempore, extemporary, impromptu, improvised, °spontaneous, unrehearsed, extemporized, °unprepared, unplanned, unscripted, °offhand, ad-lib, extempore, *Colloq* off the cuff: *Though extraordinarily well-organized, Charlotte's speech was entirely extemporaneous.*

extend *v.* **1** stretch *or* spread (out), outstretch, outspread, °open (out), unroll, °unfold; °reach, °range; °carry on, °draw out, °continue, °develop: *The carpet extends from wall to wall. Her reputation extends worldwide. The city walls extended to the river.* **2** °lengthen, elongate, °widen, °continue; broaden, °enlarge, °add to, augment; °increase, °stretch (out), °supplement: *We are planning to extend the sitting room. The definitions were extended to include metaphoric uses. Will the bank extend the time to repay the loan?* **3** °last, °stretch, °continue, go *or* carry on; °perpetuate, drag on *or* out, keep up *or* on, °prolong: *Their visit extended till the following week.* **4** °offer, proffer, °give, °present, °hold out, stretch forth, °tender; °bestow; °grant, °impart, °confer, °accord, °advance: *She extended her hand. The shop extends credit to qualified customers.*

extension *n.* **1** stretching, °expansion, °increase, enlargement, augmentation, °development, amplification, broadening, widening, lengthening, °spread, spreading: *The extension of the plan is scheduled for next year.* **2** °range, extensiveness, °scope, °extent, °magnitude, °gauge, compass, °sweep, °reach, °size, °volume, dimension(s), °proportions, °capacity, °span; °breadth, °width, °height, °length, °spread, °stretch: *The extension of the mind seems almost limitless.* **3** addendum, °addition, annex, wing, adjunct, ell, appendage; appendix, °supplement: *We have built an extension* *to the house. The extension adds sixteen pages to the book.*

extensive *adj.* **1** °broad, °wide, °expansive, far-reaching, far-ranging, wide-ranging, far-flung, °sweeping, widespread, °comprehensive, all-embracing; °national, nationwide, °international, intercontinental, cosmopolitan, worldwide, °global, °universal, °vast; cosmic; °catholic: *The storm is responsible for extensive crop damage. He has extensive business connections.* **2** °large, °big, °great, °huge, °substantial, °considerable, sizable, °immense, °enormous, °vast, °gigantic, °massive; °voluminous, °spacious, commodious, capacious: *The extensive gold deposits were soon depleted. The cave contains several extensive chambers.*

extent *n.* **1** °magnitude, dimensions, compass, °size, °range, °scale, °sweep, °scope, °expanse, immensity, °enormousness, capaciousness, spaciousness, °space, amplitude: *The extent of space is unimaginably vast.* **2** °limit, °bounds, limitation, lengths; °range, °scope: *To what extent will he go to see justice done?* **3** °area, °region, °tract, °territory, compass: *In the whole extent of Europe you could find no better mushroom.*

extenuating *adj.* °mitigating, lessening, tempering, palliating, moderating, diminishing, qualifying: *Owing to extenuating circumstances, he could not be tried for murder.*

exterior *adj.* **1** outer, °outside, °external, °outward, °superficial, °surface: *The exterior covering becomes very hot on reentry into the atmosphere.* **2** °external, °extrinsic, °extraneous, °foreign, °alien, °exotic, °outside: *The problem can be overcome without exterior help.* —*n.* **3** °outside, °surface, covering, coating, °facing, °face, °front, °skin, °shell, façade: *The exterior is of pink stucco.*

exterminate *v.* °destroy, °root out, eradicate, extirpate, annihilate, °eliminate, weed out, °get rid of, °wipe out, °obliterate, put an end to, °terminate, liquidate, °massacre, °murder, °kill (off), °butcher, °slaughter, *Slang* °bump off, *US* °rub out, °waste: *It is his aim to exterminate all the vermin on earth.*

external *adj.* **1** outer, °outside, °outward, °exterior: *This medication is for external use only.* **2** °outside, °exterior, °extrinsic, °extraneous, °alien, °foreign, °exotic: *Do not ignore the external influences on the nation's economy.* **3** °apparent, °visible, °perceptible, °superficial, °surface: *The external features of the planet barely suggest what lies within.*

extinct *adj.* **1** °defunct, °dead, died out, gone, departed, vanished: *These are the bones of an extinct species of flying mammal.* **2** dated, outmoded, °old-fashioned, °antiquated, °obsolete, archaic, out-of-date, antediluvian, °ancient, old hat, °passé, *démodé*: *High-button shoes and bustles have been extinct for almost a century.* **3** °out, extinguished, quenched, burnt *or* put *or* snuffed out; °inactive, °dormant: *The lake is in the caldera of an extinct volcano.*

extinguish *v.* **1** put *or* snuff *or* blow out, °quench; turn off *or* out: *We are landing shortly, so please extinguish all smoking materials. Extinguish the lights before leaving.* **2** °kill (off), annihilate, °destroy, °obliterate, °abolish, °exterminate, °eliminate, do away with, nullify, eradicate, °remove, °banish, wipe *or* blot out: *The potion contained a drug that extinguished all memory of unhappiness.* **3** °obscure, °eclipse, °dim, °outdo, put in the shade, °overshadow, adumbrate, *Colloq* °show up: *His personality is completely extinguished by the dazzle of his wife.*

extol *v.* °exalt, elevate, uplift, °glorify, °praise, °laud, °applaud, commend, acclaim, °cheer, °celebrate, pay tribute *or* homage to, sing the praises of, °make much of, °honor, °congratulate, °compliment: *Waybright is always extolling the talents of some dancer or other.*

extort *v.* °exact, °extract, °blackmail, °bully, coerce, °force, wring, wrest, *Colloq* °milk, bleed, put the arm on (someone): *Threatening to swear that it was my fault, she extorted a promise from me to keep silent about the body.*

extra *adj.* **1** additional, added, °auxiliary, accessory *or* accessary, °supplementary, supplemental, °further, ancillary, °subsidiary, collateral, adventitious: *We had an extra person for dinner. Zenobia has an extra toe on each foot.* **2** left-over, °excess, °spare, °surplus, °unused, °superfluous, supernumerary, °reserve: *After all were served, we had two extra desserts.* —*n.* **3** °addition, addendum, accessory *or* accessary, appurtenance, °supplement, °bonus, °premium, dividend: *As an extra, each customer receives a free ballpoint pen.* **4** °supplement, *US* markup, surcharge, *Slang US and Canadian* kicker: *The company lost money on the basic contract but made it up on the extras.* **5** supernumerary, walk-on, *Colloq* °super, spear carrier: *One could scarcely call being an extra a "Hollywood career."* —*adv.* **6** uncommonly, unusually, exceptionally, unexpectedly, extraordinarily, remarkably, °notably, surprisingly, amazingly, °very, °particularly, °especially, °extremely, strikingly: *Teacher said that Philip was extra good today.* **7** additionally, again, more, °in addition: *They charge extra for room service.*

extract *v.* **1** draw *or* pull (out), °remove, °withdraw, pluck *or* take out, draw forth, °extricate: *The dentist extracted two teeth. I extracted some important papers from your wastepaper basket.* **2** °draw, °derive, °deduce, °develop, glean, °extricate, distill, °get, °obtain: *I could extract little wisdom from the old crone's chattering.* **3** °wrench, wring, wrest, °extort, °draw (forth), °evoke, °elicit, °extricate, winkle out, worm (out), prise (out), °force (out): *Using torture, they finally extracted a confession. She extracted money from her victims by threatening to tell their wives.* **4** °copy, °quote, cite, °abstract, °select, °choose, glean, cull: *We extracted useful material from Legman's works on the limerick.* —*n.* **5** °concentrate, distillate, °essence, distillation, °quintessence, concentration, °extraction, decoction: *We boiled off the water and used the extract as a salve.* **6** °excerpt, °abstract, °quotation, citation, clipping, °cutting, °passage, °selection: *Extracts from other works are printed in smaller type.*

extraction *n.* **1** °removal, extrication, withdrawal, uprooting, eradication, extirpation, deracination: *The extraction of my tooth was painless.* **2** °extract, °concentrate, distillate, °essence, distillation, °quintessence, concentration, decoction, °separation, °derivation: *This syrup is an extraction from coffee beans.* **3** °origin, °birth, ancestry, descent, °lineage, °derivation, blood, °parentage, °breed, °strain, °race, °stock, °pedigree: *The family is of Scandinavian extraction.*

extraneous *adj.* **1** unessential, °nonessential, inessential, °peripheral, °superfluous, °unnecessary, unneeded, °extra, added, additional, adventitious, supernumerary, °incidental, °needless: *Filter out all the extraneous substances.* **2** not pertinent, °impertinent, °inapplicable, inapt, unapt, unfitting, °inappropriate, °unrelated, °irrelevant, inapposite, unconnected, °remote, °alien, °foreign, °exotic, °strange, °outlandish, °external, °extrinsic, °external, °out of place, off the mark *or* point *or* subject, beside the point *or* mark: *Why must you constantly bring up extraneous matters?*

extraordinary *adj.* **1** °unusual, uncommon, °remarkable, °exceptional, °particular, °outstanding, °special, °rare, °unique, °singular, °signal, °unheard-of, °curious, °peculiar, °odd, °bizarre, °queer, °strange,

°abnormal, unprecedented, °unparalleled: *She showed extraordinary courage in coming here.* **2** °amazing, surprising, astonishing, astounding, °remarkable, °notable, °noteworthy, °marvelous, °fantastic, °incredible, °unbelievable, °impressive, °fabulous, °miraculous, °unparalleled, *Colloq* °super, smashing, °lovely, °gorgeous, far-out, °unreal: *This is an extraordinary example of the bookbinder's art.*

extravagance *n.* **1** wastefulness, °waste, lavishness, °profligacy, °prodigality, squandering, °dissipation, improvidence, exorbitance, recklessness, overspending, °excess: *Our present debts can be traced to extravagance.* **2** immoderation, immoderateness, excessiveness, outrageousness, unrestraint, superfluity, superfluousness, oversufficiency, preposterousness, unreasonableness, irrationality, °absurdity; capriciousness, whim, °fantasy, flightiness: *His tawdry life belied the extravagance of his dreams.*

extravagant *adj.* **1** °wasteful, °lavish, °profligate, °prodigal, °improvident, °reckless, °excessive, °spendthrift, °profuse, °extreme, °immoderate: *She cannot afford to maintain her extravagant way of life for long.* **2** unrestrained, uncontained, °wild, °outrageous, °preposterous, °immoderate, °ridiculous, °foolish, °fanciful, °unreasonable, °absurd, °impractical; undeserved, unjustified, unjustifiable: *The speaker lavished extravagant praise on the award winners.* **3** °expensive, costly, extortionate, °unreasonable, overpriced, °exorbitant, °high; °dear; *Colloq* °steep: *That trip to Istanbul by private jet was quite extravagant.* **4** °gaudy, °garish, °ostentatious, °showy, °ornate, °flashy, °loud, °flamboyant; exaggerated, high-sounding: *She appears at parties wearing the most extravagant costumes.*

extravaganza *n.* spectacular, °spectacle, °pageant, °production, °show, °exposition: *An extravaganza was staged to celebrate the opening of the Channel Tunnel.*

extreme *adj.* **1** °unusual, uncommon, °exceptional, °outstanding, °notable, °noteworthy, °abnormal, °different, °extraordinary, °remarkable: *We experienced extreme difficulty driving home.* **2** °immoderate, °excessive, °severe, °intense, °acute, °maximum, worst: *This equipment is made to withstand extreme arctic conditions.* **3** outermost, endmost, farthest, °ultimate, utmost, uttermost, remotest, °last, far-off, °faraway, °distant, °very: *She walked to the extreme end of the pier.* **4** °rigid, °stern, °severe, °strict, °conservative, °hidebound, °stiff, stringent, restrictive, constrictive, uncompromising, Draconian, °harsh, °drastic: *Grandfather favored taking extreme measures against terrorists.* **5** unconventional, °radical, °outrageous, °wild, °weird, °bizarre, °queer, °outrageous, °offbeat, °exotic, °eccentric, °different, *outré*, *Slang* far-out, °way-out, *US and Canadian* kooky: *I find your outfit too extreme to wear in court.* **6** beyond the pale *or* limits *or* bounds, °extravagant, °inordinate, °excessive, °disproportionate, °outrageous: *Because of your extreme behavior, you will be confined to the house for a week.* —*n.* **7** Often, ***extremes.*** limit(s), °bounds, utmost, °maximum, *Colloq* °swing: *Try to be more moderate and to avoid extremes.* **8** Often, ***go to extremes.*** limit(s), °bounds, °maximum, °acme, °zenith, °pinnacle, °summit, °height, apex, apogee, °peak, °extremity; °depth, nadir: *Thea's moods varied between the extremes of joy and grief.* **9** ***in the extreme.*** °extremely, °very, exceptionally, °exceedingly, extraordinarily, unusually: *His table manners are rude in the extreme.*

extremely *adv.* °very, °exceedingly, outrageously, extraordinarily, unusually, uncommonly, exceptionally, damned, hellishly, to the nth degree, *Colloq Brit* bloody, *US* darned: *The service was extremely bad. They are extremely stupid.*

extremity *n.* **1** °end, °termination, °limit, °edge, °boundary, °bound, °border, °margin; °periphery; °frontier: *A fence marks the eastern extremity of our land.* **2** ***extremities.*** fingers, fingertips, toes; °hands, feet; arms, legs, limbs; paws, trotters, hooves, wings: *His extremities were numb from frostbite.* **3** °extreme, utmost, °maximum, limit(s), °bounds: *The situation tested the extremity of my patience. We were driven to extremities in our search for food.*

extricate *v.* unravel, disentangle, untangle, °disengage, °(set) free, °turn loose, °release, °liberate, °rescue, °save, °deliver: *You helped extricate me from a terrible predicament.*

extrinsic *adj.* °external, °extraneous, °irrelevant, °exterior, °unrelated, °outside; outer, °outward: *Those extrinsic matters are not germane to the problem.*

exuberance *n.* **1** cheerfulness, °joy, joyfulness, ebullience, effervescence, exhilaration, buoyancy, °animation, °spirit, spiritedness, sprightliness, liveliness, °vitality, vivacity, °enthusiasm, °excitement, zeal, °zest, °energy, °vigor: *It is hard to imagine the dog's exuberance when he saw his master.* **2** °abundance, lavishness, effusiveness, flamboyance, copiousness, superabundance, °superfluity, °excess, °profusion, °prodigality, bounteousness, bountifulness: *Everything attests to the exuberance of the author's genius.*

exuberant *adj.* **1** °cheerful, °joyful, °ebullient, °effervescent, °buoyant, °animated, °spirited, spry, °sprightly, °lively, °vivacious, °enthusiastic, zealous, °energetic, °vigorous: *She enjoys exuberant good health.* **2** °happy, °glad, °delighted, °overjoyed, °joyful, °ecstatic, *Brit* in the seventh heaven, *US* in seventh heaven, *Colloq* on cloud nine: *She was exuberant at the news of Lyle's return.*

exult *v.* °rejoice, °revel, °glory (in), jump for joy, °delight, °celebrate, °make merry: *We exulted in our new-found freedom.*

exultant *adj.* °delighted, jubilant, °overjoyed, °elated, °joyful, °gleeful, °glad, °ecstatic, °exuberant, in seventh heaven, cock-a-hoop, *Colloq* on cloud nine, *Brit* over the moon: *We were exultant to learn that the war was over.*

eye *n.* **1** eyeball, °orb, *Colloq* optic: *The eye of the eagle is nearly as large as that of an elephant.* **2** °vision, (eye)sight, visual acuity, °perception: *Her eyes are weakened by so much reading.* **3** discernment, °perception, °taste, °judgment, °discrimination, percipience, perspicacity, °appreciation, °sensitivity; °knowledge, °recognition, comprehension: *Shirley has a good eye for Chinese antiques. Walmsley has the eye of an artist.* **4** °liking, °affection, fondness, °partiality, °appreciation; lustfulness: *Old Bisley still has an eye for the girls.* **5** °ogle, °leer, °look, wink, glad eye, sidelong glance: *The barmaid gave me the eye as soon as I walked in.* **6** °view, °respect, °regard, °aim, °intention, °purpose, °design, °plan, °idea, °notion: *He lent me the money with an eye to asking a favor in return.* **7** °attention, °regard, °look, °scrutiny, °view, °examination, °observation; supervision: *All our comings and goings were under the watchful eye of Scotland Yard.* **8** °guard, °lookout, °watch, vigil: *The class monitor will keep an eye on the little children.* —*v.* **9** °examine, °scrutinize, °look at, °regard: *My future mother-in-law eyed me up and down.* **10** °behold, gaze *or* look *or* peer at *or* upon, °contemplate, °study, °regard, °view, °inspect; °watch, °observe: *The doctor eyed the wound with concern.*

eyewitness *n.* °witness, °observer, °spectator, viewer, watcher; °bystander, °onlooker, passerby: *The police sought eyewitnesses to the accident.*

F

fabric *n.* **1** °cloth, textile, °material, °stuff: *Do you have enough of this fabric for a dress?* **2** construction, constitution, °core, °heart, °foundation, °structure, framework, °organization, configuration, makeup: *Extensive strikes threatened the very fabric of our society.*

fabricate *v.* **1** °erect, °build, °construct, °frame, °raise, put *or* set up, °assemble, °fashion, °form, °make, °manufacture, °produce: *The basic structure was fabricated of steel.* **2** °invent, °create, °originate, °make up, °manufacture, concoct, °think up, °imagine, °hatch, °devise, °design: *Numerous lies, fabricated by politicians, were already in circulation.* **3** °forge, °falsify, °counterfeit, °fake, feign, °manufacture; trump up, *Colloq* cook up, *Brit* cook: *He admitted to having fabricated the data in his application.*

fabrication *n.* **1** construction, °assembly, assemblage, making, fashioning, °production, °manufacture, putting together, °building, erection, °formation, formulation, structuring, constructing, °organization, forming, framing, architecture: *The fabrication of thousands of parts took only a month.* **2** °invention, °creation, origination, make-up, °manufacture, hatching, concoction, contrivance, °design: *Only Vanessa could have been responsible for the fabrication of such a diabolical plot.* **3** °falsehood, °lie, °fib, prevarication, °story, °tale, untruth, fiction, °yarn, fable; falsification, °forgery, °fake, °sham, *Colloq* cock-and-bull story, *Brit* fairy story, fairy tale: *His war record is a complete fabrication.*

fabulous *adj.* **1** fabled, mythic(al), °celebrated, °legendary, storied, °fictitious, °fictional, °unreal, °fanciful, °imaginary, storybook, fairytale: *Greek mythology tells us of a fabulous winged horse called Pegasus.* **2** °fantastic, °marvelous, °incredible, °unbelievable, °inconceivable, wonderful, astounding, astonishing, °amazing, wondrous, °extraordinary, °miraculous, °phenomenal: *Houses in London were selling at fabulous prices.* **3** °superb, °marvelous, °terrific, °great, wonderful, *Colloq* °great, °super, smashing, °thumping, °whopping, thundering, rattling, howling, *US* °neat, °keen, *Slang* fab, °hot, far-out, °cool, *Old-fashioned* fantabulous, in the groove, groovy, ace, *Brit* °magic, *US and Canadian* copacetic: *I bought a perfectly fabulous music video today.*

face *n.* **1** visage, countenance, physiognomy, °features, lineaments, *Slang* °mug, mush, kisser, °pan, puss, *Brit* phiz, phizog, dial, clock: *I don't like the expression on your face.* **2** °look, °appearance, °aspect, °expression, mien: *He has the face of someone who is very proud of himself. Modern farming has changed the face of the countryside.* **3** °mask, °veneer, façade, °front, °camouflage, °pretense, °disguise, °(false) impression, °semblance, °masquerade: *She puts on a bold face, but we know she was deeply hurt by your remarks.* **4** °dignity, °image, °self-respect, °standing, °reputation, repute, °name, °honor, °status: *You might lose face if you admit you were wrong.* **5** boldness, °daring, audacity, °effrontery, °impudence, °impertinence, °presumption, brashness, *Colloq* °gall, °brass, °nerve, cheek, °guts, gutsiness, *Brit* brass neck, *Slang US* balls: *Who would have the face to name such a hotel "The Palace"?* **6** °surface, °exterior, °front, °outside, °cover, °facing, façade, °skin: *The building will have a face of white marble.* **7** right side, obverse, °front; dial: *The card landed face up. The face of the clock is enameled.* **8** ***face to face.*** confronting, °facing, °opposite, *en face*, vis-à-vis, °tête-à-tête, à deux, eye to eye, head to head, *Colloq* eyeball to eyeball: *We met face to face for the first time in the courtroom.* **9** ***in the face of.*** in defiance of, °notwithstanding, °despite, °in spite of, confronting, in opposition to: *Some experts attacked the report for flying in the face of received wisdom.* **10** ***make a face.*** grimace, change one's expression, *Brit* girn, *Rare* murgeon: *When I said I would be there, she made a face.* **11** ***on the face of it.*** to all *or* outward appearances, °seemingly, °apparently, superficially, °evidently: *On the face of it, I assumed that he was guilty.* **12** ***show one's face.*** put in *or* make an appearance, °appear, °arrive, be seen, °turn up, *Colloq* °show up: *I suppose I ought to show my face at the office party.* **13** ***to one's face.*** °directly, brazenly, eye to eye, °face to face, candidly, °openly, frankly: *I told him to his face that he was a liar.* —*v.* **14** confront, °brave, °meet (with), °encounter, °experience, deal *or* cope with, come *or* go up against; appear before: *In the jungle we faced grave danger from man-eating tigers. She faces her first audience tonight.* **15** give (out) *or* front on *or* onto, front toward, °overlook, look out on *or* over; be opposite: *Our rooms face the lake. Facing page 22 is a map of the area.* **16** °coat, °surface, °cover, clad, °dress, sheathe, overlay, °finish; °veneer: *The collar is faced with velvet.* **17** ***face down.*** confront, °intimidate, cow, °subdue, °overawe, °browbeat: *I'll not be faced down by that impudent clerk.* **18** ***face up to.*** **(a)** °admit, °accept, °acknowledge, °allow, °confess: *Everyone must face up to his own shortcomings.* **(b)** confront, deal *or* cope with, come *or* go up against, °brave, come to terms with; brazen through *or* out, bite (on) the bullet, grasp the nettle: *You wouldn't be able to face up to your wife if you'd forgotten her birthday.*

facilitate *v.* °ease, °expedite, °smooth, °further, °promote, °advance; °assist, °aid, °help: *It would facilitate transport if you would bring your own car.*

facility *n.* **1** °ease, smoothness, °fluency, effortlessness, °readiness, easiness, °skill, skillfulness, deftness, °dexterity, adroitness, °ability, °aptitude, °expertise, expertness, °proficiency, mastery, masterfulness, masterliness, °efficiency; quickness, alacrity, celerity, swiftness, °speed: *They praised her new-found facility with the Italian language.* **2** Often, ***facilities.*** **(a)** °plant, °system, building(s), °structure, complex: *The company is building a new facility in France.* **(b)** convenience(s), °privy, °equipment, °lavatory, °toilet, powder room, *Nautical* °head, *Brit* water closet, WC, loo, *US and Canadian* rest room, men's room, ladies' room, *Colloq Brit* gents, ladies, *Slang Brit* °bog, karzy, *US and Canadian* john: *An American asked where he could find the facilities.*

facing *n.* façade, °front, cladding, °surface, overlay, °skin; coating: *The houses all have a colorful stucco facing.*

facsimile *n.* °copy, °reproduction, °print, carbon (copy), °replica, °duplicate, photocopy, fax, *Trademark* Xerox (copy), Photostat, *Colloq US* °dupe: *Send a facsimile of the report to each field office.*

fact *n.* **1** °reality, actuality, °truth, °certainty: *The fact of the matter is that she didn't do it. Is there a basis in fact for your allegations?* **2** °accomplishment, *fait accompli*, fact of life; °occurrence, °event, °happening, °incident, °episode, °experience, °act, °deed: *Supersonic travel has been a fact for many years. The facts described in the testimony are doubtful.* **3** Often, ***facts.*** °data, °information, particular(s), detail(s), point(s), item(s), factor(s), *Colloq* °lowdown, (inside) info, the score, *Brit* the gen, *US and Canadian* the poop: *If I am going to defend you, I need all the facts of the case.* **4** ***in fact.*** °indeed, to be sure, as a matter of (actual) fact, °in truth, °truly, truthfully, °actually, °really, °in reality, in point of (actual) fact, factually: *He didn't come; in fact, he had died a week before.*

faction *n.* **1** °group, °cabal, bloc, cadre, °camp, splinter group, °circle, camarilla, °clique, °set, coterie, °lobby, pressure group, junta *or* junto, °ring, °gang, *Brit* ginger group, *Colloq* °crowd: *There is a small faction plotting to assassinate the finance minister.* **2** °dissension, °intrigue, °strife, °sedition, disharmony, °discord, °disagreement, quarreling, contention, °controversy, infighting, °rupture, °split, °rift, °schism, °clash: *Faction within the party has no regard for national interests.*

factious *adj.* contentious, disputatious, litigious, refractory, divisive, conflicting, °discordant, °argumentative, °at odds, at loggerheads, °quarrelsome, °seditious, °mutinous, °rebellious: *A factious Congress can spell trouble for the president.*

factitious *adj.* fake(d), °bogus, °false, °mock, falsified, °artificial, °insincere, °unreal, °synthetic, fabricated, engineered, manufactured, °spurious, °counterfeit, °sham, simulated, °imitation, unauthentic, set *or* got up, rigged, *Colloq* phoney *or US also* phony: *We are all victims of factitious desires by which luxuries have become necessities.*

factor *n.* **1** constituent, °ingredient, °element, °part, °particular, °piece, component, catalyst; °circumstance, °consideration, °aspect, °fact, °influence, determinant, °cause: *Which factors contributed to the decline of Mayan civilization?* **2** °agent, °representative, °proxy, middleman, °intermediary, °deputy, °go-between: *Our company has a factor handling all our exports.* **3** banker, °financier, °backer, moneylender, lender: *The factors lent us money to buy the stock needed to fill the orders.*

factory *n.* °works, °mill, °plant: *She has a job at the piano factory.*

factual *adj.* **1** °actual, °real, °true, °authentic, verifiable, °realistic, true-to-life, °genuine, valid, °bona fide: *The report is a forgery, but the letters are factual.* **2** °accurate, °correct, °true, °correct, °faithful, °precise, unbiased, undistorted, °unvarnished, unexaggerated, °objective, °unprejudiced, straightforward: *This is a factual account of the Battle of Marathon.*

faculty *n.* **1** °ability, °capacity, °skill, °aptitude, °potential, °talent, °flair, °knack, °gift, °genius; °dexterity, adroitness, cleverness, °capability: *She has a faculty for making people feel at home.* **2** °school, °department, °discipline: *Her graduate studies were in the Faculty of Philosophy.* **3** °staff, personnel, members, *Brit* dons: *Each member of the faculty has a doctorate.* **4** °power, authorization, dispensation, °sanction, °license, °prerogative, °privilege, °right, °permission, °liberty: *The government has the faculty to judge anticivil acts.*

fad *n.* °craze, °mania, °rage, °fashion, °trend, °fancy, °vogue: *Do you remember the hula-hoop fad?*

fade *v.* **1** (grow) dim *or* pale, grow faint, cloud (over), °dull; °bleach, whiten, etiolate, °wash out, blanch *or* blench, discolor: *The scene faded in the distance. Look how the sun faded the curtains!* **2** °droop, wither, °decline, die out *or* away, °perish, °ebb, °flag, °wane, °wilt, °waste away, °sag, °diminish, °dwindle, languish, °deteriorate, °decay, °shrivel, peter out *or* away: *As he aged, his lust for life faded.*

fag *v.* **1** Often, ***fag out.*** °exhaust, °weary, °tire (out), °fatigue, °wear out, °jade, *Colloq Brit* knacker, *US* poop: *I am completely fagged out from studying all night.* —*n.* **2** *Brit* °bore, °nuisance, °drag, chore, °pain: *It's a bit of a fag having to fetch water from the garden.* **3** *Brit* °servant, °menial, °flunky, drudge, lackey, underling: *The seniors used to have fags to clean their shoes.* **4** See **homosexual 1,** below. **5** cigarette, smoke, *Colloq* °butt, cig(gy), weed, coffin nail, cancer stick, *Old-fashioned* gasper: *Got a fag?*

fail *v.* **1** not succeed, be unsuccessful, °miss, °miscarry, °misfire, °fall short (of), °fall flat, °fall through, falter, be (found) lacking *or* wanting, be defective, be deficient, be *or* prove inadequate, come to grief *or* naught *or* nothing, °go wrong, abort, meet with disaster, °founder, run aground, *Colloq* °flop, °fizzle (out), go up in smoke, *US* °strike out, *US, Canadian, and New Zealand* flunk: *Guy Fawkes's plot failed utterly. Gloria failed her history examination.* **2** °let down, °disappoint, °forsake, °desert, °abandon, °neglect, °ignore, °slight: *I had counted on Marcella, but she failed me.* **3** °decline, °peter out, °dwindle, °diminish, °wane, °deteriorate, °weaken, °decay, fade *or* die (out *or* away), °disappear, °flag, °ebb, °sink, languish, °give out; gutter, °go out: *His health is failing. The light failed, leaving us in darkness.* **4** go bankrupt, go out of business, °go under, go into receivership, become insolvent, close up shop, close up *or* down, cease operation(s), *Brit* °go to the wall, *US* file for Chapter 11, *Colloq* °fold (up), go bust *or* broke, *US* drown in red ink: *According to the statistics, hundreds of businesses fail every week.*

failing *n.* **1** °weakness, °shortcoming, °foible, °flaw, °fault, °defect, weak spot, blind spot, °blemish, °imperfection: *Bigotry and prejudice are her most serious failings.* —*prep.* **2** lacking, °wanting, in default of, without, sans, in the absence of: *Failing a favorable decision, we shall appeal the case.*

failure *n.* **1** °failing, °default, nonperformance, remissness; °neglect, °omission, dereliction, deficiency: *Failure to do one's duty results in a court martial.* **2** °breakdown, °collapse, discontinuance, °miscarriage, °loss; °decline, °decay, deterioration: *Power disruptions are caused by the failure of the national grid.* **3** °loser, nonstarter, misfire, °incompetent, also-ran, nonentity, *Colloq* °flop, °fizzle, damp squib, °dud, lemon, °washout, dead duck, *US* lead balloon: *He was an utter failure as a violinist.* **4** bankruptcy, °ruin, insolvency, °downfall, °crash, *Colloq* folding: *Bank failures increased owing to bad loans and other poor investments.*

faint *adj.* **1** °dim, °dull, °pale, faded, °indistinct, °vague, °hazy, °imperceptible, indiscernible, unclear, blurred, blurry, muzzy, wavering, faltering, ill-defined, °weak, °feeble, flickering, °subdued; °low, °soft, °slight, hushed, muffled, muted, °inaudible, stifled: *A faint light burned in the corridor. I heard a faint noise.* **2** °dizzy, lightheaded, unsteady, vertiginous, °giddy, *Colloq* woozy: *I felt faint after climbing the stairs.* —*v.* **3** black out, °pass out, lose consciousness, swoon, °drop, °collapse, *Colloq* keel over: *She fainted when they told her the news.* —*n.* **4** loss of consciousness, blackout, unconsciousness, °collapse, swoon, *Medical* syncope: *He dropped in a dead faint.*

fainthearted *adj.* **1** °cowardly, timorous, °afraid, frightened, °scared, °faint, lily-livered, white-livered, pusillanimous; °timid, °shy, diffident; *Colloq* yellow (-bellied), chicken-hearted, chicken-livered, chicken: *He's too fainthearted to ask her to marry him.* **2** °irresolute, °weak, °ineffectual, °feeble, °puny, feckless: *She made only a fainthearted attempt at reconciliation.*

fair[1] *adj.* **1** °impartial, evenhanded, °disinterested, °equitable, °just, °unprejudiced, unbiased, °objective, *Colloq* °square: *Judge Leaver is known for his fair decisions. We are counting on your sense of fair play.* **2** °honest, °aboveboard, °honorable, °lawful, °trustworthy, °legitimate, °proper, °upright, straightforward: *He won the trophy in a fair fight.* **3** °light, blond(e), fair-haired, flaxen-haired, tow-headed, tow-haired; light-complexioned, peaches-and-cream, °rosy; unblemished, °clear, °spotless, °immaculate: *She has fair hair* *and fair skin.* **4** °satisfactory, °adequate, °respectable, pretty good, °tolerable, °passable, all right, °average, °decent, middling, °reasonable, *comme ci, comme ça,* not bad; °mediocre, °indifferent, *Colloq* °so-so, OK *or* okay: *The performance was fair but not outstanding.* **5** °favorable, °clear, °sunny, °fine, °dry, °bright, cloudless, °pleasant, halcyon, °benign: *Fair weather is promised for tomorrow's picnic.* **6** unobstructed, °open, °clear, °free: *Every spectator has a fair view of the football field.* **7** °attractive, good-looking, °handsome, °comely, °pretty, °beautiful, pulchritudinous, °lovely, beauteous: *Faint heart never won fair lady.* **8** °civil, °courteous, °polite, °gracious, °agreeable: *She was not deceived by his fair words.*

fair[2] *n.* °fête, °festival, kermis *or* kirmess, °exhibition, °exposition, °show; °market, bazaar, mart, *US* °exhibit: *Ashby will judge the sheep at the annual fair.*

fairly *adv.* **1** °quite, °rather, °pretty, °somewhat, tolerably, adequately, sufficiently, passably, °moderately, *Colloq* °sort of, kind of: *I thought that the singing was fairly good.* **2** equitably, impartially, justly, °properly, °honestly, objectively: *Please make certain that everyone is treated fairly.* **3** °absolutely, °totally, °utterly, °completely, °positively, °really, °actually, veritably, °virtually: *The crowd fairly cheered themselves hoarse.*

fairyland *n.* dreamland, wonderland, never-never land, happy valley, °paradise, cloudland, enchanted forest, Cloud-cuckoo-land, Nephelococcygia, Shangri-La: *Zuleika's thoughts are always off somewhere in fairyland.*

faith *n.* **1** °belief, credence, °confidence, °conviction, °trust, °certainty, certitude, °assurance, assuredness, sureness, °reliance, dependence: *His faith in God is unassailable. I have faith in her ability.* **2** °belief, °religion, °creed, °persuasion, dogma, teaching, °doctrine, °denomination, °sect: *To which faith do you belong? She is of the Jewish faith.* **3** °duty, allegiance, °obligation, °promise, faithfulness, °loyalty, fidelity, °devotion, consecration, °dedication, fealty, °obedience: *Don't break faith with your electorate by supporting the bill.*

faithful *adj.* **1** °true, °loyal, °devoted, °steadfast, dedicated, °attached, unswerving, °firm, °staunch, unwavering, °constant: *He has always remained faithful to his wife.* **2** °close, °exact, °accurate, °true, °correct, °precise, °perfect, valid; °literal: *Jowett's is a faithful translation from the Greek.* **3** °conscientious, °dutiful, °scrupulous, °careful, °meticulous, °thorough, punctilious, finicky *or* finical, °detailed, °fastidious, rigorous, °rigid, °severe, °particular: *He received a gold watch for fifty years of faithful attendance to his duties.* **4** °reliable, dependable, trusted, °trustworthy, trusty, °honest, °true, °truthful, °righteous, °right, °moral, °virtuous, °upright, veracious: *He has remained my faithful friend for many years.*

faithless *adj.* **1** °skeptical, doubting, °unbelieving, disbelieving, agnostic, atheistic, freethinking: *Faithless wretches are called Doubting Thomases.* **2** unfaithful, °disloyal, treacherous, °traitorous, °perfidious, shifting, °shifty, °fickle, °inconstant, untrustworthy, °unreliable, °false, °hypocritical, °insincere, °dishonest, °false, °crooked, °unscrupulous, conscienceless, recreant: *With faithless friends like her, who needs enemies?*

fake *v.* **1** °falsify, °doctor, °alter, °modify, °counterfeit, °fabricate, °manufacture, °forge: *He faked the evidence to implicate his own sister.* **2** °pretend, make a pretense of, dissemble, feign, °sham, °make believe, simulate, °affect: *She faked a headache to avoid gym classes.* —*n.* **3** °hoax, °counterfeit, °sham, °forgery, °imitation, *Colloq* phoney *or US also* phony: *The experts agree that the painting is a fake.* **4** faker, °impostor, charlatan, °fraud, hoaxer, mountebank, °cheat, humbug, °quack, °pretender, *Colloq* phoney *or US also* phony: *He isn't a doctor—he's a fake!* —*adj.* **5** °false, °counterfeit, forged, °sham, °fraudulent, °imitation, pinchbeck, °bogus, °spurious, °factitious, *Colloq* phoney *or US also* phony: *He escaped the country using a fake passport.*

fall *v.* **1** °descend, °sink, °subside, °settle, drop *or* come (down), °plummet, °plunge, °dive, (take a) nose dive; cascade: *The bucket fell to the bottom of the well. A meteorite fell on my house. The water falls a hundred feet over the cliff at this point.* **2** °tumble, °trip, °stumble, °slump, °collapse, keel over, °topple, °crumple: *Mother fell in the kitchen and hurt her knee.* **3** °diminish, °(become) lower, °sink, °decline, fall *or* drop off, °drop, °decrease, °dwindle, °subside, come *or* go down: *The price of oil fell to a new low last week.* **4** °slope, fall away, °decline: *Beyond the copse, the meadow falls toward the river.* **5** °succumb, °surrender, °yield, give up *or* in, °capitulate, be defeated *or* conquered, be captured, be taken (captive *or* prisoner), be overthrown, come *or* go to ruin, be destroyed, be lost: *The castle fell after a year's siege.* **6** °die, °perish, drop dead, be slain *or* killed: *His grandfather fell at the Battle of the Marne.* **7** ***fall apart.*** °disintegrate, °crumble, °collapse, fall *or* come *or* go to pieces, °break up, be destroyed; break apart, °fragment, °shatter: *She would fall apart on the witness stand. After thirty years of hard wear, my car finally just fell apart.* **8** ***fall back.*** °retreat, °retire, °withdraw, °draw back; °recede: *As the enemy advanced, we fell back.* **9** ***fall back on*** or ***upon.*** have recourse to, rely *or* depend on *or* upon, return to, count on *or* upon, °resort to, call on *or* upon, make use of, °use, °employ: *All the ready ammunition was gone and we had to fall back on our reserves.* **10** ***fall behind.*** drop back, °trail, °lag; be in arrears: *If you fall behind, I'll wait for you. We fell behind in our mortgage payments.* **11** ***fall down.*** **(a)** °collapse, °drop: *She hit me so hard that I fell down.* **(b)** °fail, be (found) wanting *or* lacking, be unsuccessful, be *or* prove inadequate *or* disappointing: *He had to be replaced because he fell down on the job.* **12** ***fall flat.*** °collapse, °fail, *Colloq* °flop, *US* °bomb (out), lay an egg, go over like a lead balloon: *A hit in London might fall flat in New York.* **13** ***fall for.*** **(a)** fall in love with, be infatuated with: *Some people will fall for anyone who tells them the right time.* **(b)** be fooled *or* duped *or* taken in *or* deceived by, °accept, °swallow, succumb to, *Slang* be a sucker for, *US and Canadian* be a patsy for: *Did Beaseley really fall for that old confidence trick?* **14** ***fall in.*** °cave in, °collapse, sink inward: *The walls were about to fall in on us.* **15** ***fall in with.*** °join, °associate with, become associated *or* allied with, befriend; cooperate with, °go along with, concur with, °support, °accept: *He fell in with a gang of thieves and spent the next few years avoiding the police.* **16** ***fall off.*** °diminish, °decrease, °decline, °deteriorate: *Business falls off immediately after Christmas.* **17** ***fall on*** or ***upon.*** °attack, °assault, assail, °set upon: *Three muggers fell on me and stole my wallet.* **18** ***fall out.*** °disagree, °differ, °quarrel, °clash, squabble, wrangle, °dispute, °fight: *We fell out over politics.* **19** ***fall short.*** prove *or* (turn out to) be inadequate *or* insufficient *or* deficient *or* lacking *or* wanting *or* disappointing, °miss, °fail, °disappoint: *The results of the sales campaign fell short of expectations.* **20** ***fall through.*** °fail, come to nothing *or* naught, °miscarry, °die, *Colloq* °fizzle (out), °flop: *The deal to buy the company fell through.* **21** ***fall to.*** °start, °begin, °commence, set *or* go about, get under way, °undertake, °tackle, °take on; get moving, °attack, *Colloq* get the show on the road, get cracking, *US* get a wiggle on, move it: *The washing-up had to be done so I fell to.* —*n.* **22** °drop, descent, °dive, nose dive, °plunge, °tumble, dropping, falling: *How could he have survived a fall from such a height?* **23** *Chiefly US and*

Canadian autumn: *We turn the clocks back one hour in the fall.* **24** °decline, °decay, °collapse, °downfall, °failure, °destruction, °ruin, °failure, deterioration, °eclipse: *Have you read Poe's classic,* Fall of the House of Usher? **25** Usually, ***falls.*** cascade, cataract, °waterfall; rapids: *How many falls are there along the Limpopo River?* **26** depreciation, °sinking, diminution, °decrease, °decline, °lapse, downturn, downswing, °drop, drop-off, lowering, abatement, °slump, °collapse: *On the stock exchange today, investors experienced steep falls in share prices.* **27** °slope, declivity, descent, °decline, °drop, downhill, *Chiefly US and Canadian* °downgrade: *Note the smooth rise and fall of the land.* **28** °surrender, capitulation, °submission, °taking, °seizure, °capture, °overthrow, °defeat, °conquest, °downfall: *The fall of Khartoum in 1898 marked the reestablishment of British rule in the Anglo-Egyptian Sudan.*

fallacy *n.* °misconception, miscalculation, misjudgment, °mistake, °error, *non sequitur,* °solecism, °delusion; paralogism; sophism: *It is a fallacy to think that you could ever learn to play the violin as well as she.*

false *adj.* **1** °untrue, unfactual, untruthful, °wrong, °amiss, °mistaken, °erroneous, °incorrect, °inaccurate, °inexact, °imprecise, °faulty, °flawed, °invalid, °unsound, °unreal, °imaginary, °fictitious, °spurious: *The explorers gave a completely false picture of the local inhabitants.* **2** °untrue, untruthful, °lying, misleading, fallacious, fabricated, made-up, concocted, mendacious, untrustworthy, °fraudulent, meretricious, °deceptive, °deceitful, treacherous, *Colloq* phoney *or US also* phony: *The testimony of this witness is completely false.* **3** °counterfeit, °imitation, simulated, °sham, forged, °fraudulent, °fake, °artificial, °synthetic, manufactured, °unnatural, °spurious, °bogus, ersatz, °factitious, °mock, pseudo, *Colloq* phoney *or US also* phony: *Your false teeth look almost real.* **4** °sham, feigned, °affected, °insincere, faked, manufactured, counterfeit(ed): *Don't shed any false tears over me when I'm gone.* **5** illogical, fallacious, °unsound, °invalid, °flawed, °faulty: *That conclusion could come only from false reasoning.*

falsehood *n.* °lie, °fib, prevarication, untruth, °fabrication, °misstatement, fiction, °(fairy) tale, °story, distortion, *Colloq* cock-and-bull story, *Slang Brit* load of codswallop: *That was a complete falsehood about the goings-on in the attic, wasn't it?*

falsify *v.* °fake, °alter, °distort, misstate, °misrepresent, °twist, *Colloq* fudge, trump up, *Brit* cook: *The tax inspector found that the accounts had been falsified.*

falsity *n.* **1** untruthfulness, mendaciousness, fraudulence, deceptiveness, °deceit, deceitfulness, dishonesty, spuriousness, speciousness, casuistry, °hypocrisy, insincerity, falseness: *The prosecution was unable to prove the falsity of her statement.*

fame *n.* °renown, repute, °reputation, °celebrity, illustriousness, °superiority, °preeminence, stardom, °prominence, eminence, °glory, °name, °notoriety, acclaim: *Randolph's fame has spread far and wide.*

familiar *adj.* **1** °well-known, °common, commonplace, °everyday, °ordinary, °current: *That's a familiar melody.* **2** °frequent, °usual, °customary, °habitual, °routine, °traditional: *We walked along the familiar path through the forest to Grandmother's house.* **3** °friendly, affable, °close, °intimate, °sociable, °social, °free, free-and-easy, °relaxed; overfriendly, overfree, overfamiliar, °bold, °forward, °insolent, °impudent, °presumptuous, presuming, °disrespectful, unreserved, unrestrained; °informal, °casual, °cordial, unceremonious; *Colloq* °chummy, *Slang US and Canadian* buddy-buddy, palsy-walsy: *She began to get entirely too familiar.* **4** ***familiar with.*** aware *or* conscious *or* cognizant of, knowledgeable about *or* of *or* in, conversant *or* acquainted with, no stranger to, on speaking terms with, up on *or* in, (well) versed in, informed of *or* about, °privy to, in the know about, *au courant, au fait*: *Are you familiar with the latest theories in particle physics?*

familiarity *n.* **1** °knowledge, acquaintance(ship), °grasp, °understanding, comprehension, °cognizance, awareness, conversance, °experience: *I have no familiarity at all with particle physics.* **2** friendliness, affability, sociability, neighborliness, °fellowship, intimacy, intimateness, closeness, openness, naturalness, °ease, informality, unceremoniousness: *He prided himself on his familiarity with celebrities.* **3** boldness, presumptuousness, overfamiliarity, °presumption, °impudence, insolence, °impertinence, °impropriety: *He put his arm around her waist with offensive familiarity.*

familiarize *v.* Usually, **familiarize with.** accustom (to), make familiar *or* acquaint (with), °initiate (in), inform (about *or* on), enlighten (about *or* as to), °teach (about), educate *or* instruct *or* tutor (in): *I am trying to familiarize myself with the music of John Cage.*

family *n.* **1** °(kith and) kin, kinsmen, °kindred, kinfolk *or Brit* kinsfolk, next of kin, relatives, °relations, household, °people, °one's own flesh and blood, one's nearest and dearest, ménage, *Colloq* folks: *We usually spend the holidays with my family.* **2** children, °offspring, °progeny, °issue, °brood, *Colloq* kids: *Large families were much more common in the nineteenth century.* **3** ancestors, forebears, forefathers, progenitors; ancestry, °parentage, descent, °extraction, °derivation, °lineage, °pedigree, genealogy, family tree, °house, °line, bloodline, °dynasty; blood, °stock, °strain: *Most of his family were humble folk, but she comes from a noble family.* **4** °group, °set, °division, subdivision, classification, °type, °kind, °class, °genre, °order, species, genus: *English belongs to the Indo-European family of languages.*

famine *n.* starvation; °shortage, °dearth, °scarcity, deficiency, paucity, exiguity, barrenness, °lack: *In days of abundance, no one should die of famine.*

famished *adj.* starving, °starved, °voracious, °ravenous, ravening, craving, °hungry: *The survivors were famished after a fortnight in the lifeboat.*

famous *adj.* °renowned, °celebrated, °popular, famed, °well-known, °noted, °eminent, °preeminent, °conspicuous, °prominent, °illustrious, °notable, acclaimed, °venerable, °legendary, °distinguished, °praiseworthy, honored, lionized: *A famous architect has been invited to address the convention.*

famously *adv.* excellently, °(very) well, superbly, marvelously, splendidly, capitally, spectacularly, superlatively: *The prince and I get on famously.*

fan *n.* °admirer, °enthusiast, adherent, °devotee, aficionado, °follower, °supporter, °lover, °zealot, *Colloq* buff, °fiend, °hound, °bug, °addict, nut, *US* booster, *Slang* junkie, °freak, groupie: *Avid fans of the Dodgers, we go to all the games.*

fanatic *n.* °maniac, extremist, °zealot, *Colloq* °fiend, nut, *Slang* °freak: *Religious fanatics killed "heathens" or "infidels" by the thousands.*

fanatical *adj.* °fanatic, °extreme, distracted, °maniacal, °mad, °rabid, zealous, frenzied, °feverish, °burning, °frantic, frenetic, °obsessive, °fervent, °compulsive, monomaniacal, fervid, perfervid, °passionate, °enthusiastic, °agog, °immoderate, °excessive: *The entire community is fanatical about its fundamentalism.*

fanaticism *n.* **1** °devotion, °dedication, devotedness; infatuation, °enthusiasm, °fervor, zeal, obsessiveness, franticness, °frenzy, hysteria: *Her fanaticism for rock musicians is getting a bit out of hand.* **2** monomania, single-mindedness, °mania, °madness, extremism, °intolerance, °bigotry, °bias, °partiality, °prejudice, narrow-mindedness, close-mindedness: *Some religious sects are characterized by virulent fanaticism.*

fancied *adj.* °imaginary, °unreal, °fanciful, imagined, °illusory, make-believe, °mythical, fairy-tale: *They support their king in the pursuit of his fancied rights.*

fanciful *adj.* **1** °whimsical, °capricious, °impulsive, °inconstant, °fickle, °changeable, °variable: *Those graffiti are products of a fanciful mind.* **2** °extravagant, chimerical, °fantastic, °fabulous, °mythical, fairy-tale, °imaginative, °fancied, make-believe, °unreal, °illusory, imagined, °visionary, °imaginary: *Some of the fanciful ideas of science fiction have become realities.* **3** °curious, °odd, °peculiar, °bizarre, °unusual, °original: *The chalice was decorated with fanciful curlicues.*

fancy *adj.* **1** °ornate, decorative, decorated, °ornamental, ornamented, °elaborate, embellished, embroidered, °fanciful, °extravagant, rococo, baroque, gingerbread, Byzantine, °complicated, °intricate, complex: *The modern trend has been away from fancy architecture.* **2** °illusory, °capricious, °fanciful, °extravagant, °fantastic, °far-fetched, delusive, °whimsical, °visionary, °unrealistic, °grandiose: *He has some fancy ideas about building an undersea city.* **3** deluxe, °luxury, °luxurious, °choice, °select, °prime, °special, °elegant, °superior, °quality, °high-class; °posh: *They stock only fancy fruits and vegetables. She has a fancy suite at the Cardigan Hotel.* **4** °high, °exorbitant, °inflated, °outrageous: *One has to pay very fancy prices for haute couture.* —*n.* **5** °imagination, °creation, °conception, inventiveness, creativeness, creativity: *These chimeras are entirely a product of his fancy.* **6** °imagination, °fantasy, °hallucination, °delusion, °illusion, unreality, make-believe, °dream, °daydream, pipe dream, mirage, phantasm, °phantom, figment (of the imagination), °impression: *Our plans for the future must be based on fact, not fancy.* **7** °liking, °inclination, fondness, °taste, °penchant, °attraction, °preference, °partiality, predilection, yearning, craving, hankering, °wish, °desire, °longing: *Miss Crow's fancy for younger men often causes her some embarrassment.* **8** °idea, whim, caprice, whimsy, °urge, impulse, °notion, vagary, °quirk, crotchet, °peculiarity: *His fancy today is that he invented electricity.* —*v.* **9** °imagine, °conceive, °picture, visualize, °envisage, think *or* make up, conjure up, *US* °envision, *Colloq* dream up: *He fancies himself on a big yacht in the Mediterranean.* **10** °think, °imagine, °understand, °believe, °suspect, °guess, conjecture, °presume, °surmise, °assume, °take it, °suppose, °infer, °reckon: *From his costume I fancy he must be Superman. Fancy Kim winning the Nobel prize!* **11** °like, be attracted to, take (a liking) to, °desire, °want, crave, long *or* pine for, have a yen *or* craving for, have an eye for, wish for, hunger for *or* after, °favor, °prefer, lust for *or* after: *Twombly has always fancied tall women. I wouldn't fancy being forty feet up a swaying ladder like that.*

fanfare *n.* **1** °flourish, fanfaron, fanfaronade, (trumpet) blast *or* blare: *Following a loud fanfare, the toreador strutted into the bullring.* **2** hullabaloo, hubbub, brouhaha, commotion, °stir, ado, °show, °fuss, *Colloq* to-do, ballyhoo: *Despite the enormous fanfare, the film was a failure.*

fantasize *v.* °dream, °imagine, °daydream, °muse, °mull (over), build castles in the air *or* in Spain, °speculate, °envisage, stargaze; hallucinate, *US* °envision: *She often fantasized about the kind of man she would marry.*

fantastic *adj.* **1** °fanciful, °strange, °weird, °peculiar, °odd, °eccentric, °queer, °bizarre, °quaint, °outlandish, °exotic, °extravagant, °grotesque, °nightmarish, °alien, °remarkable: *She wore the most fantastic costume to the fancy-dress ball.* **2** °imaginary, °illusory, illusive, °unreal, °visionary, °fanciful, °unrealistic, imagined, irrational: *His books are inhabited by fantastic creatures.* **3** °unbelievable, °incredible, °preposterous, °extraordinary, °implausible, °absurd, °unlikely: *For a man of eighty-five, he made a fantastic showing in the marathon.* **4** °marvelous, spectacular, °splendid, wonderful, tremendous, °overwhelming, *Colloq* °great, °fabulous, °terrific: *The Picasso exhibition is simply fantastic.*

fantasy *n.* **1** °imagination, °fancy, creativity, inventiveness, creativity, °originality: *We encourage the children to give free rein to their fantasy.* **2** °vision, °hallucination, °illusion, mirage, °delusion, chimera, °dream, °daydream, °(flight of) fancy, pipe dream: *Her fantasy is to become prime minister.* **3** make-believe, °invention, °fabrication, fiction, °masquerade, fable, concoction, °pretense: *His story about being an orphan is pure fantasy.*

far *adv.* **1** afar, far away *or* off, a good *or* great *or* long way *or* distance off *or* away: *We caught sight of a sail far to the south.* **2** (very) much, considerably, decidedly, incomparably: *She is a far better swimmer than George.* **3** ***by far.*** (very) much, considerably, decidedly, incomparably, (im)measurably, by a long shot, far and away, °clearly, plainly, °obviously, doubtless(ly), indubitably, °undoubtedly, °definitely, beyond (the shadow of a) doubt, without a doubt, *Colloq Brit* by a long chalk: *She's a better swimmer than George by far. He is by far the wealthiest person I know.* **4** ***far and wide.*** °everywhere, near and far *or* far and near, extensively, °widely, high and low; here, there, and everywhere: *We searched far and wide to find these specimens.* **5** ***far gone.*** **(a)** beyond *or* past help, advanced, deteriorated, worn-out, °dilapidated, near the end: *That shirt is too far gone to send to the laundry. My house is so far gone it's beyond repair.* **(b)** °drunk, besotted, *Slang* °loaded, pissed, paralytic, paralyzed: *He's too far gone to walk.* **6** ***go far.*** **(a)** °progress, °advance, °succeed, °go places, °get ahead, °rise (in the world), make a name for oneself, become successful, set the world on fire, *Brit* set the Thames on fire, *US* cut a swath: *He is a very bright youngster and will go far in whatever profession he chooses.* **(b)** °help, °aid, °contribute, play a part: *The new law will go far towards inhibiting child abuse.* **7** ***go too far.*** go overboard *or* over the top, not know when to stop, °go to extremes; °exceed, °overdo, °overstep, °transcend, go beyond: *Ambition is one thing, but he went too far when he tried to get his boss's job.* **8** ***so far.*** **(a)** thus far, (up) to *or* till *or* until now *or* the present *or* this point, to date, to this point in time: *So far, we have been able to keep up with the mortgage payments.* **(b)** to a certain extent *or* limit *or* point: *She said she would go just so far and no further.* —*adj.* **9** (more) remote *or* distant, far-away, far-off; °extreme, farther, farthest: *She claimed the ability to see into the far future. He kicked the ball to the far end of the field.*

faraway *adj.* **1** °distant, °remote, far-off, °outlying, far-flung: *People came from faraway places as news of the miracle spread.* **2** °dreamy, °detached, °absent, °absent-minded, abstracted: *When you have that faraway expression, I know you don't hear a word I say.*

farcical *adj.* °ludicrous, laughable, risible, °funny, °nonsensical, °ridiculous, °silly, °preposterous, °absurd, °foolish; comical, °humorous, droll, amusing: *His farcical attempts at surfing while fully dressed had us in hysterics.*

fare *n.* **1** °passenger, °traveler: *The taxi driver deposited his fare at the hotel.* **2** °charge, °price, °cost: *What is the fare from Oxford to London?* **3** °food, °diet, victuals, meals, viands, eatables, °provisions: *The prison fare consisted of bread and water.* —*v.* **4** do, make one's way, °manage, get on *or* along, °make out, °survive: *The children didn't fare very well on their own.*

farewell *n.* **1** adieu, °goodbye: *We said our farewells and left.* **2** departure, leave-taking, congé, °parting, *Colloq* sendoff: *I want to avoid a tearful farewell.* —*interj.* **3** Adieu!, Goodbye!, So long!, Godspeed!, Adios!, Hasta luego!, Hasta la vista!, Auf Wiedersehen!, Ciao!, Sayonara!, Aloha!, Vaya con Dios!, *Colloq Brit* God bless!, *Old-fashioned* Toodle-oo!, Pip! Pip!, Ta-ta!, *US old-fashioned* See you later (Alligator)!, Don't take any wooden nickels!: *I said farewell and we went our separate ways.*

far-fetched *adj.* °strained, stretched, °improbable, °implausible, °unlikely, °doubtful, dubious, °questionable, °forced, unconvincing, °unrealistic, °fantastic, °preposterous, hard to believe, °unbelievable, °incredible, *Colloq* hard to swallow, °fishy: *They told some far-fetched tale about being robbed by a gang of midgets.*

farm *n.* **1** farmstead, farmhouse, grange, homestead, holding; °land, farmland, acreage, arable; *Brit* steading, smallholding, °allotment, *Scots* farm-toun, croft: *Grandfather owns a farm in Nebraska.* **2** ***buy the farm.*** °die, be killed: *A MIG caught Johnson and he bought the farm.* —*v.* **3** °cultivate, work the land, till the soil: *His family has been farming this land for centuries.* **4** ***farm out.*** °contract, subcontract, °lease, °delegate, °let (out): *They cut staff and now farm out much of the work.*

farmer *n.* husbandman, agriculturist, agronomist, yeoman, *Brit* smallholder, *US dialectal* granger: *The farmers here were hard hit by the drought.*

farming *n.* agriculture, agronomy, husbandry, agribusiness, cultivation: *Less than eight thousand years ago, human beings turned from hunting and gathering to farming.*

farsighted *adj.* **1** farseeing, foresighted, prescient, °provident, °prudent, °shrewd, °perceptive, discerning, insightful, °wise, sagacious, °acute, °sharp, °astute, °sensible, °imaginative: *The committee drew up a far-sighted plan for the redevelopment of the town center.* **2** long-sighted, hyperopic *or* hypermetropic, presbyopic: *I'm farsighted, so I wear glasses only for reading.*

fascinate *v.* °bewitch, °enchant, cast a spell on *or* over, ensorcell, spellbind, hold spellbound, put *or* have under a spell, °charm, °captivate, °intrigue, °beguile, °hypnotize, mesmerize, °transfix, °entrance, engross, enthrall, °enrapture, absorb, allure, °attract: *Desmond is utterly fascinated by Elizabeth.*

fascination *n.* °enchantment, °sorcery, °magic, attractiveness, °attraction, °draw, °pull, °(animal) magnetism, °charm, allure, captivation, °influence, witchcraft, entrancement: *Blondes always held a strange fascination for him.*

fashion *n.* **1** °style, °mode, °vogue, °trend, °look, °taste: *That year there was a fashion for stiletto heels.* **2** the fad, °mania, the craze, the rage, the latest (thing), *le dernier cri, Colloq Brit* the go: *He remembers when upswept hair-dos were the fashion.* **3** °manner, °mode, °way, °approach, °attitude: *I'll always be true to you, darling, in my fashion.* **4** ***in fashion.*** See **fashionable,** below. —*v.* **5** °make, °model, °style, °shape, °form, °frame, °mold, °create, °construct, °forge, °work, °manufacture: *She fashions the most beautiful vases out of shapeless lumps of clay.*

fashionable *adj.* °in fashion, °chic, à la mode, modish, °stylish, °smart, in vogue, °up-to-the-minute, up-to-date, *Colloq* °trendy, in, with it, *Colloq Brit* all the go: *The couturiers insist that black will be fashionable this year.*

fast[1] *adj.* **1** °quick, °swift, °fleet, °speedy, °brisk; °brief, °hurried, °hasty, high-speed, accelerated, °expeditious, °rapid, °express: *She is very fast on her feet. I have to make a fast visit to the bank.* **2** °loose, °profligate, °wild, °extravagant, dissipated, intemperate, °irresponsible, sybaritic, °self-indulgent, °dissolute, unrestrained, indecorous, °rakish, licentious, °promiscuous, °immoral, °wanton, °lecherous, °lustful: *They led quite a fast life till their divorce.* **3** °firm, fastened, secure(d), °fixed, tied, °bound, connected, °attached: *The boat was fast to the pier.* **4** °firm, °fixed, settled, °stable, °solid, °immovable, unshakable *or* unshakeable, °tight: *The sword was fast in the stone.* **5** °firm, °stable, °steadfast, °staunch, unwavering, °constant, °lasting, °close, °loyal, °devoted, °faithful, °lasting, °permanent: *We maintained a fast friendship over the years.* —*adv.* **6** °quickly, °swiftly, °rapidly, speedily, briskly, presto, °hastily, hurriedly, with all speed *or* haste, expeditiously, apace, °posthaste, like a flash, in the blink of an eye, in a wink, before one can *or* could say "Jack Robinson," in no time (at all), *Colloq* like a bat out of hell, °like a shot, PDQ *or* P.D.Q. (= 'pretty damn(ed) quick'), *Brit* like the clappers (of hell), *US and Canadian* quick like a bunny *or* rabbit, lickety-split: *Don't talk so fast. He ran out of there very fast. I'll be back very fast.* **7** °firmly, fixedly, immovably, solidly, unshakably *or* unshakeably, °tightly, securely, soundly: *The rope held fast. He is fast asleep.* **8** closely, °close to, °immediately, °near, (close) on, °right: *Fast on the heels of the fugitive came the police.* **9** loosely, wildly, recklessly, intemperately, irresponsibly, fecklessly, extravagantly, intemperately, sybaritically, self-indulgently, dissolutely, unrestrainedly, indecorously, rakishly, licentiously, promiscuously, immorally, wantonly, lecherously, lustfully: *He's been living fast since inheriting that fortune.*

fast[2] *v.* **1** abstain, go hungry, deny oneself, °diet, starve (oneself): *Do you fast during Lent?* —*n.* **2** abstention, abstinence, fasting, °self-denial, °diet; hunger strike: *She went on a two-week fast.*

fasten *v.* **1** °attach, °tie, °bind, °bond, °stick, affix, °anchor, °fix, °lock, °hook (up), °secure, °join, °connect, °link, °fuse, °cement, °clamp: *Fasten your seat belts. The mussels fasten themselves by their byssi to underwater piles.* **2** °fix, rivet, °focus, °concentrate, °direct, °aim, °point: *He fastened his gaze on Kitty. She fastened her attention on the ceiling.*

fastening *n.* fastener, °catch, °clasp, latch, °lock, °tie, °bond: *Can you see what kind of fastening is holding the cover?*

fastidious *adj.* °squeamish, °delicate, overnice, °fussy, °meticulous, °finicky, finical, pernickety *or US also* persnickety, °particular, °difficult, °critical, hypercritical, supercritical, overprecise, punctilious, *Colloq* nit-picking, picky: *The Prioress was too fastidious to allow a morsel drop from her lip.*

fat *adj.* **1** °obese, °stout, overweight, °heavy, °plump, °rotund, corpulent, portly, °well-fed, °chubby, podgy *or chiefly US* pudgy, roly-poly, tubby, °bulky, fleshy, paunchy, potbellied, overfed, °flabby, elephantine, *Colloq* broad in the beam, °beamy, beefy, *Slang US* five-by-five: *A fat man squeezed in beside me.* **2** °oily, oleaginous, unctuous, °greasy, fatty, pinguid, sebaceous, adipose: *His complexion is bad because he eats too much fat food.* **3** °rich, °wealthy, °prosperous, affluent, well-to-do, °well-off, *Colloq* well-heeled, °loaded: *They grew fat on their profits from the black market.* **4** °profitable, lucrative, °fruitful, remunerative, *Slang* cushy: *He has a fat job as a purchasing agent for the government.* —*n.* **5** °obesity, corpulence, stoutness, overweight, heaviness, plumpness, rotundity, portliness, chubbiness, podginess *or chiefly US* pudginess, tubbiness, fleshiness, paunchiness, flabbiness: *He leans a little towards fat because of lack of exercise.* **6** °riches, °wealth, °prosperity, fertility, °yield, °abundance, °plenty, plenteousness: *He's living off the fat of the land.*

fatal *adj.* **1** °fateful, °deadly, °murderous, °lethal, °mortal, toxic, °terminal, °final; baneful, °poisonous: *She drank the fatal potion without a word. Who dealt the fatal blow?* **2** °destructive, °fateful, °ruinous, °calamitous, °dreadful, °disastrous, °devastating, cataclysmic, catastrophic, °harmful, °mischievous, damaging: *Discovery by the police would prove fatal to our plan.* **3** °fateful, °fated, °destined, predestined, decreed, ordained, foreordained, preordained, °predetermined, °inevitable, °unavoidable, °necessary, °essential, inescapable, ineluctable: *The events fell into their fatal sequence.*

fatality *n.* **1** °catastrophe, °disaster, °calamity, cataclysm: *The eruption of Vesuvius was a relatively recent fatality.* **2** °death, °casualty: *Traffic fatalities are on the increase.*

fate *n.* **1** °fortune, °lot, °luck, °chance, °life, °destiny, God's will, °providence, °doom, karma, kismet, toss *or* throw of the dice, *Colloq US and Canadian* the breaks, the way the cookie crumbles, the way the ball bounces: *Fate has brought us together. Our meeting this way was just fate.* **2** °doom, °destruction, °downfall, °undoing, °ruin, °disaster, °collapse, °death, nemesis, °end, °finish: *The defenders of Masada met their fate bravely.* **3** °end, °outcome, °future, °destination, °disposition: *There was no trace of the fate of the explorers. What is to be the fate of this obsolete equipment?*

fated *adj.* **1** °destined, predestined, °predetermined, decreed, °doomed, °fateful, ordained, foreordained, preordained, °decided: *The fated day of my trial arrived.* **2** °sure, °certain, °doomed, damned, cursed: *Was I fated to spend the rest of my life in this fetid dungeon?* **3** °fatal, °fateful, °unavoidable, inescapable, °inevitable, ineluctable: *Their fated punishment was to bail out the sea using sieves.*

fateful *adj.* **1** °significant, °momentous, °ominous, °major, consequential, °important, °critical, °crucial, decisive, °weighty, °portentous, earthshaking, °pivotal: *The minister is faced with having to make a fateful decision.* **2** °deadly, °lethal, °fatal, °destructive, °ruinous, °disastrous, catastrophic, cataclysmic: *Failure to preserve the environment could have fateful consequences.*

father *n.* **1** sire, paterfamilias, *Colloq* dad, daddy, pa, papa, °pop, poppa, old man, °old boy, *Brit* governor, pater: *His father is a chemist.* **2** forebear, °ancestor, forefather, °progenitor, primogenitor: *The father of the dynasty fought at Hastings.* **3** °creator, °founder, originator, inventor, °author, architect, framer, initiator: *He regards Lavoisier as the father of modern chemistry.* **4** °priest, confessor, curé, abbé, °minister, °pastor, °shepherd, parson, °clergyman, chaplain, *Colloq* padre, *Military slang* sky pilot: *We should see the Father about the funeral service.* —*v.* **5** sire, beget, °get, engender, procreate, °generate: *The sultan fathered more than forty children.* **6** °originate, °establish, °found, °invent, °author, °frame, °initiate, °institute, °create: *Galileo fathered modern astronomy by inventing the telescope.*

fatherland *n.* motherland, native land, mother country, homeland, °(old) country, birthplace: *They return to their fatherland for an annual visit.*

fatherly *adj.* °paternal, °kindly, °kind, °warm, °friendly, °affectionate, °protective, °amiable, °benevolent, well-meaning, °benign, caring, °sympathetic, °indulgent, °understanding; parental: *My nephew used to come to me for fatherly advice.*

fathom *v.* °probe, °sound, °plumb, °penetrate, °search (out), °investigate, °measure, °gauge, °determine, ascertain, °work out, get to the bottom of, delve into, °understand, °grasp, °divine: *I never quite fathomed the purpose of this device.*

fatigue *n.* **1** weariness, tiredness, °weakness, °exhaustion, lassitude, listlessness, °lethargy, languor, °sluggishness, enervation: *She kept on tending the wounded till fatigue finally overcame her.* —*v.* **2** °weary, °tire, °weaken, °exhaust, °drain, °enervate, *Colloq* °fag (out): *My morning's work had fatigued me.*

fatigued *adj.* °weary, wearied, °tired, overtired, dead tired, °weak, weakened, °exhausted, °listless, °lethargic, languorous, sluggish, enervated, °strained, wasted, *Colloq* whacked (out), knocked out, °dead, °beat, dead beat, all in, *Brit* knackered, jiggered, buggered, *US and Canadian* pooped, bushed: *The firemen were quite fatigued from having been on duty around the clock.*

fatness *n.* °obesity, stoutness, corpulence, *embonpoint,* portliness, plumpness, chubbiness, rotundity, podginess *or chiefly US* pudginess: *His fatness is owing to a glandular condition, not overeating.*

fault *n.* **1** °imperfection, °flaw, °defect, °blemish, deficiency, °shortcoming, °failing, °weakness; °frailty, °foible, °peccadillo: *The fault lies in this circuit. Using filthy language is one of his minor faults.* **2** °mistake, °error, °blunder, °lapse, °failure, °offense, °oversight, slip(-up), °indiscretion, gaffe, gaucherie, faux pas, *Slang* boner, °howler, goof, boo-boo, *Brit* boob: *People should be responsible for their faults. There are many faults in this manuscript.* **3** °responsibility, °liability, culpability; °blame, °accountability, answerability: *It wasn't my fault that the tree fell down. The fault lies with you, Gavin.* **4** °sin, °transgression, trespass, °misdeed, °offense, misdemeanor, °vice, °indiscretion, misconduct, °misbehavior: *He will confess his faults to anyone who will listen.* **5** ***at fault.*** to blame, blamable, blameworthy, in the wrong, °responsible, answerable, °accountable, °liable, culpable, °guilty: *You were at fault for failing to report the crime.* **6** ***find fault.*** °criticize, censure, take exception (to), °carp (at), °cavil (at), °pick at, °pick on, pick apart, pick holes in, °niggle, °fuss, *Colloq* nitpick, °knock: *She constantly finds fault with everything I do.* **7** ***to a fault.*** excessively, °extremely, to an extreme, °in the extreme, unreasonably, °exceedingly, °unduly, disproportionately, immoderately, irrationally, *US* °overly: *She is modest to a fault.* —*v.* **8** find fault with, censure, °blame, °criticize, call to account, impugn, call into question, hold (someone) responsible *or* accountable *or* to blame, lay at (someone's) door, °accuse: *You cannot be faulted for not knowing the fuel tank was almost empty.*

faultfinding *n.* **1** °criticism, censure, carping, caviling, captiousness, hypercriticism, quibbling, fussiness, hair-splitting, pettifogging, *Colloq* nitpicking, pickiness: *Why continue to put up with the foreman's faultfinding?* —*adj.* **2** °critical, censorious, carping, caviling, captious, hypercritical, °fussy, hair-splitting, pettifogging, °niggling, contentious, °querulous, quibbling, *Colloq* nitpicking, picky: *She could no longer abide his faultfinding attitude.*

faultless *adj.* °perfect, °flawless, °immaculate, °ideal, °exemplary, °correct, °accurate, °foolproof, °irreproachable, unimpeachable, *Colloq Brit* bang on, spot on: *The run of the new model on the test track was faultless.*

faulty *adj.* °defective, °unsound, °imperfect, °flawed, impaired, °out of order, malfunctioning, °broken, °bad; damaged; *Slang* °on the blink, *Chiefly US* on the fritz: *The carburetor is faulty.*

favor *n.* **1** goodwill, °approval, °support, approbation: *Father looked with favor on my marriage to Leslie.* **2** °courtesy, good *or* kind deed, °good turn, °kindness, °gesture, *beau geste;* °service: *As a favor, I'll walk your* *dog while you're away.* **3** °favoritism, °partiality, °prejudice, °bias, °preference, °patronage: *They show favor in repeatedly giving the contract to the same company.* **4** °grace, °esteem, °consideration, °view, °regard, °opinion, °account, (good *or* bad) books; °disposition, °taste, predisposition: *How does Rochester stand in the favor of the king?* **5** ***in favor (of).*** °for, pro, on the side of, in support (of), °at the back of, °backing, behind, on *or US and Canadian* in behalf (of), in back of: *Please vote in favor of the resolution. Are you in favor or against?* —*v.* **6** °approve, °prefer, °like, have a liking *or* preference for, be partial to, °advocate, espouse, °back, °endorse, °support, °champion, °recommend, incline to *or* towards, °side with, take the side *or* part of, °go for, opt for, °fancy, °select, °choose, °elect, °single out, °sponsor, °adopt, °go in for: *Which candidate do you favor?* **7** °pamper, °coddle, °baby, °protect, °indulge, be partial to: *Notice how Billie favors his left leg when walking?* **8** °advance, °forward, °promote, °facilitate, °expedite, °help, °benefit, °assist, °aid, °encourage, °accommodate, smile upon: *The warm, moist climate favors the growth of mold.* **9** °resemble, look like, °take after: *I disagree and think the baby favors his father.*

favorable *adj.* **1** °advantageous, °promising, auspicious, °fair, °beneficial, °suitable, °fitting, °appropriate, encouraging, facilitative, °helpful, °helping, °supportive, supporting, °convenient, °useful, °opportune, °propitious, °accommodating, accommodative; facultative: *A favorable breeze sent our craft toward Cyprus.* **2** °good, °promising, °positive, encouraging, reassuring, affirmative, well-disposed, °sympathetic; commendatory, °laudatory, °enthusiastic, °eager, °ardent, zealous: *We concluded a very favorable deal to deliver widgets to the army. My latest book received favorable reviews.*

favorably *adv.* **1** graciously, indulgently, sympathetically, genially, °positively, agreeably, enthusiastically, with favor: *They reacted favorably to my suggestion.* **2** advantageously, affirmatively, °positively, in favorable terms: *Your application has been considered favorably.*

favored *adj.* **1** preferred, chosen, °choice, selected, °popular, °favorite, °pet: *The favored explanation is that he is the boss's son.* **2** advantaged, °privileged, blessed, °prosperous, °wealthy, °rich, affluent, °well-off: *She is one of the favored few who can afford to travel.*

favorite *n.* **1** °choice, °pick, °preference, preferred, esteemed, °darling, °pet, °ideal, apple of (someone's) eye, *Colloq Brit* blue-eyed *or* white-haired *or* white-headed boy *or* girl, flavor of the month, *US* fair-haired boy *or* girl: *My favorite flavor is chocolate. Carstairs is the favorite to win the next election.* —*adj.* **2** °beloved, chosen, picked, selected, preferred, best-liked, most-liked, °favored, °choice, °pet, °ideal: *Which is your favorite flavor?*

favoritism *n.* °partiality, °bias, predisposition, prepossession, °prejudice, °bent, partisanship, nepotism, preferment; °preference, °leaning, °inclination, proclivity: *He openly admits to favoritism in selecting employees.*

fear *n.* **1** °dread, °terror, °fright, °panic, °horror, °alarm, trepidation, apprehension, fearfulness, apprehensiveness, °dismay, consternation, hesitation, qualms, diffidence, timidity, cravenness, °cowardice, second thoughts: *His fear of going into the jungle alone is understandable.* **2** awe, °respect, °reverence, °veneration: *That'll put the fear of God into him!* **3** °horror, °specter, nightmare, bogey *or* bogy, °phobia, bugbear, *bête noire,* misgiving(s), foreboding(s): *When they came to break the news, her worst fears were realized.* **4** °solicitude, °anxiety, angst, foreboding(s), °distress, misgiving(s), °concern, apprehension, °worry, uneasiness, unease: *Fear for her safety is shared by all of us.* —*v.* **5** be afraid *or* scared *or* fearful *or* frightened (of), °dread, quail *or* tremble *or* quake at, shudder at, °shrink from, °quiver: *What man cannot understand he often fears.* **6** °revere, °respect, °venerate, be *or* stand in awe of: *If you fear the Lord as I do, you won't take his name in vain.* **7** °expect, °anticipate, °imagine, °suspect, °foresee: *I fear we are too late for dinner.*

fearful *adj.* **1** °afraid, °scared, frightened, terrified, alarmed, °panic-stricken, terror-stricken: *Of all disasters, I am most fearful of fire.* **2** °hesitant, °timid, timorous, °shy, diffident, unwilling, intimidated, °jumpy, °nervous, edgy, panicky, °anxious, apprehensive, °cowardly, pusillanimous, *Colloq* yellow, jittery: *Why is she so fearful of meeting her prospective in-laws?* **3** °awful, dire, °dreadful, °frightful, °frightening, °terrifying, °terrible, appalling, °ghastly, °atrocious, °horrible, horrifying, horrendous, horrific, °hideous, °gruesome, °grisly, °grim, baleful, °monstrous, unspeakable, °loathsome, heinous, °repugnant, °repulsive, °revolting, °disgusting, nauseating, °nauseous: *The fearful monster blocked our only way out.*

fearfully *adv.* **1** hesitantly, timidly, timorously, shyly, diffidently, unwillingly, nervously, edgily, anxiously, apprehensively: *Fearfully, she approached the altar.* **2** °very, °awfully, °terribly, °extremely, °exceedingly, °frightfully, tremendously: *It was fearfully considerate of you to come.*

fearless *adj.* °brave, °bold, °intrepid, valorous, °dauntless, °courageous, valiant, plucky, °daring, °audacious, °heroic, °venturesome, °gallant, °chivalrous: *These fearless commandos cared nothing for their personal safety.*

fearsome *adj.* °dreadful, °awesome, appalling, °formidable, daunting, °frightening, °frightful, awe-inspiring, °menacing, °terrible, °terrifying, intimidating: *Last week's typhoon was the most fearsome storm of the century.*

feasibility *n.* practicability, workability, applicability, viability, practicality: *It's a good idea, but I cannot see its feasibility.*

feasible *adj.* °practicable, workable, doable, °applicable, °viable, °practical, °possible, realizable, achievable, attainable, °sensible, usable, °realistic: *The world needs a feasible method for disposing of nuclear waste.*

feast *n.* **1** °banquet, (lavish) dinner, (sumptuous) repast, (Lucullan *or* Epicurean) treat, *Colloq* °spread, *Brit* beanfeast; *Slang* blowout, *Brit* beano: *Dinner at Patrick's is no mere meal—it is a feast.* **2** °observance, °celebration, °rite, °ritual, solemnization, commemoration, memorialization, anniversary, birthday, jubilee, feast day, °festival, °fête, holy day, °holiday, red-letter day, °occasion, °event, °gala: *Church calendars list both movable and immovable feasts.* **3** °treat, °delight, °pleasure, °gratification: *Seeing you again was a feast for the eyes.* —*v.* **4** °dine, wine (and dine), fare well *or* sumptuously, (over)indulge, °gorge (oneself), gormandize, eat one's fill: *We feasted on roast turkey with all the trimmings.* **5** °entertain, °feed, wine and dine, °treat, °regale: *Visitors to Arthur's court were feasted with the best viands available.* **6** °delight, °gratify, please, °cheer, °gladden: *Ali Baba feasted his eyes on the vast treasure in the cave.*

feat *n.* °exploit, °deed, °act, attainment, °achievement, °accomplishment, tour de force: *Persuading your wife to let you go fishing was quite a feat.*

feature *n.* **1** °characteristic, °attribute, °trait, °mark, °hallmark, earmark, °property, °character, °quality, °aspect, facet, °peculiarity, °quirk, idiosyncrasy: *What are the identifying features of traditional Chinese architecture?* **2** °(main) attraction, °draw, special attraction, high point *or* spot, best *or* memorable part; main film; *US and Canadian* drawing card; column, °piece, article, °item: *The feature of the show was the appearance of Bob Hope. He writes a gardening feature for the* Sunday Herald. **3 *features.*** °face, visage, physiognomy, countenance, looks, *Slang* °mug, kisser, chips and plaice: *Her features twisted into a frown.* —*v.* **4** °present, °promote, °publicize, advertise, highlight, °spotlight, put into the limelight, °star, °stress, °emphasize, call attention to, °play up, °puff up, *Colloq* headline, hype: *He was featured in many films.* **5** be, °act, °perform, take *or* have a role *or* part, °participate, be involved *or* drawn in: *Novels feature prominently in our publishing program.*

federation *n.* °combination, confederacy, confederation, °amalgamation, coalition, °league, °alliance, °union, °association, *Bund,* °society: *The central government consists of a federation of states.*

fee *n.* **1** °charge, °price, °cost, °fare, °bill, °tariff, °toll, *Colloq* damage(s): *The membership fee is twenty-five dollars.* **2** °pay, °payment, emolument, compensation, recompense, °honorarium, °remuneration, °rate, wage(s), °stipend, °salary: *What is your fee per hour for proofreading?*

feeble *adj.* **1** °weak, °infirm, °frail, °puny, °slight, feckless; debilitated, enervated, °decrepit, enfeebled, °exhausted, weakened; effete, °delicate, °fragile, °powerless, °impotent, impuissant, languid, spiritless, °sickly, ailing, °unsound, °faint, °dizzy; *Colloq* woozy, *Brit* wonky: *The survivors, suffering from exposure, were too feeble to walk.* **2** °flimsy, °weak, °ineffectual, °ineffective, namby-pamby, half-baked, °lame, unconvincing, °shoddy, °thin, °insubstantial, °poor, °unsatisfactory, °insufficient, °inadequate, unavailing, °meager, °paltry, °insignificant, *Colloq* °wishy-washy, *Brit* °wet: *She gave her usual feeble excuse for being late.* **3** °weak, °obscure, °dim, °imperceptible, °faint, °distant, °indistinct, unclear: *He could see little in the feeble light from the one candle. We heard a feeble cry from under the rubble.*

feebleminded *adj.* °stupid, dull(-witted), witless, moronic, idiotic, imbecilic, °simple, °slow (on the uptake), slow-witted, weak-minded, dimwitted, addlepated, °half-witted, °deficient, subnormal, mentally defective, retarded, attocerebral, *Brit* ESN (= 'educationally subnormal'), *Colloq* °dumb, °thick, °soft in the head, bone-headed, empty-headed, °vacant, *Brit* gormless, *US* °exceptional: *I cannot see your spending your time with that feebleminded idiot.*

feed *v.* **1** °provision, cater *or* provide (for), victual, purvey, °provender, °supply, °maintain, nurture, °nourish, °board, °support, °sustain, wine and dine: *Is there enough to feed everyone?* **2** °eat, °devour, graze, °pasture: *Cattle feed most of the day.* **3 *feed on*** *or* ***upon.*** subsist *or* survive *or* depend *or* thrive on *or* upon, be nourished *or* gratified *or* supported by: *Swindlers feed on others' gullibility.* —*n.* **4** fodder, forage, pasturage, silage, °food, °provender: *What kind of feed are you giving the sheep?*

feel *v.* **1** °touch, °see, °note, °sense, °perceive, °experience, °determine; °handle, °manipulate, °finger: *Feel how rough the skin is at this spot.* **2** °stroke, °caress, °pet, °fondle: *I like to feel a bare carpet with my toes.* **3** °sense, be conscious of, °perceive, be aware *or* sensible of, °experience: *I feel the heat of the fire across the room.* **4** °sense, °believe, °think, °perceive, °judge, °consider, deem, °know, discern, intuit, *Colloq* get *or* have the impression, have a hunch, °guess, have a (funny) feeling, feel in one's bones: *I feel I have outstayed my welcome.* **5** °undergo, °sense, °suffer, °bear, °endure, °withstand, °stand, °abide, °brook, °tolerate, °go through: *Let the other team feel the agony of defeat for a change.* **6** °seem, °appear, strike one, give the impression, have a *or* the feeling: *It felt as if a great weight has been lifted from me. It feels like old times. That feels silky.* **7** seem to be, be, regard *or* characterize oneself as, take oneself to be: *I feel certain that we've met before. I felt a perfect fool.* **8 *feel for.*** sympathize *or* empathize with, commiserate with, bleed for, be sorry for, °pity, have compassion for: *She feels for all mothers whose sons have died.* **9 *feel like.*** incline *or* lean to *or* towards, °prefer, °fancy, °want, °desire, crave: *I feel like a cup of hot tea.* —*n.* **10** °texture, °surface, °sensation, °touch, °finish: *This wood has the feel of satin.* **11** °feeling, °air, °atmosphere, °climate, ambiance *or* ambience, °sense, °note, °tone, °quality: *It's so cozy here—it has the feel of home.*

feeler *n.* **1** antenna, tentacle, palp, sensor: *Those moths have long, feathery feelers.* **2** °overture, °hint, °suggestion, foretaste, °probe, tester, sensor: *The questionnaire is a feeler put out to test public opinion.*

feeling *n.* **1** °(sense of) touch, °sensitivity, °sense, °perception, °sensation, °sensibility: *I had no feeling in my left side.* **2** °intuition, °idea, °notion, °inkling, °suspicion, °belief, °hunch, theory, °sense; °premonition, °hint, presentiment, sense of foreboding, °sensation, °impression, °opinion, °view; °instinct, consciousness, awareness: *I have a feeling you're fooling. Do you ever have the feeling of being watched?* **3** °regard, °sympathy, empathy, °identification, compassion, tenderness, °appreciation, °concern, °understanding, °sensitivity, °sensibility: *He has a genuine feeling for animals.* **4** °ardor, °warmth, °passion, fervency, °fervor, ardency, °intensity, °heat, °sentiment, °emotion, vehemence: *She said she loved me with so much feeling that I thought she meant it.* **5 *feelings.*** emotions, °sensitivity, sympathies, °sensibilities, susceptibilities: *She hurt my feelings.* **6** °feel, °mood, °atmosphere, °climate, °sense, °air, ambiance *or* ambience: *There is a feeling of impending doom about this place.* —*adj.* **7** sentient, °sensitive, °tender, tenderhearted, compassionate, °sympathetic: *Their behavior did the soldiers honor as feeling men.*

feint *n.* °distraction, mock attack, °bluff, °dodge, °maneuver, false move, °pretense, °ruse, ploy, °subterfuge, °deception, °tactic, °stratagem, gambit, °artifice: *The attack on the flanks is merely a feint—the main thrust will be at the center.*

fell *v.* cut *or* knock *or* strike down, °floor, °prostrate, °level, hew (down), °flatten, °demolish, °mow (down); °kill: *He felled his opponent with one blow. All the trees in this area must be felled.*

fellow *n.* **1** °man, °boy, °person, °individual, gentleman, °one, *Colloq* °guy, °chap, °customer, kid, *Brit* bloke, *Slang* geezer, gink, *US old-fashioned* gazabo: *There was a fellow here asking for you today.* **2** °colleague, °associate, °comrade, °companion, °ally, °peer, compeer: *He is going to meet some of his fellows at the club.* **3** °mate, °match, counterpart, °partner, °complement, concomitant, °accessory: *I lost one glove last week and now have lost its fellow.* **4** boyfriend, °man, °sweetheart, °love, young man; °lover, °paramour, *Formal* °suitor, *Old-fashioned* beau, *Archaic* swain, *Slang* °guy: *Bailey has been Sue's fellow for years.* —*adj.* **5** associate(d), affiliate(d), allied, °auxiliary, °related: *Ladies, not accorded membership in the men's club, belonged to a fellow organization. She shared her sandwiches with a fellow passenger.*

fellowship *n.* **1** °companionship, camaraderie, comradeship, amity, °brotherhood, fraternization, togetherness, °association, °friendship, amity, companionability, sociability, intimacy: *I always have enjoyed the fellowship of my university colleagues.* **2** °society, °club,

°association, °alliance, guild, °league, °union, sisterhood, sorority, °brotherhood, °fraternity, congregation, °circle, community, °order, °organization, consortium, partnership; °lodge, °clan, °company, coterie, °set, °clique, coalition, bloc, cartel, °trust: *They belong to the worldwide fellowship of artists.* **3** friendliness, clubbiness, sociability, intimacy, amicability, affability, kindliness, cordiality, °familiarity, °affinity, °belonging, congeniality, °warmth, °hospitality: *Being a club member gives him a feeling of fellowship.*

felon *n.* °criminal, °outlaw, lawbreaker, °offender, °culprit, °miscreant, malefactor, wrongdoer: *Only if convicted of a certain class of crime is a person technically a felon.*

feminine *adj.* **1** female, womanlike, womanly, °ladylike, °submissive, deferential, amenable, °gentle, docile, °tender, °soft, °delicate: *She is much more feminine than her sister.* **2** °effeminate, womanish, unmanly, unmasculine, sissyish, sissified, effete, °affected: *Bruce's manners and behavior seem to become more feminine every day.*

fence *n.* **1** °barrier, °enclosure, barricade, confine, °wall, °rampart; railing(s), palisade: *Maurice had a fence put up around his house.* **2** ***on the fence.*** undecided, °indecisive, vacillating, uncommitted, °uncertain, °irresolute; °impartial, °neutral, °nonpartisan, unbiased, °unprejudiced, unaligned, °nonaligned, °independent: *She prefers to remain on the fence where most moral issues are concered.* —*v.* **3** °enclose, °encircle, °surround, circumscribe, °bound, coop, °restrict, hedge, confine, °fortify, °protect, °separate: *The pasture is fenced both to keep the sheep in and the wolves out. Fence off this area for a garden.* **4** parry, °avoid, °fend off, °sidestep, °dodge, °evade, hedge, stonewall, °equivocate, palter, tergiversate, vacillate, °shilly-shally, °quibble, °cavil, beat about *or* around the bush, °qualify, prevaricate, *Colloq Brit* °waffle: *Awkward questions were adroitly fenced by the chairman. Give a straight answer and stop fencing.*

fend *v.* **1** ***fend for oneself.*** get *or* scrape along (on one's own), °make out, °get by, °make do, °shift for oneself, take care of *or* provide for *or* support oneself: *Mother didn't believe I could fend for myself.* **2** ***fend off.*** °discourage, parry, keep *or* hold at bay, °resist, °repel, stave *or* ward *or* fight off, °deflect, °turn aside, avert, °divert: *We fended off our attackers for three days.*

ferment *v.* **1** °boil, effervesce, °seethe, °bubble, °foam, °froth, °brew; °rise, °raise, °work, leaven: *A rebellion is fermenting in the northern counties. Adding yeast causes the bread to ferment.* **2** °excite, °stir up, °incite, instigate, °agitate, °foment, °inflame, °rouse, °provoke: *He helped to ferment a riot.*

ferocious *adj.* °fierce, °savage, °cruel, °vicious, feral, °fell, °brutal, bestial, °merciless, °ruthless, pitiless, °inhuman, barbaric, barbarous, °violent, °destructive, °murderous, °bloodthirsty, °sanguinary, °predatory, °fiendish, diabolical, °devilish, hellish, °monstrous: *Cerberus, a ferocious beast with three heads, guards the gates of hell.*

fertile *adj.* °fruitful, °prolific, fecund, °productive, bounteous, °profuse, °abundant, copious, fructuous, plenteous, generative, teeming, °rich, °luxuriant: *The land was fertile enough to grow anything. This book is a fertile source of ideas.*

fertilize *v.* **1** °impregnate, inseminate, pollinate, fecundate, fructify: *The flower should be fertilized by pollen from a different stock.* **2** manure, mulch, °feed, °nourish, °enrich, °dress, compost: *The earth can be fertilized at any time.*

fervent *adj.* **1** fervid, °fiery, °burning, °glowing, °hot, °intense, °passionate, °impassioned, °ardent, °hot-headed, °inflamed, fanatic(al), °excited, °frantic, frenzied: *Only the most fervent acolytes were ordained as ministers of the faith.* **2** °eager, °earnest, °enthusiastic, zealous, °animated, °intense, °heartfelt, °emotional: *It is my fervent desire to see you happy.* **3** °ecstatic, transported, °rapturous, °rapt, enrapt, enraptured, captivated: *The fervent crowds cheered wildly when Caesar arrived.*

fervor *n.* fervency, °ardor, °warmth, °passion, vehemence, °glow, °intensity, zeal, °eagerness, earnestness, °enthusiasm, °animation, °gusto, ebullience, °spirit, °verve: *She went about her studies with a fervor we had not seen before.*

fester *v.* **1** ulcerate, suppurate, °run, °ooze, °putrefy, putresce, necrose, °mortify, °rot, °decay, °decompose: *His entire body was covered with festering sores.* **2** °rankle, °smolder, °gall, °chafe, °inflame: *Bitterness over the way he had been treated festered inside him for years.*

festival *n.* **1** °holiday, holy day, °fête, °feast, commemoration, anniversary: *He wanted to be with his family for the Christmas festival.* **2** °celebration, °fête, °festivities, carnival, °entertainment, red-letter day, gala day, anniversary, birthday: *The town organizes a festival every autumn.*

festivity *n.* **1** rejoicing, °gaiety, °mirth, jubilation, conviviality, joyfulness, °merriment, °revelry, merrymaking, °glee, jollity, jollification, felicity, *Colloq Brit* mafficking: *There were laughter and festivity in the air.* **2** ***festivities.*** celebration(s), °festival, °party, fun and games, °entertainment, °amusement, °hilarity, boisterousness, frivolity: *Come along and join the festivities at our house tonight.*

fetch *v.* **1** °get, go after *or* for, °bring (back), °retrieve, °deliver, °obtain, carry *or* convey (back): *Would you please fetch my slippers?* **2** °summon, bring *or* draw forth, °call, °elicit: *The doorbell fetched me from my bath.* **3** sell for, °bring (in), °go for, °yield, °earn, °make, °cost, °realize: *These days a house near London fetches a high price.* —*n.* **4** °reach, °stretch, °range, °span, °extent: *It is quite a long fetch across the bay.*

fetching *adj.* °attractive, alluring, °taking, winsome, °winning, °cute, °enchanting, charming, captivating, intriguing: *Fingal's sister is a fetching young lass.*

fête *n.* **1** °festival, °entertainment, °reception, levee, °gala, rejoicing, °celebration, °party, °festivities, °get-together, °social, °amusement, °revel, °fair, ball, °frolic, °spree, °jamboree, carnival, °event, °occasion, *Colloq* blowout, shindig, bash, do, bust, *Slang US and Canadian* wingding, °blast: *You are invited to our annual fête on New Year's Eve.* —*v.* **2** °entertain, °feast, roll *or* bring out the red carpet for, wine and dine, °celebrate, °honor, lionize, fuss over, kill the fatted calf for: *Fiona's book was a great success and she was fêted all over Europe.*

fetish *n.* **1** °charm, °amulet, °talisman, totem, *Rare* periapt: *Around his neck he wore the dreaded fetish of the fish god.* **2** °obsession, compulsion, °fixation, °mania, *idée fixe*: *It's a fetish of hers to kiss only under the mistletoe.*

feud *n.* **1** °dispute, °conflict, °vendetta, °hostility, °strife, enmity, °animosity, hatred, °antagonism, °rivalry, °ill will, bad blood, hard feelings, contention, °discord, °grudge, °dissension, °disagreement, °argument, °quarrel, bickering, squabble, falling-out, estrangement: *A silly feud broke out over whose turn it was to bring the Christmas goose.* —*v.* **2** °dispute, °quarrel, °bicker, °disagree, °conflict, °row, °fight, °fall out, °clash, be at odds, be at daggers drawn: *The Hatfields and McCoys feuded for decades.*

feverish *adj.* °inflamed, flushed, °burning, °fiery, °hot, °ardent, °fervent, hot-blooded, °passionate, frenzied, °frantic, °excited, frenetic, zealous; *Pathology* febrile, pyretic, pyrexic: *I was looking forward to our rendezvous with feverish excitement.*

few *adj.* **1** hardly *or* scarcely any, not many, °insufficient; infrequent, °occasional: *Few people came to my party. He is a man of few words.* —*n.* **2** °handful, some, °scattering: *I invited a lot of people, but only a few came.* —*pron.* **3** (only) one or two, not many: *Many apply but few are chosen.*

fiancé *n.* fiancée, betrothed, wife- *or* bride- *or* husband-to-be, intended: *He forgot to buy the engagement ring for his fiancée.*

fiasco *n.* °failure, °disaster, °muddle, °mess, abortion, °botch, *Colloq* °fizzle, °flop: *My effort to help Donald with his math homework was a complete fiasco.*

fib *n.* **1** °falsehood, (little) white lie, untruth, prevarication, °fabrication, °invention, misrepresentation, °story, °(fairy) tale, fiction; °lie; *Colloq* tall story, *US* tall tale, cock-and-bull story, whopper: *I told a little fib when I said my grandmother was ill.* —*v.* **2** prevaricate, °misrepresent, fudge, °falsify, misspeak, palter; °lie; *Colloq* °waffle: *They admitted they had fibbed about who had thrown the egg.*

fiber *n.* **1** filament, °thread, strand, fibril, tendril: *This fabric is woven from cotton fiber.* **2** °texture, °structure, °material, °fabric: *The fiber of this rock is granular.* **3** °essence, °character, °nature, °mold, °composition, constitution, °substance, °quality, °stripe, °cast, °make-up: *Love for him is wrapped up in the very fiber of her being.*

fickle *adj.* °flighty, °capricious, °frivolous, unpredictable, °moody, °giddy, °fanciful, °whimsical, °fitful, mercurial, °volatile, °unstable, °changeable, mutable, °inconstant, changeful, unsteady, unsteadfast, °indecisive, undecided, vacillating *or rarely* vacillant, unsure, °uncertain, °irresolute, wavering, °erratic, °unreliable, undependable, °irresponsible, untrustworthy, °faithless, unfaithful, °disloyal, *Colloq* °wishy-washy: *That fickle woman has already taken up with someone else. You should know how fickle the tastes of the public are.*

fictional *adj.* °unreal, °imaginary, invented, made-up, °mythical, °fanciful; °legendary, °fabulous: *The newspaper article was a hoax—completely fictional. Lorna Doone is Janet's favorite fictional heroine.*

fictitious *adj.* **1** imagined, °imaginary, °nonexistent, °unreal, made-up, invented, fabricated, °mythical, °fancied, °fanciful, fictive, °untrue, apocryphal: *The claim that an article had appeared was completely fictitious.* **2** °false, °counterfeit, °bogus, °artificial, °spurious; °assumed, improvised, made-up, invented, make-believe, °imaginary, *Colloq* phoney *or US also* phony: *When arrested they gave fictitious names.*

fiddle *v.* **1** Often, ***fiddle with.*** °meddle (with), °tamper (with), °interfere (with), °alter, °falsify, °fix; °cheat, °swindle, flimflam; *Colloq* finagle, *Brit* °cook: *The treasurer had been fiddling the accounts for years.* **2** Often, ***fiddle with** or **about** or **around.*** toy *or* trifle *or* fidget (with), twiddle *or* play *or* tinker (with), fool *or* fuss (with), *Colloq* mess *or* muck (about *or* around) (with), frivol (away), monkey (around) (with): *He sat there nervously fiddling with his ear. Stop fiddling about and get on with your work.* —*n.* **3** violin, viola, viol, cello, violoncello, °kit: *He played fiddle for the tsar.*

fiddlesticks *interj.* Nonsense!, Rubbish!, *Colloq* Fiddle-de-dee!, Balderdash!, Stuff and Nonsense!, Poppycock!, Pish and tush!, Tommyrot!, Hogwash!, Eyewash!, Baloney!, Bilgewater!, Moonshine!, Humbug!, Bosh!, Fiddle-faddle!, Bull!, Rot!, *Brit* Codswallop!, *Taboo Slang* Bullshit!, Balls!, Crap!, Horseshit!, *Brit* Bollocks! *or Brit* Ballocks!: *When he said he'd pay what he owed, I just said, "Fiddlesticks!"*

fidget *v.* **1** °squirm, twitch, °shuffle, °wriggle, wiggle, °fiddle, °fuss; °fret, °chafe: *Stop fidgeting and tell me the truth!* —*n.* **2** cat on hot bricks, *US* cat on a hot tin roof, *Colloq* fuss-pot, fussbudget, *US* nervous Nellie: *If her conscience is clear, why is she such a fidget?* ***the fidgets.*** restlessness, fidgetiness, dither, uneasiness, the jitters, nervousness, itchiness, *Colloq* heebie-jeebies, jimjams, ants in one's pants: *Whenever he tells a lie he gets a bad case of the fidgets.*

field *n.* **1** °ground, °land, arable, °pasture, grassland, °meadow, °green, °lawn, °common, clearing, °tract, °area, acreage, *Literary* greensward, lea, sward; *Archaic* mead: *The house looks out over the fields.* **2** battlefield, battleground, airfield; °(cricket) pitch, football *or* hockey *or* soccer field, *American football* gridiron: *We marched onto the field prepared for anything.* **3** °competition, competitors, players, entrants, contestants, participants, candidates, °possibilities, applicants: *Who can match you among today's field?* **4** °area, °domain, °realm, °department, °territory, °province, °sphere, °scope, °division, °interest, °line, métier, °discipline, bailiwick, °specialty, specialization, °expertise, °forte, °strength: *She is certainly the leading expert in her field. My field is biochemistry.* —*v.* **5** °catch, °stop, °return, °retrieve, °pick up: *Did you see how Gaithers fielded that ball?* **6** °answer, reply to, °respond to, °handle, °manipulate, °deal with, react to, °cope with: *The candidate adroitly fielded questions from the press.*

fiend *n.* **1** See **devil.** **2** °addict, °maniac, °fanatic; °fan, aficionado, °enthusiast, °devotee, °hound, °follower, *Colloq* buff, nut, *Slang* °freak: *He's a Welsh rarebit fiend.*

fiendish *adj.* °cruel, °savage, °inhuman, °monstrous, °ghoulish, °malignant, malevolent, malicious, °wicked, °evil, °bad, black-hearted, °satanic, °devilish, Mephistophelian, demonic, demoniac(al), diabolic(al), cacodemonic, hellish, °infernal: *He sniggered with fiendish glee as he tightened the thumbscrew.*

fierce *adj.* **1** °ferocious, °savage, °wild, °truculent, brutish, feral, bestial, tigerish, °brutal, °cruel, °fell, °murderous, °bloodthirsty, °sanguinary, °homicidal, barbaric, barbarous, °inhuman, °dangerous: *The fierce beasts will tear you limb from limb.* **2** intractable, °angry, °furious, °hostile, °aggressive, vehement, frenzied, °stormy, °violent, turbulent, °wild, °tempestuous, °tumultuous, raging, °merciless, uncontrollable: *The old curmudgeon is in a fierce mood again today. A fierce storm raged for three days.* **3** °severe, °awful, °dreadful, °intense, °keen, dire, °bitter, °biting, racking: *Wear warm clothes to protect you from the fierce cold.*

fiercely *adv.* °very, °extremely, °exceedingly, vehemently, intensely, impetuously, violently, furiously, ferociously, viciously, savagely: *She refuses to deal with anyone who is so fiercely competitive.*

fiery *adj.* **1** °burning, °flaming, blazing, °hot, red-hot, white-hot, overheated; afire, °on fire, in flames, °ablaze: *Fiery lava burst from the volcano.* **2** °glowing, red, °incandescent, °brilliant, °luminous, °glaring, gleaming, °radiant; aglow, afire: *We watched the fiery sunset fade into azure hues.* **3** °ardent, °eager, °spirited, °passionate, °excited, °excitable, peppery, irascible, °touchy, °irritable, edgy, °hot-headed, °fierce: *I had best not be late as my boss has a fiery temper.*

fight *v.* **1** contend (with), °battle, °conflict (with), °encounter, °war (against), °engage, °clash, take up arms (against), °feud (with), °combat, bear *or* take up arms (against), °brawl, struggle *or* strive (with *or* against), cross swords (with), °close (with), come to *or* exchange blows (with), go to *or* wage war (with *or* against), joust (with), grapple *or* wrestle (with), °skirmish (with), tussle *or* scuffle (with *or* against); °box, °spar; *Old-fashioned* °broil: *He offered to fight anyone in the place. Why are the boys always fighting?* **2** °dispute, °question, confront, °contest, °oppose, °contradict, °defy, confute, °protest, °resist, rail *or* struggle, against, °withstand, refute, oppugn, make *or* take a stand against, contravene, confound: *He has sworn to fight prejudice wherever he finds it.* **3** rise up, make *or* take a stand, °struggle, take up arms: *It is better to fight for the good than to rail at the ill.* **4** °argue, °dispute, °bicker, °quarrel, wrangle, squabble, spat, °tiff, °fall out (over), have words, °disagree, °row, altercate, °debate: *I hear the neighbors fight every night.* **5** ***fight off.*** °repel, °repulse: *Prepare to fight off anyone trying to come over the side!* **6** ***fight shy of.*** °avoid, keep *or* remain aloof from *or* of, keep away from: *She usually fights shy of going to cocktail parties.* —*n.* **7** °battle, °conflict, °bout, duel, °(single) combat, monomachy, one-on-one, °action, warfare, °clash, °hostilities, °war, °match, °struggle, °engagement, °meeting, °encounter, °contest, fighting, °brawl, donnybrook, affray, °fray, °fracas, °disturbance, °riot, °row, melee *or* mêlée, tussle, scuffle, °scrimmage, °skirmish, °brush, *Colloq* free-for-all, set-to, °scrap, *Brit* scrum, bovver, *Slang US* rumble, *Old-fashioned* °broil: *Three policemen were injured in the hour-long fight.* **8** altercation, °argument, °quarrel, °feud, °dispute, °run-in, °disagreement, °dissension, dissidence, dissent, °difference (of opinion), squabble, bickering, spat, °misunderstanding, °row, °discord, *Colloq* ruckus: *Such fights between husband and wife happen all the time.* **9** pugnacity, mettle, militancy, belligerence, truculence, °spirit, °pluck, zeal, °enthusiasm, °zest: *She had enough fight left in her to go on to win.*

figure *n.* **1** °physique, °build, °shape, °form, configuration, conformation, °build, °body, °outline, °silhouette; °cut, °cast; *Slang* bod, chassis: *It's exercise that gives Kathy such a fine figure.* **2** °appearance, °image, °likeness, °representation, °semblance: *The vision came to him in the figure of a dragon.* **3** °person, °individual, °human (being): *A burly figure loomed up out of the fog.* **4** °statue, effigy, °sculpture, bust, °mold, °cast, °image, °representation, °idol, icon: *Hundreds of figures are carved into the stone of the temple façade.* **5** °picture, °illustration, diagram, °drawing, sketch, °plate: *See Figure 12 for a graph showing the rate of inflation for the past decade.* **6** °personality, °celebrity, °somebody, °leader, °personage, °worthy, °notable, °individual, °presence, °force, °character: *Grimaldi looms large as a figure in the history of comedy.* **7** °number, numeral, cipher, digit; °character, °symbol, °device, °sign; °design, °pattern, °motif, °emblem: *His salary is in the six figures. She uses many unfamiliar figures in her textile designs.* —*v.* **8** Often, ***figure up.*** °calculate, °compute, °reckon, °work out; °count, °enumerate, numerate, °total, tot up, °tally, °sum: *Please figure up how much I owe you.* **9** °picture, °imagine, °think, °take, °reckon, °consider, °calculate, °judge, °believe; °assume, °presume, °suppose; °accept, °acknowledge, °concede: *I never figured her to be the guilty one.* **10** °act, °participate, take *or* play a part *or* role, °appear, °feature, have a place; be included *or* mentioned, be featured *or* conspicuous: *He does not figure in my plans.* **11** ***figure on*** *or* ***upon.*** **(a)** rely *or* depend on *or* upon, count on *or* upon, trust in, put faith in: *You mustn't figure on his help.* **(b)** plan on *or* upon, take into consideration *or* account, °consider, make allowance for: *They are figuring on huge consumer demand for a better mousetrap.* **12** ***figure out.*** **(a)** °calculate, °reckon, °compute, °work out: *I haven't yet figured out last year's profits and losses.* **(b)** °decipher, °translate, °understand, °interpret, °solve, °grasp, °get, °fathom, °see, °perceive, *Colloq* °dig, make head(s) or tail(s) of, get through one's head, get the hang *or* drift of, °catch on (to), get a fix on, *Brit* °twig, suss out: *Can you figure out what this article is trying to say?*

figurehead *n.* °puppet, °dummy, marionette, °mouthpiece, *Brit* man of straw, *US and Canadian* straw man, *Colloq* front man, *Chiefly US* °front: *The king is merely a figurehead with no power at all.*

file *n.* **1** °document, documentation, dossier, °case, °data, folder, portfolio, °information: *Interpol has files on Cowdery's activities going back ten years.* **2** °line, °queue, column, °row, °rank: *The file of armored vehicles stretched for miles.* —*v.* **3** classify, °organize, systematize, °categorize, alphabetize, chronologize, °order, °arrange, pigeonhole, interfile, put *or* place in order, °register, °record, °enter: *Must all this correspondence be filed today?* **4** °submit, send in, °complete, fill in *or US and Canadian also* fill out, °enter: *I have already filed my application for a transfer.* **5** °walk, °march, troop, °parade: *Please file into the classroom and take your seats.*

fill *v.* **1** °crowd, °stuff, °cram, °jam, °load, °burden, °pack, °squeeze: *A huge number of people filled the stadium.* **2** °top (off), °top up, °fill up, make full; °inflate, °swell, °stretch, °blow up, distend, °expand: *Pour in just enough to fill the bottle. See if you can fill the balloon. Wind filled the sails.* **3** Sometimes, ***fill up.*** °supply, °provide, °furnish: *Irena has filled her house with paintings.* **4** °meet, °satisfy, °fulfill, °answer: *Will a dozen fill your needs for a while?* **5** °satisfy, °satiate, bloat, sate, °gorge, °stuff; °stock: *The guests filled themselves with choice foods at the reception.* **6** °abound in, overflow, be abundant *or* plentiful in: *Trout filled the lake.* **7** °close, °stop (up), °block, °stuff, °plug, caulk, °seal: *The dentist is to fill two cavities tomorrow.* **8** °occupy, °take over, °discharge, °carry out, do, °execute: *Can you fill her job while she's away?* **9** ***fill in.*** **(a)** *US and Canadian also* °fill out, °make out, °complete, °answer: *Please fill in this application.* **(b)** take the place, °stand in, °substitute: *Could you fill in during my secretary's absence?* **(c)** °inform, °tell, °advise, let in on, °notify, bring up to date, °share, let in on: *Don will fill you in on the details later.* **10** ***fill out.*** **(a)** °swell, °expand, °grow, distend, °stretch; fatten, °increase: *The sails filled out in the freshening breeze.* **(b)** *US* fill in. See **9(a).** —*n.* **11** ***one's fill.*** °plenty, °enough; °surfeit, sufficiency: *We ate our fill and left.*

filling *n.* **1** filler, stuffing, padding, wadding; °contents, components, *Colloq* innards: *What kind of filling do you like in a pie?*

film *n.* **1** coating, °skin, °coat, membrane, °peel, integument, layer, overlay, covering, °cover, °sheet, dusting, °veil, *Technical* pellicle: *There is a film of grease over everything.* **2** motion picture, °movie, °picture; videotape; *Colloq* flick, pic; video: *Why do they show the same films every Christmas?* **3** °veil, °screen, murkiness, °blur, smoke screen, haze, °mist, haziness, mistiness, cloud, °vapor, °fog, steam: *I could barely make out the road through the film on the windshield.* —*v.* **4** Usually, ***film over.*** °coat, °cover, °dim, °obscure, °fade (away), °veil, °screen, °blur, blear, °mist (over), cloud (over), °glaze (over): *The stagnant pond is filmed over with decayed bacteria. Her eyes filmed with tears.* **5** °photograph, °shoot, °take, (video)tape: *If the director insists, then the scene must be filmed again.*

filmy *adj.* **1** gauzy, gossamer(-like), °sheer, °delicate, diaphanous, °transparent, translucent, °flimsy, °light, cobwebby, °insubstantial, °see-through, peekaboo: *Phoebe looked very seductive in her filmy negligee.* **2** °murky, blurry, cloudy, °hazy, °misty, bleary, blurred,

°dim, clouded, beclouded, milky, °pearly, °opalescent: *We inched down the cliff in the filmy light of the moon.*

filter *n.* **1** sieve, colander, °riddle, °screen, strainer, gauze, °cloth, cheesecloth, membrane: *To allow the liquid to pass through, a clean filter must be used.* —*v.* **2** Sometimes, ***filter out.*** °strain, °screen, °sift, winnow, °clarify, °refine, °purify, °clean; °separate, weed out, °exclude, °eliminate: *Filter the coffee to remove the grounds. This company would be improved if we could filter out the undesirable elements.* **3** leach, °percolate, °drip, seep, dribble, °trickle, °drain, run *or* pass through, °ooze: *The rainwater is filtered through fine sand to purify it.*

filth *n.* **1** sewage *or* sewerage, °dirt, slime, °muck, °sludge, sullage, effluent, °pollution, °trash, °rubbish, °garbage, °refuse, ordure, °(night) soil, excrement, feces, excreta, manure, droppings, guano, °dung, foul matter, filthiness, offal, leavings, carrion, °decay, putrescence, putrefaction, *Slang* crud, *Brit* gunge, *US* grunge, *Taboo slang* shit: *Cleaning the filth from the Augean stables was the fifth labor of Hercules.* **2** corruption, vileness, baseness, foulness, rottenness, debasement, defilement, °taint, °pollution, adulteration, °perversion, °degradation, °taint; sullying, besmirchment: *Sodom and Gomorrah were destroyed along with their filth.* **3** indecency, obscenity, smut, °pornography, corruption, nastiness, °vulgarity, grossness: *The censors forbade the showing of films with any hint of filth.*

filthy *adj.* **1** defiled, polluted, tainted, °foul, °nasty, °dirty, unclean, °vile, °putrid, °rotten, fetid *or* foetid, maggoty, flyblown, purulent, feculent, fecal, scummy, °slimy, mucky, *Slang* cruddy, *Brit* gungy, *US* grungy: *That city's public lavatories are the filthiest I have ever seen.* **2** °dirty, °unwashed, begrimed, squalid, °sordid, °shabby, soiled, °low, stained, grimy, °bedraggled, °unkempt, slovenly, °sloppy, °mean, °scurvy, °disgusting, °miserable, °wretched, *Slang Brit* gungy, *US* grungy: *The filthy beggar turned out to be Holmes in disguise.* **3** °immoral, taboo *or* tabu, °indecent, °impure, °obscene, smutty, °pornographic, X-rated, depraved, °corrupt, °dirty, °lewd, °lascivious, licentious, °gross, °offensive, °coarse, °bawdy, ribald, °blue, °suggestive, foulmouthed, dirty-minded, filthy-minded: *Customs officers confiscated thousands of filthy books, magazines, films, and videos.*

final *adj.* **1** ending, concluding, terminating, finishing, closing, °last, °terminal, °ultimate, °end: *The final sentence needs rewriting.* **2** conclusive, decisive, unalterable, °irreversible, °irrevocable, °incontrovertible, irrefutable, °indisputable, unchangeable, immutable, °definitive; settled, °fixed, °absolute, °certain, °sure: *Those are my final words on the subject. We made a final series of measurements.*

finality *n.* conclusiveness, decisiveness, unalterability, irreversibility, irrevocableness, incontrovertibility, irrefutability, indisputability, unchangeability, immutability, definitiveness, °certainty, certitude, sureness, [illegible] *There is an air of finality in the way she gives orders.*

finalize *v.* conclude, °settle, °complete, °decide, *Colloq* °wrap up, °clinch, sew up: *We finalized the agreement yesterday by signing the contract.*

finally *adv.* **1** at (long) last, lastly, °eventually, in the end, °ultimately, °at length, when all is said and done, °in the long run, at the end of the day, at the last moment: *We finally arrived in port.* **2** conclusively, °once and for all, decisively, irrevocably, in fine, °completely, inexorably, °absolutely, definitively, °definitely, for good, °forever, for all time: *I didn't want to commit myself finally till I'd seen the plans.*

finance *n.* **1** resource(s), banking, accounting, economics, °money (management), °business, °commerce, (financial) affairs, investment: *Bendick is the officer in charge of finance for the company.* **2** ***finances.*** °capital, °money, °cash, °funds, °resources, °assets, holdings, °wealth, wherewithal: *Have you the finances needed to buy the house?* —*v.* **3** °fund, subvene, invest in, °back, capitalize, °underwrite, °subsidize, °pay for, *Colloq US* bankroll: *The new enterprise is being financed entirely by small investors.*

financial *adj.* °monetary, pecuniary, °fiscal, °economic: *We must get the financial affairs of the country in order.*

financier *n.* capitalist, banker, plutocrat, investor, °backer, *US* moneyman, *Colloq* angel: *The financiers suggest switching investment from shares to bonds.*

find *v.* **1** °discover, °come across, happen *or* come on *or* upon, hit upon *or* on, chance *or* stumble on *or* upon; °encounter, °bump into: *We found a tiny bistro on the Left Bank where we liked to go.* **2** Often, ***find out.*** uncover, °discover, °unearth, lay one's hand(s) on, °turn up, come up with, °reveal, °bring to light, light upon *or* on, °catch sight of, °see, espy, descry, °detect, °learn, °spot, °locate, °track down; °identify, become aware of, °determine, ascertain, °put one's finger on, °point to, *Colloq* °tumble to, *Brit* °twig; *Slang* °finger, rumble, *Brit* suss out: *The police are trying to find the murderer. Can you find who did it?* **3** °discover, °perceive, °see, °understand, °notice, °mark, °remark, °note, °distinguish, discern; °realize: *I find nothing odd about her attire.* **4** °consider, °think, °regard, °view, feel *or* discover to be: *I have always found Lady Sharpless exceedingly dull. He finds it impossible to refuse her demands.* **5** °get, °obtain, °procure, °secure, °acquire, °win, °gain; °experience: *We despair of finding customers for our service except among dogowners. He found relief only in painkillers.* **6** °recover, °locate, °get back; repossess, °recoup: *I hope you find the earring you lost.* **7** °summon (up), °call up, °command, °gather (up), °muster (up), °rouse, °arouse, °awaken: *I tried to find the courage to ask her to marry me.* **8** °set aside, °allot, °assign, °manage, °get: *Have you found the time to read Connie's new book?* **9** °judge, decide *or* determine to be, °pronounce, °declare: *A jury found her guilty. The judge found in favor of the plaintiff.* —*n.* **10** °discovery, °catch, °bargain, °deal; °boon, °windfall: *The gold doubloon was the find of a lifetime.*

finding *n.* **1** °discovery, °find: *The findings from Pompeii are in the Museum of Portici.* **2** °judgment, °decree, verdict, °decision, °determination, °pronouncement, °declaration, conclusion: *The findings of the court are final.*

fine[1] *adj.* **1** °superior, °excellent, °superb, °magnificent, °supreme, °marvelous, °exceptional, °splendid, °exquisite, °elegant, first-class, °first-rate, °prime, °choice, °select, top-grade, high-grade, top-drawer, °quality, °admirable, °great, °good, °satisfactory, *Colloq* °out of this world, °great, OK *or* okay, peachy, °neat, °keen, *Brit* tickety-boo, *US* °neat, *Australian* bonzer, *Slang* °swell, °cool, *Brit old-fashioned* °ripping, *Dialectal* °champion: *He is a connoisseur of fine paintings. She turned in a fine performance. Spending our vacation in Barbados was a fine idea.* **2** °sunny, °fair, °bright, °clear, cloudless, balmy, °pleasant, °dry, °nice: *It was a fine day for a picnic.* **3** enjoyable, °satisfying, °entertaining, amusing, °good, °interesting, °pleasant, °nice: *Thank you for inviting us to such a fine party.* **4** °accomplished, °skillful, consummate, outstanding, masterly, °brilliant, °virtuoso: *Ann-Sofie is a fine pianist.* **5** °delicate, °subtle, °exquisite, well-made, °dainty, °elegant; °superior, °excellent, °outstanding: *What a fine example of miniature painting!* **6** °delicate, °dainty, °thin, gossamer, diaphanous, gauzy, °sheer, °slender, °frail, °flimsy, filamentous, threadlike: *The sword hung over Damocles' head by a fine thread.* **7** powdered, powdery, pulverized, comminuted, crushed, fine-grained: *As soon as I touched the skull, it disintegrated into a fine powder.* **8** °sharp, °keen, keen-edged, razor-sharp, °pointed, °acute: *The points of the sea urchin's spines are so fine you don't feel them enter.* **9** °subtle, °delicate, °refined, °acute, °keen, °discriminating, °critical, °precise, °nice, °hairsplitting: *There is sometimes a fine distinction between pleasure and pain.* **10** good-looking, °handsome, °attractive, °striking, °beautiful, °pretty, °lovely, °seemly, °fair, °comely, *Colloq US* °cute, *Scots* °bonny: *Jeanette married a fine young man.* **11** °meritorious, °worthy, commendable, °admirable, °excellent, °superb, °splendid, °good, *Colloq* °great: *She has displayed fine courage in the face of terrible family problems.* **12** °healthy, °well, healthful, °robust, all right, *Colloq* OK *or Colloq* O.K. *or Colloq* okay: *I told the doctor that I was feeling fine.* **13** °close: *Arriving at the airport with only minutes to spare is cutting it pretty fine.*

fine[2] *n.* **1** °penalty, °charge, °fee, mulct, °forfeit, amercement; forfeiture: *He could not afford to pay the $100 parking fine.* —*v.* **2** °penalize, °charge, mulct, amerce: *She was fined for parking at a bus stop.*

finery *n.* decoration(s), ornaments, °trappings, trinkets, frippery, showy dress, *Colloq* best bib and tucker, Sunday best, Sunday clothes, °gear, glad rags: *They were all dressed in their finery at the reception.*

finesse *n.* **1** artfulness, °subtlety, cunning, craftiness, cleverness, °strategy, shrewdness, °skill, °style, °dash, élan, °panache, °knack, skillfulness, °talent, adroitness, expertness, °expertise, adeptness, °proficiency, °ability, °facility: *I admire her finesse in getting him to give the painting to the museum.* **2** °trick, °artifice, °stratagem, °wile, °ruse, °scheme, °machination, °intrigue, °device, °expedient, °maneuver, °deception, °deceit: *He used every finesse in the book to persuade her to become a double agent.* **3** °tact, °diplomacy, °discretion, °grace, °taste, °delicacy, °polish, °refinement, °elegance: *With infinite finesse, Lord Lacey suggested that the duchess might like to stay the night.* —*v.* **4** °maneuver, °manipulate, °bluff, °trick, delude, °deceive, °fool, °outwit, °hoodwink, *Colloq* finagle, *Slang* con: *She was finessed into making him heir to her fortune.*

finger *n.* **1** digit: *How many fingers am I holding up?* **2** ***have a finger in.*** be *or* become *or* get involved in, figure in, have a hand in, °influence, interfere in, tamper *or* meddle *or* tinker *or* monkey with: *I knew that Lightfoot must have had a finger in the deal.* **3** ***keep one's fingers crossed.*** hope *or* pray for the best, touch *or esp. US* knock wood: *I'm sure you'll win, but I'll keep my fingers crossed just the same.* **4** ***lay*** *or* ***put a finger on.*** °(so much as) touch, °strike, °hit, °punch: *If you lay a finger on her, you'll have to answer to me!* **5** ***lay*** *or* ***put one's finger on.*** **(a)** °recall, °remember, °recollect, bring *or* call to mind, °think of, °pin down: *I can't quite put my finger on the date when we met.* **(b)** °locate, °find, °discover, °unearth, lay *or* put one's hands on, °track down, get hold of, °come by, °acquire; °buy, °purchase: *I was finally able to put my finger on the book you asked for.* **(c)** °indicate, °identify, °point to, °pin down, *Colloq* zero (in) on: *Frank put his finger on the problem at once.* **6** **(*not*) *lift*** *or* ***raise*** **(*even*) *a*** **(*little*) *finger.*** make an attempt *or* effort, °offer, make a move, °contribute, do one's part, do anything *or* something: *She didn't so much as lift a finger when I needed her.* **7** ***pull*** *or* ***get one's finger out.*** get on with it, stop delaying *or* procrastinating, *Colloq* get cracking: *He should have finished by now, and I wish he'd pull his finger out.* **8** ***put the finger on.*** °accuse, inform on *or* against, tell *or* tattle on, °betray, °bear witness, *Slang* snitch *or* squeal on, peach on: *After Louise put the finger on the kingpin, the cops arrested him.* **9** ***slip through one's fingers.*** °elude, °escape, °get away, vanish, °disappear: *Despite the cordon around the house, the fugitive slipped through our fingers.* **10** ***twist*** *or* ***wrap around one's little finger.*** °control, °dominate, °lord it over, have under control, °manipulate, °maneuver, wield power *or* authority over, have under one's thumb, have the upper hand over, be master of, °influence, make subservient: *She has her father twisted around her little finger and can do no wrong in his eyes.* —*v.* **11** °touch, °handle, °feel; toy *or* play *or* fiddle with: *He was quite nervous and kept fingering his tie.* **12** °identify, °point out, °put the finger on: *It was too late: the mob had already fingered Slats for assassination.*

finicky *adj.* **1** finical, °fussy, °fastidious, °critical, °difficult, °meticulous, hard to please, (over)delicate, (over)dainty, (over)particular, overnice, overprecise, niminy-piminy, punctilious, (over)scrupulous, *Colloq* pernickety *or US also* persnickety, °choosy, nitpicking, picky: *No one can work for him—he's too finicky.* **2** °fussy, °elaborate, °detailed, °fine, °delicate: *Notice the tiny, finicky engraving on this watchcase.*

finish *v.* **1** °stop, conclude, °end, °cease: *When you have finished eating, please clear the table.* **2** °complete, °accomplish, °perfect, °achieve, °carry out, °fulfill, consummate, °clinch, write "finis" to, *Colloq* °wrap up: *We finished the work on time.* **3** Sometimes, ***finish off.*** °dispose of, dispatch *or* despatch, °exhaust, °consume, eat *or* drink (up), °use (up), °devour, °drain, *Colloq* °polish off: *We finished the last of the beer last night.* **4** Sometimes, ***finish off.*** °kill, °exterminate, annihilate, °destroy, °get rid of, °dispose of, dispatch *or* despatch, put an end to, administer *or* deliver *or* give the coup de grâce, bring down, °overcome, °beat, °defeat, °conquer, °best, worst, *Colloq* °polish off, °terminate, *Slang* °bump off, °rub out, *US* °waste, ice: *The cavalry finished off the last of the enemy. The contender was finished in the third round.* **5** Sometimes, ***finish up.*** conclude, °close, °terminate, °wind up, °end, *Colloq* °knock off: *He goes straight home when he finishes work. When will you finish?* **6** Sometimes, ***finish up.*** end up, °settle: *Where did you finish up after the party?* **7** Sometimes, ***finish off.*** °perfect, put the final touches on, °polish, put a finish on: *He finishes fine furniture for a living.* **8** ***finish with.*** °release, let go, have *or* be done with, °let loose, set free: *I'll finish with you later, young man!* —*n.* **9** conclusion, °termination, °end, °close, closing, °completion, culmination, ending, finale, *Colloq* winding-up, *US* windup: *Tomorrow is the finish of the hunting season.* **10** °death, °killing, annihilation, extermination, °downfall, °destruction, °defeat: *A major catastrophe spelled the finish of the dinosaurs.* **11** °polish, °surface, °texture: *Just feel the finish on this gunstock!*

finite *adj.* °limited, bounded, °bound, restricted, delimited, numerable, countable: *It has been shown that there are a finite number of grains of sand in the universe.*

fire *n.* **1** flame(s), °blaze; conflagration, °holocaust, inferno: *We gathered round the roaring fire.* **2** °feeling, °passion, °ardor, ardency, °fervor, fervency, °intensity, °vigor, °spirit, °energy, vim, vivacity, °sparkle, °animation, liveliness, °verve, °pep, élan, éclat, °dash, °vitality, °eagerness, °enthusiasm, fever, feverishness: *I have never seen the role of Carmen done with such fire.* **3** firing, fusillade, °volley, barrage, bombardment, salvo, cannonade, shelling, broadside, °flak: *The fire from the guns shook the house.* **4** ***hang fire.*** °delay, be delayed, be in abeyance, be suspended, be shelved, be put off, be postponed, be up in the air, *Colloq* be put on hold, be (put) on the back burner: *The decision is hanging fire till tomorrow.* **5** ***on fire.*** afire, °burning, blazing, alight, aflame, °flaming; °ardent, °passionate, °fervent, fervid, hot-blooded, °intense, aroused, stirred, stimulated, °enthusiastic, fired up, °eager, inspired, °excited: *The building is on fire, so you must leave at once! My very soul is on fire when I think of you.* **6** ***play with fire.*** undertake a risk *or* hazard *or* peril, run a risk, risk (something *or* everything), imperil *or* endanger (something), tempt fate, live dangerously: *Those who try to stir up racial hatred are playing with fire.* **7** ***set fire to.*** See **8,** below. —*v.* **8 (a)** °set fire to, set afire, set on fire, ignite, set alight, °kindle, °spark (off), put to the torch, °burn, *Slang US* torch: *Arsonists fired our headquarters last night.* **(b)** Sometimes, **fire up.** inflame, impassion, °incite, °excite, °provoke, °foment, °whip up, °arouse, °rouse, °work up, fire up, enkindle, light a fire under, °stimulate, inspirit, °motivate, °move, °stir, °animate, °inspire, °awaken, °energize, °vitalize, vivify: *She fired the crowd's emotions with her impassioned speech.* **9** °discharge, °shoot, let go (with), °launch, °propel, °throw, catapult, °hurl: *He fired the missile towards the enemy.* **10** detonate, °set off, ignite, °set fire to, °light, °let off: *It is against the law to fire any explosives in this area.* **11** °discharge, °dismiss, oust, let go, cashier, give (someone) notice, *Brit* make *or* declare redundant, *Colloq Brit* ask for *or* get (someone's) cards, *US* give (someone) a pink slip; *Slang* °bounce, give (someone) the bounce, ax, give (someone) the ax, show (someone) the door, can, give (someone) the (old) heave-ho, give (someone) his *or* her (*Brit*) marching *or* (*US*) walking papers, *Brit* °sack, give (someone) the sack, boot (someone) out, give (someone) the boot: *They fired me last month and I haven't been able to find work.*

firm *adj.* **1** °compact, °solid, °dense, compressed, condensed, concentrated, °rigid, °stiff, °hard, unyielding, °inflexible, inelastic: *I prefer a firm mattress.* **2** °stable, °fixed, °fast, °secure, °steady, °solid, stationary, anchored, moored, unmovable, °immovable: *This pole is firmer than that one.* **3** °steady, °strong, °sturdy, °tight, unwavering, unshakable *or* unshakeable, unswerving: *What we need is a firm hand at the controls.* **4** °resolute, °determined, dogged, °definite, resolved, °positive, decisive, °decided, set on *or* upon, °steadfast, °constant, unflinching, °staunch, unshaken, unshakable *or* unshakeable, °immovable, °inflexible, °rigid, unwavering, undeviating, unchanging, unchangeable, °obstinate, obdurate, °stubborn, °strict, unyielding, unbending, unalterable: *Father was firm about my not staying out past midnight.* —*v.* **5** Often ***firm up.*** consolidate, °establish, °settle (down), °solidify, °resolve, °determine, °set up: *We firmed up the terms of the contract. Sugar prices firmed today.* —*n.* **6** °company, °organization, corporation, limited company, public limited company, plc, partnership, proprietorship, °business, °enterprise, °concern, °house, conglomerate, multinational (company), cartel, *Colloq* °outfit; *Jargon US* CIA, Central Intelligence Agency: *He works for a firm in the City.*

firmament *n.* °heaven, the heavens, °sky, the skies, arch *or* vault of heaven, *Literary* welkin, empyrean: *Sirius is one of the brightest stars in the firmament.*

firmly *adv.* **1** solidly, strongly, securely, °tightly, rigidly, °fast, immovably: *The column is set firmly into the concrete.* **2** resolutely, steadfastly, determinedly, staunchly, unwaveringly, decisively, unhesitatingly, constantly: *Jacquelyn firmly supports John in anything he says.*

first *adj.* **1** °foremost, °leading, °chief, °head, °premier, °prime, °primary, °principal, °preeminent: *The Kennedys were regarded as among "the first people of Boston." Is it true that your son is now first violinist in the philharmonic?* **2** earliest, °original, °senior, oldest; °initial, °beginning, °maiden, °opening: *The first clock in England was made about 1288. The first assault resulted in great loss of life.* **3** °fundamental, °elementary, °basic, °primary, °cardinal, °key, °essential: *It is necessary to return to first principles.* —*adv.* **4** °before, °in front, earliest, beforehand, °ahead, sooner, °foremost: *Bannister came in first.* **5** in the first place, firstly, before all *or* anything else, initially, at the outset *or* beginning, to begin *or* start with, from the start, *Colloq* first off: *First, I want to thank you all for coming to my party.* —*n.* **6** °beginning, °start, inception, commencement, °outset, *Colloq* word "go": *I knew you'd win from the first.* **7** first place, blue ribbon, gold (medal), °triumph, °victory; °win: *A Swede took a first in the Olympic high hurdles.* **8** ***at first.*** initially, in the beginning, at the start *or* outset, *Colloq* from the word "go": *I didn't believe him at first.*

first-rate *adj.* first-class, high-grade, °prime, °excellent, °superior, °superb, °great, °remarkable, °admirable, °fine, °exceptional, °outstanding, °extraordinary, °unparalleled, °matchless, unsurpassed, *Colloq* A-1 *or* A-one, topnotch, tiptop, °crack, °top, ace, *Brit* whiz-bang *or* whizz-bang, wizard: *Samuel is a first-rate pianist.*

fiscal *adj.* °financial, °economic, budgetary, pecuniary, °monetary: *Only through the establishment of sound fiscal policies will the country's economy being to stabilize.*

fishy *adj.* **1** dubious, °doubtful, °questionable, °unlikely, °far-fetched, °suspicious, not kosher, °peculiar, °odd, °queer, °strange, °suspect, °improbable, °implausible, *Colloq* °shady, °funny: *There is something fishy about this deal.* **2** piscine, fishlike, piscatory, piscatorial: *The fishy smell grew stronger as we neared the harbor.*

fit[1] *adj.* **1** °fitting, °appropriate, °fitted, °suitable, suited, adapted, apt, °meet, apropos, °applicable; °befitting, °becoming, °convenient, °proper, °right, °correct, °fitting: *Those books are not fit for children. It's not a fit night out for man or beast.* **2** °prepared, °ready, °able, °capable, °qualified, °worthy, °right, °adequate: *Is Cordwain fit to lead men into battle?* **3** °healthy, °well, °hale, °hearty, °stalwart, °vigorous, °strong, °sturdy, °robust, strapping, able-bodied, in good shape *or* trim *or* condition, in fine fettle: *Is she fit enough to run in the marathon?* **4** °ready, °angry, troubled, °upset, °inclined, °disposed, ready *or* likely *or* about to: *He was fit to be tied when he learned of the change in plans. I worked till I was fit to drop.* —*v.* **5** befit, °suit, °become, be suited to, be suitable *or* appropriate for, °answer, °satisfy: *This computer program does not fit my needs.* **6** °join, °conform, °go (together), °match, °correspond, dovetail, °tally: *I cannot make the parts fit.* **7** °adjust, °modify, °change, °adapt, °alter, °accommodate, °shape, °fashion: *You will just have to fit your schedule to conform to ours.* **8** °equip, °supply, °furnish, °provide, °outfit, fit out *or* up, °install, °rig out, gear up: *I am having my boat fitted with radar.*

fit[2] *n.* **1** °attack, °seizure, convulsion, °spasm, °spell, °paroxysm, °bout, °throe: *She has periodic fits of severe depression.* **2** °outburst, outbreak, °paroxysm, °spell, °period: *Occasionally, Gurning would go into a fit of hysterical laughter.* **3** °tantrum; °eruption, °explosion: *She throws a fit if I so much as mention Marla's name.* **4** ***by fits and starts.*** sporadically, °occasionally, fitfully, spasmodically, intermittently, erratically, haphazardly, °now and then, irregularly, unsystematically; unreliably: *The fridge works only by fits and starts.*

fitful *adj.* °irregular, °sporadic, °intermittent, °occasional, °periodic, °erratic, °spasmodic, °haphazard, unsystematic, °changeable, °unstable, °capricious, varying, fluctuating, °variable, uneven: *The dying torch emitted a fitful light. No true patriot pays merely fitful allegiance to his country.*

fitness *n.* **1** aptness, appropriateness, suitability, suitableness, competence, pertinence, seemliness; eligibility, adequacy, qualification(s): *There is some doubt about his fitness as a leader.* **2** °health, healthiness, (good) (physical) condition, °vigor, well-being, °(good) shape, (fine) fettle, °tone, wholesomeness, salubriousness *or* salubrity: *A fanatic about physical fitness, Tony exercises daily.*

fitted *adj.* custom-made, °tailor-made; tailored, bespoke: *He is proud to say that he owns both fitted suits and fitted furniture.*

fitting *adj.* **1** °fit, °befitting, °suitable, °appropriate, °meet, °becoming, °proper, *comme il faut,* °seemly, apt, apropos, apposite, germane, °relevant: *Is it fitting to send flowers? It was a fitting end to the story.* —*n.* **2** ***fittings.*** fitments, attachments, accessories, °elements, pieces, parts, units, fixtures; appointments, extras, installations, furnishings, °trappings, °furniture, °equipment, accoutrements, °paraphernalia, trimmings: *Gold-plated bathroom fittings are available at extra cost.*

fix *v.* **1** affix, °pin, °fasten, make fast, °attach, °secure, °stick, °connect, °link, °tie, °couple, °clasp, °clamp, rivet, °cement, °weld, °fuse: *Fix this sign to the door.* **2** °establish, °set, °settle, °agree to, °determine, °organize, stabilize, °firm up, °solidify, °decide, conclude, °arrive at, °define, °specify, °resolve, °arrange, °install, °institute: *The new limits must be fixed by tomorrow.* **3** °repair, °mend, °fix up, °remedy, °rectify, °correct, emend, °adjust, °patch (up), °regulate, put *or* set to rights, °doctor, °straighten out: *My watch is at the jeweler's being fixed.* **4** °hold, °fasten, °focus, °direct, °level, rivet, °concentrate, °freeze; fixate: *He fixed his attention on the ice cream.* **5** °hold, rivet, spellbind, mesmerize, °hypnotize, °fascinate, °freeze, immobilize: *She fixed him with a baleful stare.* **6** °concentrate, °focus: *Fix your thoughts on how we're going to get out of here.* **7** °harden, congeal, °thicken, °set, consolidate, °solidify, rigidify, become rigid, °stiffen, °freeze: *Mold the clay before it becomes fixed.* **8** °establish, °set, °settle, °organize, °install, °situate, °locate, °position, °place: *Her family is comfortably fixed.* **9** °impose, °assign, allocate, °attribute, ascribe, °specify, °pin, °attach, °fasten, °establish: *They tried to fix the responsibility for the accident.* **10** °settle, °set, stabilize, °freeze, °solidify; conventionalize: *Many spellings were fixed by the spread of printed books and other materials.* **11** °bribe, suborn, °buy (off), °corrupt, °influence, °manipulate, *Colloq* grease (someone's) palm: *He attempted to fix a judge.* **12** °arrange, prearrange, predetermine, °set up, contrive, *Colloq* °fiddle, °pull strings, °rig: *We knew the winner as they had fixed the race.* **13** desexualize, desex, °alter, °cut; castrate, emasculate, °see to, eunuchize, geld, caponize; spay, oophorectimize, ovariectomize: *Animals that have been fixed are usually more tractable.* **14** retaliate against, wreak vengeance on, hit *or* strike *or* get back at, °get even with, even the score with, make reprisal against, avenge oneself against, take revenge *or* retribution on *or* against, °repay, °pay back, *Colloq* settle (someone's) hash, cook (someone's) goose, sort (someone) out: *I'll fix him for tattling to the teacher!* **15** ***fix on*** *or* ***upon.*** decide (on *or* upon), °set, agree (on *or* upon), °choose, °select, °settle (on), °determine, °finalize: *We can't fix on a date till we know David's schedule.* **16** ***fix up.*** **(a)** °furnish, °supply, °provide, °accommodate, °set up, *Brit* °lay on: *Her mother fixed her up with a nice young man. Can you fix me up with a room for the night?* **(b)** (re)decorate, °furnish, °renovate, °restore, furbish, °refurbish; straighten out *or* up, °organize, do up, °set up: *I like the way you've fixed up the basement. Please fix up your room before the guests arrive.* **(c)** °clear up, °resolve, °reconcile, °sort out, °settle: *Have they fixed up their differences yet?* **(d)** °repair, °patch (up): *Fix up my car so it will run at least till I get home.* —*n.* **17** °dilemma, °predicament, °difficulty, corner, double bind, catch-22, °quandary, °mess, °(bad) situation, strait(s), *Colloq* pickle, °jam, °hole, (tight *or* tough) spot, °pinch, *US* °bind: *You could help me get out of a terrible fix.* **18** °arrangement, prearrangement, °fiddle; bribery, subornation; *Slang chiefly US and Canadian* °setup: *The fix is on for her to win at roulette.*

fixation *n.* °mania, °obsession, compulsion, fixed idea, *idée fixe,* °fetish, monomania, preoccupation, infatuation, *Colloq* hang-up, °thing, °kick: *He has a health-food fixation.*

fixed *adj.* **1** fastened, °attached, anchored, °set, secure(d), °firm, °stable, settled, °immovable, immobile, stationary, °rigid, °rooted, °solid; immobilized, stuck: *While this part rolls, the other remains fixed. There was a fixed smile on her face.* **2** established, °secure, unalterable, °steadfast, °set, °firm, unchangeable, unchanging, °persistent, unfluctuating, unflagging, unwavering, °inflexible, undeviating, unflinching, unblinking, °rigid, °rooted, immutable, °definite, °resolute, resolved, °determined, °intent; °obstinate, °stubborn: *I admire Philip's fixed determination to persevere.* **3** settled, resolved, agreed, °regular, °habitual, °decided, arranged, prearranged, °definite, established: *We have a fixed date on which we meet each month.* **4** arranged, prearranged, set-up, framed; °crooked, °dishonest, °bent, *Colloq* rigged, °put-up: *He lost everything in a fixed poker game.*

fixture *n.* **1** °meet, °meeting, °event, °match, °occasion, °occurrence: *The Davis Cup tennis championship is an annual fixture.* **2** appliance, accessory *or* accessary, °fitting, °equipment, °apparatus, °device, °instrument, °tool, °gadget, contrivance, appendage, fitment: *The landlord forbids the removal of any fixture.*

fizz *v.* **1** °bubble, effervesce, °sparkle, °froth, °fizzle; °hiss, sputter, sizzle: *The champagne fizzed in my glass.* —*n.* **2** effervescence, °sparkle, carbonation, bubbling, °froth, °fizzle, fizziness: *All the fizz is gone from my drink!* **3** °hiss, hissing, sibilance: *The steam escaped with a loud fizz.* **4** *US* soda, soda water, club soda, seltzer, *Dialect* °plain; soft drink, °tonic: *May I have a little fizz with my scotch?* **5** champagne, *Colloq Brit* champers: *We celebrated with oysters and a bottle of fizz.*

fizzle *v.* **1** °fizz: *This headache powder makes the water fizzle.* **2** Often, ***fizzle out.*** die (out *or* away), fizz out, °expire, °peter out, come to nothing *or* naught, °fail, °fall through, °miscarry, abort, °come to grief, °misfire, °collapse, °cave in: *Her grandiose plans for staging an extravaganza quickly fizzled out.* —*n.* **3** failure, misfire, *Colloq* damp squib, washout, flop, dud : *Her debut as an opera singer was a complete fizzle.*

flabby *adj.* **1** °limp, °loose, °lax, flaccid, °slack, floppy, sagging, drooping, baggy, °pendulous, quaggy, °soft: *After forty, he began to get a bit flabby around the middle.* **2** °weak, °spineless, °feeble, °impotent, °ineffective, °ineffectual: *The critics found the characterizations flabby and the plot nonexistent.*

flag[1] *n.* **1** °banner, ensign, °standard, °pennant, banneret, pennon, °streamer, bunting, jack, gonfalon, vexillum: *Our flag was raised over the fort.* —*v.* **2** Often ***flag down.*** °warn, °signal, °hail, °inform, °stop: *A workman flagged us down because the bridge was out.* **3** °mark, °tag, °label, °tab, °identify, *Brit* tick (off), *US* *and Canadian* °check (off): *Flag the supplies you need that are listed here.*

flag² *v.* **1** °droop, °sag, °dangle, hang down, swag, festoon: *In the doldrums the sails flagged soggily about the masts.* **2** °weaken, languish, falter, °fail, °dwindle, °fade, °deteriorate, °waste away, °degenerate, °die, °decline, °diminish, °decrease, lessen, abate, °peter out, °taper off, °let up, °ease (up), °subside, °slump, °fall off, °wane, °ebb, °sink, °lag: *When he reached eighty, he sometimes complained that his energy was flagging.*

flagrant *adj.* °blatant, °brazen, °bold, °barefaced, °audacious, arrant, °glaring, °outrageous, °shocking, °shameless, °scandalous, °atrocious, °infamous, °notorious, °defiant, egregious, °obvious, °conspicuous, °open, °complete, °out-and-out, utter, flagitious, °monstrous, heinous, °cruel, °villainous, treacherous, nefarious, °awful, °gross, °rank, °inconsiderate, °scornful, °contemptuous; reprehensible; °contemptible: *Speeding shows a flagrant disregard for the safety of others.*

flair *n.* **1** °talent, °ability, °aptitude, °feel, °knack, °genius, °skill, °mind, °gift, °faculty, propensity, °bent, proclivity, °facility: *He showed a flair for music at an early age.* **2** °chic, °panache, °dash, élan, éclat, °style, stylishness, °glamour, °verve, °sparkle, °vitality, °elegance, °taste, *Colloq* savvy, pizzazz *or* pizazz, *Old-fashioned* oomph: *She dresses with great flair.*

flak *n.* flack, °criticism, °disapproval, censure, °abuse, °blame, °aspersion, complaint(s), disapprobation, condemnation, *Colloq* brickbats: *He took a lot of flak from the press for his comments.*

flake *n.* **1** snowflake; °scale, °chip, °bit, °piece, °scrap, °particle, tuft, °flock, scurf, °fragment, shaving, °sliver; wafer, lamina, *Technical* squama: *A flake of slate caught him in the eye.* —*v.* **2** Often, ***flake off.*** °scale, °chip, °fragment; *Technical* desquamate, exfoliate: *The paint is flaking off on this side.* **3** ***flake out.*** **(a)** °collapse, go to *or* fall asleep, drop off (to sleep), °pass out, keel over: *After the game, I flaked out on the couch for ten hours.* **(b)** become flaky, act crazy: *He flaked out when he heard the new record.*

flamboyant *adj.* **1** °elaborate, ornamented, °ornate, decorated, embellished, baroque, rococo, florid: *We chose a flamboyant wallpaper with purple peacocks.* **2** °extravagant, °ostentatious, °showy, °gaudy, °flashy, °dazzling, °brilliant, °splendid, °dashing, °rakish, °swashbuckling, °jaunty; °high, °wide, and handsome: *With a flamboyant display of swordsmanship, d'Artagnan dispatched his attacker.*

flame *n.* **1** °fire, °blaze; conflagration: *The flame began to lick about my feet.* **2** °passion, °fervor, °ardor, °intensity, °warmth, °fire, zeal, feverishness, °enthusiasm, °eagerness: *The flame of love is still burning.* **3** boyfriend, girlfriend, °lover, heartthrob, °sweetheart, beau: *Margo is an old flame of Larry's.* —*v.* **4** °burn, °blaze, °glow, °flare: *Flaming embers from the chimney set fire to the roof. Here and there civil unrest flamed up in the countryside.*

flaming *adj.* °obvious, °conspicuous, °blatant, °flagrant, egregious, °extravagant, *Slang* bloody, bleeding, blasted, damned, blooming: *That flaming idiot was appointed to the Department of Defense!*

flammable *adj.* °inflammable, combustible, burnable: *The warning on the tin says the contents are flammable.*

flank *n.* **1** °side, °quarter; loin, haunch: *The enemy was about to attack our flank. The pony drew back with trembling flanks.* —*v.* **2** °edge, °border, °line: *Armed guards flanked the entrance.* **3** skirt, outmaneuver, outflank, °circle, go (a)round: *They have enough troops to flank our forces to the east.*

flannel *n.* **1** °flattery, humbug, °nonsense, blarney, °rubbish, prevarication, *Colloq* eyewash, hogwash, baloney *or* boloney, soft soap, weasel words, sweet talk, bull, crap, *Brit* °waffle, cock, *US* bushwa; *Taboo slang* bullshit, horseshit, (load of) shit, *Brit* codswallop, bollocks *or* ballocks: *Don't believe a thing Fordyce says—it's all flannel.* —*v.* **2** °flatter, hedge, °mislead, pull the wool over (someone's) eyes, *Colloq* soft-soap, sweet-talk, *Taboo slang* bullshit, horseshit, shit: *They aren't sincere—they're just flanneling you.*

flap *v.* **1** °slap, slat, °beat, flail, °wave, °wag, waggle, °flutter, thresh, thrash, °oscillate, °vibrate: *An ornithopter is an airplane that flaps its wings like a bird. The Union Jack was flapping in the breeze.* —*n.* **2** flapping, °beat, °wave, °wag, waggle, °flutter, oscillation: *With a flap of its wings, the bird flew away.* **3** °fold, °fly, lappet, °lap, °tail, °tab: *Leave a flap of cloth to cover the pocket.* **4** °upset, °agitation, to-do, ado, commotion, °panic, °flurry, °fuss, °distress, *Colloq* °state, tizzy, °sweat, *Brit* kerfuffle: *She's in a flap about what to wear to the dance.* **5** °quarrel, °argument, °dispute: *There was a big flap over which car model to buy.*

flare *v.* **1** Often, ***flare up.*** blaze *or* flame (up), °flash, °erupt, °break out; °dazzle, °flicker, glimmer, °shimmer, °flutter: *The fire flared up briefly when he threw in the paper.* **2** Often, ***flare out.*** spread (out *or* outwards), °widen, broaden, °expand, °increase, °enlarge, °bulge, °swell: *The sides of the vase flare out toward the rim.* **3** Often, ***flare up.*** °anger, lose one's temper, °chafe, °seethe, °fume, °rage, throw a tantrum, become incensed *or* angry (etc.); °blow up, burst forth, °erupt, °explode; *Colloq* get one's back up, get one's Irish *or* dander up, see red, get worked up, °fly off the handle, lose one's cool, go out of *or* lose control, get hot under the collar, blow one's top: *He flares up at the very mention of her name. Tempers flared when we touched on the subject of politics.* —*n.* **4** °blaze, °flame, °burst, °outburst, °flash, °glare, °dazzle, incandescence, °brilliance, luminosity: *A bright orange flare erupted from the mouth of the cave.* **5** °beacon, °light, °signal, torch, flambeau, °link: *We prayed that the rescue party would see our flares.* **6** °spread, broadening, widening, °expansion, °swelling, °bulge, °increase, enlargement: *I like that slight flare at the hem of the skirt.*

flash *n.* **1** °blaze, °flame, °flare, °burst, °dazzle, °spark, °sparkle, coruscation, fulguration, °glitter, °twinkle, °twinkling, °flicker, flickering, scintilla, scintillation, glint, °shimmer, glimmer, °gleam, °beam, °ray, °shaft: *There was a blinding flash of blue light and the little man was gone.* **2** (sudden *or* momentary) display, °stroke, °show, °manifestation, °outburst, outbreak, °sign, °indication, °exhibition; °touch, °hint, °suggestion: *She has occasional flashes of insight bordering on genius.* **3** °moment, °(split) second, °instant, °twinkling (of an eye), trice, °minute, *Colloq* two shakes (of a lamb's tail), jiffy, °shake, before one can *or* could say Jack Robinson: *In a flash he was at her side.* —*v.* **4** °burn, °blaze, °flame, °flare, °burst (out), °dazzle, °spark, °sparkle, coruscate, fulgurate, °glitter, °twinkle, °flicker, scintillate, °shimmer, glimmer, °gleam, °beam, °glare, °shine: *The light flashed every eight seconds.* **5** °race, °speed, °dash, °streak, flick, °tear, °rush, °hurry, °hasten, °fly, zoom, °shoot, °bolt, whistle; °run, sprint, dart, scuttle, scamper, *Colloq* scoot, skedaddle, °take off, whizz *or* whiz: *The train flashed by, going about a hundred.* —*adj.* **6** °dazzling, °showy, °ostentatious, °smart, °chic, *Colloq* °swish, classy, ritzy, snazzy: *Barbara's new car looks very flash.* **7** See **flashy, 1.**

flashy *adj.* **1** °flash, °gaudy, °flamboyant, °glaring, fulgurous, °showy, °ostentatious, °loud, °garish, °vulgar, °cheap, meretricious, °pretentious, °tawdry, °tasteless, *Colloq* °tacky, *Slang* jazzy, *US* glitzy: *He was attracted by her flashy appearance.* **2** °superficial, cosmetic, °skin-deep, °surface, °shallow, °glib, °slick, facile, °insubstantial, °thin: *The reviewer called it a flashy interpretation of the Bard.*

flat *adj.* **1** °level, °horizontal, °even, °smooth, °plane, unbroken, uninterrupted: *I looked out over the flat surface of the frozen bay.* **2** °prostrate, °prone, °supine, °lying (down), stretched out, °recumbent, outstretched, reclining, spread-eagle(d), °spread out, outspread: *I lay flat on my back staring up at the sky.* **3** collapsed, leveled, overthrown, laid low: *The air raid had left all the buildings completely flat.* **4** °downright, °outright, °unqualified, unreserved, unconditional, °absolute, °categorical, °explicit, unconditional, °definite, °firm, °positive, °out-and-out, unequivocal, °peremptory, unambiguous, unmistakable, °direct, °complete, °total: *The request for clemency was met with the judge's flat refusal.* **5** featureless, °monotonous, °dull, °dead, uninteresting, unexciting, °vapid, °bland, °empty, two-dimensional, insipid, °boring, °tiresome, °lifeless, spiritless, °lackluster, °prosaic, °stale, °tired, °dry, jejune: *The critics wrote that she turned in a very flat performance.* **6** deflated, collapsed, punctured, ruptured, blown out: *We had a flat tire on the way.* **7** unchangeable, unchanging, °invariable, unvaried, unvarying, °standard, °fixed, unmodified, unmodifiable, *Colloq US* cookie-cutter: *They charge the same flat rate for children, the elderly, and all between.* **8** °dead, insipid, °stale, °tasteless, flavorless, °unpalatable; decarbonated, noneffervescent: *My beer has gone flat.* **9** °exact, °precise: *It's a flat ten minutes from here to the railway station.* **10** °definite, °certain, °sure, °irrevocable: *I said I'm not going, and that's flat.* **11** °dull, °slow, sluggish, °inactive, depressed: *Business has been a bit flat since Christmas.* **12** °dull, mat *or* matt *or* matte, unshiny, nongloss(y), nonreflective, nonglare, unpolished: *The table looks better with a flat finish.* **13** lacking perspective, two-dimensional, °lifeless, °unrealistic: *Some of his paintings seem pretty flat to me.* —*n.* **14** Often, ***flats.*** **(a)** *US* low shoes, loafers, sandals, *Colloq* flatties: *Cynthia said that wearing flats makes her feet hurt.* **(b)** lowland(s), plain(s), tundra, steppe(s), prairie(s), savannah, heath, °moor, pampas; mudflat(s); shallow(s), shoal, strand; °marsh, °bog, fen, °swamp: *Before us was a large flat of barren ground.* **15** apartment, room(s), °suite (of rooms), *Chiefly US and Canadian* garden apartment, maisonette, penthouse, studio (apartment), walk-up, duplex, triplex, *Brit* bed-sitter, bedsit, *Colloq Brit* digs: *We rent a two-bedroom flat in a good neighborhood.* —*adv.* **16** °absolutely, °completely, categorically, °utterly, °wholly, uncompromisingly, irrevocably, °positively, °definitely, °directly; °exactly, °precisely, flatly: *He has come out flat in favor of the incumbent. When he started drinking, she left him flat.* **17** ***flat out.*** **(a)** at maximum *or* top *or* full *or* breakneck speed, speedily, °quickly, apace, °on the run, °rapidly, °swiftly, at full speed *or* gallop, °posthaste, hell-for-leather, like a bat out of hell, °like a shot, like (greased) lightning, like the wind: *Someone shouted "Fire!" and we headed flat out for the exits.* **(b)** flatly, unhesitatingly, °directly, °at once, °immediately, forthwith, without delay; plainly, °openly, baldly, brazenly, brashly: *He asked to borrow some money and she told him "No," flat out.*

flatten *v.* **1** level *or* even (off *or* out); °smooth (out), press *or* iron (out), °roll: *The children's noses were flattened against the window of the toy shop. You will have to flatten the lawn if you want to play croquet on it.* **2** knock down *or* over, °knock out, °floor, °prostrate: *Menges flattened the champion in the first round.* **3** °raze, tear down, °demolish, °level: *The bomb flattened every house in the area.*

flatter *v.* **1** butter up, °play up to, °compliment, °praise, fawn (on *or* upon), toady to, truckle to, court, curry favor with, *Colloq* °flannel, soft-soap, °oil; *Slang* shine *or* suck up to, bootlick, *Taboo* kiss (someone's) *Brit* arse *or* *US* ass, brown-nose: *Flattering the boss won't get you a salary increase.* **2** °enhance, °complement, °suit, show to advantage: *The color of the scarf flatters her complexion.* **3** °cajole, °wheedle, °coax, inveigle, °beguile, sweet-talk: *He was flattered into signing that contract.*

flatterer *n.* toady, sycophant, fawner, wheedler, time-server, courtier, back-scratcher, sponge, °parasite, leech, °hanger-on, sweet-talker, backslapper, truckler, lickspittle, *Colloq* °yes man; *Slang* bootlicker, *Taboo* brown-noser, *Brit* arse-kisser, arse-licker, bumsucker, *US* ass-kisser, ass-licker: *Don't believe anything that flatterer tells you.*

flattering *adj.* **1** °complimentary, °becoming, °kind, enhancing: *That is quite a flattering portrait of Dorian.* **2** adulatory, °laudatory, gratifying, fulsome, honeyed, sugary, fawning, °ingratiating, unctuous, °slimy, *Chiefly Brit* smarmy: *Authors once wrote the most flattering dedications to their patrons.*

flattery *n.* adulation, °cajolery, blandishment, sweet talk, beguilement, wheedling, gloze, *Colloq* soft soap, *Slang* bootlicking, *Taboo* brown-nosing, *Brit* arse-kissing, arse-licking, bumsucking, *US* ass-kissing, ass-licking: *Judicious flattery moved him swiftly up the corporate ladder.*

flaunt *v.* °display, °show (off), °parade, °flourish, °exhibit, °sport, disport, °spotlight: *They were like all the other nouveaux riches who flaunt their wealth in the stylish continental resorts.*

flavor *n.* **1** °taste, °savor, °tang, piquancy, °zest; tastiness, savoriness; °essence, °extract, °seasoning, flavoring, °seasoning, °spice; °aroma, °odor, °scent; *Rare* sapor: *Overcooking destroys the flavor of food. This sweet has a minty flavor.* **2** °character, °spirit, °nature, °quality, °property, °mark, °stamp, °essence, °characteristic, °style, °taste, °feel, °feeling, ambiance *or* ambience, °sense, tinge, °aroma, °air, °atmosphere, °hint, °suggestion, °touch, soupçon: *We all agreed that the new play has a Shavian flavor.* —*v.* **3** °season, °spice: *She flavors her teacakes with ginger.*

flaw *n.* **1** °fault, °defect, °imperfection, °error, °mistake, °blemish, °blot, °stain, °taint, °(black) mark, °damage, disfigurement, °failing, °weakness, weak spot; °loophole: *Flaws in the casting have given this bell a dull sound. I can see flaws in your argument.* **2** °crack, °break, °breach, °chink, °chip, °fracture, °rupture, fissure, cleft, °split, °slit, °cut, °gash, °rent, °rift, °tear, °rip, °puncture, °hole, perforation: *The tiniest flaw may reduce the value of a teapot to a collector.* —*v.* **3** °damage, °harm, °spoil, °ruin, °mark, °weaken, disfigure: *If he cleaves the diamond improperly, he might flaw it irreparably.* **4** °discredit, °stigmatize, °damage, °hurt, °harm; °taint, °mar, °stain, °blot: *Her reputation has already been flawed.*

flawed *adj.* damaged, harmed, marred, weakened, tainted, stained, tarnished, °defective, °imperfect, °unsound, °faulty: *His flawed record prevented him from getting work.*

flawless *adj.* **1** °perfect, °pristine, °pure, uncorrupted, °chaste, virgin, °intact, °whole, °clean, °immaculate, unsullied, unspoiled *or* unspoilt, unsoiled, °impeccable, unblemished, °faultless, undamaged, unimpaired, °spotless, °untarnished: *Two flawless artifacts were found in the tomb. Many envy her flawless reputation.* **2** undeniable, unassailable, unimpeachable, °unquestionable, irrefutable, °foolproof, °sound, °demonstrable: *The prosecutor's evidence appears to be flawless.*

flecked *adj.* spotted, dappled, pied, (be)speckled, (be)sprinkled, dotted, °marked, stippled, dusted, specked, spattered, freckled: *Bananas are best to eat when they are flecked with brown.*

flee *v.* **1** °quit, run away *or* off, °escape, °get away, °fly, °take flight, °bolt, °take flight, °go (away), decamp, abscond, seek safety, °avoid, °make off, make an exit, make (good) one's escape, make a (clean) getaway, beat a (hasty) retreat, °take to one's heels, show a clean pair of heels, °turn tail, make tracks, make a run for it, cut and run, vanish, °disappear, *Brit* levant, *Colloq* °take off, scoot, make oneself scarce, °beat it, °clear out, fly the coop, skedaddle, scram, *Brit* scarper, *Australian and New Zealand* shoot through, *US and Canadian* take a (runout) powder, °skip (town), °cut out, hightail it, *Old-fashioned* skiddoo; *Slang* °split, *Brit* °bugger off, do a moonlight flit, do a bunk, *US and Canadian* vamoose, lam out, take it on the lam, °blow, bail out: *He fled the state and was never seen again. When they saw what they had done, they fled. She fled to Brazil to avoid the tax authorities.* **2** °avoid, °evade, °shun, escape from, eschew: *She fled the responsibilities of caring for her children.*

fleece *v.* °cheat, overcharge, °swindle, bilk, °defraud, °victimize, °plunder, °strip, °milk, °rob, *Colloq* °take, flimflam, gyp, diddle, bleed, take for a ride *or* to the cleaners, *Slang* °rip off, °chisel, °pluck, rook, °clip, °soak: *The gang stole credit cards and fleeced the companies out of millions.*

fleet[1] *n.* armada, flotilla, °navy, naval (task) force, task force, squadron, convoy, °division: *A fleet of submarines was sent to harass shipping in the Atlantic.*

fleet[2] *adj.* °swift, °rapid, °fast, °speedy, °quick, °nimble, °expeditious, °agile: *Nowadays, the fleetest transportation in a large city is by bicycle.*

fleeting *adj.* transitory, °fugitive, °transient, °temporary, °passing, ephemeral, fugacious, evanescent, °momentary, °short-lived, °fly-by-night, °short, °brief: *I caught only a fleeting glimpse of the car as it sped by.*

flesh *n.* **1** °meat; °tissue, muscle: *We ate the flesh and threw away the fat, bones, and skin.* **2** °body, corporeality, °flesh and blood, human nature, physicality; mortality: *The spirit is willing but the flesh is weak.* **3** ***flesh and blood.*** °real, °physical, corporeal, °human, °natural: *He pointed out that he, too, was flesh and blood, not a robot.* **4** ***in the flesh.*** °personally, °in person, °really, physically, bodily, °alive, living, in life: *She stood before me, in the flesh.* **5** ***one's (own) flesh and blood.*** °kin, kinfolk *or Brit* kinsfolk, °family, °stock, blood, kith and kin, relatives, °relations: *How could she treat her own flesh and blood that way?* —*v.* **6** ***flesh out.*** °substantiate, fill (in *or* out), give *or* lend substance *or* dimension to, °incorporate, °embody, °color: *The idea is good, but it needs fleshing out.*

flex *n.* **1** wire, °lead, °cord, °cable, °extension: *Replace this flex before you have a short circuit.* —*v.* **2** °bend, °give, °stretch, curve: *The plastic tube can be flexed many times without breaking.* **3** °exercise, °tense, °tighten, °contract: *He flexed his muscles before lifting the weight.*

flexibility *n.* **1** pliability, pliancy *or* pliantness, °elasticity, resilience *or* resiliency, suppleness, flexibleness, bendability, limberness, °stretch, °give, °spring, springiness, ductility: *Chemists improved the flexibility of the substance by adding more plasticizer.* **2** conformability *or* conformableness, adaptability, versatility, adjustability *or* adjustableness, compliance *or* compliancy, manageability, tractability *or* tractableness, malleability, °obedience, submissiveness, docility, agreeableness, conformity: *Fletcher has the flexibility needed to move from one department to another.*

flexible *adj.* **1** °pliable, pliant, °elastic, resilient, °supple, bendable, limber, lithe, stretchy, stretchable, springy, extensible *or* extensile, ductile, flexile, tensile, °yielding, °willowy: *The foil is a thin, flexible dueling sword.* **2** modifiable, °adaptable, conformable, compliant, malleable, °obedient, °tractable, °manageable, cooperative, amenable, persuadable *or* persuasible: *English seems to be an infinitely flexible language. Nicole is flexible in her attitude towards politics.* **3** °easy, facile, °submissive, complaisant, docile: *Discipline at the school might be a little too flexible.*

flicker *v.* **1** °twinkle, °blink, waver, glimmer, glint, °sparkle, °shimmer, °flare, gutter: *A breath of air made the candles flicker.* **2** °flap, °flutter, °quiver, °twitter, °fluctuate, °oscillate, °shake, °tremble, °vibrate: *Dozens of tiny flags flickered in the wind.* —*n.* **3** glimmer, glint, glimmering, °sparkle, °spark, °twinkle, °twinkling, °gleam, °flare, °glare: *With a dying flicker the torch went out.* **4** °hint, °suggestion, °trace, glimmer, °vestige, scintilla, °spark: *I thought I detected a flicker of recognition when our eyes met.*

flight[1] *n.* **1** flying, soaring, winging, °excursion: *The flight above the mountains was exhilarating. He is subject to wild flights of the imagination.* **2** (air) voyage *or* journey *or* trip: *We had a fine flight from New Delhi.* **3** airplane, airliner, °plane, aircraft: *Our flight was delayed in Istanbul.* **4** °flock, °swarm, cloud, covey (of grouse *or* partridge), bevy (of quail), skein (of geese), exaltation (of larks): *A flight of migrating birds passed overhead.* **5** feather: *The flights on this dart are broken.*

flight[2] *n.* **1** °escape, °retreat, departure, °exit, exodus, °getaway, fleeing, bolting, *Slang* °split: *Mohammed's flight from Mecca to Medina in 622 is called a "hegira."* **2** ***put to flight.*** chase *or* drive (off *or* away), °disperse, send off *or* away, send packing, °dismiss, °rout, °stampede: *They put the Saracens to flight in a decisive battle.* **3** ***take flight.*** °flee, go *or* run away *or* off, abscond, °desert, °depart, °(beat a) retreat, °exit, °bolt, decamp, °withdraw, °take to one's heels, show a clean pair of heels, *Colloq* light out, shove off, *Brit* scarper, *US* take a (runout) powder, take it on the lam, lam out; *Slang* °split, *Brit* do a bunk, do a moonlight flit, °bugger off: *The constable blew his whistle and the thief took flight.*

flighty *adj.* **1** °fickle, °frivolous, °inconstant, °capricious, °fanciful, °changeable, °variable, mercurial, skittish, °volatile, °unstable, unsteady, °giddy, °wild: *These flighty girls never stay in a job more than a month.* **2** °irresponsible, lightheaded, rattlebrained, °silly, °harebrained, °dizzy, °crazy, °mad, °reckless, °thoughtless, *Colloq* nutty, screwy, dotty, dippy: *It was flighty of Carolyn to pour brandy into the well.*

flimsy *adj.* **1** °slight, °frail, °weak, insubstantial *or* unsubstantial, °feeble, °makeshift, °fragile, frangible, breakable, °rickety, °ramshackle, °dilapidated, jerry-built, gimcrack, °slight, °delicate: *That flimsy ladder will never support me. He lives in a flimsy shack near the railway.* **2** °slight, trivial, °paltry, °feeble, unconvincing, °weak, °makeshift, °implausible, °unbelievable, °unsatisfactory, insubstantial *or* unsubstantial, °poor, °inadequate: *Her flimsy excuse was that she had a headache.* **3** °sheer, °light, gauzy, °transparent, °thin, °filmy, diaphanous, gossamer, °delicate: *That flimsy coat can't keep you warm.*

flinch *v.* wince, °draw back, °withdraw, cower, °cringe, °recoil, °start, quail, blench, °shrink (from), shy (away) (from), °dodge, °duck, *Colloq* bat an eye: *Each time he raised the whip, I flinched.*

fling *v.* **1** °toss, °pitch, °throw, °cast, °hurl, °heave, °sling, °propel, °fire, °let fly, °send, *Colloq* °lob, chuck:

The smugglers had flung the contraband into the sea. —*n.* **2** °indulgence, debauch, binge, °spree, °party, *Colloq* blow-off: *This will be my last fling before going into the army.* **3** °gamble, °risk, °venture, °attempt, °try, °go, °shot, *Colloq* °crack, whirl, bash: *Her brief fling at an acting career was unsuccessful.*

flip *v.* **1** °toss, flick, °snap, °flop, °turn, °twist, °spin: *Flip a coin to see who goes first.* **2** °anger, become angry *or* furious, go mad, go crazy, go berserk, *US also* flip out, *Colloq* go off the deep end, lose one's cool, *Slang* flip one's lid *or Brit* top, °freak, *US* freak out, lose it: *Father flipped when I said I was quitting college.*

flippancy *n.* **1** frivolousness, facetiousness, °levity, lightheartedness, frivolity, jocularity, offhandedness, unseriousness: *I was chagrined that she should treat my proposal of marriage with such flippancy.* **2** °disrespect, disrespectfulness, °impudence, °impertinence, irreverence, sauciness, rudeness, discourtesy, brazenness, brashness, pertness, insolence, *Colloq* cheek, cheekiness, lip, °mouth, *Slang Brit* °side: *The headmaster will not condone flippancy during prayers.*

flippant *adj.* **1** °frivolous, facetious, lighthearted, jocular, offhand(ed), unserious, °shallow, °thoughtless, °superficial; °supercilious, belittling, °scornful, dismissive: *I don't appreciate your flippant remarks when I'm trying to be serious.* **2** °disrespectful, °impudent, °impertinent, °irreverent, saucy, °rude, °pert, °discourteous, °brazen, °brash, °insolent, *Colloq* °cheeky, °flip: *If Charlie doesn't change his flippant attitude, he might be dismissed.*

flirt *v.* **1** coquette, play *or* act the coquette, °tease, °tantalize, °toy, °lead on, dally, philander, *Colloq Brit* °chat up; *Slang US* come on to: *Amanda enjoys flirting with the boys.* **2 *flirt with.*** trifle *or* toy *or* play *or* tinker with, °contemplate, °consider, °entertain, give a thought to, think about *or* of, *Colloq Brit* try on: *Many teenagers flirt with the idea of leaving home.* —*n.* **3** coquette, °tease, vamp, hussy, playgirl, minx; °philanderer, °playboy, *Slang* lady-killer, *Old-fashioned* wolf, sheik, masher, lounge lizard, *Taboo slang* cockteaser, prickteaser: *They are both such flirts, I can't tell if their advances are genuine.*

flirtatious *adj.* coquettish, vampish, °seductive, flirty, °coy, philandering, °provocative, enticing, alluring, amorous, come-hither: *They exchanged many flirtatious looks across the table.*

flit *v.* °move, °go, °fly, °flee, dart, °skip, °skim, flick, °hop, °whisk, °flutter, °flash: *As a bee flits from flower to flower, she flits from man to man.*

float *v.* **1** °hover, °poise, bob, °waft, be suspended, °hang; °sail, °drift, °glide, swim: *In the clear Caribbean waters, the boat seemed to float in midair.* **2** °launch, °establish, °set up, °organize, °found, °initiate, get going *or* moving: *The financiers are trying to float a new company.* **3** °negotiate, °arrange, °transact, bring *or* carry off, °get, °effect, consummate, *Colloq* °pull off, °swing: *Were you able to float a loan on your house?* —*n.* **4** raft, *Brit* °platform: *I can swim all the way out to the float, Dad!* **5** °buoy; pontoon: *The bridge is supported by floats.* **6** (parade) exhibit *or* display: *Our parade float won first prize.*

flock *n.* **1** °body, °company, °group, °band, °pack, °bunch, troop, °set, °collection, °assembly, convoy, °gathering, congregation, °crowd, °mass, °mob, °throng, °gang, multitude, °number, °quantity, °host, horde, °swarm, drove; °herd, °flight, troupe, °fleet, °school; bevy: *A huge flock of buyers sought bargains at the post-Christmas sales.* —*v.* **2** °assemble, °meet, °collect, °gather, come *or* go together, congregate, °crowd, °mass, °mob, °throng, °pour, °flood, °swarm, °herd (together), °band together, °go; *Colloq* gang up: *People flocked around the speaker.*

flog *v.* **1** °whip, °lash, horsewhip, strap, flagellate, flay, °scourge, thrash, thresh, °beat; °chastise, °castigate, °punish: *He will be flogged, then drawn and quartered. He was flogged for stealing a loaf of bread.* **2** °sell; °promote, °publicize: *Her latest job is selling encyclopedias door-to-door.*

flood *n.* **1** inundation, deluge, overflow(ing), debacle: *The hurricane was accompanied by a devastating flood.* **2** °torrent, cataract, freshet, overflow, °stream, °spate: *A flood of lava poured down the mountain.* **3** °abundance, deluge, overflowing, °surge, °outpouring, °torrent, tide, tidal wave, °stream, °rush, °flow, °glut, °surfeit, °satiety, °profusion, overabundance, superabundance, nimiety, plethora, °excess, °surplus, °superfluity: *A flood of invective poured from her lips.* —*v.* **4** inundate, °submerge, overflow, °swamp, °immerse, deluge, pour over, °drown: *The water flooded everything in the basement.* **5** °overwhelm, °glut, oversupply, °saturate, °choke: *The market is being flooded with cheap imitations.* **6** °sweep, °flow, °swarm, °surge, °rush, °crowd, °pour: *As soon as the doors opened, the people flooded in.* **7** °permeate, °fill, engulf, °cover, pour into *or* throughout *or* over: *I open the curtain and sunlight floods the room.*

floor *n.* **1** flooring, parquet, boarding, planking, *Nautical or colloq* deck: *The floor is mopped once a week.* **2** °story, °level; deck: *On which floor is your flat?* **3** °minimum, °bottom, °base, lower limit, lowest (level): *Owing to inflation, the wage floor was raised. Share prices went through the floor in today's trading.* —*v.* **4** knock over *or* down, bowl over, °prostrate, °fell, °overthrow, bring down, °(make) fall; °beat, °defeat, °conquer, °destroy, °rout, °overwhelm, °crush, °whip, trounce, thrash, drub, °best, worst: *The champion was floored by a blow to the head.* **5** °stump, °bewilder, baffle, dumfound *or* dumbfound, °confuse, confound, disconcert, °nonplus, °perplex, °puzzle, °astound, °astonish, °amaze, °surprise, °shock: *Alexandra was floored when she received a gift from her ex-husband.*

flop *v.* **1** °collapse, °drop (down), °fall (down), °tumble, °topple, plump down, plop down, flounce down: *I was so exhausted that I flopped into bed at once.* **2** °flap, °wave, °swing: *The door, unfastened, flopped back and forth in the wind.* **3** °fail, °fall flat, come to naught *or* nothing, °founder, *Colloq* °fold, *US* °bomb: *The musical flopped on the first night.* —*n.* **4** °failure, °fiasco, °disaster, nonstarter, debacle, *US* °fizzle, *Colloq* °dud, °washout, clanger, *US* lead balloon, °bomb; *Slang* lemon, *Brit* cock-up, damp squib: *His idea for a new corkscrew was a complete flop.*

flounce *n.* **1** °frill, furbelow, peplum, °ruffle, °ornament, valance, trimming: *The dress has decorative flounces round the skirt.* —*v.* **2** °fling, °toss, °bounce, °twist, °strut, °toss, °parade, °march, °storm, °stamp, *US* sashay: *She flounced out of the room in high dudgeon.*

flounder *v.* °grope, °blunder, °stumble, °tumble, °struggle, °stagger, plunge about: *We floundered about in the dark till someone found the light switch.*

flourish *v.* **1** °prosper, °thrive, °grow, °develop, °luxuriate, bloom, blossom, °flower, bear fruit, fructify, °boom, burgeon *or* bourgeon, °mature, °ripen, °increase, °succeed, °get ahead, do *or* fare well, °make good; go up *or* rise in the world; *Slang* go great guns: *We understand that Amy's medical practice has flourished.* **2** °wave, °wield, brandish, °wag, °swing, °twirl, °flaunt, vaunt, °swagger, °swish, °shake: *The bandleader flourished his baton.* —*n.* **3** °decoration, °embellishment, floridness, floweriness, ornamentation, °elaboration, adornment, °frill, embroidery, curlicue, furbelow: *His signature is characterized by all these flourishes.* **4** °fanfare, °display, °show, showiness, °dash, gesturing, °wave: *With a flourish of his hand, the king signaled the tourney to begin.*

flourishing *adj.* °luxuriant, °lush, thriving, prospering, blooming, blossoming, °fruitful, flowering, burgeoning *or* bourgeoning, °successful, °prosperous, booming, growing, increasing: *Patel runs a flourishing flower shop in Bicester.*

flout *v.* °deride, °scorn, disdain, contemn, °spurn, decry, °denounce, misprize, °blaspheme, °depreciate, °degrade, abase, °belittle, °minimize, deprecate, °disparage, denigrate, °mock, °jeer, °guy, °ridicule, °scoff, °sneer, gibe *or* jibe (at), °taunt, °twit, °insult, affront, *Archaic* fleer; *Colloq* °put down, °chaff, °rag, °knock: *When introduced in the sixties, miniskirts brazenly flouted convention.*

flow *v.* **1** °stream, °pour, °run, °rush, °course, °surge, °move, °go, °proceed, °progress, °drift; °gush, °glide, purl, °roll, °ripple, °trickle, °gurgle, °bubble; °swirl, whirl, °circulate: *People continue to flow past the window. The Thames flows silently to the sea.* **2** °rush, °gush, °surge, °well (forth), °stream, °spring, °issue, °spout, °spurt, squirt, °spew, °flood, cascade, °fall, °rain; °brim, overflow, °spill, °teem: *Open the gates to allow water to flow into the lock. A stream of insults flowed from his lips. Her tears flowed like wine.* **3** °issue, °originate, °come, °emanate, °rise, °begin: *With the dam repaired, water will once again flow from the reservoir. The authority of the state constitutions does not flow from Congress.* **4** °spread, overspread, °cover: *The paint flowed effortlessly onto the canvas.* —*n.* **5** °rush, °gush, °surge; °current, °course, °stream, °run, °movement, °drift: *A dam blocks the flow of water here. The sound of a shot interrupted the flow of conversation.* **6** °abundance, superabundance, °flood, plethora, °excess, overflow(ing), deluge, tide, °supply; °plenty: *London experiences an unending flow of tourists.*

flower *n.* **1** blossom, bloom, floret *or* floweret, bud, *Technical* efflorescence: *He brought me flowers for my birthday.* **2** cream, °best, °pick, elite *or* élite, crème de la crème, finest, choicest: *The flower of the city's youth were put to death by the wicked caliph.* —*v.* **3** blossom, bloom, bud, burgeon *or* bourgeon, °come out, °open, °unfold, *Technical* effloresce: *The century plant was so called because it was thought to flower only once every hundred years.*

flowery *adj.* florid, °ornate, °fancy, elaborate(d), decorated, ornamented, °overwrought, embellished, rococo, baroque, arabesque, euphuistic, Gongoristic, Ossianic, grandiloquent, °bombastic, °inflated, °pompous, °affected, °artificial, high-flown, °showy: *Many find his flowery style of writing difficult to stomach.*

fluctuate *v.* °vary, °change, °alternate, °seesaw, °swing, vacillate, undulate, waver, °oscillate, °shift: *Prices fluctuate according to supply and demand.*

fluctuation *n.* Sometimes, **fluctuations** variation(s), change(s), alternation(s), swing(s), vacillation(s), wavering(s), oscillation(s), undulation(s), ups and downs, instability, unsteadiness, °inconstancy: *We can expect fluctuation in temperatures during the coming week.*

fluency *n.* articulateness, eloquence, °control, °command, °ease, °grace, effortlessness, °facility, felicity, smoothness, °polish, slickness, glibness, volubility: *His extemporaneous speeches are marked by an enviable fluency.*

fluent *adj.* articulate, °eloquent, well-spoken, felicitous, °graceful, facile, °easy, °natural, °effortless, °ready, °smooth, °polished, flowing, °voluble, °glib, °slick; °expressive: *She is a fluent speaker of Spanish.*

fluff *n.* **1** down, fuzz, feather(s), thistledown, floss, lint, dust, dustball, fuzzball: *Be sure you clean the fluff from under the beds.* **2** ***bit of fluff.*** poppet, °girl (friend), °mistress, *Slang* bit of all right, (bit of) crumpet, (bit of) skirt *or* stuff, *Old-fashioned Brit* popsy: *Bickerley keeps some fluff on the side.* **3** °blunder, °error, °slip, °mistake, *Colloq* °howler, *Brit* bloomer, *US and Canadian* blooper: *They showed a videotape of the fluffs actors make during recordings.* —*v.* **4** °muddle, °spoil, °ruin, make a mess of, °bungle, °botch, *Colloq* °foul up, °screw up, °mess up, *US* snafu, *Slang Brit* cock up, *US* ball up, *Taboo* fuck up: *I have a funny story about how Barrymore fluffed Hamlet's soliloquy.* **5** ***fluff up.*** °puff up, shake out *or* up, aerate: *He doesn't like the pillows on his bed fluffed up.*

fluffy *adj.* **1** °soft, downy, puffy, whipped up, °light, airy, puffy, feathery, wispy: *Beat the egg whites till they are fluffy.* **2** °frivolous, °superficial, trivial, unimportant, airy, °thin, lightweight, °light, °insubstantial, gossamer, *Brit* airy-fairy: *Mr. Piffle is too fluffy-headed to offer anything useful.*

fluid *n.* **1** °liquid, °solution, °liquor, ichor; gas, °vapor: *It might surprise some to learn that physicists consider gases to be fluids.* —*adj.* **2** °liquid, flowing, °running, runny, °watery, aqueous: *Once the metal has become fluid, it is poured into the molds.* **3** °changeable, mutable, °flexible, adjustable, °variable, pliant, unformed, formless, unfixed, unstatic *or* nonstatic, °plastic, °protean, mercurial, °mobile, °unstable, shifting, °uncertain, °indefinite, °unsettled: *The situation is fluid and the decision could go either way.*

fluke *n.* lucky *or* successful stroke, stroke of (good) luck, lucky *or* big break, °(happy) accident, quirk *or* twist of fate, °windfall, fortuity, serendipity: *If he won first prize, it was by a fluke.*

flummox *v.* °confuse, baffle, °perplex, °bewilder, confound, throw into confusion, °stymie, °stump, °puzzle, °mystify, fox, °deceive, °hoodwink, °nonplus: *We were completely flummoxed by the licensing requirements.*

flunky *n.* **1** °servant, retainer, lackey, footman, °menial, minion, hireling, underling, °subordinate, °inferior; °slave; *Colloq* dogsbody, *Slang US and Canadian* gofer: *A company flunky drove me to the airport.* **2** toady, °hanger-on, yesman, jackal, doormat, stooge, lick-spittle, sycophant, *Colloq US and Canadian* apple-polisher, *Taboo* brown-noser: *He has a collection of flunkies waiting about for handouts.*

flurry *n.* **1** °activity, commotion, ado, to-do, °fuss, °upset, hubbub, pother, °stir, °excitement, °disturbance, °agitation, °tumult, whirl, °furor, bustle, °hurry, °hustle, °flutter, °fluster; °burst, °outburst; *Colloq* tizzy: *I was happy to leave the worry and flurry of the city. There was a brief flurry of activity at the stock exchange today.* —*v.* **2** °confuse, °bewilder, °agitate, °put out, °disturb, °excite, °fluster, disconcert, °upset, °perturb, unsettle, °shake (up), *Colloq* °rattle: *He was quite flurried by the arrival of the police.*

flush[1] *v.* **1** °blush, redden, crimson, °glow, °burn, °color (up): *When he told her she was pretty she flushed a bit.* **2** Often, ***flush out.*** °rinse, wash (out *or* away), douse, douche, hose down, °flood, °drench, °clean out, °cleanse, °purge, °discharge, °empty: *Oil tankers are flushed out far from shore.* **3** °animate, °stir, °inflame, impassion, °quicken, °arouse, °excite, elate, °encourage, °cheer, °delight, °thrill, °gladden: *Suzie was flushed with the success of having got the job.* —*n.* **4** °blush, redness, bloom, rosiness, °color, blood, °glow, °radiance: *When he was asked what had happened to the money, a flush rose to his cheek.* **5** °flood, deluge, drenching, °soaking, overflow, inundation, °rush,

°gush, °surge, °stream, °flow: *The flush of water washed away all the debris.* **6** °thrill, °excitement, °passion, quickening, arousal, °stir, °stirring, °animation, elation, euphoria, °delight, °encouragement, °thrill, tingle: *In the first flush of success, they were overwhelmed with orders. He felt a flush come over him when he won the race.*

flush[2] *adj.* **1** (on the same) plane *or* level (with), °even, °smooth, °flat, °square, °true, °continuous; adjacent, next to: *See that the tabletop is flush with the countertop. The mirror should be flush against the wall.* **2** °full, overflowing, °replete, °abundant: *Turn off the water when it is almost flush to the top of the tub.* **3** °solvent, well-supplied, °comfortable; well-to-do, °well-off, well-found, °wealthy, °rich, °prosperous, affluent, moneyed, *Colloq* well-heeled, on Easy street, °in the money, *US* in the chips; *Slang* °loaded, rolling (in money *or* it): *He's flush enough to take us all out to dinner.* —*adv.* **4** even(ly), square(ly), levelly, °plumb, °directly: *Cut this branch off flush.*

fluster *v.* **1** °agitate, °stir (up), discompose, °discomfit, °discomfort, disconcert, °shake (up), °upset, disquiet, discommode, °bother, put out *or* off, °disturb, °perturb, °flurry, °flutter, make nervous, °throw off, °distract, °confuse, baffle, confound, °puzzle, °perplex, befuddle, °bewilder, °daze, °dazzle, *Colloq* °rattle, °throw, hassle, faze, *US* discombobulate: *Don't be flustered by her rude manner.* —*n.* **2** °agitation, °upset, °discomfort, disquiet, °bother, °disturbance, commotion, perturbation, dither, °flurry, °flutter, nervousness, °distraction, °confusion, bafflement, befuddlement, °perplexity, bewilderment: *All this fluster came about through the absence of a coherent plan.*

flutter *v.* **1** °flap, °flop, °fluctuate, vacillate, °wave, waver, °oscillate: *The paper streamers fluttered in the breeze.* **2** °flit, °flicker, flitter, °hover, °dance; °fuss: *We shall never get done if you keep fluttering about.* **3** °tremble, °shake, °quiver, dither, °jump, °vibrate, twitch: *She is fluttering with anticipation waiting for the postman.* —*n.* **4** fluttering, flapping, flopping, °fluctuation, fluctuating, vacillation, vacillating, °wave, waving, oscillation, oscillating, trembling, °quiver, quivering: *The slightest flutter of her fan communicates volumes.* **5** See **fluster, 2.**

flux *n.* instability, °change, °mutation, modification, °fluctuation, °unrest, °swing, °swinging, wavering, °movement, °motion, oscillation, °indecision, indecisiveness: *The economic indicators are in a constant state of flux.*

fly *v.* **1** take wing, °take flight, take to the air, wing, °soar, °sail, °hover; °flutter, °flit, °flap: *A stork flew overhead. Daedalus taught Icarus to fly.* **2** Also, ***fly away or off.*** °take flight, °depart, °leave, °flee, decamp, °bolt, run away *or* off, °escape, make (good) one's escape, °take to one's heels, show a clean pair of heels, rush *or* nip off *or* out *or* away, °(make a) run (for it), go *or* get away *or* off, abscond, make a getaway, cut and run, beat a (hasty) retreat, °take off, scoot, *Colloq* light out, °beat it, clear out *or* off, skedaddle, scram, *US* °cut out, hightail (it), take it on the lam, take a (runout) powder, lam out, take to the hills, take to the woods; *Slang Brit* scarper, °bugger off, *US* vamoose, °blow: *She flew off before I could get her telephone number.* **3** °hasten, °hurry, °rush, °run, °race, °dash, sprint, °tear, scamper, scoot, *Colloq* make tracks: *I must fly if I am to catch the train.* **4** °pass (by), °elapse, °go (by), °expire, run its course, slip *or* slide by *or* away: *Time flies quickly when you're having fun.* **5** aviate, °pilot; jet: *Parker flies his own plane. I flew down to Paris with him.* **6** ***fly in the face of.*** °flout, °defy, go against, contemn, °scorn, scoff at, °oppose, go *or* run counter to, °counter, °counteract, countervail, countermine, °contradict, contravene, °thumb one's nose at, *Colloq Brit* cock a snook at: *His actions fly in the face of conventional behavior.* **7** ***fly off the handle.*** fly into a rage *or* fury *or* temper *or* passion, lose one's temper, have a fit *or* tantrum, be fit to be tied, go berserk, go crazy *or* mad, °explode, *Colloq* lose *or* blow one's cool, blow one's top, hit *or* go through the roof, *US* hit the ceiling, °blow one's stack, get worked up (over); *Slang* blow a fuse *or* a gasket, have a hemorrhage, flip one's lid *or Brit* one's top: *Rudolf flies off the handle at the slightest provocation.* **8** ***let fly.*** **(a)** °throw, °toss, °cast, °hurl, °fling, °heave, °pitch, °lob, °sling, chuck, °shoot, °fire (off), let rip, °discharge, °launch, °propel, let go *or* off: *They let fly a volley of arrows.* **(b)** let go with, let (someone) have it, °lash out, vent one's spleen, lose one's temper, burst out *or* forth, burst into, *Colloq* pull no punches, tear into: *That was the last straw, and she really let fly at him.* —*n.* **9** Often, *Brit* ***flies.*** °flap, fly front, *US* zipper, *Brit* zip: *He was embarrassed to find his fly was open.* **10** ***fly in the ointment.*** °hitch, °snag, °impediment, °obstacle, °obstruction, °problem, °difficulty, °drawback, detraction, °rub, °hindrance, bugbear, bogey, bugaboo, *Offensive* nigger in the woodpile: *The fly in the ointment is that my mother won't give me permission to go.*

fly-by-night *adj.* **1** °temporary, °short-lived, transitory, °fugitive, ephemeral, °transient, °fleeting, °passing, °brief, impermanent, here today, gone tomorrow: *That fly-by-night operation was in business for less than a month.* **2** °unreliable, untrustworthy, °disreputable, °irresponsible, dubious, °questionable; °shifty, °dishonest, °sharp, °crooked, *Colloq* °shady, *Brit* cowboy: *Guarantees on double glazing are worthless if installed by some fly-by-night company.*

foam *n.* **1** °bubbles, °froth, spume, °lather, suds; effervescence, °sparkle, carbonation, °fizz: *The foam leaves spots when it dries.* —*v.* **2** °bubble, °froth, spume, °lather, suds up, soap up: *Watch the solution foam when I drop this tablet into it.*

focal *adj.* °central, focused, concentrated, convergent, centered, centralized: *The light rays come together at this focal point. The focal point of the conference was Harris's keynote address.*

focus *n.* **1** °center, concentration, focal point, °heart, °core, °target, convergence, °hub, nave; cynosure: *Why should Anita always be the focus of attention?* **2** ***in focus.*** °clear, °distinct, well- *or* sharply defined: *With the telescope in focus, you feel as if you can almost touch the distant shore.* **3** ***out of focus.*** unclear, °indistinct, blurred, blurry, °fuzzy, °woolly: *These photos are so out of focus that I can't tell who's who.* —*v.* **4** °concentrate, °center, °converge, °meet, pinpoint, °spotlight; bring into focus; *Colloq* zero in: *Without distractions, we can focus on the problem at hand.*

fog *n.* **1** °mist, haze, smog, °vapor, cloud, *Colloq* peasouper: *Our flight was grounded because of fog.* **2** °trance, °daze, °stupor, brown study, confused state; coma: *He's in a fog and didn't hear a word you said.* —*v.* **3** °dim, °obscure, cloud, bedim, becloud, °blind, °stifle: *The purpose of all that verbiage is merely to fog the issue.* **4** Usually, ***befog.*** °bewilder, °mystify, °perplex, °confuse, °muddle, °puzzle, °nonplus: *I was completely befogged by the technical language he used.* **5** Also, ***fog up or over.*** mist over *or* up, cloud up *or* over, °shroud: *My bathroom mirror fogs when I take a hot shower.*

fogy *n.* Usually, **old fogy** *or* **fogey.** fogey, °conservative, °relic, *Colloq* fuddy-duddy, fossil, °antique, °stick-in-the-mud, back number, °square: *That old fogy still dances the foxtrot.*

foible *n.* °weakness, °imperfection, °weak point, °fault, °frailty, °shortcoming, °flaw, °defect, °failing,

°blemish, °infirmity; °peculiarity, idiosyncrasy, °quirk, crotchet, °eccentricity, preoccupation, °kink, *Colloq* hang-up, °bug: *Her foible is that she loves junk food.*

foil[1] *v.* °thwart, °offset, °defeat, baffle, balk, parry, °frustrate, °counter, °check, °impede, °hamper, °outwit, circumvent, checkmate, °daunt, disconcert, °discomfit, °disappoint, pull the rug out from under (someone), cut the ground from under someone's feet, nullify, °nip in the bud, countervail (against), *Brit* put a spoke in (someone's) wheel, *Colloq* clip (someone's) wings, cut (someone) down to size, spoil (someone's) game, *Chiefly US and Canadian* faze: *The villain and his evil plan were again foiled by the arrival of the sheriff.*

foil[2] *n.* layer, lamina, lamination, °sheet, membrane, °film, coating, °flake, °scale, wafer: *Cigarettes once came packaged in tin foil.*

foist *v.* palm *or* pass off, °impose, °unload, °put (off), °push (off), *Colloq* °lay (off), *Brit* fob off: *Don't try to foist those fake paintings on me!*

fold *v.* **1** °bend, °ply, double (over *or* up), °overlap, crease, pleat, °gather, crimp: *Fold the letter before putting it into the envelope.* **2** °enclose, °envelop, enfold, °wrap, enwrap, °clasp, °clip, °embrace, °hug, °gather: *He folded her in his arms.* **3** °give way, °collapse, °fail, close (up *or* down), °shut down, go out of business, go bankrupt, *Colloq* °go under, go broke, go bust, *Brit* °go to the wall: *The company folded and ten employees were made redundant.* —*n.* **4** crease, °wrinkle, crinkle, °pucker, pleat, °gather, crimp, °overlap, °lap: *The folds will remain if you use starch when ironing.*

folk *n.* °people, °tribe, °(ethnic) group, °clan, °race; °society, °nation, °(general) public, °populace, °population, citizenry: *Archaeologists refer to these prehistoric people as the "Beaker Folk." Some folks disliked the performance.*

follow *v.* **1** go *or* come after *or* next; go *or* walk *or* tread *or* move behind, bring up the rear, *Colloq* string *or Colloq* tag along: *You go on ahead and I'll follow shortly. Please see the comments that follow.* **2** adhere to, cleave to, °support, °believe in, °accept, °adopt, conform to, comply with, °obey, be guided by, be modeled after *or* on, °observe, °heed, °mind, °go along with, °reflect, °mirror, °echo, °imitate, °copy, °mimic, ape: *This newspaper follows strict party policy.* **3** °attend, °accompany, °escort, go (along) with; °serve: *Mary's little lamb followed her everywhere.* **4** °chase, °pursue, dog, °hunt (down), °stalk, °run down, °track, °trail, °tail, °trace, °shadow: *The detective swore to follow him to the ends of the earth.* **5** °succeed, °supersede, step into the shoes of, take the place of, °replace, °supplant: *Who followed Henry VIII on the throne?* **6** °practice, °pursue, °engage in, °carry on, occupy oneself with, apply *or* dedicate *or* devote oneself to, °cultivate: *After the war he followed a career as a teacher.* **7** °result from, ensue, °issue, °flow, °arise, °develop: *The conclusion follows logically from the premise.* **8** °understand, °fathom, °comprehend, °get, °grasp, °see, °catch, °appreciate, °take in, keep up with, *Colloq* °dig: *Were you able to follow everything in that lecture?* **9** °watch, be a fan *or* aficionado of, °pursue, take an interest in, keep up with, keep abreast of, cheer *or* root for, °admire: *He follows international yacht racing.* **10** ***follow through*** **(*on*).** persist *or* persevere, °continue, °perform, conclude, °realize, consummate, °pursue, °carry out, °see through, °make good, °discharge, adhere to, °keep: *I hope she follows through on her promise.* **11** ***follow up*** **(*on*).** **(a)** °pursue, go after, °track, °investigate, °check (out), check up (on), °inquire, make inquiries, °look into: *MacGregor is very good at following up on the most obscure clues.* **(b)** °pursue, °prosecute, °reinforce, consolidate, °support, °buttress, augment, °bolster, °ensure: *The Romans followed up their success with an attack on Olbia.*

follower *n.* **1** °disciple, adherent, °student, °pupil, °apprentice, protégé(e): *They were followers of Buddha.* **2** °attendant, °henchman, °servant, retainer, bodyguard, myrmidon: *Louie walked in with two of his followers.* **3** °supporter, °devotee, °fan, aficionado, promoter, °enthusiast, booster, °advocate, °proponent, *US* rooter, *Colloq* groupie: *She is a faithful follower of The Dirty Knees, a rock group.*

follow-through *n.* °perseverance, diligence, °persistence, steadfastness, °endurance, °stamina, indefatigableness, sedulousness, sedulity, pursuance, °tenacity, °resolve, °determination, *Colloq US* stick-to-itiveness: *One needs imagination for new ideas but follow-through to make them succeed.*

follow-up *n.* °reinforcement, °support, backup, bolstering; consolidation: *Failure of the sales campaign was attributed to lack of follow-up.*

folly *n.* **1** foolishness, °nonsense, °absurdity, daftness, silliness, preposterousness, absurdness, senselessness, fatuousness, fatuity, rashness, °stupidity, asininity, inanity, nonsensicality, nonsensicalness, idiocy, imbecility, irrationality, °lunacy, °insanity, °madness, craziness, °eccentricity, weak-mindedness, feeblemindedness, simple-mindedness, muddle-headedness, thickheadedness, stolidity, stolidness, obtuseness, brainlessness, *Colloq* dumbness, dopiness, nuttiness, *US and Canadian* kookiness: *They saw the folly of fighting against such enormous odds. It would be folly to leave in this storm.* **2** °absurdity, °mistake, °blunder, °error, faux pas, gaffe, *Colloq* goof, °fluff: *What follies have been committed in the name of honor!*

foment *v.* °rouse, stir *or* whip up, °awaken, waken, °provoke, °incite, instigate, °initiate, °prompt, °start, °motivate, °inspire, °work up, °inflame, fan the flames (of), °kindle, galvanize, °rally, °excite, °stimulate, °encourage, °promote, °foster, °forward, °further, °advance, °cultivate, sow the seeds of, °spur, goad, egg on, °urge: *He is always fomenting ill will against the management.*

fond *adj.* **1** °tender, loving, °affectionate, °warm, adoring, caring: *She clasped him in a fond embrace.* **2** °foolish, °bootless, °empty, °vain, naive *or* naïve *or* naïf: *She had fond hopes of his being elected.* **3** ***fond of.*** °partial to, °(having a) liking (for), soft on *or* about, affectionate towards, °attached to, having a fancy *or* taste for, fancying, predisposed *or* inclined to *or* towards; addicted to, *Colloq* hooked on: *She is inordinately fond of chocolate.*

fondle *v.* °caress, °stroke, °pet, °pat, °touch, °cuddle, °snuggle; °handle, °touch: *See that couple over there fondling one another? He fondled his dagger.*

fondly *adv.* affectionately, lovingly, tenderly, °warmly, adoringly, caressingly: *Martin regarded his wife fondly.*

food *n.* °nourishment, nutriment, aliment, °sustenance, °subsistence; foodstuffs, edibles, eatables, viands, bread, victuals, °rations, °provisions, comestibles, *Brit* commons, *Colloq* grub, eats, chow, *Brit* °scoff, prog: *Without any food for days, the survivors were near starvation.*

fool *n.* **1** simpleton, ninny, ninnyhammer, nincompoop, ass, jackass, dunce, °dolt, °half-wit, numskull *or* numbskull, blockhead, bonehead, pinhead, °silly, featherbrain, loon, goose, booby, jay, goon, mooncalf, idiot, ignoramus, dimwit, nitwit, halfwit, imbecile, moron, °clod, clodpole, clod poll, clodpate, oaf, *Psychology* retardate, *Scots* gomeril, *Colloq* birdbrain, dumbbell, fathead, chump, schmuck, °twit, knucklehead, chucklehead, nit, twerp *or* twirp, *Brit* pillock, *US and Canadian* °jerk, °retard; *Slang* °sap, °dope, *Brit* git, *Australian* boofhead: *He's such a fool he'll believe anything.* **2** (court) jester, °clown, °comic, comedian *or* comedienne, entertainer, °zany, buffoon, merry-andrew, *farceur,* °joker, jokester, droll, Punch, Punchinello, pierrot, harlequin: *In Shakespeare's plays, the fool is often the wisest character.* **3** °butt, °dupe, gull, °victim, cat's-paw, *Colloq* chump, °greenhorn, °(easy) mark, *US* fall guy; *Slang* pigeon, °sucker, stooge, *Brit* °mug: *His co-workers liked to make a fool out of him.* —*v.* **4** °trick, °deceive, °take in, °swindle, °defraud, °hoax, °cheat, °fleece, cozen, °hoodwink, °bluff, °dupe, gull, humbug, delude, °mislead, make a fool of; pull the wool over (someone's) eyes, have (someone) on, pull (someone's) leg, °tease, °twit, *Archaic* chouse; *Colloq* kid, con, snow, do a snow job on, bamboozle, put one *or* something over on, pull something *or* a fast one on, *Brit* °twist, *Chiefly US and Canadian* josh; *Slang* °pluck, *Brit* nobble: *When he said the diamond was fake, he certainly fooled me. I think she was fooling when she said she'd inherited millions.* **5** °joke, jest, °banter, °tease, °twit, feign, °fake, °pretend, °make believe, *Colloq* kid: *She insists that she is not fooling but dead serious.* **6** ***fool with*** *or* ***around*** *or* ***about*** **(*with*).** play (around *or* about) (with), toy *or* trifle (with), mess *or* fiddle (around *or* about) (with), monkey (around *or* about) with, °meddle (with), tamper with, fribble (with): *Please stop fooling with the dials on the radio.* **7** ***fool around*** *or* ***about.*** **(a)** play *or* mess around *or* about, gambol, °frolic, romp, °cavort: *Stop fooling around and get serious.* **(b)** waste *or* fritter away *or* squander *or* kill time, fribble, loiter, °dawdle, °idle, putter *or Brit* potter (about *or* around), lark *or* muck about *or* around, *Colloq* footle: *She's always fooling around instead of getting on with her work.*

foolhardy *adj.* °rash, °imprudent, °impetuous, °reckless, °brash, °venturesome, °bold, °cheeky, °daring, °audacious, temerarious, °adventurous, °daredevil, incautious, °hot-headed, °careless, °heedless, devil-may-care, °hasty, °thoughtless, °unthinking, °irresponsible, °wild, madcap, *Colloq US and Canadian* nervy, *Slang* gutsy: *It is foolhardy to go scuba diving without a companion.*

foolish *adj.* **1** °senseless, incautious, °imprudent, impolitic, °indiscreet, unwise, injudicious, ill-considered, °ill-advised, °misguided, °shortsighted, °impetuous, headlong, °rash, °brash, °reckless, °hasty, °heedless, °unwary, °foolhardy, °thoughtless, °mindless: *Standing under a tree in a thunderstorm is foolish.* **2** °nonsensical, °stupid, asinine, °inane, °silly, fatuous, fatuitous, dimwitted, °scatterbrained, °harebrained, °crazy, °mad, °insane, demented, irrational, °erratic, °unstable, crackbrained, featherbrained, birdbrained, simple-minded, lightheaded, muddle-headed, numskulled *or* numbskulled, addlepated, rattlebrained, bemused, °confused, °feebleminded, moronic, idiotic, imbecilic, °half-witted, dull-witted, slow-witted, witless, brainless, empty-headed, blockheaded, boneheaded, thickheaded, °obtuse, °stolid, *Colloq* °dumb, balmy *or Brit* barmy, loony, nuts, nutty, batty, dopey *or* dopy, °soft (in the head), °dim, °thick, dippy, dotty, °dizzy, *Brit* potty, °daft, *Slang* cuckoo, goofy, screwy, wacky: *Tracy was foolish to believe him when he said he loved her.* **3** °preposterous, °ridiculous, °absurd, irrational, illogical, °unreasonable, °ludicrous, °wild: *They have some foolish idea about building a hotel in Antarctica.*

foolproof *adj.* °safe, °certain, °sure, °trustworthy, dependable, °reliable, °infallible, unfailing, guaranteed, warranted, *Colloq* sure-fire: *She claims to have a foolproof way of preventing unwanted pregnancies.*

footing *n.* **1** °foundation, °basis, °base, ground(s): *The events put our entire relationship on a new footing.* **2** °standing, °status, °level, °condition, °position, °terms, °state, °rank: *The two departments are on an equal footing as far as funding is concerned.* **3** foothold, toehold; °balance, °stability: *She lost her footing on the icy pavement and down she went.*

footstep *n.* **1** °step, footfall, tread: *I recognized her footstep.* **2** Usually, ***footsteps.*** footprint(s), °track, °trail, °trace, spoor, footmark(s); °tradition, °example, way of life: *He followed in his master's footsteps.*

for *prep.* **1** representing, championing; °in favor of, *Brit* on *or US and Canadian also* in behalf of, on the side of, in support of, in the service of, as a service to, for the benefit of, pro: *Are you for or against the candidate? She did it for herself.* **2** in search *or* quest of, seeking, looking for *or* after, after, with a view *or* an eye to: *We're just here for the beer.* **3** °instead of, in place of, representing, as a replacement for, *Brit* on *or US and Canadian* in behalf of; in return *or* exchange for, in compensation *or* recompense *or* payment *or* repayment for, in requital for: *I am acting for my father. This check is for the toaster. Will you accept nothing for your pains?* **4** for the treatment of, as a remedy for, against; for the purpose *or* object of: *This medication is for skin infections only. It's for your own good.* **5** for the sake of, *Brit* on *or US and Canadian also* in behalf of, in the interest of: *Would you really do that for me?* **6** in the direction of, to, °toward, into: *Head for the hills!* **7** to save, in the interest *or* benefit of, for the sake *or* benefit of, conducive to; because of, on account of, °by reason of: *They go to Montecatini for their health. She hated him for what he represented.* **8** to go to, destined for: *I am leaving for Tierra del Fuego tomorrow.* **9** suitable *or* suited for, fit *or* fitted *or* fitting for, appropriate for, proper for: *I am not sure that this color is for you. Is it right for me to speak out now?* **10** for the duration of; °over (the extent of), during, in the course of, °throughout, °through: *She stayed for a week.* **11** °despite, °in spite of, °notwithstanding, allowing for: *For all his complaining, he seemed satisfied.* **12** ***as for.*** °regarding, in regard to, as regards, respecting, °concerning, as far as (something *or* someone) is concerned: *As for Betty, let her do as she likes.* —*conj.* **13** because, since, as, inasmuch as, seeing that, owing *or* due to the fact that: *I was unable to see him, for he was bedridden.*

forbid *v.* °prohibit, °ban, °hinder, °stop, °exclude, debar, °preclude, °prevent; °outlaw, interdict, disallow, proscribe, °taboo; °veto: *I was forbidden from entering. They forbid smoking in the classrooms.*

forbidding *adj.* **1** °repellent, °repulsive, °offensive, odious, °abhorrent: *In his long cloak and mask, he presented a forbidding appearance.* **2** °hostile, unfriendly, °stern, °harsh, °menacing, °threatening, °ominous, °dangerous, °bad, °nasty, °ugly, unpleasant: *The coast here is rocky and forbidding. The old crone gave him a forbidding look and we cringed.*

force *n.* **1** °power, °might, °energy, °strength, potency, °vigor, °intensity, °violence, °impact; °dynamism, °pressure: *The force of the blow knocked me down.* **2** coercion, °pressure, constraint, °duress, compulsion, arm-twisting: *When persuasion failed, the guards resorted to force.* **3** troops, soldiers, army: *NATO forces are on maneuvers.* **4** °weight, persuasiveness, °persistence, cogency, effectiveness, efficacy, °strength, validity, °significance, value: *The force of his argument was sufficient to convince us.* **5** °meaning, °import, °significance: *Are they able to cope with the full force of the word "God"?* **6** ***in force.*** °in effect, °effective, °in operation, °operative, valid, binding, °current: *The ancient customs still remain in force. The law came into force at midnight.* —*v.* **7** °make, °oblige, °require, compel, coerce, °exact, constrain, °enforce, impel, °intimidate, pressure *or Brit also* pressurize, °press, dragoon, twist (someone's) arm, *Colloq* bulldoze, °put the squeeze on (someone): *Each* *of us was forced to dig his own grave.* **8** °push, °drive, °thrust, °propel; prise *or* prize, °break, °wrench, °crack, jemmy, *US* °pry: *The tornado forced the chair through the wall. He forced open the safe and took the money.* **9** °exact, °extort, °extract, wrest, wring, °drag: *The CIA tried to force a confession from her.*

forced *adj.* °artificial, °unnatural, contrived, °stilted, °calculated, °studied, °labored, °strained, °stiff, °false, feigned, fake(d), °mannered, °affected, °self-conscious, *Colloq* phoney *or US also* phony: *Her forced joviality failed to conceal her contempt for those present.*

forceful *adj.* **1** °vigorous, °energetic, compelling, °dynamic, °intense, °potent, °strong, °mighty, °powerful, °aggressive, °weighty, °effective, convincing, °persuasive: *The colonel was known as a forceful leader of men.* **2** °effective, °efficacious, cogent, °logical, °impressive, °telling, convincing, °persuasive, °strong, °mighty, °forcible, °powerful, compelling, °irresistible; pithy, meaty: *The most forceful arguments emerged in the last stages of the debate.*

forcible *adj.* **1** See **forceful, 2.** **2** °drastic, °forceful, °violent, °aggressive, coercive, °severe, stringent: *We may have to use forcible measures to get his confession.*

foreboding *n.* **1** apprehension, apprehensiveness, °feeling, °sense, °misgiving, °dread, °suspicion, °intuition, °anxiety, °fear: *An atmosphere of foreboding filled the crypt.* **2** °premonition, augury, °prophecy, °prediction, prognostication, °warning, foretoken, foreshadowing, presentiment, °omen, °sign, portent, intimation, forewarning, presage, advance word: *I had a strange foreboding of horror as I entered the room.*

forecast *v.* **1** °predict, foretell, °prophesy, °prognosticate, °foresee, augur, presage, vaticinate, °anticipate; forewarn, °calculate: *Seers forecast coming events; meteorologists forecast the weather.* *—n.* **2** °prediction, °prophecy, °prognosis, prognostication, °foresight, augury, vaticination, °anticipation; forewarning, °calculation: *Have you heard the weather forecast for tomorrow?*

foregoing *adj.* °preceding, °above, °former, °previous, °precedent, °prior, antecedent; earlier, °preliminary, anterior; aforementioned, aforesaid: *The foregoing paragraph set forth the principles.*

foregone *adj.* ***Usually in foregone conclusion.*** °assumed, established, preestablished, °predetermined, °fixed, °inevitable, °set, accepted, °cut and dried: *His incarceration for the crime was a foregone conclusion.*

foreign *adj.* **1** °alien, imported, nonnative; °overseas, °distant, tramontane, transalpine, transatlantic, transpacific: *Duty is payable on foreign goods. We holiday in a different foreign country each year.* **2** °strange, °outlandish, °exotic, °unfamiliar, °peculiar, °odd, °curious: *She regards anything except beans and toast as foreign food.* **3** °unknown, °unfamiliar, °strange, °inappropriate, °unrelated, unconnected, unassimilable, °remote, °extrinsic, °extraneous: *Their philosophy is completely foreign to my way of thinking.*

foreigner *n.* °alien, nonnative, °immigrant, °newcomer, new arrival, °outsider, outlander, °stranger: *The word* xenophobia *means 'dread of foreigners.'*

foreman *n.* °superintendent, °manager, °boss, °supervisor, °overseer, *Brit* shopwalker, *US* floorwalker, *Colloq* °super, *Brit* gaffer, *US* straw boss: *The foreman insists we use the time clocks.*

foremost *adj.* **1** °first, °primary, °prime, °leading, °preeminent, °supreme; °prominent, °notable, °noteworthy, °noted, °chief, °paramount, °main, °best, °superior: *She is one of the foremost eye surgeons in her country.* *—adv.* **2** °first, firstly, °primarily, in (the) first place, before anything else: *First and foremost, we must consider the matter of safety.*

forerunner *n.* **1** °predecessor, °precursor, foregoer; forebear, °ancestor, forefather, °progenitor; herald, °harbinger, °envoy: *This dishwasher is better designed than its forerunners. The cuckoo is thought by many to be the forerunner of spring.* **2** °omen, portent, foretoken, °premonition, °sign, augury, °token: *A drop in the barometer is a forerunner of bad weather.*

foresee *v.* presage, foretell, °envisage, °picture, °forecast, °predict, °prophesy, augur, *US* °envision: *The only problem I foresee is that she doesn't do windows.*

foreshadow *v.* presage, foretoken, portend, augur, °indicate, °prophesy, °predict, °bode, °signal, °signify, betoken: *The surrender of Ghent foreshadowed the fate of Flanders.*

foresight *n.* **1** °providence, °prudence, °care, farsightedness, watchfulness, °caution, °precaution, long-sightedness, perspicacity, °wisdom, sagacity, °insight, circumspection: *The commander demonstrated foresight in ordering a retreat.* **2** prevision, °perception, °prospect, °vision, foreknowledge, prescience; °expectation: *A little foresight could have prevented the calamity.*

forestall *v.* °anticipate, °prevent, °obstruct, °hinder, obviate, °thwart, °preclude, °frustrate, avert, ward *or* stave *or* fend off, °intercept, parry, °stop, °delay: *The frigate effectively forestalled our plan to sail away secretly.*

forethought *n.* premeditation, planning, plotting, farsightedness, long-sightedness: *Much forethought went into the assassination scheme.*

forever *adv.* **1** °always, for good, °ever, (for) evermore, forevermore, eternally, everlastingly, for ever and a day, undyingly, for eternity, till Doomsday, till the end of time, *Colloq* till the cows come home, till hell freezes over: *They swore to be true forever.* **2** constantly, continually, continuously, °all the time, unceasingly, incessantly, without cease *or* surcease, endlessly, °consistently, persistently, interminably, perpetually: *She was forever watching soap operas on television.*

foreword *n.* °preface, prologue, prelude, prolegomenon, °preamble, *Literary* proem; introduction: *He set forth the purpose of the book in the foreword.*

forfeit *n.* **1** °penalty, °fine, °fee, °charge, °damages, forfeiture, sequestration, amercement, mulct: *You will have to give me a kiss as a forfeit.* *—v.* **2** °lose, °yield (up), give up *or* over, °relinquish, °surrender, be stripped *or* deprived of, °forgo, °renounce, °waive: *You have forfeited your right to trial and you will be hanged at dawn.* *—adj.* **3** surrendered, °lost, yielded, relinquished, forgone, waived, renounced: *His life was forfeit the moment he volunteered to defuse the bomb.*

forge *v.* **1** °make, °construct, °fashion, °fabricate, °manufacture, °shape, °mold, °cast, hammer out: *This plant forges heavy steel tools. He forged a new life for himself in the American West.* **2** °create, °invent, °conceive (of), °coin, °devise, °think up, °frame: *I have forged a new plan of escape.* **3** °counterfeit, °copy, °reproduce, °imitate, °falsify, °fake, *Slang US* hang paper: *She forged her employer's name on the check. The gang was caught because they forged three-dollar bills.*

forgery *n.* **1** counterfeiting, falsification, °fraud, fraudulence: *He was sent to prison for ten years for forgery.* **2** °counterfeit, °fake, °sham, °imitation, *Colloq* phoney *or US also* phony: *The police have found three forgeries of Dali paintings.*

forget *v.* **1** fail *or* cease to remember *or* recall *or* think of, °lose, draw a blank: *I forgot what I was supposed to buy at the market.* **2** °leave (behind), omit *or* neglect (doing *or* taking): *I keep forgetting my umbrella at your house.* **3** °ignore, dismiss from (one's) mind *or* thoughts, °disregard, °overlook, consign to oblivion: *Forget the fact that you ever met me.*

forgetful *adj.* amnesiac; °absent-minded, distracted, abstracted, °inattentive, °preoccupied, neglectful, negligent, °careless, °lax, °dreamy, dreaming, in dreamland, in the clouds, in cloud-cuckoo-land *or* cloudland *or* Nephelococcygia, °remote, distrait(e), *Colloq* not turned on, turned off, out of it: *Franklin is so forgetful he'd lose his head if it weren't stitched on.*

forgive *v.* **1** °pardon, °excuse, °allow, make allowance(s) for, °indulge, condone, °vindicate; °overlook, °ignore, °disregard, pay no attention to, °pass over, *US* slough over: *Please forgive my curiosity, but where did you get that hat?* **2** °clear, acquit, absolve, exculpate, exonerate; °spare; *Colloq* °let off: *Father, forgive them, for they know not what they do.* **3** °cancel, °waive, °abolish, °void, nullify, °erase, °delete; *Colloq* °let off (the hook): *My dream was to awake one day to find all my debts forgiven.*

forgiveness *n.* **1** °pardon, °reprieve, absolution, °remission, acquittal, acquittance, amnesty, °allowance, vindication, exculpation, exoneration, *Archaic* shrift: *Oh, Lord, I ask forgiveness for my sins.* **2** °mercy, mercifulness, compassion, °grace, leniency, clemency, °indulgence, °tolerance: *He begged her forgiveness for the way he had treated her.*

forgiving *adj.* °tolerant, °lenient, °sparing, forbearing, °merciful, compassionate, conciliatory, magnanimous, humane, soft-hearted, clement: *In a forgiving mood, the judge gave her a light sentence.*

forgo *v.* **1** °give up, °renounce, forswear, °forsake, °abandon, do *or* go without, °sacrifice, °eliminate, °omit, leave out *or* alone, °cede, °waive; °avoid, °shun, eschew, abstain from, °turn down, °pass up, °deny (oneself): *He has to forgo all meats in his new diet.* **2** °resign, °give up, °yield, °surrender, °relinquish, °cede, °waive, °renounce, forswear, °abdicate, °abandon: *Did he not forgo all right to the throne by marrying a commoner?*

forlorn *adj.* **1** °miserable, °wretched, °desolate, °woebegone, °lost, °pitiable, °pitiful, °pathetic, woeful, cheerless, °joyless, °unhappy, depressed, °sad, °desolate, disconsolate, °gloomy, lugubrious, °glum, °despondent, °dismal, °dejected, dispirited, low-spirited, comfortless, down, °melancholy, dolorous, °sorrowful, °mournful, °inconsolable: *She looked terribly fragile and forlorn as she struggled to hold back her tears.* **2** °abandoned, forsaken, °deserted, neglected, shunned, °outcast, °alone, °lonely, °lonesome, friendless, bereft: *We came upon a forlorn little village.*

form *n.* **1** °shape, configuration, conformation, °order, °organization, °arrangement, °formation, construction, °structure, °construct, °frame, °cut, °cast, °mold, °pattern, °appearance; °manifestation: *The roadside cafe was in the form of a milk bottle. In what form will the genie appear next?.* **2** °figure, °body, °build, °shape, °physique, anatomy; °silhouette, °aspect, °look, °appearance, °profile, contour; °carriage, °bearing, °attitude, °pose, *Slang US* bod, built: *She has the form of a wrestler.* **3** °type, °kind, °variety, °sort, °breed, species, genus, °genre, °character, °make, °brand, °color, °tone, °tint, °texture, °fabric, °material, feather, °description, °manner, °way, °nature, °style, °stamp, °manifestation: *What forms of life are not of divine origin? He reveled in music in all its forms.* **4** °blank; °model, °format, °frame, framework, °course, °procedure, °order, regimen, °method, °system, °ritual, °formula, rule(s), °practice, °technique, °way, °means, °approach, °mode, °fashion, °manner, °style: *Fill in the application form in ink. Please follow the proper form when submitting articles.* **5** °condition, °state, °shape, °trim, fettle: *He seems in unusually good form tonight.* **6** °decorum, °behavior, deportment, °formality, °ceremony, °convention, °etiquette, °manners, °conduct, °custom, °protocol, °propriety, °ritual: *Follow correct form when writing to the king. Spitting is considered bad form.* —*v.* **7** °make, °fabricate, °forge, °shape, °mold, °fashion, °produce, °turn out, °manufacture, °construct, °assemble, put together, °set up, °put up, °build, °erect, elevate, °raise, °put up; °organize, codify; °develop: *The architect takes the many elements and forms them into a coherent whole. Have you formed an opinion of the book?* **8** °create, °originate, °devise, °invent, °compose, °formulate, give form *or* shape, °coin, concoct, °conceive, contrive, dream up, °imagine, visualize, °envisage, *US* °envision: *His ideas of religion were formed when he was very young. A picture of the battle was formed in my mind.* **9** °make up, constitute, be made up of, comprise, °be composed of; serve as: *This compound is formed of many elements. Six fugues formed the main part of the program.* **10** °acquire, °develop, °cultivate, °contract; °get: *In prison I formed the habit of rolling my own cigarettes. We formed a lasting friendship.* **11** °develop, °grow, °arise, °appear, °materialize, °show up, take shape *or* form, °accumulate: *When the yeast is added, a barm forms on the surface.*

formal *adj.* **1** °standard, °conventional, °customary, established, prescribed, °regular, °normal, °correct, °proper; °strict, formulaic, °inflexible, punctilious, °exacting, unchanging, °methodical, °orderly, °systematic, °set, pro forma, °ritual, ritualistic, °ceremonial, °proper, °official, °routine, °fixed, °rigid, °stiff, °stilted, °stately, starched, unbending, °solemn; confining, °straitened, °limited; *Colloq* °straight, °square: *Formal rules of behavior are observed.* **2** °explicit, °express, °definite, spelled-out, formalized, °express, authorized, °official, °solemn, °legal, °lawful: *Formal demands have been made for the withdrawal of our ambassador.* **3** prim, °ceremonious, °dignified, °stuffy, °strait-laced, °stiff, °precise, °exact: *He made a formal bow and left our company.*

formality *n.* Often, **the formalities.** **1** °form, °convention, conventionality, °practice, °procedure, °custom, wont, °observance, °protocol, °ceremony, °rite, °ritual: *Observe the legal formalities by having this document witnessed. Formality requires that you walk backwards out of the room.* **2** strictness, punctilio, exactness, °precision, correctness, rigidity, stiffness, inflexibility: *We must maintain the formality of Sunday service.* **3** °etiquette, politesse, °decorum, punctilio, conformity, °propriety: *Formality prohibits casual conversation.*

format *n.* **1** °appearance, look(s), °aspect, layout, °plan, °design, °style, °form, dimension(s), °shape, °size: *In its present format, the magazine resembles an academic journal.* **2** °composition, content(s), °makeup, constitution, °arrangement, °plan, °organization, °order, °setup: *Her TV program has the format of a chat show.*

formation *n.* **1** °development, °appearance, materialization, °shape, °accumulation, °generation, °creation, crystallization, forming, genesis, °production: *The formation of bubbles indicates that the acid is working.* **2** °establishment, °institution, °creation, founding, °setup, organizing, °organization, °development, °composition: *We met to discuss the formation of a new company.* **3** array, °display, °pattern, °arrangement, °structure, grouping, °organization, configuration, °disposition: *The vast military formation covered the entire valley.*

former *adj.* **1** °previous, earlier, °prior, ex-, one-time, °preceding, erstwhile, °late, °latest, °last, °recent, *ci-devant,* quondam, *Archaic* whilom: *I ran into a former girlfriend at the art show.* **2** °old, °past, °bygone; °ancient, (pre)historic, departed, antediluvian: *In former times, one could demand—and get—decent service.*

formerly *adv.* °once, °before, °previously, hitherto, long ago, °at one time, in the old days, once upon a time, in days gone by, in days *or* time past, time was, back then, when the world was young(er), *Colloq* way back *or US also* way back when: *Formerly, there were no tall buildings to block one's view.*

formidable *adj.* **1** alarming, appalling, °dreadful, °frightful, °awesome, awe-inspiring, °menacing, horrifying, °frightening, intimidating, daunting, petrifying, °terrifying: *A formidable, fire-breathing dragon blocked our exit.* **2** °terrific, °fantastic, °unbelievable, °incredible, °impressive, °prodigious, mind-boggling, °awesome, *Colloq* mind-blowing, freaky: *They are a formidable talent and will sell a million records this year.* **3** °arduous, °indomitable, °overwhelming, staggering, °powerful, °mighty, °difficult, challenging, °burdensome, onerous: *She has formidable obstacles to overcome before becoming a doctor.*

formula *n.* °recipe, rubric, formulary; rule(s), °prescription, °directions, °instructions, blueprint, °method, °technique, °means, °way, °pattern, °procedure, modus operandi, *Colloq US* MO (='modus operandi'): *Ali Baba uttered the formula, "Open sesame!", and the rock opened. The formula for making gold from lead is known only to me.*

formulate *v.* **1** systematize, codify, °define, °specify, articulate, particularize, °denote: *You must try to formulate your ideas more clearly.* **2** °devise, °originate, °create, think up *or* of, dream up, °conceive, concoct, °invent, °improvise, *Colloq* cook up: *I have formulated a plan in my mind to deal with the problem.* **3** °develop, °forge, evolve, °work out, °design, map out, °block out, °draw up: *You have three days to formulate a workable procedure.*

forsake *v.* **1** °abandon, °desert, °quit, °leave, °flee, °depart, °vacate: *He forsook balmy California to dig for gold in the Klondike.* **2** °abandon, °desert, °leave, °jilt, °reject, °throw over, jettison, °cast off: *She has forsaken Michael for another man.* **3** °give up, °yield, °renounce, °repudiate, °relinquish, °forgo, forswear, °surrender, °resign, °abdicate, °recant, °deny, have *or* be done with, °turn one's back on: *Would you forsake a throne for the love of a woman?*

forte *n.* °talent, strong point, °gift, °specialty, °strength, °aptitude, °genius, *Colloq* long suit: *His forte is carving ivory miniatures.*

forthcoming *adj.* **1** approaching, nearing, °impending, °imminent, coming, °(close) at hand, upcoming; near *or* close (by), (near *or* close) at hand, in the offing, on the horizon, *Colloq Brit* on the cards, *US* °in the cards: *The forthcoming tax increase will affect everyone. A new regulation is forthcoming.* **2** awaited, expected, anticipated, looked-for, watched-for, °prospective, foreseen: *The forthcoming payment will be a little late.* **3** °outgoing, °friendly, °amiable, affable, °sociable, °accessible, °expansive, chatty, °talkative, communicative, informative, °open, °free, revealing, unreserved: *Barry was most forthcoming when questioned about his finances.*

forthright *adj.* straightforward, °direct, °blunt, °candid, °frank, °aboveboard, unambiguous, unequivocal, °open, °outspoken, °uninhibited, unreserved, unconstrained, unrestrained: *She has always been forthright and honest in her dealings with me.*

fortify *v.* **1** °strengthen, °reinforce, shore up, °buttress, °brace, °bolster, °secure: *Steel plates were used to fortify the walls.* **2** °cheer, °encourage, hearten, °buoy, invigorate, °energize, embolden, °reassure, °brace: *Her enthusiasm fortified him to face what was coming.* **3** °supplement, °enhance, °enrich, °boost, augment: *The drink was fortified by the addition of whisky.*

fortitude *n.* °strength, mettle, °backbone, °courage, °nerve, °resolution, resoluteness, °perseverance, °endurance, °tenacity, pertinacity, °grit, °determination, willpower, *Colloq* °guts: *His encouragement gave me the fortitude to carry on.*

fortunate *adj.* **1** °lucky, in luck, fortuitous, blessed: *You were fortunate to have survived the crash.* **2** °favored, °advantageous, °propitious, auspicious, °providential, °favorable, °opportune, °timely, °well-timed: *It was a fortunate time to be buying a house.*

fortune *n.* **1** °position, °worth, °means, °assets, holdings, °wealth, °property, °estate, °possessions; °wealth, °riches, affluence, opulence, °treasure, °money, °prosperity: *Each member is liable to the full extent of his fortune. He acquired his fortune from oil.* **2** °luck, °chance, fortuity; °lot, °fate, kismet, °destiny, karma; °future; *US* happenstance: *It was just fortune that put me in the right place at the right time. She'll tell your fortune if you cross her palm with silver.* **3** Usually, ***fortunes.*** circumstance(s), experience(s), adventures, expectation(s), °lot: *My fortunes of late have been poor.*

fortuneteller *n.* °oracle, soothsayer, °prophet, diviner, augur, °seer, clairvoyant, prognosticator, sibyl, haruspex, crystal gazer, tealeaf-reader, palmist, palmreader, stargazer; futurologist: *The fortuneteller said to beware the Ides of March.*

forward *adj.* **1** °advance, °leading, °foremost, °front, °head, °first: *The forward contingents of the army moved into the town.* **2** °bold, °pert, °impudent, °brash, °insolent, °impertinent, °disrespectful, °brazen, °audacious, °rash, unashamed, °unabashed, saucy, °flippant, °presumptuous, °cheeky, *Colloq* °flip, °fresh, °pushy: *It was a bit forward of you to call the chairman by his nickname.* **3** (well-)advanced, (well-)developed, °progressive, °precocious, forward-looking: *She was quite a forward girl at fifteen.* —*adv.* **4** forwards, °ahead, °onward, along; clockwise, deasil: *I moved forward to the head of the queue. Set the clock forward an hour tonight.* **5** up, °out, forth, to the fore, into consideration, into view, into the open, to the surface, onto the table: *Worthington brought forward an interesting proposal.* —*v.* **6** °advance, °further, °promote, °back, °foster, °support, °aid, °assist, °help; °patronize, °encourage, °nourish, nurse along: *He did his best to forward her career.* **7** dispatch *or* despatch, °send, °ship, °deliver, °transmit, °express, °post, °mail, consign, °remit; send on: *The shipment will be forwarded as soon as payment is received. The post office forwards mail to my new address.* **8** °speed (up), accelerate, °advance, °hasten, °expedite, °hurry, °quicken, °step up: *This mixture is said to forward the flowering of plants.*

foster *v.* **1** °promote, °encourage, °stimulate, °further, °forward, °advance, °cultivate, nurture, °nourish, °support, °back, °assist, °help, °aid, succor, °patronize: *Cutting back these offshoots fosters growth of the main stem. Representatives have fostered acceptance of our products throughout Europe.* **2** °bring up, °rear, °raise, °take care of, °maintain, °care for: *The Cartwrights have fostered eight orphans in their home.*

foul *adj.* **1** °offensive, °loathsome, °disgusting, °obnoxious, °revolting, °repulsive, °repellent, °repugnant, sickening, °nauseous, nauseating, °nasty, °beastly, *Archaic* fulsome: *The bartender makes a foul concoction he swears will cure a hangover.* **2** °filthy, unclean, °nasty, polluted, °putrid, putrescent, putrefactive *or* putrefacient, defiled, soiled, spoiled, °rotten, decayed, decomposed, °rancid, soured, turned, tainted, °moldy, °impure, adulterated, contaminated, °stale, °bad, *Brit* °off: *That fast-food place was closed when they found* *foul food in the freezer.* **3** °smelly, °stinking, noisome, fetid *or* foetid, °rank, evil-smelling, foulsmelling, malodorous, °musty, mephitic, graveolent: *A foul odor emanated from the cabinet.* **4** °wicked, °vile, °bad, °base, °abominable, °low, °sordid, iniquitous, °evil, flagitious, °atrocious, °monstrous, nefarious, °sinful, °immoral, amoral, °vicious, °villainous, °scandalous, °infamous, °dishonorable, °shameful, °disgraceful, ignominious; detestable: *They were finally brought to justice for their foul deeds.* **5** °dirty, °obscene, °filthy, °profane, scatological, °gross, smutty, foulmouthed, °blue, licentious, salacious, °lewd, °indecent, °improper, °coarse, uncouth, °vulgar, °rude, °scurrilous, °rough, indelicate, °immodest, °risqué, °off-color, °suggestive, °bawdy, ribald, Rabelaisian, Fescennine, *US* °raw, *Slang* raunchy: *My mother forbids my reading books containing foul language.* **6** °abusive, °offensive, affronting, insulting, disparaging, maligning, thersitical, calumnious *or* calumniatory, aspersive, °slanderous, defamatory, libelous, denigrating, °derogatory, deprecatory *or* deprecative, depreciatory *or* depreciative, denunciatory *or* denunciative, °derisory, derisive, °scornful, belittling, fulminous, objurgatory *or* objurgative, °vituperative, invective: *Keep your foul tongue to yourself.* **7** °dishonest, unfair, unjust, unsportsmanlike, °dishonorable, °fraudulent, underhand(ed), double-dealing, °two-faced, °corrupt, °venal, °dirty, treacherous, °perfidious, °traitorous, °unscrupulous, *Colloq* °crooked, °shady, *Slang, chiefly Brit* bent: *Thrupp will get his way by fair means or foul.* **8** °nasty, °dangerous, °rough, °disagreeable, unfavorable, °sloppy, °stormy,. adverse; °windy, blustery; snowy, sleety, °wet, rainy: *We ran into a spot of foul weather at Dover.* **9** obstructed, blocked, choked, stopped(-up), plugged(-up), clogged(-up): *The drain is foul with all that rubbish.* **10** tangled, entangled, caught, ensnared, enmeshed, snarled: *A foul anchor is a common nautical symbol.* **11** °illegal, prohibited, forbidden, interdicted, not fair; °dirty: *In boxing a hit below the belt is a foul blow.* —*v.* **12** °dirty, °pollute, °sully, befoul, defile, °soil, °contaminate, °adulterate, °taint: *The effluent from the plant is fouling our water supply.* **13** °tangle, °entangle, °catch, °snare, ensnare, enmesh, °snag, °snarl, °jam, °twist: *We can't hoist the mainsail because the halyard is fouled.* **14** °disgrace, °dishonor, °sully, °taint, besmirch, defile, °soil, °stain, °smear, °tarnish, °blacken, denigrate, °debase, °degrade, abase, demean, °disparage, defame, derogate, asperse, devaluate, °depreciate, °vitiate, °belittle, °discredit, bring *or* call into disrepute: *That act of treachery will foul the family name for generations.* **15** °obstruct, °block, °choke, stop *or* plug *or* clog (up): *Dead leaves fouled the downpipe.* **16** ***foul up.*** **(a)** See **14,** above. **(b)** mismanage, °mishandle, °botch, °bungle, make a mess (of), °mess up, °spoil, °ruin, *Colloq* muff, *Brit* throw a spanner into (the works), *US* throw a monkey wrench into (the machinery); *Slang* °muck up, goof (up), °blow, °screw up, louse up, *Chiefly Brit* °bugger (up), *US and Canadian* snafu: *Give him a chance and he's sure to foul up. She fouled up my hi-fi.* —*n.* **17** °violation, °infringement, infraction, illegality: *The Rangers' forward has already been charged with two fouls.* —*adv.* **18** °afoul, in conflict, °in trouble, in violation: *Curshaw has fallen foul of the law again.*

foul play *n.* treachery, °chicanery, °perfidy, perfidiousness, duplicity, double-dealing, °deception, guile, °crime, sharp practice, skulduggery, dirty work *or* business, dirty trick(s); °murder, homicide, manslaughter; *Colloq* °hanky-panky: *The travelers suspected foul play when the airplane failed to arrive. The hacked up corpse suggested foul play to Detective Lemaître.*

found *v.* **1** °establish, °originate, °institute, °set up, °organize, °inaugurate, °start, °initiate, °create, °bring about, °develop: *The society was founded a hundred years ago.* **2** °base, °ground, °establish, °set, °build; °rest: *This charity was founded on love and concern for children.*

foundation *n.* **1** °basis, °base, substructure, understructure, underpinning, °bottom, foot, basement, °cellar: *This foundation is of stone.* **2** °basis, °base, °fundamental, (underlying *or* fundamental) principle, °grounds, °groundwork, °rationale, raison d'être, °purpose: *Their morality has its foundations in Judeo-Christian culture.* **3** founding, °establishment, instituting, °institution, °creation, origination, setting up, organizing, °organization, inauguration, °endowment: *We voted for the foundation of an institute to study the English language.*

founder[1] *n.* originator, °creator, °progenitor, °author, framer, °father, architect, °designer, builder, initiator, establisher, institutor: *Today we honor the founder of this great university.*

founder[2] *v.* **1** °sink, go down *or* under, go to Davy Jones's locker, be wrecked *or* destroyed: *Many a proud vessel has foundered on this rock.* **2** °fail, °miscarry, °collapse, come to nothing *or* naught, °fall through, abort, falter, °break down, °come to grief, °die: *After Alice left to get married, the business foundered.* **3** °trip, °stumble, °stagger, °lurch, °fall, topple (over *or* down), °collapse; go lame: *He drove the horse too hard and caused it to founder.*

foundling *n.* orphan, waif; °stray, °outcast: *We are trying to raise money for a hospital for foundlings.*

fountain *n.* **1** fount, °spring, font, jet, °spout, °spray, °well, wellspring, wellhead, fountainhead: *Bathing in the city fountains is forbidden.* **2** fount, °source, °origin, genesis: *Miss Corbell was a fountain of information on the subject of basketweaving.*

foxy *adj.* **1** °clever, °sly, cunning, °wily, °crafty, °tricky, guileful, °shifty, °devious, °slippery, °smooth, °slick, °artful, °resourceful, °ingenious, °calculating, °designing, plotting, °scheming, °disingenuous, °knowing, °shrewd, °sharp, °astute, °wise; foxlike, vulpine; *Colloq* cagey *or Colloq* cagy: *Be careful dealing with him, for he can be foxy.* **2** °attractive, alluring, °seductive, vampish, *Colloq* °sexy: *Margo is a very foxy lady.*

fracas *n.* **1** °trouble, °disturbance, commotion, °rumpus, °fuss, hubbub, °pandemonium, hullabaloo, °uproar, °disorder, °scramble, scuffle, °brawl, °roughhouse, rough-and-tumble, turmoil, °tumult, free-for-all, °riot, °fray, brouhaha, melee *or* mêlée, *Law* affray; *Brit* scrum, *US* brannigan; *Colloq* ruckus, punch-up, *Slang Brit* bovver: *It was football hooligans who caused the fracas after the game.* **2** °argument, °disagreement, °quarrel, °dispute, °discord, wrangle, altercation, squabble, spat, °tiff, °fight, °row, tussle, donnybrook, °brawl, *Colloq* barney, °scrap: *After the referee's ruling, a fracas broke out.*

fracture *n.* **1** °break, breakage, breaking: *An old fracture makes him walk with a limp.* **2** °break, °crack, °split, °rupture, °breach, °separation, cleavage, °division, °rift: *We can permit no fracture in the united front we present to the voters.* —*v.* **3** °break, °rupture, °crack, °split, °breach, °separate, °cleave: *He fractured three vertebrae in the accident.*

fragile *adj.* °frail, breakable, °brittle, frangible, °delicate, °dainty, °thin, °light, °slight, °weak, °feeble, °infirm, °decrepit; °tenuous, °shaky, °flimsy, °rickety, unsubstantial *or* insubstantial: *Because of her great age, her bones had become fragile. The argument in favor of acquittal is very fragile, indeed.*

fragment *n.* **1** °piece, °portion, °part, °chip, shard *or* sherd, °splinter, °sliver, °scrap, °bit, °speck, snippet,

°morsel, °crumb, °particle, °remnant, °shred, °snatch: *A fragment of the airliner was found twenty miles from the crash.* **2** ***fragments.*** smithereens; debris, *Literary disjecta membra*: *The car was blown to smithereens. I tried to pick up the fragments of my former life.* —*v.* **3** °shatter, °splinter, break *or* split (up), °explode, °disintegrate, come *or* go to pieces, °come apart: *I picked up the skull and it fragmented in my fingers.*

fragmentary *adj.* °disconnected, °piecemeal, °incomplete, scattered, °disjointed, °incoherent, °sketchy: *Accounts of the disaster are still fragmentary.*

fragrance *n.* fragrancy, °scent, °aroma, °smell, °odor, redolence, °perfume, °bouquet, balm: *The fragrance of orange blossoms filled the room.*

fragrant *adj.* °aromatic, odorous, °redolent, perfumed, balmy, odoriferous, ambrosial, sweet-scented, sweet-smelling: *A fragrant breeze wafted over the meadow.*

frail *adj.* **1** See **fragile.** **2** ailing, unwell, °ill, °sick, °sickly, °poorly, °thin, °skinny, °slight, °puny, °scrawny, wasting *or* fading away, languishing, °infirm, °feeble; °crippled, consumptive, phthisic: *She is too frail to see visitors.*

frailty *n.* **1** °weakness, °infirmity, delicate condition, feebleness, fragility, °delicacy: *The doctors are concerned about his frailty.* **2** susceptibility, °liability, suggestibility, impressionability, vulnerability; fallibility, °foible, °flaw, °defect, °imperfection, °fault: *You just said that to take advantage of my frailty. Her vanity is her greatest frailty.*

frame *n.* **1** framework, °shell, °form, skeleton, °support, chassis, framing, °structure, °fabric, scaffolding, construction: *We made a frame of sticks over which the canvas was stretched.* **2** °border, casing, case mounting, °mount, °edge, edging; °setting: *This picture would look best in a gold frame.* **3** °system, °form, °pattern, °scheme, schema, °plan, °order, °organization, framework, °structure, °construct, construction, °arrangement, blueprint, °design, layout, °composition, °context, °makeup, configuration: *The proposed new department does not fit into the present frame of the company.* **4** °physique, °build, bone structure, °body, skeleton, °figure: *He has an unusually large frame for a dancer.* **5** ***frame of mind.*** °mood, °humor, °state, °condition, °attitude, °bent, °disposition: *I am not in the right frame of mind to put up with your nonsense at the moment.* —*v.* **6** °construct, °build, put together, °assemble, °set up, °put up, °erect, °raise, elevate: *We framed the entire house in two days.* **7** °make, °fashion, °form, °mold, °carve out, °forge, °originate, °create, °devise, °compose, °formulate, put together, °conceive, °draw up, °draft, °shape, °block out, give form *or* shape to; contrive: *The founding fathers met to frame a new constitution.* **8** °enclose, °box (in); °set off: *I like the way you've framed that painting.* **9** °set up, °incriminate (fraudulently), °trap, entrap: *Did Dr. Crippen kill his wife or was he framed?*

frank *adj.* **1** °open, °free, °candid, °direct, °outspoken, unreserved, °uninhibited, °honest, °sincere, °genuine, °truthful, plain-spoken, °forthright, °downright, °explicit, unrestrained, unchecked, unconstrained, unrestricted, °unabashed: *I asked for a frank appraisal of my work and, unfortunately, he gave it.* **2** °candid, naive *or* naïve, guileless, °artless, °ingenuous, °innocent, °(open and) aboveboard, on the up and up, *Colloq* upfront, °on the level: *The boy's description of the events was completely frank.*

frantic *adj.* frenzied, °excited, frenetic, °nervous, °overwrought, °excitable, wrought-up, distracted, °distraught, °beside oneself, °hysterical, °wild, °berserk, °mad, running amok *or* amuck; °upset, °agitated, perturbed, at one's wit's end, °disconcerted, °confused; °hectic; *Colloq* in a state, in a tizzy, up the wall, in a dither, °out of one's mind, *Chiefly US and Canadian* discombobulated: *They are frantic because they haven't heard from Edmund in a week.*

fraternal *adj.* °brotherly, °platonic, °friendly, comradely, °idealistic, °intellectual: *He insists that his interest in her is purely fraternal.*

fraternity *n.* **1** community, °brotherhood, °crowd, °set, °clique, coterie, °circle, °society, *US* °club: *Rumors have been circulating in the academic fraternity for months.* **2** brotherliness, °kinship, °fellowship, camaraderie, sodality, comradeship, °friendship, °companionship, relatedness, closeness, °association, affiliation, °solidarity, °unity, esprit de corps, clannishness: *A sense of fraternity prevents them from betraying one another.* **3** °company, guild, °clan, °league, °club, °union, °society, °association: *She has joined the fraternity of legal clerks.*

fraternize *v.* consort (with), °associate (with), °socialize (with), go (around *or* round) with *or* together, spend time with *or* together, keep company (with), hobnob with, °mingle (with), °mix (with), take up with *or* together, keep up (with), °fall in with, rub shoulders (with), *Colloq* hang out (with *or* together), hang about *or* around with *or* together: *Soldiers are forbidden to fraternize, which means they are not allowed to fraternize with local people.*

fraud *n.* **1** °deception, °trickery, cheating, sharp practice, °chicanery, °deceit, swindling, double-dealing, duplicity, °artifice, °craft, guile, humbug, humbuggery, treachery, *Colloq* monkey business, funny business, °hanky-panky: *The company directors have been convicted of fraud.* **2** °trick, °hoax, °swindle, °deception, °cheat, °wile, °stratagem, °dodge, bilk, °ruse, °sham, °fake, °subterfuge, *Colloq* flimflam, *Slang* gyp, °rip-off, scam: *The investigation exposed extensive fraud in the handling of local government funds.* **3** deceiver, trickster, cheat(er), °impostor, °swindler, charlatan, humbug, sharper, shark, bilk(er), °quack, mountebank, fake(r), °pretender, bluff(er), confidence man, inveigler, defrauder; °scoundrel, °rogue, *Archaic* knave; *Colloq* con man *or* artist, phoney *or US also* phony, flimflam artist, flimflammer, *US and Canadian* fourflusher; *Slang US* barracuda: *He is a fraud who extracts protection money from the elderly.*

fraudulent *n.* **1** °fake, °counterfeit, forged, °false, falsified, °spurious, °imitation, °sham, pinchbeck, *Colloq* phoney *or US also* phony: *These banknotes are fraudulent.* **2** °deceitful, °dishonest, °criminal, °deceptive, °tricky, °artful, °crafty, double-dealing, duplicitous, °shifty, guileful, °sharp, *Colloq* °shady, °crooked, °bent: *Substituting paste for diamonds is a fraudulent act.*

fraught *adj.* **1** Usually, ***fraught with.*** filled *or* charged *or* packed with, loaded with, teeming *or* replete *or* overflowing with, oversupplied with, abounding *or* abundant in, attended *or* accompanied by: *The scene was fraught with emotion.* **2** °tense, °taut, stressful, °trying, °anxious, distressing, distressful, upsetting, °nerve-racking, °fretful, °strained, °traumatic: *Relations between the two countries became even more fraught.*

fray[1] *n.* °disturbance, °skirmish, °fight, °battle, °brawl, tussle, scuffle, °fracas, melee *or* mêlée, donnybrook, wrangle, °rumpus, °row, °quarrel, °dispute, altercation, *Law* affray, *Colloq* ruckus, punch-up: *He leapt into the fray and fought like a tiger.*

fray[2] *v.* °shred, wear (thin *or* threadbare), °wear out, °rub, abrade, °chafe, ravel, unravel, frazzle: *He saves his frayed shirts to wear at home.*

freak *n.* **1** °monstrosity, °monster, mutant, deformity: *The show features freaks like a two-headed calf and a bearded lady.* **2** anomaly, °rarity, °abnormality, irregularity, °oddity, °curiosity, rara avis, rare bird, *Brit* one-off, *Colloq* one-shot: *Biologists regard the albino giraffe as a freak of nature.* **3** whim, caprice, vagary, crotchet, °quirk, °eccentricity, °fancy, idiosyncrasy, °peculiarity: *The snow in New York in July was a freak.* **4** °enthusiast, °fan, °devotee, aficionado; °fanatic, °addict; *Colloq* buff, °fiend, nut: *She's been a jazz freak for years.* —*adj.* **5** freakish, freaky, °abnormal, anomalous, °extraordinary, °unique, °rare, atypical, °unusual, °odd, °queer, °strange, °exceptional, °bizarre, °weird, °unparalleled, °unforeseen, unexpected, unpredicted, unpredictable, *Brit* one-off, *Colloq* one-shot: *He claims he had a freak experience with a lion that had a thorn in its foot.*

free *adj.* **1** °at liberty, unfettered, unchained, unshackled, unconfined, untrammeled, unencumbered, unrestrained, unrestricted, unconstrained, °uncontrolled, freeborn, °independent, self-governing, self-governed, self-ruling, autonomous, °democratic, °sovereign: *These people want to be free. It's a free country.* **2** liberated, °at large, let go, °let off, emancipated, delivered, manumitted, set free, unshackled, unfettered, released, freed, °loose, °out, *Colloq* sprung, on the loose: *After ten years he was free at last.* **3** unoccupied, unengaged, °at liberty, not busy, °available, °accessible; °unused, °vacant, °empty, °spare, °extra, °uninhabited, untenanted: *Are you free for dinner tonight? Here is a free room we can use for the meeting.* **4** cost-free, free of charge, °complimentary, gratis, for nothing, without cost (or obligation), unencumbered, *Colloq* for free, °on the house: *The food was free but we paid for our drinks.* **5** °unattached, unfastened, untied, °loose: *Tie the free end of the rope round your waist.* **6** unasked-for, °unsolicited, °gratuitous, unbidden, °voluntary, °spontaneous, unconditioned, unconditional: *Let me give you some free advice.* **7** °generous, °lavish, °open, °liberal, munificent, unstinting, °bountiful, openhanded, unsparing; °charitable: *Fred is quite free with his donations to good causes.* **8** °relaxed, °casual, °informal, free-and-easy, °easy, °natural, unceremonious, *Colloq* laid-back: *He is quite free about letting me use his car.* **9** °open, °aboveboard, °honest, °direct, °outspoken, °uninhibited, unconstrained, unrestrained, °relaxed: *You can be free in your criticism.* **10** unhindered, °unimpeded, unencumbered, unhampered, unobstructed, allowed, permitted, °able, °clear, unrestricted, unregulated: *She's free to do as she pleases. The two countries have signed a free-trade agreement.* **11** ***free of.*** °rid of, exempt(ed) from, relieved of, safe from, not liable *or* subject to, immune from, unaffected by, °above, without, untouched by: *He thinks he is free of her influence.* —*adv.* **12** °freely, °openly, °at will, unrestrictedly, °loose; loosely: *Our dog runs free on the farm. Let the rope hang free.* **13** gratis, at no cost, free of charge, without charge: *They give new businesses advice free.* —*v.* **14** set free, set at liberty, enfranchise, °release, let go, °liberate, °let out, °let loose, unloose, unchain, unfetter, uncage; °emancipate, disenthrall, manumit; °pardon, parole, furlough: *Bail was paid and we were freed. Lincoln freed the slaves. He was freed after six years in prison.* **15** °disengage, untie, unbind, °loose, unfasten, °undo, unshackle, unlock, °open, °release, disentangle, °loosen, °detach, °extricate: *Free the end of that rope.* **16** °relieve, °rid, unburden, disburden, disencumber, unbosom; °rescue, °redeem: *Free yourself of any preconceived notions on the subject.*

freedom *n.* **1** °liberty, °independence, °self-government, self-determination, self-direction, autonomy: *Democracy is based on the freedom of the individual.* **2** °release, deliverance, °liberation, emancipation, manumission: *Will these people ever get their freedom?* **3** °exemption, °immunity, deliverance, °liberation, °relief: *All citizens should enjoy freedom from want.* **4** °range, latitude, °scope, °play, deregulation, noninterference, °discretion, °margin, free hand, °discretion; °facility, °ease, °license, °permission, °right, °privilege, °authority, authorization, °power, °carte blanche: *He must have the freedom to make decisions. I have the freedom to do as I wish.* **5** free time, °leisure, spare time: *Have you the freedom to join me for lunch?* **6** °candor, °honesty, openness, frankness, candidness, unrestraint, unconstraint, naturalness: *Freedom of speech is practiced here.* **7** boldness, overfamiliarity, audacity, audaciousness, forwardness, brazenness, °brass, °impertinence, °impudence, °disrespect, °arrogance, °presumption, presumptuousness, °nerve, °gall: *Where does she get the freedom to talk to you that way?*

freely *adv.* **1** candidly, frankly, °openly, unreservedly, without reserve, unrestrainedly, without restraint, unconstrainedly, without constraint, unceremoniously, plainly: *Please speak freely.* **2** °willingly, spontaneously, °readily, °voluntarily, on (one's) own, independently, of (one's) own accord, of (one's) own volition *or* free will: *I didn't send for her—she came to me freely.* **3** unrestrainedly, unrestrictedly, without restriction, without let or hindrance, without interference: *He was allowed to move freely about the island.* **4** liberally, lavishly, unreservedly, generously, unstintingly, openhandedly, ungrudgingly, munificently, °amply, plentifully, abundantly: *He has no money to donate but he gives freely of his time.* **5** °readily, °easily, smoothly, cleanly, unobstructedly: *With the obstacle removed, the water ran freely through the pipes.*

freeze *v.* **1** °chill, °refrigerate, ice, deepfreeze, flash-freeze, frost: *If you freeze the leftovers, they keep better.* **2** °solidify, congeal, °harden, °stiffen, ice up *or* over: *When the lake freezes, we can go skating.* **3** °fix, immobilize, °paralyze, stop (dead (in one's tracks)), °stay, °pin, °transfix, Gorgonize; become fixed, stand (stock-)still *or* motionless; °peg, °stick, °set: *She froze him with an icy stare. He froze to the spot in horror. Rates of exchange are no longer frozen.* **4** ***freeze out.*** °exclude, debar, °ban, °reject, °ostracize; °eject, drive away *or* out, °expel, force out: *When he tried to join the club he was frozen out.* —*n.* **5** frost, *Brit* freeze-up, *US* ice-up, deepfreeze: *There will be a freeze tonight in northern counties.* **6** °fix, immobilization: *The government has put a freeze on wages.*

freezing *adj.* °frigid, °icy, arctic, frosty, boreal, hyperboreal, numbing, Siberian, °polar, glacial, ice-cold, °wintry, bone-chilling, °bitter, °biting, bitter(ly) cold, perishing, °cold, °chill, tooth-chattering; chilled to the bone, frozen, shivering, *Archaic* frore: *The rescue was performed under freezing conditions. We were freezing in our thin jackets.*

freight *n.* **1** °transport, transportation, °carriage, conveyance, shipping, shipment, freightage, °delivery: *The charges for freight are included.* **2** °goods, °cargo, tonnage, freightage; °load, boatload, shipload, lorryload, °haul, consignment, payload: *How much freight passes through here?*

frenzy *n.* **1** °excitement, °agitation, °fury, fever, °furor, °passion, turmoil, °transport: *The crowd was whipped into a frenzy at the match.* **2** °distraction, °paroxysm, °seizure, °outburst, °bout, °fit: *She went into a frenzy of despair after the death of her family.*

frequent *adj.* **1** °recurrent, recurring, °habitual, °regular, °familiar, °everyday, °customary, °usual, °ordinary, °normal, °common, repeated, iterative, reiterative, °persistent, continuing, °continual, °constant; °many, numerous, countless, innumerable: *She was a* *frequent visitor at our house. She paid us frequent visits.* —*v.* **2** °haunt, °patronize, °visit, °resort to, go to *or* attend regularly, *Colloq* hang out *or Colloq* around at: *Yes, Inspector, I used to frequent the pub called The Saracen's Head.*

frequently *adv.* **1** °often, regularly, continually, °repeatedly, over and over (again), again and again, °a lot, many times, many a time, time after time, time and (time) again, *Archaic* oftentimes, ofttimes: *She has visited me frequently during the past weeks.* **2** °often, habitually, customarily, regularly, °usually, °ordinarily, °generally, commonly, every so often, many a time, as often as not, *Archaic* oftentimes, ofttimes: *He frequently stops at The Golden Hind after work.*

fresh *adj.* **1** °new, today's, °brand-new; °(most) recent, late(st): *Don't you love the smell of fresh bread? This is the product of fresh research.* **2** °new, °modern, up-to-date, °novel, °original, newfangled, °unusual, unconventional, °different, °alternative, °unorthodox: *We are seeking a fresh approach.* **3** °raw, °inexperienced, untested, °unsophisticated, °green, °untried, °unfledged, °immature, untrained, naive *or* naïve, °callow, (still) wet behind the ears, *Brit* still in nappies; *US* still in diapers: *These troops are too fresh to send into battle.* **4** additional, °new, °further, renewed, °extra, °supplementary: *We need a fresh supply of paper towels.* **5** °alert, refreshed, °vigorous, °energetic, invigorated, spry, °lively, full of vim and vigor, fresh as a daisy, °keen, °bright, °brisk, °active, *Colloq* bright-eyed and bushy-tailed: *Put the horses through the dressage while they are fresh.* **6** °wholesome, °healthy, °well, refreshed, °glowing, °fair, °rosy, ruddy, blooming, °flourishing: *She came back to work fresh from a week's rest.* **7** °moderate, °brisk, °strong; °cool, °clean, °pure, °clear, unpolluted: *With a fresh breeze and good visibility, sailing was ideal all weekend. Open the window for some fresh air.* **8** °bold, °impudent, °impertinent, °brazen, °brassy, °forward, °disrespectful, saucy, °pert, °cheeky, °presumptuous, °insolent, °rude, *Colloq* smart-aleck *or* smart-alecky, *US* sassy, °flip: *That fresh kid called her an old bag.*

freshen *v.* **1** °strengthen, °increase, blow harder: *The wind began to freshen as we sailed past the headland.* **2** Sometimes, ***freshen up.*** invigorate, °revive, °refresh, °enliven, (re)vitalize, °stimulate, titivate *or* tittivate, °rouse, liven up: *I'll just freshen up before dinner. Freshen these flowers by changing the water.* **3** ventilate, air out, deodorize, °purify: *We freshen the rooms by opening all the windows.* **4** °strengthen, °spike, °lace, °fortify: *May I freshen your drink?*

fret *v.* **1** °worry, be concerned, agonize, lose sleep, be upset *or* distressed *or* anxious *or* disturbed, °grieve, °brood, whine, °fuss, °complain, *Colloq* °stew, tear one's hair: *Your kitten is safe now, so stop fretting.* **2** °worry, °concern, °distress, vex, °annoy, °irritate, °torment, °provoke, *US* °rankle: *She fretted herself about Henry's health.*

fretful *adj.* °irritable, vexed, ill-tempered, bad tempered, °peevish, edgy, °cross, °petulant, °testy, °touchy, tetchy, splenetic, irascible, choleric, crabby, fractious, °short-tempered, grumpy, sulky, °moody, °faultfinding, carping, °querulous, whining, complaining, captious, ill-natured, °disagreeable, °impatient, °snappish, °waspish, °short, °abrupt, °curt, *US and Canadian* °cranky: *Are you always so fretful before breakfast?*

friction *n.* **1** abrasion, rubbing, abrading, chafing, fretting, attrition, scraping, °grating, °erosion: *Friction is always accompanied by heat.* **2** °disagreement, °discord, °conflict, contention, °dispute, °dissension, disharmony, °controversy, dissent, bickering, °argument, wrangling, ill feeling, °ill will, bad blood, °animosity, °rivalry, °hostility, °antagonism, °strife: *Politics have always been a source of friction between them.*

friend *n.* **1** °(boon) companion, °partner, °comrade, crony, °familiar, confidant(e), °intimate, Achates, alter ego, °ally, compeer; °acquaintance, °playmate, pen pal *or Brit also* pen friend; *Colloq* °chum, °pal, *Brit* cocker, *Chiefly Brit and Australian and New Zealand* °mate, *Australian* cobber, *US and Canadian* (bosom) buddy, *SW US* compadre; *Slang Brit* (old) china: *She is spending the weekend with some friends.* **2** °ally, °associate, °fellow, confederate, °colleague, co-worker, confrère, compatriot, consociate, *US* °cohort: *A friend from the office is coming to dinner.* **3** roommate, bunkmate, flatmate, soul mate, bedfellow, °lover, °sweetheart, °escort; °girl, °woman, girl friend, concubine, °mistress, *Old-fashioned* doxy; °man, boyfriend, *Old-fashioned* beau; *Colloq US* alternative other, POSSLQ (= 'Person of the Opposite Sex Sharing Living Quarters'), roomie; *Slang* °baby, moll, sugar daddy, *US* °squeeze, °twist, *Chiefly Brit* bird: *Chris, why don't you introduce us to your new friend?* **4** °benefactor, benefactress, °patron, °supporter, °advocate, adherent, °backer, °financier, Maecenas; angel, investor: *He has long been associated with the Friends of the Library.*

friendly *adj.* **1** °amicable, congenial, °sociable, companionable, comradely, convivial, °familiar, well-disposed, °close, on good terms, simpatico, °comfortable, °at home, °neighborly, clubby, °fraternal, °brotherly, sisterly, *Colloq* °chummy, pally, °thick, *Brit* matey, *US* palsy-walsy, buddy-buddy: *We've always been friendly with the people next door.* **2** °amiable, °affectionate, loving, °demonstrative, °cordial, warmhearted, °warm, °genial, °agreeable, °good-natured, °pleasant, °kind, °kindly, kindhearted, °agreeable, affable, approachable, °accessible, unreserved, °open, *Brit* clubbable: *Clara can be very friendly when she wants to be.*

friendship *n.* **1** amity, congeniality, sociability, companionability, comradeship, °fellowship, conviviality, °familiarity, closeness, neighborliness, °harmony, clubbiness, °fraternity, °brotherhood, sisterhood, °alliance: *I hope that this affair will not affect the friendship between our countries.* **2** friendliness, amiability, amicability, °esteem, °warmth, °devotion, °affection, fondness, °attachment, °(deep) regard, °rapport, intimacy, °love: *What can I do to regain your friendship?*

fright *n.* **1** °fear, °alarm, °terror, °dread, °horror, °panic, trepidation, °dismay, consternation, apprehension, *Colloq* (blue) funk: *He almost dies of fright at the sight of blood.* **2** °scare, °shock: *Here, sip this brandy; you've had a terrible fright.* **3** °specter, °monster, eyesore, *Colloq* °sight, °mess: *He looked a perfect fright when I saw him, right after the accident.*

frighten *v.* °terrify, °scare, °alarm, °panic, °startle, °shock, °petrify, °horrify, °dismay, °appall, °unnerve, °distress, °daunt, cow, °intimidate, *Colloq* scare out of one's wits, make one's hair stand on end, scare the (living) daylights [etc.] out of, scare stiff, *Brit* °put the wind up (someone), put the frighteners on (someone): *She was frightened by the neighbors' vicious dog. Are you frightened of airplanes?*

frightening *adj.* °terrifying, alarming, °startling, °shocking, petrifying, horrifying, dismaying, appalling, unnerving, dire, distressing, daunting, intimidating, °formidable, °frightful, °fearful, hair-raising, °harrowing, °dreadful, *Colloq* °scary, spooky: *Skydiving can be a frightening experience. The sight of their mangled bodies was frightening.*

frightful *adj.* **1** See **frightening.** **2** °awful, °dreadful, °terrible, °disagreeable, °atrocious, °abhorrent, °loathsome, °grisly, °ghastly, °lurid, °macabre, °horrible, horrifying, horrid, horrendous, °nasty, °hideous, °vile, unspeakable, nauseating, °nauseous, °repugnant, °repulsive, °shocking, °revolting, °abominable, °offensive, °ugly: *Brixton police report a frightful crime of decapitation.*

frightfully *adv.* °awfully, °very, °extremely; amazingly, surprisingly: *We saw a frightfully good play last night.*

frigid *adj.* **1** °cold, arctic, frosty, frozen, glacial, °icy, hyperboreal, °polar, bone-chilling, boreal, Siberian, °freezing, °wintry, °chilly, °chill, *Archaic* frore: *We were huddled in that frigid alpine hut for two days.* **2** °cold, °cool, °coldhearted, °forbidding, austere, unemotional, unfeeling, °stiff, °rigid, prim, °strait-laced, °stony, °callous, °steely, obdurate, °thick-skinned, impervious, °inaccessible, °remote, °unapproachable, unfriendly, °standoffish, °haughty, °aloof, °reserved: *His behavior to his ex-wife has always been frigid.* **3** °unapproachable, unresponsive, °impassive, °passive, °indifferent, °cold; °impotent: *She says he's impotent and he insists that she's frigid.*

frill *n.* **1** trimming, °decoration, °ornament, furbelow, °flounce, °ruffle: *This dress would be suitable without the frill.* **2** ornamentation, frippery, falderal *or* falderol *or* folderol, frou-frou, showiness, °ostentation, °embellishment, °luxury, trimming, °extra, °addition, °superfluity, °gewgaw, °(bit of) paraphernalia, *Colloq US* foofaraw, bells and whistles, *Slang* jazz: *They ordered a computer system that has every imaginable frill.*

fringe *n.* **1** trimming, °edge, edging, °border, °frill, °flounce, °ruffle, purfle, purfling, ruff, ruche *or* rouche, rickrack *or* ricrac, °ornament, °decoration, furbelow: *The curtains would look better without that pink fringe.* **2** °border, °perimeter, °edge, °boundary, °bounds, °periphery, °limits, °margin, °outskirts, march(es); *Technical* fimbria: *Most of the people who live on the fringe of the city commute to work.* —*v.* **3** °edge, °border, °trim, °surround: *The grounds are fringed with trees.*

frisk *v.* **1** °caper, gambol, °cavort, °frolic, °skip, °trip, romp, curvet, °leap, °dance, °prance, °play, rollick: *She was frisking about like a lamb.* **2** °search, °inspect, °examine, °check (out), °go over: *At the airport, we were frisked for weapons.*

frisky *adj.* °lively, °frolicsome, rollicking, °playful, °active, °animated, (high-)spirited, coltish: *Tell Frances to calm down—she's getting a bit too frisky.*

fritter *v.* ***fritter away.*** squander, °waste, °idle away, misspend, °dissipate: *Stop frittering away your time watching television. She frittered away every penny of her inheritance.*

frivolous *adj.* **1** °trifling, °inconsequential, unimportant, trivial, nugatory, °insignificant, °minor, °petty, °paltry, °niggling, °peripheral, °superficial, °worthless, *Colloq* °small-time, *Brit* twopenny *or* tuppenny, two-a-penny, *US* two-bit, penny-ante, nitty-gritty: *Don't waste my time with frivolous details.* **2** °scatterbrained, bird-brained, °silly, featherbrained, °irresponsible, °flippant, °casual, °flighty, °giddy, °foolish, °childish, °puerile; airy, °light, °slight, *Brit* airy-fairy, *Colloq* °flip: *Try to be serious and less frivolous about such important matters.*

frolic *n.* **1** °merriment, merrymaking, °gaiety, °sport, °fun (and games), high jinks, jollity, °mirth, jollification, °festivity, °celebration, °revelry, °play, horseplay, *Colloq* skylarking, partying: *After we won the cup there was great frolic in the town.* **2** romp, °party, °spree, °revel, gambol, °caper, gambado, antic; escapade, °prank: *The noise of our frolics echoed across the square.* —*v.* **3** °frisk, °cavort, °caper, skylark, gambol, rollick, romp, cut capers, curvet, °play, °skip, °sport, have fun, *Colloq* °party, make whoopee, horse around *or* about: *Stop frolicking about and get down to work.*

frolicsome *adj.* °playful, °merry, °frisky, °sportive, °gay, °lively, °sprightly, °animated, °spirited, coltish: *Kathy may seem frolicsome but she has her sober side as well.*

front *n.* **1** °face, façade, °facing, forepart, anterior; obverse: *The front of the door has a painting on it. This dress buttons up the front.* **2** frontage, forefront: *The front of the property measures only forty feet.* **3** °beginning, °head, fore, vanguard, forefront, van: *At the front of the parade marched the mayor.* **4** °bearing, demeanor, mien, °air, °face, countenance, façade, °mask, °expression, °show, °appearance, °aspect, °look, °exterior: *Despite her grief, she put on a brave front at the wake.* **5** °disguise, °cover, °guise, °mask, cover-up, °show, °pretext, façade: *The restaurant was merely a front for a narcotics operation.* **6** °movement, °organization, °league, bloc, °party, °group, °faction, wing: *The new popular front was formed out of a coalition of several opposition groups.* **7** haughtiness, overconfidence, °effrontery: *That salesman frightens away business by showing so much front.* **8** ***in front.*** °first, °leading, °ahead, to the fore, in the forefront, in the vanguard *or* van, °in advance, in the lead, °before; °winning: *In this picture the man in front is my father. My horse was in front all the way.* **9** ***up front.*** **(a)** See **8,** above. **(b)** °open, straightforward, °honest, °direct, °forthright, °frank, °candid: *Why can't you be up front instead of conspiring against me?* —*adj.* **10** °first, °advance, °foremost, °leading, °head; °main: *The front carriage was smashed in the train wreck. Enter by the front door.* —*v.* **11** °overlook, °face, look out on *or* toward, be opposite: *Our house fronts the river. The flat fronts on the street.* **12** ***front for.*** act for, °represent; °substitute for, °replace: *I hate formal affairs and hoped that you might front for me.*

frontier *n.* front line; °border, °boundary, bound(s), marches, (far) reaches, limit(s), °pale, extreme(s), bourn: *We'll need our passports to cross the frontier into Italy.*

froth *n.* **1** °foam, spume, suds, °lather, °bubbles; °head: *The froth tickles my nose.* **2** trivia, °rubbish, °nonsense, twaddle, °babble, °gibberish, °drivel, *Colloq* °hot air, gas, °gab, piffle: *His sermons were just so much froth.* —*v.* **3** °foam, spume, °bubble, °fizz, effervesce, aerate: *You are supposed to drink the medicine while it is frothing.* **4** °foam, salivate; °lather: *He was so angry he began to froth at the mouth.*

frown *v.* **1** °scowl, glower, °glare, knit one's brows, grimace, give a dirty look, *Brit* lour *or* lower, *US* °lower: *Don't frown so much or you'll get wrinkles.* **2** ***frown on or upon.*** °disapprove (of), (look on *or* regard *or* view with) disfavor, discountenance, look down on *or* upon, look askance at, not take kindly to, not think much of, look disapprovingly upon, *Colloq* take a dim view of, be turned off by: *My parents frown on my seeing you.* —*n.* **3** °scowl, glower, °glare, grimace, dirty look, *Brit* lour *or* *Brit* lower, *US* °lower: *Her forehead is always furrowed by a frown.*

frugal *adj.* **1** °thrifty, °sparing, economic(al), °careful, °prudent, °provident, °saving, °conservative, conservational, °moderate: *Mother had to become quite frugal when father fell ill.* **2** parsimonious, °penurious, penny-pinching, cheeseparing, °mean, °miserly, stingy, niggardly, tight(-fisted), close(fisted), hand-to-mouth: *He led a frugal existence on his earnings as an artist.* **3** °meager, °paltry, °poor, skimpy, scant(y), °small, °negligible, piddling: *His frugal meal consisted of crusts of bread.*

fruit *n.* Often, **fruits.** product(s), result(s), revenue(s), outgrowth, °outcome, consequence(s), return(s), advantage(s), benefit(s), profit(s), emolument, °payment, °income, compensation, recompense, desert(s): *He didn't live to enjoy the fruit of his work.*

fruitful *adj.* **1** °productive, °fertile, °prolific, fecund; fructiferous, frugiferous, fructuous: *The soil in this valley is extremely fruitful.* **2** °effective, °worthwhile, well-spent, °profitable, °successful, °useful, °rewarding, °advantageous, °beneficial, °productive, °fertile: *We had a fruitful meeting and accomplished a great deal.* **3** °plentiful, °abundant, bounteous, °bountiful, °prolific, plenteous, copious, °luxurious, °rich, °flourishing: *We expect another fruitful harvest this year.*

fruition *n.* °realization, °fulfillment, °consummation, °achievement, °success, materialization, °maturity, ripeness, maturation, °completion; °perfection: *I saw my plans for the town brought to fruition.*

fruitless *adj.* °barren, unfruitful, unproductive, °worthless, °bootless, °futile, °pointless, °useless, °vain, °idle, unavailing, °ineffectual, °ineffective, °unprofitable, for naught, to no avail, °unsuccessful, unrewarding, abortive: *Our search for a new manager has so far been fruitless.*

frustrate *v.* **1** °thwart, °foil, °stymie, °block, baffle, °check, balk *or* baulk, °disappoint, °discourage, °forestall, °prevent, °stop, °halt, °cripple, °hinder, °hamper, °impede, hamstring, °defeat, °counteract, °neutralize, nullify, °counter, °fight off, °repel, °repulse: *So far, we have been able to frustrate takeover bids for the company.* **2** °discourage, °disappoint, °upset, °exasperate: *I feel frustrated because my mother won't let me help pay her bills.*

fuel *n.* **1** tinder, combustible, kindling; fossil *or* nuclear fuel: *You must pay a lot for fuel to heat that big house.* **2** ammunition, °encouragement, stimulus, °incitement, °provocation: *Her teasing him only provided more fuel for his passion.* **3** °nourishment, nutriment, °sustenance, °food, nutrition: *The body needs more fuel on a cold day.* —*v.* **4** °nourish, °feed, °sustain; °stimulate, °encourage, °incite, °provoke, °inflame, exacerbate, °excite: *U-235 is used to fuel the reactor. Frustration fuels the flame of desire.*

fugitive *n.* **1** °runaway, escapee, °deserter, °refugee, *Archaic* runagate: *Many fugitives from political oppression seek asylum in the UK.* —*adj.* **2** fleeing, escaped, running away, °runaway: *The police are able to devote little time to tracing fugitive children.* **3** °fleeting, °passing, °brief, °short-lived, transitory, °transient, ephemeral, evanescent, °momentary, °volatile, fugacious: *She once entertained the fugitive notion of becoming an opera singer.*

fulfill *v.* **1** °bring about, °achieve, °accomplish, bring *or* carry to completion, °carry out, °complete, consummate, °discharge, live up to, °abide by, °observe, °realize, °effect, bring *or* carry off, °carry through, °keep, °satisfy, do, °perform, °execute, °effectuate, °achieve: *I trust that you will fulfill all your obligations. Will she ever fulfill her ambition to be a virtuoso pianist?.* **2** °answer, °satisfy, °meet, °implement, look *or* see to, conform to *or* with, comply with, °obey: *This hammer will fulfill my needs for the moment. Will your new assistant be able to fulfill your requirements?*

fulfillment *n.* °completion, °consummation, °performance, carrying out *or* through, °discharge, °realization, implementation, °execution, °accomplishment, compliance, conformity *or* conformance, making good, °meeting, °satisfaction, answering, °achievement: *You will be paid upon fulfillment of the contract.*

full *adj.* **1** filled, °replete, brimming, brimful, °packed, jampacked, °congested, °loaded, bursting, chockablock, chock-full *or* choke-full *or* chuck-full, jammed, crammed, °solid, well-supplied, crowded, stuffed; gorged, saturated, sated, satiated: *The tank is full of gas. Her Christmas stocking was full of toys. I'm full and cannot manage another bite.* **2** °complete, °thorough, °detailed, °comprehensive, °total, (all-)inclusive, °broad, °extensive, all-encompassing, °exhaustive, plenary: *The police are demanding a full investigation.* **3** °complete, °entire, °whole: *The recipe calls for a full dozen egg yolks.* **4** utmost, greatest, °maximum, highest, °top; °extreme: *Full speed ahead.* **5** °wide, °ample, °generous, °broad, copious, loose(-fitting): *The bolero jacket is offset by a full skirt.* **6** occupied, engrossed, °absorbed, immersed, °preoccupied, obsessed, consumed, °engaged, °concerned: *She's entirely too full of herself to pay any attention to us.* **7** filled *or* rounded out, round(ish), well-rounded, °plump; °robust, °shapely, well-shaped, curvaceous, °buxom, busty, °voluptuous, full-bodied, well-proportioned, well-built, *Slang* stacked, *Brit* well-stacked, *US* zaftig, built: *His face is rather full, so wrinkles don't show. Her figure is what I'd call full.* **8** unrestricted, nonrestricted, unconditional, °unqualified: *Payment of dues entitles you to full membership privileges.* **9** °sentimental, °emotional, overflowing: *His heart was so full he could say no more.* **10** unobscured, unshaded, undimmed, °open, °broad, °bright, shining, °brilliant, °dazzling, °glaring, °intense, blazing, blinding; °harsh, °vivid, revealing: *We emerged into full daylight. In the full light of day, the plan seems unworkable.* **11** °powerful, °resonant, °rich, °deep, °loud: *His full bass voice is unmistakable.* **12** °complete, °whole, °entire; °comprehensive, uncut, °unabridged, °intact, unshortened, unbowdlerized, uncensored: *We were to receive full pay for a half day's work. Is this a full deck of cards? I have a full set of the original engravings.* —*adv.* **13** fully, °completely, °entirely, °wholly, °thoroughly, °altogether: *That is not a full-grown horse.* **14** squarely, °directly, °right, °precisely, °exactly, bang, *Colloq* °slap, smack: *The ball hit him full in the eye.* **15** °very, °perfectly, °exceedingly, °quite, *Slang* damned: *You know full well why I am here.* —*n.* **16** °maximum, greatest degree, fullest: *The moon will be at its full tonight.* **17** ***in full.*** °completely, fully, °entirely, °wholly, °thoroughly, in its entirety, °totally, *in toto*: *Copy this report in full.* **18** ***to the full*** *or* ***fullest.*** °completely, fully, °quite, °thoroughly, to the utmost, to the greatest *or* fullest extent; a great deal, greatly, hugely, enormously: *Gordon really enjoys life to the fullest.*

fumble *v.* **1** °grope, °feel, °stumble: *She fumbled about for the light switch.* **2** °mishandle, °drop, muff, °bungle, °botch, *Colloq US* bobble, flub: *Bosworth fumbled the ball and Fernpath scored again.*

fume *v.* **1** °seethe, °smolder, °chafe, °boil, °rage, °storm, °rant, °flare up, °bluster, lose one's temper, °explode, *Colloq* get steamed (up) (over *or* about), lose one's cool, flip one's lid, °flip (out), °fly off the handle, hit the roof, raise the roof, blow one's top *or* stack, get hot under the collar, blow a gasket, go off the deep end: *She was fuming because I was five minutes late.* **2** smoke: *The mixture fumes if you add these ingredients.* —*n.* **3** Usually, ***fumes.*** smoke, °vapor, effluvium, gas, °exhalation, °exhaust; °smell, °odor, °aroma, °fragrance, °stench, stink, miasma; °pollution, smog: *The fumes from O. C. Cabot's cigar spread through the building.*

fumigate *v.* °disinfect, °cleanse, °purify, sanitize, °sterilize, decontaminate: *There have been no insects since the kitchen was fumigated.*

fun *n.* **1** °merriment, merrymaking, °gaiety, °glee, jollity, °mirth, °cheer, high spirits, °delight, °frolic, °festivity, high jinks; °amusement, °diversion, °sport, °enjoyment, °recreation, °entertainment, °pastime, °joy, °pleasure, *Colloq* (making) whoopee: *Your party was great fun. We always have fun when we're together. Want to have some fun?* **2** tomfoolery, horseplay, joking, playfulness, clowning, pranks, °sport, jesting, jocularity, °nonsense, fooling around *or* about, *Colloq* skylarking: *This is no time for fun—we have to catch a train.* **3** ***in*** *or* ***for fun.*** jokingly, teasingly, in jest, facetiously, with tongue in cheek, playfully, as a lark, for a joke *or* gag; not seriously: *In fun, we told him that he had missed the last ferry.* **4** ***Like fun!*** Under no circumstances!, No way!, *Colloq* Like hell!, *US* No way, Jose!: *Like fun will I go swimming in the nude!* **5** ***make fun of.*** °poke fun at, °tease, °deride, °(hold up to) ridicule, scoff at, °lampoon, °parody, °satirize, make sport *or* game of, °taunt, °gibe, °rag, *Colloq* kid, rib, *Brit* °send up: *Kevin is always making fun of people by imitating them.*

function *n.* **1** °purpose, °aim, °use, role *or* rôle, raison d'être, °responsibility, °mission, °charge, °concern, °business, °province, °duty, °job, °occupation, °work, °office, °task, chore, °assignment, commission, °activity: *A knife's function is to cut; a guard's is to stand watch.* **2** °reception, °gathering, °affair, °party, dinner, °banquet, °gala, °ceremony, °formality, °rite, °ritual; °occasion, °event: *We are attending a function at the embassy tonight.* —*v.* **3** °act, °operate, °perform, °behave, °work, °go, °run: *Are you sure the engine is functioning properly?* **4** °serve, take the role *or* rôle of, act the part of, act as, work as: *Dunmow will function as host in my absence.*

functional *adj.* **1** utilitarian, °useful, °serviceable, °practical, °practicable, functioning, working; °essential, °important, °effective: *This gear is a functional part, not mere decoration.* **2** working, operating, operational, °running, °going: *Is this telephone functional?*

functionary *n.* °official, commissioner, bureaucrat, officeholder, °officer: *She is some sort of functionary at the Ministry of Information.*

fund *n.* **1** °supply, °stock, °reserve, °store, °pool, °cache, reservoir, repository, °mine: *Mallison is a veritable fund of information about art.* **2** Often, ***funds.*** °money, °(hard) cash, ready money, °assets, °means, °wealth, °resources, wherewithal, °savings, °capital, nest egg, °endowment, *Colloq* °loot, lucre, pelf, °green, bread, dough, *Brit* °ready, °readies, lolly, *US* bucks, °scratch: *He has the funds to buy out his partners. Have you contributed to the fund for indigent lexicographers?* —*v.* **3** °finance, °back, capitalize, °stake, °support, °pay for, endow, °grant, °subsidize: *The company has funded a number of new businesses.*

fundamental *adj.* **1** °basic, °rudimentary, °elementary, °primary, °main, °prime, °first, °principal, underlying, °cardinal, °central, °essential, quintessential, constitutional, °inherent, °intrinsic, °important, °crucial, °critical, °organic, °vital: *Our fundamental aim is to provide employment.* —*n.* **2** °principle, °law, °rule, axiom, °essential, °element, sine qua non, cornerstone, °keystone: *Freedom of speech is a fundamental in a democracy.*

funeral *n.* obsequies, exequies; °burial, interment, sepulture, entombment, inhumation; cremation: *It is fitting to wear black when attending a funeral.*

funereal *adj.* °grave, °solemn, °sad, °unhappy, morose, °somber, °mournful, °doleful, °sorrowful, °melancholy, °gloomy, lugubrious, °dismal, °grievous, depressing, °dreary, woeful, °dark, sepulchral: *Why are you in such a funereal mood?*

funny *adj.* **1** comical, °humorous, °comic, °ludicrous, laughable, °ridiculous, risible, waggish, sidesplitting, °hilarious, °uproarious, jocular, jocose, °merry, droll, facetious, °witty, °farcical, slapstick, °zany; amusing, °entertaining, diverting, *Colloq* °hysterical: *A comedian's material may be funny, but timing is everything.* **2** °peculiar, °odd, °unusual, °curious, °strange, °mysterious, mystifying, °puzzling, °queer, °weird, °bizarre, °remarkable, unconventional, °eccentric, *Slang* off the wall: *She had a funny expression on her face. If you do love me, you have a funny way of showing it.*

furious *adj.* **1** °angry, enraged, raging, infuriated, fuming, incensed, irate, maddened, °mad, boiling, wrathful, provoked, °beside oneself, up in arms, in high dudgeon, on the warpath, foaming at the mouth, *Literary* wroth, *Colloq* steaming, livid, in a tizzy, *Slang* up the wall: *Mark was furious at Betty for leaving without him.* **2** °fierce, °wild, °violent, °savage, °intense, unrestrained, °frantic, frenzied: *A few made a furious attempt to jump off the burning ship.*

furnish *v.* **1** °supply, °provide, °afford, °equip, °outfit, fit (out *or* up), rig (out *or* up), °provision, °give, stock up, *Colloq Brit* kit out *or Colloq Brit* up: *The trading post furnished what we needed for a month in the wilderness. Can you furnish me with a blanket?* **2** °decorate, °equip: *She received a commission to furnish a ten-room house.*

furniture *n.* **1** furnishings, household (goods); movables, chattels, °paraphernalia, °effects, °possessions, °belongings, *Colloq* °gear, °things, °stuff, *Slang* shit: *There wasn't any furniture in the room, not even a chair.* **2** °fittings, fitments, °equipment, fixtures, °apparatus, °devices, °tackle, °tack, °trappings, °gear, accoutrements *or US also* accouterments, accessories, appliances, *Colloq Brit* clobber: *Our new front door has some fine brass furniture.*

furor *n.* **1** °uproar, °outburst, °tumult, commotion, turmoil, brouhaha, ado, hurly-burly, to-do, hubbub, °stir, °fuss, °disturbance, °excitement: *The deportation of the student caused a furor among his supporters.* **2** °rage, °craze, °mania, °vogue, °enthusiasm, °obsession, °fad: *The furor for antiques has driven up prices enormously.*

furrow *n.* **1** °groove, °channel, °rut, trench, °track, ditch, gutter, trough, fosse, fissure, sulcus, sulcation, flute, °score, °cut, °gash, °scratch, °line; °wrinkle, crease, corrugation, crow's-feet, *Technical* sulcus: *He can't even plow a straight furrow. The furrows in his brow come from worry.* —*v.* **2** °groove, °channel, flute, °score, °cut, °gash, °scratch; °plow, harrow: *This rock was furrowed by glacial action.* **3** °wrinkle, crease, corrugate, °knit, °pucker, crinkle: *A frown furrowed his forehead.*

further *adj.* **1** more, additional, another, other, °new, supplemental, °supplementary, °accessory, °auxiliary, °extra, °spare, °fresh: *Further investment by our group is out of the question.* **2** farther, more distant *or* remote: *Some day, we may travel to the further reaches of the galaxy.* —*adv.* **3** furthermore, °besides, °moreover, too, also, additionally, °in addition, over and above, beyond, °above, what is more, °to boot, °yet, then (again), again: *She told him he was rude and, further, ugly as well.* **4** farther, at *or* to a greater distance *or* depth: *She has looked into the matter much further than I.* —*v.* **5** °advance, °promote, °favor, push *or* urge onward(s) *or* forward(s), °forward, °foster, °back, °patronize, °support, °help, °assist, °aid: *What can I do to further your efforts?*

furtherance *n.* °promotion, advancement, °pursuit, °backing, boosting, °boost, fostering, championing, championship, advocating, advocacy, °patronage, °support, °help, °aid, °assistance, succor: *The furtherance of antipollution laws is all to the good.*

furtive *adj.* **1** °secret, °private, °secretive, clandestine, °surreptitious, °stealthy, underhand(ed), covert, °hidden, conspiratorial, skulking, °deceitful, under-the-table, under-the-counter, huggermugger, *Colloq* °sneaky: *She gave me a furtive look to let me know she* *was on to me.* **2** °sly, °foxy, cunning, °crafty, °wily, °shifty, untrustworthy, *Colloq* °sneaky: *His furtive maneuvers mark him as one not to be trusted.*

fury *n.* **1** °anger, °rage, wrath, °indignation, ire, choler, °rancor: *Hell hath no fury like a woman wrong'd.* **2** impetuosity, ferocity, savagery, vehemence, fierceness, tempestuousness, turbulence, °violence: *How can we stem the fury of his attack?* **3** virago, °shrew, spitfire, hellcat, termagant, vixen, she-devil, °hag, °witch, °bitch, *Archaic* beldam: *Some of the furies from the women's liberation movement descended on him.*

fuse *v.* °blend, °merge, °unite, °combine, °compound, °mix, commingle, coalesce, flow *or* come together, consolidate, °amalgamate; °melt: *The metals fused at a relatively low temperature.*

fuss *n.* **1** °bother, pother, dither, °fluster, °flurry, °fret, commotion, ado, bustle, to-do, °excitement, furor *or Brit* furore, °unrest, °(deal of) trouble, disquiet, °upset, °stir, °uproar, °disturbance, °stir, hubbub, °agitation, brouhaha, *Colloq* hoo-ha, °flap, stink, *Brit* kerfuffle, *Slang US* hoopla: *Stop making such a big fuss over nothing.* —*v.* **2** make a fuss, rush about *or* around, °flutter, *Colloq* kick up a fuss: *I do wish she'd stop fussing—the party is going very well.*

fussy *adj.* **1** °particular, °finicky, finical, °dainty, °discriminating, °difficult, °fastidious, °exacting, °demanding, *Colloq* picky, °choosy, nit-picking, pernickety *or US also* persnickety: *He's so fussy that he'll wear only custom-made shoes.* **2** °fancy, °elaborate, overdecorated, gingerbread, rococo, °ornate, °detailed, Byzantine: *The style is a bit too fussy for my taste.*

futile *adj.* °unsuccessful, unavailing, °useless, °unprofitable, °vain, abortive, profitless, °bootless, °worthless, °empty, °sterile, °barren, unproductive, °impotent, °ineffective, °ineffectual: *After several futile attempts to get in touch with you, I gave up.*

future *n.* **1** days *or* time to come; tomorrow: *Who knows what the future may bring?* —*adj.* **2** coming, tomorrow's, later, °prospective, following, unborn, °subsequent, expected, approaching; to be, to come: *Future generations will appreciate what we have done here.*

fuzzy *adj.* **1** °woolly, downy, linty, fleecy, furry, °fluffy, frizzy, flossy, flocculent, floccose, floccus; feathery: *Tiny bits from her fuzzy pullover kept coming off on my suit.* **2** °dim, °faint, °hazy, foggy, °misty, blurred, blurry, °indistinct, unclear, °vague, °shadowy, °indefinite, °obscure, ill-defined, °woolly, distorted: *The picture was too fuzzy to make out any details.*

G

gab *v.* **1** °jabber, gabble, °chatter, gibber, blather *or* blether, prate, °prattle, °blab, °gossip, *Colloq Brit* natter, witter, *Slang* jaw, yak, *US* run (off) at the mouth: *Stop gabbing and get on with your work!* —*n.* **2** °chatter, chitchat, °prattle, gibber-jabber, °jabber, °gossip, blarney, blather *or* blether, tittle-tattle; cackle, °moonshine, °nonsense, °drivel, twaddle, °rubbish, gobbledegook *or* gobbledygook, °mumbo jumbo, poppycock, bunk *or* bunkum, balderdash, stuff and nonsense, hogwash, eyewash, *Colloq* piffle, flummery, °rot, bull, codswallop, *Slang* crap, °garbage, *Chiefly Brit* tosh, *Brit* (a load of old) cobbler's, *Taboo slang US* bullshit, horseshit, *Brit* shit, balls: *He had nothing important to say—just a lot of gab.*

gad *v.* Usually, **gad about** *or* **around.** gallivant, °run around, flit about, traipse: *He's never home, always gadding about, from one party to another.*

gadget *n.* contrivance, °device, appliance, °creation, °invention, °machine, °tool, utensil, °implement, °instrument, °mechanism, °apparatus, *Colloq* °contraption, widget, thingumabob *or* thingamabob *or* thingumbob, thingamajig *or* thingumajig *or* thingummy, whatchamacallit, whatitsname, *Brit* doodah, *US* hickey, doodad, doohickey, whosis, whatsis, dingus, *Chiefly US and Canadian* gizmo *or* gismo: *He invented a gadget for resealing soft drink bottles.*

gag[1] *v.* **1** °silence, °stifle, °still, °muffle, °stop (up), muzzle, °quiet, °curb, °suppress, °repress, °restrain, throttle, strangle, °check, °inhibit, °discourage: *They attempted to gag the media to prevent a public uproar.* **2** retch, °choke, °heave; gasp for air, struggle for breath; *US* keck: *That drink made me gag.* —*n.* **3** °restraint, °curb, muzzle, °check: *The CIA tried to put a gag on the former agent so he couldn't write about his career.*

gag[2] *n.* **1** °joke, °witticism, jest, °quip, °pun, °gibe, *Colloq* °wisecrack, *Slang* °crack: *He told some gags but nobody found them funny.* **2** practical joke, °hoax, °prank, °trick, *Colloq* fast one: *For a gag, we put a snake in the sergeant's bed.*

gaiety *n.* **1** cheerfulness, exhilaration, elation, °glee, *joie de vivre,* buoyancy, lightheartedness, blitheness, °happiness, felicity, °pleasure, °delight, °joy, joyfulness, joyousness, exultation, °merriment, °mirth, mirthfulness, jubilation, good *or* high spirits, sprightliness, liveliness, joviality, jollity, °hilarity: *The gaiety of the celebration was becoming infectious.* **2** Often, ***gaieties.*** merrymaking, °festivity, °festivities, °celebration, °revelry, revels, rejoicing, conviviality, *Old-fashioned Brit* mafficking: *We all joined in the gaiety of Mardi Gras.* **3** colorfulness, brightness, gaudiness, garishness, °brilliance, brightness, cheeriness: *The gaiety of her costume was offset by her lugubrious expression.*

gaily *adv.* **1** showily, gaudily, brightly, splendidly, brilliantly, colorfully, flashily, flamboyantly, garishly: *The dancers appeared in their gaily colored dresses.* **2** °happily, cheerfully, cheerily, gleefully, joyously, joyfully, jubilantly, merrily, blithely, lightheartedly, airily, jauntily, insouciantly: *The revelers came cavorting gaily down the street.*

gain *v.* **1** °get, °obtain, °acquire, °procure, attain, °achieve, °secure, °earn, °win, °capture, °bag, °net, °harvest, °reap, °garner, glean, °collect, °gather, °come by, °pick up: *Any advantage we gain today may be lost tomorrow.* **2** °make, °get, °profit, gain ground, °earn, °benefit, °realize, °clear, °bring in, °produce, °yield: *Have you gained from the transaction?* **3** °improve, °recuperate, °progress, °rally, get better, °advance, gain ground: *Her health has gained steadily since the cyst was removed.* **4** catch up (to *or* on *or* with), °approach, get nearer (to), °overtake, close with, close in (on), narrow the gap, **gain ground:** *As we came to the finish line, Tom was gaining on me. Though he kept gaining, I won.* **5** leave behind, outdistance, draw *or* pull away (from), widen the gap, get *or* go *or* move farther ahead, get farther away (from), increase the lead: *I kept gaining on him till I was a mile ahead.* **6** °reach, °arrive at, °get to, °come to: *After paddling for hours, we finally gained the shore.* **7** °increase, move ahead, °improve, °advance, °progress, gain ground: *Shares gained again on the exchange.* —*n.* **8** °profit, °advantage, °margin, °yield, °return, °revenue, °income, dividend, °benefit, emolument, °payment, °pay, °money; °proceeds, °earnings, winnings; *Colloq chiefly US* °take, payout, °payoff: *What was your gain on the sale of the house? He has seen gains of 20 percent on his investment.* **9** °increase,

°increment, °improvement, °rise, °addition, enhancement, °elevation, augmentation, upward *or* forward movement, °advance, °progress: *Each month has seen a gain in share prices.* **10** °acquisition, °achievement, attainment: *My gain was at the expense of their sacrifice.*

gainful *adj.* °advantageous, °profitable, °productive, °fruitful, °beneficial, °useful, valuable, °worthwhile, °rewarding, remunerative, lucrative, moneymaking: *Your father thinks it about time you sought gainful employment.*

gala *n.* **1** °fête, °festival, °festivity, °feast, °celebration, °event, red-letter day, °holiday, holy day, carnival, °occasion, °happening, °event, °pageant, °party, ball; field day: *The annual gala is planned for Midsummer's Eve.* —*adj.* **2** °merry, festive, °joyful, joyous, °gleeful, jovial, °gay, celebratory, °jolly, convivial, °happy, °cheerful, cheery: *Her homecoming was a gala occasion.*

gale *n.* **1** windstorm, strong wind, (big *or* hard) blow, °blast, turbulence, °storm, °tempest: *I refuse to take the boat out in that gale.* **2** °outburst, °burst, °explosion, °eruption; °peal, °roar, °scream, °shout, °howl, °shriek: *Bea Lillie needed merely to raise an eyebrow to send her audience into gales of laughter.*

gall[1] *n.* **1** °bitterness, acerbity, acrimony, harshness, vitriol, asperity, bile, spleen, causticness *or* causticity, °bite, mordacity *or* mordaciousness, sharpness, rancidness *or* rancidity; °venom, °poison, °rancor: *Once more he experienced the gall of disappointment.* **2** °impudence, insolence, °impertinence, audacity, brashness, brazenness, sauciness, °effrontery, temerity, overconfidence, °front, *Colloq* °brass, °nerve, °guts, cheek, lip, crust, °sauce, *US and Canadian* chutzpah, moxie; *Taboo slang* balls: *One student had the gall to suggest that he would be a better teacher than I.*

gall[2] *n.* **1** °sore (spot), abrasion, °scrape, graze, °scratch, °chafe: *This gall on my leg is from the saddle.* **2** irritation, °annoyance, °nuisance, °bother, exasperation, vexation, *Colloq* aggravation: *One has to tolerate a lot of gall in this job.* —*v.* **3** °irritate, °chafe, abrade, °fret, °scrape, °rub, °grate, °scratch: *This shoe galls my foot at the heel.* **4** °irritate, °annoy, °bother, vex, °irk, °exasperate, °harass, harry, °plague, °provoke, goad, °nag, °pester, hector, badger, nettle, needle, °ruffle, °fret, °anger, °enrage, °inflame, °infuriate, incense, °arouse, *US* °rankle: *It galls me to think of all the animals killed just to make fur coats.*

gallant *adj.* **1** °brave, °courageous, °bold, valiant, °daring, °dauntless, °intrepid, plucky, °fearless, valorous, unafraid, undaunted, °manly, manful, plucky, mettlesome, stouthearted, lionhearted, °heroic, °dashing, (high-)spirited: *Three gallant soldiers held off the enemy attack.* **2** °chivalrous, courtly, °courteous, °polite, °attentive, gentlemanly, °noble, °gracious, °considerate, °thoughtful, °well-bred, mannerly: *What man is so gallant today as to give up his seat on a bus to a lady?* **3** °dignified, °elegant, °imposing, °grand, °noble, °glorious, °fine, °splendid, °majestic, °stately, °magnificent: *Overwhelmed by four enemy men o' war, the gallant ship sank beneath the waves.* —*n.* **4** °champion, °hero, knight, cavalier, paladin: *Which gallant wears her scarf on his sleeve in today's jousts?* **5** °lover, Romeo, °sweetheart, °beloved, °paramour, boyfriend, beau, °escort, °suitor, °admirer, *Literary* swain: *In those days, gallants continually sent their ladies gifts and flowers.*

gallows *n.* gibbet: *Another name for gallows was "Tyburn tree," after that at Tyburn, in London.*

galore *adv.* in abundance, in large quantity *or* numbers *or* amounts, in excess, °everywhere, aplenty, in profusion: *There were prizes galore for everyone who competed in the games.*

gamble *v.* **1** °risk, °venture, °hazard, °bet, wager, °stake, °chance, °speculate; °play, °game, *Brit* punt: *I wouldn't gamble that he'll be on time. I go to Monte Carlo to gamble.* **2 *gamble on.*** °back, bet *or* wager on, stake *or* put money on, take a chance *or* flier on, try one's luck *or* fortune on, lay *or* place *or* make a wager *or* bet on; °count on, °rely on: *We gambled on the horse to win. I was gambling on his having forgotten the debt.* —*n.* **3** °chance, °risk, °venture; uncertainty, °speculation, *Colloq US* crapshoot: *I had to take the gamble that the rope would hold. All of life is just a gamble.* **4** °bet, wager, °stake, *Brit* punt: *She took a gamble on number 14.*

game *n.* **1** °amusement, °pastime, °diversion, °distraction, °recreation, °play; °sport: *She regards love as just a game.* **2** °contest, °competition, meeting *or US also* meet, tournament *or also* tourney, °match, °encounter, °engagement, °event, °round; regatta: *Would you like to go to a football game? The games are held annually.* **3** °scheme, °plan, °plot, °design, °stratagem, °strategy, °tactic, °artifice, °trick, device(s), ploy: *Aha! I am on to your little game!* **4** °occupation, °line (of work), °job, °position, °field, °business, °trade, °profession, *Slang* °racket: *She is in the advertising game.* **5** °quarry, °prey; °victim, °target: *The game has been plentiful this year, especially pheasant and quail. He is fair game for any unscrupulous swindler.* **6 *play games.*** dissemble, °dissimulate, be deceitful *or* underhand, °misrepresent, °pretend, practice deceit *or* deception, °fake, feign: *She wasn't serious about selling, she was just playing games.* —*adj.* **7** °ready, °willing, °prepared; plucky *or* spirited *or* daring *or* adventurous enough: *He's game for wrestling alligators—anything to prove himself.* **8** plucky, °spirited, high-spirited, °daring, devil-may-care, °adventurous, unflinching, °courageous, °brave, °bold, °heroic, *Colloq* nervy, gutsy: *Entering the cave to rescue the dog was a game thing for him to do.* —*v.* **9** °gamble: *He has taken up gaming as a hobby.*

gamut *n.* °range, °scale, spectrum, compass, °spread, °sweep, °field, series: *Parker said that Hepburn's acting ran the whole gamut of emotions from A to B.*

gang *n.* **1** °group, °band, °crowd, °company, °pack, °mob, °ring: *The police are looking for the gang that robbed the mail train.* **2** °clique, °set, coterie, °circle, °party, °company, °team, troupe: *A whole gang of my friends came over yesterday evening.* —*v.* **3 *gang up* (*on* *or against*).** conspire *or* plot against, combine *or* unite *or* unify *or* join (against), league *or* ally *or* club *or* band (against), join forces (against), °overwhelm: *When they all ganged up on me, I decided to change my mind.*

gangster *n.* °criminal, gang member, °racketeer, mafioso, °soldier, hooligan, gunman, °thug, mugger, °robber, °tough, ruffian, desperado, brigand, bandit, *Brit* skinhead, *Australian and New Zealand* larrikin, *Chiefly US and Canadian* °hoodlum, gunslinger, *Colloq* crook; *Slang* hood, hit man, *US* gunsel, torpedo, mobster, goon: *The gangsters were arrested for smuggling cocaine.*

gap *n.* **1** °opening, °space, °aperture, °distance, °hole, °void, °gulf, °cavity, °break, °breach, °crevice, °chink, °crack, °split, °division, cleft, °rift, °rip, °tear, °rent; °interruption, °interval, lacuna, hiatus, discontinuity, disruption; °lull, °pause, °rest, °recess, °halt, °stop, °suspension, °delay, °wait, °intermission, °respite: *The gap between the walls is just big enough for a child. I can see through a gap in the curtain. There is a gap in the series. After a gap of an hour, we resumed.* **2** °difference, divergence, °disparity, °disagreement, inconsistency, °discrepancy; °division, °distinction: *A gap exists between your ideas of justice and mine.*

gape *v.* 1 °stare, °gawk, goggle, *Slang* °rubberneck, *Brit* gawp *or Brit* gaup: *She gaped when they brought in the boar's head.* 2 yawn, open wide, °part; °split: *The jaws of the huge shark gaped to swallow the diver.*

garbage *n.* °rubbish, °refuse, °waste, °muck, offal, sweepings, °swill, °filth, slops, dross, °scraps, °junk, °litter, debris, detritus, *Chiefly US and Canadian* °trash, *Slang* crap: *The garbage in the river is a health hazard.*

garble *v.* 1 °warp, °distort, °twist, °corrupt, °adulterate, °slant, °color, °mangle, °mutilate, °pervert, °doctor, °falsify, °misrepresent, belie, misstate, misquote, misreport, mistranslate, misrender; °misunderstand, misconstrue, misread: *She has garbled the story and it makes no sense. I've garbled your message completely.* 2 °confuse, °mix up, °jumble, °mumble, °mutter: *He garbles his words so badly that I cannot understand him.*

garish *adj.* °gaudy, °flashy, °glaring, °cheap, °tawdry, florid, raffish, °vulgar, °harsh, °loud, meretricious, brummagem, °showy, °crude, °tasteless, *Colloq* °flash, *Slang US* glitzy: *Their flat is filled with the most garish decorations.*

garland *n.* 1 wreath, festoon, °crown, chaplet, circlet: *Diana crowned the sleeping Endymion with a garland of flowers.* —*v.* 2 wreathe, festoon, °crown, °decorate, °coil, °spiral, °encircle, °ring, °circle: *The columns were all garlanded with brightly colored paper decorations.*

garments *n.* pl. garb, °clothes, clothing, raiment, °dress, attire, °apparel, °costume, °outfit, °habit, vestments, habiliment; °wardrobe; °uniform, livery; *Colloq* duds, °rig, togs, *Slang* °rags, threads, °gear: *All his garments are custom made in Savile Row.*

garner *v.* °gather, °collect, °accumulate, °assemble, °amass, °store (up), °stock (up), °husband, lay in *or* up *or* down *or* by, heap *or* pile up, put away *or* by, °stow (away), °cache, °store, °save, °reserve: *He garnered what he could from the family estate.*

gash *n.* 1 °cut, °slash, °wound, °score, cleft, °incision, laceration, °slit, °groove, °split: *The chisel slipped, cutting a deep gash in the table.* —*v.* 2 °cut, °slash, °wound, °score, °cleave, incise, °lacerate, °slit, °groove, °split: *I've gashed my hand on a rusty nail.*

gasp *v.* 1 °pant, gulp for air, fight for air *or* breath; catch one's breath, snort, °huff, °puff: *I gasped when I saw the condition of the room.* —*n.* 2 snort, °puff, °blow, °gulp, wheeze: *He gave a gasp of horror.*

gate *n.* 1 gateway, °barrier, doorway, door, access, °entrance, °exit, °passage, °opening: *The garden gate sagged on its rusty hinges.* 2 admissions, °attendance, °crowd, audience, assemblage: *We had the biggest gate of the year at yesterday's game.*

gather *v.* 1 °collect, °assemble, °accumulate, °amass, °muster, heap *or* pile (up), °garner, °pick, °harvest, glean, get *or* bring together, stockpile, °stock: *I gathered all the firewood I could find.* 2 °collect, °assemble, convene, °meet, °huddle, forgather *or* foregather, get *or* come together, congregate, °turn out, flock *or* herd (together), °group, °cluster, °throng, °crowd, °swarm, °rally: *We gathered round the old man's bed.* 3 °purse, shirr, °pucker, tuck, °ruffle, pleat, draw *or* pull together, °contract, constrict: *The fabric is gathered at the waist.* 4 °draw, conclude, °infer, °assume, °deduce, °understand, °learn, °hear, be led to believe: *I gather that you'd prefer not to collect the award in person.* 5 °increase, °grow, °enlarge, °expand, °extend, wax, °heighten, °deepen, °intensify, °build, °rise: *He stood, menacingly, in the gathering shadows.*

gathering *n.* °assembly, convocation, °convention, congress, assemblage, °meeting, °get-together, °turnout, conclave, °rally, aggregation: *There is a gathering of the clan once a year.*

gaudy *adj.* °garish, °flashy, °glaring, °tawdry, °loud, °cheap, florid, °showy, °ostentatious, raffish, °vulgar, °crude, °tasteless, brummagem, meretricious, tinselly, gimcrack, °shoddy, trashy, *Brit* tatty, *US and Canadian* honky-tonk, *Colloq US* °tacky, chintzy: *The carnival stands were painted pink, purple, and other gaudy colors.*

gauge *v.* 1 °measure, °determine, °weigh, °calculate, °compute, °reckon, °figure, *US technical also* gage: *Try to gauge the height of that pyramid.* 2 °judge, °evaluate, appraise, assess, °rate, °estimate, °guess, *US technical also* gage: *How can we gauge the strength of the opposition?* —*n.* 3 °standard, °yardstick, criterion, benchmark, °basis, °measure, °norm, °model, °example, °pattern, °rule, °touchstone, °test, °guide, guideline, *US* litmus test, *Technical also* gage: *You cannot use your compulsive work habits as a gauge for others.* 4 °scope, °capacity, °amount, °extent, °measure, °size, dimension(s), °magnitude, °degree, °limit: *What will be the gauge of your involvement in the enterprise?*

gaunt *adj.* 1 °emaciated, °scrawny, °rawboned, bony, angular, °haggard, skeletal, wasted, starved-looking, cadaverous, scraggy, spindly, °meager, hollow-cheeked, °spare, °skinny, lank(y), pinched, °thin, underweight: *We were shocked by the pictures of the gaunt faces of the victims of the famine.* 2 °dreary, °dismal, °bleak, °bare, °barren, °deserted, °desolate, °harsh, °hostile, unfriendly, inimical, °stern, °forbidding, °stark, °grim, °forlorn: *Under the dour clouds, the moor looked even more gaunt than usual.*

gawk *n.* 1 oaf, lout, bumpkin, °clod, °boor, churl, °dolt, dunderhead *or* dunderpate, ninny, gormless ninnyhammer, ignoramus, °fool, simpleton, ass, bungler, bumbler, *Colloq* clodhopper, lummox, *Slang chiefly US old-fashioned* galoot *or* galloot, °lug: *You big gawk, get out of my way!* —*v.* 2 °stare, goggle, °gape, *Colloq* °rubberneck, *Slang Brit* gawp *or* gaup: *Stop gawking and help me get him to a hospital.*

gay *adj.* 1 See **homosexual, 1,** below. 2 °happy, °blithe, jovial, lighthearted, °carefree, debonair *or Brit also* debonnaire, °cheerful, °gleeful, °bright, °joyful, joyous, jubilant, high-spirited, °merry, °lively, °vivacious, °buoyant, °effervescent, °bubbly, bubbling, sparkling, *US* chipper: *In those days our hearts were young and gay.* 3 °garish, °gaudy, °flashy, °bright, °brilliant, °vivid, many-colored: *The gay decorations added to the atmosphere to the party.* —*n.* 4 See **homosexual, 2,** below.

gaze *v.* 1 look at *or* on *or* upon *or* over, °view, °regard, °contemplate, °stare; °wonder, °gape: *She stood gazing out to sea. I gazed at the screen in disbelief.* —*n.* 2 fixed *or* steady *or* intent look, °stare, °look: *His gaze faltered and he averted his eyes.*

gear *n.* 1 °cog, cogwheel; gearbox, °mechanism, machinery; °works: *This gear operates the automatic doors.* 2 °equipment, °apparatus, appliances, **implements,** tools, °tackle, utensils, supplies, material(s), accessories, accouterments *or* accoutrements, appurtenances, °paraphernalia, panoply, °outfit, °trappings, fixtures, materiel *or* matériel: *Have you taken all the camping gear?* 3 clothing, °apparel, attire, °clothes, °garments, °habit, habiliments, vestments, raiment, °regalia, °uniform, livery, *Colloq* duds, togs, *Brit* clobber: *You should keep your gear in your locker.* 4 °belongings, °things, °stuff, °effects, °kit, °effects, chattels, °goods, impedimenta, (bag and) baggage, accouterments *or* accoutrements, *Colloq* °junk, *Slang* shit: *Get your gear together and get out of here!* —*v.* 5 °adjust, °adapt, °fit, °suit, °tailor, °accommodate: *You will have to gear your schedule according to our plans.*

gem *n.* **1** gemstone, °jewel, stone, precious *or* semi-precious stone: *The crown was set with rubies and other precious gems.* **2** °ideal, °quintessence, perfect example; °pearl (of great price), °marvel, °flower, °elite, cream, *crème de la crème,* °pick, °nonpareil, °treasure, °prize, °masterpiece, *chef d'oeuvre*: *This is a perfect gem of a painting.*

general *adj.* **1** °common, °prevailing, accepted, °popular, °public, communal, community, widespread, shared, °extensive, °prevalent, °universal, worldwide, °global; °comprehensive, °inclusive, (all-)inclusive, nonexclusive, °overall, unrestricted: *The general feeling is that she should resign. We demanded a general revision of the law.* **2** °ordinary, °common, °normal, °customary, °habitual, °regular, °usual, °run-of-the-mill, °everyday, °familiar, °accustomed; nonspecialized, unspecialized, nonspecific, unspecific: *Indoor plumbing became general only in the twentieth century. Within the general area of ceramics there are scores of specialities.* **3** °mixed, assorted, °miscellaneous, heterogeneous, °encyclopedic, diversified, extended, °broad, °comprehensive, (all-)inclusive, nonexclusive, °overall, blanket, across-the-board, °sweeping, °panoramic, °catholic, composite, combined, blended, °hybrid, °mongrel: *The plan is to prepare a general work of reference.* **4** °vague, °indefinite, °broad, ill-defined, °loose, °inexact, °imprecise, undetailed, nonspecific, unspecific, generalized, °overall; °approximate: *This kind of general report fails to pinpoint problems.*

generality *n.* **1** generalization, abstraction, °abstract, vague *or* loose *or* sweeping *or* indefinite statement, imprecise *or* vague notion: *His talks are characterized by empty generality, without detailed facts.* **2** Often, ***generalities.*** principle(s), law(s), abstraction(s), generalization(s), universality *or* universalities: *You have given us the generalities, now we want the particulars.*

generally *adv.* **1** °usually, commonly, °ordinarily, in general, customarily, habitually, conventionally, normally, typically, °on average, °as a rule, by and large, °for the most part, mostly, °mainly, °on the whole, predominantly: *Generally, people do not have a healthy diet. He is generally in the office at noon.* **2** superficially, nonspecifically, unspecifically, °roughly, broadly, in a general way, loosely, °approximately, °largely, °in the main, °mainly, °principally: *He covered the subject generally, not dwelling on details.*

generate *v.* **1** °produce, °create, °originate, °make, °manufacture: *At these stations we can generate a thousand megawatts of electricity.* **2** °spawn, °father, sire, bring into being, procreate, °breed, beget, engender, °propagate: *This one fish would generate millions of offspring if all survived.* **3** °produce, °create, °give rise to, °inspire, °whip up, °cause, °initiate: *Her good fortune generated in him a feeling of envy.* **4** °produce, °create, °invent, °coin, °make up, °fabricate, °fashion, °devise, °develop, °form, °forge, °mold, contrive, °construct, put together: *We must generate a new theory to account for these phenomena.*

generation *n.* **1** °production, °reproduction, propagation, procreation, begetting, fathering, siring: *In the lower animals, the generation of offspring is asexual.* **2** time(s), day(s), °age, °period, °era, epoch: *The trait did not appear till later generations. The 1950s and '60s spawned the yuppie generation of the '80s.* **3** origination, °creation, genesis, inception, °initiation, °start, °beginning, °institution, °establishment, °formation, formulation: *The generation of legal systems can be traced to the oldest civilizations.* **4** crop; °age (group); contemporaries: *This generation of Americans has a rendezvous with destiny. He prefers to be with people of his own generation.*

generous *adj.* **1** bounteous, °bountiful, magnanimous, °charitable, eleemosynary, °philanthropic, °lavish, openhanded, °free, °liberal, unstinting, ungrudging, beneficent, °benevolent, bighearted, munificent: *Not everyone who has money is generous with it.* **2** magnanimous, °benevolent, °charitable, °unselfish, °humanitarian, humane, °kindly, °noble, high-minded, °lofty, °good, °disinterested, °unprejudiced: *He was too generous in acknowledging the accomplishments of his political foes.* **3** °plentiful, °full, °lavish, overflowing, °abundant, bounteous, °handsome, copious, °ample: *They certainly serve generous portions at that restaurant.*

genial *adj.* affable, °amiable, °cordial, °warm, °friendly, congenial, °agreeable, °good-natured, good-humored, well-disposed, °neighborly, °sociable, °kindly, °kind, °hospitable, easygoing, °relaxed, °pleasant, °nice, °cheerful, cheery, convivial: *And here is your genial host, Sonny Gegenschein!*

genitals *n.* pl. genitalia, sexual *or* reproductive organs, sex organs, organs of procreation *or* generation, °private parts, °privates: *They wore loincloths to cover their genitals.*

genius *n.* **1** °mastermind, °master, °virtuoso, °intellect; maestro, °expert, °adept; *Colloq* °brain, Einstein: *At the age of nine, this genius entered university.* **2** °intelligence, °brilliance, °wit, °ingenuity, brains, °ability, °aptitude: *She is a woman of incredible genius.* **3** °talent, °gift, °knack, °faculty, °flair, °aptitude, °forte, °capacity, °ability, °capability: *He has a genius for saying the wrong thing.*

genre *n.* °kind, °sort, °type, °class, °style, °brand, °character, °category, genus, species, °variety, °fashion: *Her paintings are in the classicist genre.*

genteel *adj.* **1** overpolite, °unnatural, °pretentious, °affected, °mannered, putting on airs, pompous, overdone, *Colloq* °posh, la-di-da *or* lah-di-dah, phoney *or US also* phony, *Brit* county: *The Brumfits are a bit too genteel for my taste.* **2** courtly, °polite, °civil, well-mannered, °well-bred, °courteous, mannerly, °gracious, °proper, °respectable, °decorous, °ladylike, gentlemanly, °chivalrous, cavalier, °debonair, °suave, patrician, °high-class, °upper-class, aristocratic, thoroughbred, blue-blooded, °noble, °royal, *Colloq* classy, tony, upper-crust, *US* silk-stocking: *His son married a woman from an ancient, genteel house.* **3** °refined, °polished, °sophisticated, debonair *or Brit also* debonnaire, °suave, urbane, cosmopolitan, °cultivated, cultured, °elegant, *Colloq* ritzy: *Why small feet are considered genteel, I cannot say.*

gentle *adj.* **1** °kind, °kindly, °mild, °tender, °benign, °moderate, °easy, °quiet, °calm, °still, °temperate, unruffled, untroubled, undisturbed, °tranquil, °restful, °peaceful, pacific, placid, °smooth, °lenient, °patient, °indulgent, °soothing, °thoughtful, °gracious, compassionate, humane, tenderhearted, °merciful; °soft, °light, balmy: *Amy is very gentle with the patients. A gentle breeze barely stirred the leaves. The gentle waves belied the terror that lurked below.* **2** °tame, °tractable, docile, °manageable, controllable; °broken: *As he's a beginner, he's been given a gentle mare to ride.* **3** °gradual, °easy, °moderate: *The car rolled down the gentle slope and stopped.*

gentry *n.* ladies and gentlemen, °elite, aristocracy, landed gentry, gentlefolk, upper class(es) *or* strata, cream, *crème de la crème;* landowners, *Brit* squirearchy, *Colloq* °upper crust: *He thinks he is privileged because he belongs to the gentry.*

genuine *adj.* **1** °authentic, °real, °bona fide, °veritable, °legitimate, °true, °original, °proper, not counterfeit *or* fake; pukka *or* pucka: *Is that a genuine diamond? I know his feelings for her are genuine. Armstrong is a* *genuine leader of men.* **2** °candid, °frank, °open, °sincere, °earnest, °honest, unfeigned: *His art is much more genuine than the pretentious output of his contemporaries.*

germ *n.* **1** microorganism, °microbe, bacterium, virus, *Colloq* °bug: *Germs were not conceived of before Leeuwenhoek's invention of the microscope.* **2** °source, °origin, fount, embryo, °seed, °root, rudiment, °beginning, °start, °base, °basis: *The germ of romantic literature lies in folk tales.*

gesture *n.* **1** °movement, °motion, gesticulation, °signal, °indication, °action, *Colloq US* high sign: *She made a gesture of dismissal.* **2** °formality, °move; gambit, ploy; °token, °indication, °symbol: *Sending her flowers was a friendly gesture. Their burning the flag was a gesture of defiance.* —*v.* **3** °motion, gesticulate, °signal, °sign, °indicate, *Colloq US* give (someone) the high sign: *He gestured for me to step forward.*

get *v.* **1** °obtain, °secure, °acquire, come by *or* into (the) possession of, °procure, °pick up; °collect; °buy, °purchase, °book, °retain, °hire, °engage, °rent, °lease; accede to, °inherit, fall heir to, °succeed to: *Where did you get these beautiful prints? You had better get yourself a good lawyer. They got the furniture from his mother's estate.* **2** °receive; be given, °come by: *He got a bicycle from his grandma. I got a job at the carwash.* **3** °earn, °receive, °realize, °make, °take, °gross, °clear, °net, °pocket, be paid; °win, *Colloq* take home, °pull down: *He gets a lot of money for just standing there and asking contestants questions.* **4** °fetch, go (to *or* and) get, go for *or* after, °pick up; °bring (back), °retrieve: *Please get me a cup of tea.* **5** °catch, °contract, °have, °suffer from, °come down with, fall ill *or* sick with, be afflicted with, become infected with, °acquire: *You'll get pneumonia if you go out without your coat.* **6** °become, °fall: *We got ill from eating too much chocolate.* **7** °become, °turn, °grow: *It got very cold last night. He got rich almost overnight.* **8** °capture, °seize, °arrest, °take, apprehend, °grab, °pick up, °lay hold of, °bag, *Colloq* collar, °nab, °pinch: *Are the police sure that they got the right person?* **9** °manage, °arrange, °come, °succeed; contrive, fix it, °maneuver, °manipulate, *Colloq* °wangle: *How did you get to be chairman?* **10** °reach, °arrive (at), °come, °go, °travel, °journey: *She got home by taxi.* **11** °catch, °take, °enter, °make, come *or* go by, travel *or* journey by: *He got the midnight train.* **12** °reach, get in touch with, °communicate with, get onto *or* through to, *Colloq* °contact: *We finally got him just before he left for the day.* **13** °receive, °pick up, tune in to *or* on, listen to *or* watch: *I cannot get that program.* **14** °persuade, °prevail upon, °coax, °induce, °influence, °cajole, °wheedle, talk (someone) into, °sway, bring (someone) round, °make, °cause: *How did you ever get them to let you go?* **15** °put, °place, °set, °fit, °maneuver, °manipulate, wiggle, °wriggle: *I can't get this knob back on.* **16** °affect, °stir, °move, °touch, °arouse, °stimulate, °excite, have an impact *or* effect on, make an impression on, °impress, leave a mark on, °get to, *Colloq* turn (someone) on: *Those old romantic songs really get me.* **17** °get at, °irritate, °annoy, vex, °irk, nettle, pique, °provoke, °anger, °exasperate, °bother, °perturb, rile, *Colloq* °bug, get (someone's) goat, rub (someone) (up) the wrong way, *Sometimes nonstandard* °aggravate: *Doesn't it get you to see people going unpunished for their crimes?* **18** baffle, °confuse, confound, °puzzle, °perplex, °bewilder: *You've got me! I don't know where it is!* **19** °understand, °appreciate, °fathom, °see, °grasp, apprehend, °perceive, °follow, °comprehend, °take in, °work out, make head(s) or tail(s) of: *You just don't get the joke, do you?* **20** °catch, °hear: *I didn't get the name; please repeat it.* **21** °derive, °learn, glean, absorb, °take in: *Did you get anything from yesterday's lecture?* **22** °have, °place, °put, °fix, pinpoint: *Eventually, I'll get him where I want him!* **23** °get even with, °revenge oneself on, take vengeance on, pay (someone) back, settle *or* even the score with, °get back at: *I'll get him for betraying me if it's the last thing I do!* **24** °strike, °hit, °punch, smack; °shoot; °hurt, °harm, °damage, °injure; *Slang* sock, slug: *She got him right between the eyes.* **25** ***get about.*** See **29,** below. **26** ***get across.*** get *or* put over, °put across, °get through, °communicate, make clear, °impart: *She is trying to get across to you the fact that she despises television.* **27** ***get ahead.*** °succeed, °prosper, be *or* become successful, do well, °flourish, °thrive, °make good, °progress, rise (up) in the world, *Colloq* °go places: *With his contacts, he's bound to get ahead.* **28** ***get along.*** **(a)** be friendly *or* compatible (with), °associate (with), °agree (with), be agreeable, °get on (with), *Colloq* hit it off (with): *She gets along with most people quite well. Do you think they'll get along?* **(b)** °manage, °cope, °shift, °fare, °survive, make both ends meet, keep the wolf from the door, keep one's head above water, °get on, *Colloq* °get by, °make out, °make do: *He has enough money to get along by himself.* **(c)** °leave, °depart, go *or* move away, get going, °get on, go along, °proceed: *I have to get along now or I'll miss my train.* **(d)** °progress, °proceed, °get on, °advance, move ahead *or* along *or* on: *How are you getting along with your new novel?* **(e)** °get on, °age, get *or* become *or* grow older, °advance: *They are getting along but you wouldn't guess that they're eighty.* **29** ***get around*** *or* ***about*** *or* ***round.*** **(a)** °spread, become known, °leak (out), °circulate, be bruited about *or* around, be noised abroad, go about *or* around: *He let the news get around that he needed a secretary.* **(b)** be socially active, °socialize, go *or* get out; runaround *or* about, gad about: *Since his accident, he doesn't get around much any more.* **30** ***get around*** *or* ***round.*** **(a)** °cajole, °wheedle, °flatter, °persuade, °coax, °win over: *I didn't want to propose, but she managed to get around me.* **(b)** °bypass, circumvent, skirt, °avoid, °evade, °elude, °outsmart, °outwit, outmaneuver, outflank, steal a march on, *Colloq* give (someone) the runaround: *Chris is always trying to get around the parking restrictions.* **(c)** get *or* come to, °reach, °arrive at (finally), find time for: *He said he would do the repairs if he got around to it.* **31** ***get at.*** **(a)** gain access to, access, °reach, put *or* lay one's hands on, °get to: *I can't get at the lock from here.* **(b)** °intend, °mean, °suggest, °hint (at), °insinuate, °imply, have in mind *or* view, °contemplate: *She just didn't understand what I was getting at.* **(c)** °tease, °taunt, °criticize, find fault with, °carp, °nag, °pick on, *US* °get to: *She is really beginning to get at me with her persistent criticism.* See also **17,** above. **(d)** °get to, °influence, °intimidate, °corrupt, °bribe, °undermine, °subvert, suborn: *They tried to get at the judge.* **(e)** °learn, °find out, ascertain, °determine: *The police said they were only trying to get at the facts.* **32** ***get away.*** **(a)** °escape, °leave, break out *or* away, °flee, °depart, make good one's escape, elude one's captors, break free, °disappear: *He will never be able to get away now!* **(b)** °escape, take a holiday *or US also* vacation, get *or* take a rest *or* respite: *I must get away for a few days.* **(c)** °start, get *or* take off: *Black Flash got away from the starting gate at top speed.* **33** ***get back.*** **(a)** °return, come *or* go back: *I'll try to get back before midnight.* **(b)** reacquire, °recover, regain, °retrieve, °recoup, repossess: *Can you get back your money for that hat?* **34** ***get back at.*** See **23,** above. **35** ***get behind.*** °back, °support, °promote, °finance, °fund, °push, *Colloq* °plug, hype: *If you get behind the enterprise, I know it will succeed.* **36** ***get by.*** See **28 (b),** above. **37** ***get down.*** **(a)** dismount, alight, °descend, come *or* go down, climb *or* step down, °get off: *He got down from his horse and drew his gun.* **(b)** °write (down), °note (down), °record, make a note of: *She wasn't able to get down everything*

Shaw said. **(c)** °depress, dispirit, °sadden, dishearten, °discourage, deject: *Don't let a little thing like that get you down.* **(d)** °swallow, °eat: *No sooner does he get some food down than it comes right back up again.* **38 *get down to.*** concentrate *or* focus on, turn attention to, °attend to: *Let's get down to business.* **39 *get in.*** **(a)** °enter, get into *or* on *or* onto, °embark, entrain, emplane *or* enplane, get *or* go aboard: *We got in the car and were driven away. You can get in here.* **(b)** °enter, °arrive, °return, come *or* go in: *I got in at eight today.* **(c)** fit *or* squeeze in, °insert, °slip in, °include: *He got in my comment. The doctor got me in before anyone else.* **(d)** °arrive, °come in, °land: *My flight got in at dawn.* **40 *get into.*** **(a)** °put on, don, dress in, get dressed in: *Here, get into this uniform.* **(b)** °go into, °discuss, become involved in, °pursue, °treat, °describe, delineate, °detail, °follow up on, °penetrate: *Do you really want me to get into the nasty details?* **(c)** be into, be *or* become involved in, °take up, °pursue; °enjoy, °like, become enthusiastic about, *Slang* °get off on, get *or* become high on: *I could really get into farming.* **(d)** See **39 (a),** above. **41 *get off.*** **(a)** alight, °disembark, get down from, °get out of, dismount, °descend (from), climb *or* step down off *or* from, deplane, detrain: *She got off her bike and came towards me.* **(b)** °leave, °depart, °go (off), set out *or* off: *I hope to get off tomorrow morning.* **(c)** °remove, °take off, °shed, doff: *Can you get those boots off easily?* **(d)** cause to be set *or* go free, be *or* cause to be acquitted *or* liberated *or* released *or* set free: *A clever lawyer could have got him off. He got off on a technicality.* **42 *get off on.*** See **40 (c),** above. **43 *get on.*** **(a)** See **28,** above. **(b)** See **39 (a),** above. **(c)** grow *or* become late: *It's getting on and they'll be waiting.* **44 *get onto.*** **(a)** See **12,** above. **(b)** See **48 (b),** below. **(c)** °discover, learn about, become aware of, find out about, *Colloq Brit* °twig, *US* cotton (on) to: *If they get onto the missing stores we're in for it.* **45 *get out (of).*** **(a)** °leave, °depart, go out *or* away, be off, °retire: *We got out as quickly as we could.* **(b)** °escape, extricate oneself; be released: *She thought she'd never get out alive.* **(c)** °extract, °draw, wrest, °force, drag *or* pry out, wring *or* get from: *They'll never get anything out of me!* **(d)** °gain, °profit: *Ask yourself what you will get out of the deal.* **(e)** °avoid, °evade, °sidestep, °escape: *Gabriela was somehow always able to get out of gym class.* **(f)** See **29 (b),** above. **46 *get over.*** **(a)** surmount, °cross, °climb, °pass, °traverse: *A dog can't get over a fence that high.* **(b)** recover *or* recuperate from, °survive: *Which takes longer, getting over a love affair or the measles?* **(c)** °finish, °complete, bring to an end: *Let's get this over with as soon as possible.* **(d)** See **26,** above. **47 *get round.*** See **29, 30,** above. **48 *get through.*** **(a)** (help to) succeed *or* complete: *She got me through the ordeal. Without her help, I never could have got through.* **(b)** °reach, °contact: *The line is engaged and I cannot get through to mother.* **(c)** °finish, conclude: *When you get through, please put away the dishes.* **(d)** °communicate (with): *Parents sometimes have trouble getting through to their children.* **49 *get (to).*** **(a)** °arrive at, °come to; °near, °approach: *What time did you get to the office? How did you get home?* **(b)** See **31 (a), (c), (d),** above. **50 *get together.*** **(a)** °gather, °accumulate, °collect, °assemble: *Get your things together and leave at once!* **(b)** °assemble, convene, °gather, °meet, congregate, °socialize: *Do you have time to get together with me on Friday?* **(c)** arrive at *or* reach an agreement *or* settlement, °come to terms, come to an understanding: *I think we can get together on the question of wages.* **51 *get up.*** **(a)** °arise, °awaken, °wake (up): *I get up at dawn.* **(b)** °stand (up): *Get up and you'll be able to reach it.* **(c)** °mount, °climb (up), ascend: *Can we get up the next hill? She got up on the table and started to dance.* **(d)** °create, °devise, °organize, °arrange, °prepare: *We got up a skit for the after-dinner entertainment.* **(e)** °dress, °clothe, °apparel, °outfit, attire, °turn out, deck out, °rig out, °dress up, fit out *or* up: *She got herself up as a cabaret dancer.* **(f)** °study, °learn, *US* get up on: *You'd best get up your history before the exam.* **52 *get up to.*** become *or* be involved in, be up to: *If you don't watch her, she'll get up to no good.*

getaway *n.* °escape, °flight, °retreat: *The prisoners made their getaway in a van. The dinner party was so boring I couldn't wait to make my getaway.*

get-together *n.* °gathering; °meeting, °conference, °convention: *We had a get-together at my house. Can we arrange a get-together for all the members?*

getup *n.* **1** °costume, °outfit; °rig: *He came in a diver's getup.* **2** °format, layout, °arrangement, °structure, °look, °style: *Who is responsible for the magazine's new getup?*

gewgaw *n.* trinket, °bauble, gimcrack, °trifle, knickknack *or Brit also* nicknack, bagatelle, kickshaw, °toy, °novelty, bijou, °vanity; bric-a-brac *or* bric-à-brac: *She has a huge collection of worthless gewgaws.*

ghastly *adj.* **1** °dreadful, °awful, °terrible, °terrifying, °frightful, °hideous, °horrible, horrendous, horrid, horrifying, °grim, °grisly, °loathsome, °gruesome, °ugly, °repellent, °repulsive, °shocking, appalling, *Colloq* °gross, °scary: *He was involved in a ghastly accident.* **2** °grim, cadaverous, °haggard, ashen, °wan, °pale, pallid, pasty(-faced), °drawn, livid, °ghostly, °spectral, °macabre: *She looked ghastly after the surgery.* **3** °awful, °bad, °terrible, °ill, ailing, °sick: *We felt ghastly after eating the underdone chicken.*

ghost *n.* **1** apparition, °phantom, °specter, phantasm, °shade, °spirit, wraith, poltergeist, banshee, doubleganger, doppelgänger, ghoul, manes, *No. Eng. dialect* boggart; °hallucination, °illusion, °vision; *Colloq* spook: *The ghost of his uncle appeared before him.* **2** °hint, °suggestion, °shadow, °trace, scintilla, glimmer: *Vera hasn't the ghost of a chance of meeting anyone suitable at the office.*

ghostly *adj.* **1** °spectral, ghostlike, wraithlike, phantasmal, °phantom, °eerie, °unreal, °unnatural, °supernatural, preternatural, °unearthly, °sinister, °strange, uncanny, °weird, *Colloq* spooky, °scary, creepy: *I heard a ghostly voice calling my name.* **2** See **ghastly, 2,** above.

ghoulish *adj.* **1** °devilish, demonic, °satanic, diabolic(al), °fiendish, demoniac(al), cacodemonic, Mephistophelian; °infernal, hellish, malign: *The coven engages in the most ghoulish practices.* **2** °macabre, °grisly, °morbid, °gruesome, °disgusting, °monstrous, °abominable, °hideous, horrendous, °horrible, horrifying, horrid, °brutal, barbaric, °savage, °ruthless, pitiless, °merciless, °cruel, °vicious, feral, °inhuman, °bloodthirsty, °ferocious, *Colloq* °sick: *She has some ghoulish notions about being eaten alive by rats.*

giant *n.* **1** °superhuman, titan, colossus, Goliath; giantess, Amazon, °ogre; behemoth, °monster, leviathan, mammoth: *Blocking the road was a giant nearly thirty feet tall.* —*adj.* **2** See **gigantic,** below.

gibberish *n.* °drivel, tripe, °nonsense, °rubbish, gibber, °prattle, twaddle, gabble, °jabber, balderdash, jibber-jabber, blather *or* blether, jabberwocky, gobbledygook *or* gobbledegook, °mumbo jumbo, rodomontade, Gongorism, cackle, °chatter, °patter, °chatter, °jargon, °babble, claptrap, poppycock, *Colloq* tripe, codswallop, bunk, piffle, *US* °garbage, horsefeathers; *Slang* crap, *Brit* (a load of old) cobblers, *Taboo slang* balls, *US* bull(shit), crock (of shit): *If one listens to such gibberish long enough, one begins to believe it.*

gibe *v.* **1** jibe (at), °jeer (at), °scoff (at), °flout, °mock, °deride, °make fun of, °poke fun at, °ridicule, °twit, °taunt, °sneer (at), °chaff, °tease, °rag, °heckle, *No. Eng. dialect* gird, *Colloq* kid, rib, *US and Canadian* razz: *All of us at school used to gibe at David because he wore fancy clothes.* —*n.* **2** jibe, °jeer, °taunt, °sneer, °dig, cutting remark, °thrust, °chaff, °raillery, scoffing, °derision, °ridicule, °mockery, *Rare* mycterism, *Slang* °crack, °wisecrack: *David paid no attention to his classmates' gibes.*

giddy *adj.* **1** °dizzy, °faint, unsteady, lightheaded, vertiginous, reeling, *Colloq* woozy: *This medication may make you feel a bit giddy.* **2** °silly, °frivolous, °scatterbrained, °flighty, °capricious, °irresponsible, °erratic, °fickle, °volatile, °impulsive, °reckless, °whimsical: *You can't marry a giddy young girl like Peggy.*

gift *n.* **1** °present, °donation, °favor, °grant, largess *or* largesse, °bounty, benefaction, °offering, °honorarium, contribution, giveaway, °premium, °bonus, °prize; alms, handout, °dole, °charity, °benefit; °tip, gratuity, baksheesh *or* backsheesh, *pourboire,* cumshaw: *Before we left, our neighbor gave us a gift of a suitcase.* **2** °talent, °ability, °aptitude, °genius, °flair, °knack, °facility, °forte, °strength, strong point, °bent, °capability, °capacity, °power: *Her family never recognized her special gifts.*

gifted *adj.* °talented, °able, skilled, °capable, °skillful, °outstanding, °excellent, °superior, °superb, °brilliant, °expert, °master, °masterful, masterly, °virtuoso, first-class, °first-rate, top-drawer, top-flight, °good, *Colloq* topnotch, ace, crackerjack, °crack: *Grants are available for gifted, indigent artists.*

gigantic *adj.* °big, °large, °huge, °enormous, °massive, °giant, °colossal, °immense, mammoth, tremendous, stupendous, °towering, staggering, °vast, titanic, gargantuan, elephantine, cyclopean, herculean, Brobdingnagian; kingsize, extra-large, *Colloq* °jumbo, walloping, °whopping, °thumping, thundering, strapping, super-duper, *US* humongous: *The gigantic tower, reaching to the heavens, appeared before us.*

giggle *v.* **1** °titter, °snicker, snigger, °chuckle, °laugh, chortle, cackle, °twitter: *Stop that giggling at the rear of the classroom!* —*n.* **2** °titter, °snicker, snigger, °chuckle, °laugh, chortle, cackle, °twitter: *Did I hear a giggle?* **3** °joke, °prank, °laugh: *Just for a giggle, we tied his shoelaces together.*

gimmick *n.* **1** °device, °strategy, °stratagem, ploy, °trick, °ruse, °wile, °subterfuge, °maneuver, °artifice, °deception, °trap, °snare, *US* °hook, *Colloq* °dodge: *We need some sort of gimmick to attract buyers.* **2** °device, contrivance, °gadget, °invention, *Colloq* °contraption, widget, thingumbob *or* thingumabob *or* thingamabob, thingamajig *or* thingumajig *or* thingummy, whatchamacallit, whatitsname, *Brit* doodah, *US* doohickey, doodad, hickey, whosis, whatsis, dingus, Rube Goldberg, *Chiefly US and Canadian* gizmo *or* gismo: *He uses some kind of gimmick to break into cars.*

gingerly *adv.* **1** warily, cautiously, charily, carefully, delicately, °fastidious, daintily, squeamishly, tentatively, nervously, cannily, circumspectly, guardedly, watchfully, timidly, timorously, shyly: *Handle that nitroglycerine very gingerly.* —*adj.* **2** °wary, °cautious, chary, °careful, °fastidious, °delicate, °dainty, °squeamish, °tentative, °nervous, canny, circumspect, °guarded, watchful, °timid, timorous, °shy: *Note the gingerly way he handles the nitroglycerine.*

girl *n.* **1** female, °woman, °lass, (young) lady, °miss, mademoiselle, wench, *Fräulein,* °maid, °maiden, damsel, demoiselle, *Irish* colleen, *Australian and New Zealand* sheila, *Colloq* chick, filly, gal, *Slang* bird, °frail, skirt, °piece, mouse, *Brit* crumpet, bit of skirt *or* stuff, *Old-fashioned Brit* popsy, *US* dame, °broad, (bit of) San Quentin quail *or* jailbait: *His mother wants to introduce him to a nice girl.* **2** girlfriend, °sweetheart; betrothed, fiancée; °mistress, °lover, °friend, live-in lover, inamorata, °tally, *US* POSSLQ (= 'Person of the Opposite Sex Sharing Living Quarters'), *Colloq Brit* popsy, *Slang* moll, *US* °twist, °squeeze: *She's his girl and she'll stick by him.*

girth *n.* **1** circumference, °perimeter, ambit, °periphery, °circuit: *His girth steadily expanded as he approached fifty.* **2** °belt, girdle, °border, cincture, waistband, cestus, cummerbund, *Archaic* °zone, *US and Canadian* cinch: *Tighten that girth or the saddle will come off.*

gist *n.* °essence, °core, °heart, °substance, °point, °theme, °quintessence, °pith, °meat, marrow, °focus, °nub, °significance, (main *or* basic) idea; °direction, °drift: *It will save time if I tell you the gist of the argument.*

give *v.* **1** °present, °deliver, °pass (over), turn *or* hand over, °confer, °vouchsafe, °provide, °supply, °furnish, °bestow, °donate, °accord, °afford, °award, °hand out, °contribute, °distribute, °grant, °allow, °transfer, °make over, °entrust: *Please give me the envelope. I have given the files to the tax inspector. Give freely to charity.* **2** °exchange, °pay, °transfer, °trade, barter, swap *or* swop: *Philippa gave a week's salary for that dress.* **3** °impart, °pass on, °communicate, °transmit, °send, convey, °express: *Give my regards to Broadway.* **4** afflict with, °cause, °occasion: *Acid rock gives me a headache.* **5** °sacrifice, °devote, °dedicate, °yield (up), °surrender, °give up, °cede, °concede, consign, apply (oneself) to: *Those monks have given their lives to the study of Scripture.* **6** °present, °offer; °announce, °introduce: *Ladies and gentlemen, I give you Madja Kiziamainska, the Polish Nightingale!* **7** °present, °announce, °offer, °recite, °sing, °act, °perform, °read, °put on: *We were asked to give a performance for charity.* **8** utter, °emit, °give out (with), °issue: *He gave a shout to warn us.* **9** °yield, °relax, °give way, °fail, °collapse, °buckle, °break down, fall *or* come apart: *If that pillar gives, the balcony will come down.* **10** °cause, °lead, °induce, °persuade, °make, °prompt, °move, °dispose: *I was given to understand that I was not welcome.* **11** °cede, °relinquish, °concede, °allow, °surrender, °yield: *He gave me the first game but I lost the second.* **12** ***give away.*** **(a)** See **1,** above. **(b)** °reveal, °betray, °let out, divulge, °disclose, °expose, °inform on, uncover, °leak, °let slip; let the cat out of the bag; *Colloq* blow the whistle on, *Slang* rat on, *US* fink on: *The traitor gave me away, and I was arrested.* **13** ***give in.*** °yield, °submit, °give up, give ground, back away (from), back off, °capitulate, °surrender, admit defeat: *After trying for hours to make him change his mind, he finally gave in.* **14** ***give off.*** °give out, °emit, exude, °exhale, °discharge, send *or* throw out, °release, smell of: *This clothing gives off the scent of lavender.* **15** ***give out.*** **(a)** See **14,** above. **(b)** °distribute, °hand out, °give, °deal (out), pass out *or* around, disseminate, °dispense, allot, apportion, allocate, °assign, °distribute, °issue, mete *or* hand out, °ration (out), °dole (out), °pay, *Colloq* dish *or* fork out, °shell out: *The prizes will be given out to the winners at a special ceremony. She was giving out handbills to passersby.* **(c)** °publish, °announce, °make known, °broadcast, °impart, °issue, °make public, °reveal: *You shouldn't give out that kind of information.* **(d)** become exhausted, be reduced *or* depleted, °fail, °run out: *What will we do when the food gives out?* **(e)** See **8,** above. **16** ***give over.*** °assign, °resign, °hand over, °surrender, °relinquish, pass *or* over, °give up; °assign, °entrust: *We gave our dog over to Lambert's for training.* **17** ***give up.*** **(a)** °abandon, °stop, °cease, °quit, °leave off, °forgo,

°forsake, °renounce, desist from, °swear off, abstain from; °reject; *Colloq* °cut out, chuck: *I am trying to give up smoking.* **(b)** °surrender, °capitulate, °yield, °cede, °concede, °give in (to defeat), throw in the towel *or* sponge; °despair: *Surrounded by the enemy, the entire platoon gave up. I have tried to convince him but finally gave up.* **(c)** See **4,** above. **(d)** See **13,** above. **(e)** See **16,** above. —*n.* **18** °slack, °play, °leeway, °excess; °flexibility, °stretch: *There is very little give in the rules governing working hours. Nylon is used for boat anchor rodes because it has some give.*

give-and-take *n.* compromise, °cooperation, reciprocity, interaction, fair exchange, teamwork, joint effort, synergy: *There is a lot of give-and-take in our relationship—I give and she takes.*

given *adj.* **1** stated, accepted, agreed(-upon), delineated, confirmed, °noted, affirmed, specified, settled, °set, (pre)arranged, preordained, foreordained: *You must solve the problem using the given information.* **2** presupposed, °assumed, °understood, postulated, premised, conceded, acknowledged, allowed: *I think we can take her honesty as given.* **3** °prone, °accustomed, dedicated, addicted, °inclined, °disposed: *Our headmaster is not exactly given to flights of fancy.* —*n.* **4** assumption, donnée; °fact, °certainty, °reality, actuality, °gospel, the truth: *Can we accept her interest in the job as a given?*

glad *adj.* **1** °happy, pleased, contented, gratified, satisfied; °delighted, thrilled, °joyful, °overjoyed, tickled, *Colloq* tickled pink *or* to death, pleased as punch, *Slang Brit* chuffed: *I am genuinely glad to see you.* **2** °(ready and) willing, °happy, °keen, °eager, (well-)disposed, °inclined, °ready: *I would be glad to help you, madam.*

gladden *v.* °cheer, °enliven, °brighten, °delight, hearten, exhilarate, elate, °buoy (up), °animate: *It certainly gladdened me to be home again.*

gladly *adv.* cheerfully, °happily, °readily, °willingly, with pleasure: *I'd go with you gladly if I had the time.*

glamorous *adj.* **1** alluring, fascinating, charming, °attractive, °magnetic, captivating, °enthralling, °desirable, appealing, °enchanting, entrancing, intriguing, beguiling, bewitching, magical: *Esmé designs clothes for the most glamorous women in the world.* **2** °chic, °smart, °stylish, °fashionable, *Colloq* °trendy: *She always wears the most glamorous clothes.*

glamour *n.* allure, °fascination, °charm, attractiveness, °brilliance, °glitter, °attraction, °magnetism, charisma, captivation, desirability, °appeal, °enchantment, bewitchment, witchcraft, °sorcery, °magic: *Who today can match the glamour of the movie stars of the '30s?*

glance *v.* **1** glimpse, °peek, °peep, °scan, °look, *Colloq Brit* have a shufty *or* shufti at, take a dekko at: *I had a chance only to glance at the newspaper today.* **2** °reflect, glint, °glisten, °shimmer, °twinkle, °gleam, °shimmer, °flicker, glimmer, °sparkle, scintillate, °glitter, °flash: *A ray of sunlight glanced off the windows opposite.* **3** °bounce (off), °reflect, ricochet, °rebound, carom: *The stone glanced off the pavement and struck my leg.* —*n.* **4** glimpse, °peek, °peep, °look, *coup d'oeil, Colloq* gander, *Brit* shufty *or* shufti, dekko: *She turned and gave him a questioning glance, As he fingered the tip of his fer-de-lance.* **5** °gleam, glint, glimmer, °shimmer, °twinkle, °sparkle, scintillation, °glitter, °flicker, °flash: *The silver light, with quivering glance, / Played on the water's still expanse.*

glare *n.* **1** °dazzle, °brilliance, brightness, °splendor, resplendence, °radiance, effulgence, °luster, °shine, °flame, °flare, °blaze: *We had to shade our eyes from the glare.* **2** °frown, dirty *or* nasty *or* black look, °scowl, °stare, glower, lower *or Brit also* lour: *She shrank from his fierce glare.* **3** garishness, gaudiness, floridity *or* floridness, flashiness, tawdriness, showiness, °ostentation, meretriciousness: *His writing is filled with the pomp and glare of rhetoric.* —*v.* **4** °frown, give a dirty *or* nasty *or* black look, °scowl, °stare, glower, lower *or Brit also* lour, look daggers (at): *She simply glared at him and he withered.*

glaring *adj.* **1** °blatant, °flagrant, egregious, °conspicuous, °obtrusive, °prominent, °evident, °obvious, °manifest, °patent, °overt, °clear, °visible, unconcealed, °outrageous, °gross, flagitious, °atrocious, heinous, °shameless, °disgraceful, °shocking, °scandalous: *They pointed out a glaring omission in the proposal.* **2** °garish, °dazzling, °brilliant, blinding, blazing, °vivid, °harsh, °strong: *I was blinded when we first walked out into the glaring sunlight.*

glass *n.* **1** glassware, crystal: *Use the best glass for the party.* **2** °mirror, looking glass: *He constantly looks at himself in the glass.* **3** window, °pane, windowpane, plate glass: *The children pressed their noses against the glass of the toy shop.* **4** tumbler, drinking glass, beaker, goblet: *May I have a glass of beer?* **5** barometer: *The glass fell as the storm approached.* **6** lens, magnifying glass, °telescope, spyglass, microscope: *Look at this drop of water through the glass.* **7** ***glasses.*** °spectacles, eyeglasses, lorgnon, lorgnette, opera glasses, binoculars, field glasses, bifocals, trifocals, goggles, sunglasses, *Colloq* specs: *These are my reading glasses.*

glassy *adj.* **1** shining, °shiny, gleaming, °smooth, °slippery, °slick, °glossy, °icy, mirrorlike, *US* °slick: *We skidded on the glassy surface and crashed into a wall.* **2** °fixed, staring, trancelike, hypnotic, °vacant, °empty, expressionless, °blank, °void, vacuous, dazed, °dull, glazed, °cold, °lifeless: *His glassy stare made me very uncomfortable.*

glaze *v.* **1** varnish, lacquer, shellac, enamel, °coat, °cover; °polish, burnish, °shine, °gloss: *Many of the works of art had been glazed for protection.* —*n.* **2** varnish, lacquer, shellac, enamel, coating, covering; °polish, °shine, °gloss, °luster, patina: *The glaze distorts the design beneath.*

gleam *n.* **1** °light, glimmer, glint, °glow, °flicker, °shine, °shimmer, °glitter, °twinkle, °spark, °flare, °glare, °flash; °beam, °ray, °shaft: *I could just make out the faint gleam of a candle.* **2** °hint, °suggestion, °indication, °vestige, °trace, scintilla, °inkling, glimmer, °ray, °spark, °flicker: *There wasn't a gleam of hope of rescue.* **3** °look, glint: *He got a strange gleam in his eye when he talked about his inventions.* —*v.* **4** glimmer, glint, °shimmer, °shine, °twinkle, °glitter, °glisten, °beam, °sparkle: *I could see the beast's eyes, gleaming in the dark. The shore was dotted with gleaming white houses.*

glee *n.* °delight, °exuberance, cheerfulness, high *or* good spirits, °cheer, exhilaration, elation, exultation, °joy, °happiness, °rapture, gladness, felicity, °pleasure, joyfulness, °merriment, jubilation, joyousness, joviality, jollity, °gaiety, mirthfulness; °enjoyment, °satisfaction, *Schadenfreude*: *We watched with great glee as the villain was thwarted.*

gleeful *adj.* °happy, °merry, °joyful, °delighted, °exuberant, °ecstatic, °cheerful, in high *or* good spirits, exhilarated, °elated, °exultant, °rapturous, °overjoyed, pleased, jubilant, joyous, jovial, °jolly, °gay, mirthful: *He wrote a gleeful report of the chairman's conviction for insider trading.*

glib *adj.* °ready, °fluent, °smooth, °slick, facile, smooth-spoken, smooth-tongued, smooth-talking, fast-talking, °fluid, °easy, unctuous, °suave, °nonchalant, °superficial: *Why must the obituary notices always be so glib?*

glide *v.* °slide, °slip, °coast, skate, °soar, °float, °sail, glissade, °stream, °flow: *The skiff glided peacefully down the river.*

glisten *v.* °shine, °reflect, glint, glimmer, °gleam, °sparkle, °glitter, wink, °blink; °glow, °gleam, °twinkle: *The lamplight glistened on his wet coat. A tear glistened on her cheek.*

glitter *v.* **1** See **glisten,** above. —*n.* **2** See **gleam, 1,** above. **3** See **glamour,** above. **4** showiness, gaudiness, garishness, °flash, flashiness, °ostentation, floridity *or* floridness, °spectacle, °pageantry, °splendor, refulgence, °brilliance, *Colloq* pizazz, razzle-dazzle, razzmatazz, *Slang US* glitz: *In those days, Hollywood was all glitter.*

gloat *v.* Often, **gloat over.** exult (in), °glory (in), °relish (in), °revel (in), crow (over *or* about), °delight (in): *He is still gloating over the misery he caused her.*

global *adj.* worldwide, °international, °broad, °extensive, wide-ranging, far-reaching, °epidemic, pandemic, °universal: *Protection of the atmosphere is a global responsibility.*

globe *n.* **1** °earth, °world, planet, Terra: *Our family is scattered all around the globe.* **2** °sphere, ball, °orb; globule: *On the table was a lamp with a green glass globe.*

gloom *n.* **1** shadowiness, gloominess, °shade, °shadow, murkiness, murk, dimness, °dusk, dullness, °dark, darkness, cloudiness, blackness, °obscurity: *We arose in the gloom of a midwinter's morning.* **2** despondency, °depression, °sadness, dejection, downheartedness, °melancholy, °woe, °sorrow, moroseness, °desolation, low spirits, blues, doldrums, °despair, dolor, °misery, *Colloq* dumps: *The team suffered the gloom of defeat.*

gloomy *adj.* **1** °shadowy, shaded, °shady, °murky, °dim, °dusky, °dull, °dark, cloudy, °overcast, °obscure, °black, inky, *Literary* Stygian: *It is too gloomy a day to have a picnic.* **2** depressed, °melancholy, °sad, °dejected, morose, °glum, lugubrious, °unhappy, cheerless, °dismal, °moody, down, downcast, °desolate, °doleful, °sorrowful, crestfallen, chapfallen, °downhearted, °forlorn, °despondent, °miserable, °joyless, dispirited, despairing, °dreary, °sullen, °blue, distressed, °down in the mouth, in the doldrums, saturnine, *Colloq* (down) in the dumps: *Both of them have been very gloomy since the divorce.* **3** depressing, cheerless, °dreary, °dismal, dispiriting, °sad, disheartening: *The decor is much too gloomy for a doctor's waiting room.*

glorified *adj.* **1** overrated, °pretentious, overdone, high-flown, high-sounding, °affected, °pompous, °exalted, *Colloq* jumped-up: *You say he is a scholar, but he behaves more like a glorified schoolboy.* **2** °sham, °pretend, °imitation, °counterfeit, °fake, °substitute, ersatz, *Colloq* phoney *or US also* phony: *In that outfit, she looks like a glorified chorus girl.*

glorify *v.* **1** elevate, °exalt, °raise (up), upgrade, °promote, °advance, °boost, °enhance, °dignify, ennoble, °immortalize: *Winning first prize glorified his reputation considerably. In her book she glorifies motherhood. These men are glorified by their heroism.* **2** canonize, deify, °idolize, °revere, °venerate, °sanctify, °worship, pay tribute *or* homage to, ennoble, °idealize, apotheosize, °eulogize, panegyrize, °adore, °honor, °look up to, °celebrate, °extol, °praise, °laud, commend, °hail, lionize, °applaud, acclaim: *The world glorified Lindbergh for the first solo flight across the Atlantic.*

glorious *adj.* **1** °illustrious, famed, °famous, °renowned, °celebrated, °distinguished, honored, °eminent, °excellent: *England may be proud of her glorious literary heritage.* **2** °outstanding, °splendid, °magnificent, °marvelous, wonderful, spectacular, °fabulous, °dazzling: *1942 marked a glorious victory over the Germans at El Alamein.* **3** enjoyable, °delightful, °fine, °great, °excellent, °pleasurable, °superb, *Colloq* °heavenly: *We had a glorious holiday in the Greek islands.* **4** °beautiful, °splendid, °brilliant, °gorgeous, resplendent, °admirable, °superior, °excellent, °estimable: *The walls are covered with frescoes in glorious color.*

glory *n.* **1** °honor, °fame, repute, °reputation, exaltation, °celebrity, °renown, eminence, °distinction, illustriousness, °prestige, °dignity, immortality: *Our soldiers fought for glory not for gain. Even today we sense the glory that was Rome.* **2** °honor, °veneration, °reverence, °homage, °gratitude, glorification, exaltation, °worship, adoration, °praise, laudation, thanksgiving; benediction, °blessing: *Glory be to God in the highest.* **3** °splendor, °pomp, magnificence, °grandeur, °beauty, °brilliance, °radiance, effulgence, refulgence, °excellence, °pageantry, °nobility, °triumph, greatness: *Her photographs depict the Himalayas in all their glory.* **4** aureole, nimbus, °halo; °crown, circlet, corona: *A glory surrounds the saint's head in the painting.* —*v.* **5** °revel, °relish, °delight, °exult, pride oneself, crow, °rejoice, °gloat; °show off, °boast: *Let him glory in his temporary conquest.*

gloss[1] *n.* **1** °sheen, °luster, °polish, °glow, °glaze, °shine, °gleam, burnish, brightness: *I prefer a dull gloss to a high polish on furniture.* **2** °show, façade, °mask, °front, °surface, °veneer, °disguise, °camouflage, false appearance, °semblance: *She soon saw through the gloss, and the honeymoon was over.* —*v.* **3** °glaze, °polish, burnish, °shine: *Gloss up your shoes a bit.* **4** Usually, ***gloss over.*** °veil, °cover up, °smooth over, °conceal, °hide, °disguise, °camouflage, °mask, *Colloq* °whitewash: *He tried to gloss over his voting record.*

gloss[2] *n.* **1** °explanation, °interpretation, exegesis, explication, °definition, elucidation, °comment, commentary, annotation, critique, °criticism, °analysis, footnote; °translation: *This edition of Chaucer contains glosses of Middle English words.* —*v.* **2** comment on *or* upon, °explain, °interpret, explicate, °define, elucidate, annotate, °criticize, °analyze, °review, *US* critique; °translate: *Johnson was not the first to gloss the word "pastern."* **3** See **gloss**[1]**,** 4, above.

glossary *n.* °gloss, (specialized *or* special-subject) °dictionary, wordbook, word list: *A useful glossary of architectural terms appears at the end of the book.*

glossy *adj.* **1** shining, °shiny, °smooth, °polished, glazed, °lustrous, burnished, °smooth, °sleek, waxed, °glassy, glistening: *Our magazine is printed on glossy paper.* **2** °slick, °specious, °put-on, °artificial, meretricious, contrived, °pretended, simulated, feigned, °insincere, pseudo, °false, °unreal; °bogus, °counterfeit, °fraudulent, °imitation, *Colloq* phoney *or US also* phony: *It is only a glossy remake of the original film.*

glow *n.* **1** luminosity, phosphorescence, incandescence, °light, lambency, °luster: *The surface of the flying saucer emitted a faint green glow.* **2** °light, brightness, °gleam, luminousness, °brilliance, °radiance, resplendence, °splendor, effulgence: *The glow in the east is from a forest fire.* **3** °flush, °blush, redness, ruddiness, °burning, °excitement, °warmth, °fervor, fervency, °enthusiasm, feverishness, °thrill, *Colloq* °rush: *I can feel a glow just thinking of you.* —*v.* **4** °shine, °radiate, incandesce, phosphoresce, glimmer, °gleam, °light up: *The numerals on this clock glow in the dark.* **5** °heat, overheat, °burn; ablate: *The spaceship began to glow as it entered the atmosphere.* **6** °flush, bloom, °color, °blush: *As you can see from her complexion, she simply glows with good health.* **7** °blush, °flush, redden, °color, turn red *or* scarlet: *My cheeks glowed with embarrassment.*

glowing *adj.* **1** aglow, °incandescent, °burning, lambent, °luminous, candent; smoldering: *I stirred the glowing embers.* **2** °rich, °warm, vibrant, °bright, °brilliant: *Banners in glowing colors enlivened the banquet hall. He is in glowing health.* **3** °laudatory, °complimentary, °enthusiastic, eulogistic, °rhapsodic, °favorable, encomiastic, panegyrical: *The critics described her performance in glowing terms.*

glue *n.* **1** °cement, adhesive, mucilage, gum, paste: *A special glue is needed to put together these plastic models.* —*v.* **2** °cement, paste, °stick, affix, °fix, °seal: *Let's glue this picture into the album.*

glum *adj.* °gloomy, °sullen, morose, dispirited, °woebegone, °dismal, °sad, sulky, °dour, °moody, °sour, crestfallen, °doleful, down, °low, °pessimistic, lugubrious, saturnine: *After Irena left, I was feeling rather glum.*

glut *n.* **1** °excess, °surplus, overabundance, superabundance, °surfeit, oversupply, overflow, °superfluity, nimiety: *The glut of razor blades in the market is owing to increased production.* **2** saturation, glutting, satiation: *Glut leads to a lowering of prices.* —*v.* **3** oversupply, °flood, °saturate, °swamp, inundate, deluge, °overload, overstock, °clog, °stuff, °gorge: *The markets will soon be glutted with mobile telephones.* **4** °satiate, sate, °choke, °cram, °overload, overfeed, °gorge, °surfeit, °pall, cloy, °jade, °sicken, °weary: *Everyone was thoroughly glutted before the reception was over.*

glutton *n.* trencherman, gormandizer, gourmand *or* gormand, overeater, hog, pig, *grangousier, Colloq* greedy-guts, *Slang Brit* gannet, *US* chowhound: *Like the glutton that he is, he asked for more after eating an enormous meal.*

gluttonous *adj.* °voracious, gormandizing, edacious, °greedy, °ravenous, insatiable, esurient, piggish, °hoggish, swinish: *She was so gluttonous that she ate my dinner after finishing her own.*

gluttony *n.* overeating, gormandizing, gormandism, °greed, hoggishness, piggishness, °rapacity, voraciousness, greediness, voracity, insatiability, edacity, crapulence, crapulousness, intemperance, immoderation, *Archaic* gulosity: *The Bible categorizes gluttony among the seven deadly sins.*

gnarled *adj.* twisted, knotty, °lumpy, °bumpy, knotted, °bent, °crooked, distorted, contorted, warped; , arthritic: *I concealed myself in the gnarled branches of the old oak. The beggar reached out to me with her gnarled hand.*

gnaw *v.* **1** °chew, nibble, °eat, °bite, champ: *The marks were made by deer gnawing the bark.* **2** °erode, eat away, corrode, wear down *or* away, °fret, °consume, °devour: *The acid continues to gnaw away at the metal till it is gone.* **3** °fret, °irritate, harry, hector, °pester, °worry, °bother, °plague, °trouble, °torment, torture, °distress, badger, °harass, °haunt, °nag, vex, °gall, nettle, °irk, peeve, °annoy: *The feeling that something is very wrong continues to gnaw at me.*

go *v.* **1** move (ahead *or* forward *or* onward), °proceed, °advance, °pass, °make headway, °travel, voyage, °set off, °tour, trek, wend, °stir, budge: *Would you go to the market for me?* **2** °leave, °depart, °go out, move (out *or* away), decamp, °make off, °withdraw, °repair, °retire, °retreat, *Colloq* °take off: *I wish he would go at once. She went to her country cottage for the weekend.* **3** °function, °operate, °work, °run, °perform: *I cannot get the engine to go properly.* **4** °lead, open to, give access to, communicate to *or* with, connect with *or* to: *Where does this door go?* **5** °lead, °communicate with, °run: *Does this road go to Oxford?* **6** °fit, °belong (together), agree *or* conform (with each other), harmonize, °blend, °match, be appropriate *or* suitable (for *or* to), complement each other: *These colors don't go.* **7** °become: *He went mad when he learned about the accident.* **8** °fit, °extend, °reach, °span, °stretch: *My belt would never go around your waist!* **9** be disposed of *or* discarded *or* thrown away, be dismissed, be got rid of *or* abolished, be given up, be cast *or* set *or* put aside, be done with: *That outdated computer has to go.* **10** °disappear (without a trace), vanish (into thin air), °evaporate: *Where has all the money gone?* **11** °pass, °elapse, slip *or* tick away, °fly: *Time goes quickly when you're having fun.* **12** °fail, °fade, °decline, °flag, °weaken, °degenerate, °wear out, °give (out); °give way, °collapse, fall *or* come *or* go to pieces, °disintegrate, °crack: *I'm afraid the tires are going. When that last support goes, the roof will come down.* **13** °die, °expire, be gone, meet one's maker, pass on *or* away, shuffle off this mortal coil, go to one's reward, go to the happy hunting ground, go to that great cricket pitch in the sky, *Slang* kick the bucket, snuff it: *By the time the doctor arrived Graham had gone.* **14** °sound, °pronounce, °enunciate, articulate, °say, utter: *And this little piggy goes "Wee, wee, wee," all the way home.* **15** °survive, °last (out), °endure, °live, °continue: *How long can we go without water?* **16** be used up *or* consumed *or* finished: *The last of our food was gone.* **17** go to the toilet *or* the lavatory *or* the bathroom, move (one's) bowels, °urinate, °defecate, *Slang* pee, take a leak *or* a crap, *chiefly Brit* go to the loo, *Chiefly US* go to the john, *Taboo slang* (take a) piss *or Taboo slang* shit: *We stopped at a turnpike filling station because Jane had to go.* **18** ***go about.*** °approach, °tackle, °set about, °undertake, °begin, °start: *I don't like the way she goes about her work. How does one go about establishing a business?* **19** ***go ahead.*** °proceed, °continue, move *or* go forward, °advance, progress, °go on: *She told me I could go ahead with the scheme. The policeman motioned to go ahead.* **20** ***go along (with).*** **(a)** °escort, °accompany: *We asked if we could go along with them to the movies.* **(b)** °agree (to), concur (with), acquiesce (to), assent (to), °support: *Beverly would never go along with a plan like yours.* **21** ***go around*** *or* ***about*** *or Brit usually* ***round (with).*** **(a)** move *or* go around, °circulate: *I wish he'd stop going around telling everyone about me. There's a lot of flu going around.* **(b)** °revolve, °rotate, °spin, whirl, °twirl: *The earth takes a year to go around the sun.* **(c)** °socialize (with), frequent *or* seek the company of, spend time with, °associate with, *Colloq* hang around *or* about (with), °hang out (with): *That Grant boy has started going around with the Collins girl.* **(d)** wander *or* move around *or* about: *He goes about picking through rubbish bins.* **(e)** °suffice, be sufficient *or* adequate *or* enough, °satisfy: *Are there enough life jackets to go around?* **22** ***go at.*** °attack, °assault, assail: *We went at the enemy with all the fire power we could muster.* **23** ***go away.*** °go (off), °leave, °depart, °withdraw, °exit; °retreat, °recede, decamp: *The clouds went away and the sun came out. We are going away for the weekend.* **24** ***go back (to).*** **(a)** °return (to); °revert (to), change back (to): *He went back to his old job after the war. Can we go back to the way things were before we were married?* **(b)** °originate (in), begin *or* start (with), date back (to): *Our friendship goes back to our childhood.* **25** ***go back on.*** °renege (on), °break, °retract, °repudiate, °forsake: *She's gone back on our agreement.* **26** ***go by.*** **(a)** °pass (by), go past, move by; °elapse: *We used to watch the goods trains go by. The months went by quickly since our last meeting.* **(b)** rely *or* count *or* depend *or* bank on, put faith in(to), be guided by, judge from: *You cannot go by what Atherton tells you.* **27** ***go down.*** **(a)** °sink, °go under, °founder, °submerge: *The ship went down within minutes of striking the mine.* **(b)** °decrease, °decline, °drop, °sink, °drop: *The Nikkei Index went down two hundred points.* **(c)** °fall, be defeated *or* beaten, suffer defeat, °lose, °collapse: *Our forces went down under an onslaught from the attacking armies.* **(d)** be remembered *or* memorialized, memorializedor recalled *or* commemorated *or* recorded: *That day of infamy will go down in history.* **(e)** find favor *or* acceptance *or* approval, be accepted: *His ideas have not gone down well with the council.* **28 *go for.*** **(a)** °fetch, °obtain, °get: *Please go for help.* **(b)** apply *or* relate to, °concern, °involve: *The rule against smoking goes for you, too, Smedley.* **(c)** °fancy, °favor, °like, °admire, be attracted to, °prefer, °choose, *Slang* °dig: *I can tell that Lord Percy really goes for Matilda. I could go for a pint of beer right now.* **(d)** °attack, °assault, assail, °set upon: *The dog went for him as soon as he opened the gate.* **(e)** set one's sights on, aim for, focus attention *or* effort(s) on: *I have little chance, but I am going to go for the first prize.* **29 *go in for.*** **(a)** °go into, °enter, enroll in, °start, °begin, °embark on, °pursue, °take up, °embrace, espouse, °undertake, °follow, °adopt, °go into, *US* go out for: *He is going in for a career in boxing.* **(b)** °like, °fancy, °favor, °practice, do, °engage in: *I don't go in for mountain climbing.* **30 *go into.*** **(a)** See **29 (a),** above. **(b)** delve into, °examine, °pursue, °investigate, °analyze, °probe, °scrutinize, °inquire into, °study: *I want to go into the subject of your absences with you, Fanshawe.* **(c)** °touch on, °discuss, °mention: *I should avoid going into the subject of money with Pauline if I were you.* **31 *go off.*** **(a)** °go out, cease to function: *I saw the lights go off at nine.* **(b)** °explode, °blow up, detonate, °erupt; °fire, be discharged: *The bomb is set to go off in an hour. The gun went off, killing the mouse.* **(c)** °occur, °happen, °take place: *The conference went off as planned.* **(d)** °depart, °leave, °go (away), °set out, °exit, decamp, °quit: *She went off without another word.* **(e)** °deteriorate, °rot, molder, go stale, go bad, °spoil, °sour, °turn: *Two days outside the fridge and the milk goes off.* **(f)** Usually, **go off into.** break into *or* **out in, start off into**: *He goes off into gales of laughter whenever I mention your name.* **32 *go on.*** **(a)** °continue, °proceed, °keep on, °carry on; °persist, °last, °endure, °persevere: *He went on whistling after I asked him to stop. The party went on into the wee hours.* **(b)** °happen, °occur, °take place, °come about, *Colloq* °come off: *I have always wondered what went on in there.* **(c)** come on, begin *or* resume functioning: *The lights went on at midnight.* **(d)** °enter, make an entrance: *She doesn't go on till the third act.* **(e)** gabble, °chatter, drone on, *Brit* natter, *Colloq Brit* witter (on), rabbit on: *He goes on endlessly about his cars.* **(f)** rely *or* depend on, °use: *The detective had very little to go on.* **33 *go out.*** **(a)** fade *or* die (out), °expire, cease functioning, °go off, be extinguished: *The lights went out, throwing the room into Stygian blackness.* **(b)** °depart, °leave, °exit: *He went out at six and has not been seen since.* **(c)** °socialize, °associate; court, °go together, *Brit* °walk out, *US* °date: *Harry is going out with Annabel.* **34 *go over.*** **(a)** °review, skim (through *or* over), °go through, °scan, °look at, °read, °study; °inspect, °examine, °scrutinize, °investigate: *I went over your report last night. They are going over everyone's luggage with a fine tooth comb.* **(b)** be received: *The first song went over very well.* **(c)** °clean, tidy *or* neaten (up): *I've just gone over the entire apartment.* **(d)** °rehearse, °repeat, °reiterate, °review, °go through; retrace: *We keep going over the same things, again and again.* **35 *go through.*** **(a)** °experience, °suffer, °undergo, °bear, °take, °stand, °tolerate, °put up with, °brook, °submit to, °endure, live through, °brave: *I don't think I could go through another war.* **(b)** be accepted *or* approved, °pass (muster): *The bill went through without a hitch.* **(c)** See **34 (a),** above. **36 *go together.*** **(a)** harmonize, °accord, °agree, °fit, °go, suit each other, °belong (with each other): *I don't think that puce and vermilion go together.* **(b)** See **33 (c),** above. **37 *go under.*** **(a)** See **27 (a),** above. **(b)** °fail, go bankrupt, *Slang* go belly-up: *The statistics show that more than five hundred companies go under every week in the US.* **38 *go up.*** **(a)** °rise, °increase: *If inflation goes up, interest rates will probably be raised again.* **(b)** °explode, °blow up: *The munitions factory went up, destroying several houses in the area.* **39 *go with.*** **(a)** go together with, harmonize with, blend with, be suitable *or* suited for, fit (in) with, accord *or* agree with: *That scarf does not go with the dress.* **(b)** socialize with, °associate with, °accompany, court, *Brit* walk out with, *US* °date: *I hear that Connie is going with Don.* **40 *go without.*** do *or* manage *or* get by without, °lack, be deprived of, °need; abstain from, survive *or* live *or* continue without: *In the old days, if you could not afford something, you went without. She cannot go without a cigarette for more than an hour.* —*n.* **41** °chance, °turn, °opportunity, °try, °attempt, *Colloq* whack, °crack, whirl, °shot, °stab: *I don't expect much, but I'll have a go anyway.*

go-ahead *n.* **1** °permission, °approval, °leave, authorization, °sanction, *Colloq* °say-so, okay *or* OK, green light, *US* the nod: *I have the go-ahead to proceed with the project.* —*adj.* **2** °ambitious, °enterprising, °progressive, forward-looking, °resourceful: *The directors, all under thirty, make it a real go-ahead company.*

goal *n.* °object, °aim, °purpose, °end, °objective, °target, °ambition, °ideal, °aspiration: *Fletcher's goal is to be head of the company.*

gob *n.* chunk, °piece, °blob, °lump, gobbet, °morsel, °fragment, °bite: *She took a gob of peanut butter and spread it on the bread.*

gobbledegook *n.* **1** gobbledygook, °jargon, °nonsense, °gibberish, °moonshine, °rubbish, tommyrot, °mumbo jumbo, humbug, balderdash, eyewash, hogwash, poppycock, °drivel, *Colloq* bunk, °rot, °garbage, bosh, pish and tush, piffle, bilge (water), codswallop, *Slang* crap, malarkey *or* malarky, bull, *Brit* (load of old) cobbler's, *Taboo slang US* bullshit: *Can you make sense out of all that computer gobbledegook?* **2** gobbledygook, equivocation, double-talk, °deception, deceptiveness, vagueness, quibbling, circumlocution, obscurantism, obfuscation, ambagiousness, shiftiness: *People often find the gobbledegook in contracts very confusing.*

go-between *n.* °intermediary, °agent, middleman, °medium, °mediator, °negotiator, °messenger, internuncio, °liaison; intercessor, interceder: *David served as a diplomatic go-between in our negotiations with the rebels.*

goblin *n.* elf, gnome, hobgoblin, °imp, kobold, leprechaun, °demon, brownie, pixie, nix *or* nixie: *On Halloween the ghouls and goblins will get you if you don't watch out!*

god *n.* °deity, demigod, demiurge, divinity, °spirit, °immortal, °genius, °power, tutelary, numen: *Throughout man's sojourn on earth he has worshiped many gods.*

godless *adj.* **1** °wicked, °evil, iniquitous, °sinful, unrighteous, unholy, hellish; °impious, °blasphemous, °profane, °sacrilegious, °ungodly: *The prison was a godless place, where one was murdered for a crust of bread.* **2** atheistic, nullifidian, agnostic, °unbelieving, °skeptical: *In theory, communism is a godless ideology.*

godlike *adj.* **1** °divine, °godly, °sacred, °holy, °saintly, angelic, °seraphic, blest, blessed, sainted: *The savages thought her a goddess and treated her with godlike reverence.* **2** °heavenly, °celestial, blissful, °rapturous, °ecstatic, beatific, ethereal: *For a decade, she lived a godlike existence among the savages.*

godly *adj.* °religious, °pious, °devout, God-fearing, °good, °righteous, °holy, °virtuous, °moral, °pure,

°saintly, reverent, pietistic, °devoted, °faithful: *The monks pursue a godly life in their mountain fastness.*

godsend *n.* °gift, °blessing, benediction, °boon, °windfall, bonanza, stroke of (good) fortune, piece *or* bit of (good) luck: *The Red Cross parcels were a godsend to the prisoners of war.*

goggle-eyed *adj.* °agog, awe-struck, wide-eyed, °thunderstruck, agape, open-mouthed, gawking, staring, dumfounded *or* dumbfounded, astonished, astounded, amazed, stupefied, dazed, surprised: *His first time in the city, Eugene stood goggle-eyed at the skyscrapers.*

going *adj.* **1** thriving, °successful, succeeding, °prosperous, °wealthy, affluent, booming, prospering, °flourishing, growing: *The Davises took over a failing business and turned it into a going concern.* **2** °current, °present, °contemporary, °active, °effective, accepted, °prevailing, °prevalent, °universal, °common, °usual, °customary: *What is the going rate for a skilled cabinetmaker?* **3** ***going on.*** approaching, nearing, °nearly, °almost, not quite: *He's six going on seven. It's going on eight o'clock.*

golden *adj.* **1** yellow, yellowish, gold, blond *or* blonde, flaxen, aureate; tow(-haired): *Her golden tresses fell about her shoulders.* **2** gold, auriferous; gilded, gilt, aureate; *Technical* auric, aurous: *They drank together from a golden bowl.* **3** °bright, shining, °brilliant, °sunny, gleaming, °lustrous, °shiny, glittering, °dazzling, resplendent, °radiant, °glowing, sparkling: *We emerged from the black tunnel into the golden sunshine.* **4** °happy, blissful, °delightful, °joyful, °glorious, joyous, °exuberant: *We failed trying to relive the golden days of our youth.* **5** °flourishing, halcyon, °prosperous, thriving, °favorable, °excellent, °outstanding, °productive, °fertile, blessed, blest, °good, °successful, palmy: *After the golden age of Greece and Rome, Europe was plunged into the dismal gloom of the Dark Ages.* **6** °gifted, °talented, °favored, °special, °exceptional, °favorite, cherished, °pet, *Brit* blue-eyed, white-headed, white-haired, *US* fair-haired: *As long as the company's profits were increasing, Ian was the golden boy.* **7** °advantageous, °propitious, auspicious, °promising, °rosy, °opportune, °optimistic, °favorable: *She missed a golden opportunity by turning down the job.*

good *adj.* **1** °agreeable, °satisfactory, commendable, °acceptable, °fair, °adequate, admissible, °tolerable, all right, °passable, *Colloq* okay *or Colloq* OK: *According to the reviews, the new opera is good, but not great.* **2** °admirable, °outstanding, °first-rate, first-class, °fine, °superb, °superior, tiptop, °extraordinary, °exemplary, °choice, °excellent, °capital, °marvelous, wonderful, °splendid, °sterling, *Colloq* super(-duper), °great, smashing, cracking, °fantastic, °terrific, °unbelievable, groovy, °fabulous, fab, °serious, A-1 *or* A-one, *Old-fashioned Brit* tickety-boo, *US* fab, fantabulous, °bad, °major, *Brit* °brilliant, brill, ace, °crucial, *North England* °champion, *Chiefly US* A-OK: *I thought that was a really good dinner.* **3** °correct, °proper, °decorous, °orderly, °right, °seemly, °fit, °fitting, °suitable, °meet, °appropriate, allowable, °permissible, admissible, °passable, °satisfactory, °tolerable: *He should get time off for good behavior.* **4** °obedient, well-behaved, °proper, well-mannered: *Why can't you be a good boy when we go out?* **5** °moral, high-minded, °righteous, °noble, °wholesome, °chaste, °pure, °honorable, °ethical, upstanding, °upright, °virtuous, °worthy, °lofty, °elevated, °saintly, angelic, °godly, °godlike: *Think only good thoughts and still you may not get to heaven.* **6** °kind, °benevolent, beneficent, °gracious, °gentle, °kindly, °nice, °considerate, °friendly, °solicitous, goodhearted, °sympathetic, °benign, °charitable, humane, kindhearted, well-disposed: *Her parents have always been good to me.* **7** °fresh, °unspoiled, °edible, consumable, palatable: *These eggs are good but the milk has gone sour.* **8** °genuine, valid, °legitimate, °authentic, °honest, °proper, °reliable, °secure, dependable, °safe, creditable, °sound, °solid, °substantial, well-founded, °trustworthy, °honest, °actual, °real; credible, believable, convincing, compelling, cogent: *Have you a good reason for saying that?* **9** °honorable, esteemed, respected, °respectable, °well-thought-of, °reputable, established, °solid: *They say that he comes from a good family.* **10** well-proportioned, °shapely, °attractive: *Yes, I'd say that Marilyn has a good figure.* **11** °thorough, °complete, °penetrating, °careful: *Has she had a good look at the defendant?* **12** °gifted, °talented, °competent, °capable, °skillful, °clever, °accomplished, °proficient, °adept, adroit, skilled: *Am I good enough to audition for the part?* **13** °advantageous, °propitious, °opportune, °beneficial, °profitable, °favorable; °safe, °secure, °reliable, °sound, °sensible: *Are utilities a good investment? Is this a good time to buy shares?* **14** °healthy, salubrious, salutary, °beneficial, °wholesome: *We'll have to put you on a good diet.* **15** °best, °company, Sunday, special-occasion, most luxurious: *Should we use the good glasses tonight?* **16** °ample, °sufficient, °adequate, °considerable, °full, °extensive, sizable, °large, °substantial: *We have a good supply of food available.* **17** approving, °complimentary, °flattering, °positive, °favorable, °enthusiastic, °laudatory, eulogistic, encomiastic: *His book received a very good review in the Sunday supplement.* **18** °great, °considerable, sizable, °substantial, °fair: *We are still a good distance away from land.* —*n.* **19** °benefit, °advantage, °profit, °use, °usefulness, °gain, °worth, avail: *What good does it do to complain?* **20** goodness, °morality, °virtue, °merit, righteousness, °right, °rectitude, °worth, °probity, virtuousness, °integrity, °nobility, high-mindedness, honorableness, °honesty: *The evil that men do lives after them, / The good is oft interred with their bones.* **21** ***goods.*** **(a)** °possessions, °(personal) property, chattels, °things, °gear, °belongings, °effects, °paraphernalia, movables, °stuff: *He's gone and he's taken all his goods with him.* **(b)** °merchandise, commodities, °wares, °stock, °produce, tangibles, °assets: *We shall pay on delivery of the goods.* **(c)** *US and Canadian* (incriminating) evidence *or US and Canadian* proof *or US and Canadian* information *or US and Canadian* documentation *or US and Canadian* facts *or US and Canadian* data: *The police have the goods on the murderer.* **(d)** °fabric, °cloth, textile, °material, yard *or* piece goods: *My tailor said he had made it from the best goods available.*

goodbye *interj.* Good-bye! *or* Goodby!, Farewell!, *Hawaiian* Aloha!; *Italian* Arrivederci!, Ciao!; *German* Auf Wiedersehen!; *French* Au revoir!, Adieu!; *Japanese* Sayonara!; *Spanish* ¡Adios!, ¡Hasta la vista!, ¡Hasta luego!; *Latin* Vale!; *Colloq* Bye! *or* 'Bye!, Bye-bye!, Toodle-oo!, So long!, *Brit* Tata!, Cheerio!, Cheers!, Cheery pip!, Toodle pip!, *Old-fashioned* Pip! Pip!, *US* See you later (Alligator)!: *As the sun sets over the casino, we say goodbye to Biarritz.*

good-for-nothing *adj.* **1** °worthless, °useless: *"You are asking too much for that good-for-nothing old lamp," Aladdin told the peddler.* —*n.* **2** ne'er-do-well, °wastrel, waster, °idler, °loafer, layabout, lazybones, slugabed, sluggard, black sheep, *Colloq US* goldbrick, goof-off: *Why doesn't that lazy good-for-nothing get a job?*

goodly *adj.* °considerable, sizable, °substantial, °ample, °great, °large, °significant, consequential: *We were paid a goodly sum for the house.*

good-natured *adj.* good-humored, °friendly, °agreeable, °genial, °gracious, good-hearted, °pleasant, °mellow, easygoing, °considerate, °nice, °kind, °kindly, kindhearted, tenderhearted, °charitable, °tolerant, °generous, °courteous, °cordial, °warm, warmhearted, °amiable, °amicable, cooperative: *He seems a good-natured fellow, but not too bright.*

goody-goody *adj.* °smug, °sanctimonious, °self-righteous, °priggish, prim, holier-than-thou, Pecksniffian, °hypocritical: *She is one of those goody-goody people who never do anything wrong.*

gooey *adj.* **1** gluey, °sticky, °tacky, glutinous, mucilaginous, gummy: *The meatloaf was smothered in a gooey, brown gravy.* **2** °sweet, sugary, saccharine, °sentimental, unctuous, cloying, syrupy, mawkish, °maudlin, *Colloq* °mushy, slushy, °sloppy: *They like to watch those gooey soap operas.*

gore[1] *n.* blood, °carnage, butchery, °slaughter, °bloodshed: *The current horror films have too much gore for my taste.*

gore[2] *v.* °pierce, °stab, °poke, horn, °penetrate, °puncture, spear, °gouge, °spit, °stick, °impale, disembowel: *The matador, severely gored by the bull, was carried out.*

gorge *n.* **1** °ravine, °canyon, defile, °pass, chasm, fissure, °crevasse, gully *or* gulley, wadi *or* wady, °gap, *Brit* gill *or* ghyll, *US and Canadian* °notch: *Water runs in this gorge only during the rainy season.* **2** °vomit, vomitus: *Injustice makes my gorge rise.* —*v.* **3** °fill, °stuff, °glut, °cram; °gulp, gobble (down), °devour, °bolt (down), wolf (down), gormandize, °swallow: *She acquired the habit of gorging on cream cakes.*

gorgeous *adj.* **1** resplendent, °splendid, °magnificent, °glorious, °exquisite, °sumptuous, °dazzling, °grand, °beautiful, splendrous, breathtaking, °radiant, refulgent, °brilliant, °showy, colorful, *Colloq* splendiferous: *The males are known for their gorgeous plumage.* **2** °great, °terrific, °fantastic, wonderful, °marvelous, °glorious, spectacular, °superb, °excellent, *Colloq* fantabulous, marvy, smashing, °super, °nifty, °neat, °swell: *He married a gorgeous fashion model.*

gory *adj.* bloody, °sanguinary, bloodsoaked, bloodstained; °gruesome, °grisly, horrific, bloodcurdling: *The scene was so gory as to turn the strongest stomach.*

gospel *n.* °truth, °fact, °certainty: *He swears that his story is gospel.*

gossip *n.* **1** °chat, °conversation, °talk, chitchat, small talk, °palaver; tittle-tattle, °prattle; *Scot* clishmaclaver; gup; *Colloq Brit* natter, chinwag: *I just had a gossip with my neighbor.* **2** °rumor, °scandal, hearsay, °information, *on dit,* °word, *Colloq* °grapevine, (inside) info, tittle-tattle, *Slang* °dope, *Chiefly US* scuttlebutt, *US and Canadian* poop: *The latest gossip is that she left him for another woman.* **3** rumor mill, rumormonger, scandalmonger, gossipmonger, newsmonger, °busybody, tattletale, quidnunc, blabber, blatherskite, telltale, °talebearer, flibbertigibbet, *Colloq* bigmouth, chatterbox, °blabbermouth, Nosy Parker: *That old gossip will tell everybody your business.* —*v.* **4** Sometimes, ***gossip about.*** bruit, °tattle, °rumor, °whisper, blether *or US* blather, gabble, *Colloq* °blab, *Brit* natter, *Slang* jaw: *I was just gossiping to my friend about your new job.*

gouge *v.* **1** °chisel, °gash, incise, °scratch, °groove, °dig; scoop *or* hollow (out): *The screwdriver slipped and gouged a piece out of the table. She's so mad she wants to gouge out your eyes.* **2** °extort, °extract, bilk, °defraud, wrest, °wrench, °squeeze, °blackmail, *Colloq* °milk, bleed, °skin, °fleece, °cheat, °swindle: *The boys gouged money from their classmates by threatening to tell the teacher.* —*n.* **3** °groove, °furrow, °scratch, °gash, °hollow; trench, ditch: *The retreating glacier cut a deep gouge in the landscape.*

gourmet *n.* °epicure, connoisseur, Lucullus, gastronome, *gourmand, bon vivant,* bon viveur: *My dear, that was a dinner fit for a gourmet!*

govern *v.* **1** °rule, °control, °direct, °reign, hold sway (over), °lead, °conduct, °guide, °manage, °regulate, °run, °supervise, superintend, °oversee, °steer, captain, °pilot, °command, °head (up), °look after, sit on the throne, wield the scepter, wear the crown, run the show, be in power, be in charge (of), exercise *or* wield power *or* control (over), have *or* hold the whip hand, *Colloq* wear the pants, be in the saddle *or* driver's seat: *Catherine governed with an iron hand.* **2** °control, °check, °bridle, °curb, °master, °subdue, °restrain, °contain, °hold in, °suppress, °repress: *You must govern the urge to scream out.*

government *n.* **1** °rule, °command, °authority, °regulation, °control, °management, °direction, °administration, °sway, superintendence, supervision, °oversight, °guidance, °domination: *His government of the island has led to disorder.* **2** °administration, °ministry, °regime: *Under the new government, smoking will be banned in public places.*

grab *v.* **1** °snatch, lay *or* catch hold of, fasten upon, °grasp, °seize; °catch, °grip, °clutch; *Colloq* latch on to, get one's hands *or* fingers on, °nab, *US* °snag: *The thief grabbed my purse and ran. She grabbed the railing to break her fall.* **2** °appropriate, expropriate, °seize, commandeer, °take over, usurp, arrogate: *The general grabbed the reins of government in the coup.* **3** °arrest, °capture, °catch, *Colloq* °nab, °pinch, collar: *The police grabbed him as he was leaving the scene of the crime.* —*n.* **4** °snatch, °clutch; °grasp, °grip: *I made a grab for the rope.*

grace *n.* **1** °elegance, gracefulness, suppleness, °finesse, °refinement, °ease, °polish, °poise: *She rides well, with an unconscious grace.* **2** tastefulness, °(good) taste, cultivation, suavity *or* suaveness, °culture, °savoir-faire, discernment, °discrimination, °(good) manners, politeness, °breeding, °consideration, decency, °etiquette, °tact, °propriety, °decorum, mannerliness: *They had the grace to ignore my rough attire.* **3** °indulgence, °forgiveness, °mercy, mercifulness, leniency, compassion, clemency, °charity, goodwill, goodness: *We survived the fire only by the grace of God. The company gives you thirty days' grace to pay the bill.* **4** °kindness, °favor, kindliness, °benevolence, generosity, goodness, graciousness, becomingness, seemliness; °excellence, °virtue, strength of character, considerateness: *At least he had the good grace to admit being wrong.* **5** °blessing, thanksgiving, °prayer, benediction: *Grace was said before each meal.* —*v.* **6** adorn, °embellish, °set off, °decorate, °ornament, °beautify, °enhance, garnish: *The table was graced by a huge silver centerpiece.* **7** °dignify, °enhance, °distinguish, °enrich, °honor, °favor: *The dinner party is graced by your presence.*

graceful *adj.* **1** °fluid, flowing, °supple, lissom *or* lissome, lithe, facile, °smooth, °nimble, °agile, deft: *How graceful a dancer she is!* **2** °tactful, well-mannered, °polite, °courteous, mannerly, °refined, °tasteful, °elegant, courtly, urbane, °polished, °refined, °suave: *In return for certain concessions, the chieftain was graceful enough to release the hostages.*

gracious *adj.* °kind, °courteous, °polite, well-mannered, °kindly, °benevolent, beneficent, °indulgent, kindhearted, warmhearted, °cordial, °warm, °friendly, °sociable, °good-natured, °amiable, affable, °benign, °accommodating, °obliging, °agreeable, °considerate: *It was gracious of you to grant me an audience.*

grade *n.* **1** °degree, °position, °rank, °status, °stage, °standing, °station, gradation, echelon, °class, °level, °category, °condition, °state, °estate, °situation, rung:

She is a singer of the first grade. **2** rating, °mark, °score: *What grade did you get in physics?* **3** °class, °form, year: *My daughter is in the third grade.* **4** °hill, °slope, °rise, gradient, acclivity, declivity, °incline, °decline, ascent, descent, upgrade, °downgrade: *The car coasted down a gentle grade into the lake.* **5** ***make the grade.*** °pass, measure up, °succeed, °qualify, *Colloq US* °make it, *Slang US and Canadian and New Zealand* hack *or Slang US and Canadian and New Zealand* cut it: *Do you think that Simpson can make the grade?* —*v.* **6** classify, °class, °order, °organize, °rank, °sort, °size, °group, °categorize, °rate: *Her job is to grade pearls according to their size.* **7** °mark, °rate, °correct, °evaluate: *My teacher was up late grading our exams.*

gradual *adj.* °easy, °gentle, °even, °moderate, °slow, °piecemeal, inchmeal, °regular, °steady: *There is a gradual slope down to the garden wall. The acceleration was gradual. He became aware of a gradual erosion of his authority.*

gradually *adv.* slowly, evenly, °piecemeal, inchmeal, drop by drop, °step by step, bit by bit, little by little, piece by piece, *gradatim*: *The water gradually seeped through the filter. We advanced gradually through the forest.*

graduate *n.* **1** bachelor, postgraduate, *US* alumnus (*pl.* alumni) *or US* alumna (*pl.* alumnae): *Even college graduates found it difficult to get a job.* —*v.* **2** gradate, °mark, °calibrate, °grade, °scale: *This thermometer is graduated in increments of one tenth of a degree.*

graft[1] *n.* **1** bud, scion, °shoot, °splice, implantation *or* implant, °transplant: *New varieties were created by grafting on strains that have desirable qualities.* —*v.* **2** °implant, °splice, °insert, °join: *Fruit-bearing stems are grafted onto a sturdy rootstock.*

graft[2] *n.* corruption, jobbery; bribery, extortion, *Colloq* payola, °kickback: *After only five years, he retired on the takings from graft.*

grain *n.* **1** °seed, °kernel, stone, pip, °pit: *Grains from each crop are stored for planting.* **2** cereal, corn, grist: *The grain harvest was sparse.* **3** °particle, °bit, °fragment, °crumb, °speck, granule, °morsel, mote, molecule, atom, fleck, iota, ounce, °scrap, °trace, scintilla, °hint, °suggestion, whit, °jot (or tittle), °dab, soupçon, °taste, *Colloq US and Canadian* smidgen *or Colloq US and Canadian* smidgin: *There's not a grain of truth in anything that's been said.* **4** °texture, °pattern, °fiber, °weave, °nap: *Use a different kind of saw for cutting across the grain.*

grand *adj.* **1** °large, °great, °huge, °immense, °enormous, °impressive, °imposing, °splendid, °fine, °majestic, °stately, °lofty, °monumental, °lavish, °magnificent, °opulent, °luxurious, °palatial, °sumptuous, *Colloq* °posh: *We were led into a grand hall where thousands awaited the emperor.* **2** °dignified, °distinguished, august, respected, °eminent, °preeminent, °outstanding, °celebrated, °illustrious, °renowned, °notable, °legendary, °exalted, revered, °venerable, °immortal: *He is regarded as the grand old man of chemistry.* **3** °flamboyant, overdone, histrionic, °ostentatious, °pretentious, °grandiose, lordly: *The abysmal performance was punctuated by the actors' grand strutting and gesturing.* **4** °complete, °total, °sum, °comprehensive, (all-)inclusive; bottom-line: *The grand total of expenses came to $12,467.22.* **5** °marvelous, wonderful, °outstanding, first-class, °first-rate, °splendid, °excellent, °superb, °admirable, *Colloq* °great, marvy, smashing, °terrific, °fantastic, °fabulous, fantabulous, °super: *Isn't it grand that Frieda will be able to go!* **6** °principal, °chief, °main, °head, °leading, °foremost, highest: *The grand vizier was second only to the caliph.*

grandeur *n.* **1** °splendor, magnificence, majesty, sublimity, luxuriousness, °pomp: *The grandeur of the French court was never to be equaled.* **2** °nobility, augustness, nobleness, eminence, majesty: *The emperor received us with a grandeur befitting his station.*

grandiose *adj.* **1** °pretentious, °ostentatious, °showy, °flamboyant, °bombastic, histrionic, °extravagant, °pompous, fustian, high-flown, high-flying, overambitious, overdone, overdramatic, °melodramatic, Ossianic, °inflated, °affected, florid, °flashy, *Colloq* highfalutin *or* hifalutin, °flash: *The council is unimpressed by your grandiose plans to redevelop the area, Mr. Wren.* **2** °imposing, °impressive, °ambitious, °grand, °monumental, °magnificent, °lofty: *She has some grandiose ideas for a national symphony orchestra.*

grant *v.* **1** °give, °confer, °bestow, °present, °award, °offer; °supply, °furnish, °distribute, °donate; allocate, °assign: *She has been granted a sum that will enable her to complete the book.* **2** °concede, accede (to), °cede, °give (up), °agree (to), °consent (to), °allow, °permit, °admit; °let: *They granted my request for clemency.* —*n.* **3** °gift, °present, °endowment, °bequest, subvention, °subsidy, °award, grant-in-aid, °donation, contribution, concession, °allowance: *He received a government grant to study abroad.*

granular *adj.* grainy, granulated, particulate, comminuted, gravelly, sandy, °gritty: *This scouring powder is too scratchy and granular.*

grapevine *n.* rumor mill, jungle telegraph, grapevine telegraph; °rumor, °gossip: *According to the grapevine, you two are going to be married.*

graphic *adj.* **1** °vivid, °distinct, well-defined, °detailed, °explicit, °particular, °clear, lucid, °plain, °manifest, crystal-clear, unmistakable *or* unmistakeable, unambiguous, °accurate, °precise, well-drawn, °photographic, descriptive, °telling, °picturesque, °pictorial, °realistic, °lifelike, true-to-life, *Colloq* °gory: *He described his injuries in graphic detail.* **2** written, °drawn, diagrammatic, delineated, °visible: *The magazine won a prize for graphic design.*

grapple *v.* **1** °grasp, °grab, °grip, °seize, °clasp, °catch, °wrestle; °hold, °clutch, °snatch: *They grappled each other at the edge of the Reichenbach Falls.* **2** ***grapple with.*** °come to grips with, °cope with, contend with, °deal with, struggle with, °tackle, °face, °take on: *The government is grappling with the problem of unemployment.*

grasp *v.* **1** °grip, °grab, °seize, °clasp, °clutch, °snatch, °hold, take *or* lay *or* catch hold of, *Colloq* °nab: *I grasped the rope and was pulled to safety.* **2** °understand, °comprehend, °appreciate, °catch (on), °get, get the drift *or* point of, °follow, °see, °realize, apprehend, °learn, *Colloq* make heads or tails of, *Slang* °dig: *I am trying to grasp the full import of what you are saying.* —*n.* **3** °hold, °grip, °clutches, °clasp, °embrace, °lock: *He loosened his grasp on my throat and I could breathe again.* **4** °possession, °control, °power, mastery, °sovereignty, suzerainty, °hold: *By then even the outer islands had fallen within his grasp.* **5** °understanding, comprehension, apprehension, awareness, °perception, °sense: *He has a poor grasp of the basics of the subject.*

grasping *adj.* °greedy, °avaricious, acquisitive, °rapacious, °mean, °miserly, stingy, °penurious, parsimonious, niggardly, °tight, tightfisted, penny-pinching, closefisted, °mercenary: *If the company weren't so grasping, they'd have less difficulty hiring staff.*

grate *v.* **1** °shred, °rasp, °scrape, °rub, triturate: *Peel six raw potatoes and grate them fine.* **2** °scrape, °rasp, °rub, °grind, °scratch, screech, stridulate: *The cricket grates its legs against its wing covers to produce the "singing" sound.* **3** Often, ***grate on or upon.*** °annoy, vex, °irk, °irritate, °pester, set one's teeth on edge, °jar, °fret, °chafe, rub one (up) the wrong way, go against the grain, *Colloq* get on one's nerves: *That announcer's voice just grates on me.*

grateful *adj.* °thankful, appreciative: *I am grateful for any help I can get.*

gratification *n.* °satisfaction, °fulfillment, °enjoyment, °pleasure, °delight, compensation, recompense, °reward, °return, °requital: *She does it only for the gratification she gets from making the children happier.*

gratify *v.* please, °satisfy, °fulfill, °delight, °compensate, recompense, °reward, °requite, °cheer, °gladden, °favor: *Albert would be gratified to see the good coming from his charitable bequests.*

grating *adj.* **1** jarring, °strident, °raucous, °harsh, °discordant, dissonant, unharmonious, °offensive, irritating, °irksome, annoying, vexatious, galling: *That noise from the chalk on the blackboard is very grating.* **2** grinding, °gritty, squeaky, jangling, screeching, creaking, °shrill, °piercing, squawking, croaking, rasping: *The grating noise you hear is my fridge motor.* —*n.* **3** °grate, grid, reticle *or* reticule, grille, lattice, trellis, °screen, °network, reticulation: *When a beam of light is directed through the fine grating, odd patterns are formed.*

gratitude *n.* thankfulness, °appreciation, gratefulness; °thanks, °return, compensation, thanksgiving: *Here is a bottle of wine in gratitude for looking after my cat.*

gratuitous *adj.* **1** gratis, °free, °complimentary, °spontaneous: *The earth is a machine which has, till now, yielded gratuitous service.* **2** unasked-for, unrequested, unsought-for, °wanton, unprovoked, °unsolicited, unlooked-for, uncalled-for, °unwelcome, unjustified, °unwarranted, baseless, °groundless, °needless, °unfounded, ungrounded, unjustifiable, irrational: *He seems to derive some twisted pleasure from spreading gratuitous lies about me.*

grave[1] *n.* °crypt, °sepulcher, °tomb, vault, mausoleum, last *or* final resting place, eternal rest: *A sobbing was heard in the crowd as the coffin was lowered into the grave.*

grave[2] *adj.* **1** °serious, °somber, °solemn, °earnest, unsmiling, °staid, °sedate, °sober, °dour, °gloomy, °grim, grim-faced, grim-visaged, *Brit* po-faced: *The doctor, looking very grave, said that he had done all he could.* **2** °serious, °critical, °vital, °dangerous, matter of life and death, °crucial, °urgent, °weighty, °important, °pressing, °pivotal, °perilous: *Fenlow is in intensive care in a grave condition.*

graveyard *n.* burial ground, churchyard, cemetery, God's acre, necropolis, potter's field, *Western US* boot hill, *Rare* golgotha, *Slang* boneyard: *Uncle Valentine is buried in the graveyard on the hill.*

gravity *n.* **1** gravitation; °attraction: *After middle age, the body obeys the laws of gravity but not of mutual attraction.* **2** seriousness, acuteness, immediacy, °importance, °significance, °weight, °magnitude, °severity, °urgency, exigency, momentousness, weightiness: *When I saw mother's expression I realized the gravity of the situation.* **3** °solemnity, °dignity, somberness, staidness, sedateness, °sobriety, gravitas, soberness, °reserve; gloominess, grimness: *The ceremony was conducted with suitable gravity.*

gray *adj.* **1** grey, ashen, °leaden, °colorless, °pale, pallid, °wan, livid, °pearly, griseous, smoky, sooty, bloodless: *The cadaver's skin was a dead, whitish gray.* **2** grey, °gloomy, °dismal, °dull, depressing, °glum, °dreary, °somber, °drab, cheerless, °dark, °murky, foggy, °misty, cloudy, °overcast, °sunless: *As if to mold her temperament, Kathleen was born on a gray December day in 1791.* **3** grey, °aged, °elderly, hoary, °old, °venerable, °ancient: *A stooped, gray crone inched along, muttering to herself.* **4** grey, °mature, °wise, °experienced: *Seventy years have spread their gray experience over his hoary head.*

greasy *adj.* **1** °oily, sebaceous, fatty, °fat, buttery, lardy, soapy, oleaginous, pinguid, butyraceous, saponaceous, waxy: *Before curing, the hide is greasy with natural oils.* **2** unctuous, °oily, °slippery, slithery, °smooth, °glib, fawning, °slick, toadying, sycophantic, *Slang Brit* smarmy: *Edna is so greasy, always trying to maneuver you into doing something for her.*

great *adj.* **1** °big, °large, °huge, °immense, °enormous, °gigantic, °giant, °grand, °extensive, °prodigious, °colossal, °massive, °vast, tremendous; °spacious, capacious, mammoth, gargantuan, °monstrous, titanic, cyclopean, Brobdingnagian: *Europe was covered by a great forest. A great castle loomed before me.* **2** °large, °huge, °immense, °enormous, °gigantic, °prodigious, °vast, tremendous, °abundant, countless: *Plague killed a great number of people.* **3** °extreme, °considerable, °marked, °pronounced, °inordinate, °extraordinary, °significant; °excess, °excessive: *I have something of great importance to tell you. Your news created great confusion.* **4** °critical, °important, °crucial, °momentous, °significant, °serious, °weighty, consequential: *Those were great days for the history of England.* **5** °important, °prominent, °major, °eminent, °celebrated, °distinguished, °famous, famed, °renowned, °notable, °noteworthy, °illustrious, °outstanding, °well-known, °weighty, °influential, *Rare* eximious: *What of the great universities, like Oxford and Cambridge? Our mayor thinks of himself as a great man.* **6** °talented, °gifted, °excellent, °outstanding, °exceptional, °major, °superlative, °superior, °leading, °best, °incomparable, °matchless, °peerless, °skillful, artistic, °brilliant, °first-rate, °remarkable, °top, °accomplished: *Flaubert was one of the greatest writers of his age. Tamara is a good, not a great, dancer.* **7** °lofty, °elevated, °exalted, °noble, high-minded, °grand: *Great thoughts come from great minds.* **8** °talented, skilled, °skillful, adroit, °clever, °adept, °able, °proficient, °expert: *Daphne is really great at playing bridge.* **9** °keen, zealous, °eager, °active, °enthusiastic, °devoted, °ardent, °passionate: *Frank is a great stamp collector.* **10** °close, °devoted, dedicated, °fast, °faithful, °true, °loyal, °intimate, loving: *Graham and Pembroke quickly became great friends.* **11** °terrible, °bad, °awful, unforgivable, horrendous, heinous, °grievous, horrific, °horrible, °terrific, °huge, °colossal, °enormous, °gigantic, °significant, °cardinal, egregious, °basic, °profound, °flagrant, °glaring, arrant, consummate, °out-and-out: *Inviting her at the same time as her ex-husband was a great mistake.* **12** spectacular, °marvelous, °outstanding, °excellent, °superb, °grand, wonderful, °fine, *Colloq* °fantastic, °terrific, stupendous, marvy, smashing, fantabulous, *Old-fashioned Brit* tickety-boo: *He says that we missed a truly great show.*

greed *n.* **1** greediness, °avarice, avariciousness, covetousness, acquisitiveness, cupidity, avidity, craving, yearning: *Greed accounts for most of man's dishonesty.* **2** meanness, stinginess, miserliness, selfishness, niggardliness, penuriousness, parsimony, closefistedness, penny-pinching, tightfistedness: *From a life of greed he had accumulated enormous wealth.* **3** °gluttony, voraciousness, edacity, esurience, voracity, overeating, gormandizing, ravenousness, insatiableness: *Obesity owes more to greed than to glandular disorder.*

greedy *adj.* **1** °ravenous, °voracious, °gluttonous, piggish, °hoggish, swinish, cormorant, edacious, esurient, insatiable, °unquenchable: *I'd love another slice, but I mustn't be greedy.* **2** °avaricious, acquisitive, covetous, °grasping, craving; °materialistic, money-hungry: *If some people were less greedy, there would be plenty for everyone.* **3** stingy, °miserly, °mean, °selfish, niggardly, parsimonious, °penurious, penny-pinching, °mercenary, closefisted, tightfisted, °close, *Colloq* °near, °tight, *Brit* mingy: *He's so greedy that he asks beggars for receipts.*

green *adj.* **1** verdant, grassy, °fresh, °leafy; °rural, country-like: *Plans call for a green belt to be created around every major city.* **2** °immature, unripe, unripened; naive *or* naïve *or* naïf, °callow, untested, untrained, unversed, °inexperienced, °new, °raw, unseasoned, °unsophisticated, °gullible, °amateur, unskilled, unskillful, amateurish, nonprofessional, inexpert, *Colloq* wet behind the ears: *This banana is still too green to eat. Isn't Winks a bit green to be given so much responsibility?* **3** environmental, conservationist: *The green activists appear to be gaining influence in governmental circles.* —*n.* **4** °lawn, sward, °common, grassland: *Let us meet on the village green at noon.* **5** °environmentalist, conservationist, preservationist: *The greens are opposed to draining the marshes.*

greenhorn *n.* °newcomer, beginner, °novice, tyro, neophyte, °novice, °initiate, °learner, tenderfoot, *Colloq* rookie: *They send all the greenhorns off to fetch a bucket of steam and a can of striped paint.*

greet *v.* **1** °welcome, °receive, usher in, °meet: *We were greeted warmly by the family.* **2** °hail, accost, °address, °salute: *People greeted him in the street with a wave and a smile.*

greeting *n.* **1** salutation, °hail, hello, °welcome, °reception: *Even though he hated me, he gave me a most friendly greeting.* **2** greeting card, °card, °message, °note: *The president sent me a greeting on my hundredth birthday.* **3** ***greetings.*** °regards, °respects, best *or* good wishes, devoirs, °compliments: *I sent you greetings from Barbados, where I took my vacation.*

grief *n.* **1** °anguish, °suffering, °agony, °misery, wretchedness, °pain, °hurt, °sadness, °sorrow, dejection, °depression, despondency, °melancholy, unhappiness, °woe, °torment, °desolation, heartbreak, °remorse, °regret, ruth, heartache: *Nothing equaled the grief I felt at the death of my dog, Whiffler.* **2** °distress, °trouble, °difficulty, tribulation, °trial, °burden, °load, onus, °ordeal, travail, °affliction, °worry, °bitterness, °curse; adversity, °misfortune, evil days, bad *or* ill fortune *or* luck, °calamity, °disaster, °catastrophe, trauma: *The death of her only child was a lasting grief to Millie.* **3** ***come to grief.*** °fail, go to rack and ruin, meet with disaster, °miscarry, fall *or* come apart, *Colloq* come unstuck: *Our plans for retirement came to grief when my husband went bankrupt.*

grievance *adj.* **1** °wrong, °ill, °injustice, °disservice, unfairness, °injury, °damage, °harm, °outrage, affront, °indignity, °hardship, °calamity: *I had to listen to details of every grievance she had ever suffered.* **2** °complaint, °objection, °charge, plaint, °allegation, °grudge, *Colloq* °gripe, bone to pick, *Brit* crow to pluck, *Slang* beef: *Please register all your grievances at the office next door.*

grieve *v.* **1** °mourn, °bemoan, °lament, °regret, rue, deplore, °bewail, mope, eat one's heart out: *Sylvia is still grieving over the loss of her husband.* **2** °weep, °cry, °mourn, °moan, °keen, °suffer, °sorrow; shed tears, °complain: *Go in and try to comfort the grieving widow.*

grievous *adj.* **1** °severe, °heavy, °painful, °grave, °serious, distressing, °harmful; damaging, °hurtful, °acute, wounding: *The church buildings suffered grievous damage from the storm. He was charged with causing grievous bodily harm.* **2** egregious, °awful, °flagrant, °terrible, °outrageous, heinous, °dreadful, °atrocious, °monstrous, appalling, °shocking, °deplorable, °calamitous, °lamentable, intolerable, °shameful, °unbearable: *Missing that catch was a grievous error.*

grim *adj.* **1** °stern, °severe, unrelenting, °resolute, uncompromising, unyielding, °inflexible, adamant, °stony, iron, unbending, °firm, intractable, unflinching, unmoving, °unmoved, °implacable, inexorable, °determined, °steadfast, °(dead) set, °fixed, °decided, °obstinate, headstrong, °stubborn, obdurate, dogged, unwavering: *By controlling the purse strings, she kept a grim hold over the family.* **2** °forbidding, °formidable, °harsh, °ferocious, °fierce, °cruel, °savage, °merciless, °heartless, °ruthless, pitiless, °vicious, °brutal, brutish, feral, °inhuman, °fiendish, °violent, °bloodthirsty, °murderous, °homicidal, °fell: *A grim pack of howling wolves loped after our sledge.* **3** dire, °dreadful, °ghastly, °awful, °frightful, °frightening, °sinister, °hideous, horrid, horrific, °horrible, horrendous, °terrible, °terrifying, °terrific, °harrowing, °dread, alarming, appalling, °grotesque, °gruesome, °eerie, °macabre, flagitious, heinous, °evil, °wicked, iniquitous, °atrocious, °monstrous: *Before me appeared the ghostly apparition of the Grim Reaper.* **4** stern-visaged, austere, °dour, disapproving, frowning, °grave, saturnine, °solemn, °somber, humorless, °forbidding, °severe, °strait-laced, uncompromising, puritanical, prim, °prudish, °gruff, °bluff, curmudgeonly, *Brit* po-faced: *Despite a grim approach to work, Williamson has a keen sense of humor.*

grime *n.* °dirt, °filth, soot, °mud, °muck, slime, scum: *The lad began to rub the grime from the old oil lamp.*

grind *v.* **1** °pound, °powder, °pulverize, abrade, °crush, granulate, °mill, °grate, °rasp, °crumble, kibble, mash, triturate, bray, comminute: *The rock was soon ground to a fine powder.* **2** °sharpen, °whet; °file, °smooth, °polish: *He is grinding the edge to be razor sharp.* **3** gnash, °grit, °grate: *He grinds his teeth when he's angry.* **4** Also, ***grind away.*** °labor, toil, °slave (away); °study, lucubrate, burn the midnight oil, *Colloq* °cram, *Brit* swot: *Gridley has been grinding away at the same job all his life. Languages came easily, but he had to grind away at the sciences.* **5** ***grind down.*** wear down *or* away, °crush, °oppress, °subdue, °suppress, °tyrannize, °persecute, maltreat, °ill-treat, harry, °harass, °hound, hector, °plague, badger: *Dictators first grind down all opposition, then destroy it.* **6** ***grind out.*** °produce, °generate, crank out, churn out, °turn out: *For twenty years he has been grinding out the daily gossip column.* —*n.* **7** toil, °labor, °drudgery, travail, °exertion, °task, chore: *Working at the checkout counter hour after hour is a terrible grind.*

grip *n.* **1** °hold, °grasp, °clutch, handgrip, °clasp, handclasp: *I lost my grip on the rope and dropped into the abyss.* **2** °control, °command, °hold, mastery; °authority, °influence, °power, °rule, °domination, °sovereignty, °tenure, °dominion, suzerainty, °custody: *Cardozo is losing his grip on reality. The dictator kept a tight grip on the people right up till his death.* **3** °grasp, °understanding, apprehension, comprehension, °sense, °sensitivity, °feel, °feeling, awareness, °perception, °view, *Slang US* °handle: *Skerry is having trouble getting a grip on what you plan to do.* **4** handgrip, valise, (traveling *or* overnight) bag, °case, satchel, °suitcase, *Brit* holdall, *US and Canadian* carryall: *I got off the plane in New York, but my grip went on to Los Angeles.* **5** ***come*** or ***get to grips with.*** °tackle, confront, °approach, °handle, °meet (head on), °undertake, grapple *or* contend with, cope *or* deal with, °face: *We must come to grips with the problem of Cassandra.* —*v.* **6** °grasp, °clutch, °clasp, °hold, °seize: *He gripped the handle of the revolver more tightly.* **7** engross, °engage, °hold, °fascinate, enthrall, °entrance, absorb, mesmerize,

°hypnotize, spellbind, rivet: *Just at the gripping climax, the film tore.*

gripe *v.* **1** °complain, °moan, grumble, whimper, whine, bleat, °nag, °cavil, °carp, grouse, *Colloq* beef, *Brit* whinge, *Slang* °bitch, bellyache: *It's natural to gripe about the food in the army.* —*n.* **2** °complaint, °grievance, °objection, °protest; complaining, moaning, grumbling, whimpering, whining, °nagging, caviling, carping, grousing, *Colloq* beef, *Brit* whinging *Slang* bitching, bellyaching: *I don't want to hear any more gripes about the food!* **3** Usually, ***gripes.*** cramp, °twinge, °pang, °pain, °ache, colic, °distress, *Colloq* bellyache: *The gripes became so painful that he was sure he had appendicitis.*

grisly *adj.* °gruesome, °gory, °abhorrent, °abominable, °awful, appalling, °hideous, °shocking, °nasty, °dreadful, °repulsive, °repellent, °repugnant, °disgusting, sickening, nauseating, horrific, horrid, horrendous, horrifying, °terrible, °terrifying, °terrific: *A grisly scene greeted the soldiers who liberated the concentration camp.*

grit *n.* °courage, courageousness, valor, °bravery, °fortitude, °resolution, resoluteness, °resolve, toughness, mettle, °pluck, °spirit, °backbone, °nerve, gameness, intrepidity *or* intrepidness, dauntlessness, °tenacity, °determination, firmness, hardiness, hardihood, staunchness, stalwartness, doughtiness, fearlessness, *Colloq* °guts, gutsiness, °spunk, spunkiness, starch, *Brit* °bottle, *US and Canadian* chutzpah, moxie, stick-to-itiveness: *Mountain climbing requires more grit than you might think.*

gritty *adj.* **1** sandy, gravelly, °granular, grainy, °rough, abrasive, rasping, arenose: *This mud-pack feels a little gritty.* **2** °courageous, valorous, °brave, °resolute, °tough, mettlesome, plucky, °spirited, °game, °intrepid, °dauntless, °tenacious, °determined, °persistent, °firm, °hardy, °staunch, °stalwart, doughty, °fearless, *Colloq* gutsy, spunky: *The band of gritty frontiersman was led by Sheriff Boswell.*

groan *v.* **1** °moan, °sigh, °murmur, wail, whimper, whine: *Unattended patients were groaning in agony.* **2** °complain, grumble, grouse, °object, °protest, *Colloq* °gripe, beef, yammer, *Brit* whinge, *Slang* °bitch: *Now that he's rich, he's groaning about taxes.* —*n.* **3** °moan, °sigh, °murmur, wail, whimper, whine: *I thought I heard a groan coming from the attic.* **4** °complaint, grumble, grousing, muttering, *Colloq* °gripe, griping, beef, yammering, *Slang* bitching: *Announcement of the forthcoming exams was met by groans.*

groggy *adj.* unsteady, °shaky, wobbly, weak-kneed, °weak, staggering, stupefied, dazed, stunned, reeling, punch-drunk, °numb, numbed, benumbed, °faint, in a trance *or* stupor, muddled, addled, °confused, bewildered, confounded, puzzled, baffled, befuddled, *Colloq* dopey, punchy, woozy, *Brit* muzzy: *I was still groggy from the blow on the head.*

groom *n.* **1** stableboy, stableman, *Brit* stable lad, *Archaic* ostler *or* hostler, equerry: *After unsaddling the horses the groom curried them.* **2** bridegroom: *Let's drink to the bride and groom!* —*v.* **3** °spruce up, °dress, tidy *or* neaten up, smarten up, titivate *or* tittivate, °preen, °primp, °refresh: *Every time she passes a mirror Vivian stops to groom herself.* **4** °fit, °train, °prepare, °coach, °tutor, °brief, °drill, °prime, (get *or* make) ready, °adapt, °shape: *McCusker is being groomed for a directorship.*

groove *n.* °slot, °cut, °channel, °furrow, °gouge, trough; flute, °scratch, striation *or* stria, rifling, °rifle, *Architecture* glyph, *Technical* sulcus: *The door slides in these grooves at the sides.*

grope *v.* °feel, °fumble, fish, °probe: *I groped for the switch in the dark.*

gross *adj.* **1** °fat, °obese, corpulent, overweight, °big, °large, °bulky, °great, °heavy, °ponderous, °massive, cumbersome, °unwieldy: *The gross detective eased his bulk onto a tiny chair.* **2** °total, aggregate, °entire, pretax, (all-)inclusive, °overall, °whole: *The gross profit was up by 15 percent this year.* **3** °coarse, °vulgar, °crude, °unsophisticated, uncultured, uncultivated, °earthy, crass, indelicate, °indecent, °inappropriate, °unseemly, °improper, °unrefined, °bawdy, ribald, Rabelaisian, °raw, °rude, °offensive, °obscene, °lewd, °dirty, smutty, °pornographic, °filthy: *That was too gross a story for mixed company.* **4** °outrageous, °flagrant, °obvious, °plain, °glaring, °shameful, °blatant, °monstrous, heinous, °manifest, °evident: *Her conviction was a gross miscarriage of justice.* **5** °disgusting, °repulsive, °repellent, °revolting, nauseating: *His table manners are truly gross.* —*v.* **6** °earn, bring *or* take in, °make: *We grossed over a million but netted only fifty thousand after expenses.* —*n.* **7** °(overall) total, °take, intake, takings, °receipts, °gate: *As I said, the gross was over a million.*

grotesque *adj.* **1** distorted, °bizarre, freakish, twisted, °misshapen, malformed, °deformed, °gruesome, °gnarled: *Victor Hugo created Quasimodo, the grotesque bellringer of Notre Dame.* **2** °absurd, °incongruous, °weird, °odd, °fantastic, °strange, °queer, °peculiar, °curious, °outlandish, °offbeat, °abnormal, aberrant, anomalous, °ludicrous, °ridiculous, °preposterous: *The idea of marrying Leonard is too grotesque for even Gladys to contemplate.*

ground *n.* **1** °earth, °soil, °turf, sod, °dirt, loam, clay; °land, °terrain: *There is a great hole in the ground behind my house.* **2** °territory, °area, °range, °scope, compass: *We covered the ground quite thoroughly at our meeting.* **3** Often, ***grounds.*** °basis, °foundation, °base, °reason, °footing, justification, °rationale, °argument, °cause, °motive, °excuse: *What are her grounds for suing for divorce?* **4** ***grounds.*** °sediment, °dregs, lees, °deposit, settlings, *Brit* grouts: *Filter out the grounds before drinking the coffee.* —*v.* **5** °base, °establish, °organize, °found; °settle, °set: *The school's philosophy is grounded on the principles of Felix Adler.* **6** °instruct, °teach, °train, °coach, °tutor, °inform, °prepare, °initiate: *The purpose of the course is to ground students in basic mathematics.*

groundless *adj.* baseless, without foundation, °unsound, °unfounded, unsupported, unjustified, unjustifiable, °unwarranted, uncalled-for, °gratuitous, unreasoned, °unreasonable, °speculative, suppositional, °hypothetical, °tenuous, °flimsy, °illusory, °imaginary, chimerical: *Assumptions that the boy is guilty are entirely groundless.*

groundwork *n.* °basis, spadework, preparation(s), °base, °foundation, underpinning(s), cornerstone: *The report provided the groundwork for the research.*

group *n.* **1** °assembly, assemblage, °gathering, congregation, °company, °number, alliance, union, °association, °organization, °league, °society, coterie, °clique, °set, °band, °circle, °club, °party, °body, °faction, °crowd, °team, °corps, guild, troupe, °unit, troop, °platoon, °squad, °gang: *A group of revolutionaries meets nightly in a basement in Krakowskaya Street.* **2** °batch, aggregation, °set, grouping, °collection, assemblage, °bunch, °accumulation, conglomeration, agglomeration, °assortment, series; °pile, °heap, °bundle: *The entire group of abstract paintings was auctioned at double their estimates. This group of weapons was taken from the partisans.* —*v.* **3** classify, °class, °sort, °bracket, °organize, °order, °rank, assort, °categorize, catalog: *Books are grouped by subject.* **4** °collect, °assemble,

°arrange, °place, °dispose, °gather, °organize, bring *or* put together, °set apart: *Group oversized books together on these shelves.*

groveling *adj.* °obsequious, fawning, toadying, toadeating, sycophantish, subservient, slavish, °servile, °submissive, kowtowing, cringing, cowering, truckling, sniveling, scraping, tugging the forelock, abject, crawling, °base, °low, °mean, °sordid, *Colloq* boot-licking, *US* apple-polishing; *Slang* brown-nosing, *Brit* arse-kissing, arse-licking, *US* ass-kissing, ass-licking: *The groveling coward threw himself to the ground, begging forgiveness.*

grow *v.* **1** °flourish, °develop, °increase, become larger *or* greater, °enlarge, wax, °swell, °expand, broaden, °thicken, °spread, °lengthen, multiply, burgeon *or* bourgeon, °thrive, °luxuriate, °prosper, °mature, °ripen, bloom, °flower, blossom, fructify, bear *or* yield fruit: *The seeds he had planted grew abundantly. The population continues to grow.* **2** °develop, evolve, °arise, °issue, °stem, °spring (up), °originate: *A great friendship grew out of their association.* **3** °plant, °cultivate, °breed, nurture, °raise, °propagate, °produce; °sow: *Cathcart grows sorghum where he used to grow alfalfa.* **4** °become, °get: *I am growing fonder of you every day, Abbie.* **5** ***grow on.*** get *or* become accepted by, come *or* begin to be liked by, to gain *or* increase in interest *or* attraction to, become more pleasing to: *I didn't like her at first, but she grows on you.* **6** ***grow up.*** °mature, reach *or* attain maturity *or* adulthood, come of age, reach one's majority: *Those who grew up in the Great Depression knew real poverty.*

growth *n.* **1** °development, °evolution, evolvement, cultivation, nurturing, °increase, °expansion, broadening, °extension, enlargement, °spread, °proliferation, flowering: *The growth of education was a slow and painful process.* **2** vegetation, crop: *The area is covered with a dense growth of timber.* **3** °advance, advancement, °success, °improvement, °expansion, °rise, °progress: *The growth of the economy was favorable in the last quarter.* **4** wen, excrescence, wart, °lump, °tumor, °swelling, intumescence: *He developed an ugly growth on his nose.*

grudge *n.* **1** °bitterness, °resentment, °rancor, °ill will, hard feelings, °spite, °grievance, pique, °dislike, °aversion, antipathy, animus, °animosity, enmity, °venom, malice, malevolence, hatred: *I have harbored a grudge against him since his promotion.* —*v.* **2** °begrudge, °resent, °envy, °mind, covet: *She grudges others their simple pleasures.*

gruesome *adj.* °ghastly, °repugnant, °horrible, horrid, horrific, horrendous, °grisly, °hideous, °revolting, °repellent, °repulsive, °loathsome, °grim, °grotesque, °macabre, °abominable, °frightful, °frightening, °fearsome, °shocking, °terrible, °awful: *The old crone wove a tale too gruesome to repeat here.*

gruff *adj.* **1** °surly, crusty, grumpy, curmudgeonly, °cantankerous, °sour, °peevish, churlish, °rude, uncivil, bearish, °testy, °querulous, °irritable, °cross, °petulant, crabbed, irascible, °sullen, sulky, °bluff, °abrupt, °curt, °blunt, °brusque, °short, °short-tempered, ill-humored, ill-natured, bad-tempered, stinging, °cutting, °biting, acerb, acrimonious, acid, °caustic, *Colloq* grouchy, crotchety: *I refuse to continue working for someone so gruff.* **2** throaty, °deep, °rough, guttural, rasping, °low, °husky, hoarse, harsh(-sounding): *Instantly I recognized father's gruff voice on the telephone.*

guarantee *n.* **1** guaranty, °warranty, °assurance, °pledge, °bond, °obligation, °promise; °word (of honor), °oath, °undertaking: *The guarantee expires one month from the sell-by date. Have I your guarantee that this battery will last a year?* —*v.* **2** guaranty, °warranty, °assure, °ensure, °pledge, °promise, °undertake, stand behind, vouch for, °certify, make sure *or* certain, swear to, attest to: *The manufacturer guarantees every car for three years or fifty thousand miles.*

guard *v.* **1** °protect, °shield, °safeguard, (keep *or* stand) watch (over), °defend, convoy, °escort, °police, °look after, °tend, °mind: *Only three men will be guarding the bullion when we make our move.* **2** °control, °mind: *Guard your tongue—the walls have ears.* —*n.* **3** °sentinel, °watchman, sentry, security guard, custodian, °guardian, °protector, °picket, °watch, bodyguard; evzone, Bashibazouk; *Brit* warder, wardress, *Slang* °screw, *Brit* °minder: *The guards go to eat at noon.* **4** °protection, convoy, °escort, °patrol: *A guard of six men will accompany the van along this route.* **5** °defense, °protection, °safety, °safeguard, °security, °shield: *This bandage will serve as a guard against infection.*

guarded *adj.* °careful, °cautious, heedful, °prudent, circumspect, °wary, °noncommittal, restrained, °mindful, °suspicious, leery *or Brit also* leary, apprehensive; loath *or* loth, °reticent, °reluctant, *Colloq* cagey: *He became very guarded under close questioning. When I asked where she had been, she gave a guarded answer.*

guardian *n.* °protector, defender, paladin, °champion; trustee, custodian, °keeper, preserver: *She fancies him her guardian, her knight in shining armor. Do we really need guardians of the "purity" of the language?*

guerrilla *n.* guerilla, partisan *or* partizan, resistance *or* freedom *or* underground fighter, °irregular; insurgent, saboteur, °terrorist; *US history* Jayhawker, *French history* Maquis: *The guerrillas continue to attack the government's supply convoys.*

guess *v.* **1** conjecture, °estimate, hypothesize, °speculate, postulate, *Slang* guesstimate: *We guessed that he might try to come in through the window.* **2** °think, °suppose, conclude, °assume, °believe, daresay, °surmise, °judge, deem, °reckon, °imagine, °fancy, °feel, °suspect, °divine: *I guess you were right about her.* —*n.* **3** conjecture, °estimate, °hypothesis, °speculation, °surmise, assumption, °judgment, °feeling, °suspicion, °supposition, postulate, theory; guesswork; *Colloq* shot in the dark, *Slang* guesstimate: *My guess as to which card would turn up was wrong.*

guest *n.* °visitor, °company, caller; °patron, °customer, lodger, boarder, roomer: *Our guests get clean linen daily.*

guidance *n.* **1** °leadership, °direction, °management, °government, °conduct, °control, °regulation, °charge, handling, °rule, °auspices: *The country prospered under her guidance.* **2** °counsel, °advice, counseling, advisement, °instruction, teaching: *I have turned to you for guidance in the matter.*

guide *v.* **1** °lead, show *or* lead the way, °conduct, °shepherd, °direct, usher, °steer, orient *or Brit* orientate: *She guided me to the proper office. What will guide your judgment in dealing with this situation?* **2** °steer, °pilot, °maneuver, °navigate, °direct: *He carefully guided the boat between the shoals.* **3** °counsel, °advise, °influence, °sway; °supervise, °oversee, °handle, °manage, superintend, °direct, °control, °regulate, °govern: *Mr. Thrall guided my artistic development from the beginning.* **4** °instruct, °teach, °tutor, °train: *From early youth we were guided by the elders of the community.* —*n.* **5** °leader, conductor, °director, cicerone, chaperon, mentor, °counsel, °counselor, adviser *or* advisor, guru, °master: *He served as my guide through the financial jungle. Let your conscience be your guide.* **6** °model, criterion, exemplar, °standard, °ideal, °example, °inspiration: *She looked upon the life of Mother Theresa as a guide.* **7** °beacon, °light, °signal, guiding light, °landmark, lodestar, °sign, marker: *The Southern Cross was our constant guide sailing round the Horn.* **8** handbook, °manual, enchiridion, °vade mecum, guidebook, Baedeker: *Can you recommend a good guide to English usage?*

guilt *n.* **1** culpability, guiltiness, criminality, °blame, °responsibility, blameworthiness; °crime, sinfulness, feloniousness, wrongdoing, misconduct: *Confronted by the evidence, Shillingworth admitted his guilt.* **2** °remorse, self-reproach, °regret, °sorrow, contrition, repentance, °shame, contriteness, self-condemnation, bad conscience: *That feeling of guilt would plague me for the rest of my life.*

guilty *adj.* **1** °responsible, culpable, answerable, blameworthy, °at fault, °delinquent, °wrong; offending, reprehensible: *Whoever was guilty will be prosecuted. First we must find the guilty party.* **2** °remorseful, contrite, °regretful, °sorry, °apologetic, °repentant, °sorrowful, conscience-stricken, rueful, °penitent; °ashamed, °shamefaced, °sheepish, °embarrassed, red-faced: *Gammadge felt terribly guilty for having insulted Tina.*

guise *n.* **1** °appearance, °aspect, °semblance, °look, °image, °likeness, mien; °air, °behavior, °conduct, deportment, comportment, °bearing, demeanor: *This is nothing but an old political concept in a new guise.* **2** °semblance, °disguise, façade, °front, °pretense: *This treacherous lothario assumed the guise of a knight in shining armor.*

gulf *n.* **1** bay, bight, cove, inlet, °sound, loch *or* sea loch, firth *or* frith, fjord, *Irish* lough, *Brit* °creek: *We sailed into the bay to shelter from the wind behind the cliffs.* **2** chasm, °deep, °depth, °abyss, abysm, °void, °space, °opening, °rift, °breach, °gap, °separation, °rift, °split: *After the divorce, the gulf between us widened even further.*

gullible *adj.* °innocent, °green, °simple, credulous, °unsophisticated, naive *or* naïve *or* naïf, °unsuspecting, °unwary, unsuspicious, wide-eyed, born yesterday, °inexperienced, °immature: *He was gullible enough to fall for the old con game, three-card monte.*

gully *n.* gulley, °channel, riverbed, watercourse, °gorge, °ravine, °canyon, °notch, °cut, °pass, defile, °valley, °corridor, wadi, *Brit* gill *or* ghyll, *Western US* arroyo, *US and Canadian* gulch: *We rode along in the gully so as not to be seen.*

gulp *v.* **1** °bolt, gobble, wolf (down), °devour, °gorge, °swallow, throw down, toss off, quaff, guzzle, °swill, *Colloq* knock back, swig, *US* chug-a-lug: *The boy gulped his dinner and ran out to play.* **2** °swallow, °suppress, °stifle, °choke (back), °smother, strangle: *She gulped back her tears as the coffin was lowered.* —*n.* **3** °mouthful, °swallow, °draft, °swill, *Colloq* swig: *I had a gulp of beer before placing another bet.*

gumption *n.* **1** resourcefulness, shrewdness, cleverness, °(mother) wit, °(common) sense, astuteness, °judgment, *Colloq* horse sense, brains, *Slang Brit* nous: *It takes a lot of gumption to run a good cattle auction.* **2** °backbone, °grit, °pluck, mettle, °enterprise, °initiative, °courage, °spirit, gameness, °nerve, °daring, °vigor, °energy, boldness, audacity, °stamina, *Colloq* °spunk, °guts, get-up-and-go, *US* moxie, *Slang Brit* °bottle, *Taboo slang* balls: *He hasn't the gumption to go into business for himself.*

gurgle *v.* **1** °bubble, burble, °babble, °ripple, °splash, plash, °lap, °murmur, purl: *The cool brook gurgled merrily through the meadow.* —*n.* **2** °babble, burble, bubbling, babbling, burbling, °splash, gurgling, splashing, plashing, murmuring, purl, purling: *I lay back, listening to the gurgle of the spring among the rocks.*

gush *v.* **1** cascade, °rush, °flood, °stream, °spurt, jet, °spout, °burst; °run, °flow: *The water gushed out of the pipe, soaking us all.* **2** bubble over, overflow, be ebullient *or* effusive *or* effervescent, effervesce, °make much of, fuss over, °prattle, °chatter, °babble, °jabber, blather *or* blether, *Colloq Brit* natter, witter: *It was impossible to stop them from gushing on about the performance.* —*n.* **3** cascade, °rush, °flood, °flow, °stream, °spurt, jet, °spout, °burst, °torrent: *A gush of water hit me right in the face.* **4** °exuberance, effusion, bubbling over, °outburst: *After the first gush of enthusiasm, everything quieted down.*

gushy *adj.* gushing, fulsome, cloying, mawkish, °excessive, °effusive, overdone, (over)sentimental, (over)enthusiastic, *Colloq* °sloppy, slushy: *It's another gushy film about a little girl and a lost kitten.*

gust *n.* **1** °puff, °blow, °wind, °breeze, °blast: *A gust of wind blew my hat off.* —*v.* **2** °puff, °blow, °blast, °surge, °increase: *The wind gusted up to gale force overnight.*

gusto *n.* °enthusiasm, °relish, °zest, °appetite, zeal, zealousness, avidity, °eagerness, °enjoyment, °appreciation, °pleasure, °delight, °satisfaction: *We attacked the meal with great gusto and soon dispatched every scrap.*

gut *n.* **1** Often, ***guts.*** °bowels, intestines, entrails, viscera, °stomach, offal, vitals, vital parts, (of a fish) gurry, *Brit* (of a deer) gralloch, *Colloq* °insides, innards *or* inwards: *As soon as the game is killed, the guts must be removed.* **2** °stomach, abdomen, belly; beer belly, bay window, corporation: *His enormous gut hung over his belt.* **3** ***guts.*** **(a)** °backbone, °bravery, boldness, audacity, °pluck, °courage, °determination, °daring, °spirit, °grit, mettle, °gumption, °nerve, intestinal fortitude, *Colloq* °spunk, gutsiness, *Slang Brit* °bottle, *Taboo* balls: *He hasn't the guts to tell the boss what he thinks of him.* **(b)** °integrity, willpower, °stamina, °endurance, forcefulness, °dynamism: *It takes real guts to stand up for your rights.* —*v.* **4** disembowel, eviscerate, °draw, °dress, °clean: *It will take hours to gut all these fish.* **5** °ransack, °pillage, °plunder, °sack, despoil, °strip, °ravage, °loot, °rifle, *Rare* depredate; °clean out, °devastate, °empty: *The building was completely gutted by the fire.* —*adj.* **6** °basic, °heartfelt, °instinctive, instinctual, intuitive, visceral, deep-seated, °emotional: *My gut reaction is to refuse the offer.*

guttersnipe *n.* waif, street arab, (street) urchin, °ragamuffin, brat, gamin, °rogue, *Colloq Brit rare* mudlark: *That little guttersnipe stole my wallet!*

guy *n.* **1** °man, °lad, °youth, °boy, °fellow, °person, *Colloq* °chap, geezer, *Brit* bloke, *Slang* gink, cat, °customer, *US* °dude, *Old-fashioned* gazabo *or Old-fashioned* gazebo: *Clarence is a pretty nice guy.* —*v.* **2** °mock, °ridicule, °make fun of, °caricature, °satirize, °poke fun at, °lampoon, *Colloq* rib, °take off, *Brit* °send up: *We put on a show at school in which we guyed the teachers.*

gyrate *v.* °rotate, °spin, °revolve, turn (round *or* about), whirl, °twirl, °swirl, °pirouette; °swivel: *The dancers were gyrating to the deafening music.*

H

habit *n.* **1** °custom, °routine, °practice, °convention, °policy, °pattern, °usage, °mode, °rule, wont, praxis: *Jogging every morning had become a habit. I am not in the habit of revealing others' secrets. She is a creature of habit.* **2** °tendency, °disposition, °manner, °practice,

°way, °custom, °inclination, °bent, predisposition, second nature, °frame of mind, °attitude, °penchant, propensity, proclivity; addiction, compulsion: *This kettle has a habit of leaking. Try to control your bad habits. I was irritated by his habit of drumming on the tabletop with his fingers.* **3** attire, clothing, °dress, °apparel, °clothes, garb, °costume, °garments, vestments, °uniform, raiment, livery, °regalia, habiliment(s), *Colloq* °gear: *She was disguised in a monk's habit.*

habitable *adj.* livable, inhabitable: *With some effort, two rooms were made habitable.*

habitat *n.* °abode, °home, °haunt, °domain, °range, °territory, bailiwick, °realm, °terrain, °element, °environment, surroundings, *Colloq* stamping ground: *The winter habitat of the monarch butterfly is in northern Mexico. Curtis doesn't function well outside his natural habitat.*

habitual *adj.* **1** settled, °fixed, °customary, °usual, °conventional, °accustomed, °set, °rooted, established, °traditional, °standard, °routine, °ritual, °regular, °normal, wonted, °common, °ordinary, °natural: *He follows his habitual practice of arising at dawn.* **2** inveterate, established, °chronic, confirmed, hardened, °ingrained, °frequent, °persistent, °constant, °continual, °perpetual: *Many English people are habitual tea drinkers.*

habitué *n.* frequenter, °patron, regular customer, *Colloq* °regular: *Larry has been a habitué of The Bell Inn for twenty years.*

hack[1] *v.* **1** °chop, hew, °lacerate, °gash, °slash, °cut; °mangle, °butcher, °mutilate, °ruin, °destroy, smash, °batter, °damage, °deface: *He hacked the furniture to pieces with an ax.* **2** bark, cough: *She was disgusted when Garvey kept on hacking and spitting.* *—n.* **3** °cut, °gash, °slash, °chop: *He took a hack at the log with his hatchet.*

hack[2] *n.* **1** drudge, penny-a-liner, scribbler, Grubstreet writer: *We made a mistake hiring that hack as a feature writer.* **2** plodder, drudge, toiler, °menial, °flunky, lackey, °slave, *Brit* °fag, *Slang* °grind, *Brit* swot: *Fernledge gets good grades because he's such a hack.* **3** saddle horse, riding horse, hackney, *Archaic* palfrey: *The hack was a fine bay mare.* *—adj.* **4** hackneyed, trite, °banal, overdone, commonplace, °routine, stereotyped, °stock, °tired, °tedious, °mediocre, overworked, °stale, unoriginal, °run-of-the-mill, °humdrum, motheaten, °moldy, *Colloq* old hat: *Haverstock keeps publishing hack romances.*

hag *n.* crone, °fury, °witch, ogress, gorgon, harpy, fishwife, harridan, °shrew, virago, termagant, vixen, hellcat, maenad *or* menad, Xanthippe, *Archaic* beldam; dog, °beast, °monster; *Colloq* battle-ax, *Slang* °bitch, °bag, *US* two-bagger: *Perseus met the Graeae, three old hags with one eye that they passed among them.*

haggard *adj.* °gaunt, °drawn, wasted, °emaciated, hollow-eyed, hollow-cheeked, °scrawny, scraggy, °ghastly, cadaverous, °run-down, wearied, °weary, careworn, °spent, played-out, °exhausted, toilworn, °worn, shrunken, withered: *She was haggard from her ordeal in the prison camp.*

haggle *v.* wrangle, °bargain, higgle, °bicker, chaffer, palter, °dispute, squabble, °quibble, °negotiate; barter, °deal; *Colloq US* °dicker: *However little you ask for the lamp, she's sure to haggle over the price.*

hail[1] *v.* **1** °greet, accost, °address, °signal, °call: *Robert hailed us from across the road.* **2** °cheer, °salute, °applaud, °approve, °glorify, °praise, °laud, °honor, acclaim, °congratulate, felicitate, °acknowledge: *He was hailed by all for his charitable work.*

hail[2] *v.* **1** rain *or* beat *or* shower (down) on, °bombard, °pelt, °volley, barrage: *Rocks and debris hailed down on us from the cliffs above.* *—n.* **2** °volley, °storm, °shower, °torrent, bombardment, barrage: *They were greeted by a hail of abuse when they entered the meeting hall.*

hair *n.* **1** tresses, locks, mane, curls, ringlets; braids, plaits: *You will recognize her by her red hair.* **2** hairsbreadth, whisker, °trifle, fraction, skin of one's teeth: *He won the race by a hair.*

hairdo *n.* coiffure, hairstyle, °cut, coif: *Your hairdo is very becoming.*

hairless *adj.* bald, bald-headed, bald-pated, glabrous, calvous: *Hairless men have a reputation for virility.*

hairsplitting *adj.* quibbling, (over)fussy, hypercritical, °petty, captious, carping, °faultfinding, °finicky, (over)nice, °fastidious, caviling, °niggling, *Colloq* nitpicking: *His hairsplitting approach to details gets on my nerves.*

hairy *adj.* **1** hirsute, °shaggy, downy, fleecy, °fluffy, °woolly, lanate *or* lanose, lanuginous *or* lanuginose, bristly, setaceous, setal, hispid, comate *or* comose, fringy, crinite, trichoid, strigose *or* strigous, strigillose; whiskered, bewhiskered, bearded, barbate, unshaven: *The creature's face was very hairy.* **2** °tricky, °dangerous, °perilous, °risky, °uncertain, °precarious, °hazardous, °frightening, worrying, °nerve-racking, *Colloq* °scary: *The situation at the office has become very hairy.* **3** tangled, °intricate, knotty, complex, °complicated, °difficult, °problematic, °confused, confusing: *The exam contained some hairy questions.*

hale *adj.* °healthy, °hearty, °fit (as a fiddle), °sound, able-bodied, °hardy, °wholesome, °robust, °flourishing, in good *or* fine fettle, °in the pink: *The doctor said I was as hale and hearty as ever.*

halfhearted *adj.* °indifferent, uncaring, unconcerned, °lukewarm, uninterested, °dispassionate, °cool, °unenthusiastic, half-baked, °nonchalant, °phlegmatic, °lackadaisical, insouciant: *The council has made only a halfhearted effort to solve the problem.*

half-wit *n.* dunce, °fool, idiot, simpleton, ninny, ass, ninnyhammer, moron, imbecile, °dolt, dunderhead *or* dunderpate, rattlebrain, nincompoop, dullard, *Colloq* numskull *or* numbskull, nitwit, dimwit, birdbrain, *Brit* nit, °twit: *He's such a half-wit, he can hardly expect me to give him a job!*

half-witted *adj.* °stupid, °foolish, °silly, °simple, °inane, asinine, moronic, imbecilic, doltish, rattlebrained, °feebleminded, attocerebral, cretinous, °thick, *non compos mentis,* dimwitted, weak-minded, *Colloq* °dumb, *Brit* dotty, barmy (in the crumpet): *She had some half-witted idea that she would win the lottery.*

hall *n.* **1** °corridor, hallway, passageway, °passage; foyer, °entry, entryway, °lobby, vestibule: *Let us step out into the hall for a brief word.* **2** auditorium, assembly *or* meeting *or* convention hall, °theater, amphitheater, hired hall; lecture room *or* hall, classroom: *We hired a hall for the wedding reception.*

hallmark *n.* **1** authentication, verification, seal *or* stamp (of authenticity *or* approval), °mark, °device, °sign, °symbol; plate mark; assay mark: *From this hallmark we know when and where the piece was made, by whom, and its degree of purity. Surprise endings are the hallmark of O. Henry's short stories.* **2** °feature, °stamp, °mark, earmark, trademark, °characteristic, °identification: *From his hallmark, a tiny axe, you can tell it's a genuine Fellworthy painting.*

hallow *v.* **1** consecrate, °bless, °sanctify, °dedicate, °honor, enshrine, °glorify: *Englishmen look upon this battlefield as hallowed ground.* **2** °venerate, °worship, °revere, °reverence, °respect, °honor, pay homage *or* respect *or* honor to, °exalt: *We must observe the hallowed traditions of the university.*

hallucination *n.* °fantasy, mirage, °daydream, °illusion, °delusion, °vision, °dream, aberration, chimera, phantasm, °phantom, figment of the imagination, apparition, °specter, °ghost; paresthesia: *Alone at sea for a fortnight, I began having hallucinations.*

halo *n.* nimbus, °aura, aureole *or* aureola, corona, °radiance, *Painting* vesica, mandorla; °ring, disk *or* disc, °circle, annulation, annulus: *The saints can be identified by the golden halos round their heads.*

halt *n.* **1** °stop, °standstill, °end, °termination, °close, stoppage, cessation: *We must call a halt to absenteeism in the factory.* —*v.* **2** °stop, °quit, °end, °terminate, °cease, °check, °curb, °stem, °discontinue, desist, bring *or* come *or* draw to an end *or* close, put an end *or* stop to, conclude, shut *or* close down *or* up: *We halted when we came to the river. The guerrillas halted the armored column at the pass.*

halting *adj.* °hesitant, hesitating, wavering, shifting, uneven, faltering, stumbling, °faulty, unsteady, °awkward, stammering, stuttering: *In halting English he had difficulty in making himself understood.*

hamper[1] *v.* °slow, balk *or* baulk, °delay, °hold up, °retard, °inhibit, °encumber, °hinder, °obstruct, °block, °impede, °prevent, interfere with, °frustrate, °restrict, °curb, °limit, °handicap, °restrain, °trammel, °bar, barricade, °shackle, °clog, °curtail, lessen, °reduce, °diminish: *The ice floes hampered our further progress.*

hamper[2] *n.* basket, pannier, creel, *Brit* punnet; hanaper, *Dialectal* skep: *We bought a wicker hamper for the laundry.*

hand *n.* **1** *Slang* mitt, paw, *US* lunch-hook: *Keep your hands off me, you oaf!* **2** °help, °aid, °assistance, helping hand, °relief, °boost; leg up: *Please give me a hand with this trunk.* **3** °influence, °agency, participation, involvement, °part, °share: *Did you have a hand in my getting the appointment?* **4** °(manual) laborer, °worker, workman, °man, °help, °employee: *It's getting harder to find hands for the farm.* **5** °pointer, °indicator, °index: *The minute hand covers the hour hand at noon and midnight.* **6** °(round of) applause, °ovation, °clap: *Give the lad a hand for trying.* **7** handwriting, °penmanship, °script; calligraphy: *The writing was in a hand she could not recognize.* **8** Often, ***hands.*** °control, °hold, °grasp, °possession, °custody, °clutches, keeping, °power, disposal, °jurisdiction, °authority, supervision, °management, guardianship, °care: *At last, the government is in the hands of the people.* **9** ***at hand.*** °nearby, °close, °near, °close by, °handy, °(readily) available, to *or* on hand, at one's fingertips, °convenient, within (arm's) reach, °accessible, °present; approaching, °imminent, around the corner: *I always kept a pistol at hand. The Day of Judgment is at hand.* **10** ***hand in glove.*** °hand in hand, in league, together, in collusion, collusively, conniving, conspiringly, intimately, closely, jointly, *Colloq* in cahoots: *An informer is working hand in glove with the police.* **11** ***hand in hand.*** together, °side by side, °hand in glove: *They walked hand in hand down the road.* **12** ***hand over fist.*** °quickly, speedily, °rapidly, °swiftly, steadily, °like mad: *He makes money hand over fist in the stock market.* **13** ***hands down.*** °easily, °readily, effortlessly: *He can win the marathon hands down.* —*v.* **14** °give, °pass, °deliver, present to *or* with: *Please hand me the hammer.* **15** ***hand down*** *or* ***on*** *or* ***over.*** **(a)** °bequeath, °will, °pass on; °transfer, °turn over: *The farm has been handed down from father to son for generations.* **(b)** See **18 (a),** below. **16** ***hand in.*** °submit, °give in, °tender, proffer, °offer: *I handed in my resignation.* **17** ***hand out.*** °distribute, disseminate, pass out *or* round *or* around, °give out, °deal (out), mete *or* dole out, °dispense; disburse: *She was handing out leaflets to passersby.* **18** ***hand over.*** **(a)** °deliver, °submit, °yield, °give up, °surrender, °turn over; °transfer: *The masked man told the cashier to hand over all the money.* **(b)** See **15 (a),** above.

handcuffs *n.* pl. manacles, °shackles, *Colloq* cuffs, bracelets, *Slang Brit* darbies: *The police put handcuffs on us all.*

handful *n.* **1** °few, °couple, sprinkling, small number; fistful: *Only a handful of restaurants are open that late. He gave the beggar a handful of change.* **2** (behavior *or* disciplinary) problem, °bother, mischief-maker, °troublemaker, °nuisance: *Timothy was a real handful when he was five.*

handicap *n.* **1** °hindrance, °restraint, °encumbrance, °restriction, limitation, °impediment, °barrier, °bar, °obstacle, °(stumbling) block, constraint, °check, °curb, °trammel, °disability, °disadvantage: *Did Douglas Bader view the loss of his legs as a handicap?* —*v.* **2** °hinder, °hamper, °restrain, °encumber, °restrict, °limit, °impede, °bar, °block, °check, °curb, °trammel, disable, °disadvantage: *I was severely handicapped by the loss of my sword.*

handily *adv.* **1** °readily, °easily, effortlessly, without strain, comfortably, with both hands tied (behind one's back): *Donald handily won the first prize in the poetry competition.* **2** skillfully, capably, deftly, cleverly, dexterously, adroitly, expertly, proficiently, masterfully: *I could never do that intricate work as handily as she.*

handle *n.* **1** °grip, hilt, handgrip, haft, helve: *Hold it by the handle, not the blade.* —*v.* **2** °feel, °touch, °finger, °hold; °caress, °fondle, °pat: *I noticed that he was handling a large switchblade knife.* **3** °manage, °run, °operate, °direct, °administer, °supervise, °oversee, °control, °command, °guide: *At the age of twenty-six, she was handling all foreign business for the company.* **4** °steer, °control, °manage, °cope with, °maneuver, °manipulate: *Are you sure he can handle that horse?* **5** deal *or* trade *or* traffic in, °(buy and) sell, °market: *The gang was found to be handling stolen goods worth millions every month.* **6** °treat, °control, °deal with, °cope with: *Who can handle him when he behaves that way?* **7** °treat, °employ, °use, utilize; °deal with, °wield, °tackle, °manipulate: *Don't you admire how she handled the perspective in this painting?*

handsome *adj.* **1** good-looking, fine-looking, °attractive, °fair, °comely: *He's handsome enough to be a movie star.* **2** °generous, sizable, °large, °big, °substantial, °considerable, °good, °goodly, °ample, °abundant: *She has made a handsome profit on the painting.*

handy *adj.* **1** °nearby, °accessible, °available, at *or* on *or* to hand, °close (by), °convenient, at one's fingertips, within (easy) reach, °(at the) ready: *Because of the recent break-ins, she keeps a pistol handy.* **2** usable, °serviceable, maneuverable, °clever, °useful, °helpful, °practical: *A pair of pliers and a screwdriver are the handiest tools.* **3** deft, °clever, °dexterous, adroit, °adept, skilled, °skillful, °proficient, °expert: *Aunt Sara is very handy with a needle and thread.*

hang *v.* **1** °suspend, °depend, °dangle; be poised *or* suspended, °hover, °swing: *Hang the lantern from this branch. The laundry was hanging on the line.* **2** gibbet, send to the gallows, lynch, °execute, °kill, *Colloq* °string up, °stretch: *They hanged two murderers this morning.* **3** °drape, °fall: *The skirt isn't hanging straight.* **4** ***hang about*** *or* ***around.*** **(a)** loiter, °wait, °linger, dally, °idle, °tarry, *Colloq* °hang out: *He hangs about after school every day, waiting for Susan.* **(b)** Also, **hang about** *or* **around (with).** frequent, °haunt, °visit, spend time at; °associate with, socialize with, hobnob with, rub elbows with, consort with, fraternize with, mix *or* mingle with,

Colloq °hang out (with): *He hangs about the Golden Crown. I saw him hanging around with some pretty unsavory characters.* **5** ***hang back (from).*** be reluctant, °recoil (from), °shrink (from), °hesitate, falter, stay away (from): *Move to the front of the queue and don't hang back.* **6** ***hang fire.*** be delayed, remain unsettled *or* unfinished, be in suspense *or* abeyance; °stall, °hold up, °delay: *The decision will hang fire till after the election.* **7** ***hang on.*** **(a)** Also, **hang on to.** hold on (to), °cling (to), °clutch, °grip, °grasp, °grab: *Hang on to me when crossing the street.* **(b)** °wait, °stay, °stop: *Hang on a minute—what did you call me?* **(c)** °wait, °persist, °remain, °carry on, °persevere, °go on, °hold out, °endure, hold the phone, *Colloq US* hang in there: *Hang on a minute, I have to put down the phone to get a pencil.* **(d)** Also, **hang onto** *or* **upon.** listen carefully *or* attentively, give one's undivided attention, be rapt: *She hung on each word the guru uttered.* **(e)** depend *or* rely (on), be dependent *or* contingent (on), be subject (to), be conditioned *or* conditional (on): *The entire project now hangs on their approval of the budget.* **8** ***hang one's head.*** be ashamed *or* humiliated *or* abashed *or* humbled *or* embarrassed: *She hung her head when the store detective found her with the stolen goods.* **9** ***hang out.*** See **5 (a),** above. **10** ***hang over.*** be put off *or* postponed *or* delayed: *We must first deal with matters hanging over from our last meeting.* **11** ***hang together.*** **(a)** °unite, be united, be as one, °stick together, join forces, °cooperate, act in concert *or* harmony: *We must hang together or we shall hang separately.* **(b)** make sense, be logical, be consistent, °correspond, °match (up), cohere, be coherent: *The statements issued yesterday and today just don't hang together.* **12** ***hang up.*** break the connection, °disconnect, °cut off, put down the receiver: *When a salesman phones, I simply hang up.*

hanger-on *n.* °follower, dependent, leech, °parasite, toady, sycophant, °yes man, *Colloq* scrounger, *US* freeloader, *Slang* groupie, sponger *or* sponge: *The rock star came in, surrounded by his hangers-on.*

hangman *n.* executioner, *Archaic Brit* Jack Ketch: *The hangman, a black hood concealing his identity, slipped the noose over the prisoner's head.*

hanker *v.* Usually, ***hanker after*** *or* ***for.*** yearn for, long for, thirst after *or* for, hunger after *or* for, itch for, pine for, lust after *or* for, covet, crave, have a hankering for, °want, °desire, °fancy, *Colloq* have a yen for: *She hankered after a big box of chocolates.*

hanky-panky *n.* °mischief, °trickery, double-dealing, legerdemain, °deception, duplicity, °chicanery, naughtiness, foolishness, tomfoolery, *Colloq* funny business, jiggery-pokery, monkey business, shenanigans, goings-on, antics: *He'd best not try any of his hanky-panky with the tax inspector.*

haphazard *adj.* **1** °random, °arbitrary, °chance, fortuitous, aleatory, °accidental, °unforeseen, unlooked-for, unexpected, adventitious, serendipitous: *Her fate was to be decided by a haphazard throw of the dice.* **2** °casual, °offhand, hit-or-miss, unsystematic, slapdash, °slipshod, °careless, disorganized, °disorderly: *He took some haphazard shots at the target.*

happen *v.* **1** °occur, °take place, °come about, °go on, °come to pass, °develop; betide, °chance, °prove, °materialize, *Colloq* °transpire, °come off, *Slang* cook: *What is happening? As it happens, everyone has already left.* **2** befall, °become of: *What will happen to me if I refuse?* **3** ***happen on*** *or* ***upon.*** come upon, chance *or* hit on *or* upon, stumble on *or* upon, °find, °turn up, °encounter, °meet with: *It was then that I happened on an ancient book of magic.*

happening *n.* °event, °incident, °occurrence, °occasion, taking place, °circumstance, °chance, °episode, °phenomenon: *Such a conjunction of the planets is a very rare happening.*

happily *adv.* **1** fortunately, luckily, propitiously, providentially, opportunely: *Happily, I was able to identify myself as the dog's owner.* **2** joyfully, joyously, delightedly, gleefully, cheerily, cheerfully, °gaily, merrily, blithely; enthusiastically, heartily: *We happily celebrate Maggie's ninety-seventh birthday today.* **3** °gladly, with pleasure, agreeably, contentedly, °willingly, peaceably: *And they both lived happily ever after.*

happiness *n.* °pleasure, °delight, felicity, °enjoyment, °joy, joyousness, joyfulness, jubilation, cheerfulness, cheeriness, °cheer, blithesomeness, gladness, lightheartedness, exhilaration, elation, °exuberance, high spirits, °glee, °ecstasy: *My happiness at seeing you again was unbounded.*

happy *adj.* **1** pleased, °delighted, °glad, joyous, °joyful, °overjoyed, jubilant, °cheerful, cheery, °blithe, blithesome, °glad, light-hearted, contented, exhilarated, °exultant, °elated, °exuberant, thrilled, °gleeful, euphoric, °ecstatic, satisfied, gratified, *Colloq* °on top of the world, on cloud nine, pleased as Punch, tickled pink, *Brit* in the seventh heaven, over the moon, *US* in seventh heaven: *Her parents were happy that Grendel was married at last.* **2** °lucky, fortuitous, °propitious, °fortunate, auspicious, °advantageous, °beneficial, °favorable, felicitous, °opportune, °timely, °well-timed, apt, °appropriate: *By a happy chance, we were in Bermuda at the same time.*

harangue *n.* **1** diatribe, °tirade, °oration, peroration, declamation, philippic, screed, exhortation, vituperation, rodomontade, °speech, °address, *Colloq* spiel: *This morning he delivered a ten-minute harangue on the weakness of the coffee.* —*v.* **2** declaim, °hold forth, °preach, °lecture, sermonize, pontificate, °vituperate, rant and rave: *Our neighbor is always haranguing her husband about trimming the hedges.*

harass *v.* badger, harry, hector, °trouble, °torment, °bother, °exasperate, °hound, °plague, °persecute, vex, °annoy, °irritate, °pester, °worry, °beset, bait, °nag, pick on *or* at, °tease, torture, *Brit* chivy *or* chivvy *or* chevy, *Slang* give (someone) a hard time: *The police harass me by turning up at all hours for "help with their inquiries."*

harbinger *n.* °forerunner, herald, °precursor, °omen, foretoken, °sign, portent, augury: *The crowing cock is the harbinger of dawn.*

harbor *n.* **1** °port, (safe) haven, anchorage, mooring: *We sailed into the harbor just as the storm broke.* —*v.* **2** °shelter, keep safe, °protect, °shield, °guard, °safeguard, °conceal, °hide: *They have been accused of harboring a known criminal.* **3** °cherish, °foster, nurture, °nurse, °keep, °retain, °maintain, °hold, cling to: *She harbors a grudge against whoever set the fire.*

hard *adj.* **1** °rigid, °stiff, °solid, °inflexible, °firm, °dense, condensed, compressed, °close, solidified, hardened; °stony, rocklike, °concrete, °petrified, granite(like), flinty, °steely; °tough, °rugged, leathery, °callous; unyielding, adamant(ine), impenetrable, obdurate, impervious, °impregnable: *The cement gets hard in an hour. This steak is as hard as shoe leather. This metal is so hard I broke three drills trying to make a hole in it.* **2** °difficult, °laborious, °arduous, backbreaking, °burdensome, onerous, fatiguing, tiring, °exhausting, wearying, °strenuous, °tough, °toilsome: *Laying track for the railway is a very hard job.* **3** °difficult, °perplexing, knotty, °puzzling, baffling, enigmatic, °intricate, °complicated, complex, tangled, °involved, °thorny, °incomprehensible, inscrutable, unsolvable, insoluble, *Colloq* °tough: *There were a lot of hard questions in the exam.* **4** °stern, °cold, °callous, intractable, °exacting, °strict, °demanding, hardhearted, stonyhearted, °severe, °tyrannical, °despotic, °dictatorial, magisterial, °oppressive, °cruel, °ruthless, pitiless, °merciless, °savage, °brutal, brutish, °inhuman, °heartless, °harsh, °unkind, °implacable, °unsympathetic, °dispassionate, uncompassionate, unfeeling, obdurate, indurate; unsentimental, insensitive, °thick-skinned, °tough, hard-boiled, °stony, hard-bitten, unfeeling, unsparing: *Hemel is a hard taskmaster. Of the prison warders, each was harder than the next. He advocates taking a hard line against white-collar crime.* **5** °bad, °difficult, °grievous, °calamitous, wracking, °disastrous, °dark, °grim, distressing, °devastating, °agonizing, °painful, unpleasant, °severe, austere, *Colloq* °tough, °rough: *The years of the Great Depression were hard for everyone.* **6** °cool, unemotional, °calculating, uncompromising, °methodical, °critical, °systematic, °practical, pragmatic, businesslike, °realistic, °penetrating, searching, hardheaded, *Colloq* °tough, hard-nosed: *Shareholders should take a hard look at the annual report. Chapelle drives a hard bargain.* **7** sedulous, assiduous, °devoted, °conscientious, °industrious, indefatigable, °untiring, °persistent, dogged, °intent, °eager, zealous, °ardent, °energetic, °keen, avid: *Galpin is a very hard worker who gets a lot done.* **8** °cold, °bare, °plain, °straight, straightforward, °blunt, °unvarnished, °unquestionable, verifiable, °real, °indisputable, undeniable, incontestable, °incontrovertible, °strict, inescapable, ineluctable, °unavoidable, unalterable, immutable: *The hard fact is that the bill has never been paid.* **9** °angry, °bitter, acrimonious, °hostile, antagonistic, °harsh, unpleasant, unfriendly: *I'm afraid there were some hard words between us.* **10** spirituous, °alcoholic, °strong: *She won't touch hard liquor.* **11** addictive, habit-forming: *He later changed from marijuana to hard drugs.* **12** °sharp, well-defined, °clear, °distinct, °stark, °definite: *Note the hard edges of objects in these realistic paintings.* —*adv.* **13** °vigorously, forcefully, forcibly, energetically, mightily, arduously, laboriously, strenuously, earnestly, actively, dynamically, eagerly, intensely, ardently, heartily, zealously, °intently, spiritedly, diligently, assiduously, sedulously, studiously, determinedly, steadfastly, conscientiously, industriously, devotedly, urgently, persistently, untiringly, indefatigably, perseveringly, unfalteringly, relentlessly, doggedly: *They always had to work very hard just to scrape by. He's hard at work writing his new book.* **14** violently, °deeply, intensely, °badly, distressingly, °painfully, °severely, agonizingly: *Failing the examination hit him quite hard.* **15** °intently, carefully, earnestly: *The judge thought long and hard before passing sentence.* **16** harshly, °severely, °badly, °ill: *It's going to go hard with her if she doesn't change her ways.* **17** ***hard up.*** °poor, °indigent, poverty-stricken, °impoverished, penniless, impecunious, bankrupt, *Colloq* in the red, °broke, bust(ed), °on one's uppers, *Slang Brit* skint: *He's so hard up he can't afford a decent meal.*

harden *v.* **1** °set, °solidify, °stiffen, °freeze: *The concrete will harden overnight.* **2** °intensify, °strengthen, °brace, °fortify, toughen, °reinforce, °stiffen: *The opposition have hardened their stand against privatization.*

hardly *adv.* °scarcely, °barely, °only, °just, only just; not quite, °by no means; °seldom, °rarely: *I hardly knew her. There was hardly enough to eat. He hardly ever visits his old mum any more.*

hardship *n.* °want, °privation, deprivation, °suffering, °misery, °distress, °affliction, adversity, austerity, °misfortune, unhappiness, ill fortune, bad luck, °difficulty, °trouble: *The hardship of frontier life has never been fully described.*

hardware *n.* **1** tools, metal goods, *Brit* ironmongery: *Go down to the hardware shop for some nails.* **2** °(computer) equipment, components, °devices, machinery; arms, munitions, armament(s), materiel: *We need the personnel to operate the hardware.*

hardy *adj.* **1** °robust, °sturdy, °strong, °rugged, °tough, °durable, °sound, °stalwart, °stout, °vigorous, able-bodied, red-blooded, °fit, °hale, °healthy, *Colloq* °husky: *The pioneers were hardy souls.* **2** °bold, °courageous, °daring, valorous, valiant, °brave, °manly, °intrepid, °fearless, °heroic, plucky: *He was among the hardy sailors who went with Magellan.*

harebrained *adj.* **1** °rash, °foolhardy, °wild, madcap, °reckless, °heedless, °improvident, °visionary, °fanciful, airy, *Colloq* crackpot: *Don't put money into any of Zenobia's harebrained schemes.* **2** °foolish, °silly, °inane, asinine, °flighty, witless, brainless, °mindless, °giddy, °frivolous, °scatterbrained: *Such harebrained behavior is not expected in two sober adults.*

harm *n.* **1** °injury, °damage, °mischief, °hurt, °abuse, °misfortune: *A bodyguard was hired to see that no harm would come to her.* **2** °evil, wrongdoing, wickedness, iniquity, °wrong, badness: *I meant no harm when I told Phoebe that her husband was having dinner with Kathy.* —*v.* **3** °hurt, °damage, °injure, °abuse, maltreat, °wound: *She insists that her pet viper would never harm her.*

harmful *adj.* °dangerous, pernicious, deleterious, °destructive, damaging, °bad, °detrimental, °injurious; °unhealthy, noxious, baleful, toxic, °poisonous, °venomous, *Archaic* baneful: *Do you really think that Barbara has had a harmful influence on him? That substance can be harmful if swallowed.*

harmless *adj.* °benign, innocuous, °inoffensive, °gentle, °mild, °innocent, °safe; nontoxic, nonpoisonous, nonvenomous: *Pay no attention to such harmless gossip. It is impossible to identify harmless snakes by sight.*

harmonious *adj.* °agreeable, compatible, congruous, consonant, in accord, congenial, complementary, °sympathetic, concordant, *Colloq* simpatico: *Our careers have always been harmonious.*

harmony *n.* **1** °agreement, °accord, concord, compatibility, °rapport, unanimity, °unity: *The purpose is to promote harmony among the warring factions.* **2** consonance, congruity, °balance, orderliness, closeness, togetherness, consistency, °fitness, parallelism: *A sense of harmony can be felt among the European nations.* **3** melodiousness, euphony, tunefulness: *Their voices are in perfect harmony.*

harrowing *adj.* distressing, vexing, alarming, unnerving, °frightening, °terrifying, horrifying, °horrible, torturous, chilling, °heart-rending, °nerve-racking, °traumatic, °agonizing, °painful; °disturbing, upsetting, worrying, worrisome, °disconcerting, daunting, dismaying, disquieting: *Waiting for the rescue team to release me was a harrowing experience.*

harsh *adj.* **1** °rough, °coarse, bristly, °scratchy, °hairy, °crude; hoarse, °grating, °raucous, rasping, °husky, guttural; clashing, inharmonious *or* unharmonious, °discordant, atonal, dissonant, cacophonous, °strident, °shrill, grinding, °sour; °bitter, °sour, acrid: *He found harsh the sights, sounds, tastes, and smells of the Casbah.* **2** °stern, austere, °bleak, °dour, °unkind, unfeeling, comfortless, uncompassionate, unfriendly, °grim, °hard, °Spartan, stringent, overexacting, Draconian, °tyrannical, °stark, °severe, °cruel, °abusive, °punishing, °punitive, °brutal, brutish, °inhuman, °merciless, °ruthless, pitiless: *Simon is a very harsh taskmaster.* **3** unpleasant, °disagreeable, °impolite, °discourteous, uncivil, °rude, °nasty, °curt, °abrupt, °brusque, °bluff, °gruff, curmudgeonly, choleric, splenetic, °surly, °sullen, irascible, °short-tempered, °petulant, °peevish, °waspish, grouchy, °bilious, °cross, acrimonious, °sarcastic, acerbic: *Why is Maria so harsh to Alan?*

harvest *n.* **1** crop, °yield, °produce, °output, °fruit; °vintage: *If we get enough rain, the harvest should be good this year.* —*v.* **2** °reap, °gather, °pick, glean, °collect: *The oranges are harvested while they are greenish.* **3** °earn, °make, °take in, °collect, °garner, °get, °receive, °obtain, °procure, °net: *They harvest huge profits from slot machines.*

hash *n.* **1** °mixture, °confusion, hotchpotch *or US and Canadian* hodgepodge, °potpourri, gallimaufry, farrago, °mishmash, °jumble, °mess, °shambles, olla podrida, mélange, °medley: *The building is a hash of a dozen architectural styles.* **2** °fiasco, °disaster, °botch, °mess, *Slang Brit* balls-up, *US* snafu: *The council really made a hash of housing for the elderly.* —*v.* **3** Often, ***hash up.*** °mangle, mess *or* mix up, make a hash *or* mess *or* jumble of, °muddle, °bungle, °botch, °mishandle, mismanage, °ruin, °spoil, °butcher, *Colloq* foul *or* louse up, °screw up, muff, *Brit* °bugger up: *Our plans for a holiday have been all hashed up.*

haste *n.* **1** swiftness, °rapidity, quickness, °speed, °velocity, °expedition, °urgency, dispatch *or* despatch, alacrity, celerity, briskness: *We must return to Baker Street with all haste, Watson.* **2** °hurry, °rush, rashness, hastiness, °hustle, bustle, impetuousness *or* impetuosity, recklessness, precipitancy: *Haste makes waste.*

hasten *v.* **1** °hurry, °rush, make haste, °fly, °run, sprint, °race, °bolt, °dash, °scurry, scamper, scuttle, °speed: *I hastened forward to greet her.* **2** °hurry (up), °speed (up), dispatch *or* despatch, °send, °move, °quicken, accelerate, °expedite, °rush, impel, °urge: *Is there any way to hasten delivery of the mail?*

hastily *adv.* **1** °quickly, speedily, °swiftly, °rapidly, °at once, °immediately, °instantaneously, °promptly, without delay, right away, straightaway, °posthaste, hurriedly, °directly, °suddenly, in haste, precipitately, °on the spur of the moment, in a flash *or* wink, before one can *or* could say "Jack Robinson", *Colloq* pronto, °like a shot, like greased lightning, *US* lickety-split, *Slang* pdq *or* PDQ (= 'pretty damn(ed) quick'): *She left hastily when I asked for the return of the loan.* **2** impetuously, impulsively, rashly, recklessly, unthinkingly, thoughtlessly, heedlessly, incautiously: *This is a trick question, so don't answer hastily.*

hasty *adj.* **1** °quick, °speedy, °swift, °rapid, °fast, °brisk, °prompt, °immediate, instantaneous: *He made a hasty departure so as not to be late.* **2** °careless, °rash, °precipitate, °impetuous, °impulsive, °reckless, °thoughtless, °unthinking, incautious, °heedless, ill-considered, °inconsiderate: *His was a hasty decision.* **3** °quick, °speedy, °cursory, °superficial, °fleeting, °passing, slapdash, °perfunctory, °momentary, °brief: *I had a hasty look at the contract, which seemed all right.* **4** °irritable, °quick-tempered, irascible, °testy, °passionate, °impatient, hot-tempered, °petulant, °waspish, °volatile, contentious, choleric, splenetic, bearish, °short-tempered, *US, Canadian, and Irish* cranky: *Grandfather needn't have been so hasty with the poor child.*

hatch *v.* **1** °breed, °brood, °incubate, °bring forth: *The normal clutch of three eggs is hatched in a fortnight.* **2** °devise, contrive, concoct, °design, °formulate, °originate, °invent, dream up, *Colloq* cook up: *Two of them hatched a scheme to trap me.*

hate *v.* **1** °loathe, °abhor, °detest, have an aversion to, be averse to, abominate, °dislike, execrate, °despise, °scorn: *Surprisingly, not everyone hates injustice.* **2** be loath, be reluctant *or* unwilling *or* disinclined; °resist, shrink *or* flinch from, °dislike: *I hate to tell you what I really think. She hated revealing Bill's affair to his wife.* —*n.* **3** hatred, abhorrence, °loathing, odium, °animosity, animus, antipathy, °aversion, °hostility, °antagonism, malice, enmity, detestation: *He didn't believe he could feel so much hate for one person.*

hateful *adj.* **1** °loathsome, detestable, °abhorrent, horrid, °horrible, °abominable, odious, execrable, °despicable, °scurvy, °obnoxious, heinous, °foul, °contemptible, °repugnant, °repulsive, °repellent, °revolting, °vile: *No vice is universally so hateful as ingratitude.* **2** °malignant, malefic, malevolent, malicious, °evil, °mean, °spiteful, °contemptuous: *She gave me a hateful glance.*

haughty *adj.* °arrogant, °proud, °superior, °self-important, °smug, self-satisfied, complacent, °pretentious, °conceited, °egotistical, °snobbish, °overbearing, °lofty, °presumptuous, overweening, patronizing, °supercilious, °vain, °condescending, °contemptuous, belittling, derisive, °disdainful, °scornful, *Colloq* highfalutin *or* hifalutin, °hoity-toity, stuck-up, swellheaded *or* swelled-headed *or* swollen-headed, high and mighty, on one's high horse, snooty, la-di-da *or* lah-di-dah *or* la-de-da, *Slang* snotty, °uppish, uppity: *She's too haughty to have made any friends here.*

haul *v.* **1** °drag, °pull, °tug, °tow, °trail, °lug, °heave, °draw: *Are you sure that one horse can haul that load?* **2** °cart, °transport, °carry, convey, °truck, °move: *The new vehicles are equipped to haul bulk dry goods.* —*n.* **3** °pull, °tug, °drag, °draw; °heave; °attraction: *The tides are greater under the haul of both sun and moon.* **4** °catch, °take, °yield, °harvest, °bag: *We returned with quite a good haul of cod.*

haunt *v.* **1** °visit, °frequent, hang about *or* around, spend time at, *US* habituate: *She haunts the yacht club, waiting for her ship to come in.* **2** °beset, °obsess, °plague, °torment, °trouble, °possess, °prey on: *He is haunted by the fear that she told the police everything.* —*n.* **3** gathering place, meeting place, stamping ground, *Colloq* hangout: *I returned to Westport to visit the haunts of my youth.*

have *v.* **1** °possess, own, °keep; °maintain: *Imelda had more shoes than you could count.* **2** °receive, °take, °accept, °get, °obtain, °acquire, °procure, °secure, °take: *How many gifts did you say she had from him?* **3** °entertain, be struck by: *I have an idea!* **4** °possess, °bear, °contain, °include, comprise: *The night has a thousand eyes.* **5** suffer with *or* from, be suffering with *or* from, be experiencing, be undergoing, be enduring, be subjected to: *My daughter was really ill when she had measles.* **6** °arrange, °organize, °set up, °prepare; °hold: *If I have a party, will you come? We shall have our next meeting on Saturday.* **7** °partake of, participate in, °experience, °enjoy; °eat; °drink: *I hope you have a good time.* **8** give birth to, °bear, °deliver, °bring into the world; beget, sire, °father: *She had three children in as many years. How many children has he had?* **9** °make, °demand; °force, °oblige, °cause, °induce, °press, °require, compel: *We'll have the caterers come at noon.* **10** ***had better*** *or* ***best.*** ought to, °must, should: *I had better not drink as I'm driving.* **11** ***had rather*** *or* ***sooner.*** prefer to, would rather *or* sooner: *I'd rather be with you.* **12** ***have on.*** **(a)** be wearing, be dressed *or* clothed *or* attired in: *I'll have on the silk nightgown you like so much.* **(b)** be committed to, have planned, have in the offing, have on the agenda: *I have something on next Tuesday evening.* **(c)** °trick, °tease, °deceive, pull (someone's) leg, play a joke on, °fool: *Do you really love me or are you having me on?*

havoc *n.* **1** °ruin, devastation, °destruction, °desolation, rack *or* wrack and ruin, despoliation, spoliation, °damage: *The havoc caused by the storm will cost millions to repair. The hurricane wreaked havoc throughout the island.* **2** °confusion, °chaos, °upset, °disorder,

°mayhem, °shambles, disruption: *The railway slowdown created havoc with commuter schedules.*

hazard *n.* **1** °peril, °danger, °risk, endangerment, °threat, °jeopardy: *The greatest hazard in sailing single-handed round the world is the loneliness.* **2** °chance, °gamble, uncertainty, °luck, °fortune: *They banned all games that depended on hazard.* —*v.* **3** °venture, °dare; °gamble, °risk, °jeopardize, °endanger, °threaten, imperil, °stake: *May I hazard a guess as to the origin of the word? He hazarded his entire fortune on the turn of a card.*

hazardous *adj.* unsafe, °risky, °perilous, fraught with danger, °questionable, °shaky, °dangerous, °precarious, °uncertain, unpredictable, °ticklish, °tricky, *Dialect* °parlous, *Colloq chiefly Brit* °dicey, dicky *or* dickey, *Slang* °hairy: *It took three months to make the hazardous overland journey in those days.*

hazy *adj.* **1** °misty, foggy, smoggy, cloudy, °overcast: *It was so hazy that we couldn't see the town, only a mile away.* **2** °indistinct, blurred, blurry, °dull, °dim, °faint, °nebulous, °vague, unclear, °fuzzy, °indefinite, muddled: *He has only a hazy idea of the plot.*

head *n.* **1** skull, pate, cranium, *Colloq* dome, *Slang* coco(nut), belfry, noggin, bean, nut, rocker, noodle, gourd, *Brit* conk, crumpet, noddle, °loaf: *She was sound asleep as soon as her head touched the pillow.* **2** °chief, °leader, administrator, chief executive officer, CEO, °(managing) director, MD, president, chairman, chairwoman, chairlady, chairperson, °chair, °employer, °principal, °superintendent, °supervisor, governor, prime minister, headmaster, headmistress, *Colloq* °boss, head man, the man, *Brit* guv'nor, guv, *US* (chief) honcho; *Slang* big cheese, *US* Mr. Big: *The new head has called a meeting of the board of directors.* **3** °front, vanguard, forefront, van, forepart: *At the head of the column marched the general himself.* **4** °aptitude, °intellect, °intelligence, °talent, °perception, perceptiveness, °mentality, °faculty, °flair, °genius, °brain, °mind, °wit, *Colloq* brains, gray matter: *I have no head for figures.* **5** °crisis, apex, (critical *or* turning) point, °peak, °crest, °(fever) pitch, °climax, culmination, conclusion, crescendo: *Matters have been brought to a head because of the coming elections.* **6** °source, °origin, fount, font, fountainhead, wellspring: *We were trying to reach the head of the stream nightfall.* **7** °top, first place, leading position, °leadership, forefront: *Albert is at the head of his class in mathematics.* **8** ***head over heels.*** °completely, °entirely, °deeply, °utterly, °wholly, fully, *Colloq* °madly, wildly: *The two of them are head over heels in love.* —*adj.* **9** °first, °chief, °main, °principal, °leading, °premier, °foremost, °prime, °preeminent, °cardinal, °paramount, °supreme, °superior, °senior: *Alphonse is our new head chef.* —*v.* **10** °go, °move, °proceed, °turn, °steer, °aim, °point, head for, make a beeline for: *I shall head home when I leave here.* **11** head up, be in *or* take charge (of), °direct, °supervise, °oversee, °control, °govern, °run, °(take the) lead, °guide, °manage, °command, °rule, °administer, °conduct: *Who will head the organization if you resign?* **12** °lead, °precede, °top: *Charlotte heads the list of candidates.* **13** ***head off.*** **(a)** °intercept, °divert; °cut off, °stop, °block: *The cavalry will head them off at the pass.* **(b)** °stop, °forestall, °prevent, °inhibit, avert, ward *or* fend off: *What can we do to head off inflation?*

headache *n.* **1** migraine, *Technical* cephalalgia: *Hazel says she has a headache and cannot come to work today.* **2** °worry, °bother, vexation, °inconvenience, °nuisance, °annoyance, °problem, °difficulty, °trouble, bane, *Colloq* °pain (in the neck), *Slang* pain in the *Brit* arse *or US* ass: *The balance of payments deficit is a perpetual headache.*

headway *n.* **1** °progress, forward motion, °improvement: *Headway against that current was almost impossible. Any headway to report?* **2** ***make headway.*** °advance, °progress, move forward, °go, °gain (ground), get *or* go ahead, °proceed, get going: *I tried to persuade her but haven't made much headway.*

heal *v.* **1** °cure, °repair, °renew, revitalize, °rejuvenate, °restore; °mend, °recuperate, °recover, °improve: *The wounds have healed. This medicine is guaranteed to heal athlete's foot. I heal quickly.* **2** °reconcile, °settle, °patch up, put *or* set straight *or* right, °remedy, °repair, °mend: *We met to discuss healing the rift between us.*

health *n.* **1** °condition, °fitness, °trim, fettle, °form, constitution: *My health has improved enough for me to return to work.* **2** healthiness, haleness, healthfulness, robustness, °vigor, vigorousness, salubrity, salubriousness, well-being, °strength: *The health of the economy is good.*

healthy *adj.* **1** °well, °fit, °trim, in good *or* fine fettle *or* shape, in good health, °robust, °hale (and hearty), °sturdy, °strong, °vigorous, thriving, °flourishing, *Colloq* °in the pink: *How come he is so healthy if he smokes?* **2** °wholesome, healthful, salubrious, salutary, °beneficial, nourishing, °nutritious, °tonic, °bracing: *The answer lies in eating healthy food and living in a healthy climate.*

heap *n.* **1** °collection, °pile, °mound, °stack, °accumulation, aggregation, agglomeration, congeries, conglomeration, °hoard, °mass, °store, °mountain, stockpile, °supply, *Colloq US and Canadian* stash: *Donors contributed a huge heap of clothing for the sale.* **2** Often, ***heaps.*** °abundance, plethora, superabundance, lot(s), °plenty, °great deal, °scores, peck, °sea, *Colloq* lashings, load(s), °piles, ton(s), raft(s), pots, oodles, scad(s), *US and Canadian* slew: *It was served with heaps of chocolate sauce. Heaps of people bought your record. You're in for a heap of trouble.* —*v.* **3** °collect, °gather, °harvest, °reap, glean, °garner, °pile (up), °accumulate, cumulate, aggregate, °amass, stockpile, °save (up), bank, lay by *or* up *or* in, °set aside, *Colloq* stash (away): *The squirrels heap up their supply of nuts for the winter.* **4** °shower, °load, °bestow, °give, °provide, °burden: *The grateful prince heaped us with treasure beyond imagination.*

hear *v.* **1** °perceive, °understand, °listen (to), °attend (to), pay attention (to), °catch, °heed, heark *or* hearken (to) *or US also* harken (to): *Please hear what I have to say first.* **2** °understand, °learn, °discover, °find out, °gather, °get wind of, °pick up, ascertain, be told *or* advised *or* informed: *I hear you're thinking of resigning.* **3** ***hear of.*** °entertain, °consider; °approve (of), °sanction, condone, agree *or* consent *or* assent to: *I won't hear of your leaving.*

heart *n.* **1** *Colloq* ticker, °pump: *The doctor says my heart is sound as can be.* **2** °stomach, °nerve, °courage, °bravery, mettle, °will, boldness, °pluck, °resolution, °determination; callousness, insensitivity, heartlessness; *Colloq* °guts, °spunk: *I haven't the heart to tell him that his hamster died.* **3** °(basic) nature, °core, °center, °focus, °hub, °middle, marrow, °pith, °essence, °quintessence, °nucleus, °nub, crux, basics, fundamentals, *Colloq* nitty-gritty: *Let's get down to the heart of the matter.* **4** °sincerity, sentiment(s), feeling(s), °spirit, °verve, °enthusiasm: *She's going through the motions, but her heart isn't in her work.* **5** °humanity, humanitarianism, °sympathy, °understanding, °kindness, kindliness, compassion, empathy, goodness, °consideration, °concern, °soul, tenderness, magnanimity, generosity, °sensitivity, °sensibility, °sentiment, °pity, °(brotherly) love, °affection: *Mother Theresa is known for having a lot of heart. Have you no heart, Genghis?*

heartbroken *adj.* °brokenhearted, °downhearted, dispirited, °unhappy, °miserable, grief-stricken, °upset, °dejected, heartsick, crestfallen, °despondent, depressed, disconsolate, distressed, °woebegone, °doleful, °sorrowful, °mournful, morose, disheartened, °disappointed, crushed: *Lucy was heartbroken over the loss of her locket.*

heartfelt *adj.* °sincere, °honest, °genuine, unfeigned, °earnest, °serious, °wholehearted, °deep, °profound, dedicated, °devoted, °ardent, committed, °fervent, fervid, °hearty, °passionate: *It is my heartfelt wish that you should succeed.*

heartless *adj.* °cruel, hardhearted, °callous, unconcerned, °inhuman, inhumane, °unkind, unfeeling, °unsympathetic, °brutal, °cold, °merciless, pitiless, °ruthless, °coldblooded: *Taking away the boy's ice cream was a heartless thing to do.*

heart-rending *adj.* °agonizing, distressing, °excruciating, °bitter, °painful, hearthwdbreaking, °harrowing, °piteous, °tragic, depressing, °poignant: *As the dogs closed in, the deer gave a heart-rending scream and died.*

heartwarming *adj.* **1** °moving, °touching, warming, affecting, uplifting, inspiriting, cheering, encouraging: *I heard the heartwarming news about Alan's recovery.* **2** °satisfying, gratifying, °pleasing, comforting, °pleasurable, °rewarding: *It was heartwarming to see the family together at Christmas.*

hearty *adj.* **1** °genial, °warm, kindhearted, °affectionate, °amiable, °amicable, °friendly, affable, °cordial, °open, convivial: *I was greeted by a hearty welcome when I returned.* **2** °genuine, unfeigned, °authentic, °sincere, °heartfelt, °warm, °wholehearted, °honest, °earnest, °devout, °stalwart, °stout: *Please give Desmond my hearty congratulations.* **3** °enthusiastic, °vigorous, °energetic, °eager, zealous, °exuberant, °robust, °active, °animated, °strong: *They are hearty supporters of our cause.* **4** °abundant, °ample, °substantial, °solid, sizable, °satisfying, °square; nourishing, °invigorating, strengthening: *On Sundays, we had a hearty meal at midday when we could afford it.* **5** °healthy, °hale, °vigorous, °robust, °strong, °sound: *Well, Eric, now that you're over your malaria you appear to be quite hearty.*

heat *n.* **1** °warmth, warmness, hotness, fever, fieriness, torridity *or* torridness: *The heat of the sun feels good after that cold swim.* **2** °passion, °ardor, °fervor, fervidness, °intensity, °fury, zeal, zealousness, earnestness, vehemence, °eagerness, °enthusiasm, °excitement, tenseness, °tension, °stress, °agitation, arousal, impetuosity, stimulation, exhilaration: *If I shouted at you in the heat of the moment, I apologize.* —*v.* **3** °warm (up); °boil: *Would you heat some water for tea? It must cost a lot to heat this house.* **4** Often, ***heat up.*** °excite, °intensify, impassion, °inflame, °kindle, ignite, °quicken, inspirit, °rouse, awaken *or* waken, °stir, °animate, °stimulate, °warm (up), °activate, *Colloq Brit* hot up: *It took a while for the debate to heat up.*

heated *adj.* °impassioned, °excited, intensified, aroused, quickened, stimulated, °inflamed, vehement, °fiery, frenzied, °frantic, frenetic, °passionate, °fervent, fervid, °ardent, °intense, °furious, °stormy, °tempestuous, °violent; °angry, °bitter: *The heated dispute over animal rights finally led to blows.*

heathen *n.* **1** unbeliever, °infidel, °pagan, idolater *or* idolatress, polytheist, atheist, nullifidian, °skeptic, agnostic, heretic: *One often finds more mercy in a heathen than in a brother Christian.* —*adj.* **2** °infidel, °pagan, atheist(ic), °godless, nullifidian, skeptic(al), doubting, agnostic, °heretical, irreligious: *After the missionaries left, the tribe returned to their heathen practices.* **3** °savage, °barbarian, barbaric, °uncivilized, °primitive, unenlightened, uncultured, Philistine; polytheistic, pantheistic: *The people were told not to worship heathen gods.*

heave *v.* **1** °raise, °lift, °hoist, °haul, °pull, °draw, °tug; °move: *We heaved in the anchor and made sail.* **2** °throw, °toss, °hurl, °fling, °cast, °sling, °pitch, °let fly, °send, °launch, *Colloq* °peg, chuck: *When he heaved the sword into the lake a hand rose from the waters and caught it.* **3** °breathe, utter, °sigh, °groan, °moan, °gasp: *We heaved a sigh of relief as the last survivor was rescued.* **4** °gag, retch, °vomit, be sick, °regurgitate, disgorge, *Colloq* °throw up, *Slang* puke, lose one's lunch, return one's dinner, *US* upchuck, spiff one's biscuits: *Pendrake got drunk and heaved into a nearby potted palm.*

heaven *n.* **1** °paradise, °bliss, hereafter, nirvana, Abraham's bosom, Elysian Fields *or* Elysium, Valhalla, Zion, happy hunting ground, Avalon, Isles of the Blessed, the Blessed *or* Fortunate *or* Happy Isles *or* Islands: *Will I go to heaven when I die?* **2** ***heavens.*** °sky, °firmament, *Literary* welkin, empyrean: *Thousands of multicolored balloons were released into the heavens. The heavens opened and it started to pour.* **3** °happiness, °bliss, °joy, °rapture, °ecstasy, °paradise, contentment, seventh heaven, Eden, °utopia, heaven on earth: *Darby said that his years with Joan had been sheer heaven.*

heavenly *adj.* **1** °divine, angelic, °seraphic, °celestial, °holy, °immortal, blessed, beatific, beatified, °spiritual, °saintly; supernal, °unearthly, otherworldly, ultramundane, extramundane, extraterrestrial: *Alone on the mountain, she heard heavenly voices calling her name.* **2** °delightful, wonderful, °marvelous, °sublime, paradisiac(al) *or* paradisaic(al), °glorious, °splendid, °superb, °exquisite, °perfect, °ideal, °excellent, °fantastic, °rapturous, entrancing, blissful, *Colloq* °gorgeous, °divine, smashing, °great: *André's serves a perfectly heavenly lobster thermidor.*

heavy *adj.* **1** °weighty, °ponderous, °massive, *Literary* massy; °compact, °dense: *This box is much too heavy for me to lift.* **2** °abundant, overflowing, °excessive, copious, °profuse, °prodigious, °ample, unmanageable: *We just crawled along in the heavy traffic.* **3** °serious, °grave, °important, °crucial, °critical, °acute: *You always bring up heavy topics at the end of the meeting.* **4** °burdensome, onerous, °oppressive, °weighty, °unbearable, °severe, °grievous, distressful, °sore, intolerable, insupportable *or* unsupportable, °awful: *Teenagers are often weighed down with heavy responsibilities.* **5** °sad, °sorrowful, distressing, °grievous, upsetting, depressing, °gloomy, °somber, °melancholy: *Some heavy news has been received from the front.* **6** °unhappy, °miserable, depressed, °melancholy, grieving, °sad, °dejected, °downhearted, disconsolate, downcast, °despondent, °gloomy, heavy-hearted, morose, crestfallen, cheerless: *I am heavy at heart to tell you that your husband has been taken hostage.* **7** °ponderous, °tedious, °monotonous, °boring, uninteresting, °leaden, °dull, °prosaic, °dry, dryasdust, °stodgy, °staid, °stuffy, stifling, stultifying: *The critics found the style much too heavy for the subject being treated.* **8** °thick, °coarse, °broad, °blunt, °clumsy, °ungraceful: *These heavy brushstrokes prove that the painting is not a Turner.* **9** °gloomy, cloudy, °overcast, °bleak, °dismal, °dreary, °leaden, °gray, °dark, lowering *or chiefly Brit* louring, °threatening: *It was another of Scotland's heavy February days.* **10** °intense, concentrated, °severe, °forceful, °violent, °torrential: *The snowfall was the heaviest of the year.* **11** overweight, °fat, °obese, °stout, °chubby, °plump, corpulent, portly, paunchy, tubby, *Brit* podgy *or US* pudgy, *Colloq* beer-bellied: *The doctor did not say I was too heavy, only that I ought to be six inches taller.* **12** °weighty, °difficult, complex, °recondite, arcane,

°deep, °profound, esoteric, °incomprehensible, impenetrable, unfathomable: *His specialty is some heavy subject, like lunar archaeology.* **13** burdened, laden, encumbered, °loaded, overloaded, weighed down: *We spied two Indiamen, heavy with cargoes of spices, sailing toward us.*

heavy-handed *adj.* **1** °awkward, °clumsy, °inept, maladroit, unskillful, °ungraceful, graceless, bungling: *She made a heavy-handed attempt to apologize.* **2** autocratic, imperious, magisterial, °overbearing, °despotic, °dictatorial, °tyrannical, °oppressive, °domineering, ironhanded, °harsh, °severe: *For more than thirty years he exercised his heavy-handed control over the country.*

heckle *v.* badger, °pester, °annoy, °irritate, °bother, nettle, bait, °harass, harry, °plague, hector, °taunt, °jeer, *Colloq* hassle, °bug, *Brit, Australian, and New Zealand* barrack: *The speaker was constantly heckled by some rowdies at the back of the hall.*

hectic *adj.* °feverish, °excited, °agitated, °busy, bustling, rushed, hyperactive, overactive, frenzied, °frantic, °chaotic, °wild, °mad, frenetic, °riotous: *What with the new baby and everyone else having the flu, I've had a pretty hectic week.*

heed *v.* **1** pay attention to, °attend, (take *or* make) note (of), listen to, °mark, °consider, °bear in mind; °take, °follow, °obey, °mind, °respect, °accept, °abide by: *I wish I had heeded her suggestions for redecorating the house. Heeding his advice, I joined the army.* —*n.* **2** °attention, °notice, °ear, °mind, °notice, °respect, °consideration, °thought: *Let us give heed to the speaker's admonitions.*

heedless *adj.* °inattentive, uncaring, unmindful, neglectful, unobservant, °regardless; °oblivious, °deaf, °blind: *Heedless of tradition, they replaced the older buildings with a public housing project.*

heel[1] *n.* **1** °end, butt *or* tail *or* fag (end), °stump, °remainder, °remnant, °rind, crust: *At Dotheboys Hall he always got the heel of the loaf.* **2** cad, °scoundrel, swine, °rogue, scamp, °philanderer, *Old-fashioned* worm, knave, *Chiefly Brit* blackguard, *Colloq Brit* rotter, *Old-fashioned* bounder, *Slang* bastard, *Brit* sod: *That heel got her pregnant and then refused to marry her!* **3** ***down at the heel.*** °poor, °destitute, °impoverished, down-and-out, on (one's) uppers, in straitened circumstances; °shabby, °seedy, °dowdy, out at the elbows, °run-down, slovenly, *Brit* out at elbows, down at heel, *US* out at the elbows, *Colloq* °broke, strapped: *Farrington seemed to be down at the heel, so I slipped him a five-dollar bill.* **4** ***take to (one's) heels.*** °take flight, °flee, °escape, run off *or* away, show a clean pair of heels, *Colloq* °split, *Brit* do a moonlight flit, *US* take a (runout) powder, fly the coop, *Australian and New Zealand* shoot through: *When I mentioned payment, she took to her heels.* —*v.* **5** dog, °follow (closely), °shadow, °chase, °pursue: *We have taught our puppy to heel.*

heel[2] *v.* °list, °lean (over), °tilt, °tip, °incline: *The boat heeled as we turned to beat into the wind.*

hefty *adj.* **1** °big, °large, °bulky, cumbersome, °awkward, °unwieldy, °clumsy, °substantial, °massive: *That's a pretty hefty box for one person to carry.* **2** °brawny, °strong, °powerful, °burly, °muscular, strapping, °rugged, °robust, *Colloq* °husky, beefy: *That hefty man over there is the Olympic weight-lifting champion.* **3** °substantial, °considerable, sizable, °impressive, °enormous, °huge, *Colloq* thumping *or Brit* socking great: *I've just received a hefty increase in salary.*

height *n.* **1** altitude, °elevation, °level; tallness: *At what height is the second camp on Mount Everest? The height of the skyscrapers impressed me.* **2** °acme, °crest, °pinnacle, °top, °zenith, apogee, °peak, apex, °maximum, high point, °summit, °climax, culmination, °extreme: *Popularity of that style reached its height in the 1930s. His remark reflected the height of arrogance.* **3** Often, ***heights.*** °elevation, °mound, °hill, eminence, °prominence, °mountain, °peak, °crag, °summit; tor, °cliff, °bluff, promontory, escarpment, scarp, headland, *Northern Brit* °fell: *As we gazed down from the heights, the villages seemed tiny.*

heighten *v.* **1** °raise, elevate, °build up, °increase, °lift (up), upraise: *We have to heighten the screen to cover the top of the cabinets.* **2** °intensify, °deepen, °strengthen, °reinforce, °amplify, °magnify, °increase, °enhance, augment, °add to, °supplement: *Adding some blue heightens the effect.*

heir *n.* heiress, beneficiary, inheritor, legatee, successor: *What if you are not the rightful heir to the estate?*

hell *n.* **1** Erebus, Hades, Acheron, Tartarus, Gehenna, Tophet, Abaddon, Pandemonium, Dis, Sheol, Avernus, °underworld, infernal regions, °abyss, abode of the damned, inferno, hellfire, lower world, nether regions, bottomless pit, other place: *For his sins he was condemned to everlasting hell.* **2** °chaos, °misery, °torment, °agony, torture, °ordeal, nightmare, °trial: *The office has been an absolute hell since you left.* **3** °anguish, °pain, °agony, °torment, torture, °misery, °suffering, °affliction: *We have been through hell since Grenville's heart attack.* **4** °criticism, censure; scolding, castigation, °reprimand, upbraiding: *My mother gave me hell for getting home late.*

helm *n.* **1** tiller, °wheel, rudder, steering gear *or* apparatus: *I had to put the helm over hard to avoid hitting the pier.* **2** directorship, presidency, chairmanship, °leadership, °control, °rule, °command, *Colloq* driver's seat, saddle: *Arthur remained at the helm of the company for forty years.*

help *v.* **1** °aid, °assist, lend a hand, °support, °serve; succor: *Let me help you carry that package, Mrs. Joiner. His charitable efforts have long helped the poor.* **2** °relieve, alleviate, °mitigate, °improve, °facilitate, °ease, °better, °remedy, °cure: *Complaining about it won't help the situation.* **3** °stop, °refrain from, °avoid, eschew, °resist, °keep from, forbear, °escape: *I know it was a secret but I couldn't help telling Sam.* **4** °assist, °serve, °advise, °inform: *May I help you, sir?* **5** ***help oneself.*** °appropriate, °take, arrogate, commandeer, expropriate; °steal, purloin, usurp, plagiarize, °pirate, *Colloq* °pinch, °lift, *Brit* °nick, *US* °boost: *She helped herself to some money from the till.* —*n.* **6** °aid, °support, succor, °assistance: *I need help in solving this problem.* **7** employee(s), worker(s), °staff, helper(s), hand(s), assistant(s), laborer(s), domestic(s), servant(s), *Brit* °daily (help): *We have to hire more help to finish on time.* **8** °supporter, °aide, °assistant, helper: *You have been a great help to me.* **9** °relief, °remedy, °cure, balm: *This ointment has been a great help in stopping the itching.*

helpful *adj.* °useful, °serviceable, °practical, pragmatic, utilitarian, °beneficial, valuable, °profitable, °advantageous, °constructive, °productive; °supportive, reassuring, °sympathetic, °considerate, caring, °accommodating, °kind, °considerate, cooperative, °neighborly, °friendly, °benevolent: *A reference book might be helpful in answering that question. Mrs. Carlyle gave me such helpful advice.*

helping *n.* serving, °portion, °ration, plateful, *Brit* °help, *Colloq* dollop: *He has already asked for a third helping of potatoes.*

helpless *adj.* **1** dependent, °vulnerable, °weak, °helpless, °feeble, °infirm, °lame, °crippled, °disabled: *The progress of the disease has left him completely helpless.* **2** °confused, baffled, mystified, bewildered, perplexed, at sea, confounded, muddled, nonplussed: *I am com-*

pletely helpless when it comes to balancing my checkbook. **3** weakened, °weak, debilitated, °faint, enfeebled, °feeble, worn-out, °spent, °exhausted, °prostrate, enervated: *All of us were simply helpless from laughing so much.* **4** °worthless, °incapable, °incompetent, °useless, unavailing, °inefficient, °inept, unfit, °unqualified: *Dali was said to be helpless without Gala to manage his affairs.*

helter-skelter *adj.* **1** °disorderly, disorganized, °confused, muddled, °haphazard, °careless, jumbled, °random, °topsy-turvy, *Colloq* higgledy-piggledy: *How can one find anything in this helter-skelter arrangement of books?* —*adv.* **2** confusedly, °pell-mell, in all directions, recklessly, unsystematically, chaotically, erratically, aimlessly, *US* every which way, *Colloq* higgledy-piggledy: *When the giant appeared, the children ran away, helter-skelter.*

hence *adv.* **1** °therefore, °consequently, °thus, °accordingly, ergo, as a result, for that *or* this reason: *She has homework to do, hence cannot go to the cinema.* **2** away, from here *or* this place: *Get thee hence, you wicked witch!* **3** from now, in the future: *Where will I be two years hence?*

henceforth *adv.* hereafter, henceforward *or* henceforwards, from now on, *Colloq US* from here on out: *Henceforth, make no more personal remarks.*

henchman *n.* (fellow) mobster *or* gangster *or* hoodlum, °gangster, bodyguard, myrmidon, right-hand man, °associate, °attendant, °follower, °supporter, confidant, crony, *Colloq* sidekick, hooligan, *Brit* °minder, *US* buddy, °cohort; *Slang* °heavy, *US* torpedo, gunsel, goon: *Mr. Big strutted in with his henchmen and wrecked the place.*

henpeck *v.* °nag, °harass, hector, °pester, °torment, °bully, °carp, °cavil: *She henpecked him so much that he finally left her.*

herd *n.* **1** °group, °pack, °bunch, °cluster, °flock, °crowd, multitude, °host, horde, °throng, °mass, °swarm, °press, °crush; assemblage, °collection: *A herd of protesters gathered before the embassy.* **2** common herd, °rabble, °hoi polloi, great unwashed, °riffraff, masses: *How can you expect the herd to understand these principles?* —*v.* **3** °gather, congregate, °flock, °assemble, °collect: *The reporters herded round to interview the minister.* **4** °round up, °gather (together), °shepherd, °drive, *Western US and Canadian* wrangle, corral: *Herd those cattle into the pen.*

hereditary *adj.* **1** heritable, inheritable, transmissible, transferable, inherited, genetic, congenital, °inborn, innate; atavistic: *Green eyes and red hair are hereditary in our family.* **2** °traditional, handed down, inherited, bequeathed, willed; ancestral: *Priesthood was hereditary among Aaron's descendants. These are the hereditary lands that go with the castle.*

heretical *adj.* °unorthodox, heterodox, °impious, freethinking, heretic, apostate *or* apostatical, iconoclastic, °schismatic, °skeptic, agnostic, atheist(ic), idolatrous, °heathen, °pagan, °infidel, °godless: *The purpose of the Inquisition was to discover and suppress heretical views.*

heritage *n.* **1** °estate, °inheritance, legacy, patrimony, birthright, legacy: *The manor house was included in the heritage that was passed on to me.* **2** °tradition: *The documentary examines our cultural heritage.*

hermetic *adj.* hermetical, airtight, sealed; impervious: *If the hermetic seal is broken, sterility is not guaranteed.*

hermit *n.* °recluse, eremite, anchorite *or* anchoret *or* anchoress, °solitary, stylite: *He has retired from the world to become a hermit.*

hero *n.* **1** °heroine, °champion, exemplar, °star, superstar, °idol, °ideal, man of the hour, luminary, °notable, °celebrity; knight, paladin, warrior: *She was a hero to every aspiring woman lawyer. She expects a hero in shining armor on a white horse to sweep her off her feet.* **2** °protagonist, (male) lead *or* star, leading man *or* actor, °principal: *He always plays the hero in our local repertory company.*

heroic *adj.* **1** °brave, °courageous, °bold, valiant, valorous, undaunted, °dauntless, stouthearted, °noble, °intrepid, °gallant, °chivalrous, °daring, plucky, °audacious, °fearless, °manly, virile, manful: *We published a book about Aylesworthy's heroic exploits.* **2** °noble, altruistic, magnanimous, °generous, upstanding, °honorable, °virtuous, °staunch, °steadfast, °stalwart, °determined: *This plaque commemorates Manzanilla's heroic work in curbing religious persecution.* **3** °desperate, °drastic, °extreme: *Despite heroic efforts, it was impossible to rescue the entire crew.* **4** °grand, larger than life, exaggerated, magniloquent, °grandiose, °extravagant; °giant, °gigantic, °enormous, °huge, titanic, °colossal, stupendous: *The heroic phrases of the demagogue still ring in my ears. Statues of heroic proportions have been moved from Abu Simbel.* **5** mythological, epic, Homeric, °legendary, °classical, °fabulous, wonderful, °miraculous: *He enjoyed tales of heroic demigods like Hercules.* **6** °majestic, °lofty, °elevated, °grand, august, °towering, °eminent, °distinguished, °prominent: *The entire court was silent as the heroic figure of the emperor entered.*

heroine *n.* °(female) lead, leading actress *or* lady, prima donna *or* ballerina, *première danseuse*, diva: *The heroine was played by Mme. Kropeczka.*

hesitant *adj.* **1** hesitating, undecided, °uncertain, °unsettled, °irresolute, vacillating, shilly-shallying, dithering, fluctuating, wavering, °unresolved, ambivalent, in *or* of two minds, °indefinite, *Brit* havering: *He was hesitant about buying the blue or the green model.* **2** °halting, stammering, stuttering, faltering: *From her hesitant speech I'd say she was nervous.*

hesitate *v.* **1** °delay, hold *or* hang back, °pause, dillydally, °wait, temporize, think twice, balk, boggle at, °shrink from, demur, °scruple, *Brit* haver, jib, *Colloq* °stall: *He who hesitates is lost.* **2** °fumble, °equivocate, tergiversate, °fluctuate, °alternate, waver, dither, vacillate, °shilly-shally: *He hesitated between the doors—did this one conceal the lady or the tiger?* **3** °stammer, stutter, falter, sputter, splutter, °stumble, hem and haw: *I hesitated throughout the speech purely from stagefright.*

hidden *adj.* concealed, °secret, obscure(d), °occult, °veiled, °cryptic, °recondite, arcane, covert, esoteric, unseen, °private: *Was this an ordinary shopping list or did it have some hidden meaning?*

hide[1] *v.* **1** °conceal, °secrete, °cache, squirrel away; go underground, take cover, °lie low, go into hiding, °lurk, go to ground, drop out of sight, go into hiding, hibernate, latibulize, *Colloq* hide out, hole up, °lie low, *Brit* lie doggo: *Hide the sweets where the children won't find them. I found a place to hide under the stair.* **2** °conceal, °cover, °mask, °camouflage, °disguise, °veil, °shroud, °screen, °cover up, keep secret: *He managed to hide his embarrassment.* **3** °eclipse, °blot out, °obscure, °block: *The clouds hid the moon as we made good our escape.* **4** °suppress, °hush (up), °repress, °silence, keep quiet *or* secret: *She hid all knowledge of the crime.*

hide[2] *n.* **1** °pelt, °skin, °fell, leather, fur, °fleece: *Once tanned and dyed, the hide was made into clothing.* —*v.* **2** °flog, °whip, °lash, flail, °beat, thrash: *The masters took great satisfaction in hiding the boys.*

hideaway *n.* °refuge, °retreat, °sanctuary, hide-out, hiding place, °lair, (safe) haven, *Colloq* °hole, hidey-hole: *My brother and I would crawl into our hideaway and pretend we were being held prisoner by pirates.*

hidebound *adj.* °strait-laced, °conventional, ultraconventional, °conservative, °reactionary, °rigid, °set (in one's ways), °narrow-minded, close-minded, °inflexible, intractable, uncompromising, restricted, °cramped, °bigoted, °intolerant: *She's too much of a hidebound disciplinarian to rescind the punishment.*

hideous *adj.* **1** °grotesque, °ugly, °repulsive, °revolting, °repellent, °monstrous, °beastly, gorgonian, °unsightly, °ghastly, °disgusting, °grisly, nauseating, °nauseous, sickening, °gruesome: *A hideous face appeared suddenly at the kitchen window.* **2** °foul, °abhorrent, heinous, horrifying, appalling, °outrageous, °abominable, °vile, °shocking, °loathsome, °contemptible, °hateful, odious, °atrocious, horrific, °beastly, °damnable, execrable: *Two hideous crimes were committed here last night.*

high *adj.* **1** °tall, °lofty, °elevated, °towering: *The buildings seemed high enough to touch the clouds.* **2** °extreme, °excessive, °extraordinary, °exorbitant, °outrageous, *Colloq* °steep, °stiff: *Houses in that district are fetching high prices.* **3** costly, °dear, °expensive, high-priced: *Shares are not as high as they were before the crash.* **4** °great, °huge, °enormous, °considerable, °strong; °violent, turbulent: *These wires carry high voltages. The high winds blew down trees.* **5** °exalted, °elevated, °lofty, °superior, °high-class: *With insider trading, theft became a high art in the financial world.* **6** consequential, °important, °grave, °serious, °weighty, °momentous, heinous; °capital: *His activities constitute high treason.* **7** high-pitched, high-frequency, squeaky, °acute, treble, soprano; °shrill, °strident, °sharp, °penetrating, °piercing, ear-splitting: *The higher notes make my radio speaker vibrate. She has an irritatingly high voice.* **8** °cheerful, °exuberant, °elated, °boisterous, exhilarated, °hilarious, °merry, °excited: *I find his persistent high spirits rather depressing.* **9** euphoric, intoxicated, inebriated, °drunk, drugged, *Colloq* °loaded, tipsy, turned on, on a trip, *Slang* stoned, spaced out, *Brit* squiffy, *US* spacy, squiffed: *They got high sniffing glue.* **10** gamy, tainted, °aged, °ripe, *Slang Brit* pongy: *He likes to hang venison till it is quite high.* **11** °chief, °leading, °important, °principal, °foremost: *He regards himself as the high priest of women's fashion.* **12** °elaborate, °luxurious, °grand, °extravagant, °lavish, °rich, °prodigal, sybaritic: *With her millions she can now enjoy the high life.* **13** °considerable, °favorable, °great: *She is held in high esteem by her colleagues.* —*adv.* **14** far up; great in extent: *Daedalus warned Icarus not to fly so high. I was willing to go as high as twenty thousand for the painting.* —*n.* **15** °peak, °record, °height, °maximum, °acme, apex: *Employment is expected to reach a high next summer.* **16** intoxication, altered consciousness: *The high lasts till the drug wears off.* **17** anticyclone: *The high passing over the country will bring fair weather*

highbrow *n.* **1** °scholar, °intellectual, savant, °sage, °mastermind, °genius; °aesthete, connoisseur, *Colloq* egghead, °brain: *Many who pretend to deride highbrows secretly envy them.* —*adj.* **2** °scholarly, °intellectual, °learned, erudite, °deep, bookish, cultured, °sophisticated, °cultivated, °aesthetic, *Colloq* brainy: *Books on philosophy are too highbrow for the market we cater to.*

high-class *adj.* **1** °first-rate, °superior, °better, top-drawer, *Colloq* tops, tip-top, A-1 *or* A-one, °super, °great: *He has had the benefit of a high-class education.* **2** aristocratic, °upper-class, °elite, °select, °exclusive; upper-crust, °fancy, *Brit* county, *US and Canadian* tony, *Slang* classy: *She says she comes from a high-class family.*

highly *adv.* **1** greatly, much, tremendously, °well, enthusiastically, °warmly, immensely, hugely: *Her play was highly praised by all the critics.* **2** °very, °extremely, °quite, exceptionally, extraordinarily, incomparably, decidedly: *It is highly unlikely that I shall come to your party.* **3** °favorably, °well, enthusiastically, approvingly, °warmly, praisefully: *They seem to think highly of him at the office.* **4** °well, influentially, powerfully, strongly, authoritatively, effectively, importantly: *I have the information from a highly placed government official.*

hilarious *adj.* °funny, sidesplitting, °humorous, comical, amusing, °entertaining, mirthful; °merry, °gay, °jolly, jovial, °cheerful, cheery, joyous, °joyful, rollicking, °uproarious; *Colloq* °hysterical: *Binky told me a hilarious story about his aunt's pet boa constrictor. We had a hilarious time at your party.*

hilarity *n.* °laughter, °gaiety, joviality, jollity, °merriment, °mirth, °exuberance, °glee, boisterousness, cheerfulness, joyfulness, jubilation, elation, °revelry, conviviality, high spirits, vivacity, exhilaration: *Thurgood did not join in the hilarity of his retirement party.*

hill *n.* **1** °elevation, °rise, highland, °mound, °prominence, promontory, eminence, °knoll, hillock, hummock, °height, foothill, tor, °mount, upland, downs *or* downland, *Scots* brae, *Northern Brit* °fell, *Western Brit* tump, *US and Canadian* butte: *The house is on a hill overlooking the valley.* **2** °heap, °pile, °mound, °stack; °mountain: *By the autumn, the hill of compost had reached six feet in height.* **3** °slope, incline *or* decline, acclivity *or* declivity, gradient *or esp. US* °grade, upgrade *or* °downgrade: *Are you sure this car can make it up the next hill?*

hinder *v.* **1** °hamper, °delay, °interrupt, °impede, interfere with, °foil, °thwart, °frustrate, °forestall, °bar, °stymie, °check, balk *or* baulk, °encumber, °obstruct, °handicap, set *or* keep *or* put *or* hold back, °defer, °retard, °restrain, °slow, °postpone: *The difficulty of the task should not hinder the attempt.* **2** °stop, °prevent, °check, °preclude, °arrest; °discourage, °deter, °inhibit, obviate: *Does the threat of capital punishment hinder people from committing murder?*

hindrance *n.* **1** °obstruction, °impediment, °snag, °check, °obstruction, °barrier, °obstacle, °restraint, °drawback, °hitch, °stumbling block, °deterrent, °encumbrance: *The only hindrance to the plan is Phyllis's disapproval of it.* **2** °prevention, °curb, limitation: *The presence of police cars serves as a hindrance to speeders.*

hint *n.* **1** °suggestion, °clue, °implication, °inkling, °indication, °tip, tip-off, intimation, allusion, °innuendo, insinuation; °pointer, °help, °advice: *The quizmaster gave me a hint and I got the answer at once.* **2** °trace, °suggestion, °touch, °taste, °breath, °dash, soupçon, whiff, undertone, tinge, °whisper: *Which government has ever served without any hint of scandal?* —*v.* **3** °suggest, °imply, °indicate, °tip (off), °intimate, allude, °insinuate, °mention, °clue, °cue, °signal, °refer, advert: *She hinted that I might appeal to her sister.*

hip *adj.* °informed, °aware, °knowledgeable, °knowing, °perceptive, °alert, in *or* up on, onto, *Colloq* °wise (to), with it, °cool, *Old-fashioned* hep: *He's not hip to what they're saying about him and Carrie.*

hippie *n.* bohemian, *Old-fashioned* dropout, beatnik, °beat, longhair, flower child *or* person, hipster: *A few hippies were playing their guitars at the street corner.*

hire *v.* **1** °engage, °employ, °take on, °appoint, °enlist, °sign on: *Alexandra had to hire more people to get all the work done.* **2** °rent, °lease, °engage, °charter: *We hired a* *car for the day.* **3** ***hire out.*** °rent (out), °lease (out), °let (out), °charter (out): *I hire out my boat by the day.* —*n.* **4** °rent, °lease, °charter, letting: *Do you have bicycles for hire by the day?* **5** °(hire) charge, °cost, °fee, °price, °rate, °rent, rental: *How much is the hire of a horse by the hour?*

hiss *n.* **1** hissing, sibilance: *Serpents and geese make a sound like a hiss.* **2** catcall, °jeer, boo, hoot, *Slang* raspberry, *US* Bronx cheer: *The villain was greeted by hisses from the audience.* —*v.* **3** boo, hoot, °jeer, °deride, °mock, °taunt, decry, °disparage: *The workers hissed and booed the speaker.*

historic *adj.* °momentous, °important, °noteworthy, °significant, red-letter, °notable, °celebrated, °distinguished, °prominent, °great, consequential, °signal, unforgettable, °memorable: *We are gathered here to commemorate a historic event.*

historical *adj.* °factual, °true, verifiable, °reliable, °real, °authentic, recorded, documented: *The historical truth of what actually happened on that day has been obscured by legends.*

history *n.* **1** °account, °story, record, °description, depiction, portrayal, °representation, °telling, retelling, °recital, °narration, °narrative, °relation, retailing: *Washington Irving's* Knickerbocker's History of New York *begins with the creation of the world.* **2** °news, °summary, recapitulation, °report, °intelligence, °information: *The history of these events is recounted in a book by Frostly.* **3** °past, °background, °life; experiences, adventures, °story, biography: *This woman appears to have had a rather curious history. The history of your years in Ouagadougou would make an interesting book.* **4** record, °experience, °information, biography, CV *or* curriculum vitae, *US* °résumé: *Your entire work history should be included in your application.* **5** °chronicle, annals, °record, °account: *The history of the debate on the bill can be traced through the* Congressional Record. **6** ancient history, the past, yesterday, the (good old) days, days of yore, olden days, yesteryear, antiquity: *What can history tell us about the future?* **7** dead letter, yesterday's news, old hat: *Any animosity I might have felt toward him is now history.*

hit *v.* **1** °strike, cuff, smack, °knock, whack, bash, bang, thump, thwack, °punch, buffet, °slap, swat, bludgeon, °club, smite; °spank, thrash, °beat, pummel, °batter, °flog, °scourge, birch, cane, °lash, °belabor, flagellate, °whip, horsewhip, cudgel, *Archaic* fustigate; *Colloq* °belt, wallop, clobber, clout, sock, °clip, °crown, bop, conk, paste, °lambaste, °zap: *She hit him on the jaw and he went down.* **2** °strike, bat, swat, °knock, °drive, °propel: *He hit the ball over the heads of the fielders.* **3** °strike, collide *or* impact with, run *or* smash *or* crash into, bump *or* bang into: *The car went off the road and hit a tree.* **4** °affect, °touch, °stir, °move, °wound, °hurt, strike *or* hit home, make *or* leave an impression *or* a mark on, °(make an) impact (on): *The new taxes hit the wealthy more than the poor.* **5** °dawn on, enter one's mind, °occur to, °strike: *It finally hit Graham that he had been insulted.* **6** °reach, attain, °arrive at, °gain, °achieve: *Those who have hit eighty know the meaning of old age. She hit the jackpot this week.* **7** °experience, °encounter, °meet (with): *It was at that point that we hit a snag in the negotiations.* **8** Also, ***hit up.*** importune, °beseech, °petition, °beg, implore, entreat, °ask for: *As usual, Guthrie hit me for a loan as soon as we met.* **9** ***hit on*** *or* ***upon.*** **(a)** come *or* happen *or* chance *or* light on *or* upon, °discover, °find, uncover, °unearth, stumble *or* blunder on *or* upon, °arrive at: *After years of experimentation, the Curies hit upon pitchblende as a source of radium.* **(b)** °devise, think of *or* up, °invent, dream up, come up with, °work out, °see, °perceive, °detect, discern, °find: *I have hit upon a way to counteract the force of gravity.* —*n.* **10** °impact, °collision; °blow, °punch, °knock, °strike, swat, °shot, smack, °bump, bang, *Colloq* whack, thwack, conk, bop, sock: *The weakness of the hit caused only a slight dent in my car door. The boxer reeled after a hard hit to the midriff.* **11** °success, °triumph, coup, °winner, °sensation, *Colloq* smash (hit), sellout: *After opening in the West End the musical became a hit on Broadway.* **12** °kick, °jolt, °thrill, *Slang* °charge, *US* °rush, bang: *Give me a hit off that reefer.*

hitch *v.* **1** °connect, °couple, °fasten, °attach, °join, harness, °tie, °unite, °hook (up), °link, °fix: *When I arrived, she was hitching the horses to the wagon.* **2** Often, ***hitch up.*** °raise, °pull up, hike (up), °tug (up), °hoist, °yank, °jerk, *Brit* hoick: *Hitch up your trousers so the cuffs won't drag in the mud.* **3** hitchhike, thumb a lift *or* ride, *Colloq US* bum a ride: *I had no money and no car, so I hitched here from Newcastle.* —*n.* **4** °snag, °catch, °difficulty, °trouble, °problem, °mishap, °handicap, entanglement, °interference, °impediment, °hindrance, °obstruction, °obstacle: *The entire plan went off without any hitch.*

hoard *n.* **1** °supply, °stock, °store, stockpile, °reserve, °fund, reservoir, °accumulation, °collection, °cache: *They kept a hoard of food in the shelter in case of attack. Occasionally, farmers turn up a hoard of gold coins buried by the ancient Romans.* —*v.* **2** °amass, °collect, °accumulate, °pile (up), °assemble, °gather, °put away, stockpile, °store, °reserve, °set aside, °save (up), squirrel away, lay in *or* away *or* aside *or* up, *Colloq* stash away: *Hoarding food was against the law during rationing.*

hoax *n.* **1** °deception, °fraud, °swindle, °trick, flam *or* flimflam, imposture, °cheat, humbug, mare's nest, *Slang* con (game), gyp, scam, °game, *US* snow job: *They perpetrated a hoax on you and there is no way of getting your money back.* —*v.* **2** °deceive, °defraud, °swindle, °trick, °fool, °dupe, °take in, cozen, °hoodwink, gull, °bluff, *Slang* con, gyp, bamboozle: *She was hoaxed into investing in Sicilian gold mines.*

hobble *v.* **1** °limp, falter, dodder, °totter, °stagger, °reel, °weave, °stumble, °shuffle, shamble: *I hobbled about on crutches for weeks.* **2** °shackle, fetter, °restrain, °restrict, °hamper, °hinder, °impede, °trammel: *As hobbled horses do not stray, she tried to think of some way to hobble Clarence's errant ways.* —*n.* **3** °limp, °shuffle, shamble, claudication, °stagger: *With the leg-irons on, I could walk only with a jerking hobble.*

hobby *n.* °pastime, avocation, sideline, °recreation, °diversion, °relaxation: *Her hobby is collecting wedding rings.*

hobnob *v.* °associate, °fraternize, °socialize, consort, °mingle, rub elbows *or* shoulders, °mix, hang about *or* around, keep company: *He hobnobs with the aristocrats.*

hocus-pocus *n.* **1** °trickery, °chicanery, °deceit, °deception, °artifice, °cheat, duplicity, °mischief, °hoax, humbug, °trick, °swindle, °pretense, *Colloq* con (game), jiggery-pokery, flimflam, °hanky-panky: *He tried some hocus-pocus on the company books but was caught.* **2** °mumbo jumbo, abracadabra, incantation, °nonsense, rigamarole *or* rigmarole, °gibberish, *Colloq* gobbledygook: *The medicine man muttered some hocus-pocus over the body, which began to rise into the air.* **3** sleight of hand, legerdemain, prestidigitation, °magic, conjuring, jugglery: *From the earliest days religion has had its share of hocus-pocus.*

hodgepodge *n.* °miscellany, °mixture, gallimaufry, °jumble, farrago, mélange, °mishmash, °mess, °tangle, °medley, °hash, conglomeration, agglomeration, olio, olla podrida, °potpourri, ragbag, °welter, *Chiefly Brit* hotchpotch, *Colloq* omnium-gatherum, mixed bag:

That hodgepodge of junk is scarcely what I should call an "antiques collection."

hoggish *adj.* piggish, °greedy, °avaricious, insatiable, °gluttonous, °voracious, edacious, acquisitive, °possessive, self-seeking, °selfish: *Don't be so hoggish and share with others.*

hoi polloi *n.* °riffraff, °rabble, °mob, common herd, proletariat, °populace, common people, °crowd, masses, multitude, °rank and file, plebeians, multitude, bourgeoisie, man in the street, *Brit* admass, man on the Clapham omnibus, *US* John Q. Public, *Colloq* great unwashed, proles, plebs, *US* silent majority: *Much of what goes on in government is incomprehensible to hoi polloi.*

hoist *v.* **1** °lift (up), elevate, °raise, °heave, uplift, winch: *The lifeboat with the survivors aboard is now being hoisted onto the rescue vessel.* —*n.* **2** crane, °lift, elevator, davit, winch, °tackle: *The cable of the hoist broke, and the container dropped onto the pier.*

hoity-toity *adj.* °haughty, °arrogant, overweening, °snobbish, °disdainful, °supercilious, °conceited, °lofty, °superior, °self-important, *Colloq* high and mighty, stuck-up, snooty, uppity *or chiefly Brit* uppish, *Brit* toffee-nosed, *Slang* snotty: *Eleanor became very hoity-toity after her husband got his knighthood.*

hold *v.* **1** °grasp, °grip, °clasp, °seize, °clutch, °keep; °carry, *Colloq* hang onto: *She asked me to hold the baby for just a minute while she bought her railway ticket.* **2** °hug, °embrace, °clasp, cradle, clench, °clutch, enfold: *He held me in his arms briefly before the guards led him into the quad.* **3** °maintain, °keep, °put: *Hold up your hands and kick the gun over here to me.* **4** °maintain, °keep, °sustain, absorb, °occupy, °engage, °involve, engross, °monopolize: *You hold his attention while I try to get round behind him.* **5** confine, °restrain, detain, °contain, coop up: *Even a straitjacket and chains couldn't hold Houdini.* **6** °imprison, detain, confine, place into custody, put behind bars, °jail: *He is being held overnight for questioning.* **7** °believe, deem, °judge, °consider, °regard, look on *or* upon, °maintain, °think, °esteem, °take, °assume: *What do you hold to be important in life? Father holds me responsible for every little dent in his car.* **8** °accommodate, °support, °carry: *That little nail won't hold this picture.* **9** °contain, °include, comprise: *This suitcase holds everything I own in the world.* **10** °call, convene, °assemble, convoke; °run, °conduct, °engage in, participate in, °have, °carry on, preside over, officiate at: *The next meeting will be held on Tuesday, at noon.* **11** °apply, hold good, be in effect *or* in force, stand *or* hold up, hold *or* prove *or* be true, be the case, °function, °operate, be *or* remain *or* prove valid *or* relevant *or* applicable *or* operative, *Colloq* hold water, °wash: *What may be in order for Manchester may not necessarily hold for another city.* **12** °have, °possess: *She holds two engineering degrees. He was holding four aces.* **13** remain *or* keep (fast), °stay, °stick: *Screws are needed here—nails won't hold.* **14** ***hold back.*** **(a)** °restrain, °repress, °suppress, °curb, °inhibit, °control, °check, keep back, °hinder: *Many reasons hold me back from telling you what I think.* **(b)** °withhold, °reserve, °deny, keep back, °refuse: *We ought to hold back payment till the work is completed.* **15** ***hold down.*** **(a)** °control, °restrain, °check; °reduce, °diminish: *We must hold down inflation.* **(b)** °keep, °maintain, °manage: *He has to hold down two jobs to pay all the bills.* **16** ***hold forth.*** **(a)** Often, **hold forth on** *or* **upon.** lecture (on), declaim, °harangue, preach (on *or* about), orate, sermonize (on), discourse (on), speechify (on *or* about), expatiate *or* expand on *or* upon, *Colloq* °go on (about), *Brit* rabbit *or Brit* natter *or Brit* witter on (about): *As usual, Pinckley endlessly held forth on his pet subject, fishing.* **(b)** °hold out, °offer, proffer, °tender, °submit, °advance, °propose, °propound, °hold out, °extend: *The company has held forth a profit-sharing plan that we cannot refuse.* **17** ***hold in.*** **(a)** °control, °curb, °check, °hold back, °restrain, °contain: *I could hold myself in no longer and a scream escaped my lips.* **(b)** °conceal, °hide, °suppress: *How can I hold in my feelings for you?* **18** ***hold off.*** **(a)** °delay, °defer, °put off, °refrain from, °postpone, °avoid: *We held off buying till we had the money saved up.* **(b)** °repel, keep off, °repulse, °fend off, °rebuff, °resist, °withstand: *We held off the attackers till help came.* **19** ***hold on.*** **(a)** °grip, °grasp, °hold, °clutch, °cling: *Hold on to the rope and I'll pull you up.* **(b)** °keep, °maintain, °cling, °hang on, °retain: *Don't try to hold on to yesterday's dreams.* **(c)** °stop, °wait, °hold off, *Colloq* °hang on: *Hold on a minute! I'm not finished.* **20** ***hold out.*** **(a)** °last, °carry on, °persist, °persevere, °continue, °hang on, stand firm *or chiefly US* pat, °endure: *I hope that the good weather holds out for our trip. Can we hold out till reinforcements arrive?* **(b)** °offer, proffer, °extend, °hold forth, °present: *I grasped the hand he held out.* **21** ***hold over.*** **(a)** °postpone, °delay, °defer, °put off, °hold off, °suspend: *The decision is to be held over till next year.* **(b)** °continue, °retain, °extend, °prolong: *The singer was popular enough to be held over a month.* **22** ***hold up.*** **(a)** °rob, °waylay, *Colloq* °mug, °stick up; knock off *or US* over: *Two men held up the bank courier last night.* **(b)** °delay, °impede, °hinder, slow (down *or* up), °set back, detain: *I was held up by the infernal traffic again.* **(c)** °last, °survive, °fare, °bear up, °endure: *I am not sure that my car will hold up through another winter.* **(d)** °present, °show, °exhibit, °display: *Grindling has been held up to ridicule since the scandal.* **23** ***hold with.*** °support, °sustain, agree to *or* with, °favor, countenance, °approve (of), °subscribe to, condone, concur with: *Being married to her doesn't mean you have to hold with all her ideas.* —*n.* **24** °grasp, °grip, °clasp, °clutch: *Take hold of the rope!* **25** foothold, toehold, °purchase: *She lost her hold and fell. He has a good hold on the subject.* **26** °power, dominance, mastery, °control, ascendancy, °authority, °influence, leverage, °sway, *Colloq* °pull, clout: *She has a hold over him that makes him do her bidding.*

holdup *n.* **1** °(armed) robbery, *Colloq* stickup, mugging, *US* heist: *The robbers escaped with my gold watch in the holdup.* **2** °delay, °setback, °hitch, °snag, °interruption, lacuna, °gap, hiatus, °break, stoppage: *The cause of the holdup was an overturned van.*

hole *n.* **1** °cavity, °pit, °hollow, °excavation, °burrow, crater, cavern, °cave, °recess, °niche, °nook, °pocket, °depression, °indentation, dent, °impression: *The snake disappeared into a hole in the rock.* **2** °opening, °aperture, orifice, perforation, °puncture, °slit, °slot, °breach, °rip, °tear, °rent, °break, °crack, fissure: *The water poured through a hole in the pipe.* **3** hole in the wall, °shack, °hut, shanty, °slum, °hovel; *Slang* °dump, °dive, °joint: *How can anyone live in such a hole? She finally got a job dancing in some hole downtown.* **4** °cell, °prison, °dungeon, donjon, °keep, °jail, oubliette, brig, °cage: *When he refused to talk, they put him in the hole for a week.* **5** °difficulty, °trouble, °dilemma, °predicament, °situation, °fix, corner, *Colloq* °(tight) spot, hot water, °scrape, °box, °bind, pickle, catch-22, °mess, °muddle: *She really got herself into a hole with the tax man.* **6** °flaw, °shortcoming, inconsistency, °fault, °error, °mistake, °fallacy, °discrepancy, °loophole: *He never offers anything original but is always ready to pick holes in any suggestion you make.* —*v.* **7** °puncture, °pierce, °perforate: *A floating log holed the hull and the boat went down with all aboard.*

holiday *n.* **1** time off, °break, °recess, °respite, °leave (of absence), furlough, sabbatical, *Chiefly US* vacation: *We spent our holiday in Ibiza this year.* **2** °festival,

°feast, °celebration, °fête; °gala, °fair, red-letter day, °event: *Where are you going over the Christmas holiday?*

hollow *adj.* **1** °vacant, °empty, °void, unfilled: *A hollow space in the wall concealed a secret passage.* **2** °sunken, concave, indented, dented, °sunken, recessed, depressed: *Dust gathers in the hollow places of the floor.* **3** °hungry, °ravenous, °starved, °empty, °famished: *I'm feeling hollow and should prefer to eat now, not later.* **4** °insincere, °false, °hypocritical, °sham, °artificial, °counterfeit, feigned, °fraudulent, °spurious, °deceitful, mendacious, °deceptive, cynical: *Politicians campaigning for office often make hollow promises they do not intend to keep.* **5** °empty, °futile, costly, Pyrrhic, °worthless, °vain, unavailing, °bootless, °fruitless, profitless, °unprofitable, valueless, °ineffective, °pointless, °senseless, °meaningless: *Winning the lawsuit was a hollow victory, for the man was bankrupt.* **6** muffled, °dull, °flat, °low, sepulchral, toneless: *His voice sounded hollow, as if he were speaking into a metal bowl.* —*n.* **7** °hole, °cavity, cavern, crater, basin, °depression, °excavation, °pit, trough, °furrow, °indentation, dent, °impression; °valley, dale, dell, glen, °dip: *We dug ourselves in a hollow in the ground hoping not to be seen. The dog herded the sheep into the hollow near the stream.* —*v.* **8** °excavate, dig (out *or* up), °gouge, °scoop, °furrow, dredge: *Huskies hollow out places for themselves to sleep in the snow.*

holocaust *n.* **1** conflagration, firestorm, inferno, °fire; °destruction, devastation: *When the volcano exploded, few escaped the holocaust.* **2** genocide, mass murder, °massacre, blood bath, pogrom, butchery, °carnage, annihilation, extinction, extermination, eradication, elimination: *Survivors of the Nazi holocaust hold periodic memorial services.*

holy *adj.* **1** °sacred, °religious, consecrated, sanctified, blessed, hallowed, venerated, °divine, °heavenly, supernal, °celestial: *The holy relics are kept in a silver casket.* **2** °godly, °godlike, °saintly, saintlike, °pious, °devout, reverent, reverential, °faithful, God-fearing, °chaste, °pure, unsullied, °clean, sinless, °spotless, °immaculate, undefiled, uncorrupted, untainted: *The holy men, in their saffron robes, sat in a circle, contemplating the master of the universe.*

homage *n.* °obeisance, °respect, °deference, °honor, °esteem, °admiration; °loyalty, allegiance, fidelity, °tribute: *Today we pay homage to those who fought and died so that we might be free.*

home *n.* **1** dwelling place, °residence, °domicile, °abode, °dwelling, °house, °(living) quarters, habitation, lodging(s), *Brit* accommodation *or US* accommodations, *Colloq* °place, *Chiefly Brit* digs, diggings: *He has been a guest in my home on many occasions.* **2** °(home) base, residency, °territory, °haunt, home ground, bailiwick, *Colloq* stamping ground: *As you travel so much, what do you call home these days?* **3** hospice, °retreat, nursing home, old folks' *or* people's home, retirement community, almshouse, poorhouse, °refuge, haven, °institution, °shelter, rest home, *US* snug harbor: *His parents, who are very old and indigent, have been sent to a home.* **4** ***at home.*** **(a)** °comfortable, at ease, °relaxed, °cozy, composed, °tranquil, placid, °peaceful, °serene, untroubled: *The Harrises certainly do make one feel at home.* **(b)** in, °accessible, °available, welcoming: *You know that we are always at home to you, Frances.* **5** ***at home with*** *or* ***in.*** comfortable with, conversant with, knowledgeable in *or* about, °familiar with, well-versed in, competent in, expert in, proficient in, skilled in, up on, current in, adept in, adroit in, qualified in, (well-)informed in *or* on *or* about: *Widely read, she is at home in almost any subject you can name.* —*adj.* **6** °domestic, °native, °national, °internal: *Buying more foreign than home goods upsets the balance of trade.* **7** °domestic, household: *She now sells home appliances for a big manufacturer.* **8** °family, °domestic: *What kind of home life has he had?* —*adv.* **9** homeward(s): *When will you come home?* **10** to the heart *or* core, to the quick; effectively, tellingly, °profoundly, °deeply, stingingly, cuttingly, harshly, °severely, *Colloq* where it hurts, where one lives: *That remark really hit home.* **11** ***bring*** *or* ***drive home.*** °stress, °emphasize, °impress upon, make clear: *I am trying to bring home to you the hardships people suffered during the war.*

homeless *adj.* **1** dispossessed, °outcast, exiled, °vagabond, °derelict, °unsettled; unhoused: *Their houses destroyed in the war, homeless people wandered everywhere.* —*n.* **2** ***the homeless.*** knights of the road, vagrants, vagabonds, tramps, *US* bums, hoboes: *There are not enough shelters for the homeless.*

homely *adj.* **1** homy *or chiefly US* homey, homelike, unpretentious, °modest, unassuming, °simple, °unaffected, °informal, °plain, °natural, °everyday, °unsophisticated, °homespun, commonplace, °ordinary, °familiar, °friendly, °amiable, °neighborly, affable, congenial, *Colloq chiefly US* folksy: *Her success as a doctor is traceable in part to her homely approach.* **2** homy *or chiefly US* homey, homelike, °warm, °cozy, °snug, °domestic, °comfortable, °easy, °serene, °peaceful, °restful, °tranquil: *I like staying here for the homely atmosphere.* **3** °ugly, °plain, uncomely, unattractive, unlovely, ill-favored: *He is homely enough to turn a Gorgon to stone.*

homesick *adj.* nostalgic, °longing, pining, °lonely, °lonesome; °wistful, °reminiscent: *I was homesick for the smell and taste of a good haggis.*

homespun *adj.* °rustic, °plain, °simple, °unrefined, unpolished, °unsophisticated, down-to-earth, °coarse, °rough, °rude, °crude, inelegant, °amateur, amateurish, nonprofessional, °unprofessional; handmade: *The furniture is not of museum quality, but it has a homespun quality.*

homicidal *adj.* °murderous, °lethal, °deadly, death-dealing, °mortal, blood-thirsty, °sanguinary, °ferocious, °maniacal, °berserk, amuck *or* amok, °mad, °insane: *The doctors say he is homicidal and has to be hospitalized.*

homogeneous *adj.* °uniform, °consistent, unvarying, °identical, °constant; °similar, comparable, °alike, °akin: *The rock is ground up to make homogeneous powder.*

homosexual *n.* **1** °gay, homophile; lesbian, tribade, sapphist; *All the following are offensive and derogatory* °pervert, invert, *Slang* °queer, fairy, pansy, nancy, nance, °queen, drag queen, homo; butch, (bull) dyke; *Brit* poof, poofter, ginger (beer), *US* °fruit, auntie, °fag, faggot: *We suspected that he was a closet homosexual.* —*adj.* **2** °(of either sex) gay, homoerotic, homophile; (of a female) lesbian, tribadic, sapphic; °(of a male) effeminate; *All the following are offensive and derogatory* °perverted, inverted, *Colloq chiefly Brit* °bent, *Slang* °queer, °camp, campy, faggy, °kinky, *Chiefly US* fruity, limp-wristed, faggy, °swish, swishy: *We never knew of her homosexual tendencies.*

honest *adj.* **1** °trustworthy, °truthful, veracious, trusty, °honorable, creditable, °decent, law-abiding, uncorrupted, uncorrupt, incorruptible, °ethical, °moral, °virtuous, °principled, °upright, high-minded, dependable, °reliable, °reputable, on the up and up: *How often does one meet an honest person in politics?* **2** °aboveboard, °straight, °square, square-dealing, °fair, °just, on the up and up, straightforward, °proper, °genuine, °bona fide, °real, °authentic, *Colloq* °on the level, *US* square-shooting: *I write because I cannot earn an honest* *living. Would Felix give me an honest deal on a used car?* **3** °candid, °frank, °open, °plain, straightforward, °forthright, °direct, °sincere, °ingenuous, °explicit, °uninhibited, unreserved, unrestrained, unconstrained, °aboveboard, plain-spoken, unambiguous, unequivocal, *Colloq* °up front: *To be completely honest, I cannot stand the sight of you.* **4** °fair, °just, °equitable, °legitimate, valid, °rightful, °sound, °proper: *He gets an honest day's pay for an honest day's work.*

honestly *adv.* **1** truthfully, honorably, creditably, decently, ethically, morally, uprightly, dependably, reliably, in good faith, justly, °fairly, equitably, evenhandedly, disinterestedly, objectively, impartially: *In general, the police deal honestly with suspects.* **2** candidly, frankly, °openly, straightforwardly, forthrightly, °sincerely, °truly, ingenuously, unreservedly, °aboveboard, unambiguously, unequivocally, plainly, °simply, °straight (out), °to one's face, in plain words *or* English, bluntly: *I tell you honestly that I never heard of the man.*

honesty *n.* **1** trustworthiness, uprightness, °rectitude, °probity, °integrity, °virtue, virtuousness, °honor: *As he will deal with large amounts of money, his honesty cannot be in question.* **2** truthfulness, veracity, °candor, openness, frankness, forthrightness, directness, straightforwardness, outspokenness, °sincerity, guilelessness, ingenuousness, bluntness: *You might not like what he says, but you have to admire his honesty.* **3** fairness, °equity, equitableness, evenhandedness, °objectivity, impartiality, disinterestedness, justness, °justice: *We rely on the honesty of judges in meting out punishment to convicted felons.*

honor *n.* **1** °integrity, °honesty, fairness, justness, °probity, uprightness, decency, goodness, righteousness, °rectitude, °justice, °morality, °principles, virtuousness, °virtue: *Doing the right thing is a matter of honor.* **2** °respect, °esteem, °reverence, °veneration, approbation, °deference, °admiration, °homage, °regard, accolade, °praise, °kudos, °fame, glory, °celebrity, °distinction, °prestige, illustriousness: *Great honor accompanies the award of a Nobel prize.* **3** °privilege, °distinction, °pleasure, °joy, °delight; °credit, °blessing: *I have the honor to introduce tonight's speaker. It has been an honor to serve under you, Admiral.* **4** virginity, °chastity, °virtue, °purity, innocence: *How did she manage to preserve her honor when fighting with the partisans?* —*v.* **5** °respect, °esteem, °revere, °venerate, adulate, °adore, °worship, °approve, °prize, value, defer to, °admire, pay homage to: *Albert Schweitzer was much honored in the latter part of his life.* **6** °praise, °laud, °glorify, °celebrate, °eulogize, °salute, °hail, acclaim, ennoble, °dignify, °exalt: *We have come together to honor those who have died for their country.* **7** °keep, °maintain, °carry out, live up to, °discharge, °fulfill, °observe, °meet: *She has done her best to honor her obligations.* **8** °pay, °redeem, °accept, °clear, °cash: *The bank refused to honor his check because of "insufficient funds."*

honorable *adj.* **1** °upright, upstanding, °trustworthy, trusty, °honest, °just, °fair, °moral, °principled, uncorrupt, uncorrupted, incorruptible, high-minded, °noble, °virtuous: *Mark Antony referred to Caesar as an honorable man.* **2** °right, °correct, °proper, °fitting, °appropriate, °virtuous, °ethical, °worthy, °respectable, °reputable, °decent, °square: *I think you can count on Cedric to do the honorable thing.* **3** °fair (and square), °impartial, °equitable, °just, °honest, unbiased, °unprejudiced, nonprejudicial, evenhanded, °straight, °disinterested, guileless, °ingenuous, °artless, °open, °sincere, °aboveboard, on the up and up, undeceiving, undeceitful, *Colloq* °up front, °on the level: *Eleanor is too honorable to accept payment for work she did not do.* **4** °distinguished, °prestigious, °eminent, °notable, °noteworthy, °noted, °illustrious, °famous, famed, honored, °exalted, respected, °celebrated, °renowned, acclaimed, °well-thought-of: *Ian enjoys an honorable reputation as a publisher.*

honorarium *n.* °(token) fee, compensation, recompense, °pay, °payment, °remuneration, emolument: *Would you be willing to accept an honorarium in lieu of your usual fee?*

honorary *adj.* °nominal, °titular, in name *or* title only, ex officio: *She has several honorary doctoral degrees, which is not bad for someone who never went to college.*

hoodlum *n.* °gangster, °thug, °racketeer, mobster, desperado, °terrorist, ruffian, °tough, °rowdy, knave, *Colloq* hooligan, baddie *or* baddy, crook, *US* plug-ugly, *Slang* goon, *Brit* yob, yobbo, *US* °mug, bad actor, roughneck, hood, gunsel, hit man, torpedo, *French* apache, *Australian* larrikin: *He has a few hoodlums on the payroll who do the dirty work.*

hoodwink *v.* °fool, °trick, °deceive, delude, °dupe, gull, °hoax, °defraud, °mislead, humbug, °outwit, *Colloq* bamboozle, pull the wool over (someone's) eyes, pull a fast one (on someone), lead (someone) up *or* down the garden path, put one over on (someone), throw dust in (someone's) eyes, take (someone) for a ride, string (someone) along, *Slang* rook, con, suck in, *US* sucker in, snow: *Chauncy was again hoodwinked into betting on which shell concealed the pea.*

hook *n.* **1** hanger, °peg, holder; fastener, °catch, °clasp, °clip, °pin: *Hang your hat on that hook. I'm using a safety pin because the hook on my dress is gone.* **2** °snare, °trap; fishhook: *What can we use as a hook to catch the fish?* **3** ***by hook or by crook.*** °somehow (or other), someway, come what may, by fair means or foul, (by) one way or another: *I have to get out of this place by hook or by crook.* **4** ***hook, line, and sinker.*** °completely, °entirely, all the way, through and through, °thoroughly, °entirely, °totally, °utterly, °wholly: *She actually fell for that old routine hook, line, and sinker.* **5** ***off the hook.*** °(set) free, °(in the) clear, out of it; out of trouble, acquitted, exonerated, cleared, °let off, vindicated, °off: *After paying them the extortion money, he still wasn't off the hook.* —*v.* **6** °catch, °trap, entrap, °snare, ensnare; °grab, °capture, collar, °nab, °seize; *Chiefly US and Canadian* °snag, *Colloq* °pinch: *The petty crooks have been caught but we now want to hook the big fish himself.* **7** °steal, °pilfer, filch, palm, shoplift, °rob, *Slang* snitch, °rip off, *Euphemistic* °liberate, °remove, °borrow, °appropriate, *Brit* °nick, *Chiefly Brit* °pinch: *They used to hook sweets from Woolies'.*

hop *v.* **1** °jump, °leap, °bound, °spring, vault; °skip, °caper, gambol, °dance: *Here comes Peter Cottontail, Hopping down the bunny trail.* **2** take a (short) trip *or* voyage, °travel, °come, °go, °proceed; °fly: *Those yuppies think nothing of hopping over to Paris just for dinner.* —*n.* **3** °jump, °leap, °bound, °spring, vault; °skip, °caper, °dance: *In a few hops the bird was off the branch and onto my finger.* **4** (short) trip *or* flight *or* journey *or* voyage: *The Bahamas are just a hop from Miami.*

hope *n.* **1** °desire, °wish, °expectation, yearning, hankering, craving, °longing, °fancy; °ambition, (day)dream: *It was always my hope that you would go to university.* **2** °prospect, °promise, °expectation, expectancy, °confidence, °anticipation, assumption, °security, °faith, °conviction, °belief, °trust: *We had given up hope of being rescued.* —*v.* **3** °aspire, count *or* rely on *or* upon, °anticipate, °contemplate, °foresee, °look forward to, °expect, await, °wait: *After all these years, he is still hoping to meet the girl of his dreams.* **4** °trust; °wish, °want, °desire; *Dialect* °expect: *I hope her ship* *comes in soon—and I hope she gets on it and doesn't come back.*

hopeful *adj.* **1** °expectant, anticipating, °optimistic, °sanguine, °confident, assured: *We are hopeful of victory in the forthcoming election.* **2** °promising, °bright, °rosy, reassuring, heartening, encouraging, auspicious, °propitious, inspiriting: *Today was a disaster, but tomorrow looks more hopeful.*

hopefully *adv.* **1** expectantly, optimistically, sanguinely, confidently: *Jocelyn set off hopefully for the party.* **2** with (any) luck, if things go well, all being well, it is hoped, expectedly: *Hopefully, we shall arrive in time to catch the train.*

hopeless *adj.* **1** °desperate, beyond hope *or* saving, irreparable, beyond repair, irremediable, °lost, gone, °irretrievable; °incurable, °terminal, °deadly, °fatal, °lethal: *As our ship drifted into the intergalactic void, we saw that our situation was hopeless. They told me that her condition was hopeless.* **2** °bad, °poor, °incompetent, °inferior, °inadequate, °inept, °unqualified, unfit, unskillful, °deficient: *You might make a good surgeon, but as a judge of human nature you're hopeless.* **3** despairing, °despondent, °forlorn, °woebegone, disconsolate, °inconsolable, depressed, °dejected, °melancholy, downcast, °gloomy, °miserable, discouraged, °wretched, lugubrious, °funereal, °sorrowful, °sad, °unhappy: *Utterly hopeless after his script was rejected by Hollywood, he moped in his room.* **4** °futile, °vain, °bootless, unavailing, °impossible, °impracticable, unworkable, °pointless, °worthless, °useless: *It would be hopeless to try to send out a lifeboat in this storm.*

horizon *n.* °view, purview, °range, °scope, vista, compass, °perspective, °prospect, ken, field of vision, limit(s): *His philosophy went far beyond the horizon of contemporary thought.*

horizontal *adj.* °level, °flat, °plane; °prone, °supine: *Make sure that the plank is horizontal before fastening it down. They say that dancing is merely a vertical expression of a horizontal desire.*

horrible *adj.* **1** °awful, horrendous, horrid, horrifying, horrific, °terrible, °terrifying, °dreadful, °abominable, °abhorrent, appalling, °frightening, °frightful, °ghastly, °grim, °grisly, °ghoulish, °gruesome, °loathsome, °hideous, °repulsive, °revolting, °disgusting, sickening, nauseating, °nauseous, °harrowing, bloodcurdling, °macabre, unspeakable, °shocking: *The horrible sight of the victim's mangled body haunted her for the rest of her days.* **2** °awful, °nasty, unpleasant, °disagreeable, horrid, °terrible, °dreadful, °obnoxious, °offensive, °atrocious, °monstrous, °contemptible, detestable, °despicable, *Colloq Brit* °beastly: *The food was perfectly horrible at our hotel. Take that horrible little dog away.*

horrify *v.* **1** °terrify, °frighten, °scare, °alarm, °intimidate, °panic, scare *or* frighten to death, °petrify, *Colloq* scare *or* frighten the living daylights out of, scare stiff, make (someone's) hair stand on end, make (someone's *or* the) blood run cold, curl (someone's) hair, scare the pants off: *I was horrified to see the attack dogs racing toward me.* **2** °shock, °startle, °upset, °put off, °outrage, °dismay, °appall, °distress, discountenance, disconcert: *I was horrified to hear that you weren't coming to my party.*

horror *n.* **1** fear and loathing, repugnance, °terror, °dread, hatred, °revulsion, detestation, abhorrence, °distaste, °dislike; °aversion, antipathy, °hostility, °animosity, animus, °rancor; odium, execration: *She has a horror of bats.* **2** °fear, °dismay, °distress, °dread, °fright, °alarm, °upset, perturbation, °panic, °terror, fear and trembling, trepidation, °anxiety, angst, apprehension, uneasiness, queasiness, nervousness, awe: *The doctor helped me overcome my horror of flying.*

hors d'oeuvre *n.* appetizer, apéritif, antipasto, smorgasbord *or* smörgåsbord, °relish; *Chiefly Brit* starter; *Archaic* warner: *A Sauvignon was served with the hors d'oeuvre, which was gravlax.*

hospitable *adj.* **1** welcoming, °gracious, °courteous, °genial, °friendly, °agreeable, °amicable, °cordial, °warm, congenial, °sociable, °generous: *It was most hospitable of you to invite me.* **2** open-minded, °receptive, amenable, approachable, °tolerant: *The director is always hospitable to suggestions for improving sales.*

hospital *n.* medical center, health center, °infirmary, clinic, polyclinic, dispensary, sick bay; asylum, sanatorium, nursing home, convalescent home *or* facility, *US* °sanitarium: *With that wound, you ought to be in a hospital.*

hospitality *n.* graciousness, °courtesy, courteousness, friendliness, amicability, cordiality, °warmth, congeniality, sociability, generosity: *I am grateful for the hospitality you showed my sister during her visit.*

host[1] *n.* **1** hostess, innkeeper, hotelier, hotelkeeper, hotelman, landlord *or* landlady, manager *or* manageress, proprietor *or* proprietress, *Brit* publican: *As host, it is my responsibility to greet the guests.* **2** entertainer, master *or* mistress of ceremonies, emcee, MC, °announcer, *Brit* presenter, compere, *US* tummler: *For twenty years he has been the host of popular TV quiz shows.* —*v.* **3** °entertain, act *or* play the host *or* hostess, °have: *She hosts a dinner party for close friends every Wednesday.*

host[2] *n.* army, °swarm, °crowd, horde, multitude, °throng, °mob, °pack, °herd, troop, legion, °body, °assembly, assemblage, drove: *These rock groups have a host of followers, called "groupies."*

hostage *n.* °pledge, °security, surety, °pawn, °captive, °prisoner, gage: *The terrorists threatened that if their demands were not met they would kill the hostages.*

hostile *adj.* **1** °opposed, antagonistic, °contrary, against, anti, adverse; °averse, °loath: *Was he hostile to visiting your relatives before your marriage?* **2** unfriendly, inimical, °unsympathetic, °cold, °inhospitable; unfavorable: *Why do many social workers seem hostile to the people they are supposed to help? The polar regions can be very hostile environments to man.* **3** warring, °belligerent, bellicose, °warlike, combative, °militant, °aggressive: *The UN units function as a buffer between the hostile forces.*

hostility *n.* **1** °antagonism, °opposition, enmity, °animosity, antipathy, animus, °ill will, malevolence, malice, °aversion, unfriendliness: *Do you still harbor the feelings of hostility toward them that you had during the war? The petty quarrels broke out into open hostility.* **2** ***hostilities.*** °war, warfare, fighting, °combat, °action, state of war, °bloodshed: *Hostilities ceased when the factions agreed to parley.*

hot *adj.* **1** °fiery, white hot, redhot, piping hot, °burning, blistering, °scorching, roasting, frying, sizzling, searing, boiling, scalding, steaming, simmering, °torrid, °sweltering, °sultry, °heated: *The hot, molten steel is cast into ingots here. During the day it is hot, but it cools down at night.* **2** °spicy, peppery, °sharp, piquant, °pungent, °biting, acrid: *She likes very hot spaghetti sauce.* **3** °intense, °fervent, zealous, °ardent, °enthusiastic, °passionate, fervid, vehement, °excited, °animated; °impetuous, °fiery, °fierce, °inflamed, °sharp, °violent: *He had some hot words with the foreman who had dismissed him.* **4** °eager, °keen, avid, °anxious, °burning; °intense, °fervent, zealous, °ardent, °enthusiastic, °passionate, fervid, vehement, °excited, °animated, °earnest, *Slang US* gung ho: *Desmond set out in hot pursuit of the thief.* **5** °recent, °fresh, °new, °latest, °brand-new: *The hottest gossip is that she is marrying for the seventh* *time.* **6** °popular, sought-after, commercial, saleable, marketable: *The publisher believes* Zenobia's Memoirs *to be the hot property of the year.* **7** °lustful, °lecherous, libidinous, lubricous *or* lubricious, °sensual, concupiscent, °prurient, licentious, oversexed, sex-crazed, sex-mad, *Archaic* lickerish, horn-mad, *Slang* horny, *Chiefly Brit* °randy, *US* °hard up: *They were really hot, not having been with anyone of the opposite sex for months.* **8** °intense, °vivid, °striking, °bright, °brilliant, °dazzling, °loud: *That hot pink lipstick looks good with your suntan.* **9** electrified, °live, charged, powered: *The hot wire connects to the other terminal, I think.* **10** °dangerous, °precarious, °risky, °sensitive, °delicate, °unstable, °touchy, unpredictable: *The situation is getting a bit too hot to handle.*

hot air *n.* blather *or* blether, bunkum, verbiage, °talk, °wind, pretentiousness, pomposity, °bombast, grandiloquence, magniloquence, flatulence, gasconade, rodomontade, *Colloq* claptrap, bosh, gas, guff: *Management's promises of pay increases were nothing but hot air.*

hotbed *n.* breeding ground, fertile source: *Slums are a hotbed of crime.*

hotel *n.* hostelry, inn, °lodging, caravanserai; motel, motor hotel, bed and breakfast *or* B & B, guest house, °pension, *Australian and New Zealand* °pub, *US* tourist house: *We are staying at a small hotel just outside of town.*

hotheaded *adj.* °impetuous, headlong, hot-tempered, °quick-tempered, °volatile, °rash, °hasty, °wild, °foolhardy, °reckless, °precipitate, °thoughtless, °heedless, madcap, °daredevil, devil-may-care: *He is too hotheaded to succeed in the diplomatic corps.*

hothouse *n.* **1** °hotbed, greenhouse, glasshouse, conservatory: *These tomatoes were grown during the winter in our hothouse.* —*adj.* **2** °dainty, °delicate, °sensitive, °fragile, °frail, pampered, overprotected, sheltered, shielded, spoiled, coddled, babied: *She wanted to change the hothouse atmosphere of the university for the real world.*

hotly *adv.* intensively, energetically, doggedly, persistently, zealously, fervently, fervidly, ardently, °warmly, enthusiastically: *Hotly pursued by the police, I managed to hide in a culvert.*

hound *v.* °bully, °browbeat, °persecute, °nag, °harass, °annoy, °pester, harry, badger: *Although he was found innocent, his neighbors hounded him so much that he had to leave town.*

house *n.* **1** °residence, °dwelling, dwelling place, °home, °abode, household, homestead, °domicile, lodging(s), °quarters, °building, edifice: *We are playing bridge at my house next Saturday.* **2** °family, °line, °lineage, °dynasty, °clan, ancestry, °strain, °race, blood, descendants, forebears: *She is a member of the royal house of Sweden.* **3** legislature, legislative body, congress, °parliament, °assembly, °council, °diet: *Both houses of the legislature passed the bill.* **4** °establishment, °firm, °concern, °company, °business, °organization, °enterprise, °undertaking, *Colloq* °outfit: *He has gone from one publishing house to another with his manuscript.* **5** auditorium, °theater, concert hall: *The house is sold out.* **6** house of ill repute *or* ill fame *or* prostitution, °brothel, whorehouse, bagnio, bordello, *Archaic* bawdyhouse, *Colloq* sporting house, crib, *Slang US* cathouse: *She used to run a house in Lambeth.* **7** ***on the house.*** °free, gratis, for nothing, as a gift: *To celebrate the twentieth anniversary of our opening, drinks will be on the house tonight.* —*v.* **8** °shelter, °accommodate, °domicile, °lodge, °quarter, °put up, °take in, °board, billet, °harbor: *We have housed as many as ten people at a time in the cottage.* **9** °contain, °accommodate, °quarter: *This building houses our computer operations.*

housing *n.* **1** homes, houses, lodging(s), °quarters, °accommodations, habitation, °dwelling; °shelter, °protection: *The Council is trying to provide housing for the elderly.* **2** °case, casing, °cover, covering, °enclosure, container, °box, °shield: *We keep the sensitive equipment in a dust-proof housing.*

hovel *n.* °hole, °shack, shanty, (pig)sty, pigpen, coop, crib, °hut, *Colloq US* °dump: *The beggar lives in a hovel near the railway tracks.*

hover *v.* **1** °drift, °poise, °float, °hang, be *or* hang suspended, hang in the air: *The humming bird hovers over a flower, drinking its nectar.* **2** °linger, loiter, °wait, hang about *or* around: *The waiter's constant hovering about is making me nervous.*

however *adv.* **1** °notwithstanding, °regardless, °nevertheless, °nonetheless, °despite (that), °in spite of (that), °still, but, °though, °yet, °even so, be that as it may, come what may, no matter what; °at any rate, anyway, anyhow, on the other hand, in all events, °in any event, °in any case, after all: *He insisted the council should give their approval; however, that is not their responsibility. I was ready to move to Cardiff; she didn't offer me the job, however.* **2** to whatever manner *or* extent *or* degree, howsoever, no matter how, in any way *or* manner *or* respect, anyhow, how, in whatever way *or* manner: *However you view it, he is still the boss. Spend the money however you see fit.* —*conj.* **3** how, how on earth, how in the world, in what way *or* manner: *However do you manage with only three to help, you poor thing?* **4** no matter how, regardless how, putting *or* setting aside how, notwithstanding how: *However much she earns, she is always short of money.*

howl *v.* **1** yowl, °cry, wail, ululate, bay; °shout, °yell, °bellow, °scream, °roar, *Colloq* holler: *The wolves were howling at the moon last night. He howled with pain when he caught his finger in the door.* —*n.* **2** yowl, yowling, ululation, ululating, wail, wailing, yelp, yelping, °cry; °shout, °yell, °bellow, °scream, °roar, *Colloq* holler: *The wolf's howl is an eerie, chilling sound. I kicked him and he let out a howl.*

howler *n.* °blunder, °mistake, °error, gaffe; malapropism, Irish bull; *Brit* bloomer, *US* clinker, *Colloq Brit* clanger, *US* boner: *The howler, "Money is the route of all evil," the editor explained, should be corrected to "The* love *of money is the* root *of all evil."*

hub *n.* °center, °focus, focal point, °pivot, °heart, °core, °nucleus, nave: *This is the hub of the city, around which everything else moves.*

huddle *n.* **1** °cluster, °group, °bunch, °clump, °pack, °herd, °crowd, °throng, °mass: *Everyone got into a huddle, trying to keep warm.* **2** °meeting, °conference, °discussion, consultation: *They were in a huddle, deciding what to do next.* —*v.* **3** °cluster, °gather, crowd *or* press together, throng *or* flock together, °nestle, jam *or* cram together, squeeze together: *We huddled in basements during air raids.* **4** °meet, °discuss, °confer, °consult: *Let's huddle on this question of the advertising schedule.*

hue *n.* °color, °tint, °shade, tinge, °tone, °cast, tincture, *Technical* chroma: *The dust in the atmosphere gives the sunsets a reddish hue.*

huff *n.* **1** ***in a huff.*** piqued, peeved, °testy, irritated, angered, vexed, annoyed, in high dudgeon, provoked, exasperated, °petulant, in a pet, *Colloq* (all) het up: *She went off in a huff because I didn't praise her painting.* —*v.* **2** °puff, °blow, °bluster: *When I complained about the food, the manager simply huffed and puffed and walked away.*

hug *v.* **1** °embrace, °clasp, °squeeze, °cuddle, °snuggle, *Archaic or literary* °clip: *They hugged each other warmly, then kissed goodbye.* **2** follow closely, cling to, stay *or* keep near *or* close to: *We hugged the ground as the bullets whizzed by overhead.* —*n.* **3** °embrace, °clasp, °squeeze, *Colloq* °clinch: *She saw me, ran over, and gave me a big hug.*

huge *adj.* °large, °great, °enormous, °gigantic, °giant, °immense, °massive, tremendous, gargantuan, °prodigious, mammoth, °colossal, °monumental, Brobdingnagian, titanic, stupendous, elephantine, leviathan, °mountainous, °vast, *Colloq* °jumbo, °whopping: *The red spot on Jupiter is so huge it could swallow the earth several times over.*

hulk *n.* **1** shipwreck, °wreck, °derelict, °shell, skeleton: *The rusting hulk of the once-proud ship is tied up at the dock.* **2** oaf, °clod, lout, ox, *Slang US* galoot *or* galloot, klutz: *Don't tell me that hulk of a man painted these delicate miniatures!*

hulking *adj.* °clumsy, °awkward, ungainly, lubberly, oafish, loutish; °unwieldy, cumbersome, °bulky, °ponderous, °massive, °ungraceful, inelegant: *Their great hulking son could no longer fit into his clothes. That hulking monolith of an office building dominates the entire neighborhood.*

hull *n.* **1** framework, skeleton, °frame, °structure, °body: *The hull of the ship completed, we added the superstructure.* **2** °shell, pod, °case, husk, °skin, °peel, °rind, *US* shuck: *Squirrels carefully nibble through the hull of the nut to get at the kernel inside.* —*v.* **3** °shell, °peel, °skin, husk, *US* shuck: *We hulled all the nuts and they are now ready for the fruitcake.*

hum *v.* **1** °buzz, drone, thrum, °murmur, whirr, purr, °vibrate, *Technical* bombinate *or Technical* bombilate: *I want to smell the wild flowers and hear the bees hum once again.* **2** bustle, °stir, be active, move briskly, *Colloq* tick (over): *Within three months of completion, the new plant was humming.* **3** intone: *I said, "Do you know your rock guitar is keeping me awake?" and he replied, "No, man, but if you hum it I'll try to play it."* —*n.* **4** °buzz, buzzing, drone, droning, thrum, thrumming, °murmur, murmuring, murmuration, whirr, whirring, purr, purring, vibration: *From the other room came the hum of conversation.*

human *adj.* **1** °mortal, anthropoid, hominoid, android; hominid; *Possibly offensive* manlike: *The aliens from the spaceship had few human characteristics. It is human nature to think.* **2** °sensitive, °defenseless, °weak, fallible, °vulnerable: *She is human, too, and can be easily hurt.* **3** °kind, °kindly, kindhearted, °considerate, °charitable, compassionate, °merciful, °benign, benignant, °tender, °gentle, °forgiving, °lenient, °benevolent, beneficent, °generous, magnanimous, °humanitarian, °understanding, °accommodating, °sympathetic, °good-natured, humane, °sensitive: *Human concern for others is in short supply these days.* —*n.* **4** human being, °person, °individual, °woman, °man, °child, °mortal, °one, °soul, someone, °somebody: *Humans have long maltreated many of the animals on which they depend.*

humanitarian *adj.* **1** See **human, 3.** —*n.* **2** Good Samaritan, °benefactor, benefactress, °philanthropist, altruist: *Because of her charitable works, Lady Pendleton is thought of as a great humanitarian.*

humanity *n.* **1** human race, °people, °society, humankind, Homo sapiens; the public, °the masses, community; *Possibly offensive* °man, mankind: *From what we are told, humanity took millions of years to evolve.* **2** humanness, human nature, mortality: *On his first voyage to Mars, Kollworth began to doubt his own humanity.* **3** °kindness, kindliness, kindheartedness, °consideration, helpfulness, charitableness, openheartedness, warmheartedness, goodwill, °benevolence, compassion, mercifulness, °mercy, benignity, tenderness, °warmth, gentleness, leniency *or* lenience *or* lenity, beneficence, generosity, unselfishness, magnanimity, °understanding, °sympathy, °sensitivity: *It was owing to the humanity of our neighbors that we survived the bombing.*

humble *adj.* **1** °modest, °reserved, unpretentious, unostentatious, self-effacing, unassuming, unpresuming: *For someone who has accomplished so much, she is quite humble.* **2** °submissive, °meek, °servile, °obsequious, deferential, °mild, °respectful, subservient, °subdued: *Usually pompous, Nigel became humble in the presence of the chairman.* **3** °lowly, °low, °inferior, °mean, ignoble, °ordinary, °plebeian, °common, °simple, °obscure, unprepossessing, unimportant, °undistinguished, °insignificant; lowborn, °base, baseborn: *He lives in a humble cottage. Though she came from a humble background, she rose to become prime minister.* —*v.* **4** °chasten, bring *or* pull down, °subdue, abase, °debase, demean, °lower, °degrade, °downgrade, °reduce, make (someone) eat humble pie, lose face, °shame, °humiliate, °crush, °break, °mortify, chagrin, *Colloq* °put down, take (someone) down a peg *or* notch: *He found army discipline a humbling experience.*

humdrum *adj.* °dull, °boring, °tedious, °tiresome, wearisome, °monotonous, unvaried, unvarying, °routine, undiversified, unchanging, °repetitious, uneventful, unexciting, uninteresting, °prosaic, mundane, °ordinary, commonplace, °common, °banal, °dry, insipid, jejune: *They saw no way of escaping from their humdrum lives.*

humid *adj.* °damp, °moist, °muggy, °clammy, °sticky, °steamy, soggy, °sultry, °wet: *The weather has been oppressively hot and humid all week.*

humiliate *v.* See **humble, 4.**

humiliation *n.* °disgrace, °shame, mortification, °dishonor, ignominy, °indignity, °discredit, loss of face, obloquy, abasement, depreciation, detraction, °degradation, derogation, belittlement, disparagement, shaming, °embarrassment, humbling: *He had to suffer the humiliation of being drummed out of the corps.*

humility *n.* modesty, meekness, self-effacement, shyness, diffidence, timidity, timorousness, meekness, bashfulness, mildness, unpretentiousness, submissiveness, °servility, self-abasement, lowliness: *She sees humility as dissatisfaction with oneself on account of some defect or infirmity.*

humor *n.* **1** funniness, comedy, °wit, facetiousness, ludicrousness, drollery, jocoseness *or* jocosity, jocularity, waggishness, °raillery, °banter: *One writer defined humor as "the happy compound of pathos and playfulness," while another referred to the sense of humor as a "modulating and restraining balance wheel."* **2** comedy, farce, jokes, jests, witticisms, °wit, *Slang* wisecracks, gags: *Milton was scarcely known as a writer of humor.* **3** °mood, °frame of mind, °temper; spirit(s); °disposition, °nature, temperament: *She is in a bad humor till after breakfast.* —*v.* **4** soothe, °gratify, placate, please, mollify, °indulge, appease, °pamper, cosset, °coddle, mollycoddle, °jolly, °baby, °spoil: *They are trying to humor him because he's in a bad mood.*

humorous *adj.* °funny, comical, facetious, laughable, risible, °ludicrous, °farcical, sidesplitting, °hilarious, °merry; droll, °whimsical, amusing, °witty, waggish, jocular, jocose, °playful, °pleasant, *Colloq* °hysterical: *The humorous parts of the play sent the audience into gales of laughter.*

hump *n.* **1** °bulge, °lump, °bump, protuberance, °protrusion, °projection, °knob, node, °mass, °hunch, enlargement, °swelling, °growth, excrescence, tumefaction, tumescence; °mound, barrow, °tell, hummock, hillock, tumulus, *Brit dialect* tump: *Digging into the hump near the trees, we uncovered an ancient burial ground.* —*v.* **2** °hunch, °arch, curve, crook, °bend: *Sulking in the corner, he humped his back and refused to speak with anyone.* **3** °drag, °lug, °haul, °carry, °heave: *She had to hump two heavy suitcases from the railway station all by herself.*

hunch *n.* **1** °(intuitive) guess, °intuition, °feeling, °impression, °suspicion, °premonition, presentiment: *I had a hunch they would be late.* **2** See **hump, 1.** —*v.* **3** See **hump, 2.**

hunger *n.* **1** hungriness, °emptiness, °appetite, ravenousness, voraciousness, voracity; °famine, starvation: *We ate leaves to stave off the pangs of hunger. Some optimists hope to eliminate hunger from the planet by the year 2000.* **2** yearning, °desire, craving, °itch, °thirst, °longing, hankering, °mania, cupidity, *Formal* cacoëthes, *Colloq* yen: *She felt an insatiable hunger for intellectual companionship.* —*v.* **3** Usually, ***hunger for** or **after.*** crave, °yearn, °desire, °thirst, °want, °hanker, *Colloq* yen, have a yen: *Blessed are they who hunger and thirst after righteousness.*

hungry *adj.* **1** °famished, °starved, starving, °ravenous, °voracious, °empty, °hollow, *Colloq chiefly Brit* peckish: *I'm hungry enough to eat a horse.* **2** craving, covetous, °eager, avid, °greedy, °keen, yearning, °desirous, °longing, hungering, thirsting, starving, °dying, *Colloq* hankering: *Marooned for years, Crusoe was hungry for the sight of another human being.* **3** acquisitive, °greedy, °thirsty, insatiable, °deprived: *The parched, hungry earth drank up the rain.*

hunt *v.* **1** °chase, °pursue, dog, °hound, °stalk, °trail, °track (down), °trace; °course: *The jewel thieves were hunted across three continents.* **2** Also, ***hunt for** or **up** or **out** or **through.*** °seek (out), °search (for), go in search of *or* for, look (high and low) for, °quest after, go in quest of, °scour, °ransack, °investigate, pry into, go over *or* through with a fine-tooth comb, °examine, °explore, *Colloq US* °check out: *Detectives hunted the carpet for clues. She is hunting a job in publishing. I have hunted for the ring but cannot find it. Can you hunt up someone to fill the vacancy? Hunt through your pockets again for the key.* —*n.* **3** °chase, °pursuit, tracking (down), stalking, hunting; °course: *The ten-year hunt for the thieves continues.* **4** °search, °quest: *My hunt ended when I found the ring.*

hunter *n.* huntsman, huntswoman, stalker, tracker, Nimrod, Orion; huntress: *An expert hunter, she bagged three tigers last year.*

hurdle *n.* **1** °barrier, °obstacle, °impediment, °hindrance, °obstruction, °bar, °handicap, °restraint, °snag, °(stumbling) block, °check, °difficulty, °complication, °interference: *Can she overcome the hurdle of prejudice against women?* —*v.* **2** °leap (over), vault (over), °jump (over): *The fugitive easily hurdled the fence around the compound.*

hurl *v.* °throw, °toss, °shy, °sling, °fling, °pitch, °cast, °send, °fire, °heave, °propel, °let fly, *Colloq* chuck: *The fast bowler hurled the ball past the batsman.*

hurricane *n.* cyclone, tornado, typhoon, °whirlwind, °twister, windstorm, °storm, °gale, °blow: *The 1985 hurricane destroyed millions of trees.*

hurried *adj.* **1** °hasty, °feverish, °frantic, °hectic, °breakneck, frenetic, °impetuous, rushed, °precipitate, °swift, °quick, °speedy; °brief, °short: *He had a hurried snack at home before running to catch the train.* **2** °superficial, °cursory, °offhand, °perfunctory, slapdash: *Airport security guards made only a hurried examination of the hand luggage.*

hurry *v.* **1** °rush, °hasten, make haste, °speed, °race, °dash, °hustle, °scurry, °tear, °fly, °run, °shoot, scoot, scamper, scuttle, hotfoot (it), *Colloq* °shake a leg, get cracking, °get a move on, go hell for leather, skedaddle, °step on it, step on the gas, *Chiefly US* hightail (it), go like greased lightning, get a wiggle on: *You'll have to hurry to catch your bus.* **2** °speed up, accelerate, °hasten, °rush, °push, °press, °expedite; °urge, egg: *Such changes take time and can't be hurried. She hurried the children along so they wouldn't miss the bus.* —*n.* **3** °haste, °rush, °urgency, °eagerness; °agitation, disquiet, °upset, dither, °fuss, bustle, ado, to-do, °furor, commotion, turmoil, °stir, pother; *Colloq* °stew, °sweat: *Why all the hurry to finish before dark? He's in an awful hurry to get home.*

hurt *v.* **1** °harm, °injure, °wound; °damage, °impair, °mar, °spoil, °vitiate, °ruin: *She's so gentle she wouldn't hurt a fly. The scandal hurt the candidate's chances of election.* **2** °ache, °smart, °pain, °pinch, °sting, °burn, °torment, °gripe: *My elbow really hurts where I hit it.* **3** °distress, °grieve, °affect, °afflict, aggrieve, °depress, °upset, °disappoint, °pain, cut to the quick, affront, °offend: *Sticks and stones may break my bones but names will never hurt me. You always hurt the one you love.* **4** °injure, °maim, °wound, °cripple, °lame, disable, °incapacitate, °damage, °mutilate, °mangle: *Three people have been badly hurt in a car crash on the interstate.* —*n.* **5** °harm, °injury, °damage, °detriment, °disadvantage: *How much more hurt can you cause beyond what you have done already?* **6** °ache, °pain, °pang, °distress, °discomfort, °suffering, °torment, torture, °agony; °anguish, °misery, °woe, dolor, °sadness, °depression: *If you rub on this ointment, the hurt will go away. You cannot imagine the hurt we felt when he failed his examinations.* —*adj.* **7** injured, wronged, pained, rueful, grieved, °unhappy, aggrieved, °sad, °wretched, °woebegone, °sorrowful, °mournful, depressed, °dejected, °dismal, °gloomy, °melancholy: *Whenever I mention getting a job she gets that hurt expression on her face.* **8** damaged, °defective, marred, impaired, °broken, °worn, °dilapidated, shopworn, scratched, bruised, scarred: *They are holding a sale of hurt merchandise at the warehouse tomorrow.*

hurtful *adj.* **1** °harmful, °injurious, °detrimental, pernicious, °prejudicial, disadvantageous, damaging, deleterious, °destructive, noisome, noxious, baneful, °mischievous: *Efficient government is advantageous to many and hurtful to none.* **2** °nasty, °cruel, °cutting, malicious, °mean, °unkind, wounding, °spiteful: *He made some hurtful accusations about the librarian's efficiency.*

hurtle *v.* °rush (headlong), °tear, °shoot, °race, °speed; °plunge: *The car hurtled round the corner and crashed into a tree. Two bodies hurtled past me into the abyss below.*

husband *n.* **1** °mate, spouse, °groom, bridegroom, °partner, *Colloq* old man, hubby: *My husband and I take turns cleaning the house.* —*v.* **2** °save, °keep, °retain, °hoard, °conserve, °preserve, °store; budget, °economize (on), °manage: *If we husband our resources, we shall have enough for a rainy day.*

hush *interj.* **1** Shush!, Quiet!, Be *or* Keep quiet *or* silent *or* still!, Hold your tongue!, Mum's the word!, *Slang* Shut up!, Clam up!, Shut your trap!, Button your lip!, Shut your gob!, *Brit* Belt up!, *US* Hush up!, Shut your face!, *US dialect* Hush your mouth!: *Hush! You're in a library.* —*v.* **2** shush, °silence, °still, °quiet: *I wish she'd hush the child's wailing.* **3** °suppress, °mute, °soften, soft-pedal, °whisper: *They spoke in hushed tones.* **4** Usually, ***hush up.*** °suppress, °repress, °quash, °cover up, °hide, °conceal, keep quiet, *Colloq* °squelch: *They tried to hush up the news about his extramarital affairs.*

5 soothe, allay, °calm, °quiet, mollify, pacify, placate, °tranquilize: *Our worst fears were hushed by father's comforting words.* —*n.* 6 °silence, °quiet, stillness, °peace, tranquillity: *A hush came over the crowd as she rose to speak.*

husky *adj.* 1 °brawny, strapping, °sturdy, °burly, well-built, °robust, °hefty, °rugged, °powerful, °strong, °stout, thickset, °muscular, °tough, *Colloq* beefy: *Two husky men arrived to move the piano.* 2 hoarse, °gruff, °dry, °harsh, rasping, °rough, °raucous: *She has a deep, husky voice I would recognize anywhere.*

hustle *v.* 1 °rush, °push, °hurry, °hasten, °run, °dash, scamper, scuttle, °scurry, sprint: *Mrs. Grumble hustles off to the market every morning. He said that one must really hustle to make a living these days.* 2 shove, °push, °drive, °force, °hasten, °expedite, °press: *The bill was hustled through the legislature in just one day.* 3 shove, °crowd, °push, jostle, elbow, °thrust, °force: *She tried to hustle her way to the front of the queue.* 4 °push, °eject, °force, coerce, °drive, *Colloq* °bounce: *He was hustled out the back door.* —*n.* 5 pushing, jostling, buffeting, jarring, elbowing, shoving, nudging: *We always take taxis to avoid the hustle in the underground.* 6 °activity, °action, °stir, °movement: *I can't concentrate with all the hustle and bustle going on in the office.*

hut *n.* °cabin, °shack, shanty, °shed, lean-to, °shelter, cote, *Literary* °cot, *Australian* gunyah: *We found a tiny hut where we could keep out of the storm.*

hybrid *n.* °mixture, crossbreed, half-breed, °mongrel, °cross, composite, °combination, °compound: *The best wheat for this region is a hybrid developed in our laboratory.*

hygienic *adj.* °clean, °sanitary, °sterile, disinfected, germ-free, aseptic, °pure: *It is essential that hygienic surroundings be maintained in hospitals.*

hypnotize *v.* °fascinate, mesmerize, °entrance, cast a spell over *or* on, °captivate, °enchant, °charm, spellbind, °bewitch, °enrapture, ensorcell, °transport: *Greg is completely hypnotized by that singer at Felipe's nightclub.*

hypocrisy *n.* °deceit, deceitfulness, duplicity, double-dealing, °deception, °chicanery, guile, quackery, charlatanism *or* charlatanry, falseness, fakery, falseness, °pretense, pretentiousness, °lying, mendacity, pharisaism *or* phariseeism, Tartuffery, insincerity, °dissimulation, two-facedness, double standard(s), sanctimony, sanctimoniousness, *Colloq* phoneyness *or US also* phoniness: *She says she loves him, but that's just so much hypocrisy. I loathe hypocrisy.*

hypocrite *n.* deceiver, double-dealer, °quack, charlatan, impostor *or* imposter, mountebank, confidence man *or* trickster, faker, °pretender, °liar, °pharisee, whited sepulcher, Tartuffe, flimflammer, *Colloq* phoney *or US also* phony, con man, flimflam man *or* artist, two-face: *It is incredible that so many were duped by that hypocrite.*

hypocritical *adj.* °deceptive, °deceitful, deceiving, °insincere, dissembling, feigning, dissimulating, double-dealing, °false, °fake, faking, °two-faced, pretending, °pretentious, °lying, mendacious, pharisaic(al), °sanctimonious, °dishonest, underhand, treacherous, °perfidious, untrustworthy: *Don't believe anything that hypocritical confidence man tells you.*

hypothesis *n.* theory, °theorem, postulate, premise *or* premiss, °proposition, assumption, °supposition, °speculation: *Cadwallader's Hypothesis is that television sets watch viewers while viewers are watching them.*

hypothetical *adj.* °assumed, °supposed, conjectural, conjectured, hypothesized, putative, surmised, °assumed, presumed, suspected, imagined, guessed, °speculative, speculated, °theoretical, suppositional, supposititious *or* suppositious: *Let us take a hypothetical case of one who refuses to pay income tax.*

hysterical *adj.* 1 °raving, °mad, °beside oneself, crazed, irrational, distracted, °rabid, °frantic, frenzied, °wild, °berserk, °uncontrolled, uncontrollable, unrestrained, unrestrainable: *They are subject to violent, hysterical outbursts.* 2 °hilarious, sidesplitting, °uproarious, °farcical, comical, °funny: *The absolutely hysterical sign in our Japanese hotel read, "You are invited to take advantage of the chambermaid."*

I

icing *n.* 1 frosting, °glaze, coating: *Our favorite was chocolate cake with chocolate icing.* 2 °bonus, °(fringe) benefit, °reward, (extra) added attraction, °extra, °reward, dividend: *As icing on the labor contract, all employees were given their birthdays off.*

icy *adj.* 1 ice-cold, °frigid, arctic, °bitter, glacial, °freezing, frozen, °chill, glacial, hyperborean *or* hyperboreal, °polar, Siberian, °wintry, °raw, °cold, chilling, °chilly: *The icy wind cut through to the marrow of my bones.* 2 °cool, °chill, °chilly, °frigid, °distant, °aloof, °remote, °freezing, ice-cold, unemotional, unimpassioned, °stony, °steely, °callous, flinty, °formal, °reserved, °forbidding, unfriendly, °hostile: *She fixed him with an icy stare.*

idea *n.* 1 concept, °conception, °construct, °thought, °notion, °plan, °design, °scheme, °suggestion, °recommendation: *Is that your idea of a good singing voice?! Grover has an excellent idea for increasing sales.* 2 °notion, °fancy, °impression, °picture, °(mental) image, concept, °conception, °perception, °understanding, awareness, apprehension, °inkling, °suspicion, °hint, °suggestion, approximation, °clue, intimation, °guess, °estimate, °estimation, °impression: *I haven't the slightest idea what you are talking about. Can you give us any idea of the cause of the delay?* 3 °belief, °opinion, °sentiment, °feeling, teaching(s), °doctrine, °tenet, °principle, °philosophy, °view, °viewpoint, °outlook, °notion, °conviction, °position, °stance: *Western and Eastern ideas differ as to the role of women in society.* 4 °aim, °goal, °purpose, °objective, °object, °end, °point, °reason, raison d'être: *What was the idea of telling the teacher? The idea behind junk bonds is to raise money for leveraged buyouts. The idea of the game is to capture your opponent's king.* 5 °hypothesis, theory, °notion, °dream, fantasy *or* phantasy: *His invention is based on the idea of an anti-gravity device.*

ideal *n.* 1 °model, °paragon, °standard, criterion, paradigm, exemplar, °pattern, °example, °epitome: *She regarded Florence Nightingale as her ideal.* 2 °acme, °(standard of) perfection, °nonpareil: *The ideal can rarely be achieved, so settle for reality.* 3 ***ideals.*** °principles, °morals, standards: *Everyone's ideals are compromised sooner or later.* —*adj.* 4 °perfect, °excellent, °supreme, consummate, °complete, °model, °idyllic: *This is an ideal home for a young couple.* 5 conceptual, imagined, °imaginary, °unreal, °visionary, °idealistic, °fictitious, Utopian, notional, mythical *or* mythic, °fantasy, °dream, °romantic, chimeric(al), °illusory, °fanciful, °fancied: *In his ideal world as there is no crime, there is no need for police.*

idealistic *adj.* °visionary, °romantic, romanticized, °optimistic, starry-eyed, °quixotic, Panglossian, °impractical, °unrealistic: *He is idealistic enough to believe that people are basically good.*

idealize *v.* °exalt, elevate, °glorify, °worship, ennoble, deify, apotheosize, °put on a pedestal, romanticize: *Because she idealizes everyone, she cannot see their true natures.*

ideally *adv.* **1** under *or* in the best of circumstances, at best, in a perfect world, all things being equal: *Ideally, February is the time to vacation in the Caribbean.* **2** theoretically, in theory, °in principle: *Ideally, people ought to have money before they are too old to enjoy it.* **3** °perfectly: *John and Marsha are ideally suited to each other.*

identical *adj.* **1** °same, °twin, °duplicate, °indistinguishable, interchangeable; selfsame: *The two leaves look identical to me. The duchess and I had the identical suite at the hotel, but a week apart.* **2** °similar, °matching, °like, °alike, comparable, °equal, °equivalent, corresponding: *The children are dressed in identical clothing.*

identification *n.* **1** °connection, °recognition, distinguishing, °indication, °perception, detection, °selection, naming, labeling, pinpointing, designation, characterization, °denomination; authentication, verification, °establishment, certification, substantiation, corroboration, *Colloq* fingering: *Her identification of him as her attacker will not stand up in court.* **2** classification, classifying, cataloging, categorization, categorizing, pigeonholing: *How much time is needed for the identification of the minerals in the moon rocks?* **3** ID, ID card, identity card, badge, credentials: *Only up-to-date identification will be accepted as proof of ownership.* **4** °connection, °association, affiliation, empathy, °sympathy, °rapport, °relationship: *His continued identification with his mother is immature.*

identify *v.* **1** classify, °categorize, catalog, pigeonhole, °sort (out), °specify, pinpoint, home (in) on, °name, °label, °tag, °recognize, °place, °mark, °label, °tag, pinpoint, °single out, °point out, *Colloq* °put one's finger on: *We are unable to identify the butterfly you caught.* **2** °connect, °associate, °relate, °ally: *She is closely identified with the success of the enterprise.* **3** °diagnose, °specify, °name, °recognize: *From the symptoms, the doctor identified the disease as bubonic plague.* **4** Usually, ***identify with.*** empathize (with), °sympathize (with), °relate (to), *Colloq* °dig: *His problem is that he identifies too closely with his dog.*

identity *n.* **1** sameness, oneness, unanimity, indistinguishability, °agreement, °accord, congruence: *They were united by identity of conviction.* **2** °personality, individuality, distinctiveness, uniqueness, particularity, °singularity: *Many who join the army lose their identity.*

ideology *n.* belief(s), convictions, tenets, credo, °philosophy, °principles, °creed, dogma, teachings, °doctrine: *Buddhism had a great influence on his ideology.*

idiom *n.* **1** °language, °tongue, °speech, °vernacular, °dialect, argot, patois, °jargon, °cant, idiolect, °parlance, *façon de parler,* phraseology: *The play is written in a rather old-fashioned idiom.* **2** °expression, °(set) phrase, phrasing, locution, °cliché: *The term* red herring, *an idiom meaning 'false trail,' is used of something that is neither red nor a herring.*

idle *adj.* **1** °unused, °inactive, unoccupied, nonoperative, stationary: *The looms were idle for months. The devil finds work for idle hands.* **2** °unemployed, °out of work, °redundant, jobless, workless, *Colloq* °at leisure, °at liberty, between assignments, resting, *US* on the beach: *Unable to find work, Gilbert has been idle for a year.* **3** °indolent, °lazy, °listless, °lethargic, loafing, °slothful, °shiftless, °lackadaisical, loitering, fainéant: *Since winning the lottery, Crouch has become one of the idle rich.* **4** °bootless, °fruitless, unproductive, abortive, unfruitful, °pointless, °vain, °trifling, trivial, °shallow, nugatory, °superficial, °insignificant, °meaningless, °senseless, unimportant, °frivolous, °worthless, °useless, otiose, unavailing, °futile: *A few of us were passing the time in some idle chatter when Mycroft walked into the bar. He had no factual information to offer, only idle speculation.* —*v.* **5** Often, ***idle away.*** °waste, °fritter away, while away, °kill: *They idle away the hours lying by the swimming pool.* **6** laze (about), loiter, kill time, °loaf, loll, °lounge, take it easy, putter *or Brit* potter about, mess about, fool away, fool around *or* about, *Colloq Brit* °muck about, °bugger about, *US* lallygag *or* lollygag, goof off *or* around, *Military slang US* goldbrick: *Stop idling and get down to work.*

idleness *n.* **1** °inactivity, inaction, °lethargy, °torpor, °indolence, laziness, °sluggishness, °sloth, slothfulness, shiftlessness, °inertia, lassitude, °torpor, *flânerie, dolce far niente*; unemployment, *Colloq US* lallygagging *or* lollygagging, *Military slang US* goldbricking: *The idleness enforced on nonunion workers by the strike didn't help matters.* **2** shirking, malingering, dawdling, loafing, time-wasting, lazing, *Colloq* dillydallying, shilly-shallying, *Brit* skiving: *The foreman warned that he would not tolerate idleness.*

idler *n.* °loafer, layabout, °slacker, shirker, sluggard, lazybones, slugabed, °laggard, dawdler, clock-watcher, drone, °slouch, ne'er-do-well, fainéant, *Colloq* lounge lizard, *Military slang US* goldbrick *or Military slang US* goldbricker: *To fill out a crew we often recruited men from among waterfront idlers.*

idly *adv.* **1** unproductively, lazily, indolently: *She wanders about the shops, idly whiling away the hours.* **2** offhandedly, unconsciously, mechanically, thoughtlessly, unthinkingly, obliviously, insensibly, indifferently: *He sat quietly, his fingers idly drumming a tattoo on the tabletop.*

idol *n.* **1** °(graven) image, icon *or* ikon, effigy, °fetish, tiki, °symbol: *Nebuchadnezzar's people worshiped golden idols.* **2** hero *or* heroine, superstar, °celebrity, luminary, matinee idol, °favorite, °pet, °darling: *When he was a lad his idol was Quatermain, from the Rider Haggard novels.*

idolize *v.* °adore, °admire, adulate, °worship, °revere, °reverence, °venerate, °put on a pedestal, °exalt, °glorify, deify, lionize, °look up to, apotheosize: *She idolized her father, and his death was a terrible blow.*

idyllic *adj.* Arcadian, paradisaic(al) *or* paradisiac(al), °heavenly, Edenic, halcyon, °ideal, idealized, °pastoral, °rustic, bucolic, °picturesque, charming, °unspoiled, °peaceful, pacific: *He returned to the island to live out his life in idyllic repose.*

ignorance *n.* **1** unfamiliarity, unawareness, unconsciousness, benightedness, unenlightenment, °inexperience, greenness: *Ignorance of the law is no excuse. Mistakes are often caused by ignorance, not stupidity.*

ignorant *adj.* **1** unknowing, °uninformed, untaught, °uneducated, unschooled, unread, unlearned, unlettered, °illiterate: *Is there anyone so ignorant as not to know who the Duke of Wellington was?* **2** °unaware, °unfamiliar, °unconscious, °benighted, unenlightened, unwitting, in the dark, °oblivious, *Formal* nescient: *She knows many things about which we are ignorant.* **3** °inexperienced, °green, °naive, °innocent, °unsophisticated: *The new apprentice was ignorant enough to be sent to find a left-handed hammer.* **4** uncouth, °ill-mannered, °discourteous, °impolite, uncivil, °boorish: *Don't be so ignorant and open the door for the lady!*

ignore *v.* **1** °disregard, °overlook, pass over *or* by, turn a blind eye to, be blind to, °turn one's back on, turn a deaf ear to, wink at, brush off *or* aside: *If you are over sixty-five, you may ignore this paragraph.* **2** snub, give (someone) the cold shoulder, °reject, send to Coventry, °turn one's back on, *Colloq* give (someone) the brushoff *or* go-by, °cut, turn one's nose up at: *I don't care if you are cruel to me, just don't ignore me.*

ill *adj.* **1** ailing, °unsound, °sick, °indisposed, °infirm, °unhealthy, in a bad way, °diseased, afflicted, in bad health, °sickly, unwell, not well, out of commission; invalided, valetudinarian; *Colloq* °under the weather, in a bad way, °poorly, not up to snuff, °out of sorts, on the sick list, off one's feed, *Slang Brit* dicky, °seedy: *Call a doctor—this man is ill. I was ill last night after the fish mousse.* **2** °bad, °wicked, °sinful, °evil, iniquitous, °immoral, depraved, °vicious, °vile, °wrong, °corrupt: *Police raided several houses of ill repute.* **3** °hostile, unfriendly, antagonistic, °belligerent, malevolent, malicious, ill-wishing, unkind(ly), °harsh, °cruel: *The reading of the bequests sparked ill will among the heirs.* **4** °harmful, °hurtful, °injurious, °detrimental, damaging, pernicious, °dangerous, adverse, deleterious, baleful, °bad, unfavorable, °destructive, °disastrous, catastrophic, °ruinous, cataclysmic: *He suffered no lasting ill effects from the accident.* **5** °bad, °miserable, °wretched, °disastrous, unfavorable, unpropitious, °untoward, °disturbing, °unfortunate, unlucky, °inauspicious, °ominous, °unpromising, °sinister, °unwholesome: *Ill fortune led Ulysses to our isle. It is an ill wind that blows nobody good.* **6 *ill at ease.*** uncomfortable, discomfited, uneasy, edgy, °on edge, fidgety, °nervous, °anxious, °disturbed, distressed, troubled, °awkward, unsure, °uncertain: *I felt ill at ease in the presence of so august a personage.* —*n.* **7** °evil, °abuse: *Speak no ill of her in my home!* **8** °harm, °damage, °injury, °hurt, °mischief, °trouble, °misfortune, °misery, °affliction, °pain, °distress, °woe, woefulness, °discomfort, unpleasantness, °disaster, °catastrophe, cataclysm, °calamity, adversity, °damage, °suffering, °ruin, °destruction: *You will be held responsible if any ill befalls the children.* **9** °injustice, inequity, °wrong, °evil, °sin, °transgression, °abuse, °mistreatment, maltreatment: *He entered the law hoping to cure some of society's ills.* —*adv.* **10** °badly, adversely, unfavorably, °poorly, inauspiciously, unfortunately, unluckily: *Don't speak ill of the dead.* **11** °badly, adversely, unfavorably, critically, harshly, unkindly: *Please don't think ill of me for failing to attend your wedding.* **12** unkindly, harshly, unfairly, unjustly, improperly, °badly, wrongly, wrongfully, unsatisfactorily, °poorly, malevolently, maliciously: *He insists that he was ill treated when in prison.* **13** °scarcely, °hardly, °by no means, in no way: *It ill behooves you to treat me in this way.*

ill-advised *adj.* **1** inadvisable, ill-judged, injudicious, ill-considered, °misguided, unwise, °imprudent, °inappropriate, unpropitious, inexpedient, impolitic, wrongheaded, °thoughtless; °indiscreet: *It would be ill-advised to sell your house at this time.* **2** °hasty, °rash, °reckless, °impetuous, °rash, °foolhardy, incautious, °shortsighted, °improvident: *Driving without a seat belt is ill-advised.*

illegal *adj.* °unlawful, °illegitimate, °criminal, felonious, outlawed, prohibited, interdicted, forbidden, proscribed, wrongful, °unauthorized, *verboten,* °illicit, *Law* actionable: *The bookkeeper was indicted for illegal appropriation of funds.*

illegible *adj.* unreadable, unintelligible, indecipherable *or* undecipherable, °incomprehensible: *An illegible message was scrawled on the wall.*

illegitimate *adj.* **1** See **illegal. 2** bastard, °natural, fatherless, born out of wedlock, born on the wrong side of the blanket, misbegotten: *Illegitimate children were often offered for adoption in those days.* **3** °irregular, °improper, °incorrect, nonstandard, °invalid, °unauthorized, °spurious: *Purists consider* the reason is because *an illegitimate usage.*

ill-founded *adj.* °groundless, baseless, without foundation, unsupported, unsubstantiated, °empty, unjustified, unproven, uncorroborated, °unsound, °erroneous: *He wasted a great deal of time refuting her ill-founded accusations.*

illicit *adj.* **1** See **illegal. 2** °wrong, °improper, underhand(ed), °secret, furtive, clandestine, backdoor, *Colloq US* °sneaky: *He is the offspring of illicit love.*

illiterate *adj.* unlettered, analphabetic; unschooled, untaught, °uneducated, °benighted, °ignorant, unenlightened: *She teaches illiterate people to read.*

ill-mannered *adj.* °rude, °discourteous, °impolite, ill-bred, uncivil, °disrespectful, uncourtly, ungallant, °ungracious, indecorous, ungentlemanly, unladylike, °impudent, °insolent, insulting, °impertinent, °brazen: *That ill-mannered oaf didn't even apologize for knocking me down.*

illness *n.* sickness, °disease, °disorder, °affliction, °ailment, malady, °complaint, °infirmity, °disability, indisposition, °affection, *Colloq* °bug: *His illness is serious but not contagious.*

ill-treat *v.* °mistreat, maltreat, °abuse, °misuse, °harm, °hurt, °injure, °persecute, °mishandle: *He is at last being punished for those years he ill-treated his wife.*

illuminate *v.* **1** °light (up), °brighten, °lighten, throw *or* cast *or* shed light on *or* upon: *She had only a single candle to illuminate her sewing.* **2** °clarify, throw *or* cast *or* shed light on *or* upon, °enlighten, °clear up, elucidate, °explain, explicate, °reveal: *The lectures on Heidegger illuminated his philosophy to some extent.* **3** rubricate, °decorate, adorn, °embellish, °ornament: *The museum has the best collection of illuminated manuscripts.*

illumination *n.* **1** lighting, °light, brightness, °radiance, luminosity, incandescence, fluorescence, phosphorescence: *The illumination from the fire was barely enough to read by.* **2** enlightenment, °insight, °information, °learning, °revelation, °edification, °instruction, awareness, °understanding, clarification: *It is astonishing what illumination his lectures brought to his audiences.*

illusion *n.* **1** °deception, °delusion, °fancy, °misconception, misapprehension, °fallacy, °error, °mistake, mistaken *or* false impression: *He is laboring under the illusion that he is a great pianist.* **2** °fantasy, °daydream, °hallucination, phantasm, °phantom, chimera, phantasmagoria, mirage, aberration, °vision, °specter, figment of the imagination, will o' the wisp, ignis fatuus: *For years after he died, she had the illusion of seeing her father everywhere.*

illusory *adj.* illusive, °imaginary, °fictional, °unreal, °untrue, fallacious, °false, °mistaken, imagined, °fanciful, °fancied, hallucinatory, °deceptive, misleading, °apparent: *Any perceived financial gains from this scheme are purely illusory.*

illustrate *v.* **1** °instance, °exemplify, °demonstrate: *Let me illustrate my point with the following example.* **2** °picture, °illuminate, grangerize; °decorate, °embellish, emblazon, °ornament, adorn: *Now we have to find an artist to illustrate the book.*

illustration *n.* **1** °example, °case (in point), °instance, °sample, °specimen, exemplar, *Colloq* °for instance: *This film is an illustration of the* film noir *genre.* **2** °picture, depiction, °representation, °figure: *An illustration of a jet engine appears on page 32.*

illustrious *adj.* °distinguished, °famous, °noted, °renowned, famed, °eminent, °well-known, °prominent,

°important, °notable, respected, esteemed, °venerable, honored, acclaimed, °celebrated, °great: *He belongs in the same category as other illustrious generals, like Kitchener.*

ill will *n.* °dislike, °animosity, hatred, °hate, °loathing, abhorrence, detestation, malevolence, malice, °hostility, enmity, animus, antipathy, °aversion, °rancor, acrimony, °spite, °venom, vitriol, acerbity: *I bear her no ill will for the way she treated me. She incurred his ill will by refusing him access to the children.*

image *n.* **1** °likeness, °representation, °picture, °sculpture, °statue, effigy, °figure, °portrait, simulacrum; icon *or* ikon, °idol, graven image, °fetish, tiki: *Law required that an image of der Führer be present in every public office. The prophets warned the people not to worship graven images.* **2** °epitome, °duplicate, °copy, counterpart, °facsimile, °replica, °double, °twin, doppelgänger, clone, *Colloq* spit and image *or* spitting image, (dead) ringer: *He's the image of father. He stared at his image in the mirror.* **3** °impression, concept, °conception, °perception, °idea, °perception, °notion, mental picture: *You are quite different from my image of you.* **4** °epitome, °representative, °model, °(typical) example, °essence, archetype, °embodiment, incarnation, personification, materialization, reification, corporealization: *For us, she was the very image of what a leader should be.* **5** °figure (of speech), trope, °metaphor, allusion, simile, °symbol: *The poem contains images that are obscure unless you know Greek myths.* **6** °form, °appearance, °likeness, °guise, °semblance, °aspect, °mold, °cast: *Man claims to be created in God's image.*

imagery *n.* figurativeness, allusion, symbolism: *Constance's poems are filled with ornate imagery.*

imaginary *adj.* °fictitious, °fanciful, °fancied, chimerical, imagined, fictive, illusory *or* illusive, °visionary, made-up, °unreal, °untrue, mythical *or* mythic, notional, °abstract; °legendary, mythological: *When he was a boy, he had an imaginary friend called Carbonario.*

imagination *n.* **1** mind's eye, °fancy; creativity, inventiveness, °ingenuity, °insight, °inspiration, °vision, imaginativeness, creative power(s): *The description of the aliens from Neptune was drawn from his fertile imagination.* **2** °thought, °thinking, (mental) acuity, °intelligence, °wit: *It doesn't take much imagination to see why you dislike raw turnip salad.*

imaginative *adj.* **1** °creative, °original, °clever, °ingenious, inventive, innovative, inspired, inspiring, °enterprising, °resourceful: *Rigby is a highly imaginative computer programmer.* **2** °fanciful, °fantastic, °visionary, poetic(al), °whimsical, contrived, °fictitious, °fictional: *The imaginative tale of Don Quixote sprang from the mind of Cervantes.*

imagine *v.* **1** °think of, °contemplate, °picture, °envisage, °consider, °ponder, °meditate on, °envision, visualize, °conceive (of), conceptualize, °create, °think up, concoct, °devise, *Colloq* dream up, cook up: *I cannot imagine what you are referring to. Imagine, if you can, a world without conflict.* **2** °suppose, °guess, conjecture, °assume, °presume, °take it, °infer, take (it) for granted, take it as given, °think, °fancy, °believe, °gather, °surmise, °suspect, °judge, deem: *I imagine that you won't want a drink because you are driving. You are imagining things if you think she would let me get away with that.*

imitate *v.* **1** °mimic, °copy, ape, °parrot, °monkey, emulate, impersonate, do an impression of; °echo, simulate: *Cruikshank was imitating the boss, who walked in at that very moment. Fernwake's style imitates that of Carlyle.* **2** °copy, °mimic, °mock, °parody, °satirize, °burlesque, °caricature, travesty, *Colloq* spoof, °take off, *Brit* °send up: *In* Hudibras, *Samuel Butler imitated the style of the heroic epic.*

imitation *n.* **1** copying, mimicking, mimicry, aping, parroting, emulating, emulation, impersonating, impersonation, °impression: *Imitation is the sincerest form of flattery.* **2** impersonation, °parody, satirization, °burlesque, °caricature, °mockery, travesty, *Colloq* °takeoff, *Brit* sendup: *In his nightclub act, Farquhar does scathing imitations of celebrities.* **3** °copy, °fake, °counterfeit, °forgery: *These are modern imitations of ancient coins.* **4** °copy, °replica, replication, °reproduction, simulation, °facsimile, °duplicate, duplication, simulacrum: *We keep the original in a safe place and put the imitations on display.* —*adj.* **5** °fake, °synthetic, °artificial, simulated, °sham, ersatz, °mock, °factitious, °reproduction, man-made, *Colloq* phoney *or US also* phony: *It doesn't take an expert to see that this is an imitation diamond.*

immaculate *adj.* **1** °spotless, stainless, unblemished, °pure, °clean, °untarnished, unsullied, unsoiled, snow-white, spick-and-span *or* spic-and- span, °dapper, °spruce; °tidy, °neat: *Each morning he donned an immaculate uniform.* **2** °pure, °chaste, °innocent, virginal, °virtuous, vestal, °pristine, undefiled, untainted, °unspoiled, unblemished, stainless, unadulterated: *He remained convinced of his wife's immaculate reputation.* **3** °faultless, °flawless, °perfect, errorless, °impeccable: *His essay on Gibbon is an example of immaculate scholarship.*

immaterial *adj.* **1** unimportant, °inconsequential, nugatory, trivial, °trifling, °petty, °slight, °insignificant, °slight, °flimsy, °light, unessential, °nonessential, of little account *or* value: *Whether he stays or goes is immaterial to me.* **2** airy, incorporeal, °disembodied, ethereal, ephemeral, evanescent, unsubstantial: *The concept of immaterial matter, like a gas, was difficult to comprehend.*

immature *adj.* **1** °premature, °undeveloped, unripe, °rudimentary, half-grown, unformed, °unfledged, fledgling, unfinished, °young, °new, °fresh, °incomplete: *The immature growth in the spring is not hardy enough to survive a frost.* **2** °green, °callow, °unsophisticated, °naive, jejune, °inexperienced, babyish, °childish, °childlike, °puerile, °juvenile, °raw, *Colloq* wet behind the ears: *Granby is a bit immature to assume so many responsibilities.*

immeasurable *adj.* °vast, °infinite, °immense, °huge, °great, °limitless, °boundless, °endless, interminable, unbounded, °unlimited, measureless, °inestimable, measureless, unfathomable; innumerable, °numberless, uncountable, uncounted, incalculable: *The tiny ship hurtled through the immeasurable vastness of space. An immeasurable number of angels can dance on the head of a pin. He accomplished immeasurable good during his lifetime.*

immediate *adj.* **1** instantaneous, °instant, °abrupt, °sudden, °swift; °spontaneous, °instinctive, triggered, °unhesitating, °unthinking, °automatic, reflex, knee-jerk: *Why is your immediate reaction always to say, "It cannot be done"?* **2** °direct, nearest, next, closest, adjacent, proximate; °nearby: *Although he had cancer, the immediate cause of death was poison. Is there a taxi in the immediate vicinity?* **3** existing, °present, °current, °actual, °pressing, °urgent: *The immediate problem is to find a way of getting out of here.*

immediately *adv.* **1** °at once, instantly, °instantaneously, °promptly, right away, right now, without delay, unhesitatingly, without hesitation, forthwith, this instant, °directly, in a wink, in a second, in a minute, *tout de suite,* instanter, *Chiefly Brit* straightaway *or* straight away, *Colloq* pronto, in a jiffy, in two shakes of a lamb's tail, before one can *or* could say "Jack Robinson," at the drop of a hat: *Come here immediately. I shall be there immediately.* **2** °directly, closely, intimately: *These reactions are immediately concerned with the temperature.* —*conj.* **3** when, as soon as, the moment (that), *Brit* °directly: *Immediately he learned that he was wanted by the police, he left the country.*

immense *adj.* °enormous, °gigantic, °extensive, °vast, °huge, °massive, °voluminous, tremendous, staggering, stupendous, mammoth, °colossal, °giant, titanic, cyclopean, °jumbo, elephantine, Brobdingnagian, *Slang US* humongous: *Our exit from the cave was blocked by an immense boulder.*

immerse *v.* **1** °plunge, °sink, °submerge, °dip, dunk, °duck, inundate: *Immerse the red-hot blade in water to temper it.* **2** °plunge, °sink, °submerge, °bury, absorb, engross, °engage, °occupy, °involve: *I immersed myself in my work and never noticed the time.*

immigrant *n.* °newcomer, °arrival, °settler, *Australian* °migrant; °alien, °foreigner, outlander, °outsider: *Her parents were immigrants from the Ukraine.*

imminent *adj.* °impending, looming, °threatening, °menacing, °at hand, nigh, °immediate, close (by *or* at hand), (forth)coming, drawing near *or* close *or* nigh, °momentary: *An imminent disaster was averted by defusing the bomb.*

immoderate *adj.* °excessive, °extreme, °exorbitant, °unreasonable, °inordinate, °extravagant, intemperate; °outrageous, °preposterous, exaggerated, unrestrained, undue: *Their immoderate demands for reparations cannot be met.*

immodest *adj.* **1** °indecent, °shameless, °shameful, indecorous, titillating, revealing; indelicate, °improper, °wanton, °loose, unrestrained, °provocative, °obscene, °lewd, smutty, °dirty, °lascivious, °bawdy, °coarse, *Colloq* °sexy: *The striptease was, to say the least, immodest. The streakers will be punished for their immodest behavior.* **2** °brazen, °forward, °bold, °impudent, °impertinent, °brash, °arrogant, °insolent, °presumptuous, °disrespectful, *Colloq* °fresh, °cheeky: *Carl's immodest demand for a private office was ignored.*

immoral *adj.* **1** °corrupt, °bad, °wicked, °evil, iniquitous, °sinful, °impure, unethical, unprincipled, °abandoned, °base, °wrong, °vile, depraved, °dissolute, °degenerate, °reprobate, unregenerate, nefarious, flagitious, °villainous, treacherous, °unscrupulous, °dishonest: *Is it immoral to avoid paying one's taxes?* **2** °immodest, debauched, °indecent, °wanton, °libertine, °lecherous, °lustful, libidinous, °carnal, concupiscent, salacious, licentious, °lascivious, °lewd, °obscene, °pornographic, °dirty, smutty, °filthy: *He was accused of living off immoral earnings.*

immortal *adj.* **1** undying, °eternal, °deathless, °everlasting, °eternal, imperishable, sempiternal, never-ending, °endless, ceaseless, °perpetual, °timeless, °constant, °permanent, °indestructible: *Did the Egyptian pharaohs believe themselves immortal?* **2** °divine, °heavenly, °godlike: *The immortal wisdom of God guides man.* **3** remembered, °celebrated, unfading, °famous, °renowned, °classic, lauded, praised, honored, °timeless: *The immortal writings of Shakespeare inspire us still.* —*n.* **4** Olympian, god *or* goddess; hero *or* heroine, °legend, °genius, °great: *To be sure, we must consider Bach as one of the immortals.*

immortalize *v.* °celebrate, °honor, °glorify, °memorialize, °commemorate, apotheosize, canonize, beatify, °exalt, ennoble, °extol: *Dickens's father was immortalized as Mr. Micawber.*

immovable *adj.* **1** unmovable, °fixed, °fast, °rooted, °set, immobile, stationary, motionless, °stable, riveted, anchored, frozen: *What happens when an irresistible force meets an immovable object?* **2** immutable, unchangeable, unalterable, settled, °set, unmovable, °fixed, °inflexible; unshakable *or* unshakeable, unswerving, °firm, °determined, °steadfast, °staunch, °rigid, dogged, obdurate, unyielding, unwavering, °resolute, unflinching, adamant(ine), °stony, °impassive, unbending, °impassive, emotionless, °unmoved: *Christmas day, which always falls on December 25th, is an immovable feast. Despite pleas for mercy, the judge was immovable and she was hanged the next day.*

immune *adj.* inoculated, vaccinated; °exempt, °safe, protected, insusceptible *or* unsusceptible, invulnerable, untouched, °unaffected: *She never caught smallpox because a mild case of cowpox had made her immune. Rick mistakenly thought himself immune to Sally's seductive charms.*

immunity *n.* **1** °exemption, nonliability, invulnerability, °protection, °excuse, °release, °exclusion, °privilege, °freedom, °indemnity, amnesty, exoneration, absolution: *In return for his testimony he was granted immunity from prosecution.* **2** insusceptibility *or* unsusceptibility, °protection, inoculation, vaccination: *Immunity to a number of diseases is conferred by this one injection.*

imp *n.* devil(kin), °demon, sprite, evil spirit, hobgoblin, °goblin, elf, pixie *or* pixy, leprechaun, puck, brownie, fairy; scamp, urchin, gamin, °rogue, °rascal, mischief-maker, brat: *Sitting on my shoulder as I write is a tiny imp that inserts the misspellings. Give me a hug, you little imp!*

impact *n.* **1** °collision, °contact, °striking, °crash, smash, °bump, colliding, crashing, smashing, bumping: *The impact of the car against the stone wall was heard a mile away.* **2** °effect, °impression, °influence, °import, °meaning, °bearing, °force, °thrust, °weight, °burden, °brunt, repercussions, results, consequences: *It is becoming clear that technology is having an adverse impact on the environment.* —*v.* **3** °strike, °hit, °collide with: *When its orbit decays, the satellite will impact the surface of the moon.* **4** °affect, °modify, °change: *I wonder how the news of the bankruptcy will impact share prices.*

impair *v.* °weaken, °cripple, °damage, °harm, °mar, °injure, °spoil, °ruin: *"I understand that leading a wild life can impair your hearing." "What did you say?"*

impairment *n.* lessening, weakening, °damage, °harm, °injury, °flaw, °imperfection, reduction, vitiation, deterioration, °decrease, diminution, enfeeblement, debilitation, undermining, worsening, marring: *The impairment to my eyesight was caused by reading in the dark.*

impale *v.* spear, °stab, °pierce, skewer, °spit, °stick, °transfix, °spike: *They impaled the heads of their enemies on poles as a warning.*

impart *v.* **1** °give, °cede, °lend, °bestow, convey, °confer, °grant, °afford, °accord, °contribute: *The dyes impart different colors to the ink.* **2** °communicate, °tell, °relate, °transmit, °reveal, divulge, °disclose, °pass on, °intimate, confide: *His mentor imparted much wisdom to him.*

impartial *adj.* °fair, °just, evenhanded, °disinterested, °neutral, °unprejudiced, unbiased, °objective, °equitable: *Judge Leaver can be relied on to render an impartial verdict.*

impasse *n.* °deadlock, dead end, °stalemate, standoff, °block, blockage, *Colloq* blind alley: *Negotiations reached an impasse and the meeting broke up.*

impassioned *adj.* °passionate, inspired, °spirited, °stirring, °fervent, °emotional, fervid, °ardent, °heated,

°warm, °rousing, aroused, vehement, zealous, °eager, °earnest, °enthusiastic, °vigorous, °animated, °fiery, °inflamed, °glowing: *The barrister made an impassioned plea for the accused.*

impassive *adj.* °cool, apathetic, °calm, °serene, composed, °unmoved, °cold, °coldblooded, °reserved, imperturbable, unimpressionable, unruffled, controlled, contained, °phlegmatic, °lackadaisical, °stoical, unemotional, °taciturn, unfeeling, °stolid, emotionless, uncaring, °indifferent, undisturbed, °callous, °unsympathetic, °stony, °dispassionate, °detached, °nonchalant, unconcerned, insouciant, °remote: *I could see nothing in the judge's impassive expression to hint at his decision.*

impatient *adj.* **1** uneasy, °nervous, fidgety, °agitated, °restless, °restive, unquiet, °eager, °fretful, °agog, chafing, °impetuous, athirst, *Slang* itchy, *US* antsy: *There is a short wait for seats, so please don't get impatient.* **2** °irritable, irascible, °testy, °short-tempered, °querulous, °waspish, °brusque, °curt, °short, hot-tempered, °snappish, °indignant, °demanding: *I don't mean to be impatient, but I have been waiting for three hours.*

impeach *v.* **1** °charge, °accuse, arraign, °indict, °incriminate, °implicate, inculpate, °blame, censure: *They were impeached for crimes against the state.* **2** °(call into) question, °challenge, °attack, °disparage, °discredit, impugn, deprecate, °belittle, asperse, cast aspersions on, declaim, °slander, malign, °vilify: *My daughter's character has been impeached by the witness.*

impeccable *adj.* °faultless, °flawless, °perfect, °ideal, °pure, °correct, °proper, °spotless, °immaculate, unblemished, °spotless, unimpeachable, °blameless: *Nothing that they say will mar his impeccable reputation.*

impede *v.* °bar, °obstruct, °block, °thwart, °check, °hinder, °hamper, °slow, °retard, °restrain, °brake, °hold up, °delay, °foil, confound, °inhibit, °curb, °spike, °stop: *You can do nothing to impede the relentless march of time.*

impediment *n.* °bar, °barrier, °obstruction, °block, °check, °hindrance, °encumbrance, °restraint, °holdup, °hitch, °snag, °restriction, °stricture, bottleneck, °delay, hang-up, °inhibition, °curb: *What are the impediments to a happy marriage?*

impending *adj.* °imminent, approaching, (close *or* near) at hand, °close, nearing, °forthcoming, brewing, to come, in view, °in prospect, in store, in the offing, on the horizon, in the air; looming, °threatening, °menacing; *Colloq Brit* on the cards, *US* °in the cards: *The impending deadline made us work all the faster. We had a feeling of impending doom.*

imperative *adj.* **1** °mandatory, compulsory, °necessary, required, requisite, demanded, °obligatory, °indispensable, °essential, °crucial, °vital, °urgent, °pressing, exigent: *It is imperative that we catch the 5:04 to Boston.* **2** imperious, commanding, °authoritarian, °overbearing, °peremptory, autocratic, °domineering, magisterial, lordly, °arbitrary, °dictatorial, °dogmatic, °tyrannical, °despotic, *Colloq* °bossy: *How dare you use such an imperative tone with me!*

imperceptible *adj.* **1** °invisible, indiscernible, °indistinguishable, undetectable, °obscure, °vague, ill-defined; °inaudible: *The differences between the twins are imperceptible.* **2** °indistinct, unclear, unnoticeable, °slight, °subtle, inconsiderable, inappreciable, °minute, °tiny, °slight, minuscule, infinitesimal, microscopic: *Her only response was an imperceptible flicker of her eyelid.*

imperfect *adj.* °wanting, unfinished, °undeveloped, °incomplete, °deficient, °defective, °faulty, °flawed, patchy: *This is an imperfect translation of the poem.*

imperfection *n.* °flaw, °failing, °fault, °error, °defect, °blemish, °damage; inadequacy, insufficiency, deficiency, °frailty, °weakness, °foible, °shortcoming, °peccadillo, shortfall, fallibility, °infirmity: *We have reduced the price because of tiny imperfections in the weave. Imperfections in her character make her unsuitable for the job.*

imperial *adj.* **1** kingly, kinglike, queenly, queenlike, °princely, princelike, °regal, °royal, °sovereign: *The imperial court was rich and splendid.* **2** °majestic, °royal, °regal, °lofty, °exalted, °supreme, august, °noble, °superior, °imposing, °splendid, °magnificent, °grand, °excellent: *Her imperial presence outshone all at the ball.*

impermeable *adj.* impenetrable, impassable, impervious, closed, sealed, °hermetic: *Certain classes of colloids pass through impermeable membranes.*

impersonal *adj.* **1** °detached, °objective, °disinterested, °fair, °equitable, °dispassionate, °unprejudiced, unbiased: *The enforcement of the law should be entirely impersonal.* **2** °formal, °stiff, °strait-laced, °wooden, °rigid, prim, °stuffy, °cool, °detached, unfriendly, °cold, °mechanical: *Why does my bank manager have to be so impersonal?*

impertinence *n.* insolence, boldness, brazenness, °impudence, °presumption, presumptuousness, brashness, sauciness, pertness, °incivility, forwardness, impoliteness, discourtesy, °disrespect, audacity, rudeness, °effrontery, *Colloq* cheek, °brass, brassiness, °nerve, °gall, *Slang* chutzpah: *Why, the impertinence of that waitress to call you by your Christian name!*

impertinent *adj.* °presumptuous, °insolent, °bold, °brazen, °impudent, °brash, saucy, °pert, uncivil, °forward, °impolite, °discourteous, °disrespectful, °audacious, °rude, *Colloq* °cheeky, °fresh, °brassy, *US* nervy: *That impertinent little upstart had the nerve to ask the lady's age!*

impetuous *adj.* °spontaneous, °unpremeditated, °impulsive, unplanned, °hasty, °abrupt, °precipitate, °quick, °unthinking, unreasoned, °offhand, °rash, °reckless, spur-of-the-moment, unreflective, headlong: *Don't be impetuous and accept the first offer that comes along.*

impetus *n.* °drive, stimulus, °push, impulse, goad, °thrust, °energy, °momentum, stimulation, °incentive, motivation, °encouragement, °inspiration: *Roger's innovative ideas provided the impetus to develop new products.*

impious *adj.* irreligious, °irreverent, °ungodly, °sacrilegious, °blasphemous, °profane, unholy, °wicked, °sinful, iniquitous: *The priest said that he would be punished for his impious remarks.*

implacable *adj.* unappeasable, unmollifiable, unpacifiable, unforgiving, intractable, uncompromising, °inflexible, inexorable, unyielding, unrelenting, °ruthless, °cruel, pitiless, °merciless, °hard, °rigid, °unsympathetic, uncompassionate: *Ever since the incident at school he has been my implacable enemy.*

implant *v.* **1** °introduce, °instill, °insinuate, °inject; °indoctrinate, inculcate, °teach, °impress, imprint: *They spent months implanting that notion in his mind.* **2** °graft, °root, embed, inlay: *You must implant the scion nearer the main stem.* —*n.* **3** °graft, scion, ingraft; °insert: *Are they really experimenting with brain implants to improve the memory?*

implausible *adj.* °improbable, °unlikely, °doubtful, dubious, °questionable, °unbelievable, °incredible, °far-fetched, unconvincing, °debatable, °unreasonable: *He told an implausible tale of having been raised by apes.*

implement *n.* 1 utensil, °tool, °instrument, °apparatus, °device, appliance, contrivance, °mechanism, °(piece of) equipment, *Colloq* °gadget, °contraption: *You need the proper implement for measuring inside diameters.* —*v.* 2 °carry out, °execute, °accomplish, °perform, °achieve, °(put into) effect, °bring about, °cause, °fulfill, °realize: *You will need our help to implement the plan.*

implicate *v.* 1 °involve, °include, °associate, embroil, ensnare, entrap, enmesh, °entangle: *Don't implicate me in your harebrained schemes!* 2 °incriminate, inculpate, °connect, °involve, °associate, °suspect, °concern: *Wasn't he implicated in that fraud case last year?*

implication *n.* 1 involvement, °connection, °inclusion, °association, entanglement: *My implication in that affair was minimal.* 2 °suggestion, °hint, insinuation, °innuendo, intimation: *I resent the implication that I had anything to do with the murder.* 3 °significance, purport, °drift, °meaning, denotation, conclusion, °inference, °import, connotation, °sense, °burden, °substance, °essence, °pith: *The implication of the article is that Basque is unrelated to English.*

implicit *adj.* 1 implied, °indirect, inferable, °understood, unspoken, undeclared, °tacit, °inherent, inferential, latent: *Certain provisions are implicit in every legal contract and need not be expressed.* 2 °absolute, unquestioning, unquestioned, °unqualified, °total, °sheer, °complete, °unmitigated, unalloyed, °undiluted, °unlimited, unconditional, unreserved, utter, °full, °wholehearted: *I have implicit faith in Nicole's judgment.*

imply *v.* 1 °suggest, °hint (at), °intimate, °insinuate: *Are you implying that I don't know what I'm talking about?* 2 connote, allude to, °refer to, advert to, °signify, °signal, betoken, °denote, °indicate, °mean, °express; °involve, °include, °evidence, °assume, °presume, °entail: *Silence sometimes implies consent. Discovery of the tools implies a more advanced culture.*

impolite *adj.* °discourteous, °ill-mannered, uncivil, °rude, °ungracious, ungentlemanly, unladylike, °pert, °disrespectful, saucy, °boorish, churlish, °crude, indecorous, indelicate, °unrefined, ill-bred, °vulgar, °coarse: *The waiter was so impolite that I refused him a tip.*

imponderable *adj.* unmeasurable, °inestimable, °inconceivable, °incomprehensible, °subtle: *There are too many imponderable questions to give you a definite answer right away.*

import *v.* 1 °introduce, °bring in: *We import the raw materials and export the finished product.* 2 convey, °mean, °signify, °denote, °imply, betoken: *He was a dictator, in all the senses that word imports.* —*n.* 3 °meaning, °sense, denotation, signification, °gist, °drift, °thrust, °intention, °implication, purport, connotation, °suggestion, allusion, intimation: *It was difficult for me to catch the import of her words.* 4 °importance, °significance, °weight, consequence, °moment, °substance: *One could see at once that she was a personage of some import.*

importance *n.* 1 °significance, consequence, °import, value, °worth, °weight, °account, °concern, °moment, °substance, °matter: *How chemicals affect the ozone layer is of great importance to everyone.* 2 eminence, °distinction, °esteem, °standing, °status, °position, °rank, °prominence, °preeminence, °prestige, °power, °influence, °note: *His family is of considerable importance in the community.*

important *adj.* 1 °significant, consequential, °critical, °material, °vital, °urgent, °portentous, °weighty, °grave, °substantial, °momentous, °signal: *If it is important, put it in writing.* 2 °leading, °prominent, °notable, °noted, °noteworthy, °worthy, °eminent, °distinguished, respected, high-ranking, top-level, high-level, °superior, °outstanding, °foremost, °conspicuous, °impressive: *She is the most important member of the committee.* 3 °influential, °effective, well-connected, °powerful, °formidable, °mighty, °impressive: *With those important politicians supporting him, he might win.*

impose *v.* 1 °inflict, °force, °foist: *Stop imposing your standards on others.* 2 °interrupt, °interfere, °intrude, interpose, °insinuate: *I hope I am not imposing?* 3 levy, °place, °put, °exact: *A tax has been imposed on entertainment.* 4 ***impose on** or **upon.*** (a) saddle, °burden: *He imposes on everyone with his demands for charity.* (b) °exploit, take advantage (of), °misuse: *I hope I am not imposing on your good will in what I ask.*

imposing *adj.* °grand, °magnificent, °impressive, °stately, august, °majestic, °effective, commanding: *In his uniform, the general was an imposing presence at the reception.*

imposition *n.* 1 inflicting, infliction; applying, °application, enforcing, enforcement, levy, levying, promulgating, promulgation, introducing, introduction, placing, °placement, laying on: *People resent the imposition of a curfew.* 2 °burden, onus, °weight; intrusion, °misuse: *A sales tax is a terrible imposition on the poor. Walking the dog is an imposition on my time.*

impossible *adj.* 1 °hopeless, °impracticable, °inconceivable, unimaginable, °unthinkable, unattainable, unsuitable, °out of the question, unachievable, unrealizable, unworkable, unresolvable, unsolvable: *Reading of the cipher is impossible without a key.* 2 °absurd, °ludicrous, °preposterous, °ridiculous, illogical, unworkable, °outrageous, °farcical, °outlandish, °crazy, °weird: *Inventors came up with the most impossible designs for airplanes.*

impostor *n.* imposter, impersonator, °pretender, deceiver, °cheat, °fraud, °swindler, trickster, confidence man, shark, charlatan, mountebank, °hypocrite, *Colloq* con man, phoney *or US also* phony, tricky Dick, flimflam man, *US* fourflusher: *It wasn't the bank manager but an impostor.*

impotence *n.* 1 °weakness, powerlessness, helplessness, °frailty, feebleness, enervation, debilitation: *The doctor said that his impotence was caused by exhaustion.* 2 impotency, inadequacy, inefficacy, ineffectualness, ineffectiveness, ineptness, °incompetence: *Georgiana was frustrated by her impotence in dealing with the situation.* 3 sterility, infertility: *Eating raw eggs and oysters does not cure impotence.*

impotent *adj.* 1 °weak, °powerless, °helpless, °frail, °feeble, enervated, debilitated, °infirm: *When it came to political influence, I was impotent.* 2 °inadequate, °ineffective, °ineffectual, °inept, °incompetent: *The attempted robbery was a halfhearted, impotent effort.* 3 °sterile, °barren, °infertile, infecund: *They never had any children because Dixon was impotent.*

impoverished *adj.* 1 °destitute, °poor, poverty-stricken, °penurious, beggared, °needy, necessitous, impecunious, in sore *or* bad straits, °straitened, in distress, badly off, bankrupt, °insolvent, ruined, *Colloq* (dead *or* flat *or* stony) broke, bad off, pinched, up against it, °on one's uppers, °short, *US* strapped, wiped out, *Slang Brit* skint: *After paying the bills and the mortgage, Donald awoke to find himself impoverished.* 2 stripped, °barren, °desolate, wasted, °empty, depleted, denuded, drained, °exhausted: *Two years of drought left the land impoverished.*

impracticable *adj.* 1 unworkable, infeasible *or* unfeasible, °impossible, unattainable, unachievable: *The new engine design proved impracticable because of its high fuel consumption.* 2 unsuitable, unfit, unusable,

°useless, °inapplicable: *As we have no electricity, air conditioners are impracticable.*

impractical *adj.* **1** °visionary, starry-eyed, °unrealistic, °romantic, °quixotic, °wild: *Rob's solutions are very creative but quite impractical.* **2** °useless, °ineffective, °ineffectual, unworkable, unavailing, °impracticable: *The idea looked good on paper but proved impractical when we tried it.*

imprecise *adj.* °inexact, °inaccurate, °wrong, inexplicit, °indefinite, ill-defined, °indistinct, °vague, °hazy, cloudy, blurred, °fuzzy, °woolly, °ambiguous: *The readings are imprecise because the needle wavers so much.*

impregnable *adj.* invulnerable, impenetrable, inviolable, unconquerable, °invincible, °unbeatable, °indomitable, °secure, °safe, °mighty, well-fortified: *The walls of the castle, sixteen feet thick, were impregnable.*

impregnate *v.* **1** °fertilize, inseminate, fecundate, make pregnant: *The female can be impregnated only during one ten-day interval each year.* **2** imbue, °suffuse, °permeate, °penetrate, pervade, infuse, °saturate, °drench, °soak, °steep, °fill: *The preservative must be allowed to impregnate the wood.*

impress *v.* **1** °affect, °touch, °move, °reach, °stir, °strike, °sway, °influence, °persuade, *Colloq* °grab, get under one's skin: *She was genuinely impressed by his sincerity.* **2** °print, imprint, °stamp, °mark, °engrave, emboss: *Before the bowl was placed in the kiln, the potter's seal was impressed in its bottom.* **3** Often, ***impress on*** *or* ***upon.*** °stress, °emphasize, °urge, °bring home (to): *They tried to impress on him the gravity of the situation.*

impression *n.* **1** °sensation, °feeling, °sense, °suspicion, awareness, consciousness, °idea, °belief, °fancy, °notion: *I have the impression that I have been here before. She gave me the impression that she didn't like me.* **2** °impact, °effect, °influence: *Are you concerned about the impression of his speech on the crowd?* **3** dent, °indentation, °depression, °hollow, °mark; °stamp, °impress, °brand: *Look here, where the chair has made an impression in the carpet.* **4** printing, °issue, °print, °copy, °run: *This is the third impression from the original plates.* **5** impersonation, °imitation; °parody, °satire, *Colloq* °takeoff, *Brit* sendup: *After dinner, Gerry did his impression of Churchill.*

impressionable *adj.* °suggestible, °susceptible, persuadable *or* persuasible, impressible, °receptive, °responsive: *They seek to enroll students who are at an impressionable age.*

impressive *adj.* evocative, °moving, affecting, °stimulating, °exciting, °stirring, °powerful, °provocative, arousing, awe-inspiring, °awesome, °imposing, °formidable, °portentous, redoubtable: *She has many impressive talents.*

imprison *v.* incarcerate, confine, detain, remand, °jail, °lock up, intern, °shut up, put behind bars, put in *or* throw into irons, °put away, *Colloq Brit* °send down *US* °send up (the river): *He was imprisoned for a crime he didn't commit.*

imprisonment *n.* incarceration, confinement, °detention, remand, °custody, *Literary* durance (vile): *The sentence is imprisonment for not less than a year.*

improbable *adj.* °doubtful, dubious, °unlikely, °questionable, °unrealistic, °far-fetched, °remote, °implausible, °unthinkable, hard to believe *or* imagine, °inconceivable, °unbelievable, °fanciful, °incredible, °ridiculous, °ludicrous, °absurd, °crazy, °mad, °insane, °wild, °weird, °peculiar, °strange; °impossible; *Colloq* °fishy: *It is improbable that they will marry. He gave the most improbable excuses.*

improper *adj.* **1** °wrong, °mistaken, °erroneous, °false, °incorrect, °inaccurate, °inexact, °imprecise, °amiss, °faulty, °untrue, °irregular, °abnormal: *Sloppy research has led to improper conclusions.* **2** unfit, unsuitable, °inappropriate, inapt, °inapplicable, infelicitous, °incongruous, unsuited, °unseemly, unbefitting, unfitting, uncalled-for, inapposite, malapropos, out of keeping, °out of place, °incompatible, °inopportune: *Do you think it improper for shops to open on Sundays? His wedding was an improper occasion for comments about his philandering.* **3** indecorous, °indecent, indelicate, °immodest, °unseemly, °untoward, °unbecoming, °impolite, °suggestive, °risqué, °off-color, °obscene, °corrupt, °immoral, °sinful, °wicked, °lewd, °lascivious: *Do you see anything improper in sex education for children?*

impropriety *n.* **1** improperness, erroneousness, incorrectness, °falsity, falseness, inaccuracy, inaccurateness, inexactitude, inexactness, imprecision, impreciseness, irregularity, °abnormality: *I fail to see the impropriety in refusing to support capital punishment.* **2** unfitness, unsuitableness, inappropriateness, inaptness, inapplicability, infelicity, infelicitousness, incongruity, incongruousness, unseemliness, incompatibility, inopportuneness: *The judge criticized the impropriety of allowing the child to testify.* **3** indecorousness, bad *or* poor taste, indecency, °indelicacy, immodesty, unseemliness, suggestiveness, immorality, sinfulness, wickedness, lewdness, lasciviousness: *In those days, the impropriety in wearing a brief bathing costume was thought shocking.* **4** °slip, °blunder, °mistake, °error, gaffe, gaucherie, faux pas: *She admitted to many improprieties during her term of office.*

improve *v.* **1** °better, ameliorate, upgrade, uplift, °amend, °enhance, °fix up, °reform, °redeem, °rehabilitate, redress, °repair, °correct, °rectify, put *or* set right, emend; °modernize, update, °refurbish, recondition, °renovate, °repair, °overhaul, remodel: *We must improve procedures for catching drug dealers. The landlord refuses to spend any money to improve the property.* **2** °develop, °increase, °advance, °promote, °further, °benefit; °look up, °recover, °pick up, *Colloq* give a new lease of *or US* on life, take a turn for the better, *US* take a new lease on life: *How can I improve my chances of winning at tennis? Business has improved considerably.* **3** °convalesce, °recuperate, °recover, get better, °mend, °rally, °(make) progress, gain (strength *or* ground), °revive: *Costello passed the crisis and is improving daily.*

improvement *n.* **1** betterment, amelioration; °reform, rehabilitation, upgrading, enhancement, °repair: *The committee is working toward the improvement of the neighborhood.* **2** °recovery, °rise, °increase, °gain, °advance, upswing, °progress; recuperation, convalescence: *There was an improvement in share prices after release of the trade figures. The patient shows considerable improvement.* **3** °advance, change for the better: *This year's crop is an improvement over last year's.*

improvident *adj.* **1** °shortsighted, °imprudent, °wasteful, °profligate, prodigal, °spendthrift, °extravagant, °lavish, °profuse, happy-go-lucky, penny-wise and pound-foolish, uneconomic(al), thriftless: *He who is improvident in his youth plans for a short life.* **2** incautious, °unwary, °rash, °reckless, °impulsive, °impetuous, headlong, °heedless, °careless, °unthinking, unthoughtful, unmindful: *The club's present financial straits are the result of having an improvident treasurer.*

improvise *n.* **1** ad-lib, extemporize, *Colloq* play (it) by ear, fake it, wing it: *I like a comedian who improvises as he goes along.* **2** °invent, concoct, °devise, contrive, jury-rig; °make do: *We had to improvise a mast from parts of the boom. When you have nothing to work with, you have to improvise.*

imprudent *adj.* °indiscreet, °impulsive, °rash, °reckless, °hasty, incautious, °impetuous, °improvident, °careless, °heedless, °foolhardy, °irresponsible, injudicious, ill-judged, °thoughtless, ill-considered, °ill-advised, inadvisable, unwise, inexpedient, °foolish, °mad, °crazy, °insane, °inane, °silly, °perverse, °wrong, wrong-headed: *It would be imprudent to go out during the storm.*

impudence *n.* °impertinence, °effrontery, insolence, °disrespect, °presumption, presumptuousness, audacity, shamelessness, pertness, sauciness, boldness, brazenness, °incivility, rudeness, impoliteness, *Colloq* lip, °gall, guff, °sauce, °mouth, *Brit* backchat, °side, *US* back talk, *Slang* chutzpah: *He had the impudence to suggest that I was too old for the job!*

impudent *adj.* °shameless, °impertinent, °insolent, °disrespectful, °forward, °presumptuous, °audacious, °pert, saucy, °bold, °brazen, °cocky, cocksure, °arrogant, uncivil, °ill-mannered, °rude, °impolite, *Colloq* °fresh, °brassy, °cheeky: *Asked the simplest question, she is always ready with an impudent response.*

impulsive *n.* °impetuous, °emotional, unpredictable, °unpremeditated, unplanned, spur-of-the-moment, °extemporaneous, unconsidered, °offhand, °instinctive, °involuntary, °spontaneous, °quick, °sudden, °precipitate, °immediate, °snap, °rash, headlong, °rash, °reckless, devil-may-care, °foolhardy, madcap, °wild: *It was impulsive of me to give her all that money, but I love her.*

impure *adj.* **1** °dirty, soiled, unclean, sullied, tainted, polluted, defiled, °foul, °filthy, feculent, infected, scummy, °putrid, putrescent, °rotten: *The water is impure and not fit to drink.* **2** °mixed, alloyed, contaminated, adulterated, debased, °unrefined: *Impure ore contains other substances which are removed during processing.* **3** unclean, unhallowed, forbidden, disallowed, *Judaism* tref: *Certain foods are considered impure in some religions.* **4** °unchaste, °immoral, °sinful, °wicked, °evil, °vile, unvirtuous, unvirginal, corrupted, defiled, debased, vitiated, °degenerate, depraved, °loose, °wanton, °lustful, °promiscuous, libidinous, °dissolute, licentious, °obscene, °prurient, °dirty, °filthy, lubricous *or* lubricious, salacious, °lascivious, °lewd, °lecherous: *The book was condemned for fostering "impure thoughts."*

impurity *n.* **1** °pollution, contamination, defilement, adulteration, dirtiness, uncleanness, foulness: *The impurity of the atmosphere is a cause for concern.* **2** Often, ***impurities.*** °dirt, contaminant, pollutant, °pollution, smut, °filth, foreign matter *or* body: *This filter is guaranteed to remove impurities from the water.* **3** unchastity, immorality, sinfulness, wickedness, °evil, vileness, corruption, degeneration, depravity, looseness, wantonness, °lust, lustfulness, promiscuity, promiscuousness, libidinousness, dissoluteness, licentiousness, obscenity, prurience, dirtiness, filthiness, lubricity, salaciousness, lasciviousness, lewdness, lecherousness: *Children ought not be exposed to the impurity encountered on everyday television.*

imputation *n.* °charge, indictment, °accusation, °allegation, °challenge, censure, °blame, reproach, °slur, °aspersion, °attribution, ascription, insinuation, °implication, °innuendo: *I resent the imputation that I broke the vase on purpose.*

impute *v.* ascribe, °assign, °attribute, °credit, °charge, put *or* set down to; °insinuate, °imply, °suggest, hint at: *The critic imputed meanings to the book that the author had never intended.*

inability *n.* incapacity, incapability, °incompetence, unfitness, ineptness, °ineptitude, unqualifiedness, °impotence: *The poor marks result from the students' inability to perform simple arithmetical tasks.*

inaccessible *adj.* **1** unavailable, unobtainable, unattainable, unreachable, °unapproachable: *The minister has made himself inaccessible for interviews.* **2** impenetrable, impassable, °out-of-the-way: *The tomb lies in a part of the jungle inaccessible except by air.*

inaccurate *adj.* °wrong, °incorrect, °erroneous, °mistaken, °inexact, °imprecise, °faulty, °flawed, °imperfect, °amiss, awry, °false, fallacious, illogical, °unsound, *Colloq* off the mark, off the beam, °cold, *Chiefly US* all wet, *US* (way) off base, *Slang* cockeyed, full of hot air: *Reports of my death were obviously inaccurate.*

inactive *adj.* **1** °passive, placid, °quiet, °tranquil, quiescent, °serene, °peaceful, pacific, resting, unmoving, motionless, immobile, immobilized, °still, °inert, °lifeless, °inanimate, sluggish, °listless, °lethargic, °lackadaisical, languid, °indolent, °lazy, °torpid, somnolent, °idle, °slothful, °supine: *Although there was plenty to do round the house, he remained inactive all summer long.* **2** nonfunctioning, inoperative, °dormant; unoccupied, °idle, °unemployed, jobless, °out of work, out of a job: *The volcano was inactive for a thousand years. These miners have been inactive for six months.*

inactivity *n.* **1** passiveness *or* passivity, placidity *or* placidness, °quiet, tranquillity, °serenity, °peace, peacefulness; motionlessness, immobility, stillness, °inertia, inertness, lifelessness, °sluggishness, listlessness, °lethargy, languidness, °indolence, laziness, °torpor, somnolence, °idleness, slothfulness, °sloth: *She alternates between periods of activity and inactivity.* **2** inaction, quiescence; hibernation, estivation *or esp. Brit* aestivation: *During repairs, the reactor will be in a state of inactivity. Some animals go into a period of inactivity at certain times of the year.*

inadequate *adj.* **1** °insufficient, °deficient, not enough, too little, °scarce, °meager, °scanty, skimpy, °sparse, °(in) short (supply); °unsatisfactory, °imperfect, °incomplete, °defective, unsuitable, °disappointing, °flawed, °faulty: *Funds are inadequate for the payment of the bills. The terms of the agreement are inadequate.* **2** ***inadequate to.*** unsuited to *or* for, unfit for, unqualified for, not up to, unequal to, °unworthy of, inapt for, inept for, °incapable of: *He proved inadequate to the demands of the job.*

inadmissible *adj.* disallowed, unallowable, unallowed, forbidden, °unacceptable, prohibited, unsuitable, unsuited, °unqualified, °inappropriate, °inapplicable, °improper, objectionable, °exceptionable, °incorrect, °wrong: *Such evidence is inadmissible in court.*

inadvertent *adj.* **1** unintentional, unintended, °unpremeditated, °accidental, °unthinking, unwitting, °chance; °unconscious, unplanned, unstudied, undesigned, uncalculated: *She was an inadvertent witness to the murder. I assure you that the error was entirely inadvertent.* **2** °careless, °inattentive, negligent, °heedless, unobservant: *Inadvertent commentators failed to see the significance of her words.*

inalienable *adj.* untransferable, intransferable, °absolute, °inherent, unconsignable, sacrosanct, inviolable, nonnegotiable, unnegotiable, unchallengeable, *Law* imprescriptible, entailed, indefeasible: *People have an inalienable right to life, liberty, and the pursuit of happiness.*

inane *adj.* °silly, asinine, °vapid, °vacant, vacuous, °absurd, fatuous, °foolish, °senseless, °nonsensical, °unreasonable, °preposterous, °ludicrous, °ridiculous, laughable, risible, °mad, lunatic, °crazy, °stupid, °dumb, idiotic, moronic, imbecilic, *Colloq* nutty, nuts, °daft, daffy, screwy, batty, dippy, wacky, cuckoo, loony, goofy, *Brit* bonkers, dotty: *Students seem especially prone to inane behavior.*

inanimate *adj.* °lifeless, motionless, immobile, unmoving, °inactive, °inert, °still, spiritless, soulless, °cold, °dead, °defunct: *He just sits there, like an inanimate piece of furniture.*

inapplicable *adj.* °inappropriate, unsuitable, unsuited, inapt, °irrelevant, °unrelated, unconnected, inapposite, beside the point *or* question, °extraneous, off the mark, *Colloq US* off base: *That law is inapplicable in your case.*

inappropriate *adj.* °improper, unfitting, unfit, unsuitable, unsuited, unbefitting, °incompatible, malapropos, ungermane, inapt, inapposite, out of keeping, °incongruous, infelicitous, °inopportune, untimely, °irrelevant, °inapplicable: *Trying to be funny is inappropriate at a time like this.*

inarticulate *adj.* **1** °disjointed, unconnected, °incoherent, °incomprehensible, jumbled, unintelligible, unclear, illogical, °discursive, °rambling, scrambled, °wild, irrational, muddled, mixed-up, °confused, digressive: *She sat through an inarticulate lecture on Joyce's use of the subjunctive.* **2** mumbled, garbled, blurred, muffled, muttered, faltering, °halting, °indistinct, unclear, unintelligible: *Inarticulate radio messages can be filtered to make them understandable.* **3** °speechless, °tongue-tied, °(struck) dumb, °mute, voiceless: *Ed is inarticulate when it comes to describing Vera.*

inattentive *n.* unobservant, °heedless, °careless, negligent, neglectful, °indifferent, uncaring, apathetic, °slack, °remiss, unconcerned, °detached, unmindful, distracted, distrait(e), °absent-minded, abstracted, in a brown study, daydreaming, oscitant, woolgathering, musing, °oblivious, (with one's head) in the clouds, in a world of one's own: *It was late and the children were getting restless and inattentive.*

inaudible *adj.* unheard, °imperceptible, °indistinct, °low, °faint, muted, °quiet, °soft, muffled, stifled: *The film was good, although some of the dialogue was inaudible.*

inaugurate *v.* **1** °initiate, °begin, °commence, enter upon, °start, °introduce, usher in, °institute, °launch, °originate, °set up, get under way, get going: *A new program of health services has been inaugurated.* **2** °install, °induct, °invest, °establish, instate: *A new president will be inaugurated in January.*

inauspicious *adj.* °unpromising, unlucky, °unfortunate, unfavorable, unpropitious, °untoward, ill-starred, °ominous, ill-omened, ill-fated, °portentous, °menacing, °doomed, °sinister, °dark, °gloomy, cloudy, clouded, °black: *The middle of a recession is an inauspicious time to set up business.*

inborn *adj.* innate, congenital, °inherent, inherited, °hereditary, inbred, °natural, °native, constitutional, deep-seated, deep-rooted, °ingrained, instinctive *or* instinctual, *Technical* connate: *She has an inborn aversion to turnips.*

incandescent *adj.* °glowing, red-hot, white-hot, alight, aflame, °flaming, °burning, °fiery, candent, flaring: *The iron was heated till it became incandescent, lighting up the room.*

incapable *adj.* **1** Often, ***incapable of.*** °unable (to), °powerless (to), °incompetent (of), unfit (to), °unqualified (to), °impotent (to), unequal to, not up to: *Pelton seems incapable of completing his assignment.* **2** ***incapable of.*** insusceptible to, °resistant to, impervious to, ill-disposed to, disinclined to, not open to: *He is incapable of dishonesty.*

incapacitate *v.* disable, °cripple, °paralyze, °lame, °wound, °maim, °impair, °weaken, enfeeble, °enervate, °exhaust, devitalize; immobilize, inactivate, deactivate, put out of action, indispose: *Severely incapacitated in the crash, John cannot walk without the aid of crutches. The power plant was incapacitated by one well-placed grenade.*

incentive *n.* °incitement, °impetus, °encouragement, goad, °prod, °provocation, °spur, impulse, °enticement, °lure, °inducement, stimulus, motivation, *Colloq* carrot: *The discount is offered as an incentive to buy a new car.*

incidence *n.* frequency, °rate, °degree, °extent, °occurrence, °prevalence; °quantity, °amount, °number: *The incidence of heart disease is high in Britain.*

incident *n.* **1** °event, °occasion, °occurrence, °proceeding, °circumstance, °fact, °happening, °experience, °episode: *A curious incident led to the discovery.* **2** °disturbance, °scene, °affair, °upset, commotion, °fracas, °skirmish, set-to, *Colloq* to-do, do: *The police were called because of an incident at Maloney's Bar last night.*

incidental *adj.* **1** °casual, °chance, fortuitous, aleatory, °random, °haphazard, serendipitous, unpredictable, °accidental, adventitious, unplanned, unlooked-for, *Colloq* fluky: *Any good that came from the war was incidental.* **2** °subordinate, °secondary, ancillary, °minor, lesser, °nonessential, unimportant, trivial, °negligible, °inconsequential, °insignificant, °petty, °trifling, °paltry: *The incidental events were not covered by the press.*

incidentally *adv.* **1** °by the way, by the bye, apropos (of), parenthetically: *Incidentally, I won't be able to go the party tonight.* **2** casually, as luck would have it, accidentally, °by chance, perchance: *In addition to being held for murder, he was incidentally charged with possession of an illegal weapon.*

incision *n.* °cut, °gash, °slit, °slash: *The surgeon made an incision in the abdominal wall.*

incisive *adj.* **1** °keen, °sharp, °acute, °piercing, perspicacious, °perceptive, percipient, °penetrating, °trenchant, canny, °shrewd: *Shirley offered some incisive observations on the matter.* **2** °sarcastic, °biting, mordant, °cutting, °caustic, °sardonic, ironic(al), °sharp, acid, °tart, acrid, acrimonious, °bitter, acerbic, cynical, stinging, °critical: *Your incisive comments are not really appreciated, George.*

incite *v.* °stimulate, °inspire, °prompt, °move, °stir, stir *or* whip *or* work up, bestir, °excite, °fire, exhort, °agitate, °foment, °inflame, °provoke, °rally, goad, °spur, °prick, °prod, °drive, °push, egg on, °encourage, °urge, °influence, °wake, waken, °awaken, °rouse, °arouse: *Who was it that incited the crowd to violence?*

incitement *n.* **1** stimulation, instigating, spurring, urging, influencing, awakening, wakening, arousing, prodding, prompting, °stirring, °whipping, °exciting, firing, exhorting, agitating, fomenting, inflaming, firing, provoking, rallying, goading, prodding, pricking, needling: *Certain rabble-rousers have been accused of incitement of the crowds at the football stadiums.* **2** stimulus, °incentive, °inducement, °enticement, °temptation, impulse, motivation, °influence, instigation, °provocation, °inspiration, °persuasion, exhortation, °agitation, fomentation, °inflammation, arousal, °encouragement, °excitement: *Greed is a powerful incitement to dishonesty.*

incivility *n.* rudeness, boorishness, coarseness, discourtesy, uncourtliness, unmannerliness, indecorum, indecorousness, discourteousness, impoliteness, tactlessness, ungentlemanliness, bad breeding, ill breeding, bad manners, °misbehavior: *The incivility with which the ambassador was treated is unforgivable.*

inclement *adj.* °extreme, intemperate, °severe, °harsh, rigorous; °stormy, °violent, rainy, squally, blustery, °raw, °bad, °tempestuous: *We are expecting inclement weather for the weekend.*

inclination *n.* **1** °bow, bowing, °bend, bending, °nod, °nodding, °tilt, tilting: *His tacit sign of recognition was an inclination of his head.* **2** °slope, °slant, °angle, °bend, °incline, °tilt: *The degree of inclination of the rails is far too steep for this engine.* **3** °disposition, predisposition, °tendency, °bent, °bias, °leaning, °preference, °turn, °cast, proclivity, propensity, °attitude, proneness, susceptibility, predilection, °partiality, °affection, °taste, °liking, °desire, velleity: *His inclination in favor of the offer is largely instinctive.* **4** °desire, °longing, craving, °appetite, °taste, °stomach, °sympathy, predilection, °penchant, °fancy, °eagerness, °enthusiasm, zeal, °fervor, °ardor: *I have no inclination to watch open-heart surgery.*

incline *v.* **1** °tend, °lean, °bend, °bow, °slant, °tilt, °angle, bank, °slope, ascend, °rise, °descend: *The road inclines to the horizontal after a curve.* **2** °dispose, °influence, °persuade, predispose, °make, °lead, °prejudice, °bias: *Their accents might incline you to think of them as foreigners.* **3** °tend, °lean, gravitate, show favor *or* preference, be attracted to, be biased *or* prejudiced, have a mind; be disposed *or* predisposed, °lean, °tend: *I incline toward the other candidate. He inclines toward corpulence.* —*n.* **4** °slope, °pitch, °grade, gradient, °slant, °ramp, °hill, °dip, descent, declivity, °rise, ascent, acclivity: *The car picked up speed going down one incline and slowed going up the next.*

inclined *adj.* **1** tending, °disposed, predisposed, °prone, °willing, °keen, °eager, *Colloq* of a mind: *Though she is inclined to believe, she has doubts.* **2** °likely, apt, °liable, minded, °prone: *He is equally inclined to kiss you as kill you.* **3** sloping, slanting, °leaning, bending, tilting, gravitating, °bearing, verging: *This wall is slightly inclined toward the river.*

include *v.* **1** °incorporate, °embody, comprise, °embrace, °cover, encompass, °take in, subsume, °comprehend, °contain: *Staff changes are included in the plan for reorganization.* **2** classify, °categorize, °group, °file, °list, catalog, °tabulate, °register: *Did you include psychology among the social sciences?* **3** °involve, °count, °number, allow for: *Please don't forget to include me in the invitation list.*

inclusion *n.* incorporation, involvement, counting, numbering, grouping, classification: *The inclusion of her name among the culprits was an error.*

inclusive *adj.* **1** °comprehensive, °general, all-encompassing, °wide, °broad, °extensive, °full, °umbrella, blanket, across-the-board, all-in-one, unified: *We have an inclusive insurance policy that covers the entire family.* **2** ***inclusive of.*** including, embracing, comprising, taking in, covering, incorporating, embodying: *Is this bill inclusive of service?*

incognito *adj.* **1** °unknown, disguised, concealed, °unidentified, in disguise, unrecognizable, unrecognized: *I didn't tell them who I was, as I wished to remain incognito.* —*adv.* **2** unrecognizably, in disguise, °secretly, °on the sly, under cover, clandestinely: *Why travel incognito if no one knows or cares who you really are?*

incoherent *adj.* °confused, garbled, mixed up, disordered, jumbled, muddled, scrambled, °rambling, illogical, irrational, °wild, unstructured, disjoined, °disconnected, °disjointed, °loose, unconnected, uncoordinated, unintelligible, °inarticulate: *He arrived late with an incoherent excuse involving measles, his car, and a policeman.*

incombustible *adj.* °nonflammable, noninflammable, fireproof, noncombustible; flameproof: *She has to wear incombustible clothing at her job.*

income *n.* °return, revenue(s), °receipts, °proceeds, profit(s), gain(s), takings: *Most of his income is from laundromats.*

incoming *adj.* **1** arriving, entering: *Incoming passengers must pass through customs.* **2** °new, entering: *The incoming president must contend with next year's budget.*

incomparable *adj.* beyond compare, unequaled, °matchless, °peerless, inimitable, °unparalleled, unrivaled, °nonpareil, °transcendent, °surpassing, °supreme, °superior, °superlative, unsurpassed, unsurpassable: *Her performance as Lady Macbeth was incomparable.*

incompatible *adj.* °mismatched, unsuited, °discordant, clashing, jarring, °inconsistent, °contradictory, conflicting, uncongenial, irreconcilable, °incongruous; antithetic, °opposed, °opposite, °contrary, antipathetic, antagonistic, °hostile: *The colors in the carpet are incompatible with those in the curtains. The brothers are totally incompatible and can never meet without arguing.*

incompetence *n.* °inability, incapacity, inadequacy, insufficiency, deficiency, °ineptitude, inefficiency, uselessness, faultiness: *The students' failure to perform well may be attributable to the incompetence of their teachers.*

incompetent *adj.* °unqualified, unfit, °unable, °incapable, unskilled, unskillful, °inept, maladroit, inexpert, °awkward, floundering, °clumsy, bungling, gauche, °useless, °inadequate, °insufficient, °ineffective, °ineffectual, °inefficient: *What do you expect if you give your watch for repair to some incompetent oaf?*

incomplete *adj.* unfinished, °undone, °imperfect, °undeveloped, °deficient, °defective, unaccomplished, °partial, °sketchy, °crude, °rough, °fragmentary, fragmented, °piecemeal: *An incomplete manuscript of a new novel was found in his desk.*

incomprehensible *adj.* unintelligible, unfathomable, impenetrable, °deep, abstruse, arcane, °recondite, indecipherable, undecipherable, inscrutable, °cryptic, °obscure, °opaque, °dark, °occult, °perplexing, °mysterious, mystifying, enigmatic, °puzzling, over (someone's) head, baffling, unimaginable, °inconceivable: *They spoke in Swahili, an incomprehensible language to me.*

inconceivable *adj.* °incredible, °unbelievable, °unthinkable, unimaginable, °incomprehensible, °unheard-of, undreamt-of, unthought-of, °impossible, °overwhelming, staggering, *Colloq* mind-boggling: *Putting a man on Mars before the year 2000 is inconceivable.*

inconclusive *adj.* °indecisive, °unresolved, °indefinite, °unsettled, °open, indeterminate, °in limbo, *Colloq* up in the air: *As the test results are inconclusive, we cannot be sure exactly what is wrong with you.*

incongruous *adj.* °inconsistent, inharmonious, disharmonious, °discordant, dissonant, disconsonant, °incoherent, °incompatible, incongruent, conflicting, °unbecoming, °unseemly, unsuited, unsuitable, unapt, °inappropriate, misallied, unfitting, unfit, °improper, malapropos, unmeet, °absurd, discrepant, disparate, °different, °divergent, disagreeing, °contrary, °contradictory, °paradoxical, out of step, out of keeping, out of line: *That jacket is incongruous with those trousers. It was incongruous to see two octogenarians dancing at the disco.*

inconsequential *adj.* unimportant, °insignificant, trivial, °trifling, nugatory, inconsiderable, inappreciable, °negligible, °minor, °paltry, °petty, °immaterial, °slight, lightweight, °worthless, *Colloq* piddling: *Why dwell on inconsequential details?*

inconsiderate *adj.* °thoughtless, unthoughtful, °unthinking, unconcerned, uncaring, unmindful, °heedless, unheeding, insensitive, °unsympathetic, °tactless, °intolerant, °rude, °ungracious: *How can you be so inconsiderate as to work late on my birthday?*

inconsistent *adj.* **1** See **incongruous. 2** °irregular, °capricious, °fickle, °erratic, °inconstant, uneven, unpredictable, °unreliable, undependable, °unstable, unsteady, °changeable, °variable: *Their behavior is so inconsistent that I can't tell what they'll do next.*

inconsolable *adj.* disconsolate, broken-hearted, °heartbroken, °desolate, °forlorn, despairing, °miserable, °wretched, grief-stricken: *They are inconsolable over the death of their parakeet.*

inconspicuous *adj.* unnoticeable, °unnoticed, °unobtrusive, unostentatious, °insignificant, °indefinite, °indistinguishable, °undistinguished; °modest, unassuming, °discreet: *Wearing a gray coat and hat, he was quite inconspicuous in the crowd. He played an inconspicuous role in the entire affair.*

inconstancy *n.* changeableness, fickleness, irregularity, mutability, variability, unsteadiness, unsteadfastness, capriciousness, volatility, mercurialness *or* mercuriality, inconsistency, unreliability; faithlessness, unfaithfulness: *How do you deal with inconstancy in one you thought a friend?*

inconstant *adj.* °changeable, mutable, °fickle, °capricious, mercurial, °volatile, vacillating, unsteady, unsteadfast, °irresolute, °unreliable, undependable, fluctuating, wavering, °erratic, °inconsistent, °flighty, °unstable, °unsettled, °fitful, °vague, °indefinite, °variable, °moody: *His supporters proved inconstant, and the caliph fell from power. The inconstant wind blew our craft hither and thither.*

incontinent *adj.* **1** unrestrained, unconstrained, unrestricted, °uncontrolled, uncontrollable, ungoverned, °ungovernable, unbridled, uncurbed: *She must learn to suppress her incontinent tongue.* **2** °lecherous, libidinous, °lascivious, °libertine, °lustful, °lewd, debauched, °wanton, °dissolute, °loose, lubricous *or* lubricious, salacious, °profligate, °obscene, °dirty, °filthy: *Members of the Hellfire Club were well known for their incontinent behavior.* **3** self-soiling, bed-wetting, *Medical* enuretic: *Some old people lose control and become incontinent.*

incontrovertible *adj.* irrefutable, °indisputable, indubitable, undeniable, incontestable, °unquestionable, °sure, °certain, °definite, °definitive, established, °absolute, °positive: *The fingerprint was incontrovertible evidence that he had been there.*

inconvenience *n.* **1** °disadvantage, °discomfort, °pain, °trouble, °bother, °annoyance, °nuisance, awkwardness, °disturbance, disruption, °burden, °drawback, °hindrance, °impediment, °difficulty, °upset: *How are you coping with the inconvenience of walking with crutches?* **2** cumbersomeness, unwieldiness, burdensomeness, onerousness, troublesomeness, disadvantageousness, awkwardness, inappropriateness, untimeliness: *The delay has led to considerable public inconvenience.* —*v.* **3** discommode, °trouble, incommode, °disturb, °disrupt, °upset, °put out, °bother, °annoy, °irritate, °irk: *Would it inconvenience you to remove your parcel from my chair?*

inconvenient *adj.* cumbersome, °unwieldy, °burdensome, onerous, °troublesome, bothersome, annoying, irritating, °irksome, °unsettling, °disturbing, upsetting, disrupting, disadvantageous, °awkward, °inappropriate; inexpedient, °inopportune, untimely, ill-timed: *Those large boxes of laundry soap are inconvenient because they're so heavy. It is inconvenient for me to go to the door now.*

incorporate *v.* °embody, °include, °combine, comprise, °embrace, °integrate, consolidate, °unite, °amalgamate, assimilate, coalesce, °unify; °merge, °mix, °blend: *The university incorporates several independent colleges. Gently fold in the egg whites, incorporating them into the mixture.*

incorrect *adj.* °wrong, °mistaken, °inaccurate, °untrue, °imprecise, °inexact, °erroneous, fallacious, °specious; °improper, °false, °faulty: *Your method is right but the answer is incorrect. That is the incorrect key for this lock.*

incorrigible *adj.* °bad, °naughty, °villainous, °wicked, °sinful, °hopeless; intractable, unchangeable, unalterable, °habitual, inveterate, °incurable, °stubborn, hardened, obdurate, °inflexible, uncontrollable: *When a teenager, Giles was absolutely incorrigible. Among other things, he was an incorrigible liar.*

incorrupt *adj.* **1** °moral, °upright, °righteous, °pure, upstanding, °honorable, °good, °virtuous, °honest, straightforward, °straight, unimpeachable, incorruptible, undefiled, °impeccable, °spotless, °immaculate, °faultless, °flawless: *Show me one nation run by a totally incorrupt government.* **2** error-free, °correct, uncorrupted: *He possesses the only incorrupt manuscript of the play.*

increase *v.* **1** °grow, °swell, °enlarge, dilate, wax, °expand, °extend, distend, °inflate, augment, snowball, °enhance, °heighten, °raise, °develop, multiply, burgeon *or* bourgeon, °flourish, °proliferate, °spread, broaden, °widen, °lengthen, *Colloq* jackup: *Since I added the plant food, the aspidistra has increased twofold.* **2** °prolong, °advance, °further, °improve, °better, °strengthen: *The more lottery tickets you buy, the more you increase your chances of winning.* —*n.* **3** °growth, enlargement, °expansion, °extension, augmentation, enhancement, °development, multiplication, °proliferation, °spread: *The increase in scientific knowledge has been phenomenal.* **4** °addition, °increment, escalation, inflation, °gain, °rise, °boost: *Increases in costs are reflected in increases in prices.* **5 *on the increase.*** waxing, developing, growing, expanding, increasing, escalating, on the rise, proliferating, spreading: *Trade with the Far East will be on the increase next year.*

incredible *adj.* **1** °unbelievable, beyond belief, °inconceivable, unimaginable, °unthinkable, °improbable, °implausible, °far-fetched, °absurd, °preposterous, °ridiculous, °unlikely, °impossible, °unrealistic, °unreal, °fictitious, mythic(al): *Incredible as it may appear, continents move a few inches a year.* **2** °extraordinary, astounding, astonishing, °amazing, wonderful, awe-inspiring, °awesome, tremendous, °marvelous, °prodigious, *Colloq* far-out, *US* humongous: *The incredible amounts spent on arms could be better spent on medical research.*

incredulous *adj.* disbelieving, °unbelieving, dubious, °doubtful, °skeptical, mistrustful, °distrustful, °suspicious: *We were incredulous when we learned who had won the award.*

increment *n.* °increase, °addition, °gain, accrual *or chiefly Brit* accrument, augmentation: *The increment last year barely compensated for inflation.*

incriminate *v.* °accuse, °charge, °blame, °implicate, inculpate, °indict, °impeach, °involve, °entangle, *Colloq* point the finger at, *Chiefly US* °put the finger on,

Slang US °finger: *The suspect confessed and incriminated two others.*

incubate *v.* °hatch, °brood; °nurse, nurture, °develop: *The plan for a police raid has been incubating for a few weeks.*

incumbent *adj.* **1** °obligatory, °necessary, required, °mandatory, compulsory, binding, °demanding, commanding, prescribed: *As their father, you ought to feel it incumbent to talk to the children's teachers.* **2** officeholding: *The incumbent chairman is not responsible for his predecessor's peccadillos.* —*n.* **3** officeholder, °official, °occupant: *As the incumbent, you have a better chance of being reelected.*

incur *v.* bring upon *or* on (oneself), °draw, °attract, °arouse, °provoke, invite, expose (oneself) to, lay (oneself) open to: *For such acts of cruelty he has incurred the wrath of his subjects.*

incurable *adj.* **1** irremediable, °terminal, inoperable, °fatal, °hopeless: *The doctor told me yesterday that the condition is incurable.* **2** °hopeless, inveterate, °habitual, °incorrigible, dyed-in-the-wool, unflagging, °relentless, irredeemable; irreparable, unrectifiable: *You are an incurable optimist if you think you will complete this book by the deadline.*

indebted *adj.* obligated, owing, °obliged, °beholden, °bound, °liable, °responsible: *I am indebted to the Salvation Army for their help.*

indecent *adj.* **1** indecorous, indelicate, °immodest, °improper, °unbecoming, unsuitable, unfit, °inappropriate; in bad taste: *When he proposed marriage she accepted with indecent haste.* **2** °unseemly, °shameless, °shameful, °offensive, °outrageous, °repellent, °repulsive, °distasteful, °ill-mannered, °rude, °suggestive, °coarse, °risqué, °vulgar, °blue, °obscene, °gross, °rank, °prurient, °dirty, °foul, °filthy, °pornographic, ithyphallic, scatological, salacious, °lascivious, licentious, °lewd, lubricous *or* lubricious, smutty, °vile, °degenerate, debauched: *There are always complaints about indecent language on television.*

indecision *n.* hesitation, wavering, indecisiveness, vacillation, irresolution, uncertainty, ambivalence, shilly-shally *or* shilly-shallying, °fluctuation, tergiversation: *Her indecision stems from caution, not ignorance of the issues.*

indecisive *adj.* **1** hesitating, °hesitant, wavering, °doubtful, vacillating, undecided, °irresolute, °uncertain, °of two minds, ambivalent, shilly-shallying, °wishy-washy, namby-pamby, fluctuating, tergiversating: *Why are you so assertive at the office yet so indecisive at home?* **2** °indefinite, indeterminate, undecided, °inconclusive, °open, °unsettled, °moot, °doubtful: *The battle was indecisive, with heavy losses on both sides.*

indeed *adv.* **1** certainly, °surely, to be sure, doubtless(ly), °undoubtedly, undeniably, °definitely, °positively, °absolutely, °exactly, just so, °actually, °truly, truthfully, °seriously, (all) joking aside, in (point of) fact, °of course, °really, °in reality, to be realistic, °naturally, upon my word, on my honor, on my oath, *Brit* °rather, *Colloq* no kidding: *Indeed, that is the news he has been expecting.* **2** what is more, °still, not to say, as a matter of fact, if the truth be known, to say the least: *His had been a depressing, indeed miserable childhood.* —*interj.* **3** Is that so!, You don't say!, Really!, By George!, By Jove!, (Upon) my word!, My goodness!, Goodness!, Gracious!, Mercy!, Good Lord!, Good heavens!, My stars!, Fancy that!, Imagine (that)!, Well, I'll be (damned)!, *Colloq Brit* Blimey!, Cor!, Crikey!: *"I've torn up the winning lottery ticket." "Indeed!"*

indefinite *adj.* **1** °uncertain, undetermined, undefined, °imprecise, °inexact, inexplicit, unspecified, °unsettled, unfixed, unspecific, nonspecific, °vague, °general, indeterminate, undecided, sub judice: *The time of the meeting was indefinite.* **2** °vague, unclear, °obscure, °confused, confusing, °puzzling, baffling, °cryptic, bewildering, mystifying, °equivocal, °ambiguous, unspecific, nonspecific, inexplicit, °inexact, °imprecise: *Your directions for getting here were too indefinite.* **3** ill-defined, undefined, blurred, blurry, °hazy, °indistinct, °obscure, °dim, °fuzzy, unrecognizable, °indistinguishable: *Indefinite shapes loomed in the fog.* **4** °unlimited, °unknown, uncounted, uncountable, undefinable, indeterminate, indeterminable, unbounded, °boundless, °immeasurable, incalculable, °limitless, °endless, °infinite: *He ponders the indefinite capacities of the mind.* **5** °hesitant, °vague, shilly-shallying, vacillating, °indecisive, undecided, °inconstant, wavering, unsure, °uncertain, *Colloq* °wishy-washy: *She is still indefinite about going to the party.*

indelible *adj.* ineradicable *or* uneradicable *or* noneradicable, inerasable *or* unerasable *or* nonerasable, ineffaceable, inexpungible, °indestructible, uncancelable *or* noncancelable, °enduring, °permanent, °lasting, °fixed, °ingrained, inextirpable: *The laundry marks are in indelible ink. Irena makes an indelible impression on everyone she meets.*

indelicacy *n.* coarseness, crudeness, roughness, °vulgarity, boorishness, churlishness, offensiveness, rudeness, immodesty, indecency, shamelessness; °incivility, indecorum, inelegance, uncourtliness, unmannerliness, impoliteness, unrefinement, unseemliness, tastelessness, bad *or* poor taste, grossness: *She refuses to comment on the indelicacy of his proposal.*

indemnity *n.* **1** compensation, repayment, reimbursement, °remuneration, recompense, °consideration, °restitution, reparation(s), redress, indemnification, °return, quid pro quo, °restoration, °award, °reward, °payment, disbursement, °amends, °requital, °atonement, °reckoning, quittance: *The conquerors demanded indemnity for the losses they had suffered.* **2** °insurance, °protection, °security, °safety, °guarantee, °assurance, underwriting, °warrant, °endorsement, certification; °exemption, impunity, °privilege: *Am I covered for indemnity against third-party claims?*

indentation *n.* °notch, dent, °nick, °cut, °score, °mark, °depression, °impression, °hollow, dimple, °pit; *Typography* indention: *The chair legs make indentations in the carpet. The indentation of each paragraph should be five em spaces.*

independence *n.* **1** °freedom, °liberty, autonomy, °sovereignty, self-rule, home rule, self-determination, °self-government, self-direction, autonomy, autarchy: *The colony gained independence from Portugal.* **2** °confidence, °self-confidence, self-sufficiency, self-reliance, self-assurance: *Edwina asserted her independence by taking over the company herself.*

independent *adj.* **1** °free, self-governing, autonomous, °sovereign: *In 1829 Greece was acknowledged as an independent state.* **2** unrestrained, unrestricted, unfettered, untrammeled, unregulated, °uncontrolled, separate(d), unconnected, °disconnected, °unrelated, °distinct: *Air is made up of independent particles.* **3** self-reliant, °self-sufficient, self-assured, (self-)confident, °bold, individualistic, °competent: *For a six-year-old, he is very independent.* **4** °voluntary, °nonpartisan, °spontaneous, °unsolicited, unbidden, °unprejudiced, unbiased, °nonaligned, unaligned, °disinterested, °neutral: *Four independent witnesses testified to seeing him at the scene of the crime.* **5** °unlimited, unrestricted, affluent, °self-sufficient; unearned: *She is a lady of independent means.* **6** unallied, unaffiliated, uncommitted, individualistic, undecided: *The candidate hopes to at-*

tract the support of independent voters. 7 °outside, °external, unaffiliated, nonaffiliated, unconnected, °disinterested: *An independent auditor was engaged to go over the company's books.* **8** ***independent of.*** °irrespective of, disregarding, °notwithstanding, ignoring, excluding; °exclusive of, °except for, °barring, °apart from, °besides, beyond, *US* aside from: *He insists on doing things his way, independent of others' feelings. Independent of you and me, no one cares what he does.* —*n.* 9 °individual, °nonconformist, maverick, *Colloq* loner: *He is an independent, unaffiliated with any political party.*

indestructible *adj.* °durable, long-lasting, °everlasting, °eternal, °endless, °perennial, °permanent, °fixed, unchanging, °changeless, unchangeable, °indelible, ineradicable, inextirpable, immutable, unalterable, °constant, undying, nonperishable, imperishable; unbreakable, nonbreakable, shatterproof: *The problem is that many plastics are virtually indestructible.*

index *n.* 1 °guide, directory, °list, listing, table of contents, catalogue, °key, °thesaurus: *To find synonyms for a word in this book, look first in the index.* 2 °mark, °sign, °clue, °token, °hint, °pointer, °indicator, °indication, °guide: *An index to his feelings can be seen in the way he treats people.* 3 index finger, forefinger, first finger; °pointer, marker, needle, *Chiefly Brit typography* °hand, *Chiefly US typography* fist: *Use an index to emphasize important paragraphs.* 4 °ratio, °measure, °formula, °factor: *The cephalic index is the ratio of the width of a skull to its length multiplied by one hundred.*

indicate *v.* 1 °point out, °point to, °mark, °specify, °designate, °indicate; call *or* direct attention to, °show: *The thermometer indicates that the temperature is twenty-two degrees. Sally indicated the place where the dam would be built.* 2 °imply, °suggest, betoken, °manifest, °signify, bespeak, °reveal, evince, °evidence, °denote: *Measles is indicated by the presence of Koplik's spots.* 3 °suggest, °hint, °imply, °intimate; °say, °express, °state, °disclose, °tell, °make known, make clear, °register, °show, °display, °exhibit: *His frown indicated his doubts. She indicated her displeasure in a few, well-chosen words.* 4 °call for, °require, °demand, °need, °recommend: *The doctor advised that a surgical procedure is indicated in such cases.*

indication *n.* 1 °sign, °signal, °token, °suggestion, °hint, intimation, °inkling, °clue, °implication, °symptom: *A bullet-riddled corpse, said the detective, is usually a good indication of foul play.* 2 reading, °measure, °degree: *The indication on the pressure gauge is reaching the danger point.* 3 °omen, portent, forewarning, °warning, augury, foreshadowing, foretoken: *Those black clouds are an indication that a storm is brewing. Can't you give us any indication of what will become of us?* **4** ***indications.*** °evidence, °data, clues, signs: *Indications are that she left on the last train.*

indicative of *adj.* signifying, indicating, indicatory of, suggesting, °suggestive of, hinting (at), symptomatic of, denotative of, characteristic of, typical of, typifying: *Stalling is indicative of some fault in the fuel.*

indicator *n.* °pointer, needle; °gauge, meter, °display: *The indicator reads exactly sixty-seven degrees.*

indict *v.* Often, **indict for.** charge (with), accuse (of *or* with), arraign (for), incriminate (in *or* for), inculpate (in *or* for), cite (for *or* with), °denounce (for), blame (for *or* with); °summon, summons, subpoena: *She has been indicted for murder in the first degree.*

indifference *n.* 1 unconcern, apathy, listlessness, disinterest, coolness, nonchalance, insouciance, aloofness, °detachment, °disregard, inattention, pococuranteism *or* pococurantism, coldness, phlegm, stolidity, callousness, insensibility, impassiveness *or* impassivity: *I cannot believe the indifference she showed on learning about the tragedy.* 2 unimportance, insignificance, irrelevance, unconcern, inconsequence, °triviality: *Whether I go or stay is a matter of indifference to me.* 3 dispassion, disinterestedness, impartiality, neutrality, °objectivity, fairness, equitableness, evenhandedness: *The judges demonstrated their indifference by awarding first prize to a novice.*

indifferent *adj.* 1 unconcerned, apathetic, uncaring, °listless, °disinterested, uninterested, °cool, °nonchalant, °lukewarm, °lackadaisical, laodicean, °dispassionate, insouciant, °aloof, °detached, °distant, removed, °inattentive, pococurante, °cold, °phlegmatic, °stolid, °callous, unemotional, uncompassionate, insensitive, unfeeling, °inconsiderate, °unsympathetic, °insensible, °impassive: *Joan remained totally indifferent to Charles's attentions.* 2 °impartial, °neutral, °just, evenhanded, °objective, °fair, °equitable, unbiased, °unprejudiced, °nonpartisan, nondiscriminatory, °dispassionate, °disinterested: *The problem is in selecting an indifferent judgment.* 3 °neutral, °average, °mediocre, °fair, °undistinguished, uninspired, lightweight, °passable, middling, °ordinary, commonplace, °everyday, °so-so, not bad; °poor, °inferior, not very *or* particularly *or* especially good, *Slang Brit* naff: *He is a writer of indifferent quality. I held indifferent cards but lost little money.* 4 unimportant, °insignificant, trivial, °trifling, nugatory, °immaterial, °inconsequential, °minor, inappreciable, °slight, neither here nor there: *It is indifferent to him where she spends her money.*

indigenous *adj.* 1 °native, °local, autochthonous, endemic, °natural, °aboriginal, °original: *The indigenous inhabitants lived a Stone Age existence.* 2 innate, °inborn, inbred, °inherent: *It is often difficult to distinguish between man's indigenous and his acquired characteristics.*

indigent *adj.* °needy, °poor, °destitute, poverty-stricken, in want, penniless, °impoverished, °penurious, impecunious, necessitous, *Colloq* (dead *or* flat *or* stony) broke, °hard up, °short, °on one's uppers, *Brit* skint, *US* strapped: *The two brothers made certain that their indigent parents were well looked after.*

indigestion *n.* dyspepsia, upset stomach, stomach distress; stomach ache, gastralgia: *A bit of bicarbonate of soda cured my indigestion at once.*

indignant *adj.* provoked, exasperated, °disgruntled, piqued, irked, irritated, annoyed, vexed, °angry, °furious, irate, angered, enraged, incensed, wrathful, in high dudgeon, in a temper, in a rage, in a pet, *Literary* wroth, *Colloq* peeved, riled, °in a huff, huffy, miffed, °mad, livid, °sore, *US* teed-off, *Slang* pissed off: *Peter was quite indignant to learn that the publisher had lost his manuscript.*

indignation *n.* °anger, °fury, °rage, wrath, exasperation, irritation, °annoyance, vexation, °resentment, *Literary* ire, choler: *She felt righteous indignation at being treated as a child.*

indignity *n.* °insult, affront, °outrage, °injury, °offense, °humiliation, °disrespect, °slight, °dishonor, snub, obloquy, contumely, °scorn, reproach, °abuse, discourtesy, °aspersion, *Colloq* °slap (in the face): *The duchess had to suffer the indignity of arrest for shoplifting.*

indirect *adj.* 1 °roundabout, circuitous, circumambient, °devious, °tortuous, zigzag, winding, °rambling, roving, wandering, °erratic, °meandering, ambagious, °crooked, °oblique, twisted, twisting; circumlocutory, periphrastic: *In an indirect way she told me that I was fired.* 2 °secondary, °incidental, ancillary, collateral, °accidental, °side, °subordinate, °subsidiary, °accessory, additional, °accidental, adscititious, adventitious:

An indirect effect of winning the prize was paying tax on it.

indiscreet *adj.* °imprudent, °tactless, incautious, impolitic, undiplomatic, °improvident, injudicious, °ill-advised, ill-judged, ill-considered, °rash, °reckless, °audacious, °bold, temerarious, °impulsive, °hasty, °impetuous, °thoughtless, insensitive, °heedless, °careless, °unthinking, °mindless, unwise, naive *or* naïve, °foolish, °foolhardy: *Your reference to her plastic surgery was indiscreet to say the least.*

indiscretion *n.* **1** imprudence, tactlessness, improvidence, injudiciousness, rashness, recklessness, audacity, boldness, temerity, impulsiveness, hastiness, °haste, impetuousness, impetuosity, thoughtlessness, insensitivity, heedlessness, carelessness, naiveté *or* naïveté *or* naivety, foolishness, foolhardiness, °folly: *He might not have committed treason but he was certainly guilty of indiscretion.* **2** °blunder, °error, °mistake, °slip, °lapse, °misstep, gaffe, faux pas, °peccadillo, *Colloq* boner, *Brit* bloomer: *He spent much of his adult life paying for the indiscretions of his youth.*

indiscriminate *adj.* **1** undiscriminating, unselective, unparticular, uncritical, undiscerning, °careless, °promiscuous, °random: *Viola seems indiscriminate in her choice of companions.* **2** °confused, °haphazard, unorganized, °chaotic, disorganized, jumbled, disordered, disarranged, scrambled, mixed-up, °casual, °random, unsystematic, unsystematized, uncoordinated, unmethodical, wholesale, °erratic, *Colloq* higgledy-piggledy: *The Aztecs were victims of indiscriminate slaughter at the hands of the conquistadors.*

indispensable *adj.* **1** °crucial, °vital, °essential, °urgent, °imperative, °necessary, needed, needful, required, requisite, demanded, °in demand, called for, °of the essence, °important, compelling: *Today, charisma, not character, is indispensable in a national leader.* **2** °key, °unavoidable, inescapable, ineluctable, compulsory, °mandatory, sine qua non, °obligatory: *A rifle is indispensable if you are going camping in the Canadian Rockies.*

indisposed *adj.* **1** °ill, ailing, unwell, °sick, °sickly, °unsound, °unhealthy, in bad health, valetudinarian, out of commission, *Colloq* on the sick list, °(doing) poorly, laid up, in a bad way, not up to snuff, off one's feed *or Brit also* grub, °under the weather: *Madam is indisposed and cannot have tea with you.* **2** °averse, °disinclined, °loath, unwilling, °reluctant, °resistant, °hesitant: *The judge was indisposed to dispense a light sentence owing to the severity of the crime.*

indisputable *adj.* °unquestionable, °incontrovertible, incontestable, irrefutable, undeniable, indubitable, beyond doubt, °sure, °certain, °positive, °definite, °definitive, °absolute, °fixed: *Her role in the affair is indisputable. The show was an indisputable success.*

indistinct *adj.* **1** °faint, °dim, °obscure, °vague, blurred, blurry, °filmy, °hazy, °misty, bleary, °shadowy, °fuzzy, foggy, °murky, °muddy, unclear, indiscernible, °illegible, muffled, unintelligible, °indistinguishable, indeterminate, °confused, °indefinite: *I could barely make out an indistinct figure in the dark. The writing on the matchbox was indistinct.* **2** °indistinguishable, °ambiguous, not distinct, °equivocal, inseparable, ill-defined, undefined, °nebulous, °vague, °confused: *The identity of the three brothers is indistinct in my mind.*

indistinguishable *adj.* **1** Often, ***indistinguishable from.*** undifferentiated (from), °identical (to), °alike, like two peas in a pod, °twin, inseparable: *He is indistinguishable from his brother. Her fingerprints and those found on the dagger are indistinguishable.* **2** indiscernible, °imperceptible, °indefinite, unclear, °indistinct: *Whoever was the older was indistinguishable to me.*

individual *adj.* **1** °single, °sole, °particular, °separate, °distinct, °discrete, °solitary, lone: *Consider the entire utterance, not just the individual words.* **2** °singular, °special, °specific, idiosyncratic, own, °characteristic, °distinctive, individualistic, °unique, °peculiar, °personal, °proper; unitary: *She has her individual style of dressing.* —*n.* **3** °person, °human (being), °(living) soul, °mortal; °one, °party: *In a democracy each individual is entitled to one vote.*

individualist *n.* °independent, freethinker, °nonconformist, maverick, loner, lone wolf: *Nick is an individualist and seldom agrees with the majority.*

individually *adv.* one at a time, °singly, one by one, °separately, severally, °apart: *We must study each of these specimens individually.*

indoctrinate *v.* °train, °teach, °instruct, °school, °discipline, °drill, brainwash, propagandize; inculcate, imbue, °instill, °implant: *The children were indoctrinated to believe that they were the master race. He indoctrinated us with the idea that we are capable of anything.*

indolence *n.* laziness, slothfulness, °sloth, °sluggishness, °idleness, °lethargy, shiftlessness, languor, languidness, lassitude, listlessness, °torpor, torpidity, °inertia, inaction, °inactivity, faineance, *dolce far niente,* oscitancy: *After winning the lottery, he spent his days in indolence.*

indolent *adj.* °lazy, °slothful, sluggish, °idle, °lethargic, °shiftless, languorous, languid, °torpid, °inert, °inactive, °stagnant, fainéant, °listless: *"Peel me a grape," she said in her most indolent manner.*

indomitable *adj.* °resolute, resolved, °determined, °steadfast, °staunch, °persistent, unstoppable, °irrepressible, indefatigable, °untiring, °tireless, unflagging, unyielding, unswerving, unwavering, unflinching, undaunted, °dauntless, °fearless, unafraid, °intrepid, °brave, °courageous, plucky, mettlesome; unconquerable, °unbeatable, °invincible: *She survived the prison camp by dint of her indomitable strength of character.*

induce *v.* **1** °lead, °persuade, °influence, prevail on *or* upon, °sway, °move, °convince, °get, °talk into, °prompt, °incite, instigate, actuate, °motivate, impel, °encourage, °inspire, °stimulate, °nudge, °push, °press, °urge, °prod, goad, °spur, egg on, °coax, °cajole, °lure, °entice, inveigle, °seduce: *He was induced to sell the secret plans to the enemy for a large sum.* **2** °cause, bring about *or* on, °produce, °give rise to, engender, °create, °generate, lead to; °effect, °occasion, set in motion: *Some drugs induce a state of euphoria.*

inducement *n.* °attraction, °lure, °incentive, stimulus, °enticement, bait, °encouragement, °incitement, °provocation, °spur, °premium, °consideration, °reward, *Colloq* carrot, *Chiefly US* °come-on: *What inducement were you offered to buy that soap?*

induct *v.* **1** install *or* instal, °inaugurate, °invest, instate, °establish, swear in: *Crocker was inducted as president of the lodge last night.* **2** °call up, °enlist, conscript, °enroll, °register, *US* °draft: *He was inducted into the army at eighteen.*

indulge *v.* **1** Often, ***indulge in.*** °yield (to), °succumb (to), °favor, °gratify, °humor, °oblige (with), °comply (with), °minister to, °cater to, °pander to, °treat (to), °pamper (with): *When it comes to chocolates, I indulge myself to the fullest. We indulge in an occasional beer. She indulges his every whim.* **2** °coddle, °baby, °pamper, cosset, mollycoddle, °spoil: *Do you believe in indulging children?*

indulgence *n.* **1** °tolerance, sufferance, °understanding, °patience, goodwill, °allowance, forbearance; acceptance, overlooking: *Genevieve's behavior is testing the limits of Tom's indulgence.* **2** self-indulgence, °luxury, °extravagance, °profligacy, self-gratification, self-satisfaction: *Carbuckle's capacity for indulgence is legendary.* **3** °treat, °luxury, °extravagance: *My sole indulgence is vintage port.*

indulgent *adj.* °tolerant, °permissive, °patient, °understanding, forbearing, °lenient, easygoing, °relaxed, °liberal, °lax, °kind, °kindly, well-disposed, °agreeable: *Tony and Hazel are extremely indulgent parents.*

industrious *adj.* sedulous, assiduous, hard-working, °diligent, °intense, °conscientious, °energetic, °dynamic, °aggressive, °vigorous, °untiring, °tireless, indefatigable, unflagging, °persistent, pertinacious, dogged, °tenacious, *US* hustling: *The business was built up only because she was so industrious.*

industry *n.* **1** °production, °manufacture, °trade, °commerce, °business: *Competition in the car industry helps keep prices lower.* **2** diligence, assiduity, sedulousness, °energy, °exertion, °effort, °application, °perseverance, °persistence, °work, °labor, toil: *We were complimented on our industry by the teacher.* **3** industriousness, °energy, °activity, °vigor, °hustle, bustle, °dynamism, °enterprise, earnestness, °determination: *The sales office is a beehive of industry.*

ineffable *adj.* **1** unutterable, unspeakable, °unmentionable, °taboo: *In certain religions, the name of the deity is ineffable.* **2** °inexpressible, indescribable, indefinable *or* undefinable, indescribable *or* undescribable, beyond description, beyond words: *Jeannette has an ineffable, je-ne-sais-quoi air about her.*

ineffective *adj.* **1** unproductive, unfruitful, °bootless, °idle, °vain, °useless, °ineffectual, °inefficacious, inoperative, nonfunctioning, °inadequate, °insufficient, °worthless: *The anticrime measures appear to have been ineffective.* **2** °inefficient, °incompetent, °incapable, unskilled, unskillful, °inept, unfit, unproficient: *The new office manager is totally ineffective.*

ineffectual *adj.* **1** °unsuccessful, unavailing, °futile, °bootless, °sterile, °barren, °fruitless, unproductive, °ineffective, °inefficacious, inoperative: *Our efforts to deal with pollution have been largely ineffectual.* **2** °weak, °feeble, effete, °impotent, °tame, °lame, °powerless, °inefficient, °incompetent, °inadequate: *We tried to read the inscription by the ineffectual light of a match.*

inefficacious *adj.* See **ineffectual** and **ineffective.**

inefficient *adj.* **1** °ineffective, °incompetent, °incapable, °unqualified, inexpert, unskilled, unskillful, unfit, °inept, °ineffectual, °deficient: *Poorly paid labor is inefficient labor.* **2** uneconomic(al), °wasteful, disorganized, °slipshod: *For our purposes, the layout of the factory is quite inefficient.*

ineligible *adj.* °unqualified, °unacceptable, unfit, unsuited, unsuitable, °inappropriate, °improper: *He is ineligible for unemployment insurance if he left work voluntarily.*

inept *adj.* **1** °clumsy, °awkward, bungling, maladroit, ungainly, bumbling, gauche, inexpert, unskilled, unskillful, °incompetent, °inefficient: *If he is inept at sailing, he'd best become proficient at swimming.* **2** °inappropriate, inapt, °unseemly, °improper, unfitting, unfit, unsuitable, impolitic, undiplomatic, injudicious, °imprudent, °indiscreet, ill-considered, ill-judged, °ill-advised, unadvised, unadvisable *or* inadvised, °misguided, inexpedient, °out of place, unwise: *Zifkovich, as usual, made some inept comment which shocked everyone.*

ineptitude *n.* **1** ineptness, clumsiness, awkwardness, maladroitness, ungainliness, gaucherie, °incompetence, inefficiency, unfitness, unsuitableness: *He displays a remarkable ineptitude for the pronunciation of French.* **2** inappropriateness, inaptness, °absurdity, unsuitability, unseemliness: *The ineptitude of her comments embarrasses everyone.*

inequality *n.* **1** °disparity, °difference, °discrepancy, unevenness, nonconformity, incongruence, incongruity, inconsistency, °dissimilarity, imbalance: *The inequality between their incomes made him resentful.* **2** °bias, °prejudice, °partiality, unfairness, °injustice, inequity: *Why should they have to tolerate any inequality of treatment?*

inert *adj.* **1** °inactive, unreactive, unresponsive, °neutral: *Argon, xenon, and neon are inert gases.* **2** motionless, immobile, °inanimate, °lifeless, °still, °quiet, quiescent, stationary, °static: *Speaking to him was like addressing a mute, inert stone.* **3** sluggish, °slow, °torpid, °dull, °inactive, °idle, °indolent, °lazy, °slothful, °leaden, °slack, °passive, °supine, °dormant, otiose, °listless, languid *or* languorous: *I contemplated the inert bodies sunbathing around the swimming pool.*

inertia *n.* inertness, °inactivity, °sloth, °sluggishness, °torpor, dullness, °idleness, °indolence, laziness, slothfulness, passivity, apathy, lassitude, dormancy, listlessness, languor, immobility, motionlessness: *Adele finds it difficult to overcome her feeling of inertia in the morning.*

inestimable *adj.* **1** incalculable, °immeasurable, measureless, °untold, incomputable; °priceless, °invaluable, °precious: *You have been of inestimable help in completing the project on time.* **2** countless, innumerable, °vast, °immense, °prodigious, °boundless, unfathomable, °infinite, incalculable, °immeasurable, measureless, °untold, incomputable: *An inestimable number of organisms have inhabited the earth.*

inevitable *adj.* °unavoidable, inescapable, ineluctable, unpreventable, °certain, °sure, °destined, °fated, assured, inexorable, °irrevocable, unchangeable, °absolute, ordained, decreed, °authoritative, incontestable: *The inevitable outcome of war is misery for all participants.*

inexact *adj.* °imprecise, °inaccurate, °erroneous, °incorrect, °wrong, °false, °faulty, °indefinite, fallacious, °fuzzy, muddled: *This gauge is giving an inexact reading.*

inexcusable *adj.* unjustifiable, unjustified, indefensible, unpardonable, unforgivable, intolerable, inexpiable: *Insulting your hostess was inexcusable.*

inexhaustible *adj.* **1** °limitless, °boundless, °unlimited, unbounded, unrestricted, °endless, measureless, indeterminate, °infinite, incalculable: *The supply of tasteless television programs seems virtually inexhaustible.* **2** °untiring, °tireless, indefatigable, unflagging, unfailing, unfaltering, unwearying, unwearied: *Bartell is an inexhaustible practical joker.*

inexpensive *adj.* °cheap, °economical, low-priced, low-cost, °reasonable, budget(-priced): *What might be inexpensive for you is very costly for me.*

inexperience *n.* immaturity, innocence, naiveté *or* naïveté *or* naivety, greenness, callowness, unsophistication: *I attribute his mistakes to inexperience.*

inexperienced *adj.* °immature, °innocent, naive *or* naïve *or* naïf, °green, °callow, °unsophisticated, unworldly, °unfledged, °raw, uninitiated, untrained, unschooled, °uninformed, unseasoned, *Colloq* (still) wet behind the ears, born yesterday: *Isn't she too inexperienced to take on such a great responsibility?*

inexplicable *adj.* unexplainable, °unaccountable, unintelligible, inscrutable, °puzzling, enigmatic, mystifying, °perplexing, confounding, baffling, bewildering, °incomprehensible: *I find inexplicable their refusal to allow freedom of religion.*

inexpressible *adj.* unutterable, °ineffable, unspeakable, indescribable, indefinable: *I felt inexpressible joy at seeing my children again.*

inextinguishable *adj.* °unquenchable, °irrepressible, unsuppressible; °enduring, undying, imperishable, °eternal, °everlasting: *The teacher had to cope with inextinguishable giggling. Our hopes for freedom were inextinguishable.*

inextricably *adv.* inescapably, ineluctably, unavoidably, irretrievably, °completely, inseparably, indissolubly, °totally, complicatedly, intricately: *I found myself inextricably caught up in her emotional problems.*

infallible *adj.* **1** unerring, °faultless, °flawless, °perfect, oracular, unmistaken: *I judge the accuracy of her calculations to be infallible.* **2** unfailing, dependable, °sure, °secure, °certain, °reliable, °foolproof: *The vaccination provides infallible protection against the virus.*

infamous *adj.* **1** °notorious, °disreputable, ill-famed, stigmatized, °scandalous, discreditable, °dishonorable, ignominious: *So this is the infamous house where the murders took place!* **2** °bad, °awful, °wicked, °evil, iniquitous, °villainous, heinous, °vile, °abominable, °outrageous, execrable, °abhorrent, opprobrious, °despicable, °loathsome, detestable, odious, °foul, °scurvy, °rotten, °atrocious, flagitious, °revolting, °monstrous, egregious, °base, °low, °shameful, °disgraceful: *A number of infamous criminals are portrayed in the Chamber of Horrors.*

infamy *n.* **1** °notoriety, ill repute, ill fame, disrepute, °shame, ignominy, obloquy, °disgrace, °dishonor, °stigma, °discredit: *She has to suffer the infamy of her husband's treachery.* **2** wickedness, °evil, iniquity, villainy, heinousness, vileness, abomination, °outrage, abhorrence, opprobrium, loathsomeness, detestation, hatred, odium, °atrocity, °revulsion, °monstrosity, egregiousness, °shame, shamefulness, °disgrace, disgracefulness: *The infamy of his profligate behavior is widely known.*

infancy *n.* **1** babyhood, early childhood: *His limp was a result of polio, which he had contracted in infancy.* **2** beginning(s), inception, early *or* initial, stage(s), commencement, °start, °emergence, °dawn, °rise: *When psychiatry was in its infancy it was marked by a notable lack of success.*

infantile *adj.* °childish, °immature, °puerile, babyish, °juvenile: *Ben behaves in the most infantile way whenever there are girls about.*

infatuated *adj.* fascinated, beguiled, °enchanted, bewitched, spellbound, °charmed, ensorcelled, enraptured, °possessed, hypnotized, mesmerized, captivated, besotted, taken with, obsessed, °smitten, enamored, °fond: *Ferdinand is completely infatuated with the girl who moved in next door.*

infectious *adj.* contagious, °catching, communicable, transmissible: *There is also a highly infectious variety of the disease.*

infer *v.* °deduce, °derive, °draw, conclude, °understand, °gather, °surmise, °guess, °assume: *From the evidence, I infer that you were at home last night.*

inference *n.* °deduction, conclusion, °understanding, °surmise, assumption, °presumption: *If this man's father is my father's son, the inference is that this man is my nephew.*

inferior *adj.* **1** °lower, nether; *Typography* subscript: *We descended into the inferior regions of the cavern. The numeral in* H_2O *is called an inferior character.* **2** lesser, °lower, °subordinate, °secondary, °junior, °minor, unimportant, °insignificant, °lowly, subservient, °humble, °servile: *He occupies an inferior position in the company.* **3** °poor, °bad, low-quality, °mediocre, °indifferent, °imperfect, °defective, second-rate, second-class, substandard, low-grade, °shoddy, gimcrack, *Colloq* grotty, junky, crummy *or* crumby, °lousy, *Slang* crappy, *Brit* naff: *They tried to market an inferior product, but consumers rejected it.* —*n.* **4** °subordinate, underling, °junior, °menial, lackey, flunky *or* flunkey, dogsbody, cat's-paw, doormat, stooge, °yes man, lickspittle, bootlicker: *She is tired of being treated as an inferior.*

inferiority *n.* **1** unimportance, insignificance, lowliness: *The inferiority of his position does not entitle him to use the executive lavatory.* **2** inadequacy, deficiency, insignificance, unimportance, worthlessness: *Do not confuse modesty with a sense of inferiority.* **3** shoddiness, °imperfection, mediocrity: *Consumers are not fooled by the inferiority of a product for long.*

infernal *adj.* **1** hellish, °underworld, nether, Hadean: *Vergil was Dante's guide in the infernal regions.* **2** °devilish, diabolic(al), demonic(al), demoniac(al), °fiendish, °satanic, Mephistophelian; °damnable, damned, execrable, malicious, malevolent, maleficent, °wicked, °evil, iniquitous, flagitious, °villainous, heinous, dire, °sinister, °dreadful: *She tells everyone that there is an infernal plot against her.*

infertile *adj.* °sterile, °barren, infecund, unproductive, unfruitful, °nonproductive: *The stony, infertile earth yielded only cactus and scrub grass.*

infest *v.* invade, °plague, °beset, °overrun, overspread, °flood, swarm over, inundate, pervade, °permeate, °penetrate, infiltrate: *Mosquitoes infested this area before the swamp was drained.*

infidel *n.* unbeliever, °heathen, disbeliever, heretic, °pagan, agnostic, atheist, nullifidian, freethinker: *Thousands of the fanatics swarmed over the infidels, slaughtering them mercilessly.*

infidelity *n.* **1** unfaithfulness, faithlessness, treachery, traitorousness, disloyalty, °perfidy, falseness, apostasy, heresy: *The infidelity of his supporters brought about his downfall.* **2** adultery, cuckoldry; °affair, °liaison, amour, *Colloq* cheating: *Infidelity is a frequent cause of the breakdown of marriages. She forgave him his many infidelities.*

infinite *adj.* **1** °boundless, °vast, °limitless, °unlimited, illimitable, °endless, interminable, indeterminable, indeterminate, °inestimable, astronomical, °numberless, multitudinous, uncountable, uncounted, innumerable, incalculable, °inexhaustible, °immense, °enormous, °immeasurable, measureless, °bottomless, unfathomable: *We hurtled through infinite space at the speed of light. There seem to be infinite stars in the sky. In his infinite wisdom, God is right.* **2** °eternal, °everlasting, °perpetual, °endless, unending, °inexhaustible, undying, never-ending: *It is difficult to conceive of time as being anything but infinite, without beginning or end. With infinite patience we put the vase back together again.*

infirm *adj.* **1** °ill, °indisposed, debilitated, °frail, °fragile, °weak, °feeble, weakened, ailing, °decrepit, enfeebled, °failing, wasted, on the decline, °sick, °sickly, unwell, °lame, °crippled: *Campbell is infirm and unable to travel.* **2** °shaky, °flimsy, wobbly, wobbling, °doddering, °unstable, faltering, vacillating, wavering, unsteady, unsteadfast, °inconstant, °changeable, °irresolute: *We must find stalwart recruits, recruits who are not infirm of purpose.*

infirmary *n.* clinic, °hospital, sick bay, first-aid station; dispensary, *Brit* nursing home, surgery: *The infirmary is able to deal only with emergencies, not with long-term care.*

infirmity *n.* **1** °weakness, feebleness, frailness, °frailty, debility, °decrepitude, sickliness: *At one hundred, she finally succumbed to infirmity and stopped going out* *altogether.* **2** sickness, °ailment, °disease, malady, °affliction, °disorder, °defect, °complaint: *What infirmity struck him down?*

inflame *v.* **1** °arouse, °incite, °touch off, ignite, enkindle, °provoke, °rouse, °excite, impassion, °foment, incense, °agitate, °stir (up), °fire (up), °heat, °anger, °enrage, °madden, °infuriate, whip *or* lash up, °work up, rile, °exasperate, °stimulate, °animate, °move, °motivate, °urge, °prod, goad, °spur (on), °rally, °drive: *The speakers inflamed the crowd to riot against the police.* **2** °aggravate, °intensify, °deepen, °heighten, °fan, exacerbate, °increase, augment, °fuel: *She inflamed his dislike of his brother into hatred.*

inflamed *adj.* irritated, °sore, °angry, chafing, chafed, red, °swollen, °heated, °hot, fevered, °feverish, infected, septic: *The wound became inflamed and we had no medicine to reduce the infection.*

inflammable *adj.* burnable, combustible, °flammable: *The plastic foam formerly used to stuff furniture is highly inflammable.*

inflammation *n.* irritation, redness, °swelling, °sore, infection: *The inflammation subsided as the injection took effect.*

inflammatory *adj.* incendiary, °fiery, inflaming, °explosive, °rousing, °provocative, °rabid, rabble-rousing, °passionate, °fervent, fervid, °frantic, frenzied, fomenting, demagogic, insurgent, °riotous, °mutinous, °seditious, °rebellious, °revolutionary, °traitorous, treacherous: *His inflammatory speeches at the university created tension between the students and the administration.*

inflate *v.* **1** °blow up, balloon, °expand, dilate, °enlarge, °swell, °pump up, puff up *or* out, distend: *We inflated two plastic bags to use as floats.* **2** °boost, °increase: *The price of the company's shares was inflated owing to takeover rumors.* **3** °exaggerate, °amplify, °magnify, °blow up: *He tends to inflate to huge proportions the small contribution he made to the project.*

inflated *adj.* **1** exaggerated, °conceited, overblown, °grandiose, puffed-up, overstated, magnified, amplified, overdrawn, °smug, egotistic, °immodest, °cocky, °vain, °self-important, *Colloq Brit* swelled-headed, *US* swellheaded: *He has an inflated idea of the importance of his work.* **2** grandiloquent, °bombastic, orotund, high-flown, °pompous, °pretentious, °extravagant, magniloquent: *She always talks about her husband in such inflated terms.*

inflexible *adj.* unbending, °stiff, °rigid, °firm, rigorous, unyielding, adamant(ine), °severe, Rhadamanthine, inelastic, °hard (and fast), determined, °fixed, obdurate, °immovable, intractable, unvaried, unvarying, °invariable, unchangeable, immutable, °obstinate, °stubborn, °obstinate, pigheaded, mulish, dyed-in-the-wool, headstrong, refractory, °steely, °stony, °resolute, resolved, unadaptable, unaccommodating, uncompliant, uncompromising, unshakable *or* unshakeable: *The committee remains inflexible about changing the rules of the club to admit women.*

inflict *v.* °impose, °apply, °visit, °administer, levy, force on *or* upon; °trouble, °afflict: *The new government is seeking ways of inflicting more taxes on us. You ought not inflict punishment on such small children.*

influence *n.* **1** °power, °pressure, °weight, °sway, °impact, °force, °effect, leverage, potency; °hold, °control, mastery, ascendancy, *Colloq* °pull, clout: *Would you actually use your influence to fix a parking ticket? The boy is under the influence of his tutor.* **2** °connections, °favor, °favoritism, *Colloq* °pull: *He got his position through influence, not merit.* —*v.* **3** °affect, °move, °change, °modify, °alter, °bias, °sway, °persuade, °induce, °work on, °impress (upon), play *or* act upon *or* on, °incline; bring pressure to bear on *or* upon, °move, °motivate, °manipulate, *Brit* pressurize, *US* °pressure, *Colloq* pull strings *or US also* wires: *She was able to influence the minister to drop the plan.*

influential *adj.* °powerful, °weighty, °strong, °forceful, °effective, °effectual, °efficacious, °instrumental, °telling, °significant, °persuasive, °dominant, °leading, guiding, °authoritative, °predominant, °important, °substantial, °prestigious, °significant, controlling: *He comes from an influential family. What were the factors influential in your reaching this decision?*

inform *v.* **1** °tell, apprise, °advise, °communicate, °enlighten, °notify, °acquaint, °brief; °impart, °disclose, divulge, °reveal, °report, *Colloq* °tip (off): *I have informed everyone about the storm warning.* **2** Usually, ***inform against*** *or* ***on.*** °betray, °give away, °incriminate, °implicate, °identify, °tattle (on), *Colloq* °tell (on), °blab (on), rat (on), *US* blow the whistle (on); *Slang* peach (on), snitch (on), squeal (on), °put the finger on, °sing, °name names, *Brit* nark (on), grass (on), split on, blow the gaff, *US* °finger: *He informed on his accomplices in return for a lighter sentence.*

informal *adj.* **1** unceremonious, °casual, °natural, unstilted, °familiar, °ordinary, °everyday, °unaffected, unassuming, unpretentious, °simple, °relaxed, °free, free-and-easy, *Brit* common-or-garden, *US* garden-variety: *We had an informal dinner in the kitchen.* **2** °unofficial, unconventional, unconstrained, °casual, °everyday, °simple: *On duty he wore informal clothes—blue jeans and a denim jacket.* **3** °vernacular, colloquial, °simple, °unaffected, °ordinary, unpretentious, °everyday: *He uses informal language devoid of scientific jargon.*

information *n.* °knowledge, °data, °facts, °intelligence, °message, °word, °advice, °news, tidings, °report, communication, *Colloq* info, °lowdown, *Slang* °dirt, °(inside) dope, *Brit* gen, bumf, *US* poop: *I have information regarding her divorce.*

informative *adj.* communicative, °instructive, °educational, edifying, revealing, illuminating: *We found his report about missiles very informative.*

informed *adj.* **1** °intelligent, °knowledgeable, °enlightened, °educated, °learned, cultured, °cultivated: *An informed public is democracy's greatest asset.* **2** °alert (to), °aware (of), advised, in touch, *au fait,* briefed, conversant (with), (well-)versed, up (on), up-to-date, *Colloq* in the know, °wise: *Keep me informed as to your whereabouts.*

informer *n.* °traitor, betrayer, tattletale, taleteller, informant, *Slang* stool pigeon, snitch, squealer, weasel, *Brit* grass, nark, *Chiefly US* rat, *US* stoolie, canary, shoofly: *According to the informer, the secret meeting is scheduled for tonight.*

infrequently *adv.* °rarely, °seldom, sporadically, °occasionally, °now and then, irregularly, exceptionally: *Only infrequently have they been seen together.*

infringe *v.* **1** °violate, contravene, °break, °disobey, °transgress, °overstep: *Publication of this article has infringed her right to a fair trial.* **2** ***infringe on*** *or* ***upon.*** intrude on, impinge on, trespass on, encroach on, invade: *Never let pleasure infringe on the domain of duty.*

infringement *n.* °violation, °breach, contravention, infraction, disobedience, infraction, °noncompliance, breaking, °transgression: *Publication without permission constitutes infringement of copyright. The inventor sued the company for patent infringement.*

infuriate *v.* °enrage, °anger, °madden, incense, make (someone's) blood boil, °provoke, °inflame, work *or* stir *or* fire up, rile, °arouse, vex, pique, °gall, °annoy, °irritate, °bother, °chafe, °agitate, °irk, nettle, °exasperate, raise (someone's) hackles, make (someone's) hackles rise, *Brit* have *or* get (someone's) blood up, *Colloq* miff, °bug, peeve, get under (someone's) skin, get *or Brit also* put (someone's) back up, make (someone) see red, *Chiefly US* get (someone's) Irish *or* dander up, *US* burn up, *Slang* piss (someone) off: *It really infuriated him to see injustice done to anyone.*

ingenious *adj.* °clever, °skillful, skilled, °shrewd, cunning, °crafty, canny, °dexterous, adroit, °acute, °sharp, °keen, °resourceful, °adept, apt, °smart, °gifted, °bright, °brilliant, °talented, deft, °handy, inventive, Daedalian, °creative, °imaginative, °original, *Colloq* °neat, °keen, *US* crackerjack, *Slang* on the ball: *Gridley has come up with an ingenious solution to our problem.*

ingenuity *n.* ingeniousness, °genius, °brilliance, cleverness, °skill, shrewdness, cunning, °craft, °art, °knack, °flair, °dexterity, dexterousness, adroitness, acuteness, sharpness, keenness, resourcefulness, adeptness, aptness, smartness, canniness, °gift, °talent, °ability, °capability, °faculty, deftness, handiness, inventiveness, creativity, creativeness, °imagination, imaginativeness, °originality: *She displayed extraordinary ingenuity in preparing the case against the plaintiff.*

ingenuous *adj.* **1** naive *or* naïve, °simple, °innocent, °unsophisticated, °childlike, °suggestible, °artless, °sincere, °genuine, °trusting, guileless, °natural, °straight, uncomplicated, °(fair and) square, °honest, °fair, °just, °open, undeceitful, °unaffected, undeceptive, undissembling, unfeigning, °(open and) aboveboard, *Colloq* °on the level: *What ideas have you been implanting into the mind of this ingenuous young girl?* **2** °frank, °candid, °open, °trustworthy, °honorable, °forthright, °direct, straightforward, foursquare, °honest, °outspoken, °blunt, °bluff, °bold, unreserved, °free, °uninhibited, °unabashed: *I shall be ingenuous enough to confess that what you heard about me is true.*

ingrained *adj.* engrained, deep-rooted, °fixed, inveterate, deep-seated, °fundamental, °basic, °essential, °inherent, °inborn, innate, inbred, inherited, °hereditary, °organic, °intrinsic, °native, °natural: *He has an ingrained fear of heights.*

ingratiating *adj.* fawning, °groveling, sycophantic, toadying, toadeating, °servile, °obsequious, °flattering, °timeserving, wheedling, cajoling, unctuous, °oily, buttery, sweet-talking, sugary, saccharine, *US* blandiloquent, *Colloq* bootlicking, °slimy, *Brit* smarmy, *US* apple-polishing, *Slang US* brown-nosing: *It is Crawley's ingratiating manner that none of us can abide.*

ingratitude *n.* unthankfulness, ungratefulness, thanklessness, unappreciativeness, nonrecognition: *Despite the favors they had received, they showed their ingratitude by refusing to contribute to the fund drive.*

ingredient *n.* constituent, °element, °part, component, °factor; (*plural*) makings: *The elixir of life is wisdom, and its mystic ingredient is not knowledge but understanding.*

inhabit *v.* dwell in, reside in, live in, abide in, °occupy, °settle; locate in, °populate, °people; colonize: *This bird inhabits northern Africa. North America was inhabited by emigrants from Europe.*

inhabitant *n.* °resident, dweller, °denizen, °citizen, °tenant, °occupant, occupier: *The inhabitants of the village refused to pay taxes.*

inhale *v.* breathe in, °inspire, draw *or* suck in: *I stepped outside and inhaled the fresh spring air.*

inherent *adj.* °intrinsic, indwelling, °essential, °basic, innate, connate, ingrained *or* engrained, °native, congenital, inherited, °hereditary, °inborn, inbred, °indigenous, immanent, built-in: *It is difficult to tell which characteristics are derived and which are inherent.*

inherit *v.* come into, fall *or* be *or* become heir to, be bequeathed, °succeed to, be left, °receive, °acquire, *Colloq* °come by: *She inherited the property when her father died.*

inheritance *n.* patrimony, °heritage, legacy, °bequest, birthright; °property: *He came into his inheritance when he was twenty-one.*

inhibit *v.* °discourage, °repress, °frustrate, °hold back, °bridle, °curb, °control, °govern, °hinder, °restrain, °impede, °obstruct, interfere with, °check, °prevent, °bar, °stop: *Her sad look inhibited my expression of delight at seeing her again.*

inhibited *adj.* °reticent, restrained, repressed, °reserved, °self-conscious, °shy, abashed, °embarrassed, *Colloq* uptight: *He is much too inhibited to appear on television.*

inhibition *n.* self-consciousness, °restraint, constraint, °impediment, °hindrance, °bar, °barrier, °defense, defense mechanism, blockage, °interference, °check, °curb, °stricture: *Grover overcame his inhibitions and went over to talk with Anne. Gypsy Rose Lee had no inhibitions about doing a striptease.*

inhospitable *adj.* **1** unwelcoming, unreceptive, °uninviting, unsociable, °unsocial, °aloof, °cold, °cool, °standoffish, unfriendly, inimical, antisocial, °hostile, xenophobic: *They were most inhospitable and refused to let me in.* **2** unfavorable, °forbidding, °hostile, °barren, °desert, °uninviting, uninhabitable: *We were marooned on an inhospitable reef in the middle of the Pacific.*

inhuman *adj.* **1** inhumane, °merciless, °cruel, pitiless, °ruthless, °heartless, °severe, °unsympathetic, unfeeling, °unkind, unkindly, uncompassionate, °cold-blooded, °vicious, stonyhearted, hardhearted, °callous, insensitive, barbaric, barbarous, °savage: *Concentration camp inmates suffered unbelievable torture and other inhuman treatment.* **2** °animal, bestial, °brutal, brutish, °fiendish, diabolical, demonic: *The specter, assuming an inhuman form, loomed over us.*

initial *adj.* **1** °original, °primary, °first; °prime, °beginning, incipient, inaugural, °opening, °introductory, commencing: *His initial plan, to take the children, was vetoed by his wife. It is best to tread cautiously during the initial stages of the program.* —*v.* **2** °sign, °approve, °endorse: *Please initial the clauses of the contract that we have changed.* —*n.* **3** monogram: *The initial "M" is on all their towels.*

initiate *v.* **1** °begin, °commence, enter upon *or* on, °originate, °introduce, set in motion, °start, °give rise to, get under way, °launch, get *or* set going, trigger, °set off, actuate, °activate, instigate, °institute, °inaugurate: *The program was initiated last spring with much fanfare.* **2** °admit, °accept, °introduce: *The new members were initiated last night, with due ceremony.* **3** °teach, °instruct, °train, °tutor, °drill, °coach: *His responsibility is to initiate recruits in the rudiments of jungle warfare.* —*n.* **4** °novice, beginner, new boy *or* girl, °greenhorn, rookie, neophyte, tyro *or* tiro, °newcomer, tenderfoot, fledgling, °apprentice, °(raw) recruit, abecedarian, novitiate, catechumen, *Brit* fresher *or US only* freshman, *Australian* new chum: *In the first weeks, initiates are drilled in the basics.*

initiation *n.* **1** °beginning, commencement, inauguration, origination, °debut, introduction, inception, °establishment: *April 1 is the date for initiation of the new law.* **2** °admittance, °admission, °entrance, induction, enrollment, instatement, investiture, ordination, °installation; °ceremony, °rite, °ritual: *The initiation of new members is scheduled for tomorrow at midnight.*

initiative *n.* **1** first move *or* step, °lead, opening move: *Hogan took the initiative by winning the first set.* **2** °enterprise, aggressiveness, °drive, °ambition, ambitiousness, resourcefulness, °leadership, °dynamism, °energy, °vigor, °hustle, élan, *Colloq* get-up-and-go, °pep, °snap, zip, zing: *At least she had the initiative to get the company back on a profitable basis.*

inject *v.* **1** °introduce, °insert, drive *or* force (in), °shoot (in), intromit; inoculate: *The serum is injected into the tissue of the upper arm. The doctor injected me with antibiotics.* **2** °introduce, °insert, imbue, °instill, °bring in, interject, throw in: *Can't you inject a little more enthusiasm into your work?*

injunction *n.* **1** °prohibition, interdict, interdiction, °restriction, °restraint, *US Law* restraining order: *There is an injunction against picketing.* **2** °order, mandate, directive, °command, °direction, °instruction, ruling, °dictate, exhortation; °warning, admonition: *Mother's parting injunction was always to dress warmly.*

injure *v.* **1** °harm, °wound, °hurt, °damage, °impair: *He's sure to injure himself if he's allowed to play with a knife.* **2** °wrong, °offend, °abuse, °hurt, °wound, °outrage, °slight, °insult, affront, °mistreat, °misuse, °ill-treat, maltreat: *She starts the arguments yet she invariably takes the role of the injured party.*

injurious *adj.* **1** damaging, °harmful, °hurtful, °bad, deleterious, unfavorable, °detrimental, °unhealthy, insalubrious, pernicious, °destructive; adverse, °ruinous: *Smoking is said to be injurious to your health.* **2** °abusive, insulting, °scornful, °slanderous, libelous, defamatory, °scandalous, °scurrilous, °harsh, calumnious, disparaging, °derogatory, deprecatory, °contemptuous, denigrating, °offensive: *She speaks of her ex-husband in the most injurious terms.*

injury *n.* °damage, °hurt, °harm, °wound, °impairment; °wrong, °abuse, maltreatment, °mistreatment, °mischief, °offense, °outrage; °mayhem: *The injury was less painful to his foot than to his self-esteem.*

injustice *n.* **1** unfairness, °favoritism, °discrimination, °bias, °inequality, °partiality, partisanship, °prejudice, °bigotry, onesidedness, unjustness, inequity: *Class privilege has always condoned injustice.* **2** °wrong, °injury: *You do me an injustice if you think me capable of dishonesty.*

inkling *n.* °hint, °suggestion, glimmering, °suspicion, °whisper, intimation, °indication, soupçon, °clue, °notion, (faintest *or* foggiest) idea, °tip, tip-off: *Can you give me an inkling of what you are getting at?*

inmate *n.* °prisoner, °convict, °captive, jailbird, *Slang Brit* °lag; °patient, °case; °inhabitant, °occupant, °resident: *The inmates were fed on bread and water in those days. Inmates at the sanatorium are well treated.*

innocent *adj.* **1** not guilty, guiltless, °blameless, °honest, °(in the) clear, unimpeachable, above suspicion, above reproach, °faultless: *We all believe her innocent of extortion.* **2** °pure, sinless, °virtuous, °chaste, virgin(al), undefiled, untainted, unstained, unsullied, °pristine, °incorrupt, uncorrupted, °immaculate, °spotless, unblemished, unpolluted: *She is still a sweet child, innocent as the day she was born.* **3** °harmless, well-intentioned, °safe, innocuous, °inoffensive, unobjectionable: *What began as an innocent stroll became a dangerous expedition.* **4** naive *or* naïve *or* naïf, °unsuspecting, unsuspicious, °unsuspecting, °ingenuous, °unsophisticated, °trusting, trustful, °gullible, credulous, °green, °inexperienced, °childlike, unworldly, guileless, °artless, °simple, °open, unartificial, °sincere: *Youngsters would arrive directly from the farm, entirely innocent and unaware of the perils of city life.* —*n.* **5** infant, babe (in arms *or* in the wood(s)), °child; ingénue, °novice, beginner, °newcomer, *Colloq* °greenhorn: *She's certainly no innocent when it comes to men.*

innovation *n.* **1** °novelty; °invention: *The flush toilet was a nineteenth-century innovation.* **2** modernization, °alteration: *Consumer product development is marked by a spirit of tireless innovation.*

innuendo *n.* insinuation, °imputation, °suggestion, °hint, intimation, °implication, allusion, °overtone; °reference, animadversion: *I resent your innuendo that I know more about the affair than I am letting on.*

inoffensive *adj.* °harmless, unobjectionable, innocuous, unoffending, °neutral, °retiring, °mild, °tame: *Bligh is so inoffensive, how could you possibly have misinterpreted his intentions?*

inopportune *adj.* °inappropriate, malapropos, ill-timed, untimely, °inconvenient, unsuited, unsuitable, °out of place, unpropitious, unfavorable, °inauspicious, ill-chosen, °unseasonable, °unseemly, °untoward, °unfortunate: *As we are just sitting down to dinner, your call has come at a most inopportune moment.*

inordinate *adj.* **1** °immoderate, unrestrained, intemperate, °excessive, °disproportionate, °extravagant, overdone, °extreme, °exorbitant, °outrageous, °preposterous, °unconscionable, °unreasonable, undue, uncalled-for, °unwarranted: *Her inordinate appetite for sweets has caused all that dental decay.* **2** °irregular, °disorderly, disordered, °uncontrolled, °unlimited, unregulated, unsystematic, °erratic, °haphazard: *Laws have been passed to limit the inordinate hunting of deer.*

inquire *v.* **1** Usually, ***inquire into.*** °search, °investigate, °probe, °examine, °research, °look into, °inspect, °study, °explore, °survey, °scrutinize: *Scientists are inquiring into the causes of the greenhouse effect.* **2** See **enquire.**

inquiry *n.* **1** enquiry, °investigation, °probe, °examination, °research, °search, inspection, °study, °exploration, °survey, °scrutiny, inquest; questioning, querying, °interrogation, inquest, cross-examination, inquisition: *The police have asked me to help them with their inquiry. They are conducting an inquiry into the cause of the accident.* **2** enquiry, °question, °query, °interrogation: *Would you please repeat the inquiry?*

inquisitive *adj.* **1** inquiring, °curious, probing, questioning, °interested, investigative, searching, exploring, analytical: *Scientists have a naturally inquisitive nature.* **2** prying, °intrusive, *Colloq* snooping, snoopy, nosy *or* nosey: *Don't be so inquisitive about things that don't concern you!*

inroad *n.* **1** incursion, °raid, °attack, °invasion, °penetration, foray, encroachment, forced entry, intrusion: *Our armored units have made an inroad into enemy territory.* **2** Often, ***inroads.*** advance(s), °progress, breakthrough: *Have researchers made significant inroads into the problem of AIDS?*

insane *adj.* **1** °mad, demented, °psychotic, schizophrenic, schizoid, *non compos mentis,* manic, °maniacal, lunatic, °deranged, °unbalanced, psychoneurotic, °neurotic, °eccentric, °crazy, of unsound mind, crazed, unhinged, out of one's mind *or* head, mad as a hatter *or* a March hare, quirky, *Colloq* round the bend *or* twist, off one's rocker *or* chump, loopy, loony, certifiable, °mental, screwy, dotty, cuckoo, not all there, not have all one's marbles, off the wall, out of it, *Brit* potty, *Slang* °daft, nutty (as a fruitcake), nuts, spaced out, spacy, batty, have bats in one's belfry, not right upstairs, balmy, crackers, have a screw loose, schizo, *Brit* barmy (in the crumpet), bonkers, *US* bugs, bughouse, loco, crazy as a bedbug, (gone) off the deep end, kooky, °kinky, out to lunch: *Witney, certified as insane, has been institutionalized. You're insane if you think I am* *going out on that ledge.* **2** °stupid, °dumb, °dull, °silly, asinine, °ridiculous, idiotic, irrational, °absurd, fatuous, °ludicrous, °foolish, °nonsensical, °irresponsible, °reckless, °wild, imbecilic, moronic, °feebleminded, °harebrained, addlepated, °scatterbrained, thimble-witted, attocerebral, *Brit* gormless, *Colloq* nutty, screwy, °crazy: *People in love sometimes do insane things.*

insanity *n.* **1** °madness, °lunacy, mental illness *or* disorder, dementia (praecox), psychosis, schizophrenia, (mental) derangement, °mania; psychoneurosis, neurosis: *Because of the plea of temporary insanity he never stood trial for the murders.* **2** °folly, foolishness, °stupidity, idiocy, imbecility, °lunacy, °absurdity, fatuity, fatuousness, °nonsense, senselessness, irresponsibility, irrationality, inanity: *What insanity it was to think I could win the lottery!*

insecure *adj.* **1** °uncertain, °afraid, unsure, unconfident, °nervous, °worried, °anxious, °disconcerted, apprehensive, uncomfortable, °shaky, °jumpy, unnerved, °fearful: *Harry is beginning to feel insecure about his job.* **2** unsafe, °dangerous, unprotected, °vulnerable, °unguarded, °defenseless, undefended, exposed, °open: *The machine-gun position is insecure and will soon be taken.* **3** °unstable, °shaky, wobbly, °precarious, °infirm, °weak, °flimsy, °frail, °rickety, unsubstantial, °rocky, °unsound, unsteady, °unreliable, °uncertain: *The structure collapsed because the footings were insecure.*

insensible *adj.* **1** insensate, °unconscious, insentient, °numb, benumbed, °senseless, °torpid, anesthetized, *Colloq* °out, out of it: *She was totally insensible for an hour after the blow on the head.* **2** Often, ***insensible to*** *or* ***of.*** insensitive, °callous, °indifferent, impervious, °impassive, apathetic, °cool, unsusceptible; °unaffected, °unmoved, untouched; unaware, °deaf, °inconsiderate; hardhearted, °thick-skinned, unfeeling, emotionless, °dispassionate, °thoughtless, *Colloq* clotheared: *He remained calmly insensible to her desires and needs.*

insert *v.* **1** °introduce, place *or* put *or* stick in, intercalate; interpolate, interject, interpose: *Please do not insert comments of your own.* —*n.* **2** insertion, °addition, addendum, °supplement, °advertisement, broadside, °brochure, tip-in, handbill, °circular, *Colloq Brit* advert, *US* ad, flier *or* flyer; outsert, wraparound *or* wrapround: *The postage will be higher with this eight-page insert.*

inside *n.* **1** °interior, °center, °core, °middle, °heart; °contents; lining, °backing: *From the outside one could never guess what the inside looks like. Please paint the inside of the cabinet blue.* **2** Usually, ***insides.*** °bowels, entrails, viscera, gut(s), °stomach, *Colloq* innards: *He is seeing the doctor about some trouble with his insides.* **3** ***inside out.*** everted, reversed; backwards: *He is wearing his pullover inside out.* —*adj.* **4** °favored, °advantageous, °favorable, advantaged, °privileged, preferred, °preferential, °propitious, °exclusive; °internal, °private, °secret, °confidential, °privy, clandestine: *She has an inside track to the chairman. He claims to have inside information about the successful bidder.* **5** °internal, °interior; arranged, prearranged: *The police think it was an inside job.* —*adv.* **6** fundamentally, basically, °at bottom, °by nature: *Despite her gruff exterior, inside she is quite sentimental.* **7** in prison, in jail, imprisoned, incarcerated, *Slang Brit* in quod, *US* up the river: *He'll be inside for another five years.* —*prep.* **8** within, *US* inside of: *Because of rain, the party was held inside the building.*

insight *n.* °perception, percipience, °sensitivity, perspicacity, perceptiveness, perspicaciousness, discernment, acuteness, acuity, acumen, sharpness, °understanding, °judgment, comprehension, °vision: *Chambers has brought his considerable insight to bear on this complex problem.*

insignificant *adj.* °paltry, °trifling, °petty, inconsiderable, °inconsequential, trivial, unimportant, °nonessential, °minor, °negligible, nugatory, unessential, °niggling, °puny, °insubstantial, unsubstantial, *Colloq* piddling: *The union rejected management's insignificant offer. Insignificant differences can be ignored. What you and I might do with our lives is, in the larger sense, insignificant.*

insincere *adj.* °hypocritical, °dishonest, dissembling, °deceptive, °disingenuous, °deceitful, untruthful, °false, °artificial, ungenuine, treacherous, °perfidious, °faithless, double-dealing, duplicitous, °two-faced, Janus-faced, °lying, mendacious, °sly, Machiavellian, cunning, °crafty, °slick, °foxy, vulpine, °wily, °artful, °evasive, °shifty, time-serving, unctuous, °slimy, °slippery, °tricky, underhanded, °crooked, *Colloq* phoney *or US also* phony: *From the first act I thought Iago insincere.*

insinuate *v.* **1** °impute, °suggest, °hint, °intimate, °imply, °whisper, °indicate; convey, °signify: *They insinuated that she married me for my money.* **2** °insert (oneself), °inject (oneself), worm *or* work *or* inveigle *or* maneuver (oneself *or* one's way); infiltrate, °intrude: *She insinuated herself into our group uninvited.* **3** °inject, infuse, °instill, °introduce: *Wild imaginings began to insinuate themselves into my mind.*

insist *v.* **1** °demand, °require, °call for, °command, importune, °urge, exhort, °argue, remonstrate, expostulate: *The chairman insisted that proper parliamentary procedure be followed.* **2** assert, °state, °declare, °emphasize, °stress, °maintain, °persist, °hold, °dwell on, °vow, avow, aver, asseverate: *He insisted that he wanted to see the last act but she insisted on leaving.*

insistent *adj.* °firm, dogged, °emphatic, °persistent, °tenacious, °resolute, °determined, °assertive, uncompromising, unfaltering, unwavering, persevering, perseverant, unrelenting, inexorable, °stubborn, °obstinate, unyielding, compelling, °urgent, importunate: *The judges were insistent that we should follow the rules to the letter.*

insolent *adj.* °impertinent, °impudent, °pert, saucy, °bold, °presumptuous, °brazen, °brash, °disrespectful, insulting, °contemptuous, °offensive, °rude, °crude, uncivil, °insubordinate, *Colloq* °fresh, °brassy, °cheeky, *Slang Brit* smart-arsed *or US* smart-ass(ed), °wise: *That insolent boor had the gall to call her "Queenie"!*

insolvent *adj.* bankrupt, ruined, in receivership, penniless, °impoverished, °destitute, *US* in Chapter 11, *Colloq* °broke, wiped out, in the red, °on the rocks, (gone) bust, gone to the wall, *Brit* in Queer street, skint: *A business is insolvent if its liabilities exceed its assets.*

inspect *v.* °examine, °scrutinize, °study, °scan, °survey, °vet, check (up (on) *or* out), °investigate, °pore over; °peruse: *The customs man inspected every last inch of our luggage.*

inspiration *n.* **1** awakening, arousal, stimulus, °revelation, impulse, °feeling, afflatus, enlightenment, °insight, °spur, °incitement, °incentive; °spirit, °energy, élan, °passion, °ardor, zeal, °enthusiasm, °vigor, °gusto, ebullience, °sparkle: *Her extraordinary eyes gave me the inspiration I needed and I finished writing the song in an hour.* **2** °influence, stimulus, stimulation, °encouragement, °provocation, °suggestion, °guide, °education: *Eisenhower's life was an inspiration to many a career soldier.*

inspire *v.* **1** °animate, °activate, actuate, °stimulate, invigorate, °stir, °move, waken, °awaken, °arouse,

°rouse, instigate, °prompt, °rally, °energize, °enliven, °vitalize, vivify, galvanize, inspirit, °excite, °spark (off), °quicken, °kindle, °fire, °provoke: *It was the inflammatory speeches that inspired the crowd to riot. His enthusiasm inspired her to take up writing as a profession.* **2** °encourage, °strengthen, °support, °reinforce, °buoy (up), uplift, °boost, affirm, °confirm, °fortify, °buttress: *Your example has inspired many young people.*

install *v.* **1** instal, °invest, instate, °initiate, °establish, °inaugurate, °induct, °institute; °place, °put, °position, °introduce, °settle: *The new bishop will be installed in office next week.* **2** instal, °fit, °set up, °connect, °fix (in place): *I installed our new air conditioner myself.*

installation *n.* **1** investiture, instatement, °initiation, °establishment, inauguration, induction, °institution, °placement, introduction, solemnization, swearing-in, consecration, *Ecclesiastical* ordination; crowning, coronation: *Will you attend the installation of the new vice-chancellor?* **2** °fitting, °placement, °connection, positioning: *The plumber will supervise the installation of the new heating system.* **3** °base, °post, °station, depot, °camp, °establishment: *Military installations all over the country are being picketed.*

instance *n.* **1** °case (in point), °example, exemplar, °illustration, °precedent, exemplification, °occurrence, °event: *The damage to the bus shelter is another instance of hooliganism.* **2** ***for instance.*** °for example, as an example, °say, e.g.; °in the event, as it happens *or* happened: *Consider her, for instance—she managed to work and raise a family.*

instant *n.* **1** °moment, °point, °second, °time: *At the very instant the phone rang I was thinking of you.* **2** °moment, °second, °minute, °flash, °twinkling (of an eye), trice, *Colloq* jiffy: *I'll be with you in an instant, sir.* —*adj.* **3** instantaneous, °immediate, on-the-spot, overnight: *His novel was an instant bestseller.* **4** °urgent, crying, °pressing, °earnest, °imperative, °critical, exigent; split-second, °direct: *We have an instant need for a computer specialist.* **5** °ready-made, ready-mixed, °prepared, ready-to-serve, precooked: *This is an instant soup—just add hot water to the powder.*

instantaneously *adv.* instantly, °immediately, °at once, °(right) now, °directly, forthwith, °promptly, this minute *or* second *or* instant, without delay, *tout de suite,* instanter, *Brit* straightaway *or US* right away; *Colloq* pronto, *US* momentarily: *If you rub this lamp, the genii will appear instantaneously.*

instead *adv.* **1** as an alternative *or* a substitute: *She wanted me to go but I stayed instead.* **2** ***instead of.*** alternatively, preferably, in preference to, in place of, in lieu of, rather than, as a substitute for; as contrasted with, as opposed to: *Instead of going to the movies we went to the theater.*

instill *v.* imbue, inculcate, infuse, ingrain *or* engrain, °implant; °insinuate, °impart: *From childhood a sense of justice was instilled in us.*

instinct *n.* intuition, °feel, °feeling, empathy, °sensitivity, °tendency, propensity, °leaning, °bent, °skill, °talent, °faculty, sixth sense, °knack, predisposition, °capacity, °aptitude; °subconscious: *She has an uncanny instinct for finding the most interesting person in a crowd.*

instinctive *adj.* **1** instinctual, intuitive, °natural, innate, °native, °inborn, inbred, congenital, constitutional, reflex, visceral, intestinal, °intrinsic, intuitional, °subconscious, *Colloq* °gut: *She has an instinctive flair for design.* **2** °immediate, °involuntary, irrational, °mechanical, °automatic, °spontaneous: *Harvey took an instinctive dislike to Percy when they met.*

institute *n.* **1** °society, °organization, °association, °league, °alliance, guild: *They tried to establish an institute for research on badgers.* **2** See **institution, 2.** —*v.* **3** °establish, °found, °launch, °organize: *The principles of sportsmanship were instituted on the playing fields of England.* **4** °inaugurate, °introduce, °initiate, °set up, °start, °begin, °originate, °commence, °pioneer: *It was she who instituted the practice of wearing white gloves at school dances.*

institution *n.* **1** establishing, °establishment, forming, °formation, origination, founding, °foundation, °installation, introduction, °creation, °organization: *Who was responsible for the institution of baseball as a national pastime?* **2** °establishment, °institute, academy, °foundation, university, college, °school: *He attended an institution of higher learning after leaving the army.* **3** °hospital, medical center, sanitarium *or Brit* sanatorium, °home, asylum: *They spent their last days in an institution.* **4** °custom, °tradition, °habit, °practice, °routine, °rule, °order (of the day), °code (of practice); °doctrine, dogma: *Throwing the coxswain into the river after the race has become an institution.*

instruct *v.* **1** °teach, °train, °tutor, °drill, °educate, °coach, °inform, °guide, edify, °prepare, °indoctrinate, inculcate: *My parents employed a tutor to instruct me in Latin.* **2** °direct, °order, °bid, °require, °tell, enjoin, °command, importune, °charge: *We were instructed to tell no one where we were going.*

instruction *n.* **1** Often, ***instructions.*** °order, °direction, °brief, briefing, directive, guideline, °advice, °recommendation, °rule; °information; *Colloq* °drill: *The next instruction is to insert the plug into the back of the receiver. Follow the instructions carefully.* **2** teaching, °education, °schooling, training, °drill, drilling, °tuition, °guidance, indoctrination, °preparation, lessons, classes, coaching, tutelage; tutorial: *Where did you receive instruction in demolition work?*

instructive *adj.* °informative, informational, informatory, °educational, instructional, °helpful, revealing, edifying, enlightening, illuminating: *The book on computers that you lent me is highly instructive.*

instructor *n.* °teacher, trainer, °tutor, °coach, mentor, adviser *or* advisor, educator, pedagogue, scholastic, academe, academician, °doctor, docent, lecturer, professor, *Brit* °master, °mistress, don, preceptor, *US* docent: *Charlotte was not spending time with the tennis instructor just to improve her backhand.*

instrument *n.* **1** °implement, °tool, °device, °apparatus, utensil, appliance, contrivance, °mechanism, °gadget, *Colloq* °contraption, thingumabob, thingumajig, thingummy, thingy, whatsit, whatitsname, whatnot, what-d'you-call-it, *Brit* gubbins, *US and Canadian* gizmo *or* gismo: *Which instrument should I use to loosen this bolt? The doctor extracted the splinter with some sort of instrument.* **2** °agency, °means, °way, °factor, °mechanism, instrumentality, wherewithal, (prime) mover, catalyst, °agent: *Summers refused to be the instrument of bringing free people into slavery.* **3** °contract, °(legal) document, °(written) agreement, °pact, °compact, °paper: *After the signing, copies of the instruments were exchanged.*

instrumental *adj.* °helpful, °useful, utilitarian, contributory, of service, °supportive, supporting, °advantageous, catalytic, conducive, °beneficial, valuable, °significant, °important; °accessory, ancillary: *Your presence will be instrumental in effecting a favorable settlement.*

insubordinate *adj.* °disobedient, °rebellious, °defiant, refractory, °mutinous, insurgent, insurrectional, insurrectionist, °revolutionary, °seditious, incompliant *or* uncompliant, uncooperative, °recalcitrant, contumacious, fractious, °unruly, °perverse, °contrary, °obstreperous, *Colloq Brit* stroppy: *The insubordinate officers have been court-martialed.*

insubstantial *adj.* **1** unsubstantial, °insignificant, °meager, diaphanous, °small, °flimsy, °frail, °weak, °feeble, °paltry, °puny, °slight, °thin, °tenuous, °fragile, °light, gossamer, wispy, wisplike, °fine: *Though the spider's thread appears insubstantial, for its weight it is enormously strong.* **2** °illusory, °unreal, illusive, °imaginary, imagined, °fanciful, °fancied, °visionary, °immaterial, °intangible, impalpable, incorporeal, airy, ethereal, °spiritual, chimerical, °phantom, phantasmal, phantasmagorical, hallucinatory, °fantastic; °false: *Through the mist London seemed so insubstantial, a dream city of vapors rising from the Thames.*

insufferable *adj.* °unbearable, insupportable *or* unsupportable, intolerable, unendurable, °impossible: *The baby sitter said she refused ever again to look after such an insufferable brat.*

insufficient *adj.* °inadequate, °deficient, °unsatisfactory, °meager, °scanty, scant, °scarce; too little, not enough: *This light is insufficient for me to read by. We have insufficient skill to play the Brahms concerto.*

insulate *v.* **1** °detach, °separate, °isolate, °segregate, °shelter, °preserve, set *or* keep apart, sequester, sequestrate, quarantine: *The rock star's aides did their best to insulate her from her screaming fans.* **2** °lag, °protect, °shield, °cushion, °wrap, °cover: *Insulate the pipes to prevent heat loss. Insulate the wire with this tape. Nitroglycerine must be insulated from shock.*

insult *v.* **1** °offend, affront, °slight, °outrage; °abuse, °dishonor, defame, °injure; asperse, °slander, °libel: *Don't be insulted if I arrive late and leave early. You insult her by suggesting that she has never heard of Keats.* —*n.* **2** °offense, affront, °indignity, °slight, °outrage, barb, °dig, °slur, °dishonor, °abuse, defamation, discourtesy; °aspersion, °slander, °libel; *Colloq* °slap (in the face), °put-down: *Refusing to bow at a Japanese funeral is taken as an insult.*

insurance *n.* °assurance, surety, °indemnity, indemnification, guarantee *or* guaranty, °warranty, °bond, °security, °protection, °cover: *The cost of insurance is higher owing to the increase in crime.*

intact *adj.* °whole, °entire, °perfect, °complete, °integral, °sound, unbroken, °solid, (all) in one piece, °undivided, uncut, together, untouched, unreduced, undiminished, unimpaired, inviolate, unblemished, °unscathed, uninjured, unharmed, undamaged, unsullied, undefiled, untainted: *The ancient city has been preserved virtually intact.*

intangible *adj.* impalpable, unperceivable, °imperceptible, incorporeal, unsubstantial, °insubstantial, °imponderable, °immaterial, ethereal, vaporous, airy, evanescent, °vague, °obscure, °dim, °imprecise, °indefinite, °shadowy, °fleeting, °elusive: *The architect must deal with intangible ideas of design, how people work and live, etc.*

integral *adj.* **1** °basic, °elementary, °elemental, °fundamental, °essential, °intrinsic: *The rhythm is an integral part of the music.* **2** See **intact.**

integrate *v.* °combine, °unite, °blend, bring *or* put together, °assemble, °merge, °amalgamate, °join, °knit, °mesh, consolidate, coalesce, °fuse; *US* desegregate: *We must integrate all the parts into a coherent whole. Several cultures have been well integrated into our community.*

integrity *n.* **1** °honesty, °probity, veracity, uprightness, °honor, °rectitude, °principle, °morality, goodness, trustworthiness, decency, °virtue, incorruptibility, righteousness: *He is a person of unimpeachable integrity.* **2** wholeness, °entirety, °unity, togetherness, soundness, completeness, coherence, oneness, °totality: *Care should be taken not to disturb the integrity of the protective film.*

intellect *n.* **1** rationality, °reason, reasonableness, °(common) sense, °understanding, °judgment, cleverness, °intelligence, °mind, *Colloq* brains: *You have the intellect, my boy, but it needs developing.* **2** See **intellectual, 3.**

intellectual *adj.* **1** °mental, cerebral: *The greatest minds have brought their intellectual powers to bear on the problem.* **2** °thoughtful, thought-provoking, °highbrow, °academic, bookish, °scholarly, *Colloq* brainy: *The professor did not find the subject intellectual enough for a dissertation.* —*n.* **3** °thinker, °intellect, °highbrow, °mastermind, °genius, *Colloq* °brain, egghead: *He enjoys associating with intellectuals at the university.* **4** °scholar, academician, professor, savant, °sage, wise man, guru, polymath, pundit, °authority: *The editor of the journal is an intellectual who leaves the details of administration to a deputy.*

intelligence *n.* **1** °intellect, °understanding, °aptitude, °capacity, brainpower, cleverness, astuteness, quickness, alertness, keenness, brightness, shrewdness, °wit, mother wit, °(common) sense, °insight, perspicacity, °perception, discernment, °discretion, percipience, perspicaciousness, °wisdom, sagacity, *Colloq* brains, savvy, gray matter, *Slang Brit* nous: *He may not know computer programming, but he has the intelligence to learn it.* **2** °information, °knowledge, °word, °data, °facts, °advice, °news, tidings, findings, *Colloq* °dope, °lowdown, info, *Brit* gen, *US* °inside, poop: *Our agents are trying to gather intelligence on the Bulgarian situation.*

intelligent *adj.* °bright, °smart, °clever, discerning, perspicacious, °perceptive, percipient, °understanding, °rational, apt, °astute, °quick, °quick-witted, °keen, °sharp, °alert, °shrewd, canny, insightful, °gifted, °sensible, °wise, °sage, sagacious, °enlightened, °knowing, °aware, °knowledgeable, erudite, *au fait, Colloq* brainy, *Chiefly US* savvy: *She is certainly intelligent enough to be the first in her class; if only she would study harder.*

intelligentsia *n.* pl. intellectuals, literati, savants, illuminati, masterminds, highbrows, *Colloq* brains, eggheads, brains trust: *The regime stifled free speech and persecuted the intelligentsia.*

intelligible *adj.* understandable, comprehensible, fathomable, decipherable, °legible, °clear, °plain, lucid, unambiguous: *Crowther's notes are barely intelligible.*

intend *v.* °mean, have in mind *or* in view, °propose, °contemplate, °design, °plan, °aim, °purpose, °resolve, °determine: *I intend to give him a piece of my mind. She is intending to go, but I don't know if she will be able to.*

intense *adj.* **1** °extreme, °excessive, °severe, °strong, °great, °fierce, °harsh, °acute, °powerful, °profound, °deep: *The intense heat kept the firemen at bay.* **2** °eager, °enthusiastic, °keen, °earnest, °sincere, °heartfelt, °deep, °passionate, °impassioned, °ardent, zealous, °animated, °burning, consuming, °fervent, fervid, perfervid, vehement, °frantic, °fanatical, frenzied: *I had an intense desire to see the culprit brought to justice.* **3** highly strung *or Chiefly US* high-strung, °emotional, °temperamental, °tense, °touchy, °testy, °volatile, °hysterical, °hot-headed, °feverish, °nervous, (high-)spirited, °impetuous, °impulsive, *Colloq* uptight: *I appreciate his enthusiasm, but I wish Henshawe were a bit less intense about his politics.*

intensify *v.* °concentrate, °focus, °sharpen, °whet, °strengthen, °reinforce, °heighten, escalate, °deepen, °quicken, °emphasize, °magnify, °increase, augment, °double, redouble, °heat up, *Colloq* °step up, *Brit* hot up: *We must intensify our efforts to effect a settlement of the crisis. The war is intensifying.*

intensity *n.* concentration, °focus, °strength, forcefulness, °force, °power; °vigor, °energy, vehemence, °fervor, zeal, °ardor, °passion, °sincerity: *The storm is diminishing in intensity. Political partisanship is often attended by great intensity of emotion.*

intensive *adj.* concentrated, focused, intensified, °comprehensive, °exhaustive, thorough(going), all-out: *The police have launched an intensive manhunt.*

intent *n.* **1** °intention, °aim, °goal, °target, °purpose, °object, °objective, °end, °design, °plan, °idea: *Was it your intent that I should go with you? The charge is assault with intent to kill.* **2** °inclination, °tendency, °desire, intending: *The mere intent to commit treason is often tantamount to treason.* **3** ***to all intents and purposes.*** °virtually, °practically, for all practical purposes, (almost) as good as, (almost) the same as, more or less, °in effect: *Telling you that the position was being discontinued is, to all intents and purposes, the same as dismissing you.* —*adj.* **4** concentrated, focused, °fixed, °steady, °intense, °determined, °earnest, engrossed, °absorbed, °rapt, °steadfast, °resolute, °attentive: *Her intent stare unnerved many of her suitors.* **5** °bent, °set, °resolute, committed, °decided, °firm, °keen; resolved, °eager, °firm, °determined, zealous, avid, °enthusiastic: *If you're intent on going, we won't try to change your mind.*

intention *n.* °aim, °purpose, °intent, °design, °goal, °end, °object, °objective, °target, °ambition: *Was it your intention to stay till the fat lady sings?*

intentional *adj.* °deliberate, intended, °premeditated, meant, °willful, designed, planned, °preconceived, °studied, considered, contrived; °purposeful, °on purpose: *That slur was intentional and not just a slip of the tongue.*

intently *adv.* closely, attentively, concentratedly, earnestly, fixedly, unflinchingly, determinedly, °searchingly, steadily, steadfastly, continuously, assiduously, doggedly, unremittingly, eagerly, keenly, studiously: *They regarded one another intently across the table.*

intercept *v.* °interrupt, °deflect, °stop, °arrest, °check, °interfere (with), °head off, °block, °impede, °cut off, °seize, °grab, °catch, °trap: *He was intercepted as he was about to board a plane with the loot. Kevin intercepted the ball, preventing a goal.*

intercourse *n.* **1** °commerce, °traffic, °trade, °dealings, °exchange, communication, °contact, interaction: *In normal business intercourse our paths cross often.* **2** sexual intercourse, coitus, coition, sexual congress *or* union, mating, copulation, sexual relations, carnal knowledge, making love, lovemaking, intimacy, sexual connection, *Colloq* °sex: *The plaintiff admits engaging in intercourse with the defendant.*

interest *n.* **1** °attention, attentiveness, °concern, °regard, °curiosity, °scrutiny, °notice, °engagement: *The way the puzzle fitted together drew my interest. She examined the books with interest. She shows interest in taking a writing course.* **2** °concern, °significance, °importance, °weight, °moment, °note, consequence: *Of what interest is a newly found painting by Tiepolo?* **3** Often, ***interests.*** °profit, °advantage, °benefit, °good, avail, °worth, value, °consideration, °behalf, behoof: *Is it in our interest to sell the business?* **4** °share, °portion, °stake, investment, °piece, °cut, °percentage, participation, involvement: *I have a small interest in an emerald mine.* **5** Often, ***interests.*** °business, °concern, °affair, °property; °hobby, °pastime, °diversion, avocation, °amusement, °entertainment, °pursuit, °relaxation, °occupation: *She travels abroad to look after her interests. Caroline has many interests besides business.* **6** (lending) fee *or* charge, °percentage, °rate, *Slang US* vigorish: *How much interest does your bank charge for a twenty-year mortgage at 50 percent of value?* —*v.* **7** °engage, absorb, engross, °attract, °draw, °catch, °capture, °captivate, °hold, °fascinate, °intrigue, °excite, °incite, °provoke, °arouse, °affect, °quicken, infect, °animate, °kindle, °fire: *At that time, our daughter was interested only in boys. There is something about astronomy that interests me.* **8** °influence, °induce, °persuade, °move, °tempt, °involve, °enroll, °enlist, °dispose, °incline, °prevail upon, °talk into, °concern: *Could I interest you in investing in my company?*

interested *adj.* **1** Also, ***interested in.*** °engaged, °absorbed, engrossed, °drawn (to), attracted (by), °involved (in), °curious (about), fascinated (by), °keen (on), stimulated (by), °responsive (to), °concerned (about): *We talked about investing in my plastics company, and he seemed interested. She has become interested in designing jewelry.* **2** °concerned, °involved, nonobjective, °partial, °biased, °prejudiced, °prejudicial, °partisan, predisposed: *You cannot get an honest appraisal of the painting's value from an interested party.*

interesting *adj.* °absorbing, °engaging, gripping, °riveting, engrossing, °attractive, compelling, intriguing, °provocative, °stimulating, °exciting, °inviting, fascinating, °enchanting, °spellbinding, captivating: *Nick has just told me the most interesting story about Tony.*

interfere *v.* **1** °meddle, °intrude, °butt in, °intervene, intercede, interpose, °interrupt, *Colloq* horn in, put *or* stick in one's oar *or* one's oar in, poke one's nose in, *US* kibitz: *Stop interfering in things that do not concern you.* **2** °hinder, °impede, °hamper, °block, °obstruct, °encumber, °slow, °retard, °handicap, °set back, get in the way of, °frustrate, °conflict, °inhibit, °trammel, °subvert, °sabotage: *Allow nothing to interfere with the course of true love.*

interference *n.* **1** meddling, intrusion, intruding, intervention, interceding, intercession: *This interference in our personal affairs has gone far enough.* **2** °hindrance, °impediment, °block, °obstruction, °encumbrance, impedance, °difficulty, °snag, °hitch, °handicap, °setback; frustration, °inhibition, °conflict, °opposition: *The lawsuit became an unwelcome interference in the smooth flow of our lives.*

interior *adj.* **1** °inside, °internal, inner, inward: *The interior surfaces are to be painted white.* **2** °internal, °domestic, °civil, °national, °local, °home: *Pollution of water sources is an interior, not an international problem.* **3** inner, °private, °intimate, °personal, °individual, °secret, °hidden, °veiled: *Some of Hamlet's great speeches are in the form of interior monologues, or soliloquies.* **4** upland, inland, upcountry, landlocked: *The most beautiful scenery is in the interior sections, away from the marshy coast.* —*n.* **5** °inside: *The interior is coated with Teflon.* **6** °heart, °center, °middle, °core, °depths: *This is the story of my journey to the interior of the earth.* **7** uplands, upcountry, heartland, hinterland: *The interior is covered with rocks and scrub pine.*

interjection *n.* °exclamation, ejaculation, °cry, interpolation, utterance: *Grammarians regard expressions like "Hello" and "Goodbye" as interjections.*

interlude *n.* °interval, entr'acte, °intermission, °pause, °stop, stoppage, °respite, °interruption, °break, hiatus, lacuna, °gap, °halt, °wait, breathing space, °recess, °rest, °spell, °lull, *Colloq* °letup: *There is a ten-minute interlude between the acts.*

intermediary *n.* °go-between, middleman, °agent, °representative, °broker, °intermediate, third party, °mediator, arbitrator, arbiter, referee, °umpire, °judge: *Donaldson acted as intermediary and arranged the deal.*

intermediate *adj.* **1** °middle, in-between, medial, midway, halfway, transitional, intervening, °intermediary: *In sublimation a substance changes from a gas to a solid (or vice versa), skipping the intermediate liquid state.* **2** See **intermediary.**

intermission *n.* See **interlude.**

intermittent *adj.* °irregular, discontinuous, °disconnected, °sporadic, °occasional, °random, °spasmodic, °fitful, °broken, °periodic, alternating, cyclic(al), rhythmic(al), pulsating, seasonal, on-and-off, on-again-off-again, stop-and-go, stop-go: *Tomorrow, intermittent showers in the morning will yield to sunshine.*

internal *adj.* See **interior, 1, 2, 3.**

international *adj.* supranational, °global, worldwide, °universal, intercontinental, cosmopolitan, ecumenic(al) *or* oecumenic(al); °foreign: *An international conference is being held on disarmament. My father is engaged in international trade.*

interpret *v.* **1** °explain, explicate, °clear up, °clarify, elucidate, °illuminate, throw *or* shed light on, °simplify, °decipher, decode, °define, °spell out, make sense (out) of, °translate, °paraphrase: *Would you interpret this clause of the agreement for me?* **2** °understand, construe, °take (to mean), °read, figure *or* work out, °sort out, unravel: *I haven't any idea how to interpret this poem.*

interpretation *n.* **1** °explanation, clarification, elucidation, simplification, decipherment, °solution, working out, unraveling, sorting out, decoding, °definition, °illustration, °translation, paraphrasing: *Sharon's detailed interpretation of the data is quite easy to understand.* **2** °analysis, diagnosis, °examination, exegesis, explication, reading, construal, °inference, °understanding: *These figures are subject to individual interpretation.*

interrogation *n.* questioning, °examination, cross-examination, inquisition, °investigation, *Colloq* third degree, grilling: *Caught by the rebels, we were subjected to hours of interrogation.*

interrupt *v.* **1** °break in, °cut in, intrude in, °butt in, interfere in, °punctuate, °disturb, *Colloq* barge in, °chime in, horn in: *He keeps interrupting the discussion with his silly remarks.* **2** °discontinue, °break off, °cut off, °cut short, interfere with, °disrupt, °suspend, °hold up, °halt, °stop, °end, °terminate, °cease: *We interrupt this program to bring you a news bulletin.*

interruption *n.* **1** °break, intrusion, °disturbance, °interference, disruption: *Please forgive this interruption, but you're wanted on the phone.* **2** °break, °gap, °interval, lacuna, hiatus, °respite, °rest, °pause, pausing, °intermission, stopping, °stop, °suspension, cessation, °cease, ceasing, surcease, hesitation, *Prosody* caesura, *Colloq* °letup: *She talked for an hour without interruption.*

interval *n.* **1** °intermission, °interlude, entr'acte, °break, °pause; °recess, °rest (period), °period, °time, °wait, °spell, °delay, °lapse: *The play was so bad we left in the interval. After a brief interval, the lights came on again.* **2** °meanwhile, °meantime, interim: *He returned later but found that in the interval she had changed her mind.* **3** °gap, °opening, °space, °hole, °void, lacuna, °distance, interstice, *Architecture* intercolumniation: *The interval between the columns is exactly twenty feet.*

intervene *v.* **1** °interfere, °intrude, °break in, °interrupt, intercede, °meddle, interpose, °butt in, *Colloq* poke one's nose in, horn in, put in one's oar, °step in: *If I had not intervened, you might have been killed. She intervened on my behalf.* **2** come *or* go (between), °pass, °elapse: *A week intervened before we saw each other again.*

interview *n.* **1** °meeting, °(press) conference, °discussion, °conversation, °talk, question period, audience: *The reporter phoned to ask for an interview.* **2** °evaluation, appraisal, vetting, assessment: *I have a job interview scheduled for tomorrow.* —*v.* **3** °question, °examine, interrogate, °sound out, talk with *or* to: *She interviewed the prime minister on television last night.* **4** appraise, °evaluate, °check (out), °vet: *The headmaster himself interviews the teachers.*

intimate[1] *adj.* **1** °close, °personal, °warm, °affectionate, loving, °dear, °bosom, cherished, °familiar, *intime*: *She had a party for intimate friends.* **2** °secret, °confidential, °private, °personal, °privy, °hidden, *intime;* °detailed, °penetrating, °deep, °profound, °thorough, °exhaustive: *In his autobiography he reveals intimate particulars of his marriage.* **3** °sexual; °carnal: *It is well known that Millie was on intimate terms with her chauffeur.* **4** °cozy, °informal, °snug, °friendly, °warm, °comfortable; *intime, à deux,* °tête-à-tête, *Colloq* comfy: *The two of us had an intimate dinner at home last night. Mariuccia is singing at an intimate club in Soho.* —*n.* **5** °friend, °associate, °comrade, crony, °familiar, confidant(e), °(constant) companion, Achates, alter ego, °colleague, confrère, *Colloq* sidekick, °chum, °pal, *Brit and Australian* °mate, *US* buddy, *Slang Brit* china (plate), mucker: *He and a few of his intimates like to get together now and then for a game of poker.*

intimate[2] *v.* °hint, °imply, °suggest, °insinuate, °indicate, °refer to, allude to, °communicate, °make known, give (someone) to understand, °warn, °caution, *Colloq* °tip (off): *He intimates that my wife will be a widow if I tell the police.*

intimidate *v.* °frighten, °scare, °alarm, cow, °daunt, °dismay, abash, °appall, awe, °overawe, °browbeat, °menace, °threaten, °terrify, °petrify, terrorize, °tyrannize; *Slang* have *or Slang* get (someone) by the short and curlies: *You can't intimidate me with your threats!*

intolerance *n.* °bias, °prejudice, °bigotry, °discrimination, °partiality, illiberality, narrow-mindedness, dogmatism; racism, racialism, sexism, classism, ageism, xenophobia: *The government has legislated against intolerance of minorities.*

intolerant *adj.* **1** °unsympathetic, unforbearing, unindulgent, °impatient, °inconsiderate, °inhospitable, uncharitable: *The present system seems intolerant of the aged.* **2** °biased, °prejudiced, °bigoted, discriminatory, °partial, illiberal, °narrow-minded, °parochial, °provincial, °jaundiced, warped, twisted, °one-sided, °opinionated, close-minded; racist, racialist, sexist, classist, ageist, xenophobic: *How can any religion be so intolerant of other people's beliefs?*

intonation *n.* °accent, accentuation, speech *or* sound pattern, °delivery, modulation, articulation, °pronunciation, vocalization, °pitch, °tone, inflection: *You can tell the speaker's mood by his intonation.*

intoxicate *v.* **1** inebriate, make drunk, addle, stupefy, °muddle, befuddle: *He tried to intoxicate me with his homemade brew.* **2** °stimulate, °excite, °overwhelm, elate, exhilarate, °animate, °enliven, invigorate, inspirit, °thrill, galvanize, °electrify, make one's head spin, °take one's breath away, infatuate, °entrance, °enchant, °enrapture, °fascinate, °bewitch, cast a spell on, ensorcell: *You intoxicate my soul with your eyes.*

intoxicating *adj.* **1** °alcoholic, spirituous, inebriant: *Do not drink intoxicating beverages if you plan to drive.* **2** °exhilarating, °invigorating, °thrilling, °exciting, heady, °stimulating, electrifying, entrancing, fascinating: *Making a film of the life of Lola Montez is an intoxicating idea.*

intrepid *adj.* °fearless, °brave, °bold, °daring, °dauntless, undaunted, °steadfast, °resolute, °courageous, unafraid, plucky, °gallant, valiant, valorous, doughty, °audacious, °heroic, °manly, manful, °dashing, °adventurous, °venturesome, stouthearted, lionhearted, °game: *Quatermain was known to the enemy as an intrepid soldier.*

intricate *adj.* **1** °involved, °complicated, convoluted, entangled, tangled, knotty, complex, twisted, winding, °tortuous, sinuous, anfractuous, °labyrinthine, °elaborate, Byzantine, °fancy, °ornate, rococo, Daedalian *or* Daedalean *or* Daedalic, *Literary* daedal *or Literary* dedal: *She related to me a plot that was far too intricate for children to follow.* **2** °perplexing, °puzzling, mystifying, enigmatic: *The code was so intricate that even the computer needed hours to decipher the message.*

intrigue *v.* **1** °fascinate, °beguile, °captivate, °attract, absorb, °charm, pique, °interest, titillate, arouse *or* excite the curiosity (of): *It intrigues me to watch them put those ships into bottles.* **2** conspire, °plot, connive, °scheme, °maneuver: *She was sure that everyone was intriguing against her.* —*n.* **3** °conspiracy, °plot, °scheme, °maneuver, collusion, °stratagem, °trickery, °chicanery, double-dealing, guile, °subterfuge, °artifice, °machination, °deception: *Many of those close to the government are engaged in intrigue of some kind.* **4** °affair, °liaison, amour, °romance, °affaire, *affaire d'amour* , *affaire de coeur*, intimacy; adultery: *We all knew about the intrigue he was carrying on with the duke's wife.*

intrinsic *adj.* °inherent, °basic, °fundamental, °essential, °proper, °elemental, °organic, °natural, °native, inbred, congenital, inherited, °hereditary, innate, °inborn, immanent, indwelling, underlying, constitutional; °real, °true, °actual, °genuine: *As no painting has intrinsic worth, its value is arbitrary.*

introduce *v.* **1** °acquaint, °present, °make known: *Sandy introduced Gerald and Daphne to each other.* **2** bring in *or* up, °advance, °present, °broach, put *or* set forth, °put forward, °suggest, °offer, °propose, °mention: *It was she who introduced the issue of bacteria in canned goods.* **3** °announce, °present: *Please introduce the next speaker.* **4** °start, °begin, °originate, °launch, °establish, °set up, °pioneer, °initiate, usher in, °institute, bring out *or* in, °set up, °organize: *When was paper money first introduced?* **5** °insert, °add, interpose, °inject, °put in, °inject, interpolate: *Why introduce irrelevant matters into the discussion?*

introductory *adj.* **1** °opening, °prefatory, °preliminary, °preparatory, °beginning, inaugural, °initial: *His "introductory" remarks lasted longer than the speeches!* **2** °primary, °basic, °fundamental, °elementary, °first, °rudimentary: *Take an introductory course before enrolling for more advanced study.*

intrude *v.* °interfere, °break in, °interrupt, °intervene, push in, interpose, °butt in, °infringe, °encroach, °obtrude, *Colloq* horn in, barge in: *They want to be alone and you are intruding. Forgive me for intruding into your conversation.*

intruder *n.* **1** interloper, gate-crasher, uninvited guest, unwelcome visitor; trespasser, encroacher, invader, infiltrator, squatter; °burglar, °thief: *The intruders were quickly expelled by the guards.* **2** meddler, °busybody, *Colloq* snoop(er), Nosy Parker, *US* kibitzer, buttinsky: *You are an intruder into my private affairs.*

intrusive *adj.* intruding, interfering, meddlesome, invasive, meddling, prying, °inquisitive, °obtrusive, importunate, °officious, °presumptuous, °forward; °unwelcome, uncalled-for, unwanted, unsought, *Colloq* °nosy, °pushy, snoopy: *He said he apologized if he was being intrusive.*

intuition *n.* °instinct, °insight, °hunch, sixth sense, presentiment, °premonition, °foreboding; °perception, perceptiveness, percipience, perspicacity, common sense, mother wit: *Intuition told me I would find the answer here.*

invalid[1] *adj.* **1** ailing, °sick, °sickly, °ill, °infirm, valetudinarian, °disabled: *His invalid sister is confined to a wheelchair.* —*n.* **2** °patient, valetudinarian, °victim, sufferer, °incurable, °cripple, *Chiefly US and Canadian* shut-in: *Many invalids have claimed complete recovery after a visit to Lourdes.*

invalid[2] *adj.* °void, null (and void), nullified, annulled, repudiated, °untrue, °false, °faulty, °erroneous, °wrong, °spurious; °incorrect, °imperfect, impaired, °unsound, °untenable, °ineffective: *He was driving with an invalid license. Your reasons for quitting school are invalid.*

invaluable *adj.* °priceless, valuable, °precious, of inestimable *or* incalculable value; irreplaceable, irredeemable; costly, °expensive, high-priced, °dear: *The thieves made off with several invaluable paintings.*

invariable *adj.* **1** unchanging, °changeless, unvarying, invariant, unwavering, °constant, °steady, °stable, °regular; °fixed, °fast, °set, °rigid, °uniform, unfailing, unexceptional: *She is invariable in her opinion of a woman's right to abortion. His invariable routine calls for a dawn swim, regardless of the weather.* **2** immutable, unchangeable, unalterable, unmodifiable: *The law of supply and demand is invariable.* **3** °permanent, °fixed, °enduring, °abiding, °eternal, unaltered, unvarying, unchanged, unvaried, unaltered, unmodified: *The positions of the stars appear to be invariable.*

invasion *n.* **1** incursion, °raid, foray, intrusion, °inroad, encroachment, trespass, infiltration; °infringement, infraction, °transgression, °violation: *The poachers were charged with invasion of private property. Unauthorized publication of that material is an invasion of your rights.* **2** °attack, °assault, onslaught, °aggression, °offensive, °drive, storming, blitzkrieg: *The armored divisions succeeded in stopping the invasion.*

invent *v.* **1** °create, °devise, contrive, °originate, °think up, dream up, °conceive, concoct, °make up, °imagine, °formulate, °improvise, °design, °hit upon; °coin: *He claims to have invented the toothpaste tube.* **2** °fabricate, °make up, concoct, *Colloq* cook up: *She invented that story about having been a lion tamer.*

invention *n.* **1** °creation, origination, contriving, devising, °conception, contrivance, introduction, °development: *The invention of the screwdriver has spared many a broken fingernail.* **2** °creation, contrivance, °device, °gadget, *Colloq* °contraption, *US* gizmo: *Thomas Edison held patents on a huge number of inventions.* **3** fiction, figment, °story, °fantasy, °fabrication, °tale, fable, °yarn, °fib, tall story *or* tale, falsification, °fake, °sham, °falsehood, °lie, prevarication: *Her claim that the accident was my fault is a flagrant invention.*

invest *v.* **1** °venture, °lay out, °put in, °sink: *She was persuaded to invest her life savings in government bonds and municipals.* **2** °devote, °allot, °spend, °contribute, °supply, °provide: *We have invested a lot of time in cleaning up local government.* **3** °install, °inaugurate, °induct, °initiate, instate, °establish, ordain, swear in, °seat: *Dmitri was invested as archimandrite at yesterday's ceremony.*

investigate *v.* enquire *or* inquire into, °examine, °study, °consider, °explore, °probe, °look into, °research, °scrutinize, °analyze, °sift (through), winnow: *The laboratory is investigating the nature of the strange phenomenon.*

investigation *n.* enquiry *or* inquiry, °examination, °study, °review, °exploration, °quest, °search, °probe,

°research, discovery procedure, °scrutiny, °analysis, inquest, inquisition, °interrogation, questioning: *Has the investigation turned up any evidence of collusion?*

invigorating *adj.* °stimulating, °bracing, rejuvenating, °tonic, vitalizing, restorative, energizing, vivifying, enlivening, °exhilarating; °fresh, healthful, °healthy, salubrious, salutary: *Each morning I go for an invigorating walk. The doctor recommended the invigorating mountain air.*

invincible *adj.* **1** unconquerable, °unbeatable, °indomitable, insuperable, undefeated, unstoppable: *Crusher McGuhan reigned as invincible boxing champion for ten years.* **2** °impregnable, invulnerable, impenetrable, °indestructible, unassailable: *The Romans believed the fortress to be invincible.*

invisible *adj.* **1** unseeable, °imperceptible, undetectable, imperceivable; unseen: *The air we breathe is invisible.* **2** concealed, °hidden, disguised, camouflaged, masked, covered, unperceived, °veiled, indiscernible: *Once the part has been painted, the damage will be invisible.*

invitation *n.* **1** summons, °request, °call, °bidding, *Colloq* invite: *I am still waiting for my invitation to the White House.* **2** °attraction, °inducement, allure, allurement, °enticement, °temptation, °magnetism, bait, °lure, °draw, °pull: *The possibility of going where no man had gone before was too great an invitation to ignore.*

inviting *adj.* alluring, °tempting, enticing, °attractive, beckoning, appealing, captivating, °engaging, intriguing, °irresistible, winsome, beguiling, bewitching, entrancing, fascinating, tantalizing, °seductive: *She gave him an inviting smile but he still approached cautiously.*

involuntary *adj.* °unconscious, unintentional, °unthinking, °impulsive, °spontaneous, °unpremeditated, °instinctive, instinctual, unwitting; °automatic, reflex, °mechanical, conditioned, °uncontrolled, uncontrollable: *When I was struck, my involuntary reaction was to strike back at once. She is not winking at you, Bostick, she has an involuntary tic in her eye.*

involve *v.* **1** °include, °contain, comprise, °cover, °embrace, °incorporate, encompass, °take in, subsume, °embody, °comprehend, number among, count in: *The survey involved many people from all walks of life.* **2** °imply, °entail, °suggest, °mean, betoken, °require, necessitate, °presuppose: *Enrolling for a course involves doing homework as well as attending classes.* **3** Often, ***involve in*** *or* ***with.*** °implicate, °concern, °affect, °touch, °entangle, °draw in; °incriminate, inculpate; °associate with, connect with, catch (up) in: *I didn't know that Annette was involved. Are you involved in that murder investigation?*

involved *adj.* **1** implicated, °concerned, °affected, °interested, °active: *This hearing was attended by all involved citizens of the community.* **2** tangled, °complicated, complex, twisted, snarled, convoluted, °confused, confusing, °intricate, °tortuous, °elaborate, knotty, Byzantine, °labyrinthine: *The plot is too involved to be followed easily. The involved problems of adolescence cannot be treated in a one-day conference.* **3** ***involved with.*** associated with, entangled with, embroiled with, enmeshed with, *Colloq* °mixed up with: *Dennis is still very much involved with that singer from the Green Dragon.*

irk *v.* °irritate, °annoy, vex, °pester, °provoke, °chafe, nettle, °exasperate, *Colloq* needle, miff, °aggravate, °bug, peeve, rub (someone) the wrong way, °put out: *It really irks me to know that we lost because we didn't practice.*

irksome *adj.* irritating, annoying, vexing, vexatious, chafing, nettling, exasperating, bothersome, °troublesome, °burdensome, °tiresome, °tedious, °boring, wearisome, uninteresting, *Colloq* aggravating, pestiferous: *The mosquitoes are particularly irksome on muggy, windless evenings.*

irregular *adj.* **1** uneven, °bumpy, °lumpy, °coarse, °rough, unequal, unsymmetrical, asymmetric(al), °pitted, potholed, °jagged, craggy, °lopsided: *The irregular surface of the road bounced us about in the car.* **2** °sporadic, uneven, °random, °erratic, unequal, °fitful, °haphazard, unsystematic, unsystematized, °disorderly, °uncertain, unmethodical; °occasional, °casual: *In the distance I could hear the staccato of irregular machine-gun fire.* **3** °extraordinary, °unusual, °eccentric, °abnormal, anomalous, aberrant, °unnatural, °peculiar, °queer, °odd, °weird, °bizarre, °strange, °singular, nonconforming, °nonconformist, °exceptional, unconventional, °offbeat, uncommon, freakish, *Colloq* freaky: *Don't you think that keeping a Komodo dragon as a pet is a bit irregular?*

irrelevant *adj.* °inappropriate, °inapplicable, °impertinent, °unrelated, °alien, inapposite, malapropos, °beside the point, inapt, nongermane, unconnected, °extraneous, neither here nor there, °out of place, °gratuitous, uncalled-for, *Colloq* °out of the blue, off the beam, *Slang* off the wall: *The name of the person who asked the question is entirely irrelevant.*

irrepressible *adj.* unrestrainable, irrestrainable, uncontainable, uncontrollable, unmanageable, insuppressible *or* unsuppressible, unstoppable, °ebullient, °buoyant, °effervescent, bubbling, °boisterous: *Nothing could dampen our irrepressible high spirits after winning the game.*

irreproachable *adj.* °blameless, unimpeachable, beyond reproach, unreprovable, °faultless, °innocent, above suspicion, °impeccable, inculpable, °honest, °pure: *Till now, Forsyth's record in the army has been irreproachable.*

irresistible *adj.* **1** °irrepressible, unconquerable, °indomitable, °overpowering, °unbearable, °overwhelming, °overriding, unmanageable, °ungovernable, uncontrollable: *I had an irresistible desire to punch him in the nose.* **2** unstoppable, inexorable, °relentless, °unavoidable, ineluctable, inescapable: *What happens when an irresistible force meets an immovable object?*

irresolute *adj.* vacillating, wavering, faltering, °indecisive, infirm of purpose, in *or US only* of two minds, undecided, °hesitant, hesitating, shifting, changing, °erratic, °uncertain, unsure, undetermined, °unresolved, °halfhearted, *Colloq* °wishy-washy: *You must act; this is no time to be irresolute.*

irrespective of *prep.* °regardless of, °notwithstanding, °despite, °apart from, °in spite of, without regard to, ignoring, discounting: *We shall carry on irrespective of public opinion.*

irresponsible *adj.* °careless, °reckless, devil-may-care, unanswerable, °unaccountable, nonliable, °rash, °unruly, °wild; °unreliable, undependable, untrustworthy, °weak, feckless, °ineffectual: *Cyril is too irresponsible to take care of the children by himself.*

irretrievable *adj.* **1** nonretrievable, unretrievable, unrecoverable, irrecoverable, unsalvageable, unsavable, °lost, irreclaimable: *The data deleted in the computer is totally irretrievable.* **2** irreparable, irremediable, uncorrectable, unrectifiable, irredeemable, °irreversible, °irrevocable: *The radiation from the atomic blast did irretrievable damage.*

irreverent *adj.* **1** °blasphemous, °impious, °profane, °sacrilegious, unholy, °ungodly, irreligious: *He was rep-*

rimanded for his irreverent attitude at morning prayers. 2 °disrespectful, insulting, °insolent, °rude, °discourteous, uncivil, derisive, °impudent, °impertinent, saucy, °flippant, mocking, tongue-in-cheek, *Colloq* °flip, °cheeky: *The prince did not appreciate being the butt of the irreverent skit.*

irreversible *adj.* unreversible, nonreversible, °irrevocable, unchangeable, unalterable, °permanent, °fixed, °final, unrepealable, irredeemable, °irretrievable: *Burning, essentially the chemical process of rapid oxidation, is irreversible.*

irrevocable *adj.* °irreversible, unchangeable, immutable, °changeless, °fixed, unalterable, settled, unrecallable, °irretrievable, irrepealable, undoable; irreparable, °permanent, °enduring, °everlasting: *The colonel says that his was an irrevocable order. That speech did you irrevocable harm.*

irritable *adj.* °impatient, °excitable, °testy, °touchy, °quarrelsome, grouchy, °fretful, °peevish, °cross, crabby, crusty, °short-tempered, °petulant, °prickly, irascible, °moody, °temperamental, °gruff, °cantankerous, curmudgeonly, dyspeptic, bad-tempered, ill-tempered, ill-humored, snappy *or* snappish, grumpy *or Brit also* grumpish, *Colloq* crotchety, *US and Canadian and Irish* °cranky: *Why are you always so irritable before breakfast?*

irritate *v.* °annoy, vex, nettle, °pester, °provoke, °bother, °anger, °enrage, °chafe, pique, °exasperate, °ruffle, hector, °harass, harry, °nag, °plague, °worry, °fret, °fluster, °trouble, pick at *or* on, *Colloq* needle, get under (someone's) skin, get in (someone's) hair, hassle, peeve, get on (someone's) nerves, drive (someone) up the wall, get (someone's) hackles up, get (someone's) back up, drive (someone) crazy *or* mad, *Brit* rub (someone) up the wrong way, get up (someone's) nose, *US* rub (someone) the wrong way, burn (someone) up: *All these stupid questions are beginning to irritate me.*

island *n.* isle, islet, ait, cay, °key; atoll; archipelago; *Brit dialect* eyot, holm: *There are actually about fifteen hundred islands in the Thousand Islands in the St. Lawrence river.*

isolate *v.* °separate, °segregate, sequester, cloister, °detach, °cut off, send to Coventry, °ostracize, °maroon, °exclude, °shut out, °bar, debar, °banish, deport, °transport, °exile, °reject, °eject, °throw out, °expel, °shun, °spurn, °avoid, °ignore, snub; quarantine; *Colloq* °cut, send (someone) to Siberia, give (someone) the cold shoulder: *If they disapproved of a person's actions, they would isolate him.*

isolated *adj.* 1 lone, °solitary, °single, °singular, °unique, anomalous, °separate, °special, °particular, °individual, °exceptional, °unrelated: *In one isolated case they recommended a suspended sentence.* 2 °alone, separated, segregated, °secluded, sequestered, cloistered, unconnected, °detached, °(set) apart, removed, °cut off, excluded; °forlorn, °lonely, hermitic(al), eremitic(al), anchoretic(al), troglodytic(al), monastic: *Away from his friends and family, he felt totally isolated. After his wife died, he led an isolated existence.* 3 °secluded, °remote, °out-of-the-way, off-the-beaten-track, unfrequented, °lonely; °secret, °hidden: *For twenty years we lived in that isolated shack in the wilderness.*

issue *n.* 1 outflow, °outgoing, °exit, egress, issuance, emanation, efflux, debouchment, °emergence, °outlet: *The river's color changed abruptly at its point of issue into the sea.* 2 °outcome, conclusion, consequence, culmination, °result, °end, °effect, °consummation, °event, °climax, *Colloq* °payoff: *Whatever the issue, we can be satisfied that we made a brave effort.* 3 °point, °topic, °subject, °matter, °affair, °problem, °question: *That is an issue you should take up with the mayor.* 4 Usually, ***major issue.*** (major *or* big) problem *or* difficulty, °controversy, °fight, °dispute, *cause célèbre*: *He turns the question of who is to walk the dog into a major issue.* 5 printing, °edition, °version; °copy, °number: *I have a copy of the Sunday issue. See if you can buy an issue of today's paper.* 6 °publication, promulgation, issuance, issuing, °distribution, °delivery, dissemination, broadcasting, °proclamation, °circulation: *The issue of the warning came only minutes before the explosion.* 7 °offspring, child *or* children, descendant(s), °progeny, °young, scion(s), son(s), daughter(s): *According to the records, your uncle died without issue, making you his sole heir.* 8 ***at issue.*** in contention, in dispute, °unresolved, °unsettled, °uncertain, up in the air, to be decided: *The point at issue is which shell has the pea under it.* 9 ***take issue.*** °disagree, °argue, contend, °dispute, °oppose, take exception: *I feel that I must take issue with your conclusion.* —*v.* 10 °proclaim, promulgate, °declare, °publish, °put out, put *or* set forth, °announce, °circulate, °distribute, get out, °release, °deliver, °broadcast, disseminate, get out: *The kidnappers have issued an ultimatum.* 11 °emerge, come *or* go forth, °exit, °emanate, °discharge, °stream, °flow, °pour; °appear, °originate, °spring, °stem, °arise: *The play over, people issued from the theater. Where Pegasus stamped his foot the Pierian spring issued forth.*

itch *v.* 1 °tickle, tingle, °prickle: *These mosquito bites itch terribly.* 2 °desire, crave, °hanker, °hunger, °thirst, °yearn, pine, °wish, °want, °die: *I am itching to get my hands on whoever told you that I was dead.* —*n.* 3 °tickle, tickling, tingle, tingling, °prickle, prickling, irritation: *My frustration is like having an itch I can't scratch.* 4 °desire, craving, hankering, °hunger, °thirst, yearning, °longing, *Colloq* yen: *I have a sudden itch to visit mother for the weekend.*

item *n.* 1 °detail, article, °point, °particular, °matter, °thing, °element, component, °ingredient: *There's one item I'd like you to keep in mind.* 2 °piece, °mention, °notice, °note, °memorandum, memo, filler, jotting: *We often publish short items to fill out a column.*

itemize *v.* °enumerate, °list, °specify, particularize, °detail, °document, °number, °record, °count, °tabulate: *Must I itemize every single book in the inventory?*

J

jab *v.* 1 °stab, °thrust, °poke, °dig, °prod; °plunge; °nudge; °tap: *The doctor jabbed the needle into my arm.* 2 °punch, °hit, °strike, °belt, smack, °rap, whack, thwack, cuff, thump, wallop; elbow; *Colloq* °clip, sock, slug, biff: *I jabbed him in the jaw with a quick left.* —*n.* 3 °stab, °thrust, °poke, °dig, °prod, °nudge: *I felt the jab of her elbow signaling me to be quiet.* 4 °punch, °belt, smack, °rap, whack, thwack, cuff, thump, wallop, *Colloq* °clip, sock, slug, biff: *A hard jab to the midsection made the bully turn to pudding.*

jabber *v.* 1 blether *or US only* blather, °chatter, °babble, gibber, gabble, prate, °prattle, °patter, °drivel, °rattle, *Brit* natter, *Scots* yatter, *Colloq* °gab, gas, °yap, witter: *The couple behind me jabbered throughout the entire film.* —*n.* 2 See **jargon, 2.**

jade *n.* 1 °nag, °hack, *Slang Brit* °screw, *US* °plug: *That old jade hasn't won a race in his last ten times out.* 2 °shrew, harridan, °nag, °hag, °drab, °witch, crone, hussy, minx, vixen, virago, termagant, beldam, slut, °slattern, trull, trollop, baggage, °tart, *Slang* battle-ax,

°broad, °bitch, old bag, floozy *or* floozie *or* floosie: *He was married to an expensive jade of a wife.*

jaded *adj.* **1** °exhausted, °weary, °tired, dead tired, bone-tired, bone-weary, dog-tired, °fatigued, enervated, °spent, *Colloq* °(dead) beat, °dead, bushed, fagged, *US and Canadian* pooped: *She charmed away the weariness of his jaded mind.* **2** sated, satiated, cloyed, °surfeited, glutted, gorged, fed up, sick (and tired) of, slaked; °dull, bored: *You need a little champagne and caviar to reawaken your jaded palate.*

jag *n.* °spree, °carouse, °orgy, °bout, *Colloq* binge, *US and Canadian* toot: *She was terribly hungover after last night's jag.*

jagged *adj.* °rough, uneven, °notched, sawtooth, °ragged, toothed, spiked, indented, denticulate, °serrated, chipped: *I cut myself on the jagged edge of that broken window.*

jail *n.* **1** °prison, lockup, reformatory, *Brit* borstal, *Brit also* gaol, *US* penitentiary, °reform school, *Nautical* brig, *Slang* cooler, clink, can, °jug, °stir, slammer, *Brit* °nick, quod, choky *or* chokey, *US* calaboose, big house, °pen, coop, hoosegow, pokey: *They were sent to jail for life.* —*v.* **2** °imprison, °lock up, incarcerate, detain, confine, *Brit* °send down, *Brit also* gaol, *US* °send up (the river): *He was jailed for thirty days.*

jailer *n.* turnkey, °guard, *Brit* warder, governor, *Brit also* gaoler, *US* warden, *Slang* °screw: *The jailers let us out for exercise for an hour each day.*

jam *v.* **1** °cram, °force, °push, °wedge, °stuff, °press, °ram, °squeeze, shove, °pack, °crowd: *We were jammed in so tightly that we couldn't move.* **2** °block, °obstruct, congest, °fill up, °clog, °plug, °stop up: *The toilet is jammed with paper again.* **3** °slam, °activate, actuate: *I jammed on the brakes.* —*n.* **4** °obstruction, blockage, blocking, °block, congestion, °tie-up, bottleneck, stoppage: *She was stuck in a traffic jam for an hour.* **5** °crush, °squeeze, °crowd, °mob, °swarm, multitude, °throng, °mass, horde, °pack, °press: *You cannot believe the jam of football fans at the playoffs!* **6** °trouble, °difficulty, °predicament, °quandary, °dilemma, *Colloq* °bind, °fix, °hole, pickle, hot water, °(tight) spot, °scrape: *Harry helped me out of a jam once, and I won't forget it.*

jamboree *n.* °gathering, °get-together, °party, °celebration, °fête, °festival, °festivity, carnival, °frolic, °revelry, °spree, °carouse, jubilee, revels, charivari *or US and Canadian also* shivaree: *Everyone is invited to the annual jamboree in the village square.*

jangle *v.* **1** clatter, °clash, °rattle, clang, clank, °crash, °ring, °jingle: *The chains jangled as the prisoners marched to their cells.* **2** °jar, °upset, °irritate: *The continuous screaming of the sirens jangled my nerves.* —*n.* **3** jangling, clatter, °clash, °rattle, jarring, clang, clanging, clank, clanking, °crash, clangor, °noise, °din, °racket, clamor, dissonance, cacophony, reverberation, *Literary* stridor: *I heard the jangle of the rag-and-bone man's cart in the next street.*

jar[1] *n.* crock; °receptacle, °vessel, container, urn, °pot, vase; °jug, pitcher, ewer, flagon, carafe, °bottle, amphora: *We always keep some small change in that blue jar.*

jar[2] *v.* **1** °shake, °agitate, °disturb, °stir, °shock, °jolt, jounce, °bounce, °jog, °jerk, °jiggle, joggle: *Don't jar the oven or the cake will collapse.* **2** °disagree, °conflict, °clash, °bicker, °quarrel, wrangle, °oppose, °discord: *She finds that her emotions about her ex-husband are jarring.* **3** °disturb, °upset, disconcert, unsettle, disquiet, °bother, °trouble, vex, °gall, °offend, °take aback, °irritate, °grate, °irk, nettle, °annoy: *It jars me to think that they got off with light sentences.* —*n.* **4** °shock, °start, °jolt, °surprise: *Seeing Sam after all those years gave me quite a jar.*

jargon *n.* **1** °cant, argot, °parlance, °idiom, °vernacular, slang; patois, creole, °dialect, pidgin; *Colloq* °lingo: *In the jargon of philately, that is known as a "first day cover."* **2** blether *or US also* blather, °chatter, °babble, °gibberish, °jabber, gabble, gobbledygook, °prattle, °patter, °drivel, cackle, jabberwocky, twaddle, °(stuff and) nonsense, °rubbish, codswallop, balderdash, bunk, humbug, °palaver, *bavardage, Colloq* rot, °garbage, hogwash, bosh, piffle, flapdoodle, chitchat, °gab, claptrap, *Slang* bull, crap: *When I questioned my bank manager about the fee, he just gave me a lot of jargon.*

jaundiced *adj.* **1** colored, tainted, distorted, twisted, °prejudiced, °opinionated, °biased, °preconceived, untrustworthy, °bigoted, °partial, unfair, °perverted; °dishonest, °corrupt: *Even the most jaundiced view must acknowledge the merits of the plan.* **2** splenetic, cynical, °bitter, °envious, °resentful, °jealous, °hostile, °spiteful, unfriendly, disapproving, °critical, unfavorable, disparaging, denigrating: *I can't say that I agree with Cartwright's jaundiced review of the play.*

jaunty *adj.* **1** °spirited, °lively, high-spirited, °buoyant, °brisk, °frisky, °sprightly, °free (and easy), °blithe, jovial, °happy, jubilant, °jolly, °merry, °cheerful, °gay: *It is heartening to see those pensioners in such a jaunty mood.* **2** °chic, °smart, °stylish, °dashing, °debonair, °elegant, colorful, °spruce, °flashy, °flash, °showy, °flamboyant, *Colloq* °sporty, natty: *Tipping his hat at a jaunty angle, the old boulevardier strolled off, twirling his walking stick.*

jealous *adj.* **1** °resentful, °bitter, grudging, °envious, covetous, green with envy, green-eyed: *Brian is jealous of attention paid to anyone but himself.* **2** °distrustful, distrusting, mistrustful, mistrusting, °suspicious; °anxious, °insecure, threatened, imperiled, °vulnerable: *Karen is very jealous of Kathleen. If anyone so much as looks at her, he feels jealous.*

jealously *adv.* watchfully, carefully, guardedly, protectively, warily, vigilantly, scrupulously, zealously, eagerly, attentively, anxiously, suspiciously: *Victor jealously keeps all details of his business to himself.*

jeer *v.* **1** Often, ***jeer at.*** °mock, laugh *or* scoff *or* sneer (at), °flout, °deride, °ridicule, °make fun of, °thumb one's nose at, gibe *or* jibe, °chaff, decry, °twit, °taunt, *Colloq* °rag, bullyrag, roast, *Brit* cock a snook at, *Brit and Australian* barrack, *Slang* °knock: *Don't jeer at aroma therapy till you've tried it.* —*n.* **2** °taunt, gibe *or* jibe, °aspersion, hoot, °hiss, boo, catcall; °derision, °ridicule, obloquy: *Just because he was fat, Charlie had to suffer the jeers of his classmates.*

jell *v.* **1** °set, congeal, °solidify, °harden, °coagulate, °thicken, °stiffen, gelatinize: *The mixture won't jell till you add hot water.* **2** °(take) form, take shape, crystallize, °materialize, come together, be set: *Our plans for the shopping center have not yet jelled.*

jeopardize *v.* °endanger, imperil, °threaten, °menace, °risk, °hazard, °venture: *You might jeopardize your freedom if you stand up for your rights.*

jeopardy *n.* Usually, **in** sometimes **at jeopardy.** danger, °peril; °threat, °menace, °risk, °hazard, °chance, uncertainty, vulnerability, °exposure, °liability: *Enter that lion's cage and you put your life in jeopardy.*

jerk *v.* **1** °yank, °wrench, °pluck, °nip, °tug, °twist, °tweak: *I jerked the dagger out of his hand, leaving him defenseless.* **2** twitch, °lurch, °jolt, °jump, °start, jig, °jiggle, °wriggle, wiggle: *The creature jerked about convulsively, screaming, then lay still.* —*n.* **3** °yank, °pull, °wrench, °tug, °twist, °tweak: *With one sharp jerk, he pulled off the adhesive tape.* **4** °lurch, °jolt, °start,

°bump: *The train stopped with a jerk, throwing me off balance.* **5** idiot, °fool, moron, imbecile, *Slang US* °dope, °creep, yoyo, nerd, dweeb: *Why would she want to go out with a jerk like that?*

jewel *n.* **1** °gem, gemstone, °brilliant, °ornament, bijou, *Colloq* °rock, sparkler: *Thieves stole a diamond necklace and an heirloom brooch set with precious jewels.* **2** °treasure, °marvel, °find, °godsend, °gem, °pearl, °prize, °boon, *Colloq* °catch: *What would you do without your secretary—she's an absolute jewel!*

jewelry *n.* gems, precious stones, jewels, ornaments, °finery, bijouterie: *Alexandra keeps her jewelry in a bank vault.*

jiggle *v.* **1** °jog, joggle, jig, °shake, °agitate, wiggle, °wriggle, °jerk: *It's odd to see grown people jiggling about on the dance floor like that. Jiggle the key up and down—maybe then you can turn it.* *—n.* **2** °jog, joggle, jig, °shake, wiggle, °jerk: *I gave the line a few jiggles, hoping to attract a fish.*

jilt *v.* °throw over, °reject, °dismiss, °drop, °discard, °desert, break (up) with, °forsake, °abandon, *Colloq* ditch, °dump, brush off *or* give (someone) the brushoff, *Chiefly US and Canadian* give (someone) his *or Chiefly US and Canadian* her walking papers: *Angela met Tony and promptly jilted Merton.*

jingle *v.* **1** tinkle, °ring, tintinnabulate, clink, °chink, °chime: *She wore a dozen bracelets, which jingled when she walked.* *—n.* **2** tinkle, tinkling, °ring, ringing, tintinnabulation, clink, clinking, °chink, chinking, °chime, chiming: *I like to feel the jingle of change in my pocket.* **3** °tune, ditty, °melody, °song, °rhyme, verse, doggerel: *The only thing he ever wrote was a jingle for a dog food commercial.*

jingoism *n.* chauvinism, flag waving, superpatriotism, nationalism; hawkishness, warmongering, belligerence, bellicosity: *"Might makes right" is a basic tenet of jingoism.*

jinx *n.* **1** °(evil) spell, °curse, evil eye, malediction, voodoo, *US and Canadian* hex: *I think I lost at roulette because she put a jinx on me.* **2** nemesis, Jonah: *If we don't throw that jinx overboard we shall all die.* *—v.* **3** °curse, °bewitch, °damn, °doom, °sabotage, °condemn, *US and Canadian* hex: *My career was jinxed from the start.*

jitters *n.* pl. shakes, fidgets, °nerves, uneasiness, queasiness, nervousness, skittishness, restlessness, apprehension, apprehensiveness, *Slang* heebie-jeebies, willies, *US* whim-whams: *He always gets an attack of the jitters before an exam.*

job *n.* **1** °work, °employment, °position, berth, livelihood; °career, °occupation, °calling, °vocation, °appointment, °pursuit, °field, °trade, °craft, °profession, métier, °area: *What kind of job is she looking for? Harry has a new job.* **2** °assignment, °responsibility, °concern, chore, °task, °undertaking, °function, °duty, °role, °mission, °province, contribution, °charge: *It is my job to see that the machines run properly.* **3** °task, °undertaking, °procedure, °proceeding, °affair, °operation, °project, °activity, °business, °matter, chore: *The job of changing the gasket may take a while.* **4** °problem, °difficulty, °burden, °nuisance, °bother; toil, °grind, °drudgery; *Colloq* °headache, °pain (in the neck), hassle, *Slang* pain in the *Brit* arse *or US* ass: *It was a real job getting them to pay for the damage.* **5** °crime, felony; °robbery, burglary, *Slang US and Canadian* °caper: *From the modus operandi, I'd say that the same gang did that job in Manchester.* *—v.* **6** Often, ***job out.*** °let out, °assign, apportion, °allot, share out, °contract, °hire, °employ, subcontract, °farm out, consign, commission: *We undertake to do the work, then we job it out to others.*

jog *v.* **1** °trot, lope, dogtrot, °run: *I jog around the reservoir every morning for exercise.* **2** °jar, °prod, °nudge, °arouse, °stir, °stimulate, °prompt, °activate, °shake: *I jogged his memory by referring to the time the dog bit him.* **3** °bounce, °shake, °jolt, joggle, jounce, °jerk: *I was being jogged about in the back of the van as we sped over the rocky terrain.*

join *v.* **1** °unite, °connect, °couple, °link, °marry, yoke, °combine, fasten *or* tie *or* glue *or* weld *or* solder (together), °unify: *These two pieces should be joined for greater strength.* **2** ally *or* league with, associate (oneself) with, team up with, throw (one's lot) in with, °enlist (in), sign (up) (with), °enroll (in), °enter: *She was invited to join the Carpathian Society.* **3** go *or* be with, °associate with, °accompany, attach (oneself) to, participate with: *Would you care to join us for a game of bridge?* **4** border (on *or* upon), °meet, °touch, abut, °butt, adjoin, be adjacent (to), extend to, °verge on, °coincide (with), juxtapose, be contiguous *or* conterminous (with), be coextensive (with): *The two properties join at the top of the ridge.*

joint *n.* **1** °seam, °union, °juncture, °connection, °junction, intersection: *Once painted, the joint won't show.* **2** *Slang* °dive, °dump, *US and Canadian* honkytonk: *We went into a joint that Looie knew to look for some action.* **3** roast: *Who carves the Sunday joint at your house?* *—adj.* **4** shared, °mutual, combined, collective, cooperative, °common, communal, collaborative: *The result will be achieved only by joint effort.*

jointed *v.* articulated, segmented, sectioned, sectionalized, hinged: *The stick is jointed so that it can be folded for carrying in the pocket.*

joke *n.* **1** jest, °witticism, °quip, bon mot, °laugh, wordplay, °pun, °story, anecdote, *Colloq* °gag, °wisecrack, one-liner, °crack: *Ronnie comes up with the funniest jokes I have ever heard.* **2** laughingstock, °butt, °(fair) game, buffoon: *After that incident, he became the joke of the regiment.* **3** farce, °mockery, °absurdity, travesty, °caricature: *My efforts to play the piano became a joke.* *—v.* **4** jest, °quip, °pun, °frolic, °wisecrack, °tease, °taunt, °banter, °chaff, °fool, *Colloq* kid, *US* crack wise: *They joked about our predicament. You must be joking if you think I'm letting you shoot an apple off my head!*

joker *n.* **1** jokester, °comedian, comedienne, funnyman, humorist, jester, °comic, °clown, °wag, °wit, punster, droll, °zany, merry-andrew, buffoon, trickster, prankster, *Colloq* °card, gagster, gagman, kidder: *Give him a drink and a funny hat and Roger thinks he's the greatest joker in the world.* **2** °catch, °hitch, °snag, °drawback, °trap, °twist, °pitfall, fine *or* small print, *Colloq* catch-22, no-win situation, *Taboo* nigger in the woodpile: *The joker is that whoever treats the patients catches the disease.*

jolly *adj.* **1** °merry, °cheerful, °frolicsome, °gay, jovial, °joyful, °sportive, convivial, jocund, jocose, jocular, °frisky, coltish, °playful, festive, jubilant, cheery, °exuberant, high-spirited, °animated: *Everyone was in a jolly mood at Harriet's birthday party.* *—v.* **2** Often, ***jolly along.*** °humor, appease, °deceive, °string along, °fool, °hoax: *They're just jollying him along because they want him to invest in their scheme.*

jolt *v.* **1** °jar, °shake (up), jostle, °bump, °bounce, °jerk: *The cart jolted over the rough terrain.* **2** °butt, °strike, °hit, °push, °nudge, elbow, °knock, °jab: *He jolted me so hard he actually cracked a rib.* **3** °shock, °astonish, °astound, °amaze, °surprise, °startle, °stun, dumbfound *or* dumfound, stupefy, strike dumb, °daze, °shake (up): *I was jolted to learn that my husband had been arrested for murder.* *—n.* **4** °lurch, °jar, °jerk, °bump, °jump, °bounce, °start: *The train started with a* *jolt that almost knocked me over.* **5** °blow, °shock, °surprise, bolt from the blue, °bombshell: *It was certainly a jolt to discover I had failed three exams.*

jot *v.* **1** Usually, ***jot down.*** make a note of, write *or* note (down), put *or* set *or* take down, °record: *Jot down this telephone number.* —*n.* **2** °scrap, °grain, °(wee) bit, °speck, mite, iota, whit, °particle, tittle, *Colloq* slightest, *US and Canadian* tad, smidgen *or* smidgin: *I don't care a jot what she thinks about the situation in Central America.*

journal *n.* **1** °periodical, °magazine, gazette, newspaper, °paper, newsletter, °review, tabloid; °daily, weekly, monthly, fortnightly, °quarterly, annual: *The journal contains information about every building permit awarded in the entire country.* **2** °diary, °chronicle, dossier, °record, °register, log, logbook, minute book, °minutes, documentation, album, °scrapbook, °memoir, almanac, annal, °history, yearbook, record book; °roll, catalog, °list: *He kept a detailed journal of every event in his twenty-year exile.*

journalist *n.* °reporter, newspaperman, newspaperwoman, °correspondent, newsman, newswoman, member of the fourth estate, gentleman *or* lady of the press, stringer; columnist; °hack; newscaster, anchorman, anchorwoman, °anchor, anchorperson, commentator, broadcaster, *Brit* pressman, paragraphist; newsreader, *Colloq* °scribe, newsmonger, *US and Canadian* legman, newshawk, newshound, newshen: *Journalists crowded around the ambassador, urging her to make a statement.*

journey *n.* **1** °trip, voyage, °excursion, °tour, °travel, °outing, °expedition, junket, °cruise, jaunt, °pilgrimage, peregrination, odyssey, trek: *Did your wife accompany you on your journey to Tierra del Fuego?* **2** °way, °passage, °passing, °transit, °transition, °progress, °course, °way, °trip, °route, °career: *On your journey through this life, / Whatever be your goal, / Keep your eye upon the doughnut, / And not upon the hole.* —*v.* **3** °travel, °tour, voyage, go (abroad *or* overseas), make *or* take a trip, make *or* wend one's way, make a pilgrimage, peregrinate, trek, rove, °range, °wander, °roam, °tour, °cruise, °gad (about), gallivant *or* galivant *or* galavant: *He journeyed to the far corners of the earth seeking an answer to life's riddle.*

joy *n.* **1** °pleasure, °gratification, °satisfaction, °happiness, contentment, °enjoyment, gladness, °delight, felicity, elation, exaltation, °ecstasy, °bliss, exhilaration, exultation, °rapture: *We felt indescribable joy at seeing the children safe and sound.* **2** °gaiety, cheerfulness, °cheer, °glee, buoyancy, joviality, jollity, jocundity, joyfulness, joyousness, jubilation, °merriment, lightheartedness, blithesomeness: *Let me wish you joy in this holiday season.* **3** °delight, °pleasure, °treat, °blessing, °gratification, °satisfaction, °prize: *A thing of beauty is a joy forever.*

joyful *adj.* **1** °cheerful, °happy, °buoyant, °gleeful, °merry, jovial, °jolly, jocund, joyous, jubilant, °gay, lighthearted, °blithe, blithesome, °sunny: *We are pleased that you came to help us celebrate this joyful occasion.* **2** °glad, pleased, gratified, °delighted, °happy, °elated, °ecstatic, exhilarated, °exultant, °overjoyed, jubilant, in heaven, *Brit* in the seventh heaven, *US* in seventh heaven, *Colloq* on cloud nine, tickled (pink), *Brit* over the moon: *Mark was joyful at the news that he was father of a boy.*

joyless *adj.* **1** °sad, °unhappy, °miserable, depressed, °dejected, °mournful, °downhearted, downcast, down, °despondent, dispirited, °melancholy, heavy-hearted, cheerless, °doleful, grief-stricken, disheartened, saddened, crestfallen, °wretched, disconsolate, °inconsolable, morose, heartsick, °sorrowful, woeful, °woebegone: *It was a joyless company that stood at the graveside. The cat died, the dog died, and my husband was ill—all in all, a joyless time.* **2** °gloomy, depressing, dispiriting, disheartening, °dreary, lugubrious, cheerless, °dismal, °bleak, °inhospitable, °desolate, °grim, austere, °severe: *The shuttered, joyless house loomed out of the misty moor ahead.*

judge *n.* **1** °justice, magistrate, jurist, *Isle of Man* deemster *or Isle of Man* dempster, *Slang Brit* beak: *The judge demanded order in the court.* **2** arbitrator, arbiter, °umpire, referee, adjudicator, judicator, °mediator, °moderator: *She served as a judge at Crufts dog show last year.* **3** connoisseur, °expert, °authority, arbiter, appraiser, evaluator, reviewer, critic, *arbiter elegantiarum* or *elegantiae*: *Let me be the judge of which wine should be served.* —*v.* **4** adjudicate, adjudge, arbitrate, °decide, °find, conclude, °settle, °determine, °decree, pass judgment, deem, °rule, pronounce *or* pass sentence: *Do you think the jury will judge in my favor?* **5** assess, °evaluate, appraise, °estimate, °rate, value, °weigh, °measure, °review, °consider, °size up, °appreciate: *A ballistics expert is required to judge this evidence.* **6** referee, °umpire, mediate, °moderate, arbitrate: *Farnsworth agreed to judge the essay competition.* **7** °believe, °suspect, °think, °consider, °suppose, °guess, conjecture, °surmise, conclude, °infer: *Paleontologists judge the age of the specimens to be four hundred million years.*

judgment *n.* **1** judgement, °discretion, discernment, °discrimination, judiciousness, °prudence, °wisdom, °wit, sagacity, perspicacity, clearheadedness, °perception, perspicuousness, percipience, acumen, °intelligence, °(good) sense, common sense, levelheadedness, °understanding, shrewdness: *Charlotte's judgment is often sought in such matters.* **2** judgement, °decision, ruling, verdict, conclusion, °determination, °opinion, adjudication, °finding, °decree, °order; °outcome, °result, °upshot: *The judgment of the court is final. It was the judgment of Paris to award the golden apple to Aphrodite.* **3** judgement, °criticism, censure, °disapproval, °reproof, condemnation: *They offered a moral, not a legal judgment.* **4** judgement, °opinion, °view, °belief, °(way of) thinking, °mind, °perception; °sentiment: *In my judgment, she is innocent.* **5** judgement, °evaluation, valuation, appraisal, °estimation, assessment: *One critic's unfavorable judgment of a play can spell its doom.*

judicial *adj.* **1** °legal, judiciary, judicatory, juridic(al); °official, forensic: *A formal judicial procedure can be quite costly.* **2** °critical, analytical, °discriminating, distinguishing, discerning, °keen, °sharp, °perceptive, percipient, perspicacious, differentiating, discriminatory, discriminative, °judicious: *Her decisions have always been judicial.* **3** judgelike, magisterial, °impartial, °fair: *He brought judicial procedures to bear on the handling of the problem.*

judicious *adj.* °sensible, common-sensical, °sound, °sober, °intelligent, °aware, °enlightened, °wise, °sage, sapient, °thoughtful, °reasonable, °rational, °sane, °logical, discerning, °discriminating, discriminative, °astute, °perceptive, percipient, perspicacious, °well-advised, (well-)informed, °prudent, °discreet, °tactful, °diplomatic, °politic, °careful, considered, circumspect: *There is some question whether the treasurer has made judicious use of the club's funds.*

jug *n.* pitcher, ewer, urn, carafe, °bottle, flask, decanter, °jar: *She came in from the barn carrying a jug of fresh milk.*

juggle *v.* °manipulate, tamper with, °falsify, °fix, °rig, °distort, misstate, °misrepresent, °alter, °arrange, *Col-*

loq °doctor, cook: *The accountant refused a bribe to juggle the company's books.*

juice *n.* **1** °extract, °liquid, °fluid: *The recipe calls for the juice of one lemon.* **2** °essence, °pith, °extract, °vigor, °force, °vitality, °spirit, °strength, °power: *He really squeezed the juice out of my argument.*

juicy *adj.* **1** °succulent, °moist, °lush: *This is a very juicy pear.* **2** °interesting, °sensational, °lurid, colorful, °vivid, °exciting, °stirring, °thrilling, intriguing, fascinating, °provocative, °suggestive, °racy, °spicy, °risqué: *I've got such a juicy piece of gossip for you!*

jumble *v.* **1** °disorder, °mix (up), °mingle, °confuse, confound, °muddle, °shuffle, disarrange, disorganize, °tangle, °entangle: *I found my belongings all jumbled together.* —*n.* **2** °muddle, °tangle, °medley, °mess; °disorder, °confusion, disarray, °chaos, °clutter: *My clothes were in a jumble on the bed.*

jumbo *adj.* °huge, °gigantic, °enormous, elephantine, °immense, oversized, king-sized, *Colloq US* humongous: *Grandad brought us a jumbo box of chocolates.*

jump *v.* **1** °leap, °bound, °spring, °pounce, °hurdle, vault, °hop, °skip; °caper, °cavort, gambol: *Jack, jump over the candlestick! Lambs were jumping around in the meadow.* **2** °start, °jerk, wince, °flinch, °recoil: *A sudden noise made me jump.* **3** Sometimes, ***jump over.*** °skip (over), °omit, pass over *or* by, °bypass, °avoid, °leave out, °ignore, °disregard, °overlook, °gloss over: *Jump over the boring parts and read me the sexy bits.* **4** °pass, °move, °leap, °skip: *She jumped from one subject to another so quickly that I couldn't keep track.* **5** °advance, °increase, °rise, °gain, °surge, escalate: *The cost of living jumped again this month, causing fear of inflation.* **6** ***jump at.*** °accept, °grab, °snatch, swoop up, °leap at, °pounce on: *Most people would jump at the chance to better themselves.* **7** ***jump on.*** °attack, swoop down on; °reprimand, °rebuke: *She jumps on anyone who suggests that she used influence to get her job.* —*n.* **8** °leap, °bound, °spring, °pounce, °hurdle, vault, °hop, °skip: *With one jump the cheetah was upon the gazelle.* **9** °rise, °increase, °boost, hike, °advance, °gain, °surge, escalation, upsurge, °increment, °elevation: *A jump in the Consumer Price Index drove share prices lower.* **10** barricade, °obstacle, °hurdle, °fence, °rail, °obstruction: *My horse cleared the jump easily, but I didn't quite make it.* **11** °start, °jerk, °spasm, twitch, °recoil, °lurch, °jolt: *When they called his name, he gave a little jump.* **12** °break, °gap, hiatus, lacuna, °space, °hole, °breach, °rift, °interruption: *There's a jump in continuity at the end of the fourth chapter.*

jumpy *adj.* °nervous, °agitated, °anxious, jittery, fidgety, °restless, edgy, °on edge, °tense, °shaky, skittish, °fretful, uneasy, °queasy, °restive, panicky: *Do you think he was jumpy because we were approaching Count Dracula's castle?*

junction *n.* °juncture, °union, °combination, joining, conjunction, °meeting, linking, °connection, conjoining, intersection, confluence; crossroads, interchange: *The train robbery took place at the junction of the two railways.*

juncture *n.* **1** See **junction.** **2** °point, °time, °moment, °stage, °period: *At this juncture, making an offer for the company might be premature.*

junior *adj.* °secondary, lesser, °lower, °minor, °subordinate, °inferior; younger: *Thompson has been offered a junior partnership in his firm.*

junk *n.* **1** °rubbish, °waste, °refuse, °litter, debris, °scrap; *US* °garbage, °trash: *The man came to cart away the junk we cleared out of the garage.* —*v.* **2** *Colloq* °discard, °throw away, °scrap, °cast aside, jettison, *US* °trash: *We bought a new washing machine and junked the old one.*

junta *n.* junto, °cabal, °clique, °faction, °gang, coterie, °band, °set, camarilla: *After the coup, the country was run by a military junta.*

jurisdiction *n.* °authority, °power, °prerogative, °dominion, °sovereignty, °say, °control, °rule, ascendancy, hegemony, °influence; °province, °district, °area, bailiwick, compass, °realm, °sphere (of influence), °reach, °clutches, °range, °orbit: *Tierra del Fuego seems a bit far to go just to be outside the jurisdiction of the Internal Revenue.*

just *adj.* **1** °fair, °equitable, °impartial, unbiased, °unprejudiced, °reasonable, fair-minded, evenhanded, °neutral, °objective: *Do you think you can expect a just trial after all that publicity?* **2** °upright, °righteous, right-minded, °honorable, °honest, °ethical, °moral, °principled, °straight, °decent, °good, upstanding, °virtuous, °lawful: *In our system of law, one must believe that juries are basically just.* **3** justified, justifiable, well-founded, well-grounded, °legitimate, valid, °reasonable, °rightful, (well-)deserved, °due, °fitting, °proper; condign: *She has a just claim to her father's estate. His punishment was just.* —*adv.* **4** °only, °merely, nothing but, solely, °simply, at best, at most, no more than: *She said just that and nothing else.* **5** °exactly, °precisely, °perfectly; °barely, only just, °hardly, °scarcely, by a hair's breadth, *Colloq* by the skin of one's teeth: *My new car barely fits into my garage.* **6** (only *or* just) now, a moment ago, (very) recently, °lately: *I have just returned from a vacation in Tegucigalpa.*

justice *n.* **1** fairness, impartiality, °objectivity, objectiveness, °equity, equitableness, fair-mindedness, justness, evenhandedness, neutrality, fair play: *Justice triumphed on this occasion, and he was convicted of fraud.* **2** the law, the police; °punishment, °prison, °imprisonment, incarceration, °detention: *He is a fugitive from justice.* **3** °law, °right, °morality, lawfulness, rightfulness, legitimacy, judiciousness: *His conviction was a miscarriage of justice.* **4** See **judge, 1.**

justify *v.* °vindicate, °legitimate, legitimatize *or* legitimize, legalize, °rationalize, °substantiate, °defend, °support, °uphold, °sustain, validate, °warrant, °confirm; °excuse, °explain, absolve, acquit, exculpate: *My worst fears were justified. How can you justify the theft of the icon?*

jut *v.* °extend, °overhang, °project, °protrude, °stick out, beetle: *The balcony juts out over the lake.*

juvenile *adj.* **1** °young, youthful, underage, °minor, teenage(d), °immature, °adolescent, °childish, °infantile, babyish, °puerile, °unsophisticated: *What could be more juvenile than painting graffiti on lavatory walls?* —*n.* **2** °youth, °boy, °girl, °adolescent, °minor, *Law* infant: *The police report arresting two juveniles for attacking an old lady.*

K

keen[1] *adj.* **1** °enthusiastic, avid, zealous, °devoted, °ardent, °fervent, fervid, °earnest, °impassioned, °passionate, °intense, °active; °agog, °eager, itching, °anxious: *They are keen fans of TV soap operas. I was keen to go swimming.* **2** °sharp, sharpened, razor-sharp, razorlike, knife-edged; °trenchant, °incisive, °cutting, rapierlike, °pointed, mordant, acid, vitriolic, acerbic, astringent, °biting, acrid, acrimonious, stinging,

°scorching, °caustic, searing, °withering, °virulent, °pungent, °sarcastic, °sardonic: *This ax is very keen, so be careful. With his keen wit, the playwright answered his severest critics. His play is a keen satire on the government.* **3** °painful, °bitter, °acute, °poignant, °fierce, °grievous, °severe, distressing, distressful, °strong, °deep, °profound, °intense, °extreme, °heartfelt: *She feels keen resentment at the way she had been treated.* **4** °vivid, °detailed, °specific, unmistaken, unmistakable *or* unmistakeable, °distinct: *He has a keen recollection of seeing the suspect on the bus.* **5** °sharp, °acute, °sensitive, °penetrating, °discriminating, °fine: *The bloodhound has an extremely keen sense of smell.* **6** °intelligent, °sharp, °acute, °perceptive, perspicacious, percipient, °sensitive, discerning, °astute, °smart, °bright, °discriminating, discriminative, quick(-witted), °shrewd, °clever, canny, cunning, °crafty, °wise: *Philip has a keen understanding of what is required of him.* **7** ***keen on*** *or* ***about.*** °fond of, enamored of, devoted to, °interested in: *Alan is almost as keen as Martha on stamp-collecting.*

keen[2] *v.* **1** °weep, wail, °moan, °lament, °mourn, °grieve; °bewail, °bemoan: *Finnegan's widow still keens over him.* —*n.* **2** dirge, elegy, knell, °lament, °lamentation, requiem, monody, threnody, thanatopsis, epicedium, *Scots and Irish* coronach: *The keens could be heard throughout the neighborhood.*

keep *v.* **1** °retain, °hold, hang on to, °preserve, °conserve, °have, °save, °maintain, °control: *The difficulty is not making money but keeping it.* **2** °hold, °have, take care *or* charge of, °mind, °tend, °care for, °look after, °guard, keep an eye on, watch over, °protect, °safeguard; °maintain, °feed, °nourish, victual, °board, nurture, °provide for, °provision: *Would you keep this parcel for me until I return? They keep chickens and ducks.* **3** °accumulate, °save (up), °amass, °hoard (up), °husband, °retain, °preserve, put *or* stow away: *My brother has kept all the toys he ever had.* **4** °maintain, °store, °preserve: *I keep woolen things in a cedar chest.* **5** Often, ***keep on*** *or* ***at.*** °continue, °carry on, °persist (in), °persevere (in); °prolong, °sustain: *How did you keep going after Bryony's death? She kept at it all night long. Keep on working till I tell you to stop.* **6** keep to, °abide by, °follow, °obey, °mind, adhere to, °attend to, pay attention to, °heed, °regard, °observe, °respect, °acknowledge, defer to, accede (to), °agree (to): *Only a stickler would keep the letter of such a law.* **7** °stay, °remain: *Keep off the grass. Keep to the left.* **8** °support, °finance, °provide for, °subsidize, °maintain: *He keeps a mistress in Chelsea.* **9** confine, detain; °imprison, incarcerate, jail *or Brit also* gaol: *The police have kept him overnight.* **10** °celebrate, °observe, solemnize, °memorialize, °commemorate: *We always kept the Sabbath at mother's house.* **11** °last, be preserved, °survive, °stand up, stay fresh: *Eggs keep longer in the refrigerator.* **12** °harbor, °maintain, °safeguard, keep dark: *Can you keep a secret?* **13** ***keep from.*** °prevent, keep *or* hold back, °restrain, °(hold in) check, °restrict, °prohibit, °forbid, °inhibit, disallow, °block, °obstruct, °deny, °curb, °deter, °discourage: *How can I keep my dog from biting the postman?* **14** ***keep in.*** **(a)** keep *or* hold back, °repress, °suppress, °stifle, °smother, muzzle, °bottle up, °withhold, °conceal, °hide, °shroud, °mask, °camouflage: *She keeps in her true feelings. He could hardly keep in the tears.* **(b)** confine, shut in *or* up, coop up, detain; fence in: *The teacher kept him in after school to punish him.* —*n.* **15** °upkeep, °maintenance, °support, room and board, °subsistence, °food, °sustenance, living: *He works hard to earn his keep.* **16** donjon, °tower, °dungeon: *He was imprisoned in the keep for twenty years.*

keeper *n.* custodian, °guardian, °guard, warden, caretaker; warder, °nurse, °attendant, *Brit* °minder: *The keepers at the zoo forbid your feeding the animals. How come they let you out without your keeper?*

keepsake *n.* °memento, souvenir, °token, °reminder, °remembrance, °relic: *The locket is a keepsake from those happy days in St. Agnes.*

keg *n.* cask, barrel, °butt, hogshead, tun, puncheon: *We bought a keg of beer for the party.*

kernel *n.* **1** °grain, °seed, pip, stone; nut, °meat, *US* nutmeat: *Try to extract the walnut kernel without breaking it.* **2** °center, °core, °nucleus, °heart, °essence, °quintessence, °substance, °gist, °pith, °nub, quiddity: *Let's get down to the kernel of the problem.*

key *n.* **1** latchkey, skeleton key, passkey, opener: *Have you lost your car keys again? You will need a key to open the air valve on the radiator.* **2** °clue, °cue, °guide, °indication, °indicator, °explanation: *A key to his problem can be found in the way his mother treated him when he was young.* **3** °pitch, °tone, °timbre, °level, tonality, frequency: *The song was originally written in the key of C.* **4** °legend, °explanation, °description, explication, clarification, °translation: *A key to the symbols appears on every other page.* **5** °mood, °tenor, °tone, °humor, °style: *In this passage the change is to a somber key.* —*adj.* **6** °important, °essential, °vital, °necessary, °crucial, °critical, °main, °pivotal: *Automatic reversal is a key feature of this tape player. Roderick is a key man in the company.*

keystone *n.* °necessity, crux, linchpin, °basis, °principle, °foundation, cornerstone: *Predestination was the keystone of his religion.*

kick *v.* **1** °boot, punt: *Satterthwaite kicked the ball over the goalpost.* **2** °recoil, °backlash, °rebound: *The shotgun kicked when I fired it.* —*n.* **3** punt, drop-kick: *His kick sent the ball down the field.* **4** °recoil, °backlash, °rebound: *That gun has quite a kick!*

kickback *n.* °rebate, refund, °share, compensation, commission, °percentage, °reward; °bribe, °payoff, *Colloq chiefly US* payola, *US* plugola: *We get a kickback on every computer sold. Some disk jockeys were accepting kickbacks for playing certain records.*

kidnap *v.* °abduct, °capture, °seize, °carry off, *Slang* °snatch: *He was kidnapped in Beirut and held for two years.*

kill *v.* **1** °execute, slay, °murder, assassinate, do away with, put to death, cause the death of, liquidate, dispatch *or* despatch, take (someone's) life, °finish (off), put an end to, write "finis" to, °silence, kill off, administer the *coup de grâce,* °eliminate, put (someone) out of (his *or* her) misery, °exterminate, °extinguish, °obliterate, eradicate, °destroy, annihilate, °massacre, °slaughter, decimate, °butcher, °(of animals) put down, put to sleep, *Slang* do in, bump *or* knock off, °hit, °polish off, snuff (out), °take for a ride, *US* °waste, °rub out, ice, fit with concrete overshoes *or* a wooden kimono: *He was the third police officer to be killed this year.* **2** °destroy, °ruin, °devastate, °ravage, wreak *or* work havoc on, kill off: *The entire orange crop was killed by the sudden frost.* **3** °muffle, °neutralize, °deaden, °damp, °silence, nullify, °dull, absorb, °smother, °stifle, °suppress, °still: *This padding should kill the noise of the motor.* **4** °exhaust, °tire (out), °fatigue, °weary, *Colloq* °fag (out): *Dragging that pushcart around Delancey Street all those years is what killed grandad.* **5** °hurt, °pain, °torment, torture: *These shoes are killing me.* **6** °quash, °suppress, °defeat, °veto, °cancel: *The tobacco interests campaigned to kill the bill to ban smoking in public places.* **7** °consume, °use up, °spend, while away, °occupy, °fill, °pass, °idle: *While waiting, I killed time doing a crossword puzzle.* —*n.* **8** °game, °prey; °quarry: *The lioness* *allowed her cubs to eat part of the kill.* **9** °death, °killing, °end, °finish, deathblow, *coup de grâce;* °termination, denouement *or* dénouement, conclusion: *She wants to be in at the kill.*

killer *n.* **1** °murderer, assassin, slayer, °cutthroat, °butcher, exterminator, Bluebeard, (Jack the) ripper, *Slang US* torpedo, hit man, triggerman, gunsel, hooligan, gunfighter, iceman, hatchet man: *What are the statistics on the number of killers who are not caught?* **2** *Slang old-fashioned* bee's knees, *US* humdinger, doozy, killer-diller (from Manila), lollapalooza *or* lallapalooza, lulu, daisy, dilly: *The new show at the Odeon is a killer.*

killing *n.* **1** °murder, °carnage, butchery, °execution, °slaughter, °bloodshed, °death, °massacre, genocide, liquidation, mass murder *or* destruction, decimation, extermination, blood bath, manslaughter; slaying, homicide, °fatality: *The killing of dissidents must stop. There has been another killing in the park.* **2** coup, bonanza, °success, °windfall, stroke of luck, °gain, °profit, *Colloq Brit* °bomb: *They've made a killing on the stock market.* *—adj.* **3** °devastating, °ruinous, °destructive, °punishing, °exhausting, debilitating, fatiguing, tiring, enervating, °difficult, °arduous: *How can you keep up the killing pace of having two jobs?*

killjoy *n.* °spoilsport, damper, dampener, grouch, grump, malcontent, pessimist, cynic, prophet of doom, Cassandra, *Colloq* wet blanket, sourpuss, *US* party pooper, gloomy Gus, picklepuss: *Come to the dance and stop being such a kill-joy!*

kin *n.* **1** °family, relative(s), relation(s), °kindred, *US and Canadian* kinfolk *or Brit* kinsfolk, kinsman, kinswoman, °stock, °clan, blood relation(s), blood relative(s): *We might have the same name, but he's no kin of mine. Too many people are thoughtless of their kin.* *—adj.* **2** °related, °akin (to), °kindred, consanguineous, consanguine, cognate, agnate: *One often forgets that all men are kin.*

kind[1] *adj.* °friendly, °kindly, °nice, congenial, affable, approachable, °amiable, °obliging, °accommodating, °amicable, well-disposed, °courteous, °good, °good-natured, °benevolent, well-meaning, well-wishing, °thoughtful, well-intentioned, °generous, bighearted, °humanitarian, °charitable, °philanthropic, °gentle, °understanding, °sympathetic, °considerate, °lenient, °tolerant, °indulgent, compassionate, kindhearted, °gracious, °warm, warmhearted, °cordial, tenderhearted, °affectionate: *It was kind of you to stop and help us. I never thought of him as a kind man.*

kind[2] *n.* **1** °sort, °type, °variety, °style, °genre, species, °class, °breed; °brand, °make: *You brought a new kind of love to me. What kind of bird is the peewit? Do we always have to have the same kind of breakfast cereal?* **2** °nature, °character, °manner, °description, °sort, °persuasion, °stripe, feather, kidney: *What kind of person are you? Today's students are of a totally different kind.*

kindle *v.* ignite, °light, set alight, °set fire to, set afire, °inflame, °fire, °foment, °incite, instigate, °provoke, °prompt, °prick, goad, °spur, °whip up, °stir (up), °work up, °excite, °agitate, °shake up, °jolt, °arouse, °rouse, (a)waken, °inspire, inspirit, °stimulate, °animate, °enliven, °energize, innervate, galvanize: *Kindling a fire in the rain is not easy. The wholesale pillage kindled a feeling of deep resentment.*

kindly *adj.* **1** See **kind,** above. *—adv.* **2** cordially, graciously, obligingly, amiably, amicably, politely, genially, courteously, thoughtfully, considerately, hospitably, agreeably, pleasantly: *He very kindly invited me in for a cup of tea.* **3** please, be so kind as to, be good enough to: *Would you kindly pass the salt?*

kindness *n.* **1** friendliness, kindheartedness, warmheartedness, graciousness, goodness, goodnaturedness, goodheartedness, goodwill, °benevolence, benignity, humaneness, °humanity, decency, tenderness, gentleness, kindliness, °charity, charitableness, generosity, °philanthropy, beneficence, compassion, °sympathy, °understanding, thoughtfulness, °consideration, cordiality, °hospitality, °warmth, geniality, °indulgence, °tolerance, °patience: *We shall always appreciate the kindness shown us by our adoptive parents.* **2** °favor, good deed *or* turn, °service, act of kindness; generosity, °assistance, °aid: *She did me a great kindness in introducing me to you. How can I repay your kindness?*

kindred *adj.* **1** °close, associated, °united, allied, analogous, °like, °similar, °matching, °parallel, °common, °related; °akin: *Because we both like sports, Samantha's interests are kindred to mine.* **2** °related, consanguineous, consanguine, cognate, agnate: *Linguists regard the Romance languages as kindred tongues.* *—n.* **3** See **kin, 1,** above.

king *n.* prince, crowned head, majesty, °sovereign, °monarch, ruler, regent, emperor, *Colloq Brit* °royal: *The king was in the countinghouse, counting out his wealth.*

kingdom *n.* **1** °realm, empire, °sovereignty, principality, °monarchy: *He sought a suitable wife throughout the kingdom.* **2** °field, °area, °domain, °province, °sphere (of influence), °territory, bailiwick, *Colloq* °turf: *The operating theater is the surgeon's kingdom.*

kink *n.* **1** °twist, crimp, °tangle, °knot, °wrinkle, curl, °coil, curlicue, crinkle: *Those kinks wouldn't occur if you had coiled the rope properly.* **2** °pang, °twinge, °stab, °spasm, cramp, stitch, °tweak, crick: *I have a kink in my neck from constantly looking down.* **3** °difficulty, °complication, °flaw, °hitch, °snag, °defect, °imperfection, distortion, deformity: *Iron out the kinks before presenting the plan to the board.* **4** crotchet, °quirk, whim, caprice, °fancy, vagary, °eccentricity, idiosyncrasy: *He would be more believable if he rid himself of certain kinks in his thinking.*

kinky *adj.* **1** °outlandish, °peculiar, °odd, °queer, quirky, °bizarre, crotchety, °eccentric, °strange, idiosyncratic, °different, °offbeat, °unorthodox, °capricious, °irregular, °erratic, unconventional, °unique, freakish, °weird, °fantastic, °whimsical: *This school is not interested in kinky notions of education.* **2** °perverted, °unnatural, °deviant, °degenerate, warped, °abnormal, depraved: *There is a rumor that they engage in kinky sex.* **3** °crisp, frizzy, frizzed, frizzled, curly, crimped, °wiry; knotted, tangled, twisted: *He has kinky red hair.*

kinship *n.* **1** consanguinity, °(blood) relationship, (family) ties, (common) descent, °lineage, °flesh and blood: *The royal houses of several European countries were connected by kinship.* **2** °connection, correspondence, parallelism, °relationship, similarity, °association, °agreement, °alliance; °affinity: *Don't you see the kinship between the Germanic languages? I feel a kinship with other phillumenists.*

kiss *v.* **1** osculate, peck, *Colloq* smack, smooch, neck, *Old-fashioned* spoon, canoodle: *I kissed her on the cheek. Two young people were kissing on the park bench.* **2** °touch, °brush, graze: *The cue ball barely kissed the eightball, toppling it into the pocket.* **3** ***kiss goodbye.*** bid adieu, say farewell to, °give up, °relinquish, °abandon, °forsake, °desert, °renounce, °repudiate, °forget (about), °dismiss, °disregard, °ignore: *With your marks, you might as well kiss goodbye any thought of winning a scholarship.* *—n.* **4** osculation, peck, *Colloq* smack, smooch, *US and Canadian* buss, *Slang Brit* smacker: *He ran over and gave me a big hug and a kiss.*

kit *n.* °apparatus, °gear, °equipment, °paraphernalia, appurtenances, °rig, accoutrements *or US also* accouterments, °tackle, °trappings, supplies, furnishings; instruments, tools, utensils, implements: *Did you bring along your toilet kit? The model airplanes were built from kits. The plumber forgot his tool kit.*

kitchen *n.* kitchenette *or Brit also* kitchenet, cookhouse; scullery, pantry, larder; *Nautical* galley, *Brit* caboose: *I am in the kitchen getting dinner ready.*

kittenish *adj.* °coy, °seductive, °flirtatious, coquettish, °sportive, °playful: *Barbara tends to get a bit kittenish when she's with men.*

kitty *n.* °pot, °pool, °collection: *I already contributed to the kitty.*

knack *n.* °genius, °intuition, °talent, °gift, °talent, °facility, °skill, °aptitude, °bent; °ability, °flair, °dexterity, °capacity, adroitness, °proficiency, skillfulness: *He has an uncanny knack for saying the wrong thing. When it comes to gourmet cooking, Peggy certainly has the knack.*

knife *n.* **1** °blade: *This knife couldn't cut through warm butter.* —*v.* **2** °stab, °pierce, °slash, °cut, °wound: *We struggled, and I was knifed in the leg.*

knit *v.* **1** join *or* fasten *or* weave (together), interweave, interlace, interconnect, intertwine, °link, °bind, °unite, tie (up *or* together), consolidate, °combine, °compact: *He was the first king to succeed in knitting together the diverse elements of the empire.* **2** °grow (together), °heal, °mend, °join: *The broken bones will knit in about a month.* **3** °furrow, °contract, °wrinkle, °knot, crease: *He knits his brow and shows an angry eye.*

knob *n.* °boss, stud, protuberance, °projection, °protrusion, °handle: *Turn the knob, then pull the drawer open.*

knock *v.* **1** °strike, °hit, °rap, thwack, whack, thump, bang, °tap: *Knock on the door. He knocked me on the head with his cudgel.* **2** *Colloq* °criticize, deprecate, carp *or* cavil at, °disparage, °put down, °run down: *Don't knock something till you've tried it.* **3** ***knock around*** *or* ***about.*** **(a)** °wander, °roam, °ramble, rove, °travel, gad about: *She's going to knock about the world for a bit before settling down.* **(b)** °associate with, consort with: *He was only knocking about with some of the boys.* **(c)** °discuss, °debate, °talk over, *Colloq* kick around: *I have a new idea for a carwash business to knock around with you.* **(d)** °beat (up), maltreat, °mistreat, maul, °manhandle, °batter, °abuse, °hit, °strike: *He's been known to knock his wife about.* **4** ***knock down.*** **(a)** °raze, °demolish, °destroy, °level, °wreck, lay in ruins, throw *or* pull down: *They knocked down those beautiful old houses and erected the ugly county office building in their place.* **(b)** °fell, °floor, °cut down: *As soon as he got up, McCloskey would knock him down again.* **5** ***knock off.*** **(a)** *Colloq* stop work(ing), °quit, go home, clock off *or* out, °terminate, °lock up, °close down: *I think I'll knock off for a few hours' rest.* **(b)** °steal, °pilfer, thieve, °rob, *Colloq* °lift, *Brit* °pinch, *Slang US* knock over, *Brit* °nick: *Two men in balaclavas knocked off the Second National bank.* **(c)** See **kill, 1,** above. **(d)** make quick *or* short work of, °complete, °finish, bring to an end, *Colloq* °polish off: *He knocked off that book in a week.* **(e)** °copy, °imitate: *They knock off expensive items, then sell them for much less than the originals.* **6** ***knock out.*** **(a)** knock *or* render unconscious, °floor, °prostrate, trounce, °whip, *Slang* °flatten, K.O. *or* kayo: *The smart money says that the challenger will knock out the champion in the third round.* **(b)** °overwhelm, °overcome, °daze, °stagger, °astound, °astonish, °bewilder, °stun, *Colloq* bowl over, blow (someone's) mind, *Slang Brit* knock for six: *She'll really knock them out in that dress!* **7** ***knock up.*** **(a)** knock *or* put together, °improvise: *I think I can knock up something quickly that will pass muster.* **(b)** °arouse, (a)waken, °wake up: *They knocked me up at dawn to go to work.* **(c)** *Slang* °impregnate, get with child, make pregnant: *Her boyfriend knocked her up and then refused to marry her.* —*n.* **8** °blow, °rap, °tap, thump, pounding, hammering: *My neighbor's knock on the wall awoke me.* **9** °blow, °punch, °jab, smack, thwack, whack, °right, °left, cuff, *Colloq* clout, bop, biff, conk: *He gave me a knock on the nose, and it started to bleed.* **10** °slap (in the face), censure, °criticism, condemnation, °slur, °insult: *My latest novel took quite a few knocks from the reviewers.*

knockoff *n.* °imitation, °copy, simulation, °replica, °facsimile, duplication: *This cheap knockoff of a thousand-dollar watch is selling for fifty dollars.*

knockout *n.* **1** *coup de grâce, Slang* K.O., kayo: *Carnera won by a knockout in the first round.* **2** °success, °sensation, °triumph, *Colloq* °hit, °winner, smash, smash-hit, stunner: *She's a knockout since her plastic surgery!*

knoll *n.* hillock, hummock, °mound, barrow, °hill, °elevation, °rise: *We climbed to the top of the knoll where we had a better view of the house.*

knot *n.* **1** °snarl, gnarl, °tangle; °tie, °bond: *I cannot loosen this knot.* **2** °collection, assemblage, aggregation, congregation, °crowd, °cluster, °bunch, °gathering, °company, °band, °gang, °crowd, °throng: *A small knot of people surrounded my sculpture.* —*v.* **3** °fasten, °tie, °bind, °secure, °lash, °tether, affix, °fix, °attach: *I knotted the rope around his neck.*

know *v.* **1** °understand, °comprehend, be familiar with, °grasp, be acquainted with, be versed *or* skilled in: *Do you know anything about nuclear physics? I know Albanian fairly well. She knows how to do the tango.* **2** °recognize, °identify, °recall, °remember, °recollect: *I know him from somewhere.* **3** be sure *or* certain *or* positive: *I knew I was right! I just know I'm going to win the first prize.* **4** °distinguish, °separate, discern, °differentiate, °recognize, °identify: *Charles doesn't know right from wrong.* **5** be aware *or* conscious *or* cognizant of, be informed *or* advised of, have knowledge of: *He knows that he can't fire her because she knows too much.*

knowing *adj.* **1** conspiratorial *or* conspiratory, °secret, °private; °significant, °meaningful, °eloquent, °expressive; °shrewd, canny, °artful, °sly, °wily, °crafty: *She gave him a knowing wink.* **2** °wise, °clever, °shrewd, (well-)informed, °knowledgeable, °aware, °expert, °qualified, °astute, °perceptive, °intelligent, sagacious: *Harold is quite knowing about horses, why not ask him?*

knowledge *n.* **1** °knowing, awareness, apprehension, cognition, °grasp, °understanding, discernment, consciousness, °conception, °insight: *Miles's knowledge of history is extensive.* **2** °facts, °information, °data, °intelligence: *They had no knowledge of her defection to the enemy.* **3** acquaintance(ship), °familiarity, °appreciation, conversance, °expertise, °experience, adeptness, °proficiency: *Has he any knowledge of their work on the DNA molecule?* **4** °schooling, °education, °scholarship, °instruction, °learning, erudition: *A little knowledge is a dangerous thing.*

knowledgeable *adj.* **1** knowledgable, °aware, *au fait, au courant,* °up to date, (well-)informed, (well-)acquainted, cognizant, °familiar, °enlightened, °expert, °knowing, *Colloq* in the know: *She is knowledgeable about the events in eastern Europe.* **2** well-educated, erudite, °learned, cultured, well-read, °intelligent, °sophisticated, °worldly, °wise, °sage, sagacious: *Fenstermacher is one of the most knowledgeable people I have ever met.*

kowtow *v.* Often, **kowtow to.** genuflect (before), salaam (to), °prostrate oneself (before), bow (down) (to *or* before), pay court to, scrape before, cringe before, fawn (before), grovel (before), toady ((up) to), °pander to, truckle ((up) to), dance attendance on, *Colloq* butter up, *Slang* suck up to, °play up to, shine up to, *Taboo slang US* brown-nose: *He kowtows to anyone who he thinks can be of use to him.*

kudos *n.* °praise, acclaim, °glory, °fame, °renown, °honor, plaudits, °applause, laudation, acclamation, accolade: *She enjoys the kudos of the entire scientific community for her discovery.*

L

label *n.* **1** °identification, identifier, ID, °mark, marker, earmark, °tag, ticket, sticker, °stamp, imprint, °hallmark, °brand, *Brit* docket: *The label shows the weight, composition, and price of the contents. A proper shipping label should show both source and destination.* **2** °name, °denomination, designation, appellation, °nickname, epithet, sobriquet, classification, characterization, °description: *It is difficult to put a label on the kind of novels she writes.* **3** trademark, trade name, °brand, logo, °mark: *Many supermarket chains market products under their own label.* —*v.* **4** °identify (as), °mark, °tag, earmark, ticket, °stamp, °hallmark, imprint, °brand, *Brit* docket: *In those days, every liberal was labeled a communist. Prepared food packages must be labeled to show their ingredients.* **5** °name, denominate, °designate, °call, °term, dub, classify, °categorize, pigeonhole, °class, °characterize, °describe, °portray, °identify, *Colloq US* °peg: *We cannot label Voltaire either spiritualist or materialist.*

labor *n.* **1** toil, °(hard) work, travail, °exertion, °effort, laboriousness, °strain, °drudgery, °pains, °industry, °slavery, donkey-work, *Colloq* °sweat, °grind, elbow grease, *Brit* swot: *They know the labor involved in building a road.* **2** employees, workers, wage earners, laborers: *A meeting between labor and management is scheduled for today.* **3** °effort, °task, °job, chore, °undertaking: *Raising funds for the museum was a labor of love.* **4** travail, childbirth, parturition, labor pains, contractions, °delivery: *She went into labor at midnight.* —*v.* **5** °work, toil, travail, drudge, °strain, °strive, °struggle, °slave, *Colloq* °sweat, °grind, *Brit* °peg away (at), swot: *We labored hard to get where we are today.* **6** °dwell on, °overdo, overemphasize, harp on, overstress, °strain, *Colloq* °belabor: *I heard you the first time, so don't labor the point.* **7** ***labor under.*** be burdened *or* troubled *or* distressed by, be deluded *or* deceived by, be disadvantaged by, °suffer, °endure: *He is laboring under a misapprehension. She labors under the impression that she will not be affected by a lack of education.*

labored *adj.* **1** °strained, °forced, °difficult, °hard, °laborious, °heavy: *His breathing became labored.* **2** overdone, °excessive, °overwrought, °ornate, °elaborate, overworked, overembellished, contrived, °affected, °artificial, °unnatural: *Her writing style is very labored and much too stiff.*

laborer *n.* °worker, workman, °hand, blue-collar worker, working man, manual worker, drudge, *Colloq Brit* navvy: *Get some laborers to clear away that rubble.*

laborious *adj.* **1** °arduous, °burdensome, onerous, °strenuous, grueling, backbreaking, herculean, °exhausting, taxing, tiring, fatiguing, wearying, wearisome, °toilsome, °difficult, °tough, °hard, uphill, °stiff: *Few realize how laborious farm work can be.* **2** °painstaking, °detailed, °careful, °thorough, °diligent, °scrupulous, °exhaustive, °steady, °steadfast, °relentless, unrelenting, dogged, assiduous, sedulous, persevering, °persistent, °untiring, °tireless, indefatigable, unremitting: *After laborious research the virus was finally identified.* **3** °industrious, hard-working, dogged, °determined, unwavering, °obstinate, °stubborn, unflagging, obdurate: *He made a laborious effort to get ahead.* **4** °labored, °strained, °forced, °ponderous, overworked: *The book contains some of the most laborious prose ever written.*

labyrinthine *adj.* labyrinthian, mazelike, mazy, °tortuous, sinuous, winding, convoluted, °complicated, confusing, °perplexing, °puzzling, enigmatic, baffling, confounding, complex, Daedalian, daedal, °intricate, Byzantine, twisted, °gnarled, snarled, tangled, knotted, knotty, Gordian: *The plot is so labyrinthine that the audience was unable to follow it.*

lace *n.* **1** lacework, tatting, openwork, filigree, °mesh, °web, webbing, °net, netting, °network: *She wore a collar of handmade lace.* **2** shoelace, shoestring, bootlace, °cord, °string, thong, °tie, lacing: *They made me remove my belt, tie, and laces before locking me in a cell.* —*v.* **3** °thread, °weave, °string, °twine, interweave, intertwine: *First lace the cord through the eyelets, then draw it tight.* **4** °spike, °fortify, °strengthen: *I saw her lace my punch with vodka, so I didn't drink it.* **5** ***lace into.*** **(a)** °attack, °assault, °beat, assail, thrash, °belabor, fall on *or* upon, °set upon, pounce on *or* upon, °lay into, *Colloq* °light into: *The brothers laced into one another with a fury.* **(b)** °berate, °scold, revile, °attack, °upbraid, °castigate, rant *or* rave at: *She really laced into me for being an hour late for dinner.*

lacerate *v.* °gash, °cut, °slash, °tear, °rip, °claw, °mangle; °wound, °rend, °hurt: *My feet were lacerated by the sharp stones. She suffered a lacerating attack from him for forgetting to pass on the message.*

lack *n.* **1** °want, deficiency, °dearth, °absence, °scarcity, °shortage, °need, insufficiency, paucity, °deficit, inadequacy: *We suffered from a severe lack of water. The orchestra is experiencing a lack of talent.* —*v.* **2** °want, °need, °require, be deficient in, be *or* fall short of, be without: *He lacks the votes needed to win. Simeon lacks the ability to paint any better.*

lackadaisical *adj.* **1** °lethargic, languorous, languid, °listless, °lazy, sluggish, spiritless, °idle, °indolent, °inactive, °slothful, fainéant: *She's much too lackadaisical to go out and get a job.* **2** °unenthusiastic, °dull, apathetic, insouciant, uncaring, unconcerned, °indifferent, °blasé, °cold, °cool, °lukewarm, °tepid, unexcited, °phlegmatic, unemotional, unexcitable, uninterested, unimpressed, uninspired, °unmoved, pococurante: *How can you explain the government's lackadaisical attitude toward the greenhouse effect?*

lackluster *adj.* °drab, °dull, lusterless, °flat, °dingy, °colorless, °dismal, °dreary, unexciting, °boring, °prosaic, °tiresome, °tedious, wearisome, uninteresting, two-dimensional, insipid, °vapid, °bland, unimaginative, °thick, °slow, °dense, *Colloq* °wishy-washy, blah: *He gave a very lackluster performance as Shylock. She has been leading a lackluster life since John left her.*

lad *n.* °boy, young man, °fellow, schoolboy, °youth, °juvenile, youngster, hobbledehoy, °stripling, (street) urchin, (street) arab, gamin, *Colloq* °guy, (little) shaver, kid, *US* °sprout: *Round up a few of the lads for a game of football.*

ladylike *adj.* °well-bred, well-born, aristocratic, °noble; °refined, °respectable, cultured, °polished, °elegant, mannerly, °gracious, °genteel, °courteous, °polite, courtly, °dignified, °proper, °correct, °decorous: *I cannot believe that Mrs. Gibbings did not behave in a ladylike way.*

lag *v.* **1** °fall behind, °linger, loiter, °delay, °straggle, °trail, hang back, dally, °dawdle, inch *or* poke along: *If you lag behind too far, you'll never catch up.* **2** °diminish, °decrease, slacken, abate, slow (down *or* up), °flag, °wane, °ebb, falter, °fail, °let up, fall (off *or* away), °ease (up), °lighten: *Contributions to the fund are lagging. Military preparations are lagging.*

laggard *n.* straggler, °idler, dawdler, loiterer, °slouch, sluggard, °loafer, snail, loller, *Colloq Brit* slowcoach, *US* slowpoke: *Why must you always be the laggard, dragging along behind everyone else?*

lair *n.* **1** den, °burrow, °hole, °nest, °tunnel, °cave, °hollow, covert: *We cautiously approached the lair of the man-eating tiger.* **2** hide-out, °hideaway, °retreat, hiding place, °refuge, asylum, °sanctuary, *Colloq* hideyhole: *Orville seldom emerges from his shop before dinner.*

laissez faire *n.* laisser faire, laissez-faireism, free enterprise, nonintervention, noninterference, °freedom, free trade, individualism, *laissez* or *laisser aller*, deregulation, decontrol, laxness: *The government's policy of laissez faire has spurred business.*

lambaste *v.* **1** °beat, °whip, °scourge, °flog, °lash, maul, horsewhip, cane, birch, thrash, bludgeon, trounce, drub, pummel, °batter, °belabor, cudgel: *Terry threatened to lambaste anyone who interfered.* **2** censure, °rebuke, °scold, °reprimand, chide, admonish, °reprove, °upbraid, °berate, °scold, revile, °attack, °castigate, flay, *Colloq* °dress down, get on (someone's) back, rake *or* haul over the coals, *US* °call down, *Slang* chew *or Slang* bawl out: *The press lambasted the company for its unfair employment policies.*

lame *adj.* **1** °crippled, °disabled, handicapped, hobbled, limping, incapacitated, impaired, °halting, °halt, spavined, *Colloq US* gimpy: *He's twisted his ankle playing tennis and has gone lame.* **2** °feeble, °weak, °flimsy, °thin, unconvincing, unpersuasive, °awkward, °clumsy, °poor, °ineffective, *Colloq* half-baked: *Refusing to go because you have a headache sounds like the usual lame excuse to me.*

lament *v.* **1** °mourn, °bemoan, °bewail, wail, °weep (over), grieve (for *or* over), °keen (over), sorrow (for *or* over): *A year later, and she's still lamenting the death of her hamster?* —*n.* **2** °lamentation, moaning, °mourning; °keen, dirge, elegy, knell, requiem, monody, threnody, thanatopsis, epicedium, *Scots and Irish* coronach: *Solemn laments for Kitty's late husband were heard throughout the village.*

lamentable *adj.* °deplorable, °wretched, °miserable, °terrible, distressing, °awful, °regrettable, °pitiful, °despicable, intolerable, °unfortunate: *The hall was left in a most lamentable state after the party. His death was presaged by a lamentable series of events.*

lamentation *n.* °mourning, grieving, moaning, weeping, wailing, crying, sobbing, lamenting: *The lamentation went on far into the night.*

lampoon *n.* **1** °burlesque, °caricature, °satire, °parody, pasquinade *or* pasquil, squib, *Colloq* °takeoff, *Brit* sendup: *The students published a hilarious lampoon of the teaching staff.* —*v.* **2** °burlesque, °caricature, °satirize, °parody, pasquinade *or* pasquil, °mock, °ridicule, squib, *US* skewer, *Colloq* °take off, put *or* run down, *Brit* °send up: *The cows we hold most sacred are often lampooned by irreverent cartoonists.*

lance *n.* **1** spear, pike, javelin, assegai; lancet: *He transfixed the knight with his lance, and claimed victory in the joust.* —*v.* **2** °pierce, °stab, °puncture, °prick, incise, °open, °slit: *The boil must be lanced at once to draw out the poison.*

land *n.* **1** °earth, °(solid) ground, terra firma: *After weeks in an open boat, Captain Bligh was relieved to have land under his feet again.* **2** °dirt, °earth, °soil, °turf, sod, loam, °mold: *Trumble tilled the land as his ancestors had for generations.* **3** °property, °grounds, real property *or* realty *or US and Canadian also* real estate, acreage; °estate: *How much land goes with the house?* **4** °fatherland, motherland, °nation, °country, homeland, native land: *The land of our fathers flowed with milk and honey.* —*v.* **5** °arrive, alight, °light, touch *or* come *or* go down, splash down, settle on *or* upon, °come to rest, berth, °dock, °disembark, debark, go ashore, deplane, dismount: *The plane landed on the water. We land at dawn.* **6** °catch, °capture, take captive, *Colloq* °bag: *Will he be able to land that shark without help? Sadie finally landed herself a millionaire!* **7** °get, °secure, °obtain, °win, °acquire: *He finally landed a job with his father-in-law.*

landing *n.* **1** touchdown, alighting, splashdown, docking: *Owing to the strong wind, the landing was quite dangerous.* **2** disembarkation, deplaning, °arrival: *After landing, passengers proceed through passport control.* **3** landing place, °dock, °pier, jetty, wharf, quay: *We walked onto the landing to watch the ship come in.*

landlady *n.* **1** proprietor *or* proprietress, lady of the house, °mistress, manager *or Brit* manageress, hostess: *My landlady owns three bed and breakfasts.* **2** See **landlord, 2,** below.

landlord *n.* **1** °host, publican, °proprietor, innkeeper, hotelier, °manager, restaurateur, Boniface: *Landlord! Two pints of your best bitter!* **2** °landlady, landowner, householder, °(property) owner, lessor, *Brit* freeholder: *My landlord has raised my rent claiming that his expenses have increased.*

landmark *n.* **1** °feature, °guide, guidepost; °identification: *The tower serves as a landmark for those seeking the castle.* **2** turning point, watershed, milestone, °monument: *His essay remains a landmark in philosophical thought.* —*attributive* **3** °critical, °crucial, °pivotal, °important, °historic, °significant, precedent-setting, °momentous, °notable, °noteworthy, °major: *This stands out as a landmark decision in the annals of criminal law.*

landscape *n.* °prospect, °view, °scene, °aspect, vista, countryside: *Here and there the landscape is dotted with solitary trees.*

language *n.* **1** °speech, °tongue, °idiom, °parlance, °dialect, idiolect, patois, °jargon, °cant, argot, °vernacular, *Colloq* °lingo: *How many languages do you speak?* **2** communication, °intercourse, interaction: *Karl von Frisch studied the language of the bees.* **3** °jargon, lingua franca, vocabulary, °terminology, °vernacular, *Colloq* °lingo: *In medical language, a head cold is called "coryza."* **4** °wording, °words, phrasing, phraseology, °style, °diction: *Guarantees, insurance policies, and so forth should be written in simple language.*

lanky *adj.* °thin, loose-jointed, lank, °lean, °gaunt, gangling, rangy, long-legged: *They recruited tall, lanky youths for the basketball team.*

lap[1] *v.* **1** Often, ***lap up.*** lick up, °tongue; °sip, °drink: *In a moment, the cat had lapped the cream out of the bowl.* **2** °wash, °splash, °ripple, plash, purl: *The water rose till it was lapping at the doorstep.* **3** ***lap up.*** **(a)** °consume, °drink, °eat: *The dog lapped up every last scrap of food.* **(b)** °accept, °believe, °credit, °fall for, *Colloq* °swallow (whole), °buy: *They lapped up Felix's story about being an orphan.*

lap[2] *n.* **1** °circuit, °orbit, ambit, °circle, °tour, °trip, °revolution: *She runs four laps around the track every day for exercise.* **2** °flap, °fold, lappet, °projection, lapel, °overlap: *This lap hides the zip.* —*v.* **3** °overlap, °fold, enfold, °envelop, °wrap: *Lap the fabric so as to cover the seam.*

lapse *n.* **1** °slip, °error, °mistake, °fault, °failing, °oversight, °blunder, °shortcoming, °omission, *Formal* lapsus, *Colloq* slip-up, °fluff, goof: *Apart from a brief lapse, the child behaved very well. Owing to a lapse, your payment was not credited.* **2** °gap, °break, °interval, °intermission, °interruption, °pause, lacuna, hiatus, *Prosody* caesura, *Colloq* °holdup: *Listeners in Scotland experienced a lapse in transmission, for which we apologize.* **3** °decline, lowering, °fall, deterioration, °drop, diminution, descent: *There has been a serious lapse in church attendance.* —*v.* **4** °decline, °lower, °fall, °drop, °diminish, °sink, °slip, °slump, °subside, °deteriorate: *Despite the doctors' best efforts, he lapsed into a coma.* **5** °run out, °expire, be discontinued, become void, °terminate, °end, °cease, °stop: *Your option to sell the shares at the offering price lapses on Friday.* **6** °pass, °elapse, °go by, °slip away: *Three days have lapsed since I last saw you.*

large *adj.* **1** °big, °great, °broad, °stout, °heavy, thickset, chunky, °stocky, heavyset, °brawny, °husky, °sturdy, °hefty, °muscular, strapping, °burly, °solid, °weighty, corpulent, °fat, °obese, rotund, portly, adipose, °beamy, overweight: *Two large gentlemen offered to help me find the door.* **2** °big, °generous, °bountiful, °charitable, eleemosynary, °philanthropic, openhanded, magnanimous, munificent, °unselfish, bighearted, largehearted, °substantial, °considerable, °ample, beneficent, °liberal; °goodly, °kind, °good; *Colloq* °tidy: *We have to thank Mr. Wilson for large donations to our Christmas fund. It was large of you to donate the use of a car.* **3** °big, °huge, °ample, °enormous, °gigantic, °immense, °colossal, °monumental, °massive, mammoth, Brobdingnagian, gargantuan, elephantine, °monstrous, staggering, sizable *or* sizeable, °substantial, °wide, °broad, capacious, °extensive, *Colloq* °jumbo, *US* humongous, ginormous: *The house has a large dining room. Supertankers are the largest ships afloat.* —*adv.* **4** °big, prominently, overwhelmingly, imposingly, °eminently, °preeminently: *The problem looms large in the history of biology.* —*n.* **5** ***at large.*** **(a)** °free, unfettered, °at liberty, on the loose, unconfined, unrestrained: *One of the robbers is still at large.* **(b)** °generally, °mainly, in general, °chiefly, as a whole, in a body, °altogether, °in the main: *The people at large support reductions in taxes.*

largely *adv.* °chiefly, °mainly, °as a rule, by and large, °generally, in general, to a great extent, mostly, in great part, in great measure, °in the main, °on the whole, pretty much, essentially, °at bottom, basically, fundamentally: *The voters are largely in favor of keeping the present government.*

largess *n.* largesse, gifts, alms, grants, bonuses, °endowments, presents, contributions, donations, handouts; °support, subvention, °aid, °subsidy, °charity, °philanthropy, generosity, munificence, °bounty, liberality, openhandedness: *The church depends on the largess of its members.*

lark *n.* **1** °frolic, °spree, escapade, °caper, °fling, romp, °adventure, °revel, jape, °game, antic, horseplay, shenanigans, °mischief, °prank, practical joke: *The boys say they took the horse just as a lark.* —*v.* **2** Often, ***lark about.*** °frolic, °caper, romp, °revel, °play, °sport, °cavort, gambol, *Colloq* skylark: *At the end of exams, students spend a lot of their time larking about.*

lascivious *adj.* **1** °lustful, °randy, °lecherous, licentious, °lewd, °prurient, salacious, libidinous, °erotic, °sensual, lubricious *or* lubricous, ruttish, goatish, hircine, satyrlike, °wanton, Cyprian, debauched, *Slang* horny, °hot: *I wouldn't let that lascivious old man near my daughter.* **2** °pornographic, °obscene, °blue, °lurid, °indecent, smutty, °dirty, Fescennine, °filthy, °vile, ribald, °bawdy, °gross, °coarse, °offensive: *The shipment of lascivious magazines was confiscated by customs officials.*

lash[1] *n.* **1** °whip, °scourge, cat-o'-nine-tails, cat, quirt, knout, bullwhip, thong; rope's end: *The bosun gave him twenty of the best with his lash.* **2** °stroke, °blow, °strike, °slash, °cut: *Few could stand up after five lashes.* —*v.* **3** °flog, °beat, thrash, °switch, °whip, °scourge, horsewhip, °lambaste, flail, smite, thwack, *Colloq* whack: *The men were lashed for the most trivial infraction.* **4** ***lash out.*** °attack, flay, °lambaste, °belabor, °punish; °criticize, °berate, °scold: *She often lashes out at her critics quite virulently.*

lash[2] *v.* °fasten, °tie, °bind, °secure, °rope, °fix, strap, make fast: *His crew lashed Odysseus to the mast as they sailed past the land of the Sirens.*

lass *n.* °girl, young woman, °miss, mademoiselle, schoolgirl, *Old-fashioned* °maiden, °maid, damsel, demoiselle; *Scots* lassie, *Irish* colleen: *I married the lass with the delicate air.*

lasso *n.* **1** lariat, °rope, *Southwestern US* reata *or Southwestern US* riata: *The cowboy tossed the lasso round the calf's neck.* —*v.* **2** °rope: *He lassoed steers to practise for the rodeo.*

last[1] *adj.* **1** hindmost, rearmost, aftermost; °final: *I was the last person to board the plane.* **2** °latest, newest, most recent *or* up-to-date: *What was the subject of your last novel? During the last week I have seen seven films.* **3** °final, concluding, °terminal, °ultimate, terminating: *That was the last flight till tomorrow. She was the last speaker of Cornish.* **4** °definitive, conclusive, decisive, closing: *That is my last word on the subject.* —*adv.* **5** behind, at *or* in the end, in *or* at the rear, after: *My horse always comes in last.* —*n.* **6** °end, °termination, °finish: *That was the last we ever saw of Charlie.* **7** ***at last.*** °finally, °eventually, °ultimately, °at length: *At last we shall see the results of our work.*

last[2] *v.* **1** °continue, °endure, °survive, °keep on, °persist, °remain, °stay, °abide, °carry on, °hold out, *Colloq* go the distance: *He wouldn't last a day at my job.* **2** °wear, °stand up, °endure, °survive: *This shirt has lasted through sixty-two washings.*

last[3] *n.* °mold, matrix, °form, °model, °pattern: *These shoes were not made on the same last as those.*

lasting *adj.* °permanent, °enduring, °durable, °everlasting, long-term, undying, °eternal: *She made a lasting contribution to ophthalmology.*

late *adj.* **1** °tardy, delayed, °overdue, behindhand, dilatory, unpunctual; °belated, *US* past due: *I cannot believe that they are late for their own wedding. Better late than never. Your mortgage payment is late again.* **2** °recent, °last, °new, °fresh, °current, up-to-date: *Do you have a copy of the late edition of yesterday's newspaper?* **3** deceased, departed, °dead; °former, °past, ex-, °recent, °previous, °preceding, °old (See also **5**, below.): *Here is a song by the late, the great Bing Crosby. I shall be meeting with Lord Pamby, late chairman of United Oxidation.* **4** ***of late.*** See **5**, below. —*adv.* **5** recently, °lately, °previously, °formerly, °once, heretofore, in recent *or* former time(s), °of late, latterly (See also **3**, above.): *We are dining tonight with Sir Keith, late of Scotland Yard.* **6** till *or* at an advanced hour *or* time: *I was up late last night. She married late.* **7** tardily, unpunctually, belatedly: *We arrived too late for the first* *act.* **8** recently, °lately: *As late as an hour ago we still could have caught the ferry.*

lately *adv.* See **late, 5,** above.

latest *adj.* **1** See **last**[1], 2, above. **2** °fashionable, °current, °modern, °up-to-the-minute, *Colloq* in: *She drives the latest model.* —*n.* **3** most recent *or* up-to-date *or* modern development *or* news *or* example: *What's the latest? Have you seen the latest in scuba gear?*

lather *n.* **1** suds, °froth, °foam: *Spread the lather on before shaving.* **2** °sweat; dither, °fuss, pother, °flutter, *Colloq* tizzy, °state, °flap: *He got into a real lather when he missed the train.* —*v.* **3** soap (up); °foam, °froth: *The barber lathered my chin thoroughly.* **4** thrash, °beat, °belabor, maul, drub, °flog, whack, pummel, buffet, °whip, flail, °lash: *If he dared to swear, his father would lather him.*

laud *v.* °praise, commend, °celebrate, sing *or* speak *or* sound the praises of, °honor, acclaim, °extol, °glorify, °promote, °advance, °recommend, °exalt: *Bramble's paintings have been lauded by leading art critics the world over.*

laudable *adj.* °praiseworthy, °meritorious, creditable, °admirable, °outstanding, °excellent, °noteworthy, °notable, commendable, °estimable, creditable: *He made a laudable effort to keep up with her.*

laudatory *adj.* laudative, praiseful, eulogistic, panegyric(al), encomiastic(al), °complimentary, °favorable: *None of the laudatory comments made in the introduction was true.*

laugh *v.* **1** °titter, °giggle, snigger, °snicker, °chuckle, chortle, guffaw, split one's sides, *Colloq* break *or* crack up, roll on the floor, go into hysterics, roar with laughter, hoot: *If you say something funny, maybe I'll laugh.* **2** Often, ***laugh at.*** **(a)** °deride, °ridicule, °mock (at), °jeer (at), make a mockery of, poke fun (at), make fun *or* sport (of), make an ass *or* a fool (out) of, °tease, °taunt, pull (someone's) leg, °satirize, pasquinade *or* pasquil, °parody, °lampoon, jest (at), °joke (about), °scoff (at), °scorn, *Colloq Brit* take the mickey out of, *US* roast: *They laughed when I sat down—I didn't know the chair was broken. Why did the other children laugh at you?* **(b)** laugh away *or* off, °spurn, °dismiss, °brush aside, °minimize, shrug off, °reject, °ignore, °disregard, °deny, °belittle, *Colloq* pooh-pooh: *He laughed at the wound and insisted on continuing to fight.* —*n.* **3** °titter, °giggle, snigger, °snicker, °chuckle, chortle, guffaw, horse laugh: *She would do anything for a laugh.*

laughter *n.* laughing, tittering, giggling, sniggering, snickering, chuckling, chortling, guffawing: *Children's laughter filled the air.*

launch *v.* **1** °start (off), set in motion, set *or* get going, °begin, embark upon *or* on, °initiate, °inaugurate: *We are about to launch a scheme for dockside development. The attack will be launched at dawn.* **2** °originate, °establish, °organize, °set up, °found, °open, °start: *Robson is talking about launching a new company.* **3** °shoot, °fire, °discharge, °hurl, °throw, °sling, °pitch, °fling, catapult, °send, dispatch *or* despatch: *The enemy launched ground-to-air missiles against our planes.* **4** °float, set afloat: *Two frigates will be launched today.* —*n.* **5** inauguration, °start, °initiation, °opening: *The launch of the offensive is scheduled for noon tomorrow.* **6** °boat, skiff, °tender, motorboat, runabout, gig, dinghy: *The launch will shuttle passengers between the pier and the cruise ship.*

launder *v.* **1** °wash, °clean, °scrub, °cleanse: *The chambermaid laundered three shirts for me.* **2** legitimize, legitimatize, °legitimate, legalize: *The money from heroin was being laundered through several car dealerships.*

laurels *n.* pl. honor(s), distinction(s), °fame, awards, tributes, rewards, acclaim, acclamation, °glory, °renown, °celebrity, °popularity, °reputation; successes, accomplishments: *With no new fields to conquer, he is resting on his laurels.*

lavatory *n.* water closet, °toilet, bathroom, gents' (room), ladies' (room), men's (room), restroom, °privy, *Chiefly military* or *institutional* latrine; *Nautical* °head; *Chiefly Brit* WC *or* wc, *Brit* (public) convenience, *US* °equipment, outhouse; *Colloq* (little) boys' *or* girls' room, powder room, *Brit* loo, lav, *US* can; *Slang Brit* °bog: *My new house has two upstairs bathrooms and a downstairs lavatory.*

lavish *adj.* **1** °profuse, °abundant, °liberal, copious, °plentiful, °prolific, °opulent: *He wore a silk tunic with lavish gold embroidery.* **2** °generous, °liberal, openhanded, unstinting, °bountiful, unsparing, °unselfish, °effusive, °free: *She bestowed on the children the most lavish gifts.* **3** °extravagant, °wasteful, exaggerated, °prodigal, °improvident, °excessive, °unreasonable, °immoderate, °profligate, uncurbed, unrestrained, intemperate: *The council was accused of lavish overspending on the civic center.* —*v.* **4** squander, °waste, °throw away, °dissipate, °spend, °expend, °sink: *We lavished a fortune on the gardens.* **5** °shower, °bestow, °thrust, °heap, °pour: *Granny always lavished gifts on us.*

law *n.* **1** °rule, °regulation, ordinance, statute, °act, enactment, bylaw, °measure, edict, °decree, °order, directive, °injunction, °command, commandment, canon, mandate, ukase: *They have a law that forbids smoking on airplanes.* **2** corpus juris, °(legal) code, constitution, rules and regulations, °charter, *Law* °equity: *The law must be applied equally to all citizens.* **3** °principle, °proposition, theory, °theorem, °formula, axiom, °deduction, corollary, postulate, conclusion, °inference: *Not every observable phenomenon obeys the laws of physics.*

lawful *adj.* **1** °legal, licit, °legitimate, de jure, constitutional, °just, °rightful, valid, °proper: *I am the lawful heir.* **2** °permissible, allowable, justifiable, authorized, allowed, permitted: *It is not lawful to let your dog foul the footpath.*

lawless *adj.* **1** anarchic(al), anarchistic, °chaotic, °unruly, unregulated: *A lawless interval came between the revolution and the establishment of government.* **2** °illegal, °illicit, °unlawful, °criminal, felonious, larcenous, °dishonest, °corrupt, °venal, *Colloq* °crooked: *Assembly without a permit is a lawless act.* **3** °villainous, nefarious, °wicked, °sinful, flagitious, iniquitous, treacherous: *He was once a member of a lawless motorcycle gang.*

lawn *n.* sward, greensward, °turf, sod, °green: *I lay back on the lawn and gazed at the sky.*

lawyer *n.* °counsel, °advocate, member of the bar, legal practitioner, *Brit* solicitor, barrister, Queen's *or* King's counsel, bencher, *US* attorney(-at-law), counselor(-at-law), *Derogatory* shyster, ambulance-chaser, pettifogger, *Slang US* °mouthpiece: *We have turned the case over to a lawyer.*

lax *adj.* **1** °loose, °slack, °casual, °slipshod, easygoing, °careless, negligent, °permissive, °weak, °indulgent, °flexible, °relaxed: *Law enforcement has always been lax in that city.* **2** °imprecise, undefined, °indefinite, nonspecific, unspecific, °vague, °shapeless, amorphous, °general, °broad, °inexact, hit-or-miss, °careless, °untidy, *Colloq* °sloppy: *His ideas about tactics are too lax for a military man.*

lay[1] *v.* **1** °place, °put (down), °set (down), °position, °deposit; °spread: *Lay the cloth on the table, then lay the book on top of it.* **2** °set, °arrange: *The table was laid for four.* **3** Often, ***lay down.*** °establish, °build, °construct: *Before I can argue my case, I must lay a firm foundation of trust.* **4** °stake, °bet, wager, °gamble, °hazard, °risk: *I'll lay ten to one that you can't swim across the lake.* **5** °destroy, °suppress, exorcise: *Did Edward ever succeed in laying that ghost of a rumor that haunted his career for so long?* **6** °present, °offer, °submit, set *or* put forth, °advance, bring *or* put forward, °set out: *We have laid our proposal before the board.* **7** °charge, °impute, °direct, °lodge, °prefer, °aim, °attribute, ascribe: *There is no justification for laying the blame on Farmer.* **8** copulate (with), °couple (with), have (sexual) intercourse (with), °sleep (with), °lie (with), bed, go to bed (with), °mate (with), have sex (with), go all the way (with), °service, *Slang* °screw, shack up (with), *Brit* roger, *Taboo* fuck. **9** ***lay bare.*** °expose, °reveal, uncover, °disclose, divulge, °bring to light, °show, °unveil, lift the veil from: *She laid bare the most intimate details of their relationship.* **10** ***lay down.*** °stipulate, °require, °demand, insist on, °dictate: *She laid down the conditions on which he could return home.* **11** ***lay hold of.*** °seize, °grab, °snatch, °nab, catch *or* get hold of, °get: *I have been trying to lay hold of that book for weeks.* **12** ***lay in.*** See **19 (a)**, below. **13** ***lay into.*** °attack, °assault, °set about, assail, °lambaste, °belabor: *We laid into the gang and gave them a sound drubbing.* **14** ***lay it on.*** °exaggerate, °overstate, embroider: *Don't believe what Frank tells you—he really lays it on pretty thick sometimes.* **15** ***lay low.*** See **18 (c)**, below. **16** ***lay off.*** **(a)** °suspend; °dismiss, °discharge, °fire, let go, cashier, drum out of the corps, *Colloq* °(give the) sack, (give the) ax, kick out, °(give the) boot, boot out, give (someone) his *or* her walking papers: *Fifty of us were laid off because business was slow.* **(b)** °let up, °quit (it), °stop (it), °cease, desist, °leave off, leave alone, *Colloq* knock (it) off, cut (it) out, °come off (it): *I wish you'd lay off for a while and stop bothering me. You must lay off phoning your boyfriend twice a day.* **17** ***lay on.*** **(a)** °provide, °cater (for), °supply: *The company laid on quite a Christmas party.* **(b)** °impose, °charge, assess; °demand, °require: *In addition to making us pay the back taxes, they laid on a penalty of 20 percent.* **18** ***lay out.*** **(a)** °design, °plan, °outline, sketch, °arrange, °set up: *The art department laid out the advertisements for the new campaign.* **(b)** °advance, disburse, °spend, °expend, °pay, °give, °contribute, *Colloq* °shell out, ante up, kick in with, fork out: *The salesmen often lay out the money for their expenses and are then reimbursed.* **(c)** °lay low, °floor, °prostrate, knock down *or* out, strike *or* cut down, °flatten, °fell, *Colloq* knock for six, kayo *or* K.O.: *He laid out the challenger with a blow to the solar plexus.* **19** ***lay up.*** **(a)** °lay in, °amass, °accumulate, °save (up), °hoard, °preserve, °store, °keep, °put away, °put by: *We laid up a supply of canned goods for the winter.* **(b)** hospitalize, °incapacitate, disable, confine to bed, keep indoors: *She has been laid up by a virus all week.*

lay[2] *adj.* **1** °secular, nonclerical, laic, nonecclesiastical: *He served as a lay preacher for many years.* **2** °amateur, nonprofessional, nonspecialist: *Members of the lay public did not immediately understand the dangers of the greenhouse effect.*

lay[3] *n.* ballad, °song, °air, °refrain, °strain, °melody; °poem, ode, °lyric, °rhyme, ballade: *Macaulay wrote a book entitled* The Lays of Ancient Rome.

lazy *adj.* **1** °indolent, °slothful, dilatory, °idle, °shiftless, °inactive, °listless, fainéant, otiose, °slack, °lax, °lethargic: *He's so lazy and so rich that he has a barber come to his house to shave him every day.* **2** °slow, languid, °easy, easygoing, sluggish, slow-moving, languorous: *The hot lazy summer days were whiled away fishing.*

lead *v.* **1** °conduct, °escort, usher, °guide, show the way, °pilot, °steer: *If you lead, I'll follow.* **2** °cause, °influence, °prompt, °bring, °incline, °induce, °persuade, °move, °dispose, °convince: *What led you to suspect the butler?* **3** °head (up), °direct, °govern, °command, °supervise, superintend, °preside (over), take the lead, take *or* assume command (of), °manage, captain, *Colloq* °skipper: *The orchestra would like you to lead. Who is going to lead the men into battle? Nicole leads a weekly discussion group on alcoholism.* **4** come *or* be *or* go first, °excel, °surpass, °exceed, °precede, be ahead (of), °outstrip, °distance, outrun, °outdo: *They lead the world in the production of sugar.* **5** °live, °experience, °spend, °pass; while away: *He is leading a life of ease on the Riviera.* **6** be conducive to, °create, engender, °cause, °contribute to, °result in, bring on *or* about, °produce: *Your stubbornness can lead only to frustration.* **7** ***lead astray.*** °lead on, °mislead, misguide, °misdirect, °deceive; °fool, °decoy, °hoodwink, *Colloq* bamboozle: *She might have been led astray by that wolf in sheep's clothing.* **8** ***lead off.*** start (off *or* in *or* out *or* up), °begin, °commence, get going *or* moving, get under way, °initiate, °inaugurate, *Colloq* kick off: *The proceedings will lead off with the national anthem. Who is going to lead off the singing?* **9** ***lead on.*** **(a)** See **7**, above. **(b)** °lure, °entice, °seduce, °beguile, inveigle, °tempt: *They led me on with offers of a huge salary.* **10** ***lead up to.*** **(a)** prepare *or* pave *or* clear (the way), do the groundwork *or* spadework, °precede: *The events leading up to the overthrow of the government may never be known.* **(b)** °approach, °broach, °bring up, °present, °introduce, work up *or* round *or* around to, get (up) to: *I'd lead up to the subject delicately, if I were you.* —*n.* **11** °front, vanguard, van, lead *or* leading position *or* place, advance *or* advanced position *or* place: *My horse was in the lead coming down the stretch.* **12** °advantage, °edge, °advance, °supremacy, °margin, °priority, primacy, °preeminence: *She has the lead over all her competitors. Our team had a two-point lead at the half.* **13** °direction, °guidance, °leadership, °precedent, °example, °model, exemplar, °pattern, °standard: *We decided to follow his lead.* **14** °tip, °clue, °hint, °suggestion, °cue, intimation; °prospect, °possibility, °potential; *Colloq* tip-off: *Barton is following up some leads for a new job. The advertisement produced some new sales leads.* **15** leash, °tether, °restraint, °cord, °chain: *The dog fetches his lead when he wants to go out.* **16** °protagonist, hero *or* heroine, leading *or* starring role *or* part, leading *or* lead actor *or* actress, leading lady *or* man, male *or* female lead, °principal; prima donna, diva, prima ballerina, *première danseuse, premier danseur*: *She has the lead in the new production of* Cassandra. **17** wire, °cable, *Brit* °flex: *Connect this lead to the power source.* —*adj.* **18** °leading, °foremost, °first; °main, °chief, °principal, °premier, °paramount: *The lead climber fell when the rope broke. The lead story in today's paper is about Universal's takeover bid for United Basket.*

leaden *adj.* **1** °heavy, onerous, °ponderous, °dense, °burdensome: *Exhausted, he dragged his leaden feet to the next meeting.* **2** °heavy, °dull, numbing, °oppressive: *I was happy to escape the leaden atmosphere of the theater.* **3** °gray, °dull, °dingy, °gloomy, glowering, lowering, °dreary, °dismal, °oppressive, °dark, °sullen, °somber: *The leaden sky seemed to forebode disaster as he fled across the moor.* **4** °inert, °lifeless, °listless, sluggish, °inanimate, °inactive, °lethargic, languid, languorous, °torpid, spiritless, °stagnant, °static, °dormant, soporific, °sleepy: *The refugees trudged along at a leaden pace.*

leader *n.* **1** °chief, °head, commander, ruler, °superior, °director, chairman, chairwoman, chairlady, chairperson, chieftain, captain, commandant, °principal, *Colloq* °boss, bossman, kingpin, big cheese, number one, numero uno, *Brit* gaffer, *Chiefly US* Mr. Big, *Slang US* the man: *The police can identify the leader of the gang. The leader of the assembly is elected for a year.* **2** bandmaster, °director, *US and Canadian* conductor, bandleader, concertmaster: *The leader tapped his baton for silence.*

leadership *n.* °direction, °guidance, °management, directorship, °administration, supervision, °command, °regulation, °control, °operation, °influence, °initiative; governorship, superintendence, °running: *The company made huge profits under Katherine's leadership. The leadership of the country was in the hands of the privileged class.*

leading *adj.* **1** °important, °influential, °chief, °prime, °cardinal, °foremost, °paramount, °primary: *A leading politician hinted today that the prime rate might go down soon.* **2** °best, °outstanding, °preeminent, greatest, °supreme, °peerless, °matchless, unequaled, unrivaled, unsurpassed: *He was the leading writer of his time.*

leaflet *n.* folder, °circular, °brochure, handbill, °bill, booklet, °advertisement, *US and Canadian* throwaway, flier *or* flyer, *Colloq Brit* advert: *The candidate gave out leaflets urging us to vote for him.*

leafy *adj.* °green, verdant, bosky, woody, °shady, shaded, arborescent: *We lay beneath the leafy bowers and dreamed of summers past.*

league *n.* **1** confederation, °association, °alliance, °combination, coalition, °federation, confederacy, guild, °society, °fraternity, °union, °band, °fellowship, °club: *The teams in the south formed a football league of their own.* **2** ***in league (with).*** allied (with), °united (with), associated (with), leagued (with), federated (with), collaborating (with), conspiring (with), in collusion (with), *Colloq* in cahoots (with): *The suspects are said to be in league with saboteurs.* —*v.* **3** °ally, °unite, °associate, °band, °combine, °collaborate, °join (forces), conspire, collude: *Did most people league with the rebels or the official government?*

leak *n.* **1** leakage, leaking, °discharge, °trickle, °escape, seepage, seeping, oozing, exudation: *We have to stop leaks of radioactive waste from power plants.* **2** °hole, fissure, °crack, °chink, °crevice, °aperture, °opening, °puncture, °cut, °break, °split, °gash, °rent, °tear, °gap, °flaw: *There was a leak in the balloon and it began to descend.* **3** disclosure, °revelation: *The press knew about the defection because of a security leak.* —*v.* **4** °escape, °discharge, °spill, °trickle, °drip, seep, °ooze, exude, extravasate: *The cream was leaking out through a hole in the container.* **5** °disclose, divulge, °let slip, °release, °give away (the game), make known *or* public, let (something *or* it) be known, *Colloq* °spill the beans (about); let the cat out of the bag: *Someone as yet unidentified leaked the scandal to a reporter.* **6** ***leak out.*** °transpire, become known, °come out, be revealed: *Their secret relationship has leaked out.*

lean[1] *adj.* **1** °thin, °slim, °slender, rangy, °spare, °wiry, °lanky, lank, °skinny, angular, bony, °rawboned, °gaunt, gangling, gangly, °spare, °meager, skeletal, scraggy, °scrawny, °haggard, °emaciated, pinched, wasted, shrunken, macilent: *She has the lean, leggy build typical of a runner.* **2** unfruitful, unproductive, °barren, °infertile, °poor, °meager, °scanty, °bare, arid, °sparse, °impoverished: *The Kanes tried for years to eke a living from the lean soil.* **3** °impoverished, °destitute, °needy, poverty-stricken, °penurious, °indigent, necessitous, °hard, °bad, °difficult: *The Depression was a lean period for most people.*

lean[2] *v.* **1** rest (against *or* on *or* upon), be held up *or* supported by: *"Lean on me, Father William," the young man said.* **2** °incline, °slant, °tilt, °bend, °tip: *Isn't that pole leaning over too much?* **3** Often, ***lean toward(s).*** °favor, gravitate toward(s), tend toward(s), be disposed toward(s), °prefer, show a preference for, incline toward(s), be *or* lean on the side of, be biased toward(s), be (prejudiced) in favor of: *She leans toward conservatism when it comes to morality.* **4** ***lean on.*** **(a)** °rely on, °depend on, °count on, believe *or* trust in, pin one's hopes *or* faith on *or* upon: *You can lean on me if you need help.* **(b)** °pressure, bring pressure to bear on, °intimidate, °threaten, cow, terrorize, °terrify, °scare, °frighten, °warn, °menace, °endanger, imperil, *Brit* pressurize: *Sometimes, the gang would lean on a shopkeeper who refused to pay for "protection."*

leaning *n.* °bent, °inclination, °bias, °prejudice, °favoritism, °partiality, predilection, °liking, °taste, °preference, °penchant, °sympathy, °tendency, tendentiousness: *The judges' leanings in favor of short-haired dogs are well known.*

leap *n.* **1** °spring, °bound, °jump, vault (into); °hurdle, °clear, hop over, skip over, °negotiate: *He leapt into the saddle. Superman leaps tall buildings at a single bound.* **2** °jump, °cavort, gambol, °dance, °frisk, °caper, °frolic, romp, °prance, curvet: *He leapt for joy at winning first prize.* **3** °jump, °rush, °hasten, form hastily, accept (prematurely *or* without question): *The speaker kept leaping from one subject to another. Don't leap to conclusions about her strange behavior.* **4** ***leap at.*** °jump at, °accept, be eager for, move quickly, °take: *I leapt at the chance to apply for a scholarship.* —*n.* **5** °spring, °bound, °jump, vault, °hurdle, °hop, °skip: *With one leap I was on the other side.* **6** °jump, °increase, (up)surge, °rise, upswing, °growth, escalation: *The figures show a leap in inflation for the month.* **7** ***by leaps and bounds.*** °rapidly, °quickly, °swiftly, speedily: *The plans for the new theater are progressing by leaps and bounds.*

learn *v.* **1** Often, ***learn of.*** °find out, °discover, °hear (of), chance *or* hit upon, °understand, °gather, have revealed to one; °determine, ascertain, uncover: *I learned that you are leaving. I learned of your plans yesterday. I learned the particulars today.* **2** be taught, be instructed in, °master, become proficient (in), acquire knowledge (of): *I learned German at school.* **3** °understand, see the light, get the picture, *Colloq* °catch on, get the idea, get it, *Brit* °twig: *No matter how often they are told, some people never learn.* **4** °memorize, commit to memory, learn by heart: *Bolton was told to learn his part by the next day.*

learned *adj.* (well-)informed, erudite, °knowledgeable, *au fait*, cultured, °intellectual, °highbrow, (well-)educated, °scholarly, °academic, scholastic, °lettered, °experienced, skilled, practised, °accomplished, °expert, literate, well-read, (well-)versed, (well-)trained, (well-)grounded: *The consultants to the encyclopedia are all learned people.*

learner *n.* °student, °pupil, °scholar, trainee, °apprentice, °novice, tyro, abecedarian, beginner, °initiate, neophyte: *Learners of English as a second language will benefit from this book.*

learning *n.* °knowledge, erudition, °culture, °scholarship, °lore, °information, °wisdom: *You have the advantage of being able to draw on centuries of learning.*

lease *n.* **1** rental agreement *or* contract, sublease, sublet: *A new ten-year lease will be signed tomorrow.* —*v.* **2** °rent (out), °let (out), sublet (out), sublease (out),

°charter (out), °hire (out): *They leased their home to us for the summer. We arranged to lease their boat for the weekend.*

leave[1] *v.* **1** go (away *or* off), °depart, °set off, be off, get away *or* off, °retire, °retreat, °withdraw, decamp, °(make an) exit, °run, be gone, bid (someone *or* something) goodbye, say goodbye *or* adieu (to), take (one's) leave (of), °quit, °desert, °pull out, °check out, °fly, *Colloq* push *or* shove off, °take off, skedaddle, °flit, °disappear, do a disappearing act, °pull up stakes, up-anchor, *Slang* °beat it, scram, vamoose, °split, *Brit* beetle off, do a bunk, do a moonlight flit, hop it, *Taboo slang Brit* °bugger off: *Please leave at once. May I leave the room? She left him alone.* **2** go *or* run off, deviate from: *The car left the road and hit a tree.* **3** °forget, °mislay, °lose: *I left my car keys somewhere in the house.* **4** °abandon, °desert, take leave of, °wash one's hands of, °turn one's back on; °quit, resign from, °give up, °renounce, °drop (out of): *She left him because he mistreated her. I left school at sixteen.* **5** °make, °render, cause to be *or* become *or* remain: *The beauty of the place left me speechless.* **6** °bequeath, °will, °hand down, °devise, demise, °transfer: *Her aunt left some valuable paintings to my wife.* **7** °entrust, °commit, °assign, °cede, °relinquish, give over *or* up, consign, °resign: *They are leaving all the responsibility to me.* **8** cause *or* allow to remain, have as a remainder, °yield, °give: *Seven from eleven leaves four.* **9** ***leave off.*** °stop, °cease, desist, forbear, °give up, °refrain from, °discontinue, abstain (from), °renounce: *He's left off gossiping about her.* **10** ***leave out.*** °omit, °disregard, °ignore, °neglect; count out, °reject, °exclude, °eliminate, °bar, °except: *You left out the best part. Don't leave me out of the fun!*

leave[2] *n.* **1** °permission, authorization, °consent, °freedom, °liberty, °license, °sanction, dispensation: *Who gave you leave to borrow my car?* **2** furlough, leave of absence, time off, sabbatical, °recess, °holiday, vacation: *I am overworked and badly in need of some leave.* **3** departure: *He took his leave of us at last.*

lecherous *adj.* °lewd, °lascivious, salacious, libidinous, °sensual, lubricious *or* lubricous, °lustful, concupiscent, licentious, °prurient, dirty-minded, filthy-minded, °carnal, goatish, hircine, °randy, °libertine, °wanton, °profligate, depraved, °degenerate, °decadent, °dissolute, *Slang* horny: *The lecherous villain dared to proposition my wife.*

lecture *n.* **1** °speech, °address, °talk, discourse, disquisition, treatise, dissertation, °paper, °lesson, °instruction, °sermon, declamation, °harangue, diatribe, philippic, screed: *The preacher delivered a lecture on the evils of drink.* **2** °reproof, °reprimand, °rebuke, °criticism, censure, reproach, scolding, upbraiding, chiding, berating, °tongue-lashing, remonstration, *Colloq* dressing-down, telling-off, *Slang Brit* wigging, *US and Canadian* chewing-out: *I had to listen to another lecture about staying out late.* —*v.* **3** make *or* deliver *or* give a speech *or* address *or* talk, discourse, sermonize, °hold forth, moralize, pontificate, orate, °preach, declaim, expound, go on about, °harangue: *She lectures on Chinese art of the eighteenth century. He insists on lecturing to everyone he meets.* **4** °reprove, °reprimand, °rebuke, reproach, °scold, °upbraid, °berate, chide, tongue-lash, remonstrate with, rail at, fulminate against; admonish, °warn; *Colloq* °dress down, °tell off, send (someone) off with a flea in (his *or* her) ear, *Slang Brit* wig, *US and Canadian* °chew out: *She is always lecturing me about keeping my room tidy.*

ledge *n.* shelf, °projection, °step, mantel *or* mantle, mantelpiece *or* mantlepiece, °overhang, sill: *We stood on a small ledge of rock unable to move.*

leer *v.* **1** °ogle, °eye: *He leers at every pretty girl.* —*n.* **2** °ogle, the eye, *Colloq* once-over, *Slang* glad eye: *She was repelled when he gave her a leer.*

leery *adj.* °suspicious, °skeptical, dubious, °doubtful, doubting, °distrustful, °wary, °cautious, chary, °careful: *I was very leery of his story.*

leeway *n.* °space, elbowroom, °room, °play, °scope, °slack, latitude, wide berth; °freedom: *We allowed him some leeway to make changes in the manuscript.*

left *adj.* **1** left-hand, sinistral, *Nautical* °port, (formerly) larboard, *Heraldry* °sinister: *In most countries, the driver sits on the left side.* **2** leftist, left-wing, °progressive, °liberal, socialist(ic), °pink; °radical, communist(ic), red: *The party's platform is too left for my taste.* —*n.* **3** left side *or* hand *or* fist, *Nautical* °port, portside, (formerly) larboard: *The fort is on the left as you enter the bay. The champion has a powerful left.*

left-handed *adj.* **1** °clumsy, °awkward, gauche, fumbling, maladroit, cack-handed: *He has a left-handed way of doing things.* **2** insulting, disparaging, derisive, uncomplimentary, insulting, °paradoxical, °ambiguous, °questionable, dubious, °doubtful: *Calling my piano playing "not bad" was a left-handed compliment.*

leftover *n.* **1** Usually, **leftovers.** remainder(s), remnant(s), °rest, °residue, residuum, °balance, °surplus, °excess, °superfluity, overage; scrap(s), leavings, crumbs, °odds and ends, debris *or* débris, °refuse, °waste, °rubbish, rubble, detritus, *Archaic* orts: *I'm having leftovers for dinner tonight.* —*adj.* **2** °remaining, °residual, °extra, °excess, °unused, uneaten: *Save some of the leftover ice cream for later.*

leg *n.* **1** limb, °member, *Colloq* °pin, °peg, °stump, *Slang* gam: *I have a touch of arthritis in my left leg.* **2** °support, °brace, °prop, °upright, °standard, column, °pillar: *One of the legs of this table is about to collapse.* **3** °part, °portion, °segment, °stretch, °stage, °section, °length, °lap: *The first leg of my journey passed without incident.* **4** ***a leg up.*** °boost, °assistance, °push, °help, helping hand, °advance, °support, *US and Canadian* °assist: *Let me give you a leg up over the wall. Gorton might give you a leg up in getting a job.* **5** ***leg it.*** °run, °hurry, °hasten, °scurry, *Colloq* scoot, skedaddle: *He legged it down the street after the pickpocket.* **6** ***not a leg to stand on.*** °defenseless, unsupported, insupportable *or* unsupportable, indefensible, unjustifiable, °untenable, °invalid: *She insists she's right, but she hasn't a leg to stand on.* **7** ***on one's*** *or* ***its last legs.*** °decrepit, °failing, °exhausted, °dying, worn-out, °run-down, falling apart *or* to pieces, broken-down, the worse for wear; °dilapidated, °rickety, °shabby, °ramshackle, crumbling, tumbledown: *George looked as if he was on his last legs. My car is on its last legs.* **8** ***pull (someone's) leg.*** °tease, °mock, °jeer at, °taunt, °gibe, °make fun of, °chaff, °guy, °fool, °deceive, *Chiefly Brit* °twit, *Colloq* rib, kid, °rag: *We were just pulling his leg, but he took it seriously.* **9** ***shake a leg.*** **(a)** °hurry (up), °hasten, °rush, *Colloq* get going *or* moving *or* cracking, look alive *or* lively: *You'll have to shake a leg to catch that train.* **(b)** °dance, trip the light fantastic (toe), *Slang* hoof it, *US* cut a rug: *What do you say we go out and shake a leg tonight?* **10** ***stretch one's legs.*** (take *or* go for a) walk, °(take some) exercise: *After sitting for so long, I have to stretch my legs.*

legal *adj.* **1** °lawful, licit, statutory, °acceptable, °permissible, permitted, admissible, authorized: *The police officer was found to have exceeded his legal authority.* **2** °legitimate, °proper, °right, °rightful, °sound, constitutional, authorized, *Slang* legit: *I question whether his business is legal. Who is the legal heir?* **3** °judicial, juridical, judiciary, forensic: *The case is without legal precedent.*

legalistic *adj.* narrow(-minded), disputatious, contentious, litigious, °literal, °strict, °niggling, °hairsplitting, caviling, quibbling, jesuitical, pettifogging, °nice, °fine, °subtle, *Colloq* nitpicking: *Her argument is legalistic and ignores the spirit of the law.*

legend *n.* **1** epic, °saga, °myth, °story, °(folk) tale, °romance, °narrative, fable, °tradition, fiction, *Scandinavian* edda: *Scores of legends are woven into the cultural fabric of our lives.* **2** °celebrity, °phenomenon, °tradition, °wonder, luminary, °personage, °somebody: *Someone said that Rudolph had become a legend in his own mind.* **3** °key, table of symbols, °code, explanatory note: *Refer to the legend for explanations of symbols used in the map.* **4** °motto, °slogan, inscription, caption, °title: *The legend reads, "Dieu et mon droit."*

legendary *adj.* **1** fabled, storied, °traditional; °heroic, epic, mythic: *Legendary tales have been identified in every culture.* **2** °fanciful, °imaginary, °fabulous, °mythical, °romantic, °fictional: *Was King Arthur a real person or a legendary figure?* **3** °famous, °celebrated, °noted, famed, °well-known, °renowned, °illustrious, °immortal, °prominent, °eminent, °great, acclaimed, °noteworthy: *Irena became known for her legendary dinner parties.*

legible *adj.* °readable, decipherable, °clear, understandable, °plain, °distinct: *The handwriting is legible, but I don't know what language it's in.*

legitimate *adj.* **1** valid, °proper, °right, °rightful, authorized, °legal, °genuine, °real, °true, °authentic: *Which one of them was the legitimate Anastasia?* **2** °lawful, licit, °legal, by law, de jure, statutory: *Bonnie Prince Charlie claimed to be the legitimate ruler of Scotland.* **3** common-sensical, °sensible, °reasonable, °proper, °correct, °acceptable, valid, °logical, justifiable, °just, °fair: *Hers is a legitimate question that merits an answer.* —*v.* **4** legitimize, legitimatize, legalize, °authorize, °sanction, °warrant, validate, °certify: *By signing before two witnesses, she legitimated her will.*

leisure *n.* **1** spare *or* free time, °time (off), °liberty, °freedom, °opportunity: *How many people have the leisure to do what they like?* **2** °holiday, vacation, °respite, °relief, °rest, °recreation, °relaxation, °ease, breathing space, °quiet, tranquillity, °repose, *US military* rest and recreation, R and R: *Everyone needs a few days of leisure now and then.* **3** ***at leisure.*** **(a)** unoccupied, °inactive, retired, resting, °free, on holiday *or chiefly US and Canadian* vacation; °at liberty, °available, °unemployed: *I have been at leisure for a year now.* **(b)** *Often,* ***at one's leisure.*** °at one's convenience, °when convenient, °unhurriedly, °in one's own time: *I shall mow the lawn at my leisure.* —*adj.* **4** recreational: *The "leisure industry" is booming, with the popularity of theme parks.* **5** °free, unoccupied, nonworking, unencumbered, °idle, °holiday, vacation: *How do you usually spend your leisure time?*

leisured *adj.* °wealthy, °rich, affluent, °prosperous, moneyed *or* monied, well-to-do: *Yachts of that size are strictly for the leisured classes.*

lend *v.* **1** make a loan of, °loan, °advance: *The bank agreed to lend me the money for a boat.* **2** °impart, °furnish, °bestow, °give, °confer, °contribute, °add: *The silk scarf lends cachet to the whole outfit.* **3** ***lend itself to.*** °suit, °fit, be fitted *or* appropriate *or* suitable (to *or* for), be applicable *or* adaptable (to *or* for): *The book lent itself well to treatment as a musical.*

length *n.* **1** °extent, °measure, °span, °reach, °size, °magnitude, dimension, °measurement: *What is the length of a piece of string? The pool is fifty meters in length and fifteen meters in width.* **2** duration, °stretch, °term, °period, °space: *What length of time did he spend in prison?* **3** ***at full length.*** fully, °completely, to the fullest extent: *They were stretched out at full length on the floor.* **4** ***at length.*** **(a)** °finally, at (long) last, °eventually, after a (long) time *or* while, °ultimately; °in the long run: *At length, the doctor emerged from his office.* **(b)** for a long time, interminably, for ages: *He read to her at length till she dozed off.* **(c)** °in depth, °thoroughly, °completely, exhaustively, extensively, to the fullest *or* greatest extent, °in detail: *I haven't the time to go into the story at length just now.*

lengthen *v.* make longer, °extend, elongate; °stretch, drag out, °draw out, °prolong, protract, °expand, °continue: *Please lengthen the skirt an inch. The ten-minute interview lengthened into an hour. They lengthened the playing field.*

lengthy *adj.* °long, overlong, long-drawn(-out), °protracted; °endless, interminable, prolonged, long-winded, °wordy, prolix, verbose, garrulous, °talkative, loquacious, °boring, °dull, °tedious: *He underwent a lengthy operation for removal of the clot. She launched into a lengthy description of the life cycle of the frog.*

lenient *adj.* °gentle, °kind, °kindly, °easy, °sparing, °merciful, °tender, humane, tenderhearted, kindhearted, °indulgent, °permissive, °forgiving, easygoing, °tolerant, °patient, compassionate, forbearing, °understanding, magnanimous, °generous, °charitable: *Do you think they were lenient in their treatment of prisoners caught while escaping?*

lesson *n.* **1** °exercise, °drill, reading, °lecture, °recitation; °assignment, homework, °task: *The pupils are having an English lesson at the moment. You must finish your lessons before you can go out and play.* **2** °class, °session; °instruction, teaching, tutoring, °schooling; °practice: *I take piano lessons in the afternoons.* **3** °example, exemplar, °model, °guide, °maxim, °paragon, °message, °deterrent, discouragement; °warning, admonition; °moral, °precept: *Norbert's punishment should be a lesson to you.* **4** °punishment, chastisement, chastening, castigation, scolding, chiding, °rebuke, °reprimand, °reproof: *If she does that again, I'll give her a lesson she'll not soon forget!*

let[1] *v.* **1** °allow (to), °permit (to), °sanction (to), give permission *or* leave (to), °authorize (to), °license (to), °suffer (to): *Don't let him go alone. Let George do it.* **2** °cause (to), arrange for, °enable (to): *Let me know if you have any difficulty. Let me help you. Let the police know of any crime.* **3** Sometimes, ***let out.*** °rent (out), °hire (out), °lease (out), °charter (out); °contract (out), subcontract (out), °farm (out), °job (out): *Mrs. Finney lets rooms by the week. Much of the work is let out to people who work at home.* **4** ***let down.*** °disappoint, °fail, °frustrate; disenchant, dissatisfy, °disillusion: *He promised to be here at nine, but he let me down.* **5** ***let in.*** °admit, allow in; °include, °take in, °receive, °welcome, °induct, °install: *Open the window and let in some fresh air. I wouldn't want to be a member of any club that would let in someone like me.* **6** ***let off.*** **(a)** °pardon, °forgive, °excuse, °release, °discharge, let go: *They let him off with only a severe reprimand.* **(b)** exonerate, absolve, °clear, acquit, °vindicate, *Slang* let off the hook: *When the real criminal confessed, she was let off.* **(c)** detonate, °explode, detonate, °discharge, °fire, °set off: *Don't let off those fireworks so near the haystack.* **(d)** °emit, give out *or* off, °release, throw off *or* out, °let loose, exude: *Some plastics let off an awful odor when they burn.* **7** ***let on.*** **(a)** °confess, °admit, °disclose, divulge, °reveal, °expose, let it be known, °let out, °say, °tell, °give away, °let slip, °betray; °leak: *I'd never let on that you are the guilty one.* **(b)** feign, °affect, °pretend, °fake, °(put on an) act, simulate, dissemble, °dissimulate: *He let on that he had a bad leg, but he just didn't feel like going to the office.* **8** ***let out.*** **(a)** See 7 (a), above. **(b)** °(let) loose, °liberate, °(set) free, °release, let go, °discharge: *He served three years before being let out. Please let the cat out.* **(c)** °emit, °give vent to, °produce: *She let out a scream that could have woken the dead.* **(d)** °end, °stop, °break up, °finish, °close, °terminate: *When does school let out for the Christmas vacation?* **9** ***let up.*** °decrease, abate, °ease (up), slacken, °diminish, lessen, °mitigate, °subside, °moderate: *Perhaps we can go if the rain lets up.* **10** ***let up on.*** ease up on, slack off on: *They didn't let up on haranguing us for hours.*

let[2] *n.* °hindrance, °impediment, °obstruction, °obstacle: *People can now cross the border without let or hindrance.*

letdown *n.* °disappointment, disillusionment, disenchantment: *She was so eager to win that losing was all the more of a letdown.*

lethal *adj.* °deadly, °fatal, °mortal: *The lethal blow was delivered by a blunt instrument.*

lethargic *adj.* **1** lethargical, sluggish, °slow, °dull, °heavy, °lazy, °indolent, °phlegmatic, °slothful, °idle, languid, languorous, °listless, fainéant, °inactive, °torpid, stuporous, comatose; °indifferent, apathetic: *They were so lethargic that they did absolutely nothing about it.* **2** °weary, °tired, fagged out, °fatigued, enervated, °weak, °exhausted, °drowsy, °sleepy, somnolent: *I have a bad cold and feel very lethargic.*

lethargy *n.* **1** °sluggishness, °sloth, dullness, heaviness, laziness, °indolence, phlegm, °idleness, languidness, languor, listlessness, *dolce far niente,* °inactivity, °inertia, °torpor, °stupor; °indifference, apathy: *After taking the medicine, a feeling of lethargy overtook me.* **2** weariness, tiredness, °fatigue, °weakness, °exhaustion, drowsiness, sleepiness, somnolence: *The strident persistence of my alarm clock dragged me out of my lethargy.*

letter *n.* **1** °character, °symbol, °sign: *Epsilon is a letter in the Greek alphabet.* **2** °missive, epistle, communication, °note, °line, °message, dispatch *or* despatch; correspondence: *Write me a letter when you get there.* **3** ***letters.*** °literature, the humanities, belles-lettres, the classics; erudition, °culture, the world of letters, °learning, °scholarship: *He is a widely known man of letters.* **4** ***to the letter.*** °precisely, °literally, °exactly, accurately, strictly, *sic,* °thus, letter-for-letter, *literatim,* word-for-word, °verbatim: *The document has been copied to the letter.* —*v.* **5** inscribe, °write, °spell (out): *She carefully lettered the caption under the picture.*

lettered *adj.* literate, °literary, (well-)educated, erudite, °scholarly, °learned, °well-informed, °enlightened, °knowledgeable, (well-)versed, well-read, cultured, °cultivated: *He was greatly influenced by the lettered men of his time.*

letup *n.* °cease, °stop, stopping, ceasing, cessation, surcease, °break, °interruption, °pause, °intermission, °suspension, time out *or* off; moderation, lessening, abatement, diminution, °relief, hesitation, °respite, °relaxation: *I was nagged without letup about getting the work done by the end of the year.*

level *adj.* **1** °even, °smooth, °plane, °uniform, °plain, °flat, °flush, °straight, °true: *The plaster is level enough to apply the wallpaper.* **2** °horizontal; °prone, °supine: *Make sure the playing field is level. Try to keep the body level when doing this exercise.* **3** °uniform, °constant, °steady, °consistent, °invariable, unvarying, unalterable, unchanging, unfluctuating; °very: *I did my level best to warn you of the dangers of taking drugs.* **4** °parallel, °even, °equal, °equivalent, °consistent: *The top floor of my house is level with the roof of the one across the street. The problem arises because wages are not level with inflation.* **5** up (on), informed (about *or* on), up to date (on *or* with), *au fait* (with), *au courant* (with): *He does his best to keep level with what is going on in the world.* **6** °even, tied, °equal, neck and neck: *The scores are level at the half.* **7** See **levelheaded,** below. —*v.* **8** level off, °even, °smooth (out), °flatten (out): *The concrete is leveled by dragging a straightedge over it.* **9** °destroy, °demolish, °raze, °lay waste, °devastate, °knock down, tear down, °pull down, °wreck, bulldoze: *The building was leveled in a few hours.* **10** °aim, °point, draw a bead, °direct, °train, °focus: *I looked him straight in the eye when he leveled his gun at me.* **11** ***level with.*** be *or* play fair with, be honest *or* straight with, be open *or* frank *or* straightforward with, *Colloq* be up front with: *I am going to level with you—I find you very hard to work with.* —*n.* **12** °plane, °horizontal, °elevation, °height; altitude: *The windows are at the same level.* **13** °floor, °story: *At which level is Mr. Stone's office?* **14** °plane, °position, °status, °standing, °rank, °stage: *She functions at a higher level, both administratively and intellectually.* **15** ***on the level.*** °straight, straightforward, °honest, °direct, °sincere, °square, °open, °aboveboard, *Colloq* °up front, on the up and up, *US* straight-shooting: *You haven't been on the level with me about where you went last night.*

levelheaded *adj.* (well-)balanced, °sensible, °sane, °reasonable, common-sensical, °level, unruffled, undisturbed, unperturbed, imperturbable, even-tempered, composed, °calm, °cool, °collected, °tranquil, °serene, unflappable, °poised, °relaxed, °self-possessed: *Even when she was a teenager, Fern was quite levelheaded.*

levity *n.* lightheartedness, lightness, frivolity, frivolousness, °flippancy, trivialization, °triviality, facetiousness: *Politics is a weighty business and should not be treated with levity.*

lewd *adj.* °lascivious, salacious, °lecherous, °lustful, licentious, °carnal, goatish, hircine, satyric(al), ruttish, concupiscent, libidinous, lubricious *or* lubricous, °indecent, °offensive, °wild, debauched, °obscene, smutty, °crude, °dirty, °foul, °filthy, °rude, °pornographic, °prurient, °gross, °dissolute, Fescennine, °bawdy, ribald, °scurrilous, °raw, °blue, °erotic, °suggestive, °unchaste, unvirtuous, *Colloq* °randy, *Slang* °hot, horny: *He is said to have chronicled the lewd goings-on at the Hellfire Club.*

liability *n.* **1** answerability, °responsibility, °burden, onus, °accountability: *Who bears the liability for the company's losses?* **2** °obligation, °debt, indebtedness, arrear(s), debit: *Outstanding shares are treated as a liability of the corporation.* **3** °disadvantage, °drawback, °hindrance, °impediment, °encumbrance, °snag, °hitch, °barrier, °obstacle, °obstruction, °burden, onus: *Her biggest liability is her inability to read.* **4** °exposure, susceptibility, vulnerability: *The insurer's liability is spread among many underwriters.*

liable *adj.* **1** °likely, apt, °prone, °inclined, °disposed: *She is liable to want her necklace back.* **2** answerable, °responsible, °accountable, obligated, blamable, blameworthy, *Law* actionable: *An owner is liable for damage done by his dog.* **3** exposed, °susceptible, °vulnerable, °open, °subject: *Certain kinds of plants are more liable to disease.*

liaison *n.* **1** °connection, communication, contact, linkage, affiliation, °relationship, °relations: *Liaison between the ministries is improving.* **2** °contact, °intermediary, °link, °tie, °medium, °go-between, °agent: *She acts as liaison between the heads of research and manufacturing.* **3** °(love) affair, amour, °relationship, *affaire d'amour, affaire de coeur,* °intrigue, °romance, entanglement, flirtation: *Fred's wife found out about his liaison with the girl in the typing pool.*

liar *n.* fabricator, prevaricator, perjurer, falsifier, teller of tales, false witness, Ananias, Baron von Münchhausen, *Colloq* fibber: *Why do you believe me when I say I love you when you know I've been a liar all my life? He's a liar if he says he saw me at the theater that night.*

libel *n.* **1** defamation, vilification, denigration, denunciation, deprecation, depreciation, belittlement, disparagement, derogation, °disgrace, ill repute, °dishonor, obloquy, °shame, °humiliation, mortification: *His entire conversation is a perpetual libel on his acquaintances.* **2** °slander, calumny, °lie, °falsehood, prevarication, untruth, misrepresentation, °aspersion, °innuendo, insinuation, °slur, °smear, °blot, °stain, smirch, °stigma: *For years he suffered the libels circulated by his ex-mistress.* —*v.* **3** defame, °vilify, denigrate, °denounce, deprecate, °depreciate, °belittle, °disparage, derogate, °disgrace, °dishonor, °shame, °humiliate, °mortify: *He claims to have been libeled by the article in the newspaper.* **4** °slander, calumniate, lie about, °misrepresent, asperse, °insinuate, °slur, °smear, malign, °stain, °blacken, °discredit, besmirch, °stigmatize, traduce, °vilify, *Rare* vilipend, *Colloq chiefly US* badmouth: *There is often a fine line between libeling a person and exercising freedom of speech.*

liberal *adj.* **1** °generous, °bountiful, °free, openhearted, °open, openhanded, bounteous, °charitable, °philanthropic, munificent, magnanimous, °big, bighearted, unstinting, °unselfish, unsparing; °lavish, °abundant, °ample, °large, °handsome, °plentiful, copious: *Lady Farthing is a liberal benefactor of our cause. Thank you for your liberal contributions.* **2** °progressive, libertarian, reformist, humanistic, °left (of center), latitudinarian, °nonpartisan, unaligned *or* nonaligned, individualistic: *The liberal voters found little to attract them in the latest election.* **3** °free, not literal, °flexible, °lenient, °loose, °broad, °open, °disinterested, °impartial, °dispassionate, °fair, broad-minded, openminded, °unprejudiced, unbigoted, unjaundiced, unopinionated, °tolerant: *Judge Revere is known for his liberal interpretation of the law.* —*n.* **4** °progressive, libertarian, reformer, progressivist, latitudinarian, °independent, freethinker, leftist, left-winger: *They think of themselves as liberals, not as conservatives.*

liberalize *v.* **1** broaden, °widen, °extend, °expand, °stretch, °enlarge: *The purpose of the law is to liberalize the functions of solicitors.* **2** °loosen, °ease, slacken, °relax, °modify, °change, °moderate, °soften: *On holidays, the police ought to liberalize the parking regulations.*

liberate *v.* **1** °(set) free, °release, set at liberty, disenthrall, °emancipate, manumit, °deliver, enfranchise, °(let) loose, let go, °let out, °let off: *When the enemy retreated, the army liberated the prisoners of war. After her divorce, she felt like a liberated woman.* **2** °steal, °pilfer, purloin, °take, °appropriate: *She was caught liberating some tights from the clothing store.*

liberation *n.* freeing, liberating, deliverance, emancipation, enfranchisement, enfranchising, °delivery, °rescue, rescuing, °release, releasing, loosing, unfettering, unshackling, unchaining: *The liberation of the prisoners occurred on May Day, 1945.*

libertine *n.* **1** lecher, °reprobate, °profligate, °rake, rakehell, °roué, debaucher, womanizer, °seducer, fornicator, adulterer, debauchee, whoremonger, °philanderer, Don Juan, Lothario, Casanova, *Colloq* wolf, ladykiller, (old) goat, dirty old man: *He fancied himself a libertine, but women shunned him like the plague.* —*adj.* **2** licentious, °lecherous, °reprobate, °profligate, °rakish, rakehell, philandering, °dissolute, °immoral, °degenerate, depraved, debauched, °decadent, °dirty, °filthy, amoral, °wanton, °lewd, °lascivious, °prurient, lubricious *or* lubricous, salacious, Paphian, libidinous, ruttish, goatish, hircine, satyric(al), °carnal, bestial, *Colloq* °randy, *Slang* horny: *A paragon of prudery to his family, Viscount Guarnier secretly led a life of wild, libertine abandon.*

liberty *n.* **1** °freedom, °independence, self-determination, autonomy, self-rule, °self-government, self-direction, °sovereignty: *People have always been willing to die for their liberty.* **2** °right, °freedom, franchise, °carte blanche, °privilege, °prerogative, °license, °leave, °permission, authorization: *I was given the liberty to go where I pleased.* **3** °freedom, °license, °initiative, °exemption, °exception, °privilege: *On a strict diet, she allowed herself no liberties whatever.* **4** ***at liberty.*** °free, °uninhibited, unfettered, unconstrained, unrestricted, unrestrained, liberated: *I was at liberty to do as I pleased.* **5** ***take a liberty** or **the liberty** or, often, **liberties.*** be unrestrained *or* presumptuous *or* bold *or* uninhibited *or* overfamiliar *or* forward *or* aggressive *or* impudent *or* impertinent *or* audacious *or* improper; display *or* exercise boldness *or* impropriety *or* presumption *or* presumptuousness *or* indecorum *or* unseemliness *or* boldness *or* arrogance: *She thought he was taking liberties in asking her to dance. I took the liberty of asking her to go out with me.*

license *n.* **1** °leave, °permission, °liberty, authorization, °authority, entitlement, dispensation, °right, °carte blanche, °freedom, latitude, free choice, °privilege, °charter: *Merely being an adult does not give you license to do exactly as you please.* **2** °permit, certificate, credential(s), paper(s): *Bring along your driving license.* **3** °disregard, deviation, departure, nonconformity, °noncompliance, divergence: *Such grammatical abnormalities are an exercise of poetic license.* —*v.* **4** °authorize, °allow, °permit, °certify, °sanction, °approve, commission: *The government once had the power to license books for publication.* **5** °certify, °document, accredit, °entitle, validate, °enable, empower: *Whoever licensed him to drive a car made a grave mistake.*

lie[1] *v.* **1** prevaricate, °fabricate, °misrepresent, °invent, commit perjury, perjure *or* forswear oneself, *Colloq* °fib: *If he says that I stole the tarts, he's lying.* —*n.* **2** °falsehood, untruth, falsification, misrepresentation, fiction, °invention, prevarication, °fib, °fabrication, *Colloq* °story, cock-and-bull story, °(tall) tale, whopper, *US* fish story, fishtale: *His claim of having been attacked was a lie.*

lie[2] *v.* **1** °recline, stretch out, be prostrate *or* recumbent *or* prone *or* supine: *I'm going to lie down for a nap before dinner.* **2** °rest, °repose; can be found, be, be situated: *The book is lying on the table. The land lies on the Vermont-New York border.* **3** °rest, °repose, be, reside, °dwell, °abide, °remain, °belong: *The responsibility lies with you.* **4** °press, °burden, °weigh, °rest, be: *The onus for the crime lies heavily on me.* **5** ***lie low.*** °hide, remain concealed *or* in hiding, keep out of sight, *Colloq Brit* lie doggo: *You'd best lie low till the affair is forgotten.* —*n.* **6** ***lie of the land.*** °state, °status, °condition, °situation, °atmosphere, °mood, °spirit, °temper, °character: *You had best determine the lie of the land before submitting your proposal.*

life *n.* **1** °existence, °entity, being; sentience, viability: *Who can imagine the possible forms of life in the universe?* **2** °existence, survival, living, °subsistence, °sustenance: *Certain nutriments are necessary to life on earth.* **3** °existence, living, way of life, life style: *What kind of life does he lead? I enjoy country life.* **4** °existence, lifetime, °time; duration: *She did much during her short life. The life of a light bulb depends on how many hours it is on.* **5** °person, °mortal, °human (being), °individual, °soul: *The new drug might save millions of lives.* **6** biography, autobiography, memoir(s), °(life) story: *He has written a life of Gibbon.* **7** °soul,

°spirit, spark of life, vital spark, moving spirit, life force, *élan vital;* lifeblood; °animation, °vitality, liveliness, sprightliness, vivacity, °sparkle, °dazzle, °dash, élan, °vigor, °verve, °zest, °flavor, pungency, freshness, effervescence, *brio,* °flair, vim, °energy, °exuberance, °enthusiasm, *Colloq* °pep, zing, get-up-and-go: *We must breathe some life into these people. Eva is always the life of the party.* **8** °obsession, preoccupation, °passion, °fixation, compulsion: *Model railways are his life.* **9** °bounce, °resilience, °spring, °elasticity: *There's no life in this old tennis ball.*

lifeless *adj.* **1** °dead: *I stared at the lifeless body of the man.* **2** °unconscious, °inanimate, insensate, °inert, unmoving, °dead, °insensible: *At the news of the accident, she sank, lifeless, to the floor.* **3** °dull, °boring, °tiresome, °heavy, °lackluster, °torpid, °tedious, °flat, °stale, uninteresting, °colorless, uninspiring, °vapid, °wooden: *Mrs. Gabling gave a lifeless performance as Portia.* **4** °barren, °desert, °desolate, °bare, °sterile, °bleak, °empty, °uninhabited, unoccupied, °dreary, °waste: *We gazed at the lifeless landscape of the moon.*

lifelike *adj.* °authentic, °realistic, °natural, true-to-life, °real, °faithful, °graphic, °vivid: *She painted a very lifelike picture of her cat.*

lift *v.* **1** °raise, elevate; °hoist, °heave (up): *She lifted her eyes to the heavens. The pallbearers lifted the coffin to their shoulders.* **2** Often, ***lift up.*** °exalt, °raise, elevate, uplift, °boost, upgrade, °promote, °advance; °improve, ameliorate, °better, °dignify, °enhance, ennoble, enshrine, deify, °immortalize: *His unusual talents lifted him up from the ranks of ordinary mortals. Whom they most lift up they soonest forsake.* **3** °discontinue, °end, °terminate, °stop: *The siege was lifted after six months.* **4** °withdraw, °cancel, °confiscate, take away, rescind, °void, annul: *They charged him with driving while intoxicated and lifted his license.* **5** °rise, °disappear, °dissipate, vanish: *When the fog lifted, I saw we were about to run aground.* **6** °steal, °appropriate, °pilfer, °pocket, thieve, °take, purloin; plagiarize, °copy, *Colloq* °pinch, crib, °liberate, *Slang Brit* °nick, *Old-fashioned* half-inch: *We went into the supermarket and lifted a loaf of bread and a sausage. This passage was lifted from my book in its entirety.* —*n.* **7** °ride: *It is against the law to give hitchhikers a lift on a highway.* **8** °encouragement, °boost, stimulus, °inducement, °inspiration, reassurance, cheering up, *Colloq* °shot in the arm: *Winning first prize gave Susan the lift she needed to continue with her writing.*

light[1] *n.* **1** °illumination, brightness, °daylight, lamplight, candlelight, firelight, gaslight, torchlight, starlight, moonlight, sunlight; gegenschein, counterglow: *There is enough light for me to read the label.* **2** lamp, light bulb, torch, °beacon, lantern, candle, °flare, headlight *or* headlamp, streetlight *or* streetlamp, *US and Canadian* flashlight: *I saw a light in the distance. Someone turned off the lights.* **3** °radiance, °radiation, luminescence, °glare, °gleam, °glow, °reflection, luminosity, °shine, °sparkle, scintillation, incandescence, phosphorescence, fluorescence: *The light of the full moon shone through the window.* **4** daybreak, °dawn, sunrise, *US* sunup: *We leave at first light.* **5** clarification, enlightenment, °insight, °understanding, elucidation, simplification, °explanation: *After an hour's talk with his lawyer, he finally saw the light and confessed. Can you shed a little light on this problem, Robyn?* **6** window, windowpane: *The west wall has three windows of six lights each.* **7** °match, lighter, °spill, °taper, °fire, °flame, ignition: *Can I borrow a light for my cigarette?* **8** highlight: *Note the chiaroscuro effect from the lights and shadows in the drawing.* **9** ***bring to light.*** °reveal, °unearth, °find, uncover, °unveil, °discover, °expose, °disclose, °make known: *A vital fact has been brought to light.* **10** ***come to light.*** be revealed, be unearthed, be uncovered, be unveiled, be discovered, be exposed, be disclosed, °appear, °come out, °turn up, °transpire, °develop, evolve, °emerge: *The truth about her parents has finally come to light.* **11** *US* ***in light of*** *or Brit* ***in the light of.*** °considering, °in view of, in consideration of, taking into account, keeping *or* bearing in mind: *In the light of his condition, he won't be competing in this year's Olympics.* **12** ***shed*** *or* ***throw (some*** *or* ***a little) light on.*** °explain, elucidate, °simplify, °clarify: *Perhaps you could shed some light on the reason why she refuses to leave the house.* —*adj.* **13** (well-)illuminated, °bright, alight, (well-)lit, (well-)lighted, shining, °luminous, effulgent, °brilliant, beaming, °incandescent, phosphorescent, fluorescent: *The lightest room in the house is the kitchen.* **14** °pale, light-hued: *Put the light blue chair over here.* —*v.* **15** ignite, set alight, set *or* put a match to, °kindle; °burn, °touch off, °set fire to, °fire: *As it's getting chilly, you'd better light the fire. When the bomb was in place, they lit the fuze.* **16** °illuminate, °light up, °lighten, °brighten: *Take this candle to light your way to the bedroom.* **17** °turn on, switch on, °put on: *Light the lamps.* **18** ***light up.*** °lighten, °brighten, cheer up, liven up: *The moment she walked in, the whole room lit up. Her face lit up when she saw him.*

light[2] *adj.* **1** lightweight, °portable: *I bought some light garden furniture.* **2** underweight, °skinny, °slight: *Isn't Ted a bit light for his height?* **3** °faint, °dim, °obscure, °indistinct, unclear, faded, °imperceptible: *There is a very light pencil mark on the paper.* **4** °faint, °gentle, °mild, °slight, °delicate, °insignificant: *A light breeze rippled the surface of Fran's martini.* **5** °dainty, °graceful, °delicate, °gentle, °slight: *He felt the light brush of her lips on his.* **6** not weighty, °frivolous, unimportant, °insignificant, °inconsequential, inconsiderable, trivial, °trifling, evanescent, unsubstantial, °slight, °superficial: *The party was a very light affair.* **7** °nimble, °agile, °active, °swift, spry, lithe, °sprightly, lightsome, light-footed, limber, lissom *or* lissome: *When dancing, Greg may be light on his feet, but he's not very light on mine.* **8** simple-minded, lightheaded, °scatterbrained, rattlebrained, bird-brained, featherbrained, °harebrained, °flighty, °giddy, °dizzy, °silly, °inane, °foolish, °frivolous, empty-headed, °vacant, vacuous, °shallow, °superficial: *Featherstonehaugh is a bit too light in the head to be a good agent.* **9** °cheerful, °happy, °gay, °sunny, °merry, lighthearted, happy-go-lucky, easygoing, °joyful, jovial, °jolly: *Noël has written another light drawing room comedy.* **10** °easy, not burdensome, endurable, °bearable, °tolerable, °supportable, undemanding, °effortless, untaxing, °moderate: *After my surgery, I was able to resume light duties at the office. The tax on that little income is very light. You should take some light exercise every day.* **11** amusing, °entertaining, °witty, diverting: *The book hasn't much substance, but it makes good light reading.* **12** ***make light of.*** °dismiss, °write off, shrug off; °trivialize; °ridicule: *The teacher made light of my attempts to play the concerto.* —*v.* **13** alight, °land, come *or* go down, °descend, °settle; deplane, disembark *or* debark, detrain, dismount: *The bird was so tame it lighted on my finger.* **14** ***light into.*** °attack, assail, °lambaste, °assault, pounce *or* fall on *or* upon, °beat, °belabor; °abuse, tongue-lash, °harangue, °upbraid, °scold, °berate, *Colloq* °lace into; *Slang* clobber: *He really lit into the challenger in the third round. Her mother lit into her for not doing her homework.* **15** ***light on*** *or* ***upon.*** chance *or* happen *or* stumble *or* hit on *or* upon, °come across, °encounter, °find, meet up with: *I lit upon the formula when reading some ancient hieroglyphics.*

lighten[1] *v.* **1** °illuminate, °brighten, °light up: *Opening the curtains would lighten up the room.* **2** °cheer (up), °brighten, °gladden, °shine; °smile: *Her face lightened when he said he would be home for the holidays.*

lighten[2] *v.* disencumber, disburden, °relieve, alleviate, °reduce, lessen, °mitigate: *To lighten my load, I had left my suitcase at the hotel.*

likable *adj.* likeable, °genial, °amiable, congenial, °pleasant, simpatico, °agreeable, °pleasing, °attractive, appealing, °nice, °friendly, °winning, charming, °engaging, °good-natured, winsome: *He is a likable enough fellow, but not for a son-in-law.*

like[1] *adv.* **1** °similar (to), °akin (to), allied (to), parallel (to *or* with), comparable (to *or* with), °equivalent (to), °equal (to), °identical (to), cognate (with), analogous (to), corresponding (to), °correspondent (to), °close (to), homologous (to *or* with), of a piece (with), (much) the same (as), along the same lines (as), not unlike: *We hold like opinions. Her opinions are like mine. Her pies are like those mother used to bake. Problems like this give me a pain. Opposite charges attract, like charges repel.* **2** in the mood for, disposed to: *Do you feel like a walk in the park? He felt like going with me.* —*adv.* **3** as if, as though: *He ran like mad.* —*prep.* **4** similar to, identical to *or* with: *Her daughter looks like her. He makes her feel like a perfect fool.* **5** in the same way as, in the manner of, similarly to: *Sometimes she acts like a maniac. The coat fits him like a glove. He laughed like a drain. He treats his mother like a child.* **6** such as, °for example, °for instance, e.g., that is (to say), i.e., in other words, °namely, to wit, viz.: *Mistletoe grows on various trees, like oaks and birches.* —*n.* **7** °match, °equal, °peer, °fellow, opposite number, counterpart, °twin: *It is doubtful that we shall see his like again.* **8** same *or* similar kind *or* sort *or* ilk *or* type *or* kidney *or* breed *or* mold *or* cast *or* strain: *The book deals with music, literature, theater, painting, and the like.*

like[2] *v.* **1** be fond of, °approve of, °appreciate, be partial to, have a fondness *or* liking for, have a weakness for, °take to, °delight in, take pleasure in, derive *or* get pleasure from, °delight in, find agreeable *or* congenial, feel attracted to, be *or* feel favorably impressed by, °relish, °love, °adore, adulate, *Colloq* °take a shine to, *Slang* °go for, °dig, get a kick out of, *US* °get off on, groove on, get a bang *or* a charge out of: *She likes a day at the seaside. He likes chocolate ice cream. I think she likes me.* **2** Usually, ***would** or **should like.*** °prefer, °want, °wish, °ask: *Mr. Legree would like you to get to work on time in the future. I should like a drink. We'd like to have you come to dinner.* —*n.* **3** Usually, ***likes.*** °preference, °partiality, predilection, °liking: *All those people have their likes and dislikes.*

likelihood *n.* °probability, strong *or* distinct possibility, good chance: *We must consider the likelihood that she will refuse to go.*

likely *adj.* **1** °probable, °liable, expected: *Three o'clock is the likely arrival time.* **2** °probable, conceivable, °reasonable, credible, °plausible, °tenable: *It seemed likely that Holmes would solve the case.* **3** °fitting, °able, °suitable, °probable, °seemly, °meet, °right, °proper, °qualified, °acceptable, °appropriate, apposite; °favorite, odds-on, °favored, °promising: *Penhaligon is the likely man for the job.* **4** disposed to, apt to, inclined to, liable to: *It's likely to rain. You're likely to think me mad.* —*adv.* **5** °probably, °undoubtedly, indubitably, no doubt, in all probability, *Colloq* like as not: *You would most likely say that I am wrong. He will very likely win the election.*

liken *v.* °compare, equate, °match, juxtapose: *How can you liken him to Mahatma Gandhi?*

likeness *n.* **1** similarity, °resemblance, correspondence, analogy, °agreement, parallelism: *The artist caught your likeness in the second painting.* **2** °copy, °replica, °facsimile, °duplicate, °reproduction, °model, °representation, °portrait, painting, °picture, °drawing, °photograph, °sculpture, °statue, statuette, °image, simulacrum, icon *or* ikon: *He sells miniature likenesses of the Eiffel Tower to tourists.* **3** °appearance, °face, °figure, °image: *The premier's likeness appears on posters throughout the country.*

likewise *adv.* **1** similarly, in the same *or* like manner *or* way: *Penny is sitting quietly and you should do likewise.* **2** as well, too, also, furthermore, °further, °besides, °in addition, °moreover, °to boot: *There are many ways to do exercise and, likewise, many ways to avoid it.*

liking *n.* **1** °affinity, fondness, °affection, °love, °partiality, °bias, °preference, °bent, predilection, predisposition, °inclination, °appreciation, °penchant; °eye, °appetite, soft spot, °weakness: *He has a liking for good claret and beautiful women.* **2** °taste, °pleasure, °fancy, °preference: *The chef will prepare the steak to your liking.*

limbo *n.* ***in limbo.*** up in the air, consigned to oblivion, °in abeyance, suspended, hanging (fire), neither here nor there, *Colloq* on hold, treading water, holding one's breath, *US* in a holding pattern, on the shelf, on the back burner: *My career is in limbo till I learn the result of the interview.*

limit *n.* **1** Sometimes, ***limits.*** °extent, bound(s), °end, limitation, °check, °curb, °restriction, °restraint: *Is there no limit to how far you will go to get your own way?* **2** Often, ***limits.*** °border, °edge, °end, °extent, °boundary, bound(s), (boundary *or* border *or* partition) line, °frontier, °perimeter, °periphery: *That row of poplars marks the limit of the property to the east.* **3** Often, ***limits.*** °area, °territory, confines, °zone, °region, °quarter, °district, precinct(s): *Stay outside the three-mile limit. We have jurisdiction only within city limits.* **4** ***the limit.*** **(a)** °the end, the last straw, the straw that broke the camel's back, all (that) one can take, °enough, too much, *Colloq* it: *After murdering his parents, throwing himself on the mercy of the court because he was an orphan was the limit.* **(b)** °outrage, °joke, °surprise, *Colloq* °caution: *Wasn't Nigel the limit in that hat?* —*v.* **5** °check, °curb, °bridle, °restrict, °restrain, hold in check: *We must limit spending on arms. The shackles limited my movements.* **6** °restrict, confine, delimit, °narrow, °focus, °guide, °channel: *Please limit your questions to the subject at hand.* **7** °set, °define, °determine, °fix: *The bank manager has limited my overdraft to half what it was.*

limited *adj.* **1** circumscribed, restricted, °fixed, °predetermined; °small, °little, reduced, °minimal: *The theater seats a limited number of people. She was awarded limited access to the children.* **2** °narrow, restricted, restrictive, °meager: *Few original ideas presented themselves to his limited imagination.*

limitless *adj.* unrestricted, unrestrained, unconfined, unbounded, °boundless, °extensive, °vast, °immense, °enormous, °unlimited, illimitable; interminable, unceasing, incessant, undefined, °immeasurable, innumerable, °numberless, countless, myriad, unending, °perpetual, °everlasting, °eternal: *Dickens possessed a seemingly limitless capacity for characterization. In the depths of space exist limitless numbers of stars. God's love is limitless.*

limp[1] *v.* **1** °hobble, °stagger, °totter, dodder, falter: *He limped painfully back to his car.* —*n.* **2** °hobble, hobbling, °stagger, staggering, °totter, tottering, dodder, °doddering, falter, faltering, claudication, *Slang US* gimp: *He was easy to recognize at a distance because of his limp.*

limp[2] *adj.* **1** flaccid, °flabby, °lax, °slack, °soft, drooping, °relaxed, unstarched, unstiffened, °soft, °flexible, °pliable, floppy, °loose: *The banners hung limp in the still, sultry air.* **2** °exhausted, °tired, °fatigued, worn-out, °spent, enervated, wasted, debilitated, °weak, °feeble, °frail: *I was limp after the day's hike.* **3** °weak, °feeble, °ineffective, °ineffectual, °flimsy, half-hearted, °lukewarm, °spineless, thewless, namby-pamby, *Colloq* wishy-washy *Slang* gutless: *Having to visit her grandmother seemed a pretty limp excuse to me.*

line[1] *n.* **1** °mark, pencilmark, penmark, °rule, °score; °stroke, underline, underscore; diagonal, °slash, virgule, shilling mark, solidus, separatrix, °oblique: *The lines on this paper are very faint. Draw a red line under your name. A line separates the numerator from the denominator.* **2** °strip, °belt, °stripe, °band, °border, °edge, edging: *Do you see that line of trees? Put a heavy black line round the obituary notices.* **3** °wrinkle, crease, crinkle, °furrow, crow's-foot: *He has many lines around his eyes.* **4** °border, borderline, °frontier, °limit, °boundary; demarcation, °threshold: *We crossed the line into Italy. I don't mind cooking dinner, but I draw the line at washing the dishes. There is a fine line between genius and insanity.* **5** °outline, °silhouette, contour, °figure, °profile: *The line of the skirt is too straight.* **6** °row, °rank, column, °file, °string, °chain, concatenation, °train, °parade, cortege *or* cortège, cordon, °retinue, °procession, °succession, *Brit* °queue, *Colloq Brit* crocodile, tailback: *I stood in line for six hours for tickets. The line of people wound all the way around the block.* **7** °field, °area, °activity, °forte, °specialty, specialization, °business, °profession, °occupation, °(line of) work, °job, °vocation, °pursuit, °trade, °calling, °employment, *Colloq* °racket, °game: *Saying that he was a computer programmer, he asked me my line.* **8** °note, °word, °card, postcard, °letter, *US* postal card: *Drop me a line when you get there.* **9** °course, °direction, °path, °way, °route, °road, °track, °procedure, °tack, °policy, °strategy, tactic(s), °approach, °plan: *What line will our competitors take to win the account? I adopted the line of least resistance.* **10** °information, °data, °word, °lead, °clue, °hint: *Interpol is trying to get a line on where he might be found.* **11** °cord, °string, °thread, °twine, °yarn, strand, filament, °rope, °cable, hawser: *The anchor line became entangled. Hang the clothes out on the line.* **12** °track, railway *or US and Canadian also* railroad: *The accident on the southern line is delaying all the trains.* **13** °telephone, wire, °cable: *We have three lines at the office. He tried all day to get you on the line.* **14** °front (line), vanguard, °formation: *These men have been in the line for weeks. Captain Hatch was given command of a ship of the line.* **15** ancestry, descent, °stock, °lineage, °succession, °family, °parentage, °extraction, °heritage, genealogy: *She comes from a long line of horsewomen.* **16** °assortment, °stock, °merchandise, offerings, °goods, °brand, °make, °type, °kind, °variety: *What line of lawnmower do you distribute?* **17** Often, ***lines.*** °part, °role, °speech, °script, °words, *Theater US* °sides: *I know my lines by heart for the school play. She has one line in the third act.* **18** °story, °(sales) pitch, blarney, *Colloq* spiel, song and dance, *Slang* con: *He handed her a line about being lonely, but she soon found out why.* **19** ***in*** *or* ***into line.*** **(a)** aligned, in alignment, °true, °straight, in a row, °plumb: *Bring the balusters for the railing in line.* **(b)** in agreement, in accord, in accordance, in conformity, in step, in harmony, *US* lined up: *We are bringing the staffing requirements into line with the council's recommendations.* **(c)** Usually, **into line.** under *or* **in control**: *Were you able to bring the other board members into line?* **20** ***in line for.*** ready for, short-listed for, on the short list for, up for, being considered for, under consideration for, a candidate for, in the running for: *We all think she's in line for a promotion.* —*v.* **21** °rule, inscribe, °score, underline, underscore: *If you use lined paper, your writing wouldn't be so wavy.* **22** °edge, °border, °fringe: *Millions lined the streets for the parade.* **23** ***line up.*** **(a)** °organize, °prepare, °ready, °assemble, °set up, put *or* set in place, °develop, °formulate, °arrange (for), °coordinate: *Everything is all lined up for the big event.* **(b)** arrange for, °secure, °get (hold of), °obtain, contract for; uncover, °dig up, °acquire, °engage, °hire, °sign (up), contract with, °hire, °employ: *We lined up an excellent after-dinner speaker.* **(c)** form a line *or* file, get in *or* into line, form ranks *or* columns, *Military* °fall in, *Brit* queue up: *Hundreds of us lined up for meals three times a day.* **(d)** align, array, °straighten, °order: *These paragraphs should line up. Line up the troops for inspection at 0600.*

line[2] *v.* **1** interline, °cover, °face; ceil: *The coat is lined with fur.* **2** ***line one's pockets.*** accept bribes, °graft, *US* °sell out, *Colloq US* be on the take: *He was secretly lining his pockets during his years on the council.*

lineage *n.* **1** °extraction, ancestry, family tree, °pedigree, descent, °stock, bloodline, °parentage, genealogy: *Jennifer's young man is pleasant enough, but what is his lineage?* **2** forebears, forefathers, foremothers, °family, °people, °clan; descendants, °succession, °progeny, °offspring: *His lineage is listed in* Debrett's Peerage. *Peerages were confined to the lineage of the person ennobled.*

linen *n.* Often, **linens.** bedclothes, bedding, bed linen(s), sheets and pillow cases; table linen(s), napery, tablecloths and napkins; bath linen(s), towels and washcloths: *The soiled linen is sent to the laundry.*

linger *v.* **1** Sometimes, ***linger on.*** °stay (behind), °remain, °tarry, loiter, °persist, °hang on, °endure, °persevere, °survive, *Colloq* hang *or Colloq* stick about *or Colloq* around: *The unmistakable odor of burning sulfur lingered after the stranger disappeared. He lingered on for a day, then died of the King's evil.* **2** °pause, °dawdle, dally, °lag, °idle: *She lingered for a moment at the jeweler's window.* **3** Often, ***linger on*** *or* ***over.*** °dwell on, °elaborate, harp on, °labor: *I shall not linger over the details of the crime.* **4** °procrastinate, dither, °shilly-shally, temporize: *You must stop lingering and decide today.*

lingering *adj.* **1** °long, °persistent, °protracted, °remaining: *I was left with the lingering feeling that I had forgotten to tell her something.* **2** °slow, long-drawn-out, °gradual: *She is dying a lingering, painful death.*

lingo *n.* °jargon, argot, °cant, patois, pidgin, creole, °parlance, °vernacular, °dialect, °idiom, °language, °talk, °speech; gobbledygook, °gibberish, °mumbo jumbo: *In medical lingo, he is suffering from coryza—in other words, a cold.*

link *n.* **1** °tie, °bond, coupling, connector, vinculum; °element, constituent, component: *A chain is no stronger than its weakest link.* **2** °connection, °tie-up, °tie-in, °relation, °relationship, °association, affiliation, interdependence: *What is the link between her arrival and his sudden departure? He thinks he's identified the missing link.* —*v.* **3** Often, ***link up.*** °couple, °join, °fasten (together), °unite; concatenate: *They linked arms to form a barrier.* **4** °connect, tie (up *or* in *or* together), °associate, °relate, °identify with: *I cannot help feeling that this murder links up with the others.*

liquid *n.* **1** °fluid, °liquor, °juice, °solution: *Mixing these two liquids could be dangerous.* —*adj.* **2** °fluid, flowing, °running, runny, °fluent, liquefied, °watery, molten, melted: *Her hair was the color of liquid gold.* **3** °bright, shining, °brilliant, °clear, °transparent, translucent, limpid: *The surface of the lake gleamed like liquid sunshine.* **4** convertible; °solvent, °profitable:

Government bonds are regarded as liquid assets. The company remains liquid despite losses.

liquor *n.* **1** °spirits, °alcohol, °(strong) drink, intoxicants, John Barleycorn, schnapps, *US and Irish* whiskey *or Brit* whisky, demon rum, °moonshine, *US* white lightning, white mule; *Colloq* °booze, pick-me-up, hard stuff, firewater, °juice, mother's ruin (= 'gin'), *US* Kickapoo mountain joy juice; *Slang* rotgut, *Chiefly US and Canadian* hooch *or* hootch, *US* °sauce, red-eye, mountain dew: *I don't drink liquor, but I will have some tea, please.* **2** °liquid, °fluid, °extract, °broth, °stock, distillate, °concentrate, infusion: *After poaching the fish, pour off the liquor and save it.*

list[1] *n.* **1** listing, roster, °roll, rota, catalog, directory, °register, °index, °record, °schedule, °muster, slate, beadroll, laundry list, shopping list, inventory, °file, tabulation; bibliography, *liber veritatis, catalogue raisonné*: *Select what you need from this list.* —*v.* **2** catalog, °register, °index, °record, °note, °itemize, °enumerate, °schedule, °tabulate, °chronicle, °book, °enter, °enroll: *Those eligible to vote are listed in this computer printout.*

list[2] *v.* **1** °lean (over), °tilt, °slant, °heel (over), °tip, °careen, °cant, °incline: *The ship suddenly listed to starboard and I was thrown to the deck.* —*n.* **2** °lean, °tilt, °slant, °heel, °tip, °cant, °slope, °inclination: *The leak caused a severe list to port.*

listen *v.* **1** °hear, pay attention (to), °attend, lend an ear (to), prick up one's ears, keep one's ears open, *Archaic* heark *or US also* hark (to),. hearken *or also* harken (to): *I was listening to the radio when you phoned.* **2** °obey, °heed, °mind, pay attention (to), do as one is told: *You are in trouble because you refused to listen to me.*

listless *adj.* sluggish, °lethargic, °weary, °weak, enervated, °spent, languid, °lifeless, °heavy, °phlegmatic, unemotional, °impassive; °unenthusiastic, °indifferent, apathetic, unconcerned, °lukewarm, °tepid, °cool, uncaring, insouciant; pococurante, laodicean: *She was just over an illness and feeling quite listless.*

litany *n.* **1** °prayer, invocation, °supplication, °petition: *The monks trudged through the streets, chanting their litany.* **2** °recitation, °recital, enumeration, listing, °list, cataloging, catalog, inventorying, inventory: *We had to sit through an interminable litany of the names of contributors to the fund.*

literal *adj.* **1** word-for-word, °verbatim, line-for-line, letter-for-letter, *literatim*, °exact, °precise, °faithful, °strict: *In a literal translation the spirit of the original is often lost.* **2** denotative, etymological, semantic, °dictionary, lexical, °basic, °essential, °pure, °simple, simplistic, °real, °objective, °true, °genuine, °bona fide, °unvarnished, unadulterated, unembellished, simple-minded, uncomplicated, unbiased, °unprejudiced, *Colloq* honest-to-goodness, honest-to-God: *"Female parent" may be the literal definition of* mother, *but it doesn't convey the emotional and connotative senses of the word.* **3** °prosaic, °matter-of-fact, °colorless, °dull, down-to-earth, literal-minded, unimaginative, °humdrum, °boring, °tedious: *His literal approach to everything spoils the conversation for those who have any spirit.*

literally *adv.* **1** word for word, °verbatim, line for line, letter for letter, *literatim*, faithfully, strictly, °exactly, °precisely, closely; °thus, *sic*: *Generally* cordon sanitaire *means 'buffer zone,' but literally it means 'sanitary line.'* **2** °actually, °truly, °in fact, °really: *When I said he'd spilled the beans, I meant it figuratively, not literally.*

literary *adj.* **1** erudite, well-read, cultured, °learned, bookish, °scholarly, °lettered, °cultivated, °refined, °educated; literate: *The literary elite have taken the author of "One Day" to their bosom.* **2** written, °formal, °scholarly, °pedantic, °learned, °academic, scholastic, schoolmarmish: *The writing contains literary words rarely heard in ordinary conversation.*

literature *n.* **1** writing(s), °letters, belles-lettres, creative writing(s): *These works are representative of the literature of the period.* **2** °information, °facts, °data, publicity; °propaganda; brochures, pamphlets, handouts, handbills, leaflets, circulars: *As a member, you will be sent literature about future events. They print subversive literature.*

litigant *n.* litigator, °party, plaintiff, appellant, °suitor, petitioner, suer, defendant, appellee, accused: *The judge suggested that the litigants settle out of court.*

litigation *n.* lawsuit, °suit, °action, °case, legal remedy: *The costs of litigation are extremely high.*

litter *n.* **1** °rubbish, debris, °refuse, °fragments, °odds and ends, *US and Canadian* °trash, *Colloq* °junk: *Food wrappers, cigarette butts, and other litter was lying about.* **2** °brood, °issue, °offspring, °young: *Note the differences in the young of the same litter.* **3** stretcher, palanquin *or* palankeen, sedan chair: *The potentate was borne in on a litter carried by four huge Nubians.* —*v.* **4** °clutter, °strew, °scatter: *Rubbish was littered about the streets.*

little *adj.* **1** °small, °short, °slight, °tiny, °minute, °diminutive, °miniature, mini, °baby, doll-sized, °undersized, °dwarf, midget, pygmy, °elfin, °toy, bantam, °petite, °wee, infinitesimal, minuscule, Lilliputian, teeny, teeny-weeny, teensy-weensy, itty-bitty, itsy-bitsy, microscopic: *I live in a little house. A little green being suddenly appeared. This bed is too little for me.* **2** °young, °small, youthful: *You are asking a lot from a little boy.* **3** °small, °sparse, skimpy, °meager, scant, °insufficient, °inadequate, not enough, °scanty, barely *or* hardly any: *A little crowd gathered around us. You will get little help from her. There's too little milk for all of you.* **4** °short, °brief: *We had only a little time together before he was sent overseas.* **5** °trifling, trivial, °small, °minor, °petty, °paltry, °insignificant, inconsiderable, unimportant, °inconsequential, °negligible: *My safety is of little importance. He does a little business on the side. You are too concerned about little things.* **6** small(-minded), °petty, picayune, °mean, ungenerous, illiberal, °cheap, nar- row(-minded), unimaginative, °shallow: *What is going on in his cunning little brain?* —*adv.* **7** °seldom, °rarely, hardly ever, °scarcely: *I travel little these days.* **8** (but *or* only) slightly, °barely, °hardly, °scarcely, no, not any, not much: *She possessed little more than what she was wearing.* **9** °scarcely, °hardly: *Little does he care whether he wins his wager.* —*n.* **10** °bit, °dab, dollop, °particle, spoonful, °taste, thimbleful, °speck, °spot, °scrap, °crumb, °particle, *Colloq* smidgen: *If you could spare me a little of your time, I'd be grateful.*

live *adj.* **1** living, breathing, °animate, °viable, existent; °material, °physical, °tangible, °real, °actual, palpable: *She said she'd rather be a live beggar than a dead countess. My cat brought a live mouse into the house yesterday.* **2** °energetic, °lively, °spirited, °vigorous, °active, °dynamic, °busy; °current, °contemporary: *Our new advertising agency seems to have some live ideas. The other party made nuclear disarmament a live issue in the election.* **3** °burning, °glowing, °flaming, alight, red-hot, white-hot: *A live coal popped out of the grate onto the carpet.* **4** °loaded, °explosive, unexploded, combustible: *Builders have dug up a live bomb in London.* **5** charged, electrified: *Don't touch a live wire or you'll get a shock.* —*v.* **6** °breathe, °exist; °function: *A more gifted scholar never lived.* **7** °survive, °persist,

°last, °persevere, °endure; °spend, °continue, live out, °complete, °end, conclude, °finish: *He lived out his days happily in Torquay.* **8** reside, °dwell, be; °abide, °stay, °remain, °lodge, °room: *He normally lives in Acton, but at the moment he's living with his mother in Kent.* **9** subsist, °get along, °survive, °fare: *Many retired people complain that they have too little to live on.*

lively *adj.* **1** full of life, °vigorous, °energetic, °vivacious, °spirited, °brisk, spry, °active, °animated, °frisky, °sprightly, °agile, °nimble, °perky, chirpy, bouncy, °buoyant, °gay, cheery, °cheerful, *Colloq* chipper, full of pep, peppy: *A lively discussion followed the speech. The carriage was drawn by a pair of lively horses.* **2** °strong, °intense, °vivid, °keen, °pointed, °eager, °energetic, °active: *She takes a lively interest in current events.* **3** °active, °busy, bustling, °stirring, °eventful, swarming, teeming; astir, °alive: *We joined the lively crowd at the antiques fair. The streets near the square were lively with people rushing to and fro.* **4** °vivid, °bright, °gay, °cheerful, °glowing, °brilliant, °gorgeous, °rich: *Some lively colors in the upholstery and curtains would brighten up the living room.*

liven *v.* **1** Often, **liven up.** brighten, °cheer, °enliven, °perk up: *You need some fresh paint to liven up the outside of the house.* **2** invigorate, °stimulate, °energize, inspirit, °activate, °animate, °fire, °stir (up), put (some) life into, °enliven, °perk up, *Colloq* °pep up: *We need a good dance band to liven up the place in the evenings.*

load *n.* **1** °weight, °burden; onus, °pressure, °encumbrance, millstone, °cross, albatross, °responsibility, °care, °anxiety, °worry, °trouble: *He put down his load and rested. Seeing her safe is a load off my mind.* **2** shipment, consignment, °cargo, °weight: *That load is too heavy for the van.* —*v.* **3** °pack, °pile, °stack, °heap, °fill, lade, °stuff, °cram, °jam, °squeeze: *Load all the boxes you can in the truck.* **4** Often, ***load down.*** °weigh down, °burden, °encumber, saddle with, °overwhelm: *I was loaded down with too many responsibilities.*

loaded *adj.* **1** overloaded, overwhelmed, burdened, laden, weighted (down), filled (up), chock-full, chockablock, stuffed, jammed, °packed, crowded, brimming, brimful, crammed: *A fully loaded truck overturned near exit 5. The train was loaded with commuters.* **2** charged, primed, °ready: *Be careful—the gun is loaded.* **3** charged, °tricky, manipulative, insidious, °prejudiced, °prejudicial, °trap, °devious: *"When did you stop beating your wife?" is what anybody would call a loaded question.* **4** °rich, °wealthy, affluent, moneyed, well-to-do, well-off, *Colloq* °flush, well-heeled, *US* in the chips, *Slang* rolling in it: *I hear that her new husband is really loaded.* **5** °drunk: *He was loaded before he even went to the party.*

loaf[1] *n.* **1** °brick, °cake, °block, chunk; °lump, cube: *We could use a loaf of bread.* **2** °head, °sense, brains, *Colloq* noggin, noodle, °block, bean, *Brit* noddle, *Slang Brit* chump: *Use your loaf!*

loaf[2] *v.* **1** lounge (about *or* around), loiter, °idle, laze, lie about *or* around, take it easy, vegetate, watch the grass grow, *Colloq US* lallygag *or* lollygag, *Slang Brit* skive, *Military* scrimshank, *US* goof off, *US military* goldbrick, *Taboo US* fuck off, fuck the dog: *Stop loafing and get a job!* **2** ***loaf away.*** °waste, °fritter away, °idle away: *He loafs away all his time watching the television.*

loafer *n.* °idler, layabout, °wastrel, shirker, *flâneur,* ne'er-do-well, °tramp, vagrant, *Old-fashioned* lounge lizard, slugabed, *Colloq* lazybones, *Brit* drone, *US* °bum, (ski *or* tennis *or* surf) bum, *Slang Brit* skiver, *Military* scrimshanker, *US military* goldbrick(er), *Taboo US* fuck-off: *Why did she marry that good-for-nothing loafer?*

loan *n.* **1** °advance, °allowance, °credit, °accommodation: *We needed a loan to pay for Melissa's education.* —*v.* **2** °lend, °advance, °allow, °credit: *The bank loaned us enough money to buy a car.*

loath *adj.* loth, unwilling, °reluctant, °averse, °disinclined, °indisposed: *I was loath to tell my wife where I had been.*

loathe *v.* °detest, °hate, °despise, °abhor, abominate, execrate, shrink *or* recoil from, shudder at: *I love broccoli but loathe cauliflower.*

loathing *n.* hatred, abhorrence, °aversion, °hate, odium, detestation, antipathy, repugnance, °horror, °revulsion: *She treated her critics with loathing.*

loathsome *adj.* detestable, °abhorrent, odious, °hateful, °disgusting, execrable, °abominable, °despicable, °contemptible, noisome, °offensive, °horrible, °repulsive, °repugnant, nauseating, sickening, °revolting, °nasty, °vile: *Without warning, the loathsome beast attacked me. She died shortly of a loathsome disease.*

lob *v.* **1** loft, °toss, °pitch, °shy, °heave, °fling, chuck, °hurl, °throw: *She lobbed the ball high in the air over the net.* —*n.* **2** °toss, °throw, °bowl, °pitch, °hit, *US* °fly: *I misjudged the slow lob and swung too soon.*

lobby *n.* **1** foyer, entrance hall, vestibule, °entry, *US* entryway; reception (room *or* area), waiting room, °corridor, °hall, hallway: *We were to meet in the main lobby of the office building.* **2** special-interest group, pressure group, lobbyist: *A powerful lobby has been formed to influence farming legislation.* —*v.* **3** °(bring) pressure (to bear), °(exert) influence, °persuade, pull strings *or* wires, put one's weight behind, °sway, °press, °push, °promote, °urge, *Brit* pressurize: *Environmentalists are lobbying for a ban on fluorocarbons.*

local *adj.* **1** °neighborhood, °neighboring, °nearby, °close by; °adjoining: *She works for a local butcher. The law affects only the local area.* **2** °provincial, regional, °district, °state, county, shire, °municipal, °city, °town, village, °neighborhood; restricted, °limited, °specific, °particular, °peculiar: *Clean water is a general, not a local problem.* —*n.* **3** °resident, °native, townsman, townswoman, townsperson: *The locals all voted against the proposition.* **4** neighborhood pub: *He stopped in at his local on the way home.*

locale *n.* °site, °location, °spot, °place, °setting, venue, °neighborhood, °situation, locality: *We moved the locale of the film to Spain.*

locate *v.* **1** °situate, °place, °site, °position, °set (up), °fix, pinpoint, °establish, °settle, °base: *Where is your office located? We located our warehouse close to the railway.* **2** °find, °come across, °discover, °unearth, lay *or* put *or* get one's hand(s) *or* finger(s) on, chance *or* hit upon, °turn up, °track down: *Where can I locate a good player piano?*

location *n.* **1** °site, °place, °locale, °spot, °setting, °situation, °position: *We have just moved to a new location.* **2** °finding, °discovery, laying *or* putting *or* getting one's hand(s) *or* finger(s) on, unearthing, turning up, tracking down: *The location of the other half of the coin became an obsession with him.*

lock[1] *n.* **1** padlock, hasp, °bolt, latch, °bar, °hook, °clasp, °catch: *As you have nothing worth stealing, why put a lock on the door?* **2** °hold; °control, °power, °sway, °authority, °command, supervision: *The bureau has a complete lock on the issuance of licenses.* —*v.* **3** Often, ***lock up.*** padlock, °bolt, latch, °bar, °secure, °seal: *Lock the doors before leaving for the day.* **4** °clasp, °entangle, °engage, °join, °entwine, °close; °clutch, °grasp, °grapple: *For six hours the knights were locked in battle.* **5** ***lock away.*** See **lock up** *or* **away,** below. **6** ***lock in.*** **(a)** °secure, °retain, °fix, °plant, °implant, stabilize:

What does the maker mean by "the flavor is locked in"? **(b)** °commit, °oblige, constrain, °bind: *We are locked in to the promises we made to our employees.* **(c)** lock up *or* away, confine, °restrain, coop up: *You'd best lock in the dogs before you leave.* **7** ***lock on.*** fix on *or* upon, °track, °follow, °pursue, °keep track of: *Our sights are locked on the target.* **8** ***lock out.*** °exclude, °shut out, close out, keep out, °bar, debar: *During the strike, the management locked out all employees.* **9** ***lock up*** *or* ***away.*** confine, jail *or Brit also* gaol, °imprison, coop up, incarcerate, detain, impound, put behind bars, °restrict, °cage: *They threatened to lock him up and throw away the key.*

lock[2] *n.* tress, curl, ringlet: *As a love token, she sent him a lock of her hair.*

lodge *n.* **1** hunting lodge, °cabin, chalet, °cottage, °house; gatehouse, °shelter, °hut: *I have a small lodge in the mountains.* **2** °branch, chapter: *He belongs to the local lodge of the Society of Odd Fellows.* —*v.* **3** reside, °live, °dwell, °abide, °stay, °stop, °room, occupy rooms: *For years they lodged together at 221B Baker Street.* **4** °accommodate, °board, °put up, billet, °quarter, °house, °take in; °shelter, °harbor: *Mrs. Mulrooney is lodging two gentlemen at her house. Her house cannot lodge more than three.* **5** °stick, °wedge, °catch, °deposit, become stuck *or* wedged *or* fixed *or* caught *or* deposited, embed itself, become embedded: *An apple seed has lodged itself between my front teeth.* **6** °register, °enter, °record, °submit, bring forward, set forth *or* out, °file: *We lodged a complaint against our neighbors because of the noise.*

lodging *n.* Often, **lodgings.** accommodation(s), °shelter, °quarters, °rooms, apartment, °housing, °house, °dwelling, dwelling place, °residence: *I found a night's lodging in Cranberry Street.*

lofty *adj.* **1** °tall, °high, °elevated, °towering, soaring: *On first sight, the lofty skyscrapers of New York impress the visitor.* **2** °exalted, °majestic, °imposing, °grand, °magnificent, °noble, °regal, °imperial, blue-blooded, thoroughbred, aristocratic, magisterial, august, °stately, °venerable, °distinguished, °dignified, °elevated, °eminent, °celebrated, honored, °honorable, respected, °renowned, °famous, °prominent, °illustrious, °notable, °leading, °preeminent, °sublime, °immortal: *He is heir to one of England's loftiest peerages.* **3** °elevated, °honorable, °superior, °exalted, °noble: *Public servants should adhere to lofty principles of morality.* **4** °grand, °grandiose, °haughty, °arrogant, °disdainful, °condescending, °contemptuous, °scornful, °supercilious, contumelious, patronizing, °superior, overweening, vainglorious, °pompous, °snobbish, *Colloq* high and mighty, snooty, uppity, *Brit* °uppish, *Slang* snotty, *Brit* toffee-nosed: *She's much too lofty to have anything to do with the likes of us.*

logic *n.* **1** °reasoning, °deduction, dialectics, ratiocination, inferential *or* scientific reasoning: *Mill defined logic as the science of proof, or evidence.* **2** (good *or* common) sense, sound judgment, °wisdom, °presence of mind: *According to Disraeli, England was not governed by logic.* **3** reasonableness, °intelligence, judiciousness, practicality, rationality: *Your logic is valid, but you are omitting the human factor.*

logical *adj.* **1** syllogistic(al), inferential, deductive, inductive: *Her argument follows basic logical criteria.* **2** °reasonable, °plausible, °sensible, °sound, valid, °intelligent, °judicious, °practical, °rational, °wise, °proper: *Is it logical to assume that people will believe you?* **3** well-organized, °sound, °coherent, °consistent, °sensible, °rational, °reasonable, well-reasoned, well-thought-out: *The proposal is set forth in a logical fashion, from premise to conclusion.*

lonely *adj.* **1** °single, °solitary, °sole, lone, °one; °unaccompanied, °alone: *The lonely survivor of the crash struggled out of the jungle. She walked for miles, lonely and desolate.* **2** °desolate, °uninhabited, °deserted, °barren: *Robinson Crusoe's world was a lonely place.* **3** friendless, °lonesome, °abandoned, °outcast, forsaken; °solo, hermitlike, eremitic(al), °reclusive, °secluded, °retiring, °withdrawn, °unsocial: *He felt lonely in the strange city. She leads a lonely existence with only her cat.*

lonesome *adj.* **1** °alone, forsaken, friendless, rejected, °unpopular, °unwelcome, °outcast, °deserted, °abandoned, °estranged: *She was terribly lonesome and homesick before meeting Patrick.* **2** See **lonely, 2.**

long[1] *adj.* **1** °extensive, extended, elongate(d), °large, °great, °big: *That is a very long snake!* **2** °lengthy, prolonged, °protracted; °extensive, extended, °sustained: *I waited for a long time.*

long[2] *v.* °wish, crave, °want, °yearn, °desire, °hunger, °fancy, covet, dream of, °hanker, eat one's heart out: *I am longing for a good roast beef dinner. I long to see you once more.*

longing *n.* craving, °wish, yearning, °hunger, °fancy, °desire, hankering, *Colloq* yen: *I have a longing to be home again.*

look *v.* **1** Usually, ***look at.*** °see, °observe, °consider, °contemplate, °regard, °view, °survey, °inspect, °scrutinize, °study, °scan; pay attention, °attend, °notice, °watch, °witness; *Literary* °behold, *Slang US* eyeball: *Look at what you've done! Look at the way I do it.* **2** °seem (to be), °appear (to be): *That looks like a useful utensil. You certainly look healthy.* **3** °face, °front (on), °overlook, look out on: *The house looks west.* **4** ***look after.*** °care for, °take care of, be responsible for, °attend, °mind, °watch, °serve, °wait on, °nurse, °protect: *Rose will look after your house while you're away. I need someone to look after me.* **5** ***look down on*** *or* ***upon*** *or US* ***at.*** disdain, °despise, contemn, °scorn, °disparage, derogate, °spurn, °sneer, misprize, *Colloq* turn one's nose up at, look down one's nose at: *They look down on anyone with less money.* **6** ***look for.*** **(a)** °seek, °demand, °require: *We aren't looking for the same things in a person.* **(b)** °hunt for, forage for, °search for: *I am looking for a cup to match my set.* **(c)** °expect, °hope, °anticipate, °count on, °reckon on: *I am looking for a better school report next term.* **7** ***look forward to.*** **(a)** °anticipate, await, wait for: *I am looking forward to Sunday's match.* **(b)** °expect, count *or* rely on *or* upon: *We were looking forward to your help.* **8** ***look into.*** °examine, °study, °investigate, °inspect, delve into, °dig into, °probe, °scrutinize, °explore, °go into, °research, °check (out), °check into: *I shall have to look into the matter.* **9** ***look out.*** be careful, be alert, be vigilant, be on the qui vive, be watchful, watch out, °beware, pay attention, be on guard: *If you don't look out, you'll bang your head.* **10** ***look over.*** °look at, °examine, °read, °scan, °study, check (out *or* over), *Slang US* eyeball: *I haven't had a chance to look over the material you gave me.* **11** ***look up.*** **(a)** °seek, °search for, °hunt for, try to find, track *or* run down: *I meant to look up the etymology of "picnic."* **(b)** get in touch with, (tele)phone, °ring (up), °visit, °call on, °call up, look *or* drop in on, go to see: *Look up a friend of mine when you are in Chicago.* **(c)** °improve, get better, °pick up, show improvement, °progress, °gain, make headway *or* progress: *He was pleased to see that business is looking up.* **12** ***look up to.*** °admire, regard highly, °respect, °esteem, °honor, °revere, °extol, °worship, °idolize, °venerate: *The boy really looks up to you.* —*n.* **13** °gaze, °glance: *Her look was inviting. He gave me a dirty look. Let me have a look at you.* **14** looks, °appearance, °aspect, °bearing, °manner, °air, demeanor; °expression, countenance, °face, mien:

I just didn't like the look of the man. He had a nasty look.

lookalike *n.* °twin, °double, exact *or* perfect likeness *or* match, clone, *Colloq* spit and image *or Colloq* spitting image; doppelgänger; *Slang* (dead) ringer: *She is such a perfect lookalike of Princess Di that people follow her in the street.*

lookout *n.* **1** °guard, sentry, °sentinel, °watchman: *The lookout reported that the fort was surrounded by Cochise's braves.* **2** °alert, *qui vive*; °guard, °watch: *Be on the lookout for shoplifters.* **3** °responsibility, °worry, °concern, °problem, °difficulty, *Colloq* °headache: *How he wastes his free time is his lookout, how he spends his working time is ours.*

loom *v.* **1** °appear, °emerge, take shape *or* form, °materialize, °surface, °arise: *A huge figure loomed out of the fog.* **2** °menace, impend, °threaten, °overshadow, °tower, °dominate, hang *or* hover over: *Count Dracula's dark castle loomed over us.* **3** ***loom large.*** °dominate, °predominate, play a big *or* an important role *or* part: *Sales to children loom large in your predictions.*

loop *n.* **1** hoop, noose, °ring, °circle, °bow, °eye, eyelet, °coil, whorl, *Nautical* °bend: *I caught my toe in a loop of rope and fell on the deck.* —*v.* **2** °twist, °coil, °wind, °tie, °circle, curl, °entwine, °turn, °ring, *Nautical* °bend: *Loop the thread round the crochet hook and pull it through.*

loophole *n.* °outlet, way out, means of escape, °escape, °subterfuge, °pretext, °evasion, °quibble, *Colloq* °dodge: *They couldn't prosecute him because he found a loophole in the law.*

loose *adj.* **1** °unattached, unconnected, °disconnected, °detached, °free, unsecured, unfastened, °movable: *I have a loose tooth. The vacuum cleaner picks up any loose bits.* **2** unconfined, untied, unfettered, released, freed, unshackled, unchained; °free, °at liberty, °at large, on the loose, untrammeled: *In Pamplona, they allow loose bulls to run through the streets to the arena.* **3** unconfining, free-flowing, flowing, baggy, °slack, hanging: *That skirt is loose on you since you lost weight.* **4** disordered, disorganized, unbound, untied, messy; strewn *or* spread *or* tossed *or* thrown about *or* around, scattered (about *or* around), in disorder, in disarray, dispersed: *On the desk was a loose pile of papers.* **5** °rambling, °disconnected, unstructured, unconnected, discontinuous, nonspecific, unspecific, °indefinite, °imprecise, °inexact, °inaccurate, °free, °broad, °rough, °offhand, °casual, °careless, °untidy, °sloppy, slapdash, °general, °vague: *He gave a loose description of his assailant. A loose translation does not convey the flavor of the original.* **6** °lax, °relaxed, negligent, °careless, °sloppy: *Discipline has been quite loose around here lately.* **7** °wanton, °dissolute, debauched, °immoral, °promiscuous, °abandoned, °fast, °libertine, °profligate, licentious, °lewd, °perverted, °corrupt: *She was quickly gaining a reputation for being a loose woman.* —*adv.* **8** ***break loose.*** °escape, °flee: *The boy broke loose from my grasp and ran to greet his mother.* **9** ***hang*** *or* ***stay loose.*** °relax, stay *or* keep calm *or* cool, cool off *or* down, sit back, take it easy: *Hang loose and everything will be all right.* **10** ***let*** *or* ***set*** *or* ***turn loose.*** °discharge, let go (with); °emit, °give out (with), °fire: *They let him loose after questioning. She let loose a stream of invective like a fishwife.* —*v.* **11** let go, °(set) free, °release, let *or* set *or* turn loose; °liberate, °deliver: *Ten thousand balloons were loosed in the celebration.* **12** untie, °undo, unfasten, let go, °disengage, °relax, °ease, °loosen, slacken; , °cast off: *Loose these ropes, please.* **13** let go, °let fly, °fire, °discharge, °shoot, unleash, °deliver: *We loosed a devastating barrage at the enemy.*

loosen *v.* **1** °loose; °undo, unfasten, unhook, unbutton, unlace, untie, unbind, unbuckle; unscrew: *Quick! Loosen his collar!* **2** °weaken, °detach, °separate, °sever, break *or* cut (apart): *The vibration has loosened the rivets.*

loot *n.* **1** °booty, °spoils, °plunder, °prize, °haul, *Slang* swag, boodle: *They divided up the loot amongst the gang.* —*v.* **2** °plunder, °sack, °ransack, °rob, °pillage, despoil, °raid, °ravage, maraud, *Rare* depredate: *The soldiers looted every city of its art.*

lop *v.* Often, **lop off.** chop off, °trim, °top, °head, crop, °prune, °dock, °clip, °trim, snip off, shear off, °cut off, °pare, °shorten, hack off, amputate: *Lop the lower branches off close to the trunk.*

lopsided *adj.* **1** uneven, askew, °one-sided, awry, unsymmetrical, asymmetrical, unequal, °crooked, °unbalanced, °irregular, *Colloq* cockeyed: *One lopsided bookcase is, fortunately, the sole product of Ed's cabinetmaking class.* **2** uneven, unequal, °one-sided, °biased, °disproportionate, unfair, warped, twisted: *Four against two—I'd call that lopsided, wouldn't you?*

lord *n.* **1** °master, °monarch, ruler, °sovereign: *As Harry considers himself lord of all he surveys, it's a good thing he's so shortsighted.* **2** °noble, nobleman, °peer, aristocrat; earl, duke, °count, viscount, baron: *In gratitude for his heroism, the nation made him a lord.* **3** ***The*** *or* ***Our Lord.*** God, the Almighty, God Almighty, the Creator, the Supreme Being, Christ, Jesus, Jehovah: *The Lord moves in mysterious ways his miracles to perform.* —*v.* **4** ***lord it over.*** domineer, °swagger, be overbearing, *Colloq* °boss (around), act big, pull rank: *He lords it over everyone in the office.*

lore *n.* **1** folklore, beliefs, °culture, tradition(s), °mythology, myths, mythos, ethos, teaching(s), °doctrine, °wisdom: *According to their lore, the American Indians go to the Happy Hunting Ground when they die.* **2** °knowledge, °learning, erudition: *He was acquainted with medical lore through his reading.*

lose *v.* **1** °mislay, misplace, °displace, °part with; suffer the loss of, be deprived of: *I lost my pen. He lost a leg in the war.* **2** °forfeit, °yield: *He says he lost the rent money gambling.* **3** °give up, °yield, °capitulate, admit defeat, °succumb, bow to, be defeated *or* conquered, suffer defeat, be beaten *or* overcome *or* worsted *or* bested, *Colloq* lose out: *It was a good game, but we lost.* **4** °waste, °let slip, squander, °dissipate, fritter *or* trifle away, °run out of; °consume, °use (up), °expend, °spend: *We lost a lot of time waiting for him to decide.* **5** °elude, °evade, °escape, throw *or* shake off, give the slip: *I managed to lose the sinister-looking man who was shadowing me.*

loser *n.* also-ran, °misfit, °failure, °fiasco, nonstarter, *Colloq* °flop, °dud, °washout, bummer, lead balloon, lemon, born loser, *Brit* damp squib, *Brit and Australian* no-hoper, *US* clinker, nebbish, schlemiel *or* schlemihl *or* shlemiel, schlimazel *or* shlimazel *or* shlimazl, schnook, sad sack: *Glenn is a real loser and will never amount to anything. His idea for a company newsletter was a real loser.*

loss *n.* **1** deprivation, bereavement, °privation, °denial, °sacrifice, forfeiture, disappearance: *We must be constantly vigilant to watch for any loss of our liberty.* **2** diminution, °erosion, reduction, impoverishment, depletion, shrinkage: *New policies are in place to stem the loss of parkland.* **3** °disadvantage, °detriment, °harm, °impairment, °injury, °damage: *I don't regard Wentworth's resignation as any great loss.* **4** °waste, wastage, wasting, squandering: *The loss of time was owing to a train delay.* **5** °defeat, °setback, °disadvantage, °disappointment, °failure, °downfall, °collapse, °breakdown, °ruin; drubbing, trouncing: *His loss is my* *gain.* 6 Often, ***losses.*** debit(s), °liability (liabilities), negative cash flow: *The company wrote off the losses.* 7 Often, ***losses.*** °death, °dying, °passing, demise, °destruction, extermination, extinction, annihilation: *The loss from disease was overwhelming. What losses did we suffer in the battle?*

lost *adj.* 1 gone, departed, vanished, strayed; missing, mislaid, misplaced, irrecoverable: *They found my lost dog. The airline told me that my bag was lost.* 2 wasted, °misspent, gone by the board, squandered, °down the drain, °spent, °exhausted, *Colloq* out of the window: *The new arrangement is resulting in a lot of time lost. I watched another lost opportunity slip by.* 3 °confused, baffled, perplexed, puzzled, mystified, bewildered, confounded, adrift, °helpless, disoriented, at sea, astray: *I am totally lost when it comes to high finance.* 4 forgotten, °bygone, °extinct, °past, °obsolete, vanished, buried: *Good writing needn't be a lost art.* 5 °dead, °extinct, departed, fallen, °late: *We held a memorial service for our lost comrades.* 6 destroyed, demolished, devastated, ruined, wrecked, irreparable, unsalvageable, irreclaimable, irremediable: *After the exposé in the newspaper, his name was consigned to the limbo of lost reputations.* 7 damned, cursed, accursed, °abandoned, °corrupt, fallen, °wanton, °unchaste, °dissolute: *Unrepentant sinners are lost souls.* 8 °hopeless, °distraught, distracted, °desperate, °frantic, frenzied: *He had the lost look of a cornered fugitive.*

lot *n.* 1 °collection, °batch, consignment, °assortment, °group, °portion, °set, °quantity, grouping, apportionment: *They are expecting a new lot of furniture this afternoon.* 2 °luck, °fortune, °destiny, °fate, kismet, °plight, °doom, °end: *If it is my lot to die for my principles, then so be it.* 3 °lottery, °drawing, °raffle, drawing lots *or* straws: *The winner will be decided by lot.* 4 °share, °portion, °division, °interest, °part, °allotment, °assignment, apportionment, °ration, °allowance: *Marguerite was bequeathed the house, while my lot included the paintings.* 5 ***a lot*** *or* ***lots.*** **(a)** a good *or* great deal: *You have a lot of nerve! I'd give a lot to be able to paint like you. He got into a lot of trouble.* **(b)** much, loads *or* a load, mountains *or* a mountain, tons *or* a ton, barrels *or* a barrel, stacks *or* a stack, piles *or* a pile, heaps *or* a heap, masses *or* a mass, °oceans, *Colloq* oodles, scads, *US* gobs: *She has lots of money.* **(c)** °many, myriad, numerous, countless, reams, infinite *or* an infinity, quantities *or* a quantity, enormous numbers *or* an enormous number, *Colloq* oodles, scads, loads, tons, masses, : *There are lots of fish in the sea.* 6 ***the lot.*** °everything, *Colloq* the whole kit and caboodle, all: *I'll give you ten dollars for the lot.*

lotion *n.* cream, liniment, balm, °salve, °ointment, embrocation, unguent, pomade: *The doctor said to rub in some of this lotion twice a day.*

lottery *n.* °raffle, sweepstake, °drawing, °pool, *Brit* tombola: *I've never known anyone who won anything in the national lottery.*

loud *adj.* 1 deafening, earsplitting, booming, blaring, stentorian, thundering, °thunderous, sonorous, °noisy, clamorous, °piercing, fortissimo: *The hi-fi is so loud I can't hear myself think. A loud crash came from the kitchen.* 2 °tawdry, °garish, °flashy, °gaudy, °tasteless, °extravagant, °showy, °ostentatious, *Colloq* splashy, snazzy, jazzy: *He was wearing a plaid suit and the loudest tie and shirt I had ever seen.*

lounge *v.* 1 °idle, °loaf, laze, loll, languish, vegetate: *After dinner, we lounged till midnight, sipping port.* —*n.* 2 °lobby, foyer, waiting room, °reception (room), vestibule: *I was waiting in the lounge when the doctor returned.* 3 cocktail lounge, *Brit* (lounge *or* saloon) bar: *We went into the lounge and ordered a drink.* 4 sofa, °couch, divan, studio couch, daybed, settee, °settle, love seat, chaise longue; causeuse, °tête-à-tête, °serpentine; *US and Canadian* davenport: *The psychoanalyst has a lounge on which you lie while talking to him.*

lour *v.* See **lower**[2], below.

lousy *adj.* 1 °awful, °terrible, °mean, °contemptible, °low, °base, °hateful, detestable, °despicable, °vile, °wretched, °miserable, °scurvy, °dirty, °vicious, *Colloq* °rotten: *Telling tales to the teacher was a lousy thing to do.* 2 °bad, °poor, °awful, °terrible, °inferior; low-quality, °shoddy, °shabby, °inferior, °miserable, second-rate, °wretched: *We had lousy seats for the theater last week. That dealer sold you a really lousy car.* 3 pedicular, pediculous, *Brit* lice-infested, lice-ridden, *US* louse-infested, louse-ridden: *Even lice are lousy—they have their own parasites.* 4 ***lousy with.*** alive with, overloaded with, swarming with, teeming with, *Colloq* crawling with, knee-deep in: *In five minutes the place was lousy with cops.*

lovable *adj.* loveable, °adorable, °darling, °dear, cherished, °likable, °attractive, °engaging, °cute, °fetching, °taking, alluring, °endearing, appealing, winsome, °sweet, °tender, cuddly, °affectionate, charming, °enchanting: *She has the most lovable little baby boy.*

love *n.* 1 °warmth, °affection, °attachment, fondness, tenderness, °devotion, °attraction, °friendship, amity, °regard, °admiration, °fancy, adoration, adulation, °ardor, °passion, °fervor, °rapture, infatuation: *Her love for him grew over the years.* 2 °liking, °delight, °enjoyment, °pleasure, fondness, °attraction, predilection, °bent, °leaning, proclivity, °inclination, °disposition, °weakness, °partiality, °preference, °taste, °relish, °passion: *In his retirement, Brett has developed a love for golf.* 3 °darling, °beloved, °sweetheart, sweetie, °sweet, honey, dear one, dearest, angel, turtledove, truelove, light of one's life, °lover, °paramour, °mate, intended, betrothed; girlfriend, inamorata, ladylove, young lady, fiancée; boyfriend, beau, inamorato, °suitor, swain, young man, °fiancé, *Archaic* leman, °tally, *US* POSSLQ (= 'Person of the Opposite Sex Sharing Living Quarters'), *Colloq* °girl, °woman, °guy, °man: *Let me tell you something, my love, "'Tis best to be off with the old love before you be on with the new." Come live with me and be my love.* 4 °sympathy, tenderness, °concern, °charity, °care, °solicitude, °affinity, °rapport, °harmony, °brotherhood, sisterhood, fellow feeling: *He has great love for his fellow human beings.* 5 ***love affair.*** **(a)** amour, °liaison, °affair, °romance, °relationship, *affaire de coeur, affaire d'amour,* °intrigue: *He's been having a love affair with his secretary.* **(b)** °passion, °mania, *Colloq* °thing: *George allows nothing to interfere with his love affair with tennis.* 6 ***make love*** **(*to*)** *or* **(*with*).** °embrace, °cuddle, °caress, °fondle, have sexual intercourse, °take, *Archaic* °know, *Colloq* neck, °pet, canoodle, °romance, have sex, make the beast with two backs, *US and Canadian* °make out; *Taboo slang* °screw, fuck, °hump, bang, *Brit* roger, bonk: *He still makes love to his wife though they're in their eighties.* —*v.* 7 °cherish, °admire, °adore, be in love with, lose one's heart to, °worship, °idolize, dote on, °treasure, be infatuated with, think the world of, adulate, hold dear, °like, *Colloq* be hung up on, be crazy *or* nuts *or* wild *or* mad about, have a crush on: *Only after thirty-five years of marriage did she discover that she loved him.* 8 °delight in, take pleasure in, derive pleasure *or* enjoyment from, °relish, be partial to, have a passion *or* preference *or* taste for, be attracted to, be captivated by, be fond of, °like, °enjoy, °appreciate, value, *Colloq* get a kick from *or* out of, be wild about, be thrilled by, *US* get a bang *or* charge from *or* out of: *She loves chocolates. I just love your new dress! Love me, love my dog. A man may love his house and yet not ride upon the ridgepole.*

love letter *n.* billet-doux, *Archaic* mash note: *She saved all his love letters.*

lovely *adj.* **1** good-looking, °pretty, °handsome, °attractive, °comely, °fair, °fetching, °engaging, captivating, alluring, enticing, bewitching, °ravishing, °gorgeous, °beautiful, beauteous, pulchritudinous: *Larry has two lovely daughters. The house has a lovely view of Lake Windermere.* **2** °satisfying, °satisfactory, °agreeable, enjoyable, gratifying, °nice, °pleasing, °pleasant, °pleasurable, °engaging, °delightful: *What a lovely way to spend an evening!*

lover *n.* See **love, 3,** above.

low[1] *adj.* **1** °short, squat, °little, °small, stubby, stumpy, °stunted; low-lying: *The terrain has many low shrubs, not more than two feet tall.* **2** °inadequate, °insufficient, °deficient, down, °short, °sparse, °scanty, scant, °limited: *The air in the tires is a bit low. Our water supply ran low.* **3** °coarse, °unrefined, indelicate, °improper, °naughty, °risqué, °indecent, °unseemly, °vulgar, °crude, °common, °rude, °offensive, °gross, ill-bred, °lewd, °obscene, ribald, °bawdy, °scurrilous, smutty, °pornographic, °dirty: *The old burlesque acts were characterized by their low humor.* **4** °weak, °frail, °feeble, debilitated, enervated, °sickly, °unhealthy, °infirm, °shaky, °decrepit, °ill, °sick: *He's very low and sinking fast.* **5** °ineffectual, °ineffective, °weak: *Her resistance to disease is low. I have a very low pain threshold.* **6** °miserable, °dismal, °wretched, °abysmal, °sorry, abject, °destitute: *I had reached the low point in my career.* **7** °humble, °poor, lowborn, °lowly, °base, °inferior, baseborn, °plebeian, proletariat, ignoble: *He rose from those low beginnings to a peerage.* **8** °unhappy, depressed, °dejected, °sad, °gloomy, °melancholy, °miserable, °despondent, disconsolate, °blue, downcast, down, °glum, °wretched, morose, crestfallen, °brokenhearted, °heartbroken, °tearful, lachrymose, °sorrowful, °mournful, heavy-hearted: *I felt low for weeks after the death of my kitten.* **9** °inferior, second-rate, °poor, °bad, not up to par, °worthless, °shoddy, °shabby, °mediocre, substandard: *Their products are of low quality.* **10** °inferior, °lower, lesser, °small, smaller: *Play a low card.* **11** low-cut, décolleté, revealing, *Colloq US* low and behold in the front and vie de Bohème in the back: *Gilda wore a low, strapless black satin evening gown.* **12** °base, °vile, abject, °contemptible, °despicable, °mean, °menial, °servile, ignoble, degraded, °vulgar, °foul, dastardly, depraved, °nasty, °sordid: *Telling her husband about our rendezvous was a low trick. I didn't know that anyone could sink to such a low level.* **13** °quiet, hushed, °soft, °subdued, °gentle, muted, muffled, stifled, °indistinct, whispered, murmured, murmurous: *Her low voice was sweet.* **14** unfavorable, °critical, adverse: *She has a rather low opinion of his singing.*

low[2] *v.* moo, °bellow; °bawl: *The lowing herd wound slowly to the pasture.*

lowdown *n.* °information, °intelligence, °data, the facts, inside story, *Colloq* info, °dope, °dirt, *Brit* bumf: *I have the lowdown on the scandal.*

lower[1] *v.* **1** °drop, °reduce, °decrease, °mark down, °discount, lessen, °diminish, °downgrade, °cut, °slash: *Prices on all goods have been lowered.* **2** let *or* move *or* bring *or* put down, °drop: *The drawbridge was lowered and we rode across. They lowered the coffin into the grave.* **3** cut *or* lop off, cut *or* take down, °reduce, °diminish, crop, °trim: *We lowered the hedge by a foot.* **4** abase, °debase, °degrade, °discredit, °shame, °disgrace, demean, °belittle, °humble, °humiliate; °stoop, °deign, °condescend: *I wouldn't lower myself so much as to speak to him.* **5** °turn down, quieten, °moderate, °modulate, °soften, tone *or* tune down: *He lowered his voice to a whisper.* —*adj.* **6** farther down: *My mother has an apartment on a lower floor of the building.* **7** earlier: *These primitive amphibians are from the Lower Carboniferous.* **8** ***lowercase.*** °small, minuscule: *Use capitals for the first letters and lowercase for the rest in these names.*

lower[2] *v.* **1** °lour, darken, °threaten, °menace, °loom: *"Between the dark and the daylight, / When the night is beginning to lower, / Comes a pause in the day's occupation, / Which is known as the children's hour."* **2** °lour, °frown, °scowl, glower; °sulk, °pout, mope: *He said nothing, but lowered at me from beneath his beetled brow.*

lowly *adj.* See **low**[1]**,** 7, above.

loyal *adj.* °faithful, °true, dependable, °devoted, °trustworthy, trusty, °steady, °steadfast, °staunch, trusted, °reliable, °stable, unswerving, unwavering, dedicated, °constant, °patriotic: *Give me twelve men loyal and true and we shall rout the enemy.*

loyalty *n.* faithfulness, fidelity, dependability, devotedness, °devotion, allegiance, patriotism, trustworthiness, steadfastness, staunchness, firmness, °resolution, °resolve, reliability, °stability, °dedication, constancy: *I hope that we can rely on the loyalty of everyone in your unit, Colonel.*

luck *n.* **1** °fortune, °chance, °destiny, °fate, °accident, fortuity, serendipity; °fluke, stroke of luck, *US* happenstance: *Luck brought us together.* **2** good fortune, °(good) break: *It takes a lot of luck to get the kinds of roles you want as an actor.* **3** chance(s), success rate, fortune(s): *I hope that my luck improves soon.*

lucky *adj.* **1** °fortunate, blessed, °favored, °charmed: *She was very lucky to get the job.* **2** °providential, °timely, °opportune, °propitious, °favorable, auspicious, °advantageous, °convenient, fortuitous: *It was lucky that you were at the swimming pool to save her.*

ludicrous *adj.* °ridiculous, laughable, °absurd, °farcical, °nonsensical, °preposterous, °incongruous, asinine, °foolish, °silly, °zany, °crazy, comical, risible; °funny, facetious, droll, waggish, jocular, °witty, jocose: *The explanations of the new tax laws have been carried to ludicrous extremes. Count on Bosworth to come up with some ludicrous prank on April Fools' Day.*

lug *v.* °drag, °tug, °tow, °haul, °heave; °carry, tote, °transport: *You're mistaken if you think I'm going to lug that heavy suitcase everywhere.*

luggage *n.* baggage, bags, °gear, impedimenta, °paraphernalia, °things, °belongings: *I found my luggage was missing when I arrived in Torre del Greco.*

lukewarm *adj.* **1** °tepid, room temperature, °warm: *I like my bathwater lukewarm.* **2** °cool, °indifferent, °halfhearted, °chill, °chilly, °phlegmatic, unresponsive, °unenthusiastic, °nonchalant, °lackadaisical, apathetic, insouciant, laodicean, °unmoved, *US* half-baked, *Colloq* laid-back: *My ideas for improving efficiency continued to receive a lukewarm reception from the board.*

lull *n.* **1** °pause, °respite, °interlude, °intermission, °interval, °break, hiatus, °interruption, °stop, °halt, °lapse, °delay, *Literary* caesura, *Colloq* °letup: *After a brief lull, the hurricane resumed in all its ferocity.* **2** °quiet, quiescence, °hush, °calm, calmness, stillness, °silence, °peace, peacefulness, tranquillity: *There was a lull, then the noise of the artillery again shattered the night.* —*v.* **3** soothe, °calm, °quiet, °hush, pacify, mollify, °tranquilize: *Do not let her sweet words lull you into a sense of false security. I was lulled to sleep by the crickets and the bees.*

lumber *n.* **1** °odds and ends, °junk, °clutter, °jumble, rejects, white elephants; °rubbish, °litter, *Chiefly US* °trash: *The small room off the kitchen is for lumber.*

2 °timber, wood, beams, planks, boards, wood: *I have ordered the lumber for building the garage.*

luminous *adj.* 1 °shiny, shining, °bright, °brilliant, lighted (up), lit (up), illuminated, °radiant, alight, resplendent, °lustrous, gleaming, shimmering, glistening, sparkling, °dazzling, refulgent, effulgent: *The birches shone with a luminous beauty.* 2 °glowing, aglow, luminescent, °incandescent, phosphorescent, fluorescent: *I have a watch with luminous numbers and hands.* 3 °clear, lucid, perspicuous, percipient, perspicacious, °penetrating, discerning, °perceptive, cleareyed, clearheaded, °keen, °acute, °sharp, °explicit, °incisive, °specific, °express; understandable, °intelligible: *His novel provides another example of his luminous style.*

lump[1] *n.* 1 °mass, °piece, °gob, gobbet, °clod, chunk, clot, °wad, °clump, hunk, nugget; cube, °wedge, °cake: *A lump of dirt was clogging the drain. May I have two lumps of sugar, please?* 2 °bump, °growth, °swelling, protuberance, °protrusion, °prominence, °bulge, excrescence, tumescence, nodule, °knob; wen, cyst, °boil, carbuncle, blister, wart, corn: *You ought to see a doctor about that lump on your nose.* —*v.* 3 Often, ***lump together.*** °combine, °join, consolidate, °collect, °bunch, °group, °unite, °mass, aggregate, °blend, °mix, throw *or* put together: *Treat me as an individual and don't lump me together with everyone else.*

lump[2] *v.* Usually, **lump it.** allow, °tolerate, °suffer, °put up with, °bear, °stand, °brook, °endure: *Whether you like it or not, you'll have to lump it.*

lumpy *adj.* chunky, °bumpy, uneven, °granular, grainy: *Mix the batter till it is no longer lumpy.*

lunacy *n.* 1 °madness, °insanity, dementia, craziness, derangement, psychosis, °mania: *Formerly, it was believed that lunacy fluctuated in accordance with the phases of the moon.* 2 °folly, foolishness, bad *or* poor judgment, illogicality, illogic, senselessness, ridiculousness, irrationality, foolhardiness, °stupidity: *It would be sheer lunacy to confess to something you did not do.*

lunge *n.* 1 °thrust, °jab, , °strike: *He was impaled by the first lunge of the sword.* 2 °dive, °plunge, °rush, °leap, °jump, °spring, °pounce: *I made a lunge for the knife but missed.* —*v.* 3 °dive, °plunge, °charge, °pounce, °dash, °bound, °jump; °thrust, °stab, °strike, °hit, °jab, °cut: *He lunged at me with his machete.*

lurch[1] *n.* ***leave in the lurch.*** °desert, °abandon, °forsake; °drop, °jilt: *He left her in the lurch, waiting at the church.*

lurch[2] *n.* 1 °stagger, °sway, °pitch; °list, °tilt, °toss: *The man gave a sudden lurch into me, knocking the vase from my hands.* —*v.* 2 °stagger, °sway, °stumble; °roll, °tilt, veer, °pitch, °list, °heel, °wallow: *He lurched into the room, a dagger protruding from his back. The ship lurched dangerously in a heavy sea.*

lure *v.* 1 °tempt, °attract, °induce, °coax, inveigle, °seduce, °draw in, °entice, °lead on, °decoy, °charm, °persuade, allure, °catch: *She lured me into the trap of believing that she loved me.* —*n.* 2 bait, °decoy, °attraction, °temptation, °inducement, magnet, siren song, °charm, *US* drawing card, *Slang* °come-on: *Too many naive people fall for the lure of easy money.*

lurid *adj.* 1 °sensational, °vivid, °shocking, °startling, °graphic, °melodramatic: *The tabloids delight in a lurid scandal.* 2 °ghastly, horrid, horrifying, horrendous, °gory, °grisly, °gruesome, °macabre, °revolting, °disgusting, appalling, °frightful, °terrible, °awful: *Every lurid detail of the massacre was shown on TV.* 3 °pale, ashen, sallow, °wan, pallid, °ghastly, baleful: *From the lurid shade of his skin, I could see he had some horrible affliction.* 4 °glaring, °fiery, °flaming, °burning, aglow, °glowing, glowering: *The burning city cast a lurid light in the sky.*

lurk *v.* skulk, °slink, °prowl, °steal, °sneak, °hide, °(lie in) wait, °lie low: *The muggers were lurking in the shadows, waiting for a victim.*

luscious *adj.* delectable, °delicious, mouthwatering, °tasty, toothsome, °savory, appetizing, °rich, °sweet, °epicurean, ambrosial, palatable, °pleasant; °succulent, °juicy, *Colloq* scrumptious, °yummy: *Mother used to make the most luscious pies and tarts.*

lush *adj.* 1 °luxuriant, °thick, °lavish, °flourishing, verdant, °green, °dense, °overgrown, °thick, °exuberant: *The walls were covered with a lush growth of ivy.* 2 °juicy, °succulent, mouthwatering, °fresh, °moist, °ripe: *We ate lush pears that we picked right off the tree.* 3 °palatial, °extravagant, °elaborate, °luxurious, °opulent, °sumptuous, *Colloq* ritzy, °plush: *The bridal suite is the lushest accommodation they offer.*

lust *n.* 1 sensuality, libido, libidinousness, sexuality, lustfulness, concupiscence, sexual appetite, *Slang* horniness: *The nouveaux riches spend much of their time and money satisfying their lust.* 2 °desire, °drive, °energy, voracity, avidity, avidness, °ambition, ravenousness: *Buonvolio has an infectious enthusiasm and lust for life.* —*v.* 3 ***lust after.*** °desire, crave, hunger *or* thirst *or* hanker for *or* after, ache for: *Brumble was lusting after the barmaid at The Two Magpies.*

luster *n.* 1 °sheen, °gleam, °glow, °gloss, luminosity, luminousness, °radiance: *A good waxing should restore the luster to the table.* 2 °glory, °renown, °brilliance, °celebrity, °honor, °distinction, °fame, illustriousness: *Winning first prize lent a little luster to his tarnished reputation.*

lustful *adj.* libidinous, °carnal, concupiscent, licentious, °lewd, °prurient, °lascivious, salacious, *Colloq* horny, °randy: *Any girl passing the construction site has to put up with the lustful comments from the workers there.*

lustrous *adj.* °glossy, °shiny, shined, °polished, burnished: *French polishing gives the furniture a lustrous finish.*

lusty *adj.* 1 °vigorous, °healthy, °strong, °energetic, °robust, hale and hearty, °lively; °buxom: *The sailors were accompanied by a couple of lusty young women.* 2 °vigorous, °substantial, °strong, °husky, °powerful: *He sang in a lusty voice.*

luxuriant *adj.* 1 °abundant, °profuse, copious, °lush, °rich, bounteous, overflowing, °full, °luxurious: *Her luxuriant hair hung down to her waist.* 2 °lavish, °full, °rank, °prolific, thriving, rife, °exuberant, °lush, abounding, plenteous, °abundant, superabundant, °dense, °fruitful, teeming: *The luxuriant orange groves are yielding a bumper crop this year.* 3 °ornate, °elaborate, decorated, °fancy, rococo, baroque, °flowery, frilly, florid, overdone, °flamboyant, °showy, °ostentatious, °gaudy, °garish, *Colloq* °flashy: *The elders disapproved of the luxuriant ornamentation of the churches.*

luxuriate *v.* 1 Often, ***luxuriate in.*** °wallow in, swim in, bask in, °indulge in, °delight in, °relish, °revel in, °enjoy oneself, °savor, °appreciate, °like, °love: *She luxuriates in her new-found wealth.* 2 live in luxury *or* comfort, be in the lap of luxury, have a good *or* great *or* marvelous time, take it easy, °enjoy oneself, live the life of Riley, live off the fat of the land, *Colloq* have the time of one's life, have a ball, *US* live high off the hog: *Their postcard says that they are luxuriating on Capri for a week.*

luxurious *adj.* 1 °opulent, °sumptuous, °grand, °extravagant, °lavish, °magnificent, °splendid, de luxe, °fancy; °epicurean, °gourmet; *Colloq* °swanky, swank, ritzy, °plush, °posh: *We had a luxurious room overlooking the sea and ate the most luxurious meals.* **2** °self-indulgent, °voluptuous, voluptuary, sybaritic, hedonistic, pampered: *We lived a truly luxurious life till the money ran out.*

luxury *n.* **1** opulence, °splendor, sumptuousness, °grandeur, °extravagance, magnificence, richness, luxuriousness: *I cannot describe the overwhelming luxury of the maharajah's palace.* **2** °indulgence, self-indulgence, hedonism, sybaritism, voluptuousness: *He had been living a life of luxury in Tahiti.* **3** °security, °confidence; °gratification, °satisfaction, °enjoyment, °pleasure, °delight, °comfort: *He has the luxury of knowing that his dog will be well looked after.* **4** °frill, °extravagance, °extra, °indulgence, °nonessential, °expendable, °treat: *He showered her with luxuries from the finest shops.*

lying *n.* **1** prevarication, fibbing, mendacity, mendaciousness, falsification, untruthfulness, °perjury; dishonesty, °deceit, duplicity: *She was accused of lying while under oath.* —*adj.* **2** untruthful, °false, mendacious, °hypocritical, °dishonest, °deceitful, °deceptive, duplicitous, treacherous, °perfidious: *Whoever told you that your latest novel is marvelous is a lying hypocrite.*

lyric *adj.* **1** melodic, songlike, °musical, °melodious, °lyrical: *Lyric drama is no longer fashionable.* **2** °personal, °subjective, °individual, idiosyncratic; °sentimental, °rhapsodic: *He expresses his own feelings in the lyric poems.* **3** °sweet, dulcet, °graceful, silvery, lilting, mellifluous, °mellow, °light: *This song sounds best when sung by a lyric tenor.* —*n.* **4** ***lyrics.*** libretto, °book, °words: *Ira Gershwin wrote the lyrics for many of George Gershwin's songs.*

lyrical *adj.* **1** See **lyric, 1,** above. **2** °enthusiastic, °ecstatic, encomiastic, °rapturous, °rhapsodic, °effusive, °impassioned, °emotional, °ebullient, °exuberant, panegyrical: *He waxed lyrical whenever he spoke of his children.*

M

macabre *adj.* °grim, °ghastly, °grisly, °gory, °gruesome, °grotesque, °ghoulish, °fiendish, °dread, °eerie, °fearsome, °frightful, °frightening, °terrifying, °terrible, °dreadful, dire, °morbid; deathly, °deadly, deathlike, °ghostly, cadaverous: *He told a macabre story of how they survived by resorting to cannibalism. The crypt had a macabre eeriness about it.*

machiavellian *adj.* °deceitful, cunning, °shrewd, °crafty, °wily, °foxy, °scheming, °tricky, °perfidious, nefarious, treacherous, °sneaky: *Mr. Sidwell has concocted a truly machiavellian plan for getting the pensioners to move out.*

machination *n.* plotting, °scheming, intriguing, maneuvering, °designing, manipulating; °plot, °scheme, °intrigue, °maneuver, °design, °stratagem, °ruse, °trick, °trickery, °artifice, dirty trick(s), °wile, manipulation, ploy, tactic(s), °move, gambit: *He escaped from the machinations of his enemies. Must you resort to such machinations merely to get them to listen?*

machine *n.* **1** °mechanism, °device, °apparatus, contrivance, appliance, °instrument, °implement, °tool, utensil, °gadget, *Colloq* °contraption, *US and Canadian* gizmo *or* gismo: *What kind of machine is used to make a corkscrew?* **2** °engine, motor, prime mover, °vehicle; °car, automobile, motorcar, *US* auto: *We used to get into the machine and go for Sunday picnics.* **3** °organization, °system, °ring, °gang, °cabal, °clique, °party, °faction: *The entire council is run by a political machine.* —*v.* **4** °shape, °make, °manufacture: *In this department we machine the castings to a tolerance of one ten-thousandth of an inch.*

machismo *n.* masculine pride *or* arrogance, manliness, virility, masculinity, °grit, *Colloq* °guts, *Slang* balls: *He is just trying to impress you with his machismo.*

macho *adj.* °manly, masculine, virile, °proud, °arrogant: *He's afraid his macho image will be destroyed if he admits to enjoying ballet.*

mad *adj.* **1** °insane, °deranged, °crazy, crazed, demented, lunatic, unhinged, °delirious, out of one's mind *or* head, °psychotic, °maniacal, °(mentally) unbalanced, mentally ill, of unsound mind, *non compos mentis, Chiefly Brit* daft, *Colloq* out of one's head, touched (in the head), screwy, cuckoo, °mental, balmy, certifiable, having a screw loose, dotty, cracked, mad as a March hare, mad as a hatter, not all there, off the wall, stark raving mad, *US* nutty as a fruitcake, *Chiefly Brit* potty; *Slang* nuts, loony, goofy, loopy, crackers, batty, off one's rocker *or* trolley, out of one's tree, bananas, *US* out to lunch, bughouse, bugs, crazy as a bedbug *or* a coot, loco, wacky, out of one's tree, meshuga, *Brit* round the bend *or* twist, twisted, off one's chump, barmy, bonkers: *You're mad if you think I'm going in there with that lion.* **2** °foolish, °silly, °childish, °immature, °puerile, °wild, °nonsensical, °foolhardy, madcap, °heedless, °senseless, °absurd, °imprudent, unwise, °indiscreet, °rash, °ill-advised, ill-considered, °reckless, °extravagant, irrational, fatuous: *I did many mad things when I was a student. Hitchhiking to California is a mad idea.* **3** °wild, °ferocious; °rabid: *Have they caught that mad dog yet?* **4** °furious, °angry, infuriated, incensed, enraged, irate, fuming, °berserk, irritated, provoked, wrathful, exasperated, *Literary* wroth: *Will you get mad if I ask you a question? Please don't be mad at me—I wrecked your car.* **5** ***like mad.*** °madly, feverishly, in a frenzy, frenziedly, desperately, excitedly, violently, wildly, hysterically, furiously; enthusiastically, fervently, ardently; *Colloq* like crazy: *He's been running about like mad trying to find her.* **6** ***mad (about or for).*** °crazy, °infatuated, °ardent, °enthusiastic, °eager, avid, zealous, °passionate, °fervent, fervid, °keen, °fanatical, °wild, *Colloq* hooked, *Brit* dotty, *Slang* nuts: *She's absolutely mad about cricket.*

madden *v.* **1** °infuriate, °anger, °enrage, incense, °provoke, °inflame, excite (someone) to (a) frenzy *or* rage, make (someone's) blood boil, raise (someone's) hackles, make (someone) see red, get (someone's) back up, drive someone crazy, *Literary* raise (someone's) ire, *Colloq* drive (someone) up the wall, *Brit* drive (someone) round the bend *or* twist, *US* tick (someone) off, burn (someone) up, tee (someone) off, °bug, drive (someone) up the wall: *Their getting away with something like that maddens us all.* **2** °irk, vex, pique, °exasperate, °irritate: *He has the maddening habit of affecting a lisp.* **3** bait, badger, °torment, °plague, bedevil, *US* rile, hassle: *The dog was maddened into attacking anything that moved.*

madly *adv.* **1** insanely, hysterically, dementedly, wildly, distractedly, frenziedly: *The patient keeps screaming madly, day and night.* **2** foolishly, stupidly, inanely, ridiculously, ludicrously, idiotically, absurdly, irrationally, senselessly: *The king madly spurned the proffered aid and was soundly defeated.* **3** furiously, wildly, ferociously, °fiercely, energetically, desperately, °like mad, vehemently, feverishly, excitedly, fanatically, violently, impetuously: *He dashed about madly trying to get help.* **4** excessively, °extremely, desperately, intensely, passionately, wildly, ardently, fervently, fervidly, °exceedingly: *I am madly in love with you, darling.*

madman *n.* madwoman; lunatic, psychopath, °psychotic, °maniac, *Colloq* crackpot, psycho, loony, screwball, *US* kook, *Slang* nut, nutcase, *Brit* nutter: *I am working like a madman to finish the book by the deadline.*

madness *n.* **1** °insanity, °lunacy, °mania, dementia, psychosis, mental illness: *That form of madness is called schizophrenia.* **2** craziness, °lunacy, °folly, foolishness, °nonsense, senselessness, ridiculousness, pointlessness, illogicality, illogic, illogicalness, impracticality, preposterousness, futility: *He soon realized the madness of trying to stem the tide of public opinion.*

magazine *n.* **1** °periodical, °journal, °publication: *We publish a quarterly magazine about language.* **2** arsenal, ammunition *or* munitions dump, armory: *The saboteurs infiltrated the defenses and blew up the magazine.*

magic *n.* **1** witchcraft, °sorcery, wizardry, black magic, necromancy, black art, voodoo, obeahism, devilry *or* deviltry, diabolism, demonolatry, occultism; sortilege, theurgy, white magic; °spell: *As the magic began to work, he slowly changed into a horrible ghoul.* **2** legerdemain, conjuring, prestidigitation, sleight of hand, °illusion, °hocus-pocus, °trickery: *He performs tricks of magic, including sawing a woman in half.* **3** °enchantment, allure, allurement, °charm, bewitchment, °spell, witchery, witchcraft, wizardry, °glamour, °fascination, °magnetism, ensorcellment: *She worked her magic on every man she met.* —*adj.* **4** magical, °miraculous: *The magic ointment made her young again.* **5** necromantic, °occult, mystic, shamanistic, theurgical: *Using a magic spell, she turned him into a toad.* **6** magical, °enchanting, entrancing, bewitching, fascinating, hypnotic, mesmerizing, entrancing, °spellbinding, charming, °magnetic, ensorcelling: *The music had a magic effect on them.*

magician *n.* **1** conjuror *or* conjurer, illusionist, wizard, °sorcerer, sorceress, magus, necromancer, enchanter, enchantress, Merlin, Houdini, Circe, °witch, warlock; thaumaturge, theurgist: *The court magician changed the pebbles into precious stones by a wave of his wand.* **2** °marvel, miracle worker, °virtuoso, wizard, °genius, °master, *Colloq* whiz: *Mary Lou Williams is a magician on the boogie-woogie piano.*

magnetic *adj.* °attractive, attracting, °engaging, captivating, °enthralling, °seductive, alluring, entrancing, bewitching, beguiling, °arresting, °spellbinding, °irresistible, charismatic, °winning, winsome, °inviting: *Because of Amy's magnetic personality, men flock about her.*

magnetism *n.* °attraction, °draw, °appeal, allure, °magic, °lure, attractiveness, °charm, °pull, seductiveness, irresistibility, drawing power, charisma, *duende,* likableness, sex appeal: *JFK's personal magnetism won him millions of votes.*

magnification *n.* enlargement, amplification; buildup, strengthening, enhancement, aggrandizement, raising, °elevation, °increase, °expansion, heightening, glorification, ennoblement: *The microdot can be read only at a magnification of 260X. The advertising agency worked on the magnification of the candidate's record of achievement.*

magnificent *adj.* °great, °excellent, °splendid, °superior, °superb, °marvelous, °glorious, °grand, °fine, °impressive, °imposing, awe-inspiring, °impressive, °brilliant, commanding, august, °noble, °majestic, °regal, °distinguished, °elegant, °exalted, °sublime, °outstanding; °sumptuous, resplendent, °opulent, °rich, °luxurious, °lavish: *She lives in a magnificent country house.*

magnify *v.* **1** °enlarge, °expand, °amplify, °inflate, °increase, augment, °exaggerate, °heighten, °build up, °boost, °dramatize, °aggravate, °worsen, exacerbate; °overstate, *Colloq* °blow up, make a mountain out of a molehill: *Tiny errors in the beginning are magnified later on. I fear that he has magnified his importance out of all proportion.* **2** °enlarge, °blow up: *We had to magnify the photograph to read the registration number of the stolen car.*

magnitude *n.* **1** greatness, °size, °extent, bigness, immensity, °enormousness, dimensions: *You cannot imagine the magnitude of the unemployment problem in our area.* **2** °importance, °significance, consequence, °note: *This is a matter of sufficient magnitude to warrant a board meeting.*

maid *n.* **1** °girl, °maiden, °lass, °miss, nymphet, nymph, wench, damsel, mademoiselle, demoiselle, *Scots* lassie; virgin, *virgo intacta*: *A lovely young maid was milking the cow.* **2** housemaid, maidservant, °domestic, chambermaid, lady's maid, *Literary or archaic* abigail, *Brit* °daily, *Archaic colloq Brit* tweeny: *When she rang, the maid brought in the tea.* **3** ***old maid.*** spinster, bachelor girl: *She never married and is now an old maid.*

maiden *n.* **1** See **maid, 1,** above. —*adj.* **2** virgin, virginal, undefiled, °intact, °chaste, (*virgo*) *intacta*; °unmarried, unwed: *My maiden aunt is visiting for the weekend.* **3** inaugural, °first, °initial, *Colloq US* shakedown: *The* Titanic *sank on her maiden voyage.*

mail *n.* **1** °post, correspondence; °letters: *Has the mail arrived?* —*v.* **2** °post, °send, dispatch *or* despatch: *Please mail this letter tonight.*

maim *v.* °cripple, °mutilate, °lame, disable, °incapacitate, °wound, wing, °impair, hamstring, put out of action *or* commission; °injure, °harm, °damage: *Mr. Crampton was badly maimed in the automobile accident.*

main *adj.* **1** °chief, °primary, °prime, °(most) important, °principal, °cardinal, °paramount, °first, °foremost, °leading, °preeminent, °predominant, predominating, °dominant, ranking, °major; °outstanding: *The main cause of traffic deaths is drunken driving. In Britain, the main street of a town is called the high street. We arrived late and missed the main entertainment.* **2** largest, biggest, greatest, strongest: *The main part of the strike force landed in Normandy.* **3** °necessary, °essential, °basic, °particular, °fundamental, °critical, °crucial, °vital: *Economic recovery is the main thrust of our program for reform.* **4** °sheer, °brute, utter, °pure, °out-and-out, °absolute, °mere, °plain: *He lifted that block by main force!* —*n.* **5** °pipe, duct, °channel, °line, °pipeline, water *or* gas main, *Brit* °(electric) cable, mains, °power (supply), conduit: *Water reaches the city through a huge underground main.* **6** °strength, °power, °might, °effort, °energy, °vigor: *With all his might and main he tried to move the stone blocking the cave.* **7** ***in the main.*** See **mainly,** below.

mainly *adv.* °in the main, °chiefly, °principally, predominantly, °generally, °above all, °on the whole, in general, mostly, most of all, effectively, essentially, °at bottom, first and foremost, °for the most part, °largely, by and large, °primarily, °as a rule, °usually, all in all, on balance, for all practical purposes, °in the long run: *We are concerned mainly with safety. The tourists are mainly from Italy.*

mainstay *n.* main *or* chief *or* principal support, °anchor (to windward), sheet anchor, °bulwark, °buttress, linchpin, main *or* greatest strength: *David remains the mainstay of the sales force.*

maintain *v.* **1** °continue, °preserve, °persevere in, keep going, °persist in, °keep (up), °carry on, °retain,

°perpetuate, °prolong, °sustain, °uphold: *Sarah maintained friendly relations with her ex-husband. Try to maintain your composure.* **2** °look after, °take care of, °care for, °preserve, °(keep in) service, keep up, keep in repair; nurture, °support: *It is expensive to maintain a vintage car. Allison maintains an ancient aunt in Piddlington.* **3** °hold, °state, °say, °declare, °claim, assert, °allege, °testify, contend, aver, avow, °announce, °proclaim, °vouchsafe, °profess, °insist (on), affirm: *Despite evidence to the contrary, he maintains that he is innocent.* **4** °defend, °stand by, °keep, fight for; take up the cudgels for, make a case for, °advocate, °champion, take *or* make a stand for, °plead for, °back (up), °support, °vindicate, °justify, *Colloq* go to bat for: *He maintained his ground in the face of virulent attacks.*

maintenance *n.* **1** °upkeep, °care, °preservation, °conservation, °support, sustention, sustentation: *I can no longer afford the maintenance of a large country estate.* **2** continuation, continuance, perpetuation, prolongation, °persistence, maintaining: *His maintenance of an untenable position will lose him votes.* **3** °upkeep, livelihood, °subsistence, °support, °allowance, living, °sustenance, °stipend, subvention, contribution, alimony, °keep: *How much is the monthly maintenance?*

majestic *adj.* **1** °regal, °dignified, °grand, °imperial, °royal, °noble, lordly, °lofty, °elevated, °exalted, °glorious, °magnificent, °monumental, °impressive, °striking, °imposing, °awesome, °splendid, °marvelous, kingly, queenly, °princely: *With majestic ceremony, the procession entered the cathedral.* **2** °pompous, °supercilious, °disdainful, °superior, °arrogant, °haughty, magisterial, imperious, °grandiose, °affected: *She dismissed him with a majestic wave of her hand.*

major *adj.* **1** larger, greater, bigger, °main, °chief, °important: *Leave it to Lloyd to grab the major portion of the credit.* **2** °vital, °important, °critical, °crucial, °principal, °foremost, °paramount, °primary, °prime, °main, °big, biggest, °preeminent, °notable, °noteworthy, °significant, °outstanding, °dominant, dominating; °serious, °grave, worst: *One of Throssell's major contributions was the restoration of Hartwell House. The major problem in commercial kitchens is sanitation.*

majority *n.* **1** °bulk, °preponderance, °mass, more than half, the better *or* best part, the greater part *or* number, lion's share: *The majority of shares were bought by existing shareholders.* **2** adulthood, °maturity, seniority, womanhood, °manhood: *They reach their majority in 1997.*

make *v.* **1** °build, °assemble, °construct, °erect, put together, °set up, °fashion, °form, °mold, °shape, °frame, °create, °originate, °fabricate, °manufacture, °produce, °put out, °forge, contrive, °devise: *She makes her own dresses. They make TV sets here. The Colossus of Rhodes was said to be made of bronze.* **2** °cause, compel, °force, impel, coerce, °provoke, °urge, exhort, °press, °pressure, °require, °command, °order, °induce, °persuade, °prevail upon, insist upon, °oblige, *Brit* pressurize: *The devil made me do it.* **3** °bring about, °occasion, °cause, °give rise to: *The new regulation is going to make trouble for you.* **4** make out *or* up, °draw (up), °create, °write, °sign, °frame: *I made a new will leaving everything to my children.* **5** °produce, °cause, °create, °generate: *The engine made a funny noise, then died.* **6** °enact, °pass, °frame, °establish, °institute: *He thinks that laws were made to be broken.* **7** °earn, °return, °reap, °garner, °take in, °get, °procure, °gather, °clear, °realize, °gross, °net, °pocket, °get, °acquire, °obtain, °receive; °win, °gain, *Slang US* °pull down: *Has her invention made money? He makes a good living out of his shop. He made twenty-five dollars an hour playing poker.* **8** °amount to, constitute, °represent, add up to, °total, °come to: *He knows how many beans make five. Three and two do not make four. One singer does not make an opera.* **9** °change, °turn, °alter, °modify, °transform, °convert; transmute, mutate, metamorphose: *He made her into a star. You cannot make a silk purse out of a sow's ear. The alchemists tried to make base metal into gold.* **10** °become, be, change *or* turn *or* grow into, perform as: *I think that Quentin will make a good surgeon one day.* **11** serve as *or* for, be suitable for, be, prove to be, turn out to be, °turn into, °become: *This cut of meat will not make a good roast.* **12** °fetch, °realize, °bring, °earn, °return: *The locket made a thousand dollars at the auction.* **13** °score, °earn, °secure: *Our team made six goals against Maine in the first six minutes of the game.* **14** °reach, °arrive at, attain, °get (to), °win, °achieve, °accomplish; °come in, *Brit* be placed, *US* °place: *Fanny might make first place in the marathon. As it was, she made second.* **15** °prepare, °arrange, rearrange, °tidy (up), °neaten (up): *You have made your bed, now you will have to lie on it.* **16** °record, °arrange, °fix, decide (on *or* upon), °agree (to): *I made an appointment to see the doctor.* **17** °prepare, °fix, cook: *I made what you like for dinner.* **18** °deliver, °present: *Janet made a good speech.* **19** °traverse, °cover, do, °travel, °navigate: *We cannot make more than a hundred miles a day over this terrain.* **20** do, °go, travel *or* move at, °move: *You think that this car will make a hundred miles an hour?* **21** °judge, °think, °calculate, °estimate, °reckon, °gauge, °suppose: *What do you make of Sidney's new book?* **22** °establish, °set up, °organize: *We made our headquarters in the farmhouse.* **23** °appoint, °name, °select, °choose, °elect, °vote (in as), °designate, °authorize, commission, °delegate, depute, deputize, °assign, °sanction, °approve, affirm, °certify, °confirm: *They made him their leader.* **24** °seduce, make it with: *Kenneth tried to make Sharon last night.* **25** ***make as if*** *or* ***as though.*** °pretend, feign, act as if *or* as though, °affect, make a show *or* pretense of, give the impression of: *He made as if to strike me.* **26** ***make away.*** run off *or* away, °flee, °fly, °make off, abscond, °take to one's heels, decamp, beat a (hasty) retreat, *Colloq* run for it, make a run for it, °beat it, °clear out, cut and run, skedaddle, °take off, °cut out, °skip (town), make tracks, *US* fly the coop, *Slang* scram, vamoose, *US* hightail it, take a (runout) powder: *Taking the jewels, he made away as fast as he could run.* **27** ***make away with.*** °steal, °rob, filch, °pilfer, purloin, walk away *or* off with, *Colloq* °borrow, °liberate, °boost, *Slang* °pinch, °hook, °swipe, °rip off, °lift, *US* °boost: *That boy has made away with my cherry tarts!* **28** ***make believe.*** °pretend, °fancy, playact, °dream, °fantasize, °imagine, act as if: *We used to make believe we were grown-ups.* **29** ***make do.*** get by *or* along, °cope, scrape by *or* along, °manage, °muddle through, °survive, *Colloq* °make out: *We have to make do on the pittance Randolph gets from the university.* **30** ***make for.*** **(a)** head for *or* toward(s), aim for, steer (a course) for, proceed toward(s), be bound for: *After this is done, I am making for the nearest pub.* **(b)** °assault, °attack, °set upon, °charge, °rush (at), °pounce upon, fall upon *or* on, °go for, lunge at, °storm, assail: *The big fellow was making for me with a knife when the lights went out.* **(c)** °promote, °contribute to, be conducive to, °favor, °facilitate: *Good fences make for good neighbors.* **31** ***make good.*** **(a)** °make up (for), °pay (for), compensate for, recompense (for), °repay, °offset, make restitution for, °settle, °square, °rectify, put to rights, set right, °remedy, °correct, °restore: *He agreed to make good any losses.* **(b)** °succeed, °prosper, °flourish, °thrive, *Colloq* °make it: *Cranshaw made good in the commodities market.* **(c)** °fulfill, °carry out, *Colloq* °deliver (the goods): *She made good on her promise to return my book.* **32** ***make it.*** **(a)** °succeed, °prosper, °triumph, °win, °make good, *Colloq* °make the grade: *Do you think she will make it as* *a doctor?* **(b)** °arrive, °get (somewhere), °show up, °appear, °turn up: *He was supposed to be interviewed on the nine o'clock news, but he never made it to the studio.* **33 *make known.*** °tell of, °impart, °disclose, °reveal, divulge, °mention, °communicate, °announce, °declare, promulgate, °publish, °let slip, *Colloq* °tip off: *She made known her demands.* **34 *make much of.*** **(a)** °exaggerate, °overstate, °color, hyperbolize, *Colloq* make a big deal of, °blow up: *He made much of his new title.* **(b)** °coddle, cosset, °baby, °pamper, dote on, °flatter, toady (up) to, °cajole, °humor, °indulge, *Colloq* butter up: *Henry makes much of his grandchildren.* **35 *make off.*** See **make, 26,** above. **36 *make off with.*** See **make, 27,** above. **37 *make out.*** **(a)** °see, discern, descry, espy, °detect, °discover, °distinguish, °perceive: *I made out a dim figure in the gloom.* **(b)** °complete, °fill in, *Brit* °fill up, *US and Canadian* °fill out: *I made out an application for the job.* **(c)** °draw (up), write (out *or* down), °record, *Colloq US* °cut: *Please make out a list of your complaints. Make out the check to me, personally.* **(d)** °understand, °fathom, °comprehend, °figure out, °perceive, °follow, °grasp, °see, °decipher, °read: *She mumbles so, I cannot make out what she's saying. Can you make out this name?* **(e)** °suggest, °imply, °hint, °insinuate, °indicate, °impute, °intimate, make to appear, °pretend, make as if *or* as though, °represent; °present, °show, °demonstrate, °establish: *She tried to make out that I was a fool. He made out a strong case for dog licensing.* **(f)** °get on, °survive, °manage, °fare, °thrive, °succeed: *How are you making out in your new house?* **38 *make over.*** **(a)** do over, remodel, redecorate, °alter: *We are making over our kitchen.* **(b)** °transfer, °hand over, °sign over, convey, °assign, °turn over: *The property has been made over to me.* **39 *make up.*** **(a)** °complete, °fill out, °finish (out), °flesh out: *We need you to make up the team.* **(b)** °compose, °form, constitute, be comprised of: *The gang is made up of ex-convicts.* **(c)** °hatch, °invent, concoct, °devise, °create, °construct, dream up, °originate, °coin, °compose, *Colloq* cook up: *He made up that story about the murder.* **(d)** be reconciled, make peace, settle amicably, °come to terms, bury the hatchet: *The litigants have kissed and made up.* **(e)** °construct, °build: *The shack is made up of scrap boards.* **40 *make up for.*** °compensate, redress, °make good, °atone, °make amends: *How can I make up for all the bad things I said about you?* **41 *make way.*** move aside, clear the way, allow to pass, make room *or* space: *Make way for the Lord High Executioner!* —*n.* **42** °kind, °brand, °style, °sort, °type, °mark: *That make of computer comes from Japan.* **43 *on the make.*** °aggressive, °assertive, °go-ahead, °enterprising, °vigorous, °energetic, *Colloq* °pushy: *The book is about a young man on the make in today's financial world.*

makeshift *adj.* **1** °temporary, °stopgap, °expedient, °emergency, °temporary, jury-rigged, improvised, °tentative, °standby, slapdash: *We were able to fashion a makeshift rudder out of an oar.* —*n.* **2** °stopgap, °expedient, improvisation, °substitute: *The heater broke down and we had to get along with a makeshift.*

makeup *n.* **1** cosmetics, *maquillage,* greasepaint, *Colloq* war paint: *Do you think I am wearing too much makeup?* **2** constitution, °character, °cast, °disposition, °personality: *Sympathy is not a quality in Witlow's makeup.* **3** constitution, °arrangement, construction, °composition, °format, configuration, °build, °form: *What is the chemical makeup of the plastic?*

male *adj.* masculine, man's; virile, °manly, manful; *Archaic* spear: *His is bound to be a male point of view.*

malignant *adj.* **1** °virulent, pernicious, °deadly, °fatal, toxic, °poisonous, °harmful, life-threatening: *The tumor proved to be malignant.* **2** malign, malevolent, °evil, malicious, pernicious, °vicious, invidious, °spiteful, °bitter, °hateful, °venomous: *The exposé was full of malignant insinuations.*

man *n.* **1** gentleman, °male, °fellow, *Colloq* °guy, °chap, *Brit* bloke, °squire, *Slang* gink, geezer, *US* gazabo: *Have you met the man she is going to marry?* **2** °people, human beings, mankind, mortals, Homo sapiens, °humanity, humankind, the human race: *Man wants but little here below.* **3** valet, manservant, gentleman's gentleman, °servant, retainer, houseboy, houseman: *Humberson sent his man round with a note.* —*v.* **4** °staff, °people, °crew; °cover: *Who will man the office while I am away?*

manacle *n.* **1 *manacles.*** °shackles, fetters, °handcuffs, gyves, chains, irons, *Colloq* cuffs, *Slang* bracelets, *Brit* darbies: *The manacles are cutting into his wrists.* —*v.* **2** °shackle, fetter, handcuff, °restrain, put *or* throw *or* clap in irons, °chain, *Colloq US* cuff: *The manacled prisoner was led away.* **3** confine, °inhibit, °restrain, °curb, °check, °control, °hamper: *We felt manacled by the nine o'clock curfew.*

manage *v.* **1** °handle, °administer, °run, °supervise, °look after, watch over, °direct, °head, °oversee, superintend, °direct, preside over, be in charge (of), °take care of, °control; °rule (over), °govern, °regulate: *Mr. Gant manages the glove department.* **2** °handle, cope *or* deal with, °control, °govern, °manipulate: *Will you be able to manage such a frisky horse?* **3** °conduct, °carry on, °carry out, °bring off, °control, °undertake, °take care of, °look after, °handle: *Is she old enough to manage her own affairs?* **4** °succeed, °function, °make do, °make it, °shift (for oneself), get along *or* by *or* on, °make out, °muddle through, °survive: *Will you be able to manage from here on?*

manageable *adj.* controllable; °tractable, compliant, amenable, docile, tamable, °tame, trainable, teachable, manipulable, °submissive: *Keep your spending within manageable limits. The dog is quite manageable.*

management *n.* **1** managing, °control, supervision, manipulation, handling, °direction, directing, directorship, °administration, °government, °conduct, governance, °operation, °running, superintendence, °command, °guidance, stewardship: *The management of the company is in your hands, Thewless.* **2** °administration, executive(s), bosses, directors, °board (of directors), directorate, *Colloq* °(top) brass: *The management believes that you are the right person for the job.*

manager *n.* °supervisor, °superintendent, °director, °executive, °head, °proprietor, °overseer, °foreman, forewoman, administrator, *Chiefly Brit* manageress, *US* straw boss, *Colloq* °boss, °chief: *The position of manager entails a lot of responsibility.*

mandatory *adj.* compulsory, °obligatory, requisite, required; °essential, commanded, demanded, °necessary, needed: *The wearing of safety belts in cars is mandatory.*

mangle *v.* °destroy, °mutilate, °butcher, deform, disfigure, °spoil, °mar, °ruin, °wreck; °cut, °hack, °lacerate, °chop (up), °crush, °damage, °cripple, °maim: *Don't you hate the way they mangle the English language? His hand was badly mangled in the accident.*

mangy *adj.* scruffy, °dirty, °sleazy, °wretched, °miserable, °repulsive, °sorry, squalid, slovenly, °unkempt, °filthy, °dingy, °seedy, °poor, °shabby, °mean, °low, ignominious, °base, abject, odious, °disreputable, moth-eaten, °contemptible, °despicable, °nasty, °scurvy: *The mangy beggar turned out to be our agent in Calcutta.*

manhandle *v.* maul, paw, °rough up, °batter, °beat (up), pummel, °abuse, °mistreat, maltreat, °ill-treat, trounce, °belabor, *Slang* knock about *or* around, clobber: *His captors manhandled him into a car and drove off.*

man-hater *n.* °misanthrope, misanthropist: *He says he prefers to live alone because he's a man-hater.*

manhood *n.* **1** masculinity, manliness, manfulness, virility, *Colloq* °machismo: *He felt that looking after the children compromised his manhood.* **2** °bravery, °pluck, boldness, °determination, °resolution, °fortitude, °grit, °spirit, °force, *US* intestinal fortitude, *Colloq* °guts: *He joined the army to test his manhood.*

mania *n.* **1** °rage, °craze, °passion, °obsession, compulsion, °urge, °fascination, preoccupation, °furor, yearning, craving, °desire, cacoëthes, *Colloq* °fad, yen: *By that time, the hula hoop mania had died down. He indulges her mania for collecting apostle spoons.* **2** °madness, °lunacy, °insanity, dementia, derangement, hysteria, *Technical* manic disorder: *The psychiatrists diagnosed his problem as a mania.*

maniac *n.* **1** °madman, madwoman, lunatic, psychopath, °psychotic, *Colloq* crackpot, *Slang* nut, *Brit* nutter, loony, *US* kook: *In former times, maniacs were scorned and maltreated.* **2** °fanatic, °fan, °enthusiast, °zealot, *Slang* °freak, °fiend: *Since his retirement, he has become a golf maniac.*

maniacal *adj.* **1** manic, °maniac, °insane, lunatic, °mad, demented, °deranged, °hysterical, mentally ill, of unsound mind, *non compos mentis,* °psychotic: *We hospitalized him because of his maniacal behavior.* **2** °hysterical, °berserk, °wild, °crazy, *Slang* loony: *His maniacal outbursts frightened us.*

manifest *adj.* **1** °apparent, °clear, °evident, °obvious, °plain, °patent, °blatant, °conspicuous, unmistakable *or* unmistakeable, °discernible, recognizable, comprehensible, °distinct, palpable, °definite, °explicit, unambiguous, °unquestionable, indubitable, °indisputable: *The bank had to deal with the manifest dishonesty of that one employee.* —*v.* **2** °show, °demonstrate, °exhibit, evince, °reveal, °disclose, °display, °betray; °express, °declare: *Fitzsimmons has been manifesting signs of dissatisfaction with his assignment.* **3** °prove, corroborate, °substantiate, °attest: *His distaste for cauliflower is manifested by the look on his face.*

manifestation *n.* °display, °exhibition, °demonstration, °show, disclosure, °appearance, °exposure, °presentation, °sign, °indication, °mark, °expression, °example, °instance; °declaration, avowal, °publication, °announcement: *The police feared that the crime wave was a manifestation of civil unrest. The first manifestation of thought is speech.*

manifestly *adv.* °evidently, °clearly, °obviously, plainly, °apparently, patently, unmistakably *or* unmistakeably, palpably, unquestionably, indubitably, °undoubtedly, indisputably: *The treaty is manifestly to Canada's advantage.*

manifold *adj.* °diverse, diversified, multifarious, °varied, °various, assorted, multiplex, °miscellaneous, °sundry, many-sided, many different, *Literary* °divers; °many, numerous, multitudinous: *I have manifold reasons for wishing you to stay.*

manipulate *v.* **1** °manage, °handle, °control, °maneuver, orchestrate, choreograph, °influence, °use, °exploit, °play on, utilize: *He knows how to manipulate people to make them do his bidding.* **2** °handle, °control, °operate, °direct, °work, °use, °employ, °negotiate: *The controls can be manipulated to make the robot imitate human motion.* **3** °rig, °falsify, °juggle, tamper with, °doctor, *Colloq* cook, *chiefly Brit* °fiddle: *He's been manipulating his expense account for years.*

manly *adj.* manful, virile, °courageous, °bold, °brave, °intrepid, valorous, valiant, °dauntless, °fearless, plucky, °daring, °venturesome, stouthearted, °resolute, °stable, °steadfast, unflinching, unwavering, unshrinking, °chivalrous, °gallant, °noble, °heroic; masculine, °male, *Colloq* °macho, red-blooded: *Emerson characterized the English as manly rather than warlike.*

manner *n.* **1** °way, °mode, °style, °technique, °procedure, °method, °fashion; °means, °approach: *Her incisive manner of reporting is widely admired.* **2** °air, °behavior, mien, demeanor, °bearing, deportment, comportment, °conduct, °attitude, °aspect: *His manner is overbearing and dictatorial.* **3** ***manners.*** °etiquette, °decorum, °(good) form, politeness, proprieties, °protocol, politesse, °civility, °ceremony, social code, social graces, formalities, niceties, amenities, social conventions; °behavior, °conduct: *Can't you teach him some manners?*

mannered *adj.* °artificial, contrived, °stilted, °stiff, °affected, °insincere, °pompous, °pretentious, posed, °unnatural, °hypocritical, *Colloq* phoney *or US also* phony, pseudo, highfalutin *or* hifalutin *or* hifaluting, la-di-da *or* lah-di-dah *or* la-de-da, °hoity-toity, on one's high horse, high-hat, uppity *or Brit* uppish: *Alice's speech is too mannered for anyone to feel close to her.*

mannerism *n.* °quirk, °peculiarity, idiosyncrasy, °trait, °characteristic, °habit: *She has the irritating mannerism of pulling on her ear lobe.*

maneuver *n.* **1** °move, °stratagem, °tactic, °trick, gambit, ploy, °subterfuge, °ruse, °dodge, °artifice, °device, °wile, *démarche;* °strategy, °plan, °plot, °scheme, °intrigue, °machination: *That maneuver will never work on a clever woman.* **2** °exercise, °operation, °drill, war game, °operation, kriegspiel, training: *We were out on maneuvers for a week.* —*v.* **3** °manipulate, contrive, °plot, °scheme, machinate, °intrigue, °trick, °devise, °engineer, °finesse, °manage, *Colloq* finagle, °wangle: *We maneuvered our way into going on that mission.* **4** °manipulate, °operate, °run, °drive, °guide, °navigate, jockey: *Can you maneuver the car into that tiny space?*

mantle *n.* **1** °cloak, °cape, °wrap, shawl, pelisse, pelerine: *Over his armor he wore a mantle embroidered with his bearing.* **2** covering, °cover, °sheet, °veil, blanket, °screen, °cloak, °shroud, °pall, canopy, curtain: *A mantle of snow lay on the ground.* —*v.* **3** °cover, °clothe, °envelop, °surround, °encircle, °shroud, °veil, °screen, °obscure, °cloak, °conceal, °hide, °mask, °wrap, °disguise: *The mountains were brilliantly mantled and capped with snow.*

manual *n.* handbook, °vade mecum, enchiridion; °directions, °instructions, °guide: *Follow the manual carefully to avoid mistakes.*

manufacture *v.* **1** °make, (mass-)produce, °construct, °build, °assemble, °fabricate, put together, °turn out, °create, °originate: *That company manufactures windshield wipers.* **2** concoct, °create, contrive, °invent, °make up, °fabricate, °think up, *US and Canadian* create out of *or US and Canadian* from whole cloth, *Colloq* cook up: *She manufactured that tale about being an orphan.* —*n.* **3** making, (mass-)production, construction, °building, °assembly, °fabrication, turning *or* putting out, putting together, °creation, origination: *The manufacture of durable goods increased .02 percent in the last quarter.*

manufacturer *n.* maker, °producer, industrialist, fabricator: *Manufacturers of computers usually buy components from specialized firms.*

many *adj.* **1** numerous, multitudinous, myriad, °profuse, innumerable, °numberless, uncountable: *Professor Gerard has earned the respect of many students.* **2** °diverse, multifarious, °varied, °various, assorted, °sundry, *Literary* °divers: *There are many stories in the* *naked city.* —*pron.,* (*n.*) **3** horde(s), crowd(s), lot(s), swarm(s), throng(s), mass(es), °profusion, multitude(s), °abundance, °plenty, shoal(s), flock(s), drove(s), torrent(s), flood(s), number(s), score(s), hundred(s), thousand(s), etc.; *Colloq* ton(s), scads: *A great many will turn out to vote next week. Many are called but few are chosen.*

mar *v.* **1** °damage, °ruin, °mutilate, °deface, °spoil, °scar, disfigure: *The surface of the table was marred by a deep scratch.* **2** °damage, °wreck, °ruin, °impair, °harm, °hurt, °blight, °blot, °taint, °stain, °tarnish: *She claims that her reputation was marred by the article.*

march *v.* **1** °parade, °step, stride, °strut, tread, °pace, °walk: *We marched from Baxter Street to Avenue A.* —*n.* **2** °parade, °procession, °demonstration, cortege *or* cortège, °walk: *The march was organized to protest against the government's nuclear arms policy.* **3** °walk, trek, slog, hike: *We had a long march ahead of us to get to Hastings.*

margin *n.* **1** °edge, °border, °perimeter, °periphery; °rim, lip, °side, °brink, °verge: *Leave a one-inch margin at the left side of the page. Trees grew at the margin of the pond.* **2** limit(s), bound(s), °boundary (line), °border, °frontier, °line, partition line: *These streets form the margin of the inner city.* **3** °allowance, °play, °leeway, latitude, °scope, °freedom, °room, °space; compass: *These calculations do not provide much of a margin for error.*

marginal *adj.* **1** borderline, °minimal, °small, °slight, °negligible, °insignificant, °tiny, infinitesimal: *There are marginal differences in policy on this issue.* **2** borderline, on the edge, °disputable, °questionable, °doubtful, dubious: *His qualifications for the job are marginal.*

marine *adj.* **1** maritime, °nautical, naval, °seafaring, seagoing, oceangoing, °sea: *Marine commerce has declined in recent decades.* **2** maritime, °sea, °oceanic, aquatic, saltwater, pelagic, thalassic: *Many groups are concerned about the conservation of marine life.*

mark *n.* **1** °spot, °stain, °blemish, °smear, smudge, °trace, °impression, dent, °nick, °scratch, pockmark *or* pock, °streak, °line, *Brit* splodge, *US* splotch: *You've made a mark on the newly painted wall.* **2** °sign, °symbol, insigne, °emblem, °device, °hallmark, earmark, fingerprint, badge, °characteristic, °token, °brand, °stamp, °label, °identification, °indication, °feature, °attribute, °trait, °quality, °property: *This work bears the mark of a true genius.* **3** °standard, criterion, °norm, °yardstick, °level, °measure: *Your son's work is not up to the mark.* **4** rating, °grade, grading: *I'd give her a low mark for originality but high marks for technique.* **5** °influence, °impression, °effect: *Her innovations have left their mark on all music composed since 1900.* **6** °target, °goal, °objective, °aim, °purpose, °end, °object: *Your criticism has fallen wide of the mark.* **7** marker, °indicator, °guide, signpost, °landmark: *The next mark, a bell buoy, should be left to port.* **8** consequence, °importance, °note, noteworthiness, notability, °distinction, eminence, °dignity, °prestige, °standing, °account: *Granada was a place of little mark as compared with Cordova.* **9** ***make one's mark.*** °succeed, °get ahead, °triumph, distinguish oneself, attain distinction, bring honor upon oneself, acquit oneself, bring credit to oneself, have an effect, *Colloq* make it big, °make the grade: *Randolph has made his mark in the world of scholarship.* —*v.* **10** Sometimes, ***mark up.*** °spot, °stain, °blemish, °smear, smudge, °streak, dent, °trace, pockmark, °nick, °scratch, °cut, °chip, °pit, °bruise: *That glass has marked the finish on the table.* **11** °signify, °specify, °indicate, °designate, °identify, tick, °label: *Mark your choice with an "X" next to the candidate's name.* **12** pay attention to, °attend (to), pay heed to, °note, °notice, take notice of, °watch, °see, °look at, °observe; °respect, °mind, °heed, °obey: *Mark the way she swings the golf club. Mark my words—Don't go near the water!* **13** °brand, °stamp, °identify, °characterize, °distinguish: *His music is marked by long atonal passages.* **14** °correct; °grade, °evaluate, assess, appraise: *The teachers are busy marking exam papers.* **15** ***mark down.*** **(a)** °write (down), °record, °register, make (a) note of, °note (down): *Mark down these numbers in your book.* **(b)** °decrease, °reduce, devalue, devaluate, °cut, °slash, °discount: *We marked down the prices for a quick sale.* **16** ***mark up.*** **(a)** See **10,** above. **(b)** °increase, °raise, hike, up: *Prices were marked up to keep pace with inflation.*

marked *adj.* °noticeable, °conspicuous, °decided, °pronounced, °considerable, °remarkable, °significant, °signal, unmistakable *or* unmistakeable, °prominent, °obvious, °patent, °evident, °apparent: *There is a marked improvement in your work.*

market *n.* **1** marketplace, °exchange, stock exchange: *There are some excellent buys on the market.* **2** °shop, °store, bazaar, supermarket, *Chiefly US* superstore: *If you are going to the market, please buy some milk.* **3** °demand, customer base, °call: *The market for computers continues to grow.* —*v.* **4** °sell, °merchandise, retail, vend, °peddle, hawk, make available, °furnish; °trade (in), buy and sell, deal in: *We market these computers worldwide.*

maroon *v.* °abandon, °cast away, °desert, strand, °forsake; °isolate, seclude: *One is marooned at Mont-Saint-Michel when the tide comes in.*

marriage *n.* **1** matrimony, wedlock: *They were joined in marriage forty years ago todayy.* **2** nuptials, °wedding: *The marriage took place at noon.* **3** °association, °alliance, confederation, °federation, affiliation, °connection, coupling, °union, °merger, °amalgamation, integration, *Colloq* hookup: *Would a marriage of the companies benefit the shareholders?*

marry *v.* **1** °wed, join in matrimony *or* wedlock, become man and wife, *Colloq* get hitched *or* spliced, tie the knot: *They married and lived happily ever after.* **2** °match (up), go *or* fit together, °fit; °unite, °unify, °bond, °weld, °fuse, put together, °couple, °join, °link; °league, affiliate, °ally, °amalgamate, °combine: *The marble head, found in London, married perfectly with the torso, found in Crete. The best qualities of each earlier model are married in this new product.*

marsh *n.* °swamp, °bog, fen, slough, quagmire: *The marsh teems with new life in the spring.*

martial *adj.* **1** °warlike, °belligerent, bellicose, °pugnacious, °militant: *We had no suspicion of their martial intentions. He is an expert in the martial arts.* **2** °military, soldierly, °courageous, °brave, valorous, valiant, °stalwart, °staunch, stouthearted: *The speech was intended to rouse the soldiers' martial spirit.*

marvel *v.* **1** °wonder (at), °gape (at), be awed *or* amazed (by), be agog *or* astonished (at), °gape (at): *We marveled at how homing pigeons find their way home.* —*n.* **2** °wonder, miracle, °phenomenon: *Tammy was a marvel at cooking paella.*

marvelous *adj.* wonderful, astonishing, °amazing, astounding, surprising, °remarkable, °extraordinary, °phenomenal, °glorious, °splendid, °superb, °excellent, spectacular, breathtaking, °miraculous, °unbelievable, °incredible, °sensational, mind-boggling, °unparalleled, *Colloq* °terrific, °great, °fantastic, °fabulous, smashing, °crazy, far-out, *Slang* °wild, groovy, °super, °out of this world, fantabulous, *Brit* spot *or* bang on, *US* marvy: *It was a marvelous show, and Renfrew is a marvelous actor.*

mask *n.* **1** false face, domino: *She wore a mask, and I don't know who she was.* **2** °disguise, °guise, °camouflage, °show, °semblance, °pretense, °cover, coverup, false colors, false flag, concealment, °cloak, façade, °veil: *He wormed his way into her confidence under the mask of friendship.* —*v.* **3** °disguise, °camouflage, °cover (up), °conceal, °hide, °obscure, °veil, °screen, °shroud: *She was unable to mask her true feelings for him.*

masquerade *n.* **1** masked ball, masquerade ball, costume party, *ballo in maschera,* bal masqué: *He went to the masquerade as Lorenzo di Medici.* **2** °disguise, °deception, °pose, °dissimulation, °bluff, °subterfuge, false show, outward show, fakery, imposture, play-acting, false front, cover-up, °camouflage, *Colloq* °act, °front, *Slang* °put-on: *How long will she continue her masquerade as a wealthy widow?* —*v.* **3** Usually, ***masquerade as.*** °pretend (to be), pass oneself off (as), impersonate, simulate, °pose (as), °imitate, °mimic: *He is a conservative masquerading as a liberal to garner votes.*

mass *n.* **1** °pile, °heap, °mountain, °load, °stack, °mound, °bunch, °bundle, °lot, °batch, °quantity, °hoard, °store, °collection, °accumulation, aggregation, agglomeration, congeries, °assortment, °miscellany, assemblage, conglomeration: *There was a mass of boulders at the bottom of the cliff.* **2** °abundance, °quantity, °profusion, °volume, multitude, horde, °host, °mob, °crowd, °throng, drove(s), herd(s), swarm(s), legion(s), score(s), number(s), *Colloq* bunch(es), ton(s), °mountain, °piles, bags, barrels, oodles, lots, °oceans, loads, scads, *US* °mess, slew(s), motherlode *or facetious* motherload: *A mass of bills awaited my return from a trip abroad. Masses of people greeted us at the airport.* **3** °block, concretion, chunk, °lump, hunk, nugget: *A mass of sludge was clogging the valve.* **4** °majority, best *or* better *or* greater part, °bulk, °body, °preponderance, almost all, lion's share: *The mass of the votes was still uncounted.* **5** dimension, °size, °magnitude, °bulk, bigness, massiveness, °enormousness, immensity: *The very mass of the elephant makes it a target hard to miss.* **6** ***the masses.*** the common people, the (common) herd, the proletariat, the plebeians, °hoi polloi, the lower class(es), the man *or* woman in the street, *Brit* the man *or* woman on the Clapham omnibus, A. N. Other, *US* John Q. Public, John *or* Jane Doe, Richard Roe: *The taxes became burdensome and the masses rebeled.* —*v.* **7** °amass, pile *or* heap up, °gather: *Thunderheads began to mass over the mountains.* **8** aggregate, °accumulate, °collect, °assemble, congregate, °meet, get *or* come together, °gather, forgather *or* foregather, °throng, convene, flock together, °rally, °group, °cluster, marshal, °muster, °mobilize: *Thousands of people massed in front of the embassy. The massed bands made a tremendous noise.*

massacre *n.* **1** °slaughter, slaughtering, °carnage, annihilation, annihilating, blood bath, °killing, °execution, extermination, exterminating, butchery, butchering, °(mass) murder, murdering, slaying, liquidation, pogrom, genocide: *The massacre of millions of innocent people followed the coup.* —*v.* **2** °slaughter, annihilate, °kill, °execute, °exterminate, °butcher, °murder, slay, liquidate, °destroy, °eliminate, °obliterate, eradicate, put to the sword, decimate, *Colloq* °mow down, *Slang* °bump off: *It was their policy to massacre all the men and take captive all the women and children.*

massage *n.* **1** rubdown, °rub, manipulation, kneading: *I was really relaxed after the massage and steam bath.* —*v.* **2** rub down, °manipulate, knead, palpate: *She gently massaged my temples with her fingertips.* **3** °manipulate, °handle, °maneuver, *Colloq* finagle, *Brit* °fiddle: *He massaged the figures to make it look as if the company was solvent.*

massive *adj.* °big, °large, oversized, °huge, °bulky, °enormous, °hulking, °immense, °gigantic, °towering, mammoth, °colossal, titanic, °vast, tremendous, °prodigious, °mountainous, gargantuan, Brobdingnagian, cyclopean, elephantine, °jumbo, stupendous, °mighty, °weighty, °ponderous, *Literary* massy, *Colloq* walloping, °whopping, °monster, *Brit* whacking (great), *US* ginormous, humongous: *They built a massive wall round the city. Massive beams support the roof.*

master *n.* **1** °owner, °head, °chief, °leader, chieftain, commander, °lord, governor, °director, controller, °employer, °manager, °overseer, °supervisor, °superintendent, taskmaster, slave driver, °principal, °sovereign, °monarch, ruler, *Colloq* lord high muck-a-muck, Pooh-Bah, kingpin, big fish, °boss, °skipper, *Brit* gaffer, *US* kingfish, the man, big fish, big boss, bossman; *Slang US* big cheese, big wheel, Mr. Big, chief *or* head honcho: *He is master of all he surveys. Decisions about punishment are referred to the master.* **2** °expert, °authority, °genius, craftsman, °adept, maestro, °mastermind, past master, old hand, °virtuoso, ace, °professional, *Colloq* pro, wizard, *Chiefly Brit* °dab hand, *US* crackerjack; *Slang US* maven *or Slang US* mavin: *He is a master in the art of chicanery.* **3** °teacher, °tutor, °instructor, °guide, °leader, guru, swami: *She studied under the master for years.* —*adj.* **4** °adept, °ingenious, °expert, °masterful, masterly, °skillful, skilled, °proficient: *After twenty years, he became a master craftsman.* **5** °overall, controlling, °principal, °main, °prime, °basic, °chief: *Only the leader knows the master plan for conquering the planet. This is the master switch that turns off all the power.* **6** °main, biggest, °principal, owner's: *We slept in the master bedroom.* —*v.* **7** °learn, °grasp, become expert in, know inside out and backwards, °know, °understand: *He claims to have mastered telekinesis.* **8** °control, °overcome, °repress, °suppress, °subdue, °bridle, °check, °quell, get the better of, *Colloq* get a grip on: *You must learn to master your emotions.*

masterful *adj.* **1** masterly, °adept, °expert, °excellent, °superior, °superb, adroit, °exquisite, °superlative, °supreme, consummate, °accomplished, °peerless, °matchless, °first-rate, °proficient, °dexterous, deft, °skillful, skilled, *Colloq* °crack: *She gave a masterful performance of the violin concerto.* **2** °authoritarian, °dictatorial, °tyrannical, °despotic, °arbitrary, °domineering, imperious, °overbearing, °arrogant, dominating, autocratic, highhanded, magisterial, overweening, °self-willed, *Colloq* °bossy: *I dislike Ian's masterful attitude, always telling everyone what to do.*

mastermind *v.* **1** °plan, °devise, °conceive, °think up, °engineer, °design, °generate, °create, °manage, °organize, °develop, work up *or* out: *It was probably Fu Manchu who masterminded this diabolical plot.* —*n.* **2** planner, contriver, conceiver, °creator, architect, °genius, °mind, °intellect, *Colloq* brain(s): *It would take a mastermind to get even a passing mark on that exam.*

masterpiece *n.* masterwork, *magnum opus, chef d'oeuvre,* tour de force, °jewel, work of art, work of genius, *pièce de résistance: Among his many fine paintings, I consider this his masterpiece.*

match *n.* **1** °equal, °equivalent, °peer, °equivalent, °fellow, °mate; °parallel, °replica, °copy, °double, °twin, °lookalike, °facsimile, counterpart: *When it comes to punning, Crosby has finally met his match.* **2** °contest, °competition, °game, °meet, tourney, °tournament, °bout, duel, °rivalry, °trial: *The match was between the top-seeded players.* **3** °marriage, betrothal, °alliance, °combination, °compact, °contract, partnership, °union, affiliation: *It was a match made in heaven.* **4** °prospect, °candidate: *With his inheritance, he's considered a good match.* —*v.* **5** match up, °join, °marry, °unite, °link, °combine, put together, pair up *or* off, juxtapose, conjoin: *However do you match the names with the faces in such a crowd? Glynis matched them up at the dinner table.* **6** °equal, be equivalent (to), °resemble, °compare (with), °tie, measure up (to), °compete (with), °vie (with), °rival: *His skill on the flute will never match hers.* **7** °fit, °go with, °suit, °accord, °agree, harmonize, °go (together), °coordinate, °blend, °correspond: *Who will notice that your socks match your tie? Your eyes don't match.*

matching *adj.* **1** corresponding, homologous, comparable, °equivalent, complementary: *Birds' wings and fishes' fins are matching parts.* **2** analogous, °like, corresponding, °identical: *He wore a pink jacket and matching shoes. She bought a set of matching luggage.*

matchless *adj.* °unique, °original, °peerless, unequaled, without equal, inimitable, unmatched, °incomparable, °unparalleled, beyond compare: *Stefania has a contralto voice of matchless quality.*

mate *n.* **1** °companion, °associate, °colleague, °fellow, °chap, co-worker, °comrade, crony, °ally, °friend, alter ego, *Colloq* °chum, °pal, *US* buddy, °cohort, *Slang Brit* cully, china: *After work, my mates and I stop off for a beer or two (or three).* **2** spouse, °partner, helpmeet, helpmate, consort, husband *or* wife, better half, *Colloq* hubby, old man *or* lady *or* woman, lord and master, *US* bride, *Slang Brit* trouble and strife (= 'wife'): *He doesn't seem a likely mate for the winner of a Miss World contest.* **3** °fellow, °twin, counterpart, °parallel, one of a pair: *Have you seen the mate to this sock anywhere?* —*v.* **4** °pair (up), °match (up), °marry, °wed, °join, °unite, °couple, °link (up): *People should be free to mate with whom they wish.* **5** °breed, °couple, copulate, °pair (up): *The zoo was able to get the rhinoceroses to mate. A lion was mated with a tiger to produce a tigon.* **6** °match (up), °fit (together), synchronize, °join: *I cannot get these gears to mate.*

material *n.* **1** °substance, °matter, °stuff, °fabric: *It is hard to tell what material this is made of.* **2** °cloth, °fabric, textile, °stuff: *This material is too sheer for a dress.* **3** constituents, °elements, components: *We import the raw materials and make the finished product here.* **4** °information, °data, °facts, statistics, figures, documents, documentation, °papers, °notes, °resources, °means, °research, °apparatus, supplies: *I haven't yet gathered enough material for the article.* —*adj.* **5** °physical, °tangible, °concrete, °solid, °real, substantive, °substantial, palpable, corporeal, bodily: *It was many years before scientists could persuade people that air is a material substance.* **6** consequential, °important, °significant, °serious, °substantial: *How could one horseshoe nail be material in the outcome of a battle?* **7** °worldly, °earthly, mundane, °temporal, nonspiritual, °secular, °lay, °materialistic: *He ponders the abstract, not the base, material things of everyday life.*

materialistic *adj.* °expedient, money-oriented, possession-oriented, °greedy, *Slang* yuppie: *He's very materialistic and married her only for her money.*

materialize *v.* **1** °appear, °turn up, become visible, become manifest, be revealed, take shape *or* form, °form, °emerge: *A hulking shape began to materialize out of the fog.* **2** °happen, °come to pass, °take place, °occur, become manifest *or* real, be realized, become an actuality, be actualized: *The dream of a united Europe was finally materializing.*

materially *adv.* °substantially, palpably, significantly, °seriously, essentially, basically, considerably, greatly, much, °in the long run, °at bottom: *I doubt that his criticism will materially affect the success of the play.*

maternal *adj.* motherly, °warm, nurturing, caring, °understanding, °affectionate, °tender, °kind, °kindly, °devoted, °fond, doting; maternalistic: *Her interest in him is strictly maternal.*

maternity *n.* **1** motherhood, parenthood, pregnancy: *The prospect of maternity frightened her.* **2** parenthood, motherhood: *It never occurred to us to question the maternity of the child.*

mathematical *adj.* arithmetical; °precise, °exact, rigorous: *He went about his calculations with mathematical precision.*

matrimonial *adj.* marital, °marriage, °wedding, conjugal, °nuptial; married, °wedded, connubial: *They retired to the matrimonial bed. They enjoy matrimonial bliss.*

matter *n.* **1** °material, °substance, °stuff, sum and substance: *Some yellowish matter oozed out of the wound. It is a question of mind over matter.* **2** °situation, °issue, °question, °affair, °business, °subject, °topic, °condition, °thing, °fact, °concern; °occurrence, °episode, °incident, °event, °occasion, °proceeding: *It was a matter of life and death. His loss of face is not a matter that need involve you.* **3** °problem, °difficulty, °trouble, °complication, °worry, °upset, °dilemma, °quandary, °enigma, °puzzle: *What is the matter?* **4** °content, essentials, °pith, °context, °theme, °argument, purport, °implication; signification, °meaning, meaningfulness, °import, °importance, °significance, °moment, °weight, consequence: *Pay attention to the matter in his speech, not his manner. It is of little matter to me whether you go.* **5** °amount, °sum, °quantity, °question: *It was only a matter of a few pennies' difference.* —*v.* **6** be important *or* of importance, °count, be of consequence, make a difference, mean something: *What you think matters a great deal.*

matter-of-fact *adj.* straightforward, °direct, °forthright, °sober, °factual, unimaginative, unartistic, °prosaic, unpoetic, °dry, dryasdust, °dull, °boring, °tiresome, °flat, mundane, °lifeless, featureless, °unvarnished, °colorless, unembellished, °unadorned: *The report contained a matter-of-fact description of the events.*

mature *adj.* **1** °adult, grown (up), full-grown, fully grown, of age, fully fledged, full-fledged, fully developed, matured, °experienced, °knowledgeable, °sophisticated: *At twenty-two, she is mature enough to be on her own.* **2** °ripe, °ready, ripened, °mellow, °aged, °seasoned: *A good wine, like fruit, should not be consumed till it is mature.* **3** °complete, matured, °perfect, perfected, °polished, °refined, °ready, fully developed, consummated: *Our plans are not yet mature.* —*v.* **4** °grow up, °age, °develop, come of age; *Brit* be one's age, *US* act one's age: *Has he matured sufficiently to be left alone in the house for the weekend? I wish he'd mature a bit more.* **5** °ripen, °mellow, °age, °season; maturate: *Fruit that has matured on the plant tastes best.* **6** °develop, °perfect, °refine, °polish, maturate, bring to fruition: *The felon's maturing his felonious little plans.*

maturity *n.* **1** adulthood, °majority, full growth *or* development: *Maturity is as much a state of mind as a matter of age.* **2** ripeness, °readiness, mellowness; maturation: *Different wines need aging for different periods to attain maturity.* **3** °readiness, °perfection, °completion, fullness, °consummation, operability, applicability; maturation: *We have brought our plans to maturity.*

maudlin *adj.* °sentimental, (over)emotional, mawkish, °romantic; °tearful, lachrymose, weepy, teary(-eyed); *Colloq* °mushy, slushy, *Brit* soppy, *US* soupy: *He gets maudlin thinking about his childhood.*

maxim *n.* saying, °proverb, axiom, aphorism, adage, °byword, °saw, apophthegm *or* apothegm, °epigram, °motto, °slogan; mot, °witticism; °cliché, °truism: *Her maxim is "What you don't know, can't hurt you."*

maximize *n.* **1** °increase, broaden, °improve, °magnify, augment, °add to, °expand, °build up, °enlarge: *He maximizes his chances for winning by betting on several numbers. To compete, you must maximize your productivity.* **2** °inflate, overplay, °overdo, °overstate, °exaggerate, oversell, °make much of, overstress, (over) color, °enhance, embroider (on), °embellish, °elaborate, °magnify: *Cliff tends to maximize his role while minimizing others'.*

maximum *n.* **1** most, utmost, uttermost, greatest, °peak, °extreme, °extremity, °pinnacle, °top, highest, °(upper) limit: *Turn the volume up to the maximum.* **2** °zenith, °pinnacle, °peak, °limit, apex, °acme, apogee, °climax, °crest, high(est) point, °top, °summit: *Her career reached its maximum in the 1930s.* —*adj.* **3** maximal, greatest, most, utmost, uttermost, °superlative, °supreme, °paramount, °extreme, highest, °top, topmost, climactic, crowning: *She viewed winning an Oscar as the maximum achievement of a lifetime.*

maybe *adv.* °perhaps, °possibly, *Literary* perchance, *Archaic* or *dialectal* mayhap, *Archaic* peradventure: *Maybe you'd prefer to stay home, as we went out only last night.*

mayhem *n.* maihem, °violence, °havoc, °destruction, °disorder, devastation, °chaos; °fracas, commotion, °confusion: *Someone screamed "Fire!", and the ensuing mayhem was unbelievable.*

maze *n.* labyrinth, complex, intricacy, twistings and turnings, convolutions: *His application has been lost in the maze of bureaucratic bungling.*

meadow *n.* °field, meadowland, °pasture, pastureland, *Literary or archaic* lea, mead: *The entire meadow was dotted with little puffy clouds of white sheep.*

meager *adj.* **1** °scanty, scant, °poor, °paltry, °inadequate, skimpy, scrimpy, °sparse, °spare, °insufficient, °bare, °puny, piddling, °trifling, °pathetic, exiguous, *Colloq* °measly: *Angela scraped out a meager living selling cosmetics door-to-door. The meager supplies wouldn't last another day.* **2** °spare, °skinny, °scrawny, bony, °emaciated, °gaunt, °thin, °lean, bare-boned, (half-)starved, underfed, undernourished, starving: *The meager faces of the survivors told of their ordeal.* **3** °spare, °plain, bare-boned, °unadorned, unembellished, unelaborate, unelaborated, °simple, simplified, oversimplified, °bare, °inadequate, °deficient, undetailed, °indefinite, nonspecific, °general, °broad, °loose, °vague: *The police have only a meager description to go on.* **4** unfruitful, °infertile, °barren, °deficient, °poor, unproductive: *This land is too meager to provide us with a living.*

meal *n.* **1** °food, repast, victuals, °nourishment, °spread, collation, refection; dinner, supper, breakfast, lunch, luncheon, *Brit* tea: *Meals will be served in the main dining room.* **2** ***make a meal of.*** °overdo, overplay, go overboard, go *or* carry to extremes, carry *or* go too far, do to excess: *Perkins is bound to make a meal of the slightest thing.*

mealy-mouthed *adj.* °mincing, °reticent, °reluctant, °hesitant, °equivocal, equivocating, °ambiguous, °indirect, unwilling to call a spade a spade, euphemistic, °roundabout, °vague, circumlocutory, periphrastic, °hypocritical, °deceitful, °artful, °slick, °oily, unctuous: *All you'll get from that mealy-mouthed hypocrite is platitudinous twaddle.*

mean[1] *v.* **1** °intend, °design, °purpose, °plan, °aim, have in mind, °contemplate, have in view; °want, °wish, °expect, °hope; be motivated by, have as justification: *I did mean to phone you but had no time. I am sure she meant nothing by her remark.* **2** °denote, °signify, °indicate, °note, °specify, °designate, °represent, betoken, °signal, °carry, convey, °drive at, °refer to, allude to, °communicate, °express, °bring out, get over *or* across; °imply, °suggest, connote, °intimate, °hint (at): *What does "canicular" mean? It depends on what you mean by "liberty." The footprints meant that someone had already been here.* **3** portend, °show, foretell, °foreshadow, °promise, presage, augur, herald: *Those clouds mean that it will soon rain.* **4** carry *or* have the weight *or* significance *or* importance of: *Money means little to someone in her circumstances.*

mean[2] *adj.* **1** stingy, °miserly, °tight, °close, °cheap, parsimonious, °penurious, stinting, niggardly, penny-pinching, tightfisted, closefisted, °mercenary, uncharitable, ungenerous, mean-spirited, unaccommodating, °small, °petty, *Colloq* °near, money-grubbing, °measly, *Brit* mingy: *She is far too mean to buy anyone a Christmas present. It's rather mean of him to refuse to contribute, considering his income.* **2** °lowly, °low, °base, °inferior, abject, °menial, °servile, degraded, °degenerate, undignified, ignoble, °plebeian, proletarian, °modest, °humble, °common: *He rose to a position of great power despite his mean background.* **3** °disgraceful, °rundown, °poor, °sorry, °miserable, scruffy, °seedy, °scurvy, °shabby, squalid, °wretched, °vile, °mangy, °sordid, °contemptible, °dismal, °dreary, °abysmal: *They live in very mean conditions.* **4** °unkind, malicious, °cruel, unaccommodating, disobliging: *It was very mean of you to steal that toy from the baby.* **5** °cantankerous, churlish, °nasty, °hostile, ill-tempered, bad-tempered, °sour, unpleasant: *That mean old lady would never even talk to us.* **6** °excellent, wonderful, °marvelous, °great, °exceptional, °effective, °skillful, skilled, *Colloq* far-out, *Slang US* bad: *The saxophonist in that combo is one mean cat.*

mean[3] *n.* **1** °average, °middle, °norm, °(happy) medium: *We are attempting to achieve the mean between the two extremes.* **2** ***by all means.*** **(a)** °absolutely, °definitely, certainly, °surely, assuredly, °of course, °positively: *By all means, do tell us what you think is wrong.* **(b)** °in any event, °at all events, no matter what, without fail, at any cost, °in any case: *By all means, you must do what you think is right.* **3** ***by means of.*** by dint of, via, °through, °by way of, with the help *or* aid of, employing, using, utilizing: *She gained access to her victims by means of the telephone directory. We entered by means of the front door.* **4** ***means.*** **(a)** °instrument, °agency, °method, °process, °technique, °mode, °manner, way(s), °approach, °course, °procedure, avenue, °medium, °vehicle: *If you can find the means, then get the job done at once.* **(b)** °resources, °funds, °money, wherewithal, °capital, °finances, °backing, °support: *Where will we find the means to launch the advertising campaign?* **(c)** °substance, °wealth, °property, °position, financial stability: *She is a woman of considerable means.* **5** ***by no means.*** by no manner of means, in no way, not at all, definitely *or* absolutely not, on no account, not conceivably, not in one's wildest dreams *or* fantasies, not by any stretch of the imagination, *Colloq* no way, *US* no way José: *By no means can I let you leave here alive. He is by no means a great artist.* —*adj.* **6** °middle, °center, °intermediate, medial, °medium, median, °average, middling: *The mean temperature for the year has increased.*

meander *v.* **1** °wander, °ramble, zigzag, °snake, °wind, °twist, °turn; °stroll, amble, rove, *Colloq* mosey, *Brit* swan around *or Brit* about: *The river meandered with a mazy motion down to a sunless sea. We meandered through the forest.* —*n.* **2** Often, ***meanders.*** meandering(s), turn(s), turning(s), twist(s), twisting(s), winding(s), curve(s), curving(s), loop(s), looping(s), bend(s), coil(s), zigzag(s), convolutions; tortuosities, flexuosities, anfractuosities: *We followed the river's meanders till we came to the falls.*

meandering *adj.* wandering, °roundabout, circuitous, sinuous, °tortuous, winding, °serpentine, °indirect, flexuous, curvy, °crooked, convoluted, °labyrinthine, mazy, anfractuous: *The meandering road followed the river through the valley.*

meaning *n.* **1** °sense, °import, °content, signification, denotation, °message, °substance, °gist: *The meaning of the word "lucid" is perfectly clear to me.* **2** purport, °implication, °drift, °spirit, connotation, °significance, °intention: *You understand my meaning, so I need not explain.* **3** °interpretation, °explanation: *What is the meaning of my dream about being attacked by my philodendron?*

meaningful *adj.* **1** °significant, °important, consequential, °serious, °sober, °deep, °substantial, pithy, substantive, °telling, °weighty, valid, °relevant: *Tired of casual encounters, he was ready for a meaningful relationship.* **2** °suggestive, °pregnant, telltale, °pointed, sententious, °significant, °expressive, °eloquent: *She gave him a meaningful look.*

meaningless *adj.* **1** °empty, °hollow, vacuous, unsubstantial, °absurd, °silly, °foolish, fatuous, asinine, °ridiculous, °preposterous, °nonsensical: *He gave some meaningless excuse for having forgotten the appointment. This gibberish is entirely meaningless to me.* **2** °ineffective, °ineffectual, °inefficacious, °bootless, unavailing, to no avail, trivial, nugatory, °trifling, °puny, °paltry, °worthless, not worth anything *or* a straw *or* a rap, valueless, °inconsequential, unimportant, of no moment, °insubstantial, °vain, °pointless, °senseless, °purposeless, undirected, °irrelevant, °insignificant: *Considering the enemy forces, anything we could do would be meaningless.*

means *n.* See **mean**[3], 4, above.

meantime *n., adv. See* **meanwhile,** below.

meanwhile *n.* **1** interim, °meantime, °interval: *In the meanwhile, you could study for the exam.* —*adv.* **2** in the meanwhile, °meantime, in the meantime, in the interim, for the moment, °temporarily, for now, °for the time being, during the interval, in the intervening time: *Jennie was in the shower; meanwhile, Abby finished doing the dishes.*

measly *adj.* °sparse, scant, °scanty, °meager, °paltry, °pathetic, skimpy, °puny, piddling, °miserly, niggardly, °miserable, beggarly, stingy, *Colloq Brit* mingy: *"Nouvelle cuisine" is the name given to a restaurateur's design to serve decorative, but measly, portions.*

measure *n.* **1** °amount, °quantity, °magnitude, amplitude, °size, °bulk, °mass, °extent, °reach, dimension, °scope, °proportions, °range, °spread; °capacity, °volume; °width, °length, °breadth, °height; °weight: *It is difficult to calculate the measure of an irregularly shaped object.* **2** °scale, gauge *or Technical* gage, °yardstick, °rule, *US* litmus test: *I cannot determine the length without a proper measure.* **3** °system, °standard, criterion, °rule, °method; barometer, Richter scale: *The measure by which ice cream is sold ought to be weight, not volume. What sort of measure do you apply when judging character?* **4** assessment, °evaluation, valuation, appraisal, value, gauge *or* gage, °rank, rating, °measurement, °stamp, °estimation: *It is not easy to get the measure of a man who wants to marry your daughter.* **5** °quota, °allotment, °ration, °share, °amount, °degree, °proportion, °quantity, °allowance; °portion, °part: *He was accorded a measure of freedom in choosing the method for getting the job done. She has had her measure of praise.* **6** Often, ***measures.*** step(s), °procedure, °proceeding, °action, °course (of action), °plan, °method, °means, avenue, tactic(s), °way, °direction, °approach, °technique: *Measures are in place to ensure that such a catastrophe never happens again.* **7** °bill, °resolution, legislation, °act, statute, °law; °plan, °proposal: *A measure to change the speed limit is before the council.* **8** °bound, °limit, °end, °extreme, °extent, limitation, moderation, °control, constraint, °restraint: *Drug trafficking has gone beyond all measure in some countries.* **9** °beat, °rhythm, °cadence, meter, °time; °melody, °tune, °bar, °theme, °motif: *The ambassador entered, and the band played in stately measures.* **10** ***for good measure.*** °to boot, °in addition, additionally, as a dividend, into the bargain, °besides, as *or* for a bonus, °moreover, furthermore: *The job includes a company car and, for good measure, use of the company ski chalet.* —*v.* **11** °rank, °rate, gauge *or* gage, meter, °weigh, °calculate, °reckon, °compute, °calibrate, °determine, ascertain, figure out *or* up, assess, appraise, °estimate, °evaluate, °judge, value; °survey, °find out: *We must measure the effect of the new safety regulations on our equipment budget. Public acceptance of the new product may be difficult to measure.* **12** °proportion, °pace, °adapt, °gauge, °relate, °tailor, °fit, °adjust, °regulate, °control: *Teachers should measure homework assignments according to the pupils' available time. Measure your words when speaking to young children.* **13** ***measure off or out.*** mark off *or* out, °limit, delimit, °fix, pace off *or* out, °lay off: *Now measure off exactly 122.5 meters due north.* **14** ***measure out.*** mete out, dole out, °ration (out), °parcel out, apportion, °allot, share out, °assign, allocate; °give out, deal out, °distribute, °issue, °pass out, °hand out, °dispense, °disperse, spread around *or* about: *The relief workers measured out food and water to the victims of the famine.* **15** ***measure up (to).*** **(a)** °meet, °equal, °fulfill, °match, °reach, attain: *The team has not measured up to the coach's expectations. Will they ever measure up?* **(b)** °qualify (for), be suitable (for), be equal to, be fit *or* fitted for, be adequate (to), be up to, *Colloq* °make the grade, come up to scratch, be up to snuff, *US* cut the mustard: *If Wiswell gets the assignment, are you sure that he'll measure up? No one else measures up to the job.*

measured *adj.* **1** °slow, regulated, °unhurried, leisurely, °stately, °majestic, °dignified, °sedate, °solemn: *The silent procession moved through the streets at a measured pace.* **2** °careful, °cautious, °prudent, °calculated, °studied, considered, °deliberate, °systematic, °sober, °intentional, planned, regulated, °premeditated, well-thought-out, reasoned: *He planned the murder with the same measured deliberation he used in writing computer programs.* **3** rhythmic(al), °regular, cadenced, °steady, °uniform, °even, °monotonous: *The measured marching of the soldiers echoed through the night.* **4** °precise, regulated, °exact, °predetermined, modulated, quantified; clockwork: *A measured amount of each ingredient is automatically added at each stage.*

measurement *n.* **1** measuring, °reckoning, gauging *or* gaging, ascertainment, °determination, assessment, °estimation, appraisal, °evaluation, valuation, °judgment, °calculation, computation, mensuration, commensuration; metage: *The measurement of astronomical distances is approximate at best.* **2** dimension, °extent, °size, °amount, °magnitude, amplitude; °length, °breadth, °height, °width, °depth; °area; °volume, °capacity; °weight, tonnage; °(elapsed) time, °period; (square *or* cubic) footage, (square) yardage, mileage, acreage: *I need the measurements of the bedroom if I am to order the carpet.*

meat *n.* **1** °flesh; °food, °nourishment, °sustenance, viands, victuals, nutriment, °provisions, °provender, comestibles, edibles, eatables, *Colloq* eats, chow, grub: *I was starving and they gave me meat and drink.* **2** °pith, °core, °heart, marrow, °kernel, vital part, °essence, °gist, °substance, basics, essentials, crux: *The meat of his argument can be found in the closing chapters.*

mechanical *adj.* **1** °automatic, automated, machine-driven; machine-made: *Early mechanical toys are sometimes quite intricate. This is a mechanical, not an electronic device.* **2** °automatic, reflex, °involuntary, °instinctive, °routine, °habitual, °unconscious, °perfunctory, machinelike, robotlike: *Good manners ought to be mechanical. Blinking is a mechanical reaction to bright light.* **3** °impersonal, °distant, °cold, °matter-of-fact, unfeeling, °insensible, ritualistic, °lifeless, spiritless, °dead, °inanimate, unanimated, unemotional, unartistic, mechanistic, °colorless, uninspired, businesslike: *The girl at the checkout gave us a mechanical smile.*

mechanism *n.* **1** °device, appliance, contrivance, °apparatus, °instrument, °machine: *This mechanism will detonate the bomb at the right moment.* **2** machinery, workings, °works, °structure, °system, °organization, °arrangement: *Physicists offer only an imperfect account of the mechanism of the universe.* **3** °way, °means, °method, °procedure, °approach, °technique, °medium, °process, °agency: *We shall use every mechanism at our disposal to have the case heard by a different judge.* **4** materialism, mechanicalism, physicalism, logical positivism, identity theory, monism: *His notions of mechanism reject the idea of free-thinking individualism.*

meddle *v.* °interfere, °intrude, °butt in, °pry, °intervene, interlope, °tamper, °snoop, *Rare* intermeddle, *Colloq* stick *or* poke one's nose in, have a *or* one's finger in the pie, kibitz: *Why are you always meddling in things that don't concern you?*

mediator *n.* arbitrator, arbiter, referee, °umpire, °judge, °negotiator, °intermediary, °go-between, middleman, °moderator, °liaison, intercessor, interceder, conciliator, appeaser, °peacemaker: *To avoid a lawsuit, we agreed to allow a mediator to settle our differences.*

medicinal *adj.* healing, remedial, °therapeutic, curative, restorative, sanative; medical, iatric(al), *Medicine* roborant, analeptic, alexipharmic: *The doctor prescribed a medicinal ointment for the rash.*

medicine *n.* medication, medicament, °remedy, °drug, pharmaceutical, °prescription, *Archaic* physic; nostrum, panacea, cure-all: *There is no medicine that can cure your affliction.*

mediocre *adj.* middling, °indifferent, °ordinary, commonplace, °average, °medium, °everyday, garden variety, °run-of-the-mill, °pedestrian, °undistinguished, uninspired, unimaginative, unexceptional, °tolerable, °fair, not (that *or* too) good, not bad, second-rate, third-rate, °inferior, °poor, *Brit* common-or-garden variety, *US* garden-variety, *Colloq* °so-so, fair to middling, nothing to brag *or* to write home about, no great shakes: *We sat through a mediocre play acted by mediocre performers.*

meditate *v.* **1** °reflect, °think, °muse, °ponder, °study, ruminate, cogitate, °contemplate, cerebrate, be lost in thought, be in a brown study: *Please don't interrupt me while I'm meditating.* **2** Often, ***meditate on*** *or* ***upon.*** °consider, °contemplate, mull over, reflect on *or* upon, ponder on *or* over, °chew over, °plan, °scheme, °devise, °design, °conceive, °frame, °think up, have in mind: *She was meditating revenge, while I was meditating on how to discourage her.*

meditative *adj.* °thoughtful, °pensive, contemplative, °reflective, °studious, cogitative, excogitative, abstracted, °rapt, engrossed, lost *or* deep in thought, ruminative, brooding: *She suddenly stopped talking and became meditative, looking at me with a curious expression.*

medium *adj.* **1** °average, °middle, mid, medial, median, °normal, °standard, °usual, °everyday, °ordinary; mid-sized: *His wife has a medium-sized station wagon.* **2** See **mediocre,** above. —*n.* **3** °average, °middle, midpoint, compromise, °center, °mean, °norm, mediocrity: *Instead of going to extremes, try for the happy medium.* **4** °atmosphere, °environment, ambiance *or* ambience, °milieu: *The air serves as a medium of sound. The light is refracted as it passes into a denser medium.* **5** °means, °method, °mode, °approach, instrumentality, °device, °mechanism, intermediation, °technique, contrivance, °agency, °expedient, °way, °course, °route, °road, avenue, °channel, conveyance, °vehicle: *Latin and Greek were the media of the scholarship, the science, and the theology of the Middle Ages. Television is a powerful advertising medium.*

medley *n.* °mixture, °assortment, °combination, °miscellany, mélange, °collection, conglomeration, agglomeration, hotchpotch *or US and Canadian also* hodgepodge, olio, °blend, gallimaufry, omnium-gatherum, °pastiche, °potpourri, salmagundi, olla podrida, °mishmash, °jumble, °mess, farrago, °stew, goulash, *Colloq* mixed bag: *The program consisted of a medley of show tunes.*

meek *adj.* **1** °modest, °humble, °submissive, unassuming, unambitious, unpretentious, °mild, °bland, °patient, deferential, °shy, °retiring, °lowly: *The meek shall inherit the earth.* **2** °tame, °timid, °weak, docile, compliant, °submissive, °yielding, acquiescent, unaggressive, nonmilitant, °tractable, °manageable, °subdued, repressed, spiritless, suppressed, °broken, *Colloq* wimpish: *The people are meek and will not object to being ruled by a dictator.*

meet[1] *v.* **1** °encounter, °come across, chance on *or* upon, happen on *or* upon, stumble on *or* into, °see, *Colloq* run across *or* into, °bump into: *Guess where Donald and Clare are moving next week!* **2** Often, ***meet with.*** rendezvous (with), °get together (with); convene, °assemble, °gather, °collect, forgather *or* foregather, congregate: *I met with the mysterious stranger as arranged. The board meets at nine in the morning.* **3** make the acquaintance of, be introduced to, first encounter, °come across, °find: *Where did you meet your wife?* **4** °link up, °join, come together, °unite, adjoin, abut, °touch, intersect: *The horizon is where the sky meets earth.* **5** °answer, °deal with, °handle, °satisfy, °fulfill, °take care of, °dispose of, °heed, °observe, °carry out; °gratify, °pay, °settle, °defray, liquidate: *Will these precautions meet the requirements of the Department of the Environment? The police advised against meeting the kidnapper's demands.* **6** ***meet with.*** °encounter, be met by, °experience; °undergo, °endure, °suffer, °have, °go through: *All our efforts met with little or no response. Gary met with an accident yesterday.* —*n.* **7** °competition, °contest, °meeting, °match, tourney, °tournament; °muster, °rally: *The track meet was won by our team.*

meet[2] *adj.* °fitting, °suitable, °appropriate, °proper, °fit, congruous: *It is meet to offer praises to God.*

meeting *n.* **1** °appointment, °engagement, rendezvous, °encounter, assignation, tryst, *Slang US* °meet: *She arranged a meeting with her ex-husband.* **2** °assembly, °convention, °conference, °gathering, congress, conclave, °session, congregation, convocation, *US* caucus, *Colloq* °get-together: *Nominations for next year's officers will be presented at the next meeting.* **3** convergence, converging, confluence, joining, °union, °junction, conjunction, intersection: *The earliest settlement was at the meeting of the two rivers.* **4** See **meet, 7,** above.

melancholy *adj.* **1** °sad, morose, depressed, unhappy, °dejected, °despondent, °blue, °downhearted, °glum, °gloomy, woeful, °woebegone, lugubrious, disconsolate, downcast, dispirited, low-spirited, cheerless, crestfallen, chapfallen, °forlorn, °heartbroken, °mournful, °sorrowful, °miserable, °dismal, *Colloq* °down in the mouth, (down) in the dumps, °low: *Harold has been melancholy ever since Lucinda left.* —*n.* **2** °sadness, °sorrow, °misery, °woe, °gloom, unhappiness, blues, moroseness, °depression, dejection, dejectedness, despondence, despondency, downheartedness, glumness, gloominess, woefulness, lugubriousness, disconsolateness, dispiritedness, cheerlessness, mournfulness, sorrowfulness, miserableness, dolor, °anguish: *Getting Harold's mind off his problems may relieve his melancholy.*

mellow *adj.* **1** °soft, °juicy, °luscious, °delicious, °rich, °sweet, flavorful, full-flavored, °ready, °ripe, °mature, ripened, °aged: *She set before us a dish of mellow peaches. From its mellow taste, I'd say that this is an amontillado sherry.* **2** °musical, °melodious, °full, °pure, °rich, °sweet, dulcet, mellifluous, euphonious, vibrant: *Her voice was low and mellow and suggestive.* **3** °soft, softened, °subtle, muted, pastel: *The mellow shades of this painting would fit well into the decor.* **4** easygoing, °genial, °gentle, °good-natured, °easy, °cordial, °friendly, °warm, °amiable, °agreeable, °pleasant, °cheerful, °happy, jovial, felicitous; °suave, mellowed out, *Colloq* °cool: *Patrick seems to have become more mellow with age.* —*v.* **5** °mature, °ripen, °age, °season, °sweeten, °develop, °improve (with age), °soften: *Frank has mellowed since taking up golf.*

melodious *adj.* sweet(-sounding), dulcet, °tuneful, euphonious, °harmonious, melodic, °lyrical, °musical, mellifluous, mellisonant, silvery, °golden: *I agreed with her father that Sylvia has a most melodious voice.*

melodramatic *adj.* °sensational, sensationalistic, °dramatic, stagy, °theatrical, emotionalistic, (over)sentimental, (over)sentimentalized, overdrawn, overworked, °overwrought, overdone, exaggerated, blood-and-thunder, histrionic, *Colloq* hammy, *US* schmaltzy, hokey: *The slightest thing that happened to him he blew up with melodramatic effect to elicit Maria's sympathy.*

melody *n.* **1** °song, °tune, °air, °strain, °measure, °theme, °refrain: *Isn't that the melody from Mimi's aria in* La Bohème? **2** tunefulness, melodiousness, euphoniousness, euphony, °harmony, musicality, sweetness: *The older music seems to be marked by so much more melody than rock 'n' roll.*

melt *v.* **1** °soften, °thaw, liquefy, °fuse, °dissolve, liquidize, deliquesce: *The fridge is out of order, and all the ice cream has melted.* **2** °soften, °thaw, mollify, assuage, °touch, °move, °disarm, °mellow: *The look on the child's face melted his icy reserve.* **3** Usually, ***melt into.*** °blend, °fade, °merge, °disappear, °dissolve, °shrink: *As soon as the police arrived, Phyllis tried to melt into the background.* **4** ***melt away.*** °disappear, °dissolve, vanish, °evaporate, °go away, °fade, °pass, °decline, °decrease, °shrink, °dwindle, °diminish: *Our objections to Carrie's wearing her best dress melted away when we saw her walk in.*

member *n.* °colleague, °associate, °fellow: *I am going to propose Fred as a member of the club.*

memento *n.* **1** souvenir, °keepsake, °remembrance, °relic, °trophy, °token: *I kept his medals as a memento of our time together.* **2** ***mementos.*** memorabilia: *There will be an auction of war mementos next week.*

memoir *n.* **1** °account, °report, reportage, °narrative, °essay, dissertation, disquisition, °paper, °journal, °record, biography, °life: *They found her memoir about life with Livingstone very exciting.* **2** ***memoirs.*** autobiography, reminiscences, recollections, memories, °diary, confessions, °letters, life story; annals, °history, account(s), record(s), °chronology: *Perhaps they will publish my memoirs after I am gone.*

memorable *adj.* unforgettable, catchy, never-to-be-forgotten, °noteworthy, °notable, °remarkable, °significant, °important, °worthy, °momentous, °eventful, °historic, °illustrious, °celebrated, °great: *His first book contains many memorable and quotable passages.*

memorandum *n.* °note, °record, °minute, °reminder, °message, *Brit* chit *or* chitty, *Colloq* memo: *Send a memorandum reminding them of the meeting.*

memorial *adj.* **1** commemorative: *A memorial plaque will be dedicated to him tomorrow.* —*n.* **2** °monument, marker, °plaque, cenotaph, °statue, °memento, °remembrance, °reminder, souvenir: *Let this serve as a memorial to those who died in the wars. Wreaths were laid at the foot of the memorial.*

memorialize *v.* °honor, °commemorate, pay homage *or* respect *or* tribute to, °remember, °eulogize, °celebrate, °mark: *His life, devoted to botany, is memorialized in these beautiful gardens.*

memorize *v.* learn by heart *or* rote, commit to memory, learn word for word, °retain; °remember: *He memorized the entire* Rime of the Ancient Mariner.

memory *n.* **1** °recall, °recollection, retention: *My memory of the incident is very vivid. Her memory for names is very poor.* **2** °recollection, °reminiscence, °thought: *The interviewer was drawing on grandfather's memories of the 1920s.* **3** °remembrance, °honor, °homage, °respect, °tribute, °celebration: *The poem was written in memory of a dear friend.*

menace *v.* **1** °threaten, °intimidate, °daunt, terrorize, °terrify, cow, °bully, °frighten, °scare, °alarm; bare one's teeth: *The loan sharks menaced him when he refused to pay.* —*n.* **2** °threat, °danger, °peril, °hazard, °risk: *The sunken wreck is a menace to navigation.* **3** intimidation, °scare, °threat, °warning, commination: *According to law, a menace alone does not constitute an injury.*

menacing *adj.* °threatening, looming, °impending, °ominous, °frightening, °terrifying, intimidating, minacious, minatory *or* minatorial, lowering *or Brit also* louring; °dangerous, °perilous, °hazardous, °risky, chancy: *Menacing storm clouds scudded across the darkening sky.*

mend *v.* **1** °repair, °fix, °patch (up), °rectify, °correct, °remedy, °restore, °rehabilitate; °heal: *When will you get around to mending the roof?* **2** °correct, °improve, °better, ameliorate, °reform, °revise, °rectify, set *or* put right, emend: *When are you going to mend your ways?* **3** °heal, °improve, °recover, °convalesce, °recuperate, get better: *It took me a few months to mend after the accident.* —*n.* **4** °repair, °patch: *You can hardly see where the mend was made.* **5** ***on the mend.*** recovering, recuperating, convalescing, convalescent, improving: *The good news is that father is on the mend after his attack.*

menial *adj.* **1** °lowly, °servile, °humble, subservient, °base, °low, °mean, slavish, demeaning, °degrading, ignoble; °routine, unskilled: *He took a number of menial jobs to pay for his daughter's wedding.* **2** °servile, fawning, °groveling, toadying, sycophantic, °obsequious, cringing, °timeserving, °flattering, sniveling, *Colloq* bootlicking, *Slang taboo* brown-nosing: *His insecurity makes him surround himself with menial hangers-on.* —*n.* **3** lackey, flunky *or* flunkey, serf, °slave, underling, minion, *Brit* °fag, *Colloq Brit* dogsbody, skivvy, *US* gofer: *He works as one of the menials in the hotel kitchen.* **4** toady, sycophant, °yes man, lickspittle, leech, °parasite, timeserver, *Colloq* bootlicker, *Slang taboo* brown-nose(r): *The film star was attended by a retinue of menials ready to jump at her command.*

mental *adj.* **1** °intellectual, cognitive, cerebral, perceptual, °rational, conceptual, °theoretical, noetic, °abstract: *The boy staggered us with his ability to perform rapid mental arithmetical calculations.* **2** lunatic, °mad, °crazy, °psychotic, demented, mentally ill, °unstable, °unbalanced, °deranged, °disturbed, °daft, certifiable, *Slang* off one's rocker, nutty, batty, balmy, loony, screwy, bonkers, crackers, nuts, bananas, *US* loco, *Brit* barmy: *Uncle Ronnie has gone a bit mental in his old age.*

mentality *n.* **1** °intelligence, °brain, °capacity, °intellect, °wit, °sense, °judgment, acuity, acumen, I.Q., rationality, °understanding: *I'd say that Kurt is a man of average mentality.* **2** °inclination, °attitude, °bent, mind-set, °disposition, °frame of mind, temperament, °outlook, °view: *What can be the mentality of people who focus only on corporate mergers, takeovers, and the like?*

mention *v.* **1** speak *or* write about, °refer to, allude to, touch on *or* upon, make mention (of), bring up *or* in, °introduce, °broach, call *or* direct attention to, °note, °name, cite, °acknowledge; °point out, °indicate, °make known, adduce, °report, °quote: *Sergeant Patton was mentioned in dispatches from the front. I should mention the works of Corbell in support of my thesis.* **2** divulge, °reveal, °intimate, °disclose, °impart, °suggest, advert (to), °hint (at), °imply, °insinuate: *Did he mention that Liz was present?* —*n.* **3** °reference, allusion, °note, naming, citation, mentioning, *Colloq* cite: *There was a mention of you in the newspaper today.* **4** °recognition, °tribute, °acknowledgment, °kudos, °praise: *She won no prize but came in for an honorable mention.* **5** °announcement, °reference, referral, °remark: *The speaker's mention of the conference reminded me of something.*

mercantile *adj.* commercial, °business, °trade, marketing, °market: *The book dealt with the mercantile affairs of the Hanseatic League.*

mercenary *adj.* **1** money-oriented, °grasping, °greedy, acquisitive, covetous, °predatory, °avaricious, °venal, *Colloq* money-grubbing: *She objected that the "financial world" ought to be renamed the "mercenary world."* **2** °venal, bribable, °corrupt, bought, *Colloq US* on the take: *They found themselves a mercenary judge and the indictment was quashed.* —*n.* **3** hireling, legionnaire, soldier of fortune: *The highly trained task force was made up entirely of mercenaries.*

merchandise *n.* **1** °goods, commodities, products, °stock, °staples, °produce: *The merchandise you ordered was shipped yesterday.* —*v.* **2** °trade, deal in, traffic in, °market, °distribute, retail, °(buy and) sell, °promote, advertise: *We are merchandising the new product line via direct mail.*

merchant *n.* **1** °dealer, retailer, °seller, shopkeeper, store owner, °trader, tradesman *or* tradeswoman, vendor: *High Street merchants are complaining about a decline in sales.* **2** °peddler, hawker; (traveling) salesman, °(sales) representative, commercial traveler, huckster, door-to-door salesman, *US old-fashioned* drummer, *Colloq* (sales) rep: *His work as an itinerant merchant took him away from home most of the week.* **3** distributor, wholesaler, jobber, °broker, °agent, forwarder; businessman, merchant prince, °mogul, °tycoon, magnate, industrialist, *US* baron: *Generations of her family have been steel merchants.*

merciful *adj.* compassionate, °sympathetic, °forgiving, °kind, °kindly, clement, kindhearted, forbearing, °sparing, °lenient, °tender, humane, °liberal, °mild, tenderhearted, °softhearted, °gracious, °generous, magnanimous, benignant, beneficent, °charitable, °thoughtful, °considerate, °indulgent, °big: *Which of the conquerors in history could be called merciful?*

merciless *adj.* °cruel, pitiless, °ruthless, °heartless, °unmerciful, inhumane, °inhuman, °brutal, °savage, barbarous, barbaric, °barbarian, °crude, °rude, °rough, °harsh, °tough, °callous, °hard, hardhearted, illiberal, °tyrannical, stonyhearted, °cold, °severe, unsparing, insensitive, °indifferent, °unsympathetic, unforgiving, °ungracious, malevolent, °thoughtless, uncharitable, °inconsiderate, °unmoved, unbending, °inflexible, °relentless, unrelenting, inexorable: *Our captors proved to be utterly merciless. The merciless sun beat down on the desert outpost.*

mercy *n.* compassion, °pity, forbearance, °quarter, °tolerance, °sympathy, °favor, °forgiveness, °kindness, kindliness, leniency, tenderness, °humanity, humaneness, liberality, kindheartedness, tenderheartedness, softheartedness, graciousness, generosity, magnanimity, benignity, beneficence, °charity, thoughtfulness, °consideration, °indulgence: *May the Lord have mercy on your soul. We could expect no mercy from that slave driver.*

mere *adj.* °bare, °basic, scant, °stark, °sheer; °absolute, unmixed, °only, °just, nothing but, °pure (and simple), °unmitigated, °undiluted: *She was a mere slip of a girl. We need something more to go on than mere hearsay evidence.*

merely *adv.* **1** °only, °simply; basically, purely, essentially, fundamentally, °at bottom: *I was merely going to ask if you'd like to go to the cinema.* **2** °only, no more than; °barely, °scarcely, °simply, solely, °entirely: *She was merely the scullery maid. We can't unlock the safe merely on your authority.*

merge *v.* °combine, coalesce, °unite, °join, °amalgamate, consolidate, °pool, °blend, °mix, °mingle, commingle, °fuse: *If we merge our companies, we shall save on overhead costs.*

merger *n.* °combination, coalescence, °union, merging, °amalgamation, consolidation, coalition, pooling, blending, mixing, mingling, commingling, fusing, fusion: *A merger of the two firms might result in a monopoly.*

merit *n.* **1** °worth, worthiness, value, °excellence, °quality, °virtue, °good, goodness: *Alan's plan has a good deal of merit.* **2** Often, **merits.** °assets, strong point, °advantage, rights and wrongs: *What exactly are the merits of his approach over mine? The case should be tried on its own merits.* —*v.* **3** °earn, °deserve, °warrant, °rate, have a right *or* claim to, be entitled to, be qualified for, be worthy of: *What gave you the idea that you merited any special consideration?*

meritorious *adj.* °honorable, °laudable, °praiseworthy, commendable, creditable, °admirable, °estimable, °excellent, °exemplary, °outstanding: *Penny has won the company award for meritorious service five months in a row.*

merriment *n.* jollity, joviality, merrymaking, °revelry, °gaiety, high *or* good spirits, °mirth, mirthfulness, joyfulness, felicity, jubilation, °festivity, exhilaration, buoyancy, °exuberance, °cheer, cheerfulness, °glee, °fun, °hilarity, °enjoyment, °happiness, blithefulness, blithesomeness, frolicking: *Even old Stickley joined in the merriment at his grandson's wedding party.*

merry *adj.* **1** °cheerful, °happy, °gay, cheery, °jolly, jovial, in high *or* good spirits, mirthful, °joyful, joyous, °hilarious, jubilant, rejoicing, festive, °exhilarating, °exuberant, °vivacious, convivial, °buoyant, °gleeful, °blithe, blithesome, °carefree, lighthearted, °delighted: *Didn't we have a merry old time at the Christmas party!*

2 *make merry.* °revel, °celebrate, °carouse, °frolic: *The guests made merry till the wee hours of the morning.*

mesh *n.* **1** meshwork, °network, netting, °net, °web, webbing, lattice, latticework, °screen, screening, interlacing, lacework, grid, °grate, °grating, grater, sieve, strainer, trellis, trelliswork, decussation, *Technical* rete, reticle *or* reticule *or* graticule, reticulation, plexus, plexure, reticulum; interstice: *The mesh of the hammock broke and I fell to the ground.* **2** Often, ***meshes.*** °grip, °clutches, °grasp, toils, °web, °trap, entanglement, °tangle, complex, °complexity, intricacy: *Her papers were lost in the meshes of the filing system.* —*v.* **3** °catch, °entangle, enmesh, °grab, °trap, entrap, °snare, ensnare, °involve: *I became meshed inextricably in the complexities of her life.* **4** °engage, °fit (together), dovetail, °knit, enmesh, °match, interlock: *After they were aligned, the gears meshed perfectly.*

mess *n.* **1** °chaos, °disorder, disarray, disorganization, °shambles, °muddle, disarrangement, °clutter, hotchpotch *or US also* hodgepodge, °litter, °tangle, dog's breakfast *or* lunch *or* dinner, mare's nest, °jumble, °confusion, °mishmash; untidiness: *Would you help me clean up this mess? Drugs have made a mess of her life.* **2** concoction, °mixture, °medley, °miscellany, °hash, gallimaufry, farrago, olio, olla podrida, °potpourri, smorgasbord *or* smörgåsbord, kedgeree: *They presented us with a huge mess of things from which we must select what we want.* **3** °predicament, °difficulty, °plight, °pinch, °trouble, °dilemma, °quandary, imbroglio, *Colloq* foul-up, °stew, °fix, hot water, (pretty *or* fine) kettle of fish, pickle, °jam, *Slang* screw-up, *Brit* balls-up, can of worms, *US* snafu: *You have got yourself into a fine mess this time!* —*v.* **4 *mess about*** or ***around (with).*** **(a)** potter, °fool (around), dally, busy oneself, fiddle about *or* around, °play: *He enjoys messing about in boats.* **(b)** philander, °trifle, °toy, °flirt, °seduce, sleep around, °fool around, °run around: *She doesn't want him to mess about with other women.* **5 *mess up.*** **(a)** disarrange, disarray, dishevel, °tousle, *Colloq US* muss (up): *He always messes up my hair when he kisses me.* **(b)** °ruin, °destroy, make a shambles of, °wreck, °bungle, °botch, °foul up, *Colloq* make a hash of, *Slang* °muck up: *Making that one little mistake messed up his whole career.* **(c)** °dirty, °clutter up, make untidy, turn upside down, pull to pieces, °upset: *The police messed up my whole house looking for the pistol.* **6 *mess with.*** interfere in *or* with, °intervene, meddle with *or* in, intrude in, butt in *or* into, tinker with, tamper with, get involved in *or* with: *They ought to learn not to mess with things that don't concern them.*

message *n.* **1** communication, °bulletin, °report, communiqué, °news, dispatch *or* despatch, °information, °word, °intelligence, tidings; °note, °missive, °letter, °memorandum: *The message reached us by carrier pigeon.* **2** °speech, °address, °presentation, °statement, °declaration: *His message was delivered via television to people everywhere.* **3** °idea, °point, °import, °meaning, °essence, °implication: *The message that he isn't liked around here still hasn't penetrated his thick skull.*

messenger *n.* °envoy, emissary, legate, nuncio, °intermediary, °go-between; °page, errand boy *or* °girl, messenger boy *or* girl, courier, °runner, dispatch rider, Pheidippides, Mercury, Hermes; herald, °harbinger; *Colloq US* gofer: *If you don't like the message, don't take it out on the messenger.*

Messiah *n.* deliverer, liberator, emancipator, savior *or* saviour, rescuer: *They believe in a Messiah who will release them from earthly bondage.*

metaphor *n.* °figure (of speech), allusion, analogy, analogue, °reference, °image, trope, °symbol; simile, parabole; metonymy, symbolism, °imagery: *When she said she was blue, she was using "blue" as a metaphor. Poetry is rife with metaphor.*

metaphoric *adj.* metaphorical, nonliteral, allusive, analogic(al), analogous, figurative, °symbolic, referential, parabolic(al), metonymic, metonymous, tropological: *I meant "filthy" in a metaphoric, not a literal, sense.*

mete *v.* Usually, **mete out.** deal (out), apportion, °distribute, °dole (out), °allot, °assign, allocate, °parcel out, °share (out), °ration (out), °measure out, °dispense, °hand out, °give out, °pass out, *Colloq* dish out: *Punishment was meted out to those who deserved it.*

meteoric *adj.* **1** °brief, °short-lived, °temporary, transitory, °transient, ephemeral, evanescent, impermanent, °fleeting, °momentary, °swift, overnight: *She enjoyed a meteoric rise as a singer but, after a year, was never heard from again.* **2** °brilliant, °dazzling, flashing, spectacular, °sensational: *His meteoric career was launched in 1974.*

method *n.* **1** °way, °means, °procedure, °approach, °route, avenue, °road, °mode, °manner, °technique, °process, °routine, modus operandi; °plan, °scheme, °program, °course, °routine, °practice, °pattern, °system, methodology; *Colloq US* MO (= 'modus operandi'): *What method was used to arrive at the answer?* **2** °arrangement, °order, °system, °structure, °organization, °design, °pattern, orderliness, neatness, °regularity, °discipline: *There is a discernible, underlying method to what seems superficially chaotic.*

methodical *adj.* organized, ordered, °systematic, structured, businesslike, °orderly, °neat, °tidy, °regular, °routine, balanced, disciplined, °painstaking, °meticulous, °deliberate, paced, °laborious, plodding, °labored: *The weaver resumed his methodical throwing of the shuttle, back and forth, back and forth.*

meticulous *adj.* °careful, °precise, °accurate, °exact, °fastidious, °scrupulous, °thorough, °particular, °painstaking, punctilious, °fussy, °finicky, °demanding, °strict, °critical, °exacting, °perfectionist: *It is amazing how meticulous the craftsmen had to be to create such detailed mosaics.*

metropolis *n.* °capital, °(capital) city; metropolitan area, urban sprawl, megalopolis, °municipality: *Street crime has become a problem in every major metropolis.*

microbe *n.* microorganism, microzoon, °germ; bacterium, virus, *Colloq* °bug: *Only recently has science learned that diseases are caused by microbes.*

midday *n.* °noon, noontime, twelve (o'clock) noon, *US* high noon: *By midday, the temperature was more than 100° in the shade.*

middle *adj.* **1** °central, °center, halfway, mid, midway, °mean, medial, *Technical* mesial: *We had reached the middle stage of our journey. Why do you wear your wedding ring on your middle finger?* —*n.* **2** °center, midpoint, °midst, halfway point; °heart, bull's-eye: *He lives in the middle of the city. The arrow found the very middle of the target.* **3** midriff, waist, midsection, °stomach: *She wore a red sash round her middle.*

midst *n.* °middle, °center, midpoint, halfway point: *In the midst of life we are in death.*

midwife *n.* accoucheur *or* accoucheuse: *Most babies in the world are delivered by midwives.*

might *n.* **1** °strength, °power, °energy, °force, muscle, potency, *Literary* puissance: *I hit my attacker with all my might. Might makes right.* **2** °influence, °authority, °weight, °sway, °dominion, ascendancy, °superiority, mightiness, °capability, °capacity, °power, °effect, effectiveness, *Colloq* clout: *It was not till the twentieth century that the might of the United States was felt outside its borders.*

mighty *adj.* **1** °powerful, °strong, °potent, °influential, °dominant, °predominant, ascendant, °weighty, doughty, °authoritarian, autocratic, °indomitable: *The pen is mightier than the sword. He should have thought twice before challenging such a mighty opponent.* **2** °strong, °muscular, °powerful, °robust, strapping, °sturdy, °brawny, °burly, well-built, able-bodied, °hardy, *Colloq* °husky, °hefty: *"The smith a mighty man is he, / With large and sinewy hands."* **3** °big, °large, °huge, °grand, °great, °enormous, °gigantic, tremendous, °towering, °monumental, °prodigious, °massive, °bulky: *The giant plucked a mighty tree from the earth and flung it across the sea.* *—adv.* **4** °very, °extremely: *It was mighty nice of you to visit me here in the Ozarks.*

migrant *n.* **1** wanderer, °rover, °drifter, gypsy, nomad, itinerant, °transient, migrator, wayfarer, bird of passage, peregrinator, °traveler; vagrant; *Colloq US* wetback: *The townspeople were having difficulty finding accommodation for the migrants.* *—adj.* **2** °transient, migratory, itinerant, peripatetic, drifting, nomadic, °traveling, gypsy, floating; vagrant: *The farmers depend on migrant workers to bring in the harvest.*

migrate *v.* **1** °go, °move, °travel, °settle, resettle, relocate, move house; °emigrate, immigrate, expatriate: *During the potato famine, many migrated from Ireland to America.* **2** °wander, °roam, voyage, rove, °drift, °range: *Some say that the American Indians migrated across the Bering land bridge from Asia.*

mild *adj.* **1** placid, °peaceful, °calm, °tranquil, °bland, °mellow, °inoffensive, °gentle, °serene, °good-natured, affable, °amiable, °kind, °kindly, °equable, easygoing, °temperate, nonviolent, conciliatory, °indulgent, °merciful, °forgiving, compassionate, °lenient, forbearing, °peaceable, pacific, °passive, °submissive, °yielding, °tractable, °meek, unassuming, °modest, °quiet, °subdued: *Judge Leaver is known for his mild disposition.* **2** clement, balmy, °warm, °fair, °pleasant, °temperate, placid, °moderate: *We've had a succession of mild winters here.* **3** °bland, °soothing, lenitive, mollifying, demulcent, emollient, °gentle, calming, softening: *The doctor recommended a mild laxative.*

milieu *n.* °environment, °climate, surroundings, environs, °background, ambiance *or* ambience, °sphere, °setting, °context, °atmosphere, °medium, °element, °precincts: *He sought a more peaceful milieu in whcih to pursue his research and writing.*

militant *adj.* **1** °aggressive, combative, °pugnacious, °belligerent, °hostile, contentious, antagonistic, °offensive, °truculent, °fierce, °ferocious, °warlike, bellicose, °martial, jingoistic, hawkish: *Those who have experienced war are not likely to be militant.* **2** warring, fighting, combatant, combating, embattled; °at war, up in arms: *The militant members of the council refused to discuss peace.* *—n.* **3** fighter, °aggressor, combatant, °belligerent, warrior, °soldier: *Anyone who experiences the hardships of a long strike may think twice before being a militant.*

military *adj.* **1** °martial, soldierly, naval, army, fighting, °service: *The military machine was poised to strike.* *—n.* **2** (armed) services *or* forces, army, °navy, air force, military establishment, soldiery: *The military will oppose any reductions in spending for arms.*

militate *v.* **1** Usually, ***militate against.*** °discourage, work *or* go *or* operate against, °foil, °counter, countervail, °cancel (out), °reduce (possibility of), °prevent, °hinder, °resist, °oppose: *His poor prison record militates against his early parole.* **2** Usually, ***militate for*** or ***in favor of.*** be on the side of, °favor, °further, °promote, °help, °aid: *The same reasoning that militated in favor of lower taxes then persists today.*

milk *v.* °drain, bleed, °extract, °tap, °exploit, wring, draw off *or* out, °withdraw: *The comic milked the joke for yet another laugh. The taxman is milking me dry.*

milksop *n.* sissy *or Brit also* cissy, °coward, °weakling, namby-pamby, mollycoddle, crybaby, nancy (boy), dastard, poltroon, caitiff, (little) Lord Fauntleroy, poltroon, *Archaic* caitiff, *US* milquetoast, mama's boy, *Colloq* chinless wonder, *Brit* mother's *or* mummy's boy, *US* mama's boy, pantywaist, *Slang* pansy: *We want men, not milksops, in the Marines.*

mill *n.* **1** grinder, quern, crusher, °roller: *In many parts of the world, corn is still ground by hand in a stone mill.* **2** °plant, °factory, °works, workshop, °shop, foundry: *This mill turns out rolled steel.* **3** ***been through the mill.*** °experienced, °knowledgeable, °sophisticated, toughened, hardened, °seasoned, battle-scarred, *Colloq* been through the wringer, in the know: *He's been through the mill and knows what to expect.* **4** ***run-of-the-mill.*** °average, unexceptional, °ordinary, °common, unremarkable, °everyday, °basic, °simple, *Brit* common-or-garden variety, *US* garden-variety: *Critics found her voice pleasant but rather run-of-the-mill.* *—v.* **5** °grind, °crush, comminute, °powder, °pulverize, °grate, granulate, °pound, triturate, masticate, bray; °crunch, mince: *Flour milled to a very fine consistency is more suitable for pastries than for breads.* **6** ***mill about*** or ***around.*** °meander, °wander, °walk, °stroll, amble, move about *or* around, °crowd, °throng, °swarm: *There were quite a few people milling about our stand at the fair.*

mimic *v.* **1** °imitate, ape, °copy, simulate, °mirror, °echo: *The image in the mirror mimicked every move I made.* **2** °reproduce, °duplicate, °copy: *This frieze mimics one on the Parthenon.* **3** °mock, °ridicule, °satirize, °caricature, °parody, °make fun of, °lampoon, impersonate, *Colloq* °take off: *It is rude to mimic your elders.* *—n.* **4** impersonator, imitator, impressionist, caricaturist, parodist, *Colloq* copycat: *Professor Eidolon, the greatest mimic in the world, is here to entertain you!* *—adj.* **5** imitative, °imitation, °mock, simulated, mimetic, °sham, make-believe, pretend(ed); °fake, °counterfeit, feigned: *Naumachia is the name given to the mimic sea battle staged in ancient Rome.*

mincing *adj.* °effeminate, °dainty, °delicate, niminy-piminy, foppish, dandyish, overdainty, °affected, °put-on, °pretentious, °precious, *Brit* twee, *Colloq* la-di-da *or Colloq* lah-di-dah *or Colloq* la-de-da: *His mincing speech sets my teeth on edge.*

mind *n.* **1** °intelligence, °intellect, °wit, wits, °mentality, °brain, brains, brainpower, °sense, sagacity, °wisdom, °perception, percipience, °reason, astuteness, °insight, shrewdness, sapience, *Colloq* gray matter: *There is nothing wrong with your son's mind, Mr. Fell; he just doesn't want to apply it to school work.* **2** °memory, °recollection; °remembrance: *Keep in mind what I am about to tell you.* **3** °aptitude, °head, °perception, °capacity, °brain: *She has a great mind for dates—for figures of all kinds.* **4** °intellect, °intellectual, °sage, °genius, °thinker, *Colloq* °brain: *There is no doubt that your daughter is one of the great minds of all time.* **5** °intention, °disposition, °temper, temperament, °humor, °fancy, °tendency, °bent, °inclination, °bias, °persuasion: *I was of a mind to let him have his own way.* **6** °opinion, °sentiment, °attitude, °(point of) view, °feeling, °judgment, °belief, °viewpoint, °position: *She has a mind of her own. To my mind, the entire process could be speeded up considerably.* **7** °feeling, °position, °will, °wish, °desire, plan(s): *Won't you change your mind and stay to dinner?* **8** °attention, °thoughts, concentration, °thinking: *Try to keep your mind on your work.* **9** ***bear*** or ***keep in mind.*** °remember, do not forget *or* overlook, °recall, °retain, be aware *or* cognizant *or* mindful of, °consider: *Bear in mind that the post office is closed tomorrow.* **10** ***give someone a piece of one's mind.*** °castigate, °scold, °rebuke, °reprimand, rail at, °reprove, reproach, °chastise, °upbraid, °berate, read (someone) the riot act, *Colloq* °tell off, °dress down, haul *or* rake over the coals, skin alive, *US* °bawl out; *Slang* give someone hell, *US* °chew out: *This time Mrs. Overholt really gave the maître d'hôtel a piece of her mind.* **11** ***know one's (own) mind.*** be decided *or* resolved, be firm *or* resolute, be sure *or* certain *or* positive, be (self-)assured *or* (self-)confident, be in touch with oneself: *He knows his own mind when it comes to his career.* **12** ***make up one's (own) mind.*** °decide, °choose, conclude, form an opinion; °determine, °consider, °weigh, °judge, deem: *Sally hasn't yet made up her mind about marrying Jan. Let her make up her own mind.* **13** ***of*** *or* ***in two minds.*** vacillating, undecided, ambivalent, °uncertain, shilly-shallying, unsure, wavering: *Sam was of two minds about ordering a new computer.* **14** ***out of one's mind.*** °insane, °mad, °crazy: *You're out of your mind if you think I'm swimming in that polluted pond!* —*v.* **15** object to, °resent, take offense at, be offended by, °dislike, be troubled *or* annoyed by, °care, have any objection to, disapprove of, be bothered *or* affronted by: *Do you mind not smoking? Yes, I mind very much. Would you mind moving your car so I can get out?* **16** °heed, °attend to, pay attention to, °obey, listen to, make *or* take note of, °mark, °note: *Please mind what your mother says.* **17** °watch, be careful of, take care with, be cautious of: *Mind your head at that low doorway!* **18** watch over, °take care of, °care for, °look after, °sit with, baby-sit, °guard, keep an eye on *or* out for, have *or* take charge of, °attend: *Suzie is minding the children so that we can get some work done.* **19** ***never mind.*** °ignore, °disregard, °forget, pay no attention to, do not think twice about, do not give a second thought to, erase *or* obliterate *or* cancel from the mind, slough off: *Never mind what that naughty boy calls you!*

minder *n.* **1** (baby) sitter, nanny, °nurse, governess, *Chiefly Brit* child minder: *The minder will put the children to bed.* **2** bodyguard, °escort, °protector: *Arthur hired an ex-fighter as his minder.*

mindful *adj.* Often, **mindful of.** aware, °alert, attentive to, °alive, °conscious, heedful, °conscientious, watchful, °vigilant, on the qui vive, on the lookout, circumspect, °cautious: *I am ever mindful of the way you like things done. She was always mindful of keeping the plants well watered.*

mindless *adj.* **1** °stupid, asinine, °thick, thickheaded, °obtuse, idiotic, imbecilic, moronic, °thoughtless, witless, °senseless, brainless, °feebleminded, fatuous, addlebrained, addlepated, featherbrained, *Colloq Brit* gormless: *Many of the programs on television are mindless twaddle.* **2** °inattentive, °unthinking, °thoughtless, °unaware: *He plunged into the pool, mindless of any of the perils that awaited him.*

mine *n.* **1** °pit, °excavation, lode, °vein; colliery, coalfield: *My father worked in a mine from the age of fourteen.* **2** °source, motherlode, °vein, °store, °storehouse, °supply, °deposit, depository *or* depositary, repository, °reserve, °hoard, treasure trove, reservoir, wellspring; °abundance, °fund, gold mine, °wealth, °treasury: *This thesaurus is a rich mine of information about the English language.* —*v.* **3** °excavate, °dig, °quarry, °extract, scoop out *or* up, °remove, °unearth; °derive, °extract, °draw: *He claims to have mined coal in Wales and gold in South Africa. Analysis shows this ore was mined in Pennsylvania.* **4** °ransack, °search, °rake through, °scour, °scan, °read, °survey, look through, °probe: *Several directories were mined to compile our list of sales prospects.*

mingle *v.* **1** °mix, °blend, intermingle, commingle, intermix, °combine, °amalgamate, °merge, °compound, °marry, °join, °unite: *When these substances are mingled, the result is gunpowder. The loud applause was mingled with a few chuckles and gasps.* **2** °mix, °socialize, °associate, °join, °circulate, °fraternize, °hobnob, consort, °go, spend time, *Colloq* hang about *or* around *or* out, rub shoulders, *Brit* pal up, *US* pal around: *I asked Fred to mingle with the guests and make them feel at home.*

miniature *adj.* °small, small-scale, °little, °tiny, °diminutive, °minute, °wee, minuscule, mini, microscopic, micro, midget, °dwarf, bantam, °baby, pygmy, °pocket, Lilliputian, *US* vest-pocket, *Colloq* °minimal: *This miniature TV fits in the palm of your hand.*

minimal *adj.* least, smallest, minutest, littlest, tiniest, slightest; °minimum, °nominal, °token: *As a house plant, the cactus requires minimal attention. There is a minimal charge for service.*

minimize *v.* **1** °reduce, °shrink, lessen, °diminish, °prune, °abbreviate, °pare (down), °cut (down), °curtail, °abridge, °shorten, °decrease, minify: *Our stringent safety measures minimize the chance of an accident.* **2** °belittle, de-emphasize, downplay, °play down, make little *or* light of, °disparage, decry, deprecate, °depreciate, misprize, devalue, devaluate, undervalue, underrate, °underestimate, *US* °talk down: *Being modest, she minimizes her role in the organization.*

minimum *n.* **1** least, lowest, nadir: *The hotel offered the minimum of comfort for the maximum price.* —*adj.* **2** °minimal, °nominal, reduced, minutest, littlest, least, slightest, lowest: *What is the minimum deposit the travel agent will accept? He refused to make even the minimum effort needed to save himself.*

minister *n.* **1** cleric, °clergyman, clergywoman, ecclesiastic, °pastor, vicar, °priest, °father, reverend, churchman, °divine, parson, °preacher, man *or* woman of the cloth, evangelist, °missionary, dean, curate, curé, abbé, chaplain, *Colloq* padre, *Slang US military* sky pilot, Holy Joe: *The marriage was performed by a visiting minister.* **2** °envoy, °delegate, legate, diplomat, °ambassador, emissary, plenipotentiary, minister plenipotentiary, envoy extraordinary, minister resident, consul, °agent, chargé d'affaires, *Brit* cabinet officer *or Brit* member: *The minister himself signed the treaty.* —*v.* **3** Usually, ***minister to.*** attend (to *or* on *or* upon), °wait on, °care for, °look after, °see to, °accommodate; °serve, °supply, °aid, °help, °assist, °support: *In his last days, she ministered to him in every way.*

ministry *n.* **1** priesthood, sacred calling, the church, the pulpit, °the cloth; °religion, holy orders: *Burt retired from the ministry to write tracts.* **2** clergy, clergymen *or* clergywomen, clericals, °the cloth, church elders *or* elders of the church: *During his regime, the ministry exercised great influence on the government.* **3** °department, °office, °bureau, °agency: *Military affairs are the responsibility of the Department of Defense.*

minor *adj.* **1** lesser, smaller, °secondary, °subordinate, °subsidiary: *The minor planets are bodies orbiting the sun in the asteroid belt.* **2** °insignificant, °obscure, °inconsequential, unimportant, °trifling, trivial, °negligible, inconsiderable, °slight, °petty, °paltry, °small, *Colloq* °small-time, one-horse, *Brit* two-a-penny, *US* minor-league, bush-league, penny-ante, two-bit, picayune: *It is a minor matter to me whether you stay or go. She is only a minor player in this game.* —*n.* **3** °child, youngster, °youth, °stripling, °teenager, °adolescent, schoolboy, schoolgirl, °boy, °girl, °lad, laddie, °lass, lassie, *Law* °ward, infant: *At your age, you are still a minor in the eyes of the law.*

minstrel *n.* bard, troubadour, balladeer, jongleur, skald *or* scald, minnesinger, Meistersinger: *The medieval minstrels wandered the countryside, entertaining as they went.*

mint *n.* **1** °(small) fortune, °lot, king's ransom, millions, billions, *Colloq* °bundle, °pile, °heap, wad(s), °packet, pot(s), loads, ton, *Slang Brit* °bomb, *US* (big) bucks: *He lost a mint at roulette last night. She made a mint on her inventions.* —*v.* **2** °make, °coin, °produce, °earn: *Special coins were minted to commemorate the Coronation.*

minute[1] *n.* **1** °instant, °second, split second, °flash, °moment, before one can *or* could say "Jack Robinson," before you can say "knife," blink *or* wink *or* twinkling of an eye, *coup d'oeil,* trice, *Colloq* one sec, two secs, bat of an eye, °shake, jiffy, tick, *Brit* half a mo, two shakes (of a lamb's tail): *I'll be with you in a minute.* **2** ***minutes.*** log, °record, °journal, °transcript, °notes, °summary, °résumé, °proceedings, °transactions, °memorandum: *Have you the minutes of the last meeting?* **3** ***up-to-the-minute.*** °latest, newest, °modern, up-to-date, °trendy, °fashionable, °smart, all the rage, in vogue, °stylish, °in style, °in fashion, à la mode, *Colloq* in, with it, hep *or* hip, °hot, °cool, *US* °now: *She always wears the most up-to-the-minute clothes.* —*v.* **4** °record, °transcribe, °take down, °write down, °note, make (a) note of, °document, log: *We need someone to minute the meeting.*

minute[2] *adj.* **1** °small, °little, °tiny, tiniest, minuscule, °miniature, °wee, infinitesimal, microscopic, micro, °diminutive, mini, °baby, pint-sized, bantam, Lilliputian, *Colloq* teeny, teensy(-weensy), itty-bitty, eensy-weensy, itsy-bitsy: *The minute furniture in the doll's house was made entirely by hand.* **2** unimportant, °petty, °insignificant, least, °slight, °mere, °meager, °trifling, trivial, °minor, °small, °little, *Colloq* piddling, *US* picayune: *The painting was a perfect copy, down to the minutest detail.*

miraculous *adj.* °marvelous, wonderful, wondrous, °incredible, °unbelievable, °inexplicable, unexplainable, °extraordinary, spectacular, °amazing, astounding, astonishing, mind-boggling, °remarkable, °phenomenal, °fantastic, °fabulous; magical, °supernatural, preternatural, °superhuman, *Colloq* °out of this world, *Slang* far-out, °crazy: *Langley has made a miraculous recovery from malaria.*

mire *n.* **1** °swamp, °bog, fen, °marsh, quagmire, °morass, slough, *Brit dialect* sump: *The horse stumbled into the mire and slowly sank to its death.* **2** °mud, °ooze, °muck, slime, °dirt: *I had to wade through muck and mire to get to your front door.* —*v.* **3** enmire, °bog down, become entangled *or* tangled, become enmeshed *or* meshed, become involved: *Sorry I'm late; I got mired in a problem at the office.* **4** °dirty, °soil, begrime, °muddy, befoul, besmirch, °sully, °tarnish, °smear, °blacken, defile, smudge: *My boots were badly mired. The scandal badly mired his reputation.*

mirror *n.* **1** looking glass, °glass, speculum, reflector: *The face that looked back at me from the mirror seemed not to be my own.* **2** °reflection, °reproduction, °picture, °representation, replication, °(mirror) image: *Language is the mirror of society, reflecting every social change.* —*v.* **3** °reflect, °reproduce, °represent, depict, °repeat, °echo, send back: *The calm lake mirrored the moonlight. The people's misery was mirrored in their faces.*

mirth *n.* °merriment, merrymaking, jollity, °gaiety, °fun, °laughter, °amusement, °frolic, frolicking, joviality, joyousness, °revelry, rejoicing, °glee, high spirits, mirthfulness, °hilarity, buoyancy, *Formal* jocundity: *Life without mirth is a lamp without oil. Christmas is a time for mirth.*

misalliance *n.* mésalliance, mismarriage, mismatch, mismatchment, bad match, mismating: *No one is surprised that John and Marsha's misalliance has ended in divorce.*

misanthrope *n.* misanthropist, mankind-hater; °man-hater, woman-hater, misogynist; loner, °hermit, °recluse, anchorite *or* anchoret, *Colloq* lone wolf: *The longer he lived and the more he learned of his fellow man, the more a misanthrope he became.*

misanthropic *adj.* man-hating; antisocial, °unsocial, unfriendly, egocentric, °egoistic: *In his misanthropic moods, Joel disdains human company.*

misappropriate *v.* **1** °embezzle, °steal, filch, expropriate, °pocket, *Formal* peculate, defalcate: *He was found guilty of misappropriating club funds.* **2** misapply, °misuse, °pervert, misemploy: *The money intended for the health service was misappropriated for the war effort.*

misbehave *v.* °disobey, behave badly *or* improperly, be bad *or* naughty *or* mischievous, *Colloq* °carry on, act up, *Slang* raise hell, raise Cain: *The teacher sent for me because Robyn misbehaved in class.*

misbehavior *n.* naughtiness, badness, misconduct, misdemeanor(s), disorderliness, disobedience, delinquency, disorderly conduct, °rowdyism: *Wendy has been reprimanded again for misbehavior in class.*

miscalculate *v.* misjudge, °err, misevaluate, misestimate, misreckon, miscompute, miscount, misappreciate, misread; °underestimate, undervalue, underrate; overestimate, overvalue, °overrate: *She miscalculated the effect of her resignation.*

miscarriage *n.* °failure, abortion, °collapse, °breakdown, °failing, mismanagement, nonfulfillment, °defeat, nonsuccess, frustration: *Her acquittal was a miscarriage of justice.*

miscarry *v.* abort, °fail, °fall through, °break down, °go wrong, °founder, come to nothing *or* naught *or* nought, go awry, °come to grief, go amiss, °misfire, go up *or* end up in smoke, °perish, °die: *The plan to steal the crown jewels miscarried.*

miscellaneous *adj.* °varied, heterogeneous, °diverse, °mixed, diversified, °divers, motley, °sundry, assorted, °various, varying, multifarious, multiform, many-sided, multiplex, °manifold: *Today's auction includes miscellaneous items.*

miscellany *n.* °mixture, °assortment, °variety, °medley, °diversity, mixed bag, job lot, ragbag, mélange, °potpourri, gallimaufry, motley, hotchpotch *or US also* hodgepodge, salmagundi, olio, olla podrida, smorgasbord *or* smörgåsbord, °odds and ends, omnium-gatherum, °hash, °mess, *Brit* lucky dip, °jumble, *US* grab bag: *A miscellany of household goods was contributed for the rummage sale.*

mischief *n.* **1** °misbehavior, naughtiness, impishness, elfishness *or* elvishness, roguishness, rascality, devilry *or* deviltry, mischievousness, playfulness, devilment, badness, *Colloq* monkey business, shenanigans, *Brit* monkey tricks *or US* monkeyshines: *That boy is always up to some mischief.* **2** °harm, °injury, °damage, °detriment, °trouble, °hurt, °wrong, °difficulty, disruption, °destruction, °misfortune, °evil: *He didn't realize the mischief a casual remark could do. She was accused of malicious mischief for puncturing his car tires.*

mischievous *adj.* **1** °naughty, impish, roguish, rascally, °devilish, elfish *or* elvish, puckish, scampish, °frolicsome, °playful, °sportive: *Tying his brother's shoelaces together was a mischievous prank and not malicious.* **2** °harmful, °injurious, °hurtful, damaging, pernicious, °detrimental, °destructive, deleterious, °dangerous, °spiteful, malicious, °vicious, malign, baleful, baneful, noxious, °wicked, °evil, °bad: *The government regards as extremely mischievous the holding captive of any Foreign Office personnel.*

misconceive *v.* °misunderstand, misconstrue, misjudge, °mistake, misapprehend, °misinterpret, misread, get *or* have the wrong idea, get *or* have (hold of) the wrong end of the stick: *He has misconceived the purpose in banning firearms.*

misconception *n.* false *or* wrong notion *or* idea, °misunderstanding, misconstruction, misconstrual, misjudgment, miscalculation, misapprehension, mistaken belief, °error, °mistake, °delusion: *The confused wording of the warranty could result in a misconception of one's rights.*

miscreant *n.* **1** °villain, °wretch, mischief-maker, scamp, °rascal, °criminal, evildoer, °felon, malefactor, °rogue, °reprobate, °scoundrel, wrongdoer, °good-for-nothing, ne'er-do-well, blackguard, hooligan, ruffian, °hoodlum, °thug, °rowdy, *Archaic* knave, caitiff, varlet, rapscallion, *Colloq* crook, roughneck, scallywag *or* scalawag *or* scalawag, *Slang* hood, *Brit* °mug, °rough, *Australian* larrikin, *US* baddie *or* baddy, bad actor: *The police have apprehended the miscreant who had been attacking passersby.* —*adj.* **2** °villainous, °wretched, °mischievous, rascally, °criminal, felonious, °corrupt, malefic, malevolent, °evil, depraved, °base, nefarious, iniquitous, °vicious, unprincipled, ne'er-do-well, °reprobate, scoundrelly, °wicked: *The miscreant terrorists may be hiding in Greece.*

misdeed *n.* Often, **misdeeds.** offense, °crime, felony, wrongdoing, misdoing, °transgression, misdemeanor, °fault, misconduct, °sin, trespass, °wrong, °peccadillo: *Who will punish him for his misdeeds now that he is safe in Argentina?*

misdirect *v.* misguide, misadvise; misaddress: *He misdirected us by saying the lavatory was downstairs. The letter, misdirected, was returned.*

miser *n.* skinflint, hoarder, niggard, pennypincher, pinchpenny, cheeseparer, Scrooge, *Colloq* cheapskate, *US* tightwad: *For that old miser, charity begins and ends at home.*

miserable *adj.* **1** °wretched, °unhappy, depressed, woeful, °woebegone, °sad, °dejected, °forlorn, disconsolate, °despondent, °heartbroken, °sorrowful, °brokenhearted, °mournful, °desolate, °desperate, despairing, °downhearted, °melancholy, °glum, low-spirited, °gloomy, °dismal, lachrymose, °tearful, *Colloq Brit* °cut up: *Tanya is miserable because no one invited her to go to the dance.* **2** unpleasant, °inclement, °inconvenient, °untoward, °bad, unfavorable, °awful, °terrible, adverse, *Colloq* °rotten, °lousy: *The weather was miserable throughout the weekend.* **3** °inadequate, °unworthy, °poor, °deplorable, °contemptible, °bad, °despicable, °sorry, °pitiful, °pathetic, °lamentable, *Colloq* °rotten, °lousy: *That was the most miserable play I have ever had the misfortune to sit through.* **4** squalid, °wretched, °bad, abject, °deplorable, °shabby, °mean, °vile, °shameful, °scurvy, °awful, °disgraceful, °contemptible: *He lives in a cellar in the most miserable conditions.*

miserly *adj.* stingy, °penurious, niggardly, pennypinching, parsimonious, °mean, °cheap, cheeseparing, °tight, tightfisted, °close, closefisted, °mercenary, °avaricious, °greedy, covetous, *Colloq* moneygrubbing, *Brit* mingy, *US* chintzy: *Flynt is too miserly to lend you a penny.*

misery *n.* **1** unhappiness, °distress, °discomfort, wretchedness, °woe, °sadness, °melancholy, °sorrow, dolor, heartache, °grief, °anguish, °anxiety, angst, °depression, °despair, °desperation, °desolation, despondency, °gloom: *Worn out by grief and misery, she collapsed into a chair.* **2** squalor, °poverty, destitution, °privation, indigence, penury, wretchedness, sordidness: *The misery of those slums is unbelievable.* **3** °hardship, °suffering, °calamity, °disaster, °curse, °misfortune, °ordeal, °woe, °trouble, °catastrophe, °trial, tribulation, adversity, °burden, °affliction: *She cannot bear the twin miseries of illness and poverty.* **4** °spoilsport, damper, °killjoy, dampener, Job's comforter, grouch, grump, malcontent, pessimist, cynic, prophet of doom, Cassandra, *Colloq* wet blanket, sourpuss, *US* party pooper, gloomy Gus, picklepuss: *How can you stand living with such a misery?*

misfire *v.* **1** °fail, abort, °miscarry, °go wrong, °fizzle (out), °fall through, *Colloq* °flop, come a cropper, *Brit* go phut, *US* go pfft *or US* phft: *The plan to take over the conglomerate misfired when the shares soared in price.* —*n.* **2** °miscarriage, °failure, °fizzle, °dud, *Colloq* abort, °flop: *After two misfires, we gave up trying to buy a timeshare in the Algarve.*

misfit *n.* oner, °eccentric, °individual, °nonconformist, maverick, square peg in a round hole: *People regard him as one of society's misfits and tend to steer clear.*

misfortune *n.* **1** bad luck, ill luck, ill fortune, hard luck, infelicity, adversity, °loss: *He had the misfortune to marry the wrong woman.* **2** °accident, misadventure, °mishap, °calamity, °catastrophe, mischance, °disaster, contretemps, °tragedy, °blow, °shock; °reverse, stroke of bad luck, *Colloq* bad news: *With the death of his father, then his illness, then the loss of his business, one misfortune followed another.*

misgiving *n.* apprehension, °mistrust, °worry, °concern, °anxiety, °qualm, °scruple, disquiet, hesitation, °doubt, °question, uncertainty, °suspicion, unease, uneasiness, °discomfort; °dread, °premonition, °foreboding, *Colloq* funny feeling: *I had grave misgivings about driving all that way on my own.*

misguided *adj.* misled, °wrong, misdirected, °foolish, °unreasonable, °erroneous, °mistaken, misplaced, °imprudent, unwise, impolitic, °ill-advised, fallacious, uncalled-for, laboring under a misapprehension, °wide of the mark, *Colloq* °off (the mark), barking up the wrong tree: *Her helpfulness is often misguided, ending in disaster for all concerned.*

mishandle *v.* **1** °abuse, °mistreat, maltreat, °ill-treat, °beat (up), brutalize, maul, °molest, °injure, °hurt, °harm, handle *or* treat roughly, °manhandle, *Colloq* knock about *or Colloq* around: *The poor child was mishandled by her parents.* **2** mismanage, °bungle, °botch, misconduct, °mangle, °mess up, °muddle, °wreck, °ruin, °destroy, *Colloq* muff, make a mess *or* hash of, *Slang* °screw up, °bugger up, *Taboo* fuck up: *Gerber mishandled the deal so badly that it cost the company millions.*

mishap *n.* See **misfortune, 2,** above.

mishmash *n.* °mess, °medley, °hash, gallimaufry, farrago, °potpourri, °jumble, °pastiche, °mixture, salmagundi, hotchpotch *or US also* hodgepodge, °tangle, omnium-gatherum, mélange, olio, olla podrida, goulash, °stew: *His paper is nothing more than a mishmash of other people's ideas.*

misinform *v.* misguide, °mislead, misadvise, °misdirect, delude, °deceive, °dupe, °defraud, °fool, gull, °lead astray, throw someone off the scent, *Colloq* con, slip *or* put something over on someone, pull a fast one on someone, *US* give (someone) a bum steer, throw someone a curve: *If you think you are entitled to compensation, you've been misinformed.*

misinformation *n.* disinformation, misintelligence; red herring, false trail, false scent: *When we uncover an enemy agent in our midst, we allow him to collect misinformation.*

misinterpret *v.* °misunderstand, °mistake, misconstrue, °misconceive, misread, misjudge, misapprehend, *Slang* °screw up, °bugger up: *He misinterpreted what I said and stepped off the ledge.*

mislay *v.* misplace, °lose, mislocate; misfile: *I have mislaid my reading spectacles.*

mislead *v.* °misinform, °lead astray, misguide, °misdirect, throw off the scent *or* track, pull the wool over someone's eyes, °fool, °outwit, °bluff, °hoodwink, °trick, humbug, bamboozle, °deceive, °dupe, gull, cozen, *Colloq* con, °take in, lead up the garden path, flimflam, slip *or* put *or* pass one over on, *Slang* °take, *US* give someone a bum steer: *I was misled into believing that she loved only me. He is misleading you if he says that the shares will increase in value.*

mismatched *adj.* mismated, ill-matched, ill-mated, °incompatible, unfit, °inappropriate, unsuited, unsuitable, °incongruous, misallied, disparate, uncongenial, °inconsistent, inharmonious, °discordant: *With her interest in art and his in football, they made a mismatched pair.*

misprint *n.* °error, °mistake, erratum, typographical error, printer's *or* printing error, *Brit* °literal, *Colloq* typo: *We cannot publish a book containing so many misprints.*

misrepresent *v.* °distort, °twist, °pervert, °garble, misstate, °mangle, °falsify, belie, °disguise, °color: *In his report, he misrepresented everything I said.*

miss[1] *v.* **1** °skip, °forgo, °absent oneself from, be absent from, fail to keep; °avoid, °evade, °escape, °dodge, *Colloq* °pass up: *I missed my dentist's appointment. I missed school because my grandmother was ill.* **2** long for, yearn for, pine for, feel nostalgia for, be nostalgic for *or* about, °want, °need, wish for: *I miss those long summer evenings by the lake.* **3** °misunderstand, °misinterpret, misconstrue, misapprehend, fail to understand *or* perceive, °mistake: *You have completely missed the point.* **4** Sometimes, ***miss out (on).*** °pass up, °omit, °leave out, °slip up (on), °overlook, °let slip (by), let pass, °pass over, °disregard, °ignore: *You can be sure that Frobisher would never miss a free meal.* —*n.* **5** °omission, °oversight, °slip, °failure, °mistake, °error, °blunder, *Colloq* slip-up: *A miss is as good as a mile.*

miss[2] *n.* Ms., °girl, °lass, lassie, °maid, °maiden, young lady, young woman, schoolgirl, mademoiselle, nymphet, °teenager, virgin, spinster, °old maid, *Brit* bachelor girl, *Irish English* colleen, *Literary* nymph, *Archaic* demoiselle, damsel, *Colloq old-fashioned* gal, *US* coed, Valley girl, bachlorette, *Slang* teeny-bopper, groupie, *Chiefly Brit* bird, *Old-fashioned US* bobby-soxer: *She's a pretty young miss, isn't she?*

misshapen *adj.* distorted, twisted, contorted, °crooked, °deformed, °crippled, malformed, °grotesque, awry, warped, °gnarled, ill-proportioned, ill-made, °monstrous, *Technical* acromegalic: *The misshapen bell-ringer in* The Hunchback of Notre Dame *was named Quasimodo.*

missile *n.* °projectile, brickbat; guided missile, ballistic missile: *Bottles, stones, and other missiles were thrown at the police during the riot.*

mission *n.* **1** °task, °duty, °function, °purpose, °job, °office, °work, °assignment, °errand, °charge, °business, commission, °undertaking, °pursuit, °activity, °aim, °objective: *He accepted the mission to convert the natives to Christianity.* **2** °calling, °occupation, °vocation, °trade, °line (of work), °profession, métier: *It has become my mission in life to see that you learn how to do sums.* **3** delegation, legation, deputation, commission, °committee, °group, °ministry: *She has been sent on a governmental trade mission to Botswana.*

missionary *n.* evangelist, °preacher, °minister, proselytizer: *During the nineteenth century, many missionaries were dispatched to spread the faith in Africa.*

missive *n.* °letter, communication, °message, °dispatch, °note, °line, postcard, °card, epistle: *I just received a missive from an old flame that she will be visiting here next week.*

misspent *adj.* wasted, squandered, °idle, dissipated, thrown away, profitless, °prodigal: *Proficiency at billiards is a sign of a misspent youth.*

misstatement *n.* **1** falsification, misreport, misquotation, miscitation, distortion, misrepresentation, misconstruction, misinterpretation, °perversion, °lie, °falsehood, untruth, °fabrication: *The prosecutor's indictment is a misstatement of the charges against the defendant.* **2** °solecism, °error, °mistake, gaffe, faux pas, °slip of the tongue, *lapsus linguae,* °blunder, *Slang* °howler, *Brit* bloomer, *US and Canadian* blooper: *For an educated man his speech is sprinkled with too many grammatical misstatements.*

misstep *n.* **1** false step, °blunder, °mistake, °error, bad *or* wrong *or* false move, °trip, °stumble, °slip: *One misstep meant a fall into the abyss.* **2** °indiscretion, °mistake, °lapse, faux pas, °oversight, °error, gaffe, *Colloq* slip-up, *Slang* °howler, *Brit* bloomer, *US and Canadian* blooper, goof: *Her misstep was telling a journalist that she had been an artist's model.*

mist *n.* **1** °fog, haze, smog, (low-hanging) cloud, °vapor; drizzle, *Brit dialect* mizzle: *I could scarcely make out the road in the mist.* —*v.* **2** Usually, ***mist up*** *or* ***over.*** cloud (up *or* over), becloud, °fog, °befog, °dim, °blur, °film, steam up: *The mirror in the bathroom was all misted over after my shower.*

mistake *n.* **1** °misconception, misapprehension, °error, °fault, miscalculation, misjudgment, °blunder, °botch, °fumble, bad move, °misstep, °slip, erratum, gaffe, faux pas, *Colloq* boo-boo, clanger, muff, °howler, *Brit* boob, bloomer, *US* blooper, goof, goof-up, flub: *Giving Aston a raise was a mistake.* **2** °indiscretion, °misstep, false step, wrong move: *It was a mistake not to stand up when she entered.* —*v.* **3** °misunderstand, °misinterpret, misjudge, misconstrue, take the wrong way, get wrong, misread, misapprehend: *The fool mistook my remark as a compliment.* **4** ***mistake for.*** °mix up with, misidentify as, confuse with, take for: *In the dark, she mistook him for her husband.*

mistaken *adj.* **1** (all *or* completely) wrong, °amiss, °incorrect, °in error, °wide of the mark, in the wrong, °inaccurate, °out of order, *Colloq* barking up the wrong tree, °off, (way) off the beam, on the wrong track, *US* full of hot air, *Slang* full of it (= *Taboo* 'full of shit'), *US* all wet: *I may be mistaken, but I thought I saw your wife with Lord Cholmondley.* **2** °erroneous, °faulty, °false, fallacious, misinformed, °incorrect, °wrong, °inaccurate, °flawed, warped, distorted, twisted, °misguided, *Slang* cockeyed: *You have a mistaken impression of me.*

mistreat *v.* °abuse, maltreat, ill-use, °ill-treat, °misuse, °damage, °manhandle, °harm; °hurt, °injure, °molest, maul, °rough up, brutalize: *The book has no value because you mistreated it. She was mistreated by her husband for years.*

mistreatment *n.* °abuse, maltreatment, ill use, ill treatment, brutalization, °misuse; manhandling, molestation, mauling, roughing-up, rough handling, battery, °assault: *This mistreatment of employees must stop at once. The children complained of their parents' mistreatment.*

mistress *n.* **1** °lover, girlfriend, live-in lover, kept woman, concubine, inamorata, °paramour, *Literary* odalisque, *Archaic* doxy, *Colloq US* alternative other,

POSSLQ (Person of the Opposite Sex Sharing Living Quarters): *How can Nigel afford a mistress on his civil service salary?* **2** schoolmistress, instructress, governess; headmistress: *She had a rewarding career as mistress of a girls' school in Lancaster.*

mistrust *v.* **1** °suspect, °distrust, be suspicious of, °doubt, be *or* feel wary *or* suspicious *or* doubtful of *or* about, have (one's) doubts about, °question, have reservations; °beware; *Slang* be *or Slang* feel leery of *or Slang* about: *He has not done anything specifically bad, but I mistrust his motives.* —*n.* **2** °suspicion, °distrust, °doubt, °skepticism, wariness, °reservation, chariness, misgiving(s), uncertainty, unsureness, apprehension, apprehensiveness: *Is there nothing I can say to dispel your mistrust of me?*

misty *adj.* cloudy, foggy, °hazy, °murky; °fuzzy, °dim, blurred, blurry, unclear, °indistinct, °vague, °dark, °opaque, °shadowy, °obscure, unintelligible: *It is very misty, so drive carefully. I have only the mistiest idea of what you mean.*

misunderstand *v.* °misconceive, misconstrue, °misinterpret, misapprehend, get (it (all)) wrong, get the wrong idea (about), misread, misjudge, °miscalculate, miss the point (of): *He understood your words but misunderstood your meaning.*

misunderstanding *n.* **1** °misconception, misconstruction, misinterpretation, misapprehension, misreading, misjudgment, miscalculation, wrong idea, wrong *or* false impression, *malentendu,* mistaking, *Technical* parasynesis: *Misunderstanding of the law is no excuse for flouting it.* **2** °disagreement, °discord, °dispute, °argument, °difference, °dissension, °controversy, °quarrel, °rift, *Colloq* falling-out: *We had a slight misunderstanding about who was to pay the restaurant bill.*

misuse *n.* **1** misapplication, misusage, misappropriation, misemployment, diverting, °diversion, perverting, °perversion: *He was accused of the misuse of public funds for his personal gain.* **2** misusage, °abuse, corruption, °solecism, malapropism, barbarism, catachresis, ungrammaticality, infelicity: *His misuses of English have made him a laughingstock.* **3** See **mistreatment,** above. —*v.* **4** °abuse, misapply, misemploy, °misappropriate; °pervert: *I broke the knife blade misusing it as a screwdriver. He was accused of misusing the courts to his own ends.* **5** See **mistreat,** above.

mitigate *v.* °moderate, °temper, °reduce, abate, lessen, °decrease, °relieve, °ease, ease up (on), °relax, alleviate, °remit, assuage, allay, °let up (on), slacken, °slacken up (on), °tone down, °lighten, appease, palliate, mollify, °calm, °tranquilize, soothe, placate, quiet *or chiefly Brit* quieten, °still, °soften, °dull, °blunt, take the edge off, *US* lighten up (on): *The pain was mitigated by taking aspirin. The relief organizations did their best to mitigate the effects of the disaster.*

mitigating *adj.* °extenuating, justifying, excusatory, palliating, vindicating, qualifying: *Were there mitigating circumstances in her taking the money?*

mix *v.* **1** °mingle, °combine, intermingle, °blend, °incorporate, put together, °merge, °unite, °alloy, commingle, °amalgamate, coalesce: *Mix the ingredients to a creamy consistency.* **2** °socialize, °fraternize, consort, °hobnob, go round *or* around *or* about (together), °get together, keep company, °join (with), °associate (with), *Colloq* hang out *or Colloq* about *or Colloq* around (with): *The Hatfields and the McCoys do not mix very well, so you'd best invite them separately.* **3** ***mix in.*** °add, stir in, °combine, fold in: *Mix in the chocolate at the very end.* **4** ***mix up.*** **(a)** See **mix, 1,** above. **(b)** °confuse, confound, °bewilder, °muddle, °perplex, °puzzle, °fluster, °upset; addle, °disturb; *Colloq US and Canadian* discombobulate: *She mixed me up so that I cannot decide what to do.* **(c)** °snarl, ensnarl, °tangle, °entangle, °scramble, °jumble: *The treasures were all mixed up with the rubbish at the church sale.* **(d)** °confuse, interchange, °exchange: *Have they mixed up our room keys again?* **5** ***mix up in.*** °involve, °implicate, °include, °connect, draw *or* drag into: *Wasn't Wilcox the one who was mixed up in that scandal last year?* **6** ***mix up with.*** °confuse, °mistake, misidentify, confound; interchange, °exchange: *I think you must have me mixed up with somebody else. The porter mixed up my bag with someone else's.* —*n.* **7** °mixture, °blend, °compound; °amalgam, °combination, °alloy, °assortment, °distribution: *Was this cake made from a mix or from scratch? The center party has a good mix of liberals and conservatives.*

mixed *adj.* **1** °hybrid, halfbred, °mongrel, interbred, crossbred; °impure, tainted, adulterated: *This horse has mixed blood.* **2** °confused, muddled; °varied, °various, °diverse; conflicting, °contradictory, °opposing, clashing, °opposite: *He had mixed feelings about ending his relationship with her.* **3** ***mixed up in*** *or* ***with.*** °involved, connected, associated: *Wasn't she mixed up in that scandal last year? He doesn't want his children mixed up with people who take drugs.*

mixture *n.* **1** °assortment, °amalgam, °amalgamation, °medley, °combination, mingling, intermingling, composite, °blend, °jumble, °mix, °miscellany, mélange, °mess, °mishmash, hotchpotch *or US also* hodgepodge, gallimaufry, farrago, olio, olla podrida, °hash, °potpourri, ragout, goulash, omnium-gatherum, salmagundi: *The design is an unfortunate mixture of many incompatible styles.* **2** mixing, °amalgamation, amalgamating, combining, mingling, intermingling, °combination, °blend, blending, °association, associating, °compound, compounding, °synthesis, interweaving, merging, °merger, fusion, fusing, °alloy, alloying: *The mixture of household ammonia and bleach creates noxious fumes.*

mix-up *n.* °confusion, °mess, °muddle, hotchpotch *or US also* hodgepodge, °tangle, °jumble, *Colloq* °botch, °mishmash, foul-up, *Slang US* screw-up, snafu, *Taboo slang Brit* balls-up: *The mix-up in the files is owing to a computer malfunction.*

moan *n.* **1** °complaint, °lament, °lamentation, °groan, wail, moaning, °grievance: *Disregard his moans about money—he has plenty.* —*v.* **2** °complain, °lament, °groan, wail, °bewail, grumble, °bemoan, deplore, whine, whimper, *Colloq* grouse, °gripe, beef, °bitch, *Brit* whinge: *She's always moaning about the way people take advantage of her.* **3** °sigh, °mourn, °weep, °sorrow, °cry, wail, °keen, °grieve, °sob, °snivel, °bawl, mewl, pule, ululate: *The women, all in black, moaned and tore their hair when they neared the coffin.*

mob *n.* **1** horde, °host, °press, °throng, °crowd, °pack, °herd, °swarm, °crush, °jam, multitude, °mass, °body, assemblage, °collection, °group: *The mob surged forward, hurling sticks and stones at anyone in uniform.* **2** °rabble, °riffraff, proletariat, °populace, °the masses, great unwashed, °hoi polloi, *canaille,* bourgeoisie, lower classes, scum (of the earth), dregs of society: *The mob cheered when the king was beheaded.* —*v.* **3** °crowd (around), jostle, °throng, °surround, °beset, clamor over, swoop down on: *Thousands of screaming teenagers mobbed the singer.*

mobbed *adj.* crowded, °packed, thronged, °congested, teeming, swarming, °full, filled: *I tried to leave by the emergency exits, but they were mobbed, too.*

mobile *adj.* **1** °movable, nonstationary, unstationary, unfixed, °traveling, °portable, transportable: *A mobile* *display was designed for industrial fairs.* **2** motorized, °mechanical, transportable, °movable: *They live in a mobile home. The mobile library comes around every Wednesday.* **3** °expressive, °sensitive, °animated, °plastic, °flexible, *US* facile: *Her mobile face at once reflected her thoughts.* **4** °agile, °versatile, °nimble, °quick, °alert, °active, °responsive: *You need your wits about you to keep up with Randolph's mobile mind.* **5** ambulatory, ambulant: *After the artificial joint surgery, Keith was again completely mobile.*

mobilize *v.* °assemble, marshal, conscript, °enroll, °enlist, °organize, °muster, levy, °rally, °activate, °call up, °prepare, °ready, *US* °draft: *A huge army was mobilized, virtually overnight.*

mock *v.* **1** °deride, °ridicule, °make fun of, °tease, °taunt, °tantalize, °jeer (at), gibe *or* jibe (at), °thumb one's nose at, °chaff, °laugh at, °poke fun at, °make sport of, °guy, °scorn, °flout, °abuse, °defy, °scoff (at), °sneer (at), disdain, °disparage, decry, *Archaic* fleer (at), *Colloq* °rag, rib, kid, put (someone) on, *Brit* take the mickey out of, cock a snook at: *He was taken in for questioning only because he mocked a police officer. Jealousy is the green-eyed monster that mocks the meat it feeds on.* **2** ape, °mimic, °imitate, °caricature, °lampoon, °satirize, °parody, °burlesque, travesty, *Colloq* spoof, °take off, *Brit* °send up: *The political cartoonist mocks governments daily on the pillory of his pen.* —*adj.* **3** °substitute, °artificial, simulated, °fake, °synthetic, °imitation, °false, forged, ersatz, °sham, feigned, °counterfeit, °fraudulent, °bogus, make-believe, °pretend, *Colloq* phoney *or US also* phony, pseudo: *Mock turtle soup, though good, does not taste like real turtle soup.*

mockery *n.* **1** °ridicule, °derision, disdain, taunting, disparagement, °abuse, °scorn, °contempt, contumely, decrial: *Vladimir was expelled for his mockery of the headmaster.* **2** °semblance, °imitation, impersonation; °caricature, °parody, °burlesque, travesty, °lampoon, °satire, pasquinade, farce; °miscarriage; *Colloq* spoof, °takeoff, *Brit* sendup: *That wasn't a trial, it was a mockery of justice.* **3** °disappointment, °joke, °laugh, °absurdity: *It is a mockery to speak of that sadist and humanity in the same breath.*

mode[1] *n.* **1** °way, °manner, °method, °approach, °form, °course, °fashion, °procedure, °technique, °system, °wise, modus operandi, methodology, standard operating procedure, SOP: *What mode of teaching foreign languages do you use?* **2** °status, °condition, °state, configuration, °setup: *The computer is now in printing mode.*

mode[2] *n.* °fashion, °style, °look, °vogue; °trend, °rage, °craze, *Colloq* °fad: *We went to the ball dressed in the mode of the 1920s.*

model *n.* **1** °representation, °replica, mock-up, maquette, scale model, working model, °miniature, °dummy, °image, °likeness, °facsimile, °copy: *This is a model of the house he is planning.* **2** °original, °mold, archetype, °prototype, °pattern, °paragon, °ideal, exemplar, °example, °standard: *When the wax model is finished, it is cast in bronze using the* cire perdue *process. Their government was established after the Roman model.* **3** °ideal, °paragon, exemplar, °epitome, *beau idéal,* cream, *crème de la crème, ne plus ultra,* °nonpareil, nonesuch *or* nonsuch: *"He is the very model of a modern major general."* **4** °subject, sitter, poser: *Dali's wife served as his model for many paintings.* **5** mannequin; °dummy: *She is tall and thin enough to be a high-fashion model. May I see the hat that is on that model in the window?* **6** °design, °kind, °type, °style, °version; °variety, °sort, °form, °fashion, configuration; °brand, °mark: *My next car will be a convertible model.* —*v.* **7** °fashion, °mold, °shape, °form, sculpt, °carve (out), °make, °fabricate, °produce: *She modeled these figures out of fine clay.* **8** pose in, °display, °show (off), °wear, *Colloq* °sport: *She makes a fortune modeling swimwear.* **9** ***model after*** *or* ***on.*** °imitate, °copy, pattern on *or* after, emulate, °follow: *Architects often model public buildings after the temples on the Acropolis.* —*adj.* **10** °copy, °imitation, °facsimile, °representative, °miniature: *He enjoys building model airplanes.* **11** °ideal, °exemplary, °perfect, archetypal, unequaled, consummate, inimitable: *Wortley has been a model student.*

moderate *adj.* **1** °temperate, °calm, °reasonable, °cool, °judicious, °rational, balanced, unexcessive, °modest, °sober, °sensible, common-sensical, controlled, °deliberate, °steady: *In view of the conflicting interests expressed, the council has adopted a moderate policy regarding parking.* **2** °center, middle-of-the-road, nonradical, nonreactionary: *Bella's political outlook is said by some to be moderate.* **3** °fair, middling, °average, °ordinary, °medium, °middle, °modest, °mediocre, unexceptional, *Colloq* fair to middling: *They make clothing of moderate quality and sell it at a moderate price.* —*n.* **4** middle-of-the-roader, nonradical, nonreactionary, centrist: *Where do the moderates stand on the issue of free speech?* —*v.* **5** abate, °calm, mollify, soothe, °ease, °relax, alleviate, °mitigate, °soften, °dull, °blunt, °cushion, °relieve, °reduce, lessen, °remit, slacken, °diminish, °decrease, defuse, °temper, *Colloq* °let up (on): *He would get along better if he moderated the virulence of his reactions.* **6** mediate, arbitrate, referee, °judge, °chair, °supervise, °preside (over), °coordinate, °run, °regulate, °manage, °direct: *Professor Girning will moderate the panel discussion.*

moderately *adv.* °somewhat, °rather, °quite, °fairly, °pretty, comparatively, °slightly, passably, more or less; to some extent, °within reason, to a certain extent, °to a degree, to some degree, in some measure, in moderation, within limits; temperately; *Colloq* °sort of, kind of: *Clifford is moderately well off. It was a moderately hot day.*

moderator *n.* °mediator, arbiter, arbitrator, °judge, referee, °umpire; *Chiefly US* °chair, chairperson, chairman, chairwoman, chairlady, presiding officer, president, coordinator, °(discussion) leader; anchorman, anchorwoman, anchorperson; master of ceremonies, toastmaster, *Brit* compere, *Colloq* emcee, MC: *Lady Felicia was chosen as the moderator of next year's panel.*

modern *adj.* up-to-date, °current, °contemporary, today's, °new, °fresh, °novel, °brand-new, °up-to-the-minute, present-day, °latest, new-fashioned, newfangled; à la mode, modish, in vogue, °fashionable, °in fashion, °stylish, °in style, °chic, *Chiefly Brit* flavor of the month, *Slang* °trendy, in, with it, mod, °hip, °hot: *Stern has little patience for modern electronic gadgets.*

modernize *v.* °renovate, streamline, redo, redecorate, °refurbish, refurnish, update, do over, °rejuvenate, °refresh, °revamp, redesign, remodel, refashion, remake: *We installed air conditioning when we modernized our offices.*

modest *adj.* **1** unassuming, unpresuming, °humble, unpretentious, °unobtrusive, °reserved, retiring, diffident, °shy, °bashful, demure, °coy, shamefaced, self-effacing, °self-conscious, °reticent, °reluctant, °timid, °meek, timorous, *Rare* verecund: *Caroline is quite modest about her accomplishments.* **2** °humble, °simple, °plain, °ordinary, unpretentious, °homely, °lowly, unexceptional, unostentatious; °inconspicuous, °unobtrusive: *They live in a modest little cottage.* **3** °moderate, °limited, °understated, unimportunate, unexaggerated, °reasonable, °sensible, constrained, restricted, restrained: *The workers' demands seem modest enough.*

modicum *n.* °bit, °trifle, °jot, jot *or* tittle, atom, scintilla, °spark, °particle, iota, °speck, °grain, whit, °scrap, °shred, snippet, °sliver, °fragment, °splinter, °morsel, °crumb, ounce, dram, °drop, °dash, °spot, °touch, tinge, °hint, °suggestion, *Colloq* smidgen: *One must exercise a modicum of judgment in affairs of the heart.*

modify *v.* **1** °adjust, °adapt, °change, °transform, °alter, °revise, °amend, redo, remake, remold, reshape, reconstruct, °reform, °revamp, refashion, remodel, rework, °reword, reorient, reorganize: *They are modifying the car to improve its performance. Ought we to modify our views about free speech?* **2** °reduce, °decrease, °diminish, lessen, °moderate, °temper, °soften, °lower, abate, °tone down, °modulate; °qualify, °limit, °restrict: *The severity of the winter is modified by our proximity to the sea. We shall have to modify our demands.*

modulate *v.* °adjust, °regulate, °set, °tune, °balance, °temper, °moderate, °modify; °lower, tune *or* tone *or* turn down, °soften: *Modulate the volume to suit the size of the auditorium. Please modulate your voice—everyone can hear you!*

mogul *n.* magnate, °tycoon, baron, mandarin, *Colloq* big shot, big gun, big cheese, Pooh-Bah, °bigwig, big wheel, big (White) Chief, Big Daddy, hotshot, VIP, big noise, nabob, *Slang US* Mr. Big: *In the old days, he was one of the moguls in Hollywood.*

moist *adj.* **1** °damp, wettish, dampish, dewy, dank, °humid, °clammy, °muggy, °steamy, °misty, foggy: *During the monsoon everything becomes so moist that your clothes never seem to dry.* **2** °damp, °wet, rainy, drizzly, soggy, moisture-laden: *The moist climate keeps Ireland green.* **3** °tearful, teary, °misty, lachrymose: *Every eye in the theater was moist at the end of that scene.*

mold[1] *n.* **1** °form, °cast, matrix, °die; template *or* templet, °pattern, °form: *He broke the mold to prevent the casting of a duplicate statuette.* **2** °form, °shape, °pattern, °format, °structure, °build, construction, °design, °arrangement, °organization, configuration, °kind, °brand, °make, °line, °type, °stamp, °cut: *Was man made in the same mold as his God, or vice versa?* **3** °character, °nature, °stamp, °type, °kind, kidney, ilk, °sort: *Grandfather said that he wouldn't want someone of her mold in the family.* —*v.* **4** °shape, °form, °make, °work, °fashion, configure, sculpture *or* sculpt, °model, knead, °construct, °carve, °cut: *The artist molds the original sculpture out of clay.* **5** °forge, °cast, °stamp, die cast: *These beads were molded of solid gold.* **6** °influence, °shape, °form, °affect, °make, °control, °direct, °guide, °lead: *The speechwriter's job is to mold the language to fit his client's ideas. The candidate is trying to mold public opinion in his favor.*

mold[2] *n.* mildew, fungus, °blight, smut: *A mild solution of bleach will clean that mold off the bathroom tiles.*

mold[3] *n.* °soil, °earth, loam, topsoil, °dirt, humus: *Add some rich mold to the houseplants periodically and they will flourish.*

moldy *adj.* °aged, °ancient, outdated, °old-fashioned, antediluvian, °unused, °stale, decayed, decaying, carious, mildewed, moldering, °musty; spoilt *or* spoiled, °rotten, rotting, °putrid, putrescent, putrefying, °rancid, °rank, decomposed, decomposing, mucid: *Digging about in the files, he unearthed some moldy old plans for reorganizing the company. Owing to the power cut, all the food in the fridge became moldy.*

molest *v.* **1** °annoy, °irritate, vex, °disturb, °pester, badger, needle, °provoke, nettle, °tease, °harass, harry, °worry, hector, °irk, °bother, °gall, °chafe, roil, °torment, °plague, beleaguer: *For weeks people on the beach have been molested by green flies.* **2** accost, meddle with, interfere with, °annoy, °abuse, °bother, °attack, °ill-treat, maltreat, °manhandle; paw: *The police are looking for a man who was molesting girls in the park.*

moment *n.* **1** °instant, °second, °minute, half a second, two seconds, °flash, °twinkling, blink *or* wink of an eye, twinkling of an eye, trice, *Colloq* jiffy, °shake, two shakes (of a lamb's tail), before one can *or* could say "Jack Robinson," before you can say "knife," *Brit* mo, half a mo, tick: *I'll be with you in a moment, Madam.* **2** °instant, °time, °second, °minute, hour, °point (in time), °juncture, °stage: *At that very moment, I was trying to phone you.* **3** °importance, °weight, consequence, °significance, °import, °gravity, seriousness, °prominence, °concern, °note, °interest, °consideration: *What you decide to do is of no great moment to me.*

momentary *adj.* °fleeting, °temporary, ephemeral, evanescent, impermanent, °fugitive, °passing, transitory, °brief, °short-lived, °quick, °short, °hasty: *The medicine afforded only momentary relief from the pain.*

momentous *adj.* °important, °weighty, consequential, °significant, °grave, °serious, decisive, °crucial, °critical, °vital, °pivotal, °portentous, charged, laden, °fraught, of concern: *The Augustan Age was a momentous period in Latin literature.*

momentum *n.* °energy, °force, °drive, °strength, °impetus, °power, °inertia, impulse, °thrust, °push: *The momentum of the car carried it over the cliff. Enthusiasm has given me the momentum needed to carry on.*

monarch *n.* **1** ruler, °sovereign, potentate, crowned head; °queen, °king, empress, emperor, tsar *or* czar: *Mary was the monarch of Scotland from 1542 to 1567.* **2** ruler, °sovereign, °chief, °lord, °master, °owner, *Colloq* °boss: *He is monarch of all he surveys.*

monarchy *n.* **1** °kingdom, empire, °domain, °dominion, principality; °state, °nation, °country: *According to Gibbon, a hereditary monarchy presents the fairest scope for ridicule.* **2** monocracy, autocracy, absolutism, royalism, monarchism, °sovereignty, totalitarianism, authoritarianism; °despotism, °tyranny: *Those who are discontented with monarchy call it tyranny.*

monastery *n.* abbey, cloister, priory, friary, charterhouse, hospice, *Buddhism* vihara, *Hinduism* ashram, *Tibetan Buddhism* lamasery: *In later life he retired to a monastery to contemplate the world.*

monetary *adj.* pecuniary, °cash, °money, °fiscal, °financial, °capital; numismatic, *Technical* nummular, nummary: *Monetary shortages result from the government's tightening its hold on the economy by raising interest rates.*

money *n.* **1** currency, legal tender, medium of exchange, specie, °(hard) cash, ready money, banknotes, paper money, *Brit* °notes, *US* bills, coin(s), °change, small change, *Derogatory* (filthy) lucre, pelf, *Colloq* shekels, *US* folding money, cold (hard) cash, *US and Canadian and Australian* shinplasters, *Brit* lolly; *Slang* °loot, dough, bread, spondulix *or* spondulicks, boodle, readies *or* (the) ready, moolah, *US* °(long) green, greenbacks, mazuma, wampum, simoleons, bucks, °scratch, gelt, kale, cabbage, lettuce, spinach, *Old-fashioned* jack, *Brit* rhino, Bugs Bunny: *He makes daily rounds to collect the money from the laundromats he owns.* **2** °resources, °wealth, °fortune, °funds, °capital, wherewithal, affluence, °means, °(liquid) assets, °riches; *Slang* °bundle: *All the family money is invested in long-term bonds.* **3** °gain, °profit, °net, *Colloq* °take, °percentage, *Slang* °rake-off: *How much money does the shop take in at the end of the day?* **4** ***in the money.*** °rich, °wealthy, affluent, moneyed *or* monied, °well-off, well-to-do, °prosperous, *Colloq* °flush, in clover, in *or* on Easy Street, *Slang* °loaded, well-heeled, rolling in it *or* in money *or* in dough, filthy rich, stinking rich, °fat: *They have been in the money since Uncle Bernard died.*

mongrel *n.* cur, mutt, crossbreed, mixed breed, °hybrid, halfbreed, *Technical* bigener; lurcher: *There is little chance of a mongrel's winning a first at Crufts dog show.*

monitor *n.* **1** watchdog, °supervisor, °sentinel, °guard, °guardian, custodian; *Brit* invigilator, prefect, *Rare* prepositor *or* prepostor; *US* proctor: *The government has appointed a committee to monitor oil prices.* **2** °(television) screen, cathode ray tube screen, CRT, °display, *Chiefly Brit* visual display unit, VDU: *Do you prefer an amber, green, or black-and-white monitor?* —*v.* **3** °watch, °oversee, °observe, check (out *or* up on), audit, °supervise, superintend, °scan, °examine, °study, °follow, keep an eye on, °survey, °keep track of, °track, °trace, °record, *Brit* °vet: *Before a drug is approved, it must be monitored through thousands of tests.*

monk *n.* °brother, °religious, cenobite, monastic, *Loosely* friar: *He was accepted into the Benedictine order, also called the black monks.*

monkey *n.* **1** simian, ape, primate, *Colloq* °monk: *The monkeys' antics forever delight the children.* **2** °fool, ass, laughingstock, °butt, °victim, °target, °(fair) game, *Colloq* goat, *Slang* °sucker: *The teacher should not have made a monkey out of Hubert in front of his classmates.* **3** °imp, °devil, mischief-maker, °rascal, scamp, rapscallion: *She's really a little monkey—you wouldn't believe what she got up to this morning.* —*v.* **4** °mimic, mime, °imitate, impersonate, °copy, ape, °duplicate: *He monkeys everything you do.* **5** Usually, ***monkey around*** *or* ***about*** **(*with*).** °fool around (with), °play (with), fiddle (about *or* around) with, meddle (with *or* in), interfere (with *or* in), mess (about *or* around) (with), °tinker (with), °tamper (with), *Colloq US* screw around (with): *Stop monkeying around with that button or the bomb will go off. Barton is always monkeying around in class. If the hi-fi works, don't monkey with it.*

monograph *n.* treatise, dissertation, disquisition, °essay, °paper: *Entwhistle's monograph is the definitive work on Hawthorne.*

monolithic *adj.* °massive, °huge, °enormous, °monumental, °imposing, °colossal, °gigantic, °giant; featureless, °uniform, undifferentiated, characterless; °rigid, impenetrable, invulnerable, unbending, °inflexible, °solid, °stolid, intractable, °immovable: *These monolithic international conglomerates, assembled with the aid of junk bonds, staffed by yuppies, and controlling markets worth hundreds of billions, resemble the monolithic glass towers that house them.*

monopolize *v.* corner (the market in), °control, °dominate, own, *Slang* hog: *Till recently, one company monopolized the telephone business in the US.*

monotonous *adj.* °boring, °tedious, °dull, °tiresome, °humdrum, sleep-inducing, soporific, wearisome, wearying, tiring, monotonic, °repetitious, °prosaic, °banal, °dry, dryasdust, uninteresting, °dreary, °colorless, unexciting, °run-of-the-mill, °ordinary, commonplace, °routine, uneventful, °everyday, °mechanical, banausic, *Colloq* ho-hum: *Working in an assembly line became monotonous, so I took a job in advertising.*

monster *n.* **1** °beast, °fiend, °ogre, °giant, dragon, °brute, °demon, troll, bogeyman: *Why are all children's fairy tales populated by monsters?* **2** °monstrosity, (living) abortion, mutant, °mutation, °freak, deformity, *lusus naturae*, eyesore, °horror, miscreation, missing link: *Monsters of every description were the main features of circus sideshows for centuries.* —*adj.* **3** See **monstrous, 3,** below.

monstrosity *n.* **1** See **monster, 2,** above. **2** monstrousness, heinousness, horribleness, horridness, hideousness, awfulness, nightmarishness, dreadfulness, frightfulness, °horror, hellishness, ghoulishness, fiendishness, °barbarity: *The monstrosity of the Nazi concentration camps was not widely known till after the war.*

monstrous *adj.* **1** °awful, °horrible, horrid, horrific, horrendous, horrifying, °hideous, °ugly, °nightmarish, °dreadful, heinous, °grisly, °gruesome, °disgusting, °nauseous, nauseating, °repulsive, °repellent, °revolting, °frightful, °grotesque, hellish, °ghoulish, freakish, °fiendish, barbaric, barbarous, °savage, °inhuman, °merciless, °ruthless, °brutal, brutish, °beastly: *How could anyone treat another human being with such monstrous cruelty!* **2** °outrageous, °shocking, °scandalous, °atrocious, appalling, °wicked, °villainous, °evil, °vile, insensitive, °cruel, °base, debased, °shameful, °shameless, °infamous, °disgraceful, nefarious, egregious, °foul, °vicious, flagitious, °loathsome, depraved: *It is monstrous to suggest that money could compensate for the loss of a child. It was a monstrous trick to make him think he had failed the exam when he'd got the top mark.* **3** °gigantic, °giant, °huge, °vast, °enormous, °colossal, °monster, gargantuan, °jumbo, °immense, tremendous, titanic, °prodigious, °massive, °towering, elephantine, mammoth: *They are planning to build a monstrous shopping mall in the conservation area.*

monument *n.* **1** marker, cairn, °memorial, °tablet, shrine, commemoration; °sepulcher, gravestone, °tombstone, headstone, °tomb, mausoleum, cenotaph: *Many monuments have been established in memory of Julio Casteña.* **2** °testimony, °testimonial, testament, °token, °witness, °record, °evidence, °example, exemplar: *The book will survive as a lasting monument to her scholarship.*

monumental *adj.* **1** staggering, awe-inspiring, °outstanding, °prominent, stupendous, °vast, °awesome, epoch-making, °historic, history-making, °memorable, °lasting, °permanent, unforgettable, °significant, °notable, °noteworthy, °impressive, °marvelous, °prodigious, wonderful, wondrous, spectacular, °magnificent, °grand, °striking, °glorious, °enduring, °classic: *This work is a monumental contribution to linguistic theory.* **2** °massive, °huge, °gigantic, °enormous, °prodigious, °colossal, °immense, °vast, tremendous: *These monumental stones, each as big as a house, were dragged from a quarry hundreds of miles away.* **3** commemorative, °memorial: *A monumental obelisk will be raised to mark the battle.* **4** egregious, catastrophic, °calamitous, °huge, °enormous, °awful, abject, °terrible, unforgivable, °unbelievable, °monstrous, *Colloq* °whopping: *Announcing our plans in advance could prove to be a monumental blunder.*

mood *n.* **1** °humor, °attitude, °inclination, °disposition, °nature, °temper, °frame of mind, °spirit, °atmosphere, °sense, °feeling: *The mood of the people is rebellious. She's in a bad mood.* **2** ***in the mood.*** °ready, °willing, °eager, °keen, (well)disposed, °inclined, °sympathetic, minded: *I'm not in the mood to go to the movies, but I am in the mood to go dancing.*

moody *adj.* **1** °sullen, °melancholy, °blue, °sad, °unhappy, °dejected, depressed, crestfallen, downcast, °despondent, chapfallen, in the doldrums, °downhearted, °gloomy, °glum, moping, mopy, mopish, sulky, sulking, morose, brooding, broody, heavy-hearted, °dour, cheerless, °dismal, °desolate, disconsolate, lugubrious, disheartened, saturnine, *Colloq* °down in the mouth, (down) in the dumps, °out of sorts, *US* off (one's) feed: *He's been moody the past few days, and I don't know how to cheer him up.* **2** °testy, crotchety, °short-tempered, °abrupt, °short, °curt, °impatient, crabby, crusty, huffy, huffish, crabbed, °cantankerous, curmudgeonly, ill-humored, ill-tempered, °cranky, °petulant, °waspish, °temperamental, °snappish, °snappy, °irritable, °peevish, °cranky, °touchy, piqued; in a (fit of) pique, in high dudgeon: *If I were you, I shouldn't ask her for a favor today as she's very moody.* **3** °fickle, °volatile, °capricious, mercurial, °unstable, °fitful, °flighty, unsteady, °changeable, °erratic, uneven, °inconstant, undependable, °unreliable, unpredictable: *Being artists, they are moody, and there's no telling if they will want to be interviewed today.*

moonshine *n.* **1** moonlight, moonbeams: *Her hair reflected silver in the moonshine.* **2** °(stuff and) nonsense, °rubbish, tarradiddle, humbug, °drivel, twaddle, balderdash, blather *or* blether, *Colloq* °hot air, claptrap, pack of lies, con, bosh, gas, eyewash, hogwash, bunk, guff, piffle, hokum, °rot, malarkey, bilge (water), tripe, (the old) song and dance, °line, *Brit* tommyrot, *US* applesauce, razzmatazz, jive, *Slang* crap, bull, hooey, *Brit* tosh, *US* °garbage, BS, *Taboo slang* horseshit, bullshit: *He fed her a load of moonshine about owning a big yacht.* **3** poteen, *Colloq chiefly US and Canadian* hooch *or* hootch, white lightning, white mule, home brew, bootleg, *US* Kickapoo (Mountain) Joy Juice: *Government agents raided a still that produced one hundred gallons of moonshine a day.*

moor[1] *n.* heath, moorland, wasteland, *Northern English and Scots* °fell: *The police chased the escaped prisoner across the moor, where they lost him.*

moor[2] *v.* °secure, °tie up, make fast, °dock, berth, °(drop) anchor; °fix: *The vessel was soon moored and the passengers disembarked. The huge aerial is moored by cables.*

moot *adj.* **1** °debatable, arguable, undecided, undetermined, °controversial, °doubtful, °disputable, open to debate, °at issue, °indefinite, problematic(al), °questionable, open (to question *or* to discussion), confutable, confuted, contestable, contested, °unsettled, °unresolved, up in the air, unconcluded: *Whether he would be found innocent was still a moot point.* —*v.* **2** bring up *or* forward, °introduce, °broach, °put forward, proffer, °posit, °propound, °advance, °submit, °suggest: *It has been mooted that she was with the victim just before the murder.*

moral *adj.* **1** °ethical; °right, °good, °pure, °honest, °proper, °upright, °honorable, °decent, moralistic, °respectable, high-minded, °virtuous, upstanding, °righteous, °principled, °scrupulous, incorruptible, °noble, °just: *She is one of the most moral people I have ever met.* **2** °ethical, moralizing, moralistic: *Questions of right and wrong are a subject for moral philosophy. It is a moral, not a legal, issue.* —*n.* **3** °lesson, homily, teaching, °point, °message; aphorism, °maxim, °precept, apophthegm *or* apothegm, adage, °saw, °proverb, °epigram, °motto, °slogan: *The moral of the story is "Do unto others before they do unto you."* **4** ***morals.*** °behavior, °conduct, mores, °belief, habit(s), custom(s), practice(s), principle(s), scruples, ethics, °ideals, standards; °probity, °morality, °rectitude, °integrity: *For this appointment, we require a person of the highest morals.*

morale *n.* °dedication, spirit(s), °unity, esprit de corps, °disposition, °attitude, °confidence, °self-confidence, °self-esteem: *Morale would be improved if only the team could win a game.*

morality *n.* **1** ethics, °morals, ethicalness, moralness, principle(s), mores, °integrity, °propriety, standards, °ideals; °honesty, °right, rightness, righteousness, °rectitude, °justice, fair play, fairness, decency, uprightness, °integrity: *Morality should not be conditioned by expediency.* **2** °behavior, °conduct, habit(s), custom(s): *Are you suggesting that utilitarianism is a guide for morality?*

morass *n.* **1** °bog, °marsh, °swamp, fen, quagmire, slough, marshland, moorland: *The land not under cultivation was morass or forest.* **2** entanglement, °confusion, °muddle, °mess, quagmire, °tangle, quicksand: *It took years to extricate the company from the legal morass of the bankruptcy courts.*

moratorium *n.* °halt, hiatus, °suspension, °stay, °respite, °freeze, °delay, waiting period, °postponement: *The arts commission requests a six-month moratorium on sales to foreign purchasers.*

morbid *adj.* **1** °unhealthy, °unwholesome, disordered, °unsound, °sick, pathological, pathogenic: *The eating of peeling paint, which contains lead, has led to this morbid condition.* **2** °grim, °ghoulish, °macabre, °monstrous, °ghastly, °grotesque, °grisly, °gruesome: *At an early age, Glynis exhibited a morbid interest in cadavers. Spare me the morbid details—please!* **3** °gloomy, lugubrious, °glum, morose, °somber, °blue, °sad, °melancholy, °despondent, depressed, °dejected, downcast: *With Charley again in one of his morbid moods, we can anticipate another cheerless visit.*

moreover *adv.* furthermore, °further, °besides, not only that, more than that, what is more; °to boot, into the bargain, °in addition, additionally, as well, too: *She was pleased that I had brought the wine—it was, moreover, her favorite burgundy.*

moribund *adj.* **1** °dying, *in extremis,* at death's door, °failing, fading, with one foot in the grave, half dead, breathing one's last, expiring, °on one's last legs, on one's deathbed: *Apparently moribund for weeks, he suddenly revived when he saw her.* **2** ending, declining, °obsolescent, °weak, on the way out, waning, °on the wane, dying out; stagnating, °stagnant: *At the time, the Ottoman Empire was already moribund.*

morning *n.* **1** forenoon, (the) a.m., °dawn, daybreak, sunrise, *Literary* morn, *Archaic* cockcrow, dayspring, morrow, *Chiefly US* sunup: *The meeting will be at nine o'clock in the morning.* —*adj.* **2** matutinal, matinal, a.m., forenoon: *I prefer a morning meeting.*

morsel *n.* **1** °mouthful, °bite, gobbet, spoonful, forkful, °chew, °taste, °sample, nibble, °bit, °drop, dollop, soupçon: *She delicately ate tiny morsels of the banana.* **2** °bit, °crumb, °fragment, °scrap, °sliver, °splinter, shard *or* sherd, °shred, °remnant, °particle, atom, °speck, whit, fraction, °grain, granule, °pinch, °piece, *Colloq* smidgen: *There wasn't the slightest morsel of evidence to connect me with the crime.*

mortal *adj.* **1** °human; transitory, °temporal, °transient, ephemeral: *It finally dawned on him that the king was mortal, like other men.* **2** °physical, bodily, corporeal, corporal, fleshly, °earthly, °worldly, perishable: *Her mortal remains are buried at Bognor Regis.* **3** °deadly, °fatal, °lethal, °terminal, °destructive, °disastrous: *The admiral lay dying from a mortal wound.* **4** °relentless, °implacable, unrelenting, °bitter, sworn, °deadly, unremitting, unappeasable, unceasing: *Halifax was the mortal enemy of despotism.* **5** abject, °extreme, °awful, °great, °enormous, °intense, °terrible, °inordinate, dire: *Gill is in mortal fear of offending his mother.* —*n.* **6** °human (being), °man, °woman, °person, °soul, °individual, °creature, earthling: *We mortals have nothing to fear from intergalactic aliens.*

mortify *v.* **1** °humiliate, °shame, °humble, °embarrass, abash, chagrin, °rebuff, °crush, °discomfit, deflate, bring down, °degrade, °downgrade, °reduce, °chasten, °subdue, °suppress, make someone eat humble pie, teach someone his *or* her place, *Colloq* °put down: *She was mortified to learn that she had been disqualified.* **2** °punish, °castigate, °discipline, °control, °subdue, °subjugate: *Repentance was considered insufficient penance, so the flagellants mortified the flesh with scourges.*

3 gangrene, °fester, necrose, °putrefy, °rot, °decompose, °decay, putresce: *After a fortnight, the flesh mortifies and falls away.*

mother *n.* 1 dam, materfamilias, °(female) parent, *Old-fashioned or formal or jocular* mater, *Formal* progenitrix; matriarch, *Colloq* ma, old lady, old woman, *Brit* mummy, °mum, *US* mom, mommy, mama, mamma, maw, mammy, mam: *He runs crying to his mother when someone teases him.* 2 °source, °origin, genesis: *Necessity is the mother of invention.* 3 nourisher, nurturer, °nurse: *Poverty was the mother of anarchy.* —*adj.* 4 °native, °natural, innate: *She has the mother wit to know better.* —*v.* 5 nurture, °nourish, °nurse, °care for, °look after, °protect, °shelter, watch over, °take care of: *She mothered me when I needed her, and I won't desert her now.* 6 °pamper, °baby, °coddle, °spoil, °indulge, fuss over, overprotect: *You mother the child too much.*

motif *n.* °theme, °idea, °topic, °subject, concept, leitmotif; °pattern, °figure, °refrain, °device, °ornament, °decoration, °element, °convention: *The painter repeats the motif throughout in much the same way that a composer repeats a bar of music.*

motion *n.* 1 °movement, °moving, °change, °shift, shifting, °action, °going, °traveling, °travel, °progress, °passage, °transit; °activity, commotion, °stir, °agitation, turmoil, turbulence: *One could barely sense the slow motion of the train. The leaves of the aspen appear to be in constant motion.* 2 mobility, movability, motility: *I could feel the motion gradually returning to my numbed limbs.* 3 gait, °bearing, °carriage, tread, °walk, °step: *Different motions are involved in walking upstairs and walking downstairs.* 4 °gesture, gesticulation, °signal, °sign: *She seemed to make an encouraging motion with her head.* 5 °proposal, °suggestion, °proposition, °recommendation, °offering, °submission: *A motion must be seconded by a member in good standing.* —*v.* 6 °gesture, gesticulate, °beckon, °signal, °sign, °wave: *The cashier motioned to me and I went over to the window.*

motivate *v.* °prompt, °activate, °move, °inspire, °incite, °induce, actuate, °stimulate, °provoke, °influence, °encourage, °occasion, °bring about, °cause; °excite, egg (on), °urge, °prod, °spur, galvanize, goad, °rouse, °arouse, °stir (up), °wheedle, °coax, °persuade, °cajole, °tempt, °push, impel, °drive, instigate: *The biggest problem is motivating students to apply themselves to academic subjects.*

motive *n.* 1 °inducement, °incentive, motivation, stimulus, motivating force, stimulation, °incitement, °influence, °cause, °reason, °rationale, °grounds; °attraction, °lure, °enticement, goad, °spur, °urge, °prod: *Jealousy has frequently been a motive for murder.* 2 °purpose, °aim, °intention, °intent, °object, °objective, °goal, °end, *arrière pensée;* °ambition, °desire; *Colloq* °angle: *I asked the president what was his motive in expanding the company.* —*adj.* 3 driving, impelling, propelling, propulsive, °moving, kinetic, activating, °operative: *The motive power used in the ultralight aircraft is provided entirely by the pilot's pedaling.*

mottled *adj.* dappled, brindled, marbled, streaked, splodgy *or US* splotchy, blotched, blotchy, freckled, spotted, °spotty, patchy, °speckled, flecked, sprinkled, spattered, splashed, streaky, stippled, pied, piebald; multicolored, °variegated, parti-colored, *Colloq* splodged *or US* splotched: *Do you like that mottled effect in the wallpaper?*

motto *n.* °maxim, °proverb, saying, adage, °saw, aphorism, apophthegm *or* apothegm, gnome, °slogan, °byword, catchword, battle cry, °guide, °moral, °principle, °rule, °precept: *The motto of the Order of the Garter is* Honi soit qui mal y pense, *"Shame on whoever thinks ill of it."*

mound *n.* 1 hillock, °rise, hummock, °hill, °hump, bank, °elevation, °plateau, °knoll, °knob, °swell, dune, °slope, tor, *Chiefly W US* mesa, *Chiefly W US and Canadian* butte: *We stood on a high mound from which we could see for miles.* 2 °heap, °pile, °stack, *Archaeology* tumulus, °tell, barrow, (kitchen) midden: *Excavation of the mound yielded scores of Iron Age artifacts.*

mount[1] *n.* See **mountain, 1,** below.

mount[2] *v.* 1 °climb (up), °go up, ascend, °scale, clamber up, make one's way up: *We mounted the ladder to the roof. The speaker mounted the dais.* 2 °rise (up), °arise, °soar, °fly (up), rocket (upward(s)): *With the salmon in its grasp, the eagle mounted to the sky.* 3 climb *or* get *or* clamber up on, bestride, straddle, bestraddle: *The gunfighter mounted his horse and rode off.* 4 °(put on) display, °(put on) exhibit, put on exhibition, °present, install *or* instal, °stage, °prepare, °ready, °put on, put in place, °set up; °arrange, °coordinate, °compose, °organize, set in motion, °launch: *The gallery will mount a show of John's sculpture next spring. We have mounted a major advertising campaign for our face cream.* 5 °frame, mat *or* matt, °set off: *This print should be mounted using an olive green paper.* 6 °increase, wax, °rise, escalate, °intensify, °swell, °expand, °grow, mount up, multiply, °pile up, °build up, °accumulate: *Complaints have been mounting against the vulgar language heard on prime-time TV.* —*n.* 7 °setting, mounting, °support; °backing, °background, °set, °arrangement, backdrop, °scene: *This style of mount will show the ruby to its best advantage. Will this serve as a suitable mount for the jewelry display?* 8 horse, steed, charger, palfrey: *Her mount was a cream-colored mare.*

mountain *n.* 1 °height, °elevation, °mount, eminence, °prominence, °peak, alp, tor, °summit, *No. Eng. and Scots* °fell, *Scots and Irish Eng.* ben: *Her favorite sport is climbing mountains.* 2 °heap, °pile, °stack, °mound, °accumulation, °abundance, °mass, *Colloq* ton(s), °heaps, °piles, stacks: *I have a mountain of laundry to do.*

mountainous *adj.* 1 craggy, alpine, Himalayan: *It was very slow going through the mountainous parts of the country.* 2 °huge, °towering, °high, °steep, °enormous, °immense, °formidable, °mighty, °monumental, °prodigious, staggering: *Our tiny craft was almost engulfed by the mountainous seas.*

mourn *v.* °grieve (over), °lament, °sorrow (over), °bemoan, °bewail, °keen, weep for *or* over, °regret, rue, deplore: *We all mourned the loss of a great leader. After his death, the nation mourned for a month.*

mournful *adj.* 1 °sad, °sorrowful, °dismal, °melancholy, °blue, afflicted, °doleful, dolorous, grief-stricken, rueful, °forlorn, °woebegone, °somber, lugubrious, °funereal, °joyless, dispirited, cheerless, °unhappy, °downhearted, heavy-hearted, disconsolate, °heartbroken, °inconsolable, °despondent, °desolate, despairing, heartsick, °overcome, °prostrate: *The mournful crowd filed past the bier to pay their last respects.* 2 °deplorable, °sorrowful, °grievous, distressing, upsetting, °tragic, saddening, disheartening, depressing, °lamentable, catastrophic, °calamitous, °disastrous: *The mournful news of the loss of life in the explosion was broadcast round the world.*

mourning *n.* 1 °grief, °lament, grieving, °lamentation, sorrowing, keening, weeping, wailing: *The Highlanders are wont to mingle a degree of solemn mirth with their mourning.* 2 bereavement, °loss, °anguish, °sorrow, °misery, °grief, °sadness, °woe, woefulness, °melancholy, heartache, despondency, °despair, °desolation: *In that country they wear white as a sign of mourning.* **3** °black, widow's weeds, sackcloth and ashes: *It was customary to spend at least one month in mourning.*

mousy *adj.* **1** mousey, mouse-colored, °dun, °gray, grayish-brown, brownish-gray, brownish, brown, °dull, lusterless, °lackluster, °drab, °flat, °plain, °colorless: *Her mousy hair hung to her shoulders.* **2** °timid, cowering, timorous, °shy, self-effacing, diffident: *Adrian is far too mousy to ask for an increase in salary.*

mouth *n.* **1** lips; maw, jaws, oral cavity, *Technical* stoma, *Slang* °trap, kisser, muzzle, °gob, chops, °yap, *US* bazoo: *I want to hear from your own mouth that you don't love me.* **2** °opening, °aperture, doorway, door, gateway, °gate, access, °entrance, inlet, °entry, entryway, way in, entrée; °passage, passageway, °way, orifice; °exit, way out, °vent, °outlet, outfall, *Technical* debouchment *or* debouchure, debouch *or* débouché, embouchure: *We stood at the mouth of the cave. An enormous delta formed at the mouth of the river.* **3** bragging, boasting, braggadocio, empty *or* idle talk, °bombast, rodomontade, fustian, *Slang* claptrap, °hot air, gas: *The fellow is all mouth and no action.* **4** °disrespect, °impudence, insolence, sauciness, rudeness, °impertinence, pertness, boldness, audacity, presumptuousness, brashness, °flippancy, *Colloq* lip, cheek, backchat, °sauce, freshness, *US* sass, back talk: *He'd better not give me any of his mouth or I'll have his guts for garters.* **5** grimace, °pout, *moue,* °face: *She made a mouth when I said she couldn't go.* **6** ***down in*** *or* ***at the mouth.*** °dejected, °despondent, °sad, °sorrowful, °unhappy, °melancholy, °blue, crestfallen, dispirited, disheartened, downcast, *Colloq* (down) in the dumps, broken up: *He was very down in the mouth after seeing the results of the laboratory tests.* —*v.* **7** utter, °say, °speak, °pronounce, °announce, °enunciate, articulate, °voice, °sound, °express, vocalize; declaim, orate: *He mouthed each word with excruciating care. Who wants to hear some pompous fool mouthing platitudes?*

mouthful *n.* °morsel, °bite, spoonful, forkful, °lump, chunk, °gob, hunk: *He took a mouthful of food.*

mouthpiece *n.* **1** embouchure; °bit: *I needed a new mouthpiece for my trumpet.* **2** spokesman, spokeswoman, spokesperson, °agent, °representative, intermediator, °mediator, °delegate: *He is always quoted through his mouthpiece, never directly.* **3** °lawyer, attorney, *Slang US* shyster: *A good mouthpiece could get me out of this fix.*

movable *adj.* moveable, floating, °variable, °changeable, unfixed; °portable, transportable, transferable: *Easter Sunday, a movable feast, has no fixed date on the calendar. Recent developments have produced a computer that is smaller than an attaché case and easily movable.*

move *v.* **1** °shift, °stir, budge, make a move, °go; °proceed, °advance, °progress: *It is difficult to move in this straitjacket. We are moving at two hundred miles a day. Don't move—I have a gun.* **2** move house, move out, °remove, move away, relocate, decamp, °depart, change residence, °emigrate, go *or* make off, °transfer, *Colloq* °take off (for), °pull up stakes, *Brit* up sticks, *Slang US* °split (for): *When did you say you were moving?* **3** °shake (up), °disturb, °stir (up), °agitate, °affect, °touch: *We moved heaven and earth to find a proper place for her to stay.* **4** °affect, °touch, °stir, °shake up, °agitate, °hit (hard), °upset, °strike, smite, °disturb, °ruffle, disquiet, have an (*or* a profound) effect (on), make a (deep) impression (on): *The pictures of starving children moved everyone.* **5** °provoke, °arouse, °excite, °stir up, °lead, °rouse, °stimulate: *At the end of the film, when she left him, I was moved to tears.* **6** °arouse, °rouse, °provoke, actuate, °lead, °prompt, °spur, °motivate, °influence, impel, °prod, °remind, °inspire, °make: *His mention of families moved me to ask when he had last seen his father.* **7** °propose, put forward *or* forth, °forward, °advance, °submit, °suggest, °advocate, °propound: *The question of budget was moved at the meeting.* —*n.* **8** °change, changeover, relocation, °transfer, °shift, °removal: *My supervisor suggested a move to a different department.* **9** °maneuver, °device, °trick, °caper, °dodge, ploy, °stratagem, °artifice, °ruse, °action, °act, °deed, *Colloq* °gimmick: *Getting Daphne to invite Oscar was a very good move.* **10** °turn, °time, °opportunity: *I have had my turn, now it's your move.* **11** °gesture, gesticulation, °action, °motion, °stirring: *One move and you're a dead man!* **12** ***get a move on.*** **(a)** get moving, °begin, °start, °commence, get going, get under way, get started, stir *or* bestir oneself, *Colloq* get *or* start the ball rolling, get the show on the road, °break the ice, get cracking, step on it *or* the gas: *He doesn't usually get a move on till noon.* **(b)** °hurry, °hasten, make haste, °rush, °run: *Get a move on down to the supermarket before it closes.* **13** ***on the move.*** **(a)** °traveling, in transit, on the way, on one's way, on the road, on the go, °moving: *She is always on the move and it is difficult to catch her.* **(b)** on the go, working, °on the run, °busy, occupied: *I have been on the move since six this morning.* **(c)** °proceeding, progressing, advancing, moving ahead, succeeding, on the go: *It looks as if the industry is again on the move after a brief decline.*

movement *n.* **1** repositioning, °move, °motion, relocation, °moving, migration, °shift, °transfer, °flow, displacement: *The specialist's term for mass population movement is* Wölkerwanderung. **2** °action, °activity, °move, °moving, °stir, °stirring: *There was a sudden movement in the bushes.* **3** °gesture, gesticulation, °move, °flicker, °sign, °signal; °maneuver, change of attitude *or* position: *A slight upward movement of his eyebrow means you should swallow the microdot. Every little movement has a meaning all its own.* **4** °mechanism, °works, workings, moving parts, machinery, °action, gears, *Colloq* innards: *Technically, the movement of a watch does not include the escapement.* **5** °campaign, °crusade, °drive; °front, °faction, °party, °group, wing: *The 1960s saw many successes of the civil rights' movement in America.* **6** °change, °activity, °action, °shift, advance *or* decline, increase *or* decrease, upward *or* downward movement, °stirring; °development, °progress: *Shares displayed little movement in today's trading.* **7** °drift, °trend, °tendency, °course, °swing: *They have noted a growing movement toward religious fundamentalism.*

movie *n.* **1** motion picture, °film, moving picture, °silent (picture), talking picture, *Colloq* talkie, flick: *Her grandfather was one of the first actors in silent movies.* **2** Usually, ***movies.*** picture show, cinema, flicks, *Colloq* big *or* large screen, silver screen: *Why don't we go to the movies instead of watching television tonight?*

moving *adj.* **1** °touching, °poignant, emotive, affecting, °stirring, °heart-rending, °emotional, °telling, °effective, °impressive, °striking, compelling; °pathetic, °exciting, °thrilling, inspiring, inspirational, impelling, °persuasive: *He related a moving story of his years in various concentration camps.* **2** °active, °mobile, unfixed, unstationary, motile, °going, operating, working, in motion, °on the move: *Be careful not to touch any of the moving parts of the machine.*

mow *v.* **1** °cut (down), scythe, °trim, shear: *Do not mow the lawn so much during a dry spell.* **2** ***mow down.*** annihilate, °kill, °massacre, °butcher, °slaughter, °exterminate, liquidate, eradicate, °wipe out, °cut down, cut to pieces, °destroy, decimate: *The first platoon was mowed down by machine-gun fire.*

muck *n.* **1** ordure, manure, °dung, excrement, feces, droppings; guano: *The muck is saved for use as fertilizer.* **2** °dirt, °filth, bilge, slime, °sludge, °ooze, scum, sewage, °mire, °mud, feculence, *Colloq* gunge, gunk, *US* grunge: *Give me a chance to get the muck off my shoes.* —*v.* **3** ***muck about.*** °fool around, waste time, °idle, loiter, mess around *or* about: *She told me to stop mucking about and get a job.* **4** ***muck up.*** °ruin, °wreck, °destroy, make a mess of, °botch, °mess up, °bungle, *Colloq* °screw up, *Slang* °bugger up, make a muck of: *She's mucked up her own life, and, given the chance, she'll muck up mine.*

mud *n.* °muck, °ooze, slime, °mire, clay, °sludge, °silt, °dirt, *US and Canadian* gumbo *or US and Canadian* gombo: *The mud came up to my ankles.*

muddle *v.* **1** °bewilder, °confuse, confound, °mystify, baffle, °mix up, disorient, befuddle, °perplex, °bemuse, °puzzle, °befog: *I am totally muddled by your explanation of differential calculus.* **2** °confuse, °mix up, °jumble, °scramble, °entangle, °tangle, °mess up, °disorder, disarrange, disorganize, °bungle, mismanage, *Colloq* muff: *My financial affairs got hopelessly muddled when I changed banks.* **3** ***muddle through.*** °(barely) manage, °cope, °make it, scrape through *or* along, contrive, °make do, °get by, °get along: *We were just muddling through when James was offered this wonderful job.* —*n.* **4** °mess, °confusion, °mix-up, °jumble, °tangle, °disorder, hotchpotch *or US also* hodgepodge, °mishmash, °chaos, °disaster, *Colloq* °stew, *Slang US* screwup, snafu, *Taboo slang Brit* balls-up: *A lot of muddles and misunderstandings ensued.*

muddy *adj.* **1** fouled, befouled, muddied, mud-spattered, °dirty, grubby, grimy, soiled, mud-caked, °slimy, mucky, miry; oozy, squelchy, squashy, boggy, fenny, marshy, swampy; *Formal* feculent; *Colloq* squishy, squushy: *Take off those muddy boots before you come into the house. The walk up to the house has become all muddy in the rain.* **2** °confused, unclear, °vague, °obscure, °dull, °dim, °fuzzy, muddled, addled, mixed-up: *Gregory has only a muddy idea of what you're talking about.* **3** °drab, °subdued, blurred, °dingy, °flat, °murky, mat, °washed-out: *The colors in her paintings look very muddy to me.* —*v.* **4** °obscure, °dull, °dim, °confuse, °mix up, °befog, cloud: *His explanation muddied the issue rather than clarifying it.* **5** °dirty, °soil, begrime, smirch, besmirch, °spatter, bespatter: *A passing car muddied my new skirt.*

muffle *v.* **1** Often, ***muffle up.*** °wrap, °swathe, swaddle, °cloak, °envelop, °cover (up), enfold, °shroud, enshroud, °conceal, °protect: *Make sure your neck is muffled against that icy wind.* **2** °deaden, °silence, °suppress, °stifle, °subdue, °damp, °dampen, °mute, °hush, °quiet, °tone down, °still: *A few more thicknesses of flannel will muffle the noise of the motor.*

muffler *n.* scarf, boa; shawl, °wrap: *Wrap this muffler round your neck for added warmth.*

mug *n.* **1** °jug, tankard, stein, toby (jug), °pot, beaker, cup: *The relief workers passed round mugs of hot coffee to the homeless.* **2** °face, visage, °features, countenance, *Slang* puss, kisser, mush, *Brit* clock, dial, *Archaic* phiz *or* phizog, *US* °pan: *The ex-boxer has a mug that only a mother could love.* **3** °fool, °duffer, simpleton, °dupe, gull, °innocent, *Colloq* chump, °mark, soft *or* easy touch, *Brit* muggins, *Slang* °sucker: *The mug thought he had found a full wallet.* —*v.* **4** make a face *or* faces, grimace: *Here are some pictures of Alexa mugging for the camera.* **5** °attack, °set upon, °rob, °assault; garrote, throttle: *Walk through the park at night and you risk getting mugged.* **6** ***mug up (on).*** °study, *Formal* lucubrate, *Colloq* °cram, burn the midnight oil, °get up (on), *Brit* swot, bone up (on): *I have to mug up on my Latin for tomorrow's examination.*

muggy *adj.* °humid, °damp, °sticky, °sultry, °oppressive, °clammy, °steamy, °close, °stuffy, °moist, soggy: *It was so muggy that every night the bedlinen was completely soaked.*

mull *v.* Usually, **mull over.** ponder, °consider, °study, think (over *or* about), cogitate (on *or* over *or* about), con, °evaluate, °turn over, °weigh, deliberate (on *or* over), °reflect (on), °muse (on), °review, °examine, °contemplate, °meditate (on), °chew over, ruminate (on *or* over): *I have been mulling over your suggestion that I should join the navy.*

mum *adj.* **1** °silent, °mute, close-mouthed, °quiet, °tight-lipped: *She swore she would keep mum about where the money was hidden.* —*n.* **2** ***Mum's the word.*** Don't tell a soul, Keep silent, Keep secret, Keep quiet, Keep (something) to oneself, Keep (something) under one's hat, Say nothing, Tell no one, Play dumb, *Brit* Keep schtum: *He told who stole the bell, but mum's the word.*

mumble *v.* °murmur, °mutter, say inarticulately, utter indistinctly, swallow one's words: *He mumbled the name of the person who had shot him, but I didn't catch it.*

mumbo jumbo *n.* **1** °gibberish, °nonsense, °rubbish, gobbledegook *or* gobbledygook, °drivel, humbug, bunkum, double-talk, rigmarole *or* rigamarole, jabberwocky, blather *or* blether, poppycock, *Colloq* eyewash, °rot, tommyrot, hogwash, bilge, bosh, bull, malarkey *or* malarky, claptrap, piffle, *US* hooey, °moonshine, bunk *Slang* bull, crap, *Brit* tosh, *Taboo slang* bullshit: *Before he would give me the check, the lawyer made me sign something full of legal mumbo jumbo.* **2** °spell, incantation, °chant, °formula, °charm, abracadabra, °hocus-pocus, °rite, °ritual, conjuration, °magic: *She had to sit through all the mumbo jumbo of her husband's investiture.*

munch *v.* °chew, °crunch, masticate, champ, chomp, scrunch: *He just sat there, munching peanuts.*

municipal *adj.* civic, °civil, metropolitan, urban, °city, °town, village, borough, parish, *Brit* °council: *Voting takes place in the municipal elections next week.*

municipality *n.* °city, °metropolis, °town, village, borough, °district, township; suburb, exurb: *The recreational facilities offered by the municipality are in need of refurbishing.*

murder *n.* **1** homicide, manslaughter, regicide, patricide, matricide, parricide, fratricide, sororicide, uxoricide, infanticide; °killing, slaying, assassination: *In many countries, murder is a crime punishable by death.* **2** °slaughter, butchery, genocide, °massacre, liquidation, decimation, extermination, eradication, wiping out, murdering, slaying, °killing, °bloodshed, °carnage: *Their wanton murder of thousands of prisoners of war has been proved.* —*v.* **3** °kill, slay, assassinate, put to death, end the life of, put away *or* down, °put out of one's misery, °wipe out, °destroy, °butcher, °massacre, liquidate, °exterminate, eradicate, annihilate, °extinguish, °slaughter, °lay low, *Slang* °eliminate, °bump off, °knock off, do in, °polish off, blow away, *US* °rub out, °waste, ice, °take for a ride, fit with a concrete overcoat *or* with concrete overshoes, snuff (out): *The drug dealers were murdered one by one by the rival gang.* **4** °spoil, °ruin, °mar, °destroy, °wreck, °kill, °mangle, °butcher, °mutilate: *When she finally did sing, she murdered the song.*

murderer *n.* murderess, °killer, slayer, assassin, homicide, °cutthroat, liquidater, executioner, °butcher,

Slang hit man: *The police have arrested someone they are certain is the murderer.*

murderous *adj.* **1** °fatal, °lethal, °deadly, deathly, °mortal, °destructive, °devastating, °sanguinary, bloody, °brutal, °savage, °bloodthirsty, barbarous, °fell, °cruel, °inhuman: *The tolling of the bell signaled that the murderous deed had been done.* **2** °killing, °strenuous, stressful, °difficult, °arduous, °exhausting, °punishing, hellish, °harrowing, rigorous, intolerable, °unbearable: *You cannot keep up such a murderous pace, without any rest, and remain healthy.*

murky *adj.* °dark, °gloomy, °threatening, °dim, clouded, cloudy, °overcast, °gray, °dismal, °dreary, °bleak, °somber, °grim, °funereal, °shady, °shadowy: *I could barely make out her face in the murky corner of the chapel.*

murmur *n.* **1** °undercurrent, undertone, background noise *or* sound, rumble, rumbling, °mumble, mumbling, drone, droning, °buzz, buzzing, murmuration, murmuring, °hum, humming, whispering, *Formal* susurration *or Formal* susurrus: *They spoke in low voices, barely audible above the murmur of the mourning doves.* **2** muttering, complaining, °complaint, grumble, grumbling, *Colloq* grousing, grouse: *We have heard not any murmurs since everyone had an increase in salary.* —*v.* **3** °mumble, °mutter, °whisper: *"But I came especially to see you," he murmured.* **4** °complain, grumble, °mutter, °moan, °lament, wail, *Colloq* grouse: *Employees will always find something to murmur about.*

muscular *adj.* °sinewy, °brawny, °burly, °powerful, powerfully built, strapping, °rugged, °husky, °robust, athletic, °sturdy, well-muscled, broad-shouldered: *We all envied Nick's muscular physique.*

muse *v.* cogitate, °meditate, °reflect, °contemplate, ruminate, think over, °think about, °consider, °chew over, °deliberate, °revolve, °weigh, °evaluate, °study, mull over, °brood (over), °ponder; be absorbed (in thought), be in a brown study, °dream, °daydream, be in a trance *or* reverie: *Asked for a decision, Brinkton mused for a bit before answering.*

mushy *adj.* **1** °soft, pulpy, squashy, doughy, squidgy, spongy; swampy, boggy, miry; *Colloq* squishy, squushy, squashy: *This apple is overripe and mushy. The bottom of the pond is all mushy and feels awful when you stand up.* **2** mawkish, °maudlin, °sentimental, °romantic, saccharine, sugary, syrupy, *Colloq* corny, °sloppy, °gooey, slushy, *Brit* °wet, *Slang* schmaltzy: *She cried three handkerchiefs-full over the mushy parts in* Gone With The Wind.

musical *adj.* °tuneful, melodic, °harmonious, lilting, °lyrical, °melodious, mellifluous, dulcet, euphonious: *Many modern compositions, while technically competent, are not as musical as the old-fashioned pieces.*

must *v.* **1** ought (to), should, have to, be obliged *or* obligated to, be compelled *or* forced to, be required to: *I must go down to the sea again. Must you? Yes, I must.* —*n.* **2** °necessity, requisite, °requirement, °obligation, sine qua non, °essential: *Honesty is an absolute must in this job.*

muster *v.* **1** call *or* come together, °assemble, convoke, convene, °collect, °mobilize, °rally, °round up, °gather, marshal, °summon (up): *Within a month he had mustered a force of thousands to fight the invaders. I mustered up all my courage to ask Joan out to dinner.* —*n.* **2** °rally, °assembly, assemblage, convocation, °meet, °meeting, °convention, congress, °roundup, °turnout, °gathering, congregation, aggregation, aggregate: *We attended the annual muster of marching bands in Deep River, Connecticut.* **3 *pass muster.*** come up to scratch, °make the grade, measure up, be acceptable, *Colloq* come *or Colloq* be up to snuff: *Are you sure that your new design will pass muster with the art director?*

musty *adj.* **1** °moldy, °damp, mildewed, mildewy, °sour, °rancid, spoiled, decayed, °rotten, °putrid, fetid *or* foetid, fusty, °stale: *Open the windows to get rid of that musty odor.* **2** °stale, °old-fashioned, °antiquated, antediluvian, °ancient, out-of-date, °bygone, °passé, old hat, °obsolete, archaic, °tired, hoary, worn-out, trite, clichéd, stereotypical: *Some of those musty jokes of his are no longer funny.*

mutation *n.* **1** °change, changing, °alteration, altering, modification, modifying, °transformation, transforming, metamorphosis, metamorphosing, transmutation, transmuting, transfiguration, transfiguring, °evolution, evolving, °variation, varying: *For years he studied the mutation of the sweet pea.* **2** °variant, °variation, deviation, °deviant, mutant, anomaly, departure: *The mutations exhibited marked differences over the generations.*

mute *adj.* **1** °silent, °dumb, °speechless, voiceless, wordless, °tight-lipped, °taciturn, °tacit, °reserved, °quiet, *Colloq* °mum: *Though I persisted in asking his name, he remained mute.* **2** unspoken, unsaid, °tacit, °silent: *The stain on the floor was a mute reminder of the crime.* —*v.* **3** °deaden, °silence, °muffle, °stifle, °dampen, °damp, °subdue, °suppress, quiet *or Brit also* quieten, °hush, soft-pedal, °turn down, °tone down: *The carpet muted her footsteps as she crept up behind him.*

mutilate *v.* **1** °maim, disfigure, °mangle, °cripple, °lame, °butcher, disable; dismember, amputate, hack off, °cut off, lop off, tear off, °rip off: *Formerly, the hands of thieves were mutilated as an example to others.* **2** °spoil, °mar, °ruin, °damage, °deface, vandalize, °destroy: *As the book had been mutilated, he had to buy the library a replacement.*

mutinous *adj.* **1** °rebellious, °revolutionary, °subversive, °seditious, insurgent, insurrectionary: *The mutinous crew was finally subdued and clapped in irons.* **2** °recalcitrant, refractory, contumacious, °obstinate, °defiant, °insubordinate, °disobedient, °unruly, unmanageable, °ungovernable, uncontrollable: *Whenever we told Clara she had to do her homework before going out she became mutinous.*

mutiny *n.* **1** °revolt, rebellion, °revolution, °subversion, subversiveness, insurgency, insurgence, insurrection, °uprising: *The crew are threatening a mutiny.* —*v.* **2** °rebel, rise up (against), °strike, °revolt; °disobey, °subvert, °agitate (against): *The crew of the* Bounty *mutinied and set Captain Bligh adrift.*

mutter *v.* **1** °mumble, °murmur, grunt: *He was muttering something to himself that I didn't catch.* **2** grumble, °complain, *Colloq* grouch, grouse, *Brit* chunter: *She has been muttering about cuts in the postal service ever since I can remember.*

mutual *adj.* **1** °reciprocal, reciprocated, requited, interactive, complementary: *I love her and I hope that the feeling is mutual.* **2** °common, communal, °joint, shared: *Gerald is our mutual friend.*

mysterious *adj.* **1** °puzzling, enigmatic, baffling, insoluble, unsolvable, bewildering, confounding, confusing, °perplexing, mystifying, °weird, °bizarre, °strange, uncanny, °curious: *A mysterious force makes the saucer rise off the ground and fly at incredible speeds.* **2** °cryptic, arcane, °secret, inscrutable, covert, °hidden, furtive, unclear, °dark, concealed, °occult, °inexplicable, °incomprehensible, mystic(al), °unknown, unfathomable, °recondite, abstruse: *It is mysterious how the agency accomplishes its ends.*

mystery *n.* **1** °puzzle, °enigma, conundrum, °riddle, °question: *How the murder was committed in a room locked on the inside is a mystery.* **2** °obscurity, °secrecy, indefiniteness, vagueness, nebulousness, °ambiguity, ambiguousness, inscrutability, inscrutableness: *The identity of the beautiful lady in the green dress is shrouded in mystery.* **3** detective story *or* novel, °murder (story), *Colloq* whodunit: *Joel, who fancied himself an intellectual, read nothing but mysteries.*

mystical *adj.* **1** allegorical, symbolic(al), mystic, cabalistic, arcane, unrevealed, °secret, °occult, °supernatural, esoteric, otherworldly, preternatural, °cryptic, concealed, °hidden, clandestine, °private, °veiled, °ineffable, °mysterious: *A mystical spirit dwells in the raven that sits on his shoulder.* **2** See **mysterious, 2,** above.

mystify *v.* °fool, °hoax, humbug, °confuse, confound, °mix up, °bewilder, °stump, °puzzle, baffle, *Colloq* bamboozle, °stump, °flummox, *Slang* °beat: *We were totally mystified by the unfamiliar symbols on the wall.*

mystique *n.* °mystery, °magic, charisma, °aura, inscrutability, supernaturalism, preternaturalism, strangeness: *She has a certain mystique about her.*

myth *n.* **1** °legend, fable, allegory, °parable, °tradition, °saga, epic, °(folk) tale, °story, mythos; °history: *The ancient myths have provided sources for much of modern literature.* **2** fable, °lie, °(tall) tale, °fib, prevarication, fiction, untruth, °falsehood, °fabrication, cock-and-bull story, *Colloq* whopper: *He told her some myth about his being a multimillionaire and she believed it.*

mythical *adj.* **1** mythic, mythological, fabled, °legendary, °traditional, folkloric, storied, °romantic, fairytale, storybook; allegorical, °symbolic, parabolic(al): *Campbell has demonstrated astounding similarities amongst the mythical tales of divers cultures.* **2** mythic, °fanciful, °imaginary, °fictitious, make-believe, made-up, chimerical, °untrue: *When she was about seven, my daughter had a mythical friend named Theodosia.*

mythology *n.* (body of) myths, folklore, °tradition, °lore, stories, mythos: *After only one generation, an extraordinary mythology has been created about Marilyn Monroe.*

N

nab *v.* °catch, °capture, °arrest, put *or* place under arrest, °seize, apprehend, °pick up, °bring in, take into custody, *Colloq* °pinch, collar, °run in, °nail, *Brit* °nick: *They finally nabbed that cat burglar in Hampstead.*

nag[1] *v.* **1** Sometimes, ***nag at.*** °annoy, °irritate, °irk, °pester, °criticize, °ride, °scold, °carp (at), °upbraid, badger, °harass, harry, vex, °henpeck, °torment, hector, °pick at, goad, °pick on, find fault with, berate, nettle, °bully, °provoke, °plague, °worry, °bother, *Brit* chivy *or* chivvy *or* chevy, *Colloq* needle: *He nags her day and night about going on a diet.* —*n.* **2** °scold, harpy, °pest, °shrew, virago, termagant, fishwife: *You certainly can be a terrible nag, can't you?*

nag[2] *n.* °jade, Rosinante; horse, °hack, pony, dobbin, racehorse, thoroughbred, *Slang* gee-gee, *US* hayburner, °plug, bangtail: *He lost his day's winnings on an old nag in the last race.*

nagging *adj.* distressing, °chronic, °continuous, °continual, °persistent, unrelenting, °relentless, recurring: *He complains of a nagging pain in his shoulder. I have a nagging feeling that I have an appointment to be somewhere.*

nail *n.* **1** fastener, °fastening, °spike, °pin: *The pieces were held together by several nails.* **2** fingernail, toenail, °claw, talon: *The detective found some fibers under the victim's nails.* **3** ***bite one's nails.*** °worry, agonize, °fret, lose sleep (over), °chafe, °suffer, *Colloq* stew (over *or Colloq* about): *Carl is biting his nails over the lab report on his cholesterol.* **4** ***hard*** or ***tough as nails.*** **(a)** °tough, °vigorous, °hardy: *After years of mountain-climbing, René is as hard as nails.* **(b)** °cold, unsentimental, °unsympathetic, unfeeling: *The boss, as hard as nails, doesn't care what you sacrifice to get the job done.* **5** ***hit the nail on the head.*** be accurate, be correct, be precise, be right, put (one's) finger on it: *When you said they were fools, you really hit the nail on the head.* **6** ***on the nail.*** °immediately, °at once, *Brit* °straightaway, *US* right away, °promptly, without delay, on the spot, *Colloq US* on the barrelhead: *He has always paid his bills right on the nail.* —*v.* **7** °attach, °secure, °join, °pin, °tack, clinch *or* clench; °fasten, °fix, °focus, rivet, °glue: *The door to the mysterious room was nailed shut. His eyes were nailed to the pressure gauge.* **8** See **nab,** above. **9** °hit, °strike; °punch; °shoot: *She nailed him with a left hook and he sank like a stone.* **10** ***nail down.*** °settle, °secure, °resolve, °complete, conclude, make final; °finalize: *Let's celebrate—I nailed down the order for ten thousand air conditioners.*

naive *adj.* naïve *or* naïf, °ingenuous, °innocent, credulous, °childlike, born yesterday, °unaffected, °unsophisticated, °inexperienced, °green, unworldly, °unsuspecting, unenlightened, unsuspicious, °trusting, trustful, °gullible, °artless, guileless, °simple, simplistic, simple-minded, unpretentious, unpretending, °candid, °natural: *Is Chatterley naive enough to believe that the gamekeeper's meetings with his wife were to discuss fox-hunting?*

naiveté *n.* naïveté *or naivety or* naïvety, ingenuousness, innocence, credulity, credulousness, °inexperience, °(blind) trust, gullibility, artlessness, callowness, guilelessness, °simplicity, unpretentiousness, °candor, naturalness, frankness, openness, °sincerity: *I hope that no one takes advantage of her naivety.*

naked *adj.* **1** stark-naked, unclothed, undraped, °bare, exposed, stripped, undressed, unclad, uncovered, bared, °nude, in the nude, *Colloq* in the altogether, in one's birthday suit, in the buff, °in the raw, *au naturel,* in a state of nature, *Brit* starkers, in the nuddy: *The two of them stood there, completely naked.* **2** unaided, unassisted: *Stars of the seventh magnitude or brighter are visible to the naked eye.* **3** °plain, °unadorned, unembellished, °stark, °overt, °patent, °obvious, °conspicuous, °manifest, °sheer, °undisguised, °unvarnished, °unmitigated, °evident, palpable, unconcealed, in plain sight *or* view, °blatant, °barefaced, undeniable, °glaring, °flagrant, unmistakable *or* unmistakeable, unalloyed, unmixed, °blunt, unadulterated, °pure: *He told her the naked truth about how he felt.* **4** unsheathed, unprotected, °bare, exposed: *How dare he show a naked sword in the presence of the Emperor!*

name *n.* **1** designation, °label, appellation, °term, °tag, °style, *Colloq* moniker *or* monicker, °handle: *His name is Chauncy but they call him Rusty.* **2** °reputation; repute, °honor, °esteem, °(high) regard, °respect, °rank, °standing, rating, °preeminence, °superiority, eminence, notability, °prominence, °prestige, °favor, °distinction, °renown, °fame, °popularity, °celebrity: *She has made a name for herself as a clever business executive. He thinks he has to protect his name as a ladies' man.* **3** °personage, °somebody, °celebrity, °star, superstar, °hero, VIP, °dignitary, luminary, *Colloq* big shot, °bigwig, big cheese, big name: *The program in-*

cluded some well-known names in the entertainment world. —*v.* **4** °label, °tag, °style, °entitle; °call, dub, °christen, baptize: *They named their book* The Alien Tongue. *Why would anybody want to name a child Kerdel?* **5** °choose, °elect, °select, °delegate, °nominate, °designate, °appoint; °identify, denominate, pinpoint, °specify: *She has been named "Woman of the Year." I asked her to name our wedding day. She refuses to name her attacker.* **6** ***name names.*** °identify, °specify, °mention, cite: *In exchange for a lighter sentence, the witness agreed to name names.*

nameless *adj.* **1** unnamed, innominate, °unidentified, anonymous, pseudonymous, °incognito, °unknown, °unheard-of, °unsung: *How much we owe the nameless inventor of the wheel!* **2** °inexpressible, indefinable, unidentifiable, unspecified, unspecifiable: *A nameless dread seized him as he entered the cave.* **3** °ineffable, unutterable, unspeakable, °unmentionable, °abominable, °horrible, indescribable, °repulsive: *Paganism allowed man to sink beneath a flood of nameless sensualities.*

namely *adv.* specifically, to wit, that is (to say), *id est,* i.e., *videlicet,* viz., *scilicet,* sc.; °for example, °for instance, *exempli gratia,* e.g. *or* eg *or* eg.: *We visited three countries, namely, Lichtenstein, Andorra, and San Marino.*

nap[1] *v.* **1** °doze, °nod (off), catnap, *Colloq* catch forty winks, drop off (to sleep), get some shuteye, snooze, zizz, *US* catch a few Z's: *I nap every afternoon.* —*adv.* **2** ***napping.*** °unawares, off guard, unexpectedly, in an unguarded moment: *The ball, hit to his backhand, caught him napping.* —*n.* **3** °doze, catnap, siesta, *Colloq* forty winks, shuteye, snooze, zizz, *Brit* lie-down: *Take a short nap before dinner.*

nap[2] *n.* °pile, °fiber, °texture, °weave, down, shag: *Install a carpet with a short nap for areas of heavy wear.*

narcotic *adj.* **1** soporific, stuporific, hypnotic, °sedative, somnolent, sleep-inducing, opiate, dulling, numbing, anesthetic, stupefacient, stupefying, stupefactive, tranquilizing, Lethean: *Most narcotic drugs may be sold only with a doctor's prescription.* —*n.* **2** °drug, soporific, stuporific, hypnotic, °sedative, opiate, anesthetic, stupefacient, °tranquilizer: *Many doctors are reluctant to prescribe narcotics.*

narrate *v.* °relate, °tell, °recount, °report, give an account (of), °recite, °rehearse, °repeat, °review, °unfold, °chronicle, °describe, °detail, °reveal, retail: *She narrated a bone-chilling story of intrigue and murder. Please narrate the events leading up to your finding the body, Mrs. Estoril.*

narration *n.* **1** °telling, relating, unfolding, recounting, chronicling, recording, describing; °report, °recital, °recitation, °rehearsal, °relation, °chronicle, °description, portrayal, detailing, °revelation, °story, °tale, °narrative: *His narration was accompanied by nervous gestures. Her narration disagrees with her husband's in certain essential respects.* **2** reading, voice-over: *The narration did not seem to be coordinated with the pictures on the screen.*

narrative *n.* **1** °story, °tale, °chronicle, °description, °revelation, portrayal, °account, °report, °record, °history, °recital, °statement: *The characterizations were poor, but the narrative was fascinating.* —*adj.* **2** storytelling, chronicling, anecdotal: *One of her long, narrative poems has been published in a collection.*

narrator *n.* °reporter, storyteller, °raconteur, taleteller, teller of tales, anecdotist *or* anecdotalist, relator, annalist, chronicler, describer, °author; voice-over: *We sat spellbound waiting for the narrator to continue.*

narrow *adj.* **1** constricted, °slender, °slim, °thin, restricted, °straitened, attenuated, narrowed; narrowing, tapering: *We squeezed through the narrow passage to freedom.* **2** confined, confining, °limited, °cramped, °close, °meager, pinched, °tight, incommodious: *I awoke in a cell so narrow I could scarcely move.* **3** °strict, °careful, °close, °precise, °exact, °exacting, °demanding, °finicky, finical, °sharp, °meticulous, °scrupulous, °fussy, °rigid, searching, °critical: *The suspected forgeries were submitted to the narrow scrutiny of several experts.* **4** restricted, °limited, circumscribed, proscribed, denotative: *I learned the meaning of charity in its narrowest sense.* **5** See **narrow-minded,** below. **6** °close, hairsbreadth, °lucky: *I'd had a very narrow escape, for the bullet just grazed my scalp.* **7** stingy, niggardly, parsimonious, °miserly, °tight, °sparing, tightfisted, °mean, mercenary *Brit* mingy, *Dialectal* °near, *Colloq* °close: *He was so narrow he barely allowed himself the essentials.* —*v.* **8** constrict, °limit, °qualify, °reduce, lessen, °diminish, °decrease: *She narrowed her chances of winning by buying only one lottery ticket.* **9** °limit, °restrict, °focus, confine, °concentrate, narrow down: *They have narrowed the search for the boy to the area near Chester.* —*n.* **10** ***narrows.*** strait(s), °channel, °passage: *The vessel approached the narrows at dead slow speed.*

narrowly *adv.* **1** °barely, °(only) just, °scarcely, °hardly, by a hair's breadth; by the skin of one's teeth, *Colloq* by a whisker: *The speeding lorry narrowly missed those children.* **2** closely, carefully, meticulously, scrupulously, °searchingly, critically: *She regarded him narrowly before speaking.*

narrow-minded *adj.* °bigoted, °prejudiced, illiberal, °narrow, °biased, °opinionated, °one-sided, °intolerant, nonobjective, °conservative, °reactionary, °parochial, ultraconservative, stiff-necked, °conventional, °hidebound, fundamentalist, literal-minded, narrow-spirited, mean-minded, mean-spirited, °petty, pettifogging, °small-minded, puritanical, unprogressive, °old-fashioned, old-fogyish, °strait-laced, *Colloq* °stuffy, *US* close-minded, °square, screed-bound, rednecked: *The vote reflected a reasonable balance between broad-minded and narrow-minded factions.*

nasty *adj.* **1** °foul, °filthy, °dirty, unclean, °offensive, °disgusting, nauseating, °revolting, °horrible, °loathsome, °repugnant, °repellent, °vile, odious, °obnoxious, objectionable, °nauseous, sickening, vomit-provoking, fetid *or* foetid, noisome, mephitic, °rank, malodorous, °rancid, noxious: *The nasty stench of rotting vegetation assailed our nostrils.* **2** unpleasant, °disagreeable, °unsavory, °painful, objectionable, annoying, °untoward, °awkward, °difficult, °bad, °serious: *Lord Petherbridge had some very nasty experiences in the war.* **3** °obscene, °dirty, °pornographic, °blue, smutty, °lewd, °vulgar, °sordid, °indecent, licentious, °gross, °coarse, °crude, °rude, ribald, °bawdy, °risqué, °off-color, °suggestive, *Colloq* X-rated, raunchy: *The shops are selling some rather nasty videotapes that I don't want the children to watch.* **4** unpleasant, °disagreeable, °ugly, bad-tempered, °vicious, currish, °surly, °abusive, °spiteful, irascible, ill-natured, ill-tempered, °cruel, °inconsiderate, °rude, churlish, °obnoxious, crotchety, curmudgeonly, °cantankerous, crabbed, °cranky, *US and Canadian* °mean: *Why is your father so nasty to everyone who wants to go out with you?* **5** °bad, °severe, °acute, °painful, °serious; °dangerous, °critical: *I got a nasty shock when I opened the cupboard door. That's a nasty wound you have there, Philip.*

nation *n.* °country, °state, °land, political entity, polity, °domain, °realm: *The countries of Europe are unlikely to become one nation.*

national *adj.* **1** nationwide, countrywide, °state, governmental, °civil; °public, °popular, *US* federal: *It took years to enact national clean air laws.* **2** °nationalistic, nationalist, °patriotic, jingoistic, chauvinistic: *During the war, national feelings ran high.* —*n.* **3** °citizen, °subject, °inhabitant, °resident; °native: *Stephenson left England years ago and is now an Australian national.*

nationalistic *adj.* nationalist, °patriotic, jingoist(ic), chauvinist(ic), xenophobic, isolationist: *As communications improved, nationalistic feelings were eroded.*

nationality *n.* **1** citizenship: *Some countries allow their citizens dual nationality.* **2** °race, °nation, ethnic group, ethnos, °clan, °tribe; °strain, °stock, °pedigree, °heritage, °roots, °extraction, bloodline, °breed: *The country became a melting pot of myriad nationalities.*

native *adj.* **1** innate, natal, °inborn, °natural, °inherent, congenital, indwelling, inherited, °hereditary, in the blood, °intrinsic, constitutional: *Early in life, Carla demonstrated a native ability for music.* **2** °domestic, °local, homegrown; °indigenous, autochthonous, °aboriginal: *The native oysters in this area are superb.* **3** °basic, °first, °best, °original, °exclusive: *Can you tell that my native language is Turkish?* **4** °national, ethnic, °clan, tribal: *We visited a Dutch town where the people wear traditional native dress.* **5** °aboriginal, °provincial, °local: *We soon fell in with the native custom of taking a siesta.* **6** born; by birth: *Are you a native Glaswegian?* —*n.* **7** aborigine, indigene, autochthon; °national, °citizen, °resident, °inhabitant: *It is not difficult to distinguish the natives from the tourists in London.*

natural *adj.* **1** °ordinary, °common, commonplace, °normal, °standard, °regular, °usual, °customary, unexceptional, °routine, °habitual, °typical, °everyday; °reasonable, °logical, °reasonable, °sensible, accepted: *They say that Count Dracula could not die a natural death. The natural thing to do in case of attack is to defend oneself.* **2** °normal, °ordinary, °regular, expected; °spontaneous: *The natural motion of the waves carried the bottle out to sea.* **3** °simple, °basic, °fundamental, °real, unartificial, °genuine, unembellished, °unadorned, unpretentious: *She has great natural beauty and needs no cosmetics.* **4** unstudied, unconstrained, °candid, °frank, °spontaneous, °unaffected, °easy, °honest, °straight, straightforward, °artless, guileless, °impulsive, °unpremeditated, °unaffected, °ingenuous, °unsophisticated, unsophistic(al): *His kindness is quite natural.* **5** See **native, 1,** above: *She has a natural gift for painting.* **6** °true, °real, °genuine, °actual, °authentic, °bona fide: *That, believe it or not, is his natural hair.* **7** °lifelike, true-to-life, °realistic: *Note the natural colors of the sea in this painting by Whistler.* **8** °illegitimate, bastard: *He was the duke's natural son and had no claim on the estate.* **9** °consistent, consonant, consequent, °logical, °reasonable, °fitting, °appropriate, °proper, expected, not incongruous, understandable: *In the circumstances, it would have been natural for her to despise Monty.* **10** °organic, organically grown, nonchemical, °health: *They eat only natural foods, which they grow themselves.* —*n.* **11** °genius, artist, °talent: *When it comes to chess, Boris is a natural.* **12** *Archaic* idiot, imbecile, simpleton, °fool, °half-wit: *One would have to be a natural to give his money to that mountebank.*

naturally *adv.* **1** °(as a matter) of course, °needless to say, to be sure, certainly, °surely, not unexpectedly, as expected *or* anticipated, °obviously, °clearly, logically, °consequently, as a consequence *or* result: *He treated her badly, so, naturally, she refuses to see him again.* **2** normally, °by nature, by character, °really, °actually, genuinely; inherently, instinctively, innately, congenitally: *My hair is naturally curly. He is not aloof, just naturally shy.* **3** unaffectedly, unpretentiously, °easily, candidly, °openly, °simply, plainly, °honestly, straightforwardly, uncomplicatedly: *If only politicians expressed themselves naturally and not pompously.*

nature *n.* **1** °quality, properties, °features, °character, °personality, °makeup, °essence, constitution, °identity, attributes, °disposition, temperament, complexion: *Only detailed analysis will reveal the nature of this compound. It is not in his nature to be envious.* **2** °universe, cosmos, °world, °creation, °environment: *Science fiction deals with phenomena and contrivances that contradict known laws of physics.* **3** scenery, countryside, wildness, primitiveness, °simplicity: *I often enjoyed sitting by the river, communing with nature.* **4** °kind, °variety, °description, °sort, °class, °category, °type, °genre, species; °stamp, °cast, °mold, feather, kidney, °color, °stripe: *The duties of the position are largely of a ceremonial nature.* **5** ***by nature.*** See **naturally, 2,** above.

naught *n.* nought, °nothing, °nil, °zero, aught *or* ought; °ruin, °destruction, °disaster, °collapse, °failure: *All my efforts had come to naught.*

naughty *adj.* **1** °mischievous, impish, puckish, roguish, scampish, °devilish; °frolicsome, °playful: *The children would get naughty the minute the teacher turned her back to the class.* **2** °disobedient, refractory, °insubordinate, °bad, °perverse, °wicked, fractious, °unruly, wayward, unmanageable, °ungovernable, °undisciplined, °defiant, °obstreperous: *Naughty pupils in this school are birched.* **3** °improper, °offensive, °vulgar, °indecent, °immoral, °risqué, °off-color, ribald, °bawdy, °blue, °pornographic, smutty, °lewd, °obscene, °dirty, *Colloq* X-rated, raunchy: *The teacher caught Kenneth reading a naughty book.*

nauseate *v.* °sicken, °disgust, °repel, °revolt, °offend: *Your hypocrisy nauseates me. When you get used to it, eating insects is no longer nauseating.*

nauseated *adj.* sickened, °disgusted, repelled, revolted, offended, °sick (to one's stomach), °queasy, °squeamish; seasick, carsick, airsick: *They were nauseated when they saw the cadavers. I became nauseated by the rocking of the boat.*

nauseous *adj.* nauseating, °loathsome, sickening, °disgusting, °repellent, vomit-provoking, °offensive, °revolting, °repugnant, °repulsive, °abhorrent, °nasty, °foul, unpleasant, stomach-turning, *Technical* emetic: *A nauseous odor emanated from the crypt.*

nautical *adj.* maritime, °marine, °seafaring, seagoing; naval; boating, yachting, sailing; navigational: *Britain has always been a nautical nation. One must observe the nautical rules of the road.*

navel *n. Technical* umbilicus, omphalos, *Colloq* bellybutton: *The dancer wore a ruby in her navel.*

navigable *adj.* **1** °passable, traversable, negotiable, unblocked, unobstructed, °clear: *The Thames is not navigable above Lechlade.* **2** maneuverable, sailable, controllable, steerable, yare: *My boat is navigable when under way at about four knots.*

navigate *v.* **1** °sail, voyage, °cruise, °journey; °cross, °traverse: *After navigating the Indian Ocean for a month, we reached the Seychelles.* **2** °maneuver, °handle, °sail, °guide, °pilot, °steer, °direct, °skipper, captain, *Nautical* con: *It was tricky navigating through the shoals. Ordway has navigated the company through rough waters over the years.*

navigation *n.* pilotage, helmsmanship, seamanship, steersmanship, steering, sailing: *The navigation of a small vessel in such a storm is no mean task.*

navigator *n.* °pilot, helmsman, seaman, tillerman, wheelman, steersman, °skipper: *The navigator without knowledge of local waterways must engage a pilot.*

navy *n.* °fleet, flotilla, naval force(s), armada, *Literary* argosy: *The entire navy sailed across the Channel to Le Havre.*

naysayer *n.* denier, refuser, disdainer, rejecter *or* rejector; prophet of doom, pessimist, °skeptic, dissenter, defeatist: *Don't expect any encouragement from a naysayer like Raymond.*

near *adv.* **1** close (by *or* at hand), not far (off *or* away), °nearby, nigh, in *or* into the vicinity *or* neighborhood, within (easy) reach: *Draw near and listen to my tale.* **2** °close to, next to: *Don't go near the edge!* **3** °nearly, °almost, just about, well-nigh, close to being; not quite, °virtually: *She was damn near killed in the car crash.* —*adj.* **4** °close, °imminent, °immediate, °impending, looming, coming, approaching, °forthcoming; in the offing, °at hand: *We hope to settle the pollution problem in the near future. The time is near for me to go.* **5** near by, °close, adjacent, next-door, close-by, °adjoining, abutting, °neighboring, contiguous: *My nearest neighbors live a mile away.* **6** stingy, °mean, niggardly, °miserly, parsimonious, °penurious, °cheap, penny-pinching, cheeseparing, °selfish, °close, tightfisted, closefisted: *He is so near he begrudged me even a candle to read by.* **7** °close, °intimate, connected, °related, °attached: *In case of emergency, list the name and address of a near relative.* **8** °close, °narrow, hairsbreadth: *Although I escaped, it was a near thing.* —*prep.* **9** °close to, in the vicinity *or* neighborhood of, next to, adjacent to, within reach of, within a mile of; a stone's throw from, not far (away) from: *She wouldn't allow the doctor to come near her. I live near the railway.* —*v.* **10** °approach, draw near *or* nigh, come close *or* closer, °verge on, approximate on, lean toward(s): *As summer nears, I think of going on holiday. The ship neared port. His estimate is beginning to near mine.*

nearby *adv.* **1** °close by, close at hand, not far off *or* away, in the vicinity *or* neighborhood, within (easy) reach, °about, °around: *Vultures and hyenas loitered nearby waiting for the lions to leave the kill.* —*adj.* **2** °close, within reach, °handy, °accessible, at *or* to hand, adjacent: *Nearby villagers helped put out the fire. We always kept a gun nearby because of pirates.*

nearly *adv.* **1** °almost, not quite, °about, °approximately, all but, just about, °virtually, well-nigh, °practically, as good as, more or less; °around, approaching, nearing, °barely, °hardly, °scarcely, °close to: *We were nearly there but couldn't make it. She was nearly ninety when her first book was published.* **2** closely, identically, °exactly, °precisely: *Her opinions agree most nearly with his. My paintings most nearly resemble Picasso's.*

nearsighted *adj.* **1** myopic, °shortsighted: *Being nearsighted, I must wear spectacles for driving.* **2** *Chiefly US* °shortsighted, °narrow-minded, blinkered, °narrow, close-minded, illiberal, °unthinking, °heedless, insular, °partial, °one-sided, °parochial, °unsophisticated, unimaginative, °biased, unobjective, °opinionated, °dogmatic, °prejudiced, °intolerant, °bigoted: *They maintain a nearsighted attitude toward associating with anyone outside their immediate clique.*

neat *adj.* **1** °tidy, °orderly, °clean, uncluttered, °trim, °spruce, natty, °fastidious, spick-and-span, °shipshape (and Bristol fashion), organized, well-organized, well-ordered, °systematic, *Brit dialect* trig, *Colloq* neat as a pin, *Brit* dinky: *His room was always neat—very odd for a teenager.* **2** °straight, unadulterated, unmixed, °undiluted, uncut, unblended, °pure; °on the rocks: *He drinks his whiskey neat.* **3** unembellished, °unadorned, unornamented, °simple, °elegant, °graceful, °smart, uncomplicated, °regular, °precise, copperplate; calligraphic: *An invitation in her neat handwriting awaited my return from abroad.* **4** deft, adroit, °clever, °efficient, °ingenious, °expert, °practiced, °skillful, °dexterous: *He contrived a neat plan to avoid paying taxes.* **5** °fine, wonderful, °marvelous, °great, °splendid, °excellent, °exceptional, °capital, °grand, first-class, *Colloq* °cool, smashing, °keen, °nifty, topnotch, A-1 *or* A-one, *Brit* top-hole, *Chiefly US* A-OK, *Slang* °swell, far-out, °boss, *Brit* topping, *US and Canadian* spiffy: *She found a really neat way to get boys to ask her out. Gordon has a neat new car.*

neaten *v.* Often, **neaten up.** tidy (up), straighten (up *or* out), °clean (up), °spruce up, °(put in) order, *Brit dialect* trig: *If you don't neaten up your room, Kiley, you may not borrow the car.*

nebulous *adj.* °vague, °hazy, clouded, unclear, °obscure, °indistinct, °fuzzy, °muddy, ill-defined, °shapeless, amorphous, blurred, indeterminate, °murky, °opaque, turbid, °dim, foggy, °faint, °pale: *He has only the most nebulous idea of what the lecture was about.*

necessarily *adv.* inevitably, unavoidably, inescapably, axiomatically, inexorably, ineluctably, irresistibly, incontrovertibly, automatically, °naturally, °(as a matter) of course, as a result, certainly, °surely, to be sure, like it or not, °willy-nilly, perforce, of necessity, by definition: *If you accept his premise, then you must necessarily accept his conclusion. You don't necessarily have to be rich to be happy, or so Jean-Paul Getty told me.*

necessary *adj.* **1** °indispensable, °essential, required, needed, compulsory, requisite, °vital, demanded, °imperative, °obligatory, needful, °of the essence, °important, of the utmost importance, top priority, high priority, °urgent, exigent, compelling, life-and-death *or* life-or-death: *A good diet is necessary for good health. It is necessary that you come at once. Take the necessary steps to get the job done.* **2** °inevitable, °unavoidable, inescapable, ineluctable: *In the circumstances, we find it necessary to ask for your resignation.* **3** °sure, °certain, °predetermined, predestined, °fated, inexorable; resulting, resultant: *The necessary outcome of the affair was that the child was returned to her natural parents.* —*n.* **4** See **necessity, 1,** below.

necessity *n.* **1** °requirement, °essential, °necessary, requisite, °need, °prerequisite, °basic, °fundamental, sine qua non, desideratum, constraint: *I was marooned for a year with only the bare necessities of life. We regard honesty as a necessity in a bank manager.* **2** indispensability, unavoidability, needfulness, inexorability: *The necessity for exercise is often met by private health clubs.* **3** °poverty, °want, indigence, °need, destitution, penury, °straits, °difficulty, °difficulties, pauperism, neediness: *The extreme necessity of the unemployed demands that the government act quickly.* **4** °urgency, °emergency, °crisis, °misfortune, exigency, °pinch, °extreme, matter of life and death: *She has made a virtue of necessity. Electric power is no longer a matter of choice but of necessity.*

need *v.* **1** °require, °demand, °want, be in want of, °call for, have need of *or* for; °lack, °miss, have occasion for: *This railing needs a coat of paint. Although she may want more money, she doesn't need it. Do you need anything to make you more comfortable?* —*n.* **2** °necessity, °requirement; °call, °demand, constraint: *There's no need to shout—I can hear you. There is a need to keep this matter confidential. This facility will meet our electricity needs for decades.* **3** °essential, °necessary, requisite, °prerequisite, °necessity, °basic, °fundamental, sine qua non, °necessary, desideratum: *I am perfectly capable of taking care of my family's needs.* **4** °distress, °difficulty, °trouble, (dire *or* desperate) straits, °stress, °emergency, exigency, °extremity, neediness, needfulness; °poverty, penury, impecuniousness, destitution, °privation, deprivation, indigence, beggary: *She was very supportive in his hour of need. The need of the people in that district is heart-rending.* **5** °want, °lack, °dearth, °shortage, paucity, °scarcity, insufficiency, desideratum: *The need for medical supplies was most sharply felt in areas that were already ravaged by famine.*

needless *adj.* **1** °unnecessary, °nonessential, unessential, unneeded, unwanted, °useless, uncalled-for, °gratuitous, °superfluous, °redundant, °excess, °excessive, tautological, °dispensable, °expendable, supererogatory, *de trop,* pleonastic: *He went to a lot of needless trouble to change the tire. It is needless to raise further questions.* **2** ***needless to say.*** °naturally, °(as a matter) of course, °obviously, it goes without saying: *Needless to say, they will have to get married.*

needy *adj.* °poor, °indigent, poverty-stricken, °destitute, °impoverished, penniless, impecunious, necessitous, underprivileged, °deprived, disadvantaged, below the poverty level, in dire straits, in *or* on the way to the poorhouse, in reduced circumstances, °down-and-out, °insolvent, *Colloq* °on one's uppers, flat *or* stony broke, °hard up, strapped, pinched, on the breadline, up against it, *Brit* on the dole, *US* dead broke, on welfare, on relief: *It is not enough to look after needy families only at Christmas time.*

negative *adj.* **1** °contradictory, anti, °contrary, dissenting, dissentious, disputing, disputatious, °argumentative, adversarial, adversative, antagonistic, antipathetic, adverse, *US* °adversary: *He has adopted a very negative attitude toward his job. Few negative voices were heard on the issue.* **2** °pessimistic, °unenthusiastic, °cool, °cold, uninterested, unresponsive: *The reaction to our offer has been largely negative.* **3** nullifying, annulling, neutralizing, voiding, canceling: *The laws are mainly negative, listing only things one must not do.* **4** negating, refusing, denying, gainsaying, °opposing: *The judge came to a negative decision regarding bail.* **5** ***in the negative.*** negatively, "No": *Asked if she wanted to go, she replied in the negative.*

neglect *v.* **1** °disregard, °ignore, °slight, pay no attention to, be inattentive to, °overlook, °pass by, °spurn, °rebuff, °scorn, disdain, contemn, *Colloq* cold-shoulder: *Scholars neglected his work for years.* **2** °fail (in), °omit; °disregard, let slide *or* pass, be remiss (about *or* in *or* regarding), °abandon, lose sight of, °forget, °shirk: *Have I neglected telling you how much I love you? Sybil has neglected her obligations.* —*n.* **3** °disregard, °disrespect, inattention, °indifference, slighting, unconcern, °oversight, heedlessness, neglectfulness, carelessness, inadvertence: *We lost business to our competitor through simple neglect.* **4** °negligence, laxity, laxness, slackness, neglectfulness, passivity, passiveness, °inactivity, inaction, dereliction, °default, °failure, °failing, remissness: *She has been accused of neglect in looking after her children properly.*

negligence *n.* inattention, inattentiveness, °indifference, carelessness, unconcern, dereliction, °failure, °failing, heedlessness, laxity, laxness, °disregard, °oversight, °omission, inadvertence, °neglect, remissness, forgetfulness, oscitancy *or* oscitance: *The car crash was attributed to the truck driver's negligence.*

negligible *adj.* °insignificant, °minor, unimportant, °trifling, trivial, °inconsequential, piddling, inappreciable, °small, °slight, °paltry, nugatory, °worthless, °petty, °niggling, not worth mentioning *or* talking about: *The differences between the two plans are negligible.*

negotiate *v.* **1** °deal, °bargain, °dicker, °haggle, chaffer, palter; °discuss, °debate, mediate, °consult, °parley, °speak, °talk, °transact, °come to terms: *A conglomerate is negotiating to buy our company. The company asked me to negotiate on our behalf.* **2** °arrange (for), °organize, orchestrate, °conduct, °handle, °maneuver, °manage, °engineer, °work out, °settle, °get, °obtain, bring off *or* about, °carry off, °accomplish, do, °execute, °effect, °complete, conclude, *Colloq* °pull off: *Will you be able to negotiate a loan for the car?* **3** °maneuver, °clear, get through *or* past *or* round *or* over, °pass, °cross, *Colloq* make (it (through *or Colloq* past *or Colloq* around *or Colloq* over)): *Will you be able to negotiate the barbed-wire fence?*

negotiation *n.* **1** °discussion, mediation, arbitration, bargaining, °parley, parleying, °talk, coming to terms: *The disarmament negotiations have dragged on for years.* **2** °deal, °bargain, °transaction, °agreement, °arrangement, °understanding, °determination, °decision, °settlement; °contract, °pact, °compact, covenant, concordat, °treaty: *All parties seem pleased with the final negotiation.*

negotiator *n.* arbitrator, arbiter, °mediator, °moderator, diplomat, °ambassador, °go-between, middleman, intercessor, interceder, intervener, °agent, °broker: *An independent negotiator was invited to the bargaining table with the union and management representatives.*

neighborhood *n.* **1** locality, °area, °region, °vicinity, vicinage, environs, °quarter, °district, precinct(s), purlieus, °locale; surroundings, confines: *Houses in the neighborhood of the blast were leveled.* **2** ***in the neighborhood of.*** °approximately, °about, °around, °nearly, °practically, °close to, °almost, more or less, *Colloq* in the ballpark of, in the region of, *Brit* getting on for, not far off, *US* within an eyelash of, *Slang* as near as dammit to: *The playground will cost in the neighborhood of five thousand dollars.*

neighboring *adj.* °nearby, °near, °around, adjacent (to), °surrounding, °adjoining, contiguous (to), °touching, bordering (on), next (to), nearest: *Owners of properties neighboring the nuclear power plant worry about radiation. The neighboring towns will participate in the fair.*

neighborly *adj.* °friendly, °cordial, °warm, °amiable, °agreeable, affable, companionable, well-disposed, °kindly, °kind, well-wishing, °genial, °sociable, °social, °harmonious, °considerate, °thoughtful, °helpful, °gracious, °courteous, °civil: *It was quite neighborly of the Constables to look after our cat.*

neologism *n.* neoterism, coinage, neology, nonce word; °blend, portmanteau word: *Lexicographers must decide on the neologisms to be added to their dictionaries.*

nerve *n.* **1** °courage, coolness, boldness, °bravery, intrepidity, °determination, valor, °daring, fearlessness, dauntlessness, °pluck, mettle, °spirit, °fortitude, °will, °tenacity, steadfastness, staunchness, firmness, doughtiness, resoluteness, *Colloq* °guts, °grit, °gumption, °spunk, *US* sand, *Brit* °bottle, *US* moxie, *Taboo slang* balls: *It took a lot of nerve to go back into that burning building.* **2** °effrontery, brazenness, °gall, °impertinence, °brass, °impudence, insolence, audacity, brashness, °presumption, presumptuousness, temerity, *Colloq* cheek, °sauce, chutzpah, *Slang* crust: *You have a lot of nerve, talking to your mother that way!* **3** ***get on someone's nerves.*** °annoy, °irritate, °upset: *That loud rock music gets on my nerves.* **4** ***nerves.*** °tension, nervousness, hysteria, °anxiety, fretfulness, °stress, °worry, apprehension, °fright, *Colloq* the jitters, *Slang* the willies, the heebie-jeebies, *US* the whim-whams: *I had a bad case of nerves before learning the doctor's diagnosis.*

nerve-racking *adj.* nerve-wracking, °harrowing, °agonizing, distressing, °trying, vexing, vexatious, °troublesome, worrisome, °irksome, irritating: *Waiting *for the names of the survivors was the most nerve-racking experience of my life.*

nervous *adj.* **1** highly strung, °excitable, °sensitive, °tense, °agitated, wrought up, worked up, °upset, flustered, ruffled, °disturbed, perturbed, distressed, °worried, °anxious, troubled, °concerned, disquieted, edgy, °on edge, on tenterhooks, fidgety, °fretful, uneasy, apprehensive, frightened, °fearful, °shaky, °scared, skittish, *US* on a tightrope, *Colloq* °jumpy, jittery, flappable, in a stew, in a dither, in a sweat, in a tizzy, in a flap, uptight, *Brit* nervy, *US* on pins and needles, *Slang* strung out: *Tom is nervous because he has to give a speech.* **2** °difficult, °tense, °critical: *There were a few nervous moments before we knew if the rope would hold.*

nest *n.* **1** roost, °perch, eyrie *or US also* aerie, den, °lair: *The birds, which mate for life, return to the same nest each year.* **2** snuggery, °retreat, °refuge, °haunt, °hideaway, hide-out; °resort: *We have a little nest in the country which we escape to on weekends.*

nestle *v.* °cuddle (up), °snuggle (up), °huddle, curl up, nuzzle (up): *They nestled close together to keep warm.*

net[1] *n.* **1** °network, netting, °mesh, meshwork, °web, webbing, openwork, lattice, latticework, trellis, trelliswork, lacework, reticulum, reticle, rete, plexus, grid, gridwork, grille, °grate, °grating, fretwork; sieve, °screen, strainer, sifter: *They watched the fishermen mending their nets.* —*v.* **2** °catch, °capture, °trap, entrap, °snare, ensnare, °bag: *As I bring the fish close to the boat, you net it with this.*

net[2] *n.* **1** nett, °(net) profit, °gain, °earnings, return(s), *Colloq US* °take: *This year's net was 15 percent higher than last year's.* —*adj.* **2** °clear, after deductions, after taxes, take-home, °final, bottom-line: *The tax increase was greater than my salary increase, so my net income was lower this year.* **3** °final, °end, closing, concluding, conclusive, °effective, °ultimate: *The net result of the advertising campaign was a sales increase of 18 percent.* —*v.* **4** °make, °realize, °clear, take home, °bring in, °earn, °pocket, °take in, °get: *How much did you net last year—after taxes, that is.*

network *n.* **1** See **net**[1], 1, above. **2** °system, °arrangement, °structure, °organization, complex, grid, crisscross, °web, plexus; °maze, labyrinth, jungle, °tangle: *She has developed a worldwide network of contacts for her business. Before transistors, the back of a radio was a network of multi-colored wires.*

neurotic *adj.* psychoneurotic, °unstable, °disturbed, °confused, irrational, disordered, maladjusted, °distraught, °overwrought, °anxious, °nervous, °obsessive, °deviant, °abnormal: *Most people are neurotic about something.*

neuter *adj.* **1** asexual, sexless, epicene: *Worker bees are neuter, neither male nor female.* —*v.* **2** desex *or* desexualize, °doctor; castrate, emasculate, geld, capon *or* caponize, eunuchize; spay, ovariectomize, oophorectomize; *Colloq* °fix, °cut, *US* °alter: *We had our cat neutered after her first litter.*

neutral *adj.* **1** nonbelligerent, noncombatant, unaligned, °nonaligned, unaffiliated, uninvolved, unallied, nonallied, °nonpartisan, °impartial, °disinterested, °indifferent, °dispassionate, unbiased, uncommitted, °noncommittal, °aloof, °withdrawn, °detached, °remote, removed: *Switzerland has remained neutral during both world wars.* **2** °dull, °drab, °colorless, achromatic, toneless, indeterminate, °washed-out, °pale, °indefinite, °indistinct, °indistinguishable, indeterminate, °vague, °drab, beige, ecru: *A neutral wallpaper color won't clash with the paintings.*

neutralize *v.* °void, annul, °cancel (out), nullify, invalidate, negate, °delete, °undo, make *or* render ineffective, counterbalance, °counteract, °offset, °equalize, °even, °square, compensate for, °make up for: *The forces on both sides of the sail, being equal, neutralize each other.*

never *adv.* **1** at no time, not ever, not at any time, on no occasion, under no circumstances *or* condition(s), on no account, not at all, *Colloq* not in a million years, not till hell freezes over: *You are never to use such language again!* **2** in no way, not in any way, not in the least, not in any degree, not under any condition; (*postpositive*), not: *Never fear, for I am near.*

nevertheless *adv.* °still, °notwithstanding, °yet, in spite of that, despite that, °nonetheless, °regardless, be that as it may, for all that, °even so, but, °however, just *or* all the same, everything considered, °at any rate, anyway, °in any case, °in any event, °at all events, *Literary* withal: *He said he dislikes sweet things; nevertheless, he ate all the chocolates.*

new *adj.* **1** °novel, °original, °unique, °unusual, °different, °fresh, °creative, °imaginative, °brand-new: *Ramona has a new idea for the sales campaign.* **2** °latest, °late, °modern, °contemporary, modish, °stylish, °fashionable, °chic, °recent, advanced, up-to-date, °brand-new, late-model, *Colloq* °trendy, *Slang* mod, °hip: *MacGregor buys a new car every year—it has to be equipped with the newest gadgets and set off his wife's new wardrobe.* **3** °fresh, °further, additional, supplemental, °supplementary: *Has the new issue of* Verbatim *come out yet?* **4** °unfamiliar, °unknown, °strange, °different; °unique, unheard of: *I hear there's a new girl in town. I want to introduce my new friend, Dmitri Theotocopoulis. Every year they discover a new virus.* **5** revitalized, reborn, renewed, rejuvenated, changed, altered, redone, restored, redesigned, remodeled: *I saw before me a new Marie. They published a new version of the Bible.* **6** °inexperienced, °green, °fresh, °callow, °unfledged, budding, °immature, unripe, untrained: *Let us look over the new recruits, Sergeant.* **7** °late, °young, °recent: *We found newer fossils at higher levels.* **8** °uncharted, unexplored, untrodden, °unknown, °experimental: *Astronomers are breaking new ground in the analysis of pulsars.*

newcomer *n.* **1** °alien, °immigrant, °foreigner, outlander, °stranger, °settler, colonist, °outsider: *The newcomers quickly established themselves and became self-sufficient.* **2** beginner, °amateur, °novice, proselyte, neophyte, tiro *or* tyro, °initiate, trainee, °learner, fledgling *or Brit also* fledgeling, *US* freshman, *Colloq* °greenhorn: *Though experienced in book publishing, he was a newcomer to magazine publishing.*

news *n.* **1** tidings, °word, °information, °advice, °intelligence; °rumor, °talk, °gossip, hearsay, °dirt, °scandal, exposé, *Colloq* info, °lowdown, °scoop, *US* scuttlebutt, *Slang* °dope: *What's the latest news about the situation in the Middle East?* **2** dispatch *or* despatch, °report, °account, °story, communication, °bulletin, communiqué, °announcement, °information, °message, °word, °statement, °(press) release, °(news) flash: *The news from our correspondent contains no mention of finding a cache of arms.* **3** newscast, news broadcast *or* telecast, news program: *Please be quiet so that I can listen to the news.* **4** °(good) copy, front-page news, °(hot) item: *The royal family is always news.*

nice *adj.* **1** °pleasant, °agreeable, °amiable, °amicable, °friendly, °cordial, °warm, °gracious, warmhearted, °kind, °kindly, °outgoing, charming, °genial, °delightful, °courteous, °polite, °refined, gentlemanly, °ladylike, winsome, °likable, °attractive: *They are one of the nicest couples I have ever met.* **2** °good, °satisfactory, commendable, °worthy, °worthwhile: *The manager said that I had done a nice job in preparing the specifica-*

tions. **3** °good, °fine, °superb, °superior, °attentive, °sharp, °acute, °keen, °careful, °exact, °exacting, rigorous; °precise, °accurate, unerring, °scrupulous, °meticulous, punctilious, °discriminating, discriminative, °perceptive, °delicate, °fastidious, °flawless, °faultless, °subtle, °strict, °close, °small, °slight, °minute, complex, °complicated, °intricate: *A diamond cutter must have a nice eye for detail. It is difficult to maintain nice shades of meaning among certain words.* **4** °delicate, °subtle, °sensitive, °exquisite, hair-splitting, overnice, °fine, °critical, °ticklish, °dangerous, °precarious, °perilous, *Colloq* °hairy: *The matter of Hong Kong requires some nice political negotiations.* **5** °trim, well-turned-out, °tidy, °neat, °fine: *Don't you want to look nice for your birthday party?* **6** ***nice and ...*** pleasantly, delightfully, pleasingly, agreeably, enjoyably, gratifyingly, satisfyingly, comfortably: *It's nice and warm by the fire.*

niche *n.* **1** °recess, °hollow, alcove, °nook: *A small statue of Buddha stood in a niche in the wall.* **2** °place, °position, *Colloq* °slot, pigeonhole: *Humphrey has finally found a niche for himself working for the FBI.*

nick *n.* **1** °cut, °notch, °chip, °gouge, °gash, °scratch, dent, °indentation, °flaw, °mark, °blemish, °defect: *Be careful of that nick in the rim of the glass.* —*v.* **2** °cut, °notch, °chip, °gouge, °gash, °scratch, dent: *You nicked the top of the desk when you dropped the chisel on it.*

nickname *n.* **1** pet name, sobriquet, epithet, agnomen, *Colloq* moniker *or* monicker, °handle: *Her real name is Josephine, but her nickname is Dusty.* **2** °diminutive, shortening: *A common nickname for Terence is Terry.*

nifty *adj.* **1** °smart, °stylish, modish, °chic, °spruce: *I borrowed a nifty outfit from Grandma for the flapper's costume ball.* **2** °healthy, in good form, spry, °energetic, °agile, °quick: *I'm not as nifty as I was in 1950.* **3** °excellent, °neat, °great, °splendid, °fine, °clever, °skillful, apt, °suitable: *Having a picnic was a nifty idea. That was as nifty a bit of bargaining as I've ever seen, Captain.* **4** °satisfactory, °satisfying, °good, °profitable, °substantial, °considerable: *He made a nifty profit on the sale of his house in Chelsea.*

niggle *v.* °find fault, °nag, °carp, °fuss, °cavil, °criticize; °complain, *Colloq* grouse, *Slang* °bitch, *US* kvetch: *I do wish she would stop niggling when we cannot do anything about what is already done.*

niggling *adj.* **1** irritating, worrying, worrisome, °irksome, vexing, vexatious, annoying, °troublesome: *There are a few niggling matters that I must see my accountant about.* **2** °petty, nugatory, °trifling, trivial, °fussy, °insignificant, unimportant, °inconsequential, °frivolous, *Colloq* piddling, nit-picking, *US and Canadian* picayune: *Billings always ignores the core of a problem and occupies himself with the niggling details.*

night *n.* **1** (Stygian *or* Cimmerian) dark *or* darkness *or* blackness *or* gloom; nighttime, shades of night, *Formal* tenebrosity *or Formal* tenebrousness *or Formal* tenebriousness: *The strange creature slunk off into the night.* **2** nightfall, gloaming, °twilight, °dusk, eventide, °evening, evensong, edge of night, sunset, sundown, end of day, vespers: *When night comes, one can hear the frogs calling from the pond.* **3** ***night and day.*** °all the time, continually, incessantly, unceasingly, continuously, unendingly, endlessly, round-the-clock, ceaselessly, °nonstop: *Those beasts keep up their caterwauling night and day.*

nightly *adj.* **1** every night, each night, each and every night, night after night: *The watchman does his nightly rounds.* **2** nighttime, nocturnal, bedtime: *A nightly glass of warm milk promotes sound sleep.* —*adv.* **3** every night, each night, nights, after dark, after sunset; nocturnally: *The bird sings nightly outside my window.*

nightmarish *adj.* °frightening, °terrifying, alarming, horrific, °horrible, °dreadful, °awful, °ghastly, dismaying, °agonizing, worrisome, exasperating, frustrating, Kafkaesque, *Colloq* creepy, °scary: *Dealing with officialdom can be a nightmarish experience when abroad.*

nil *n.* °nothing, °zero, °nought, ought *or* aught, *Tennis, table tennis, etc.* love, *Brit cricket* °duck, *US* goose egg, *Slang US* zip, zilch: *Cleveland 4; Boston Nil.*

nimble *adj.* **1** °agile, °lively, °active, °light, lithe, limber, spry, °sprightly, °brisk, °smart, °energetic, °rapid, °quick, °swift, adroit, deft, °dexterous; nimble-fingered; nimble-footed: *She's as nimble as a cat on a hot tin roof.* **2** °agile, °alert, °acute, nimble-witted, °quick-witted, ready-witted, °intelligent, °keen, °sharp; °smart, °brilliant, sparkling, °scintillating, coruscating: *Despite his inability to make decisions, Desmond has quite a nimble mind.*

nip[1] *v.* **1** °bite, nibble; °pinch, °snip, °clip, °cut, °snap, °tweak, twitch, °trim, °lop, crop, shear; °grip, °squeeze: *The dog nipped the postman's ankle. Nip off the suckers to promote healthy growth of the tomatoes.* **2** ***nip in the bud.*** °stop, °arrest, °check, °thwart, °frustrate, °stymie, °forestall; °quash, °squelch, °suppress, °extinguish, °put down: *The revolt of the army officers was nipped in the bud.* —*n.* **3** °bite, nibble, °morsel, °pinch, °snip: *The deer had taken nips off the tips of the shrubbery.* **4** °chill, coldness, iciness, sharpness, °tang, °bite: *There's a definite wintry nip in the air tonight.*

nip[2] *n.* °taste, °drop, °sip, *soupçon,* °portion, °swallow, °gulp, °mouthful, °finger, draft *or Brit* draught, °peg, °tot, *Scots* dram, *Colloq* snort, °shot: *I had a few nips of brandy to ward off the cold.*

nobility *n.* **1** nobleness, °dignity, °grandeur, illustriousness, greatness, °glory, °influence, °authority, °leadership, °distinction, °probity, °integrity, °excellence, goodness, °character, °rectitude, righteousness, ethics, °honesty, honorableness, decency, justness, high-mindedness, magnanimity, °prestige, loftiness, primacy, °significance: *The man's nobility was evident from his mien and bearing.* **2** °rank, °position, °class, °birth, blue blood: *Their nobility is recognized only at court and in fashionable society.* **3** ***the nobility.*** the gentry, the elite, the aristocracy, *Colloq* the upper crust, the ruling class(es), °the Establishment, *US* the Four Hundred: *With a name like Hohenzollern, her family must be from the European nobility.*

noble *n.* **1** nobleman, noblewoman, aristocrat, patrician, °lord, lady, °peer; gentleman, gentlewoman, *Colloq* blue blood: *The nobles united and forced King John to sign the Magna Carta.* —*adj.* **2** highborn, °high-class, upperclass, aristocratic, titled, high-ranking, lordly, patrician, *Colloq* blueblood(ed): *She comes from a noble Austrian family.* **3** °dignified, °eminent, °distinguished, august, °grand, °lofty, °elevated, °illustrious, °prestigious, °preeminent, °noted, honored, esteemed, °celebrated, °renowned, acclaimed, respected, venerated: *The noble Knights of the Round Table have become legend throughout the world.* **4** °upright, °righteous, °honorable, °honest, °virtuous, incorruptible, °chivalrous, °staunch, °steadfast, °true, °loyal, °faithful, °trustworthy, °true, °principled, °moral, °good, °decent, self-sacrificing, magnanimous, °generous: *The qualities that make a noble friend make a formidable enemy.* **5** °splendid, °magnificent, °imposing, °impressive, °stately, °exquisite, °sublime, °grand, °striking, °stunning, °superb, °admirable, °elegant, °rich, °sumptuous, °luxurious: *The Taj Mahal is probably one of the noblest works of man.*

nobody *pron.* **1** no one, not anyone, no person: *We had to wait because nobody was in the shop when we entered.* —*n.* **2** nonentity, °unknown, °zero, cipher,

Colloq °nothing: *Overnight, Cosgrove went from being a celebrity to being a nobody.*

nod *v.* **1** °greet, °acknowledge, °recognize: *The maître d'hôtel nodded to me as I entered the restaurant.* **2** say yes; °consent, assent, °agree, concur, acquiesce: *Asked if she had seen Nick, the barmaid nodded. I asked permission to leave the room and the teacher nodded.* **3** °doze (off), °nap, drowse, drop off, fall asleep: *Exhausted travelers nodded in their chairs waiting for their flights to be announced.* **4** °slip (up), °err, make a mistake, be mistaken *or* wrong; be careless *or* negligent *or* lax *or* inattentive: *Sometimes, even Homer nods.* —*n.* **5** °signal, °sign, °cue, °indication, °gesture: *I saw him give a nod to the barber, who suddenly held a cutthroat razor to my throat.* **6** °approval; °consent, acquiescence, concurrence, assent, °agreement, *Colloq* OK *or Colloq* O.K. *or Colloq* o.k. *or Colloq* okay: *The company has given me the nod to open an office in Acapulco.*

nodding *adj.* °casual, °slight, °superficial, °distant; °incomplete: *I know Featherstonehaugh slightly—he's a nodding acquaintance. It was clear that the violinist had only a nodding acquaintance with the finer points of Mozart's concerto.*

noise *n.* **1** °sound, clamor, °crash, °clap, °clash, clangor, °din, °thunder, thundering, rumble, rumbling, °outcry, hubbub, °uproar, hullabaloo, °racket, charivari *or US and Canadian also* shivaree, °rattle, caterwauling, °rumpus, °blare, °blast, blasting, bawling, babel; commotion, °bedlam, °fracas, °tumult, °pandemonium, turmoil; discordance, dissonance, cacophony; *Archaic* alarms *or* alarums and excursions, *Colloq* ruckus, ruction, ballyhoo: *I couldn't sleep because of the unbearable noise from the party next door. You may call acid rock music, but she calls it noise.* **2** °sound, °disturbance: *Did you just hear that strange noise? It's only the noise of the crickets.* —*v.* **3** Often, ***noise about*** or ***around.*** °circulate, °spread, °rumor, bruit (about): *It is being noised about that John and Marsha are getting a divorce.*

noiseless *adj.* muted, °quiet, °soft, hushed, muffled, deadened, dampened, damped; °silent, °mute, °still, °inaudible, soundless: *We watched the noiseless boats gliding past.*

noisy *adj.* °loud, deafening, earsplitting, jarring, °grating, °harsh, °piercing, °shrill, °discordant, unmusical, dissonant, cacophonous *or* cacophonic, resounding, clarion, clamorous, clangorous, °thunderous, °uproarious, blaring, blasting, °obstreperous, vociferous, °boisterous, °tumultuous, °riotous: *I could hear nothing over the noisy airplane engines. When she arose to speak, the noisy crowd fell silent.*

nominal *adj.* **1** °titular, in name only, °formal, °pretended, °so-called, °self-styled, *soi-disant,* °professed, purported, °supposed, would-be, representational, represented, supposititious *or* suppositious; proposed, propositional; °puppet: *White is the nominal chairman, but Burlington actually runs the company.* **2** °insignificant, trivial, °trifling, °minor, minuscule, °tiny, °small, °insubstantial, °minimal, inconsiderable, °token: *We charge a nominal fee for the analysis if you refuse our recommendation.*

nominate *v.* °choose, °select, °name, °appoint, °designate, °suggest, °offer, °submit, °recommend, °propose, °present, put up *or* forward; °forward; *Formal* °put forth: *Barker has been nominated for the presidency.*

nominee *n.* °candidate, officeseeker, designee, selectee, appointee, assignee: *We must choose a nominee to run in the next election.*

nonaligned *adj.* uncommitted, nonallied, nonaffiliated, unaligned, unaffiliated, unallied; °neutral, °impartial: *There are several nonaligned nations that remain independent of the influence of the superpowers.*

nonbeliever *n.* unbeliever, disbeliever, cynic, doubting Thomas, doubter, °skeptic, freethinker, agnostic, atheist, nullifidian; °infidel, °heathen, °pagan: *During his reign, thousands of nonbelievers were put to the sword.*

nonchalant *adj.* °cool, unexcited, unexcitable, unperturbed, imperturbable, undisturbed, untroubled, unflappable, unruffled, °dispassionate, unemotional, °detached, °distant, unconcerned, °indifferent, pococurante, insouciant, uninterested, °aloof, °blasé, °offhand, °calm, °collected, composed, easygoing, free and easy, happy-go-lucky, °casual, °relaxed, at ease; °unenthusiastic, apathetic; *Colloq* laid-back, together: *How can you be so nonchalant about important issues that affect all our lives?!*

noncommittal *adj.* °wary, °cautious, °careful, °gingerly, °guarded, °(playing it) safe, circumspect, watchful, °prudent, canny, °tentative, on guard, °reserved, °cool; precautionary *or* precautional, precautious; *Colloq* playing it cool, playing it safe, playing it *or* one's cards close to the chest: *The company is still noncommittal on the takeover of Continental, Ltd.*

noncompletion *n.* nonfulfillment, unfulfillment, nonperformance, incompleteness, deficiency: *We regret the noncompletion of the work on your house and will refund in full the amount already paid.*

noncompliance *n.* disobedience, nonconformity, nonobservance, °disregard, disregarding, °failure, noncooperation, uncooperativeness, unresponsiveness, °rejection, °refusal, °denial: *If you fail to provide the records requested by the court, you can be charged with noncompliance.*

nonconformist *n.* **1** nonconformer, °renegade, maverick, °rebel, °radical, °individualist, heretic, dissenter, °dissident, iconoclast, loner, °exception, anomaly: *In the 1960s, Alastair counted himself among the nonconformists who wore unconventional clothes and flouted conventional behavior.* —*adj.* **2** nonconforming, °renegade, maverick, °rebellious, °radical, individualist(ic), °heretical, dissentient, °dissident, iconoclastic: *Suffragists were jailed for their nonconformist ideas.*

nondescript *adj.* indescribable, unclassifiable, unclassified, °ordinary, common-or-garden variety, °common, commonplace, unremarkable, °colorless, °drab, °everyday, °bland, uninteresting, insipid, characterless, undistinctive, unexceptional: *He was wearing nondescript clothing and blended into the crowd.*

none *pron.* no one, not anyone, °nobody, no person; not one; not any; °nil: *Of all those I met, none impressed me more than Kathy. I'd give you an ice-cream cone if I had one, but I have none. I'd even give you money, but I have none.*

nonessential *adj.* **1** nonvital, unessential, unneeded, °unnecessary, °needless, °inconsequential, °insignificant, unimportant, °superfluous, °dispensable, °expendable, °gratuitous, uncalled-for, °extraneous, °peripheral, °extra, *de trop,* adventitious, additional, supplemental, adscititious, °redundant, °accessory, °subordinate, °secondary, °subsidiary: *The nonessential industries were converted to the war effort during the 1940s.* —*n.* **2** unessential, inessential, nonentity, cipher, °zero, °nobody; °extra, supernumerary, spear carrier, *Colloq* °nothing, *Slang US* nebbish: *She used him as a lackey, a nonessential whom she could dispose of at will.*

nonetheless *adv.* See **nevertheless,** above.

nonevent *n.* anticlimax, *Colloq* nonstarter, lead balloon, °dud, *Brit* damp squib: *Martin's party turned out to be the nonevent of the social season.*

nonexistent *adj.* °unreal, °imaginary, imagined, °fictional, fictive, °fanciful, °fancied, °mythical, °fabulous, fabled, °illusory, chimerical, delusive: *Although the entire Greco-Roman pantheon became nonexistent overnight, its gods and goddesses continue to pervade our culture.*

nonflammable *adj.* noncombustible, °incombustible, noninflammable, unburnable; fire-retardant: *The government recently passed regulations requiring that nonflammable materials be used in upholstered furniture.*

no-nonsense *adj.* °serious, unfrivolous, businesslike, °practical, nontrivial, untrivial: *Customs officials take a no-nonsense approach to drug trafficking.*

nonpareil *n.* °paragon, °model, °standard, *ne plus ultra,* exemplar, °ideal, *Literary* nonesuch, *Colloq* oner, one in a million, *Brit* one-off: *Annabelle was a nonpareil among the artists of her day.*

nonpartisan *adj.* **1** nonpartizan, °nonaligned, unaligned, unaffiliated, °independent, noncommitted, uncommitted, °neutral, uninvolved, °free, °(sitting) on the fence: *Choose between nonpartisan and coalition candidates.* **2** °impartial, evenhanded, °fair, °just, °objective, unbiased, °unprejudiced, °equitable, °dispassionate, °disinterested: *One could not find a more nonpartisan judge than Sir Ronald.* —*n.* **3** °independent, °neutral, mugwump: *I am a nonpartisan on the subject of privatization of public utilities.*

nonplus *v.* confound, °perplex, °puzzle, °confuse, °dismay, baffle, °stop, °check, °stun, °shock, dumbfound *or* dumfound, °take aback, °astonish, °astound, *US* faze, *Colloq* bring up short, °flummox, °stump: *She was nonplussed to learn that her husband had been arrested.*

nonproductive *adj.* **1** unproductive, °barren, °sterile, °infertile, unfertile, unfruitful, infecund: *Nonproductive land should be left uncultivated.* **2** °ineffectual, °bootless, °ineffective, °impractical, unavailing, °pointless, °useless, °worthless, °wasteful, time-consuming, time-wasting: *Executives should spend more time working and less in nonproductive meetings discussing the work to be done.*

nonsense *n.* **1** °rubbish, °drivel, °gibberish, gobbledegook *or* gobbledygook, twaddle, °trash, °babble, balderdash, °moonshine, °bombast, fustian, rodomontade, puffery, flummery, blather *or* blether, bunkum, poppycock, stuff and nonsense, double-talk, °jargon, °mumbo jumbo, jabberwocky, cackle, gas, °palaver, *Colloq* bunk, piffle, °rot, bosh, eyewash, hogwash, malarkey *or* malarky, bilge (water), baloney *or* boloney, claptrap, °hot air, *Brit* tosh, *Old-fashioned Brit* gammon, °waffle, *US* applesauce, horsefeathers, °garbage, bushwa; *Slang* crap, tripe, bull, hooey, double Dutch, *Brit* (a load of old) cobbler's, *Taboo slang* bullshit, horseshit: *Pinkleton's speech on the benefits of tobacco was unmitigated nonsense.* **2** °mischief, clowning, antics, capering, horseplay, pranks, tricks, jokes, silliness, foolishness, inanity, frivolity, tomfoolery, joking, jesting, waggishness, buffoonery, shenanigans, *Colloq* monkeybusiness, *Brit* monkey tricks, *US* monkeyshines: *Keep up this nonsense and you'll get into a lot of trouble!*

nonsensical *adj.* °senseless, °meaningless, °absurd, °ridiculous, °ludicrous, laughable, °preposterous, irrational, warped, askew, °crazy, °mad, °silly, °foolish, °harebrained, asinine, idiotic, moronic, imbecilic, °stupid, °dumb, *Colloq* nutty, screwy, cockeyed, °fool, screwball, *Slang* loony: *What is your nonsensical excuse for being late this time?*

nonstop *adj.* **1** uninterrupted, °continuous, unbroken, °direct: *Is this a nonstop flight to New York?* **2** unending, °endless, interminable, unceasing, ceaseless, °continual, °continuous, uninterrupted, unbroken, °persistent, °relentless, °constant, unremitting, °steady, round-the-clock, °ongoing, continuing, °unhesitating, unfaltering, °tireless; °regular, °habitual: *I told my neighbor that his nonstop rock 'n' roll music is driving me mad.* —*adv.* **3** unendingly, endlessly, interminably, unceasingly, ceaselessly, continually, continuously, uninterruptedly, persistently, relentlessly, constantly, unremittingly, steadily, round-the-clock, day-in and day-out, tirelessly; regularly, habitually: *He continues to play his hi-fi nonstop.*

nook *n.* **1** °cranny, °recess, °niche, alcove, corner, °cavity, °crevice, °crack, °opening: *Flowers grew from the nooks in the wall.* **2** °retreat, hide-out, °hideaway, °nest; inglenook: *I curled up with a book in my nook next to the fireplace.*

noon *n.* twelve o'clock (noon), °midday, 1200 hours, noontime, high noon, *Archaic* noontide; noonday: *We sit down to lunch promptly at noon. The noon sun is terribly hot in the tropics.*

norm *n.* **1** °usual, °average, °mean, °normal: *The norm for the day is sixty-eight degrees.* **2** °model, °standard, °type, °pattern, criterion, °rule, °measure, °gauge, °yardstick, bench mark: *That style of window became the norm for many years.*

normal *adj.* **1** °standard, °regular, °average, °conventional, °usual, °run-of-the-mill, °ordinary, °routine, °universal, °general, °common, °customary, °natural, °typical, conformist, °orthodox; °healthy: *If living in such squalor seems normal to you, we clearly have different standards. His temperature and pulse are normal.* **2** °sane, °stable, °rational, °reasonable, well-adjusted: *The psychologists say that, though his behavior is sometimes odd, Teddy is quite normal for a teenager.*

normalize *v.* regularize, °standardize, °regulate, °control; °conform: *The program is directed at normalizing the code patterns used in keyboarding text.*

nosy *adj.* nosey, °curious, °inquisitive, prying, meddlesome, °spying, peeping, eavesdropping, *Colloq* snooping, snoopy: *Mind your own business and don't be so nosy.*

notable *adj.* **1** °noteworthy, °noted, °famous, famed, °well-known, °renowned, °illustrious, °important, °prominent, °eminent, °outstanding, °great, °distinguished, °celebrated, acclaimed: *Many notable people attended my college.* **2** °remarkable, °different, °distinctive, °singular, °unusual, uncommon, °preeminent, °peerless, °matchless, unmatched, unequaled, °unparalleled, °extraordinary, °conspicuous, °outstanding, °memorable, unforgettable, °striking: *She enjoyed a notable reputation as a cellist. Last night's notable performance was enjoyed by a capacity audience.* —*n.* **3** °dignitary, °personage, °worthy, VIP; °celebrity, luminary, *Colloq* °(big) name, big shot: *Many notables attended the charity ball.*

notably *adv.* **1** °particularly, °especially, markedly, noticeably, signally, distinctly, remarkably, unusually, uncommonly, outstandingly, conspicuously, °clearly, °obviously, °evidently, °manifestly, specifically, distinctly, curiously, oddly, uniquely, strangely, strikingly, shockingly, surprisingly, stunningly: *In the temperate zones, the seasons vary notably in length.* **2** meaningfully, significantly, importantly, prominently: *The viruses changed in notably different ways.*

notation *n.* **1** °note, °memorandum, jotting, °record, °reminder, minute(s), °abstract, *Colloq* memo: *I shall make a notation of that in my diary.* **2** symbols, signs,

°code, characters, symbolism: *In chemical notation, water is H_2O and table salt is NaCl.*

notch *n.* 1 °nick, °cut, dent, °indentation, °groove, cleft, °score, °mark, °gouge, °gash: *For every man he killed, the gunfighter cut a notch in the barrel of his gun.* 2 °step, °grade, °level, rung, °peg, °degree, °stage, gradation: *Gillian's promotion moves her up another notch toward the chairmanship.* —*v.* 3 °nick, °cut, dent, indent, °groove, °score, °mark, °gash, °gouge: *You notch the end of the arrow to admit the bowstring.* 4 ***notch up.*** °gain, °win, °accomplish, °achieve, °score, °register, °mark (up): *The US Davis Cup team has notched up another win.*

notched *adj.* serrate(d), saw-tooth(ed), crenelate(d), crenate, serriform, pinked, scalloped, zigzag, toothed, dentate, denticulate(d), dentiform: *The notched parapet of the castle protected defending bowmen. The flower has notched yellow petals.*

note *n.* 1 See **notation, 1,** above. 2 °message, °letter, communication, (piece of) correspondence, °memorandum, epistle, postcard *or* (postal) card, fan letter, °love letter, billet-doux, bread-and-butter letter, °word, °line, thank-you note, *Colloq* memo, *US old-fashioned* mash note: *She sent Rob a note only last week about the meeting.* 3 °comment, commentary, °remark, °observation, °explanation, annotation, footnote, side note, marginalia (*pl.*), °gloss, critique, °criticism, *Literary* scholium, exegesis, eisegesis, *Technical* shoulder note: *Her notes on insect behavior are unique.* 4 banknote, °money, °bill, currency, treasury note; promissory note, demand note, bill of exchange, letter of credit, °(bank) draft, note of hand; *Colloq* folding money: *I found a packet of notes dropped by the bank robbers. The bank is holding my note for ten thousand dollars.* 5 °theme, °characteristic, °motif, °element, °quality, °mood, °tone, °tenor: *There is a note of angry frustration that runs through her writing.* 6 °signal, °cue, intimation, °hint, °inkling, °suspicion, °clue, °suggestion, °idea, °tip, *Slang* tip-off: *Her greeting, though warm, contained a note of terror. On that note, I decided to leave.* 7 °heed, °attention, °notice, °regard, °respect, °thought, *Colloq US* °mind: *Note of his objection to that line of questioning has been taken.* 8 °mark, consequence, °substance, °importance, °moment, °weight, °distinction, °merit, °prestige, (high) rank *or* standing, eminence, °prominence, repute, °reputation, °renown: *People of note have been entertained at our table.* 9 °tone, °sound; °key: *He knows the music but gets the notes wrong.* 10 ***notes.*** jottings, impressions, record(s), °report, (thumbnail) sketch, °(rough) draft, °outline, °synopsis: *The entire account is based on the notes she made during the trial.* —*v.* 11 °notice, °observe, °perceive, °see, °mark, °think about, give thought to, °consider, °contemplate, °study, pay attention to, °attend to; °look into, °investigate, °check out: *Have you ever noted how people try to avoid you because of your bad breath? Note how quickly the days seem to grow shorter toward the end of summer. A detective has been assigned to note every move the suspect makes.* 12 °record, °register, °write down, put *or* set down, put on record, °jot down, put in writing, °chronicle: *The policeman noted down the license plate of the car.* 13 call attention to, remark on *or* about, °mention, °report, °touch on, comment on *or* about: *The report failed to note the disappearance of the murder weapon.*

noted *adj.* respected, °eminent, °distinguished, °illustrious, esteemed, acclaimed; °well-known, °famous, famed, °prominent, °celebrated, °notable, °popular; °notorious: *We are pleased to welcome the noted violinist, Beau Stringfellow. Was it his aunt or his mother who ran the noted house of ill repute near Brighton?*

noteworthy *adj.* °notable, of note, °exceptional, °extraordinary, out-of-the-ordinary, °unusual, °rare, uncommon, °singular, out-of-the-ordinary, °unique, °different: *Miss Byrne has made many a noteworthy contribution to our knowledge of flatworms.*

nothing *n.* 1 °nought, nothing at all *or* whatsoever, no thing, not anything, *Taboo slang Brit* bugger-all, (sweet) fanny adams *or* f.a. *or* FA, SFA, fuck-all: *Nothing that you tell me can be held against you in court. You may think it important, but I tell you it is nothing.* 2 cipher, °zero, °nobody, nonentity: *If it weren't for her, he would be nothing.* 3 °trifle, bagatelle, *Colloq* peanuts: *A million is nothing to him.*

notice *v.* 1 °note, take *or* make note (of), take notice (of), pay *or* give attention to, °attend (to), °heed, take heed (of), give heed to, °mark, °remark, °mind, °observe, °perceive, °see: *Notice how quickly he retracted his remarks when challenged. I waved to them, but they didn't notice me.* 2 °mind, °observe, °perceive, discern, °see, °detect, °make out, °identify, °recognize, *Colloq* °spot: *I noticed signs of illness when I visited her.* —*n.* 3 °attention, awareness, consciousness, °perception, °observation, °cognizance: *Let me bring to your notice the footnote on page six.* 4 °regard, °consideration, °respect, °observation, °attention, °note, °heed: *They have published many books worthy of notice. He considers matters of money beneath his notice.* 5 notification, °announcement, °information, °advice; °warning, °bulletin, °poster, intimation: *A notice showing the currency exchange rates is posted in the bank. The company let Corbett go without notice.* 6 °criticism, critique, °review, °comment, commentary: *The play has enjoyed excellent notices in the London newspapers.* 7 ***give notice.*** °warn, admonish, °notify, °announce, °advise, °inform: *Amtrak gave notice of a curtailment of service to this station. Jepson gave notice today of his resignation.*

noticeable *adj.* 1 °discernible, perceivable, °observable, °perceptible, recognizable, distinguishable, °visible, palpable, °manifest, °distinct, °evident, °clear, clear-cut, °conspicuous, °obvious; °patent, unmistakable *or* unmistakeable, °undisguised, unconcealed: *Is the scratch noticeable? Those wrinkles around the eyes are the first noticeable signs of ageing.* 2 °noteworthy, °notable, °significant, °signal, °remarkable, °important, °singular, °exceptional, °pronounced, °distinct, especial, °considerable, °major: *After the reprimand, there was a noticeable improvement in his work.*

notify *v.* 1 °inform, °tell, °advise, °alert, apprise, °warn: *She notified us that she might come in late today. They wrote to notify shareholders of the new share offer.* 2 °announce, °publish, °declare, °proclaim, give notice of; °intimate, °hint: *The arrival of the first salmon notified to us the change of season.*

notion *n.* 1 °idea, °thought, concept, °conception, °image, °impression, general idea, °(mental) picture, °inkling: *She has a pretty good notion of who did it. I haven't the slightest notion of what you are talking about.* 2 °fancy, whim, crotchet, whimsy, caprice, impulse, °inclination, vagary, °conceit, °quirk, °kink: *She suddenly took a notion to fly to New York and left.*

notoriety *n.* notoriousness, disrepute, °dishonor, °disgrace, °infamy, °shame, °discredit, °scandal, °stain, °blot, obloquy, ignominy, opprobrium: *The notoriety attending his latest escapades displeased the prime minister.*

notorious *adj.* 1 °disreputable, °dishonorable, °disgraceful, °infamous, °shameful, shaming, °embarrassing, discreditable, °scandalous, °naughty, °flagrant, ignominious, opprobrious: *Was Cogglesworth again seen in public with that notorious dancer?* 2 °celebrated,

°renowned, °famous, °well-known, fabled, °legendary, °memorable: *Annie was notorious for riding her horse into hotel lobbies.*

notwithstanding *adv.* **1** °nevertheless, °nonetheless, despite that, in spite of that, °yet, anyway: *He was refused permission to go but he left notwithstanding.* —*prep.* **2** °despite, °in spite of, °regardless of, °in the face of, against: *Notwithstanding his mother's objections, George married Marsha.* —*conj.* **3** although, °though, °even though, despite the fact that: *The product was almost totally unknown, notwithstanding it had been on the market for years.*

nought See **naught.**

nourish *v.* **1** °feed, °sustain, °support, °maintain, °keep, °provide for, °care for, °take care of, °look after, nurture, °nurse: *The child seems to be thriving and well nourished.* **2** °foster, °cherish, °nurse, °maintain, °harbor, °keep, nurture, °sustain: *Iago nourished a terrible hatred for Othello.* **3** °strengthen, °fortify, °encourage, °promote, °stimulate, °cultivate, °help, °advance, °aid: *These malcontents continue to nourish trouble in the party ranks.*

nourishment *n.* °food, °sustenance, nutriment, nutrition, victuals: *You must take nourishment to maintain your strength.*

novel *adj.* **1** °new, °unusual, °unfamiliar, unconventional, °fresh, °different, °original, °creative; untested, °untried: *This is quite a novel design for an ultralight aircraft.* —*n.* **2** °story, °tale, °narrative, °romance; novella, novelette, bestseller, *Colloq* blockbuster: *The members of our company board behave like characters out of a novel.*

novelty *n.* **1** °originality, newness, uniqueness, freshness, innovativeness: *We are enthusiastic about the novelty of the new sales campaign.* **2** °gimmick, gimcrack, °trifle, °gewgaw, °bauble, knickknack, °toy, trinket, °ornament, °plaything, brummagem, kickshaw: *New subscribers receive as a premium some novelty, like a ballpoint pen or a cheap clock.*

novice *n.* beginner, neophyte, °newcomer, proselyte, tiro *or* tyro, novitiate *or* noviciate, °learner, °amateur, °initiate, °learner, °apprentice, trainee, probationer, fledgling *or Brit also* fledgeling, *US* freshman, *Colloq* °greenhorn, rookie: *She is a mere novice at scuba diving.*

now *adv.* **1** °at present, just now, right now, at the present time *or* moment, at this (very) moment *or* minute *or* second *or* instant: *He's in the shower and cannot come to the phone now.* **2** these days, °nowadays, today, in these times, at the moment, in this day and age, under *or* in the present circumstances *or* conditions, in the present climate, things being what they are, contemporarily, any more, any longer; °for the time being, for the nonce: *What makes you say that the minimum wage is enough to live on now? Selling now might bring the highest price.* **3** °at once, °immediately, right away, without delay, instantly, °promptly, *Chiefly law* instanter, *Chiefly Brit* °straightaway: *I want you here now—not in five minutes, but now!* **4** ***now and then*** or ***again.*** °occasionally, from time to time, °at times, °on occasion, °sometimes, sporadically, °once in a while, every now and then *or* again, randomly, intermittently; °infrequently, °seldom, °rarely, once in a blue moon: *There was a power cut now and then. He visits his mother only now and then.* —*adj.* **5** °contemporary, up-to-date, °modern, °stylish, °fashionable, °trendy, *Colloq* in, with it: *Advertisers must appeal to the yuppies of the now generation.*

nowadays *adv.* See **now, 2,** above.

nub *n.* **1** °projection, protuberance, °knob, °boss, °lump, °bump, knop, °protrusion, °bulge, node, °knot; excrescence, °swelling, tumescence: *Press the small nub on the side to open the door.* **2** °essence, °core, °heart, °nucleus, crux, °point, °gist, °pith, °kernel, °nucleus, °meat, °(sum and) substance, main issue, gravamen: *Let's get to the nub of the argument.*

nuclear *adj.* atomic: *He joined the protest march against nuclear weapons.*

nucleus *n.* °core, °heart, °center, °kernel, °pith, °focus, °nub: *We already have the nucleus of a very good team. The factory was the nucleus around which many towns were built.*

nude *adj.* unclothed, undressed, uncovered, *au naturel,* °bare, °naked, in the nude, stark-naked, undraped, without a stitch (on), *Colloq* in the buff, in the altogether, in one's birthday suit, mother naked, *Brit* starkers, *Brit and Australian* in the nuddy: *A nude man streaked across the football field.*

nudge *v.* **1** °jog, °poke, elbow, °jab, °dig, °bump, °prompt, shove; °prod, °push, *US* °encourage: *I had to nudge him to stay awake for the film's thrilling climax.* —*n.* **2** °jog, °poke, elbow, °jab, °dig, °bump, shove; °prod, °push, °encouragement: *Albert just needs a nudge in the right direction.*

nuisance *n.* **1** °annoyance, °inconvenience, °trial, °ordeal, °burden, irritation, irritant, thorn in the flesh *or* side, °difficulty, °bother, *US* bur under the saddle, *Colloq* pain (in the neck *or* rear), °headache, hassle, *Slang* pain in the *US* ass *or Brit* arse, *US and Canadian* pain in the butt: *Having to paint the room again was a terrible nuisance.* **2** °bore, °pest, °nag, °tease, tormentor: *Brendan has made a nuisance of himself by telephoning every fifteen minutes.*

numb *adj.* **1** numbed, benumbed, °insensible, insensate, °dead, deadened, without feeling, sensationless, °senseless; asleep: *My feet are numb from standing up so long in the cold.* —*v.* **2** benumb, anesthetize, °drug, °deaden, °dull, °freeze, °paralyze, immobilize, °stun: *The doctor numbed my hand before removing the wart. Fear numbed her, and she felt nothing as the monster approached.*

number *n.* **1** numeral, integer, °figure, digit: *The columns of numbers were entered in a neat hand.* **2** °few, °handful, °crowd, slew, °gang, °bunch, °party, bevy, covey, troop, °company, °platoon, °swarm, horde, multitude, °mob, °host, army, °mass, hundred, thousand, million, billion; °several, °many, numbers, legions, *US and Canadian* slew(s) *or* slue(s), *Colloq* loads, tons: *A number of people attended the gallery opening. An enormous number of viruses could fit on the head of a pin.* **3** °issue; °edition, °copy: *The fourth number of the quarterly is published at the end of the year.* —*v.* **4** °count, °enumerate, °compute, °calculate, °tally, °figure (up), °add (up), °include, °total, °tot (up), °reckon, °sum (up): *Who can number the stars?*

numberless *adj.* uncountable, uncounted, countless, innumerable, incalculable, °immeasurable, numerous, °untold, myriad, °infinite: *Although seemingly numberless, the number of grains of sand in the universe is calculable.*

nuptial *adj.* °bridal, °matrimonial, °wedding, spousal, °wedded, marital; connubial, conjugal, *Literary* hymeneal: *The nuptial arrangements have been made.*

nurse *n.* **1** angel of mercy, Florence Nightingale, *Brit* sister: *I awoke to see two nurses bending over me.* —*v.* **2** °care for, °look after, °tend, °attend, °minister to, °treat; nurture, °foster, °coddle, °baby, °pamper, °cherish, °preserve, keep alive, °cultivate, °develop: *During his illness, she nursed him night and day. We nursed the company along through the first year.* **3** wet-nurse, suckle, breast-feed, °nourish: *Nursing mothers must be* *careful about what they eat.* **4** °preserve, °harbor, keep alive, nurture, °foster: *She has nursed a grudge against him for ten years.*

nutritious *adj.* healthful, °healthy, nutritive, °wholesome, life-giving, °beneficial, salutary, nourishing, alimentary, nutrimental: *Be sure you eat a nutritious breakfast every day.*

O

oar *n.* **1** °paddle, scull: *The ancient galleys sometimes had six men on each oar.* **2** oarsman, oarswoman, bencher, sculler, rower, paddler: *With Hanson out because of his back, we'll need a new oar for tomorrow's race.*

oasis *n.* **1** fertile patch, watering hole: *In the desert, you cannot be sure whether you are looking at an oasis or a mirage.* **2** haven, °refuge, °(safe) harbor, °sanctuary, °retreat, asylum, °resort, °sanctum: *We escaped to the cottage, a tiny oasis away from the city's frenetic activity.*

oath *n.* **1** °vow, avowal, °pledge, °promise, °word (of honor), °promise, °plight, guarantee *or* guaranty, warrant *or* warranty, °(sworn) statement, *Archaic* troth: *She has taken an oath to tell the whole truth.* **2** °curse, °profanity, blasphemous language *or* expression *or* word, imprecation, malediction, swearword, °expletive, four-letter word, obscenity, dirty word: *The door slammed on his finger and he muttered a foul oath.*

obedience *n.* compliance, dutifulness, °observance, °respect, respectfulness, tractability, conformity *or* conformance, °yielding, conformability, adaptability, °agreement, agreeability, agreeableness, acquiescence, submissiveness, °submission, subservience, docility, passiveness, passivity: *The abbot demanded unquestioning obedience to every rule of the monastic order.*

obedient *adj.* compliant, °dutiful, duteous, °observant, °respectful, °tractable, °yielding, conformable, °adaptable, °agreeable, amenable, acquiescent, °submissive, subservient, docile, °passive, °timid, biddable, pliant: *Prunella was always an obedient child. All matter and energy is obedient to the laws of physics.*

obeisance *n.* °deference, °respect, respectfulness, °homage, °submission, °reverence, °honor: *As she entered the chamber, she made obeisance to the king.*

obese *adj.* °fat, overweight, °stout, fleshy, °gross, corpulent, °heavy, °plump, portly, tubby, pudgy, °chubby, paunchy, °rotund, potbellied, *Rare* abdominous: *I sat down next to an obese person who occupied nearly two chairs.*

obesity *n.* corpulence, plumpness, tubbiness, chubbiness, grossness, *embonpoint,* rotundity, portliness, paunchiness, °size, °bulk, °weight, avoirdupois: *Has your obesity made it difficult to find attractive clothing that fits?*

obey *v.* **1** °comply (with), °agree (to), °consent (to), °submit (to), °abide (by), °observe, °respect, adhere to, °follow, conform (to *or* with), acquiesce (to *or* in), °mind, °accept, °heed, defer to, °yield (to), knuckle under (to), °give way (to), °surrender (to), °succumb (to), °give in (to), truckle to, bow to, bend to, take *or* accept orders from: *Unfortunately, Ogilvy has obeyed his baser instincts in making the punishment fit the crime. Everything must obey the laws of nature. Harold obeys Millie's slightest whim.* **2** °discharge, °execute, °effect, °carry out, °fulfill, °meet, °satisfy, do, °perform; °serve, °act: *We obeyed the colonel's orders to the letter. It is your function to command, mine to obey.*

obituary *n.* necrology, death notice, °eulogy, necrologue, *Colloq* obit: *Mark Twain is one of the few people ever to have read his own obituary.*

object *n.* **1** °thing, °tangible, °item; °reality, °entity, °fact, °phenomenon: *A number of objects lay on the table. Thoughts may be considered as objects of the imagination.* **2** °focus, °target, °butt, °aim, °destination, °quarry, °goal: *The object of my affection has married someone else.* **3** °purpose, °end, °intention, °objective, °reason, °intent, °idea, °goal: *The object of our visit is to ask you a few questions.* —*v.* **4** protest (to *or* against), °interfere (with), raise objections (to), °argue (against), °oppose, be against, take exception (to), °disapprove (of), draw the line (at), °complain (about), remonstrate (over *or* about), take a stand (against), °refuse: *I won't object if you want to bring the wine. Would you object to rereading that passage? If they ask me, I cannot object.*

objection *n.* °protest, °opposition, °exception, °argument, °challenge, °interference, demur *or* demurral *or* demurrer, °question, °doubt, °disapproval, °complaint, remonstration, remonstrance, °stand, °refusal, °dislike, antipathy: *The meeting proceeded without further objection from the audience. If you have no objection, I'd like to leave now. The secretary has raised an objection to the method of procedure.*

objective *adj.* **1** °fair, °impartial, °just, °judicious, °equitable, °neutral, °disinterested, °dispassionate, openhanded, open-minded, °detached, unbiased, °unprejudiced, unbigoted, evenhanded, uncolored, unjaundiced: *How can you be objective about the guilt or innocence of your own child?* —*n.* **2** °target, °goal, °object, °aim, °purpose, °end (in view), °intent, °intention, °design, °aspiration, °ambition, °hope: *If we capture the flag, we shall have gained our objective. My objective is to win the pentathlon.*

objectivity *n.* impartiality, fairness, fair-mindedness, equitableness, equitability, evenhandedness, neutrality, disinterest, °detachment, °indifference, dispassion: *The jury's objectivity was never in doubt.*

obligate *n.* °oblige, °pledge, °commit, °bind; °require, compel, constrain, °force: *I feel deeply obligated to her for her kindness to our children. We are obligated to do what we are told.*

obligation *n.* **1** °responsibility, °duty, °charge, °burden, onus; °accountability, °liability, °trust; °demand, °requirement, compulsion, *Literary* devoir: *It was Frank's obligation to get the children home safely. Civil servants have an obligation to serve the people. I could never fulfill all my obligations.* **2** constraint, °requirement, °contract, °promise, °pledge, °bond, °agreement, covenant: *The company is under no obligation to replace a product because the customer dislikes its color. I am under an obligation to her for introducing us.* **3** °debt, °liability: *Denby may be unable to meet all his obligations.*

obligatory *adj.* required, demanded, °necessary, requisite, compulsory, °mandatory; °incumbent; °indispensable, °essential: *Has she been able to meet all the obligatory qualifications?*

oblige *v.* **1** °accommodate, °indulge, °favor, serve, please, °cater to, °gratify: *The hotelier obliged us with every luxury he had to offer. Please oblige us by keeping your dog on a leash.* **2** °make, °require, °demand, °force, compel, coerce, °bind, °obligate: *What hold has she over you that obliges you to do her housework?*

obliged *adj.* **1** °thankful, °grateful, appreciative, °beholden, °indebted, obligated: *We are deeply obliged to you for asking us with the children.* **2** °bound, required, compelled, °forced, made, obligated: *Under the terms of the agreement, I am obliged to repay the debt by May.*

obliging *adj.* °accommodating, °willing, °indulgent, °gracious, °courteous, °civil, °considerate, °polite, °agreeable, amenable, °kind, °kindly, °helpful, °friendly, °amiable, °neighborly, °supportive: *It was very obliging of you to look after my cat while I was gone.*

oblique *adj.* **1** slanting, slanted, sloping, aslant, °inclined, diagonal, inclining, angled, angling, canted, canting, banked, banking, cambered, °crooked, askew, °divergent, diverging, tilted, atilt, tilting: *The roof joins the wall at an oblique angle.* **2** awry, °devious, °roundabout, °indirect, circuitous, circumlocutionary, °evasive, °sly, °sidelong, °offhand, °surreptitious, furtive, implied, clandestine, underhand(ed), °deceitful, °devious, °deceptive, °false: *She made some oblique comments about the candidate's wife.*

obliterate *v.* **1** °erase, expunge, °rub out, efface, eradicate, °wipe out, °delete, dele, strike off *or* out, °strike from, °rule out, °eliminate, °write off: *After the scandal, that name was obliterated from the roll of honor.* **2** annihilate, °destroy, °kill, °exterminate, °wipe out, °eliminate, °blot out, eradicate, extirpate: *The entrance to the cave was completely obliterated by the explosion.*

oblivion *n.* **1** blankness, blackness, darkness, °obscurity, nothingness, nihility, anonymity, extinction, nonexistence, °void, °limbo: *The rock band enjoyed brief fame, then sank into oblivion.* **2** unawareness, obliviousness, forgetfulness, heedlessness, °disregard, unconsciousness, insensibility: *I sank back into the sweet oblivion of deep sleep.*

oblivious *adj.* °unaware, °unconscious, unmindful, disregardful, °insensible, insensitive, °distant, unconcerned, °detached, removed, unfeeling, abstracted, °absent-minded, °forgetful, Lethean: *Many are completely oblivious to the plight of the starving multitudes in Africa.*

obnoxious *adj.* °revolting, °repulsive, °repugnant, °disgusting, °offensive, objectionable, fulsome, noisome, °vile, °repellent, °nauseous, nauseating, sickening, °foul, noxious, mephitic, °unsavory, execrable, °abominable, °abhorrent, °loathsome, detestable, °hateful, odious, °scurvy, °base, °obscene, °despicable, °awful, °terrible, °unpalatable, °distasteful, unlikable, unpleasant, °nasty, *Colloq chiefly Brit* °beastly: *Bleddoe's obnoxious sister even has table manners that are obnoxious.*

obscene *adj.* **1** inelegant, °improper, °rude, °impure, °unchaste, °shameless, °shameful, °indecent, °immodest, °off-color, indecorous, indelicate, °risqué, °vulgar, °immoral, °degenerate, amoral, °dissolute, °broad, °suggestive, °erotic, °sensual, ribald, debauched, °wanton, °loose, °libertine, °bawdy, °blue, scabrous, °coarse, °dirty, °filthy, smutty, °pornographic, libidinous, °lewd, licentious, °lecherous, °lustful, goatish, °carnal, ruttish, °lascivious, °filthy, salacious, °prurient, °disgusting, °offensive, °repulsive, °foul, °abominable, °vile, °loathsome, °gross, foulmouthed, °scurrilous, scatological, °vile, *Literary* Cyprian, Paphian, Fescennine, thersitical: *In the cinema business, "adult" and "obscene" seem to be synonyms.* **2** °evil, °wicked, heinous, °atrocious, °awful, °outrageous, °repulsive, °shocking, °repellent, °obnoxious, off-putting, objectionable, °beastly, intolerable, °insufferable, °unpalatable, °distasteful, °nauseous, nauseating, sickening, execrable, °despicable, °nasty: *The obscene monster dashed the brave warriors to the rocks far below.*

obscure *adj.* **1** °dark, unlit, °gloomy, °somber, °dismal, °murky, °dusky, °black, Cimmerian, tenebrous, °dim, °faint, blurred, °veiled, °shadowy, subfusc, subfuscous, umbral, °shady, °hazy, foggy, befogged, clouded, °nebulous, °overcast, cloudy: *The traveler's lantern was barely seen in the obscure reaches of the wood.* **2** unclear, °uncertain, °ambiguous, °vague, °hazy, °doubtful, dubious, °equivocal, °indefinite, °indistinct, °fuzzy, blurred, °confused, confusing, delphic, °puzzling, enigmatic, °perplexing, baffling, mystifying, °mysterious, °cryptic, °incomprehensible, °unfamiliar, °foreign, °strange: *The sorcerer muttered some obscure words, and a golden horse stood prancing before them.* **3** °secret, concealed, °hidden, °remote, °out-of-the-way, °inconspicuous, unnoticeable, °secluded, °unnoticed: *The caped figure scurried down the alley and disappeared into some obscure doorway.* **4** °unknown, °unheard-of, anonymous, unnamed, °insignificant, unimportant, °inconsequential, °humble, °lowly, °mean, inglorious, °inconspicuous, °undistinguished, °unnoticed, °unsung, °minor, little-known: *Though extremely popular, the song was written by an obscure composer.* **5** abstruse, arcane, °recondite, esoteric, °intricate, complex, °occult, out-of-the-ordinary, °unfamiliar, *Colloq* far-out: *He is an authority on some obscure subject like Coptic calligraphy.* —*v.* **6** °cover, °conceal, °hide, °veil, °shroud, °cloak, °mask, °screen, °disguise, °keep from: *Her link with military intelligence was obscured from her family.* **7** °dim, bedim, cloud, becloud, °dull, °shroud, °shade, adumbrate, °overshadow, darken, obfuscate, °block, °eclipse: *The street lamp was obscured by trees.*

obscurity *n.* **1** dimness, darkness, °gloom, murk, murkiness, duskiness, °dusk, blackness, faintness, blurriness, °shade, °shadow, haze, °fog, cloudiness, nebulousness: *The two of them vanished into the obscurity of the night.* **2** abstruseness, ambiguousness, intricacy, °complexity, unintelligibility; °mystery, arcanum, °secret, esoterica (*pl.*): *Can he truly believe that he has fathomed all the obscurities of Scripture?* **3** insignificance, unimportance, ingloriousness, inconspicuousness, anonymity, namelessness, °limbo: *After a fleeting surge of popularity, punk rock sank into obscurity.*

obsequious *adj.* °low, cringing, toadying, toadyish, sycophantic(al), sycophantish, unctuous, truckling, °groveling, crawling, fawning, deferential, °ingratiating, °menial, °flattering, °servile, slavish, subservient, °submissive, abject, °mealy-mouthed, °slimy, *Colloq* bootlicking, *Chiefly Brit* smarmy, *Taboo slang* brown-nosing, *Brit* arse-kissing, arse-licking, *US* ass-licking, ass-kissing: *He is surrounded with obsequious followers who cater to his every whim.*

observable *adj.* °perceptible, perceivable, °noticeable, °discernible, recognizable, detectable, °visible, °apparent, °distinct, °evident, °manifest, °plain, °obvious, °clear, °explicit, °transparent, °patent, °tangible, unmistakable *or* unmistakeable: *A marked change in public sentiment became at once observable.*

observance *n.* **1** °observation, observing, °obedience, obeying, compliance, complying, conformity, conforming, adherence, adhering, keeping, accordance, °regard, °recognition, recognizing, °respect, respecting, °heed, heeding, °attention: *Observance of the rules by everyone makes for a happier community.* **2** °ceremony, °celebration, °ceremonial, °practice, °rite, °ritual, °service, °performance, °form, °custom, °convention, °tradition, °formality, °usage, °habit, wont, °institution: *Almost all the fine arts derived their origin from religious observances.* **3** °observation, °examination, inspection, °scrutiny, looking, watching: *His observance of the passing scene was chronicled in his diary.*

observant *adj.* **1** watchful, °alert, °attentive, °vigilant, on the lookout, on the qui vive, on guard, wide-awake, regardful, °mindful, °aware, °keen, keen-eyed, °sharp-eyed, °eagle-eyed, °perceptive, °sharp, °shrewd: *Hannay was observant of all who passed him on the way to the train. How very observant of you to spot the man in the crowd!* **2** Usually, ***observant of.*** °obedient (to), compliant (with), °respectful (of), heedful (of), attentive (to *or* of), conformist (to), adherent (to): *One must always be observant of the rules of the road.*

observation *n.* **1** watching, °examination, °scrutiny, inspection, viewing, °survey, °surveillance; °notice, °discovery, °attention, awareness: *The police put the house under twenty-four-hour observation. The smuggled weapon escaped the guard's observation.* **2** °comment, °remark, °note, °reflection, °opinion, °sentiment, °point of view, °impression, °feeling, commentary, °criticism; utterance, °word, °announcement, °pronouncement, °proclamation, °declaration: *Have they published their observations regarding terrorism in the Middle East?*

observe *v.* **1** °obey, °abide by, comply with, be heedful of, °attend to, conform to, °regard, °keep, °follow, adhere to, °respect, pay attention to: *We observed the prohibition against swimming.* **2** °watch, °look at, °examine, °monitor, °scrutinize, °study, °regard, °view, °inspect, °pore over, °contemplate, °consider, *Colloq* check (out *or* up on), check over, °size up, *Slang* °case: *The naturalists went to the Arctic to observe the polar bears.* **3** °see, °mark, °notice, °look, °perceive: *Observe how lithe the penguin is in the water!* **4** Sometimes, ***observe on*** *or* ***upon.*** comment (on *or* upon), remark (on *or* upon), °mention, °say, °note, °refer (to), make reference to, animadvert on *or* upon *or* to; °state, °declare: *It is impolite to observe on others' manners. He couldn't help observing to his cellmate how easy it would be to escape.* **5** °celebrate, °keep, solemnize, °respect, keep holy, °mark, °commemorate, °memorialize, °remember, °recognize: *We always observe the sabbath.*

observer *n.* °witness, °eyewitness, °spectator, viewer, °onlooker, beholder, watcher, looker-on; nonparticipant: *UN observers reported that the battle was over.*

obsess *v.* °haunt, °harass, °plague, bedevil, °torment, °take over, preoccupy, °dominate, °control, °grip, °possess, °hold: *He was obsessed by the conviction that he could design a successful flying machine.*

obsession *n.* fixed idea, *idée fixe,* °fixation, °conviction, preoccupation, prepossession, °passion, °mania, °phobia, *Colloq* hang-up, °thing: *Thoughts of death became his constant obsession.*

obsessive *adj.* haunting, harassing, tormenting, dominating, controlling, possessing, all-encompassing, °passionate, unshakable *or* unshakeable: *She has an obsessive fear of heights.*

obsolescent *adj.* fading, waning, °on the wane, declining, °dying, on the way out, on the decline, going *or* passing out of use *or* fashion *or* style: *Much of yesterday's newest slang is obsolescent today.*

obsolete *adj.* °out of date, out of fashion, out-dated, °passé, °out, °dead, outmoded, °old, °antiquated, antediluvian, °ancient, superannuated, dated, archaic, °old-fashioned, *démodé;* °unused, °disused, discarded, superseded, °extinct, *Colloq* old hat: *The Briticism "tickety-boo" is obsolete. He plays his 78-rpm records on an obsolete phonograph.*

obstacle *n.* °impediment, °hindrance, °obstruction, °hurdle, °hitch, °catch, °snag, °stumbling block, °barrier, °bar, °check: *The obstacles in the road prevented our proceeding further. One must often overcome many obstacles before achieving success.*

obstinacy *n.* obstinateness, stubbornness, doggedness, °tenacity, persistence *or* persistency, mulishness, pigheadedness, willfulness, contrariness, perverseness, perversity, cantankerousness, recalcitrance, uncooperativeness, rebelliousness, contumacy, contumaciousness, refractoriness, intractability, intransigence, pertinacity, pertinaciousness, obduracy, fixedness, stolidity, inflexibility, firmness, *Archaic* frowardness, *Colloq Brit* bloody-mindedness: *His obstinacy makes it difficult to work with him.*

obstinate *adj.* °stubborn, dogged, °tenacious, °persistent, mulish, °perverse, headstrong, pigheaded, °single-minded, °willful, strong-willed, °self-willed, °contrary, °recalcitrant, uncooperative, °rebellious, contumacious, refractory, intransigent, pertinacious, obdurate, °fixed, °inflexible, °stony, adamant, °set, unmoving, °immovable, inexorable, intractable, unchangeable, °resolute, °steadfast, unyielding, persevering, °stiff, °rigid, °hard, *Archaic* froward, *Colloq Brit* bloody-minded: *The obstinate man does not hold opinions—they hold him.*

obstreperous *adj.* vociferous, clamorous, °noisy, °loud, °raucous, °riotous, °uproarious, °tumultuous, °boisterous, °rowdy, rumbustious, °tempestuous, °unruly, °disorderly, unmanageable, uncontrollable, °uncontrolled, unrestrained, irrepressible, out of control, °undisciplined, roisterous, °wild, turbulent, *Colloq* rambunctious, *Brit* mafficking: *The party was getting rough and some of the guests got a bit too obstreperous.*

obstruct *v.* **1** °block, °bar, °check, °prevent, °stop (up), °arrest, °halt, °clog, make impassable; bring to a standstill: *The vein is obstructed by a large blood clot. A clogged drain is obstructing the water.* **2** °hamper, °slow, °impede, interfere with, °retard, °hinder, °interrupt, °delay, °stay, °stall: *An overturned truck obstructed traffic on the interstate.* **3** °preclude, °prevent, debar, °block, °prohibit, °forbid, °stop, stand in the way of: *They are maneuvering to obstruct her from taking over the company.*

obstruction *n.* **1** °obstacle, °barrier, °bar, °check, °stumbling block, °hindrance, °impediment, °hurdle, °hitch, °snag, °catch, bottleneck, limitation, constraint, °restriction: *The fallen trees created an almost impassable obstruction.* **2** checking, stopping, cessation, proscription, °forbidding, forbiddance; hindering, impeding, limiting, °halting, slowing: *The obstruction of the bill's passage can be blamed on the beef lobby.*

obtain *v.* **1** °get, °procure, °acquire, °come by, come into (the) possession of, °secure, get hold of *or* one's hands on, °grasp, °capture, °take possession of, °seize; °buy, °purchase: *She has been unable to obtain the job she wants. You can obtain that kind of soap at the supermarket.* **2** °earn, °gain: *We talked to the manager about obtaining an increase in wages.* **3** °prevail, be in force, be in vogue, °exist, subsist, have (a) place, be prevalent, be established, be customary, °apply, be relevant, °relate: *A different set of regulations obtains here.*

obtrude *v.* thrust (oneself) forward *or* forth, °intrude, °impose (oneself), °force (oneself): *The best writers never obtrude between the reader and the story.*

obtrusive *adj.* interfering, °intrusive, meddling, °officious, meddlesome, importunate, °forward, °presumptuous, °forceful, *Colloq* °pushy: *Parker's inclined to be even more obtrusive when he's had a drink or two.*

obtuse *adj.* **1** rounded, unpointed, °blunt: *When mature, the leaves become more obtuse.* **2** °dull, insensitive, unfeeling, imperceptive, °thick-skinned, °stolid, °thick, °dense, doltish, cloddish, thickheaded, dull-witted, dimwitted, slow-witted, (mentally) retarded, boneheaded, lumpish, loutish, oafish, °simple, simple-minded: *Luke is a bit too obtuse to get the point of the story.*

obvious *adj.* °clear, °plain, °apparent, °patent, °perceptible, °evident, °self-evident, °barefaced, clear-cut, °manifest, palpable, (much) in evidence, °conspicuous, °open, °visible, °overt, ostensible, °pronounced, °prominent, °glaring, undeniable, unconcealed, unhidden, unsubtle, °distinct, °simple, bald, bald-faced, straightforward, °direct, self-explanatory, °indisputable, unmistakable *or* unmistakeable: *There are obvious flaws in the fabric. The reason you were refused seemed obvious to me.*

obviously *adv.* °clearly, plainly, °apparently, patently, °evidently, °simply, certainly, °of course, undeniable, unmistakably *or* unmistakeably, indubitably, doubtless(ly): *You are obviously the right person for the assignment.*

occasion *n.* **1** °time, °moment, °circumstance, °incident, °occurrence, °opportunity, °chance, °opening, °advantage: *I took the occasion of the inquiry to leave town.* **2** °reason, °cause, °call, justification, ground(s), °warrant, °provocation, prompting, impulse, stimulus, °incitement, °inducement: *Tom gave David no occasion to doubt his honesty.* **3** °event, °function, °happening, °affair, °observance, commemoration, °ceremony, °celebration, °gala, °party: *Grandpapa's ninetieth birthday was a great occasion. This hall is saved for important occasions.* **4** ***on occasion.*** See **occasionally,** below. —*v.* **5** °give rise to, °bring about, °cause, °bring on, °effect, °prompt, °provoke, °evoke, °call forth, °elicit, °call up, °induce, impel, °create, °generate, engender, °produce, °make (for): *An increase in the inflation rate occasions a decrease in the value of money.*

occasional *adj.* **1** °intermittent, °irregular, °periodic, °random, °sporadic, infrequent, °casual, °incidental: *They staged occasional raids on the arsenal. He works as an occasional farm hand.* **2** additional, °extra, °spare, °supplementary, °incidental, °auxiliary, °accessory: *We bought a few occasional chairs in case we have company.* **3** °special, °particular, °ceremonial, °ritual: *She writes occasional verses for memorial services.*

occasionally *adv.* °sometimes, °on occasion, °(every) now and then, from time to time, °at times, °(every) now and again, °once in a while, every so often, periodically, intermittently, sporadically, irregularly, off and on: *We take our vacation in the Vermont occasionally. Trevor occasionally drops in at the local bar for a beer.*

occult *adj.* **1** °secret, °dark, concealed, °private, °privy, °hidden, °obscure, °veiled, obscured, shrouded, °vague, abstruse, °shadowy, °mystical, °mysterious, cabalistic, esoteric, °recondite, arcane: *Printing was kept an occult art for generations.* **2** magical, °mystical, alchemic(al), unexplained, unexplainable, °inexplicable, °puzzling, baffling, °perplexing, mystifying, °mysterious, °incomprehensible, inscrutable, indecipherable, impenetrable, unfathomable, transcendental, °supernatural, preternatural, mystic: *They dress in odd clothes and participate in occult rituals at peculiar times of the day and night.* —*n.* **3** Usually, ***the occult.*** the supernatural, the unknown, the black arts; arcana, cabbala *or* cabala *or* kabbala *or* kabala; cabbalism *or* cabala *or* kabbala *or* kabala, occultism, °sorcery, witchcraft, black magic: *Gerard has studied the occult for many years.*

occupant *n.* °resident, °inhabitant, occupier, °tenant, lessee, leaseholder, renter, °owner, householder, indweller, dweller, °denizen, lodger, roomer, boarder; addressee; °incumbent: *The occupant of the apartment upstairs is rarely home.*

occupation *n.* **1** °job, °position, °post, °situation, °appointment, °employment, °vocation, °line (of work), °career, °field, °calling, °trade, métier, °craft, °skill, °profession, °business, °work: *Claverton pursued his occupation as a miniaturist for some fifty years.* **2** °possession, °tenure, occupancy, °rule, °control, suzerainty, subjugation, °subjection, °oppression, °bondage: *Terrible atrocities were committed while the land was under the occupation of the Mongols.* **3** °conquest, °seizure, appropriation, takeover: *Francisco Pizarro was responsible for the occupation of Peru.*

occupy *v.* **1** °capture, °seize, °take possession of, °conquer, invade, °take over, °overrun, garrison, °dominate, °hold: *Rebel forces had occupied the capital.* **2** live *or* reside *or* dwell in, °tenant, be established *or* ensconced *or* situated in, establish *or* ensconce *or* situate oneself in, °inhabit, be settled in *or* into, settle in *or* into, take up residence in, make one's home in, move in *or* into; be located in: *She occupies a luxurious house on Lake Shore Drive.* **3** °engage, °busy, absorb, °monopolize, °hold, take up *or* over, °catch, °grab, °seize, °grip; °divert, °amuse, °entertain, °distract, °beguile, preoccupy, hold (someone's) attention, °interest, engross, °involve: *Other matters occupied my attention last Sunday. While one man was keeping the shopkeeper occupied, the other man was robbing the till.* **4** fill (in *or* up), °take up, °cover, extend over, °consume, °use (up), *Colloq* eat up: *The spaceship occupies more garage space than I thought it would. Singing lessons occupy very little of my time.*

occur *v.* **1** °happen, °take place, °arise, °come about, befall, °come to pass, °chance, °appear, °surface, °materialize, °develop, become manifest, manifest itself, *Colloq* °transpire, crop up, °come off, °turn up: *We reported to the police all that had occurred. What occurred to make you late this time?* **2** ***occur to.*** °dawn on, °strike, °hit, °come to, suggest itself to, cross (someone's) mind, enter (someone's) head, be brought to (someone's) attention: *Has it occurred to you that she might not like diamonds?*

occurrence *n.* **1** °happening, °event, °incident, °phenomenon, °affair, °matter, °experience: *Earthquakes are frequent occurrences in California.* **2** °existence, °instance, °manifestation, materialization, °appearance, °development: *The occurrence of mutations diversifies the species.* **3** frequency, °incidence, °rate; °likelihood, °chance: *What is the recorded occurrence of typhoons in the South China Sea?*

ocean *n.* **1** °(deep blue) sea, °(bounding) main, high seas, °the deep, Davy Jones's locker, the depths, *Colloq* the briny, °the drink: *The sails filled, and our tiny craft was swept out into the open ocean.* **2** Often, ***oceans.*** °flood, °abundance, multitude, °profusion, plethora, *Colloq* scads, loads, tons, lots, oodles, gobs, zillions: *The direct mail campaign yielded oceans of responses.*

oceanic *adj.* °marine, pelagic, thalassic; saltwater, deep-water, aquatic, maritime, °sea, °ocean: *These creatures are chiefly oceanic, coming ashore only to breed.*

odd *adj.* **1** °strange, °peculiar, °unusual, uncommon, °different, unexpected, °unfamiliar, °extraordinary, °remarkable, atypical, untypical, °exotic, °out of the ordinary, °unparalleled, unconventional, °exceptional, °unique, °singular, °individual, anomalous, idiosyncratic, °rare, °deviant, °outlandish, uncanny, °queer, °curious, °bizarre, °weird, °eccentric, °funny, °quaint, °fantastic, °freak, °abnormal, freakish, *Colloq* °offbeat, screwy, °kinky, freaky, *Slang Brit* °bent, rum, *US and Canadian* kooky or *US and Canadian* kookie: *Didn't you think it odd coming down with prickly heat in the Antarctic? Yes, Ebenezer is an odd name for a dog. Where did you get that odd hat? I cannot account for his odd behavior.* **2** °occasional, °casual, part-time, °irregular, °random, °sporadic, discontinuous, °disconnected,

°various, °varied, °miscellaneous, °sundry, °incidental: *After being cashiered from the army, he worked at odd jobs for a year or so. The odd shower can be expected during the afternoon.* **3** °leftover, °surplus, °remaining, °unused, °spare, °superfluous, °extra: *After the patterns were cut, we were allowed to take the odd scraps of fabric.* **4** uneven, unmatched, unpaired: *This gallery has an odd number of columns.*

oddity *n.* **1** °peculiarity, strangeness, unnaturalness, curiousness, incongruity, incongruousness, °eccentricity, outlandishness, extraordinariness, unconventionality, bizarreness, weirdness, queerness, oddness, unusualness, individuality, °singularity, distinctiveness, anomalousness, anomaly, *Colloq* kinkiness, *US and Canadian* kookiness: *What caught my attention was the oddity of the clothes worn by the students.* **2** °peculiarity, °curiosity, °rarity, °freak, °original, °phenomenon, °character, °eccentric, °nonconformist, fish out of water, odd bird, rara avis, °misfit, square peg in a round hole, maverick, *Colloq* °card, °crank, weirdie *or* weirdo, oner, *Brit* odd fish, *US and Canadian* kook, oddball, screwball: *The townspeople thought Albert an oddity, but we knew he was a genius.* **3** °peculiarity, irregularity, anomaly, idiosyncrasy, °eccentricity, deviation, °quirk, °mannerism, °twist, °kink, crotchet: *I suppose people all have their own oddities when it comes to food.*

odds *n.pl.* **1** °chances, °likelihood, °probability: *The odds are that Janet will finish the job in time.* **2** °edge, °advantage, °lead, °superiority: *We have won before against greater odds.* **3** °difference, °inequality, °disparity, unevenness, °discrepancy, °dissimilarity, °distinction: *It makes no odds who you are, you may not go in there.* **4** ***at odds.*** °at variance, at loggerheads, at daggers drawn, at sixes and sevens, at cross purposes, at each other's throats, in disagreement, °in opposition, on bad terms, not in keeping, out of line, inharmonious, conflicting, clashing, disagreeing, differing: *Teenagers have been at odds with their parents since time immemorial.* **5** ***odds and ends.*** oddments, °fragments, debris, leftovers, leavings, remnants, bits (and pieces), particles, shreds, snippets, °scraps, °rubbish, °litter, *Colloq Brit* odds and sods: *The children ate everything except for a few odds and ends of the birthday cake.*

odor *n.* **1** °smell, °scent, °aroma, °bouquet, °fragrance, °perfume, redolence; °stench, stink, fetor *or* foetor: *The air was filled with the odor of orange blossoms. The odor of rotting vegetation assailed our noses.* **2** °air, °breath, °hint, °suggestion, °atmosphere, °spirit, °quality, °flavor, °savor, °aura, °tone: *She would never allow the odor of scandal to touch her family.*

off *adv.* **1** away, °out, °elsewhere: *His secretary said that he'd gone off somewhere for the weekend.* **2** °distant, away, afar, far off: *The U-boat was a mile off. Christmas is only a month off.* —*adj.* **3** °incorrect, °wrong, °inaccurate, °in error, °mistaken, °misguided, misled, off the mark: *I'm afraid you're off on the question of the best way to approach him.* **4** °mad, °insane, °crazy, °eccentric, touched (in the head), *Colloq* dotty, dippy, nutty, potty: *Underwood's aunt is slightly off.* **5** °remote, °distant, °improbable, °unlikely: *He went to the station on the off chance that she would be on the noon train.* **6** off work, °at leisure, °idle, °free, °open: *Can you get the day off tomorrow to go on a picnic with me?* **7** °sour, °moldy, °bad, °rotten, °rancid, turned, °high: *The cream smells a bit off.* **8** °bad, unpropitious, °disappointing, °unsatisfactory, disheartening, displeasing, °slack, °slow, substandard, °below par, below average, °quiet: *It has been an off year for the Seychelles chess team.* **9** canceled, postponed: *The meeting is off till next week.* **10** situated, °fixed, supplied: *Is he really that well off? She was much worse off when they were married.*

offbeat *adj.* °strange, °eccentric, °bizarre, °weird, °peculiar, °odd, °queer, unconventional, °unorthodox, bohemian, idiosyncratic, °unusual, unexpected, °outré, °outlandish, °deviant, °novel, innovative, *Colloq* °kinky, °way-out, far-out, off the wall, freaky, °weirdo: *Jasper's offbeat, satirical humor has made him a popular comedian.*

off-color *adj.* **1** unwell, °ill, off form, °out of sorts, °queasy, °sick, °run down, °awful, °seedy, *Colloq* °under the weather, °poorly, *Slang* °lousy, °rotten: *I have been feeling off-color since eating that fish.* **2** indelicate, °risqué, ribald, °bawdy, °indecent, °suggestive, °broad, indelicate, inelegant, °improper, °inappropriate, °unseemly, °blue: *The countess does not countenance off-color remarks at the dinner table.*

offend *v.* **1** hurt (someone's) feelings, affront, °insult, °slight, snub, °give offense, °hurt, °pain, displease, disgruntle, chagrin, °humiliate, °embarrass; pique, °fret, °gall, vex, °annoy, °irritate, nettle, needle, °rankle, °provoke, °ruffle, °outrage, rile, °anger, *Colloq* miff, put (someone's) back up, put (someone's) nose out of joint, tread *or* step on (someone's) toes, put (someone) out, °rattle: *I hope you weren't offended at my saying that you have the Parthian pox.* **2** °disgust, °sicken, turn (someone's) stomach, °nauseate, °repel, °repulse, °revolt, *Colloq* turn (someone) off: *I, for one, am offended by seeing open-heart surgery on television.*

offender *n.* °criminal, malefactor, lawbreaker, °outlaw, wrongdoer, °culprit, °miscreant, °transgressor, °sinner, evildoer, *Slang* crook: *I don't know if they apprehended the offender.*

offense *n.* **1** °violation, °breach, °crime, felony, misdemeanor, infraction, °transgression, trespass, °wrong, wrongdoing, °sin, °peccadillo, °misdeed, °fault, °infringement, malefaction; dereliction, °lapse, °slip, °error: *He was accused of offenses against the rights of others. Some regard the splitting of an infinitive an offense against the purity of the language.* **2** ***give offense.*** incur displeasure, create annoyance *or* irritation *or* resentment *or* pique, evoke indignation *or* anger; °slight, °injure, °hurt, °harm, °offend, °insult, °outrage, *Colloq* put (someone) down: *He denied that he meant to give offense in his criticism of the play.* **3** ***take offense.*** take umbrage, feel displeasure *or* annoyance *or* resentment *or* pique *or* indignation, be angered *or* enraged: *Why should you take offense at what a fool says?*

offensive *adj.* **1** antagonistic, °hostile, contentious, °quarrelsome, attacking, °aggressive, °threatening, °provocative, combative, °martial, °belligerent, °warlike, bellicose: *The minute the enemy made an offensive move, we attacked.* **2** insulting, °rude, °disrespectful, uncivil, °insolent, °discourteous, °impolite, unmannerly, °impertinent, °impudent, objectionable, displeasing: *Nigel has been asked to leave because of his offensive behavior.* **3** °disgusting, °unsavory, °unpalatable, nauseating, °nauseous, noisome, noxious, °obnoxious, °repugnant, °repulsive, °repellent, °revolting, °abominable, °foul, °loathsome, °vile, sickening, fetid *or* foetid, °rank, malodorous, mephitic, °putrid, putrescent, putrefying, °rancid, °rotten: *An offensive stench emanated from the gorgon's mouth as she drew near to kiss him.* —*n.* **4** °attack, °offense: *At last, our team was on the offensive.* **5** °attack, onslaught, °drive, °assault, °offense, °push: *The offensive to capture the arsenal will be launched at dawn tomorrow.*

offer *v.* **1** proffer, °propose, °tender, °bid: *They offered twice what I had paid for it. She offered to buy my Vermeer.* **2** make available, °present, °tender, put on the market, °sell, put up for sale, °put up, °furnish: *The supermarket is offering lettuce at half price.* **3** proffer, °provide, °submit, put forward *or* forth, °advance, °tender, °extend, °make; °suggest: *Can you offer a suggestion for improving office efficiency? He offered to forget the whole thing if I paid him a thousand dollars.* **4** volunteer, present oneself, step *or* come forward: *I offered to help with her packages.* —*n.* **5** °proposal, °bid, °tender, °offering: *She said she would double any offer we have already had for the sword.* **6** °proposal, °presentation, proffer, °proposition: *The company said they could entertain no offers past the deadline. My offer to lend the money was contingent on being repaid.*

offering *n.* °sacrifice, oblation, contribution, °donation, °gift, °present: *They made offerings to the gods in order to propitiate them.*

offhand *adj.* **1** offhanded, °casual, °informal, °nonchalant, °cool, °distant, °aloof, easygoing, °blasé, unceremonious, °relaxed, °easy, °smooth, unconcerned, insouciant, lighthearted, uninterested, °superficial, °cursory, cavalier, °careless: *His offhand reaction shows that he doesn't care as much about her as we thought.* **2** °curt, °brusque, °abrupt, °perfunctory, °ungracious, °glib, °smooth: *When asked when he expected to pay, he made an offhand reply.* **3** extempore, impromptu, °unpremeditated, unstudied, °extemporaneous, °informal, off the cuff, ad-lib: *She rose to make some offhand comments about the accomplishments of the guest of honor.* —*adv.* **4** extempore, impromptu, extemporaneously, informally, off the cuff, ad-lib, °on the spur of the moment, at the drop of a hat: *That was a pretty good speech considering it was made offhand.* **5** casually, informally, °incidentally, °by the way, offhandedly, by the bye, parenthetically, °in passing, *en passant*, cursorily, superficially: *Offhand, I'd say that the two of them deserve each other.*

office *n.* **1** °business, °organization, °department, °firm, °house, °establishment, °company, corporation: *When I travel, I make certain to phone the office once a day.* **2** commission, °department, °branch; °section, °division: *He was with the overseas office for years.* **3** workplace, °offices; °room, °area: *Our new office is completely air-conditioned. My office is next to the board room.* **4** °duty, °obligation, °responsibility, °charge, commission, °service, °employment, °occupation, °position, °post, °appointment, °assignment, chore, °task, °job, °place, berth, °work, °role, °function, °purpose, °part, °bit, *Colloq* °thing, *Slang* shtick: *He was appointed to the office of Minister of Health. In her office as Minister of Finance, she wields great power.* **5** ***offices.*** °indulgence, intermediation, °auspices, °support, advocacy, aegis, °help, °aid, intercession, mediation, °patronage, °favor, °backing, back-up: *She appealed to the police chief's good offices to allow her to visit her son.*

officer *n.* **1** °(public) official, °dignitary, officeholder, public servant, officebearer, (political) appointee, °(government) agent, bureaucrat, °functionary, commissioner, administrator, °manager, °director; apparatchik: *Pepys followed a career as a customs officer. The bailiff is an officer of the court.* **2** policeman, policewoman, °police officer, officer of the law, °constable, *Old-fashioned* catchpole, *US* lawman, peace officer, G-man, T-Man, *Colloq* gendarme, *Slang* cop, copper, fuzz, *US* dick, narc, *Brit* Old Bill, tec: *The officer standing at the door was there to serve a writ.*

official *adj.* **1** authorized, °legitimate, °lawful, °legal, °authentic, °bona fide, °proper, °true, accredited, valid, documented, licensed, sanctioned, endorsed, certified, verified, recognized, accepted: *I won't believe I've won till I hold the official notification in my own hands.* **2** °ceremonial, °formal, °solemn, ritualistic, °ceremonious, °pompous, °stiff, °proper, °seemly, °decorous: *She has to make an acceptance speech at the official dinner.* —*n.* **3** See **officer, 1,** above.

officiate *v.* °preside, °direct, °manage, °chair, °conduct, °oversee, °head (up), °run, °lead, °supervise, superintend; °umpire, referee, °judge, adjudicate, °moderate, mediate: *Who will officiate at the annual meeting? Dennis has been invited to officiate at the football game on Saturday.*

officious *adj.* °dictatorial, °intrusive, intruding, meddlesome, meddling, °obtrusive, °forward, °bold, interfering, °aggressive, °insistent, °persistent, °demanding, importunate: *Skervy is one of those officious little men who are always ready to give unasked-for advice.*

offset *v.* **1** °compensate, counterbalance, countervail, counterpoise, °counteract, °balance (out), °equalize, even (out *or* up), °square, °cancel (out), °neutralize, nullify, °make up (for), °atone (for), redress; recompense, °repay, make amends *or* restitution, °make good, °reimburse, indemnify: *The votes from the center offset those lost to the far left. How are you going to offset losses resulting from pilferage by shop assistants?* —*n.* **2** compensation, counterbalance, counteraction, °check, equalizer, neutralizer: *The bank manager considered the money owed to the company as sufficient offset for the money it owed.*

offshoot *n.* **1** °branch, °spur; °shoot, limb, bough, °twig, °stem, appendage, °sucker, °sprout, sprig, tendril, scion: *There is an offshoot of this road that goes up the hill. The offshoots are trained to grow along the arms of the espalier.* **2** °descendant, °relation, °relative, °kin, °kindred, °offspring, scion, °heir: *One offshoot of the family later emigrated to the United States.* **3** outgrowth, °development, °branch, spinoff; by-product, °derivative: *An offshoot of the company manufactures optical instruments. The sale of gravel for concrete is an offshoot of our mining operations.*

offspring *n.* °(*Often used as plural.*) child, °progeny, °issue, °seed, youngster, °brood, °young, successor, °heir: *None of the earl's offspring ever amounted to much.*

often *adv.* °frequently, regularly, much, many times, °usually, habitually, commonly; °ordinarily, again and again, over and over again, time after time, °repeatedly, time and (time) again, in many cases *or* instances, on numerous occasions, day in (and) day out, continually, *Literary* oftentimes, oft: *How often do you visit your mother? We often went to the seaside for our summer holiday. She was often warned not to go too near the edge.*

ogle *v.* **1** °leer, °eye, make eyes at, *Colloq* give (someone) the glad eye, give (someone) the once-over, make sheep's eyes at: *Colonel Bumper is always ogling the pretty girls.* **2** °gape, °gaze, goggle, °gawk, °stare, *Slang Brit* gawp *or Slang Brit* gaup: *We took turns ogling the rings of Saturn through the telescope.* —*n.* **3** °leer, °stare, °gape, goggle, oeillade, *Colloq* once-over, glad eye: *One frowning glance from that virago was like the ogle of a Gorgon.*

ogre *n.* ogress, °monster, °giant, °fiend, °demon, troll, man-eater, bogey, bogeyman, bugbear, °specter, Minotaur, Cyclops, Gorgon, Caliban; °brute, sadist, °villain, cad, °scoundrel: *The ogre chased Jack to the beanstalk. The persistent ogre of poverty threatened him all his life.*

oil *n.* **1** lubricant, grease, lubricator, unguent: *A little oil will stop that squeak.* **2** °fuel: *Miraculously, the lamp burned for eight days with only one day's supply of oil.* —*v.* **3** lubricate, grease: *Oil the bearings or they will burn out.*

oily *adj.* **1** °greasy, oleaginous, °fat, fatty, adipose, pinguid, sebaceous, soapy, saponaceous, buttery, butyraceous, lardaceous; °slippery, °slimy, slithery, °smooth, unctuous: *The cars slid about as the tires failed to grip the oily surface.* **2** °glib, °smooth, unctuous,

°servile, °obsequious, sycophantic, °ingratiating, °flattering, °hypocritical; °suave, urbane, °sophisticated, *Colloq* smarmy: *Sarah found Curtis's approach sickeningly oily.*

ointment *n.* unguent, balm, °salve, emollient, embrocation, demulcent, pomade, pomatum, petrolatum; °lotion, cream: *A little ointment will keep the sore moist till it heals.*

OK *interj.* **1** O.K.!, Okay!, Fine!, Yes!, Definitely!, Agreed!, Very well!, All right!: *"Would you have dinner with me?" "OK!" "I think you ought to leave." "OK!"* —*adj.* **2** °satisfactory, °acceptable, °correct, °suitable, all right, °fine, °good, °in order: *Is it OK if I go to the movies tonight? That dress is OK to wear to the dance.* **3** °adequate, °mediocre, °fair, middling, °passable, °tolerable, *Colloq* °so-so, pretty good, not bad, not great: *The play was OK, I suppose.* **4** °well, °healthy; °sound, in good condition, in fine fettle, °fine, all right: *Now that he's on the proper medication, Sam is OK. The mechanic assured me that my car would be OK.* —*v.* **5** °approve, °sanction, °ratify, °authorize, °endorse, °support, °agree to, °allow, °consent to, °agree to, *Colloq* give the go-ahead *or* green light to, give the thumbs-up *or* the nod to *or* on, rubber-stamp: *A department head must OK your expense account before you can be reimbursed.* —*n.* **6** °approval, °sanction, ratification, authorization, °endorsement, °agreement, °support, °permission, °consent: *You need an OK from the security guard to enter the restricted area.* —*adv.* **7** all right, satisfactorily, °well (enough), adequately: *She can get along OK without me.*

old *adj.* **1** °elderly, aging, °aged, advanced in years *or* age, long-lived, past one's prime, °gray, full of years, getting on (in years), hoary, superannuated, *Colloq* over the hill, past it: *Bill is too old to continue working in the mine.* **2** °ancient, °antiquated, antediluvian, fossil, °prehistoric, Noachian, °obsolete, °antique, outdated, out-of-date, old-time, dated, archaic, °stale, outmoded, °passé, *Literary* Ogygian: *The archeological dig has turned up some interesting old artifacts.* **3** °timeworn, decayed, °dilapidated, °ramshackle, disintegrated, crumbling, °shabby, worn-out, dusty, broken-down, tumbledown, °disused, °unused, °cast off, °cast aside: *They have torn down the old mill near the river.* **4** long-standing, °well-established, °enduring, °lasting, age-old, °time-honored: *It is hard to see an old friendship die.* **5** °former, olden, °bygone, °early, primordial, °primitive: *In the old days, it took a week to travel from London to Edinburgh.* **6** °previous, °preceding, °prior, °former, quondam, erstwhile, one-time, ex-: *The West End was my old stamping ground when I lived in London.* **7** °experienced, °veteran, °practiced, (well-)versed, °knowledgeable, °proficient, °accomplished, °adept, skilled, °expert, old-time: *Charles is an old hand at steam engines.* **8** °dear, °beloved, loved, esteemed, valued, °precious, °well-known, °intimate, °close, °familiar: *Penelope is an old friend of the family's.*

old-fashioned *adj.* °antiquated, °antique, °passé, outmoded, outdated, unfashionable, °stale, dated, out-of-date, °tired, old-time, °obsolete, °obsolescent, °dead, superseded, replaced, °disused, °out, oldfangled, old hat: *Whoever thought we would see the day when miniskirts would be old-fashioned?*

omen *n.* portent, augury, °sign, °token, foretoken, °indication, °harbinger, forewarning, °premonition, foreshadowing, handwriting on the wall, prognostic, presage: *Solar eclipses were once regarded as omens, sometimes good, sometimes bad.*

ominous *adj.* **1** °foreboding, °threatening, °fateful, °dark, °black, °gloomy, lowering *or* louring, °menacing, °sinister; unpropitious, unfavorable, ill-omened, ill-starred, °unpromising, star-crossed, °inauspicious: *With ominous solemnity, the judge placed a black cloth square on his head before passing sentence.* **2** minatory, °warning, admonitory, cautionary: *The whispering had taken on ominous overtones.* **3** °portentous, °prophetic, oracular, vaticinal, predictive, prognostic, augural, mantic, sibylline *or* sibyllic *or* sibylic, °meaningful, premonitory, foreshadowing, foretelling, foretokening, indicative: *Virtually everything was regarded as ominous in ancient times.*

omission *n.* **1** noninclusion, omitting, leaving out *or* off, excluding, eliminating, dropping, skipping; °exclusion, °exception, deletion, elimination, excision: *The omission of your name from the list was a mistake. Allowing for inadvertent omissions, the inventory is complete.* **2** °failure, °default, °neglect, dereliction, °oversight, °shortcoming, °negligence: *She is being punished for her innocent omission in failing to notify the police while he is at liberty despite his deliberate commission of a crime.*

omit *v.* **1** °leave out, °exclude, °skip, °except, °pass over; °delete, °erase, °cancel, eradicate, °edit out, °strike (out), dele, °cut (out), °cross out, °obliterate: *She was offended because he omitted any mention of all that she had contributed.* **2** °neglect, °disregard, °fail, °forget, °overlook, °let slide, °ignore: *I omitted to tell you that your sister telephoned yesterday.*

once *adv.* **1** once upon a time, °formerly, (at) one time, on a former occasion, °previously, °before, in days gone by, in olden days, in the (good) old days, long ago, some time ago, years *or* ages *or* eons ago, in days of yore: *Your hair is as long as mine once was. He was once a famous film star. That once revered leader has fallen.* **2** one time, on one occasion, a single time: *He has visited his family only once in all these years.* **3** ***once and for all.*** °finally, °positively, °definitely, decidedly, conclusively, for good: *We must settle the itinerary once and for all before we can make the bookings.* **4** ***once in a while.*** °occasionally, °(every) now and then, °now and again, °at times, °sometimes, periodically, from time to time, at intervals, sporadically: *We go to the theater once in a while.* —*conj.* **5** °(if) ever, as soon as, at any time: *Once the bus comes, you'd best get on it straightaway.* —*n.* **6** ***at once.*** **(a)** °immediately, straightaway, right away, °directly, without delay, °promptly, instantly, °posthaste; in a wink, in the twinkling of an eye, in a minute *or* moment *or* second *or* split second, °in no time (at all), before one can turn around, before one can *or* could say "Jack Robinson," in a trice, *Colloq* in a jiffy, in two shakes of a lamb's tail: *Watson, come here at once. I'll be there at once.* **(b)** together, °at the same time, simultaneously, at a stroke, in the same instant, in the same breath, *Colloq* at one go, at a go, in one go: *You cannot be in two places at once.*

oncoming *adj.* **1** advancing, arriving, coming, nearing, approaching, onrushing, °imminent: *We barely avoided the oncoming train.* —*n.* **2** °onset, °beginning, nearing, °arrival, °advance, °approach: *With the oncoming of spring, the birds returned.*

one *adj.* **1** °single, lone, °solitary, °individual, °sole, °only: *The one time I kissed Margie it was heaven.* **2** unified, °united, inseparable, joined, °undivided, one and the same, °identical, °equal, at one, °harmonious, in unison, °whole, °entire, °complete: *When he went into a trance, he felt one with his God.* **3** a particular, a certain, a given, a specific: *I recall one time when we met she brought her dog along.* —*pron.* **4** a person, an individual, a man *or* a woman, everybody, °everyone, anybody, anyone; °people; *Possibly offensive* °man: *One ought to treat others as one would like to be treated. One cannot be too careful these days.* —*n.* **5** °joke, °story, anecdote, chestnut, one-liner; limerick, °rhyme, ditty, °song; bromide: *Have you heard the one that begins, "There was a young man from Loch Ness"?*

one-sided *adj.* **1** °partial, °biased, °partisan, °prejudiced, °bigoted, unfair, unjust, inequitable, close-minded, °narrow-minded, °intolerant: *His is a one-sided view of the problem.* **2** °lopsided, °unbalanced, unequal, unequalized, uneven, °disproportionate, *Slang* cockeyed: *The swelling on his left cheek made Tom's face look very one-sided.* **3** unilateral, °independent, exclusionary, °exclusive: *They made a one-sided decision to halt production of nuclear weapons.*

ongoing *adj.* **1** continuing, continued, °continuous, °continual, ceaseless, unbroken, uninterrupted, °constant, °perpetual, °nonstop, °relentless, °persistent, unending, °endless, interminable, °running: *There has been an ongoing dispute with the museum over the authenticity of the sculpture.* **2** developing, evolving, growing, °successive, unfolding, progressing, °progressive: *Rather than come to a hasty decision, we decided to monitor ongoing developments.*

onlooker *n.* °spectator, °observer, looker-on, °eyewitness, °witness, watcher, viewer; °bystander, passer-by: *She was merely an onlooker, not a participant. Onlookers reported that the driver had run away from the accident.*

only *adj.* **1** °sole, °single, °solitary, lone, one and only, °exclusive: *He is the only one who can identify the murderer.* —*adv.* **2** solely, °just, exclusively, °alone: *He has a face that only a mother could love. They have been here only twice. She is right only because she wields the power. Harry was correct in one respect only.* **3** °merely, °simply, °barely, at best, at worst, at most, °just, purely, not *or* no more than, not *or* no greater than: *She is only seventeen. I received your note only today. Don't get excited, it's only your sister.* —*conj.* **4** but, °however, on the other hand, on the contrary, contrariwise: *The flowers are lovely, only they have no scent.*

onset *n.* **1** °attack, °assault, onrush, onslaught, °charge, °strike, °hit, °raid, storming, sally, sortie: *These troops had to bear the brunt of the onset.* **2** °beginning, °start, °outset, °initiation, inauguration, commencement, inception, °dawn, °birth, °origin, genesis, °appearance, °debut: *We must leave before the onset of the monsoon season. The sudden onset of a new policy will throw the ministers into a panic.*

onward *adj.* **1** °forward, advancing, °progressive, progressing, moving onward *or* forward: *They resumed their onward march, laying waste to the countryside as they went.* —*adv.* **2** onwards, forward(s), on, henceforward, °henceforth, °ahead, °in front, on, forth: *Onward, Christian soldiers, / Marching as to war. From this day onward afternoon visiting hours will be from two to four o'clock.*

ooze *n.* **1** slime, °muck, °mud, °mire, °silt, °sludge, °sediment, slush, *Colloq* goo, gunk, guck, *Slang US* glop, goop: *I stepped into the bog and the ooze rose over the tops of my shoes.* —*v.* **2** exude, °weep, seep, °secrete, bleed, °leak, °drain, °trickle; °emit, °discharge: *Sap continues to ooze from the gash in the tree.*

opacity *n.* **1** opaqueness, darkness, murkiness, dimness, °obscurity, impermeability, impenetrability: *The opacity of the lens increases automatically in the presence of sunlight.* **2** °obscurity, density, impenetrability, unintelligibility, indefiniteness, vagueness, reconditeness, abstruseness, °ambiguity, equivocation, mystification: *For centuries scientists were unable to penetrate the opacity of the question of what occurred when substances burned.* **3** °stupidity, dullness, denseness, thickness, obtuseness: *A light finally dawned through the thick opacity of his brain.*

opalescent *adj.* opaline, iridescent, nacreous, °pearly, °lustrous: *The gem seemed to breathe with an inner opalescent fire.*

opaque *adj.* **1** °dark, °murky, °dim, turbid, °muddy, cloudy, °obscure, obscured, obfuscated, °black, °impermeable, impenetrable, clouded, nontransparent, untransparent, nontranslucent, °hazy, blurred, blurry, smoky: *Solar eclipses should be viewed directly only through special opaque glass.* **2** unclear, °vague, °indefinite, °obscure, unfathomable, unplumbable, baffling, mystifying, °ambiguous, °equivocal, impenetrable, °cryptic, enigmatic, °puzzling, °perplexing, °mysterious, °elusive, abstruse, arcane, °recondite: *Despite years of study, the inscriptions on the tomb have remained opaque to scholars.* **3** unintelligent, °dense, °thick, °dull, °obtuse, °stupid, dull-witted, °stolid, thickheaded, dunderheaded, dunderpated, °slow, doltish, °backward, cloddish: *Some people are opaque, able only to borrow and reflect the light of others.*

open *adj.* **1** ajar, gaping, agape, unfastened, unlocked, unbarred, unbolted, unlatched, unclosed: *Come on in—the door is open.* **2** yawning, agape, uncovered, revealed, unsealed, exposed, °bare: *At midnight on Halloween they found three graves open which were covered again the next morning.* **3** unwrapped, unsealed, unfastened: *The package was open and the contents gone.* **4** °free, °accessible, °public, °available; obtainable; unrestricted, unobstructed, unencumbered, °unimpeded, unhindered, unhampered, unregulated, unconditional, °unqualified: *The parks are open to all. We were allowed open access to the library stacks.* **5** unprotected, unenclosed, unsheltered, °bare; uncovered, exposed: *They spent a week in an open boat before being rescued. The roof can be retracted, leaving the interior completely open to the sky.* **6** °unsettled, unagreed, unsigned, unsealed, unclinched, unestablished, unconcluded, undecided, °pending: *As far as I am concerned, the deal is open till the contract is signed.* **7** undecided, °unsettled, °unresolved, °debatable, arguable, °problematic, °moot, *US* up in the air: *Whether they should get married is a question that will remain open until he returns from abroad.* **8** unscheduled, unbooked, unspoken-for, unreserved, uncommitted, °free, unpromised: *The doctor has an hour open at noon on Friday.* **9** °clear, unobstructed, wide-open, uncluttered, °roomy, °spacious, °extensive, °expansive; treeless, uncrowded, unfenced, unenclosed; ice-free, °navigable, unblocked, °passable: *We traveled through open country for days. When are the shipping lanes open again in the spring?* **10** °available, unfilled, °vacant, untaken: *There are not many jobs open in this part of the country.* **11** °receptive, open-minded, °flexible, amenable, persuasible *or* persuadable, pliant, °willing, °responsive: *The management is open to suggestions for improving its products and services.* **12** exposed, °public, °well-known, widely known, unconcealed: *That they are living together is an open secret.* **13** °evident, °obvious, °conspicuous, °manifest, °clear, unconcealed, unequivocal, °plain, palpable, °apparent, °patent, °downright, °out-and-out, °blatant, °flagrant, °glaring, °brazen: *He operates with open disregard for the law.* **14** °generous, °liberal, °charitable, unreserved, openhanded, °liberal, munificent, magnanimous, bighearted, beneficent, bounteous, °unselfish, unstinting, °humanitarian, altruistic: *They are quite open in their gifts to charities.* **15** unreserved, °candid, °frank, °outspoken, straightforward, °forthright, °direct, °honest, °sincere, guileless, °artless, °fair: *That philandering wretch has never been completely open with his wife.* **16** °free, unrestrained, unconstrained, °uninhibited, unreserved, unrestricted: *They have an open marriage, each aware of the other's affairs.* **17** unfolded, extended, °spread (out), outstretched, outspread: *She ran into my open arms.* **18** °liable, °subject, °susceptible, exposed, °inclined, predisposed, °disposed: *These fraudulent shipping documents may well leave the captain open to prosecution for barratry.* **19** unprotected, undefended, unfortified, exposed: *With the invaders at the gates, Paris was declared an open city.* —*v.* **20** °begin, °start, °initiate, °commence, get under way, °inaugurate, °launch, put in *or* into operation, °activate, get going, set in motion; °establish, °set up; *Colloq* get *or* start the ball rolling, get *or* put the show on the road, kick off: *The minister opened the proceedings with an interminable speech. Jeremy is planning to open a restaurant in Pebble Lane. The show opens in Manchester next week.* **21** unlock, unbar, unlatch, unbolt, unfasten; uncover; uncork, unseal; °undo, untie, unwrap; °pull out: *Open the door. Open the box. Open the bottle. Open your present. He opened the drawer.* **22** unblock, °clear, unobstruct, unclog, unstop: *The drain was opened by taking it apart. The new law has opened the way for increased exports.* **23** °disclose, °unveil, uncover, °expose, °display, °show, °exhibit, °reveal, divulge, °bring to light, °communicate, °bring out, unbosom, °explain, °present, °announce, °release, °publish, °air, °make known, advertise: *The wonders of the language were opened to me by my first dictionary.* **24** °expand, °spread (out), stretch out, open up *or* out, unfurl, °extend: *The flag opened to the breeze.* **25** °present, °offer, °furnish, °provide, °afford, °yield, °reveal, uncover, °raise, °contribute, °introduce: *Expansion of technology opens new business opportunities every day.*

opening *n.* **1** °break, °breach, °rent, °rift, cleft, °crack, °crevice, fissure, °cranny, °chink, °pit, °gap, °split, °slit, °slot, °aperture, °hole, orifice, °separation: *Flowers grew from openings in the wall.* **2** °opportunity, °chance, °occasion, toehold, foothold, *Colloq* °break, toe *or* foot in the door, *Brit* look-in: *I was waiting for an appropriate opening to make my presentation.* **3** °job, °position, °opportunity, °vacancy: *Is there likely to be an opening in the art department of your company?* **4** °beginning, commencement, °start, °birth, °origin, °outset, °onset, inauguration, °launch, send-off, °initiation, °presentation, °debut; vernissage, *US* start-off, start-up: *The opening of the autumn social season was marked by Malcolm's birthday party. Aren't you going to the opening at the museum tonight?*

openly *adv.* **1** brazenly, brashly, flagrantly, unabashedly, unashamedly, unreservedly, boldly, audaciously, flauntingly: *She has openly defied the direct orders of her mother.* **2** frankly, unreservedly, plainly, forthrightly, candidly, °directly, °outright, °freely, outspokenly: *Count Knoedelkopf admitted openly that he had stolen the plans for the missile.*

operable *adj.* workable, °practicable, °serviceable, usable, °functional, °fit, operational, in working order *or* condition: *Aircraft engines that rely on oxygen for burning fuel are not operable where the air is too thin.*

operate *v.* **1** °go, °run, °perform; °work, °function, °serve, °act: *This watch operates under water. The drug operates to reduce the blood pressure.* **2** °manage, °run, °direct, °conduct, °control, °carry on, °ply, °manipulate, °handle; *US* °drive: *Katherine has been operating as an antiques dealer for years. It is unsafe to operate machinery without goggles.*

operation *n.* **1** °function, functioning, working, °running, °performance, °action, °motion, °movement: *The operation of my car's engine is very erratic.* **2** manipulation, handling, °direction, °running, °control, °management, managing; maneuvering: *The operation of the aircraft is under the control of the captain.* **3** °undertaking, °enterprise, °venture, °project, °affair, °deal, °procedure, °proceeding, °(day-to-day) business, °transaction: *Who will be in charge of the operation while the president is abroad?* **4** Often, ***operations.*** °action, °maneuver, °mission, °task, °campaign, °exercise: *The generals directed military operations from positions close to enemy lines.* **5** ***in*** *or* ***into operation.*** functioning, °operative, °in effect, °in force, operating, operational, °functional, °effective, °efficacious: *Is the factory in operation yet? The new regulations went into operation last week.*

operative *adj.* **1** See **operation, 5,** above. —*n.* **2** °worker, °hand, °employee; craftsman, craftswoman, artisan, mechanic, machinist: *We hired two more lathe operatives today.* **3** private detective, (private) investigator, *Colloq* private eye, °sleuth, *Brit* sleuthhound, *US* P.I., gumshoe, *Slang* (private) dick, *US* shamus, °eye: *Our operatives reported that Jones had been seen with an enemy agent.* **4** espionage *or* intelligence agent, counterespionage *or* counterintelligence agent, °spy, counterspy, undercover agent *or* man, (FBI *or* CIA) agent, *US* G-man, *Colloq US* company man, member of the firm: *We had an operative at the top level of the NKVD.*

operator *n.* **1** (bus *or* taxi *or* train) driver; °worker, °operative, manipulator, practitioner: *These operators are required to take safety courses.* **2** °director, administrator, °manager, °supervisor, °superintendent: *Shaughnessey is the operator of a roofing business in Tring.* **3** machinator, faker, °fraud, manipulator, maneuverer, *Colloq* finagler, °wise guy, *Slang* smooth *or* slick operator, smoothie, wheeler-dealer, big shot, big-time operator, *Chiefly US and Canadian* big wheel: *Claude is an operator who always gets what he goes after.*

opinion *n.* **1** °belief, °judgment, °thought, °sentiment, °(point of) view, °viewpoint, °conviction, way of thinking, °perception, °idea, °impression, °notion, °conception, theory, *idée reçu;* °mind: *It is my opinion that sickness benefits ought to be increased. In her opinion all men are chauvinists.* **2** °evaluation, °estimation, °estimate, appraisal, °appreciation, °impression: *Myra has a very low opinion of Ray's taste in architecture.*

opinionated *adj.* **1** °stubborn, pigheaded, °obstinate, doctrinaire, °inflexible, °dogmatic, °single-minded, cocksure, obdurate, °dictatorial, dogged, mulish, bullheaded, °overbearing: *Felix is too opinionated to change his mind even if he knows he is wrong.* **2** °prejudiced, °biased, °bigoted, °one-sided, °jaundiced, colored, °partial, °partisan: *You can count on him for an opinionated view of social values.*

opponent *n.* °antagonist, °adversary, disputant, °contestant, °competitor, contender, °rival, foe, °enemy; the opposition: *He may be my opponent in the chess competition, but we are the best of friends.*

opportune *adj.* **1** °favorable, °advantageous, auspicious, °good, felicitous, °happy, °propitious, °beneficial, °helpful, °fortunate, °lucky, °profitable: *As I need money, and you have it to invest, our meeting is most opportune.* **2** °timely, °well-timed, °seasonable, apt, °appropriate, germane, °pertinent, °convenient, °fitting, °suitable, °becoming: *If this is not an opportune time to bring up the matter of the money you owe me, just say so.*

opportunistic *adj.* °expedient, °selfish, taking advantage, exploitive *or* exploitative, unprincipled, Machiavellian, opportunist: *Don't you agree that it was opportunistic of him to inform on his own brother for a reward?*

opportunity *n.* °chance, °occasion, °opening, °possibility, °moment, °time, *Slang* °break: *She has taken advantage of every opportunity to vilify her ex-husband.*

oppose *v.* **1** °resist, °counter, °object (to), °defy, take a stand against, °withstand, °resist, °combat, °contest, °attack, counterattack, °fight, °grapple with, contend with *or* against: *If it comes to that, we must oppose force with force.* **2** °check, °bar, °obstruct, °block, °hinder, °impede, °stop, °slow, °curb, °restrain, °inhibit, interfere with, °restrict, °prevent, obviate, °preclude, °thwart, °foil, °frustrate: *In England, the Labour party opposes the privatization of industry.* **3** °match, °offset, counterbalance, °contrast, pit *or* set against, play off (against), °set off: *What can they call upon to oppose the power of the Evil One?*

opposed *adj.* Often, **opposed to.** against, °in opposition (to), °opposing, in conflict (with), antipathetic, conflicting, °contrary (to), °at variance (with), antithetical (to), °hostile (to), inimical (to), °opposite (to), contrasting: *Harvey is among those opposed to abortion. All those who are opposed to the motion raise your hands.*

opposing *adj.* °opposite, conflicting, °contrary, antithetical, antagonistic, antipathetic, °hostile, inimical, contrasting, °rival, °contradictory, °incompatible, irreconcilable, °dissident, discrepant: *Those with opposing views will be heard from later.*

opposite *adj.* **1** °facing, vis-à-vis, *en face*: *The sniper was on the roof of the building opposite.* **2** °opposing, conflicting, °contrary, contrasting, °contradictory, antithetical, differing, °different, °divergent, °diverse, antagonistic, °inconsistent, irreconcilable: *Pitmarsh and his wife hold diametrically opposite political views.* —*n.* **3** °reverse, °converse, °contrary, antithesis: *Whatever you tell teenagers to do, their first reaction is to do the opposite.*

opposition *n.* **1** °hostility, °antagonism, unfriendliness, °resistance, counteraction, °disapproval, °objection, °conflict, defiance, °contrast, antipathy, adversity, *Colloq* °flak: *There is strong opposition to plans for an amusement arcade.* **2** °competition, °opponent, °adversary, °competitor, °antagonist, °enemy, foe, °rival, other side: *We must overcome the opposition in order to win.* **3 *in opposition.*** competing, competitive, antagonistic, °hostile, conflicting, in conflict, antithetic(al), °opposed, at daggers drawn, in deadly embrace: *Although they agree on some points, the parties are in opposition on others.*

oppress *v.* **1** °burden, °afflict, °trouble, °weigh down, °overload, °encumber, °wear (down), °press, °weary, overburden, °overwhelm, *Brit* pressurize, *US* °pressure: *He was oppressed by the heavy burden of responsibility.* **2** °crush, °repress, °put down, °suppress, °subjugate, °tyrannize (over), °subdue, °overpower, °enslave, °persecute, maltreat, °abuse, harry, °harass, trample underfoot, ride roughshod over: *The Russian serfs had been oppressed for centuries before they finally rose up against tyranny.*

oppression *n.* °repression, °suppression, subjugation, °subjection, °tyranny, °despotism, enslavement, °persecution, maltreatment, °abuse, °torment, torture, °hardship, °injury, °pain, °anguish, °injustice: *People who have not known oppression cannot imagine the agonies suffered by the oppressed who once were free.*

oppressive *adj.* **1** °burdensome, °overpowering, °overwhelming, onerous, °heavy, cumbersome, °exhausting, racking, °unbearable, intolerable, °agonizing, unendurable, °harsh, °brutal, °severe, °tyrannical, °repressive; dispiriting, depressing, disheartening, discouraging, °grievous, distressing, dolorous, °miserable, °harrowing, °wretched: *The conquerors resorted to oppressive measures to keep the people subjugated.* **2** suffocating, stifling, °stuffy, °close, airless, unventilated, uncomfortable: *The atmosphere in the tiny cell quickly became oppressive, and several people fainted.*

oppressor *n.* °bully, °tyrant, taskmaster, taskmistress, °despot, autocrat, persecutor, slave driver, °dictator, overlord, iron hand, °scourge, tormentor, torturer, intimidator: *The citizens finally banded together and overthrew their oppressors.*

optimistic *adj.* °sanguine, °positive, °cheerful, °buoyant, °bright, °hopeful, °expectant, °confident, bullish, °idealistic, Pollyannaish: *We have every reason to be optimistic that the venture will succeed.*

optimum *n.* **1** °best, finest, most favorable, °ideal, °perfection, °model, °paragon, exemplar: *In all work, the optimum is difficult to achieve.* —*adj.* **2** °best, finest, most favorable, °ideal, °perfect, choicest, optimal, °first-rate, first-class, °sterling, °prime, °capital, °excellent, °exceptional, °superlative, °extraordinary, °unique, °peerless, unequaled, unexcelled, unsurpassed: *These instruments keep the chamber at the optimum temperature. After months of training, Guy is in optimum condition to win the marathon.*

option *n.* **1** °choice, °selection, °alternative, °recourse, °opportunity, way out: *There are fewer employment options open to the uneducated.* **2** °choice, °privilege, °election, °opportunity, °chance: *Investors pay for the option to buy at a fixed figure if the price of the shares goes up.*

optional *adj.* °voluntary, discretionary *or* discretional, elective, facultative, °free, °spontaneous, uncoerced, unforced, noncompulsory, uncompulsory, nonmandatory, unmandatory, nonrequisite, unrequisite: *Life insurance is optional for those who have our hospitalization policy.*

opulent *adj.* **1** °wealthy, affluent, °rich, °prosperous, well-to-do, °well-off, °comfortable, *Colloq* °flush, well-heeled, °loaded, rolling in it, made of money, in clover, on Easy Street, *Brit* on velvet, *US* in velvet, in the chips: *Timothy was fortunate in having met and wed an opulent widow.* **2** °luxurious, °lavish, °sumptuous: *That poor little rich girl was raised in the most opulent surroundings.* **3** °abundant, copious, °bountiful, °plentiful, °prolific, °profuse, plenteous: *We enjoyed a most opulent harvest this year.*

opus *n.* °work, °composition, °production, *oeuvre*, °creation; *magnum opus*: *Her most important opus will be performed at the Hollywood Bowl next week.*

oracle *n.* **1** °prophet, sibyl, °seer, soothsayer, augur, °fortuneteller, diviner, prognosticator, *US* reader (and adviser *or US* advisor), Cassandra, Nostradamus; °authority, guru, °mastermind, mentor, wizard: *He insists on consulting his oracle before making any final decision.* **2** °prophecy, augury, °prediction, divination, °advice, prognostication, °answer, °message, divine utterance: *According to the oracle, the travelers would survive the perils of the journey.*

oral *adj.* °spoken, said, °verbal, uttered, voiced, vocal, vocalized, enunciated, °pronounced, articulated, word of mouth, viva voce: *Tomorrow James must make an oral presentation of his plan to the entire staff.*

oration *n.* °speech, °declaration, °address, °lecture, °recitation, discourse, monologue, declamation; valedictory, °eulogy, homily, panegyric; *Colloq* spiel: *Bentley delivered himself of a two-hour oration on the state of the economy.*

oratory *n.* public speaking, speechmaking, eloquence, °rhetoric, way with words, command of the language, °fluency, glibness, grandiloquence, magniloquence, declamation; elocution, °diction, enunciation, articulation, °address; *Colloq* gift of the gab: *The crowds who thronged to hear Churchill's oratory were seldom disappointed.*

orb *n.* °sphere, ball, °globe: *The golden orb of the sun sank into the sea.*

orbit *n.* **1** °circuit, °course, °path, °track, °revolution, °circle, °round, °cycle: *The earth's orbit round the sun is elliptical.* —*v.* **2** °revolve, go *or* turn around, °circle, °encircle: *The earth orbits the sun in a year. Electrons orbit the nucleus of an atom.*

ordeal *n.* °trial, °test, tribulation(s), °hardship, °affliction, trouble(s), °suffering, °distress, °anguish, nightmare, °misery, °grief, °misfortune, adversity, °tragedy, °disaster: *After his ordeal as a hostage, Gibson was in no mood to talk to a platitudinous television interviewer.*

order *n.* **1** °organization, °arrangement, grouping, °disposition, °form, °structure, categorization, systematization *or* systemization, classification, codification, disposal, layout, array, °sequence, *Colloq* °setup: *The order of the library has to be restored if we are to find anything.* **2** °organization, °uniformity, °regularity, °system, °pattern, °symmetry, °harmony, tidiness, orderliness, neatness: *Some believe that there is an order of things in the universe, others that the universe tends to chaos.* **3** °category, °class, °caste, °level, °kind, °sort, °rank, °group, °scale, °importance, hierarchy, °position, °status, °degree, *Colloq* pecking order: *Gregory's musical talents are of a very high order.* **4** °command, °direction, directive, °instruction, commandment, °dictate, mandate, edict, behest, °request, °demand, ukase, °decree, fiat, °proclamation, °pronouncement, pronunciamento; °rule, °regulation, °law, ordinance, statute, °requirement: *The police have issued an order to surrender all hand guns.* **5** °procedure, proceeding(s), °discipline, °conduct: *The order of the meeting was breached by some rowdies.* **6** °condition, °state (of affairs): *Please leave everything in the order in which you found it.* **7** purchase order, °request, °requisition, commitment, commission, °instruction: *We received a large order for office furniture.* **8** °calm, °peace, peacefulness, tranquillity, °quiet, °serenity, law and order, °discipline, lawfulness: *After a brief commotion, order was restored.* **9** °brotherhood, °fraternity, sisterhood, sorority, °fellowship, sodality, °association, °organization, °society, guild, °sect, °company, community, °lodge, °body, knighthood: *One of her ancestors was a Knight of the Teutonic Order.* **10 *in order.*** **(a)** °neat, °clean, °tidy, °shipshape, °orderly, (well-)organized, °ready, °prepared, arranged: *Is everything in order for the wedding tomorrow?* **(b)** °fitting, °suitable, °appropriate, °correct, °right, apt, called for; required, demanded, needed: *I think that an apology is in order for the way you behaved.* **11 *in order that.*** so (that), with the aim *or* purpose that, to the end that: *We invited him in order that you might meet him.* **12 *in order to.*** to, for the purpose of: *In order to get there, you have to drive up the hill.* **13 *out of order.*** **(a)** disordered, nonsequential, out of sequence, nonalphabetical, disorganized, unorganized, in disorder: *The cards in this catalog are out of order and I cannot find anything.* **(b)** °unseemly, °out of place, °improper, uncalled-for, unsuitable, indecorous, *Colloq chiefly Brit* not cricket: *Your remark about her religion was completely out of order.* **(c)** out of commission, °broken, in disrepair, nonfunctioning, nonfunctional, not working, °broken down, inoperative, out of kilter *or Brit also* kelter, *Colloq* (gone) haywire, kaput, bust(ed), *US* out of whack, on the fritz, °shot; *Slang* °on the blink, *Brit* wonky, gone phut: *The television is out of order again.* —*v.* **14** °direct, °command, °instruct, °charge, °tell, °bid, °require, enjoin; °demand, ordain; °force, °make: *The sergeant ordered the men to run around the drill field with full packs. The king ordered that bread be given to the poor.* **15** °requisition, °ask for, send (away) for, °call for, apply for, °reserve, °engage, commission, contract for; °purchase, °buy: *Have you ordered a room for the weekend? Let's order something from the Chinese restaurant.* **16** °organize, systematize, °arrange, classify, °categorize, codify, °lay out, °sort (out), straighten (out *or* up): *The bottles were ordered in neat rows along the wall.*

orderly *adj.* **1** in (good) order, (well-)organized, °neat, °shipshape, °tidy, arranged, °methodical, °systematic, systematized *or* systemized, °harmonious, °symmetrical, °regular, °uniform: *Before you leave, make sure that your room is orderly.* **2** well-behaved, disciplined, °decorous, law-abiding, well-mannered, °peaceable, °tranquil, mannerly, °polite, °courteous, °civil, civilized, nonviolent: *Everyone left the burning theater in orderly fashion.* —*n.* **3** °assistant, adjutant, °attendant, °messenger; °menial, °servant; nurse's aide; *Brit military* batman; *US* candy-striper; *Slang US and Canadian* gofer: *An orderly arrived with dispatches from the general. She has a job as a hospital orderly.*

ordinarily *adv.* °usually, normally, °as a rule, commonly, °generally, in general, customarily, routinely, typically, habitually, by and large, °for the most part: *Ellie is ordinarily at her desk by nine o'clock.*

ordinary *adj.* **1** °usual, °normal, expected, °common, °general, °customary, °routine, °typical, °habitual, °accustomed, °traditional, °regular, °everyday, °familiar, °set, °humdrum: *This wine is quite good for ordinary drinking. Just display ordinary good manners when you meet the queen.* **2** °common, °conventional, °modest, °plain, °simple, °prosaic, °homespun, commonplace, °run-of-the-mill, °everyday, °average, unpretentious, workaday, °mediocre, °fair, °passable, °so-so, °undistinguished, unexceptional, unremarkable, uninspired, °pedestrian, °bourgeois, °peasant, °provincial, °unrefined, *Colloq Brit* common or garden, *US* (common-or-)garden variety: *They bought a rather ordinary house in an inferior neighborhood.* —*n.* **3** °standard, °norm, °average, status quo, °convention, expected: *Saint-Gaudens' architectural designs are far from the ordinary.* **4 *out of the ordinary.*** °extraordinary, °unusual, uncommon, °strange, °unfamiliar, °different, unexpected, unconventional, °curious, °eccentric, °peculiar, °rare, °exceptional, °original, °singular, °unique, °odd, °bizarre, °weird, °offbeat, °outlandish, °striking, °quaint, °picturesque: *She was looking for a gift that was a little out of the ordinary, so I suggested a pet tarantula.*

organ *n.* **1** °device, °instrument, °implement, °tool; °member, °part, °element, °unit, component, °structure, *Technical* °process: *The eye, come to think of it, is a truly miraculous organ.* **2** °medium, °vehicle, °voice, °mouthpiece, forum, °publication, °paper, °magazine, newsletter, house organ, newspaper, annual, semiannual, °quarterly, monthly, fortnightly, weekly, hebdomadal, °daily, °journal, °periodical: *The official organ of the society is published in Abergavenny.*

organic *adj.* **1** living, °natural, biological, biotic, °animate, breathing: *Though coal may seem to be a mineral, it is organic, for it was formed from plants.* **2** °basic, °elementary, °essential, innate, °inborn, °natural, °native, °ingrained, °primary, °fundamental, visceral, constitutional, °inherent, structural, °integral: *The organic differences between the styles of writing are obvious.* **3** organized, °systematic, °coherent, coordinated, integrated, structured, °methodical, °orderly, °consistent: *The various elements of the painting blend into an organic whole.*

organism *n.* living thing, °structure, °body; being, °creature: *It is the work of natural scientists to classify all kinds of organisms.*

organization *n.* **1** organizing, structuring, assembling, °assembly, putting together, coordination, systematizing, systematization, classifying, classification, categorizing, categorization, codifying, codification: *The organization of the teaching schedule took hours to complete.* **2** °structure, °pattern, configuration, °design, °plan, °scheme, °order, °system, °organism, °composition, °arrangement, constitution, °makeup, grouping, framework, °format, °form, °shape: *One must consider the organization as a whole, not merely its constituent elements.* **3** °body, °system, °institution, °federation, confederacy, confederation, °society, °group, °league, coalition, conglomerate, °combine, consortium, °syndicate, °organism: *The organization is a coherent structure made up of an enormous number of disparate elements.*

organize *v.* **1** °structure, °coordinate, systematize, systemize, °order, °arrange, °sort (out), classify, °categorize, codify, catalogue, °group, °tabulate, pigeonhole, °standardize: *These files ought to be organized so that you can find something when you need it.* **2** °form, °found, °set up, °establish, °institute, °start, °begin, °create, °originate, °initiate, put together, °build, °develop, US pull together: *In 1969, we organized a company to publish reference books.*

orgy *n.* **1** bacchanalia, bacchanal, saturnalia, Dionysia, debauch, carousal, °carouse, °spree, °revel, °party, *Colloq* binge, °bender, °drunk, bust, *Slang* °jag, *US and Canadian* toot, °tear: *The journalist represented the party as a wild orgy.* **2** overindulgence, °splurge, °spree, °fling, *Slang US* °bender: *Trying to lift himself out of depression, Roger went on a spending orgy.*

orient *n.* **1** east: *Harriet is in the orient on business.* —*adj.* **2** *Literary* oriental, eastern: *The grass was sown with orient pearls.* —*v.* **3** °adjust, °adapt, acclimatize *or* acclimate, habituate, °accommodate, °condition, °accustom, °familiarize, feel one's way, assess, get one's bearings, *Colloq* orientate: *It is a new job and she needs a few days to orient herself.*

orientation *n.* **1** °placement, °bearings, °attitude, alignment, °lie, placing, °situation, layout, °location, °position, positioning, °arrangement, °setup: *The orientation of the buildings is such that the windows face south.* **2** introduction, training, °initiation, briefing, familiarization, assimilation, acclimatization, °preparation, °instruction: *The orientation of the new employees is scheduled for next week. We were given orientation lectures.*

origin *n.* **1** °source, °derivation, °rise, fountainhead, °foundation, °basis, °base, wellspring, fount, provenance, *Chiefly US* provenience: *The origins of many English words are unknown.* **2** °creation, genesis, °birth, birthplace, cradle, dawning, °dawn, origination, °start, °beginning, commencement, °outset, °launch, launching, inception, inauguration: *The origin of the notion of democracy can be traced to ancient Greece.* **3** Often, ***origins.*** °parentage, ancestry, °extraction, descent, °lineage, °pedigree, genealogy, °stock, °heritage: *We have traced our family's origins back to the Middle Ages.*

original *adj.* **1** °initial, °first, earliest, °primary, °beginning, starting, °basic: *The original report made no mention of any missing jewelry.* **2** °native, °indigenous, autochthonous, °aboriginal, primordial, primeval, °primitive: *At first, we could find only slight traces of the original inhabitants.* **3** °master, °actual, °primary, °authentic, °true, °genuine, °real, °basic; prototypic(al), archetypal, °source: *I have the original document and my lawyer has a copy.* **4** °creative, °novel, innovative, °unique, °imaginative, °unusual, inventive, °ingenious; first-hand, °fresh, underived, unprecedented: *The film is based on a highly original story by Claurène duGran. The author has some original insights into Hamlet's relationship with Ophelia.* —*n.* **5** °prototype, archetype, °source, °model, °pattern; °master: *The original hangs in the National Gallery.* **6** °eccentric, °nonconformist, °individualist, *Colloq* °case, °card, °character, *Brit* queer fish: *Regarded as an original, Wilde sauntered down the Strand with a lily in his hand.*

originality *n.* creativeness, creativity, inventiveness, °ingenuity, innovativeness, °innovation, °novelty, newness, unorthodoxy, unconventionality, cleverness, °daring, resourcefulness, °independence, individuality, uniqueness, nonconformity: *One must admire Dali for his originality.*

originally *adv.* in *or* at *or* from the beginning, (at) first, from the first, initially, to begin with, at *or* from the outset, at *or* from the start, in the first place *or* instance, *Colloq* from the word "go," from day one: *Originally, we were to have gone in Patrick's car.*

originate *v.* **1** °create, °bring about, engender, give birth to, beget, °conceive, °initiate, °inaugurate, °start, °begin, °introduce, °launch, °found, °set up, °institute, °establish, °invent, °coin, °devise, °pioneer, °design, contrive, concoct, °mastermind, °compose, °organize, °formulate, °form, °generate, °produce, °develop, evolve: *Wasn't it the Chinese who originated free public health schemes?* **2** °arise, °rise, °begin, °start, °come, °spring, °stem, °flow, °issue, °emerge, °emanate, °proceed, °grow, °develop, evolve, °derive, °result: *Where did the idea of the democratic form of government originate?*

ornament *n.* **1** enhancement, °embellishment, adornment, °decoration, ornamentation, gingerbread, trimming, garnish, garnishment, °frill, embroidery, beautification, °accessory; frippery; knickknack, furbelow, °bauble, °gewgaw, *Slang US* tchotchke: *We spent a pleasant afternoon putting ornaments on the Christmas tree. The ornaments on the mantelpiece need dusting.* —*v.* **2** °decorate, °embellish, °enhance, adorn, °trim, garnish, embroider, °elaborate, °beautify, accessorize, deck (out), °dress up: *The cabinet is ornamented with ormolu fittings in the Empire style.*

ornamental *adj.* decorative, beautifying, adorning, garnishing, embellishing: *Nothing in the house is ornamental, everything is functional.*

ornate *adj.* °elaborate, florid, overdone, °labored, rococo, baroque, gingerbread, arabesque, °fancy, °lavish, °rich, °flowery, °busy, °fussy, frilly, °intricate; high-flown, euphuistic, Ossianic, °bombastic, °pompous, °pretentious, °affected, °grandiose, fulsome, highfalutin *or* hifalutin, grandiloquent, °flamboyant: *Louis Quinze style is far too ornate for my tastes. We find his writing, with its contorted artificialities, far too ornate to appeal to the modern reader.*

orthodox *adj.* conformist, accepted, °authoritative, authorized, recognized, received, °official, °standard, °prevailing, °prevalent, °common, °regular, °popular, °ordinary, doctrinal, established, °traditional, traditionalist, °accustomed, °conventional, °customary, °conservative: *The orthodox view is that he was killed by an assassin acting on his own.*

oscillate *v.* °fluctuate, °vibrate, waver, °seesaw, °swing, °sway; vacillate, °equivocate, °shilly-shally, hem and haw, tergiversate: *The needle is oscillating between the "Safe" and "Danger" marks. I wish he'd stop oscillating and make up his mind.*

ostensibly *adv.* °outwardly, externally, superficially, patently, ostensively, demonstrably, °apparently, °evidently, °seemingly; °clearly, plainly, °manifestly, conspicuously, °obviously, patently, noticeably, prominently: *Ostensibly, he was visiting his aunt, but we all know he went to see Stella.*

ostentation *n.* °show, °display, °exhibition, exhibitionism, showing off, °pretension, pretentiousness, flaunting, flashiness, °flourish, flamboyance, °parade, window-dressing: *His clothes are elegant without ostentation.*

ostentatious *adj.* °showy, °boastful, °braggart, vaunting, °vain, vainglorious, flaunting, °pretentious, °flamboyant, °theatrical, *Colloq* °flash: *It's terribly ostentatious of Lady Penny to wear her diamond tiara to the disco.*

ostracize *v.* blackball, blacklist, °banish, °exile, °boycott, °isolate, °segregate, °exclude, excommunicate, snub, °shun, °avoid, *Chiefly Brit* send to Coventry, *Colloq* °cut, cold-shoulder, give (someone) the cold shoulder: *Wetherby has been ostracized at the club since that nasty business in the scullery.*

otherwise *adv.* **1** if not, or else, under other circumstances, in another situation, on the other hand: *I learned something about you tonight that otherwise I should never have guessed.* **2** differently, in another manner *or* way: *Mother wanted you to mow the lawn, and you'd best not do otherwise.*

out *adv.* **1** °outside, outdoors, in *or* into the open air: *If you go out take an umbrella.* **2** away (from), °abroad, °elsewhere, not (at) home, gone (from), gone away (from), °absent (from): *She was out of the house when I phoned.* **3** in *or* into the open, to *or* into public notice, for all to see, out of the closet: *Everything will be brought out at the trial.* **4** revealed, exposed, °visible, °discernible, °manifest, in sight, in view: *Are the stars out tonight?* **5** °short, minus, missing, in default, out of pocket: *He is out the two thousand he invested in a diamond mine.* **6** °free, °at liberty, °at large, °loose, unconfined: *They let Matilda out after she served her sentence.* **7** °completely, °thoroughly, effectively, °entirely: *I was tired out after that long swim.* —*adj.* **8** °unconscious, °senseless, °insensible, *Colloq* out cold, out like a light: *He has been out for almost an hour.* **9** dated, outdated, outmoded, °passé, °old-fashioned, °antiquated, old hat, *démodé*, °obsolete, unfashionable: *Longer skirts were out in the mid-1960s.* **10** °outlying, °distant, far-off, °peripheral: *He was planning to travel to the outer reaches of his empire. Let's sail to some of the out islands.* **11** °exhausted, gone, finished, ended; °over, completed: *Our food was out. Rescue came before the day was out.* **12** °inaccurate, °incorrect, °wrong, °at fault, °faulty, °off, °wide of the mark: *Your figures are out by a factor of four.* **13** °unacceptable, forbidden, prohibited, not allowed, *Colloq* not on: *Smoking is out in the dining room.* **14** extinguished, unlit; °off, doused; inoperative, nonfunctioning, out of order *or* commission, unserviceable, °broken: *Make certain that all campfires are completely out. The light is out in the corridor.* —*n.* **15** °alibi, °excuse, °escape, °loophole, °evasion: *She used your visit as an out to avoid calling on her mother.*

out-and-out *adj.* °complete, °unmitigated, unalloyed, °undiluted, °pure, utter, °perfect, consummate, °outright, °total, °downright, °unqualified, °thorough, thoroughgoing, through and through, dyed-in-the-wool: *She was an out-and-out fool to turn down his marriage proposal.*

outburst *n.* outbreak, °eruption, °explosion, blowup, flare-up, fulmination; upsurge, °surge, °outpouring, welling (forth), upwelling, outflow(ing), °rush, °flood, effusion, effluence *or* efflux; °fit, access, °attack, °spasm, °paroxysm, °seizure, °tantrum: *Another outburst like that, young man, and you'll be sent home.*

outcast *n.* pariah, °exile, °reject, *persona non grata*, leper, untouchable, expatriate, °refugee, displaced person, DP, evacuee: *In 1946, Europe swarmed with outcasts, the detritus of the war.*

outcome *n.* °result, consequence, end (result *or* product), aftereffect, °effect, °upshot, °sequel, °development, outgrowth, aftermath, °wake, °follow-up, *Medicine* sequela (usually *pl.* sequelae), *Colloq* °payoff, bottom line: *One outcome of the new safety regulation will be higher fares. We eagerly awaited the outcome of the race.*

outcry *n.* °protest, protestation, decrial, °complaint, °indignation, °uproar, vociferation, clamor, clamoring, commotion, °outburst, °noise, hullabaloo, °howl, howling, hoot, hooting, boo, booing, °hiss, hissing: *The public outcry against terrorism was heard around the world.*

outdo *v.* °exceed, °surpass, °excel, °transcend, °beat, °outstrip, outshine, °top, °cap, trump, °overcome, °defeat, °outweigh: *Their prices are lower because they outdo us in cheapness of labor.*

outdoor *adj.* °outside, out-of-door(s), alfresco, open-air: *Both of them enjoy outdoor activities like hiking and bicycling.*

outfit *n.* **1** °gear, °rig, °equipment, equipage, °apparatus, accoutrements *or US also* accouterments, °paraphernalia, °trappings, °tackle, °tack, utensils: *Our mountain-climbing outfit turned out to be extremely expensive.* **2** °clothes, °costume, °ensemble; attire, garb, clothing, °dress; *Colloq* get-up, togs: *The other guests at the wedding wore such weird outfits that we felt out of place in our jump suits.* **3** °firm, °concern, °business, °organization, °company, °(military) unit, corporation; °party, °set, °group; *Colloq* °setup: *I joined the outfit when it consisted of only a hundred people.* —*v.* **4** fit (out *or* up), °equip, kit out, °provision, °stock, accoutre *or US also* accouter, rig (out *or* up), °supply, °furnish: *The shop is prepared to outfit anyone for anything from a walk in the country to an African safari or an Arctic expedition.*

outgoing *adj.* **1** departing, °retiring, ex-, °former, °past, emeritus, leaving, withdrawing: *It is our custom to honor the outgoing president with a banquet.* **2** °genial, °friendly, °amiable, °cordial, °warm, °expansive, approachable, affable, °accessible, amenable, easygoing, °amicable, °sociable, congenial, extrovert, °familiar, °informal, communicative: *Because of his outgoing attitude, Keith gets along well with most people.*

outing *n.* jaunt, junket, °excursion, °trip, °expedition, °tour, °ride, *Colloq* °spin: *This year, the annual family outing will again be at Torquay.*

outlandish *adj.* °unfamiliar, °strange, °odd, °queer, °offbeat, °peculiar, °curious, °exotic, °foreign, °alien, °unknown, °unheard-of, °different, °exceptional, °extraordinary, °quaint, °eccentric, °bizarre, °outré, °weird, °fantastic, °unusual, °singular, °unique; freakish, °grotesque, barbarous; *Colloq* far-out, camp(y), °kinky: *Those youngsters wear the most outlandish hairdos you have ever seen.*

outlast *v.* °survive, outlive; outwear; °weather: *Considering her condition, it seems doubtful that she will outlast her husband.*

outlaw *n.* **1** °criminal, °gangster, °robber, desperado, bandit, highwayman, brigand, footpad, picaroon, °pirate, fugitive (from justice *or* the law), °renegade, *US* road agent: *In Westerns, the sheriff always wins out over the outlaws.* —*v.* **2** °forbid, disallow, °ban, interdict, °bar, °exclude, °prohibit, proscribe: *Some countries have outlawed prostitution, but with little effect.*

outlay *n.* °expense, °cost, °expenditure, spending, disbursement, °payment: *The city council refused to approve the outlay for a new swimming pool.*

outlet *n.* **1** way out, °exit, egress, °loophole, °relief, °escape, escape hatch, °vent, °opening, °release, safety valve, °discharge: *There seemed to be no outlet for his anger but to throw the cushion at her. The main outlet of the Great Lakes is the St. Lawrence River.* **2** retailer, °shop, °store, °market: *The company is having difficulty finding outlets that will stock its leather T-shirts.*

outline *n.* **1** °profile, °silhouette, contour, °periphery, °boundary, footprint: *This outline is of the desk area occupied by the keyboard and monitor.* **2** °précis, °synopsis, °résumé, °summary, °digest, °abstract, conspectus, °survey, overview, °rundown, recapitulation, °review, (thumbnail) sketch, skeleton, °(overall) plan, layout, framework, °draft, °scenario: *Osumara presented an outline of what his company planned to do after the takeover.* —*v.* **3** °trace, °draft, sketch, °rough out, °profile, °block (out), °plan (out), °lay out, °define, delineate: *None of the divers was particularly enthusiastic about the procedure outlined for collecting sharks.*

outlook *n.* **1** °view, °position, °point of view, °viewpoint, °prospect, °perspective, °slant, °angle, °standpoint, °attitude, °opinion: *His outlook on the situation in the Middle East is bound to be somewhat biased.* **2** °prospect, °forecast, expectation(s): *What is the outlook for the value of the pound sterling in the next year?*

outlying *adj.* °distant, far-off, far-flung, outer, outermost, °out-of-the-way, °remote, °faraway, °peripheral, farthest: *In those days it took weeks for the news to reach the outlying parts of the empire.*

out-of-the-way *adj.* **1** untraveled, unfrequented, °isolated, °lonely, °outlying, °obscure, °hidden, °secluded, °inaccessible: *She now lives in some out-of-the-way village in the Himalayas.* **2** °unusual, °odd, °peculiar, °extraordinary, °far-fetched, °remarkable, °outré, °exceptional, °outlandish, °strange, °rare, uncommon, °exotic, °unheard-of, unconventional, °queer, °weird, °bizarre: *His latest book is a treatise on an out-of-the-way subject.*

outpouring *n.* effusion, outflow, °flow, °outburst, °flood, deluge, °torrent, °spate, emanation, spouting, °spurt, gushing, efflux, effluence, outrush, tide, cascade, cataract, Niagara, *Technical* debouchment: *We scarcely expected such an outpouring of grief at her death. This writing appears to reflect the outpourings of his soul.*

output *n.* **1** °production, °result, °yield, crop, °harvest: *Nobody was quite ready for such a massive output.* **2** productivity, °achievement, °efficiency: *Job insecurity has diminished her output.* —*v.* **3** °put out, °produce, °generate, °create, °manufacture, °yield, °achieve: *Our new laser printer outputs about ten pages a minute.*

outrage *n.* **1** °violence, °atrocity, inhumanity, barbarism, °enormity, °evil, °barbarity, savagery, brutality, malignity, malefaction, wrongdoing, evildoing, maltreatment, °abuse, cruelty, °injury, °harm, °damage: *Wherever there is war there is misery and outrage.* **2** °resentment, affront, °bitterness, °indignation, °hurt, °shock, °anger, wrath, ire: *The minister felt outrage at being given a parking ticket.* **3** °insult, °indignity, °slight: *Contributors to the charity considered it an outrage that the fund raisers should keep so much of the money.* —*v.* **4** °offend, °insult, affront, vex, displease, °distress, nettle, °chafe, °infuriate, °anger, °enrage, °madden, make one's blood boil, raise (someone's) hackles, rile: *He was outraged to discover that the wretch had proposed to his daughter.* **5** °violate, °desecrate, defile, °do violence to, °injure, °harm, °abuse, °damage: *Such deeds outrage human feelings.* **6** °rape, °violate, °ravage, °ravish, deflower, °attack: *He seized the unhappy girl and outraged her.*

outrageous *adj.* **1** °excessive, °extravagant, °immoderate, °exorbitant, °enormous, °unreasonable, °preposterous, °shocking, °extreme, °unwarranted, exaggerated, °unconscionable, °inordinate, intolerable, °disgraceful, °shameful, °scandalous: *The prices at that restaurant are absolutely outrageous.* **2** °vicious, °cruel, heinous, °atrocious, barbaric, °inhuman, °abusive, °beastly, °horrible, horrid, horrendous, iniquitous, °villainous, °wicked, °evil, egregious, °flagrant, °grievous, °infamous, execrable, °abominable, °grisly, °hideous, °monstrous, °vile, °unthinkable, °foul, °awful, unspeakable, appalling, °offensive, °indecent: *The captives suffered the most outrageous treatment at the hands of their conquerors.* **3** °indecent, °offensive, °immoral, °rude, indelicate, °obnoxious, °profane, °obscene, °dirty, °filthy, °lewd, salacious, °foul, smutty, scatological, °pornographic, objectionable, °repellent, °repulsive, nauseating, °nauseous, °nasty, °gross, °revolting, °shocking, °repugnant, °disgusting, fulsome, °perverted, depraved, °dissolute, °degenerate, dissipated, debauched, °profligate; °explicit, unrestrained; foulmouthed, thersitical, insulting; °unseemly, °inappropriate, indecorous, °improper, °naughty, appalling, °embarrassing; *Literary* Fescennine, *US* shy-making: *The sermon denounced the outrageous films, books, magazines, and television programs to which children are exposed. Warren sometimes says the most outrageous things.*

outré *adj.* unconventional, °unusual, °extravagant, °bizarre, °weird, °strange, °odd, °peculiar, °grotesque, °outlandish, freakish, °out-of-the-way: *The attention of the Inquisition was first called to Cagliostro because of his outré behavior.*

outright *adj.* **1** °unqualified, °total, unreserved, unrestricted, °full, °complete, unconditional, unequivocal, °clear, °direct, °definite, unmistakable *or* unmistakeable: *The duke is the outright owner of the property.* **2** °undisguised, °unmitigated, utter, consummate, °pure, °out-and-out, all-out, °sheer, °absolute, °stark, bald, °thorough, arrant, thoroughgoing, through and through, °downright, °direct, °definite, unmistakable *or* unmistakeable: *Her outright refusal to provide further help was met with dismay.* —*adv.* **3** °directly, °at once, °immediately, °instantaneously, instantly, then and there *or* there and then, *Brit* °straightaway, *US* right away, on the spot, right off: *One passenger was killed outright, the other died later in the hospital.* **4** °completely, °entirely, °exactly, °precisely, °totally, *in toto*, °utterly, baldly, starkly, consummately, purely, °thoroughly, °directly, unhesitatingly, °quite, °absolutely, explicitly, categorically, straightforwardly, plainly, °openly, forthrightly, unequivocally, unambiguously, candidly: *One wishes that Henley weren't so reticent and would say outright what he means.* **5** unrestrictedly, unqualifiedly, unreservedly, unconditionally: *The duke owns the property outright.*

outset *n.* °beginning, °start, inauguration, inception, °first, *Colloq* kickoff: *Had you let them know who you were at the outset, this wouldn't have happened.*

outside *n.* **1** °exterior, °face, °facing, °shell, °skin, °case, casing, °surface, °front; façade: *What is that on the outside of the box? The outside of the house is painted white.* **2** °aspect, °appearance, °look, demeanor, °face, °front, façade, mien, °mask, °disguise, false front, °pretense: *One cannot tell what people are really like from the outside they present to the world.* **3** °extreme, °limit, most, °maximum, utmost, °best, worst, longest: *At the outside, you shouldn't pay more than half your income for housing. I'll wait for her for an hour at the outside.* **4** the world at large: *We had to bring in someone from the outside to complete the work.* —*adj.* **5** °exterior, °external, out-of-door(s), °outdoor: *They have added an outside swimming pool to the house.* **6** °maximum, maximal, highest, °best, worst, greatest, most, largest, longest, farthest: *What was their outside estimate for replacing the roof? The outside time for driving here from London is about an hour.* **7** °private, °home, °cottage, °secondary, °peripheral, °independent, freelance: *Her outside job pays more than her regular work.* **8** °unlikely, °remote, °faint, *Colloq* °slim: *He has an outside chance of beating the world record.* **9** °foreign, °alien, °outward; unconnected, excluded, uninvolved, disinvolved, °independent, °separate, °different: *I'm worried that Philbert might be subject to outside influences. An outside contractor is doing the work.* —*adv.* **10** outdoors, out-of-doors: *Perhaps you'd like to step outside to discuss the matter further?*

outsider *n.* nonmember, noninitiate, °foreigner, °alien, outlander, °stranger, °newcomer, °guest, °visitor, trespasser, interloper, °intruder, squatter, invader, *Colloq* gate-crasher: *The others always treated Peter as an outsider. Why do I feel an outsider in my own home?*

outskirts *n.* pl. periphery, °edge, environs, outer reaches, °vicinity, border(s), suburb(s), exurb(s), general area *or* neighborhood, purlieus, fringes, vicinage, faubourg(s): *The university is on the outskirts of the city.*

outsmart *v.* °outwit, outfox, outthink, outmaneuver, outmanipulate, outplay, steal a march on, get the better *or* best of, °trick, °dupe, °hoodwink, °fool, °deceive, °hoax, gull, make a fool of; °swindle, °cheat, °defraud, cozen, *Colloq* put one over on, pull a fast one on, °take in, make a monkey (out) of, bamboozle, con, *Brit* nobble, *Slang* slip *or Slang* put one *or Slang* something over on (someone): *"I have been outsmarted by bigger fools than you!" Mr. Winkle shouted.*

outspoken *adj.* °candid, °frank, °open, °free, °direct, unreserved, unreticent, straightforward, °forthright, °explicit, °specific, plain-spoken, plain-speaking, unequivocal, unceremonious, unambiguous, unsubtle, °uninhibited, unshrinking, °blunt, °bold, °brusque, °brash, undiplomatic, °tactless, °crude: *Asperia was always quite outspoken in her opinions of her neighbors. Her outspoken observations are a fruitful source of gossip.*

outstanding *adj.* **1** °prominent, °eminent, °renowned, °famous, famed, unforgettable, °memorable, °celebrated, °distinguished, °special, °choice, °noteworthy, °notable, °noted, °important, °conspicuous, °exceptional, °excellent, °superior, first-class, °first-rate, °superb, °remarkable, °extraordinary, °marvelous, °sensational, *Colloq* smashing, °super: *Liszt was the outstanding pianist-composer of his time.* **2** °unsettled, on-going, °unresolved, °unpaid, °due, owed *or* owing, receivable *or* payable; °remaining, °leftover: *The company has a few outstanding debts.*

outstrip *v.* °overcome, °surpass, °outdo, outperform, outshine, outclass, °better, °beat, °transcend, °best, worst, °exceed, °excel, outdistance, °overtake, °top, °cap, put in the shade, °eclipse: *Bannister again outstripped everyone in the race.*

outward *adj.* **1** °external, °exterior, outer, °outside, °outlying, °manifest, °obvious, °evident, °apparent, °visible, °observable; °superficial, °surface, °extrinsic, °skin-deep, °shallow, °pretended, °false, ostensible, °formal, °physical, bodily, fleshly, °carnal, mundane, °worldly, °secular, °temporal, °terrestrial, °material, nonspiritual: *She gave every outward sign of being the bereaved widow. Whatever outward trappings money may buy, A man's true wealth lies deep inside.* —*adv.* **2** outwards, °outside, away, °out, without: *Towns were formerly centered on the railway station and radiated outward from there.*

outwardly *adv.* externally, °apparently, visibly, superficially, °ostensibly, °evidently, °seemingly, °on the surface, to all appearances, °to all intents and purposes: *Though the town was outwardly quiet, we had a feeling of ominous foreboding.*

outweigh *v.* °overcome, outbalance, overbalance, overweigh, tip the scales, preponderate (over), °surpass, °prevail (over), override, take precedence (over), °compensate (for), °make up for: *His feeling for his wife outweighed all else in his life.*

outwit *v.* See **outsmart**, above.

oval *adj.* egg-shaped, ovoid, ovate, oviform, obovoid, obovate; elliptical, ellipsoid(al): *His mother's picture hung in an oval frame over the mantel.*

ovation *n.* °applause, acclamation, acclaim, plaudits, cheers, cheering, clapping, laudation, °praise, °kudos, *Colloq* °(big) hand: *At the conclusion of the concerto, the pianist was given a standing ovation.*

over *prep.* **1** °above, on, upon, on top of, atop (of): *Build a shed over the boat for protection.* **2** more than, greater than, upward(s) of, in excess of, °(over and) above, (over and) beyond; °exceeding: *Of the two thousand people questioned in our survey, over half said they think prunes are funny. The thieves took over half a million dollars' worth of paintings.* **3** across, to *or* from *or* on the other side of; beyond: *Over the river and through the woods to grandmother's house we go.* **4** °for, during, in *or* over *or* during the course of, °through, °throughout: *Over the next week she will be working in the Paris office.* **5** °(all) through, °throughout, °(all) about, all over: *We traveled over the entire country without seeing a camel. Have you gone over the manuscript I left with you?* —*adj.* **6** done (with), finished, terminated, concluded, ended, °past, settled, closed, at an end, over with: *I'm afraid it's all over between us, Carrie.* —*adv.* **7** to, onto, °past, beyond, across: *This room looks out over the sea.* **8** °remaining, as a remainder, as surplus, °outstanding: *When we finished eating, there wasn't much left over.* **9** (once) again, once more, one more time: *This pot will have to be cleaned over again.* **10** down, to the ground *or* floor: *You almost knocked over the lamp.*

overall *adj.* °total, °complete, °comprehensive, (all-)inclusive, °whole, °entire, all-embracing, blanket: *The overall cost, including materials and labor, came to more than I had expected.*

overawe *v.* °overwhelm, °intimidate, cow, °daunt, awe, °bully, hector, °browbeat, °dominate, domineer, °frighten, °scare, °terrify, disconcert, °discomfit, °upset, abash: *The children were overawed by their father's slightest sign of displeasure.*

overbearing *adj.* °repressive, °domineering, bullying, imperious, °officious, high and mighty, highhanded, overweening, magisterial, lordly, °authoritarian, °willful, °despotic, °dogmatic, autocratic, °tyrannical, °dictatorial, °peremptory, °arbitrary, °assertive, °arrogant, cavalier, °haughty, °superior, °supercilious, °pretentious, *Colloq* °bossy, °pushy, °hoity-toity, highfalutin *or* hifalutin, snooty, *Slang* snotty: *His temper was harsh and severe, his manner haughty and overbearing.*

overcast *adj.* cloudy, clouded, °sunless, moonless, starless, °murky, °gray, lowering *or* louring, °dull, °dark, darkened, °dreary, °somber, °gloomy, °dismal, °threatening, °menacing: *The sky was overcast this morning, but the sun shone through in the afternoon.*

overcome *v.* **1** °beat, °defeat, °conquer, °overpower, °subdue, worst, °best, °triumph over, win (out) (over), °prevail (over), °overthrow, °overwhelm, vanquish, get the better *or* best of, °whip, drub, °rout, °break, °subjugate, °suppress, °crush, °master, *Colloq* lick: *If we do not overcome these temptations, they will overcome us. The superior force easily overcame the tiny group of defenders.* —*adj.* **2** beaten, defeated, overwhelmed, °subdued, worsted, bested; °affected, °speechless, swept off one's feet, rendered helpless, overpowered, moved, influenced, at a loss (for words), *Colloq* bowled over: *The victim's parents were overcome when the casualty list was published. Culligan was too overcome to speak at the award presentation.*

overconfident *adj.* **1** °brash, °arrogant, cocksure, °cocky, °brazen, hubristic *or* hybristic, swaggering, °audacious, °overbearing, vainglorious, *Colloq* °pushy: *It is harmful for a sportsman's morale to be overconfident of winning.* **2** °heedless, °foolhardy, °thoughtless, °short-sighted, °hasty: *Rock climbers should not be overconfident and rely entirely on their ropes.*

overcritical *adj.* supercritical, hypercritical, captious, carping, °niggling, caviling, °querulous, °fault-finding, °finicky, °fussy, °hairsplitting, °difficult, °fastidious, °harsh, °severe, °demanding, °exacting, °small, °small-minded, *US and Canadian* picayune, *Colloq* picky, nit-picking, pernickety *or US also* persnickety: *You should not be overcritical of Maddy—she's only a child.*

overcrowded *adj.* jammed, °packed, °congested, °populous, overpopulous, jampacked, overpopulated; swarming, crawling, choked, packed to the gunwales: *The planet is becoming overcrowded. We squeezed into an already overcrowded train.*

overdo *v.* **1** carry to excess, overindulge, be intemperate, go *or* carry to extremes, overact, °exaggerate, carry *or* go too far, overreach, not know when to stop, paint the lily (or gild refined gold), out-Herod Herod, *Colloq* go overboard, do to death, lay it on thick, lay it on with a trowel; go off the deep end: *Exercise a little judgment and try not to overdo it.* **2** °overwork, do too much, overtax, °exhaust, °fatigue, °overload, overburden, *Colloq* bite off more than one can chew, burn the candle at both ends: *Don't overdo it or you'll feel the results tomorrow.*

overdue *adj.* °late, °tardy, behindhand, behind, unpunctual, °belated, *US* past due: *As it is now noon, she is already an hour overdue. I have an overdue library book to return.*

overeat *v.* °gorge, binge, gormandize, stuff oneself, overindulge, guzzle, °feast, wolf down, overfeed, do the gavage, *Colloq* pack away, *Brit* pig, *US* pig out: *People not only overeat but they tend to eat the wrong foods.*

overgrown *adj.* covered, °overrun, overspread, °luxuriant, weedy, °abundant: *The walls are all overgrown with ivy.*

overhang *v.* **1** °jut (out), beetle, °bulge (out), project (out), °protrude, °stick out, °loom (out), °extend (out), hang (out) over: *The balcony overhangs the sea.* **2** impend, °threaten, °menace, imperil, °loom: *The overhanging scent of death has pervaded the family since the suicide of Crenshawe.* —*n.* **3** °ledge, °projection, °bulge, °protrusion, °extension: *A swallow nested under the overhang of the roof.*

overhaul *v.* **1** °overtake, °pass, gain on *or* upon, draw ahead of, catch up with, get ahead of, °outstrip, outdistance, leave behind, °lap: *We overhauled the ketch and were beginning to close on the yawl when the squall hit us.* **2** °renovate, °refurbish, recondition, rebuild, °restore, °repair, °service, °adjust, °patch (up), °mend, °fix (up): *The car engine needs to be overhauled.* —*n.* **3** reconditioning, overhauling, refurbishing, rebuilding, renovation, servicing, °adjustment, mending, fixing (up): *The overhaul of the die-stamping machine will require two months to complete.*

overhead *adv.* **1** °(up) above, (up) in the air *or* sky, high up, on high, °aloft, skyward: *The airplanes passed overhead at dawn.* —*adj.* **2** °elevated, raised, °upper: *The crane travels on an overhead track.* —*n.* **3** *Brit* ***overheads,*** *US* ***overhead.*** (basic *or* fixed) costs, operating cost(s), expense(s), °outlay, disbursement(s), running cost(s), expenditure(s), °maintenance, cost(s) of doing business: *One way to increase profits is by reducing overhead.*

overjoyed *adj.* °delighted, °ecstatic, °elated, °happy, °rapturous, euphoric, jubilant, thrilled, transported, *Colloq* tickled pink, in seventh heaven, on cloud nine, *Brit* over the moon: *We were overjoyed to receive an invitation to your party.*

overlap *v.* **1** °lap (over), overlie, overlay, shingle, *Technical* imbricate, strobilate: *Note how the upper tiles overlap the lower to keep the rain out.* **2** °coincide, °correspond, intersect: *Our work shifts overlap by four hours.* —*n.* **3** °lap, °flap, overlay, fly (front) *or Brit* flies, imbrication: *The overlap conceals the buttons that close the skirt on the side.*

overload *v.* **1** °weigh down, °burden, overburden, °load (up), overtax, saddle with, °tax, °strain, °impede, °handicap, °oppress, °encumber, cumber, overcharge: *They overloaded me with so much work that I'll never finish.* —*n.* **2** surcharge, overcharge, overburden, dead weight, °oppression, °handicap, °tax, °load, °encumbrance, °impediment, °hindrance: *We had to hire extra hands at Christmas to help clear the overload.*

overlook *v.* **1** °miss, slip up on, °omit, °neglect, °slight, °disregard, fail to notice, °ignore, °pass over, °leave out, °forget, *Colloq* °pass up: *You overlooked an error on the first page.* **2** °blink at, wink at, let go (by), let pass, let ride, turn a blind eye to, shut (one's) eyes to, pretend not to notice, take no notice of, °ignore, °disregard, °forgive, °pardon, °excuse, °permit, °allow, forget about, °write off, condone, make allowances (for), let bygones be bygones, °gloss over: *I shall overlook your tardiness if you agree to stay late to compensate for it.* **3** front on (to), °face, give upon, command *or* afford a view of, look out on *or* over, have as a vista *or* view: *My room overlooks the lake.*

overly *adv.* excessively, too, °exceedingly, immoderately, disproportionately, °unduly, inordinately, extraordinarily, °very, *Colloq* damned: *She seemed overly anxious for us to leave.*

overpower *v.* **1** °overcome, °overwhelm, °beat, vanquish, °conquer, °defeat, °crush, °put down, worst, °best, °prevail, °master, °quell, °subdue, °subjugate: *The thugs easily overpowered the elderly couple and took their money.* **2** °overcome, °overwhelm, dumbfound *or* dumfound, °daze, °stagger, °amaze, °stun, stupefy, °nonplus, °strike, *Colloq* °floor: *I was quite overpowered by the grandeur of the house.*

overpowering *adj.* °overwhelming, °irresistible, °powerful, °telling, compelling, unendurable, °unbearable, °oppressive: *They presented an overpowering argument against the use of asbestos as an insulator.*

overrate *v.* overvalue, make too much of, exaggerate the worth *or* value of, attach too much importance to, overprize, assess too highly: *They insisted that his value to the company is overrated and he should be dismissed.*

overreact *v.* °exaggerate, make much ado about nothing, make too much of (something), make a mountain out of a molehill, lose all *or* one's sense of proportion, blow (up) out of (all) proportion: *I'd say that she overreacted by leaving him because he forgot her birthday.*

overriding *adj.* °dominant, dominating, °predominant, predominating, compelling, °prevailing, °primary, °prime, most important, overruling, °overwhelming, °paramount, preponderant, °principal, °cardinal,

°main, °chief: *His overriding reason for buying the larger dictionary was that it made a better doorstop.*

overrun *v.* invade, °defeat, °attack, °ravage, °destroy, °overwhelm, °conquer, harry, vandalize, °plunder, maraud, °scourge, despoil, °sack, °strip, °pillage, °storm, *Colloq* blitz: *Within a week the armies had overrun the country and captured the capital city.*

overseas *adv.* °abroad: *Did you serve overseas during the war?*

oversee *v.* °direct, °manage, °watch (over), keep an eye on, °administer, superintend, °run, °supervise, °operate, °manipulate, °handle, °control: *We employed Gertrude to oversee our offices in Frankfurt.*

overseer *n.* °superintendent, °supervisor, °manager, °foreman, forewoman, °superior, *Colloq* °boss, °chief, °super, *Brit* gaffer, *US* straw boss, (head *or* chief) honcho: *Simon is too hard a taskmaster to be overseer of that sensitive operation.*

overshadow *v.* **1** °dominate, outshine, °eclipse, °dwarf, °diminish, °minimize, put in(to) *or* throw into *or* leave in the shade, steal the limelight from, tower over *or* above, °excel: *The Crown, overshadowed by the great barons, turned to the Church for aid.* **2** °spoil, °blight, °ruin, °mar, take (all) the pleasure from, put a damper on, take the edge off, °impair, take the enjoyment out of: *The news from the east overshadowed everyone's spirits at the party.*

oversight *n.* **1** °omission, inadvertence, °neglect, laxity, laxness, °fault, °failure, dereliction, °error, °mistake, °blunder, carelessness, heedlessness: *It was owing to an oversight by a clerk that the flight booking was not made.* **2** supervision, superintendence, °surveillance, °management, °direction, °guidance, °administration; °charge, °care, °custody, keeping, °hands, °protection, °auspices: *The matter of expenses must be left to the oversight of the committee established to deal with such things.*

overstate *v.* °exaggerate, °magnify, hyperbolize, embroider, overstress, °color, make (too) much of, overdraw, overemphasize, °stretch, °enlarge, °inflate, °blow up: *They might have overstated the problem by asking for a twenty-four-hour police guard.*

overstep *v.* °exceed, °transcend, °surpass, go beyond: *Again Hortense has overstepped the bounds of propriety by inviting herself to the reception.*

overt *adj.* °apparent, °evident, °plain, °clear, °obvious, °manifest, clear-cut, unconcealed, °patent, °open, °visible, °observable, °public: *Moving the troops to the border may be considered an act of overt hostility.*

overtake *v.* **1** catch (up with *or* to), °reach, draw level *or* even with, °overhaul, gain on *or* upon, move by *or* past, °pass, leave behind, °outstrip, outdistance: *The express train overtakes the local just outside of Milwaukee.* **2** come upon, °seize, °catch (unprepared), befall, °strike, °hit, °overwhelm: *Fortunately for some, the weaknesses of old age overtook Becky before she could finish writing her memoirs.*

overthrow *v.* **1** °defeat, °beat, °rout, °conquer, °overpower, °master, bring down, depose, oust, °overwhelm, unseat, unhorse, °topple, °overturn, dethrone, thrash, worst, °best: *The new leader was overthrown by the partisans in two weeks. The Lakers overthrew the Celtics in last night's game.* —*n.* **2** °defeat, °rout, °conquest, deposing, ousting, unseating, toppling, °overturn, overturning, °downfall, °end, °ruin, °fall, °collapse, °destruction, °suppression, quashing, crushing, subjugation, *US* ouster: *The overthrow of the military regime was followed by weeks of rioting and looting.*

overtone *n.* undertone, connotation, °hint, °suggestion, °innuendo, insinuation, intimation, °indication, °implication: *Do I detect overtones of regret now that you are finally leaving?*

overture *n.* Often, overtures. approach, °advance, °offer, °proposal, °proposition, °tender: *While they were winning, they rejected all peace overtures.*

overturn *v.* **1** °turn over, knock down *or* over, °tip over, °capsize, upend, °upset, °turn turtle, turn upside down, turn topsy-turvy, invert: *The tanks overturned when they tried to drive past the huge concrete caltrops. On his first day as a waiter, he overturned the soup in a guest's lap.* **2** bring down, °overthrow, °throw over, °upset, depose, unthrone, dethrone, unseat, oust, °eject: *The dictatorship was overturned and the country returned to being a democratic republic.* —*n.* **3** overturning, °overthrow, unseating, ousting, toppling, °fall, °destruction, °ruin, °defeat, *US* ouster: *Could the overturn of the present regime be effected without force of arms?*

overwhelm *v.* **1** °overpower, °overcome, overtax, °devastate, °stagger, °crush, °defeat, °destroy, °subdue, °suppress, °quash, °quell, °conquer, °beat, bring down, °prostrate, °weigh down, °oppress: *Overwhelmed by grief, she dissolved into tears.* **2** inundate, °overcome, engulf, °submerge, °flood (over); deluge, °swamp, °bury, °immerse: *A feeling of terror suddenly overwhelmed me as I watched my safety rope fray and break. In only a few hours the rising waters had completely overwhelmed the house.* **3** °overcome, °stagger, °astound, °astonish, dumbfound *or* dumfound, °shock, °stun, °bewilder, °confuse, confound, °nonplus, °surprise, °take aback, *Colloq* bowl over, knock off one's feet *or* pins, blow one's mind, discombobulate, *Brit* knock for six: *We were overwhelmed by the friendly reception that awaited our return.*

overwhelming *adj.* **1** °overpowering, uncontrollable, °irresistible, °devastating, unendurable, °unbearable, crushing, °burdensome, °formidable: *He had an overwhelming sense of shame for what he had said.* **2** °awesome, awe-inspiring, stupefying, astounding, astonishing, staggering, bewildering, mind-shattering, °prodigious, mind-boggling, *Colloq* mind-blowing: *When we first approached Jupiter, all of us found its sheer size totally overwhelming.*

overwork *v.* **1** overexert, overstrain, overburden, °oppress, overtax, °overload, overuse: *We were so overworked that the quality had to suffer.* **2** °slave (away), burn the midnight oil, lucubrate: *All of us overworked to get the project done in time for the presentation.* —*n.* **3** overexertion, overstrain, °strain: *I was ready to drop from overwork.*

overwrought *adj.* **1** °tense, °nervous, jittery, °jumpy, fidgety, °touchy, in a dither *or* twitter, all atwitter, overexcited, °on edge, overstimulated, °frantic, frenetic, distracted, *Brit* strung up, *US* strung out, *Colloq* (all) worked up, edgy, in a tizzy, wound up, uptight: *They were quite overwrought worrying about the children's safety.* **2** overworked, °ornate, °elaborate, baroque, rococo, florid, °flowery, °fussy, °ostentatious, °busy, °gaudy, °garish: *Some of the Victorian houses were characterized by overwrought gingerbread decoration. A few of his poems are overwrought and difficult to digest.*

owe *v.* **1** be in debt to, be indebted to, be beholden to: *It must be a good feeling not to owe anyone any money at all.* **2** ***owing to.*** because of, on account of, °thanks to; °through, as a result of, resulting from, *Colloq* due to: *Owing to my schedule, I cannot stay the night. The higher tides are owing to the positions of the sun and moon.*

owner *n.* possessor, holder; °proprietor, proprietress: *The papers have been returned to their rightful owner. Who is the owner of the property?*

P

pace *n.* **1** °step, stride: *Please walk forward two paces.* **2** °rate (of speed), °tempo, °speed, °velocity, *Colloq* °clip: *We were proceeding at a pace of five miles per hour.* —*v.* **3** °walk, stride, tread; °traverse: *Alistair paced back and forth nervously, awaiting word from the surgeon.* **4** °measure, gauge *or* gage, °judge, °rate, °estimate, °determine, °reckon, °figure, °compute: *I was trying to pace my running to conserve some energy for a final burst at the end of the race.*

pack *n.* **1** °parcel, °package, °packet, °bundle, bale, backpack, knapsack, rucksack, haversack, kit-bag, °kit, duffel *or* duffle bag: *The stranger hoisted his pack to his shoulder and loped off.* **2** °load, °lot, °bunch, multitude, °heap, °pile, °accumulation, °mass, amassment, °mess, barrel, peck: *She told the jury a pack of lies.* **3** °group, °collection, °assembly, assemblage, congregation, °gathering, °crowd, °number, °throng, horde, °mass, °crew, °gang, °body, lots, loads, °band, °company, °party, °set, °flock, °herd, drove, °mob, °swarm, bevy, covey, °circle, coterie, °clique: *A huge pack of people were waiting at the stage door for the star to emerge.* **4** deck: *We ought to have two packs of cards for bridge.* —*v.* **5** Often, ***pack in*** *or* ***into.*** °package, bale, °bundle, °compact, °crowd, °cram, °jam, °fill, °stuff, °squeeze, °ram, °press, °wedge, tamp: *They packed us into the train like sardines. I packed as much as I could into the short time I had left.* **6** ***pack it in.*** °stop, °cease, °end, °finish, °quit, °wind up, °terminate, °give up, call it a day, *Colloq* chuck: *Edward finally packed it in because they were giving him too much trouble.* **7** ***pack off.*** °dismiss, send off *or* away, bundle off *or* out, hustle off *or* out *or* away, °get rid of, drive off *or* away, order off *or* away *or* out, send (someone) about his (*or* her) business: *At the beginning of the war, Sally's mother packed her off to America to stay with her aunt.* **8** ***pack up.*** **(a)** get *or* gather together, °put away, °store: *Pack up your troubles in your old kit bag and smile, smile, smile.* **(b)** °quit, °stop, °break down, °fail, °give out, °stall, °die, give up the ghost, *Colloq* conk out, have had it: *After three days crossing the desert, the engine finally packed up.*

package *n.* **1** °packet, °parcel, °box, container, °case, carton, °bundle: *I am donating a package of clothing to the relief fund.* **2** °combination, °unit, package deal: *Included in the package from the car dealer are several luxury features.* —*v.* **3** °wrap, °pack, containerize, carton, °case, encase, °enclose, °include; °combine, °unite, coupled, °incorporate: *You shouldn't package the perishables in with the canned foods. More and more merchandise comes packaged in plastic these days. A special sponge is packaged with the cleaning liquid.*

packed *adj.* filled, °full, °loaded, crowded, stuffed, jammed, crammed, brimful, chockablock, chock-full, jampacked, overloaded, overflowing, loaded *or* filled to the gunwales, up to there, bursting, groaning, °swollen, °replete, *Colloq* wall-to-wall: *The gallery was packed with the soprano's relatives, who cheered every note. The publicity promised a fully packed program of mirth and merriment.*

packet *n.* **1** °package, °parcel, °pack, °batch: *We found a packet of your old love letters in a trunk in the attic.* **2** loads, lot(s), °great deal, °fortune, °mint, *Colloq* °bundle, pretty penny, pile(s), tidy sum, king's ransom, *Brit* °bomb: *He lost a packet on that property when they built a sewage plant alongside it.*

pact *n.* °agreement, °treaty, °bargain, °alliance, °contract, °compact, concord, covenant, concordat, entente, °understanding, °arrangement, °deal: *The companies entered into an illegal pact not to compete in the same markets.*

pad *n.* **1** °cushion, pillow, °wad, wadding, stuffing, padding, °filling, filler: *Wrap a soft cotton pad over the wound to protect it.* **2** writing pad, note pad, memo pad, °block (of paper), jotter, *US* filler: *You'd best take a pad with you to make notes at the lecture.* **3** °flat, apartment, room(s), °home, °place, °quarters, *Colloq* hangout, *Brit* digs *or* diggings, *Slang US* °flop: *A few friends are crashing at my pad while they're in town.* —*v.* **4** °cushion, °wad, °stuff, °fill; upholster: *The chair seats are padded with foam rubber, but the arms are bare.* **5** Sometimes, ***pad out.*** °expand, °inflate, °stretch, dilate, °lengthen, protract, °extend, °blow up, °flesh out, °puff up, augment, °spin out, °amplify: *He pads his weekly newspaper column with trivia in order to fill the space.*

paddle *n.* **1** °oar, °sweep, scull: *A "crab" occurs when you catch the water with the paddle blade on the return stroke.* —*v.* **2** °row, scull, °oar: *Deftly she paddled the canoe over to the pier.* **3** °wade: *The children were paddling at the water's edge.* **4** °spank, paddywhack *or* paddywack, thrash, °beat, °whip, °flog: *Dad threatened to paddle me if he caught me playing hooky again.*

pagan *n.* **1** °heathen, unbeliever, idolater, polytheist, °infidel, Gentile: *He joined the religion of the pagans and worshiped the golden calf.* —*adj.* **2** °heathen, °infidel, idolatrous, polytheistic, heathenish, Gentile: *Professor Exmount has made a study of pagan rites of passage among the pygmies.*

page[1] *n.* **1** leaf, folio, °side, °sheet, verso *or* recto: *On which page does the index begin?* **2** °episode, °phase, °period, °time, °stage, °point, °era, epoch, °age, chapter: *The early 1940s were among the darkest pages in Britain's history.* —*v.* **3** paginate, folio, °number: *Roman numerals were used in paging the front matter of the book.*

page[2] *n.* **1** °attendant, pageboy, °servant, errand boy *or* girl, messenger (boy *or* girl), *Brit* foot-boy, *US* bellman, bellhop, *Offensive used of an adult* bellboy: *Please have a page deliver this message to Mr. Simmons.* —*v.* **2** °announce, °summon (forth), send for *or* after, °call, bleep, beep, °call for, call out: *They are paging Lady Vance to deliver an urgent message.*

pageant *n.* °spectacle, °display, °grandeur, °tableau, °show, °parade, °procession, °ceremony, °formality, °ritual, °event, °affair, °extravaganza, °presentation, °gala: *The children stage a medieval mystery play as a part of the annual Christmas pageant.*

pageantry *n.* °pomp, °ceremony, °display, magnificence, °extravagance, panorama, showiness, °show: *The pageantry of the Oberammergau passion play should be experienced at least once.*

pain *n.* **1** °hurt, °suffering, °discomfort, soreness, ache, aching, °pang, °spasm, smarting, cramp: *I feel the pain in my back from lifting that box.* **2** °anguish, °agony, °affliction, °distress, °grief, °woe, °suffering, °misery, travail, wretchedness, °despair, °torment, tribulation, °trial, torture, dolor, discomposure, °ordeal, disquiet: *No one who has not experienced it can imagine the pain of losing a child.* **3** irritation, vexation, °annoyance, °bother, °nuisance, °pest, *Colloq* pain in the neck, °headache, °drag, °bore, *Taboo slang* pain in the *Brit* arse *or US* ass: *What a pain it must have been to have forgotten your keys again! Boris can really be a pain when he goes on about the book he's writing.* **4** ***pains.*** °effort, °trouble, °exertion, toil, °labor: *She went to great pains to make our stay comfortable.* —*v.* **5** °hurt, °distress, °grieve, °wound, °injure; °trouble, °depress,

°sadden, °sorrow, cut to the quick: *It pained us to learn of Mrs. McArthur's illness.*

painful *adj.* **1** hurting, °grievous, °hurtful, °sore, distressing, distressful, °excruciating, torturous, °agonizing, smarting, stinging, aching, achy, throbbing, °burning, °piercing, stabbing, °sharp, °tender, °sensitive, °raw, °bitter, *Formal* nociceptive: *The bullet caused a painful wound in the knee. Mother's arthritis can be quite painful at times.* **2** vexing, vexatious, annoying, harassing, irritating, °irksome, aggravating, galling, exasperating, unpleasant, afflictive, °harrowing, worrisome, worrying, troubling, disquieting, °disturbing, distressing: *Fighting in the front lines was a painful experience for Rupert.* **3** °painstaking, °laborious, °careful, rigorous, °arduous, assiduous, sedulous, °diligent, °earnest, °conscientious, °meticulous, °scrupulous, °detailed, °thorough, thoroughgoing, °exacting, °demanding: *A great deal of painful research went into discovering the cause of yellow fever.*

painfully *adv.* agonizingly, distressingly, disagreeably, unpleasantly, unfortunately, °sadly, woefully, lamentably, ruefully, unhappily: *I became painfully aware that much of the relief destined for the poor never reached them.*

painkiller *n.* anodyne, analgesic, anesthetic, °sedative, palliative: *When he wrenched his back, the doctor gave him a painkiller so that he could walk.*

painless *adj.* trouble-free, °easy, °simple, °comfortable, °effortless, *Colloq* easy as 1, 2, 3 *or* as ABC, easy as pie, a piece of cake, °pushover, child's play, *Slang* cinch, no sweat: *The procedure for assembling the bicycle is relatively painless.*

painstaking *adj.* See **painful, 3,** above.

paint *n.* **1** °color, °tint, dye, coloring, pigment, °stain: *On their first day apprentices are always sent out for a bucket of striped paint.* **2** coating, °coat, °surface; enamel: *The rust has come through the paint here and there.* **3** °makeup, cosmetics, *maquillage,* greasepaint, *Colloq* war paint, °face: *Gladys is in her room putting on some paint before we go out.* —*v.* **4** depict, °portray, °picture, °show, °represent, delineate, °render, °draw, limn, °characterize, °describe: *Whistler painted a portrait of my grandmother. Correspondents painted a very grim picture of the plight of the hostages. I doubt that the captain was as bad as he was painted.* **5** °coat, °brush, °apply, °cover, daub: *The doctor painted the area with some medication that soothed the pain.* **6** °color, °tint, dye, °stain, °decorate: *We are planning to paint the nursery pale blue.* **7** ***paint the town red.*** °make merry, °carouse, °revel, go on a spree, go out on the town, *Colloq* whoop it up, live it up, (go on a) pub-crawl, °step out, *Slang* make whoopee, go on a bender *or* drunk *or* binge, booze it up: *Saturday nights a few of the boys would go out and paint the town red till the wee hours.*

pair *n.* **1** °couple, twosome, two of a kind, set of two, matched set, duo, dyad, twins, °double, doublet; °brace, °span, yoke, °team, tandem: *A pair of silver candelabra disappeared during the party.* —*v.* **2** °match (up), °mate, pair off *or* up, °team (up), put together, °partner, °twin, °double, °join, °unite, yoke; °wed, °marry, join in wedlock *or* in holy matrimony: *Pair these gloves for me, would you? Husbands may not pair off with wives for the next dance.*

pal *n.* **1** °friend, consort, °comrade, alter ego, crony, °companion, amigo, °playmate, classmate, *Colloq* °chum, sidekick, °mate, *Chiefly US and Canadian* buddy: *Jim left a little while ago with a few of his pals.* —*v.* **2** ***pal (up) with*** *or* ***about*** *or US* ***around (with).*** °associate (with), be *or* become friendly *or* friends (with), be *or* get *or* become on friendly *or* intimate terms (with), go (around *or* about) with, °fraternize (with), consort (with), spend time together *or* with, keep company (with), *Colloq* hang out *or* about *or* around (with), knock about *or* around (with): *Fiona and Isabel palled up last year. Do you still pal around with Timothy?*

palace *n.* mansion, °castle, stately *or* palatial home *or* residence, manor (house), °(country) estate, chateau, *palazzo,* villa: *After the wedding, they went to the prince's palace and lived happily ever after.*

palatial *adj.* °luxurious, de luxe, °magnificent, °splendid, °stately, °sumptuous, °opulent, °majestic, °magnificent, °grand, °elegant, palatine, *Slang* °posh, ritzy, °swanky, classy: *After winning the lottery, they moved into a palatial house in Palm Beach.*

palaver *n.* **1** °nuisance, °procedure, red tape, to-do, rigmarole *or* rigamarole, *Colloq* song and dance, °bother, °nonsense, °business, carry-on, °performance: *He lost the certificates, so she now has to go through all the palaver of getting new ones.* **2** °chatter, °babble, °jabber, (empty *or* small) talk, blather *or* blether, °gossip, prating, °prattle, prattling, palavering, *Brit* natter, nattering, *Scots* clishmaclaver, *Colloq* jawing, °hot air, *Colloq Brit* witter, wittering: *There is so much palaver at the annual meeting that nothing important ever gets said.* **3** °parley, °talk, °conference, °discussion, colloquy, °conversation, confabulation, °meeting, °get-together, round table, powwow, *Colloq* confab, °huddle, chinwag: *The annual palaver of the regional general managers is scheduled for next week.* —*v.* **4** °chatter, °babble, °jabber, blather *or* blether, °gossip, °prattle, prate, chitchat, gabble, *Brit* natter, witter, *Colloq* jaw, chinwag, *US and Canadian* shoot the breeze, *Slang* yak, yacketyyak: *Irmintrude and I were palavering outside the supermarket when Sophronia came along.* **5** °confer, °consult, °discuss, °parley, °talk, °converse, powwow, °meet, °get together, sit down (together), confabulate, °negotiate, *Colloq* confab, °huddle, chew the fat *or* the rag: *Management and union representatives will palaver tomorrow.*

pale[1] *adj.* **1** °colorless, °white, °wan, sallow, waxen, livid, ashen, ashy, pallid, bloodless, whitish, °pasty, °washed-out, anemic, blanched, wheyfaced, drained, °ghostly, °ghastly, peaky *or* peakish, °peaked, cadaverous: *If you had been through an ordeal like that, you'd look pale, too.* **2** °faint, °light, °dim, °washed-out, pastel: *Gunilla promised to wear that pale green evening gown tonight.* **3** °feeble, °weak, °flimsy, °meager, enfeebled, °ineffective, °ineffectual, °puny, °insignificant, °paltry, °lame, °poor, °inadequate, °halfhearted, °tame, spiritless, °empty, °sterile, °lifeless, uninspired, anemic, *Colloq* half-baked: *The sequel was only a pale imitation of the original movie.* —*v.* **4** blanch, blench, °dim, whiten: *He paled when they told him of the car crash.* **5** °diminish, lessen, °fade (away), °decrease, abate: *The works of most modern writers pale in comparison with those of the Elizabethans.*

pale[2] *n.* **1** paling, palisade, °picket, °upright, °post, °stake: *The pales are attached to horizontal rails to form a fence.* **2** °boundary, limit(s), °restriction, °bounds, border(s), confines: *Nothing is forbidden outside the pale of the imagination.* **3** ***beyond the pale.*** °improper, °irregular, °unseemly, unsuitable, °indecent, °unacceptable, °inadmissible, forbidden, anathema, disallowed, prohibited, *verboten,* interdicted; *US* °unusual, °bizarre, °peculiar, °outré, °weird, °abnormal, °strange: *The committee find your behavior to be beyond the pale, Frank, and we demand your resignation.*

pall[1] *n.* **1** °shroud, covering, °mantle, °cloth, °veil: *A black velvet pall covered the coffin.* **2** gloomy *or* melancholy *or* somber *or* grave *or* depressing air *or* mood *or* atmosphere; damper, cold water, *Colloq* wet blanket:

The widow's keening cast a pall over the memorial ceremony.

pall² *v.* **1** Often, ***pall on*** *or* ***upon.*** °bore, °tire, °weary, °jade, °irk, °irritate, °sicken: *His position lost all its charm when the work began to pall on him.* **2** sate, °satiate, cloy, °glut, °surfeit, °gorge: *I had reached the point where even the choicest foods began to pall me.*

paltry *adj.* °trifling, trivial, °petty, °small, °insignificant, °worthless, °pitiful, °pathetic, °pitiable, °puny, °sorry, °wretched, °miserable, °inconsequential, inconsiderable, unimportant, °meager, °mean, beggarly, °base, °low, °contemptible, *Colloq* piddling, *Brit* twopenny, tuppenny, twopenny-halfpenny, mingy, *US* penny-ante, *Slang* Mickey Mouse: *Despite his wealth, his charitable contributions are rather paltry. We all recognize Denison for the paltry pedant he is.*

pamper *v.* °baby, °coddle, cosset, (over)indulge, °spoil, mollycoddle, °cater to, °pet, *Rare* cocker, *Irish* cosher: *The Relphs pamper their children far too much.*

pamphlet *n.* booklet, °brochure, °tract, °essay, folder, °leaflet, °circular; handbill, °bill, °notice, °bulletin, °advertisement, handout, ad, *Brit* advert, *US* flier, throwaway: *He was arrested for printing a pamphlet that denounced the government's repression of free speech. Yesterday we received our sixteenth pamphlet promoting replacement windows.*

pan *n.* **1** saucepan, frying pan, skillet, °pot, casserole, *US* spider: *After melting a teaspoonful of butter in the pan, add the mixture and stir briskly over medium heat.* **2** °face, visage, mien, façade, *Slang* kisser, °mug, puss: *The worst part of Alf's not working is that I have to look at his ugly pan all day.* **3** °depression, °indentation, concavity, °cavity, °hollow, °pit, °hole, crater: *In a rainshower, the pans fill with water and the desert blooms.* —*v.* **4** °wash, °separate, °sift: *For years prospectors have panned for gold in these hills.* **5** °criticize, censure, °find fault, °put down, °reject, flay, excoriate, *Brit* hammer, *Colloq* °knock, roast, slate, *Slang US* °trash, *Brit* °rubbish: *The critics panned the play and it closed after a week.* **6** ***pan out.*** °succeed, °thrive, °prosper, °flourish, fare well, °make it; °work out, °turn out, °result, °come out, °end (up), conclude, culminate, eventuate: *Your grandiose plans for irrigating the Sahara didn't pan out, either! How did the election pan out?*

panache *n.* °flourish, °dash, élan, éclat, °chic, °sophistication, °savoir-faire, °savoir-vivre, flamboyance, °verve, °style, cultivation, °(good) taste, °flair, smartness, boldness, self-assurance, °swagger, °vigor, liveliness, °spirit, brio, °gusto, °zest, °animation, °enthusiasm, °energy: *Whatever needs to be done, you can count on Irena to carry it off with panache.*

pandemonium *n.* °bedlam, °chaos, turmoil, °disorder, °tumult, °frenzy, °uproar, °furor, °confusion: *Pandemonium reigned after the first bomb struck.*

pander *v.* **1** Usually, ***pander to.*** °satisfy, °gratify, °humor, °indulge, °fulfill, bow to, yield to, truckle to, °cater to: *She is tired of pandering to his every wish.* **2** °procure, °pimp, °solicit: *Adolf was hired solely to pander for Mr. Big.* —*n.* **3** panderer, °pimp, °procurer, solicitor, whoremonger, white slaver, *Slang* flesh-peddler, *Brit* ponce, mack: *The word* pander *is an eponym for* Pandarus, *who acted as a sycophantic go-between for Troilus in Chaucer's* Troilus and Criseyde.

pane *n.* panel, °sheet, °glass, windowpane, °light, °quarrel, bull's-eye: *Phone the glazier and have the pane replaced in that door.*

pang *n.* **1** °pain, °stab, °ache, °pinch, °twinge, stitch, °spasm: *I felt a sharp pang of hunger.* **2** °qualm, hesitation, °scruple, °misgiving, °remorse, °regret, contrition, contriteness, self-reproach, mortification, °guilt, °anguish, °discomfort, malaise: *Does Claire feel the slightest pangs for treating him badly?*

panic *n.* **1** °terror, °alarm, °fear, °fright, °dread, °horror, °dismay, consternation, hysteria; °anxiety, apprehension, apprehensiveness, nervousness: *As the speeding car veered toward us, a feeling of panic gripped me.* **2** source of merriment, °comedian, °clown, funnyman, jester, °comic, °wit, °wag, °zany, *farceur,* droll, *Colloq* °(laugh) riot, °scream: *It is hard to see why she considers the Gaekwar of Baroda to be such a panic.* —*v.* **3** be terrified *or* alarmed *or* fearful *or* frightened *or* terror-stricken *or* terror-struck, °dread, °fear, lose one's nerve; °frighten, °scare; *Colloq* °go to pieces, °fall apart, *Brit* lose one's bottle: *I panicked, turned the wheel the wrong way, and crashed the car into a tree. Gregory panics at the slightest sign of danger.* **4** °frighten, °scare, °alarm, °terrify, °unnerve: *Something panicked the horses and one kicked out his stall door.*

panic-stricken *adj.* panic-struck, terrified, alarmed, horrified, aghast, terror-stricken *or* terror-struck, panicky, frenzied, in a frenzy, °hysterical, beside oneself with fear *or* terror, °fearful, °afraid, °scared (stiff), °petrified, horror-struck *or* horror-stricken, frightened *or* scared out of one's wits, appalled, stunned, stupefied, perturbed, unnerved, °nervous, distressed, °upset, jittery, °jumpy, (all) worked up, *Colloq* in a cold sweat, in a flap, in a tizzy, *Slang taboo* scared shitless, shitting green: *At the news of the radioactive leak at the nuclear plant, a few panic-stricken people packed up and moved away.*

panoramic *adj.* °sweeping, commanding, °extensive, °comprehensive, °wide, °overall, °scenic, far-reaching, all-embracing, far-ranging, all-encompassing, °inclusive, bird's-eye, °general: *This room affords a panoramic view of the seacoast.*

pant *v.* **1** °gasp, °huff (and puff), °blow, °heave, breathe hard, wheeze: *He was panting when he came into the room, having run up the stairs.* **2** Usually, ***pant for*** *or* ***after.*** crave, °hanker after, hunger *or* thirst for *or* after, yearn for, ache for, °want, °desire, covet, wish for, long *or* pine *or* sigh for, have one's heart set on, die for, be dying for, *Colloq* have a yen for, give one's eyeteeth *or* right arm for: *Margery was the girl that all the boys panted after when we were ten.*

pants *n.pl.* **1** *In Britain* (men's) drawers, smallclothes, smalls, drawers, underpants, boxer shorts, trunks, undershorts, Y-fronts, briefs, *Colloq* smalls; (women's) knickers, cami-knickers, panties, drawers, bloomers, pantalettes, °tights, pantyhose, undies: *The baby was dressed in a frilly dress and matching pants.* **2** *In the US* trousers, slacks, breeches, (Oxford) bags, knickerbockers, flannels, °shorts, Bermuda shorts *or* Bermudas, pedal pushers, bellbottoms, pegtops, hiphuggers, (blue) jeans, dungarees, denims, *Trademark* Levis, *Scots* trews; *Scots and North England* breeks, *US* knickers, *Colloq* cut-offs: *Americans who go into British shops asking for a pair of pants are embarrassed to be offered only underwear.*

paper *n.* **1** newspaper, tabloid, °daily, weekly, °journal, gazette, °publication, °periodical, newsletter, °organ, *Colloq* °rag, °sheet: *She is feature editor of a paper in Manchester.* **2** Often, ***papers.*** **(a)** document(s), °instrument, legal papers, °form, certificate, °deed, ownership papers; credential(s), °identification: *Bring the papers to my office for signing. His papers are not in order. Many people lost their papers during the war.* **(b)** docket, files, dossier, record(s), archive(s): *Your papers seem to have disappeared from our office, Count Alucard.* **3** °stationery, letterhead, writing paper, letter paper, notepaper; foolscap; scrap *or US also* scratch paper; wrapping paper; gift-wrapping, gift-wrap; wallpaper: *She wrote to me on the most beautiful engraved paper. We buy our Christmas paper from a museum shop.* **4** article, °composition, °essay, °assignment, °report, °thesis, °study, °tract, °analysis, critique, exegesis, treatise, dissertation, disquisition, manuscript, Ms. *or* ms., autograph, holograph, typescript, °script, °speech: *Adelaide will present her paper at the meeting of the National Academy this year.* —*v.* **5** (hang) wallpaper, °line; °post, °distribute: *The entire area was papered with posters announcing the meeting.*

par *adj.* **1** °standard, °normal, °average, expected: *A success rate of two percent is par for these candidates.* —*n.* **2** °level, °rank, °standing, °scale, °standard: *How can you place yourself on a par with her as a tennis player?* **3** ***above par.*** above average, °superior, °outstanding, °excellent, °exceptional; °choice, °select, °prime: *Since I stopped taking that medication, I've been feeling above par. I am pleased to announce that the test results were above par for the school as a whole.* **4** ***at par.*** °average, °level, °even, °equal, °equivalent, °standard; par value: *The shares are selling at par, or below their true market value.* **5** ***below*** or ***under par.*** **(a)** below average, substandard, °inferior, second-rate, °mediocre, middling, not up to par, °poor, °inadequate, °unsatisfactory, °wanting, °bad, °wretched, °miserable, °awful, °terrible, *Colloq* °lousy, not up to snuff *or* scratch: *Ian's goal keeping has been below par all season.* **(b)** °ill, °sickly, °unhealthy, unwell, not (very) well, not oneself, not in good *or* the best shape, in bad shape, *Brit* off form, off color; *Colloq* °under the weather, °poorly, not up to snuff: *I was feeling below par.* **6** ***up to par.*** all right, °adequate, °average, °satisfactory, good enough, °passable, °acceptable, *Colloq* OK *or* O.K. *or* okay, up to scratch *or* snuff, fair to middling: *Had your game been up to par, Davis would not have won.*

parable *n.* allegory, fable, °lesson, morality tale: *When we first read Aesop's* Fables *we didn't know they were parables.*

parade *n.* **1** °procession, °march, °train, °file, °promenade, cortège, column; entourage: *On Saturday there will be a parade to commemorate those who died in Vietnam.* **2** °exhibition, °(ostentatious) display, °show, °spectacle, array, °pomp, °splash: *She often makes a parade of her knowledge of architecture.* **3** °promenade, °walk, °(pedestrian) way, mall, esplanade: *Let's meet in the Parade when you have finished shopping.* —*v.* **4** °march, pass in review, °promenade, °walk, °file: *The generals saluted smartly as the troops paraded past.* **5** °strut, °flaunt, °show (off), brandish, °wave, vaunt, °display, °air: *Why does Thea feel it necessary to parade every new beau in front of all the neighbors?*

paradise *n.* **1** °heaven, City of God, Zion, Elysium, Elysian Fields, happy hunting grounds, Abraham's bosom, heavenly kingdom, Promised Land, Celestial City, New Jerusalem, Avalon, Valhalla, Hesperides, Isles *or* Islands of the Blessed, seventh heaven: *People all have their own ideas of paradise, but all agree it's a nice place to be.* **2** heaven on earth, (Garden of) Eden, (land of) Goshen, °utopia, Shangri-La, Land of Beulah: *Early settlers of California thought it a paradise.* **3** °bliss, °happiness, °rapture, °heaven, °delight, blessedness, °ecstasy, seventh heaven, °joy, dreamland, nirvana: *Being with you has been sheer paradise.*

paradox *n.* contradiction, self-contradiction, incongruity, inconsistency, °absurdity, °ambiguity, °enigma, °puzzle, °mystery, °quandary, °problem, °dilemma: *Can you explain away Zeno's paradox of Achilles and the tortoise? The paradox was that although it was Edward's fifth birthday, he was twenty years old.*

paradoxical *adj.* °contradictory, self-contradictory, conflicting, oxymoronic, °impossible, °improbable, °incongruous, illogical, °inconsistent, °absurd, °ambiguous, confusing, °equivocal, enigmatic, °puzzling, baffling, °incomprehensible, bewildering, °perplexing, °mysterious, °problematic: *It would be true, though it might sound paradoxical, to say that the Norman Conquest made England Saxon.*

paragon *n.* °epitome, archetype, °model, °prototype, °quintessence, °pattern, °standard, exemplar, °ideal, *beau idéal,* criterion: *Before the scandal, Michael had been long regarded as a paragon of virtue.*

parallel *adj.* **1** °similar, corresponding, congruent, analogous, analogic(al), °correspondent, °like, °matching, homologous, °coordinate, °equivalent, coequal, °proportional, proportionate, *pari passu, mutatis mutandis,* in proportion, °uniform; contemporary *or* cotemporary, contemporaneous *or* cotemporaneous: *The development of the limbs during gestation is parallel.* —*n.* **2** analogue, °match, homologue, °equivalent, °complement, counterpart, °equal, coequal: *Campbell found many parallels among the legendary heroes of disparate cultures.* **3** analogy, parallelism, equivalence, complementarity, °relationship, °kinship, correspondence, °resemblance, °likeness, similarity, °symmetry, °equality, coequality, °parity, correlation; °proportion, °balance, equiponderance, equipoise, counterbalance, °offset: *One can draw parallels between the human arm, the fish's fin, and the bird's wing. By way of parallel, consider the flood in the Bible and that in the Gilgamesh epic.* —*v.* **4** correspond to *or* with, °match, equate to *or* with, be likened to, correlate to *or* with, compare with *or* to, °imitate, °repeat, °echo, iterate, °reiterate, °duplicate, °follow, °agree with; keep pace with, °conform (to), °balance, °set off, °offset, even off *or* up, be accompanied by, coincide with, *Colloq* jibe with: *In many respects, your attitudes toward life parallel mine when I was your age. A decrease in the rate of inflation supposedly parallels a rise in interest rates.*

paralyze *v.* **1** immobilize, inactivate, deactivate, °transfix; °halt, °stop: *The stroke paralyzed his entire left side. As the thing approached, the children became paralyzed by fear.* **2** °deaden, °numb, °freeze, anesthetize, benumb, render insensible: *This injection will paralyze your thumb so that we can operate on it.* **3** disable, °cripple, °incapacitate, disenable: *His left leg was paralyzed in the accident.*

paramount *adj.* °preeminent, °chief, °supreme, °dominant, °main, °predominant, °cardinal, °first, °prime, °primary, °principal, °essential, °vital, requisite, °basic: *The company feels that the independence of researchers is paramount. It is of paramount importance that you study trigonometry before calculus.*

paramour *n.* °lover, °love, inamorato *or* inamorata, amorist, °mistress, gigolo, concubine, *cicisbeo,* kept woman, *Colloq* °flame, sugar daddy, *US* POSSLQ (= 'Person of Opposite Sex Sharing Living Quarters'), *Slang* fancy man *or Slang* woman: *At eighty, Lady Paphian liked to reminisce about her many paramours.*

paraphernalia *n.* usually pl. equipment, °apparatus, accessories, °outfit, appliances, utensils, °gear, °rig, material(s), materiel, °things, °tackle, equipage, accouterments *or* accoutrements, °effects, chattels, °possessions, °belongings, appurtenances, °trappings, °property, baggage, impedimenta, supplies, °stuff, *Colloq* °junk, *Brit* °rubbish, clobber, *Slang* crap, *Taboo US* shit: *Bring along your scuba paraphernalia. It took us a month just to pack up all the paraphernalia we had in the apartment.*

paraphrase *n.* **1** rephrasing, rephrase, rewording, restatement, rewriting, rewrite, °rehash, °rendition, °rendering, °version, *Technical* paraphrasis: *The editor wanted a revision, not merely a paraphrase of the origi-*

nal. —*v.* **2** rephrase, °reword, restate, rewrite, explicate, °explain: *Please paraphrase the specialized jargon to make it understandable to nonprofessionals.*

parasite *n.* leech, °hanger-on, *Colloq* freeloader, sponger *or* sponge, °bloodsucker, cadger, scrounger *or* scrounge, barnacle, jackal, hyena: *That parasite Cadwell lives off Gemma's money and has never worked a day in his life.*

parcel *n.* **1** °package, °packet, carton, °box, container, °case: *We are sending food parcels to the needy.* **2** °portion, °plot, plat, °lot, °piece, °section, °tract: *We have bought a small parcel of land and plan to build on it next year.* **3** °lot, °group, °batch, °collection, °pack, °bundle, °set: *She gathered a parcel of drawings and published them as a travel sketchbook.* —*v.* **4** Often, ***parcel out.*** apportion, °allot, °deal (out), °dole (out), °mete (out), °hand out, °distribute, °share (out), °divide, *Colloq* divvy (up): *Before he died he parceled out his fortune among his grandchildren.*

parch *v.* dry (out *or* up), desiccate, dehydrate, exsiccate; °scorch, sear, °burn, bake; °shrivel (up), wither: *The earth was parched by the unrelenting sun.*

pardon *n.* **1** °forgiveness, °forgiving, amnesty, °remission, °release, reprieval, absolution, °indulgence, °excuse, excusal, °allowance, overlooking, condonation, exoneration, exculpation: *Should there be any pardon for crimes against humanity?* —*v.* **2** °forgive, °remit, °release, °reprieve, absolve, °indulge, °allow, °overlook, °let off, °excuse, condone, exonerate, exculpate: *There are some crimes that ought not be pardoned.*

pare *v.* **1** °trim, °peel, °skin, °shave (off), shuck; decorticate, excoriate: *It took me hours to pare the skins from that bushel of apples.* **2** Often, ***pare down.*** °reduce, °diminish, °decrease, cut (back *or* down), °curtail, °slash (back), °lower, lessen: *Since his illness, Bob has pared down his time at the office. We'll have to pare the prices to move this produce before the "sell-by" date.*

parent *n.* **1** father *or* mother, °progenitor, progenitrix, procreator, begetter, materfamilias *or* paterfamilias; foster parent, stepmother *or* stepfather, °guardian, *Colloq* old lady *or* old man, *Brit old-fashioned or facetious* mater *or* pater: *The principal must like you and Dad a lot—he said he wouldn't let me return to school unless I brought my parents.* **2** °source, °origin, originator, wellspring, fountainhead, °root: *The liturgy of St. James is undoubtedly the parent of the Armenian Rite.*

parentage *n.* °lineage, ancestry, °line, °family, °extraction, descent, °origin, °pedigree, °stock, °birth, °strain, bloodline, °heritage, °roots: *As he was separated from his mother at one, he knows nothing of his royal parentage.*

parenting *n.* (of children) rearing, °upbringing, raising, nurturing: *Brian and Betty take their parenting very seriously.*

parity *n.* **1** °equality, equivalence, consistency, °uniformity, °par, °likeness, similarity, analogy, congruity, similitude, conformity, congruence: *The ministers agreed that they would observe parity of power among themselves at the conference.* **2** °proportion, parallelism, analogy, °balance, correspondence: *We are trying to maintain parity between the Deutschemark and the dollar.*

park *n.* **1** garden, green(s), common(s), °preserve, °reserve, greensward, parkland, woodland, °estate, *Chiefly Brit* °reservation: *Twice a day he strolls through the nearby park.* **2** *Brit* car park, *US and Canadian and New Zealand* parking lot: *I shall meet you in the car park at the railway station.* —*v.* **3** °leave, °put, °deposit, °store: *It is forbidden to park your car in this street.*

parlance *n.* way *or* manner of talking *or* speaking, *façon de parler,* phrasing, phraseology, °speech, °wording, °language, °idiom, °dialect, °jargon, *Colloq* °lingo: *In the parlance of the yuppies, "That's a ballpark figure."*

parley *n.* **1** °conference, °discussion, °dialogue, °palaver, deliberation, °meeting, colloquy, colloquium, confabulation, powwow, talk(s), *Colloq* °huddle, confab: *The issue will be decided at the executive council's parley.* —*v.* **2** °confer, °discuss, °palaver, °deliberate, °talk (over), °negotiate, °deal, *Colloq* °huddle: *Would Sitting Bull parley with General Custer?*

parliament *n.* **1** In Great Britain, ***Parliament.*** Houses of Parliament, House of Lords and House of Commons, Westminster, the House, Mother of Parliaments: *The new health bill got its second reading in Parliament today.* **2** legislature, °council, congress, °diet, °council, °assembly, upper and lower house *or* chamber: *The parliament consists of an upper house and the lower house.*

parliamentary *adj.* °formal, ordered, °orderly, procedural, conforming, conformist, *US* according to *Roberts Rules of Order*: *The debate must follow parliamentary procedure.*

parlor *n.* living room, *Old-fashioned or Brit* drawing room, *Chiefly Brit* sitting room, °reception (room), °lounge: *She invited us into the parlor where tea was laid out.*

parlous *adj.* °perilous, °risky, °precarious, °uncertain, °dangerous, °hazardous, °difficult, °ticklish, °awkward, *Colloq* chancy, iffy, *Slang* °hairy: *These are parlous times in the Middle East.*

parochial *adj.* regional, °provincial, °narrow, °local, insular, °isolated, °limited, restricted, °narrow-minded, °petty, °shortsighted, °hidebound, °conservative, °conventional, illiberal, °bigoted, °prejudiced, °intolerant, °one-sided, °partial, °biased, °stubborn, °opinionated, °dogmatic, °rigid, °stiff, stiff-necked, °immovable, intractable, unchangeable, unchanging, close-minded, °unsophisticated, unworldly, uncultivated, uncultured: *Uneducated and untraveled, Mr. Shriver maintained a very parochial view of the world.*

parody *n.* **1** °burlesque, °lampoon, °satire, °caricature, °mockery, mimicry, *Colloq* take-off, spoof, *Brit* sendup: *The more serious the intent of the original author's work, the easier it is to write a parody of it.* **2** travesty, °mockery, feeble *or* poor imitation, distortion, °perversion, corruption, debasement: *The Inquisition was a parody of justice and of religion.* —*v.* **3** °burlesque, °lampoon, °satirize, °caricature, °mock, °mimic, ape, °ridicule, °deride, °laugh at, °poke fun at, °guy, scoff at, sneer at, rib, °tease, °twit, roast, pillory, make a laughingstock (of), °make sport of, °make fun of, make a monkey of, *Archaic* fleer, *Colloq* °take off, spoof, kid, *Brit* °send up: *Swift parodied English political figures and policies of the day in* Gulliver's Travels.

paroxysm *n.* °fit, convulsion, °spasm, °throe, °seizure, °spell, °outburst, °eruption, °explosion, *Colloq* flare-up: *At the sight of George dressed as Madame Pompadour, Laura rolled on the floor in a paroxysm of hysterical laughter.*

parrot *n.* **1** imitator, °mimic, *Colloq* copycat: *Never with an original idea of her own, she's a mere parrot of others' fashions and fancies.* —*v.* **2** °imitate, °mimic, ape, °copy, °echo, °repeat, °reiterate: *Mike, who doesn't understand the issues, parrots what he hears on TV.*

part *n.* **1** °piece, °portion, °division, °allotment, °share, °percentage, participation, °interest; °parcel, °fragment, °scrap, shard; some: *I bought a part of the company when it was available. I want no part of the* *deal now.* **2** °portion, component, °factor, constituent, °element, °ingredient: *A part of her problem is that she doesn't speak Japanese.* **3** °role, °function, °duty, °responsibility, °share; °say, °voice, °influence, participation, involvement, °business: *Every man must do his part. Don't thank me! I had no part in your getting the contract.* **4** °role, °character: *He plays the part of Tweedledum in the school play.* **5** °side, °interest, °cause, °behalf, °faction, °party: *Which part did you support in the argument? No explanation has been offered on either part.* **6** °neighborhood, °quarter, °section, °district, °region, °area, corner, °vicinity, vicinage, *Colloq* neck of the woods: *I come from the same part of the country as you.* **7** °piece, °portion, °segment, °section; °department, °division, component, °unit: *Which part of the turkey do you prefer? In which part of the company do you work?* **8** ***for the most part.*** mostly, °generally, °usually, °mainly, °in the main, by and large, °on the whole, °chiefly, °principally, essentially, for all practical purposes, °to all intents and purposes, in most cases *or* instances: *The shops are for the most part closed on Sunday.* **9** ***in part.*** °partly, °partially, to some extent *or* degree, in some measure, °relatively, comparatively, °somewhat: *He is himself in part responsible for the present state of affairs.* **10** ***on the part of (someone)*** *or* ***on (someone's*** *or* ***one's) part.*** °by, on *or US also* in behalf of, °(as) for, as regards, as far as (someone) is concerned, in the name of, for the sake of, in support of: *Tyrannical acts on the part of the king were not condoned. For my part, I want nothing to do with it.* **11** ***take (a) part (in).*** °participate (in), °join (in), be (a) party to, play a part *or* role (in), be involved (in *or* with), associate oneself (with), have *or* take a hand in, °partake (of), °contribute (to): *Why insist that she take part in your nefarious plot? People began to sing, but she didn't feel like taking part.* —*v.* **12** °separate, part company, °split up, go his *or* her (*or* their) separate way(s), °break up, say *or* bid goodbye (*or* adieu, etc.); °leave, °depart, go (away *or* off): *We parted on the best of terms.* **13** °separate, °divide, put *or* pull apart, put asunder: *I saw a pale hand part the curtains for a brief moment. As the saying goes, "A fool and his money are soon parted."* **14** ***part with.*** °give up, °yield, °relinquish, °release, °sacrifice, °forgo, °renounce, °forsake, let go, °surrender: *I doubt that you'll persuade Rover to part with his bone.* —*adj.* **15** °partial, fractional, °limited: *Ronald is a part owner of the company.*

partake *v.* **1** Usually, ***partake in.*** °share (in), °participate (in), take (a) part (in), enter (in *or* into): *We share each other's burdens and partake in each other's joys.* **2** Usually, ***partake of.*** **(a)** °receive, °get, have a share *or* portion *or* part (of), °share: *We were invited to partake of a meager repast.* **(b)** °evoke, °suggest, hint at, °intimate, °imply, possess *or* have the quality of: *Greater knowledge often partakes of insolence.*

partial *adj.* **1** °incomplete, °fragmentary, not total *or* whole, °imperfect: *They were able to afford only a partial restoration of the house. There will be a partial solar eclipse at noon.* **2** °prejudiced, °biased, °partisan, °inclined, influenced, °one-sided, °jaundiced, unfair, discriminatory: *It will be difficult to find a judge who is not partial.* **3** ***partial to.*** °in favor of, predisposed to, °fond of, having a soft spot *or* weakness for, having a liking *or* taste *or* predilection for, having a fondness for, feeling an attraction *or* affinity to *or* toward(s), finding enjoyment in: *She used to be partial to punk rockers. I am partial to Scotch beef.*

partiality *n.* **1** °prejudice, °bias, °inclination, °favoritism, predilection, predisposition, °leaning, °preference: *The losers of the contest accused the judges of partiality.* **2** °preference, °taste, °relish, °liking, fondness, °appreciation, °fancy, °love, °eye, °weakness, soft spot, °penchant; fetish *or* fetich: *Alas, my wife has acquired a partiality for emeralds.*

partially *adv.* °partly, °in part, to some extent *or* degree, to a limited *or* a certain extent *or* degree, not totally *or* wholly *or* entirely, restrictedly, incompletely, in some measure, °relatively, comparatively, °moderately, (up) to a (given *or* certain) point, °somewhat: *Jones found the meal only partially satisfying and ordered another banana split.*

participant *n.* **1** participator, partaker, sharer, °party, contributor, prime mover: *The chief participants in the recent terrorist attack are in custody.* —*adj.* **2** Usually, ***participant in*** *or* ***of.*** participating, partaking, sharing: *As a shareholder, he will be participant in the profits.*

participate *v.* Often, **participate in.** take part (in), °share (in), partake (in *or* of), °join (in), °engage (in), get *or* become involved (in), be *or* become associated (with), enter (in *or* into), °contribute (to): *We invited her to join the group, but she refuses to participate. I would like to participate in the venture.*

particle *n.* atom, molecule, scintilla, °spark, mote, °suggestion, °hint, °suspicion, °gleam, °bit, °crumb, °jot, tittle, whit, mite, °speck, °dot, °spot, iota, °grain, °morsel, °shred, °sliver, °scrap, *Colloq* smidgen *or Colloq* smidgin: *There isn't the slightest particle of evidence linking Lagrange with the murder.*

particular *adj.* **1** °certain, °specific, °special, °peculiar, °singular, °single, °isolated, °individual, °distinct, °discrete, °separate, °definite, °precise, °express: *These are not Florian's particular sentiments but those of thousands of his followers. In this particular instance your theory does not apply.* **2** °marked, °special, especial, °exceptional, °remarkable, °noteworthy, °notable, °outstanding, °unusual: *The commendation is for Dr. Fantod's particular contribution to nervous disorders. Vintage port is a particular favorite of mine.* **3** °detailed, itemized, item-by-item, °thorough, °minute, °precise, °exact, °exacting, °painstaking, °nice, rigorous, °close, blow-by-blow: *The inscriptions reveal the particular care taken in keeping daily records.* **4** °fussy, °meticulous, °finicky, finical, °fastidious, °discriminating, °selective, °demanding, hypercritical, °critical, *Colloq* pernickety *or US also* persnickety, °choosy, picky: *Daphne is very particular about whom she invites to dinner.* —*n.* **5** Usually, ***particulars.*** °detail, minutia, fine point, °item, °specific, °element, °fact, °circumstance, °information: *The particulars may never be revealed, but we know in general what took place that fateful night.* **6** ***in particular.*** °particularly, specifically, °precisely, °exactly, °especially, °specially; °particular, °specific, °special, °definite: *I told mother that I was just going out, not anywhere in particular.*

particularly *adv.* **1** °especially, °specially, exceptionally, peculiarly, singularly, distinctively, uniquely, unusually, uncommonly, °notably, outstandingly, markedly, extraordinarily, °very, °extremely, strikingly, surprisingly, amazingly: *Susannah went through a particularly bad patch but she's better now. Holroyd is a good writer, particularly of biography.* **2** °in particular, specifically, °especially, °principally, °mainly, exceptionally, °expressly, explicitly, °notably, markedly; °only, solely: *Maeterlinck was particularly interested in bees.*

parting *n.* **1** separating, splitting, dividing, breaking (up *or* apart), sundering, cleaving; °separation, °split, °division, breakup, °rift, °rupture: *The high point of the film was the parting of the Red Sea. The parting between Harriet and Sam after twenty turbulent years was to be expected.* **2** leave-taking, °farewell, saying goodbye, departure, leaving, °going (away), making one's adieus *or* adieux; valediction: *Parting is such sweet sorrow.* —*adj.* **3** closing, °final, concluding, °last, departing, valedictory; deathbed, °dying: *Her parting comment was, "If you're leaving, take the garbage."*

partisan *n.* **1** partizan, °devotee, °follower, °supporter, adherent, °backer, °champion, °enthusiast, °fan, °zealot, °fanatic, *Chiefly US and Canadian* booster, *Colloq US* rooter: *He has long been a partisan of surrealism.* **2** guerrilla *or* guerilla, freedom fighter, underground *or* resistance fighter, °irregular: *During the war the Yugoslav partisans were led by Marshal Tito.* —*adj.* **3** °one-sided, factional, °biased, tendentious, °sectarian, °opinionated, °partial, °bigoted, °prejudiced, °parochial, myopic, °shortsighted, °nearsighted, °narrow, °narrow-minded, °limited: *She refuses to become involved in the petty infighting of partisan politics.* **4** guerrilla *or* guerilla, °freedom, °underground, °resistance, °irregular: *Both were members of the Maquis, the French partisan group who fought against the Nazis.*

partition *n.* **1** °separation, °division, splitting (up), split-up, partitionment, breakup, breaking up, segmenting, segmentation: *One of the results of World War II was the partition of Germany.* **2** °allotment, allotting, apportionment, apportioning, °distribution, distributing, meting out, doling out, rationing (out), sharing (out), dividing (up), giving *or* handing *or* passing out, parceling out, *Colloq* divvying up: *The partition of the estate is in the hands of our solicitors.* **3** (room) divider, °(dividing) wall, °barrier, °screen, separator: *Let's put up a partition to divide this office into two.* **4** °compartment, °room, °chamber, °section, °part, °area, °division, subdivision, °cell, °stall, °booth: *In open floor planning, furniture, plantings, and other movables are used to create the partitions where people work.* —*v.* **5** °divide (up), °separate, °cut up, subdivide, °split (up): *Balkanize means to "partition off a country into smaller, often quarrelsome units."* **6** Often, ***partition off.*** °divide, °separate, subdivide, °wall off, °screen (off), fence off: *We must partition off the machine room so that the office staff are not disturbed by the noise.*

partly *adv.* See **partially,** above.

partner *n.* sharer, partaker, °associate, °colleague, °participant, °accomplice, °accessory, confederate, °comrade, °ally, collaborator, °companion, teammate, °fellow, alter ego, °friend, *Colloq* °pal, sidekick, °mate, *US and Canadian* buddy, °cohort: *His partner in the bank robbery was arrested this morning.* **1** wife *or* husband, spouse, °mate, helpmate, helpmeet, consort; °(live-in) lover, *US* significant other, POSSLQ (= 'Person of Opposite Sex Sharing Living Quarters'): *She became my life partner more than fifty years ago.* **2** °companion, fellow dancer: *Won't you change partners and dance with me?*

party *n.* **1** °(social) gathering, (cocktail *or* dinner) party, °celebration, fête *or* fete, °function, °reception, soirée, levee, °festivity, °festivities, °festival, °frolic, °spree, romp, carousal *or* carouse, saturnalia, bacchanalia, debauch, °orgy, *Colloq* °get-together, bash, bust, shindig *or* shindy, ball, at-home, do, *US* blowout, toot, *Brit* beanfeast, beano, knees-up, *Slang US* °hop, *Chiefly US and Canadian* wingding, bust-up, *Brit* rave *or Brit* rave-up: *Campbell said he had a marvelous time at your birthday party.* **2** °group, °company, °band, °body, °corps, °gang, °crew, °team, °squad, troop, °platoon, °detachment, °detail, cadre, °unit, *Colloq* °bunch, °outfit: *A small party of men is trapped on the other side of the river.* **3** °side, °interest, °faction, °league, °club, coalition, bloc, °division, °sect, °denomination, °clique, coterie, °set, °cabal, junta *or* junto, partisans, adherents, confederacy, confederation, °federation, *Chiefly US and Canadian* caucus: *The proliferation of political parties confuses the electorate.* **4** °participant, participator, confederate, °associate, °ally, °accomplice, °accessory, approver, ratifier, upholder, contributor, °supporter, °advocate, °backer, °aid, helper, seconder, promoter, °partisan, defender, exponent, °proponent, °champion: *I refuse to be a party to any illegal act.* **5** °individual, °person, °litigant, plaintiff, defendant, °side, °interest, signer, signatory, cosignatory, °participant, *US* cosigner: *The party of the first part accuses the party of the second part of infidelity. How many parties are there to this contract?*

parvenu *n.* **1** parvenue, °upstart, arriviste, nouveau riche, °intruder, °adventurer, social climber: *Usually one generation has to pass before a parvenu is accepted by the class with which his money associates him.* —*adj.* **2** nouveau riche, °upstart, °intrusive: *This parvenu industrialist tried to use his wealth to make up for his lack of refinement.*

pass *v.* **1** Often, ***pass by.*** proceed *or* move past, go by *or* past: *She passed me in the street. I didn't recognize her till she had passed. We met like ships that pass in the night.* **2** °proceed, move (onward(s)), °go (ahead), °progress, °extend, °lie, °run, °flow, °fly, °roll, °course, °stream, °drift, °sweep: *A flock of geese passed overhead.* **3** let pass, let go (by), °overlook, °disregard, °ignore, pay no heed, °omit, °skip: *I think I'll pass on making a decision till the proper time comes.* **4** °qualify (in), °pass muster (in), get *or* come through, °succeed: *Diarmuid didn't believe she would pass chemistry, but pass she did.* **5** °spend, °devote, °use (up), °expend, °employ, °occupy, °fill, while away, °take (up); °dissipate, °waste, °fritter away, °kill: *He passed his time on Devil's Island planning ways to escape.* **6** °surpass, °exceed, °outdo, °transcend, go beyond, overshoot, °outstrip, outrun, surmount, outdistance: *She passed her own expectations in winning the scholarship. How Atterburgh ever got a degree passes all comprehension.* **7** °allow, °tolerate, °permit, °approve, °sanction, °accept, °authorize, °endorse, °carry, °agree to, °confirm: *The customs inspectors passed my luggage without question. The bill passed the committee and came up for voting.* **8** °give, hand round *or* along *or* over, °transfer, pass on *or* over, °deliver, convey, *Sports US* hand off, *Colloq* °toss, °throw, °reach: *Please pass the salt. Caverton passed the ball to the goalkeeper.* **9** utter, °express, °issue, °declare, °pronounce, °deliver, °set forth, °offer: *Who is she to pass judgment on abstract paintings?* **10** °go away, °disappear, vanish, °evaporate, fade away, °melt away, evanesce, °cease (to exist), °(come to an) end, °die out, go by the board, °terminate, *Literary* evanish, *Colloq* blow over: *For an instant I was gripped by a horrible fear, but the feeling passed.* **11** °go (by), °expire, °elapse; slip by *or* away, °fly; °crawl, °creep, °drag: *Weeks have passed since we last met. Time passes quickly when you're having fun.* **12** °evacuate, °void, °eliminate, excrete, °defecate, °urinate: *The tests showed that he had passed some gravel in his urine.* **13** ***come to pass.*** befall, °happen, °occur, °take place, °come about, °arise, *Colloq* °come off: *And it came to pass that a great pestilence was abroad in the land.* **14** ***pass away.*** **(a)** °die, °expire, °perish, °succumb, breathe one's last, °pass on, go to one's reward, go to one's last resting place, (go to) meet one's maker, *Colloq* go west, give up the ghost, *Slang* croak, kick the bucket, *Chiefly US* bite the dust, turn up one's toes: *He was the sole beneficiary when she passed away.* **(b)** vanish, °disappear, °go away, °stop, °cease, °end: *That feeling of vertigo simply passed away when the plane landed.* **15** ***pass by.*** See **1,** above: *The strikers were urging people to pass by the shop.* **16** ***pass for*** *or* ***as.*** **(a)** be taken for, be mistaken for, be regarded as, be accepted as: *She wrote a book about being a grandmother who passed for twenty.* **(b)** impersonate, °imitate, °mimic, pass (oneself) off as, come *or* go as, be disguised as, disguise oneself as, assume the guise of, °masquerade as, °pose as, assume the role of, act the part of, act like, pretend to be, °play: *Using her best cockney accent, she tried to pass as a Londoner.* **17** ***pass off.*** °evaporate, °disappear, evanesce, be emitted: *The water passes off as steam, leaving the distillate in the flask.* **18** ***pass on.*** °proceed, °continue, °progress: *Let us now pass on to the next lesson.* **(a)** °bequeath, hand down *or* on, °transfer, °make over, °will, °cede, °give: *The old gambler passed on to me the secret of never losing—don't gamble.* **(b)** See **14, (a),** above. **19** ***pass (oneself) off as.*** See **16, (b),** above. **20** ***pass out.*** **(a)** °faint, °collapse, swoon, black out, °drop, *Colloq* conk out, keel over: *When they told her of the accident, she passed out on the spot.* **(b)** °distribute, dole out, mete out, °deal (out), °hand out: *After we took our seats, the cabin staff passed out glasses of champagne.* **21** ***pass over.*** See **3,** above: *I shall pass over his latest misdeeds without comment.* **22** ***pass up.*** °reject, °decline, °refuse, °waive, °turn down, °dismiss, °spurn, °renounce; °deny (oneself), °skip, °give up, °forgo, let go (by), °abandon, forswear, °forsake, let pass, °ignore, pay no heed, °disregard, pay no heed, °omit: *If you don't buy this car, you are passing up an opportunity of a lifetime.* —*n.* **23** defile, °gorge, col, cwm, °cut, °canyon, °notch, °gap, °gully, couloir; °passage, °opening, °way, °route, °road: *We had mined the pass to prevent the enemy's advance.* **24** authorization, °permit, °license, °approval, safe-conduct, green light, °go-ahead; °permission, °freedom, °liberty, °authority, °clearance; *Colloq* OK *or Colloq* O.K. *or Colloq* okay: *He bribed an official for a pass to leave the country.* **25** free pass, complimentary ticket, *Slang US* twofer, Annie Oakley: *I have two passes to the opera tonight—want to go?* **26** °state (of affairs), °condition, °situation, °stage, °juncture, °status, crux; °predicament, °crisis: *At that critical pass, it was too late to rebuild the navy virtually overnight. Things had come to a pretty pass.* **27** °attempt, °trial, °try, °effort, °endeavor: *Anita's first pass at a perfect score resulted in disaster. We made a pass at docking while under sail.* **28** (sexual *or* indecent) overture *or* advance(s), °proposition, indecent proposal: *Letchworthy has made a pass at every secretary he's ever had.* **29** °maneuver, °approach; °passage, °flight, fly-by, °transit: *The pilot's first pass over the airfield was at five hundred feet.* **30** °transfer, °toss, °throw, *US* hand-off: *A forward pass is illegal in rugby.*

passable *adj.* **1** °satisfactory, °acceptable, °tolerable, all right, °adequate, admissible, allowable, °presentable, °average, °fair (enough), fair to middling, fairly good, middling, not bad, unexceptional, °sufficient, °indifferent, *Colloq* OK *or* O.K. *or* okay, °so-so: *I thought the play passable, not up to his usual standard.* **2** traversable, °navigable, °open, unobstructed, unblocked: *The channel is passable till October, when the river freezes over.*

passage *n.* **1** °movement, °moving, °going, °transition, °transit, traversal, °traverse, °progress, crossing, °passing: *The queen's passage through the town was attended by much jubilation. We observed the passage of Mercury across the sun's disk.* **2** °extract, °excerpt, °selection, °section, °part, °portion, °text, paragraph, canto, stanza, verse, °line, °sentence, °phrase, citation, °quotation: *I recognize this passage from* The Faerie Queene. **3** °way, °route, avenue, °course, °channel; °road, thoroughfare: *Amundsen found a navigable passage between the Atlantic and Pacific across North America.* **4** °corridor, °hall, passageway, hallway, vestibule, °lobby, foyer: *As I stepped into the passage outside my room I tripped over a body.* **5** °change, °mutation, °shift, conversion, °progression, °passing: *In sublimation, the passage is directly between the solid and gaseous or vaporous states, without an intermediate liquid stage.* **6** °passing, °elapse, °progress, °progression, °flow, °march, °advance: *She may become more tractable with the passage of time.* **7** voyage, °trip, °journey, °cruise, crossing, °sail, °run, °travel, °traveling; *Brit* accommodation *or US* accommodations, arrangement(s), °facilities: *In 1620 the trans-Atlantic passage was fraught with danger. The refugees were guaranteed safe passage to the border. We have been unable to book a passage to Majorca for tomorrow.* **8** safe-conduct, °permission, °privilege, °liberty, °freedom, visa, authorization, °allowance: *Our troops were denied passage through Nepal.* **9** enactment, ratification, °sanction, °approval, acceptance, °passing, adoption, °endorsement, endorsing, legitimatization *or* legitimization, legalization, legislation, constitutionalization, ordainment: *It appears that there is enough support to ensure passage of the bill.* **10** °traffic, trafficking, dealing, shipment, shipping, °commerce, °trade, °exchange, °transaction: *Representatives of several governments discussed putting a halt to the passage of narcotics from Colombia.* **11** °aperture, °hole, orifice, °opening; °entry, access, inlet; °exit, °outlet: *We found a tiny passage through which the termites were entering and leaving.*

passé *adj.* °old-fashioned, unfashionable, dated, out-of-date, °behind the times, outmoded, °obsolete, °obsolescent, °antiquated, archaic, démodé, °quaint, °antique, superseded, *Colloq* °out, not *or* no longer in, old hat, back number: *She persisted in dressing in a style that had been passé at the turn of the century.*

passenger *n.* rider, °fare, °traveler, voyager, commuter: *The cruise ships have very comfortable accommodation for their passengers.*

passing *adj.* **1** disappearing, vanishing, ephemeral, °brief, °going, fading (away), slipping away, °short-lived, expiring, °transient, transitory, °temporary, °momentary, °fleeting, transitional, impermanent: *I hope that Anne's obsession with acid rock is a passing fancy.* **2** °hasty, °superficial, °cursory, °casual, °quick, °fleeting, °brief, °summary, °abrupt, dismissive; glancing: *The speaker paid only passing attention to those he disagreed with.* **3** ***in passing.*** °by the way, °incidentally, by the by, parenthetically, *en passant*: *I might mention in passing that my train was late owing to the strike.* —*n.* **4** °death, °dying, demise, °end, °loss, expiry, °expiration, dying out, extinction, disappearance, vanishment: *Her passing is a great sorrow to all of us. Who thought we would live to see the passing of the steam engine?*

passion *n.* **1** Often, ***passions.*** °ardor, ardency, °eagerness, °intensity, °fervor, fervency, fervidness, zeal, zealousness, avidity, avidness, °zest, zestfulness, vivacity, vivaciousness, °gusto, °verve, °emotion, °feeling, °animation, °spirit, spiritedness, °vigor, °enthusiasm, °eagerness; °zealotry, °fanaticism, feverishness: *The passions of the mob are uncontrollable. Passion and prejudice govern the world. Passions ran high at the political convention.* **2** °fit, °outburst, °frenzy, °paroxysm, °seizure, °spasm, convulsion, °eruption, °whirlwind, °tempest, °storm, °ferment, °fury, furor *or Brit* furore: *Henry fell on his knees in a passion of grief.* **3** infatuation, °mania, °obsession, °craze, craving, °lust, °(unquenchable) thirst, °(insatiable) hunger, °itch, yearning, °longing, °desire, concupiscence, °love, °affection, °enthusiasm, compulsion, fondness, predilection, keenness, °fancy, °fascination, °partiality, °liking, °interest, °weakness, *Colloq* yen: *Would the world be a better place if the passion for spiritual values were as great as for material things? She has a passion for chocolates.* **4** °love, heart's desire, °beloved, °idol, hero *or* heroine, °obsession, *Colloq* heartthrob, dreamgirl *or* dreamboy: *This year the teenagers' passion is the rock* *group called The Miseries.* **5** Usually, ***Passion.*** °pain, °suffering, °agony, martyrdom: *Christ underwent his Passion at Calvary.*

passionate *adj.* **1** °ardent, °eager, °intense, fervid, zealous, avid, °eager, °earnest, zestful, °feverish, fanatic(al), vehement, °impassioned, °emotional, °animated, °spirited, °enthusiastic, °vigorous, invigorated, °energetic: *Passionate environmentalists campaign for better governmental control of carbon dioxide emissions.* **2** aroused, °lustful, °lecherous, °erotic, °sexual, amorous, °sensual, *Colloq* °hot: *She tried to resist his passionate advances.* **3** °quick-tempered, irascible, °hotheaded, °fiery, °testy, huffy *or* huffish, °peevish, °cranky, peppery, choleric, °touchy, blious, °snappish, °volatile, °cross, °temperamental, °irritable, °quarrelsome, °pugnacious, °argumentative, contentious, °belligerent, *Rare* atrabilious *or Rare* atrabiliar: *Hortense gets into a passionate mood whenever she thinks she is being treated unjustly.*

passive *adj.* **1** °inactive, nonaggressive, °inert, motionless, unresponsive, °quiet, °calm, °tranquil, °serene, placid, °still, °idle, unmoving, °unmoved, °impassive, untouched, °cool, °indifferent, °phlegmatic, uninterested, uninvolved, °dispassionate, apathetic, °lifeless, °listless, quiescent, unperturbed, °unaffected, imperturbable, unshaken, *Colloq* laid-back, out of it: *The comic was unable to get a reaction from the passive audience.* **2** °submissive, repressed, deferential, °yielding, compliant, complaisant, °receptive, °flexible, malleable, °pliable, °tractable, docile, °subdued, °sheepish, ovine, lamblike, cowlike, bovine, °tame, °gentle, °meek, °patient, unresisting, unassertive, forbearing, °tolerant, resigned, long-suffering, *Colloq US* excuse-me-for-living: *People tend to bully those of a more passive nature.* **3** unexpressed, °tacit, unrevealed, undisclosed, °implicit, unasserted: *We can no longer take a passive position regarding crime. Your passive support, while welcome, is not as effective as active campaigning.*

password *n.* watchword, °shibboleth, open sesame, countersign: *If you don't know the password, the guard won't let you in.*

past *adj.* **1** °over, done, finished, (over and) done with, gone (and forgotten), dead (and buried *or* gone), °defunct: *The day of the horse and carriage is past.* **2** °late, °former, one-time, °sometime, °previous, °prior, erstwhile, quondam, whilom; °last, °recent: *Past owners of the house had no idea that this treasure was buried in the basement.* —*adv.* **3** on, °(close) by, °nearby: *I was standing here when he ran past.* **4** ago, °before, heretofore, since: *The Cottles moved out two years past.* —*n.* **5** °history, °background, °life, lifetime, °existence, °career, lifestyle, biography: *She reveals all about her past in her new book.* **6** days *or* years *or* times gone by, days of yore, old times, olden times *or* days, former times, (good) old days, days of old, days beyond recall: *In the past, life proceeded at a more leisurely pace.*

pastiche *n.* °mixture, °medley, °blend, °compound, composite, °patchwork, olla podrida, °potpourri, motley, °miscellany, omnium-gatherum, mélange, gallimaufry, farrago, °mishmash, hotchpotch *or US also* hodgepodge, °tangle, *Colloq* °mess: *The room was furnished in a pastiche of styles from every imaginable period.*

pastime *n.* °hobby, avocation, °recreation, °diversion, °distraction, °amusement, °entertainment, °fun, °play, leisure-time activity, °relaxation, °leisure, °sport, divertissement: *As a pastime, she collects books.*

pastor *n.* vicar, °clergyman, clergywoman, parson, °minister, churchman, churchwoman, rector, canon, reverend, °father, °divine, ecclesiastic, °priest, bishop: *The pastor makes the church available for small civic functions.*

pastoral *adj.* **1** bucolic, °idyllic, Edenic, °innocent, °simple, °tranquil, °serene, °quiet, °restful, °peaceful, °peaceable, placid, pacific, °harmonious, °simple, uncomplicated, *Literary* Arcadian, georgic: *After a week in those pastoral surroundings, she felt inspired to write poetry in the style of Theocritus.* **2** °country, °rural, °rustic, °provincial, °farming, agricultural, agrarian; °humble: *He has chosen a pastoral setting for the play, which contrasts sharply with the emotional intensity of the characters.* **3** °clerical, ministerial, ecclesiastic(al), church(ly): *Our vicar discharges his pastoral duties with energy and sincerity.* —*n.* **4** idyll, eclogue, georgic: *Flambeau did these marvelous engravings for a book of pastorals.*

pasture *n.* °meadow, meadowland, pastureland, grassland, grass, lea, °range; pasturage: *How much pasture is needed for two hundred head of cattle?*

pasty *adj.* °wan, pallid, pasty-faced, sallow, °pale, pale-faced, wheyfaced, °sickly, anemic: *His complexion is a bit pasty, but don't forget—he's still recuperating.*

pat[1] *v.* **1** °tap, °touch, °dab, °pet, °stroke, °caress: *Being short, he hated it when people affectionately patted him on the head. Pat down the coverlet to smooth out the wrinkles.* **2** ***pat on the back.*** °congratulate, commend, °praise, °compliment, °encourage, °reassure: *The sales manager patted me on the back for clinching the contract.* —*n.* **3** °tap, °touch, °dab, °stroke, °caress: *Give the dog a pat and he'll leave you alone.* **4** °(small) piece, patty, °lump, °cake, °portion; patty: *The waiter put a pat of butter on my plate.* **5** ***pat on the back.*** commendation, °praise, °compliment, °flattery, °encouragement, °credit, reassurance, °approval, °endorsement, °recognition; honeyed words: *I was expecting a big bonus, but all I got was a pat on the back.*

pat[2] *adv.* **1** °perfectly, °exactly, °precisely, faultlessly, flawlessly, just so *or* right, *Brit* off pat: *She has the new technique pat.* **2** aptly, suitably, appositely, °readily, °appropriately, fittingly, relevantly: *He hit the nail pat on the head with that remark.* —*adj.* **3** apt, °suitable, apposite, °ready, °appropriate, °fitting, °relevant: *The poet made a very pat comparison.*

patch *n.* **1** °piece, °scrap, °reinforcement; °shred, °snip, snippet, °tatter; °pad: *Mother sewed colorful patches over the holes in my jeans.* **2** °area, °section, °segment, plat, °plot, °lot, °tract, °ground, °parcel, °field: *All he ever wanted was a little house on a little patch of land.* **3** °responsibility, °area, bailiwick, °territory; °(line of) country, line of work: *The handling of customer complaints is in your patch, Gordon.* **4** °period, °interval, °spell, °stage, °episode, °time; °experience: *She went through a bad patch shortly after Starkey died.* —*v.* **5** patch up *or* over, °mend, °repair, vamp, °revamp, darn, °sew (up), °reinforce, °cover: *I had my trousers patched where the dog bit me. Bellshaw always wore the same patched jacket to the club.* **6** Often, ***patch up.*** °fix (up), °doctor, jury-rig, °improvise, knock together *or* up: *Can you patch it up so I can drive it home?* **7** ***patch up.*** °settle, set right *or* straight, °straighten out, °reconcile, °resolve, °heal; come *or* bring to terms, bury the hatchet, kiss and make up, call a truce: *Will the two girls ever patch up their differences?*

patchwork *n.* pastiche *or* pasticcio, °mixture, °confusion, hotchpotch *or US also* hodgepodge, gallimaufry, olio, olla podrida, °mishmash, °jumble, mélange, °medley, °hash, *US* crazy quilt, *Colloq* mixed bag: *The concerto by Lindslade is "derivative," which is a nice way of saying a patchwork of others' work.*

patent *n.* **1** certificate of invention, letters patent, trade name, trademark, copyright, *US* service mark;

°license, °permit, °charter, franchise, °grant; °control: *Oliver Tremble holds the patent on a new kind of comb. You think you have a patent on misery, but I have news for you.* —*adj.* **2** °obvious, °clear, °transparent, °manifest, °apparent, °plain, °evident, °self-evident, unmistakeable *or* unmistakable, unequivocal, °explicit, palpable, °tangible, °physical, °conspicuous, °flagrant, °blatant, °prominent: *The results are as patent as the fact that two and two make four.*

paternal *adj.* **1** °fatherly, °kindly, °indulgent, °solicitous, °fond, °concerned, °devoted, loving; patriarchal: *Uncle Charles takes a paternal interest in the welfare of his nieces and nephews.* **2** patrilineal *or* patrilinear, patriclinous *or* patroclinous *or* patriclinal *or* patroclinal *or* patriclinic *or* patroclinic, patrilateral, patrimonial: *She resembles her paternal grandmother. His fortune is partly paternal, partly acquired.*

paternity *n.* fatherhood, fathership; °parentage, descent, °heritage, °line, °lineage, °extraction, °family, °stock, °strain, blood, patrilineage: *The child's paternity was established through DNA tests.*

path *n.* **1** footpath, pathway, towpath, °track, °trail, °walk, walkway, *Brit* footway: *A little kitten was sitting on the garden path.* **2** °way, °course, °track, °route, °road; °orbit, °trajectory, °circuit: *She had to overcome many obstacles in her path to the directorship. The path of the missile will take it outside the atmosphere.* **3** °course, °approach, °channel, °direction, °procedure, °process, °way, avenue, °means, °method, °technique, °strategy, °scheme, °plan, *Colloq US* game plan, °scenario: *What path would you follow to accomplish your ends?*

pathetic *adj.* **1** °moving, °stirring, affecting, affective, °touching, °emotional, emotive, °poignant, °tragic, °heart-rending, heartbreaking, °pitiful, °pitiable, °piteous, plaintive, plangent, °wretched, °miserable, °sorrowful, °grievous, °sad, °doleful, dolorous, °mournful, woeful, °lamentable: *The boat people told a pathetic tale of the hardship of weeks in the open sea.* **2** °meager, °paltry, °feeble, °inadequate, °poor, °petty, °puny, °sorry, piddling, *Colloq* °measly, *Slang* crummy: *He made a pathetic effort to claw his way back to safety. A return of 3 percent a year is pathetic.*

patience *n.* **1** °tolerance, forbearance, °restraint, toleration, sufferance, leniency, °submission, °resignation, °self-control, imperturbability, even temper, unflappability, composure, calmness, °serenity, equanimity: *Fräulein Rottenmeyer lost her patience when dealing with children.* **2** diligence, °tenacity, doggedness, indefatigability, °endurance, assiduity, °perseverance, constancy, °persistence, steadfastness, pertinacity, °determination, °resolve, °resolution, firmness, °stoicism, °fortitude, *Colloq US* stick-to-itiveness: *Where do you get the patience to wait in those long queues?*

patient *adj.* **1** resigned, °submissive, °stoical, long-suffering, compliant, acquiescent, °passive, °self-possessed, °philosophical, °serene, unaggressive: *The staff who handle complaints must be extremely patient.* **2** °diligent, dogged, °tenacious, °persistent, assiduous, sedulous, °steadfast, °staunch, perseverant, unwavering, unswerving, °constant, unfaltering, unfailing, °untiring, °tireless, indefatigable, pertinacious, °determined, resolved, °resolute, °firm, unyielding: *Be patient, and don't do anything rash.* **3** forbearing, °tolerant, °forgiving, °lenient, °accommodating: *We have been patient long enough and must now put a stop to the vandalism.* —*n.* **4** °invalid, sufferer, °case, valetudinarian: *Doctors were called in to treat patients who developed the symptoms.*

patriot *n.* nationalist, loyalist; flag waver, jingo, jingoist, chauvinist: *She was among the patriots ready to do battle for their country.*

patriotic *adj.* nationalist(ic), loyalist; flag-waving, jingoist(ic), chauvinist(ic): *The fact that they criticize the government may show that they are more rather than less patriotic.*

patrol *n.* **1** °guard, sentry, °watch, °watchman, °sentinel, patrolman: *The patrol passes here once every hour.* **2** °rounds, policing, patrolling, °beat; protecting, °protection, guarding, safeguarding, defending, watchfulness, °vigilance: *The patrol must be maintained night and day.* —*v.* **3** °police, °guard, °protect, °defend, watch over, walk a beat, make (the) rounds, stand *or* keep guard *or* watch (over), keep vigil: *Some of the local residents have taken to patrolling the neighborhood at night.*

patron *n.* **1** patroness, °benefactor, benefactress, °philanthropist, Maecenas, °protector, °supporter, defender, °advocate, °champion, °guardian (angel), °sponsor, °backer, promoter, °sympathizer, °friend, *US* booster; friend at court; *Colloq* angel: *Lady Agnes is a well-known patron of the arts.* **2** °customer, °client, purchaser, °buyer, patronizer, °habitué, °regular, frequenter: *The patrons return to her shop because of the bargains.*

patronage *n.* **1** sponsorship, °support, °backing, °promotion, °encouragement, boosting, °aid, °help, °sympathy, financing, °auspices, °protection, guardianship, aegis: *The exhibition was organized with the duke's patronage.* **2** °trade, °business, °custom, trading, °traffic: *The shop could never succeed with our patronage.* **3** condescension, disdain, °scorn, °contempt, contumely, °superiority, patronizing, stooping, deigning, °humiliation: *There is an air of patronage about him when he finally does allow me a visit.* **4** °favoritism, °partiality, °preference, °bias, nepotism, political patronage, granting of indulgences, *US* °spoils (system): *Politicians have always used patronage as repayment for support.*

patronize *v.* **1** °look down on, °scorn, look down one's nose at, treat condescendingly, talk down to, treat as (an) inferior, disdain, demean, °put down, °humiliate, *Formal* contemn: *He patronizes people by patting them on the head.* **2** bring trade to, deal *or* trade *or* do *or* transact business with, buy *or* purchase from, °frequent, shop at, be a customer *or* client of: *You should patronize the local merchants in your town.* **3** °sponsor, °support, °back, °promote, °encourage, °boost, °aid, °assist, °help, °fund, contribute *or* subscribe to, °underwrite, °foster: *The fund has patronized the arts in this town for many years.*

patter[1] *v.* **1** tiptoe; °scurry, scuttle, °skip, °trip: *I could hear the squirrels pattering across the metal roof.* **2** °spatter, pitter-patter, °tap, pitapat; °beat, °pelt: *The rain pattered on the tent with increasing intensity.* —*n.* **3** °spatter, spattering, pitapat, pitter-patter, tattoo, drum, thrum, °beat, beating, °tap, rat-a-tat, tap-tap-tap: *The rhythmical patter of the rain lulled me to sleep.*

patter[2] *n.* **1** °pitch, sales talk, spiel, °line: *Anyone listening to his patter might be inclined to buy a time-share vacation home.* **2** °chatter, °prattle, prate, babbling, °babble, gabble, gabbling, cackle, cackling, °palaver, °jabber, jabbering, chitchat, small talk, °gossip, blather *or* blether, gibberish *or* gibber, *Chiefly Brit* natter, nattering, *Scots* clishmaclaver, *Colloq* °gab, gabbing, *Slang* gas, °hot air, yackety-yak, yak: *The poor man tapping our phone had to listen to the teenagers' patter for hours on end.* —*v.* **3** °chatter, °prattle, prate, °babble, gabble, cackle, °palaver, °jabber, °rattle (on), chitchat, chaffer, °gossip, blather *or* blether, gibber,

Chiefly Brit natter, witter (on), *Colloq* °gab, *Slang* gas, yackety-yak, yak, jibber-jabber: *They just patter on about nothing in particular.*

pattern *n.* **1** °model, °original, archetype, °prototype, exemplar, °paragon, °ideal, °standard, °yardstick, criterion, °gauge, °measure: *Their life together could serve as a pattern for any couple.* **2** °figure, °motif, °design, °device, °decoration, °ornament: *The pattern of the wallpaper clashes with that of the curtains.* **3** °system, °order, °arrangement, °plan, °theme; °repetition, consistency, orderliness, °regularity, °sequence, °cycle: *By carefully noting the movements of the heavenly bodies, ancient observers were able to detect a pattern.* **4** blueprint, diagram, °plan, layout, °design, °draft, °guide, template *or* templet, stencil, °mold, matrix: *These new lighting fixtures are made to a nineteenth-century pattern.* **5** °sample, °example, °instance, °specimen, °representation: *Were you able to match up your fabric pattern with one of those in the tailor's book?* **6** layout, configuration, °figure, °formation, °composition: *The geese often flew in a V-shaped pattern.* —*v.* **7** Often, ***pattern on.*** °model on, °imitate, °copy, °mimic, °duplicate, °follow, emulate, simulate: *She tried to pattern her behavior on that of her older sister.* **8** °decorate, °figure, °ornament: *Cynthia's gifts were wrapped in a beautifully patterned paper.*

paunch *n.* belly, potbelly, *Colloq* corporation, *US* bay window, *Slang* beer belly: *Capitalism is often conventionalized as a top-hatted older man in tails with a huge paunch and a cigar.*

pauper *n.* have-not, °indigent, down-and-out(er), bankrupt, °insolvent; °beggar, mendicant; °tramp, hobo, vagrant, *US* °bum: *The mortgage payments on such a lavish house soon made him a pauper.*

pause *v.* **1** °hesitate, °interrupt, °delay, °hold up, °discontinue, °break, °wait, mark time, °suspend, intermit, falter, °rest: *He paused for a moment to allow latecomers to take their seats.* —*n.* **2** hesitation, °interruption, °delay, °lull, °lapse, °moratorium, °holdup, °wait, °break, breather, breathing space, discontinuity, lacuna, hiatus, °abeyance, discontinuation, discontinuance, *Prosody* caesura, *Music* fermata, *Colloq* °letup: *There was a brief pause, then the altercation in the apartment next door resumed.*

pave *v.* **1** macadamize, tarmac, asphalt, tile, °flag, °concrete, °cover, °surface: *Europeans thought that the streets of America were paved with gold.* **2** ***pave the way for*** *or* ***to.*** prepare *or* smooth the way for, open the door for, make easy *or* easier for; °facilitate, °ease: *If I talk to him first, perhaps I can pave the way for you to make your proposal.*

pawn[1] *v.* **1** °pledge, mortgage, hypothecate, °plight, °deposit, *Formal* pignorate, *Archaic* gage, *Colloq Brit* °pop, *Chiefly US and Canadian* hock: *I had to pawn my watch to get enough money to eat.* **2** °venture, °stake, °risk, °gamble, °hazard, °chance, °jeopardize: *By agreeing to donate one of his kidneys, he pawned his life to save hers.* —*n.* **3** collateral, guaranty *or* guarantee, °pledge, surety, °security, °assurance, °bond, bail, °deposit: *Cornelia gave her jewels as pawn for the safe return of the children.*

pawn[2] *n.* °tool, cat's-paw, °puppet, °instrument, °dummy, °dupe, *Colloq* stooge: *He's not important: he's just being used as a pawn by the powerful interests involved.*

pay *v.* **1** recompense, °compensate, remunerate, °reward, indemnify; °repay, refund, °reimburse; °pay off, °pay out, pay up, °satisfy, °clear, °remit, °discharge, liquidate, °settle, °honor, °meet: *Sue insists that she is poorly paid as a lexicographer. Please pay me for the dinner expenses I advanced last night. You have thirty days to pay this invoice.* **2** °reward, °benefit, recompense, °requite, °compensate: *Thurlew has been amply paid by having his name engraved on the roll of honor.* **3** °extend, °bestow, °transmit, °pass on, °give, °deliver: *Please pay my respects to Lady Crouch.* **4** °benefit, °profit, avail, (turn out to) be *or* prove profitable *or* worthwhile, yield a return, be advantageous, produce results, °pay off: *It no longer pays to complain about the service in the shops. Honesty pays.* **5** Often, ***pay back.*** °repay, °retaliate, settle (accounts) (with), even the score *or* a score (with), °reciprocate, °requite, take *or* get revenge on, avenge oneself for *or* on, treat in kind, hit *or* strike *or* get back (at), settle *or* pay off (with), exact one's pound of flesh (from), make (someone) pay (for), °punish, °chastise, °castigate, *Brit* °pay out, *Colloq* get even (with): *His classmates threatened to pay him back for turning informer.* **6** °suffer (the consequences), °answer (for), °make amends (for), °atone (for), get one's (just) deserts, undergo punishment (for), be punished (for): *If I catch him, I'll make him pay. Society requires that he pay for his crime.* **7** produce *or* make *or* generate *or* earn money, yield a return, °pay off: *His vending machine business certainly pays well.* **8** ***pay back.*** **(a)** recompense, °compensate, remunerate, °reward, indemnify; °repay, °pay off, refund, °reimburse: *I paid back every penny I borrowed from the bank.* **(b)** See **5,** above. **9** ***pay for.*** See **6,** above. **10** ***pay off.*** **(a)** See **4** *and* **7,** *and* **8 (a),** above. **(b)** °bribe, suborn, buy off, grease (someone's) palm, give (someone) a bribe *or* a rebate, *Colloq* give (someone) a kickback, slip (someone) something: *Did he really pay off the judges to select his daughter Miss Tyneside?* **11** ***pay out.*** **(a)** °distribute, deal out, °give out, °disperse, disburse: *Mr. Kritchey paid out bonuses to the entire staff.* **(b)** disburse, °expend, °spend, °contribute, *Colloq* °shell out, °lay out, *US and Canadian and Australian and New Zealand,* kick in with, *Slang* cough up, fork out *or* over *or* up: *In the office, everyone had to pay out something today for Kitty's wedding present.* **(c)** °release, °loosen, °let out, slack *or* slacken off (on): *Pay out more rope so that he can reach it.* **(d)** See **5,** above: *We'll pay him out for everything he's done to you, never fear!* —*n.* **12** °payment, compensation, recompense, °settlement, °return; °remuneration, °consideration, °reward, °money, °wages, °salary, °fee, °honorarium, °remittance, °stipend, °income, takings, take-home (pay), °gain, °profit, *Colloq US* °take: *His sole pay for painting the kitchen was a big hug and a kiss. The work is tedious, but the pay isn't bad.*

payable *adj.* °due, owed, owing, °outstanding, °unpaid, receivable, °mature: *This bill is payable at the end of the month.*

payment *n.* **1** See **pay, 12,** above. **2** °expenditure, disbursement, °distribution, °outlay, °fee, contribution, °charge, °expense, payout: *Payment of any amount over one hundred dollars is not authorized.*

payoff *n.* **1** See **pay, 12,** above. **2** °result, °outcome, °upshot, conclusion, windup, °settlement, final reckoning, *Colloq* punch line, °crunch, grand finale, *Slang US and Canadian* kicker: *The payoff was that permission to march on Sunday was denied.* **3** °bribe, °graft, °rebate; °ransom, blood money, *Colloq* °kickback, hush money, *Chiefly US* payola, *US* plugola: *The payoff was always made in cash, in used notes.*

peace *n.* **1** °serenity, tranquillity, °calm, calmness, placidity *or* placidness, peace of mind, °quiet, peacefulness, peaceableness, stillness: *For a summer holiday, we enjoy the peace of the mountains.* **2** °harmony, °accord, harmoniousness, concord, amity, peacefulness, peacetime; cease-fire, armistice, °truce: *After the war, Europe enjoyed more than forty-five years of relative peace.*

peaceable *adj.* 1 See **peaceful, 1,** below. 2 pacific, °inoffensive, dovish, peace-loving, °mild, nonviolent, nonbelligerent, unbelligerent, unwarlike, nonwarring, noncombative, °temperate, °agreeable, compatible, congenial, °genial, °friendly, °amiable, °amicable, °cordial, °civil: *Despite cultural differences, the two countries maintained peaceable relations.*

peaceful *adj.* 1 °peaceable, °serene, placid, °calm, °quiet, quiescent, °gentle, °restful, °tranquil, untroubled, undisturbed, unruffled: *After our weekend guests left, the house was again peaceful.* 2 See **peaceable, 2,** above.

peacemaker *n.* conciliator, pacifier, reconciler, propitiator, placater, pacificator, °mediator, arbitrator, intermediator, °intermediary, diplomat, appeaser, interceder, °go-between, referee, °umpire, adjudicator; peacemonger: *The ambassador was called upon to act as peacemaker between the warring nations.*

peak *n.* 1 °top, °pinnacle, °crest, °ridge, tor, mountaintop, °summit, °mountain, eminence, °elevation, °hill: *We were just able to make out the snow-capped peaks in the distance.* 2 °top, °tip, tiptop, apex, °acme, culmination, apogee, °zenith, high point, °crown, °extreme, utmost, uttermost, °perfection, *ne plus ultra,* °consummation, °climax: *Irene has brought the office up to the peak of efficiency.* 3 visor, °brim, *US* °bill, nib: *The peak of the fisherman's cap serves as an eyeshade.* —*v.* 4 °rise, °crest, culminate, °(reach a) climax, °top (out): *Prices peaked during the Christmas shopping season.*

peaked *adj.* pinched, °unhealthy, °sickly, ailing, °ill, unwell, °infirm, °unwholesome, °pale, pallid, °wan, waxen, anemic, °pasty, sallow, wheyfaced, ashen, °washed-out, drained, °emaciated, wasted, °gaunt, hollow-eyed, °haggard, °drawn, °weak, °feeble, *Brit* peaky *or Brit* peakish: *Rudolph has been looking a bit peaked of late.*

peal *n.* 1 ringing, °ring, carillon, °chime, chiming, °toll, tolling, clang, clangor, tintinnabulation, clamor, reverberation; knell; °clap, °crash, °roar, rumble, °thunder: *The constant peal of the bells nearly drove him mad. The doleful peal rang out for yet another fisherman lost at sea. A resounding peal of thunder shook the house.* —*v.* 2 °ring, °toll, °chime, clang, tintinnabulate, reverberate, resonate, °resound; knell; °boom, °crash, °roar, °roll, rumble, °thunder: *The cowbells pealed plaintively, reminding me of my youth in the Alps. Lightning flashed, and the echoing thunder pealed through the valley.*

pearl *n.* °gem, °treasure, °prize, cream, °flower, °wonder, °nonpareil: *If I am not mistaken, Cuba was once called the Pearl of the Antilles.*

pearly *adj.* nacreous, pearllike, perlaceous, °lustrous, mother-of-pearl: *Separating the rooms was a Coromandel screen elaborately set with pearly designs.*

peasant *n.* °rustic, countryman, countrywoman, °farmer, °provincial, °(farm) worker, (country) bumpkin, bucolic; peon, fellah, muzhik *or* mouzhik *or* mujik; *Historical* esne, serf, *Archaic* swain, hind, churl, *Derogatory* yokel, hillbilly, bogtrotter, oaf, °lump, lout, °boor, churl, °clod, clodhopper; *Colloq US and Canadian* hick, galoot *or* galloot, *Derogatory* hayseed, rube, *Derogatory and offensive US* poor white (trash): *Regimes came and went, but the life of the medieval peasant endured. Always the snob, Rupert's worst term of abuse is "Peasant!"*

peccadillo *n.* °slip, °error, °lapse, °mistake, infraction, °violation, °misdeed, °shortcoming, °misstep, °blunder, faux pas, °indiscretion, gaffe, °botch, °stumble, °fault, petty sin, °(minor) transgression, trespass, *Colloq* slip-up, goof: *We have to forgive the occasional peccadillo.*

peculiar *adj.* 1 °odd, °curious, °strange, °queer, °bizarre, °weird, °unusual, °abnormal, anomalous, aberrant, deviant *or* deviate, °eccentric, uncommon, °outlandish, °exceptional, °extraordinary, out-of-the-ordinary, °offbeat, °unorthodox, atypical, idiosyncratic, unconventional, °out-of-the-way, °quaint, °unique, °singular, one-of-a-kind, sui generis, °distinct, °distinguished, °special, °particular, quirky, °funny, freakish, *Slang* far-out, freaky, *Brit* rum: *Don't you agree that anyone who continuously hums* The Marseillaise *is a bit peculiar? She enjoys a peculiar immunity to snakebite.* 2 Usually, ***peculiar to.*** typical of, characteristic of, characterized by, natural to, symptomatic of, appropriate to *or* for, distinctive of, restricted to, °specific to, °indicative of, denotative of, limited to, individual to, personal to, special to, unique to; seen *or* observed (only) in, local to, native to, indigenous to: *Such behavior is peculiar to those who have served long prison sentences.* —*n.* 3 *Typography* °arbitrary, °sort: *Characters with diacritical marks are called "peculiars" in the printing trades.*

peculiarity *n.* 1 idiosyncrasy, °oddity, °eccentricity, °abnormality, irregularity, °quirk, °kink, crotchet, caprice: *A peculiarity of Jess's was his predilection for eating chocolate-covered bees.* 2 °feature, °characteristic, °property, °quality, °trait, °attribute, earmark, °hallmark, °mark, particularity, °singularity, °specialty: *One of the peculiarities of the books is their edge printing.*

pedantic *adj.* 1 didactic, doctrinaire, donnish, pedagogic, pedantical, preachy, professorial, bookish, °ostentatious, °pretentious, sententious, °pompous, °vain, °stuffy, °stilted, °stiff, °dry: *A pedantic approach to marketing may be all right in the classroom, but it doesn't sell products.* 2 °perfectionist, °scrupulous, overscrupulous, finicky *or* finical, °fussy, punctilious, °fastidious, °meticulous, °exact, choplogic, hair-splitting, quibbling, *Colloq* nit-picking: *Carew is so pedantic that when he buys aspirin, he counts the tablets.*

peddle *v.* °sell, hawk, °market, vend, huckster, *Colloq* °push, °flog: *He is a publishing tycoon today, but he started by peddling books door-to-door.*

peddler *n.* hawker, (door-to-door) salesman *or* saleswoman *or* salesperson, vendor, huckster, °seller, colporteur, *US* drummer, *Chiefly Brit* pedlar, *Archaic* chapman, *Colloq* cheap-jack: *How can you expect the watch to work if you bought it from a street peddler?*

pedestal *n.* 1 °foundation, °base, °platform, °stand, substructure, mounting, °pier, foot, mounting, °support, *Technical* plinth, socle, dado: *The statue crumbled away and only its pedestal remained.* 2 ***put*** *or* ***place*** *or* ***set on a pedestal.*** °glorify, °exalt, °worship, deify, °revere, °idolize, °dignify, apotheosize, ennoble, elevate, °raise: *What does she see in him that merits putting him on a pedestal?*

pedestrian *n.* 1 walker, stroller, ambler, rambler, footslogger; itinerant, peripatetic: *A special crossing has been installed for pedestrians.* —*adj.* 2 °boring, °dull, °banal, °tiresome, commonplace, mundane, °tedious, unimaginative, uninteresting, °monotonous, °run-of-the-mill, °humdrum, °stock, °prosaic, insipid, °dry, °flat, jejune, °colorless, °dreary, °pale, °ordinary, hackneyed, trite, (as) dull as ditchwater *or US also* dishwater, °vapid, °stale, uninspired, uninspiring, spiritless, °lifeless, °dead: *The dean delivered his customary pedestrian lecture to the new students.* 3 walking, strolling, ambulatory, on foot, °rambling, peripatetic: *We were exhausted after taking a pedestrian tour of London.*

pedigree *n.* (line of) descent, ancestry, genealogy, blood, bloodline, °line, °extraction, °lineage, °stock, °heritage, °family, °derivation, °birth, °parentage, °strain, °roots: *What makes you think that you'll find your pedigree in* Burke's Peerage?

peek *v.* **1** °peer, °peep, glimpse, °look, squint (at), squinny (at), *Scots* keek, *Colloq* take *or* have a gander (at), *Brit* take a dekko (at): *A small boy peeked out from under the blanket.* —*n.* **2** °look, glimpse, °peep, °glance, *Scots* keek, *Colloq* gander, look-see: *I've had a peek at what Father hid in the cupboard.*

peel *v.* **1** Sometimes, ***peel off.*** °skin, °strip (off), °pare, flay, °flake off, descale, decorticate; shuck, °hull, bark, °scale; desquamate: *Peel me a grape, Beulah.* **2** °strip, undress, °disrobe; do a striptease: *In the last act, everyone peels to the bare skin.* **3** ***peel off.*** take off *or* doff, strip off: *I peeled off my coat and dived into the canal to save her.* —*n.* **4** °skin, °rind, coating, peeling: *Don't you like candied orange peel?*

peep *v.* **1** °chirp, tweet, cheep, squeak, °twitter, °pipe, chirrup: *The birds peeped excitedly when they saw the cat.* —*n.* **2** °chirp, tweet, cheep, squeak, °twitter, °pipe, chirrup, chirr *or* chirre *or* churr: *The night was filled with the peep of the frogs at the pond.* **3** °sound, °complaint, °outcry, °protest, protestation, grumble, °murmur: *The students didn't let out a peep when ordered to remain after class.*

peer[1] *n.* **1** °noble, nobleman *or* noblewoman, lord *or* lady, aristocrat; duke *or* duchess, marquess *or* marchioness, earl *or* countess, viscount *or* viscountess, baron *or* baroness: *As a peer of the realm, she had certain rights.* **2** °equal, coequal, compeer, °like, °match, confrere, °associate, °colleague: *Under law, he is entitled to trial by a jury of his peers.*

peer[2] *v.* **1** °peep, °peek, squint (at), squinny (at), °look, °examine; °spy: *Scrooge peered closely at the accounts ledger.* **2** °appear, peep through *or* out, °break through, °show, become visible, °emerge: *Now and then, the moon peered through the clouds.*

peerless *adj.* without equal, unequaled, °matchless, unmatched, unrivaled, °unique, °incomparable, beyond compare, °unparalleled, °nonpareil, inimitable, unexcelled, unsurpassed, °superior, °superb, °excellent, °supreme, °superlative, finest, °best, *ne plus ultra,* °sovereign, consummate, °preeminent, °paramount: *Pontefract's mother was a gifted, peerless beauty.*

peevish *adj.* °irritable, °testy, °touchy, °fretful, ill-humored, °waspish, °petulant, crabbed, churlish, °querulous, °short-tempered, ill-natured, tetchy, °cross, bad-tempered, ill-tempered, °faultfinding, captious, carping, caviling, crusty, curmudgeonly, crotchety, °cantankerous, grumpy *or* grumpish, pettish, acrimonious, splenetic, *Colloq* °bilious, *US and Canadian and Irish* °cranky: *He's very peevish today, so don't get on his wrong side.*

peg *n.* **1** °pin, dowel, °rod, °stick, °bolt; thole *or* tholepin; clothes peg, °hook: *Pegs and glue are often used for fastening good furniture together. Hang your coat on the peg.* **2** ***take down a peg (or two).*** °humble, °diminish, °lower, °subdue, °suppress, °downgrade, °dishonor, °mortify, °humiliate, °put down, abase, °debase, devalue *or* devaluate: *He was acting a bit high and mighty, so she took him down a peg or two.* —*v.* **3** °fasten, °secure, make fast, °fix, °attach, °pin: *They pegged the tent firmly to the ground.* **4** °fix, °attach, °pin, °set, °control, °limit, °restrict, confine, °freeze, °bind, °regulate, °govern: *In indexing, the rate of inflation is often used for pegging wages.* **5** °toss, °throw, °shy, °flip, °sling, °cast: *I'll bet a fiver you can't peg that stone across the river.* **6** ***peg away*** *or US also* ***along.*** work (away) (at), °persevere (at), apply oneself (to), persist (in *or* at), go to *or* at (it), °keep at (it), stick to *or* with *or* at (it), stay with *or* at (it), carry on (with *or* at), *Colloq* plug away (at), beaver away (at), hammer *or* bang *or* peck away (at): *Trevor pegs away at his homework every evening.*

pell-mell *adv.* **1** °helter-skelter, slapdash, rashly, feverishly, incautiously, confusedly, chaotically, wildly, impulsively, recklessly, slapbang, impetuously, °hastily, hurriedly, precipitately, spontaneously: *The ice-cream vendor pedaled down the street with the children running pell-mell after him.* —*adj.* **2** °helter-skelter, slapdash, °rash, °feverish, incautious, °confused, disordered, °disorderly, disorganized, °wild, °mad, °chaotic, °tumultuous, panicky, °impulsive, °reckless, °precipitate, °impetuous, °hasty, °hurried: *Quick action by the soldiers was responsible for the pell-mell rout of the attacking Rifs.* —*n.* **3** °confusion, °disorder, °chaos, °tumult, °pandemonium, turmoil, melee *or* mêlée, °furor, commotion, °bedlam, brouhaha, hubbub, °excitement: *In the pell-mell that followed, several bystanders were injured.*

pelt[1] *v.* **1** °bombard, °shower, °bomb, °pepper, strafe, °batter, °shell, assail, °assault, °attack, *US* pummel *or* pommel, °belabor, °pound, *Old-fashioned* lay about, *Slang US* clobber, wallop, paste, work over: *The hooligans pelted the crowd with sticks and stones.* **2** Often, ***pelt down.*** °beat, °dash, °pound, °hit; come down, °teem, °pour, *Colloq* rain cats and dogs, bucket down, *US* rain pitchforks: *The rain came pelting down just as we were ready to go out.* **3** ***pelt along*** *or* ***over.*** °hurry, °rush, °run, °dash, °shoot, scoot, °scurry: *Constable Clarendon pelted along after the escaping thief.* —*n.* **4** °stroke, °blow, whack, °hit, smack, °slap, thwack, bang, thump, *Colloq* wallop, °belt: *Chris suffered a solid pelt on the head from a rock, making the blood flow.*

pelt[2] *n.* °skin, °hide, °coat, fur, °fleece: *The trappers traded the fox pelts for supplies.*

pen[1] *n.* **1** writing instrument, fountain pen, ballpoint (pen), *Brit Trademark* Biro, *Old-fashioned* quill: *Dearest, I have finally found time to put pen to paper. This week saw the publication of still another book from the pen of Adam Isakoff. I prefer a pen with a fine point.* —*v.* **2** write (down *or* up *or* out), °jot down, °(make a) note (of), °draft, °draw up, °compose, put on paper, commit to paper, commit to writing, put in writing, scribble, °scrawl, °scratch, *Formal* indite: *While he was in prison, he penned letters to many men of influence.*

pen[2] *n.* **1** coop, °enclosure, hutch, (pig)sty, °pound, °fold, °stall, confine, *US and Canadian* corral: *We kept the geese in a pen by the barn.* —*v.* **2** Often, ***pen up.*** °enclose, confine, coop up, °shut up, impound, °round up, *US and Canadian* corral: *It took three of us to pen the sheep. During the blockade, the ships remained penned up at Gibraltar.*

penal *adj.* correctional, °punitive, disciplinary: *Crime and its punishment are covered in the penal code. He was sent to a penal colony for life.*

penalize *v.* °punish, °discipline, mulct, amerce, °fine, °handicap, impose *or* invoke a penalty against, impose a penalty on, *Formal* amerce; °sentence: *This judge penalizes speeders mercilessly.*

penalty *n.* °punishment, °discipline, °penance, °sentence; °forfeit, °fine, °handicap, °price, mulct, *Formal* amercement; °imprisonment, incarceration: *The prosecution demanded the maximum penalty allowable under law for the crimes.*

penance *n.* **1** °punishment, °penalty, reparation, °amends, °atonement, self-punishment, self-mortification, °regret, repentance, contrition, °suffering, °penitence: *A year of public service was fair penance for the offense.* **2** ***do penance.*** °pay, °suffer, make amends *or* reparation(s), °atone, wear sackcloth and ashes *or* a hair shirt: *She has done penance enough for her crime.*

penchant *n.* °inclination, °bent, proclivity, °leaning, °bias, predisposition, predilection, °partiality, proneness, propensity, °tendency, °affinity, °liking, °preference, fondness, °taste: *Agnes has a penchant for tennis and for men who play tennis.*

pendant *n.* °ornament, tassel, lavaliere, medallion, locket, necklace, rivière, eardrop, teardrop, °drop, *Old-fashioned* carcanet: *I recognized the pendant she wore at her neck as my mother's.*

pending *prep.* **1** awaiting, waiting (for), depending on, °till, until, 'til, till such time as; while, during: *Pending the outcome of the trial, he was remanded in custody.* —*adj.* **2** °unsettled, undetermined, undecided, unconfirmed, unfinished, °inconclusive, up in the air, hanging fire, in the balance, °in abeyance; °forthcoming, °imminent, °impending, in the offing, *Colloq US* in a holding pattern, on hold: *The pending negotiations on the rent will determine whether we stay or move house. While the matter is pending, we can do nothing.*

pendulous *adj.* **1** pendent, hanging, drooping, sagging, dangling, suspended, pensile: *The weaverbird's pendulous nest is a marvel of engineering.* **2** °swinging, swaying, waving, undulating, undulatory, oscillating, oscillatory: *I was hypnotized by the pendulous motion of the clapper of the ship's bell.*

penetrate *v.* **1** °enter, go *or* pass through *or* into, °pierce, °bore (into), °lance, spear, go through *or* into, °probe, °stab, °puncture, °perforate, °drill: *The shell penetrated the tank's heavy armor.* **2** °permeate, °diffuse, °suffuse, pervade, filter *or* seep through, percolate through: *The soothing balm penetrated my aching body and I relaxed into a deep sleep.* **3** °reach, °get to, °get at, °touch, °affect, °hit, °strike: *Her cruel words penetrated the darkest recesses of my soul.* **4** °sink in, be absorbed, be understood, °register, come *or* get through, become clear, °come across, be realized, *Colloq* soak in, seep in: *It took a while for it to penetrate that she did not wish to see him again.* **5** °understand, °sense, become aware *or* conscious of, °see (through), gain insight (in)to, discern, uncover, °discover, °find (out), °comprehend, °grasp, °work out, unravel, °fathom, °perceive, *Colloq* °get, °figure out, °dig, *Brit* suss out: *Will we ever penetrate all of nature's secrets?*

penetrating *adj.* **1** °incisive, °trenchant, °keen, searching, °deep, °acute, °sharp, °perceptive, perspicuous, percipient, °quick, °discriminating, °intelligent, °sensitive, °clever, °smart, discerning: *Sherman has done a penetrating analysis of the situation in Rarotonga.* **2** °piercing, °shrill, °strident, earsplitting, earshattering, °pervasive; °pungent, °harsh, °biting, mordant, °strong, stinging: *I was awakened by a penetrating scream. The penetrating odor of ammonia assailed my nostrils.*

penetration *n.* **1** °piercing, perforation, puncturing, °incision, °puncture, °penetrating; °inroad, °entry, °entrance: *Penetration of the skull required great effort. Tentatively we attempted the penetration of the jungle.* **2** °insight, keenness, °perception, percipience, °intelligence, perspicacity, perspicuity, perspicaciousness, perceptiveness, acuteness, incisiveness, °sensitivity, sentience, °understanding, acuteness, discernment, °discrimination, cleverness, shrewdness, °wit, quick-wittedness: *We all admired the penetration shown by the examiner in the questions she asked.*

penitence *n.* °penance, contrition, °regret, repentance, regretfulness, °compunction, °remorse, °sorrow, sorrowfulness, ruefulness, °grief, °sadness, °shame, self-reproach: *Brendan's penitence for his wrong-doing took the form of guiding young people in the ways of the Lord.*

penitent *adj.* contrite, °regretful, °repentant, °remorseful, °sorrowful, °sorry, rueful, grief-stricken, °sad, °shamefaced, self-reproachful, °apologetic, conscience-stricken: *He regarded himself as a penitent prodigal who had squandered God's gifts.*

penmanship *n.* calligraphy, °hand, fine Italian *or* Italic hand, handwriting, °script, °writing, longhand, chirography: *One must admire the penmanship of the medieval scribes.*

pennant *n.* °flag, °banner, pennon, °streamer, banderole, gonfalon, ensign, °colors, °standard, labarum, *Chiefly nautical* jack, *Nautical and yachting* burgee, *Technical* vexillum: *We saw from her pennant that she was a Spanish frigate.*

pension *n.* **1** °benefit, °allowance, annuity, °subsistence, superannuation, °allotment, old-age pension, *US* social security, *Colloq* golden handshake: *She finds that her pension is not enough to live on.* —*v.* **2** Usually, ***pension off.*** °(cause to) retire, superannuate; °dismiss; *Colloq* °shelve, put out to pasture: *The company cut back on staff by pensioning off everyone over sixty.*

pensioner *n.* retiree, °veteran, °senior citizen, *Brit* OAP (= 'old age pensioner'), *US* golden ager, *Colloq Brit* wrinkly: *The housing units were specially designed for pensioners' needs.*

pensive *adj.* °thoughtful, °meditative, musing, in a brown study, cogitative, contemplative, °reflective, °preoccupied, ruminative, °wistful, daydreaming, in a trance, in a reverie, brooding, °sober, °serious, °grave: *I found the professor in a pensive mood, staring out the window.*

pent-up *adj.* restrained, constrained, repressed, stifled, bottled-up, corked-up, held-in, checked, held-back, curbed, °inhibited, restricted: *After weeks of frustration, he wanted to release his pent-up emotions in a scream. They took a long swim to work off some pent-up energy.*

penurious *adj.* **1** stingy, °mean, penny-pinching, °miserly, °tight, tightfisted, closefisted, cheeseparing, niggardly, °cheap, ungenerous, parsimonious, skinflinty, °thrifty, begrudging, grudging, Scroogelike, *Colloq* °near, *Brit* mingy, *US* chintzy: *Even today, he is so penurious that he gives his children an allowance of only fifty cents a week.* **2** °poor, poverty-stricken, °destitute, °impoverished, penniless, °indigent, °needy, impecunious, necessitous, beggarly, bankrupt, *Colloq* (stony *or* flat) broke, °hard up: *Kershaw was in a penurious state, without two coins to rub together.*

people *n.pl.* **1** persons, individuals, men and women, ladies and gentlemen, males and females, living souls; mortals; bodies: *How many people can this airplane carry?* **2** °relations, relatives, °kin, kinsmen, kinfolk, °family, kith and kin; ancestors, forebears: *His people left Russia in 1917.* **3** masses, °(general) public, °hoi polloi, consumers, multitude, °populace, common people, common man, commoners, subjects, citizenry, plebeians, grass roots, proletariat, °rank and file, the crowd, commonalty *or* commonality, *mobile vulgus,* bourgeoisie; man *or* woman in the street, everyman, everywoman, Mr. *or* Mrs. Average, *Brit* A. N. Other, Joe Bloggs, man *or* woman on the Clapham omnibus, *US* John Doe, Mary Doe, Richard Roe, John Q. Public; *Colloq and often derogatory* proles, °the rabble, ragtag and bobtail, silent majority, common herd, *Brit* plebs, admass: *Politicians unable to communicate with the people are seldom elected.* —*n. sing.* **4** °race, community, °clan, °tribe, °folk, °nation, °population, °society: *On the subject of religion, they were a people divided. The anthropologists were studying the peoples south of the Sahara.* —*v.* **5** °populate, colonize, °settle, °occupy: *The area was once peopled with Berber tribesmen.*

pep *n.* **1** °vigor, vim (and vigor), °spirit, °animation, vivacity, °energy, °verve, °zest, °fire, sprightliness, °life, effervescence, °sparkle, ebullience, °dash, °enthusiasm, *brio,* élan, *Colloq* zip, zing: *He certainly has a lot of pep for an octogenarian.* —*v.* **2** ***pep up.*** °stimulate, invigorate, °animate, °enliven, °vitalize, vivify, °energize, exhilarate, °quicken, °arouse, breathe (some) life into, °inspire, °activate, actuate, °fire, cheer up, *Colloq* buck up, °spark, work *or* fire up, *US* °wind up: *After those defeats, the team needed to be pepped up.*

pepper *v.* sprinkle, °scatter, °dot, speckle, fleck, bespeckle, speckle, °spatter, stipple, mottle: *This ice cream is peppered through with bits of chocolate. Their speech is peppered with obscenities.*

perceive *v.* **1** °see, °make out, discern, °catch sight of, glimpse, °spot, espy, apprehend, °take in, °notice, °note, °discover, descry, °observe, °mark, °remark, °identify, °distinguish, °detect: *I perceived his hesitation when a solo flight was suggested. She perceived a strange odor emanating from the floor.* **2** °appreciate, °grasp, °feel, °sense, apprehend, °understand, °gather, °comprehend, °deduce, °infer, °figure out, ascertain, °determine, conclude, °decipher, *Colloq* °dig, °catch on: *She perceived that he was going to renege on his promise to take her to the Riviera.* **3** Often, ***perceive of.*** °regard, °view, look on, °consider, °contemplate, °judge, deem, °believe, °think: *Calthorpe perceives of himself as a great actor, but we perceive him to be a ham.*

percentage *n.* °share, °part, °portion, °proportion, °interest, °piece, *Colloq* °cut: *We are to get a percentage of the profits in return for our investment.*

perceptible *adj.* °discernible, detectable, °observable, perceivable, °noticeable, distinguishable, recognizable, °apparent, °evident, °notable, °obvious, °patent, °manifest, palpable, °plain, °clear, °prominent, unmistakable *or* unmistakeable: *There are perceptible differences between your playing and Heifetz's.*

perception *n.* **1** °appreciation, °grasp, apprehension, °understanding, comprehension, °knowledge, °perspective, °view: *Mrs. Hart's perception of the situation is quite different from mine.* **2** °intuition, °insight, °instinct, °feel, °feeling, °sense, °impression, awareness, °idea, °notion, consciousness, °realization: *Norton hasn't the slightest perception of what is going on behind his back at the office.*

perceptive *adj.* °astute, °alert, °attentive, °quick, °alive, °quick-witted, °intelligent, °acute, °sharp, °sensitive, °sensible, percipient, discerning, °observant, perspicacious; on the qui vive; *Colloq* on the ball: *"It was very perceptive of you to spot the two small marks on the dead girl's throat," said the count. Anne is one of the more perceptive journalists reporting on events in eastern Europe.*

perch *n.* **1** roost, °rest, °seat; °spot, °location, °position, °place, °site, vantage point, °perspective: *From his perch at the top of the cliff, Martin had a clear view of the cave entrance.* —*v.* **2** roost, °rest, °sit, °nest; °place, °put, °set, °situate, °locate, °position, °site: *The owl was perched in the tree, waiting for the vole to appear from its burrow.*

percolate *v.* seep, °steep, °transfuse, leach, °drip, °drain, °strain, °filter, pervade, infuse, °ooze, transude, filtrate, °trickle, °permeate, °suffuse, °penetrate: *The earth is too hard for the rainwater to percolate to the roots below. It slowly percolated through to me that Clare had told the police where I was hiding.*

perdition *n.* damnation, °hell, hellfire, °doom, °ruin, condemnation, °destruction, ruination, °downfall: *If what divines call lust be punished with perdition, who is pure?*

peremptory *adj.* **1** commanding, °imperative, compelling, °obligatory, °mandatory, irrefutable, °incontrovertible, decretal: *After the sound of marching came a peremptory knock at the door.* **2** decisive, °final, preclusive, °arbitrary, °categorical, unequivocal, °dogmatic, unconditional, unreserved, °flat, °out-and-out, °outright, °unqualified, °unmitigated: *The juror was subject to peremptory dismissal because he was the defendant's cousin.* **3** imperious, °authoritative, °tyrannical, °despotic, °dictatorial, autocratic, °emphatic, °positive, °firm, °insistent, *Colloq* °bossy: *How dare you take such a peremptory tone when speaking to your father!*

perennial *adj.* **1** °durable, °lasting, continuing, °enduring, °constant, °stable, lifelong, °persistent, incessant, uninterrupted, °continual, °continuous, °chronic: *Once we expected our rivers to yield a perennial supply of fresh water. Bing Crosby has been a perennial favorite since the 1930s.* **2** °permanent, unfailing, never-failing, °endless, unending, ceaseless, unceasing, imperishable, undying, °perpetual, °everlasting, °timeless, °eternal, °immortal, *Literary* sempiternal: *The artificial flowers at her grave symbolize Ingrid's perennial youth.*

perfect *adj.* **1** °complete, °absolute, finished, (fully) realized, fulfilled, consummate, °pure, °entire, °whole, perfected, °best, °ideal: *Many have called the building a perfect example of the Palladian style.* **2** °sublime, °ideal, °superb, °supreme, °superlative, °best, °flawless, °faultless, °preeminent, °excellent, °exquisite, unexcelled, unrivaled, unequaled, unmatched, °matchless, °incomparable, °nonpareil, °peerless, inimitable: *At her throat she wore the most perfect emerald I had ever seen.* **3** °blameless, °righteous, °holy, °faultless, °flawless, °spotless, °immaculate: *Nobody's perfect.* **4** °fitting, °appropriate, °(just) right, apt, °suitable, °correct, °proper, made-to-order, °best, *Brit* spot on: *Arthur would be perfect for the role of Quasimodo.* **5** °precise, °exact, °accurate, °correct, unerring, °true, °authentic, °lifelike, right on, °excellent, °superlative, °superb, °reliable, *Brit* spot on: *These copies are perfect replicas of the original.* **6** utter, °absolute, °complete, °mere, °thorough, °out-and-out, through and through; 24-karat, °categorical, °unqualified, unalloyed, °unmitigated: *We were perfect strangers at the time. By his behavior, he's shown himself to be a perfect idiot.* **7** °expert, °proficient, °accomplished, °experienced, practised, °skillful, skilled, °gifted, °talented, °adept, deft, adroit, °polished, °professional, masterly, °masterful: *The admirable Crichton was the perfect butler.* —*v.* **8** °complete, °finish, °realize, °fulfill, consummate, °accomplish, °achieve, °effect, °execute, carry out *or* through, °bring (to perfection): *In 2100, the design of the interplanetary passenger ship was not to be perfected for several years.* **9** °rectify, °correct, emend, (put *or* set) right, °improve, °refine, °polish, °cultivate, °better, ameliorate: *Viniculture has been perfected in the Bordeaux region of France.*

perfection *n.* **1** °purity, flawlessness, faultlessness, sublimity, °superiority, °excellence, °preeminence, transcendence: *Though we strive for perfection, we never can achieve it.* **2** °completion, completeness, °achievement, °fulfillment, °realization, °consummation, °accomplishment, attainment: *The building doesn't reach perfection till the last roof tile is in place.* **3** °ideal, °paragon, °model, archetype, °pattern, °mold, °standard, idealization, °essence, °quintessence, °acme, °pinnacle, °summit: *Machiavelli probably achieved the perfection of political cunning.*

perfectionist *n.* **1** °purist, pedant, precisian, precisionist, stickler, *Colloq* fuss-pot, *US* fussbudget: *The foreman is such a perfectionist that there's no satisfying him.* —*adj.* **2** °meticulous, °precise, punctilious, °scrupulous, °exacting, °particular, °demanding, °fastidious,

°fussy; °obsessive; *Colloq* picky, nit-picking: *She takes a perfectionist attitude toward everything she does.*

perfectly *adv.* **1** °completely, purely, °entirely, °absolutely, °utterly, °totally, °wholly, consummately, °thoroughly, °quite, °definitely, °positively, unambiguously, unequivocally, unmistakably *or* unmistakeably, explicitly, °truly, °very, °extremely, extraordinarily, remarkably: *Your instructions for finding the house were perfectly clear.* **2** superbly, superlatively, flawlessly, faultlessly, impeccably, inimitably, incomparably, sublimely, exquisitely, marvelously, admirably, wonderfully: *Sally plays that Chopin étude perfectly.* **3** °exactly, °precisely, flawlessly, faultlessly, accurately, °literally, line for line, word for word, °verbatim, letter for letter, °to the letter, *literatim*: *He knows the entire Koran perfectly.* **4** °very, °full, °quite, *Dialectal* °right, *Brit* °jolly, *Slang* damned, bloody: *You know perfectly well that I hate cauliflower.*

perfidious *adj.* treacherous, °deceitful, °traitorous, treasonous, treasonable, °disloyal, °faithless, °false, unfaithful, °untrue, insidious, °hypocritical, °two-faced, Janus-faced, °corrupt, °dishonest: *It was perfidious of her to tell Bill's wife of their affair.*

perfidy *n.* perfidiousness, treachery, °deceit, traitorousness, treason, disloyalty, faithlessness, falseness, °falsity, unfaithfulness, °infidelity, insidiousness, °hypocrisy, °betrayal: *The name of Judas is a byword for perfidy.*

perforate *v.* °riddle, °puncture, °pierce, honeycomb, °drill, °bore, °punch; °enter, °penetrate, pass into: *This perforated utensil is called a colander. The shrapnel perforated only the outer layer of the bulletproof vest.*

perform *v.* **1** °execute, °complete, bring off *or* about, °accomplish, °effect, °carry out, °discharge, °dispatch, °conduct, °carry on, do, °fulfill, *Colloq* °pull off, °knock off, °polish off: *Postmen perform their duties despite hazards, like vicious dogs.* **2** do, °act, °behave, °operate, °function, °run, °work, °go, °respond: *How does your new car perform?* **3** °present, °stage, °produce, °put on, °mount, do; °act, depict, °take, °play, appear as: *The repertory group performs six shows weekly. He performs the role of Scrooge every Christmas.*

performance *n.* **1** °execution, °completion, bringing off *or* about, °accomplishment, effectuation, carrying out, °discharge, °dispatch, °conduct, carrying on, doing, °fulfillment: *The soldiers acquitted themselves nobly in the performance of their duties.* **2** °show, °exhibition, °exhibit, °play, playing, °engagement, °act, °appearance, *Colloq* gig: *She does three performances nightly at the Blue Angel.* **3** playing, acting, °interpretation, °presentation, portrayal: *His performance in the courtroom scene was outstanding.* **4** °behavior, °conduct, deportment, demeanor: *Her performance in the nightclub last night was outrageous, to say the least.* **5** °scene, °show, °exhibition, °display: *The lad put on quite a performance till threatened with punishment by the headmaster.*

performer *n.* actor *or* actress, °thespian, trouper, °player, *Chiefly Brit* artiste: *They received their training as performers in a repertory company.*

perfume *n.* **1** °essence, °extract, *parfum*, eau de Cologne, toilet water, °scent, °fragrance; °aroma, °odor, °smell, °bouquet, nose: *What is that wonderful perfume you're wearing? The perfume of the wine wafted over to me.* —*v.* **2** °scent: *Orange blossoms perfumed the air.*

perfunctory *adj.* **1** °routine, °mechanical, °automatic, robotlike, °unthinking, businesslike, unspontaneous, °formal, dismissive, °inattentive, uninvolved, apathetic, °indifferent, unconcerned, removed, °distant, *dégagé*, °offhand, °heedless, uninterested, °hasty, °hurried, °superficial, °cursory, °fleeting, rushed: *I don't care for the perfunctory service in those fast-food restaurants.* **2** °careless, °slipshod, slovenly, negligent, °sketchy, °spotty: *The bill, made out in a perfunctory fashion, was incorrect.*

perhaps *adv.* °maybe, °possibly, it is possible that, conceivably, as the case may be, it may be, *Archaic or literary* perchance, peradventure, *Archaic or dialectal* mayhap: *Perhaps she'll be on the next train.*

peril *n.* °danger, °threat, °risk, °jeopardy, °exposure, vulnerability, susceptibility, uncertainty, insecurity: *You enter there at the peril of your life. It was a time of peril for us all. The child's life was in peril.*

perilous *adj.* °dangerous, °risky, °hazardous, °vulnerable, °susceptible, °uncertain, °insecure, unsafe, unsure: *Why undertake such a perilous journey alone?*

perimeter *n.* °boundary, °border, borderline, °margin, °periphery, limit(s), °bounds, ambit, circumference, °edge, °verge, fringe(s), *Archaic or literary* bourn *or Archaic or literary* bourne: *The perimeter of the military base is patrolled by sentry dogs.*

period *n.* **1** °interval, °time, °term, °span, duration, °spell, °space, °stretch; while; *Colloq chiefly Brit* °patch: *During the period of his absence, his children had grown up. I waited a short period, then phoned again. We went through a bad period last year.* **2** °era, days, epoch, eon *or* aeon, °age, years: *During the Old English period, very little was written down.* **3** full stop: *Place periods at the ends of sentences.*

periodic *adj.* °periodical, °intermittent, °regular, °recurrent, °repetitive, iterative, cyclic(al), repeated; episodic, °sporadic, °occasional: *We called in the plumber because of the periodic hammering in the pipes. The next periodic return of Halley's comet is expected in 2061–62.*

periodical *n.* °magazine, °journal, °paper, °publication, newsletter, °organ, serial, weekly, fortnightly, semimonthly, monthly, bimonthly, °quarterly, semiannual, annual, yearbook, almanac, *Rare* hebdomadal *or Rare* hebdomadary: *Nicole's story will soon be published in an important literary periodical.*

peripheral *adj.* **1** °incidental, unimportant, °minor, °secondary, inessential *or* unessential, °nonessential, °unnecessary, °superficial, °tangential, °irrelevant, °beside the point: *Whether the units are grams or ounces is peripheral to the question of how much something weighs. Try to concentrate on the important issues and leave aside the peripheral ones for now.* **2** circumferential, °external, perimetric, °outside, outer: *The peripheral measurement, or circumference, of a circle with a diameter of seven inches is approximately twenty-two inches.*

periphery *n.* **1** °perimeter, circumference, °border, °edge, °rim, °brim, ambit, °boundary, °bound, °margin: *Trees will be planted along the periphery of the car park.* **2** °surface, °edge, °superficies: *Your analysis deals with the periphery, not the core, of the problem.*

perish *v.* °die, °expire, lose (one's) life, be killed, be lost, meet (one's) death, be destroyed: *Three gnus perished in the fire at the zoo.*

perjury *n.* °lying, mendacity, mendaciousness, forswearing, prevarication, bearing false witness: *Convicted of perjury, he was remanded for sentencing a week later.*

perk *n.* See **perquisite,** below.

perk up *v.* cheer up, become jaunty, °brighten, liven up, invigorate, smarten up, °quicken, (re)vitalize, °pep up, °revive, inspirit, *Colloq* buck up: *Old Mrs. Kertens perks up when her grandchildren come to visit. To perk up the party, Edgar put on some boogie-woogie records.*

perky *adj.* °lively, cheery, °cheerful, °jaunty, bouncy, °bright, invigorated, °vigorous, vitalized, peppy, °spirited, °sprightly, °frisky, °animated, °vivacious, °effervescent, °bubbly, °buoyant, °gay, *Colloq* bright-eyed and bushy-tailed, full of pep: *Sue's been quite perky since she started going with Trevor.*

permanence *n.* permanency, °stability, durability, fixedness, changelessness, lasting quality, longevity, °endurance, °persistence, dependability, reliability, survival: *The new roofing material is being tested for permanence.*

permanent *adj.* **1** °everlasting, °eternal, unending, °endless, °perpetual, unceasing, °constant, undying, imperishable, °indestructible, °stable, °abiding, long-lasting, °lasting, °enduring, °perennial, long-lived, °durable: *The satellite is in permanent orbit around the earth.* **2** unchanging, °invariable, °changeless, °fixed, unchangeable, immutable, unalterable, °stable, °persistent: *This stain is permanent and will not come out in the wash. Bonham is a permanent member of the executive committee.*

permanently *adv.* °forever, for good, °once and for all, forevermore, °always, eternally, everlastingly; perpetually, constantly, incessantly, °nonstop, continuously, endlessly, ceaselessly, unendingly, interminably: *Is the boy permanently disabled? According to one theory, the universe will not continue to expand permanently.*

permeate *v.* imbue, °diffuse, °penetrate, pervade, infiltrate, °enter, spread through(out), °saturate, seep through(out), percolate through, soak through: *He felt the warmth permeate every limb. Greed seemed to permeate every level of society.*

permissible *adj.* allowable, admissible, °acceptable, allowed, permitted, °tolerable, °legal, licit, °lawful, °legitimate, authorized, °proper, °(all) right; pardonable, excusable, °venial, *Colloq* O.K. *or* OK *or* okay, kosher, legit: *Some things that are permissible are not necessarily honorable.*

permission *n.* °consent, assent, °leave, acquiescence, sufferance, °tolerance, laxity, leniency *or* lenience, °leave, °license, °sanction, acceptance, authorization, °approval, approbation, countenance, °allowance, °liberty, °indulgence, sufferance; franchise, enfranchisement: *Have you permission to watch television? She eats anything she likes with the doctor's permission.*

permissive *adj.* assenting, consenting, °indulgent, °lenient, latitudinarian, acquiescent, °lax, easygoing, °liberal, °tolerant, nonconstraining, nonrestrictive, libertarian: *Allison grew up in the permissive society of the 1960s.*

permit *v.* **1** Often, ***permit to.*** °allow, °agree (to), °consent (to), give permission *or* leave (to), °authorize, °sanction, °license, °tolerate, countenance, °suffer, °brook, °admit, °grant, °enable, empower, franchise, enfranchise; °let: *He permitted me to use his name as a reference. Do they permit smoking here?* —*n.* **2** °license, °authority, authorization, franchise, °warrant, °pass, passport, visa: *Your parking permit expired last week. Shirley has a permit to visit Tibet.*

perpendicular *adj.* **1** °erect, °upright, vertical, °plumb, °straight (up-and-down): *This wall isn't perpendicular.* **2** Often, ***perpendicular to.*** at right angles (to), at ninety degrees (to): *The two paths are perpendicular to one another.*

perpetrate *v.* °commit, °execute, °perform, carry out *or* through, °effect, °effectuate, °accomplish, do, be responsible for, °practice, *Colloq* °pull (off): *Atrocities were perpetrated by both sides in the war.*

perpetual *adj.* **1** °eternal, °infinite, °everlasting, never-ending, unending, °perennial, ageless, °timeless, long-lived, °permanent, unceasing, °lasting, °enduring, unvarying, unchanging, immutable, °invariable, undeviating, *Literary* sempiternal: *They declared their perpetual love for each other.* **2** °constant, uninterrupted, °continuous, unfailing, incessant, °persistent, unremitting, unending, °nonstop, °endless, °recurrent, °continual, °repetitive: *Why should we have to listen to the neighbors' perpetual bickering?*

perpetuate *v.* °continue, °maintain, °extend, keep (on *or* up), keep going, °preserve, °memorialize, °immortalize, eternalize: *We perpetuated father's memory with an endowment to the Salvation Army.*

perpetuity *n.* °permanence, constancy, timelessness; °eternity: *The estate was bequeathed to the townspeople in perpetuity.*

perplex *v.* °confuse, °bewilder, °puzzle, °mystify, °distract, baffle, befuddle, confound, °muddle, disconcert, °stump, °nonplus, °stymie, stupefy, °stun, °daze, dumbfound *or* dumfound, flabbergast, *Colloq* bamboozle, hornswoggle, *Chiefly US and Canadian* discombobulate, throw for a loop: *The more I tried to understand bathymetric semiotics, the more perplexed I became. Annemarie's reticence perplexed us all.*

perplexing *adj.* confusing, bewildering, °puzzling, mystifying, baffling, confounding, °disconcerting, stupefying, flabbergasting, enigmatic, °paradoxical, °incomprehensible, unfathomable, impenetrable, °recondite, arcane, °labyrinthine, complex, °complicated, Byzantine, °intricate, °involved, convoluted, twisted, knotty, Gordian: *The writing is filled with perplexing references to the author's personal experiences, of which the reader is told nothing.*

perplexity *n.* **1** °confusion, bewilderment, bafflement, °distress, °doubt, °difficulty: *My perplexity grew as he related his version of the event.* **2** intricacy, °complexity, complicatedness, arcaneness, reconditeness, impenetrability, impenetrableness, involvement, unfathomability, °obscurity, °difficulty: *The more deeply the enigma was probed, the greater its perplexity.* **3** °puzzle, °enigma, °mystery, °dilemma, °problem, °paradox, catch-22, °quandary, °predicament, °bind: *Because of the interlocking directorships of the companies, we faced many perplexities in trying to sort out what had happened to the funds.*

perquisite *n.* °consideration, emolument, °bonus, °(fringe) benefit, °extra, °bonus, dividend, gratuity, °tip, douceur, baksheesh, °token (of appreciation), *US* lagniappe *or* lagnappe, *Colloq* °perk: *It was traditional to furnish each director with a company car as a perquisite.*

persecute *v.* **1** °oppress, °suppress, °subjugate, maltreat, °ill-treat, °abuse, °outrage, °molest, °victimize, °tyrannize, °afflict, °punish, martyr, °torment, torture: *For years black people had been persecuted.* **2** °bother, °annoy, °pester, °plague, hector, °bully, badger, harry, °harass, °irritate, °worry, vex, °trouble, °worry, importune, °hound: *Her lawyers continually persecuted him for nonpayment of alimony.*

persecution *n.* **1** °oppression, °suppression, subjugation, maltreatment, ill-treatment, °abuse, °outrage, molestation, victimization, °tyranny, °affliction, °punishment, °torment, torture: *They had to suffer persecution only because they belonged to a different religious order.* **2** °bother, °annoyance, hectoring, bullying, badgering, harrying, harassing, irritation, °worry, vexation, °trouble: *Persecution of the animals at the zoo is strictly forbidden.*

perseverance *n.* °persistence, steadfastness, °determination, °resolution, °resolve, decisiveness, °decision, firmness, purposefulness, pertinacity, staying power,

°stamina, sedulousness, assiduity, °grit, °pluck, tirelessness, indefatigableness, indefatigability, °patience, °endurance, diligence, °devotion, °tenacity, doggedness, stubbornness, inflexibility, °obstinacy, obstinateness, obdurateness, *Colloq* °guts, *US* stick-to-itiveness: *New ideas can make their effects felt only through perseverance.*

persevere *v.* Often, ***persevere in*** *or* ***with*** *or* ***at.*** °persist, °resolve, °decide, °endure, °continue, carry on *or* through, keep at *or* on *or* up, be steadfast *or* staunch *or* constant, keep going, stand fast *or* firm, °see through, be *or* remain determined *or* resolved *or* resolute *or* stalwart *or* purposeful *or* uncompromising, be tenacious *or* persistent *or* constant *or* pertinacious *or* assiduous *or* sedulous, be tireless *or* untiring *or* indefatigable, show determination *or* pluck *or* grit, be plucky, be patient *or* diligent *or* stubborn *or* inflexible *or* adamant *or* obstinate *or* obdurate, show *or* exhibit *or* demonstrate patience *or* diligence *or* stubbornness *or* inflexibility *or* obstinacy *or* obduracy, remain dogged, pursue doggedly, be intransigent *or* intractable, cling to, stick to, °support, stop at nothing, °sustain, *Colloq* °stick with, stick (it) out: *We must persevere if we are to win. I shall persevere in my loyalty.*

persist *v.* **1** Often, ***persist in*** *or* ***at.*** °persevere, be persistent, °insist (on), stand firm *or* fast, be steadfast *or* staunch, °strive, toil, °labor, work (hard) (at): *She persists in arguing her innocence. Only those who persist will succeed.* **2** °remain, °continue, °endure, °carry on, keep up *or* on, °last, °linger, °stay: *The bad weather persisted through the weekend.*

persistence *n.* °perseverance, °resolve, °determination, °resolution, steadfastness, °tenacity, constancy, assiduity, °stamina, tirelessness, indefatigability, indefatigableness, tirelessness, °pluck, °grit, °patience, diligence, pertinacity, doggedness, stubbornness, °obstinacy, obduracy: *Father's persistence finally won the day and we again spent our vacation camping in Yosemite Park.*

persistent *adj.* **1** persisting, persevering, °tenacious, °steadfast, °firm, °fast, °fixed, °staunch, °resolute, resolved, °determined, unfaltering, unswerving, undeviating, unflagging, °tireless, °untiring, indefatigable, dogged, unwavering, °stubborn, °obstinate, obdurate, °inflexible, °rigid: *He was persistent in his demands for justice. The inspector never gave up his persistent pursuit of criminals.* **2** continuing, °constant, °continuous, °continual, unending, interminable, unremitting, unrelenting, °perpetual, incessant, unceasing, °nonstop: *The persistent rainy weather began to depress us. At last he gave in to her persistent complaints and bought a washing machine.*

person *n.* **1** °individual, °human (being), being, man *or* woman *or* child, °(living) soul; °mortal: *Not a single person knew the answer to my question.* **2** ***in person.*** physically, °personally, bodily, °actually, myself *or* yourself *or* himself *or* herself *or* ourselves *or* yourselves, *or* themselves, *Colloq* °in the flesh: *Secondhand reports being unreliable, the correspondent visited the battlefield in person. I know their records, but I have never seen them in person.*

persona *n.* °face, °front, façade, °mask, °guise, °exterior, °role, °part, °character, °identity, self: *Her office persona is quite different from the one she displays at home.*

personage *n.* °celebrity, luminary, VIP (= 'very important person'), °name, °notable, °somebody, °personality, °star, superstar, magnate, °mogul, *Colloq* big shot, big wheel, hotshot, hot stuff, *Brit* big noise, *Theater US* headliner: *Blaisdell is fast becoming a personage in the financial world.*

personal *adj.* **1** °individual, °physical, bodily, °actual, °live; °in person, °in the flesh: *The star is scheduled to make a personal appearance on Wigan's radio show.* **2** °intimate, °exclusive, °private, °special, °particular: *Would you do me a personal favor? I hear they are having personal problems.* **3** °intimate, °close, °dear, °bosom, °familiar, °special: *Wendy happens to be a personal friend of ours.* **4** °intimate, °individual; disparaging, slighting, °offensive, °derogatory, °critical, deprecating, belittling, adverse, unfriendly, insulting: *He should confine his criticism to her acting and avoid personal remarks about her bad breath.*

personality *n.* **1** °character, °nature, temperament, °disposition, °makeup, °persona; °identity, °psyche: *That fellow Cox has an extremely abrasive personality and often upsets people.* **2** °celebrity, luminary, °star, superstar, °name, headliner, °somebody: *Whom shall we get as a personality to attract the crowds?*

personalized *adj.* monogrammed, initialed, individualized; signed: *He has written to me on his personalized stationery.*

personally *adv.* **1** °in person, °alone, by oneself, on one's own, myself *or* yourself *or* himself *or* herself *or* ourselves *or* yourselves *or* themselves, *Colloq* °in the flesh: *She has not met them personally, but we have. They will see to the matter personally.* **2** in one's own view *or* opinion, for one's part, for oneself, as far as one is concerned, from one's own viewpoint, from where one stands, as one sees it *or* things, as for oneself: *Personally, I wasn't sure I would make it.* **3** as an individual, as a person, privately, °in private: *I like him personally but would never have him as my dentist.*

personify *v.* **1** °embody, °typify, °exemplify, epitomize, be the embodiment of, °manifest, °represent, °stand for, °symbolize, *Archaic* impersonate, personate: *In my view, he personifies everything that is evil.* **2** humanize, personalize: *In literature, personifying inanimate things in nature with human attributes is called the pathetic fallacy.*

perspective *n.* **1** °(point of) view, °viewpoint, °standpoint, °prospect, vantage point, °position, °angle, *Colloq* where one is coming from: *I can see that my view would seem illogical from his perspective.* **2** °attitude, °position, °angle, °approach, °sentiment, °outlook, °lookout: *Management hasn't the right perspective on what is good for workers.*

perspiration *n.* °sweat, dampness, wetness; sweating; *Technical* sudor; diaphoresis: *I could feel the perspiration stand out on my forehead. They say that perspiration makes one feel cooler on a hot day.*

persuade *v.* **1** °urge, °induce, °prevail upon, °influence, exhort, importune, °dispose, °incline, °prompt, °sway, °press: *The officer persuaded him to surrender.* **2** bring round, °convince, °win over, talk *or* argue into, °convert: *We persuaded her to open the door. He was persuaded to vote Republican.*

persuasion *n.* **1** °inducement, inducing, °influence, influencing, exhortation, exhorting, persuading: *At the bank's persuasion, the builders extended credit to buyers. She has extraordinary powers of persuasion at her command.* **2** °opinion, °belief, °creed, °faith, set of beliefs, °faith, °religion, °(religious) conviction; °sect, °denomination, °faction, °school (of thought), affiliation: *Till he met Maggie, he had always been of the Baptist persuasion.*

persuasive *adj.* convincing, °telling, °influential, °effective, °productive, °impressive, °efficacious, cogent, °weighty, compelling, °forceful, valid, °winning, °authoritative, °dynamic: *His most persuasive argument for our leaving was that if we stayed we'd be shot.*

pert *adj.* °forward, °brash, °brazen, °cheeky, °insolent, °impertinent, °flippant, saucy, °bold, °presumptuous, °impudent, °disrespectful, °audacious, °rude, °impolite, uncivil, °ill-mannered, unmannerly, *Archaic* malapert, *Colloq* °fresh, °flip, out of line, °brassy, bigmouthed, wise-guy, *Slang US* smart-ass(ed), wise-ass(ed), *Brit* smart-arsed: *The child might be clever, but I don't like his pert manner.*

pertain *v.* Often, **pertain to.** concern, °refer to, °regard, have reference *or* relation (to), °apply (to), °relate (to), °include, °cover, °affect, appertain (to), be appropriate (to), be fitting (for), befit, °bear on, have bearing (on): *The sign, "Keep Off the Grass," does not pertain to the people who mow the lawn, Morris.*

pertinent *adj.* pertaining, °appropriate, °fitting, °suitable, apt, °relevant, germane, apropos, apposite: *Try to keep your comments pertinent to the subject under discussion.*

perturb *v.* °upset, °disturb, °fluster, °ruffle, unsettle, disconcert, make uneasy, °discomfit, vex, °worry, °agitate, °shake up, °alarm, disquiet, °confuse, discompose, °unnerve, addle, disorganize: *He became quite perturbed when the police asked him to help with their inquiries.*

perusal *n.* reading, °scrutiny, °check, °examination, °study, inspection, scanning, °review: *I saw nothing blasphemous in my perusal of the text.*

peruse *v.* °read, °study, °scan, °scrutinize, °examine, °inspect, °review, °browse, run one's eye over: *As Gregory was perusing the ancient manuscript, a sudden draft blew out the candle.*

pervasive *adj.* °penetrating, pervading, omnipresent, °general, inescapable, °prevalent, °universal, widespread, ubiquitous, permeating, permeative: *A pervasive sense of doom possessed Rodolfo whenever the wolf howled.*

perverse *adj.* **1** °wrong, wrongheaded, awry, °contrary, wayward, °incorrect, °irregular, unfair, °improper, °contradictory: *It was most perverse of you to change your mind about our outing to the seashore.* **2** °cantankerous, °testy, curmudgeonly, churlish, crusty, bad-tempered, °petulant, captious, °cross, cross-grained, °peevish, °waspish, °snappish, °bilious, splenetic, fractious, ill-tempered, °quarrelsome, irascible, °sullen, contentious, °touchy, °obstreperous, crabby, crabbed, °irritable, °surly, *Colloq* grouchy, *Brit* stroppy, *US and Canadian* °cranky: *With everything going wrong, Catherine feels particularly perverse today.* **3** °stubborn, °self-willed, wrong-headed, intractable, °willful, obdurate, °obstinate, pigheaded, adamant(ine), °inflexible, unbending, refractory, unyielding: *I refuse to give in to a perverse child just because he has a tantrum.*

perversion *n.* **1** deviation, irregularity, misdirection, corruption, °subversion, distortion, twisting, falsification, misrepresentation, °diversion, sidetracking: *The conduct of this trial has been a most heinous perversion of the course of true justice.* **2** unnatural act, deviation, deviance *or* deviancy, °abnormality, depravity, °vice, aberration, debauchery, *Colloq* kinkiness, *Brit* °kink: *I forbear describing perversions that would shock your delicate sensibilities, Gentle Reader.*

pervert *v.* **1** °deflect, °divert, °sidetrack, turn aside *or* away, °subvert, °misdirect, °distort, °twist, °abuse, °falsify, misapply, misconstrue, °misrepresent, °corrupt: *By withholding evidence, you have perverted the course of justice.* **2** °seduce, °lead astray, debauch, °degrade, °corrupt, °demoralize, °subvert: *He was accused of perverting young girls.* **3** °deviant, °degenerate, debauchee, *US* °deviate, *Colloq* °weirdo: *The perverts were caught* in flagrante delicto.

perverted *adj.* °deviant, °deviate, °abnormal, amoral, unmoral, °immoral, °bad, depraved, °unnatural, warped, twisted, °profligate, °dissolute, °delinquent, °degenerate, °evil, °wicked, malign, malicious, malefic, malevolent, °evil-minded, °sinful, iniquitous, °base, °foul, °corrupt, unprincipled: *Members of the Hellfire Club yielded themselves up to the most perverted, abandoned behavior.*

pessimistic *n.* °gloomy, °negative, despairing, °hopeless, °inauspicious, depressed, °despondent, °dejected, °melancholy, °downhearted, heavy-hearted, defeatist, °glum, °sad, °blue, °unhappy, cheerless, °joyless, cynical, °bleak, °forlorn: *The bears in the stock exchange take a pessimistic view of share prices.*

pest *n.* °nuisance, °annoyance, °nag, irritant, °bother, gadfly, bane, °trial, heckler, vexation, °curse, thorn in one's flesh, *Colloq* °pain (in the neck), *Slang US* (*Yiddish*) nudge *or* noodge *or* nudzh, nudnik, *Taboo slang* pain in the *Brit* arse *or US and Canadian* ass: *Derwent is such a pest, I wish he'd leave me alone.*

pester *v.* °annoy, °nag, °irritate, °irk, °bother, get at *or* to, badger, °plague, vex, °fret, hector, °harass, harry, °heckle, nettle, °chafe, peeve, pique, °provoke, °exasperate, bedevil, get *or* grate on (someone's) nerves, get under (someone's) skin, get in (someone's) hair, °try (someone's patience), °torment, °persecute, *Brit* chivvy, *Colloq* drive (someone) up the wall, needle, give (someone) the needle, hassle, °ride, give (someone) a hard *or* bad time, °bug: *Please stop pestering me about going to the football game.*

pestilence *n.* **1** °plague, °epidemic, pandemic, Black Death, *Rare* °pest: *The pestilence raged throughout all Europe, killing fifty million people.* **2** °scourge, °blight, °curse, cancer, canker, bane, °affliction: *How are we to overcome the pestilence of greed?*

pet[1] *n.* **1** °darling, °favorite, °idol, apple of (someone's) eye, *Colloq Brit* blue-eyed boy, *US* fair-haired boy: *You know that you were always Father's pet.* —*adj.* **2** °tame, trained, domesticated: *Doesn't the landlord take a dim view of your keeping a pet alligator in the bath?* **3** °favorite, °favored, preferred, cherished, °special, °particular; indulged, prized, treasured, °precious, dearest, adored, °darling: *Building the summerhouse was Desmond's pet project. Elizabeth's pet pupil is Anne.* —*v.* **4** °caress, °fondle, °stroke, °pat; °cuddle, nuzzle, °nestle, °snuggle, *Colloq* neck, smooch *or Australian and New Zealand also* smoodge *or also* smooge, *Chiefly US and Canadian* °make out: *Please do not pet my tarantula. Two teenagers were petting in the back seat of the car.* **5** °humor, °pamper, °favor, °baby, °coddle, cosset, mollycoddle, cocker, °spoil, °indulge, dote on: *His mother pets him far too much.*

pet[2] *n.* (bad *or* ill) temper, pique, °sulk, °(bad) mood, °fume, *Colloq Brit* paddy *or Colloq Brit* paddywhack *or Colloq Brit* paddywack: *He's in a terrible pet because they forgot to cancel the milk and paper deliveries when they went to Europe.*

peter out *v.* °diminish, °evaporate, °wane, come to nothing *or* naught *or US also* nought, °die out, °disappear, °fail, fade (out *or* away), °dwindle (into nothing), °run out, °give out, °flag, °melt away: *After a while, our money petered out and we had to send home for more.*

petite *adj.* °delicate, °dainty, mignon, °diminutive, °small, °little, °slight, °tiny, small-boned, *Colloq Brit* dinky: *It was incongruous to see the basketball player with a petite blonde.*

petition *n.* **1** °request, °application, solicitation, °suit, entreaty, °supplication, °plea, °appeal: *The anti-pollution petition, signed by thousands of people, was delivered to the Environmental Protection Agency.* —*v.* **2** °request, °ask, apply to, apply for, °solicit, °sue, °call upon, entreat, supplicate, °plead, appeal (to) *or* (for), °beseech, implore, importune, *Rare* obsecrate: *The shopkeepers petitioned the council for better police protection.*

petrified *adj.* **1** horrified, horror-stricken, terrified, terror-stricken, °panic-stricken, frightened, °afraid, paralyzed, numbed, benumbed, frozen: *The maiden stood petrified as the dragon, breathing fire, approached.* **2** shocked, °speechless, dumbfounded *or* dumfounded, dumbstruck, stunned, °thunderstruck, astonished, astounded, confounded, stupefied, appalled, aghast, *Colloq* flabbergasted: *The firemen rescued three petrified children who were huddled in a corner.* **3** ossified, fossilized: *These were not stones but the petrified remains of ancient trees.*

petrify *v.* **1** °frighten, °scare, °horrify, °terrify, °paralyze, °numb, benumb: *Consuela was petrified by the noise of the explosion.* **2** °shock, dumbfound *or* dumfound, °stun, °astonish, °astound, °amaze, confound, disconcert, stupefy, °appall, *Colloq* flabbergast: *Finding a body in one's bedroom can be a petrifying experience.* **3** ossify, fossilize, turn to stone: *Over thousands of years the desert conditions petrify the wood.*

petty *adj.* **1** °insignificant, trivial, °paltry, °minor, °inferior, °niggling, °trifling, °negligible, °puny, inessential, °nonessential, °inconsequential, unimportant, °slight, nugatory, of no account, *US* dinky, *Colloq* piddling, °measly, no great shakes, no big deal, °small-time, *Brit* twopenny-halfpenny *or* tuppenny-ha'penny, *US and Canadian* picayune: *He has been convicted only of petty crimes.* **2** °miserly, °mean, mingy, stingy, cheeseparing, grudging, °small-minded, °cheap, niggardly, parsimonious, °tight, tightfisted, °close, close-fisted: *It was very petty of you to refuse the beggar a few pennies.*

petulant *adj.* °peevish, pettish, °impatient, ill-humored, °testy, °waspish, irascible, choleric, °cross, captious, ill-tempered, bad-tempered, splenetic, °moody, °sour, °bilious, crabby, crabbed, °irritable, huffish, huffy, °perverse, °snappish, crotchety, °cantankerous, curmudgeonly, grouchy, grumpy: *With a petulant gesture she hurled the rose away.*

phantom *n.* **1** apparition, °specter, °ghost, °spirit, phantasm, °shade, wraith, revenant, °vision, *Formal* eidolon, phantasma, *Colloq* spook: *The so-called phantom of the opera turned out to be a real person.* **2** figment (of the imagination), °illusion, °delusion, chimera *or* chimaera, °hallucination, °fancy, mirage: *She was a phantom of delight / When first she gleamed upon my sight.*

pharisaic *adj.* pharisaical, °hypocritical, °insincere, °self-righteous, °pretentious, holier-than-thou, °sanctimonious, pietistic(al), formalistic, canting, unctuous, °oily, °slimy, *Literary* Tartuffian, Pecksniffian, *Colloq* °goody-goody, *Chiefly Brit* smarmy: *That pharisaic gospeler never practices what he preaches.*

pharisee *n.* °hypocrite, °pretender, dissembler, humbug, °fraud, whited sepulcher, pietist, formalist, canter, *Literary* Tartuffe, Pecksniff, *Colloq* phoney *or US also* phony: *She is such a pharisee that butter wouldn't melt in her mouth.*

pharmacist *n.* pharmacologist, *Rather old-fashioned or formal* apothecary, *Brit* (pharmaceutical) chemist, *US and Canadian* °druggist, *Formal* posologist, *Colloq* pill pusher, *US* pill roller: *The pharmacist said that those pills have bad side effects.*

pharmacy *n.* **1** dispensary, *Rather formal or old-fashioned* apothecary, *Brit* chemist's (shop), *US and Canadian* drugstore, druggist's: *Stop off at the pharmacy and get me something for this headache, please.* **2** pharmaceutics, pharmacopoeia: *Research has vastly expanded modern pharmacy.*

phase *n.* **1** °stage, °period, °development, °step: *The boy is just going through a phase.* **2** °time, °moment, °juncture, °occasion: *At this phase of the discussion, I should like to introduce a new subject.* **3** °state, °form, °shape, configuration, °aspect, °appearance, °look, °condition, °status: *Here is a diagram of the phases of the moon at various points in its orbit.* **4** facet, °side, °angle, °viewpoint, °point of view, *Colloq* °slant: *Only one phase of the argument has so far been presented.* —*v.* **5 *phase in.*** °(gradually) introduce, usher in, °work in, °inject, °insert, °insinuate, °include, °incorporate: *The new work schedules will be phased in over the next month.* **6 *phase out.*** ease out *or* off, °taper off, °wind up, put a stop to, °(gradually) eliminate, °remove, °withdraw, °discontinue, °end: *The use of nonbiodegradable and nonrecyclable packaging is being phased out.*

phenomenal *adj.* °outstanding, °remarkable, °exceptional, °extraordinary, °unusual, freakish, °rare, uncommon, °singular, °unorthodox, unprecedented, °unheard-of, °unparalleled, °unbelievable, °incredible, °marvelous, wonderful, °amazing, astonishing, astounding, staggering, °stunning, °prodigious, °miraculous, °fantastic, *Colloq* mind-boggling, mind-blowing: *She made a phenomenal recovery and can walk again.*

phenomenon *n.* **1** °event, °happening, °occurrence, °incident, °occasion, °experience, °fact: *Everyone knows the phenomenon of the souring of milk.* **2** °wonder, °curiosity, °spectacle, °sight, °sensation, °marvel, °rarity, °exception, miracle, *Slang* stunner: *An eight-year-old chess champion is truly a phenomenon, even in Russia.*

philanderer *n.* °flirt, °gallant, °roué, °rake, Casanova, Lothario, Don Juan, Romeo, °lover, °playboy, gay dog, *Colloq* lady-killer, womanizer, *Old-fashioned* wolf, *Slang* stud: *In his memoirs it emerged what a philanderer he had been, often dallying with six women at once.*

philanthropic *adj.* °charitable, eleemosynary, °generous, magnanimous, munificent, °benevolent, openhanded, ungrudging, unstinting, beneficent, °humanitarian, altruistic, humane: *Finlay was always philanthropic, unlike some wealthy people I could mention.*

philanthropist *n.* contributor, °donor, benefactor *or* benefactress, patron *or* patroness, °sponsor, Maecenas, Good Samaritan, °humanitarian, altruist: *Some anonymous philanthropist provided the funds for the new school swimming pool.*

philanthropy *n.* **1** generosity, °benevolence, °charity, °patronage, magnanimity, charitableness, public-spiritedness, bigheartedness, thoughtfulness, almsgiving, kindheartedness, beneficence, benignity, liberality, openhandedness: *Carnegie's philanthropy was focused on libraries and education.* **2** °donation, contribution, °largess, °aid, °grant, °assistance, °help: *Mrs. Ander's recent philanthropy allows us to offer six new scholarships.*

philistine *n.* **1** °boor, °barbarian, °yahoo, lowbrow, Boeotian, vulgarian, ignoramus, °bourgeois, *US* Babbitt: *Charlotte never could abide the philistines who put love of money above love of culture.* —*adj.* **2** uncultured, uncultivated, °tasteless, commonplace, unenlightened, °unrefined, unread, unlettered, °uneducated, untutored, unlearned, °narrow-minded, anti-intellectual, °boorish, lowbrow, °dull, °prosaic, °boring, °bourgeois, crass, commercial, °materialistic: *The television companies were accused of pandering to the basest appetites of the philistine viewer.*

philosophical *adj.* **1** philosophic, °abstract, esoteric, °learned, °scholarly, erudite, °theoretical, °rational, °logical, °impractical: *The person who is out of a job cares nothing about philosophical reasons for unemployment.* **2** °detached, unconcerned, unemotional, unimpassioned, composed, °thoughtful, °reflective, °meditative, cogitative, contemplative, °judicious, °sober, °levelheaded, °realistic, °practical, pragmatic(al), down-to-earth, °cool, °calm, °serene, placid, °stoical, °patient, unruffled, coolheaded, °tranquil, unperturbed, even-tempered, °temperate, °moderate, °equable, equanimous, imperturbable: *Over the years, Evelyn had learned to take a philosophical attitude toward her husband's shortcomings.*

philosophy *n.* **1** metaphysics, epistemology, °logic, natural *or* moral *or* metaphysical philosophy, rationalism, °thinking, aesthetics: *He views philosophy as the attempt to describe and codify universal truths.* **2** °viewpoint, °(point of) view, °outlook, °opinion, °attitude, °feeling, °sentiment, °idea, °notion, °ideology, (set of) beliefs *or* values, tenets, *Weltanschauung,* worldview: *Harold's philosophy of life is "Live and let live."* **3** °stoicism, °sang-froid, °control, °self-control, °restraint, coolness, composure, calmness, °serenity, placidity, coolheadedness, equanimity, thoughtfulness, imperturbability, self-possession, aplomb, dispassion, °patience, °resignation: *You may be sure that Paul allows nothing to disturb his philosophy.*

phlegmatic *adj.* **1** phlegmatical, stoic(al), unemotional, °unenthusiastic, unanimated, sluggish, apathetic, uninvolved, °lethargic, unfeeling, uncaring, °cold, unresponsive, °stolid, °unmoved, insensitive, °unaffected, °insensible, °indifferent, unconcerned, uninterested, °listless, °torpid, °indolent, °inactive, °passive, *Rare* hebetudinous: *Hutton is far too phlegmatic to be stirred by the fervor of the revolutionaries.* **2** phlegmatical, °self-possessed, self-controlled, controlled, restrained, composed, °calm, °serene, °tranquil, placid, coolheaded, equanimous, °cool, undisturbed, unperturbed, unruffled, imperturbable, even-tempered, °philosophical, °temperate, °moderate: *One has to learn to be phlegmatic about things going wrong at the office.*

phobia *n.* °fear, °horror, °terror, °dread, hatred, detestation, abhorrence, °loathing, execration, °aversion, °revulsion, repugnance, °dislike, °distaste, antipathy; disquiet, nervousness, °qualm, °distrust, °suspicion, apprehension, °worry: *He suffers from claustrophobia, which means that he has a phobia for enclosed places, so he avoids elevators and telephone booths.*

phony *adj.* **1** phoney, °unreal, °fake, °synthetic, °artificial, °factitious, °false, °fraudulent, °imitation, °bogus, °spurious, °counterfeit, °mock, ersatz, paste, trumped-up; °sham, °pretended, °insincere, °hypocritical, dissimulating, °deceitful, °dishonest; *Colloq* pseudo *or Brit* pseud: *They were caught trying to collect insurance on the theft of a string of phony pearls. Every time she wanted something, she'd turn on the phony charm.* —*n.* **2** phoney, °fake, °fraud, °imitation, °counterfeit, °forgery, °hoax, °sham: *The diamond she's wearing is a phony.* **3** phoney, trickster, faker, humbug, °impostor, °pretender, charlatan, mountebank, double-dealer, °counterfeiter, °quack, deceiver, *Colloq Brit* pseud, *Slang US* paperhanger: *Madame Tatiana claims to be able to communicate with the dead, but I think she's a phony.*

photograph *n.* **1** snapshot, °print, °picture, °slide, transparency; °negative, °positive, *Colloq* photo, °snap, °shot, pic (*pl.,* pix): *Heinrich has taken some marvelous photographs of the children.* —*v.* **2** take a picture (of), °shoot, °film, °take, *Colloq* °snap: *He photographed Jennifer holding her kitten.*

photographer *n.* lensman, lenswoman, cameraman, camerawoman, cinematographer, *paparazzo* (*pl., paparazzi*), *Old-fashioned* photographist: *The photographers clustered round the prime minister.*

photographic *adj.* **1** °vivid, °natural, °realistic, °graphic, °accurate, °exact, °precise, °faithful, °detailed, °lifelike, true-to-life: *Some of the hard-edge artist's paintings are virtually photographic.* **2** cinematic, filmic, °pictorial: *Uncannily, a photographic image appeared on the shroud under certain light.*

phrase *n.* **1** clause, noun phrase, verb phrase, prepositional phrase, adverbial phrase, adjectival phrase: *When asked for an example of a verb phrase, the student responded, "Shut up."* **2** °expression, word group, collocation, locution, °idiom, idiomatic expression, collocution, °proverb, °motto, °slogan, saying, catch phrase, adage, °maxim, axiom, °saw, colloquialism, °cliché, platitude, commonplace, *Colloq* chestnut: *Do you know the source of Churchill's famous phrase, "Blood, sweat, and tears"?* **3** phraseology, phrasing, °wording, °language, °usage, way *or* manner of speaking, °diction, °parlance, *façon de parler, modus loquendi, modus scribendi,* speech habit, °style, choice of words, word choice, syntax, vocabulary: *He was to "let slip," to use Shakespeare's phrase, "the dogs of war."* —*v.* **4** °express, °term, °word, °put, °frame, °formulate, °couch, put into words, put *or* set forth, verbalize, articulate, °voice, utter, °say, °write; °describe, delineate: *I am pondering over the best way to phrase this example.*

physical *adj.* bodily, corporeal, corporal, fleshly, incarnate, °carnal, °mortal, °earthly, °natural, somatic; °material, °tangible, palpable, °real, °actual, °true, °concrete, °manifest, °solid: *Curtis doesn't have the physical strength to lift that weight. As a chemist, James deals with the physical, not the spiritual universe.*

physician *n.* °doctor, medical doctor, M.D., doctor of medicine, medical practitioner, general practitioner, G.P., medical man *or* woman, °specialist, diplomate, *Brit navy* surgeon, *Colloq* doc, medico, medic, *US* °man, *Slang* sawbones, bones: *You ought to see a physician if the pain persists.*

physique *n.* °build, °figure, °body, °frame, °shape, bodily structure, °form, *Slang* chassis, *US* bod, built: *Mark did not appreciate your comments about his physique.*

pick *v.* **1** Often, ***pick out.*** °select, °choose, cull, °sort out, hand-pick, °single out, opt for, fix *or* decide upon *or* on, °elect, settle upon *or* on, °screen (out), °sift (out): *Melanie was picked to succeed Hubert as president.* **2** °pluck, °gather, °collect, °harvest, bring *or* take in, °garner: *Scores of extra workers were brought in to pick apples.* **3** °provoke, °foment, °incite, °start, °initiate, work *or* stir up: *He tried to pick an argument with me about who is taller.* **4** ***pick at.*** **(a)** °criticize, carp at, find fault with, cavil (at *or* about), quibble (at *or* about), °pick on, °nag (at), °niggle (at), °harass, °pester, °annoy, °irritate, °bother: *Stop picking at your brother!* **(b)** nibble (at), peck at: *We can't get him to eat a thing, he just picks at his food.* **5** ***pick off.*** °shoot (down), °kill: *We picked them off one by one as they emerged from the trench.* **6** ***pick on.*** °bully, °ride, °intimidate, °abuse, °browbeat, badger, harry, hector, °harass, °tease, °taunt, needle, °torment: *Penrose must learn not to pick on boys who are bigger than he.* **7** ***pick out.*** **(a)** See **1,** above. **(b)** discern, °distinguish, tell apart, °make out, °recognize, °separate, °discriminate: *I was able to pick out a distant rider approaching swiftly. From among the forest sounds she picked out the song of the nightingale.* **8** ***pick up.*** **(a)** °raise (up), °lift (up), heft, °hoist, °take up: *The stone is too heavy to pick up. Please pick up that piece of paper.* **(b)** °gather, °collect, glean, °take up: *I* *wish you would pick up your clothes.* **(c)** °tidy (up), °neaten, straighten up *or* out, °clean (up): *She refuses to continue to pick up after him.* **(d)** °acquire, °learn, become acquainted with; °master; *Colloq* get the hang of: *We picked up a little Spanish on our vacation, which was in New York City.* **(e)** °acquire, °find, °come by, get hold of, °obtain; °buy, °purchase, °get: *Basil picked up a few bad habits in the army. Let's pick up a bottle of wine on the way.* **(f)** °improve, get better, °gain, °make headway, °recover, °perk up, °rally, °recoup, °(make) progress, move ahead, °increase, *Colloq* make a comeback: *Business usually picks up before Christmas.* **(g)** accelerate, °speed up: *The pace picked up as they approached the finish line.* **(h)** °arrest, apprehend, take into custody, *Colloq* °pinch, collar, °nab, bust, °run in, °pull in, *Brit* °nick: *Two men were picked up trying to break into the bank.* **(i)** °call for, give a lift *or* ride to, °collect, go for *or US also* after, go to get: *I'll pick you up at the railway station at noon.* **(j)** °meet, introduce oneself to, strike up an acquaintance with, accost, make advances to: *I think she picked him up in a wine bar.* **(k)** °catch, °come down with, °contract, °get: *He picked up a rare virus while in the tropics.* —*n.* **9** °selection, °choice, °option, °preference: *She could have had the pick of any man in the place. Take your pick.* **10** choicest, °best, *crème de la crème,* cream: *The early shoppers had already taken the pick of the crop.*

picket *n.* **1** °stake, °pale, °post, °peg, stanchion, °upright, vertical, palisade, paling: *The walls of the fort were of strong pickets tapered to a point at the top.* **2** demonstrator, protester, striker: *The police prevented the pickets from attacking workers who entered the factory.* **3** picquet, °sentinel, °watchman, °guard, °observer, °patrol, vedette *or* vidette *or* vedette boat: *The picket reported no unusual activity.* —*v.* **4** °enclose, °shut in, wall in, °fence (in), hem in, °box in: *We picketed the camp for the night.* **5** °protest, °demonstrate, blockade: *Strikers picketed the factory twenty-four hours a day.*

picnic *n.* **1** garden party, *fête champêtre,* meal alfresco, barbecue, *US* clambake, *US and Canadian* cookout: *The clouds had gone, and it was a great day for a picnic.* **2** child's play, *Colloq* °pushover, °snap, cinch, °piece of cake, walkover, *US and Canadian* °breeze, lead-pipe cinch: *Sailing home with a quartering wind—it was a picnic all the way.* **3** ***no picnic.*** °difficult, °arduous, torture, torturous, °agony, °agonizing, °painful, °disagreeable, discomfiting, °misfortune, *Colloq* °tough, tough luck, tough going, °rough, a pain in the neck, *US* tough sledding, *Taboo slang* pain in the *Brit* arse *or US* ass: *Being marooned on a desert island for a month was no picnic, I assure you.*

pictorial *adj.* **1** °graphic, °picturesque, °vivid, °telling, °striking, °expressive, °plain, °explicit, °clear, lucid: *Many of the poet's images are amazingly pictorial.* **2** illustrated: *They published a pictorial history of biblical lands just for children.*

picture *n.* **1** °drawing, painting, °representation, °portrait, depiction, artwork, °illustration, sketch, °photograph: *Here is a picture of our cottage in the Cotswolds.* **2** °image, (perfect *or* exact) likeness, °(carbon) copy, °double, °duplicate, °twin, °(exact) replica, look-alike, °facsimile, *Colloq* spitting image *or* spit and image, *Slang* (dead) ringer: *Isn't she just the picture of her mother?* **3** °impression, °idea, °notion, °understanding, °image: *I think I have a pretty good picture of the situation.* **4** °model, °prototype, °epitome, °essence, °embodiment, incarnation, personification, perfect example: *From her autobiography, you might believe her to be the picture of sweet innocence.* **5** ***put (someone) in*** *or* ***into the picture.*** inform *or* advise fully, *Colloq* fill (someone) in: *Put me into the picture about what went on here last night.* —*v.* **6** °envision, °envisage, visualize, °imagine, °fancy, conceive of, see in the mind's eye: *Picture me / Upon your knee.* **7** depict, °draw, °portray, °paint, °represent, °show, °illustrate, °display: *In this fresco, Salome is pictured dancing before Herod Antipas.*

picturesque *adj.* **1** colorful, °interesting, intriguing, °unusual, °unique, °original, charming, °idyllic, °fetching, °attractive, °pretty, °lovely, °quaint, °delightful, °pleasing, °scenic: *We came upon a picturesque village in which all the houses were half-timbered Tudor style with thatched roofs.* **2** colorful, °graphic, °realistic, °vivid, °striking: *Bernard has written a picturesque account of a walking trip through Bavaria.*

piece *n.* **1** °bit, °morsel, °scrap, chunk, hunk, °sliver, °lump, °portion, °particle, °fragment, °shred, shard *or* sherd, °remnant, °quantity: *All the beggar wanted was a piece of bread. A piece of shrapnel is still embedded in my arm.* **2** °wedge, °slice, serving, °helping, °portion: *You may not have a piece of pie till you've eaten your vegetables.* **3** °share, °portion, fraction, °part, °division, °segment, °section, °interest, holding, °percentage, °proportion: *It turned out that a piece of the company had been sold without shareholders' approval.* **4** °(short) story, article, °essay, °report, °theme, °draft; °poem; music, °opus, °(musical) number, °composition, °arrangement, °tune, °melody, °song, °air, °jingle, ditty; °production, °play, °drama, sketch, °show: *I read that piece about cholesterol in yesterday's paper. Destry wrote that piece for the flute. Which piece by Strindberg will you put on next?* **5** °man, °token, chessman, chesspiece, checker, *Brit* draughtsman: *Once you have touched a piece you must move it.* **6** ***go to pieces.*** °fall apart, °disintegrate, °crumble, °shatter; be shattered, be upset, be disturbed, have a nervous breakdown, go out of *or* lose control, °break down, *Colloq* crack up: *Another earthquake and this wall will go to pieces. At the news of his son's death, Heywood simply went to pieces.* **7** ***in pieces.*** smashed, destroyed, ruined, shattered, °broken, in smithereens, smashed: *The vase lay in pieces at my feet. Though I had won the case, my life was in pieces.* **8** ***of a piece (with).*** °similar, similarly constituted, °alike, of the same sort *or* kind *or* type, °uniform, the same, part and parcel (of the same thing), °identical; in harmony, in agreement, °harmonious, in keeping: *This book is of a piece with the others in the same series. All his paintings are of a piece.* **9** ***piece of cake.*** *Colloq* °snap, cinch, *US and Canadian* lead-pipe cinch, °breeze: *The French exam was a piece of cake.* **10** ***piece of (one's) mind.*** *Colloq* scolding, °rebuke, °lecture, °reprimand, °tongue-lashing, chiding, rap over *or US* on the knuckles, *Colloq* °hell, what for, dressing-down, *US* bawling-out, chewing-out: *She gave him a piece of her mind about the amount of time he spent in the pool halls.* **11** ***piece of the action.*** °share, °portion, °interest, °stake, °percentage, holding, °quota: *For a thousand dollars you can have a piece of the action.* **12** ***speak (one's) piece.*** have (one's) say, express (one's) opinion, say what is on (one's) mind; vent (one's) spleen, *Colloq* get a load off (one's) mind *or Colloq* chest: *All were given a chance to speak their piece.* —*v.* **13** ***piece together.*** °assemble, put together, °connect, °gather, °compose; °fix, °unite, °restore, °mend: *We pieced together what happened from the witnesses' accounts. You'll never be able to piece together the bits of that lamp.* **14** ***pièce de résistance.*** —*n.* highlight, (special *or* main) feature *or* attraction, *spécialité (de la maison)*, °masterpiece, *chef d'oeuvre,* °specialty: *In this restaurant, the stuffed pheasant hongroise is the* pièce de résistance.

piecemeal *adv.* **1** piece by piece, little by little, inch by inch, bit by bit, inchmeal, °gradually, °by degrees, slowly, in bits and pieces, °by fits and starts, fitfully, intermittently, sporadically, disjointedly: *One misses the perspective of the entire project if it is examined piecemeal.* 2 into fragments *or* shreds *or* pieces: *He angrily took the check and tore it up piecemeal.* —*adj.* 3 °fragmentary, bit by bit, inchmeal, °gradual, °disjointed, °sporadic: *All continuity of cause and effect is lost in Ledger's piecemeal approach to teaching history.*

pier *n.* 1 wharf, landing (stage *or* place), jetty, quay, floating dock, *Technically inaccurate* °dock: *The ship is tied up at the pier.* 2 °pile, piling, °post, °upright, column, °support, °buttress: *Owing to lack of maintenance, the piers supporting the building have crumbled.*

pierce *v.* 1 °stab, °puncture, °penetrate, thrust *or* poke into, °lance, spear, °spit, run through *or* into, skewer, °impale, °fix, °transfix: *The arrow pierced his heart and he dropped down dead.* 2 bore into *or* through, °penetrate, °drill, °perforate, °riddle, punch through, °hole, tunnel into: *The wall is pierced to provide ventilation.* 3 °penetrate, °fathom, °see, °understand, °comprehend, , °grasp, °discover, °realize: *His keen analytical mind allowed him to pierce the mysteries of nature.* 4 °affect (keenly), °touch, °move, °melt, °stir, °rouse, °pain, cut to the quick, °wound, °strike: *It pierced my heart to hear the child weep.*

piercing *adj.* 1 °strident, °shrill, °harsh, earsplitting, earshattering, high-pitched, screaming, shrieking, screeching, °loud, blaring: *The piercing sound of the police siren made me sit bolt-upright in bed.* 2 probing, searching, °penetrating, °sharp, °keen; °arresting, gripping, °spellbinding, °enthralling, fascinating, entrancing: *I was completely transfixed by her piercing green eyes.* 3 °penetrating, °icy, frosty, °frigid, chilling, °freezing, °cold, numbing, °keen, °wintry, arctic, °raw, °bitter, °fierce, °biting, nipping, nippy: *Shivering in that piercing wind I thought I'd never be warm again.* 4 stabbing, shooting, °excruciating, °exquisite, °acute, °sharp, °severe, °agonizing, °fierce, °intense, °painful, racking: *I suddenly felt a piercing pain in my left ear.*

piety *n.* 1 °devotion, devotedness, °respect, °deference, °dedication, dutifulness, °loyalty, °affection: *In filial piety he hung the painting of his parents prominently over the mantelpiece.* 2 piousness, °reverence, °veneration, devoutness, holiness, godliness, pietism, devotedness, °devotion, °observance, religiousness, °grace, °sanctity: *His life of piety had marked him out as a likely candidate for sainthood.*

pile[1] *n.* 1 °heap, °mound, °stack, °accumulation, stockpile, °mass, °supply, °deposit, °collection, assemblage, °batch, °hoard, aggregation, congeries, conglomeration, °assortment, agglomeration, concentration, amassment: *A huge pile of gravel was delivered today for the builders.* 2 °money, °fortune, °wealth, holdings, *Colloq* °bundle, °loot, °mint, *Slang* °packet, tidy sum, *US* bankroll, °roll, °wad: *She made her pile selling arms to terrorists.* 3 Usually, ***piles.*** °abundance, overabundance, superabundance, °plenty, °great deal, °quantity, ocean(s), lot(s), stack(s), plethora, *Colloq* oodles, ton(s), bag(s), heap(s), bundle(s): *He made piles of money in the black market.* 4 See **pier, 2,** above. —*v.* 5 Often, ***pile up.*** °stack (up), °heap (up), °mound, °accumulate, stockpile, °amass, °collect, °assemble, °hoard, aggregate, cumulate: *Please pile the cartons in the corner.* 6 ***pile in*** *or* ***into.*** °enter, get in *or* into, crowd in *or* into, pack in *or* into, flood in *or* into, jam in *or* into, crush in *or* into, *Colloq* jump in *or Colloq* into: *All of us piled into my car to go to the drive-in.* 7 ***pile it on.*** °exaggerate: *Ronnie was really piling it on about how much his new job pays.* 8 ***pile on*** *or* ***onto.*** **(a)** get in *or* into *or* on *or* onto, crowd on *or* onto, jump on *or* onto: *We piled on the train after the game. They piled onto the hay wagon for a ride home.* **(b)** °attack, °assault, °jump on, °overwhelm: *They all piled on me and I had to give up.* 9 ***pile out.*** °leave, get out (of) *or* down (from), °exit: *When we arrived, we all piled out of the bus. Hordes of people piled out of the theater.* 10 ***pile up.*** °accumulate, °amass, °collect: *The rubbish kept piling up during the strike.*

pile[2] *n.* °nap, shag, °plush; fuzz, bristles, °fleece: *The feet of the chairs have left marks in the carpet pile.*

piles *n.* pl. hemorrhoids.

pileup *n.* 1 °(road) accident, smash, °crash, (multiple) (rear-end) collision, *Colloq* smashup: *Thirty cars were involved in that pileup on the turnpike.* 2 °accumulation, °heap, °stack, °mass, *Colloq* °mountain: *How will you ever get through that pileup of work on your desk?*

pilfer *v.* °steal, °rob, °plunder, thieve, filch, °embezzle, °misappropriate, purloin, °take, walk off with, palm, *Colloq* °appropriate, °pinch, °snatch, °grab, °lift, °borrow, *Brit* °nick, snaffle, *US* °boost, *Slang* °hook, snitch, °swipe, °rip off: *The auditors found that he had pilfered small sums from the company for years.*

pilgrim *n.* hajji *or* hadji *or* haji, *Medieval history* palmer; crusader: *The pilgrims visited holy places in and near Jerusalem.*

pilgrimage *n.* hajj *or* hadj, holy expedition, °crusade; °expedition, °journey, trek, voyage, °tour, °trip, °excursion: *Every year the entire family made a pilgrimage to grandfather's grave.*

pill *n.* 1 °tablet, capsule, bolus, pellet, pilule; °medicine, medication, medicament, °drug, pharmaceutical, °remedy, °cure; cough drop, pastille, lozenge, troche: *Doctor, can't you give me some kind of pill for this headache?* 2 °nuisance, °bore, °pest, *Colloq* °pain (in the neck), °crank, °drag: *I don't understand what she sees in Leonard—he's such a pill.*

pillage *v.* 1 °plunder, °raid, °ravage, °sack, despoil, °rob, °loot, °ransack, °rifle, maraud, depredate, °devastate, vandalize, °ruin, °demolish, °raze, °level, °strip: *The Goths pillaged every community they conquered and carried off the booty.* —*n.* 2 °plunder, rapine, despoliation, looting, °robbery, °sack, sacking, ransacking, marauding, brigandage, piracy, freebooting, buccaneering, banditry, °depredation, devastation, vandalization, defilement, °destruction, laying waste, °destruction, razing, demolition, leveling, °ruin, stripping: *Operating from Tripoli, pirates were responsible for the pillage of one coastal town after another.* 3 °plunder, °loot, °booty, °spoils: *In the cave Ali Baba found the pillage from a thousand robberies.*

pillar *n.* 1 column, pilaster, °pile, piling, °pier, °upright, °post, °shaft, °prop; atlas, caryatid: *The roof is supported by a single pillar. Lot's wife was turned into a pillar of salt.* 2 °mainstay, °supporter, °worthy, upholder, °backbone, °(tower of) strength, °leader: *Cummings has been a pillar of strength in our community.*

pilot *n.* 1 aviator, aviatrix, flier, airman, airwoman, aeronaut, captain: *The pilot announced that we would land in five minutes.* 2 steersman, helmsman, °navigator, *US* wheelman *or* wheelsman; °guide, °leader, cicerone, conductor: *The pilot knows his way through the shoals. Cavanaugh acted as our pilot in our tour of the Casbah.* —*v.* 3 °guide, °steer, °run, °direct, °shepherd, °control, °lead, °navigate, °drive; °fly: *Only someone with local knowledge can pilot the ship into that harbor. The chairman piloted the company through turbulent times.*

pimp *n.* 1 °procurer, panderer *or* pander, white slaver, whoremonger, *Slang US* hustler: *The police rounded up the pimps and prostitutes for questioning.* —*v.* 2 °procure, °pander, °solicit, *Slang US* °hustle: *He was well known as having pimped for three generations of the nobility.*

pimple *n.* pustule, papule, °boil, °swelling, °eruption, blackhead *or Technical* comedo, excrescence, *Brit* °spot, *Scots* plouk *or* plook, *US* whitehead, *Old-fashioned US* hickey: *The old witch had a pimple at the end of her nose.*

pin *n.* **1** °peg, dowel, °bolt, thole, tholepin, °spike, rivet; *Brit* drawing pin, *US* pushpin: *The table is held together entirely by wooden pins. We need some more pins for the bulletin board.* **2** °brooch, °clip; stickpin, tiepin, scarfpin, *US* tietack: *She is wearing the cameo pin that her mother gave her. I had George's pin made into a brooch.* —*v.* **3** °attach, °fix, affix, °fasten, °secure, °tack; °hold, °staple, °clip: *Let's play "Pin the tail on the donkey." Pin these papers together.* **4** ***pin down.*** **(a)** °force, °make, compel, coerce, constrain, °press, *Brit* pressurize, *US* °pressure: *We must pin him down to give his decision by tomorrow.* **(b)** °define, °specify, pinpoint, °name, °identify, °determine, °name, put *or* lay one's finger on, home *or* zero in on, focus on: *The doctor was unable to pin down what is wrong with Matilda.* **(c)** confine, °hold (down), °fix, immobilize, °tie down, constrain: *We were pinned down by enemy fire.* **5** ***pin on.*** °blame, hold responsible *or* accountable, point the finger at, °accuse; lay at (someone's) door, *Slang* °nail: *They'll never be able to pin the murder on Drayton.*

pincers *n.* pl. pliers, nippers, tweezers: *Can you get the nail out with these pincers?*

pinch *v.* **1** °squeeze, °nip, °tweak, °press, compress, °grip, °grasp: *I pinched my finger in the drawer.* **2** °squeeze, cramp, confine, °crush, °hurt: *These shoes really pinch badly.* **3** °steal, thieve, °rob, °take, shoplift, filch, °pilfer, purloin, *Colloq* °lift, *Brit* °nick, *US* °boost, *Slang* °swipe, °knock off: *I think it was Andrew who pinched my book.* **4** °arrest, apprehend, take into custody, *Colloq* °nab, °run in, collar, bust, °run in, *Brit* °nick: *She was pinched for driving while under the influence.* **5** ***pinch pennies.*** scrimp, °save, skimp, °economize: *We're pinching pennies now so that we can afford to buy a better car next year.* —*n.* **6** °squeeze, °nip, °tweak, °twinge: *Montrose gave the girl's cheek an affectionate pinch.* **7** °touch, (tiny *or* wee) bit, soupçon, °jot, mite, °taste, *Colloq US* tad, smidgen *or* smidgin: *Add a pinch of salt to the boiling water.* **8** °predicament, °emergency, °crisis, °difficulty, °dilemma, (ticklish *or* delicate) situation, °complication, *Colloq* pickle, °jam, °scrape, *Chiefly Brit* °crunch: *Sue will help me out in a pinch.*

pink[1] *n.* **1** ***in the pink.*** at one's best, °healthy, °hearty, in the best of health, in top form, in good shape, *Colloq US* up: *I saw Rob the other day, and he's in the pink.* —*adj.* **2** °rosy, rose, rose-colored, pinkish, flesh-color(ed), salmon(-color(ed)): *The designer suggests pink curtains for the bedroom.*

pink[2] *v.* serrate, °notch, scallop; °perforate, °puncture, °prick: *Pink the edge of the fabric to prevent fraying.*

pinnacle *n.* °top, °peak, apex, °acme, °summit, °zenith, °maximum, °climax, crowning point, °consummation, utmost, °extreme, °perfection; °tip, °cap, °crest, °crown: *Being elected chairman was the pinnacle of Mark's career. The climbers reached the pinnacle, where they will rest overnight.*

pioneer *n.* **1** pathfinder, frontiersman, frontierswoman, trailblazer, explorer, colonist, (early) settler; groundbreaker, °forerunner, °precursor, predecessor, innovator, °leader, trendsetter, pacemaker, pace-setter: *The pioneers who explored and settled the American west were daring men and women. Lee De Forest was a pioneer in the development of radio.* —*v.* **2** °create, °originate, °invent, °initiate, take the first step, °introduce, °institute, actuate, trigger, °set off, °inaugurate, °start, °begin, °launch, °establish, °found, °set up, °develop, lay the groundwork *or* foundation, set *or* put in motion, take the lead, lead *or* show the way, blaze the trail, be a prime mover, open up, break new ground, *Colloq* kick off, get the ball rolling: *Our company has pioneered systems for automatic typesetting.*

pious *adj.* **1** °devout, °religious, reverent, reverential, worshipful, °dutiful, God-fearing, °godly, °faithful, °holy, dedicated, °devoted, °spiritual, °moral, °good, °virtuous, right-minded, °saintly, °holy, angelic, °seraphic, Christ-like, God-like: *Extremely pious, Francis attends church regularly and obeys all the Commandments.* **2** °hypocritical, °sanctimonious, pietistic, °self-righteous, °pharisaic, °mealy-mouthed, °pretended, °fraudulent, °goody-goody, unctuous, °oily, *Colloq Brit* smarmy: *Pious deception marked their so-called service to God.*

pipe *n.* **1** °pipeline, tube, duct, hose, °line, °main, conduit, °passage, conveyor, °channel: *The pipes are inspected regularly for corrosion.* **2** briar, meerschaum, corncob, calabash, clay pipe, water pipe, hookah, narghile, chibouk *or* chibouque, peace pipe *or* pipe of peace *or* calumet, *Colloq* Irish briar, *Brit* hubble-bubble, *US* bong: *He slowly tamped down the tobacco in his pipe, then lit it.* **3** panpipe, whistle, boatswain's pipe, tooter, horn, °wind, wind instrument, woodwind, °brass: *They play the pipes in the London Symphony.* **4** tootle, tweet, skirl, warble, whistle, °peep, cheep: *One was piping away on a flute, the other was dancing a jig.* **5** °transmit, °deliver, °channel, °conduct, convey, °supply: *The gas is piped directly into our homes.* **6** *US* °look at, °notice, °spot, °note, °look at, *Colloq* get a load of: *Pipe the guy trying to climb up the outside of that building.* **7** ***pipe down.*** become quieter, quiet(en) down, make less noise, °hush (up), shush (up), °whisper, *Colloq* belt up, °shut up: *I wish they'd pipe down, I can't sleep.* **8** ***pipe up.*** °speak (up), °say, raise one's voice, make oneself heard, °offer, volunteer: *Sally piped up with the correct answer from the back of the classroom.*

pipeline *n.* **1** °pipe, tube, duct, hose, °line, °main, conduit, °passage, conveyor, °channel: *Pipelines carry gas from the North Sea throughout Britain. Very little information was flowing through the pipeline from Beirut.* **2** ***in the pipeline.*** on the way, °under way, in the offing, °ready, °imminent, coming, *Colloq* in the works, cooking, *US* °in work: *They said that the contract was in the pipeline and should arrive soon.*

pirate *n.* **1** buccaneer, sea rover, corsair, privateer, freebooter, sea robber, filibuster, *Archaic* picaroon: *Captain Kidd and Blackbeard were actual pirates who looted shipping in the Americas in the seventeenth century.* **2** plagiarist, plagiarizer, infringer: *Some of these pirates offer unauthorized cheap reprints of expensive textbooks.* —*v.* **3** plagiarize, °infringe, °copy, °reproduce, °steal, °appropriate, poach, *Colloq* °lift, °pinch, crib: *Our government has no jurisdiction over those who pirate books in countries with which we have no treaty.*

pirouette *n.* **1** °spin, whirl, °twirl, °turn, °revolution, pivoting: *Antoinette did a beautiful pirouette followed by a pas de chat.* —*v.* **2** °spin, whirl, °turn (round), °revolve, °pivot: *Alphonse was so happy that he fairly pirouetted round the room.*

pistol *n.* gun, handgun, °revolver, °automatic, *Slang* °rod, °piece, shooting iron, *Chiefly US* gat, *US* Saturday night special, heater, roscoe: *In the United States, law enforcement officers are required to carry a pistol.*

piston *n.* plunger: *The piston in a bicycle pump is worked by hand.*

pit[1] *n.* **1** °hole, °excavation, °shaft, °cavity, °mine, mineshaft, ditch, trench, trough: *The men went down into the pit every morning at dawn.* **2** pothole, °hollow, °depression, dent, °indentation, dimple, pockmark: *The entire surface was marked by pits.* **3** °abyss, chasm,

°well, °crevasse, crater: *We found a bottomless pit that led to the center of the earth.* **4** ***the pits.*** °awful, °terrible, the worst, *Slang* °lousy: *That TV show last night was the pits. He thought that going to school was the pits.* —*v.* **5** dent, pockmark, °dig, °scar, hollow out, °gouge: *The salt air has pitted the metal parts of my car.* **6** Often, ***pit against.*** °match, °oppose, °set against; °contrast: *Shirley doesn't stand a chance if they pit her against Maria.*

pit[2] *n.* stone, °seed, pip: *I prefer seedless grapes, the ones without pits.*

pitch[1] *v.* **1** °toss, °throw, °cast, °fling, °hurl, °heave, °sling, °fire, °launch, °shoot, °send, °let fly, *Cricket* °bowl, *Colloq* chuck, °peg, °lob, *Brit* bung: *He rolled the paper into a ball and pitched it into the basket.* **2** °erect, °raise, set *or* put up, °position, °fix, °place: *We ought to pitch the tent in a level area.* **3** °plunge, °fall (headlong), °dive, °drop, °plummet, (take a) nose dive: *I caught my toe on the curb and pitched forward onto the pavement.* **4** *Chiefly nautical* toss about, °lurch, °plunge, °flounder, go head over heels, go keel over truck, *US* pitchpole *or US* pitchpoll: *The wind reached force ten as we pitched and rolled and yawed in the heavy seas.* **5** ***pitch in.*** °contribute, °cooperate, °help, °assist, *Colloq* °chip in: *Everyone pitched in to make the church fair a success.* **6** ***pitch into.*** **(a)** °attack, °lay into, assail, lash out at, °abuse, rail against, *Colloq* °lace into, tear into, jump down (someone's) throat, °jump on: *Reggie's wife really pitched into him about going out with other women.* **(b)** °attack, °assault, °set upon, °belabor, *Colloq* °light into, sail into, tear into: *They pitched into each other and fought like Kilkenny cats.* **7** ***pitch on*** *or* ***upon.*** °determine, °decide on, °select, °pick, °choose, opt for, °elect, °nominate, °name, *Colloq* °light on: *They pitched on Carrie to be the best candidate for treasurer.*

pitch[2] *n.* tar, bitumen, asphalt: *Peter patched potholes with pitch.*

pitch-black *adj.* °black, °dark, ebon(y), stygian, inky (black), unlit, unlighted, pitch-dark, coal-black, jet-black; raven, sable: *The cellar was pitch-black. Her pale complexion was in stark contrast to her pitch-black hair.*

pitched *adj.* organized, planned, °deliberate, coordinated, arranged, systematized: *The two armies fought a pitched battle on the plain.*

piteous *adj.* °pitiable, °pathetic, °pitiful, plaintive, °miserable, °heart-rending, plangent, °poignant, distressing, °grievous, heartbreaking, °mournful, °sad, °doleful, dolorous, °tearful, °lamentable, °deplorable, °regrettable, rueful, woeful, °moving, °emotional: *We heard the piteous wailing of the mothers who had lost children in the disaster.*

pitfall *n.* **1** °trap, °pit: *They often used pitfalls for capturing tigers in Burma.* **2** °danger, °peril, °hazard, °catch, °difficulty, °snag: *You might well encounter pitfalls, but don't be discouraged.*

pith *n.* **1** °core, °heart, °kernel, °nucleus, crux, °gist, °focus, focal point, °essence, °meat, marrow, °nub, °point, °spirit, °substance, °quintessence: *As usual, Randolph at once came directly to the pith of the argument.* **2** °weight, °burden, gravamen, °gravity, °force, °moment, °import, °importance, °significance, °substance, °depth, °matter: *I have something of great pith to tell you about.*

pitiable *adj.* See **piteous,** above.

pitiful *adj.* **1** See **piteous,** above. **2** °small, °little, °insignificant, °trifling, unimportant, beggarly, °sorry, °mean, °contemptible: *Burnside actually expected us to praise him for that pitiful contribution of his?*

pittance *n.* mite, shoestring, *Slang* peanuts, chickenfeed, small potatoes: *The miserable pittance that she receives does not even cover the necessities of life. He's trying to run that business on a pittance.*

pitted *adj.* eaten away, corroded, eroded, pockmarked, defaced, marred, pierced, perforated: *The chrome plating is all pitted where the acid splashed.*

pity *n.* **1** °sympathy, commiseration, °sorrow, condolence, compassion, tenderness, *Archaic* ruth: *I really feel pity for Betty, being married to such a brute.* **2** (crying *or* damned) shame, sad thing, °disgrace, °misfortune, °sin, °sacrilege, *Colloq* °crime: *It's a pity that no one can do anything.* —*v.* **3** °sympathize, °feel for, commiserate with, feel sorry for, feel *or* have compassion *or* tenderness for, bleed for, weep for: *I pity any mother whose son goes off to war.*

pivot *n.* **1** pintle, gudgeon, hinge, °swivel, °pin, kingpin, spindle, fulcrum: *Use this stone as a pivot for the lever to lift the rock.* **2** °center, °heart, focal point, °hub, nave, crux: *The finance minister regards the interest rate as the pivot on which the economy turns.* —*v.* **3** °rotate, °revolve, °turn, °spin, °twirl, whirl, °swivel: *The flywheel pivots on a bearing that requires constant lubrication.* **4** hinge, °depend, °hang, be contingent, revolve around, °rely: *The safety of the hostages pivots on the mood of the terrorists.*

pivotal *adj.* °critical, °central, °focal, °crucial, °significant, °important, °essential, °vital, °pressing, °urgent, °radical: *The attitude of the judge is pivotal in the jury's decision.*

place *n.* **1** °location, °site, °position, °point, °spot, locus, °area, °locale, °scene, °setting: *This looks like a nice place for a picnic. She likes to see a place for everything and everything in its place.* **2** °locale, °area, °neighborhood, °vicinity, °district, °section, part of the country, °quarter, °region; °city, °town, village, hamlet: *She comes from some place near Glasgow.* **3** °status, °station, °standing, °grade, °rank, °position, °niche, °slot, °situation, °estate, °state, circumstance(s): *Angela was just saying how difficult it is today to find a servant who knows his place.* **4** °function, °role, °part, °purpose, °duty, °obligation, °task, °responsibility, °charge, chore, °burden, °concern, °mission: *It is scarcely my place to remind you of your appointments with the dentist.* **5** °position, °job, °post, berth, °appointment, livelihood; °employment, °occupation, *Colloq* billet: *Is there a chance of my earning a place in your new company?* **6** °home, °house, °flat, apartment, room(s), °quarters, lodgings, *Rather formal* °residence, °domicile, °dwelling, °abode, *Colloq* digs *or* diggings, °pad: *Why not stop by my place for tea on Sunday?* **7** stead; lieu: *As I cannot go, would you go in place of me?* **8** °position, °situation, °circumstances, °condition: *Put yourself in my place and I think you would have done exactly the same.* **9** °seat, °chair, °position: *Kevin, please take your place at the head of the table.* **10** ***go places.*** °succeed, become successful, °get ahead, °advance, °prosper, °thrive, °flourish, go up in the world, °make good, strike it rich, *Colloq* °arrive, make a splash, *US and Canadian* hit pay dirt, luck out: *That boy will go places one day.* **11** ***in place.*** **(a)** °fitting, °suitable, °appropriate, °right, °proper, °correct, good form: *I don't think it in place for you to tell me what to do.* **(b)** *in situ,* in (the right *or* proper *or* correct) position, °ready, all set, °set up, °in order, all right, *Colloq* OK *or* O.K. *or* okay: *Is everything in place for tonight's party?* **12** ***out of place.*** °awkward, uncomfortable, unsuitable, °inappropriate, °wrong, °improper, misplaced: *Compassion is out of place when dealing with war criminals.* **13** ***put (someone) in his*** *or* ***her*** *or* ***their place.*** °humble, °mortify, bring down, °embarrass, °squelch, *Colloq* cut down to size, take down a peg (or two): *Aunt Agatha used to put Uncle Wilfred in his place by reminding him who held the purse strings.* **14** ***take***

place. °happen, °occur, °go on, °come about; °arise, *Colloq* °transpire: *We shall never know what took place behind those locked doors.* —*v.* **15** °put (out), °position, °situate, °locate, °dispose, °arrange, °order, °set (out), °lay, °deposit; °station, °post, °spot, pinpoint, *Colloq* °stick, *Brit* bung: *Place the forks on the left and the knives on the right. They placed guards at the door of my room.* **16** °class, classify, °sort, °order, °arrange, °rank, °group, °categorize, °bracket, °grade; °regard, °view, °see, °consider: *She places love of family above love of country. Critics place him among the best writers of the century.* **17** °identify, °put one's finger on, °recall, °remember, °recognize; °associate: *I just can't place her for the moment. He finally placed me with those who had ragged him at school.* **18** °put, °set, °assign, °give: *People place too much importance on material things.*

placement *n.* **1** °arrangement, placing, °position, °distribution, array, °disposition, deployment, positioning, stationing, °organization, °order, ordering, °location, locating, arraying, emplacement, emplacing: *The placement of the chairs is all wrong for tonight's meeting.* **2** °employment, °appointment, °engagement, hiring: *Placement of qualified engineers has not been a problem.*

plagiarism *n.* plagiarizing, plagiary, piracy, pirating, °theft, purloining, °stealing, copying, appropriating, appropriation, thievery, usurpation, infringing, °infringement, °imitation, *Euphemistic* borrowing, *Colloq* lifting, cribbing: *The similarities between the two books could be attributable only to plagiarism.*

plague *n.* **1** °scourge, °epidemic, °pestilence, °affliction, pandemic, °calamity, °curse, °evil, bane, °blight, °visitation: *The inhabitants turned a plague of locusts to advantage by eating them.* **2** irritation, °annoyance, °nuisance, °pest, vexation, °bother, thorn in one's side *or* flesh, °torment, torture, *Colloq* °pain (in the neck), °headache, aggravation, *Slang* °drag, °bitch, hassle, *Taboo slang* pain in the *US* ass *or Brit* arse: *It's a plague trying to find a place to park the car.* —*v.* **3** badger, harry, °hound, °pester, °annoy, vex, °irritate, °bother, °harass, °nag, nettle, needle, °exasperate, °gall, °annoy, °irk, °torment, torture, °anguish, °distress, *Brit* chivy *or* chivvy *or* chevy, *Colloq* °bug: *I wish the police would stop plaguing me with questions about Jonathan's whereabouts.*

plain *adj.* **1** °flat, °smooth, °even, featureless, °level, °plane: *I prefer a plain wallpaper to an embossed one.* **2** °clear, °evident, °simple, °distinct, crystal-clear, lucid, °vivid, °transparent, °apparent, °obvious, °patent, °self-evident, °manifest, °distinct, unmistakable *or* unmistakeable, unequivocal, unambiguous, understandable, °intelligible, °graphic, °direct, in black and white: *His intentions regarding my daughter are plain enough. The plain fact is that she despises him.* **3** °open, °honest, straightforward, °forthright, °direct, °frank, °candid, °blunt, °outspoken, °ingenuous, °sincere, guileless, °artless, unreserved: *I want to see some plain talk between the two of us.* **4** °simple, °unadorned, undecorated, unembellished, °basic, austere, °stark, unostentatious, °colorless, °drab, °bare, °unvarnished, °Spartan: *Don't you find the average business suit a very plain affair?* **5** °homely, unattractive, ordinary-looking, unlovely, °ugly: *Who would believe that such a plain child could become such a beautiful woman?* —*n.* **6** prairie, grassland, °pasture, meadowland, veldt *or* veld, pampas, campo, llano, savanna *or* savannah, steppe, tundra, champaign *or* campagna; heath; °moor, moorland; °plateau, flatland; down, downland, *Literary* wold, *Literary or archaic* mead: *The plain stretched out before us as far as the eye could see.*

plan *n.* **1** °design, layout, blueprint, °scheme, °method, °procedure, °system, °arrangement, °program, °project, °formula, °pattern, *Colloq* °script, °scenario: *The plan is to warm up the audience with a comedian. The best laid plans can often go wrong.* **2** °drawing, sketch, °design, layout, blueprint, °chart, map, diagram, °representation: *This plan shows where the furniture will be placed.* —*v.* **3** lay out *or* down, °design, °arrange, °devise, °outline, °organize, °plot, map out, delineate, °develop: *She has been invited to plan the new shopping center.* **4** °intend, °expect, °envisage, °envision, °foresee, °aim, °contemplate, °propose: *Why don't we plan to stay home and watch TV tonight?*

plane *n.* **1** flat *or* level (surface): *The lines meet in the same plane.* **2** airplane, aircraft, airliner, jet (plane): *I caught the next plane for Marrakesh.* —*adj.* **3** °flat, °even, °level, °smooth, °plain, °regular, unbroken, uninterrupted, °uniform, °horizontal: *We landed on the plane surface of the glacier.* —*v.* **4** °glide, °skim, skate, skid, °slip, °slide: *I swiftly planed along on my sailboard.*

plank *n.* °board, °timber, °slab: *The flooring was made up of wide planks laid side by side.*

plant *n.* **1** °flower, vegetable, herb, bush, shrub, tree, vine, weed: *Because of all the rain, the plants are flourishing this summer.* **2** °factory, °mill, °shop, °works, workshop, foundry: *The new plant in Crawley is hiring lathe operators.* **3** °equipment, machinery, °apparatus; °gear, fixtures: *The plant includes heavy cranes, earth movers, and bulldozers.* **4** °spy, (undercover *or* secret) agent, °informer, informant: *Chatham is a plant, sent in by management to report on union activities.* —*v.* **5** bed (out), °sow, °seed, °set (out), °transplant: *We planted a herbaceous border along the south wall of the garden.* **6** °implant, °establish, °root, °fix, ingrain, °lodge, °instill, °insinuate, °inject, °introduce, °impress, imprint: *Who planted the idea in your mind that you were a gifted writer?* **7** °place, °put, °position, °station, °assign, °situate: *Watchtowers are planted at fifty-foot intervals around the prison.* **8** °hide, °secrete, °conceal: *The company has planted detectives in the store to watch out for shoplifters.*

planter *n.* flowerpot, cachepot: *He poured the drugged wine into a nearby planter.*

plaque *n.* **1** °tablet, medallion, °plate, panel, marker, °slab, plaquette: *The plaque on the house marks it as the birthplace of Thomas Carlyle.* **2** badge, °brooch, °pin, °patch, medallion, medal, insignia *or* insigne: *The plaque shows him to be a member of the Royal Yacht Squadron.* **3** °prize, °honor, °award, °trophy: *At the annual dinner, she was presented with a plaque to mark her many years of service.*

plaster *v.* °smear, daub, bedaub, °spread, °coat, °cover, overlay, superimpose: *The mud was plastered all over my boots. They plastered posters all over the walls.*

plastic *adj.* **1** moldable, shapable *or* shapeable, fictile, °soft, malleable, workable, ductile, °flexible, °soft, pliant, °supple, °pliable, clayey, waxy: *The clay must be worked into the desired form while it is still plastic.* **2** °impressionable, °receptive, °open, persuadable *or* persuasible, °susceptible, °tractable, compliant, °responsive, °manageable, unformed, °inexperienced: *One encounters the children at an age when their minds and personalities are still plastic enough to be influenced.* **3** °artificial, °synthetic, °imitation, °fake, °counterfeit, ersatz, paste, °bogus, meretricious, °sham; °cheap, pinchbeck, °shoddy; *Colloq* phoney *or US also* phony, crummy, *US* chintzy: *She was wearing a tawdry plastic brooch.*

plate *n.* **1** °platter, dish, *Archaic* trencher, charger: *I left the cold mashed potato on my plate.* **2** °course, serving, °portion, dish, °platter: *I ordered their specialty, a plate of spaghetti.* **3** layer, leaf, °sheet, °pane, panel, lamina, °slab: *In the condenser, insulation separates the thin metal plates.* **4** coating, °coat, plating, layer, lamination: *The plate on our cutlery is wearing thin.* **5** °illustration, °picture, °print, *US* °cut: *The book contains many beautiful color plates of flowers.* —*v.* **6** °cover, °coat, overlay, °face, laminate: *All the serving dishes were plated with gold.*

plateau *n.* **1** tableland, upland, highland, mesa: *We climbed till we reached a grassy plateau.* **2** °level, °lull, °pause, leveling off: *At thirty-nine, Julia felt she had reached a plateau in her career.*

platform *n.* **1** °stand, dais, °stage, podium, °rostrum: *After the introduction, the speaker mounted the platform.* **2** °policy, party line, principle(s), tenet(s), °program, °plank: *The main elements of our platform will be revealed at the party conference.*

platonic *adj.* nonphysical, asexual, nonsexual, °celibate, °chaste, °dispassionate, °detached, °spiritual, °ideal, °intellectual: *Some say that their relationship has not always been purely platonic.*

platoon *n.* °company, °squad, squadron, °group, °patrol, °team, cadre, °body, °formation, °unit, *Colloq* °outfit: *A platoon of soldiers was marched to the barracks in close-order drill.*

platter *n.* serving dish, server, salver, tray, °plate, dish: *Waiters walked among the guests with platters of hot hors d'oeuvres.*

plausible *adj.* **1** °likely, believable, °reasonable, credible, °tenable, conceivable, °thinkable, °probable, imaginable, admissible, °sound, °sensible, °rational, °logical, °acceptable, °trustworthy, °presentable: *The police regarded our alibi as plausible.* **2** °specious, °deceptive, meretricious, misleading, °deceitful, casuistic, sophistical, jesuitical, °smooth, °empty: *He was a cunning, plausible sort of fellow.*

play *v.* **1** amuse oneself, °frolic, °frisk, °cavort, gambol, °caper, °sport, have fun, have a good time, °enjoy oneself, disport (oneself), °carouse: *Ken's mother won't let him go out and play.* **2** °participate (in), take part (in), °join (in), be occupied (in *or* with); °engage in, contend in, °take up, occupy oneself in *or* with, °undertake: *He was invited for a game of poker, but he refused to play. I understand that you play bridge.* **3** °engage, contend with, compete with *or* against, °challenge, vie with, pit oneself against, °take on, °rival: *I played pool with the stranger, and he beat me three games out of three.* **4** °portray, depict, °perform, act *or* take the role *or* part of, °act: *In the new production of* Spitfire *she plays Cuspidora.* **5** perform (upon *or* on); °put on: *Play "Misty" for me. She plays the piccolo very well. Could you play that Caruso record?* **6** °operate: *I wish they wouldn't play their hi-fi so loud.* **7** °gamble, °bet, wager, °stake, °place, °put: *He played his last chip on number 14.* **8** ***play along.*** **(a)** Often, **play along with.** cooperate, °participate, go along (with), do *or* play one's part, be a party to: *I agreed to play along with her charade.* **(b)** °manipulate, °jolly along: *She played him along till he made her his beneficiary.* **9** ***play around.*** **(a)** °fool around, °tease, *Colloq* monkey about *or* around, horse around *or* about: *Stop playing around and get to work.* **(b)** dally, °flirt, be unfaithful; philander, womanize; *Colloq* °fool around, °run around, sleep around, play the field: *She found out about his playing around and filed for divorce.* **10** ***play at.*** °pretend, °make believe, °fake, feign, simulate, °affect: *She's merely playing at enjoying skiing to please you.* **11** ***play ball.*** °cooperate, °agree, work together, work hand in glove, °play along: *They want him to smuggle diamonds, but he won't play ball.* **12** ***play by ear.*** °improvise, extemporize, ad-lib, *Colloq* wing it: *She can't read music and just plays by ear. With no definite plan of action, I'll just play it by ear.* **13** ***play down.*** °belittle, °minimize, °diminish, °disparage, °make light of, deprecate, decry, de-emphasize: *He has always played down his role in the affair.* **14** ***play for time.*** °delay, °procrastinate, °stall (for time), temporize, °hesitate, *Colloq* °drag one's feet: *They don't yet have the money to pay, so they are playing for time.* **15** ***play on*** *or* ***upon.*** °use, °misuse, °abuse, trade on, °exploit, take advantage of, °impose on: *He plays on women's affections then persuades them to give him money.* **16** ***play the game.*** °behave, °conduct oneself, deport oneself, °act: *It makes no difference whether you win or lose, it's how you play the game.* **17** ***play up.*** **(a)** °stress, °emphasize, underscore, underline, accentuate, call attention to, highlight, °spotlight, °dramatize, °build up: *Always try to play up your assets and play down your liabilities.* **(b)** act up, °misbehave, give *or* cause trouble, malfunction, *Colloq* go on the blink *or* *US* fritz, *Brit* be wonky: *The bloody engine started playing up again, right in the middle of a rainstorm.* **18** ***play up to.*** curry favor with, °flatter, toady to, ingratiate oneself with, butter up, truckle to, court, *Colloq* soft-soap, suck up to, bootlick, *US* apple-polish, *Slang taboo* brown-nose: *Ray is always playing up to the teacher, trying to get a better mark.* **19** ***play with.*** **(a)** °toy with, °trifle with, treat cavalierly *or* lightly, °make light of, think nothing of, dally with, amuse oneself with: *He's just playing with her till he gets what he's after.* **(b)** °consider, °think about, °toy with, not treat seriously: *We were playing with the idea of a winter holiday this year.* **(c)** °mess with, °fiddle with, °toy with, fidget with: *Stop playing with your food!* —*n.* **20** °drama, stage play, °show, °piece, °production, °entertainment: *We have tickets to a different play for every night this week.* **21** °behavior, °actions, deportment, °conduct, demeanor: *In the game of life, many have no regard for the rules of fair play. The police suspect foul play.* **22** °amusement, frivolity, °entertainment, °recreation, °fun, °pleasure, °sport, merrymaking, °revelry, tomfoolery, *Colloq* horseplay, skylarking, monkey business, *Brit* monkey tricks *or* *US* monkeyshines: *The time for play is past, and we must get down to serious business.* **23** °move, °maneuver, °action: *That last play might have won you the game.* **24** °flexibility, looseness, °freedom, °leeway, °margin, °room, °space, °movement, °motion, *Colloq* °give: *There's too much play in this gear lever.* **25** °treatment, coverage, °attention: *The newspapers gave Connie's new book a big play.*

playboy *n.* man about town, °roué, °rake, debauchee, gay dog, womanizer, Don Juan, Casanova, Lothario, Romeo, *Colloq* wolf, lady-killer: *Egbert fancies himself a playboy, but no women seem to agree.*

player *n.* **1** °contestant, °participant, °competitor, contender; athlete, sportswoman, sportsman, *Colloq US* jock: *Chess players from everywhere came together for the match. The referee sent three players off.* **2** actor *or* actress, °performer, entertainer, trouper, °thespian: *Mr. Openshaw was a renowned member of the Mighty Allen Arts Players.* **3** gambler, bettor *or esp. Brit* better, gamester, speculator, *Brit* punter: *C.B. is a big player in the commodities market, where the stakes are high.* **4** musician, instrumentalist, °performer, °virtuoso: *Gladys is one of the best tuba players in the band.*

playful *adj.* **1** (high-)spirited, °cheerful, °frisky, °frolicsome, °kittenish, °sprightly, fun-loving, °sportive, coltish, °mischievous, puckish, impish, elfish, °devilish: *The cat didn't mean to scratch you, she was just being playful.* **2** joking, facetious, teasing, roguish, waggish, jesting, °humorous, tongue-in-cheek: *Cleo's in a playful mood today, isn't she?*

playmate *n.* playfellow, °friend, °comrade, *Colloq* °pal, °chum, *US and Canadian* buddy: *Some of Molly's playmates are asking if she can come out to play.*

plaything *n.* 1 °toy, °game, knickknack *or* nicknack, °pastime: *Please pick up your playthings before supper.* 2 °tool, cat's-paw, °dupe, pigeon, °pawn, *Colloq US and Canadian* fall guy: *She had no use for Gerry except as a plaything.*

playwright *n.* °dramatist, dramaturge *or* dramaturgist, scriptwriter, screenwriter, scenarist: *The playwright has deftly caught the way people talk to each other.*

plea *n.* 1 °request, entreaty, °appeal, °petition, °supplication, °suit, °cry, solicitation: *Teenagers' mischievous behavior is an earnest plea for attention.* 2 °answer, °defense, °argument: *Her response to the indictment was a plea of "Not guilty."* 3 °excuse, °reason, °explanation, justification; °pretext: *His plea for not paying was that he had not received the bill.*

plead *v.* 1 Often, ***plead for.*** °request, °appeal (for), °cry (for), °ask (for), °seek, °beg (for), °pray (for), supplicate (for): *After an hour of persuasive questioning, we'll have him pleading for permission to tell us everything he knows.* 2 Usually, ***plead with.*** °request (of), entreat, appeal to, °petition, °ask, apply to, implore, °beseech, °beg, importune, °solicit; °demand: *Bryan pleaded with us to let him go to the rock concert.* 3 assert, °say, aver, °allege, °argue, °maintain, °offer, °put forward, °declare, affirm, avow, °swear: *She pleaded that she was unaware that these were stolen goods.*

pleasant *adj.* 1 °pleasing, °pleasurable, °nice, enjoyable, °satisfying, °good, °lovely, °attractive, °inviting, enjoyable, gratifying, °delightful, charming, °agreeable, °suitable, °fitting, °appropriate; °harmonious, euphonious, melodic, mellifluous; °delicious, delectable, palatable, °savory, toothsome: *Traveling on the Orient Express was a most pleasant experience. The bedroom is decorated in a combination of pleasant colors. He simply doesn't find acid rock particularly pleasant to listen to. That medicine has a very pleasant taste.* 2 °friendly, affable, °amiable, °amicable, gregarious, companionable, °sociable, °engaging, °attractive, °winning, °open, approachable, °outgoing, welcoming, °hospitable, °agreeable, °gracious, charming, congenial, °genial, °nice, °likable, urbane, °cultivated, °genteel, °polite, °courteous, well-mannered, °suave, °debonair, °polished, °well-bred, cultured: *Murray always has a pleasant expression on his face. He is quite a pleasant fellow, once you get to know him.* 3 °fair, °sunny, °clear, °bright, cloudless, balmy, °nice, °fine: *What a pleasant day for an outing in the country!*

please *v.* 1 °delight, °gratify, °satisfy, °suit, °humor, °content, °cheer, °gladden, °amuse, °divert, °interest, °entertain: *I cannot tell you how much it pleases me to see you again.* 2 °like, °prefer, °choose, °desire, °want, see fit, °wish, °will, °elect, opt: *She may have her birthday party wherever she pleases.*

pleased *adj.* °happy, °delighted, °glad, gratified, satisfied, contented, thrilled; *Colloq* tickled pink, pleased as punch, on cloud nine, in seventh heaven, °on top of the world, walking on air, *Brit* over the moon, chuffed: *His pleased look comes from his having won first prize.*

pleasing *adj.* 1 See **pleasant, 1,** above. 2 See **pleasant, 2,** above.

pleasurable *adj.* See **pleasant, 1,** above.

pleasure *n.* 1 °enjoyment, °happiness, °delight, °joy, °satisfaction, °fulfillment, contentment, °gratification; °comfort, °recreation, °amusement, °entertainment, °diversion: *It is a pleasure to meet you at last. Fred derives so much pleasure from tinkering with his model railways.* 2 °choice, °option, °desire, °wish, °preference, °fancy, °inclination, °discretion: *Outpatients are permitted to come and go at their pleasure.*

plebeian *adj.* 1 proletarian, working-class, blue-collar, low-class, lower-class, °lowly, lowborn, °common, °mean, °humble, °inferior, peasantlike, *Colloq* non-U: *What would you expect from someone with a plebeian background?* 2 °unrefined, °coarse, °vulgar, ignoble, lowbrow, unpolished, uncouth, crass, brutish, gauche, °provincial, °rustic, °popular, commonplace, °undistinguished: *The level of entertainment caters to the most plebeian tastes.* —*n.* 3 proletarian, common man *or* woman, commoner, man *or* woman in the street, (any *or* every) Tom, Dick, or Harry, *Brit* man *or* woman on the Clapham omnibus, *Colloq* pleb, prole: *This art exhibition is not going to mean very much to the plebeians.*

plebiscite *n.* popular vote *or* ballot, referendum, °poll: *The plebiscite revealed that the people favored joining the Common Market.*

pledge *n.* 1 °promise, °oath, °vow, °word (of honor), convenant, °assurance, guaranty, °guarantee, °warrant, °warranty: *They had our solemn pledge that we would return within the hour.* 2 bail, surety, collateral, °security, °deposit, °earnest (money), °pawn, gage, °bond, guaranty, °guarantee: *They wanted something besides my signature as a pledge for the money.* 3 °toast, °tribute, °cheer, °health: *They drank a pledge to the success of the voyage.* —*v.* 4 °swear, °vow, °promise, give one's word (of honor), °contract, °undertake, °agree, °vouch, °vouchsafe: *I pledge allegiance to the nation.* 5 °deposit, °pawn, mortgage, *Archaic* gage, *Colloq US and Canadian* hock: *He pledged his gold watch to pay for her birthday gift.* 6 °toast, °drink (to), drink (someone's) health: *Drink to me only with thine eyes / And I will pledge with mine.*

plentiful *adj.* 1 °ample, °abundant, °profuse, copious, °lavish, plenteous, °bountiful, °generous, bounteous: *We found a plentiful supply of food in the markets.* 2 °fertile, °fruitful, °productive, bumper, °luxuriant, thriving, °prolific: *We thanked the Lord for a plentiful harvest.*

plenty *n.* 1 °abundance, more than enough, great deal, mass(es), quantity *or* quantities, multitude(s), number(s), load(s), °mess, °scores, *Colloq* lot(s), mountain(s), heap(s), stack(s), pile(s), load(s), ton(s), ocean(s), oodles, *US and Canadian* slew(s): *There is plenty of food, so it is just as well that plenty of people are coming. There's plenty of time before your bus arrives.* 2 plentifulness, fertility, copiousness, °abundance, plenteousness, °wealth, °profusion, lavishness, °prodigality, plenitude, bountifulness: *It was a land of plenty, a land of milk and honey.*

pliable *adj.* 1 °flexible, pliant, °elastic, °plastic, fictile, malleable, workable, bendable, bendy, ductile, flexuous, °supple; lithe, limber: *This substance remains pliable only while it is warm.* 2 °tractable, °adaptable, °flexible, pliant, compliant, persuadable *or* persuasible, °impressionable, °susceptible, °responsive, °receptive, docile, °manageable, °yielding: *The other directors may not be as pliable when it comes to improving employee benefits.*

plight *n.* °condition, °state, °circumstances, °situation, °case; °difficulty, °predicament, °quandary, °dilemma, catch-22, °straits, °trouble, °extremity, *Colloq* °hole, °jam, pickle, °spot, °scrape, °fix, °bind, hot water, °mess, fine kettle of fish, fine state of affairs: *As soon as Elliott's sorry plight became known, we all rallied around to help.*

plod *v.* 1 Often, ***plod along.*** trudge (along), °tramp, slog, °drag, tread, °lumber, °labor, *Colloq* stomp, galumph: *We plodded wearily up the road carrying our heavy packs.* 2 Often, ***plod along or away.*** °labor, °work, drudge, toil, moil, °slave (away), grind (away *or* along), grub (on *or* along), plug (along *or* away), *Brit* peg away (at) *or Brit* along: *Plodding through the compilation of the* Oxford Thesaurus, *he was only halfway through the letter P.*

plot[1] *n.* **1** °scheme, °plan, °intrigue, °machination, °cabal; °conspiracy: *Guy Fawkes's plot to blow up the Houses of Parliament was uncovered just in time.* **2** °story (line), chain of events, °theme, °outline, °scenario, °thread, skeleton: *When Ethel learns that Oscar knows about Jane's liaison with Philip, who is Edwin's Aunt Margaret's cousin by marriage to Ernest, who died saving Richard's dog, the plot begins to thicken.* —*v.* **3** °scheme, °plan, °intrigue, machinate, °cabal, collude, conspire, °hatch, °devise, °design, °arrange, °organize, concoct, dream up, °conceive, *Colloq* cook up: *Martin's wife had been plotting to murder him since their wedding night.* **4** °draw, °plan, diagram, °lay down, °outline, °calculate, °compute, °figure, °chart, map (out), °find, °determine, depict, °show: *The navigator plotted our position as being in the Strait of Malacca.*

plot[2] *n.* °lot, plat, patch *or* parcel (of land), °tract, acreage, °area, *Brit* °allotment: *I always wanted a small plot of land of my own to cultivate with a few vegetables.*

plow *v.* **1** °till, °cultivate, °furrow, harrow, *Literary* delve, *Chiefly Brit* plough: *They ought to finish plowing the south field by tomorrow.* **2** °drive, °plunge, °push, bulldoze, °lunge, °dive, shove, °hurtle, °crash, *Chiefly Brit* plough: *Out of control, a bus plowed through the fence and into my garden.*

pluck *n.* **1** °courage, °spirit, °bravery, °grit, boldness, intrepidity, °backbone, mettle, °determination, gameness, °resolve, °resolution, steadfastness, hardiness, sturdiness, stoutheartedness, stoutness, °fortitude, °nerve, *Colloq* °guts, °spunk, *US* sand, *Slang Brit* °bottle, *US* moxie: *Wentworth showed his pluck by rescuing the children from the raging torrent while under enemy machine gun fire.* —*v.* **2** °pick, °remove, °withdraw, °draw out, °extract: *Do you remember when you bought a chicken and had to pluck it yourself?* **3** °snatch, °grab, °yank, °jerk, °tear (away): *She was plucked from the jaws of death by Tarzan, who happened to be swinging by on a vine.* **4** °tug (at), °pull (at), °catch (at), °clutch (at); vellicate: *An ancient crone plucked at his sleeve and whispered, "The key is in the egg."*

plug *n.* **1** °stopper, stopple, bung, cork: *If you pull the plug, of course the water runs out of the basin.* **2** °chew, °twist, quid, °wad, pigtail, cavendish: *As they forbade smoking, he would take snuff or chew a plug of tobacco.* **3** publicity, °mention, °promotion, °recommendation, °puff, blurb, PR; °advertisement, *Colloq* advert, hype: *Mullins's book was given a plug on yesterday's evening news.* —*v.* **4** Often, ***plug up.*** °stop (up), close (up *or* off), seal (off *or* up), cork, °stopper, bung, °block, °jam, °stuff, °clog, °obstruct, dam (up): *A piece of soap has plugged the drain.* **5** °publicize, °mention, °promote, °push, advertise, °puff, commend, *Colloq* °boost, beat the drum for: *It wasn't right to plug her brother's company in her article on double glazing.* **6** See **plod, 2,** above.

plum *n.* °find, °catch, coup, °prize, °treasure; °bonus, cream: *Harvey's new job as a tax collector is a real plum.*

plumb *n.* **1** °weight, (plumb) bob, °plummet, °lead, sinker: *The plumb holds the string taut and exactly vertical.* —*adj.* **2** vertical, °perpendicular, straight up and down: *Make sure that the jamb is plumb before hanging the door.* —*adv.* **3** vertically, perpendicularly, straight up and down: *The cord doesn't hang plumb because the shelf is in the way.* **4** °exactly, °precisely, °dead, °right, accurately, *Colloq* °slap, *Brit* bang, °spot: *The navigator checked our position and we were plumb on course.* —*v.* **5** °sound, °fathom, °measure, °probe, °explore, °measure, °gauge, delve, °penetrate: *The bathyscaphe has plumbed the deeps of the Indian Ocean. I plumb the depths of depression thinking of starving people everywhere.*

plummet *v.* See **plunge, 1,** below.

plump[1] *adj.* **1** °chubby, °stout, fleshy, °ample, full-bodied, portly, tubby, °rotund, squat, chunky, °buxom, corpulent, roly-poly, °fat, °obese, overweight, steatopygous, *Brit* podgy *or US* pudgy, *Colloq* busty, °beamy, hippy, beefy, well-upholstered: *They went on a diet because they were getting a bit plump.* —*v.* **2** puff up *or* out: *When you get up, please plump up the pillows on the sofa.*

plump[2] *v.* **1** °drop, °plummet, °fall, °plunge, °dive, °sink, °collapse, °flop: *I was so tired that all I could do was plump down in an armchair and watch TV.* **2** °deposit, set *or* put (down), plunk, plop: *She plumped down the defective mixer on the counter and asked for a refund.* **3** ***plump for.*** °support, °choose, °select, °back, °side with, campaign for: *The residents are plumping for better garbage collection.* —*n.* **4** °drop, plunk, °flop, thump, clunk, °clump, °thud, °bump: *He dropped the book on the floor with a plump.* —*adv.* **5** abruptly, °suddenly, °directly, unhesitatingly, °at once, unexpectedly, surprisingly, without warning, (all) of a sudden, plunk, bang: *As I came out of the bank, I ran plump into a policeman.* —*adj.* **6** °direct, unequivocal, unmistakable *or* unmistakeable, unambiguous, °definite, °definitive, °blunt, °simple, °plain, °forthright, °downright, °straight, °matter-of-fact: *He asked Sibyl to marry him and she gave him a plump "No."*

plunder *v.* **1** °pillage, °loot, °rob, °ravage, °ransack, °rifle, despoil, °spoil, vandalize, °sack, °strip, maraud, °devastate, °desolate, °lay waste: *The attacking tribes plundered the Roman villages.* **2** prey on *or* upon, °pirate, °capture, °seize: *The buccaneers plunder the Spanish treasure ships on the high seas.* —*n.* **3** °pillage, looting, °robbery, °depredation, rapine, despoliation, spoliation, vandalization, °sack, vandalism, vandalizing, sacking: *It is said that the plunder of Rome by the Vandals took no more than a fortnight.* **4** °booty, °loot, °spoils, prizes, *Slang* boodle: *The police displayed an Aladdin's cave of plunder recovered from the thieves.*

plunge *v.* **1** °descend, °drop, °plummet, °dive, °pitch, nose-dive, °fall (headlong): *When they lost their footing, both Holmes and Moriarty plunged into the abyss.* **2** °submerge, °sink, °immerse; engulf, °overwhelm: *I plunged the hot poker into the beaker and Voilà! mulled wine. Fully recovered from the tragedy, Humbert plunged himself ever more deeply into his work.* —*n.* **3** °dive, nose dive, °fall, °pitch, °plummet, °drop, descent; submersion, immersion: *Judge Crater's plunge into oblivion has never been satisfactorily explained.* **4** °gamble, wager, °bet, °risk: *Are you taking the plunge with an investment in Abernathy Consolidated?*

plus *prep.* **1** and, added to, increased by, with the addition of, with an increment of, (coupled) with, together with: *Two plus three equals five. Your initial deposit plus accumulated interest will be substantial.* —*adj.* **2** added, additional, °supplementary, °extra: *One must take into account the plus value of the good publicity.* —*n.* **3** °addition, °bonus, °extra, °gain, °benefit, °asset, °advantage, °profit, °return: *Improvements constitute a big plus in evaluating the building.*

plush *adj.* °luxurious, °posh, costly, (de) luxe, °palatial, °lavish, °rich, °opulent, °sumptuous, °regal, °elegant, *Colloq* ritzy, classy, *Old-fashioned* swank(y): *I had no idea that you lived in such plush surroundings.*

ply *n.* layer, leaf, thickness, °fold: *These seams are sewn through two plies of fabric.*

pocket *n.* **1** °pouch, °sack, °bag, °receptacle, reticule, satchel: *You should always have a clean handkerchief in your pocket.* **2** °cavity, °pit, °hollow, crater: *We found* *water that had collected in pockets in the surface of the rock.* **3** °area, °island, °camp, °center, °cluster, concentration: *Leaders encountered pockets of resistance to the new policies among the farmers.* —*v.* **4** °take, °appropriate, °keep; filch, °embezzle, °steal, purloin, thieve, °pilfer, help oneself to, palm, *Colloq* walk off *or* away with, °pinch, °swipe, °rip off, °hook, °lift, snitch, *Brit* °nick, snaffle: *While collecting for charity, Johnson was known to have pocketed occasional contributions.*

poem *n.* verse, °lyric, rhyme *or Archaic* rime, °song, ode, rhapsody, °jingle, ditty: *The anthology contains poems by Japanese politicians.*

poet *n.* poetess, versifier, metrist, lyricist *or* lyrist, versemaker, sonneteer, elegist, bard, °minstrel; rhymester *or* rimester *or* rhymer *or* rimer, poetaster: *A widely published poet, Constance is often invited to give readings.*

poetic *adj.* **1** poetical, lyric(al), metrical, °musical, melodic; °idyllic, elegiac, georgic, °rhapsodic, epic, dithyrambic: *Many enjoy and admire the poetic works of the Romanticists.* **2** artistic, °aesthetic, Parnassian, Hippocrenian, °melodious: *The novel is brightened here and there by poetic passages of surpassing quality.*

poetry *n.* verse, versification, metrical composition, metrics, °rhyme, *Archaic* poesy: *Modern poetry is often rather obscure.*

poignant *adj.* **1** distressing, upsetting, °agonizing, °grievous, °painful, woeful, °melancholy, °blue, °sad, °sorrowful, °tragic, °disastrous, heartrending, heartbreaking, °excruciating, °bitter, °pathetic, °pitiable, °piteous, °pitiful, °miserable, °moving, °touching: *The biography also covers the poignant early years, spent largely in orphanages and foster homes.* **2** °keen, °acute, °intense, °incisive, °sharp, stinging, °pointed, °piercing, °penetrating, barbed, °cutting, °caustic, acid, acerbic, °bitter, °biting, mordant, °sarcastic, °sardonic, °severe: *The headmaster's poignant remarks about honesty and loyalty really hurt.* **3** °sincere, °earnest, °heartfelt, °deep, °profound, °dramatic, deeply felt, °stirring, °moving, °touching, °emotional: *The reunion of the little boy and his dog was the most poignant moment in the film.*

point *n.* **1** °dot, °mark, °speck; °(full) stop, °period; decimal point: *The writing is smudged, but that looks like a point at the end of the line.* **2** °tip, °peak, apex, °spike, °spur, prong, sharp end: *Be careful of the point of that knife.* **3** °spot, °place, °stage, °position; °site, °station, °location, °locale: *We have reached the point of no return on that subject. I stopped at a few points along the way to admire the view.* **4** °time, °moment, °instant, °juncture: *At that point, we were summoned by the dinner bell.* **5** °focus, °essence, °meat, °pith, quiddity, °substance, °heart, °nucleus, crux, °nub, °core, °bottom, °details, *Colloq* °guts, nitty-gritty: *I wish he would stop all the palaver and get to the point.* **6** °purpose, °intent, °intention, °aim, °goal, °object, °objective, °sense: *What was the point of bringing that account of your surgery into the discussion?* **7** °thrust, °drift, °theme, purport, °burden, °import, °implication, °significance, signification, °sense, °meaning; °application, applicability, relevancy, appropriateness: *We found it difficult to see the point of the argument.* **8** promontory, °projection, headland, °cape, peninsula: *It may take us a day to sail round that point.* **9** °brink, °verge: *We were on the point of leaving when the guest of honor finally arrived.* **10** °detail, °particular, °item, °element, nicety, °aspect, facet, °matter, °issue, °subject, °question; specifics: *Some of the points in her speech need clarification.* **11** °pointer, °hint, °suggestion, piece of advice, °tip: *The golf pro at the club gave me some good points on improving my swing.* **12** °thought, °idea, °consideration; °notion, °view, °plan, °tactic; something: *That's a point to keep in mind if you buy another sailing boat.* **13** °unit, °tally, °score: *She made her point and went on to win the match.* **14** °attribute, °characteristic, °feature, °aspect, °trait, °quality, °side, °property: *I suppose he has his good points, too.* **15** ***beside the point.*** °irrelevant, inapt, °inappropriate, malapropos, °incidental, °immaterial, unimportant, °pointless, °inconsequential: *The fact that she's my sister is beside the point.* **16** ***in point of.*** in reference to, °regarding, as regards, in the matter of, °concerning, with respect to: *In point of date, the two events coincide perfectly.* **17** ***make a point of*** *or US also* ***make (it) a point to.*** make an effort (to), put *or* place emphasis on, go out of one's way (to); °emphasize, °single out, °stress: *He made a point of apologizing to the hostess for his behavior. She made a special point of forgiving him.* **18** ***to the point.*** °relevant, °pertinent, °appropriate, °fitting, apropos, germane, apt, °applicable, apposite: *The speech was short and to the point.* —*v.* **19** Often, ***point to.*** °indicate, call *or* direct attention: *Desirée pointed to the building and asked what it was.* **20** °direct, °level, °aim, °train: *I awoke to find her pointing a gun at my head.* **21** ***point out.*** **(a)** °designate, call *or* direct attention to, °show, °exhibit, °indicate, °identify: *The taxi driver pointed out all the sights as we drove along.* **(b)** °say, °bring up, °mention, allude to, °bring up, °emphasize, °stress, °point up, °single out; allude to, call attention to, °remind: *I'd like to point out that you have rarely won at bridge. She pointed out the fallacy in his argument.* **22** ***point up.*** °emphasize, °stress, accentuate, underline, underscore, °accent, °spotlight: *Why must you point up all my shortcomings every time we argue?*

point-blank *adj.* **1** °direct, °straight, °blunt, °flat, straightforward, °abrupt, °categorical, °explicit, uncompromising, °unmitigated, unalloyed, °downright, °outright, °to the point, straight-from-the-shoulder, (open and) above-board, unreserved: *I asked her out to dinner and got a point-blank refusal.* **2** °close, °short, °nearby: *He was shot at point-blank range.* —*adv.* **3** °directly, *Brit* °straightaway, *US* right away, bluntly, °flat, flatly, abruptly, categorically, unqualifiedly, explicitly, uncompromisingly, unmitigatedly, °outright, unreservedly, plainly, frankly, °openly, candidly: *He turned her down point-blank when she proposed marriage.* **4** °directly, °straight: *He fired point-blank at the target.*

pointed *adj.* **1** needle-shaped, °sharp, °acute, barbed, °peaked, spiked, spiky, *Technical* acuminate, cuspidate, aciform, acicular, aciculiform, aculeous, apiculate, spiculate, serrate(d), acuminate, mucroniform, mucronulate, mucronate(d), muricate, hebetate: *The bird actually uses a pointed stick to pick insects from a hole in a tree.* **2** °incisive, °piercing, °cutting, °sharp, °pungent, °keen, °penetrating, °telling, °trenchant, °biting, unmistakable *or* unmistakeable: *The critic made a few pointed remarks about the actors' reading of the lines.*

pointer *n.* **1** °indicator, °rod, °stick; °index, °sign, arrow, *Typography* fist: *The pointer is used to call attention to something important.* **2** °tip, °advice, °hint, °suggestion, °recommendation, piece of advice: *McLeod gave me a few pointers on how to improve the restaurant service.*

pointless *adj.* °purposeless, °aimless, °worthless, °ineffective, °meaningless, °ineffectual, °futile, unproductive, °fruitless, °bootless, °useless, °vain, °senseless, °absurd, °silly, °stupid, °inane, asinine, fatuous, °preposterous, °nonsensical, °ridiculous, °empty, °hollow: *After her paper, someone rose and wasted half an hour making pointless remarks.*

point of view *n.* **1** °viewpoint, °perspective, °approach, °position, °angle, °slant, °orientation, °outlook,

°stance, °standpoint, vantage point: *From his point of view, I can see why Frobisher thinks the research a waste of time.* **2** °opinion, °view, °belief, °(way of) thinking, °principle, °doctrine: *Should everyone or only those who have some qualifications be entitled to express a point of view?*

poise *n.* **1** °balance, equilibrium, equipoise, equiponderance, °parity, °par: *The chariots of the gods in even poise, obeying the rein, glide rapidly.* **2** composure, °control, self-possession, aplomb, °assurance, °dignity, equanimity, °sang-froid, coolheadedness, imperturbability, °presence of mind, coolness, staidness, °reserve, sedateness, calmness, °serenity, tranquillity, *Colloq* °cool: *Jane's poise is remarkable, despite the heckling by the audience.* —*v.* **3** °balance, be balanced, °hover, °hang, °float; make *or* be *or* get ready, °prepare: *The boulder was poised on the edge of the cliff.*

poised *adj.* **1** composed, controlled, °self-possessed, unflappable, (self-)confident, (self-)assured, °dignified, coolheaded, imperturbable, unruffled, °cool, °staid, °reserved, °sedate, °calm, °serene, °tranquil, *Colloq* together: *For a lad of eleven, Richard is quite poised.* **2** °ready, standing by, waiting, °prepared: *Our commando unit is poised to attack at your signal.* **3** teetering, hovering, tottering, wavering, suspended, trembling, wobbling, balanced: *In the present business climate, the company is poised on the brink of bankruptcy.*

poison *n.* **1** toxin, °venom, bane; miasma, mephitis: *The power station persists in spewing poisons into the atmosphere. He kills rats using poison.* **2** virus, bane, cancer, canker, corrupt *or* evil influence, °pestilence, °plague, °blight: *The drug barons continue to spread their poison throughout the world.* —*v.* **3** defile, °adulterate, infect, °taint, °pollute; °contaminate, °debase, °pervert, °vitiate, °subvert, °warp, envenom: *Chemicals from the nearby plant have poisoned the drinking water. Why have you poisoned the child's mind against eating spinach?* **4** °murder, °kill, do away with, °destroy, dispatch *or* despatch: *Lucrezia Borgia poisoned her enemies in ingenious ways.*

poisonous *adj.* **1** °lethal, °deadly, °fatal, °mortal, toxic, septic, °virulent, noxious *or rare* nocuous, °venomous, °malignant, pernicious, miasmic, mephitic: *Poisonous effluents are leaked into the rivers by the chemical plant.* **2** malicious, malevolent, °malignant, corruptive, °vicious, baleful, °evil, °foul, diabolic(al), defamatory, libelous, °slanderous, °dangerous, deleterious: *They have been spreading poisonous gossip about him again.*

poke *v.* **1** °jab, °stick, °prod, °dig, goad, °stab, °thrust, °push, elbow, °nudge, °jog, jostle, °butt, shove: *Be careful not to poke someone in the eye with your umbrella. He tells a joke, then pokes me in the ribs to make sure I've got it.* **2** °punch, °hit, °strike, °box, cuff, smite, smack: *I made a suggestive remark and she poked me in the jaw.* **3** °pry, nose (about *or* around), stick one's nose into, °intrude, °dig, °probe, °investigate; °meddle, °interfere, °butt in, °tamper; *Colloq* °snoop: *I wish they would stop poking into my affairs.* **4** ***poke fun (at).*** °tease, °ridicule, °mock, °make fun of, °jeer (at), °chaff, °taunt, °twit, °make sport of, needle, *Colloq* kid, rib, *Brit* °send up, take the mickey out of: *People who don't understand something often poke fun at it. Stop poking fun at my hat.* —*n.* **5** °jab, °prod, °dig, °stab, °thrust, °push, elbow, °finger, °nudge, °jog, jostle, °butt, shove: *Every time Rodney wants to emphasize a point, he gives you a poke.* **6** °punch, °hit, °box, °jab, cuff, smack, °blow: *If anyone says anything bad about you, I'll give him a poke in the nose.*

polar *adj.* **1** arctic, antarctic, °frigid, °icy, glacial, °freezing, frozen, numbing, Siberian, hibernal, hyperborean, brumal, °wintry: *It is positively polar in the house without the heat turned on.* **2** °opposite, °opposed, antithetical, °contrary, °contradictory, diametric, antipodal, antagonistic, °hostile: *Dante felt Good and Evil to be the two polar elements of the Creation, on which it all turns.*

pole[1] *n.* °rod, °stick, °staff, °spar, °shaft, mast, °standard, °upright; flagpole, flagstaff, jackstaff; beanpole, hoppole: *We had to use a long pole to untangle the kite from the edge of the roof.*

pole[2] *n.* **1** °extremity, °end, °limit, °extreme: *The Nominalists and the Realists maintained opposite poles of the same truth.* **2** ***from pole to pole.*** °everywhere, all over, °far and wide, high and low, leaving no stone unturned, throughout the world *or* the length and breadth of the land, *Colloq US* everyplace: *He searched for her from pole to pole.* **3** ***poles apart.*** (very *or* completely) different, worlds apart, at opposite extremes, at opposite ends of the earth, °at odds, irreconcilable: *I'm afraid that the union and management remain poles apart on the issue of working hours.*

police *n.* **1** constabulary, policemen, policewomen, police officers, *Colloq* boys in blue, the (long arm of the) law, the cops, the gendarmes, *Slang* the coppers, the fuzz, *US* the heat, *Brit* the (Old) Bill: *There was a bit of an argument, so we called the police.* —*v.* **2** °patrol, °guard, °watch, °protect: *Additional officers were assigned to police the neighborhood.* **3** °enforce, °regulate, °administer, °oversee, °control, °observe, °supervise, °monitor: *A unit was established to police the terms of the agreement.*

police officer *n.* °officer, policeman, policewoman, °constable, *Brit* police constable, PC, WPC, *Chiefly US* peace officer, patrolman, patrolwoman, *Colloq* cop, gendarme, *Brit* bobby; *Slang* copper, fuzz, flatfoot, *Brit* rozzer, Old Bill, *Offensive and derogatory* pig, *US* bull, fuzzball, *Historical Brit* peeler: *The policemen then asked if I wouldn't mind helping with their inquiries.*

policy *n.* °approach, °procedure, °(game) plan, °design, °scheme, °program, °method, °system, °management, °conduct, °behavior, °strategy, tactic(s), principle(s), °protocol, °regulation, °rule, °custom, °way, °practice, ways and means, °action: *What policy should we follow regarding interest rates?*

polish *v.* **1** °shine, °brighten, burnish, buff, furbish, wax, °clean, °smooth, °rub, °gloss: *He polished up the handles so carefully / That now he is the ruler of the Queen's navy.* **2** Often, ***polish up.*** °refine, °improve, °perfect, °finish, °cultivate, ameliorate, °enhance; °correct, emend: *If you polish up the article we might consider it for publication.* **3** ***polish off.*** **(a)** conclude, °end, °terminate, °finish: *Ned polished off his homework in less than an hour.* **(b)** °kill, slay, °murder, dispatch *or* despatch, °destroy, °dispose of, do away with, liquidate, °eliminate, *Slang* °bump off, °rub out, do in, °take for a ride: *The three guards were polished off quickly.* **(c)** °dispose of, °put away, °eat, °consume, wolf (down): *In a few minutes, MacGregor had polished off the entire meal.* **4** ***polish up.*** °study, °review, °learn, *Archaic* con, *Colloq* bone up (on), *Slang Brit* swot up (on): *If you're going to Málaga, you'd better polish up your Spanish.* —*n.* **5** °gloss, °shine, °luster, °sheen, °glaze, smoothness, °brilliance, °sparkle, °gleam, °glow, brightness, °radiance: *You cannot imagine what a fine polish she put on that old table.* **6** wax, °oil: *The new polish really put a shine on the desk.*

polished *adj.* **1** °accomplished, °adept, °proficient, °expert, °fine, °outstanding, °skillful, °gifted, °masterful, masterly, °virtuoso, °excellent, °superior, °superb,

°superlative; °flawless, °faultless, °perfect, °impeccable: *Edna is quite a polished cellist. She gave a polished performance.* **2** °refined, °elegant, °cultivated, °graceful, °debonair, °sophisticated, urbane, soigné(e), courtly, °genteel, cultured, civilized, °well-bred, well-mannered, °polite: *Under her guidance, James has become a polished gentleman.*

polite *adj.* **1** °civil, °respectful, well-mannered, mannerly, °courteous, deferential, °diplomatic, °tactful, °formal, °proper, °cordial: *We asked our new neighbors to tea but were met with a polite refusal.* **2** See **polished, 2,** above.

politic *adj.* **1** °ingenious, °shrewd, °crafty, canny, cunning, °designing, °scheming, °clever, °wily, °foxy, °tricky, °artful, Machiavellian, °evasive, °shifty, *Colloq* cagey: *Daniel is politic enough to know how to maneuver situations to his benefit.* **2** °tactful, °diplomatic, °discreet, °prudent, °judicious, °wise, °sage, sagacious, °sensible, °intelligent, percipient, °discriminating, °farsighted, °expedient, °perceptive: *It was politic of the winner to share the prize with those who had helped him most.*

political *adj.* **1** governmental, civic, °civil, °public, °state, °national, federal; administrative, bureaucratic: *The movement began as an attempt to gain political freedom.* **2** °partisan, °factious, factional: *A coalition government was formed by the major political parties.*

politician *n.* legislator, lawmaker, statesman, stateswoman; °minister, *Brit* Member of Parliament, MP, *US* public *or* civil servant, administrator, °official, bureaucrat, officebearer, senator, congressman, congresswoman, °representative, assemblyman, assemblywoman, selectman; *Colloq US derogatory* politico, (political) boss *or* hack, machine politician, ward heeler, wirepuller: *Some politicians have given a bad name to the art of politics.*

politics *n.* **1** public affairs, political science, civics, civil affairs, °government, statecraft, °diplomacy, statesmanship: *Hume viewed politics as people united in society and dependent on each other.* **2** maneuvering, manipulation, wirepulling, °machination: *Office politics are such that they wouldn't dare dismiss that department head.*

poll *n.* **1** voting, °vote, °returns, °tally, figures: *The poll shows that the incumbents are not as secure as they thought.* **2** opinion poll, °survey, °canvass, census, ballot, °count: *A poll of homeowners shows that most would like to own a washing machine.* —*v.* **3** °sample, °survey, °question, °canvass, ballot, °ask, °interview; °count, °enumerate, °tally, °register, °record: *We polled teenagers to determine their reaction to a youth center; those in favor polled 73 percent.* **4** °receive, °get, °win, °register, °tally: *Cavendish polled more than 60 percent of the votes cast.*

pollute *v.* **1** °contaminate, °adulterate, befoul, °foul, °soil, °spoil, °taint, °stain, °dirty, °sully, °blight, °poison: *These factories must be prevented from continuing to pollute the atmosphere.* **2** °corrupt, °desecrate, °profane, defile, °violate: *The altar was polluted because it had been touched by a nonbeliever.*

pollution *n.* contamination, adulteration, corruption, polluting, fouling, befouling, soiling, spoiling, tainting, staining, dirtying, sullying, blighting, poisoning, vitiation: *Laws were passed to prosecute those responsible for the pollution of the environment.*

pomp *n.* °glory, °grandeur, magnificence, °splendor, °show, °extravaganza, °pageantry, °ceremony, °spectacle, °brilliance, ceremoniousness: *The Chinese court's pomp and display of wealth had never before been seen by a westerner.*

pompous *adj.* **1** °vain, vainglorious, °proud, °arrogant, °pretentious, °ostentatious, °showy, °grandiose, °haughty, °overbearing, °conceited, °egotistical, °self-important, °boastful, °braggart, °inflated, °snobbish, magisterial, imperious, pontifical, °affected, exhibitionist, *Colloq* uppity, highfalutin *or* hifalutin, °hoity-toity, high-hat, *Slang* snooty, snotty: *Don't you despise the pompous way Marshall struts about in that ridiculous uniform?* **2** °bombastic, °flowery, grandiloquent, °pedantic, °stuffy, fustian, orotund, °ornate, embroidered, flatulent, °windy, turgid, °inflated, high-flown, euphuistic: *He's always making long-winded, pompous speeches because he loves the sound of his own voice.*

ponder *v.* °consider, muse (over *or* on), brood (over *or* upon *or* on), mull over, °deliberate (over), meditate (upon *or* on), think (over *or* on *or* about), °weigh, ruminate (over), °chew over, cogitate, excogitate, reflect (on *or* over), °contemplate: *I shall need a while to ponder the answer to that question. Give me a little more time to ponder.*

ponderous *adj.* **1** °weighty, °unwieldy, °heavy, °massive, °huge, °big, °large, °awkward, °clumsy, cumbersome *or* cumbrous: *A ponderous juggernaut bore down on us, completely out of control.* **2** °dull, °tedious, °labored, °laborious, °tiresome, turgid, °boring, °dreary, °pedestrian, °stilted, °windy, °inflated, long-winded, °wordy, verbose, prolix, elephantine, °pompous, grandiloquent, overdone: *Many students faced with textbooks written in ponderous prose become disenchanted with learning.*

pool *n.* **1** pond, lake, tarn, °mere, lagoon; swimming pool, leisure pool, wading pool, *Brit* paddling pool, *US* wading pool, *Formal* natatorium: *We found an icy mountain pool where we went for a swim.* **2** °collection, fund(s), °purse, stakes, reserve(s), bank, *Colloq* °pot, jackpot, °kitty: *So far we have too little money in the pool for an office party at Christmas.* **3** °syndicate, °trust, °group, consortium, cartel, °combine: *The object of the pool was the private regulation of market prices, which is illegal.* —*v.* **4** °accumulate, °collect, °gather, °combine, °merge, consolidate, °amalgamate, °league, bring *or* come *or* band *or* get together, team (up) with: *We would do better if we pooled our resources.*

poor *adj.* **1** °needy, °destitute, °indigent, in want, in need, penniless, poverty-stricken, °impoverished, badly off, necessitous, poor as a church mouse, °straitened, pinched, in reduced circumstances, impecunious, financially embarrassed, °down-and-out, out of pocket, ruined, °insolvent, bankrupt, *Colloq* °broke, °hard up, °on one's uppers, °short, *US* wiped out, *Brit* in Queer street, *Slang Brit* skint: *They were poor and didn't know where their next meal was coming from.* **2** °low, °bad, skimpy, °meager, scant, °scanty, °inadequate, °deficient, °insufficient, °sparse: *How can a family of five survive on such a poor salary? These days, 3 percent would be considered a poor return on investment.* **3** °barren, unproductive, unfruitful, °fruitless, °infertile, °sterile; depleted, °exhausted, °impoverished: *This is poor soil, and no amount of cultivation is likely to make it productive.* **4** °bad, °awful, °inadequate, °unsatisfactory, °unacceptable, bumbling, °inefficient, amateurish, °unprofessional, °inferior, second-rate, third-rate, low-grade, °shabby, °shoddy, °mediocre, °defective, °faulty, °flawed, substandard, °sorry, not up to par *or* snuff, °slipshod, below *or* under par, *Colloq* °rotten, °lousy: *They did a poor job repairing my car. Her latest recital was pretty poor.* **5** °insignificant, °slight, °paltry, °inconsequential, °mean, °modest, trivial, °trifling: *They made only a poor attempt to correct the problem.* **6** °unfortunate, unlucky, °pathetic, luckless, °pitiful, °pitiable, ill-fated, °miserable, °wretched, ill-starred, star-crossed, jinxed, hapless: *The poor chap lost his en-*

tire family in the disaster. 7 °bad, °ill: *Aunt Theresa has been in poor health lately.*

poorly *adv.* 1 °badly, inadequately, unsatisfactorily, incompetently, inexpertly, improperly, crudely, unprofessionally, amateurishly: *The decorators you recommended have done their work very poorly indeed.* —*adj.* 2 unwell, °indisposed, ailing, °sick, °below par, *Colloq* °rotten, °under the weather: *Charles is rather poorly, I'm afraid.*

pop *v.* 1 °burst, °explode, bang, °go off: *The balloon popped, making me jump.* 2 Often, ***pop in** or **out** or **by.*** °visit, °stop, °call, °appear, *Colloq* drop in, *Brit* nip in: *Guess who popped in to see me on his way to the airport.* 3 °bulge, °protrude, °stick out, *US* bug out: *The little boy's eyes popped when they brought in the birthday cake.* —*n.* 4 °explosion, bang, °report, °crack: *The firecracker went off with a loud pop.* 5 soft drink, soda (water); cola, *Brit* fizzy drink, lemonade, *US* soda (pop): *A bottle of pop for my daughter and a pint of ale for me, please.*

populace *n.* °people, masses, commonalty, °(general) public, commoners, multitude, °hoi polloi, °crowd, °throng, °rabble, peasantry, proletariat, common folk, °rank and file, working class, bourgeoisie, °mob, *Contemptuous* great unwashed, °riffraff, °rabble, *canaille,* ragtag and bobtail: *Some senators act on behalf of their own constituencies rather than the populace at large.*

popular *adj.* 1 °favorite, °favored, in favor, accepted, well-received, approved, (well-)liked, °fashionable, °in fashion, °stylish, in vogue, °celebrated, °renowned, acclaimed, °famous, °in demand, sought-after, all the rage, *Colloq* °trendy, in, °hot: *You would be a millionaire overnight if you could accurately predict the popular music in the coming year.* 2 °conventional, °stock, commonplace, °public, °normal, °standard, °general, °universal, °average, °everyday, °ordinary, °routine, °common, °habitual, °prevalent, °current, °prevailing, °dominant, °predominant, predominating, °customary: *His films cater to the popular taste in science fiction.* 3 °lay, nonprofessional, °amateur, understandable, °accessible, popularized, simplified: *He presents a popular astronomy program on TV.*

popularity *n.* °favor, acceptance, °approval, °esteem, °regard, repute, °reputation, °vogue, °trend, stylishness, °renown, acclaim, °fame, °celebrity, lionization, °(hero) worship, *Colloq* trendiness: *The immense popularity of Juan Ramōn Gonçalvo's novels in Tegucigalpa has not been equaled elsewhere.*

popularly *adv.* commonly, °generally; °ordinarily, °usually, °universally, °widely, regularly, customarily, prevalently, habitually: *It is a popularly held belief among the French that theirs is the best wine in the world.*

populate *v.* colonize, °settle, °people, °occupy; °inhabit, dwell in, reside in, live in: *Shiploads of families braved the voyage to populate the New World.*

population *n.* °people, °populace, inhabitants, residents, natives, denizens, citizenry, citizens, °folk: *In those days there was no one to look after the legal rights of the population.*

populous *adj.* crowded, (heavily) populated, peopled, teeming, thronged, crawling, swarming, jammed, jampacked, °packed: *The populous cities contrast with the sparsely inhabited rural areas.*

pore[1] *v.* ***pore over.*** °study, °examine, °scrutinize, °peruse, °read, °go over, *Colloq* con: *Hamilton sits in the library, day after day, poring over books of forgotten lore.*

pore[2] *n.* °opening, orifice, °hole, °aperture, °vent, perforation, *Technical* spiracle, stoma: *Water oozed through the pores of the goatskin bag.*

pornographic *adj.* °obscene, °lewd, °offensive, °indecent, °prurient, smutty, °taboo, °blue, °dirty, salacious, licentious, °nasty, X-rated, *Colloq* porno, *US* raunchy: *Some prudes consider anything concerning sex to be pornographic.*

pornography *n.* obscenity, smut, °filth, °dirt, erotica, *Colloq* porn: *The council voted to forbid the sale of pornography within the town precincts.*

porous *adj.* spongy, spongelike, permeable, pervious, penetrable: *The rainwater runs through the porous rock and collects in the pools below.*

port *n.* °harbor, haven, seaport; mooring, anchorage; °refuge: *We were bound east for the port of Cardiff.*

portable *adj.* transportable, °manageable, carryable, °handy, °light, lightweight, °compact, °pocket, pocket-sized, °little, °small, *US* carry-on, vest-pocket, shirt-pocket: *He needs a van just to transport his portable television, portable radio, portable hi-fi, portable typewriter, portable computer, and portable calculator.*

portentous *adj.* 1 °ominous, °threatening, °momentous, °sinister, °fateful, alarming, °menacing, °foreboding, ill-omened, °inauspicious, unfavorable, ill-starred, ill-fated, star-crossed, lowering *or* louring, °gloomy, °unpromising, unpropitious: *Precisely at midnight they appeared with the most portentous news.* 2 °extraordinary, °amazing, astonishing, astounding, °prodigious, °awesome, awe-inspiring, °remarkable, °marvelous, °phenomenal, °fabulous, °fantastic, mind-boggling, wondrous, °miraculous: *The coronation of a British monarch is truly a portentous event.*

porter[1] *n.* 1 bearer, (baggage) carrier *or* attendant, *US Airports* skycap, *US Railways* redcap: *I tipped the porter to carry our bags to a taxi.* 2 cleaner, *Chiefly US and Canadian* janitor, °superintendent, *Colloq US* °super: *When we are away, the porter keeps our mail for us.*

porter[2] *n.* doorkeeper, °watchman, doorman, gatekeeper, concierge, caretaker, *US* tiler: *He went to my club, but the porter would not let him in.*

portico *n.* porch, veranda *or* verandah, gallery, colonnade, galilee: *We kept dry standing under the portico while waiting for the museum to open.*

portion *n.* 1 °segment, °part, °section, °division, subdivision, °parcel, °piece, hunk, chunk, °lump, °wedge, °slice, °sliver, fraction, °fragment, °bit, °morsel, °scrap: *A portion of the masonry fell into the street in the earthquake.* 2 °share, °part, °allotment, °quota, °ration, apportionment, °allowance, allocation, °assignment, °percentage, °measure, °division, °quantity: *The size of an investor's portion depends on the amount invested.* 3 °helping, serving; °ration, °plate, °platter: *He complained that the portions in nouvelle cuisine are too small.* —*v.* 4 Often, ***portion out.*** apportion, share out, allocate, °ration, °allot, °partition, °assign, consign, dole out, °deal (out), °parcel out, °distribute, °administer, °dispense, °disperse, °divide, °split up, °carve up, °cut up, °break up, °section, *Colloq* divvy up: *The conquered territories were portioned out among the victors.*

portrait *n.* °picture, °likeness, °image, sketch, °rendering, vignette; °representation, °description, °profile, thumbnail sketch, portrayal, picturization, depiction; °account, °story, characterization, °study, °record, °file, dossier: *We have the culprit's portrait in our rogues' gallery. We need a more accurate portrait of the kidnapper.*

portray *v.* 1 °represent, °picture, °show, depict, °paint, °render, °characterize, °describe, delineate: *She is portrayed wearing the Castelli tiara. Why must I* *always be portrayed as the villain?* **2** act *or* play (the part *or* role of), take the part *or* role of, °represent, °pose as, impersonate: *She portrays an aged crone who regains her youth in the last act.*

pose *v.* **1** °sit, °model; °position, °place, °arrange, °set (up), °put: *Would you let your husband or wife pose in the nude? She posed the model standing, looking out of the window.* **2** Usually, ***pose as.*** °portray, act *or* play (the part *or* role of), take the part *or* role of, °represent, impersonate, be disguised as, °masquerade as, pretend *or* profess to be, pass (oneself off) as, °pass for, °imitate, °mimic; attitudinize, °posture, put on airs, *Colloq* °show off: *She posed as an art expert from Switzerland. She thinks him a roué, but he's only posing.* **3** °set, °put, °ask, °submit, °broach, °posit, °advance, °present, predicate, postulate: *The interviewer posed some questions that were quite embarrassing.* —*n.* **4** °position, °attitude, °posture, °stance: *Her profile is not shown to best advantage in that pose.* **5** °affectation, °act, °pretense, attitudinizing, affectedness, °display, façade, °show, °ostentation: *The Forbushes put on a pose, pretending they have a lot of money. Her interest in art is just a pose, put on to impress him.*

poseur *n.* posturer, exhibitionist, °pretender, °impostor, masquerader, attitudinizer; °fake, faker, dissembler, °fraud, *Colloq* °showoff, phony *or* phoney: *He talks about literature, but he's a poseur, spouting others' opinions.*

posh *adj.* (de) luxe, °luxurious, °elegant, °sumptuous, °lavish, °opulent, °rich, °royal, °regal, °luxury, °grand, °fashionable, *Colloq* swank(y), classy, ritzy, *Slang* snazzy: *You ought to see what a posh office my son Edgar has!*

posit *v.* postulate, hypothesize, °propound, put *or* set forth, °put forward, °advance, °propose, °pose, °offer, °submit, predicate: *In the company's projections for next year, they posit a rebirth of interest in Victorian furniture.*

position *n.* **1** °posture, °attitude, °stance, °pose; °disposition, °arrangement, disposal: *Once in a while, he would change his position so that he faced the sun. With the pieces in this position the chess game is a draw.* **2** °site, °placement, °situation, °whereabouts, placing, emplacement, °location: *Those bearings put his present position about twenty miles west of Cádiz.* **3** °viewpoint, °point of view, °outlook, °attitude, °angle, °slant, °stance, °stand, °opinion, °inclination, °leaning, °bent, °sentiment, °feeling, way of thinking: *My solicitor is pessimistic about my position in this case.* **4** °status, °condition, °state, °circumstances, °situation: *Our financial position vis-à-vis investment in gilt bonds has changed.* **5** °class, °caste, °place, °rank, °standing, °station, °status, °importance: *They insist on knowing the social position of the girl's family.* **6** °job, °occupation, °situation, °post, °office, °function, °appointment, °capacity, °place, °role, *Colloq* billet, berth, *Australian* possie *or Australian* possy: *My mother's position as managing director had nothing to do with my getting a promotion.* **7** °hypothesis, °thesis, °principle, contention, °assertion, predication, °belief, °proposition, postulate: *His position is based on the implications of the third law of thermodynamics.* —*v.* **8** °put, °place, °situate, °site, °set, °fix, °settle, °dispose, °arrange: *The pointer is again positioned at zero. Position your forces along this ridge and stay on full alert.* **9** °place, °locate, °establish, °determine, °fix, localize: *The scientists positioned the seismic activity as being along the San Andreas fault.*

positive *adj.* **1** °sure, °certain, °definite, unequivocal, °categorical, °absolute, °unqualified, unambiguous, unmistakable *or* unmistakeable, clear-cut, °clear, °explicit, °express, decisive, °indisputable, indubitable, °unquestionable, unquestioned, incontestable, uncontested, undeniable, °reliable, °persuasive, convincing, irrefutable: *The police have positive evidence that the men are innocent of any crime. Spending this evening with you has been a positive delight.* **2** °definitive, °emphatic, °decided, °forceful, °firm, °peremptory, °definite: *He gave a positive denial when asked if he had agreed to let the hostages go. She made a positive commitment to meet me here at noon.* **3** °sure, °certain, °confident, convinced, satisfied: *Are you positive that the last train is at midnight?* **4** °beneficial, °favorable, °complimentary, °productive, °useful: *I heard many positive things about the way she handled herself at the interview.* **5** °overconfident, °dogmatic, doctrinaire, pontifical, °opinionated, pigheaded, °stubborn, °obstinate, obdurate, °arbitrary, overweening, °arrogant, °assertive, °dictatorial, °despotic, imperious, *Rare* thetic(al): *His problem is that he is quite positive about things that simply are not so.* **6** °practical, °realistic, utilitarian, °productive, °functional, pragmatic(al), °realistic, °matter-of-fact, *Colloq* hard-nosed: *Corrigan has finally taken a positive approach to the study of navigation.* **7** encouraging, °supportive, °constructive, reassuring, °enthusiastic, °favorable, affirmative, yes, confirming: *Our plan to buy out the company has received a positive reaction from our bankers. Asked if they would buy our product, businessmen gave a positive response.* **8** auspicious, °promising, °propitious, encouraging; °optimistic, °favorable, °cheerful, °confident; *Colloq* bullish, °upbeat: *In the long run, the picture for homeowners looks positive. It pays to maintain a positive outlook and not get depressed.* **9** °complete, utter, °total, °perfect, °out-and-out, consummate, °unmitigated, °thorough, thoroughgoing; egregious, °glaring, °stark, °sheer, °outright, °unqualified, unequivocal, *Colloq* dyed-in-the-wool: *You were a positive idiot not to let me know you needed help. All attempts at reconciliation ended in positive disaster.*

positively *adv.* °definitely, °absolutely, unquestionably, certainly, (most) assuredly, undeniably, °undoubtedly, °surely, to be sure, emphatically, unmistakably *or* unmistakeably, unqualifiedly, categorically, indisputably, beyond *or* without (a *or* the shadow of) (a) doubt, indubitably, beyond question: *The wine list at Le Maître's is positively the best in the area.*

possess *v.* **1** be possessed *or* in possession of, °have, own, °enjoy, be blessed *or* endowed with: *In the nineteenth century, the family possessed great wealth.* **2** °have, be born *or* gifted *or* endowed with, °contain, °embody, °embrace, °include: *He possesses a talent to amuse people. Man does not possess as keen a sense of smell as many animals.* **3** °dominate, °control, °govern, °consume, take control of, preoccupy, °obsess; °charm, °captivate, °enchant, cast a spell over, °bewitch, enthrall: *What possessed her to think that I could help? He behaves as if he is possessed by the devil.* **4** ***be possessed with*** *or* ***of.*** °have, be held *or* influenced *or* dominated by, be imbued *or* inspired *or* permeated *or* affected with: *She became possessed with the notion that she could sing. Fiona believes herself to be the only person possessed of reason.* **5** ***possess oneself of.*** °acquire, °achieve, °get, °come by, °gain, come into, °win, °obtain, °procure, °secure, °take, °seize, take *or* gain possession of: *Urquhart has possessed himself of the only comfortable bed in the place.*

possessed *adj.* obsessed, driven, controlled, dominated, ridden, bedeviled, consumed, haunted, pressed, maddened, crazed, demented, frenzied, *Colloq* eaten up: *He behaved like a man possessed.*

possession *n.* **1** ownership, °title, proprietorship, °control, °hold, °tenure, keeping, °care, °custody, guardianship, °protection: *The paintings are now in my possession.* **2** holding, °territory, °province, °dominion, colony, protectorate: *How long will the Falkland Islands remain a possession of the Crown?* **3** ***possessions.*** °belongings, °property, °effects, chattels, °assets, worldly goods, °things: *The prison clerk placed all my possessions in an envelope, explaining that they would be returned when I was released.* **4** ***take*** *or* ***gain possession of.*** °seize, °capture, °take, °conquer, °occupy, °acquire, °win, °possess oneself of, °secure, °obtain; repossess: *The raiding party gained possession of the port and surrounding area. The finance company has taken possession of my car.*

possessive *adj.* **1** °greedy, unyielding, °selfish, ungiving, ungenerous, stingy, niggardly, °materialistic, covetous, acquisitive: *He refuses to see anyone, as he is very possessive of his time.* **2** overprotective, controlling, °grasping, dominating, °domineering, °overbearing: *Anyone with a possessive parent knows how difficult it is to leave home.*

possibility *n.* **1** °chance, °odds, °prospect, conceivability, °feasibility, plausibility, °likelihood, admissibility: *There is a possibility of my leaving work early to meet you. What is the possibility that he might be honest? They deny even the possibility of God's existence.* **2** Often, ***possibilities.*** °opportunity, potentiality, °potential, °capacity, °promise: *The old house has many large rooms and offers great possibilities.*

possible *adj.* **1** °feasible, °plausible, imaginable, conceivable, °thinkable, credible, °tenable, °reasonable, admissible: *It is remotely possible, though improbable, that she committed the crime.* **2** realizable, °practicable, workable, °practical, doable, achievable, attainable, reachable, accomplishable, °viable, *Colloq* on: *In the present state of the art, an antigravity device is still not possible.*

possibly *adv.* **1** °maybe, °perhaps, God willing, *Deo volente,* if possible, *Archaic or literary* perchance, mayhap, peradventure: *She is possibly the best person for the job.* **2** in any way, under any circumstances, by any chance, by any means, at all, conceivably: *Could you possibly help me find an example for this entry? The weather forecast couldn't possibly be correct.*

post[1] *n.* **1** °pole, °stake, °upright, column, °pillar, °pale, °picket, °shaft, °standard, newel, °pier, pylon, °pile, piling, °strut, shore, stanchion, °leg, °prop, °stay, °support, °brace: *A huge center post held up the roof.* —*v.* **2** advertise, °announce, °proclaim, °publish, °circulate, °propagate, promulgate; put *or* pin *or* tack *or* stick *or* hang up, affix: *The notice of the meeting has been posted for all to see.*

post[2] *n.* **1** °assignment, °appointment, °position, °situation, °job, °place, °duty, °role, °function, °employment, °work, °task, chore: *Clarke was given a post as consul in some forgotten country.* —*v.* **2** °place, °put, °station, °assign, °appoint, °position, °situate, °set, °locate: *Guards have been posted around the enclosure.*

post[3] *n.* **1** postal service, °mail; °delivery; °collection: *I am expecting an important letter in the post. It hasn't come in the first post.* —*v.* **2** °send, dispatch *or* despatch, °transmit, *Chiefly US and Canadian* °mail: *I posted your check this morning.* **3** °record, °enter, °register, °list: *We post the daily receipts in this ledger.* **4** ***keep (someone) posted.*** °inform, °advise, °brief, °notify, *Colloq* fill (someone) in on: *Our observers have kept me posted as to your whereabouts every step of the way.*

poster *n.* placard, °notice, °bill, °advertisement, °announcement, broadside, broadsheet; °circular, flier: *Who designed that striking poster for the new play?*

posterior *adj.* **1** hind, °rear, °back, after, °hinder, rearward, *Nautical* aft: *The posterior legs are somewhat longer.* **2** later, after, latter, ensuing, following, succeeding, °subsequent: *Analysis has yielded evidence of the posterior origin of the lava.* —*n.* **3** °buttocks, °bottom, rump, °seat, *derrière, Colloq* behind, °rear, backside, °tail, hinie, °butt, °bum, °duff, *Slang Brit* °bum, *Taboo slang US* ass, *Brit* arse, *Yiddish* tokhes *or Yiddish* tokus *or Yiddish* tochis *or Yiddish* tuchis: *The term describing a person with a fat posterior is "steatopygous."*

posterity *n.* descendants, successors, heirs, children, °offspring, °issue, °progeny: *These buildings must be preserved for posterity. Posterity will be the judge of our success.*

posthaste *adv.* °quickly, °at once, without delay, °immediately, °directly, °straightaway, right away, °promptly, speedily, °swiftly, instantly, before one can *or* could say "Jack Robinson," before you can say "knife," rapidly, at full tilt, in a wink, in a trice, in the twinkling of an eye, *Colloq* pronto, chop-chop, PDQ (= 'pretty damn(ed) quick'), *US and Canadian* lickety-split, like greased lightning: *He got me the report posthaste.*

post-mortem *n.* **1** autopsy, necropsy: *We'll know the cause of death after the post-mortem.* **2** °review, °analysis, *Slang US* Monday morning quarterbacking: *We can do without the post-mortem on every bridge hand.*

postpone *v.* °delay, adjourn, °defer, keep in abeyance, put off *or* aside, lay aside, °suspend, °shelve, put *or* keep on ice, temporize, dally, *Colloq* put on the back burner, *US* °table: *We ought to postpone further discussion till we have the facts.*

postponement *n.* °delay, adjournment, °abeyance, °suspension, °stay, deferment, deferral, °moratorium: *There has been a postponement of the meeting till Monday.*

posture *n.* **1** °pose, °position, °attitude, °stance, °appearance, °carriage: *She stood there in a defiant posture, with arms akimbo.* **2** °position, °condition, °situation, °state, °disposition; °arrangement, °organization, layout, array, °format: *The government adopted a conciliatory posture in the matter. The entire fleet was in the best posture to await the hurricane.* **3** °attitude, °stance, °position, °feeling, °sentiment, °outlook, °(point of) view, °viewpoint, °orientation, °disposition, °frame of mind, °mood: *The interviewer tried to determine the general's posture on disarmament.* —*v.* **4** °pose, attitudinize, °affect, put on a show, do for effect, *Colloq* °show off: *Despite all their posturing, we believe our politicians, on the whole, to be sincere.*

pot *n.* **1** °pan, saucepan, cauldron, cookpot, stewpot; kettle: *How many politicians have promised the people "a chicken in every pot"?* **2** jackpot, bank, °kitty: *You have to add five dollars to the pot to see my hand.* **3** potbelly, °paunch, °gut, *Colloq* corporation, beer belly, spare tire, *US* bay window: *If he exercised more, Patrick wouldn't have such a pot.*

potent *adj.* **1** °powerful, °strong; °mighty, °vigorous, °forceful, °formidable, °authoritative, °influential, *Literary* puissant: *A potent poison will kill those weeds. We have to reckon with an enemy that is quite potent.* **2** °effective, convincing, cogent, °persuasive, compelling, °efficacious, °telling, °sound, valid, °impressive: *Their argument is potent enough to convince the biggest skeptics.*

potential *adj.* **1** °possible, °likely, °implicit, implied, °imminent, developing, budding, embryonic, °dormant, °hidden, concealed, covert, latent, quiescent, °passive, °future, unrealized, °undeveloped: *We are dealing with a potential threat to our liberty.* —*n.* **2** °capacity, °capability, °possibility, °aptitude, potency, *Colloq* the (right) stuff, what it takes: *Teenagers have* *potential; the problem is to teach them to exploit it to the best advantage.*

potion *n.* °draft, °brew, beverage, °drink, philtre, potation, °elixir, °tonic, cup, °dose, concoction, decoction: *The old crone gave him a tiny bottle containing the love potion.*

potpourri *n.* °mixture, °medley, °miscellany, °assortment, olla podrida, smorgasbord *or* smårgøsbord, gallimaufry, salmagundi, °patchwork, °collection, hotchpotch *or US also* hodgepodge, melange *or* mélange, motley, pastiche *or* pasticcio, °mishmash, °jumble, °mess: *What a potpourri of styles is represented in that art gallery!*

pottery *n.* earthenware, ceramics, terra cotta, crockery, stoneware, porcelain, china, delftware: *Ornamented pottery has been found in pre-Columbian digs.*

pouch *n.* °pocket, °sack, °bag, °purse, reticule, *Dialect* °poke: *A Highlander's pouch is called a sporran. As I was going out, I filled my tobacco pouch.*

pounce *v.* **1** Often, ***pounce on*** *or* ***upon.*** spring (on *or* upon), leap (at *or* on), swoop down (on *or* upon), °fall upon, jump (at *or* on), °strike, take by surprise *or* unawares, °attack, °ambush, *Colloq* °mug: *As I rounded the corner, three youths pounced on me, stole my bag, and ran off.* —*n.* **2** °spring, °leap, °swoop, °jump: *The cat was on the mantel shelf in a single pounce.*

pound[1] *v.* **1** °beat, °batter, °pelt, hammer, pummel; thump, °belabor, thrash, bludgeon, cudgel, maul, °strike, *Colloq* lambaste *or* lambast, *Slang* paste, clobber, work over, give (someone) the works *or* a pasting: *She pounded on the door till someone came. It was satisfying to see him pound the class bully into submission.* **2** °crush, °powder, °pulverize, bray, comminute, triturate, mash, °pulp: *The corn must be pounded into a fine meal before use.* **3** °beat, throb, hammer, °pulse, °pulsate, palpitate: *My heart was pounding, waiting to see if the tiger would attack.* **4** ***pound into.*** °instill, din into, drill into, drub into, hammer into, beat into: *Her parents have pounded into her that she must show respect to her elders.* **5** ***pound out.*** °rid, °expel, °clear, °cleanse, °empty, °purge, beat out, hammer out: *The last bit of stubbornness was pounded out of me at school.* **6** beat out; hammer out, °produce: *The jungle drums pounded out the message that Tarzan was coming.* —*n.* **7** pounding, °beat, beating, thump, °thumping: *The pound of horses' hoofs heralded the arrival of the cavalry.*

pound[2] *n.* °enclosure, °pen, °compound, confine, yard: *My car was towed to the pound, and it cost a fortune to retrieve it.*

pour *v.* **1** °flow, °run, °gush, °rush, °flood, °stream, °course, °spout, °discharge, °spurt, °spew out, cascade: *Water was pouring from a crack in the dam.* **2** °empty, °discharge, °let out: *Pour the boiling water over the tea leaves and let them steep for a few minutes.* **3** °rain, °teem, *Colloq* come down in buckets *or* by the bucketful, bucket down, rain cats and dogs, *US* rain pitchforks: *It poured all night.* **4** °stream, °swarm, °crowd, °throng, °teem, °emerge, sally forth, °issue (forth), °go (forth): *The show over, people poured into the streets.*

pout *v.* **1** mope, °brood, °sulk, make a *moue,* pull a long face, °frown, lour *or* lower, knit one's brows: *Don't pout—I'll buy you an ice cream.* —*n.* **2** °frown, *moue,* long face: *Her pout was occasioned by her father's scolding.*

poverty *n.* **1** °want, penury, indigence, insolvency, destitution, pauperism, impecuniousness, neediness, beggary: *Her childhood was spent in poverty.* **2** °scarcity, scarceness, °want, °need, °lack, meagerness, insufficiency, °shortage, °dearth, paucity, inadequacy: *They criticized the poverty of talent among the current crop of actors.*

powder *n.* **1** dust; talc: *After centuries the wood had turned into powder. A little powder will soothe that irritation.* **2** ***take a (runout) powder.*** °run away, abscond, °escape, vanish, °disappear, *Slang Brit* scarper, do a moonlight flit, *US* take it on the lam: *The cops are coming, so I'm taking a powder.* —*v.* **3** °pulverize, bray, °grind, °crush, °pound, granulate, triturate, comminute, levigate: *The rocks must first be crushed, then powdered.* **4** sprinkle, besprinkle, dust, dredge, °cover, °coat: *Lightly powder the top of the cake with icing sugar, and it is ready to serve.*

power *n.* **1** Sometimes, ***powers.*** °capacity, °capability, °ability, °potential, °faculty, competency *or* competence, °faculty, potentiality, *Colloq* what it takes, *US* the (right) stuff, the goods: *Some fishes have the power to change their buoyancy. He has remarkable powers of observation.* **2** °control, dominance, °authority, mastery, °rule, °influence, °sway, °command, ascendancy, °sovereignty, °dominion, °weight, clout, *Colloq* °pull, *US* °drag: *He maintains a Svengalilike power over her. The prime minister has the power to appoint and dismiss cabinet ministers.* **3** °control, °command, °authority: *The party might not be in power for very long.* **4** °strength, °might, °vigor, °energy, °force, mightiness, potency, forcefulness, °brawn, muscle, *Literary* puissance: *His speech was delivered with great power and a certain wit. Has she the power it takes to be a weightlifter?* **5** °talent, °skill, °ability, °faculty, °gift, °aptitude, °genius, °knack: *They say that she has the power to see into the future.* **6** °authority, °license, °right, authorization, °privilege, °warrant, °prerogative: *By the power vested in me, I now declare you man and wife. It is within her power to grant a stay of execution.* **7** Often, ***powers.*** °activity, effectiveness, °effect, °ability, °capacity, active ingredient(s): *This mushroom has hallucinogenic powers.* **8** °energy, °momentum, °impetus, °drive, °force, °inertia: *The car rolled partway up the next hill under its own power.* **9** (mechanical *or* electrical *or* atomic) energy, °electricity, °fuel: *Yesterday there was another demonstration against the use of nuclear power.* **10** ***powers that be.*** °government, °administration, °authorities, incumbents: *It is up to the powers that be to investigate corruption.*

powerful *adj.* **1** °potent, °strong, °mighty, °vigorous, °robust, °energetic, °sturdy, °stalwart, °tough, resilient, °dynamic: *Leslie has a powerful physique. The cities of the Hanseatic League had powerful economies.* **2** °influential, °strong, compelling, °forceful, °potent, °substantial, °weighty, °authoritative, °effective; °important, °impressive, °telling, °effectual, °formidable, °persuasive: *Powerful arguments have been brought to bear against taking such action. She has many powerful friends.* **3** °strong, °potent; °intense, °substantial, °great, °high: *The drug is too powerful to be taken regularly. This wire cannot resist a powerful electrical charge.*

powerless *adj.* **1** °helpless, °incapable, °unable, unfit, °incompetent, °ineffectual, °ineffective: *Without his wand, the magician was powerless.* **2** incapacitated, °helpless, °weak, °feeble, debilitated, °crippled, paralyzed, °disabled: *He is completely powerless without the use of his right hand.*

practicable *adj.* doable, °feasible, workable, performable, achievable, attainable, accomplishable, °possible, °viable: *We ascended the glacier as far as was practicable. At last we had a practicable flying machine.*

practical *adj.* **1** pragmatic, °useful, usable *or* useable, °functional, °realistic, °reasonable, °sound, utilitarian,

°applicable, °serviceable, °empirical, °efficient: *Whether the device is practical will depend on its cost effectiveness.* **2** °sensible, °reasonable, common-sense *or* common-sensical, °everyday, °ordinary, down-to-earth, °expedient, °matter-of-fact, mundane, businesslike, hardheaded, °judicious, *Colloq* hard-nosed: *Put theory aside and focus on a practical solution to the problem.* **3** applied, °field, hands-on, °personal, °everyday: *Bagley has had practical experience in servicing jet engines.*

practically *adv.* **1** °almost, °(very) nearly, well-nigh, °virtually, °in effect, just about, essentially, fundamentally, °at bottom, basically, when all is said and done, at the end of the day, °to all intents and purposes: *We are practically there. I have practically finished the ironing.* **2** realistically, matter-of-factly, °clearly, °simply, reasonably, rationally, sensibly: *Practically, there was nothing to be done except let the fire burn itself out.*

practice *n.* **1** °custom, wont, °habit, °routine, °convention, °tradition, °rule, °procedure, °usage, °mode, °style, °way, modus operandi, technique *or* technic, *Formal* praxis, *Colloq* MO (= 'modus operandi'): *She makes a practice of swimming for an hour every day. It is our practice to get patients out of bed as soon as possible after surgery.* **2** °exercise, °discipline, °drill, practising, °repetition, °rehearsal, training, °preparation; workout, warm-up; °application, °study: *She needs more practice on the beginner's slope before going down the main piste.* **3** °pursuit, °exercise, °work, °profession, °career, °vocation, °conduct; °business, °office: *He genuinely enjoys the practice of law. I heard of a veterinary practice for sale in Yorkshire.* **4** ***in practice.*** °practically, °actually, day-to-day, realistically, in real life: *In practice no one would ever treat an injury the way this book recommends.* **5** ***out of practice.*** °inexperienced, unpracticed, °unaccustomed, rusty: *I once played a good game of chess, but I'm out of practice now.* —*v.* **6** °drill, °exercise, °work out, °train, °prepare, °rehearse, °run through, °repeat, °study, *Brit* practise: *I am practicing for Sunday's tennis tournament. She practices the piano every day. Practice your routine for the show.* **7** °carry on, make a practice of, °perform, do, °act, °carry out, put into practice, *Brit* practise: *She practices law. You should practice what you preach.*

practiced *adj.* **1** °accomplished, °proficient, °expert, skilled, °experienced, °capable, °adept, °seasoned, °able, °qualified, °gifted, °talented, °skillful, °masterful, consummate, °superb, °superior, *Brit* practised: *She is a practiced liar.* **2** trained, rehearsed, °versed, °cultivated, schooled, finished, perfected, *Brit* practised: *He greeted them with the practiced air of a veteran diplomat.*

praise *n.* **1** acclaim, °approval, approbation, °applause, plaudits, °kudos, °endorsement, acclamation, °tribute, accolade, °compliments, commendation, encomium, °eulogy, panegyric, °ovation: *Sheilah has received well-deserved praise for her work with underprivileged children. Would you expect the critics to sing the praises of the new oratorio?* **2** °honor, glorification, adoration, exaltation, °devotion, °homage, °worship, °veneration, adulation, °reverence, °glory, hymn *or* song of praise, paean, hosanna: *The king rejoiced and said, "Praise be to God."* —*v.* **3** acclaim, °approve, °laud, °applaud, °endorse, pay tribute to, °compliment, commend, °eulogize, °extol, °honor, sing the praises (of): *The speaker praised Read for his many contributions to linguistics.* **4** °worship, °revere, °reverence, °exalt, °glorify, °adore, pay homage to, °venerate, give thanks to, °hallow: *Praise the Lord and pass the ammunition.*

praiseworthy *adj.* commendable, °laudable, °admirable, creditable, °worthy, °meritorious, °deserving, °exemplary: *Despite the goalkeeper's praiseworthy attempt, the ball flew past him, making the score one to nothing.*

prance *v.* **1** °caper, °dance, gambol, °skip, °cavort, romp, °leap, °frisk, °jump, °spring, °bound, *Dressage* curvet, capriole: *Ahead of the royal coach came a dozen riders on prancing horses.*

prank *n.* °trick, °(practical) joke, °frolic, escapade, antic, °caper, °stunt, °lark, jest, jape, monkeytricks *or esp. US* monkeyshines, °mischief: *Those schoolboy pranks we once thought hilarious now seem quite silly.*

prattle *v.* **1** prate, °babble, blather *or* blether, blither, gibber, °jabber, gibber-jabber, °palaver, °tattle, twaddle, gabble, °chatter, °patter, °drivel, °twitter, °rattle on, °go on (and on), maunder, *Brit* natter, *Colloq* witter (on), gas, °gab, °spout, °gush, run (on) at the mouth, *US* run off at the mouth, *Slang* jaw, ya(c)k, ya(c)ketyya(c)k, shoot off one's mouth: *Do I have to listen to him prattle on and on about his grandchildren?* —*n.* **2** prate, prating, °babble, babbling, blather *or* blether, blathering *or* blethering, gibber, gibbering, °jabber, jabbering, °palaver, palavering, °tattle, tattling, twaddle, °chatter, chattering, gabble, gabbling, °patter, °drivel, °twitter, twittering, rattling on, °going on, maundering, cackle, *US* gibber-jabbering, twattle, clack, *Colloq* gas, °gab, running off at the mouth, *Slang* jawing, ya(c)ketyya(c)k: *We were treated to a constant barrage of prattle about eating healthy foods.* **3** jabberwocky, gobbledygook *or* gobbledegook, °mumbo jumbo, °rubbish, balderdash, °(stuff and) nonsense, humbug, bunkum, tommyrot, °trash, °rot, foolishness, *Colloq* pish and tush, °hot air, flapdoodle, rigamarole *or* rigmarole, bunk, piffle, °moonshine, poppycock, claptrap, bull, hogwash, °swill, *Brit* tosh, fiddle-faddle, baloney, *Chiefly US* °garbage, horsefeathers, *Slang* crap, hooey, guff, *Taboo slang* bullshit, *US* crock (of shit): *What they told you about guaranteeing your safety in Beirut is a lot of prattle.*

pray *v.* **1** °beseech, °ask, call upon *or* on, entreat, implore, °request, appeal to, °plead (with), °beg (for), importune, °solicit, °petition, supplicate, *Rare* obsecrate: *I pray you to find it in your heart to approve my work.* **2** say one's prayers, offer a prayer: *We prayed for her safe return.*

prayer *n.* **1** °petition, °supplication, °request, entreaty, °plea, °suit, °appeal, *Rare* obsecration: *Our prayers were answered, and she returned safely.* **2** °devotion, praying, invocation, °(divine) service, *Literary* orison: *Man ascends to God through prayer.*

preach *v.* **1** deliver a sermon, evangelize, spread the Word *or* the Gospel; catechize: *The Most Reverend Rev. Attwood will preach next week at Westport.* **2** moralize, sermonize, °advise, °counsel, admonish, °reprimand, °lecture, °harangue, pontificate; °urge, inculcate, °advocate: *Mother used to preach to us about being charitable. Father preached restraint in all things.*

preacher *n.* °minister, evangelist, °clergyman, clergywoman, cleric, ecclesiastic, reverend, °divine, *Colloq* tub-thumper: *He studied for many years to become a preacher.*

preamble *n.* introduction, °foreword, prologue, °preface, *Formal* proem, prolegomenon, exordium: *As a preamble to today's proceedings, I should like to welcome our honored guests.*

precarious *adj.* °uncertain, °unreliable, unsure, °risky, °hazardous, unpredictable, °insecure, °unstable, unsteady, °unsettled, °shaky, °doubtful, dubious, °questionable, °tricky, °delicate, °ticklish, °sensitive, °slippery, touch-and-go, (hanging) in the balance, hanging by a thread, Damoclean, °perilous, treacherous, °dangerous, °difficult, °problematic, *Colloq* chancy, *Brit* °dodgy, °dicey, *US* iffy, *Slang* °hairy: *If sales continue to* *drop, the company will be in precarious condition. We followed a precarious trail down the mountainside.*

precaution *n.* **1** °provision, preventive measure, safety measure, °safeguard, °insurance, °protection, °cover, °escape: *Unfortunately, he had failed to take any precautions against storm damage.* **2** °foresight, °prudence, °providence, °forethought, °caution, cautiousness, circumspection, °care, °attention, watchfulness, °vigilance, alertness, wariness, chariness, apprehension, farsightedness, °anticipation: *Precaution is wiser than hindsight.*

precede *v.* come *or* go *or* proceed before *or* first, go ahead *or* in advance (of), °lead (the way), pave the way (for), herald, usher in, °introduce, antecede; °foreshadow, antedate, predate: *He let his wife precede him because he knew that the area was mined.* The Decameron *preceded* The Canterbury Tales *by about fifty years.*

precedence *n.* precedency, °priority, °preeminence, °superiority, °supremacy, °preference, °privilege, °prerogative, °importance, °rank, °position, primacy: *Your homework takes precedence over football practice.*

precedent *n.* °prototype, °model, °example, exemplar, °pattern, paradigm, °yardstick, criterion, °standard, °lead: *The French considered the War of American Independence a precedent for their revolution.*

preceding *adj.* °foregoing, °former, °previous, °above, °prior, earlier, above-mentioned, aforementioned, above-stated, above-named: *The defendant in the preceding action was remanded in custody for a week.*

precept *n.* **1** °rule, °guide, °principle, unwritten law, canon, guideline, °dictate, °code, °injunction, °law, commandment, °instruction, °injunction, directive, °prescription, mandate, °charge; statute, °regulation, edict, ukase, °decree, °order, fiat, ordinance: *The Ten Commandments contain basic—and minimal—precepts of moral behavior.* **2** °maxim, °proverb, axiom, °motto, °slogan, saying, °byword, aphorism, apophthegm *or* apothegm: *Does anyone follow the precepts set forth in Scripture?*

precinct *n.* **1** Usually, ***precincts.*** °area, °territory, °region, °province, environs, purlieus, °borders, °bounds, confines: *Does he have any authority outside the precincts of the city?* **2** °sphere, °neighborhood, °zone, sector, °section, °quarter, °district, °locale: *In some US cities, the area covered by a police station is called a precinct.*

precious *adj.* **1** °dear, dearest, costly, °expensive, high-priced, valuable, °invaluable, prized, °priceless, irreplaceable, *Colloq* °pricey: *The entire cabinet was filled with precious jewels.* **2** esteemed, °choice, cherished, °beloved, idolized, adored, loved, valued, prized, revered, venerated, °venerable, hallowed: *The church keeps its most precious relics in a special vault.* **3** °precise, °exquisite, overrefined, chichi, overnice, °studied, °artificial, effete, °affected, overdone, °pretentious, euphuistic, alembicated: *His saccharine style is characterized by the elaborate embroidery of precious language.* **4** °sweet, °sentimental, °quaint, °dainty, °cute, mignon, bijou, *Colloq Brit* twee, *Slang US* cutesy: *The tea-roomy atmosphere is a bit too precious for my taste.*

precipice *n.* °cliff, escarpment, °bluff, °crag: *With a scream, he plunged over the precipice into the raging sea below.*

precipitate *v.* **1** accelerate, °hasten, °speed (up), °advance, °hurry, °quicken, °expedite, bring on *or* about, trigger, °provoke, instigate, °incite, °facilitate, °further, °press, °push forward: *Glasnost precipitated perestroika—or was it the other way around?* **2** °hurl, °fling, °cast, °launch, °project: *The force of the crash precipitated him through the windshield.* —*adj.* **3** headlong, °violent, °rapid, °swift, °quick, °speedy, °meteoric, °fast: *A powerful counterattack brought about the enemy's precipitate rout.* **4** °sudden, °abrupt, unannounced, unexpected, unanticipated: *Today's fall in share prices was as precipitate as yesterday's rise.* **5** °rash, °impetuous, °hasty, °volatile, hotheaded, °careless, °reckless, incautious, injudicious, °foolhardy, °impulsive, unrestrained: *The selection of a career should not be the result of a precipitate decision.*

precipitation *n.* showers, drizzle, °downpour, °rain, rainfall, snow, snowfall, °hail, sleet: *Precipitation can be expected in advance of the low-pressure area pushing down from the north.*

precipitous *adj.* **1** °abrupt, °steep, °perpendicular, °sheer, °bluff, acclivitous, declivitous: *I cannot see how anyone can climb up the precipitous face of that crag.* **2** See **precipitate, 5,** above.

précis *n.* °outline, °summary, °synopsis, *aperçu,* °résumé, conspectus, °survey, overview, °abstract, °abridgment, °digest, compendium, recapitulation; table of contents: *As I haven't the time to read the entire study document, let me have a précis of it.*

precise *adj.* **1** °correct, °exact, °definite, well-defined, °explicit, word-for-word, °verbatim, °literal, letter-for-letter, *literatim,* °faithful, °specific, unambiguous, unequivocal, °strict, °authentic, °perfect, °true, veracious, °truthful, unerring, error-free, °accurate: *Errors show that medieval scribes did not always succeed in making precise copies.* **2** °strict, °meticulous, °scrupulous, °careful, °conscientious, °exact, unconditional, rigorous, °rigid, puritanical, unbending, °inflexible, unyielding, °demanding, °severe, prim, °absolute: *He has been very precise about dates and facts. Margaret is precise in insisting on obedience to her orders.* **3** °fastidious, °particular, °finicky, finical, °fussy, °meticulous, °scrupulous, °careful, °conscientious, °nice, °exacting, °critical, °demanding: *Painting miniatures is very precise work.* **4** °exact, °very: *This is the precise spot where I found the body.*

precisely *adv.* **1** °exactly, °just, strictly, *Colloq* °on the nail, smack, °slap, on the nose, *Brit* bang on, spot on: *His view is precisely the opposite of hers. We left precisely at five o'clock. You phoned, and that is precisely why I came. The two paintings are not precisely the same.* **2** °exactly, exactingly, correctly, rigorously, °absolutely, punctiliously, minutely, carefully, meticulously, scrupulously, conscientiously, strictly, rigidly, inflexibly; in all respects, in every way: *This judge follows the letter of the law precisely.*

precision *n.* **1** correctness, exactness, fidelity, faithfulness, exactitude, preciseness, °accuracy, °rigor, °perfection, flawlessness, faultlessness, literalism, faithfulness, unerringness: *This copy follows the original with precision.* **2** definiteness, °care, nicety, meticulousness, rigorousness, °rigor, fastidiousness, punctiliousness, scrupulousness, unambiguousness, nicety, strictness, explicitness: *The precision of the wording leaves no doubt about what the writer intended.*

preclude *v.* obviate, °bar, °prevent, °stop, °exclude, °prohibit, °shut out, °forestall, °rule out, debar, °check, °block, °obstruct, avert, °avoid, °thwart, °frustrate, °impede, °inhibit, °hinder, interfere with: *To preclude misunderstanding, please repeat what you said. They may not engage in activities that preclude them from performing their regular duties.*

precocious *adj.* advanced, °mature, °bright, °gifted, °intelligent, °smart, °quick: *It is hard to believe that Oliver was a precocious child.*

preconceived *adj.* beforehand, predisposed, prejudged, °predetermined, °prejudiced, °biased, anticipatory: *He has many false, preconceived notions about people.*

preconception *n.* predisposition, prejudgment, predetermination, °prejudice, °bias, °presumption, °presupposition, assumption, *idée fixe,* prepossession, preconceived notion *or* idea: *From a basic moral sensitivity, we all have a preconception of what is right and wrong.*

precondition *n.* °prerequisite, °stipulation, °condition, °essential, °must, sine qua non, °imperative, °requirement, °proviso, °provision, °qualification, °necessity: *A precondition for membership is Swiss citizenship.*

precursor *n.* **1** °harbinger, herald, vanguard: *The glow on the eastern horizon is the precursor of another day.* **2** See **predecessor, 1,** below

predatory *adj.* **1** predacious *or* predaceous, carnivorous, preying, raptorial: *Despite domestication, dogs and cats are predatory animals.* **2** °rapacious, °ravenous, plundering, robbing, pillaging, marauding, despoiling, looting, piratical, vulturine, °avaricious, °greedy, °voracious, larcenous, thieving, extortionate, usurious: *Predatory pirates from Tripoli once ravaged the Mediterranean. Keep out of the hands of predatory money-lenders, my son.*

predecessor *n.* **1** °forerunner, °predecessor, antecedent: *I could not match the accomplishments of my predecessor in this post.* **2** forebear, forefather, °ancestor, antecedent: *Can you name the Tudor predecessors of Elizabeth I?*

predestination *n.* °destiny, °future, °lot, °fortune, kismet, karma; °doom, °fate; foreordainment, foreordination: *It was not my predestination to be born rich or beautiful.*

predetermined *adj.* **1** °fixed, prearranged, preestablished, °set (up), °foregone, preplanned, preset: *A predetermined amount of milk is automatically poured into each cup of coffee.* **2** °fated, °doomed, °destined, ordained, foreordained, *Colloq* °cut and dried, *Brit* on the cards, *US* °in the cards: *One gets the feeling that the outcome was predetermined.*

predicament *n.* °dilemma, °quandary, °difficulty, °trial, °situation, °state, °condition, imbroglio, °emergency, °crisis, °impasse, *Colloq* pickle, °jam, °fix, °pinch, °scrape, °spot, °bind, corner, °hole, °mess, *US* °box: *I was hoping that you might help me out of a very awkward predicament.*

predict *v.* foretell, °prophesy, °forecast, °foresee, augur, °prognosticate, forewarn, presage, vaticinate; portend, °foreshadow, foretoken, forebode; °intimate, °hint, °suggest: *My mother predicted that there would be moments like this. The term for predicting coming events from the entrails of animals is "haruspicy."*

predictable *adj.* foreseeable, foreseen, °probable, °likely, °liable, expected, anticipated, (reasonably) sure *or* certain, *Colloq Brit* on the cards, *US* °in the cards: *Her angry reaction at being dismissed was quite predictable.*

prediction *n.* °forecast, °prophecy, augury, °prognosis; intimation, °hint, °suggestion: *The weatherman's predictions are more accurate than one might think. The prediction that I might be released the next day did not come true.*

predominance *n.* predominancy, °superiority, °influence, dominance, °preeminence, °preponderance, ascendancy, °precedence, °power, °supremacy, °hold, °sway, hegemony, °leadership, mastery, °control, °dominion, °sovereignty, transcendence *or* transcendency, °authority, the upper hand, the whip hand, °advantage, the edge: *We must reestablish the predominance of intellectual vigor over crass commercialism.*

predominant *adj.* °dominant, predominating, controlling, °sovereign, ruling, °preeminent, preponderant, ascendant, °superior, °supreme, °leading, °paramount, °main, °chief, °transcendent, °important, °telling, °influential, °primary, °prevailing, °prevalent: *England was the predominant power in the world during the nineteenth century. The phlogiston theory was once predominant among scientists.*

predominate *v.* Often, **predominate over.** dominate, °control, °rule, °reign, preponderate, °outweigh, °obtain, °prevail, °overshadow, get *or* have the upper hand, °lord it over, hold sway, overrule: *Though the American president is a Republican, the Democrats predominate in Congress. Eventually, the orthodoxy predominated over the Albigensian heresy.*

preeminence *n.* **1** See **predominance,** above. **2** peerlessness, magnificence, °excellence, °distinction, eminence, inimitability, °superiority: *There is no gainsaying the preeminence of Shakespeare as a poet and playwright.*

preeminent *adj.* **1** See **predominant,** above. **2** °peerless, °excellent, °distinguished, °eminent, inimitable, °superb, unequaled, °matchless, °incomparable, °outstanding, °unique, unrivaled, unsurpassed, °supreme, °superior: *In her opinion, Craig is the preeminent authority on the subject.*

preeminently *adv.* °primarily, °principally, °by far, far and away, °manifestly, °eminently, °notably, conspicuously, prominently, signally, uniquely, extraordinarily, °supremely, superbly, matchlessly, incomparably, outstandingly: *There is no doubt that Nathalie is preeminently qualified to direct the play.*

preempt *v.* °appropriate, usurp, arrogate, °take over, °assume, °take possession of, °seize, °acquire, °take, °possess, expropriate: *All the seats on the committee had been preempted by the ruling faction.*

preen *v.* **1** °trim, °clean, plume, °groom: *The gliding swans stopped now and then to preen their feathers.* **2** °primp, °dress up, titivate *or* tittivate, prettify, °beautify, prink, °spruce up, deck (out), *Colloq* doll up, *Brit* tart up: *He made Sonia wait while he preened himself before the mirror.*

preface *n.* **1** introduction, °foreword, prologue, °preamble, *Formal* proem, prolegomenon, exordium: *By way of preface, I should like to say how happy I am to be here. The preface to the book is far too long.* —*v.* **2** °precede, °introduce, prefix, °begin, °open: *The speaker prefaced his acceptance speech with a tribute to his predecessor.*

prefatory *adj.* °opening, °introductory, °preliminary, °preparatory: *Would you like to make a few prefatory remarks introducing this evening's speaker?*

prefer *v.* **1** °favor, like better, °fancy, lean *or* incline toward(s) *or* on the side of, be inclined, be partial to, °pick, °select, opt for, °choose, °single out, take a fancy to, °embrace, espouse, °approve, °esteem: *Which flavor do you prefer, chocolate or coffee? I should prefer to take my own car.* **2** °present, °offer, °propose, proffer, °advance, °submit, °tender, °put forward, °file, °lodge, °enter: *They preferred charges against the hooligans for assault and battery.*

preference *n.* **1** °favorite, °choice, °selection, °desire, °option, °pick: *My preference is the Dover sole meunière.* **2** °partiality, proclivity, °prejudice, °favoritism, predilection, °liking, °fancy, predisposition, °bent, °inclination, °leaning: *She shows a marked preference for short men.*

preferential *adj.* °advantageous, °biased, °prejudiced, °favorable, °privileged, °partial, °better, °favored, °superior: *What entitles her to preferential treatment?*

pregnant *adj.* **1** gravid, parturient, °expectant, (heavy) with child, *enceinte*, *Colloq* °expecting, in a family way, *Brit* preggers, *Slang* having a bun in the oven, *Brit* in the (pudding) club, °up the spout: *Even though she is pregnant, no man has given her his seat on the bus.* **2** charged, °fraught, °loaded, °weighty, °significant, °meaningful, °suggestive, °eloquent, °expressive, °pointed: *Although she said nothing, her wan smile was pregnant with meaning.* **3** °fruitful, teeming, °fertile, fecund, °rich, abounding, °replete, °productive: *My mind was so pregnant with ideas that I couldn't wait to get down to work.*

prehistoric *adj.* **1** primordial, primal, primeval, °primitive, earliest, °early, antediluvian, Noachian *or* Noachic, fossil, °ancient: *The prehistoric remains of yet another skeleton have been found in the peat bogs.* **2** °antiquated, outdated, °old-fashioned, °passé: *My mother makes me wear these absolutely prehistoric clothes to school.*

prejudice *n.* **1** °partiality, °preconception, prejudgment, °bias, °leaning, °warp, °twist, preconceived notion, predisposition, predilection, jaundiced eye, jaundice: *The judge demonstrated an unfortunate prejudice against my client.* **2** °bigotry, unfairness, °bias, partisanship, °favoritism, cronyism, °discrimination, °intolerance, °inequality; racism, racialism, apartheid, Jim Crowism, sexism, (male) chauvinism: *The failure to accord everyone equal opportunity can be laid at the door of racial, religious, social, or sexual prejudice.* —*v.* **3** °bias, °influence, °warp, °twist, °distort, °slant; °color, jaundice, °poison: *Stop trying to prejudice me against the book, and let me form my own opinion. Are you prejudiced in favor of Anita's getting the job?*

prejudiced *adj.* **1** unfair, °one-sided, °biased, °jaundiced, °opinionated, predisposed, °partial, °partisan, nonobjective, unobjective: *The prosecution objected to him as a prejudiced witness.* **2** °bigoted, °intolerant, °narrow-minded, closed-minded, °parochial, sexist, racist, chauvinistic: *We have no room in our organization for people who are prejudiced.*

prejudicial *adj.* °injurious, damaging, °detrimental, °harmful, unfavorable, inimical, deleterious, disadvantageous, counterproductive, pernicious: *Such an investment decision might prove prejudicial to her financial security.*

preliminary *adj.* **1** °advance, °prior, °introductory, °beginning, °initial, °opening, °preparatory, °prefatory, °preceding, antecedent, forerunning; premonitory; *Formal and technical* prodromal *or Formal and technical* prodromic: *Eric's preliminary design for the swimming pool is ready. After some preliminary remarks, the ceremonies got under way.* —*n.* **2** introduction, °beginning, °opening, °preparation, °groundwork, prelude, °precedence; °overture: *We insist on rigorous training as a preliminary to working in the field.* **3** ***prelims.*** Rarely, **preliminaries.** introduction, °preface, °foreword, °preamble, prologue, front matter, *Formal* proem, exordium, prolegomenon, prodromus, prodrome: *The manuscript is finished and we are waiting for the prelims.*

premature *adj.* **1** °immature, °undeveloped, underdeveloped, °unfledged, untimely, unready, °early, unripe, °green: *Much premature fruit was blown down in yesterday's windstorm.* **2** °hasty, untimely, ill-timed, too early, too soon, beforehand, °unseasonable, overhasty, °impulsive, °inopportune: *The mails may be slow, but still I think September a bit premature to send Christmas cards. His realization that they knew about the missing money accounts for his premature departure.*

prematurely *adv.* **1** untimely, too soon, too early: *He died prematurely at the age of forty-six.* **2** rashly, (over)hastily, at half cock, half-cocked: *She dismissed him prematurely, before he had the time to prove himself. The gun went off prematurely, with no chance to aim it.*

premeditated *adj.* planned, °conscious, °intentional, intended, °willful, °deliberate, °studied, purposive; contrived, preplanned, °calculated, °preconceived: *From the cobra venom in the syringe, Inspector Maigre deduced that the murder was premeditated. I admit to being wrong, and I have no premeditated excuses.*

premier *n.* **1** prime minister, PM, head of state, chief executive, president, chancellor: *The premiers of the Commonwealth met at Kuala Lumpur in 1989.* —*adj.* **2** °first, °prime, °primary, °chief, °principal, °head, °main, °foremost, top-ranking, highest-ranking, ranking, °leading, °top, °preeminent: *What is the name of the premier danseur of the ballet?*

première *n.* **1** premiere, first night, °opening (night), °debut: *The premiere of* Aïda *was at the opening of the Suez Canal.* —*v.* **2** premiere, °open, °debut: *The film will première at the Barbizon Cinema tomorrow.* —*adj.* **3** premiere, °opening, °debut, °first, °original, °initial: *The première performance of* Aïda *was staged to celebrate the opening of the Suez Canal.*

premise *n.* **1** premiss, assumption, °proposition, postulate, °hypothesis, conjecture, °assertion, °supposition, °thesis, °presupposition, °proposal, °theorem, °surmise, °basis, °ground: *He started out with the premise that time had a beginning and will have an end.* —*v.* **2** °assume, °propose, postulate, hypothesize, hypothecate, conjecture, °posit, assert, °suppose, °presuppose, °theorize, °surmise, put *or* set forth, predicate, °argue: *If you accept what Einstein premised in the Special Theory of Relativity, then you accept that space is curved.*

premium *n.* **1** °bonus, °extra, dividend, °prize, °award, °reward, °perquisite: *In our slimming competition, a premium will be paid for each pound you lose beyond five.* **2** °incentive, °inducement, stimulus, °incitement, °lure, bait, °spur, goad, °reward, *Colloq* °come-on, *Slang US and Canadian* freebie: *If you buy gas there, the premium is free glassware for each transaction over ten dollars.* **3** value, °importance, °regard, °stock, °store, °appreciation: *We place a premium on the way our staff treat customers.* **4** ***at a premium.*** **(a)** °scarce, °rare, scant, °scanty, °sparse, hard to come by, °in short supply, *Colloq* scarce as hen's teeth, *Chiefly Brit* °thin on the ground: *Proficient, literate, experienced editors are at a premium these days.* **(b)** costly, °expensive, °dear, high-priced, *Colloq* °steep, °stiff: *The convertible model of this car is at a premium.*

premonition *n.* °intuition, °foreboding, presentiment, forewarning, °suspicion, °feeling, °hunch, *Colloq* funny feeling, sneaking suspicion: *I had a premonition that something terribly evil was lurking in the basement.*

preoccupied *adj.* **1** engrossed, lost in thought, °rapt, °thoughtful, °pensive, °absorbed, cogitating, cogitative, meditating, musing, reflecting, °reflective, contemplative, contemplating, pondering, brooding, ruminating, in a brown study: *John is preoccupied, thinking about the speech he has to deliver tonight.* **2** °vague, °offhand, °faraway, °absent-minded, abstracted, °oblivious, °unaware, wrapped up, immersed, °inattentive, distracted, distrait, *Colloq* turned off, *US* out of it: *From her preoccupied air it was clear that something was wrong.*

preparation *n.* **1** Often, ***preparations.*** **(a)** °groundwork, spadework, provision(s), °foundation, preparing, °measures, °proceedings: *Nothing will interfere with our preparation for the king's visit.* **(b)** plans, °arrangements: *We are making preparations to leave tomorrow.* **2** °fitness, °readiness, readying, preparing, training, °education, teaching, °instruction, instructing, °tuition, briefing, grooming, *Colloq* gearing up, prep, *US* prepping: *The preparation of the students was your responsibility.* **3** drawing up, drafting, planning, setting up, putting together, organizing, °organization, composing, making: *How is your preparation of the new proposal coming along?* **4** °work, preparing, getting ready, °study, studying, practicing, °practice, *Colloq* cramming, *Brit* swotting: *All my preparation for the exam was to no avail.* **5** °substance, °compound, concoction, °mixture, °product, °material, °stuff, °composition: *This preparation is a furniture wax, not a suntan lotion.*

preparatory *adj.* **1** preparative, °preliminary, °introductory, °prefatory, °opening: *After some preparatory warnings about safety, the scuba-diving lessons began.* **2** °elementary, °basic, °essential, °fundamental, °primary, °rudimentary: *His preparatory training left him ill equipped to deal with such a major problem.* **3** ***preparatory to.*** °before, in preparation for, °preceding: *Preparatory to the mission, we were briefed by the CIA.*

prepare *v.* **1** (get *or* make) ready, °prime, °arrange, °(put in) order, °organize, °provide for, make provision(s) for, lay the groundwork (for), °(make) fit, °fit (out), °equip, °outfit, °adapt: *Have you prepared adequately for the meeting? We prepared the house to receive guests.* **2** °train, (get *or* make) ready, °study, °practice, *Colloq* °cram, *Brit* swot, °get up: *He is preparing to take an exam.* **3** °train, °educate, °teach, (get *or* make) ready, °groom, °brief, °develop: *Her mother is preparing her for the next Olympics.* **4** cook (up), °make, do, *Colloq* °whip up, *US and Canadian* °fix: *I shall prepare dinner for eight o'clock.* **5** °manufacture, °fabricate, °produce, °make, °put out, °build, °construct, °assemble, put together, °turn out, °fashion, °forge, °mold: *Our company prepares components for the printing industry.* **6** °brace, °strengthen, °steel, °fortify, °ready: *When I saw the doctor's face, I prepared myself for the bad news.* **7** °process, °produce, °make, °treat, °change, °modify, °transform: *This fabric has been specially prepared to repel stains.*

prepared *adj.* **1** °ready, °set, °advance, prearranged, planned: *We have six prepared questions for the interview.* **2** treated, processed, modified, changed: *Using an etching needle, incise the lines on the prepared surface of the steel plate.* **3** °willing, °disposed, predisposed, °able, °inclined, of a mind; °ready, °(all) set: *Are you prepared to apologize?* **4** oven-ready, microwave-ready, microwave-able, °instant, convenience, ready-to-eat, ready-to-serve, precooked, °ready-made: *He lives alone and usually buys prepared dinners that just need reheating.*

preparedness *n.* °vigilance, alertness, °readiness, °fitness: *The armed forces were kept in a continuous state of preparedness.*

preponderance *n.* **1** °majority, greater part, °bulk, °mass, lion's share: *The preponderance of voters want a change of government.* **2** °weight, °influence, weightiness, °superiority, °supremacy, °predominance, primacy, ascendancy, °sway, °strength, °force, °power, °advantage, °control, °authority, hegemony, °leadership, °rule: *Good has always appeared to have an incontestable preponderance over evil.*

prepossessing *adj.* °attractive, appealing, °pleasing, °favorable, °engaging, charming, captivating, fascinating, winsome, °winning, °magnetic, alluring, bewitching, °taking, °fetching, °inviting, good-looking, °handsome, °lovely, °beautiful: *She is a woman of prepossessing presence and a delightful nature.*

preposterous *adj.* °absurd, °ridiculous, °ludicrous, laughable, risible, asinine, °foolish, °senseless, irrational, °nonsensical, fatuous, fatuitous, °mindless, °insane, °crazy, crackbrained, °mad, idiotic, moronic, imbecilic, °incredible, °unbelievable, °outrageous, °extravagant, °extraordinary, °extreme, °exorbitant, °outlandish, °outré, °weird, °bizarre, *Colloq* balmy, *Brit* barmy, *Slang* nutty, screwy, batty, dotty, wacky, loony, cuckoo, *US* wacko: *Today he had another preposterous excuse for coming in to work late.*

prerequisite *adj.* **1** °essential, °necessary, requisite, °imperative, °indispensable, °obligatory, required, called for, demanded: *Naval service is a prerequisite condition for joining the club.* —*n.* **2** °precondition, °requirement, °qualification, requisite, °condition, sine qua non, °proviso, °provision, °necessity: *A thorough grounding in mathematics is a prerequisite for the study of advanced physics.*

prerogative *n.* °privilege, °right, °liberty, °power, °due, °advantage, °license, franchise, °claim, °sanction, °authority, authorization: *As managing director, she exercises her prerogative to park her car closest to the door.*

prescribe *v.* ordain, °order, °direct, °dictate, °demand, °decree, °require, enjoin, °rule, °set (down), °stipulate, °command, °instruct, °define, °specify, °impose, °lay down, °exact, constrain: *The doctor prescribed rest. An independent journalist, Healey would not let anyone prescribe what he could or could not write.*

prescription *n.* **1** °formula, °recipe, °instruction, °direction: *The prescription calls for one capsule before each meal.* **2** °remedy, medication, °medicine, °drug, °preparation, medicament: *That prescription had no effect at all on my headache.*

prescriptive *adj.* °dictatorial, constrictive, didactic, restrictive, °dogmatic, °authoritarian, °overbearing, autocratic, imperious: *As grammar is a description of how language works, it cannot be prescriptive.*

presence *n.* **1** °proximity, nearness, closeness, adjacency, °vicinity, *Formal* propinquity: *The presence of a hospital near her home made Aunt Mary feel more secure.* **2** °attendance, °company, °companionship, °society, °association, °existence, °manifestation, manifestness, being: *The Queen honored us by her presence at the unveiling of the plaque.* **3** °poise, self-assurance, °bearing, self-possession, °confidence, mien, °carriage, comportment, deportment, °air, °personality, °aspect, °aura, °appearance: *When Geoffrey entered, his presence created quite a stir.* **4** °spirit, wraith, °specter, °shade: *I had the vague but pervading sense that an unidentifiable presence was with me in the empty room.* **5** ***presence of mind.*** aplomb, °sophistication, coolness, coolheadedness, composure, imperturbability, phlegm, °sang-froid, self-possession, self-assurance, °calm, equanimity, levelheadedness, quick-wittedness, alertness, *Colloq* °cool: *With great presence of mind, Grainger walked up to the man and took the gun from him.*

present[1] *adj.* **1** °current, °contemporary, present-day, existing, existent, up-to-date, *Colloq* °now: *Do you understand the present generation?* **2** °nearby, nearest, °immediate, closest, adjacent, proximate, propinquitous; °close, °remaining; accounted for: *Everybody has been awful to me—present company excepted. All those present heard what she said to me.* —*n.* **3** ***at present.*** (right *or* just) now, °for the time being, for the present, today, these days, *Nonstandard* °presently, *Colloq* at this point in time: *I am not ready to invest at present.* **4** ***the present.*** the time being, the moment, the hour, the nonce, this juncture, these days, our times; today, °(right) now, °nowadays; *Colloq* this point in time: *She lives for the present, without a thought for the consequences.*

present[2] *n.* **1** °gift: *May I open my birthday presents now?* **2** °donation, °offering, °bounty, °grant, largesse *or* largess, contribution, °endowment: *Cublington made the Society a present of Cublington Hall to house its collection.* **3** °tip, gratuity, *pourboire*, baksheesh *or* backsheesh, °bonus; alms, handout, °dole, °aid, °allowance: *She gave the maître d'hôtel a present for looking after us so well.* —*v.* **4** °introduce, °acquaint with, °make known: *I'd like to present my wife, Cordelia, who has been looking forward to meeting you.* **5** °offer, °give, °stage, °show, °exhibit, °put on (show), °mount, °produce: *The Mighty Allen Arts Players will present a new play tonight.* **6** °give (out), °award, °confer (on), °bestow (on), turn *or* hand over, °grant, °provide, °furnish; °dispense, °distribute, dole out, °pass out, deal out, mete out: *They presented her with a prize for the best poem. Is it tonight that they present the awards?* **7** °offer, bring (in *or* up), proffer, °tender, °introduce, °produce, °submit, set *or* put forth, °put forward, adduce; °register, °file, °record: *The lawyer for the defense presented evidence of his client's alibi.* **8** °introduce, °announce, *Brit* compère *or* compere, *Colloq* emcee: *And here, to present our show, is the inimitable Reginald Norton!*

presentable *adj.* **1** °fit, °fitting, °suitable, °acceptable, °satisfactory, °adequate, °passable, °tolerable, admissible, all right, allowable, up to par *or* standard *or* the mark, good enough, *Colloq* °up to scratch, OK *or* O.K. *or* okay, up to snuff: *Whether that argument is presentable in a court of law is disputable.* **2** °decent, °proper, °polite, °decorous, °respectable, °well-bred, well-mannered, fit to be seen: *After a bath, shave, and change of clothing he looked quite presentable.*

presentation *n.* **1** giving, bestowal, °offering, proffering, presenting, °award, awarding, conferral, conferring, °delivery; °donation: *The presentation of the prizes will be at the annual dinner.* **2** °appearance, °image, °display, visual(s), °spectacle, °show, °performance, °demonstration, °production: *The text of the proposal was fine, but the presentation could be improved.* **3** °debut, °launch, introduction, unveiling, disclosure: *We attended the presentation of the latest hovercraft car.*

presently *adv.* °soon, by and by, in a little while, °shortly, after a short time, in due course, after a while *or* a time, before long, in a moment *or* a minute *or* a while, *Archaic or literary* anon, *Colloq* in a jiffy, in two shakes (of a lamb's tail), *Nonstandard* °now, °at present: *The show opens presently in the West End. I shall be with you presently.*

preservation *n.* **1** °upkeep, °maintenance, °care, °conservation: *The preservation of old paintings has become an art in itself.* **2** keeping, retention, retaining, perpetuation, perpetuating, continuation, °safekeeping, °security, safeguarding, °protection, protecting, °conservation: *Several charities are dedicated to the preservation of our heritage.*

preserve *v.* **1** keep safe, °protect, °guard, °take care of, °care for, °safeguard, watch over, °shield, °shelter, °defend, °spare: *What can we do to preserve ourselves from this terrible pestilence? May God preserve me, I never expected to see you again!* **2** °keep (up), °maintain, °conserve, °spare, °perpetuate; °continue, °support, °sustain, °save: *We developed a technique to preserve antique furniture. The sea preserves much of the heat it absorbs in the summer. It is difficult to preserve one's sanity in this madhouse.* **3** °conserve, °put up, pickle, °cure, smoke, kipper, °salt, corn, marinate, can, °freeze, freeze-dry, °refrigerate, °dry, dehydrate, vacuum-pack; embalm, mummify: *The berries were preserved in jams and jellies. Some of the remains were preserved for hundreds of years.* —*n.* **4** Often, ***preserves.*** conserve(s), °jam, jelly, confiture, marmalade: *Susan likes thick toast with butter and preserves for tea.* **5** °(game) reserve, °reservation, °sanctuary, *Brit* °park: *No hunting is allowed in this preserve.*

preside *v.* °supervise, °run, °oversee, °direct, °operate, °lead, °head (up), °govern, °rule, °manage, °handle, °control, °direct, °chair, °administer, administrate, °regulate, °officiate: *Melissa presides over the meetings of the board.*

press *v.* **1** subject to *or* exert pressure *or* force, °force, °push, impel, °thrust, °bear (on), weigh on *or* upon, °jam, °cram, °crush; pressure *or Brit also* pressurize: *The crowd was pressing against the door. Although hard pressed at first, we eventually won.* **2** °squeeze, compress, °depress, °push: *When the picture is in focus, press this button to actuate the shutter.* **3** °squeeze, °crush, compress, mash: *After harvesting, the grapes are pressed to extract the juice.* **4** iron, °smooth, °flatten, put through a mangle; steam: *I shall have to press my jacket before going out to dinner.* **5** °clasp, °embrace, °hug, hold (close *or* tight(ly)), take in one's arms, throw one's arms about *or* around, cleave to, *Archaic* °clip: *She pressed the child to her with great affection.* **6** constrain, °urge, °force, °pressure, compel, °demand, °persuade, °induce, °prod, °provoke, importune, °beseech, °ask, °request, °beg, entreat: *They pressed me to tell where the money was hidden.* **7** °crowd, °flock, °gather, °mill, °swarm, °throng, °seethe, °cluster, congregate, °meet, °converge, °huddle: *The reporters pressed round the chancellor to catch every word.* —*n.* **8** crowding, °gathering, thronging, converging, convergence, °crowd, °throng, °swarm, °cluster, °huddle, °pack, °herd, °host, multitude, horde, °mob, °crush: *When the doors opened, the huge press of people pushed onto the train.* **9** °urgency, °haste, °hurry, °hustle, bustle, °pressure, °stress: *The press of business required me to postpone my trip to Paris.* **10** ***the press.*** **(a)** newspapers, the papers, Fleet Street, the fourth estate, the media, the wire *or* news services, broadcasting, °television, °radio: *The press will have a field day when this news gets out.* **(b)** newspaper people, newspapermen *or* newspaperwomen, newsmen *or* newswomen, reporters, correspondents, ladies *or* gentlemen of the press, journalists, commentators, *paparazzi, Brit* leader writers, *US* editorial writers, *Colloq* newshounds, *Brit* journos: *The doors were opened and the press was invited in.*

pressing *adj.* °urgent, compelling, °crucial, °pivotal, °burning, °grave, °serious, °major, °important, °vital, high-priority, °critical, °portentous, °momentous, °profound, °significant: *Some pressing matters kept me late at the office.*

pressure *n.* **1** °force, compression; °weight, °power, °strength: *The air pressure in your tires is low. The pressure of the water burst the pipe.* **2** compression, °pressing, squeezing, compressing, °tension, °stress, crushing: *The pressure of a tourniquet has to be loosened now and then.* **3** °affliction, °oppression, °press, °weight, °burden, °load, albatross, °strain, °stress, °urgency, demands, exigency *or* exigencies, vexation, °distress, trouble(s), adversity, difficulty *or* difficulties, °straits, constraint(s), problem(s): *The pressure of work prevents me from taking any time away from the office. As soon as interest rates increase, we all feel financial pressure.* **4** °influence, °power, °sway, constraint, insistence, coercion, intimidation, arm-twisting; °inducement, °persuasion, urging, °pressing: *Without more pressure, the road repairs will not be completed in time.* —*v.* **5** °persuade, °influence; prevail upon *or* on, °press, °urge, °sway, °intimidate, bring pressure to bear (on), apply pressure (on *or* to), coerce, °force, compel, constrain, °require, °demand, °make, insist upon *or* on, *Brit* pressurize, *Colloq* twist (someone's) arm, °lean on, turn the heat on, *Slang* put the screws on *or Slang* to: *They pressured him to take the assignment in Toronto.*

prestige *n.* °status, °reputation, °standing, °rank, stature, °importance, °significance, eminence, °esteem, °preeminence, °prominence, °predominance, primacy, °superiority, °supremacy, ascendancy, °distinction, °renown, °regard, °fame, °cachet, repute, °celebrity, °glamour, stardom: *To raise money, someone with prestige must be found to serve as director of the charity.*

prestigious *adj.* °important, °significant, °eminent, °estimable, °imposing, °impressive, °preeminent, °prominent, °distinguished, august, °dignified, °renowned, °famous, famed, °well-known, °illustrious, acclaimed, respected, °celebrated, °noted, °notable, °noteworthy, °outstanding, °glorious, honored, °glamorous: *She may be a prestigious author, but she does not live at a very prestigious address.*

presumably *adv.* °probably, in all likelihood, (very *or* most) likely, in all probability, °seemingly, doubtless(ly), indubitably, no doubt, °undoubtedly, unquestionably, without a doubt, °surely, certainly, °on the face of it, all things considered, all things being equal: *Presumably, you have heard the one about the colonel's poodle.*

presume *v.* **1** °assume, take for granted, °suppose, °surmise, °infer, °suppose, °presuppose, °take it, °gather, °understand, °think, °believe, °imagine, °suspect, °fancy, conjecture, postulate, °posit, °theorize, °speculate, hypothesize, hypothecate, *US and Canadian* °guess: *For some unaccountable reason, we presumed that your train would arrive on time. Dr. Livingstone, I presume?* **2** °dare, °take the liberty, be so presumptuous as, make (so) bold (as), have the audacity *or* effrontery, go so far as, °venture: *Who is he to presume to judge others?* **3** Often, ***presume on*** *or* ***upon.*** encroach (on *or* upon), impose (on *or* upon), °take liberties (with), intrude (on *or* upon *or* into): *I hate to presume on our friendship, but could you lend me some money?*

presumption *n.* **1** °arrogance, °pride, °effrontery, audacity, boldness, brazenness, °impudence, °impertinence, insolence, temerity, overconfidence, presumptuousness, forwardness, immodesty, *Colloq* pushiness, cheek, cheekiness, °nerve, °gall, chutzpah, °brass, *Brit* brass neck: *He had the presumption to ask my age.* **2** assumption, °supposition, °presupposition, °preconception, premise *or* premiss, °surmise, °proposition, postulation; °probability, °likelihood, plausibility, °feasibility: *The presumption of the innocence of a person accused of a crime is established in law.* **3** assumption, °stand, °position, °inference, °feeling, °deduction, conclusion, °conviction, °bias, °guess, theory, °hypothesis, conjecture, °belief, °thought; °suspicion: *Having examined the evidence, the pathologist's presumption was that the victim had died of natural causes.* **4** ground(s), °reason, °basis, °evidence: *What is your presumption for thinking that you might win the lottery?*

presumptive *adj.* **1** °likely, °reasonable, °plausible, °tenable, believable, credible, conceivable, °acceptable, justifiable, °sensible, °rational, °sound: *There is strong presumptive evidence, but we need proof.* **2** inferred, presumed, °assumed, °supposed, °understood, predicted, predicated: *As the king was childless, his brother was heir presumptive to the Crown.*

presumptuous *adj.* °arrogant, °proud, prideful, °audacious, °bold, °brazen, saucy, °impudent, °impertinent, °insolent, temerarious, °brash, °overconfident, overweening, °forward, presuming, °immodest, °egotistical, *Colloq* °pushy, °cheeky, too big for one's boots, *Brit* °uppish: *He is presumptuous enough to think he can do no wrong.*

presuppose *v.* See **presume, 1,** above.

presupposition *n.* See **presumption, 1, 2,** above.

pretend *v.* **1** feign, °profess, °represent, °allege, °make believe, °make out: *Let's pretend that we are royalty. Often, he pretends to be me.* **2** °try, °attempt, °endeavor, °venture, °presume, °undertake: *I could not pretend to guess the number of people attending last night's meeting.* **3** °make believe, act *or* play, play-act, °fake, feign, put on an act, dissemble, °sham, sail under false colors: *Is she serious about wanting you to leave or was she just pretending?*

pretended *adj.* °so-called, °alleged, asserted, °reputed, °professed, ostensible, purported, °so-called, °imaginary, make-believe, °fictitious, °fictional, °sham, °false, °fake, feigned, °bogus, °counterfeit, °spurious, *Colloq* phoney *or US also* phony, pseudo, °pretend: *So this is the pretended hero of yesterday's battle!*

pretender *n.* claimant, aspirant, °candidate, °suitor, °rival, seeker: *He was the pretender to the throne of Austria.*

pretense *n.* **1** °show, °display, °pretension, °ostentation, °airs, °front, façade, °appearance, make-believe, fiction, °hypocrisy, fakery, faking, feigning, humbuggery, humbug, °deception, °artifice, °pretext, posturing, pretentiousness, pretending, °camouflage, cover-up: *Her charming manner was all pretense, for in reality she despised him.* **2** °hoax, humbug, °artifice, °pretext, °sham, °show, °pose, façade, °front, cover-up, °cover, °cloak, °veil, °mask, °masquerade, °disguise, °guise, °deception, °ruse, °dodge, °blind, °fabrication, °invention, fiction, °story, fable, make-believe, fairy tale, figment, falsification, °impression: *His loyalty was a pretense under which he carried on many liaisons.* **3** °excuse, °pretext, °pretension: *They would ring for the butler on the slightest pretense, just to impress us.*

pretension *n.* **1** Often, ***pretensions.*** claim(s), pretense(s), aspiration(s), ambitiousness, ambition(s): *He is known to have pretensions to the chairmanship of the department.* **2** °pretext, °pretense, pretentiousness, °ostentation, pretending, °affectation, °hypocrisy: *Gladwin has behaved with complete sincerity and utterly without any pretension.*

pretentious *adj.* **1** °ostentatious, °showy, °superficial, °pompous, °arrogant, °bombastic, °inflated, high-flown, exaggerated, vainglorious, fastuous, °grandiose, grandiloquent, °extravagant, magniloquent: *The minister's pretentious language is a bit too much to take.* **2** °snobbish, °lofty, °haughty, flaunting, *Colloq* high and mighty, highfalutin *or* hifalutin, °hoity-toity, high-hat, *Slang* snotty, *Brit* toffee-nosed: *That new couple next door are so pretentious—they think that nobody is good enough to associate with them!*

pretext *n.* **1** °pretense, °excuse, °camouflage, °guise, °disguise, °cover, °veil, °cloak, °color: *They carried on the surveillance under pretext of working on the sewer.* **2** °ruse, red herring, °cover (story), °rationale, °pretense, rationalization, °explanation: *What pretext did the robber use that made you let him in?*

pretty *adj.* **1** °comely, °attractive, good-looking, nice-looking, appealing, °lovely, °cute, minion *or* mignon *or* mignonne, °graceful, °fair, °bonny, °fetching, charming, winsome, °beautiful, pulchritudinous, *Colloq* easy on the eye(s): *A pretty girl is like a melody. That's a very pretty cottage.* **2** °tuneful, melodic, °melodious, dulcet, °musical, °lyrical, °harmonious, catchy, mellifluous, euphonious: *He played a pretty tune on the piano.* —*adv.* **3** °rather, °quite, °fairly, °moderately, reasonably, tolerably; °somewhat; °very, °extremely, unbelievably, incredibly: *The children put on a pretty good performance, I thought. He had become pretty fat since we last saw him.*

prevail *v.* **1** hold sway, °win (out), °predominate, °succeed, °triumph, gain *or* achieve a victory, prove superior, gain mastery *or* control: *It is sometimes disappointing to learn that right does not always prevail.* **2** °predominate, be prevalent *or* widespread *or* current, preponderate, °dominate, be the order of the day: *As usual during Oktoberfest in Bavaria, revelry prevailed.* **3** ***prevail on*** *or* ***upon.*** °persuade, °induce, °influence, °sway, °dispose; °incline, °win over, bring round, °convince: *We prevailed on the guard to let us through the gate.*

prevailing *adj.* **1** °dominant, °predominant, °prevalent, °main, °chief, °principal, common(est), °usual, °customary, °universal: *The prevailing winds are westerlies.* **2** °influential, °effective, °effectual, dominating, affecting, °powerful, °potent, °forceful, ruling, °telling, °main, °principal: *The prevailing religion there is Buddhism.*

prevalence *n.* **1** prevalency, frequency, commonness, currency, universality, ubiquitousness, ubiquity, pervasiveness, omnipresence, extensiveness; °predominance, °practice, acceptance, °popularity: *The prevalence of disease among the population is distressing.* **2** °sway, °control, °rule, primacy, ascendancy, mastery, °predominance: *The prevalence of Bushido in Japan has diminished little since 1945.*

prevalent *adj.* °universal, °catholic, °common, °frequent, °prevailing, °current, ubiquitous, °pervasive, omnipresent, °general, °usual, °customary, commonplace, °extensive, widespread, established, ascendant, °dominant, °predominant, governing, ruling: *Cannibalism, once prevalent in Borneo and New Guinea, has not quite died out.*

prevent *v.* °anticipate, °preclude, obviate, °forestall, avert, °avoid, °prohibit, °ban, °bar, °forbid, interdict, °taboo, enjoin, proscribe, °foil, °frustrate, °obstruct, debar, °intercept, °nip in the bud, abort, °thwart, °check, °block, ward *or* fend *or* stave off, baffle, balk *or* baulk, °(put a) stop (to), °arrest, °(bring to a) halt, °hinder, °impede, °curb, °restrain, °hamper, °inhibit, °delay, °retard, °slow, °mitigate, °control: *Some diseases can be prevented by inoculation or vaccination. There is nothing to prevent us from leaving. The barrier was built to prevent flooding.*

prevention *n.* preventing, °anticipation, preclusion, obviation, forestalling, avoidance, avoiding, °prohibition, prohibiting, °ban, banning, °bar, °barring, forbiddance, °forbidding, interdiction, interdicting, °taboo, tabooing, enjoining, °injunction, proscription, proscribing, foiling, frustration, frustrating, °obstruction, obstructing, debarment, debarring, interception, intercepting, abortion, aborting, thwarting, checking, °check, blocking, °block, warding *or* fending *or* staving off, balk *or* baulk, balking *or* baulking, stopping, °arrest, °arresting, °halt, °halting, °hindrance, hindering, impedance, impeding, °curb, curbing, °restraint, restraining, hampering, °inhibition, inhibiting, °delay, delaying, retardation, retarding, slowing, mitigation, °mitigating, °control, controlling: *The first item on the agenda is the prevention of cruelty to children. As the total prevention of crime is impossible, we must at least try to curb it.*

preventive *adj.* **1** preventative, preventing, hindering, impeding, restraining, hampering, inhibitive *or* inhibitory, inhibiting, restrictive: *We must take preventive steps to ensure the stability of the rate of exchange.* **2** preventative, prophylactic, precautionary, anticipatory *or* anticipative, °protective, counteractive: *Preventive means are available to limit heart disease.* —*n.* **3** preventative, °hindrance, °curb, °inhibition, °impediment, °block, °barrier, °obstacle, °obstruction: *Caffeine is one of the most powerful preventives of sleep available.* **4** preventative, prophylactic, °protection, °shield, °safeguard, °prevention, countermeasure, counteractant, counteragent, inoculum *or* inoculant, vaccine, serum, °antidote, °remedy: *Heart specialists have recommended an aspirin every other day as a preventive to arterial blood clotting.*

preview *n.* advance showing, private showing; °opening, vernissage: *We saw the Picasso exhibition at a preview held for members of the museum.*

previous *adj.* **1** °former, °prior, °past, earlier, onetime, °foregoing, °sometime, erstwhile, °preceding, *Literary* quondam, *Archaic* whilom: *A previous owner of the house filled in the fishpond.* **2** °prior, °former, °foregoing, °above, °preceding, *Formal* antecedent, anterior, aforementioned, above-mentioned, before-mentioned, aforesaid, above-named: *Please see the previous examples in the Foreword.* **3** °premature, untimely, too soon *or* early: *Isn't putting up Christmas decorations in October being a bit previous?* **4** ***previous to.*** previously to, °before, °prior to, °preceding, anterior to, antecedent to: *Previous to the advent of automobiles, we had traffic jams of horse-drawn vehicles.*

previously *adv.* °before, °once, °formerly, earlier, °at one time, then, beforehand, heretofore, theretofore, hitherto, thitherto, in the past, in days gone by, in days of old, in days *or* time past, in the old days, some time ago, a while ago, once upon a time, yesterday, *Literary* in days of yore, in olden days *or* times: *The same thing had happened previously when I was in London. Previously, people lived more relaxed lives—or so we like to think.*

prey *n.* **1** °quarry, °kill, °game, °objective, °target: *The lioness singled out her prey from the herd of zebra.* **2** °victim, °target, °objective; °dupe, *Colloq* °mark, *Slang* fall guy, °pushover, *Brit* °mug: *A public company with huge cash reserves, United Vector seemed easy prey for a takeover bid.* —*v.* **3** ***prey on*** *or* ***upon.*** **(a)** live off, feed on *or* upon, °eat, °consume, °devour, °kill, °destroy, °stalk, °pursue, °hunt, °seize: *These snakes prey mostly upon other snakes.* **(b)** °victimize, go after, °exploit, °use, take advantage of, °intimidate, °bully, °cheat, °dupe, °swindle, gull, °trick, snooker, °defraud, °outwit, °outsmart, outfox, °hoodwink, *Literary* cozen, *Colloq* rook, bamboozle, flimflam: *An unscrupulous gang is preying on the elderly, persuading them to invest in nonexistent properties.* **(c)** °oppress, weigh on *or* upon, °burden, °depress, °distress, °strain, vex, °worry: *His wretched condition preyed very much on her mind.*

price *n.* **1** °charge, °cost, °expense, °expenditure, °outlay, °payment, °amount, °figure, °fee; °quotation, appraisal, value, valuation, °evaluation, °worth: *The price of this lamp is too high. What is the price of that box? Can he afford the price of a ticket? At current prices a house is out of reach for me.* **2** °sacrifice, °toll, °penalty, °cost, consequence: *Loss of his freedom was too high a price for standing by his principles.* **3** °reward, °bounty, °premium, °prize, °payment, °bonus, °honorarium, *Literary* guerdon: *The gunfighter had a price of a thousand dollars on his head.* **4** ***without price.*** See **priceless, 1,** below. —*v.* **5** value, °evaluate, °rate, assay, assess, °cost (out): *How would you price a piece of furniture like this chair?*

priceless *adj.* **1** costly, °dear, °expensive, high-priced, valuable, °invaluable, °precious, °inestimable, incalculable; irreplaceable, °unique: *The vaults in the Vatican contain a king's ransom in priceless jewels.* **2** °hilarious, °riotous, °(screamingly) funny, sidesplitting, °hysterical, droll, comical, amusing: *I'll never forget Vernon's priceless remark when he got the speeding ticket.*

pricey *adj.* pricy, °expensive, °dear, costly, °exorbitant, °outrageous, °excessive, extortionate, *Colloq*

°steep, *Brit* over the odds: *The restaurant where Bernice took him was certainly pricey—they even charged for a glass of water!*

prick *n.* **1** °puncture, pinhole, pinprick; °hole, perforation: *For the blood test, the doctor made a tiny prick in my finger with a needle.* **2** °sting, °pinch, °twinge, °prickle, tingle, °pain: *The teacher leapt from her chair the instant she felt the prick of the tack.* —*v.* **3** °puncture, °pierce, °stab, °jab, °punch, °perforate, °riddle; °lance: *Using a pin, prick tiny holes in the paper to let the steam escape. The doctor pricked a boil on my neck.* **4** °stab, °sting, °hurt, °prickle, °pinch, °bite, °smart: *The hypodermic needle really pricked me when it went in.*

prickle *n.* **1** °spine, °bristle, barb, °thorn, bur, needle, tine, °spike, °spur, prong: *The prickles make the brambles cling to your clothes.* **2** pricking, prickliness, °itch, itchiness, °sting, tingling, tingle: *I felt the prickle of the rough wool against my skin.* —*v.* **3** tingle, °sting, °itch, °smart: *Sloane complained that the beard he had to grow for the pirate role made his face prickle.* **4** °stick, °jab, °prick: *Be careful—the chestnut burs can prickle your fingers.*

prickly *adj.* **1** bristly, °thorny, brambly, spiny, barbed, briery, spinous, spiky, *Technical* setaceous, setose, acanthoid, aciculate, aculeate, muricate, spiculate: *Dad's face is all prickly when he hasn't shaved.* **2** tingling, stinging, pricking, prickling, itchy, crawly, crawling: *The squeak of the chalk on the blackboard gives me a prickly feeling.* **3** °touchy, °irritable, °petulant, °cantankerous, °testy, °waspish, bad-tempered, °peevish, fractious, °short-tempered, curmudgeonly, *Colloq* °cranky: *Moira gets a bit prickly if you ask her why she married Noel.* **4** nettlesome, °thorny, °ticklish, °touchy, °troublesome, °intricate, °complicated, complex, knotty, °hard, °difficult, contentious: *The prickly problem of how to pay for the university must still be faced.*

pride *n.* **1** °honor, proudness, °self-esteem, °self-respect, *amour-propre,* °dignity: *It is gratifying to be able to look with pride on one's children's achievements.* **2** °conceit, egotism, self-importance, °vanity, hubris, °arrogance, overconfidence, overweeningness, self-admiration, self-love, self-importance, smugness, haughtiness, hauteur, °snobbery, snobbishness, *Colloq* uppitiness: *Pride goeth before a fall. Her pride stems from an exaggerated notion of her own worth.* **3** °boast, °flower, °best, °prize, pride and joy, °treasure, °jewel, °gem: *Those model railway trains are Eustace's pride and joy.* —*v.* **4** Usually, ***pride oneself on.*** be proud of, take pride in, °delight in, °revel in, °celebrate, glory in: *Irena prides herself on having made her own way in life, without anyone's help.*

priest *n.* priestess, clergyman *or* clergywoman, ecclesiastic, cleric, churchman *or* churchwoman, reverend, vicar, °divine, man *or* woman of the cloth, man *or* woman of God, curate, confessor, °minister (of the gospel), servant of God, °father, °mother, holy man *or* woman, °preacher, °missionary, evangelist, abbé, abbot *or* abbess, *Colloq* padre: *The high priest muttered some incantations over the body of the sacrificial victim.*

priestly *adj.* °clerical, ecclesiastic, °pastoral, hieratic, sacerdotal; ministerial, canonical, °missionary: *She has taken her priestly vows.*

prig *n.* (ultra)conservative, °prude, °purist, pedant, schoolmarm, °puritan, (Mrs.) Grundy, Grundyite, Grundyist, precisionist, precisian, conformist, formalist, *Colloq* stuffed shirt, °stick-in-the-mud, °goody-goody: *A terrible prig, the headmaster forbade even the slightest hint of slang usage.*

priggish *adj.* (ultra)conservative, prim, demure, °prudish, °purist, puristic, °pedantic, schoolmarmish, °strait-laced, °hidebound, stiff-necked, puritanical, conformist, (Mrs.) Grundyish, punctilious, °formal, formalistic, °strict, °severe, °fastidious, °fussy, °particular; °precious, *précieux* or *précieuse,* niminy-piminy, overnice, *Colloq* °stick-in-the-mud, °goody-goody, °prissy, oldmaidish, stuffed-shirt, °stuffy, uptight, nit-picking, *Brit* twee: *Victorians were less priggish in their private behavior than in their public image.*

primarily *adv.* **1** °principally, °mainly, °chiefly, °especially, °at bottom, °particularly, first of all, °preeminently, basically, essentially, fundamentally, °on the whole, °for the most part, mostly, predominantly *or* predominately, °generally: *The rain in Spain falls primarily in the plain.* **2** initially, °originally, from *or* at the start, °first (and foremost), in the first instance, *ab initio*: *The colonists, primarily refugees from England, began to settle the New World in the seventeenth century.*

primary *adj.* **1** °first, °prime, °principal, °chief, °main, °leading, °preeminent, °cardinal, °fundamental, °basic, °essential, °predominant, °elementary, °elemental, underlying: *The primary reason I want to see you is to discuss your future with the company. The primary meaning of a word is given first.* **2** earliest, °first, °original, °initial, °primitive, primeval, primordial, embryonic, germinal, °beginning, °ultimate: *The primary source of life was possibly a sort of soup containing proteins and other molecules.* **3** first-hand, °direct, °immediate: *Bauxite is the primary source of aluminum ore.* **4** °elementary, °basic, °rudimentary, °fundamental: *One of the primary lessons we are taught is consideration for others.* **5** unmixed, unadulterated, °pure, °simple, °rudimentary, °fundamental, °principal: *The primary colors in art are red, yellow, and blue.*

prime *adj.* **1** See **primary, 1,** above. **2** °best, °foremost, °chief; °first-rate, first-class, °choice, °select, °superior, °preeminent, °leading, ranking, °predominant, °unparalleled, °matchless, °peerless, °noteworthy, °outstanding, °admirable, °worthy, °exceptional, °excellent, °extraordinary, °exceptional: *She is a prime example of the results of a modern education. Abernathy is certainly a prime candidate for the position.* **3** °original, °fundamental, °basic, °elemental, °elementary: *The prime source of scurvy is lack of vitamin C, ascorbic acid.* —*n.* **4** °youth, springtime; best years, heyday, °pinnacle, °acme, °peak, °zenith: *Some people reach the prime of life at sixty.* —*v.* **5** (make *or* get) ready, °prepare, °educate, °teach, °instruct, °coach, °train, °tutor, °drill: *Has Sonia been fully primed to take over the chairmanship when Sir William steps down?* **6** °inform, °advise, °notify, apprise, °brief: *Having read your book, I am fully primed on American history.*

primitive *adj.* **1** °first, °original, °aboriginal, earliest, primordial, primal, primeval *or Brit also* primaeval, °pristine, °prehistoric; antediluvian, Noachian *or* Noachic, °old, °ancient: *In its most primitive state, life probably originated from some random strings of molecules. The most primitive farming tools date from some ten thousand years ago.* **2** °crude, °rude, °unrefined, °raw, barbaric, uncultured, °barbarian, °coarse, °rough, °uncivilized, °savage, uncultivated, °unsophisticated, uncouth: *I cannot tolerate Nigel's primitive table manners.* **3** °simple, °basic, simplistic, naive *or* naïve *or* naïf, °childlike, °unsophisticated, uncultivated, °unrefined, unpolished, °rough, untutored, untaught, untrained, unschooled, °undeveloped: *Gary collects paintings of the primitive school and has one by Grandma Moses.*

primp *v.* °preen, prink, prettify, titivate *or* tittivate, plume, °dress up, °groom, *Colloq* doll up, get (all) dolled up, °spruce up, put on one's best bib and tucker, *Chiefly Brit* tart up, get (all) tarted up, *Slang* deck out, trick out *or* up, put on one's glad rags, *Brit* fig out, *US* gussy up, get (all) gussied up, dude up: *She was primping before the mirror, for Mr. D'Arcy was coming.*

princely *adj.* **1** °lavish, °bountiful, °generous, °liberal, °ample, °substantial, °huge, °enormous: *They paid a princely sum for their stately home in Surrey.* **2** °lavish, °magnificent, °splendid, °luxurious, °majestic, °royal, °regal, °sumptuous, °superb, *Colloq* ritzy, swank(y), °posh, °plush: *The hotel laid on princely accommodation for us with rooms overlooking the sea.* **3** °royal, °noble, °regal, °sovereign, of royal *or* noble blood *or* rank: *Who would have thought that our humble home would ever shelter a princely guest?*

principal *adj.* **1** °chief, °primary, °prime, °paramount, °main, °first, °foremost, ranking, °preeminent, °predominant, °dominant, °prevailing; °leading, starring: *The principal reason I am here is to see you. O'Hara is a principal contender for the lightweight title. The principal food of the people is corn. The principal role was sung by the diva, Maria Mofetta.* **2** °important, °prominent, °leading, °key, °cardinal: *Cuba is a principal source of sugar. Hers is one of the principal families of India.* —*n.* **3** °owner, °proprietor, chairman, chairwoman, chairperson, °(managing) director, °head, president, °chief, chief executive officer, CEO, manager *or Brit* manageress, °superintendent, °supervisor, *Colloq* °boss, *US* (head *or* chief) honcho: *We should talk to the principals about buying that company.* **4** dean, °director, *Chiefly Brit* headmaster, headmistress, °master, rector, (vice) chancellor: *His appointment as principal is for a two-year period.* **5** °(working) capital, capital funds, °resources, investment, °backing, (cash) reserve(s), °assets; °money: *She is fortunate to be able to live on the income from her investments, without touching the principal.* **6** °star, °lead, °heroine, °hero, leading lady *or* man, leading role, main part; diva, *première danseuse, premier danseur,* prima donna, prima ballerina: *The principal in the ballet company was, as you might expect, a Russian.*

principally *adv.* °chiefly, °mainly, °first (and foremost), °primarily, °above all, °in the main, mostly, °for the most part, °largely, predominantly, °on the whole, °at bottom, °in essence, essentially, basically, fundamentally; °especially, °particularly: *He seems to be interested principally in money, with little regard for anything else.*

principle *n.* **1** °truth, °given, °precept, °tenet, °fundamental, °grounds, °law, °rule, dictum, canon, °doctrine, teaching, dogma, °proposition, (basic) assumption, postulate, axiom, °maxim, °truism, °guide, °standard, criterion, °model: *The perpetual-motion machine violates a basic principle of physics.* **2** Often, ***principles.*** °philosophy, °code, °attitude, °(point of) view, °viewpoint, °sentiment, °belief, credo, °creed, °idea, °notion, ethic, sense of right and wrong: *That cynic conducts his life on the principle, "Do unto others before they do unto you." I am not sure I condone his principles.* **3** °(sense of) honor, uprightness, °honesty, °morality, °morals, °probity, °integrity, °conscience: *If you don't think him a man of principle, don't do business with him. Saying everything she thinks is a matter of principle with Doreen.* **4** ***in principle.*** on principle, in theory, theoretically, basically, fundamentally, °at bottom, °in essence, essentially, °ideally: *I like your plan in principle, but in practice it cannot be accomplished that way.*

principled *adj.* °moral, °righteous, right-minded, °virtuous, °noble, high-minded, °ethical, °honorable, °proper, °correct, °right, °just, °upright, °honest, °scrupulous: *Michael is too highly principled to submit to bribery.*

print *v.* **1** °impress, imprint, °stamp, °publish, °issue, °run off, °put out; °copy; °(pull a) proof: *We decided to print five hundred copies of the book. You may have your name printed on the cover for an additional amount.* —*n.* **2** °reproduction, °copy, °replica, °facsimile; °positive, °photograph, etching, (steel *or* wood) engraving, lithograph, woodcut, linocut, silk screen, rotogravure, *Trademark* Xerox; °picture, °illustration; *Colloq* photo, °cut, pic (*pl. pix*): *Today, a good print of a Picasso costs more than an original did fifty years ago.* **3** °text, printed matter, °type, °writing; °language, °wording, °(choice of) words, phrasing: *You are advised to read all the small print before you sign any contract.*

prior *adj.* **1** °former, °previous, earlier, onetime, ex, erstwhile; °old, °last, °late, °latest, *Literary* quondam, whilom: *The prior owner left all the curtains and carpets behind. If you overdraw your bank account without prior arrangement, you will be charged a higher rate of interest.* **2** ***prior to.*** °before, °previous to, previously to, °till, until, °preceding: *Prior to Charlotte's chairmanship, her husband held the office.*

priority *n.* °precedence, precedency, primacy, °urgency, immediacy, °predominance, °preeminence, °preference, °rank, °superiority, °prerogative, °right, seniority, °importance, °weight, immediacy: *The applications for aid will be processed in the order of priority.*

prison *n.* jail *or Brit also* gaol, °dungeon, oubliette, lockup, penal institution, house of correction, correctional institution, reformatory, house of detention; confinement, °detention; *Old-fashioned* °reform school, *Military* guardhouse; *US* penitentiary, *US Military* brig; *Brit* remand center, detention center, remand home, community home, *Brit Military* glasshouse, *Brit formal* CHE (= 'community home with education on the premises'), *Brit old-fashioned* approved school, *Brit archaic* bridewell; *Slang* clink, can, cooler, °jug, °stir, *US* °pen, calaboose, slammer, hoosegow, *US old-fashioned* big house, *US and Canadian* pokey *or* poky, *Brit* quod, chokey *or* choky: *Those convicted of a felony are sent to prison for not less than one year.*

prisoner *n.* °convict, trusty; internee, detainee; *Colloq* lifer, jailbird *or Brit also* gaolbird; *Slang* con, *Brit* °(old) lag, *Old-fashioned* ticket-of-leave man: *The prisoner in cell block 64 has a visitor.*

prissy *adj.* °fussy, °precious, overnice, finicky *or* finical, °strait-laced, schoolmarmish, prim (and proper), °prudish, °squeamish, °fastidious, *Colloq* oldmaidish: *Llewellyn is awfully prissy about changing into swimming trunks on the beach.*

pristine *adj.* **1** °original, primal, °basic, primeval *or Brit also* primaeval, °primitive, primordial, earliest, °first, °initial: *It is impractical to try to return the world to what some regard as its pristine purity.* **2** uncorrupted, °pure, unsullied, undefiled, virginal, virgin, °chaste, untouched, °unspoiled, unpolluted, °untarnished, °spotless, °immaculate, °natural: *One must travel far today to experience the pristine beauty of nature.*

privacy *n.* **1** °seclusion, retirement, °solitude, isolation, °retreat, sequestration, reclusiveness, reclusion, solitariness; monasticism: *Coleman very much enjoys the privacy of living alone.* **2** °secrecy, secretiveness, clandestineness, confidentiality, surreptitiousness, covertness, concealment: *The press agreed to respect the privacy of his arrangement with the minister from Walachia.*

private *adj.* **1** °(top) secret, °confidential, undisclosed, °hidden, clandestine, concealed, covert, °surreptitious, °off the record, not for publication, °unofficial, *Colloq* hush-hush: *I think our relationship should be kept private for the time being. What I am about to tell you is strictly private.* **2** °privileged, restrictive, restricted, °exclusive, °special, °reserved, °personal, °inaccessible, nonpublic; °hidden, °secluded, concealed, °secret, °sneaking: *The house is situated on a private road. I had a private suspicion that they would cancel their trip.* **3** °personal, °individual, own, °intimate, °particular: *My private affairs are nobody's business but mine.* **4** °solitary, seclusive, °reclusive, °withdrawn, °retiring, °reticent, ungregarious, nongregarious, °unsocial, unsociable, antisocial, °reserved, uncommunicative, hermitic(al), hermitlike, eremitic(al); sequestered, °secluded, retired: *You have to bear in mind that Edmund is a very private person.* —*n.* **5** private soldier, infantryman, foot soldier, *US* enlisted man, *Colloq Brit* tommy, Tommy Atkins, squaddie, *US* GI (Joe), *Slang US* grunt: *Before cashiering him, they reduced him from colonel to private.* **6** ***in private.*** °in secret, °secretly, privately, sub rosa, °personally, confidentially, behind closed doors, in camera, °off the record, *US* on the Q.T.; clandestinely, secretively, sneakily, sneakingly, surreptitiously, furtively, covertly, °on the sly: *Family matters should be discussed only in private. They met in private with agents of the Paludian government.* **7** ***private parts*** *or* ***privates.*** °genitals, sexual *or* sex organs, genitalia: *They wore loincloths to cover their private parts.*

privation *n.* °need, neediness, °want, deprivation, °hardship, indigence, °necessity, °poverty, penury, destitution, strait(s), pauperism, beggary; °distress, °misery: *The lower classes in Victorian England suffered terrible privation.*

privilege *n.* °benefit, °advantage, °right, °prerogative, concession, °allowance, °indulgence, °immunity, °exemption, dispensation, °freedom, °liberty, franchise, °permission, °consent, °leave, authorization, °sanction, °authority, °license, °carte blanche: *The people were given the privilege of choosing their own form of religion.*

privileged *adj.* **1** °favored, advantaged, indulged, entitled, elite *or* élite, °special, honored: *Harvey was one of the privileged few to be told her private telephone number.* **2** protected, exempt(ed), °immune; licensed, empowered, admitted, permitted, sanctioned, authorized, enfranchised, chartered: *This is a privileged institution as far as the dispensation of grants is concerned.* **3** °powerful, ruling; °wealthy, °rich: *He despised the privileged class till he had made enough to become a member.* **4** °confidential, °secret, °private, °privy, °inside, off the record, not for publication, restricted, *Colloq* hush-hush: *What I am about to tell you is privileged information.*

privy *adj.* **1** See **privileged, 4,** above. **2** ***privy to.*** aware of, in on, on to *or* onto, sharing (in), cognizant of, apprised of, informed *or* advised about *or* of, informed on, knowledgeable about, *Colloq* in the know about, *Slang* hip to, °wise to, *Old-fashioned* hep to: *Was the minister's wife privy to what was discussed at cabinet meetings?* —*n.* **3** °lavatory, (outside *or* outdoor) toilet, latrine, water closet, WC, *US* outhouse, *Colloq chiefly Brit* loo, *Slang Brit* °bog, *US* crapper, *Taboo slang US* shithouse: *Almost every home had an outside privy, as there was no inside plumbing.*

prize[1] *n.* **1** °reward, °award, °trophy, °premium; °honor, accolade, *Literary* guerdon: *The first prize was a week in Terre Haute, the second prize, two weeks in Terre Haute.* **2** winnings, jackpot, °purse, °receipts, °gain, °windfall, stakes, *Colloq* °haul, *Chiefly US* °take: *He used his prize from winning the lottery to buy a new car.* **3** °aim, °goal: *The prize they all strove for was a grant to carry on lexicographic research.* **4** °loot, °booty, spoil(s), °trophy, °plunder, pickings: *The pirates took the galleon as their prize.* —*adj.* **5** °choice, °excellent, (prize-)winning, °best, °champion, °outstanding, °select, °superior, °superlative, °first-rate: *This dairy owns a prize herd of Guernseys.*

prize[2] *v.* value, °treasure, °esteem, °cherish, °appreciate, rate highly, hold dear: *I prize your friendship above all things.*

probability *n.* °likelihood, likeliness, °odds, °expectation, chance(s), °(distinct) possibility, °presumption: *There is a good probability that it will rain.*

probable *adj.* °(most) likely, apt, °(quite) possible, presumed, °plausible, undoubted, indubitable, °apparent, °unquestionable, °evident, ostensible, odds-on, °feasible, believable, credible: *The probable cause of the flooding was that hole in the dam. Is it probable that she would come in this storm?*

probably *adv.* °(very) likely, in all likelihood, in all probability, °undoubtedly, doubtlessly, indubitably, unquestionably, °presumably, °quite, all things considered, °to all intents and purposes, °possibly, °perhaps, *Colloq* as likely as not, °quite: *She is probably the best tennis player in Sitka.*

probe *v.* **1** °explore, °examine, °scrutinize, °investigate, °search (into), °look into, °go into, °study, °dig into, delve into, poke about *or* around, *Colloq* poke into: *They probed his background thoroughly but found nothing unusual.* **2** °poke, °prod, °explore, °examine; °plumb, °dig: *The doctor probed for the bullet, deep in the wound.* —*n.* **3** °investigation, °examination, °exploration, °scrutiny, °search, °study, inquiry *or* enquiry, inquest: *There will be a governmental probe into the mismanagement of the department.*

probity *n.* °integrity, uprightness, °honesty, °morality, °rectitude, °virtue, goodness, decency, righteousness, right-mindedness, °sincerity, trustworthiness, °honor, °equity, justness, °justice, fairness: *Supreme Court justices have consistently been men of unquestioned probity.*

problem *n.* **1** °difficulty, °trouble, °question, °dilemma, °quandary, °predicament, °complication, hornet's nest, imbroglio, °mess, °muddle, °stew, *Colloq* can of worms, fine kettle of fish, (pretty) pickle, *Brit* facer: *The Secretary of the Treasury must constantly deal with the problems of the country's economy.* **2** °puzzle, conundrum, poser, °riddle, °question, °enigma, puzzler, *Colloq* mind-boggler, hard *or* tough nut to crack: *The problem is how to get the yolk out without breaking the shell.* —*adj.* **3** °unruly, unmanageable, intractable, uncontrollable, °difficult, °ungovernable, refractory, °incorrigible, °obstreperous, °delinquent, maladjusted, °disturbed, emotionally upset: *He teaches at a school where most of the students are problem children.*

problematic *adj.* problematical, °difficult, °uncertain, °questionable, questioned, °doubtful, doubted, °debatable, °disputable, disputed, °unsettled, °moot, undecided, °controversial, °tricky, °touchy, °sensitive, *Colloq* °hairy, iffy: *The value of treating colds with massive doses of vitamin C is problematic.*

procedure *n.* °way, °conduct, °course, °action, course of action, °method, methodology, °mode, °form, °system, °approach, °strategy, °plan (of action), °scheme, modus operandi, °operation, °policy, ways and means; °routine, °tradition, °practice, °custom, wont, standard operating procedure, *Colloq* MO (= 'modus operandi'), SOP (= 'standard operating procedure'), *Chiefly Brit* °drill: *What procedure will you follow to regain the stolen property? The procedure in ballroom dancing has always been for the man to lead.*

proceed *v.* **1** Sometimes, ***proceed with.*** go *or* move (on *or* ahead *or* forward), °advance, °continue, °progress, °carry on, get *or* move along, get going *or* moving *or* under way, °start, °pass on, make progress *or* headway, push *or* press on *or* onward(s), forge ahead; °resume, °renew, go on with, °pick up (where one left off): *Proceed to the fork, then turn left. If the membership* *committee has finished, let us proceed to the report of the finance committee. After the break, the performance will proceed.* 2 Often, ***proceed from*** or ***out of.*** °result from, °arise (from), °come (from), stem from, °spring from, °develop (from), issue (from *or* forth), derive from, be derived (from), descend from, °emerge (from), grow (from *or* out of), originate (in *or* from *or* with), °begin (with), start (with *or* from): *That outcropping proceeded from glaciation and erosion. They say that more than half of a person's happiness proceeds from hope.* —*n.* 3 ***proceeds.*** profit(s), °gain, °yield; °income, °receipts, return(s), °gate, box office, *US* °take: *The proceeds from the charity auction were better than ever this year.*

proceeding *n.* 1 °measure, °act, °(course of) action, °move, °step, °undertaking, °deed, °procedure, °process, °operation, °transaction, °maneuver, °feat, °accomplishment: *Ida Lewis's rescue of the shipwrecked men was truly a heroic proceeding.* 2 ***proceedings.*** **(a)** °transactions, report(s), °minutes, record(s), annals, affairs, °dealings, °business, account(s), archives, *Formal* acta: *The proceedings of the society are published annually.* **(b)** events, goings-on, doings; celebration(s); performance(s): *The proceedings were briefly interrupted while the undesirables were ejected from the hall.*

process *n.* 1 °procedure, °proceeding, °operation, °system, °method, °approach, °technique; course of action: *I am sure you understand how the process works. What process do they use to make paper out of wood?* 2 °activity, °function, °development: *The process of photosynthesis is the means by which plants synthesize carbohydrates.* —*v.* 3 °treat, °prepare, (make *or* get) ready, °change, °modify, °transform, °convert, °alter: *This is the room where the ingredients are processed before mixing.* 4 °handle, °take care of, °organize, °deal with, °manage; °dispose of, °answer, °manipulate: *After the earthquake the insurance companies processed thousands of claims.*

procession *n.* 1 °parade, °march, cavalcade, motorcade, cortege *or* cortège, column, °line, °file, °train, marchpast: *A strange procession of thousands of beggars silently wound its way through the city.* 2 °succession, °cycle, °sequence, °string, °train, °chain, series, °course, °run, °progression, cavalcade: *He began to recount the procession of events that led up to the murder.*

proclaim *v.* 1 °announce, advertise, °circulate, °declare, °broadcast, °pronounce, °make known, bruit (about), trumpet, °publish, promulgate, herald; °profess, °protest, °enunciate, articulate: *Edgar was proclaimed king. She proclaimed her innocence.* 2 °brand, accuse of being, stigmatize as, °pronounce, °rule, °decree, °characterize, °report: *She found herself proclaimed a traitor.*

proclamation *n.* °announcement, °advertisement, °declaration, °publication, promulgation, °statement, manifesto, pronunciamento, notification, °notice: *A proclamation was issued granting amnesty to all political prisoners.* 1 proclaiming, announcing, advertising, declaring, broadcasting, publishing, promulgating, heralding, making known, bruiting about: *The proclamation of his appointment as head of the secret police came at a bad time.*

procrastinate *v.* 1 temporize, act evasively, °play for time, dally, °delay, °stall; °postpone, °defer, put off *or* aside, °shelve, *US* °table: *I wish Bunter would stop procrastinating and say what he plans to do. Because the attack was procrastinated, its timely effect was lost entirely.* 2 °hesitate, °pause, waver, vacillate, be undecided, °equivocate, tergiversate, °shilly-shally: *Helen procrastinates every time she is faced with making a decision.*

procure *v.* 1 °obtain, °acquire, °get, °come by, °secure, get *or* lay one's hands on, get (a) hold of, °gain, °win, come into, °pick up, °find, °appropriate, °requisition; °buy, °purchase: *I have been trying to procure early editions of Defoe's works.* 2 °accomplish, °bring about, °effect, °cause, °produce: *Using every resource available to her, she finally procured his release.*

procurer *n.* pander *or* panderer, °pimp, white slaver, flesh-pedlar *or US also* flesh-peddler *or also* flesh-pedler, *Archaic* whoremaster, bawd; madam, procuress: *He soon became a procurer for the wealthy men in the village.*

prod *v.* 1 °jab, °dig, °poke, °nudge, elbow: *Every time the comedian said something risqué, Aunt Flora would giggle and prod me in the ribs.* 2 °spur, °urge, impel, egg on, °push, °thrust, °prompt, °rouse, °stir, °incite, °move, °motivate, actuate, °activate, °provoke, °encourage, °stimulate: *I am grateful to Vincent for prodding me to continue with my studies.* 3 °incite, goad, needle, °irritate, °irk, °annoy, °pester, °harass, hector, badger, °plague, °nag, °hound, carp at, °cavil; °henpeck: *My parents are constantly prodding me to go out and get a job.* —*n.* 4 °jab, °dig, °poke, °nudge, elbow, °push: *I was quite black and blue from those prods I was getting from Aunt Flora.* 5 goad, °spur; needle, rowel: *These days electrified cattle prods are available.* 6 stimulus, °push, °prompt, °reminder, °signal, °cue: *Calvin needed that prod to make him finally go home after the party.*

prodigal *adj.* 1 °wasteful, °extravagant, °spendthrift, °lavish, °excessive, °profligate, squandering, °immoderate, intemperate, °wanton, °improvident, °reckless: *Were the government less prodigal, it would be possible to have a balanced budget.* 2 °generous, °bountiful, copious, °profuse, bounteous, °lavish, °liberal, °luxuriant, °sumptuous, °abundant, abounding, °rich, °plentiful, plenteous, superabundant, thriving, swarming, teeming: *The pioneers who reached California could scarcely believe there was such a prodigal land. He was prodigal of both compliments and criticism.* —*n.* 3 °wastrel, °spendthrift, °profligate, squanderer, waster, big spender: *That prodigal, deHaven, went through his inheritance in less than a year.*

prodigality *n.* 1 wastefulness, °waste, °extravagance, °excess, excessiveness, immoderation, intemperateness, wantonness, recklessness, °profligacy, improvidence, °dissipation, squandering: *Surely his contemporaries must have condemned Shah Jehan's prodigality in building the Taj Mahal.* 2 lavishness, profuseness, °luxury, luxuriousness, luxuriance, °abundance, °plenty, °bounty, bountifulness, bounteousness, copiousness, °profusion, profuseness, sumptuousness, richness, plentifulness, plenteousness, superabundance, °exuberance: *The prodigality of Nature is unbounded.*

prodigious *adj.* 1 °vast, °immeasurable, °colossal, °enormous, °huge, °giant, °gigantic, °immense, mammoth, °monumental, tremendous, stupendous, titanic, Brobdingnagian, gargantuan, herculean, cyclopean, leviathan, °monstrous, °extensive, *Colloq US* ginormous, humongous: *Consider the prodigious amount—the very weight—of water vapor contained in a cloudy sky. You may accuse her of being a workaholic, but consider the prodigious amount she gets done in a day.* 2 °amazing, astonishing, astounding, °startling, °extraordinary, °exceptional, °marvelous, wonderful, wondrous, °fabulous, °miraculous, °phenomenal, spectacular, °fantastic, °sensational, °unusual, staggering, °striking, dumbfounding *or* dumfounding, °remarkable, °noteworthy, °notable, *Colloq* flabbergasting, mind-boggling, mind-blowing: *In those days, the four-minute mile was considered a prodigious feat.*

prodigy *n.* **1** (child *or* girl *or* boy) genius, wonderchild, *Wunderkind,* °mastermind, °talent, °intellect, intellectual *or* mental giant, wizard, °virtuoso, *Colloq* °brain, Einstein, whiz kid *or* whizz kid, whiz *or* whizz, walking dictionary *or* encyclopedia: *Although she might have been a prodigy at the age of six she was merely a good violinist at the age of thirty-six.* **2** °wonder, °marvel, °phenomenon, °sensation, miracle: *Some regarded the painting* Nude Descending a Staircase *a prodigy of cubist art.*

produce *v.* **1** °make, °develop, °turn out, put *or* bring out, °manufacture, °fabricate, °generate, °create; °construct, °assemble, put together, °compose; °mold, °cast; extrude: *This machine is capable of producing ten thousand units an hour.* **2** °yield, °give rise to, °cause, °bring up, °bring forth, °spark, °initiate, °occasion, °bring about, °prompt, °evoke, °start, °create, °generate, beget, °originate: *He was unprepared for the laughter produced by his slip of the tongue.* **3** °generate, beget, °create, put out *or* forth, °breed, °propagate, °bear, give birth to, °hatch, °grow: *See how the plant is producing buds at these nodes?* **4** bring forward *or* in *or* out, °introduce, °present, °offer, °show, °exhibit, °display, °stage, °put on, °mount: *She was the first to produce Japanese Noh drama in the west.* **5** °disclose, °reveal, °bring to light, °show, °display, °draw: *When I refused to give him the money, he produced a gun.* **6** °supply, °furnish, °provide, °deliver, °distribute: *They are accused of continuing to produce fluorocarbons despite the government ban.* —*n.* **7** vegetables, °fruit, *Chiefly Brit* greengrocery: *The produce is delivered to the markets from the farms at the crack of dawn.*

producer *n.* **1** maker, °manufacturer, fabricator, processor, °creator; grower, °farmer: *That company is the largest producer of microchips in the world.* **2** °(in Britain) director, auteur, impresario, *régisseur;* (in US and Canada) entrepreneur, (business *or* financial) manager, organizer, impresario: *He was the producer of more TV soap operas than anyone else.*

product *n.* **1** °result, consequence, °output, °outcome, °issue, °effect, fallout, °yield, °upshot; spinoff, °offshoot, by-product: *It is evident that this work is the product of much thought. One product of her interest in literature is the publishing company Lizard Press.* **2** artifact, good(s), °produce, commodity, °output, °merchandise, °offering, °work: *The quality of a product doesn't matter if nobody knows about it.*

production *n.* **1** producing, °manufacture, manufacturing, making, °fabrication, °preparation, origination, °creation, °output, putting out, °development; °formation, forming, forging, shaping, molding, casting, °assembly, °building, construction: *Our business is the production of insulating materials.* **2** °product, °(end) result, °work, °effort, handiwork, °output, °opus, *oeuvre*: *These paintings are clearly productions of a fertile imagination.* **3** (in Britain) artistry, °direction, staging; (in US and Canada) °display, °presentation, staging, *mise en scène,* °setting: *The critics liked the plot and raved about the production in general.* **4** °drama, °play, (stage *or* television *or* radio) show, °performance; °film, motion *or* moving picture, °movie: *Our repertory company is putting on a production of* Othello *next week.*

productive *adj.* **1** °fruitful, °fertile, °rich, fecund, °plentiful, plenteous, °abundant, °bountiful, bounteous, °prolific, °dynamic: *Owing to a perfect balance of rain and sunshine, the earth has proved productive this year.* **2** °imaginative, °creative, inventive, °resourceful, generative, °ingenious, °fertile, °vigorous: *Her latest science-fiction trilogy is certainly the creation of a highly productive mind.* **3** °profitable, remunerative, °rewarding, valuable, °worthwhile: *Years ago he made an investment in computer shares that has turned out to be very productive. Discussing sales strategy is not as productive as getting out there and selling.*

profane *adj.* **1** °irreverent, °sacrilegious, °blasphemous, idolatrous, irreligious, °infidel, °heathen, °unbelieving, disbelieving, °pagan, atheist(ic), °impious, °godless, °ungodly, °sinful, °wicked, iniquitous, °contemptuous, °disrespectful: *The faithful attacked anyone who was seen to perform a profane act.* **2** unsanctified, unholy, unconsecrated, defiled, °impure, unclean, unhallowed, nonreligious, nonsacred, unsacred; °lay, nonclerical, °secular, °temporal; *Judaism* tref *or* treif *or* treifa, nonkosher: *The missionaries persuaded them to abandon their profane rites and practices.* **3** °bad, °dirty, °filthy, smutty, °foul, foulmouthed, °obscene, °vulgar, °coarse, uncouth, °low, °taboo, °blasphemous, °bawdy, ribald, °scurrilous, °off-color, °immodest, °improper, °naughty, °indecent, unprintable, °unmentionable, indecorous, indelicate, °common; °abusive, °vituperative, °venomous, thersitical, *Literary* Fescennine, *Colloq* °blue: *The boy was sent home from school for using profane language.* —*v.* **4** °debase, °contaminate, °pollute, °taint, °vitiate, °degrade, defile, °desecrate, °violate, °pervert, °corrupt: *They swore undying retribution against those who had profaned their temple.*

profanity *n.* blasphemy, obscenity, cursing, curse word(s), swearing, swear word(s), foul *or* bad *or* dirty *or* vulgar *or* coarse *or* filthy *or* smutty *or* taboo language, four-letter word(s), billingsgate: *We could not interview her live because she uses a lot of profanity, which we have to bleep out.*

profess *v.* **1** assert, °claim, aver, asseverate, °state, affirm, °confirm, °confess, °declare, °say, °hold, °maintain, °present, °offer, proffer, °tender, °set forth, °put forward, °pronounce, enounce, °enunciate, °announce, utter, °vow, avow: *He professed himself satisfied with the judge's decision.* **2** °pretend, lay claim, make a pretense, purport, act as if, simulate: *They professed to be there to read the electricity meter, so she let them in.*

professed *adj.* **1** °pretended, ostensible, °apparent, °alleged, purported, °so-called, would-be, °self-styled, *soi-disant(e)*: *The professed wine "expert" brought a bottle of awful-tasting stuff from his "cellar," a shelf in the kitchen cupboard.* **2** confessed, avowed, sworn, admitted, acknowledged, confirmed, certified, declared: *What do you do when a professed friend betrays you?*

profession *n.* **1** °occupation, °calling, °work, °field, °vocation, °employment, métier, °trade, °business, °craft, °line, °sphere, specialty *or Brit* speciality, °job, °position, °post, °situation, *Slang* °racket: *I know that she's a doctor, but what is her husband's profession?* **2** confession, affirmation, °statement, avowal, °assertion, asseveration, °declaration, °acknowledgment, °testimony, averment, °admission, °announcement: *I was uncertain whether to believe his professions of love.*

professional *adj.* **1** trained, °educated, °practiced, °veteran, °experienced, °seasoned, °able, skilled, °skillful, °gifted, °talented, °qualified, °knowledgeable, licensed, °official, °expert, °masterful, masterly, °master, °efficient, °adept, °proficient, °competent, °polished, finished: *He is a professional ambulance driver who has to work as a plumber to make ends meet.* **2** °excellent, °proficient, °efficient, °skillful, masterly, °thorough, °prompt, °conscientious, °authoritative, businesslike: *She did a very professional job of repairing my car.* —*n.* **3** °master, °expert, maestro, °virtuoso, past master *or* mistress, °specialist, °authority, *Colloq* wizard, pro, whiz *or* whizz, *US* maven *or* mavin, *Brit* °dab hand: *You can always tell when the work has been done by a professional.*

proficiency *n.* °facility, °skill, °talent, adeptness, °expertise, expertness, skillfulness, °aptitude, °capability, °ability, °dexterity, °capability, competence *or* competency, °ingenuity, °knack, °savoir-faire, *Colloq* know-how: *The proficiency he displayed as a racing car driver was second to none.*

proficient *adj.* °skillful, skilled, °talented, °adept, °gifted, °expert, °experienced, °practiced, *au fait,* °veteran, well-versed, (highly) trained, °professional, °qualified, °capable, °able, °accomplished, °dexterous, °competent, °knowledgeable, topnotch, °first-rate, *Colloq* ace, °crack, whiz-bang *or* whizz-bang, *Brit* wizard: *She is an extremely proficient surgeon.*

profile *n.* **1** °outline, °silhouette, contour, side view: *His aquiline nose shows up prominently in this profile.* **2** biography, (biographical *or* thumbnail *or* character) sketch, °life, °portrait, vignette: *That profile of Allen in a well-known magazine did not do him justice.* **3** °analysis, °study, °survey, °examination; graph, diagram, °chart, °list, statistics: *The profile of the average newspaper reader revealed some interesting anomalies.* —*v.* **4** °describe, °draw, sketch, limn: *The editor phoned to say that they want to profile you in the Gazette's Sunday color supplement.*

profit *n.* **1** Often, ***profits.*** gross *or* net profit, net *or Brit also* nett, return(s), °gain, °yield, °revenue, °proceeds, bottom line, °surplus, °excess, *US* °take, *Slang* gravy, *US* vigorish: *The profits are down in the third quarter owing to the higher interest rate.* **2** °advantage, avail, °good, °benefit, °welfare, °gain, value, °interest, °use, °usefulness, *Archaic or Literary* behoof: *Let me recommend a book that you will find of some profit.* —*v.* **3** °advance, °further, be of profit to, °benefit, °promote, °aid, °help, be advantageous *or* of advantage, °serve, avail, °improve: *A few lessons from the club professional could profit your golf game.* **4** Often, ***profit from.*** take advantage of, °use, turn to advantage *or* account, °exploit, utilize, make (good) use of, make capital (out) of, capitalize on, °maximize, make the most of, *Slang* cash in on: *I hope you profit from the experience.* **5** °clear, °realize, °earn, °gain, *Colloq* make a killing, °clean up, °rake it in, make a bundle *or* a packet: *Ernestine profited to the tune of one hundred thousand dollars on the sale of her house.*

profitable *adj.* **1** °beneficial, °productive, lucrative, °fruitful, (well-)paying, well-paid, °worthwhile, °effective, cost-effective, °gainful, remunerative, money-making, °rewarding: *I had never thought of being a pavement artist as a profitable enterprise. Playing professional golf can be very profitable—if you win tournaments.* **2** °beneficial, °helpful, °useful, utilitarian, valuable, °worthwhile, °advantageous, °productive, °rewarding: *I found his advice profitable in the conduct of my life.*

profiteer *n.* **1** °racketeer, exploiter, extortionist, black marketeer: *The profiteers turned to selling sugar at exorbitant prices.* —*v.* **2** overcharge, °fleece, °exploit, °milk, make the most of; °extort; *US* °gouge: *The arms dealers are profiteering by selling banned materiel to the guerrillas.*

profligacy *n.* **1** debauchery, °vice, immorality, °sin, sinfulness, wickedness, °evil, °dissipation, dissoluteness, degeneracy, licentiousness, depravity, corruption, promiscuity, lechery, lasciviousness, lewdness, indecency, °perversion, carnality, libertinism, wantonness, unrestraint, eroticism, sybaritism, voluptuousness, sensuality: *The town's reputation for profligacy is much deserved.* **2** °prodigality, °extravagance, °excess, °waste, wastefulness, recklessness, exorbitance, lavishness, improvidence, squandering: *The profligacy of the French court at the end of the seventeenth century was unequaled before or since.*

profligate *adj.* **1** debauched, vice-ridden, °immoral, unprincipled, °sinful, °shameless, °evil, iniquitous, °wicked, dissipative, °dissolute, °degenerate, °loose, licentious, depraved, °corrupt, °promiscuous, °lecherous, °lascivious, °lewd, °indecent, °perverted, °carnal, °libertine, °wanton, unrestrained, °erotic, sybaritic, °voluptuous, °sensual, °wild, °abandoned: *The duke suddenly gave up his profligate life and became an ascetic monk.* **2** °extravagant, °prodigal, °wasteful, °reckless, squandering, °improvident, °spendthrift, °immoderate, °excessive: *As the family fortunes waned, Aunt Clotilde had to curb her profligate ways.* —*n.* **3** debauchee, °sinner, °degenerate, °pervert, sodomite, °reprobate, °rake, rakehell, °libertine, lecher, whoremonger, °roué, °wanton, °sybarite, voluptuary, °sensualist: *With the restraints removed, the profligate abandoned himself to every excess imaginable.* **4** °prodigal, °spendthrift, °wastrel, waster, squanderer: *Such are the extravagances indulged in by the profligates who never learned the value of money.*

profound *adj.* **1** °deep, unfathomable, abstruse, °recondite, arcane, esoteric, °intricate, knotty, °involved, °tricky, inscrutable, indecipherable, cabalistic, °incomprehensible, °obscure, °subtle, °occult, °secret, °cryptic, °puzzling, enigmatic, mystifying, °mysterious: *Some aspects of the theory of thought are even too profound for many specialists.* **2** °learned, °scholarly, °intellectual, erudite, discerning, °astute, sagacious, °sage, °wise, °penetrating, °sharp, °keen, insightful, analytical, °knowledgeable, °informed, °well-informed, well-read: *Russell was one of the most profound thinkers of his day.* **3** °deep, °great, °intense, °sincere; °heartfelt, °keen, °acute, utter, °extreme, °overpowering, °overwhelming: *He gave a profound sigh and fell asleep at once. It is with profound regret that I must tender my resignation.* **4** utter, °complete, °total, °perfect, °absolute, °thorough, thoroughgoing, °out-and-out, °downright, consummate; °awful, °terrible: *Her casual attitude stems from her profound lack of awareness of the importance of the work. I was immediately struck by the profound silence in the house. Profound changes are under way.*

profoundly *adv.* °very, °extremely, °deeply, greatly, keenly, acutely, intensely, °sincerely; °terribly, °awfully: *We are profoundly sorry to have been so much trouble to you.*

profundity *n.* **1** °depth, profoundness, abstruseness, reconditeness, arcaneness, intricacy, °subtlety, °complexity, complicatedness, °difficulty, inscrutability, inscrutableness, involvement, involvedness; indecipherability, incomprehensibility, incomprehensibleness, °obscurity: *The profundity of this article on recombinant DNA has left me completely uninformed.* **2** erudition, discernment, °scholarship, scholarliness, sagacity, °wisdom, sharpness, keenness, astuteness, acumen, insightfulness, knowledgeableness, knowledgeability: *This book reflects a profundity of information about semiotics unequaled elsewhere in the literature on the subject.*

profuse *adj.* **1** °abundant, °ample, °plentiful, copious, unstinting, unsparing, ungrudging: *My profuse apologies if I have offended you in any way.* **2** °extravagant, °lavish, °bountiful, bounteous, °prolific, °luxuriant, °abundant, °exuberant, superabundant, °lush, °thick, teeming, overflowing, bursting, thriving, °productive, °fruitful, °rich: *From the seventh year onward, the trees yielded profuse crops of pears and plums.* **3** °excessive, copious, °considerable, °prolific, °liberal, °lavish, °free, °abundant, °ample: *This stage of the fever is accompanied by profuse sweating.* **4** °generous, °abundant, °plentiful, copious, unsparing, °unselfish, unstinting, °exuberant, magnanimous, °liberal: *The audience* *was profuse in its applause and cries of "Bravo!" and "Encore!"*

profusion *n.* profuseness, °quantity, °abundance, °plenty, plentifulness, plenteousness, °bounty, copiousness, superabundance; °mass, °host, °hoard, °number, multitude, °lot, °mountain, °load, °mess, °stack, °pile, °heap, agglomeration, conglomeration, °accumulation, °wealth, °glut, °surplus, oversupply, °surfeit, plethora, °superfluity, *Formal* nimiety: *Ivy grew in profusion over the front of the house. We had never seen such a profusion of bric-a-brac as they had collected.*

progenitor *n.* **1** progenitrix, °ancestor, forefather, forebear: *They regarded the snake god as the progenitor of the human race.* **2** °predecessor, °forerunner, °precursor, antecedent, foregoer, °source, originator; °origin, °original, °prototype, archetype, °pattern, °guide: *The ancient philosophers were progenitors of many modern ideas. Some regard da Vinci's designs as the progenitors of some of today's machines.*

progeny *n.* °offspring, children, descendants, °issue, °young, °posterity, heirs, scions, successors, sons and daughters, *Colloq* kids, °spawn, fry, *US* sprouts: *He dreamed that his progeny might have a freer, healthier life than his.*

prognosis *n.* °forecast, forecasting, °prediction, °prophecy, prognostication, °projection: *The doctor has recorded his prognosis for the course of the disease.*

prognosticate *v.* **1** °predict, foretell, °prophesy, °forecast, presage, °divine, forebode: *Those prophets of doom are always prognosticating disaster.* **2** betoken, augur, herald, °foreshadow, foretoken, °announce, °harbinger, °signal, portend: *The freshening breeze from the east prognosticates some rainy periods ahead.*

program *n.* **1** °schedule, °plan, °scheme, agenda, order of the day, °routine, °protocol, slate, °list, listing, °description, °outline, °abstract, °précis, °calendar, menu, bill of fare, curriculum, syllabus, °synopsis, °summary, °prospectus: *The program shows the sequence of events. You cannot tell what is being offered without a program.* **2** °performance, °production, °show, °presentation, (radio *or* television) play, telecast, °broadcast, °recital, concert: *Television programs seem to cater to lower levels of taste and intelligence every year.* **3** °proceedings, events, affairs, activities: *The program includes an hour of exercise before breakfast every day.* —*v.* **4** °organize, °arrange, prearrange, °plan, °lay out, map (out), °design, °formulate, °set (up), °schedule, °book, *US* slate: *The solo pianist is programed to follow the string quartet.*

progress *n.* **1** °(forward) movement, going forward, °progression, °advance, °headway, advancement: *Progress through the dense underbrush was very slow.* **2** advancement, °advance, °promotion, °improvement, betterment, °elevation, °rise, °development, °furtherance: *Her progress in the company might have been impeded by the fact that she was a woman.* **3** advancement, °course, °development, °growth, °expansion, °increase, °evolution, maturation, ripening, burgeoning *or* bourgeoning, amplification, enlargement, °spread, °extension, broadening, °promotion, °furtherance, °advance, °encouragement: *Our publishing credo is that every book must contribute to the progress of culture or knowledge.* **4** ***in progress.*** °under way, °ongoing, °going on, °happening, occurring, taking place, at work, °in operation, being done, °proceeding, *Colloq* in the works: *Some of the plans for reorganization are already in progress.* —*v.* **5** °advance, move *or* go (forward(s) *or* onward(s)), °proceed, °continue, go *or* forge ahead, go *or* move along, make (one's) way, °make headway, °travel, go *or* push *or* press on: *Throughout the nineteenth century wagon trains of pioneers progressed across the prairies to settle the American West. I haven't progressed far enough in my reading to say whether I like the book or not.* **6** °advance, °improve, get well, get better, °develop, °grow, °expand, °increase, evolve, °mature, °ripen, burgeon *or* bourgeon, °amplify, °enlarge, °spread, °extend, broaden, °rise, move up, upgrade: *Sarah's condition progressed to the point where she could get out of bed. Christmas sales of the new toy are progressing nicely. In only ten years, Michael progressed to a directorship in the company.*

progression *n.* **1** movement forward *or* forward movement, °advance, advancement, (making *or* gaining) headway, °progress, ascension, °rise, °elevation: *This mode of progression requires enormous effort. It was interesting to watch Curtis's progression through the corridors of power.* **2** °progress, °development, °advance, advancement, °spread, spreading, °extension, extending, broadening, enlargement, °headway, intensification, °rise: *It seemed impossible to halt the progression of the disease.* **3** °order, °sequence, °succession, °train, °chain, concatenation, °course, °flow: *The report traces the progression of events leading up to the war.*

progressive *adj.* **1** advancing, continuing, developing, increasing, growing, °ongoing, °continuous, °step by step, °gradual: *The doctors noted the progressive deterioration in his condition. With the onset of spring, we shall see progressive improvement in the weather.* **2** reformist, revisionist, °liberal, °radical, °revolutionary, °avant-garde, advanced, °dynamic: *Carrie was soon regarded as one of the more progressive thinkers of her day.* —*n.* **3** reformist, reformer, revisionist, °liberal, leftist, left-winger: *Once a conservative, he did a complete volte-face after the war and became a progressive.*

prohibit *v.* **1** °bar, °ban, °forbid, disallow, interdict, °outlaw, °taboo, debar, proscribe: *The deprivation of others' rights is prohibited by law.* **2** °prevent, °stop, °preclude, °rule out, °obstruct, °block, °impede, °hinder, °hamper, °inhibit, °frustrate, °foil, °thwart, °restrain, °check: *A turnstile prohibits entry through the exit doors.*

prohibition *n.* **1** forbiddance, °barring, °bar, banning, °ban, disallowance, disallowing, interdiction, interdicting, outlawing, outlawry, °taboo, debarment, debarring, proscription, proscribing: *Did the members of the motorcycle club vote for the prohibition of spirits on the club premises?* **2** °bar, interdict, °injunction, debarment, °embargo, proscription, °ban: *Imposition of an excessively high tariff on imported computer components has amounted to the prohibition of their use in this country.*

prohibitive *adj.* **1** discouraging, suppressive, °repressive, restrictive, prohibitory, inhibitory, restraining: *Prohibitive legislation has forced capital out of some profitable areas and into others less productive.* **2** °excessive, °exorbitant, °outrageous, °dear, °high, °outlandish, °abusive, extortionate, insupportable, °criminal: *The prices of the new designer dresses are prohibitive.*

project *n.* **1** °proposal, °plan, °scheme, °design, layout: *The project for housing low-income families had to be modified to allow for car parking.* **2** °activity, °enterprise, °program, °undertaking, °venture, °assignment, commitment, °obligation, °contract, °engagement; °occupation, °job, °work: *What project will you move on to when you complete this one?* —*v.* **3** °plan, °scheme, °prepare, °devise, conjure up, concoct, °think up, °contemplate, contrive, °invent, work up *or* out, °propose, °present, °outline, °design, °draft, °draw up, delineate, °describe, put forth *or* forward, *Colloq* cook up: *In a new theory, scientists project a cyclical system of rejuvenation of the genes.* **4** °cast, °hurl, °fling, °throw, °toss, °launch, °propel, °discharge; °shoot, °transmit; *Colloq* chuck, °lob: *A body projected into space will keep on traveling unless it strikes something. This device projects stereoscopic pictures on a screen.* **5** jut out, °stick out, °protrude, °stand out, °bulge (out), °extend (out), poke out, beetle (out), °overhang: *The cornice projects a full six feet over the street.* **6** °estimate, °reckon, °calculate, °predict, °forecast: *The economists projected a decline in inflation over the coming months.*

projectile *n.* °missile, °shell, bullet, °rocket: *Projectiles rained down on the enemy gun emplacement.*

projection *n.* **1** °protrusion, protuberance, °bulge, °extension, °overhang, °ledge, flange; °ridge, eminence, °prominence, °spur, °crag, outcropping: *The safety line caught on a building projection and I was saved. That projection off to the north is covered with ice all year long.* **2** °proposal, outlining, mapping, mapping-out, presenting, °presentation: *The projection of the plan for the new development must be effective for the council to accept it.* **3** °plan, °scheme, blueprint, °program, °design, °proposal, °outline, diagram, map, °representation, planning: *The council has approved the projection for the development of the waterfront.* **4** °estimate, prognostication, °forecast, °prediction, °calculation, °reckoning: *These sales projections for next year are quite optimistic.*

proliferate *v.* °grow, °increase, burgeon *or* bourgeon, multiply, mushroom, snowball; °breed, °reproduce: *Fast-food restaurants are proliferating everywhere. Failure to disinfect the wound allows germs to proliferate.*

proliferation *n.* °growth, °increase, burgeoning *or* bourgeoning, °expansion, °spread, escalation, buildup, °rise: *Some attribute the increase in violent crime to the proliferation of handguns.*

prolific *adj.* **1** °fertile, fecund, °productive, °fruitful, °abundant, copious, °bountiful, bounteous, °profuse, °plentiful, plenteous, °lush, °rich, rife: *The more perilous its natural survival, the more prolific a species is likely to be.* **2** °productive, °creative, °fertile: *A prolific author, more than two hundred of his novels have been published.*

prolong *v.* °extend, °lengthen, elongate, °stretch (out), draw *or* drag out, °drag (on), keep up, °string out, protract: *Not wishing to prolong the agony, we left after the first act. Aunt Caroline prolonged her visit by yet another week.*

promenade *n.* **1** °walk, °parade, esplanade, boulevard: *We strolled along the promenade by the river.* **2** °walk, °stroll, °saunter, °ramble, °turn, constitutional, airing, °turn: *After a brief promenade around the garden, she returned to the house.* —*v.* **3** °walk, °stroll, °saunter, amble, °ramble, °parade, perambulate, take a walk *or* stroll: *Every Sunday, regardless of the weather, he promenades for an hour in the park.* **4** °flaunt, °show (off), °display, °parade, °strut: *Why he promenaded that awful person before all his friends is beyond me!*

prominence *n.* **1** °celebrity, eminence, °fame, °distinction, notability, °reputation, °preeminence, °standing, °position, °rank, °prestige, °renown, repute, °importance, °weight, °influence, °account, °name, consequence: *They are a family of considerable prominence in the county. She rose to prominence as an actress just last year.* **2** °hill, hillock, °rise, hummock, outcrop, outcropping, °spur, tor, °crag, arête, °spine, °ridge, °peak, °mount, °pinnacle; headland, °point, promontory: *We planted our flag on the prominence and constructed a small cairn.* **3** protuberance, °projection, °protrusion, °growth, excrescence, °swelling, tumefaction, tumescence, extrusion, outshoot, outgrowth, °spur, °bulge: *If that prominence on your elbow doesn't disappear in a day or so, you should see a doctor.*

prominent *adj.* **1** °conspicuous, °obvious, °evident, recognizable, °pronounced, °discernible, distinguishable, identifiable, °noticeable, °remarkable, °noteworthy, eye-catching, °striking, °outstanding, °chief, °main, °principal, °significant, °important; °apparent, unmistakable *or* unmistakeable, °patent, °glaring, °salient, °flagrant, egregious: *The most prominent peak in the region is Annapurna. Some of the prominent shortcomings of the plan can be rectified easily.* **2** °eminent, °preeminent, °distinguished, °notable, °noteworthy, °noted, °leading, °foremost, °first, °outstanding, °eminent, °well-known, famed, °illustrious, °famous, °celebrated, °renowned, acclaimed, honored, °honorable, respected, °well-thought-of, °prestigious, °reputable, creditable: *A prominent politician is to be invited to cut the ribbon opening the new bridge.* **3** °protuberant, protruding, protrusive, projecting, jutting (out); excrescent, bulging, raised, °elevated: *She has a rather prominent chin. There is one prominent spot on the tabletop that needs more sanding.*

promiscuous *adj.* **1** °indiscriminate, undiscriminating, unselective, nonselective, nondiscriminatory, unconscientious, °heedless, °haphazard, °indifferent, uncaring, uncritical, unfussy, unfastidious, disregardful, neglectful, negligent, °slipshod, slovenly, °irresponsible, °careless, °cursory, °perfunctory, °unthinking, °thoughtless, unconsidered: *General Clargis was involved in the promiscuous massacre of the Kelman sect in 1684.* **2** °lax, °loose, °unchaste, °wanton, °wild, °uninhibited, unrestrained, ungoverned, °uncontrolled, unbridled, uncurbed, °immoderate, °abandoned, amoral, °immoral, °indecent, °libertine, licentious, dissipated, °dissolute, depraved, °profligate, debauched, °fast; unfaithful, °faithless, °dishonorable: *Nothing you say could persuade me to condone your promiscuous behavior.* **3** °mixed, °miscellaneous, heterogeneous, °random, intermixed, jumbled, °disorderly, disordered, °confused, °chaotic, motley, intermixed, intermingled, scrambled, unorganized, disorganized, unsystematic, unsystematized, °helter-skelter, higgledy-piggledy, hotchpotch *or US also* hodgepodge: *Problems arose in addressing a promiscuous assembly consisting of business executives, unemployed dockworkers, and of doctors, teachers, and other professionals.*

promise *n.* **1** °assurance, °(solemn) word (of honor), °pledge, °vow, °oath, °undertaking, °engagement, °bond, commitment, guaranty, °guarantee, °warranty; °agreement, °contract, covenant, °compact: *You have my promise that the book will be completed by the deadline. She broke her promise to marry him. Unlike a contract, a promise does not imply a mutual commitment.* **2** °expectation, °potential, °capability; °likelihood, °probability: *This first novel shows a lot of promise. There is little promise that compensation will be paid to the victims by the government.* —*v.* **3** °assure, give one's word (of honor), °pledge, °swear, °vow, take an oath, °undertake, °engage, °commit oneself, guarantee *or* guaranty, °warrant, cross one's heart (and hope to die): *Harold promised to do what cannot be done. We were promised a free ride on the roundabout. She promises she will repay the loan promptly. Make him pay for the damage as he promised.* **4** give indication of, hint at, °suggest, foretell, augur, °indicate, show signs of, be in store for, look like, seem *or* appear likely *or* probable, *Brit* be on the cards, *US* be in the cards, *Literary* bid fair, betoken, bespeak: *Though the economic situation promised to improve, many were disappointed that it took so long.*

promising *adj.* °hopeful, encouraging, °favorable, auspicious, °positive, °rosy, °optimistic, °propitious, cheering, full of promise, reassuring, heartening: *I see a promising future for you as a lexicographer.*

promote *v.* **1** °help, °further, °encourage, °assist, °advance, °support, °forward, °back, °sanction, °abet, °aid, °boost, °foster, °patronize, nurture, °develop, inspirit, °strengthen, °stimulate, °inspire: *Their continual bickering did little to promote domestic tranquillity. The sale of portable appliances has done much to promote the battery business.* **2** °advance, move up, °raise, upgrade, elevate, °exalt, *Colloq* kick upstairs: *He is to be promoted to floor manager next month.* **3** °recommend, °endorse, °sponsor, °support, espouse, commend, °advocate, °advance, °champion, °speak for, °side with, °present, call attention to: *She is promoting the cause of ecology in her campaign.* **4** advertise, °publicize, °push, °sell, *Colloq* beat the drum for, °plug, *Slang* hype, *Chiefly US* ballyhoo: *They are promoting the new chewing gum by giving out free samples in Times Square.*

promotion *n.* **1** °furtherance, advancement, °advance, °encouragement, °support, °backing, °sanction, sanctioning, abetting, aiding, °helping, assisting, boosting, fostering, nurturing, cultivation, °development, developing, °improvement, improving, °inspiration, inspiriting, strengthening, stimulation, °stimulating: *The charter of the society calls for the promotion of learning in a number of disciplines.* **2** advancement, °advance, upgrading, upgrade, °rise, °elevation, preferment, exaltation: *He was envied for his promotion to the highest ranking post in the department.* **3** promoting, °recommendation, °presentation, espousal, commendation, advocacy, championing: *I am grateful for your promotion of my name for the job.* **4** advertising, publicity, public relations, °propaganda, selling, hard *or* soft sell, °fanfare, plugging, *Colloq* puffery, *Slang* (media) hype, *Chiefly US* ballyhoo, hoopla: *The company is spending millions on the promotion of a new cola.* **5** °advertisement, advertising, °circular, °brochure, handbill, °bill, handout, °leaflet, °poster, *affiche,* placard, publicity, °space, °(publicity) release, *US and Canadian* flier *or* flyer, billboard, broadside, *Colloq US* puff piece, poop sheet: *The promotions for the new detergent feature endorsements by celebrities.*

prompt *adj.* **1** °quick, °ready, °immediate, instantaneous, °unhesitating, °rapid, °fast, °swift, °speedy, °punctual, °timely, °on time, °instant, °summary, °brisk, alacritous: *No matter what the question, Gemma gave a prompt response.* **2** °alert, °eager, °ready, °quick, °expeditious, ready and willing, °disposed, predisposed, °unhesitating, °keen, avid: *Slow to praise, Alan was always prompt to criticize.* —*v.* **3** °urge, egg (on), °prod, °nudge, °spur, exhort, °incite, °induce, impel, °provoke, °rouse, °arouse, °encourage, work *or* stir *or* fire up, °move, °motivate, °cause, °influence, put (someone) up to (something), °coax, °persuade, °cajole, prevail upon *or* on, talk (someone) into (something): *If he hadn't prompted her, she never would have apologized.* **4** °cue, °remind, feed lines (to), °help: *She never could remember her lines and needed to be prompted.* **5** °bring about, °inspire, °occasion, °give rise to, °elicit, °evoke, °provoke, °call forth, °stimulate, °awaken: *The mention of hostages prompted a question about what was being done to gain their release.* —*n.* **6** °reminder, °cue, °hint, stimulus: *He always needs a prompt to remind him to send his mother flowers on her birthday.*

promptly *adv.* °quickly, °at once, straightaway *or* straight away, °directly, right away, °immediately, without delay *or* hesitation, unhesitatingly, °swiftly, speedily, °readily, instantly, °instantaneously, punctually, expeditiously, with celerity, with alacrity, *Colloq US and Canadian* momentarily: *If this bill is not paid promptly, we shall have to discontinue your credit arrangements.*

prone *adj.* **1** face down *or* downward(s), °prostrate, lying down, reclining, °recumbent, °horizontal, procumbent, *Formal or technical* decumbent, accumbent: *She was lying prone, resting between push-ups.* **2** °inclined, apt, °likely, °liable, °disposed, predisposed, of a mind, °subject, °given, tending, °leaning: *Though they are prone to petty squabbling, they agree about major issues.*

pronounce *v.* **1** °declare, utter, °say, °voice, °express, articulate, °enunciate, vocalize, put into words: *He pronounced a curse on the family. Try to pronounce it one syllable at a time.* **2** °declare, affirm, °proclaim, °announce, °decree, °judge, aver, °state, asseverate, assert, say to be: *I now pronounce you man and wife. The doctor pronounced mother to be out of danger.* **3** °announce, °proclaim, promulgate, °publicize, °publish, °deliver, °broadcast, °make known, let (something) be known, put out *or* forth, °set forth; °pass: *The judge will pronounce sentence tomorrow.*

pronounced *adj.* **1** definite, °clear, °plain, well-defined, °decided, °conspicuous, °noticeable, recognizable, identifiable, °obvious, °striking, °prominent, °notable, °distinct, unmistakable *or* unmistakeable, °marked, strong: *There was a pronounced odor of bitter almonds about the dead man. Rita spoke with a pronounced lisp.* **2** definite, °distinct, unequivocal, unambiguous, °specific, °unqualified, °undisguised, °downright, °outright, °out-and-out, °decided, °complete, °total, °unmitigated, °strong, utter, unalloyed, unmixed, °clear, clear-cut, unmistakable *or* unmistakeable: *A pronounced prejudice against fuzzy thinking emerges from his writings.*

pronouncement *n.* **1** °statement, °assertion, °observation, °comment, °opinion, °announcement, °proclamation, pronunciamento, manifesto, °declaration, avowal, affirmation, asseveration, averment, promulgation: *Occasionally he would issue a pronouncement on the sad state of English usage.* **2** °judgment, °decree, edict, °proclamation, dictum, °command, ukase, (papal) bull, °imperative, °order, ordinance: *The authorities made several official pronouncements regarding travel restrictions.*

pronunciation *n.* enunciation, articulation, elocution, °diction, °speech, speech pattern, manner of speaking, °delivery, °accent, accentuation, °intonation, inflection, modulation: *Standard pronunciations of English are so diverse that it is sometimes difficult for one native speaker to understand another. Rock singers seem always to imitate a Texan pronunciation.*

proof *n.* **1** °evidence, verification, corroboration, confirmation, validation, authentication, ratification, substantiation; documentation, °document, °facts, °data, certification, °testimony, *Colloq* ammunition: *Is there any proof that she was with you? The prosecution claims to have enough proof of his guilt to convict him of embezzlement.* **2** °test, °trial, measure, °standard, °touchstone, criterion: *Do you intend to put his loyalty to the proof? The proof of the pudding is in the eating.* —*adj.* **3** impervious, impenetrable, able to withstand *or* resist, °protective, °strong, °tough, °impregnable, °resistant, tempered: *They maintain that this armor is proof against a nine-inch shell.*

prop *v.* **1** Often, ***prop up.*** °support, °brace, °hold (up), °buttress, °stay, °bolster, °uphold, °bear, °sustain, shore up, keep up: *A longer post is needed to prop up this end of the roof.* **2** °lean, °stand, °rest: *Do you mind if I prop my crutches against the desk while I sit down?* —*n.* **3** °support, °brace, truss, °stay, °buttress, °mainstay, °upright, vertical, shore: *Can you use this broomstick as a prop to support the shelf temporarily?*

propaganda *n.* **1** agitprop, disinformation, newspeak, rumors, lies: *The state-controlled media organized a campaign of agitation and propaganda.* **2** advertising, °promotion, publicity, public relations, °puff, °fanfare, *Colloq* puffery, ballyhoo, *Slang* hype, *US* hoopla, whoop-de-do *or* whoop-de-doo: *All that propaganda was merely to announce a new laundry soap!*

propagate *v.* **1** °breed, °generate, °reproduce, multiply, °proliferate, °deliver, °bring forth, °bear, procreate: *Biologists are studying ways of making the species propagate at a faster rate.* **2** multiply, °increase, °spread, °grow, °develop: *These plants propagate very rapidly.* **3** °publicize, °promote, disseminate, °dispense, °distribute, °spread, °publish, °broadcast, °circulate, °make known, °transmit, °disperse, propagandize, °proclaim, promulgate, bruit about, noise abroad, herald: *The purpose of the organization is to propagate information about the state of the environment.*

propel *v.* °drive, impel, °move, actuate, set in motion, get moving, °move, °push, °thrust, °force, °send, °launch, °start: *Each galley was propelled by huge oars. The capsule is propelled by a rocket. Job opportunities propelled her to move to Boston.*

proper *adj.* **1** °right, °appropriate, apropos, apt, °suitable, °fit, °fitting, °befitting, °becoming, suited, apposite, *de rigueur, comme il faut,* adapted, *Literary* °meet: *Is this the proper time to ask for a salary increase?* **2** °correct, °accurate, °exact, °right, °precise, °orthodox, °formal, expected, °normal, °usual, accepted, established, *Old-fashioned Brit* tickety-boo: *Some insist that the only proper way to pronounce "affluence" is with the stress on the first syllable.* **3** °decorous, °dignified, °genteel, °fitting, °right, *de rigueur,* °appropriate, °becoming, suitable, °decent, °seemly, °due, °correct, apt, *comme il faut,* conformist; gentlemanly, °ladylike, °polite, °refined, punctilious, °respectable: *Jeans are not proper attire for dinner at La Hacienda. Do you call that proper behavior for the president of a learned society?* **4** °fitting, °suitable, °correct, °right, °satisfactory, °good, °sensible: *The kitchen, not the garage, is a proper place for the dishwasher. Proper nutrition is very important.* **5** °complete, °perfect, utter, °thorough, thoroughgoing, °out-and-out, °unmitigated: *He felt a proper fool for having lost the key to his car.* **6** own, °individual, °separate, °distinct, °correct, °specific, °special, °particular, °respective; °characteristic, °distinctive, °peculiar, °singular, °unique: *When you have finished, return each book to its proper place.* —*quasi-adv.* **7** strictly speaking *or* so called, in the strict(est) *or* narrow(est) sense, °only, solely, °alone, on (its *or* someone's) own: *Does the deed refer to the house proper or is the adjacent land included, too?*

properly *adv.* **1** °appropriately, fittingly, correctly, °well, becomingly, suitably, politely, decently, decorously, nicely: *If you cannot behave properly I shall have to ask you to leave.* **2** °duly, °appropriately, °well, suitably, rightly, correctly, aptly: *Are they properly equipped for the rescue mission?*

property *n.* **1** °possessions, °belongings, °effects, °gear, °paraphernalia, °goods, chattels: *All his worldly property fit into the small suitcase he carries everywhere.* **2** °assets, °means, °resources, holdings, °capital (goods), °fortune, °riches, °estate, °worth: *We look on all our shares, personal belongings, and real estate as part of our property.* **3** °land, acreage, realty, real estate *or* property: *That property on Main Street was recently sold to a group of investors.* **4** °characteristic, °attribute, °quality, °feature, °trait, °mark, °hallmark, idiosyncrasy, °peculiarity, °oddity, °quirk, *Formal* haecceity, quiddity: *A property of iodine is its change directly from a solid to a vapor—and vice versa—without first liquefying.*

prophecy *n.* **1** forecasting, foretelling, °prediction, fortunetelling, divination, soothsaying, augury, prognostication, crystal gazing, *Formal* vaticination: *The prophecy of future events lies far outside the realm of science.* **2** °prediction, °forecast, °prognosis, °revelation: *Her prophecy warned of ice storms that would ravage the earth.*

prophesy *v.* **1** °predict, foretell, °forecast, forewarn, °prognosticate, vaticinate: *His father prophesied that he would come to no good if he quit school.* **2** augur, foretell (of), presage, °foreshadow, portend, °bode, °harbinger, herald, °promise, vaticinate: *The early disappearance of the caterpillars prophesies a long, cold winter.*

prophet *n.* prophetess, °oracle, forecaster, °seer, soothsayer, clairvoyant, prognosticator, °fortuneteller, augur, diviner, °witch, warlock, sibyl, haruspex, vaticinator; (of doom) Cassandra: *Statistically, the weather prophets have a fair record of accuracy.*

prophetic *adj.* predictive, prognostic, divinatory, oracular, inspired, prescient, sibylline, *Literary* fatidic, vatic: *His warnings of a drought proved to be prophetic.*

propitiatory *adj.* **1** conciliatory, pacifying, appeasing, expiatory, placative, propitiative, pacificatory, placatory: *The government made a propitiatory payment to the relatives of those killed in the air disaster.* **2** deferential, °ingratiating, °obliging, obeisant, acquiescent, compliant, °tractable: *He was most propitiatory in his treatment of his stepchildren.*

propitious *adj.* °advantageous, °timely, °well-timed, °opportune, °lucky, °fortunate, °happy, °providential, °favorable, °bright, encouraging, auspicious, °promising, °rosy: *It was not a propitious moment to ask her to marry him,.*

proponent *n.* proposer, promoter, °supporter, upholder, °backer, subscriber, °patron, espouser, adherent, °enthusiast, °champion, °friend, °partisan, defender, °advocate, exponent, pleader, apologist, spokesman, spokeswoman, spokesperson: *Ashley is a staunch proponent of the literacy movement.*

proportion *n.* **1** °ratio, °(comparative) relation, °relationship, °comparison: *The proportion of men to women on the course is three to one. The demand is out of all proportion to the supply.* **2** °balance, °agreement, concord, °harmony, suitableness, °symmetry, congruity, correspondence, correlation, °arrangement, °distribution: *The beauty of the building is at once apparent in the proportion of its parts.* **3** °portion, °division, °share, °part, °percentage, °quota, °allotment, °ration, *Colloq* °cut: *Water covers a large proportion of the surface.* **4** ***proportions.*** °size, °magnitude, dimensions, measurements, °extent; °volume, °capacity, °mass, °bulk, °area, °expanse, °scope, °range, °degree: *The dragon suddenly assumed gigantic proportions.* —*v.* **5** °adjust, °modify, °change, °modulate, °poise, °balance, °shape, °fit, °match, °conform, equate: *The punishment should be proportioned to the crime.*

proportional *adj.* proportionate, proportioned, comparable, analogous, analogical, °relative, °related, correlated, balanced, °symmetrical, corresponding, compatible, °harmonious, °consistent, commensurate, in accordance with: *The zeal with which their ideals are defended is directly proportional to the fierceness with which they are attacked.*

proposal *n.* **1** °offer, °presentation, °bid, °tender, °proposition, °recommendation, °suggestion, *Literary* proffer: *Her proposal was to provide the funding if we would do the work.* **2** °plan, °scheme, °outline, °draft, °design, layout; °program, °proposition, °project: *The proposal for the new civic center was rejected as too expensive.*

propose *v.* **1** °offer, °tender, proffer; °present, °introduce, °submit, °advance, °set forth, °put forward, °propound, °bid, °recommend, °suggest, come up with, call attention to, °broach, *Brit* °table: *Management pro-*

posed a wage increase of 5 percent. The architect proposed that we consider an addition to the north wing. 2 °offer, °mean, °intend, °plan, °expect, °aim: *As it's your birthday, the children propose to do the dishes.* 3 °nominate, °name, put forward *or* forth, °suggest, °introduce, °submit, °put up: *His boss proposed him for membership in the country club.*

proposition *n.* 1 See **proposal, 1,** above. 2 See **proposal, 2,** above. —*v.* 3 accost, °solicit, make an indecent *or* sexual advance *or* proposal *or* overture, *Colloq* make a pass at: *She said she had been propositioned by him, but we never learned if she accepted or not.*

propound *v.* put *or* set forth *or* forward, °propose, °offer, proffer, °suggest, postulate: *It was she who first propounded the idea of a unified country.*

proprietor *n.* 1 proprietress, °owner, landowner, °landlady, °landlord, landholder, titleholder, deedholder, property owner: *He is the proprietor of a large house in Washington.* 2 °owner, °partner, °landlord, restaurateur, innkeeper, hotelkeeper, hotelier, licensee, °manager, *Brit* publican: *The proprietor greeted us warmly and showed us to our rooms.*

propriety *n.* 1 correctness, properness, conformity, suitableness, appropriateness, suitability, aptness, °fitness, seemliness, °decorum; advisability, °wisdom: *I question the propriety of concealing this information from the stockholders.* 2 °protocol, good *or* proper form, punctilio, °etiquette, politeness, °courtesy, politesse, °refinement, sedateness, °dignity, modesty, °decorum, decency, °breeding, respectability, gentility, °grace, mannerliness: *He always behaved with the utmost propriety in the presence of his aunt.* 3 ***the proprieties.*** the social graces, the amenities, the civilities, formality *or* the formalities, social convention *or* the social conventions, social procedure *or* codes, accepted practice, °tradition, °ceremony, °ritual: *Ramsbotham observes the proprieties only when he wants to make a good impression.*

propulsion *n.* °drive, impulse, °impetus, °thrust, °power, driving *or* propelling *or* propulsive force, °pressure, °momentum, °push: *Enormous propulsion is required to lift the space shuttle into orbit.*

prosaic *adj.* °dull, °banal, overdone, °tedious, clichéd, commonplace, stereotyped, °pedestrian, °flat, stereotypical, hackneyed, °stock, °routine, °everyday, °ordinary, °common, workaday, °mediocre, °undistinguished, °bland, characterless, °homely, °plain, trite, °stale, °threadbare, °tired, °lifeless, °dead, °dry, jejune, °boring, °tiresome, unimaginative, unpoetic, unromantic, uninspiring, uninspired, insipid, uninteresting, °humdrum, °monotonous, *Literary ennuyant, Colloq* ho-hum, °run-of-the-mill, °moldy: *She found the law too prosaic and trained to become a jockey.*

prose *n.* °(expository) writing, °text, °language: *It came as a great revelation to learn that he had been speaking prose all his life.*

prosecute *v.* 1 arraign, °indict, °charge, put on *or* bring to trial, °try, take to court, °sue, bring suit *or* action against, °accuse, *Brit* put in the dock: *She threatened to prosecute him for slander.* 2 °pursue, follow up *or* through, see *or* carry through, °persist, go on with: *Will the same policy be prosecuted in other countries?* 3 carry on *or* out, °perform, do, °exercise, °conduct, °follow, °engage in, °practice, °continue: *The government needed popular support if it was to continue to prosecute the war successfully.*

prospect *n.* 1 °view, °scene, panorama, °landscape, seascape, °outlook, vista, °sight, °spectacle, °perspective, °aspect: *The sitting-room window offered a splendid prospect of the sea and cliffs below.* 2 °anticipation, contemplation, °outlook, °promise, °plan, °design, °intention, expectancy, °expectation, °thought, °likelihood: *The prospect of spending the Christmas holidays with your family is quite daunting.* 3 Often, ***prospects.*** °future, °outlook, chance(s), hope(s), possibility *or* possibilities, °likelihood, opportunity *or* opportunities: *The prospects for her advancement are excellent.* **4 *in prospect.*** in sight *or* view, in the offing, on the horizon, in store, °in the wind, projected, in store, °likely, °probable, °possible, *Brit* on the cards, on the table, *US* °in the cards: *As a doctor, he has a number of opportunities in prospect.* —*v.* 5 Often, ***prospect for.*** °explore, °search (for), °look (for): *In 1896, his grandfather went to the Klondike to prospect for gold.*

prospective *adj.* anticipated, expected, awaited, looked-for, °future, °forthcoming, coming, approaching, °imminent, nearing, °pending, °impending, °destined, °potential, incipient: *Jane's prospective wedding to Hoccleston had excited the entire family.*

prospectus *n.* °announcement, °plan, °scheme, °program, °outline, conspectus, °description: *According to the prospectus for the fund, the minimum investment is one thousand dollars.*

prosper *v.* °flourish, °thrive, °succeed, fare well, °progress, °get ahead, °grow, °develop; °profit, °gain, become wealthy, grow rich, make one's fortune, °make good, *Colloq* °make it, make one's pile: *With hard work, thrift, and perseverance, MacIntosh prospered and was soon able to buy the house he wanted.*

prosperity *n.* °success, °(good) fortune, °wealth, °riches, affluence, °money, °luxury, °plenty, prosperousness, opulence, °bounty, *Colloq* life of Riley: *The basis of the family's prosperity was huge land holdings in Australia.*

prosperous *adj.* 1 °rich, °wealthy, moneyed *or* monied, affluent, well-to-do, °well-off, *Colloq* well-heeled, °loaded, °flush, °in the money, rolling in it *or* wealth *or* money, in clover, on Easy Street, *Slang* stinking rich: *Anyone who owns six houses and four yachts must be prosperous.* 2 °successful, thriving, °flourishing, booming, prospering: *Farnsworth owns a prosperous chain of video shops.*

prostitute *n.* 1 whore, call girl, streetwalker, strumpet, trollop, harlot, lady of the night *or US also* evening, fallen *or* loose woman, demimondaine, cocotte, *fille de joie,* painted woman, woman of ill repute, camp follower, *Archaic* catamite, *Literary* hetaira *or Literary* hetaera, courtesan *or* courtezan, *Brit* rent boy, toy boy, *US* boy toy, *Archaic* bawd, quean, trull, cotquean, *Colloq* °tart, hustler, *Slang* pro, moll, *Brit* °brass, hooker, *US* bimbo, working girl, chippy *or* chippie, roundheels: *There was a terrible scandal when that Member of Parliament was found to have consorted with prostitutes.* —*v.* 2 Often, ***prostitute oneself.*** °degrade, demean, °lower, cheapen, °debase, °profane, defile, °desecrate, °pervert, °abuse, °misuse, devalue, *Colloq* °sell out: *To ward off starvation, he prostituted his talent by drawing comic strips.*

prostitution *n.* 1 whoredom, harlotry, the oldest profession, Mrs. Warren's profession, streetwalking, °vice: *Out of desperation, some countries have experimented with the legalization of prostitution.* 2 °degradation, debasement, profanation, defilement, desecration, °misuse, °abuse, devaluation, lowering, °perversion, corruption: *Don't you regard pornography as a prostitution of the principle of freedom of expression?*

prostrate *v.* 1 Usually, ***prostrate oneself.*** lie down, °kowtow, °bow (down), °bow and scrape, grovel, kneel, fall to *or* on one's knees, °truckle, °crawl, °cringe, °submit, abase oneself: *He was forced to prostrate himself before the emperor.* 2 °overwhelm, °overcome, °overpower, °crush, lay *or* bring low, °paralyze, °fell, bowl over, °floor, bring down, °humble, make helpless, °ruin; °exhaust, °fatigue, °weary, wear down *or* out, °tire (out): *They were prostrated by grief at the loss of their son. Having been prostrated for months by glandular fever, Dalgleish finally recovered.* —*adj.* **3** °prone, °horizontal, lying down, laid low, stretched out, procumbent, °recumbent, *Formal or technical* accumbent, decumbent: *The prostrate bodies of the worshipers in their multicolored garb resembled a huge patchwork quilt.* **4** overwhelmed, °overcome, overpowered, crushed, brought *or* laid low, paralyzed, felled, bowled over, brought down, humbled, °helpless, ruined, brought to one's knees, °powerless, °impotent, °defenseless, disarmed, *Colloq* floored: *She was prostrate at the news of the car crash. After the long war, the countries were struggling to revitalize their prostrate economies.* **5** °exhausted, drained, °fatigued, °spent, worn-out, wearied, °weary, °tired (out), dead tired, dog-tired, played out, *Colloq* fagged out, knocked out, all in, °beat, bushed, *US* wiped out, *Slang* shagged out, *US and Canadian* pooped (out): *We were prostrate after the long climb.*

prostration *n.* **1** genuflection *or Brit also* genuflexion, kowtowing, °kowtow, kneeling, bowing, °bow, salaaming, salaam, °submission: *Prostration before a superior was a mark of honor.* **2** °servility, °veneration, °worship, °humiliation, °respect, adulation, °deference, °obeisance, °homage: *Their silence betokened the profound prostration they felt before her superior intellect.* **3** °despair, °misery, °desolation, °desperation, dejection, °depression, despondency, wretchedness, unhappiness, °grief, °woe, woefulness: *Years of poverty created in him a spiritual prostration from which he never recovered.* **4** weariness, °exhaustion, °weakness, debility, feebleness, enervation, lassitude, paralysis, °collapse, °breakdown: *The diagnosis was nervous prostration and the treatment was bed rest for a week or more.*

protagonist *n.* **1** °hero, °heroine, antihero, antiheroine, °principal, leading character; °lead, leading role, title role: *Mother was always the protagonist in our little domestic dramas.* **2** °leader, °supporter, °advocate, °backer, prime mover, moving spirit, °champion, °mainstay, standard-bearer, exponent: *He is considered the chief protagonist of reformist policies in the party.*

protean *adj.* °variable, ever-changing, multiform, °variable, mutable, °changeable, labile, polymorphous *or* polymorphic, kaleidoscopic: *The magic ring gave him protean powers to appear now as an eagle, now as a serpent.*

protect *v.* **1** °defend, °guard, °safeguard, keep safe, °shield, °cover, °screen: *Company rules require the wearing of safety goggles to protect the eyes when operating any machine.* **2** °care for, °preserve, °keep, °shelter, watch over, °safeguard, °take care of, °conserve, take under one's wing, °foster, nurture, °tend, °mind: *We must protect wildlife in the wild, not by bringing animals into zoos.*

protection *n.* **1** °defense, °screen, °shield, °barrier, °guard, °safeguard, °immunity, °bulwark, buffer, °shelter, °refuge, haven, °sanctuary, °security, °safekeeping, °safety, °preservation: *They have not yet developed any protection from the common cold. How can we offer these fledglings protection against predators?* **2** °care, guardianship, aegis, °custody, °charge, °safekeeping, °patronage, sponsorship, keeping: *Even under the protection of the government, elephants continue to be slaughtered.* **3** extortion, °blackmail, protection money: *If we refused to pay protection, they threatened to bomb the restaurant.*

protective *adj.* defensive, °jealous, °vigilant, watchful, heedful, °careful, °possessive; preservative, shielding, sheltering, safeguarding: *She is fiercely protective of her independence and refuses all help. Many animals rely on protective coloring as a defense against predators.*

protector *n.* protectress, defender, °benefactor, benefactress, °patron, patroness, °guardian (angel), °champion, knight in shining armor, paladin, bodyguard, *Slang Brit* °minder: *She came to regard him as her friend and protector.*

protégé *n.* protégée, °ward, °charge, dependant; °discovery, °student, °pupil: *Alphonse became her protégé and she has taught him everything he knows.*

protest *n.* **1** °objection, °opposition, °complaint, grumble, °grievance, dissent, °disapproval, protestation, °exception, °disagreement, demur *or* demurral, demurrer, disclaimer, °denial, °scruple, °compunction, °qualm, *Colloq* °gripe, grouse, °squawk, *US* °kick, *Slang* beef, °bitch: *The governor's office has received many protests against the treatment of prisoners. There was no protest from you when someone else was given a ticket for parking in this spot.* **2** ***under protest.*** unwillingly, reluctantly, involuntarily: *I paid the fine under protest.* —*v.* **3** °object, °oppose, °complain, grumble, dissent, °disapprove, take exception, take issue with, °disagree, demur, disclaim, °deny, °scruple, *Colloq* °gripe, grouse, °squawk, *Brit* °kick (against), *US* °kick, *Slang* beef, °bitch: *The bank employees protested at being required to work on Saturday mornings.* **4** assert, °confirm, °declare, aver, asseverate, affirm, °announce, °profess, insist on, avow, avouch: *The convicted man went to the gallows protesting his innocence.*

protocol *n.* **1** rule(s) *or* code(s) *or* standard(s) of behavior *or* conduct, convention(s), custom(s), °diplomacy, °formality, formalities, °form, °etiquette, politesse, °manners, °practice, °usage, °authority: *According to protocol, the lady stands at the right of the gentleman. Which protocol are you following, the British or the French?* **2** °treaty, °pact, °compact, covenant, °agreement, concordat; °memorandum, °minute, °note, °draft, °outline: *The original protocol must be checked before the ministers attach their signatures.*

prototype *n.* **1** °model, archetype, °first, °original, °pattern, exemplar, °precedent, °mold: *The prototype of elaborate tombs was that of Mausolus at Halicanarssus, from which we derive the word "mausoleum."* **2** °example, °instance, °illustration, °sample, °norm, °paragon, °epitome, °model, °standard, analogue, referent, °metaphor: *Mrs. Grundy is a common prototype for narrow-mindedness and prudishness.*

protracted *adj.* °long, long-drawn-out, interminable, prolonged, overlong, never-ending, extended, stretched-out, marathon, °endless, °everlasting, long-winded: *The protracted union negotiations delayed the start of the bridge expansion by six months.*

protrude *v.* °stick out, °jut (out), °project, °extend, poke out, °stand out, thrust out *or* forward, °start (from), exsert, *Rare* extrude; °bulge, balloon, °bag (out), belly (out); °(of the eyes) pop, goggle, *Colloq US* °bug (out): *The tops of the plants protruded through the snow. The children's eyes protruded when they saw the ice cream.*

protrusion *n.* °projection, protuberance, °prominence, °swelling, excrescence, tumescence, °bump, °lump, °knob, °bulge; (condition of the eyes) *Technical* exophthalmic goiter: *I can feel the slight protrusion of the nailheads above the smooth surface.*

protuberant *adj.* protrusive, protruding, bulging, gibbous, jutting, bulbous, °swelling, °swollen, turgid, tumescent, distended, tumid, extrusive, excrescent, extruding, projecting, beetling, overhanging, °prominent:

The bomber's protuberant gun turrets resemble blisters on the fuselage.

proud *adj.* 1 Often, ***proud of.*** pleased (with), satisfied (with), contented (with), °glad (about), happy (with *or* about), delighted (with *or* about), elated (with *or* about); honored, gratified: *He is very proud of his children and what they have achieved. I am proud to be your friend.* 2 °conceited, °boastful, self-satisfied, narcissistic, °self-important, °egotistical, °vain, vainglorious, prideful, self-centered, complacent, °snobbish, °haughty, °supercilious, °smug, °arrogant, °cocky, cocksure, °boastful, °braggart, *Colloq* high and mighty, snooty, stuck-up, *Slang* snotty, *Brit* toffee-nosed: *She's far too proud to have anything to do with the likes of us.* 3 °lofty, °dignified, lordly, °noble, °great, respected, honored, °honorable, °important, °glorious, august, °illustrious, °estimable, creditable, °eminent, °prominent, °distinguished, °reputable, °worthy, °notable, °noted, °noteworthy: *His will always be a proud name in British military history.* 4 °stately, °majestic, °magnificent, °splendid, °grand: *Proud Edinburgh earned the sobriquet, "Athens of the North."*

prove *v.* 1 °verify, °authenticate, °confirm, °make good, corroborate, °demonstrate, °show, validate, °establish, °substantiate, °certify, affirm; °support, °sustain, °back (up), °uphold: *Prove that he lied under oath, and we shall have him for perjury. If you cannot prove her guilt, she must be presumed innocent.* 2 °try, °test, °examine, °check, °analyze, assay: *Division can easily be proved by multiplication, and vice versa. The proving ground for military vehicles is off limits to the public. Come live with me and be my love, / And we shall all the pleasures prove.* 3 °turn out, be found, be shown, be established, end up; °develop, °result: *The child proved to be his long-lost grandson.* 4 °show, evince, °demonstrate: *He proved his love many times over during their fifty years of marriage.*

provender *n.* 1 °provisions, °food, supplies, victuals, °rations, foodstuffs, groceries, eatables, edibles, comestibles, aliment, °nourishment, °sustenance, *Colloq* grub, eats: *The armies relied for their provender on farms they passed on their advance.* 2 fodder, forage, °feed, hay, silage, corn, °grain: *Provender for the livestock was running low because of the long winter.*

proverb *n.* saying, °maxim, aphorism, °saw, adage, apophthegm *or* apothegm, axiom, °moral, moralism, homily, dictum, gnome, °epigram; commonplace, platitude, °truism, °cliché, chestnut, bromide: *According to the old proverb, "A fool and his money are soon parted."*

proverbial *adj.* 1 axiomatic, aphoristic, °epigrammatic, apophthegmatic *or* apothegmatic, homiletic, moralistic; acknowledged, °well-known, accepted, °time-honored, °traditional: *The language is full of proverbial sayings reflecting popular wisdom.* 2 °typical, archetypal, °exemplary: *In mentioning Dr. Albert Schweitzer, he was referring to his proverbial humanitarianism.*

provide *v.* 1 °supply, °furnish, °equip, °outfit, fix up (with) provision, °contribute, °accommodate, purvey, °cater, °stock (up), victual, °provender: *After providing us with some necessities, the pirates marooned us on the island.* 2 °produce, °yield, °afford, °lend, °give, °present, °offer, °accord: *The fertile land provided food in plentiful abundance. During those bleak years, the radio provided us not only with news but also with entertainment.* 3 °stipulate, °lay down, °require, °demand, °specify, °state: *The lease provided that the rent be reviewed every five years.* 4 ***provide for.*** °look after, °care for, °support, °take care of, take under one's wing, °minister to, °attend (to): *The bequest ensured that his widow would be amply provided for.* 5 ***provide for*** *or* ***against.*** arrange for, prepare for, °anticipate, forearm, make *or* get ready for, plan for, take precautions, take measures: *It would seem that you have provided for any eventuality.*

providence *n.* 1 °foresight, °forethought, °preparation, °anticipation, °readiness, farsightedness, °caution, °precaution, °discretion, °prudence, °care; °thrift, frugality, husbandry, thriftiness, °conservation, °economy: *Because of our providence, we were able to weather severe financial setbacks.* 2 Usually, ***(divine) Providence.*** °protection, °care, °concern, beneficence, °direction, °control, divine intervention, °guidance; °destiny, °fate, °lot, °fortune, karma, kismet: *Providence is always on the side of those who help themselves.*

provident *adj.* 1 °cautious, °wary, °discreet, canny, °prudent, °careful, °vigilant, °prepared, °farsighted, °thoughtful, °wise, °shrewd, sagacious, °sage, °judicious: *In the fable, the grasshopper learns from the ant what it means to be provident.* 2 °frugal, economic(al), °thrifty, °prudent: *Because he has never been a provident man, he will always be poor.*

providential *adj.* °fortunate, °lucky, blessed, felicitous, °happy, °opportune, °timely: *Leaving the house just before the earthquake was most providential.*

providing *conj.* Sometimes, **providing that.** provided (that), on (the) condition (that), if (only), only if, as long as, °in the event (that), with the proviso (that), °in case, with the understanding (that): *Simon is always ready to go out to dinner, providing someone else pays for it.*

province *n.* 1 °territory, °state, °zone, °region, °quarter, °area, °district, °domain, dependency *or US also* dependancy, °division, °section, °district: *Quebec is one of the administrative provinces of Canada.* 2 °country, °territory, °region, °dominion, °realm, strand, °tract: *During the war, those who escaped fled to distant provinces.* 3 sphere *or* area (of responsibility), °responsibility, bailiwick, °concern, °function, °charge, °business, °field; *Colloq* °thing, °headache, °worry: *The payroll falls within the province of my department.* 4 ***provinces.*** outlying districts, countryside, hinterland(s), *Chiefly US* exurbia, *Slang US and Canadian* boondocks, boonies, hicksville: *Once a year, people flock into London from the provinces to do their Christmas shopping.*

provincial *adj.* 1 °local, regional: *Provincial administration is the responsibility of the sheriff.* 2 uncultured, uncultivated, °unsophisticated, °limited, °uninformed, naive *or* naïve *or* naïf, °innocent, °ingenuous, unpolished, °unrefined, °homespun, °rustic, °rude, °country, °parochial, insular, °narrow-minded, °boorish, loutish, cloddish, °awkward, °ungraceful, oafish, backwoods, *Brit* parish pump, *US* small-town, *Colloq US and Canadian* hick, hick-town: *These paintings are unlikely to appeal to provincial tastes.* —*n.* 3 °rustic, country cousin, (country) bumpkin, yokel, *US and Canadian* out-of-towner, hick, hayseed: *One could tell they were provincials by the cut of their clothes.*

provincialism *n.* 1 dialectalism, localism, regionalism; °idiom, patois, °dialect: *His southern speech is peppered with provincialisms unfamiliar to those who live in the north.* 2 narrow-mindedness, insularity, parochialism, narrowness, benightedness; unsophisticatedness, °simplicity, lack of awareness, naivety *or* naiveté *or* naïveté, ingenuousness, innocence, °inexperience: *Their provincialism led the townspeople to regard any stranger as a city slicker. There was something charming about the provincialism of this rough man from the outback.*

provision *n.* 1 °providing, supplying, furnishing; catering, victualing, provisioning, purveyance, purveying, furnishing, equipping, fitting out, outfitting, accoutrement *or US also* accouterment, °equipment: *The* *school is responsible for the provision of textbooks. The provision of a cruise ship of that size requires days.* **2** °stipulation, °proviso, °condition, °restriction, °qualification, clause, °term, °exception, °demand, °requirement, °prerequisite, *Colloq* °catch, °string, *US* hooker: *This provision of the contract calls for a penalty for each day's delay beyond the guaranteed completion date.* **3** °preparation, prearrangement, °arrangement, °measures, °steps: *They had failed to make provision for so many customers and soon ran out of food.* **4** Usually, ***provisions.*** supplies, stores, stockpile, stock(s), °quantity; °food, foodstuffs, eatables, edibles, drinkables, potables, victuals, viands, comestibles, °rations, groceries, °provender, °staples: *We had enough provisions to last a year.* —*v.* **5** stockpile, °stock, °supply, victual, °cater, purvey: *We provisioned the expedition for a three-month period.*

provisional *adj.* **1** °temporary, interim, provisionary, transitional, °stopgap, *Colloq* pro tem: *The provisional government was expected to be in place for less than a month.* **2** conditional, contingent, provisory, °qualified, stipulatory, provisionary, probationary: *The appointment is provisional, subject to review after six months.*

proviso *n.* See **provision, 2.** above.

provocation *n.* **1** °grounds, °reason, °cause, justification, instigation, °initiation, °incitement, stimulus, °incentive, motivation, °motive, °inducement: *What was the provocation for that unpleasant outburst?* **2** °insult, °offense, °taunt, irritation: *After a series of provocations, the final straw was his remark about my mother.*

provocative *adj.* **1** °inviting, alluring, °tempting, charming, tantalizing, teasing, intriguing, fascinating, °seductive, °stimulating, °voluptuous, °sensual, °sensuous, °suggestive, °erotic, arousing, °exciting, entrancing, °irresistible, bewitching, *Colloq* °sexy: *The council found the film too provocative to be shown to schoolchildren.* **2** irritating, annoying, galling, °irksome, nettlesome, harassing, plaguing, exasperating, infuriating, angering, incensing, maddening, enraging, vexing, vexatious, disquieting, challenging, upsetting, distressing, °disturbing, °outrageous, wounding, stinging, °offensive, humiliating, mortifying: *She is given to making provocative remarks that drive him to distraction.*

provoke *v.* **1** °stir (up), °stimulate, °move, °motivate, °push, impel, °drive, °get, °spur (on), egg on, goad, °force, compel, °prompt, °rouse, °arouse, waken, °awaken, °enliven, °animate, °activate, °induce, °encourage: *She did her best to provoke him to start his new novel.* **2** °start, °incite, instigate, °produce, °promote, °foment, °kindle, °work up: *Are you trying to provoke an argument?* **3** °irritate, °annoy, °irk, °pester, vex, pique, °anger, °enrage, °madden, incense, °infuriate, °gall, rile, nettle, °harass, hector, °plague, badger, °exasperate, get on one's nerves, try one's patience, °frustrate, °upset, °disturb, °perturb, °distress, °outrage, °offend, °insult, affront: *If he continues to provoke me I shall punch him.*

prowess *n.* **1** °ability, °skill, skillfulness, °aptitude, adroitness, °dexterity, dexterousness, adeptness, °facility, °finesse, °expertise, mastery, °genius, °talent, know-how, °ingenuity, °capability, °proficiency: *Her prowess as a sculptor is unquestioned.* **2** °bravery, valor, °courage, boldness, °daring, intrepidity, dauntlessness, mettle, stoutheartedness, valiance, lionheartedness, fearlessness, gallantry, doughtiness, °fortitude: *He was famed for his prowess in single combat.*

prowl *v.* **1** °lurk, °sneak, skulk, °steal, °slink: *I thought I saw someone prowling about in your back garden.* **2** °scour, scavenge, range over, rove, °roam, °patrol, °cruise, °cover: *The police continued to prowl the waterfront looking for smugglers.* —*n.* **3 *on the prowl.*** lurking *or* sneaking *or* skulking *or* stealing *or* slinking about *or* around, searching, seeking, hunting, tracking, stalking: *The lioness was on the prowl for wildebeest. After a few drinks, the boys went on the prowl for some female companionship.*

proximity *n.* nearness, closeness, adjacency, °neighborhood, °vicinity, vicinage, contiguity, contiguousness, propinquity: *The proximity of a nuclear power plant kept housing prices relatively low.*

proxy *n.* °substitute, °agent, °delegate, surrogate, °representative, °factor: *As I was unable to attend the owners' meeting, I sent Jane as my proxy.*

prude *n.* °prig, °puritan, Mrs. Grundy, *Colloq* °goody-goody, *US* bluenose: *He was a bit of a prude and disapproved of anything even slightly suggestive.*

prudence *n.* **1** °discretion, °wisdom, sagacity, °judgment, °discrimination, common sense, canniness, °presence of mind, awareness, wariness, °care, °tact, carefulness, °caution, cautiousness, circumspection, watchfulness, °vigilance, heedfulness: *All prudence abandoned him when he lost his temper.* **2** planning, °preparation, °preparedness, foresightedness, °forethought, °foresight, °providence, °precaution, farsightedness; °economy, husbandry, °thrift, (good *or* careful) management: *Owing to the prudence of her investments, Clara had accumulated a substantial nest egg.*

prudent *adj.* **1** °careful, °cautious, °discreet, discerning, °wise, °sage, sagacious, °politic, °judicious, °discriminating, °sensible, °reasonable, canny, °shrewd, °cautious, circumspect, watchful, °vigilant, heedful, °wary, °attentive, °alert, °guarded: *James had always found it prudent to listen to others before speaking his mind.* **2** °provident, °thrifty, economic(al), °frugal, prudential: *Father favors prudent, conservative management of the family business.*

prudery *n.* prudishness, priggishness, puritanicalness, puritanism, squeamishness, Grundyism, primness, stuffiness, oldmaidishness, precisianism: *Until late in the nineteenth century, prudery prevented many women from pursuing a career in the theater.*

prudish *adj.* °priggish, puritanical, old-maidish, °prissy, prim, °fussy, °squeamish, °strait-laced, °stiff, °rigid, overnice, overmodest, overcoy, °proper, demure, °decorous, °formal, *Colloq* oldmaidish: *The dividing line between the prudish and the prurient Victorians was very thin indeed.*

prune *v.* °clip, °cut back, °lop, °dock, °pare (down), °trim: *If you prune fruit trees, they bear more abundantly. We have to prune the entertainment budget.*

prurient *adj.* **1** libidinous, °lecherous, °lascivious, °lewd, lubricious *or* lubricous, salacious, °lustful, concupiscent, licentious, °carnal, debauched, rakehell, °sensual, °randy, °voluptuous, °loose, goatish, ruttish, *Literary* Cyprian, Paphian, *Archaic* lickerish *or Archaic* liquorish, *Slang* horny, °hot: *Complaints were lodged that the tabloid press was taking a prurient interest in the private lives of celebrities.* **2** °dirty, °lewd, °filthy, °pornographic, smutty, °obscene, °foul, °scurrilous, °vile, °indecent, °gross, °lurid, °blue, °bawdy, ribald, titillating, °suggestive, °coarse, °vulgar, °low, °crude, *Literary* Fescennine: *The police raid uncovered a store of prurient literature, films, and videos.*

pry *v.* **1** °investigate, ferret about, °examine, °peer, °peek, be inquisitive, inquire *or* enquire: *They pried into her past but found nothing revealing.* **2** °intrude, °meddle, °interfere, *Colloq* poke *or* stick one's nose in *or* into, °snoop, be nosy, nose about *or* around, poke about *or* around: *They have no right to pry into my private affairs.*

pseudonym *n.* nom de plume, nom de guerre, alias, pen name, stage name, °incognito: *It is common knowledge that "George Eliot" was the pseudonym of Mary Ann Evans.*

psyche *n.* °soul, °spirit, °mind, *élan vital,* divine spark, life force, anima, self, °subconscious, °unconscious, °personality, °(essential) nature, inner man *or* woman *or* person, *Philosophy* pneuma: *To Homer, the psyche was more like an alter ego, or conscience.*

psychic *adj.* **1** psychical, °mental, °spiritual, psychologic(al), °subjective, psychogenic, cognitive, metaphysic(al), °intellectual, cerebral; philosophic(al): *The psychic effect of losing a loved one can have physical repercussions.* **2** psychical, extrasensory, °supernatural, °occult, magical, telepathic, telekinetic, preternatural, spiritualistic, °unearthly, extramundane, supermundane: *Many psychic phenomena have never been satisfactorily explained.* —*n.* **3** °medium, spiritualist, clairvoyant, mindreader, telepathist, °seer, seeress, crystal gazer, soothsayer, astrologer, °fortuneteller, °prophet, prophetess, sibyl: *We were disturbed to learn that our political leaders consulted psychics.*

psychological *adj.* °mental, °intellectual, cerebral, cognitive, psychic(al), °spiritual, °subjective, °subconscious, °unconscious, °subliminal, psychogenic; philosophic(al): *He has plenty of money, so there must be psychological reasons for his stealing.*

psychology *n.* °(mental) makeup, constitution, °attitude, °behavior, thought processes, °thinking, °psyche, °nature, feeling(s), emotion(s), °rationale, °reasoning: *I cannot understand the psychology behind hatred and intolerance.*

psychotic *adj.* **1** °mad, °insane, psychopathic, °deranged, demented, lunatic, paranoiac *or* paranoid, °abnormal, °unbalanced, (mentally) ill *or esp. US* sick, °disturbed, *non compos mentis,* of unsound mind, °exceptional, certifiable, °daft, unhinged, °raving, *Slang* °crazy, nuts, nutty, loony *or* looney *or* luny, off one's rocker *or* trolley *or* chump *or* head, cracked, crackbrained, °mental, out to lunch, batty, bats, having bats in one's belfry, having a screw loose, not all there, touched (in the head), bonkers: *The doctors diagnosed him as psychotic and insisted he be hospitalized.* —*n.* **2** °madman, madwoman, °maniac, psychopath, lunatic, paranoid *or* paranoiac, schizophrenic, bedlamite, *Slang* nut, nutter, screwball, crackpot, °crazy, loony *or* looney *or* luny, schizo, *US* kook: *How many psychotics do you think are wandering about loose?*

pub *n.* public house, alehouse, tavern, inn, °bar, cocktail lounge, saloon, taproom, hostelry, *Brit* saloon *or* lounge bar, *US* barroom, *Colloq Brit* °local, *Slang* boozer, watering hole, °joint, °dive, *Chiefly Brit* gin palace, *US* barrelhouse, gin mill, honky-tonk: *We went down to the pub for some beer and a game of darts.*

puberty *n.* pubescence, sexual maturity, adolescence, juvenescence, teens; nubility: *Parents' anxiety over a child's health is nothing compared with their concern when the child reaches puberty.*

public *adj.* **1** communal, community, °common, °general, collective, °universal, °catholic, °popular, worldwide: *One need only compare early and new films to see how public taste has changed.* **2** °civil, civic, °social, societal, community, communal: *Although they may seem restrictive, these laws are for the public good.* **3** °accessible, °open, °free, unrestricted, nonexclusive, communal, community, °available: *He does much of his research in the public library.* **4** °open, °manifest, exposed, °overt, projected, °plain, °obvious, °apparent, °patent, °clear, clear-cut, acknowledged, known, admitted, °visible, viewable, °conspicuous: *Her public image is quite different from her private persona.* **5** °visible, viewable, °conspicuous, unconcealed, unshrouded, °flagrant, °blatant: *Has Alfie made a public spectacle of himself again?* **6** °well-known, °prominent, °eminent, °celebrated, °famous, °renowned, °noted, °notable, °influential, °illustrious; °notorious, °disreputable, °infamous: *The American gangsters of the Prohibition era became public folk heroes.* **7** ***make public.*** See **publish.** —*n.* **8** community, people (at large *or* in general), citizenry, citizens, °nation, °populace, °population, °society, masses, multitude, °hoi polloi, bourgeoisie, plebeians, proletariat, °rank and file, middle class, third estate, commonalty, voters, man *or* woman in the street, *Brit* admass, *US* John Q. Public, Mr. *or* Mrs. Average, *Colloq* (any *or* every) Tom, Dick, and Harry: *The public has responded generously to the charity appeal.* **9** clientele *or Brit also* clientage, customers, °custom, patrons, followers, supporters, buyers, consumers, purchasers, following, °business, °trade: *How can you expect to attract the public without advertising?* **10** sector, °segment, special-interest group, °portion: *The commuting public will no longer tolerate these excessive train delays.* **11** ***in public.*** publicly, °openly, in the open, *Colloq* out of the closet: *Jarvis has finally confessed in public what we long suspected privately.*

publication *n.* **1** dissemination, promulgation, publicizing, publishing, °proclamation, issuance, reporting, °announcement, °advertisement, advertising, °pronouncement, airing, putting out, °revelation, °declaration, °appearance: *The publication of the news about the sale of the company was no surprise to us.* **2** °book, booklet, °pamphlet, °brochure, °leaflet, broadside *or* broadsheet, flier *or* flyer, handbill, handout; °periodical, °magazine, °journal, newsletter, newspaper, °paper, broadsheet, tabloid; annual, semiannual, °quarterly, bimonthly, monthly, semimonthly, fortnightly, biweekly, weekly, hebdomadal *or* hebdomedary, semiweekly, °daily: *He buys every publication he can find dealing with model railways.*

publicize *v.* °promote, advertise, give publicity to, *Colloq* beat the drum for, °plug, °puff, *US* °push, *Slang* hype: *They once hired an elephant to publicize a huge department-store sale.*

publish *v.* °make public, °put out, °broadcast, spread (about *or* around), advertise, °make known, let (something) be known, °announce, °publicize, °report, °proclaim, promulgate, bruit about, °reveal, divulge, °disclose, break the news (about), *Colloq* °leak: *The minister resigned the day the news of the scandal was published.*

pucker *v.* **1** °gather, draw together, compress, °purse, crinkle, ruck, shirr, °ruffle, corrugate, °furrow, °wrinkle, crease, °screw up, °tighten, °contract, °squeeze: *Pucker up your mouth and give me a kiss. The shirring is done by puckering up the fabric and stitching it through with decorative thread.* —*n.* **2** °gather, tuck, pleat, pleating, shirr, shirring, °ruffle, ruck, ruche, ruckle, °wrinkle, wrinkling, °fold, crinkle, crinkling: *The curtains have some puckers that need ironing.*

puerile *adj.* °childish, °immature, babyish, °infantile, °juvenile, °silly, asinine, °foolish, trivial, °ridiculous, °irresponsible, °shallow, °inconsequential, °insignificant, *US* sophomoric: *Stop that puerile horseplay and settle down.*

puff *n.* **1** °blow, °breath, °wind, whiff, °draft, °gust, °blast, °huff: *A slight puff through the open window stirred the curtains.* **2** °draft, °draw, °pull, *Colloq* °drag: *She took a puff from her cigarette and blew the smoke in his face.* **3** °praise, commendation, favorable mention *or* review *or* notice, publicity, puffery, *Colloq* °plug, blurb, *Slang* hype: *The late editions of the papers carry puffs for your new book.* —*v.* **4** °blow, °breathe, °huff,

°pant, °gasp, wheeze: *Victor came huffing and puffing up to the finishing line.* **5** °draw, pull (at *or* on), °inhale, suck, smoke, *Colloq* °drag: *Holmes was puffing silently on his pipe.* **6** Usually, ***puff up*** *or* ***out.*** °inflate, distend, bloat, swell (up *or* out), °stretch, balloon, °expand, °pump up, °enlarge: *Cavendish swaggered in, all puffed up with pride as the new father of twins.* **7** °publicize, advertise, °promote, °push, trumpet, ballyhoo, °extol, commend, °praise, *Colloq* °plug, beat the drum (for): *He has been shamelessly puffing his book on every radio and TV show.*

pugilism *n.* boxing, prizefighting; the manly art of self-defense, fisticuffs; *Colloq* the boxing *or Colloq* fight game: *His career in pugilism came to an abrupt end in the second minute of the third round.*

pugilist *n.* boxer, prizefighter, fighter, contender, °contestant, battler, combatant, *Colloq* °bruiser, scrapper, champ, *Slang* slugger, pug: *The pugilists came out of their corners and at once started jabbing away at each other.*

pugnacious *adj.* °aggressive, °belligerent, combative, °quarrelsome, bellicose, antagonistic, °argumentative, °hostile, litigious, contentious, disputatious, °disagreeable, fractious, °petulant, °testy, irascible, hot-tempered, choleric, unfriendly, curmudgeonly, °irritable, °short-tempered: *The Gothic tribes were very pugnacious, always ready to fight with little or no provocation.*

puling *adj.* whining, wailing, °querulous, whimpering, sniveling, weeping, caterwauling: *The third-class carriage was filled with chickens, pigs, and puling infants.*

pull *v.* **1** °draw, °haul, °drag, °lug, °tow, °trail: *Do you think the car is strong enough to pull that load?* **2** °tug, °jerk, °yank, °wrench, °pluck: *He pulled suddenly on the door and it opened.* **3** Sometimes, ***pull out*** *or* ***up.*** °pluck (out), °withdraw, °extract, °uproot, pick (up *or* out), snatch out *or* up, tear *or* rip out *or* up, cull, °select, °draw out, °take out, °remove: *We pulled out all the weeds and threw them on the compost heap. He has a collection of jokes pulled from his speeches.* **4** Often, ***pull apart.*** tear *or* rip (up *or* apart), °rend, pull asunder, °wrench (apart), °stretch, °strain: *This fabric is so weak it pulled apart as soon as I touched it. I think I pulled a muscle in my calf.* **5** Often, ***pull in.*** °attract, °draw, °lure, °entice, allure, °catch, °captivate, °fascinate, °capture: *We need something besides the "Sale" sign to pull the customers into the shop.* **6** ***pull apart.*** pull to pieces *or* shreds, °criticize, °attack, pick *or* take apart *or* to pieces, flay, °run down, *Colloq* °put down, °pan, °knock, °devastate, °destroy, slate, *Slang* °slam: *The critics really pulled apart her new play.* **7** ***pull away.*** °withdraw, draw *or* drive *or* go *or* move away; outrun, outpace, draw ahead of: *She pulled away abruptly when he touched her hand. The green car is pulling away from the others.* **8** ***pull back.*** **(a)** °withdraw, °draw back, back off *or* away, °recoil, shrink (away *or* back) from, °shy, °flinch (from), °jump, °start: *Cadogan pulled back when he saw the ferocious dog.* **(b)** °withdraw, °(beat a) retreat, °take flight, °flee, °turn tail, drop *or* fall back, back out: *We cheered when we saw the enemy troops pulling back.* **9** ***pull down.*** **(a)** °demolish, °raze, °level, °destroy, °wreck: *It takes only minutes to pull down a house that it has taken generations to build.* **(b)** °draw, °receive, °get, be paid, °earn: *He pulls down much more at his new job.* **(c)** °lower, °debase, °diminish, °reduce, °degrade, °dishonor, °disgrace, °discredit, °humiliate: *When his fortunes declined, he pulled down all his friends with him.* **10** ***pull for.*** hope *or* pray for, be enthusiastic for, be supportive of, °support, campaign for, cheer for, °encourage, °boost, *US* root for: *We are all pulling for Aylesbury to win the game. When Collings went into hospital, we all pulled for his quick recovery.* **11** ***pull in.*** **(a)** drive up, °arrive, °come, draw up *or* in, °reach: *The train finally pulled in at midnight. We need gas, so pull in at the next filling station.* **(b)** °arrest, apprehend, take into custody, *Colloq* °pinch, °nab, collar, °nail, *Brit* °nick, *Slang* bust: *The cops pulled him in for possession of narcotics.* **12** ***pull off.*** **(a)** °detach, rip *or* tear off, °separate, wrench off *or* away: *When he was cashiered from the army, they pulled off all his insignia and medals.* **(b)** °accomplish, do, °complete, °succeed, °carry out, °bring off, °manage, °perform: *Three men pulled off the robbery in broad daylight.* **13** ***pull oneself together.*** °recover, get a grip on oneself, get over it, °recuperate, *Colloq* °snap out of it, buck up: *Try to pull yourself together and stop crying.* **14** ***pull out.*** **(a)** °uproot, °extract, °withdraw: *In the ensuing battle royal, someone tried to pull out his hair. Two survivors were pulled out of the rubble.* **(b)** °withdraw, °retreat, beat a retreat, °recede, °draw back, °leave, °depart, go *or* run away *or* off, °evacuate, *Colloq* °beat it, do a bunk, *Brit* do a moonlight flit: *The artillery unit pulled out yesterday.* **(c)** °leave, °go, °depart, °take off: *When that train pulls out, I want you on it!* **(d)** °withdraw, °quit, °abandon, °resign (from), °give up, °relinquish: *You can still pull out of the deal if you want to.* **15** ***pull someone's leg.*** °tease, °chaff, rib, °have on, °rag, °twit, °poke fun at, °make fun of, °hoodwink, °ridicule: *He said that I'd just eaten a fly, but he was pulling my leg.* **16** ***pull strings.*** use influence *or* connections, *US* use pull, °pull wires: *His uncle pulled strings to get him the job.* **17** ***pull through.*** °survive, °recover, °improve, get better, get over (it *or* some affliction), °rally; °live: *Murphy was at death's door, but luckily he pulled through.* **18** ***pull up.*** **(a)** °stop, °halt, come to a standstill: *We pulled up alongside the road for a few minutes' rest.* **(b)** °uproot, °root out, °dig out, deracinate, eradicate: *Your dog has pulled up all the flowers in my garden.* **(c)** draw even *or* level with, come up to, °reach: *On the fifth lap, Manson pulled up to, then passed Sabbatini.* —*n.* **19** °draw, °tug; °yank, °jerk: *Give the bellrope a strong, steady pull, and try not to yank it suddenly.* **20** °attraction, °draw, °magnetism, °appeal, drawing *or* pulling power, seductiveness, seduction, °lure: *The pull that golf has on certain people is hard to explain.* **21** °influence, °authority, °connections, °prestige, °weight, leverage, *Colloq* clout, muscle: *You'd better treat her nicely, for she has a lot of pull with the boss.* **22** °puff, °draw, inhalation, *Colloq* °drag: *He took a long, meditative pull on his cigarette and blew some smoke rings.*

pulley *n.* sheave, °block: *The rope is run through a system of pulleys, called a tackle, for lifting weights.*

pulp *n.* **1** marrow, °pith, °heart, soft part, °flesh: *After removing the seeds, add two tablespoonfuls of sugar and mix into the pulp.* **2** mush, paste, mash, pap, pomace, °mass, *Technical* triturate: *Water is added to the waste paper, which the machines then reduce to a pulp.* —*v.* **3** mash, squash, °pulverize, °destroy, *Technical* levigate, triturate: *The damaged books were pulped when she threatened to sue the publisher.* —*adj.* **4** °lurid, °sensational, trashy, °cheap: *What he considers great literature you would call pulp fiction.*

pulsate *v.* °beat, °pulse, throb, °pound, thrum, drum, thump, °thud, reverberate, hammer, palpitate, °vibrate; °oscillate, °quiver: *Throughout the voyage, I could feel the ship's engines pulsating.*

pulse *n.* **1** °beat, beating, throb, throbbing, pulsing, pulsating, pulsation, pounding, thrumming, drumming, °thumping, thudding, reverberation, reverberating, hammering, palpitation, palpitating, vibration, vibrating: *The pulse of the jungle drums became louder and faster.* —*v.* **2** See **pulsate,** above.

pulverize *v.* **1** °powder, comminute, °grind, °crush, °mill, granulate, °crumble, °break up, bray, °pound, *Technical* triturate, levigate: *This machine pulverizes the rock, after which the binding agents are added.* **2** °devastate, °destroy, °demolish, °crush, smash, °shatter, °ruin, °wreck, annihilate: *The four battalions were pulverized in the attack.*

pump *v.* **1** °send, °force, °deliver, °push: *The heart pumps blood around the body.* **2** interrogate, °question, °examine, cross-examine, °quiz, °probe, *Colloq* grill, give (someone) the third degree: *They pumped her for hours, but she told them nothing about my whereabouts.* **3** ***pump out.*** pump dry *or* empty, °empty, °drain, bail out, draw *or* drive *or* force out, siphon (out): *After the flood it took two days to pump out my basement.* **4** ***pump up.*** **(a)** °inflate, °blow up; dilate, °swell, bloat, °expand, puff out *or* up: *If you don't repair the tire, pumping it up will accomplish nothing.* **(b)** °excite, °inspire, °stimulate, °animate, inspirit, °electrify, galvanize, °energize, °motivate, *Colloq* enthuse: *The coach is talking to the team, trying to pump them up before the big game.* **(c)** °intensify, °concentrate, °emphasize, °stress, °increase: *We ought to pump up our promotion campaign just before Christmas.*

pun *n.* play on words, °quip, (bon) mot, °witticism, double entendre, *Literary* equivoque, *Technical* paronomasia: *Dennis made some awful pun on "nose" and "knows" that nobody got.*

punch[1] *v.* **1** °hit, °clip, °jab, whack, thwack, °knock, smack, °box, pummel, °strike, cuff, *Colloq* clout, bop, slug, wallop, thump, lambaste *or* lambast, °slam, *Slang* sock, biff, °plug, °belt, °lace (into), *US* paste: *Anyone says anything about my girl, I'll punch him in the jaw!* —*n.* **2** °clip, °jab, whack, thwack, °knock, smack, °box, cuff, uppercut, left *or* right, *Colloq* clout, bop, slug, wallop, thump, °slam, *Slang* sock, °belt, biff, haymaker, °plug, paste: *The punch knocked him down.* **3** °effect, °impact, effectiveness, °force, forcefulness, °power, °vitality, °gusto, °vigor, °life, vim, °zest, ginger, *Colloq* it, oomph, what it takes, *Slang* zing, zip: *These advertisements are pretty, but they are lacking in punch.*

punch[2] *n.* **1** awl, auger, bodkin, perforator; °drill, brace and bit: *Use the punch to make another hole in your belt.* —*v.* **2** °pierce, °stab, °puncture, °perforate; °bore, °drill: *She uses a needle to punch a tiny hole in each end of the egg.*

punctual *adj.* °on time, °timely, °prompt, *Colloq* °on the dot: *Please be punctual, as I don't fancy waiting about in the street.*

punctuate *v.* **1** °interrupt, °break, intersperse; °pepper, sprinkle: *The speeches were punctuated by frequent shouts from the audience.* **2** °accent, accentuate, underline, underscore, °emphasize, °stress, °mark: *He punctuated each element of his argument with a sharp rap of his pencil on the lectern.*

puncture *n.* **1** °hole, perforation, °opening, °leak; °flat (tire): *I haven't the tools needed to patch the puncture in the tire.* **2** perforation, perforating, holing, puncturing, °piercing, stabbing, punching: *The puncture of the balloon means that we shall have to postpone our trip.* —*v.* **3** °perforate, °hole, °pierce, °stab, °penetrate, °go through, °prick, °nick, °rupture: *An arrow punctured the side of the tent.* **4** deflate, °disillusion, bring up short, °discourage, °humble, °dash, °destroy, °ruin: *Any hopes I had of winning the marathon were punctured when I tripped and fell, spraining my ankle.*

pungent *adj.* **1** °spicy, °hot, °sharp, °strong, °penetrating, °aromatic, °seasoned, peppery, piquant, tangy, flavorful, °tasty, sapid: *The pungent odors of stuffed roasting turkey assailed my nostrils.* **2** °sharp, °biting, stinging, °caustic, °severe, astringent, °stern, acrid, °harsh, °sour, acid, °tart, acrimonious, °bitter, °cutting, °keen, barbed, °trenchant, °scathing, °incisive, mordant, °sarcastic: *This week's* Review *carries a pungent criticism of MacReady's new novel.* **3** distressing, upsetting, °poignant, °painful, °hurtful, °penetrating, °piercing, stabbing, °intense, °severe, °acute, °agonizing, °oppressive, °excruciating, racking, consuming: *Only when surgery eased the pressure was he relieved of the pungent pain.*

punish *v.* **1** °penalize, °chastise, °castigate, °discipline, °chasten, °scold, °rebuke, °take to task, °reprove, °dress down, admonish, °correct, teach someone a lesson, give a lesson to, throw the book at, rap someone's knuckles, slap someone's wrist, have *or US* call on the carpet, *Colloq* take it out on (someone): *A child's first taste of injustice is being punished for something she did not do.* **2** °imprison, jail *or Brit also* gaol, incarcerate, °lock up; °fine, mulct, amerce; °lash, °flog, °beat, °scourge, °spank, °whip, cane, birch, put across *or US also* turn over (one's) knee, *US* °paddle; pillory, crucify; tar and feather, °exile, °banish, excommunicate, cashier, drum out of the corps; °hang, °execute, electrocute, draw and quarter, send to the gas chamber, *Colloq* °put away, *Slang Brit* °send down, *US* °send up: *They were punished—justly, some say—for their crimes.* **3** °hurt, °harm, °injure, °damage, °abuse, maltreat, °rough up, knock about *or* around, maul, thrash, °beat, trounce, °manhandle, °batter, *Slang* beat up: *The challenger punished the champion so badly that he could not come out for the tenth round.*

punishing *adj.* grueling, °hard, °arduous, °strenuous, °laborious, °exhausting, tiring, wearying, fatiguing, wearing, taxing, °demanding, °burdensome, backbreaking, torturous: *Laying railway track was the most punishing job I ever had. This punishing schedule is beginning to tell on me.*

punishment *n.* **1** chastisement, chastising, castigation, castigating, °discipline, disciplining, chastening, scolding, °rebuke, °reproof, dressing-down, admonishment, admonition, °correction, punitive measures: *Every society must continually revise its approach to the punishment of criminals.* **2** °penance, °penalty, °sentence, sentencing, just deserts; °imprisonment, incarceration, jailing *or Brit also* gaoling; lashing, flogging, beating, °whipping, scourging, °spanking, caning, birching, *US* paddling; °exile, banishment, excommunication, cashiering; hanging, °execution, electrocution, drawing and quartering: *Her punishment is to perform twenty hours a week of public service for a year. Stop whimpering and take your punishment like a man.* **3** °injury, °harm, °damage, °abuse, maltreatment, mauling, beating, °thrashing, trouncing, manhandling, battering, torture: *How are professional wrestlers able to withstand so much punishment?*

punitive *adj.* chastening, castigatory, disciplinary, retributive, punitory, retaliatory, correctional: *A punitive expedition was sent there to quell the colonists' uprising.*

punk *n.* **1** ruffian, °hoodlum, hooligan, °delinquent, °tough, °thug, vandal, °yahoo, °barbarian, *Colloq* goon, °mug: *A couple of young punks were standing at the bar, trying to look important.* —*adj.* **2** °inferior, °rotten, unimportant, °worthless, °bad, °poor, °awful, *Colloq* °lousy: *I thought that the lead in the play was a punk actor.*

puny *adj.* **1** °small, °little, °insignificant, °petty, unimportant, °inconsequential, °paltry, trivial, °trifling, °minor, °negligible, nugatory, of little *or* no account, °inferior, °worthless, °useless, *Colloq* piddling, *Slang Brit* not much cop: *Farr's contribution to the success of the business was really quite puny.* **2** °small, °little,

°diminutive, °tiny, °minute: *Mattson looks so puny alongside that sumo wrestler.* **3** °weak, °feeble, °frail, °sickly, weakly, underfed, undernourished, °undersized, underdeveloped, °stunted, °dwarf, midget, pygmy: *You cannot expect such a puny chap to lift such a great weight.*

pup *n.* puppy, whelp, °upstart, whippersnapper, popinjay, cub, jackanapes, °showoff, °braggart: *That young pup had the gall to tell me I looked ancient.*

pupil *n.* °student, °learner, °scholar, schoolchild, schoolgirl, schoolboy, °disciple, °apprentice; beginner, °novice, neophyte, tyro *or* tiro, *Chiefly ecclesiastical* catechumen: *A number of Geoffrey's former pupils have organized a dinner to honor him.*

puppet *n.* **1** hand puppet, finger puppet, glove puppet, doll; marionette, string puppet: *The children built a little theater and put on a puppet show.* **2** °figurehead, cat's-paw, °pawn, °dupe, °tool, hireling, yes man *Colloq* front (man), *Slang* stooge, °sucker, patsy: *He is without authority and is merely a puppet of the drug dealers. The army of occupation set up a puppet government to rule the country.*

purchase *v.* **1** °buy, °acquire, °procure, °obtain, °get, °secure, °pay for: *He recently purchased a new personal computer and cannot get it to work. I can remember when you could purchase a loaf of bread for ten cents.* **2** °win, °gain, °achieve, °realize, attain, °obtain: *His liberty was purchased by the betrayal of his colleagues.* —*n.* **3** °acquisition, acquiring, buying, purchasing, obtaining, securing, procurement: *The purchase of books for the university library is the responsibility of Monica Turnbull.* **4** °buy, °acquisition: *Tom returned from the shop carrying his purchases.* **5** °grip, °hold, °support, toehold, foothold, °grasp; leverage, °position, °advantage, °edge: *The climber was having trouble getting a good purchase on the icy ledge.*

pure *adj.* **1** unmixed, unadulterated, unalloyed, simon-pure; 24-karat, °sterling, °solid; °real, °genuine, °authentic, °flawless, °faultless, °perfect, °natural, °true, °simple: *We have made every effort to keep the bloodline of these horses completely pure. Pure gold is not used for coins or jewelry because it is too soft. She wore a dress of the purest white.* **2** uncontaminated, °clear, °clean, °wholesome, °sanitary, uninfected, disinfected, pasteurized, sterilized, °sterile, antiseptic, unpolluted, °spotless, °immaculate, unsullied, unbesmirched, unblemished, unmarred, unstained, untainted: *It was refreshing to breathe pure air again. Stringent tests ensure that the substances are pure before they go on the market.* **3** °chaste, virginal, virgin, °intact, maidenly, vestal, undefiled, °innocent, guileless, °virtuous, °modest, °moral, °correct, °proper, °decent, °decorous, uncorrupted, °blameless, sinless, °impeccable: *As far as we can tell, she led a totally pure life.* **4** °theoretical, °hypothetical, conjectural, °speculative, °abstract, conceptual, notional, °philosophical, academic(al): *His field is pure science, and he cares little how the results of his work might be applied to everyday life.* **5** unalloyed, °simple, °unmitigated, °sheer, utter, °absolute, °unqualified, °complete, °total, °perfect, °thorough, °outright, °downright, °out-and-out, °mere: *Many regarded the notion that man could ever fly as pure nonsense. Any resemblance to a living person is pure coincidence.* **6** °honorable, °(highly) principled, °righteous, °upright, °honest, straightforward, high-minded, °pious, °worthy, °good, °ethical, °virtuous, °sincere, above suspicion, above reproach, like Caesar's wife: *Can anyone doubt that their motives in striving for international peace are pure?*

purebred *adj.* full-blooded, thoroughbred, pedigreed: *We raise only purebred Angus cattle.*

purgative *n.* **1** laxative, cathartic, aperient, °purge, physic, depurative: *The doctor prescribed a mild purgative, and I felt much better the next day.* —*adj.* **2** laxative, cathartic, aperient, evacuant, diuretic, depurative; abstergent: *We were totally unprepared for the purgative effects of the spicy food.*

purge *v.* **1** °cleanse, °purify, °clean (out), °clear, °scour (out), depurate, deterge, °wash (out): *The tanks must be purged before filling them with the new fuel.* **2** °eject, eradicate, °expel, °eliminate, °get rid of, °dismiss, clear out *or* away, sweep away *or* out, oust, °remove, rout out, weed out, °root out; do away with, °exterminate, liquidate, °kill, °destroy: *After the coup, the old guard was purged from all the governmental agencies. The rebels purged the entire council of elders.* **3** °clear, exonerate, absolve, °forgive, °purify, °pardon, exculpate: *He was given every opportunity to purge himself of the charge of contempt of court.* —*n.* **4** °ejection, eradication, °expulsion, elimination, °dismissal, clearing out *or* away, ousting, ouster, °removal, routing out, weeding out, rooting out, unseating; defenestration; extermination, liquidation, °killing, °murder, °slaughter: *The bloody purge of dissidents was carried out by the secret police.* **5** See **purgative, 1,** above.

purify *v.* **1** °cleanse, °clean, °clarify, °wash, sanitize, depurate, decontaminate, °freshen, °disinfect; °fumigate: *The water is then purified by passing it through sand.* **2** exonerate, exculpate, absolve, °clear, °redeem, shrive, lustrate, acquit, °pardon, °forgive, °excuse: *Before the aspirants can be admitted to the church they must be purified.*

purist *n.* pedant, precisian, formalist, stickler, bluestocking, dogmatist, °pharisee, °fanatic, *Colloq* diehard, stuffed shirt: *Language purists are often totally unaware of their own grammatical and semantic blunders.*

puritan *n.* **1** moralist, pietist, religionist, °fanatic, °zealot, °purist: *All religions seem to have their puritans.* —*adj.* **2** °prudish, puritanical, prim, °proper, °straitlaced, ascetic, austere, moralistic, pietistic, °intolerant, disapproving, °bigoted, °narrow-minded, °stuffy, stiffnecked, °rigid, uncompromising, hardline, °stern, °severe, °strict, *Colloq* uptight, hard-nosed: *Puritan notions about women's clothing set fashions back by fifty years.*

purity *n.* **1** pureness, faultlessness, correctness, flawlessness, °perfection, spotlessness, cleanness, cleanliness, °clarity; healthfulness, wholesomeness, salubrity; innocuousness, harmlessness: *The purity of the drinking water has been affected by the nearby chemical plant.* **2** °chastity, chasteness, virginity, virtuousness, °virtue, °morality, °propriety, °honesty, °integrity, °rectitude, properness, innocence, guilelessness, decency, decorousness, modesty, blamelessness, sinlessness: *We never had any reason to doubt the purity of their character.*

purpose *n.* **1** °object, °intention, °intent, °end, °goal, °ambition, °objective, °target, °aim, °principle, °point, °rationale, °reason; °scheme, °plan, °design, °motive, motivation, °view: *What purpose was served by Tweed's confession? If your purpose in inviting me was to insult me, I shall leave now. I fail to understand Laura's purpose in warning me against David.* **2** °resolution, firmness, °determination, °persistence, °drive, single-mindedness, deliberation, deliberateness, purposefulness, steadfastness, °tenacity, doggedness, °will, °resolve, resoluteness, °perseverance, stubbornness: *She admired the strong purpose behind his refusal to compromise with quality.* **3** °use, practicality, avail, °effect, utility, °usefulness, °outcome, °result; °advantage, °profit, °gain, °good, °benefit: *To what purpose are you planning to put this knowledge? I cannot see the purpose in pursu-*

ing this line of questioning. **4 *on purpose.*** **(a)** purposely, intentionally, °deliberately, willfully, by design, consciously, knowingly, designedly, wittingly: *That was no accident—you kicked me on purpose!* **(b)** °especially, °specially, °expressly, °exactly, °precisely, specifically, °in particular: *He kicked me on purpose, Mom, just to make me cry!* —*v.* **5** °plan, °intend, °design, °resolve, °mean, °aim, have in mind *or* view, have a mind, °propose, °consider, °contemplate, °aspire, °long, °yearn: *They purpose to open an office of their own.*

purposeful *adj.* °intentional, intended, planned, °deliberate, resolved, settled, °determined, °resolute, °decided, confirmed, affirmed, °sure, °certain, °positive, °definite, °staunch, °steadfast, °persistent, strong-willed, dogged, °tenacious, pertinacious, unfailing, unfaltering, °firm, °fixed: *Gillian's purposeful manner inclines others to defer to her.*

purposeless *adj.* °pointless, °bootless, °meaningless, °empty, vacuous, °senseless, °aimless, °rambling, °discursive, wandering, disorganized, unorganized: *He made what we thought was a purposeless trip to Cuernavaca, where he died.*

purse *n.* **1** °pouch, (money-)bag, °wallet, °pocket, *Dialect* °poke, *Highland dress* sporran, *US and Canadian* (woman's) handbag, pocketbook, °bag: *The money goes out so fast I think there is a hole in my purse.* **2** °money, °wealth, °resources, °funds, °finances, exchequer, °treasury, °capital, °revenue, °income, °means, °cash, °riches; dollars, shekels, *Derogatory* mammon, pelf, (filthy) lucre, *US* almighty dollar, *Slang* ready *or* readies, *US and Canadian* °scratch, *Brit* lolly: *The spread of capitalism shows that the power of a man's purse can transcend his political idealism.* **3** °prize, °reward, °award, °present, °gift: *The purse for the next race will be five thousand dollars.* —*v.* **4** °pucker (up), °contract, °wrinkle, compress, press together: *Robert pursed his lips thoughtfully.*

pursue *v.* **1** °follow, °chase, go *or* run after, hunt (after *or* down *or* for *or* up), °trace, °trail, °track, °run down, take off after, dog, °stalk, °shadow, *Brit* chivy *or* chivvy *or* chevy, *Colloq* °tail: *The police pursued the escaped convicts across the moor.* **2** follow (up *or* on with), °trace, carry on with, °continue, °conduct, devote *or* dedicate oneself to, °cultivate, °undertake, practise, persist *or* persevere in, °maintain, °exercise, °proceed with, adhere to, stay with, apply oneself to, *Colloq* °stick with: *If you pursue such a course, you are certain to land in jail.* **3** °aspire to, aim for, work for *or* toward(s), try *or* strive for, °purpose, °seek, °search for, go in search of, quest after *or* for, be intent on, be bent upon *or* on: *Jonathan is pursuing a doctorate in history at Oxford.* **4** woo, (pay) court (to), seek the hand of, *Formal* press (one's) suit with, pay suit *or* court *or* (one's) addresses to, *Colloq* set one's cap for, *Slang* °chase (after): *Jack is pursuing Jacquelyn with matrimony in mind.*

pursuit *n.* **1** pursuing, chasing, following, hunting, °hunt, going *or* running after, tracing, trailing, tracking, running down, dogging, stalking, shadowing, *Brit* chivy *or* chivvy *or* chevy, *Colloq* tailing: *We all joined in the pursuit of the banditti who stole Terry's ice cream.* **2** pursuance, striving after, seeking, searching, °search, looking for: *Many think that the pursuit of wealth is the only worthwhile thing in life.* **3** °work, °line (of work), °employment, °field, °area, °specialty, specialization, °business, °profession, °trade, °vocation, °calling, °career, lifework, °activity; °hobby, °pastime, avocation, °interest; *Slang* °racket: *What pursuits are open to Jack Ketch now that hanging has been abolished?*

push *v.* **1** °thrust, shove, °drive, °move, set in motion, get moving, °propel; °press: *Push the boat away from the pier. He pushed the door shut.* **2** °press, °depress: *Push the button if you want anything.* **3** shove, °thrust, elbow, °shoulder, °force, jostle, °nudge: *We pushed our way through the crowd.* **4** °urge, °encourage, °press, °induce, °ask, °persuade, °get, egg on, °press, °prod, °spur, goad, °rouse, °prompt, °incite, °move, °motivate, °stimulate, °influence, impel, °make, compel, °force, dragoon, coerce, constrain; badger, °hound, °pester, °harass, °plague, °nag, °browbeat; °beg, importune, entreat: *His family is pushing him to find work. I'll soon have the work done, but stop pushing me.* **5** °force, °strain, overstrain, °tax, overtax, °burden, overburden: *You might have got away with it this time, but don't push your luck.* **6** °promote, °publicize, advertise, °boost, propagandize, °puff, *Colloq* °plug, *Slang* ballyhoo, hype: *Because we have too much stock, we are pushing suntan oil this week.* **7 *push about*** *or* ***around.*** °intimidate, °bully, cow, domineer, °tyrannize, bullyrag, °torment, °force, coerce: *Don't push people about if you want them to do what you want, just ask politely.* **8 *push away.*** °reject, °deny, brush off *or* aside, give (someone) the cold shoulder, °rebuff, distance oneself from: *She pushes everyone away and then wonders why she has no friends.* **9 *push off.*** **(a)** shove off, sail away *or* off: *We stepped into the dinghy and pushed off.* **(b)** °leave, °depart, °go away, *Colloq* light out, °take off, hit the road, skedaddle, scram, make oneself scarce, *Slang* °beat it: *I have to push off now if I am to catch that train.* **10 *push on*** *or* ***forward*** *or* ***along.*** move onward(s) *or* ahead *or* forward, °continue, °proceed, °advance, press on *or* onward(s): *If we do stay the night, we'll have to push on first thing in the morning.* **11 *push through.*** °force, °press, *Colloq* °railroad: *The bill was pushed through committee without delay.* **12 *push (up) daisies.*** be dead (and buried), be six feet under: *Her husband's pushing up daisies.* —*n.* **13** shove, °thrust, °nudge: *My car needs a push to get it started.* **14** °effort, °energy, °dynamism, °drive, °force, °enterprise, °ambition, °vigor, °determination, °initiative, °eagerness, °spirit, °enthusiasm, zeal, °verve, *Colloq* get-up-and-go, zing, zip, °gumption, °go: *If he had a bit more push, he might succeed.* **15** °campaign, °attack, °assault, °advance, °offensive, °charge, onslaught, foray, sortie, °invasion, incursion, °raid, sally, blitzkrieg, blitz, °strike: *The troops were massed along the coast, ready for the big push inland.* **16 *the push.*** °dismissal, °notice, *Colloq* one's marching orders, *Chiefly Brit* °the sack, the boot, *Chiefly US and Canadian* one's walking papers, a pink slip: *He'd been with the company for twenty years, but they gave him the push anyway.*

pushover *n.* **1** sure thing, *Colloq* °piece of cake, child's play, °snap, cinch, °picnic, walkover, *US* °breeze, *Slang* cinch, *Brit* doddle, *US* lead-pipe cinch: *The Latin exam was a pushover.* **2** *Colloq* walkover, chump, soft touch, soft *or* easy mark, easy prey *or* game, *Slang* patsy, °sucker, stooge, °sap, *Brit* °mug, *US and Canadian* milquetoast: *I spotted him at once as a pushover for the Spanish handkerchief con.*

pushy *adj.* °forward, (self-)assertive, °forceful, °aggressive, °obnoxious, °arrogant, bumptious, °brassy, °brazen, °impertinent, °insolent, pushing, °presumptuous, °officious, °loud, °showy, °cocky, °brash, °offensive, °bold, *Colloq* °cheeky: *I hate pushy salesmen who won't take "No" for an answer.*

pussyfoot *v.* **1** °sneak, °creep, °slink, °prowl, °steal, tiptoe: *I wish she'd stop pussyfooting about the house.* **2** beat about *or* around the bush, hem *or* hum and haw, °equivocate, °hesitate, be evasive, evade the issue, prevaricate, tergiversate, be noncommittal, be *or* sit on the fence, °blow hot and cold: *Don't pussyfoot with me—just say whether you'll come or not.*

put *v.* **1** °place, °position, °situate, °set, °lay, °station, °stand, °deposit, °rest, °settle; °locate: *Please put all books where they belong.* **2** °assign, °commit, °cause, °set, consign, °subject: *I joined the army to fight for my country, and they put me to work peeling potatoes.* **3** °subject, cause to experience *or* undergo *or* suffer, consign, °send: *He was put to death for his beliefs, not his actions.* **4** °express, °word, °phrase; °say, utter, °write : *How did they put it?—"We hold these truths to be self-evident, that all men are created equal. . ." To put it to you straight, shape up or ship out! Can you put that into French?* **5** °offer, °advance, bring forward, °present, °propose, °submit, °tender, set before: *I put it to you that you are lying when you say that you have never seen this woman before.* **6** °bet, °gamble, wager, °stake, °play, °chance, °risk, °hazard: *I put two dollars on the horse to win.* **7** °throw, °heave, °toss, °fling, °cast, °pitch, °bowl, °lob, °send, °shy, °shoot, °snap, catapult: *How far did you say you can put a ten-pound shot?* **8** °place, °assign, °attribute, °lay, °pin, °attach, °fix: *They try to put the blame on me for everything that goes wrong around here.* **9** ***put about.*** °broadcast, °publish, °make known, °publicize, °announce, spread about *or* around: *They have been putting about some story linking the two of us romantically.* **10** ***put across*** *or* ***over.*** make clear, °get across, make (something *or* oneself) understood *or* understandable, °explain, °spell out, convey, °communicate: *Thea is quite adept at putting across her ideas.* **11** ***put aside.*** set *or* lay aside, °ignore, °disregard, pay no heed to, push aside, shrug off: *They put aside their differences and discovered they had much in common. Isn't it time you put aside childish things?* **12** ***put aside*** *or* ***by*** *or* ***away.*** lay aside *or* by, °set aside; °save, °store, stow *or* store *or* salt *or* squirrel away, lay away, °cache, bank: *You ought to put a little money aside for a rainy day.* **13** ***put away.*** **(a)** See **12,** above. **(b)** jail *or Brit also* gaol, °imprison, incarcerate, *Colloq* °send, °jug, confine, °commit, institutionalize, remand, *Brit* °send down, *US* °send up: *She wants rapists and child molesters put away for a long, long time.* **(c)** See **16 (d),** below. **(d)** °consume, °gorge, gormandize *or US also* gourmandize: *It is unbelievable how much that man can put away at one sitting.* **14** ***put back.*** °return, °replace, °restore: *He insists that he put back all the money he took. Put the pictures back where they belong.* **15** ***put by.*** See **12,** above. **16** ***put down.*** **(a)** °record, °register, °write down, °set down, °enter, °list; log, °note (down), °jot down, make a note *or* notation of: *You may put me down for a contribution of five dollars. Please put down your name and address in the box provided.* **(b)** depose, put an end to, °topple, °crush, °overthrow, °subdue, °suppress, °check, °quash, °quell: *The loyalists put down the rebellion in short order.* **(c)** ascribe, °assign, °attribute: *The doctor put my feelings of fatigue and listlessness down to anemia.* **(d)** °kill, °exterminate, °destroy, put to death, put to sleep, °put away, do away with: *We had to have our cat put down, which was very upsetting for us all.* **(e)** abash, °humiliate, °crush, °silence, °mortify, °lower, take down (a peg *or* a notch), °shame, snub, deflate, °slight, °reject, °dismiss: *Head waiters have a way of putting people down if they behave presumptuously.* **(f)** take for, °reckon, °account, °count, °categorize, °regard: *Don't put Martin down as a dunce just because he acts the fool now and then.* **(g)** °belittle, °diminish, °disparage, deprecate, °depreciate, °criticize, disdain, °look down on, °despise, contemn: *They deny being prejudiced, but I see how they put down everyone who isn't exactly like them.* **17** ***put forth.*** **(a)** °propose, °offer, °set forth, °advance: *Several theories have been put forth on the function of mitochondrial DNA.* **(b)** °grow, °produce, send out *or* forth, bud, °flower: *When the plant puts forth suckers, they must be pinched back.* **(c)** °begin, °set out, °set forth, °start: *We put forth on our voyage with great enthusiasm.* **(d)** promulgate, °issue, °publish, °make known, °make public: *The committee has not put forth any suggestions for improving traffic flow.* **18** ***put forward.*** **(a)** °propose, °present, °tender, °nominate, °name; °recommend: *Your name has been put forward for next year's chairmanship.* **(b)** °suggest, °offer, °propose, °set forth, °put forth, °present, °submit, °tender, proffer, °introduce, °advance, °propound, °air, °make known, °announce: *No new solutions to the problem were put forward.* **19** ***put in.*** **(a)** °insert, °introduce: *Why did you put in a reference to your mother?* **(b)** °spend, °devote: *During the holiday season I plan to put in a lot of time to reading.* **(c)** °make: *So, you finally decided to put in an appearance.* **20** ***put in for.*** **(a)** apply for, °request, °ask for, petition for: *MacTavish has put in for a transfer twice and has been refused both times.* **(b)** °seek, apply for, °pursue, °file: *Do you really think that Jamieson will put in for reelection?* **21** ***put off.*** **(a)** °postpone, °delay, °defer, °put back, °stay, °hold off, °shelve, put *or* set aside, *Chiefly US* °put over, °table, put on hold: *Consideration of your request has been put off till tomorrow.* **(b)** °dismiss, °get rid of, send away, turn away; °discourage, *Colloq* give (someone) the brushoff, *US* °brush off: *The manager put me off when I went in to ask about a salary increase.* **(c)** °dismay, disconcert, °upset, °confuse, °disturb, °perturb, abash, °distress, *Colloq* °throw, °rattle: *Curiously, Michael didn't seem a bit put off to learn about the murder.* **(d)** °repel, °disgust, °sicken, °revolt, °nauseate: *That television program on open-heart surgery really put me off.* **(e)** °leave, °depart, °go (off), °set off: *We put off tomorrow morning for Tasmania.* **22** ***put on.*** **(a)** don, clothe *or* attire *or* dress (oneself) in, get dressed in, change *or* slip into: *We put on our best clothes to come to your party.* **(b)** °assume, °take on, °pretend, °affect, feign, °bluff, make a show of: *She puts on that air of indifference, but she really does care.* **(c)** °add, °gain: *Bill put on quite a bit of weight when he stopped smoking.* **(d)** °stage, °mount, °produce, °present, °perform: *Our local repertory group is putting on* Hedda Gabler *this weekend.* **(e)** °tease, °mock, *Colloq* kid, pull (someone's) leg, rib, °rag, *Brit* °have on: *When he told you that all holidays had been canceled he was only putting you on.* **23** ***put out.*** **(a)** °inconvenience, discommode, °disturb, °embarrass, °trouble, °bother, impose upon *or* on, *Colloq* put on the spot: *Her arriving an hour early didn't put me out one bit.* **(b)** °annoy, vex, °irritate, °anger, °exasperate, °irk, °perturb, °provoke, *Slang* °bug: *Donald was very much put out to discover that the appointment had gone to someone else.* **(c)** snuff out, °extinguish, °blow out, douse, °quench, °smother: *We put out the fire on the stove before it could spread.* **(d)** °exert, °put forth, °expend, °use, °exercise: *We put out considerable effort moving the furniture before she decided she liked it the way it had been.* **(e)** °publish, °issue, °broadcast, °make public, °circulate, °spread, °make known, °release: *The information office put out the news that the summit meeting had been a great success.* **24** ***put out of* (*one's*) *misery.*** °release, °relieve, °free, °deliver, °rescue, °save, °spare: *Just tell him what they said and put him out of his misery.* **25** ***put over.*** **(a)** put *or* get across, convey, °communicate, set *or* put forth, °relate: *I am trying to put the point over to you that I cannot leave now.* **(b)** See **21 (a),** above. **26** ***put* (*one or something*) *over on* (*someone*).** °fool (someone), pull (someone's) leg, °deceive (someone), °mislead (someone), pull the wool over (someone's) eyes, °trick (someone), °hoodwink (someone): *She really put one over on me when she said she loved me. I think that Manfred is trying to put something over on you.* **27** ***put through.*** **(a)** carry out *or* through, °execute, °(put into) effect, °bring off, °accomplish, °complete, °finish, conclude, *Colloq* °pull off: *If they succeed in putting through* *their plan, the company will be bankrupt.* **(b)** °process, °handle, °organize, °see to, °follow up on: *I will personally see that your application is put through the proper channels.* **(c)** °connect, hook up: *I asked the operator to put me through to the chairman himself.* **28** ***put up.*** **(a)** °erect, °build, °construct, °raise, °set up, put together, °fabricate: *Unbelievably, they demolished some lovely old Georgian houses and put up some ugly council offices in their place.* **(b)** °accommodate, °lodge, °board, °house, °take in, °quarter, *Chiefly military* billet: *As our house was small, we were able to put up only three people.* **(c)** °preserve, can; °cure, pickle: *There are many foods that can be put up for later consumption.* **(d)** °contribute, °pledge, °offer (as collateral), °stake, mortgage, °post: *We put up the house as security for the loan.* **(e)** °contribute, °give, °supply, °donate, ante (up), °advance, °pay, °invest: *We put up all our savings to secure our retirement pension.* **(f)** °increase, °raise, °boost, elevate: *As soon as there is a hint of an oil shortage, they put up the price of gas.* **(g)** See **18 (a),** above. **(h)** °offer, °tender, put *or* place on the market: *They put their house up for sale and are planning to move to the Riviera.* **29** ***put up to.*** °incite, °urge, goad, °spur, egg on, °encourage, °prompt, instigate: *Was it you who put Reggie up to playing that prank on the teacher?* **30** ***put up with.*** °tolerate, °abide, °take, °brook, °stand (for), °stomach, °accept, °resign oneself to, °bear, °endure, °swallow: *Why do you put up with Calthorp's outrageous behavior?*

put-down *n.* °dig, °sneer, snub, disparaging *or* denigrating remark, °slight, °offense, °insult: *The worst put-down was to tell her she looks like her sister, whom she hates.*

put-on *n.* **1** °deception, °hoax, °trick, leg-pull, jest, °(practical) joke, °prank, °pretense, *Colloq* spoof: *Witherton was completely taken in by Threep's put-on about being from the Nobel prize committee.* **2** °takeoff, °parody, °satire, °burlesque, °caricature, °act, *Brit* sendup, *Colloq* spoof: *You shouldn't be offended, it was only a put-on.*

putrefy *v.* °rot, °decompose, °decay, molder, go bad, °spoil, °deteriorate, °go off: *The battlefield was strewn with putrefying bodies.*

putrid *adj.* °rotten, rotting, decomposed, decomposing, decayed, decaying, °moldy, moldering, spoilt *or* spoiled, putrefied, putrescent, putrefying, °foul, fetid, °rank, tainted, °corrupt: *Only the vultures continued to hover about the putrid carcass.*

putter *v.* Usually, ***putter around or about.*** dabble (in *or* with), °toy with, °trifle with, fribble, fool (with *or* about *or* around), °fritter (away), mess (about *or* around *or* with), °tinker (with), °meddle (with), monkey (about *or* around *or* with), °fidget (with), *Colloq* fiddle (about *or* around *or* with), footle (around *or* about): *On Sundays I like to putter about in the garden.*

put-up *adj.* °(secretly) preconceived, prearranged, plotted, preconcerted: *If the butler had a hand in it, you know it was a put up job.*

put-upon *adj.* imposed upon, inconvenienced, put-out, taken advantage of, exploited, °abused: *I shall be happy to help, and I don't feel put-upon at all.*

puzzle *v.* **1** baffle, °bewilder, °confuse, confound, °mystify, °flummox, °perplex, °nonplus, °stymie, °stump: *I am completely puzzled by what you wrote in your last letter.* **2** Usually, ***puzzle over.*** °study, °ponder (over), mull over, °contemplate, meditate on *or* upon *or* over, °consider, muse over *or* on, reflect on *or* over, think about *or* over *or* on: *If you puzzle over it long enough, the problem may vanish of its own accord.* **3** ***puzzle out.*** °solve, °decipher, °crack (the code), unravel, °work out, °figure out, think through, °sort out, unlock: *I finally puzzled out the inscription on this old coin. It might take a while, but I can puzzle out the solution to this crossword.* —*n.* **4** °enigma, °problem, °question, °paradox, poser, °mystery, °riddle, conundrum, *Colloq* brain-teaser: *The puzzle is how the murder was committed when the door was locked from the inside.*

puzzling *adj.* mystifying, enigmatic(al), bewildering, baffling, confounding, °perplexing, confusing, °ambiguous, °contradictory, abstruse: *It is a bit puzzling to see Catherine so often in the company of the husband she recently divorced.*

Q

quack *n.* **1** charlatan, °impostor, °pretender, fake(r), °fraud, *Colloq* phoney *or US also* phony: *That quack diagnosed Franny as having appendicitis, but it turned out that she was pregnant.* —*adj.* **2** °fake, °fraudulent, °sham, °counterfeit, *Colloq* phoney *or US also* phony: *Are you still going to see that quack doctor—the one who told you to take aspirin to relieve the pain of your ulcer?*

quaint *adj.* **1** °curious, °odd, °strange, °bizarre, °peculiar, °unusual, °queer, uncommon, °singular, °unorthodox, °eccentric, °whimsical, °offbeat, °fanciful, °outlandish, unconventional, °fantastic: *Her sitting room was furnished in a quaint mixture of Gothic and Victorian styles.* **2** °old-fashioned, archaic, °antiquated, outdated, °picturesque, °antique: *A rotund little figure appeared, dressed in quaint clothing of another era.*

quake *v.* **1** °tremble, °shake, °quiver, °shudder; °vibrate, °stagger: *We were quaking with fear that the giant would return to find us in his cave.* **2** earthquake, tremor, temblor *or* trembler *or* tremblor, seismic(al) activity, *Rare* seism: *In a few moments the quake leveled three cities that had stood for thousands of years.*

qualification *n.* **1** Often, ***qualifications.*** °fitness, °ability, °aptitude, °capability, competence *or* competency, °capacity, suitableness, suitability, eligibility, °proficiency, °skill, °knowledge, *Colloq* know-how: *You cannot persuade me that Powell has the qualifications needed to do the job.* **2** limitation, °restriction, modification, °reservation, caveat, °condition, °stipulation, °proviso, °prerequisite, °requirement: *The only qualification for use of the club's facilities is that one must be a member. My endorsement of her for the position was without qualification.*

qualified *adj.* **1** °able, °suitable, °capable, °competent, °fit, °fitted, equipped, °prepared, °ready, trained, °proficient, °accomplished, °expert, °talented, °adept, °skillful, skilled, °experienced, °practiced, °knowledgeable, °well-informed: *Is Kimble qualified to become foreman?* **2** contingent, conditional, restricted, modified, °limited, °provisional: *The chairman has given his qualified approval to the scheme.*

qualify *v.* **1** °equip, °fit (out), °ready, °prepare, °condition, make eligible; °certify: *His training in the air force qualified him to be a pilot.* **2** be eligible, meet the requirements, be fit *or* suitable *or* equipped *or* ready *or* prepared, °make the grade: *Sue will qualify as a doctor if she passes just one more exam.* **3** °temper, °mitigate, °modify, °moderate, °modulate, °restrict, °limit: *In light of their later behavior, I shall have to qualify my recommendation that you accept them as students.*

quality *n.* **1** °property, °attribute, °characteristic, °mark, °distinction, *je ne sais quoi,* °trait: *Can you describe the qualities needed to be a good musician?*

2 °grade, °caliber, °rank, °status, °importance, value, °worth: *In the curiosity shop, items of high and low quality are mixed together, helter-skelter.* 3 *Old-fashioned* eminence, °prominence, °importance, °excellence, °superiority, °distinction, °standing, °supremacy, °dignity, °grandeur, °nobility, blue blood: *Despite her shabby appearance, from her bearing and speech the woman was clearly a person of quality.*

qualm *n.* second thought, °doubt, uncertainty, °misgiving, hesitation, °scruple, uneasiness, °compunction, °reluctance, disinclination, queasiness, apprehension, apprehensiveness, °twinge, °pang, °worry, °concern, *Colloq* funny feeling, sinking feeling: *He had no qualms about turning his mother in for a reward, why should he hesitate about his father?*

quandary *n.* °dilemma, °predicament, °difficulty, °plight, cleft stick, uncertainty: *Faced with two identical doors, he was in a terrible quandary—behind one waited the beautiful maiden; behind the other, a ferocious, man-eating tiger.*

quantity *n.* °amount, °extent, °volume; °sum, °number, °total; °weight, °measure: *There seemed to be no limit to the quantity of gold in the mine. The person who most accurately guesses the quantity of beans in the jar wins a prize. What quantity of flour should I buy to make enough cakes for fifty people?*

quarrel *n.* 1 °dispute, °argument, °disagreement, °debate, °controversy, °discord, °difference (of opinion), contention, °misunderstanding; wrangle, °tiff, °row, squabble, altercation, set-to, scuffle, °feud, °fight, °fray, °fracas, °brawl, donnybrook, melee *or* mêlée, °riot, °battle (royal), *Colloq* dust-up, barney, °scrap, *US* spat: *I have no quarrel with my accountant—it's the tax auditors I worry about. The quarrel began over something trivial—but ten minutes later the police had to be called in.* —*v.* 2 °argue, °disagree, °dispute, altercate, have an altercation, °differ, wrangle, be at odds *or* loggerheads, °clash, squabble, °feud, °fight, °brawl, °battle, *Colloq* °fall out, °scrap: *When we were children, my sister and I quarreled all the time, but we still loved each other.*

quarrelsome *adj.* °testy, °petulant, irascible, °irritable, °disagreeable, fractious, °querulous, °peevish, °cross, choleric, curmudgeonly, °contrary, dyspeptic, °cranky, grouchy, °argumentative, combative, squabbling, disputatious, °hostile, antagonistic, dissentious, dissentient, °dissident, °pugnacious, bellicose, °belligerent, contentious: *I have to listen to the continual bickering of my quarrelsome neighbors.*

quarry[1] *n.* °prey, °game, °prize, °object: *The hunter's dogs had run their quarry to ground. Wealthy widows were regarded as fair quarry by the swindlers.*

quarry[2] *n.* 1 °mine, °pit, °excavation: *Only one quarry yielded the statuary marble needed by those sculptors.* —*v.* 2 °mine, °extract, °obtain, °get: *Gault's ideas were all quarried from the jottings of obscure writers.*

quarter *n.* 1 fourth: *Statistics showed that a quarter of the population attended church regularly.* 2 three-month period, three months, ninety days, thirteen weeks; fifteen minutes; °phase (of the moon), quadrature: *The first quarter's rent is now due. We waited a quarter of an hour. The moon is in its third quarter.* 3 °area, °region, °part, °section, °district, °zone, °division, °territory, °place, °neighborhood, locality, °locale, °location, °point, °spot; °direction: *The army was called in to patrol the quarter of the city where looting had occurred. It was difficult to tell from which quarter the wind was blowing.* 4 °mercy, compassion, mercifulness, clemency, leniency, °forgiveness, °favor, °humanity, °pity: *The hostages expected no quarter from their captors.* 5 ***quarters.*** living quarters, lodging(s), dwelling place, °dwelling, accommodation(s), °rooms, chambers, °residence, °shelter, habitation, °domicile, °home, °house, °abode; *Military* billet, barracks, cantonment, casern *or* caserne: *We found comfortable quarters in which we remained during our stay in the area.* —*v.* 6 °lodge, °accommodate, °house, °board, °shelter, °put up; *Military* billet, °post, °station: *The tour guide stayed at the best hotel, while we were quartered in a youth hostel.*

quarterly *adj.* 1 trimonthly, three-monthly: *It is one of the best quarterly journals on language.* —*adv.* 2 every three months, every ninety days, every thirteen weeks, four times a year: *The landlady insisted on being paid quarterly, in advance.*

quash *v.* 1 annul, nullify, °void, declare *or* render null and void, invalidate, °revoke, °vacate, °set aside, rescind, °cancel, °reject, °throw out, °reverse, °overthrow, °discharge, overrule: *When the real culprit confessed, my conviction was quashed and I was set free.* 2 °suppress, °subdue, °quell, °put down, °squelch, °repress, °overthrow, °crush, °overwhelm: *The rebellion was quashed by the secret police.*

quasi *adv.* 1 Sometimes, ***quasi-.*** as if, as it were, °seemingly, °apparently, °partly, to some extent, °to all intents and purposes, more or less, °virtually, °almost: *Her apology seemed only quasi-sincere to me. His function in the proceedings was quasi-judicial.* —*adj.* 2 °so to speak, kind of, °sort of; °pretended, °fake, pseudo, °so-called, °supposed, °artificial, °mock, °sham, *Colloq* phoney *or US also* phony: *His quasi excuse does not diminish the horror of what he did.*

quaver *v.* 1 °tremble, °quiver, °shake, °shiver, °vibrate, waver, °shudder, °fluctuate, °oscillate, °flutter: *His voice quavered with emotion as he described his rescue.* —*n.* 2 trembling, °tremble, °quiver, quivering, tremor, shaking, vibration, wavering, °break, °fluctuation, oscillation: *Only a slight quaver in her voice betrayed her feeling of panic.*

queasy *adj.* 1 uncomfortable, uneasy, °nervous, apprehensive, °ill at ease, troubled, °worried, discomfited; °doubtful, °hesitant: *I had a queasy feeling that I was being watched. The manager was a bit queasy about giving Denis so much responsibility.* 2 °sick, °nauseous, °nauseated, °ill, °bilious, °queer, *Colloq* green around *or* about the gills, °groggy, woozy: *Catherine felt very queasy after the ride on the roller coaster.*

queen *n.* 1 °sovereign, °monarch, ruler; empress; queen consort; queen mother; queen dowager: *She insists that she is the queen of Romania.* 2 beauty queen, movie queen, °star, prima donna, diva; °epitome, °model, °idol, leading light, cynosure, °leader: *At the height of her career, she was known as "the Queen of the Silver Screen." "Queen of the Antilles" was once a nickname of Cuba.*

queer *adj.* 1 °odd, °strange, °different, °peculiar, °funny, °curious, uncommon, unconventional, °unorthodox, atypical, °singular, °exceptional, anomalous, °extraordinary, °unusual, °weird, °bizarre, uncanny, °unnatural, freakish, °remarkable, °offbeat, °irregular, °unparalleled, °incongruous, °outlandish, °outré, °exotic, °eccentric, fey, °quaint, °absurd, °ridiculous, °ludicrous, unexampled: *How do you explain Basil's queer behavior at the office party? I agree that a chimney sweep's costume would be queer attire for the chairman of a bank, but he was going to a costume ball.* 2 °questionable, dubious, °suspicious, °suspect, °doubtful, °puzzling, °mysterious, *Colloq* °fishy, °shady: *We had a feeling that something queer was going on when the clocks started running backward.* 3 °(slightly) ill, °queasy, °sick, unwell, °poorly, °faint, uneasy, °dizzy, °giddy, vertiginous, lightheaded: *It's no surprise you feel queer after drinking all that wine.* **4** °mad, °unbalanced, unhinged, demented, °deranged, °insane, °daft, touched, *Colloq* dotty, potty, nutty, nuts, loony, batty, cracked: *Fenby seems to have gone a bit queer—fancies himself a reincarnation of Cicero.* —*v.* **5** °ruin, °spoil, °bungle, °botch, °muddle, muff, °mar, °wreck, °destroy, *Colloq* make a hash *or* mess of, gum up (the works), *Slang* °muck up, °screw up, louse up, *US* ball up: *Grimstone queered the deal by telling the buyer the company was worth half what we were asking.*

quell *v.* **1** °suppress, °put down, °repress, °subdue, °quash, °overcome, °crush, °squelch: *The army arrived to quell the uprising at the prison.* **2** °moderate, mollify, soothe, assuage, alleviate, °mitigate, allay, °quiet, °calm; pacify, °tranquilize, °compose: *Any feelings of bitterness were quelled by my mother's comforting words.*

quench *v.* **1** °satisfy, °slake, sate, °surfeit, °satiate, allay, appease: *My thirst isn't quenched by those sweet cola drinks.* **2** °put out, °extinguish, douse, °smother, snuff out, °stifle, °kill, °destroy, °suppress, °squelch, °repress, °overcome, °subdue: *Even the most totalitarian regime cannot entirely quench the fire of freedom in the hearts of the oppressed.*

querulous *adj.* complaining, carping, °critical, criticizing, hypercritical, °faultfinding, °finicky, finical, °fussy, overparticular, censorious, °petulant, whining, murmuring, grumbling, °peevish, °testy, °touchy, °irritable, irritated, annoyed, piqued, in a pique, irascible, fractious, °perverse, °quarrelsome, ill-natured, ill-humored, °cantankerous, curmudgeonly, crusty, crotchety, °fretful, bad-tempered, ill-tempered, °waspish, crabby, °cross, splenetic, choleric, °sour, dyspeptic, grumpy, *Colloq* grousing, pernickety *or US also* persnickety, *Slang* bitching: *He objected to almost all the television programs and sent querulous letters to the broadcasters telling them so.*

query *n.* **1** °question, inquiry *or* enquiry: *The information office will answer any query you may have about social security benefits.* **2** °doubt, uncertainty, °skepticism, °reservation, °problem: *She wants to see the manager about some query concerning her bank balance.* —*v.* **3** °ask (about), inquire *or* enquire (about), °question; °challenge, °doubt, °dispute: *I am phoning to query the third item on your latest bill.*

quest *n.* **1** °search, °pursuit, °exploration, °expedition, voyage (of discovery), °pilgrimage, °mission, °crusade; °chase, °hunt: *When a child, he was inspired by the legend of Sir Launcelot's quest for the Holy Grail.* —*v.* **2** Usually, ***quest after*** *or* ***for.*** seek (after *or* for), search after *or* for, °hunt (for), °track down, °pursue, °stalk: *In 1928, joined by his father, he quested for gold along the Orinoco River.*

question *n.* **1** °query, inquiry *or* enquiry: *I asked you a simple question, "Where were you last night?" and I expect a civil answer.* **2** °problem, °difficulty, °confusion, °doubt, dubiousness, uncertainty, °query, °mystery, °puzzle: *Is there some question about the qualifications of the new employee?* **3** °matter, °issue, °point, °subject, °topic, °theme, °proposition: *Whether you stay is not the question. The question remains, How are you going to earn a living?* **4** ***beyond*** **(*all*** *or* ***any) question.*** beyond (the shadow of) a doubt, °without question, without a doubt, indubitably, °undoubtedly, doubtlessly, °definitely, certainly, assuredly: *He is guilty without question.* **5** ***call in*** *or* ***into question.*** °question, °doubt, °query, °challenge, °dispute, harbor *or* entertain *or* have doubts *or* suspicions about, °suspect, cast doubt *or* suspicion on: *I cannot see why he has called her loyalty into question.* **6** ***in question.*** **(a)** under discussion *or* consideration: *The person in question never worked for this company.* **(b)** °questionable, °debatable, °at issue, °in doubt, °doubtful, open to debate: *Her honesty was never in question.* **7** ***out of the question.*** °unthinkable, °impossible, °absurd, °ridiculous, °preposterous, °inconceivable, beyond consideration, insupportable: *An increase in salary at this time is out of the question.* **8** ***without question.*** See **4**, above. —*v.* **9** °ask, °examine, interrogate, °query, °interview, °sound out, °quiz, *Colloq* °pump, grill, *Slang* give (someone) the third degree: *He was questioned for six hours as to his whereabouts.* **10** call in *or* into question, °doubt, °query, °mistrust, °distrust, cast doubt upon, °dispute, °suspect: *Investigators have long questioned the relationship between the banks and the drug barons.*

questionable *adj.* °doubtful, dubious, °debatable, °moot, °disputable, borderline, °suspect, °suspicious, °shady; open to question, °in question, in dispute, problematic(al), °uncertain, arguable, unsure, °unreliable, °ambiguous: *The evidence shows that the defendant engaged in questionable banking practices while a director of Offshore Ltd. The advantages of quick divorce are questionable.*

queue *n.* pigtail, °braid, plait; °tail, ponytail: *Not long ago, Chinese men wore their hair in a queue.*

quibble *v.* **1** °equivocate, split hairs, °evade, be evasive, palter, chop logic, bandy words, °cavil, pettifog, *Colloq* nitpick: *While people continued to be kept in slavery, the politicians quibbled over the difference between "liberty" and "freedom."* —*n.* **2** quibbling, equivocation, hair-splitting, splitting hairs, °evasion, paltering, sophistry, *Colloq* nitpicking: *Lawyers often become involved in a quibble over the* mot juste *and forget their clients.* **3** °cavil, sophism, °subtlety, nicety: *It's a quibble whether you say "either" and "neither" to rhyme with "peek" or "pike."*

quick *adj.* **1** °rapid, °fast, °speedy, °swift, °fleet; °expeditious, °express: *We made a quick descent to reach the camp before dark. The hand is quicker than the eye. Can you arrange the quick delivery of this parcel to Aylesbury?* **2** °sudden, °precipitate, °hasty, °brisk, °short, °abrupt, °hurried, °perfunctory, °summary; °immediate, °prompt, °timely, instantaneous: *There is no quick answer to your question about the causes of World War II. Don't be so quick to criticize others. I had a quick response to my letter complaining about service.* **3** °agile, °lively, °nimble, °energetic, °vigorous, °alert, °animated, °keen, °sharp, °acute, spry, °spirited, °vivacious, °rapid, °swift: *In three quick steps she was at my side. Antonia is known for her quick mind.* **4** °intelligent, °bright, °brilliant, facile, °adept, adroit, °dexterous, apt, °able, °expert, °skillful, deft, °astute, °clever, °shrewd, °smart, °ingenious, °perceptive, perspicacious, discerning, °farsighted, °responsive; nimble-witted, °quick-witted: *You can rely on Rob's quick thinking to come up with a solution to the problem.* **5** °excitable, °touchy, °testy, °petulant, irascible, °irritable, °impatient: *She said that I have a quick temper because of my red hair.*

quicken *v.* **1** accelerate, °hasten, °speed up: *As we neared the house, our pace quickened.* **2** °expedite, °hurry, accelerate, °hasten, °speed (up): *We did everything we could to quicken her departure.* **3** °stimulate, °arouse, °kindle, °spark, invigorate, °excite, °animate, °vitalize, vivify, galvanize, °enliven, °awaken, °energize, °revive, resuscitate, reinvigorate: *Olivia's inspiring lectures quickened the imagination of everyone who heard them.*

quickly *adv.* **1** °rapidly, °swiftly, speedily, °fast: *Although share prices dropped in the morning, they recov-*

ered quickly in afternoon trading. **2** °rapidly, °swiftly, speedily, °fast, with dispatch *or* despatch, apace, °post-haste, at *or* on the double, with all speed, °quick, *Colloq US and Canadian* lickety-split: *Run quickly, and you'll get there before the shop closes.* **3** instantly, °promptly, °hastily, °at once, °immediately, *Brit* °straightaway, *US* right away, °shortly, without delay, °(very) soon, hurriedly, °quick, *Colloq* pronto: *You'll have to leave quickly to catch your train.*

quick-tempered *adj.* °excitable, °impulsive, °temperamental, hot-tempered, °waspish, choleric, splenetic, °impatient, °short-tempered, °touchy, irascible, °irritable, °snappish, °abrupt, °short, short-spoken, °quarrelsome, °testy, °volatile, hot-blooded, bad-tempered, ill-tempered, churlish, highly strung, *US* high-keyed: *Legrand is very quick-tempered these days and flies into a rage at the slightest provocation.*

quick-witted *adj.* °acute, °sharp, °clever, °smart, nimble-witted, °alert, °keen, °astute, °perceptive, perspicacious: *Hallam is known for his quick-witted responses to hecklers in the audience.*

quiet *adj.* **1** °silent, soundless, °noiseless, hushed, °still: *We have to be very quiet so as not to wake the children.* **2** °still, °serene, °silent, °peaceful, unperturbed, °calm, °tranquil, placid, pacific, °smooth, °mild, °restful, unagitated: *The sea was very quiet as we rowed toward the beach.* **3** °dormant, quiescent, °inactive, retired, °withdrawn, °unobtrusive: *His last days were spent living a quiet life in the Outer Hebrides.* **4** °still, motionless, unmoving, °fixed, stationary, at rest, °inactive, composed, °temperate, unexcited, °calm: *There was a knock at the door and everyone suddenly became quiet.* —*n.* **5** °silence, stillness, soundlessness, noiselessness, °hush, quietness, quietude, °ease, °rest, calmness, °serenity, tranquillity, °peace, peace of mind, °repose: *In the quiet of the cave I could hear my heart beating.* —*v.* **6** Often, ***quiet down.*** °still, °silence, °hush, °calm, °tranquilize, °lull, *Brit* quieten (down): *He gradually quieted down the audience, and the curtain rose.*

quietly *adv.* **1** °silently, soundlessly, noiselessly, inaudibly, in silence, softly: *She tiptoed quietly out of the room.* **2** in hushed tones, in whispers: *In a corner of the room a young couple were quietly conversing.* **3** peacefully, calmly, serenely, peaceably, meekly, mildly: *When the police explained they only wanted him to help in their inquiries, he went with them quietly.* **4** modestly, humbly, demurely, unpretentiously, unostentatiously, unobtrusively, unassumingly, sedately: *The applicants sat quietly, waiting to be called in for the interview.*

quintessence *n.* °essence, °heart, °core, quiddity, essentialness, essentiality, °pith, marrow, sum and substance, °epitome, °nonpareil, °embodiment, personification, °model, °prototype, exemplar, °ideal, *beau idéal,* °paragon: *Hernandez considered the ancient Greek temple the quintessence of classic design.*

quip *n.* **1** (bon) mot, °witticism, sally, jest, ad lib, °joke, °gibe, barb, aphorism, °epigram, apophthegm *or* apothegm, °pun, double entendre, equivoque *or* equivoke, *Colloq* °gag, one-liner, °crack, °wisecrack, wheeze, chestnut: *In his worst quip he attributed the breakdown of his computer to a slipped diskette.* —*v.* **2** °joke, jest, °gibe, *Colloq* °wisecrack, *US* crack wise: *As Mae West quipped, "It's not the men in my life that count; it's the life in my men."*

quirk *n.* °peculiarity, caprice, vagary, °eccentricity, °fancy, °twist, °warp, aberration, idiosyncrasy, °oddity, °kink, °characteristic, crotchet, whim, °trick: *By some strange quirk of fate, after weeks in an open boat I was cast ashore at Cap d'Antibes.*

quit *v.* **1** °leave, °depart from, go (away) from, get away from, decamp, °exit, °desert, °flee, °forsake, °abandon, *Colloq* °take off, °beat it, °skip: *They think that he will try to quit the country if he is released on bail.* **2** °resign, °give up, °relinquish, °leave, °renounce, retire from, withdraw from: *Would you really quit your job to sail around the world?* **3** °cease, °stop, °discontinue, °leave off, desist from: *They would always quit laughing when I walked into the room.* —*adj.* **4** °free, °clear, discharged, °rid of, released (from), °exempt (from): *When my ex-wife remarried I was quit of the burden of paying alimony.*

quite *adv.* **1** °completely, °very, °totally, °utterly, °entirely, from head to toe, from A to Z, fully, °wholly, °thoroughly, unequivocally, °absolutely, °perfectly, °altogether, unreservedly: *Her trouble is that she is quite certain of many things that are simply not so. The last time I saw him, he was quite drunk.* **2** °rather, °fairly, °moderately, °somewhat, °relatively, to some *or* a certain extent *or* degree, noticeably: *I thought that the paintings were quite well done.* **3** °rather: *We had quite a good dinner last night.* **4** very much, °totally, °entirely, °wholly, °altogether; °really, °actually, °truly, °definitely, °positively, °undoubtedly, indubitably, °absolutely, unequivocally, certainly, °surely, unreservedly, °honestly: *What you are talking about is quite another matter. Your homemade chutney is quite the best I have ever tasted.*

quiver *v.* **1** °shake, °tremble, °vibrate, °shiver, °quaver; °shudder, tremor, °oscillate, °fluctuate, wobble: *He felt his lips quivering as he told them where to find the body. The outcome of the contest quivered in the balance.* —*n.* **2** °tremble, °quaver, °shudder, °spasm, °shake, tremor, °shiver: *Pierced by the dart, the tiny bird fell to the earth, gave a little quiver, and lay still.*

quixotic *adj.* °idealistic, °impractical, °impracticable, °unrealistic, unrealizable, °visionary, °romantic, °fantastic, chimerical, °fanciful, °dreamlike, °dreamy, nephelococcygeal, starry-eyed, °optimistic, °rash, °absurd, °mad, °foolhardy, °reckless, °wild, °preposterous, °ridiculous: *When he was young, he had a quixotic notion that he could improve the world.*

quiz *n.* **1** °examination, °test, *Colloq* exam: *We have a ten-minute quiz every week on our reading assignment.* —*v.* **2** °question, interrogate, °ask, °examine, *Colloq* grill, °pump: *The police have quizzed everyone in the neighborhood about what they might have seen.*

quizzical *adj.* °curious, °queer, °odd, inquiring *or* enquiring, questioning, puzzled: *The interviewer had a quizzical expression, as if he didn't quite understand his own questions.*

quota *n.* apportionment, °portion, °allotment, allocation, °allowance, °ration, °share, °part, °proportion, °percentage, °equity, °interest, *Colloq* °cut: *Each shareholder is allowed a quota of the new shares based on the number now held.*

quotation *n.* **1** °quote, °passage, citation, °reference, allusion, °extract, °excerpt, °selection: *For many years Frances collected quotations from the speeches of politicians.* **2** °quote, (bid *or* asking *or* offer *or* market) price, °charge, fixed price, °rate, °cost; value: *The quotation for Universal General International stood at 48 at the close of the market on Friday.*

quote *v.* **1** cite, °mention, °recite, °repeat, retell, °reproduce, °duplicate, °call up, °bring up, °bring in, °instance, °exemplify, °refer to, °reference, °extract, °excerpt: *Let me quote to you the relevant passage from* Paradise Lost. —*n.* **2** See **quotation,** above.

R

rabble *n.* **1** °mob, °crowd, horde, °throng, °swarm, °gang: *We narrowly escaped the wrath of the rabble fighting the police in the square.* **2 *the rabble.*** *Contemptuous* masses, proletariat, °hoi polloi, commoners, peasantry, ragtag and bobtail, vermin, outcasts, °riff-raff, scum, °dregs (of society), lower classes, *canaille,* commonalty, *Colloq* °trash, °the great unwashed: *He referred to the pantheon of Greek and Roman gods as a "rabble of deities."*

rabble-rouser *n.* °agitator, demagogue, instigator, inciter, firebrand, incendiary, °radical, °troublemaker, *agent provocateur,* °revolutionary, insurrectionist, *Colloq* hell-raiser: *The crowd was being whipped into a frenzy by a rabble-rouser who was not identified.*

rabid *adj.* **1** °unreasonable, unreasoning, °extreme, °fanatical; raging, °furious, °violent, crazed, frenzied, °maniacal, °wild, °mad, infuriated, frenetic, °berserk: *Please don't mention politics to Nigel—he gets rabid on the subject.* **2** hydrophobic, °mad: *Animals are kept in quarantine to make sure they are not rabid.*

race[1] *n.* **1** footrace, horse race, marathon, rally *or* rallye; °competition, °contest, °rivalry, contention: *How many will be running in Saturday's race? Several countries had joined the race to complete a workable antigravity device.* **2** sluice, flume, °chute, watercourse, °course, °channel, bed, millrace, raceway, spillway; °track: *The water in the race enabled us to wash the sand in order to extract the gold. The ballbearings don't fall out because they are retained in the race.* **3 *the races.*** horse races, dog races, the dogs, the track, °the turf, *Brit* racecourse, race meeting, *US* racetrack: *We're off to the races at Newmarket.* —*v.* **4** °speed, °hurry, °hasten, °dash, sprint, °fly, °rush, °scramble, highball, step lively, *Colloq* °tear, °rip, zip, step on the gas, °step on it, hop to it, °get a move on, *Brit* hare, *US* get a wiggle on: *You're going to have to race to catch the last bus.* **5** °compete (with): *I'll race you to the gate and back.*

race[2] *n.* **1** °stock, °line, °lineage, °type, °tribe, °nation, °people, °folk, °clan, °family: *The citizens constitute a mix of all imaginable races.* **2** blood, descent, °breed, °kin, °family, °stock, °line, °lineage: *You say you are liberal, but how would you feel about your daughter's marrying someone of a different race?*

racial *adj.* ethnic, genetic, genealogical, ethnological, °folk; tribal; °national: *Bigots emphasize racial differences rather than racial similarities.*

rack *n.* **1** framework, °frame, trestle, holder, °support; °stand, scaffold, scaffolding, °structure, hatrack, hat stand, coat rack, *Technical* stretcher, tenter, *US* hat tree: *The fish are hung out on long racks in the sun to dry. Please take your luggage off the seat and put it on the overhead rack.* **2** °suffering, °torment, torture, °agony, °anguish, °pain, °misery, °distress, °affliction, °scourge, adversity: *Only those who have experienced it can know the severe rack of gout.* —*v.* **3** °distress, °torment, torture, agonize, °oppress, °pain, °persecute, °anguish, beleaguer, °plague, °harass, harrow, °hurt: *Gruber was racked by the feeling that he had been responsible for the road accident. Quentin has been racked by the pain of arthritis for ten years.* **4** °beat, °strain, °wrench, tear at, °lash (at), °batter, °shake, °damage: *Winds of hurricane force racked villages along the coast last night.*

racket *n.* **1** °noise, °din, °uproar, °disturbance, clamor, hubbub, °row, °rumpus, hullabaloo, °fuss, ado, commotion, to-do, hue and cry, °outcry, brouhaha, °tumult, babel, °pandemonium, *Archaic* alarms *or* alarums and excursions, *Colloq* ballyhoo: *How can you sleep through the racket from your neighbor's all-night parties?* **2** °(organized) crime, criminal *or* illegal activity *or* enterprise, °trickery, °trick, °dodge, °scheme, °swindle, °stratagem, °artifice, °game, °ruse, *Slang* °caper, scam, gyp: *Their racket is selling "protection" to shopowners.* **3** °business, °line, °profession, °occupation, °trade, °vocation, °calling, °job, °employment, livelihood: *I'm in commodities; what's your racket?*

racketeer *n.* mobster, °gangster, mafioso: *The Chicago racketeers of the 1930s seem tame compared with today's terrorists.*

raconteur *n.* storyteller, anecdotalist *or* anecdotist, °narrator, relater *or* relator, *Colloq* spinner of yarns: *The after-dinner speaker was a marvelous raconteur, with a wealth of stories about the theater.*

racy *adj.* **1** °fresh, °lively, °animated, °spirited, °sprightly, °buoyant, °vivacious, °energetic, °vigorous, °dynamic, zestful, °stimulating, mettlesome *or* mettled, *Colloq* full of vim and vigor, peppy, full of pep, full of beans: *Journalists are told that their writing must be racy—straightforward reporting is out of fashion.* **2** °risqué, ribald, °bawdy, °naughty, °lusty, °earthy, °gross, °off-color, salty, °suggestive, °sexual, °immodest, indelicate, °indecent, °blue, smutty, °lewd, salacious, °vulgar, °dirty, °filthy, °pornographic, °obscene, sex-oriented, °adult, °rude, °crude, °coarse, *Colloq* raunchy, °sexy, °spicy, °hot: *My name got on the mailing list of some company selling racy videos.* **3** °spicy, piquant, °tasty, flavorful, °pungent, °strong, °savory, °sharp, zesty, tangy, °tart, °hot: *That curry you served last night had quite a racy flavor.*

radiance *n.* **1** radiancy, °splendor, brightness, °brilliance, resplendence, luminosity, luminousness, °dazzle, °sparkle, coruscation, scintillation, °twinkle, effulgence, refulgence, incandescence, °glow, phosphorescence, °gleam, °luster, °shimmer, °shine: *As Aladdin rubbed the ring it began to glow with an inner radiance and then—suddenly—a genie appeared.* **2** °warmth, gladness, °joy, °pleasure, °happiness, cheeriness, °delight: *It was good to bathe again in the radiance of the family at Christmas dinner.*

radiant *adj.* **1** shining, °bright, beaming, °burning, °ablaze, blazing, °brilliant, °luminous, resplendent, °splendid, splendorous *or* splendrous, °lustrous, gleaming, °glowing, phosphorescent, shimmering, °shiny, °glossy, glistening, °incandescent, alight, effulgent, refulgent, sparkling, °dazzling, glittering, coruscating, °scintillating, °twinkling; aglow: *In her hair she wore a diamond tiara, a radiant crown diminished only by her beauty. The room was radiant with sunshine.* **2** °happy, °overjoyed, °ecstatic, °rapturous, °delighted, °joyful, °blithe, blithesome, blissful, beatific, °glad, °gleeful, joyous, °gay, bubbling, °bubbly, jubilant, °elated, °rhapsodic, °exultant, exhilarated, in seventh heaven, in heaven, *Colloq* on cloud nine, *Brit* over the moon: *Carrie was absolutely radiant when she learned she was pregnant.*

radiate *v.* **1** °shine, °beam, °burn, °blaze, °gleam, °glow, °shimmer, °glisten, °sparkle, °dazzle, °glitter, coruscate, scintillate, °twinkle: *The jewel seemed to radiate with an inner light.* **2** °emanate, disseminate, °disperse, °spread, °diffuse, °shed, °send out, °emit, give off *or* out, *Rare* irradiate: *He radiated goodwill the way a fire radiates heat and light.*

radiation *n.* emission, emanation, diffusion, dispersal, shedding: *Because the night was cloudless, unhindered radiation of the earth's heat rapidly cooled the air.*

radical *adj.* **1** °basic, °fundamental, °elementary, °inherent, constitutional, °elemental, °essential, °cardinal, °principal, °primary, °deep, deep-seated, °profound, underlying, °organic, °natural, °rudimentary: *There is a radical error in your approach to the problem. There are radical differences in the ways we regard the English language.* **2** °thorough, thoroughgoing, °complete, °entire, °total, °exhaustive, °sweeping, (all-)inclusive, °comprehensive, all-embracing, °out-and-out, °drastic, °severe, °extreme, extremist, °revolutionary: *We have made radical changes in the way we live. Every political party that is not in power promises radical reform.* **3** extremist, °revolutionary, fanatic(al), °militant, anarchist(ic), °immoderate: *A staunch conservative, Vincent regarded as radical anyone who disagreed with him.* —*n.* **4** extremist, °revolutionary, °fanatic, °zealot, °immoderate, anarchist, °militant: *The radicals favor reform, the conservatives the status quo—but there is nothing new in that.* **5** communist, leftist, left-winger, red, Bolshevik, Bolshevist, °pink, *Slang US* pinko: *Carlo, recently arrested as a spy, had been a radical since his days at university.*

radio *n.* **1** receiver, °portable (radio), *Old-fashioned* crystal set, *Brit and US old-fashioned* wireless, *Colloq* transistor, *Slang* ghetto blaster, *Brit* trannie *or* tranny, *US* boom box: *He listens to BBC World Service on his shortwave radio.* —*v.* **2** °transmit, °broadcast, °air, disseminate, °announce, °present: *Weather reports are radioed continuously to ships at sea and aircraft.*

raffle *n.* °lottery, °draw, sweepstake *or US* sweepstakes, sweep *or US* sweeps: *The church is holding a raffle for a new car next week.*

rag[1] *n.* **1** °tatter, °piece, °shred, °scrap, °fragment, °bit, *Dialect* clout: *Take a clean rag and dust the bookshelves when you get the chance.* **2** newspaper, °periodical, °magazine, °publication, °journal: *You can't believe anything you read in that rag.* **3** ***rags.*** *Facetious* °clothes, clothing, attire, °dress, °garments, *Old-fashioned* duds: *I bought some new rags at the post-Christmas sales.* **4** ***rag trade.*** garment industry, clothing business, fashion industry: *Peggy has been a model in the rag trade for years.* **5** ***chew the rag.*** **(a)** °converse, °talk, °gossip, °chat: *We bought a bottle of wine and just sat about chewing the rag all evening.* **(b)** *Brit* °argue, wrangle: *They are constantly chewing the rag over trivial nonsense.*

rag[2] *v.* °tease, °taunt, °belittle, °twit, °ridicule, °mock, °make fun of, pull (someone's) leg, *Brit* °rally, *Colloq* kid: *The boys at school were always ragging him about his lisp.*

ragamuffin *n.* (street) urchin, street arab, waif, mudlark, gamin, little lost lamb, babe in the wood, °stray, °guttersnipe, scarecrow: *We took in a poor little ragamuffin who was shaking from the cold.*

rage *n.* **1** °anger, °fury, wrath, ire, high dudgeon, exasperation, vehemence, °passion: *Imagine her rage when Simon phoned to say that he couldn't make it home for dinner!* **2** °fury, °passion, °frenzy, hysterics, °tantrum, °fit, °temper, *Brit* paddy *or* paddywhack *or* paddywack, *Colloq Brit* wax: *He flies into a rage if Grace is as little as five minutes late for an appointment.* **3** °fashion, °craze, °vogue, the (latest *or* newest) thing, last word, *dernier cri,* °mode, *Colloq* °fad: *Miniskirts were all the rage in the 1960s.* —*v.* **4** °rant, °rave, °storm, go mad *or* crazy *or* bonkers *or* wild *or* out of one's mind, go berserk, run amok *or* amuck, behave *or* act *or* be like one possessed, °fret, be beside oneself (with anger *or* fury), lose one's temper, go berserk, have a tantrum, fulminate, °explode; °fume, foam at the mouth, °stew, °smolder, °boil, °seethe, °simmer, *Colloq* have kittens, lose one's cool, °fly off the handle, go off the deep end, *Slang* get into *or* work oneself up into a lather *or* stew *or* sweat, get all worked up, blow one's top, blow a gasket, °blow up, flip one's top *or* lid, hit the ceiling *or* roof, freak out, be fit to be tied, be ready for a straitjacket, *Brit* throw a wobbly, *US and Canadian* °blow one's stack, flip one's wig, blow a fuse, have a hemorrhage, go ape, do a slow burn, have a conniption fit: *Father raged for hours when he found that Donald had borrowed the car without permission.*

ragged *adj.* **1** °rough, °shaggy, °shabby, °seedy, °tattered, °unkempt, scraggy, torn, °rent, ripped, frayed, °worn (out), °threadbare, patched, patchy, raveled, *Chiefly Brit* tatty, *Brit* down at heel, *US* down at the heel(s), *Colloq US* frazzled, beat-up: *On weekends, Charles put on his most ragged clothes to work in the garden.* **2** °rough, uneven, °irregular, nicked, °jagged; °serrated, sawtooth(ed), zigzag, °notched, toothed, denticulate(d), ridged: *Every morning on his way to work, his first sight of the factory was the ragged edge of its roof.* **3** worn-out, °tired, °exhausted, °on one's last legs, the worse for wear, dead tired, overtired, °fatigued, °weary, fagged out, °spent, *Colloq* all in, dog-tired, *Slang* °dead (on one's feet), °(dead) beat, *Brit* knackered, *US and Canadian* pooped (out): *The emergency crew was run ragged by thirty hours on duty without rest.* **4** °bad, °rough, patchy, °imperfect, °shabby, messy, disordered, °disorderly, °run-down, battered, broken-down, neglected, deteriorated, °dilapidated, *Colloq* beat-up: *From its ragged condition, it is clear that the building wasn't properly maintained.* **5** °rough, °harsh, °discordant, °grating, rasping, hoarse, °scratchy, croaking: *His voice, once a beautiful lyric tenor, was ragged after years of alcoholism.* **6** uneven, °bad, °poor, °shabby, patchy: *The guests gave a ragged rendition of "Happy Birthday" as their host entered the room.*

ragman *n.* rag dealer, ragpicker, scrap dealer, *US* junk dealer, junkman, *Brit* rag-and-bone man, knacker: *When the ragman comes round, give him those old curtains.*

raid *n.* **1** °(surprise) attack, incursion, °invasion, °onset, onslaught, blitz, sortie, sally, °expedition, *Slang* (police) bust: *The partisans made a raid on the arsenal last night and got away with guns and ammunition.* —*v.* **2** °attack, invade, °assault, °storm, °set upon, descend upon, swoop down upon, °pounce upon, *Military* forage; °sack, °plunder, °pillage, °loot, °ransack, °rifle, °strip, maraud, depredate, *Slang* bust: *The commandos raided the enemy stronghold and destroyed it.*

rail[1] *n.* **1** °bar, °rod, handrail, footrail, towel rail; railing, banisters, balustrade, baluster, °fence: *Don't lean on the rail or it will break.* **2** ***by or via rail.*** by train, by railway *or US also* railroad: *We traveled by rail from Paris to Geneva.*

rail[2] *v.* Usually, **rail at** *or* **against.** vituperate, vociferate, fulminate, be *or* become abusive, revile, °attack, °berate, °scold, °upbraid, °criticize, censure, decry, °condemn, °denounce: *She continues to rage and rail against those who would deny equal rights to women.*

raillery *n.* °banter, badinage, persiflage, °repartee, frivolity, joking, jesting, chaffing, teasing, °ridicule, *Colloq* kidding: *You shouldn't get upset over what was nothing more than some good-natured raillery.*

railroad *n.* **1** railway, °train, °rail, rolling stock, *Archaic* iron horse: *Railroad service could be improved on some suburban lines. Diamond Jim Brady, the American millionaire, made his money from the railroads.* —*v.* **2** °force, compel, °expedite, coerce, °intimidate, °push (through), *Brit* pressurize, *US* °pressure, °bully, hector, °tyrannize, dragoon, °browbeat, bullyrag, *Colloq* bulldoze, °squeeze, °lean on: *Clayton was railroaded into appointing Pinbury to head the committee to investigate corruption.*

rain *n.* **1** °precipitation, drizzle, sprinkle, °downpour, °shower, thundershower, cloudburst, rainstorm, squall, deluge, *US* sunshower: *After the rain the sun came out.* **2** rainfall, °precipitation: *The rain in Spain falls mainly in the plain.* **3** °flood, °torrent, °shower, °volley, °stream, °outpouring: *Arriving home, she could barely shield herself from the rain of children's kisses.* —*v.* **4** come down, °pour, °teem, sprinkle, drizzle, °spit, *Brit dialect* mizzle, *Colloq* come down in buckets, rain cats and dogs: *The weatherman said it would rain today, so take an umbrella.* **5** °trickle, °pour, °run, °fall: *Tears rained down her sallow cheeks.* **6** °descend, °shower: *Blows rained on him from the tiny fists of the furious child.* **7** °bestow, °lavish, °shower: *He rained kisses on her upturned face.*

raise *v.* **1** °lift (up), elevate; °hoist, °pull up, haul up, run up, *Literary* upraise: *She raised the glass to her lips. We raised the mainsail and sailed out of the harbor.* **2** °erect, °put up, °construct, °build, °produce, °create, put together, °assemble, °frame: *With the help of some neighbors, the barn was raised in less than a day.* **3** °farm, °grow, °cultivate, °plant, °bring up, nurture, °harvest, °propagate: *During that year's mild winter, we were able to raise three crops.* **4** °bring up, nurture, °rear; °mother, °father, °parent: *My mother raised ten children, six girls and four boys.* **5** °assemble, °gather, bring *or* gather *or* get together, °muster, °mobilize, °round up, °rally, °collect, convene, °recruit, *Colloq* pull together: *Do you think you could raise an army quickly enough to defend us against attack?* **6** °increase, °boost, °advance, °put up, jack up, run up, °inflate, escalate: *Why are consumer gas prices always raised at the vaguest hint of an increase in oil prices?* **7** °cultivate, °foster, nurture, °heighten, °stimulate, °buoy, °lift, uplift, °boost, °arouse, °quicken, °encourage, °develop: *The news about the new company raised our hopes of finding jobs.* **8** °open, °introduce, °initiate, °broach, °bring up, bring *or* put forward, °present, °suggest, °mention, °moot: *It is regrettable that you saw fit to raise such a delicate issue.* **9** °remove, °relieve, °lift, °abandon, °eliminate, °discontinue, °(bring to an) end, °terminate: *The ban against travel to Southeast Asia was about to be raised.* **10** °cause, °provoke, °evoke, °occasion, put *or* set in motion, °institute, °prompt, engender, °stir up, instigate, °inspire, °give rise to, °bring about, °arouse, °originate: *Objections to the relaxation of safety regulations were raised by relatives of the victims of the earlier disaster.* **11** utter, °express, °bring up, °put forward, °shout, °call: *Was it you who raised the alarm at the first sign of smoke?* **12** °assemble, °obtain, °get, °collect, °amass, °solicit: *We raised twice as much for charity as last year.* **13** invigorate, °animate, °vitalize, vivify, °buoy, °lift, uplift, °cheer (up), exhilarate, elate: *Our spirits were raised considerably by the news from Waterloo.*

rake[1] *v.* **1** Often, ***rake up.*** °scrape together, gather (together *or* up), °collect, draw together: *Your weekend chore is raking up the leaves from the lawn.* **2** °scrape, comb, °scratch, °grate, graze: *Don't you just hate the sound of a fingernail being raked across a blackboard?* **3 *rake in.*** °collect, gather (up *or* in), °pull in, °make: *Pamela is raking in the royalties from her latest thriller.* **4 *rake it in.*** °coin money, make money (hand over fist), *Brit* coin it in: *The new record shop in the mall is really raking it in.* **5 *rake out.*** °sift (out), °screen, °remove, °clear, °eliminate: *Please rake out the cinders from the grate.* **6 *rake over*** *or* ***through.*** °search, °probe, °ransack, °scour, comb, rummage through, pick through *or* over, go through *or* over (with a fine-tooth(ed) comb), °rifle (through): *The police raked through her entire flat but found nothing incriminating.* **7 *rake up.*** **(a)** See **1,** above. **(b)** °revive, resuscitate, °resurrect, °raise, °bring up, °recall: *Must you persist in raking up those bad memories?* **8 *rake up*** *or* ***together.*** °gather, scrape up *or* together, °collect, drag together, °pick up, °dig up, dredge up, °find, °unearth: *Were you able to rake up the others for a game of poker tonight?*

rake[2] *n.* °libertine, womanizer, lecher, °roué, °playboy, ladies' man, Don Juan, Casanova, Lothario, debauchee, voluptuary, °profligate, °prodigal; °scoundrel, °rascal, cad, *Archaic* rakehell, masher; *Colloq* lady-killer, *Colloq old-fashioned* wolf, *Brit* bounder: *In those days, young rakes accosted ladies in the park.*

rake-off *n.* °kickback, commission; °discount, markdown, °rebate, *Colloq* °cut, °piece: *Does Farquhar get a rake-off for every buyer he sends to the shop?*

rakish *adj.* °dashing, °jaunty, °dapper, °spruce, °debonair, raffish, °smart, °breezy, °flashy, °chic, °fashionable, °elegant; °dandy, foppish: *Adjusting his slouch fedora to a rakish angle, Peter strode down Fifth Avenue.*

rally *n.* **1** °gathering, °(mass) meeting, °meet, convocation, °convention, assemblage, °assembly, °muster: *There is to be a political rally tonight in the town square. Are you taking your Daimler to the vintage car rally?* **2** °recovery, °improvement, °revival, turn for the better, recuperation, renewal: *His doctors said that the rally might be only temporary.* —*v.* **3** Often, ***rally around.*** bring *or* call *or* get together, °round up, °assemble, convene, °group, congregate, °organize, come together, troop; marshal, °mobilize, °summon, °gather, °muster: *All Sally's friends rallied around her to be supportive. He rallied every last ounce of strength in the dash for the finish line.* **4** °revive, °rouse, °recover, °improve, get better, take a turn for the better, °recuperate, °perk up, °pick up, *Colloq* °snap out of it, make a comeback: *According to the nurse, he rallied during the night and asked for something to eat.*

ram *v.* **1** °jam, °force, °drive, °cram, °crowd, °pack, compress, °stuff, °squeeze, °thrust, tamp, °pound, hammer: *We rammed as much insulation as we could into the crevices.* **2** °butt, °bump, °strike, °hit, °collide with, °dash, °crash, °slam: *Their car rammed into mine in the garage.*

ramble *v.* **1** amble, °wander (off), °stroll, °saunter, °walk, perambulate, go (off *or* away), °travel, °drift, °range, rove, go *or* move about, hike, trek, *Colloq* mosey: *Don't ramble too far away as dinner will be ready soon. We used to ramble over those hills, but that area is now an industrial site.* **2** °meander, °wander, digress, maunder: *Fiona tends to ramble from the main topic of conversation.* **3** Sometimes, ***ramble on.*** °babble, °chatter, gibber, °rave, °go on (and on), °rattle on, *Colloq Brit* witter on, rabbit on *or* away: *If you didn't stop him, Hugh would ramble on forever about his grandchildren.* —*n.* **4** °stroll, amble, °saunter, °walk, °promenade, constitutional, walkabout, °tour, °tramp, hike, trek: *We went on a short ramble through the park.*

rambling *adj.* **1** °discursive, °roundabout, circuitous, °tortuous, °incoherent, °diffuse, °unsettled, °disconnected, °disjointed, disorganized, unorganized, illogical, maundering, °aimless, °confused, muddled, jumbled, scrambled, unintelligible, °inarticulate, periphrastic, circumlocutory, circumlocutional, circumlocutionary, ambagious, °wordy, verbose, prolix, °endless, interminable: *We couldn't make heads or tails of the professor's rambling lecture.* **2** unplanned, straggling, °irregular, sprawling, spread-out, spreading, straggly: *The Wilsons live in that big old rambling Victorian house down the road.* **3** roving, wandering, °traveling, peripatetic, itinerant, wayfaring, migratory, nomadic: *For years I led a rambling life as a folk-singing banjo player.*

ramification *n.* **1** consequence, °result, °effect, °upshot, °implication, °subtlety; °complication, °difficulty: *Have you considered the ramifications of your decision to retire?* **2** °branch, °extension, outgrowth, subdivision, °offshoot: *Courier services may be considered a ramification of services once performed by the post office.*

ramp *n.* °slope, °grade, gradient, °incline; °rise, ascent, acclivity; descent, °dip, declivity: *There is a special ramp provided for wheelchair access.*

rampage *n.* **1** °excitement, °agitation, recklessness, °riot, °tumult, °uproar, °frenzy, °fury, °rage, °furor, turmoil: *His latest rampage was prompted by their serving him a cup of cold coffee.* **2** *US* ***on a rampage,*** *Brit* ***on the rampage.*** °berserk, °mad, °crazy, amuck *or* amok, °wild, out of control: *She went on a rampage when he refused to let her visit the children on the weekend.* —*v.* **3** °storm, °rage, °rant, °rave, go berserk, run amok *or* amuck: *He rampaged like a man possessed when they took back the key to the executive washroom.*

rampant *adj.* **1** unchecked, °uninhibited, unrestrained, °wild, °uncontrolled, out of control, out of hand, frenzied, unbridled, uncontrollable, °violent: *He was a victim of rampant jealousy. Jennie could hardly be described as a rampant feminist.* **2** °profuse, unbounded, abounding, °flourishing, rife, widespread, °everywhere, °epidemic, pandemic, °prevalent, unrestrained, unchecked, running wild, °uninhibited, °indiscriminate, °wild, °uncontrolled; in control, holding sway, in full sway, °dominant, °predominant; °exuberant, °rank, °luxuriant: *Before Masterson became sheriff, vice and crime were rampant in this town.*

rampart *n.* °defense, bastion, °guard, fortification, °security, °stronghold, °bulwark, barricade, °wall; earthwork, breastwork, parados, gabion: *They regard religion as an impenetrable rampart against immorality.*

ramshackle *adj.* °dilapidated, tumbledown, crumbling, broken-down, °rickety, unsteady, jerry-built, °decrepit, °flimsy, °shaky, °unstable, tottering, unsubstantial *or* insubstantial, ruined, in disrepair, beyond repair, °run-down, neglected, °derelict: *He lives in a ramshackle hut on the outskirts of town.*

rancid *adj.* stinking *or* reeking (to high heaven), foul-smelling, ill-smelling, evil-smelling, noisome, mephitic, miasmic *or* miasmal *or* miasmatic(al), °smelly, °rank, malodorous, fusty; °nasty, °disagreeable, odious, fetid *or* foetid, °rotten, decayed, spoiled, turned, °bad, °awful, °sour, tainted, °high, gamy, °ripe, °putrid, °corrupt, °stale: *To rid your fridge of that rancid smell, wash it out with a solution of baking soda.*

rancor *n.* hatred, °hate, antipathy, °spite, °resentment, resentfulness, °antagonism, °hostility, malignity, °bitterness, malevolence, malice, venomousness, °venom, vindictiveness, vengefulness, spleen, acrimony, animus, °animosity, enmity, bad feeling, bad blood: *Owing to the rancor between them, the brothers had not exchanged a word in forty years.*

rancorous *adj.* °hateful, °spiteful, °resentful, °hostile, malign, °malignant, °bitter, malevolent, malicious, °venomous, °vindictive, vengeful, splenetic, acrimonious: *Olivia felt rancorous toward her brother-in-law because he refused to lend her money.*

random *adj.* **1** °haphazard, °chance, fortuitous, serendipitous, aleatory, °arbitrary, °casual, °stray, °occasional, °indefinite, °indiscriminate, nonspecific, unspecific, unspecified, unordered, unorganized, undirected, °unpremeditated, unplanned, °accidental, uncalculated, unsystematic, adventitious, °incidental, hit-or-miss: *He used a special computer program to generate random numbers. A random selection of slips was drawn from the drum and Madelaine's name was on one of them.* **2** ***at random.*** randomly, haphazardly, fortuitously, °by chance, serendipitously, arbitrarily, casually, °occasionally, °(every) now and then, °(every) once in a while, irregularly, indefinitely, erratically, indiscriminately, unsystematically, adventitiously, °incidentally, unpremeditatedly: *They picked names from a hat at random. Instead of publishing on a fixed schedule, they issued books at random, as they became ready.*

randy *adj.* aroused, °lustful, °lecherous, *US* in heat *or Brit* on heat, *US* in estrus *or Brit* on oestrus, *Slang* °hot, horny, in rut, rutting, at stud: *Seeing Betty always made Lonigan terribly randy.*

range *n.* **1** °scope, °sweep, °reach, °limit, °extent, °span, °area, radius, °distance, compass, latitude, °stretch, °sphere, °orbit: *The gun has a range of about two hundred yards. This radio transmitter is effective only within a thirty-mile range. The minister is trying to broaden the range of his influence.* **2** °assortment, series, °collection, °lot, °spread, °selection, °choice, °number, °variety, °kind, °sort, °scale, °gamut, °register: *The courses offered cater to a wide range of interests.* **3** °rank, °row, °tier, °line, °file, series, °string, °chain: *From my window I could see the range of hills stretching to the horizon.* **4** kitchen range, (cooking) stove, *Chiefly Brit* cooker, *Brit trademark* Aga, *US* cookstove: *In every French farmhouse there was always a pot of soup on the range.* —*v.* **5** °line up, °rank, °order, align, °order, array: *The teacher ranged the children according to their height.* **6** °vary, °fluctuate, °spread, run the gamut, °extend, °stretch, °run, °go: *The students range in age from eight to eighty.* **7** °organize, °categorize, catalogue, °arrange, classify, °sort, °class, °group, °rank, °bracket, pigeonhole, °file, °index, °break down, °grade, °distribute: *The books are ranged according to their subject matter.* **8** °cover, °traverse, °roam, rove, travel over *or* across, go *or* pass over, °drift, °migrate, °wander, °move, °extend: *The wildebeest ranges far and wide over vast tracts of grazing land in Africa.*

rank[1] *n.* **1** °status, °standing, °position, °place, °level, °stratum, °class, °caste, °circumstances, echelon, °grade: *It is virtually impossible to determine rank simply by looking at a person.* **2** °weight, °authority, °power, °superiority, seniority, ascendancy, °priority, °influence, eminence: *Caldwell uses the privilege of rank to cover up some of his questionable transactions.* **3** °nobility, °title, high birth, aristocracy, °dignity, °prestige, (blue) blood; peerage: *Gentlemen of rank behave with honor, my boy.* **4** °line, °row, column, °queue, series, °formation; °sequence: *We gazed on the serried ranks of warriors, golden shields glistening in the sun.* **5** ***ranks.*** soldiers, workers, °staff, employees: *Smithers rose from the ranks to become chief executive officer.* —*v.* **6** °grade, °rate, classify, °class, °categorize; °dispose, °organize, °order, °sort, assort, °arrange, array, align, °range, °graduate: *He was ranked among the best in the school. The applicants are ranked according to their test scores.* **7** °rate, °count, °stand, have standing *or* value *or* prestige, be important *or* distinguished: *To Janet, character ranks far above wealth. Phil is so mean that Scrooge ranks high on his list of charitable men.*

rank[2] *adj.* **1** °lush, °luxuriant, °abundant, °flourishing, °profuse, °prolific, °dense, superabundant, °exuberant, °fertile, °productive, fructuous: *The expedition could hardly make its way through the rank undergrowth to the bottom of my garden.* **2** °corrupt, °foul, °low, °base, °gross; °downright, utter, °sheer, °absolute, °complete, °out-and-out, °blatant, °flagrant, unalloyed, °unmitigated, °undiluted: *I cannot understand the rank treachery of the man after all that we have done for him. Would you admit a rank outsider to the inner circle?* **3** °offensive, °loathsome, °disgusting, °gross, °foul, °corrupt, °indecent, °shocking, °immodest, indecorous, °shameless, °risqué, °lurid, °off-color, °outrageous, °blue, °nasty, °vulgar, °vile, °dirty, °filthy, smutty, scatologic(al), °pornographic, °obscene: *Those bigots always resort to the rank vocabulary of malice and hate. In a secret cupboard was a collection of the rankest literature.* **4** °offensive, °loathsome, °disgusting, °gross, °foul, foul-smelling, °smelly, °rancid, noisome, °stinking, reeky, reeking, mephitic, miasmic *or* miasmal *or* miasmatic(al), fetid *or* foetid, noxious, °rotten, °putrid, °musty, °stale, °disagreeable, °strong, °pungent: *The rank stench from the open sewers pervades the air in those pockets of poverty.*

rank and file *n.* (general) membership, members, °majority: *The union's rank and file must now vote on whether to accept management's pay offer.*

rankle *v.* °gall, °fester, °irk, vex, °plague, °chafe, °grate, nettle, °torment, °pain, °hurt, °provoke, °anger, °exasperate, °get (to), °upset: *Although she had forgiven him, the insult still rankled.*

ransack *v.* **1** °search, °examine, go through *or* over (with a fine-tooth(ed) comb), comb, rake *or* rummage through, °scour, °explore, °scrutinize, turn inside out: *I ransacked the secondhand bookshops for a copy but had no success.* **2** °rob, °plunder, °pillage, °sack, despoil, °loot, °strip; burgle, *US and Canadian* burglarize: *The thieves who ransacked the shop apparently missed the most valuable jewel.*

ransom *n.* **1** redemption, °rescue, deliverance; °release, °liberation: *The envoy was held to ransom by a fanatical band of zealots.* **2** °payment, payout, °payoff, °price: *The police refuse to reveal the amount of the ransom given to the kidnappers.* —*v.* **3** °redeem, °rescue, °release, °deliver: *The boy was ransomed for fifty thousand dollars.*

rant *v.* **1** declaim, °hold forth, expound, expatiate, orate, perorate, pontificate, trumpet, °preach, °harangue, °lecture, deliver (oneself of) a tirade *or* diatribe *or* speech, °speak: *The minister ranted pompously about his accomplishments.* **2** vociferate, °bluster, °rave, rant and rave, °bellow, °rage: *The master kept ranting on about his newspaper having been creased.* —*n.* **3** °tirade, philippic, °bluster, flatulence, °rhetoric, °bombast, pomposity, turgidity, gasconade, rodomontade, theatrics, histrionics, °act: *As he carried on his rant, the theater gradually emptied till he was alone.*

rap *v.* **1** °knock, °strike, °tap, °hit: *The policeman rapped smartly at the door with his truncheon.* **2** °criticize, °rebuke, °scold, °reprimand, rap over the knuckles, *Colloq* °knock, *Brit* tick off: *We were severely rapped for dropping litter on the pavement.* **3** °converse, °talk, °chat, °gossip, *Colloq* °gab, *Slang* chew the fat *or Slang* the rag: *That guy was so fascinating—we could rap with him for hours.* —*n.* **4** °knock, °tap, °hit, °blow, °crack, °stroke, cuff, whack, thwack, °punch, *Colloq* °belt, clout, *Slang* sock, slug, biff: *The punishment for failing to complete the assignment was a severe rap on the knuckles.* **5** °conversation, °discussion, °chat, confabulation, °talk, °dialogue, discourse, colloquy, *Colloq* confab, *Slang chiefly US and Canadian* bull session, rap session: *We used to get together for a nice long rap now and then.* **6** °responsibility, °blame; °punishment, °sentence; °charge, °accusation, indictment: *Why should he have to take the rap for something you did? The witness failed to show up, so Baxter beat the rap.*

rapacious *adj.* °greedy, covetous, °grasping, °avaricious, °mercenary, usurious, acquisitive, °predatory, predacious, °ravenous, ravening, °voracious, insatiable, insatiate, wolfish, wolflike, lupine, vulturine, raptorial: *That rapacious swindler took everything they had in the world.*

rapacity *n.* °greed, greediness, cupidity, covetousness, °avarice, acquisitiveness, predaciousness, ravenousness, voracity, voraciousness, insatiability, insatiableness, rapaciousness: *The rapacity of the invading hordes of Mongols knew no bounds.*

rape *n.* **1** ravishment, defloration, deflowering, °violation, sexual assault, defilement: *The police are investigating reports of rape by a masked man.* **2** rapine, despoliation, spoliation, despoilment, °pillage, °depredation, ravagement, ravaging, °plunder, plundering, °sack, sacking, looting, ransacking: *The reports document the rape of the war-torn areas.* **3** abduction, carrying-off, kidnapping, °seizure, °capture, snatching: *History debates whether the ancient Romans were truly responsible for the rape of the Sabine women.* —*v.* **4** °violate, °ravish, assault sexually, deflower, defile, force to submit to sexual intercourse, have one's way with, take advantage of: *The defendant raped three women before he was caught.* **5** despoil, spoliate, °pillage, depredate, °ravage, °plunder, °sack, °loot: *Throughout history the city has been raped by marauding tribes.*

rapid *adj.* °quick, °fast, °swift, °speedy, high-speed, °brisk, °expeditious, °prompt, °express, °fleet, lightning(-fast), alacritous; °hurried, °hasty, °precipitate, °impetuous, °immediate, instantaneous, °instant, °sudden: *The company we use offers rapid delivery service. What do you suppose occasioned Flannery's rapid departure?*

rapidity *n.* quickness, swiftness, °speed, speediness, briskness, expeditiousness, promptness, promptitude, alacrity, celerity, immediateness, dispatch *or* despatch, instantaneousness: *The rapidity with which they responded to our request is truly commendable.*

rapidly *adv.* **1** °quickly, °fast, °swiftly, speedily, briskly, expeditiously, °like a shot, at the speed of light, double-quick, at full speed, like one possessed, at a gallop, tantivy, *Colloq* like blazes, like (greased) lightning, *US* lickety-split, *Slang* like a bat out of hell, °like mad: *He was cycling down Main Street as rapidly as he could go.* **2** °promptly, instantly, °instantaneously, instanter, without delay, °at once, *Brit* °straightaway, *US* right away, in a moment, in a trice, °like a shot, in a wink, in (less than) no time, double-quick, in a flash, at the speed of light, before one can turn around, *Colloq* in a jiffy, in two shakes (of a lamb's tail), like (greased) lightning, before one can *or* could say "Jack Robinson", *US and Canadian* (right) off the bat, lickety-split, *Slang* like a bat out of hell: *When I mentioned that you were coming, she left very rapidly.*

rapport *n.* empathy, °relationship, °sympathy, °harmony, °affinity, °accord, °bond, °relationship, °(mutual) understanding, camaraderie: *The minute we met we established a rapport.*

rapprochement *n.* °reconciliation, °understanding, °settlement: *Perhaps we can bring about a rapprochement between the radicals and the reactionaries.*

rapt *adj.* **1** entranced, fascinated, spellbound, mesmerized, hypnotized, engrossed, enthralled, bewitched, °absorbed, transported, captivated, °delighted: *She performed the dance of the seven veils before a rapt audience.* **2** enraptured, °rapturous, °delighted, °elated, °happy, °ecstatic, blissful, °overjoyed, joyous, °joyful, beatific: *You should have seen the rapt expression on the child's face when Santa Claus appeared.*

rapture *n.* °ecstasy, °delight, °joy, joyfulness, joyousness, °pleasure, exaltation, elation, °thrill, °enchantment, euphoria, beatitude, °transport, °enchantment: *Nothing could compare with the rapture he felt at seeing his daughter win the Nobel prize.*

rapturous *adj.* °ecstatic, °delighted, °joyful, joyous, °elated, thrilled, °enchanted, euphoric, in seventh heaven, °overjoyed, °rhapsodic, *Colloq* on cloud nine, *Brit* over the moon: *He was rapturous over qualifying for the Olympics.*

rare[1] *adj.* **1** uncommon, °unfamiliar, °unusual, °exceptional, °out of the ordinary, °extraordinary, atypical; °scarce, °unparalleled, °choice, recherché, °phenomenal, infrequent, few and far between, °sparse, °scanty, °limited, seldom encountered *or* met with *or* seen; °unique, °singular, one of a kind; *Chiefly Brit* °thin on the ground: *It was rare to see Beauchamp is such good humor. A copy of the first edition in this condition is rare.* **2** °fine, °good, °admirable, °excellent, °choice, °select, °special, °first-rate, first-class, °exquisite, °superior, °superlative, °peerless, unequaled, °matchless, °incomparable, in a class by itself *or* herself *or* himself *or* themselves, *sui generis,* °outstanding; collectible *or* collectable: *O, rare Ben Jonson! On exhibit was a black pearl of rare beauty and remarkable size. Consideration for others seems a rare quality these days.*

rare[2] *adj.* underdone, undercooked, *bleu(e), saignant(e)*: *Do you want your steak rare, medium, or well-done?*

rarefied *adj.* **1** °thin, °lean, attenuated, diluted, °sparse, scant, °scanty: *As we climbed higher, the rarefied atmosphere made breathing more difficult.* **2** °exalted, °lofty, °elevated, °high, °sublime, °noble, °high: *I feel out of place in the rarefied atmosphere of academia.* **3** cliquish, clannish, °exclusive, °private, °select, esoteric: *Clarence copes well in the rarefied atmosphere of Whitehall's corridors of power.*

rarely *adv.* °seldom, °infrequently, on rare occasions, °hardly (ever), °scarcely (ever), almost never, *Colloq* once in a blue moon: *He rarely attends formal dinners at his club.*

rarity *n.* **1** °curiosity, °oddity, curio, collector's item, °find, °treasure, conversation piece, *Brit* one-off, *Colloq Brit* oner: *Among the rarities in his collection is a stuffed dodo.* **2** unusualness, uncommonness, rareness, uniqueness, °scarcity: *The rarity of a first edition makes this a valuable book.*

rascal *n.* °imp, °devil, scamp, mischief-maker; rapscallion, °rogue, °scoundrel, cad, °villain, blackguard, knave, °good-for-nothing, ne'er-do-well, °wastrel, scapegrace, dastard, °wretch, *Colloq* scallywag *or* scalawag, *Brit* rotter, blighter, bounder: *That little rascal has eaten all the cherry tarts. They have caught the rascal who was painting graffiti on the walls.*

rash[1] *adj.* °impetuous, °impulsive, °unthinking, °thoughtless, °foolhardy, unconsidered, ill-considered, °ill-advised, injudicious, °imprudent, °indiscreet, °precipitate, °hasty, °careless, °heedless, °reckless, headlong, °wild, madcap, °harebrained, °hot-headed, °adventurous, °quixotic, °venturesome, °audacious, °bold, °dashing, °brash, devil-may-care: *It was rash of him to try to sail around the world single-handed, but he did it.*

rash[2] *n.* **1** °eruption, eczema, redness, efflorescence, dermatitis: *One of the signs of measles is a skin rash.* **2** number(s), °quantity, lot(s), multitude, °profusion, outbreak, series, °succession, °spate, °wave, °flood, deluge, °plague, °epidemic: *There has been a rash of burglaries in the neighborhood.*

rasp *n.* **1** °grating, °scrape, scraping, °scratch, scratching, grinding, stridulation: *One could hear the rasp of the file as they tried to cut through the door.* **2** °file, grater: *The coarser rasp is for use on wood.* —*v.* **3** °scrape, abrade, °grate, °file: *First, rasp the horseradish, then combine the gratings with a little vinegar.* **4** °irritate, °jar (upon), grate upon *or* against, rub (someone) (up) the wrong way, nettle, °irk, °annoy, vex, wear on, °get, *Literary* gride: *Her rasping metallic voice gets on my nerves.* **5** croak, °squawk, screech: *The sergeant rasped out another command and the company marched forward.*

rate[1] *n.* **1** °measure, °pace, gait, °speed, °velocity, *Colloq* °clip: *We moved along at a pretty fast rate. Rates of production must increase.* **2** °charge, °price, °fee, °tariff, °figure, °amount; °toll: *The rate for a hotel room has increased enormously in the past few years.* **3** °percentage, °scale, °proportion: *What is the current rate of interest on government savings bonds?* **4** Usually, ***-rate.*** °rank, °grade, °place, °standing, °status, °position, °class, classification, °kind, °sort, °type, rating, °status, °worth, value, valuation, °evaluation: *After a few years, he had improved from a second-rate typist to a first-rate secretary.* **5** ***at any rate.*** °in any case, °in any event, anyway, °at all events, anyhow, under any circumstances, °regardless, °notwithstanding: *At any rate, with their last child out of college, they have no more worries about paying tuition.* —*v.* **6** °rank, °grade, °class, classify, °evaluate, °estimate, °calculate, °compute, °count, °reckon, °judge, °gauge, assess, appraise, °measure: *How do you rate your chances of beating the champion?* **7** °merit, be entitled to, °deserve, be worthy of, have a claim to: *The play doesn't rate all the attention it's been getting in the press.* **8** °count, °place, °measure: *She rates very high among her friends.*

rate[2] *v.* °scold, °berate, °reprimand, °rebuke, reproach, °reprove, °take to task, °upbraid, censure, *Colloq* °bawl out, °dress down, *US and Canadian* °chew out: *I was rated severely for staying out after midnight.*

rather *adv.* **1** °quite, °very, °somewhat, °fairly, °moderately, to a certain extent *or* degree *or* measure, to some extent *or* degree *or* measure, more or less, °pretty, °slightly, *Colloq* °sort of, kind of: *I suppose she was rather good-looking, but at that age I didn't pay much attention to girls. This coffee tastes rather peculiar.* **2** preferably, sooner, °instead, more readily *or* willingly: *Given an option, I choose to live alone rather than marry.* **3** ***would*** or *US Colloq* ***had rather.*** °prefer, °choose: *Given a choice, I would rather live alone.*

ratify *v.* °approve, °sanction, °endorse, °support, corroborate, °uphold, °back (up), °sustain, °establish, validate, °substantiate, °verify, °authenticate, °guarantee, °warrant, °certify, affirm, °ensure, °clinch, °settle: *The pact was ratified by all members.*

ratio *n.* °proportion, °relationship, correlation, correspondence: *What is the ratio of single to married parents? In that area, the ratio of tenants to homeowners is about two to one.*

ration *n.* **1** °share, °quota, °allotment, °portion, °helping, °part, °provision, °measure, °dole, °percentage, °amount: *Each person is permitted a weekly fixed ration of butter, sugar, coffee, meat, etc. We shall have to go on short rations till we are rescued.* **2** ***rations.*** supplies, °provisions, °food, °provender, victuals, viands, eatables, edibles, comestibles, *Brit* commons: *It was not till the third week, with our rations running low, that we really began to worry.* —*v.* **3** Often, ***ration out.*** °allot, apportion, °dole (out), °give out, °distribute, deal out, mete out, °parcel out, °measure out, °hand out: *As water was in short supply, we had to ration it. One cup was rationed out to each of us every morning.* **4** budget, °schedule, °restrict, °control, °limit: *Water had to be rationed to one cup a day for each of us.*

rational *adj.* **1** °well-balanced, °sane, °sound, °normal, °reasonable, °reasoning, °logical, ratiocinative, clearheaded, cleareyed, °sober; of sound mind, *Colloq* all there: *Desmond is a rational human being who would be able to advise you about your problem.* **2** °discriminating, °intelligent, °thinking, °enlightened, °prudent, °wise, °knowledgeable, °informed: *We could not have elected a more rational representative for our cause than Reggie Blunt.* **3** °sensible, common-sense, commonsensical, °practical, pragmatic, down-to-earth, °everyday, °acceptable, °reasonable, °logical: *Fosditch thinks he has a rational explanation for the liquid's rising in the straw by itself.*

rationale *n.* °reason, °explanation, logical basis, °grounds, °logic, °reasoning, °philosophy, °principle, theory: *Do you understand the rationale of a policy for curbing inflation that sets interest rates at a level where business, initiative, and productivity are stifled?*

rationalize *v.* **1** make plausible *or* believable *or* understandable *or* acceptable *or* reasonable, make allowance(s) *or* excuses for, °vindicate, °account for, °justify, °excuse, reason away, °explain away: *How can you rationalize taking my car without my permission?* **2** think through, reason out; apply logic to, ratiocinate: *He warps the statistics to rationalize his argument. The fundamentalists make no attempt to rationalize their literal interpretation of the Scripture.*

rattle *v.* **1** clatter: *An acorn dropped from an overhanging tree and rattled down the tin roof.* **2** °shake, °vibrate, °jar, joggle, °jiggle: *The wind was strong enough to rattle the windows in their frames.* **3** °unnerve, disconcert, °discomfit, °disturb, °perturb, °shake, discountenance, °upset, °agitate, °put off, *Chiefly US and Canadian* faze: *They are only trying to rattle us by saying that we might lose our jobs if we strike.* **4** jounce, °bounce, °bump, °jolt, °speed, °hurtle: *The stagecoach came rattling down the road, swaying wildly.* **5** ***rattle off.*** °recite, °list, utter, °reel off, °run through, °enumerate, °call off: *The announcer rattled off a long list of trains canceled because of the storm.* **6** ***rattle on.*** °chatter, °babble, °jabber, gibber, prate, °prattle, blabber, cackle, blather, °ramble, *Chiefly Brit* witter, natter, *Slang US* run off at the mouth: *The speaker kept rattling on and on about the delights of collecting beer cans.* —*n.* **7** clatter, °racket, °noise; rale *or* râle, crackle, crackling; death rattle: *I heard the rattle of the beans and knew you were about to grind the coffee. The doctor said he didn't like the rattle in my chest.* **8** °clapper, sistrum, *US* noisemaker: *Rattles have been used as musical instruments for thousands of years.*

rattletrap *n.* flivver, rattler, *Colloq* jalopy *or* jaloppy, tin lizzie, *US* Model T: *This old rattletrap won't get us all the way to Hawkhurst.*

ratty *adj.* **1** °irritable, °cross, °testy, °touchy, annoyed, crabbed, irritated, °angry, °short-tempered, °impatient, °disagreeable: *I felt really ratty this morning after partying all night.* **2** °dirty, °greasy, straggly, °unkempt, matted: *His hair was all ratty and he hadn't bathed in a month.*

raucous *adj.* °harsh, rasping, °rough, °husky, hoarse, °grating, scratching, °scratchy, °discordant, dissonant, jarring; °strident, °shrill, °noisy, °loud, earsplitting, °piercing: *You could hear Ramsay's raucous voice above the others. A raucous scream came from the parrot's cage.*

ravage *v.* **1** °lay waste, °devastate, °ruin, °destroy, °demolish, °raze, °wreck, wreak havoc on, °damage: *The hurricane ravaged outlying areas but did little damage in the city.* **2** °pillage, °plunder, despoil, °ransack, °sack, °loot: *The police tried to prevent hooligans from ravaging the shops in the town center.* —*n.* **3** Usually, ***ravages.*** °destruction, °damage, depredation(s), devastation, wrecking, °ruin, demolition: *All about us we saw the ravages of war.*

rave *v.* **1** °rant, °rage, °storm, fulminate, °roar, °thunder, °howl, °yell, caterwaul, yowl, °bellow, °shout, °scream, °fly off the handle, *Slang* flip one's lid *or Brit* top, raise hell: *He kept raving on about his coffee mug having been taken.* **2** ***rave about.*** °praise, °laud, rhapsodize over, °applaud, gush over: *Bernard can't stop raving about that new Thai restaurant.* —*n.* **3** °praise, acclaim, °favor, enthusiastic reception, °tribute, °testimonial, encomium, °bouquet, plaudits, accolade, °admiration, *US* hosanna: *The play got raves in all the newspapers when it opened.* **4** °rage, °fashion, °vogue, °trend, °thing, last word, *dernier cri,* °craze, °mania, *Colloq* °fad: *The rave that year was for stiletto heels.*

ravenous *adj.* **1** °hungry, °famished, starving, °starved: *We were ravenous after the long hike.* **2** °voracious, °gluttonous, °greedy, insatiable, ravening, swinish, piggish, °hoggish, edacious, wolfish: *With my family's ravenous appetites, there are rarely any leftovers.*

ravine *n.* °gorge, canyon *or* cañon, °pass, cleft, defile, gully *or* gulley, °valley, *Dialect Brit* clough, *Scots* linn, *US* °gap, gulch, arroyo: *At the spring thaw, torrents filled the ravines.*

raving *adj.* **1** °mad, °insane, °berserk, raging, °crazy, crazed, irrational, manic, °maniacal, °frantic, frenzied, °delirious, °hysterical; out of one's mind *or* head: *When she learned that he had overdrawn their account, she became a raving lunatic.* **2** °extraordinary, °outstanding, °unusual, °rare, uncommon, °phenomenal, °great, °striking, °ravishing, *Colloq* °stunning: *Diana had been a raving beauty in her time.* —*n.* **3** Often, ***ravings.*** °rant, ranting, °bombast, pomposity, grandiloquence, magniloquence, rodomontade, °rhetoric, °bluster, blustering, claptrap, balderdash, puffery, bunkum, flatulence, hyperbole, vaporing, fustian, *Colloq* °hot air, bunk: *Why must we be subjected to the ravings of this politician?* **4** gabble, °babble, babbling, °gibberish, *Colloq* °gab: *This is nothing more than the raving of a maniac.*

ravish *v.* **1** °enrapture, °delight, °captivate, enthrall, °fascinate, °charm, °entrance, spellbind, °bewitch, °transport: *I was ravished at the prospect of seeing you again.* **2** °rape, °violate, have one's way with, deflower, defile: *The soldiers attacked the village and ravished the women.*

ravishing *adj.* °dazzling, °beautiful, °gorgeous, °striking, °radiant, charming, alluring, °attractive, entrancing, captivating, °enthralling, bewitching, °spellbinding, *Colloq* °stunning: *Daphne was even more ravishing than her sister Diana.*

raw *adj.* **1** uncooked, °unprepared, °fresh: *Many vegetables can be eaten raw.* **2** unprocessed, untreated, °unrefined, unfinished, °natural, °crude: *The raw sugar is shipped to the plant where it is refined. The raw statistics revealed that many women had refused to respond to the questionnaire.* **3** °new, inexperienced, unseasoned, °immature, °green, °untried, °fresh, untrained, unskilled, untested: *They sent us some raw recruits who had never seen battle.* **4** exposed, unprotected, uncovered, °open; °sore, °tender, °inflamed, °painful, °sensitive: *When he mentioned her name, it was as if he had touched a raw nerve.* **5** °chill, °chilly, chilling, °cold, °damp, °frigid, °freezing, °biting, stinging, nippy, nipping, °sharp, °keen, °piercing, °penetrating: *Wear a warm scarf against that raw wind.* **6** °brutal, °frank, °candid, °blunt, °direct, °unvarnished, unmollified, unembellished, °realistic, °honest, °plain, unreserved, unrestrained, °uninhibited, °bluff, straightforward: *The producer felt that no domestic audience could stomach a raw depiction of the horrors of war.* —*n.* **7 *in***

the raw. °naked, stark-naked, undressed, unclothed, °nude, in the nude, *Brit* starkers, *Colloq* in the buff, in the altogether, in one's birthday suit: *For punishment, they made us stand outside, in the freezing cold, in the raw.*

rawboned *adj.* °gaunt, °lean, gangling, °thin, °skinny, °spare, °meager, °scrawny, underfed, bony, °emaciated, half-starved, wasted, hollow-cheeked, cadaverous: *They'll take that rawboned youngster into the army and put some meat on him.*

ray *n.* **1** °beam, °shaft, °bar, °streak, pencil, °gleam, °flash: *A single ray from the setting sun illuminated the head of the statue.* **2** glimmer, °trace, °spark, scintilla, °flicker: *As he had not been refused, there was still a ray of hope that he might be accepted.*

raze *v.* tear *or* pull *or* bring *or* knock *or* throw down, °demolish, °destroy, °level, °flatten, bulldoze: *As usual, beautiful old buildings were razed to make way for a new shopping center.*

reach *v.* **1** Often, ***reach out.*** °hold out, °extend, °stretch (out), °stick out, thrust out, outstretch, outreach: *He reached out his hand and took mine warmly.* **2** °arrive at, °get to, °come to, go to, end up at *or* in; land at *or* in, *Colloq* °make (it to): *I won't reach the office till noon today. Have we reached the point of no return?* **3** °get, get in touch with, °communicate with, establish *or* make contact with, get through to, get (a) hold of: *I tried to reach her at home, but she was out.* **4** attain, °achieve, °accomplish, °make, get *or* go to, get *or* go as far as: *Do you think that McClintock will reach the semifinals?* **5** come *or* go *or* get up to, °amount to, attain, climb to, °rise to, run to, °equal, °match: *It looks as though the trade deficit might reach last year's figure.* **6** get through *or* across to, °register with, °communicate with, reach into the mind of, °impress, °influence, °sway, °move, °stir, carry weight with: *It is very difficult to reach teenagers, who have a natural suspicion of adults.* —*n.* **7** °range, ambit, °scope, °orbit, compass, °sphere, °territory: *Those outlying areas are beyond our reach. Padua is within easy reach of Venice.* **8** °capability, °capacity: *Tragedy often occurs when one's ambition exceeds one's reach.*

react *v.* **1** °act, °behave, °conduct oneself, °proceed; °retaliate, °reciprocate, get even: *I wonder how you would have reacted if someone had insulted you!* **2** °respond, °answer, °reply, °retort: *The crowd reacted to the police presence by becoming very nasty. She reacted with surprise on learning of his arrest.*

reaction *n.* **1** °response, °reply, °answer, °effect, feedback: *Did you get any reaction when you told her I had been arrested? Her reaction was to turn over and go back to sleep.* **2** repulsion, °resistance, counteraction, counterbalance, compensation: *A rocket derives forward thrust by reaction against its own exhaust, hence can operate in a vacuum.* **3** retaliation, reciprocation, °reprisal, °revenge: *In reaction to the insult, the knight ran the varlet through with his sword.*

reactionary *adj.* **1** °conservative, ultraconservative, °right, rightist, right-wing; traditionalist; *Chiefly Brit* blimpish, *South African* verkrampte: *Reactionary governments tend to reduce social services.* —*n.* **2** °conservative, ultraconservative, rightist, right-winger; traditionalist; *Brit* Colonel Blimp, *South African* verkrampte: *As a reactionary, he feels that the status quo is fine and change unnecessary.*

read *v.* **1** °peruse, °scan, °skim, °review, °study, °look over, °pore over: *I have read the article twice but still cannot understand it.* **2** °understand, °know, be familiar with, °comprehend, °interpret, °decipher: *Can you read Greek? Madame Sonja reads tea leaves.* **3** °announce, °present, °deliver: *Anna reads the news on TV at six o'clock every evening.* **4** ***read into.*** assign to, °impute (to), °infer (from), °assume (from), °presume (from), conclude (from): *You might be reading too much into that letter of commendation.*

readable *adj.* **1** °intelligible, comprehensible, understandable, easy to understand, easily understood, °plain: *Surprisingly, the instructions for operating the computer are fairly readable.* **2** °entertaining, easy to read, enjoyable, °pleasurable, °absorbing, °interesting, °engaging, °stimulating, °worthwhile: *That book on insomnia is so readable that I stayed up reading it all night.* **3** °legible, decipherable, °distinct: *The print on that microfiche is too small to be readable.*

readily *adv.* **1** cheerfully, °willingly, eagerly, ungrudgingly, unhesitatingly, °freely, °gladly, °happily, agreeably, graciously, charitably: *He always gave readily of his time to anyone who sought his help, and we shall miss him sorely.* **2** effortlessly, °easily, smoothly, without difficulty: *I was able to get the wheel off readily enough, getting it back on was the problem.* **3** °promptly, °quickly, speedily, °swiftly, apace, °at once, without delay, °in no time, °immediately, instantly, °instantaneously, instanter, *Brit* °straightaway, *US* right away, at *or* on short notice, *Colloq* pronto: *The police came readily but were unable to catch the thieves, who had escaped.*

readiness *n.* **1** willingness, cheerfulness, goodwill, °eagerness, agreeableness, graciousness: *Her readiness to help marked her as truly charitable.* **2** promptness, quickness; °facility, °ease, °skill, adroitness, expertness, °proficiency: *The readiness with which he gave expert advice on gardening made him very popular with his neighbors.* **3** ***in readiness.*** See **ready, 1,** below.

ready *adj.* **1** °prepared, °(all) set, °in readiness, in (proper) shape; up, primed, °ripe, °fit, in condition; *Colloq* psyched (up): *The charge is ready for detonation. I am ready for a good run in the park. Are these apples ready for eating? Are you ready to meet the girl of your dreams?* **2** °agreeable, consenting, acquiescent, °willing, °content, °eager, °keen, °happy, °cheerful, °genial, °gracious, °cordial, °friendly, well-disposed, °enthusiastic, *Colloq* °game: *Walter was always a ready accomplice to any mischief devised by his brother.* **3** apt, °likely, °inclined, °disposed, °given, °prone: *She was all too ready to believe anything she was told.* **4** °about, °liable, °likely, apt; °on the verge of, °subject to, °in danger of, on the brink of, on the point of, °close to: *The volcano seemed ready to erupt at any moment.* **5** °prompt, °rapid, °quick, °immediate, °speedy, °swift, °punctual, °timely: *Alistair seems to have a ready answer for everything.* **6** °clever, °keen, °sharp, °agile, deft, °skillful, adroit, °alert, °bright, °intelligent, °perceptive, °quick: *It was Carlotta's ready wit that first attracted me.* **7** on *or* at *or* to hand, °handy, °available, °accessible, at (one's) fingertips, °at the ready, close at hand, °convenient: *If you don't have the ready cash, you may pay by check.* —*n.* **8** ***readies.*** °money, °cash, wherewithal: *If you have the readies, please pay for this meal.* **9** ***at the ready.*** **(a)** waiting, °on tap, °expectant, in position, °poised: *Keep your pistol at the ready in case there's trouble.* **(b)** See **8,** above. —*v.* **10** °prepare, make *or* get ready, °set, fit out, °equip, °organize, psych up: *He readied himself as best he could for the ordeal.*

ready-made *adj.* **1** ready-to-wear, finished, prefabricated, *Brit* off-the-peg: *These ready-made clothes fit me better than custom-made.* **2** °convenient, °expedient, °serviceable, usable, °handy, °useful, °suitable, °adaptable; °plausible, credible, believable: *The storm provided a ready-made excuse for cancelling the appointment.* **3** stereotyped, stereotypic(al), hackneyed, time-worn, trite, °stale, °conventional, unoriginal, °stock, °pedestrian, °routine, °run-of-the-mill: *Count on Weekley to* *come up with one of his ready-made clichés to describe the situation.*

real *adj.* **1** °genuine, °true, °actual, °authentic, verified, verifiable, °legitimate, °right, °bona fide, °official; °legal, °legitimate, licit, °natural, valid, °veritable: *He says that his name is Smith, but what is his real name? Her diamonds are not real. It has been proved beyond the shadow of a doubt that he is the real father.* **2** °genuine, °actual, °true, existent, °authentic, °natural; °material, °physical, °tangible, palpable, corporeal: *How can one distinguish between the real object and what one sees, which is merely its image?* **3** °genuine, °sincere, °heartfelt, °true, °actual, unfeigned, °unaffected, °earnest, °honest, °truthful, °legitimate, valid: *Deirdre's face is so impassive that it is impossible to tell what her real feelings are. What is the real reason that you are here?* **4** °genuine, °actual, °true, °loyal, °trustworthy, trusted, °honest: *You are a real friend, and I know I can count on you.* **5** °intrinsic, °actual, °true, °genuine, °proper, °essential: *The real value of something is what one is willing to pay for it.* —*adv.* **6** See **really,** below.

realistic *adj.* **1** practical, °matter-of-fact, down-to-earth, pragmatic, commonsense, °sensible, reasonable, °levelheaded, °rational, °sane, hardheaded, businesslike, °no-nonsense, unromantic, unsentimental, °tough, tough-minded, *Colloq* hard-nosed, hard-boiled: *Kevin's realistic grasp of the problems may make the company profitable.* **2** °natural, °lifelike, true-to-life, naturalistic, °vivid, °genuine, °real, °graphic: *This painting is so realistic that it looks like a photograph. Harriet's description was so realistic that I felt I had actually been to Benidorm.*

reality *n.* **1** actuality, °fact, °truth, genuineness, authenticity, *Aristotelianism* entelechy: *It is important to deal with the reality not the potentiality of the situation.* **2** ***in reality.*** See **really,** below.

realization *n.* **1** °conception, °understanding, comprehension, apprehension, awareness, °appreciation, °perception, °recognition, °cognizance: *Full realization of what he had done finally dawned on him.* **2** actualization, °consummation, °accomplishment, °achievement, °establishment, °fulfillment, materialization, effectuation: *The realization of the dream of a united Europe seems possible in the last decade of the twentieth century.*

realize *v.* **1** make real, °effect, °bring about, make happen, make a reality, actualize, °accomplish, °produce, °achieve, °fulfill, °materialize, °effectuate: *Many of Leonardo da Vinci's inventions were not realized in his lifetime.* **2** °understand, °appreciate, °comprehend, be aware of, conceive of, °grasp, °perceive, discern, be *or* become conscious *or* aware *or* appreciative of, °recognize, °see, *Colloq* °catch on (to), cotton (on) to, *Brit* °twig: *Did Clarissa fully realize the consequences of her actions?* **3** °return, °gain, °clear, °profit, °make, °earn, bring *or* take in, °net, °produce, °get: *We have realized a substantial profit on the sale of the shares bought when the company was privatized.*

really *adv.* **1** genuinely, °actually, °truly, °honestly, °in reality, in actuality, in (point of) fact, as a matter of fact, °surely, °indeed, °absolutely, °definitely: *Is he really going to do the high-wire act without a net?* **2** °indeed, °actually, °absolutely, unqualifiedly, °positively, categorically, unquestionably, °definitely, undeniably: *She is really the musical one in the family.* **3** °in effect, °in reality, °actually, °in fact, de facto, in the end, °at bottom, in the final analysis, at the end of the day, *Colloq* deep down: *Greene's protests to the contrary, it was really Culligan who saved the platoon.* **4** °very, °extremely, °quite, exceptionally, remarkably, unusually, uncommonly, extraordinarily, °exceedingly, *Nonstandard* °real: *I saw a really good film last night.*

realm *n.* **1** °domain, °kingdom, empire, °monarchy, principality, palatinate, duchy *or* dukedom: *They searched the realm for a girl whose foot would fit into the glass slipper.* **2** °territory, °area, bailiwick, °department, °responsibility, °jurisdiction: *Collecting taxes is not within the realm of our office.* **3** °area, confines, °sphere, °limits: *Cooperation with foreign companies is certainly within the realm of possibility.*

ream *v.* °drill (out), °bore (out), open up, °tap: *The carpenter reamed three holes in the beam.*

reap *v.* **1** °harvest, °garner, glean, °gather (in), °mow, take in *or* up: *The land was fertile, and a bountiful crop was reaped. As ye sow, so shall ye reap.* **2** °profit, °bring in, °gain, °procure, °acquire, °get, °obtain, °take in: *It is impossible to reap much from so small a margin between cost and selling price.*

rear[1] *n.* **1** °back (part), °end, hind part, *Nautical* °stern, *Colloq* °tail (end), fag end, *US and Canadian* tag end: *The rear of the column was attacked as it marched through the defile. We took seats at the rear of the auditorium.* **2** hindquarters, °posterior, rump, °buttocks, *Colloq* °bottom, behind, backside, rear end, *Slang Brit* arse, °bum, *US and Canadian* ass, tokus, toches; *Babytalk US and Canadian* tushie *or* tushy *or* tush, hinie: *As he left, he felt someone give him a pat on the rear.* —*adj.* **3** °back, °last, °end, rearmost, *Nautical* aft, after, °stern: *The rear window is all misted over. The troop was bringing up the rear echelon.*

rear[2] *v.* **1** °raise, °bring up, °care for, nurture, °nurse; °cultivate, °educate, °train; °breed, °produce: *They reared six children in that tiny house. Is this where they rear Aberdeen Angus cattle?* **2** °erect, °raise, °build, °put up, °construct, °fabricate, °create: *The family reared an elaborate mausoleum to Ortega's memory.* **3** °raise, °lift, °put up, upraise, uplift, °hold up: *Just mention intelligence these days and espionage rears its ugly head.*

reason *n.* **1** justification, °argument, °case, °explanation, °rationale, ground(s), °pretext, vindication, °defense, why (and wherefore), *Literary* apologia, apology: *What reason did they give for confiscating our luggage?* **2** °judgment, common sense, °intelligence, °sanity, sense(s), saneness, °insight, perspicacity, percipience, °understanding, rationality, °reasoning, °mind, °intellect: *He must have lost his reason to accuse you of embezzlement.* **3** °purpose, °aim, °intention, °object, °objective, °goal, °motive, °end, °point: *His reason for leaving now is to be on time for the curtain.* **4** °excuse, rationalization: *Ignorance of the law is not an accepted reason for breaking it.* **5** ***by reason of.*** because of, on account of, °owing to, °by virtue of, as a result of; due to: *He was declared unfit to stand trial by reason of insanity.* **6** ***within*** *or* ***in reason.*** °reasonable, °sensible, justifiable, °rational, °fitting, °proper, °acceptable: *We will consider any offer for the house that is within reason.* —*v.* **7** °think, conclude, °calculate, °reckon, °estimate, °figure (out), °work out, °deduce, act *or* think rationally *or* logically, ratiocinate, use (one's) judgment *or* common sense, use (one's) head, put two and two together: *I reasoned that she would be arriving at the station at that very moment. Rudolph certainly hasn't lost his powers of reasoning.* **8** ***reason with.*** argue with, remonstrate with, debate with, discuss with, talk over with, °plead with, °convince; °persuade, dissuade, °urge, °prevail upon: *Gene is very stubborn and I cannot reason with him.*

reasonable *adj.* **1** °sensible, °rational, commonsense, common-sensical, °sane, °logical, °sober, °sound, °judicious, °wise, °intelligent, °thinking: *I had always thought Philip a reasonable man.* **2** credible, believable, °plausible, °tenable, reasoned, arguable, well-thought-out, well-grounded: *Her story about having found the* *child wandering in the street doesn't sound very reasonable to me.* **3** °moderate, °tolerable, °acceptable, °within reason, °equitable, °fair; °inexpensive, unexcessive, unextravagant, °economical, °conservative: *Considering inflation, we thought the price increases were reasonable.* **4** °appropriate, °suitable, °proper, °sensible, °right: *The car was traveling at a reasonable speed, considering conditions.*

reasoning *n.* **1** °thinking, °logic, °analysis, rationalization: *The entire argument falls to pieces because of faulty reasoning.* **2** reasons, arguments, premises, °rationale, postulate, °hypothesis, theory, °explanation, explication: *His reasoning had no effect on her--she simply refused to listen.*

reassure *v.* °comfort, °encourage, hearten, °buoy (up), °bolster, °cheer, uplift, inspirit, °brace, °support, restore confidence to, set *or* put (someone's) mind at rest, set *or* put (someone) at ease, settle (someone's) doubts: *The minister reassured the public regarding the safety of public transport.*

rebate *n.* **1** °discount, reduction, °deduction, °allowance, markdown, cutback, refund, repayment, *Colloq US* °rake-off: *This coupon entitles you to a rebate of 10 percent on the purchase of four new tires.* **2** °kickback, °percentage, °rake-off, commission, °cut, °bribe, °graft, *Colloq chiefly US* payola, *US* plugola: *He gets a rebate of 5 percent on every client he refers to his lawyer.* —*v.* **3** °discount, °reduce, °deduct, °mark down, refund, °repay; kick back: *Periodicals rebate 15 percent of their gross advertising revenue to the agencies that place ads.*

rebel *v.* **1** °revolt, °mutiny, rise up: *The people will rebel if subjected to more oppression.* **2** Often, ***rebel against.*** °defy, °flout, °dare, °challenge; °disobey, dissent: *It is in the nature of young people to rebel against authority and the status quo.* —*n.* **3** °revolutionary, revolutionist, insurgent, insurrectionist, mutineer, resister, resistance fighter, freedom fighter: *We call ourselves fighters for justice, but the unjust call us rebels and traitors.* **4** heretic, °nonconformist, apostate, dissenter, recusant, °schismatic: *James's problem was that he was a rebel without a cause.*

rebellious *adj.* **1** °insubordinate, °defiant, °mutinous, °revolutionary, contumacious, insurgent, insurrectionary, °seditious: *The more rebellious students organized mass meetings against the university authorities.* **2** unmanageable, °disobedient, °incorrigible, °ungovernable, °unruly, °difficult, refractory, °stubborn, °obstinate, °recalcitrant: *They resented being treated as rebellious children.*

rebirth *n.* renaissance *or* renascence, °revival, renewal, reawakening, °resurgence, revitalization, resurrection, regeneration, rejuvenation, °restoration, new beginning, reincarnation; metempsychosis, palingenesis: *The fifteenth century marked the rebirth of learning in Europe after the so-called Dark Ages.*

rebound *v.* **1** spring back, °bounce, °recoil, ricochet, resile: *The ball rebounded from the goal post.* —*n.* **2** °bounce, °recoil, ricochet, °return, comeback, °repercussion, °backlash, reflex: *White caught the ball on the rebound.*

rebuff *n.* **1** °rejection, snub, °check, °denial, °repulse, °refusal, °dismissal, °defeat, repudiation, °slight, discouragement, *Colloq* cold shoulder, °cut, °put-down, °brushoff, *Slang US* °brush: *Everton's friendly approach was met by an abrupt rebuff.* —*v.* **2** °reject, snub, °check, °deny, °repel, drive away, °spurn, °repulse, °refuse, °dismiss, °defeat, °repudiate, °slight, °ignore, *Brit* send (someone) to Coventry, *Colloq* give (someone) the cold shoulder, °cut, put (someone) down, brush (someone) off, give (someone) the brushoff, tell (someone) where to go *or* get off, tell (someone) where to get lost, *Slang* hand (someone) his *or* her *or* their walking papers, freeze (someone) out, *US* give (someone) the brush, show (someone) the door: *Because she rebuffed everyone who proposed marriage, Cynthia remained a spinster.*

rebuke *v.* **1** scold, reproach, admonish, °reprove, °reprimand, °lecture, censure, chide, reprehend, °berate, °castigate, °criticize, °take to task, °upbraid, revile, *Colloq* °dress down, °bawl out, give (someone) a piece of one's mind, haul (someone) over the coals, let (someone) have it, give (someone) hell *or* what for, tell (someone) off, tell (someone) where to get off, *Brit* carpet, tear (someone) off a strip, tick (someone) off, wig, blow (someone) up, give (someone) a wigging, *US and Canadian* rake (someone) over the coals, give (someone) the business, chew out: *The headmaster severely rebuked those boys who had violated the school curfew.* —*n.* **2** scolding, reproach, admonition, °reproof, °reprimand, °lecture, censure, chiding, reprehension, berating, castigation, °criticism, upbraiding, revilement, °tongue-lashing, *Colloq* dressing-down, what for, *Brit* wigging, blowup *or* blowing up, *Slang* hell: *Rawlings was let off with nothing more than a rebuke.*

rebut *v.* refute, °deny, °disprove, confute, invalidate, negate, °discredit, belie, °contradict, controvert, °puncture, °expose, °destroy, °ruin, °devastate, *Colloq* shoot full of holes, knock the bottom out of, shoot down, blow sky-high: *We have all the evidence needed to rebut the contention that our client was anywhere near the scene of the crime.*

rebuttal *n.* °answer, °reply, °retort, °response, rejoinder, counter-argument, riposte, retaliation, °denial, refutation, contradiction, confutation, *Colloq* comeback: *In rebuttal, the defense tried to prove that the witness for the prosecution had been bribed.*

recalcitrant *adj.* °stubborn, °obstinate, °willful, °defiant, refractory, headstrong, °perverse, °contrary, contumacious, °mutinous, °rebellious, fractious, °unruly, unmanageable, °ungovernable, uncontrollable, wayward, °insubordinate, intractable, unsubmissive, unyielding, unbending, adamant, °immovable, °inflexible, °stiff, °firm: *Recalcitrant students who refuse to obey the rules risk expulsion.*

recall *v.* **1** °remember, °recollect, think back to, reminisce over *or* about, call to mind: *I recall the happy days we had at Biarritz.* **2** °withdraw, °retract, call back, °summon: *The legions guarding the frontier provinces were recalled to Rome.* **3** rescind, °cancel, annul, nullify, °retract, °withdraw, °revoke, °recant, °take back, call back; disavow, disown, °deny: *Once you have moved a piece on the board, you may not recall it. Despite the fact that you might have been hasty, you cannot recall your promise.* —*n.* **4** °memory, °recollection, remembering, °remembrance: *The settlement of the area was in days beyond recall.* **5** withdrawal, recantation, °cancellation, revocation, annulment, nullification, recision, rescission, retraction, °repeal, disavowal, °denial: *The decision was made without any possibility of recall.* **6** withdrawal, retraction, °return: *The manufacturer's recall of the shipment of defective toasters was voluntary.*

recant *v.* °recall, forswear, °deny, rescind, °repudiate, disavow, disclaim, °withdraw, °revoke, °retract, °forsake, °abandon, apostasize, °renounce, abjure, °take back: *Once his accuser recanted his sworn testimony, the defendant was released.*

recapitulate *v.* summarize, °sum up; °repeat, °go over (again), °reiterate, restate, °review; °recount, °enumerate, °recite, °relate, °list, *Colloq* recap: *I shall begin by recapitulating the issues covered in my last lecture.*

recede *v.* **1** °ebb, °subside, fall *or* go *or* move back, abate, °return, °withdraw, °retreat, °back up: *As the flood waters receded, people began to go back to their homes.* **2** °diminish, lessen, °decline, °dwindle, °shrink,

°wane, °fade, become more distant *or* less likely: *The likelihood of our being rescued receded with each passing day.*

receipt *n.* **1** sales receipt, register receipt, sales slip, ticket, °stub, counterfoil, proof of purchase, voucher: *Any request for a return or exchange must be accompanied by a receipt.* **2** °delivery, acceptance, °reception, °arrival: *He has acknowledged receipt of the merchandise.* **3** ***receipts.*** °income, °proceeds, °gate, takings, gains, °return, *Colloq* °take: *The receipts from ticket sales were below expectations.*

receive *v.* **1** °get, °obtain, °come by, °collect, °take (into one's possession), °accept, be given, °acquire, come into, °inherit, °gain, °profit, °make: *She received a fortune from the sale of the house. What does he receive when his aunt dies?* **2** °earn, be paid, °make, °draw, °gross, °net, °clear, °pocket, *Colloq* take home, walk off *or* away with, *US* °pull down: *He receives a bigger weekly salary than I make in a year.* **3** °greet, °meet, °welcome; show in, give entrée, °let in, °admit: *We received our guests in the foyer.* **4** °experience, °undergo, °endure, °suffer, °bear, °sustain, be subjected to, °meet with: *In return for helping him all I received was a punch in the nose.* **5** °gather, °hear, °learn, ascertain, be told, be informed *or* notified of, °find out, °pick up: *We received the information only yesterday that you were coming to stay for a month.*

recent *adj.* °late, °latest, °new, just out, °brand-new, °fresh; °current, °modern, up-to-date, late-model: *Recent arrivals will be interviewed by the committee. This wine seems to be of recent vintage.*

receptacle *n.* container, holder, repository; °box, tin, can, °case, °casket, °chest, reliquary, °vessel, °bag, basket: *Please put the rubbish into the receptacle. She kept the diamond in a velvet-lined receptacle which she always carried with her.*

reception *n.* **1** °welcome, °greeting, °treatment, °reaction, °response: *My suggestion that we open a bottle of champagne met with a warm reception.* **2** °party, levee, °social, soirée, °function; °opening, °preview, vernissage, *Colloq* do: *Would you like to go to the reception at the new art gallery next week?*

receptive *adj.* **1** °open, °hospitable, amenable, pervious, persuasible, °tractable, °flexible, pliant, °interested, °willing, °responsive: *I have always found Peggy receptive to suggestions on improving her garden.* **2** °quick, °alert, °perceptive, °astute, °intelligent, °keen, °sharp, °bright, °sensitive: *Julia has a receptive mind.*

recess *n.* **1** alcove, °niche, °nook, °cranny, bay, °hollow: *The statuette fits perfectly in that recess outside the library.* **2** °respite, °rest, °interlude, time off, °break, °intermission, breather, breathing spell, °pause; °holiday, vacation: *A brief recess has been scheduled to allow us to relax before continuing.* **3** ***recesses.*** innermost reaches, corners, secret places, °depths, penetralia: *I had to search for her name in the recesses of my mind.*

recession *n.* °setback, (economic) downturn, °slump, °decline, dip, depression: *The slowing of the economy led to a recession which soon deepened into a depression.*

recipe *n.* **1** °formula, °prescription: *Elizabeth gave me a marvelous recipe for chocolate cake.* **2** °plan, °procedure, °method, °approach, °technique, °way, °means, °system, °program, modus operandi, *Colloq US* MO (= 'modus operandi'): *The minister insists that his recipe for conquering inflation will work.*

recipient *n.* receiver, beneficiary, heir *or* heiress, legatee: *Who is to be the recipient of her estate?*

reciprocal *adj.* °mutual, exchanged, returned, complementary, correlative, °common, shared, °joint, requited: *The reciprocal courtesies enjoyed among our neighbors benefit us all.*

reciprocate *v.* °repay, recompense, °requite, °exchange, °return, °trade, °match, °equal, °correspond: *You have always been very kind to me and I want to reciprocate in any way I can.*

recital *n.* **1** (solo) concert, °performance, musicale, °presentation, °show, °entertainment: *We have tickets to a recital for that night.* **2** °report, °narration, °account, °recitation, °description, °relation, °telling, recounting, °narrative, °rendition, °interpretation, °version, recapitulation, °rehearsal, °repetition, *Colloq* recap: *A reporter who was on the flight provided us with a recital of the events leading up to the crash.*

recitation *n.* **1** reciting, °performance, reading, monologue: *At the age of four Hubert delighted his parents with his recitation of "The Charge of the Light Brigade."* **2** See **recital, 2,** above.

recite *v.* **1** °quote, °repeat, °present: *In the old days, children were often called upon to recite poems they had learned by heart.* **2** °report, °narrate, °recount, °relate, °tell, °describe, °detail, °chronicle, °list, °share, °recapitulate, °repeat, *Colloq* recap: *Again I recited every particular of the holdup that I could recall.*

reckless *adj.* °careless, °rash, °thoughtless, incautious, °heedless, °foolhardy, °imprudent, unwise, injudicious, °impulsive, °irresponsible, negligent, unmindful, °foolish, devil-may-care, °daredevil, °wild, °breakneck, °dangerous, madcap, °mad, °harebrained: *I won't go with Olive because she's a reckless driver.*

reckon *v.* **1** Often, ***reckon up.*** °calculate, °compute, °add (up), °figure (up), °tally (up), °sum up, °total (up), work out *or* up: *If you will reckon up the bill, I'll pay it at once.* **2** °include, °count, °number, °enumerate, °list, °name, °consider, °account, °judge, deem, look upon, °regard, °view, °think of, °hold, °gauge, °estimate, appraise, value, °rank, °rate, °class: *I have always reckoned Arthur as being among my closest friends.* **3** °suppose, °think, °assume, °presume, daresay, °venture, °imagine, °fancy, °consider, conclude, be of the opinion, *US or Colloq* °guess: *She reckoned that no one could accuse her of idle chatter.* **4** ***reckon on*** *or* ***upon.*** °count on, °rely on, °depend on, °lean on, trust in, take for granted, *Colloq* bank on: *I was reckoning on your help, and I hope you won't disappoint me.* **5** ***reckon with.*** **(a)** settle (accounts) with, °take care of, °look after, see *or* attend to, °deal with, °handle, pay attention to, °think about: *I have to go now, but I'll reckon with you later.* **(b)** take into account *or* consideration, °consider, °contemplate, °account for, °remember, °bear in mind: *He thought he had got away with it, but he failed to reckon with Inspector Hawkeye of Scotland Yard.*

reckoning *n.* **1** counting, °calculating, °calculation, computation, enumeration, °addition: *The reckoning of Old Style dates used the Julian calendar.* **2** °bill, °account, invoice, *US* °check, *Colloq chiefly US and Canadian* °tab: *If you let me have the reckoning, I'll pay it with my credit card.* **3** °(last) judgment, °retribution, final account(ing) *or* settlement, °doom: *On the day of reckoning we must all face the music.*

reclaim *v.* °restore, °recover, °rescue, °redeem, °salvage, °save, regain, °retrieve, regenerate, °rejuvenate: *Much of the farmland was reclaimed from the sea.*

recline *v.* °lie (down), lie back, lean back, °lounge, °rest, °repose, °sprawl, loll, stretch out: *Just recline on the sofa and make yourself comfortable.*

recluse *n.* °hermit, anchorite *or* anchoress, monk *or* nun, eremite: *For the past years, he has been living the life of a recluse.*

reclusive *adj.* °solitary, lone, °secluded, °isolated, eremitic(al), hermitic, anchoritic, monastic, cloistered, sequestered, °retiring, °shut off: *She insists that she enjoys her reclusive existence.*

recognition *n.* **1** °identification, detection: *Her recognition of the man on TV led to his arrest.* **2** °acknowledgment, °notice, °attention, °cognizance, acceptance, awareness, °perception, °admission; °honor, °appreciation: *The award was recognition of his standing as an outstanding art expert.*

recognize *v.* **1** °identify, °detect, °place, °recall, °remember, °recollect, °know (again): *I recognized him from seeing his picture in the newspaper.* **2** °acknowledge, °perceive, °understand, °realize, °see, °admit, °accept, own, °concede, °allow, °grant, °appreciate, °respect, be aware of: *He refuses to recognize that he is responsible for his own actions.* **3** °approve, °sanction, °endorse, °accept, validate, °ratify: *A totalitarian government refuses to recognize the rights of the individual.* **4** °honor, give recognition to, °salute, show gratitude *or* appreciation, °reward, °distinguish, pay respect, do homage: *The society recognizes Dr. Jackson for his many contributions to science.*

recoil *v.* **1** jerk *or* jump *or* spring back, °start, °flinch, wince, °shrink, blench, balk *or* baulk, shy (away) (from): *He recoiled when confronted with the evidence of the gruesome experiments with live animals.* **2** °rebound, bounce back, resile, kick back: *When fired, the gun recoiled, bruising his shoulder.* —*n.* **3** °kick, °rebound, °repercussion, °backlash: *He was unprepared for the violent recoil of the old blunderbuss.*

recollect *v.* °recall, °remember, call to mind: *Now that you mention it, I do recollect your asking me to buy some milk when I went out.*

recollection *n.* °memory, °recall, °remembrance, °impression, °reminiscence: *His recollection of our grandparents is quite different from mine.*

recommend *v.* **1** °counsel, °advise, °guide, °urge, exhort, °suggest, °advocate, °propose, °(put) forward, °propound, °persuade: *He recommended that the entire system should be changed.* **2** °endorse, °praise, commend, mention favorably, vouch for, °second, °subscribe to, °back, °push, °favor, °approve, °underwrite, °stand up for, °support, °promote, *Colloq* °tout, °plug: *I wish Caswell would stop recommending his brother-in-law for a job with us.* **3** make attractive *or* advisable *or* interesting *or* acceptable: *Wendover has little to recommend it as a tourist spot.*

recommendation *n.* **1** °counsel, °advice, °guidance, urging, exhortation, °direction, °encouragement, °suggestion, prompting, advocacy, °proposal: *It was on your recommendation that we tried that restaurant.* **2** °endorsement, °praise, commendation, favorable mention, °backing, °blessing, °approval, approbation, °support, °promotion, good word, °testimonial, °say-so: *The chairman of the board gave Caswell's brother-in-law his personal recommendation.*

reconcile *v.* **1** get *or* bring (back) together, °unite, reunite, settle *or* resolve differences between, restore harmony between, make peace between, placate, make compatible: *Do you think that a marriage counselor could help reconcile John and Martha?* **2** °resign, °submit, °accommodate, °adjust: *Lady Mary Wortley Montagu wrote that she was reconciled to being a woman when she reflected that she was thus in no danger of ever marrying one.*

reconciliation *n.* **1** conciliation, appeasement, propitiation, pacification, placation, °rapprochement, reconcilement, °understanding, détente, reunion, °harmony, concord, °accord, amity, °rapport: *It was with great effort that the reconciliation between the two powers was finally arranged.* **2** compromise, °settlement, °agreement, arbitration, conformity, compatibility, °adjustment: *Reconciliation of their differences seems difficult but not impossible.*

recondite *adj.* abstruse, arcane, °obscure, esoteric, °deep, °profound, °incomprehensible, unfathomable, impenetrable, undecipherable, °opaque, °dark, °occult, cabbalistic *or* cabalistic *or* kabbalistic *or* kabalistic, °inexplicable, enigmatic: *I have difficulty following the more recondite aspects of information theory.*

reconnaissance *n.* °survey, °examination, scouting, °exploration, reconnoitering, °investigation, inspection, °scrutiny, *Slang Brit* recce: *Their reconnaissance showed that the enemy forces had withdrawn.*

reconnoiter *v.* °survey, °examine, °scout (out), °scan, °explore, °investigate, °inspect, °scrutinize, °check out, check up (on), *Slang Brit* recce: *We reconnoitered the territory and reported the enemy troop movements.*

record *v.* **1** °write (down), °transcribe, °document, °register, °note, make a notation (of), °take down, put *or* set down, log, °chronicle, °report, °itemize, °list, °enumerate, catalog: *I recorded the events of the day in my diary.* **2** °tell of, °relate, °recount, °narrate, °recite: *This ancient manuscript records an eyewitness account of the Battle of Hastings.* —*n.* **3** recording, °report, °document, log, °journal, °memorandum, °note, °notation, minute(s), °transactions, archive(s), annal(s), °chronicle, °diary, °account, dossier, °register, °list, catalog: *Portia has kept a careful record of all the men she ever knew.* **4** documentation, °data, °information, °evidence; °memorial, °memento, souvenir: *There is no record of your birth in the church archive. I took away a piece of the Berlin Wall as a record of my visit.* **5** °(best) performance, track record, °accomplishment; °time; °distance; °height: *Her records for the 110-meter hurdles and the high jump still stand.* **6** accomplishment(s), deed(s), °history, °reputation, curriculum vitae, CV, *Colloq* track record: *He has a record of being late everywhere. His criminal record is none too savory.* **7** disk *or* disc, recording, album, °release, LP (= 'long playing'), 78, 33⅓, 45, EP (= 'extended play'), maxisingle, CD (= 'compact disk *or* disc'), *Brit* gramophone record, *US* phonograph record, *Colloq* °single, *Slang US* °platter: *We put a jazz combo together and cut a few records last year.* **8 *off the record.*** confidential(ly), private(ly), not for publication, secret(ly), °in confidence, unofficial(ly), sub rosa: *We cannot publish his statement, which was off the record. Off the record, I'll tell you where Hazel is hiding.* —*adj.* **9** record-breaking, °extreme: *Last week we had a record snowfall.*

recount *v.* **1** °relate, °narrate, °tell, °recite, °report, °communicate, °impart, °unfold: *The old man recounted for us the legend of the lavender house.* **2** particularize, °review, °detail, °describe, °enumerate, °specify: *Must I recount every particular all over again?*

recoup *v.* regain, °make good, °make up, recompense, °repay, °recover; refund, °redeem, °reimburse, remunerate: *From his profits on the Culpepper deal, he recouped all the losses he had sustained when Cranston Ltd failed.*

recourse *n.* **1** °resort, access, entrée, °admittance, availability: *If Olivia has a problem, she has recourse to sources you know nothing about.* **2** °resource, backup, °reserve, °refuge, place to turn, °alternative, °remedy: *When their appeal was rejected, they were totally without recourse.*

recover *v.* **1** regain, get *or* take *or* win *or* make back (again), °recoup, repossess, retake, recapture, °restore, °retrieve, °reclaim, °redeem: *It was impossible to recover the time lost. After a fierce battle, the territory was* *recovered.* **2** get well *or* better, °recuperate, °convalesce, °mend, return to health, regain one's strength *or* health, be on the mend, °heal, °improve, °revive, °rally, take a turn for the better, get back on one's feet, °pull through: *It took me a week to recover from the operation. Browne had a virus, but she's recovering now.* **3** °save, °salvage, °deliver, °rescue, °return, bring back, °reclaim, °retrieve: *They never did recover the gold from the sunken galleon.*

recovery *n.* **1** recuperation, convalescence, °restoration, °improvement, healing, °rally, turn for the better; °rise, °revival, °increase, amelioration, bettering, betterment, °advance, °gain, °advance, advancement, *Colloq* pickup, comeback: *After so serious an illness, Bentley has made a remarkable recovery. Share prices staged a recovery after the reduction in interest rates.* **2** retrieval, recouping, repossession, retaking, °restoration, reclamation; recapture, redemption: *The recovery of the stolen gems was credited to excellent detective work. Hard fighting resulted in the recovery of the land west of the river.* **3** °salvage, °delivery, deliverance, °rescue, °return, °saving, reclamation, retrieval: *Recovery of the climbers' bodies from the ravine was very hazardous.*

recreation *n.* °entertainment, °amusement, °enjoyment, °diversion, °distraction, fun and games, °leisure (activity), °pastime, °relaxation, °sport, °play: *What do you like to do for recreation? His favorite recreation is building model airplanes.*

recrimination *n.* counter-accusation, countercharge, retaliation, counterattack, blaming, °aspersion, °reprisal: *Recriminations for real and imagined ills were continually exchanged throughout their marriage.*

recruit *v.* **1** °draft, °induct, °enlist, °enroll, °muster, °raise, °call up, conscript, °mobilize, °impress, levy: *Owing to the nefarious enemy attack, there was little time to recruit men for a strong fighting force.* —*n.* **2** conscript, trainee, beginner, °apprentice, °initiate, °novice, neophyte, tiro *or* tyro, *US* draftee, enlistee, *Colloq* °greenhorn, rookie: *The new recruits have to be issued with their uniforms and other gear.*

rectify *v.* °correct, °revise, redress, put *or* set right, °cure, °repair, °remedy, °improve, emend, °fix, °adjust, °square, ameliorate: *What are you doing to rectify the damage you have done?*

rectitude *n.* °propriety, correctness, °morality, uprightness, °probity, °virtue, decency, goodness, °honesty, °integrity, incorruptibility, righteousness, °principle, good character, respectability, unimpeachability: *Fred's unquestioned moral rectitude makes him an outstanding candidate for membership.*

recumbent *adj.* reclining, lying down, flat on one's back, °horizontal, °lying, reposing, accumbent, decumbent, °supine, stretched out; °leaning (back): *Oscar resumed his recumbent posture, sprawled out on the cushions.*

recuperate *v.* °improve, °recover, °convalesce, get better, °rally, °revive, °mend, °heal, get back on one's feet, regain one's health, °pull through, °survive, take a turn for the better: *Victor went to Ibiza to recuperate from his illness.*

recur *v.* **1** °return, happen *or* occur again, reoccur, °repeat, come (back) again, reappear: *The condition recurred after a year, and he needed another set of treatments.*

recurrent *adj.* repeated, recurring, returning, reoccurring, reappearing, °frequent, °periodic, cyclical, °regular, °repetitive, °repetitious, °persistent, incessant, °continual, °intermittent, °habitual, iterative: *She told the psychiatrist about her recurrent dream of flying.*

redeem *v.* **1** °reclaim, °recover, regain, repossess, °retrieve, °get back, buy back, repurchase; °pay off, *Brit* °clear: *He went into the pawn shop to redeem the gold watch. It is marvelous that you can afford to redeem your mortgage.* **2** °rescue, °ransom, °deliver, °free, °save, °liberate, set free, °emancipate, °release: *He was able to redeem the boy only after paying the kidnappers.* **3** °exchange, °cash (in), collect on, trade in: *These coupons must be redeemed before the end of the month.* **4** °rehabilitate, °save, °reform, °convert, absolve, restore to favor, reinstate: *She can be redeemed only by atoning for her sins.* **5** make amends for, °make up for, atone for, redress, compensate for, °offset, make restitution for: *Do you think that two hundred hours of community service was enough to redeem his crime of snatching handbags in the shopping mall?* **6** °perform, °fulfill, °keep, °make good (on), °discharge, °satisfy, °abide by, keep faith with, be faithful to, hold to, °carry out, °see through: *Hortense is here to redeem her promise to help at the charity bazaar.*

red-handed *adj.* in the (very) act, (in) *flagrante delicto*, *Colloq* with one's hand in the till, *US* with one's hand in the cookie jar: *There is no doubt that Rick did it—he was caught red-handed.*

redolent *adj.* **1** °fragrant, sweet-smelling, °aromatic, perfumed, odorous, scented, °savory: *The entire house was redolent with the odor of pine needles.* **2** ***redolent with*** *or* ***of.*** reminiscent of, °suggestive of, evocative of, remindful of, characteristic of, having the earmarks *or* hallmark of: *The style of the film is redolent of 1930s Hollywood.*

reduce *v.* **1** °cut (back), °cut down (on), °decrease, °diminish, °moderate, abate, lessen, °shorten, truncate, °abbreviate, °abridge, °curtail, crop, °trim, compress, °restrict, °limit, °stunt: *We must reduce expenses this year.* **2** °ease (up on), °let up (on), °decrease, °mitigate, °tone down, °slacken up (on): *Will it reduce your burden if you share it with me?* **3** lose weight, °slim (down), °diet, trim down, *Chiefly US* slenderize: *You'll have to reduce to fit into that dress, Bertha.* **4** °change, °modify, °adjust, °turn, °convert; break down *or* up, °grind, °rub, triturate, °pulp, °powder: *Payment of the debt reduced me to poverty. The first step is to reduce the liquid to a thick extract. Sugar reduced to a very fine consistency is used for confections.* **5** °cut, °decrease, °trim, bring down, °lower, °drop, °mark down, °slash, *Colloq* °knock down: *The prices on these shoes have been reduced for a quick sale.* **6** demote, °degrade, °lower, °downgrade, °break; °humble; *Military US and Canadian* bust: *He was reduced in rank from sergeant to private.* **7** °diminish, lessen, bring down, °depreciate, °subdue, °belittle, °minimize: *He was reduced in her estimation when she saw the deprecating way his friends treated him.* **8** *Medicine* °set, °adjust, reset: *The doctor caused little pain when he reduced my dislocated finger.*

redundant *adj.* **1** °superfluous, °unnecessary, °surplus, inessential *or* unessential, °nonessential, unneeded, unwanted, *de trop*, in excess, °excessive: *In the UK, when a worker is "declared redundant," it means he has lost his job.* **2** °wordy, verbose, prolix, overlong, long-winded, °repetitious, tautologic(al) *or* tautologous, circumlocutory, °roundabout: *The book could be improved by making the writing less redundant.*

reek *v.* **1** stink *or* smell (to high heaven), *Slang Brit* pong, °hum: *Alben reeks from cleaning out the stable.* **2** smoke, steam: *It is because it reeked so badly that Edinburgh acquired its nickname, "Auld Reekie."* —*n.* **3** stink, °stench, fetor *or* foetor, miasma, mephitis, °odor, °smell, *Slang Brit* °hum, pong: *The reek of death permeated the atmosphere.* **4** °fumes, smoke, steam, °vapor, °exhalation, cloud, °mist: *A sickening reek arose from the mephitic bog.*

reel *v.* **1** °stagger, °totter, waver, °stumble, °lurch, falter, °roll, °rock, °sway, °flounder, °pitch: *Halliburton came reeling down the lane, dead drunk.* **2 *reel off.*** °list, °recite, °rattle off, °enumerate, °review, °itemize, °name, read off, °call off, °run through, °run over: *Under questioning, the suspect began to reel off the names of his accomplices.*

refer *v.* **1** Often, ***refer to.*** allude to, make reference to, °mention, make mention of, °touch on, °bring up, advert to, °speak of, turn *or* call *or* direct attention to, direct to, °point to, °indicate, advert to, °specify, °name, °concern, °quote, cite, make a note of, take note of, °note: *A footnote refers to the author's article on netsukes.* **2** °assign, °hand over, pass on *or* over, °send, °direct, °commit: *I was referred to your office for information about rates. The question will be referred to committee.* **3** Usually, ***refer to.*** °look at, °use, °study, °check, °consult, °resort to, have recourse to, °turn to, appeal to, confer with; °talk to, °ask, inquire *or* enquire of, apply to: *For information about words, you should refer to a dictionary. Refer to the departmental office for Professor Clarke's address.* **4** Usually, ***refer to.*** °mean, °signify, °denote, °say: *What are you referring to?*

reference *n.* **1** allusion, °mention, °remark, referral, °direction, °indication, °specification, naming, °quotation, citation, °note, °notation, notification, °hint, intimation, °innuendo, insinuation: *I cannot find a reference in any footnote. Few readers make use of the bibliographical references in scholarly articles. Her persistent references to my baldness embarrass me.* **2** °regard, °concern, °connection, °respect, °relation, °relevance, pertinence: *People are selected without reference to race, creed, color, sex, or age.* **3** °endorsement, °recommendation, °testimonial, certification; credentials: *I relied on my former employer to give me a good reference. Let me see your references as an engineer.*

refine *v.* **1** °purify, °cleanse, °clear, °clarify, decontaminate: *We continued to refine the liquid till all the impurities were gone.* **2** °cultivate, °civilize, °polish, °improve, elevate, °perfect: *He is trying to refine his speech by getting rid of his country accent.* **3** hone, °sharpen, °concentrate, °focus, subtilize: *We ought to refine our argument for higher wages before approaching the management.*

refined *adj.* **1** °cultivated, cultured, civilized, °polished, °sophisticated, urbane, °elegant, °well-bred, °genteel, courtly, °ladylike, gentlemanly, polite, °courteous, mannerly, well-mannered, °gracious, °gentle, °noble, aristocratic, °dignified, °elevated, *Colloq* °posh: *It is seldom that I find myself in such refined company.* **2** °subtle, °discriminating, discerning, °sensitive, °fastidious, °nice, °precise, °exacting, cultured, °educated, °cultivated, °knowledgeable, advanced, °sophisticated: *Her taste in the visual arts is quite refined, but she knows little about music.* **3** °exact, °precise, °fine, °subtle, °sensitive, °nice, °sophisticated: *In his poetry one encounters an extremely refined use of metaphor. These new devices are far more refined than any previously available.* **4** purified, clarified, cleansed, °pure, °clean; distilled: *Only refined chemicals may be used in pharmaceuticals.*

refinement *n.* **1** °culture, °polish, °elegance, °sophistication, urbanity, urbaneness, °breeding, cultivation, gentility, °propriety, courtliness, °civility, politeness, politesse, °delicacy, °tact, °diplomacy, °finesse, suavity, suaveness, °taste, tastefulness, °discrimination, discernment, °sensitivity: *Julia and Ewan behave with a measure of refinement befitting their station in life.* **2** °subtlety, nicety, nuance, °distinction, °detail, fine point, °delicacy, minutia; fastidiousness, finickiness: *The latest version contains refinements that are beyond the appreciation of all but a few.* **3** refining, purification, purifying, clarification, clarifying, cleaning, cleansing; filtration, distillation: *The sugar must undergo refinement before being sold commercially.* **4** °improvement, betterment, bettering, enhancement, °development, °perfection: *She is working on the refinement of her technique before performing in public.*

reflect *v.* **1** °mirror, send *or* throw back, °reproduce, °return; °echo: *The smooth surface of the lake reflects the sunlight. The radio waves are reflected and focused by this parabolic dish.* **2** °show, °demonstrate, °exhibit, °illustrate, °exemplify, °reveal, °lay bare, °expose, °display, °disclose, °bring to light, uncover, °point to, °indicate, °suggest, °evidence: *They say that your true emotions are reflected in your dreams. This crime reflects the depths of depravity to which a person can sink.* **3** Often, ***reflect about*** *or* ***on.*** think (about *or* over *or* on), °contemplate, muse (about *or* on), °consider, ponder (about *or* over *or* on), deliberate (on *or* over), ruminate *or* meditate (about *or* on *or* over), cogitate (about *or* on *or* over); mull over, °weigh, °evaluate, °examine: *I was just sitting here reflecting. I was reflecting on what you said about wanting to quit your job.* **4 *reflect on*** *or* ***upon.*** °result in, end in, °bring, °attract, °cast, °throw: *The success of the team reflects credit on the coach.*

reflection *n.* **1** °image, °echo, *Brit* reflexion: *Sonar works by detecting the reflection of generated sound waves.* **2** °thought, °thinking, meditation, meditating, °consideration, cogitation, rumination, deliberation, deliberating, pondering, °study, cerebration: *After a moment's reflection, I arrived at the same conclusion as you had.* **3** °aspersion, °imputation; °effect: *Any reflection on his reputation might harm his career.* **4** °result, °sign, °token, °symbol, °mark; °evidence, °testimony, testament, °proof, substantiation, corroboration: *Her tears are a reflection of how she feels about having betrayed me.*

reflective *adj.* reflecting, °thoughtful, °pensive, contemplative, meditating, musing, °meditative, cogitating, cogitative, ruminating, deliberative, deliberating, pondering: *I found Alan in a reflective mood, staring out of the window.*

reform *v.* **1** °improve, °better, ameliorate, meliorate, emend, °rectify, °correct, °mend, °repair, °fix, °remedy, redo, °revise, revolutionize, °rehabilitate, remodel, refashion, °renovate, reorganize, rebuild, °recover: *The first thing to be done was to reform the political system.* **2** mend one's ways, turn over a new leaf, *Colloq* go straight: *If there is a chance that he would reform, he ought to get a light sentence.* —*n.* **3** °improvement, betterment, amelioration, melioration, emendation, rectification, °correction, rehabilitation, °change, modification, perestroika, reorganization, renovation, °recovery: *Many believe that prison reform would reduce crime by second offenders. Many reforms were condemned by conservative members.*

reform school *n.* *US* reformatory, *Brit* youth custody center, CHE (= 'community home (with education on the premises)'), *Brit formerly* borstal, approved school: *Did your parents threaten to send you to a reform school if you misbehaved?*

refrain[1] *v.* **1** Usually, ***refrain from.*** °keep (from), forbear, abstain (from), eschew, °avoid: *Whatever you do, refrain from whistling while aboard his boat.* **2** °stop, °cease, °give up, °discontinue, desist, °quit, °leave off, °renounce: *Refrain from calling the headmaster, "Pops," even if he is your father.*

refrain[2] *n.* °melody, °song, °tune, chorus, °burden, reprise: *Every time I hear that sweet refrain I think of you, darling.*

refresh *v.* **1** °enliven, °renew, °revive, °freshen (up), resuscitate, bring back to life, breathe new life into, invigorate, °vitalize, °energize, °brace, °fortify, exhilarate, revitalize, reinvigorate, reanimate: *I felt fully refreshed after a short nap.* **2** °arouse, °rouse, °awaken, waken, reawaken, °stimulate, °jog, °prod, °activate: *Would you mind refreshing my memory as to when and where we last met and who you are?* **3** °fix up, °repair, redo, °revamp, °overhaul, °spruce up, recondition, °renovate, °refurbish, refurnish; °renew, restock, °restore: *We called in a service to refresh all the rooms in the hotel. We took the opportunity of the ship's arrival to refresh our supplies.*

refreshing *adj.* °invigorating, °stimulating, °bracing, °exhilarating, °fresh, inspiriting, fortifying, °tonic, like a breath of fresh air, rejuvenating, enlivening, vitalizing, revitalizing; cooling, thirst-quenching, slaking: *It is refreshing to be around young people. I would love a refreshing glass of beer right now.*

refreshment *n.* **1** Usually, ***refreshments.*** °nourishment, nutriment, °sustenance, restorative, °food, drink(s), edibles, eatables, °bite, snack(s), tidbit(s), *Brit* titbit(s), *Slang* grub, eats, chow, nosh: *Will they be serving refreshments after the meeting?* **2** stimulation, invigoration, exhilaration, °tonic, rejuvenation, enlivenment, °revival, °restoration, renewal, resuscitation, fortification, °reinforcement: *He enjoys taking tea for his afternoon refreshment.*

refrigerate *v.* °cool, °chill, keep cool *or* cold *or* chilled, ice, °freeze: *The soufflé must be refrigerated for three hours to allow it to set.*

refuge *n.* **1** °sanctuary, °shelter, haven, asylum, °protection, °cover, °retreat, °harbor, °security, safe house, °stronghold, citadel, bolt-hole, hideaway, hide-out, *Colloq* hidey-hole: *The abandoned mine served as a refuge from the storm.* **2** °excuse, °pretext, °resort, °recourse, °ruse, °trick, °stratagem, °subterfuge, °dodge, °evasion, °expedient: *Patriotism is often the last refuge of a scoundrel.*

refugee *n.* °fugitive, °runaway, escapee, displaced person, DP, °exile, émigré: *After the war, the most severe problem was that of the refugees.*

refurbish *v.* °restore, refurnish, redecorate, °clean (up), °polish, °renew, °renovate, °spruce up, remodel, refit, °overhaul, °repair, recondition, °revamp, rebuild, *Colloq* do up, *US* °fix up: *The person who bought the house refurbished it completely, at enormous expense.*

refusal *n.* **1** °denial, °rejection, °disapproval, turndown: *She pleaded against the grand vizier's refusal to grant amnesty to her brother.* **2** °option, °choice, °privilege, disposal: *We were promised first refusal on the house should it come on the market.*

refuse[1] *v.* **1** °decline, °reject, °spurn, °repudiate, °turn down, °rebuff, give (something) the thumbs down, *US* turn thumbs down on, *Colloq* pass by *or Colloq* up: *They refused our application to build a hotel here. Did he actually refuse the Nobel prize?* **2** °deny, °deprive (of), °withhold, disallow, not allow *or* permit: *They refused me my right to vote.*

refuse[2] *n.* °rubbish, sweepings, °waste, °litter, dust, °dirt, °dregs, dross, °garbage, debris, detritus, castoffs, °junk, *Chiefly US* °trash: *Refuse collection is scheduled for once a week.*

regal *adj.* **1** °royal, kingly, queenly, °princely, fit for *or* befitting a king *or* queen, °noble, lordly, °sovereign, °majestic, °imperial, °stately, °splendid, °magnificent, °grand, resplendent, °palatial, °exalted: *The regal throne was decorated with gold and precious stones.* **2** °disdainful, °haughty, °proud, °scornful, contumelious, °contemptuous, °derisory, derisive, °pompous, lordly: *She was dismissed with a regal wave of the hand.*

regale *v.* °entertain, °amuse, °delight, °divert, °indulge, please, °gratify, °captivate, °fascinate, °entrance, °enchant, spellbind, °bewitch, °charm, °enrapture: *She regaled us with entertaining tales of her travels.*

regalia *n.* °finery, decorations, insignia, emblems, accouterments *or* accoutrements, furnishings, °apparatus, °gear, °paraphernalia, °trappings, °tackle, appurtenances, °equipment, equipage: *He carried with him everywhere the full regalia of a sorcerer and magician.*

regard *v.* **1** °view, look at *or* upon *or* on, °observe, °note, °notice, °watch, °eye, keep an eye on, gaze at *or* upon, stare at: *The audience regarded the magician's performance with fascination.* **2** °contemplate, °consider, °perceive, °view, look upon *or* on, °treat: *I had a feeling that the lion regarded me as his dinner. Hollings regards his appointment as treasurer as a license to steal.* **3** °respect, °esteem, value: *Gwen's friends regard her very highly indeed.* **4** °consider, °view, look upon *or* on, °think (of), °judge, deem, °rate, °believe (to be), °gauge, °see, pay heed *or* attention to, °esteem, °account, °take into account, °imagine, °reckon, °evaluate: *I don't regard her to be a likely suspect. He regards his professor as his mentor.* **5** °concern, °relate to, be relevant to, pertain to, apply to, °refer to, °affect, have (a) bearing on, bear on *or* upon, °involve, have to do with, °go for: *I don't see why this matter regards anyone but me.* —*n.* **6** °reference, °relation, °relevance, relevancy, °association, pertinence, °application, °bearing, °connection, °link, °tie-in: *With regard to your interest in this affair, how did you get involved?* **7** °respect, °consideration, °attention, °reverence, °veneration, awe, °deference, °honor, °favor, °esteem, high opinion, °approval, approbation, °appreciation, °admiration, °affection, fondness: *They have no regard for the law.* **8** °point, °particular, °respect, °aspect, °detail, °matter: *In this regard, I shall take your past record into consideration.* **9** °care, °concern, °thought, °consideration, °respect, °sympathy, °feeling, °sentiment; °heed, °attention, °notice, °mind: *The police behave as if they have no regard for the safety of the hostages. This work reflects the artisan's regard for detail.* **10** ***regards.*** best wishes, good wishes, °compliments, °greetings, °respects, salutations, *Archaic* devoirs: *Please give my regards to your father.*

regarding *adj.* °concerning, °about, respecting, with regard to, with respect to, with reference to, on *or* in the matter of, pertaining to, on the subject of, apropos, re, *Law* in re, *Archaic or Scots* anent, *Dialectal* anenst: *We have received your letter regarding poor service. Regarding your claim, processing it has been delayed because you failed to sign the form.*

regardless *adj.* **1** Sometimes, ***regardless of.*** °despite, °notwithstanding; °in spite of, heedless of, *Nonstandard* irregardless: *Regardless of her father's threats, she continued to go out with William.* —*adv.* **2** °notwithstanding, °nevertheless, no matter what, °in any event, °in any case, °at all events, anyway, anyhow, *Nonstandard* irregardless: *She heard his threats but is going out with William regardless.*

regime *n.* régime, regimen, °reign, °government, °rule, °regulation, °administration, °direction, °order, °leadership, °management, °system, °discipline: *Coffee breaks will be held to ten minutes under the new regime.*

regiment *v.* °discipline, °order, °organize, systematize, whip into shape, °standardize, °regulate, °control: *Management has been trying to regiment employees into pigeonholes, so they can ignore their individuality and treat them uniformly.*

region *n.* **1** °district, °area, °zone, °territory, °division, locality, sector, °section, °tract, °part, °dominion, °precinct, °province, °quarter, °department: *Each re-*

gion is under the suzerainty of a powerful dictator. **2** °sphere, °territory, °domain, °province, °field, ambit, °pale, °jurisdiction, bailiwick: *She is an expert in the region of climatology.*

register *n.* **1** °record, °roll, roster, rota, catalogue, annal(s), archive(s), °journal, daybook, °diary, appointment book, °calendar, °chronicle, °schedule, °program, directory, ledger, °file, °index, inventory, °list, listing, °poll, °tally, : *The town hall has a register of all houses sold since 1900 with the names of the sellers and buyers.* **2** cash register, °till, money box, cashbox: *The money in the register does not agree with the figures for receipts.* —*v.* **3** °record, write *or* jot *or* take *or* put *or* set down, °list, °enroll, sign in *or* on *or* up, °enter, catalog, log, °index, °chronicle, °note, make *or* take note (of): *All transactions must be registered.* **4** °show, °display, °express, °indicate, °manifest, °reveal, °betray, divulge, °record, °reflect: *James pretended to like my book, but his contempt for it was registered on his face.* **5** °make known, inform of, °advise, °transmit, °communicate, °record, °note, make note of, °report, °write down, °minute: *I wish to register my disapproval of the way prisoners are treated here.* **6** °check in, sign in *or* on, log in: *In the US, all aliens are required to register each year in January. We registered in the hotel as man and wife.* **7** Sometimes, ***register with*** or ***on.*** °sink in, °impress, become apparent (to), make an impression (on), come home (to), °get through (to); dawn on *or* upon, °occur to: *I heard what you said, but it just didn't register. The name didn't register with me at first, but I now know who you are.* **8** °indicate, °read, °mark, °represent, °measure, °point to, °specify, °exhibit, °show, °manifest: *The earthquake registered 7 on the Richter scale.*

regret *v.* **1** rue, °mourn, °lament, °bemoan, °bewail, °repent, be *or* feel sorry for, feel remorse for, feel *or* be remorseful over, be *or* feel upset, never forgive oneself, deplore, deprecate, weep *or* cry over: *You will regret speaking to me that way! The management regrets any inconvenience caused by the redecoration of the restaurant.* —*n.* **2** Sometimes, ***regrets.*** repentance, °guilt, °sorrow, °disappointment, contrition, °remorse, regretfulness, (pang *or* pangs of) conscience, self-reproach, self-condemnation, °qualm, second thoughts, rue, ruefulness, °grief, °woe, °sadness, mournfulness, *Literary* dolor: *He felt no regret for his evil deeds. Do you ever have any regrets over the way you treated him?* **3 *regrets.*** °refusal, nonacceptance, nonconsent, *US* turndown: *Thank you for the invitation, but I must tender my regrets.*

regretful *adj.* rueful, °mournful, °sad, °repentant, °guilty, °sorry, °sorrowful, °disappointed, contrite, °remorseful, °apologetic, °penitent: *I was most regretful for the way I had behaved.*

regrettable *adj.* °lamentable, °deplorable, woeful, °sad, distressing, upsetting, °unhappy, °unfortunate, unlucky, too bad, °awful, °terrible, execrable, reprehensible, °wrong, °shameful, shaming, *Colloq* °tough, °rough: *It is regrettable that he had the car crash, but he had been warned about drinking and driving.*

regular *adj.* **1** °routine, °ordinary, °common, °everyday, °customary, °accustomed, wonted, commonplace, °normal, °usual, °traditional, °time-honored, °conventional, °typical, °habitual, °natural, °familiar, °standard, °predictable, scheduled, °fixed, unvarying, °invariable, °methodical: *The attack occurred while Eric was taking his regular evening stroll.* **2** scheduled, °routine, °systematic, ordered, °steady, °consistent, rhythmic(al), °automatic, °uniform, °periodic, cyclic(al); hourly, °daily, weekly, hebdomadal, semiweekly, biweekly, fortnightly, semimonthly, monthly, bimonthly, seasonal, °quarterly, semiannual, annual: *The patrol made their regular rounds, as usual. The printers' strike might delay regular publication of the magazine.* **3** °symmetrical, °uniform, °even, even-sided, equal-sided, equilateral, equal-angled, equiangular; °harmonious: *Any six-sided plane figure is a hexagon, but if it has six equal angles and six sides equal in length, it is a regular hexagon.* **4** undistorted, °even, well-proportioned, °proportional, °symmetrical, °classic: *She's not beautiful, but her face has very regular features.* **5** °even, °smooth, °level, °straight, °uniform, uninterrupted, unvarying, °continuous, °flat, °plane, °plumb: *The surface of the wing is regular except for the spoilers.* **6** dependable, °methodical, (well-)regulated, (well-)ordered, °orderly; °proper, °correct, °legal, °official, °bona fide, °legitimate, established, recognized, °orthodox, approved, *Colloq* kosher, O.K. *or* OK *or* okay: *Everything about the council election was entirely regular.* **7** °usual, expected, °normal, °habitual, °accustomed, °familiar: *Our regular doctor was on vacation, and a woman was substituting for him.* **8** °acceptable, accepted, °estimable, °fine, °good, °likable, °popular, °pleasant: *We all consider Bob to be a regular fellow and a friend.* **9** °complete, utter, thoroughgoing, °unmitigated, unalloyed, °unqualified, consummate, °perfect, °thorough, °absolute, °well-known, acknowledged: *The man is a regular dunce when it comes to repairing cars.* **10** °permanent, °career: *She's in the regular navy.* **11** °conventional, °usual, °common, conformable, °ordinary, °systematic, *Colloq US* ruly: *Regular English verbs form their past by the addition of -d or -ed to the infinitive.* —*n.* **12** °fixture, °habitué, °(steady) customer, °patron, °client, frequenter: *Jan has been a regular at the Star and Garter pub for years.*

regularity *n.* **1** consistency, constancy, °uniformity, evenness, sameness, °symmetry, °balance, °harmony, harmoniousness, orderliness, °order, °stability, predictability: *To the casual observer, nothing interferes with the regularity of planetary movement.* **2** °routine, reliability, dependability, steadiness, invariability; °pace, °rhythm, °cadence: *Mr. Warbridge listened to the six o'clock evening news with unfailing regularity. Our business depends on the regularity of mail delivery. The regularity of the ticking of the clock is driving me mad.*

regulate *v.* **1** °adjust, °modify, °modulate, °control, °balance, °set, °fix, °order, °govern, °organize, °maintain, °manage: *The mixture of air and fuel in modern cars is regulated by an injection system. The responsibility of this department is to regulate prices.* **2** °control, °monitor, °govern, °run, °operate, °administer, °handle, °guide, °steer, °conduct, °direct, °supervise, superintend, °oversee, °manage: *It is the foreman's job to regulate the work on the assembly line.*

regulation *n.* **1** °adjustment, modification, modulation, °control, °balance, balancing, °setting, fixing, °organization, °maintenance: *The regulation of the temperature is accomplished by the thermostat.* **2** °rule, ruling, °law, °code, bylaw *or* byelaw, edict, °order, ordinance, statute, °decree, directive, °dictate: *There are strict regulations governing the sale, disposal, and movement of radioactive materials.* **3** edict, ukase, °pronouncement, fiat, (papal) bull, °proclamation: *A regulation was issued against meetings of more than five people.* —*adj.* **4** °standard, accepted, °official, required, prescribed, °mandatory: *That is not a regulation uniform, Soldier.* **5** °usual, °normal, °ordinary, °customary, °typical: *If you use anything but regulation accessories, the guarantee will be null and void.*

regurgitate *v.* °vomit, disgorge, return one's dinner, °spew up, °(of birds of prey) cast; retch, °gag; *Colloq* °throw up, upchuck, *Slang* puke, *US* spiff one's biscuits, barf: *Drinking even a great deal of water can make one regurgitate.*

rehabilitate *v.* **1** °restore, °save, °reclaim, °rescue, °redeem, reestablish, reinstate, reeducate, reorient, °reform, *US* °straighten out, *Colloq US* rehab: *Cordwain's has had marked success at rehabilitating criminals.* **2** °renew, redecorate, °renovate, °refurbish, °restore, °fix (up), °repair, reconstruct, rebuild, °change, °transform: *Cordelia wants to rehabilitate vintage cars for a living.*

rehash *v.* **1** rework, go over again, restate, redo, rearrange, reshuffle, reuse: *Heverton has nothing new to offer—he just keeps rehashing the same old material.* —*n.* **2** reworking, restatement, redoing, rearrangement, reshuffle, reshuffling, reuse, rewording: *Her speech was a mere rehash of what she's said scores of times already.*

rehearsal *n.* **1** °practice, °exercise, dry run, °drill, °run-through, read-through, dress rehearsal, *Technical* undress rehearsal: *Is there a rehearsal of the school play scheduled for tonight?* **2** °narration, recounting, °relation, °recital, °telling, °description, enumeration, °account, °repetition, °repeat: *Do we have to listen to another rehearsal of your great deeds of derring-do at the unemployment office?*

rehearse *v.* **1** °practice, °exercise, °run through, read through, °study, °repeat: *Sorry, Annie, but we are going to rehearse the scene till you get it right.* **2** °repeat, °relate, °recite, °tell, °describe, °recount, °review, go through *or* over, °review, °report, °recapitulate, *Colloq* recap: *I shall go mad if I hear him again rehearse the events leading up to his open-heart surgery.*

reign *n.* **1** °rule, °sovereignty, ascendancy, °power, hegemony, °influence, °command, suzerainty, °administration, °jurisdiction, °leadership, °government, °direction, °control, °domination, mastery; °kingdom, °monarchy, empire: *During her reign, the kingdom saw its greatest prosperity.* —*v.* **2** °rule, °control, °command, °govern, °lead, °direct, °dominate, °supervise, °manage, hold sway, wear the crown, wield the scepter, occupy the throne, *Colloq* run the show, rule the roost, *Slang* °call the shots: *During the sixth Egyptian dynasty, Pepi II Nefektare reigned for ninety-four years, the longest in recorded history.* **3** °prevail, be *or* become prevalent, °predominate, hold sway, °obtain, be *or* become rampant, be *or* become universal: *An argument broke out, which developed into a fist fight, and soon pandemonium reigned.*

reimburse *v.* °repay, recompense, refund, °pay back, °compensate, remunerate, indemnify: *Don't worry—we will reimburse you for any money you laid out.*

rein *n.* **1** °check, °curb, °control, °restraint, constraint, limitation, harness, °bridle, °brake: *You are going to have to put a rein on your expenses.* **2** ***reins.*** °control, °command, °control, °administration, °running, °leadership, °power, °rule, tiller, °helm: *He should hand over the reins to a younger person.* —*v.* **3** ***rein in.*** °check, °curb, °control, °restrain, °limit, harness, °bridle, °restrict, pull back on: *You will have to rein in your passions and stop behaving like that.*

reinforce *v.* °strengthen, °buttress, °bolster, °support, °fortify, °prop (up), shore up, augment, °brace, °stay, °steel: *The bookshelf will fall down if you don't reinforce it with another bracket.*

reinforcement *n.* **1** °buttress, °support, °prop, °brace, °stay, °bolster: *This iron reinforcement ought to be strong enough to secure the wall.* **2** strengthening, buttressing, bolstering, shoring (up), augmentation, °bracing: *This weak shelf is in need of reinforcement.* **3** ***reinforcements.*** °reserves, auxiliaries, men, soldiers, forces, personnel: *We cannot hold out till reinforcements arrive.*

reiterate *v.* °repeat, restate, iterate, °labor, harp on, °dwell on, °rehash, °recapitulate, *Colloq* recap: *I must reiterate my warning to be careful.*

reject *v.* **1** °decline, °refuse, disallow, °spurn, °veto, °turn down, give the thumbs-down (on *or* to), °set aside, *US* turn thumbs down (on); °deny, °repudiate, °renounce, °rebuff, °shun, °brush aside, turn a deaf ear to, will not hear of: *Cornelius has rejected the plan for a vacation in Torremolinos. She rejects any suggestion that she might be wrong.* **2** °refuse, °repel, °repulse, °spurn, °rebuff, say no to, °turn down, °decline, °brush aside; °jilt, °drop, °dismiss, °throw over, give (someone) the cold shoulder, show (someone) the door, send (someone) away *or* on his *or* her *or* their way, °turn one's back on, *Slang* give (someone) the brushoff *or US also* the brush, give (someone) his *or* her *or* their walking papers, *Brit* give (someone) the boot: *Ted proposed, but Alice rejected him.* **3** throw away *or* out, °discard, disown, jettison, °eliminate, °scrap, *Colloq* °junk, °scratch: *The inspector rejects all defective merchandise.* —*n.* **4** °second, °irregular, °discard, castoff: *The rejects are sold at reduced prices in the factory store.*

rejection *n.* °refusal, °denial, repudiation, °rebuff, °dismissal, spurning, renunciation, turndown; cold shoulder, *Slang* brushoff *or US also* brush, the (old) heave-ho: *The union voted for the rejection of management's offer of a 9.2 percent increase. Her advances were met with rejection.*

rejoice *v.* °delight, °exult, °glory, °celebrate, °revel, be happy *or* delighted *or* pleased *or* overjoyed *or* elated *or* glad, *Colloq* be tickled (pink): *We all rejoiced to be home once again after the ordeal.*

rejuvenate *v.* °restore, °refresh, reinvigorate, revitalize, revivify, °renew, reanimate, regenerate, recharge, breathe new life into: *I think a week in Barbados would rejuvenate me.*

relapse *v.* **1** °get back, backslide, °fall back, °lapse, slip back, regress, retrogress, recidivate; go back, °return, °retreat, °revert: *Lottie has relapsed into her bad habits.* **2** °decline, °deteriorate, °weaken, °degenerate, °fail, °fade, °sink, °sicken, °worsen, get *or* become worse: *Just when we thought he was improving, he relapsed into a coma and died.* —*n.* **3** backsliding, falling *or* going back, °lapse, lapsing, °return, returning, reversion, reverting, regression, regressing, retrogression, retrogressing, recidivism, apostasy: *The church's main concern was a relapse into idolatry.* **4** °decline, deterioration, weakening, degeneration, °failing, °failure, fading, °sinking, worsening: *Barford was getting better but has suffered a relapse.*

relate *v.* **1** °associate, °connect, °couple, °link, °tie, °ally, correlate, °coordinate: *Volta first enabled us to relate the forces of chemistry and electricity.* **2** °recount, °narrate, °tell, °report, °present, °describe, °recite, °detail, °set forth, °communicate, divulge, °impart, °reveal, delineate, °make known, give an account of, °rehearse: *It was not till years later that Olive related to me the strange events of that night.* **3** apply to, coordinate with, °respect, °regard, bear upon *or* on, have a bearing on, have reference to, have to do with, pertain to, °refer to, appertain to, belong with *or* to: *Does that comment relate to the matter at hand?* **4** Often, ***relate to.*** be in *or* en rapport with, °understand, empathize, °sympathize, °communicate with, °identify with, °grasp, °comprehend, be in tune with, °deal with, °handle, °cope with, *Colloq* °dig, tune into, be hip to, be turned on to, be *or* get into: *I have been having trouble relating to my new psychiatrist. Your favorite music is reggae?—Man, I can relate to that.*

related *adj.* **1** associate(d), affiliate(d), connected, coupled, linked, tied up, allied, correlated, coordinate(d), interconnected, interrelated, interdependent, °mutual, °reciprocal, °common, °joint, cognate: *She and her husband work for related companies.* **2** °kin, °kindred, consanguineous, cognate, agnate; °akin: *Just how are you two related?*

relation *n.* **1** °relationship, °connection, affiliation, °association, °bearing, °link, °tie, °tie-in, °reference, pertinence, interconnection, interdependence, correspondence, °kinship: *The relation between time and money is rarely disputed these days.* **2** °kinship, °relationship: *Just what is your relation to this lady?* **3** °relative, kinsman *or* kinswoman, blood relative, in-law, family member: *Is she a relation of yours?* **4** °narration, °narrative, °telling, recounting, °description, °report, °recital, °recitation, delineation, portrayal, °story, recapitulation: *The police say that your relation of what happened is quite different from hers.* **5** ***in relation to.*** °concerning, °about, °regarding, respecting, pertaining to, with regard to, with respect to, referring to, with reference to, on *or* in the matter *or* subject of, apropos, re, *Archaic or Scots* anent, *Dialectal* anenst: *I should like to talk to you in relation to the matter of company expenses.* **6** ***relations.*** **(a)** sexual intercourse, coitus, °sex, criminal conversation; carnal knowledge of: *It is well known that he had relations with his secretary.* **(b)** °dealings, °intercourse, link(s), association(s), °liaison, °relationship, *Colloq* doings, °truck: *We have no relations with that company.*

relationship *n.* See **relation, 1, 2,** above.

relative *adj.* **1** °related, connected, associated, allied, °affiliated, interconnected, interrelated, °pertinent, °relevant, germane, °applicable; apropos: *Your comments are not relative to the matter at hand.* **2** Sometimes, ***relative to.*** comparative, comparable, proportionate, °proportional, commensurate; analogous to, contingent on, dependent on, reliant on, conditioned by, °subject to, attendant on, correspondent *or* corresponding to, provisional on, appurtenant to; °subordinate to, ancillary to: *Well, as they say, everything is relative. Relative humidity depends on the temperature of the air. The yield is relative to the effort.* —*n.* **3** See **relation, 3,** above.

relatively *adv.* more or less, °somewhat, comparatively, °rather, to some degree *or* extent: *Business has been relatively quiet lately.*

relax *v.* **1** °loosen, let go, °release, °let up (on), °relieve, °ease, °reduce, °moderate, slacken, °remit, *Colloq* ease up on, slacken up on: *He relaxed his grip on the rope, and she fell to her death on the rocks below.* **2** °diminish, °decrease, lessen, °reduce, abate, °weaken, °mitigate, °modify, °tone down, °moderate, °modulate, °lighten (up on), °check, °temper, °curb: *We must not relax our vigilance. I hope that they don't relax the penalties against smoking at the office.* **3** ease up, °slow down, loosen up, put one's feet up, °rest, unbend, *Colloq* take it easy, unwind: *Why not come for a weekend and relax a bit?* **4** °calm down, cool down, °quiet down, stay calm, cool and collected, *Colloq* take it easy, *Slang* cool it: *You ought to relax and not get so excited.*

relaxation *n.* **1** °ease, °repose, °rest, °leisure, °recreation, °amusement, °entertainment, °fun, °pleasure, °diversion, *Colloq* R and R (= 'rest and relaxation'): *For relaxation we usually charter a boat and go sailing.* **2** easing (up *or* off), °relief, alleviation, abatement, diminution, lessening, mitigation, moderation, slackening, °remission, weakening, letting up, *Colloq* °letup: *What do you think about the proposed relaxation of the laws against pornography?*

relaxed *adj.* °nonchalant, easygoing, °calm, °peaceful, °tranquil, °serene, pacific, °carefree, insouciant, °blasé, languorous, languid, devil-may-care, free and easy, happy-go-lucky, °mellow, at ease, composed, °cool: *Richard takes a relaxed view of life.*

release *v.* **1** let go, °(set) free, °liberate, (set *or* let *or* turn) loose, unloose, untie, unchain, unfetter, unshackle, °deliver, °let out, disenthrall, °discharge, °let off, °emancipate, manumit, °rescue, °save: *To celebrate the country's centennial, all political prisoners are to be released.* **2** °issue, °publish, make available, °put out, °pass out, °hand out, come out with, °circulate, °distribute, disseminate; °launch, °unveil, °present: *The government refuses to release information that is considered harmful to its interests. The band's new album will be released tomorrow.* —*n.* **3** freeing, releasing, liberating, loosing, unloosing, delivering, emancipating, manumitting, rescuing, °saving, °freedom, °liberation, deliverance, °discharge, emancipation, manumission, °rescue, °remission, salvation: *The release of the prisoner is scheduled for noon. It would be a great release to be rid of housework.* **4** press *or* publicity release, °announcement, publicity, °notice, °story, °report: *The press office sent out a release to the media about Randolph's appointment.*

relegate *v.* **1** consign, °banish, °exile, °transfer, dispatch *or* despatch: *Owing to his disagreement with the State Department, Kendall was relegated to some outpost in the Indian Ocean.* **2** °downgrade, demote: *I fear that Corporal Schweik has been relegated to private.* **3** °assign, °commit, °hand over, °refer, °transfer, °pass on: *She relegates most of her everyday correspondence to her secretary for reply.*

relent *v.* °relax, °soften, °yield, °give, give way *or* ground, °bend, °comply, acquiesce, compromise, °capitulate, come round, be merciful, show pity *or* compassion, °melt, show mercy, °succumb: *They finally relented and allowed me to have visitors.*

relentless *adj.* **1** unyielding, inexorable, unstoppable, unrelenting, dogged, °implacable, °inflexible, unbending, unmoving, °unmoved, unrelieved, °stiff, °hard, stiff-necked, °rigid, °obstinate, adamant, obdurate, intransigent, °determined, unswerving, undeviating, intractable, persevering, °steely, °tough, intransigent, unsparing, uncompromising, pitiless, unforgiving, °ruthless, °merciless, °cruel, °unmerciful, °remorseless: *The novel is about the relentless horde of Mongols that swept across Europe in the thirteenth century.* **2** °nonstop, °persistent, incessant, unrelenting, unremitting, unstoppable, °perpetual, unfaltering, unfailing, unflagging, unrelieved, unabated, unrelieved, unbroken, °continual, °continuous, ceaseless, °constant, unceasing, °steady, °habitual, °regular: *Won't you ever stop your relentless nagging?*

relevance *n.* relevancy, appropriateness, aptness, pertinence, °bearing, °connection, °affinity, °tie-in, °relation, °significance, suitability, suitableness, applicability, °application, applicableness: *What you just said has no relevance whatsoever to the issue under discussion.*

relevant *adj.* °pertinent, °appropriate, apt, °related, °relative, °significant, suited, °applicable, °fitting, °proper, germane, °akin, allied, associated, apposite, °to the point: *Alan's suggestion, as it happens, is immediately relevant to our plans.*

reliable *adj.* dependable, °trustworthy, °honest, trusty, trusted, °principled, °conscientious, punctilious, °honorable, credible, believable, °safe, °sure, °certain, °secure, °sound, °responsible, °predictable, °stable, unfailing, °infallible, °reputable: *Would a reliable employee like Fiedler abscond with the money?*

reliance *n.* °confidence, °trust, °faith, dependence: *You may be placing too much reliance on the smoke alarm.*

relic *n.* **1** °memento, °keepsake, °memorial, °remembrance, souvenir, heirloom, °token, artifact: *It has been proved again and again that the shroud of Turin, though old, is not a genuine relic of Christ.* **2** °remains; °fragment, °trace, °scrap, shard *or* sherd, °remnant: *Deep in the jungle, archaeologists unearthed relics of a previously unknown civilization.*

relief *n.* **1** °ease, easing, abatement, easement, deliverance, °remedy, redress, alleviation, °release, °remission, assuagement, °liberation, °recess: *We would all welcome relief from this blistering heat. Contributions for the relief of famine in Africa would be most welcome.* **2** °aid, °help, °support, °assistance, succor; °comfort: *How many troops can we send for the relief of the besieged garrison?* **3** °elevation, °projection, °contrast, °prominence; bas-relief *or* basso rilievo (= 'low relief'), mezzo-rilievo (= 'medium relief'), alto-rilievo (= 'high relief'): *The Greek reliefs brought to England by Lord Elgin can be seen at the British Museum.* **4** °substitute, surrogate, replacement, °alternate, locum (tenens), °stand-in; °understudy, °double: *I wish that my relief would arrive so I could take a brief respite.*

relieve *v.* **1** °ease, lessen, °reduce, °diminish, abate, °lift, °raise, alleviate, °mitigate, palliate, °soften, soothe: *Business will suffer if nothing is done to relieve the burden of high interest rates. The doctor gave me something to relieve the pain.* **2** disburden, °free, °rid, °liberate, disencumber, unburden, °rescue, °save, °release: *Winning the lottery relieved me of any immediate worries of poverty.* **3** °help, °aid, °assist, °support, succor, °rescue, save, °deliver: *Will the cavalry arrive in time to relieve the fort, which is under attack?* **4** °stand in (for), °replace, °substitute for, take over for *or* from, *US* °spell, *Colloq* sub for: *I thought you were coming to relieve me at noon.*

religion *n.* °creed, °belief, °faith; °doctrine, dogma: *People must be treated equally, regardless of their religion.*

religious *adj.* **1** °devout, churchgoing, °pious, godfearing, °holy, spiritual-minded: *You never struck me as a particularly religious person.* **2** °scrupulous, °exact, °precise, °conscientious, rigorous, °strict, °fastidious, °meticulous, °faithful, punctilious, unerring, unswerving, undeviating: *Edmund's work is characterized by religious devotion to detail.*

relinquish *v.* **1** °yield, °give up, °cede, °waive, °leave, °quit, °abandon, °drop, °forsake, forswear, °desert, °abdicate, °resign, °renounce, let go, °surrender, °vacate, retire from: *David was forced to relinquish the throne.* **2** let go, °give up, °release, unloose, °loose, °free: *She relinquished her hold on the rope and fell into the chasm below.*

relish *n.* **1** °enjoyment, °pleasure, °delight, °gusto, °eagerness, avidity, °anticipation, °taste, °appetite, °zest, °liking, °appreciation, fondness, °fancy, °partiality, °preference: *The children attacked the birthday cake with great relish.* —*v.* **2** °enjoy, °delight in, take pleasure in, °fancy, be partial to, °appreciate, °savor, °look forward to, °anticipate: *After eating in restaurants for a month, I'd really relish a home-cooked meal. I don't relish having to tell her she has lost her job.*

reluctance *n.* unwillingness, disinclination, °aversion, °dislike, disrelish, hesitancy: *Cooper's reluctance to leave before the end of the play was understandable.*

reluctant *adj.* unwilling, °disinclined, °averse, °hesitant, °loath, °unenthusiastic, °indisposed, °opposed, antagonistic; °cautious, chary, °wary, °leery, circumspect, °careful: *I remember how reluctant you were to try raw oysters, and then you couldn't get enough of them.*

rely *v.* **1** ***rely on*** *or* ***upon.*** depend on *or* upon, lean on *or* upon, count on *or* upon, bank on *or* upon, have confidence in, bet on, trust in, °swear by, be sure *or* certain of: *You can rely on Phoebe to come through when you need a friend. I am relying on the continuation of interest rates at their present levels.*

remain *v.* **1** °stay (behind), be left, °tarry, °linger, °wait, *Colloq* stay put: *I remained to help wash the dishes.* **2** be left, be there: *How many days remain before your trip?* **3** °stay, °continue, °carry on, °abide; °endure, °persist, °last, °persevere: *She refuses to remain with him another minute. His last book will remain a monument to his scholarship.* —*n.* **4** ***remains.*** **(a)** leavings, remnants, crumbs, debris, detritus, °remainder, °balance, °residue, leftovers, °scraps, vestiges, °traces, °fragments, oddments, °odds and ends: *The remains of a half-eaten meal were on the table.* **(b)** °body, °cadaver, °corpse; carcass: *The remains of the impala were being picked over by vultures.*

remainder *n.* **1** °rest, °balance, °remains, °residue: *He spent the remainder of his life in penury.* **2** °excess, overage, °surplus, °residue, residuum, leftovers: *I used all I needed and don't know what to do with the remainder.*

remaining *adj.* **1** °left (over), extant, °outstanding: *I did as much as I could in the time remaining.* **2** °leftover, surviving, °residual; °unused, uneaten, unconsumed: *The remaining books will be sold at auction.*

remark *n.* **1** °note, °notice, °observe, °perceive, °regard, °look at, take notice *or* note of: *His friends remarked his tired and overworked appearance.* **2** comment (on *or* upon), °say, °observe, °reflect, °mention, °declare, °state, assert: *Carrie remarked that she had seen Ted the night before.*

remarkable *adj.* **1** °extraordinary, °unusual, °singular, °exceptional, °noteworthy, °notable, uncommon, °incredible, °unbelievable, °impressive, °phenomenal, astonishing, astounding, surprising: *With remarkable swiftness the falcon dropped on its prey.* **2** °striking, °distinguished, °signal, °special, wonderful, °marvelous, out-of-the-ordinary, °special, °unique, °significant, °outstanding, °rare, °memorable, unforgettable, never-to-be-forgotten: *This novel is truly a remarkable achievement.* **3** °strange, °different, °odd, °peculiar, °curious: *Contortionists can do truly remarkable things with their bodies.*

remedy *n.* **1** °cure, °treatment, °therapy, °antidote, counteractant, countermeasure, medication, medicament, °medicine, °prescription, °drug, pharmaceutical, cure-all, panacea, nostrum, restorative, °specific: *What remedy did the doctor recommend?* **2** °cure, °antidote, cure-all, panacea, nostrum, countermeasure, °relief, redress, °answer, °solution: *The only remedy for superstition is knowledge.* —*v.* **3** °cure, °treat, °heal, °mend, °restore, °relieve, soothe, °control, °ease, °mitigate, alleviate: *This ointment will remedy the itching but not its cause.* **4** °correct, °rectify, °reform, °improve, ameliorate, redress, °repair, put *or* set right, °straighten out: *How can I remedy a situation that has already gone too far?*

remember *v.* **1** call to mind, °bear in mind; °recall, °recollect: *Can you remember her funny little laugh? Yes, I remember.* **2** muse (on *or* about), reminisce over *or* about, think back on *or* about, °memorialize, °commemorate, °recognize: *He will be remembered for his many achievements.* **3** °retain, °keep in mind, °recall: *He was unable to remember his own name.* **4** °tip, °reward: *The Elliotts always remembered the servants at Christmas.*

remembrance *n.* **1** °memory, °recollection; °reminiscence: *My remembrance of her is quite different from* *yours.* **2** °memento, °reminder, souvenir, °keepsake, °memorial: *I keep this piece of shrapnel as a remembrance of being wounded in the war.*

remind *v.* °prompt, °cue, cause to remember, jog the memory, put in mind of: *Remind me to set the clock back an hour tonight.*

reminder *n.* mnemonic, refresher; °cue, °prompt: *I tied this string around my finger as a reminder, but I don't know of what.*

reminisce *v.* Sometimes, **reminisce about.** remember, °recollect, think back, look back, turn one's mind *or* thoughts back; call to mind, hark back, °return: *Most of the time we just sit about and reminisce. The other day we were reminiscing about the time Alex was arrested for shoplifting.*

reminiscence *n.* Usually, **reminiscences.** anecdote(s), memory *or* memories, reflection(s), memoir(s): *The* Sunday Clarion *has paid a six-figure sum to Charlotte for her reminiscences as a bag lady.*

reminiscent *adj.* Usually, **reminiscent of.** recalling, °redolent of, evocative of, °indicative of, °suggestive of, similar to, comparable with *or* to: *That odor of burning cabbage is reminiscent of my husband's cooking.*

remiss *adj.* °slack, °careless, negligent, neglectful, °heedless, unheeding, °inattentive, unmindful, °thoughtless, °forgetful, °unthinking, °slow, °indolent, °lazy, dilatory, °delinquent: *I was remiss in failing to remind you to turn off the bath water.*

remission *n.* **1** °forgiveness, °pardon, deliverance, amnesty, °reprieve, exoneration, °release, absolution, exculpation, °indulgence, °excuse, °exemption, acquittal: *As we were granted remission of our crimes, we were freed.* **2** diminution, abatement, °decrease, lessening, subsidence, alleviation, mitigation, assuagement, ebbing, °relaxation, easing: *Remission of the disease can occur if it is caught in time.*

remit *v.* **1** °send, °transmit, °forward, dispatch *or* despatch; °pay, °compensate, °settle, liquidate: *We have remitted in full the amount requested.* **2** abate, °diminish, slacken, °decrease, lessen, °subside, alleviate, °mitigate, assuage, °ebb, °dwindle, °reduce, °relax, ease (up *or* off), °fall off: *The fever ought to remit on the sixth day.*

remittance *n.* °payment, °settlement, °allowance, °consideration: *Remittance was made regularly on the tenth of the month. He lives on remittances he receives from home.*

remnant *n.* **1** °scrap, °shred, °fragment, °end, °bit, °piece, °trace, °vestige, °relic: *A silver button was the only remnant of the splendid uniform.* **2** °leftover, °remainder, °residue, °rest, °remains, °part: *The surviving remnant of the battalion made its way across the desert to Djibouti.*

remorse *n.* **1** °regret, repentance, ruefulness, °sorrow, °woe, °anxiety, guilty *or* bad conscience, pangs of conscience, °humiliation, °embarrassment, °guilt, self-reproach, mortification, °shame, contrition, contriteness, °penitence, °compunction, °bitterness: *It is impossible to describe the remorse he felt at having run over the little girl's puppy.*

remorseful *adj.* °regretful, °repentant, rueful, °sorry, °apologetic, °sorrowful, woeful, °anxious, °guilty, °bad, conscience-stricken, guilt-ridden, humiliated, humbled, °embarrassed, mortified, °shamefaced, °shameful, °ashamed, shamed, contrite, °penitent, °bitter: *Harold made a remorseful confession of all his sins.*

remorseless *adj.* **1** °cruel, °heartless, °callous, °harsh, hardhearted, stonyhearted, °savage, °merciless, °unmerciful, pitiless, °ruthless: *The heavyweight champion administered a remorseless beating to the challenger.* **2** °relentless, unrelenting, unremitting, unstoppable, inexorable, °implacable: *The remorseless storm thrashed at our tiny vessel, tearing away every last shred of rigging.*

remote *adj.* **1** °distant, °faraway, far-off, removed, °outlying, °inaccessible: *The more remote the transmitter, the weaker the signal.* **2** °lonely, °isolated, godforsaken, °secluded, °out-of-the-way, sequestered, tramontane, ultramontane: *He was transferred to a remote outpost of the empire.* **3** °unfamiliar, °obscure, arcane, °recondite, °subtle, °alien, °far-fetched, °unusual, °unlikely: *Her writing is interlarded with references remote to the average reader's experience.* **4** °unrelated, °irrelevant, °inappropriate, unconnected, °outside: *Her conclusions were not remote from what we now know to be the truth.* **5** °slight, °faint, foggy: *I haven't the remotest idea what you are talking about.* **6** °slight, °faint, °slender, °insignificant, °slim, °small, °meager, °outside, °poor, inconsiderable, °negligible, °improbable, °unlikely, °implausible: *There is only a remote chance that the experiment will succeed.* **7** °aloof, °detached, °withdrawn, °reserved, °indifferent, °standoffish, abstracted: *He keeps himself quite remote from the hurly-burly of everyday life.* **8** °early, °ancient, far-removed, °distant: *The archaeopteryx is a remote ancestor of modern birds.*

removal *n.* **1** elimination, removing, eradication, taking away: *The removal of the wallpaper revealed a mural of astonishing beauty.* **2** extermination, °murder, elimination, °killing, slaying, assassination, °execution, liquidation, eradication, °massacre, °slaughter, °purge, doing away with, *Slang* bumping off, rubbing out, doing in, *US* rubout, wasting: *The removal of the finance minister was accomplished by the secret police.* **3** °dismissal, °transfer, transference, transferal, shifting, °discharge, throwing over, throwing out, deposition, unseating, dethroning, dethronement, displacement, °expulsion, ouster, ousting, riddance, °purge, *Colloq* firing, sacking: *The company functions better since the removal of the former chairman.* **4** °move, °transfer, departure, °moving: *Her removal to a branch office was requested by top management.*

remove *v.* **1** °take off, doff, °shed, °cast off: *Please remove your hat in her presence.* **2** take away, °get rid of, carry away *or* off, °shift; °transfer: *When I arrived this morning, my desk had been removed. The threat of war has been removed. Kindly remove your elbows from the table.* **3** °obliterate, °delete, °erase, expunge, eradicate, efface, °eliminate, °take off, wipe *or* rub out, wipe *or* rub off, °get rid of: *Use chemicals to remove the graffiti from the wall.* **4** °murder, assassinate, °kill, slay, °execute, °exterminate, °eliminate, liquidate, eradicate, °massacre, °slaughter, do away with, °dispose of, °get rid of, °purge, *Slang* rub *or* wipe out, do in, °bump off, *US* °waste: *We brought in some of the boys to remove the competition.* **5** °discharge, °dismiss, depose, unseat, °displace, °expel, oust, °turn out, °get rid of, °purge, *Colloq* °fire, °sack, kick out: *Abernathy must be removed from his job before he does any more harm.* **6** relocate, °move, °transfer, °shift: *I removed to a quieter place where I could write in peace.* **7** °take out, unfasten, °detach, °disconnect, °separate, °undo: *They have to remove the transmission to overhaul it.* —*n.* **8** °distance, °space, °interval, °separation: *They tried to keep the girls and the boys at some remove from one another.*

remuneration *n.* **1** °payment, compensation, °salary, °wages, °earnings, emolument, °income, °pay, °stipend, °consideration, °reward: *Her remuneration is higher than her husband's.* **2** recompense, repayment, reimbursement, °restitution, reparation(s), °damages, °indemnity, indemnification, redress: *Have they received any remuneration for their loss of property during the war?*

renaissance *n.* renascence, °rebirth, °revival, reawakening, °restoration, resumption, renewal, °resurgence, °return, regeneration, rejuvenation, new dawn, new birth: *The renaissance is so called because of the return to classical tradition. Heppleworth seems to have undergone a renaissance since taking that rest cure.*

rend *v.* **1** rip *or* tear *or* pull (to pieces *or* apart *or* asunder), °wrench, °mangle, °shred: *The bully threatened to rend me limb from limb.* **2** °split, °tear, °rip, °rupture, °cleave, °split, rive, °separate, °slice, °lacerate: *When the balloon was rent from top to bottom, the basket fell to the ground.* **3** °pain, °distress, °pierce, °stab, smite, °wound, °afflict, °torment, wring, °hurt: *My heart was rent by conflicting emotions.*

render *v.* **1** depict, °picture, °represent, °reproduce, °portray, °create, °produce, do, °execute, °make, °accomplish, °achieve: *Cordillero rendered this portrait of the family from a photograph.* **2** °make, cause to be *or* become: *The news of winning the lottery rendered me speechless.* **3** °give (up), °yield (up), °surrender, °relinquish, °resign, °cede, °deliver, °hand over, °tender, °offer, proffer, °present, °furnish, °provide: *The inscription on the tomb has still not rendered up its secret. To Caesar will I render my legions and my horse.* **4** °play, °perform: *She rendered a fugue most beautifully on a harpsichord.* **5** °deliver, °return: *The jury rendered a verdict of guilty.* **6** °translate, decode, °decipher, °transcribe, °convert, °explain, °interpret, °put, restate, °reword, rephrase: *She rendered the passage into idiomatic English.* **7** °deliver, °hand in, °present, °offer, proffer, °furnish, °provide, °tender: *We shall render our invoice for the work already done.* **8** °melt, °clarify, °extract, *Brit* render down: *Fat must be rendered before it can be used for making soap.*

rendering *n.* depiction, showing, °presentation, °interpretation, °conception, °version, °rendition, °representation, delineation, portrayal, °picture: *This is an architect's rendering of what the building will look like when completed.*

rendition *n.* **1** °performance, °interpretation, °execution, °conception, concept, °understanding, construction, reading, °rendering: *I have never before heard a rendition of a Bach cantata on the zither.* **2** See **rendering,** above.

renegade *n.* **1** °deserter, °turncoat, heretic, °defector, °traitor, apostate, *Archaic* renegado: *The renegade has organized a conspiracy against us.* —*adj.* **2** °traitorous, treacherous, °perfidious, treasonous, apostate, °heretical, °disloyal: *A renegade band of deserters has turned against the military junta.*

renege *v.* **1** *Cards* °revoke: *When you renege you fail to follow suit.* **2** °go back on, back out, °default, °repudiate, go back on *or* break (one's) promise *or* word, °recant, abrogate, abjure, *Slang* welsh (on): *You swore you would keep your word, and now you're reneging.*

renew *v.* **1** °restore, °refresh, °rejuvenate, revitalize, reinvigorate, resuscitate, °revamp, redo, °rehabilitate, °transform, regenerate, °refurbish, refurnish, °renovate, refit, °overhaul, recondition, °modernize, redecorate, do over: *All the meeting halls of the hotel have been renewed to attract more conventions.* **2** °resume, °resurrect, restart, pick *or* take up again, recommence, return to, reopen: *I wish to renew my argument in favor of the teaching of Latin.* **3** °restore, °replace, restock, °replenish: *We must renew our supply of claret.* **4** °repeat, °reiterate, reaffirm, °confirm, reconfirm, restate, reassert: *Each autumn she renews her promise to return in the spring.*

renounce *v.* °give up, forswear, °surrender, °abandon, °desert, abjure, °reject, °repudiate, °spurn, °swear off, abstain from, °deny, °forgo, °forsake, eschew, disown, throw off *or* out, °shun, °avoid: *He renounced his claim to the throne and retired to a monastery.*

renovate *v.* redecorate, °modernize, do over, °refurbish, refurnish, refit, remodel; recondition, °rehabilitate, °restore, °repair, °revamp, °overhaul, °patch up, *Colloq* do up, *US* °fix up: *We are having the entire house renovated by an interior decorator. It was quite costly to have those paintings renovated.*

renown *n.* °fame, °celebrity, °glory, °distinction, °esteem, acclaim, °reputation, °prominence, eminence, °note, °mark, °honor, °prestige, repute, éclat, °luster, illustriousness, stardom: *She enjoys great renown as a rock star.*

renowned *adj.* °famous, famed, °celebrated, °distinguished, acclaimed, °prominent, °eminent, °well-known, °noted, °notable, honored, °illustrious: *He was one of the most renowned authors of the nineteenth century.*

rent¹ *v.* **1** °let (out), °lease, °hire (out), °charter (out), °farm out: *Eric owns a three-bedroom house that he rents out. I should like to rent a yacht for a cruise this summer.* —*n.* **2** rental, °hire, °lease, °fee: *The rent comes to about a quarter of my salary.*

rent² *n.* °tear, °rip, °split, °gash, °slash, °hole, °slit: *The skirt has a six-inch rent in it.*

repair *v.* **1** °mend, °patch (up); °renew, put *or* set right, °restore, °fix (up), °service, put (back) in *or* into working order, vamp, °revamp, °adjust: *Can you repair the tear in my trousers? I must remember to ask the garage to repair the brakes on my car.* —*n.* **2** °mend, °patch; °restoration, fixing (up), servicing, °improvement, °adjustment, renovation, revamping, renewal: *The cost of repair to our roof after the storm was covered by our insurance.* **3** °form, °condition, fettle, °state, working order, *Colloq* °shape, *Brit* °nick: *The engine, I'm afraid, has not been kept in good repair.*

repartee *n.* °banter, badinage, persiflage, °patter, °(witty) conversation, wordplay, °raillery, °give-and-take, *Literary* deipnosophy: *Simon is a welcome dinner guest because of his clever repartee.*

repay *v.* °pay back, recompense, °compensate, °requite, °reciprocate, return the favor *or* compliment, °reward, °square with, settle (up) with; refund, give back, °return, °reimburse, °restore: *How can I ever repay you for what you have done? Alan repaid me the money he borrowed.*

repeal *v.* **1** °revoke, °recall, rescind, °reverse, °cancel, annul, nullify, invalidate, °void, °set aside, °abolish, abrogate, *Law* °vacate: *They have repealed the rule forbidding women from wearing slacks in the restaurant.* —*n.* **2** revocation, °recall, rescission, rescindment, °reversal, °cancellation, annulment, nullification, invalidation, voiding, °abolition, abrogation: *Alcoholic beverages were forbidden in the US till the repeal of Prohibition.*

repeat *v.* **1** °reiterate, restate, °echo, retell, °recite, °quote, °rehearse, °recount, °recapitulate, *Colloq* recap: *Would you please repeat to Anne what you told me?* **2** °duplicate, °reproduce, replicate: *History repeated itself here today, when the Yankees won the World Series.* —*n.* **3** °repetition, °duplicate, °copy, duplication, replication, °reproduction, °replica, rerun, rebroadcast, replay, encore, reprise: *This is a repeat of a show we saw last year.*

repeatedly *adv.* again and again, over again, over and over, °frequently, °often, time and (time) again, time after time, recurrently, repetitively, repetitiously: *When the needle stuck, the record player kept playing the same thing repeatedly.*

repel *v.* **1** °repulse, drive back *or* away *or* off, °reject, °fend off, parry, °ward off, °hold off, °rebuff, °resist, °withstand, keep at bay *or* arm's length: *"Make ready to repel boarders" was a shout heard when new lodgers were preparing to move in to Mrs. Hudson's rooms.* **2** °revolt, °offend, °disgust, °sicken, °nauseate, turn one's stomach, make one's skin crawl, *Colloq* give one the creeps, turn one off: *Even the idea of eating sheep's eyes repels me.*

repellent *adj.* °repulsive, repelling, °revolting, °disgusting, nauseating, °nauseous, stomach-turning, sickening, °offensive, °loathsome, °repugnant, °distasteful, vomit-provoking, sick-making, °disagreeable, °obnoxious, off-putting: *He described some of the more repellent examples of vivisection he had encountered.*

repent *v.* °regret, feel contrition, °lament, °bemoan, °bewail, be sorry, rue, feel remorse, feel remorseful *or* penitent, show penitence: *Now that the evil deed is done he can but repent.*

repentant *adj.* °regretful, contrite, rueful, °remorseful, °apologetic, °sorry, °ashamed, °embarrassed, °penitent: *It was terrible that she wasn't the least bit repentant for what she had done.*

repercussion *n.* Often, **repercussions.** reaction, °response, °effect, °outcome, consequence, reverberation, °result, aftermath, aftereffect, °upshot, fallout, °backlash, °echo: *We were dealing with the repercussions of that decision long afterward.*

repertory *n.* repertoire, °store, reservoir, °collection, °hoard, °cache, repository, °stock, °supply, inventory, stockpile: *To my embarrassment, Cosgrove felt compelled to run through his repertory of dirty jokes at dinner.*

repetition *n.* **1** Often, ***repetitions.*** reiteration(s), duplication(s), redundancy *or* redundancies, repeats, tautology *or* tautologies: *This article is full of boring repetitions.* **2** reiteration, °repeat, °echo, echoing, repeating, duplication, duplicating, °rehearsal, recapitulation, restatement, replication, rereading, retelling, relisting, °recital, reprise, rerun, rerunning: *We sat through a repetition of every name for the benefit of any who might have missed it the first time.*

repetitious *adj.* °tiresome, °tedious, °boring, °redundant, prolix, °windy, long-winded, °wordy, tautological, pleonastic: *Streatham's lectures are very repetitious—he makes the same point over and over again.*

repetitive *adj.* iterative, °repetitious, incessant, °monotonous, repeated, °redundant, °humdrum, unceasing, ceaseless, °relentless, unremitting, °persistent, °recurrent, °nonstop, uninterrupted, °continual, °constant, °continuous: *The repetitive beat of that rock music is driving me mad.*

replace *v.* **1** °change, put in place of, °substitute, °supplant, °renew: *If the toy won't work, try replacing the batteries.* **2** °succeed, °supersede, °substitute for, take over from, °supplant: *Tim will be replacing David as department head.* **3** °restore, °return, °put back, °make good (on), °repay, refund, make restitution for: *Are you going to replace the money you took?*

replenish *v.* refill, restock, °restore, °renew, °replace, °fill, °top up; °furnish, °provide: *The stationer is supposed to replenish our supply of paper clips.*

replete *adj.* **1** Often, ***replete with.*** °full (of), filled up (with), overflowing (with), well-supplied *or* well-provided *or* well-stocked (with), chock-full (of), crammed *or* jammed *or* jampacked (with), brimful (with), chockablock, bursting, teeming, °loaded, overloaded (with), gorged (with), stuffed (with), *Colloq* up to the eyes *or Colloq* ears (in), up to here *or* there (in), *US* up to the old wazoo (in), *Slang* °lousy (with), *Taboo slang* up to the *Brit* arse (in) *or US* ass (in): *His speech was replete with names, dates, and precise quotations.* **2** satisfied, sated, satiated: *After a meal like that, anybody would feel replete.*

replica *n.* °copy, °duplicate, °facsimile, °reproduction, °likeness, °imitation, carbon copy, photocopy, duplication, *Colloq* °dupe, *US* °knockoff: *Do you believe he doesn't know that it's illegal to make replicas of currency?*

reply *v.* **1** °answer, °respond, rejoin, °retort, °return, come back, °acknowledge: *When I asked where he had been, he refused to reply and walked out of the room.* —*n.* **2** °answer, °response, rejoinder, °retort, comeback, riposte, °reaction, *US* °rise: *I have written to the prime minister but as yet have had no reply.*

report *n.* **1** °account, °description, °story, article, write-up, °piece, °statement, dispatch *or* despatch, communication, communiqué, °announcement, °narrative, °record; °news, °information: *Stanley hasn't yet sent in the full report of his meeting with Livingstone. Is there any report from the front?* **2** °explosion, bang, °boom, °shot, gunshot, gunfire, backfire, °discharge, °crack, °blast, detonation: *It was not till we found the body that we realized that the report we had heard last night was not a backfire.* —*v.* **3** °relate, °recount, °describe, °narrate, °tell of, °detail, give an account of, write up, °document: *Our man in Chichicastenango reported a disturbance in a remote village.* **4** °publish, promulgate, °publicize, °put out, °announce, °set forth, °reveal, °disclose, divulge, °announce, °circulate, °make public, °broadcast: *Two newspapers and a television station reported the arrest of the suspect.* **5** °arrive, °appear, °surface, °check in, sign in, clock in *or* on, °turn up, °come in, *US* report in: *We were asked to report for work at eight o'clock Friday morning.* **6** ***report on.*** °investigate, °cover, °examine, °explore, °look into, °inquire into, check into *or* on, check (up) on, °research, °study, °probe, °scrutinize, *Slang* °check out, *Brit* suss out: *Nicholas was assigned to report on allegations of bribery in the council.*

reporter *n.* °journalist, newspaperman, newspaperwoman, newsman, newswoman, °correspondent, columnist, newswriter, gentleman *or* lady of the press, gentleman *or* lady of the fourth estate, *Brit* pressman *or* presswoman; newscaster, news presenter, commentator, broadcaster, anchorman, anchorwoman, anchorperson, anchor, photojournalist, cameraman *or* -woman, *Colloq* newshound, newshawk, newshen, stringer: *Reporters swarmed around the athletes who had returned home with the Olympic gold medals.*

repose *n.* **1** °rest, °inactivity, °calm, °respite, tranquillity, °quiet, restfulness, °peace, °relaxation: *A half-hour's repose would be welcome after a rigorous day of door-to-door selling.* **2** °sleep, °nap, °doze, catnap, siesta, slumber, *Colloq* forty winks, snooze, *Slang* zizz, shuteye: *This is the time of madam's daily repose, and she cannot be disturbed.* **3** composure, calmness, °calm, °serenity, equanimity, °poise, self-possession: *He became angry in a flash, his face changing from quiet repose to livid distortion.* —*v.* **4** °lie, °abide, be, °lodge, °rest: *The power to make the modifications reposes with you.*

represent *v.* **1** °substitute for, stand (in) for, °replace, act for: *I have appointed her to represent me in all legal matters.* **2** Often, ***represent oneself as.*** °present (oneself), depict (oneself), put *or* set (oneself) forth, °masquerade as, take (on) *or* assume the guise *or* role *or* part of, characterize as, impersonate, pretend to be, °pose as, °imitate, °mimic: *He represents himself as a plumber in order to get into houses to rob them. His wife was represented to me as a social worker.* **3** °describe, delineate, °reproduce, °report, assert, °state, put *or* set forth, °show, °reflect, °mirror, °characterize, °define, °note, °outline, sketch, depict, °picture, °portray, °draw, °paint; °pretend: *What took place that fateful day is not as represented in Inspector Throttle's testimony.* **4** °symbolize, °stand for, °typify, °exemplify, °embody, epitomize; °illustrate: *He represents all that is good in the world today.*

representation *n.* **1** °reproduction, °image, °likeness, °portrait, °picture, depiction, portrayal, °semblance, °model, °manifestation: *This is an accurate representation of the town's appearance in the eighteenth century.* **2** °agency: *The sales representation for this area is under Postlethwaite's direction.* **3** °statement, °account, °exposition, °declaration, deposition, °assertion, °presentation, °undertaking: *Certain representations have been made concerning harassment by the police.* **4** °replica, °reproduction, °figure, figurine, °statue, statuette, bust, °head, °model, effigy: *This representation of the goddess Athena was found in Khios.*

representative *adj.* **1** °symbolic, °typical, °characteristic, °emblematic, archetypal, evocative, illustrative: *Is this a representative sample? These poems are not representative of your best work.* **2** elected, chosen, °democratic: *They are struggling to establish a representative form of government.* —*n.* **3** °agent, °deputy, legate, °ambassador, (papal) nuncio, spokesman *or* -woman, °proxy, °delegate, envoy, emissary, °missionary, commissioner; councillor, congressman *or* -woman, assemblyman *or* -woman, *New England* selectman *or* -woman, *Brit* member of parliament, MP: *We like to think that our representatives are acting on our behalf.* **4** °agent, (traveling) salesman *or* -woman, *Colloq* rep: *We have hired a new representative to market refrigerators in that territory.*

repress *v.* °suppress, °put down, °(keep in) check, °curb, °quash, °stifle, °overcome, °squelch, °(keep under) control, °contain, °restrain, constrain, °limit, keep back, °quell, hold back *or* in, °subdue, °inhibit, °hamper, °hinder, °deter, °frustrate, °discourage, disallow: *The government has tried to repress freedom of expression.*

repression *n.* **1** °restraint, °suppression, subjugation: *Increasing repression of their liberties finally led the people to revolt.* **2** °check, squelching, °control, °inhibition, hampering, hindering, °hindrance, stifling, deterring, frustration, frustrating: *The repression of the revolt was unsuccessful.*

repressive *adj.* °tyrannical, °oppressive, °dictatorial, °despotic, °brutal, suppressive, °authoritarian, °totalitarian; fascist(ic), Nazi: *In some countries, the people were unable to rebel against the repressive regime.*

reprieve *v.* **1** °respite, °rescue, °save; °let off, °spare: *The prisoner has been reprieved pending further investigation.* —*n.* **2** °delay, °postponement, °suspension, °remission, °respite, °stay, amnesty: *The governor granted the condemned man an eleventh-hour reprieve.*

reprimand *n.* **1** scolding, °reproof, °rebuke, admonition, upbraiding, castigation, reproach, °lecture, censure, °criticism, °disapproval, remonstrance, remonstration, reprehension, °tongue-lashing, *Colloq* dressing-down, talking-to, telling-off, ticking-off, rap on the knuckles, slap on the wrist, *Brit* slating, *US and Canadian* chewing-out, *Slang Brit* wigging: *Brown received a severe reprimand for violating the curfew.* —*v.* **2** °scold, chide, °reprove, °rebuke, admonish, °upbraid, °castigate, reproach, °berate, °lecture, censure, °criticize, find fault with, °attack, flay (alive), reprehend, read the riot act to, slap on the wrist, rap over the knuckles, °take to task, *Colloq* °bawl out, °dress down, give a dressing-down, °tell off, tick off, haul over the coals, give (someone) a piece of (one's) mind, *Brit* slate, give (someone) a row, send (someone) away with a flea in his *or* her *or* their ear, tell (someone) a thing or two, carpet, wig, °skin (alive), *US and Canadian* °chew out, rake over the coals, call on the carpet, pin (someone's) ears back: *Nigel deserves to be reprimanded for what he did.*

reprisal *n.* retaliation, °revenge, °retribution, redress, °requital, °vengeance, repayment, recompense, vindication, getting even, °indemnity: *They were going to torture us in reprisal for the way we had treated them.*

reproachful *adj.* °faultfinding, °critical, censorious, disapproving, disparaging, upbraiding, reproving, scolding, admonitory, condemnatory, hypercritical: *She was quite reproachful after seeing what I had written about her in my book.*

reprobate *adj.* **1** unprincipled, °immoral, amoral, °abandoned, depraved, °despicable, °dissolute, °low, lowlife, °base, °mean, debased, damned, accursed, cursed, °degenerate, °profligate, °shameful, °shameless, °vile, °evil, °wicked, °villainous, °sinful, irredeemable, °foul, iniquitous, reprehensible: *Such reprobate behavior soon brought him to the attention of the church fathers.* —*n.* **2** °scoundrel, blackguard, °miscreant, °rake, °profligate, °roué, °villain, °wastrel, °wretch, °degenerate, unprincipled person, evildoer, debauchee, °libertine, °good-for-nothing, ne'er-do-well, cur, rapscallion, scamp, knave, °rascal, *US* lowlife, *Colloq* scallywag *or Colloq* scalawag: *I cannot understand what she sees in that old reprobate.*

reproduce *v.* **1** °duplicate, °copy, replicate, °match, recreate, °repeat, °imitate, simulate: *The printer thinks he can reproduce the colors of the original picture in a print.* **2** °breed, multiply, °propagate, procreate, °spawn, produce *or* bring forth *or* beget young; regenerate: *These animals normally reproduce twice a year.*

reproduction *n.* **1** duplication, copying, printing: *Unlicensed reproduction of this film is against the law.* **2** °duplicate, °copy, °print, clone, carbon (copy), °facsimile, °replica, °lookalike, °double, °twin, °imitation: *These reproductions look better than the original.* **3** propagation, °breeding, spawning, °proliferation, °production: *The trick is to control the reproduction of this strain to make it free of disease.*

reproof *n.* See **reprimand, 1,** above.

reprove *v.* See **reprimand, 2,** above.

repudiate *v.* °reject, °scorn, °turn down, °renounce, °retract, rescind, °reverse, °abandon, abrogate, forswear, °forgo, °deny, disown, °discard: *The government has repudiated its earlier policy of refusing to negotiate with the strikers.*

repugnant *adj.* °repulsive, °abhorrent, °disgusting, off-putting, °offensive, °repellent, °revolting, °vile, °abominable, °loathsome, °foul, °distasteful, °unpalatable, °unsavory, execrable, intolerable, °obnoxious, noisome, nauseating, °nauseous, sickening, unpleasant, objectionable: *The very thought of seeing them again is repugnant to me.*

repulse *v.* **1** °repel, °rebuff, drive back, °ward off, fight *or* beat off, °check: *With a mighty effort, we were able to repulse the attacking forces.* **2** °refuse, °spurn, snub, °reject, °rebuff, °fend off, °resist, °turn down, give the cold shoulder to: *She repulsed the only offer of marriage she ever received.* —*n.* **3** °rejection, °rebuff, °refusal, °denial, snub, cold shoulder, spurning: *Alistair told himself he could not accept another repulse.*

repulsive *adj.* °disgusting, °revolting, °abhorrent, °loathsome, °repugnant, °repellent, °offensive, °obnoxious, objectionable, °gross, °unsavory, °distasteful, °nasty, unpleasant, displeasing, °disagreeable, °ugly, off-putting, sickening, nauseating, °nauseous, °beastly, °vile, °dreadful, °awful, °rotten, feculent, °foul, odious, °horrible, horrid, °abominable, execrable, fulsome: *I have seldom had so repulsive an experience as the tour of that abattoir. The repulsive sights and smells were more than I could bear.*

reputable *adj.* °respectable, °honorable, °well-thought-of, °estimable, respected, °trustworthy, trusted, °honest, °reliable, dependable, °principled, °virtuous, °good, °worthy: *Alan is one of Aylesbury's most reputable businessmen.*

reputation *n.* **1** repute, °name, °standing, stature, °position, °status: *Jean is a woman of excellent reputation in the community.* **2 *have a reputation for.*** be known *or* noted *or* notorious *or* famous for: *The company has a reputation for dismissing employees without notice.*

reputed *adj.* °alleged, purported, °supposed, °assumed, presumed; rumored, said, deemed, held, regarded, viewed, looked on *or* upon, judged, considered, °thought, believed: *The police arrested three reputed gang members. They are reputed to be international drug dealers.*

request *v.* **1** °ask for, °seek, °plead for, apply for, °put in for, °requisition, °call for, °demand, insist on, °solicit, °beg, entreat, °beseech, importune: *They request my cooperation. I requested to be excused. He requested her to open the door.* —*n.* **2** °plea, °petition, °application, °requisition, °call, °demand, solicitation, entreaty: *We receive many requests for information.*

require *v.* **1** °order, °command, °ask (for), °call (for), °press (for), °instruct, coerce, °force; °insist, °demand; °make: *I am required to appear in court on Monday. The teacher required that I bring my mother to school.* **2** °need, °want, °lack, be lacking, be missing, be short (of); °desire: *We require one more trainee to fill our quota.*

requirement *n.* **1** requisite, °prerequisite, °demand, °precondition, °condition, °qualification, °stipulation, sine qua non, °provision, °proviso, °demand, °necessity, °essential, desideratum, °must: *It is a requirement of the loan agreement that you maintain an account with this bank.* **2** °need, °want, °demand: *Will he be able to fill all of Mandy's requirements?*

requisition *n.* **1** °request, °order, °demand, °call, authorization, mandate, voucher: *A requisition arrived from headquarters for two hundred desks and chairs.* —*v.* **2** °request, °order, °demand, °call, °authorize, mandate: *Have you requisitioned the office furniture we need?* **3** °seize, °appropriate, commandeer, °confiscate, °take possession of, °take (over), °occupy; expropriate: *The army of occupation has requisitioned all available sleeping accommodation in the town.*

requital *n.* **1** repayment, °return, recompense, redress, °restitution, reparation, °remuneration, quittance, °amends, °satisfaction, compensation, °payment: *I look forward to an opportunity to make requital for all the favors you have done for me.* **2** °revenge, retaliation, °retribution, °reprisal, °vengeance; quid pro quo, *Brit or literary* Roland for an Oliver: *In requital for his shameful treachery, MacFallon was sent into exile.*

requite *v.* **1** °repay, °reward, °reciprocate, recompense, compensate for, °respond to: *Her years of tender care were requited with abuse and scorn.* **2** retaliate for, °revenge, avenge, make restitution for, redress, pay back for, give tit for tat for, make amends for: *Who can requite the wrongs inflicted on these poor people?*

rescue *v.* **1** °save, °deliver, °(set) free, °liberate, let go (free), °release, °(let) loose: *We rescued all the men who were being held prisoner. Jane was rescued by a very handsome lifeguard.* —*n.* **2** deliverance, °saving; freeing, liberating, °release: *The seaman's rescue was conducted by a navy helicopter team. Because of the heavy guard, the prisoners' rescue is out of the question.*

research *n.* **1** °investigation, investigating, °exploration, delving, digging, enquiry *or* inquiry, fact-finding, scrutinization, °scrutiny, °examination, inspection, probing, °analysis, experimentation: *Her research into the causes of the disease have yielded good results. The government no longer supports much research.* —*v.* **2** °investigate, °explore, delve into, °dig into, enquire *or* inquire into, °scrutinize, °examine, °study, °analyze, °inspect, check in *or* into *or* (up) on, °probe, °experiment with, *Colloq* °check out: *We are researching ways in which the virus attacks the blood cells.*

resemblance *n.* °likeness, similarity; correspondence, congruity, °coincidence, conformity, °accord, °agreement, equivalence, comparableness, comparability, °comparison: *There is a striking resemblance between your wife and mine. Any resemblance is purely a matter of chance.*

resemble *v.* look *or* sound *or* taste *or* seem *or* be like *or* similar to, bear (a) resemblance to, °approximate, smack of, correspond to, have (all) the hallmarks *or* earmarks of, °take after, *Colloq* °favor: *You are right, the taste does resemble that of licorice. Does she resemble her mother?*

resent *v.* feel embittered *or* bitter about, feel envious *or* jealous of, °begrudge, have hard feelings about, be displeased *or* disgruntled at, be angry about: *I resent having to go to work while you sit about the house all day.*

resentful *adj.* °embittered, °bitter, acrimonious, °spiteful, °envious, °jealous, begrudging, °vindictive, °indignant, displeased, °disgruntled, °dissatisfied, unsatisfied, °unhappy, peeved, irritated, irked, annoyed, provoked, riled, °angry, piqued, irate, °furious, incensed, °agitated, °upset, worked up, antagonistic, °hostile: *Ronald is resentful because you got the promotion that he considered to be rightfully his.*

resentment *n.* °bitterness, acrimony, °rancor, °envy, jealousy, °grudge, °indignation, °displeasure, °dissatisfaction, unhappiness, irritation, °annoyance, °provocation, pique, °anger, ire, °fury, °agitation, °upset, °anxiety, °ill will, malice, °antagonism, °hostility, °animosity, enmity, antipathy, °hate: *Some believe that inequality among people provokes resentment.*

reservation *n.* **1** keeping *or* holding back, withholding, reticence, °reluctance, hesitation, hesitancy, hedging: *It was not without some reservation that I gave the book my endorsement.* **2** °qualm, °scruple, °qualification, hesitancy, limitation, °exception, °objection, demur *or* demurral *or* demurrer, °condition, °proviso, °provision: *He has no reservations about revealing the most intimate details in his autobiography.* **3** booking, °appointment, °arrangement: *I'll make a reservation for dinner for the two of us.* **4** °preserve, °sanctuary, °reserve, °area, °tract, °territory, °region, °section, °plot: *In North America, many Indians still live on reservations.*

reserve *v.* **1** keep *or* hold (back), °withhold, °save, set *or* put aside, °conserve, °preserve, °retain, keep (to *or* for oneself), °hold over, °postpone, °delay, °put off, °defer: *Reserve judgment till you have heard all the facts. I shall reserve my answer to that question for another day.* **2** °hold, °keep, °book, °register, °save, put *or* set aside, °charter, °engage, °secure, contract for: *I have reserved two seats for the matinee.* —*n.* **3** Often, ***reserves.*** °store, °stock, stockpile, inventory, °supply, nest egg, reservoir, °fund, °hoard, °cache: *We prayed for rain, as our water reserves were getting low.* **4** reticence, (self-)restraint, (self-)control, taciturnity, aplomb, °formality, coolness, aloofness, guardedness, standoffishness, remoteness, °detachment: *I had to admire Bagley's reserve in the face of such insults.* **5** Often, ***reserves.*** °auxiliary, °alternate, °substitute, °reinforcements, backup, °spare: *If battle goes badly, we shall have to call out the reserves.* **6** See **reservation, 4,** above. **7** °reservation, °restriction, °restraint, hesitancy, hesitation, °limit, limitation, hedging, avoidance, °evasion, dodging, fudging: *The lawyer acknowledged without reserve that the case was weak.* **8** ***in reserve.*** °ready, °in readiness, on hand, °available, °on call, °accessible, as backup, in store, °on tap, at (one's) fingertips: *We have huge resources in reserve should we require them.*

reserved *adj.* °reticent, restrained, unresponsive, controlled, °silent, °taciturn, uncommunicative, unforthcoming, closemouthed, unresponsive, undemonstrative, unemotional, poker-faced, °cool, °formal, °aloof, °guarded, °standoffish, °unsocial, antisocial, °distant, °remote, °detached, °retiring, °withdrawn, °sedate, demure, °dignified, °guarded, prim, °rigid, °strait-laced, °icy, °frigid, ice-cold: *She was always very reserved, not showing any emotion.*

residence *n.* **1** °abode, °home, °domicile, °dwelling, dwelling place, °place, °house, habitation, °(living) quarters: *A crowd was waiting to greet them outside their residence.* **2** residency, °stay, °sojourn, °visit, °tenancy: *During his residence in Cape Province, he became bilingual in Xhosa.* **3** mansion, villa, manor (house), stately home, °estate, chateau, °castle, °palace, *Brit* °hall: *The ambassadorial residence was a very imposing structure indeed.*

resident *adj.* **1** in residence, residing, living, staying, °abiding, °dwelling, °remaining: *Is he resident in this country?* **2** °local, °neighborhood, °district, regional, °neighboring: *Our resident astronomer mentioned there would be an eclipse of the moon tonight.* —*n.* **3** °denizen, dweller, °inhabitant, °citizen, householder, homeowner, °tenant, °local: *Parking permits are available only to residents.*

residual *adj.* °remaining, °leftover, °surplus, °spare, °extra, residuary: *Residual units of the army are still bivouacked in the field.*

residue *n.* °remainder, °leftover, °surplus, °remains, °rest, °excess, °dregs, residuum: *The residue at the bottom of the victim's glass shows traces of arsenic.*

resign *v.* **1** °quit, °leave, °go, °abdicate, °give notice; °retire (from), °abandon, °give up, °forsake, °hand over, yield up, °renounce, °relinquish, let go, °release, °vacate, °surrender, deliver up, °turn over: *Amanda resigned because she was offered a better job elsewhere. Anthony refused to resign the last of his responsibilities till he was ninety.* **2** ***resign (oneself) (to).*** reconcile (oneself) (to), be *or* become resigned *or* reconciled (to), accommodate (oneself) (to), adjust (oneself) (to), adapt (oneself) (to), acclimatize *or* acclimate (oneself) (to), submit (oneself) (to): *Knowing that no one had ever escaped from the dungeon, he resigned himself to his fate.*

resignation *n.* **1** °notice; abandonment, abdication, resigning, renunciation, forgoing, relinquishment: *After that incident in the boardroom, Kim was asked for her resignation. Sidney's resignation of his responsibilities led to a severe reprimand.* **2** °reconciliation, reconcilement, °adjustment, °adaptation, acclimatization *or* acclimation, °submission, acceptance, compliance, capitulation, abandonment, acquiescence, passivity: *The saddest thing I saw was the prisoners' complete resignation to their fate.*

resilience *n.* °rebound, °recoil, °bounce, °elasticity, springiness, °spring, buoyancy, °flexibility, suppleness, ability to recover: *The resilience of youth is astonishing, and Withers was soon on his feet again.*

resist *v.* **1** °stop, °hinder, °prevent, °hold out (against), be proof (against), keep *or* hold at bay, hold the line (against), °thwart, °impede, °block, °obstruct, °inhibit, °restrain, °preclude, °check, °control, °curb, °stem, °bridle, °hold back, °withstand, °weather, °last (against), °endure, °outlast, stand up (to *or* against); °combat, °fight (against), °battle, countervail (against), °counteract, °oppose, °rebuff, °defy: *It is impossible to resist these overpowering forces of nature. Indeed, it is pointless to resist.* **2** °refuse, °deny, °turn down, °forgo: *I cannot resist a rich sweet for dessert.*

resistance *n.* **1** °opposition, defiance, °refusal, °denial, °obstruction, intransigence, rebelliousness, recalcitrance, stubbornness: *The child puts up considerable resistance to any form of discipline.* **2** defenses: *My resistance is low, I can deny you nothing.* **3** ***Resistance.*** °underground, partisans, freedom fighters, guerrilla *or* guerilla movement, guerrillas *or* guerillas, irregulars, Maquis, *US* resisters: *During the war, he was a member of the Resistance and engaged in sabotage against the invaders.*

resistant *adj.* **1** °opposed, against, °defiant, °averse, unsubmissive; impervious, °unaffected: *These organisms are resistant to any change.* **2** °recalcitrant, °stubborn, °obstinate, intransigent, °rebellious, °immovable, intractable, refractory, °willful, °ungovernable, unmanageable, °unruly, uncompliant, uncooperative: *We have run up against a highly resistant group of activists.* **3** Often, ***resistant to.*** impervious (to), impenetrable (to), °repellent (to); °proof (against); shedding: *You need a more resistant material. This fabric is resistant to rain and snow.*

resolute *adj.* resolved, °determined, °purposeful, °steadfast, °firm, °stubborn, adamant, °set, °decided, °staunch, °bold, dogged, undaunted, °dauntless, persevering, persisting, °persistent, perseverant, pertinacious, °tenacious, °single-minded, dedicated, °devoted, bulldog, purposive, °deliberate, °inflexible, unwavering, unshakable *or* unshakeable, unshaken, unflagging, °untiring, indefatigable, °tireless, unfaltering, °unhesitating, unhesitant, unswerving, °irreversible, undeviating, unchanging, °changeless, unchangeable, immutable, unalterable: *We remain resolute in our determination to end injustice.*

resolution *n.* **1** °resolve, resoluteness, °determination, °purpose, purposefulness, steadfastness, firmness, decidedness, °decision, staunchness, boldness, doggedness, dauntlessness, stubbornness, °obstinacy, °perseverance, °persistence, relentlessness, pertinacity, °tenacity, single-mindedness, °dedication, °devotion, constancy, devotedness, deliberation, deliberateness, inflexibility, inflexibleness, unshakability *or* unshakeability, fixedness, indefatigability, indefatigableness, irreversibility, changelessness, unchangeability, immutability, immutableness, unalterability, *Colloq US* stick-to-itiveness: *The boys showed extraordinary resolution in deciding to continue despite the hardships.* **2** °promise, commitment, °pledge, °word (of honor), °oath, °vow, °undertaking, °obligation; °intention: *I find that my New Year's resolutions last till about January 5th.* **3** °motion, °resolve, °proposal, °proposition, °plan, °suggestion, °idea, °notion; °determination, verdict, °decision, °judgment: *The committee votes today on the resolution to increase membership fees.* **4** °answer, answering, °solution, solving, unraveling, disentanglement, sorting out, explication; °outcome, °issue, °result, °end (result): *The resolution of a family problem is seldom easy. Can there be a final resolution of the question of a free market economy?* **5** acutance, sharpness, °precision, °accuracy, exactness, exactitude, fineness, °discrimination, detailing, distinguishability: *The new optical system provides for a much better resolution.*

resolve *v.* **1** °determine, °decide, make up one's mind, °agree, °undertake, °settle, °fix, conclude: *When did you resolve to move to Australia? Let us resolve never to let this happen again.* **2** °work out, °figure out, °solve, °clear up, °answer: *How have you resolved the problem of looking after the cats while you are away?* **3** °adopt, °pass, °approve, °decide: *It was resolved that membership fees should be raised.* **4** ***resolve into.*** °change into, convert into, alter into, transform into, transmute into, metamorphose into, be convert(ed) into, °become, dissolve into, break down into, liquefy into, disintegrate into, reduce to *or* into: *O, that this too too solid flesh would melt, / Thaw, and resolve itself into a dew!* —*n.* **5** See **resolution, 1,** above. **6** See **resolution, 2,** above: *She kept her resolve never to marry again.*

resonant *adj.* vibrating, vibrant, resounding, (re)echoing, reverberating, reverberant, pulsating, ringing, booming, thundering, °thunderous, °loud: *The resonant pealing of the bells almost drove me mad.*

resort *n.* **1** spa, °retreat, *Chiefly Brit* watering place: *We go to a lovely resort in the Alps every summer.* **2** °resource, backup, °reserve, °refuge, place to turn, °alternative, °remedy: *He will see a doctor only as a last resort.* —*v.* **3** ***resort to.*** have recourse to, °turn to, look to, °fall back on, repair to, °take to, °frequent, °patronize, °attend; °visit, °haunt, hang out in: *I have to resort to father for a small loan. At that time, she was resorting to the lowest dens in the Casbah.*

resound *v.* °boom, resonate, °ring (out), °boom (out), (re)echo, reverberate, °pulsate, °thunder: *The laughter resounded around the entire office.*

resource *n.* **1** Often, ***resources.*** °initiative, °ingenuity, °talent, inventiveness, °imagination, imaginativeness, cleverness, quick-wittedness, °capability, resourcefulness, °aptitude, °qualifications, °strength, °quality, °forte, *Colloq Brit* °gumption, *Slang* °guts: *Has he the inner resources for the job that lies ahead?* **2** Often, ***resources.*** °capital, °assets, °money, °possessions, °wealth, °property, °cash, °funds: *She has sufficient resources to buy a jet plane if she wants one.*

resourceful *adj.* °ingenious, inventive, °imaginative, °clever, Daedalian, °creative, °skillful, °smart, °slick: *Sylvia is resourceful enough to get out of any situation.*

respect *n.* **1** °regard, °consideration, °admiration, °esteem, °(high) opinion, °appreciation: *We have great respect for Samuel Sypher, our president.* **2** °regard, °consideration, °courtesy, politeness, °civility, attentiveness, thoughtfulness, °etiquette, °deference, °reverence, °veneration: *Few treat Alastair Windbourne with the respect he deserves.* **3** °reference, °relation, °connection, °comparison, °regard, °bearing: *What are they planning to do with respect to the flooding?* **4** °detail, °point, °element, °aspect, °characteristic, °feature, °quality, °trait, °particular, °matter, °attribute, °property: *Describe the respects in which the War of American Independence and the French Revolution were similar.* **5** ***respects.*** °regards, good *or* best wishes, °greetings, °compliments, *Formal* salutations, *Formal or archaic* devoirs: *I went to pay my respects to my aunt on her ninetieth birthday.* —*v.* **6** °consider, °admire, °esteem, °honor, °appreciate, value, defer to, pay homage to, think highly *or* well of, °look up to, °revere, °reverence, °venerate: *I am not sure that I like Mrs. Howling, but I certainly respect her for what she has done for the poor.* **7** °heed, °obey, show consideration *or* regard for, pay attention to, °attend to, be considerate *or* polite *or* courteous to, defer to: *Children are no longer taught to respect their elders.*

respectable *adj.* **1** °proper, demure, °decorous, °seemly, °estimable, °worthy, °dignified, °decent, °upright, °honest, respected, °genteel, °refined, °reputable, °aboveboard, unimpeachable, law-abiding: *Patricia Smyrke is a very respectable member of the community. I'll have you know that I run a respectable boarding house!* **2** °moderate, appreciable, °goodly, °reasonable, °fair, not inconsiderable, °considerable, °tolerable, °satisfactory, sizable, good-sized, °substantial, not insignificant, °significant, *Colloq* °tidy: *She earns a respectable living. A respectable number of people showed up. Climbing Annapurna, though it is not Everest, is a respectable feat.* **3** °presentable, °moral, °decent, °proper, °modest, °chaste, °innocent, °pure, °clean: *The board decided that the film is not respectable enough to be shown to schoolchildren.*

respectful *adj.* °courteous, °polite, well-mannered, well-behaved, mannerly, °civil, °cordial, gentlemanly, °ladylike, °gracious, °obliging, °accommodating, °considerate, °thoughtful: *On entering, Captain Gregory made a respectful bow to the ladies.*

respective *adj.* °separate, °individual, °particular, °pertinent, °specific, °special, °personal, own, °relevant, corresponding, °several: *Each retired to his respective room for the night. All of you know your respective duties.*

respectively *adv.* °separately, °individually, °singly, severally, *mutatis mutandis,* each to each: *Ronald and Archibald Snyde were, respectively, president and secretary of the association.*

respite *n.* **1** °interval, °intermission, °break, °interruption, °recess, breather, °rest; °holiday, *Chiefly US and Canadian* vacation: *I haven't had a moment's respite since six o'clock this morning. We are planning a brief respite from work in August.* **2** °pause, °delay, hiatus, °stay, °extension, °reprieve, °postponement: *After a ten-minute respite, the infernal noise began again.*

respond *v.* **1** °answer, °reply, come back, °return, °react, °reciprocate, °counter; rejoin, °retort: *When Sean insulted her, Eunice responded with a slap to his face. I said, "Good morning!" and Alphonse responded, "What's good about it?"* **2** Often, ***respond to.*** be responsive (to), °react (to), empathize (with), °sympathize (with), commiserate (with), °feel for, °pity, be affected *or* moved *or* touched (by): *Many respond to those charity appeals for the crippled and disabled.*

response *n.* °answer, °reply, °retort, rejoinder; °reaction, °effect, feedback, °return, *Colloq* comeback: *Whatever you say to Ronnie Farrago, he always has a snappy response. The response to our advertising has been quite good.*

responsibility *n.* **1** °accountability, °liability, chargeability, answerability, °obligation: *Responsibility is one of the burdens a parent must undertake.* **2** °charge, °duty, onus, °burden, °trust, °job, °role, °task: *Has Ted Brittle taken on more responsibilities than he can handle? It is your responsibility to see that the children are awakened in time for school.* **3** °blame, °guilt, °fault, culpability: *Professor Klein assumed full responsibility for the failure of the experiment.* **4** dependability, reliability, trustworthiness, °stability, °accountability, creditability: *Before accommodating you with a loan, Mr. Smeek, we must confirm your financial responsibility.*

responsible *adj.* **1** °accountable, answerable, °liable, chargeable: *The court determined that the defendant was not responsible for her actions.* **2** °reliable, °trustworthy, dependable, °stable, creditable, °accountable, °ethical, °honest: *If teenagers can show that they are responsible, the bank will lend them money for their enterprise.* **3** °executive, °leading, °authoritative, administrative, °important, decision-making, managerial, directorial, °principal, °chief, °top, *US* front-office: *Oliver does play a responsible role in the running of the company.* **4** °guilty, to blame, °at fault, culpable: *We never did find out who was responsible for putting the frog in the teacher's desk.*

responsive *adj.* °alert, °alive, (wide-)awake, reactive, communicative, °sharp, °keen, °receptive, °sensitive, °open, °sympathetic: *I was delighted to find students who were so responsive to the ideas put forth in my lectures.*

rest[1] *n.* **1** °repose, °sleep, °nap, °doze, siesta, slumber, *Chiefly Brit* lie-down, *Colloq* forty winks, zizz, snooze; shuteye: *I think I'll have a bit of a rest before dinner.* **2** °relaxation, °intermission, °interval, °interlude, entr'acte, rest period, cessation, (tea *or* coffee) break, °recess, breather, breathing spell, °respite, time off, °holiday, *Chiefly US and Canadian* vacation: *Why don't you take a rest for a while?* **3** °ease, °relaxation, °leisure, °indolence, °idleness, °inactivity, loafing, dozing: *My well-deserved rest was disturbed by an urgent request from Interpol to investigate a smuggling operation in Bioko.* **4** °prop, °support, holder, °brace, trestle, shelf, °bracket: *Those old guns were too heavy to hold and fire without using a rest.* **5** ***come to rest.*** °stop, end up, °turn up, °arrive: *After flying through the window, the tennis ball came to rest in my soup.* —*v.* **6** °(go to) sleep, °doze, °relax, take a rest, °(take one's) repose, lie down, °recline, go *or* take to one's bed, take one's ease, unwind, loll, languish, laze about, be idle, idle about, °lounge, °(take a) nap, put one's feet up, *Colloq* take it easy, snooze, count sheep, have a zizz, catch *or* grab some shuteye, get *or* take forty winks, *US* catch *or* log some Z's, *Slang Brit* kip, doss down, °hit the sack, hit the hay, *US* sack out: *Rest now—you'll feel better tomorrow.* **7** reside, be situated, be lodged, °lie, be placed, hinge, be found, °remain, °stay: *The responsibility for passenger safety rests with the captain and the crew. The blame for this fiasco rests on you.* **8** °place, °position, °put, °lay, °set, °lean, °prop: *Rest your head on my shoulder.* **9** °lie, °remain, °stay: *Can't you let the matter rest?* **10** allay, °calm, °quiet, °still, °stay: *Rest your fears —I shall let nothing happen to you.*

rest[2] *n.* **1** °remainder, °balance; °remains, remnants, leftovers, °residue, residuum, °excess, °surplus, overage: *If you carry these bags, I'll take the rest. Oswald bought up all the best books and left the rest for us.* —*v.* **2** (continue to) be, °remain, keep on being: *Rest assured, she'll get home safely.*

restful *adj.* **1** relaxing, °soothing, comforting, tranquilizing, °sedative, calming, sleep-inducing, hypnotic, soporific, somnolent: *Don't you find the sound of the waves restful?* **2** °tranquil, °calm, °peaceful, °quiet, °still, °serene, pacific, °comfortable, °relaxed, reposeful: *After a hectic day in the city, I was happy to return to the restful atmosphere of the country.*

restitution *n.* **1** °amends, compensation, redress, recompense, °remuneration, reparation, °requital, indemnification, °indemnity: *It was agreed that victims of the Nazis were entitled to restitution for what had been taken from them.* **2** °restoration, °return, reestablishment, reinstatement, °recovery: *A commission is to investigate the restitution of plundered property.*

restive *adj.* See **restless,** below.

restless *adj.* °restive, uneasy, edgy, °on edge, on tenterhooks, fidgety, °nervous, skittish, °excitable, highly strung, high-strung, worked up, °agitated, °fretful, °jumpy, apprehensive, itchy, *Colloq* jittery, *Slang* uptight, *US* antsy, hyper: *The crowd in the square were becoming restless because of the presence of the security police.*

restoration *n.* **1** See **restitution, 2,** above. **2** renovation, refurbishment, rehabilitation, renewal, °repair, rejuvenation, reconstruction, resurrection, reconversion, °revival: *The program for the restoration of ancient buildings suffers from lack of funding.*

restore *v.* **1** give *or* hand back, °return, make restitution, bring back: *We must restore to the people the land that is rightfully theirs.* **2** °revive, °rejuvenate, reestablish, °renew, bring back, give (someone) back, resuscitate, °resurrect, rekindle, reinvigorate, °refresh, °stimulate, revitalize, °strengthen: *Your kindness has restored my faith in mankind.* **3** °renovate, °refurbish, °renew, °repair, °rejuvenate, °resurrect, °revive, reconstruct, °rehabilitate, rebuild; °mend, °fix, °retouch, °touch up; *Colloq US* °fix up: *They live in a seventeenth-century cottage that has been lovingly restored. Antiques often lose their value entirely if they are restored.* **4** °replace, reinstate, °put back; °return, bring back: *We shall not rest till we have restored the rightful king on the throne. The doctor said I would soon be restored to good health.* **5** °replace, °reimburse, °repay, °return, pay *or* put *or* give back: *Even though he has restored all the money he took, she refuses to forgive him.*

restrain *v.* **1** (keep under *or* in) control, (keep *or* hold in) check, hold (back *or* in), °curb, °govern: *A poor rider, he was unable to restrain his horse. Something must be done to restrain the general's power.* **2** °limit, °restrict, °inhibit, °regulate, °suppress, °repress, °bar, debar, °curtail, °stifle, °hinder, interfere with, °hamper, °handicap: *Trade between the countries was restrained because of the tariff dispute.* **3** °(place under) arrest, confine, °imprison, incarcerate, detain, °hold, °lock up, jail *or Brit also* gaol, shut in *or* up: *For his most recent offense, he was restrained for two months.*

restraint *n.* **1** °control, °check, °curb, °rein, °bridle, °restriction, constraint, °limit, limitation, curtailment, °taboo, °ban, interdict *or* interdiction, proscription, delimitation, bound(s), °embargo: *The law places restraints on executives' buying and selling shares in their own companies. In 1863, a bill for the restraint of the press was brought before the House of Commons.* **2** °control, °restriction, constraint, confinement; °bondage, bonds, fetters, °shackles, °handcuffs, gyves, bilboes, pinions, °manacles, ball and chain, straitjacket, *Colloq* cuffs, bracelets: *Despite the restraints to his liberty, he felt a free man. He became violent and had to be put under restraint.* **3** °control, °reserve, °self-control, self-possession, °poise, equanimity, self-discipline, self-restraint: *Heather Gorse exhibited admirable restraint in remaining silent when teased her about her name.*

restrict *v.* °limit, confine, °bound, circumscribe, delimit, mark off, demarcate, °regulate; °qualify, °restrain, °impede: *Smoking is allowed only in restricted areas. People get upset when you restrict their freedom of movement.*

restriction *n.* **1** °condition, provision, °proviso, °qualification, °stipulation: *One restriction is that purchasers of shares in the utility must be resident in the UK.* **2** See **restraint, 1,** above.

result *n.* **1** °outcome, consequence, °effect, °end (result), °fruit; conclusion, °upshot, °issue, °development, °sequel, °follow-up, denouement *or* dénouement: *It will take years to evaluate the results of the new educational curricula. As a result of his speeding ticket, his driving license was suspended.* —*v.* **2** Often, ***result from.*** °develop, °emerge, °follow, °happen, °occur, °come (about), °come to pass, °arise, evolve, be produced: *That is what can result from allowing children to play with matches. The mixture that results may be highly volatile.* **3** ***result in.*** °end, conclude, culminate, °terminate: *The explosion resulted in a heavy loss of life. The experience gained often results in better safety devices.*

resume *v.* °continue, °carry on, take up again, °pick up (where one left off): *After prison, it is not easy to resume one's life where it left off. When the audience quieted, the speaker resumed.*

résumé *n.* **1** °summary, °digest, °abstract, °synopsis, °précis, °outline, °review, recapitulation, °epitome, *Colloq* °rundown, recap: *Give me a résumé of what went on at the board meeting.* **2** curriculum vitae, CV, °summary, biography, work *or* job history, career description, *Formal* prosopography, *Colloq* bio, *US* vita: *My résumé was enough to get me an appointment for an interview.*

resurgence *n.* °renaissance, renascence, °rebirth, °revival, reawakening, °restoration, renewal, resumption, °return, resurrection, regeneration, rejuvenation, new dawn, new birth: *Simon is banking his entire future on a resurgence of the fashions of the seventies.*

resurrect *v.* °revive, bring back, °return, reawaken, °restore (to life), reintroduce, °renew, regenerate, °rejuvenate, °raise (from the dead), resuscitate, breathe new life into, reanimate, reincarnate: *Iris Higgens has succeeded in resurrecting the Gothic novel to popularity.*

retain *v.* **1** °keep (possession of), °hold (on to), °save, °preserve, *Colloq* hang on to: *Retain the receipt in case you wish to exchange the merchandise.* **2** °engage, °hire, °employ, commission, °take on: *We have retained a caretaker to look after the estate.* **3** °hold, absorb, °contain, °soak up, °preserve: *This type of soil retains little water.* **4** °remember, keep *or* bear *or* have in mind, °recall, remain aware of, °memorize, impress on the memory, °recollect: *As he aged, he found it increasingly difficult to retain the simplest information.*

retaliate *v.* °repay, °pay back (in kind), °counter, strike back (at), take revenge (on), wreak vengeance (on), revenge oneself (on), avenge, °reciprocate, settle *or* even a score (with); give tit for tat, take an eye for an eye (and a tooth for a tooth), give as good as one gets, give (someone) a taste of his *or* her *or* their own medicine, pay (someone) back in his *or* her *or* their own coin, *Brit or literary* give a Roland for an Oliver; *Colloq* get even (with), °get back (at): *When government forces shelled their headquarters, the guerrillas retaliated by bombing bridges and rail lines.*

retard *v.* **1** slow (down *or* up), hold up *or* back, °set back, °hinder, °impede, °delay, keep back, °stall, °thwart, balk, °block, °restrict, hold in check, °frustrate, interfere with: *We must find out what is retarding the growth of these seedlings.* —*n.* **2** *Offensive and derogatory* idiot, moron, °fool, imbecile, dunce, *Slang chiefly US and Canadian* °jerk: *Why she invites that retard to her parties I cannot imagine.*

reticent *adj.* °quiet, °shy, °timid, °retiring, °reserved; °taciturn, °silent, unresponsive, °tight-lipped, unforthcoming: *If you are proud of your accomplishments, why be reticent about saying so? Crouch is reticent about how he acquired his gold earring.*

retinue *n.* entourage, °escort, convoy, cortège, °company, °train, °suite, followers, attendants, following, hangers-on, *Colloq* groupies: *The rock band arrived at the airport accompanied by a retinue of flunkies.*

retire *v.* **1** °withdraw, rusticate, go off *or* away, °take off, °retreat; hibernate, estivate *or Brit* aestivate, seclude *or* sequester *or* cloister oneself: *I think I'll retire to some mountaintop to finish my book.* **2** stop *or* give up work(ing), be pensioned off, (be) put out to grass *or* pasture, take the golden handshake, be given the gold watch, go on Social Security, go on a pension, be superannuated, go out of circulation: *Gemma Frobisher retired from her job in the bakery and now lives with her son in Indianapolis.* **3** go *or* take to (one's) bed *or* bedroom, °(go to) sleep, lie down, °(take one's) repose, °(take a) nap, put one's feet up, *Colloq* take it easy, snooze, count sheep, have a zizz, catch *or* grab some shuteye, get *or* take forty winks, *US* catch *or* log some Z's, *Slang* °hit the sack, sack out, hit the hay, *Brit* kip, doss down: *I usually retire around midnight.*

retiring *adj.* °shy, °bashful, °coy, demure, °modest, diffident, °timid, unpretentious, unassuming, °humble, self-effacing, timorous, °meek, °reticent, °reserved, °unsocial, unsociable, °aloof, removed, °standoffish, °distant, °reclusive, eremitic(al): *Hoggett is of a retiring disposition, preferring to keep to himself.*

retort *n.* **1** °response, °reply, rejoinder, °answer, riposte, °rebuttal, *Colloq* comeback: *Faced with criticism of the novel, the author's retort was a four-letter word.* —*v.* **2** fling *or* hurl back, rejoin, °answer back, riposte, °rebut, °counter, come back with, °return, °respond, °answer, °reply, °retaliate: *"And don't expect a Christmas present from me, either!" she retorted vehemently.*

retouch *v.* °touch up, °correct, °restore, °repair, recondition, °refresh, brush up, °adjust, °improve, °finish, put the finishing touches on: *Only an expert could have retouched the damaged painting so skillfully.*

retract *v.* **1** °withdraw, pull *or* draw back: *The instant he felt the heat, he retracted his hand.* **2** °take back, °withdraw, rescind, °revoke, °repeal, °deny, disavow, °recant, °renounce, abjure, °cancel, forswear, °repudiate, disclaim, disown, °reverse: *If they refuse to retract their statement, I shall sue them for libel.*

retreat *n.* **1** retirement, withdrawal, pulling *or* falling *or* drawing back, giving ground, evacuation, °flight: *The regiment's retreat in the face of overwhelming odds is complete.* **2** retirement, °seclusion, withdrawal, isolation, °solitude, rustication: *Since her retreat from public life, Amelia has devoted herself to painting.* **3** °sanctuary, °sanctum (sanctorum), °refuge, °shelter, den, haven, asylum, °resort, °hideaway, hide-out: *Antonia insisted on having her own retreat where she could occasionally escape from the children.* —*v.* **4** °withdraw, decamp, °run (away), °turn tail, °depart, give *or* lose ground, pull *or* fall *or* draw back, °retire, °evacuate, °flee, °take flight: *We must retreat to the shelter of the hills, where we can regroup.* **5** °ebb, draw *or* move back, °recede: *The wavelets, constantly advancing and retreating across the pebbly strand, lulled me to sleep.*

retribution *n.* °vengeance, °revenge, °reprisal, retaliation, °requital, redress, quid pro quo, °satisfaction, °punishment, °justice, just deserts, recompense, compensation: *It is a moral question whether society's treatment of malefactors should be an act of retribution or the setting of an example to others.*

retrieve *v.* **1** bring *or* get (back), °fetch, come back with: *Simon trained his dog to retrieve his slippers—Simon's, that is.* **2** °recover, °save, °rescue, °take back, °recoup, regain, °reclaim: *I tried using a fishing line to retrieve my hat from the lake.* **3** °make up, make amends for, °recover, °cover, °redeem, °repay, °pay for, °return, °get back, regain, be repaid *or* reimbursed for: *Will we be able to retrieve our losses when we sell the business?*

retrospect *n.* hindsight, reconsideration, °review, remembering, afterthought, °recollection, looking back: *In retrospect, I am not sure we did the right thing.*

return *v.* **1** come *or* go back, °revert, °turn back: *She is returning to her old habits. I shall return to that subject if I have time.* **2** come back, reappear, resurface, crop up again, turn *or* show up again, put in an appearance again; °recur, reoccur, *Colloq* pop up again: *Will he return before March? The symptoms returned.* **3** °replace, °put back, °restore, give back, bring *or* carry back: *The missing books have mysteriously been returned.* **4** °exchange, bring *or* carry *or* send back: *You may return any merchandise, for any reason, for refund or exchange, within ten days of its purchase.* **5** °yield, °earn, °gain, °profit, °benefit: *Where else can you find a secure investment that returns more than 15 percent?* **6** °deliver, °render, °give, °offer, °turn in, proffer, °report: *The jury returned a verdict of not guilty.* —*n.* **7** recurrence, reappearance, °repetition, renewal, recrudescence, resurfacing, reemergence: *There was a return of the symptoms when I stopped taking the medicine.* **8** replacement, replacing, °restoration, restoring, °restitution: *We shall see to the return of the stolen items to their rightful places.* **9** Sometimes, ***returns.*** °yield, °earnings, °profit, °gain, °benefit, °income, °revenue, °proceeds, °interest, takings, results: *The return on my investment was almost 100 percent!* **10** °arrival, advent, coming, homecoming: *A small crowd had gathered to await his return.* **11** reciprocity, reciprocation, repayment, recompense, reimbursement, compensation, °payment, reparation, °indemnity, indemnification, °consideration, °amends, redress, °requital: *What will I get in return for the sheep killed by your dog?*

revamp *v.* °overhaul, redo, recondition, °renovate, °repair, °fix, do up, refit, °refurbish, °restore, °rehabilitate, *US* °fix up: *To bring it into line with the twenty-first century, the constitution may need to be revamped completely.*

reveal *v.* °expose, °display, divulge, °disclose, °show; °make known, °let on, °let out, let be known, °let slip, °communicate, °give vent to, °air, ventilate, °leak (out): *The increasing light of dawn revealed a no-man's-land of utter devastation. The book is said to reveal everything you never wanted to know about Fernand Coutier.*

revel *v.* **1** Usually, ***revel in.*** °(take) delight (in), take pleasure (in), °rejoice (in), °luxuriate (in), bask (in), °wallow (in), °lap up, crow (about *or* over), °glory (in), °savor, °relish: *Gavin reveled in the news that his ex-wife was remarrying and he would no longer have to pay her alimony.* **2** °make merry, °carouse, °celebrate, cut loose, go on a spree, *Colloq* live it up, make whoopee, whoop it up, °paint the town red, °party, *Brit* push the boat out, *Slang Brit* have a rave *or Slang Brit* rave-up: *The night after exams finish, the students traditionally revel till dawn.* —*n.* **3** °spree, °party, merrymaking, debauch, °carouse, °festival, °fête, carousal, °celebration, °gala, ball, romp, °fling, carnival, °jamboree, bacchanal, saturnalia: *Everyone joined in the revels on New Year's Eve.*

revelation *n.* °news, °information, °proclamation, °publication, °bulletin, communiqué, °announcement, °pronouncement, °declaration, °statement, °leak; °admission, confession; °discovery, unveiling, uncovering, °exposure, disclosure, exposé: *We were shocked at the revelation that they had been married all these years. Scientists' revelations about the depletion of the ozone layer have everyone concerned.*

revelry *n.* merrymaking, °fun, reveling, carousal, carousing, °gaiety, °festivity, jollity, °mirth, °celebration, ball, high jinks *or* hijinks, °spree, *Colloq* partying, *Slang Brit rave or* rave-up: *The revelry went on into the wee hours of the morning.*

revenge *n.* **1** °vengeance, retaliation, °reprisal, °retribution, vindictiveness, spitefulness, repayment, °satisfaction: *They think he set fire to the place in revenge for having been dismissed.* —*v.* **2** avenge, get even for, take revenge for, make reprisal for, exact retribution *or* payment *or* repayment for: *She was seeking ways to revenge the murder of her brother.* **3** ***revenge oneself (on).*** settle a score *or* an old score (with), pay (someone) back in his *or* her *or* their own coin, give tit for tat, take an eye for an eye (and a tooth for a tooth), °punish, *Colloq* °get, get even (with), give (someone) his *or* her *or* their comeuppance, give (someone) a taste of his *or* her *or* their own medicine, *Brit or literary* give a Roland for an Oliver: *He found ways to revenge himself on them for all that they had done to him.*

revenue *n.* °(gross) income, °proceeds, °receipts, return(s), °yield, °interest, takings, °net (income), °gate; °profits, °gain, *Colloq chiefly US* °take: *The revenue from the drugs operation must run into billions.*

revere *v.* °adore, adulate, °reverence, °venerate, °worship, °idolize, enshrine, °sanctify, beatify, °glorify, °esteem, °admire, °respect, °honor: *All had heard of Rome's glories and revered the name of Caesar.*

reverence *n.* **1** °honor, °respect, °esteem, °admiration, glorification, beatification, sanctification, idolization, °worship, °veneration, adulation, adoration, °homage, fealty, °obeisance, °deference, awe: *Their totalitarian discipline was able to succeed because of the people's reverence for authority.* —*v.* **2** See **revere,** above.

reverie *n.* °daydream, °fantasy, brown study, woolgathering, absent-mindedness; meditation, °thought: *I shouldn't disturb Kevin—he's deep in a reverie.*

reversal *n.* **1** °reverse, °turnabout, turnaround, U-turn, °change, volte-face, °(complete) switch, *Brit* about-turn, *US* °about-face: *This represents a reversal of former policy, doesn't it?* **2** See **reverse, 8,** below. **3** annulment, nullification, nulling, °cancellation, revocation, °repeal, rescission: *The present case is an appeal against the reversal of an earlier judgment.*

reverse *adj.* **1** °opposite, °contrary, inverse, °converse; inverted, upside-down, °mirror, reversed, backward(s): *Now try turning the wheel in the reverse direction. What appears on the retina is a reverse image, which the brain inverts.* —*v.* **2** invert, °overturn, turn upside down, turn topsy-turvy, °turn over, upend; °exchange, °change, interchange, °transpose: *Using your finger to keep the mercury from running out, reverse the tube in the dish of mercury. Perhaps the car will start if you reverse the leads to the battery.* **3** °overturn, °overthrow, °upset, °set aside, °quash, override, annul, nullify, °vacate, °abandon, °revoke, negate, °veto, declare null and void, disaffirm, invalidate, °cancel, °repeal, rescind, overrule, countermand, °undo: *The superior tribunal reversed the decision of the lower-court judge.* **4** °alter, °change, °modify; °renounce, °recant, °take back: *Reverse the polarity of the wiring. I reversed my opinion after hearing her sing.* **5** °back up, move *or* go backward(s), backtrack, *Nautical* make sternway: *You will be in the right spot if you reverse a yard or two.* —*n.* **6** °opposite, °contrary, °converse, antithesis: *Whatever she tells you, I suggest you do the reverse.* **7** °back, °rear, wrong side, verso, underside; flip side, B-side *or* side B; tail side: *Read what it says on the reverse.* **8** °setback, °disappointment, °misfortune, °reversal, °mishap, misadventure, °trouble, problem, °difficulty, °vicissitude, adversity, °defeat; °disaster, °catastrophe, débacle, °rout, *Colloq* °washout: *We suffered a temporary reverse or two in business, but we're all right now. The charge of the Light Brigade was scarcely a reverse—more like a slaughter!*

revert *v.* °return, come *or* go back, take *or* pick up again, °lapse, backslide, regress, °relapse, °retreat: *He reverted to his old beliefs about male superiority.*

review *v.* **1** °survey, °examine, °regard, °look at, °study, con, °consider, °weigh, °inspect, °look over, °scrutinize: *Would you be good enough to review my work? She is coming to review the troops this weekend.* **2** reexamine, reconsider, go over again, look at *or* over again, reassess: *Billings reviewed once more all the bad things he had done in his lifetime.* **3** °criticize, critique, assess, °judge, °evaluate, give one's opinion of, comment on *or* upon, °discuss: *Who reviewed your latest book?* —*n.* **4** °criticism, critique, review article, assessment, °judgment, °evaluation, commentary, °study, °comment, °notice: *The play received rave reviews from all the critics but one.* **5** °survey, °examination, °study, °consideration, inspection, °scrutiny, °analysis; reviewing, reading: *After a detailed review of the circumstances, we have decided to let you off easy this time, Finnegan.* **6** reexamination, reconsideration, rehashing *or* rehash, °post-mortem, reassessment, rethinking, rethink: *Why do you always insist on a review of the bidding?* **7** °periodical, °journal, °magazine: *Have you read Stulpnagel's article in the latest* Birmingham Review? **8** °parade, °procession, array, cavalcade, marchpast, flypast *or chiefly US* flyover: *A great naval review was held at Newport for the Sultan.*

revise *v.* **1** °edit, emend, °improve, °correct, °rectify, °modify, °revamp, redact, rework, °overhaul, update; rewrite: *A revised edition of the dictionary has been published every ten years.* **2** °alter, °change, °amend: *Since hearing him sing, I have been forced to revise my taste in music.*

revision *n.* editing, revising, emendation, °improvement, °correction, rectification, modification, revamping, reappraisal, reexamination, reinterpretation, reassessment, redaction, reworking, °overhaul, overhauling, updating, update; rewrite, °edition, °version, °rendition: *Her revision of the text is based on the fourteenth-century palimpsest. The fourth revision will be available next April.*

revival *n.* **1** resurrection, resuscitation, renewal, °restoration, revitalization, resurfacing, °return, returning: *How much more often we see the revival of outmoded fashions than the introduction of new ideas!* **2** °rebirth, renaissance *or* renascence, °resurgence, awakening, quickening, reanimation: *Do you regard the current trend toward fundamentalism as a genuine revival of religion?* **3** °recovery, °improvement, °increase, upsurge, upturn, °boost, upswing, °advance, advancement, °progress, °rise, escalation, *Colloq* pickup, comeback: *This week saw a revival in the metals market.*

revive *v.* **1** (re)awaken, °wake (up), come *or* bring (a)round, waken, resuscitate; °recover, (re)gain consciousness: *Dashing cold water in his face revived him at once. I revived to find her leaning over me.* **2** reawaken, stir up again, °renew, °resume, reopen, °refresh: *You can blame Kershaw for reviving the feud between our families.* **3** bring back, reactivate, °renew, °resurrect, reestablish, °resume, reopen, revitalize, breathe life into, reinvigorate, °rejuvenate: *Perhaps we ought to revive the practice of putting hooligans, vandals, and delinquents into stocks in the market square.*

revoke *v.* °cancel, °deny, invalidate, annul, declare null and void, °void, nullify, negate, °repudiate, °renounce, rescind, °repeal, °recall, °recant, °quash, °veto, °set aside, abrogate, °abolish, °withdraw, °take back, °retract: *The health department warned that if there was one more offense, the restaurant's license would be revoked. Morgan le Fay refused to revoke the curse she had placed on Merlin.*

revolt *n.* **1** rebellion, °revolution, °uprising, °mutiny, insurrection, coup d'état, putsch, takeover: *The council ordered the secret police to suppress the revolt.* —*v.* **2** °rebel, rise up, °mutiny; °protest, dissent: *When the people revolted, the deposed leaders escaped to safety with the aid of their Swiss bank accounts.* **3** °repel, °offend, °disgust, °shock, °horrify, °repulse, °nauseate, °sicken: *I was revolted by the poverty and sickness I saw in that country.*

revolting *adj.* °disgusting, sickening, nauseating, °nauseous, stomach-turning, stomach-churning, vomit-provoking, sick-making, °foul, °loathsome, °abhorrent, horrid, °horrible, °nasty, °vile, °obnoxious, °repulsive, appalling, °abominable, °repellent, °offensive, objectionable, off-putting, °beastly, °gross, °rotten, °rancid, inedible, °disagreeable, unpleasant, *Slang* icky, yukky: *Rick, who regards himself as a great chef, served up a revolting mess for dinner.*

revolution *n.* **1** °mutiny, °revolt, rebellion, coup (d'état), °uprising, insurgency, insurrection, putsch, takeover, °overthrow: *If the government fails to ensure that the people have food, there will be a revolution.* **2** °upheaval, cataclysm, °transformation, (drastic *or* radical *or* major) change, sea change, metamorphosis: *Recent successes of capitalism have sparked a revolution in the political philosophies of communist nations.* **3** rotation, °turn, °orbit, °circuit, °spin, °lap, °round, °cycle, °circle, gyration; °wheel, whirl, °pirouette: *At each revolution of the gear, the ratchet moves one notch. The rapid revolutions of the ice-skater dissolved her in a blur of color.*

revolutionary *adj.* **1** °mutinous, °rebellious, insurgent, insurrectionist, insurrectionary, °radical, °rebel, °seditious, °subversive: *Vladimir had dreamed for years of the revolutionary overthrow of the government.* **2** °novel, innovative, °creative, °new, °different, °original, °avant-garde: *Gently has patented a revolutionary device that is sure to make his fortune.* —*n.* **3** °rebel, mutineer, insurgent, insurrectionist, insurrectionary, revolutionist, sans-culotte *or* sans-culottist, anarchist, °radical, extremist, °terrorist, °rabble-rouser: *In his younger days at college, Culbert fancied himself a revolutionary, without quite knowing what he might rebel against.*

revolve *v.* **1** °spin, °turn, °pivot, °rotate, °gyrate, whirl, °twirl, °reel, °wheel, go (a)round, °circle, °cycle, °orbit; °swivel: *As the earth revolves on its axis, the moon revolves round the earth.* **2** °turn, °depend, °pivot, °rely: *His decision to buy a new car revolves on his finances.* **3** °turn over (in one's mind), °ponder, °weigh, °consider, meditate upon *or* on, °think about, reflect upon *or* on, ruminate over *or* on, °chew over, °contemplate: *I have been revolving in my mind your most attractive proposal, but I regret I must turn it down.*

revolver *n.* °pistol, gun, sidearm, firearm, *Chiefly US and Canadian* handgun, *Colloq US* six-gun, six-shooter, *Slang US* °rod, gat, roscoe, °piece, Saturday night special, shooting iron: *How do you explain the fact that your fingerprints were found on the revolver with which she was shot?*

revulsion *n.* °loathing, detestation, °disgust, repugnance, abomination, abhorrence, °aversion, hatred, antipathy, odium, execration: *I cannot describe the revulsion I felt at seeing those elephants slaughtered for their ivory.*

reward *n.* **1** °award, °favor, recompense, compensation, °return, °payment, °pay, °requital: *For your good deeds you will have to seek your reward in heaven; dishonesty is recompensed here on earth.* **2** °prize, °award, °tribute, °honor, *Literary* guerdon: *The Victoria Cross was instituted as a reward for military valor in 1856.* **3** °retribution, °punishment, just deserts, *Colloq* comeuppance: *Hanging was the reward of treason.* —*v.* **4** recompense, °compensate, °pay, °repay, remunerate, redress, °requite, make (something *or* it) worth someone's while: *Only if they reward him will Fuller give information leading to the capture of the thief.*

rewarding *adj.* °satisfying, gratifying, °worthwhile, enriching, enriched, °fruitful, °worthwhile; °profitable, °advantageous, °productive, °gainful: *Laurence says that his most rewarding experience has been seeing his children again.*

reword *v.* °paraphrase, rephrase, put into different words, put another way, express differently, °revise, recast, rewrite: *Perhaps if you reworded your request in politer terms it might get more sympathetic consideration.*

rhapsodic *adj.* rhapsodical, °ecstatic, °enthusiastic, °elated, °overjoyed, °effusive, °rapturous, thrilled, blissful, transported, orgasmic, intoxicated, euphoric, walking on air, °delighted, intoxicated, happy as a sand boy, happy as a pig in clover, °(sitting) on top of the world, *Brit* in the seventh heaven, *US* in seventh heaven, happy as a cow in clover, *Taboo slang US* happy as a pig in shit: *What kind of society have we that is scornful of intellect and rhapsodic about celebrities?*

rhetoric *n.* **1** eloquence, expressiveness, elocution, way with words, *Colloq* gift of the gab: *His rhetoric stirred us, beguiled us, angered us, soothed us, delighted us, and now, alas, he will be heard no more.* **2** °bombast, °bluster, fustian, rodomontade, grandiloquence, magniloquence, °oratory, windiness, high-flown poppycock, wordiness, sesquipedality, verbosity, prolixity, long-windedness, turgidity, flatulence, gasconade, *Colloq* °hot air, puffery: *Her threats are mere rhetoric and I'd pay them no heed.*

rhetorical *adj.* **1** stylistic, linguistic, °poetic, °expressive, oratorical: *Zeugma, hysteron proteron, anacoluthon—these are terms for rhetorical devices.* **2** °artificial, contrived, for effect, unanswerable, not literal: *The question "Why don't you drop dead?" was purely rhetorical, Firnby.* **3** °pretentious, °bombastic, °flamboyant, °extravagant, florid, fustian, high-flown, °inflated, °grandiose, euphuistic, turgid, grandiloquent, magniloquent, long-winded, °windy, orotund, °wordy, prolix, sesquipedalian, *Colloq* highfalutin *or Colloq* hifalutin *or Colloq* highfaluting: *The rhetorical style of Crampton's parliamentary speeches appear to be doing the party more harm than good.*

rhyme *n.* **1** rime, °poem, °poetry, verse, versification, metrical composition, °song: *The art of rhyme may well be coincident with the earliest stages of language.* **2** ***rhyme or reason.*** °(common) sense, °logic, °intelligence, °meaning, °wisdom, rationality, °rationale, soundness, °organization, °structure: *There appears to be no rhyme or reason behind the bombing of the library.*

rhythm *n.* °tempo, °beat, cadence *or* cadency, throb, throbbing, °accent, accentuation, °time, timing, stress *or* rhythmic(al) pattern, °measure, meter, °pulse, lilt, *Music* downbeat, °thesis, °upbeat, arsis: *That song is in samba rhythm.*

rhythmic *adj.* rhythmical, °measured, cadenced, throbbing, pulsing, pulsating, °regular, °steady, beating: *The rhythmic sound of the waves soon sent me to sleep.*

ribaldry *n.* °vulgarity, immodesty, °indelicacy, indecency, coarseness, bawdiness, earthiness, wantonness, raciness, naughtiness, shamelessness, lustfulness, rakishness, rascality, dissoluteness, lubricity, lasciviousness, looseness, scurrility *or* scurrilousness, lewdness, salaciousness, licentiousness, grossness, offensiveness, rankness, rudeness, smuttiness, smut, °dirt, °filth, foulness, obscenity: *The ribaldry of some of Chaucer's tales provides a clue to what was on many people's minds in fourteenth-century England.*

rich *adj.* **1** °wealthy, affluent, °prosperous, well-to-do, °well-off, well-provided-for, °opulent, moneyed, in clover, on velvet, *Colloq* °flush, °loaded, on Easy Street, rolling in it *or* money *or* wealth, in the chips *or* the dough *or* the money, well-heeled, *US* well-fixed: *A year ago Montmorency was a rich man, but that was before Blue Thursday on the Stock Exchange.* **2** Sometimes, ***rich in.*** °abundant (in), overflowing (with), °fertile (in), fecund (in), °productive (of), copious (in), abounding in, well-supplied (with), well-stocked (with), rife (with), °replete (with), profuse (in *or* with): *The rich prairie land was soon yielding abundant harvests. The surrounding mountains were rich in minerals of all kinds.* **3** valuable, °precious, °invaluable, °priceless: *For centuries these people have enjoyed the rich bounty of the seas. The thieves who broke into the vaults got a rich haul.* **4** costly, °expensive, °dear, valuable, °invaluable, °precious, °priceless, °lavish, °sumptuous, °lush, °luxurious, °palatial, °elaborate, °splendid, °exquisite, °superb, °elegant: *The caravans returned laden with rich treasures of the Orient—silks, jewels, spices, rare woods, and other luxuries.* **5** °intense, °dark, °deep, °warm, vibrant, °strong, °lustrous: *At the rajah's court the rich colors of the carpets and hangings dazzle the eye.* **6** °fat, fattening, °heavy, creamy, °succulent, °savory, mouthwatering, °luscious, sapid, °delicious: *When Vickie complained that her sweet was too rich, Arthur ate hers, too.* **7** °mellow, mellifluous, °resonant, sonorous, °full: *Edgar's rich basso filled the cavernous amphitheater.* **8** °aromatic, ambrosial, °savory, °fragrant, °redolent, °pungent, °strong: *The house was filled with the rich odor of freshly made coffee.* **9** °productive, °plentiful, °abundant, °ample, °bountiful, °prolific, °fruitful, °fertile, fecund, copious, °profitable, °potent: *The letters proved a rich source of intimate details about their love life.* **10** laughable, °funny, °hilarious, comic(al), °humorous, amusing, *Colloq* sidesplitting: *He had many rich anecdotes to relate about the practical jokes they used to play on one another.* **11** °ridiculous, °preposterous, °outlandish, °ludicrous, °absurd, °nonsensical: *One particularly rich episode involved a chimpanzee.*

riches *n.* pl. wealth, affluence, opulence, °plenty, °prosperity, °abundance, °fortune, °means, °resources; lucre, pelf: *It was rumored that the family had acquired its apparently endless riches in the slave trade, in the eighteenth century.*

richly *adv.* **1** sumptuously, lavishly, luxuriously, splendidly, elaborately, exquisitely, elegantly, superbly: *The room was richly appointed with Chinese antiques, thick oriental carpets, and silk hangings.* **2** °well, °thoroughly, °amply, fully, °appropriately, fittingly, condignly: *Shirley's work was at last given the recognition it so richly deserved.*

rickety *adj.* wobbly, unsteady, broken-down, °decrepit, °shaky, tottering, teetering, °ramshackle, °flimsy, °frail, °precarious, °dilapidated, in disrepair, tumbledown, unsecure: *Don't put that heavy lamp on that rickety table!*

rid *v.* **1** ***rid of.*** deliver from, relieve of, free from *or* of, °rescue, °save: *What can we do to rid ourselves of the nuisance of mosquitoes?* —*adj.* **2** ***be*** *or* ***get rid of.*** °banish, °exile, °eject, °expel, °dispose of, throw out *or* away, °eliminate, °dispense with; °refuse, °reject, °dismiss, shrug off: *If she had wanted to be rid of him, she would have divorced him long ago. I am having trouble trying to get rid of my old car.*

-ridden *adj.* afflicted *or* harassed *or* affected *or* dominated by, infected *or* infested with: *They were shocked when they first came to this drought-ridden province. We visited a rat-infested, disease-ridden quarter of the city as part of the tour.*

riddle[1] *n.* conundrum, °puzzle, °enigma, poser, °question, °mystery, °problem, brain-teaser *or Brit* brain-twister: *The riddle is how they put eight great tomatoes in that itty-bitty can.*

riddle[2] *v.* **1** °perforate, °pepper, °puncture, °pierce, honeycomb: *Before they adjusted the blindfold, I saw that the wall behind me was riddled with the bullets of countless executions.* **2** °penetrate, °infest, infect, pervade, °permeate, °fill, °spread: *The entire department was so riddled with corruption that we had to bribe the staff to learn of it.* —*n.* **3** sieve, colander *or* cullender, strainer, °grating, °screen, sifter, °filter: *We shoveled the dirt through the riddle to sift out the larger stones.*

ride *v.* **1** sit on *or* in, travel *or* journey *or* go *or* proceed on *or* in, be borne *or* carried *or* conveyed (on *or* in), °take; propel *or* control *or* drive (a horse *or* a bicycle *or* a motorcycle): *They ride the bus to work each day. The movie begins with a witch riding a broomstick. Do you know how to ride a motorbike?* **2** °float, °lie: *The yacht is riding at anchor off Portsmouth.* **3** °tyrannize, terrorize, °intimidate, °dominate, °oppress; °bully, °rag, harry, °harass, hector, °nag, °provoke, °heckle, badger, °plague, °pester, °annoy, °irritate, *Colloq* hassle: *He has been ridden by obsessive ambition all his life.* —*n.* **4** °drive, °journey, °trip, °excursion, °tour, jaunt, °outing, °expedition, *Colloq* °spin: *Let's go for a ride in the country!* **5** ***take for a ride.*** **(a)** delude, °swindle, °trick, °deceive, °defraud, humbug, °defraud, gull, °take in, °cheat, bamboozle: *Those tricksters with the loaded dice certainly took Charlie for a ride last night!* **(b)** (kidnap and) murder *or* kill *or* execute *or* do in, *Slang* °bump off, *US* (snatch and) rub out *or* *US* waste: *They said that if he identified them, they would take him for a ride.*

ridge *n.* °crest, °line, °strip, top edge, arête: *The mountain ridge was partly obscured by clouds.*

ridicule *n.* **1** °derision, deriding, °jeer, jeering, °taunt, taunting, °mockery, mocking, gibe *or* jibe, gibing *or* jibing, °raillery, *Colloq* ribbing, *US and Canadian* razzing, joshing: *Pettigrew was often subjected to ridicule because of his outlandish clothes.* —*v.* **2** °deride, °jeer at, °taunt, °tease, °mock, gibe *or* jibe, °guy, °chaff, °laugh at, °caricature, °poke fun at, make fun *or* sport of, °lampoon, °burlesque, travesty, °parody, make a laughingstock (of), *Colloq* rib, roast, *Brit* °send up, take the mickey out of, *US and Canadian* razz, josh: *In a democracy, it is an inalienable right of the people to ridicule their leaders and government.*

ridiculous *adj.* °absurd, laughable, °preposterous, °weird, comical, °funny, °humorous, °ludicrous, °farcical, droll, amusing, mirthful, °hilarious, sidesplitting, risible; °silly, °inane, °nonsensical, °foolish, °stupid, °outlandish, °bizarre, °grotesque, °queer, °crazy, °insane, °zany, °wild, *Slang* far-out: *You look absolutely ridiculous in that outfit! Where did she ever get the ridiculous notion that she would marry a prince?*

riffraff *n.* °rabble, °hoi polloi, ragtag and bobtail, scum, *canaille,* masses, °dregs (of society), *Colloq* great unwashed: *I am shocked that you continue your membership in a club to which riffraff like Trevor Ascham belong.*

rifle *v.* **1** °rob, °loot, °ransack, °plunder, despoil, burgle, °pillage, *US* burglarize: *We caught the man who had rifled the till.* **2** °search, °ransack, °go through, rummage through: *She rifled through the files till the will was found.*

rift *n.* **1** °separation, °break, °split, °schism, cleft, °gulf, °gap; °disagreement, °conflict, disruption, °difference, °breach, breakup, °division, °distance, alienation: *The continued bitterness served only to intensify the rift between them.* **2,** tear, rent, opening, hole, crack, chink, crevice, cleavage, fracture, flaw, fault: *Water was pouring through an ever-widening rift in the dam.*

rig *v.* **1** Often, ***rig out or up.*** fit (out *or* up), °equip, °set up, °outfit, °supply, °provision, accoutre *or* *US also* accouter, caparison, °set up, *Chiefly Brit* kit out: *Good Sports was the shop that rigged us out completely for our Everest climbing expedition.* **2** °falsify, °manipulate, °doctor, °juggle, °fix, tamper with, °fake, *Colloq* °fiddle (with), cook: *The accountants rigged the books to make the company appear profitable enough to merit a bank loan.* —*n.* **3** °equipment, equipage, °gear, °tackle, °apparatus, °outfit, °kit, accoutrements *or* *US also* accouterments, °paraphernalia, appurtenances, *Colloq* °things, °stuff: *It cost a fortune to transport his diving rig by plane.*

right *adj.* **1** °just, °moral, °good, °proper, °correct, °legal, °lawful, licit, °honest, °upright, °righteous, °virtuous, °ethical, °fair, °true, °honorable, right-minded, °principled, °open, °aboveboard, °legitimate, *Colloq* legit: *There is no doubt in my mind that Simon can be relied on to do the right thing.* **2** °correct, °fitting, °suitable, °proper, °accurate, °exact, °precise, °perfect; °factual, °truthful, veracious, valid, °sound, *Colloq Brit* bang on, spot on: *Have you bought the right kind of nail to go into plaster? Jennie had nine right answers out of ten. How many of yours are right?* **3** °propitious, °convenient, °strategic, °advantageous, °beneficial, °favorable, auspicious, preferred, preferable, °promising: *It is critical for the mixture to reach the boiling point at exactly the right time.* **4** °sound, °sane, °normal, °rational, lucid, °healthy: *Is she right in the head, letting him drive in his condition?* **5** right-hand, dextral, dexter, *Nautical* starboard: *In the United Kingdom, drivers sit on the right side.* **6** rightist, right-wing, °conservative, °reactionary, Tory: *The right wing of the liberal party is not far from the left wing of the conservatives.* **7** °face, °upper, °principal, °front: *Have you laid the tablecloth right side up?* **8** utter, °complete, °perfect, °unmitigated, unalloyed, °out-and-out, °thorough, thoroughgoing, 24-karat, dyed-in-the-wool, °pure, °absolute, °real, *Brit* °proper: *I've been a right idiot not to realize that he loved only her.* —*n.* **9** °privilege, °prerogative, °license, °power, °claim, °title, °freedom, °liberty: *I allow nothing to interfere with my right to vote.* **10** °justice, °reason, °fact, °truth, fairness, °equity, °good, goodness, °integrity, °virtue, virtuousness, °honesty, honorableness, °morality, °propriety, °rectitude, right-mindedness, high-mindedness, °nobility, uprightness: *He believes that he has right on his side.* **11** right side *or* hand *or* fist, *Nautical* starboard: *She stood on his right. She gave him a quick right to the jaw.* **12** ***by rights.*** °properly, °fairly, justly, to be just, to be fair, in fairness, °honestly, in all honesty, to be honest: *By rights, because I paid for it, the house is mine.* —*adv.* **13** °directly, °straight, °straightaway, right away *or* off, in a beeline, as the crow flies, forthwith; unhesitatingly, °immediately, °promptly, °at once, instantly, without hesitating *or* hesitation, without delay, °quickly, °swiftly, speedily, *Colloq* pronto, °straight off: *The minute he walked into the room, he went right to the wall safe. Please come over right now.* **14** °exactly, °precisely, unerringly, accurately; °just: *She hung the paintings right where they belonged.* **15** °just, °only: *I stopped right now—when you walked in.* **16** °well, satisfactorily, advantageously, profitably, °favorably, opportunely: *I hope that everything turns out right for you.* **17** correctly, accurately, °properly, °precisely, °well, sensibly, fittingly, suitably, aptly: *If you can't do it right, why bother doing it at all!* —*v.* **18** straighten (up *or* out), set upright *or* aright: *I righted all the chairs the children had knocked over.* **19** put *or* set *or* make right, put *or* set to rights, °correct, °straighten out, redress, °amend, °make up for, °rectify, °sort out, °repair, °fix: *She has always felt that her mission in life was to right wrongs.* **20** avenge, retaliate for, °vindicate, °repay, °revenge, °settle, *Colloq* get even for: *Langdon swore that he would right the injury done to his father's name.*

righteous *adj.* **1** °moral, °just, °virtuous, upstanding, °upright, °good, °honest, °ethical, °honorable, °fair, °reputable, °trustworthy: *He insists that his father was a righteous man who had been unjustly accused.* **2** °right, °correct, justifiable, justified, °appropriate, condign, °fitting, apt, °self-righteous: *Her righteous indignation at being criticized stems from her overweening pride.*

rightful *adj.* **1** °legal, °lawful, °legitimate, licit, de jure, °correct, °proper, °bona fide, valid, °true, authorized, °right: *If the deed is in his name, then he is the rightful owner.* **2** °just, °fair, °equitable, °right: *Emily has supported rightful causes all her life.*

rigid *adj.* **1** °stiff, unbending, unbendable, °firm, °hard, °strong: *Are those brackets rigid enough to support such a heavy shelf?* **2** °inflexible, inelastic, unyielding, undeviating, unalterable, °set, °firm, °hard, hard-line, unbending, adamant(ine), °steely, iron, °strong, uncompromising, rigorous, unrelenting, intransigent, stringent, °severe, °strict, rigorous, °stern, °harsh, austere: *Frank is known to favor rigid discipline in the classroom.* **3** rigorous, °exact, °precise, unwavering, °unqualified, unswerving, undeviating, °demanding, °strict, hard and fast, °literal, °nice, °close, °thorough, °scrupulous, °careful, °conscientious, °painstaking, °meticulous, punctilious, °exacting, °strait-laced: *Rigid observance of the regulations produced the desired results. Pinwell's rigid parsimony met little favor from his family.* **4** °obstinate, °stubborn, pigheaded, °inflexible, °immovable, adamant, adamantine, °fixed, °set, obdurate, °willful, headstrong, dogged, persevering, °determined, °resolute, °steadfast, resolved, °tenacious, °relentless, unrelenting, uncompromising, unadaptable, mulish, close-minded: *Candida remained rigid in her opposition to birth control.*

rigmarole *n.* rigamarole, motions, complication(s), formalities, red tape, °bureaucracy, punctilio, ceremoniousness, °ceremony, °ritual, °procedure, °mess, °mumbo jumbo, gobbledygook *or* gobbledegook, bunkum *or* buncombe, °bother, balderdash, °rubbish, °nonsense, foolishness, *Colloq* hassle, bunk, *Slang* crap, *US* meshugaas *or US* mishegaas: *Why do you have to go through such a rigmarole just to get a driving license?*

rigor *n.* **1** Usually, ***rigors.*** °severity, harshness, °hardship, inhospitableness, bleakness, inclemency *or* inclementness, °bitterness, °cold: *I am not sure that Grandmama can endure the rigors of another Maine winter.* **2** strictness, rigidity, °precision, preciseness, punctilio, literalness, exactness, meticulousness, stringency, inflexibility, rigorism, harshness, °severity, °hardship, asperity, austerity, sternness: *If you work with radioactive materials, safety regulations must be observed with rigor. The rigor of monastic life has toughened him over the years.*

rim *n.* °edge, °brim, lip, °border, °periphery, °perimeter: *To make a margarita, you first coat the rim of a cocktail glass with coarse salt.*

rind *n.* °peel, °skin, husk: *Cut the thin outer part of the rind of a lemon into thin strips and soak them in brandy.*

ring[1] *n.* **1** °loop, hoop, °band, °circle, ringlet, circlet, annulus, grommet, eyelet, quoit, noose, cincture, °belt, girdle, cestus *or* cestos, °border, °halo, aureole, nimbus, corona, cuff, collar, necklace, neckband, bandeau, fillet, bracelet, armlet, torque *or* torc, °crown, coronet, tiara, diadem, (laurel) wreath, °garland, *Mechanics* gland, °bearing, *US and Canadian* bushing *or Brit also* bush, washer, O-ring, *Nautical* thimble, *Architecture and heraldry* annulet, *Literary* roundlet, *Anatomy* cingulum: *The rings brought up from the wreck of the ancient Greek trading vessel are worth a great deal.* **2** rink, °enclosure, arena, circus, bullring, *Spanish plaza de toros,* boxing ring: *I would no sooner get into the ring with him than with a bull!* **3** °circle, °organization, °gang, °band, °pack, °cell, °team, °crew, confederacy, confederation, °combination, cartel, °mob, bloc, coterie, °set, °clan, °clique, °fraternity, °brotherhood, sorority, sisterhood, guild, °(secret) society, junta *or* junto, °cabal, °faction, °group, °league, °alliance, °federation, coalition, °union, affiliation, camorra, camarilla, Bund: *She was running an international ring of jewel thieves who worked only in the most exclusive resorts.* —*v.* **4** °encircle, °surround, °bind, gird, girt, °circle; °loop, compass, °envelop, encompass, °embrace: *The swimming pool is ringed with dense shrubbery. After the vase is fired, I shall ring the base with a gold band.*

ring[2] *v.* **1** °peal, °chime, °toll, knell, tintinnabulate, ding-dong, gong, °sound, resonate, °resound, °echo, re-echo, reverberate; clang, °jangle; tinkle, clink, °jingle, ding-a-ling, ring-a-ding-ding, *Archaic and dialectal* °knoll: *The bells rang out their joyous announcement of peace on earth.* **2** °telephone, ring up, phone, °call, give (someone) a ring, *Colloq* (give (someone) a) buzz, give (someone) a tinkle, get (someone) on the blower *or* horn, *US* (give (someone) a) jingle: *I'll ring you when I am next in Krivoy Rog.* —*n.* **3** ringing, °peal, pealing, °chime, chiming, °toll, tolling, tintinnabulation, knell, sounding, resonating, °echo, echoing, reecho, reechoing, reverberation, reverberating; clang, clanging, °jangle, jangling; tinkle, tinkling, clink, clinking, °jingle, jingling, ding-a-ling, ring-a-ding-ding, *Archaic and dialectal* °knoll: *I couldn't hear the ring at the door because the water was running.* **4** (telephone *or* phone) call, *Colloq* °buzz, tinkle, *US* °jingle: *Don't forget to give your mother a ring on Sunday.*

rinse *v.* **1** wash (out *or* off), wash up, °clean, °cleanse, bathe, °drench, °flood, °flush, irrigate, *Chiefly Brit* °swill (out), *Literary* lave: *Use plenty of fresh water to rinse all the soap out of the clothes.* **2** °tint, dye, °touch up, highlight: *Martine is at the age when she rinses a little blue into her graying hair.* —*n.* **3** rinsing, °wash, washing, bath, bathing, cleaning, cleansing, ablution, drenching, °flood, flushing, irrigation, *Medical* lavage, *Literary* laving: *The softening agent should be added in the final rinse.* **4** °tint, dye: *I had no idea that she used a rinse—I thought her hair was naturally reddish.*

riot *n.* **1** rioting, riotous behavior, °disturbance, °uproar, °tumult, turmoil, °(civil) disorder, lawlessness, hubbub, °rumpus, turbulence, °fracas, °fray, affray, melee *or* mêlée, °pandemonium, donnybrook, °brawl, °row, °unrest, commotion, °bother, imbroglio, °outburst, anarchy, disruption, °violence, °strife, *Colloq* ruckus, ruction, to-do, do, *Brit* bovver, punch-up: *The government is unable to quell the food riots, which have become more frequent and violent.* **2** funny person *or* woman *or* man, comedian *or* comedienne, hilarious event *or* bit *or* shtick *or* thing *or* piece of business, *Colloq* gas, *US* °panic, laugh-riot: *Benny's routine with the model is an absolute riot.* —*v.* **3** mount the barricades, take to the streets, °rebel, revolt, create *or* cause a disturbance, °brawl, °fight, (go on the *or US also* a) rampage, run riot, °storm: *The prisoners rioted to protest against the crowded conditions.*

riotous *adj.* **1** °tumultuous, unrestrained, °wild, °noisy, °uncontrolled, uncontrollable, unmanageable, °chaotic, °disorderly, disordered, °lawless, turbulent, °violent, brawling, °obstreperous: *The headmaster warned that he would not tolerate such riotous behavior.* **2** °rowdy, °boisterous, °unruly, °uproarious, rollicking, roisterous, °wild, rumbustious, unrestrained, °uninhibited, *Archaic* robustious, *Colloq* rambunctious, no-holds-barred: *Bernard's retirement party was certainly a riotous affair.*

rip *v.* **1** tear (apart *or* asunder), °rend, be torn *or* rent, °split, °cut (apart): *I ripped my trousers on that nail. The paper ripped because you yanked it out of my hand.* **2** ***rip off.*** **(a)** °steal, purloin, °rob, °snatch, °pilfer, filch, °take, shoplift, *Colloq* °pinch, *Brit* °nick, *Slang* °lift, °swipe, *US* °boost, °promote: *They broke into my car and ripped off my tape deck.* **(b)** °cheat, °swindle, °trick, °fleece, bilk, °dupe, °deceive; °defraud, °embezzle, *Colloq* con, bamboozle, *Slang* °skin, gyp, rook: *A smooth couple ripped me off by using the old "stuffed wallet" trick.* —*n.* **3** °tear, °rent, °split, °gash, °slash; °rift, cleft, °rupture: *The vandals tore a foot-long rip in the top of my new convertible.*

ripe *adj.* **1** °mature, matured, °seasoned, (fully) grown, (well-)ripened, developed, °mellow, °ready, °fit, *US* °(fully) aged, well-aged: *The grapes are ripe for picking.* **2** °prepared, °ready, °fit, °appropriate, °experienced, °veteran, °seasoned, °sage, °wise, °sophisticated, °mature, °informed, °qualified, dependable, °reliable: *Cootes is ripe in judgment and understanding.* **3** ***ripe for.*** **(a)** °timely, °opportune, °propitious, °favorable, auspicious, °ideal, °right, °suitable, apt, °proper, suitably advanced: *Is the time yet ripe for another revival of* Grand Hotel? **(b)** °ready, °eager, °enthusiastic, °prepared, °fit, °ready, °in readiness: *The mob is ripe for revolt.*

ripen *v.* °develop, °mature, °age, °season, maturate, bring *or* come to maturity; °perfect: *This peach needs another day to ripen.*

rip-off *n.* **1** °stealing, purloining, robbing, pilfering, °taking, filching, pilferage, °theft, °robbery, larceny; shoplifting, *Colloq* pinching, *Brit* nicking, *Slang* lifting, swiping, *US* boosting: *The rip-off of books from the library is a disgrace.* **2** °swindle, confidence trick, swindling, cheating, °cheat, °fraud, °deception, defrauding, defalcation; °embezzlement, *Colloq* con (job *or Colloq* trick): *A husband-and-wife team is being sought for executing a rip-off of thousands from the Outland Bank using a forged letter of credit.* **3** overcharging, exploitation, *Colloq* highway *or chiefly Brit* daylight robbery: *It's a rip-off to charge two dollars for something that costs only three cents to make.*

ripping *adj. Somewhat archaic* °fine, °splendid, °marvelous, °excellent, °exciting, °thrilling, °stirring, spine-tingling: *Buchan and Henty and Haggard wrote ripping adventure novels.*

ripple *n.* **1** wavelet, °wave, °ruffle, ruffling, cat's-paw, purl, purling, undulation, *US* riffle, riffling: *There wasn't even the slightest breath of air to cause a ripple on the mirrorlike surface of the lake.* **2** °(slight) disturbance, °upset, perturbation, °agitation, °flurry, °flutter, °suggestion, °hint, soupçon: *The divorce occasioned only a ripple of dissent in the royal family.* —*v.* **3** °ruffle, purl, undulate, °wave; °splash, °wash, *US* riffle: *I sat musing, watching the water ripple over the stones in the brook.*

rise *v.* **1** °get up, °arise, °stand (up), get to one's feet, *Brit* be upstanding: *All rise when the judge enters the court.* **2** °get up, °arise, °awaken, waken, °wake up, start *or* begin the day, *Nautical or colloq* hit the deck, *Colloq* °turn out: *We usually rise at about six o'clock.* **3** ascend, be elevated, °arise, °climb, °lift, °go up, °mount: *The smoke slowly, almost furtively rose from the campfire into the still night air. The morning star can be seen tomorrow rising in the sky just before dawn. Some of those skyscrapers rise to a height of 110 stories.* **4** Often, ***rise (up) (against).*** °rebel, °revolt, °mutiny, kick over the traces, take up arms, mount the barricades, take to the streets: *If the food shortages persist, the people are sure to rise up and overthrow the present regime.* **5** °swell, °flood, °increase, °grow; wax: *Owing to the heavy rains, the river is rising rapidly.* **6** slant *or* incline *or* slope (upward(s)), ascend, °climb, go uphill: *The path rises steeply just beyond the cwm.* **7** °fly, °take flight, take wing, take to the air, °take off, °arise, °lift, °climb, °soar, °mount: *At the roar of the gun, tens of thousands of flamingoes rose from the marshes.* **8** °advance, improve one's lot *or* position, progress, °get ahead, go *or* get somewhere, °succeed, make something of oneself, be promoted, °prosper, °thrive, °make good, *Colloq* °make it, °make the grade, °go places: *Fields could well have risen in the company had his attitude been different.* **9** °start, °begin, °originate, °arise, °occur, °happen, °take place: *Somehow, a little feud rose between us that I cannot explain.* **10** °increase, be elevated *or* lifted *or* boosted, °grow, °go up, move upward(s), °climb, escalate, ascend, snowball: *When interest rates rise, they contribute to the inflation that raising them was supposed to reduce.* **11** °arise, be nurtured, be produced, be generated, be created, °spring up, be engendered: *The feelings that rose within him were spurred on by sheer hatred of the man.* **12** ***rise to.*** **(a)** °arise, °come up, °meet, be equal to, prove adequate to: *You can count on Farley to rise to the occasion should you need a good man.* **(b)** come to get, °take, °swallow, react to, °respond to, succumb to, be tempted by: *Sue rose to the bait of a company car and accepted our offer.* —*n.* **13** ascent, °hill, hillock, °knoll, eminence, °prominence, °elevation, upland, highland, (upward) slope *or* incline, acclivity, *US* upgrade: *As we climbed a slight rise, we saw the golden towers gleaming in the distance.* **14** °increase, °increment, °gain, °addition: *In return for only a slight rise in wages, they are being asked to shoulder much more responsibility.* **15** ascent, ascension, °elevation, °flight, °climb, °takeoff: *The balloon's rise was suddenly halted when the basket caught on a tree.* **16** ***get*** *or Brit also* ***take a rise out of (someone).*** °provoke, °stimulate, °incite, instigate, °foment, goad, °encourage, °press, °push, °shake up, waken, °awaken, °move, °motivate, °activate, °agitate, °stir (up), °inflame, impassion: *We finally got a rise out of Sidney by threatening to charge him for cleaning his room.* **17** ***give rise to.*** °start, engender, °generate, °begin, °commence, °produce, °bring out, °cause, °bring about, bring into being: *We never discovered what gave rise to Theedle's dislike of dogs.*

risk *n.* **1** °danger, °peril, °jeopardy, °hazard, °chance, °gamble: *I would take up sky diving, but my mother won't let me because of the risk.* —*v.* **2** °endanger, imperil, °jeopardize, °hazard, °chance, °gamble: *Would you risk all your money on one roll of the dice?*

risky *adj.* °dangerous, °perilous, °hazardous, chancy, touch-and-go, °precarious, *Colloq* iffy, °dicey, °dodgy, °touchy: *Futures trading is a very risky business, as many have discovered to their dismay.*

risqué *adj.* indelicate, °unrefined, indecorous, °indecent, °improper, °broad, °naughty, °spicy, salty, °off-color, °racy, °bawdy, °erotic, °suggestive, °wicked, °blue, ribald, °daring, salacious, °gross, *Colloq Brit* near the knuckle: *Some of those photographs of beauty contest queens are pretty risqué.*

rite *n.* °ceremony, °ritual, °ceremonial, °observance, °formality, °custom, °practice, °routine, °procedure, °solemnity, solemnization, liturgy: *This weekend I have to perform the annual rite of having tea with Great Aunt Elizabeth.*

ritual *adj.* **1** °ceremonial, °ceremonious, sacramental: *The celebrants practice ritual washing, called ablutions.* **2** procedural, °formal, °conventional, °customary, °habitual, °routine, prescribed, °usual, °automatic, °perfunctory: *Each time they meet, Harold and Biffy have to go through a ritual handshake—learned in Patagonia, I think.* —*n.* **3** °formality, °routine, °custom,

°practice, °convention, °protocol: *If everyone goes through this tea-drinking ritual twice a day, how do you get any work done?* **4** See **rite,** above.

rival *n.* **1** °competitor, °opponent, contender, challenger, °antagonist, °adversary; °opposition: *This year there are many formidable rivals for the championship.* —*v.* **2** compete with *or* against, contend with *or* against, °oppose, °challenge, °contest, struggle with *or* against, vie with, °combat, °compare with, °equal, °measure up to, °(be a) match (for): *Jan's prowess at weightlifting rivaled that of anyone else in Craigie.*

rivalry *n.* °competition, competitiveness, contention, vying; °dispute, °feud, feuding, °conflict, °struggle, °strife, °controversy, °dissension, °discord, difference(s): *Once again these two teams meet in a rivalry that has been a tradition for more than a hundred years. It appeared unlikely that the rivalry between the brothers would be settled amicably.*

river *n.* **1** watercourse, °branch, °tributary, °stream, waterway, estuary, rivulet, °creek, °brook, streamlet, runnel, rill, *Scots and Northern England* °burn, *Brit* beck, *US* °kill: *We used to sail our toy boats on the river here when I was lad.* **2** °stream, °flood, °torrent, °quantity, cataract, °flow, cascade: *Rivers of sweat poured down me as I toiled in the humid heat of the jungle.*

riveting *adj.* °spellbinding, engrossing, hypnotic, hypnotizing, transfixing, fascinating, °enthralling, gripping, captivating, °absorbing: *Unpredictably, alas, television producers are capable of putting on absolutely riveting programs.*

road *n.* **1** °way, °means, °approach, °route, °procedure, °technique, °method, °passage, °street, avenue, °course, °track, entrée, access, °direction: *Our researchers arrived at the identical results via a completely different road.* **2** thoroughfare, °way, byway, highway, roadway, highroad, lowroad, avenue, boulevard, °street, lane, alley(way), *Brit* motorway, carriageway, *US* turnpike, expressway, freeway, parkway, throughway *or* thruway, *German Autobahn, Italian autostrada, French autoroute, Colloq US* pike: *The roads are jammed with traffic on Sunday evenings during the summer.*

roam *v.* °wander, rove, °ramble, °range, °walk, °drift, dally, °dawdle, °cruise, °stroll, amble, °meander, °saunter, °stray, °prowl, perambulate, °travel, voyage, peregrinate, circumambulate, traipse, gallivant, jaunt, *Colloq* mosey, swan: *They spent a year roaming about the continent, staying in a place as long as they pleased.*

roar *v.* **1** °bellow, °thunder; °howl, °bawl, squall, °cry, °yell, yowl; bay, °snarl, growl: *A train roared by. The crowd roared appreciation. A lion roared once in the night.* **2** °laugh, guffaw, °howl (with laughter), hoot: *They fell off their chairs roaring when he told the joke about the egg and the taxi.* —*n.* **3** roaring, °bellow, °thunder, rumble, °boom; °howl, °bawl, squall, °cry, °yell, yowl, clamor, °outcry; °snarl, snarling, growl, growling: *The roar of the falls could be heard miles away. There was a roar of indignation at the proposal. The beast's roar was enough to strike terror into our hearts.* **4** guffaw, °outburst, °howl, hoot: *Everything she said was met by a roar of laughter from the gallery.*

rob *v.* **1** burgle, °loot, °rifle, °ransack, °plunder, depredate, °raid; hijack; °pillage, °sack; *US* burglarize, *Colloq* °hold up, *Slang* °stick up, knock off *or* over, °rip off: *They robbed a jeweler's shop last week.* **2** prey upon *or* on, *Colloq* °hold up, °mug, *Slang chiefly US* °stick up, °rip off, *Chiefly US and New Zealand* °roll: *They have taken to robbing old ladies in the street.* **3 rob (*someone*) (*of*).** °deprive (of), cheat *or* swindle (out of), °defraud (of), °strip (of), °fleece (of), bilk (of), °victimize, mulct (of), *US* euchre (out of), *Colloq* rook (out of), do *or* diddle (out of), gyp (out of), *Slang Brit* nobble (of): *I was robbed of my last penny by those confidence men. We were robbed by a taxi driver who offered to change our money. She robbed me of my self-respect.*

robber *n.* °thief, pickpocket, shoplifter, °burglar, bandit, housebreaker, sneak thief, cat burglar, safebreaker, highwayman, gentleman of the road, brigand, °pirate, freebooter, buccaneer, privateer, corsair, *Colloq* mugger, holdup man, *Slang* cracksman, rip-off artist, safecracker, safe-blower, *Chiefly US* stickup man, *US* second-story man: *The police were making every effort to catch the robbers.*

robbery *n.* robbing, °theft, thievery, thieving, burglary, burgling, pilfering, pilferage, °stealing, plundering, °plunder, looting, °sack, sacking, ransacking, °depredation, °pillage, pillaging, hijacking, hijack, larceny, breaking and entering, *Colloq* pinching, °holdup, holding-up, mugging, *Slang* °rip-off, ripping-off, *Chiefly US* stickup, sticking-up, *Brit* nicking, *US* heist: *The robbery of the paintings took place in broad daylight. The robbery of the shop was aided by a former employee. The robbery of elderly persons is on the increase.*

robe *n.* **1** °cloak, °dress, garment, vestment, °habit, frock, cassock, caftan, muumuu, surplice, bathrobe, dressing gown, lounging robe, housecoat, kimono, housedress, peignoir, °wrapper, *French robe de chambre,* °costume: *A woman wearing a green robe came to the door and asked what I wanted.* **2 *robes.*** °costume, °habit, °uniform, garb, attire, vestments, °apparel, raiment, livery, clothing, °garments, °outfit, accoutrements *or US also* accouterments, °regalia, °finery, °trappings, panoply, °gear, °paraphernalia, appurtenances, equipage, °rig, *Archaic* vesture: *She looked absolutely splendid decked out in the full robes of state.* —*v.* **3** °cloak, garb, °dress, °cover, enrobe, °clothe: *The dish consists of ladyfingers topped with fresh strawberries robed in chocolate, with whipped cream spread over the top.*

robot *n.* **1** automaton, mechanical man *or* monster, android: *In the future, much of the drudgery will be taken over by electronic robots.* **2** drudge, °clod, °tool, °puppet, cat's-paw, myrmidon, mechanical man, automaton: *Charlton has a number of robots on his payroll who do his bidding without question.*

robust *adj.* **1** °healthy, °fit, °sound, °hale (and hearty), °sturdy, °hardy, °hearty, °strong, °stout, °tough, able-bodied, strapping, °brawny, °sinewy, °rugged, °muscular, °powerful, well-knit, athletic, °staunch, °vigorous; in fine *or* good fettle, *Colloq* °husky: *Clampett was such a robust chap one might have thought he'd live forever.* **2** °pungent, °strong, flavorful, sapid, °rich, full-bodied, nutty, fruity: *He gave me a glass of an excellent robust tawny port.*

rock[1] *n.* **1** stone; boulder: *In the avalanche, the rocks thoroughly destroyed the ski lodge.* **2** °crag, tor, escarpment, scarp, *Brit* outcrop, *US* outcropping: *We used to go rock climbing when I had the strength.* **3 *on the rocks.*** **(a)** on ice: *I'd like a whiskey on the rocks, if you please.* **(b)** in (a) shambles, destroyed, in ruins, ruined, finished, °broken down, beyond repair: *He lost his job and his marriage went on the rocks.* **(c)** °destitute, poverty-stricken, °indigent, penniless, bankrupt: *Unable to find another job, he's been on the rocks for a year, now.*

rock[2] *v.* **1** °sway, °swing, °lull: *Her father used to rock her to sleep in his arms.* **2** °roll, °reel, °lurch, °toss, °swing, °sway, wobble: *The building rocked so during the earthquake that I thought it was going to fall down.* **3** °astound, °astonish, °shock, °surprise, °jar, °stagger, °amaze, °stun, dumbfound *or* dumfound, °daze, stupefy, °overwhelm, disconcert, °unnerve, *Colloq* set (someone) back on his *or* her *or* their heels, °throw, °rattle,

°shake up: *I was really rocked by the news that Sue was marrying Count Eigenblick.*

rocket *v.* zoom, °take off, skyrocket, shoot up, °climb, rise rapidly, °soar, spiral upward(s), *Colloq* go through the roof: *The day controls were lifted, prices rocketed, almost doubling overnight.*

rocky[1] *adj.* **1** °stony, pebbly, shingly, boulder-strewn, craggy; °bumpy, °difficult, °hard, uncomfortable, °arduous: *The beach at Nice might be romantic, but it's very rocky to lie on. The road to love can be very rocky indeed.* **2** °stony, adamant(ine), °firm, unyielding, rocklike, °tough, unbending, flinty, °firm, °solid, °steadfast, °steady, unfaltering, °staunch, unflinching, °determined, °resolute, unwavering, unchanging, unvarying, invariant, °invariable, °reliable, dependable, °sure, °certain: *I knew we could depend on Charles's rocky determination to see the project through.* **3** °stony, flinty, unfeeling, °unsympathetic, unemotional, emotionless, °impassive, °cold, °cool, apathetic, °indifferent, uncaring, °detached, *dégagé,* °callous, °thick-skinned, °tough, °hard: *Beneath that rocky exterior beats a tender heart.*

rocky[2] *adj.* °unstable, tottering, teetering, unsteady, °shaky, °rickety, unsure, °uncertain, °unreliable, °weak, °flimsy, wobbly, wobbling, vacillating, dubious, °doubtful, °questionable, *Colloq* iffy: *The project will be very rocky till we are assured proper funding.*

rod *n.* **1** °bar, °pole, baton, °wand, °staff, °stick, dowel, cane, °shaft: *Steel rods are used to reinforce the concrete in modern buildings.* **2** cane, birch, °switch, °scourge, °lash; °punishment, chastisement, castigation, °discipline, chastening, °correction: *As they say, spare the rod and spoil the child.*

rogue *n.* **1** trickster, °swindler, °cheat, cad, ne'er-do-well, °wastrel, °good-for-nothing, °miscreant, *Rather old-fashioned* scamp, °scoundrel, blackguard, knave, °rascal, rapscallion, scapegrace, dastard, cur, churl, °wretch, °villain, charlatan, mountebank, *Brit* bounder; *Colloq* louse, °stinker, rat, °creep, *Brit* rotter, blighter, *Chiefly US and Canadian* son of a gun, *Slang* bastard, *Chiefly US and Canadian* son of a bitch, s.o.b. *or* SOB, °bum: *She met some clever rogue in Capri who separated her from a few million lire before he was through.* —*adj.* **2** °independent, °undisciplined, uncontrollable, °ungovernable, unmanageable, unpredictable, °disobedient, °incorrigible, fractious, °self-willed, °unruly, intractable, unrestrained, °wild, °lawless, strong-willed, headstrong, refractory, contumacious, °recalcitrant, cross-grained, rampageous: *Part of the village was destroyed when a rogue elephant ran amok. Perkins quit the company to start up his own rogue concern.*

role *n.* **1** rôle, °part, °character, impersonation; °lines: *She had the role of Cuspidora in the Off-Broadway success,* Spitfire. **2** °function, °position, °situation, °place, °post, °capacity, °job, °duty, °task, °responsibility: *Hermione's role in the firm is now as managing director.*

roll *v.* **1** °rotate, °cycle, turn (over (and over)), °wheel, trundle; °revolve, go (a)round, °orbit, °tumble, somersault *or* somerset *or* summersault *or* summerset: *Roll out the barrel, / We'll have a barrel of fun! The children loved rolling about in the ride at the carnival.* **2** °pass, °go, °flow, °slip, °flit, °glide, °slide, °move (on); °expire, °elapse, °disappear, vanish, °evaporate: *The time certainly rolls by quickly when you're having fun.* **3** °move, °drive, °bowl, be carried *or* conveyed, °cruise, °sail, °coast, °ride, °float, °fly: *Here we are, just rolling along, singing a song.* **4** undulate, billow, rise and fall: *The countryside rolled away before us as far as the eye could see.* **5** °roar, °echo, reecho, rumble, reverberate, °resound, °sound, °boom, °peal, resonate, °thunder: *The roar of the explosion rolled over us as we ran for cover.* **6** °rob, steal from: *They got their money by rolling drunks.* **7** Often, ***roll out.*** °flatten, level (off *or* out), °smooth (out), °even (out), °grade: *After cutting, the lawn ought to be rolled.* **8** Usually, ***roll over.*** °turn (over), °rotate, °spin: *Roll onto your stomach so I can examine your back.* **9** Usually, ***roll up.*** furl, °coil, curl, °wind (up), °wrap (up); enwrap, °swathe, enfold, °envelop, °shroud, enshroud: *Every night he takes down the banner and carefully rolls it. Cleopatra was rolled up in a rug to be smuggled past the guards.* **10** ***roll in.*** **(a)** °arrive, °come in, pour in, flow in, °turn up, °show up: *The contributions really started to roll in after the appeal.* **(b)** °luxuriate in, °revel in, °wallow in, °savor, bask in, °delight in, take pleasure in, °indulge in, rejoice in, °relish: *Bascombe has been rolling in new-found wealth since his aunt died.* **11** ***roll out.*** unroll, unfurl, °spread (out), °unfold, uncoil, uncurl, unwind, °open (out): *Roll out the bolt of cloth so that we can measure it.* —*n.* **12** °reel, spool, cylinder, scroll; tube: *The message was written on a long roll of paper.* **13** °list, rota, °register, °record, directory, listing, roster, slate, docket, catalogue, inventory, °muster, °index, census, annal(s), °schedule, chronicle(s), *Sports* lineup: *With great ceremony, the clerk read the roll.* **14** rolling, billowing, waving, wave action, °wave, billow, °swell, undulation, pitching, rocking, tossing: *With each roll of the boat, I felt more seasick.* **15** °peal, rumble, reverberation, °boom, °echo, °thunder, °roar: *A roll of thunder drowned out her shout for help.* **16** rotation, °spin, °toss, whirl, °twirl: *I decided to bet everything on one last roll of the dice.* **17** bun; scone, croissant; *Brit* breadroll: *Waiter, would you please bring some more rolls?* **18** bankroll, °money, °wad, °bundle: *He took out a roll that could choke a horse and peeled off ten C-notes.*

roller *n.* **1** drum, cylinder, barrel, calender; tube; windlass; rolling pin; °mangle, wringer: *Make certain that the roller is clean.* **2** billow, °wave, comber, breaker, °swell: *After the storm, the heavy rollers continued to pound the sea wall.*

romance *n.* **1** °(love) affair, amour, affair of the heart, *affaire de coeur, affaire d'amour,* °liaison, °relationship, dalliance, °intrigue: *Wasn't Rick the one from Canada with whom Dora had a romance?* **2** °(true) love: *You are lucky if you find romance even once in your lifetime.* **3** °novel, °narrative, fiction, °story, °mystery, thriller, horror story, ghost story, science fiction, °fantasy, Western, melodrama, gothic *or* Gothic *or* Gothick novel *or* tale, °(fairy) tale, love story, idyll, epic, °legend: *It is probably accurate to say that the romance is the most popular form of fiction.* **4** °sentiment, nostalgia, °mystery, °intrigue, °fantasy, °imagination, imaginativeness, °adventure, °excitement, °fascination, exoticism, °glamour, °color, colorfulness: *It was the romance, hardly the beauty of Morocco that entranced me.* **5** tall tale *or* story, °fantasy, °fabrication, fairy tale, °exaggeration, prevarication, concoction, flight of fancy, °fib, °(white) lie, prevarication, balderdash, fiction, °nonsense, °imagination: *That episode with the girl on the train that David told you about was all romance.* —*v.* **6** °make love to; woo, court: *Stiller has been romancing Meara ever since they met.* **7** °pander to, °flatter, curry favor with, toady (up) to, *Colloq* butter up, soft-soap, *Taboo slang* brown-nose: *The new restaurant is romancing neighborhood office executives, luring them in for business lunches.*

romantic *adj.* **1** °imaginary, imagined, °fictitious, °fictional, °ideal, idealized, °fancied, °fabulous, made-up, dreamed-up, dreamt-up, fantasized, °fanciful, fairy-tale, °mythical, °idyllic, utopian, °illusory: *Vivian isn't a real person—she's just Howard's romantic notion of the girl he wants to marry.* **2** °impractical, °visionary, °fictitious, unpractical, °unrealistic, °ideal, °abstract,

°quixotic, chimerical, °absurd, °extravagant, °wild, crackpot, °mad: *Running a bakery is a business, incompatible with your romantic notions of giving free pastries to the children.* **3** nostalgic, °sentimental, °emotional, °sweet, °tender, °picturesque, °exotic, °glamorous; mawkish, °maudlin, saccharine, *Colloq* soppy, sugary, °mushy, °sloppy: *I can think of a lot more romantic places for our anniversary dinner than a hamburger stand! Fred hates it when I get too romantic.* **4** amorous, °affectionate, aroused, °impassioned, °passionate, libidinous, °lustful, overfriendly, *Colloq* lovey-dovey; °fresh: *After a drink or two Horace starts getting romantic—if you know what I mean.* —*n.* **5** romanticist, °dreamer, Don Quixote, °visionary, idealist, sentimentalist: *I've never lost my taste for soft lights and sweet music—I guess I'm just a romantic at heart.*

room *n.* **1** °space, °area, °scope, °extent, °allowance, latitude, elbowroom, °range, °leeway, °margin: *There is room for one more inside.* **2** °chamber, apartment, °compartment, °office, °cell, cubicle: *For years Pablo occupied a small room on the top floor.* **3** ***rooms.*** °quarters, lodgings, °accommodation, apartment, °dwelling, *Chiefly Brit or old-fashioned US* °flat: *We had rooms at 221B Baker Street—but you would not have heard of that place.* —*v.* **4** °live, °lodge, °dwell, °abide, reside, °stay: *Did you really room with the great man himself?*

roomy *adj.* °spacious, capacious, commodious, °large, sizable, °big, °ample: *For what appears to be a small car, this has a surprisingly roomy interior.*

root[1] *n.* **1** °base, °basis, °foundation, °source, °seat, °cause, fountainhead, °origin, fount, wellspring: *Love of money is the root of all evil.* **2** rootstock, rootstalk, taproot, rootlet; tuber; *Technical* radix, radicle, radicel, rhizome, rhizomorph: *In transplanting seedlings, care should be taken not to damage the root.* **3** ***root and branch.*** radically, °completely, °utterly, °entirely, °wholly, °totally: *The Romans sought to destroy Carthage root and branch.* **4** ***roots.*** °origins, °heritage, °family, °lineage, °house, antecedents, forefathers, foremothers, descent, genealogy, family tree, forebears, ancestors, predecessors, °stock, °pedigree; birthplace, motherland, °fatherland, native land *or* country *or* soil, cradle: *Carlotta spent years tracing her roots to ancient Rome. The roots of civilization first appeared in Mesopotamia.* **5** ***take root.*** become set *or* established *or* settled, germinate, °sprout, °grow, °develop, °thrive, burgeon, °flourish, °spread: *Good work habits should take root at an early age; then they will last a lifetime.* —*v.* **6** °plant, °set, °establish, °found, °fix, °settle, embed *or* imbed; entrench, °anchor: *These annuals failed to grow because they were not properly rooted. Herbert's fear of heights is rooted in a childhood fall from a tree.* **7** ***root out.*** **(a)** Sometimes, **root up.** uproot, eradicate, °eliminate, °destroy, extirpate, °exterminate: *We must root out any subversives in the organization.* **(b)** °find, uncover, °discover, dig up *or* out, °unearth, °turn up, °bring to light: *The survey of accounts is aimed at rooting out customers who are slow in paying.*

root[2] *v.* rootle, forage, °dig, °pry, nose, °poke, ferret, °burrow, °rummage, delve, °search, °ransack: *Harvey has been rooting about in the garage looking for his tennis racket.*

root[3] *v.* Usually, **root for.** cheer (for), °applaud (for); °boost, °support, °encourage, urge on: *I'm rooting for our side to win.*

rooted *adj.* °firm, established, °set, °fixed, °fast, settled, deep-rooted, deep-seated, °entrenched, ingrained *or* engrained, (firmly) embedded *or* imbedded, implanted, instilled; °chronic, inbred, °inherent, °intrinsic, °essential, °fundamental, °basic, °radical: *Have you any rooted objections to the principle of living with someone of the opposite sex without being married?*

rope *n.* **1** °line, °cord, °cable, hawser; strand, °string: *This rope is too light for mooring the boat. She wore a rope of baroque pearls.* **2** ***the ropes.*** the routine, the procedure, one's way around, the ins and outs; the truth, the (real) situation; *Colloq* what's what, the score, *Brit* the gen: *It didn't take her long to learn the ropes.* —*v.* **3** °tie, °bind, °lash, °hitch, °fasten, °secure; °tether, °attach: *Rope these two crates together. The horses were roped to the post.* **4** ***rope in.*** °attract, °draw (in), °tempt, °entice, °lure, °persuade: *They are going to use a lottery to rope the customers in to supporting our cause.*

ropy *adj.* **1** ropey, viscous, °stringy, viscid, glutinous, mucilaginous, gluey, gummy, thready, fibrous, filamentous: *The plastic resin emerges from this vat in a ropy form, like mozzarella cheese.* **2** questionable, °inadequate, °inferior, °deficient, °indifferent, °mediocre, substandard, °unsatisfactory, °poor, °sketchy: *Greenwood's plan for extracting gold from seawater looks pretty ropy to me.*

rostrum *n.* °platform, °stage, dais, podium, °stand; pulpit; lectern, reading stand: *As the crowd hooted and booed, the speaker descended from the rostrum.*

rosy *adj.* **1** °pink, rose-colored, red, roseate, reddish, pinkish, cherry, cerise, ruddy, flushed, °glowing, blushing, ruby, rubicund, florid; rose-red: *"The object of my affection has changed my complexion from white to rosy red."* **2** °optimistic, °promising, °favorable, auspicious, °hopeful, encouraging, °sunny, °bright: *Caldwell has a rosy future ahead of him.*

rot *v.* **1** °decay, °decompose, °fester, °spoil, go bad *or* off, be tainted, be ruined, °mold, molder, °putrefy; corrode, rust, °disintegrate, °deteriorate, crumble *or* go *or* fall to pieces: *When the power went off for a week, all the food in my fridge rotted. The piers holding up one end of the bridge have all rotted away.* **2** °waste away, wither away, languish, °die, molder, °decline, °deteriorate, °degenerate, °decay, atrophy: *Despite the fact that the police knew he was innocent, he's been rotting away in jail all these years.* —*n.* **3** °decay, decomposition, °mold, putrefaction, putrescence, °blight, corrosion, corruption, disintegration, deterioration: *Once the rot has set in, the job of revitalization can be overwhelming.* **4** °(stuff and) nonsense, balderdash, °rubbish, bunkum, tommyrot, twaddle, °drivel, hogwash, eyewash, °trash, *Colloq* claptrap, bunk, boloney *or* baloney, bosh, malarkey, °moonshine, poppycock, tosh, *Slang* crap, bull, codswallop, *Brit* (a load of (old)) cobblers, *Taboo* balls, bullshit: *Vogel knows nothing whatsoever about linguistics and is just talking a lot of rot.*

rotary *n.* traffic circle, *Brit* °roundabout, mini-roundabout: *When you come to the rotary, turn right till you come to a traffic light.*

rotate *v.* **1** °turn, °revolve, go round, °spin, °gyrate, °pirouette, whirl, °twirl, °wheel, °pivot, °reel: *As the gear rotates, its teeth mesh with those of the other gear, causing it to turn in the opposite direction.* **2** °change, °exchange, °alternate, interchange, °switch, trade places; °take turns, *Colloq* swap *or Colloq* swop: *Rotate the positions of the tires for more even wear. We rotate the chores in our family.*

rote *n.* **1** °routine, °ritual: *He acquired a rote knowledge of the words, but doesn't know what they mean.* **2** ***by rote.*** **(a)** by heart, from memory: *He can recite the entire Prologue of* Canterbury Tales *by rote.* **(b)** unthinkingly, automatically, mechanically: *She memorized the French dialogue by rote but doesn't understand a word of it.*

rotten *adj.* **1** rotted, decayed, decomposed, decomposing, °putrid, putrescent, putrescing, °moldy, moldering, spoilt *or* spoiled, mildewed, °rancid, fetid *or* foetid, °stale, °rank, °foul, feculent, tainted, contaminated, festered, festering, °corrupt, °bad, °off, turned, overripe, soured, °sour: *One rotten apple in the barrel can spoil the rest. The stench of rotten meat filled the room.* **2** rotted, rusted, corroded, deteriorating, disintegrating, crumbling, crumbly, falling to pieces, friable: *He stepped on a spot where the wood was rotten and his foot went right through the stair.* **3** °immoral, °corrupt, °dishonest, °deceitful, °venal, °shameless, °degenerate, °villainous, iniquitous, °evil, °wicked, °vile, debased, °base, °perverted, depraved, °unscrupulous, unprincipled, amoral, warped, *Slang* °bent: *These drug dealers are rotten at the core.* **4** heinous, °evil, °vile, °base, °miserable, °despicable, °wretched, °awful, °terrible, °horrible, horrific, °nasty, °contemptible, °filthy, °mean, °low, *Colloq* °lousy, °stinking, °lowdown, dirty-rotten: *Pairing me off with Vera at the dance was a rotten trick.* **5** °ill, unwell, °sick, °nauseated, °awful; hung over, *Colloq* °lousy, ropy *or* ropey, °rough: *I feel really rotten—maybe it was that pizza.*

rotund *adj.* **1** round(ed), °circular, orbicular, globular, °spherical: *His rotund, cherubic little face simply invited a pinch of the cheek.* **2** °full, full-toned, °deep, °resonant, reverberant, reverberating, sonorous, °rich, °round, °mellow; pear-shaped: *They thrilled to hear the rotund notes issuing from Caruso's lips as he sang "Vesti la Giubba."* **3** °chubby, podgy *or US chiefly* pudgy, °(pleasingly) plump, portly, tubby, °heavy, fleshy, corpulent, °stout, °fat, °obese, overweight, *Colloq* roly-poly: *Jack's rotund figure was easily recognizable waddling up the street.*

roué *n.* °playboy, womanizer, ladies' man, °rake, lecher, Lothario, Don Juan, Romeo, Casanova, °charmer, °flirt, °libertine, debauchee, *Old-fashioned* masher, gay dog, *Colloq* wolf, lady-killer, dirty old man, *Taboo slang* gash-hound: *Alistair was quite a roué in his youth but has settled down since turning fifty.*

rough *adj.* **1** uneven, °irregular, °coarse, °jagged, °rugged, °bumpy, °lumpy, °broken: *The next mile or so we slogged over some very rough terrain.* **2** °agitated, turbulent, choppy, °stormy, storm-tossed, °tempestuous, roiled: *We negotiated a very rough patch of water before sailing into the calm, sheltered bay.* **3** °brusque, °bluff, °curt, °short, °abrupt, unpleasant, churlish, °discourteous, °impolite, rough-spoken, °ungracious, °surly, °disrespectful, °rude, uncouth, loutish, °unrefined, uncivil, °uncivilized, uncultured, °vulgar, unladylike, ungentlemanly, °coarse, °ill-mannered, ill-bred, °inconsiderate: *Ian can be a bit rough, I know, but underneath he is a true and loyal friend.* **4** °tough, rough-and-tumble, roughneck, °rowdy: *There are some very rough characters frequenting that pub.* **5** °harsh, °violent, unfeeling, unjust, °severe, °cruel, °tough, °hard, °brutal, °extreme; ungentle: *They were subjected to very rough treatment when being questioned by the police.* **6** °dirty, °obscene, smutty, °pornographic, °vulgar, °crude, °raw, °rude: *After midnight they watch the rough shows beamed by satellite from Italy.* **7** °hard, °tough, °Spartan, °difficult, °arduous, °laborious, °rugged, unpleasant: *It was pretty rough going with almost no money and all those expenses.* **8** °harsh, °grating, cacophonous, °discordant, jarring, °strident, °raucous, rasping, unmusical, inharmonious, °gruff, °husky: *In acknowledgment he made a rough, growling noise.* **9** unfinished, °incomplete, uncompleted, °imperfect, °rudimentary, °crude, °rude, formless, unformed, °raw, rough-and-ready, rough-hewn, roughcast, °undeveloped, unshaped, unworked, unwrought, unprocessed, °unrefined; uncut, unpolished; °shapeless, unshaped, undressed: *I saw the sculpture in its rough state and cannot say how it looked when finished. The stones are rough, as they were when taken from the ground.* **10** °general, °approximate, °inexact, °cursory, °quick, °hasty, °sketchy, °vague, °hazy; foggy, *Colloq* ballpark: *Can't you give me even a rough idea of how much you want for the house?* **11** unfair, unjust, °bad, °tough; unlucky, °unfortunate: *We went through a spell of rough luck last year. This is a rough time for everyone.* **12** See **rotten, 5,** above. *—n.* **13** °rowdy, °tough, hooligan, ruffian, °thug, brawler, °yahoo, *Slang* roughneck, *Australian* larrikin, *US* °mug: *Phineas will get into trouble if he continues to associate with those roughs at the Red Lantern.* **14** sketch, °(rough) draft, mock-up, °outline: *Let me see roughs of the presentation before the meeting with the client.* *—v.* **15** ***rough out*** *or* ***in.*** sketch, °draft, mock up, °outline, mark out, °trace, °block out: *Our escape plan was roughed out by Captain Gallagher.* **16** ***rough up.*** °beat (up), thrash, °attack, °batter, °assault, pummel *or* pommel, °lay on, °knock about, °belabor, lambaste *or* lambast, *Colloq* wallop: *Some of the boys roughed him up a bit when they learned that he was an informer.* *—adv.* **17** violently, savagely, brutally, brutishly: *You'd best not mention Cassie or Henry might cut up rough.*

roughhouse *n.* **1** boisterousness, rowdiness, °rowdyism, °violence, brawling, disorderliness, disorderly conduct, ruffianism: *The police were called in to quell the roughhouse that broke out at the dance.* *—v.* **2** °brawl: *If you want to roughhouse, do it outside!*

roughly *adv.* **1** °approximately, °around; °about, °nearly: *Yes, the museum is roughly in the same area as the library. It cost roughly two weeks' wages. Roughly two years later I saw her again.* **2** harshly, unkindly, °severely, sternly, unsympathetically, brutally, violently, savagely, inhumanly, mercilessly, unmercifully, ruthlessly, pitilessly, cruelly, heartlessly: *The police were accused of dealing too roughly with those suspected of harming a fellow officer.* **3** clumsily, rudely, crudely, awkwardly, primitively, inexpertly, amateurishly, maladroitly, heavy-handedly, ineptly, inefficiently, unskillfully, inartistically: *This roughly hewn block of local granite is what the council paid a fortune for as "sculpture."*

round *adj.* **1** °circular; disk-shaped *or* disc-shaped, discoid, disklike *or* disclike: *She made a round motion to show the shape. I prefer a round table to a square one.* **2** ring-shaped, annular, hoop-shaped, hooplike: *The track is oval, not round.* **3** °spherical, ball-shaped, balllike, globular, spheroid, spheroidal, globe-shaped, globelike, globate, orb-shaped, orblike, orbicular: *You will never convince me that the earth is round—I know what I see.* **4** curved, curvilinear, rounded, arched: *The round chisel is used to cut these grooves.* **5** °exact, °precise, °complete, °entire, °full: *That makes a round hundred times you have asked me the same question.* **6** °approximate, °rough, rounded (off *or* up *or* down), °whole: *In round numbers, that makes 40 percent for and 60 percent against.* **7** rounded, °mellow, °full, vibrant, reverberant, reverberating, sonorous, °rich, mellifluous, orotund, pear-shaped: *Couldn't you listen forever to those beautiful round tones sung by Fischer-Dieskau?* **8** °plain, °honest, straightforward, °direct, °unvarnished, unembellished, unelaborated, °outspoken, °candid, °truthful, °frank, °open, °blunt, *Colloq* upfront: *They deliver what they have to say in round, assertive statements.* **9** *Chiefly Brit* °return: *We made the round trip to Bristol in less than three hours.* *—n.* **10** °circle, disk *or* disc; °ring, hoop, annulus; ball, °sphere, °globe, °orb, bead: *The molding is a repeated pattern of a round alternating with three vertical grooves.* **11** °cycle, series, °sequence, °succession, °bout, °spell: *The rounds of talks on inflation lasted for three days. We made the usual round of parties on New*

Year's Eve. **12** Often, ***rounds.*** °beat, °route, °routine, °circuit, °course, °tour, °turn, ambit: *The milkman still makes his daily rounds in England.* **13** °heat, °stage, °level, °turn: *After each round in the tournament, the losers are eliminated, reducing the competitors by half.* **14** °spell, °period, °run, °spate, °bout, °outburst, °burst, °volley: *Each new acrobat who added to the human pyramid got a round of applause.* **15** bullet, cartridge, °charge, °shell, °(single) shot: *Each man was given only ten rounds of ammunition.* —*v.* **16** °turn, go (a)round: *The car rounded the corner on two wheels.* **17** °orbit, circumnavigate, go (a)round, °circle, °encircle: *He rounded the earth in the fastest recorded time.* **18 *round off*** or ***out.*** °complete, °close, °end, bring to an end *or* completion *or* a close, °finish: *Crumpleigh suggested we round off the evening with a rubber of bridge.* **19 *round up.*** °gather, °assemble, °muster, draw *or* pull *or* get together, °collect, °herd, marshal, *US and Canadian* (of cattle *or* horses) corral, wrangle: *The police are rounding up witnesses who might have seen the shooting. More than ten thousand head have been rounded up for the drive.* —*prep.* See **around,** *prep.* —*adv.* See **around,** *adv.*

roundabout *adj.* **1** circuitous, °circular, °indirect, °long: *We had plenty of time so we came the more scenic, roundabout way.* **2** °devious, circuitous, °evasive, °indirect, °oblique: *If you wanted to borrow some money, asking me for tax advice was quite a roundabout way of asking.*

roundup *n.* **1** °gathering, °assembly, °rally, °collection, herding, *US and Canadian* (of cattle *or* horses) corralling, wrangling: *The roundup of suspects by the police was focused on the Philadelphia area.* **2** °summary, °synopsis, °digest, °outline, recapitulation, °review, °survey, *Colloq* recap: *And now, with a roundup of the news, here is Dan Murrow.*

rouse *v.* **1** °arouse, °call, waken, °awaken, °wake (up), °get up, °arise: *I was roused from a deep slumber by a dog's barking.* **2** °stir (up), °arouse, bestir, °stimulate, inspirit, °animate, invigorate, °electrify, °excite, °provoke, °prompt, goad, °prod, galvanize, °incite, °whet, °move, °work up, fire up: *Bruce's unquenchable enthusiasm roused many to become involved in the campaign.*

rousing *adj.* °stimulating, inspiriting, animating, enlivening, energizing, inspiring, °invigorating, vitalizing, electrifying; °fervent, °vigorous, °energetic, °enthusiastic, °spirited, °brisk, °lively, °animated, *Colloq* peppy: *The huge crowd gave the team a rousing cheer as they jogged out onto the field.*

rout *n.* **1** °defeat, trouncing, °ruin, °overthrow, subjugation, vanquishment, debacle, °conquest, °thrashing, drubbing, beating; dispersal, °retreat, °collapse; *Colloq* licking, hiding, *US and Canadian* shellacking: *Owing to the government's overwhelming forces, the rout of the rebels was complete.* —*v.* **2** °defeat, win (out) over, trounce, °ruin, °overthrow, bring down, °subjugate, °subdue, °suppress, vanquish, °conquer, °overwhelm, °overpower, put to rout *or* flight, worst, °best, °trample, °overrun, thrash, °trim, °whip, drub, °beat, °crush, °batter, smash, °shatter, cut to pieces *or* ribbons *or* shreds, °destroy, °devastate, °wipe out, °eliminate, °put down, seal the doom *or* the fate of, eradicate, °obliterate, *Colloq* lick, wipe the floor with, °polish off, °knock off, *Hyperbolic sports jargon* °pulverize, make mincemeat of, ride roughshod over, °demolish, °mangle, °ravage, °mutilate, °flatten, squash, °topple, °lay waste, wreak havoc on, °ravage, °massacre, °murder, °exterminate, annihilate, liquidate, °smother, °stifle, do away with, *Slang* clobber, do in, *US* skunk, *Chiefly US and Canadian* cream: *The Buffalo Bills routed the Miami Dolphins 6-0.*

route *n.* **1** °way, itinerary, °course, °direction, °path, °road, avenue: *If you have the time, take the scenic route. Heevey's route to the chairmanship is strewn with his colleagues' corpses.* —*v.* **2** °direct, convey, °carry: *Goods destined for Cleveland will be routed through Northampton.*

routine *n.* **1** °custom, °habit, °procedure, °practice, °method, °schedule, °plan, °program, °formula, °pattern, °way, °usage, wont, *Colloq chiefly Brit* °drill: *Chalmers has always been a stickler for routine. What routine does one follow to apply for a job here?* **2** °act, °piece, °bit, °performance, °number, °part, *Colloq* °thing, shtick: *The fat lady doesn't sing till after Joe Miller finishes his routine.* —*adj.* **3** °customary, °habitual, °usual, °rote, °accustomed, °familiar, °conventional, °regular, °ordinary, °everyday; programmed, assigned, designated, scheduled: *Her routine responsibilities include proofreading.* **4** °boring, °tedious, °tiresome, unimaginative, uninteresting; hackneyed, trite, stereotypic(al), clichéd, °run-of-the-mill, °ordinary; unvaried, unvarying, unchanging, °monotonous, uneventful, °rote, °automatic, °mechanical, °perfunctory: *His is a routine job on the assembly line.*

rover *n.* wanderer, bird of passage, itinerant, °traveler, rolling stone, nomad, gypsy *or chiefly Brit also* gipsy, wayfarer, gadabout, sojourner, °tourist; °drifter, °tramp, °vagabond, vagrant, *US* hobo, °bum: *He's a born rover and will never settle down.*

row[1] *n.* °line, °rank, °tier, bank, °string, series, °file: *A row of trees serves as a windbreak along one side of the farm.*

row[2] *n.* **1** altercation, °argument, °dispute, °quarrel, °disagreement, squabble, spat, °tiff, °conflict, °fracas, *Colloq* shouting match, °scrap, falling-out, *Brit* slanging match: *We had a silly row about where we should spend the Christmas holidays.* **2** commotion, °disturbance, clamor, hubbub, °racket, °din, °rumpus, °tumult, °uproar, brouhaha, °fuss, °stir, turmoil, hullabaloo; °bedlam, °pandemonium, °chaos; *US* foofaraw, *Colloq* ruckus: *There was a tremendous row over the increase in membership fees.* —*v.* **3** °dispute, °quarrel, °argue, °disagree, wrangle, cross swords, have words, °bicker, °tiff, *Colloq* °scrap, °fall out: *The Finleys are constantly rowing over something, usually something trivial.*

rowdy *adj.* **1** °boisterous, °uproarious, °disorderly, °noisy, °loud, °obstreperous, °unruly: *It only takes a couple of rowdy youths to set off a free-for-all these days.* —*n.* **2** ruffian, °tough, hooligan, °yahoo, brawler, lout, *Brit* lager lout, skinhead, *Chiefly US and Canadian* °hoodlum, hood, *Slang Brit* bovver boy: *The police arrested three of the rowdies involved in last night's fight.*

rowdyism *n.* rowdiness, ruffianism, hooliganism, roughhousing, barbarism, troublemaking, brawling, unruliness, boisterousness, *Slang Brit* bovver: *Have you any suggestions about what to do to curb rowdyism at football matches?*

royal *adj.* **1** queenly, kingly, queenlike, kinglike, °regal, °imperial, °sovereign, °princely, °majestic: *The royal Family usually spend their summer holidays at Balmoral. The local saddlery proudly boasts royal patronage.* **2** °grand, °splendid, °stately, °impressive, august, °imposing, °superior, °superb, °magnificent, °majestic: *She arrived at the ball in a royal coach drawn by six prancing white horses.* —*n.* **3** °king, °queen, prince, princess, duke, earl, duchess, viscount, viscountess, baron, baroness, °noble, nobleman, noblewoman, °peer: *She would swoon if she met a royal in the flesh.* **4 *the royals.*** °royalty, °nobility, nobles, peerage: *He writes the daily column reporting on the activities of the royals.*

royalty *n.* **1** queenship, kingship, royal house *or* line *or* family, °sovereignty: *Royalty does have its privileges, doesn't it?* **2** °percentage, commission, °share, °payment, compensation: *He receives a royalty every time his song is played or sung commercially, anywhere in the world.* **3** °nobility, nobles, peerage, *Colloq Brit* royals: *It is ironic that Americans are more impressed than the British with royalty.*

rub *v.* **1** °massage, knead, °stroke; °scour, °scrub, °scrape, abrade, °chafe, °clean; °wipe, °smooth, °polish, °shine, buff, burnish: *Rub my back, would you?—No, down a little lower. You are not supposed to rub those plastic baths with abrasive cleansers. Before the guests arrived, we rubbed the silver till you could see yourself in it.* **2** ***rub in*** *or* ***on.*** °apply, °smooth, °smear, °spread, °put: *The ache may be relieved a bit if you rub on some liniment.* **3** ***rub (it*** *or* ***something) in.*** °emphasize, °stress, make an issue of, harp on, °reiterate, °dwell on, hammer away, °dramatize: *It's bad enough that Curby did better than I—you don't have to keep rubbing it in.* **4** ***rub off*** *or* ***out.*** expunge, °erase, °remove, °delete, °cancel, °eliminate, eradicate: *Pity, Pilgrove, that you rubbed out the right reply and wrote in the wrong one. The boys were released on the condition that they would rub the graffiti off the wall.* **5** ***rub off (on).*** °affect, be transferred (to), be communicated *or* transmitted (to), be passed on *or* along (to), be imparted to: *Your mother and I had hoped that some of our love of music might have rubbed off on you.* **6** ***rub out.*** °murder, °kill, °execute, slay: *The boss ordered all witnesses to be rubbed out.* **7** ***rub shoulders with.*** rub elbows with, °associate with, socialize with, mix with, fraternize with, keep company with, consort with, *Colloq US* run *or Colloq US* pal *or Colloq US* chum around with: *Since getting his knighthood Chilsworthy no longer rubs shoulders with the likes of us.* **8** ***rub (someone) (Brit*** **up)** ***the wrong way.*** °annoy, °irritate, °irk, °anger, °provoke, go against the grain, *Colloq* °bug, get under one's *or* someone's skin, peeve: *I don't know what it is about Underbridge, but every time we meet he manages to rub me up the wrong way.* —*n.* **9** °wipe, °stroke, rubbing: *You have to give that stain a good rub to make it come out.* **10** °massage, rubdown: *After exercising, Jane always has an alcohol rub.* **11** ***the rub.*** the *or* a catch *or* hitch *or* snag *or* hindrance *or* setback, the *or* an obstacle *or* impediment, the *or* a difficulty *or* problem *or* trouble: *The only rub I can see is that Winkley may not agree to resign.*

rubberneck *v.* **1** °gape, °stare, goggle, °gawk: *They spent a weekend in New York, rubbernecking at the skyscrapers.* —*n.* **2** °tourist, °sightseer, rubbernecker, *US* out-of-towner: *Buses for the rubbernecks leave Trafalgar Square every hour or so.*

rubbish *n.* **1** °refuse, °waste, debris, rubble, detritus, °litter, °garbage, sweepings, dross, °dregs, °residue, leftovers, remnants, lees, °scraps, °fragments, leavings, residuum, °junk, rejects, *Chiefly US* °trash, *Slang chiefly US* dreck: *A private company carts away the rubbish once a week.* **2** °(stuff and) nonsense, balderdash, °moonshine, °gibberish, gobbledygook *or* gobbledegook, tommyrot, bunkum, °trash, °garbage, twaddle, *Colloq* °rot, flapdoodle, crap, hokum, codswallop, bosh, piffle, hooey, bunk, malarkey, poppycock, boloney *or* baloney, eyewash, hogwash, bilgewater, bull, *Scots* havers, *Brit* tosh, gammon, *US* a crock, horsefeathers, gurry, *Slang* °rot, *Brit* (a load of (old)) cobblers, *Taboo slang* bullshit, horseshit, *US* a crock of shit: *What she told you about how I treated her is rubbish.* —*v.* **3** °criticize, °attack, °destroy, *Colloq* clobber, °pan, *Chiefly US* °trash, *Slang* °jump on, *Chiefly US and Canadian* badmouth, jump all over: *In his latest book he rubbishes the newspaper that had given his previous book a bad review.*

rude *adj.* **1** °impolite, °discourteous, °disrespectful, °ungracious, ungallant, unmannerly, °ill-mannered, uncivil, bad-mannered, ungentlemanly, unladylike, ill-bred, °unrefined, unpolished, uncouth, °boorish, churlish, oafish, loutish, °coarse, °uncivilized, uncultured, unceremonious: *It was rude of him not to get up when she came in, but he doesn't know any better.* **2** °impertinent, °impudent, °discourteous, insulting, °insolent, °offensive, saucy, °bold, °disrespectful, uncivil, °flippant, °brusque, °curt, °gruff, °tactless, °outrageous, *Colloq* °fresh: *I simply asked the chambermaid to make up my room early and she responded with a rude remark.* **3** °naughty, °unrefined, ribald, °bawdy, °indecent, indelicate, °vulgar, °obscene, °dirty, °filthy, lubricious *or* lubricous, °lewd, °gross, smutty, °taboo, °pornographic: *Police found thousands of rude magazines in the warehouse.* **4** °crude, °rough, °clumsy, °awkward, unskillful, unskilled, °artless, inartistic, °imperfect, unpolished, °inaccurate, gauche, bumbling, °raw, inelegant, °makeshift, °homespun, °primitive, °misshapen, ill-formed, unfinished, rough-hewn, °simple, °basic, °bare: *The replica on view at Port House is only a rude approximation of the original.*

rudimentary *adj.* **1** °basic, °essential, °elementary, °fundamental, rudimental, °primary, °introductory, abecedarian, formative, °first, °initial, °elemental, primal, °seminal: *The purpose of the course is provide students with a rudimentary knowledge of computer programming.* **2** °crude, °coarse, unshaped, unfinished, °imperfect, °primitive, °undeveloped, °vestigial, embryonic, primordial, °immature: *These fishes, which spend their lives in darkness, have only a rudimentary organ for an eye.*

rudiments *n.* pl. basics, °elements, essentials, fundamentals, first principles: *She learned the rudiments of musical composition from the maestro himself.*

ruffle *n.* **1** trimming, °flounce, °frill, ruff, peplum, °flare, smocking, ruche, ruching, °gather, °gathering: *Doesn't that ruffle at the back of the skirt make it look a bit old-fashioned?* **2** °ripple, wavelet, °disturbance, °flurry, bustle, °stir, perturbation, °wrinkle: *The news of the divorce caused a slight ruffle at court.* —*v.* **3** °agitate, disconcert, °confuse, discompose, °discomfit, °upset, °disturb, °stir up, °perturb, unsettle, disorient, °unnerve, °fluster, °affect, °bother, °intimidate, unstring, °put out, vex, °trouble, °worry, *Colloq* °rattle, °throw, °shake up, *Chiefly US and Canadian* discombobulate, voodoo, hex, psych out, *Slang chiefly US and Canadian* get (someone) all shook up, spook: *She was ruffled by questions concerning her relationship with the deceased.* **4** disarrange, dishevel, °disorder, °rumple, °mix up, °tousle, °tangle, °disorder, *Colloq* mess *or Colloq* muss (up): *His unpleasant attitude ruffled quite a few feathers in the department.*

rugged *adj.* **1** °rough, uneven, °broken, °stony, °rocky, °irregular, °bumpy, °pitted, °jagged, °ragged: *The horses picked their way slowly across the rugged terrain.* **2** °tough, °rough, °severe, °hard, °harsh, °difficult, °arduous, °Spartan, rigorous, onerous, °stern, °demanding, °burdensome: *It is amazing how that small band of pioneers withstood the rugged life in the untamed West.* **3** °hardy, °durable, °strong, °sturdy, °hale, °robust, °tough, °vigorous, °hard, rough-and-ready, °stalwart; °independent, individualistic, self-reliant, °self-confident, °self-sufficient, °bold: *WANTED—Five rugged pioneering types to join an expedition to the source of the Amazon River.* **4** °rude, uncouth, uncultured, °uncivilized, °unrefined, unpolished, °crude, °ungraceful, churlish: *I admit that Gene's manners are a bit rugged, but he's one of the best men we have.*

ruin *n.* **1** °downfall, °destruction, devastation, °havoc, °breakdown, breakup, debacle, °collapse, °fall, disintegration, ruination, °dissolution, wiping out, °failure, °decay, °undoing, °end; °conquest, °defeat, °overthrow; bankruptcy, liquidation, °failure: *Alexander was responsible for the ruin of the Persian empire. Overwhelming debts spelled Tulley's financial ruin.* **2** °degradation, °dishonor, debasement, defilement, corruption, vitiation, seduction, °degrading, dishonoring, debasing, defiling, corrupting, vitiating, seducing; deflowering, defloration: *Who was the libertine who brought about the ruin of Lord Harecombe's daughter?* **3** nemesis, °curse, °end, bane, °curse: *That boy will be my ruin!* **4** gin, mother's ruin, blue ruin, *Slang* rotgut: *After eating, we sat down and polished off a bottle of mother's ruin.* **5** °hag, °witch, crone, beldam; dotard, (old) fogy *or* fogey, fossil, fuddy-duddy, °antique, *Brit* OAP (= 'old-age pensioner'), *Chiefly US* retiree, *Colloq* (old) geezer, °wreck, *Slang* dodo: *It is hard to understand what those two old ruins see in each other.* **6** ***ruins.*** debris, °wreckage, °fragments, rubble, °remains: *He stood among the ruins of the burnt-out house wondering what to do next. She was trying to piece together the ruins of her disastrous marriage.* —*v.* **7** °destroy, °devastate, °demolish, annihilate, °dissolve, °wipe out, °undo, °overthrow, °lay waste, °raze, °shatter, °wreck, °crush, °flatten, wreak havoc upon *or* on, reduce to nothing *or* naught, °pulverize, smash, bring to ruin: *The bomb ruined the entire center of the city.* **8** °spoil, disfigure, °damage, °botch, °mess up, make a mess of, °mar, uglify: *The county office building completely ruins the Aylesbury skyline.* **9** °spoil, °destroy, °wreck, nullify, °damage, °harm, °hurt, °impair, °poison, *Slang* louse up, °screw up, put the kibosh on, *US* bollix up, *Taboo slang chiefly Brit* make a balls-up of: *I think that your behavior at the Christmas party might have ruined your prospects for a promotion, Wigborough.* **10** bankrupt, pauperize, impoverish, reduce to penury *or* poverty *or* destitution *or* indigence: *Paying off the loan virtually ruined me.* **11** °violate, deflower, °ravish, °seduce, °lead astray, °dishonor, defile, °corrupt, °debase, defile: *She claimed that it was Derek Littlewick who ruined her, despite evidence to the contrary.*

ruinous *adj.* °disastrous, °destructive, catastrophic, °calamitous, deleterious, pernicious, crippling, cataclysmic, baleful, °fatal, toxic, °poisonous, noxious, °harmful, °injurious, °nasty, *Archaic* baneful: *Smoking can be ruinous to your health. Who was behind the ruinous decision to move the factory to the Brae of Moan?*

rule *n.* **1** °regulation, °order, °law, ordinance, ruling, °decree, ukase, statute, °principle, °direction, °guide, guideline, °precept: *We go by the Golden Rule here, Jarvis, "Do unto others before they do unto you."* **2** °dominion, °authority, °control, °sovereignty, °sway, °command, ascendancy, °direction, °oversight, supervision, mastery: *How could such a despised monarch maintain rule over a people for so long?* **3** °fact, °standard, °customs, °practice, °form, °routine, °convention, °policy, way things are: *Honesty and integrity among our students are the rule rather than the exception.* **4** ***as a rule.*** °generally, °usually, normally, customarily, °for the most part, mostly, °ordinarily, °mainly, °in the main, °chiefly, °on the whole, commonly, more often than not: *As a rule, we require payment in advance.* —*v.* **5** Sometimes, ***rule over.*** °reign (over), °govern, be in control *or* charge *or* command (of *or* over), be in power (over), hold sway (over), wield the scepter, wear the crown, °run; °prevail, hold sway, °dominate, °predominate, °control: *The hand that rocks the cradle rules the world. Supposedly, the majority rules in a democracy.* **6** °direct, °guide, °manage, °control, °lead, °head (up), °preside (over), superintend, °oversee, °supervise, °regulate, °govern, °run: *The company is ruled by a committee made up of employees and directors.* **7** °decide, °judge, hand down a judgment *or* decision, °decree, deem, °resolve, °settle, °determine, °find, °declare, °pronounce: *The referee ruled that Wiggins had committed a foul.* **8** ***rule out.*** °ban, °bar, °prohibit, °exclude, °eliminate, °forbid, °preclude, proscribe, negate, °dismiss, °disregard, °bypass, °ignore: *The program committee ruled out Las Vegas as the site of next year's conference.*

rummage *v.* **1** °search, °hunt, comb, °scour, scrabble, look through, sift through, turn inside out *or* upside down, °examine, *Colloq* °scrounge: *We rummaged about in four dustbins before finding the discarded receipt.* —*n.* **2** °jumble, miscellanea, °miscellany, knickknacks, °odds and ends, hotchpotch *or US also* hodgepodge: *May we borrow your car to pick up the contributions to the church rummage sale?*

rumor *n.* **1** °news, °gossip, hearsay, °information, °scoop, tidings, °chat, chitchat, tittle-tattle, *on dit;* °grapevine, jungle telegraph, *Colloq* °lowdown, info, *US and Canadian* poop, *Slang chiefly US nautical* scuttlebutt: *The rumor is that they are moving from Boston to Santa Rosa, California. Rumor has it that you are not going with them.* —*v.* **2** bruit about, noise abroad, °circulate, pass around, °intimate, °breathe, °suggest, °whisper, °leak, °reveal, °make known, °put about, °say, °report, °tell: *It has been rumored that she is going to have his baby.*

rumple *v.* Sometimes, **rumple up.** wrinkle, °crumple, °crush, crease, °fold, crinkle, dishevel, °ruffle, °tousle, scrunch (up), °pucker, muss (up), °mess (up): *Every time I drive the car, the back of my jacket gets all rumpled.*

rumpus *n.* commotion, °disturbance, °fuss, °confusion, °uproar, °tumult, to-do, ado, °mayhem, °bedlam, brouhaha, °stir, pother, affray, °fracas, °row, melee *or* mêlée, °roughhouse, °brawl: *You have no idea what a rumpus he created just because my dog went in his garden.*

run *v.* **1** sprint, °race, scamper, °scurry, °scud, dart, °bolt, °dash, °flit, °tear (along), scoot, scuttle, zip, whiz *or* whizz, gallop, °jog, °trot, lope; °rush, °hurry (up), °hasten, °scramble, °hustle, step lively, hop (to) it, °step on it, put on some speed, *Archaic* hie, *Colloq* °get a move on, hoof it, °leg it, hotfoot (it), stir one's stumps, *Brit* hare, *US* step on the gas, °hump (it), *Slang* get cracking, *US* get the lead out (of one's pants *or taboo* ass), *Chiefly US* get a wiggle on: *If you run, you might catch her before she gets on the train.* **2** run away *or* off, °flee, °escape, °take flight, °take to one's heels, °bolt, decamp, °make off, °clear out, show a clean pair of heels, abscond, cut and run, °(beat a (hasty)) retreat, °retire, make a getaway, (make a) run for it, *Colloq* °beat it, scram, skedaddle, °take off, °skip (out), take French leave, fly the coop, *Slang* head for the hills, *Brit* scarper, do a bunk, *US and Canadian* take a (runout) powder, lam out of (somewhere), take it *or* go on the lam, *US* vamoose: *The minute the enemy appeared, the entire troop ran. Cavendish ran, hoping the police wouldn't catch him.* **3** °go, °cover, °pass over, sprint, °race: *He ran the mile in just under four minutes.* **4** °wander, rove, °roam, °meander, °drift: *Some farmers allow their animals to run free rather than penning them up.* **5** Often, ***run for.*** °compete (for), be a candidate (for), °vie, °struggle, contend, °fight (for), °stand (for): *I doubt that she will run for reelection in East Hagbourne.* **6** °pass, °flow, °pour, °stream, °flood, °gush, °spill, dribble, °spurt, °issue, °move, °trickle, seep, °discharge, cascade, °spout: *You can see where the rainwater runs down* *this crack in the roof and into the room below.* **7** °flow, °diffuse: *The color ran when he put his red underwear in with some white shirts, and now his shirts match his underwear.* **8** °melt, liquefy, °dissolve, °fuse: *It was so hot that the icing ran off the side of the cake and onto the plate.* **9** °keep, °maintain, °support, °sustain, °manage: *It is very expensive to run a large house.* **10** °operate, °manage, °direct, °supervise, °oversee, °conduct, superintend, °control, °handle, °manipulate, °head, °carry on, °lead, °regulate, °take care of, °look after, °administer, be in charge of, °coordinate: *You must be terribly clever to be able to run a huge international conglomerate like that!* **11** °operate, °perform, °function, °work, tick, °go: *Even after all that abuse, the watch kept running and kept good time.* **12** °extend, °stretch, °reach; °amount, add up, total up, °come (up): *The balance of payments deficit is running to twice last year's figure. His bill at the restaurant last night ran to more than what I earn in a month!* **13** convey, °transport, give (someone) a lift, °drive, °take, °bring: *Tammy said he would run me home in his new car.* **14** bootleg, smuggle, deal *or* traffic in, *Chiefly US and Canadian* °rustle: *During Prohibition, his father ran whiskey from Canada to the US.* **15** be in effect *or* force, be effective, have force *or* effect: *This lease has another year to run.* **16** °incur, invite, °encourage, °attract, be subjected to: *Aren't you running a risk asking Sheila to handle your money?* **17** °propel, °drive; °steer, °guide, °navigate: *If you are inept enough to run a boat aground, at least have the sense do it at low tide.* **18** unravel, come undone, *Chiefly Brit* ladder: *My stocking ran just before I had to leave for the dance!* **19** Sometimes, ***run off.*** °print, °offset, lithograph, °reproduce, °publish, °display; imprint, °position, °place, °locate, °lay out: *They are going to run my story in the next issue of* The Old Review! *We'll run the photographs of the fire alongside the article on page one.* **20** ***run across.*** °meet (up with), °run into, °come across, °find, stumble on *or* upon, hit *or* chance *or* happen upon, *Colloq* °bump into: *Guess whom I ran across in the supermarket yesterday.* **21** ***run after.*** °chase, °pursue, go after, court, woo, *Colloq* set one's cap for: *Wall Street is filled with young men and women running after the almighty dollar. Muggleswick is running after this girl half his age, and he hasn't a clue that she's leading him on.* **22** ***run along.*** °go away, °leave, *Slang* get lost: *Sue wanted to get rid of Wally and told him to run along.* **23** ***run around.*** philander, be unfaithful, gallivant, *Colloq* sleep around, play the field: *How does he cope with his wife's running around?* **24** ***run around with.*** °associate with, spend time with, dally with, consort with: *Cynthia is running around with a bad crowd these days.* **25** ***run away.*** See **2**, above. **26** ***run down.*** **(a)** °trace, °track, °hunt, °stalk, °follow, °pursue, dog, °shadow; °find, °locate, °track down, °discover: *We ran down the suspect in an abandoned barn in Littlehampton.* **(b)** °criticize, decry, defame, °vilify, °disparage, deprecate, °depreciate, denigrate, *Colloq* °knock, °pan: *If you keep running him down like that he'll have an inferiority complex before he's seven.* **(c)** °weaken, °tire, °expire, play (itself) out, burn out, °run out, °fail, *Colloq* °peter out: *After the batteries run down we can no longer transmit a distress signal.* **(d)** °strike, °hit, smash *or* crash *or* slam into, °run over, knock over *or* down, °collide with: *On the way home, he almost ran down a boy on a bicycle.* **27** ***run in.*** °arrest, take into custody, °jail, apprehend, take *or* bring in, *Colloq* °pinch, °nab, °pull in, bust, collar, *Brit* °nick: *This is the third time tonight that Minette has been run in for soliciting.* **28** ***run into.*** See **20**, above. **29** ***run off.*** **(a)** See **2**, above. **(b)** °duplicate, °print, °copy, °turn out, °produce, °make, °manufacture, °generate, *Colloq* do, churn out: *How quickly can you run off a thousand copies of this brochure?* **30** ***run out.*** **(a)** be exhausted, °expire, °terminate, (come *or* draw to a) close, °end, °cease: *The time for renewing your subscription has run out.* **(b)** °finish, °go, be exhausted, be used up, *Colloq* °peter out: *The food had run out and we had no water when we were rescued.* **31** ***run out of.*** °use up, °consume, eat up, °exhaust, be out of: *Sorry, we ran out of self-raising flour and are awaiting a new shipment.* **32** ***run out on.*** °desert, °abandon, leave high and dry, °forsake, °leave in the lurch, leave holding the baby: *Despite all his promises, Edgar ran out on her when she became pregnant.* **33** ***run over.*** **(a)** See **26 (d),** above. **(b)** °read (through), °(copy) edit, °study, °scan; go over *or* through: *I promised Casey I would run over his article if I had the chance.* **(c)** overflow, °spill (over), brim over, slosh over, pour over; °extend, °reach, spread over, stretch over; °exceed, go beyond, overreach, overshoot, °surpass, °transcend: *She let the bath run over, ruining the ceiling below. This article will run over onto the next page. McClelland always runs over the time allotted for his speech.* **(d)** °rehearse, °run through, °repeat, °practice, °review, °go over, °study, °learn, °memorize: *You'd best run over your speech again before the banquet.* **34** ***run through.*** **(a)** °pierce, °stab, °transfix, °stick, °spit: *Zounds! The base varlet has run my master through with his sword!* **(b)** squander, °consume, °use up, °waste, °fritter away, °exhaust, deplete, °spend, °dissipate, °throw away, *Slang* °blow: *It didn't take Hayden more than a year to run through his inheritance.* **(c)** See **33 (b),** above. *n.* **35** sprint, °dash, °race, °jog, °trot: *They go for a two-mile run every morning before breakfast.* **36** °trip, °journey, °visit, °drive, °expedition, trek, °outing, °excursion, jaunt, junket, °sojourn, *Colloq* °spin, joy ride: *He has to make a run over to Paris for the weekend, supposedly on business.* **37** °route, °routine, °circuit, °passage, °trip, °cycle, °round; °beat: *I was on the Murmansk run during the war.* **38** °period, °spate, °interval, °time, °spell, °stretch, °course; °engagement, booking, *Colloq* °patch: *Rumbold has had a run of bad luck and could use some help. The show had a long run on Broadway.* **39** access, °freedom, °liberty: *We left to escape our guests, allowing them the run of the house for a week.* **40** °return, °satisfaction, °reward, recompense, compensation, °requital, expiation, °atonement, repayment, °remuneration: *Although the odds are never in your favor, you get a good run for your money at Diamond Lil's Gambling Emporium.* **41** series, °sequence, °stream, °spate, °string, °succession, °progression: *Witlow attributed his poor showing at the tournament to a bad run of cards.* **42** °stream, °brook, runnel, °creek, rill, rivulet, *Brit* beck, *Scots* °burn, *US* °branch, °kill: *There is a small run at the bottom of the hill.* **43** °demand, °call, °request: *Bouillon cubes seem to be enjoying quite a run lately.* **44** °type, °category, °class, °kind, °sort: *Marston seems to be a bit more refined than the usual run of guest at Lisa's parties.* **45** °trail, °track, piste, °path, °slope; °way, runway: *She skied down the north run in record time.* **46** °enclosure, yard, °pen, °compound, runway; paddock; °pound: *There is a good-sized run at the kennel where we board the dogs when we go away on vacation.* **47** *Music* roulade, cadenza, arpeggio, riff: *We were treated to a marvelous run in which each note was distinct and pure.* **48** ***in the long run.*** °eventually, °finally, after all, °ultimately, in (due) time, in due course, in fine, in the end, at the end of the day, in the final analysis, all things considered, when all is said and done: *Regardless of early troubles, things always seem to turn out all right in the long run in his plays.* **49** ***on the run.*** **(a)** °hastily, in haste, hurriedly, while under way, in a hurry, at speed, in a rush: *It isn't supposed to be good for you to eat breakfast on the run every day.* **(b)** on the loose, fleeing, escaping, in flight, °running (away), *Slang US* on the lam: *Himself on the run from* *the police, Dannay had to find Mr. Memory.* **(c)** °running, retreating, on the move *or* the go: *We kept the enemy on the run for a week.* **50 *the runs.*** diarrhea, dysentery, upset stomach, *Jocular* tummy rot, Delhi belly, Aztec hop, tourista *or* turista, Mexican fox-trot *or* two-step *or* toothache, Montezuma's revenge, curse of Montezuma, Rangoon runs, Tokyo trots, Lambeth run(s): *I came down with the runs when we got to Ankara.*

runaround *n.* evasive treatment; °slip: *Every time he applied for help he was given the runaround.*

runaway *n.* **1** °fugitive, escapee, °refugee, °deserter, °truant, absconder: *The lad is a notorious runaway who has escaped four times before.* —*adj.* **2** °wild, °uncontrolled, unchecked, °rampant, °renegade, unsuppressed; driverless, riderless, °loose; escaped: *Runaway inflation is ruining the economy of the country. The runaway horses were finally brought back.* **3** °easy, °effortless, °overwhelming, uncontested: *The incumbent enjoyed a runaway victory in yesterday's election.*

run-down *adj.* **1** wearied, °exhausted, debilitated, weakened, worn-out, °peaked, °fatigued, enervated, °tired, drained, °spent, out of shape *or* condition, °below par, in bad shape; °unhealthy, °sickly, °ill: *In a badly run-down condition when he was found, he was soon back on his feet.* **2** °ramshackle, °dilapidated, tumbledown, °decrepit, °rickety, broken-down: *They live in that run-down shack at the edge of the town dump.*

rundown *n.* °run-through, °synopsis, °summary, °survey, °précis, °résumé, (thumbnail) sketch, °outline, rough idea, °review, recapitulation, briefing; highlights, high points: *Give me a quick rundown of what happened while I was gone.*

run-in *n.* °disagreement, °argument, °dispute, altercation, °quarrel, confrontation, contretemps: *We had a run-in with our neighbor about playing his hi-fi too loud.*

runner *n.* **1** sprinter, racer, jogger, hurdler, miler: *Gordon is one of our best runners and ought to make the Olympic team.* **2** °messenger, courier, errand boy *or* girl, messenger boy *or* girl, °page, dispatch-bearer *or* despatch-bearer, dispatch-rider *or* despatch-rider, *Colloq US* gofer: *The runner said that the package of securities fell off his motorcycle and he lost it.* **3** °sucker, tendril, creeper, °shoot, °branch, °stem: *Tie these runners to the frame if you want to train the vine.* **4** °blade: *The runners on your skates need sharpening.*

running *n.* **1** °operation, °management, °direction: *Who attends to the everyday running of the business while you are away?* **2** °competition, °contest, °meet, °tournament, °race, °match; °event, °game: *If he is found to take steroids, he'll be out of the running.* —*adj.* **3** °continuous, °ongoing, °continual, °perpetual, °sustained, °constant, uninterrupted, ceaseless, unceasing: *Where do they get the energy to keep up a running argument on the same silly subject?*

runt *n.* °dwarf, pygmy, midget: *Save the runt of the litter for me.*

run-through *n.* **1** °rehearsal, °practice, °trial, °test: *Let's have one more run-through before tomorrow's dress rehearsal.* **2** See **rundown,** above.

rupture *n.* **1** °break, °rift, °split, fissure, °fracture, cleavage, bursting; breaking, splitting, breakup, °breach, °schism, disunity, breaking up, severance, °division, °separation: *The rupture was caused by the expansion of the water on freezing. There was no rupture in friendly relations between the families.* **2** hernia: *He is to undergo surgery for a rupture.* —*v.* **3** break (up *or* apart), °split, °fracture, °cleave, °divide, °breach, °separate; °disrupt, °part, sunder: *The fabric of the balloon ruptured, the gas escaped, and the basket plummeted earthward.*

rural *adj.* **1** °country, °pastoral, sylvan, bucolic, °rustic, Arcadian, exurban; agricultural, agrarian, *Literary* georgic: *They always preferred rural life to life in the city. Constable painted mainly rural scenes. He sells farm equipment, strictly a rural business.* **2** See **rustic, 2,** below.

ruse *n.* °trick, °device, °deception, °maneuver, °dodge, °pretense, °pretext, °subterfuge, °stratagem, ploy, °hoax, °wile, °artifice, imposture: *His feigned friendship was only a ruse to get them to reveal their next move.*

rush *v.* **1** °hurry (up), °hasten, °run, °race, °hustle, bustle, make haste, °dash, °speed, °scurry, °scramble, scoot, °jump, sprint, scamper, scuttle, *Colloq* °move (it), hotfoot (it), skedaddle, °step on it, make it snappy, *US* hightail (it), step on the gas, *Slang* get moving, get cracking, get a wiggle on, go like a bat out of hell, °shake a leg: *If you want the job, you'll have to rush to send in your application. Don't rush—there's plenty for everyone.* **2** °attack, °assault, °charge, °storm, blitz: *If we rush them, we might take them by surprise.* —*n.* **3** °hurry, °haste, °hustle, bustle, °dash, °speed, turmoil, turbulence, °flurry, commotion, °ferment, pother, ado, to-do, °excitement, °pell-mell, harum-scarum: *What's the rush?* **4** °surge, °sensation, °thrill, °charge: *I felt a rush of pleasure when I saw Irena coming toward me.* —*adj.* **5** °urgent, hurry-up, exigent, high-priority, top-priority, °emergency: *This is a rush job that must be completed today.*

rustic *adj.* **1** See **rural, 1,** above. **2** °peasant, °plain, °simple, uncomplicated, °unsophisticated, naive *or* naïve *or* naïf, °ingenuous, guileless, °artless, °unrefined, unpolished, countrified, uncultivated, uncultured, °boorish, °crude, °rough, unmannerly, hillbilly, backwoods, °awkward, ungainly, cloddish, plodding, oafish, gawky, lumpen, loutish: *Who could dream that such a rustic homebody would some day be famous for the most fashionable literary salon in London?* —*n.* **3** °peasant, bumpkin, °boor, yokel, hillbilly, countryman, countrywoman, country boy *or* girl, oaf, country cousin, *Colloq* clodhopper, *Brit derogatory and offensive* bogtrotter, *US and Canadian* hayseed, hick: *In those days few rustics had ever even visited a city.*

rustle *v.* **1** °whisper, °swish, sibilate, susurrate: *Outside my window, the leaves rustled in the evening breeze.* —*n.* **2** °whisper, whispering, rustling, °swish, swishing, sibilation, sibilance, susurration, susurrus: *I heard the faint rustle of satin and knew she was listening at the door.*

rut *n.* **1** °groove, °furrow, wheelmark, °track, trough: *In the winter the ruts fill with water and freeze, making driving difficult.* **2** °pattern, °habit, °routine, °groove, °grind, treadmill, dead end, *Colloq* rat race: *After forty-seven years as a horse groom, I'm beginning to think about getting out of my rut.*

ruthless *adj.* pitiless, unpitying, °cruel, °unsympathetic, °merciless, °unmerciful, °harsh, °fierce, °remorseless, uncompassionate, °vicious, °savage, °ferocious, hardhearted, °callous, unfeeling, °tough, °severe, °heartless, °inhuman, °brutal, brutish, unrelenting, °relentless, *Chiefly US and Canadian* °mean: *In all of history there was no more ruthless a tyrant.*

S

sabotage *n.* **1** °destruction, °damage, wrecking, °impairment: *Enemy infiltrators were responsible for the sabotage of our radar.* **2** °subversion, treachery, treason: *When the radar was found damaged, sabotage was suspected.* —*v.* **3** °undermine, °subvert, °disrupt, °spoil, °ruin, °cripple; °damage, °incapacitate, disable, °destroy, °wreck, *Colloq US* throw a monkey wrench into the machinery, *Brit* throw a spanner in the works, *Slang Brit* °queer (someone's pitch): *Only someone who had something to lose would have sabotaged our plan for reorganization. The engine had been sabotaged by pouring sand into the fuel tank.*

sack *n.* **1** °pouch, °bag, *Scots and US Dialectal* °poke; *Technical* sac: *You'd better get a sack of potatoes while you're at the market.* **2 *hit the sack.*** °retire, °turn in, go to bed *or* to sleep, *Slang* hit the hay, *Brit* kip (down), *US* sack out: *I need my beauty sleep so I'm going to hit the sack.* **3 *the sack.*** °dismissal, °discharge, firing, *Colloq* heave-ho, the ax, marching orders, *US* pink slip, °bounce, *Chiefly US and Canadian* walking papers, *Slang Brit* the boot, the chop, °the push: *Ten of us got the sack when the new management took over.* —*v.* **4** °dismiss, °discharge, °fire, let go, °lay off, *Brit* make *or Brit* declare redundant, *Colloq* give (someone) the ax *or* the (old) heave-ho, give (someone) his *or* her *or esp Brit* their marching orders, *Brit* give (someone) the sack, *US* °bounce, *Slang Brit* give (someone) the boot *or* the chop *or* the push: *Now that he's been sacked, Hal will be able to spend more time with the children.*

sacred *adj.* **1** consecrated, dedicated, hallowed, °holy, blessed, blest, sanctified, revered, °divine, awe-inspiring, °venerable, venerated, sainted, heaven-sent: *This is a most sacred Islamic site.* **2** inviolable, inviolate, untouchable, protected, sacrosanct: *They held nothing sacred in their depredations of the temple.* **3** °religious, °spiritual, °ceremonial, church(ly), ecclesiastical, °priestly, hieratic, °ritual, °solemn, sacramental, liturgical: *He continued to carry out his sacred duties despite the war.*

sacrifice *n.* **1** immolation, °surrender, forfeiture, forgoing, giving up, yielding up, °offering (up), °offer, *Christianity* oblation: *The Aztec religion demanded the sacrifice of human beings.* **2** forfeiture, forgoing, giving up, relinquishment, °loss: *Mrs. Clinton continues her charitable work at the sacrifice of much of her time.* —*v.* **3** immolate, °offer (up), °yield (up), °give up: *In their religion, they were required to sacrifice small animals to propitiate the gods.* **4** °give up, °forgo, °forfeit, °relinquish, °surrender, let go, °lose, °yield, °renounce, forswear; forbear, desist, °cease, °stop, °refrain from: *She was quite ready to sacrifice her inheritance to see justice done. If you sacrificed smoking you could give the money you save to charity.*

sacrificial *adj.* **1** sacrificed, immolated, surrendered, given up, yielded: *They led the sacrificial lamb to the slaughter.* **2** atoning, expiatory, °propitiatory, conciliatory: *The solemn sacrificial ceremonies were carried out by the high priestess.*

sacrilege *n.* **1** desecration, profanation, debasement, °violation, °prostitution, dishonoring, vitiation, defilement, befouling, fouling, contamination, befoulment, °misuse, °abuse, °perversion, maltreatment: *The horrendous office building is an example of architectural sacrilege.* **2** impiety, heresy, profanation, °outrage, °violation, °profanity, blasphemy, impiousness, irreverence, °disrespect, secularization: *It was an unscrupulous sacrilege to appropriate to his own use the offerings to the gods.*

sacrilegious *adj.* °profane, °impious, °heretical, °blasphemous, °irreverent, °disrespectful: *The inscription contains a curse on any who perpetrate any sacrilegious act against the remains of the pharaoh.*

sad *adj.* **1** °unhappy, °melancholy, downcast, °dejected, depressed, °low, °sorrowful, °gloomy, morose, °glum, lugubrious, °mournful, heartsick, crestfallen, chapfallen, disheartened, °downhearted, °blue, °despondent, °brokenhearted, °heartbroken, °woebegone, °miserable, °wretched: *Danny was so sad when his dog died that he wept for a week.* **2** depressing, °gloomy, disheartening, °dreary, °dismal, °funereal, °somber, lugubrious, saddening, heartbreaking, °bleak, distressing, dispiriting, °calamitous: *It was a sad day when the US team failed to win the gold medal.* **3** °unfortunate, °unsatisfactory, °awful, °bad, °shabby, °dirty, °lamentable, °miserable, °sorry, °wretched, °pathetic, °pitiful, °pitiable, °deplorable, °terrible, *Colloq* °lousy, °rotten: *After two weeks under water, the sofa was in pretty sad shape.*

sadden *v.* °depress, deject, °sorrow, dishearten, °distress, dispirit, °discourage, °grieve, aggrieve: *It saddens me to think of all the starving people.*

sadistic *adj.* °cruel, °monstrous, °brutal, brutish, °beastly, °ruthless, °perverse, *Technical* algolagnic: *She derives sadistic pleasure from reminding me of all my failures. The scars and bruises confirmed the sadistic treatment he had undergone.*

sadly *adv.* **1** unfortunately, alas, unhappily, unluckily, lamentably, regrettably, deplorably, sad to relate: *Sadly, all our oldest oaks were uprooted in the hurricane.* **2** unhappily, gloomily, morosely, mournfully, despondently, miserably, wretchedly, dejectedly, dismally, somberly, lugubriously: *He told me sadly that he did not expect to return to England again.*

sadness *n.* unhappiness, dolor, °misery, °sorrow, dispiritedness, °grief, °depression, dejection, dejectedness, sorrowfulness, despondency, °melancholy, °gloom, gloominess: *It is hard to describe the sadness we all felt when she left.*

safe *adj.* **1** unharmed, °whole, uninjured, unhurt, °(safe and) sound, °secure, protected, shielded, sheltered, out of harm's way, all right, *Colloq* O.K. *or Colloq* OK *or Colloq* okay: *You'll be safe here with me.* **2** °harmless, nontoxic, nonpoisonous, innocuous; unpolluted: *This medication is safe if you do not exceed the recommended dosage. Is the water safe to drink?* **3** °sure, °secure, °sound, protected, risk-free, riskless, safe as the Bank of England, °reliable, dependable, °solid, °bona fide, °conservative, tried and true, *Brit* safe as houses: *It was previously thought that an investment in Alfordyce Ltd was safe.* **4** all right, allowable, °permissible, °acceptable, °satisfactory, °appropriate, °suitable, °timely, °right, °correct, °proper, justifiable, justified, *Colloq* OK *or Colloq* O.K. *or Colloq* okay: *Is it safe to come out now?* **5** secured, protected: *The children are safe in bed. Your secret is safe with me.* —*adv.* **6** safely, securely: *I won't breathe easy till he's safe behind bars.* —*n.* **7** vault, °crypt, strongbox, safe-deposit *or* safety-deposit box, coffer, °chest, repository: *The necklace is kept in the safe, and I never get a chance to wear it.*

safeguard *n.* **1** °protection, °defense, °security: *I do not consider such a flimsy fence to be a sufficient safeguard.* —*v.* **2** °protect, °defend, °shield, °shelter, keep safe, °guard; °conserve, °save, °keep, °care for, °look after: *What is being done to safeguard the passengers* *from attacks by bandits? My aunt in Slowe Bourne is safeguarding our valuables while we travel abroad.*

safekeeping *n.* °charge, °protection, keeping, °custody, °care, guardianship: *The deed to the house is in the safekeeping of Mr. Williams, my solicitor.*

safety *n.* safeness, °protection, aegis, °cover, °shelter, °security, °refuge; °sanctuary, °safekeeping: *There's safety in numbers, so the more the merrier. We retreated to the safety of the castle dungeon.*

sag *v.* **1** °droop, °sink, °slump, °bend, °dip; swag, °bag: *The board sagged precariously under my weight. Without a breath stirring, the banners sagged in the humid air.* **2** °drop, °decrease, °decline, go *or* come down, °fall, °slide, °slip, °weaken, °slump, °descend, °diminish, lessen, °droop, °subside, °flag, falter, °wilt: *The value of the dollar sagged in world markets today.* —*n.* **3** sagging, °drop, °droop, °sinking, sinkage, subsidence, °dip; reduction, °decrease, °decline, °fall, °slide, weakening, °slump, lessening, flagging, faltering: *If you stand over here you can see a definite sag in the roof. The sag in profits was attributed to a poor response to the new model.*

saga *n.* °(heroic) legend, °narrative, epic, edda, °chronicle, °romance, *roman-fleuve,* °story, °tale, °adventure: *Are you really interested in the continuing saga of the inhabitants of Hydrangea Crescent?*

sage *adj.* **1** °wise, sagacious, °prudent, °sensible, perspicacious, °profound, °intelligent, discerning, °reasonable, °logical, °judicious, common-sense, common-sensical: *The old woman gave her some very sage advice about men, which she promptly forgot.* —*n.* **2** wise man, savant, °expert, °elder, guru, Nestor, pundit, Solomon, philosopher, °oracle, °authority: *The sage whose council you must seek lives deep in the forest.*

sail *v.* **1** °navigate, °pilot, °steer: *Sail as close to the shore as possible to avoid the whirlpool.* **2** go sailing *or* boating *or* yachting, °cruise, set sail, put (out) to sea: *Would you like to sail to the Isle of Wight for the weekend?* **3** °drift, move lightly, °breeze, °flow, °waft, °sweep, °coast, °float, °scud, °glide, °slide, °slip, °plane, °skim, °fly, °flit: *It was delightful to watch the tiny boats sailing, tacking, beating, and running in the brisk wind.* —*n.* **4** canvas: *The schooner was carrying all the sail she could.*

sailor *n.* seaman, seafarer, seafaring man *or* woman, seagoing man *or* woman, mariner, °(old) salt, seadog, bluejacket, shellback, yachtsman, yachtswoman, boatman, boatwoman, deck hand, captain, °skipper, *Old-fashioned or literary* Jack Tar, *Colloq old-fashioned* tar, *Naval US* swab, swabbie, °gob: *When she said she had married a sailor I didn't know he had a hundred-foot yacht.*

saintly *adj.* °holy, blessed, blest, beatific, °godly, sainted, angelic, °seraphic, °pure, °righteous, °virtuous, °blameless: *Donald was such a saintly man, it was impossible to think ill of him.*

sake *n.* **1** °benefit, °welfare, well-being, °good, °advantage, °behalf, °profit, °gain, °account: *You needn't go to school for my sake—you're the one who will benefit.* **2** purpose(s), reason(s), objective(s): *Just for the sake of comparison, let us now look at the situation in Taiwan.*

salary *n.* °income, °pay, °earnings, compensation, °remuneration, emolument; wage(s): *I can't afford a new car on my present salary.*

sale *n.* **1** selling, °traffic, vending, marketing, trafficking, trading; °trade, °exchange, °transaction: *The sale of alcoholic beverages on Sundays is prohibited. Provide a receipt for each sale.* **2** °trade, °purchase; trading, buying, purchasing: *We must do something to boost sales. The house is priced for a quick sale.* **3** sales event, markdown, white sale, jumble sale, *Brit* boot sale, *US* rummage sale, garage sale, tag sale, yard sale, *Colloq* sell-a-thon, sales marathon: *We are having a sale to clear out things we no longer want or need.* **4** °transaction: *She hasn't made a sale in a week.* **5** ***on sale.*** *US* marked down, cut-price, bargain-priced; reduced (in price): *Milk is on sale this week at half price.* **6** ***(up) for sale.*** on the market, °available, in stock, *Chiefly Brit* °on sale, on offer, *US* on the block: *My Vespa is for sale to the highest bidder.*

salesperson *n.* salesman, saleswoman, saleslady, salesgirl, salesclerk, clerk, *Chiefly Brit* shopgirl, *Brit* shop assistant, *Old-fashioned* counter-jumper: *Is it possible to find a salesperson who knows the merchandise in this shop?*

salient *adj.* °conspicuous, °outstanding, °pronounced, °noticeable, °prominent, °significant, °important, °marked, °impressive, °striking, °remarkable, distinguishing, °distinctive, °unique, °eminent, °noteworthy, °notable, °principal, °chief, °primary: *A salient feature is the privilege of converting the insurance at any time from a straight life to an endowment policy.*

salt *n.* **1** common *or* table salt, sodium chloride, sea salt, rock salt: *A little salt will bring out the flavor.* **2** °spice, spiciness, °zest, zestiness, pungency, °vigor, °vitality, liveliness, °pep, °pepper, poignancy, piquancy, °relish, °bite, °savor, °seasoning, °taste, *Colloq* zip, zing, °punch: *Yes, do invite Randolph, for he can be relied on to add a little salt to the dinner conversation.* **3** °(Attic) wit, Attic salt, dry humor, °sarcasm: *Her conversation is sprinkled with the salt of clever aphorisms.* **4** See **sailor,** above. **5** ***with a grain*** *or* ***pinch of salt.*** *cum grano salis,* warily, cautiously, qualifiedly, qualifyingly, doubtfully, skeptically, suspiciously, reservedly, with reservation(s) *or* qualification(s): *You have to take what Winkleigh says with a grain of salt.* —*v.* **6** °season, °spice, °flavor, °pepper: *The pirate's speech was well salted with expressions like "Shiver me timbers!," "Avast there!," and "Blow me down!"* **7** pickle, cure, °preserve, corn, marinate, souse: *To keep, the meat must be salted.* **8** ***salt away.*** °save (up), °hoard, put *or* lay *or* set by *or* aside, squirrel away, store up, stockpile, °amass, °accumulate, °pile up, *Colloq* stash away, *US and Canadian* sock away: *They were always terribly stingy, claiming they were salting something away for their old age.* —*adj.* **9** salty, saline, brackish, briny: *The water in your area has a salt taste.* **10** pickled, kippered, marinated, soused; corned; cured: *Do you like salt herring?*

salute *v.* **1** °greet, °hail, °address, accost: *The moment I stepped in the door I was saluted with the peremptory question, "Where have you been?"* **2** pay respects *or* homage *or* tribute to, °honor, °recognize, °acknowledge: *Wherever she went she was saluted as a heroine.* —*n.* **3** °greeting, °address, salutation: *He returned my salute with a nod.*

salvage *v.* **1** °save, °recover, °rescue, °redeem, °deliver, °retrieve, °reclaim: *Were you able to salvage anything of value after the fire?* —*n.* **2** °recovery, °rescue, retrieval, redemption, deliverance, reclamation, salvation: *The salvage of a vessel or cargo at sea invests the salvager with legal rights under certain conditions.*

salve *n.* **1** balm, °ointment, unguent, dressing, cream, °lotion, demulcent, embrocation, liniment: *The doctor applied a salve to the wound and bandaged it.* **2** emollient, balm, palliative, °tranquilizer, opiate, anodyne, °narcotic, °relief, assuagement: *The money was more a salve to his conscience than a token of his charity.* —*v.*

3 °mitigate, °relieve, °ease, alleviate, assuage, palliate, soothe, mollify, °comfort, appease: *There must be something you can do to allow them to save face and to salve their wounded pride.*

same *adj.* **1** °identical, °exact (same), selfsame; °very: *When I said they were wearing the same dress I really meant identical dresses. That's the same tie you wore when we first met.* **2** unchanged, unchanging, °changeless, unmodified, unaltered, °constant, °uniform, unvaried, unvarying; word-for-word, °verbatim: *As you can see, it's the same old Charley you used to know and love. Professor Lympette is still giving the same lecture that he has given all these years.* **3** ***all the same.*** °at the same time, °nevertheless, °nonetheless, °even so, °yet, but, anyway, anyhow, °in any case, °in any event, °at any rate, °regardless, °still (and all), in spite of *or* despite the fact, °notwithstanding, for all that, that (having been) said, having said that, after all is said and done, just the same: *Of course you have a right to go; all the same, I wish you had let me know.*

sample *n.* **1** °specimen, °example, °representative, °representation, °illustration, sampling, sampler, cross section; swatch; °bite, nibble, °taste: *I should like a sample to see whether I want to place an order.* —*v.* **2** °test, °try, °taste, °experience: *Anyone is welcome to sample the merchandise on request.* —*adj.* **3** °representative, °specimen, illustrative, representational, °trial, °test: *A sample copy of the book is available for examination.*

sanitarium *n.* rest home, convalescent home, nursing home, clinic, health farm, sanatorium: *After spending a month recuperating in a Swiss sanitarium, Gladys was as good as new.*

sanctify *v.* **1** consecrate, °hallow, make sacred *or* holy, °glorify, °exalt, canonize, enshrine; *Roman Catholic Church* beatify: *The site was sanctified and is visited by pilgrims throughout the year.* **2** °purify, °cleanse: *According to Scripture, some are sanctified before birth.* **3** °confirm, °sanction, °ratify, °justify, °legitimate, legitimatize *or* legitimize, legalize, °license: *Her family insisted that the marriage be sanctified at the church in Cockayne, where they live.*

sanctimonious *adj.* °hypocritical, °self-righteous, canting, °mealy-mouthed, holier-than-thou, pharisaical, pietistic, unctuous, Tartuffian, *Colloq* °goody-goody, *Chiefly Brit* smarmy, *Slang Brit* pi: *While that sanctimonious dog preaches the benefits of honesty to me, his accomplices are secretly buying shares in the company for a fraction of their value.*

sanction *n.* **1** confirmation, ratification, secondment, authorization, legalization, legitimatization *or* legitimization, validation, °license, certification, °approval, °permission, imprimatur, seal *or* stamp (of approval), signet: *If you get the sanction of the rest of the members, then I shall agree.* **2** °help, °aid, °encouragement, °support, advocacy, °backing, sponsorship, °favor, countenance: *You will need the sanction of the entire committee in order to win a vote of confidence.* **3** °agreement, concurrence, acceptance, affirmation, assent, acquiescence, compliance, °approval, OK *or* O.K. *or* okay: *He would never have proceeded with the invasion without the sanction of his generals.* **4** °ban, °penalty, °punishment, °retribution, °discipline, retaliation, redress: *The government voted to reduce the sanctions against their former enemy.* —*v.* **5** °confirm, °ratify, °second, °authorize, legalize, legitimatize *or* legitimize, validate, °license, °certify, °approve, °permit, °allow, notarize, °vouchsafe, °subscribe to, commission, °consent to: *If the board sanctions the purchase of the company, it is then up to the shareholders to vote.* **6** °support, °encourage, °advocate, °back, °sponsor, °favor, countenance, °help: *You know I cannot be seen to sanction your plan in preference to others'.*

sanctity *n.* °piety, holiness, saintliness, divinity, °grace, sacredness, godliness, °devotion, °dedication: *One could but admire the sanctity of his life spent as a missionary.*

sanctuary *n.* **1** °sanctum, shrine, chapel, °temple, church, house of worship, house of God; synagogue, mosque, pagoda: *All of us retire at this time to the sanctuary for evening prayer.* **2** asylum, °refuge, °retreat, °protection, °shelter, °safety: *Formerly, criminals could seek sanctuary in religious houses.* **3** (nature *or* wildlife) reserve *or* preserve, °reservation, conservation area, national park: *If housing is built here, it will destroy the bird sanctuary.*

sanctum *n.* **1** °sanctuary, holy of holies, shrine: *No one is allowed into the sanctum except the high priestess.* **2** sanctum sanctorum, den, °study, °retreat; hiding place, hide-out, °hideaway, cubbyhole: *After dinner Roger goes to his sanctum to write.*

sane *adj.* °normal, of sound mind, °rational, *compos mentis,* °well-balanced, right-minded, °levelheaded, °rational, °reasonable, °sensible, °judicious, *Colloq* right in the head, all there: *There is no justification for considering the people of Marystoe any less sane than the rest of us.*

sang-froid *n.* coldbloodedness, coolness, coolheadedness, °indifference, composure, phlegm, self-possession, °self-control, °poise, imperturbability, equanimity, *Colloq* unflappability, °cool, coolth: *What amazed us was the sang-froid displayed by even the smallest children when faced by brutality.*

sanguinary *adj.* **1** °bloodthirsty, °cruel, °brutal, brutish, °gory, °merciless, °remorseless, °ruthless, pitiless, °heartless, °savage, barbarous, slaughterous, °grim, °fell, °murderous, °homicidal: *That ax murder was the most sanguinary crime it has been my misfortune to investigate.* **2** bloody, sanguineous, sanguinolent: *I am unsure how I survived such a sanguinary battle.*

sanguine *adj.* °optimistic, °rosy, °confident, °hopeful, forward-looking, anticipatory, °expectant, °enthusiastic, fervid, zealous: *Some experts took a more sanguine view, expecting an economic upturn in the next quarter.*

sanitary *adj.* °clean, °sterile, °hygienic, antiseptic, disinfected, aseptic, germ-free, bacteria-free; °healthy, unpolluted, salubrious, healthful, salutary, °wholesome: *Without the personnel and equipment, it has been impossible to maintain a sanitary hospital environment.*

sanity *n.* saneness, °reason, mental health *or* soundness, normality, rationality, °reason, reasonableness, °stability, °balance: *Have you any reason for suspecting her sanity?*

sap[1] *n.* **1** (vital) juice *or* fluid, bodily *or US also* body fluid, lifeblood, °essence, *Literary* ichor: *The sap of youth has shrunk from my veins.* **2** °fool, idiot, nincompoop, ninny, ninnyhammer, simpleton, ignoramus, nitwit, dimwit, dunce, ass; °dupe, gull, *US* thimble-wit; *Colloq* chump, °drip, *Brit* (right) charlie, noddy, noodle, °wet, *Chiefly Brit* °twit, *Slang* patsy, °sucker, °(easy) mark, °pushover, saphead, *Brit* muggins, *US* schnook, schlemiel *or* schlemihl *or* shlemiel, fall guy: *You certainly were a sap to fall for that old line!* —*v.* **3** bleed, °drain, °draw, °tap, °rob, °milk: *I could feel the strength being sapped from my body.*

sap[2] *v.* °undermine, °sabotage, °weaken, °cripple, °wreck, devitalize, deplete, °drain, °erode, °enervate, debilitate: *They did everything they could to sap my self-confidence.*

sarcasm *n.* °scorn, contumely, °derision, °ridicule, °bitterness, acrimony, acrimoniousness, acerbity, harshness, acridity, acridness, asperity, °venom, °poison, venomousness, poisonousness, °virulence, °spite, spitefulness, malice, maliciousness, malevolence, °satire, irony, cynicism, disdain: *Kathy is too thick-skinned to feel the sarcasm of your cutting commentary on her acting.*

sarcastic *adj.* °scornful, contumelious, derisive, °derisory, ridiculing, °bitter, °biting, °cutting, °trenchant, °incisive, acrimonious, acerbic, acid, acidic, acidulous, °harsh, acrid, aspersive, °venomous, °poisonous, °virulent, °spiteful, malicious, malefic, malevolent, satiric(al), ironic(al), cynical, °disdainful, mocking, °contemptuous, °critical, censorious, captious, carping, caviling, °sardonic, °scathing, °caustic, °nasty: *Lymping, known for his sarcastic book reviews, appears to dislike everything written by anyone later than Jane Austen.*

sardonic *adj.* ironic(al), derisive, °derisory, mocking, cynical, °sarcastic: *His only response to the remark was a sardonic smile.*

satanic *adj.* **1** diabolic(al), °fiendish, °devilish, Mephistophelian, demonic, demoniac(al), °infernal, cacodemonic, °ghoulish, hellish, °infernal, °evil, °wicked, iniquitous, °corrupt, depraved, °perverted, °perverse, °godless, °ungodly, °impious, unholy, °sinister, °dark, °black, °immoral, amoral: *They practice witchcraft and other satanic inventions.* **2** dire, °monstrous, heinous, °atrocious, °hideous, °horrible, horrendous, horrid, horrifying, °loathsome, °vile, °abhorrent, unspeakable, unutterable, °damnable, °despicable, °abominable: *The miners worked under positively satanic conditions.*

satellite *n.* **1** moon, spacecraft, *Old-fashioned* sputnik: *The sky is crowded with artificial satellites orbiting the earth.* **2** °follower, °attendant, retainer, °disciple, acolyte, °aide, aide-de-camp, minion, lieutenant, °assistant, helper, °hanger-on, dependent, °shadow, right-hand man, vassal, °parasite, sycophant, *Colloq* sidekick: *Boswell became the obsequious satellite of Samuel Johnson.*

satiate *v.* **1** °stuff, °glut, °gorge, cloy, °surfeit, overfill, overstuff, °glut, °pall, overindulge, °saturate, °choke, deluge, °flood, °choke, suffocate, °weary, °exhaust, °bore, °tire, °jade: *The travelers were so satiated with food and drink that nothing could wake them.* **2** °slake, °satisfy, °quench, °content, °gratify, sate: *I felt the it would take at least a gallon of water to satiate my thirst.*

satiety *n.* °surfeit, °glut, superabundance, overindulgence, saturation, nimiety, °excess, °superfluity: *It was a case of satiety or famine.*

satire *n.* **1** °ridicule, irony, °sarcasm, °mockery, spoofing, °exaggeration, °caricature: *Often, the best way to comment on a distasteful political situation is through satire.* **2** °burlesque, °lampoon, °parody, travesty, pasquinade, spoof, cartoon, °caricature, *Colloq* °takeoff, *Chiefly Brit* sendup: *For several years he wrote mainly satires of life among the wealthy.*

satirical *adj.* satiric, ironic, °sarcastic, mocking, spoofing, °irreverent, exaggerating, Hudibrastic, derisive, disparaging, °abusive, °scornful, °flippant, ridiculing, chaffing: *After a number of satirical plays, he returned to writing novels, which, he felt, reached a wider audience.*

satirize *v.* °lampoon, °burlesque, °parody, °caricature, travesty, °poke fun at, °(hold up to) ridicule, make fun *or* sport of, pillory, °deride, °mock; °mimic, °imitate; *Colloq* °take off, °put down, *Brit* °send up: *Since ancient times writers have satirized the follies of their age.*

satisfaction *n.* **1** °gratification, °comfort, °fulfillment, contentment, °delight, °joy, °enjoyment, °pleasure, °happiness: *Wallingford certainly gets a great deal of satisfaction out of seeing his children happily married.* **2** °payment, °requital, repayment, compensation, recompense, °remuneration, reparation, °indemnity, indemnification, °restitution, vindication, °damages, °amends, redress, °atonement, expiation: *The purpose of the lawsuit is to demand satisfaction for the wrongs done to our family.*

satisfactory *adj.* °adequate, °sufficient, °acceptable, °passable, all right, not bad, good enough, °fair, *Colloq* OK *or Colloq* O.K. *or Colloq* okay: *I'd say that the food at Michelle's is satisfactory but not outstanding.*

satisfy *v.* **1** °gratify, °fulfill, °comfort, please, °content, placate, appease, pacify: *If you liked that dinner at the Grendelkellar, I would say that it doesn't take much to satisfy you.* **2** °fill, °meet, °fulfill, °provide for, look after *or* to, °serve, °answer, comply with, °resolve, °solve, °gratify, °indulge; °slake, °quench, sate, °satiate: *Your requirement for superior working conditions would be satisfied in the Newcastle area. What will it take to satisfy your thirst for adventure?* **3** °convince, °persuade, °reassure, °assure, put (someone's) mind at rest, °content: *He was satisfied that emigration to Australia was the only option open to him.* **4** °pay, °repay, redress, °make good, indemnify, °write off, liquidate: *I shall see to it that the debt is satisfied at once.*

satisfying *adj.* gratifying, °satisfactory, fulfilling, °filling, satiating; comforting, °pleasing, pacifying, °pleasurable: *That was a very satisfying meal. It is a satisfying feeling to know that you are happy.*

saturate *v.* °soak, °wet, °drench, °steep, °fill, imbue, souse, °suffuse, °impregnate, °permeate; waterlog; *Technical* ret: *Make sure to saturate the earth with water after transplanting.*

sauce *n.* **1** gravy, condiment: *That sauce on the meat was truly delicious.* **2** °impertinence, sauciness, °impudence, audacity, insolence, brazenness, pertness, °disrespect, disrespectfulness, *Colloq* cheek, cheekiness, lip, back talk, backchat, °brass, °nerve, °gall, *Slang* crust, *US and Canadian* sass, sassiness: *She had the sauce to tell me to pay at once or get out.*

saunter *v.* °walk, °stroll, amble, °meander, °ramble, °wander, *Colloq* mosey, traipse: *An oddly dressed man sauntered over and asked if I had ever bet on a horse before.*

savage *adj.* **1** °wild, °untamed, undomesticated, feral, unbroken: *These savage beasts had never seen a man before.* **2** °vicious, °ferocious, °fierce, °beastly, bestial, brutish, °bloodthirsty, °brutal, °cruel, °ruthless, pitiless, °merciless, °harsh, bloody, °unmerciful, °fell, barbarous, barbaric, °murderous, demonic, demoniac, °sadistic: *The boy was subjected to a savage attack by the defendant's dog. The king's guards were selected for their savage nature.* **3** °wild, °uncivilized, uncultivated, °primitive, °inhuman, bestial, barbaric, barbarous, °untamed, °rude: *The savage behavior of some New Guinea tribes included cannibalism.* —*n.* **4** wild man *or* woman, °brute, °barbarian; Caliban: *When the explorers landed, they were welcomed by the savages who lived there.*

save *v.* **1** °(come to someone's) rescue, °deliver; °(set) free, °liberate, °release, °redeem, bail (someone) out; °recover, °salvage, °retrieve: *She expects to be saved by a knight in shining armor. What can be done to save me from the drudgery of doing the laundry? We saved only one thing of value from the fire.* **2** °keep, °preserve, °guard, °safeguard, °protect, °conserve, °secure, °shelter, °shield: *One way to save dried flowers is to press them between the leaves of a book.* **3** lay *or* put aside, lay *or* put by, lay *or* put away, °keep, °retain, °set apart, °hold, °reserve, °preserve, °conserve; °economize, scrimp, °scrape: *I have saved enough to buy a new car. If you keep on saving, you may soon have enough for a new suit.* **4** obviate, °preclude, °spare, °prevent: *Extra care taken now will save a lot of trouble later on.*

saving *adj.* **1** redeeming, redemptory *or* redemptive *or* redemptional, compensating, °compensatory, qualifying, °extenuating, extenuatory: *His one saving grace is that he has a lot of money.* **2** parsimonious, °economical, °thrifty, °provident, °frugal, °sparing, °prudent: *It was only because of Allison's saving nature that we had enough to live on during the strike.* —*n.* **3** economizing, economization, °economy, °thrift, °providence, frugality, °prudence, scrimping, scraping, sparingness: *It was only by her saving that we survived.* **4** ***savings.*** °resources, °reserve, °cache, °hoard, nest egg: *Our savings are to pay for your education, not for having a good time.*

savior *n.* **1** rescuer, salvation, friend in need, Good Samaritan, liberator, redeemer, deliverer, emancipator, °champion, knight-errant, knight in shining armor: *Because he gave her a job when she most needed it, she regarded him as her savior.* **2** ***the*** *or* ***our Savior*** *or* ***Saviour.*** *Christian religion* Christ (the Redeemer), Jesus, the Messiah, Lamb of God, Our Lord, Son of God, King of Kings, Prince of Peace, *Islam* Mahdi.

savoir-faire *n.* °tact, tactfulness, °sophistication, °finesse, urbanity, °discretion, knowledgeability, °diplomacy, urbanity, smoothness, °polish, suavity *or* suaveness, °poise, °grace, °style, °skill, adroitness, °knowledge, comprehension, *Slang* savvy: *You can count on Clapworthy to have the savoir-faire needed to carry off the deception.*

savoir-vivre *n.* °breeding, °upbringing, comity, °knowledge, °sophistication, °polish: *He has at least enough savoir-vivre to know that one doesn't drink red wine with oysters.*

savor *n.* **1** °taste, °flavor, °zest, °tang, smack, piquancy: *These poached peaches have a savor of brandy about them.* **2** °hint, °suggestion, °odor, °scent, °fragrance, °smell, °perfume, redolence, °bouquet, °breath, °trace, °quality, *soupçon,* °dash: *He preferred talking about the savor rather than the odor of sanctity, for alliteration's sake.* —*v.* **3** °taste, °sample, °perceive, °detect, °sense, discern, °mark, descry, °observe, °notice, °note, °identify; °enjoy, °luxuriate in, °relish, °indulge in, bask in, °appreciate, °revel in, °delight in, value, °cherish, *Colloq* lick *or Colloq* smack one's lips *or Colloq* chops over: *One could savor a trace of honey in the wine. For a week we savored the delights of the Costa Smeralda.*

savory *adj.* **1** palatable, °delicious, delectable, °tasty, toothsome, appetizing, flavorful, flavorous, flavorsome, ambrosial, °luscious: *Pettigrew bit greedily into the savory flesh of the ripe melon.* **2** °tasteful, °honest, °proper, °decent, °reputable, °respectable, °honorable, creditable, °upright, °decorous, °seemly, °wholesome, °innocent: *I am not sure that Victoria is traveling in particularly savory society.* —*n.* **3** appetizer, °hors d'oeuvre; °dessert, °sweet; °morsel, °dainty, tidbit *or Brit* titbit, *Chiefly Brit* starter, *Archaic* warner: *I think I'd like the Welsh rabbit as a savory.*

saw *n.* °proverb, °maxim, (old) saying, aphorism, apophthegm *or* apothegm, axiom, adage, °epigram, gnome; °slogan, °motto, catchword, catch phrase, °byword; dictum, platitude, °truism, °cliché, commonplace: *She always quoted to him the old saw, "The grass on the other side of the fence is always greener."*

say *v.* **1** °state, affirm, °declare, °maintain, °hold, aver, °remark, assert, °claim, asseverate, °announce: *She said that I had to go at once, and I said, "I shall never leave you!"* **2** assert, °allege, °report, °mention, °rumor, °reveal, bruit about, °disclose, divulge, °bring to light, °put about, noise abroad, °suggest, °hint, °whisper: *It was said that spies had already infiltrated the party.* **3** °pronounce, articulate, utter; °phrase, rephrase, °translate: *How do you say* gemütlich? *How do you say it in English?* **4** °tell, °put, °express, verbalize, °communicate, °explain, °reveal, °bring up, °break, °impart: *I don't quite know how to say this, Ormond, but you have bad breath.* **5** °reply, °respond, °answer: *What have you to say to her allegation that it is all your fault?* **6** °guess, °estimate, conjecture, °venture, °judge, °imagine, °believe, °think: *Yes, I'd say you look about seventy years old. I say the card is the king of diamonds.* **7** mean *or* intend *or* try to say, °think, °contemplate, °imply, °suggest: *Are you saying that you would steal if you had the opportunity?* **8** °predict, °prognosticate, foretell: *The newspaper says that shares will go down.* **9** °signify, °denote, °symbolize, °communicate, °indicate, convey, °suggest, °imply, °mean: *What does a red light say to you?* **10** °order, °require, °demand, °bid, °stipulate, °command, give the word: *If I say that you are to go, then you go—and quickly.* **11** °deliver, utter, °speak: *Say your lines and exit stage left without waiting for a reply.* —*n.* **12** °voice, °authority, °influence, °power, °weight, °sway, clout: *Does she have that much say about how the money should be spent?* **13** °turn, °chance, °opportunity, °vote: *You've had your say, now let's hear from Ackroyd.* —*adv.* **14** °approximately, °about, °roughly, circa; °nearly: *The snake was, say, twenty feet long.* **15** °for example, °for instance, as *or* for an illustration, e.g.: *Take any novel, say,* Wuthering Heights, *and analyze the characters.*

say-so *n.* °authority, °word, °say, °order, dictum; authorization: *I shouldn't do it just on his say-so if I were you.*

scale[1] *n.* Often, ***scales.*** °balance: *We need a new bathroom scale—the old one gives too high a reading.*

scale[2] *n.* **1** °flake, imbrication; scurf, dandruff; *Technical* squama, °plate, scute *or* scutum, lamina, lamella: *While the fish is alive, its scales are a brilliant hue.* **2** coating, encrustation *or* incrustation, crust, overlay, layer, °cake, caking, tartar, °plaque: *The hardness of the water causes the scale inside the kettle.*

scale[3] *n.* **1** °range, compass, °rank, ranking, gradation, graduation, °register, spectrum, calibration, °progression, hierarchy, °scope, °gamut: *As we ascend in the scale of life we rise in the scale of longevity.* **2** °proportion, °ratio: *The scale of these drawings is one foot = one eighth inch.* —*v.* **3** °climb, ascend, °mount, clamber up, surmount, °go up, escalade: *Using grappling hooks and ropes, we scaled the wall in minutes.* **4** °regulate, °adjust, °proportion, *Chiefly US and Canadian* prorate: *The size of the cable is scaled to the weight it must carry.* **5** ***scale up*** *or* ***down.*** °increase, °enlarge, °raise; °decrease, °reduce, °diminish, °lower: *Depending on the expanse of the façade, the windows ought to be scaled up or down accordingly.*

scaly *adj.* **1** °rough, imbricated, shingly, flaky, *Technical* lamellar, laminar, lamellate, scutate: *The scaly covering of reptiles and fishes is related to the feathers of birds.* **2** scabby, scabrous, squamous, squamulose, squamosal, squamose, scurfy, furfuraceous, scruffy: *The medication causes the skin to dry and become temporarily scaly.*

scan *v.* **1** glance at *or* through, °look over, °skim, read over, flip *or* thumb *or* leaf through: *I didn't have time to read it thoroughly, but I did scan it.* **2** °study, °pore over, °examine, °investigate, °scrutinize, °inspect, delve into, °research, °explore (in depth), °sweep, *Archaic* con: *We scanned for traces of blood. Scan the horizon for hostile planes.* —*n.* **3** °examination, °investigation, °scrutiny, inspection, °research, °exploration: *A microscopic scan revealed no trace of blood.*

scandal *n.* **1** °shame, °disgrace, °embarrassment, °sin, °outrage: *It is a scandal how much of the money raised for charity goes into the pockets of the fundraisers.* **2** °discredit, °damage, calumny, ignominy, obloquy, °dishonor, °degradation, disrepute, °infamy: *The scandal resulting from their being found together could never be lived down. The breath of scandal never touched her.* **3** °slander, °libel, °aspersion, °innuendo, insinuation, °abuse, °dirt, defilement, defamation, °slur, °smear, °taint, °blemish, °spot, °stigma, smirch, black mark *or* spot, °blot (on the escutcheon), °(badge of) infamy, skeleton in the cupboard, *Brit* blot on one's copybook: *His cowardice brought scandal to the name that could never be lived down.*

scandalize *v.* appall *or* appal, °shock, °outrage, affront, °offend, °horrify, °upset, °disturb; °rankle, stick in (someone's) craw *or* throat, °gall: *We were scandalized to learn the truth about Cooksley's father.*

scandalous *adj.* **1** °shocking, °disgraceful, ignominious, °improper, indecorous, °unseemly, °infamous, °outrageous, °shameful, °immodest, °dishonorable, °disreputable, °sordid, °despicable, flagitious, °wicked, °sinful, °evil, iniquitous, °profligate, °immoral, °indecent, °lewd, °lascivious, °lustful, licentious, °lecherous, °atrocious, heinous, °disgusting, fulsome, °taboo, °unmentionable, unspeakable: *The scandalous goings-on at the Hellfire Club are well documented.* **2** defamatory, libelous, °slanderous, calumnious, calumniatory, aspersive, °abusive, °scurrilous, °injurious, defamatory: *Baddley is suing the newspaper over that scandalous article about his alleged defalcations.*

scanty *adj.* **1** scant, °sparse, °scarce, °little, °meager, °minimal; barely adequate *or* sufficient, °limited, restricted, *Colloq* °measly: *The scanty food would have to last us for another week. The news from the front was scanty. We received only scanty support from the Arts Council.* **2** skimpy, °short, °small, °sparse, °minimal, °meager, °in short supply, *Colloq chiefly Brit* °thin on the ground: *Investor interest in the new share offering seemed to be scanty.*

scapegoat *n.* °victim, °front, °dupe, gull, cat's-paw, whipping boy, *Brit* man of straw, Aunt Sally, *US* straw man, *Colloq* fall guy, *Slang* °sucker: *We have to find a scapegoat to take the punishment for us.*

scar *n.* **1** °blemish, °mark, °damage, disfigurement, °wound, °injury, °scratch, °mar, °cut, °burn, °brand, cicatrix: *Fortunately, it is only a bruise and there won't be any scar.* —*v.* **2** °blemish, °mark, °damage, disfigure, °wound, °injure, °scratch, °mar, °cut, °burn, °brand; dent: *The blow scarred him for life.*

scarce *adj.* °scanty, scant, °insufficient, °inadequate, °deficient, °wanting, lacking, °rare, °unusual, °at a premium, °in short supply, °meager, few and far between, seldom met with, hard to come by, *Chiefly Brit* °thin on the ground: *Good editors are very scarce.*

scarcely *adv.* **1** °hardly, °barely, °(only) just, not quite: *He had scarcely uttered the magic words when the rock split open.* **2** (probably *or* certainly *or* surely *or* definitely) not, in no way, not at all, not in the least, °by no means, on no account, under no circumstance, nowise, *Colloq US* noway: *I scarcely need remind you that you are getting married tomorrow.*

scarcity *n.* °lack, °want, °need, paucity, °dearth, insufficiency, °shortage, inadequacy, inadequateness: *The teaching of foreign languages is not being improved by the scarcity of qualified teachers.*

scare *v.* **1** °frighten, °alarm, °startle, °shock, °dismay, °daunt, °appall, give (someone) a shock *or* a fright, °terrify, terrorize, °threaten, °menace, cow, °intimidate, °horrify, *US and Canadian* spook, *Colloq* scare the pants off, scare the life *or* the living daylights *or* the hell out of, scare out of one's wits, make one's hair stand on end, make one's flesh creep *or* crawl, give one goose bumps *or* goose pimples, *US* scare the bejesus out of, *Taboo slang* scare *or* frighten the shit out of, scare shitless, *US* scare shitty: *A man's piercing scream scared me, and I clambered out of the old mine as fast as I could.* **2** ***scare up.*** scrape together *or* up, °find, °gather, °collect, °raise, °dig up, °get, °come by, °scrounge (up): *I can't scare up that much money in one day!* —*n.* **3** °fright, °shock, °surprise, °start: *I had a terrible scare when I saw what I thought was blood.*

scared *adj.* frightened, alarmed, °afraid, appalled, shocked, terrified, horrified, startled: *Don't be scared, I won't hurt you.*

scary *adj.* °frightening, °eerie, °terrifying, °frightful, hair-raising, unnerving, bloodcurdling, horrifying, spinechilling, intimidating, daunting; horrendous, horrid, °horrible, creepy, crawly, *Colloq* spooky: *Aren't you afraid to watch scary movies on TV when you're home alone?*

scathing *adj.* searing, °withering, damaging, °harmful, °severe, °harsh, °stern, °nasty, °biting, acrid, acrimonious, mordant, °incisive, °cutting, °sharp, °keen, °virulent, vitriolic, acid, °scorching, °burning, °fierce, °savage, °ferocious: *She was totally unprepared for the critic's scathing attack on her book.*

scatter *v.* **1** °spread, °diffuse, °shower, °litter, sprinkle, °strew, °circulate, °distribute, disseminate, °sow, °broadcast: *As the seeds of dissent were scattered far and wide, we became aware of a groundswell of hatred for the regime.* **2** °disperse, °separate, °dissipate, dispel, °disband, °break up, °go off: *My notes are scattered all over the floor. Immediately after dividing up the loot, the gang scattered to the four winds.*

scatterbrained *n.* °harebrained, rattle-headed, rattlebrained, °frivolous, flibbertigibbet, °giddy, dazed, °flighty, woolgathering, *Colloq* dippy, °dizzy, dopey *or* dopy, slaphappy: *It irritates me to think that our lives might depend on such a scatterbrained fool.*

scattering *n.* smattering, sprinkling, °trifle, °bit, °suggestion, soupçon, °hint: *It is true, there was a scattering of rebellious sentiment in the town.*

scenario *n.* (master *or* ground *or* floor) plan, °(grand) scheme, °plot, schema, °design, °outline, layout, framework, °structure; sequence of events, °routine: *According to my scenario, the bank provides the financing and we engineer the takeover.* **1** °(plot) summary, °précis, °résumé, °synopsis; (working *or* shooting) script, screenplay: *The scenario was written by Arthur de Vere, who had nothing to do with writing* Casablanca.

scene *n.* **1** °location, °site, °place, °area, °locale, °spot, locality, °whereabouts, °sphere, °milieu, backdrop, °background: *The scenes of my youth are always in my mind.* **2** °action, °episode, °part, chapter, °section, °segment, °(stage) setting, *mise en scène;* scenery: *The next scene takes place in Venice. Is the Venice scene ready?* **3** commotion, °upset, °exhibition, °display, °row, brouhaha, °disturbance, °furor, °tantrum, °argument, altercation, uncomfortable *or* disagreeable situation, °episode, °incident: *He was entirely responsible for that scene in the restaurant.* **4** °view, scenery, °sight, °landscape, seascape, panorama, vista, °picture, °view, °prospect: *The scene from the terrace was completely tranquil.* **5** ***behind the scenes.*** °secretly, privately, clandestinely, confidentially, surreptitiously, sub rosa, on the Q.T.: *Sigrid operated behind the scenes and her identity was never revealed to the other agents.* **6** ***make or do the scene.*** °socialize, °appear, get around *or* about,

°go out, °participate: *We used to make the scene in the Village back in the forties, but no more.*

scenic *adj.* °picturesque, °panoramic, °pretty, °beautiful, °grand, °awesome, awe-inspiring, °impressive, °striking, spectacular, breathtaking: *The train passes along incredibly scenic routes in Switzerland.*

scent *n.* **1** °fragrance, °aroma, °perfume, redolence, °smell, °odor, °bouquet, whiff, °trace: *The room was filled with the scent of wildflowers.* **2** °trail, spoor, °track: *Lagrange hasn't been seen since the day he said he was following the scent of Mack the Knife into Limehouse.* —*v.* **3** °perceive, °detect, °find out, °determine, discern, °distinguish, °recognize, °sense, °smell, °sniff (out), °get wind of, learn *or* hear about: *Scenting possible trouble, I decided I'd better take along my revolver.* **4** °perfume: *She had left behind her handkerchief, scented with Chanel No. 19.*

schedule *n.* **1** °program, °timetable, °plan, °calendar, agenda, °outline, °list, listing, °record, °register: *I'd like a complete schedule of your work for the next month on my desk tomorrow.* —*v.* **2** °program, °organize, °plan, °outline, °list, °record, °register, °arrange, °book, °time, slate, °appoint, °assign, °allot, °dedicate, earmark: *Schedule the next meeting for the 15th of the month. Alistair is scheduled to speak tomorrow.*

schematic *adj.* **1** diagrammatic(al), representational, °graphic, charted: *We have the schematic drawings showing the placement of the equipment.* —*n.* **2** diagram, blueprint, layout, (floor *or* game) plan, °scheme, °design, °representation, graph, (flow *or* PERT) chart: *Give copies of the schematics to the telephone company so that they can plan where to put the wiring.*

scheme *n.* **1** °plan, °plot, °design, °program, °system, °course (of action), schema, °outline, °exposition, °projection, °draft, °method, °technique, °approach, game plan, °scenario: *In my scheme, compensation would be dependent on productivity, merit, and length of service.* **2** °pattern, °arrangement, layout, °design, diagram, blueprint, °chart, map, °drawing, °schematic, °disposition, °order, °organization, schema: *This alternative scheme shows the executive offices on the second floor.* **3** °plot, °plan, ploy, °maneuver, °strategy, °stratagem, °tactic, °machination, °subterfuge, °trick, °device, °dodge, °wile, °ruse, °intrigue, *Colloq* °racket, °game, °move: *Ashton's scheme is to lure the security guards into the outer room, then lock them in.* —*v.* **4** °plan, °plot, °devise, contrive, °intrigue, °organize, °formulate, °hatch, conspire, machinate, °maneuver, connive, concoct, *Colloq* cook up: *Clifton has been scheming to get his revenge on me ever since the Schickard episode.*

scheming *adj.* conniving, plotting, nefarious, treacherous, °crafty, cunning, °artful, °sly, °wily, °devious, Machiavellian, intriguing, °slick, °calculating, °tricky, °foxy, °slippery, underhanded, duplicitous, °deceitful: *Wait till I lay my hands on the scheming little thief!*

schism *n.* °split, °rift, °break, °breach, °division, °rupture, °separation, disunion: *The schism was caused by the far left faction.*

schismatic *adj.* schismatical, separatist, breakaway, divisive, °dissident, °heretical: *The schismatic movement in art was marked by one faction's abandonment of realism.*

scholar *n.* **1** °academic, professor, °teacher, pedagogue, °authority, °expert, pundit, savant, bookman, bookwoman, man *or* woman of letters, °intellectual, °highbrow, °bookworm, *Colloq* egghead, °brain, *US* longhair: *Professor Read is one of the most respected linguistics scholars in the world.* **2** °student, °pupil, schoolboy, schoolgirl, undergraduate: *Cosgrove was a failure as a scholar, preferring football to physics.*

scholarly *adj.* °learned, erudite, °lettered, scholastic, °profound, °deep, °intellectual, °academic, highbrow(ed), ivory-tower(ed), *Colloq* egghead, brainy, *US* longhair(ed): *A scholarly person with scholarly interests, her scholarly achievements are not limited to scholarly articles for scholarly journals.*

scholarship *n.* **1** °learning, erudition, °knowledge, °lore, °education, °schooling, training, °preparation, *Colloq* know-how: *Elsa Cairn brings to her appointment as headmistress considerable scholarship and experience.* **2** °grant, °endowment, °award, °fellowship, *Brit* °exhibition, *Chiefly Scots and New Zealand* bursarship, bursary: *Without the scholarship, you would never have been able to continue your education.*

school *n.* **1** °(educational) institution, kindergarten, nursery school, primary *or* grammar *or* secondary *or* high school, °institute, college, university, °seminary; *US* alma mater; boarding school, day school; public school, private school, *Brit* state school; *lycée;* Lyceum, first *or* middle school, *US* junior high school: *His mother was very upset when he hinted that he might quit school.* **2** °set, coterie, °circle, °clique, °group, °denomination, °faction, °sect, followers, devotees, adherents, votaries, disciples; °style, °kind, °form, °manner, °fashion: *He belonged to the Ashcan School of painters.* **3** °philosophy, °principles, °creed, set of beliefs, way of life, °persuasion, credo, dogma, teaching, °view, °opinion, °faction, °approach: *Regarding the creation of the universe, do you support the big-bang school or the steady-state school?* —*v.* **4** °teach, °educate, °drill, inculcate, °instill, °indoctrinate, °instruct, °tutor, °train, °discipline, °coach, °prepare, °prime, °equip, °ready; °mold, °shape, °form; school in, imbue with, infuse with: *Our students are schooled intellectually, morally, and socially to take their places in society. Young ladies were schooled in all the social graces.*

schoolbook *n.* text(book), primer, grammar (book), reader, °manual, handbook, exercise book, notebook, copybook, *Rare* enchiridion, *Old-fashioned* hornbook, abecedarium: *All I did was ask if I could carry her schoolbooks!*

schooling *n.* °education, teaching, °instruction, tutelage, °tuition, °guidance, training, °preparation, indoctrination, °edification, enlightenment; °learning, °study, °research: *How many years of schooling are needed to become a doctor?*

schoolteacher *n.* °teacher, professor, °instructor, °tutor, pedagogue, schoolmaster, schoolmistress, *Scots* dominie, *Colloq* schoolmarm: *It is not often realized what a profound influence schoolteachers have on one's entire life.*

science *n.* **1** (body of) knowledge *or* information, (body of) laws *or* principles, °discipline, °study, °branch, °field, °area, °subject, °realm, °sphere: *Many new sciences have sprung up even in the past fifty years. DNA research falls properly into the science of microbiologic genetics.* **2** °skill, °art, °technique, °expertise, °proficiency, °method, °system: *It would appear that Gorbleigh has perfected beer-drinking into a science.*

scientific *adj.* (well-)organized, (well-)regulated, (well-)controlled, (well-)ordered, °orderly, °systematic, °methodical, °precise, °meticulous, °thorough, °painstaking, °detailed: *A scientific approach to the problem would start with a complete analysis of the present conditions.*

scintillating *adj.* **1** sparkling, coruscating, flashing, °dazzling, gleaming, glittering, °twinkling, shimmering, glistening, shining, °lustrous, °radiant, effulgent, °brilliant, *Literary* nitid: *Madame Irena appeared wearing a scintillating silver lamé gown.* **2** °exciting, °engaging, °lively, °effervescent, fascinating, entrancing, °stimulating, °invigorating, °dynamic, °vivacious: *I cannot remember when I last found myself in such scintillating company.*

scoff *v.* Often, **scoff at.** deride, °belittle, °dismiss, °disparage, °mock, °make light of, °sneer (at), poke fun (at), °ridicule, spoof, °lampoon, °jeer (at), °chaff, °tease, °twit, rib, kid, *Brit* °rag: *It's easy for you to scoff at Clare's tantrums—you don't have to live with her. Don't scoff—your turn might come next.*

scold *v.* **1** °reprimand, chide, °reprove, °upbraid, °criticize, censure, °find fault (with), °rebuke, reproach, °lecture, °berate, °rate, °castigate, take (someone) to task, find fault with, rap someone's knuckles, slap someone's wrist, *Colloq* °bawl out, °dress down, give (someone) hell, give (someone) what for, °jump on (someone), jump down (someone's) throat, call (someone) to account, bring (someone) to book, let (someone) have it with both barrels, give (someone) a piece of (one's) mind, give (someone) a tongue-lashing *or* a talking-to, give (someone) a hard time, rake *or* haul (someone) over the coals, tell (someone) off, tick (someone) off, skin (someone) alive, call *or* have (someone) on the carpet, light *or* rip *or* tear *or* lace *or* sail into (someone), *US* °chew out, jump all over (someone), *Brit* carpet: *When their mother scolded them, they began to cry.* —*n.* **2** °nag, °shrew, termagant, virago, fishwife, beldam, harridan, hellcat, °fury, amazon, tigress, Xanthippe, *Colloq* battle-ax: *In the old days, they used to take scolds like you, tie them in a chair at the end of a pole, and dunk them in a pond till they cooled off.*

scoop *n.* **1** ladle, dipper, bailer, spoon: *Use the scoop to skim the fat off the soup.* **2** °exclusive: *Fergus's story on the minister's illegal business dealings was a real scoop for the paper.* **3** °(latest) news, °(inside) story, °revelation, °truth, *Colloq* °latest, °lowdown, info, °dope, *Brit* gen, *US* poop: *What's the scoop on that man seen coming out of her bedroom?* —*v.* **4** Often, ***scoop up.*** bail, °dip, ladle, spoon: *Scoop up some sand and spread it on the ice for traction.* **5** ***scoop out.*** gouge out, °excavate, spoon out, hollow out, °dig, °cut: *Scoop out some melon balls for the fruit salad.* **6** ***scoop up.*** °pick up, °gather (up), sweep up *or* together, take up *or* in: *He scooped up the money and ran out of the bank.*

scope *n.* **1** °range, °reach, °field, °area, °extent, compass, °expanse, °breadth, °sphere, °orbit, °span: *The scope of her interests includes both the sciences and the arts.* **2** °leeway, °space, °room, elbowroom, °freedom, °opportunity, °capacity, °stretch, latitude, °spread: *Have you allowed enough scope for growth?*

scorch *v.* sear, °blacken, °burn, roast, °singe, char: *Though scorched in the fire, the papers were saved intact.*

scorching *adj.* **1** °hot, °torrid, searing, parching, shriveling, tropical, hellish, sizzling, broiling, boiling, °sweltering: *This scorching heat wave can be dangerous for the very young and the elderly.* **2** °critical, °caustic, °scathing, mordant, °vituperative, excoriating, °harsh, acrimonious, °bitter: *The shadow minister for finance issued a scorching condemnation of the government's tax policies.*

score *n.* **1** °record, °account, °reckoning, °register, °tally, °amount, °number, °count, °sum, °total; °mark, °grade: *What was your best golf score? Chester got a score of 100 on the French exam.* **2** °nick, °groove, °scratch, °line, °mark, °stroke, °notch, °cut, *Archery* nock: *Every time he won, he made a score on his walking stick.* **3** twenty: *And the days of our lives shall be three score years and ten.* **4** Often, ***scores.*** dozens, hundreds, (tens *or* hundreds of) thousands, millions, number(s), drove(s), horde(s), host(s), multitude(s), herd(s), legion(s), lot(s), mass(es), myriad(s), shoal(s), pack(s), covey(s), bevy *or* bevies, swarm(s), flock(s), army *or* armies, crowd(s), throng(s): *Passengers by the score angrily protested against the airline's policy of overbooking flights. Scores of animals were fleeing before the forest fire.* **5** music, accompaniment, *Technical* full *or Technical* short *or Technical* vocal score: *Fingal might have written the libretto, but the score was by Flaherty.* **6** °situation, °story, °news, °status (quo), °condition, °word, *Colloq* °latest, °scoop, *US* poop: *What's the score on your applications to medical school?* **7** °dupe, gull, °victim, °greenhorn, *Colloq* fall guy, chump, goat, sitting duck, *Slang* °sucker, patsy, °mark, *Brit* °mug: *That fellow in the Hawaiian shirt looks a likely score for our little scam.* **8** ground(s), °basis, °account, °reason, °rationale, °provocation, °cause: *On what score can you justify eliminating him from the competition?* **9** ***settle*** *or* ***pay off*** *or* ***even a score*** *or* ***the score*** *or* ***old scores.*** get revenge, °retaliate, get even, avenge, °repay, get an eye for an eye, give tit for tat, give measure for measure, pay (someone) back in his own coin, give (someone) a taste *or* a dose of his *or* her *or* their own medicine, *Colloq* get one's own back: *I wish I could find where MacTavish moved as I have an old score to settle with him.* —*v.* **10** °mark, °line, incise, °scratch, °nick, °notch, °cut, °groove, °graduate; °scrape, °deface, °mar, °gouge, °slash: *The dipstick is scored at intervals to indicate how much oil is in the crankcase. Stop dragging that chest—you're scoring the polished floor!* **11** gain *or* make a point *or* points, °record, °tally, °account for: *George scored a hat trick in the game against the Whalers.* **12** °count (for), reckon for *or* as: *The ace scores either 1 or 11 in blackjack.* **13** °succeed, be successful, °triumph, °win, make an impression, have an impact, *Colloq* make a hit: *Your plan really scored with the boss.* **14** succeed in seducing, *Slang* °make out, get laid: *Did you score with Ava last night?*

scorn *n.* **1** contumely, °contempt, contemptuousness, disdain, deprecation; °rejection, °dismissal: *He treated their demands with the utmost scorn.* **2** °mockery, °derision, derisiveness, sneering, °ridicule, scoffing, jeering, taunting: *The crowd's scorn was directed at the candidate who failed even to promise them benefits, no matter how unrealistic.* —*v.* **3** °reject, °rebuff, disown, disavow, °disregard, °ignore, °shun, snub, °flout, contemn, treat with *or* hold in contempt, °have no use for, disdain, °spurn, °despise, turn up one's nose at, curl one's lip at, look down on *or* upon, look down one's nose at, °thumb one's nose at, *Colloq* pooh-pooh, °put down, *Brit* cock a snook at: *His parents offered to finance his education, but he scorned their help.* **4** °mock (at), °deride, sneer at, °ridicule, scoff at, °jeer at, °taunt, °poke fun at, °make fun of, °laugh at: *People scorned him because he said the earth was round.*

scornful *adj.* contumelious, °contemptuous, °disdainful, deprecative, disparaging, °derisory, derisive, snide, °supercilious, mocking, sneering, scoffing, °haughty, overweening, highhanded, °superior, *Colloq* snooty, *Slang* snotty: *You needn't be so scornful of my attempts—I didn't see you do much better.*

scoundrel *n.* °villain, °rogue, °wretch, °good-for-nothing, scapegrace, blackguard, °rascal, scamp, cur, *Old-fashioned* bounder, cad, knave, *Colloq* °heel, *Slang* louse, *Brit* rotter: *The scoundrel not only ran off with my wife but then had the effrontery to desert her and send her back to me!*

scour *v.* **1** °scrub, °clean, °cleanse, °wash, °rub, abrade, °polish, burnish, buff, °shine: *The pots were scoured until they shone.* **2** scrape (about *or* around), °rake, comb, turn upside down, °search, °ransack: *I scoured the shops looking in vain for the coffee filters you wanted.*

scourge *n.* **1** °curse, °misfortune, bane, °evil, °affliction, °plague, adversity, °torment, °misery, °woe: *He questions whether AIDS should be considered the scourge of the twentieth century as the plague was that of the seventeenth. Was it Attila who proved himself the scourge of Rome?* **2** °whip, °lash, cat-o'-nine-tails, knout, quirt, horsewhip, bullwhip: *They saw fit to punish themselves with scourges.* —*v.* **3** °whip, °flog, °beat, °lash, horsewhip, whale, °belt, flagellate: *Members of some sects scourged themselves as they wandered through the medieval cities.* **4** °punish, °castigate, °chastise, °discipline, °afflict, °torment: *It is hard to believe that there are some schools that still condone the scourging of students.*

scout *v.* **1** Often, ***scout about*** *or* ***around.*** °reconnoiter, °investigate, °study, °research, °examine, °explore, °spy, search *or* look (about *or* around) (for), hunt (about *or* around) (for), cast around *or* about (for), *Colloq* check (about *or Colloq* around): *I am going to scout about to find a key that fits.* **2** ***scout up*** *or* ***out.*** °discover, °find, °locate, uncover, *Colloq* °dig up: *I was finally able to scout up a key that fits.*

scowl *v.* **1** glower, °frown, grimace, °glare, look daggers, lower *or* lour: *It's hard to tell if he's ever pleased because he's always scowling.* —*n.* **2** °frown, grimace, °glare, dirty look: *I do wish she would wipe that scowl off her face and try to look more pleasant.*

scramble *v.* **1** °climb, clamber, °crawl, scrabble, °struggle: *We scrambled up the rocky hillside with the big cat in pursuit.* **2** °rush, °hurry, scamper, °run, °hasten, °race, °scurry, scoot, scuttle, °dash, hotfoot (it), °hustle, *Chiefly US and Canadian,* hightail (it), *Colloq* skedaddle: *When the siren sounded, we scrambled to the air-raid shelter.* **3** Often, ***scramble up.*** °mix up, °confuse, °jumble, intermingle, °mingle, commingle: *Scramble up the numbered slips in a bowl, and whoever holds the number drawn is the winner.* —*n.* **4** °scrimmage, °struggle, tussle, contention, °clash, °competition, °contest, °race, °rush, °clash, °conflict, °disorder, commotion, °riot, melee *or* mêlée, °pandemonium, *Colloq* free-for-all, hassle, *Brit* scrum *or Brit* scrummage: *Thousands were involved in the scramble for tickets to the rock concert. In the scramble for a directorship Jeremy lost.* **5** °struggle, °climb: *I twisted my ankle in the scramble over the rocks.*

scrap[1] *n.* **1** mite, °bit, °shred, °bite, °morsel, °piece, °fragment, shard *or* sherd, °particle, °sliver, snippet, °snip, °crumb, whit, iota, jot *or* tittle, °snatch, °drop, °drip, °grain, °speck, molecule, atom, °dab, °trace, scintilla, °hint, °suggestion: *There wasn't a scrap of evidence to link him with the crime.* **2** ***scraps.*** remnants, °remains, leftovers, leavings, °residue, vestiges, °traces, scrapings, discards, rejections, rejects: *In the alley, two cats fought over some scraps of garbage.* **3** °waste, debris, °rubbish, *Colloq* °junk: *Any computer more than five years old isn't even worth its weight as scrap.* —*v.* **4** °discard, °throw away, °reject, °abandon, °give up, consign to the scrapheap, °forsake, °forget, °get rid of, °dispose of, °dispense with, *Colloq* °junk, *US* °trash: *If I were you, I would buy a new car and scrap this one.*

scrap[2] *n.* **1** °fight, °brawl, °fracas, °fray, affray, °rumpus, scuffle, donnybrook, °battle (royal); °row, °dispute, °argument, °quarrel, °disagreement, wrangle, squabble, °tiff, spat, *Colloq* ruckus, set-to, dust-up: *A terrible scrap broke out in the pub last and the police were called in. Our neighbors were having a scrap about who should take the dog for a walk.* —*v.* **2** °fight, °brawl, °spar, scuffle, °battle, °row, wrangle, °argue, °disagree, squabble, °bicker: *Let's not scrap over trifles—only important things.*

scrapbook *n.* album, portfolio, °collection: *She keeps a scrapbook of reviews of her stage triumphs.*

scrape *v.* **1** abrade, graze, °scratch, bark, scuff, °skin, °bruise, °damage, °injure: *I scraped my knee on the pavement when I fell.* **2** Often, ***scrape off*** *or* ***away*** *or* ***out.*** °remove, rub off *or* away, scour *or* scrub *or* clean (off *or* away), scratch off *or* away, claw (at *or* away *or* out), gouge out, scrabble (at), dig out *or* away at: *As I scraped away the grime, a beautiful mosaic was revealed. Each sled dog scraped out a hole in the snow and snuggled into it.* **3** skimp, scrimp (and scrape), °save, °stint, be frugal *or* stingy *or* parsimonious *or* thrifty, pinch and save *or* scrape, °economize; °struggle, *US* scrabble: *Only by scraping were we able to pay the increased taxes.* **4** ***bow and scrape.*** make obeisance, °kowtow, salaam, genuflect, kiss the feet *or* hem *or* ring, grovel, demean *or* lower oneself, °prostrate oneself, toady, *Colloq* bootlick: *He can bow and scrape all he likes, but Stratton will get nowhere with the foreman.* **5** ***scrape by*** *or* ***through.*** °get by, °cope, °(barely) manage, °survive, scrape *or* get along, *Colloq* squeak by, barely make it: *We are just about able to scrape by on the little that Nigel is making.* **6** ***scrape together*** *or* ***up.*** glean, °garner, scratch *or* get *or* rake together *or* up, dredge up, scrabble for, °gather, °save (up), get hold of, marshal, °amass, °muster, °accumulate, aggregate, °compile, °pile up, °stack up, °assemble, *Colloq* °scrounge (up): *By the time we finally managed to scrape up enough for a down payment on a house, the prices had gone up.* —*n.* **7** abrasion, °bruise, °scratch, graze, scuff, °damage, °injury: *Who is responsible for this scrape on my new car?* **8** °predicament, °difficulty, °quandary, °dilemma, °plight, (fine) kettle of fish, °muddle, °stew, °situation, °position, °pinch, *Colloq* pickle, °fix, °mess, the crunch, (tight *or* tough) spot: *I got into a terrible scrape by forgetting our wedding anniversary. Can you lend me some money to help me out of a scrape?*

scratch *v.* **1** °mar, °mark, °gouge (out), °gash, abrade, graze, scuff, grate against, °bruise, °damage, °injure; °claw: *If you glue some felt to the bottom, the lamp won't scratch the table.* **2** °chafe, °rub: *If you scratch those mosquito bites you might get an infection.* **3** Often, ***scratch out*** *or* ***off.*** °erase, °obliterate, rub out *or* off, °cross out, °delete, strike out *or* off, expunge; °exclude, °eliminate, *US* x out: *You can scratch my name from the list of candidates. She scratched out what she had written and started again.* —*n.* **4** °mark, °gouge, °gash, abrasion, °scrape, graze, scuff, °bruise, °damage, °injury; °line: *Where did you get that scratch on your face? "It's only a scratch," Sandy protested as they tried to remove the spear from his shoulder.* **5** ***up to scratch.*** up to standard *or* par, °adequate, °sufficient, good enough, °competent, °satisfactory, competitive, *Colloq* up to snuff: *Cynthia's performance is not yet up to scratch, so she might not make the team.* —*adj.* **6** °hasty, °hurried, impromptu, unplanned, °haphazard, °rough, °casual, °informal, °unprepared, °unpremeditated, °makeshift, extempore, *Colloq* off the cuff, *US* pickup: *We didn't have time to practice, so we are racing with a scratch crew.*

scratchy *adj.* **1** itchy, irritating, °prickly: *That pullover is as scratchy as a hair shirt.* **2** °rough, hoarse, °raw, °grating, °sore, raspy, °dry: *Eating dry biscuits made my throat a bit scratchy and I need something to drink.*

scrawl *n.* **1** scribble, *Colloq* squiggle, *US* hen-scratch, chicken-scratch: *I can't read that scrawl of his.* —*v.* **2** scribble, °scratch, doodle: *How do pharmacists read the prescriptions that doctors scrawl for you?*

scrawny *adj.* bony, °skinny, °spare, °drawn, reedy, °haggard, °lean, lank(y), scraggy, °gaunt, °rawboned, angular, °emaciated, cadaverous; anorectic *or* anorexic:

She certainly has changed from that scrawny kid I used to know.

scream *v.* **1** °shriek, screech, squeal, yowl, wail, caterwaul, °howl, °cry: *She screamed when I applied iodine to the cut.* **2** °laugh, °roar, hoot, °howl, guffaw: *The women screamed hysterically as the male stripper got down to the bare essentials.* —*n.* **3** °shriek, screech, squeal, yowl, wail, caterwaul, °howl, °cry: *When she saw what had happened, she let out a bloodcurdling scream.* **4** *Colloq* °card, °panic, °riot, thigh-slapper: *Phyllis is such a scream when she does her imitations.*

screen *n.* **1** °partition, (room) divider, paravent, °wall; *Theater* tormentor: *In our one-room apartment a bookcase serves as a screen to separate her part from mine.* **2** °shelter, °protection, °shield, °cover: *A row of poplars acts as a screen against the wind.* **3** curtain, °blind, °shroud, °cloak, °cover; concealment, °camouflage: *Gauze fabric is too transparent to act as a screen.* **4** sieve, °mesh, strainer, °filter, colander, °riddle: *The sand has to be shoveled through a screen to get rid of the stones.* **5** motion pictures, °movies, silver screen; °television, small screen, home screen, *Colloq* °box, telly, *US* boob tube: *She is a star of both the big and small screens.* —*v.* **6** °partition (off), °separate, °divide, °wall off: *Why not use a beaded curtain to screen the kitchen area from the dining room?* **7** °shelter, °protect, °shield, °cover, °guard, °conceal, °camouflage, °mask, °veil, °hide: *A decorative pierced wall screens the ladies of the harem from the prying eyes of visitors.* **8** °sift, °separate (out), °sort (out), °filter, °select, cull, °process, °interview, °evaluate, °grade, °gauge, °qualify, °examine, °scan, *Chiefly Brit* °vet: *Our agency screens all job candidates before sending them to see a client.*

screw *n.* **1** °bolt, screw bolt, machine screw, lag bolt, lag screw: *This screw is too small to hold that door hinge.* **2** helix, °spiral, corkscrew: *The spiral staircase should not be too small for safety.* **3** sexual intercourse; sexual partner, *Slang* °lay, *Taboo slang* fuck: *I heard that he was out in bars all night looking for a screw.* **4** ***put the screws on (someone).*** °pressure, °influence, °force, constrain, °press, °oblige, °require, °demand, coerce, compel, apply pressure, bring pressure to bear (on); °insist; *Chiefly Brit* pressurize, *Colloq* twist (someone's) arm, °put the squeeze on (someone): *I believe that someone put the screws on him to leave town.* —*v.* **5** °twist, °turn, °rotate: *Screw the bolt in clockwise.* **6** Often, ***screw out of.*** °defraud, °cheat, °swindle, gull, bilk, do out of, *Slang* °take, °clip, °fleece: *They screwed him by convincing him that the bonds were real. He was screwed out of all his savings.* **7** ***screw up.*** **(a)** °raise, °increase, °stretch, °strain; °summon, °call up, °call upon, °tap, draw on *or* upon: *I screwed up my courage to ask for a raise.* **(b)** °ruin, °destroy, make a mess of, °botch, °bungle, °muddle, mismanage, °mishandle, *Colloq* make a hash of, *Slang* louse up, *US* bollix up, *Brit* make a muck-up of, *Taboo slang* fuck up, *Brit* °bugger up, ballocks *or* bollocks up, make a balls-up of: *It is hard to see how they could have screwed up such a simple operation.* **(c)** contort, °twist, deform, °warp: *At the suggestion of cauliflower, Ambrose screwed up his face in displeasure.*

scribe *n.* **1** copyist, copier, transcriber, *Archaic* scrivener; amanuensis, clerk, secretary: *Before the invention of movable type printing, copies of books and documents were prepared by scribes.* **2** °writer, °author, penman, scrivener, wordsmith, scribbler, °hack, penny-a-liner; °dramatist, dramaturge, °playwright, °poet, novelist, essayist, columnist, technical writer; °journalist, gentleman *or* lady of the press, newspaperman, newspaperwoman, °reporter, rewrite man *or* woman *or* person, °editor, reviewer, commentator, newswriter, sob sister, agony aunt *or* uncle, Miss Lonelyhearts, gossip columnist, member of the fourth estate, *Brit* paragraphist, leader writer, *US* Grubstreeter, *Colloq Brit* journo: *In a general way, he refers to all writers as "scribes," but particularly those he considers a bit old-fashioned.* —*v.* **3** inscribe, incise, °etch, °engrave, °mark, °scratch, °score, °grave, scrimshaw, °carve, chase *or* enchase: *The shotgun was scribed with the most beautiful designs, some inlaid in gold.*

scrimmage *n.* °skirmish, scuffle, °fray, affray, °disturbance, brouhaha, melee *or* mêlée, °riot, °row, °brawl, °struggle, °scramble, tussle, °fracas, °rumpus, donnybrook, °battle, °fight, *Colloq* ruckus, set-to, dust-up, free-for-all, °scrap, *Brit* scrum *or* scrummage, *Slang Brit* (bit of) bovver: *The police were called when what started as a minor scrimmage began to develop into a riot.*

script *n.* **1** handwriting, °hand, °(cursive) writing, °penmanship; calligraphy: *These love letters are written in an unusually beautiful script.* **2** manuscript, °scenario, °book, °play, screenplay, teleplay, libretto, continuity: *Stick to the script—I don't want you ad-libbing!* —*v.* **3** °write, °pen, °prepare, °create: *The new series was scripted by the same writer who wrote* Back Out in the Outback. **4** °plan, °organize, °design, °arrange, °lay out, °order, configure, °pattern: *The takeover of your company has not been scripted in our business forecast scenario for the coming year.*

Scripture *n.* Scriptures, sacred writings, Bible, Good Book, Holy Writ *or* Scripture(s), Word of God, Gospel(s); Book of Mormon; Koran; Upanishad(s), Bhagavad-Gita: *Throughout his life, he did only those things approved in Scripture.*

scrounge *v.* **1** ferret out, seek out, nose *or* smell out, come up with, scrape together *or* up, scratch up; importune, cadge, °beg, (borrow, or steal), *Colloq US* freeload, °bum: *I have to scrounge up the rent money somewhere. Did Hedley try to scrounge some money from you, too?* —*n.* **2** scrounger, cadger, °parasite, *Colloq* sponger, *US* freeloader: *Hedley, that scrounge, has just asked me for a cigarette.*

scrub *v.* **1** See **scour, 1,** above. **2** °cancel, °call off, abort, °scratch, °drop, °terminate, °give up, °end, °abandon, °stop, °cease, °discontinue, do away with: *Something went wrong with the fuel computer, so they scrubbed today's satellite launch.*

scruple *n.* **1** °compunction, °qualm, °reluctance, °misgiving, second thoughts, °doubt, °(twinge of) conscience, hesitation, uneasiness, °discomfort, squeamishness: *Pendleton hadn't the slightest scruple about taking out my wife.* —*v.* **2** °pause, falter, °hesitate, vacillate, have doubts *or* compunctions (about), demur, waver, shrink from *or* at, have misgivings *or* qualms (about *or* over), be loath *or* loth (to), think twice (about), °stick at, be reluctant, balk (at), have scruples (about): *Susan didn't scruple for a moment about keeping the money she found in the street.*

scrupulous *adj.* **1** °careful, °cautious, °meticulous, °exacting, °precise, overnice, °strict, °rigid, rigorous, °severe, °critical, °fastidious, °neat, °conscientious, finicky *or* finical, °fussy, °painstaking, punctilious: *Andrew has always been scrupulous about his toys, his books, his clothes, etc.* **2** °ethical, °honorable, upstanding, °moral, °righteous, °principled, high-minded, °just: *Meg was always scrupulous in her business dealings.*

scrutinize *v.* °examine, °analyze, dissect, °investigate, °probe, °study, °inspect, °sift, go over *or* through, °check: *I have scrutinized the results of the blood tests and can find nothing abnormal.*

scrutiny *n.* °examination, °analysis, °investigation, °probe, probing, °study, inspection, sifting, inquiry *or* enquiry, °exploration, °check: *The tax inspectors have* *subjected the company books to close scrutiny but have found nothing amiss.*

scud *v.* °fly, °skim, °race, scoot, °speed, °shoot: *We scudded along in the catamaran at a good thirty knots.*

sculpture *n.* **1** °figure, figurine, °statue, statuette, °group, °head, bust, °relief; bronze, marble: *His favorite sculpture is Rodin's* Burghers of Calais. —*v.* **2** sculpt *or* sculp, °model, °chisel, °carve, °cast, °form, °fashion: *The students are learning how to sculpture in clay.*

scurrilous *adj.* foulmouthed, thersitical, °gross, °indecent, °profane, Fescennine, °vulgar, °obscene, licentious, Sotadean *or* Sotadic, °foul, °vituperative, °low, °coarse, scabrous, °vile, °nasty, defamatory, °derogatory, disparaging, vilifying, calumnious *or* calumniatory, malign, aspersive, opprobrious, °offensive, °abusive, insulting: *Scurrilous humor, no matter how hilarious, seems inappropriate when delivered by a comedienne. Clifford was an inveterate collector of scurrilous limericks.*

scurry *v.* °dash, °scramble, scamper, scoot, dart, °fly, °race, sprint, scuttle, °hurry, °hasten, °speed, °hustle, °rush, °tear, zoom, zip, °bolt, °rip, °scud: *A rat scurried across the floor. You had best scurry if you expect to get to the store before it closes.*

scurvy *adj.* °low, °miserable, °contemptible, °vile, °base, °despicable, °rotten, °sorry, °bad, ignoble, °dishonorable, °mean, °worthless, °shabby: *Get out of here, Egan, and take your scurvy crew with you. Robbing poor old ladies is a pretty scurvy activity.*

sea *n.* **1** °ocean, deep blue sea, high seas, *Literary* °(briny) deep, °(bounding) main, Neptune's *or* Poseidon's kingdom *or* domain, *Nautical* blue water, Davy Jones's locker (= 'bottom of the sea'), *Colloq* briny, °drink, pond (= 'Atlantic Ocean'): *For years he sailed the seas in search of adventure.* **2** °swell, breaker, °wave: *As the wind increased, huge seas began to wash over the boat.* **3** plethora, °quantity, °abundance, °surfeit, °profusion, °flood, multitude, °spate, legion, °mass, *Colloq* lot(s), heap(s), pile(s), ton(s), mountain(s), load(s), oodles, gobs, scads: *Hamlet's sea of troubles was nothing compared to mine if this book isn't completed soon!* **4** ***(all) at sea.*** °confused, disoriented, at sixes and sevens, bewildered, perplexed, baffled, mystified, °lost, adrift: *They were clearly not expecting a woman and were completely at sea as to what to do.*

seacoast *n.* seashore, shore, °coast, seaside, seaboard, shoreline, coastline, littoral, sand(s), °beach, strand: *The people in the villages along the seacoast depend mainly on fishing for their livelihood.*

seafaring *adj.* maritime, °nautical, naval, °marine: *Britain was at one time the largest seafaring nation in the world.*

seal *n.* **1** °symbol, °token, °mark, insigne (*plural* insignia), °sign, signet, °crest, °bearing, coat of arms, escutcheon, °emblem, badge, monogram, °identification, cartouche, °design, imprint, °stamp: *The seal on his ring showed a crouching lion.* **2** authentication, confirmation, verification, validation, affirmation, attestation, ratification, corroboration, °assurance, guarantee *or* guaranty, °endorsement, substantiation, °evidence, °notice, notification: *The company sets its seal of approval on all its goods.* —*v.* **3** Sometimes, ***seal off*** *or* ***up.*** close (off *or* up), °shut (off), zip up, °plug (up), °stop (up), °lock, °bolt, °secure, batten down, make airtight *or* waterproof; cork: *All exits have been sealed off. Don't seal up the windows so that no air can get in.* **4** °authenticate, °confirm, °verify, validate, affirm, °attest, °ratify, °clinch, corroborate, °assure, °ensure, °guarantee, °endorse: *The peace treaty was sealed by the marriage of the king to the emperor's daughter.*

seam *n.* **1** °junction, °juncture, °joint, suture, *Technical* commissure; °scar, °ridge, °line, cicatrix: *It is easy enough to see the seam where the parts meet.* **2** lode, °vein, °stratum, bed, layer, thickness: *The miners have been working on the new seam discovered at the end of Tunnel 4.*

seamy *adj.* °sordid, °nasty, °dark, °disreputable, °shameful, °unwholesome, °unpalatable, °unsavory, °distasteful, °unseemly, squalid, °low, depraved, °degenerate, degraded, °foul, °vile, odious, °abhorrent, °contemptible, °scurvy, °rotten, unattractive, °ugly, °repulsive, °repellent: *During his later years, he preferred to paint the seamy side of life in the slums.*

search *v.* **1** Often, ***search through.*** °examine, °scrutinize, °check, comb (through), °explore, °go through, °investigate, °scout out, °inspect, look at *or* into, °probe, °scour, sift through, pry into, hunt *or* rummage through; inquire *or* enquire of, *Colloq* plow through: *I searched every book on the shelf, but there was no will. Search your conscience to determine the best course.* **2** Often, ***search for.*** look (about *or* around), cast about, °seek, leave no stone unturned: *I searched high and low but couldn't find my key.* —*n.* **3** °hunt, °pursuit, °quest: *The search for the killer went on for years.* **4** researching, °analysis; °exploration, °examination, °scrutiny, °probe, °study, °perusal, sifting, inspection, scouring, inquiry *or* enquiry: *A thorough search of the documents failed to turn up anything useful. The police made a house-to-house search of the neighborhood.*

searchingly *adv.* penetratingly, piercingly, °intently, °deeply, fixedly, concentratedly, eagerly: *She looked searchingly into my eyes for some glimmer of hope.*

season *n.* **1** °time, °period, °occasion, °opportunity: *This is the season when the birds migrate north. Is the silly season upon us again so soon?* **2** ***in season.*** °ripe, °ready, °edible, °seasoned, °seasonable, °available: *Fresh strawberries will be in season soon.* —*v.* **3** °spice, °salt, °flavor, °pep up, °enliven: *One of the great things about chicken is that you can season it in many different ways.* **4** °ripen, °mature, °age, °condition, °mellow: *The wood should be well seasoned before being made into furniture.*

seasonable *adj.* °appropriate, °opportune, °suitable, apt, °timely, °fitting, °providential, °well-timed, °proper, °fit, °propitious, °welcome, well-suited, °happy, °lucky, °fortunate, °convenient, auspicious, °favorable, °advantageous, °expedient: *The success of the book was owing largely to its seasonable publication, just at the end of the cold war.*

seasoned *adj.* **1** °experienced, trained, long-standing, long-serving, practised, well-versed, habituated, acclimatized *or* acclimated, °accustomed, familiarized, °prepared, established, °veteran, tempered, hardened, toughened, inured: *It was a pleasure to be working with so seasoned a performer as Margie.*

seasoning *n.* °spice, °zest, °flavor, °relish, °sauce: *The fact that a film star was running for office lent just the right seasoning to the campaign.*

seat *n.* **1** °place, °chair, bench, sofa, settee, °settle, stool, throne: *I found a seat and waited to be called.* **2** °focus, °base, °center, °heart, °hub, °site, °capital, cradle, headquarters, fountainhead: *In those days, the city was the seat of learning in western Europe. Mehmed established Constantinople as the seat of the Turkish empire.* **3** membership, °position, incumbency: *His grandfather held a seat in the Senate.* **4** °bottom, °buttocks, posterior(s), rump, hindquarters, fundament, derrière, *Colloq* behind, °butt, backside, °rear (end), *Brit* °bum, *US* fanny, tushie, tush, tokus, hinie, *Slang* °tail, *Brit* arse, *US* ass: *He needs a swift kick in the seat to get him to move.* **5** °abode, °residence, °home, °domicile, °estate, mansion: *Lady Blaunder spends the summer at her country seat near Shortley.* —*v.* **6** °hold, °accommodate, have room *or* space *or* capacity for, °contain, °sit: *The new auditorium will seat more than four thousand.* **7** install *or* instal, enthrone, ensconce, instate, °invest, °establish, °place, swear in: *Once Karpoff has been seated, it will be difficult to get rid of him.*

seating *n.* °accommodation, °capacity, °space, °room: *At present, the theater has seating for only five hundred.*

secede *v.* withdraw *or* resign *or* retire (from), °abandon, °forsake, apostasize, break with *or* away (from), drop *or* pull out (of), turn one's back to *or* on, °quit, separate from, °leave, °wash one's hands of, have nothing further to do with: *The City of New York occasionally threatens to secede from the United States.*

secession *n.* withdrawal, seceding, defection, °break, breaking, disaffiliation, retirement, °separation, splitting off *or* away, apostasy: *The secession of the splinter group was welcomed by virtually all the other members.*

secluded *adj.* **1** °private, °separate, °isolated, °lonely, cloistered, sequestered, °detached, °solitary, retired, eremitic, monastic: *He lives a very secluded life on an island in the Outer Hebrides.* **2** off-the-beaten-track, °out-of-the-way, °remote, °faraway, far-off, °separate, segregated, °private: *He chose a secluded island in the Outer Hebrides to live.*

seclusion *n.* °privacy, °private, °separation, isolation, loneliness: *He prefers to live in complete seclusion.*

second[1] *adj.* **1** °subsequent, following, next: *Let the first one go by and aim for the second target.* **2** °subordinate, next: *If the colonel isn't here, who is the second officer?* **3** °alternative, second-best: *I refuse to settle for second choice.* **4** °alternate, other: *The laundry is picked up and delivered every second Tuesday. His second language is Choctaw.* **5** other, later, younger, newer, more recent: *After his first wife died, he took her nurse as his second wife.* **6** another, °duplicate: *The boy is a second Einstein.* —*n.* **7** defective *or* imperfect *or* damaged *or* faulty *or* deficient *or* flawed *or* impaired *or* marred *or* blemished *or* bruised *or* inferior merchandise: *The prices of these shirts have been reduced because they are seconds.* **8** °subordinate, °assistant, number two, lieutenant, aide-de-camp, man Friday, girl Friday, right hand; °understudy, °stand-in, °substitute, surrogate, °double, °alternate, °backer, °supporter; *Colloq US* gal *or Colloq US* guy Friday: *Not being ambitious, Helen would prefer to be a second to the person who runs things.* —*v.* **9** °support, °back, °aid, °help, °assist, °approve (of), °advance, °promote, °subscribe to, espouse, °sponsor, °patronize, °favor, °encourage, °go along with: *Clara is inclined to second most of her husband's decisions.* **10** °transfer, °move, °assign, °shift, relocate: *After the Gulf War, he was seconded to fleet headquarters for a year.* —*adv.* **11** secondly, in the second place, secondarily, (number) two, b *or* B: *First, you look great; second, I love you.*

second[2] *n.* °moment, °instant, °flash, °minute, twinkling *or* wink *or* bat (of an eye), split second, *Colloq* sec, jiffy, two shakes (of a lamb's tail), *Brit* tick, half a mo: *I'll be with you in a second, as soon as I finish writing this letter.*

secondary *adj.* **1** less important, unimportant, inessential *or* unessential, °nonessential, noncritical, °subsidiary, ancillary, °minor, °inferior, °subordinate: *These secondary matters can be dealt with after the important ones.* **2** °derivative, derived, °indirect, °secondhand, unoriginal, not original; copied, imitated: *The biography was based entirely on secondary sources.* **3** °auxiliary, second-line, backup, °extra, °reserve, °spare, °provisional, supporting, °supportive, °alternate, °alternative: *We were forced to rely on our secondary line of defenses after their cavalry broke through.*

secondhand *adj.* °used, °old, °worn, *Colloq* hand-me-down: *When we were children our parents could only afford to dress us in secondhand clothes. All of Clumpett's inspirations are based on secondhand ideas.*

secrecy *n.* **1** °mystery, concealment, confidentiality, °stealth, secretiveness, surreptitiousness, °privacy, furtiveness, covertness, clandestineness: *Why was there so much secrecy surrounding the publication date of Jennie's book?* **2 *in secrecy.*** °secretly, mysteriously, confidentially, stealthily, secretively, surreptitiously, privately, furtively, covertly, clandestinely, sneakily: *They carry on much of their business in secrecy.*

secret *adj.* **1** concealed, °hidden, °private, covert, shrouded, clandestine; °confidential, covert, °quiet, under cover, °secretive, unpublishable, unpublished, *Colloq* hush-hush: *There is a secret passageway leading from the cave to the cove. Keep secret what I shall tell you. She revealed to me her most secret desires.* **2** °cryptic, °private, arcane, °mysterious, °incomprehensible, esoteric, °recondite, abstruse; cryptographic, encrypted, encoded: *We used to communicate by secret code, which one needed a special decoder ring to interpret. Did you receive my secret message?* —*n.* **3** private *or* confidential matter *or* affair, °mystery: *If I tell you my secret will you tell me yours?* **4 *in secret.*** privately, confidentially, °secretly, on the Q.T.; surreptitiously, under cover, by stealth, stealthily, furtively, °quietly, °on the sly, clandestinely: *The office manager told me in secret that she is the boss's girl friend. They see one another in secret.*

secrete[1] *v.* °hide, °conceal, °cache, °bury, °cloak, °shroud, enshroud, °camouflage, °mask, °disguise, *Slang* stash away: *They secreted the haul from the robbery in a tomb.*

secrete[2] *v.* °yield, excrete, °pass, °generate, °release, °ooze, seep, exude, °discharge, °leak, °drip, °drop, dribble, °trickle, °run, °drain, °emit, °give off, °emanate, transude, *Technical* extravasate: *The ants keep aphids for the substance they secrete. The wound will secrete serum for a while.*

secretion *n.* secreting, °release, °escape, oozing, seeping, seepage, °discharge, discharging, °leak, leaking, leakage, °drip, dripping, °drop, dropping, dribbling, trickling, °trickle, °running, °drain, draining, emission, emitting, giving off, exudation, transudation, excretion, excreting, emanation, emanating, °generation, *Technical* extravasation; transudate, excreta, *Technical* extravasate: *The secretion of perspiration may be profuse under such circumstances. The secretion is highly acid in content.*

secretive *adj.* °reticent, °silent, closemouthed, °taciturn, uncommunicative, °reserved, °tight-lipped, °close, *Colloq* °mum: *Robert is very secretive about where he goes every day at lunchtime.*

secretly *adv.* surreptitiously, °quietly, privately, covertly, on the Q.T., furtively, stealthily, mysteriously, clandestinely, °in secret, confidentially, °on the sly, slyly, sub rosa, sub sigillo: *I can tell you secretly that she is wanted by the police. Henry secretly rents videotape cartoons to watch on weekends.*

sect *n.* **1** religious order *or* group *or* denomination *or* body *or* cult *or* persuasion *or* subdivision: *At times it seems as if there are as many sects as worshipers.* **2** °school (of thought), °faction, ism, °set, °clique, °cabal: *He belonged to a small sect that promulgated the Ptolemaic cosmogony.*

sectarian *adj.* **1** cultist, cultish, clannish, cliquish, °partisan, °partial, °dogmatic, doctrinaire, factional: *They have seldom mingled with others, maintaining their sectarian existence.* **2** °parochial, °narrow, °narrow-minded, °limited, insular, °provincial, °rigid, fanatic(al), °prejudicial, °prejudiced, °bigoted: *Because they refused to consider ideas other than their own, they gradually became more and more sectarian in their views.* —*n.* **3** adherent, °member, sectary, votary, cultist, °partisan: *They tried to banish the Anabaptists, Lutherans, Calvinists, and other sectarians.* **4** (true) believer, dogmatist, °fanatic, bigot, °zealot, extremist, *Slang* nut, °bug, °fiend: *Some of the sectarians engage in the ritualistic handling of live rattlesnakes.*

section *n.* **1** °part, °division, °department, °branch, sector, °group, °detachment, °segment, °portion, subdivision, component, °element: *Sadler has been assigned to work in the research section. Sorry, the first thing I do is throw away the sports section. Which sections of the country will vote for you? The string section needs practice before tonight's concert.* **2** °sample, °slice, cross section, fraction: *It can be faulty to assume that the characteristics of the whole are necessarily represented in a small section of it. I examined a section of tissue under the microscope.* **3** °part, °stage, °segment, °portion, °leg: *The first section of the journey was more comfortable than the last.* —*v.* **4** °cut (up), °divide (up), °segment, °split, °cleave, °measure out, apportion, °allot, allocate: *The land was sectioned into four-acre parcels.*

secular *adj.* °worldly, °terrestrial, mundane, °temporal, °material, °lay, laic *or* laical, nonclerical, nonecclesiastic(al), nonspiritual, nonreligious, °civil, °state: *Once he had taken his vows, he put aside secular matters.*

secure *adj.* **1** °safe, shielded, sheltered, protected, °immune, unthreatened, unexposed, unimperiled, °snug, °cozy: *Fiona feels quite secure in her new house. Considering the success of the company, Bill's job looks secure.* **2** °firm, °steady, °stable, °fixed, °fast, moored, anchored, °immovable, closed, °shut, fastened, locked (up), °tight, °sound, °solid, °sturdy, °strong: *That button on your jacket doesn't look very secure. Make sure that the house is secure before you go to sleep.* **3** °reliable, °safe, °good, °profitable, °healthy, °solid: *Telephone shares looked like a secure investment at the time.* **4** °sure, °certain, assured, ensured, °definite, °inevitable, °assumed, °evident, °obvious, °unquestionable, established, °probable, °easy: *With only two minutes left to play, the Rangers' victory seemed secure.* —*v.* **5** °obtain, °get (hold of), °come by, °acquire, °procure, °win; °gain, get *or* take possession of, arrogate: *Tim has secured a responsible position at the bank. She always manages to secure the sympathy of some men.* **6** °guarantee, °underwrite, hypothecate, collateralize: *We used our house to secure the loan.* **7** °protect, °shelter, °shield, °defend, °guard, °safeguard, °preserve: *It seems impossible to secure the walls of public buildings from the ravages of graffiti artists.* **8** °fasten, make fast, °fix, affix, °attach, °anchor: *The sconce is firmly secured to the wall now and will not fall off again.*

security *n.* **1** °safety, °shelter, °protection, fastness, °refuge, °safekeeping, °sanctuary, asylum: *During the air raids, we retired to the security of the basement.* **2** °confidence, °certainty, surety, °assurance, °conviction: *He has the security of knowing that he is right in this instance.* **3** guarantee *or* guaranty, collateral, °deposit, gage, °pledge, °insurance: *What are you going to offer as security for the loan?* **4** °surveillance, safeguarding, guarding, °safekeeping, °protection, °custody, custodianship, °care: *What plans does the company have for the security of the office building?*

sedate *adj.* **1** composed, °serene, °peaceful, °calm, °tranquil, °cool, °collected, even-tempered, °detached, imperturbable, unruffled, undisturbed, unperturbed, controlled, placid, °grave, °serious, °sober, °solemn, *Colloq* unflappable: *Despite all the turmoil round her, Shari remained quite sedate.* **2** °dignified, °decorous, °refined, °formal, °stiff, °staid, °proper, °strait-laced, °prudish, °fussy, prim, °conventional, °old-fashioned: *Donald preferred a sedate waltz to boogieing and jiving.*

sedative *n.* **1** °narcotic, °tranquilizer, opiate, sleeping pill, soporific, calmative, anodyne, depressant, hypnotic, barbiturate, lenitive, *Colloq* downer, knockout drop, *Slang* Mickey (Finn): *The doctor has given him a sedative and he should sleep now.* —*adj.* **2** °narcotic, tranquilizing, relaxing, °soothing, calming, allaying, opiate, soporific, sleep-inducing, calmative, anodyne, lenitive, depressing, hypnotic: *I found that listening to a recording of the surf splashing on a beach has a sedative effect.*

sedentary *adj.* seated, sitting, stationary, °fixed, immobile, unmoving, house-bound, desk-bound: *There are sedentary occupations you could follow while you recover. As a writer, I lead a sedentary existence and get little exercise.*

sediment *n.* lees, °dregs, °deposit, °grounds, °precipitate, °remains, °residue, settlings, residuum, detritus: *Periodically, the sediment must be cleaned out of the filters.*

sedition *n.* °agitation, °incitement (to riot), rabble-rousing, fomentation, instigation, firing-up, stirring-up, whipping-up; °mutiny, insurrection, insurgency *or* insurgence, rebellion; treason, treachery: *Because he had organized the strike in the munitions plant during wartime, he was accused of sedition.*

seditious *adj.* °rebellious, °mutinous, °revolutionary, insurgent, °inflammatory, rabble-rousing, insurrectionist, insurrectionary, refractory, °subversive, treacherous, °dissident, °disloyal, °turncoat, unfaithful: *The colonists' most recent seditious action was to dump tea into the harbor at Boston, Massachusetts.*

seduce *v.* **1** °lure, °entice, °attract, allure, °tempt, °mislead, °beguile, °deceive, °decoy, °draw on, °charm, °captivate, vamp, entrap, ensnare, °trap, *Colloq* sweet-talk: *He was seduced into giving her his life's savings.* **2** °dishonor, °ruin, °corrupt, °lead astray, defile, debauch, deflower, °violate, °ravish: *How many girls have been seduced in the name of undying love?*

seducer *n.* See also **seductress,** below. °rake, °libertine, °roué, °playboy, lady-killer, lecher, debauchee, debaucher, °lover, *cicisbeo,* Don Juan, Lothario, Casanova, *Colloq* wolf: *That wretched seducer is going out with my daughter!*

seductive *adj.* alluring, °attractive, °tempting, tantalizing, enticing, °inviting, seducing, °enchanting, entrancing, bewitching, fascinating, °flirtatious, coquettish, captivating, beguiling, °provocative, °siren, °irresistible, °winning, appealing, °prepossessing, *Colloq* °sexy: *He was lured into the whirl of the financial world by the seductive five-letter word—money. Never had Hortense looked more seductive than on that night in Rio.*

seductress *n.* See also **seducer,** above. °temptress, °siren, *femme fatale,* enchantress, Circe, Lorelei, Jezebel, vamp: *That clever seductress will soon have him eating out of her hand.*

see *v.* **1** °perceive, °note, °notice, °mark, °spot, °watch, °witness, °recognize, °behold, discern, °distinguish, °observe, °look at, °regard, °sight, °catch sight of, descry, espy, °spy, °make out, look upon, °view, glimpse, catch a glimpse of, *Slang* get a load of, *US* glom: *I saw him buy a bottle of whiskey. We saw two yellow-bellied sapsuckers in one day! Can you see the sea from your suite? Did you see the Houses of Parliament when you were in London?* **2** °understand, °comprehend, apprehend, °perceive, °appreciate, °fathom, °grasp, °take in, °realize, °know, be aware *or* conscious of, get the idea *or* meaning of, *Colloq* °dig, °get, get the drift *or* the hang of: *I see what you are saying, but I don't agree. After her speech, I saw Gladys in a new light.* **3** °foresee, foretell, °imagine, °envisage, °envision, visualize, °picture, °divine, °conceive (of), dream of, conjure up, °accept: *I can see a day when warming from the greenhouse effect will make the sea level rise. Whitney said he can't see you as a married man.* **4** °determine, ascertain, °find out, °investigate, °discover, °learn: *See if the bakery has any macaroons. I'll see what she thinks.* **5** Often, ***see to it.*** °ensure, °assure, make sure *or* certain, °mind, be vigilant: *It's cold outside, so see that you dress warmly.* **6** °accompany, °escort, °show, °lead, °conduct, usher, °take, convoy, °bring, °walk, °drive: *It was from Aunt Dinah's quilting party, I was seeing Nellie home.* **7** go out with, socialize with, keep company with, consort with, °associate with; court, woo; *Colloq* go steady with, *Chiefly US* °date: *Are you still seeing that boy you met at the school dance?* **8** make up one's mind, think over, mull over, °consider, ponder (on *or* over), °contemplate, °decide, °reflect (on), meditate (on *or* over *or* about), ruminate (on *or* over), °brood over: *Mother said she'd see whether I should go. Then I asked father, and he said he'd see.* **9** °receive, °meet (with), talk *or* speak with, confer with, °consult (with), have a word with, sit down with, visit with, °interview; °welcome, °greet: *The ambassador will see you now.* **10** °undergo, °experience, °go through, °endure, °survive: *He saw service overseas during the war.* **11** °help, °aid, °assist, °support, °finance, pay the way for; °guide, °shepherd: *She saw three children through university without anyone's help.* **12** ***see about.*** **(a)** °see to, °attend to, °look after, take care *or* charge of, look to, °organize, °manage, do, °undertake, °sort out; °think about, °consider, give some thought to, pay attention *or* heed to: *Could you please see about feeding the horses while I am away? I asked Martin and he said he'd see about it.* **(b)** °investigate, °study, °probe, °look into, make enquiries *or* inquiries, enquire *or* inquire about: *The teacher said she would see about letting us out early before the holiday.* **13** ***see off.*** bid adieu *or* bon voyage: *We went to the airport to see them off.* **14** ***see through.*** **(a)** °penetrate, °detect, °perceive, *Slang* be wise to: *She saw through your subterfuge at once.* **(b)** ***see (something) through.*** °persevere, °persist, °manage, °survive, °last, ride out, *Colloq* °stick out: *Once you start on something, I wish you'd see it through for a change.* **(c)** ***see (someone) through.*** provide with help *or* aid *or* assistance, °help, °aid, °assist, °last: *We'll give you enough money to see you through.* **15** ***see to.*** See 12 **(a)**, above.

seed *n.* **1** °grain, spore, °kernel, °pit, tuber, bulb, corm, *Technical* ovum, ovule, embryo, egg, °germ: *Add water and the seeds will soon germinate.* **2** °origin, °source, °cause, °root, °provocation, °reason, °basis, °grounds; °motive, motivation, motivating factor: *The seeds of suspicion were sown by the continuous police presence in the neighborhood.* **3** °offspring, children, °progeny, °young, °issue, descendants, heirs, successors: *Yea, and verily I say unto you that his seed shall populate the earth.* **4** ***go*** *or* ***run to seed.*** °run down, become dilapidated *or* worn out *or* shabby, °decay, go downhill, °decline, °degenerate, °deteriorate, go to rack and ruin, *Colloq* go to pot: *If he doesn't start to look after it, the entire place will go to seed.* —*v.* **5** °scatter, °sow, °distribute: *Don't walk on the freshly seeded lawn.*

seedy *adj.* **1** °shabby, °dilapidated, °worn (out), decayed, deteriorated, °run-down, broken-down, °mangy, grubby, decaying, tatty, scruffy, squalid, °sleazy, *Colloq* °ratty: *They live in a seedy little shack on the other side of town.* **2** °tired, °weary, wearied, °run-down, worn-out, unwell, °out of sorts, ailing, °ill, °sickly, *Colloq* °poorly, °under the weather, off one's feed: *I begged off at dinner time because I was feeling a bit seedy.*

seeing *conj.* °in view of (the fact that), whereas, in (the) light of, inasmuch as, since, °considering: *Seeing that you are here, you might as well tell her yourself.*

seek *v.* **1** °look (for), °search (for), °hunt (for), go *or* be after, °quest after, °pursue: *All of us are seeking the bluebird of happiness. Seek and ye shall find.* **2** °hope, °aim, °aspire, °try, °essay, °endeavor, °undertake: *I am seeking to determine who committed these crimes.* **3** °ask for, °request, °beg, °solicit, invite; °demand: *He sought her help and she refused him.*

seem *v.* °appear, look (as if *or nonstandard in US* like), °sound, °feel, have (all) the hallmarks *or* earmarks of, give every indication *or* appearance of: *He seems all right to me. She seems to have forgotten her key again. It seems as if I've never been away at all. He seemed frightened.*

seeming *adj.* °apparent, °evident, ostensible, °outward, °superficial, °surface, °assumed, feigned, °pretended, °false, °so-called, °alleged, °specious, purported, °professed: *She was shocked by his seeming indifference to her problems.*

seemingly *adv.* °apparently, °evidently, °ostensibly, °outwardly, superficially, falsely, allegedly, speciously, purportedly, professedly, °on the face of it, °possibly, feasibly, conceivably, plausibly, believably: *The purpose of this seemingly honest confession was to throw the detectives off the scent.*

seemly *adj.* **1** °proper, °fitting, °appropriate, °becoming, suited, °suitable, °fit, °befitting, apt, *comme il faut,* °right, apropos, apposite, °characteristic, °meet, °reasonable, °sensible: *You can count on Felix to do the seemly thing in such cases.* **2** °decent, °decorous, °proper, °dignified, °genteel, gentlemanly, °ladylike, °diplomatic, °discreet, °prudent, °politic: *She behaved in a seemly manner, in keeping with what was expected of a princess.*

seer *n.* soothsayer, °fortuneteller, sibyl, °oracle, °prophet, prophetess, augur, vaticinator, prophesier, clairvoyant, °psychic, crystal gazer, stargazer: *In ancient times, many consulted seers before taking an important decision.*

seesaw *n.* **1** °teeter: *The children were playing on the seesaw.* —*v.* **2** °teeter, °totter, waver, °vary, vacillate, °oscillate, °alternate, °fluctuate, °swing, °switch: *He couldn't make up his mind and kept seesawing between staying and going.*

seethe *v.* **1** °boil, °stew, °simmer, °foam: *The devils danced round the cauldron, which seethed on the bonfire.* **2** °stew, °simmer, °foam (at the mouth), °fume, °smolder, °burn, °rage, °rant, °rave, become livid *or* feverish, be in ferment, be furious *or* incensed, *Colloq* blow one's stack *or* top, °carry on, °take on, get hot under the collar, get red in the face, get all steamed up: *The ball sailed through Mr. Garamond's new greenhouse and he came out seething with rage.*

see-through *adj.* °sheer, diaphanous, gauzy, °transparent, translucent, gossamer, °filmy, peekaboo: *She was wearing a see-through negligee when she appeared at the door.*

segment *n.* **1** °section, °part, °division, °portion, component, °element; °piece, fraction, °fragment, °length, °joint, °slice, °wedge: *A different tutor was en-*

gaged for each segment of her education. The orange was divided into three equal segments. —*v.* **2** °divide, °partition, °section, °separate, °part, °cleave, °split, subdivide, °fragment: *The department was then further segmented into smaller units, each with its own manager.*

segregate *v.* °separate, °segment, °partition, °isolate, seclude, sequester, °set apart, compartmentalize, °exclude, °ostracize, discriminate against: *In the sorting process, each size must be segregated into its own compartment. Their policy was to segregate the Blacks from the rest of the population.*

segregation *n.* °separation, segmentation, °partition, isolation, °seclusion, sequestration, setting apart, compartmentalization, °exclusion, ostracism, °discrimination, apartheid, *US* Jim Crowism: *No modern culture can long survive a segregation of the races.*

seize *v.* **1** Sometimes, ***seize on.*** °grab, °grasp, °clutch, °take (hold of), °grip, °snatch: *Robin felt that he had to seize every opportunity that presented itself. The dog seized the robber by the seat of the pants.* **2** °capture, °catch, °arrest, take into custody, take prisoner, apprehend, °round up, *Colloq* °pinch, °nab, collar, °pick up, *Brit* °nick, *Slang* bust: *The police seized him as he was trying to leave the country.* **3** °catch, °transfix, °stop, °hold, °possess, °take possession of, °afflict, °beset, °visit, °subject: *I was seized by a sudden desire to kiss those quivering, pouting lips.* **4** take advantage of, make good use of: *When their attention was diverted, I seized the opportunity to escape.* **5** °confiscate, °take (away), commandeer, °appropriate, °capture, °take possession of, impound: *Ten tons of cannabis was seized in the raid.* **6** seize up, °bind, °jam, °stop, °lock (up), °stick, °freeze (up): *Because of the excessive heat, the gears seized and the engine stalled.*

seizure *n.* **1** seizing, confiscating, confiscation, appropriation, impounding, commandeering, °capture, °taking, °possession, annexation, sequestration, usurpation: *Police today announced the seizure of a shipment of two tons pounds of marijuana.* **2** °spasm, °attack, °fit, °paroxysm, convulsion, *Technical* ictus: *He should not be out alone as he could have a seizure at any time.*

seldom *adv.* °rarely, °infrequently, not often, hardly ever, very occasionally: *Since they moved away, we seldom see the Pattersons any more.*

select *v.* **1** °choose, °pick, show (a) preference for, °prefer, opt for, °single out, handpick, °distinguish: *Select any two free books from this pile. Anne easily selected the real antiques from among the reproductions.* —*adj.* **2** selected, chosen, handpicked, °choice, °special, preferred, preferable, °favored, °favorite, °exceptional, °excellent, °first-rate, first-class, °superior, °supreme, °prime, °better, °best, finest, tiptop: *We have set aside some of our select jewels for you to consider.* **3** °limited, restricted, restrictive, °exclusive, °privileged, élite *or* elite, closed: *He belongs to one of those select clubs that accept you only if your family can be traced back five hundred years.*

selection *n.* **1** °choice, °pick, °preference, °option: *Please make your selection from the merchandise in this display.* **2** °assortment, °variety, °collection, °range, °batch, °number, °set, series, °group: *Tonight I shall play for you a selection of jazz recordings made at the Savoy—the one in Harlem, not London.* **3** selecting, choosing, picking, singling out, electing, settling on, voting for, opting for, °choice, °pick, °election: *Selection has been delayed till tomorrow.* **4** °extract, °quotation, °excerpt, °abstract, °passage, °piece, °quote: *I shall read you a selection from Carmody's treatise on the life of the flea.*

selective *adj.* °particular, discerning, discriminative, °discriminating, discriminatory, eclectic, °exacting, °demanding, °choosy, *Colloq* picky: *Thea has been very selective in her choice of running mate in the election.*

self-abuse *n.* masturbation, onanism, self-gratification, autoeroticism *or* autoerotism, self-stimulation, self-manipulation, self-pollution, self-defilement, self-contamination, *Technical* manustrupation: *Psychologists differ widely in their opinions regarding the harm, if any, from self-abuse.*

self-confidence *n.* °confidence, self-assurance, °self-respect, °self-esteem, °assurance, °poise, *armour-propre*, aplomb, self-reliance, self-sufficiency: *Harrigan has demonstrated that he has the self-confidence to run the department.*

self-confident *adj.* °confident, self-assured, assured, °poised, self-reliant, °secure, sure of oneself, °positive, °definite, °assertive, °independent: *We need someone who is self-confident but not conceited or overconfident.*

self-conscious *adj.* °embarrassed, °coy, diffident, °shy, °modest, self-effacing, °sheepish, shrinking, °retiring, unsure, apprehensive, °reserved, insecure, °affected, °awkward, °nervous, uncomfortable, °hesitant, °timid, timorous: *Phil felt a bit self-conscious about asking Maria to the dance.*

self-contained *adj.* **1** °self-possessed, unemotional, self-controlled, in control, composed, °serene, °peaceful, °calm, °tranquil, °cool, °collected, even-tempered, °detached, imperturbable, unruffled, undisturbed, unperturbed, controlled, placid, °grave, °serious, °sober, °solemn, *Colloq* unflappable: *Sometimes I think Forslund a little too self-contained.* **2** °reserved, controlled, °distant, °aloof, °formal, °withdrawn, °reticent, °standoffish: *Greg is so self-contained that he didn't tell anyone he was terminally ill, not even his wife.* **3** °whole, °entire, °complete, stand-alone, unitary: *In this self-contained bathroom you have basin, toilet, bidet, and shower all molded together in one fiberglass element.*

self-control *n.* **1** self-discipline, self-restraint, °restraint, °self-denial, °control, willpower, strength (of character *or* of mind *or* of will), mettle, °fortitude, moral fiber, °determination, self-possession, resoluteness, °resolve, °will, constancy, steadfastness, °perseverance, doggedness, obduracy, °persistence, *Facetious US* won't-power: *Can I exercise the self-control needed to stop smoking?* **2** calmness, tranquillity, °serenity, placidity, imperturbability, coolheadedness, coolness, °poise, levelheadedness, °patience, aplomb, °dignity, equanimity, forbearance, °control, °restraint, self-restraint, even temper: *You exhibited marvelous self-control by not losing your temper when you were insulted.*

self-denial *n.* **1** self-sacrifice, self-abnegation, renunciation, selflessness, °altruism, unselfishness, magnanimity: *His contributions must have been made at the cost of some self-denial.* **2** °hardship, °suffering, self-mortification, asceticism, °privation, renunciation, renouncing, abstemiousness, abstinence, abstention, self-deprivation, keeping away from, °refusal, refusing, giving up, desisting, *Colloq* swearing off: *As far as certain foods are concerned, a little self-denial is good for one's health.* **3** See **self-control, 1,** above.

self-esteem *n.* **1** °conceit, °vanity, egoism, narcissism, self-centeredness, egotism, *amour-propre,* self-approbation, self-satisfaction, self-admiration, self-love, self-adulation, self-idolatry, smugness, self-importance, self-regard: *The arrogant, overweening self-esteem of some of the club members is unbearable.* **2** See **self-confidence,** above.

self-evident *adj.* °evident, °obvious, °patent, °clear, °incontrovertible, °definite, °express, °distinct, clear-cut, °apparent, unmistakable *or* unmistakeable, undeniable, inescapable, incontestable, °plain, axiomatic, °proverbial, °manifest, °true, palpable, °tangible: *Her guilt seemed self-evident at first. The gradual destruction of the beaches is self-evident.*

self-government *n.* 1 self-rule, °independence, self-determination, home rule, autonomy, °freedom: *The satellite countries, one by one, voted for self-government.* 2 See **self-control, 1** above.

self-important *adj.* °conceited, self-centered, self-seeking, self-absorbed, °vain, egotistic(al), self-satisfied, °smug, °pompous, swollen-headed, swellheaded, °arrogant, overweening, °overbearing, vainglorious, self-glorifying, self-engrossed, °presumptuous, °snobbish, °haughty, *Colloq* snooty, *Slang* snotty, stuck-up: *That self-important little cockalorum ought to be taken down a few pegs.*

self-indulgent *adj.* self-gratifying, °selfish, self-gratifying, °extravagant, °sensual, intemperate, overindulgent, °greedy, °immoderate, hedonistic, sybaritic, °epicurean, °gluttonous, gormandizing, pleasure-bound, pleasure-seeking, °dissolute, dissipating, licentious, °profligate, debauching: *Dorell's self-indulgent habits take no account of his wife and children.*

selfish *adj.* 1 °greedy, covetous, °grasping, °avaricious, °self-indulgent, self-aggrandizing, acquisitive, self-seeking, self-loving, self-centered, self-absorbed, self-interested, self-serving, egotistic(al), egoistic(al): *They pursue success for their own selfish ends.* 2 stingy, °mean, °mercenary, °tight, tightfisted, °narrow, °penurious, parsimonious, °miserly, niggardly, penny-pinching, cheeseparing, ungenerous, illiberal, grudging, uncharitable, °possessive, °inconsiderate, °thoughtless: *He is so selfish that he won't even give you the right time.*

selfless *adj.* °open, °charitable, °unselfish, self-denying, °generous, altruistic, ungrudging, magnanimous, °considerate, °thoughtful; self-sacrificing: *Always ready to help anyone with a need, Archer is the most selfless man I know.*

self-made *adj.* °independent, self-reliant, entrepreneurial, °self-sufficient: *A self-made woman, she accomplished it all without anyone else's help.*

self-possessed *adj.* composed, °cool, °serene, placid, °collected, self-assured, °peaceful, °calm, °tranquil, even-tempered, °detached, imperturbable, unruffled, undisturbed, unperturbed, controlled, °dignified, °refined, *Colloq* unflappable: *Considering what she has been through, Tanya is quite self-possessed for a teenager.*

self-respect *n.* °honor, °dignity, °integrity, self-regard, °self-esteem, °pride, *amour-propre,* °morale: *He managed to maintain a semblance of his self-respect despite the accusations hurled at him.*

self-righteous *adj.* pharisaic(al), °sanctimonious, holier-than-thou, pietistic, °mealy-mouthed, °hypocritical, complacent, °smug, self-satisfied, °priggish, °superior, tartuffian, canting, *Colloq* °goody-goody, *Slang Brit* pi: *Butter would not melt in the mouth of that self-righteous hypocrite!*

self-styled *adj.* would-be, self-called, *soi-disant,* °professed, self-appointed, self-christened, °so-called, °quasi: *Everyone seems to be a self-styled television critic.*

self-sufficient *adj.* °independent, self-reliant, self-supporting, self-sustaining: *Agriculture will reach the point where the country is self-sufficient for food.*

self-willed *adj.* headstrong, °determined, °forceful, refractory, °stubborn, °obstinate, pigheaded, °willful, °ungovernable, uncontrollable, °unruly, unmanageable, intractable, °contrary, °perverse, uncooperative, contumacious, °recalcitrant, stiff-necked, vexatious, °difficult, °incorrigible, °disobedient: *That self-willed, over-indulged little brat needs a good spanking.*

sell *v.* 1 vend, °transfer, convey (title), °trade, barter, °exchange, °dispose of: *Sorry, but I sold that car last week.* 2 °market, deal in, °merchandise, trade in, traffic in, °peddle, vend, hawk, °handle, retail, °carry, °stock, °furnish, °supply, °offer, *Colloq* °push, *Slang* °flog: *We no longer sell that model. She sells seashells by the seashore.* 3 Often, ***sell out.*** °betray, °inform against, deliver up, °give away, *Slang* rat on, grass on, °tell on, tattle on, sell down the river, blow the whistle on, °double-cross, *Brit* °shop: *He swore he'd get the man who sold him.* 4 °promote, °push, put across *or* over: *He couldn't sell sand to a beach flea.* 5 ***be sold on.*** persuaded, convinced, won over: *After seeing those videos, I was sold on taking my vacation in Mexico.*

seller *n.* °dealer, vendor, °merchant, retailer, shopkeeper, °salesperson, salesman, saleswoman, saleslady, sales agent, °representative, °traveler, traveling salesman, °peddler, hawker, colporteur, sutler, chandler, *Old-fashioned* counter-jumper, *Brit* shop assistant, -monger (*as in* costermonger, ironmonger, fishmonger, etc.), *US and Canadian* salesclerk, clerk, shopgirl, *Colloq* rep, *US* drummer: *Time ran out on his job as a seller of water clocks and hourglasses.*

semblance *n.* 1 °appearance, °image, °likeness, °resemblance, °form, °figure, °bearing, °aspect, °air, °look, mien, °exterior, °mask, façade, °front, °face, °show, °veneer: *Beneath that semblance of tranquillity lies a turbulent, tortured mind.* 2 °guise, °face, °front, façade, °air, °show, °veneer, °look, °pretense, °show, °cloak, simulation, °impression, °affectation, *Rare* °superficies: *You might try to give the sculpture at least a semblance of reality.*

seminal *adj.* 1 °original, °basic, °creative, °primary, °prime, formative, innovative, °imaginative, °new, unprecedented, precedent-setting, °landmark, bench mark, °important, °influential, °telling: *Michaelson's paper is considered the seminal work on the subject. It sets forth many seminal ideas.* 2 embryonic, germinal, °potential, °undeveloped, incipient: *Each seed contains a seminal plant.*

seminary *n.* academy, °school, °institute, °institution, college, university, training ground: *Three little maids who, all unwary, Come from a ladies' seminary, / Freed from a genius tutelary / Three little maids from school!*

send *v.* 1 Sometimes, ***send off.*** dispatch *or* despatch, commission, °charge, depute, °delegate, °assign: *The pope sent an emissary to London.* 2 °communicate, °transmit, convey, °deliver, consign, address to, °mail, °post, fax, °remit, °ship, °forward; °broadcast, telecast, televise, °radio, telegraph: *The message is being sent to all concerned.* 3 °release, °discharge, °shoot, °propel, °fire, °fling, °project, °hurl; °cast, °throw, °toss, °let fly: *With a great burst of flame the satellite was sent into orbit.* 4 °delight, please, °charm, °enrapture, °stir, °thrill, °move, °electrify, *Slang* turn (someone) on: *That boogie-woogie piano really sends me.* 5 ***send down.*** °imprison, incarcerate, send away, jail *or Brit also* gaol, *Slang US* °send up (the river): *If she doesn't mend her ways, she'll be sent down.* 6 ***send for.*** °call for, °summon, °order, °request, °ask for: *I have sent for a taxi which should be here shortly.* 7 ***send forth*** *or* ***out.*** °emit, °radiate, °discharge, °give off, exude, °grow: *This piece of rotten wood sends forth a phosphorescent glow. In the spring, the vine sends forth tendrils.* 8 ***send off.*** send (someone) away (with a flea in his *or* her *or* their ear), °dismiss, °discharge, send (someone) packing, send (someone) (on) about his *or* her *or* their business, *Colloq* give (someone) his *or* her *or* their walking papers,

Slang give (someone) the brushoff, *US* give (someone) the brush: *I went to the office as you instructed, but they just sent me off, with no explanation.* **9** ***send up.*** **(a)** °lampoon, °satirize, °burlesque, °parody, °make fun of, *Colloq* °take off, spoof, *Brit* take the mickey out of: *In Britain one of the enduring forms of humor is in sending up the government.* **(b)** °imprison, incarcerate, send away, jail *or Brit also* gaol, *Slang US* °send up (the river): *He was sent up for insider trading.*

senile *adj.* (of a woman) anile; senescent, °decrepit, declining, °failing, in one's dotage, doting, °doddering, in one's second childhood, dotty, °simple, °feebleminded; °forgetful, *Colloq* past it: *He's senile and needs constant care.*

senility *n.* (of a woman) anility; senile dementia, Alzheimer's disease, senescence, °decrepitude, °decline, dotage, second childhood, loss of one's faculties: *An increased lifespan has created increased cases of senility.*

senior *adj.* °elder, older, *Brit* °(*postpositive*) major; (higher) ranking, °superior, °chief: *The senior members of the club are impossibly stuffy. You must not address your senior officer in such a manner.*

senior citizen *n.* elderly person, retired person, °pensioner, *Chiefly US* retiree, golden ager, *Brit* OAP (= 'old age pensioner'), gray panther: *Special tax provisions have been made for senior citizens.*

sensation *n.* **1** °feeling, °sense, °impression, °perception, °foreboding, presentiment, prescience, awareness, °suspicion, *Colloq* sneaking suspicion, funny feeling, °hunch: *I had the distinct sensation of having been there before.* **2** commotion, °stir, °thrill, °furor; °excitement: *The news of the wedding caused a sensation.* **3** °hit, *coup de théâtre,* °success, *Colloq* show-stopper, crowd-puller, crowd-pleaser: *Edith Piaf was a sensation at Le Théâtre de la verdure in Nice.*

sensational *adj.* **1** °exciting, °stimulating, electrifying, galvanizing, °shocking, hair-raising, spine-tingling, °thrilling, °stirring, breathtaking, °amazing, astonishing, astounding, staggering, mind-boggling, °unbelievable, °incredible, spectacular, *Slang* mind-blowing: *They certainly put on a sensational show—not a good show, mind you, but a sensational one.* **2** °lurid, °vivid, overdone, overdrawn, °extreme, °melodramatic, exaggerated, °dramatic, °extravagant: *The weekly I worked for insisted that each issue's first page carry a sensational story—whether it was true or not.* **3** °(very) good, °great, °marvelous, wonderful, °superior, °superb, °matchless, °peerless, unequaled, °nonpareil, °extraordinary, °terrific, °phenomenal, °splendid, °fabulous, °fantastic, stupendous, *Colloq* °super, smashing, *Slang* far-out: *Merly Peters's performance was sensational in* La Canaresa.

sense *n.* **1** °faculty: *It's not always easy to tell if Belfin's senses are all functioning.* **2** common sense, °intelligence, °perception, quick-wittedness, quickness, °(mother) wit, °judgment, °reason, °wisdom, sagacity, °discrimination, discernment; °sanity, *Colloq* brains, *Slang Brit* nous: *Imagine doing something like that!—she hasn't the sense she was born with!* **3** See **sensation, 1,** above. **4** °meaning, intelligibility, coherence, head or tail, °drift, °gist, °import, purport, nuance, °significance, °message, °substance: *I was unable to get any sense out of what she was saying. Yes, I wasn't able to make sense out of it either.* **5** °sentiment, °atmosphere, °impression, °intuition, °sensation: *I had a sense of being* de trop, *that I was not wanted.* —*v.* **6** °feel, °perceive, °detect, °divine, intuit, have a hunch *or* feeling, have *or* get *or* be under the impression that, °suspect, *Colloq* have a funny feeling that, feel (something) in one's bones, °pick up: *I sensed a certain antagonism in that last remark. I sensed that he was getting bored with me.*

senseless *adj.* **1** °insensible, °unconscious, °(knocked) out (cold), stunned, insensate, comatose: *Agatha's head hit the andiron and she was borne senseless to her room.* **2** numb(ed), insensate, unfeeling, benumbed, unfeeling, °insensible, anesthetized, °dead, deadened, insentient: *My feet are senseless from the cold.* **3** °pointless, °purposeless, °ridiculous, °ludicrous, unintelligent, illogical, irrational, °incongruous, °meaningless, °absurd, °wild, °mad, °crazy, demented, °insane, asinine, °nonsensical, imbecilic *or* imbecile, idiotic, moronic, simple-minded, fatuous, °stupid, °foolish, °silly, °dizzy, °half-witted, °mindless, brainless, witless, empty-headed, thimble-witted, pea-brained, bird-brained, °harebrained, featherheaded, rattlebrained, addlepated, muddle-headed, *Colloq* °daft, *US* daffy, nutty, batty, *Slang* wacky, dippy: *Many of the government's objections to the plan are so much senseless posturing. He is promoting some senseless plan for double-decker highways.*

sensibility *n.* **1** See **sensitivity, 3,** below. **2** Often, ***sensibilities.*** °feelings, responsiveness, responsivity, emotions, °sentiments: *Personal events are more likely to affect our sensibilities than major disasters in which we are not involved.*

sensible *adj.* **1** °reasonable, °realistic, °logical, common-sense, common-sensical, °rational, reasoned, °sound, °practical, °prudent, °judicious, °discreet, °intelligent, °sage, °wise, °sane; down-to-earth, °matter-of-fact, well-thought-out: *Do you really think it sensible to go out in this storm? Arthur has devised a sensible way of dealing with junk mail.* **2** perceivable, °perceptible, detectable, °evident, °discernible, recognizable, ascertainable, apprehensible, cognizable, °manifest, palpable, °physical, °tangible, corporeal, substantive, °material, °visible, °observable, seeable: *Some maintain that in addition to the sensible world there is another world, an antiworld, that we do not normally encounter.* **3** sentient, °feeling, °sensitive, °live, °conscious, °aware: *Who among us ever regards microbes as sensible creatures?* **4** Usually, ***sensible of*** *or* ***to.*** °conscious (of), °aware (of), °acquainted (with), cognizant (of), °sensitive (to), °alive to, °mindful (of), °understanding (of), in touch (with), °observant (of), °awake (to), °alert (to), *Slang* °wise (to), hip *or old-fashioned* hep (to): *The people of Graustark are ever sensible to the needs of their neighbors. There are some things to which we are not at all sensible.* **5** appreciable, °significant, °considerable, °substantial, substantive, °noticeable: *The theory is that an increased interest rate will have a sensible effect on the rate of inflation.*

sensitive *adj.* **1** °delicate, °tender, °sore, °susceptible: *My arm is still sensitive where I got the injection.* **2** °touchy, °susceptible, susceptive, reactive, °responsive, attuned, °impressionable, °emotional, thin-skinned, °vulnerable, supersensitive, hypersensitive, °testy, irascible, °quarrelsome, °irritable, °volatile, °excitable, °temperamental, °petulant, hot-tempered, °quick-tempered: *Philomela is very sensitive on the subject of being unmarried, and I wouldn't bring it up.* **3** finely tuned, °delicate, °responsive, °subtle, °acute, reactive, °receptive: *We need a more sensitive test to detect neutrinos. This radio is not sensitive enough to receive distant stations.*

sensitivity *n.* **1** sensitiveness, °delicacy, touchiness, oversensitivity, hypersensitivity, supersensitivity; soreness, irritability: *Their sensitivity to criticism is well known to our diplomatic corps. The sensitivity of the bruised area is likely to continue for a few days.* **2** compassion, °concern, °sympathy, tenderness, tenderheartedness, kindheartedness, kindliness, °warmth, °feeling:

The nurses exhibit a genuine sensitivity for the problems of the handicapped. **3** awareness, consciousness, acuteness, °perception, °understanding, °intuition, °feeling, °sense, °sensitivity, sensitiveness, receptivity *or* receptiveness, receptibility, °appreciation, appreciativeness, susceptibility, susceptivity *or* susceptiveness: *Harvey has a keen sensitivity for the feelings of others. Sometimes I think the duke lacks the sensitivity needed to sympathize with the needy.*

sensual *adj.* °physical, appetitive, °voluptuous, °carnal, bodily, fleshly, °erotic, °sexual, °lustful, °unchaste, °abandoned, °dissolute, °profligate, dissipated, licentious, °lewd, °lascivious, lubricious *or* lubricous, goatish, hircine, °lecherous, libidinous, salacious, °prurient, °rakish, °wanton, debauched, Cyprian, °loose, °dirty, *Slang* °randy: *Cranston pursued his sensual pleasures with no regard for morality.*

sensualist *n.* lecher, °profligate, °wanton, debauchee, °roué, °rake, Romeo, Don Juan, Casanova, Lothario, °libertine; voluptuary, hedonist, °sybarite, bon viveur, *bon vivant,* °epicure, °epicurean, °gourmet, gourmand, gastronome, pleasure-seeker: *For the sensualist, the senses are mere implements of fantasy.*

sensuous *adj.* sensory, sybaritic, °epicurean, hedonist(ic), °sumptuous, °luxurious, °rich: *Jim O'Shea was known to revel in sensuous pleasures on his isle.*

sentence *n.* °judgment, °decision, ruling, verdict, °decree, °determination; °punishment, *Slang* °rap: *It is up to the judge to announce the sentence of the court. Asked by his teacher for an example of a short sentence, Johnny replied, "Thirty days."*

sentiment *n.* **1** °attitude, °feeling, °sensibility, °emotion, susceptibility, tenderness, tender-heartedness, °sentimentality, sentimentalism: *He wanted to preserve the old village square more for sentiment than historical importance.* **2** Often, ***sentiments.*** °view, °outlook, °opinion, °position, °attitude, °judgment, °thought, °belief, °feeling, °emotion: *His sentiments regarding apartheid are better left unexpressed.*

sentimental *adj.* **1** °emotional, °sympathetic, compassionate, °tender, warmhearted, tenderhearted: *Has he returned for sentimental or for practical, mercenary reasons?* **2** °romantic, nostalgic, °emotional, °maudlin, mawkish, overemotional, °tender, °tearful, weepy, sickening, nauseating, simpering, °sweet, saccharine, *Colloq* °sloppy, °gooey, °sticky, °tacky, °mushy, slushy, °gushy, soppy, drippy, tear-jerking, ill-making, sick-making, *Slang* corny, schmaltzy, icky, yucky *or* yukky: *His last tune is always some sentimental ballad like "Good Night, Sweetheart."*

sentimentality *n.* romanticism, nostalgia, pathos, emotionalism, maudlinism, bathos, mawkishness, overemotionalism, tenderness, tearfulness, weepiness, sweetness, *Colloq* sloppiness, gooeyness, mushiness, slushiness, gushiness, soppiness, drippiness, *US* stickiness, tackiness, *Slang* corn, corniness, schmaltz, schmaltziness, ickiness, yuckiness *or* yukkiness: *The role of the bereaved mother was overplayed with enough gushing sentimentality to make one sick.*

sentinel *n.* sentry, °guard, °watchman, °watch, °picket, °lookout, °patrol: *The sentinels measured out the ramparts of the fortress with their stately paces.*

separable *adj.* distinguishable, segregable, detachable, divisible, severable, removable, fissile, scissile: *Unfortunately, as history demonstrates, morality and religion are separable.*

separate *v.* **1** disjoin, pull *or* take *or* break apart, come *or* fall apart, fall *or* take *or* break to pieces, split *or* divide *or* break (up), split *or* break (off *or* away), °disconnect, °disengage, °part, °partition, °sort (out), uncouple, disarticulate, disassemble, unhook, °detach, disunite, unyoke, disentangle, unravel: *In order to clean it, I first have to separate all the individual elements of the motor. The glue didn't hold, and the vase separated into a dozen fragments. Separate this yarn into different piles.* **2** °distinguish, °discriminate, °analyze, °sort, °break down, classify, °segregate, °single out, sequester, °type, codify, °organize, °split up; °group, collate: *What criteria are used to separate the men from the boys, the sheep from the goats, or the wheat from the chaff?* **3** split *or* break up, °part (company), °divide (up), °disband, °divorce: *After ten years we separated, and we each went our own way.* **4** fork, split (up *or* off), bifurcate, °diverge, °branch: *Beyond the river, the road to Norton separates from the main road.* —*adj.* **5** divided, separated, disjoined, °disconnected, °detached, °isolated, °discrete, °distinct, °individual, °independent, °solitary, °different: *Sort these books into separate piles by subject.* **6** °different, °independent, °unrelated, other: *How you behave when you are not at home is a separate matter.* **7** °withdrawn, °solitary, °alone, shut *or* closed off *or* away, °apart, °detached, removed, cloistered, °secluded, sequestered, °isolated, separated: *He prefers a separate existence, apart from the rest of the world.*

separately *adv.* °individually, independently, °singly, one by one, one at a time, °personally, °alone, severally: *Each child approached separately to wish her bon voyage.*

separation *n.* **1** °rift, °break, °split, split-up, °divorce, breakup, disunion, estrangement: *Their separation, after forty years, was a surprise.* **2** °partition, °division, °split, °schism, dividing line, dissociation, disassociation, severance: *The United States Constitution provides for the separation of church and state.* **3** disintegration, shattering, breakup, fragmentation, dismemberment, taking *or* keeping apart, °segregation, °division, disjoining, disjunction, fission, scission, °rupture, °schism, splitting, °split, fracturing, °fracture, °break: *After a crash, investigators supervise the separation of the airplane into its tens of thousands of pieces and then its reassembly.*

sepulcher *n.* °tomb, mausoleum, burial vault, °grave, °crypt, pyramid, burial-place: *The sepulchers of important people are beneath the floor of the church.*

sequel *n.* °follow-up, °upshot, °issue, °result, consequence, °development, °supplement: *We must wait for the sequel to see the effect of your work.*

sequence *n.* °succession, °progression, °order, series, °chain, °string, °course, °cycle, °arrangement, °organization, °train, °line, °set, °run, concatenation, °system: *His description conflicts with the sequence of events as recorded by the video camera. Some of these dates are out of sequence.*

sequential *adj.* °successive, ordered, °orderly, serial, °progressive, organized, °systematic, cyclic, °continuous: *Would you please put these periodicals into sequential order?*

seraphic *adj.* angelic, °celestial, °divine, °heavenly, blissful, °sublime, empyrean, elysian, ethereal, °holy, °saintly, °godly: *In her last years, Ariadne's beauty took on a detached, seraphic quality.*

serene *adj.* **1** °peaceful, °tranquil, °calm, pacific, °peaceable, °restful, halcyon, °idyllic, bucolic, °pastoral, undisturbed, unruffled, imperturbable, unperturbed, untroubled, °quiet, °still: *With the turmoil of the love affair behind her, Shari looked ahead to a more serene life.* **2** °calm, °cool, °collected, placid, composed, °self-possessed, °poised, unexcitable, even-tempered, °temperate, °nonchalant, easygoing, coolheaded, °easy, *Colloq* unflappable: *Beneath William's serene exterior lay a seething, churning miasma of violent lust.*

serenity *n.* **1** peacefulness, °peace, tranquillity, °calm, calmness, restfulness, °quiet, stillness: *The serenity of the warm summer's evening was shattered by an enormous explosion.* **2** tranquillity, peacefulness, peaceableness, unexcitability, calmness, °calm, composure, self-possession, °poise, aplomb, even-temperedness, °temperance, nonchalance, coolheadedness, placidity, *Colloq* unflappability: *The serenity of William's nature was like a safe harbor to the friends who sought comfort from the turbulence of their lives.*

serious *adj.* **1** °grave, °solemn, °earnest, unsmiling, poker-faced, straight-faced, °sedate, °sober, °pensive, °thoughtful; humorless, °somber, °grim, °dour, °severe: *One should be wary of the person who is always serious, who cannot see the ludicrous side of life.* **2** °grave, °important, °vital, °dangerous, °weighty, °significant, °momentous, °crucial, consequential, life-and-death, °urgent, °pressing; no laughing matter, of consequence *or* moment *or* importance: *Things like the greenhouse effect and the hole in the ozone layer are serious concerns for all mankind.* **3** °sincere, straightforward, not joking *or* fooling, °genuine, °honest: *Is he serious about wanting me to fetch him a left-handed wrench?* **4** °acute, °critical, life-threatening, °bad, °dangerous, °nasty, °perilous, alarming, °grave, °severe, °precarious: *His condition is serious and we might have to operate tonight. She had no idea that measles could be such a serious illness.*

seriously *adv.* **1** gravely, °badly, °severely, critically: *Two men were seriously injured in the explosion.* **2** °really, °honestly, scout's honor, °sincerely, °truly, candidly, °openly, *Colloq* joking *or* kidding aside, no joking *or* kidding *or* fooling, cross one's heart (and hope to die): *Seriously, though, I meant what I said about going to bed on time, or no treats tomorrow at the beach.* **3** soberly, earnestly, unquestioningly, without a doubt, at face value: *What teacher takes seriously that old story about a grandmother's illness as an excuse for absence?*

sermon *n.* **1** °lecture, °lesson, preaching, °reprimand, reproach, °reproof, remonstration, remonstrance, scolding, °harangue, *Colloq* talking-to, dressing-down: *Mother gave me a long sermon last night about getting home late.* **2** homily, °address, exhortation, °lesson, °lecture, °speech, °talk, discourse: *We thought that Father Keiller's sermon on vanity was especially appropriate.*

serpentine *adj.* **1** °evil, °bad, diabolical, °satanic, Mephistophelian, reptilian, °devilish, °wily, cunning, conniving, °sneaky, °shrewd, °artful, °crafty, °slick, °sly, insidious, °shifty, °tricky, °scheming, plotting, Machiavellian: *With serpentine deceit she swore that she had returned the ring.* **2** twisting, winding, °tortuous, snaking, snakelike, sinuous, anfractuous, °roundabout, °meandering, ambagious, °indirect, °devious, °crooked, °labyrinthine, vermicular, vermiculate, complex, °complicated, Byzantine: *The amulet was traced through its serpentine trail of ownership back to Xerxes.*

serrated *adj.* sawlike, sawshaped, sawtoothed, toothed, °notched, zigzag, °jagged, serrate, serriform, serratiform, serrulate(d), crenelated, crenulate, crenate, denticulate: *The knife with the serrated edge would cut the meat more easily.*

serried *adj.* ranked, tiered, row on row, ranged, assembled, °packed, °close, compacted, °compact: *The serried army of brightly clad warriors stretched as far as the eye could see.*

servant *n.* **1** °domestic (servant), °help, retainer, servitor; °maid, maidservant, parlor maid, upstairs maid, lady's maid, cleaner, cleaning man, cleaning woman, amah; housekeeper, chatelaine, major-domo, factotum, steward, seneschal, butler, houseman, houseboy, °boy, °page; valet, °man, gentleman's gentleman, manservant, serving-man, footman, footboy, chauffeur, driver, °attendant, °groom; governess, au pair (girl), °nurse, nursemaid, ayah; cook, chef, °waiter, waitress, stewardess, wine steward, sommelier (des vins); °menial, lackey, dogsbody, *Archaic* servitor, *Historical* coachman, postillion, serving-woman, serving-girl, servant-girl, serving-wench, scullery maid, scullion, *Literary* cupboy, Ganymede, *Chiefly Brit* nanny, *Brit* boots, charwoman, charlady, °daily, tweeny, *US* scrubwoman: *Who can afford to keep domestic servants these days?* **2** ***civil servant.*** civil-service employee *or* worker, public servant, (government *or* state) official, officeholder, government worker: *In his capacity as a civil servant, Richard has worked for many different governments.*

serve *v.* **1** °attend (to), wait on *or* upon, °minister to, °look after (the needs of), be at (someone's) beck and call, °assist, °help, be of assistance *or* help, be in the service of, °oblige, °accommodate, °gratify: *After a patron is seated and settled, approach him and ask, "May I serve you, sir (or madam, as the case may be)?" We serve all customers with equal consideration.* **2** fulfill *or* carry out *or* perform *or* discharge (a function *or* a duty *or* one's duty), °work (for), do (duty)(as *or* for), do one's part, °suffice, be used *or* of use *or* useful (to *or* for), function (as *or* to), act (as *or* to), fill the bill, be serviceable (as *or* for), be available (for), °answer (for), be sufficient *or* adequate *or* suitable, °suit, be advantageous *or* of advantage (to); °fight (for), be obedient (to), take one's part: *This catch will serve to keep the door closed. I had no screwdriver, but a coin served just as well. They also serve who only stand and wait.* **3** °distribute, deal out, dole out, °give out, °present, °set out, °provide, °supply, °offer, pass out *or* about *or* around, make available, come (a)round with, dish up *or* out: *They will serve hors d'oeuvres at seven-thirty. Dinner is served, Madam.* **4** be convenient *or* opportune *or* favorable (to): *If the opportunity serves me, I shall repay you many times over.* **5** not fail, not play tricks (on), work *or* function (for), be accurate *or* correct: *If memory serves, he was last here about ten years ago.* **6** °go through, °complete, °spend, °last, °endure, °survive: *Hepbourne served ten years in Leavenworth for that crime.*

service *n.* **1** °help, °assistance, °aid, °use, °usefulness, utility, °benefit, °advantage: *May I be of service to you? You could do me a great service by saying nothing of this to my mother.* **2** °maintenance, °overhaul, servicing, checking, °repair, mending: *Must we send the copier out for service or can they do it in the office?* **3** serving, °accommodation, amenities, waiting, °care: *The food is good but the service leaves much to be desired.* **4** °employment, °employ: *How long have you been in my service, Jiggs?* **5** °use, utilization, °usage, handling: *This old teapot has seen much service.* **6** °assignment, °post, °appointment, *Military Brit* secondment: *Oscar was in service in the Far East during the war.* **7** °rite, °ceremony, °ritual, °worship: *I shall be attending a memorial service for Grimes on Saturday.* **8** Often, ***services.*** army, °navy, air force, marines; armed forces *or* services, °military: *He was a pianist before joining the service.* **9** Often, ***services.*** talents, °help, professional care, °work, °advice: *You will need the services of a good accountant.* **10** °serve, serving; putting into play: *Clarke's tennis service is faster and more accurate than before.*

serviceable *adj.* **1** workable, working, °functional, functioning, usable, °useful, °operative, operating: *It seems unlikely that the old tractor would still be serviceable.* **2** hard-wearing, long-wearing, °durable, utilitarian, long-lasting, °tough, wear-resistant: *I really must buy a serviceable pair of hiking boots.*

servile *adj.* °submissive, subservient, °menial, craven, acquiescent, abject, cringing, slavish, °mean, fawning, deferential, mean-spirited, °vile, °low, °base, °ingratiating, °groveling, °obsequious, toadying, toadyish, sycophantish, sycophantic, truckling, wheedling, unctuous, °slimy, °flattering, °timeserving, bootlicking, *Colloq chiefly Brit* smarmy, *US and Canadian* apple-polishing, *Slang Brit* arse-kissing, *US* ass-kissing, *Taboo slang* brown-nosing, *Brit* arse-licking, *US* ass-licking: *Vespertillo is constantly trying to gain favor with the directors in the most servile way.*

servility *n.* submissiveness, °submission, subservience, servileness, acquiescence, abjectness, abjection, cringing, slavishness, meanness, fawning, mean-spiritedness, vileness, baseness, °groveling, obsequiousness, obsequence, toadying, toadyism, sycophancy, sycophantism, truckling, wheedling, unctuousness, sliminess, °flattery, bootlicking, *Colloq chiefly Brit* smarminess, *US and Canadian* apple-polishing, *Slang Brit* arse-kissing, *US* ass-kissing, *Taboo slang* brown-nosing, *Brit* arse-licking, *US* ass-licking: *We are sick of the servility with which he treats the directors.*

servitude *n.* °bondage, °slavery, thralldom, serfdom, subjugation, enslavement, °subjection, vassalage: *If captured, they were sentenced to penal servitude for their entire lives.*

session *n.* **1** sitting, °seating, °assembly, °conference, °meeting, hearing: *Congress is in session as this is being written.* **2** °term, °period: *At the next session of Congress the bill will be introduced again.*

set *v.* **1** °set down, °place, °put, °situate, °locate, °site, °plant, °position, °station, °stand, °lay, install *or* instal, °lodge, °mount, °park, °deposit, °plump, °drop, plunk *or* plonk (down): *Please set that blue box over here.* **2** °go down, °sink, °decline, °subside: *The sun is setting.* **3** jell *or* gel, congeal, °freeze, °solidify, °stiffen, °harden, clot, °coagulate, °thicken, °cake: *The concrete takes a long time to set in this weather.* **4** °establish, °fix, fasten on, °appoint: *The date of the wedding has been set for May 23.* **5** focus on, home *or* zero in on, pinpoint, °pin down: *When Ivor sets his mind on doing something it usually gets done.* **6** °adjust, °regulate, °turn, synchronize, °fix, °calibrate, °coordinate: *Set your watches to Greenwich Mean Time. Set the counter back to zero.* **7** °present, °introduce, °establish, °determine, °stipulate, °lay down, °define, °indicate, °designate, °specify; set *or* mark off, delineate: *Who set the pattern for such behavior? You'll have to set the rules.* **8** °prepare, °set up, concoct, °lay, °arrange, °fix: *I think we ought to set a trap for the burglar.* **9** °set forth, °propound, °present, °devise, work out *or* up, °make up, °select, °choose, °decide, °settle, °establish: *Who is going to set the questions for the examination?* **10** °arrange, °lay, °spread: *If you set the table, I'll serve dinner.* **11** °adjust, °move, °tilt, °fix, °place, °position, °lodge: *He set his hat at a jaunty angle and strode out of the restaurant.* **12** ***set about.*** **(a)** get *or* make ready, °start, °begin, get to work, set in motion, get under way, start the ball rolling, °break the ice, °undertake, °launch, °tackle, °address oneself to, enter upon, *Colloq* get cracking: *It's time you set about deciding what you want to do for a living.* **(b)** °attack, assail, °assault, beat up: *The gang set about him with clubs.* **13** ***set against.*** **(a)** °compare, °evaluate, °rate, °balance, °weigh, juxtapose, °contrast: *If you set his assets against his liabilities, he comes off pretty well.* **(b)** antagonize, set at odds, °alienate, °divide, disunite: *They succeeded in setting brother against brother.* **14** ***set apart.*** **(a)** °distinguish, °separate, °differentiate: *His talent for language sets him apart from the others.* **(b)** °reserve, put *or* set aside, °store, earmark, °put away, lay away, set by, °save, keep back: *I set apart this piece of chocolate just for you.* **15** ***set aside.*** **(a)** See **14 (b),** above. **(b)** annul, °cancel, nullify, declare *or* render null and void, °reverse, °repudiate, abrogate, °quash, °overturn, overrule, °discard: *The judge set aside the ruling of the lower court.* **16** ***set back.*** °put back, °hinder, °slow, °hold up, °retard, °delay, °impede, °obstruct, °stay, °thwart, °frustrate, °inhibit: *Completion has been set back because the materials failed to arrive.* **17** ***set down.*** **(a)** °write (down), put in writing, °put down, °record, °register, mark *or* jot down, °list: *The secretary laboriously set down every word the official said.* **(b)** ascribe, °assign, °attribute, °impute, °charge: *She set down her failure to marry to her mother's choosiness.* **(c)** °put down, °land: *See if you can set the plane down in that field over there.* **18** ***set forth.*** **(a)** °express, °voice, °propose, °propound, °state, °offer, °submit, °suggest, °broach, make a motion, °move: *Inez set forth the proposal that we break for dinner, a most welcome idea.* **(b)** set out *or* off, put forth *or* out, °begin, °start (out), get under way, °go, °embark, sally forth, °push off, °depart, °leave: *After much hemming and hawing, Clive finally set forth on his big adventure.* **(c)** °set out, °present, °declare, °describe, °propose, °state, articulate, °enunciate: *Hermione's theory is set forth in a lengthy report.* **19** ***set in.*** °begin, become established, °arrive, °come: *Spring has set in quite early this year.* **20** ***set off.*** **(a)** ignite, °kindle, detonate, °light, °touch off, trigger, °trip: *That little spark set off an explosion that destroyed the entire plant.* **(b)** °dramatize, °enhance, highlight, throw into relief, °show (off), °display: *The engraving is beautifully set off by the new mat and frame.* **(c)** See **18 (b),** above. **21** ***set on.*** °set upon, °attack, °assault, pounce on *or* upon, fall on *or* upon, fly at: *You were set on by a pack of dogs in the park?* **22** ***set out.*** **(a)** See **17 (b)** and **(c),** above. **(b)** °put out, °lay out, °arrange, °dispose, °display: *The stallholders are all setting out their wares for the jewelry fair this morning.* **23** ***set up.*** **(a)** °build, °put up, °erect, °assemble, °construct, °raise, elevate, put together, °arrange, °prepare: *They are setting up the Christmas displays over the weekend.* **(b)** °start, °begin, °initiate, °organize, °establish, °found; °finance, °fund, invest in, °back, °subsidize: *The Crockers are setting up their third shop in Halsted. His father set him up in business.* **24** ***set upon.*** °attack, °assault, °ambush, beat up, °fall upon, set about *Colloq* mug: *The gang set upon him as he was getting home from work.* —*n.* **25** °collection, °group, °combination, °number, grouping, °assortment, °selection, °arrangement, series: *Brackenhall bought a set of Hogarth etchings from the gallery.* **26** °clique, coterie, °group, °company, °circle, °ring, °crowd, °faction, °sect, °gang: *The set she associates with think nothing of flying to Rio for the weekend.* **27** °kit, °outfit, °rig; °equipment, °apparatus: *When I was ten, my father bought me a chemistry set.* **28** °setting, stage set *or* setting, °scene, *mise en scène,* mounting, scenery: *Hunter's set for* Waiting for Godot *won a prize.* —*adj.* **29** °fixed, established, °determined, °predetermined, arranged, prearranged, °decided, °customary, °usual, °normal, °regular, agreed, °conventional, °habitual, °definite, defined, °firm, unvarying, unvaried, unchanging, wonted, °rigid, °strict, settled, scheduled: *There is a set order in which these things are done.* **30** stereotyped, trite, hackneyed, °routine, °standard, °traditional, unchanged, unvaried, °invariable: *He made his set welcoming speech, as usual dripping with condescension, to the incoming classes.* **31** °prepared, °ready, °fit, primed: *Are you all set to go?*

setback *n.* °hindrance, °hitch, °check, °reverse, °reversal, °impediment, °block, °obstruction, °defeat, °holdup, °delay, °check, °rebuff, °upset; °relapse; *Colloq* hiccup: *The reduction in arts funding has resulted in some severe setbacks in certain programs. Derek has* *suffered a setback and will remain in the hospital for a while.*

setting *n.* mounting, scenery, °background, backdrop, °locale, °location, surroundings, °habitat, °home, environs, °environment, °milieu, °frame, °context, °site, °placement; stage set *or* setting, *mise en scène,* °scene: *It is quite interesting to see how the animals live in their natural setting. The setting of Hardy's novels is in the West Country.*

settle *v.* **1** °arrange, °order, °dispose, °organize, °straighten out, put in *or* into order, °compose, °sort out, classify, °coordinate, °resolve, set to rights, °reconcile: *Alben settled his affairs before undergoing surgery.* **2** Often, ***settle on*** *or* ***upon.*** °fix (on), °decide (on), °establish, °appoint, °set, °confirm, affirm, conclude, make sure *or* certain (of), °determine, agree (upon *or* on), °pick, °choose, °select: *Have you been able to settle on a suitable time to visit Max and Maddy?* **3** °decide, °reconcile, °resolve, put an end to, conclude, °clear up, °patch up, °adjust, °negotiate, mediate: *We have settled our differences.* **4** Often, ***settle down.*** take up residence, go *or* come *or* move to, °dwell, reside, make one's home, °abide, °remain, °stay, °live, set up home *or* house, put down roots, °locate, °inhabit, *US* set up housekeeping, °locate: *They have decided to settle in Australia after George retires.* **5** Sometimes, ***settle down.*** °light, alight, °land, come down, °put down, °set down, (come to) rest *or* (roost), °descend, °perch: *The butterfly settled on the tip of Alice's nose.* **6** °populate, °people, colonize, °pioneer: *That part of the country wasn't settled till late in the nineteenth century.* **7** Usually, ***settle down.*** °calm down, °subside, °quiet down, be *or* become tranquil, *Chiefly Brit* quieten (down): *The class settled down and we began the lesson. I must settle down and get back to work.* **8** °calm, °quiet, soothe, °tranquilize, °relax, *Chiefly Brit* quieten: *He took a bromide to settle his nerves.* **9** °subside, °sink, °decline, °fall: *A great crack appeared in the wall when the house settled a bit.* **10** Often, ***settle up.*** °pay, °square, °dispose of, °clear, °balance, liquidate, °discharge: *If you settle the bill, we can leave.* **11** Often, ***settle out.*** gravitate, °sink, °fall, °precipitate (out): *Stop shaking the container and let the sand settle to the bottom.* **12** °clarify, °clear: *After about an hour, the particles in the suspension will settle.*

settlement *n.* **1** colony, outpost, °post, °camp, community, encampment, village, hamlet: *The first colonists established a small settlement on the bank of the river.* **2** colonization, settling, populating, settling: *The settlement of the valley was mainly by Scandinavians.* **3** °agreement, °rapprochement, °resolution, °adjustment, elimination, °reconciliation, working out, °accommodation, arbitration, °arrangement: *Settlement of the dispute was supervised by an impartial judge.* **4** °payment, defrayal, °discharge, liquidation, °satisfaction, settling, quittance, clearing, °clearance: *The settlement of our debts is of paramount importance.* **5** deciding, settling, °setting, °decision, conclusion, confirmation, affirmation, °establishment, stabilization, °determination, °agreement, °choice, °selection: *The only thing on his mind was the settlement of regular days for the class to meet.*

settler *n.* colonist, frontiersman, frontierswoman, °pioneer, °immigrant: *The early settlers cleared the land for farming.*

setup *n.* **1** °arrangement, °system, °organization, layout, °regime, °structure, °makeup, °composition, framework, °frame, construction; °conditions, °circumstances: *This setup of interlocking directorships makes it difficult to see who answers to whom and who owns what. What, exactly, is the setup between you and Eustace?* **2** prearrangement, °trap, *Slang* put-up job: *As soon as I saw the police car, I knew we had walked into a setup.*

sever *v.* **1** cut off *or* apart *or* in two, lop *or* chop *or* hew *or* hack off, slice *or* shear off, °cleave, °dock, bob, dissever, °split, °separate, °divide, disjoin, °detach, °disconnect: *The animal's claw was severed from its toe by the trap.* **2** °separate, disunite, °dissolve, break off *or* up, °terminate, °end, °cease, °stop, °discontinue, °suspend, °abandon, put an end to: *The final divorce decree severed us from one another once and for all.*

several *adj.* **1** some, a few, not too *or* very many, a handful *or* a sprinkling *or* a number (of): *Several people asked for you at the party last night.* **2** °various, °sundry, a variety of, °diverse, °divers, °different, °respective, °individual, °distinct, disparate, °particular, °certain, °specific, °discrete, °dissimilar: *There are several stages one must go through before reaching the end.*

severe *adj.* **1** °strict, °harsh, rigorous, austere, °hard, °stony, stonyhearted, hardhearted, flinty, inexorable, ironhanded, °oppressive, unbending, °rigid, uncompromising, °relentless, unyielding, obdurate, pitiless, °punitive, °merciless, °unsympathetic, unfeeling, °cruel, °brutal, °mean, °savage, °inhuman, °beastly, °ruthless, °despotic, °dictatorial, °tyrannical, autocratic, °demanding, °exacting, °painstaking, °fastidious, exigent, taxing: *We were subjected to extremely severe treatment at Dotheboys Hall. Mr. Bradshaw was a severe taskmaster.* **2** °stern, °forbidding, °dour, glowering, °grave, °grim, °stiff, °strait-laced, °serious, unsmiling, °sober, °cold, °frigid, °aloof, austere: *If he so much as dropped a spoon the master would give Nicholas a severe look.* **3** °dangerous, °critical, °dreadful, °awful, life-threatening, °acute, dire; °mortal, °fatal, °terminal: *She has had a severe attack of the flu.* **4** stringent, °punitive, °harsh, °punishing, °burdensome, °tough, onerous, °grievous, °painful, Draconian *or* Draconic: *There are severe penalties for breaking the parking laws in this city.* **5** °harsh, °bitter, °cold, °inclement, °keen, °violent, °stormy, °intense, turbulent, °fierce, °wicked: *Granny couldn't live through another severe winter at Fort William.* **6** °stark, °bare, °plain, austere, °Spartan, ascetic, °primitive, °simple, °sparse, °spare, monastic, °modest, °unadorned, unembellished, °crude, undecorated, unembroidered: *The severe decor in the servants' quarters was almost prisonlike.*

severely *adv.* **1** acutely, °seriously, °badly, °dangerously, dreadfully; °permanently, fully, °entirely: *The accident left him severely handicapped.* **2** strictly, harshly, rigorously, austerely, oppressively, relentlessly, mercilessly, cruelly, brutally, savagely, inhumanly, tyrannically: *The boys were severely maltreated at the school.* **3** sternly, forbiddingly, dourly, gloweringly, gravely, °seriously, grimly, unsmilingly, soberly, coldly, coolly, austerely: *Mrs. Gladwin regarded me severely over the top of her pince-nez.* **4** stringently, punitively, harshly, onerously, grievously, °painfully: *One is punished severely for the slightest infraction of the law.* **5** °dangerously, acutely, critically, dreadfully, °awfully; mortally, fatally, terminally: *Your mother is severely ill, child.* **6** starkly, plainly, °barely, modestly, austerely, ascetically, monastically, primitively, °simply, crudely, sparsely, sparely: *The room is furnished a bit severely for my taste.*

severity *n.* **1** strictness, harshness, °rigor, rigorousness, austerity, hardness, flintiness, inexorability, inexorableness, stringency, °oppression, oppressiveness, rigidity, inflexibility, relentlessness, obduracy, obdurateness, pitilessness, mercilessness, coldbloodedness, abusiveness, cruelty, brutality, meanness, savagery, inhumanity, beastliness, ruthlessness, °despotism, °tyranny, fastidiousness, exigency: *The severity of the discipline was almost unbearable.* **2** coldness, aloofness, sternness, °gravity, grimness, frigidity, austerity, °solemnity: *The severity of his look froze me to the spot.* **3** dangerousness, acuteness, seriousness, °gravity, ferocity, fierceness, °virulence, °violence, °intensity: *The severity of the fits has diminished since we put the patient on the new medication.* **4** stringency, punitiveness, °punishment, harshness, onerousness, grievousness, painfulness, burdensomeness, oppressiveness: *The severity of the sentence is not inappropriate to the heinousness of the crime.* **5** harshness, inclemency, °violence, storminess, °intensity, ferocity, fierceness, °fury, furiousness, tempestuousness: *The severity of the weather has interfered with our maintenance of the electrical power lines.* **6** plainness, starkness, austerity, asceticism, bareness, modesty, °simplicity, primitiveness, spareness, sparseness, monasticism, crudeness: *The severity of the cells at the monastery was equaled by the austerity of the regimen.*

sew *v.* sew up, stitch, darn, °mend, °repair; sew on, °attach, °fasten; °tack, baste, hem: *Would you sew up a hole in my sweater for me and, while you're at it, would you mind sewing a button on my shirt?*

sex *n.* **1** gender: *It is difficult to determine the sex of some of the lower animals.* **2** sexual intercourse *or* relations, coitus, coition, mating, copulation, (sexual) congress *or* union, intimacy, lovemaking, making love, coupling, *Colloq* making out, going to bed, shacking up, *Slang* having it away, *Taboo slang* screwing, shafting, shagging, fucking, *Chiefly Brit* bonking: *They engaged in sex only after they were married.*

sexual *adj.* **1** °sex, reproductive, genital, procreative *or* procreant, progenitive, propagative: *The project is an investigation of sexual behavior in young adults.* **2** °sensual, °sensuous, °erotic, °carnal, fleshly, °voluptuous, libidinous, °earthy, bodily, °physical, °lustful, °animal, *Colloq* °sexy: *They have a deep sexual attraction to one another.*

sexy *adj.* **1** °erotic, arousing, °exciting, °sensual, °sensuous, °seductive, °suggestive, °provocative, °inviting, alluring, bedroom, °flirtatious, appealing, fascinating, °striking, °tempting, captivating, °enchanting, °stunning, *Colloq* come-hither: *She thought he had a sexy smile.* **2** °sex, °dirty, °pornographic, °obscene, °filthy, smutty, °lewd, °foul, °lascivious, °indecent, °explicit, °gross, X-rated, °vulgar, °rude, °coarse, °off-color, °risqué, titillating, °bawdy, ribald, °lusty, °immodest, °rough, indelicate, °suggestive, °unseemly, °improper, indecorous, °naughty, °shameless, *Slang* raunchy: *They sat there leering over a sexy magazine.*

shabby *adj.* **1** °worn, °dingy, faded, worn-out, °threadbare, tatty, °tattered, frayed, raggedy, °ragged, scruffy, °dilapidated, °ratty, °dirty, °bedraggled, °mangy, °run-down, °seedy, (much) the worse for wear, *Brit* down at heel, *US* down at the heel(s), *Colloq* grubby, scrubby, gungy, °tacky, *US* grungy: *The best clothes he could find to wear to the interview were very shabby.* **2** °poor, °peremptory, unpleasant, °nasty, °disagreeable, °mean, °contemptuous, demeaning, grudging, ungenerous, °impolite, °rude, unfriendly, unhelpful, °shoddy, ungentlemanly, unladylike, °dishonorable, °unworthy, °scurvy, *Colloq* °rotten: *We were subjected to very shabby treatment at the office where we applied for work.* **3** °dilapidated, tumbledown, broken-down, shattered, battered, °run-down, °ramshackle, °seedy, neglected, °dirty, squalid, slumlike, slummy, *Colloq* beat-up, crummy: *Sam lives in a shabby little shack near the railway yard.* **4** °mean, stingy, ungenerous, niggardly, °contemptible, °low, °lowly, °base, mean-spirited, °despicable, °vile, uncouth, discreditable, °inferior, °disreputable, °infamous, °abominable, °dishonorable, ignoble, °atrocious, ignominious, odious, detestable, opprobrious: *The shabby apology for their behavior fell on deaf ears.*

shack *n.* °hut, °hovel, shanty, °cabin, lean-to, *Colloq* °dump: *Sam's shack is built out of scrap lumber and cardboard boxes.*

shackle *n.* **1** Often, ***shackles.*** fetter(s), leg iron(s), chains, iron(s), bilboes, gyve(s), ball and chain, manacle(s), handcuff(s), restraint(s), bond(s), trammel(s), *Colloq* cuff(s), *Slang* bracelet(s), *Brit* darbies: *When those shackles went around my ankles I thought I'd never see them taken off.* **2** Usually, ***shackles.*** °restriction, °restraint, °deterrent, °impediment, °check, °obstacle, °obstruction, °barrier, °hindrance, °bar, °encumbrance: *When shall we throw off the shackles of ignorant prejudice and see a truly free world?* —*v.* **3** °chain, fetter, °manacle, handcuff, °bind, °restrain, °tie, °secure, truss, pinion, °tether: *Shackle that scurvy varlet to the mast! In this glorious spring weather I resent being shackled to a desk all day long.* **4** °restrain, °hold back, °check, °deter, °hinder, °discourage, °hobble, °handicap, °restrict, °curb, °rein, °bridle, °control, fetter, °inhibit, °limit: *Freedom of expression has been shackled during all those years of oppression.*

shade *n.* **1** °shadow, shadiness, dimness, duskiness, semidarkness, gloominess, murkiness, °dusk, °gloom, murk, darkness, °obscurity, *Literary* °shades: *We rested in the shade of an old oak to get out of the hot sun.* **2** °tint, tinge, °tone, °color, °hue, °intensity: *The curtains should be of a lighter shade to match the wallpaper.* **3** °hint, intimation, tinge, °suggestion, °modicum, sprinkling, soupçon, °trace, °suspicion, undertone, °overtone, °touch, °speck, °dash, nuance, atom, °grain, scintilla, iota, jot *or* tittle: *Did I detect a shade of sarcasm in your voice, Aloysius?* **4** °ghost, °specter, apparition, °phantom, phantasm, °spirit, wraith, °vision, banshee, *Colloq* spook: *As night came on, horrifying shades took shape among the leafless trees.* **5** °blind, window blind, curtain, Venetian blind; lampshade; °screen, °cover, covering, °protection, °veil, awning, canopy, °shield, °shelter, °umbrella, parasol: *We need shades to reduce the glare of the light. We had only a light tent as a shade against the broiling sun.* **6** fraction, hairsbreadth, °bit, °hair, *Colloq* smidgen: *Try moving the picture a shade to the left.* **7** °variation, °variety, nuance, °degree: *Do you really think his approval would make a shade of difference to me?* **8** **put (someone) in** *or* **into the shade.** °overshadow, °exceed, °surpass, °outstrip, outclass, °eclipse, outshine, °best, °better, °beat, °put to shame, outplay, outperform, °outdo, *Colloq* run rings *or* circles around, °show up: *Your record time in the thousand-meter put the other contenders in the shade.* **9** ***shades.*** sunglasses: *Take off those shades so I can see what you are thinking.* —*v.* **10** °screen, °protect, °shield, °shelter, °cover: *Shade your eyes from the direct sunlight.* **11** darken, °opaque, black out, °blacken: *The chiaroscuro effect would be heightened by more shading.* **12** °dim, °shadow, °veil, °blot out, cloud, °conceal, °hide, °obscure, °shroud, °screen, °mask, °camouflage, °disguise: *The house is shaded from view by trees on the south side.*

shadow *n.* **1** darkness, °gloom, dimness, °dusk, °obscurity: *In the shadow I could make out the dim outline of a figure.* **2** °cover, °screen, covering, °shield, °veil, curtain, °protection, concealment: *We stole past the sentry under the shadow of darkness.* **3** °hint, intimation, °suggestion, °suspicion, °trace, °vestige; °remnant: *That the event took place in Briantspuddle in 1901 is beyond any shadow of a doubt. Martin is merely a shadow of his former, robust self.* **4** cloud, °blight, °curse: *A terrible shadow has fallen over the family name.* **5** See **shade, 4,** above. **6** °companion, alter ego, °comrade, *Colloq* sidekick, crony, °chum, °(bosom) pal, *US* (bosom) buddy: *Boswell was Johnson's shadow for almost* *twenty years.* —*v.* **7** °follow, °trail, °track, dog, °stalk, °pursue, °trace, *Colloq* °tail, *US and Canadian* bird-dog: *For a month detectives shadowed the suspect.*

shadowy *adj.* **1** °dark, °shady, bowery, °leafy, shaded, °gloomy, °dusky, °dim, *Literary* bosky: *We opened our bottle lounging in a shadowy glen far from the rampaging city.* **2** °vague, °dim, °dark, °obscure, °faint, °indistinct, °indefinite, °hazy, ill-defined, unclear, indeterminate: *Perhaps we can get this shadowy part of the photograph enlarged to clarify it. The years before he came to Piddlington are part of his shadowy past.* **3** °spectral, °ghostly, °phantom, phantasmal, wraithlike, phantasmagoric(al), °illusory, °dreamlike, °imaginary, °visionary, chimerical, hallucinatory, °unreal, unsubstantial, °fleeting, impalpable, transitory, ethereal, °immaterial: *He has seen shadowy figures lurking near the tomb. The plaintiff's case is of a shadowy character.* **4** See **shady, 2,** below.

shady *adj.* **1** See **shadowy, 1,** above. **2** °indistinct, °indefinite, °questionable, °doubtful, °uncertain, °unreliable, °suspicious, °suspect, dubious, °shifty, °disreputable; °devious, °tricky, °slippery, underhand(ed), unethical, °unscrupulous, °dishonorable, °dishonest, *Colloq* °fishy, °crooked, °bent, not (strictly) kosher: *Hagny's shady past will be revealed in the next* Tribune. *I'll never forget the morning / That Grandpa ate the awning / To impress a pretty lady / Who went for men that were shady.*

shaft *n.* **1** °pole, °rod, °staff, °stick, °stem, shank, °handle, helve; °pillar, column, °post, stanchion, °upright: *The shaft of the harpoon splintered as the behemoth sounded to the depths. These shafts help support the floor above.* **2** °beam, °ray, °gleam, °streak, pencil: *At exactly noon on the winter solstice a shaft of light struck the jewel and it glowed like fire.* **3** °thrust, barb, °sting, dart, gibe *or* jibe, barb, *Colloq* °slap (in the face), °knock, °put-down: *After the episode in the restaurant, Barlow had to suffer the shafts of his co-workers' teasing attacks.* **4** mineshaft, °tunnel, adit, °well, °pit; airshaft, duct, flue: *How many men went down the shaft in the first shift? The screen over the ventilating shaft must be kept clean.*

shaggy *adj.* °hairy, °woolly, °unkempt, unshorn, uncut, hirsute, disheveled, matted, °untidy: *They have a large shaggy brown dog.*

shake *v.* **1** °quiver, °quake, °shudder, waver, wobble, °tremble, °shiver: *The poor child is shaking from the cold.* **2** wiggle, °wriggle, °squirm, shimmy, twitch, joggle, °jiggle, waggle, °sway, °swing, °roll, °bump, °grind, °vibrate, °oscillate, °pulsate, °gyrate: *When Lola shakes her hips the strongest men have been known to weaken.* **3** °weaken, undermined, °impair, °harm, °damage, °discourage; disenchant, °disappoint, disaffect: *The incident shook my faith in the political party I had supported all my voting life.* **4** °wave, brandish, °flourish, °display, °show off, °parade, °exhibit, vaunt, waggle, °flap, °flutter: *It was a thrill to see the once-banned flag being shaken under the noses of the deposed leaders.* **5** Often, ***shake up.*** °agitate, °stir (up), °mix (up); °upset, °distress, °frighten, °scare, °shock, °disturb, °unnerve, unsettle, disconcert, °discomfit, °worry, °fluster, disquiet, confound, °confuse, °perplex, °puzzle, *Colloq* °rattle, °get to, *US* °throw (for a loop): *Shake the medicine before using it. I was really shaken by the news that Sheila had quit.* **6** ***shake down.*** **(a)** °break in, °condition, °test, °prove, *Colloq* debug: *We shook down the submarine on a thirty-day cruise.* **(b)** °blackmail, extort *or* extract *or* squeeze *or* wrest money from, °hold up, °squeeze, °threaten: *The shopkeepers were so frightened that the gang easily shook them down.* **7** ***shake off.*** °get rid of, °discard, dislodge, °drop, °brush off, °elude, °evade, °lose, °throw off, rid oneself of, give the slip to: *We finally shook off our pursuers and felt safe at last.* —*n.* **8** shaking, quivering, quaking, shuddering, wavering, wobbling, trembling, shivering, °quiver, °quake, °shudder, waver, wobble, °tremble, °shiver, wiggle, °wriggle, twitch, joggle, °jiggle, °sway, °swing, °roll, gyration: *The dying beast gave a shake of its massive head and expired.* **9** agitating, °agitation, shaking, °stirring (up), °jolt, °jar, jarring, jounce, jolting, jouncing: *Give him a good shake and he'll wake up.* **10** ***the shakes.*** trembling, tremors, delirium tremens, *Colloq* D.T.'s: *She's had so much to drink that she's got the shakes.*

shake-up *n.* reorganization, rearrangement, °overhaul, °revamp, restructuring, rehabilitation, makeover, realignment: *With new owners you can expect a complete shake-up of top management.*

shaky *adj.* **1** °uncertain, wobbly, °unstable, °precarious, °unsound, °flimsy, °weak, unsteady, unsupported, unsubstantiated, undependable, °unreliable, °tenuous, untrustworthy, dubious, °questionable, °doubtful, *Colloq* iffy: *The evidence against him is very shaky indeed.* **2** wobbly, wobbling, °unstable, °precarious, °dilapidated, °ramshackle, °on its last legs, °decrepit, falling down *or* apart, °rickety, °flimsy, unsteady, °insecure, °unsound, unsubstantial, °insubstantial, °feeble: *Surely, you can't expect that shaky old chair to support your weight!*

shallow *adj.* **1** °surface, °skin-deep, °superficial, °thin, °empty, °flimsy, trivial, unimportant, °slight, °frivolous, °idle, °foolish: *If you read only comic books, your thoughts are bound to be shallow.* **2** Often, ***shallows.*** shoal(s), sandbar, sandbank, bank, shelf: *This boat draws too much water to go fishing in the shallows.*

sham *n.* **1** °fake, °fraud, °counterfeit, °imitation, °hoax, humbug, °pretense, °forgery, °copy, imposture, *Colloq* phoney *or US also* phony: *There was no robbery—it was all a sham to add a bit of excitement to the party.* —*adj.* **2** °fake, °fraudulent, °counterfeit, °imitation, paste, simulated, °false, make-believe, °fictitious, made-up, °bogus, °spurious, °mock, ersatz, °artificial, °synthetic, *Colloq* phoney *or US also* phony, pseudo: *That was not only a sham Dutch Master, it was a sham Van Meegeren!*

shambles *n.* °chaos, devastation, °mess, °disaster, Augean stables, pigsty, °muddle, pigpen: *Why do teenagers have an innate compulsion to create a shambles in their bedrooms?*

shambling *adj.* shuffling, lumbering, dragging, scuttling, °awkward, °clumsy, lurching, unsteady, faltering: *The old beggar hobbled off with a shambling gait.*

shame *n.* **1** °embarrassment, °humiliation, mortification, chagrin, ignominy, shamefacedness, loss of face, abashment: *Watts felt abysmal shame at being discovered stealing from the petty cash.* **2** °disgrace, ignominy, °dishonor, disrepute, °degradation, opprobrium, vilification, calumniation, °infamy, obloquy, odium, °contempt, °scandal, denigration, loss of face, defamation, °discredit, disesteem, °disfavor, derogation, disparagement: *To think that one knight's single act of cowardice in 1249 could have brought such shame to all his descendants.* **3** °pity, °calamity, °disaster, °catastrophe; °outrage: *What a shame that Gerda failed the exam!* **4** °humility, modesty, (sense of) decency *or* decorum *or* propriety, respectability, decorousness, diffidence, shyness, coyness, prudishness, timidity, shamefacedness: *The people who wear bathing suits like that obviously have no shame at all.* **5** ***put to shame.*** **(a)** °surpass, °eclipse, outclass, °overshadow, cast into the shade, °outdo, °outstrip, outshine, °show up: *Cochrane put the other tennis players to shame when they failed to score a point against him.* **(b)** See **8,** below. —*v.* **6** °embarrass, °humiliate, °mortify, °humble, chagrin, disconcert, discountenance, °put down, bring down, abash, °chasten, *Colloq* bring (someone) down a peg, °suppress, °subdue: *He hated the teacher because she had shamed him in front of his friends.* **7** coerce, °force, °drive, °bully, °push; °embarrass, °humiliate, °mortify, °humble: *The other boys shamed him into admitting that he had done it.* **8** °disgrace, °embarrass, °dishonor, °scandalize, calumniate, °degrade, °debase, defame, °discredit, °stigmatize; °smear, °blacken, °stain, °taint, besmirch, °tarnish: *I am sure, son, that nothing you ever do would shame the family or me.*

shamefaced *adj.* **1** °bashful, °shy, °modest, self-effacing, diffident, °timid, °meek, °coy, °sheepish, timorous: *He felt shamefaced as a schoolboy when they awarded him the medal for saving the girl's life.* **2** °ashamed, shamed, abashed, °embarrassed, humiliated, dishonored, mortified, humbled, chastened, chagrined, uncomfortable, discomfited, °remorseful, red-faced: *Shamefaced, she finally admitted the truth about where she had been.*

shameful *adj.* °disgraceful, °dishonorable, °base, °low, °mean, °vile, °degrading, °indecent, inglorious, °deplorable, discreditable, °corrupt, °disreputable, °infamous, ignominious, humiliating, °embarrassing, mortifying, humbling, chastening, discomfiting, shaming, blameworthy, °scandalous, °outrageous, unprincipled: *Quentin was known for his shameful behavior both at school and in the army.*

shameless *adj.* °wild, °flagrant, unreserved, °uncontrolled, °immodest, °wanton, indecorous, °indecent, °rude, °improper, °forward, °bold, unembarrassed, unblushing, °audacious, °brazen, °brash, °unabashed, unashamed, °impudent, °shocking, °outrageous: *That dancing involves some shameless actions seldom witnessed in politer venues.*

shape *n.* **1** °form, °figure, °build, °body, °physique; °lines, °profile, °silhouette, contours: *I've never heard anyone complain about Marilyn Monroe's shape.* **2** °form, °pattern, configuration, °structure, °aspect: *The discussion is on the shape of things to come in the plumbing trade. Our new swimming pool will be in the shape of an oval.* **3** °form: *Gradually, the shopping mall began to take shape.* **4** °state, °condition, fettle, °status, °(state of) health, °order, °trim: *When I saw Willis yesterday, he seemed in very good shape indeed.* **5** °guise, °disguise, °form, °appearance, °likeness, °image: *Zeus visited her in the shape of a golden shower—or so she claims.* —*v.* **6** °form, °fashion, °mold, °cast, °make, °model, °sculpture, sculpt; °cut, °carve, hew, °hack, °trim: *Using primitive tools, they were able to shape the clay into bowls. The topiary shrubs were shaped into animal forms.* **7** °determine, give form to, °control, °govern, °regulate, °affect, °condition, °influence, °decree, °frame, °define: *There are many factors that shape our future.* **8** °word, °express, embody in words, °put, °formulate, °form: *Perhaps you could shape your question differently.* **9** °change, °modify, remodel, °accommodate, °fit, °adapt, °adjust: *The box was shaped especially for this bottle.* **10** ***shape up.*** **(a)** take form, take shape, °develop, evolve, °proceed: *The plan is shaping up very nicely.* **(b)** °conform, °improve, °progress, go *or* move *or* come along, show improvement, come up to snuff; behave better: *Those who cannot shape up will have to ship out.*

shapeless *adj.* **1** amorphous, formless, °nebulous, unformed, °indefinite, unstructured, °vague: *Sculptors start with a shapeless mass of clay.* **2** unshapely, °deformed, °misshapen, distorted, twisted, °bent, battered: *The wreckage was a shapeless tangle of steel and mortar.*

shapely *adj.* curvaceous, °comely, well-proportioned, °graceful, °neat, well-turned-out, good-looking, °pleasing; °voluptuous, *Colloq* °sexy: *That slinky black dress certainly sets off her shapely figure.*

share *n.* **1** °portion, °allotment, °division, apportionment, allocation, °ration, appropriation, dispensation, °allowance, °part, °due, °percentage, °interest, dividend, °quota, °portion, °helping, serving, *Colloq* °cut: *Each of us is entitled to a share in the profits. We made sure that all participants got their shares.* **2** °interest, °piece, °part, °stake, °equity, °slice: *Do you own any shares in the stock market? My share in United Telephone is worth twice what I paid for it.* —*v.* **3** share out, °divide up, °allot, apportion, allocate, °ration, °appropriate, share in, °split, °partition, parcel *or* deal *or* dole out, °pay out: *Ownership in the company is shared equally among the employees. Let's share the car expenses this weekend.*

sharp *adj.* **1** °acute, °keen, edged, razor-sharp, knife-like, knife-edged, sharpened; °pointed, needle-sharp: *How do you keep the telescope in sharp focus? You need an extremely sharp knife for boning fish. This pencil isn't sharp enough.* **2** °abrupt, °sudden, °precipitous, °sheer, vertical, °marked: *Just round the corner is a sharp dip in the road. There was a sharp rise in prices on the London Stock Exchange today.* **3** °keen, keen-witted, keen-minded, sharp-witted, °shrewd, °intelligent, °smart, °alert, °bright, °quick, °agile, °astute, °clever, on the qui vive, °penetrating, °observant: *Penny has a sharp mind and good eye for detail.* **4** °caustic, °bitter, °biting, acrid, °hot, °spicy, °pungent, piquant, tangy, °harsh, °sour, acid, acidulous, °tart: *That chili sauce is a little too sharp for my taste.* **5** acid, acidulous, acerbic, vitriolic, acrimonious, °cutting, piquant, °biting, °bitter, °unkind, °strict, °hurtful, °spiteful, °virulent, °sarcastic, °sardonic, °trenchant, °severe, °scathing, malicious, °nasty, °malignant, °venomous, °poisonous: *How does one escape the sharp comments of the critics? I was unprepared for the sharp exchange between the guests on my talk show.* **6** °clever, °shrewd, °artful, °crafty, °sly, cunning, °foxy, °calculating, °unscrupulous, °dishonest, *Colloq* °sneaky, °fly: *There is evidence that the stock manipulators were engaged in very sharp practice.* **7** high-pitched, °shrill, °penetrating, °piercing, °strident, °harsh, earsplitting, °loud: *Jane gave a sharp cry when she felt the ice break under her feet.* **8** °poignant, °severe, °cutting, °intense, °sudden, °piercing, °extreme, °acute, °fierce: *Complaining of a sharp pain in his chest, he fell down.* **9** °chic, °dapper, °spruce, °stylish, °smart, °fashionable, °dressy, *Colloq* °snappy, natty, classy, °nifty, °swell, °swanky: *If you want to see the latest collection of sharp clothes, go to the disco.* —*adv.* **10** °sharply, °precisely, °exactly, punctually, °on the dot, *Colloq* on the button, *US* on the nose: *Our appointment was for seven o'clock sharp.* **11** °sharply, °suddenly, abruptly: *I was pulled up sharp by her mention of the Black Rose.* **12** °sharply, alertly, attentively, vigilantly, watchfully, carefully: *Listen sharp to what I tell you.*

sharpen *v.* hone, °grind, strop, °whet: *Of what use is a knife if you don't sharpen it periodically?*

sharp-eyed *adj.* sharp-sighted, eagle-eyed, hawk-eyed, lynx-eyed, gimlet-eyed, keen-sighted, wide-awake, °wakeful, watchful, °observant, °(on the) alert, on the qui vive, °wary, circumspect, Argus-eyed, °cautious, °careful: *It takes quite a sharp-eyed person to check the accuracy of these circuits.*

sharply *adv.* **1** °severely, sternly, harshly, cuttingly, acerbically, peremptorily, angrily, strictly, °firmly: *Mrs. Rottenmayer sharply rebuked Heidi for smiling.* **2** °suddenly, °quickly, abruptly, precipitously, precipitately: *Prices advanced sharply for the third day in succession.* **3** acutely, distinctly, °definitely, definitively: *These new cameras automatically remain sharply focused on the object.*

shatter *v.* **1** °disintegrate, °burst, °pulverize, °shiver, smash, °demolish, °break (to smithereens), °splinter, °fragment, °fracture, dash to pieces: *If you hadn't caught that Ming vase, it would have shattered into a million pieces.* **2** °destroy, °ruin, °devastate, °wreck, °dash, °crush, °demolish, torpedo, °undermine, °blast: *Any dreams he might have had of being published were shattered by the rejection letter.* **3** °upset, °disturb, °perturb, °trouble, °unnerve, °overcome, °overwhelm, °crush, °devastate, °depress, deject, °rattle, °shake (up), unsettle, °agitate, confound, °confuse, stupefy, °daze, °stun, °paralyze, *Colloq* °throw: *He was shattered by the news of Mill's death.*

shave *v.* **1** shear (off), °cut (off), °trim, °clip, crop, snip off, °remove: *Don't shave off your beard—it hides more of your face.* **2** °pare, °scrape, °plane, °whittle: *If you shave just a hair off this side, it will fit into the hole.* —*n.* **3** ***close shave.*** narrow escape, *Colloq* narrow *or* near squeak, *US* squeaker: *Was that ever a close shave!—You almost fell out the window!*

shed[1] *n.* lean-to, °shelter, °structure, °addition, penthouse, °hut, °shack, °stall, °booth, °pen, cote, hutch: *The tools are in the shed alongside the barn.*

shed[2] *v.* **1** °spill: *Much blood has been shed in the name of freedom.* **2** °shine, °spread, °scatter, °throw, °cast, let fall, °impart, °release, °focus, pour forth, let fall, °radiate: *Can you shed a little light on a matter that has puzzled me for years?* **3** pour *or* stream *or* flow *or* surge *or* spill (out *or* forth), °discharge, °emanate, °emit, °drop; exude, °ooze, °weep: *The gentle rain sheds its life-giving moisture on the world below. Don't shed any tears over me.* **4** °cast off, doff, °drop, °abandon; molt, defoliate, desquamate, °peel (off), °flake (off): *The trees shed their leaves. The ecdysiast sheds her clothing as the snake sheds its skin.*

sheen *n.* °shine, °gleam, °polish, °luster, shininess, burnish, brightness, °gloss, °glow, glimmer, °shimmer, °radiance, glint, °dazzle: *The sheen of the horse's coat glistened in the sunlight.*

sheepish *adj.* **1** °timid, °withdrawn, °passive, docile, °obedient, compliant, sheeplike, manipulable, °tractable, °pliable, °meek, amenable: *Dashworth might seem sheepish but he doesn't like to be bullied.* **2** See **shamefaced,** above.

sheer *adj.* **1** °steep, °precipitous, °abrupt, °perpendicular, °bluff, vertical: *There is a sheer drop of two thousand feet to the valley below.* **2** °absolute, °unmitigated, °unqualified, °downright, °out-and-out, unalloyed, unadulterated, °pure, unmixed, °plain, °simple, °rank, °total, °complete, arrant, °thorough, thoroughgoing, utter: *Johnson said that he had got his definition of "pastern" wrong out of "sheer ignorance."* **3** °transparent, °see-through, °thin, diaphanous, °filmy, gauzy, gossamer, translucent, peekaboo: *Kathryn was wearing the sheer nightgown I had given her for her birthday.*

sheet *n.* **1** bedsheet, cribsheet, fitted sheet, flat sheet, *US* contour sheet: *How often do you change the sheets on your bed?* **2** leaf, folio, °page: *He tore a sheet of paper from the notebook.* **3** °pane, panel, °plate, °slab: *The desk has a sheet of glass on top for protection.* **4** lamina, lamination, layer, °stratum, °veneer, membrane: *Plywood is made up of a number of thin sheets of pressed wood bonded together for strength.* **5** °area, °expanse, °stretch, layer, °film, °coat, coating, covering, blanket, °cover, °surface, °skin, °veneer: *The sheet of ice on the road made driving hazardous.* **6** newspaper, °journal, °paper, tabloid, gazette, °daily, weekly, monthly, *Colloq* °rag: *Have you an extra copy of the sheet that published the article about you?*

shell *n.* **1** cartridge, °projectile, °shot: *We huddled in the foxhole as the shells whistled overhead.* **2** °exterior, °outside, façade, framework, °frame, chassis, externals, skeleton, °hull: *After the earthquake, only the shell of the building was left standing.* —*v.* **3** shuck, husk, °peel, °hull, excorticate, decorticate: *After the harvest, we all helped to shell the pecans.* **4** fire on *or* upon, °bombard, barrage, °attack, °bomb, blitz, cannonade, *Slang Brit* prang: *The enemy shelled our position relentlessly.* **5** ***shell out.*** °pay out, °give out, disburse, pend, °expend, °hand over, °hand out, *Colloq* °lay out, fork out, dish out, *Chiefly US* ante up: *I had to shell out a month's salary for that suit.*

shelter *n.* **1** °protection, °cover, °refuge, asylum, °sanctuary, haven, °safety, °security: *During the hailstorm, we sought shelter in a cave.* **2** °cover, covering, concealment, °screen, °umbrella: *Under the shelter of a moonless night, we stole quietly past the guards.* **3** dwelling place, habitation, °home, °dwelling, °housing, °accommodations: *He felt he was entitled to food, clothing, and shelter.* —*v.* **4** °protect, °screen, °shield, °safeguard, °guard, °keep, °secure, °harbor: *The parapet sheltered the sentinel from the icy wind.* **5** seek *or* take refuge *or* shelter, hole up, *Colloq* lie *or Colloq* lay low: *Muggleswick sheltered in West Hagbourne till the hue and cry died down.*

shelve *v.* °postpone, °defer, put off *or* aside *or* on ice *or* on the shelf, pigeonhole, lay aside, hold in abeyance, *US* °table: *We must shelve this matter till we have more information.*

shepherd *v.* °lead, convoy, °escort, °conduct, °guide, usher, °take, °pursue: *I want you, personally, to shepherd Miss Jones through the plant. As this project requires special care, I shall shepherd it through personally.*

shibboleth *n.* °password, catchword, catch phrase, buzzword, °byword, watchword, °jargon: *As Shep knew none of the shibboleths of teenagers' slang, he remained an outsider to his classmates.*

shield *n.* **1** °protection, °guard, °safeguard, °defense, °screen, °bulwark, °shelter: *Education is a poor shield against poverty.* —*v.* **2** °protect, °guard, °safeguard, °keep, °defend, °screen, °shelter: *The mother bear will shield the cubs from harm or die in the attempt.*

shift *v.* **1** °move, change position; °edge, budge, relocate, rearrange, °transpose, °switch: *Shift the picture slightly to the left. Marla shifted over closer to me on the bench. I almost broke my leg because you shifted the chair as I was about to sit on it.* **2** Usually, ***shift for (oneself).*** °manage, °succeed, °make do, °look after, °take care of, get *or* scrape by *or* along, fend for (oneself), °make it, paddle one's own canoe: *Lorenzo prefers to live alone and shift for himself.* **3** °sell, °market, °move: *The object of advertising is to shift merchandise.* —*n.* **4** hours, °stint, °schedule; work force, relay, °crew, cadre, °staff, workers, °squad, °team, °corps, °group, °party, °gang: *I prefer to work the morning shift. The night shift comes on at four o'clock.* **5** °change, °movement, °switch, °transfer, °swerve, deflection, veer: *Owing to a wind shift, it became impossible to maneuver among the rocks.* **6** smock, chemise, muumuu; kaftan *or* caftan: *In this weather a light cotton shift is most comfortable.*

shiftless *adj.* unambitious, °lazy, °indolent, uninspired, unmotivated, °idle, °lackadaisical, °irresponsible, uncaring, unenterprising, °aimless, °slothful, otiose, °ineffective, ne'er-do-well, °good-for-nothing, fainéant, pococurante: *He calls Mr. Porbeagle a shiftless vagabond, virtually a tautology.*

shifty *adj.* °tricky, °artful, °shrewd, canny, cunning, °foxy, °wily, °sharp, °devious, °slick, °evasive, °smooth, °slippery, °scheming, °designing, conniving, °calculating, underhand(ed), conspiratorial, treacherous, °trai-

S

torous, °deceitful, deceiving, duplicitous, °two-faced, Janus-faced, °dishonest, °crooked, untrustworthy, *Colloq* °bent: *The shifty scoundrel gave me a bad check and then left town.*

shilly-shally *v.* hem and haw, dillydally, teeter-totter, °seesaw, yo-yo, vacillate, waver, °alternate, °fluctuate, dither, falter, tergiversate, *Brit* haver, hum and haw, *Scots* swither; °delay, °hesitate, °dawdle: *Stop shilly-shallying and decide whether you want chocolate or strawberry!*

shimmer *v.* **1** °shine, °gleam, °glow, glimmer, glint, °glisten, °ripple, °flicker: *The desert shimmered like a lake under the relentless sun.* —*n.* **2** shimmering, °shine, °gleam, °glow, glimmer, glint, °gloss, °flicker, °light: *As Julia slunk up to the microphone, the spotlight picked up the shimmer from her black satin gown.*

shin *v.* Usually, **shin up.** climb, clamber up, °scramble up, °scale, *US* shinny up: *The lad shinned up the flagpole carrying the banner.*

shine *v.* **1** °gleam, °glow, °shimmer, °radiate, °beam, °glare, °flare, °glisten, °glitter, coruscate, °twinkle, °sparkle, scintillate, glint, °flash, °flicker: *The light of a candle shone from her window. The stars are shining brightly. I want that brass polished till it shines.* **2** °polish, burnish, rub (up *or* down), buff, °brush, °brighten: *You'd best shine your shoes before going for the job interview.* **3** °excel, °surpass, °stand out, outshine, be outstanding *or* preeminent *or* excellent *or* prominent *or* conspicuous: *Athletes who shine at tennis are unlikely to be weightlifters.* —*n.* **4** °gleam, °glow, °shimmer, °sparkle, brightness, °radiance, °gloss, °luster, °sheen, °glaze, patina: *That shine was achieved with a little wax.* **5** ***take a shine to.*** °like, be attracted to, take a fancy to, °fancy: *Her Great Dane has taken a shine to me and likes to sit on my lap.*

shiny *adj.* gleaming, °glowing, shimmering, °glossy, shimmery, °lustrous, °glassy, °radiant, °bright, beaming, glistening, °polished, burnished, glittering, °dazzling, coruscating, °twinkling, sparkling, °scintillating, glinting, flashing, °flashy, flickering, lambent, fulgent: *She loves riding about in her shiny new car.*

ship *n.* **1** °vessel, (ocean *or* passenger) liner, steamer, windjammer, cutter: *The statue is dedicated to the men who go down to the sea in ships.* —*v.* **2** °send, °move, ferry, °transport, °deliver, °carry, dispatch *or* despatch, °freight, °haul, °truck, °cart: *You need a specialized company that ships computers and other delicate equipment.* **3** ***ship out.*** °leave, °depart, °embark, set sail, °take off, get out, °quit, *Slang* scram: *You'll ship out if you know what's good for you.*

shipshape *adj.* °neat, °trim, °spotless, °orderly, Bristol fashion, spick-and-span *or* spic-and-span, °tidy: *Pick up your clothes and make this room shipshape at once!*

shirk *v.* °avoid, °evade, °shun, °dodge, °get out of, °shrink from, *Colloq* °duck (out of), *Brit* skive, *Military Brit* scrimshank, *US* goldbrick: *Stop shirking your responsibilities and get to work! Is Calliger shirking again?*

shiver[1] *v.* **1** °shake, °quake, °tremble, °shudder, °quiver: *We sat in our wet clothes, shivering in the cold. The old shack shivered in the wind.* **2** °vibrate, luff, °flap, °flutter, °chatter, °rattle, °shake, wallop: *The sails shivered violently in the stiff breeze.* —*n.* **3** °shake, °quake, °tremble, °shudder, °quiver, °thrill, *frisson,* trembling, tremor, °flutter: *She gave a little shiver and then snuggled into my arms.* **4** ***the shivers.*** trembling, shivering, goose pimples *or* bumps, *Colloq* °the shakes, *US* the chills: *He's had malaria and occasionally gets the shivers. Those horror films give me the shivers.*

shiver[2] *v.* °shatter, °fragment, °splinter, °disintegrate, °rupture, °explode, implode, smash (to smithereens), °crash: *The light bulb fell and shivered into a million shards.*

shock *v.* **1** °startle, °surprise, °stagger, °jar, °jolt, °shake (up), °stun, °numb, °paralyze, °daze, stupefy, dumbfound *or* dumfound, bowl over, °appall, °astonish, °astound, °frighten, °scare, °petrify, traumatize, °horrify, °outrage, °disgust, °nauseate, °repel, °revolt, °sicken, °upset, disquiet, °disturb, °perturb, discompose, unsettle, *Colloq* °throw, *US* throw for a loop, flabbergast, give (someone) a turn: *He shocked us when he announced his resignation. We were shocked to learn of the train wreck.* —*n.* **2** trauma, °stupor, paralysis, °prostration, °breakdown, °collapse, nervous exhaustion: *He is in a state of shock after the accident.* **3** °surprise, thunderbolt, bolt from the blue, °bombshell, °revelation, shocker, eye-opener, °jolt: *It came as a shock to learn whom I had been talking with so casually. The death of her husband was a terrible shock.* **4** tingle, °jolt, °impact: *You'll get a nasty shock if you touch those wires.*

shocking *adj.* **1** surprising, astounding, astonishing, °amazing, °striking, stupefying, numbing, °sudden, unexpected, electrifying, °startling, *Colloq* mind-boggling, mind-blowing: *It was shocking to see the way the neighborhood had deteriorated.* **2** °disgusting, °revolting, nauseating, °nauseous, sickening, °repulsive, °abominable, °hideous, °horrible, horrifying, horrific, horrid, °foul, °loathsome, °abhorrent, °ghastly, °hideous, unspeakable, °dreadful, distressing, °outrageous, appalling, °monstrous, °scandalous: *The conditions in the concentration camps were shocking.*

shoddy *adj.* °shabby, tatty, °inferior, °poor, rubbishy, °cheap, pinchbeck, meretricious, °tawdry, °gaudy, brummagem, °plastic, °artificial, tinsel, tinselly, second-rate, trashy, junky, *Colloq* crappy, cheap-jack, °tacky, *US* chintzy: *They can't get away for long with selling such shoddy merchandise.*

shoo *interj.* **1** Scat!, Go away!, Away with you!, Be off!, Get out!, Begone!, *Colloq* Get lost!, Beat it!, Scram!: *I shouted "Shoo!" and the chickens scattered.* —*v.* **2** Often, ***shoo away*** *or* ***off.*** scare off, frighten away, drive away, force to leave: *She shoos away anyone who disturbs her while she is painting.*

shoot *v.* **1** scoot, dart, °whisk, °speed, °bolt, °run, °race, °rush, °flash, °fly, °dash, °hurtle, °streak, scuttle, °bound, °leap, °spring, *Colloq* zip, whiz: *The instant the baby seal cried out, the mother shot over to protect it.* **2** °discharge, °fire, open fire; °let fly, °launch, °propel, °project, °fling, °hurl, °throw, °toss: *Don't shoot till you see the whites of their eyes! With all his strength he shot the arrow into the beast's neck.* **3** °wound, °hurt, °harm, °injure; °kill, slay, assassinate, °execute, *Slang* fill *or* pump full of lead, °plug, °blast, °zap, °knock off, snuff (out): *He insulted her, so she shot him. He's to be shot at sunrise.* **4** °sprout, germinate, burgeon, °flourish, °grow, °spring up, mushroom, °develop: *Owing to the perfect weather, the corn shot up very quickly.* —*n.* **5** °sprout, °stem, bud, °branch, °offshoot, °slip, scion, °sucker: *If you want fruit, the new shoots have to be pinched back.*

shop *n.* **1** °store, boutique, department store: *Would you stop at the shop and buy some lettuce?* **2** workshop, machine shop: *Henry gained experience in operating a lathe at my father's shop.* —*v.* **3** °betray, inform on *or* against, *Slang* peach on, rat on, snitch on, blow the whistle on: *The police persuaded Lucchese to shop his accomplices in exchange for a lighter sentence.* **4** ***shop for.*** °buy, °purchase, °seek, °look for, °research: *I've been shopping for a new dishwasher.*

short *adj.* **1** °small, °little, °slight, °petite, °diminutive, °wee, °tiny, °elfin, minuscule; midget, dwarfish, squat, °dumpy, runty, stubby, °stunted, *Colloq* pint-sized, knee-high to a grasshopper, sawn-off: *Did you know that Queen Victoria was quite short?* **2** shortened, °brief, °concise, compressed, compendious, °compact, °pocket, *US* vest-pocket; °abbreviated, abridged, °cut: *A short version of the book was published in paperback.* **3** laconic, °terse, °succinct, pithy, sententious, °epigrammatic: *He made a few short remarks that were very much to the point.* **4** °abrupt, °curt, °terse, °sharp, °blunt, °bluff, °brusque, °sharp, °offhand, °gruff, °testy, °snappish, °discourteous, uncivil, °impolite: *It is unnecessary to be so short with the staff.* **5** °direct, °straight, straightforward, short and sweet: *In reply to your request for permission to leave early, the short answer is "No."* **6** Usually, ***short of.*** °deficient (in), lacking (in), needful (of), °wanting, °inadequate, °shy (of), °low (on): *The hotel is short of clean linen because the laundry failed to deliver.* **7** °brief, °limited; transitory, °temporary, °short-lived, °momentary, °quick, °transient: *HW had a short life but a happy one. I'll just make a short stop in here and will join you in a moment.* **8** impecunious, °straitened, pinched, underfunded, °poor, penniless, °deficient: *I'm a bit short today and wonder if you could lend me some money?* **9** ***in short supply.*** °rare, °scarce, °scanty, unplentiful, °meager, °sparse, *Colloq chiefly Brit* °thin on the ground: *Good editors are in short supply these days, Grimthorpe.* **10** ***short of.*** °before, °failing, excluding, °exclusive of, °barring, eliminating, precluding, excepting, °except for, leaving out, °apart from, setting aside: *Short of killing him, I am not sure what she could have done in the circumstances.* —*adv.* **11** abruptly, °suddenly, peremptorily, without warning, instantly, unexpectedly, hurriedly, °hastily, °out of the blue: *She stopped short in the middle of the road and made me get out.* **12** ***cut short.*** **(a)** °trim, °curtail, °shorten, °abbreviate, °cut: *I must cut my speech short because I have a train to catch.* **(b)** °stop, °cut off, °terminate, cut in on, break in on, °interrupt; °butt in: *I am sorry to cut you short, but it is getting on to midnight.* **13** ***fall*** *or* ***come short.*** °fail, be *or* prove inadequate *or* insufficient: *These grades fall short of our expectations, Bobby.* —*n.* **14** ***in short.*** °briefly, °in a word, all in all, to make a long story short, in a nutshell: *He told me, in short, that there was no job and there never had been one.* **15** ***shorts.*** Bermuda shorts, knee breeches, knee pants, hot pants: *I usually wear shorts to play tennis.*

shortage *n.* °deficit, deficiency, shortfall, °dearth, °scarcity, °lack, °want, paucity: *They are predicting a shortage of water again this summer.*

shortcoming *n.* °failure, °defect, deficiency, °weakness, °frailty, °drawback, °liability, °imperfection, °weak point, °flaw: *One might justifiably regard color-blindness as a shortcoming in an interior decorator.*

shorten *v.* **1** °cut, °curtail, cut off *or* down *or* short, °reduce, °diminish, °trim; lop off, °dock, °prune; hem: *Please shorten the time you spend in the bathroom in the mornings. Do you think I'd look better with my hair shortened? This skirt has already been shortened.* **2** condense, °abridge, °abbreviate, °digest, compress: *Can one shorten the book without losing too much of the characterization?*

short-lived *adj.* ephemeral, evanescent, °temporary, °fleeting, transitory, °transient, °passing, fugacious, °volatile: *He enjoyed a short-lived reputation that died with him.*

shortly *adv.* **1** °soon, °presently, anon, before long, in a (little) while, by and by, *Archaic* ere long: *I'll see you shortly.* **2** °just, °immediately, °soon, °right: *We had dinner with him shortly before he went abroad.* **3** abruptly, °briefly, peremptorily, curtly, brusquely, °sharply, tersely, testily, gruffly, rudely, tartly: *"Leave me alone," said Bill shortly.*

shortsighted *adj.* **1** °nearsighted, myopic, dim-sighted: *He wears thick spectacles because he's so shortsighted.* **2** unimaginative, unprogressive, °improvident, °imprudent, injudicious, °rash, °brash, °impulsive, °reckless, impolitic, °limited, °unwary, incautious, °careless, °thoughtless, unmindful: *Haverstake's shortsighted view makes no provision for the future.*

short-staffed *adj.* undermanned, short-handed, understaffed: *We are usually short-staffed during the summer because of the holiday schedules.*

short-tempered *adj.* °testy, irascible, °short, °curt, °abrupt, °gruff, °peremptory, °bluff, °rude, °tart, acid, acidulous, °terse, °brusque, crabbed, crabby, °irritable, °touchy, °petulant, °peevish, bearish, °snappish, °waspish, shrewish, curmudgeonly, crusty, °surly, °discourteous, grouchy, °disagreeable, °caustic, acrimonious, acerbic: *Henry is often short-tempered if he has slept badly.*

short-winded *adj.* short of *or* out of breath, winded, °breathless, panting, huffing (and puffing), gasping (for air *or* for breath), *Technical* dyspnoeal: *He is very short-winded because he smokes too much.*

shot *n.* **1** °discharge, shooting: *The town was captured without a shot being fired.* **2** bullet, ball, slug, cannon-ball, buckshot, pellet, °projectile, °missile: *A dumdum bullet is a piece of soft shot with a hollowed-out nose.* **3** °attempt, °try, °opportunity, °chance, °go, °essay, °endeavor, °guess, conjecture, *Colloq* °stab, °crack, whack: *You are allowed three shots at the answer.* **4** marksman, markswoman, sharpshooter, sniper, rifleman: *Geraghty, an excellent shot, has won a gold in the Olympics.* **5** °photograph, snapshot, °picture, *Colloq* °snap, photo: *Mili took some excellent action shots using stroboscopic lighting.* **6** injection, inoculation, vaccination: *The doctor gave me a flu shot this morning.* **7** °drink, jigger, °tot, dram, °nip, °spot, °swallow, *Colloq* °finger, swig, slug, snort: *Will you have a shot of whiskey with me, Muldoon?* **8** (space) launch *or* launching: *The moon shot scheduled for today has been postponed owing to technical difficulties.* **9** ***call the shots.*** run *or* direct *or* manage *or* administer *or* control things *or* affairs *or* matters, run the show, rule the roost, be in command *or* the driver's seat: *Ravelli calls the shots around here, and don't you forget it.* **10** ***like a shot.*** °quickly, °swiftly, °rapidly, speedily, hurriedly, °hastily, °at once, like a flash, °immediately, instantly, °instantaneously, *Colloq* in two shakes of a lamb's tail, like greased lightning, before you can say Jack Robinson: *All I had to do was mention ice cream, and the children were out of here like a shot.* **11** ***(not) by a long shot.*** no way, under no circumstances, °by no means, on no account, by no chance, °never: *You ask if he can get elected and my response is, "Not by a long shot."* **12** ***shot in the arm.*** °boost, stimulus, °encouragement, °incentive, °inducement, °provocation, motivation: *The team needs a shot in the arm to improve morale.*

shoulder *n.* **1** °side, °edge, °verge, °margin; breakdown lane: *If you have a flat tire, drive onto the shoulder and phone for help.* **2** ***give (someone) the cold shoulder.*** °rebuff, snub, °ostracize, *Chiefly Brit* send (someone) to Coventry, cold-shoulder, put (someone) down, °reject, °exclude, freeze (someone) out, °shun, °avoid, *Colloq* °cut (dead): *After that episode with the call girl, they gave Flambeau the cold shoulder at his club.* **3** ***put (one's*** *or* ***the) shoulder to the wheel.*** make every effort, make an effort, °strive, work hard, °pitch in, apply oneself, roll up one's sleeves, set *or* get to work, *Colloq* knuckle down, buckle down: *If we put our shoulders to the wheel, we ought to be able to get our candidate elected.* **4** ***rub***

shoulders (with). °associate (with), °hobnob (with), °socialize (with), consort (with), °mix (with), °fraternize (with), keep company (with): *Once in a while, she condescends to come down and rub shoulders with the common people.* **5** ***shoulder to shoulder.*** °side by side, °united, as one, cooperatively, jointly, arm in arm, °hand in hand, in partnership: *If we fight shoulder to shoulder, we can defeat this menace and regain our freedom.* **6** ***straight from the shoulder.*** °directly, straightforwardly, candidly, frankly, °honestly, °openly, unabashedly, unashamedly, unambiguously, unequivocally, plainly, bluntly, man to man, (with) no holds barred, °outright, *Colloq* without beating about the bush, without pulling (any) punches: *I want to tell you, straight from the shoulder, Willie, you're the best pal in the world.* —*v.* **7** °push, shove, jostle, °hustle, thrust aside, elbow, °force: *She shouldered everyone aside roughly to fight her way to the bargain counter.* **8** °support, °carry, °bear, take upon oneself, °take on, °accept, °assume: *Are you ready to shoulder the responsibilities of caring for a family?*

shout *v.* **1** °yell, °scream, °bellow, °bawl, °howl, °roar, °cry (out), °call (out), °whoop, *Colloq* holler: *We lined up when the sergeant shouted out our names.* —*n.* **2** °yell, °scream, °bellow, °howl, yelp, °roar, °cry, °call, °whoop, *Colloq* holler: *Nobody on shore could hear the boys' shouts when their boat capsized.*

show *v.* **1** °display, °present, °expose, °demonstrate, °indicate, °exhibit, °manifest, °(lay) bare, °disclose, °reveal, °betray, °make known, divulge, °register, °express, make clear *or* plain *or* manifest, elucidate, °clarify, °explain: *Show your identification as you enter. Your persistent refusal to associate with Dominic shows how you feel. For the first time, they showed how the perpetual motion machine works.* **2** °escort, °accompany, °conduct, usher, °lead, °guide, °direct; °steer: *Please show Mr. Sorenson to his table.* **3** °prove, °demonstrate, °confirm, corroborate, °verify, °substantiate, °bear out, °certify, °authenticate: *This signature shows that the will was signed personally by your father.* **4** °teach, °instruct, °tell, °inform, let (someone) in on, give (someone) an idea of: *Would you please show me how to use this washing machine?* **5** °appear, become *or* be visible, peek through, can *or* may be seen: *The floor shows through that hole in the carpet.* **6** °exhibit, °reveal, °indicate, °display, °register: *A thermometer shows the temperature, a barometer the atmospheric pressure.* **7** make an appearance, °appear, °show up, °arrive, °come, *Colloq* °surface: *Do you really think that the prince will show?* **8** °represent, °symbolize, depict, °portray, °picture, °illustrate: *This drawing shows what happens to children who bite their fingernails.* **9** °present, °play, °put on, °stage, °screen; be (being) presented *or* played *or* playing *or* put on *or* staged *or* screened: *They are showing* The Maltese Falcon *at the local cinema. What is showing there tomorrow?* **10** °grant, °accord, °bestow: *They finally showed mercy and let the hostages go.* **11** ***show off.*** make an exhibit *or* a spectacle of, °flaunt, advertise, °display, °parade, °pose, °swagger, °posture, °boast, °brag, *US and Canadian* grandstand: *They invited us to show off their new house. Quilty shows off too much for my taste.* **12** ***show up.*** **(a)** °expose, °give away, °reveal: *His collection of animal trophies shows him up for what he really is.* **(b)** °stand out, be conspicuous, be noticeable, °contrast: *The white dress shows up nicely against your suntan.* **(c)** °embarrass, °(put to) shame, °mortify, upstage, °overshadow, outshine, °eclipse: *Candace runs in the marathon just to show me up.* **(d)** See **7**, above. —*n.* **13** °display, °demonstration, °exhibition, °exposition, °fair, °presentation, *Colloq* expo: *There will be a computer show at the convention center next week.* **14** °production, °presentation, °drama, °musical, °entertainment: *We flew to New York for a week to see a few Broadway shows.* **15** °ostentation, °display, °appearance, pretentiousness, °pretension, °affectation: *Those paintings are there just for show, not because they enjoy art.*

showdown *n.* confrontation, °climax, moment of truth, final settlement, moment of decision, *US* face-off: *It's come to the point where we must have a showdown about who's boss.*

shower *n.* **1** sprinkle, sprinkling, drizzle: *Scattered showers are predicted for this afternoon.* **2** deluge, °torrent, °flood, °stream, barrage, overflow, °abundance, °profusion: *Showers of confetti rained on the parade from the offices.* —*v.* **3** sprinkle, °rain, °pour, °spray, °bombard, °fall, °descend, °drop: *Debris and ash from the volcano showered down on the village.* **4** °lavish, inundate, °overwhelm, °heap, °load (down): *On return from her triumphant tour, Sylvia was showered with gifts and praise.*

showman *n.* °producer, impresario, °director: *P. T. Barnum was possibly the greatest showman in modern history.*

showoff *n.* °braggart, exhibitionist, swaggerer, egotist, boaster, *Colloq* blowhard, windbag: *Sit down and stop being such a showoff, Tetsworth.*

showy *adj.* °flashy, °garish, °flamboyant, °conspicuous, °ostentatious, °pretentious, bravura, °gaudy, *US* °showoff; °elaborate, °fancy, florid, °ornate, °fussy, °intricate, baroque, rococo, Byzantine, arabesque: *Their new house is very showy, with gargoyles and lions and unicorns everywhere you look.*

shred *n.* **1** °scrap, °fragment, °bit, °tatter, °sliver, snippet, °rag, °remnant, °chip, °piece; atom, °trace, whit, °grain, jot *or* tittle, scintilla, °trace, °hint, °suggestion, iota, °speck: *We found shreds from his shirt caught on the thornbushes. There isn't a shred of evidence to suggest that she killed him.* —*v.* **2** °fragment, °tear (up), °tatter, °rip (up); °destroy, °demolish; °throw away, °dispose of, °scrap, *US* °trash: *At the news of the break-in, he dashed to his office and shredded the incriminating evidence.*

shrew *n.* harridan, virago, termagant, vixen, °scold, fishwife, °nag, °fury, spitfire, maenad, harpy, °witch, °hag, crone, hellcat, beldam, °bitch, banshee, Xanthippe, Thyiad *or* Thyad, *Colloq* battle-ax, dragon: *His mother is an old shrew who does nothing but complain all the time.*

shrewd *adj.* °clever, °smart, °astute, cunning, canny, °acute, °sharp, keen-minded, °keen, °quick-witted, °crafty, °artful, manipulative, °calculating, °calculated, °foxy, °sly, °wily, °perceptive, percipient, perspicacious, discerning, °wise, °sage, sagacious, long-headed, °intelligent, °ingenious, Daedalian, inventive, °resourceful: *He is a shrewd politician, very good at knowing exactly what people want to hear him say.*

shriek *n.* **1** °scream, °cry, screech, squeal, °squawk, squall: *The girls collapsed in shrieks of laughter.* —*v.* **2** °scream, °cry, screech, squeal, °squawk, squall: *It is very boring to listen to one's neighbors shrieking at each other day and night.*

shrill *adj.* high-pitched, earsplitting, °piercing, earpiercing, °sharp, piping, screeching, screechy, °penetrating: *She let out a shrill scream when she saw what had happened.*

shrink *v.* **1** wither, °shrivel (up), °contract: *My woolen pullover shrank because you washed it in hot water.* **2** Often, ***shrink from.*** °withdraw (from), °draw back, °recoil (from), back away (from), °retreat (from), cower, °cringe, °flinch, shy away (from), wince, balk (at): *Once* *committed to a responsibility, nothing could persuade Ella to shrink from it.*

shrivel *v.* Often, **shrivel up.** shrink, °wrinkle, °pucker (up), curl (up), wizen, °contract; wither, °wilt, dry up, desiccate, dehydrate: *The soil had dried up and all the plants had shriveled.*

shroud *v.* **1** °screen, °veil, °mask, °disguise, °camouflage, °cover, °shield, blanket, °shade, °hide, °conceal, °protect, °cloak, °swathe, °wrap, °envelop: *The tanks were shrouded from view in the forest. His activities are shrouded in mystery.* —*n.* **2** winding sheet, cerement, cerecloth, graveclothes: *They wrapped his body in a shroud and buried him in the churchyard.* **3** °veil, °cover, °shield, °cloak, blanket, °mask, °mantle, °pall, °screen, covering, cloud: *A shroud of secrecy descended over the Morocco affair.*

shrubbery *n.* shrubs, planting(s), hedge(s), hedging, hedgerow, °thicket, underbrush, °brake, bracken, undergrowth, coppice *or* copse: *The fox lay low in the shrubbery till the dogs flushed him out.*

shudder *v.* **1** °quiver, °shake, °tremble, °shiver, °jerk, convulse, °quaver, °quake; °vibrate, °rattle: *I shudder to think what he might do next. Olive slammed on the brakes and the ancient car shuddered to a stop.* —*n.* **2** °quiver, °shake, °tremble, twitch, °shiver, convulsion, °paroxysm, °spasm, °quaver, °quake; vibration, °rattle: *Hallcroft gave a shudder at the prospect of having to clean out the stable. With a shudder, the dilapidated shack fell apart.*

shuffle *v.* **1** °mix (up), intermix, disarrange, rearrange, interfile, intersperse, °jumble, °confuse; °shift (about), °mess up, turn topsy-turvy, °scatter, disorganize: *I shall cut the cards after you have shuffled them. Osbert shuffled the papers on his desk.* **2** scuff *or* drag (one's feet), scrape along, shamble: *The boy shuffled into the headmaster's office, guilt written all over his face.* **3** °equivocate, hem and haw, bumble, °shift, °cavil, °fence, be evasive *or* shifty, °dodge, °niggle, split hairs, °quibble, prevaricate, *Brit* hum and haw, *Colloq* °waffle: *The suspect shuffled when asked the names of his accomplices.* —*n.* **4** shamble, °shambling, scuffling, scraping: *Whenever he was lying, he did a funny little shuffle with his feet.* **5** °sidestep, °evasion, °subterfuge, °trick, °dodge, °shift, prevarication, °quibble, shuffling: *Tell me what you've done with the money—and no shuffle!*

shun *v.* °avoid, keep *or* shy away from, °steer clear of, eschew, °shrink from, °fight shy of, run *or* turn (away) from, flee *or* escape from; °forgo, °give up; disdain, °spurn, °rebuff, °reject, cold-shoulder, give the cold shoulder to: *After her experience, Penelope shunned the company of men for several years.*

shut *v.* **1** °close, °fasten, °secure, °shut up; °lock, °bolt, °seal: *Come in and shut the door.* **2** ***shut down.*** °close down, °discontinue, °cease, °suspend, °halt, °leave off, °shut up; switch *or* turn *or* shut off, °stop: *We shut down operations at the Enfield factory last week. Make sure you shut down the machinery before leaving.* **3** ***shut in.*** **(a)** confine, seclude, °keep in, °pen, fence in, °secure: *We shut the dog in before we went out.* **(b)** See **6 (a),** below. **4** ***shut off.*** **(a)** switch or turn off, °shut down, *Colloq* °kill, douse, °cut (off): *They shut off the lights while I was shaving.* **(b)** °separate, °isolate, seclude, °segregate, sequester, °bar, °shut out, °cut off, *Chiefly Brit* send to Coventry: *The more violent cases are shut off in another wing.* **(c)** °close (off), °shut (down): *To save on heating, we shut off part of the house in the winter.* **5** ***shut out.*** **(a)** °exclude, °eliminate, °bar, debar, °lock out, °ban, keep out *or* away, disallow, °prohibit: *Aliens without work permits are shut out from employment. We shut out all hooligans and rowdies.* **(b)** keep out, °screen, °exclude, °block out, °cut out: *This windowshade shuts out the direct sunlight.* **(c)** °screen, °mask, °hide, °conceal, °veil, °cover: *The house is shut out from view by the hedge.* **6** ***shut up.*** **(a)** confine, °shut in, coop (up), °cage (in), °bottle up, °box in; °imprison, °jail, incarcerate, intern, immure: *People suspected of being subversives were shut up during the war.* **(b)** °silence, keep quiet, °stifle, °mute, °gag, shush, *Chiefly Brit* quieten: *As soon as Bert wants to say something, Sidonie shuts him up.* **(c)** See **1,** above. **(d)** See **2,** above. —*adj.* **7** closed (up), sealed (up), locked (up), bolted, fastened: *There is not much hope of penetrating their shut minds on this issue.*

shuttle *v.* commute, °alternate: *He shuttles daily between Los Angeles and San Diego.*

shy *adj.* **1** diffident, °coy, °bashful, °retiring, °withdrawn, withdrawing, °reserved, °timid, °meek, °modest, °sheepish, unconfident, °self-conscious, introverted, °nervous, apprehensive, timorous, *Rare* verecund: *Casey is quite shy and you have to draw him out gently.* **2** timorous, °cautious, °wary, chary, °leery, °guarded, °afraid, °fearful, frightened, °anxious, °worried, °suspicious, °distrustful, °cowardly, craven, uncourageous: *The local people are shy of anyone or anything unfamiliar.* **3** missing, lacking, deficient in, °short of: *The shipment is shy a few dozen shoelaces we ordered.* —*adv.* **4** ***fight shy of.*** °avoid, be unwilling *or* reluctant *or* averse *or* loath *or* loth *or* disinclined *or* not disposed; be wary *or* cautious *or* watchful: *The children fight shy of talking to strangers.*

sick *adj.* **1** °nauseated, °queasy, sick to one's stomach, °squeamish, qualmish; seasick, carsick, airsick, *Colloq* green around the gills: *Patrick felt a bit sick after the rough ferry ride.* **2** °ill, unwell, °unhealthy, °sickly, °indisposed, °infirm, ailing, °diseased, afflicted, *Colloq* °under the weather, on the sick list, °poorly, laid up, not (feeling) up to snuff: *If you feel sick, you'd best see a doctor. I've been sick for a week.* **3** °affected, troubled, °stricken, heartsick, °wretched, °miserable, burdened, weighed down: *I was sick with worry to learn of the problems you have been having with your business.* **4** °mad, °crazy, °insane, °deranged, °disturbed, °neurotic, °unbalanced, psychoneurotic, °psychotic, *Colloq* °mental, *US* off ((one's) rocker *or US* trolley): *Ashley must be sick to have committed such a crime.* **5** °peculiar, unconventional, °strange, °weird, °odd, °bizarre, °grotesque, °macabre, °shocking, °ghoulish, °morbid, °gruesome, stomach-turning, °sadistic, masochistic, sadomasochistic, *Colloq* °kinky, °bent, far-out, *US* °off: *In thirty years as a policeman, I've seen sick things, but this is the end.* **6** sickened, shocked, °put out, °upset, appalled, °disgusted, revolted, repulsed, offended, repelled, °nauseated; annoyed, chagrined, irritated: *She was sick at what she saw in the army field hospital. The way you do things sometimes makes me sick.* **7** Sometimes, ***sick of.*** °(sick and) tired, bored, °weary, *Colloq* fed up with: *Don't you get sick of telling them the same thing so many times?*

sicken *v.* **1** fall ill, take sick, °contract, be stricken by, come down with *or* catch something *or* a bug, °fail, °weaken: *After he seemed to be recuperating so well, he suddenly sickened and died.* **2** make ill *or* sick, °afflict, °affect, °disgust, °nauseate, turn one's stomach, °upset, appall *or* appal, °shock, °repel, °revolt, °repulse, °offend, make one's gorge rise; °put out: *The scene in the abattoir sickened me.*

sickly *adj.* **1** See **sick, 2,** above. **2** ailing, °feeble, °delicate, °wan, °weak, pallid, °pale, °drawn, °peaked, peaky, peakish: *Jennie looks sickly and must build up her resistance.* **3** °mushy, mawkish, °maudlin, cloying, insipid, °weak, °watery: *The sickly sentimentality of the* *episode about the boy and his lost puppy was unbearable.*

side *n.* **1** Sometimes, ***sides.*** flank(s), edge(s), °verge, margin(s), °rim, °brim, °brink, border(s); bank; boundary *or* boundaries, °perimeter, °periphery, limit(s): *This side of the road is in West Sussex, the other in East Sussex. The sides of the swimming pool are tiled.* **2** °surface, °face, °plane; facet: *The smallest possible number of sides of a pyramid is four.* **3** °faction, °interest, °party, °part, °sect, °camp, °(point of) view, °viewpoint, °aspect, °opinion, °standpoint, °stand, °cause, °angle, °position, °attitude, °school, °philosophy: *Which side are you on? I know the side that Charlotte supports.* **4** °team; °string, °squad; *American football* and *Association Football* or *Soccer* and *Cricket* and *Field Hockey* eleven, *Australian Rules football* eighteen, *Baseball* nine, *Basketball* five, *Gaelic football* and *rugby union* fifteen, *Rugby league football* thirteen, *Ice hockey* six: *Whichever side Craig is on seems likely to win. The Ranger's side is now coming out onto the field.* **5** °affectation, °pretension, haughtiness, °arrogance, insolence, pretentiousness, °airs: *Chauncy puts on a bit too much side for my taste.* **6** ***side by side.*** together, jointly, cheek by jowl: *These paintings should not have been hung side by side.* **7** ***take sides.*** show preference, be partial, show favoritism: *A judge must avoid taking sides.* —*adj.* **8** °secondary, °incidental, °subordinate, °tangential, °subsidiary, °indirect, ancillary, °marginal, lesser, °minor, unimportant, °inconsequential, inconsiderable, °insignificant: *The main point is that he refuses to cooperate, and his reasons are entirely a side issue.* **9** °auxiliary, °secondary: *After leaving the highway, drive along the side road for a mile.* —*v.* **10** ***side with.*** take sides with, show preference for, be partial to, show favoritism to *or* for, °support, °favor, °prefer, go in *or* along with, join ((up) with), ally with, be *or* become allied with, *Colloq US* throw in with, team up with: *You can be sure that Kenneth will always side with the favorite candidate.*

sidelong *adj.* °oblique, °indirect, °sideways, covert, °surreptitious: *He marched straight ahead, without so much as a sidelong glance in her direction.*

sidestep *v.* °avoid, °dodge, circumvent, skirt, °evade, °shun, °steer clear of, *Colloq* °duck: *The minister neatly sidestepped all questions about her retirement.*

sidetrack *v.* °divert, °deflect, draw off *or* away, °distract, °turn aside; shunt: *Where was I? I got sidetracked by your comment about being overweight.*

sidewalk *n.* °walk, *Chiefly Brit* pavement, footpath, footway: *You may play on the sidewalk, but don't you dare step into the street.*

sideways *adv.* obliquely, laterally, edgeways, edge on, °sidelong, crabwise, indirectly, *US and Canadian* edgewise: *The dresser will go through the door if you turn it sideways.*

sidle *v.* °edge: *She sidled toward the door, opened it, and was gone.*

siege *n.* **1** blockade, encirclement, beleaguerment, besiegement: *The siege of Troy lasted ten years.* —*v.* **2** lay siege to, °besiege, blockade, beleaguer, cordon off, °encircle, box *or* pen *or* shut in, °bottle up: *They sieged the city and the inhabitants surrendered when they ran out of food and water.*

sift *v.* **1** °strain, sieve, °riddle, °filter, °screen, °bolt; winnow, °separate, weed out, °sort out, °select, °choose, °pick: *It is not easy to sift out the promising manuscripts.* **2** °examine, °analyze, °study, °probe, °screen, °scrutinize, °investigate: *The police are sifting the evidence of the witnesses.*

sigh *v.* **1** °breathe, sough; suspire: *Troilus sighed forth his soul to the tent where sleeping Cressid lay.* **2** ***sigh for.*** °bemoan, lament *or* mourn *or* grieve *or* weep for, °bewail; yearn *or* pine for: *Now and then, all of us sigh for the good old days.* —*n.* **3** °murmur, °exhalation, °sound; suspiration: *I gave a sigh of relief as the train left the station.*

sight *n.* **1** eyesight, °vision, eyes: *His sight was completely restored by the operation.* **2** field of view *or* of vision, range of vision, ken, °perception, °view, eyeshot, °gaze: *As we turned the corner, the welcoming lights of the city came into sight.* **3** °spectacle, °scene, °show; °rarity, °marvel, °wonder, °phenomenon; °pageant: *I shall never forget the sight of Biffy, stark naked, standing on that table in the officers' club. We went to Rome to see the sights.* **4** °mess, °disaster, eyesore, °monstrosity, *Colloq* °fright, °atrocity: *The flat looked a sight after the boys had been home for only a week.* **5** ***catch sight of.*** °spot, °notice, descry, °spy, espy, glance at, (catch a) glimpse (of), (get a) look *or* peep *or* peek at, *Colloq* take a gander at, get a look-see at, *Slang* get a load of, *US* glom, °pipe: *We caught sight of a sinister-looking person slipping behind a pillar.* **6** ***out of sight.*** **(a)** °remote, °distant, far away, unseeable, °imperceptible, °invisible: *She soon walked out of sight down the road. The mortar emplacement is out of sight, over the hill.* **(b)** °unusual, °rare, °extraordinary, °outrageous, °imaginative, awe-inspiring, °incredible, °shocking, °unreal, °moving, jolting, *Colloq* °neat, °cool, *Brit* °brilliant, brill: *The jazz concert last night was out of sight, man.* —*v.* **7** °look, °view, °(take) aim, °peer, °peek, °peep, draw a bead: *Sight along this stick and you'll see the cluster of stars called the Pleiades.* **8** °spot, °see, °catch sight of, °mark, °observe, °behold, °view, °distinguish, discern, °identify, °note, °notice, °remark, glimpse, descry, espy, °spy: *Three birdwatchers actually sighted a golden eagle in the Adirondacks.*

sightseer *n.* °tourist, °traveler, globe-trotter, *Colloq* rubberneck(er), *Brit* tripper, day-tripper: *This parking area was built for sightseers to stop and view the mountains.*

sign *n.* **1** °token, °symbol, °indication, °mark, °signal, °indicator; °notice: *I interpreted the shouts of the crowd as a sign that I had won the race.* **2** °movement, °gesture, °motion, °signal, °cue, gesticulation: *She made a sign motioning me to come over.* **3** °trace, °indication, °evidence, °mark, °clue, °hint, °suggestion, °vestige: *She waited for an hour but there was no sign of him.* **4** °device, °mark, °symbol, °representation, °emblem, trademark, badge, °brand, °stamp, °seal, ideogram, ideograph, lexigram, phonogram, grapheme, hieroglyph, cartouche, rebus, logo(type), colophon, ensign, °standard, °banner, °flag; monogram, initials, cipher *or* cypher: *Sir Roger fought under the sign of a rampant lion.* **5** signboard, °advertisement, placard, °poster, *US* broadside; shingle, °notice, °announcement: *You can't miss the huge supermarket sign at the corner.* **6** °omen, augury, °warning, forewarning, °foreboding, portent, °indication, (hand)writing on the wall, °prophecy, prognostication, foreshadowing: *We took the rumbling as a sign of an imminent earthquake.* —*v.* **7** autograph, put one's signature on *or* to, inscribe, countersign, °endorse, °witness, put *or* set one's hand to, °mark; sign on the dotted line; *Colloq US* put one's John Hancock on *or Colloq US* to: *I have already signed the petition, have you? The painting is valueless unless signed by the artist.* **8** ***sign away.*** °forgo, °relinquish, °give up, °abandon, abandon *or* quit claim to, °waive, °release, °surrender, °dispose of, °sacrifice, °get rid of: *With one stroke of the pen, he signed away most of his fortune.* **9** ***sign off.*** °close down, °discontinue (broadcasting, writing a letter, etc.): *It is midnight, time for us to sign off till* *tomorrow.* **10** ***sign on*** *or* ***up.*** **(a)** °enroll, °enlist, °sign up (for), °register, volunteer, °join (up), °contract: *I signed on to be a bodyguard, not a nursemaid.* **(b)** °enroll, °enlist, °hire, °employ, put under contract, °retain, °engage, °take on, *Colloq* take on board, bring aboard: *We signed up three new engineers this week.* **11** ***sign over.*** °assign, consign, °transfer, °make over, °deliver, °give, °donate, present, °dispose of, °turn over: *Aunt June was persuaded to sign over the ownership in the business to my cousin Gurney.*

signal *n.* **1** See **sign, 1,** above. **2** °incitement, stimulus, °spur, °impetus, goad, °prick: *The waving of the flag will be the signal to advance.* —*adj.* **3** °remarkable, °conspicuous, °striking, °extraordinary, °unusual, °unique, °singular, °special, °noteworthy, °notable, °exceptional, °significant, °important, °outstanding, °momentous, consequential, °weighty: *Her Majesty's forces won a signal victory in Khartoum that day. Winning the competition is a signal accomplishment.* —*v.* **4** °motion, °indicate, °gesture, gesticulate, °communicate, °announce, °notify; whistle, wink, °blink, °nod, °beckon, °wave, °sign: *When your turn comes, a clerk will signal you to approach the counter. Peter raised his hand to leave the room and the teacher signaled his assent.*

significance *n.* **1** °meaning, °sense, signification, denotation, °message, °idea, °point, °import, purport, °implication, portent, °content, °pith, °essence; °gist, °drift, °vein, °impression, connotation: *The significance of words like* God, peace, *and* hunger *is different for almost everyone who uses them.* **2** °importance, °weight, weightiness, consequence, °moment, °relevance, value: *The significance of the agreement lay not in its wording but in the fact that it was drawn up at all.*

significant *adj.* **1** °important, °weighty, °momentous, consequential, °critical, °substantial, substantive, °noteworthy, °notable, valuable, valued, °meritorious, °outstanding, °impressive, °historic, °relevant, °signal: *Appleton made significant contributions to knowledge about the ionosphere.* **2** °meaningful, °eloquent, pithy, °expressive, °pregnant, °suggestive, °informative: *It is significant that they prefer the term* Britain *to* England, Scotland, and Wales.

signify *v.* **1** °sign, °signal, °indicate, °communicate, °make known, convey, °symbolize, betoken, °represent, °express, °announce, °declare, °denote, °say, °mean, °specify; connote, °intimate, °suggest, °reveal, °disclose, °impart: *She signified her agreement by a nod. What does the phrase 'pursuit of happiness' signify to you?* **2** °matter, °count, be significant *or* important *or* consequential, be of significance *or* of importance *or* of consequence, carry weight, °impress, °stand out, deserve *or* merit consideration: *It signifies little that they were found innocent of a crime we all believe they committed.*

silence *n.* **1** °quiet, quietness, stillness, soundlessness, noiselessness, °calm, calmness, °hush, quietude, tranquillity, °peace, peacefulness, °serenity: *The silence was so complete that I could hear my heart beating.* **2** speechlessness, muteness, dumbness, reticence, taciturnity, uncommunicativeness: *Do not construe his silence as agreement.* —*v.* **3** °quiet, °mute, °hush, °still, shush, °calm, °tranquilize, soothe, *Chiefly Brit* quieten: *The audience was silenced by the start of the overture.* **4** °mitigate, °temper, mollify, take the sting out of, propitiate, pacify, °blunt, °suppress, °repress, °restrain, °subdue, draw the fangs *or* teeth of, °inhibit, °put down, °damp, °mute, °squelch, °quash, emasculate, muzzle, °muffle, °shut off, °gag, °stifle, °smother, °deaden (the effect of): *Only a full exposure of the events will help to silence criticism in the media.*

silent *adj.* **1** °quiet, °still, soundless, °noiseless, °tranquil, hushed, shushed, °mute; °calm, °serene, placid, °peaceful, pacific, unagitated, unruffled, untroubled, undisturbed, *Literary* stilly: *After the children left the house fell silent. The earth in silent stillness lay to hear the angels sing.* **2** uncommunicative, °mute, close-mouthed, °taciturn, °reticent, °reserved, °mum, °tight-lipped, °secretive: *Millie told me a great deal about her children but was silent on the subject of her own activities.* **3** unspeaking, unspoken, °mute, unexpressed, °tacit, °understood, °implicit, implied, unstated, unsaid: *There is a silent agreement among the islanders not to reveal where the whisky is hidden.* **4** °inactive, nonparticipating, °passive, quiescent, *Brit* sleeping: *Although John runs the company, he has several silent partners.* **5** unpronounced, unuttered, not sounded, *Technical* aphonic: *The* g *in* sign *is silent, like the* k *in* know *and the* l *in* would.

silently *adv.* °quietly, soundlessly, noiselessly, with catlike tread, as quietly as a mouse, stealthily; wordlessly, speechlessly, mutely: *They fold their tents like the Arabs, / And silently steal away. He stood by silently, allowing his brother to do all the talking.*

silhouette *n.* °outline, °profile, contour, °form, °figure, °shape, °shadow, configuration, °periphery, °perimeter: *I could see the silhouette of a person on the window blind.*

silky *adj.* silken, silklike, °delicate, °sleek, °soft, °smooth, satiny, °shiny, °glossy, °lustrous, *Technical* sericeous: *Her long, silky black hair hung below her shoulders.*

silly *adj.* **1** °senseless, °nonsensical, °absurd, °ridiculous, °ludicrous, laughable, risible, asinine, °inane, °preposterous, idiotic, °childish, °puerile, °foolish, °foolhardy, °irresponsible, °unreasonable, illogical, irrational, °pointless, fatuous, °stupid, unwise, imbecilic, °crazy, °mad, °insane: *It would be silly to pay for it if you can get it for nothing.* **2** stunned, stupefied, °daze, °giddy, °dizzy, muzzy, benumbed: *Culley said something rude and the girl knocked him silly with a karate chop.* —*n.* **3** °fool, nincompoop, idiot, dunce, ninny, °half-wit, simpleton, numskull *or* numbskull, dimwit, booby, °dolt, jackass, °twit, °dolt, blockhead, bonehead, nitwit, ignoramus, °clod, *US* thimble-wit, *Colloq* °dope, °dummy, knucklehead, goose, °drip, silly-billy, *Brit* clot, *Slang US and Canadian* °jerk, nerd: *Warren is such a silly to spend so much money on a girl who cares nothing for him.*

silt *n.* **1** °deposit, °sediment, alluvium, °ooze, °sludge: *The irrigation pumps get clogged up with silt.* —*v.* **2** Usually, ***silt up*** *or* ***over.*** become clogged *or* choked *or* obstructed *or* dammed *or* congested: *The channel silts up and has to be dredged yearly so that the boats can pass through.*

silver *n.* **1** silverware, °sterling, °(silver) plate; cutlery, *US* flatware; holloware: *Guests are coming so I'll have to polish the silver.* **2** °white, grayish, whitish-gray, grayish-white, °gray: *When your hair has turned to silver, / I will love you just the same.* —*adj.* **3** silvery, °shiny, shining, °polished, burnished, °lustrous, °pearly, nacreous, °bright, gleaming, *Literary and heraldry* argent; °white: *The moon's silver crescent hung low in the sky. Her hair is more silver than gray.* **4** silvery, silver-toned, silver-tongued, °sweet, °pretty, euphonious, °melodious, mellifluous, mellifluent, dulcet, °musical: *The silver tones of the muezzin echoed in the streets, calling the faithful to prayer.*

similar *adj.* **1** °like, almost identical, comparable, °equivalent, nearly the same; °alike: *That brooch is similar to the one I bought for my mother. The two may be similar, but they are not the same.* **2** ***be similar to.***

°resemble, be like, correspond to, compare favorably with: *That ten-dollar bill is very similar to the one I had yesterday.*

simmer *v.* **1** °seethe, °stew, cook, °boil, °bubble: *A huge caldron of soup simmered on the fire.* **2** °chafe, °seethe, °stew, steam, °smolder, °fume, °rage, °burn, *Colloq US* do a slow burn: *Father is still simmering because you damaged his car.* **3** ***simmer down.*** calm *or* cool down, cool off, calm oneself, become quiet, control oneself, get control of *or* over oneself, *Chiefly Brit* quieten down, *Slang* cool it: *Yesterday he was in a towering rage, but he has simmered down a bit today.*

simple *adj.* **1** uncomplicated, °plain, uninvolved, °unsophisticated, understandable, °intelligible, °(easily) understood, comprehensible, °clear, lucid, straightforward, °easy (as ABC), °elementary, °basic: *At the age of four, Jan was able to do simple mathematical calculations.* **2** °plain, °unadorned, undecorated, unembellished, °basic, °fundamental, °elementary, °elemental, °mere, unostentatious, unassuming, unpretentious, °modest, °classic, uncluttered, °stark, °clean, °severe, austere, °Spartan, °homely; °unvarnished, °naked, °honest: *Thea was stunning in a simple black dress with a colorful scarf and gold jewelry. The simple truth is that he came home from work to find the house broken into and his wife gone.* **3** °sincere, °frank, °candid, °open, °unaffected, uncomplicated, unpretentious, straightforward, °aboveboard, simple-hearted, uncontrived, °direct, °upright, °square, °forthright, foursquare, °righteous, °honest, naive *or* naïve *or* naïf, guileless, °artless, undesigning, °childlike, °ingenuous, °unsophisticated, °innocent, °green: *The colonel was a simple man who was respected because he always said just what he meant. A simple heart may be its own best guide.* **4** °unsophisticated, naive *or* naïve *or* naïf, °slow, slow-witted, °stupid, °thick, thickheaded, simple-minded, °feebleminded, oafish, bovine, °dense, °obtuse, °dull, dull-witted, witless, °half-witted, brainless, °backward, imbecilic *or chiefly Brit* imbecile, cretinous, *Colloq* °dumb, moronic: *That chap Dawson is far too simple to understand what you are telling him.* **5** °lowly, °humble, °inferior, °mean, °base, subservient, °common, °subordinate: *The lord of the manor regarded his tenants as simple folk, a breed apart from himself and his family.*

simplicity *n.* **1** uncomplicatedness; understandability, comprehensibility, lucidity, straightforwardness, °clarity, intelligibility, decipherability: *Consider the simplicity of some life forms in contrast to the complexity of others. The simplicity of the Bible can be deceptive.* **2** plainness, cleanness, °clarity, °severity, starkness, austereness, asceticism, °restraint, bareness, °purity: *She prefers the simplicity of classical or very modern design to baroque and rococo clutter.* **3** °sincerity, openness, artlessness, °candor, guilelessness, frankness, unsophisticatedness, ingenuousness, straightforwardness, forthrightness, unaffectedness, unpretentiousness, modesty, naivety *or* naiveté *or* naïveté; plainness, directness, inelegance, rusticity, pastoralism: *At the time, many were attracted by the charming simplicity of bucolic poetry.* **4** °stupidity, slow-wittedness, thickheadedness, simple-mindedness, feeblemindedness, oafishness, cloddishness, obtuseness, dullness, dull-wittedness, witlessness, half-wittedness, imbecility, brainlessness: *The simplicity of some of those yokels is incredible.*

simplify *v.* °clarify, °clear up, make easy, °paraphrase, °explain, explicate, disentangle, untangle, unravel, streamline: *Simplify your request and we shall be able to do what you ask.*

simply *adv.* **1** °merely, °barely, purely, °only, solely, °just, °entirely, fully, °totally, °wholly, °altogether, °absolutely, °really: *Is it true that they met regularly simply to discuss politics?* **2** °totally, °completely, °altogether, °entirely, °just, plainly, °obviously, °really, unreservedly, unqualifiedly: *If you believe that, you're simply too stupid for words!* **3** naively *or* naïvely, artlessly, guilelessly, °openly, innocently, ingenuously, unaffectedly, unpretentiously, plainly, °naturally: *He asked her very simply what she did as a "working girl."* **4** plainly, modestly, starkly, °severely, sparely, sparsely, austerely, ascetically: *His home is simply furnished, without paintings or any other decoration.* **5** distinctly, unambiguously, plainly, °obviously, unmistakably *or* unmistakeably: *The instructions must be written simply so that they are understandable to anyone.*

simultaneous *adj.* coincident, coinciding, concurrent, contemporaneous, synchronous; °contemporary: *The simultaneous appearance of two books on such an obscure subject was very unusual.*

sin *n.* **1** trespass, °transgression, °offense, °wrong, impiety, °misdeed, profanation, desecration, iniquity, °evil, °devilry, °sacrilege, °crime, infraction, °misdeed, dereliction, °infringement, °violation, misdemeanor, °fault, °foible, °peccadillo: *How many sins have been committed in the name of righteousness! It's a sin to tell a lie. The seven deadly sins are anger, covetousness, envy, gluttony, lust, pride, and sloth.* **2** wickedness, sinfulness, °vice, corruption, ungodliness, badness, °evil, wrongfulness, iniquity, iniquitousness, immorality, depravity, impiety, irreverence, impiousness, °sacrilege: *Her brother Edmund was conceived in sin, and sin ruled his entire life.* —*v.* **3** °transgress, °offend, °fall (from grace), °lapse, °go wrong, °stray, go astray, °err, *Biblical or archaic* trespass: *Strictly speaking, we can sin only against God.*

sincere *adj.* °honest, °truthful, °true, veracious, °genuine, °heartfelt, true-hearted, undissembling, unfeigned, °open, °(open and) aboveboard, straightforward, °direct, °frank, °candid, guileless, °artless, *Colloq* upfront, °on the level, on the up and up: *I took her criticism as a sincere desire to see improvement in my work. You have always been sincere in your emotional involvements.*

sincerely *adv.* °truly, °honestly, °really, wholeheartedly, candidly, frankly, unequivocally, °seriously, earnestly, genuinely, °deeply, fervently: *When they say they love each other, they mean it sincerely.*

sincerity *n.* °honesty, truthfulness, straightforwardness, openness, forthrightness, frankness, °candor, candidness, seriousness, genuineness, uprightness: *Everyone appreciates sincerity in dealings with others.*

sinew *n.* **1** ligament, tendon; muscle, thew: *After the climb, I ached in every sinew.* **2** Usually, ***sinews.*** °strength, °force, °power, °energy, °brawn, °vigor, °might, °stamina, °vitality: *The sinews of his argument were considerably weakened by her disarming candor.*

sinewy *adj.* °strong, °powerful, °muscular, °mighty, °stout, °wiry, °robust, °tough; strapping, °brawny, °burly: *The smith a mighty man was he, / With large and sinewy hands.*

sinful *adj.* °corrupt, °evil, °wicked, °bad, °wrong, wrongful, iniquitous, °vile, °base, °profane, °immoral, °profligate, depraved, °criminal, °sacrilegious, °ungodly, unholy, demonic, irreligious, piacular, °impious, °irreverent: *Anything that you like as much as you like chocolate must be sinful. They would never condone such sinful behavior in our church.*

sing *v.* **1** °chant, intone, carol, vocalize, trill, croon, °pipe, °chirp, warble; chorus; yodel: *It was a wonderful spring day and Barbara felt like singing. Can you sing* On the Road to Mandalay? **2** whistle, °pipe, °peep: *I have just bought a singing kettle.* **3** °tell, °tattle, °name names, *Slang* rat, snitch, squeal, blow the whistle, peach, °spill the beans: *We learned that Johnny sang to the cops about last week's bank job.*

singe *v.* char, °blacken, sear, °scorch, °burn: *I singed the tail of my shirt ironing it.*

singer *n.* °vocalist, soloist, songster, crooner, chanteuse, nightingale, °minstrel, troubadour, balladeer, caroler, chorister, choir boy *or* girl *or* member, chorus boy *or* girl *or* member, *Colloq* songbird, canary, thrush: *She used to be a professional singer.*

single *adj.* **1** °unmarried, unwed, °unattached, °free; °celibate: *She remained single till she was twenty-two.* **2** °singular, °individual, °distinct, °solitary; °one, °only, °sole, lone, °unique, °isolated: *There wasn't a single person there who could think for himself. She was the single survivor whom we questioned about the disaster.* **3** °separate, °distinct, °individual, °solitary: *Every single hair is rooted in its follicle. A single tree grew in the center of the lawn.* —*v.* **4** ***single out.*** °select, °choose, °pick, °separate, take *or* put *or* set aside *or* apart, °distinguish, cull, °segregate, fix *or* fasten on: *Why was she singled out for special treatment?*

single-handed *adj.* **1** °solo, lone, °solitary, °independent, unaided, unassisted: *Chichester was the first winner of the single-handed transatlantic sailing race.* —*adv.* **2** single-handedly, by oneself, °alone, °solo, on one's own, independently: *He sailed around the world single-handed.*

single-minded *adj.* dedicated, °devoted, °resolute, °steadfast, persevering, °firm, °determined, dogged, unswerving, unwavering, °tireless, °purposeful: *Single-minded perseverance saw him through the completion of the work in eight years.*

singly *adv.* one at a time, °separately, °individually, one by one, successively, one after the other, seriatim: *Taken singly, the facts are not hard to assimilate.*

singular *adj.* **1** °unusual, °different, atypical, °eccentric, °extraordinary, °remarkable, °special, uncommon, °strange, °odd, °peculiar, °bizarre, °outlandish, °curious, °queer, *outré,* °offbeat, *Colloq* far-out: *Ideas that yesterday were considered singular have today become quite commonplace.* **2** °outstanding, °prominent, °eminent, °preeminent, °noteworthy, °significant, °important, °conspicuous, °particular, °notable, °signal, °exceptional, °superior: *Stempel made a singular contribution to the development of microscopy.* **3** lone, °isolated, °single, °separate, uncommon, °rare, °unique, °distinct, one-of-a-kind: *This church is a singular example of an architectural style that did not catch on.*

singularity *n.* **1** individuality, distinctiveness, uniqueness, idiosyncrasy: *Personality demonstrates its singularity even in handwriting.* **2** °eccentricity, °peculiarity, strangeness, oddness, queerness, outlandishness, uncommonness: *The singularity of the marriage ceremony among these tribes was often remarked on by anthropologists.* **3** *Technical* black hole: *The ultimate collapse of the star results, theoretically, in a singularity.*

sinister *adj.* **1** °fateful, °inauspicious, unfavorable, °foreboding, °threatening, °menacing, minacious, minatory *or* minatorial, °portentous, °ominous, unpropitious, °disastrous, °dark, °gloomy: *There was a sinister meaning in the look he gave her.* **2** °evil, °bad, °corrupt, °base, malevolent, °malignant, malign, °harmful, pernicious, treacherous, nefarious, °wicked, diabolic(al), baleful, °villainous, insidious, °sneaky, furtive, underhand(ed): *The poor girl fell under the sinister influences of a real-life Svengali.*

sink *v.* **1** °founder, °submerge, °go down, °go under, °plunge, °descend, be engulfed: *Struck below the waterline, the ship sank in minutes.* **2** °subside, °cave in, °collapse, °subside, °settle, °drop, °fall in, °go down, °slip away: *The earth suddenly sank beneath our feet.* **3** °descend, °go down, °drop, °fall, move down *or* downward(s), go down to *or* on: *The parachute slowly sank to the ground. When he saw the statue move, he sank to his knees, terrified.* **4** °decline, °weaken, °worsen, °degenerate, °subside, °deteriorate, °flag, °fail, °diminish, °die, °expire; languish; *Colloq* go downhill: *Grandfather, sinking fast, summoned the family to his bedside.* **5** °disappear, vanish, fade away, °evaporate; °set, °go down, °descend, °drop: *After a meteoric rise, many rock stars sink into oblivion. As the sun sinks slowly in the west, we return to our hotel to reminisce about our day at the pyramids.* **6** °settle, °precipitate, °descend, °drop: *After a few minutes, the sand sinks to the bottom.* **7** °bore, °put down, °drill, °dig, °excavate, °drive: *A special shaft will be sunk just for ventilating the mine.* **8** °submerge, °immerse, °plunge: *He sank his hand deep into the mud and came up with a small box.* **9** °stoop, °bend, °get, °go, lower *or* humble oneself: *I never dreamt that he would sink so low as to beg in the streets.* **10** °invest, °venture, °risk, °put: *He sank his life's savings into his son's business, only to see it go bankrupt.* **11** ***sink in.*** be understood, °penetrate, °register, make an impression on, get through to: *How many times do you have to say it in order for it to sink in?* —*n.* **12** basin, washbasin, washbowl, lavabo; *Church* font, stoup, piscina: *The kitchen has a large double sink.* **13** cesspool, cesspit, °pit, hellhole, den of iniquity, sinkhole, *Colloq* °dive: *We found him playing piano in some sink frequented by the worst element.*

sinking *adj.* **1** °queasy, °nervous, uneasy, °tense, apprehensive, unquiet, °fretful, °shaky, jittery, °jumpy, °anxious: *Just thinking about the way they treat hostages gave me a sinking feeling in the pit of my stomach.* **2** depressed, °dejected, °miserable, dolorous, °doleful, °mournful, °forlorn, woeful, °desolate, despairing, °stricken, heavy-laden: *With a sinking heart, I knew that no one would think of looking for me at the bottom of the well.*

sinner *n.* °transgressor, wrongdoer, °miscreant, °offender, evildoer, malefactor, °reprobate, *Biblical or archaic* trespasser: *How many sinners are found out?*

sip *v.* **1** °taste, °sample, sup: *Sip some of this soup and tell me if you like it.* —*n.* **2** °taste, °sample, *soupçon,* °drop, °bit, °swallow, °mouthful, spoonful, thimbleful, °nip, dram, *Colloq* swig: *She took a sip, savored it, then turned to me with a smile.*

siren *n.* **1** whistle, warble, wailer, horn, foghorn; °signal, °alarm, °warning, °alert, tocsin: *The air raid siren sounded and we ran for shelter.* **2** °temptress, °seductress, enchantress, °charmer, sorceress, *femme fatale,* Circe, Lorelei, *Colloq* vamp, *US* mantrap: *Have you ever seen the latest siren to whom John has fallen prey?*

sissy *n.* °milksop, mama's boy, mummy's boy, namby-pamby, °weakling, °baby, crybaby, mollycoddle, *US* milquetoast, *Colloq* softy, *Brit* °wet: *The boy you remember from school as a sissy is now a football star.*

sit *n.* **1** be seated, °settle, sit down, take a seat, °rest, *Colloq* take the weight *or Colloq* a load off one's feet: *Come in and sit with me for a while. If you sit on that chair it will break.* **2** hold a session, be in session, °assemble, °meet, convene; °gather, °get together: *The Supreme Court of the United States will be sitting next week.* **3** Often, ***sit on.*** have *or* hold *or* occupy a seat (on), °participate (in), be a member (of): *Kathy sits on the board of directors.* **4** °remain, °stay, °lie, °rest; °relax, mark time, °abide, °dwell: *Let the dough sit a while to rise. Tina sat in Vienna awaiting instructions.* **5** °seat, °contain, °hold, °accommodate, have seats *or* seating for, have room *or* space *or* capacity for seating: *The auditorium sits only six hundred.* **6** ***sit in.*** **(a)** Often, **sit in on.** play, °join (in), °participate (in), take part (in); °observe, °watch: *Mind if I sit in on your poker game? We invited Mary Lou to sit in during our discussion.* **(b)** °substitute, °fill in, °stand in, °double, *Colloq* °cover, sub, *US* pinch-hit: *Mary Lou often sat in for our pianist if he had another gig.* **7** ***sit out.*** wait out, outwait, outstay, °outlast, outlive, last through, live through: *Will sat out endless dances. He sat out the whole evening merely waiting to walk home with Lily.* **8** ***sit tight.*** °wait, hang back, °hold back, be patient, bide (one's) time, play a waiting game, take no action, °delay, temporize, *Colloq* hold (one's) horses: *Just sit tight till you hear from them—don't rock the boat.* **9** ***sit up.*** °awaken, pay attention, °notice, become alert *or* interested *or* concerned: *Jennie's performance in the hurdles made many track coaches sit up.* **10** ***sit (with).*** Often, **sit well** *or* **right (with).** agree with, be agreeable to; °seem, °appear, °look: *The way Ashton handled the matter did not sit well with the directors.*

site *n.* **1** °location, °place, °plot, plat, °spot, °locale, °area, °milieu, °neighborhood, locality, purlieus, °placement, °position; °situation, °orientation: *A site near that of ancient Pergamum has been acquired for the building. The site alongside the river has much to recommend it.* —*v.* **2** °locate, °position, °place, °put, °situate, install *or* instal: *The building has been sited to take maximum advantage of the sun.*

situate *v.* place in a position *or* situation *or* location, °place, °position, °locate, °set, °spot, °put, install *or* instal: *Harwood is very well situated to learn what the high command is planning. The greenhouse should be situated on the south side of the house.*

situation *n.* **1** °place, °position, °location, °spot, °site, °locale, °setting: *The situation of the monastery, high on the mountain, makes it almost inaccessible.* **2** °state (of affairs), °condition, °circumstances, °case, °status (quo), lay of the land, °picture; °plight, °predicament; *Colloq* ball game; kettle of fish: *The present situation calls for careful planning. The chancellor's policies have done little to improve the economic situation.* **3** °position, °place, °job, °employment, °post, *Colloq* berth: *Jenkins likes his new situation as Lord Jellinek's valet.*

size *n.* **1** °magnitude, largeness, bigness, °bulk, °extent, °scope, °range, dimensions, °proportions, measurement(s), °expanse, °area, square footage, °volume, °mass, °weight; hugeness, immensity, greatness, vastness, °enormousness: *How do astronomers determine the distance and size of stellar objects? A shape of extraordinary size suddenly loomed up in the darkness.* —*v.* **2** dimension, °measure: *The furnishings are sized in proportion to these miniature room displays.* **3** ***size up.*** assess, °judge, °evaluate, °measure, take the measure of, appraise, assay, make an estimate of, °estimate, value, °gauge, °rate: *She looked him up and down, sizing him up as a prospective husband.*

skeptic *n.* doubter, questioner, doubting Thomas, disbeliever, nullifidian, agnostic, scoffer, cynic: *You'll always find some skeptics who don't believe that a woman could be a good prime minister.*

skeptical *adj.* doubting, dubious, °doubtful, questioning, disbelieving, °incredulous, agnostic, scoffing, cynical, mistrustful, °distrustful: *We were skeptical at first that anything would come of it, but then things began to happen.*

skepticism *n.* °doubt, dubiety, dubiousness, doubtfulness, disbelief, incredulity, incredulousness, agnosticism, cynicism, °mistrust, °distrust, mistrustfulness, distrustfulness: *At first, the entire matter was greeted with some skepticism; but then the Berlin Wall came tumbling down.*

sketchily *adv.* cursorily, superficially, incompletely, patchily, °roughly, perfunctorily, skimpily, °vaguely, imperfectly, crudely, °hastily, hurriedly: *His account of the event was sketchily written.*

sketchy *adj.* °cursory, °superficial, °incomplete, patchy, °rough, °perfunctory, skimpy, °imperfect, °crude, °hasty, °hurried, °vague, ill-defined, °fuzzy, °indistinct, °inexact, °imprecise, °unrefined, unpolished, rough-hewn, unfinished: *We received only sketchy reports of what was going on in the capital. This will do as a sketchy outline, but now the details must be filled in.*

skill *n.* **1** °talent, °ability, °aptitude, expertness, °expertise, °facility, skillfulness, °art, artistry, cleverness, adeptness, adroitness, mastery, °dexterity, handiness, °ingenuity, °experience, °proficiency, °finesse, °knack, quickness, deftness, °technique: *It requires great skill to operate this machine.* **2** °accomplishment, °forte, °strength, °gift, °capability, know-how, °faculty: *Her skill is in teaching others how to be skillful sales people.*

skillful *adj.* skilled, °accomplished, °adept, adroit, °dexterous, °expert, °proficient, masterly, °masterful, °gifted, apt, °able, °clever, °talented, °capable, °professional, trained, °qualified, °experienced, °practiced: *Julio is a skillful enough driver to race tomorrow. I have to admire the skillful way he handles people.*

skim *v.* **1** Often, ***skim off.*** °separate, cream, scoop *or* ladle off, °take off, °remove: *After the water has come to a boil, skim off the scum that has collected on top.* **2** Often, ***skim through*** *or* ***over.*** °scan, flip *or* thumb *or* leaf through, skip through, glance at *or* through, °dip into: *I had a chance only to skim through your report, but at a glance it looks good.* **3** °soar, °glide, skate, °slide, °sail, °fly: *Along came Calabro on his sailboard, skimming along the tops of the waves.*

skin *n.* **1** epidermis, derma, integument, °hide, °pelt, °fleece, °fell: *Her skin reddens in the sun. How many skins are needed to make a coat?* **2** °coat, °film, coating, crust, incrustation, husk, °peel, °rind, °outside, °shell, pellicle, °veneer, outer layer, lamina, overlay: *The frame is first covered with a tough plastic skin to make it waterproof.* —*v.* **3** flay, °strip, decorticate, excoriate: *I shall skin that boy alive if I catch him!* **4** °peel, °hull, husk, °shell: *This machine skins the fruit automatically.* **5** abrade, °scrape, graze, bark: *I always skin my shin on the edge of your coffee table.*

skin-deep *adj.* °superficial, °shallow, °surface, °slight, °external, unimportant, trivial, unprofound, °insubstantial: *The impression he makes on people is only skin-deep.*

skinny *adj.* °thin, underweight, °gaunt, bony, scraggy, lank, °lanky, gangly, gangling, °rawboned, °meager, °spare, °emaciated, half-starved, undernourished, °spare, pinched, hollow-cheeked, wasted, shrunken: *Two skinny children were clinging to their mother's skirts.*

skip *v.* **1** °leap, °cavort, °caper, gambol, °frisk, °prance, °jump, °hop, romp, °bound, °dance: *Eleanor came skipping down the walk to the house.* **2** °omit, °leave out, °pass by, °overlook, °pass over, °avoid, °ignore, °disregard, °steer clear of, °cut: *In my haste, I skipped over your name. Please skip the reading of the roll today.* —*n.* **3** °leap, °cavort, °caper, gambol, °frisk, °prance, °jump, °bound, °dance, °hop, romp: *He had a curious way of walking, giving a little skip before each step.* **4** lacuna, °gap, °omission, avoidance, °disregard; °miss, *Colloq* go-by: *There was a skip of thirty-two pages after page 64. If you take my advice, you will give that restaurant a skip.*

skipper *n.* captain, °master, commander; °boss, °leader, °chief: *The skipper gave orders to put him in irons.*

skirmish *n.* **1** °fight, °encounter, °fray, °brush, °clash, °engagement, confrontation, °showdown, °combat, °battle, °conflict, °struggle, set-to, °contest, °scrimmage, °fracas, tussle, melee *or* mêlée, *Law* affray, *Colloq* °scrap, dust-up, *Brit* scrum: *There was a brief skirmish when the troops met a band of partisans.* —*v.* **2** °fight, °clash, °struggle, °battle, tussle, °clash: *At the edge of the convoy a destroyer skirmished with a submarine.*

sky *n.* **1** heaven(s), skies, arch *or* vault of heaven, °firmament, °(wild) blue (yonder), ether, *Literary or archaic* welkin, empyrean, azure: *The dour, overcast sky gave the bleak moor a chilling sense of foreboding.* **2 *to the skies.*** °overly, excessively, extravagantly, fulsomely, profusely, inordinately, °highly: *If I praise him to the skies people will get the idea that I benefit from his success.*

slab *n.* °slice, °wedge, °piece, hunk, chunk, tranche, *Colloq Brit* wodge: *The keeper threw a large slab of meat into the lion's cage.*

slack *adj.* **1** °remiss, °careless, °indolent, negligent, °lax, °lazy, °idle, neglectful, °delinquent, °inattentive, otiose, dilatory, cunctatory, °laggard, easygoing, °slothful, sluggish, °lethargic, °shiftless, do-nothing, fainéant, *Colloq* asleep at the switch *or* the wheel, asleep on the job: *Production has fallen off because the workers are getting slack.* **2** °loose, °flabby, flaccid, °soft, °limp, baggy, drooping, droopy, bagging, sagging, floppy: *The flag hung down, slack in the still air.* —*v.* **3** Often, ***slack** or **slacken off** or **up.*** **(a)** let go, let run, °let loose, °release, slacken, °loose, °loosen, °relax, ease (out *or* off), °let up (on): *Slack off the stern line a bit.* **(b)** slow (down *or* up), °delay, reduce speed, °tire, °decline, °decrease, °diminish, °moderate, abate, °weaken: *Barnes could not keep up his terrific pace and is beginning to slack off. Business has slackened off since Christmas.* **4** °neglect, °shirk, *Colloq Brit* skive (off), *US* goof off, *Chiefly military* goldbrick, *Taboo slang US* fuck the dog: *Don't let the foreman find you slacking.* —*n.* **5** °lull, °pause, °inactivity, cutback, lessening, reduction, abatement, drop-off, downturn, diminution, °decline, falloff, °decrease, dwindling: *How do you compensate for the slack in sales of ski equipment during the summer?* **6** °room, looseness, slackness, °play, °give: *There's too much slack in that mooring line.*

slacker *n.* shirker, °loafer, °idler, *Slang Brit* skiver, *Military* scrimshanker, *US* goldbrick, goof-off: *Because of the tight deadlines, there is no room for slackers on the team.*

slake *v.* °satisfy, °quench, °gratify, allay, assuage, °ease, °relieve: *Nothing would slake my thirst better right now than a pint of ice-cold lager.*

slam *v.* **1** °shut, fling closed, bang: *The woman said she hated encyclopedias and slammed the door in the salesman's face.* **2** °crash, smash, smack, °dash, °ram, bang, °slap: *Not looking where he was going, Newland slammed his car into a street lamp.* **3** °criticize, °attack, °vilify, pillory, °run down, °disparage, denigrate, °denounce, °put down, flay, pounce on *or* upon, *Colloq* shoot down, °pan, *Chiefly Brit* slate: *The critics slammed his play because of the way it portrayed women.*

slander *n.* **1** defamation (of character), calumny, obloquy, misrepresentation, °slur, vilification; °libel: *He spread lies about me and I am suing him for slander.* —*v.* **2** defame, calumniate, °disparage, °slur, traduce, malign, °smear, °vilify, decry; °libel: *He must be stopped from slandering people and ruining their reputations.*

slanderous *adj.* defamatory, calumnious, disparaging, °smear, deprecatory, depreciative, discrediting, decrying; libellous *or* libelous: *I understand that she made some slanderous remarks about your relationship with your ex-wife.*

slant *n.* **1** °angle, °viewpoint, °(point of) view, °standpoint, °approach, °twist, °idea, °aspect, °attitude: *The article reflects a new slant on why governments are sometimes out of touch with the electorate.* **2** °bias, °prejudice, °partiality, one-sidedness, °turn, °bent: *Carla's reporting has a feminist slant which occasionally distorts the facts.* **3** °slope, °incline, °tilt, °ramp, gradient, °pitch, °lean, °leaning, deflection, °angle, °rake, °cant, camber: *The window sill ought to have an outward, downward slant. The road has a slant downward to the right on right-hand curves.* —*v.* **4** °tilt, °angle, °incline, °pitch, °cant, °slope, °bend, °lean, °list, °tip, bevel, °shelve: *The land slants downward near the lake. Notice how his writing slants upward at the ends of the lines. Cut the edges to slant outward.* **5** °bend, °distort, °deviate, °twist, °warp, °color, °weight, °bias: *The editor slanted the story to put the minister in a favorable light.*

slap *v.* **1** smack, cuff, whack, °rap; °spank; *Colloq* clout, wallop: *He said something extremely rude so she slapped him.* **2** °flap, slat, °whip, °beat, bat: *Can't you stop that blind from slapping in the wind?* **3** °fling, °toss, °splash, °hurl, °throw, °sling: *If you slap some paint on it, no one will know the difference.* —*n.* **4** smack, °blow, cuff, whack, °rap, *Colloq* clout, wallop: *He got a hard slap on the cheek for using foul language.* **5** Often, ***slap in the face.*** °reprimand, °reproof, °rebuff, °criticism, censure, °rebuke, °shot, °thrust, °attack, °put-down, °insult, °offense, smack in the eye: *The speaker's reference to Anne's paper as 'trivial' was a severe slap in the face.* —*adv.* **6 *slap on.*** °exactly, °directly, °precisely, °straight, °point-blank, °right, squarely, °plumb, smack, bang: *As usual, Barry's comments were slap on the mark.*

slash *v.* **1** °cut, °gash, °hack, °score, °slit, °knife, °lacerate; °wound; °scar: *The guide slashed away at the undergrowth with his machete.* **2** °lash, °whip, °scourge, °flog, °beat, horsewhip, flail, flagellate, flay, thrash, °beat, *Colloq* lambaste *or Colloq* lambast: *In those days, a convicted felon was beaten and slashed in front of a crowd in the market-place.* **3** °cut, °reduce, °decrease, °drop, °mark down, °trim, °lower: *Prices were slashed to clear out last season's styles.* —*n.* **4** °cut, °gash, °incision, °slit, °slice, °gouge, °rent, °rip, °score, laceration: *There is a slash in each sleeve that reveals the colorful fabric underneath.* **5** °cut, reduction, °decrease, markdown: *The department stores continued their price slashes to the end of January.*

slattern *n.* slut, °tramp, sloven, trollop, hussy, °wanton, whore, °prostitute, harlot, streetwalker, lady of the evening, woman of ill repute, loose *or* fallen woman, *trottoise, Colloq* call girl, pro, *Slang* °tart, hooker, hustler, *US* roundheel(s), bimbo: *His reputation will not be enhanced if he associates with slatterns.*

slaughter *n.* **1** butchery, butchering, *Rare* abattage: *We arranged the slaughter of these cattle for next week's market.* **2** °massacre, °killing, °bloodshed, blood bath, °murder, homicide, manslaughter, °carnage, extermination, °execution, liquidation, slaying, bloodletting, butchery, pogrom, genocide, mass murder *or* execution *or* extermination, °sacrifice, hecatomb: *1930 to 1945 saw the slaughter of millions in Europe and the Far East.* —*v.* **3** °butcher, °kill, °murder, slay, °execute, °exterminate, °massacre, put to the sword, put to death, liquidate, °destroy: *How many more must be slaughtered before war is made obsolete?* **4** °defeat, °beat, win (out) over, vanquish, °overcome, °overwhelm, smash, °crush, thrash, °destroy, °rout, °upset, trounce, *Colloq* clobber: *Our school soccer team slaughtered the visitors 10–0.*

slave *n.* 1 lackey *or* lacquey, scullion, serf, slave girl, slaveling, odalisque *or* odalisk, bondservant, bondslave, bondsman *or* bondman, bondswoman *or* bondwoman, bondmaid, vassal, thrall, *Disparaging chiefly Brit* skivvy, *Archaic* esne, helot, hierodule, *Colloq Brit* slavey: *In ancient times, captured peoples and those of lower social status were kept as slaves.* 2 drudge, workhorse, °hack, °grind, toiler, °laborer, *Chiefly Brit* °fag, dogsbody, *Colloq US* gofer: *Susan was a slave to her job for years, only to be sacked without a pension by the new owners.* —*v.* 3 °labor, toil, moil, °grind, grub, drudge, °sweat, burn the midnight oil, lucubrate, work one's fingers to the bone, work like a Trojan *or* a horse, *Brit* skivvy: *I have been slaving over a paper due tomorrow.*

slaver[1] *v.* 1 drool, salivate, slobber, °drivel, dribble, °spit, *Dialect* slabber: *Why go to a restaurant where you have to watch slavering infants?* —*n.* 2 drool, saliva, °drivel, dribble, °spit, spittle, *Dialect* slabber: *The slaver formed on his lips as he raged on.* 3 °nonsense, °drivel, °rubbish, twaddle, piffle: *The slaver spewed forth by the critics set her teeth on edge.*

slaver[2] *n.* 1 slave ship, slave trader: *Conditions were so bad aboard the slavers that many of the poor souls died before reaching their destination.* 2 blackbirder, slave trader; white slaver, °pimp, panderer: *The cruelest man I ever met was Dunvey, the South Sea slaver.*

slavery *n.* 1 enslavement, °bondage, thralldom, thrall, enthrallment, °servitude, serfdom, vassalage, yoke; subjugation, °captivity, *Historical US* peculiar institution: *All of those ancient cultures that we venerate practiced slavery.* 2 slave trade, blackbirding: *No one talks about the family fortune having been made from slavery.* 3 toil, moil, °drudgery, travail, °grind, °strain, °(hard) labor: *She thought of the years she had spent in slavery at the kitchen sink.*

sleazy *adj.* 1 unsubstantial *or* insubstantial, °flimsy, °slight, °shabby, °poor, gimcrack, jerry-built, °tawdry, °cheap, tatty, °ramshackle, °rickety, °slipshod, *Colloq US* chintzy: *Her garish dress was made from some sleazy synthetic material.* 2 °disreputable, low-class, low-grade, squalid, °dirty, °base, °seedy, °sordid, °contemptible, trashy, °run-down, °mean, °cheap, *Colloq* crummy, slummy, *Slang* crappy, cheesy: *He was lying low in a sleazy hotel while the police searched for him.*

sleek *adj.* 1 °smooth, °slick, velvety, °lustrous, °shiny, shining, °glossy, °silky, silken: *The seals' sleek fur glistened in the sunlight.* 2 °graceful, °trim, °streamlined: *For her birthday Edmund gave her a sleek convertible.* 3 °suave, unctuous, °slimy, fawning, °oily, °specious, °hypocritical, *Chiefly Brit* smarmy: *The sleek villain has separated many an unsuspecting widow from her savings.*

sleep *v.* 1 °doze, °(take a) nap, catnap, °rest, °repose, slumber, drowse, drop *or* nod off, be in the Land of Nod, be in the arms of Morpheus, snore, *Colloq* snooze, saw wood, catch a few Z's, take *or* have a zizz, catch forty winks: *I was so tired, I slept for a full ten hours.* —*n.* 2 °nap, °doze, slumber, °rest, siesta, *Colloq* forty winks, snooze, zizz, beauty sleep: *As she grew older, she found she needed less sleep.*

sleepless *adj.* 1 °restless, °wakeful, insomniac, °disturbed: *Your mother and I spent a sleepless night worrying about you.* 2 °alert, watchful, °vigilant, unsleeping: *Nugent's wife kept a sleepless vigil at his bedside.*

sleepwalking *n.* 1 noctambulism, somnambulism, noctambulation, somnambulation: *Because of his sleepwalking, this patient is usually strapped in at night.* —*adj.* 2 noctambulant, somnambulant: *Proper drugs will control her sleepwalking activities.*

sleepy *adj.* 1 °drowsy, somnolent, °tired, °nodding, dozy, °lethargic, °torpid, slumberous, sluggish, oscitant; °weary, °fatigued, °exhausted, *Colloq* dead on one's feet, °(knocked) out, °beat, *US and Canadian* pooped: *All that exercise has made me sleepy.* 2 °boring, °inactive, °dull, °quiet, soporific, °slow, sluggish: *He grew up in a sleepy little village in the Outer Hebrides.*

slender *adj.* 1 °slim, °lean, °willowy, sylphlike, svelte, lissom *or* lissome, lithe, °graceful, snake-hipped, °thin, °spare, °slight, °lanky: *By exercising regularly, she has retained her slender figure.* 2 °slim, °narrow, °slight, °poor, °unlikely, °small, °little, °scanty, °remote, °meager, °weak, °feeble: *Prospects for a quick recovery of the stock market seemed slender.* 3 °slim, °slight, °little, °scanty, °inadequate, °insufficient, °insignificant, °trifling: *The evidence against the defendant is too slender for conviction.*

sleuth *n.* °(private) detective, (private) investigator, *US* P.I., *Colloq* private eye, sherlock, °snoop, *Brit* tec *or* 'tec, *US* hawkshaw, *Slang US* dick, shamus, *US and Canadian* gumshoe: *Dissatisfied with the progress of the police, we hired a private sleuth to investigate the theft.*

slice *n.* 1 °slab, °piece, rasher, collop, shaving, layer, *Cookery* scallop, escalope, scaloppine (*pl. of scaloppina*) *or* scaloppini (*pl.*): *May I have another slice of ham?* 2 °portion, °piece, °part, °wedge, °share, °sliver, °helping: *He wants to make certain he gets his slice of the pie.* 3 spatula; slicer: *The Cabots gave us a silver fish slice for a wedding present.* —*v.* 4 °cut, °carve, °divide: *Father always slices the roast for Sunday dinner.*

slick *adj.* 1 °smooth, °sleek, °glossy, °silky, silken, °shiny, shining, °glassy, °slippery: *His slick hair looked as if it had been greased.* 2 °smooth, urbane, °suave, smooth-spoken, °glib, °smug, °plausible; sycophantic, unctuous, *Colloq* smarmy: *Thomas was slick enough to worm his way into a position of power.* 3 °smooth, °clever, °skillful, adroit, °dexterous, °professional, °ingenious, °imaginative, inventive, °creative, *Colloq* °neat: *Fiona came up with a truly slick plan for handling customer complaints.* 4 °superficial, °shallow, meretricious, °specious, °glib: *They have made a lot of money publishing slick magazines for the yuppie market.* —*v.* 5 Often, ***slick down.*** °smooth, plaster down, grease, °oil: *He likes his hair slicked down to look like Valentino's.*

slicker *n.* 1 confidence man *or* woman, °cheat, °swindler, mountebank, *Colloq* con man, city slicker: *That slicker she met in Paris separated Harriet from a lot of money before he disappeared.* 2 oilskin (raincoat): *Best wear your slicker in this storm, Clemson.*

slide *v.* 1 °glide, °slip; °coast, °skim, glissade, skate, °plane, skid, toboggan, °slither: *Slide the door back into the wall, out of the way. Terry came sliding down the icy hill, arms and legs flailing.* 2 °creep, °steal, °slip, °slink, °move: *My contact slid into the seat beside me and slipped me a note.* 3 °decline, °decrease, °drop, °fall: *Shares slid to an all-time low on this morning's market.* **4 *let slide.*** °forget, °ignore, °neglect, gloss *or* pass over, pay no heed *or* mind (to): *When Bartlett borrowed my lawnmower, I let the matter slide till he began to think of it as his own, and offered to lend it to me!* —*n.* 5 landslide, earthslip, avalanche, mudslide: *Slides, caused by the recent torrential rains, destroyed or buried a dozen houses.*

slight *adj.* 1 °small, °little, °minor, °negligible, °unlikely, °insignificant, °inconsequential: *There is always the slight chance that you might be wrong, you know.* 2 °trifling, °tiny, °slender, °minute, infinitesimal; °trace, °perceptible: *Don't attach the slightest importance to the rumors you hear. You'd best wash it again, as there is still a slight odor of garlic.* 3 °small, °short, °petite, °thin, °slim, °slender, °delicate, °diminutive, °tiny,

°miniature, bantam, °wee, °pocket, pocket-sized, *US* vest-pocket, *Colloq* pint-sized: *Charlotte's slight build seems totally unaffected by the amount she eats.* **4** insubstantial *or* unsubstantial, °weak, °feeble, °delicate, °dainty, °frail, °unstable, °fragile, °flimsy, lightly made *or* built, °precarious, °inadequate, °rickety, °insecure: *This table is too slight to support both the computer and the printer.* —*v.* **5** °disregard, °ignore, disdain, °scorn, snub, °rebuff, °cut, °disrespect, cold-shoulder: *Did you slight me deliberately at the dance last night?* **6** °insult, °offend, affront, °mortify, °diminish, °minimize, °depreciate, °disparage: *Siobhan felt slighted by Harry's inattention.* —*n.* **7** °insult, affront, °slur, °indignity, °outrage, offense, °disrespect: *One finds it hard to understand why she tolerates his persistent slights.* **8** inattention, °neglect, °disregard, °indifference, snub, cold shoulder, coldness, ill-treatment: *Perhaps she misinterprets his obsessive occupation in his work as a slight to her.*

slightly *adv.* a little, °somewhat, to a certain *or* slight *or* minor extent *or* degree *or* measure, marginally: *Yes, I do feel slightly better, thank you.*

slim *adj.* **1** See **slender, 1,** above. **2** See **slender, 2, 3,** above. —*v.* **3** °reduce, lose *or* shed weight, °diet, *Chiefly US* slenderize: *I really must slim if I am to fit into my summer clothes.*

slimy *adj.* **1** oozy, °slippery, mucky, squashy, squishy, viscous, °sticky, gluey, mucilaginous, uliginous, °oily, oleaginous, glutinous, mucous, °clammy, °mushy, *US* squushy *or* squooshy, *Colloq* °gooey, gunky, *US* gloppy: *He had cleverly buried the bag of diamonds in a mass of slimy waste where no one would want to look.* **2** °slippery, unctuous, °obsequious, sycophantic, toadying, °servile, creeping, °groveling, abject, *Colloq* smarmy: *I don't mind his trying to sell me insurance, but why does he have to be so slimy about it?*

sling *v.* **1** °toss, °throw, °cast, °propel, °hurl, °shy, °fling, °fire, °shoot, °pitch, °let fly, °launch, °heave, °lob, *Colloq* chuck: *On release of the enormous tension, the missile was slung over the wall of the besieged city.* —*n.* **2** slingshot, catapult, trebuchet *or* trebucket: *It is said that David slew Goliath with a stone from his sling.* **3** °support, strap, °band; °belt: *The doctor ordered him to keep his sprained arm in a sling for a few days. The military rifles are supplied with leather slings.*

slink *v.* °sneak, °creep, °steal, °prowl, skulk: *Whenever I visited Blanche, her mother was always slinking about in a nearby room, keeping an eye on us.*

slip[1] *v.* **1** °slide, skid, °glide, °slither: *Waxing the runners helps the sleigh slip along more easily.* **2** °stumble, lose one's footing *or* balance, miss one's footing, °trip; °fall, °tumble: *Mother slipped on the ice but luckily didn't fall down.* **3** Often, ***slip up.*** °err, °blunder, make a mistake, °miscalculate, °go wrong, °botch (up), *Slang* °screw up: *He slipped up again by failing to give you your telephone messages.* **4** ***let slip.*** °reveal, divulge, °blurt out, °leak, °let out, °disclose, °expose, *Colloq* come out with, °blab: *Oops! I let slip the news that you were back in town.* **5** ***slip away*** *or* ***by.*** °pass, °elapse, vanish, °go by: *The hours just slipped away when I was with Irena.* **6** ***slip away*** *or* ***off*** *or* ***out.*** °escape, °disappear, °leave, vanish, °steal, go *or* run away *or* off *or* out, °break away, °get away, give (someone) the slip; sneak away *or* off *or* out: *After the speech, he slipped away before we could interview him.* **7** ***slip in.*** °enter, °get in, sneak in; °put in: *A field mouse slipped in past the wire netting. I managed to slip in my suggestion before the meeting was adjourned.* —*n.* **8** °blunder, °error, °mistake, °fault, °oversight, slip of the tongue *or* pen, inadvertence, °indiscretion, °impropriety, °transgression, °peccadillo, faux pas, *Colloq* slip-up, *Chiefly US* blooper, *Slang Brit* boob, bloomer: *Revealing the date of the ship's arrival was a serious slip, Yamamoto.*

slip[2] *n.* **1** °piece, °scrap, °strip, °sliver; °paper, °note, chit, °permit, °permission, °pass, °document: *I'll put my phone number on this slip of paper. To go on the outing, each of you will have to bring in a slip signed by a parent.* **2** °shoot, scion, °cutting, sprig, °twig, °sprout, °runner, °offshoot: *After the slip is inserted, the branch must be bandaged up firmly.*

slippery *adj.* **1** °slick, °sleek, °slimy, °icy, °glassy, °smooth, °greasy, °oily, lubricated, *Colloq* skiddy: *Be careful! Those stairs can be slippery when they're wet.* **2** °evasive, °devious, °shifty, °unreliable, undependable, °questionable, untrustworthy, °dishonest, treacherous, °disloyal, °perfidious, °slick, °crafty, °sly, °foxy, cunning, °tricky, °sneaky, °false, reptilian, °faithless, *Colloq* °shady: *I have dealt with some very slippery characters but none as bad as Dan Writher.*

slipshod *adj.* °careless, slovenly, slapdash, °haphazard, messy, °untidy, disorganized, °lax, unorganized, *Colloq* °sloppy: *I don't much care for the slipshod way they did the repairs.*

slit *v.* **1** °split, °cut, °slash, °gash, °knife, °slice: *The bark of the log is first slit open, then carefully peeled back.* —*n.* **2** °split, °cut, °gash, °incision, fissure, °groove, °slash, cleft, °aperture, °opening: *We actors could watch the audience through a slit in the curtain.*

slither *v.* °slide, worm, °snake, °slip, °slink, °glide, skitter, °creep, °crawl: *After regarding me for a moment, the cobra turned and slithered away.*

sliver *n.* °fragment, °piece, shard, °shred, °splinter, °slip, shaving, paring, °flake, °chip, °bit, °scrap, snippet, °snip: *Two weeks later, we were still finding tiny slivers from the broken vase.*

slob *n.* oaf, °boor, pig, lout, churl, °yahoo, *Archaic* slubberdegullion, *Slang Brit* yob, yobbo, *Chiefly US* galoot *or* galloot, slobbovian: *He was a terrible slob—seldom bathed and kept his room like a pigsty.*

slogan *n.* war cry, battle cry, rallying cry, catchword, watchword; °motto: *The magazine's slogan was "If you love words, you'll love* Verbatim*."*

slope *v.* **1** °incline, °decline, ascend, °descend, °rise, °fall, °dip, °sink, °drop (off), °angle, °slant, °pitch, °tilt, °tip: *The lawn slopes downward from the house, then upward toward that grove of trees.* —*n.* **2** °incline, °decline, ascent, descent, acclivity, declivity, °rise, °fall, °ramp, °dip, °sink, °drop, °angle, °slant, °pitch, °tilt, °rake, °tip, camber, °cant, °grade, bevel, °hill, bank, °mount, gradient, *US* °grade, upgrade, °downgrade: *Sheep can be seen grazing on the grassy slopes.*

sloppy *adj.* **1** messy, °dirty, slovenly, °careless, °slipshod, °untidy, disordered, °disorderly; draggletailed, °bedraggled, disheveled, °unkempt, °dowdy, frumpish, °shabby, scruffy, *Colloq US* grungy: *She is certainly a sloppy housekeeper. After working in the garden all day I am too sloppy to receive visitors.* **2** °wet, slushy, °watery, soggy, soppy, sopping, sodden, sloshy, °muddy, rainy: *The roads were sloppy after all the rain. You're lucky if you don't have to go out in this sloppy weather.* **3** °sentimental, °gushy, gushing, mawkish, °maudlin, °mushy, overemotional, *Colloq* slushy, *Brit* °wet, soppy: *She gets so sloppy when she attends weddings that she weeps the entire time.*

slot *n.* **1** °groove, fissure, °notch, °slit, °opening, °hollow, °depression, °channel, sulcus: *Each of these pieces fits into its own slot. Drop a coin in the slot and watch what happens.* **2** °opening, °position, °vacancy, °job, °place, °assignment, °niche, °space, °spot, pigeonhole: *We filled that slot in the sales department. Schedule the new program into the half-hour slot after the six* *o'clock news.* —*v.* **3** °groove, fissure, °notch, °slit, hollow out: *We slotted the stanchions to accommodate the shelf brackets.* **4** °assign, °schedule, °place, °position, pigeonhole, °fit: *Can we slot this interview into the documentary on China?*

sloth *n.* °idleness, laziness, °indolence, slothfulness, °inertia, apathy, °indifference, accidie, °torpor, faineance, pococurantism, torpidity, °sluggishness, languor, languidness, °lethargy, phlegm, *Rare* hebetude: *The maharajahs kept themselves in sloth and luxury at the expense of their poverty-stricken subjects.*

slothful *adj.* °idle, °lazy, °indolent, apathetic, °indifferent, °torpid, °inert, pococurante, °slack, °lax, °shiftless, fainéant, °inactive, do-nothing, sluggish, sluggard(ly), °slow, °laggard, languorous, languid, °lethargic, °lackadaisical, °phlegmatic, hebetudinous: *One cannot pursue a slothful attitude and expect to get anywhere in life.*

slouch *v.* **1** °droop, °sag, °stoop, loll, °slump, °hunch: *Sit up straight and stop slouching.* —*n.* **2** °stoop, °sag, °droop, °slump, °hunch: *He seems to have developed a slouch from the burden of his responsibilities.* **3** Usually, ***no slouch.*** sloven, °loafer, sluggard, °laggard, °loafer, °idler, malingerer, lazybones: *Nellie is certainly no slouch when it comes to exercising to keep fit.*

slow *adj.* **1** lagging, °laggard, dawdling, sluggish, sluggard(ly), slow-moving, °leaden, °ponderous, °unhurried, plodding, snaillike, tortoiselike, °torpid, leaden-footed, creeping, crawling; °deliberate, slow-paced, leisurely, °gradual, °easy, °relaxed, °lax, °lackadaisical, °lazy, *US* lallygagging *or US* lollygagging: *They walked at a slow pace back to the house. Phileas has become a little slower in his old age.* **2** °gradual, °progressive, °moderate, °perceptible, almost imperceptible, measurable: *The church elders have noted a slow decline in moral standards.* **3** °unhurried, slow-moving, slow-paced: *The slow funeral cortege crept through the streets.* **4** behindhand, unpunctual: *I missed the train because my watch is five minutes slow.* **5** °late, °tardy, behindhand, dilatory, delayed, unpunctual: *You were so slow in getting here that everyone has gone.* **6** °slack, °inactive, °quiet, sluggish, unproductive: *Business is always a bit slow after the holidays.* **7** °dense, °dull, slow-witted, dull-witted, °obtuse, °backward, bovine, °dim, dimwitted, °stupid, unresponsive, blockish, cloddish, unintelligent, doltish, °simple, °stolid, unimaginative, Boeotian, *Colloq* slow on the uptake, °thick, °dumb: *I'm afraid I'm a bit slow when it comes to particle physics.* **8** °conservative, unprogressive, °old-fashioned, out-of-date, °backward, old-fogyish, °behind the times, *Colloq* °square, not with it, past it, *US* out of it: *The council has been rather slow in realizing the needs of the community.* **9** °boring, °dull, °tiresome, ennuyant, °tedious, °dead, °sleepy, somnolent, °torpid, soporific, wearisome, dryasdust, uninteresting, °monotonous, °tame, uneventful, °humdrum, *Colloq* ho-hum, °dead, *Brit* dead-and-alive: *This town becomes really slow in the autumn, till the tourists reappear in the spring.* **10** °reluctant, not quick, unwilling, °hesitant, °disinclined, °averse, loath *or* loth, °indisposed: *Even when provoked, Cassie tends to be slow to anger.* —*adv.* **11** slowly, °unhurriedly, cautiously, carefully, circumspectly: *Drive slow where you see children playing.* **12** behindhand, tardily, °late, unpunctually: *The trains seem to be running a bit slow for a change.* **13** slowly, °easy, leisurely, °easily: *They told him to take things slower or he'd have another heart attack.* —*v.* **14** Often, ***slow down*** *or* ***up.*** slack *or* slacken off, reduce speed, °hold back, put on the brakes, take it easy: *You'd better slow down before you come to Deadman's Hill.* **15** °relax, take it easy, *Colloq* ease up: *The doctor suggested that at my age I ought to slow down a little.*

sludge *n.* °muck, °mire, °ooze, °mud, slime, °dregs, °silt, °residue, °precipitate, *Colloq* goo: *The mechanic said that there was a lot of sludge in the engine.*

sluggishness *n.* °sloth, laziness, slothfulness, languor, lassitude, °lethargy, languidness, laggardness, °torpor, phlegm, lifelessness, stagnation, shiftlessness, pococurantism, faineance, accidie, *Rare* hebetude: *A few days in the Inverness climate will help to shake off the sluggishness of a February Caribbean vacation.*

slum *n.* Often, **slums.** ghetto, warren, shantytown, *US* skid row *or US* Skid Road: *Many very famous men emerged from the slums of the Lower East Side in New York.*

slump *n.* **1** °dip, trough, depreciation, °decline, downturn, downslide, °recession, °depression, falling-off, falloff, °fall, °drop, °plunge, descent, °crash, °collapse, °failure; nose dive, tailspin: *A slump in the price index was taken as a good sign by the economists concerned about inflation. The market went into a slump this morning from which it barely recovered before closing.* —*v.* **2** °decline, °slip, °recede, °fall (off), °drop, °plunge, °descend, °sink, °crash, °collapse, °dive, °plummet, take *or* go into a nosedive *or* tailspin: *Prices on the stock market slumped on the news of a rise in the interest rate.* **3** See **slouch, 1,** above.

slur *n.* **1** °smear, °insult, calumny, °aspersion, affront, °stigma, °stain, °blot, °spot, °(black) mark, °discredit, insinuation, °innuendo, °imputation, °slander, °libel, °slight, *Colloq* °put-down: *Mrs. Cassidy regards any question about her husband's military service as a slur on his character.* —*v.* **2** °mumble, misarticulate, °garble, stutter, lisp: *Some youngsters slur their speech so badly I cannot understand them.* **3** ***slur over.*** °gloss over, °pass over, °disregard, give short shrift to, °ignore: *The eulogy slurred over his faults and focused on his accomplishments.*

sly *adj.* **1** cunning, °artful, °crafty, °clever, °wily, guileful, underhand(ed), °deceitful, treacherous, °foxy, °scheming, plotting, °designing, conniving, furtive, °shrewd, °sneaky, °stealthy, insidious, °devious, °disingenuous, °tricky, °shifty, °sharp, canny, *Colloq* °shady: *It was devilishly sly of him to maneuver you into paying for his party.* **2** impish, elfish, roguish, °mischievous, puckish, °devilish, scampish, °naughty, °arch, waggish: *Even as he was flirting with Allison he was giving Antonia sly, conspiratorial winks.* —*n.* **3** ***on the sly.*** slyly *or* slily, °quietly, surreptitiously, covertly, stealthily, furtively, sneakily, underhandedly, clandestinely, *Colloq* on the Q.T., on the side: *While Matilda was engaged to be married to Walter, she continued to go out with Carlton on the sly.*

small *adj.* **1** °little, °tiny, °short, °diminutive, °petite, mignon(ne), °wee, teeny, °elfin, Lilliputian, midget, °miniature, °minute, minuscule, °baby, bantam, pocket(-sized), mini; °undersized, °immature, °young, underage; *Colloq* pint-sized, *US* peewee: *She was once small enough to fit into a size 7. When I was very small, my father took me to my first cricket match.* **2** °slight, °secondary, °insignificant, trivial, °inconsequential, lesser, °puny, °negligible, °minor, °trifling, unimportant, °paltry, nugatory: *What she says about me is of only the smallest moment to me.* **3** unimaginative, °shallow, unoriginal, mundane, °everyday, °limited, unprofound, uninspired, commonplace, °matter-of-fact, °flat, two-dimensional: *With their small minds they were incapable of fully appreciating the scope of the project.* **4** skimpy, niggardly, stingy, uncharitable, ungenerous, °scanty, °meager, °cheap, °petty, parsimonious, grudging, stinting, °selfish, °miserly, °tight, tight-fisted, closefisted, °close; °poor, °insignificant, °inadequate, °insufficient, °unsatisfactory, °negligible,

°trifling, °humble, small-scale, °modest, unpretentious, *Colloq* piddling, °measly: *With all your money, it was very small of you not to contribute more to the scholarship fund. We all thought your check was too small.* **5** °insignificant, °limited, °negligible, °trifling, °tiny; °little, °minor, diminished, reduced, °slight: *Only a small number of people came to the opening night. They derived small consolation from the knowledge that their daughter had placed second in the competition.* **6** ***feel small.*** feel embarrassed *or* ashamed *or* shamed *or* humiliated *or* foolish, feel discomfited *or* disconcerted *or* uncomfortable, feel mortified *or* chagrined, *Colloq* feel put down: *Doreen always managed to make some remark that made Andrew feel small.*

small-minded *adj.* °small, °petty, °selfish, stingy, grudging, niggardly, ungenerous, °mean, °narrow-minded, °narrow, close-minded, uncharitable, °hidebound, °rigid, °intolerant, unimaginative, °shortsighted, °nearsighted, myopic: *How very small-minded of you it is to consider only your own immediate interests!*

small-time *adj.* °small, small-scale, unimportant, °petty, piddling, °minor, °insignificant, °trifling, trivial: *In less than ten years he succeeded in building a small-time car dealership into a national chain.*

smart *adj.* **1** °intelligent, °clever, °bright, °brilliant, °quick-witted, °sharp, °acute, °astute, °capable, °adept, apt, °quick, °ingenious: *Emily is one of the smartest children in the class.* **2** canny, perspicacious, °perceptive, percipient, discerning, °knowledgeable, *au fait,* well-educated, well-read, erudite, °learned, well-versed, °aware, °shrewd, streetwise, *Slang* °hip, tuned-in, *US* savvy: *Brendan is smart enough to avoid trouble.* **3** °elegant, °chic, °fashionable, °stylish, modish, à la mode, °well-groomed, °trim, °neat, °dapper, °spruce, soigné(e), *Colloq* °snappy, natty: *You must admit that both Della and Paul are smart dressers.* **4** °pert, °pointed, saucy, °witty, nimble-witted, °poignant, °trenchant, °effective: *Stop making such smart remarks when I am speaking to you.* **5** °brisk, °vigorous, °animated, °active, °energetic, °spirited, °lively; °quick, °alert, °jaunty, °perky, °breezy: *The enemy launched a smart counterattack at dawn. He set off down the road at a smart pace.* **6** °quick, °swift, °stiff, smarting, stinging, °sharp, °severe: *The mugger gave him a smart blow on the back of the neck.* —*v.* **7** °sting, °hurt, °pinch, °pain, °ache, tingle, °prickle, °burn, throb, °stab, °pierce: *The antiseptic may smart when it touches those cuts. I have a smarting pain in my leg from the tick bite.* —*n.* **8** °injury, °harm, °pain, °pang, °twinge, °affliction, °suffering, smarting: *For years he silently endured the smart from the calumnies Buford repeatedly heaped upon him.*

smear *v.* **1** daub, °rub, anoint, °spread, °cover, °coat, °wipe, °plaster, bedaub; besmirch, °dirty, smudge, °stain, °soil, begrime: *Lydia smears on a mudpack to improve her complexion. The car windows were all smeared up by the children.* **2** °blacken, besmirch, °soil, °sully, calumniate, °slander, °discredit, °tarnish, defile, °vilify, °scandalize, °stigmatize, *Colloq* drag through the mud: *We must not allow our fearless leader's reputation to be smeared by the tabloids.* —*n.* **3** smudge, daub, °stain, splodge *or chiefly US* splotch, °blot, °taint, °spot: *These outrageous lies constitute a smear on his reputation. There's a smear of blood on your collar.* **4** °slander, °scandal, °libel, vilification, mudslinging, defamation, calumny, °aspersion, °reflection: *The article contains several smears on the character of a Senator.*

smell *n.* **1** °odor, °scent, °aroma, °perfume, °fragrance, °bouquet, °breath, whiff: *Don't you love the smell of freshly brewed coffee in the morning?* **2** stink, °stench, fetor *or* foetor, fetidness, mephitis, effluvium, *Colloq Brit* pong: *The smell from the rotting garbage was overpowering.* —*v.* **3** °scent, °sniff, *Colloq* get a whiff of: *The moment I smelled that perfume I knew that Nicole had been there.* **4** stink, °reek, *Colloq Brit* pong, °hum: *The milk had gone off and smelled to high heaven.*

smelly *adj.* malodorous, evil-smelling, foul-smelling, °foul, mephitic, fetid, °putrid, reeky, °stinking, noisome, °rank, °offensive, miasmic *or* miasmatic *or* miasmatical *or* miasmal, odoriferous, °rancid, °high, gamy, *Slang Brit* whiffy: *Wash out those smelly jogging clothes right away.*

smile *v.* **1** grin, °beam: *He smiled and said he was pleased to meet me.* —*n.* **2** grin: *Georgina bestowed on me a sweet, wistful smile.*

smirk *n.* **1** °leer, °sneer, grin, grimace, simpering smile: *Wipe that conceited smirk off your face.* —*v.* **2** °sneer, grimace, °leer: *Instead of saying, "I told you so," he just stood there smirking.*

smitten *adj.* **1** °affected, afflicted, °beset, °stricken, troubled, distressed, burdened, crushed, plagued, haunted, °worried, bothered, vexed: *We were smitten by the news that Peter had married Vespasia despite our warnings.* **2** captivated, enthralled, struck, bewitched, °enchanted, beguiled, °charmed, enraptured, °infatuated, enamored, ensorcelled, swept off one's feet, *Colloq* bowled over, gaga: *Anthea met some rock star and was totally smitten. She was what you might call a "smitten kitten."*

smolder *v.* °burn; °seethe, °simmer, °chafe, °rage, °fume, °foam, °boil, °stew, °fester, *Colloq* get hot under the collar, get (all) steamed up, see red, *US* do a slow burn, get (all) burnt up: *The fire continued to smolder, then flared up again a week later. He has been smoldering ever since his wife confessed she was going out with another man.*

smooth *adj.* **1** °regular, °even, °flush, °flat, °level, °plane, unruffled, unbroken, unwrinkled, undisturbed, °tranquil, °peaceful, °calm, °serene, °glassy: *A light breeze rose to ripple the smooth surface of Tony's martini.* **2** °slick, °sleek, °shiny, °glossy, °glassy, mirrorlike, °uniform, °polished, burnished; °silky, silken, velvety, satiny: *Optical mirrors for telescopes must be as smooth as modern technology can made them. He ran his fingers over the smooth fabric.* **3** unobstructed, °easy, °effortless, °free, uncluttered, °even, °orderly, well-ordered, uneventful, flowing, °fluent, unconstrained, uninterrupted: *There were no obstacles in the way of a smooth return to normal operations. The road to ruin is short and smooth.* **4** °hairless, bald, °bare, °naked, cleanshaven, smooth-shaven, depilated, glabrous: *His smooth pate shone in the moonlight.* **5** °soothing, °mellow, °pleasant, °bland, °mild, °soft: *That is a very smooth whiskey indeed.* **6** °suave, °slick, °slippery, unctuous, silken, °silky, °glib, urbane, soigné(e), °agreeable, °winning, °plausible, facile, °nonchalant, courtly, °eloquent, honey-tongued, smooth-spoken, °persuasive, °oily, °slimy, syrupy, *Colloq chiefly Brit* smarmy: *With his smooth approach, Edward was able to beat the store's record for appliance sales.* **7** °sweet, dulcet, pear-shaped, °mellow, well-modulated, silver-tongued: *The smooth tones of his serenade wafted through the evening air.* **8** °slick, °scheming, conniving, °crafty, °shrewd, cunning, °tricky, °shifty, °sly, °foxy, Machiavellian, sophistic(al), °plausible, credible, believable, *Colloq* cagey: *Muggins came up with a very smooth plan to bilk wealthy widows out of their money.* —*v.* **9** Often, ***smooth out*** *or* ***away.*** °flatten, °even, °level, iron, °press, °mangle, calender: *Please smooth the wrinkles out of this shirt so that I can wear it tonight.* **10** °prepare, °lay, °pave, °ease, °ready, °clear, °open, °prime, lubricate, °facilitate: *What is your policy for smoothing the way for* *employees joining the firm?* **11** sand, °plane, °polish, buff, burnish: *The wood is smoothed to a glassy finish before being used for the cabinets.* **12** Often, ***smooth over.*** ameliorate, assuage, allay, °calm, °gloss over, °minimize, °mitigate, lessen, soothe, °reduce, °temper, mollify, smoothen, °soften, palliate, appease: *An otherwise uncomfortable situation was smoothed over by her quiet words.*

smother *v.* **1** suffocate, °stifle, °choke, asphyxiate; throttle, strangle, snuff (out), °kill: *She was smothered by her pillow when she had an epileptic fit.* **2** be suffocated *or* stifled *or* asphyxiated, be choked *or* strangled, be killed: *The infant apparently smothered in the blanket.* **3** °overwhelm, °overcome, blanket, inundate, °cover, °shower; °envelop, °wrap, enshroud, °surround: *The children crowded around and smothered her with kisses. The banquet table was smothered in white roses.* **4** °repress, °subdue, °suppress, °conceal, °hide, keep *or* hold back, °cover up, °mask, choke back *or* down, °check; °stifle, °muffle, blanket, blank out: *He managed to smother any display of grief and put on a brave face. Our whispers were smothered by the noise of the hi-fi.* **5** °extinguish, °put out, snuff out: *The foam from the fire extinguisher smothered the blaze in a few moments.*

smug *adj.* self-satisfied, complacent, holier-than-thou, °self-important, °overconfident, °conceited: *She is a bit too smug about the security of her job and might get a nasty shock some day soon.*

snack *n.* **1** °bite, nibble, °morsel, tidbit *or Brit also* titbit, refreshment(s), *Brit* elevenses, *Anglo-Indian* tiffin, *Colloq* nosh: *Have a little snack or you'll be hungry.* —*v.* **2** °bite, nibble, *Colloq* nosh: *He gained all that weight just from snacking between meals.*

snag *n.* **1** °hitch, °catch, °problem, °(stumbling) block, °stricture, bottleneck, °complication, °obstacle, °impediment, °obstruction, °hindrance, °difficulty, *US* hang-up: *A snag developed that prevented my keeping my promise to you.* —*v.* **2** °catch, °tear, °rip: *He snagged his new jacket on a nail.*

snake *n.* **1** reptile, serpent, ophidian, viper: *A specialist in snakes is called a herpetologist.* **2** snake in the grass, °traitor, °turncoat, Judas, quisling, betrayer, double-crosser, °informer, rat, *US* Benedict Arnold, *Slang chiefly US and Canadian* fink, rat fink: *Eventually they will catch the snake who gave the plans to the enemy.* —*v.* **3** °slither, °glide, °creep, °crawl, worm: *To avoid being seen, he snaked along on his stomach for a few yards.* **4** °twist, °wind, curve, °bend, °turn, zigzag, worm, °wander, °loop, crook, °meander: *The road snakes through the jungle for ten miles or more.*

snap *v.* **1** °break (off), °separate, °crack; °cleave, °split, °fracture, °give way, °part: *The wind snapped off tree branches like matchsticks. The back leg snapped in two when she sat down on the chair.* **2** click; °pop; °crack: *The door snapped shut.* **3** Often, ***snap at.*** **(a)** °bite (at), °nip, gnash *or* snatch at: *The postman ran down the street, the dog snapping at his heels.* **(b)** °attack, lunge at, °lash out (at), snarl at, growl (at), bark (at), be brusque *or* short *or* curt (with), *Colloq* jump down (someone's) throat, °fly off the handle (at): *Robin is in a bad mood today, snapping at everyone.* **4** Usually, ***snap up*** *or US also* ***off.*** °grab (up), °snatch (up), °seize, °pluck, pounce on *or* upon, °make off with, °take (away), °capture, °catch, °get, °secure: *The people who arrived early at the sale had snapped up all the bargains.* **5** °shoot, snapshoot, snapshot, °photograph, click, °catch: *He always carried his camera with him so that he could snap anything interesting.* **6** ***snap one's fingers at.*** disdain, °scorn, °flout, °dismiss, contemn, °disregard, °ignore, °defy, °mock, °deride, °thumb one's nose at, *Brit* cock a snook at: *He merely snapped his fingers at the dangers when I mentioned them.* **7** ***snap out of it.*** °recover, come round *or* around, °revive, °awaken, °wake up, °perk up, liven up, cheer up; get a grip *or* (a) hold on *or* of oneself, °pull oneself together, (re)gain control of oneself: *I was feeling very blue yesterday but managed to snap out of it.* —*n.* **8** °crack, °pop, click: *The lid shut with a snap.* **9** °spell, °period, °interval, °wave: *During that cold snap we thought we were in for a terrible winter.* **10** °catch, spring-catch, (snap) fastener, °fastening, °clasp: *Do you want snaps or buttons sewn on this shirt?* **11** °energy, °vigor, °animation, liveliness, °vitality, °bounce, alertness, sprightliness, élan, °dash, °sparkle, °verve, *Colloq* zip, zing, get-up-and-go, °pep, pizazz: *Granny certainly has a lot of snap left in her.* **12** easy job, *Slang* °picnic, *US and Canadian* °breeze: *There had been no need to worry, for the exam turned out to be a snap.* —*adj.* **13** °abrupt, °sudden, °precipitate, °hurried, °hasty, incautious, °rash, °unpremeditated, unplanned, not well-thought-out, °quick, instantaneous, °instant: *This is too important a matter for a snap decision.*

snappish *adj.* **1** °short-tempered, °testy, °petulant, °peevish, °irritable, °prickly, °touchy, irascible, quick to anger, °quick-tempered, hot-tempered, °waspish, *Brit* °snappy, *US* on a short string *or US* tether: *He fancies that being snappish goes with his "artistic" temperament.* **2** °curt, °short, °abrupt, °brusque, curmudgeonly, °cantankerous, °sharp, °cross, grouchy, °gruff, °cranky, crusty, crabby, crabbed, acid, °tart, acerbic, churlish, dyspeptic, choleric, splenetic, ill-humored, ill-tempered, °temperamental, °moody, *Brit* °snappy: *Grandad tended to be snappish when he became impatient with our lack of experience.*

snappy *adj.* **1** °quick, °sharp, °brisk, °smart, °crisp, °lively, °rapid, °speedy: *If you want to come with us, you'd better make it snappy.* **2** °fashionable, °chic, °sharp, °smart, °stylish, °dapper, modish, *Colloq* natty, *Brit* °trendy: *Your new friend is a very snappy dresser, isn't he?*

snare *n.* **1** °trap, °net, springe, noose, gin: *He had caught only a pigeon in the snare.* —*v.* **2** °trap, °catch, entrap, °seize, °capture, ensnare: *Using a different bait, William snared some partridges.*

snarl[1] *v.* **1** growl; °snap: *I admit I found three snarling Dobermans a bit off-putting. The clerk snarled at me when I asked for a form.* —*n.* **2** growl: *With a snarl the cat leapt at his throat.*

snarl[2] *v.* **1** Often, ***snarl up.*** °tangle, °entangle, °complicate, °confuse, °scramble, °muddle, °twist, mix *or* mess up, *Colloq* ball up, °screw up: *This situation is so snarled up that we'll never straighten it out.* **2** °tangle, °entangle, °knot, °twist, ravel, °jam, °kink: *The rope won't feed through the pulley because it's all snarled.* —*n.* **3** °tangle, entanglement, °complexity, °snag, °problem, °difficulty, °complication, °muddle, °mess, °predicament, °fix, °quandary, °dilemma, *Colloq* snarl-up, tight spot, pickle: *Will they ever get out of this snarl?* **4** jungle, °maze, labyrinth, °knot: *The drain was clogged by a snarl of hair.*

snatch *v.* **1** °grab, °grasp, °seize, °clasp, °clutch, °pluck, °take (hold of), °catch, °lay hold of, wrest, latch onto, °capture, °snap up, °win, °get, lay *or* get one's hands on: *The thief snatched her purse and ran away. One must snatch every available opportunity for happiness.* **2** *Chiefly US* °kidnap, °abduct: *They snatched the kid and are holding him for ransom.* **3** °save, °rescue, °deliver, °remove: *At the very last moment, we were snatched from the jaws of death.* —*n.* **4** °grab, °clutch, °grasp: *He made a snatch for her necklace, but she avoided him.* **5** °scrap, °bit, °fragment, snippet, °segment, °morsel, °specimen, °sample: *We were able to* *hear only brief snatches of their conversation through the closed door.*

sneak *v.* **1** °lurk, °slink, °steal, °creep, skulk, cower, °lurk, °pad, °prowl, °sidle, *Colloq* °pussyfoot: *We caught Francis sneaking about the house last night.* —*n.* **2** °informer, *Colloq* tattletale, *Brit* grass, *Slang* stool pigeon, snitch, *Brit and Australian* nark, *US* stoolie, *Chiefly US and Canadian* fink, rat fink, *US* shoofly: *Frank picked up a little money acting as a sneak for the police.*

sneaking *adj.* **1** °persistent, °lingering, lurking, °nagging, worrying, worrisome, °niggling, intuitive, deep-rooted, deep-seated, *Slang* °gut: *She had a sneaking suspicion that Stephen was a police informer.* **2** innate, intuitive, °inherent, °private, °secret, suppressed, °hidden, unexpressed, undeclared, unvoiced, unavowed, unconfessed, unrevealed, unadmitted, undivulged, undisclosed, covert: *I have a sneaking sympathy with the underdog.*

sneaky *adj.* underhand(ed), °devious, furtive, °sly, °slippery, °disingenuous, °deceitful, °dishonest, °unscrupulous, °shifty: *Going behind your back to tell your mother was a sneaky thing for Gaffer to do.*

sneer *v.* **1** °smirk, curl one's lip, °sniff: *Don't sneer—you might be the next to lose your job.* **2** °scorn, disdain, °despise, contemn, turn up one's nose (at), °sniff (at), °jeer (at), °laugh (at), °deride, °mock, °ridicule; underrate; *Colloq* °sneeze at, *Slang* °knock: *They sneered at her attempts to better the record for the hundred-meter backstroke.* —*n.* **3** °scorn, °jeer, disdain, °contempt, °derision, °mockery, °ridicule; sneering, jeering: *Blaydon endured his classmates' sneers.*

sneeze *v.* **1** sternutate: *About to sneeze, I grabbed a tissue.* **2** ***sneeze at.*** See **sneer, 2,** above. —*n.* **3** sternutation; sneezing: *The sneeze is an involuntary convulsive action.*

snicker *v.* **1** snigger, °chuckle, °giggle, °titter, laugh up one's sleeve, °mock, °scorn, °laugh (at), °jeer (at): *It was embarrassing when the audience began to snicker at the inept juggler.* —*n.* **2** snigger, °chuckle, °giggle, °titter: *The villain gave a little snicker when he thought he had her in his clutches.*

sniff *n.* **1** whiff, °breath, °odor, °scent: *I got a sniff of her perfume as she walked by.* **2** °hint, °spirit, °feeling, °suggestion: *There is the sniff of spring in the air.* —*v.* **3** °smell, snuffle, snuff: *Why is your dog always sniffing at my shoes?* **4** ***sniff (at).*** See **sneer, 2,** above.

snip *v.* **1** °nip, °clip, crop, °cut, °lop, °prune, °dock: *When the blossoms begin to fade, snip them.* —*n.* **2** °cut, °slit, °gash, °slash, °incision, °nick: *We could watch the audience through a small snip in the curtain.* **3** °bit, °scrap, °shred, snippet, °fragment, °cutting, clipping, °remnant, °morsel: *You are welcome to the leftover snips of fabric.* **4** ***snips.*** scissors, shears, tin snips: *This duct can be cut with the snips.*

snipe *v.* Usually, **snipe at.** shoot at, fire at; °attack, °criticize, °deride, find fault with, carp at, pick apart: *Newby has been sniping at the organization for years and we simply ignore him.*

snivel *v.* sniffle, snuffle, blubber, whimper, whine, mewl, pule; °cry, *Colloq Brit* whinge: *I'll buy some of your matches, little girl, but only if you stop sniveling.*

snobbery *n.* snobbism, snobbishness, pretentiousness, °pretension, hauteur, haughtiness, superciliousness, condescension, loftiness, contemptuousness, presumptuousness, lordliness, disdainfulness, disdain, pompousness, pomposity, °affectation, inflatedness, self-importance, °conceit, vainness, °vanity, narcissism, self-admiration, self-centeredness, egotism, smugness, *Colloq* uppishness, uppitiness, snootiness, snottiness: *Snobbery is more characteristic of the nouveau riche than of the "riche."*

snobbish *adj.* °condescending, °superior, patronizing, °arrogant, °haughty, lordly, °lofty, putting on airs, °disdainful, °supercilious, °contemptuous, °pretentious, °smug, °scornful, °self-important, °affected, °conceited, egotistic(al), °vain, self-satisfied, complacent, °pompous, *Colloq* snooty, snotty, highfalutin *or* hifalutin *or* highfaluting, on one's high horse, uppity, °hoity-toity, high and mighty, stuck-up, *Brit* °uppish, *Chiefly US* high-hat, *Slang Brit* toffee-nosed: *Since he got his knighthood, Cathcart has become so snobbish he'll have nothing to do with us.*

snoop *v.* **1** °pry, °spy, °interfere, °meddle, °intrude, butt in(to), *Colloq* stick *or* poke one's nose; be nosy, nose around *or* about: *She couldn't stop snooping into other people's affairs.* —*n.* **2** °busybody, meddler, °spy, °intruder, snooper, peeper; private detective *or* investigator, *Brit* Paul Pry; *Colloq* Nosy Parker, *US* buttinsky *or* buttinski; private eye, shamus: *There was a snoop here today, asking a lot of personal questions.*

snug *adj.* °cozy, °comfortable, °intimate, relaxing, °restful, °warm, sheltered, °friendly, °easy, °homely, °casual, *Colloq* comfy: *In his favorite armchair, Chris looks as snug as a bug in a rug.*

snuggle *v.* °cuddle, snug down, °nestle, nuzzle: *The puppies snuggled up close to their mother for warmth and protection.*

soak *v.* **1** °drench, °saturate, °wet, °immerse, souse, douse *or* dowse, bathe, °steep, inundate, ret: *You'd better let the tablecloth soak for a while to get out those wine stains.* **2** ***soak up.*** absorb, °take in, sponge up; assimilate, °learn: *The parched earth soaked up the water like a sponge. Let me just lie here and soak up some sunshine. She spent years soaking up the culture of the Cameroons.* —*n.* **3** °alcoholic, drunkard, °drunk, dipsomaniac, drinker, tippler, toper, sot, *Slang* sponge, souse, boozer, *US* °dip, °lush, juicer: *You'll never get a straight answer out of that old soak.*

soaking *n.* **1** drenching, wetting, dousing *or* dowsing, immersing, saturating: *We got a thorough soaking when we got caught in the storm.* —*adj.* **2** °wet, sopping, drenched, dripping, saturated, soaked, wringing wet, streaming, sodden, waterlogged: *Take off your soaking clothes and sit by the fire in this robe.*

soar *v.* **1** °rise, °fly, °hover, °float, °hang: *The glider soared over the hills, catching every updraft.* **2** °rise, °increase, escalate, °climb, spiral upward(s), shoot up *or* upward(s), °rocket, skyrocket: *To counteract inflation, interest rates began to soar.*

sob *v.* °cry, °weep, blubber, shed tears, °snivel, whimper, °sniff, snuffle, pule, wail, °moan, boohoo, mewl, pule, °bawl, °howl, yowl: *The poor, lonely child, suddenly an orphan, sobbed quietly.*

sober *adj.* **1** teetotal, °temperate, *US* °dry, *Colloq* on the (water) wagon: *Of course I am sober, officer—I don't drink.* **2** °serious, °solemn, °earnest, °dispassionate, unruffled, unflustered, unexcited, unperturbed, °steady, °sedate, °staid, composed, °dignified, °cool, °calm, °serene, °tranquil, °collected, coolheaded, °levelheaded, °sane, balanced, °practical, °realistic, °rational, clearheaded, *Slang* together: *We invited you because we respect your sober judgment.* **3** °sedate, °somber, °plain, °simple, °subdued, °quiet, repressed, °dreary, °dark, °drab, °colorless, °neutral: *At the funeral the widow wore a sober black costume.* —*v.* **4** ***sober up.*** detoxify, °recover, *Colloq* dry out: *When she finally sobered up she awoke to find herself in custody.*

sobriety *n.* **1** teetotalism, abstemiousness, abstention, abstinence, nonindulgence, °temperance: *Since* *taking the pledge, Oscar has been a model of sobriety.* **2** seriousness, soberness, °solemnity, staidness, °gravity, temperateness, sedateness, °formality, °dignity: *They made her promise to do nothing to upset the sobriety of the memorial service.*

so-called *adj.* **1** styled, °self-styled, designated, *soi-disant,* called, °professed: *I took lessons from Madam LaZonga, the so-called "Queen of the Dance."* **2** °alleged, °pretended, °supposed, ostensible; misnamed, misdesignated; °suspect: *She bought it through the so-called Honest Used Car Dealers Association.*

sociable *adj.* °friendly, affable, approachable, °social, gregarious, °outgoing, extrovert(ed) *or* extravert(ed), companionable, °accessible, °amiable, °amicable, °genial, congenial, convivial, °warm, °cordial, °neighborly, hail-fellow-well-met, *Colloq* °chummy, °cozy: *The people in this area are quite sociable, and we get together often.*

social *adj.* **1** communal, community, °common, collective, °group, °public, °popular, societal: *In many New England towns, the business of government is a social activity. Crime is a social, not an individual problem.* **2** °sexual, sexually transmitted, °venereal: *Sweet Eloise has a social disease.* **3** See **sociable,** above.

socialize *v.* °mix, °get together, °fraternize, keep company, °go out, get out; °associate: *Since contracting the disease, Peter hasn't cared much about socializing.*

society *n.* **1** °fellowship, °brotherhood, °association, °intercourse, °companionship, °company, camaraderie, °friendship: *Few people could be happy for long without the society of others.* **2** mankind, °people, the public: *I am not sure that Rob is yet ready to mingle with society. Do you believe that society demands too much?* **3** °culture, °civilization, community, way of life, °world; °organization, °system: *The hunting society was replaced largely by the agricultural some 10,000 years ago.* **4** high society, *haut monde, beau monde,* upper classes, polite society, elite *or* élite, °gentry, *Colloq* °upper crust: *The purpose of the debutante ball is to introduce young ladies of about eighteen to society.* **5** °organization, °club, °association, °circle, °league, °institute, academy, °alliance, guild, °group, °fraternity, sorority, °brotherhood, sisterhood, °fellowship, °union, consociation, sodality, *Verein,* bund *or* Bund: *The society, founded to advance and support linguistic scholarship, is now a hundred years old.*

soft *adj.* **1** °yielding, cushiony, plushy, spongy, squeezable, compressible, squashy, squashable, °flexible, °plastic, °pliable, pliant, °supple, flexile, flexuous, unstarched: *This mattress is much too soft for my back.* **2** °easy, °comfortable, undemanding, *Colloq* cushy: *His uncle got him a soft job at Cit Hall.* **3** °gentle, °mild, balmy, °pleasant, °moderate, °warm, halcyon, springlike, summery, °restful, °tranquil, relaxing, °lazy: *Alexandra was looking forward to a week in the soft Caribbean climate.* **4** °subdued, toned *or* turned down, muted, °low, °quiet, °melodious, mellifluous *or* mellifluent, °mellow, °gentle, °faint, softened, °soothing, °smooth: *Riley thought she might succumb to the influences of the wine and the soft music and lights.* **5** easygoing, °tolerant, °kind, compassionate, °gentle, °merciful, °lenient, °indulgent, °permissive, °liberal, °lax, °easy, docile, °tame, °submissive, deferential, °benign, tenderhearted, °sympathetic, kindhearted, °kind: *Some believe that the jailors were too soft with the criminals. There was a concern that he might have been a bit soft on communists.* **6** Usually, ***soft in the head.*** °foolish, °silly, °simple, *Colloq chiefly Brit* °daft, *US* °off: *Kevin must have gone a bit soft in the head if he thinks that Clara still loves him.* **7** depressed, declining, in decline, in recession, °slow, °unprofitable, borderline, °questionable, °weak: *Owing to the stormy, cold summer, the resort business has been soft this year.* **8** downy, °silky, silken, satiny, furry, °fluffy, feathery, fleecy, °fuzzy, velvety, °smooth (as a baby's bottom): *The wool from the lambs is much softer.* **9** pastel, °pale, °delicate, °fine, °subdued, °light, matte *or* matt, °quiet, diffuse(d), °soothing: *He wanted the bedroom painted in soft shades of greenish blue.* **10** °harmless, nonaddictive: *Some consider cannabis a soft drug.* **11** °fuzzy, °woolly, blurred, blurry, foggy, diffuse(d): *They use a soft focus lens when filming romantic scenes.* **12** °weak, °feeble, °frail, effete, °delicate, nonphysical, nonmuscular, °puny, °flabby, out of training *or* condition *or* shape, pampered; namby-pamby, °effeminate, unmanly, unmanful, *Colloq* sissified, °sissy: *He seemed a little too soft to qualify for the marines.* **13** °easy, °luxurious, pampered, °rich, °opulent, °plush, °posh, *Colloq* ritzy, swank(y): *Since selling his business, Aubrey has been living the soft life in Costa Smeralda.*

soften *v.* **1** Often, ***soften up.*** °melt, °affect, mollify, °mellow, palliate, soothe, °relax, appease: *Nothing would soften the heart of that cruel tyrant.* **2** °mitigate, assuage, °diminish, °moderate, °reduce, °cushion, lessen, °weaken, allay, °ease, °lighten, abate, °temper, °relieve: *Can't you think of some way to soften the blow of such bad news?* **3** °muffle, °deaden, °damp, soft-pedal, °lower, °still, °quiet, °tone down, lessen, °diminish, °lighten, °turn down, °quell, *Chiefly Brit* quieten: *Try to soften the high notes and bring up the bass.* **4** °give in, °succumb, °surrender, °yield, °agree, °consent, concur, assent, °give way, °relax, °ease (up), °let up: *The committee finally softened and allowed the park to be used for games.*

softhearted *adj.* tenderhearted, compassionate, °tender, warmhearted, °sentimental, °charitable, °generous, giving, °sympathetic, °indulgent, °kind, kindhearted, °responsive: *Lady Barking is quite softhearted when it comes to animal causes and contributed willingly.*

soil[1] *v.* **1** °dirty, °stain, begrime, °muddy, °smear, °spot: *This shirt was returned by the laundry still soiled.* **2** °pollute, °contaminate, °sully, defile, °foul, befoul, °tarnish, besmirch, °disgrace, °muddy, °smear, °blacken; °blot: *Dealing with such blackguards could soil your reputation, Simon.* —*n.* **3** °dirt, °filth, °muck, °mire, °mud, °sludge, °dregs, °refuse; excrement, °waste (matter): *The soil is carried away by these pipes.*

soil[2] *n.* °earth, loam, °dirt, °ground, °turf, humus; clay: *The men who dug the swimming pool took away the soil from the pit.*

sojourn *n.* **1** °stay, °stop, stopover, °visit, °rest, °holiday, vacation: *We had a very pleasant sojourn in Toronto on our way home.* —*v.* **2** °stay, °stop (over), °visit, °rest, °holiday, vacation, °tarry: *Next year we plan to sojourn in Bermuda.*

solace *n.* **1** °comfort, consolation, condolence, °relief, balm, °support, °help, succor; reassurance, °cheer: *Only the children could bring her solace in her bereavement.* —*v.* **2** °comfort, °console, condole, °support, °help, succor, soothe, allay, alleviate, ameliorate, °mitigate, assuage, °relieve; °cheer (up), °reassure, hearten: *There was little to solace James's misery.*

soldier *n.* **1** serviceman, servicewoman, °recruit, fighter, infantryman, foot soldier, trooper, warrior, military man, man-at-arms, *Brit* Tommy (Atkins), *US* enlisted man *or* woman, *Colloq Brit* squaddie, *US* GI *or* G.I. (Joe), *Old-fashioned* (*WWI*) doughboy: *Enemy soldiers were reconnoitering our position during the night.* **2** fighter, °stalwart, °supporter, °militant: *Clampett has been a soldier in the fight against poverty all his life.* —*v.* **3** °serve (in the army): *In his family the* *men have been soldiering for generations.* **4 *soldier on.*** °continue, °persist, °persevere, °endure, °drive, keep going, keep on *or* at, °grind, drudge: *Despite setbacks, they soldiered on to complete the project in time.*

sole *adj.* lone, °only, °singular, °unique, °solitary; °particular, °exclusive, °individual, °personal: *As Serena is the sole survivor, she has the sole right to sell the house if she wishes.*

solecism *n.* °error, °slip, °impropriety, °fault, °breach, °violation, °lapse, °mistake, misusage, incongruity, inconsistency, barbarism, °blunder, gaffe, °bungle, °fumble, gaucherie, faux pas, botch *or* botch-up, *Colloq* boo-boo, *US* flub, *Slang* boner, *Brit* boob, bloomer, *Chiefly US and Canadian* blooper: *Solecisms include certain instances of failure of subject and verb to agree in number.*

solemn *adj.* **1** °serious, °sober, °reserved, °grave, °earnest, °sedate, °staid, °taciturn; morose, °morbid, mirthless, unsmiling, °gloomy, °somber, °grim; °glum, long-faced, saturnine: *We observed a moment of solemn silence in memory of departed friends. From his solemn expression, we knew that something was wrong.* **2** °ceremonial, ritualistic, liturgical, °religious, ecclesiastical, °holy, °divine, °sacred, hallowed, sacramental, reverential, devotional: *We attended a solemn service to celebrate the return of the hostages.* **3** °ceremonious, °ritual, °formal, °dignified, °stately, °grand, august, °imposing, °impressive, awe-inspiring, °awesome, °important, °momentous: *The Trobriand islanders regarded the rites of passage as most solemn.*

solemnity *n.* solemnness, °gravity, seriousness, soberness, °reserve, sedateness, taciturnity, staidness, earnestness, impressiveness, °grandeur, °importance, momentousness, consequence: *All who attended the service were affected by the solemnity of the lighting of the eternal flame.*

solicit *v.* **1** entreat, °beseech, °ask (for), implore, °petition, importune, appeal for *or* to, call on *or* upon, °beg, supplicate, °pray, crave: *How many times have you been solicited for help? He solicited my help on more than one occasion.* **2** accost, °approach, °entice, °lure, °pander to, *Slang* °hustle: *You can't go out in that neighborhood without being solicited.*

solicitous *adj.* **1** °concerned, caring, °considerate, uneasy, troubled, °anxious, apprehensive, °worried: *She seemed genuinely solicitous over the state of my health.* **2** °eager, °earnest, zealous, °keen, °anxious, °desirous, °ardent, avid: *Sylvia seems sincerely solicitous to see that the customers are satisfied.*

solicitude *n.* °concern, °consideration, °regard, disquiet, disquietude, uneasiness, °anxiety, apprehension, °worry, nervousness, °fear, fearfulness, °alarm: *As she had not yet made him her sole heir, his solicitude over her health was understandable.*

solid *adj.* **1** three-dimensional, cubic: *The perspective drawing is a representation of a solid object.* **2** filled (in *or* up), °packed, jammed, crowded, teeming, °congested, crammed, swarming, compressed, concentrated, *Colloq* chockablock, jampacked, chock-full: *By the time I arrived, the gallery was solid with people.* **3** °compact, °firm, °hard, °stable; unshakable *or* unshakeable, unshaky, °substantial, °concrete, °sturdy, °sound, °stout, °strong: *The vessel floated through the air before coming to rest on solid ground. His theories rest on a solid foundation.* **4** °consistent, °homogeneous, °uniform, unalloyed, unmixed, °pure, °continuous, unbroken, °real, °authentic, °true, °genuine, 24-karat, unadulterated, *Slang* honest-to-God: *This mountain seems to be solid iron all the way through. John gave me a solid gold necklace for my birthday.* **5** law-abiding, upstanding, °upright, °decent, °stout, °substantial, °powerful, °reliable, °regular, °steady, °steadfast, °stalwart, °straight, °estimable, °sure, trusty, °trustworthy, trueblue, °worthy, dependable, °sober: *Mr. Hart is one of the solid citizens of this town and a solid defender of the underdog.* **6** °steady, °stable, °stalwart, dependable, °sturdy, °strong, °substantial, °sound, °firm, well-built, well-constructed, well-made, °tough, °durable, °rugged, °stout: *That chair is not solid enough for you to stand on.* **7** cogent, °sound, °concrete, °weighty, proved, provable, valid, °reasonable, °sensible, °rational, °sober, well-founded, °authoritative, °indisputable, °incontrovertible, irrefutable, incontestable, °good, °powerful, °potent, °forceful, convincing, °persuasive: *He presents a solid argument for a sales tax.* **8** °firm, °downright, °vigorous, °telling, °effective, °forceful, °potent, °powerful, °mighty, °dynamic, °thorough, through and through, °intensive: *Jan struck a solid blow against the supporters of abortion.* **9** See **solvent, 1,** below. **10** °entire, °complete, °whole, °continuous; uninterrupted, °undivided, unbroken, unrelieved, °blank, windowless: *I had to wait at the dentist's for a solid hour. Along the park drive they erected a solid wall.*

solidarity *n.* °unity, unanimity, unification, °accord, concord, concordance, °harmony, concurrence, like-mindedness, °agreement, mutuality, single-mindedness, singleness (of purpose), community of interest, esprit de corps, camaraderie, comradeship, sodality, *Solidarnos'c'*: *It was through solidarity, not the pursuit of individual interest, that freedom was achieved.*

solidify *v.* **1** °harden, °freeze, °set, °cake, °compact, compress, crystallize; jell *or* gel, clot, congeal, °coagulate, °thicken, *Technical* inspissate; °sublime: *When the lava solidified it molded the inundated figures, preserving them in rocky tombs.* **2** consolidate, °unite, °unify, pull *or* draw together: *Studying the speeches of politicians enables me to solidify my own views.*

solitary *adj.* **1** lone, °single, °sole, °individual; unattended, °solo, companionless, friendless, °lonesome, °lonely, °unsocial, cloistered, °secluded, °reclusive, °separate, eremitic(al), hermitic(al), °remote, °withdrawn, °distant, °out-of-the-way, unfrequented, °desolate: *There is not a solitary exception to the rule. Flora's solitary style of living allows her to concentrate on her writing. There is a solitary little inn where we go for weekends.* —*n.* **2** solitary confinement: *The prisoner has been in solitary for striking a guard.*

solitude *n.* **1** solitariness, aloneness, isolation, °seclusion, °privacy: *Peter enjoys the solitude of living on an island.* **2** loneliness, remoteness; °emptiness, wilderness: *How long could one person survive in the solitude of interstellar space?*

solo *adv.* **1** °alone, °unaccompanied, on one's own: *Jack flew solo for the first time yesterday.* —*adj.* **2** °individual, °unaccompanied, °solitary: *Let Eugene play the solo trombone part.*

solution *n.* **1** solving, working *or* figuring out, °discovery, finding out, unraveling, explication, deciphering, decipherment, elucidation, °revelation, clarification, °explanation; °answer, °key: *The solution of the problem should be left up to the experts. Several solutions have been found through computer analysis.* **2** °settlement, settling, °resolution, °result, denouement *or* dénouement, °outcome, conclusion: *It seems unlikely that we shall be able to bring this matter to a solution in a one-day meeting.* **3** °mixture, °blend, °compound, infusion; °liquid, °fluid; *Technical* emulsion, °suspension, colloid *or* colloidal solution *or* colloidal suspension: *After mixing the powder with water, use the solution to bathe the wound.* **4** dissolving, °dissolution, mixing, °mixture: *The solution of these ingredients in alcohol was first accomplished in ancient times.*

solve *v.* work *or* figure out, unravel, disentangle, untangle, °clarify, °clear up, make plain *or* clear, °interpret, explicate, °decipher, °crack, °explain, elucidate, °reveal, °answer, °resolve: *Oedipus' problems began in earnest when he solved the Riddle of the Sphinx.*

solvent *adj.* creditworthy, °(financially) sound, °solid, °reliable; debt-free; °profitable: *The auditors rate the business as solvent.*

somber *adj.* **1** °gloomy, morose, lugubrious, °funereal, °morbid, lowering *or* louring, °melancholy, °sad, °dismal, °unhappy, cheerless, °joyless, °serious, °sober, °doleful, dolorous, °mournful, depressed, depressing, °grave, °grim, grim-faced, grim-visaged, *Literary* melancholic, darksome: *Despite his somber appearance, Sid was one of the funniest men I've ever met.* **2** °dark, °gloomy, °foreboding, °bleak, depressing, °shadowy, °murky, °leaden, °gray, °black, °dismal, °dreary, °overcast, °dusky, °dim, °dingy, darkling, °dull, subfusc *or* subfuscous: *The somber sky set the mood for the proceedings that were to follow.* **3** °staid, °sedate, °sober, °solemn, °dark, °dull, subfusc: *A young person like you ought to dress in less somber colors.*

somebody *pron.* **1** °one, someone, some person: *There must be somebody who will take you to the dance.* —*n.* **2** °personage, °celebrity, °dignitary, VIP, luminary, °notable, °star, superstar, *Colloq* hotshot, °bigwig, big wheel, big gun, big noise, big White Chief, big Daddy, big Chief, big-timer; hot stuff; *Old-fashioned* big cheese; *US* Mr. Big: *He must be a somebody to merit such treatment.*

somehow *adv.* someway, in one way or another, in some way, somehow or other, °by hook or by crook, by fair means or foul, *Colloq* come hell or high water: *Somehow, we have to get into the vault for those papers.*

sometime *adj.* **1** °former, erstwhile, °past, °recent, one-time, quondam: *Crawford was a sometime student of the occult.* —*adv.* **2** at some time or other, someday, one day, any time, on a future occasion, when *or* if the opportunity arises, °soon, by and by, one of these days: *Come up and see me sometime.* **3** °sooner or later, in (due) time, in the fullness of time, °in the long run, one fine day, *Un bel di,* °eventually, when all is said and done, before long, before you know it: *You can be sure that the tax man will find out about it sometime.*

sometimes *adv.* °occasionally, °on occasion, °(every) now and then, °now and again, off and on, °at times, from time to time, every so often, °(every) once in a while: *Sometimes Ambrose likes to visit his grandchildren.*

somewhat *adv.* °rather, °quite, °relatively, more or less, °moderately, °pretty, °fairly, to some *or* a certain extent *or* degree *or* measure, °slightly, a bit, a little, *Colloq* °sort of, kind of: *Ingrid was somewhat put out by my refusal.*

song *n.* **1** °tune, °air, °melody, ditty, °number: *Together, they wrote some of the most popular songs of this century.* **2** ***for a song.*** cheaply, inexpensively, at a bargain price: *That old book you wanted went for a song at the auction.* **3** **(*old*) *song and dance.*** **(a)** °fuss, to-do, commotion, °bother, ado, *Colloq* °flap, °performance, *Brit* kerfuffle: *I cannot see why they made such a song and dance out of telling us they couldn't come.* **(b)** °evasion, °tale, prevarication, °(long) story, °(long) explanation: *To excuse his absence he gave me the old song and dance about his grandmother being ill.*

soon *adv.* **1** before long, °presently, ere long; in the near future, any minute (now), before you know it, in good time, in a little while, in a minute *or* a moment, momentarily, °shortly, anon, in a second, *Colloq* in a jiffy: *Soon another year will have passed. The doctor will be with you soon.* **2** °quickly, speedily, °at once, °promptly, °immediately, °directly, without delay, straightway, *Brit* °straightaway, *US* right away, forthwith, in short order, °on the double, in two shakes (of a lamb's tail), in a wink, *tout de suite,* without delay, *Colloq* pronto, *US and Canadian* lickety-split: *If he doesn't come soon, dinner will be spoilt.* **3** °quickly, speedily, °promptly, °swiftly: *I'll be there as soon as I can.* **4** °willingly, lief, °gladly, °happily, °readily: *I'd just as soon be left alone, if you don't mind.* **5** ***sooner or later.*** at some time or other, some time, one day, °in time, in due course, °eventually, °ultimately, in the end, when all is said and done, at the end of the day, in the last *or* final analysis, °at bottom: *Don't worry, she'll come home sooner or later. Sooner or later they will have to yield.*

soothing *adj.* **1** relaxing, °restful, °serene, °peaceful, pacifying, °calm, calming, °quiet, °soft, quieting: *I put on some soothing music and offered her a cocktail.* **2** mollifying, comforting, palliative, lenitive, demulcent, balsamic, emollient: *The doctor prescribed a soothing embrocation which helped immeasurably.*

sophistic *adj.* sophistical, °specious, fallacious, °deceptive, °hypocritical, °false, °unsound, baseless, °groundless, casuistic(al), jesuitic(al), captious, misleading, °bogus, °sham, °untenable: *The politicians advanced their usual sophistic arguments for continuing a war that no one wanted to support.*

sophisticated *adj.* **1** °cultivated, cultured, °refined, °experienced, °worldly, cosmopolitan, °polished, °elegant, urbane, worldly-wise, °knowledgeable, °knowing, °suave, soigné(e), °blasé, chi-chi, °slick, *Slang* hip *or* hep, °cool, with it: *The Gordons travel in a pretty sophisticated crowd.* **2** advanced, complex, °complicated, °intricate, °elaborate, °subtle, °refined, multifaceted: *The desktop computers of today are much more sophisticated than most older mainframe systems.*

sophistication *n.* **1** worldliness, urbanity, °culture, °refinement, °knowledge, knowledgeability, cosmopolitanism, °polish, °elegance, °poise, suavity *or suaveté,* °savoir-faire, °savoir-vivre, °finesse, °discrimination, discernment, awareness, °taste, tastefulness, °style: *They want him on the board of directors of the museum because he lends it sophistication.* **2** °complexity, intricacy, °subtlety, °refinement: *There is a staggering amount of sophistication built into modern integrated circuits.*

sorcerer *n.* sorceress, magus, necromancer, wizard, °witch, warlock, enchanter, enchantress, °magician, thaumaturgist, shaman, witch doctor, medicine man: *The law demanded death for the sorcerers and for those who dealt with them.*

sorcery *n.* witchcraft, °enchantment, sortilege, necromancy, wizardry, (black *or* white) magic, shamanism, black art, diabolism: *Once Vera works her sorcery on you, you obey her every whim.*

sordid *adj.* **1** °base, °vile, °corrupt, °low, ignoble, debased, degraded, abased, °mean, ignominious, °dishonorable, °despicable, °disreputable, °shabby, °shameful, °scurvy, °rotten, execrable: *He forced his family to do the most sordid things to raise money.* **2** °avaricious, °greedy, °grasping, °mercenary, piggish, °hoggish, °selfish, °rapacious, money-grubbing, stingy, parsimonious: *His most sordid plan was to pretend he was collecting money for the handicapped.* **3** °dirty, °foul, °filthy, squalid, unclean, °untidy, °mean, slummy, °seamy, °seedy, °wretched, unsanitary, insanitary, °offensive, defiled, polluted, fetid, feculent, mucky, maggoty, °putrid, flyblown, °slimy: *The conditions of some of the homes the social workers visit are indescribably sordid.* **4** °wretched, °miserable, °poor, poverty-stricken, °down-and-out, °impoverished, °ramshackle, hovel-like, tumbledown, °dingy, °seamy, °seedy, slummy, deteriorated, °sleazy, back-alley: *When he first came to town, he lived in a sordid shack lent him by a farmer.*

sore *adj.* **1** °painful, °sensitive, °tender, °raw, °angry, °burning, stinging, smarting, hurting; irritated, °inflamed, chafed: *My finger is still sore around the cut.* **2** °sensitive, °delicate, °tender, °embarrassing, °awkward, °ticklish, °touchy, °thorny, °prickly: *Failing to complete medical school is a sore point with her.* **3** dire, °serious, °acute, °extreme, °critical, °urgent, °pressing, °desperate: *Camberly is in sore need of psychiatric help, I fear.* **4** °angry, angered, annoyed, irritated, vexed, irked, °upset, *Colloq* peeved: *Don't get sore at me because you got a low mark on the exam.* **5** °painful, °troublesome, °grievous, aggrieved, distressing, distressful, °harrowing, °severe, °agonizing, °bitter, °fierce, °burdensome, onerous, °heavy, °oppressive: *His son's divorce proved a sore trial for Bodley.* **6** ***sore straits.*** °difficulty, °trouble, °distress, °danger, dangerous *or* precarious condition: *The business is in sore straits, I fear.* —*n.* **7** °injury, °damage, °swelling, rawness, infection, °inflammation, °bruise, abrasion, °cut, laceration, °scrape, °burn, canker, °ulcer: *If that sore on your arm doesn't heal, you'd better see a doctor.*

sorrow *n.* **1** °sadness, heartbreak, °grief, unhappiness, dolor, °misery, °woe, °anguish, °distress, °suffering, °torment, °agony, wretchedness, heartache, °desolation, desolateness: *How can one deal with the sorrow of the loss of a child?* **2** °affliction, °trouble, °trial, tribulation, °misfortune, °hardship, adversity, bad *or* hard luck, cares, °pressure, °strain, travail: *It is unbelievable that anyone could survive a life filled with such sorrow.* —*v.* **3** °grieve, °lament, °mourn, °regret, °keen, °bemoan, agonize, °moan, °bewail: *She is sorrowing over the loss of her nephew.*

sorrowful *adj.* **1** °sad, °unhappy, °regretful, °sorry, depressed, °dejected, crestfallen, chapfallen, °gloomy, downcast, °blue, dispirited, °melancholy, in the doldrums, °wretched, °woebegone, °miserable, heartsick, disheartened, °piteous, heavy-hearted, °brokenhearted, rueful, woeful, °tearful, disconsolate, °inconsolable, grief-stricken, *Colloq* °down in the mouth, down in the dumps: *The entire family was sorrowful to learn of Carriston's death.* **2** distressing, °lamentable, °doleful, °unfortunate, °bitter, distressful, troublous, °grievous, unlucky, hapless, afflictive: *What a sorrowful life some of those miners' families have led!*

sorry *adj.* **1** °regretful, °penitent, °remorseful, contrite, conscience-stricken, guilt-ridden, °repentant, °apologetic, penitential: *I am sorry if I offended you.* **2** abject, °miserable, depressing, °wretched, °pitiful, °pitiable, °pathetic, °deplorable, °stark, °grim, °sordid, °dismal, °base, star-crossed, ill-starred: *Crackham has led a rather sorry life, most of it as a vagrant. Agnes again made a sorry spectacle of herself at the Christmas party.* **3** See **sorrowful, 1,** above: *I was indeed sorry to learn that you are moving away, Peggy.*

sort *n.* **1** °kind, °variety, °type, °class, classification, °group, °category, °brand, °make, °mark, °stamp, °description, °mold, °stripe, ilk, feather, kidney, °character, °nature; °manner, species: *A Reo coupé might not be right sort of car for a family. Lewis is a person of a different sort. What sort of person is he?* **2** °kind, °type, °manner: *This sort of behavior must stop.* **3** species, genus, °family, phylum, subgenus, subspecies, °race, °breed, °strain, °stock, °kind, °variety, °type: *Aberdeen Angus is a better sort of beef for steaks.* **4** °person, °individual, °lot; °thing: *They are not really such a bad sort.* **5** ***of sorts.*** of a sort, of a mediocre *or* passable *or* admissible *or* not (too) bad *or* fair *or* sufficiently good *or* adequate *or* undistinguished *or* indifferent kind *or* quality *or* proficiency: *Yes, one might say that he is a violinist of sorts.* **6** ***out of sorts.*** not oneself, not up to snuff, unwell, ailing, °indisposed, °(slightly) ill, °low, *Colloq* off one's feed, °under the weather: *He is out of sorts after last night's revelries.* **7** ***sort of.*** See **somewhat,** above. —*v.* **8** assort, classify, °file, °order, °rank, °grade, °class, °categorize, °separate, °divide, °combine, °merge, °arrange, °organize, systemize, systematize, catalogue, °group, °sort out: *Sort these into alphabetical order.* **9** °describe, °characterize, °categorize, °cast, °throw, °combine, °mold, °type: *I did not enjoy being sorted together with thieves and murderers.* **10** ***sort out.*** **(a)** °organize, set *or* put straight, °straighten out, °resolve, °tidy (up), °clarify, °clear up, °solve; °decide: *Someone will have to sort out how we are to proceed.* **(b)** °choose, °select, °separate, °divide: *Sort out all the books that you want to keep from those to be donated to the library.*

so-so *adj.* °mediocre, all right, °average, °undistinguished, °passable, not (too) bad *or* good, °adequate, °fair (to middling), middling, °indifferent, °ordinary, °tolerable, *comme ci, comme ça,* °modest: *Sabrina, whose singing is just so-so, will never be a star.*

soul *n.* **1** (vital) spirit *or* force, being, (inner *or* true) self, °essence, °psyche, °heart, °mind, °intellect, °reason, anima: *He finally surrendered, body and soul.* **2** °person, °individual, °man, °woman, °mortal, (human) being: *I won't tell a soul about our conversation. What a sweet old soul she is!* **3** incarnation, °embodiment, personification, typification, °essence, °quintessence: *You can count on me as the soul of discretion.* **4** °emotion, °feeling, °sentiment, °sincerity, °fervor, °ardor, °warmth, °dynamism, vivacity, °energy, °spirit, °vitality, °force: *Alice puts a lot of soul into her singing, and that comes across to the audience.*

soulful *adj.* °sincere, °deep, °profound, °moving, °emotional, °warm, °ardent, °intense, °fervent, °expressive: *The lovers exchanged soulful looks.*

sound[1] *n.* **1** °tone; °noise; °din; cacophony; °report: *This piano has a tinny sound. We could hear the sound of laughter from below. I heard nothing over the sound of the crowd. Sounds of gunfire echoed in the valley.* **2** °ring, °tone, °impression, °characteristic, °quality, °effect, °aspect, °look: *I didn't like the sound of that remark.* **3** hearing, °range, earshot: *Anyone within the sound of his voice could tell he was angry.* —*v.* **4** °resound, reverberate, °echo, resonate: *The noises sound very close.* **5** °seem, °appear, °look; strike one, give one the impression *or* feeling *or* sense (that); °resemble, sound *or* seem *or* look like: *It sounds as if you don't want to come to my party. That sounds like Choctaw to me.* **6** °ring, (be) activate(d), (be) set *or* touch(ed) off, °signal: *Anyone walking into the room sounds the alarm. The alarm sounds when the beam is broken.* **7** Sometimes, ***sound out.*** articulate, °pronounce, °enunciate, utter; °voice, vocalize: *I was taught to speak clearly, to sound each consonant and vowel.* **8** Sometimes, ***sound out*** *or* ***off.*** °shout (out), cry out, °yell (out): *Sound out those numbers loud and clear!* **9** ***sound off.*** °vituperate, °complain, °bluster, grumble, *Slang* °bitch: *Today Morrison was again sounding off about the service in the cafe.*

sound[2] *adj.* **1** undamaged, uninjured, °whole, unmarred, in good condition *or* shape, °intact, unimpaired, °unscathed: *When the vase was sold to you it was sound.* **2** °healthy, °hale (and hearty), °fit (as a fiddle), °robust, °vigorous, blooming, °rosy, ruddy: *Samantha is as sound as ever.* **3** °firm, °solid, °substantial, °strong, °sturdy, °tough, °rugged, °durable, well-built, well-constructed, dependable: *A business, like a building, must be built on sound foundations if it is to last.* **4** °sane, balanced, °normal, °rational, °wholesome, °reasoning,

°reasonable, clearheaded, lucid, right-minded, °responsible, °practical, °prudent, °politic, °wise, °sensible, °logical, common-sense, common-sensical, °astute, °farsighted, °perceptive, perspicacious, percipient; valid, °good, °judicious, °reliable, °useful: *As Juvenal said, "Mens sana in corpore sano," or "A sound mind in a sound body." He provided a lot of other sound advice, too.* **5** °safe, °secure, °good, °conservative, nonspeculative, °solid, riskless; °profitable: *Are you sure that utilities are a sound investment now?* **6** unbroken, uninterrupted, undisturbed, untroubled, °peaceful, °deep: *I drank the potion and at once fell into a sound sleep.*

sound[3] *v.* **1** Often, ***sound out.*** °plumb, °probe, °test, check (out *or* into), °fathom, inquire *or* enquire of, °question, °poll, °canvass, °investigate, °examine, °survey: *Sound out the membership on their feelings about increasing the dues.* **2** °dive, °plunge, °submerge: *The great whale sounded, almost taking the tiny boat down with it.*

sound[4] *n.* inlet, strait(s), fjord, bight, (sea) loch, bay, arm of the sea, cove, *Scots* firth: *The breeze freshened as we sailed across the Sound of Mull.*

sour *adj.* **1** acid, acidic, °tart, vinegary, lemony, acidulous *or* acidulent, acidulated, acescent, acerbic: *This apple has a pleasant, slightly sour taste. Do you like sour dill pickles?* **2** turned, °bad, °(gone) off, fermented, curdled, °rancid, spoilt *or* spoiled: *If the food is sour, throw it away.* **3** °disagreeable, unpleasant, °distasteful, °bad, °nasty, °bitter, °terrible: *The poor service and facilities made the entire vacation a very sour experience.* **4** acrimonious, °bitter, °embittered, unpleasant, churlish, ill-natured, ill-tempered, bad-tempered, crusty, curmudgeonly, crabbed, crabby, grouchy, °cross, °cranky, °testy, °petulant, °impatient, °abrupt, °nasty, °curt, °caustic, °brusque, °peevish, °snappish, edgy, °sullen, morose, °gloomy, °discontented: *The waitress's sour expression matched her sour disposition.* —*v.* **5** °turn, °spoil, curdle, go bad *or* off, °ferment: *Milk sours if you don't refrigerate it.* **6** embitter, acerbate, disenchant, °exasperate, vex, *Colloq* peeve: *Being forced to practice for hours when a child was what soured me on playing the piano.*

source *n.* **1** fountainhead, wellspring, °origin, provenance, provenience, inception, °start, °outset, °beginning, root(s), commencement, °rise: *We set out to find the source of the Blue Nile. The mineral bauxite is the source of aluminum.* **2** originator, °author, °creator, begetter: *Who is credited as the source of this quotation?* **3** °authority, documentation; informant, *Colloq* horse's mouth: *To verify the information, go to the source.*

sovereign *n.* **1** °monarch, ruler, emperor, empress, °king, °queen, prince, princess, potentate, °chief, °master, °mistress, shah, sultan; Akund (of Swat), Gaekwar (of Baroda), Nizam (of Hyderabad), Mehtar (of Chitral), Nucifrage of Nuremberg, Sheik of Araby, *Colloq* supremo: *She is the sovereign and must be obeyed.* —*adj.* **2** supreme, paramount, highest, °principal, °foremost, greatest, °predominant, °dominant, ranking, °leading, °chief, °superior, °preeminent, ruling, regnant, reigning, governing, all-powerful, °absolute, °unlimited: *The firstborn of the monarch is vested with sovereign authority upon the monarch's death.* **3** °royal, °regal, °majestic, °noble, lordly, aristocratic, kingly, queenly: *He was hailed as the sovereign ruler of all the islands.*

sovereignty *n.* suzerainty, hegemony, °dominion, °rule, °preeminence, °power, °jurisdiction, °authority, °leadership, °command, °sway, °supremacy, ascendancy, primacy: *Anyone who doubted his sovereignty was seized and thrown into a dungeon.*

sow *v.* °seed, disseminate, °broadcast, °plant: *The grand vizier sowed the seeds of dissension amongst the subjects of the caliph.*

space *n.* **1** spaciousness, °room, °place, °expanse, elbowroom, °leeway, °margin, latitude, °play: *There was no space for another desk.* **2** °blank: *When the spaces are filled in, the puzzle is done.* **3** °interval, °lapse, °period, °time, hiatus, lacuna, °span, while, duration, °extent, °spell, °stretch, °pause, °wait, °intermission, °gap, °break, °interruption: *After a short space we were home again.* **4** °accommodation, °seat, berth, °room, °place: *Luckily, I got the last available space on the plane.* —*v.* **5** °arrange, °organize, array, °set out, align, °range, °order, °rank, °lay out, °measure (out): *Space the trees about eight feet apart.*

spacious *adj.* °vast, °large, °extensive, °enormous, °wide, °broad, commodious, °ample, °expansive, °roomy, °huge, sizable, °large, capacious, °great, °immense, outsize(d), °voluminous, oversize(d): *Asia offers spacious territory, ripe for habitation over the next century. The inside of the house was much more spacious than it seemed from outside.*

span *n.* **1** °bridge, °link, °stretch, overpass, *Chiefly Brit* flyover: *We drove across the entire span in less than five minutes.* **2** °course, °extent, °interval, °stretch, °period, °time, °term, °spell: *In the short span of six hours we had crossed the Atlantic.* —*v.* **3** °cross, stretch over, reach over, extend over, °go over, °bridge: *A rude bridge spans the river at Lexington. Their lives spanned more than a century.*

spank *v.* °slap, smack, put *or* take over one's knee, thrash, °paddle; °chastise, °punish, °castigate; *Colloq* wallop, tan (someone's) hide, paddywhack *or* paddywack, whack, give (someone) a (good) licking *or* hiding: *In those days when you misbehaved you were spanked.*

spanking *adj.* **1** spick and span *or* spic and span, °smart, °bright, °snappy, gleaming, °brand-new, °fine, °remarkable, °outstanding, °big, °large, °great: *You ought to see Trowbridge's spanking new yacht.* **2** °brisk, °lively, °crisp, °bracing, °fresh, freshening, rattling, °strong, °invigorating, blustery: *A spanking breeze blew up from the west.* **3** °quick, °rapid, °swift, °lively, °snappy, °fast, °smart, °energetic, °vigorous, °brisk: *The pony drew the dogcart at a spanking pace.*

spar[1] *n.* Nautical mast, yard, yardarm, °boom, boomkin, gaff, jigger, mizzen, *Colloq* °stick, °pole: *The first gusts of the storm tore the remaining canvas from the spars.*

spar[2] *v.* **1** °fight, °box, exchange blows; shadowbox: *He was once the champion's sparring partner.* **2** °dispute, °argue, °bicker, squabble, wrangle, bandy words, have words; °fight, *Colloq* °scrap: *I have better things to do than spar with you over breakfast every day.*

spare *adj.* **1** °extra, °surplus, supernumerary, °auxiliary, °supplementary, additional; °odd, °leftover; °in reserve, °in addition: *You'll have to put on the spare tire. Give them the spare room. Have you a spare cigarette?* **2** unoccupied, °leftover, °leisure, °free, °surplus, °extra; not spoken for: *If you have some spare time, I could use some help moving my piano.* **3** °thin, °skinny, °scrawny, cadaverous, °gaunt, °rawboned, °meager, gangling, lank(y), °wiry, °slim, °slender; all skin and bones: *The spare, ragged figure crouched in the shadows, his hand outstretched in supplication.* **4** See **sparing,** below. **5** °meager, °frugal, °small, skimpy, °modest, °scanty: *Even Sunday dinner was for us a very spare meal, usually consisting of a boiled turnip or parsnip.* —*v.* **6** °save, °rescue, °deliver, °redeem: *Bill spared me from a fate worse than death. Spare me the embarrassment of having to ask for my money back.* **7** °pardon, let go, °release, have mercy on, °let off, °free, °liberate: *At the last moment, the judge spared him.* **8** °allow, °relinquish, let go (of), °give, °award, °bestow, let have, °donate, °part with, °give, °yield: *Brother, can you spare a dime?* **9** °avoid, °dispense with, manage *or* do without, °give up, °forgo, °forsake, °surrender, °sacrifice: *We spared a great deal of trouble by settling the dispute. Spare the rod and spoil the child.*

sparing *adj.* **1** °thrifty, °saving, °frugal, °spare, °careful, °prudent, parsimonious, °economical, °penurious, °mean; penny-pinching, stingy, niggardly, °miserly, °close, closefisted, tightfisted, °cheap, *Colloq* °tight, *Brit* mingy: *As sparing in his praise as he was in his purse, General Waller rarely complimented his troops.* **2** See **sparse, 2** below.

spark *n.* **1** scintilla, °flicker, glimmer, glint, °sparkle, °speck, °hint, °suggestion, °vestige, atom, whit, °jot (or tittle), iota: *As long as a spark of life remained in his body he fought for freedom and justice.* —*v.* **2** Often, ***spark off.*** set *or* touch off, ignite, °kindle, enkindle, °electrify, °animate, trigger, °energize, galvanize, °activate, °excite, °stimulate, set in motion, °bring about, °start (up), °begin, °initiate, °inspire, inspirit, °provoke, °precipitate: *Was there any single factor that sparked the Renaissance?*

sparkle *v.* **1** °glitter, scintillate, glint, °flicker, °shine, °twinkle, wink, °blink, glimmer, °flash, coruscate, °blaze, °burn, °flame: *The setting sun sparkled on the sea. Brian's eyes sparkled with an inner fire.* **2** effervesce, °fizz, °bubble: *We ordered a jeroboam of sparkling Burgundy.* —*n.* **3** °glitter, scintillation, °twinkle, coruscation, °dazzle, °spark, °gleam, brightness, °brilliance, °radiance: *I fear that the sparkle has gone out of her smile.* **4** vivacity, liveliness, °fire, brightness, wittiness, effervescence, ebullience, °excitement, °animation, °vigor, °energy, °spirit, °cheer, °joy, lightheartedness, élan, zeal, °gusto, °dash, °life, °gaiety, °cheer, cheerfulness; certain something; *Colloq* vim, zip, zing, pizazz, oomph: *It is easy to see that she gets her sparkle from her mother.*

sparse *adj.* **1** °thin (on the ground), °few (and far between), °meager, °scanty, (widely) dispersed *or* scattered, °spread out, °spotty, °in short supply, °scarce: *Tourists are sparse in Eastport, Maine, during January. Sparse blooms appeared on the rosebushes this year.* **2** °little, °limited, °meager, scant, °sparing, inappreciable, not much, °insignificant: *Sparse use was made of the new tennis court because of the rain.*

Spartan *adj.* austere, °strict, °severe, °harsh, °hard, °stern, rigorous, °rigid, ascetic, stringent, controlled, disciplined, self-denying, abstinent, abstemious: *He gave up all luxuries and today lives a Spartan life in the forest.*

spasm *n.* **1** convulsion, °throe, °fit, twitch, °paroxysm: *Deprived of his medication, Breen might go into spasms.* **2** °fit, °seizure, convulsion, °paroxysm, °spell, °outburst, °burst, °eruption: *After that, every time he said "furniture," the crowd went into spasms of laughter.*

spasmodic *adj.* **1** spasmodical, paroxysmal, convulsive, jerky, jerking, °sudden, *Technical* spastic: *John was quite uneasy, unaware that Cynthia's spasmodic winking was due to a tic.* **2** °fitful, °irregular, °intermittent, arrhythmic, °random, interrupted, °sporadic, °erratic, °occasional, °periodic, unsustained, discontinuous, pulsating, cyclic(al), °broken: *Spasmodic peals of thunder shook the house.*

spate *n.* °flood, inundation, onrush, °onset, °rush, deluge, °outpouring, outflow, outflowing, °flow: *What did I say to provoke that spate of foul language?*

spatter *v.* °splash, splatter, speckle, bespatter, °spray, °dabble, daub, bedaub, sprinkle, besprinkle, *Brit* splodge, *US* splotch: *When she dropped the pan the grease spattered all over my new white flannels.*

spawn *v.* give birth to, °yield, °bear, °bring forth, °breed, beget, °create, °father, sire, °produce, °generate, engender, °give rise to, °bring about, °cause: *Donald's arrogance over his wealth spawns a great deal of resentment.*

speak *v.* **1** °talk, °converse, discourse: *We were not allowed to speak during class.* **2** °talk to, converse *or* discourse with, °address, say (something *or*) anything to: *That is no way to speak to your elders. Never speak to me again, you oaf!* **3** °talk, communicate in, discourse *or* converse in, communicate in, utter in, articulate in, °use: *Can you really speak Chinese?* **4** °express, utter, °say, °state, °tell, °pronounce, °enunciate, °voice; articulate, °make known, °communicate, °reveal, °indicate: *He is speaking the truth.* **5** °symbolize, betoken, °signify, °communicate, convey, °indicate: *Her refusal to return your love letters speaks volumes.* **6** ***so to speak.*** as it were, in a manner of speaking, figuratively *or* metaphorically (speaking): *When it comes to employee benefits, the boss is Scrooge himself, so to speak.* **7** ***speak for.*** **(a)** °support, °uphold, °defend, °stand up for, °plead for, make a plea for, °recommend, *Colloq* °stick up for: *Be quiet and let Simon speak for himself.* **(b)** act on *or* in behalf of, act for, °represent, act as agent for: *The Senator rose and announced, "I am speaking for Buncombe!"* **(c)** °demand, °require, °beg, °request, °ask for: *The situation speaks urgently for an effective drug rehabilitation program.* **8** ***speak for itself.*** be self-evident, be obvious, be significant: *The fact that both your son and the ring are missing speaks for itself.* **9** ***speak of.*** °mention, advert to, allude to, °refer to, make reference to, °comment on, speak *or* talk about: *I begged you not to speak of her again in my presence.* **10** ***speak on.*** °discuss, °address, discourse upon *or* on, °speak to, °treat (of), °deal with, °examine, touch upon *or* on: *In his presentation Dr. Abernagle spoke on the uses of disease in population control.* **11** ***speak out*** *or* ***up.*** **(a)** talk (more) loudly *or* clearly, make oneself heard: *I cannot hear you unless you speak up a bit.* **(b)** talk freely *or* unreservedly, express one's opinion, speak one's mind, °declare, °come out, state one's position, take a stand: *I heard a mother of ten speak out against birth control.* **12** ***speak to.*** **(a)** °reprove, °scold, °reprimand, °rebuke, admonish, °warn, °lecture: *After his father spoke to him, Nicholas stopped staying out so late.* **(b)** be meaningful to, appeal to, °influence, °affect, °touch: *Constance's poetry effectively speaks to the heart.* **(c)** accost, °address, °talk to, *Formal* apostrophize: *I have to speak to the teacher after class.* **(d)** See **10,** above. **13** ***spoken for.*** °reserved, °engaged, bespoke, °set aside, accounted for, chosen, selected: *I'm afraid that the purple dress has already been spoken for, madam.*

speaker *n.* orator, lecturer; keynoter; °rabble-rouser, demagogue, *Colloq* tub-thumper, spieler: *It was my function to introduce the speaker.*

spearhead *v.* **1** °launch, °initiate, °lead (the way), take the initiative, °pioneer, blaze the trail, °break the ice, take the lead, be in the van *or* vanguard: *Who was it that spearheaded medical research into arthritis?* —*n.* **2** vanguard, advance guard, van, forefront, cutting edge: *They claim to be in the spearhead of space technology.*

special *adj.* **1** °particular, °specific, °exceptional, uncommon, especial, °rare, °unusual, out-of-the-ordinary, °extraordinary, °different, °unorthodox, unconventional, °unique, °precise, °individual, °singular, °distinctive, specialized, °certain, °remarkable, inimitable, idiosyncratic, °curious, °peculiar, °odd, °strange, °bizarre, °weird, one-of-a-kind; °distinguished, °notable, °noteworthy: *They have a special way of communicating with* *each other. I bought something very special for your birthday. A special relationship sprang up between us.* **2** °significant, °important, °momentous, earthshaking, °memorable, red-letter; °gala, festive, celebratory: *The fifth of September is a special day in our family.* **3** °pointed, concerted, °deliberate, °particular, °extra, °determined: *After I had made a special effort to get there, nobody was home.* **4** °exclusive, °express, °individual, °extra: *Send the packet by special messenger.* **5** °dear, °intimate, °particular, °good, °close, °bosom, °staunch, °loyal, °faithful, °devoted, °steadfast; dearest, °best, closest; esteemed, valued: *I wasn't aware that Spencer was a special friend of yours.* **6** °prime, °primary, °major, °prominent, °paramount: *I shouldn't attach special importance to a love letter addressed to "Occupant."* —*n.* **7** See **specialty, 2** below.

specialist *n.* °expert, °authority, °professional, °master, connoisseur, maestro, artist, °adept, *Chiefly Brit* artiste: *Legrand is a specialist in matching antique furniture finishes. Such work requires specialist knowledge.*

specially *adv.* °especially, °particularly, °custom, °expressly, exclusively: *I had this suit specially made for me.*

specialty *n.* **1** °expertise, °talent, °genius, °gift, °skill, °aptitude, °trade, °craft, °accomplishment, °ability, °strength, °forte, strong point, °capability, adeptness, °art, °sphere, °field, °area, °subject, concentration, specialization, métier, *Chiefly Brit* speciality, *Colloq* °bag, °thing, cup of tea, °baby, claim to fame: *Mona's specialty is tropical diseases.* **2** *pièce de résistance, spécialité de la maison,* °special, *Chiefly Brit* speciality, *US* blueplate special: *Today's specialty is calves' brains au beurre noir.*

specific *adj.* **1** °definite, °precise, °exact, °particular, °explicit, °express, unambiguous, °definitive, clear-cut, unequivocal, (well-)defined, °determined, specified, °individual, °peculiar, °certain, °limited, indicated, °predetermined, established, spelled out, delineated, °set, °distinct, °fixed, circumscribed, restricted: *Each part has its specific function.* **2** Often, ***specific to.*** °characteristic (of), unique to, °individual (to), sui generis, °proper (to), °typical (of), °peculiar to, identified with, °personal (to), °discrete (to), °special (to), associated with: *The disease is specific to one area only. Are these problems group-specific?*

specification *n.* **1** °identification, identifying, °description, describing, particularization, particularizing, specifying, naming: *The police are relying on the accomplice's specification of those involved in the crime.* **2** itemization, itemizing, °list, listing, checklist, inventory, list of particulars, °detail, enumeration: *When can you let me have the specifications for the new building?* **3** °requirement, qualification, °condition, °restriction, °stipulation, °consideration: *The specification called for verde antique marble and no substitute.*

specify *v.* particularize, °enumerate, °itemize, °name, denominate, °list, °indicate, °mention, °identify, cite, °define, °detail, °stipulate, °spell out, set out *or* forth, individualize, be specific about, delineate, °determine, disambiguate, °establish: *You must specify the source of each quotation.*

specimen *n.* °sample, °example, °instance, exemplar, °representative, °representation; °illustration, °case (in point), °type, °model, °pattern: *This is a specimen taken from the Precambrian strata. Using this as a specimen, can you reproduce the entire fabric?*

specious *adj.* °deceptive, °superficial, casuistic, ostensible, misleading, °apparent, °seeming, fallacious, sophistic(al), °plausible, °likely, conceivable, °possible, °supposed, purported, presumed, presumable, °alleged, °so-called: *The candidate affected a specious show of liberality merely to attract votes. Do not be misled by specious reasoning.*

speck *n.* °spot, °dot, fleck, mote, speckle, °mark, °bit, °particle; °crumb, iota, °jot (or tittle), whit, atom, molecule, °touch, °hint, °suggestion, °suspicion, tinge, °modicum, °amount, °grain, smidgen: *There's a speck of soot on your collar. Is there a speck of truth in what she says about you?*

speckled *adj.* spotted, °mottled, dotted, sprinkled, °flecked, stippled, dapple(d), freckled, brindle(d); discolored, spattered, bespattered: *I can see by looking at it that it's a speckled trout.*

spectacle *n.* **1** °show, °display, °sight, °performance, °event, °presentation, °exhibition, °exhibit, °exposition, °demonstration, °extravaganza, °marvel, °wonder, °sensation: *Instead of criticizing everything in life, why not just sit back and enjoy the spectacle?* **2** °fool, laughingstock, °curiosity: *Mona made a spectacle of herself by her solo dancing when she got a little tipsy at the wedding.* **3** ***spectacles.*** eyeglasses, °glasses, *Colloq* specs: *He wears spectacles only for reading.*

spectator *n.* °witness, °eyewitness, °observer, viewer, °onlooker, looker-on, watcher, beholder: *There were more spectators than participants at the spring dance.*

specter *n.* **1** °ghost, °phantom, wraith, apparition, °vision, °spirit, °shade, revenant, doppelgänger, chimera, *Colloq* spook, bogeyman *or* bogyman: *Each year the specter appeared to her on Christmas Eve.* **2** °image, °vision, °(mental) picture: *Again the horrible specter of war and pestilence arose.*

spectral *adj.* °ghostly, ghostlike, °phantom, °eerie, wraithlike, incorporeal, °unearthly, °supernatural, °weird, *Colloq* spooky: *I could have sworn there was a spectral figure, dressed in white, standing near the window.*

speculate *v.* **1** Often, ***speculate on*** *or* ***upon*** *or* ***about*** *or* ***over.*** reflect (on *or* about *or* over), °consider, muse (on *or* about *or* over), meditate (on *or* over *or* about), contemplate (on *or* about), cogitate (on *or* about), think (about *or* over *or* on), ponder (over *or* about), mull over, chew on *or* over, ruminate (on *or* over *or* about), °wonder (about), deliberate (over *or* on *or* about), °weigh, °judge, °evaluate, theorize (on *or* about), conjecture (on *or* about), postulate, hypothesize: *I have been speculating about whether I should take a vacation in the Canadian Rockies. Why not go and enjoy yourself instead of speculating on the world's injustices?* **2** °gamble, wager, take a chance, *Colloq* have a flutter, play the market, take a plunge: *She is too conservative to speculate in the stock market.*

speculation *n.* **1** conjecture, °guess, °hypothesis, theory, guesswork, postulation, °surmise, °supposition, °opinion: *My speculations have often proved true. Both the steady-state and the big-bang theories are pure speculation.* **2** °thinking, rumination, cogitation, °reflection, meditation, contemplation, °consideration, cerebration, pondering, wondering, deliberation, °evaluation: *All serious philosophers engage in idle speculation.* **3** gambling, °gamble, wagering, wager, taking (a) chance(s) *or* risk(s), chance-taking: *Her speculation on the market paid off handsomely.*

speculative *adj.* **1** °intellectual, ideational, °abstract, cogitative, notional, °theoretical, °hypothetical, conjectural, suppositional, supposititious *or* suppositious, suppositive, °rational, ratiocinative, °ideal, idealized, °idealistic, °unrealistic, unpractical, °impractical, analytical: *Frank's theories of sailing was entirely speculative and substantially wrong.* **2** °risky, °hazardous, °uncertain, °unreliable, untrustworthy, °doubtful, dubious, untested, unproven, unproved, *Colloq* iffy, chancy,

Slang °dicey: *You ought to invest your money in less speculative enterprises.*

speech *n.* **1** speaking, talking, articulation, °diction, °language, °expression, enunciation, elocution, speech pattern; communication: *His speech is marred by a lisp. Martha studied speech at a well-known drama school.* **2** °oration, °address, °lecture, °talk, °lecture, discourse, disquisition, °sermon, homily; °tirade, °harangue, philippic; °(sales) pitch, °line, song and dance; *Colloq* spiel, °blast: *Her speech left her audience awed by her knowledge of the subject. When the householder opens the door, you launch into your speech about the encyclopedia.* **3** °dialect, idiolect, °jargon, °parlance, °idiom, *façon de parler,* °language, °tongue, *Colloq* °lingo: *English speech varieties range from Scots to Cockney to Texan to Jamaican, etc.*

speechless *adj.* **1** °mute, °dumb, voiceless: *He was speechless and in shock for two days following the accident.* **2** dumbfounded *or* dumfounded, dumbstruck *or* dumbstricken, wordless, struck dumb, °tongue-tied, °thunderstruck, shocked, dazed, °inarticulate, paralyzed, nonplussed: *He was speechless with fury to find his car missing.*

speed *n.* **1** °rapidity, fleetness, quickness, speediness, swiftness, °velocity, dispatch *or* despatch, °hurry, hurriedness, °haste, hastiness, celerity, alacrity, expeditiousness, °expedition, briskness, promptness, timeliness; suddenness, precipitateness, precipitousness, abruptness: *We reached a speed of Mach 2. What is the speed of your new computer? The speed of her reactions will be recorded during the experiment.* —*v.* **2** Often, ***speed up.*** accelerate, °move, °expedite, °forward, °advance, °facilitate, °boost, °further, °promote, °help, °assist, °aid, *Colloq* give a leg up: *He vowed to do anything he could to speed the troops' return from the battlefield.* **3** °hasten, make haste, °hurry, °rush, °charge, dart, °bolt, °shoot, °run, °race, sprint, °fly, °streak, °scurry, °tear, °hustle, °scramble, scamper, °career, highball, bowl along, go *or* fly like the wind, *Colloq* go hell for leather, go like a bat out of hell, belt along, °step on it, put one's foot down, zip, zoom, skedaddle, go like a shot, *US* hightail it, step on the gas, go like greased lightning, make tracks, *Slang US* burn rubber: *Where the road is straight, cars speed along at 120 miles per hour.*

speedy *adj.* **1** °quick, °rapid, °swift, °brisk, °expeditious, °fast, °immediate, °prompt: *What happened to the form of justice under which the accused were guaranteed a speedy trial?* **2** °hasty, °precipitate, °precipitous, °hurried, °summary: *Some of the townspeople favored the wrong kind of speedy justice—by lynching.* **3** °fleet, °nimble, wing-footed, winged, °fast, °quick, °rapid, °swift: *Carol is one of the speediest sprinters on our team.*

spell[1] *n.* **1** °period, °interval, °time, °term, °season; °stint, °turn, °run, °course, °shift, °tour (of duty), °watch, °round: *We had a spell of bad weather. Each of us had a spell at the tiller during the crossing.* —*v.* **2** °relieve, °replace, °substitute for, take over for: *After thirty-six hours of uninterrupted duty, I needed someone to spell me for a while.*

spell[2] *n.* **1** °enchantment, allure, °charm, °magic, witchcraft, witchery, °fascination, captivation, enthrallment: *She had us all under her spell.* **2** incantation, °formula, °charm: *The priest uttered some spell and I found myself unable to move.* **3** °attraction, °lure, allure, °appeal, °draw, °pull, °magnetism, °fascination, °influence, mesmerism, hypnotic effect: *We were drawn ever onward by the spell of the jungle.*

spell[3] *v.* **1** augur, portend, presage, °promise, hold promise of, °signify, °point to, °indicate, °omen, °bode, look like, °amount to, °mean: *Harry's conviction will spell the end of his career as a bank robber.* **2** ***spell out.*** °specify, delineate, make clear *or* plain *or* explicit, °clarify, elucidate: *She despises you—I can't spell it out for you more plainly than that.*

spellbinding *adj.* fascinating, °enchanting, °enthralling, captivating, enrapturing, bewitching, mesmerizing, charming, °overpowering: *The preacher put on one of his most spellbinding performances.*

spend *v.* **1** °pay out, disburse, °expend, °lay out, *Colloq* fork out, dish out, °shell out, *Brit* splash out: *That man spends more for a suit than you earn in a year.* **2** squander, °throw away, °fritter away, °waste, °go through, °splurge, °lavish, °dissipate: *He spends money as if it's going out of style.* **3** °devote, °allot, °assign, °invest, °put in, °pass: *If you don't spend more time with the children they'll forget they have a father.*

spendthrift *n.* **1** °profligate, °wastrel, (big) spender, squanderer, °prodigal: *Because her nephew is such a spendthrift, she put his inheritance in a trust fund.* —*adj.* **2** °wasteful, free-spending, °prodigal, °profligate, squandering, °extravagant, °improvident: *The chancellor's spendthrift budget will increase inflation.*

spent *adj.* **1** drained, °exhausted, °prostrate, °tired, °fatigued, fagged out, °weary, wearied, worn-out, *Colloq* °(dead) beat, done in, done for, all in, dog-tired, played out, burnt *or* burned out, used up, *Brit* knackered, done up, *US* pooped: *I was totally spent from sitting in the airport all night during the storm.* **2** °exhausted, used up, emptied, gone, expended, finished, *fini,* consumed, depleted: *At seventy, he was wrong to feel that his life was spent.*

spew *v.* Often, ***spew forth*** or ***out*** or ***up.*** belch (up *or* out *or* forth), vomit (up *or* forth), °regurgitate, spit up *or* out, °spout, °discharge, °emit, °eject, °send forth, °spurt, °gush, throw up *or* out, disgorge, *Slang* puke: *The volcano spewed forth lava for six days and nights.*

sphere *n.* **1** °globe, °orb, globule, spherule; °drop, droplet, °bubble: *Flotation is provided by thousands of hollow plastic spheres.* **2** °society, °class, °level, °caste, °rank, °domain, walk of life, °station, °stratum, °position: *We do not belong to the same social sphere.* **3** °area, °field, °province, °subject, °discipline, °range, °specialty, °forte, *Colloq* bailiwick, °territory, °department, °thing, °bag: *Are Gregorian chants within Stanley's sphere of interest?*

spherical *adj.* spheric, globular, °round, ball-shaped, ball-like, globelike, globe-shaped, globose *or* globous, globoid, globate(d), spheroid(al): *The radiolarians are spherical with radial spikes.*

spice *n.* **1** condiment, °relish, °seasoning, flavor(ing); herb: *Much of the purpose of early explorers was to search for spices.* **2** °zest, spiciness, piquancy, °tang, pungency, °bite, sharpness, poignancy, °gusto, °excitement, °seasoning, °dash, élan, °color, °life, °vigor, °interest, stimulation, °stimulant, °spirit, *Colloq* vim, zip, °pep, °kick, pizazz, °punch, ginger, °pepper: *Carruthers' notion of adding a little spice to his life consists of reading a tabloid newspaper.* —*v.* **3** °season, °flavor: *Highly spiced food does not agree with many people.* **4** Often, ***spice up.*** °enliven, inspirit, °stimulate, invigorate: *I doubt that the magazine would be spiced up much by a centerfold of puppies or babies.*

spicy *adj.* **1** zesty, zestful, piquant, tangy, (well-)spiced, (well-)seasoned, °hot, peppery, °sharp, °pungent, °snappy, °biting, full-bodied, °aromatic, °savory, flavorsome, flavorful: *This curry is a little too spicy for my taste.* **2** °off-color, indelicate, °suggestive, °risqué, °improper, °indecent, indecorous, ribald, °racy, °bawdy, °unseemly, °offensive, titillating, °sexy, *Colloq* °hot: *This magazine often publishes spicy nude* *photos.* **3** °scandalous, °sensational, °outrageous, °notorious, revealing, revelatory, °intimate: *She likes to read about the spicy goings-on in Hollywood.*

spike *n.* **1** skewer, °stake, prong, treenail, °nail, °peg, °picket, °pin, pike, °spine: *To subdue a vampire, drive a wooden spike through its heart.* —*v.* **2** °impale, °stab, °stick, spear, °pierce, °spit, °lance: *He was spiked through the thigh by a bayonet.* **3** disable, °thwart, nullify, °disarm, °block, °frustrate, °foil, °void, balk, °check, °cancel, annul: *Our plans for the picnic were spiked by the weather.* **4** °strengthen; °drug, °poison, *Slang* slip in a Mickey (Finn): *I passed out because my drink was spiked.*

spill *v.* **1** pour (out *or* over), overflow, slop *or* run *or* brim over: *The milk spilled all over the floor.* **2** °waste, °throw out, °lose: *Don't cry over spilt milk.* **3** ***spill the beans.*** reveal *or* tell *or* disclose *or* divulge all *or* everything, °blab, °tattle, let the cat out of the bag, °confess, *Slang* squeal, be a stool pigeon *or* stoolie, spill one's guts, °sing (like a canary), *Brit* blow the gaff: *Finnegan spilled the beans to the cops.* —*n.* **4** °outpouring, °flood, °leak, leakage: *Don't tell me there's been another oil spill!* **5** °fall, °tumble, °accident, *Colloq* cropper, header: *Crutchley had a bad spill at the third jump in the Grand National.*

spin *v.* **1** °revolve, °turn, °rotate, °gyrate, °twirl, whirl, °twist, °reel, °pirouette, °pivot: *The car spun out of control on the ice. Fred spun Ginger around the dance floor.* **2** °invent, concoct, °make up, °devise, °produce, °fabricate; °weave, °relate, retail, °recount, °narrate, °tell, °unfold: *Each night, father would spin a new chapter in the yarn of the vagabond princess.* **3** be dizzy, suffer vertigo, swim, whirl, be giddy: *My head was spinning from so much attention by the media.* **4** ***spin off.*** °separate, °derive: *The manufacturer spun off two new companies specializing in parts and service.* **5** ***spin out.*** °prolong, protract, drag *or* draw out, stretch out, °perpetuate, °continue, °extend, keep alive, keep going: *The case was spun out for years in the courts.* —*n.* **6** whirl, whirling, °twirl, twirling, °turn, turning, gyration, °reel, °pirouette, °revolution, revolving, rotation, rotating: *He puts a spin on the ball when he throws it. Is the spin of all planets in the same direction?* **7** °drive, whirl, joy ride, °ride, °tour, °excursion, °outing, jaunt: *On Sunday afternoons we sometimes went for a spin in the car.*

spine *n.* **1** °backbone, spinal column, vertebrae: *He can't stand up straight—he has curvature of the spine.* **2** °thorn, needle, barb, °spike, °spur, prong, quill, °ray, barbel, °bristle, °prickle, *Technical* barbule, spicule *or* spiculum *or* spicula: *The surface of the sea urchin is covered with needle-like spines.*

spineless *adj.* **1** *Technical* invertebrate: *There are more species of spineless creatures than those with internal backbones.* **2** °weak, °feeble, °flabby, °irresolute, weak-willed, °indecisive, °ineffectual, °ineffective, °impotent, °powerless: *After her reign, any king would appear spineless.* **3** °cowardly, dastardly, pusillanimous, timorous, lily-livered, white-livered, craven, °fearful, °timid, spiritless, °squeamish, *Colloq* yellow, chicken-hearted, chicken, yellow-bellied, wimpish: *She said you were spineless because you refused to go into the lions' cage.*

spiral *n.* **1** helix, °coil, corkscrew, °screw, scroll; whorl, volute, °turn, curl: *In earlier Greek motifs, the spiral is commoner than the rectangular fret.* —*adj.* **2** helical, coiled, °screw, corkscrew, cochlear *or* cochleate; scrolled, volute(d), whorled: *A spiral staircase is attractive and space-saving, but hard to use.*

spire *n.* **1** column, °tower, belfry; steeple, flèche: *The spire of the church was repeatedly struck by lightning that night.* **2** °top, °pinnacle, apex, °peak, °summit, °acme, °tip, °crest, °crown, °vertex: *We climbed till we stood on the highest spire above the valley.*

spirit *n.* **1** anima, °breath, °life, °vitality, vital spirit, pneuma, °soul, consciousness, °psyche, self, °heart, °essence: *Though past eighty, Winifred has a great deal of spirit. Even though she is gone, Annabel's spirit is still with us.* **2** °character, temperament, °temper, °persona, °disposition, °mind, °will, willpower, °attitude, °bent, °inclination, °energy, °ardor, °desire, °impetus, °drive, °urge, °eagerness, °zest, zeal, zealousness, °fire, passion(s), °enthusiasm, motivation, mettle, °resolution, °resolve, °intention, °enterprise: *The computer translated "The spirit is willing but the flesh is weak" as "The vodka is good but the meat is awful." Trumbull has shown a lot of the right spirit.* **3** °zest, pungency, piquancy, °warmth, °fire, °animation, °life, liveliness, vivacity, vivaciousness, °panache, élan, °dash, °spice, *Colloq* °sauce, °pepper: *He needs to put more spirit into his sales presentation.* **4** See **specter, 1,** above. **5** °bravery, °courage, °grit, °backbone, valor, °pluck, °daring, stoutheartedness, manfulness, manliness, gameness, resoluteness, °will, willpower, *Colloq* vim, °spunk, get-up-and-go, °(right) stuff, °guts, *US* sand: *Johnston has the spirit to be a marine.* **6** °meaning, °sense, °tenor, signification, purport, °intent, °intention, °purpose, °aim, °implication, °message, °essence, °quintessence, °core, °heart, °meat, °pith, °substance, marrow: *The spirit of the agreement is different from the way the lawyers worded the contract. Is it in keeping with the spirit of the law?* **7** °attitude, °principle, °thought, °idea, °inspiration, °notion, °feeling, °inclination, impulse: *Christmas is associated with the spirit of giving.* **8** Often, ***spirits.*** °temper, °mood, °sentiments, °feelings, °cheer, °humor, °frame of mind; °morale: *I hope he took my criticism in good spirit. You seem to be in pretty good spirits today. My spirits are quite low.* **9** ***spirits.*** **(a)** °feelings, °mood, °temper, °sentiments; °morale, esprit de corps, team spirit: *Spirits were high at the annual boat races.* **(b)** °alcohol, °liquor, °whiskey, strong drink, *Colloq* °booze, firewater, *Slang chiefly US and Canadian* hooch *or Slang chiefly US and Canadian* hootch: *Is it legal to sell spirits to minors?* —*v.* **10** ***spirit away*** *or* ***off.*** °abduct, make off *or* away with, °carry off, °transport, take away, °kidnap, steal (off *or* away with), whisk away, abscond with; make disappear: *A thief spirited away a painting from the exhibition. She was spirited away by a band of elves.*

spirited *adj.* °lively, °sprightly, °energetic, °vigorous, °animated, sparkling, °dynamic, °buoyant, °effervescent, °vivacious, °ardent, mettlesome: *The horse I chose was a bit too spirited. The barrister presented a spirited defense for his client. The 1970s saw spirited revivals of art nouveau and art deco styles.*

spiritual *adj.* **1** °sacred, ecclesiastic(al), churchly, °clerical, °priestly, devotional, °holy, °divine, sacerdotal, °religious, nonsecular: *Father Cratter was in charge of the spiritual aspects of our lives.* **2** nonmaterial, incorporeal, psychic(al), °mental, °psychological, inner: *Spiritual needs are less easily satisfied than physical needs.*

spit *v.* **1** expectorate; dribble, salivate, drool, °slaver, sputter, splutter; °discharge, °spew (forth), °eject: *Spitting is forbidden. Don't you hate people who spit when they talk? The volcano spat huge boulders into the sky.* **2** ***spitting image*** *or* ***spit and image.*** °twin, °duplicate, clone, °image, counterpart, °likeness, °copy: *She is the spitting image of her mother.* —*n.* **3** spittle, saliva, drool, *Technical* sputum: *A large gobbet of spit clung to his beard.*

spite *n.* **1** spitefulness, maliciousness, malice, malevolence, malignity, °ill will, °venom, spleen, °rancor,

°animosity, °gall (and wormwood), °resentment, °bitterness, °hostility, °antagonism, hatred, °hate, *Colloq* bitchiness: *Marian saw to it that he didn't get the job just out of spite.* **2** ***in spite of.*** °despite, °notwithstanding, °regardless of, ignoring, in defiance of: *In spite of his efforts to be helpful, she despises him.* —*v.* **3** °annoy, °irritate, vex, °upset, disconcert, °offend, °provoke, °discomfit, pique, °put out, °hurt, °injure, °wound, *Colloq* peeve, get under (someone's) skin, needle, *US* do a number on: *He'd cut off his nose to spite his face. He told her he was happy only to spite her for having divorced him.*

spiteful *adj.* °rancorous, °bitter, acrimonious, malevolent, malicious, °venomous, °hateful, invidious, °hostile, antagonistic, unfriendly, unforgiving, retaliative *or* retaliatory, °punitive, retributive *or* retributory: *Cutting off Casper's allowance because he forgot your birthday was a spiteful thing to do.*

splash *v.* **1** °spatter, bespatter, splatter, °shower, °spray, sprinkle, besprinkle, *Brit* splodge *or US also* splotch; mottle, °spot: *The painters splashed paint all over the floor. My freshly washed car is splashed with mud.* **2** blazon, °spread, °plaster: *Was that your picture splashed across the front page of yesterday's paper?* —*n.* **3** °spatter, °spray, splatter, sprinkle, °spot, °stain, °smear, smudge, *Brit* splodge *or US also* splotch: *There's a splash of tomato sauce on the tablecloth.* **4** °impression, °show, °uproar, ado, brouhaha, °sensation, commotion, °excitement, *US* foofaraw, *Colloq* to-do: *After all that advance publicity, he was expected to make a big splash in the theater.*

splendid *adj.* **1** splendorous, °magnificent, resplendent, °dazzling, °gorgeous, °showy, °dashing, °marvelous, spectacular, °grand, °glorious, °lavish, °ornate, °sumptuous, °majestic, °brilliant, °extraordinary, °exceptional, °superb, °supreme, °imposing, °impressive, awe-inspiring, °awesome, °lush, °plush, °rich, °luxurious, *Colloq* splendiferous, °posh, swank(y), ritzy: *The mausoleum built by Shah Jehan for his wife was far more splendid than his own palace.* **2** °impressive, °marvelous, °brilliant, °eminent, °prominent, °superior, °noteworthy, °notable, °celebrated, °illustrious, °famous, °distinguished, °exemplary, °remarkable, °admirable, °conspicuous, °outstanding, °sublime, °striking, °extraordinary, °successful, °admirable, °meritorious, creditable: *His splendid victory at Semolina climaxed an enviable career.* **3** °excellent, °superior, °preeminent, °fine, °marvelous, °extraordinary, °exceptional, °unbelievable, °incredible, first-class, unequaled, unsurpassed, °fabulous, °peerless, °matchless, °nonpareil, °superlative, °praiseworthy, °laudable, *Brit* °brilliant, *Colloq* °great, °colossal, supercolossal, stupendous, fab, °fantastic, °super, smashing, A-1 *or* A-one, tiptop, °capital, *Brit* brill, *Slang* far-out, °way-out, °dandy, °cool, °keen, *US* °solid, out-of-sight, fantabulous, °boss, °neat, °major: *Bill is a splendid dancer.*

splendor *n.* **1** magnificence, °grandeur, °brilliance, °display, °radiance, resplendence, sumptuousness, stateliness, majesty, panoply, °spectacle, °show, °glory, °pomp, gorgeousness, °dazzle, refulgence, °beauty, splendidness, exquisiteness, luxuriousness, richness, lavishness, °luxury, *Colloq* swankiness, poshness, swank, ritziness: *There was nothing to match the splendor of the court of Louis XIV.* **2** °brilliance, °shine, °luster, °light, effulgence, brightness, °glitter, °dazzle, refulgence, luminosity, luminousness, °gloss: *The splendor of the city's golden domes was reflected in the setting sun.*

splice *v.* **1** °join, °unite, °marry, °bind, conjoin; °knit, °entwine, intertwine, °braid, plait, °twist, interlace: *After the rope ends are spliced together, they should be seized with marline.* —*n.* **2** joining, °union, splicing, °joint, connection *or Brit* connexion, °tie, °bond, binding, °fastening, linking, linkage: *A proper splice will probably be stronger than the single cable.*

splinter *n.* **1** °sliver, °fragment, °piece; °scrap, shard *or* sherd, °shred, °chip: *A splinter of glass is stuck in my finger.* —*v.* **2** °shatter, °break, °fragment, °split, °disintegrate, smash into smithereens: *The lamp hit the floor and splintered into a million pieces.*

split *v.* **1** Often, ***split up*** *or* ***apart.*** °divide, °separate, °cleave, cut *or* chop apart, cut *or* chop in two, pull *or* tear apart, °rend, break *or* snap apart *or* in two, °break up, °come apart, °rupture, °partition, °detach, become detached; bisect, dichotomize: *He split the log with one blow of his ax. The ship split apart on the reef. The party may split on the tax issue.* **2** Often, ***split up.*** °divorce, °separate, go separate ways, °break up, part company: *After thirty years, the Hitchcocks have decided to split up.* **3** Often, ***split up.*** °branch, fork, °diverge, °separate: *The road splits to pass round the pond.* **4** Often, ***split up.*** °divide (up), apportion, deal out, dole out, °distribute, °allot, share *or* parcel out, °carve up: *Why don't we split the dinner bill five ways?* **5** °burst, crack *or* break up, fall apart *or* about, *Slang* bust: *I laughed so much I thought I'd split!* **6** °leave, °depart, °go, *Slang* take a (runout) powder, take it on the lam, °beat it, scram, skedaddle: *When they heard the sirens of the police cars, they split.* —*n.* **7** °crack, cleft, fissure, °chink, °cranny, °slit, °slot, °crevice, °groove, °furrow, °channel, sulcus; °gap, hiatus, lacuna, °opening, °separation, °division, chasm; °rift, °break, °rupture, °fracture; °slash, °gash, °tear, °rip, °rent: *The edges of the tectonic plates mark splits in the earth's crust. Pressure caused the splits in these plaster columns. I had to borrow a tablecloth to cover up the split in my trousers.* **8** °division, dichotomy, °schism, °breach, °rupture, °partition, disunion, °discord; °break, °separation: *Don't you think that the issue will cause a split within the party?* —*adj.* **9** divided, separated; halved, bisected, cleft; °cut, °broken, fractured: *The result was a split vote along party lines. Their first containers were of split bamboo.*

splurge *n.* **1** °display, °show, ostentatiousness, °extravagance, °indulgence, access, °splash, °burst, °outburst, °spree: *In his final splurge he bet everything on the number 14.* —*v.* **2** Often, ***splurge on.*** squander *or* dissipate *or* waste *or* burn (up) *or* throw away money (on), show off *or* flaunt one's money, *Slang* blow everything (on): *Victor always splurges when it comes to the children.*

spoil *v.* **1** °ruin, °destroy, °wreck, °queer, °mess up, °upset, °demolish, °harm, °damage, *Colloq* °kill: *You really spoiled everything by telling Mum about my pet frog. Spare the rod and spoil the child.* **2** °damage, °mar, °injure, °harm, °deface, disfigure, °scar, °blemish: *The painting was spoiled by some fanatic who slashed it.* **3** °baby, mollycoddle, °coddle, °indulge, °pamper, dote on, spoon-feed, *Rare* cocker: *Barbara really spoils her family terribly by indulging their every whim.* **4** °turn, go off *or* bad, curdle, molder, °decay, °decompose, become addle(d), °rot, °putrefy, mildew: *The milk will spoil if not refrigerated.* **5** ***be spoiling for.*** itch (for *or* after), °yearn (for), be eager (for), be keen (for), °look for, be bent on, be desirous of, crave, be after: *If you give Tom a drink he'll be spoiling for a fight. Oh, I thought he was always spoiling for a fight.* —*n.* **6** ***spoils.*** °loot, °booty, °plunder, °pillage, prizes, pickings, *Slang* swag, °take, °goods, boodle: *The Nazis took the paintings as spoils of war.*

spoilsport *n.* °killjoy, damper, dog in the manger, *Colloq* wet blanket, *US* party pooper: *Some spoilsport tipped off Jane about her surprise birthday party.*

spoken *adj.* 1 °oral, vocal, °verbal, viva voce: *My spoken commentary was recorded for later broadcast. She has an excellent command of spoken French.* 2 ***spoken for.*** See **speak, 13,** above.

sponsor *n.* 1 °backer, °supporter, promoter, angel, °patron, Maecenas, subsidizer: *Make a contribution and we will list you as a sponsor of the games.* 2 (radio *or* television) advertiser: *"A word from our sponsor" is not to be taken literally.* —*v.* 3 °back, °support, °promote, °fund, °patronize, °subsidize, °finance, °underwrite: *Hughes agreed to sponsor a campaign to clean up the environment.*

spontaneous *adj.* 1 unannounced, °unpremeditated, unplanned, impromptu, °extemporaneous, extempore, °unprepared, unrehearsed, °offhand, ad-lib, spur-of-the-moment, *Colloq* off the cuff: *Miss Malkin's spontaneous speech could not have been better had it been written out.* 2 °natural, unforced, unbidden, °instinctive, instinctual, °unconscious, reflex, °automatic, °mechanical, °immediate, °offhand, °unguarded, °unthinking, unwitting, °involuntary, °impetuous, °impulsive, *Slang* knee-jerk: *How could you doubt the sincerity of such an outburst of spontaneous affection? My spontaneous reaction is to refuse permission for you to take up sky diving.*

sporadic *adj.* °occasional, °intermittent, °random, °irregular, uneven, °erratic, °chance, unexpected; spasmodic(al), °fitful, periodic(al): *There has been sporadic unrest in that area of the city.*

sport *n.* 1 °recreation, °diversion, °pastime, °amusement, °entertainment, °play, °distraction, °relaxation, divertissement, °pleasure, °enjoyment, °fun: *I fish for sport, not to earn a living.* 2 jest, °humor, °fun, °mockery: *Surely you called me that in sport?* 3 ***make sport of.*** °tease, °deride, make a laughingstock, °(hold up to) ridicule, make a fool of: *She doesn't like it when you make sport of her husband.* —*v.* 4 °frolic, gambol, °cavort, romp, °caper, °play, °frisk, °lark, rollick, skip about: *These octogenarians are sporting about like young foals.* 5 °show off, °exhibit, °flaunt, °display, °wear: *Kingsley is always sporting the latest in Italian footwear.*

sportive *adj.* °frisky, gamboling, cavorting, frolicking, romping, capering, rollicking, °sprightly, coltish, °spirited, °frolicsome, °buoyant, gamesome, °gay, °kittenish, °merry, °playful, °gleeful, lighthearted, °blithe, prankish, waggish: *He watched a sportive group of children playing hide-and-seek.*

sportsmanship *n.* fair play, sportsmanliness, fairness, honorableness, °honesty, °honor, °probity, scrupulousness, °integrity, uprightness, °justice, justness: *Why does good sportsmanship require that the loser congratulate the winner?*

sporty *adj.* °informal, °casual; °stylish, °chic, °smart, °trendy, °fashionable, modish, à la mode, up-to-date, °showy, °rakish, *Colloq* swank(y), °loud, *Slang* classy, °swell, °flashy, snazzy, °sharp, *US and Canadian* spiffy: *He came to dinner wearing sporty clothes. Yes, I suppose you might call Fred a sporty dresser.*

spot *n.* 1 °mark, °patch, °speck, °blot, blotch, °blemish, speckle, fleck, °particle, mote, macula, smudge, °stain, °stigma, discoloration, *Brit* splodge *or US also* splotch: *There's a spot of ink on your shirt.* 2 °site, °place, °locale, °location, locality, °scene, °setting, °section, °area, °neighborhood, °quarter: *The Costers picked out a lovely spot to build their new house.* 3 °morsel, °bit, °bite, *Colloq* smidgen: *Let's stop in here for a spot of tea.* 4 °predicament, °situation, °quandary, °mess: *After promising to get Ronnie Barker to entertain at my boss's birthday, I am in a terrible spot if he refuses.* 5 ***spots.*** eruptions, pimples, acne, pustules, blackheads, comedos *or* comedones, whiteheads; boils, blains, wens; pockmarks; *US and Canadian old-fashioned* zits, hickeys: *The British word for acne is "spots."* —*v.* 6 °see, °catch sight of, glimpse, discern, °identify, °pick out, °distinguish, °single out, °detect, °sight, °recognize, °make out, descry: *I think I spotted Molly in the crowd.* 7 °mark, °stain, fleck, speckle, °spray, °splash, °spatter, bespatter, °sully, °soil, °dirty, °taint, besmirch, smudge: *We spotted the beige wall with lavender paint because we liked the effect.*

spotless *adj.* 1 °immaculate, °clean, gleaming, °shiny, °polished, unspotted, spick-and-span *or* spic-and-span: *The floor is so spotless you could eat off it.* 2 °pure, unsullied, unassailable, °flawless, °faultless, °untarnished, °blameless, °irreproachable: *Samantha has a spotless reputation.*

spotlight *n.* 1 arc light, searchlight, *US* pin spotlight, *Colloq US* °(pin) spot: *I couldn't see the audience with the spotlight in my eyes.* 2 °focus (of attention), limelight, public eye: *How does it feel to have been in the spotlight most of your life, Miss Hallward?* —*v.* 3 °light (up), °illuminate, focus (light) upon *or* on, shine *or* shed *or* throw *or* cast light upon *or* on, °emphasize, highlight, draw attention to, °feature, give prominence to, °stress, accentuate, °accent, °point up, underscore, underline: *The exam required us to spotlight the most important battles of the war.*

spotty *adj.* 1 spotted, dotted, °speckled, freckled, °flecked, blotched, blotchy, stained, °marked, pied, piebald, brindle(d), skewbald, °mottled, motley, dapple(d), macular, foxed; soiled, °dirty; *Brit* splodgy, splodged *or US also* splotchy, splotched: *Pages of some old books are spotty with brown stains.* 2 pimply, pimpled, blotched, blotchy, acned, pockmarked, pocky, °bad, *Scots* plouky *or* plooky: *Ian used to have a spotty complexion, but it's clear now.* 3 patchy, °irregular, uneven, °erratic, °sporadic, °capricious, °fitful: *Radio reception in our area is spotty during electrical storms.*

spout *v.* 1 °discharge, °emit, squirt, °spurt, jet, °shoot, °gush, °erupt, spew (up *or* out *or* forth), °spit, °eject, disgorge, vomit (up *or* forth), pour (out *or* forth), °flow, °stream: *Steam spouts from many vents in Iceland. The water came spouting out of the hose and drenched me.* 2 °ramble on, °rant, °rave, °carry on, pontificate, orate, declaim, °hold forth, maunder (on), witter on, °hold forth, expatiate, speechify, °talk, *Colloq* °go on, *Brit* rabbit on: *Harry, as usual, was spouting about his golf handicap.* —*n.* 3 waterspout, gargoyle, downspout, duct, °drain, °outlet, conduit: *The rainwater drains into the gutters, then into the spouts.* 4 ***up the spout.*** gone, °lost, destroyed, beyond hope *or* recovery, to be written off *or* abandoned: *Our dream of finding the treasure was now up the spout.*

sprawl *v.* 1 °spread (out), °stretch (out), straddle, °ramble, °meander, °wander, °straggle, branch out: *Greater Los Angeles sprawls over a huge area of some 450 square miles.* 2 °spread out, stretch out, loll, °lounge, °slouch, °slump, °recline, lie about *or* around: *A dozen guests were sprawled on blankets at the beach party.* —*n.* 3 °spread, °stretch, °expansion, °extension: *Some suburbs of large cities have been swallowed up in urban sprawl.*

spray[1] *v.* 1 sprinkle, °spatter, °scatter, °shower, °disperse, °diffuse, atomize, °spread: *Try to spray the insecticide evenly over the plants.* —*n.* 2 °shower, sprinkling, drizzle, °mist, sprinkle, spindrift *or* spoondrift: *The strong wind carried the spray from the whitecaps right up into the streets of the town.* 3 atomizer, sprayer, sprinkler, vaporizer, aerosol: *Richard bought me a cologne spray for Mother's Day.*

spray[2] *n.* flower *or* floral arrangement, nosegay, posy, °bouquet, sprig, °branch, bough: *The scores of tiny roses were interspersed with sprays of baby's-breath.*

spread *v.* **1** Often, ***spread out.*** °diffuse, °distribute, °disperse, disseminate, °broadcast, °sow, °scatter, °strew, °shed, dispel, °dissipate: *Spread the mulch as evenly as possible to a depth of several inches.* **2** Often, ***spread about*** *or* ***around.*** °broadcast, °publicize, °make known, bruit about, °air, televise, °circulate, °publish, °distribute, disseminate, trumpet, °announce, °pronounce, promulgate, advertise, enounce, °make public, tell the world, herald, °repeat, °recite: *Start spreading the news—I'm leaving today.* **3** Often, ***spread out.*** °unfold, °draw out, °display, stretch out, open out, °extend, °lay out, fan out, unroll, unfurl: *Please spread the cloth on the table.* **4** Often, ***spread out.*** °stretch (out), °extend, protract, °prolong, drag out, °distribute, °disperse: *The bank said I could spread the payments over twenty years.* **5** Often, ***spread out.*** °stretch, °extend, °separate, put apart *or* out, °part: *He spread his arms and Lorna rushed into them.* **6** °grow, °develop, °increase, broaden, °expand, °extend, °widen, °enlarge, mushroom, °proliferate, °sprawl, branch out; metastasize: *To meet the demand for the pies, bakeries began to spread throughout the country. The cancer, unchecked, was bound to spread.* **7** °smear, °apply, °smooth, °put, °rub, °cover, layer, °plaster, °plate, °coat, °wash, °glaze, °paint, varnish, overlay, overspread; °cloak, °mantle, swaddle, °wrap, blanket: *I was careful to spread the paint evenly, avoiding streaks. When the cake has cooled, spread the chocolate icing over it.* —*n.* **8** spreading, °extension, extending, °expansion, expanding, enlargement, enlarging, °development, developing, °increase, increasing, °proliferation, proliferating, broadening, °growth, widening, mushrooming, dispersion, dispersal, dispersing, dissemination, disseminating, °distribution, distributing, dispensation, dispensing: *The spread of the disease is uncontrolled. We must encourage the spread of knowledge.* **9** °extent, °expanse, °area, °span, °sweep, vastness, °stretch, °reach, °breadth, °depth, °size, dimensions, compass, °limits, °bounds, boundary *or* boundaries: *As our craft accelerated past the moon, we were awed by the spread of empty space that lay before us.* **10** °range, °extent, °scope, °span, °difference: *There was too much of a spread between the cost and the selling price.* **11** °feast, °banquet, °meal, dinner, repast, barbecue; °table; *Colloq* °feed: *I have never seen a spread like the one laid on for the returning champions.* **12** butter, margarine, °jam, jelly, °preserve, °conserve, confiture, paste, *US old-fashioned* oleo: *What kind of spread do you like on your toast?* **13** ranch, landholding, holding, °property, °place, plantation, °farm, homestead, °place: *His spread in Texas eventually amounted to ten thousand acres.* **14** bedspread, counterpane, coverlet, bedcover, °cover, quilt, eiderdown, duvet, afghan, *US* comforter, *US and Canadian* °throw: *Leave the spread off the bed, Carlotta, as I want to take a nap.*

spree *n.* °frolic, romp, °lark, °outing, escapade, °revel, wild party, °fling, debauch, °orgy, bacchanalia; drinking bout, carousal, *Colloq* °bender, binge, °jag: *Bruce has gone off on another of his sprees.*

sprightly *adj.* °lively, chipper, spry, °vivacious, °cheerful, °gay, °brisk, °animated, °sportive, °active, °alert, °nimble, °agile, °energetic, °jaunty, °perky, °playful, °spirited: *She is quite sprightly for a grandmother of seventy-eight years.*

spring *v.* **1** °leap, °bound, °jump, °hop, vault, dart, °fly, °bounce: *He was brought up to spring to his feet when a woman enters the room. When Jim came to the stream, he sprang over it. The leprechaun sprang out from behind a tree.* **2** °arise, °appear, °grow, °come up, °rise, come into being *or* existence, be born, °emerge, °sprout, shoot up, burst forth: *Where the dragon's teeth were sown, an entire army sprang up.* **3** Often, ***spring up or from.*** °originate, °begin, °start, evolve; °proceed from, stem from, descend from, derive from, come from, develop from: *He learned how the practice of scratching the head in perplexity had sprung up. I wondered where the term* horse latitudes *sprang from.* **4** start *or* begin *or* experience *or* cause to occur *or* appear *or* happen suddenly *or* unexpectedly, °broach, °pop, introduce *or* divulge *or* reveal *or* disclose suddenly *or* unexpectedly: *The canoe sprang a leak and started to sink. He likes to spring trick questions at you in an exam.* **5** ***spring for.*** °pay for, treat (someone) to, assume the expense(s) of: *If you go to a place I choose, I'll spring for dinner.* —*n.* **6** °leap, °bound, °jump, °hop, vault, °bounce, °skip: *In one spring the cat cleared the garden wall and was off.* **7** bounciness, °bounce, resiliency, °resilience, springiness, buoyancy, °elasticity, sprightliness, airiness, °flexibility: *Despite her age, there's still a lot of spring in her step.* **8** °source, fount, fountainhead, wellspring, °well, °origin, °beginning, °root: *From what secret spring does his knowledge arise?* **9** springtime, Eastertide, Maytime: *At thirty-five, James is still in the spring of life.* —*adj.* **10** vernal: *We exulted in the spring sunshine.*

sprout *v.* bud, germinate, °come up, °arise, °begin, bloom, blossom, °flower: *With this warm weather, everything in my garden is beginning to sprout.*

spruce *adj.* **1** °neat, °dapper, °smart, °trim, well-turned-out, °well-groomed, °elegant, *Colloq* natty: *Thornton turned up for dinner looking as spruce as can be.* —*v.* **2** ***spruce up.*** °tidy (up), °neaten (up), °primp, °clean (up), straighten out *or* up, smarten (up), titivate *or* tittivate: *Company is coming, so first spruce up your room and then spruce yourself up.*

spunk *n.* °nerve, °courage, °pluck, °spirit, gameness, °resolve, °resolution, mettle, °heart, °grit, spunkiness, °backbone, marrow, *Colloq* °guts, °gumption, *Brit* °bottle, *US* sand: *Frances has shown she has tremendous spunk to have got this far despite her handicap.*

spur *n.* **1** goad, °prod, urging, impulse, °incitement, instigation, prompting, °pressure, stimulus, stimulation, °incentive, °provocation, °inducement, °encouragement, °motive, motivation: *The company pays workers by the piece as a spur to productivity.* **2** °projection, prong, °spike, °spine, gaff, barb, quill, tine, barbel, barbule, °process: *The hook has a spur that prevents the fish from disengaging.* **3** ***on the spur of the moment.*** impetuously, impulsively, unthinkingly, unpremeditatedly, impromptu, on the spot; rashly, thoughtlessly, recklessly, °hastily, brashly, incautiously, unexpectedly, °suddenly: *On the spur of the moment, we decided to fly to Cap d'Antibes for the weekend.* —*v.* **4** goad, °prod, °urge, egg on, impel, °incite, °prompt, press, °push, pressure *or Brit* pressurize, °stimulate, °drive, °provoke, °induce, °encourage, °motivate, °excite, °animate: *The sales force are spurred on by the huge bonuses the company offers.*

spurious *adj.* °false, °counterfeit, °sham, °fake, °fraudulent, °bogus, °mock, °imitation, simulated, unauthentic, ungenuine, forged, feigned, °pretended, °deceitful, meretricious, contrived, °factitious, °artificial, ersatz, °synthetic, *Colloq* pseudo, *Brit* phoney, *Chiefly US* °phony: *His brother was jailed for peddling spurious Egyptian burial artifacts to archaeologists.*

spurn *v.* °reject, disdain, °scorn, contemn, °despise, °rebuff, °repudiate, °refuse, sneer at, snub, °brush off, °turn down, turn one's back on *or* upon, look down on *or* upon, *Colloq* cold-shoulder, turn one's nose up at, °sneeze at: *Nell curtly spurned my offer of a loan.*

spurt *n.* **1** °burst, access, °effort, outbreak, °spell, °interval, °spate, °moment, °instant: *With a sudden spurt of speed he caught up with me.* **2** °increase, °advance, acceleration, °rise, °improvement: *After a brief spurt in business in January, everything stopped.* —*v.* **3** °gush, °spew, squirt, jet, °shoot, °erupt, °burst, °surge: *The minute the plumber left, water began spurting out of the drain again.*

spy *n.* **1** double agent, foreign agent, secret (service) agent, intelligence agent, undercover agent, mole, fifth columnist, CIA man *or* woman *or* agent, *Brit* MI5 *or* MI6 man *or* woman *or* agent; °informer, informant, *Colloq* mole, *Slang* stool pigeon, stoolie, fink, rat fink: *In those days, everyone was a spy and you could trust no one.* —*v.* **2** Usually, ***spy on*** *or* ***upon.*** °follow, °shadow, °trail, °watch, °observe, °reconnoiter, keep under surveillance, *US* surveil, *Colloq* °tail, °check out, °case: *Why should anyone from a foreign government be spying on my mother?* **3** espy, glimpse, °spot, catch sight *or* a glimpse of, descry, °note, °notice, °see, discern: *I spied him getting into a yellow convertible.*

spying *n.* espionage, undercover work, secret service; detection, °intelligence, °surveillance: *Virtually every government engages in spying.*

squad *n.* °unit, °team, °band, °company, °crew, °force, troop, cadre, °gang, °section, °group, squadron, °platoon, °party: *Has the work squad returned to base yet?*

square *adj.* **1** equilateral, quadrangular, rectangular, right-angled, quadrilateral, four-sided, cubic, cubed, six-sided, boxy: *The work he paid ten thousand dollars for consists of three square pieces of red plastic on a mauve ground. He keeps trying to put the square pegs into the round holes.* **2** °equal, on a par, °even, on equal terms, settled, balanced: *With this payment, our accounts are now all square.* **3** °even, °true, °exact, °straight, °accurate, °precise, °correct: *Make sure that the corners are perfectly square.* **4** °honorable, °upright, °honest, straightforward, °fair (and square), °decent, °ethical, °open, (open and) above board, °right, °(right and) proper, °clean, °just, °equitable, *Colloq* °on the level, on the up and up: *I have never had anything but square dealings with Fred.* **5** healthful, °healthy, °nutritious, °substantial, °solid, °full, °generous, °satisfying, °filling, unstinting: *You could always count on three square meals a day if you stayed at Adele's.* **6** naive *or* naïve *or* naïf, °innocent; °bourgeois, °conservative, °conventional, °unsophisticated, °provincial, °old-fashioned, conformist, °strait-laced, unimaginative, °predictable, *Colloq* antediluvian, up tight, out of it, not with it, not in the know, not hip *or* hep, unhip, °stuffy, °behind the times, °straight: *We were at the age when anyone over twenty-five was considered square.* —*n.* **7** rectilinear figure, rectangle; cube, °block: *He took a square of wood and quickly carved it into the shape of a duck.* **8** plaza, piazza, °place, °park, °(village) green, marketplace, °market (square), agora, quadrangle: *We like to sit on the benches in the square outside my house and chat with the neighbors.* **9** °bourgeois, °conservative, conformist, traditionalist, (old) fogy *or* fogey, diehard; °outsider; *Colloq* stuffed shirt, fuddy-duddy, *US* longhair, *Slang US* nerd, dweeb: *My parents will never understand me—they're real squares.* —*v.* **10** °stiffen, throw back, °straighten (up), °tense: *I squared my shoulders and prepared for the worst.* **11** Usually, ***square with.*** °meet, °match (with), conform to *or* with, °obey, correspond to *or* with, tally with, accord with, °agree with, reconcile with *or* to: *These procedures do not square with the plans laid out by the directors.* **12** °adapt, °adjust, °change, °modify, harmonize, °accommodate, °arrange, comply with, °fit: *Ronald was unable to square his beliefs with what they are teaching him at the seminary.* **13** °settle, °arrange, °come to terms, °patch up, °clear up, °satisfy, °fix: *Don't worry about the customs officials—I'll square it with them later on.*

squawk *v.* **1** cackle, screech, °shriek, °yell, yowl, °whoop, hoot, °scream, °call, °cry: *The parrot kept squawking "Where is the treasure? Where is the treasure?"* **2** °complain, grumble, whine, grouse, °protest, °object, °(make a) fuss, °yap, yowl, *Slang* bellyache, °bitch, °kick, beef, °gripe, kick up a fuss: *Oh, stop squawking and get on with your work!* —*n.* **3** °complaint, grouse, grumble, °protest, *Colloq* °kick, beef, °gripe: *If you have a squawk, take it up with the sergeant.*

squeamish *adj.* **1** °dainty, °delicate, °prudish, punctilious, °demanding, °critical, °exacting, °difficult, °fussy, °scrupulous, °fastidious, °meticulous, °painstaking, finicky *or* finical, *Colloq* persnickety *or* pernickety, fuddy-duddy: *The museum directors are too squeamish to put on an exhibition of nudes.* **2** °nauseous, qualmish, easily disgusted *or* revolted *or* nauseated: *Sam was too squeamish to watch open-heart surgery on the TV.*

squeeze *v.* **1** °press, compress, °compact, °crush, squash, wring, °pinch, °nip, °grip, °tweak: *Each morning Anne squeezes an orange to make fresh juice for Edgar. Everyone knows the symbolism of squeezing the nose between the thumb and forefinger.* **2** °extract, wrest, °exact, °extort, °milk, °wrench, °pry (out), °tear, *Colloq* bleed: *It is doubtful that they will be able to squeeze a contribution out of old Scrooge.* **3** °milk, *Colloq* °shake down, bleed, °lean on, put the screws to, °put the squeeze on, twist (someone's) arm, *US* put the arm on: *You'll have to squeeze Fletcher if you want to get paid.* **4** °ram, °jam, °pack, °stuff, °cram, °crowd, °force, °press, °wedge: *We were squeezed into the train so tightly that I could hardly breathe.* **5** °clasp, clench, °embrace, °hug, °hold, enfold, °fold, °clutch, *Archaic* °clip: *She squeezed him close as they bade goodbye.* **6** ***squeeze through*** *or* ***by.*** get through *or* by, °pass, °(barely) succeed, *Colloq* squeak through *or* by, °(barely) make it: *Cassie managed to squeeze through with a passing grade, but it was touch and go.* —*n.* **7** °clasp, °embrace, °hug, °clutch, *Colloq* °clinch: *She gave me a little squeeze to signal her affection.* **8** °pressure: *I was beginning to feel the squeeze between inflation and high interest rates.* **9** °crush, °jam, °crowd, squash, °press: *There was such a squeeze of people at the sale that I couldn't get in.* **10** girl friend, °mistress, °sweetheart, *Colloq* sweetie, *Slang* moll, °broad, *Archaic* doxy: *Willie showed up with his current squeeze.* **11** ***put the squeeze on.*** °press, bring pressure to bear on, °urge, °influence, *Brit* pressurize, *US* °pressure: *The country club was beginning to put the squeeze on me for the annual membership dues.*

squelch *v.* **1** °suppress, °subdue, °put down, °quell, °quash, °defeat, °overcome, °outdo, °humiliate, *Colloq* shoot *or* slap down, take down a peg (or two), take the wind out of (someone's) sails, settle (someone's) hash: *The slightest sign of disapproval from Noël was enough to squelch the most arrogant toady.* —*n.* **2** riposte, °retort, comeback, °quip, sally, jibe, barb, *Colloq* °wisecrack, °put-down: *One of the best squelches was when Oscar Wilde said, referring to a bon mot, "I wish I'd said that," to which Whistler said, "You will, Oscar. You will."*

squire *v.* **1** °escort, °accompany, °conduct, °go with, °take; convoy: *Who will squire my sisters to the prince's ball?* —*n.* **2** esquire, gentleman, landowner, landholder, landed proprietor: *The old squire held his head high among the county aristocracy.*

squirm *v.* °wriggle, writhe, °twist, °flounder, °shift, °fidget, be (very) uncomfortable, agonize, *Colloq*

°sweat: *The boy squirmed under the stern gaze of the headmaster.*

stab *v.* **1** °stick, °puncture, °prick, °lance, °jab, °pierce, °run through, °impale, °gore, °transfix, °knife, bayonet, skewer, °spike, °spit, spear, °pin; °plunge, °poke, °thrust: *In the melee, George was stabbed in the arm. You say she stabbed a hatpin into his hand?* **2** ***stab in the back.*** °harm, °betray, °sell out, °double-cross, give the Judas kiss, play false with: *As soon as he had left, they stabbed him in the back by phoning the authorities.* —*n.* **3** °puncture, °jab, °thrust, °(stab) wound: *The coroner's office reported that the body had six stabs in the abdomen.* **4** °attempt, °try, °essay; °guess, conjecture: *Even if you are wrong, you lose nothing by making a stab at the right answer.* **5** °pang, °twinge, °pain, °ache, °hurt, stitch: *Doctor, I get a terrible stab of pain in my side when I turn that way.* **6** ***stab in the back.*** treachery, °betrayal, °double-cross, Judas kiss, kiss of death, duplicity: *Informing on anyone to the tax collector is the worst kind of stab in the back.*

stability *n.* **1** steadiness, solidity, firmness, soundness, sturdiness, °strength: *These flimsy structures are not known for their stability.* **2** steadfastness, constancy, dependability, reliability, °tenacity, °resolve, resoluteness, °perseverance, °determination, °persistence, durability, lasting quality, solidity, °permanence: *The stability of the government was in some doubt as the crisis continued.*

stable *adj.* **1** °steady, °solid, °firm, °sound, °sturdy, °strong, °durable, well-founded, °fast, °sure, established, deep-rooted, °stout: *The building is designed to remain stable even in a severe earthquake.* **2** °lasting, °enduring, long-lasting, longstanding, °secure, °steadfast, °steady, °strong, unchanging, unchanged, unchangeable, unalterable, °fixed, °invariable, unwavering, immutable, °permanent: *A stable relationship has existed between our countries for generations. The relationship you and I enjoy has remained stable all our lives.* **3** °sane, (well-)balanced, °responsible, °reasonable, °sensible, °competent, °accountable: *If he is released from our care, what guarantee is there that he will continue to behave in a stable fashion?*

stack *n.* **1** °pile, °heap, °mound, °mass, °accumulation, °hill, °mountain, °store, °stock, bank, °deposit, °supply, stockpile, °hoard, °load, °bundle, bale, *Colloq US and Canadian* stash: *This stack of paper ought to be enough to last through the next printing.* **2** haystack, cock, haycock, rick, rickle, hayrick, *Brit* °clamp: *The stacks must be moved into the barn before the next rain.* **3** °collection, aggregation, °accumulation, agglomeration, amassment, °mass, °load, °pack, °amount, °abundance, °plenty, °profusion, °volume, array, °sea, °throng, multitude, °swarm, °host, °number, °quantity, °pileup: *We have a huge stack of orders to process this morning.* **4** smoke stack, chimney, chimney stack, funnel; *Building* soil stack: *They built the stack very tall to carry the fumes away from the town below.* **5** ***blow one's stack.*** °anger, become angry, become furious *or* infuriated, °rage, °rant, lose one's temper, *Slang* blow *or* lose one's cool, get hot under the collar, blow one's top: *The boss will blow his stack if we miss the deadline.* —*v.* **6** Often, ***stack up.*** °pile (up), °heap, °accumulate, °amass, °store, °stock, stockpile, °hoard, °collect, aggregate, agglomerate, *Colloq* stash (away), squirrel away: *Stack those boxes neatly in the corner. Has he stacked up enough points to qualify?* **7** ***stack up.*** **(a)** make sense, add up, °agree, jibe, be verifiable, *Colloq* °check out: *The ledgers don't stack up with the checkbook. The two sets of figures simply don't stack up.* **(b)** °compare, measure up, hold a candle to, be on a par (with), be as good as: *He could never stack up to Olivier.*

stadium *n.* arena, °ground, amphitheater, hippodrome, coliseum *or* colosseum, circus: *This football stadium can accommodate sixty thousand spectators.*

staff *n.* **1** °stick, °pole, °standard, baton, °rod, pikestaff, pike, °stake, cane, stave, °shaft, alpenstock, shillelagh, °club, truncheon, mace, crook, crozier, scepter, °wand, caduceus: *Carrying the ceremonial staff and dressed in full regalia, the officer impressively opened the royal court proceedings.* **2** personnel, employees, °help, work force, °crew, °team, °organization: *Before computers, we had a staff of fifty to do nothing but filing.*

stage *n.* **1** °position, °situation, °grade, °level, °stratum, °tier, echelon, °step, °station, °place, °point, °spot, °juncture, °division, °phase, °lap; °status, °condition: *We have to reach the third stage before going on to the fourth.* **2** °platform, dais, podium; °rostrum: *After the performance, the audience threw flowers onto the stage.* **3** ***the stage.*** show business, °the theater, the boards, the footlights, Broadway, *Chiefly Brit* the West End; acting, thespianism; *Colloq* show biz: *That first burst of applause, just for her, decided her to make the stage her career.* —*v.* **4** °put on, °produce, °present, °mount, °exhibit: *If we can't get the theater, why don't we stage the show in Mr. Kimble's barn?* **5** °put on, contrive, °organize, °originate, °devise, °make up, concoct, °fake, trump up, stage-manage, °manipulate, °maneuver: *It seems that the enthusiastic reception given his wife had been staged solely for his benefit.*

stagger *v.* **1** °totter, °reel, °lurch, °teeter, °sway, walk unsteadily *or* shakily, °pitch, °rock, wobble: *When he staggered in, we thought he was drunk till we saw the knife in his back. I was already staggering under the burden of a heavy mortgage and three children and two elderly parents to care for.* **2** °surprise, °amaze, °astound, °astonish, °overwhelm, °overcome, dumbfound *or* dumfound, °shock, stupefy, °stun, °nonplus, °floor, confound, °bewilder, °startle, °jolt, °shake (up), °take one's breath away, make one's head swim, take (someone) aback, throw (someone) off balance, °tax, °burden, *Colloq* flabbergast, °flummox, bowl over, *Slang* blow (someone's) mind: *The cost of cleaning up the oil spill will be staggering. Your excuses for being tardy stagger the imagination.* **3** °alternate, °space (out), °vary, rearrange, zigzag, *US* change off: *Working hours will be staggered during the holiday period. We could strengthen the structure by staggering the positions of the columns.*

stagnant *adj.* motionless, °standing, °still, °quiet, sluggish, unmoving, immobile, °flat; °stale, °foul, °putrid, putrescent, putrefied, polluted, °dirty, contaminated, °filthy: *The malarial mosquitoes thrive in pools of stagnant water in the tropical swamps.*

stagnate *v.* languish, °idle, vegetate, °deteriorate, °degenerate, °decline, go to seed *or* pot, °decay, rust, molder, °decompose, °spoil, °rot: *He felt that he had stagnated in that backwater of civilization far too long.*

staid *adj.* °sedate, °rigid, °stiff, prim, °dignified, °sober, °calm, composed, °quiet, restrained, °solemn, °serious, serious-minded, °grave, sober-sided: *One expects a judge to be rather staid in his behavior.*

stain *n.* **1** °blot, °mark, °spot, discoloration, blotch, smutch, smirch, °speck, *Brit* splodge *or US also* splotch: *There's a stain on your tie.* **2** °mark, °blot (on the escutcheon), °stigma, °blemish, *Brit* blot on one's copybook, *Colloq US* black eye: *His cowardice in battle was a permanent stain on his reputation.* **3** dye, °color, coloring, °tint, tinge, pigment: *An indigo stain was used to bring up certain features in microscopy.* —*v.* **4** °blot, °mark, °spot, discolor, blotch, speckle, dye, °spatter, splatter, tinge, smudge, smutch, °splash: *The tablecloth was stained red from the spilled wine.* **5** °spoil, defile,

°ruin, smirch, besmirch, °taint, °tarnish, °stigmatize, °shame, °disgrace, °sully, °contaminate, °soil, °corrupt: *Her reign was stained with the blood of the thousands she had tortured and executed.*

stake[1] *n.* **1** °stick, °post, °spike, °picket, paling, °pale, °pole, pike, stave; palisade, °pole, °upright, °pillar, column: *We tied the goat to a stake and hid, waiting for the tiger to catch the scent.* **2** ***pull up stakes.*** °move (house), resettle, move on, °migrate, °emigrate, °leave, °depart: *It was time to pull up stakes and find a new campsite.* —*v.* **3** °tether, °tie (up), °secure, °fasten, °picket, °lash, leash, °hitch, °chain: *The ponies were staked out in the pasture.* **4** Usually, ***stake out.*** **(a)** fence (in *or* off), confine, °pen, °enclose, close in *or* off, hem in, °shut in, impound, °enclose, °cage, wall in: *We hired a man to stake out a large enough area to keep the horses.* **(b)** mark off *or* out, °define, delimit, °outline, demarcate, delineate, circumscribe: *The pioneers had already staked out the land they planned to cultivate.*

stake[2] *n.* **1** °bet, wager, ante, °risk, °hazard: *If you want to make the game interesting, you'll have to raise the stakes.* **2** investment, °interest, °share, involvement, °concern: *Myra has a considerable stake in the outcome of the shareholders' meeting.* **3** ***at stake.*** at hazard, hazarded, at risk, risked, on the table, in jeopardy, jeopardized, °concerned, °involved: *Don't take the matter lightly, as Tom's reputation is at stake.* —*v.* **4** °risk, °jeopardize, °venture, put (money) on, °chance, °hazard, °gamble, wager, °bet: *I had staked my last penny on that gold mine.*

stale *adj.* **1** °old, past its prime, unfresh, °dry, dried-out, hardened, °limp, wilted, withered, °flat, °sour, turned, °(gone) off, °moldy, °musty, spoiled, °rotten: *All the food is stale because you failed to wrap it carefully.* **2** °old, °banal, overused, °antiquated, °old-fashioned, °threadbare, trite, clichéd, unoriginal, hackneyed, stereotyped, °tired, °weary, °boring, °tiresome, warmed-over, shopworn, °familiar, °stock, °well-known, *Colloq* hand-me-down, *Brit* reach-me-down: *These comedians come up with the same old stale jokes year after year.*

stalemate *n.* °impasse, °deadlock, standoff, °standstill, (dead *or* full) stop, °tie; °check, checkmate, °mate; *US* Mexican standoff: *As both parties refused to back down, the situation was a stalemate.*

stalk[1] *v.* °follow, dog, °haunt, °shadow, °trail, °track (down), °hunt (down), °pursue, °hound, °chase, *Colloq* °tail: *The leopard stalks its prey in complete silence.*

stalk[2] *n.* °stem, °trunk, cane, main axis, leafstalk, °shaft, °spike: *The branches cluster at the very top, leaving the entire stalk of the plant bare.*

stall[1] *v.* **1** °stop, °halt, °die, °quit, °shut down, °fail, cease operating, come to a standstill, *Colloq* conk out: *My car stalled in the middle of rush-hour traffic—isn't that typical?* —*n.* **2** °compartment, °stand, °booth, cubicle, alcove, °section, °space, °area, °slot, °enclosure, °quarters; °counter, °table: *The antiques dealers hire stalls here on Saturdays to display and sell their wares.* **3** °shed, °pen, cote, °fold, coop, sty, corral, °enclosure, cowshed, barn, °stable: *The animals belong in their stalls at night.*

stall[2] *v.* **1** °delay, °dawdle, dillydally, dally, loiter, °linger, temporize, °equivocate, °hesitate, prevaricate, °play for time, waste time, stonewall, be obstructive, put (someone *or* something) off; vacillate, dither, hedge, °procrastinate; *Brit* haver, *Colloq* beat about *or* around the bush, °drag one's feet, give (someone) the runaround: *Stop stalling and give me the money or I'll shoot.* —*n.* **2** stalling, °delay, hedge, hedging, °pretext, °subterfuge, °wile, °trick, °ruse, °artifice, °stratagem, °maneuver, °move, stonewalling, obstructionism, playing for time, procrastination, procrastinating, *Colloq* beating about *or* around the bush, °runaround, foot-dragging: *Her claim that she was dressing was nothing but a stall to give her time to hide the pistol.*

stalwart *adj.* **1** °robust, °stout, °strong, °mighty, °powerful, °rugged, °staunch, °hardy, °sturdy, °vigorous, °lusty, °indomitable, °solid, able-bodied, °brawny, °sinewy, °muscular, °fit, °healthy, °hale, °(hale and) hearty, *Colloq* °husky, °hefty, beefy: *We shall need a dozen stalwart men to carry out the raid.* **2** redoubtable, °intrepid, undaunted, °resolute, °firm, °determined, unbending, °steadfast, °tenacious, unswerving, unwavering, unfaltering, unflinching, uncompromising, unyielding, persevering, °persistent, unflagging, °relentless, °tireless, °untiring, indefatigable: *Although they lost, their lawyers put up a stalwart defense.* **3** °brave, °courageous, °daring, °intrepid, valiant, °heroic, °manly, manful, °fearless, °indomitable, stout-hearted, °bold, °audacious, °game, red-blooded, plucky, mettlesome, lionhearted, °spirited: *These stalwart knights were to meet at dawn on the battlefield.* —*n.* **4** °supporter, upholder, sustainer, °partisan, loyalist, °(party) faithful, trouper, °hero, °heroine: *And most of all I want to thank those stalwarts who have supported me throughout this victorious campaign.*

stamina *n.* ruggedness, °vigor, vigorousness, °(intestinal) fortitude, robustness, indefatigability, staying power, °endurance, °energy, °power, °might, mettle, °(inner) strength, staunchness, stalwartness, °courage, indomitability, *Colloq* °grit, °guts, starch, *US* stick-to-itiveness, sand, *Taboo slang* balls: *Even at the age of seventy, Vernal had the stamina to run in the marathon.*

stammer *v.* **1** stutter, °hesitate, hem and haw, °stumble, falter, °pause, *Brit* hum and haw: *Is he stammering from embarrassment or because he has a speech defect?* —*n.* **2** stutter: *Many young people who have a stammer lose it when they get out of their teens.*

stamp *v.* **1** °trample, bring down one's foot; tread, °step, °tramp; *Colloq* stomp (on): *He flew into a rage and stamped on the floor, like a petulant child.* **2** °impress, °mark, imprint, °print, °record, °document, °register, log; °engrave, emboss, inscribe; °sign, °initial: *That last night in Rio is stamped forever in my memory. Our passports were stamped at the border.* **3** °brand, °label, °mark, °tag, °term, °name, °style, °identify, °categorize, classify, °characterize, °designate, denominate, show to be: *After the episode on Friday, he was stamped* persona non grata *in Juan Fernández.* **4** ***stamp out.*** °eliminate, eradicate, °abolish, °get rid of, annihilate, °exterminate, °kill, snuff out, °terminate, °end, put an end to, °destroy, °put down, °put out, °extinguish, extirpate; °quell, °subdue, °suppress, °squelch, °repress: *Something must be done to stamp out the drug traffic.* —*n.* **5** °mark, °sign, °hallmark, earmarks, traits, °features, characteristics: *Her work bears the stamp of genius.* **6** °die, °block, °punch, °seal, matrix, °plate, die stamp, stereotype, °mold; signet (ring): *The stamps for the binding are not quite ready.* **7** °seal, (trade *or* service) mark, °brand, logo, logotype, °symbol, °representation, colophon, imprint, °emblem, insigne (*singular of* insignia), °label, monogram, °sign, °crest, coat of arms, escutcheon, cartouche, signature, initials: *As it bears his stamp, it must belong to him.* **8** °character, °kind, °sort, °make, °fashion, °type, °cast, °mold, °grade, °style, °cut, °genre, °class, °level, kidney, feather, °stripe, classification, species, genus, °variety, °description: *Men of Randolph's stamp are hard to find among gentlemen and scholars alike.*

stampede *n.* **1** °rout, °flight, °scattering, °panic, °rush, °dash: *When the store opened its doors for its annual sale, we were nearly trampled in the stampede.* —*v.* **2** °panic, °frighten, °rush, °scatter, °rout: *Millions of buffalo were killed by stampeding them off cliffs.*

3 °rush, °run, °race, °charge, °take to one's heels, °flee, °take flight: *At the cry of "Fire!" the audience stampeded for the exits.*

stance *n.* °carriage, °bearing, deportment; °position, °posture, °attitude, °standpoint, °stand, °viewpoint, °point of view: *Legs apart and arms akimbo, she assumed a defiant stance at the door. Her stance was that she must protect the children at all costs.*

stanch *v.* °staunch, °stop, °stem, °halt, °check, °arrest, °stay, °end, °cease; °prevent: *Apply pressure with a pad to stanch the flow of blood from the wound.*

stand *v.* **1** °rise, °arise, °get up, *Brit* be upstanding: *We all stood when the master of ceremonies proposed a toast to McArthur.* **2** Sometimes, ***stand up.*** **(a)** °set, °place (upright), °position, °put, °move; upend: *Please stand the cabinet in the corner.* **(b)** °stay, °remain (standing): *Just for that you can go and stand in the corner!* **3** °endure, °survive, °tolerate, °brook, countenance, °face, confront, last through, °abide, °allow, °accept, °take, °suffer, °bear, °withstand, °undergo, °experience, °cope with, °brave, stand *or* bear up under, °stand for, °withstand, °stomach, °weather, °handle, *Colloq* °put up with: *I could not stand another year in the tropics. She can stand neither hypocrites nor vanilla ice cream.* **4** °continue, °remain, °persist, be *or* remain in effect *or* in force, °prevail, °obtain, °apply, °exist: *The law will stand as it is till it is rescinded or modified by the court.* **5** ***stand by.*** **(a)** °support, °defend, °back, stand *or* stick up for, stand behind, be *or* remain loyal *or* faithful to, °uphold, take the side of, °side with, °sympathize with, *US* go to bat for: *No matter what he did, she always stood by him.* **(b)** °wait (in the wings), stand *or* wait *or* stay *or* remain on the sidelines, be *or* stand ready *or* available *or* accessible, be *or* stand in readiness: *Would you mind standing by in case we need your help?* **(c)** stick to, adhere to, °support, °maintain, °persist in, affirm, reaffirm, °confirm, °abide by: *We shall stand by every word of the agreement.* **6** ***stand down.*** °resign, °quit, step aside, °withdraw: *When it was revealed that she was taking steroids, she agreed to stand down from the team.* **7** ***stand for.*** **(a)** °symbolize, betoken, °represent, °signify, °mean, be emblematic of, °exemplify, epitomize, °illustrate, °typify, °refer to, allude to: *The initials in P. G. Wodehouse's name stand for "Pelham Grenville."* **(b)** °support, °advocate, °favor, °sponsor, °promote, espouse (the cause of), °subscribe to, °back, °champion, lend support *or* one's name to, °second: *Norma has always stood for the principles of democratic freedom.* **(c)** campaign for, be *or* present (oneself) as a candidate for, *US* stump for, °run for: *Who will stand for chairman this year?* **(d)** See **3**, above. **8** ***stand in.*** °substitute (for), °understudy (for), °replace, °relieve, double for, take the place of, *US and Canadian* pinch-hit (for), *Colloq* cover for: *If you can't get to the rehearsal, I'll stand in for you.* **9** ***stand out.*** **(a)** be prominent *or* conspicuous *or* noticeable, be notable *or* noteworthy: *With her green spike hairdo and a safety pin through her cheek, Kiley really stands out in a crowd. His farewell speech stands out in my mind as one of his best.* **(b)** °protrude, °project, °stick out, jut out, °bulge, °obtrude, beetle, °overhang, °extend: *The balcony stands out over the village square.* **10** ***stand up.*** **(a)** °stand, °rise, °arise, get to one's feet, °get up: *Children should stand up when an adult enters the room.* **(b)** °endure, °last, °wear (well), °survive: *Today's products, with their built-in obsolescence, don't stand up to long use.* **(c)** °jilt, break *or* fail to keep an appointment with: *That wretch has stood me up for the last time!* **11** ***stand up for.*** °support, °defend, take the side of, °side with, °champion, °uphold, *Colloq* °stick up for: *I shall never forget how you stood up for me when they accused me of cheating.* **12** ***stand up to.*** **(a)** confront, °brave, °challenge, °encounter, °dispute, °question, °resist, °defy, °withstand: *He's a coward when it comes to standing up to authority.* **(b)** °resist, °defy, °withstand, °endure, °outlast, last through, °suffer: *This material will stand up to years of abuse.* —*n.* **13** °position, °attitude, °stance, °posture, °policy, °philosophy, °point of view, °viewpoint, °standpoint, °belief, °opinion, °sentiment, °feeling, °line: *I disagree completely with the Chancellor's stand on the economy.* **14** °defense, °resistance, °effort: *This campaign will be our last stand in trying to retrieve some of the market we lost.* **15** °stop, stopover, °halt, °stay; °performance, °show: *We do some one-night stands in the country before returning to Manchester.* **16** °counter, °booth, °stall, °table; wagon *or Brit also* waggon, barrow, °cart: *He sells fruit from a stand in the street.* **17** °rack, °frame, °bracket; hat stand, coat rack: *This stand is for displaying a china plate. I left my hat on the stand in your office.* **18** staging, °platform, dais, °stage, staging, bandstand, summerhouse: *We like to go to the stand in the park to hear the band play.* **19** copse, grove, wood, °thicket, °brake, wood, *Brit* spinney, coppice: *Near the house was a small stand of beeches.*

standard *n.* **1** criterion, °measure, bench mark, °model, °pattern, archetype, °touchstone, °yardstick, °gauge, °guide, guideline, paradigm, °paragon, exemplar, °example, °sample, °type, °ideal, *beau idéal,* °rule, canon, °law, °requirement, °precept, °principle: *The metric system has become the standard in many countries. People resent having imposed on them the standards of another culture.* **2** °mean, °average, °norm, °par, °level, rating: *With many luxuries now necessities, the standard of living has improved enormously. The accepted standard for this class is a grade of C.* **3** °flag, °banner, ensign, °emblem, °pennant, burgee, insigne (*singular of* insignia), guidon, gonfalon *or* gonfanon, labarum: *The black knight bore a curiously colored standard into battle.* **4** °pole, °post, stanchion, lamppost, column, °pillar, °support, °pedestal, °pier, °footing, (upright) bar *or* rod *or* timber: *The car went out of control and knocked down two lighting standards.* —*adj.* **5** accepted, approved, °definitive, defined, °authoritative, °official, required, regulative, regulatory, textbook: *Must we follow standard procedure? The tests were conducted according to the standard methods.* **6** recognized, °prevailing, °prevalent, °usual, °customary, °habitual, °orthodox, °set, established, °regular, °familiar, °ordinary, °traditional, °classic, °stock, °typical, °normal, °staple, °conventional, °universal: *To some, standard English is that used by the largest percentage of its native speakers.*

standardize *v.* °regiment, systematize, codify, °normalize, homogenize: *We could save money if we standardized the dimensions of the books we publish.*

standby *n.* **1** °supporter, defender, °backer, upholder, °partisan, °sympathizer, adherent, °stalwart: *His mother had always been his standby against the severity of his father.* **2** °substitute, surrogate, replacement, backup, °understudy, °second, *US and Canadian* °alternate: *An ambulance is kept as a standby in case the other breaks down or needs servicing.* **3** °resource, °support, replacement: *I knew I could always count on you as a standby in case I had trouble.*

stand-in *n.* °double, °substitute, stunt man *or* woman; surrogate, replacement, °standby, backup, °understudy, °second, *US and Canadian* °alternate: *Any time they had to shoot a dangerous scene, my stand-in went through the motions.*

standing *adj.* **1** established, °set, °standard, °conventional, °customary, °usual, °normal, °regular, °fixed, °permanent, continued, continuing, °regular: *The club has standing rules concerning use of the cardroom.*

2 °stagnant, motionless, unmoving, stationary, °still, °static: *Insects bred in the standing pools of rainwater.* 3 °continuous, °fixed, °ongoing, °perpetual, unbroken: *Michael placed a standing order for six cases of Beaujolais Nouveau every November.* 4 °erect, °upright, on one's feet, vertical, unseated: *Please remain standing till the signal is given to be seated.* —*n.* 5 °status, °rank, °station, °position, °place, °grade, °order, °level, °stratum: *Considering her standing in the party, Mrs. Miller ought to win the nomination easily.* 6 eminence, °prominence, repute, °reputation: *Farquhar is a man of considerable standing in the field of speleology.* 7 Usually, ***long standing.*** (considerable) age *or* longevity *or* experience *or* seniority *or* duration: *Ivan is an art expert of long standing.*

standoffish *adj.* °aloof, °haughty, °unsocial, °reserved, °cool, frosty, °withdrawn, °remote, removed, °distant, °detached, °unapproachable, °inaccessible, uncongenial, unfriendly, unsociable; Olympian, lordly, °pompous, *Colloq* highfalutin *or* hifalutin *or* highfaluting, snooty: *Oliver is much too standoffish to deign to come to my party.*

standpoint *n.* °viewpoint, °point of view, vantage point, °perspective, °position, °angle, °view: *From the standpoint of mice, cats do not make good pets.*

standstill *n.* (dead *or* full) stop, °halt: *The car came to a standstill with the front wheels overhanging a cliff.*

staple *adj.* 1 °basic, °elementary, °essential, °necessary, requisite, required, °vital, °indispensable, °critical, °fundamental, °primary, °principal, °main, °chief: *These people cannot afford even staple commodities like flour and salt.* 2 °standard, °usual, °habitual, °ordinary, °customary, °prevailing, °normal, °conventional, °universal: *The staple fare on television seems to consist of old and new sitcoms.* —*n.* 3 Often, ***staples.*** necessities, essentials, basics, fundamentals: *Could you pick up a few staples for me in the supermarket? Children ought to read the staples of the literature of their country.*

star *n.* 1 celestial *or* heavenly body; evening star, morning star, falling star, shooting star, comet; nova, supernova: *It was a beautiful clear night and the sky was full of stars.* 2 °celebrity, °personage, °dignitary, VIP, °name, °somebody, luminary, leading light, leading man *or* woman *or* lady, °lead, °principal, diva, prima donna, °hero, °heroine, °idol, superstar, *Technical slang* top banana, headliner, *Colloq* big shot, °(big) draw, celeb, big name: *Clancy knows many of the big stars personally.* —*adj.* 3 °principal, °major, °leading, °important, °celebrated, °famous, famed, °prominent, °eminent, °preeminent, °distinguished, °brilliant, °illustrious, unequaled, °peerless, °matchless, °incomparable, unrivaled, inimitable, unmatched, °unparalleled, °top, °foremost: *Tonight's star attraction at the hippodrome will be Washout, The Talking Horse.* —*v.* 4 feature *or* be featured; play *or* act *or* take the lead *or* the leading part *or* role: *The film stars Cary Stewart, who began his career starring in spaghetti westerns.*

stare *v.* 1 °gaze, °gape, goggle, °gawk, °watch, *Colloq* °rubberneck, *Slang Brit* gawp: *The people were powerless to do anything but stare as the molten lava engulfed the city.* —*n.* 2 fixed *or* blank look; goggle, °gaze: *Monica has a stare that I find vapid but Frank finds very engaging.*

stark *adv.* 1 °completely, °utterly, unqualifiedly, °wholly, °absolutely, °entirely, °totally, fully, °altogether, plainly, °obviously, °clearly, certifiably: *Is that an elephant on the beach, or am I going stark raving mad? "Those horses are stark naked!" exclaimed Mrs. Whitelouse.* —*adj.* 2 °plain, °simple, °Spartan, °severe, unembellished, °unadorned, °cold, °bare, °harsh, °hard, °grim, bald, °blunt: *The room looks less stark with curtains. Just give me the stark facts without all the elaboration.* 3 °harsh, °severe, °bleak, austere, °barren, °desolate, °dreary, °gray, depressing, ravaged, °empty, °vacant, *Literary* drear, *Colloq US* spooky: *As day dawned, I looked out over the stark landscape of the nuclear winter.* 4 °sheer, °complete, utter, °absolute, °perfect, °pure, °thorough, thoroughgoing, arrant, °unmitigated, °out-and-out, °downright, °outright, °total, unconditional, °unqualified, °clear, °plain, °evident, °obvious, °patent, °flagrant, °gross, °rank: *They were shocked by the stark poverty they encountered in Appalachia.*

start *v.* 1 Often, ***start off*** *or* ***up.*** °begin, °commence, get (something) going, get off the ground, °originate, °initiate, °open, set in motion, °activate, °embark on; set in motion, turn *or* switch on, crank up, *Colloq* kick off: *Start operations at once. I can't start my car.* 2 Often, ***start off*** *or* ***up*** *or* ***in.*** °arise, °come up, come to be *or* into being, °emerge, crop up, °develop, °begin, °commence, get under way, °originate: *How did this situation start?* 3 Often, ***start off*** *or* ***up*** *or* ***in.*** °go, °leave, °depart, get going, move (off *or* out *or* on), get under way, set off *or* out *or* forth, *Colloq* hit the road, get the show on the road: *You start and I'll follow shortly.* 4 Often, ***start in.*** °begin, °commence, get *or* start the ball rolling, get things under way, be on one's way, get going: *When I went in I found that she had started without me.* 5 °jump, °flinch, blench, quail, °shy, °recoil, wince, °shrink, °draw back: *He started when he saw her suddenly behind him in the mirror.* 6 cause to spring *or* leap *or* dart *or* jump *or* bound: *Any abrupt movement will start the plovers from cover.* 7 °establish, °found, °begin, °set up, °initiate, °institute, °create, °father, give birth to, beget: *It was her ideas that started the current trend in mystery novels.* 8 °bulge, °protrude, °stick out; *Colloq US* bug out: *Warping caused these boards to start from the floor. When he saw what was in his swimming pool, his eyes started from their sockets.* —*n.* 9 °beginning, °opening, °move: *This is not very good, but it's a start.* 10 °beginning, commencement, °opening, °outset, °onset, inception, startup: *The start of the race is set for ten o'clock.* 11 beginning(s), inception, °birth, °initiation, °onset, °rise, genesis, °creation, °emergence, °origin: *The blues had their start in New Orleans.* 12 °opportunity, °chance, °beginning; °help, °assistance, °aid, °backing, financing, sponsorship, °encouragement, *Colloq* °break: *Who gave you your fresh start when you were down and out?* 13 head start, °advantage, °edge, °lead, *Colloq* °(the) jump, *US and New Zealand* °drop (on someone): *Having the inside track gave her a start on the competition. If he has a ten-minute start, we'll never catch him.* 14 inauguration, °opening, °beginning, °initiation, *Colloq* kickoff: *When is the start of the salmon fishing season this year?* 15 beginning(s), founding, °foundation, °establishment, inception, °birth, °origin: *This company had its start in 1781.*

startle *v.* °frighten, °surprise, °scare, °disturb, unsettle, °upset, discompose, make (someone) jump, °jolt, °jar, °dismay, °perturb, °stun, take (someone) aback, °shock, °astound, °astonish, *Colloq* °shake up, give (someone) a turn, *US* discombobulate: *He jumped out from behind the door and startled me. She was startled when told she had to vacate the house at once.*

startling *adj.* °shocking, °terrifying, °frightening, astounding, astonishing, °awesome, staggering, jarring, °disturbing, °unsettling, upsetting, °amazing, surprising: *Miriam fainted when told the startling news.*

starved *adj.* 1 starving, °(extremely) hungry, °famished, °ravenous: *Humbert came in from the fields, starved and ready to eat a horse.* 2 ***starved*** *or* ***starving for.*** yearning for, dying for, hankering for, hungry *or* hungering for, pining for, longing for, burning for, craving, thirsting for *or* after, desirous of, aching for, *Colloq* hurting for: *I was starved for a pint of cold lager.* **3** ***starved of.*** deprived of, in need *or* want of, lacking, bereft of: *The poor orphans seemed starved of affection.*

state *n.* **1** condition(s), circumstance(s), °situation, state of affairs, °status, °shape, °position: *He was concerned about the state of the company's finances.* **2** °structure, °form, constitution, °shape, °phase, °stage: *One does not see iodine in its gaseous state outside the laboratory.* **3** °grandeur, °pomp, °style, °splendor, magnificence, °glory, °brilliance: *The prince was met at the airport and escorted in state to the palace.* **4** °nation, °country, °land, body politic: *No longer a colony, it became an independent state in 1952.* *—adj.* **5** governmental, °government, °national, federal: *That is a local community problem, not a state issue.* **6** °ceremonial, °formal, °dignified, °stately, °solemn, °official; °royal, °regal, °imperial, °majestic: *We attended a state dinner at the White House. The Queen traveled in the state coach from Buckingham Palace to Westminster.* *—v.* **7** aver, assert, asseverate, °declare, affirm, °express, °report, articulate, °voice, °specify, delineate, °claim, °maintain, °allege, °submit, °confirm; °say, °testify, °hold, °have: *He has often stated his firm belief in democracy. The defendant stated that she was not in the house on the night of the murder.*

stately *adj.* °dignified, august, °solemn, °distinguished, °impressive, °striking, °imposing, °awesome, °grand, °lofty, °elevated, °noble, °majestic, °regal, °royal, °imperial: *The stately procession wound through the streets of the capital.*

statement *n.* °assertion, °allegation, °declaration, °expression, °report, °account, affirmation, asseveration, averral, °announcement, annunciation, °proclamation, utterance, communication, communiqué, disclosure: *Your statement is subject to interpretation.*

static *adj.* **1** °immovable, immobile, unmoving, motionless, stationary, °fixed, °stagnant, °inert, °still, unchanging, unchanged, °changeless, unvarying, °invariable, °constant: *These brackets are static but those pivot freely. The situation has remained static for several years.* *—n.* **2** °interference, °noise, atmospherics; difficulty *or* difficulties, °trouble, problem(s), *Colloq* °flak: *With all the static, I couldn't understand what she was saying. You can expect some static about being late for your wedding.*

station *n.* **1** °place, °position, °spot, °post, °site, °location: *The sentry is not to leave his station till relieved.* **2** °position, °place, °status, °rank, °caste, °standing, °class, °level: *As a doctor, he attained a station in life far above that of his father.* **3** railway station, train station, passenger station, bus station, *US and Canadian* depot: *Let's meet in the waiting room in the station.* *—v.* **4** °position, °place, °spot, °post, °site, °locate, °assign, °appoint, garrison, °install, *Colloq* billet: *The State Department stationed me in Poona for six years.*

stationery *n.* writing paper, letterhead(s), paper and envelopes, writing implements *or* supplies; office supplies *or* equipment: *We consider stationery costs as part of overhead.*

statue *n.* °sculpture, °figure, figurine, statuette, carving, casting, °model, bronze, °image, icon *or* ikon, effigy, °representation; bust, atlas, caryatid, colossus, °figurehead, *Bible* graven image: *A statue of Disraeli stands in the market square of Aylesbury.*

statuesque *adj.* °imposing, °impressive, °majestic, °regal, °stately, °magnificent, °noble, °dignified, august, °grand, well-proportioned, °comely, °handsome, queenly, Junoesque: *Julie's statuesque figure was well set off by the draped silk gown.*

status *n.* **1** eminence, °prominence, °preeminence, °standing, stature, °importance, °significance, repute, °reputation, °rank, °station: *Few could approach Keith's status as a pathologist.* **2** See **standing, 5,** above.

staunch *adj.* **1** °steadfast, °loyal, °firm, unflinching, °steady, unshrinking, unswerving, dependable, °reliable, °(tried and) true, °devoted, trueblue, trusty, trusted, °faithful, unfaltering, undeviating, unwavering: *Cargrove has always been a staunch party supporter.* **2** °strong, °solid, °sturdy, °sound, well-built, °stout, °substantial, well-constructed, well-made, °tough, °rugged, long-lasting; °watertight, seaworthy: *That staunch oaken hull has withstood much abuse over the years.*

stay[1] *v.* **1** °remain, °stop, °continue, °tarry, °wait, °stand, *Colloq* °freeze: *Stay where you are or I'll shoot!* **2** °remain, °stop, °lodge, °sojourn, °abide, reside, °dwell, °live, °visit: *I heard that Sheila was back in town, staying at her aunt's.* **3** °keep, °remain, continue to be: *I was having trouble staying awake.* **4** °stop, °arrest, °thwart, °prevent, put an end to, °halt, °interrupt, °block, °check; °curb, °retard, °slow, °impede, °foil, °obstruct, °hamper, °hinder, °discourage, °deter; °delay, °postpone, °put off, °discontinue, °defer, *Technical* prorogue: *Only one man has the authority to stay the execution. What can be done to stay the advance of the killer bees?* **5** °linger, loiter, °wait, °tarry, °stop, °remain, *Archaic* bide: *I like this part of the world and plan to stay here a while.* *—n.* **6** °stop, stoppage, °arrest, °setback, °check, °halt, °prevention, discontinuance, discontinuation, °interruption, blockage, °delay, °postponement, deferment, deferral, °reprieve: *Have you been able to arrange a stay in carrying out the sentence?* **7** stopover, °sojourn, °visit, °stop: *We really enjoyed our stay at Fred's house in Fort Lauderdale.*

stay[2] *n.* **1** °guy, °line, °rope, °cable, °chain, °support, °brace, °reinforcement; *Technical* headstay, (running) backstay, forestay, °mainstay, mizzenstay: *If one of those stays gives way, the entire structure may fall.* *—v.* **2** °support, °strengthen, °secure, °reinforce, °brace, °buttress, gird, shore (up): *The mast is stayed by two steel cables, fore and aft.*

steadfast *adj.* °resolute, °determined, persevering, resolved, °single-minded, °steady, unflinching, unfaltering, unwavering, unswerving, indefatigable, dependable, °immovable, °stable, °firm, °fixed, °constant, °persistent, unflagging, °tireless, °enduring, dedicated, deep-rooted, °faithful, °true, °loyal, °staunch: *We admire how steadfast Janet has been in supporting our cause. Sheldon was a steadfast friend and will never be forgotten.*

steady *adj.* **1** °stable, °firm, °solid, °substantial, °sound, °stout, °strong: *Is this chair steady enough to stand on?* **2** °even, °regular, °uniform, °habitual, °invariable, unvarying, unfluctuating, unwavering, undeviating, °changeless, unchanging, °continuous, °constant; °perpetual, °nonstop, around-the-clock *or* round-the-clock, °persistent, uninterrupted, unbroken, unrelieved, unceasing, ceaseless, incessant, °relentless, unremitting, never-ending, unending, °endless: *Steady trade winds could be relied on to carry vessels to the Caribbean. Inflation has remained steady for a year. How do the children manage to survive on a steady diet of junk food? The economists fear a steady rise in inflation.* **3** unflinching, unblinking, °fixed, °constant, unfaltering, °continuous, °direct: *The boy began to quail under the headmaster's steady gaze.* **4** °calm, °cool, balanced, °equable, controlled: *Steady nerves are needed to handle this new breed of fighter plane.* **5** °devoted, °firm, °staunch, °faithful, °loyal, longstanding, inveterate, °consistent, confirmed, °persistent: *The Pendergasts have always been steady supporters of the museum.* **6** °staid, °sedate, °sober, °dignified, °poised, °sophisticated, civilized, °sensible, down-to-earth, settled, °serious, °levelheaded, °reliable, *Colloq* unflappable: *WANTED—Steady person as housekeeper to eccentric editor.* —*adv.* 7 °firmly, solidly: *Hold the ladder steady while I climb up to the roof.* **8 *go steady.*** keep company, °date, °socialize: *Is Jane still going steady with Hubert?* —*n.* 9 boyfriend, girlfriend, (regular) fellow *or* girl, °sweetheart, *Colloq* °guy, gal, °woman, °man: *Yes, Hubert is still Jane's steady.* 10 °regular, °habitué, °customer, frequenter, familiar face: *Gil has been a steady here ever since we opened.* —*v.* 11 stabilize, hold fast; °brace, °secure, °support, °strengthen: *Prices steadied after the first hour of trading. To steady the table, merely tighten the screws holding the legs.*

steal *v.* 1 °take (away), °appropriate, filch, shoplift, °pilfer, purloin, make *or* walk off *or* away with, get away with; °embezzle, °misappropriate, peculate; *Colloq* °lift, °pinch, °hook, snitch, °borrow, *US* °boost, °liberate, heist, hijack *or* highjack, *Slang* °swipe, *Brit* °nick, °prig, *US* °hoist: *The thieves stole only the emeralds, leaving the diamonds. The bookkeeper stole the money by diverting it into his own account.* 2 plagiarize, °pirate, °copy, °imitate, °appropriate, usurp, °take: *Howalls claims that the story of the film was stolen from his short story.* 3 °sneak, °creep, °slip, tiptoe, °prowl, °lurk, skulk, *Colloq* °pussyfoot: *Silently, we stole into the garden at the rear of the house.* —*n.* 4 °bargain, °(good) buy, *Colloq* giveaway: *At that price, the rug was a steal!*

stealing *n.* °theft, °robbery, robbing, larceny, pilferage, shoplifting, poaching, °embezzlement, peculation, thievery, thieving, filching, burglary, °plagiarism, plagiarizing, piracy, pirating: *Stealing from the church poor box must be one of the lowest things a person can do.*

stealth *n.* furtiveness, °secrecy, clandestineness, surreptitiousness, sneakiness, slyness, underhandedness: *What the thieves lacked in stealth they compensated for in knowledge of art.*

stealthy *adj.* stealthful, furtive; °secretive, °secret, °sly, clandestine, °surreptitious, °sneaky, °sneaking, skulking, covert, °undercover, underhand(ed), backstairs, huggermugger, closet: *In her movements she was stealthy as a cat. They were stealthy collectors of pornography.*

steamy *adj.* 1 °humid, steaming, °damp, °moist, °muggy, °sticky, dank, sweaty, °sweltering, sodden, °sultry, boiling, °wet: *We were not prepared for the steamy jungle of equatorial Africa.* 2 steamed (up), fogged (up), befogged, °misty, misted, °hazy, clouded, cloudy, beclouded, °dim, blurred: *Every time I take a hot shower, the bathroom mirror gets all steamy.* 3 °erotic, °passionate, °(sexually) exciting, arousing, °hot, *Colloq* °sexy, *Slang* horny: *The film was notorious for its steamy scenes.*

steel *n.* 1 sword, °dagger, °blade, °knife, dirk, stiletto: *When the foeman bares his steel—Tarantara! Tarantara! / We uncomfortable feel.* —*v.* 2 °brace, °nerve, °stiffen, °fortify, grit one's teeth, °bear up, bite the bullet, screw up one's courage (to the sticking point); inure, °insulate, °protect: *People must learn to steel themselves against criticism in this business.*

steely *adj.* 1 grayish, °gray: *His steely blue eyes pierced deep into her soul.* 2 iron, °tough, indurate, adamant(ine), °hard, °strong, °rugged, unyielding, flinty, °sturdy: *Her children battled her steely determination to leave her entire fortune to her cat.*

steep[1] *adj.* 1 °sheer, °abrupt, °precipitous, °bluff, °sharp, nearly vertical *or* perpendicular *or* upright: *In those days, a car had to be in first gear to climb such a steep hill.* 2 °expensive, °dear, °high, overpriced, °exorbitant, °excessive, °extravagant, extortionate, *Colloq* °stiff: *The house prices in that area were much too steep for all but the wealthiest people.*

steep[2] *v.* 1 °soak, °submerge, souse, °drench, °immerse, °saturate, douse, °wet, ret; pickle, marinate: *Before cooking, the meat has to be steeped in brine for at least six hours to tenderize it.* 2 imbue, °fill, °saturate, °immerse, inundate; °bury: *He learned Japanese by steeping himself in the language for six months.*

steer *v.* 1 °guide, °pilot, °conduct, °direct; °manage, °control, °channel: *Steer the boat closer to the pier. David has steered the company to greater profits than ever before.* **2 *steer clear of.*** °avoid, °dodge, keep away from, °shun, circumvent, give (something *or* someone) a wide berth: *You'd best steer clear of Melissa when she's angry.* —*n.* 3 Usually, ***bum steer.*** (bad *or* poor) tip *or* suggestion *or* hint; (bad *or* poor) guidance *or* advice *or* information: *He avoids me because I once gave him a bum steer on a horse.*

stellar *adj.* 1 astral, °star, sidereal: *The sidereal year is based on stellar calculations.* 2 °chief, starring, °principal, °leading, °main, headlining: *For years Lee has been a stellar performer on the golf circuit.*

stem[1] *n.* 1 °trunk, °stalk, °stock; *Technical* peduncle, pedicel, petiole, °shoot: *Three white blossoms are borne on each stem of the plant.* 2 bows, prow, stempost: *The ship shook from stem to stern.* —*v.* 3 °come, °arise, °develop, °derive, °issue, °flow, °generate, °originate, °spring, °emanate, °sprout, °grow, °descend, °result, °proceed: *The dispute stems from basic differences in the ways the parties regard property.*

stem[2] *v.* 1 °check, °stop, °halt, stanch *or* staunch, °arrest, °stay, °curb, °control, °quell, °suppress; °retard, °slow, lessen, °diminish, °reduce, cut (back (on)): *The government introduced legislation to stem immigration. I was able to stem the bleeding by applying the tourniquet.* **2 *stem the tide (of).*** °resist, °withstand, go *or* make headway *or* advance *or* make progress against, prevail over *or* against: *It is exceedingly difficult to stem the tide of public opinion.*

stench *n.* stink, °reek, noisomeness, mephitis, fetor *or* foetor, foul odor, effluvium, *Colloq Brit* pong: *A terrible stench emanated from beneath the floorboards in the barn. For decades their government has wallowed in the stench of corruption.*

stenographer *n.* secretary, amanuensis, stenotypist, tachygrapher, phonographer: *The court stenographer read back part of the testimony.*

stenography *n.* shorthand, stenotypy, tachygraphy, speed writing: *I suppose that people do, still, study stenography.*

step *n.* 1 °movement, °move: *One step forward and two steps back is not what we call "Ballin' the Jack."* 2 footfall, °footstep, tread: *I think I hear father's step on the stair.* 3 °footstep, footprint, °trace, spoor, °track, °trace, °mark, °impression; imprint, °vestige: *In his master's steps he trod, where the snow lay dinted.* 4 °action, °initiative, °measure, °activity, °procedure, °move, °motion: *What steps are needed to improve the situation?* 5 °stage, °move, gradation, °degree, °progression: *Can we not proceed from one to the other in easy steps?* 6 °pace, °footstep, stride: *My mother's cottage is just a few steps away, at the bottom of the garden.* **7 *in step (with).*** in keeping (with), in harmony *or* agreement (with), °harmonious (with), °agreeable (with), according (with *or* to), concordant (with), attuned (to), in tune (with), consonant (with), °consistent (with), °appropriate (to), °fitting (for); °conventional, °traditional, °routine: *Do you think that her ideas of discipline are in step with the times? I am not sure they are in step at all.* **8 *out of step (with).*** out of keeping (with), out of *or* not in harmony *or* agreement (with), not harmonious (with), not agreeable (with), not according (with *or* to), °discordant (with), not concordant (with), not attuned (to), out of tune (with), not consonant (with), °inconsistent (with), °inappropriate (to), not fitting (for); °offbeat, unconventional, °eccentric, *Slang* °kinky: *You must admit that her views are out of step with the committee's.* **9** ***step by step.*** °gradually, a step at a time, slowly, steadily: *The way to unravel the problem is to analyze it step by step.* **10** ***steps.*** **(a)** °course, °way, °route, °direction, °path, °movement, °passage; °journey, journeying, °travels, °traveling: *After leaving the village, I directed my steps southward, toward the coast.* **(b)** stairway, stairs, stair, staircase, stepladder, *US and Canadian* °stoop: *Claire walked down the steps to greet me.* **11** ***take steps.*** °proceed, °move, begin *or* start *or* commence to act *or* to take action, do something: *We must take steps to prevent this from happening again.* **12** ***watch one's step.*** tread carefully *or* cautiously, be cautious *or* careful, exercise care *or* caution, be wary *or* discreet, be on the qui vive, be *or* remain alert, be on one's guard, have *or* keep one's wits about one, take care *or* heed, *Colloq* pussyfoot about: *You must really watch your step with Marnie to avoid upsetting her.* —*v.* **13** °move, °walk, °look; °pace, stride: *Step lively or you'll miss your last chance to see the elephants.* **14** ***step down.*** **(a)** °resign, °abdicate, °quit, bow out, °retire: *Don't you think it time you stepped down from the chairmanship?* **(b)** °decrease, °diminish, °reduce: *Using this transformer, we can step down the voltage gradually.* **15** ***step in.*** °intervene, °interfere, intercede, become involved: *It seemed the right time for us to step in and take over the company.* **16** ***step on it.*** °hurry (up), make haste, °hasten, °speed up: *He'd better step on it if he is going to catch his plane.* **17** ***step out.*** **(a)** go outside *or* out of doors, °leave: *Would you mind stepping out for a few minutes while we settle this in private?* **(b)** °go out, °socialize: *Notice how quickly Genevieve has started stepping out again after her bereavement.* **(c)** become disinvolved, °withdraw, °secede: *Our firm stepped out of the negotiations when we saw who was bidding.* **18** ***step up.*** **(a)** °improve, °progress: *Gurrey has certainly stepped up in the world since I knew him at college.* **(b)** °increase, accelerate, °raise, °intensify, °boost, escalate, up, °speed up: *They stepped up the pace till she could no longer keep up with them.*

sterile *adj.* **1** °barren, °fruitless, unfruitful, childless, unproductive, °infertile, infecund: *The first great disappointment of Napoleon's life was that Josephine was sterile.* **2** °pure, aseptic, uninfected, unpolluted, uncontaminated, disinfected, °sanitary, sterilized, germ-free, antiseptic: *Make sure that you always use a sterile bandage for a wound.* **3** °barren, unproductive, °stale, effete: *Vanessa's greatest fear was that her mind would become sterile and she would run out of ideas.*

sterilize *v.* **1** °purify, °disinfect, °cleanse, °clean, °fumigate, depurate, *Technical* autoclave: *Sterilize the instruments before using them.* **2** castrate (males), emasculate (males), geld (horses), spay (female animals), °alter (animals), °neuter (animals), caponize (male fowl), eunuchize (males), *Technical* ovariectomize (females), vasectomize (males), *Colloq* °fix (animals), °cut (male animals), *Slang* tie (someone's) tubes: *Steers—that is, sterilized bulls—yield tenderer meat.*

sterling *adj.* **1** °genuine, °authentic, °real, °true, °pure: *Is this candle-snuffer sterling silver?* **2** °excellent, °superior, °superb, °superlative, first-class, °exceptional, °matchless, °peerless, unequaled, °nonpareil, °incomparable, °fine, very good, °worthy, °estimable, °admirable: *Commander Ian Johnston has acquitted himself as an officer of sterling character.*

stern *adj.* **1** austere, °severe, °strict, stringent, °demanding, °critical, °rigid, rigorous, flinty, °steely, °authoritarian, uncompromising, °hard, °tough, °inflexible, °firm, °immovable, °unmoved, unrelenting, unremitting, °steadfast, °resolute, °determined, unyielding, adamant, adamantine, obdurate, hard-hearted, °stony, stonyhearted, unsparing, unforgiving, °unsympathetic, °harsh: *Discipline in the French Foreign Legion is said to be quite stern.* **2** °serious, frowning, °grim, °forbidding, °grave, °gloomy, °dour, °somber, saturnine, lugubrious, °gruff, °taciturn, crabby, crabbed, crusty, churlish, °sour: *Beneath that terribly stern exterior he really is a pussycat.*

stew *n.* **1** gallimaufry, goulash, salmagundi, °hash, °mess, olla podrida, olio, °mixture, °mishmash, *Brit* hotchpotch, *US also* °hodgepodge: *His book is a stew of many different opinions, none of them his own.* **2** state of excitement *or* alarm *or* anxiety, dither, pother, °bother, °lather, °sweat, *Colloq* tizzy, °state: *She really worked herself up into a stew over the boy Paula is engaged to.* —*v.* **3** °simmer, °seethe, agonize, °fret, dither, °chafe, °burn, °smolder, *Colloq* get steamed (up) (about *or* over), work (oneself) (up) into a sweat *or* lather *or* state (over): *Paula's father is all stewed up over her leaving school to get married.*

stick[1] *v.* **1** °pierce, °thrust, °stab, °transfix, °pin, °spike, °impale, spear, °spit, °run through, °poke, °gore, °jab, °prick, °puncture, °punch, °penetrate, °drill, °bore, °riddle, °perforate: *He stuck the wild boar with his spear.* **2** °put, °drop, °place, °deposit, *Colloq* shove, plonk, plunk, plop: *Stick another ice cube in my drink, would you?* **3** °put, °poke, °push, °thrust, °prod, °dig; °insert: *If you stick your head through that hole, you'll get hit in the face with a custard pie. Stop sticking your finger in my ribs!* **4** °attach, °fasten, affix, °fix, °nail, °pin, °tack; °glue, °cement, paste, gum, °weld, solder, °bind, °tie, °tape, wire; °bond, °melt, °fuse, °unite, °join: *What shall we used to stick the poster to the wall? How can I stick the pieces of the vase together again?* **5** Often, ***stick together.*** cohere, adhere, stay *or* remain *or* cleave *or* cling together: *I cannot make these parts stick together.* **6** °hold, °last, °endure, °go through, be upheld, be *or* remain effective, remain attached: *The prosecutor was unable to make a charge of murder stick.* **7** °linger, °dwell, °remain (fixed), °continue, °stay; be *or* become lodged *or* stopped *or* fixed *or* fast *or* immovable *or* stationary, be *or* become entangled *or* enmired *or* bogged down: *Something sticks in my mind about your leaving next week. We were stuck in the Sunday traffic for hours. The wheel is stuck in the sand.* **8** °burden, °weigh down, °encumber, saddle with, °charge, °impose on, force on: *We stuck Tony with the nasty job of changing the tire.* **9** baffle, °puzzle, °bewilder, °perplex, °confuse, °stump, °stop, °nonplus: *I was totally stuck for a solution.* **10** °stand, °abide, °tolerate, °endure, °bear: *I can't stick people watching me while I am painting.* **11** ***stick around*** *or* ***about.*** °wait, °tarry, °linger, °stay, °stand by, °remain, *Colloq* hang around *or Colloq* about *or Colloq* on: *Can you stick around for a few minutes after the meeting?* **12** ***stick at.*** stop at, hesitate at, pause at, scruple at, be deterred *or* put off by, °take exception to, shrink from *or* at, balk at: *Barnes sticks at nothing to get his way.* **13** ***stick by.*** °support, be loyal *or* faithful to, °stand by: *Arnold will stick by you, come what may.* **14** ***stick it (out).*** °persevere, °persist, stand fast, bear it, be resolute, °soldier on, hold (one's) ground, grin and bear it, see it through, weather it, *Colloq US* tough it out: *It was a very hard job, but I stuck it out to the very end.* **15** ***stick out*** *or* ***up.*** °protrude, °jut (out), °extend, °project, °poke (out); °bulge, °obtrude, °stand out, °overhang, beetle: *Stick out your tongue. What is sticking out of your ear? Balconies stick out from all sides of the building.* **16** ***stick together.*** **(a)** °unite, °unify, °join (forces), consolidate, °merge, confederate, °amalgamate, °cooperate, work together: *We must stick together if we are to accomplish anything. The family sticks together in time of crisis.* **(b)** See **5,** above. **17** ***stick up.*** **(a)** °rob, °mug, *Colloq* °hold up, *US* heist: *They stuck up a bank courier this morning, in broad daylight!* **(b)** °put up, °post, affix, °display: *We went round town sticking up posters for our candidate.* **18** ***stick up for.*** rally to the support of, °support, stand by *or* up for, °defend, speak for *or* in behalf of, take up the cudgels for: *A person must stick up for what he thinks is right.* **19** ***stick with.*** °persevere, °persist, stay *or* remain *or* continue with, not change one's mind about: *Stick with me, kid, and you'll wear diamonds. I'll stick with the smoked eel as a starter.*

stick[2] *n.* **1** °stake, °twig, °branch, baton, °wand, °staff, °rod, cane, °pole, pike, walking stick: *We put a stick in the ground to mark the place.* **2** °person, °man, °fellow, °chap, *Colloq* °guy, *Brit* geezer, bloke: *Desmond isn't such a bad old stick after all.* **3** ***the sticks.*** the country, the provinces, the countryside, the backwoods, the bush, *Brit* the hinterland *or US* the hinterlands, *Australian* the outback, *US* the boondocks, the boonies: *He hates the city and now lives somewhere in the sticks.* **4** ***wrong end of the stick.*** °misunderstanding, misreading, misconstruction, misinterpretation: *When it comes to understanding a regulation, Ed always seems to get hold of the wrong end of the stick.*

stick-in-the-mud *n.* (old) fogy *or* fogey, °conservative, °anachronism, *Colloq* fuddy-duddy, fossil, °square, back number: *Get up and boogie, Man; don't be such a stick-in-the-mud!*

sticky *adj.* **1** gluey, gummy, viscous, °tacky, glutinous, viscid, *Colloq* °gooey: *Children, please keep your sticky fingers off the car windows.* **2** °awkward, °ticklish, °tricky, °sensitive, °delicate, uncomfortable, discomfiting, discomforting, °embarrassing, *Slang* °hairy: *Inviting Steve with his ex-wife might be a bit sticky, don't you think?* **3** °humid, °clammy, dank, °damp, °muggy, °close, °sultry, °oppressive, °sweltering: *It's so hot and sticky, why not turn on the fan?*

stiff *adj.* **1** °firm, °rigid, inelastic, unbending, °inflexible, °hard, unbendable, °tough, °solid, solidified, stiffened, unyielding, °brittle: *The table napkins were so stiff with starch that I almost cracked one.* **2** °severe, °harsh, °punitive, °hurtful, °punishing, °abusive, torturous, distressing, afflictive, °painful, °overwhelming, °unbearable, tormenting, °merciless, °excruciating, °cruel, °drastic, *US* cruel and unusual: *The government here has stiff penalties for drug traffickers.* **3** °strong, °potent, °powerful, °overpowering, °alcoholic: *After that kind of an ordeal, you could use a stiff drink, I'm sure.* **4** °vigorous, °energetic, °staunch, dogged, °tenacious, °resolute, resolved, °determined, °stubborn, °obstinate, unyielding, °indomitable, °relentless: *They met with stiff opposition in trying to capture the fort.* **5** °strong, °steady, °powerful, °fresh, °brisk, °spanking, gusty, °forceful, howling: *We had to reduce sail because of a stiff westerly wind.* **6** °excessive, °exorbitant, °high, °steep, °expensive, °dear: *They are asking a pretty stiff price these days for a bottle of good vintage port.* **7** °cool, °haughty, °rigid, °wooden, °stuffy, °aloof, °tense, °intense, unrelaxed, °forced, °pompous, °stilted, °mannered, °ceremonious, austere, °formal, °chilly, °cold, unfriendly, °standoffish, °reserved, °snobbish, *Colloq* snooty, *Slang* uptight: *Vince is warm and friendly, but his wife is as stiff as a poker.* **8** °stilted, unrelaxed, °wooden, °forced, °artificial, °labored, °pedantic, turgid, °formal, prim, *Colloq* °stuffy: *I have always found her writing to be rather stiff.* **9** °difficult, °hard, °steep, uphill, °laborious, °arduous, tiring, fatiguing, °exhausting, °harrowing, °toilsome, rigorous, challenging, *Colloq* °rough, °tough: *It is quite a stiff climb to the top of the pyramid of Cheops. That was a stiff homework assignment.* **10** °solid, semisolid, °firm, °hard, °thick, °dense, °compact: *If you add a bit more water the next time, the jelly won't get quite so stiff.* —*n.* **11** °corpse, °body, °cadaver: *Bancroft had to go down to the morgue to identify some stiff they hauled out of the river.* **12** skinflint, °miser, *Colloq* cheapskate, *Slang* piker, *US and Canadian* tightwad: *A stiff is a customer who fails to leave a tip.*

stiffen *v.* **1** °thicken, °coagulate, clot, °harden, °jell, °set, °solidify, congeal, crystallize: *Beat the egg whites with a whisk till they stiffen.* **2** °brace, °reinforce, tauten, rigidify, toughen, °strengthen: *The weak support needs stiffening with a steel bar.*

stifle *v.* **1** suffocate, °smother, °choke, strangle, throttle, asphyxiate: *The firemen were almost stifled by the smoke from the chemical fire.* **2** °choke back, keep *or* hold back, °withhold, °repress, °suppress, °hold in, °restrain, °prevent, °curb, °cover up, °control: *I stifled a yawn as Cromarty went on about his grandchildren.* **3** °destroy, °crush, °demolish, °extinguish, °stamp out, °kill, °quash, °silence, °stop, °check: *Under his tyrannical rule all artistic creativity was stifled for fifty years.*

stigma *n.* °brand, °(bad) mark, °blot, smirch, °stain, °spot, °taint, °blemish, demerit, blot on the escutcheon, *Brit* blot in one's copybook: *That one small misstep remained a stigma on his reputation till he died.*

stigmatize *v.* °brand, °mark, °scar, °blemish, besmirch, °sully, °disparage, °depreciate, °denounce, °condemn, calumniate, defame, pillory, °slander: *His foul treachery stigmatized his entire family.*

still *adj.* **1** °quiet, °serene, placid, °calm, °tranquil, motionless, unmoving, °peaceful, pacific, at rest, quiescent, °even, °flat, °smooth, °inert, stationary, undisturbed, unruffled: *In the moonlight, the lake lay still and dark.* **2** °silent, °quiet, °noiseless, soundless; hushed, °restful, *Literary* stilly: *Be still or they'll hear you! In the still night I heard the distant sound of an owl.* —*n.* **3** stillness, °hush, °quiet, °silence, tranquillity, noiselessness, peacefulness, °calm: *In the still of the evening came the call of the nightingale.* —*adv.* **4** even now, to *or* till *or* until this *or* that time, (up) till *or* until now, °yet: *And still he gazed, and still the wonder grew, / That one small head could carry all he knew. Henry came for dinner five years ago, and he's here still. Do you still smoke?* **5** °even, °in addition: *Hobart weighed three hundred pounds when you last saw him, but he's heavier still today.* **6** °notwithstanding, °yet, even then: *Are you still going to Sue's party, despite what she said?.* **7** motionless(ly), quiet(ly), silent(ly), stock-still: *Lie still while I put on this bandage.* —*conj.* **8** °however, but, °notwithstanding, °nevertheless, °even so, °in any event, °in any case: *She said she'd be here; still, you never know.* —*v.* **9** °calm, allay, assuage, alleviate, °relieve, °silence, °lull, quiet *or chiefly Brit* quieten, pacify, soothe, mollify, appease, °subdue, °suppress: *A few kind words and she had stilled my fears.*

stilted *adj.* °awkward, °ungraceful, graceless, °clumsy, °wooden, °stiff, turgid, °affected, °artificial, °unnatural, °mannered, °labored; °pretentious, °formal, °pompous, °lofty, °bombastic, grandiloquent, high-flown, °inflated: *Latham has a stilted way of expressing himself.*

stimulant *n.* **1** stimulus, °incentive, °provocation, °spur, °prompt, goad, °urge, °prod, fillip, °impetus, °incitement, °drive, impulse, °push, °pull, °draw: *She yielded readily to the stimulants of literature, science, and the fine arts.* **2** energizer, antidepressant, °tonic, restorative, *Colloq* bracer, pick-me-up, °shot in the arm, *Slang* pep pill, °upper, bennie, °speed: *He found he couldn't get through the day without a stimulant.*

stimulate *v.* **1** °rouse, °arouse, waken, °awaken, °wake up, °excite, °incite, °inspire, °encourage, °spur, °quicken, °fire, °fuel, °nourish, °activate, whip *or* stir up, goad, galvanize, °jolt, inspirit: *What can be done to stimulate students to study history? Teachers must find ways to stimulate students' interest.* **2** °increase, °encourage, °prompt, °provoke, °quicken: *This drug stimulates the circulation of the blood.*

stimulating *adj.* °exciting, inspirational, inspiring, arousing, °stirring, animating, °exhilarating, °provocative, thought-provoking: *I cannot recall having spent a more stimulating evening in the theater.*

sting *v.* **1** °prick, °stab, °pierce, °stick; °bite: *Bees, wasps, scorpions, etc. sting you, but mosquitoes, spiders, and snakes bite.* **2** °hurt, °wound, °pain, °injure, °distress, nettle, cut to the quick: *David was really stung by Maria's callous remarks.* **3** See **stimulate, 1,** above: *We were stung into action by the minister's inspiring speech.* **4** °cheat, overcharge, °swindle, °fleece, °defraud, *Slang* °rob, °soak, °rip off, °take for a ride: *You really were stung if you paid that much for such an old piano.*

stinker *n.* °wretch, °villain, °scoundrel, cad, °heel, °beast, cur, viper, snake in the grass, skunk, swine, polecat, *Somewhat old-fashioned* blackguard, °rogue, *Archaic* knave, varlet, dastard, (base) caitiff, *Colloq* stinkpot, louse, °creep, rat, *Brit* nasty piece of work, sod, *Old-fashioned* rotter, bounder, blighter, *Slang* (rotten) bastard, son of a bitch, *Brit* toerag, °bugger, *US* SOB, °bum, stinkeroo *or* stinkaroo, *Taboo slang* shit, *Brit* arse-hole, *US* asshole: *That stinker told the teacher on us.*

stinking *adj.* **1** foul-smelling, °smelly, fetid *or* foetid, mephitic, °rank, noisome, malodorous, reeking, °putrid, miasmal *or* miasmatic(al) *or* miasmic, °rancid, gamy, *Colloq Brit* pongy, whiffy: *They were kept in that stinking dungeon with little food for a month.* **2** °wretched, °villainous, °beastly, °vile, °contemptible, °low, °despicable, °mean, °nasty, °disgusting, °rotten, °terrible, °awful, *Old-fashioned* dastardly, *Colloq* °lousy, *Taboo slang* shitty: *Taking my clothes while I was swimming was a stinking thing to do.* **3** drunken, drunk (as a lord *or US also* a skunk), intoxicated, inebriated, (be)sotted, °under the influence, over the limit, °high, °maudlin, tipsy, woozy, *Colloq* pie-eyed, °loaded, in one's cups, °under the weather, three sheets to the wind, *Slang* sozzled, soused (to the gills), potted, plastered, smashed, bombed, pissed, boozed, boozy, tanked, stoned, canned, *US* in the bag: *The last time I saw Bob, he was so stinking I had to pour him onto his train.*

stint *n.* **1** °share, °quota, °allotment, °bit, °assignment, °stretch, °shift, °term, °time, °job, chore, °task, °routine, °turn, °tour, °duty, °responsibility, °obligation, °charge: *Officer Muldoon completed her daily stint of awarding parking tickets to deserving motorists.* **2** °control, °curb, °limit, limitation, °restriction, °check, °restraint, constraint, °condition, °qualification, °reservation: *The committee is free to exercise without stint its power to raise membership fees.* —*v.* **3** °control, °curb, °limit, °restrict: *Don't stint yourself, lunch is on the firm.* **4** skimp, scrimp, be stingy *or* cheap *or* penurious *or* parsimonious *or* sparing *or* cheeseparing *or* frugal, °hold back (on), °withhold, °economize, °pinch (pennies), cut corners, *Colloq Brit* be mingy: *They certainly didn't stint on the food and drink at Alexa's party.*

stipend *n.* °pay, °salary, °payment, °remuneration, °remittance, recompense, compensation, °reward, emolument, °earnings, °income; °grant, subvention, °scholarship, °subsidy, °allowance, °allotment, °(financial) support: *His stipend barely covers his basic expenses.*

stipulate *v.* °specify, °demand, °require, covenant, °set forth, °agree (to), °provide (for), °guarantee, °warrant, °promise, insist (upon *or* on); °call for: *The agreement stipulates that the goods be received prior to payment. The compensation stipulated in the contract has been paid.*

stipulation *n.* °condition, °demand, °essential, °given, °requirement, requisite, °prerequisite, °specification, °undertaking, °obligation, covenant, clause, °proviso, °term, °agreement, °provision, °guarantee, °warranty, °promise: *This stipulation calls for payment on delivery in full and in sterling.*

stir *v.* **1** Often, ***stir up.*** °agitate, °shake (up), °mix (up), °scramble, °amalgamate, °mingle, commingle, intermingle, °merge, °blend, °fold (in), churn (up), °beat, °whip (up): *Stir all the ingredients together.* **2** °move, °rise, °arise, °get up, bestir (oneself), be up and about, *Colloq* °get a move on, get moving, get a wiggle on, °shake a leg, look *or* step lively, look alive, stir one's stumps: *It is about time you were up and stirring.* **3** °disturb, °trouble, °affect, °upset, °stimulate, °activate: *The play stirred the conscience of the king. Mention "diet" to Roberta and you'll stir up a wasp's nest.* **4** Often, ***stir up.*** °motivate, °encourage, °stimulate, °energize, galvanize, °electrify, °animate, °excite, °inspire, °provoke, °move, °rouse, °arouse, °get, °prompt, °urge, °incite, °spur, °prod, °induce, °persuade, °convince: *What will it take to stir the council to take action? The usual rabble-rousers were there, stirring up the crowd.* **5** Often, ***stir up.*** °awaken, °rouse, (cause to) recall *or* call to mind, °revive, resuscitate: *Those photographs stir up so many old memories!* —*n.* **6** bustle, °activity, °movement, °stirring, °action, commotion, °flurry, °confusion, °tumult, ado, to-do, °fuss, °disturbance, °excitement, hubbub, *Colloq Brit* kerfuffle: *There was a stir at the door, which was flung wide to allow Kitty to sweep in. The news of her marriage caused quite a stir.* **7** °prison, jail *or Brit also* gaol, jailhouse, clink, penitentiary, lockup, *Military Brit* glasshouse, *US* brig, *Slang chiefly Brit* quod, *US* big house, °pen, slammer, can, calaboose: *He claims he's in the stir for something he didn't do.*

stirring *adj.* °moving, °telling, °emotional, emotive, emotion-charged, °rousing, °stimulating, inspiring, gripping, evocative, °exciting, °thrilling, °melodramatic, °dramatic, heady, °intoxicating, °spirited, inspiriting, °exhilarating, awe-inspiring: *He was treated to the stirring experience of hearing fifty thousand people shouting his name in unison, again and again.*

stock *n.* **1** °supply, °store, inventory, stockpile, °reserve, reservoir, °cache, °hoard; °wares, °merchandise, °goods, °selection, °assortment, °range, °variety, array: *One can select something suitable from the stock at the local wine shop.* **2** °pedigree, bloodline, °house, °dynasty, (line of) descent, genealogy, °extraction, °roots, °lineage, °family, ancestry, °parentage, °breeding, °heritage: *Her mother was concerned that she should marry someone of good stock.* **3** °source, °progenitor, °creator, °father, begetter, forefather, °ancestor, °precursor, °forerunner, forebear; °founder: *He might well provide the stock for a new line of kings.* **4** livestock, (domestic *or* farm) animals, °cattle, beasts; horses, cows, oxen, sheep, goats: *Don't forget to water the stock before driving to the village.* **5** °share, ownership, investment, °capital, °funds; °property, °assets: *He owns ten shares of stock in our corporation. The company's stock is worth next to nothing today.* **6** ***take stock.*** °weigh (up), °estimate, °review, appraise, °look at, *Colloq* °size up: *We must take stock of the situation and decide what to do.* —*adj.* **7** °routine, stereotyped, °banal, clichéd, commonplace, °usual, hackneyed, °ordinary, °stale, °staple, °run-of-the-mill, °tired, °old, °everyday, °customary, °set, °standard, °traditional, trite, worn-out,

Colloq corny: *Ask him how he is and you always get his stock reply, "Fair to middling."* **8** °standard, °ordinary, °regular, °routine, °staple: *The store doesn't carry "extra tall" or "extra short," only stock sizes.* —*v.* **9** °carry, °have, have *or* make available, °handle, deal in, °market, °sell, °supply, °furnish, °provide, °offer, trade in, °keep: *Do you stock any other brand of pantyhose?* **10** Often, ***stock up (on).*** °accumulate, °amass, °pile up, stockpile, °hoard, °store (up), °cache, °lay in, inventory: *We have no plans for stocking large quantities of buggy whips.*

stocky *adj.* thickset, °sturdy, chunky, °dumpy, °solid, stumpy, °burly, beefy, heavyset, squat, pyknic, mesomorphic: *He is stocky and powerfully built.*

stodgy *adj.* °stuffy, °dull, °heavy, °ponderous, elephantine, °boring, °tedious, °humdrum, °tiresome, turgid, uninteresting, unimaginative, dryasdust, jejune, °vapid, °dreary, °flat, °colorless, °bland, *Colloq* ho-hum, blah, °deadly: *That was the stodgiest talk I have ever heard on such a fascinating subject.*

stoical *adj.* stoic, °impassive, resigned, apathetic, °cool, unemotional, emotionless, °frigid, imperturbable, °calm, °dispassionate, °indifferent, °phlegmatic, long-suffering, °stolid, disciplined, °self-possessed, (self-)controlled, *Colloq* unflappable: *Bertram remained stoical as the judge read out the sentence of death.*

stoicism *n.* °indifference, self-possession, austerity, °self-control, °fortitude, calmness, °calm, coolness, imperturbability, longanimity, forbearance, °patience, fatalism, °resignation, *Colloq* unflappability: *The people accepted the news of one defeat after another with increasing stoicism.*

stole *n.* tippet, scarf, boa, shawl: *Mindy was too embarrassed to wear her mink stole to the Animal Rights League dinner.*

stolid *adj.* °impassive, °dull, doltish, °obtuse, °thick, °dense, bovine, °wooden, °slow, lumpish, unemotional, clodlike, °phlegmatic, °lethargic, apathetic, °indifferent, uninterested: *He is a stolid character, very different from his dynamic younger brother.*

stomach *n.* **1** abdomen, belly, °gut, potbelly, °pot, °paunch, *Colloq* corporation, bay window, tummy, breadbasket, spare tire: *His huge stomach hung over and concealed his belt.* **2** °tolerance; °taste, °appetite, °desire, °hunger, °thirst, craving, °need, °inclination, °relish, °longing, yearning, hankering: *I have no stomach for those TV sitcoms featuring precocious four-year-olds.* —*v.* **3** °abide, °tolerate, °endure, °stand, °bear, °suffer, °take, °accept, °swallow, resign *or* reconcile oneself to, °put up with, countenance, °brook, *Brit* °stick: *He walked out when he could no longer stomach her continual criticism.*

stony *adj.* **1** stoney, °rocky, pebbly, shingly, shingled: *Some of the beaches along the Riviera are too stony to lie on directly.* **2** stoney, °hard, obdurate, adamant(ine), °heartless, stonyhearted, hardhearted, °indifferent, °unsympathetic, °implacable, intractable, °heartless, insensitive, °insensible, unfeeling, unsentimental, °merciless, pitiless, °cold, °coldhearted, °chilly, °frigid, °icy, °tough, °callous, °steely, °inflexible, unresponsive, *Colloq* hard-boiled: *He listened to her pleadings in stony silence.*

stoop *v.* **1** Sometimes, ***stoop down.*** °bend (down), °bow, °duck (down), °lean (down), °hunch (down), hunker (down), °crouch (down), scrunch down: *She had to stoop down to talk to the child.* **2** Often, ***stoop low.*** °condescend, °deign, lower *or* abase *or* degrade oneself, °sink, humble oneself; be demeaned *or* diminished: *She had to stoop pretty low to accept a job paying only half of what she had demanded.* —*n.* **3** °hunch, °slouch, scrunch, °crouch, stooping, slouching, *Technical* lordosis, curvature of the spine, torticollis, wryneck: *The orthopedist thinks your stoop could be corrected by a back brace.*

stop *v.* **1** °discontinue, °halt, °terminate, °cease, °break off, °end, put an end *or* a stop to, bring to a stop *or* a halt *or* an end *or* a close, °give up, °quit, °leave off, °finish, conclude, desist (from), °refrain (from), °abandon; draw to a close, be over, come to a stop *or* a halt *or* an end *or* a close; *Colloq* °cut (out), °lay off, *Brit* °pack in: *Stop shouting, I'm not deaf. Stop the car. It won't stop! Will the fighting ever stop?* **2** bring to a stop *or* a halt *or* a standstill, °check, °cut off; °arrest, °suppress, °restrain, °thwart; °block, °bar, °obstruct, dam, keep *or* hold back, °prevent, °hinder; °slow, °impede, °stem, stanch *or* staunch: *How do we stop the drug traffic? Can we stop the river from flooding? This tourniquet will stop the bleeding.* **3** Often, ***stop up.*** °obstruct, °block (up), °jam (up), °plug (up), °clog (up), °choke (up), °stuff (up), °fill (up), close (up *or* off): *The drain is all stopped up with paper.* **4** °peter out, be over, °end: *He gave me an injection and the pain finally stopped.* **5** °pause, °break, take a break, °interrupt, °tarry; °sojourn, °rest, °stay, °put up, °lodge, °visit, stop off *or* in *or* over; pull over, °pull up: *I'd like to stop for a cup of coffee. I stopped at The Bell Inn on my way here. We stopped short at the sign.* —*n.* **6** °halt, °end, cessation, °termination, °ban, °prohibition; °close, °standstill, conclusion: *The new law put a stop to door-to-door selling. We must bring that practice to a stop.* **7** °stay, °sojourn, °visit, °break, °rest, stopover, *US* layover: *Did you enjoy your stop at Tegucigalpa as much as that at Chichicastenango?* **8** stopping place, °station, °terminal, °stage, terminus, *US and Canadian* depot: *There is a bus stop in the next street.* **9** blockage, blocking, stopping(-up), stoppage, closing up, °obstruction, °block: *A stop in the water supply was soon cleared up.*

stopgap *n.* **1** °makeshift, improvisation, °substitute: *This wire will serve as a stopgap till we can get the proper part.* —*adj.* **2** °makeshift, °temporary, improvised, impromptu, °substitute, °emergency, °provisional, °standby; jury-rigged: *A broomstick served as a stopgap mast for the dinghy's sail.*

stopper *n.* stopple, cork, °plug, bung: *Until we could get the stopper out, we wouldn't be serving any beer.*

store *v.* **1** °stock, °collect, °accumulate, °put by, lay away, °set aside, °pile (up), aggregate, °amass, cumulate; °hoard; °assemble: *The corn is stored in huge silos.* **2** °keep, °hold, °stow (away), °preserve, °warehouse, stockpile: *Apples should be stored in a cool, dry place.* —*n.* **3** °supply, inventory, °collection, °accumulation, °stock, stockpile, reservoir, °cache, °fund: *We keep a large store of dry foods for an emergency. Grandfather has an endless store of tales about the sea.* **4** °shop, department store, °market, retailer, °outlet, cooperative (store), *Colloq* co-op: *You might get a xylophone at a department store.* **5** ***set*** or ***lay store by.*** give credence to, °believe (in), have faith *or* trust in, °trust (in), bank *or* rely on, depend on *or* upon, °count on, value: *I shouldn't set much store by what that old gossip tells you.*

storehouse *n.* °warehouse, depository *or* depositary, repository, storeroom, bank, °store, (*in Asia*) godown; arsenal, °magazine, armory: *The company maintains storehouses in a dozen countries to better supply their customers.*

storm *n.* **1** °tempest, °disturbance, turbulence; windstorm, mistral, °gale, °whirlwind, °hurricane, tornado, typhoon, cyclone, *US and Canadian* williwaw; °shower, cloudburst, °downpour, rainstorm, deluge, monsoon, thundershower, thunderstorm, electrical storm; dust storm, sandstorm, simoom *or* simoon *or* samiel, harmattan, khamsin, sirocco; snowstorm, blizzard; hailstorm, ice storm: *People were evacuated from the area where it was thought the storm would strike.* **2** °outburst, °outcry, °explosion, °eruption, °outpouring, furor *or Brit* furore: *The proposed taxes were met by a storm of protest.* **3** °disturbance, °rumpus, °stir, commotion, °agitation, furor *or US* furor; turbulence, °strife, turmoil, °disorder: *I am not sure the company could weather another serious financial storm.* —*v.* **4** °rage, °rant, °rave, °bluster, °fume, °explode, °thunder, °roar, raise the roof, raise hell, raise Cain, *Colloq* °fly off the handle, blow one's top, *US* °blow one's stack: *When she found out where Leonora had gone, mother stormed around the house.* **5** °blow, °rain, °hail, snow, sleet, °rage, °bluster, squall, °howl: *It stormed for a week, and much property was destroyed.* **6** °attack, °assault, assail, °raid, blitz, blitzkrieg, °bombard, barrage, fire upon *or* on, °shell; °besiege, lay siege to, °siege: *The enemy stormed the castle again and again without effect.*

stormy *adj.* **1** °violent, °tempestuous, blustery, turbulent, °wild, howling, raging, roaring, °foul, °nasty, °bad, not fit for man or beast, °inclement: *It was always stormy along the eastern coast of the island.* **2** °violent, °tempestuous, turbulent, °fierce, °fiery, °frantic, frenetic, °nerve-racking, frenzied, °feverish, °raving, °wild: *He and Kate enjoyed—if that is the right word—a stormy relationship for thirty years.*

story[1] *n.* **1** °narrative, °tale, recounting, anecdote, °yarn; °account, °recital, °chronicle, °record, °history; °legend, °myth, fairy tale *or* story, °romance, gest *or* geste, fable, fabliau; epic, °saga, edda; °joke, *Colloq* °gag; °mystery, detective story, whodunit, thriller; horror story; allegory, °parable; °piece, article: *George really knows how to tell a story. Have any of your stories been published?* **2** contention, °testimony, °assertion, °version, °statement, °representation, °allegation: *That's his story and he's sticking to it.* **3** °fib, confabulation, (white *or* black) lie, °alibi, °excuse, untruth, °falsehood; tall tale, fishing *or* fish story: *He told us some story about having sailed round the world single-handed in a dinghy. Did you believe her story about being of royal blood?* **4** article, °item, °report, °dispatch, °news, tidings, °release, °information, °copy, °feature; °scoop, °exclusive: *A story about China appears on page 2.* **5** story line, °plot, °scenario, °(plot) outline, °summary, °book: *The story needs some revision before the dialogue can be rewritten.* **6** biography, curriculum vitae, °life (story); °facts, experiences, adventures, °fortunes: *His story is going to be made into a film.*

story[2] *n.* storey, °floor, °level, °tier: *The new building is five stories high.*

stout *adj.* **1** °fat, °obese, tubby, overweight, thickset, heavyset, °big, °burly, corpulent, fleshy, °heavy, °plump, portly: *A stout person ought to wear clothing one size too large to make him look smaller.* **2** valiant, °brave, undaunted, °dauntless, °hardy, °courageous, °gallant, plucky, valorous, °staunch, °resolute, doughty, °bold, °gallant: *"Stout fellow!" exclaimed the colonel, the ultimate compliment he was ever heard to utter.* **3** °strong, °tough, °substantial, °durable: *Tie stout cord round the box before posting it.* **4** °brawny, °sturdy, °healthy, °robust, strapping, °stalwart, °lusty, °hulking, athletic, *Colloq* beefy, °husky: *Two stout bouncers flanked the entrance to the club.*

stow *v.* °pack, °store, °load, °deposit, °put (away), °place; °cram, °stuff, °wedge, °bundle, °jam; °hide, °secrete, °conceal, °cache, *Colloq* stash (away): *We stowed our gear the moment we went aboard. They couldn't stow another thing in the basement. Where did he stow the pistol when the police searched the place?*

straggle *v.* °stray, °ramble, loiter, rove, °prowl, °range, °drift, °wander, °meander, °(be) spread, *Colloq* mosey: *Thousands of refugees straggled along the road away from the bombed village.*

straight *adj.* **1** °direct, unbending, undeviating, uncurved, °regular, linear: *The tracks are straight for as far as the eye can see.* **2** °erect, vertical, °upright, upstanding, °perpendicular; °plumb: *Stop slouching and stand up straight. He stood straight as a ramrod.* **3** °even, °square, °true, °right, °flat, °smooth, °horizontal, °level: *Are you sure that the billiard table is straight?* **4** °honest, °frank, straightforward, °direct, °forthright, °legitimate, °(fair and) square, °fair, °equitable, °just, °aboveboard, °upright, °respectable, °decent, °trustworthy, °honorable, dependable, °reliable, *Colloq* °up front: *Sternway seems to be perfectly straight in his business dealings.* **5** unequivocal, unambiguous, straightforward, °candid, °plain, °simple, °explicit, °blunt, unembellished, unelaborated, °unqualified, °outright, °accurate: *The straight facts are given in this article. The police cannot get a straight story out of him as to his whereabouts last night.* **6** °direct, °point-blank, straightforward, straight-from-the-shoulder, °candid, °outright, °plain, °frank, °no-nonsense: *Do you mind answering a straight question?* **7** °shipshape, °orderly, °neat, °tidy, °in order, arranged, organized, sorted out, °spruce, straightened out: *I wish Leonard would put his room straight before the guests arrive.* **8** °sober, °staid, °sedate, °serious, unsmiling, unemotional, °impassive, emotionless, °taciturn, composed, masklike: *He tried to keep a straight face when he told her what happened to the cherry tart.* **9** °undiluted, °neat, unmixed, °pure, unadulterated, uncut, unmodified, unaltered, unalloyed: *May I have a straight whiskey?* **10** °even, °square, settled, straightened out, agreed: *The account will be straight when Dick makes the last payment.* **11** heterosexual, *Slang* hetero: *Till Evelyn came out of the closet, I thought he was straight.* —*adv.* **12** °directly, °right, undeviatingly, unswervingly; as the crow flies, in a beeline: *I should have known she would go straight to the police.* **13** °(straight) ahead: *Go straight for a mile and you'll see it on the right.* **14** Sometimes, ***straight out.*** °directly, unequivocally, unambiguously, forthrightly, straightforwardly, °point-blank, candidly, plainly, °simply, in plain *or* simple English, explicitly, °outright, °honestly, accurately: *Just give me the story straight, without beating about the bush.* **15** Often, *Brit* ***straightaway*** *or* ***straight off.*** °immediately, °at once, without delay, instantly, °summarily, °directly, right off the bat, °promptly, *US* right away, *Brit* right off, *Colloq* PDQ (= 'pretty damned quick'): *She went back to work straight after having a baby. I told him straight off to leave town.* **16** ***straight up.*** without ice: *I like my gin straight up.*

straighten *v.* **1** Often, ***straighten out.*** uncurl, untangle, disentangle, unsnarl, unravel, unkink; °clear (up), °settle, °resolve, °sort out, set *or* put straight *or* right *or* to rights, °correct, °adjust, °rectify: *Straighten this skein of yarn for me, would you? They straightened out the matter of his father's will.* **2** Often, ***straighten out.*** °reform, °rehabilitate, °organize, reorganize: *Perhaps psychotherapy will straighten him out.* **3** Often, ***straighten out*** *or* ***up.*** °tidy (up), °arrange, rearrange, °neaten, °spruce up, put in order, °clean (up): *Would you help me straighten up the place before you leave?*

strain[1] *v.* **1** °stretch, °force, °tax, overtax, °burden, overburden, °overwork, °push; °exceed, °surpass: *His story that a bushy-haired intruder had shot her strained the jury's credulity.* **2** °push, °pull, °tug, °heave, °stretch, °twist, °wrench, °struggle: *As he strained at his bonds the rope cut more deeply into his wrists.* **3** °injure, °hurt, °harm, °impair, °damage, °overwork, °tax, °pull, °tear, °twist, °wrench: *I strained my back lifting that* *box of books.* **4** °stretch, crane, °twist; °try (hard), °struggle, °strive, °labor, toil, °push, make an effort, °exert oneself: *She strained to catch a glimpse of the prince. They strained to hear every word.* **5** °filter, °sift, °drain, °screen, sieve; winnow, °draw off, °separate; °purify, seep, °percolate: *Strain the soup to remove the bones. Strain the dregs from the wine. The water is strained through sand at this stage.* —*n.* **6** sprain, °injury, °damage, °harm, °wrench: *Strains often hurt more than broken bones.* **7** °anxiety, °worry; °effort, °exertion, °stress, °tension, °pressure, °burden; °tax, °demand, °obligation: *The strain was beginning to tell on all of them. Another rope is needed to relieve the strain on the first. This advertising budget will put a great strain on the company's resources.* **8** Often, ***strains.*** °air, °melody, °tune, °song, °sound, music: *In the distance I made out the faint strains of skirling bagpipes.* **9** °tenor, °tone, °drift, °inclination, °tendency, °quality, °spirit, °mood, °humor, °character, complexion, °cast, °impression, °thread, °vein, °theme: *I detected a mellowing strain in her letters.*

strain[2] *n.* **1** °family, °stock, ancestry, °roots, °extraction, °derivation, °(family) background, °heritage, descent, °parentage, °lineage, °pedigree, bloodline, °race, °line, descendants: *This sample of sweet pea came from an entirely different strain.* **2** °trace, °hint, °suggestion, °suspicion, *soupçon,* °streak, °trait, °mark, °indication, °vestige, °evidence, °sign: *There seems to be a strain of ruthlessness in her character.*

strained *adj.* °labored, °forced, °artificial, °stiff, °tense, °awkward, uneasy, uncomfortable, °difficult, tension-ridden, °self-conscious, °unnatural, °insincere, °put-on: *Philip's already strained relations with his wife scarcely improved when he met Lisa.*

strait *adj.* **1** °narrow, °tight, constricted, constricting, confining, confined, restricting, restricted, °limited, limiting, rigorous, °demanding, °exacting; °difficult, °straitened: *Strait is the gate and narrow is the way which leadeth unto life, and few there be that find it.* —*n.* **2** Usually (except in gazetteers), ***straits.*** °narrows, °channel: *We coasted through the Straits of Magellan with all sails flying.* **3** ***dire*** *or* ***desperate*** *or* ***sore straits.*** bad *or* poor state *or* condition, °trouble, °predicament, °plight, °mess, °dilemma, tight spot, hot water, *US* bind *Colloq* pickle, °jam, °scrape, pretty *or* fine kettle of fish, *US* °box: *We shall be in dire straits if father's check doesn't come soon.*

straitened *adj.* °inadequate, °insufficient, reduced, oppressed, distressed, °needy, necessitous, °poor, poverty-stricken, °indigent, °impoverished, °destitute, penniless, °insolvent, *Colloq* °hard up, *US* strapped: *As a writer, he was accustomed to living in straitened circumstances between sales of his stories.*

strait-laced *adj.* °priggish, prim, °conservative, °old-fashioned, Victorian, oldmaidish, °proper, °prudish, puritanical, moralistic, °strict, °narrow-minded, (over)scrupulous, °fussy, *Colloq* pernickety *or US also* persnickety, °stuffy, °goody-goody: *Her aunt has the strait-laced notion that girls ought to be home by ten o'clock.*

strange *adj.* **1** °odd, °peculiar, °bizarre, °weird, °curious, uncommon, °unusual, °rare, °singular, exceptional, °eccentric, °weird, °funny, °quaint, °fantastic, out-of-the-ordinary, °extraordinary, °out-of-the-way, °queer, °outlandish, °unheard-of, °grotesque, °abnormal, °remarkable, surprising, °inexplicable, °unaccountable, uncanny, *Colloq* °offbeat, far-out, *Slang* °kinky, *Brit* rum, *US* kooky: *I had a strange dream last night.* **2** °unfamiliar, °unknown, °unaccustomed: *The man who entered was strange to me. The children were warned not to get into strange cars.*

stranger *n.* °foreigner, outlander, °alien, °newcomer, °visitor: *People in these parts don't take kindly to strangers, Mister.*

stratagem *n.* °trick, °artifice, °device, °dodge, °subterfuge, °lure, °wile, °ruse, °plan, °scheme, °plot, °intrigue, °maneuver, ploy, °tactic: *What stratagem did they use to tempt you to attend this boring affair?*

strategic *adj.* °tactical, °key, °crucial, °principal, °cardinal, °critical, °vital, °key: *That line of supply is of strategic importance. Sending in the landing force was a strategic maneuver.*

strategy *n.* °plan, tactic(s), °design, °policy, °procedure, °scheme, blueprint, *Colloq* game *or* master plan, °scenario: *The enemy's strategy calls for quick air strikes by low-flying bombers.*

stratum *n.* **1** layer, °level, stratification, °table, °vein, °seam; °plane: *These coal strata have been mined for hundreds of years.* **2** °level, °caste, °class, °rank, °station, °standing, °status, °bracket, °group, °estate: *She married someone from a higher stratum of society.*

stray *v.* **1** °wander, °roam, rove, °range, °straggle, °drift, °meander: *You must not let your cattle stray about the countryside any longer.* **2** °deviate, °diverge, °wander, digress, °ramble, divagate, get *or* go off the track *or* subject, go off on *or* at a tangent, get sidetracked: *Occasionally she let her mind stray from the subject at hand.* —*n.* **3** straggler, vagrant, waif, *US* dogie: *They have always taken in and cared for strays.* —*adj.* **4** vagrant, °lost, roving, roaming, wandering, °homeless, °derelict, °abandoned: *The problem with stray pets increases when people abandon them on returning home after the summer.* **5** °random, °casual, °chance, °accidental, °haphazard, °singular, °freak, °accidental, unexpected: *A stray bullet caught him in the leg.* **6** °isolated, separate(d), lone, °odd, °single: *She brushed a stray hair from her face.*

streak *n.* **1** °stripe, striation, °strip, °stroke, °bar, °band, °line, °mark, °smear, °slash, °dash, °touch, daub, fleck, °trace; °vein, layer, °seam, °stratum: *The male bird is marked with a streak of bright reddish-orange on its throat. A red streak marks the iron deposit.* **2** °flash, °bolt: *I saw the dark figure in the light from a streak of lightning.* **3** °spell, °spate, °period, °stretch, °run: *Let's celebrate last night's streak of good luck at roulette.* —*v.* **4** °stripe, striate, °line, °bar, °mark, °smear, daub, °slash: *Dawn was beginning to streak the sky when she finally left the party.* **5** °race, °run, °rush, °dash, sprint, dart, °hurtle, °fly, scoot, °speed, °hasten, °hurry, °tear, whistle, zip, zoom, *Colloq* whiz *or Colloq* whizz: *The gang streaked through the village on their motorcycles and were gone.*

stream *n.* **1** °brook, brooklet, streamlet, rivulet, °tributary, °river, freshet, °run, watercourse, waterway, °channel, *Chiefly literary* rill, runnel, *Literary or N Brit dialect* beck, °burn, *Archaic NE US except in place names* °kill, *US* °creek, °branch: *If the streams continue to swell, flooding can be expected in low-lying areas.* **2** °flow, °current, °outpouring, effluence *or* efflux, effusion, °rush, °spurt, °surge, °fountain, geyser, °torrent, °flood, deluge, cataract, cascade: *A stream of steamy bubbles emanated from the underwater mofette. You have never heard such a stream of obscenities in all your life!* **3** °flow, °rush, °swarm, tide, °flood, deluge, °succession, series, °row, °line, °string, °chain, barrage, *Brit* °queue: *Simon has had a constant stream of visitors all morning.* —*v.* **4** °run, °flow, °course, °glide, °rush, °slide, °slip, °surge; °pour, °issue, °emanate, °gush, °flood, °spout, well up *or* out *or* forth, squirt, °spurt, °shoot, jet; cascade: *The Danube streamed past in the moonlight. She stabbed the statue and blood streamed out.* **5** °issue, °emanate; °rush, °surge, °pour, °flood,

°file, °proceed, °march, °walk, °move: *People stream in and out of the building all day long.*

streamer *n.* °pennant, °banner, pennon, °flag, bannerette *or* banneret, banderole *or* bannerol, gonfalon *or* gonfanon, jack, burgee: *The long streamer fluttering from the mast truck carried the company's colors.*

streamlined *adj.* **1** aerodynamic, hydrodynamic, curved, curvilinear; °smooth, flowing: *Streamlined design began to appear in cars and boats in the 1930s.* **2** °modern, ultramodern, modernistic, modernized, up-to-date, timesaving, labor-saving, °compact, (well-)organized, °efficient, automated: *Mother always dreamed about having a streamlined kitchen.* **3** well-run, °smooth, °efficient, automated, labor-saving, timesaving, °profitable, °productive, simplified: *He designed a fully streamlined manufacturing process for us.*

street *n.* **1** thoroughfare, °way, °road, roadway, highroad, avenue, concourse, boulevard, lane, °drive, terrace, °circle, °row, °passage, alley, byway: *I don't know the name of the street where she lives.* **2** ***up (someone's)*** *Brit* ***street*** *or US* ***alley.*** (someone's) cup of tea, in (someone's) bailiwick, suiting (someone) to a T: *Crossword puzzles are very much up Eugene's street.*

strength *n.* **1** °power, °might, °force, mightiness, robustness, toughness, stoutness, sturdiness, °brawn, brawniness, muscle, °sinew: *Do you have the strength to lift that by yourself?* **2** °fortitude, °backbone, °stamina, °tenacity, tenaciousness, willpower, °perseverance, °persistence, resoluteness, °resolution, pertinacity, °nerve, °grit, °pluck, °determination, gameness, intrepidity, firmness, °stability, *Colloq* °guts, gutsiness, °spunk, *US* intestinal fortitude, stick-to-itiveness: *Sally was a tower of strength when I needed her the most.* **3** °talent, °ability, °aptitude, °gift, strong point, °asset: *One of his strengths is being able to sight-read music.* **4** concentration, concentratedness, °intensity, potency: *Coffee of that strength would keep me awake all night.* **5** °vigor, °force, °energy, °power, potency, °intensity: *The strength of the wind was great enough to pick up cars and toss them about.* **6** durability, °power, toughness, °stability, reliability, °resistance, solidity, °stamina, ruggedness, °endurance, soundness: *The manufacturer claims that his product has the strength to resist wear for a hundred years.* **7** persuasiveness, cogency, °weight, °force, convincingness, incisiveness, soundness: *There is no disputing the strength of the argument against smoking.*

strengthen *v.* **1** °reinforce, °renew, °bolster, °fortify, °support, °confirm, corroborate, °substantiate, °buttress, °step up, °boost: *A thousand troops were sent to strengthen the garrison at Fort Old.* **2** °encourage, hearten, invigorate, °fortify, °rejuvenate, °nourish, °energize, °vitalize, toughen, °brace (up), °steel, innervate, °stiffen: *Whatever she said seems to have strengthened his determination to succeed.*

strenuous *adj.* **1** °demanding, taxing, °tough, °arduous, °laborious, °toilsome, °burdensome, tiring, °exhausting, °difficult, °hard, uphill: *Hauling in fishing nets is strenuous work.* **2** °energetic, °active, °vigorous, °enthusiastic, zealous, °earnest, °dynamic, °intense, indefatigable, °tireless, °persistent, dogged, °determined, °tenacious, pertinacious, °resolute, °sincere, °eager: *He has made a strenuous effort to be more accommodating.*

stress *n.* **1** °emphasis, °force, °pressure, forcefulness, °accent, accentuation, °prominence, *Technical* ictus: *Place the stress on the first syllable of "controversy."* **2** °emphasis, °significance, °importance, °weight, °force, insistence, °urgency: *She lays too much stress on looks.* **3** °(stress and) strain, °burden, °anxiety, °worry, °distress, °pain, °grief, °suffering, °anguish, °pressure, tenseness, °tension: *She had four elderly people to care for, and the stress was beginning to show.* —*v.* **4** °emphasize, °accent, accentuate, lay stress *or* emphasis on, underscore, underline, °mark, °note, °make a point of, °bring home, focus on, bring into prominence, °spotlight, °feature, highlight: *The conference succeeded in stressing the importance of a good home life.* **5** °strain, put under strain *or* stress, °upset, °disturb, °burden, °worry, °distress, *Brit* pressurize *or US* pressure: *One could hardly say that Duncan ever stressed himself with overwork.*

stretch *v.* **1** °extend, °reach; °span, °spread: *Her memory stretches back to the First World War. His Texas ranch stretches as far as the eye can see.* **2** distend, °lengthen, elongate, °widen, broaden, °swell, draw *or* pull out, balloon, °inflate, °enlarge, °expand, °increase, dilate, °blow up: *Stretch the netting to cover the frame.* **3** overtax, overextend; °warp, °strain, °distort, °bend, °break: *Don't stretch the capacity of people to believe all you say. He is stretching the rules by allowing undergraduates to take his course.* —*n.* **4** °elasticity, °give, °resilience, resiliency, stretchability, stretchiness: *There isn't much stretch left in this old rubber band.* **5** °extent, °reach, °span, °spread, °expanse, °sweep, °area, °tract, *US* °section: *There is a stretch of desert just south of the coast.* **6** °time, °stint, °period, °spell, °term, °tour (of duty), *Colloq US and Canadian* °hitch: *How many sign up for a two-year stretch in the navy? He once did a six-month stretch in prison for assault.*

strew *v.* °scatter, bestrew, sprinkle, °disperse, °spread, °toss, °distribute; °litter: *Stop strewing your rubbish all over the floor!*

stricken *adj.* Usually, **stricken by. 1** struck (down) (by), °hit (by), laid low (by *or* with), affected (by *or* with), afflicted (with), wracked (by *or* with): *He was stricken by pneumonia but is recovering.* **2** Often, ***struck.*** °affected (by), °smitten (by), overwhelmed (by *or* with), overcome (by *or* with), plagued (by *or* with), tormented (by); °broken, crushed, demoralized, °brokenhearted, grief-stricken: *He was stricken by her beauty. During the plague year, commerce virtually ceased with the stricken city of London.*

strict *adj.* **1** rigorous, °narrow, °close, undeviating, confining, constricting, constrictive, °rigid, defined, °precise, °exact, °exacting, stringent, °meticulous, °compulsive, punctilious, finicky *or* finical, °scrupulous, °attentive, °conscientious, °faithful, °thorough, °complete: *The judge's strict interpretation of the law left no room for leniency or sympathy.* **2** °severe, austere, °authoritarian, autocratic, °stern, °firm, °hard, °tough, uncompromising, °inflexible, °coldblooded, ironfisted, °tyrannical, °harsh, °ruthless, pitiless, °unsympathetic: *Miss Weed is remembered by all her students as a strict disciplinarian.*

stricture *n.* **1** interdiction, blockage, °restriction, °restraint, constraint, °deterrent, °impediment: *The strictures imposed on credit have throttled trade.* **2** °criticism, censure: *He deplored conservative strictures against anything new in the arts.*

strident *adj.* °shrill, °raucous, °harsh, °loud, °grating, stridulous, stridulant, scraping, scratching, °scratchy, grinding, hoarse, °rough, guttural, °husky, gravelly, rasping, jarring, °discordant, unharmonious, unmelodious, unmusical, cacophonous, croaking, creaking: *Cassie's strident voice could be heard all over the house.*

strife *n.* **1** °discord, disharmony, °disagreement, °difference, °conflict, °rivalry, °competition, contention, °dispute, °dissension, °struggle, squabbling, bickering, arguing, quarreling: *The nation is torn by the strife of perpetual labor disputes.* **2** °animosity, °friction, hard feelings, bad feeling(s), bad blood, °antagonism, °ill will, hatred, enmity, °hostility, unfriendliness: *The* *strife between us will continue because our philosophies differ so basically.*

strike *v.* **1** °hit, deal a blow to, °knock, smack, thump, thwack, °crown, cuff, °punch, smite; °beat, hammer, °belabor, °batter, pummel *or* pommel, °pelt, buffet, thrash; cudgel, bludgeon, °club, °whip, horsewhip, °scourge, °lash, cane, °flog, birch, °slap, *Colloq* wallop, slug, whack, clout, sock, conk, °belt, bash, lambaste *or* lambast, bop: *Small wonder he struck you after what you called his wife.* **2** °deliver, °deal, °aim, °direct: *He struck a blow for freedom.* **3** °hit, °collide with, land on *or* in *or* against, smash *or* bump *or* bang *or* crash *or* dash into, go *or* run into, °impact: *Mrs. Humphrey's car went out of control and struck a tree.* **4** °remove, take away, take apart, dismantle, °knock down; take *or* pull *or* haul down: *Stagehands struck the set. As the wind piped up we struck the mainsail.* **5** Usually, ***strike off*** *or* ***from*** *or* ***out.*** °obliterate, expunge, °erase, eradicate, °blot out, °delete, °scratch, °eliminate, °rub out, °cross (out), °cancel, °wipe out, *US* x out: *Owing to the scandal, they want to strike his name from the register. Strike out any references that might be thought libelous.* **6** °light, ignite: *She struck a match to see where she was going.* **7** °affect, °impress, °influence, °afflict, *Colloq* °hit: *The death of her kitten struck Mandy to the heart.* **8** °make, °reach, attain, conclude; agree *or* settle (on *or* upon), °ratify, °confirm: *We struck a bargain on the sale of the car.* **9** occur *or* come to, dawn on *or* upon, *Colloq* °hit, °register (with): *It suddenly struck me that you are leaving tomorrow.* **10** °impress, °print, °stamp, °punch, °mint, °make: *They have struck a new gold coin for collectors.* **11** °instill, °implant, °induce: *Horror films struck fear into our hearts when we were children.* **12** °assume, °adopt, °put on, °display, °affect, °take on, feign: *It was so annoying when he struck that supercilious pose of insouciant hauteur.* **13** Often, ***strike down.*** °afflict, °affect, °attack, indispose, °incapacitate, disable, °cripple, °invalid: *Polio struck her down in the prime of life.* **14** °encounter, come *or* happen *or* hit upon, °come across, chance upon, °discover, °stumble on, °find: *The news is that they've struck oil in the North Sea.* **15** °revolt, °rebel, °mutiny, walk out (of *or* off the job): *The machinists struck at midnight, bringing the industry to a halt.* **16** ***strike on*** *or* ***upon.*** dream up, °devise, conjure up, °improvise, °work out, °invent, contrive, come up with, hit on *or* upon, °arrive at: *The alchemist never did strike upon a way to turn lead into gold.* **17** ***strike out.*** °fail, get nowhere, *Colloq US* miss the boat, °flop, come a cropper, *Slang US* blow it, blow the gaff, come to nothing *or* naught *or* nought: *Wally always strikes out when it comes to girls.* **(a)** See **5**, above. **18** ***strike up.*** (cause to) begin *or* start *or* commence: *Strike up the band. We struck up an acquaintance aboard ship.* —*n.* **19** °attack, °assault: *A dawn air strike is planned.* **20** walkout, sit-down (strike), job action, slowdown, go-slow, work-to-rule: *They will call a strike if negotiations break down.*

striking *adj.* °remarkable, astounding, astonishing, °amazing, wondrous, awe-inspiring, °awesome, °stunning, °impressive, °imposing, °fabulous, out-of-the-ordinary, °unusual, °rare, °exceptional, °marvelous, °extraordinary, °magnificent, °superb, °splendid, stupendous, *Colloq* °great, smashing, *Slang old-fashioned Brit* °ripping, rip-snorting, top-hole, topping: *I met the most striking girl at the golf club.*

string *n.* **1** °line, °cord, °thread, °twine, °fiber, °rope, °cable, ligament, strand, filament: *To fly a kite one needs the right weight of string.* **2** leash, °lead, °leader: *Why does he let her drag him around on the end of a string?* **3** °line, °row, series, °sequence, °succession, °chain, °procession, °stream, °train, °file, *Chiefly Brit* °queue: *A string of coincidences led to their meeting again after twenty years.* **4** necklace, °chain, °loop, strand, dog-collar, choker, chaplet, wreath, rivière, *Archaic* carcanet: *In her hair was a string of pearls.* **5** ***pull strings*** *or* ***wires.*** use *or* exert influence, *Colloq* throw one's weight around: *Furness had to pull strings to get his son a job on the paper.* **6** ***pull the strings.*** be in control, °control, °run, °operate, °dominate, be in command, be in the driver's seat, hold the reins, °manipulate: *She is the one who pulls the strings when it comes to new investments.* **7** ***strings.*** °conditions, stipulations, provisos, °qualifications, requirements, prerequisites, °terms, obligations, limitations, °provisions, musts, *Colloq* catches: *I might have known that there would be strings attached to their offer of a free vacation in Corfu.* —*v.* **8** °thread, °join: *Would you help me string these beads?* **9** Often, ***string together*** *or* ***up.*** °loop, festoon, °link, °drape, °suspend, °sling, °hang, array, concatenate, chain together: *At holiday time, colored lights are strung round the town square.* **10** ***string along.*** **(a)** °follow, go along (with), °agree, concur, °collaborate: *I said I would string along with the plan for the time being.* **(b)** keep waiting *or* dangling, keep on a string, keep on tenterhooks, *Colloq* play fast and loose with (someone): *She's just stringing you along till her boyfriend comes back.* **(c)** °fool, °deceive, °bluff, °dupe, °cheat, °trick, °hoax, *Colloq* take someone for a ride, put one *or* something over on (someone): *They strung Harold along for months before they made off with his life savings.* **11** ***string out.*** **(a)** °stretch, °reach, °extend: *The line of refugees strung out as far as one could see.* **(b)** °delay, °postpone, drag out, protract, °spin out: *He strung out the repayment for as long as possible.* **12** ***string up.*** °hang, lynch: *He knew that if the jury didn't convict him the townspeople would string him up.*

stringy *adj.* fibrous, chewy, °sinewy, gristly, °ropy, leathery, °tough: *Unless it is Aberdeen Angus beef, the roast always seems stringy to me.*

strip[1] *n.* **1** °band, ribbon, fillet, °belt, swath *or* swathe, °stripe: *She wore a strip of embroidered silk tied round her head. They have begun farming the strip of land near the lake.*

strip[2] *v.* **1** °peel, °skin, °bare, uncover, denude, °lay bare, decorticate, excoriate, flay: *Strip the paint off this chair before repainting it.* **2** °disrobe, undress, get undressed, unclothe, strip down to nothing *or* to the skin *or* to the buff *or* to (one's) birthday suit, take off *or* peel off *or* divest (oneself) of *or* shed (one's) clothes *or* clothing, get naked: *The doctor told me to go behind the screen and strip.* **3** (do a) striptease, *US* work the runway: *Sugar Caine says she enjoys stripping for an appreciative audience.* **4** °remove, take away, °confiscate, °seize, expropriate, *Slang* °rip off: *He was accused of stripping the assets from the companies he purchased.* **5** °rob, °pillage, despoil, °plunder, °ransack, °loot, °sack: *Armies could be raised only by letting them strip the cities they conquered.*

stripe *n.* **1** °band, °bar, striation, °strip, °streak, °line, °stroke, °slash, °length: *Each railway car has a bright orange stripe along each side.* **2** °style, °kind, °sort, °class, °type, complexion, °character, °nature, °description, °persuasion, kidney, feather: *These new students are of a completely different stripe.*

striped *adj.* streaked, lined, striated: *Why not wear your striped dress to the party?*

stripling *n.* °lad, °boy, °adolescent, °juvenile, °minor, schoolboy, youngster, °teenager, °youth, young fellow *or* man, fledgling, *Dialect* gossoon, young 'un, *Archaic* hobbledehoy: *Artisans would take in striplings as apprentices and teach them their craft.*

strive *v.* **1** °endeavor, °strain, °struggle, make every effort, °attempt, °try (hard), do one's best *or* utmost,

°exert oneself, work at, *Colloq* give (it) one's all, go all out: *Jim strove to keep his self-control. Jennie strives for perfection in all she does.* **2** °compete, contend, °fight: *One must often strive against unbelievable odds to achieve a goal.*

stroke *n.* **1** °blow, °rap, °tap, thump, °knock, smack, whack, °swipe, °slam, °strike, *Colloq* wallop: *One stroke of the hammer and the nail went in.* **2** °action, °motion, °go, °move, °movement, °feat, °achievement: *She won the leading role and defeated her shyness at one stroke.* **3** °flourish, °movement, °gesture; °mark, °dash, *Colloq* °splash: *With a stroke of the pen the man's fate was sealed.* **4** °beat, throb, °pulse, pulsation, thump: *His heart was beating at exactly seventy-two strokes a minute.* **5** °attack, °seizure, °fit, apoplexy, apoplectic fit, °spasm, paralytic attack *or* fit; *Technical* embolism, thrombosis, cerebrovascular accident, aneurysm: *Gerry has been incapacitated since he had a stroke.* **6** °pat, °touch, °caress: *A few strokes of her fingers on my brow and the headache was gone.* **7** °achievement, °accomplishment, °feat, °act, °action, °work; °example; °touch: *A jazz concert to raise money for musicians is a stroke of genius.* **8** °bit, jot or tittle, °scrap, iota, °touch, stitch, °hint, °suggestion: *You haven't done a stroke of work all morning.* **9** °occurrence, °happening, °matter: *It was just a stroke of luck that I came along when I did.* —*v.* **10** °caress, °pet, °pat, °fondle; °massage, °rub, soothe: *His heart beat faster when she stroked his hand lightly with her fingertips.*

stroll *v.* **1** amble, °saunter, °ramble, °walk, °wander, °promenade, °meander, °stray, *Colloq* mosey: *I think I'll stroll over and buy a newspaper.* —*n.* **2** amble, °ramble, °saunter, °walk, °wander, °promenade, °meander, constitutional: *Father rarely missed his Sunday stroll in the park.*

strong *adj.* **1** °powerful, °muscular, °mighty, °brawny, strapping, °robust, °sturdy, °stalwart, °burly, °stout, °sinewy, athletic, °wiry, *Colloq* beefy, °hefty, °husky: *He was so strong he carried both wounded men five hundred yards to safety.* **2** °powerful, concentrated, °intense, °pungent, °potent, °sharp, piquant, acrid, heady, °penetrating, °aromatic, °fragrant, °hot, °spicy: *There was a strong odor of ammonia. The spicing is a little strong for my taste.* **3** °smelly, odoriferous, noisome, °stinking, °foul, mephitic, miasmic, °putrid, putrescent, °rotten: *There is a strong smell of rotting flesh.* **4** concentrated, °undiluted, °potent, intensified: *How can you drink such strong coffee and still sleep?* **5** °vigorous, °active, °dynamic, °energetic, °eager, unflagging, °tireless, unfailing, °diligent, indefatigable, °staunch, true-blue, °steadfast, dedicated, °enthusiastic, °ardent, °fervent, fervid, vehement, °rabid, zealous, °resolute, °determined, unwavering, unswerving, °firm, uncompromising, °regular, °persistent, °tenacious, sedulous, assiduous, hard-working: *We could use other strong party supporters like Sandy.* **6** °competent, °talented, skilled, °qualified, °knowledgeable, °able, °experienced, well-versed, trained, °efficient, °capable: *We are looking for someone who is particularly strong in writing advertising copy.* **7** °influential, °persuasive, convincing, compelling, °trenchant, unmistakable *or* unmistakeable, °telling, °great, °profound; °effective, °efficacious, °effectual, °powerful, °formidable: *His evidence had a strong effect on the jury. She takes very strong medicine for her migraines.* **8** well-supported, irrefutable, well-substantiated, cogent, °forceful, °substantial, convincing, conclusive: *The prosecution made out a strong case against her for premeditated murder.* **9** °well-established, well-founded, redoubtable, °substantial, °powerful, °formidable: *The position inside the fortress was quite strong.* **10** °likely, °definite, °substantial, °good, better than average, °reasonable, sizable: *There is a strong possibility that she will be elected. Sales of cars were strong during the first quarter.* **11** °stable, °sound, °solvent, °prosperous, °flourishing, thriving, affluent: *The economists think that the economy will remain strong.* **12** °solid, °sturdy, °substantial, °tough, well-built, reinforced, heavy-duty, °durable; hard-wearing: *Is the cage strong enough to hold the beast?* **13** °drastic, °extreme, Draconian, highhanded, °severe, °forceful, rigorous, °harsh, stringent, °aggressive, °strenuous, °stiff, °tough, *Colloq* hard-nosed: *We had to resort to strong measures.* **14** numerous, °large, °considerable, °great; numerically, in number, in strength: *They arrived on the battlefield with a strong force of ten thousand men. Their army was ten thousand strong.* **15** °vivid, °graphic, etched, engraved, imprinted, impressed; °definite, clear-cut, °clear, °pronounced, °distinct, °striking, °marked: *The memory of her embrace was still strong in my mind. He bears a strong resemblance to his brother.* **16** °willful, °aggressive, combative, defensive, °difficult, °assertive, °incisive, °dogmatic, doctrinaire, °opinionated, °self-willed, hard-headed, °strong-minded, °recalcitrant, °stubborn, °obstinate, °emphatic, *Colloq* °pushy: *One cannot deny that Caroline has a strong personality.* **17** °vigorous, °forceful, °powerful, °heavy: *Strong winds lashed the west coast last night.* **18** °rugged, craggy, °rough, weather-beaten: *She was attracted by Merton's strong, suntanned good looks.* **19** °dazzling, °glaring, °bright, °garish, °brilliant, °vivid, °bold, blinding: *The strong sunlight made her blink. Shocking pink may be too strong a color for a small room.* **20** °urgent, strongly worded, °emphatic, °assertive: *A proposal to repeal the tax was met by strong protests.* **21** unvarying, the same, °steady, °stable, °firm, balanced: *The market has remained strong all day.* **22** °emotional, deep-felt, deep-rooted, °basic, °intense, °fervent, °passionate, °deep, °earnest: *Doris has very strong feelings on the subject of feminism.* —*adv.* **23** overbearingly, overenthusiastically, offensively, aggressively, antagonistically, truculently: *He always comes on a bit strong when he's enthusiastic about something.*

strong-arm *adj.* °threatening, °menacing, bullying, high-pressure, thuggish, °violent, °brutal, brutish, °aggressive, terrorizing, °terrorist, intimidating, minacious: *The gang uses strong-arm tactics to force shopkeepers to buy protection.*

stronghold *n.* fortress, °bulwark, bastion, fastness, fortification, citadel: *The eastern district was one of the last strongholds of the Democratic Party.*

strong-minded *adj.* strong-willed, °obstinate, °firm, °determined, uncompromising, °resolute, resolved, °independent: *He realized he was dealing with a very strong-minded young woman.*

structure *n.* **1** °form, °shape, configuration, °organization, °arrangement, °makeup, framework, °order, °design, °formation, °system, °nature, °character: *The career structure of the company has served as a model for similar firms.* **2** °building, edifice, °house, construction: *Planning regulations allow only a certain kind of structure in the residential areas.* —*v.* **3** °construct, °build, °organize, °design, °form, °shape, °arrange, systematize: *Their language seems to be structured in an unusual way.*

struggle *v.* **1** °strive, °strain, expend energy, °exert oneself, °endeavor, °try, °attempt: *He struggled to keep his head above water.* **2** contend, °fight, °wrestle, °battle: *They struggled against the twin odds of poverty and sickness.* **3** °wriggle, wiggle, °squirm, writhe, °twist, worm: *She finally struggled free of the ropes.* —*n.* **4** °effort, °exertion, °strain; toil, °work, travail, °labor, °drudgery, striving, struggling: *Earning even a humble living was a struggle in those days. Was the outcome worth the struggle?* **5** contention, °competition, °contest, °battle, °fight, tussle, °match, °clash, °encounter, °strife: *The struggle between good and evil continues.*

strut *v.* °swagger, °parade, °promenade, peacock, °prance: *She goes strutting about the office as if she were the boss.*

stub *n.* **1** °butt, °end, °stump, °tail (end), °remnant, *Colloq Brit* fag end: *He produced a stub of pencil and a scrap of paper from an inside pocket.* **2** counterfoil; °receipt: *You will need your ticket stub for any claim.*

stubborn *adj.* °obstinate, unyielding, °inflexible, intransigent, intractable, uncompromising, mulish, pigheaded, refractory, wayward, adamant(ine), °recalcitrant, bullheaded, °persistent, °tenacious, pertinacious, unrelenting, dogged, °determined: *He remains stubborn in his refusal to join the group.*

student *n.* **1** °pupil, °learner, °scholar, undergraduate, schoolboy, schoolgirl, schoolchild, trainee, °apprentice, °disciple; *Colloq Brit* swot *or* swotter *or* swat, *US* °grind: *Only students of advanced standing are admitted to this class.* **2** °devotee, °follower, °admirer, °observer, evaluator, commentator, critic: *Nicole is a perceptive student of human nature.*

studied *adj.* °premeditated, °deliberate, °calculated, planned, °intentional, °willful, well-thought-out, °conscious, contrived, feigned, °forced, °labored: *He treated even the greatest crisis with a studied air of insouciance.*

studious *adj.* **1** assiduous, sedulous, °diligent, °industrious, °attentive, °careful, °painstaking, °thorough, °tireless: *Miniaturists must pay studious attention to detail.* **2** °scholarly, bookish, °academic: *She refused to allow her athletic achievements to interfere with her studious pursuits.*

study *v.* **1** °learn (about), °read, con, °memorize, burn the midnight oil, lucubrate, *Colloq* bone up (on), °cram, *Brit* swot *or* swat, mug up: *I want to study music. She has to study for an exam.* **2** °contemplate, °consider, °reflect on, think over *or* about, ruminate on, °chew over, °turn over, °weigh, °ponder, deliberate over *or* on *or* about, muse about *or* on, mull over, meditate on *or* about *or* over: *I want to study the problem before deciding what to do.* **3** look *or* go into *or* over, °look at, °scan, °examine, °analyze, °inspect, °investigate, °scrutinize, °survey, °observe: *The crime squad studied every square inch of the murder scene. Avoiding the issue as usual, Joyce silently studied her fingernails.* —*n.* **4** °analysis, °review, °examination, °survey, inquiry *or* enquiry, °investigation, °scrutiny, °research, °exploration: *Government funds are paying for a study of the ecosystem of Baja California.* **5** °learning, lessons, bookwork, °work, reading, contemplation, °investigation, *Colloq* boning up, cramming, *Brit* swotting *or* swatting: *The study of anthropology is what led him to his theory of language universals.* **6** library, reading *or* writing room, °sanctum (sanctorum), °haunt, studio, °retreat, den, workroom, °office: *Let's go into the study for a glass of port.*

stuff *n.* **1** °substance, °material, °matter, °fabric, ingredients, °essence, essentials, fundamentals, building blocks, makings: *This is the stuff that dreams are made on. Is hydrogen the basic stuff of the universe?* **2** °equipment, °goods, °gear, °trappings, °kit, °tackle, accessories, °paraphernalia, accoutrements *or US also* accouterments, °effects, °belongings, °possessions, °things, bits and pieces, impedimenta, baggage, °property, chattels, °furniture, *Brit* °lumber, *Colloq* °junk, °rubbish, crap, *Brit* clobber, *Taboo slang* shit: *Greeley left all his stuff with his mother when he went west.* **3** °spirit, °attitude, °grit, °substance, makings, talent(s), °abilities, capabilities, qualities, attributes: *Is Speckler the stuff of which commandos are made?* **4** °nonsense, °trash, °rubbish, stuff and nonsense, twaddle, humbug, bunkum, tommyrot, balderdash, *Colloq* °rot, °garbage, bunk, tripe, poppycock, crap, malarkey, baloney *or* boloney, bosh, hogwash, °swill, claptrap, piffle, °hot air, flapdoodle, fiddle-faddle, codswallop, bull, *US* horsefeathers, *Taboo slang* bullshit, horseshit: *You don't believe all that stuff you've been reading about me, do you?* **5** creations, accomplishments, °things, °works, materials, °matter: *Without the attributions, I'd have trouble telling their stuff apart.* —*v.* **6** °jam, °ram, °cram, °crowd, compress, °pack, °press, °squeeze, squash, shove, °thrust, °force: *You couldn't stuff another handkerchief into that suitcase.* **7** °line, °fill, °pack: *Stuff the chicken with the following mixture.* **8** °overeat, °gorge, overindulge, gormandize *or US also* gourmandize, gluttonize, *Colloq* make a pig *or* a hog of oneself: *They really stuffed merrily away at the wedding reception.* **9** ***stuff up.*** °clog, °plug, °obstruct, °choke, °block (up), stop *or US also* pack up: *Phone the plumber—the toilet is stuffed up again. He had a stuffed-up nose from a cold.*

stuffy *adj.* **1** °close, airless, unventilated, °oppressive, stifling, suffocating, °stale, °musty, fusty, °moldy, mildewy, °muggy, fetid *or* foetid, frowzy *or* frowsy, *Brit* frowsty: *It was stuffy in the ancient tomb.* **2** °pompous, °pedantic, self-important, self-centered, °stodgy, old-fogyish *or* old-fogeyish, °old-fashioned, °strait-laced, °staid, °conventional, prim (and proper), °priggish, niminy-piminy, °stilted, °stiff, °rigid, *Colloq* fuddy-duddy, uptight: *What a bore to be trapped for an entire evening with those stuffy old codgers!*

stumble *v.* **1** falter, °blunder, °slip, °trip, miss one's footing, °stagger, °lurch, °flounder: *Minnie caught her heel in the grating and stumbled.* **2** falter, °pause, °hesitate, °trip, °slip, °blunder: *I used to stumble over the pronunciation of the word "floccinaucinihilipilification."* **3** ***stumble on or upon.*** chance *or* come *or* happen on *or* upon, °hit upon, come *or* run across, °find, °discover, °encounter, *Colloq* °bump into: *He stumbled on a new lime drink while experimenting with acaricides.*

stumbling block *n.* °impediment, °obstacle, °bar, °block, °obstruction, °hurdle, °hindrance, °barrier, °difficulty, °snag: *It was often the labor unions that proved the stumbling blocks to technological progress.*

stump *n.* **1** °stub, °butt, °end: *The branches were lopped off, leaving only the stumps.* —*v.* **2** °mystify, °confuse, °perplex, °bewilder, °flummox, °foil, °puzzle, baffle, confound, dumbfound *or* dumfound, °stop, °stymie, °nonplus, bring up short: *I am completely stumped as to how the trick was done.* **3** °campaign, °electioneer, °canvass, *US and Canadian* barnstorm: *The candidates are out stumping in the farm areas in Iowa.* **4** ***stump up.*** pay up *or* out, °contribute, °donate, *Colloq* cough up, °chip in, shell *or* fork out: *Everyone managed to stump up some loose change at the office collection for Jeffrey's birthday.*

stun *n.* **1** °daze, °numb, benumb, °knock out: *I was stunned by an unexpected blow to the solar plexus.* **2** °astonish, °daze, °paralyze, °stagger, stupefy, °overcome, °overwhelm, °astound, °jar, °shock, °jolt, strike dumb, °amaze, confound, °bewilder, take (someone's) breath away, *Colloq* °shake up, bowl over, discombobulate, flabbergast: *I was stunned to hear that mother had been arrested again.*

stunning *adj.* **1** stupefying, paralyzing, staggering, benumbing, numbing; °knockout: *Another stunning punch to the head and McGinty went down.* **2** °beautiful, °dazzling, °brilliant, °gorgeous, spectacular, °ravishing, °sensational, °extraordinary, °remarkable, °marvelous, stupendous, °fabulous, wonderful, °superb, °grand, °divine, °heavenly, °sublime, °lovely, °exquisite, °glorious, astonishing, astounding, °amazing, °striking, °splendid, staggering, °overpowering, mind-boggling, earthshaking, °magnificent: *There stood the* *most stunning creature I had ever laid eyes on. The TV brought us the news of our team's stunning victory in the Olympics.*

stunt[1] *n.* °caper, °act, °deed, °feat, tour de force, °exploit, °trick, *US* dido: *The dangerous stunts are done by a stand-in for the star.*

stunt[2] *v.* °stop, °limit, delimit, °restrict, °check, °arrest, put an end to, °end; °impede, °hamper, °hinder, °slow, °retard: *Whatever they didn't want me to do would stunt my growth, according to my parents.*

stunted *adj.* dwarfed, shrunken, °undersized, °small, °tiny, °diminutive, °little, °wee: *We saw a bonsai—a stunted Japanese potted tree that was hundreds of years old.*

stupid *adj.* **1** unintelligent, fatuous, °obtuse, bovine, °dull, °dense, lumpish, doltish, °simple, simple-minded, moronic, imbecilic, cretinous, Boeotian, subnormal, °feebleminded, weak-minded, °stolid, dull-witted, °dim, dimwitted, °half-witted, °thick, thick-witted, thickheaded, slow-witted, witless, brainless, °mindless, empty-headed, birdbrained, featherbrained, featherheaded, rattlebrained, rattleheaded, oxlike, boneheaded, addlepated, addle-headed, addled, *Chiefly Brit* imbecile, *Chiefly US* °dumb, jerky, thimble-witted, *Colloq* dopey, *Brit* dozy: *He must be really stupid to believe that Gertie loves him for himself.* **2** °foolish, °silly, °frivolous, asinine, °harebrained, °crazy, °insane, °mad, crackbrained, °scatterbrained, °absurd, °inane, idiotic, °ridiculous, risible, laughable, °ludicrous, °nonsensical, °senseless, °bootless, °irresponsible, irrational, °ill-advised, °foolhardy, half-baked, *Colloq* cuckoo, balmy, cockeyed, damnfool, *Chiefly Brit* °daft, barmy, *US* cockamamie *or US* cockamamy: *His next stupid plan was to corner the escargot market.* **3** insipid, °dull, °tedious, °boring, °tiresome, °humdrum, °prosaic, °monotonous, unimaginative, uninspired, uninteresting, °vapid, vacuous, *Colloq* ho-hum: *Most sitcoms feature stupid characters doing stupid things.*

stupidity *n.* **1** fatuity, obtuseness, dullness, denseness, lumpishness, doltishness, °simplicity, simple-mindedness, imbecility, cretinism, feeblemindedness, weak-mindedness, stolidity, dull-wittedness, dimness, dimwittedness, half-wittedness, thick-wittedness, slow-wittedness, thimble-wittedness, witlessness, brainlessness, mindlessness, empty-headedness, featherheadedness, rattleheadedness, boneheadedness: *In public affairs stupidity is more dangerous than knavery.* **2** foolishness, °folly, asininity, craziness, °insanity, °madness, °absurdity, absurdness, inanity, idiocy, ridiculousness, risibility, ludicrousness, °nonsense, senselessness, bootlessness, irresponsibility, irrationality, foolhardiness: *The stupidity of such a venture is obvious when you consider the enormous risk and the minuscule return.*

stupor *n.* insensibility, stupefaction, °torpor, °lethargy, listlessness, languor, laziness, lassitude, lifelessness, supineness, °inertia; inertness, coma, °trance, unconsciousness, numbness: *His stupor might be caused by the medications he is taking for his cold.*

sturdy *adj.* **1** °strong, °solid, °stout, °rugged, °tough, well-built, °substantial; strapping, °muscular, °powerful, °brawny, °burly, °robust, well-muscled, athletic, °hardy, *Colloq* °husky, °hefty: *This ladder should be sturdy enough to hold you. James got that sturdy build from exercise.* **2** °stalwart, °staunch, °steadfast, °resolute, °firm, °vigorous, °determined, uncompromising, unyielding, unwavering, unswerving, unfaltering, °enduring, °indomitable: *One must admire her sturdy independence in the face of all that criticism.*

style *n.* **1** °type, °kind, °variety, °category, °genre, °sort, °manner, °mode, °make, °design, °fashion, °look, °period, °pattern, configuration, °line, °cut, °shape, °form: *That style of jacket looks very good on you. They are planning to build the house in the pointed Gothic style.* **2** °fashion, °trend, °vogue, °mode, °look, °rage, °craze, *Colloq* °fad, °(latest) thing: *The current style is for shorter skirts.* **3** °luxury, high style, °comfort, opulence, °splendor, °elegance: *Now that he's won the lottery, they live in style on the Costa Smeralda.* **4** °chic, stylishness, °taste, smartness, °flair, °dash, élan, °panache, °cachet, tastefulness, fashionableness, °elegance, °refinement, °polish, °sophistication, sophisticatedness, cosmopolitanism, *Colloq* pizazz; ritziness: *Irena has more style in her little finger than you have in your whole body.* **5** °quality, °character, mode of expression, °approach, °treatment, °vein, coloring, °spirit, °mood, °form, °technique; °tenor, °tone, °wording, phraseology, phrasing, °language, vocabulary, word choice, °diction, sentence structure: *The pointillist style appeals to many. His style of writing is reminiscent of Stevenson's.* **6** ***in style.*** See **stylish,** below. —*v.* **7** °characterize, °designate, denominate, °call, °name, °term, °label, °tag, °brand: *The use of the indicative for the subjunctive is no longer styled a solecism in British English.* **8** °fashion, °design, °arrange, °set, do, °cut, °tailor, °shape, °form: *Antoine styled my hair in a pageboy for the reception.*

stylish *adj.* °chic, °fashionable, °smart, à la mode, modish, in style *or* fashion *or* vogue, °elegant; chi-chi; *Colloq* in, with it, °swanky, *Chiefly Brit* °trendy, *Slang* °swell, °neat, classy, snazzy, *US* spiffy: *Nicole always looks so stylish.*

stymie *v.* °thwart, °obstruct, °block, °frustrate, snooker, °defeat, °spike, °ruin, °foil, confound, °stump, °nonplus, °hinder, °impede, *Colloq* °flummox: *The government has stymied all efforts to have him extradited.*

styptic *adj.* astringent: *The styptic effect of alum stops bleeding.*

suave *adj.* °debonair, °sophisticated, urbane, cosmopolitan, °worldly, °smooth, °gracious, °nonchalant, civilized, °cultivated, °courteous, °diplomatic, °polite, charming, °agreeable, affable, °bland: *Fictional detectives range from the suave Simon Templar to the earthy Philip Marlowe.*

subconscious *adj.* **1** °subliminal, °unconscious, suppressed, °hidden, latent, repressed, inner, innermost, underlying, deep-rooted, *Colloq* Freudian: *Despite his belligerence, he has a subconscious desire to be loved.* —*n.* **2** °(collective) unconscious, inner self; °heart: *Her subconscious tells her that all confined areas are dangerous.*

subdue *v.* **1** put *or* beat down, °quell, °repress, °suppress, °quash, °crush, °control, °master, °overpower, gain mastery *or* control *or* the upper hand over, get the better of, °dominate, °triumph over, hold *or* keep in check, °bridle, °tame: *Having subdued the uprising, government forces are again in control.* **2** °conquer, vanquish, °defeat, °overcome: *The Mongol horde subdued all the people in its path.* **3** quiet *or* tone down, *Chiefly Brit* quieten down, °moderate, °mellow, °temper, °soften, soft-pedal, °check, °curb, °control: *Suppress your animation and subdue your excess effervescence.*

subdued *adj.* **1** °quiet, mellow(ed), toned-down, moderate(d), tempered, hushed, muted, low-key, °unenthusiastic, repressed, restrained, °peaceful, °tranquil, placid, calm(ed), °temperate, °reserved: *There was a little subdued criticism at first, but the bill was passed.* **2** chastened, °sober, sobered, °solemn, saddened, °dejected, °sad, °down in the mouth, crestfallen, downcast, °grave, °serious: *He looked subdued when he emerged from the headmaster's office.*

subject *n.* **1** °(subject) matter, °topic; °issue, °theme, °angle, °thesis, °gist, °substance, °business, °affair, °point: *What is the subject of conversation today? The subject before the committee was critical to the people's welfare.* **2** °course (of study), °field, °area, °discipline, branch of knowledge: *In which subject did Frank take his doctorate?* **3** °cause, ground(s), °motive, °reason, °basis, °source, °rationale; °excuse: *Increased taxes are always a subject of complaint.* **4** °participant, °case, guinea pig, testee: *The subjects of the experiment were coal miners.* **5** °citizen, °national; taxpayer, voter; liegeman, vassal: *She became a Peruvian subject on her marriage to Gonçalves.* —*adj.* **6** Usually, ***subject to.*** exposed (to), °open (to), °vulnerable (to), °susceptible (to), °prone (to), °disposed (to), at the mercy (of), liable (to suffer *or* undergo): *She is subject to asthma attacks. This kind of wood is subject to worm infestation.* **7** discussed, under discussion, referred to, °above: *The subject book was not returned before the due date.* **8** ***subject to.*** **(a)** answerable to, responsible for, bound by, obedient to, subservient to, submissive to, controlled by, under the control of: *You are subject to the same laws as everyone else.* **(b)** dependent on, conditional on, contingent on: *Permission to leave is subject to my approval.* —*v.* **9** ***subject to.*** °expose, lay open, °submit, °put through, °impose on, cause to undergo: *How could anyone subject another human being to such cruelty?* **10** °conquer, °subjugate, °dominate, °subdue, °enslave, °crush, °humble, *Archaic* enthrall: *The peoples subjected by the Romans sometimes fared better than when independent.*

subjection *n.* subordination, °domination, °conquest, subjugation, enslavement, enthrallment, humbling, °humiliation: *Their goal was the subjection of all peoples in the Mediterranean area.*

subjective *adj.* **1** °personal, °individual, idiosyncratic; °prejudiced, °biased: *He imposes his subjective criteria on others.* **2** self-centered, °egoistic, egocentric, °selfish, self-serving: *Your approach is much too subjective to be of interest to others.* —*n.* **3** *Technical* nominative: *The subjective of "me" is "I."*

subjugate *v.* °dominate, °enslave, °crush, °humble, °subject, °oppress, °suppress, °put down, °tyrannize, °subdue, °reduce, °quell, °overcome, °overpower, make subservient *or* submissive, °humble, °humiliate, *Archaic* enthrall: *Few peoples have been subjugated so ignominiously as the American Indians.*

sublimate *v.* transmute, °alter, °transform; °channel, °divert: *He sublimates his aggressions by jogging and marathon running.*

sublime *adj.* **1** °lofty, °high, °supreme, °exalted, °elevated, empyrean *or* empyreal, °heavenly, °noble, °glorious, °grand, high-minded; °honorable, ennobled, °eminent, °glorified, beatified, canonized, sanctified, °great, °good: *Connie's poetry evokes sublime emotions.* **2** °awesome, °overwhelming, inspiring, mind-boggling, °overpowering, humbling, awe-inspiring, °majestic, °splendid, empyrean: *Only astronauts have experienced a truly sublime perspective of the universe.*

subliminal *adj.* °subconscious, °unconscious, °suggestive: *Advertisers know that subliminal techniques insinuated into TV commercials can be effective.*

submerge *v.* **1** °plunge, submerse, °immerse, inundate, °dip, °wash, °soak, °drench, °saturate, °wet, douse, *Colloq* dunk: *To cleanse, submerge the garment in a basin of warm water for ten minutes.* **2** °dive, °plunge, °go down, °descend, °sink, °sound, °plummet: *The order to submerge was given as soon as the planes were spotted.* **3** °flood, °immerse, inundate, °swamp, °bury, engulf, °overwhelm, deluge, °drown; °conceal, °hide, °camouflage, °obscure, °cloak, °veil, °shroud: *Submerged under a sea of information, how can the truth be rescued?*

submission *n.* **1** concession, acquiescence, capitulation, °surrender, °yielding, °deference, giving in, °obedience, compliance, °resignation, submissiveness, tractability; meekness, docility, passivity, timidity, unassertiveness: *After years of oppression their abject submission to a new tyrant was not altogether surprising.* **2** submittal, °offering, °tender, contribution, °entry: *Submissions for the essay competition should be sent to the address given below.*

submissive *adj.* **1** °yielding, acquiescent, deferential, compliant, °obedient, °tractable, amenable, °agreeable, °accommodating, °passive, unresisting, pliant, °flexible, °manageable, unassertive, docile, °meek, °timid, resigned, uncomplaining: *Feeble sovereigns and a submissive people could not advance into national greatness.* **2** °obsequious, abject, subservient, °servile, °humble, deferential, slavish, °ingratiating, truckling, biddable, sycophantic, toadying, *Colloq* bootlicking, *Taboo slang* brown-nosing, *Brit* arse-kissing, arse-licking, *US* ass-kissing, ass-licking: *She keeps a few submissive lackeys about to do her bidding.*

submit *v.* **1** Often, ***submit to.*** °surrender (to), °yield (to), °capitulate (to), give in *or* up (to), °comply (with), °agree (to), °concede (to), °consent (to), accede (to), °defer (to), bow *or* bend (to), °succumb (to), °truckle (to), knuckle under (to), resign (oneself) (to), be *or* become resigned (to); °respect, °accept, *Colloq* °put up with: *Despite the threats, he refused to submit. The company finally submitted to the workers' demands.* **2** °offer, proffer, °tender, °enter, °propose, °present: *Many of the suggestions that Alan submitted have been accepted.*

subordinate *adj.* **1** Often, ***subordinate to.*** °minor; °inferior (to), °lower (than), lesser (than), °secondary (to), °second (to), °junior (to), °subsidiary (to); next to, °below, °beneath, °under: *He had to accept a subordinate position in the new company. His new job is subordinate to that of purchasing director.* —*n.* **2** °assistant, °aide, °junior, subaltern, staff member; underling, hireling, °inferior, lackey, °servant, °slave, vassal; *Colloq US* staffer: *She is a subordinate to the editor in chief.* —*v.* **3** make (something) secondary: *Too many subordinate community interest to personal greed.*

subscribe *v.* **1** Often, ***subscribe to.*** °endorse, °support, °underwrite, °advocate, °back (up), °approve (of), agree (with *or* to), °accept, °consent (to), assent (to), countenance, °tolerate, condone, °allow, °permit, °brook: *I sympathize with many of their arguments but I cannot subscribe to terrorism.* **2** Often, ***subscribe to.*** °contribute (to), °support, °give (to), °donate (to), °pledge, °promise, sign (up) (for), *Colloq* chip in (to *or Colloq* for): *Mrs. Donaldson has subscribed thousands to Amnesty International.*

subscription *n.* **1** °payment, °remittance, investment; commitment, °dues, °fee, °price, °cost: *The subscription is now a hundred dollars a year.* **2** °obligation, °pledge, °promise, underwriting: *If we get enough subscriptions, we shall be able to print the book.*

subsequent *adj.* **1** succeeding, following, ensuing, next, °future, later, °successive; resultant, resulting, consequent: *Subsequent problems, which she did not have to cope with, did not surface until after her resignation.* —*prep.* **2** ***subsequent to.*** after, following, succeeding, in the wake *or* aftermath of: *Other problems arose subsequent to her term of office.*

subsequently *adv.* later (on), afterward(s): *Subsequently, the law was changed to protect witnesses.*

subside *v.* **1** °sink (down), °drop (down), °go down, °recede, °descend, °decline; °lower, °settle: *The waters* *of the swollen river subsided. The volcano's center subsided, leaving a caldera.* **2** abate, quiet *or chiefly Brit* quieten (down), °calm (down), °moderate, °let up, °decrease, °diminish, lessen, die (down *or* off *or* out), °pass (away), °wear off: *My enthusiasm for sky diving subsided a bit when I reached seventy. When the clamor had subsided, the president rose to speak.*

subsidiary *adj.* Often, **subsidiary to.** ancillary (to), °secondary (to), °auxiliary (to), lesser (than), additional (to), supplementary *or* supplemental (to), complementary (to), °accessory (to), °subordinate (to), adjuvant (to): *The company offers removal of hazardous waste as a subsidiary service.*

subsidize *v.* °fund, °finance, °support, °aid, °sponsor, subvene, °maintain, °underwrite; capitalize, *Slang US and Canadian* bankroll: *Bill and Sue can afford to live in a big house only because they're subsidized by her parents.*

subsidy *n.* funding, financing, subsidizing, sponsoring, sponsorship, °assistance, °aid, contribution, °support, °grant, subvention, °maintenance, underwriting, capitalization: *Farmers are being threatened by reduced subsidies.*

subsistence *n.* **1** °existence, living, survival, subsisting, being: *You are worried about buying beer when I feel my very subsistence threatened!* **2** °food, °rations, victuals, °provision, °sustenance, °board, °nourishment, nutriment, aliment; °maintenance, °keep, °upkeep: *We depend on your salary for our subsistence, so forget about quitting your job.*

substance *n.* **1** °material, °matter, °stuff; °fabric, °composition, °makeup: *What is this substance? The comet's substance is mainly ice and dirt.* **2** °essence, °pith, °heart, °core, °gist, °burden, °theme, °meat, °kernel, °nub, crux, sum total, sum and substance, °point, gravamen, haecceity, °quintessence, quiddity: *Explain, in five hundred or fewer words, the substance of Hegel's dispute with Kant.* **3** °meaning, °import, °significance, purport, signification, °point: *Our visit to San Felipe gave substance to all we had read about it.* **4** °reality, corporeality, solidity, actuality, concreteness: *You must learn to deal with the substance, not the shadows.* **5** °means, °wealth, °property, °possessions, °riches, °resources, affluence, °assets: *Culver is a citizen of some substance in this town.*

substantial *adj.* **1** °material, °considerable, °significant, °great, °worthwhile, consequential, °ample, °goodly, °respectable, °abundant, °generous, °big, °large, sizable, °major, *Colloq* °tidy, °healthy: *We understand that he made a substantial payment to keep his name out of the papers.* **2** °strong, °solid, well-built, °durable, °sound, °stout, °sturdy; °big, °large, °massive, °huge, sizable, °impressive, °vast; numerous, °numberless: *A substantial mausoleum was constructed for his family. A substantial crowd had gathered.* **3** well-founded, °sound, °weighty, °solid, °well-established, °telling, °good, valid, °actual: *Many economists produced substantial arguments supporting further cuts in interest rates.* **4** °wealthy, well-to-do, °rich, affluent, °prosperous, °profitable, °successful; landed, propertied: *The Trembal Ironworks is a substantial business, and Mr. Trembal is one of our more substantial businessmen.*

substantially *adv.* in substance, essentially, °at bottom, fundamentally, basically, °in essence, intrinsically, °in reality, at heart, °sincerely, °truly, °actually, °in truth, veritably, °indeed, °in fact, as a matter of fact; °largely, to a large extent, in large measure, °materially, °practically, °in the main, °for the most part, mostly, °virtually, °to all intents and purposes; *Archaic* verily: *I think that Merriwether is substantially an honest man. Do you agree substantially with the ruling of the court?*

substantiate *v.* °confirm, affirm, corroborate, °support, °sustain, °back up, °bear out, °authenticate, °show (clearly), °prove, °document, °verify, °certify, validate: *Do not make accusations that cannot be substantiated by evidence.*

substitute *v.* **1** Sometimes, ***substitute for.*** °replace, °exchange, °displace, °relieve, °supplant; °switch; take the place of, stand in for, double for, *Colloq* sub for, cover for, swap *or* swop, *US and Canadian* pinch-hit for: *She substituted the real diamond with a paste replica. I will substitute for you while you are away.* —*n.* **2** °substitution, replacement, °alternative, °relief, °representative, °deputy, °delegate, °stand-in, °standby, °understudy, surrogate, succedaneum, *Brit* locum (tenens), *US and Canadian* °alternate: *In those days a man could hire a substitute to serve in his place in the militia.*

substitution *n.* **1** °exchange, exchanging, °change, changing, replacement, replacing, supplanting, °switch, switching, interchange, interchanging, *Colloq* swap *or* swop, swapping *or* swopping: *The substitution of certain words distorted the sense.* **2** See **substitute, 2,** above.

substratum *n.* substrate, °foundation, underlayer, °basis, fundament, °base, substructure, °groundwork: *The agreement rests on a substratum of mutual trust and respect.*

subterfuge *n.* °artifice, °trick, °device, °stratagem, °maneuver, ploy, °evasion, °deception, °dodge, °feint, °shift, °excuse, °expedient, contrivance, °intrigue: *Although he told no lies, he used every subterfuge to avoid telling the truth.*

subtle *adj.* **1** °delicate, °fine, °refined, °exquisite, °nice, *Archaic* subtile: *I became aware, almost subconsciously, of the subtle odor of jasmine. These subtle shades of green are so peaceful.* **2** abstruse, arcane, °recondite, °remote, °deep, °profound, concealed, °hidden, °shadowy, °nebulous, °vague, °obscure, °veiled, °thin, airy, °insubstantial, °elusive, °faint; sophistic(al): *The symbolism is so subtle as to be meaningful to the poet alone. A subtle hint gave me to understand that I ought to go.* **3** °tricky, °shrewd, cunning, °wily, °sly, °devious, °crafty, °smart, °clever, °foxy, °artful, °scheming, °designing, underhand(ed), °deceptive, jesuitical, Machiavellian, °ingenious, °skillful, °strategic, insidious, casuistic, °shifty, °slick, °slimy, *Chiefly Brit* smarmy: *You were a fool to allow that subtle serpent to insinuate himself into your confidence.*

subtlety *n.* **1** °refinement, nicety, °delicacy, exquisiteness, intricacy, fineness, acuteness, °elegance, °sophistication: *The subtlety of the detail in this work is quite unique. One must admire the subtlety of expression in her writing.* **2** treachery, guile, insidiousness, casuistry, cunning, artfulness, craftiness, deviousness, slyness, deceptiveness: *The subtlety of her deception was revealed only after the war.*

subtract *v.* **1** °deduct, take away, °take off, take (something) from: *First calculate the discount, then subtract it from the price.* **2** Sometimes, ***subtract from.*** °detract (from), °diminish, take away (from): *Nothing could subtract from the exhilaration of that moment.*

subversion *n.* °overthrow, °ruin, °destruction, undermining, °upheaval, displacement: *The dictator used extreme tactics to avoid the subversion of his leadership.*

subversive *adj.* **1** subversionary, °seditious, seditionary, treasonous, treacherous, °traitorous, °revolutionary, insurrectionary: *The government regarded as subversive anything and anyone disagreeing with their policies.* —*n.* **2** °traitor, insurgent, saboteur, fifth columnist, collaborator, collaborationist, quisling, °radical, °revolutionary, insurrectionist, insurrectionary; °dissident, °defector: *The security agencies were weeding subversives out of sensitive jobs.*

subvert *v.* °overthrow, °ruin, °destroy, °undermine, °topple, °demolish, °wreck, °sabotage: *The seventh dynasty was subverted from within, not by outside forces.*

subway *n.* **1** *Brit* °underground (railway), tube: *She takes the subway to work.* **2** *US* °tunnel, underpass: *Use the subway to cross the road in safety.*

succeed *v.* **1** °follow, come after, supervene: *The end of the war in the Pacific succeeded the surrender in Europe.* **2** Often, ***succeed to.*** be successor (to), °follow, be heir (to), °replace, take the place of, inherit *or* take over from: *Who succeeded Queen Nefertiti?* **3** Often, ***succeed in*** *or* ***at.*** °make good, °thrive, °prosper, °flourish, be a success, be successful, °progress, °advance, get ahead *or* on, attain *or* gain *or* achieve success, °win, °triumph, *Colloq* °make it, °arrive, get to the top: *Has she succeeded in persuading you to sing? He always wanted to make a lot of money and now he has succeeded. They have succeeded at whatever they have tried.*

success *n.* **1** good *or* happy result *or* outcome, good fortune, °achievement, °triumph, attainment, ascendancy, °prosperity: *The success of the fast food business is evident everywhere.* **2** °star, °celebrity, °(big) name, °sensation: *She was a great success as a singer, dancer, and actress.*

successful *adj.* **1** °wealthy, °rich, °prosperous, °fortunate, °lucky, °flourishing, thriving, prospering, well-to-do, affluent, *Colloq* °loaded, well-heeled, °flush, °in the money, *US* well-fixed: *He is a very successful investor.* **2** lucrative, booming, °profitable, °fruitful, money-making, remunerative: *Burbank Boiler Company has been a successful investment. Was last year successful, too?* **3** °famous, °well-known, famed, °celebrated, °renowned, °eminent; °prominent, °preeminent, °popular, °leading, °top, best-selling: *Donald is one of our most successful poets. That book of his was also successful.* **4** °victorious, °triumphant; °first; °winning: *Henry was successful in his bid for the chairmanship. I want to see a successful conclusion to the tour.*

succession *n.* **1** °passing (on), handing down *or* on, transmittal, °transmission, °transfer, transferal, °shift, conveyance, conveyancing: *According to her will, the succession of the property is via her first-born.* **2** °sequence, °progression, °order, series, °turn, °course, °flow, °chain, °train, °procession: *The succession of events was as described.* **3** accession, assumption, attainment, °elevation, °promotion; °inheritance: *What was the year of Charles II's succession to the Spanish throne?* **4** °lineage, descent, birthright, °dynasty, ancestry, descendants, bloodline: *The eighteenth century began with the War of the Spanish Succession.* **5** ***in succession.*** one after *or* behind the other, at intervals, successively, consecutively, in a row, °running, without interruption, uninterruptedly, °in order, °in line: *They published several historical novels in quick succession.*

successive *adj.* uninterrupted, °continuous, unbroken, °continual, consecutive, succeeding: *Each successive success succeeded as a* succès d'estime.

succinct *adj.* °compact, °brief, °concise, pithy, °terse, °short, compressed, condensed, °epigrammatic: *He prefers a succinct style of writing, one that gets to the point.*

succulent *adj.* °juicy, °rich, °luscious, mouthwatering, toothsome: *A shipment of the most succulent peaches has arrived at the fruit shop.*

succumb *v.* °yield, °give up, °give way, °surrender, accede, °submit, °capitulate: *I finally succumbed and agreed to their request for help.*

sucker *n.* °dupe, goat, gull, °victim, °butt, cat's-paw, °fool, *Colloq* °(easy) mark, easy *or* fair game, chump, °pushover, soft touch, *Chiefly US and Canadian* fall guy, *Slang* °sap, pigeon, *Brit* °mug, *Chiefly US and Canadian* patsy: *If Barnaby won't buy Dick's car at that price, Dick will find some other sucker who will.*

sudden *adj.* unexpected, unannounced, unanticipated, °unforeseen; °unwonted, surprising, °startling; °precipitate, °abrupt, °quick, °immediate, °rapid, °swift, °brisk; °impetuous, °hasty, °rash, °impulsive: *His sudden turn caught us off balance. Your change of heart was quite sudden.*

suddenly *adv.* **1** in a flash *or* a moment *or* a split second, all at once, instantly, °instantaneously, momentarily, fleetingly, in the twinkling of an eye, in a trice; °quickly, abruptly, °swiftly, speedily, °rapidly: *Suddenly, my whole life flashed before me. She suddenly realized why the head of the department had summoned her.* **2** all of a sudden, °out of the blue, unexpectedly, without warning, °on the spur of the moment, °hastily, hurriedly, feverishly: *Suddenly, she turned on her heel and stormed out.*

sue *v.* **1** proceed *or* move *or* act (against), take (legal) action *or* bring suit *or* prefer charges (against); summon(s), °charge, °accuse: *He threatened to sue. They are suing him for the injuries they sustained.* **2** °petition, °beg, °plead, entreat, °pray, °request, °solicit, °apply, °beseech, implore, supplicate: *After five years of constant warfare, the country is suing for peace.*

suffer *v.* **1** Sometimes, ***suffer from*** *or* ***with.*** agonize, °smart, °hurt, writhe, °sweat, °ache: *He is suffering with a stomach virus. How terribly they suffered when their child was kidnapped!* **2** °endure, °undergo, °experience, °bear, live *or* go through, °tolerate, °withstand, °sustain, °take, °submit to, °abide, °submit to, *Colloq* °put up with: *You have no idea the punishment the survivors of the concentration camps suffered.* **3** °allow, °tolerate, °permit, °let, °admit, °humor, °indulge: *They suffer the storks to build their nests on their chimneys.* **4** °deteriorate, °diminish, °decline, °go down, °fall off, be reduced *or* diminished: *If you discontinue advertising, sales will suffer.*

suffering *n.* °pain, °agony, °distress, °misery, °affliction, °hardship, °torment, torture, tribulation, °trial: *The man's suffering is written in his face.*

suffice *v.* °satisfy, °serve, do, be sufficient *or* enough *or* adequate, °answer, sate, °satiate, °quench: *If you are merely thirsty, water should suffice.*

sufficient *adj.* °adequate, °enough: *Have we sufficient food to last through the winter?*

suffix *n.* **1** ending, desinence, °addition; affix: *The suffix* -ness *is added to many English adjectives to form nouns.* —*v.* **2** °add (on), °join, fasten to, subjoin, °amend, *Colloq* °tack on: *One can suffix* -ness *to a huge number of adjectives to form nouns.*

suffrage *n.* °(right to) vote, voting right(s), franchise, °voice, °say, ballot, °option, °choice: *They campaigned for universal adult suffrage.*

suffuse *v.* overspread, imbue, pour *or* spread over, bathe, °cover, °permeate, pervade, °flood, °flush, °penetrate, °saturate, °mantle, infuse, °transfuse, imbrue *or* embrue: *A rosy glow suffused her face.*

suggest *v.* **1** °propose, °advance, °recommend, °urge, °advocate, °support, °offer, proffer, put *or* set forward, °present, °mention, °introduce: *Wasn't it Whitfield who suggested coming here for dinner? Did you suggest an alternative?* **2** call to mind, °bring up, °hint (at), °imply, °insinuate, °intimate, make one think, lead one to believe, °indicate: *These trophies suggest that she is interested in racing.*

suggestible *adj.* °impressionable, °susceptible, °receptive, impressible, susceptive, °open, moldable, fictile: *Octavia was very suggestible at that age and readily did everything she was bidden.*

suggestion *n.* **1** °proposal, °proposition, °recommendation, °plan, °advice, °counsel, °idea, °notion, °opinion; prompting, urging: *The membership thought little of your suggestion that the dues be increased.* **2** °indication, °trace, °whisper, insinuation, °innuendo, °implication, intimation, °hint, *soupçon,* °touch, tinge, °suspicion, °breath, iota, jot or tittle: *There wasn't the slightest suggestion of animosity in her criticism.*

suggestive *adj.* **1** Often, ***suggestive of.*** °reminiscent (of), evocative (of), indicative (of): *To me, the smell of orange blossoms is always suggestive of southern California.* **2** °provocative, °naughty, °risqué, ribald, °off-color, °racy, °bawdy, °earthy, °lusty, °rude, indelicate, °unseemly, °immodest, °improper, °indecent, °prurient, °blue, °offensive, °vulgar, smutty, °dirty, °pornographic, °lewd, salacious, *Colloq* °sexy, °spicy, *Slang* raunchy: *He used to sing his suggestive songs in night clubs.*

suit *v.* **1** °adapt, °accommodate, °fit, °adjust, °tailor, make appropriate *or* suitable: *It will require some reorganization of the space to suit the offices to our needs.* **2** please, °satisfy, fill (someone's) needs, °gratify, be acceptable *or* suitable *or* convenient to *or* for, befit; conform to: *Choose whichever room suits you. The schedule does not suit my personal plans.* —*n.* **3** jacket and trousers *or* skirt, °outfit, °uniform, °ensemble, °costume, °habit; garb, clothing, °clothes, livery: *He showed up wearing a new silk suit.* **4** lawsuit, °action, °case, °proceeding, °process, °cause, °trial; °litigation: *I do not want to enter into a suit over the damage to my car.* **5** °petition, °plea, °request, entreaty, °prayer, solicitation, °application, °appeal, °supplication; courtship: *The queen has rejected all suits for her daughter's hand in marriage.*

suitable *adj.* °appropriate, apt, °fit, °fitting, °befitting, °becoming, °right, °proper, °correct, °acceptable, °satisfactory, °applicable, °meet, °seemly; °timely, °opportune: *Ferdinand has at last found suitable employment. That chandelier is not suitable for your decor. Is this a suitable time for me to ask a favor?*

suitcase *n.* °bag, valise, overnight bag, holdall, grip *or Brit* handgrip, *Brit formerly* portmanteau: *You traveled all over Europe with only one suitcase?*

suite *n.* **1** °set, series, °collection, °number: *They gave us a beautiful suite of rooms at the hotel.* **2** °set: *She bought a new suite of dining room furniture.* **3** following, °retinue, entourage, °train, cortège, convoy, °escort; followers, attendants, retainers: *Her suite includes advisers, secretaries, hairdressers, bodyguards, servants—you name it.*

suitor *n.* °admirer, beau, wooer; boyfriend, °paramour, °lover, inamorato, *cicisbeo,* °escort, *Archaic* swain: *Madelaine has more suitors than I can count.*

sulk *v.* mope, °brood, °pout, be sullen *or* moody *or* ill-humored: *I wish you'd get a job instead of sulking about the house all day.*

sullen *adj.* sulky, sulking, morose, brooding, pouting, °gloomy, °moody, °temperamental, °dour, lugubrious, °funereal, °dismal, °dreary, °grim, depressing, depressed, churlish, ill-humored, °glum, grumpy, °somber, out of humor, antisocial, unsociable, °cross, °petulant, °perverse, crusty, crotchety, choleric, crabby, ill-natured, ill-tempered, bad-tempered, splenetic, °peevish, dyspeptic, °out of sorts, *US* °cranky: *He remained quite sullen and refused to associate with anyone.*

sully *v.* besmirch, °stain, smirch, °blemish, °mar, defile, °soil, °disgrace, °dirty, °tarnish, °pollute, °spoil, °ruin, °destroy, °wreck: *Crumleigh's reputation was sullied by his association with that debauched profligate, Tremlow.*

sultry *adj.* **1** °hot, °humid, °sticky, °stuffy, stifling, °oppressive, °close, °muggy, °steamy, steaming, °moist, °damp, °sweltering, suffocating: *I was happy to be languishing again in the sultry warmth of St. Lucia.* **2** °lusty, °lustful, °passionate, °erotic, °seductive, °voluptuous, °provocative, °sensual, *Colloq* °sexy, °hot: *Tondelayo continued her sultry, slow, sinuous dance.*

sum *n.* **1** °total, aggregate, grand total, sum total, °whole, °totality; °amount, °quantity: *The sum of money found in the satchel was not revealed by the police. He would give the neighborhood children small sums to spend on sweets.* —*v.* **2** ***sum up.*** **(a)** °recapitulate, summarize, encapsulate, synopsize, °digest, °abridge, condense, consolidate, epitomize, °review: *He summed up by reminding the children not to accept gifts from strangers.* **(b)** °reckon, add up, °calculate, °total, tot up, °measure (up), take the measure of: *We summed up the chances of escaping.* **(c)** °estimate, °evaluate, °size up, assess: *She had summed him up the minute she saw him.*

summarily *adv.* **1** °immediately, °at once, °straightaway, °directly, °quickly, without delay, unhesitatingly, without hesitation, forthwith, °promptly, °swiftly, speedily, expeditiously, instantly, *Colloq* PDQ (= 'pretty damn(ed) quick(ly)'): *Any dog caught without a muzzle will be impounded summarily.* **2** °suddenly, without warning, abruptly, peremptorily, precipitately: *He was summarily dismissed, with no advance notice.*

summary *n.* **1** summarization, recapitulation, encapsulation, compendium, °synopsis, °digest, °abridgment, condensation, shortening, consolidation, °epitome, epitomization, °review, distillate, conspectus, °brief, °outline, °précis, °résumé: *I want a summary of the meeting on my desk in the morning.* —*adj.* **2** °abrupt, °peremptory, °short, °quick, °brief, laconic, °perfunctory, °curt, °terse: *My simple request for some time off was denied with a summary, "No!"*

summit *n.* °peak, °top, apex, °acme, °pinnacle, °zenith, °crown; culmination, °climax: *The winds at the summit of Annapurna reached hurricane force. There was no doubt that I had reached the summit of my achievements.*

summon *v.* **1** °call, °assemble, convoke, convene, °send for, invite, °muster, °get together, °arouse, °rouse: *Why would he summon us to the office at midnight?* **2** Often, ***summon up.*** call *or* draw on *or* upon, °draw up, °mobilize, °muster (up), °work up, °gather, invoke: *I had to summon up all my courage to ask Eve to go out.*

sumptuous *adj.* °expensive, costly, °extravagant, °exorbitant, °dear, °rich; °lavish, °luxurious, de luxe, °opulent, °palatial, °royal, °majestic, °regal, °magnificent, °dazzling, °splendid, °showy, *Colloq* °posh, °plush, ritzy: *The family treated us to a sumptuous repast to celebrate our fiftieth wedding anniversary.*

sun *n.* **1** (old) Sol, Helios, Phoebus (Apollo), Ra, Sunna, daystar: *That lucky old sun just rolls around heaven all day.* —*v.* **2** tan, suntan, sunbathe, bask, bake, brown, bronze: *I was sunning myself at Ipanema Beach when a girl walked by.*

sundries *n.* pl. knickknacks *or* nicknacks, trinkets, small items, notions, miscellanea, °miscellany, kickshaws, brummagem, frippery, °bric-a-brac, °odds and ends: *We might find a button for your shirt in a shop that sells sundries.*

sundry *adj.* °various, °varied, °miscellaneous, assorted, °different, °mixed, diversified, °divers: *In addition to food, supermarkets stock sundry other items.*

sunken *adj.* **1** °hollow, hollowed-out, °haggard, °drawn: *From her sunken cheeks the child looked as if she hadn't eaten for weeks.* **2** submerged, undersea, underwater, submersed: *The boat hit a sunken wreck and foundered.* **3** buried, °underground, in-ground, below-ground, settled, lowered: *The old term for a sunken fence is "ha-ha."*

sunless *adj.* °dark, °grim, cheerless, °unhappy, °joyless, °funereal, depressing, °dreary, drear, °somber, °gloomy, °gray, Stygian, °black, pitchy, inky, °shadowy, tenebrous, unlit, unlighted, °dusky, subfusc *or* subfuscous, darkling: *Just where did Alph, the sacred river, run into a sunless sea?*

sunny *adj.* **1** sunlit, sunshiny, °brilliant, °bright, °radiant, °fair, °fine, cloudless, °clear, unclouded: *The people there tell you it is never sunny because they hate tourists.* **2** °cheerful, cheery, °happy, joyous, °joyful, lighthearted, smiling, beaming, °buoyant, °blithe, °gay, mirthful, °jolly, °bubbly, °ebullient, °genial, °warm, °friendly, °outgoing: *Being with someone who has a sunny disposition was a pleasant change.*

super *adj.* wonderful: *A super collection of synonyms for this word appears under* **superb,** *below.*

superb *adj.* wonderful, °marvelous, °excellent, °superior, °gorgeous, °glorious, °magnificent, °outstanding, °exquisite, °fine, °splendid, unequaled, °sensational, °noteworthy, °admirable, °peerless, °matchless, unrivaled, °first-rate, °superlative, °perfect, °classic, °exceptional, °extraordinary, °striking, °brilliant, °dazzling, °miraculous, °incredible, °unbelievable, °fantastic, °fabulous, stupendous, staggering, mind-boggling, breathtaking, *Woman's dialect* °divine, *Colloq* °great, °super, smashing, °magic, °terrific, fantabulous, °unreal, °out of this world, mind-blowing, °super, far-out, *Slang* °out of sight, °boss, °solid, °cool, °hot, °bad: *The curator said he had never seen such a superb collection. She does a superb* Hedda Gabler. *We were treated to a superb dinner.*

supercilious *adj.* °haughty, °contemptuous, °superior, °snobbish, °disdainful, °arrogant, °condescending, patronizing, °overbearing, °scornful, lordly, high and mighty, °pompous, °lofty, °stuffy, °pretentious, *Colloq* °hoity-toity, highfalutin *or* hifalutin *or* hifaluting, uppity, snooty, stuck-up, *Brit* toffee-nosed, °uppish, la-di-da *or* lah-di-dah *or* la-de-da: *She looked upon her employees' personal lives with supercilious disdain.*

superficial *adj.* **1** °surface, °external, °exterior, °shallow, °skin-deep, °slight, °outside: *I suffered a superficial wound to the forehead.* **2** °surface, °slight, °external, °apparent, °skin-deep, °outward, °cursory, °insignificant, °passing, unimportant, trivial, °empty, °insubstantial; paying lip-service, for appearances' sake, cosmetic: *She bears a superficial resemblance to your sister. The empire's interest in the welfare of its colonies was purely superficial.* **3** °cursory, slapdash, °quick, °hurried, °hasty, °perfunctory, °nominal, °meaningless, °passing: *The machinery is subjected to a superficial inspection before shipping.*

superficies *n.* °(outer) surface, façade, °face, externals, °outside: *The superficies may appeal to children, but adults appreciate the book's profundities.*

superfluity *n.* °excess, superabundance, overabundance, °surplus, oversupply, °surfeit, °glut, superfluousness, °profusion, plethora, oversupply, supersaturation: *The middle and upper classes wallow in a superfluity of goods.*

superfluous *adj.* °excessive, °excess, superabundant, overabundant, supererogatory, °surplus, unneeded, uncalled-for, °unnecessary, °redundant, °extra; °needless, °dispensable, °gratuitous: *The superfluous grain that remains unsold should be donated to third world countries. His widow certainly shed a few superfluous tears over him.*

superhuman *adj.* **1** °heroic, herculean, °godlike, °legendary, valiant, °courageous, °brave, °daring, °dangerous, death-defying, °extraordinary, °miraculous, °phenomenal, °incredible, °fabulous, °fantastic, °unbelievable, °amazing: *His superhuman feats would never be duplicated.* —*n.* **2** superman, °hero, superhero, *Übermensch,* Hercules: *They still believe that one day a superhuman will come along to lead them.*

superintendent *n.* °supervisor, °foreman, °overseer, °manager, administrator, °chief, °head, °boss; governor, controller, °director, conductor: *Alan has been appointed superintendent of the new project.*

superior *adj.* **1** higher, higher-ranking, higher-level, higher-class, higher caliber, °upper, upper-level, °upper-class, loftier, nobler, °better; of a higher order *or* status *or* standing, *Colloq* classier, tonier: *Report to your superior officer for reassignment. It is evident that she comes from a superior tier of society to his.* **2** °high-class, °elevated, °first-rate, °distinguished, °exceptional, °excellent, preferred, choice, °select, élitist, °outstanding, °superlative, °matchless, unequaled, °peerless, °nonpareil, °sterling, °supreme, °fine, °noteworthy, °notable, °worthy, °estimable: *Kermit has had the benefit of a superior education.* **3** See **supercilious,** above. —*n.* **4** See **supervisor,** below.

superiority *n.* **1** ascendancy, °preeminence, °supremacy, °leadership, °lead, dominance, °predominance, primacy, hegemony: *Her superiority in the field is unchallenged.* **2** °excellence, greatness, peerlessness, matchlessness, inimitability, superlativeness, °prominence, eminence, °importance, °distinction, °prestige, °renown: *All commented on the superiority of Mark's contribution.*

superlative *adj.* unsurpassed, °paramount, °supreme, consummate, °superior, °best, choicest, finest, °matchless, °peerless, unequaled, unrivaled, °singular, °unique, °incomparable, °excellent, °superb, °sterling, °dazzling, °first-rate, °exceptional, °extraordinary, °marvelous, spectacular, °capital, *Colloq* tip-top, °super, smashing, °great, ace, °terrific, °fantastic, *Slang* °crack: *Max turned in a superlative performance in the tournament.*

supernatural *adj.* preternatural, °unusual, °extraordinary, °exceptional, °unnatural, °miraculous, °remarkable, °fabulous, preterhuman, °ghostly, °spectral, °abnormal, °inexplicable, unexplainable; metaphysical, otherworldly, °unearthly, ultramundane, supramundane, extramundane, °occult, mystic, paranormal, °psychic, uncanny, °weird, °mysterious, arcane, °unreal, magical, °dark: *Some claim to hear supernatural voices speaking to them. He says that he avoids using his supernatural powers, which would enable him to lift this building.*

supersede *v.* °replace, °succeed, °displace, °supplant, oust, take the place of, °substitute for: *The 286 computers have been superseded by 386 models.*

supervise *v.* °oversee, °overlook, °watch (over), °manage, °run, °control, superintend, °govern, °direct, be in *or* have charge (of), °handle, keep an eye on, °administer: *They are moving you up to supervise the production department.*

supervisor *n.* °overseer, °foreman, °manager, controller, °superintendent, °superior, governor, °director, °boss, °chief, °head, administrator: *As supervisor, you are responsible for the work done in your department.*

supervisory *adj.* managerial, administrative, °executive: *Do you think that Legree is ready for supervisory responsibilities?*

supine *adj.* **1** °flat (on one's back), °lying (down), °prostrate, °recumbent, *Formal or technical* procumbent, accumbent, decumbent: *When I found him, he was supine, staring up at the stars.* **2** °indolent, °lazy, °lethargic, °idle, °listless, °indifferent, apathetic, unconcerned, uninterested, °torpid, languid, languorous, sluggish, °slothful, °phlegmatic, lymphatic, °lackadaisical, °inert, °inactive, °passive, motionless, °inanimate, spiritless, abject: *During the entire crisis, the town council was completely supine and uncaring.*

supplant *v.* °replace, °displace, oust, °turn out, °eject, °remove, °expel, °dismiss, unseat, °supersede, °substitute, °exchange: *The council will be supplanted by a new group more sympathetic to our needs.*

supple *adj.* **1** °flexible, flexile, pliant, bendable, °elastic, resilient, °pliable, tractile, fictile: *He made a rugbeater out of some supple lengths of bamboo.* **2** °willowy, lithe, limber, °nimble, pliant, lissom *or* lissome, °graceful, athletic: *She has the supple body of a dancer.* **3** °tractable, compliant, °yielding, °accommodating, °obliging, complaisant, acquiescent, °submissive, unresistant, unresisting, °servile, °obsequious, °ingratiating, fawning, toadying: *With supple words the miserable sycophant wormed his way into father's confidence.*

supplement *n.* **1** addendum, °addition, appendix, epilogue, endpiece, postscript, appendage, °extension, continuation, adjunct, annex, appurtenance, °accessory, codicil, °insert, °sequel; supplementation; *Technical* suppletion: *The later version includes the 1909 supplement.* —*v.* **2** add (on *or* to), °extend, augment; °complement: *Various scholars supplemented the earlier work with their own notes and comments. He supplements his income by working at a restaurant.*

supplementary *adj.* **1** additional, added, annexed, adjunct, °new: *The supplementary buildings increased the size of the hospital greatly.* **2** supplemental, °supportive, contributory, ancillary, °secondary, °subordinate, annexed, additional, °attached, added, appended, °subsidiary, adscititious; °extraneous, adventitious, supervenient, °extra, °excess; *Technical* suppletive: *The supplementary material expanded the original work fivefold.*

supplicant *adj.* **1** suppliant, supplicating, entreating, petitioning, supplicatory, beseeching, praying, imploring, °solicitous, importunate, begging, mendicant: *In 1897, he carved a supplicant group in white marble, now in the Fraser Museum.* —*n.* **2** suppliant, applicant, petitioner, beseecher, °suitor, pleader, aspirant, appellant, plaintiff, °beggar, mendicant: *He came to me as a supplicant, begging forgiveness for what he had done.*

supplication *n.* **1** entreaty, °petition, °prayer, °appeal, pleading, °plea, °suit, solicitation, obsecration, obtestation, impetration: *Heeding the pitiful supplications of the beggars, we gave them some money.* **2** supplicating, begging, pleading, soliciting, petitioning, entreating, beseeching: *All his supplication could not wring a farthing from the old skinflint.*

supply *v.* **1** °furnish, °provide, °give, endow, present, purvey, °deliver, come up with, °contribute, °distribute, °sell; °stock, °accommodate, °afford, °equip, °outfit, °gear (up), °rig (out), °fit (out), °provision, °cater to, *Chiefly Brit* kit out *or Chiefly Brit* up; victual: *Her company supplies radios to the army. Her husband's firm supplies the navy with antifouling paint.* **2** °yield, °give, °contribute, come up with, °deliver, °provide, °furnish: *The farm supplies our basic foods.* **3** °satisfy, °fulfill, °replenish, °fill: *Can you supply the demand for clothing?* —*n.* **4** °stock, stockpile, °store, inventory, °quantity, reservoir, °reserve, °cache, °hoard, °accumulation, °fund: *Our supply is big enough to serve the entire area.* **5** furnishing, °provision, °providing, purveying, supplying, °distribution, equipping, outfitting, provisioning, °delivery, stocking, stockpiling: *The supply of microchips on a large scale is beyond our capacity.*

support *v.* **1** °back (up), °stand by, °help, °bolster, °uphold, °brace, °strengthen, °fortify, °buttress, °prop (up), shore up, °reinforce, °boost, °champion, °assist, take up the cudgels for, °aid, °promote, °forward, °second, °advance, °advocate, °stand up for, be supportive (of *or* in), *Colloq* °stick up for: *Peterson agreed to support you in your bid for the presidency.* **2** °brace, °hold up, °carry, °prop (up); °strengthen, shore up, °reinforce, °fortify, °buttress: *You need a column to support this beam.* **3** °tolerate, °bear, °stand (for), °suffer, °submit to, °undergo, °brook, °stomach, °endure, °abide, countenance, °face, *Brit* °stick, *Colloq* °put up with: *He cannot support the notion that she might be guilty.* **4** °pay for, °fund, °maintain, °keep, °finance, °subsidize, °underwrite, °sponsor, *Colloq US* bankroll: *I can no longer support myself or my extravagances.* **5** °sustain, °withstand, °stand, °take, °bear, °tolerate, hold up under, °weather: *That bracket looks too weak to support so much pressure.* **6** °verify, corroborate, °authenticate, vouch for, °endorse, °confirm, affirm, °bear out, attest to, °certify, °substantiate, validate, °ratify: *Can anyone support his alibi?* —*n.* **7** °help, °backing, backup, °reinforcement, bolstering, °encouragement, reinforcing, fortifying, °assistance, °aid, succor, °sustenance: *Thank you for your support over the years.* **8** °brace, °prop, °stay, °frame, °foundation, underpinning, substructure, truss, °beam, column, °pillar, °strut, °guy, guywire, °mainstay, °buttress, °bolster, °reinforcement, °supporter: *The supports failed under the grandstand and it fell, injuring a few. Aloysius was a tremendous support during my illness.* **9** °sustenance, °(living) expenses, °keep, °maintenance, °subsistence, °upkeep; °finances, funding: *Mona is suing David for the support of their child. He lives mainly on government support.*

supportable *adj.* **1** °tolerable, °bearable, endurable, °acceptable, sufferable: *Thoughts of death are more supportable when it seems remote.* **2** defensible, confirmable, verifiable, °demonstrable, °tenable, believable: *The argument that we should keep the old tax structure is simply not supportable.*

supporter *n.* **1** °enthusiast, °champion, promoter, °fan, aficionado, °devotee, °admirer, °backer, °follower, °support, °advocate, exponent, adherent, °aid, °assistant, helper: *I wish to thank my many supporters, who were there when I needed them.* **2** See **support, 8,** above.

supportive *adj.* °helpful, sustaining, supporting, encouraging, °sympathetic, °understanding, reassuring: *Among our many friends, Cleo was the most supportive during the recent trouble.*

suppose *v.* **1** °assume, °presume, °presuppose, °surmise, °take, take as given *or* as read, take for granted; °believe, °think, °fancy, °imagine; *Colloq* °take it: *Don't people usually suppose that civil servants are honest? I supposed her to be his wife. Do you suppose you could return my key today?* **2** hypothesize, °theorize, postulate, °posit, °assume: *Supposing a reduced rate of inflation, will they reduce interest rates? Suppose that you have lost the election—what will you do?*

supposed *adj.* **1** °alleged, °assumed, putative, °reputed, presumed, °hypothetical, °theoretical, theorized, imagined, suppositious, supposititious: *Her supposed drug addiction turned out to be a myth.* **2** °obliged, expected, required; meant, intended: *You were supposed to return my key. What is that supposed to mean?*

supposedly *adv.* allegedly, reputedly, theoretically, hypothetically, °presumably; rumor has it: *The furniture is to be delivered, supposedly today.*

supposing *conj.* if, even if, in the event that, despite the fact that, although, °though: *Supposing he says no, what will you do?*

supposition *n.* assumption, °presumption, °surmise, °belief, °thought, °fancy, theory, °hypothesis, postulate, °proposal, °proposition: *On the supposition that your calculations are correct, the earth should have exploded yesterday.*

suppress *v.* **1** °end, °discontinue, °cut off, °cease, °stop, °terminate, put an end to, °halt, °prohibit, °preclude, °prevent, °repress, censor, °forbid, interdict, °block, °obstruct, °withhold, °stifle, °inhibit, °hinder, °arrest: *They have been unable to suppress the publication of the book.* **2** °put down, °quell, °crush, °squelch, °quash, °subdue, °check, °stamp out, snuff out, °smother, °extinguish, °quench, crack down on: *The uprising was suppressed with the shooting of the students.* **3** keep down, °control, keep under control, keep *or* hold in check, °restrain, hold in *or* back, °repress, °cover up, °conceal, °hide, keep quiet *or* secret, °mute, °muffle, °quiet, °silence: *Objections to the guest speaker were suppressed at the request of the university's chancellor.*

suppression *n.* suppressing, ending, °end, discontinuation, discontinuing, cutting off, cutoff, cessation, ceasing, surcease, stopping, °stop, terminating, °termination, °halting, °halt, prohibiting, °prohibition, preclusion, precluding, preventing, °prevention, repressing, °repression, censoring, censorship, °forbidding, forbiddance, interdicting, interdiction, blocking, obstructing, °obstruction, withholding, stifling, hindering; putting down, °put-down, quelling, crushing, squelching, quashing, subduing, checking, °check, stamping out, smothering, snuffing out, extinguishing, extinction, elimination, quenching, cracking down on, crackdown; °control, controlling, restraining, °restraint, concealing, concealment, hiding, muting, muffling, quieting, silencing: *That constitutes suppression of the freedom of speech. The suppression of heresy is a perennial church problem. Dictators confuse the suppression of ideas with the oppression of people.*

supremacist *n.* supremist, bigot, racist, racialist, dogmatist, °zealot, °fanatic: *Supremacists' philosophies are based on hatred, not love.*

supremacy *n.* **1** transcendency, °preeminence, °superiority, ascendancy, °excellence, primacy, peerlessness, matchlessness, incomparability, inimitability: *Colonialism was aided by the supremacy of British gunpowder.* **2** °sovereignty, °dominion, °sway, mastery, °control, dominance, (supreme *or* absolute) rule *or* authority, autarchy, omnipotence, hegemony: *They maintained their supremacy over the Iberian peninsula for centuries.*

supreme *adj.* **1** highest, loftiest, topmost, greatest, °first, °foremost, °principal, unsurpassed, °top, °uppermost, °chief, °paramount, °sovereign: *There is no appeal to a decision of the supreme tribunal.* **2** greatest, °maximum, °extreme, uttermost, utmost, °ultimate: *I fear that your son has made the supreme sacrifice, Mrs. Atkins.* **3** °best, greatest, °first, °outstanding, °preeminent, °first-rate, °prime, °primary, unexcelled, °leading, crowning, consummate: *Many regard her as the supreme artist of her day.* **4** °superb, °marvelous, °excellent, °outstanding, °superlative, °matchless, °peerless, °incomparable, °unparalleled, °masterful, masterly, °sublime, °brilliant, °transcendent, inimitable, °choice: *There is no doubt that she is a supreme artist.*

supremely *adv.* °very, °extremely, °completely, °perfectly, superlatively, sublimely, transcendently: *I have never been so supremely happy as I was in Capri in '38.*

sure *adj.* **1** °certain, assured, convinced, persuaded, °positive, °definite, unwavering, unswerving, unflinching, °steadfast, °steady, unshakable *or* unshakeable, °confident, satisfied, undeviating, unfaltering: *How can you be so sure in your opinion that Whimsey did it? I'm sure I've met you before.* **2** established, °firm, °solid, trusty, °stable, °steadfast, °secure, °safe, °trustworthy, °reliable: *I never travel without my sure companions, courage and caution.* **3** °accurate, °reliable, dependable, tried and true, unfailing, °infallible, °foolproof, °effective, *Colloq* sure-fire: *They say that chicken soup is a sure cure for what ails you.* **4** °certain, °inevitable, indubitable, °unavoidable, ineluctable, inescapable, guaranteed: *He who enters the Devil's Cave meets sure death.*

surely *adv.* **1** certainly, to be sure, °positively, °absolutely, °definitely, °undoubtedly, indubitably, unquestionably, beyond the shadow of a doubt, beyond question, °doubtless, doubtlessly, assuredly, *Colloq* °sure, *US* absotively-posolutely: *He is surely one of the finest riders I've ever seen.* **2** °firmly, solidly, confidently, unfalteringly, steadily, unswervingly, unhesitatingly, determinedly, doggedly, securely: *Slowly but surely, the Great Bamboni stepped out onto the tightrope stretched across the Niagara Falls.*

surface *n.* **1** °exterior, covering, °outside, °top, °skin, integument, façade, °face, °boundary, interface, °superficies; °side, °plane: *Most of the earth's surface is covered by water. How many surfaces on an icosahedron?* **2** ***on the surface.*** superficially, to all appearances, at first glance, °outwardly, to the casual observer, extrinsically, °ostensibly: *It looks pretty enough on the surface, but it is rotten underneath.* —*v.* **3** °appear, °show up, °emerge, °materialize, °arise, °rise, °come up, *Colloq* pop up, crop up: *Guess who surfaced after ten years in Tierra del Fuego! An enormous turtle surfaced near our boat.* **4** °pave, °concrete, tarmac: *They are surfacing the road in front of my house.*

surfeit *n.* overabundance, superabundance, plethora, °glut, °excess, °surplus, oversupply, overdose, °satiety, overflow, °flood, deluge, °superfluity, nimiety: *There has been a surfeit of those dolls on the market. She is suffering from a surfeit of sushi and sashimi.*

surfeited *adj.* gorged, overfed, satiated, sated, stuffed, glutted, °jaded: *He passed the sleeping surfeited diners and stole into the harem unseen.*

surge *v.* **1** °swell, °wave, billow, °bulge, °heave, °roll, undulate, well forth *or* up, rise and fall, ebb and flow, °pulsate; °rush, °gush, °pour, °flood, °stream, °flow: *The sea surged through the narrow gorge with a roar. The crowd surged around the hero.* —*n.* **2** °swell, °wave, billow, °roller, whitecap, white horse, breaker, comber, upsurge, °eddy, °rush, °gush, °flood, °stream, °flow: *The raft was caught by a surge and tossed high up on the beach.*

surly *adj.* unpleasant, °rude, crusty, °cantankerous, curmudgeonly, churlish, crabby, crabbed, choleric, splenetic, dyspeptic, °bilious, °temperamental, °cross, crotchety, grouchy, grumpy, bearish, °testy, °touchy, °short-tempered, ill-tempered, bad-tempered, ill-natured, bad-natured, ill-humored, °peevish, °quarrelsome, °argumentative, °obnoxious, uncivil, °rough, °obstreperous: *I see no reason to shop where the salespeople are surly.*

surmise *v.* **1** °imagine, °guess, conjecture, °speculate, °suppose, hypothesize, °theorize, °assume, °presume, conclude, °gather, °infer, °understand, °fancy, °suspect, °feel, °sense: *I surmised that the treaty terms were arranged by deputies long before the summit.* —*n.* **2** °guess, conjecture, °speculation, °notion, °hypothesis, theory, °supposition, assumption, °presumption, conclusion, °understanding, °fancy, °suspicion, °feeling,

°sense: *It was an early surmise of the experts that man could not survive passing through the Van Allen belts.*

surpass *v.* °exceed, °excel, go *or* pass beyond, °outdo, °beat, worst, °better, °best, °outstrip, outdistance, outperform, outclass, outshine, °eclipse, °overshadow, °top, °cap, °transcend, prevail over, leave behind: *He easily surpassed the statesmen of his time.*

surpassing *adj.* °excessive, °extraordinary, °great, °enormous, unrivaled, °matchless, °peerless, unmatched, unequaled, unsurpassed: *I could not believe the surpassing gall of the man.*

surpassingly *adv.* °exceedingly, extraordinarily, incomparably, *Literary* °surpassing: *The girls at Poggibonsi were surpassingly beautiful.*

surplus *n.* **1** surplusage, overage, °excess, leftover(s), °surfeit, overabundance, oversupply, overdose, °glut: *The price of oil dropped owing to the surplus in the market.* —*adj.* **2** °excess, °leftover, °extra, °spare, overabundant, °superfluous, °unused, °redundant: *The surplus grain is sold abroad.*

surprise *v.* **1** °shock, °astound, °astonish, °amaze, disconcert, °nonplus, dumbfound *or* dumfound, °stagger, °take aback, °strike, °hit, *Colloq* °floor, bowl over, flabbergast, rock *or* set (someone) back on his *or* her *or chiefly Brit* their heels, *Brit* knock (someone) for six, *US* knock (someone) for a loop: *What surprised me was her coolness in the face of danger.* **2** take *or* catch unawares, catch red-handed *or* in the act *or* in flagrante delicto, catch napping *or* off-guard, °discover: *When Samuel Johnson's wife found him with a maid, she said, "Dr. Johnson! I am surprised!" to which he replied, "No, madam—you are amazed. I am surprised."* —*n.* **3** °shock, °astonishment, °amazement, stupefaction, °wonder, incredulity: *Imagine my surprise to learn that I was being arrested.* **4** °blow, °jolt, shocker, bolt from *or US also* out of the blue, °bombshell, eyeopener: *It came as a surprise to me that Mrs. R— was pregnant.*

surrender *v.* **1** °give up, °yield, let go (of), °relinquish, °deliver (up), °hand over, °forgo, °forsake, °turn over, °turn in, °part with, °cede, °concede: *We were forced at gunpoint to surrender our valuables.* **2** °give up, °yield, °quit, cry quits, °capitulate, throw in the sponge *or* the towel, raise the white flag, throw up one's hands, °succumb, °submit, °give way, acquiesce, °comply, °give in, °concede, °crumble: *The argument is so overwhelming that I must surrender.* —*n.* **3** °submission, capitulation, °yielding, renunciation, relinquishment, °yielding, transferal, °transfer, transference, handing *or* turning over, conveyancing, ceding, cession, concession: *The official surrender of the disputed territory took place in the following year.*

surreptitious *adj.* furtive, °secret, clandestine, °stealthy, underhand(ed), covert, °(on the) sly, °secretive, °private, concealed, °hidden, °veiled, *Colloq* °sneaky: *She stole surreptitious glances at him.*

surround *v.* **1** encompass, °encircle, °envelop, °enclose, hem in, °ring: *She likes being surrounded by flowers. Troops surrounded the building.* —*n.* **2** environs, °environment, surroundings, °atmosphere, ambiance *or* ambience, °setting: *The surround of formal gardens is an excellent setting for an art gallery.*

surrounding *adj.* °nearby, °neighboring, °local, °adjoining, °neighborhood, adjacent, bordering, abutting, circumambient, circumjacent: *The surrounding countryside was bursting into bloom.*

surveillance *n.* °observation, °watch, °scrutiny, °reconnaissance: *Was this entire activity carried on while you were under police surveillance?*

survey *v.* **1** °examine, appraise, °evaluate, take the measure of, °inspect, °study, °scan, °scrutinize, °measure, °size up, assess, °investigate, look into *or* over, °review: *She surveyed the situation and found nothing wrong.* **2** °view, °look at, get a bird's eye view of, °contemplate: *From the top of the tower I surveyed the surrounding countryside.* —*n.* **3** surveying, °examination, appraisal, °evaluation, °measure, °study, °scan, scanning, °scrutiny, inquiry *or* enquiry, °measurement, °investigation, inspection: *Our survey of the company finances yielded some very interesting information.*

survive *v.* **1** °continue, °last, °live (on), °persist, subsist, °pull through, °endure; remain solvent, keep one's head above water: *The village was destroyed, but its people survived. Companies found it difficult to survive during the last recession.* **2** °outlast, outlive: *At the age of 114, MacMurtagh has survived all his children and many of his grandchildren.*

susceptible *adj.* **1** Often, ***susceptible of*** or ***to.*** °open (to), °prone (to), °subject (to), °disposed (to), predisposed (to), °receptive (to), affected by, °responsive (to): *Brass is susceptible of a high shine. Are you still susceptible to her blandishments?* **2** °impressionable, influenceable, °vulnerable, reachable, °accessible, credulous, °suggestible, °gullible, naive *or* naïve *or* naïf: *Those peoples need more protection from western exploitation because they are more susceptible.*

suspect *v.* **1** disbelieve, °doubt, °mistrust, °distrust, harbor *or* have suspicions about *or* of, be suspicious of: *Do you suspect the butler?* **2** °feel, °think, °believe, °sense, have a feeling, °fancy, °imagine, °theorize, °guess, °surmise, have a sneaking suspicion, think it likely *or* probable, *Colloq* °expect: *I suspect that the butler might have done it.* —*adj.* **3** °suspicious, °questionable, °doubtful, dubious, °shady, °shadowy; suspected: *If you ask me, his behavior since boarding the Orient Express has been highly suspect.*

suspend *v.* **1** hold up *or* off (on), °withhold, °put off, put in(to) *or* hold *or* keep in abeyance, °shelve, °postpone, °delay, °defer, °interrupt, stop *or* check *or* cease *or* discontinue temporarily, *US* °table: *We suspended payment pending inspection of the work already done. Train service on this route will be suspended till further notice. Suspend your disbelief for a moment to consider the possible motive for stealing a prune.* **2** °hang, °attach, °fasten, °dangle, °swing: *She suspended the cameo from a gold chain which she wore as a necklace.* **3** debar, °exclude, °eliminate, °reject, °expel, °eject, °evict; deprive the rights of, deny the privileges of; blackball: *A member may be suspended if his dues are six months or more in arrears.*

suspense *n.* **1** uncertainty, indefiniteness, insecurity, °doubt, irresolution, expectancy, °indecision, not knowing: *May I have the envelope please? I cannot stand the suspense!* **2** °anxiety, °tension, apprehension, nervousness, °agitation, anxiousness, °anticipation, °expectation, °excitement: *The story that creates suspense for the reader is often suitable for film treatment.*

suspension *n.* **1** debarring, disbarment, °exclusion, elimination, °rejection, °expulsion, °ejection, °eviction, deprivation, °denial: *The suspension of privileges cannot last for more than a fortnight.* **2** °intermission, °moratorium, deferment, °holdup, °delay, delaying, °interruption, °postponement, postponing, discontinuing, discontinuation, °stay: *Luckily, the suspension of hostilities lasted till the summer.*

suspicion *n.* **1** °doubt, dubiousness, dubiety, °misgiving, °mistrust, °distrust, °skepticism, °qualm, wariness, apprehension, apprehensiveness, cautiousness, hesitation, second thought(s), uncertainty, leeriness, *Colloq* funny feeling, bad vibes: *We all had our suspicions about Edgar's genuine intentions. My worst suspicions have now been confirmed.* **2** °notion, °inkling,

°suggestion, °hint, °trace, °flavor, soupçon, °taste, °dash, glimmer, tinge, °touch, °shadow, °shade, scintilla, *Colloq chiefly US and Canadian* tad: *If you have even a suspicion of doubt, do not find the defendant guilty. Did I detect a suspicion of a sneer?*

suspicious *adj.* **1** °doubtful, °in doubt, dubious, °questionable, °debatable, suspect(ed), under suspicion, open to doubt *or* question *or* misconstruction, *Colloq* °shady, °fishy: *There is something very suspicious about her behavior.* **2** mistrustful, °distrustful, °doubtful, °in doubt, °skeptical, suspecting, disbelieving, °unbelieving, °leery, apprehensive, °wary, °uncertain, uneasy: *I was suspicious of Douglas from the very beginning.*

sustain *v.* **1** °uphold, °support, keep up, °maintain, °continue, keep (someone *or* something) going, keep alive, °preserve; °prolong, °persist in: *He has been sustained on a life-support system for weeks. If Georgette can sustain the pretense of loving him, he won't change his will.* **2** °support, °carry, °bear, °bolster, °buoy (up), °reinforce, keep (someone) going, °strengthen, shore up, underpin, °prop up, °buttress: *During those months as a hostage, only prayer sustained me.* **3** °endure, °stand, °withstand, bear up under, °put up with, °suffer, °undergo, °experience, °tolerate, °weather, °brave: *It is amazing how much punishment the human body can sustain.* **4** °bear, °carry, °support: *This column is supposed to sustain the weight of half the building.* **5** °uphold, °recognize, °allow, °admit, °approve, °ratify, °sanction, °authorize, °endorse, validate: *The judge sustained the objection by the defense.*

sustained *adj.* continued, °continuous, °continual, prolonged, unremitting, °steady, ceaseless, unceasing, incessant, interminable; °uniform, °even, °level, unchanged, unchanging: *The neighbors keep up that sustained caterwauling every night. Five minutes of sustained middle C on the zither is not my idea of music.*

sustenance *n.* **1** nutriment, °nourishment, °food (and drink), daily bread, °rations, victuals, °provisions, °provender, groceries, aliment, edibles, eatables, foodstuff(s), viands, °meat, *Colloq* grub, eats, chow, nosh, *Slang Brit* prog, °scoff: *My first concern is sustenance for my family.* **2** livelihood, °support, °maintenance, °upkeep, °keep, °subsistence, living: *The fur trade provided the sustenance of early Canadian settlers.*

swagger *v.* **1** °strut, °prance, °parade, *Archaic* swash, *Colloq US* sashay, cut a swath: *Look at that Charlie swaggering down the street in his new suit!* **2** °boast, °brag, °show off, vaunt, crow, *Colloq Brit* swank: *He is always swaggering on about his war experiences.* —*n.* **3** °strut, °prance, strutting, swaggering, °show, °display, showing off, °ostentation, braggadocio, °arrogance, boastfulness: *Plentworth's outrageous swagger developed only after his son was admitted to Yale.*

swallow *v.* **1** °eat, °consume, °devour, ingest, dispatch *or* despatch; °drink, °gulp, guzzle, down, *Colloq* put *or* pack away, swig, °swill: *Snakes swallow their prey whole. The amount that man can swallow at one meal is incredible.* **2** °accept, °allow, °credit, °believe, °take, *Colloq* °buy, °fall for: *She swallowed that tale about an airplane hook, line, and sinker.* **3** Often, ***swallow up.*** absorb, make disappear, engulf, °consume, assimilate: *The fugitive was swallowed up by the crowd.* **4** Sometimes, ***swallow back.*** keep *or* choke back *or* down, °repress, °suppress, °control, °stifle, °smother, °overcome, °conquer: *Haldane swallowed back the tears. Try swallowing your pride just this once.* —*n.* **5** °bite, nibble, °morsel, °mouthful; °drink, °gulp, guzzle, *Colloq* swig: *Have a swallow of this to warm you up a little.*

swamp *n.* **1** °bog, fen, °marsh, quagmire, °morass, °moor, *Chiefly literary* slough, *Scots and northern English dialect* moss, *Southern US* everglade: *The draining of swamps destroys the breeding places of fishes.* —*v.* **2** °overwhelm, °overcome, °flood, inundate, °submerge, °immerse, deluge, °overload, overtax, overburden, *Colloq* snow under: *I am so swamped with work I'm sleeping at the office.* **3** scuttle, °sink, °founder: *A huge wave swamped the boat and she sank quickly.*

swanky *adj.* °smart, °stylish, °fashionable, °chic, chi-chi, °fancy, °luxurious, °grand, °elegant, *Colloq* swank, snazzy, °neat, °nifty, °plush, °posh, ritzy, *Brit* °swish: *Living in a swanky house and wearing swanky clothes—what else would anyone want?*

swarm *n.* **1** °throng, horde, army, °host, multitude, hive, °herd, °mob, °mass, drove, °flood, °stream, cloud, °flock, °pack, shoal, °bunch: *Swarms of people showed up for our garden party.* —*v.* **2** °throng, °mass, °crowd, congregate, °flock, °gather, °flood, °stream, °flow: *The fans swarmed around the rock group, The What.* **3** ***swarm with.*** Often, ***be swarming with.*** crawl with, abound in *or* with, throng with, °teem with, burst with, bristle with, be overrun with: *In two minutes the place was swarming with police.*

swarthy *adj.* swart, °black, ebon, ebony, sable, °pitch-black, jet-black, coal-black, raven, °dark: *His swarthy complexion contrasted with her chinalike pallor.*

swashbuckling *adj.* °adventurous, °daring, °daredevil, swaggering, roisterous, °bold, °dashing, °flamboyant: *Those old swashbuckling films with Errol Flynn, Richard Green, and Douglas Fairbanks formed a genre all their own.*

swath *n.* °swathe, °path, °belt, °strip: *A flock of birds was feeding along the swath cut by the harvester.*

swathe *v.* °tie, °bind, bandage, °wrap, enwrap, swaddle, °bundle (up), °envelop, °shroud, °muffle (up): *I can't tell who he is as his head is swathed in bandages.*

sway *v.* **1** °wave, waver, °swing, °sweep, °oscillate, undulate, °reel, °totter, swing *or* move to and fro *or* back and forth *or* from side to side *or* backward(s) and forward(s), °rock, °fluctuate; °bend, °lean: *The reeds swayed in the breeze.* **2** °move, °incline, °divert, °tend, veer, °tilt, °lean, °slant, °bias; °influence, °persuade, °impress, °win over, °bring around, °convince, °talk into: *The wind is causing the trees to sway toward the east. We tried to sway him to our way of thinking.* —*n.* **3** °sweep, °wave, °swing, (period of) oscillation, libration: *There must be a sway of twenty feet in the top of that skyscraper.* **4** °influence, °control, °power, °command, °authority, °dominion, °rule, °sovereignty, °leadership, mastery; °grip, °clutches, °grasp: *His ambition was to get the mountain people under his sway.*

swear *v.* **1** depose, aver, asseverate, °declare, °insist, assert, solemnly affirm *or* state, °testify, °promise, take an oath, °undertake, °vow, avow, °vouchsafe, °warrant, °pledge, give one's word, °agree: *He swore he had not done any such thing.* **2** °curse, °blaspheme, imprecate, use profanity, utter profanities, execrate, *Colloq* cuss: *Priscilla stopped her ears at the men's swearing.* **3** ***swear by.*** °trust (in), °believe in, °rely on, have confidence in, °count on: *The colonel swears by the same old moustache wax and will use no other.* **4** ***swear off.*** forswear, °renounce, abjure, °go off, °forgo, °shun, °avoid, °give up, eschew, °forsake, °throw over: *Bernie tells me he's sworn off Pimms Cup—at least till the summer.*

sweat *v.* **1** perspire, °glow: *That workout made me sweat.* **2** Often, ***sweat out.*** °worry, be anxious, agonize, °anguish, bite (one's) nails, be on pins and needles, °fret, °fuss, °stew, torture *or* torment oneself, lose sleep (over), *Colloq* sweat blood, be in a tizzy, *US* sweat bullets: *The professors really like to make students sweat, don't they? Fiona is sweating it out till the results* *are posted.* **3** °slave (away), °labor, drudge, °grind, toil and moil, slog, work like a Trojan *or* a horse, *Slang Brit* swot *or Slang Brit* swat: *I sweat over a hot stove all day while you're out and about town.* **4** °ooze, exude, squeeze out, transude: *When a boy, he believed that dew came from flowers sweating water.* —*n.* **5** °perspiration, *Technical* diaphoresis, sudor: *In addition to body moisture, salt and other minerals are lost in sweat.* **6** °(hard) work, °labor, laboriousness, °grind, toil, °drudgery, slogging, sweating, *Slang Brit* swotting *or Slang Brit* swatting: *Nothing worthwhile is gained without some sweat.* **7** state of confusion *or* upset *or* excitement *or* distraction *or* agitation *or* anxiety *or* distress *or* worry; pother, *Colloq* dither, tizzy, °lather: *Herman has really worked himself up into a sweat over this problem.* **8** No sweat! No problem!, Don't worry!, Everything is taken care of!, All is well!, That presents no difficulty!: *When asked to change a flat tire, the stranger's response was, "No sweat!"*

sweep *v.* **1** °brush, °whisk, °clean, °clear, °tidy up: *This weekend I have to sweep out the basement.* **2** Often, ***sweep away.*** carry *or* take (away *or* off), °destroy, °wipe out, °demolish, °remove, °wash (away); °blow (away): *The raging torrent swept away the bridge. The wind swept the debris before it.* **3** °swoop, °flounce, °glide, °sail, °march, °parade, °skim, °tear, °dash, zoom: *Hattie swept into the room majestically.* **4** curve, arc, °arch, °bend, °bow, °circle, °turn: *The long drive to the house sweeps round the pond and gardens.* —*n.* **5** °pass, °clearance, °stroke; °purge: *In one clean sweep the police rounded up all the suspects.* **6** curve, arc, °arch, °bow, °bend, curvature, flexure: *Note the sweep of those main arches holding up the dome.* **7** °range, °extent, compass, °reach, °stretch, °scope, °swing, °span: *The Hubbell space telescope has a farther sweep than any on earth.*

sweeping *adj.* **1** °comprehensive, (all-)inclusive, °general, °extensive, °universal, all-embracing, °broad, widespread, wide(-ranging), far-ranging, blanket, °umbrella, °catholic, °exhaustive, °radical, thorough(going), °out-and-out, across-the-board, wholesale, *Colloq* wall-to-wall: *We must institute sweeping reforms of the party before the election.* **2** °complete, °total, °overwhelming, decisive: *The team enjoyed a sweeping victory in track and field events.*

sweet *adj.* **1** sugary, honeylike, honeyed, sweetened: *Sweet breakfast cereals are bad for your teeth.* **2** °fragrant, perfumed, scented, °aromatic, ambrosial, sweet-smelling, sweet-scented, balmy, °redolent: *The night air was sweet with jasmine.* **3** °harmonious, °melodious, sweet-sounding, euphonious, dulcet, °musical, °tuneful, euphonic, mellifluous, °mellow, °lyric, silvery, bell-like, °golden: *I could hear McCormick's sweet voice singing "Danny Boy."* **4** °gentle, °amiable, °agreeable, °genial, °warm, °friendly, °kind, °nice, unassuming, easygoing; °attractive, appealing, charming, °winning, °pleasant, °pleasing, °lovely; °cute, °pretty: *Jane normally has such a sweet disposition. Old Mrs. Hughes was always sweet to me when I was little. Amy looks so sweet in her new dress. What a sweet little kitten!* **5** °dear, °beloved, °precious, prized, treasured, wonderful, °marvelous, °splendid, *Colloq* °great: *How sweet life can be!* **6** °considerate, °attentive, °solicitous, °thoughtful, °sympathetic, compassionate, °kind, kindhearted, °generous, °gracious, °accommodating: *How sweet it was of you to bring flowers!* **7** cloying, °sentimental, syrupy, saccharine, treacly, °precious, honeyed, sickening, *Colloq* gushing, °gushy, °sloppy, soppy, °maudlin, °sticky, *Brit* twee, *Colloq* icky: *Claire is so sweet that she really overdoes it sometimes.* **8** ***sweet on.*** °fond of, taken with, °keen on, devoted to, enamored of, infatuated with, (head over heels) in love with, *Colloq* wild *or* mad *or* crazy about, nuts about *or* over, *Slang* gone on, stuck on, batty about: *We were teasing Rick about being sweet on Anne.* —*n.* **9** Often, ***sweets.*** bonbon, chocolate, confection, sweetmeat, *Old-fashioned* comfit, *US* °candy: *Hernando always brings a box of sweets for my mother when he comes to see me.* **10** °dessert, *Brit* pudding, afters: *For a sweet we can offer you a choice between chocolate cake and rhubarb pie.*

sweeten *v.* **1** sugar, sugar-coat: *Can you think of nothing that would sweeten the bad news?* **2** °dress up, make more attractive *or* agreeable, sugar-coat, °embellish, embroider; make less painful, °mitigate, alleviate, assuage, °lighten, °soften, palliate, mollify, °ease, allay, °moderate, °temper: *Perhaps the shareholders would be more amenable to the takeover were the buyers to sweeten the offer.*

sweetheart *n.* girlfriend, boyfriend, °friend, °admirer, beau, °darling, °dear, °love, °beloved, °lover, °paramour, inamorato, inamorata, ladylove, betrothed, intended, fiancé(e), *Archaic* swain, *Colloq* heartthrob, °flame, sweetie, °steady: *She goes out so seldom, her sweetheart might as well be the Man in the Moon.*

swell *v.* **1** Often, ***swell out*** *or* ***up.*** °grow, °increase, °enlarge, °expand, blow *or* puff up *or* out, distend, °inflate, dilate, wax; mushroom, belly, balloon, bloat, °bulge, billow, fatten, °rise, tumefy: *After cooking, the mixture swells to twice its original volume.* **2** °grow, °increase, mushroom, snowball, °accumulate, °mount: *Forgotten for years, her savings had swelled to become a small fortune.* **3** °increase, °raise, augment, °enlarge, °boost, °step up: *Something must be done to swell the number of recruits.* —*n.* **4** enlargement, broadening, °increase, °extension, °spread, °swelling, inflation, °expansion, °rise, °surge: *Have you noted the swell in popularity of health foods?* **5** °wave, °surge, billow: *As we rose to the top of each swell, we could see the shore.* **6** fop, °dandy, gay blade, fashion plate, Beau Brummell, *Archaic* coxcomb, *Historical* macaroni, *Colloq* clotheshorse, *US* fancy Dan, *Slang* nob, *Archaic* lounge lizard, *Brit* toff: *He frequented the nightclubs populated by the other city swells.* —*adj.* **7** °smart, °chic, °stylish, °fashionable, modish, °grand, °luxurious, deluxe, °elegant, °first-rate, first-class, top-grade, *Colloq* °posh, swank, °swanky, ritzy: *Charlie always took Diane to the swellest places.* **8** °marvelous, °thrilling, °splendid, spectacular, °first-rate, °fine, *Colloq* °great, °super, °terrific: *I really had a swell time at your party.*

swelling *n.* enlargement, distension, tumescence, protuberance, °bump, °prominence, °bulge, °lump, excrescence, °protrusion, °tumor, node, nodule: *It is a good idea to have any unexplained swelling examined by the doctor.*

sweltering *adj.* °hot, °torrid, steaming, °sultry, °steamy, °muggy, °sticky, °oppressive, stifling, °stuffy, suffocating, °clammy, °humid, °wet, broiling, boiling, °scorching, roasting, baking, wilting, melting, tropical, *Colloq* °close: *I cannot live through another sweltering summer in the Amazon jungle.*

swerve *v.* veer, °career, °swing, °diverge, °deviate, sheer off, skew, °stray, °turn (aside): *The car swerved wildly off the road and into a ditch.*

swift *adj.* °fleet, °fast, °rapid, °speedy, °hasty, °lively, °nimble, °expeditious; °quick, °brisk, °sudden, °abrupt: *The postman was bent on the swift completion of his rounds. What Bill needs is a swift kick to jolt him out of his lethargy.*

swiftly *adv.* °fast, °quickly, speedily, °rapidly, expeditiously; briskly, hurriedly, °hastily, °suddenly, abruptly, in a flash, in a trice, in the wink of an eye, before one can *or* could say "Jack Robinson," before you can say "knife," like a shot, in an instant, in (less than) no time, precipitately, unexpectedly, *Colloq* like greased lightning, *US* lickety-split, in a jiffy, *Slang* pronto, like a bat out of hell: *The assembly is done much more swiftly by machine. Swiftly, I ducked into a doorway to avoid being seen.*

swill *n.* **1** hogwash, pigswill, °refuse, pigwash, slop(s), °garbage, °waste: *If you saw the swill that pigs eat, you might not eat pork.* **2** °nonsense, °rot, °rubbish, *Slang* crap: *That novel of hers was the worst swill I've read in a long time.* —*v.* **3** °drink, guzzle, quaff, °swallow, *Colloq* swig, toss off *or* down, throw down, °polish off, knock back *or* off, *US* chug-a-lug: *They waste every evening swilling beer.*

swimmingly *adv.* smoothly, °easily, effortlessly, °well, successfully, without a hitch *or* a problem, like a dream, cozily, like clockwork, without difficulty, °handily, °readily: *The plan has gone swimmingly thus far.*

swindle *v.* **1** °cheat, cozen, bilk, °defraud, °deceive, °hoodwink, °take in, °fleece, °dupe, °fool, mulct, gull, make a fool *or* sucker (out) of, °victimize, °exploit, °trick, *Old-fashioned or literary* euchre, *Archaic* chouse, *Colloq* bamboozle, °chisel, diddle, pull a fast one on, flimflam, °pluck, °burn, take (someone) for a ride, *Brit* °fiddle, *US* buffalo, *Slang* con, °sting, °screw, rook, gyp, rip (someone) off: *The gang has swindled several elderly people out of their life's savings.* —*n.* **2** °fraud, confidence game *or* trick, cheating, swindling, defrauding, °deception, °racket, °trickery, sharp practice, thimblerigging, °chicanery, knavery, *US* shell game, three-card monte, *Colloq Brit* °fiddle, swizzle *or* swizz, *Slang* °rip-off, scam, con (game), gyp, *US* bunco: *Those old-time swindles are nothing compared with the activities of unscrupulous stock brokers.*

swindler *n.* °cheat, confidence man *or* woman, hoaxer, mountebank, charlatan, knave, °scoundrel, sharper, °fraud, trickster, thimblerigger, °villain, *Technical* defalcator, *Colloq* flimflam man, *Slang* con man *or* woman, *US* bunco artist, fourflusher: *Swindlers bilked them out of all their cash and traveler's checks.*

swing *v.* **1** °sway, move *or* go to and fro *or* back and forth *or* backward(s) and forward(s), come and go, °wave, °fluctuate, °flap, °oscillate, °vibrate, librate, waver, wobble, waggle, zigzag, wigwag, °flourish: *The branches swung in the wind. The signalman is swinging the semaphore flags.* **2** °hang, °dangle; be hanged, be suspended: *They watched the monkeys swinging from tree to tree. He will swing for his crime.* —*n.* **3** °swinging, °sway, swaying, to-ing and fro-ing, coming and going, waving, °fluctuation, fluctuating, flapping, °flap, oscillation, oscillating, vibration, vibrating, libration, waver, wavering, wobble, wobbling, waggle, waggling, zigzag, zigzagging, wigwag, wigwagging, °flourish, °flourishing, °stroke: *Catch the rope on its next swing. He was caught on the shoulder by a wild swing of the bat.* **4** °sweep, °scope, °range, °trend, limit(s); °change, °switch, °shift: *The swing of public opinion is unpredictable in these matters. There has been a small percentage swing to the Conservatives.* **5** °pace, °routine, °groove, °pattern: *Don't you find it hard to get into the normal swing of things after a long holiday weekend?* **6** ***in full swing.*** in (full) operation, °under way, in business, °animated, °lively, °on the move, °moving, °going, *Colloq* on the hop, cooking: *The party was in full swing when we arrived.*

swinging *adj.* °fashionable, °chic, up-to-date, °modern, *Colloq* à-go-go, in the swim, with it, *Chiefly Brit* °trendy, *Slang* °hip, groovy, in the groove: *That's what the swinging '60s were like in London.*

swipe *v.* **1** Usually, ***swipe at.*** swing at, strike at, hit at, lash out at: *He swiped away at the golf ball but kept missing it.* **2** °steal, filch, °pilfer, purloin, *Colloq* °pinch, °lift, snitch, *Chiefly Brit* °nick, °whip, snaffle: *They caught Freddie swiping a bag of mints at Woolworth's.* —*n.* **3** °swing, °strike, °clip: *Ronald took a swipe at the ball with his racket.*

swirl *v.* **1** whirl, °spin, °eddy, churn, °circulate, °gyrate, °surge, °boil, °seethe: *Faster and faster the water swirled around in the maelstrom.* **2** °twist, whirl, whorl, curl, °roll, furl, °spin, curve, °spiral, °twirl, °wind (round): *She nervously swirled ringlets of hair round her fingers.* —*n.* **3** °twist, whirl, curl, °roll, °twirl, °spiral: *At the fabric exhibition swirls of textiles were decoratively displayed.*

swish *v.* **1** °hiss, °whisk, °rustle, °whisper, susurrate: *I heard Elena swish up behind me in her red silk taffeta dress.* —*n.* **2** °hiss, hissing sound, whoosh, swoosh, °rustle, whistle: *With a quick swish the machete knifed through the underbrush.* —*adj.* **3** °elegant, °fashionable, °stylish, de rigueur, °smart, *Colloq* °posh, °plush, ritzy, °swell, swank(y): *Liza bought a swish new outfit for the wedding.* **4** °homosexual, °effeminate, camp(y), °gay, *All of the following are offensive and derogatory* swishy, *US* fruity, limp-wristed, *Colloq* °queer, *Chiefly Brit* °bent, *US* faggy, *Slang* °kinky: *He used to bring his swish friends here, but not lately.*

switch *n.* **1** twitch, °lash, °rod, °whip, birch (rod), °scourge: *He picked up the switch as if to hit me with it.* **2** °change, °alteration, °exchange, °shift, changeover, °reversal, deflection, °trade, swap *or* swop: *A switch of costume was done swiftly backstage, yielding the expression "quick-change artist."* —*v.* **3** twitch, °lash, °whip, birch, °beat, °strike, thrash, °scourge, °flog: *The headmaster switched the boys for the slightest infraction.* **4** °change, °shift, °exchange, °divert, °deviate: *The colonel switched to another subject. They switched the two stones, leaving the fake emerald.* **5** °divert, °turn, rechannel, redirect, °direct: *The funds have been switched to another project. It was a teacher who switched me on to becoming a biologist.*

swivel *v.* **1** °pivot, °turn, °rotate, °spin, °revolve, °pirouette, move freely: *This joint allows the arm to swivel in all directions.* —*n.* **2** °pivot, elbow joint, gimbal, ball-and-socket joint: *The stern chaser is fitted with a swivel, allowing it to be fired at any angle.*

swollen *adj.* enlarged, distended, °inflated, °bloated, bulging, puffed-up *or* -out, tumid, tumescent, expanded, turgid, puffy, oversized, outsized, *Technical* dropsical, hypertrophied, °proud: *My knee was swollen for three days after that knock.*

swoop *v.* **1** °descend, °dive, sweep down, °pounce, °stoop: *The eagle swooped down across the stream, catching a salmon in its talons.* —*n.* **2** descent, °dive, °sweep, °pounce, °stoop, °stroke, °blow, °rush: *The police raid caught several drug dealers in one fell swoop.*

sybarite *n.* °epicure, °epicurean, hedonist, voluptuary, °sensualist, °aesthete, gastronome, °gourmet, *bon vivant, bon viveur,* pleasure-seeker, °playboy, jet-setter: *These sybarites spend all their time and their parents' money in the fleshpots of the world.*

symbol *n.* °representation, °figure, °metaphor, allegory, insigne (*singular; plural is* insignia), °token, °sign, °emblem, badge, °image, logotype, °mark, trademark, colophon, °brand, °code, °abbreviation, phonogram, initialism, cryptogram, acronym, monogram, °password, °shibboleth, watchword, codeword; arms, °bearing, armorial bearing, °crest, escutcheon, coat of arms, °banner, °flag, °pennant, °standard, *Colloq* logo: *This jewel is a mere symbol of my devotion and love for you.*

symbolic *adj.* Often, **symbolic of.** symbolical (of), tokening, betokening, °emblematic (of), figurative, allegoric(al), °typical (of), °representative (of), °symptomatic (of), °characteristic (of), metaphoric(al), allusive (of), denotative (of), connotative (of), mnemonic (of):

From ancient times the laurel, or bay, has been regarded as symbolic of praiseworthiness.

symbolize *v.* °represent, °stand for, °denote, connote, °suggest, °express, °imply, °signify, °mean, °typify, °exemplify, betoken, °illustrate, °embody, epitomize: *Green symbolizes envy and jealousy, red anger, blue sadness, white purity, and so on.*

symmetrical *adj.* symmetric, (well-)balanced, proportionate, °proportional, well-proportioned, °orderly, (well-)ordered, in proportion, °even, °regular, congruous, congruent, °uniform, °harmonious; °equal, mirror-image, mirrorlike: *The temple was exactly symmetrical, with a wing on each side. The two wings are symmetrical.*

symmetry *n.* °balance, °proportion, evenness, °order, orderliness, °regularity, °uniformity, congruity, congruousness, correspondence, °agreement, °harmony, consistency, °equality: *With the crudest of tools, the pyramid builders achieved incredible symmetry. What immortal hand or eye / Could frame thy fearful symmetry?*

sympathetic *adj.* **1** Often, ***sympathetic to** or **toward(s).*** compassionate (to *or* toward(s)), commiserating (with), commiserative (with), °understanding (of), °supportive (of), caring (to *or* toward(s)), concerned (about *or* with), solicitous (of *or* to *or* (toward(s)), warmhearted (to *or* toward(s)), kind-hearted (to *or* toward(s)), responsive (to *or* toward(s)), well-meaning, well-intentioned, good-natured (to *or* toward(s)), considerate (of *or* to *or* toward(s)), empathetic *or* empathic (with *or* to *or* toward(s)); sympathizing, °kindly, comforting, consoling: *The social workers are sympathetic to the people's problems. One should be more sympathetic. Julie takes a sympathetic outlook.* **2** Often, ***sympathetic to** or **toward(s).*** °agreeable, °pleasant, °friendly, well-disposed, favorably disposed, encouraging, like-minded, °responsive, congenial, *en rapport,* simpatico: *You will find a sympathetic ear at the council. The people are not sympathetic to a tax increase.*

sympathize *v.* **1** Often, ***sympathize with.*** suffer *or* grieve *or* mourn (with), feel (sorry) (for), have pity (for), empathize (with), condole (with), commiserate (with): *I sympathized with you over the loss of your grandfather. Now that mine is gone, it is your turn to sympathize.* **2** Often, ***sympathize with.*** harmonize (with), °get along (with), °relate (to), °identify (with), go along (with), see eye to eye (with), °agree (with), °side (with), °understand, be *en rapport* (with), be in sympathy (with), be *simpatico* (with), have (a) rapport (with), *Colloq* be *or* vibrate on the same frequency *or* wavelength (with), *Slang* °dig: *I can sympathize with anyone who despises terrorism.*

sympathizer *n.* condoner, approver, conspirator, coconspirator, collaborator, °accomplice, °accessory, °supporter, fellow traveler, °ally: *At one time, they tried to incarcerate all suspected of being Fascist sympathizers.*

sympathy *n.* **1** compassion, commiseration, °pity, °concern, tenderness, empathy, °understanding, solicitousness, °warmth, tenderheartedness, warmheartedness, *Archaic* ruth: *We feel great sympathy for any child who loses a pet.* **2** °agreement, °harmony, compatibility, °rapport, concord, °accord, fellow feeling, congeniality, °affinity, closeness, °unity, communion, °fellowship, camaraderie: *It is amazing that there can be so much sympathy between two such different people.*

symptom *n.* °manifestation, °evidence, syndrome, °mark, °token, °indication, °cue, °clue, °(warning) sign, °characteristic, °trait, °feature, earmark, marker: *Esmeralda is extremely jealous—I recognize the symptoms.*

symptomatic *adj.* Often, **symptomatic of.** indicative (of), °representative (of), °suggestive (of), °characteristic (of), °emblematic (of), °symbolic (of), °peculiar (to), °specific (to), idiosyncratic (of); indicating, suggesting: *A high fever is symptomatic of numerous afflictions. Epilepsy bears the symptomatic name of the "falling sickness."*

syndicate *n.* **1** °trust, monopoly, bloc, cartel, syndication: *They were part of a syndicate that controlled the worldwide sale of nickel.* **2** °(crime) family, cosa nostra, mafia: *Guido used to be an enforcer for the syndicate.* —*v.* **3** affiliate, °ally, °associate, °amalgamate, consolidate, °league, confederate, synthesize: *Mrs. Cartwright was successful in syndicating the milk producers in the entire county.* **4** serialize, °distribute: *His business is syndicating old movies to television stations.*

synonymous *adj.* Often, ***synonymous with** or **to.*** °equal (to), °equivalent (to), tantamount (to), identified (with), corresponding (to *or* with); transposable (with), exchangeable (with), identical (to *or* with), interchangeable (with), the same (as): *A word is almost never truly synonymous with another in all respects. Even terms like "salt" and "sodium chloride" are not synonymous.*

synopsis *n.* °summary, condensation, °abridgment, epitomization, °outline, °abstract, °digest, °précis, °epitome, compendium, conspectus, *aperçu,* °résumé: *He was assigned to write a ten-page synopsis of a six-hundred-page book.*

synthesis *n.* °blend, °compound, °merge, °union, °amalgamation, coalescence, integration, unifying, unification, composite, °composition, °mixture, °combination; compounding, combining, blending, merging, °union, °amalgamation, coalescence, integrating, mixing, fusing, fusion, unifying, unification: *This mixture is a synthesis of many different ingredients. How do they effect the synthesis of coal, water, and air to make nylon?*

synthetic *adj.* °artificial, man-made, manufactured, ersatz; °fake, °false, °counterfeit, °sham, °bogus, °spurious, °mock, °imitation, pseudo, °plastic, *Colloq* phoney *or US also* phony: *Because of the animal rights activists, many women today wear synthetic fur.*

system *n.* **1** organized whole, °organization, °set, °group, °combination; °structure, °arrangement, °pattern, °setup: *Some believe that the universe is a vast ordered system, others that it is chaos in which only man sees order. It took years to establish the banking system in this country.* **2** °scheme, °method, °approach, modus operandi, °way, °procedure, methodology, °technique, °plan, °process, °practice, °routine: *There is a proper system for doing the job and that is not it.*

systematic *adj.* organized, systematized, planned, °methodical, businesslike, °orderly, well-organized, well-ordered, °regular, °routine, standardized, °standard: *There are several systematic ways for cataloging books.*

T

tab *n.* **1** °flap, °tag, °loop, ticket, sticker, °label, °flag, lappet, strap, °handle: *Pull on the red tab and the life jacket inflates automatically.* **2** °charge, °bill, °account, °reckoning, *Chiefly US* °check: *They left me to pick up the tab for everyone.*

table *n.* **1** °food, victuals, °provender, comestible, edibles, eatables, °fare, °board, °provisions: *One can rely on a fine table at Mrs. McGuire's.* **2** °plain, flatland, mesa, tableland, °plateau, steppe: *The area is barren lava table from horizon to horizon.* **3** (tabular *or* columnar) list *or* listing, °register, °record, tabulation, °chart, catalogue, °index, inventory, itemization, °précis, table of contents: *The figures are easier to understand when arranged in a table.* —*v.* **4** °submit, °present, °offer, proffer, bring forward, °bring up, °propose: *The bill will be tabled for action today.* **5** °shelve, °postpone, °defer, °suspend, °put off, °stay, pigeonhole, mothball, *Colloq* put on ice: *When the bill came up we tabled it for six months.*

tableau *n.* °scene, °sight, °spectacle, °picture, °image; °composition, °arrangement, grouping, °effect: *The children raptly listening to father telling a story—what a charming tableau!*

tablet *n.* **1** (scribbling *or* writing *or* note *or* memo) pad, (spiral(-bound)) notebook, *US* scratch pad: *I write all my dreams in a little tablet lest I forget to relate them to my psychiatrist.* **2** °slab, °plaque, °plate, panel, plaquette: *The tablet on the base of the statue says it is of Disraeli.* **3** stone, gravestone, headstone, °tombstone, °memorial: *We each contributed to the tablet that marks Croombender's grave.* **4** °pill, capsule, troche, pellet, pastille, °drop, lozenge, bolus: *If you are required to take a tablet in water, you needn't get into the bath.*

taboo *adj.* **1** tabu, anathema, forbidden, interdicted, off limits, out of bounds, *verboten,* proscribed, banned, prohibited, restricted, °unmentionable, unspeakable; censored, censorable, °unacceptable, °rude, °impolite, indecorous, °dirty, °explicit; outlawed, °illegal, °illicit, °unlawful: *Our vacation in a camper at Lake Tahoe is a taboo subject in my house. Many people dislike hearing taboo language used on radio or television.* —*n.* **2** tabu, anathema, interdict, interdiction, proscription, °ban, °prohibition, °restriction: *There is a strict taboo against mentioning bodily functions in public.* —*v.* **3** tabu, °forbid, interdict, proscribe, °ban, °prohibit: *The hotel taboos the sharing of a room by unmarried couples.*

tabulate *v.* systematize, °organize, °order, °group, °list, °arrange, classify, °categorize, °rate, °grade, catalog, codify, pigeonhole, °sort, assort, °index, °itemize; °record, °note: *We shall report the results of the poll as soon as they have been tabulated.*

tacit *adj.* unspoken, undeclared, unsaid, unstated, unvoiced, unuttered, °silent, °understood, unexpressed, implied, °implicit: *We have a tacit agreement not to steal each other's boyfriends. He gave me a look of tacit gratitude.*

taciturn *adj.* °silent, uncommunicative, °mum, °mute, °reticent, °reserved, uncommunicative, unforthcoming, °tight-lipped, close-lipped, untalkative, °quiet: *Even when he was small, Chris was quite taciturn, saying little but evidently thinking much.*

tack *n.* **1** °pin, pushpin, °nail, *Brit* drawing pin, tintack, *US* thumbtack: *Someone keeps stealing the tacks from the bulletin board.* **2** °fastening, stitch, baste: *Could you put some tacks in the hem so I can try on the skirt?* **3** °direction, °bearing, heading, °course, °approach; °way, °path, °procedure, °method, °technique, °attack, °line: *Our new tack was due east. We might have to take a different tack to win the contract.* **4** °tackle, °gear, °equipment, equipage, harness, saddlery, °fittings, fitments, °kit, °outfit, °rig, rigging, accoutrements *or US also* accouterments: *He has ordered all new tack for his horse.* —*v.* **5** °pin, °attach, °fasten, °secure, °join, °couple, °unite, °combine, °stick, °fix, affix, °staple, °nail, skewer, °peg, °screw, °bolt, rivet; baste, stitch, °sew, °bind, °tie; paste, °glue, °cement, solder, braze, °weld: *You were supposed to tack these together lightly, but I can't get them apart.* **6** change direction *or* heading *or* course, *Nautical* go *or Nautical* come about; zigzag, veer off *or* away, *Nautical* °beat: *We tacked quickly to avoid striking the buoy.* **7** ***tack on.*** °add (on), append, annex, °attach, tag on: *When they tacked on a rider to the contract, we refused to sign it.*

tackle *n.* **1** °gear, °rig, °fittings, °equipment, equipage, rigging, °paraphernalia, °outfit, tools, °apparatus, °trappings, accoutrements *or US also* accouterments, *Colloq Brit* clobber: *The steeplejack brought along all his tackle and made ready for his climb.* **2** °block (and tackle), °fall, hoisting gear, °pulley, sheave: *You'll need heavier tackle to lift this cargo.* —*v.* **3** °come to grips with, °grapple with, °approach, °take on, try to solve, (try to) deal *or* cope with, stand *or* face up to, °face, confront, °address oneself to, °attend to, °set about, °pursue, *Colloq* take a crack at, have a go at: *Perhaps you'd better tackle the parking problem yourself. I'll tackle the boss on the matter as soon as I see him.* **4** °attack, °fall upon, °devour, °consume, °demolish, °destroy: *You've seen nothing till you've seen Gruber tackle a roast turkey.*

tacky[1] *adj.* °sticky, gluey, gummy, adhesive, °ropy, viscous, viscid, *Colloq* °gooey: *The varnish is still tacky and must dry before sanding.*

tacky[2] *adj.* °tawdry, °cheap, brummagem, °gaudy, °tasteless, °vulgar, °shabby, tatty, °sleazy, chintzy, °shoddy, °seedy: *Richard consistently buys the tackiest suits on the market.*

tact *n.* °discretion, °diplomacy, °sensitivity, °savoir-faire, °judgment, politesse, °delicacy, °finesse, cleverness, °prudence, °care, carefulness, °dexterity, dexterousness, discernment, judiciousness, adroitness, °skill, acumen, acuteness, °perception, °understanding, °consideration, thoughtfulness, politeness: *You can count on Robert to handle such matters with consummate tact.*

tactful *adj.* °discreet, °diplomatic, °sensitive, °politic, °judicious, °delicate, °clever, °prudent, °careful, °dexterous, discerning, adroit, °skillful, °acute, °perceptive, °considerate, °understanding, °considerate, °thoughtful, °polite: *Even the people working in the complaints department of that store are tactful.*

tactic *n.* **1** °move, °maneuver, ploy, °caper, °plan, °strategy, °stratagem, °device, °ruse, °plot, °scheme, °design: *Inviting her husband along was a tactic merely to avert suspicion.* **2** ***tactics.*** maneuvers, °strategy, plans, °campaign, generalship, military science, military operation(s), orchestration, engineering, masterminding: *Entirely different tactics must be employed in jungle fighting.*

tactical *adj.* °artful, °clever, cunning, °shrewd, adroit, °strategic, °skillful, °adept, °politic, °smart, °tactful: *General Schweik thinks that disguising his men as trees is the ultimate in tactical genius.*

tactician *n.* strategist, campaigner, °mastermind, intriguer, plotter, planner, schemer, manipulator, maneuverer, orchestrator, *Colloq* °operator: *Montgomery was regarded by some as a master tactician.*

tactless *adj.* °coarse, °boorish, °uncivilized, °unsophisticated, °rough, °rude, uncouth, °discourteous, ungentlemanly, unladylike, °crude, °gruff, °bluff, °abrupt, °blunt, °brusque, °impertinent, °disrespectful, uncivil, °impolite, insensitive, °awkward, bungling, °clumsy, maladroit, °inept, undiplomatic, °thoughtless, gauche, unskillful, impolitic, °imprudent, °inconsiderate, injudicious, °indiscreet, unwise: *It was pretty tactless of him to ask her for a date right in front of her husband!*

tag *n.* **1** °label, name *or* price tag, °mark, marker, °tab, ticket, sticker, °stub, docket: *The tag was clearly marked with a bar code.* **2** °name, epithet, °label, designation, °title, appellation, °nickname, *Slang* °handle, moniker: *Called "Eddie the dip," he got his tag "the* *dip" because he's a pickpocket.* —*v.* **3** °label, °mark, ticket, °identify, earmark: *Please tag all these cartons so we'll know what's in them without opening them.* **4** °label, °name, °call, dub, °nickname, °style, °entitle, °christen, baptize: *I have already said that Eddie was tagged "the dip" because he was a pickpocket.* **5** ***tag along.*** °follow, trail (along) after, °tail, °shadow, °attend, °accompany, drag along with *or* after: *Does your little brother have to tag along wherever we go?*

tail *n.* **1** appendage, °brush (of a fox), scut (of a hare, rabbit, deer), °dock, caudal fin (of a fish), uropygium (of a bird), pope's *or* parson's nose, tailpiece, °flag: *I just caught a glimpse of the fox's tail as it disappeared into the thicket.* **2** rear end, tail end, backside, °buttocks, croup, rump, posterior(s), °bottom, *Colloq* behind, hinie, *Taboo slang Brit* °bum, arse, *US* ass: *When I catch you, I'll kick your tail!* **3** °reverse: *The obverse of a coin is the head, the reverse the tail.* —*v.* **4** dog, °follow, °trail, °stalk, °shadow, °track: *We tailed the suspect from Newcastle to New Mexico.*

tailor *n.* **1** couturier, couturière, costumier, °dressmaker, modiste, clothier, garment maker, outfitter, seamstress: *She has all her clothing made by her tailor.* —*v.* **2** °fit, °adapt, °suit, °adjust, °alter, °accommodate, °modify, °change, °convert, °cut, °fashion, °mold, °stretch, °accustom: *Your telephone system can be tailored to your most exacting demands.*

tailor-made *adj.* **1** °fitted, custom-made, made-to-order, bespoke; made-to-measure: *He would not wear anything but tailor-made clothes.* **2** °ideal, °perfect, customized, made-to-order, custom-made, suited, °suitable, °(just) right, *Colloq* right up one's *Brit* street *or US* alley: *The job of lighthousekeeper was tailor-made for a misanthrope.*

taint *n.* **1** °stain, °blot, °blemish, °slur, tinge, tincture, (black *or* bad) mark, °stigma, °imperfection, °flaw, °scar, °defect; °discredit, °dishonor: *His writing suffers from the taint of pedantry.* —*v.* **2** °sully, °tarnish, °stain, °stigmatize, °smear, °harm, °hurt, °damage, °debase, °vitiate, °blacken, °foul, °contaminate, °pollute, °dirty, °muddy, smirch, besmirch, °blemish, °soil, °corrupt, °spoil, defile, °ruin, °destroy: *It takes very little to taint one's reputation in a small village.*

take *v.* **1** °grip, °seize, °grasp, °clasp, °get, get *or* take hold of, °grab, °snatch, °clutch, °catch, °capture, °obtain, °lay hold of, lay (one's) hands on, °procure, °acquire, °gain (possession of), °take possession of, °secure, °win, °carry off, °abduct, *Colloq* °nab: *The police took him into custody. He takes what he can get. When it comes to ineptitude, Sue certainly takes the prize.* **2** °pick, °select, °choose, opt for, settle *or* decide *or* fasten on *or* upon: *When you get to the fork, take the road to the left.* **3** °appropriate, arrogate, °extract, carry off *or* away, °steal, purloin, °pilfer, filch, palm, °rob, shoplift, °pocket, °remove, walk off *or* away with, run *or* make off *or* away with; °embezzle, °misappropriate, peculate; plagiarize, °pirate; *Colloq* °lift, °swipe, snitch, *Chiefly Brit* °pinch, °nick, *Slang* °knock off, °hook, °rip off, °liberate, *US* °boost, crook: *They took what didn't belong to them.* **4** °reserve, °book, °engage; °hire, °rent, °lease: *He took a room in a sleazy hotel, going out only at night.* **5** °acquire, °get, °adopt; °assume, °derive, °obtain, °draw, °receive, °inherit: *He has taken his bad manners from you. The film takes its title from the book.* **6** °accept, °receive, °bear, °withstand, °stand, °endure, °weather, °tolerate, °abide, °brave, °go through, °undergo, °suffer, °submit to, °swallow, *Colloq* °put up with, °brook, °stomach, *Brit* °stick: *She took the news about Leon's relapse quite well. I have taken quite enough from you already.* **7** °assume, °bear, °undertake, °adopt, arrogate; °acknowledge, °accept: *Kevin took full responsibility for the mistake.* **8** °believe, °think, °judge, deem, °hold, °feel; take for, assess (as), °consider (as), °regard (as), °view (as), °accept (for): *I take people to be honest till proven otherwise. She took him for a fool when they first met.* **9** °carry, convey, °bear, °transport, °bring, °deliver, ferry; °haul, °cart: *Will this train take me to Aylesbury? They took the bicycle to Old Lyme in a van.* **10** °take up, °study, be involved *or* occupied in *or* with, apply oneself to, °learn; °read, *Colloq* °tackle: *I cannot believe that Doreen is taking home economics.* **11** prove *or* be effective *or* efficacious *or* operative *or* functional, °take effect, take hold, °operate, °function, °work, °perform, *Colloq* °do the trick: *If the transplant fails to take, the doctors will have to operate again.* **12** °exact, °extract, °get: *She took revenge by denying him certain privileges.* **13** °swallow, °eat, °consume, ingest, °devour, gulp down, gobble up *or* down, wolf, °bolt; °drink, imbibe, quaff; °inhale: *Rudolf takes pills of every conceivable color. The doctor said I could take a wee nip now and then. Sonya stepped out to take a breath of fresh air.* **14** °subtract, °deduct, °remove, take away, take from, °take off: *Take three from five. Five, take away two, leaves three. It's cheaper when you've taken off the discount.* **15** °end, °terminate, annihilate, °wipe out; °kill: *The eruption at Thera took tens of thousands of lives. The Lord giveth and the Lord taketh away.* **16** °require, °demand, °need, necessitate, °call for: *Remember, it takes two to tango. It will take two years to complete the fresco. It takes six to sail his yawl.* **17** °hold, °contain, °accommodate, °accept, fit in: *This storage bin cannot take another stick of furniture.* **18** convey, °lead, °conduct; °escort, convoy, °guide, °accompany: *This road takes you directly to the yacht club. He asked to be taken to the manager's office.* **19** °understand, °gather, °interpret, °perceive, apprehend, °deduce, conclude, °infer, °judge, deem, °assume, °suppose, °imagine, °see: *I take it from your expression that you've had bad news.* **20** °charm, °captivate, °capture, °attract, °lure, allure: *There was something about him that took her fancy.* **21** °use, °employ, make use of, °establish, put in(to) place, °adopt, put into effect, °effect, °apply; °resort to, have recourse to, °turn to: *The police have taken measures to ensure that it doesn't happen again.* **22** °clear, get *or* go over *or* past *or* round *or* through: *Whipround took the last jump easily. He must have taken that corner at ninety!* **23** °experience, °entertain, °feel: *She seems to have taken a great liking to you.* **24** °express, °voice, °raise, °put forth: *I hope you will not take objection or exception to what I am about to tell you.* **25** cause *or* make *or* induce *or* drive *or* persuade (someone) (to) go *or* be: *What takes you to Málaga in August?* **26** °act, °assume, °play, °perform: *They asked me to take the part of Yorick in the next production of* Hamlet. **27** bilk, °cheat, °swindle, °defraud, *Colloq* con, *Brit* °fiddle: *When he examined his wallet he realized he'd been taken.* **28** ***take aback.*** °astound, °astonish, °surprise, °startle, °shock: *She was really taken aback at the news about Edmund.* **29** ***take after.*** **(a)** °resemble, look like, be the spit and image *or* the spitting image of, °favor, remind one of, *Colloq* be a chip off the old block: *He takes after his grandfather.* **(b)** Sometimes, **take off after.** chase, °follow, °run after, °pursue: *When the man stole the newspaper and ran, the newsvendor took after him at a gallop.* **30** ***take back.*** °retract, °withdraw, °recant, disavow, °repudiate: *He now wants to take back what he said about you.* **31** ***take down.*** **(a)** °note, make a note *or* memo *or* memorandum of, °write down, °record, put *or* set down, put in writing, °document, °transcribe, °chronicle: *Please take down what I am about to tell you.* **(b)** °debase, deflate, °lower, °diminish, °belittle, °depreciate, deprecate, °humble, °humiliate, °shame, °disparage, °degrade, °disgrace: *She certainly took that pompous ass down a peg or two.* **32** ***take in.*** **(a)** °accommodate, °receive, °let in, °quarter, °board, °lodge: *Mrs. Hervey has a B & B and takes in lodgers.* **(b)** °deceive, °fool, °trick, °impose upon, overcharge, °cheat, mulct, °defraud, cozen, bilk, °dupe, gull, °hoodwink, °swindle, *Colloq* bamboozle, con, pull the wool over (someone's) eyes, *Slang Brit* do: *He was really taken in by that time-share deal.* **(c)** °include, subsume, °embrace, comprise, °cover, encompass, °contain: *Our sales figures take in all of North America, not just Canada.* **33** ***take it.*** **(a)** withstand *or* tolerate *or* survive punishment *or* abuse, °survive: *The Marines are extremely tough and can take it.* **(b)** See **20,** above. **34** ***take off.*** **(a)** °remove, doff, strip *or* peel off, °discard, divest (oneself) of: *Take off your hat in the house.* **(b)** °satirize, °lampoon, °caricature, °mock, °parody, travesty, °burlesque, °mimic, °imitate, *Colloq* spoof, *Brit* sendup: *It is not always easy to take off members of the cabinet.* **(c)** °depart, °leave, °go (away), decamp; °fly off, become airborne, lift off, blast off; *Colloq* skedaddle, make (oneself) scarce, *Slang* hit the road, scram, °beat it, °split: *You'd better take off before they find you here. The plane takes off whether you're aboard or not.* **35** ***take on.*** **(a)** °hire, °engage, °employ, °enroll, °enlist, °retain: *We are so busy that we took on ten new people.* **(b)** °challenge, °rival, °face, contend against, °oppose, match *or* pit (oneself) against, vie with, °fight: *Are you in any condition to take on the champion?* **(c)** °assume, °accept, °undertake, °tackle: *You might be taking on more than you bargained for.* **36** ***take out.*** °entertain, °escort, invite out; court, woo: *You allow Billy Spodge to take your daughter out!* **37** ***take over.*** assume *or* take *or* usurp *or* gain control *or* possession *or* command of: *A huge conglomerate is trying to take over our company.* **38** ***take to.*** **(a)** °like, find pleasant *or* pleasing, feel affection *or* liking *or* affinity for, find suitable: *We took to each other the very first time we met. He took to computers as a duck takes to water.* **(b)** leave *or* depart *or* take off for, °run for, head for, flee to, °make for: *Take to the hills!* **39** ***take up.*** **(a)** °pick up, °accept, °agree to, acquiesce to, accede to: *She took up his offer, and they've been married for forty years.* **(b)** °assume, °resume, °carry on, °continue, go on with, follow on with, °pick up: *After a pause, he took up the thread of the story once more.* **(c)** espouse, °embrace, become interested *or* involved in, °support, °sponsor, °advocate: *She took up the cause of women's rights at her age?* **(d)** °occupy, °cover, °use (up), °fill (up): *Bernard's exercise equipment now takes up half the living room.* **(e)** °deal with, °treat, °consider, °bring up, °raise: *Next week our panelists will take up the question, "Who benefits from education?"* —*n.* **40** °revenue, takings, °yield, °return, °receipts, °proceeds, °gain, profit(s); °gate, box office: *How much was the take from the first day?*

taken *adj.* captivated, entranced, °enchanted, °charmed, bewitched, °infatuated: *I was so taken with Lucinda that I could scarcely breathe.*

takeoff *n.* **1** °flight, flying, taking off, departure, leaving, °going; °launch, liftoff: *The takeoff was delayed by bad weather.* **2** °satire, °lampoon, °caricature, °mockery, °parody, travesty, °burlesque, °imitation, *Colloq* spoof, *Brit* sendup: *The annual students' entertainment, as usual, was a takeoff of the teachers.*

taking *adj.* °attractive, alluring, °engaging, captivating, °winning, winsome, charming, entrancing, °enchanting, bewitching, °fetching, fascinating, °delightful, °irresistible, compelling, intriguing, °prepossessing: *Millie had taking ways that won her many friends.*

tale *n.* **1** °story, °narrative, °report, °account, °record, °chronicle, °history, °narration, °recital, anecdote: *We were brought up on tales of the supernatural. Her tale of what happened does not agree with the facts.* **2** °falsehood, °lie, fiction, °fib, °fabrication, untruth, falsification, °exaggeration, *Colloq* tall tale *or* story, °(cock-and-bull) story, *US* fish story: *He told some tale about having been captured by cannibals in Kansas.* **3** °rumor, °gossip, °slander, °allegation, tittle-tattle, °libel, °story, *US chiefly naval* scuttlebutt: *I have been hearing tales about wild parties at the neighbors' again.*

talebearer *n.* °gossip, rumormonger, gossipmonger, taleteller, talemonger, scandalmonger, telltale, °troublemaker, quidnunc, tattler, °informer, sieve, *Chiefly US and Canadian* tattletale, *Slang* bigmouth, °blabbermouth, squealer, stool pigeon, stoolie, rat, *Brit* °sneak, nark, *US* fink, rat fink: *The words, spoken in private, were repeated to him by some talebearer.*

talent *n.* **1** °ability, °power, °gift, °faculty, °flair, °genius, °facility, °aptitude, °capacity, °knack, °ingenuity, °forte, °strength; °endowment: *Bill's extraordinary talent for playing the trombone may be unmatched anywhere.* **2** °tendency, proclivity, propensity, °penchant, predilection, predisposition, °bent, °inclination: *Anastasia certainly has a talent for saying the wrong thing.*

talented *adj.* °gifted, °accomplished, °brilliant, skilled, °skillful, °masterful, °expert, °adept, adroit, °dexterous, deft, °clever, °good, °polished, °proficient, °first-rate, top-drawer, °excellent, *Colloq* ace, °crack, topnotch, *Brit* wizard, whizzo, *US* crackerjack: *But Leslie is the most talented baton twirler we have.*

talisman *n.* °amulet, °charm, tiki, fetish *or* fetich, juju, periapt, abraxas; wishbone, rabbit's foot, *Brit* merrythought: *He was given a talisman that made its wearer invisible or visible on command, but he was never told the command.*

talk *v.* **1** Sometimes, ***talk in.*** °speak (in), °use, communicate in, converse in, express (oneself) in, discourse in: *If you talk English, I might understand.* **2** °confer, °consult, °parley, have a (little) talk, °(have a) chat, confabulate, *Colloq* confab, *Slang US* °rap: *I think we ought to talk before you decide which offer to accept.* **3** °chatter, prate, °prattle, °jabber, blather *or* blether, gibber, gibber-jabber, cackle, °babble, °patter, °rattle on, °go on, *Brit* natter, witter, rabbit on, *Colloq* °gab, *Slang* gas, jaw, *US* run off at the mouth: *They talk continuously, mostly about nothing.* **4** °chat, °gossip, °palaver, *Slang* chew the fat *or* the rag, shoot the breeze, *US* chin, schmooze, °rap, have a bull session, bat the breeze, shoot the bull: *We were just sitting around talking when Tom suggested we go for a swim.* **5** °inform, °confess, give the game away, °blab, *Colloq* °come clean, *Slang* rat, squeal, °sing, °spill the beans, *Brit* grass: *If they give him the third degree, you can be sure he'll talk.* **6** °speak, give *or* deliver a speech *or* a talk *or* an address, °lecture: *I believe that Professor Bardolph is talking at Mingley College tonight.* **7** ***talk about*** *or* ***over*** *or* ***of.*** °discuss, confer about *or* on, parley about: *I don't want to talk about football all the time. Can't we talk it over? She talks only of you.* **8** ***talk big.*** °boast, °brag, vaunt, crow, °bluster, °exaggerate, blow *or* toot (one's) own horn: *He was never in the war—he just talks big.* **9** ***talk down.*** **(a)** °depreciate, deprecate, denigrate, °disparage, °belittle, °minimize, °diminish, °criticize, *Colloq* °knock, °pan, °put down: *It's the only thing he's ever done, and you shouldn't talk it down.* **(b)** Usually, **talk down to.** condescend to, °patronize: *I resent your talking down to me that way.* **10** ***talk into.*** °convince, °bring around, °sway, °persuade: *We have been trying to talk him into running in the marathon.* **11** ***talk over.*** °discuss: *We talked over the best way to prepare the advertising proposal.* **12** ***talk to*** *or* ***with.*** speak to *or* with, °communicate with, converse with: *Since the episode with the monkey, she refuses to talk to me. I talked with him yesterday.* **13** ***talk up.*** °promote, °support, °sponsor, advertise, °publicize, °push, *Colloq* °plug, hype, ballyhoo: *She has been talking up the book on her* *radio show.* —*n.* **14** °oration, °lecture, °address, °presentation, °speech, discourse, °report, disquisition, dissertation; °sermon; °harangue, °tirade, *Colloq* spiel: *Claire delivered a talk on computational linguistics. Every week we have to listen to the same talk about sin.* **15** °conversation, °conference, °discussion, °meeting, consultation, °dialogue, colloquy, °parley, °palaver, °chat, °tête-à-tête, confabulation, *Colloq* confab, powwow, *Slang* chinwag, head-to-head, *US* one-on-one, rap session: *We must have a little talk about your school grades.* **16** subject *or* topic of conversation *or* gossip *or* rumor: *Her behavior at the wedding was the talk of the town for weeks.* **17** °gossip, °rumor, hearsay, °information, °news, °report, *Colloq* info, *Slang* °dope: *There is some talk that you might be leaving.* **18** °palaver, °gossip, claptrap, °prattle, prattling, °chatter, verbiage, cackle, bunk, °nonsense, °rubbish, balderdash, poppycock, °hot air, stuff and nonsense, twaddle, *Colloq* malarkey, piffle, hooey, hokum, bunkum, bosh, hogwash, horsefeathers, *Slang* bilge (water), crap, bull, tripe, *Brit* tosh, balls, *US* applesauce, *Taboo slang* horseshit, bullshit: *That's just a lot of talk.* **19** °dialect, °speech, way *or* manner of speaking, *façon de parler,* °language, °jargon, argot, °cant, patois, °accent, *Colloq* °lingo: *Can you understand musicians' talk? My parents' talk is peppered with words long obsolete in standard English.*

talkative *adj.* garrulous, loquacious, verbose, long-winded, °voluble, prolix, °wordy, chatty, gossipy, °effusive, talky, logorrhoeic *or* logorrhoeal, *Colloq* gabby, blabby, *Slang* big-mouthed; *US* running off at the mouth: *The children are so talkative I can't get a word in edgewise.*

talker *n.* **1** °speaker, lecturer, orator, speechmaker, keynoter, spellbinder, tub-thumper, °rabble-rouser, demagogue, haranguer, ranter, speechifier: *You might disagree with what he says, but you must admit he's a great talker.* **2** blusterer, blatherskite, swaggerer, °show-off, *Slang* windbag, gasbag, lot of hot air, blowhard: *He is a big talker, but he never gets anything done.*

tall *adj.* **1** °high, °towering, °big, soaring, °lofty, °giant, °gigantic; multistory: *The giant sequoias of California are the tallest trees in the world. There are many tall buildings in most major cities.* **2** °lanky, gangling, rangy, leggy, long-legged, °big, °giant, °huge, °gigantic, °large: *Some of those basketball players are unbelievably tall.* **3** exaggerated, overblown, °far-fetched, °improbable, °unbelievable, °incredible, °preposterous, °outrageous, overdone, °absurd, *Colloq Brit* °steep: *He told us some amusing tall tales about his fishing adventures.*

tally *v.* **1** °agree, °coincide, °accord, °correspond, °fit, °compare, °match (up), °square, °conform, concur, harmonize, *Colloq US* jibe: *Nothing I could do would make my figures and the bank's tally.* **2** Sometimes, ***tally up.*** count (up *or* out), °enumerate, °record, °register, °reckon, °add (up), °total (up), °tabulate, °itemize, °list, °calculate, °compute: *Please tally up what I owe you.* —*n.* **3** °count, enumeration, °record, °register, °reckoning, °addition, °total, tabulation, itemization, listing, °calculation, computation: *According to my tally, you owe $156.72.* **4** ticket, °label, °mark, marker, °tag, °tab: *Each plant can be identified by its own tally.* **5** counterfoil, °stub, counterpart, °duplicate, °mate: *If you lose your tally you cannot get back your luggage.*

tame *adj.* **1** tamed, docile, disciplined, °obedient, domesticated, housebroken, trained, °broken: *He used to keep a tame ocelot in his spare room.* **2** °mild, °gentle, °fearless, unafraid: *It is amazing how tame animals can be when they do not perceive man as a threat.* **3** °tractable, pliant, compliant, °meek, °submissive, °mild, under (someone's) control *or* thumb, °subdued, suppressed; unassertive, °feeble, °ineffectual, °timid, timorous, °cowardly, pusillanimous, chicken-hearted, °fainthearted, white-livered, lily-livered, yellow, *Colloq* wimpish: *Casper is far too tame to defy Katinka.* **4** °boring, °tedious, °tiresome, °dull, insipid, °bland, °lifeless, °flat, °vapid, °prosaic, °humdrum, °bland, unexciting, uninspired, uninspiring, °run-of-the-mill, °ordinary, uninteresting, °dead, *Colloq* °wishy-washy: *As a safari park, the place is too tame for my tastes.* —*v.* **5** °break, domesticate, °train, house-train, °gentle, °master, °subdue, °subjugate: *The brown bear is a most difficult animal to tame.* **6** °calm, °subdue, °control, mollify, pacify, °mute, °temper, °soften, °curb, °tone down, °moderate, °mitigate, °tranquilize: *After a week the Lord tamed the violent sea and our tiny craft was allowed to continue.*

tamper *v.* °interfere, °meddle, intermeddle, °intrude, °tinker, mess (about *or* around), *Colloq* fiddle *or* fool (about *or* around), °monkey (around), °muck (about): *Anyone caught tampering with the machinery will be prosecuted.*

tang *n.* **1** pungency, piquancy, °bite, °zest, zestiness, sharpness, poignancy, spiciness, °nip, °edge, °spice, °taste, °flavor, °savor, °aroma, °smell, °odor, *Colloq* zip, °kick: *Ginger root gives the dish just the right tang.* **2** tinge, °flavor, °hint, °suggestion, *soupçon,* °trace, °dab, smack, °touch, smattering: *The new version has the tang of the original without its obsolete words.* **3** prong, °tab, °projection, °tongue, °strip, tine, shank, °pin, °spike: *The tang of a good knife goes right through to the end of the handle.*

tangential *adj.* °divergent, digressive, off *or* beside the point, °peripheral, °irrelevant, °extraneous, °unrelated: *The matter of cost is tangential to the main issue.*

tangible *adj.* °material, °real, °physical, corporeal, bodily, somatic, °solid, °concrete, touchable, tactile, °manifest, palpable, °evident, °actual, °substantial, °visible, seeable, °discernible, °perceptible, ponderable, °objective, ostensive: *Leaving aside intangibles like taste, smell, and color, the important tangible characteristic of food is texture.*

tangle *n.* **1** °confusion, °knot, gnarl, °mesh, °snarl, °twist, °kink, entanglement, °jam, °snag, °jumble, °mess, skein, °web, °coil: *It took hours to straighten out the tangle of ribbons.* **2** °muddle, °complication, °jumble, °puzzle, °medley, °complexity, °complication, °scramble, °mishmash, °mix-up, hotchpotch *or US and Canadian also* hodgepodge, jungle, °maze, labyrinth: *Reference books help organize the vast tangle of information.* —*v.* **3** Often, ***tangle up.*** °confuse, °knot, °mesh, °snarl, gnarl, °twist, °kink, ravel, °entangle, °jam, °snag, intertwist, intertwine, interlace, interweave, °jumble, °mess up, °scramble, °shuffle, °muddle: *Now you've tangled all the ribbons I carefully straightened out earlier.* **4** Often, ***tangle with.*** wrangle (with), contend (with), fight (with *or* against), °(come into) conflict (with), come *or* go up against, lock horns (with), °dispute, cross swords (with), °disagree (with): *The sheriff is the last person I want to tangle with.*

tantalize *v.* °tease, °taunt, °provoke, °torment, torture, bait, °tempt, °plague, °frustrate: *They tantalized him by hanging the keys to his cell just outside his reach.*

tantamount to *adj.* amounting to, as good as, virtually the same as, (pretty) much the same as, °equal to, equivalent to, °like, °of a piece with, comparable to, commensurate with: *Coming from Felix, "I picked up a little something for your birthday" was tantamount to a confession of larceny.*

tantrum *n.* fit (of anger *or* of passion), °outburst, °eruption, blowup, °explosion, flare-up, °storm, °rage, °fury, *Colloq Brit* paddy, wax: *He has a tantrum every time he fails to get his own way.*

tap[1] *v.* **1** °rap, °knock, °dab, °strike, peck; drum, °beat: *Must you continually tap on the table with that blasted pencil!* —*n.* **2** °rap, °knock, °dab, °strike, peck, °pat; tapping, tap-tap, rapping, knocking, pecking, °beat, beating, °patter, pattering: *I felt a tap on my shoulder. The tap, tap, tap of the dripping water was driving me mad.*

tap[2] *n.* **1** cock, stopcock, petcock, sillcock, seacock, spigot, °spout, valve, *US* faucet: *Replace the washer in the tap to stop the leak.* **2** bung, °stopper, cork, spile, °plug, stopple, °peg: *Move the tap from the empty barrel to the new one.* **3** wiretap, °bug, listening device, electronic eavesdropper: *I think there's a tap on my phone.* **4** ***on tap.*** **(a)** on draft, out of the barrel *or* keg: *They have real ale on tap.* **(b)** °ready, °available, on *or* at hand, waiting, °in reserve, °on call: *He promised to be on tap if I needed him for anything.* —*v.* **5** °drain, °draw (off), siphon off *or* out, °extract, °withdraw: *We tapped the wine directly from the casks.* **6** °open, °drain, unplug, °sap, bleed, °milk, °broach, °mine, °use, utilize, make use of, put to use, draw on *or* upon, turn to account: *Whenever she needed money, she tapped her savings.* **7** °bug, eavesdrop on, wiretap: *The FBI still taps her phone.*

tape *n.* **1** °strip, °band, fillet, °stripe, strap, °belt, ribbon: *She wore a colorful woven tape round her head to hold her hair.* **2** (tape) recording, °reel, spool, cassette, video: *He played us the tape of the whole interview.* —*v.* **3** strap, °band, °bind; °seal, °stick: *If you tape the packet, customs cannot open it for inspection.* **4** °record; tape-record, video: *Programs can be taped automatically when we are out.*

taper *v.* **1** °narrow (down), °thin, °diminish, come *or* go down: *Note how the trousers taper towards the ankles.* **2** Often, ***taper off.*** °diminish, °reduce, °thin out, °wind down, °decrease, °fade, lessen, °peter out, °wane, °subside, °let up, slacken, die away *or* down *or* off *or* out, °decline, slow (down *or* up), °weaken, abate, °ebb, °slump, °drop (off), °fall (off), °plummet: *Sales of ice skates can be expected to taper off during the summer.*

tardy *adj.* **1** °late, unpunctual, behind schedule, °overdue, behindhand: *You will be given an extra assignment because you were tardy again.* **2** °slow, dilatory, °belated, °slack, retarded, sluggish, °reluctant, °indolent, °lackadaisical, °listless, °phlegmatic, °slothful, °lethargic, languid: *They have been tardy in acknowledging the importance of Allen's work.*

target *n.* °goal, °object, °objective, °aim, °end; °butt, °quarry: *The target of the charity drive is to raise fifty thousand dollars. Why make Peter the target of your ridicule?*

tariff *n.* **1** °tax, assessment, °duty, excise, levy, impost, °toll, *Brit* °rate: *The tariff on imported clothing may be increased.* **2** °schedule (of charges), price list; bill of fare, menu: *I stopped by the hotel to ask about their tariff.*

tarnish *v.* °sully, °disgrace, °taint, °blacken, °blemish, °stain, °blot, °soil, °spot, °dirty, °contaminate, defame, °injure, °spoil, °ruin, °damage, °harm, °hurt, °stigmatize, °debase, °degrade, denigrate, °dishonor, asperse, calumniate: *By his behavior, he has tarnished the reputation of his entire family.*

tarry *v.* **1** °delay, °pause, °wait, °linger, loiter, °stall, °procrastinate, °dawdle, bide one's time, temporize, hang back, *Colloq* hang on *or Colloq* about *or Colloq* around: *Don't tarry any longer or you will miss your train.* **2** °remain, °sojourn, °stay, °stop, °rest, °dwell, bide (one's) time, °settle: *She plans to tarry at Eastbourne for a while.*

tart[1] *adj.* **1** °sour, acidic, acidulous, acidulated, lemony, citrusy, vinegary, acetous, acescent; °sharp, tangy, astringent, acerb, acerbic, acrid, °bitter, °pungent, piquant, °harsh: *Those grapes are a bit tart for my taste.* **2** °biting, °bitter, °caustic, acid, corrosive, mordant, astringent, acrimonious, °trenchant, °harsh, °scathing, stinging, acerbic, °incisive, °cutting, °keen, barbed, °nasty, curmudgeonly, °testy, crusty, °abusive, °virulent, °sarcastic, °sardonic, satiric(al), °vicious, cynical: *You can rely on Henry for a tart rebuff every time.*

tart[2] *n.* **1** pie, tartlet, pastry, °turnover, flan, quiche, patty, *Brit* °pasty: *I'll give you three guesses who stole the tarts from the Queen of Hearts.* **2** strumpet, streetwalker, °prostitute, whore, harlot, fallen woman, trollop, °wanton, working girl, *fille de joie,* call girl, loose woman, slut, °drab, °jade, demimondaine, courtesan, woman of ill repute, hussy, doxy, camp follower, piece of baggage, lady of the evening *or* the night, woman of easy virtue, *Slang* floozy, hooker, *US* chippy *or* chippie, roundheel(s), bimbo: *The tabloid claimed that the minister was found with a woman it described as "a tart."*

task *n.* **1** °duty, °assignment, °business, °job, °charge, °stint, °mission, °work, chore, °undertaking: *It will be your task to keep the others working.* **2** °(major) effort, °test (of strength), piece of work, °struggle, °strain: *Getting him to do anything at all was quite a task.* **3** ***take to task.*** °scold, °reprimand, call to account, °blame, censure, recriminate, reproach, °reprove, °rebuke, °criticize, °lecture, °upbraid, chide, reprehend: *I was taken to task for failing to notify the police of the break-in.*

taste *n.* **1** °drop, *soupçon,* °dash, °pinch, °touch, °hint, °suggestion, °grain, °trace, °bit; °flavor, °savor, °relish, °tang: *I added just the tiniest taste of coriander to the sauce.* **2** °sample, °morsel, °bite, °mouthful, °bite, °sip, °nip, °swallow: *I took a taste of the sauce and it is superb.* **3** palate, °desire, °inclination, °leaning, °partiality, °disposition, °penchant, °liking, °fancy, °preference, fondness, °appetite, °relish, °stomach, °tolerance: *Unfortunately, she developed a taste for caviar and champagne while on a fish and chips and beer budget.* **4** discernment, °discrimination, °perception, °judgment, cultivation, °refinement, stylishness, °grace, °polish, °elegance: *Her unerring taste leads her to select only the best.* **5** °style, °mode, °fashion, °manner, °form, °design, °motif: *The room was decorated in Moorish taste.* **6** °decorum, °discretion, tactfulness, °delicacy, °refinement, politesse, politeness, correctness, °propriety, tastefulness: *Evan's taste would never allow him to say anything rude.* —*v.* **7** °savor, °sample, °examine, °try, °test: *Taste this and see if it's too salty.* **8** °experience, °sample, °know, have knowledge of, °undergo, °encounter, °meet (with), come up against: *As a young man he had tasted the pleasures of Paris.*

tasteful *adj.* in good taste, °decorous, °refined, finished, °tactful, °polite, °polished, restrained, °correct, °harmonious, °fitting, °fit, °proper, °discriminating, °aesthetic, discriminative, °fastidious, °cultivated, *comme il faut,* °elegant, °graceful, charming: *Though a bit shabby, her clothes were demure and tasteful.*

tasteless *adj.* **1** in bad *or* poor taste, °garish, °gaudy, °loud, °tawdry, meretricious, °cheap, °flashy, °unrefined, inelegant, unesthetic; °improper, °wrong, indecorous, indelicate, uncultivated, uncouth, uncultured, gauche, °boorish, maladroit, °distasteful, °unsavory, °coarse, °crude, °gross, °vulgar, °base, °low: *That is the most tasteless wardrobe I have ever seen. Leave it to him to make a tasteless remark about her table manners.* **2** insipid, °bland, °dull, °flat, °watery, °vapid, flavorless, °unsavory, *Colloq* °wishy-washy: *The tea served there is completely tasteless.*

tasty *adj.* °delicious, delectable, °luscious, flavorous, flavorsome, flavorful, °savory, toothsome, palatable, appetizing, sapid, mouthwatering, ambrosial, *Colloq*

°yummy, scrumptious: *They serve a very tasty curry at that Indian restaurant.*

tatter *n.* **1** Often, ***tatters.*** scrap(s), rag(s), shred(s), bit(s), piece(s): *Tatters from the culprit's clothing were found on a thorn bush.* **2** ***in tatters.*** in ruins, in shreds, destroyed, ruined, shattered, in disarray, demolished: *My life has been in tatters ever since she left me.*

tattered *adj.* °ragged, torn, shredded, °rent, °threadbare: *Oh, tear her tattered ensign down, / Long has it waved on high.*

tattle *v.* **1** °blab, °tell, reveal *or* divulge *or* give away secrets, *Slang* squeal: *Though sworn to silence, it was Davis who tattled.* **2** °gossip, °prattle, prate, °babble, °chatter, °jabber, blather *or* blether, *Brit* natter, witter, *Slang* yak: *The old crones were tattling away in the kitchen.*

taunt *v.* **1** °tease, °jeer (at), °flout, °twit, °mock, °torment, °annoy, make fun *or* sport of, °poke fun at, °deride, °sneer (at), °scoff (at), °insult, °ridicule, °burlesque, °lampoon, *US* °ride, *Colloq* kid, rib, roast, °put down, *Brit* °guy, *Slang* °bug, °rag, hassle, *US* get on (someone's) case: *The boys used to taunt him, calling him "Shrimpo" because he was so short.* —*n.* **2** °jeer, °gibe, brickbat, °insult, °scoff, °derision, °sneer, °slap (in the face), raspberry, *Colloq* °dig, *US* Bronx cheer: *At seventeen, he was six feet tall and no longer had to tolerate his classmates' taunts.*

taut *adj.* **1** °tight, °tense, °strained, stretched, °rigid, °stiff: *When his jaw goes taut like that I know he's angry.* **2** *Nautical* °neat, °tidy, Bristol fashion, °shipshape, °spruce, °(in) trim, °smart, °orderly, well-organized; well-disciplined: *Captain Scriven runs a taut ship, no doubt about it.*

tautology *n.* °repetition, redundancy, battology, pleonasm, iteration, tautologism; repetitiousness, repetitiveness, wordiness, prolixity, verbiage, verbosity, long-windedness: *Clichés like "null and void" and "cease and desist" are tautologies. Tautology, when called "repetition," is often used for rhetorical effect.*

tawdry *adj.* °gaudy, °cheap, °flashy, brummagem, °showy, meretricious, °garish, °loud, tatty, tinsel, tinselly, °plastic, °tinny, °shabby, *US* °tacky, *Colloq* cheapjack: *She was wearing a tawdry outfit picked up at a church rummage sale.*

tax *n.* **1** levy, impost, °duty, °tariff, assessment, °tribute, °toll, excise, °customs, °charge, contribution, *Archaic* scot, tithe, *Brit* octroi, cess, rate(s), °dues: *Most people try to pay the lowest taxes they can get away with.* **2** onus, °burden, °weight, °load, °encumbrance, °strain, °pressure: *The admission of twenty new children put a heavy tax on the school's resources.* —*v.* **3** assess, °exact, °demand, °charge, impose *or* levy a tax (on), *Archaic* tithe: *If the government doesn't tax citizens, where will the money come from?* **4** °burden, °strain, put a strain on, °try; °load, °overload, °stretch, °exhaust; °encumber, °weigh down, saddle, *Brit* pressurize, *US* °pressure: *This misbehavior is really taxing my patience. Our meager resources were taxed to the limit. Mother was taxed with the responsibility of caring for us.*

taxi *n.* **1** taxicab, °cab, hackney, *Colloq* °hack: *Please hurry, as I have a taxi waiting.* —*v.* **2** °drive, °ride (on the ground): *After landing, the plane taxied to the gate.*

teach *v.* °instruct (in), °inform (about), °communicate (to), °educate, °guide, °train, °tutor, °coach, °enlighten, edify, °indoctrinate, inculcate, °instill, school in, °demonstrate, °show, familiarize *or* acquaint with, give lessons (in) (to); °drill, °discipline: *Plinkton is qualified to teach chemistry. Captain Ross taught many youngsters to sail. She wants to teach when she completes her training.*

teacher *n.* °schoolteacher, educator, °instructor, professor, °doctor, °tutor, °fellow, lecturer, °master, °mistress, schoolmaster, schoolmistress, °coach, trainer, °guide, mentor, guru, cicerone, °counselor, °adviser; educationist; *Brit* don, *Scots* dominie, *US* docent, *Colloq* schoolmarm: *One can be counted extremely lucky to have inspiring teachers.*

team *n.* **1** °side, line-up, °group, °band, °gang, °body, °crew, °party, troupe: *Which team does Carswell play for?* **2** °pair, yoke, °span, duo, °set, °rig, tandem: *They hitched a fresh team of horses to the coach and we were off again in an hour.* —*v.* **3** Often, ***team up.*** join (up *or* together), band *or* get *or* work together, °unite, °combine, °link (up), °cooperate, °collaborate; conspire: *If we team up, we'll get the job done better and in half the time.*

tear *v.* **1** °rip, °rend, rive, °rupture, °pull apart, °shred, °mutilate, °mangle, °claw, °split, °divide, °separate, °sever: *The tiger is tearing its quarry to shreds. He tore the contract in half and stormed out.* **2** °pull, °snatch, °wrench, : *The force of the blast tore the attaché case from my grasp. The child was torn from her mother's arms by the soldier.* **3** °dash, °fly, °run, gallop, °race, sprint, °rush, scoot, °shoot, sprint, °speed, °bolt, dart, °flit, °scurry, scuttle, °career, zoom, °hurry, °hasten, *Colloq* zip: *He is tearing around trying to find the money to cover his loan.* —*n.* **4** °rip, °rent, °rupture, °hole, °split, °slash, °gore, °cut, °score, °slit, °gash, fissure, °rift, laceration: *Nobody will see the tear in your trousers if you remain seated.*

tearful *adj.* weeping, crying, in tears, sobbing, whimpering, dewy-eyed, blubbering, sniveling, lachrymose, *Colloq* weepy: *She was tearful as she told him they could not meet again.*

tease *v.* **1** bait, °taunt, °torment, °harass, bedevil, °bother, nettle, °plague, °chaff, °pester, °annoy, °irritate, needle, goad, badger, °provoke, vex, °twit, °tantalize, °frustrate, *Nonstandard* °aggravate, *Colloq* °guy, °pick on, rib, drive mad *or* crazy, °drive up the wall, *Brit* take the mickey out of, *Slang* °rag: *Stop teasing the animals! Frances would tease me by agreeing to go out and then begging off with a headache.* **2** °coax, °worry, winkle, °work, °manipulate: *He was finally able to tease the broken piece of key out of the lock.*

technical *adj.* **1** complex, °complicated, °detailed, °intricate, specialized: *This is a technical matter, not easily understood by the layman.* **2** °mechanical, applied, industrial, polytechnic, technologic(al): *We thought he would be happier attending a vocational school.*

technique *n.* **1** technic, °method, °approach, °manner, °mode, °fashion, °style, °procedure, °system, °tack, °line, modus operandi, standard operating procedure, *Colloq* MO (= 'modus operandi'), SOP (= 'standard operating procedure'): *After many years, he has developed his own technique for building hot-air balloons.* **2** technic, °art, craftsmanship, artistry, °craft, °knack, °touch, °skill, skillfulness, adroitness, adeptness, dexterousness, °facility, competence, °faculty, °ability, °aptitude, °performance, °proficiency, °talent, °gift, °genius, know-how, °knowledge, °expertise: *Jascha's artistry remains unsurpassed.*

tedious *adj.* overlong, long-drawn-out, prolonged, °endless, unending, °monotonous, unchanging, °changeless, unvarying, °laborious, long-winded, wearing, wearying, wearisome, tiring, °exhausting, fatiguing, °tiresome, °boring, °dreary, °dull, dryasdust, °drab, °colorless, °vapid, insipid, °flat, uninteresting, °banal, unexciting, °prosaic, prosy, soporific, °humdrum, °routine, °repetitious, °repetitive, °mechanical, automaton-like, °automatic, *Colloq* ho-hum, *Slang* °dead, *US* blah:

For almost fifty years my father had a tedious job on an assembly line.

tedium *n.* tediousness, monotony, changelessness, invariability, long-windedness, wearisomeness, tiresomeness, °boredom, ennui, dreariness, dullness, drabness, colorlessness, vapidity, insipidity, insipidness, two-dimensionality, banality, °routine, repetitiousness: *How do you survive the tedium of getting up and going to an office every day?*

teem[1] *v.* ***Usually teem with.*** °proliferate (with), be prolific (with), °abound, be abundant, °swarm (with), be alive (with), °crawl (with), °bristle (with), overflow (with), °overrun (with), be full (of), °brim (with): *When the panel was removed, the entire foundation was seen to be teeming with termites.*

teem[2] *v.* °pour, °rain, °stream (down), *Colloq* come down (in buckets), bucket down, rain *or* pour cats and dogs: *It teemed with rain continuously all day, ruining our picnic.*

teenager *n.* °adolescent, °youth, °boy, °girl, young man, young lady, °juvenile, °minor, *Colloq* kid: *The town ought to provide a place where teenagers can get together.*

teeter *v.* °balance, wobble, °rock, °sway, °totter, waver, °tremble, °stagger: *She screamed when she saw Bruce teetering on the edge of the precipice.*

telegram *n.* °cable, cablegram, radiogram, radiotelegram, wire, telex, (In France) *bleu, pneu, US trademark* Mailgram, *Brit trademark* Telemessage: *She received a telegram saying that he would be arriving on the midnight flight.*

telephone *n.* **1** handset, phone, *Colloq* blower, *Chiefly US* horn, *Slang Brit* dog (and bone): *Mr. Fleeb, your wife is on the telephone for you.* —*v.* **2** phone, °ring (up), °call (up), give (someone) a ring *or* a call, *Colloq* get (someone) on the blower *or chiefly US* the horn, give (someone) a tinkle *or* a buzz, °buzz: *Telephone when you have a chance.*

telescope *n.* **1** spyglass, *Old-fashioned* °glass; refracting telescope, reflecting telescope, radiotelescope: *The term* radiotelescope *is a misnomer, since* telescope *properly means "far seeing."* —*v.* **2** °shorten, compress, °abbreviate, °curtail, condense, summarize, °précis, °digest, °tighten (up), boil down, °abridge, truncate, °abstract: *Telescope the report into two columns for page 2.* **3** concertina, squash, °crush: *The first three carriages of the train were telescoped in the crash.*

television *n.* TV, video (receiver), small screen, *Colloq* °box, idiot box, *Brit* telly, *US* boob tube, *Slang* tube, *Brit* gogglebox: *My television was not working over the weekend.*

tell[1] *v.* **1** °relate, °narrate, °recount, °recite: *Tell me a story.* **2** Sometimes, ***tell of.*** °say, °mention, hint at, °refer to, °touch on, utter, °state, °declare, °proclaim, °announce, °publish, °broadcast, °communicate, °make known, °report, °impart, °indicate, °release, °break, let (something) be known, advertise, trumpet, herald, °bring to light, °disclose, divulge, °intimate, °leak, °admit, °betray, °acknowledge, °confess, disbosom, get (something) off (one's) chest, unburden *or* disburden (oneself), °blab, °tattle, *Colloq* °talk, let the cat out of the bag, °spill the beans, °let out, °let slip, blow the whistle on, give away the (whole) show, *US* pull the plug on, *Slang* squeal, squeak, rat, peach, *US* spill (one's) guts: *Tell it not in Gath. She told of her fears. He refuses to tell where the money is hidden. She talked a lot but told very little.* **3** apprise, °advise, °inform, let (someone) know, °notify, acquaint (someone) with (something): *He has told me everything. Tell them that I shall be late. He told them nothing of his illness.* **4** °recount, °describe, delineate, °outline, °portray, depict, °express, °put, °word, °explain: *I am not sure how to tell this to you.* **5** °order, °command, °require, °demand (that), °charge, °direct, °dictate (that), °instruct, °bid: *Tell them I want no visitors today. She told him to go.* **6** carry weight, be influential, be effective, have (an) effect: *What we accomplished here today is bound to tell in our favor.* **7** °determine, °say, °confirm, aver, assert, asseverate, °swear, take an oath, be sure *or* certain *or* positive, know (for sure *or* for certain): *It just shows you never can tell.* **8** ascertain, °determine, °perceive, °understand, °make out, discern, °identify, °recognize, °distinguish, °discriminate, °differentiate: *I cannot tell if that is George or not. I can't tell George from John in this photograph.* **9** °predict, °prophesy, °forecast, foretell, °foresee, °forecast, °determine, ascertain, °know: *Who can tell what tomorrow will bring?* **10** ***tell off.*** °scold, °reprimand, °berate, chide, °castigate, censure, °take to task, °rebuke, °lecture, reproach, °reprove, *Colloq* give (someone) a tongue-lashing, rake *or* haul (someone) over the coals, give (someone) a piece of (one's) mind, tick off, *Slang* tear a strip off, *US* °chew out: *He was told off for being late for dinner again.* **11** ***tell on.*** tattle on, blab about, *Brit* grass on: *Charlotte never forgave Tim for telling on her to Aunt Jane.*

tell[2] *n.* tumulus, °mound, barrow, hillock: *Excavation of the tell exposed ancient skeletons.*

telling *adj.* **1** °effective, °effectual, °influential, °weighty, °important, °powerful, °forceful, °potent, °significant, °considerable, °striking: *The new color scheme in the office has had a telling effect on morale.* —*n.* **2** tattling, (too) revealing, *Colloq* letting the cat out of the bag, giving away the whole show: *I cannot say why her key was found in his pocket—that would be telling.*

temper *n.* **1** °mood, °disposition, temperament, °humor, state *or* frame of mind, °character, °personality, °nature, °makeup, constitution: *Practicing law did not suit his temper, so he became an editor.* **2** composure, °self-control, self-possession, calmness, equanimity, °balance, °sang-froid, coolness, *Colloq* °cool: *He lost his temper every time she mentioned another man she had known.* **3** ill humor, ill temper, foul temper, irascibility, irritability, petulance, volatility, peevishness, huffishness, surliness, churlishness, hotheadedness, hotbloodedness: *At the first sign of temper, just ignore him.* **4** °(temper) tantrum, °fury, °fit (of pique), °rage, °passion, *Colloq Brit* wax, paddy: *She went off into a temper at the mere mention of his name.* —*v.* **5** °modify, °moderate, assuage, mollify, °soften, °cushion, °tone down, allay, soothe, °mitigate, palliate, °reduce, °relax, slacken, °lighten, appease: *The cold currents from the poles are tempered by the equatorial waters. She used all her wiles to temper his fury.* **6** anneal, toughen, °strengthen, °harden: *Plunge the heated blade into water to temper it.*

temperamental *adj.* **1** °moody, °sensitive, °touchy, hypersensitive, °volatile, irascible, °irritable, °petulant, °testy, °short tempered, hot-tempered, hotheaded, hot-blooded, °excitable, °explosive, on a short fuze *or Brit* fuse, °capricious, °impatient, bad-humored, °curt, °brusque, °short, °gruff, °bluff, curmudgeonly, °waspish, °snappish, °peevish, crabby, crabbed, grumpy, huffish, huffy, crotchety, *US* °cranky, *Colloq* grouchy: *He's not usually temperamental—must be something he ate.* **2** °erratic, uneven, °unreliable, °inconsistent, undependable, unpredictable: *That car of mine is getting temperamental in its old age.*

temperance *n.* **1** (self-)restraint, moderation, (self-)control, forbearance, (self-)discipline, continence: *Temperance is one of the four cardinal virtues, alongside justice, prudence, and fortitude.* **2** abstemiousness, teetotalism, abstinence, °sobriety, Rechabitism; °prohibition: *We have always preached temperance to our children.*

temperate *adj.* **1** °moderate, °reasonable, (self-)restrained, disciplined, controlled, forbearing, °reasonable, °sensible, °sane, °rational, not excessive, composed, °steady, °stable, even-tempered, °equable, °sober, sober-sided, sober-minded, °mild, °dispassionate, unimpassioned, °cool, coolheaded, unexcited, °calm, unruffled, °tranquil, imperturbable, unperturbed, °self-possessed, °quiet, °serene: *Even the most temperate person would get angry dealing with a man like Fred.* **2** abstemious, teetotal, abstinent, continent, °moderate, °sober; °chaste, °celibate, austere, ascetic, self-denying, puritanical: *Those who lead temperate lives seem to live no longer than those who yield to dissipation.*

tempest *n.* **1** °storm, windstorm, hailstorm, rainstorm, °hurricane, typhoon, tornado, cyclone, squall, thunderstorm: *The Red Spot on Jupiter is the center of a perpetual, violent tempest.* **2** °storm, commotion, °disturbance, °upheaval, disruption, °furor, turbulence, °ferment, °tumult, °agitation, perturbation, hurly-burly, °disorder, outbreak, °unrest, °riot, °chaos, °uproar, brouhaha, Colloq hoo-ha: *After the tempest of protest died down, a vote was taken and saner heads prevailed.*

tempestuous *adj.* °stormy, °wild, °uncontrolled, uncontrollable, disrupting, disruptive, turbulent, °tumultuous, °riotous, °chaotic, °uproarious, °boisterous, °frantic, frenzied, frenetic, °furious, wrathful, vehement, °fiery, °impassioned, °fierce: *In their tempestuous relationship, they couldn't live together or apart.*

template *n.* templet, °pattern, °mold, °guide, °model, °die: *The work on the lathe is matched to the template.*

temple *n.* place *or* house of worship, holy place, house of God, church, synagogue, mosque, pagoda, cathedral, °sanctuary, chapel, shrine, *Yiddish* shul *or Yiddish* schul: *Temples to gods are found throughout the world.*

tempo *n.* °cadence, °rhythm, °beat, °time, °pulse, meter, °measure; °pace, °speed, °rate: *The melody should be played in waltz tempo. Living in the country, Tom is unaccustomed to the tempo of big city life.*

temporal *adj.* **1** °earthly, °terrestrial, terrene, mundane, °worldly, nonspiritual, nonclerical, °lay, laic(al), °secular, nonreligious, nonecclesiastic, °material, °civil, °profane, fleshly, °mortal: *He believed that his temporal miseries would be compensated for by an eternity in the seventh heaven.* **2** See **temporary,** below.

temporarily *adv.* **1** °for the time being, in the interim, pro tem, in *or* for the meantime *or* the meanwhile, for now: *Why was he insulted when I expressed sympathy at his wife's being temporarily* hors de combat? **2** °briefly, fleetingly, for a (short *or* little) while *or* time, for the moment: *Your size is temporarily out of stock, madam.*

temporary *adj.* impermanent, °makeshift, °stopgap, °standby, °provisional; pro tem, transitory, °transient, °fleeting, °fugitive, °passing, ephemeral, °temporal, evanescent, °short-lived, °momentary: *I suppose it will have to do as a temporary substitute.*

tempt *v.* **1** °attract, °entice, °lure, allure, °draw (in), invite, °lead on, whet (one's) appetite, °seduce, °captivate, °persuade, °coax, °cajole: *Ought we to offer a premium to tempt customers?* **2** °lead, °induce, °persuade, °prompt, °move, °incline, °dispose: *I would be tempted to resign if they treated me that way.* **3** °provoke, °dare, °(put to the) test: *I shouldn't tempt Providence if I were you.*

temptation *n.* **1** °tempting, enticing, leading on, seducing, captivating, persuading, coaxing, cajoling: *The temptation by advertisers never stops, does it?* **2** °enticement, seduction, captivation, °persuasion, allurement, °invitation, °attraction, °draw, °lure, °inducement, °snare, *Colloq* °pull, °come-on: *I could never resist a temptation like that.*

tempting *adj.* **1** °seductive, enticing, °inviting, alluring, captivating, °attractive, tantalizing, appealing, °irresistible, titillating; °fetching, winsome, °prepossessing, °ravishing, °voluptuous, °sensuous, *Colloq* °sexy, *US* °foxy: *A free weekend in Pittsville is a tempting offer. Later, without all the make-up, she wasn't so tempting any more.* **2** appetizing, mouthwatering, °delicious, °savory, delectable, °succulent, °luscious, toothsome: *Why does food always look most tempting when I am on a diet?*

temptress *n.* °seductress, vamp, °siren, *femme fatale,* coquette, °flirt, enchantress, sorceress, Circe, *Slang* sexpot, man-eater, *US* foxy lady, fox, mantrap, : *You warned me against that temptress because you wanted her for yourself.*

tenable *adj.* defensible, °supportable, justifiable, maintainable, workable, °viable, defendable, °plausible, °reasonable, °rational, arguable, believable, credible, creditable, imaginable, conceivable, °possible: *Copernicus showed that Ptolemaic cosmology was no longer tenable.*

tenacious *adj.* **1** °persistent, dogged, unfaltering, pertinacious, unswerving, °determined, °diligent, °resolute, °staunch, °stalwart, °steadfast, °strong, °sturdy, unwavering, strong-willed, °strong-minded, unshaken, unshakable *or* unshakeable, °obstinate, intransigent, °stubborn, adamant, obdurate, refractory, °immovable, °inflexible, °rigid, °firm, unyielding, uncompromising: *Despite his age, Comstock maintains a tenacious grip on reality.* **2** cohesive, °strong, °tough; adhesive, °sticky, clinging; gummy, gluey, mucilaginous, glutinous, viscous, viscid: *The bricks are then smeared over with a tenacious mud, which dries hard.* **3** Often, ***tenacious of.*** clinging (to), °grasping, maintaining, keeping (up), staying with, retentive (of), persisting *or* persistent (in), retaining: *She is tenacious of the old ways of doing things.* **4** retentive, °good: *Aunt Agatha has a very tenacious memory.*

tenacity *n.* **1** tenaciousness, °persistence, doggedness, °perseverance, pertinacity, °determination, °grit, diligence, resoluteness, °resolution, purposefulness, °resolve, staunchness, steadfastness, °stamina, assiduity, sedulousness, °strength, strong-mindedness, unshakability *or* unshakeability, °obstinacy, intransigence, stubbornness, obduracy, inflexibility, rigidity, firmness, uncompromisingness, *Colloq US* sand, stick-to-itiveness: *He persisted with the tenacity of the English bulldog.* **2** tenaciousness, cohesiveness, °strength, °power, toughness, °resilience; adhesiveness, stickiness, gumminess, glueyness, mucilaginousness, glutinousness, viscousness, viscidity, *US* °cling: *The texture of the surfaces affects the tenacity of the bonding cement.*

tenancy *n.* occupancy, °occupation, °possession, °tenure: *Their tenancy of the house expires next month. The tenancy of this position is a year.*

tenant *n.* °occupant, lessee, renter, leaseholder, occupier, °resident, °inhabitant: *We could not redecorate the premises while the tenants were there.*

tend[1] *v.* be inclined *or* disposed, be liable *or* apt *or* likely, °incline, °lean, have *or* show *or* exhibit *or* demonstrate a tendency, °favor, °verge, gravitate, be biased; be prone: *The judge might tend toward leniency in your case. I tend to agree. Tree growth here tends toward the horizontal because of the wind.*

tend[2] *v.* °care for, °take care of, °look after, look out for, watch over, °see to, keep an eye on, °attend (to), °wait on, °cater to, °minister to, °serve, °nurse, nurture: *Marie tended her father throughout his long illness.*

tendency *n.* °inclination, °bent, °leaning, °disposition, propensity, predisposition, proclivity, predilection, susceptibility, proneness, °readiness, °partiality, °affinity, °bias, °drift, °direction, °trend, °movement: *Bagby has always had a tendency to being overweight. The particles exhibit a tendency to align themselves north and south.*

tender[1] *adj.* **1** °sensitive, °delicate, °fragile, °frail, °infirm, °unstable, °shaky, °weak, °feeble, unwell, °sickly, ailing, °unsound: *His condition is still a bit too tender for him to go outside.* **2** chewable, °edible, eatable, °soft: *The steak will become more tender if it is aged.* **3** °young, youthful, °immature, °juvenile, °inexperienced, °impressionable, °vulnerable, °green, °new, °raw, °undeveloped, untrained, uninitiated, °callow: *At his tender age he could not have known about such things.* **4** °sensitive, °touchy, °ticklish, °dangerous, °troublesome, °provocative, °difficult, °tricky: *Please avoid mentioning the wedding, it's a tender subject around here.* **5** °gentle, °soft, °delicate, °light, °sensitive, °soothing: *Oh, how he yearned for her tender touch at his fevered brow.* **6** °kind, kindhearted, loving, °affectionate, °fond, °gentle, °mild, compassionate, °considerate, humane, °benevolent, °sympathetic, °feeling, °thoughtful, °softhearted, °warm, caring, °merciful, °solicitous, tenderhearted, warmhearted, °good-natured: *Wounded and helpless, he succumbed to the tender care of the nurses.* **7** °touching, °emotional, °moving, °stirring, soul-stirring, °heart-rending, °heartfelt, °passionate, °impassioned, impassionate, °poignant, °sentimental, mawkish, °maudlin: *Who could forget the tender scene as Cedric bade farewell to his mother and rode off to war?* **8** °sore, °raw, °painful, °sensitive, °inflamed; smarting, °burning, hurting, aching, °agonizing: *This spot, right here, is so tender it hurts when I so much as think about it.* **9** loving, °affectionate, amatory, amorous, adoring, °romantic: *They exchanged tender, knowing looks.*

tender[2] *v.* **1** °offer, proffer, °present, °propose, °put forward, °extend, °hold out, °submit, °advance, °put up, set before: *Donald tendered his resignation yesterday. She has tendered the committee an excellent proposal.* —*n.* **2** °offer, °bid, °presentation, °proposal, °proposition: *The city is accepting tenders for the construction of a new bridge.* **3** currency, °money, specie, °(bank) notes, °cash, bills; °payment, compensation: *Today, checks are considered legal tender subject to collection.*

tender[3] *n.* **1** dinghy, gig, skiff, °launch, °boat, rowboat *or* rowing boat, jollyboat: *The yacht moved easily through the water towing a tender behind.* **2** wagon, °truck, °vehicle: *The tender of Felix's toy train was filled with sweets for his birthday.*

tenet *n.* °belief, credo, °creed, article of faith, °ideology, °precept, °conviction, °principle, dogma, °idea, °opinion, °position, °view, °viewpoint, °maxim, axiom, canon, teaching, °doctrine: *They accept the Golden Rule as a fundamental tenet of their organization.*

tenor *n.* °drift, °tone, °spirit, °essence, °character, °gist, °bias, °import, °substance, °effect, °significance, °meaning, °sense, connotation, °theme, °thread, °implication, °inference, °intent, °purpose, °tendency, purport, °direction: *The general tenor of Malvern's letters was unfriendly.*

tense *adj.* **1** °taut, °strained, °stiff, under tension, °rigid: *One could see how tense the muscles were under the skin.* **2** °intense, °nervous, °anxious, under (a) strain, highly strung, high-strung, °strained, °on edge, wrought up, keyed up, worked up, °taut, on tenterhooks, apprehensive, distressed, °upset, °disturbed, °worried, edgy, on pins and needles, °jumpy, fidgety, °overwrought, *Colloq* wound up, jittery, having a case of the jitters, *Brit* strung up, *US* strung out, *Slang* uptight, *US* antsy: *Try to relax and not be so tense.* **3** °nervous, °anxious, worrying, worrisome, distressing, °disturbing, stressful, °nerve-racking, °fraught, disquieting: *Those were tense moments while we waited for the winner's name to be posted.* —*v.* **4** °tighten, °stretch, °strain, tauten, °tension: *The mast will be secure once the stays have been tensed.*

tension *n.* **1** °stress, tightness, tautness, °strain, °pull, °traction, °pressure, tenseness, °force: *Can this cable withstand so much tension?* **2** nervousness, °anxiety, anxiousness, °strain, edginess, apprehension, °suspense, tautness, °distress, °upset, °worry, jumpiness, fidgetiness, *Colloq* jitteriness, (a case of) the jitters: *I could feel the tension building among the audience.*

tentative *adj.* **1** °experimental, °speculative, exploratory, probative, °trial, °provisional: *We have tentative plans for sectioning the fly's DNA.* **2** unsure, °hesitant, °uncertain, °indecisive, °cautious, °timid, °shy, diffident, uneasy, apprehensive: *The child gave him a tentative smile, not knowing what to expect.*

tenuous *adj.* **1** °thin, °slender, °thin, °fine, attenuated, °delicate, gossamer, diaphanous, °fragile: *The spider descended slowly on its tenuous thread.* **2** °flimsy, insubstantial *or* unsubstantial, °paltry, °weak, °feeble, °frail, °meager, °vague, °negligible, °insignificant, °trifling, °sketchy, °hazy, °nebulous, dubious, °doubtful, °shaky: *Your case rests on very tenuous evidence.*

tenure *n.* **1** °possession, holding, occupancy, incumbency, tenantry, °tenancy, °occupation, residency, °residence: *The English laws on tenure are very complex.* **2** °(job) security, °permanence, permanency: *After five years he was automatically guaranteed tenure.*

tepid *adj.* **1** °lukewarm, warmish: *The gelatin should be tepid before refrigerating it.* **2** °lukewarm, °unenthusiastic, °cool, °indifferent, apathetic, uninterested, unconcerned, °nonchalant, uncaring, °neutral, °blasé: *How can he remain tepid on an issue like apartheid?*

term *n.* **1** °name, °title, designation, appellation; °word, °expression, locution, °phrase: *The term for that is* venturi tube. Habeas corpus *is a term of art in law.* **2** semester; °time, °period (of time), °interval, length of time, °span (of time), duration, °spell, °stretch, while: *During the spring term we trained for the baseball team every Saturday. He was sent to prison for a five-year term.* **3** sitting, °stint, °session, °course; incumbency, °administration: *During his latest term of office interest rates were raised twice.* **4** Often, ***terms.*** condition(s), provision(s), article(s), clause(s), proviso(s); stipulation(s), qualification(s), assumption(s): *According to the terms of the will, proceeds from the patent go into a trust fund.* **5** ***come to terms.*** °agree, come to *or* reach an agreement *or* an arrangement *or* an understanding, °reconcile, °arrange, °settle, compromise: *Can we come to terms on the price? He will simply have to come to terms with his conscience.* **6** ***in terms of.*** °concerning, °regarding, as regards, with regard to, °in relation to, °relative to, relating to, in the matter of: *The psychiatrist offered little in terms of direct solutions to my problems.* **7** ***terms.*** **(a)** °payment, °schedule, rates: *What terms have they offered to settle the debt?* **(b)** °standing, °position, °basis, °relationship, °relations, °footing: *I always thought you two were on good terms.* —*v.* **8** °call, °name, °label, °designate, denominate, °entitle, °title, °style, dub; °nickname: *Abraham*

Lincoln was termed, among other things, "The Great Emancipator."

terminal *adj.* **1** closing, concluding, terminating, ending, °final, °ultimate, °extreme; °maximum, greatest: *Spencer wrote that the human being is at once the terminal problem of biology and the initial factor of sociology. When falling through a medium, a body eventually reaches its terminal velocity.* **2** °deadly, °mortal, °fatal, °lethal, °incurable: *The cancer was diagnosed as terminal.* —*n.* **3** terminus, °(terminal) station, end of the line, depot: *We arrived at the air terminal with five minutes to catch our plane.* **4** keyboard, °monitor, °position, °station, VDU (= 'visual display unit'), VDT (= 'visual display terminal'), PC (= 'personal computer'), module, CRT (= 'cathode ray tube'), °screen, (control) panel: *The operator controls the devices from his terminal.* **5** °connection, wire, connector, coupler, coupling, conductor: *Do I attach the red cable to the positive or the negative terminal?*

terminate *v.* °stop, °end, come to an end, °finish; put an end to, °cease, conclude, °discontinue, °drop, abort, bring to an end *or* a close, wind up *or* down, °sign off, °cut off: *The railway line once terminated at the port. She says she wants to terminate our relationship.*

termination *n.* **1** °end, ending, °stop, stopping, stoppage, ceasing, cessation, discontinuation, abortion, *Colloq* windup, winding-up, °close, °finish, finishing, conclusion: *The rights revert to us at the termination of the agreement. Because of her condition, termination of the pregnancy was recommended.* **2** °suffix, desinence, ending: *The English termination* -ly *is usually adverbial but might be adjectival, as in* kindly.

terminology *n.* nomenclature, vocabulary, °language, °words, locutions, °wording, °terms, phraseology, phrasing, °jargon, shoptalk, argot, °cant, *Colloq* °lingo: *Medical terminology requires specialized dictionaries.*

terrain *n.* topography, °landscape, °ground, °territory: *These vehicles are specially designed for rough terrain.*

terrestrial *adj.* **1** °earthly, earthbound, °worldly, terrene, tellurian *or* telluric, °global, sublunary, subastral; mundane: *He insists that we tackle our terrestrial problems before spending a fortune on space travel.* —*n.* **2** earthman, earthwoman, earthperson, earthling, °mortal, °human: *My experiences are beyond the imaginations of you mere terrestrials.*

terrible *adj.* **1** °bad, °serious, °grave, °severe, °acute, distressing, °disagreeable, °nasty, °foul, °unbearable, °dreadful, °loathsome, °hideous, °vile, intolerable, °awful, *Colloq* °rotten, °lousy, °beastly: *I've had a terrible virus.* **2** °bad, °remorseful, °regretful, rueful, °sorry, contrite, °ashamed, conscience-stricken, °guilty, distressed, °dreadful, °awful, *Colloq* °rotten, °lousy, °beastly: *I felt terrible having to tell Mrs. Barrington about her cat.* **3** °unhappy, unpleasant, °disagreeable, °awful, °miserable, °joyless, °wretched, °unfortunate, *Colloq* °rotten, °lousy, °beastly: *We spent an absolutely terrible evening at the theater.* **4** °gruesome, °grisly, °macabre, °gory, °grotesque, °brutal, °savage, °horrible, horrendous, °terrifying, °terrific, °harrowing, horrid, horrifying, °ghastly, °frightening, °frightful, unspeakable, °monstrous, °dread, °terrible, appalling, °shocking, alarming, °awful, °foul: *Casualties being brought in had the most terrible wounds.* **5** °disgusting, °revolting, nauseating, °nauseous, °offensive, vomit-provoking, °obnoxious, stomach-turning, stomach-churning, °abominable, mephitic, noisome, noxious, °loathsome, °horrible, °hideous, °terrific, °awful, °evil, °vile, °rotten, °awful: *The decaying cadavers gave off the most terrible stench.* **6** °terrifying, °frightening, °frightful, °fearsome, °formidable, redoubtable, °awesome, awe-inspiring, °terrific: *Catching sight of me, the giant roared a terrible laugh which reverberated throughout the cave.*

terribly *adv.* °very, °extremely, °exceedingly, °thoroughly, decidedly, unbelievably, incredibly, monumentally, outrageously, °awfully, fabulously, *Colloq* °frightfully: *The acting was sometimes terribly bad, sometimes terribly good.*

terrific *adj.* **1** See **terrible, 4, 5, 6,** above. **2** wonderful, °marvelous, °splendid, breathtaking, °extraordinary, °outstanding, °magnificent, °exceptional, °unbelievable, °incredible, mind-boggling, stupendous, °superb, °excellent, first-class, °superior, *Colloq* °great, ace, °fantastic, °fabulous, °sensational, smashing, °super: *They really put on a terrific show.*

terrify *v.* °alarm, °frighten, °scare, terrorize, °shock, make one's flesh crawl *or* creep, °horrify, make one's blood run cold, make one's hair stand on end, °stun, °paralyze, °petrify: *Are you trying to terrify the child with those horror stories?*

terrifying *adj.* alarming, °frightening, °scary, °shocking, horrifying, paralyzing, petrifying: *Being chased by a great white shark is a terrifying experience.*

territory *n.* **1** °area, °region, °district, °neighborhood, °zone, sector, °tract, °land, °precinct, °quarter, vicinage, °vicinity, purlieu: *Each one of these animals stakes out its own territory and protects it.* **2** °area, bailiwick, °domain, °province, haunts, °patch, *Colloq* stamping ground, *US* °turf: *Rimbeau would be best advised to stick to his own territory.*

terror *n.* **1** °fright, °dread, °fear, °horror, °panic, °shock, °alarm, °anxiety, °dismay, consternation, intimidation, awe: *I cannot tell you the terror I felt when I saw who—or what—was at the door.* **2** °scourge, °demon, °brute, °monster, °fiend, °devil, *US* mad dog: *Is this little fellow the bandito who was known as the Terror of the Pampas?*

terrorist *n.* °subversive, °radical, insurgent, °revolutionary, anarchist, nihilist; bomber, arsonist, incendiary; desperado, gunman, °thug, °felon, °criminal: *The shooting was carried out by a previously unknown gang of terrorists.*

terse *adj.* **1** °concise, °brief, °short, °compact, pithy, °succinct, °summary, laconic, short and sweet, °to the point, sententious, °crisp, °epigrammatic, aphoristic; distilled, condensed, compendious, °abbreviated, abridged, shortened, concentrated: *A terse presentation would save us a lot of time.* **2** °abrupt, °curt, °short, °brusque, °blunt, °gruff, °bluff, °ungracious, °petulant, °tart, °rude: *His answer was terse and impolite.*

test *n.* **1** °trial, °examination, exam, °proof, °evaluation, assay, °check, checkup, °investigation, °study, °analysis: *My physics test is tomorrow. Have you had a blood test lately?* —*v.* **2** °try (out), check (up) (on), °examine, °evaluate, assess, assay, °prove, °probe: *Someone ought to test the water for purity.*

testify *v.* °state, aver, assert, °attest, °swear, °say, affirm, °declare, give evidence *or* testimony, °bear witness, avow, °vouchsafe, °proclaim, °announce: *He testified that he saw you there. Is your mother going to testify at the trial? The fingerprint testifies to the fact that she was here.*

testimonial *n.* °endorsement, certification, commendation, °(letter of) recommendation, °reference, *Colloq* blurb: *I am happy to write a testimonial for you.*

testimony *n.* °evidence, attestation, affirmation, confirmation, verification, authentication, corroboration, avowal, deposition, °statement, affidavit, °declaration, °assertion, °claim, averral, asseveration, °information: *We have only his testimony that she caused the accident.*

testy *adj.* °irritable, bad-tempered, irascible, °short-tempered, °petulant, °touchy, tetchy, °querulous, °peevish, hot-tempered, crusty, °cross, grumpy, grouchy, bearish, crabby, crabbed, °fretful, captious, °waspish, °snappish, °quarrelsome, fractious, contentious, choleric, splenetic, ill-humored, °disagreeable, ill-tempered, edgy, °on edge, °quick-tempered, crotchety, °cantankerous, *US* °cranky, *Colloq or dialectal US and Canadian* ornery: *The old colonel made a few testy remarks about the boys' haircuts. Why make excuses for his being so testy?*

tête-à-tête *n.* **1** (cozy *or* personal) chat, °dialogue, causerie, pillow talk, private talk *or* word, °parley, °interview, *Colloq* confab, *US* one-on-one: *Will we have time for a brief tête-à-tête after dinner?* —*adv.* **2** intimately, privately, °in private, °face to face, confidentially, °secretly, *à deux,* °in secret, *Colloq US* one on one: *Could we meet tête-à-tête some time?* —*adj.* **3** °intimate, °private, intime, °cozy: *I arranged a little tête-à-tête dinner at my place.*

tether *n.* **1** °lead, leash, °rope, °cord, fetter, °restraint, halter, °tie, °chain: *Despite the weather, he tied the horse in the heather by a leather tether.* —*v.* **2** tie (up *or* down), °restraint, fetter, chain (up *or* down), leash, °manacle, °secure, °shackle, °fasten, °picket, °stake: *Why don't you tether the goat where the grass is more plentiful?*

text *n.* **1** °wording, °words, °content, °(subject) matter; printed matter, °(main) body (text), °contents: *The editor had no right to make changes in the text without consulting the author. The text is in 10-point roman type, the footnotes in 8-point.* **2** °extract, °abstract, °section, °quotation, °part, paragraph, °passage, verse, °line: *What text from the Scripture was selected for next week's sermon?* **3** °subject (matter), °topic, °theme, °motif, °issue, °focus: *For today's talk I have chosen for my text the problem of alcoholism.* **4** textbook, °schoolbook, reader, °manual, primer, workbook, exercise book, *Archaic* hornbook: *Today's science texts are much more advanced than those of twenty years ago.*

texture *n.* °feel, °surface, °character, °grain, °features, consistency, °weave; configuration, °nature, °structure, °fabric, constitution, °substance: *This material has a rough texture. The texture of the plot is reinforced by the dialogue.*

thank *v.* **1** express *or* show (one's) gratitude *or* thanks *or* appreciation, say "Thank you" *or* "Thanks," give *or* offer *or* tender thanks: *Don't forget to thank your uncle for the gift, William.* **2** °blame, hold responsible, °credit, °acknowledge: *You have your mother to thank for the present state of affairs.* —*n.* **3** ***thanks.*** °gratitude, °appreciation, gratefulness, °acknowledgment, °recognition, thanksgiving: *I gave thanks that I was again on terra firma.* **4** ***thanks to.*** °owing to, because of, as a result of, thanks be given to, in consequence of, as a consequence of, °by reason of, °through, *Sometimes nonstandard* due to: *Thanks to you, I have to commute to work by train.*

thankful *adj.* °grateful, appreciative, °indebted, pleased, °glad, °obliged, obligated, under obligation, beholden to: *I am most thankful that you came, doctor.*

thankless *adj.* unappreciated, unacknowledged, °useless, unrewarding, °fruitless, °unprofitable, profitless, unrequited, °vain, °futile, °bootless: *She always regarded housekeeping as a thankless task.*

thaw *v.* **1** Sometimes, ***thaw out.*** °melt, deice, liquefy, defrost, °warm (up), °heat (up), unfreeze: *We cannot leave till the river thaws. Why not thaw out the shepherd's pie for dinner?* **2** °soften, °warm, become (more) cordial *or* friendly, °relax, °yield, °relent, °bend, unbend, let (oneself) go: *After he got to know some of us better he began to thaw.*

theater *n.* **1** playhouse, °(opera) house, °(music) hall, auditorium, amphitheater, theater in the round, coliseum, hippodrome, arena (theater): *A long run in a West End or Broadway theater is every playwright's dream.* **2** °drama, stagecraft, dramaturgy, melodrama, theatrics, histrionics, staginess, acting, performing, °performance: *Even the way Mary serves dinner is pure theater.* **3** ***the theater.*** °drama, °the stage, dramaturgy, dramatic *or* thespian *or* histrionic art, the boards, show business, *Colloq* show biz: *The theater is a risky profession.* **4** °area, arena, °scene, sphere *or* place *or* field of action, °setting: *Her father served in the Pacific theater during the war.*

theatrical *adj.* **1** theatric, °dramatic, °stage, histrionic, °thespian; °repertory: *He joined a theatrical company and his family disowned him.* **2** stagy, overdone, °camp, campy, °melodramatic, °overwrought, exaggerated, °forced, overacted, overacting, °sensational, sensationalistic, °fake, °false, °mannered, °affected, °unnatural, °artificial, °showy, °ostentatious, spectacular, °extravagant, *Colloq* phoney *or US also* phony, ham *or* hammy, grandstand: *Doesn't Fern's theatrical behavior sometimes embarrass you?*

theft *n.* °robbery, °stealing, pilferage, pilfering, filching, shoplifting, thievery, purloining, °embezzlement, hijacking, larceny, *Colloq* lifting, appropriation, pocketing, pinching, swiping, snitching, *Chiefly Brit* nicking, *US* boosting, *Slang* heist, knocking off, °rip-off: *Raffles's theft of the Costanza emerald has never been proved.*

theme *n.* **1** °subject (matter), °topic, °idea, °notion, concept, °thesis, °text, °thread, keynote, °gist, °core, °substance, °point, °essence: *What is the central theme of your article?* **2** °essay, °paper, °composition, °review, article, °story, °piece, °exposition, °study, °exercise, °monograph, °tract, °thesis, dissertation, disquisition, treatise: *I have to hand in two themes next week.*

theorem *n.* **1** °hypothesis, °proposition, assumption, conjecture, °thesis, postulate: *He has set forth a clever proof of the binomial theorem.* **2** °statement, dictum, °rule, °deduction, °formula, axiom, °principle: *After a lifetime of experience, his theorem is that honesty is the best policy.*

theoretical *adj.* **1** °hypothetical, conjectural, °speculative, untested, unproved, unproven, °moot, putative, °debatable, supposititious, suppositional: *The theoretical trajectory will take it beyond Uranus.* **2** °impractical, °unrealistic, °pure, °ideal, °abstract, °academic: *As the business doesn't open its doors till tomorrow, what we do with the profits is purely theoretical.*

theorist *n.* theoretician, speculator, hypothecator, hypothesizer, theorizer, philosopher, °dreamer: *Leave guesswork to the theorists—we need hard facts.*

theorize *v.* °guess, hypothesize, conjecture, °speculate: *It is logically illogical to theorize about articles of faith.*

therapeutic *adj.* therapeutical, healing, curative, remedial, restorative, salutary, health-giving, °healthy, °beneficial, corrective, salubrious, medical, °medicinal: *Primitive medicine men knew of the therapeutic powers of these herbs.*

therapist *n.* psychotherapist, psychologist, analyst, therapeutist, psychiatrist, psychoanalyst, °counselor, °adviser, *Colloq* °shrink: *For some problems it is best to seek the advice of a trained counselor or therapist.*

therapy *n.* **1** °remedy, °treatment, remedial program; °cure: *As therapy, he suggested I swim a mile every day.*

2 psychotherapy, psychoanalysis, °analysis, group therapy: *If it weren't for my weekly therapy sessions, I'd never see anyone.*

therefore *adv.* °consequently, so, °thus, as a result *or* consequence, °hence, ergo, for that reason, wherefore, °accordingly, that being so *or* the case: *The Franks were stronger and therefore the masters.*

thesaurus *n.* **1** °treasury, treasure trove, °storehouse, armory, arsenal, repository, °cache: *It is hoped that this book will be regarded as a thesaurus of linguistic treasures.* **2** synonym dictionary, synonymy, °dictionary, lexicon: *He checked in his thesaurus for a synonym of "idea."*

thesis *n.* **1** °argument, theory, °proposition, °point, contention, °belief, °idea, premise *or* premiss, assumption, °view, °assertion, °precept, °opinion, °notion, °theorem, axiom, postulate: *Think how many centuries it took man to come to grips with the thesis that all men are created equal.* **2** See **theme, 2,** above.

thespian *adj.* **1** °dramatic, theatrical *or* theatric, histrionic, acting, performing; *Colloq* ham, hammy: *An assessment of your thespian talents, Ferncope, leads me to suggest a career in librarianship.* —*n.* **2** actor, actress, °performer, trouper, °player; supernumerary; matinee idol, °star; *Colloq* ham: *For many years Dame Maida was one of the most beloved thespians who ever trod the boards.*

thick *adj.* **1** °broad, °wide, °solid, thickset, °burly, °ample, °solid, °bulky, °substantial, °beamy: *She is a bit thicker in the midsection than I had remembered, but a yare ketch nonetheless.* **2** Usually, ***thick with.*** °dense, °solid, °compact, concentrated, condensed, °packed, close-packed, compressed, crowded, choked, filled, °full, °deep, clotted, chock-full *or* choke-full *or* chuck-full, chockablock, teeming, swarming, °alive, bristling, crawling, bursting, crammed, jammed, brimming, *Colloq* °lousy with: *The gallery was thick with visitors.* **3** °compact, condensed, compressed, choking, °packed, impenetrable, impassable, °dense; pea-soup, soupy, °murky, °misty, foggy, smoggy, smoky, °opaque, °obscure, obscuring, °hazy: *We plunged deeper into the thick jungle. A thick fog blanketed the valley.* **4** °abundant, °plentiful, bushy, °luxuriant: *When he loosed the pins her thick red hair tumbled down over her shoulders.* **5** °dense, viscid, viscous, gelatinous, mucilaginous, gluey, glutinous, °ropy, coagulated, clotted, congealed, jelled, jellied, inspissated, stiffish; °stiff, °firm, °rigid, °solid: *For a thicker sauce, use cornstarch.* **6** thickheaded, thick-witted, thick-skulled, °dense, °stupid, °slow, slow-witted, °dull, dull-witted, °stolid, °obtuse, gormless, boneheaded, fat-headed, pinheaded, woodenheaded, addlepated, °half-witted, blockheaded, doltish, Boeotian, cretinous, imbecilic, moronic, *US* thimble-witted; insensitive, °thick-skinned; *Colloq* dimwitted, *Slang* dopey: *How could anyone so thick have become an executive?* **7** guttural, hoarse, throaty, raspy, rasping, °rough, °husky, °grating, gravelly, °indistinct, distorted, °inarticulate; °gruff, °raucous: *"I need you," he moaned, his voice thick with emotion.* **8** °close, °friendly, like that, inseparable, °devoted, °hand in glove, on good terms, on the best (of) terms, °intimate, *Colloq* °chummy, pally, (as) thick as thieves, *Brit* matey, well in, *US* palsy-walsy: *We were thick in the old days, going everywhere together.* **9** °marked, °pronounced, °strong, °decided, °obvious, °typical: *She speaks English with a thick Polish accent.* —*n.* **10** °core, °heart, °center, °middle, °focus, °midst: *If it's a fight, you can be sure that Timothy will be in the thick of it.*

thicken *v.* °coagulate, clot, congeal, °jell, gel, °set, °solidify, °stiffen, °harden, °firm up, °cake, incrassate, inspissate: *The rennet causes the mixture to thicken.*

thicket *n.* copse, °brake, grove, covert, wood, *Brit* spinney: *We concealed our horses in the thicket and stole ahead on foot.*

thick-skinned *n.* insensitive, insensate, °dull, °obtuse, °stolid, °callous, numb(ed), steeled, hardened, toughened, °tough, unsusceptible, inured, unfeeling, case-hardened, impervious, pachydermatous, *Colloq* hard-boiled: *After a few years in the emergency ward, one becomes too thick-skinned to mind the gore.*

thief *n.* **1** °robber, °burglar, cat burglar, housebreaker, picklock, sneak thief, safecracker, pilferer, shoplifter, purloiner; embezzler, peculator; pickpocket, cutpurse, purse snatcher, mugger, highwayman, footpad, brigand, bandit, °thug, dacoit, ruffian, °outlaw, desperado, hijacker, gunman, plunderer; poacher; *Technical* kleptomaniac, *Australian* bushranger, *US* road agent, *Colloq* holdup man, crook, *US* second-story man, bandito *or* bandido, *Slang* cracksman, box man, °dip, stickup man: *They caught the thief who took the paintings.* **2** °cheat, °swindler, confidence man, mountebank, charlatan, sharper, trickster, flimflam artist *or* man, thimble-rigger, *Colloq* con man, con artist, shell game artist, *US* highbinder: *That thief cheated them out of their savings.* **3** °pirate, °(sea) rover, picaroon *or* pickaroon, corsair, freebooter, buccaneer, marauder, filibuster, privateer: *They might seem romantic in retrospect, but at the time, pirates were no more than common thieves.*

thin *adj.* **1** °slim, °slender, °lean, °spare, °slight, °lanky, spindly, °skinny, thin as a rail *or* reed *or* rake, wispy, twiggy, skeletal, °gaunt, gangling, bony, °emaciated, cadaverous, °meager, °scrawny, all skin and bones, scraggy, undernourished, underfed, underweight, °undersized, °puny, °sparse, hollow-cheeked, (half-)starved, pinched, withered, shrunken, shriveled (up): *Douglas is quite thin as a result of his illness.* **2** °sparse, unsubstantial, °poor, scant, °insufficient, °inadequate, °slight, °worthless, unimportant, °deficient, skimpy, unplentiful, °paltry, piddling: *This year's harvest has been very thin.* **3** attenuated, threadlike, stringlike, pencil-thin, °fine; °narrow: *Draw a thin line between the columns. Please slice the bread thin.* **4** °flimsy, °weak, °feeble, °slight, unsubstantial, °insubstantial, °fragile, °frail, °poor, °lame; °unbelievable, unconvincing: *Harry gave some thin excuse for being late.* **5** airy, °filmy, diaphanous, gossamer, °sheer, °light, °delicate, chiffon, °silky, silken, gauzy, translucent, °see-through, °transparent: *She had nothing but a thin negligee to protect her from the cold.* **6** °watery, watered-down, dilute(d), °weak, unsatisfying: *My dinner, as usual, consisted of thin gruel and a dry crust.* **7** ***thin on the ground.*** °rare, uncommon, °scarce, °few (and far between), °unusual, hard to come by *or* find, scant, °scanty: *Good lexicographers are thin on the ground.* —*v.* **8** Often, ***thin down.*** °draw out, attenuate, °reduce, °trim, °cut down, °prune; °sharpen: *At one end, thin the dowel to a point.* **9** Often, ***thin down*** *or* ***out.*** °dilute, °water (down), °decrease, °reduce, °diminish: *They always thin the wine with some water. The crowd thinned out after a while.*

thing *n.* **1** °item, °(inanimate) object, article, °possession: *They went without leaving a thing behind.* **2** °item, °subject, °matter, °detail, °feature, °aspect, °affair, constituent, °element, °factor, °point: *She added one thing that I forgot to mention. It's a small thing, but I like coffee with breakfast.* **3** °fad, °trend, °fashion: *Let me show you the latest thing in sports cars.* **4** °feeling, °reaction, °attitude, °sentiment, emotional attachment; °quirk, °fixation, preoccupation, °obsession; fetish *or* fetich, *idée fixe,* °affection, °liking, °partiality, predilection, °fancy, °love, °passion, °mania; °phobia, °fear, °terror, °aversion, °loathing, °horror, detestation, °dislike, *Colloq* hang-up: *A chocoholic has a thing about chocolate. Pet snakes are not Patricia's thing. Roger has a thing about going out in public.* **5** °device, °item, °gadget, °object, °entity, °mechanism, contrivance, °apparatus, °instrument, utensil, *Colloq* dingus, doodad, whatchamacallit, thingumajig, whosis, whatsis, thingummy, thingumabob, *Chiefly US and Canadian* gizmo *or* gismo: *Where's the thing that is supposed to hold the flywheel in place?* **6** chore, °task, °responsibility, °matter; °act, °action, °deed, °activity, °proceeding: *The first thing I must take care of is to phone home.* **7** °opportunity, °chance, °possibility: *See me first thing tomorrow.* **8** ***things.*** **(a)** affairs, matters, °business, concerns: *I have to take care of some things at the office today.* **(b)** °circumstances, events, happenings: *Things are rarely what they seem.* **(c)** °belongings, °luggage, baggage, impedimenta, °possessions, °paraphernalia, °effects, °clothes, clothing, °goods; °equipment, tools, utensils, implements, °apparatus, *Colloq* °gear, °stuff, *Slang Brit* clobber, *US* crap, °junk, *Taboo slang US* shit: *Take your things and leave at once! The plumber left his things at the shop.*

think *v.* **1** °believe, °imagine, °expect, °dream, °fantasize, °suppose: *Do you think that you will ever retire?* **2** °judge, °reckon, °consider, deem, °regard (as), °characterize (as), °believe, °assume, °mark: *They thought him a fool for trying to build a flying machine.* **3** °contemplate, cogitate (on *or* over *or* about), ruminate (over *or* about), °reflect (on), meditate (on *or* over *or* about), muse (on *or* over *or* about), deliberate (on *or* over *or* about), think about *or* of *or* over: *He likes to sit in the tower room and think. I need time to think over your proposal.* **4** Often, ***think of.*** °recall, °remember, °recollect, call to mind: *I simply cannot think of the name of that movie we saw last week.* **5** ***think of*** or ***about.*** °consider, °ponder, °weigh, °contemplate, muse over, have in mind, mull over, entertain the idea *or* notion of, °intend, have in mind, °propose: *Do you ever think of retiring?* **(a)** assess, °evaluate, value, °judge: *What do you think about the situation in Lower Badoeng?* **6** ***think up*** or ***of.*** °devise, concoct, contrive, come up with, °invent, °conceive (of), dream up, °create, °make up, °improvise: *Who thinks up the plots for some of those films?*

thinkable *adj.* conceivable, °possible, imaginable, °feasible, °reasonable, °tenable, not unlikely, °plausible, believable, credible: *I suppose it is thinkable that experience may be as important as training in some fields.*

thinker *n.* °sage, wise man, savant, Nestor, Solomon, pundit, °mastermind, philosopher, °scholar, learned person, mentor, °expert: *She has been looked on as one of the great thinkers of her time.*

thinking *adj.* **1** °rational, °sensible, ratiocinative, °intelligent, °reasoning, °reasonable; °meditative, contemplative, °reflective, °philosophical, cogitative, °pensive, °thoughtful, °intellectual: *Any thinking person would have done the same as you.* —*n.* **2** °opinion, °judgment, °belief, °thought, °point of view, °viewpoint, assessment, °evaluation, theory, °reasoning, conclusion, °idea, °philosophy, °outlook: *There is nothing either good or bad, but thinking makes it so. Do you concur with the current thinking on education?*

thirst *n.* **1** craving, °desire, °appetite, °hunger, °eagerness, avidity, ravenousness, voracity, voraciousness, °lust, °passion, °enthusiasm, °fancy, hankering, °longing, yearning, *Colloq* °itch, yen: *It was the thirst for gold that drove them mad. She has an insatiable thirst for knowledge.* —*v.* **2** Often, ***thirst for*** or ***after.*** crave, °desire, hunger for *or* after, lust for *or* after, °fancy, hanker for *or* after, long for, yearn for, wish for: *These men are thirsting for power.*

thirsty *adj.* **1** parched, °dry, dehydrated; arid: *Be careful not to drink too much if you are very thirsty. The thirsty land welcomes the rain.* **2** °desirous, °hungry, avid, °eager, °ravenous, °voracious, °burning, °greedy, °avaricious, hankering, yearning, craving, *Colloq* itching: *The students are thirsty for knowledge and very hard-working.*

thorn *n.* **1** barb, °spine, °spike, °prickle, °bristle, brier, bur *or* burr, °point, bramble, cocklebur: *My sleeve caught on the thorn and tore.* **2** Often, ***thorn in (one's) side.*** °bother, irritation, °annoyance, °nuisance, vexation, °torment, torture, °scourge, °plague, °affliction, irritant, bane, *Colloq* pain in the neck, *Taboo slang* pain in the *Brit* arse *or US* ass: *That nagging brother has been a thorn in his side for years.*

thorny *adj.* **1** °prickly, barbed, spiny, spiked, brambly, spinous, *Technical* spinose, acanthoid, spiculose, spiculate, spinulose, aciculate, muricate, barbellate, setigerous, setaceous, setiferous, setose: *The insect's feet have tiny thorny projections.* **2** °difficult, °hard, °tough, °prickly, nettlesome, °painful, °ticklish, °delicate, °intricate, °critical, complex, °complicated, °problematic, vexatious, knotty, tangled, °involved, °troublesome, °controversial, °nasty, worrying, *Colloq* °sticky, *Slang* °hairy: *Many thorny questions have vexed us throughout the ages.*

thorough *adj.* **1** thoroughgoing, °complete, °downright, °perfect, through and through, °total, °unmitigated, °undiluted, unmixed, unalloyed, °out-and-out, °unqualified, °sheer, utter, arrant, °absolute, °proper: *I have rarely encountered such thorough stupidity.* **2** °exhaustive, °extensive, °painstaking, °meticulous, assiduous, °careful, °scrupulous, °particular, °conscientious, °methodical: *We conducted a thorough search but found nothing.* **3** °extensive, °exhaustive, °detailed, in-depth, °comprehensive, °full, °complete, (all-)inclusive, °total, all-embracing, °encyclopedic, °universal, A-to-Z, *Colloq* all-out: *We need someone who has a thorough knowledge of the terrain.*

thoroughly *adv.* **1** °completely, °downright, °perfectly, °totally, unqualifiedly, °utterly, °absolutely, °entirely, °extremely, unreservedly, °wholly, fully, °positively, °definitely, °quite: *He is thoroughly stupid, as we all know.* **2** carefully, painstakingly, exhaustively, extensively, assiduously, sedulously, methodically, conscientiously, scrupulously, meticulously, intensively, comprehensively, °completely, °throughout, from top to bottom, from stem to stern, backward(s) and forward(s), in every nook and cranny: *We searched thoroughly again, but still no murder weapon.*

though *conj.* **1** although, °even though, while, in spite of *or* despite the fact that, notwithstanding that, albeit, granted, granting *or* conceding that, allowing *or* admitting that, even if, °supposing: *Though it is the right size, I hate the color.* —*adv.* **2** °however, °nonetheless, °nevertheless, °yet, but, °still, °even so, be that as it may, °all the same, °notwithstanding, for all that: *She lost the original; I kept a copy though.*

thought *n.* **1** °thinking, °reflection, reflecting, meditation, meditating, contemplation, contemplating, cogitation, cogitating, musing, pondering, rumination, ruminating, brooding, mental activity, mentation, brown study; brainwork, cerebration, deliberation, deliberating, °consideration, °considering: *She was deep in thought. The solution may require considerable thought.* **2** °idea, °notion, brainstorm, °observation: *I had a thought that I'd like to share with you.* **3** °consideration, contemplation, planning, °plan, °scheme, °design, °intention, °expectation, °hope, °prospect, °anticipation, °dream, °vision: *Any thought of taking a few days off had to be abandoned.* **4** thoughtfulness, °consideration, kindliness, kindheartedness, °concern, compassion, tenderness, °kindness, °sympathy, attentiveness, °regard, °solicitude: *I don't need any help, but I appreciate the thought.* **5** Often, ***thoughts.*** recollection(s), memory *or* memories, remembrance(s), reminiscence(s): *Now and then he comforted himself with thoughts of his happy childhood.* **6** °intellect, °intelligence, °reasoning, rationality, ratiocination, °reason: *Some believe that only humans are capable of thought.* **7** °bit, °trifle, °touch, small amount, °trace, *soupçon,* °little, tinge: *This champagne is a thought too sweet.*

thoughtful *adj.* **1** °considerate, °kind, °kindly, kindhearted, compassionate, °tender, °sympathetic, °attentive, °solicitous, °helpful, °charitable: *Wasn't it thoughtful of Simon to offer his help?* **2** contemplative, °pensive, °reflective, musing, in a brown study, pondering, °meditative, engrossed, introspective, °rapt, °wistful, brooding, woolgathering, daydreaming: *Every time I mentioned Clarissa, he would become thoughtful.* **3** °prudent, °wary, °cautious, °mindful, heedful, °thinking, °attentive, circumspect, °careful, caring: *The crew are ever thoughtful of the passengers' safety.*

thoughtless *adj.* **1** °inconsiderate, °rude, °impolite, insensitive, °tactless, undiplomatic, untactful, °unthinking: *It was thoughtless of you to mention her son, who is serving time for possession of drugs.* **2** °rash, °imprudent, negligent, °foolish, °stupid, °careless, neglectful, °reckless, °silly, °unthinking, unreflective, °absent-minded, °forgetful, °remiss, ill-considered, °heedless, °inadvertent, °inattentive: *How thoughtless of me to forget to lock up the house!*

thrashing *n.* **1** beating, drubbing, °whipping, flogging, °assault, caning, belting, mauling, lashing, trouncing, basting, battering, pounding, *Colloq* hiding, tanning, lambasting, hammering, pasting: *Every time he stepped into the ring he got a thrashing.* **2** °punishment, chastisement, disciplining, °discipline, castigation: *He dreaded the thrashing his father administered at any sign of disobedience.*

thread *n.* **1** °fiber, filament, strand, °(piece of) yarn; °string, °line, °cord, °twine: *Threads from his jacket were found clutched in the victim's hand. The glass is spun out into fine threads.* **2** °theme, °plot, story line, °subject, °motif, °thesis, °course, °drift, °direction, °tenor, °train (of thought), sequence *or* train *or* chain of events: *The thread of the story resumes after a scientific disquisition on whales.* —*v.* **3** °string: *Carefully thread the beads onto the wire.* **4** °file, °wind, °pass, °squeeze (through), pick *or* make (one's) way (through), inch, °ease: *He threaded his way through the crowd to reach the dais.*

threadbare *adj.* **1** frayed, °worn (out), worn to a frazzle, °ragged, moth-eaten, °tattered, tatty, scruffy, °shabby, °seedy, torn, °wretched, °sorry, slovenly: *I did not recognize the vagrant in threadbare rags as my brother.* **2** trite, hackneyed, overused, overworked, reworked, °stale, °tired, stereotyped, commonplace, clichéd, cliché-ridden, °banal, °prosaic, °dull, °monotonous, °tedious, °tiresome, °boring, played out, *Colloq* old hat: *You'll never make it as a comedian by relying on that threadbare material of yours.*

threat *n.* **1** intimidation, °menace, commination, °warning, °peril, °risk, °danger, Damoclean sword: *It is not honesty but the threat of punishment that deters many from crime.* **2** °omen, presage, portent, °foreboding, forewarning, intimation: *The villagers lived under the constant the threat from the smoking volcano.*

threaten *v.* **1** °intimidate, °menace, terrorize, °daunt, cow, °bully, °browbeat, °warn, °caution: *The headmaster threatened those who misbehaved with the most dire punishment.* **2** imperil, put at risk, °endanger, °jeopardize, put in jeopardy: *All life is threatened if the environment is not better looked after.* **3** impend, °loom; augur, portend, presage, forebode: *They crawl into their burrows if danger threatens. The early appearance of the woolly caterpillar threatens a cold winter.*

threatening *adj.* °ominous, °menacing, °portentous, °sinister, looming, °inauspicious, minatory, minacious, comminatory, intimidating, °foreboding, °imminent, °impending: *Threatening storm clouds began to build in the western sky.*

threshold *n.* **1** sill, doorsill, doorstep; doorway, °entrance: *He stood at the threshold and peered into the room.* **2** °brink, °verge, °edge, °beginning, commencement, °outset, °start, °dawn: *The emigrants were at the threshold of a new life.*

thrift *n.* °economy, husbandry, °care, carefulness, °prudence, parsimony, frugality, thriftiness, sparingness, scrimping, skimping; penuriousness, closefistedness, tightfistedness, niggardliness, stinginess, miserliness: *Only through our thrift were we able to save enough for a new car.*

thrifty *adj.* °economical, °careful, °prudent, parsimonious, °frugal, °sparing, scrimping, skimping; °penurious, closefisted, tightfisted, niggardly, stingy, °miserly, penny-pinching, °cheap: *He talked about the need to save money and be thrifty in every way.*

thrill *n.* **1** °excitement, titillation, frisson, tingle, tingling (sensation), stimulation, *Colloq* °kick, bang, °charge, °buzz: *Victor feels a tremendous thrill when his daughter wins a tennis match.* **2** tremor, °quiver, quivering, °shudder, shuddering, °tremble, trembling, °flutter, throb, throbbing, tremor, pulsation, vibration: *The only warning was a slight thrill in the earth just before the main quake.* —*v.* **3** °excite, °stimulate, °animate, °electrify, galvanize, enliven, °stir, titillate, °touch, °strike, °move, impassion, °arouse, *Slang* °send, give (someone) a kick: *The good news from the front thrilled us all.*

thrilling *adj.* °exciting, °stimulating, animating, electrifying, galvanizing, enlivening, °stirring, titillating, °striking, °moving, arousing, °rousing, gripping, °sensational, °riveting, spine-tingling, soul-stirring: *He reads those thrilling adventure yarns in* Boy Mercenary. *They have really thrilling rides at Disneyland.*

thrive *v.* °succeed, °prosper, °boom, °advance, °flourish, °grow, bloom, burgeon, °develop, wax, °increase, fructify, °ripen: *They thrived in the invigorating mountain air.*

throe *n.* Usually, **throes.** pang, °anguish, °struggle, °chaos, turmoil, °tumult, °paroxysm, °spasm, °fit, °seizure, convulsion, *Technical* ictus: *I was in the throes of despair at the sad news from home.*

throng *n.* **1** horde, °crowd, °host, assemblage, °assembly, °gathering, °mass, °crush, °jam, multitude, congregation, °press, °swarm, °herd, °flock, bevy, drove: *Throngs of people packed the shopping mall on Saturday.* —*v.* **2** °crowd (into), °fill, °pack (into), °cram (into), °crush (into), °jam (into), °press (into), °swarm (into), °herd (into), flock (into *or* to); assemble (in *or* at), gather (in *or* at), mass (in *or* at), congregate (in *or* at): *Thousands thronged the stadium to watch the tennis matches.*

through *prep.* **1** because of, on account of, °owing to, as a consequence *or* result of, °by virtue of, via, °by means of, °by way of, with the aid *or* help of, under the aegis *or* auspices of, *Sometimes nonstandard* due to: *It was through his good graces that I got the appointment.* **2** during, °throughout, in the course *or* middle of: *The dog barked all through the night.* **3** °inclusive of, including: *When I say "the third through the sixth" I mean* *"including the sixth."* **4** to; into: *I went through a lot of trouble to find this book.* —*adj.* **5** Often, ***through with.*** done (with), finished (with); at the end of one's tether (with), washing (one's) hands (of): *Let me know when you are through. How many times has she sworn she was through with him forever?* —*adv.* **6** °by, °past: *I am just passing through.* **7** °entirely, through and through, °completely, °thoroughly, °totally, °wholly, °utterly, fully, to the core, from head to foot *or* toe, from top to bottom, from stem to stern, from one end to the other, in every way, in all respects: *By the time I got home, I was soaked through.*

throughout *prep.* **1** during, all (the way) through, from the beginning to the end of: *We worked throughout the day and night. Throughout his life he regretted having told me his secret.* **2** everywhere in, all over, in every part of, in every nook and cranny of, from one end to the other of: *I searched throughout the house for my keys.* —*adv.* **3** all (the way) through, °everywhere, from one end to the other, °wholly, °entirely, °completely, fully: *We always preferred the cupcakes that were pink throughout.*

throw *v.* **1** °toss, °cast, °hurl, °fling, °sling, °pitch, °dash, °propel, °project, °shy, °bowl, °send, °launch, *Colloq* chuck: *He threw the ball to the batter with all his might.* **2** °cast, °shed, °project: *I was hoping you might throw some light on the subject. Coming events throw their shadow before them.* **3** throw *or* bring down, °floor, °fell, knock down *or* over, °overthrow, °upset, °overturn: *He threw the vase on the floor.* **4** °dismay, confound, °confuse, dumbfound *or* dumfound, baffle, disconcert, °unnerve, throw off *or* out, unsettle, °put off, put (someone) off his *or* her *or* their stride *or* pace *or* stroke, *Colloq* discombobulate: *Her frank questions about sex really threw me.* **5** ***throw away.*** **(a)** °discard, °cast off, °dispose of, jettison, °get rid of, °scrap, °throw out, °dispense with, *Colloq* °dump, °trash, chuck out, *Slang* ditch: *He is a hoarder, refusing to throw anything away.* **(b)** °waste, squander, °lose, °forgo, °fritter away, fail to exploit *or* take advantage of, *Slang* °blow: *He threw away his last chance for a decent life when he left her.* **6** ***throw off.*** **(a)** °eject, °expel, °emit, throw up *or* out: *The volcano throws off rocks every so often.* **(b)** °shake off, rid *or* free (oneself) of, °get rid of, °reject, °renounce, °repudiate: *He finally managed to throw off his mother's apron strings.* **(c)** °deceive, °mislead, °decoy, misguide, °misdirect, °distract, °divert, °bewilder, confound, °confuse, *Colloq* °flummox, bamboozle: *He doubled back to throw the dogs off.* **(d)** See **3**, above. **7** ***throw out.*** **(a)** °radiate, °emit, °send forth, give out *or* off, °diffuse, put out *or* forth, disseminate: *The substance throws out a greenish light.* **(b)** °expel, °eject, force out, °evict, *Colloq Brit* turf out, *Slang* °bounce: *Henry was thrown out of the pub for rowdy behavior.* **(c)** See **4 (a),** above. **(d)** See **5 (a),** above. **8** ***throw over.*** °jilt, °leave, °abandon, °desert, °forsake, break *or* split up with, *Colloq* walk out on, chuck, °drop: *She was heartbroken when he threw her over and married her sister.* **9** ***throw up.*** **(a)** °vomit, spit up, puke, °spew up, be sick; °regurgitate, disgorge, *Colloq* °heave (up): *He felt as if he was going to throw up.* **(b)** °abandon, °quit, °leave, °throw over, °give up, °relinquish, °resign, °renounce, *Colloq* chuck: *He threw up a perfectly good job to go and live on a boat.* **(c)** °reveal, bring out *or* up, bring to the surface *or* the top, bring forward *or* forth, bring to light *or* to notice: *Her treachery was thrown up in high relief by the events that followed.* **(d)** throw *or* slap *or* knock together, jerry-build: *The house was thrown up overnight—and looks it.* **(e)** See **5 (a),** above.

thrust *v.* **1** °push, shove, °drive, °force, impel, °ram, °jam, °butt, °propel, °prod, °urge, °press; °shoulder, jostle, elbow: *A stranger thrust this note into my hand. She thrust her way through the crowd.* **2** °stab, °plunge, °stick, °jab, °poke; °lunge: *The hand that held the dagger has thrust it into the back of its neighbor. She thrust at me with an ice pick.* **3** Usually, ***thrust upon.*** press (upon *or* on), impose (upon *or* on), force (upon *or* on), urge (upon *or* on): *Some men are born great, some achieve greatness, and some have greatness thrust upon them.* —*n.* **4** shove, °push, °drive, °lunge, °poke, °prod, °stab: *He depended on his agility to avoid the thrusts of the rapier.* **5** °propulsion, °force, °power, °energy: *The thrust of the photon engines drives the space ship at nearly the speed of light.*

thud *n.* clunk, thump, whomp, wham, clonk, °bump: *His head hit the wall with a thud.*

thug *n.* hooligan, °gangster, desperado, gunman, °terrorist, °hoodlum, °robber, assassin, °murderer, °killer, °cutthroat, ruffian, mafioso, (*in Paris*) *apache*, °tough, mugger, *Technical* p'hansigar, *Brit* °rough, *Slang* °tough, hood, crook, hit man, °heavy, *US* goon, *Australian* larrikin: *Some thugs roughed him up and took his wallet.*

thumb *n.* **1** *Technical* pollex. **2** ***all thumbs.*** °awkward, °clumsy, maladroit, *Colloq* butterfingered, ham-fisted, cack-handed: *When it comes to threading needles, I am all thumbs.* **3** ***turn*** *or* ***give thumbs down (to).*** °disapprove (of), °reject, °rebuff, °turn down: *We turned thumbs down on the idea of a picnic.* **4** ***turn*** *or* ***give thumbs up (to).*** °approve (of), °accept, °welcome, *Colloq* O.K. *or Colloq* OK *or Colloq* okay: *The boss gave thumbs up to our request for a Christmas party.* **5** ***under (one's) thumb.*** under (one's) control, wrapped (a)round (one's) little finger, in the palm of (one's) hand, eating out of (one's hand), at (one's) beck and call: *She has the directors under her thumb.* —*v.* **6** hitchhike, *Colloq* °hitch, *US* hook a ride: *They thumbed their way across the country.* **7** Often, ***thumb through.*** leaf (through), flick *or* flip (through), riffle (through), °skim (through), °browse (through): *I was thumbing through* City Life *and came across your picture, Bernard!* **8** ***thumb (one's) nose at.*** scoff at, °deride, °jeer at, °mock, °dismiss, °scorn, °flout, be contemptuous of, show contempt for, exhibit defiance for, be defiant of, contemn, *Brit* cock a snook at: *He has thumbed his nose at authority all his life.*

thumbnail *adj.* °rough, undetailed, °cursory, °sketchy, °superficial; °brief, °short, °quick; °compact, °concise, pithy, °succinct: *It was nothing elaborate, merely a thumbnail description of the highlights of the event.*

thumping *adj.* **1** °great, °huge, °colossal, stupendous, °gigantic, °enormous, °immense, °monumental, °massive, titanic, elephantine, behemoth, gargantuan, mammoth, °jumbo, *Colloq* °whopping, thundering, walloping: *That is the biggest thumping beach ball I have ever seen! That was a thumping lie.* **2** °complete, utter, °unmitigated, 24-karat, °perfect: *Anyone who would turn down a salary increase must be a thumping idiot.*

thunder *n.* **1** °roll, reverberation, °boom, booming, °roar, roaring, pealing, rumble, rumbling; °crash, crashing, °crack, cracking, °explosion, °blast: *We were deafened by the thunder of the tanks crossing the bridge above us.* —*v.* **2** °roll, reverberate, °boom, °roar, rumble, °resound; °explode, °crash, °crack, °blast: *The noise of the jets thundered through the valley.* **3** °shout, °yell, °scream, °bellow, bark, °roar; °denounce, fulminate against, °swear (at), °rail (at), °curse (at), execrate; °threaten, °intimidate, °menace: *"Fee, fie, foe, fum!" thundered the giant. No secretary will tolerate his thundering at her.*

thunderous *adj.* roaring, booming, thundering, °tumultuous, °noisy, °loud, earsplitting, deafening: *They took curtain calls to thunderous applause.*

thunderstruck *adj.* dumbfounded *or* dumfounded, astonished, astounded, awe-struck, awed, °speechless, struck dumb, amazed, taken aback, staggered, stunned, shocked, dazed, °numb, paralyzed, aghast, open-mouthed, nonplussed, *Colloq* flabbergasted, floored, bowled over, *Brit* knocked for six: *We were thunderstruck to learn that our team had been disqualified.*

thus *adv.* **1** so, in this manner *or* way *or* fashion *or* wise, as follows, *Nonstandard* thusly: *On receipt of the payment, I wrote thus to him.* **2** °therefore, ergo, °consequently, as a consequence, as a result, °accordingly, (and) so, then, for this *or* that reason, °hence, in which case *or* event, that being the case, that being so: *He has already decided; thus, your efforts to persuade him are useless.*

thwart *v.* **1** °frustrate, °impede, °check, °stymie, baffle, °stop, °foil, °stump, °hinder, °obstruct, balk, °block, °oppose, negate, nullify, *Colloq* short-circuit: *They have thwarted all our efforts to become friendly.* —*n.* **2** °brace, cross brace; °(rowing) seat, bench: *These canoes are fitted with two thwarts.*

tickle *v.* titillate, °delight, please, °gratify, °amuse, °entertain, °divert, °captivate, °thrill, tickle pink *or* to death: *The show we saw last night really tickled my fancy. She was tickled to receive the flowers.*

ticklish *adj.* **1** °uncertain, unsteady, unsure, °unstable, °unsettled, °fickle, touch-and-go, °touchy: *Power that relies on the sword has a ticklish basis.* **2** °delicate, °precarious, °risky, °hazardous, °dangerous, °critical, °thorny, °fragile, °awkward: *As the canoe drifted toward the waterfall, we found ourselves in a very ticklish predicament.* **3** °delicate, °sensitive, hypersensitive, °difficult, °touchy, °prickly: *Cornelia is a bit ticklish on the subject of who the father is.*

tidbit *n.* °delicacy, °(dainty) morsel, °treat, choice item, *bonne bouche, Chiefly Brit* titbit, *Colloq* goody: *Sally likes to steal little tidbits off my plate. I heard a juicy tidbit of gossip this morning.*

tidy *adj.* **1** °neat, °orderly, °trim, °shipshape, °spruce, spick-and-span *or* spic-and-span, °clean, well-kept, °well-groomed: *Change your clothes and come back when you're tidy.* **2** well-organized, organized, well-ordered, °methodical, °systematic, °trim: *They refused to clean his office till he had made it tidy.* **3** °respectable, sizable, °significant, °considerable, °substantial, °good, °goodly, good-sized, °ample, °large, °big, °fair, °generous, not insignificant; *Colloq* not to be sneezed at: *They settled a tidy sum on their daughter when she married.* —*v.* **4** Often, ***tidy up.*** °neaten (up), straighten (out *or* up), °clean (up), put in order, *Colloq* °fix (up), °spruce up, °organize, reorganize, °arrange, rearrange: *Don't come down to dinner till you've tidied up your room. Please tidy up the files by the end of the week.*

tie *v.* **1** °bind, °fasten, make fast, °tie up, °lash, °secure, truss, °attach, °tether, °rope, °chain, °moor; °connect, °join, °knot, °link, °couple, °splice, °unite: *He keeps a vicious dog tied to a stake in his garden. Tie the ends of the rope together.* **2** °bind, truss (up), °tie up, °lash, pinion, °restrict, confine, °restrain; °limit, °tie down, °curtail, °curb, cramp, °hamper, °hinder: *His hands are tied behind his back. According to the terms of the agreement, my hands are tied and I can do nothing.* **3** °connect, °associate, °unite, °join, °link, °bind (up), affiliate, °ally, °league, °team (up): *Our business interests are tied to those of the major banks.* **4** °equal, °even, be equal *or* even (with), °match, be neck and neck (with): *We tied them for first place. Yesterday, we were tied for second.* **5** ***tie down.*** **(a)** °clinch, °secure, °confirm, *Colloq* °nail down: *I tied down the PBT contract this morning.* **(b)** °restrict, °restrain, constrain, confine, °curtail: *He won't marry because he doesn't want to feel tied down.* **6** ***tie in.*** **(a)** be consistent, make sense, °correspond, °coincide, °fit (in), be logical, °coordinate: *His alibi ties in with Mrs. Gladstock's testimony.* **(b)** °relate, °connect, °link, °associate, °coordinate: *They cleverly tie in concern for the environment with their product.* **7** ***tie up.*** **(a)** °occupy, °engage, °(keep) busy: *This affair will tie me up till Tuesday.* **(b)** °use, °take up, encroach on, °impose on: *I won't tie up your time any longer.* **(c)** °stop, °halt, bring to a standstill: *The traffic was tied up for hours.* **(d)** °commit, °oblige, °obligate, °bind: *We have tied up all their output for a year.* **(e)** See **1,** above. **(f)** See **2,** above. —*n.* **8** °link, °fastening, °bond, °band, °connection, °tie-up, °relationship, affiliation, °liaison, involvement, entanglement: *Isn't there some tie between those two companies?* **9** °string, °cord, °lace, °rope, thong, ribbon, °band, ligature, shoelace, °line, leash, °stop: *The tie of her dressing gown came loose.* **10** °equality, dead heat, °deadlock, °draw, °stalemate: *When there is a tie, the game ought to go into "sudden death" overtime.* **11** cravat, *US* necktie: *Please put on a tie for dinner.* **12** railway tie, sleeper: *Ties in Europe are now made of concrete.*

tie-in *n.* °tie-up, °relationship, °relation, °association, °connection, °link, linkage: *They say that there is a tie-in between smoking and lung cancer.*

tier *n.* °row, °line, °level, °order, °range, °course, series, °stratum, layer, echelon, °file, °rank, °story: *The trireme is said to have had three tiers of oarsmen.*

tie-up *n.* **1** slowdown, slowup, entanglement, stoppage, °jam, logjam, traffic jam, °delay, congestion: *The tie-up was caused by fog on the highway.* **2** See **tie-in,** above.

tiff *n.* °(petty) quarrel, °disagreement, °misunderstanding, °dispute, °argument, °difference (of opinion), squabble, °bicker, °row, wrangle, *US* spat: *Cecil and Gertrude have had their first tiff.*

tight *adj.* **1** °secure, °firm, °fast, °fixed, secured, close-fitting, °snug, sealed, hermetically sealed, leak-proof, °hermetic, impervious, impenetrable, °impermeable, airtight, °watertight, waterproof, : *Make sure that the lid is tight.* **2** °taut, stretched, °tense, constricting, °(too) small, ill-fitting: *She was wearing a very tight dress.* **3** °strict, binding, restrictive, stringent, °severe, °tough, uncompromising, unyielding, rigorous, °stern, austere, autocratic, °harsh, hard and fast, °inflexible: *You can't get away with anything under such tight regulations. I have a tight schedule next week.* **4** *Chiefly nautical* °taut, (well-)disciplined, °orderly, °neat, well-organized, °trim, °tidy, °smart: *He runs a tight ship.* **5** stingy, niggardly, °mean, °penurious, °miserly, parsimonious, penny-pinching, tightfisted, closefisted, *Colloq Brit* mingy: *Gurney is much too tight to pay for a round of drinks.* **6** °close, °(almost) even, (highly) competitive, neck and neck, evenly matched: *It is going to be a tight race.* **7** °difficult, °trying, °dangerous, °perilous, °risky, °hazardous, °touchy, °problematic, °sticky, °tricky, °ticklish, °precarious, touch-and-go: *The strike has created a tight situation at the office.* **8** tipsy, °drunk, intoxicated, *Colloq* °high, woozy, °under the influence, *Brit* tiddly: *When I saw her, she was too tight to drive.* **9** °scarce, °scanty, hard to find *or* come by, °rare; °dear, °expensive: *Entrepreneurs have found investment money to be pretty tight this year.* —*adv.* **10** °tightly, securely, °firmly; closely: *Keep this door closed tight. Hold me tight. Her hands were clasped tight in prayer.* **11** compactly, densely, solidly, °firmly, closely: *We were packed so tight in the bus that no one could move.*

tighten *v.* 1 Sometimes, ***tighten down*** *or* ***up.*** °anchor, °fasten, °fix, °tense, °secure: *Tighten those last two screws and you're finished.* 2 make tighter *or* tenser *or* stronger, °strengthen: *He tightened his grip on my hand.* 3 Sometimes, ***tighten up.*** make more rigorous *or* strict *or* stringent *or* severe *or* restrictive, close gaps in: *We ought to tighten security at the warehouse.* 4 Sometimes, ***tighten up.*** tauten, °stiffen, °tense, °close: *I felt the noose begin to tighten around my neck.*

tight-lipped *adj.* close-mouthed, °silent, °quiet, °mum, °mute, close-lipped, °noncommittal, °reticent, °secretive, °taciturn, unforthcoming, uncommunicative, °reserved: *He remained tight-lipped about his activities during the war.*

tightly *adv.* closely, tensely, °vigorously, rigorously: *It was a tightly fought race.*

tights *n.* pl. *US and Canadian and New Zealand* pantyhose, *Australian also* pantihose: *It was much too hot to wear tights.*

till[1] *v.* plow *or* plough, °cultivate, °farm, °work, °dig, hoe, harrow, manure, *Literary* delve: *My family has tilled this land for twenty generations.*

till[2] *n.* money *or* cash drawer, cash box *or* register: *He insists that the money was in the till when he left last night.*

tilt *v.* 1 °lean, °slant, °incline, °slope, °angle, °tip, heel over, °pitch, °list, °cant: *She caught the lamp just as the table tilted.* 2 ***tilt at.*** joust with, compete with, °battle against, contend with, spar with, cross swords with, °attack: *In tilting at the media, he has chosen some formidable opponents.* —*n.* 3 °lean, °slant, °incline, °slope, °angle, °tip, °heel, °list, °pitch, °cant, °inclination: *The wall is at a perilous tilt and could easily fall.* 4 joust, tourney, °tournament, °meeting, tilting, °engagement, °encounter, °match, °contest, °test, °trial, °fight, °combat; °dispute, °argument, °difference, °quarrel, altercation, squabble, °tiff, *US* spat, *Colloq* set-to: *I have enjoyed the various tilts I have had with the editor over the years.*

timber *n.* 1 trees, forest, woodland: *The standing timber has been seriously depleted in Brazil.* 2 wood, beams, boards, planks, *US and Canadian* °lumber: *The timber for our new house arrived today.* 3 °material, °potential, °stuff, °character, °quality, °talent, °prospect: *Agatha seems to be good management timber.*

timbre *n.* tone (color *or* quality), tonality, °color, resonance: *Kurt's voice has a rich, full timbre.*

time *n.* 1 °period, °interval, °stretch, °spell, °patch: *I was going through a bad time when we last met.* 2 °period, °interval, °stretch, while, °span, °space, °term, duration: *It is a long time since I have seen you. In the time I took to write this, she had vanished.* 3 hour; °point, °moment: *What time is it? At what time is our appointment?* 4 °age, °period, epoch, °era, lifetime, heyday, day(s): *The novel is set in the time of the Caesars.* 5 °opportunity, °chance, °occasion: *You must make time to visit your mother.* 6 °experience: *I had a wonderful time at your party.* 7 °tempo, °beat, °rhythm, meter, °measure: *They are playing "Teddy Bears' Picnic" in march time.* 8 °ease, °leisure; convenience: *He'll be there in his own time.* 9 Often, ***times.*** °life, °things, °circumstance, °conditions, °everything, °culture, mores, habits, values: *How times have changed since we were young!* 10 ***ahead of time.*** °(bright and) early, °prematurely, beforehand, in good time: *He arrived ahead of time and had to wait.* 11 ***all the time.*** °always, °ever, constantly, continuously, continually, perpetually, at all times, without surcease, unceasingly: *Mother is after me all the time to do my homework.* 12 ***at one time.*** **(a)** °once, once upon a time, on one occasion, °previously, in days of yore, °formerly, heretofore, in the (good) old days: *At one time you thought me beautiful.* **(b)** simultaneously, °(all) at once, °at the same time, together, all together, in unison: *The car can only carry five people at one time.* 13 ***at the same time.*** **(a)** °all the same, °nonetheless, °yet, °even so, but, °however, be that as it may, °nevertheless, °notwithstanding, just the same: *I love her; at the same time, I cannot live with her.* **(b)** See 12 **(b)**, above. 14 ***at times.*** from time to time, °occasionally, °(every) now and then, °once in a while, °on occasion, every so often, at intervals, °sometimes, *Colloq* every so often: *At times, Ingrid would come over and cook dinner for me.* 15 ***behind the times.*** °old-fashioned, outdated, dated, outmoded, °antiquated, °passé, °obsolescent, °obsolete, *Colloq* old hat, °dead: *Isn't "The Lambeth Walk" a bit behind the times?* 16 ***for the time being.*** for now, for the present, for the moment, °meanwhile, °temporarily, pro tempore, pro tem, *Archaic* for the nonce: *He has been appointed chairman for the time being.* 17 ***in no time.*** °at once, forthwith, °straightaway, °immediately, °quickly, speedily, without delay, °swiftly: *The pizza we ordered was delivered in no time.* 18 ***in time.*** **(a)** in timely fashion, °early, in good time, in the nick of time: *The doctor arrived in time to save the baby.* **(b)** °soon, one of these days, °sometime, someday, one day, °eventually, °sooner or later, anon: *In time, people might be living on the moon.* 19 ***on time.*** **(a)** punctually, °on the dot, in good time: *He is never on time for his appointments.* **(b)** in installments, on terms, on account, on credit, *Colloq Brit* on the never-never, on hire-purchase *or* h.p.: *We bought the car on time.* 20 ***take (one's) time.*** °dawdle, dillydally, °shilly-shally, °delay, °linger, loiter: *He's certainly taking his time in the lavatory.* 21 ***time and again.*** again (and again), °repeatedly, (over and) over again, time and time again, time after time, °frequently, °often, many times, on many occasions: *Time and again I warned him he'd get a ticket if he parked there.* —*v.* 22 °schedule, °set, °regulate, °control: *The trains are timed to arrive five minutes apart.* 23 °schedule, °set, °organize, °adjust, °fix: *She timed her entrance to coincide exactly with the crash of the cymbals.*

time-honored *adj.* established, °traditional, °habitual, °customary, °rooted, °conventional, age-old, °set, °fixed; °venerable, venerated, respected, revered, honored: *We observed the time-honored custom of kissing the Blarney Stone.*

timeless *adj.* °eternal, °everlasting, °immortal, undying, °endless, unending, ceaseless, °abiding, °deathless, ageless, °changeless, unchanged, immutable, unchanging, °permanent, °indestructible: *He was enraptured by the timeless beauty of the heavens.*

timely *adj.* °punctual, °prompt, °well-timed, °propitious, °opportune, °convenient, °favorable, auspicious: *We welcomed the timely arrival of our rescuers.*

timeserving *adj.* self-seeking, self-serving, °selfish, °self-indulgent, °ambitious, °mercenary, °venal, °greedy, profit-oriented, fortunehunting, gold-digging, °opportunistic, °hypocritical, °obsequious, sycophantic, toadying, toadeating, bootlicking, subservient, *Colloq* °on the make, on the take, *Slang US* out for numero uno, *Taboo slang* brown-nosing: *Members of the cult were known for their timeserving morality.*

timetable *n.* °schedule, °calendar, curriculum, °program, agenda, *Chiefly Brit* °diary: *My timetable doesn't allow for long lunches.*

timeworn *adj.* aging *or* ageing, °old, °tired, °worn, time-scarred, °decrepit, °dilapidated, tumbledown, °ramshackle, °run-down, dog-eared, °ragged, moth-eaten, °threadbare, °seedy, °shabby, archaic, °antique, well-worn, worn-out, °passé, broken-down, °old-fashioned, out-dated, dated, °antiquated, °ancient, °obsolescent, °obsolete, stereotyped, stereotypic(al), hackneyed, °stale, trite, overused, *Colloq* old hat: *She went into her timeworn routine about two living as cheaply as one.*

timid *adj.* °shy, °retiring, °modest, °coy, °bashful, diffident, timorous, °fearful, apprehensive, °mousy, °scared, frightened, °nervous, °cowardly, pusillanimous, craven, *Colloq* chicken-hearted, yellow, yellow-bellied, chicken, chicken-livered, lily-livered, gutless: *I knew this strapping marine when he was a timid little boy.*

tinker *v.* °trifle, °dabble, °meddle, mess (around *or* about), °toy, fool *or* play (around *or* about), putter *or Brit* potter (about *or* around), *Colloq* fiddle *or* monkey *or* muck (about *or* around): *Gaylord has always enjoyed tinkering with engines.*

tinny *adj.* **1** °shabby, °flimsy, flimsily *or* poorly made, °shoddy, °inferior, °cheap, °tawdry: *He bought himself a tinny old rattletrap of a car.* **2** metallic, °harsh, twangy: *She plays that tinny old trumpet day and night.*

tint *n.* **1** tincture, °wash, °hue, °color, °cast; tinge, °touch, °hint, °trace, °dash, coloring, °shade, °tone, °suggestion: *In the west a tint of pink lingered in the sky. I thought I detected a tint of anger in his voice.* **2** dye, °rinse, °wash, °stain, tincture, colorant, coloring, touchup: *What's the harm in granny's using blue hair tint?* —*v.* **3** dye, °stain, °color, °rinse, tinge, °touch up: *Sandra tints her hair to give it a coppery sheen.* **4** tinge, °color, °influence, °affect, °taint, °stain: *A deep pessimism tints all his writing.*

tiny *adj.* microscopic, infinitesimal, °minute, minuscule, °diminutive, °wee, °small, °little, °miniature, micro, mini, °pocket, pocket-sized, bantam, pygmy *or* pigmy, midget, Lilliputian, °petite, °delicate, °dainty, °elfin, °slight, °insignificant, °negligible, °trifling, °paltry, °inconsequential, °puny, *Colloq* pint-sized, teeny, teeny-weeny, teensy-weensy, itty-bitty, itsy-bitsy: *She wore the tiniest bikini I had ever seen. When I saw them together I realized how tiny she is. Forgetting my birthday was only a tiny thing.*

tip[1] *n.* **1** °end, °extremity, °peak, apex, °summit, °vertex, °cap, °top, °pinnacle, tiptop, °crown, °head, °terminal, ferrule *or* ferule, finial, nib *or* neb, °point: *The tip of the mountain showed above the clouds. He broke off the tip of the billiard cue.* —*v.* **2** °top, °cap, °crown, surmount: *The foil is tipped with a rubber button to prevent injury.*

tip[2] *v.* **1** Often, ***tip over.*** °upset, °overthrow, knock *or* cast *or* throw down, upend, knock over, °overturn, °topple (over), °capsize: *His elbow tipped the lamp, which fell with a crash.* **2** °slant, °lean, °incline, °slant, °tilt, °cant: *The statue is tipped a bit off the vertical.* **3** °empty, °unload, °dump, °deposit, *Slang Brit* ditch: *They tipped their load of gravel all over my driveway.*

tip[3] *n.* **1** gratuity, baksheesh, *pourboire, douceur,* lagniappe *or* lagnappe, °present, °gift, *Colloq* little something: *That waiter didn't deserve a tip.* **2** tip-off, °(inside) information, °warning, °advice, °suggestion, °clue, °hint, °pointer, °forecast, °prediction, *Colloq Brit* gen: *Louie had a tip that the police were coming. Her tip was "Flapdoodle" to win the fifth race.* —*v.* **3** °reward: *Why did you tip the barmaid?* **4** Usually, ***tip off.*** °advise, °warn, °caution, °alert, forewarn, °notify, let (someone) know, *Colloq* let (someone) in on: *The thieves, who had been tipped off, never arrived.*

tirade *n.* declamation, °harangue, diatribe, philippic, °outburst, onslaught, screed, jeremiad, denunciation, stream of abuse, invective: *He let loose with such a vituperative tirade that I walked out.*

tire *v.* **1** °weary, tire out, °fatigue, °exhaust, °wear out, °drain, °sap, °enervate, °drain, °wear out, debilitate, °weaken, *Colloq* take it out of, °fag (out), take it out of: *The long climb tired me.* **2** °bore, °exasperate, °weary, °irk, °irritate, °annoy, °bother: *I wish you wouldn't tire me with your tales of your supermarket adventures.*

tired *adj.* **1** °exhausted, tired out, worn-out, °weary, °fatigued, °spent, drained, *Colloq* all in, °(dead) beat, knocked out, fagged (out), dog-tired, ready to drop, dead tired, done in, *Brit* knackered, whacked, *US* bushed, pooped, wiped out: *I was really tired after that ten-hour flight.* **2** Usually, ***tired of.*** bored with, exasperated by, °weary of, irked *or* irritated *or* annoyed *or* bothered by, sick (and tired) of, *Colloq* fed up (to here) with: *Perhaps she's tired of your constant nagging.* **3** overworked, overused, clichéd, stereotyped, stereotypic(al), hackneyed, unimaginative, trite, °stale, worn-out, unoriginal, commonplace, *Colloq* bromidic: *The gossip columns are filled with the same old tired rubbish day after day.*

tireless *adj.* °energetic, °vital, °vigorous, °dynamic, °spirited, °lively, indefatigable, hardworking, °industrious, °untiring, unflagging, unfaltering, unfailing, °persistent, dogged, °tenacious, pertinacious, persevering, °staunch, sedulous, unwavering, unswerving, undeviating, °steady, °steadfast, °resolute, °determined: *We all appreciate the honorary secretary's tireless efforts in behalf of our fund raising this year.*

tiresome *adj.* **1** °boring, °dull, fatiguing, °humdrum, °monotonous, °flat, °tedious, wearisome, tiring, uninteresting, insipid, °bland, dryasdust, fatiguing, soporific, hypnotic: *Ibsen wrote plays that actors love and audiences find tiresome.* **2** irritating, °irksome, vexing, vexatious, annoying, bothersome, exasperating, °trying, °disagreeable, °troublesome, unpleasant: *I wish she would take her tiresome problems elsewhere.*

tissue *n.* °fabric, °network, °web, interweaving, °combination, °chain, series, °accumulation, conglomeration, concatenation, °pile, °mass, °pack: *Their entire testimony was a tissue of lies.*

title *n.* **1** °name: *You cannot tell much from a book's title.* **2** designation, appellation, epithet: *Aubrey is now an earl, but he rarely uses his title.* **3** caption, inscription, headline, °head, subtitle, °legend, subhead, rubric: *The title on this picture has nothing to do with its subject.* **4** championship, °crown: *He holds the world heavyweight boxing title.* **5** °right, °interest, °privilege, entitlement, ownership, °possession, °tenure; °(title) deed, documentation of ownership: *My aunt has sole title to these lands.* —*v.* **6** °name, °call, °designate, °style, °label, °term, °entitle, °christen, baptize, °nickname, denominate, °tag, dub: *Harold was titled King of the Revels.*

titter *v.* **1** °chuckle, °snicker, chortle, °giggle; snigger: *He told the most awful jokes at which his staff would titter politely.* —*n.* **2** °chuckle, °snicker, °giggle, °(suppressed) laughter, chortle, snigger: *A titter ran through the audience.*

titular *adj.* °nominal, °so-called, so-designated, so-styled, °self-styled, *soi-disant,* °token, putative, °theoretical: *Curtin may be the titular head of the company, but his brother is really the boss.*

toast *n.* **1** °tribute, °pledge, salutation(s), greeting(s), felicitations, °honor, good wishes, °appreciation, remembrance(s), cheers: *The speaker proposed a toast to the editor in chief.* **2** °heroine, °hero, °favorite, °darling, °idol: *Jenny was the toast of the town.* —*v.* **3** pay tribute to, °salute, °drink to, °honor, °greet, °congratulate, felicitate: *We toasted our fallen comrades.*

toilet *n.* **1** (water) closet, W.C. *or* WC, men's (room), ladies' (room), (public) convenience, facility *or* facilities, washroom, bathroom, °lavatory, °privy, outhouse, urinal, (in France) *pissoir, vespasienne*; *Nautical* °head, *Chiefly military* latrine, *Chiefly US* rest room, *New England* °necessary, *Colloq* gents', ladies', powder room, little girls' room, little boys' room, *Brit* loo, *Military* ablutions, *Slang Brit* °bog, karzy, *US and Canadian* john, can, crapper: *They excused themselves to go to the toilet.* **2** *Formal or literary* grooming, dressing, making up, *Brit* toilette: *She was still at her toilet when we arrived.*

toilsome *adj.* °arduous, °laborious, °tough, hard, °difficult, °strenuous, °burdensome, onerous, backbreaking, °exhausting, fatiguing, tiring, enervating, wearying, draining: *After a toilsome ascent we finally reached the first camp.*

token *n.* **1** °coin, disk *or* disc: *This token is good for one admission.* **2** °symbol, °sign, °mark, marker, badge, °emblem, °indication, °proof, °evidence: *The purple trim on a Roman toga was a token of its wearer's elevated status.* **3** souvenir, °memento, °keepsake, °reminder, °remembrance, *Archaic* remembrancer: *The miniature Statue of Liberty and Eiffel Tower are tokens of our trip to Japan, where they were made.* —*adj.* **4** °symbolic, °emblematic, °representative: *The panel was made up of three men and a token woman.* **5** °superficial, cosmetic, °surface, °perfunctory, °minimal, °slight, °nominal: *We encountered only token resistance when taking over the communications center.*

tolerable *adj.* **1** °bearable, °supportable, allowable, endurable, °acceptable, sufferable: *The heat and humidity are barely tolerable.* **2** °acceptable, unexceptional, °common, °fair, common-or-garden variety, middling, °ordinary, °average, °so-so, °mediocre, °adequate, °run-of-the-mill, °passable, °indifferent, *Colloq* O.K. *or* OK *or* okay, not (too) bad, pretty *or* fairly good: *They serve a tolerable lunch in business class on Urgent Airways.*

tolerance *n.* **1** open-mindedness, toleration, forbearance, broad-mindedness, permissiveness, magnanimity, °indulgence, sufferance, °patience, freedom from bigotry *or* prejudice: *You must exercise more tolerance in dealing with people who are poorer than you.* **2** °play, °clearance, °allowance, °variation: *The tolerance between these parts is less than a ten-thousandth of a millimeter.* **3** toleration, °resistance, °endurance, imperviousness; °immunity, insensitivity: *His tolerance of the outrages of stupidity seems limitless. She has a low tolerance for sugar.*

tolerant *adj.* open-minded, °objective, forbearing, °unprejudiced, unbigoted, °dispassionate, broad-minded, °indulgent, magnanimous, °patient, °generous, °charitable, °catholic, latitudinarian, °permissive, °liberal, bighearted, °fair, evenhanded, °considerate: *I am more tolerant of people than of their ideas.*

tolerate *v.* **1** °stand (for), °allow, °permit, °bear, °suffer, °brook, countenance, °abide, °admit, °indulge, °concede, °sanction, °swallow, °stomach, turn a blind eye to, *Colloq* °put up with, *Brit* °stick: *The teacher refused to tolerate any more misbehavior.* **2** °bear, °stand, °submit to, °endure, °weather, °take, °accept, °undergo: *Tests show that the product can tolerate a lot of abuse.*

toll[1] *v.* **1** °ring, °peal, °chime, °strike, °sound: *The clock tower bell tolled three.* —*n.* **2** °ring, ringing, °peal, pealing, °chime, chiming, °striking, °sound, sounding, tolling, knell: *Each toll of the bell marked the death of another victim.*

toll[2] *n.* **1** °charge, °fee, °dues, assessment, °tariff; excise, °duty, impost, levy, °tribute: *The highest toll, ninety-nine thousand dollars, was charged the* Queen Elizabeth *for passing through the Panama Canal. Halliburton paid a toll of thirty-six cents for swimming through.* **2** °loss, °penalty, °cost, damage(s); exaction: *The death toll from the earthquake reached fifty thousand.*

tomb *n.* °sepulcher, °crypt, vault, mausoleum, °grave, catacomb, burial chamber, last resting place: *Every year his tomb is visited by a mysterious woman in black.*

tombstone *n.* gravestone, headstone, marker, °monument, cenotaph: *The tombstones in the military cemetery stood in serried ranks.*

tone *n.* **1** °sound, °note: *Odd tones emanated from the electronic music-maker.* **2** °stress, °emphasis, °force, °accent, °intonation, modulation, phrasing, inflection, °pitch, tonality, °timbre, °sound (color), tone color *or* quality, color *or* coloring, resonance, sonorousness, sonority, fullness, richness: *The tone of his voice is extraordinarily melodic.* **3** °manner, °style, °attitude, °air, °aspect, °approach, °note, °tenor, tone of voice, mode of expression, °temper, °vein, °spirit, °air: *Do not take that imperious tone with me!* **4** °tint, tinge, °shade, °hue, °color, coloring, °cast: *The fabric has a silvery tone that makes it shimmer.* —*v.* **5** ***tone down.*** °temper, °modify, °reduce, °moderate, °modulate, °soften, quiet *or Brit* quieten (down), °dampen, °dull, °subdue, °mute, soft-pedal: *Please tone down your voice or the neighbors will hear you.* **6** ***tone up.*** (re)invigorate, °tune (up), °brighten (up), (re)vitalize, °freshen (up), limber up, get into condition *or* shape: *I really have to tone up my body before I put on a swimsuit this summer.*

tongue *n.* **1** °language, °speech; °dialect, patois, creole, °idiom, °parlance, argot, °talk, °vernacular, *façon de parler*: *The people in that area speak an odd tongue.* **2** °(verbal) expression, utterance, °voice, articulation: *Michael is reluctant to give tongue to his innermost thoughts.* **3** ***hold (one's) tongue.*** be *or* remain *or* keep silent, keep mum, say nothing *or* nought, not breathe a word, keep (one's) counsel, not say a word, *Slang* °shut up: *Hold your tongue till you are spoken to!* **4** ***slip of the tongue.*** °slip, °mistake, gaffe, °blunder, faux pas, Freudian slip, *Colloq Brit* boob: *Saying "bitch" when I meant "witch" was a slip of the tongue.* **5** ***(with (one's) tongue in (one's) cheek.*** facetiously, whimsically, ironically, jocularly, jokingly, not seriously, in jest, jestingly, °in fun, to be funny, *Colloq* kiddingly: *As he is my older brother, I call him "Dad" with tongue in cheek, of course.*

tongue-lashing *n.* scolding, berating, °reproof, °rebuke, °reprimand; °(verbal) abuse, castigation, chastisement, vituperation, revilement, *Colloq* dressing-down, telling-off, talking-to, *Brit* slating, ticking-off, wigging: *For all his efforts to please, all he got was a tongue-lashing for interfering.*

tongue-tied *adj.* °speechless, at a loss for words, struck dumb, dumbfounded *or* dumfounded, °mute, °inarticulate: *I was so nervous when the Prince of Wales spoke to me that I got tongue-tied.*

tonic *n.* **1** °stimulant, restorative, invigorant, °boost, refresher; *Obsolete or literary* ptisan *or* tisane, *Technical* roborant, analeptic, *Colloq* bracer, pick-me-up, pickup, °shot in the arm, *US* picker-upper: *His father made his money selling snake-oil tonic.* —*adj.* **2** °stimulant, °stimulating, restorative, °invigorating, fortifying, °bracing, strengthening, reviving, enlivening, °refreshing, *Technical* analeptic, roborant: *Seeing Maddy and Max again had a tonic effect.*

tool *n.* **1** utensil, °implement, °instrument, °device, °apparatus, appliance, contrivance, °aid, °machine, °mechanism, °gadget, *Colloq* °contraption, °gimmick, *Chiefly US and Canadian* gizmo *or* gismo: *The work would go faster if you had the proper tools.* **2** °means, °way, °agency, weapon, °medium, °vehicle, instrumentality, avenue, °road: *Education is the tool you need to* *get anywhere in life.* 3 °puppet, cat's-paw, °pawn, °dupe, *Slang* stooge, °sucker: *I had been used merely as a tool in her ambitious rise to the top.* —*v.* 4 °work, °carve, °cut, °embellish, °decorate, °ornament, °dress, °shape: *This is the shop where the leather is tooled and gold-stamped.*

top *n.* 1 °summit, apex, °peak, °acme, °crest, °head, °pinnacle, °vertex, °zenith, meridian, °crown, culmination, high point, °height, apogee: *They reached the top of the mountain at dawn. For a change, I should like to be at the top looking down. Samantha remains at the top of her profession.* 2 lid, °cover, °cap, covering, °stopper, cork: *Put the top back on the bottle when you're through.* —*v.* 3 °surpass, °better, °best, °outstrip, °exceed, °outdo, °excel, °beat, °transcend: *It would be impossible to top some of his achievements.* 4 surmount, °cover, °cap, °crown, °tip; °finish, °complete, garnish: *The entire concoction was topped by a maraschino cherry.* 5 °trim, crop, lop *or* cut off, °clip, °prune, °nip, °pinch (back): *All the trees in this area have been topped to let in more sunlight.* 6 °scale, °climb, ascend, surmount: *The party topped Annapurna the next day.* 7 ***top up.*** °fill (up), °refresh, refill, °replenish, *US* °freshen (up): *He ordered the barman to top up our drinks.* —*adj.* 8 °best, greatest, °foremost, °leading, °preeminent, °eminent, °first, °first-rate, °principal, °prime, finest, choicest, topmost; °excellent, °superior, °superb, top-drawer, top-grade, topnotch, °supreme, °peerless, unequaled, °incomparable, *Colloq* °crack, ace, A-1 *or* A-one: *Is he the top man in his field? She is a top economics adviser to the government.* 9 °uppermost, topmost, highest: *The top fruit is the best.*

topic *n.* °subject (matter), °matter, °issue, °question, °point, °thesis, °theme, °text, keynote, field *or* area of study *or* of inquiry: *We need to decide on a topic for tonight's discussion.*

topical *adj.* 1 °contemporary, °current, up-to-date, °timely: *Drug addiction is certainly a topical subject for discussion.* 2 °local, °superficial: *For a minor injury, we give a topical, not a general anesthetic.*

topple *v.* 1 °upset, upend, knock down *or* over, bring down, °fell, °capsize, °collapse: *Scores of buildings were toppled by the tornado.* 2 bring *or* throw down, °overthrow, °defeat, vanquish, °overcome, °overturn, unseat, oust: *The leaders of the coup succeeded in toppling the government in a week.* 3 fall (over *or* down), °drop, °collapse, keel over, tumble down: *Seven guardsmen toppled over because of the heat on the parade ground.*

topsy-turvy *adj.* 1 upside down, wrong side up, °head over heels, inverted, reversed, backward(s), °vice versa: *They had always believed that everything was topsy-turvy in Australia.* 2 °chaotic, muddled, jumbled, °disorderly, disordered, disorganized, °confused, mixed-up, messy, °untidy, in a muddle, higgledy-piggledy, *Colloq* arsy-varsy, every which way: *The room was topsy-turvy when the police finished looking for the gun.*

torment *v.* 1 torture, °abuse, maltreat, °mistreat, °distress, agonize, excruciate, crucify, harrow, °rack, °pain: *Hunger tormented his weary body.* 2 °worry, °trouble, °plague, °annoy, bedevil, vex, harry, badger, hector, °harass, °pester, °nag, °persecute, needle, nettle, °irk, °irritate, °bother, torture, °afflict, *Brit* chivy *or Brit* chivvy *or Brit* chevy: *I was tormented by the suspicion that something was wrong.* —*n.* 3 °agony, wretchedness, °anguish, °distress, °misery, °pain, °woe, painfulness, torture, °suffering, °curse, °hell: *How can parents endure the torment of losing a child?* 4 °worry, vexation, °annoyance, harassment, °ordeal, °persecution, needling, °nuisance, bane, irritation, °bother, °affliction, °scourge, torture: *For yet another day I had to suffer the torment of her company.*

torpid *adj.* sluggish, °slow, slow-moving, slow-paced, tortoiselike, °lethargic, apathetic, °indolent, °passive, °slothful, °dull, stupefied, °sleepy, somnolent, °inactive, °inert, languid, languorous, °phlegmatic, spiritless, °lifeless, °listless, fainéant, °lackadaisical, pococurante, °indifferent, uncaring, unconcerned, insouciant: *Johnson said that it is a man's own fault if his mind grows torpid in old age.*

torpor *n.* °sluggishness, °sloth, °lethargy, apathy, °indolence, passivity, slothfulness, dullness, stupefaction, drowsiness, sleepiness, somnolence, °inactivity, °inertia, inertness, languor, laziness, phlegm, lifelessness, listlessness, °idleness, faineance, pococurantism, °indifference, unconcern, insouciance: *It is refreshing to note the new man's zeal in contrast to Cartyn's torpor.*

torrent *n.* °stream, °rush, °flood, deluge, effusion, gushing, °outburst, °outpouring, °spate, inundation, °flow, overflow, tide, cascade: *In minutes the brook became a rushing torrent. They assailed us with a torrent of abuse.*

torrential *adj.* rushing, streaming, copious, °profuse, teeming, °relentless, °violent; °fierce, vehement, vociferous, °ferocious: *The torrential monsoons come to inundate the land. I had never heard such a torrential spate of blasphemy.*

torrid *adj.* 1 °hot, °fiery, °sultry, stifling, °sweltering, broiling, sizzling, roasting, blazing, °burning, baking, cooking, boiling, blistering, blistery, °scorching, scorched, parched, parching, arid; °humid, °steamy, steaming, °muggy; tropical: *It seemed impossible that any creature could survive in such a torrid climate.* 2 °fervent, fervid, °passionate, °intense, °ardent, °inflamed, °impassioned, °lustful, amorous, °erotic, *Colloq* °sexy, °hot: *No one does those torrid love scenes the way she did.*

tortuous *adj.* 1 twisted, twisting, winding, wandering, °serpentine, turning, °crooked, sinuous, °bent, curled, curling, curved, curvy, curvilinear, flexuous, anfractuous, convoluted, involuted, zigzag, mazelike, mazy, °labyrinthine: *We followed a tortuous trail through the jungle.* 2 °roundabout, °indirect, °devious, °intricate, °involved, unstraightforward, °complicated, °ambiguous, ambagious, circuitous, warped, °crooked, °tricky, misleading, °deceptive: *We could not help being intrigued by the tortuous Machiavellianism of the scheme.*

toss *v.* 1 °throw, °cast, °lob, °pitch, °fling, °hurl, °heave, °shy, °launch, °send, °let fly, °propel, catapult, °sling, °bowl, *Colloq* chuck: *The goalkeeper tossed the ball into the middle of the field.* 2 °shake, °jerk, °stir up, °agitate, °fling: *Lucy disdainfully tossed her head.* 3 °shake (up), °stir (up), °agitate, °jiggle, °tumble, joggle; °wave, °lash, thrash: *Don't you put the dressing on before you toss the salad? The branches of the huge oak were being tossed about in the storm.* 4 writhe, °wriggle, °squirm, toss and turn, thrash: *I tossed all night, unable to get a moment's sleep.* 5 °pitch, yaw, °wallow, °roll, °lurch, undulate, °plunge: *Our little boat was tossed this way and that in the heavy seas.* —*n.* 6 °throw, °lob, °pitch, °heave, °shy: *His toss went wild and the ball rolled over the cliff.*

tot *n.* °child, toddler, infant, °baby: *Can you recall when you were a tiny tot?*

total *n.* 1 °sum (total), °totality, aggregate, °whole, °amount, total number: *The total of wounded came to only fifteen.* —*adj.* 2 °whole, °entire, °complete, °full, °gross, °overall, °comprehensive: *The total rainfall there is only three inches a year.* 3 °complete, unalloyed, °unmitigated, °unqualified, unconditional, utter,

°out-and-out, °thorough, thoroughgoing, °perfect, °outright, °downright, all-out, °absolute: *My son is a total failure in everything he attempts.* —*v.* **4** °add (up), tot up, °sum up, °reckon, °compute: *Please total my bill.* **5** °amount to, add up to, °come to, mount up to: *Your bill totals twice the amount I paid yesterday.*

totalitarian *adj.* °absolute, absolutist, °arbitrary, °authoritarian, autocratic, °dictatorial, fascist(ic), undemocratic, illiberal, °monolithic, Nazi, °oppressive, °despotic, °tyrannical: *Sooner or later, totalitarian forms of government have been seen to fail.*

totality *n.* °total, aggregate, °sum (total), °whole, °entirety, beginning and end, alpha and omega, be-all and end-all: *The totality of his interests lies in watching soap operas on TV.*

totally *adv.* °completely, °utterly, °entirely, fully, unqualifiedly, unconditionally, °perfectly, °absolutely, °thoroughly, °wholly, consummately: *She is the only person I have ever known who is totally good, in all senses of the word.*

totter *v.* waver, °topple, falter, °tremble, °teeter, °sway, °rock, °stagger, °stumble, wobble, °quiver, °shake, °quake, °shiver, dodder: *He seemed to totter a bit as he left the pub.*

touch *v.* **1** put (one's) hand on, °feel, °handle: *I became excited when she touched my hand.* **2** bring into contact with, °apply, °put, °set: *I touched a match to the fuse and ran.* **3** Sometimes, ***touch (up) against.*** be in contact (with), °border, adjoin, °meet, come up *or* be (up) against, push *or* press *or* lean (up) against, brush *or* rub (up) against, come *or* be together, abut: *The ladder is touching the freshly painted wall.* **4** lay a hand *or* finger on; meddle with, have to do with, interfere with, come near, °approach: *If you touch me, I'll scream.* **5** °drink, °eat, °consume, °partake of, °take, °use, °taste, have to do with: *Lips that touch liquor will never touch mine.* **6** °affect, °impress, °influence, °disturb, °move, °stir, °arouse, °excite, impassion, °stimulate, °strike, *Colloq* °get to: *We were touched by your sympathetic note.* **7** °rival, °match, °equal, °compare with, come up to, be on a par with, be a match for, be in the same league *or* class as *or* with, be on an equal footing with, °reach, come *or* get near *or* close to, hold a candle to, measure up to *or* against, *Colloq US* stack up to *or* *Colloq US* with *or* *Colloq US* against: *The competition couldn't touch our new model in speed and safety.* **8** Usually, ***touch on*** *or* ***upon.*** °refer to, have reference to, pertain to, °relate to, have a bearing on, °regard, °mention, allude to, speak *or* write of, °tell of, bring up *or* in, °raise, °deal with, °cover: *I must now touch upon a delicate matter.* **9** have access to, access, °use, °employ, make use of, put to use, avail (oneself) of, °take, °get, take advantage of: *I do not touch the principal and try to live off the interest.* **10** ***touch down.*** °land, alight, come to earth: *We touched down for fuel at Gander, Newfoundland.* **11** ***touch off.*** **(a)** detonate, °spark (off), set alight, °set off, ignite, °light, °fire, put a match to: *We touched off the gunpowder and blew up the arsenal.* **(b)** instigate, °initiate, °begin, °start, set in motion, ignite, °set off, trigger, °provoke, °foment, °cause, °give rise to: *Her offhand remark touched off a family row that lasted for decades.* **12** ***touch up.*** °retouch, °patch up; °beautify, °enhance, titivate *or* tittivate, °renovate, °spruce up: *I refuse to buy a painting that's been touched up. Mandy is upstairs touching up her makeup.* —*n.* **13** °feeling, °feel, °texture: *She wants nothing but the touch of silk next to her body.* **14** °pat, °tap, °blow, °hit, °stroke, °brush, °caress: *She despised him so that she cringed at his slightest touch.* **15** °dash, °hint, intimation, °suggestion, *soupçon,* °bit, intimation, °pinch, °jot, °spot, °trace, tinge, °taste, °suspicion, smattering, coloring, smack, °speck, °drop, whiff, °odor, °scent, °smell: *Cynthia has a touch of hypocrisy about her. The salad dressing could use a touch more garlic.* **16** °ability, deftness, °expertise, °dexterity, adroitness, °facility, °skill, skillfulness, °knack, °capability, °genius, °talent, °gift, °flair: *Richard has a master's touch in everything he does.* **17** °response, °feel, responsiveness, °feeling, °movement, °operation, performance level: *This piano action has an excellent touch.* **18** signature, trademark, °characteristic, °influence, °approach, °style, °manner, °technique, °execution, °method: *I thought I recognized your touch in the furniture selection. A woman's touch might have helped.*

touching *adj.* °moving, °stirring, °emotional, °tender, °poignant, °pathetic, soul-stirring, °heart-rending, heartbreaking, °sad, °pitiful, distressing, distressful: *The most touching moment was when he returned his Victoria Cross.*

touchstone *n.* °standard, °yardstick, criterion, °reference, benchmark, °test, °norm, °measure: *His craftsmanship is the touchstone by which the others are judged.*

touchy *adj.* **1** (over)sensitive, supersensitive, hypersensitive, highly strung, °tense, thin-skinned, crabby, crabbed, °testy, irascible, °irritable, tetchy, °temperamental, grouchy, °peevish, °querulous, °petulant, pettish, splenetic, captious, bad-tempered, °short-tempered, hot-tempered, °quick-tempered, crusty, °cross, curmudgeonly, grouchy, °cantankerous, choleric, dyspeptic, °waspish, bearish, snarling, °snappish, °argumentative, disputatious, contentious, *US* high-strung, °cranky: *She's very touchy on the subject of who the father is. He's always touchy in the morning.* **2** °critical, touch-and-go, °sensitive, °ticklish, °risky, °precarious, °hazardous, chancy, unsure, °uncertain, °close, hairsbreadth *or* hairbreadth, °dangerous, hair-raising, °frightening, °terrifying, °nerve-racking, *Jocular* °parlous, *Colloq* °hairy: *We have a touchy situation with the fuel leaking into the astronauts' cabin.*

tough *adj.* **1** °hard, °firm, °durable, long-lasting, wear-resistant, °substantial, °strong, °stout, °rugged, °sturdy, °sound, well-built: *Only one kind of material is tough enough to endure such conditions for so long.* **2** °stiff, °hard, leathery, °inflexible, chewy, fibrous, cartilaginous, °sinewy, °ropy, °wiry, °stringy: *That steak was as tough as shoe leather.* **3** °strong, °stalwart, °brawny, °burly, °muscular, °powerful, virile, °manly, °sturdy, doughty, °intrepid, °stout, °rough, °vigorous, strapping, athletic: *Two tough-looking characters moved toward us, so we left.* **4** °difficult, °demanding, °exacting, °hard, °troublesome; °laborious, taxing, °strenuous: *We deal with tough complaints every day. It's a tough job.* **5** baffling, °thorny, °puzzling, °perplexing, mystifying, knotty, °irksome, °difficult: *The examiners asked us some very tough questions.* **6** °stubborn, hardened, inured, °obstinate, obdurate, °hard, °harsh, °severe, °stern, °inflexible, refractory, intractable, adamant, unyielding, ungiving, °rigid, unbending, unsentimental, unfeeling, °unsympathetic, °callous, hard-boiled, uncaring, °cold, °cool, °icy, °stony, *Colloq* hard-nosed: *He's a tough man to work for, but he's fair.* —*interj.* **7** *Colloq* Too bad!, Tough luck!, Hard luck!, *Brit* Hard cheese!, *Slang* Tough titty!, *Taboo slang* Tough shit!: *When I told him that my money had been stolen, he said "Tough!"* —*n.* **8** °bruiser, hooligan, °bully (boy), °rowdy, °thug, ruffian, *Colloq* roughneck, °bruiser, tough guy, gorilla: *A couple of toughs who had been threatening the cashier disappeared when they saw our uniforms.*

tour *n.* **1** °journey, °trip, °excursion, °outing, °expedition, voyage, trek, peregrination, jaunt, junket: *We went on a guided tour because we were short of time and wanted to see everything.* **2** °stroll, perambulation, walkabout, °ramble, °walk, °drive; °round, °circuit, ambit: *The government representatives are on a tour of all nuclear facilities.* **3** °spell, °shift, °assignment, °turn, *Military* period of service *or Military* enlistment: *His tour of duty is not finished till midnight.* —*v.* **4** °journey, °travel, voyage, °visit, °trip, trek, sight-see, °cruise; *Colloq* globe-trot: *On retirement we spent a year just touring around Europe.*

tourist *n.* °traveler, voyager, °visitor, °sightseer, *Colloq* rubberneck(er), out-of-towner, *Brit* tripper, day-tripper, holiday-maker: *During the summer, this place is inundated with tourists.*

tournament *n.* tourney, °competition, °contest, °match, °meeting, °event, °meet: *The tennis tournament was delayed by rain.*

tousle *v.* dishevel, °disorder, °ruffle, disarrange, °tangle (up), °mess (up), °rumple, disarray, *US* muss (up): *Her hair was all tousled by the wind.*

tout *v.* **1** hawk, °peddle, °sell, °promote, °talk up, *Colloq* °push, °plug: *They tout these new headache tablets as miraculous.* —*n.* **2** tipster: *He is often seen with racetrack touts.*

tow *v.* °pull, °drag, °draw, °haul, °lug, °trail, °tug, trawl: *The boat was towing a long purse seine. I arrived just as the police were towing away my car.*

toward *prep.* **1** towards, in the direction of, to; °for, so as to approach *or* near, on the way *or* road to: *She turned toward me. I'd appreciate a lift if you're going toward Aston Clinton.* **2** towards, to, °for, as a help to, supporting, promoting, assisting: *We welcome any contribution toward the charitable works of the order.* **3** towards, °near, nearing, °close to, approaching, shortly before: *Toward the close of day, the cattle wander back to the barns.*

tower *n.* **1** belltower, campanile, minaret, pagoda, obelisk; belfry, °spire, turret, steeple, flèche: *There was a time when the church tower was the tallest building in the town.* **2** fortress, citadel, °stronghold, °castle, fastness; °keep, °dungeon, °prison: *The princess was imprisoned in the tower for a year and a day.* —*v.* **3** Often, ***tower over*** *or* ***above.*** °loom, °soar, °rise, ascend, °rear: *These ugly high-rise buildings tower over everything, blocking out the sun.*

towering *adj.* **1** °lofty, °tall, °high, soaring, °outstanding, °elevated, skyscraping, sky-high, °great, °impressive, °imposing, °huge, °gigantic, °supreme, °superior, °paramount, °extraordinary, unmatched, unequally, unrivaled, °unparalleled, unsurpassed: *The towering American economy dominated the world for decades after World War II.* **2** °violent, °fiery, °burning, °passionate, °excessive, vehement, °intense, consuming, °mighty, °overwhelming, unrestrained, °immoderate, °inordinate, intemperate, °extreme, °colossal, °enormous: *Every time she mentioned Valentino, he went into a towering rage.*

town *n.* township, village, hamlet, community; °municipality, °city, °metropolis, borough, burgh: *She was born in a small town in the Outer Hebrides. We have to drive into the town to do our shopping.*

toy *n.* **1** °plaything: *Whenever he visited, Uncle Jed brought toys for the children.* **2** °trifle, trinket, °bauble, °gewgaw, gimcrack, knickknack *or* nicknack, bagatelle, kickshaw, bit of frippery: *After marrying her, he treated her as a rich man's toy.* —*v.* **3** Usually, ***toy with.*** °trifle (with), dally (with), °play (with), °sport (with), °fool (with), °fiddle (with), °tinker (with): *He kept toying with his watch chain.* **4** Usually, ***toy with.*** °flirt (with), dally (with), dillydally (with), °play (with), deal with carelessly, amuse oneself with: *The cad was only toying with her affections.* —*adj.* **5** °miniature, °tiny, °diminutive, °small, °dwarf: *She breeds toy poodles.* **6** °imitation, °fake, phoney *or US also* phony, simulated, °artificial: *This is only a toy watch and doesn't work.*

trace *n.* **1** °hint, intimation, °sign, °token, °suggestion, °touch, °vestige, °indication, °mark, °record, °evidence, °clue: *There wasn't any trace of anyone's having been there.* **2** °bit, °spot, °speck, °jot, °drop, °dash, °suspicion, °remnant, tinge, *soupçon,* iota, whiff, °suggestion, °trifle: *The traces of mud on the suspect's shoes proved he had been there.* **3** Often, ***traces.*** track(s), °trail, spoor, footprint(s), print(s), footmark(s): *They followed the buffalo's traces for miles.* —*v.* **4** dog, °pursue, °follow (in the footsteps of), °stalk, °track (down), °shadow, °trail, *Colloq* °tail: *We traced him to a seedy hotel in Cuernavaca.* **5** °investigate, °discover, ascertain, °detect, °determine, °find, °seek, °search for, hunt down *or* up, °unearth, °track: *Have you traced her whereabouts? I am able to trace my ancestry back to the time of the First Crusade.* **6** delineate, °outline, °copy, °draw, map, °chart, °mark (out), °record, °reproduce, sketch: *We traced Marco Polo's route to China and back.*

track *n.* **1** °line, rail(s), °way, railway, *US* °railroad: *New track was laid between Washington and Boston.* **2** °path, °trail, °route, footpath, °course, °road, °street, alley: *A rough track leads over Sty Head Pass. Are you sure you're on the right track?* **3** spoor, °trail, footprint(s), print(s), trace(s), footmark(s), °scent, °slot, °wake: *The last time I saw him he was following the track of a yeti.* **4** ***keep track of.*** °trace, keep an eye on, °follow, °pursue, °monitor, °supervise, °oversee, keep up with *or* on, °watch, keep a record of *or* on, °record: *Using a radio implant, we were able to keep track of the animal's movements.* **5** ***lose track of.*** °lose, misplace, °mislay, lose sight of, °forget: *She lost track of some important papers. I lost track of what I was saying.* —*v.* **6** °follow, dog, °pursue, °trace, °stalk, °shadow, °trail, hunt down, °chase, *Colloq* °tail: *She was tracked to Istanbul, where they lost her.* **7** See **4,** above. **8** ***track down.*** °find, seek out, ferret out, hunt down, °trace, °catch, apprehend, °capture, smell *or* sniff out, run to earth *or* ground, *Colloq* °run down: *We finally tracked him down at his mother's house in Putney.*

trackless *adj.* °empty, pathless, untrodden, unexplored, °uncharted, virgin, untrod: *We found ourselves in the trackless wilderness of central Australia.*

tract[1] *n.* °region, °area, °stretch, °territory, °expanse, °zone, °portion, °section, sector, °quarter, °district, °patch, °plot, °parcel, *US* °lot: *The developers own a tract of land where they will build a shopping mall.*

tract[2] *n.* treatise, °monograph, °essay, article, °paper, dissertation, disquisition, homily, °sermon, critique; °pamphlet, booklet, °brochure, °leaflet: *Drusilla Clack distributed tracts to passersby in front of the church.*

tractable *adj.* **1** docile, amenable, °tame, °manageable, biddable, persuadable *or* persuasible, compliant, easygoing, °willing, °submissive, °obedient, governable, °yielding: *It was a pleasure to teach a student who was so tractable.* **2** °manageable, handleable, workable, °adaptable, malleable, °pliable, °plastic, ductile, fictile: *The new materials are quite tractable and can be drawn or molded with ease.*

traction *n.* °grip, gripping power, °drag, °purchase, °friction, adhesion: *Unable to get any traction on the ice, the car skidded off the road.*

trade *n.* **1** °commerce, °business, °traffic, °exchange, barter, dealing(s), buying and selling, merchandising, marketing, mercantilism, °truck: *Do we have any trade in shoes with South Korea?* **2** °calling, °occupation, °pursuit, °work, °business, °employment, °line (of work), métier, °job, °vocation, °craft, °career, °profession: *Just what trade are you engaged in?* **3** swap *or* swop, °exchange, interchange, barter: *Was exchanging your microwave cooker for a hi-fi set a good trade?* **4** customers, °clientele, °custom, patrons, following, °patronage, shoppers: *Her shop caters chiefly to the carriage trade.* —*v.* **5** transact *or* do business, °buy, °sell, °deal, °traffic, °merchandise, have dealings: *There are laws against Sunday trading in some areas.* **6** °exchange, swap *or* swop, interchange, °switch, barter; °return: *Having traded his car for the boat, Sam has no way of reaching the marina.*

trader *n.* °dealer, °merchant, businessman, °broker, merchandiser, distributor, °seller, salesman, saleswoman, °salesperson, vendor, °buyer, purchaser, retailer, wholesaler: *A smuggler prefers to be known by the vague designation of "trader."*

tradesman *n.* **1** °merchant, °dealer, shopkeeper, retailer, vendor, °seller: *Till Penny married a tradesman, the men in her family had always been in the army.* **2** artisan, craftsman, journeyman, handicraftsman: *He is a tradesman, a skilled cabinetmaker by profession.*

tradition *n.* °custom, °practice, °habit, °usage, °convention, °ritual, °rite, unwritten law, °institution, °form, praxis, °lore: *Exchanging gifts on Christmas Eve has long been a tradition in our family.*

traditional *adj.* °customary, °usual, °routine, °habitual, °standard, household, °stock, °time-honored, established, °well-known, °conventional, °ritual, unwritten, °accustomed, °historic, °old, ancestral: *Mazetta mocha tarts were the traditional birthday cakes at home.*

traffic *n.* **1** °movement, conveyance, shipping, °transport, °freight, *Chiefly US* transportation: *Traffic across the border has resumed.* **2** See **trade, 1,** above: *Drug traffic has increased substantially over the years.* —*v.* **3** See **trade, 5,** above: *The police think that Arthur Daly is trafficking in stolen goods.*

tragedy *n.* °catastrophe, °calamity, °disaster, °misfortune, adversity, °blow: *The death of their dog was a terrible tragedy, especially for the children.*

tragic *adj.* °sad, depressing, °lamentable, °unhappy, °funereal, °forlorn, °melancholy, cheerless, °mournful, lachrymose, dolorous, °grievous, morose, lugubrious, °dismal, °piteous, °pitiable, °pitiful, pathetic(al), appalling, °wretched, °dreadful, °awful, °terrible, °horrible, °deplorable, °miserable, distressing, °disturbing, upsetting, °shocking, unlucky, °unfortunate, hapless, ill-fated, °inauspicious, star-crossed, ill-omened, ill-starred, °calamitous, catastrophic, crushing, °disastrous; tragical: *It was terribly tragic that their new house burned down on the day before they were to move in.*

trail *n.* **1** °(beaten) path, °way, footpath, °route, °way, °track, °course: *There is a well-worn trail through the wood.* **2** °track, spoor, °scent, °smell, °trace, °footsteps, footprints, °path, °wake: *The trail of the elephant herd was quite easy to follow.* **3** See **train, 2,** below: *She was always followed by a trail of admirers.* —*v.* **4** °tow, °draw, °drag (along), °haul, °pull, °tag along, trawl, bring along (behind), carry along (behind): *We were moving so slowly because we were trailing a dinghy.* **5** °drag, °pull, °move, be drawn, °stream, °sweep, °dangle: *I heard the whisper of silken gowns trailing across Persian carpets.* **6** °lag (behind), °dawdle, loiter, °linger, °follow, °straggle, bring up the rear, hang back, fall *or* drop behind: *The Yankees were trailing in the World Series, three games to two.* **7** °follow, °pursue, dog, °trace, °shadow, °stalk, °track, °chase, °hunt, *Colloq* °tail: *We trailed the suspect to Victoria Station, where we lost him.* **8** ***trail off*** *or* ***away.*** °diminish, °decrease, fade away *or* out, °disappear, °dwindle, lessen, die out *or* away, °peter out, °subside, °taper off, °weaken, grow faint *or* dim: *As they drove away, the noise of their blasting radio trailed off.*

train *n.* **1** °carriage, °coach, *Babytalk* choo-choo: *Will this train take me to Newcastle?* **2** °retinue, entourage, cortège, °suite, following, °escort, °guard, attendants, retainers, followers, °trail; °staff, court, household: *After the duke's coffin came a train of several hundred hangers-on.* **3** °line, °queue, °procession, °succession, °string, °set, °sequence, °chain, °progression, caravan, cavalcade, °parade, column, °file: *The baggage train of the advancing army stretched for miles.* —*v.* **4** °discipline, °exercise, °tutor, °teach, °coach, °drill, °school, °instruct, °prepare, °educate, edify, °guide, °bring up, °indoctrinate, °rear, °raise: *We had been trained to put things away and avoid clutter.* **5** °work out, °exercise, practise: *Hortense is training for the next Olympics.*

trait *n.* °feature, °characteristic, °attribute, °quality, °peculiarity, idiosyncrasy, °quirk, lineament, °mark, °property: *Some of his traits, like spitting when he talks, are quite unsavory.*

traitor *n.* °turncoat, Judas, quisling, betrayer, °renegade, fifth columnist, *US* Benedict Arnold, *Colloq* double-crosser, snake in the grass, double-dealer, two-timer: *Some traitor in our midst had revealed our plans to the enemy.*

traitorous *adj.* treacherous, °perfidious, °seditious, °subversive, insurrectionist, °renegade, insurgent, °disloyal, °deceitful, °untrue, unfaithful, °faithless; treasonable, *Colloq* double-crossing, double-dealing, two-timing: *We identified the traitorous wretch and hanged him. Consorting with the enemy is a traitorous act.*

trajectory *n.* flight path, °course, °track: *The howitzer has a high trajectory.*

tram *n.* tramcar, trolleybus, *US and Canadian* streetcar, trolley (car): *Most cities with a traffic problem have replaced trams with buses.*

trammel *n.* **1** Usually, ***trammels.*** impediment(s), hindrance(s), shackle(s), handicap(s), check(s), restriction(s), restraint(s), curb(s), deterrent(s), constraint(s), hitch(es), snag(s), (stumbling) block(s), obstacle(s), °bar: *He said that a lack of education was not a trammel to his career as a silver prospector.* —*v.* **2** °impede, °hinder, °handicap, °check, °restrain, °curb, °deter, constrain, °block, °obstruct, fetter, confine: *He found himself unable to work when trammeled with a deadline.*

tramp *v.* **1** °march, hike, trudge, °plod, slog, plough, tread, trek, °walk, *US* mush: *I must have tramped across half of England looking for work.* **2** Usually, ***tramp on*** *or* ***upon.*** See **trample, 1,** below. —*n.* **3** °march, trudge, °plod, slog, trek, hike, °walk: *Every day we had to make the five-mile tramp into the village for water.* **4** °derelict, °vagabond, vagrant, °drifter, °rover, gypsy *or* gipsy, beachcomber, *Brit* dosser, °down-and-out, *Australian* swagman, *US* hobo, °bum, down-and-outer: *The police did not allow tramps to sleep in the public parks.* **5** °step, tread, footfall, °footstep: *I couldn't fall asleep till I heard my husband's tramp on the stair.*

trample *v.* **1** trample on *or* upon, tramp (on *or* upon), °stamp (on), tread (on), step on, °crush, press, squash, °flatten, *Colloq* stomp (on *or* upon), squish, squush *or* squoosh: *After the harvest, we would take turns trampling the grapes.* **2** Often, ***trample on*** *or* ***upon.*** °violate, °damage, °harm, °hurt, infringe *or* encroach on, ride roughshod over, set at naught, °scorn, contemn, disdain, °defy, °disregard, °ignore, °fly in the face of, fling *or* cast *or* throw to the winds: *The military regime trampled on the people's civil rights.* **3** Usually, ***trample out.*** trample down, trample under foot, °stamp out, °extinguish, °put out, °destroy, °crush, °break down: *The party was trampled in a military coup.*

trance *n.* °daze, °stupor, semiconscious *or* half-conscious *or* hypnotic *or* cataleptic *or* dream state, state of semiconsciousness *or* half-consciousness *or* catalepsy *or* suspended animation *or* stupefaction *or* abstraction *or* (complete) absorption *or* exaltation *or* rapture *or* ecstasy; brown study: *He was concentrating so hard that he seemed to be in a trance.*

tranquil *adj.* °calm, °serene, placid, °quiet, °peaceful, °still, °smooth, unagitated, halcyon, °relaxed; unruffled, °sedate, °steady, °regular, °even, °dispassionate, °self-possessed, °cool, self-controlled, coolheaded, unexcited, undisturbed, untroubled, unperturbed: *He often thought back to the tranquil, sultry summer days of his youth. How can Eleanor be so tranquil when everything about her is so chaotic?*

tranquilize *v.* °calm, soothe, pacify, °still, °quiet, °relax, °lull, °compose, °sedate: *The medication is supposed to tranquilize patients, allowing them time to gain a perspective on their problems.*

tranquilizer *n.* bromide, barbiturate, opiate, °sedative, antipsychotic, antianxiety drug, *Slang* downer, red: *Instead of trying to cope, he takes tranquilizers and his problems seem to vanish.*

transact *v.* do, carry on *or* out, °conduct, °manage, °handle, °negotiate, °transact, °administer, °discharge, °perform, °enact, °settle, conclude, °complete, °finish: *We are flying to Frankfurt today to transact some business.*

transaction *n.* **1** °deal, dealing, °negotiation, °matter, °affair, °business, °action, °proceeding, °agreement, °arrangement, °bargain: *This transaction must not be revealed to the stock market.* **2** ***transactions.*** °proceedings, record(s), acta, °minutes, annals, *Colloq* goings-on, doings: *The society's transactions are published annually.*

transcend *v.* °surpass, °outstrip, °exceed, go beyond, outdistance, °overstep, °outdo, °excel, °overshadow, °top, outdistance, outvie, rise above, outshine, °beat: *Her performance at La Scala transcended that of every Mimi I have heard.*

transcendent *adj.* °peerless, °incomparable, unequaled, °matchless, unrivaled, °unparalleled, °unique, consummate, °paramount, °superior, °surpassing, °supreme, °preeminent, °sublime, °excellent, °superb, °magnificent, °marvelous; transcendental: *Can there be any doubt of Einstein's transcendent genius?*

transcribe *v.* **1** °copy, °reproduce, replicate, °duplicate: *I am busy transcribing my lecture notes.* **2** °translate, transliterate, write out, °render, °represent, °show, °interpret: *The Rosetta Stone has the same text transcribed in Egyptian hieroglyphics and in demotic as well as in Greek.*

transcript *n.* **1** transcription, °translation, transliteration, °rendering, °interpretation, °representation: *Most linguists regarded writing as a mere transcript of language.* **2** (carbon *or* machine *or* Xerox *or* photostatic *or* xerographic) copy, carbon, °duplicate, duplication, photocopy, °reproduction, Photostat, *Colloq* °dupe: *I have requested a transcript of the court proceedings.*

transfer *v.* **1** °move, °transport, convey, °remove, °carry, °take, °deliver, °bring, °transmit, °cart, °haul, °shift, hand (on *or* over), °turn over, °give, pass (on *or* along *or* over): *The documents will be transferred to you today.* —*n.* **2** °move, conveyance, transmittal, °transmission, °delivery, °change: *The papers documenting the transfer of the property are here.*

transfix *v.* **1** °pin, °fix, °impale, skewer, °nail, °pierce, spear, °spike, °spit, °stick: *The shrike, or butcherbird, transfixes its prey on a thorn, then picks its bones clean.* **2** °enrapture, galvanize, °electrify, °hypnotize, mesmerize, rivet, °fascinate, °bewitch, °enchant, ensorcell, engross, root to the spot, °stun, °paralyze, *Colloq* stop dead (in one's tracks): *The felon stood transfixed with terror as the judge pronounced sentence.*

transform *v.* °change, °modify, transfigure, °alter, transmute, metamorphose, °turn into, °convert, transmogrify, mutate, permute: *What will it take to transform these students into civilized human beings?*

transformation *n.* °change, modification, transfiguration, transfigurement, °alteration, transmutation, metamorphosis, conversion, transmogrification, °mutation, permutation: *The transformation in her appearance over a few short months was miraculous.*

transfuse *v.* **1** °instill, °transmit, °transfer, °inject: *Her greatest success was to transfuse a sense of history to her students.* **2** infuse, °permeate: *An embarrassed blush transfused his face.*

transgress *v.* **1** °sin, trespass, °offend, °err, °lapse, fall from grace, °disobey, °misbehave, go wrong *or* astray, do wrong: *She knew she had transgressed and was ready to atone.* **2** break *or* violate *or* contravene *or* go beyond *or* exceed *or* overstep *or* infringe *or* defy *or* disobey (the law): *He had blatantly transgressed the laws of decency.*

transgression *n.* °sin, trespass, °offense, °error, °lapse, fall from grace, disobedience, °misbehavior, °wrong, °violation, °fault, °misdeed, misdemeanor, °crime, wrongdoing, infraction: *How should he be punished for his transgressions?*

transgressor *n.* °sinner, °offender, °criminal, °felon, °culprit, lawbreaker, trespasser, wrongdoer, evildoer, °villain, °miscreant, malefactor, °delinquent: *The majority of transgressors are apprehended within twenty-four hours of the crime.*

transient *adj.* transitory, °temporary, °brief, °fleeting, °momentary, °passing, ephemeral, fugacious, °fugitive, evanescent, °short-lived, short-term, impermanent, °fly-by-night, °volatile: *With Ananias, the urge to tell the truth was purely transient. A souvenir shop obviously sells more to the transient trade than to residents.*

transit *n.* **1** °moving, °movement, °travel, °traveling, °motion, °passing, °progress, °progression, °transition; °passage, °traverse, traversal, traversing: *We were unable to mail any letters while we were in transit. The transit of the entire island took only three days.* **2** °transport, transportation, °carriage, haulage, cartage, conveyance, °transfer, transference, transferal, transmittal: *The transit of the merchandise will be handled by our regular shipping company.* —*v.* **3** °cross, °traverse, go *or* move *or* pass *or* travel across *or* over *or* through: *We watched as the satellite transited the night sky.*

transition *n.* **1** °change, °alteration, metamorphosis, changeover, °transformation, transmutation, °mutation, °development, °evolution, conversion, modification, metastasis: *Tracing the transition from tadpole to frog is exciting for my pupils.* **2** See **transit, 1,** above.

translate *v.* **1** °convert, °paraphrase, °change, rewrite, °interpret, °transcribe, °render, decode, °decipher, metaphrase: *Can you translate this German document?* **2** °transform, °convert, °change, mutate, °turn, transmute, metamorphose, transubstantiate, °alter, transmogrify: *Why are people always trying to translate the dross of reality into the gold of dreams?* **3** °interpret, rewrite, °explain, °reword, elucidate, °spell out: *Can you translate this technical legal gobbledygook into plain English?* **4** °transfer, convey, °carry, °move, °transport, °forward, °ship, °send, dispatch *or* despatch: *Soon after being made bishop of Worcester, he was translated to Winchester.*

translation *n.* 1 conversion, °paraphrase, °interpretation, transcription, transliteration, °rendering, °rendition, metaphrase, °gloss, decipherment, decoding: *Wasn't it Michael Ventris who was responsible for the translation of Linear B?* 2 metamorphosis, °change, °alteration, transmutation, transfiguration, °transformation, transmogrification, transubstantiation, conversion: *The translation from the caterpillar to the butterfly is miraculous.* 3 °interpretation, rewriting, rewrite, °explanation, rewording, elucidation: *Infeld is well known for his translation of Einstein's ideas into terms an intelligent layman could cope with.* 4 °transfer, transference, transferal, conveyance, carrying, °moving, °movement, transportation, °transport, forwarding, shipping, shipment, sending, °transmission, dispatch *or* despatch: *His translation to a minor consular post in the south Pacific was expected.*

transmission *n.* 1 °transfer, transference, transferal, transferring, conveyance, carrying, °moving, °movement, transportation, °transport, transporting, forwarding, shipping, shipment, sending, transmittal, transmitting, dispatch *or* despatch, dispatching *or* despatching: *We were unable to arrange for the transmission of the papers in time for the meeting.* 2 °broadcast, broadcasting, sending, telecasting, dissemination, communication: *Transmission of the new programs begins next year.*

transmit *v.* 1 °send, °transfer, convey, °communicate, °pass on, °deliver, °forward, dispatch *or* despatch; °post, °ship, °cable, °radio, telegraph, fax, telex, °telephone, phone, *Chiefly US and Canadian* °mail, *Colloq* wire: *Your message was transmitted last night.* 2 pass *or* go through, °pass on, °send, °put, °direct, °conduct, °channel: *The mirror transmits the light through this filter.*

transparent *adj.* 1 °(crystal) clear, pellucid, diaphanous, °see-through, limpid, crystalline, °sheer, transpicuous: *I want completely transparent glass here and frosted glass in the lavatory window.* 2 °plain, °apparent, °obvious, °evident, unambiguous, °patent, °manifest, unmistakable *or* unmistakeable, °(crystal) clear, as plain as day, as plain as the nose on (one's) face, °undisguised, recognizable, understandable, transpicuous: *The origin of many English words is transparent because of their spelling.* 3 °candid, °open, °frank, plainspoken, °direct, unambiguous, unequivocal, straightforward, °ingenuous, °forthright, °aboveboard, °artless, guileless, °simple, naive *or* naïve *or* naïf, undissembling, *Colloq* °on the level, °up front: *His transparent honesty makes Crockett unsuited to a career in diplomacy.*

transpire *v.* 1 become known, be rumored, be revealed, °come to light: *It transpired that she had been seeing him while she was still married.* 2 *Sometimes nonstandard* °happen, °occur, °take place, °come about, °come to pass, °materialize, °arise, °turn out: *No improvement has yet transpired in their living conditions.*

transplant *v.* °displace, °move, °remove, relocate, °shift, °uproot, resettle, °transfer: *This is the third time the company has transplanted Harry in a year!*

transport *v.* 1 °carry, °bear, convey, °move, °remove, °transfer, °deliver, °fetch, °bring, °get, °take, °ship, °haul, °transmit, °send, °forward: *The goods were transported by ship.* 2 °exile, °banish, deport, send away: *Australia was settled mainly by people who had been transported from England.* 3 °carry away, °enrapture, °captivate, °delight, °charm, spellbind, °bewitch, °fascinate, °enchant, °entrance, °hypnotize, mesmerize, °electrify, °ravish: *Flavia's parents were totally transported by her winning the decathlon.* —*n.* 4 transportation, °carrier, conveyance, shipping, °transfer, transferal, shipment, haulage, cartage, °carriage, °moving: *Can't you give me any idea of what the transport costs might come to?* 5 Usually, ***transports.*** °rapture, °ecstasy, exaltation, exultation, euphoria, °delight, °(seventh) heaven, °happiness, °bliss, elation, exhilaration, °thrill, Elysium, °paradise; Elysian Fields, *Colloq* cloud nine: *Helen succumbed to transports of delight at seeing her son receive such an important award.*

transpose *v.* °exchange, interchange, metathesize, °switch, swap *or* swop, °trade, commute, °transfer: *The bank transposed two numbers on my statement, reducing my balance by five hundred dollars.*

trap *n.* 1 °snare, °pitfall, gin, springe, deadfall, booby trap: *After the Pied Piper left, Hamelin had no further need for rat traps.* 2 °trick, °subterfuge, °wile, °ruse, °stratagem, °ambush, °deception, °device, °artifice, ploy: *Oscar was the man for whom Esther set a neat trap.* 3 °mouth, *Slang* °yap, °gob, mush, °face: *Shut your trap or get out.* —*v.* 4 °snare, ensnare, entrap, °catch, °net: *We dug a pit to trap the marauding lion.* 5 °imprison, confine, °lock, °hold, °keep: *The boy was trapped in the cave without a means of escape.* 6 °trick, °deceive, °fool, °dupe, °beguile, inveigle: *Esther finally trapped Oscar into marrying her.*

trappings *n.* pl. accoutrements *or US also* accouterments, panoply, caparison, equipage, °apparatus, °equipment, °paraphernalia, appointments, furnishings, °furniture, °gear, °rig, habiliments, decoration(s), embellishment(s), accessories, frippery *or* fripperies, adornment(s), trimmings, raiment, °fittings, °finery: *Many respect the trappings of office more than the officeholder.*

trash *n.* 1 °rubbish, °(stuff and) nonsense, balderdash, °moonshine, °gibberish, gobbledygook *or* gobbledegook, tommyrot, bunkum, °garbage, twaddle, *Colloq* °rot, flapdoodle, crap, codswallop, bosh, piffle, hooey, bunk, malarkey, poppycock, boloney *or* baloney, eyewash, hogwash, bilgewater, bull, *Scots* havers, *Brit* tosh, gammon, *US* a crock, hokum, gurry, horsefeathers, *Slang Brit* (a load of (old)) cobblers, *Taboo slang* bullshit, horseshit, *Brit* balls, *US* a crock of shit: *Don't believe him—he is talking trash.* 2 °junk, brummagem, knickknacks *or* nicknacks, gewgaws, trifles, °bric-a-brac, frippery *or* fripperies, bits and pieces, °odds and ends, trinkets, tinsel: *One can sometimes find a treasure amidst the trash in those flea markets.* 3 °(*In the US and Canada*) rubbish, °litter, °garbage, °waste, °refuse, °junk, debris, rubble, °dregs, dross, scoria, slag, offscourings, °dirt, sweepings, *Slang* crap: *The service comes to remove the trash once a week.* —*v.* 4 *Slang chiefly US* °destroy, °ruin, °wreck, vandalize, °deface: *The lodgers trashed the flat before doing a moonlight flit.*

traumatic *adj.* °shocking, upsetting, °disturbing, °painful, °agonizing, distressing, °harmful, °hurtful, °injurious, damaging, wounding, traumatizing: *Surviving a plane crash may be a traumatic experience, but it can't compare with being killed.*

travel *n.* 1 °traveling, tourism, touring, globe-trotting: *Travel is broadening. Travel by air is the commonest means of long-distance transport.* 2 ***travels.*** trips, expeditions, journeys, excursions, tours, voyages, touring, treks, trekking, °traveling, wanderings, peregrinations, junkets, pilgrimages: *In all your travels, have you ever encountered an honest man?* —*v.* 3 °journey, °go, °move, °proceed, °roam, rove, °traverse, °tour, take *or* make a trip *or* tour *or* excursion *or* junket *or* journey, trek, voyage: *As one who has traveled far and wide, what is your favorite country?* 4 °go, °move, °socialize, °fraternize, °associate, *Colloq* hang around *or Colloq* about: *Cordelia and I do not travel in the same circles.*

traveler *n.* **1** °traveler, °tourist, voyager, °sightseer, globe-trotter, gypsy, wanderer, hiker, °rover, wayfarer, *Jocular* bird of passage, *Colloq* rubberneck(er), jet-setter, *Chiefly Brit* tripper, day-tripper, holiday-maker: *Travelers have been greatly inconvenienced by the strike of customs officials.*

traveling *adj.* °traveling, itinerant, wandering, peripatetic, roving, °mobile, nomadic, touring, wayfaring, migratory, °restless: *At fourteen, he joined a traveling band of jugglers, musicians, and actors.*

traverse *v.* **1** °cross, crisscross, travel *or* pass *or* move over *or* through, °walk, °cover, °roam, °wander, °range, °tramp, °tour: *He has traversed the country from end to end innumerable times.* **2** °cross, crisscross, go across; lie *or* extend across *or* athwart, °bridge, intersect: *The road traverses the river several times at Newtown.* **3** °oppose, °cross, °thwart, go *or* act against, go *or* act in opposition *or* counter to, °conflict (with), controvert, contravene, °counter, °obstruct, °contradict, gainsay, °deny: *The policies of today seem to traverse those set forth only last year.* **4** °examine, °look into, °scrutinize, °inspect, °investigate, °review, °study, °look at, °consider, °contemplate, °scan, °look over, °check, °survey, °reconnoiter, °observe: *Certain areas of knowledge are seldom traversed by scholars.*

treasure *n.* **1** °wealth, °riches, °money, °fortune, valuables, °cash, °cache, °hoard: *The existence of the treasure came to light only last week.* **2** °pride (and joy), °delight, °joy, °darling, °ideal, apple of (someone's) eye, *Colloq* °jewel, °gem, °prize, °find, °catch: *Kathy is a treasure and I don't know what we did before we hired her.* —*v.* **3** hold dear, °cherish, value, °prize, °esteem, rate *or* value highly: *We treasure the signed letter that Churchill wrote to my father. I treasure every moment we can spend together.*

treasury *n.* exchequer, bank, °cache, °resources, °funds, money(s): *I had to dig into my treasury for the money to take Susan to dinner. Using money in the treasury the company bought back a huge block of its shares.*

treat *v.* **1** °handle, °manage, behave *or* act toward(s), °deal with; °use: *Why should we treat female employees any differently?* **2** °handle, °manage, °deal with, °discuss, touch on *or* upon, °consider, °take up, °study, °examine, °explore, °investigate, °scrutinize, °analyze, °go into, °probe, °survey, expound (on), °criticize, °review, critique: *That subject is treated in Chapter VI.* **3** °nurse, °doctor, °attend, °care for, °look after, prescribe for, medicate: *He is being treated for a persistent case of the Palermo pip.* **4** °entertain, °take out, °pay for, °regale, play host to; wine and dine: *Is Philip treating us or are we taking him to dinner?* **5** ***treat (someone) to (something).*** pay (the bill) for, buy (something) for: *Edward treated us all to ice cream.* —*n.* **6** °favor, °gift, °present, °boon, °bonus, °premium, *Colloq US and Canadian* freebie: *Put your money away—this is my treat.*

treatment *n.* **1** Often, ***treatment of.*** behavior (toward(s)), conduct (toward(s)), action (toward(s)), handling (of), °care (of), °management (of), dealing(s) (with), manipulation (of), °reception (of); °usage (of): *I am not accustomed to such rude treatment. Your treatment of our customers must be more polite, Miss Diggs.* **2** °therapy, °care, curing, remedying, healing: *She went to Dr. Cutler for treatment of her condition.*

treaty *n.* °pact, °agreement, °alliance, concordat, entente, covenant, °deal, °contract, °compact, °accord: *We entered into a treaty not to violate each other's borders.*

tremble *v.* **1** °quiver, °shake, °quake, °shiver, °shudder, °quaver, quail; °vibrate, °rock: *Her first big role and she was trembling like a leaf! The earth trembled as the tanks rolled past.* —*n.* **2** °quiver, °shake, °quake, °shiver, °shudder, °quaver, tremor; vibration: *After giving a little tremble, the building collapsed.*

tremulous *adj.* **1** trembling, atremble, quivering, shaking, quaking, shivering, shuddering, quavering, °hesitant, wavering, unsure, unsteady, faltering, °doubtful, °nervous, °shaky, palpitating, °jumpy, *Colloq* jittery: *His tremulous hands revealed just how apprehensive he was.* **2** °timid, °shy, °bashful, °anxious, °worried, timorous, °fearful, °afraid, frightened, °scared: *I cannot remember when I felt so tremulous before meeting someone.*

trenchant *adj.* °cutting, °keen, °acute, °sharp, °pointed, °poignant, °penetrating, °incisive, °biting, mordant, mordacious, °sarcastic, °bitter, acerbic, acid, vitriolic, °tart, acrid, acrimonious, acidulous, corrosive, °caustic: *The critics dismissed the play with a few trenchant remarks.*

trend *n.* **1** °tendency, °leaning, °bias, °bent, °drift, °course, °inclination, °direction: *The trend seems to be toward shorter skirts.* **2** °fashion, °style, °vogue, °mode, °look, °rage, *Colloq* °fad, °craze, °thing: *Why is she so compulsive about keeping up with the latest trends?* —*v.* **3** °tend, °lean, be biased, °bend, °drift, °incline, veer, °turn, °swing, °shift, °head: *At the convention, the party leaders seemed to trend more to the left of center.*

trendy *adj.* **1** °fashionable, °stylish, à la mode, °modern, °up to date, up to the minute, in vogue, voguish, all the rage, *Slang* °hot, °now, with it, groovy, in the groove, in, °flash: *Sibyl travels with that trendy set from Belgravia.* —*n.* **2** °showoff, clotheshorse, coxcomb, exhibitionist, *Slang Brit* pseud, grandstander: *Don't you just hate those trendies down at Cole's wine bar?*

trial *n.* **1** °test, testing, °experiment, °proof, tryout, trying out, trial run, °examination, °check, checking, *Colloq* dry run: *The trials of the new life jackets are to be conducted soon.* **2** hearing, enquiry *or* inquiry, °examination, inquisition, °litigation, judicial proceeding, lawsuit, °contest: *Throughout the trial, the accused protested his innocence.* **3** °try, °attempt, °endeavor, °effort, °venture, °essay, *Colloq* °go, °shot, °stab, °fling, whirl, °crack, whack: *If he failed in his first trial at sky diving onto the Matterhorn, how did he get another chance?* **4** °trouble, °affliction, tribulation, °hardship, adversity, °suffering, °grief, °woe, °misery, °distress, bad *or* hard luck, °misfortune, hard times: *Mona acknowledged the trial of having ten children and no husband.* **5** °nuisance, irritation, °bother, bane, °annoyance, °pest, irritant, thorn in the flesh *or* side, *US* bur *or US* burr under the saddle, *Colloq* °plague, hassle, °pain (in the neck), °headache, *Taboo slang* pain in the *Brit* arse *or US* ass: *William, who is full of mischief, is a constant trial to his mother.* —*adj.* **6** °sample, °experimental, exploratory, °provisional, probationary, °tentative, conditional, °pilot: *Will you consider a trial subscription to* Verbatim, The Language Quarterly?

tribe *n.* °race, °stock, °strain, °nation, °breed, °people, °seed, °(ethnic) group, gens, °clan, blood, °pedigree, °family, sept, °dynasty, °house; °caste, °class: *It was important in that society to marry someone from the same tribe.*

tribunal *n.* court (of justice), °bar, bench, judiciary, Inquisition, Star Chamber: *Should he be tried before a judicial tribunal or pilloried by the tribunal of public opinion?*

tributary *n.* °branch, °offshoot, streamlet, feeder, °brook, rivulet, °run, rill, runnel, runlet, streamlet, *Scots and Northern England* °burn, *Northern England* beck, *US* °creek, *Northeastern US* °kill: *The Teviot is one of the tributaries of the Tweed.*

tribute *n.* 1 °honor, °homage, °recognition, °celebration, °respect, °esteem, °testimonial, °compliment, encomium, °acknowledgment, acclaim, acclamation, commendation, °praise, °kudos, laudation, panegyric, °eulogy, glorification, exaltation: *No greater tribute could be bestowed than recognition by one's fellows.* 2 °tax, exaction, impost, °duty, excise, levy, °dues, assessment, °tariff, °charge, surcharge, °payment, contribution, °offering, °gift; °ransom; tithe, Peter's *or* Peter pence: *In exchange for their freedom, the king demanded an annual tribute of a thousand oxen.*

trick *n.* 1 °ruse, °artifice, °device, °stratagem, °wile, °deception, °maneuver, °deceit, °fraud, °hoax, imposture, °intrigue, °machination, °conspiracy, °subterfuge, °dodge, confidence trick, °sham, *Slang* con: *The government's "dirty tricks squad" perpetrated crimes against their political adversaries.* 2 °prank, °frolic, antic, °(practical) joke, °hoax, tomfoolery, antic, °caper, jape; °sport, horseplay, °mischief; *Scots* cantrip, *Colloq* leg-pull, °gag, shenanigans, *US* dido: *The boys meant no harm, they're just up to their tricks.* 3 °art, °knack, °technique, °skill, °secret, °gift, °ability, *Colloq* °hang: *He has developed the trick of persuading people to buy life insurance.* 4 Usually, ***no + (adj.) + trick.*** °feat, °accomplishment, °deed: *It was no mean trick to train a cat to fetch his slippers.* 5 sleight of hand, legerdemain, °magic, °stunt: *I have taught him all the tricks I know.* 6 °trait, °characteristic, °peculiarity, idiosyncrasy, °eccentricity, °quirk, °practice, °habit, °mannerism, crotchet, °weakness, °foible: *He has an odd trick of winking while giving a twitching smile.* 7 ***do the trick.*** °work, °answer, fulfill the need, °suffice, be effective, solve *or* take care of the problem, do *or* accomplish the necessary, *US* turn the trick, *Colloq* fill the bill: *Replacing the battery cable did the trick.* —*v.* 8 °fool, °hoodwink, °dupe, °mislead, °outwit, outmaneuver, °deceive, misguide, °misinform, gull, bilk, °cheat, °defraud, cozen, °take in, °swindle, humbug, *Colloq* bamboozle, °take, put something over on (someone), pull the wool over (someone's) eyes, *Brit* gammon, *Slang* rook: *I knew I had been tricked when I missed my wallet. For years the couple made a living tricking tourists out of their money. She tricked me into taking her to dinner.* —*adj.* 9 See **tricky, 3,** below.

trickery *n.* °chicanery, °deception, °deceit, guile, shrewdness, craftiness, slyness, shiftiness, evasiveness, artfulness, °artifice, °craft, imposture, swindling, knavery, duplicity, double-dealing, °fraud, cheating, *Colloq* °hanky-panky, skulduggery, funny *or* monkey business, jiggery-pokery: *He separated her from her money by trickery.*

trickle *v.* 1 °drip, °drop, dribble, drizzle, °run, °flow, °spill; °ooze, seep, °leak, exude: *The water trickled onto the floor. Blood is trickling from the wound.* —*n.* 2 °drip, seepage, °spill, dribble, runnel, runlet, rivulet: *A tiny trickle of saliva appeared at the corner of his mouth.*

tricky *adj.* 1 °deceitful, °shady, °deceptive, °shifty, °dodgy, °artful, guileful, °crafty, duplicitous, °shrewd, cunning, °dishonest, °devious, °sly, °wily, °slippery, °foxy, double-dealing, cheating: *Malvolio is a tricky chap and I shouldn't trust him.* 2 °ticklish, °risky, °hazardous, °sensitive, °delicate, touch-and-go, °thorny, °difficult, °awkward, complex, °complicated, knotty, °uncertain, °debatable, *Colloq* iffy, °sticky: *It is a tricky decision whether you tell a patient how ill he really is.* 3 unfair, unjust, unsportsmanlike, °deceptive, *Colloq* °trick: *There were some tricky questions in that exam.*

trifle *n.* 1 knickknack *or* nicknack, trinket, °bauble, bagatelle, °toy, °gewgaw, °nothing, °plaything, bêtise, *Colloq* doodah: *Oh, it's nothing, just a trifle I picked up in the Seychelles.* 2 °little, °bit, °drop, iota, scintilla, °suggestion, °dash, °dab, °pinch, whiff, mite, whit, °jot, tittle, *Colloq* smidgen, *US* tad: *I'd like a trifle more sugar in my coffee, if you don't mind.* —*v.* 3 Usually, ***trifle with.*** dally (with), °flirt (with), °wanton (with), mess about (with), °toy (with); °play (with), °fiddle (with), dandle, °tinker (with), °fidget (with): *That blackguard trifled with my sister's affections! While trifling with this knob, I was able to get Radio Moscow.*

trifling *adj.* trivial, °insignificant, unimportant, °puny, °minor, °paltry, °slight, °petty, °inconsequential, °frivolous, °superficial, °incidental, °negligible, commonplace, inconsiderable, °shallow, valueless, °worthless, *US and Canadian* picayune, *Colloq* piddling: *Crapton's contribution to musical scholarship has been trifling.*

trim *adj.* 1 °neat, °tidy, °orderly, well-ordered, °well-groomed, well-turned-out, well-kempt, °smart, °crisp, °dapper, spick and span *or* spic and span, °spruce, °shipshape (and Bristol fashion), *Archaic or dialectal* trig, *Colloq* natty, *US* spiffy: *Nancy arrived wearing her trim new flight attendant's uniform.* 2 in good *or* fine fettle, °fit (as a fiddle), athletic, °slim, °slender, clean-cut, °shapely, °streamlined, °compact: *Larry looks so trim that I asked him if he had lost weight. The new convertible model is a very trim little car.* —*v.* 3 °curtail, °shorten, °prune, °pare, °lop (off), crop, bob, °clip, °cut, °shave, shear, °snip, °dock; barber: *They trimmed the article by cutting two paragraphs from the end. Mother wept when I went into the navy only because they trimmed off my curls.* 4 °decorate, °embellish, °dress up, embroider, adorn, °ornament, deck out, caparison, °beautify: *When should we trim the Christmas tree?* —*n.* 5 trimming, edging, piping, purfling, rickrack, embroidery, °border, hem, °frill, °fringe, °ornament, ornamentation, °decoration, °embellishment, adornment: *The fabric is nice for a skirt, but I don't like that trim around the edge.* 6 °condition, °state, fettle, °health, °form, °order, °fitness, °repair, *Colloq* °shape: *Sid's car seemed to be in good trim when I saw it yesterday.*

trio *n.* threesome, trilogy, triad, triplex, triple, troika, triptych, triumvirate, triplet, trine, triune, trinity, three: *An interesting trio showed up for dinner.*

trip *n.* 1 °stumble, °slip, °blunder, false step, °misstep, °fall: *He sprained his ankle in that trip on the stair.* 2 °stumble, °slip, °blunder, false step, °misstep, faux pas, °error, °mistake, °indiscretion, °lapse, °slip of the tongue, *lapsus linguae,* erratum, °oversight; Freudian slip; *Slang Brit* boob: *If it hadn't been for that one trip, we would have had a perfect score.* 3 °tour, °journey, °excursion, °outing, °expedition, voyage, trek, peregrination, jaunt, junket, °drive: *We took a short side trip to visit Khios.* —*v.* 4 °dance, °caper, °skip, °cavort, gambol, °frisk, °hop, °spring: *Joanne came tripping gaily down the Champs Élysées.* 5 °stumble, °slip, °blunder, °misstep, °fall (down), °tumble, °topple, °dive, °plunge, °sprawl, °lurch, °flounder, °stagger, falter: *I tripped on the doorstep and went head over heels.* 6 Often, ***trip up.*** °trap, °trick, catch out, unsettle, °throw off, disconcert: *She has been trying to trip me up and confess to something I didn't do.* 7 °journey, °travel, voyage, °visit, °tour, trek, sightsee, °cruise; *Colloq* globe-trot: *They have been tripping all over Canada this summer.* 8 detonate, °set off, trigger, °operate, °release, °explode, °spark off: *When he touched that wire, he tripped the charge.* 9 Often, ***trip out.*** hallucinate, *Slang* freak out, °turn on: *There's no talking to him when he's tripping on drugs.*

triumph *n.* 1 °victory, °conquest, °success, °achievement, °accomplishment, attainment, coup, ascendancy: *How comforting it must be to see the triumphs of democracy around the world!* 2 exultation, rejoicing, exulting, elation, °delight, °rapture, exhilaration, jubilation, °happiness, °joy, °celebration, °glory: *There was great triumph on winning the World Cup.* —*v.* **3** Often, ***triumph over.*** °win, °succeed, carry the day, be victorious, gain a victory, take the honors, °thrive, °dominate, °prevail; °defeat, °beat, °rout, vanquish, °best, °conquer, °overcome, °overwhelm, °subdue: *Does justice always triumph? The book is about how man triumphed over pain.*

triumphal *adj.* celebratory, °rapturous, jubilant, °joyful, °glorious, °exultant; commemorative: *A holiday was declared to celebrate her triumphal entry into the city. This triumphal arch commemorates Trajan's victory over the Dacians.*

triumphant *adj.* °victorious, °successful, conquering, °winning; undefeated: *The triumphant hero returns tonight!*

triviality *n.* **1** smallness, unimportance, insignificance, meaninglessness, inconsequentiality *or* inconsequentialness *or* inconsequence *or* inconsequentness, trivialness, pettiness, paltriness: *I have difficulty coping with the triviality of some of my boss's requests.* **2** °trifle, technicality, °nonessential, small matter, unimportant *or* insignificant *or* inconsequential *or* trivial *or* petty detail, bêtise: *He tends to get bogged down in trivialities, unable to see what is important.*

trivialize *v.* °belittle, denigrate, lessen, °minimize, undervalue, °depreciate, °underestimate, underrate, °make light of, laugh off, underplay, °dismiss, °disparage, misprize, °beggar, deprecate, °slight, scoff at, °scorn, °run down, decry, *Colloq* °put down, °play down, pooh-pooh: *Buckner tends to trivialize the work of others.*

trophy *n.* **1** °prize, laurel(s), wreath, cup, °award, °reward, honor(s), medal, citation, palm, bays; °booty, °spoils, *Colloq* gold, °silver, silverware: *He has won trophies for more boat races than I have participated in.* **2** °memento, souvenir, °token, °record, °reminder, °remembrance, °keepsake: *That scar is a trophy of a hand-to-hand fight near El Alamein.*

trot *v.* **1** °jog, °run; bustle, °hustle, °hurry, °hasten, scamper, scoot, *Colloq* skedaddle: *I trot round the park every morning for exercise. As I need some butter, I'd better trot down to the shop before it closes.* **2** ***trot out.*** °bring out, °show, °display, °exhibit, °flaunt, come out with; dredge up, drag out; °recite, °repeat: *Our neighbor trotted out his new lawnmower for us to admire.* —*n.* **3** °jog, lope, single-foot, °pace; °run: *It was a delight to watch Dimbletonian in a fast trot round the track.* **4** °translation, °gloss, °interpretation, crib, *Colloq US* pony, horse: *He couldn't read Homer without a trot.*

trouble *v.* **1** °bother, °upset, °anguish, °alarm, °worry, °afflict, °agitate, disquiet, °discomfit, make uncomfortable, °grieve, °perturb, discommode, °inconvenience, discompose, discountenance, °put out, °burden, °encumber, °weigh down: *I don't mean to trouble you with my problems, but I have no one else to turn to.* **2** °annoy, °irritate, °irk, vex, °bother, °plague, °pester, °torment, °harass, hector, harry, °provoke, nettle, °exasperate, °ruffle, *Colloq* get *or* grate on (someone's) nerves, give (someone) a hard time, get under (someone's) skin: *Vincent keeps troubling me for advice on starting a new business.* **3** discommode, incommode, °impose on, °inconvenience, °put out, °thank: *I'll trouble you to turn off the light when you leave the room.* **4** °care, be concerned, take the trouble *or* the time, go to the trouble, °bother, °exert (oneself), °concern (oneself), take pains: *He never troubled to find out if his family was safe.* —*n.* **5** °distress, °worry, °concern, °difficulty, °discomfort, unpleasantness, °inconvenience, vexation, °grief, °woe, °affliction, disquiet, °suffering, tribulation, °anxiety, °torment, °anguish, °strife: *Her trouble began when her ex-husband stopped paying for child support. How can someone so insignificant cause so much trouble?* **6** °annoyance, °bother, tormentor, irritation, °nuisance, °nag, heckler, °pest, *Slang US* nudnik: *Ever since she lost her job, she's been a trouble to her family.* **7** °disorder, °agitation, °row, °disturbance, turbulence, °tumult, °upset, °dissatisfaction, °unrest, °discord, °dispute, turmoil, rebellion, °revolt, °uprising, outbreak, fighting, °fight, skirmishing, °skirmish: *The trouble began when workers refused to allow management to hire replacements.* **8** °affliction, °defect, °disability, °disease, °ailment, °illness, sickness, °disorder, °complaint: *With her trouble she ought to see a doctor.* **9** ***in trouble.*** **(a)** in deep trouble, in a mess, in a predicament, in dire straits, *Colloq* in a pickle, in hot water, on the spot, in a scrape, *Slang Brit* in shtuk *or* shtuck *or* schtuck, *Taboo slang* in deep shit, up shit creek (without (the vestige of) a paddle): *They are in terrible trouble if the bank forecloses on the mortgage.* **(b)** unmarried *or* unwed and impregnated *or* pregnant *or* with child *or* expecting *or* in a delicate condition *or Colloq* in a family way: *Most of the girls who are in trouble are teenagers.*

troublemaker *n.* mischief-maker, °rabble-rouser, gadfly, firebrand, *agent provocateur,* stormy petrel, incendiary, gossipmonger, scandalmonger, malcontent, instigator, meddler, °agitator: *As far as the police were concerned, any protester was, by definition, a troublemaker.*

troublesome *adj.* worrisome, worrying, annoying, °irksome, irritating, vexatious, bothersome, distressing, °difficult, °burdensome, *Colloq* pestiferous, *US and Canadian* pesky: *We sometimes have to put up with troublesome motorcycle gangs.*

truant *n.* **1** malingerer, °runaway, absentee, °delinquent, dodger, shirker, °idler, °loafer, layabout, *Slang Brit* skiver, *Brit military* scrimshanker: *Truants were warned that their parents would be required to visit the school.* —*adj.* **2** malingering, °runaway, °absent, absentee, °delinquent, shirking, loafing, *Slang Brit* skiving: *The officer brought in three truant boys found fishing at the lake.*

truce *n.* **1** armistice, cease-fire, suspension of hostilities, °lull, °moratorium, °respite, °letup, °intermission, °interval, °interlude: *It looked as if the truce might last.* **2** °pact, °treaty, °compact, °agreement, cease-fire, armistice: *If both sides abide by the truce the war might be over.*

truck *n.* **1** °merchandise, commodities, °goods, °stock, °wares, °stuff, °odds and ends, °sundries, °junk, °rubbish, *US* °trash: *There was no one in the shop and all the truck was stacked in the corner.* **2** dealing(s), °traffic, °business, °transaction, °trade, °commerce, communication, °contact, °connection, (business *or* social) relations: *She refuses to have any truck with the likes of you.*

truckle *v.* °kowtow, be obsequious, toady, °defer, °bow, °scrape, genuflect, salaam, drop to the ground *or* to (one's) knees *or* down on (one's) knees, °submit, °yield, cower, °cringe, grovel, °crawl, quail, fawn (on *or* upon), *Colloq* butter up, fall all over, lick (someone's) boots, bootlick, *US* apple-polish, *Slang* suck up to, *Taboo slang* brown-nose, kiss (someone's) *Brit* arse *or US* ass: *Just look how Carruthers is always truckling to the boss, hoping for favors.*

truculent *adj.* °surly, °sullen, bad-tempered, ill-tempered, unpleasant, °nasty, °obstreperous, °rude, unpleasant, °ferocious, °fierce, °savage, feral, barbarous, °harsh, °scathing, °virulent, combative, °belligerent, antagonistic, bellicose, °hostile, contentious, °warlike, °violent, °pugnacious, *Colloq* scrappy: *I don't care* *enough for this job to endure the boss's truculent attitude a moment longer.*

true *adj.* **1** °accurate, °correct, °truthful, °faithful, °literal, °authentic, veracious, °actual, °factual, °realistic, °genuine, °right, valid, unelaborated, °unvarnished, unadulterated, verified, verifiable: *Do you swear that this is a true account of what actually took place?* **2** °staunch, °faithful, °devoted, dedicated, °loyal, °fast, °firm, unswerving, °steady, °steadfast, °trustworthy, trusty, °dutiful, °upright, °honorable, °constant, unwavering, °stable, dependable, °sincere, °reliable, true-blue: *D'Artagnan proved himself to be a true friend.* **3** °proper, °exact, °accurate, unerring, °correct, °precise, °right, *Slang Brit* spot on: *It is important that these matters be seen in their true perspective. If this is a true copy of the original, please sign it.* —*adv.* **4** °truly, truthfully, °honestly, accurately, candidly, frankly, °sincerely, straightforwardly: *Tell me true, do you love me? If the report speaks true, then we must find the culprit.* **5** °exactly, correctly; geographically: *We sail true north to Iceland.* **6** ***come true.*** °come to pass, °occur, °take, °place, °happen, be realized, become a reality, be fulfilled: *Her dreams finally came true when she bought a little house in Kent.*

truism *n.* commonplace, platitude, bromide, axiom, °cliché, °maxim: *You can always rely on Vera to utter a truism like, "It's a nice day," on a warm, sunny day.*

truly *adv.* **1** truthfully, °actually, °really, °honestly, °in fact, °in truth, in actuality, °in reality, in all honesty, °sincerely, genuinely: *Are you truly giving up your job to get married? I truly believed her to be guilty.* **2** °properly, rightly, rightfully, justly, legitimately, justifiably, °duly, well and truly, accurately: *In the circumstances, can we truly condemn him for behaving as he did?* **3** °definitely, °really, °actually, °undoubtedly, indubitably, beyond (the shadow of) a doubt, beyond question, without a doubt, °indeed, unquestionably, °absolutely, °positively, decidedly, certainly, °surely: *I believed her to be truly guilty.* **4** °in truth, °indeed, °really, °honestly, °sincerely, genuinely, *Archaic* (yea), verily, *Usually ironic* forsooth: *I love you Cleo, truly I do.*

trunk *n.* **1** main stem, °stalk, °stock, *Technical* bole: *After the tornado, only the trunks of the trees remained upright.* **2** torso, °body: *Only the trunk of the corpse was found.* **3** °chest, locker, footlocker, °box, °case, bin, coffer, °casket: *We found a trunk full of old books in the attic.* **4** snout, proboscis: *The elephant reached for the food with its trunk.* **5** (*In the US and Canada*) luggage compartment, *Brit* °boot: *We cannot get all the luggage into the trunk of the car.*

trust *n.* **1** °confidence, °reliance, °faith, °conviction, certitude, °certainty, sureness, positiveness, °assurance, °belief: *Place your trust in me.* **2** °credit, reliability, dependability, credibility, trustworthiness: *The company will sell you the piano on trust.* **3** °custody, °care, keeping, °charge, guardianship, °protection, °safekeeping, trusteeship: *The money is in trust for Gillian's grandchildren.* **4** monopoly, cartel; °group, corporation, conglomerate: *An international trust controls all aspects of the world market in diamonds.* —*v.* **5** rely (on *or* upon), have faith *or* confidence (in), confide (in), depend *or* bank *or* count (on *or* upon), pin (one's) faith *or* hopes on *or* upon: *I trust that you will attend the meeting. In God we trust; others must pay cash. Can I trust you to keep a secret? Don't trust to luck.* **6** °entrust, °commit, °give, °delegate, make *or* turn *or* sign *or* hand over, depute, °assign, empower, consign: *I shouldn't trust my money to her.*

trusting *adj.* trustful, unsuspicious, confiding, °confident, °unsuspecting; naive *or* naïve *or* naïf, °innocent, °gullible, incautious, credulous: *It is a good thing that her husband has a trusting nature. Samantha might be a little too trusting and could easily be deceived.*

trustworthy *adj.* °reliable, trusty, dependable, °accurate; °responsible, °steady, °steadfast, °loyal, °faithful, °(tried and) true, °honorable, °honest, °ethical, °principled, °moral, incorruptible: *Is this thermometer trustworthy? Isaac's former employer said that he is completely trustworthy.*

truth *n.* **1** genuineness, °reality, actuality, correctness, °accuracy, °fact: *The truth of the matter is that he does love you, if anything, too much.* **2** fact(s): *To tell the truth, I came here to kill you.* **3** ***in truth.*** °in fact, °truly, °actually, °really: *In truth, his name is not Jack Armstrong at all but Ebenezer McCuddigillie.*

truthful *adj.* °true, °accurate, °factual, veracious, true-to-life, °honest, °realistic, °reliable, °faithful, °trustworthy, straightforward, °candid, °frank, °sincere, °earnest, °forthright, °unvarnished, unembellished: *He gave a truthful account of the landing of the flying saucer.*

try *v.* **1** °attempt, °endeavor, °essay, °seek, °undertake, °venture, °strive, °struggle, make an effort, try (one's) hand at, *Colloq* have a stab *or* go *or* whack (at), take a shot *or* crack (at): *He tried to help me with my homework.* **2** °test, try out, °prove, °evaluate, °examine, °inspect, °check out, °sample, appraise, assay, °look over, °analyze, °scrutinize, assess, °judge: *I'll try your way of solving the problem. You won't know if it works till you try it.* **3** °test, °prove, °strain, °tax: *You are trying my patience with your silly questions.* **4** °hear, °sit on, adjudicate, °judge, adjudge: *There are three more cases to try on this month's calendar.* —*n.* **5** °attempt, °endeavor, °essay, °undertaking, °venture, °struggle, °effort, °turn, *Colloq* °go, °stab, whack, °fling, °shot, °crack: *You have three tries to pin the tail on the donkey.*

trying *adj.* irritating, exasperating, frustrating, annoying, °irksome, infuriating, maddening, bothersome, °tiresome, vexing, °troublesome, worrying, worrisome, distressing, disquieting, upsetting, dispiriting, taxing, °demanding, °tough, stressful, °difficult, tiring, fatiguing: *This must be a trying time for you, caring for eight small children.*

tug *v.* **1** °pull, °tow, °yank, °jerk, °draw, °drag, °haul, °wrench: *The boy was tugging a little puppy along on a lead.* —*n.* **2** °pull, °tow, °yank, °jerk, °drag, °haul, °wrench: *I gave a tug and the doorknob came away in my hand.*

tuition *n.* °education, teaching, tutelage, training, °schooling, °instruction, °guidance, °preparation: *The tuition of the students is more important than sports in this school.*

tumble *v.* **1** °fall (down), °pitch, turn end over end *or* head over heels, °roll, °drop: *Giggling hysterically, we tumbled in a heap on the lawn.* **2** °drop, °toss, °dump, °jumble: *The waiter tumbled some spoonfuls of berries onto my plate.* **3** ***tumble to.*** °understand, apprehend, °perceive, °comprehend, see the light, *Colloq* get the signal *or* message, °catch on, *Brit* twig to, *Slang* get wise, wise up, °dig: *I finally tumbled to what she was trying to tell me.* —*n.* **4** °fall, °slip, °stumble, *Colloq* header, °spill: *Joshua took a bad tumble on the stairs yesterday.*

tumble-down *adj.* °ramshackle, °dilapidated, ruined, in ruins, °decrepit, °rickety, °shaky, falling apart *or* to pieces, disintegrating, tottering, broken-down, crumbling, gone to rack and ruin: *He lived for years in a tumble-down shack by the railroad track.*

tumor *n.* neoplasm, cancer, melanoma, sarcoma, malignancy, carcinoma, °growth, °lump, °swelling, protuberance, excrescence: *The doctor found a tumor that he said ought to be removed.*

tumult *n.* commotion, °disturbance, °upset, °uproar, °riot, °disorder, disquiet, insurrection, °agitation, °bedlam, °chaos, brouhaha, °fracas, hubbub, °stir, °pandemonium, hullabaloo, °furor, °brawl, donnybrook, affray, °row, melee *or* mêlée, turbulence, °ferment, ado, turmoil, °confusion, °rampage, °frenzy, °rage, °excitement, °rumpus, *Colloq US* ruckus: *The tumult caused by the football hooligans spread through the city.*

tumultuous *adj.* clamorous, °noisy, °boisterous, °disorderly, turbulent, °violent, °uproarious, °chaotic, frenzied, °furious, °excited, °agitated, °hectic, °riotous, °rowdy, °unruly, unrestrained, °fierce, °savage, °wild, °hysterical, °frantic, rumbustious, °boisterous, °obstreperous, °tempestuous, °stormy: *The heroes received a tumultuous welcome on their return.*

tune *n.* **1** °melody, °air, °song, °strain, °motif, °theme: *David presents a marvelous half-hour radio program of show tunes every week.* **2** euphony, °pitch, °harmony, °accord, accordance, consonance, °unison, correspondence, conformity: *She cannot sing in tune. The guitar is out of tune with the piano. Wilberforce is out of tune with today's fashion.* —*v.* **3** tune up, °calibrate, °adjust, °regulate, °coordinate, °adapt, attune, align, °set: *That garage charges too much for tuning an engine.* **4** ***tune in (on).*** °attend (to), pay attention (to), °listen (to), °understand, be aware (of), be on the qui vive, be alert (to), *Slang* be on the same wavelength *or Slang* frequency (with): *I am not sure that Bunkley is tuned in on what his sister does for a living.* **5** ***tune out.*** °ignore, °disregard, turn a blind eye to, be blind to, °turn one's back on, turn a deaf ear to: *Sperber is able to tune out anything she doesn't like to hear.*

tuneful *adj.* melodic, °musical, sweet-sounding, °melodious, euphonious, dulcet, mellifluent, mellifluous, harmonic, catchy, °mellow, °smooth, °rich, °rhythmic, *Colloq* easy on the ear(s): *Irving Berlin wrote some of the most tuneful music that we have.*

tunnel *n.* **1** °shaft, °subway, (underground) passage(way), underpass; °burrow, °hole; *Colloq* Chunnel (= 'Channel Tunnel'): *The cat got out through this tunnel.* —*v.* **2** °burrow, °dig, °hole, °excavate, °penetrate, °mine: *The prisoners tunneled under the wall and escaped.*

turf *n.* **1** sod, sward, °green, grass, greensward, °lawn: *We now buy turf and have a ready-made lawn.* **2** °territory, bailiwick, °area, °neighborhood, back yard, *Colloq* stamping ground, home ground, °(personal) space: *You're on my turf now, so you'll do as I say.* **3** ***the turf.*** horse-racing, racing, the racing world, racecourse, racetrack: *The attractions of the turf keep them from other pursuits.*

turn *v.* **1** °rotate, °revolve, °spin, °roll, °reel, °circle, °gyrate, whirl, °wheel, go (a)round *or* about, °pivot, °swivel: *The earth turns on its axis. Turn the crank to raise the bucket.* **2** °move, °shift, °wheel, veer, °swing, °face: *As she turned I noticed that horrible scar.* **3** °reverse, turn (a)round, °alter, °change, °adapt, reorganize, remodel, °modify, refashion, reshape, °reform, °transform, °make over, °convert, bring over: *He has been trying to turn the business into a profit-making enterprise. He has turned defeat into advance. She managed to turn one of the most loyal Panoverian agents.* **4** go *or* pass *or* move (a)round, veer, °drive, °walk: *Turn left at the corner.* **5** go bad, become rancid, °spoil, curdle, addle, °sour, °decay, molder, °rot, °putrefy, *Colloq* °go off: *All the milk in the fridge turned because of the power cut.* **6** °apply, °put, °use, °employ: *Is there any way we can turn this situation to our advantage?* **7** Sometimes, ***turn aside*** *or* ***away.*** °block, avert, °thwart, °prevent, balk *or* baulk, parry, °deflect, °fend off, °check: *He deftly turned aside the thrust of the dagger.* **8** °form, °make up, °fashion, °formulate, °construct, °cast, °create, °coin, concoct, °express: *Donald certainly knows how to turn a felicitous phrase.* **9** °direct, °aim, °point: *Turning the gun on himself, he pulled the trigger.* **10** °twist, sprain, °wrench: *I have turned my ankle and cannot walk.* **11** °twist, °wind, °snake, curve, °bend, arc, °coil, °loop, °meander, zigzag: *The road turned this way and that, following the river bank.* **12** ***turn against.*** °defy, °mutiny, °rebel, °revolt, rise (up) against: *The captain had not expected the first mate to turn against him, too.* **13** ***turn back.*** **(a)** °reverse, °repulse, °repel, °rebuff, drive back, beat back: *At last we turned back the enemy's advance.* **(b)** go back, retrace (one's) steps, °return: *We must turn back before it is too late.* **14** ***turn down.*** **(a)** °refuse, °reject, °rebuff, °decline, °deny: *My request for help was turned down.* **(b)** decrease *or* diminish *or* lessen *or* lower *or* soften the sound of: *Turn down the radio, I can't hear on the phone.* **15** ***turn in.*** **(a)** go to bed *or* sleep, °retire, °withdraw, call it a day, *Slang* hit the sack *or Slang* the hay: *I usually turn in by eleven o'clock.* **(b)** hand in *or* over, °turn over, °deliver, °give in, °submit, °offer, proffer, °tender, give back, °return, °surrender, °yield: *Please turn in your visitors' badges before you leave.* **(c)** °turn over, °deliver (up), °inform on, °betray, *Colloq* squeal on, rat on, °finger, °tell on: *For enough money, he'd turn in his own mother.* **16** ***turn into.*** **(a)** °turn to, °become, change into *or* to, metamorphose into *or* to: *Right before her, the prince turned into a frog again.* **(b)** go *or* come into, drive into, pull into, walk into: *The man I was following turned into a sidestreet and vanished from view.* **17** ***turn off.*** **(a)** °stop, switch off, deactivate, °discontinue; °extinguish: *First turn off the water, then the light.* **(b)** °disillusion, °depress, °cool (off), disenchant, disaffect, °alienate, °repel, °repulse, °bore, °offend, °put off, displease, °sicken, °nauseate, °disgust: *People who don't brush their teeth turn me off.* **(c)** °deviate, °diverge: *When you come to the fork, turn off to the right.* **18** ***turn on.*** **(a)** °start (up), switch on, °energize, °activate, set in motion, cause to function *or* operate: *Turn on the light.* **(b)** depend on *or* upon, be contingent upon *or* on, hinge on *or* upon, be subject to: *The success of the venture turns on our ability to capitalize it.* **(c)** °excite, °thrill, °arouse, °stimulate, titillate, °work up, impassion: *Dan finally confessed that Joan turns him on.* **19** ***turn on*** *or* ***upon.*** **(a)** °concern, revolve about, °relate to: *The discussion turned on his ability to write music.* **(b)** be hostile to, °attack, assail, °set upon, *Colloq* tear into: *Oliver is so unpopular that his own dog turned on him and bit him.* **20** ***turn out.*** **(a)** °make, °form, °shape, °construct, °build, °fabricate, put together, °assemble, °manufacture, °produce, °put out, °bring out: *The plant turns out a thousand cars a week.* **(b)** °develop, evolve, eventuate, °happen, °result, °prove, °occur, end up, °arise: *As it turned out, he lost anyway. It turns out that he knows my sister.* **(c)** °eject, °evict, °throw out, °expel, oust, °dismiss, °terminate, cashier, *Colloq* °fire, °sack, kick out, ax, *Brit* turf out: *When they found I wasn't a member, they turned me out.* **(d)** °dress, fit out, °equip, °rig out, accoutre *or US also* accouter: *She was well turned out in a beautiful ball gown.* **(e)** °come, °arrive, °appear, °attend, °assemble, °meet, *Colloq* °show (up), °surface: *Fifty-five thousand turned out for the rock concert.* **21** ***turn over.*** **(a)** °consider, muse *or* ruminate over *or* about, °revolve, °ponder (over): *I needed a while to turn over the job offer in my mind.* **(b)** °reverse, invert, turn upside down: *Turn over the clock and read the inscription on the bottom.* **(c)** °overturn, °upset, knock over: *In my haste, I turned over the punch bowl.* **(d)** °sell, °merchandise: *A shop in that location ought to turn over a million a year.* **22** ***turn tail.*** °run away, °flee, °bolt, scoot, show a clean pair of heels, cut and run, take to (one's) heels, beat a hasty retreat, *Colloq* °take off, °beat it, scram, skedaddle: *He turned tail when I shouted for help.* **23** ***turn to.*** **(a)** appeal to, apply to, °resort to: *She turned to me for help.* **(b)** advert to, °refer to, pick *or* take up, have recourse to: *Please turn to your exercise books now.* **(c)** get to work, °pitch in, buckle *or* knuckle down: *The neighbors turned to in helping clean up the mess after the storm.* **(d)** °turn into, °change to, convert to, °become: *Lot's wife was turned to salt.* **24** ***turn turtle.*** °capsize, °overturn, keel over, °upset, upend, *Colloq* go bottoms up: *The overloaded barge turned turtle and sank in the river.* **25** ***turn up.*** **(a)** °surface, °appear, °arrive, *Colloq* °show (up), °show one's face: *Guess who turned up at our wedding?* **(b)** °come up, °arise, *Colloq* crop up, pop up: *Something will turn up soon for you.* **(c)** uncover, °discover, °find, °unearth, °come across, °hit upon, °dig up, °expose, °disclose, °reveal, °bring to light: *We turned up a formerly unknown fact about the shipwreck.* **(d)** increase *or* raise *or* amplify *or* intensify the sound of: *Turn up the TV—I can't hear what they're saying.* —*n.* **26** °revolution, rotation, °cycle, °spin, whirl, °circuit, °round, °roll, °twirl; °pirouette: *Just a few more turns and the roast was done.* **27** curve, °bend, turning, corner, sinuosity, dogleg, hairpin bend *or* curve, irregularity, °meander, °twist, zigzag, *Colloq* to-ing and fro-ing: *There are many dangerous turns on that road.* **28** °loop, °coil, °spiral, °twist: *Take two turns of this rope round your waist, then knot it.* **29** deviation, turning, °detour, °shift, change of direction *or* course: *A turn to the right is not permitted at this corner.* **30** opportunity, °chance, °say, °round, °spell, °time, °watch, °shift, °stint, °tour (of duty), °move, °trick, *Colloq* whack, °crack, °shot, go: *You have had your turn, now let someone else go.* **31** °drive, °spin, °ride; airing, constitutional, °ramble, °saunter, °stroll, °walk, °promenade, amble: *Let's take a short turn around the park.* **32** °trend, °direction, °drift: *The conversation took a new turn.* **33** °change, °alteration, °switch: *The doctor says that Vladimir has taken a turn for the better.* **34** Usually, ***bad turn.*** °disservice, °harm, °injury, °wrong: *If you do someone a bad turn, what can you expect?* **35** Usually, ***good turn.*** °favor, °(good) deed, °act (of kindness), °courtesy, °boon, °mercy: *One good turn deserves another.* **36** °shock, °fright, °surprise, °start, °scare: *You really gave me a turn, jumping out like that!* **37** °form, °style, °manner, °mode: *Each turn of phrase in her writing seems original and refreshing.* **38** °disposition, °inclination, °bent, °bias, °leaning, °tendency: *Flanagan is of a rather dour turn of mind tonight.* **39** ***at every turn.*** °everywhere, constantly, °always, °all the time: *In Scotland, we met with kindness and hospitality at every turn.* **40** ***by turns.*** alternately, reciprocally, in rotation, successively, °in succession: *The book is fascinating and frustrating by turns.* **41** ***in turn.*** sequentially, one after the other, °in succession, successively, in (proper) order: *Each patient will be treated in turn.* **42** ***out of turn.*** **(a)** out of sequence, °out of order: *I don't want you answering questions out of proper turn, Jonathan.* **(b)** imprudently, indiscreetly, improperly, disobediently, inappropriately: *I apologize if I am speaking out of turn on this issue.* **43** ***take turns.*** °alternate, °vary, °rotate, °exchange: *Let's take turns looking through the telescope.*

turnabout *n.* reciprocity, °exchange: *Turnabout is fair play.*

turncoat *n.* °renegade, °traitor, betrayer, °deserter, fifth columnist, double agent, apostate, tergiversator, °defector, backslider, Vicar of Bray, *US* Benedict Arnold, *Colloq* snake in the grass: *Democrats who voted with the Republicans on the issue were branded turncoats.*

turnoff *n.* **1** °exit, side road, feeder (road), °auxiliary (road), °ramp, *Brit* slip road, *US* (exit *or US* entrance) ramp: *Our shop is at the first turnoff after the traffic light.* **2** damper, °killjoy, *Colloq* wet blanket, *Slang US* freeze-out: *Watching open heart surgery on TV was a real turnoff during dinner.*

turnout *n.* **1** assemblage, °muster, °attendance, audience, °crowd, °gate, °throng, °gathering: *The turnout for the first day of the sale was enormous.* **2** °output, °production, outturn, °volume; gross national product, GNP, gross domestic product, GDP: *Turnout has improved since the settlement of the labor dispute.* **3** °gear, °outfit, clothing, °apparel, °apparatus, °equipment, °trappings, °fittings, equipage: *Have you seen Crawleigh in his mountain-climbing turnout?*

turnover *n.* °gross (revenue), °(total) business, °volume: *Although turnover increased by 10 percent, profits were down by 2 percent.*

tutor *n.* **1** °teacher, °instructor, educator, °coach, mentor, guru: *Bernard engaged a tutor to coach him for the examinations.* —*v.* **2** °teach, °instruct, °coach, °educate, °school, °train, °indoctrinate, °drill, °enlighten, °advise, °direct, °guide, °prepare, °ground: *Twitchell needs someone to tutor him in the fine art of going out with girls.*

tweak *v.* **1** °pinch, °nip, twitch, °squeeze, °jerk, °grip: *As a sign of affection she tweaked my nose.* —*n.* **2** °pinch, °nip, twitch, °squeeze, °jerk, °grip: *He gave her nose an affectionate little tweak.*

twiddle *v.* **1** °play with, °twirl, °fiddle (with), wiggle, °juggle, °toy with, fidget with, *Colloq* °fool with, °mess with, monkey with: *Stop twiddling with the dial on that radio!* **2** ***twiddle (one's) thumbs.*** do nothing, be idle, idle *or* while away (the) time, waste time, bide (one's) time: *I sat there, twiddling my thumbs, while you were being entertained royally.*

twig[1] *n.* sprig, °stem, °shoot, °offshoot, branchlet, °stick, °sucker, °sprout, withe *or* withy, tendril: *Gather up some dry twigs for kindling.*

twig[2] *v.* °understand, °grasp, °fathom, °get, °comprehend, °see, °know, °sense, °divine, *Colloq* °catch on, be *or* get *or* become wise to, °tumble to, *Slang* rumble, °dig: *She twigged the situation at once but didn't let on she knew.*

twilight *n.* **1** °dusk, sunset, gloaming, sundown, half-light, crepuscule *or* crepuscle: *We enjoyed dinner at twilight on the terrace overlooking the sea.* **2** °decline, °wane, waning, °ebb, downturn, downswing, °slump, °decay, weakening, declination, diminution: *Even at the twilight of his career, Jonas enjoyed the respect of his colleagues.* **3** ***Twilight of the Gods.*** Götterdämmerung, Ragnarök *or* Ragnarok: *The world ends at the Twilight of the Gods, only to be born anew.* —*adj.* **4** °evening, crepuscular, dimming, darkening, darkish, darksome, °shadowy, °shady, °dim, °dark, °obscure, °somber, °gloomy, *Literary* darkling: *The nymphs in twilight shade of tangled thickets mourn.* **5** ***twilight zone.*** °limbo: *He lives in the twilight zone, unable to distinguish reality from fantasy.*

twin *n.* **1** °double, clone, °duplicate, °lookalike, counterpart, *Slang* ringer: *The car he drives is a twin to mine.* —*adj.* **2** °identical, °matching, matched, °duplicate, corresponding, °lookalike: *My friend Donald and I have twin houses and twin cars.* —*v.* **3** °pair, °match, yoke, °join, °link, °couple, °combine, °connect, °associate: *Many towns in Britain are twinned with similar towns on the Continent.*

twine *n.* **1** °cord, °string; °rope, °cable, °yarn: *Have you some twine for tying up this package?* —*v.* **2** °entwine, °braid, °twist, intertwine, curl, wreathe, °spiral, °wind, °weave, interweave, °encircle, °wrap: *Annette's front door has roses twined all around it.*

twinge *n.* **1** °stab, °pang, cramp, °spasm, °pinch, stitch, °(sharp) pain, °prick, °bite, °gripe: *I get a terrible twinge in my back when I lift something heavy.* **2** °pang, °pain: *I felt a twinge of remorse at leaving.*

twinkle *v.* **1** scintillate, °sparkle, coruscate, °glitter, °shimmer, wink, °flicker, °glisten, glint, °flash, fulgurate, °spark, °dance, °blink, °shine, °gleam: *The stars were twinkling in the icy black sky. Nicole's eyes twinkled as she told me about Max's latest success.* —*n.* **2** °twinkling, scintillation, °scintillating, °sparkle, sparkling, coruscation, coruscating, °glitter, glittering, °shimmer, shimmering, winking, °flicker, flickering, glistening, glint, °flash, flashing, fulguration, °spark, sparking, dancing, blinking, °shine, shining, °gleam, gleaming, °dazzle, °dazzling: *From far off, we caught the twinkle of the sun on their golden armor.*

twinkling *n.* **1** °(split) second, °flash, twinkling *or* wink of an eye, °instant, trice, *Colloq* jiffy, two shakes (of a lamb's tail), tick: *Liza called out, and in a twinkling, Joseph was at her side.* **2** See **twinkle, 2,** above.

twirl *v.* **1** °spin, whirl, °rotate, °revolve, °wheel, °turn, °gyrate, °twist, wind (about *or* around): *The windmills twirled in the breeze. Katherine absently twirled a lock of hair around her finger.* —*n.* **2** twirling, °spin, spinning, whirl, whirling, °turn, turning, °revolution: *He was hypnotized by each twirl of the roulette wheel.* **3** whorl, winding, convolution, °spiral, helix, °coil, volute: *The pattern consists of interlocking twirls of green and brown.*

twist *v.* **1** plait, °braid, °weave, °entwine, intertwine, °twine, interweave, pleach, °splice, wreathe, interlace: *She twisted together some daisies to make a garland for her hair.* **2** °distort, °warp, contort, °pervert, °alter, °change, °slant, °bias, °color, °falsify, misquote, misstate, °garble, miscite, °misrepresent, °violate; °misinterpret, mistranslate, °misunderstand, misconstrue: *He twisted her words so that she seemed to be saying the opposite of what she intended.* **3** °wriggle, worm, °squirm, writhe, wiggle: *The little beggar twisted out of my grasp and ran for his life.* **4** °wind, °snake, °meander, °turn, zigzag, worm, °bend, curve: *I lost sight of the stream where it twisted through the undergrowth.* **5** °wrench, °turn, sprain, rick *or* wrick: *I twisted my ankle on that faulty step.* **6** ***twist (one's*** or ***someone's) arm.*** °force, coerce, °make, °persuade, °bully, *Brit* pressurize, *US* °pressure: *I didn't want to go, but she twisted my arm.* —*n.* **7** °coil, °spiral, skew, zigzag, dogleg, °turn, curve, °angle, °bend, °bow, °meander: *The road is full of twists and turns.* **8** °interpretation, °analysis, °understanding, °slant, °angle, construction, construal; °treatment, °approach, °version, °variation: *This book puts a new twist on the fall of the Roman Empire.* **9** distortion, misinterpretation, contortion, °perversion, warping, °alteration, °change, departure, °bias, coloring, falsification, misquotation, °misstatement, garbling, misrepresentation; mistranslation, °misunderstanding, misconstrual, misconstruction: *It was a twist of fate that brought us together again. She gave everything he said such a twist that he scarcely recognized his own ideas.* **10** °quirk, idiosyncrasy, crotchet, °peculiarity, °oddity, °trick, °eccentricity, incongruity, inconsistency, irregularity; °weakness, °flaw, °fault, °foible, °failing: *Owing to a twist in his character, he has developed a hatred of women.* **11** ***round the twist.*** °mad, °crazy, °insane, °eccentric, *Colloq* °daft, balmy, *Brit* round the bend, *Slang* nuts, nutty, bonkers, cuckoo, batty, off (one's) rocker, *Brit* barmy: *Kevin's gone round the twist if he thinks I am going out with his sister.*

twister *n.* tornado, cyclone, typhoon, °hurricane, °whirlwind; waterspout: *The twister picked up my car and dropped it in my neighbor's garden.*

twit[1] *v.* °tease, °cajole, °taunt, °jeer (at), °make fun of, °banter, °tweak, gibe *or* jibe, °chaff, °ridicule, °mock; °blame, °berate, °deride, °scorn, contemn, censure, revile, reproach, °upbraid; *Colloq* kid, pull (someone's) leg: *She never seemed to mind being twitted about her height. At that time, one could not twit the monarch and escape unscathed.*

twit[2] *n.* nitwit, nincompoop, ass, ninny, ninnyhammer, °fool, imbecile, blockhead, °half-wit, idiot, simpleton, *Colloq* chump, moron, *Brit* silly-billy, *Slang* °dope, *US and Canadian* °jerk: *That colossal twit told the teacher what we were planning.*

twitter *v.* **1** °peep, cheep, tweet, °chirp, warble, trill, chirrup, °chatter: *The birds twitter so loudly that they wake me every morning.* **2** °chatter, °prattle, °gossip, °giggle, prate, °titter, °snicker, snigger, simper: *I was embarrassed because my classmates always twittered when I did a recitation.* —*n.* **3** °peep, peeping, cheep, cheeping, twittering, tweet, tweeting, chirrup, chirruping, °chirp, chirping, warble, warbling, trill, trilling: *The twitter of the birds was driving me mad.* **4** ado, bustle, °excitement, °flutter, dither, whirl, °agitation, *Colloq* °stew, tizzy: *The girls were in a twitter of apprehension.*

two-faced *adj.* double-dealing, °hypocritical, duplicitous, dissembling, °deceitful, Janus-faced, treacherous, °dishonest, untrustworthy, °insincere, °scheming, °designing, °crafty, Machiavellian, °sly, °perfidious, °lying, mendacious: *That two-faced liar told you one story and me another!*

tycoon *n.* °mogul, magnate, baron, °financier, (multi)millionaire, billionaire, merchant prince, potentate, *Colloq* big shot, °(big-time) operator, wheeler-dealer, big-timer, *US* big wheel, big cheese: *What makes you think that Castenado is an oil tycoon?*

type *n.* **1** °class, °category, classification, °kind, °sort, °genre, °order, °variety, °breed, species, °strain, °group, genus, ilk, kidney: *Just what type of person would you say the president is?* **2** typeface, *Brit* fount, *US* font: *The body text ought to be set in Bodoni type.* **3** °prototype, paradigm, archetype, °epitome, exemplar, °model, °specimen, °pattern, personification, °standard, °quintessence: *She doesn't consider him to be the usual type of businessman.* —*v.* **4** typewrite; keyboard; °transcribe: *Miss Cantrell types at about a hundred words a minute.*

typical *adj.* **1** °representative, °characteristic, °conventional, °normal, °standard, °ordinary, °regular: *On a typical day I arise at 6:30.* **2** °orthodox, °classic, °conventional, °in character, in keeping, °usual, commonplace, °run-of-the-mill, °natural, °customary, °common, to be expected, °ordinary: *His way of looking at life is typical for someone of his educational and social background.*

typify *v.* °exemplify, °instance, epitomize, °personify, °represent, °characterize, °embody, evince, °symbolize, °suggest: *John's views typify the conservative approach.*

tyrannical *adj.* tyrannous, °oppressive, °dictatorial, fascistic, °despotic, autocratic, °authoritarian, °arbitrary, imperious, °overbearing, unjust, highhanded, °severe, °harsh, ironhanded, °heavy-handed: *The people suffered under one tyrannical form of government after another for generations.*

tyrannize *v.* Often, **tyrannize over.** domineer over, °bully, °subjugate, °enslave, °dominate, °intimidate, dictate to, order about *or* around, ride roughshod over, °browbeat, keep under (one's) thumb, °oppress, °subdue, °suppress, keep down, *Archaic* enthrall: *The people have been tyrannized long enough.*

tyranny *n.* autocracy, fascism, authoritarianism, absolutism, °despotism, dictatorship, Stalinism, Nazism; arbitrariness, °oppression, °suppression, subjugation, enslavement, enthrallment, °domination: *Tyranny would not survive long were it not expedient for its supporters.*

tyrant *n.* °dictator, °despot, autocrat, martinet, Hitler, °bully, °oppressor, °authoritarian, hard taskmaster, slave driver, Simon Legree, overlord: *It seems that no people or part of the world is immune to tyrants.*

U

ugly *adj.* **1** unattractive, unlovely, unprepossessing, °unsightly, °hideous, °grotesque, °gruesome, °ghastly, °offensive, repulsive-looking, °plain, plain-looking, plain-featured, bad-featured, ill-favored, dreadful-looking, awful-looking, terrible-looking, horrible-looking, frightful-looking, monstrous-looking, *US and Canadian* °homely: *The old houses were demolished to make way for an ugly skyscraper.* **2** objectionable, °disagreeable, unpleasant, °offensive, °nasty, °loathsome, °repellent, °repugnant, °repulsive, noisome, nauseating, nauseous, °revolting, sickening, °disgusting, °obnoxious, mephitic, °rotten, °corrupt, °filthy, °vile, heinous, °bad, °sordid, °evil, °foul, °perverted, °immoral, depraved, °degenerate, °base, debased, detestable, °hateful, °abominable, execrable, °despicable, odious: *The murders were among the ugliest crimes of the century. He sprang from his seat with an ugly curse on his lips.* **3** disquieting, uncomfortable, discomforting, °troublesome, °awkward, disadvantageous, °ominous, °dangerous, °perilous, °hazardous: *Accused of treason, Stupeleigh found himself in a very ugly position.* **4** unpleasant, °disagreeable, °surly, °hostile, °nasty, °spiteful, bad-tempered, ill-tempered, currish, irascible, curmudgeonly, °cantankerous, crabby, crabbed, crotchety, °cross, °cranky, °mean: *He's in an ugly mood till he's had his coffee.*

ulcer *n.* **1** °sore, lesion, abscess, ulceration, canker, chancre, °boil, gumboil, °eruption, carbuncle, °inflammation: *That ulcer should be treated before it gets any worse.* **2** cancer, canker, festering spot, °blight, °scourge, °poison, °disease, °pestilence, °curse, bane, °plague: *We must purge this ulcer before it corrupts the entire organization.*

ulcerous *adj.* ulcerative, cancerous, cankerous, festering, ulcerated, ulcerative, suppurating, suppurative, gangrenous, septic, *Technical* furuncular, furunculous, necrotic, necrosed, sphacelated: *The wound is in an ulcerous condition and requires immediate surgery.*

ulterior *adj.* **1** °hidden, concealed, covert, °secret, unrevealed, undisclosed, unexpressed, °private, °personal, underlying, °surreptitious, underhand(ed): *She had an ulterior motive for leaving her fingerprints on the glass.* **2** °outside, beyond, °further, °remote, remoter: *What you propose is ulterior to our immediate plan.*

ultimate *adj.* **1** °final, °last, terminating, °terminal, °end, °eventual, conclusive, concluding, decisive, deciding: *The ultimate outcome will not be known till next week. Who is the ultimate authority in such matters?* **2** °final, °maximum, highest, greatest, °supreme, utmost, °paramount: *Her ultimate goal is to win the gold medal.* **3** °elemental, °basic, °fundamental, underlying, °primary, °essential, °final: *He believes that the ultimate truths were set down in Scripture.* **4** remotest, farthest, °extreme, uttermost, °last, °final: *The ultimate stage of the experiment will be at a distance of five light-years.*

ultimately *adv.* °finally, at long last, in the final *or* last analysis, in the end, at the end of the day, after all is said and done, at (the) last, °in the long run; fundamentally, essentially, basically, °at bottom: *I thought it might ultimately come to this. He came to understand that people must ultimately depend on each other.*

ultimatum *n.* demand(s), term(s), condition(s), stipulation(s), requirement(s): *If we refuse to comply with his ultimatum, he'll kill the hostage.*

ultra- *adj.* °extreme, °immoderate, °excessive, °drastic, °radical, fanatic(al), °unmitigated, °outrageous, °unqualified, °sheer, °blatant, °out-and-out, °complete, °thorough, thoroughgoing, dyed-in-the-wool, diehard, °rabid, °opinionated, unregenerate, °unrepentant, unreformed, fundamentalist, °prejudiced, °bigoted, *Colloq* hard-nosed: *Ultraconservatism marked his views till the day he died.*

umbrage *n.* Usually, **take umbrage.** feel *or* be offended, °take offense, feel displeasure *or* annoyance *or* exasperation *or* indignation *or* vexation *or* bitterness *or* resentment, be piqued *or* displeased *or* annoyed *or* exasperated *or* indignant *or* vexed *or* resentful, harbor a grudge: *She took umbrage at what she perceived was his discourteous treatment of her.*

umbrella *n.* **1** parasol; *Colloq chiefly Brit* gamp, *Brit* brolly, *US* bumbershoot: *You will need an umbrella today.* **2** °protection, °cover, coverage, aegis, °shield, °screen, °patronage, °agency: *What benefits are included under the umbrella of this policy?*

umpire *n.* **1** referee, arbiter, °judge, °moderator, adjudicator, arbitrator; °official; *Colloq* ref, *Australian* umpy, *US* ump: *The sides agree to abide by the decision of the umpire.* —*v.* **2** referee, arbitrate, °judge, °moderate, adjudicate; °officiate: *When Caswell umpires the game, you know you get a fair decision.*

umpteen *adj.* a lot of, °many, innumerable, unnumbered, countless, a huge number of, very many, numerous, hundreds of, thousands of, millions of, billions of, trillions of: *You've been told umpteen times, "Don't go near the water."*

unabashed *adj.* unashamed, unblushing, unembarrassed, °brazen, °blatant, °bold, undaunted, unawed, undismayed, unconcerned: *I couldn't put up with his unabashed conceit for another moment.*

unable *adj.* not able, °powerless, unfit, °unqualified, °impotent: *Was Cosgrove unable to beat your golf score?*

unabridged *adj.* **1** uncut, °whole, full-length, °entire, °complete, °intact, uncondensed, unshortened; unbowdlerized, unexpurgated: *This is the original, unabridged edition, with nothing removed.* **2** °extensive, °thorough, °comprehensive, °exhaustive, all-encompassing, (all-)inclusive: *That word is too rare to be listed in anything but an unabridged dictionary.*

unaccented *adj.* unstressed, unemphasized, unaccentuated, °weak, *Technical* lenis: *The first syllable of "before" is unaccented.*

unacceptable *adj.* °unsatisfactory, objectionable, °wrong, °bad, °improper, unallowable, °undesirable, not *de rigueur,* °distasteful, °disagreeable, unsuitable, °inappropriate, unpleasant, °tasteless: *Murder is generally regarded as an unacceptable way to relieve oneself of unwanted company.*

unaccompanied *adj.* °alone, °solo, on (one's) own, unescorted, unchaperoned, unattended, *Music* a cappella, *Colloq* stag: *We went to the dance unaccompanied.*

unaccountable *adj.* **1** unexplained, °inexplicable, unexplainable, °mysterious, inscrutable, °incomprehensible, unintelligible, °strange, °puzzling, baffling, °peculiar, °odd, °bizarre, unfathomable: *I found it unaccountable that anyone would arrive for an appointment four hours early.* **2** not answerable, not responsible: *How can you have a governmental committee that is unaccountable to anyone?* **3** °weird, °unheard-of, °extraordinary, °unusual, °unorthodox, uncanny: *Sophie has unaccountable powers of perception.*

unaccustomed *adj.* **1** °unfamiliar, °unusual, °rare, unexpected, uncommon, unprecedented, unanticipated, °curious, °peculiar: *We encountered unaccustomed hostility from local people.* **2** ***unaccustomed to.*** °unused to, inexperienced in *or* at, amateurish at, unpracticed in *or* at, °unfamiliar with, uninitiated in: *Unaccustomed as I am to public speaking, I had better say nothing.*

unadorned *adj.* °plain, °simple, unembellished, undecorated, unornamented, °stark, °bare, austere: *She prefers an unadorned straightforward style.*

unaffected[1] *adj.* °genuine, °real, °sincere, °natural, °simple, °plain, unpretentious, unassuming, °ingenuous, °unsophisticated, unstudied, °honest, guileless, °artless, unartificial, straightforward, unfeigned: *Being with someone as unaffected as Ambrose was a relief after that Machiavellian Anthony.*

unaffected[2] *adj.* Usually, **unaffected by.** impervious (to), °immune (to), untouched (by), °unmoved (by), unresponsive (to), aloof (to *or* from), uninfluenced (by), unimpressed (by), remote (to *or* from), cool *or* cold (to), unconcerned (by), unstirred (by): *Minnie remained totally unaffected by the children's appeals for help.*

unapproachable *adj.* **1** °distant, °remote, °aloof, °reserved, stand-offish, austere, °withdrawn, unfriendly, °forbidding, °chilly, °cool, °cold, °frigid: *I know that Ursula seems unapproachable, but she is actually a very warm person.* **2** °inaccessible, °remote, unreachable, °out-of-the-way, out of reach, beyond reach: *Her house is unapproachable except from the sea.*

unarmed *adj.* unprotected, °defenseless, weaponless: *The army was shooting unarmed civilians.*

unasked *adj.* uninvited, unrequested, undemanded, °unsolicited, unsought, unwanted, unprompted, °gratuitous, unbidden, °spontaneous, °unwelcome, unasked-for: *I wish he would keep his unasked opinions to himself.*

unattached *adj.* **1** °separate, unconnected, °detached, °independent, unaffiliated, self-governing, self-regulating, self-regulated, autonomous, self-reliant, self-sustaining, self-sustained: *The committee was established as an unattached body, not associated with any organization.* **2** °single, °unmarried, uncommitted, unengaged, on (one's) own, unspoken for: *I joined the club hoping to meet some unattached people.*

unauthorized *adj.* unsanctioned, unapproved, °unofficial, °unlawful, °illegal, °illicit, °illegitimate: *Unauthorized use of the company's parking spaces is prohibited.*

unavoidable *adj.* inescapable, ineluctable, °inevitable, °irresistible, inexorable, °sure, °certain, °fated, °destined, predestined, °determined, °predetermined, unchangeable, unalterable, settled, °fixed, °definite: *Punishment is an unavoidable consequence of getting caught for something you shouldn't have done.*

unaware *adj.* °ignorant, °oblivious, unknowing, °unsuspecting, °unconscious, °uninformed, unenlightened, incognizant, inobservant, °insensible, °heedless, unmindful, °unsuspecting: *She was totally unaware of the huge spider dangling directly over her.*

unawares *adv.* **1** unexpectedly, abruptly, by surprise, °suddenly, off (one's) guard: *That sentry caught me unawares.* **2** inadvertently, unconsciously, unintentionally, unknowingly, unwittingly, by mistake, mistakenly, by accident, accidentally, in an unguarded moment: *She had shot her husband unawares.*

unbalanced *adj.* **1** uneven, asymmetric(al), unsymmetric(al), °lopsided, unequal, overbalanced, °unstable, wobbly, °shaky, unsteady: *That stack of chairs is unbalanced and could fall any minute.* **2** °mad, demented, certifiable, °crazy, °insane, °eccentric, *non compos mentis,* touched (in the head), °unstable, unhinged, °deranged, °disturbed, of unsound mind, out of (one's) head, *Colloq* daffy, °dizzy, *Chiefly Brit* °daft, *Slang* nuts, batty, off (one's) rocker, *Chiefly Brit* bonkers, *US* out of one's gourd, loco: *The boy is clearly unbalanced, but that doesn't excuse his appalling behavior.*

unbearable *adj.* intolerable, unsupportable, unendurable, °insufferable, °unacceptable, too much: *The pain from the bullet wound was almost unbearable.*

unbeatable *adj.* unsurpassable, undefeatable, °excellent, unexcelled, °incomparable, °matchless, unrivaled, °peerless, °unparalleled, °superlative, °supreme: *She'll go far with such an unbeatable combination of looks and brains.*

unbecoming *adj.* **1** unsuited, unsuitable, °inappropriate, ill-suited, unfitting, unfit, inapt, unapt, °out of character, °out of place: *It is unbecoming to wear your mink coat to the Save the Animals fund-raising dinner.* **2** indecorous, °unseemly, indelicate, °improper, ungentlemanly, unladylike, °offensive, °tasteless: *Father said it was unbecoming for her to use foul language.*

unbelievable *adj.* °incredible, °preposterous, °inconceivable, unimaginable, mind-boggling, °implausible, °unthinkable: *They spend an unbelievable amount of money on clothes. He told an unbelievable tale about having been picked up by a flying saucer.*

unbelieving *adj.* °incredulous, disbelieving, nonbelieving, doubting, mistrusting, distrusting, mistrustful, °distrustful, °suspicious, °skeptical, unpersuaded, unconvinced: *The unbelieving world thinks that the age of miracles is past.*

uncertain *adj.* **1** unsure, indeterminate, unpredictable, undeterminable, unforeseeable, unascertainable, °haphazard, °chance, °arbitrary, °random, aleatory, serendipitous, hit-or-miss, °casual: *It is uncertain when interest rates will go down again.* **2** unsure, in *or* of two minds, vacillating, undecided, unclear, ambivalent, °irresolute, °indecisive, °hesitant, hesitating, undetermined, shilly-shallying, *Brit* °at a loose end, *US* °at loose ends: *Julia is uncertain about what to do next.* **3** unsure, indeterminate, up in the air, °indefinite, unpredictable, °unresolved, °unsettled, in the balance, conjectural, °speculative, °debatable, touch-and-go, °unreliable, °doubtful, dubious, °questionable, °vague, °hazy: *The future of the company is uncertain. The results of the election are still uncertain.* **4** °variable, °changeable, °inconstant, unfixed, °unsettled, °irregular, °fickle, °erratic, °fitful, unsteady, wavering, °unreliable, °sporadic, °occasional; unmethodical, unsystematic: *The pictures in the cave danced in the uncertain light of the torch. How can we make an appointment if your plans are so uncertain?*

uncharted *adj.* unmapped, °unknown, unexplored, undiscovered, °unfamiliar, °strange, virgin, °trackless: *For weeks he wandered lost in the uncharted jungle.*

unchaste *adj.* °impure, °wanton, °immoral, unvirtuous, °promiscuous, °immodest, Cyprian, debased, °lecherous, °lewd, °lascivious: *Her unchaste conduct was a topic of conversation throughout the school.*

uncivilized *adj.* **1** barbarous, °savage, °wild, uncultivated, °barbarian, barbaric, °crude, °primitive, brutish: *He was an uncivilized man living in uncivilized surroundings.* **2** °unrefined, uncultured, uncouth, loutish, °coarse, °uneducated, untutored, unpolished, churlish, °boorish, °philistine, °provincial, °rough, °rude, unlearned, °ill-mannered, incondite, unmannerly, °unsophisticated, inelegant, °gross, gauche: *Anna's father forbade her going out with the uncivilized boors in the town.*

unconscionable *adj.* **1** conscienceless, °unscrupulous, amoral, unprincipled, °immoral, unethical, °evil, °criminal, unjust, °wicked, arrant: *This unconscionable thief stole from his own mother.* **2** °excessive, extortionate, egregious, °extreme, °unwarranted, °unreasonable, °outrageous, °inordinate, °immoderate, °exorbitant, indefensible, unpardonable, °inexcusable, unforgivable: *When I questioned the unconscionable size of his bill, the plumber explained that he charges extra for house calls.*

unconscious *adj.* **1** °insensible, °out (cold), knocked out, °senseless, °numb, stunned, comatose, dead to the world, *Colloq* blacked-out: *The patient was unconscious for hours.* **2** Often, ***unconscious of.*** heedless (of *or* to), unheeding, unheedful (of), insensitive (to), °mindless, unmindful (of), reflex, °automatic, °involuntary, unintentional, °instinctive, °subliminal, °unthinking, °unpremeditated, °subconscious, unwitting; °blind (to), °unaware (of), oblivious (to *or* of), °deaf (to): *For Tim, scratching his ear is an unconscious act. Tim is totally unconscious of how irritating his mannerisms are.*

uncontrolled *adj.* unrestrained, ungoverned, unchecked, untrammeled, °undisciplined, °wild, °unruly, °boisterous, °riotous, out of hand *or* of control, °rampant, frenzied, °frantic; going berserk, running amok *or* amuck: *She confessed to an uncontrolled appetite for chocolates. You cannot leave him uncontrolled.*

under *prep.* **1** °beneath, °below, underneath, covered by: *What is under that blanket? The wreck lies a mile under the surface.* **2** °subordinate to, answerable to, inferior to, second to, secondary to, subservient to, °below, °beneath, underneath, junior to, directed *or* supervised *or* controlled by, under (the) control of, at the mercy of, at the beck and call of: *The bashibazouks were under the sultan's top officers.* **3** included *or* comprised in *or* under, subsumed under: *Under which category is "software"?* **4** under the aegis *or* protection *or* eye *or* guardianship *or* care of: *The prince was trained under the grand vizier.* **5** less than, lower than: *I paid under the going price for my VCR.* **6** ***under the influence.*** °drunk, tipsy, °high, impaired: *The police charged him with driving while under the influence.* *—adv.* **7** °below, underneath, °beneath: *You looked on top of the bed, but have you looked under?* **8** underwater, beneath the waves, down, °out of sight: *We watched helpless as he went under for the third time.*

underclothes *n.* underclothing, underwear, undergarments, lingerie, *Old-fashioned* °unmentionables, *Colloq* underthings, undies, *Brit* smalls, *Old-fashioned Brit* small clothes, *US* skivvies: *When a salesman she met on the train said he was in ladies' underclothes, she moved to another carriage.*

undercover *adj.* °secret, °private, clandestine, °confidential: *He doesn't look much like an undercover agent.*

undercurrent *n.* **1** undertow, crosscurrent, riptide, °rip (current), underflow: *Caught in the undercurrent, he was carried out to sea.* **2** undertone, subcurrent, °trend, °tendency, °overtone, °tenor, °suggestion, °murmur, °implication, connotation, °sense, °feeling, °aura, tinge, °flavor, °atmosphere, ambiance *or* ambience; vibrations, *Colloq* °vibes: *There is a sinister undercurrent of gloom about this place.*

undercut *v.* **1** °undermine, °excavate, hollow out, cut out *or* away, gouge out: *If you undercut the roadway it will collapse.* **2** underprice, undercharge, °sacrifice, sell cheaply *or* at a loss, undersell: *The supermarkets undercut prices and put the independent grocers out of business.*

underdog *n.* °loser, °scapegoat, °victim; vanquished, °defenseless; *Colloq* fall guy, little fellow *or* guy: *It has always been our policy to fight for the underdog.*

underestimate *v.* undervalue, underrate, °discount, misjudge, °miscalculate, misprize, °minimize, °depreciate, °belittle, not do justice to, fail to appreciate, set (too) little store by, think (too) little of: *You might not win, as I think you have underestimated his strength.*

undergo *v.* °suffer, °bear, °endure, °experience, live *or* go through, be subjected to, subject oneself to, °sustain, °submit to, °weather, °stand, °withstand: *It is unbelievable what they have undergone for their children.*

underground *adj.* **1** subterranean, buried, belowground, °sunken, covered: *We crept through the underground passage into the treasure room.* **2** °secret, clandestine, concealed, °hidden, covert, °undercover, °surreptitious, °stealthy, °private: *The secret service has a worldwide underground network of agents.* **3** °alternative, °radical, °experimental, °avant-garde, °nonconformist, °revolutionary: *Some underground newspapers that sprang up in the 1960s are still being published.* *—n.* **4** tube, metro, underground railway, *US* °subway: *Two more stops on the underground and we'll be there.* **5** °resistance, partisans *or* partizans, freedom fighters, (in France) Maquis, insurgents, seditionaries *or* seditionists, insurrectionists, guerrillas *or* guerillas, extremists, revolutionaries; fifth columnists, fifth column, saboteurs, subversives: *The underground helped the family to escape to England through Portugal. The government blames the underground for the bombings.*

undermine *v.* **1** °sap, °drain, disable, °weaken, debilitate, °threaten, °sabotage, °subvert, °damage, °hurt, °harm, °impair, °ruin, °dash, °wreck, °spoil, *Slang* °queer, °bugger (up): *He said that the entire campaign would be undermined if the events of January 17 were ever revealed.* **2** See **undercut, 1,** above.

undersized *adj.* undersize, °little, °short, °small, °petite, °tiny, °elfin, bantam, °slight, mignon; °stunted, underdeveloped, runty, runtish, dwarfish, dwarfed, pygmy, squat; underweight, °undeveloped: *These cattle are a bit undersized for market.*

understand *v.* **1** °grasp, °comprehend, °see, °perceive, discern, °make out, get the drift *or* the hang of, °appreciate, °interpret, take cognizance of, °recognize, be aware *or* conscious of, be conversant with, °know, °realize, conceive of, be aware (of), apprehend, °penetrate, *Colloq* °get (it), °dig, °catch on (to), °tumble to, cotton on (to), *Brit* °twig: *I can understand what you are saying. I didn't think you understood.* **2** °accept, °agree, °arrange, covenant, °take: *I understand the terms of the agreement.* **3** °interpret, °take, °read, gather from, construe, surmise from, assume from, infer from, °view, °see: *As I understand the regulation, only residents may use the facilities.* **4** °hear (of), °gather, get wind (of), °take it, be told *or* informed *or* advised, have found out *or* learned, *Colloq* hear tell: *I understand that you have accepted assignments in Azusa and Cucamonga.* **5** sympathize *or* empathize (with), be in sympathy (with), show compassion (for), commiserate (with); °accept, °tolerate, °allow, °forgive: *Surely your employer will understand your being late because of the funeral.*

understanding *n.* **1** °agreement, °contract, °arrangement, °bargain, covenant, concession, °pact, °compact,

°accord, °treaty, concordat, entente, °alliance, °truce, armistice, °reconciliation, °settlement: *They reached an understanding covering nuclear weapons.* 2 discernment, °sensitivity, sensitiveness, °sympathy, empathy, °perception, °insight, good sense, °intuition, enlightenment, percipience, sagacity, sageness, sapience, °wisdom, *Colloq* savvy: *Georgianna brings so much understanding to her treatment of the aged.* 3 Usually, ***understanding of.*** comprehension *or* awareness *or* grasp *or* control *or* idea *or* conception *or* knowledge *or* mastery (of), acquaintance *or* familiarity *or* intimacy *or* dexterity *or* skillfulness *or* deftness *or* adroitness *or* adeptness (with), competence *or* skill *or* expertness *or* know-how *or* proficiency *or* expertise (in), *Colloq US* °fix (on), °handle (on): *I know no one with a better understanding of medieval Welsh.* 4 reading, °interpretation, °opinion, °judgment, °estimation, °notion, °view, °perception, apperception, apprehension: *My understanding of what took place is at odds with yours.* 5 °intellect, °intelligence, °mind, °brain, brainpower, °sense, °reason, reasoning power, °wisdom, *Colloq* brains: *I'm afraid that he lacks the understanding needed to cope with the subtleties of the situation.*

understated *adj.* °subtle, restrained, low-key, °simple, °basic, unembellished, °unadorned: *Alexandra was wearing an understated black dress with a string of pearls.*

understood *adj.* accepted, agreed, arranged, °given, covenanted, settled, conceded: *I thought it was understood that we would be dining together tonight. The terms of the agreement are fully understood.*

understudy *n.* 1 °second, °substitute, °stand-in, °alternate, backup, °double, sub, °reserve, *US* pinch hitter: *Understudy to the great Undine Seagrave, Myra got her big break one night when the star fell ill.* —*v.* 2 °substitute for, stand in for, °back up, double for, °second, °replace, *US* pinch-hit for: *A professional at understudying actors who always appeared for performances, Gabriel spent his entire professional life offstage.*

undertake *v.* 1 °assume, take on *or* upon (oneself), °accept, take *or* assume *or* bear the responsibility for, enter upon, °begin, °start, °set about, °embark on, °tackle, °try, °attempt: *Gates might be undertaking a bit too much, considering his condition.* 2 °promise, covenant, °agree, °contract, °pledge, °vow, °swear, °warrant, °guarantee, °bargain, °commit (oneself), °stipulate, °engage: *Is it not true that you undertook to complete the work in six months?*

undertaker *n.* mortician, funeral director: *The body is available for viewing at the undertaker's tonight.*

undertaking *n.* 1 °enterprise, °affair, °business, °project, °task, °effort, °venture, °work, °feat: *The scope of the undertaking was far beyond her capacity.* 2 doing, performing, °performance, °realization, °achievement: *Giles's undertaking of the work makes me feel more confident.* 3 °promise, °pledge, commitment, °assurance, °contract, °agreement, °vow, guarantee *or* guaranty, °warranty: *You gave an undertaking to complete the work by noon.*

underworld *n.* 1 Usually, ***the underworld.*** organized crime, the syndicate, the Mafia, the mob, Cosa Nostra, criminals, the criminal element, *Colloq* gangland: *Lorenzo was fast becoming a well-known figure in the underworld.* 2 nether regions, abode of the dead, Hades, Hell, Avernus, Dis, Orcus, *Facetious* Egyptian underground: *I think I prefer contemplating the Happy Hunting Ground of the American Indian to the underworlds of Christianity and classical Mediterranean culture.*

underwrite *v.* 1 °back (up), °finance, °support, invest in, °subsidize, subvene, °sponsor, °uphold, °approve, insure, °guarantee, *US* subvene: *The company has agreed to underwrite the development of your invention.* 2 °subscribe to, endorse *or* indorse, °sign, countersign, °consent to, °agree to, °confirm, accede to, °sanction, °ratify, °approve, validate, *Colloq* OK *or Colloq* O.K. *or Colloq* okay: *The government underwrote the action one day, then denied knowledge of it the next.*

undesirable *n.* 1 *persona non grata,* pariah, °outcast, °exile, °reject, leper: *The police had many requests to run the undesirables out of town.* —*adj.* 2 unwanted, objectionable, °offensive, °unacceptable, °obnoxious, °unsavory, °unwelcome, disliked, °distasteful, °repugnant, unfit, °unbecoming, unsuitable: *The parks are filled with tramps, drug addicts, and other undesirable elements.*

undeveloped *adj.* embryonic, °premature, °immature, incipient, inchoate, °potential, latent: *His sense of beauty remained undeveloped till he visited Rome.*

undiluted *adj.* °pure, °neat, °straight, unmixed, uncut, unblended, unadulterated, unwatered, unalloyed: *No ice, please—I like my whiskey undiluted. How do you like the undiluted gall of the fellow walking in here uninvited!*

undisciplined *adj.* untrained, unschooled, °unprepared, untutored, °uneducated, untaught, unpracticed, °uncontrolled, °disobedient, °naughty, °bad, °willful, wayward, unrestrained, °erratic, unpredictable, °unruly, °wild: *The sergeant has taken quite a few undisciplined recruits and knocked them into shape.*

undisguised *adj.* °open, °out-and-out, unmistakable *or* unmistakeable, °overt, unconcealed, unreserved, unrestrained, unfeigned, unpretended, obvious, °evident, °patent, °clear, °explicit, °transparent, °sincere, °heartfelt, unalloyed, °unmitigated: *They regarded the traitor with undisguised contempt.*

undisputed *adj.* unquestioned, °unquestionable, beyond question, accepted, acknowledged, admitted, °indisputable, indubitable, undoubted, °certain, °sure, unmistakable *or* unmistakeable, °definite, °explicit, °clear, (self-)evident, °obvious, uncontested, unchallenged, incontestable, irrefutable, °incontrovertible, undeniable, conclusive: *He proved once again that he is undisputed heavyweight champion of the world.*

undistinguished *adj.* °ordinary, commonplace, °common, °everyday, °run-of-the-mill, °pedestrian, unexceptional, °plain, °homespun, °simple, °prosaic, unremarkable; °mediocre, middling, °indifferent, unexciting, unimpressive, unpretentious, *Brit* °homely, *Colloq* °so-so, no great shakes, no big deal, nothing to write home about, nothing special *or* unusual *or* extraordinary: *Digby had enjoyed an undistinguished career as a civil servant before writing his novel.*

undivided *adj.* 1 °whole, °entire, unbroken, uncut, °intact, unseparated, °complete, unsplit: *The family lands remained undivided for generations.* 2 undiverted, °whole, °entire, °devoted, concentrated, °full, °complete, °exclusive, undistracted: *You have my undivided attention.*

undo *v.* 1 °loosen, °loose, °open, unfasten, unhook, unlace, unzip, unsnap, unbutton, untie, unpin; unlock, unbolt: *Help me undo my blouse. Don't undo the gate—you'll let the dogs out.* 2 unwrap, uncover, °open, untie, unbind: *Why have you allowed the children to undo all their Christmas gifts a week early?* 3 °cancel, annul, rescind, nullify, °void, declare null and void, °reverse, invalidate: *If one could undo things there would be no such thing as regret.*

undoing *n.* 1 °ruin, ruination, °destruction, devastation, °defeat, °downfall, °overthrow, °fall, °collapse, descent, debasement, °degradation, abasement, mortification, °humiliation, °shame, °disgrace: *It is not easy to* *identify the factors that brought about the demagogue's undoing.* **2** °curse, °misfortune, bane, °affliction, °trouble, °blight: *As the novelists would have it, a beautiful woman was his undoing.*

undone[1] *adj.* **1** ruined, °lost, wrecked, crushed, destroyed, devastated, shattered, brought to ruin, defeated, prostrated, °overcome: *Binky was completely undone by that gold-digging heartbreaker he met in Caracas. If the headmaster finds out about the glue, we are undone!* **2** °open, °loose, loosened, untied, unfastened, °detached, unhooked, unlaced, unzipped, unsnapped, unbuttoned, unpinned, unstuck: *Your shoelaces are undone. The knot came undone.*

undone[2] *adj.* unaccomplished, uncompleted, °incomplete, unfinished, omitted, neglected, °left (out), skipped, missed, passed over, forgotten, unattended to: *From this list of chores, how many are still undone?*

undoubtedly *adv.* indubitably, without (a) doubt, indisputably, unquestionably, beyond *or* without (a *or* the shadow of) (a) doubt, certainly, °definitely, °surely, assuredly, unmistakably *or* unmistakeably, explicitly, °clearly, °obviously, incontestably, irrefutably, incontrovertibly, undeniably: *If they win, they will undoubtedly go to the nightclub to celebrate.*

unduly *adv.* **1** disproportionately, excessively, °overly, unnecessarily, inordinately, unreasonably, irrationally, unjustifiably, improperly, inappropriately: *He is unduly sensitive to everything that is written about him.* **2** immoderately, lavishly, profusely, extravagantly: *Don't be unduly generous in tipping our waiter—he did spill the soup in your lap.*

unearth *v.* °dig up, disinter, exhume; °excavate, dredge up, °mine, °quarry, °find, pull *or* root out, °come across, °discover, °turn up, °expose, uncover: *The artifacts were unearthed at Boghazköy, ancient Hattusas. We unearthed some evidence that will put you away for years.*

unearthly *adj.* **1** °supernatural, °unnatural, preternatural, unworldly, otherworldly, psychic(al), extramundane, extraterrestrial, extrasensory, supersensory, out-of-(the)-body, asomatous, incorporeal, °sublime, °celestial, astral: *They had an unearthly experience while in the flying saucer.* **2** °weird, °bizarre, °macabre, °nightmarish, uncanny, °eerie, °strange, °ghostly, °spectral, °unreal, *Literary* eldritch, *Colloq* spooky, creepy: *An unearthly creature appeared in a mist before her.* **3** °strange, °odd, °peculiar, °unusual, °abnormal, °absurd, out-of-the-ordinary, °extraordinary, °outrageous; °unheard-of, °unreasonable, *Colloq* °ungodly: *The baby starts crying at the unearthly hour of three o'clock in the morning.*

uneducated *adj.* unschooled, untaught, uncultivated, unread, uncultured, °illiterate, unlettered, °ignorant, unenlightened: *It is debatable whether the fault for being uneducated lies with the student, the teacher, or the system.*

unemployed *adj.* °out of work, jobless, °idle, laid off, out of a job, °out of work, unoccupied, °inactive, *Facetious* resting, °at liberty, °at leisure, *Brit* °redundant, *Colloq Brit* on the dole: *He has been unemployed since the factory closed.*

unenthusiastic *adj.* °lukewarm, °cool, °cold, uninterested, °indifferent, °blasé, unresponsive, apathetic, unexcited, unimpressed: *Bentley presented the new design, but the directors were unenthusiastic and turned it down.*

unenviable *adj.* uncoveted, °undesirable, unwished-for, unattractive: *He was given the unenviable task of informing the victim's family.*

unfamiliar *adj.* **1** °new, °novel, °unknown, unconventional, °unusual, °different, uncommon, °strange, °odd, °peculiar, °bizarre: *The man spoke in a tongue that was totally unfamiliar to me.* **2** Usually, ***unfamiliar with.*** unacquainted with, °unaccustomed to, inexperienced in *or* with, °unused to, unconversant with, uninformed about, ignorant of, unpracticed *or* unskilled in, unskilled at, uninitiated in, unversed in: *The Americans were as unfamiliar with cricket as the Englishmen were with baseball.*

unflattering *adj.* **1** °harsh, °unsympathetic; °realistic, °stark, °candid: *An upswept hairdo is an unflattering style for your face.* **2** uncomplimentary, insulting, unfavorable: *Saying that I look just like my grandmother is quite unflattering.*

unfledged *adj.* °undeveloped, °immature, unmatured, °inexperienced, °green, °callow, °young, °raw, ungrown: *Take these unfledged recruits and turn them into men.*

unfold *v.* **1** open (out *or* up), °spread (out), unfurl, stretch out, °expand, uncoil, unwind, °straighten out: *Emerging from the pupa, it unfolds its wings and becomes a beautiful butterfly.* **2** °develop, evolve, °happen, °take place, °occur, be divulged, be disclosed *or* revealed: *What will unfold next in our little drama?*

unforeseen *adj.* unexpected, surprising, unanticipated, unpredicted, unlooked-for, unsought, unhoped-for, undreamt-of *or* undreamed-of, unthought-of, °startling, °surprise, °chance, fortuitous: *The sales manager's resignation was an unforeseen event.*

unfortunate *adj.* **1** unlucky, luckless; cursed, out of luck, unblessed, *Colloq* down on (one's) luck: *Claurène was unfortunate enough to have married early and divorced late.* **2** °poor, °miserable, °wretched, °woebegone, °pathetic, °dismal, °unhappy, °forlorn, °pitiable, °doomed, ill-starred, star-crossed, ill-fated: *These unfortunate refugees have nowhere to go if we refuse them asylum.* **3** °deplorable, °terrible, °awful, catastrophic, °disastrous, °calamitous, °tragic, °lamentable, °regrettable, distressing, upsetting, °disturbing, °inauspicious, °grievous, °ruinous: *Taking the children with us on our vacation was a most unfortunate decision. The earthquake was a most unfortunate sequel to the flood.*

unfounded *adj.* baseless, °groundless, °unwarranted, unjustified, unsupported, unsupportable, °unsound, unjustifiable, unattested, unproven: *Withdraw your unfounded accusations or you may expect my seconds to call in the morning. Reports of a typhoid epidemic are completely unfounded.*

ungodly *adj.* **1** °wicked, °sinful, °impious, °blasphemous, °heretical, irreligious, iconoclastic, atheist(ic), antireligious, °sacrilegious, demonic, demoniac(al), diabolic(al), °satanic, °fiendish, hellish, °infernal; depraved, °godless, °corrupt, °immoral, °evil, iniquitous, °bad, °villainous, heinous, flagitious, °profane, °vile: *The crusaders believed they had a mission to deliver Jerusalem from the ungodly heathens.* **2** °awful, °outrageous, °indecent, °monstrous, °nasty, °dreadful, °terrible, appalling, °frightful, °shocking, *Colloq* god-awful, °unearthly, *Brit* °beastly: *Margo and I were subjected to another ungodly evening at the Browns'.*

ungovernable *adj.* °unruly, refractory, intractable, unmanageable, uncontrollable, °rebellious, °wild, °disobedient, unrestrainable, °incorrigible: *The nursemaid was saddled with two ungovernable children whom she had to cosset.*

ungraceful *adj.* **1** °awkward, °clumsy, ungainly, lubberly, *Colloq* °all thumbs, butterfingered, *Slang US and Canadian* klutzy: *People in the aerobics class are always ungraceful to start off with.* **2** inelegant, graceless, °coarse, °crude, inartistic, °vulgar, °tasteless, unesthetic, °unrefined, barbarous, unlovely, °ugly, unharmonious, unattractive, ill-proportioned, unsymmetric(al), asymmetric(al): *Some regard the Albert Memorial as a paragon of ungraceful Victorian design.*

ungracious *adj.* °discourteous, °overbearing, churlish, gauche, °rude, uncivil, °impolite, ill-bred, bad-mannered, unmannerly, ungentlemanly, unladylike, °gruff, °bluff, °brusque, °abrupt, °surly, curmudgeonly: *It was ungracious of him not to acknowledge the help of his staff.*

ungrateful *adj.* unthankful, unappreciative, °rude; °selfish, °heedless: *That ungrateful wretch never thanked me for all my help.*

unguarded *adj.* **1** incautious, °unwary, °careless, °inattentive, °heedless, inobservant, °inadvertent, °unthinking, unwatchful, unvigilant: *He let the intruder past him in an unguarded moment.* **2** °defenseless, unprotected, undefended, unfortified, °open, uncovered, exposed, °vulnerable: *The sentry insists that he left his post unguarded only for a few minutes.* **3** °indiscreet, °careless, °imprudent, unwise, °hasty, °unthinking, °thoughtless; guileless, incautious: *He apologized for unguarded remarks he had made about his associates.*

unhappy *adj.* **1** °sad, depressed, °blue, °dejected, °melancholy, °despondent, downcast, °gloomy, °downhearted, dispirited, heavy-hearted, long-faced, disconsolate, °sorrowful, °miserable, crestfallen, cheerless, °forlorn, low-spirited, °glum, distressed, °tearful, *Formal* lachrymose, *Colloq* down, *Slang US* bummed out: *Mary is very unhappy that John is to be away a whole month.* **2** unlucky, °unfortunate, unpropitious, °inauspicious, unfavorable, luckless, hapless, cursed, °wretched, ill-omened, ill-fated, ill-starred, star-crossed, jinxed, °disastrous: *Let me tell you how this unhappy state of affairs came about.* **3** infelicitous, unfitting, °inappropriate, unsuitable, unsuited, °wrong, inexpedient, °ill-advised, °poor, °unfortunate: *Vivian was an unhappy choice as his successor.*

unhealthy *adj.* **1** ailing, unwell, °ill, °sickly, °infirm, °feeble, °frail, debilitated, °unsound, °sick, in poor *or* delicate health *or* condition, °indisposed, °invalid, valetudinary: *Gormley has been unhealthy ever since I have known him.* **2** °unwholesome, °harmful, noxious, °detrimental, insalubrious, damaging, °injurious, °destructive, malign: *The doctor said that this climate is unhealthy for people with respiratory disorders.* **3** °risky, °dangerous, °perilous, life-threatening, touch-and-go: *You might find it unhealthy to criticize the government here.*

unheard-of *adj.* **1** °unknown, °unfamiliar, °obscure, °unidentified, °nameless, °unsung: *She is busy trying to promote some unheard-of sculptor.* **2** unimaginable, undreamed-of *or* undreamt-of, unprecedented, unimagined, °unbelievable, °inconceivable, °unusual: *They took the unheard-of step of requiring that refunds be paid. Her starting salary is an unheard-of $250,000.* **3** °shocking, °offensive, °outrageous, °disgraceful, °extreme, °unthinkable, °outlandish: *The victims were subjected to unheard-of tortures.*

unheralded *adj.* unannounced, unpublicized, unadvertised; unexpected, °surprise, unanticipated, °unforeseen, unpredicted: *Published by some obscure company, the book turned out to be an unheralded success.*

unhesitating *adj.* **1** °swift, °rapid, °quick, °immediate, instantaneous, °prompt, °ready, unhesitant: *The response was an unhesitating "No!"* **2** unfaltering, unwavering, °wholehearted, °unqualified, unswerving, undeviating, °staunch, °steadfast, °implicit, °resolute: *We appreciate your unhesitating devotion to duty.*

unhurried *adj.* leisurely, unrushed, °easy, easygoing, °casual, °gradual, °deliberate, °steady, °sedate, °calm: *We proceeded at an unhurried pace, eventually arriving at dusk at the castle gates.*

unidentified *adj.* °nameless, anonymous, °unknown, unmarked, unnamed, °unknown, °unfamiliar, unrecognized, °mysterious: *The police found the body of an unidentified woman in the park.*

uniform *adj.* **1** °homogeneous, °consistent, unvaried, unchanged, unaltered; unvarying, unchanging; °invariable, unchangeable, unalterable, regimented, °standard; ordered, °orderly, °equal, °even, °like, °identical; °alike: *The mixture is uniform throughout. The rules are uniform for everyone. Line up in ten uniform rows of ten each.* **2** °even, unbroken, °smooth, °regular, °flat: *Not one irregularity marred the uniform surface of the plain.* —*n.* **3** livery, °habit, °regalia, °costume, °outfit; regimentals: *When I next saw him, he was in the uniform of a Gestapo officer.*

uniformity *n.* **1** °regularity, similarity, sameness, homogeneity, consistency, °symmetry, evenness, invariability, unchangeability, similitude, conformity, °agreement, concord, °accord, harmoniousness; °harmony, concordance, accordance, conformance, correspondence: *One can easily see the lack of uniformity even in what is supposed to be a classless society.* **2** dullness, monotony, drabness, sameness, °tedium, featurelessness, flatness, invariability, lack of variety, changelessness: *After a few days, the uniformity of the routine began to get on my nerves.*

unify *v.* consolidate, °unite, °combine, °amalgamate, coalesce, bring together, °fuse, °join, °weld, °merge, confederate, °integrate: *The separate elements were unified into a coherent whole. Successive Reform Bills unified the nation.*

unimpeded *adj.* unblocked, unchecked, °free, unconstrained, unrestrained, unhindered, unhampered, unencumbered, °open, untrammeled, unrestricted: *They insist on having unimpeded access to the files.*

unimposing *adj.* unimpressive, nugatory, trivial, °trifling, °minor, unimportant, °puny, inconsiderable, °negligible: *The offices were in an unimposing little building squeezed between two skyscrapers. A more unimposing little cockalorum has seldom assailed my sensibilities.*

uninformed *adj.* °ignorant, nescient, unknowledgeable, unenlightened, °uneducated, unschooled, untutored, untaught, uninstructed, °unaware, incognizant: *The problem is not that they are stupid, merely uninformed and misinformed.*

uninhabited *adj.* °desolate, °empty, °abandoned, °deserted, unoccupied, °vacant, vacated, tenantless, untenanted; °desert, unpopulated, unpeopled, °trackless, depopulated, °waste, °barren: *Tristan da Cunha, in the south Atlantic, is an uninhabited island.*

uninhibited *adj.* °wild, unchecked, unbridled, uncurbed, intemperate, °boisterous, unrepressed, unconstrained, unrestrained, °uncontrolled, unself-conscious, unreserved, °relaxed, °casual, easygoing, °free (and easy), °open, °frank, °candid, °outspoken, *Colloq* °up front: *Their behavior at the party was totally uninhibited. I will tell you what I think only if I can be totally uninhibited about it.*

uninviting *adj.* °repulsive, °repellent, °offensive, unappealing, unattractive, unpleasant, °disagreeable, °distasteful, unappetizing, °unsavory, sickening, °revolting, °obnoxious, °nasty, °disgusting, *Brit* off-putting: *The food on the table was cold and uninviting.*

union *n.* **1** uniting, °unity, combining, °combination, joining, °junction, conjoining, conjunction, allying, °alliance, associating, °association, coalition, amalgamating, °amalgamation, fusing, fusion, marrying, °marriage, confederating, confederation, confederacy, synthesizing, °synthesis, mixing, °mixture, °federation, togetherness: *Monarchies favor the union of Church and State.* **2** °alliance, °association, °organization, °society, °circle, °fraternity, °club, °fellowship, °team, °ring, °gang, °syndicate, coalition, °party, confederation, confederacy, °federation, Bund, °league, consortium, bloc, cartel, °trust: *The countries formed a union to control their joint economies.* **3** °joint, °seam, °splice, °junction, conjunction, °graft, °weld; coupling: *The union between the two pieces is often stronger than the materials themselves.* **4** °agreement, °accord, °harmony, harmoniousness, congruity, coherence, compatibility, unanimity, °unity: *The union of the various elements in the painting was skillfully effected.*

unique *adj.* **1** °single, lone, °(one and) only, °solitary, one of a kind, sui generis: *Natural forces conspired to create this unique situation.* **2** unequaled, °unparalleled, unrivaled, °incomparable, inimitable, °peerless, unmatched, unsurpassed, unexcelled, second to none: *Stimble's genius is unique in the annals of proctology.*

unison *n.* ***in unison (with).*** in harmony, together, corresponding exactly, in (perfect) accord, consonant, °harmonious: *We all responded in unison with a rousing "No!" Why is Crashawe's singing never in unison with ours?*

unit *n.* °element, component, °entity, °part, °item, constituent, °piece, °portion, °segment, °section, module: *The units fit together like Lego blocks.*

unite *v.* **1** °combine, °unify, °merge, coalesce, °amalgamate, °mix, °mingle, commingle, intermix, °blend, consolidate, °fuse: *The splinter parties have united to form a coalition.* **2** °join (forces), °unify, °wed, °marry, °link, °connect, °merge: *The two families are united by marriage.* **3** °bond, fuse *or* weld *or* solder *or* glue *or* stick *or* tie *or* bind *or* fasten *or* fix *or* fit (together): *Unite these pieces and the job is done.*

united *adj.* **1** unified, °common, °mutual, combined, merged, coalesced, pooled, shared, collective, °joint, amalgamated, connected: *The new law will serve our united interests.* **2** °joint, cooperative, °common, communal, collaborative, synergetic *or* synergistic, collective, concerted, coordinated, allied; partnership: *We must make a united effort if we are to survive.* **3** agreed, unanimous, in agreement, of one mind, of like mind *or* opinion, like-minded, in accord, in harmony, °harmonious: *We are united in our assessment of last night's performance.*

unity *n.* **1** consistency, unanimity, constancy, °uniformity, sameness, consensus, °agreement, concord, concordance, °accord, °solidarity, compatibility, concurrence, continuity, consentaneousness, °rapport, °sympathy, like-mindedness: *All members of the association enjoy a unity of purpose.* **2** oneness, °singularity, °integrity, singleness, congruity, °uniformity, congruousness, homogeneity, °identity, sameness, °resemblance, °likeness, similarity, similitude: *If there is little unity between your ideas and those of the organization, perhaps you should resign.* **3** unification, uniting, °combination: *In unity is strength.*

universal *adj.* **1** °prevalent, °prevailing, °general, worldwide, widespread, ubiquitous, omnipresent, °limitless, °unlimited, °common, pandemic, °epidemic: *The end of the twentieth century is marked by a universal preoccupation with greed.* **2** cosmic, infinite, °boundless, °limitless, °unlimited, measureless, °endless, uncircumscribed, (all-)inclusive, all-embracing, all-encompassing, wide-ranging, °comprehensive: *Renaissance man was regarded as possessing universal knowledge.*

universally *adv.* in every case *or* instance, in all cases *or* instances, unexceptionally, without exception, uniformly, °always, invariably: *In those days, it was universally accepted that the earth was flat.*

universe *n.* **1** cosmos, °creation, macrocosm: *What arrogance makes man believe that in all the vast universe God pays him any heed?* **2** °world, bailiwick, °sphere, °province, °domain, °circle, °milieu, °territory, corner, °quarter, microcosm: *Such behavior may be acceptable in her tiny universe.*

unkempt *adj.* disheveled, uncombed, tousled, disarranged, ungroomed, windblown, °untidy, disordered, mussed (up), messy, messed up, °bedraggled, °shaggy, scruffy, rumpled, slovenly, frowzy, blowzy, *Archaic* draggletailed, *Colloq* °sloppy: *Mother would refuse to allow us to the table if we were unkempt.*

unkind *adj.* °inconsiderate, unthoughtful, °thoughtless, unfeeling, unconcerned, insensitive, unkindly, °unsympathetic, uncharitable, unchristian, uncaring, hardhearted, °heartless, flinty, hard, °rigid, °callous, °tough, °inflexible, unyielding, unbending, °severe, °harsh, °stern, °cruel, °mean, °inhuman: *It was very unkind of you to refuse the beggar.*

unknown *adj.* **1** unrecognized, °unfamiliar, °strange, unnamed, anonymous, °nameless, °unidentified; °obscure, °unheard-of, little-known, °humble, °undistinguished, °unsung: *Some unknown person left this package for you. Bill is in some unknown place for the weekend. An unknown poet won the competition.* **2** °unfamiliar, unexplored, uninvestigated, unresearched, unrevealed, °mysterious, °uncharted, °unidentified, °dark: *We were now going into unknown territory.* **3** unbeknownst, °untold, unrevealed: *Unknown to me, the thieves were at that moment breaking in at a rear window.*

unlamented *adj.* unmissed, unmourned, unbemoaned, unbewailed, unloved: *They destroyed a statue of the late, unlamented Stalin.*

unlawful *adj.* °illegal, °illicit, against the law, °illegitimate, under-the-table, under-the-counter, °criminal, felonious; outlawed, banned, prohibited, forbidden, interdicted, disallowed, proscribed, *verboten;* °unauthorized, unlicensed, unsanctioned: *Unlawful payments had been made to councillors. Gambling has been declared unlawful in that county. Unlawful sales of the recording are hard to police.*

unlike *adj.* **1** °different (from), °dissimilar (to), unalike, °distinct (from), opposite (from *or* to), contrasting *or* contrastive (with *or* to), °separate (from), °divergent (from), °incompatible (with), distinguishable (from), far apart (from), °far (from), °distant (from), ill-matched (with), unequal (to), unequivalent (to): *His notions of right and wrong are unlike those of other people. Your car and mine are as unlike as can be. Why should we receive unlike amounts for the same work?* **2** atypical, uncharacteristic, untypical: *It is unlike you to complain.* —*prep.* **3** different from, differing from, in contradistinction to, in contrast with *or* to, dissimilar to, distinct from, opposite from *or* to, contrasting with *or* to, divergent from, incompatible with, distinguishable from, ill-matched with, unequal to, unequivalent to: *Unlike you, I have been in the army and know what it's like.*

unlikely *adj.* **1** °improbable, °doubtful, dubious, °remote, °unthinkable, unimaginable, °inconceivable, °implausible: *Glenn was an unlikely candidate, but he won. It is unlikely that he will come even if invited. That's a pretty unlikely story.* **2** °unseemly, °inappropriate, unfit, unfitting, unsuitable, uncongenial, objectionable, °unbecoming, °unacceptable, unattractive, °distasteful: *Claire had seldom seen a more unlikely* *prospect for a husband.* **3** unpropitious, °unpromising, °inauspicious: *The tiny shop seemed an unlikely place to find a great treasure.*

unlimited *adj.* **1** unrestricted, unrestrained, °limitless, unconstrained, °unqualified, °full, °absolute, unconditional, far-reaching, unchecked, °uncontrolled: *He has unlimited authority to issue all the paper money he wishes.* **2** °limitless, °boundless, °endless, °vast, unbounded, °immense, °immeasurable, measureless, °numberless, innumerable, °inexhaustible, interminable, never-ending, °infinite, myriad, °extensive: *As the wealthiest woman in the world, she has unlimited resources to do as she pleases.*

unload *v.* °empty, °dump, unpack, off-load, °discharge; disburden, unburden: *We unloaded the cargo at Sulawesi. The camels must be unloaded every night.*

unmarried *adj.* °single, unwed(ded), bachelor, spinster, old-maid, °maiden, °unattached, unengaged, unbetrothed, unplighted, unpromised, °free, uncommitted: *She was thirty-five, unmarried, and lived in a small apartment in Chicago.*

unmentionable *adj.* **1** unspeakable, unutterable, °ineffable, °taboo, °scandalous, forbidden, interdicted; °inexpressible: *I will not have Karen or her unmentionable boyfriend in this house!* **2** °disgraceful, °indecent, °immodest, °shameful, °shocking, appalling, °dishonorable, indescribable, °obscene, °filthy: *In carnivals, geeks do unmentionable things with live chickens.* —*n.* **3** ***unmentionables.*** °underclothes, underclothing, underwear, undergarments, lingerie, *Archaic* small clothes, *Colloq* underthings, undies, *Brit* smalls, *US* skivvies: *Polly would hang out her unmentionables for all the neighborhood to see.*

unmerciful *adj.* °merciless, pitiless, unsparing, °unkind, °relentless, unpitying, °heartless, stonyhearted, hardhearted, flinty, unfeeling, °unsympathetic, unforgiving, °mean, °cruel, °savage, °brutal, brutish, °vicious, barbarous: *The boys gave him an unmerciful beating for cheating.*

unmitigated *adj.* °undiluted, unalloyed, unmixed, untempered, unmoderated, unmodified, unabated, unlessened, undiminished, unreduced, unrelieved, °oppressive, unalleviated, unmollified, unsoftened, °relentless, °unqualified, °out-and-out, °thorough, thoroughgoing, °outright, °downright, °categorical, °absolute, °immoderate, °sheer, °complete, consummate, °total, °perfect, °true, °pure, arrant, utter, °plain: *That man is an unmitigated bore. His unmitigated pessimism was beginning to depress her.*

unmoved *adj.* °cool, °aloof, °calm, °collected, °unaffected, untouched, °unsympathetic, unstirred, undisturbed, apathetic, stoic(al), °impassive, °dispassionate, unemotional, unfeeling, unconcerned, °indifferent, unreactive, unresponsive, °stolid, °stony, adamant, stonyhearted, hardhearted: *Despite her child's tears, she remained totally unmoved.*

unnatural *adj.* **1** uncharacteristic, °out of character, °odd, °peculiar, °strange, unexpected, °abnormal, unusual: *Isn't it unnatural for someone who has won the sweepstakes to look so unhappy?* **2** °labored, °forced, °stilted, °stiff, restrained, °artificial, °false, °insincere, feigned, contrived, °affected, °mannered, °self-conscious, °theatrical, stagy: *She has an irritating, unnatural laugh.* **3** °outlandish, °weird, uncanny, °strange, °odd, °unaccountable, °supernatural, preternatural, °queer, °grotesque, °bizarre, °extraordinary, °eccentric, freakish: *Let me tell you about the unnatural events that took place in that house.*

unnecessary *adj.* unneeded, °needless, unrequired, °dispensable, °disposable, °expendable, unwanted, °surplus, °superfluous, supererogatory, inessential, unessential, °nonessential: *As Grandma was asleep, we asked the workmen to avoid making any unnecessary noise.*

unnerve *v.* °upset, °agitate, °perturb, °ruffle, °fluster, °rattle, °discomfit, unsettle, disconcert, °dismay, °intimidate, °stun, stupefy, *Colloq* °shake (up), faze: *Unnerved by the accident he'd just witnessed, Peter drove home slowly.*

unnoticed *adj.* unnoted, overlooked, unobserved, undiscovered, unremarked, unmarked, unperceived; unseen, unheard: *Fortunately, Martha's nasty aside went unnoticed.*

unobtrusive *adj.* °inconspicuous, unostentatious, low-key, °retiring, °modest, self-effacing, unpresuming, unpretentious, unassuming, °quiet, °humble, unaggressive, unassertive, nonassertive, °subdued, °reserved, °reticent, suppressed: *Agatha may seem unobtrusive, but she's a tigress when aroused.*

unofficial *adj.* °informal, °unauthorized, undocumented, °off the record, °private, °secret, unpublicized, unannounced: *The builder acknowledged having had unofficial meetings with the council.*

unopened *adj.* closed, °shut: *The book lay, unopened, on the bedside table.*

unorthodox *adj.* °irregular, unconventional, °nonconformist, unconforming, nonconforming, aberrant, aberrational, °deviant, heteroclite, °unusual, °abnormal, uncustomary, uncommon: *Roebuck's unorthodox behavior at the office party almost lost him his job.*

unpaid *adj.* **1** °payable, °outstanding, owed, owing, °due, °unsettled: *Mrs. Cudleigh absconded, leaving a pile of unpaid bills.* **2** unsalaried, °voluntary, volunteer, °honorary, *US* dollar-a-year: *On retirement Catworth did unpaid work for the local hospitals.*

unpalatable *adj.* °distasteful, °disagreeable, unpleasant, °unsavory, unappetizing, unattractive, °repugnant, °nasty, °offensive; °rancid, °sour, °off, turned, °bitter, inedible, uneatable: *Without an education, they are often offered the unpalatable jobs. We found heavy claret totally unpalatable with the sole.*

unparalleled *adj.* unequaled, °incomparable, °matchless, °peerless, unrivaled, unmatched, inimitable, unexcelled, °superior, °supreme, °superlative, unsurpassed, °unusual, °special, °singular, °rare, °unique, °exceptional, consummate: *In our experience, the food and service aboard the* Normandie *were unparalleled.*

unperfumed *adj.* unscented, °plain, °natural: *This deodorant is available in sandalwood, forest pine, mint, or unperfumed.*

unpopular *adj.* out of favor, in bad odor, unliked, disliked, shunned, avoided, snubbed, ignored, unsought-after, unaccepted, unwanted, rejected, °unwelcome, °undesirable; unloved, friendless: *New taxes are always unpopular. Adolf had always been an unpopular boy at school.*

unprejudiced *adj.* unbigoted, unbiased, °impartial, unjaundiced, °just, °fair, °objective, °disinterested, fair-minded, °nonpartisan, °liberal, open-minded, undogmatic: *An unprejudiced opinion might be hard to find, since everyone seems to have an ax to grind.*

unpremeditated *adj.* °unprepared, unplanned, unarranged, uncontrived, unstudied, °coincidental, °spontaneous, spur-of-the-moment, last-minute, impromptu, extemporaneous *or* extemporary, extempore, ad-lib, °offhand, °casual, °impulsive, °natural, °involuntary, °automatic, °unconscious, *Colloq* off the cuff: *I had no inkling of her plans, and my running into her was entirely unpremeditated. Her immediate, unpremeditated reaction was to run away.*

unprepared *adj.* **1** unready, surprised, taken aback, (caught) napping *or* off guard, dumbfounded *or* dumfounded, at sixes and sevens, *Colloq* (caught) with (one's) pants down, *Brit* caught on the hop, *US* asleep at the switch: *The sudden squall found us unprepared and we lost a sail.* **2** unfinished, °incomplete, uncompleted: *Dinner was still unprepared at eight-thirty.* **3** unwarned, unreadied, not set up, not forewarned: *I was unprepared for the strange events that followed.* **4** See **unpremeditated, above.**

unprofessional *adj.* **1** °unbecoming, °improper, unethical, unprincipled, °unseemly, undignified, unfitting, unbefitting, °unworthy, unscholarly, negligent, °lax: *The Medical Association is quick to act against doctors whose conduct might be judged unprofessional.* **2** amateurish, °amateur, inexpert, °inexperienced, untrained, untutored, unschooled, °incompetent, unskilled, unskillful, °inferior, second-rate, °inefficient, °poor, °shoddy, low-quality, °sloppy: *This unprofessional job could not have been done by a journeyman cabinetmaker.* **3** nontechnical, unspecialized, nonspecialized, nonspecialist, °lay, °everyday, °ordinary, °plain (English), understandable: *What does all that gobbledygook mean when translated into unprofessional language?*

unprofitable *adj.* **1** profitless, ungainful, unremunerative, unfruitful, non-profit-making; breaking even; losing, loss-making: *The last quarter was unprofitable for car manufacturers.* **2** °bootless, °pointless, °purposeless, unavailing, °futile, °useless, unproductive, °worthless, °ineffective, °inefficient: *Having employees sitting about waiting for the telephone to ring is unprofitable.*

unpromising *adj.* °inauspicious, unpropitious, unfavorable, °gloomy, °ominous, adverse, °portentous, baleful, °hopeless: *The forecast for the economy looked unpromising yesterday but has suddenly brightened.*

unqualified *adj.* **1** °ineligible, unfit, untrained, ill-equipped, unsuited, unequipped, °unprepared: *He was rejected as unqualified for the job of repairing computers.* **2** unrestricted, unreserved, unconditional, °categorical, °outright, °unmitigated, °downright, °out-and-out, °pure (and simple), °true, °perfect, utter, °absolute, consummate: *If he told you that he flew jets in the RAF, he is an unqualified liar. Cato's unqualified "Delenda est Cartago" was echoed in every speech he made in the Roman senate.*

unquenchable *adj.* insatiable, unslakable *or* unslakeable, unsatisfiable; °inextinguishable, unsuppressible, °irrepressible, °indestructible: *Derwent sought ways to allay his unquenchable thirst for knowledge. Those principles have kindled an unquenchable fire in the hearts of the oppressed.*

unquestionable *adj.* unexceptionable, indubitable, undoubted, °indisputable, incontestable, unimpeachable, undeniable, °certain, °sure, °positive, irrefutable, °manifest, °obvious, °patent, °clear, °definite, °incontrovertible, unequivocal, unmistakable *or* unmistakeable, conclusive: *Their integrity is unquestionable, and I have complete confidence in them.*

unreal *adj.* **1** °imaginary, °fantastic, chimeric(al), °fanciful, °fancied, °illusory, make-believe, phantasmagoric(al), phantasmal, °spectral, figmental, °unrealistic, °nonexistent: *The monster in my nightmare was unreal, not like the creature that attacked me as I awoke.* **2** °theoretical, °hypothetical, °mythical, °imaginary, made-up, °fictitious, make-believe, °fanciful: *Although the example is unreal, it will serve its purpose in my argument.* **3** °artificial, °synthetic, synthesized, °mock, °false, fake(d), °counterfeit, °fraudulent, °dummy, °spurious, falsified, pretend(ed), °sham, pseudo, make-believe: *The landscape looked a bit unreal, like the painted backdrop in a theater.*

unrealistic *adj.* **1** °impractical, illogical, °unreasonable, unworkable, unrealizable, °quixotic, °romantic, °fanciful, °visionary, delusional, delusive, delusory: *However engaging the idea of traveling faster than light, the entire notion is unrealistic.* **2** °unreal, unlifelike, °unnatural, unauthentic, nonrepresentational, unrepresentative, °inaccurate: *The background paintings of the dioramas are poorly done and unrealistic.*

unreasonable *adj.* **1** irrational, illogical, °unthinking, °absurd, °foolish, °senseless, °nonsensical, °mindless, brainless, °thoughtless, °silly, °mad, °crazy, °insane, idiotic, moronic, imbecilic, °stupid, fatuous, °ridiculous, °ludicrous, laughable, °preposterous, °far-fetched, °shortsighted, unperceptive, unperceiving, undiscerning, myopic, °blind: *Some people have an unreasonable aversion to anything scientific. Is it unreasonable to expect you to spend a little time with your mother?* **2** °excessive, °outrageous, °exorbitant, °extravagant, °immoderate, extortionate, °inordinate, °unconscionable, unjust, °unwarranted, inequitable, unfair, unequal, °improper, unjustified, unjustifiable, uncalled-for: *The new tax puts an unreasonable demand on those with lower income.* **3** °inappropriate, unapt *or* inapt, unsuitable, unbefitting, °impractical, °unrealistic: *His conduct was quite unreasonable for a man of the cloth.*

unrefined *adj.* **1** °coarse, °rude, °rough, °unsophisticated, uncultured, °uncivilized, uncultivated, unpolished, inelegant, ill-bred, °impolite, °discourteous, unmannerly, °ill-mannered, bad-mannered, ignoble, °plebeian, undignified, unladylike, ungentlemanlike, ungentlemanly, uncourtly, °ungracious, °boorish, loutish, °gross, °vulgar, uncouth, cloddish, bumbling, °awkward, gauche: *Consider their background before criticizing them for being unrefined.* **2** °impure, unpurified, unclarified, °raw, °crude, °coarse, untreated, unfinished, °natural, unprocessed: *The unrefined ore is first washed thoroughly.*

unrelated *adj.* °independent, °separate, °distinct, °different, °dissimilar, °incompatible, °inappropriate, °foreign, °alien, unassociated, unaffiliated, unconnected, uncoupled, unlinked, unallied, uncoordinated: *Why raise a totally unrelated issue?*

unreliable *adj.* °irresponsible, °disreputable; untrustworthy, undependable, °uncertain, °unstable, treacherous, °flimsy, °weak: *Fordyce is unreliable—unlikely to be there when you need him. That stepladder is too unreliable—Watch out!*

unrepentant *adj.* unrepenting, unremorseful, impenitent, unapologetic, unregretful, unashamed, unembarrassed, unself-conscious, °remorseless, unreformed, unrehabilitated, unregenerate, recidivist *or* recidivistic *or* recidivous: *When caught, the offenders were totally unrepentant for what they had done.*

unresolved *adj.* °unsettled, °open, up in the air, °moot, °pending, °debatable, arguable, problematic(al), °indefinite, °vague, open to question, °questionable, unanswered, unsolved; undetermined, undecided, °uncertain, unsure, ambivalent, wavering, vacillating, °irresolute: *The matter of joining the monetary union remained unresolved. I am still unresolved on the issue.*

unrest *n.* disquiet, uneasiness, °distress, °anxiety, anxiousness, nervousness, °anguish, unease, °worry, °concern, °agitation, turmoil, °disturbance, °trouble, °strife, °agony: *The proximity of the huge army on our borders caused considerable unrest among the populace.*

unruly *adj.* unmanageable, °ungovernable, uncontrollable, °undisciplined, unregulated, °lawless, °disobedient, °insubordinate, °rebellious, °mutinous, fractious, refractory, contumacious, °obstreperous, °willful, headstrong, °stubborn, °recalcitrant, intractable, °defiant, uncooperative, wayward, °disorderly, turbulent, °riotous, °tumultuous, °violent, °stormy, °tempestuous: *She had no idea of how to handle a capricious and unruly teenager. The first reading of the bill led to an unruly session of the House.*

unsatisfactory *adj.* °insufficient, °inadequate, °inferior, °poor, °unacceptable, displeasing, °disappointing, °unworthy, °inappropriate, °deficient, °weak, °wanting, lacking, unsuitable, °imperfect, °flawed, °defective, °faulty: *The report was unsatisfactory in that it failed to meet the committee's demands. Why put up with unsatisfactory workmanship?*

unsavory *adj.* °distasteful, objectionable, unpleasant, °disagreeable, unappetizing, °unpalatable, °offensive, °repugnant, °obnoxious, °repellent, °nasty, °repulsive, °revolting, °disgusting, nauseating, sickening: *Her ex-husband is a totally unsavory character. Even the most unsavory swill tasted like ambrosia to the starving inmates.*

unscathed *adj.* unharmed, unhurt, uninjured, unmarked, untouched, undamaged, unscarred, unscratched, safe and sound, in one piece, as new, *Archaic* scatheless, *Colloq* like new: *Miraculously, he came out of the battle unscathed.*

unscrupulous *adj.* °unconscionable, conscienceless, unprincipled, amoral, unethical, °immoral, °dishonorable, °corrupt, °dishonest, °deceitful, °sly, cunning, °artful, insidious, °shifty, °sneaky, °slippery, roguish, knavish, °disingenuous, treacherous, °perfidious, °faithless, °false, untrustworthy, °wicked, °evil, *Colloq* °crooked: *That wretch is totally unscrupulous in the pursuit of his ambition.*

unseasonable *adj.* unsuitable, °inopportune, °inappropriate, untimely, ill-timed, inexpedient: *The arrival of the pack ice in September was most unseasonable.*

unseemly *adj.* **1** °improper, °unrefined, °unbecoming, indecorous, indelicate, °unbecoming, unladylike, ungentlemanly, undignified, in poor *or* bad taste, °disreputable, discreditable, °risqué, °naughty, °indecent, °shameful, °offensive, °lewd, °lascivious, °obscene, °rude, °coarse: *Out of desperation, they might have engaged in some unseemly activities.* **2** impolitic, unwise, °imprudent, inapt, °inappropriate, °inopportune, °inconvenient, uncalled-for, unsuitable, °improper, inadvisable, °ill-advised, unbefitting, unfitting, out of place *or* keeping, °awkward, °inauspicious, inexpedient, °unfortunate, ill-timed, untimely: *It might be unseemly for you to become involved with such a notorious individual.*

unselfish *adj.* °generous, °charitable, openhanded, ungrudging, unstinting, unsparing, giving, magnanimous, °philanthropic, °humanitarian, °free, °liberal, altruistic, °selfless, self-sacrificing: *Timothy has always been most unselfish should a friend be in need of help of any kind.*

unsettled *adj.* **1** unfixed, °unstable, changing, varying, °variable, °changeable, °inconstant, ever-changing, °protean, unpredictable, °inconstant, °uncertain: *The weather remained unsettled all week. I have told you of my unsettled life as a gypsy.* **2** °disturbed, turbulent, riled, °agitated, disquieted, °upset, perturbed, ruffled, rattled, flustered, °restive, °restless, unnerved, *US* roiled: *It was being kept in the dark about Father that made us feel unsettled.* **3** disoriented, °confused, mixed up, unorganized, disorganized, °disorderly, disordered, °tumultuous: *The unsettled situation in the east worried us.* **4** See **unresolved,** above: *Some questions about Ian remain unsettled.*

unsettling *adj.* unnerving, upsetting, °disturbing, perturbing, discomfiting, °disconcerting: *Sorry for being a wet blanket, but I have had some unsettling news from home.*

unsightly *adj.* °ugly, °hideous, awful-looking, °horrible, frightful-looking, unattractive, unprepossessing, unlovely, unpretty, °plain, *US and Canadian* °homely: *The property borders an unsightly junkyard.*

unsocial *adj.* unsociable, unfriendly, °cool, °cold, °chilly, °aloof, uncongenial, unamiable, unforthcoming, °standoffish, °inhospitable, °withdrawn, °reserved, °solitary, °retiring, °distant, °detached, °reclusive, hermitic(al), eremitic(al), anchoritic *or* anchoretic; antisocial, °misanthropic, °hostile: *Erwin, who prefers to be alone, is definitely unsocial. Donna, who despises other people, is unsocial for different reasons.*

unsolicited *adj.* unlooked-for, unsought, unsought-after, unrequested, unasked-for, uncalled-for, °gratuitous, uninvited; *Colloq US* over-the-transom: *Tanya is always offering unsolicited advice on every conceivable subject. The publishers receive scores of unsolicited book manuscripts each week.*

unsophisticated *adj.* **1** naive *or* naïve *or* naïf, °inexperienced, °simple, °childlike, unworldly, °innocent, °ingenuous, °artless, guileless: *Is the unsophisticated consumer sufficiently protected against unscrupulous traders?* **2** °simple, °plain, uncomplicated, undetailed, uninvolved, °unrefined: *Earlier computers were unsophisticated compared with those of today.*

unsound *adj.* **1** °weak, °feeble, °frail, °rickety, °shaky, °ramshackle, °infirm, °unstable, wobbly, tottering, unsteady, broken-down, crumbling, disintegrating, °dilapidated, °defective, °imperfect, °faulty, decayed, °rotten: *The building inspector condemned the structure as unsound and unsafe.* **2** °unhealthy, °diseased, °ill, afflicted, in poor health, ailing, °sickly, °sick, unwell, °delicate, injured, wounded: *That attack of the flu left him unsound of wind and limb.* **3** °insane, °mad, °psychotic, °unbalanced, °unstable, demented, °deranged: *The doctors said that she was still of unsound mind and ought not be released.* **4** illogical, °faulty, °flawed, fallacious, °untenable, °invalid, °groundless, °unfounded, °erroneous, °defective, °specious: *We thought his arguments unsound, so we rejected his recommendations.*

unspoiled *adj.* unspoilt, unsullied, °pristine, virgin, °whole, untainted, unstained, °immaculate, uncorrupted, unpolluted, °spotless, stainless: *The house has been preserved in its original condition, unspoiled by modern improvements.*

unstable *adj.* **1** °changeable, °variable, unsteady, °inconstant, °inconsistent, °insecure, °capricious, °fickle, °irregular, unpredictable, °unreliable, °erratic, °volatile, fluctuating, °flighty, mercurial, vacillating, tergiversating, °indecisive, undecided, °irresolute, °indefinite, °unsettled: *Hurley is too unstable to maintain a long-standing relationship.* **2** See **unsound, 1,** above.

unsuccessful *adj.* **1** °unfortunate, unavailing, °vain, abortive, °useless, °bootless, °fruitless, unfruitful, unproductive, °ineffective, °ineffectual, °inefficacious, °worthless, °unprofitable, °sterile: *We made an unsuccessful attempt to regain control of the company.* **2** unlucky, hapless, °unfortunate, luckless, defeated, beaten, jinxed, cursed, foiled, frustrated, balked: *Trevor complained that he had been unsuccessful in business all his life.*

unsung *adj.* uncelebrated, unrecognized, unglorified, unexalted, unpraised, unhonored, °unnoticed, disregarded, °unknown, anonymous, °unidentified, °nameless, °obscure, °insignificant, °inconspicuous: *Purvey joined the long list of unsung authors of the twentieth century.*

unsuspecting *adj.* unsuspicious, °unwary, unknowing, °ignorant, °unconscious, °gullible, credulous, naive *or* naïve *or* naïf, °ingenuous, °innocent, °trusting; °unaware, off guard: *Martin became an unsuspecting dupe in Lambert's plot to embezzle money from the bank.*

unsympathetic *adj.* uncaring, unconcerned, °callous, unfeeling, °unaffected, untouched, °unmoved, °indifferent, unemotional, °dispassionate, uncompassionate, unreactive, unresponsive, °impassive, °stolid, °cold, °cool, °aloof, unstirred, apathetic, insensitive, stoic(al), °stony, adamant, stonyhearted, hardhearted, unpitying, pitiless, °ruthless: *We were unable to understand Taylor's unsympathetic attitude to the suffering around him.*

untamed *adj.* undomesticated, °wild, unbroken, unsubdued, uncontrollable, °savage, °fierce, feral, °ferocious: *The wolf boy behaves like an untamed beast.*

untarnished *adj.* unsoiled, unsullied, °immaculate, °spotless, unspotted, untainted, °faultless, uncorrupted, unfouled, °chaste, lily-white, undefiled, virginal: *Despite the obloquy heaped on his colleagues, his own reputation remained untarnished.*

untenable *adj.* insupportable *or* unsupportable, indefensible, unsustainable, unmaintainable, unjustified, unjustifiable, baseless, °groundless, °unfounded, °flawed, °faulty, °weak, illogical, °specious, °implausible, °unreasonable, °unsound: *It is beyond me why he holds so tenaciously to such untenable tenets.*

unthinkable *adj.* **1** °inconceivable, °unbelievable, unimaginable, °incredible, °incomprehensible, beyond belief, °extraordinary, *Colloq* mind-boggling, *Slang* mind-blowing: *It was unthinkable that mild-mannered Dr. Cubbleigh could have been the mass murderer.* **2** °unacceptable, °absurd, illogical, °impossible, °improbable, °unlikely, °out of the question, °preposterous, °ridiculous, laughable, °ludicrous, *Colloq* not on: *It is simply unthinkable for her to go to the ball unattended.*

unthinking *adj.* **1** °thoughtless, °mindless, undiscriminating, unconsidered, unwitting, unreflecting, unthoughtful, irrational, °unreasonable, illogical, unperceptive, unperceiving, undiscerning, witless, brainless, °foolish, °senseless, °nonsensical, °rash, °stupid, °silly, °mad, °crazy, °insane, idiotic, moronic, imbecilic, °hasty, °shortsighted: *It was unthinking of me to ask her mother to go out to dinner with us.* **2** °inconsiderate, °impolite, °tactless, °rude, °thoughtless, undiplomatic, °discourteous, uncivil, °imprudent, unwise, °indiscreet, neglectful: *On the other hand, it would have been unthinking of me not to ask her mother.*

untidy *adj.* °disorderly, messy, disheveled, °unkempt, slovenly, slatternly, °bedraggled, rumpled, frowzy, °sloppy, °dirty; littered, cluttered, °chaotic, °helter-skelter, jumbled, *Archaic* draggletailed, *Colloq US* mussy, mussed up: *You should never go to the door looking so untidy. Ken's papers are always so untidy.*

untiring *adj.* unflagging, °determined, indefatigable, dogged, persevering, perseverant, °tireless, unwearying, unwearied, dogged, dedicated, unfailing, unfaltering, °steady: *Kim has always been an untiring perfectionist.*

untold *adj.* **1** countless, uncounted, uncountable, unnumbered, °numberless, innumerable, myriad, incalculable; °immeasurable, measureless, °unlimited: *The mute stones speak to the archaeologist through untold centuries.* **2** unrecounted, unnarrated, undescribed, unpublished, unrevealed, undisclosed, undivulged, unreported, °private, °hidden, °secret: *How many untold stories are there in the naked city?* **3** °inexpressible, unutterable, indescribable, unimaginable, °inconceivable, °unthinkable, unspeakable: *The untold agony that man went through because of his family's illnesses!*

untoward *adj.* **1** adverse, unfavorable, unpropitious, discouraging, °inopportune, °unpromising, °bleak, °inauspicious, °bad, °unfortunate: *The rainstorm created untoward conditions, causing the river to rise dangerously.* **2** °unbecoming, unfitting, °awkward, °inappropriate, unapt, unsuitable, °improper, °impolite, °rude, °boorish, ungentlemanly, unladylike, indecorous, indelicate, °unwarranted, uncalled-for, °unrefined, °unseemly, unwise, °imprudent, undiplomatic, °tactless, untactful, ill-conceived, °silly, °foolish, °stupid, ill-timed, vexatious, vexing, irritating, annoying: *Cheevey made some untoward remark about the size of Sandra's feet.*

untried *adj.* untested, unproved *or* unproven, °new: *He may be good at homes, but he is as yet untried as the designer of a library.*

untrue *adj.* **1** unfaithful, °faithless, °disloyal, °fickle, °capricious, undependable, °unreliable, °dishonorable, untrustworthy, °false, °hypocritical, °dishonest, °insincere, °two-faced, duplicitous, °devious, °deceitful, treacherous, °perfidious: *Only when I saw her with Henry did I realize that Pauline had been untrue to me.* **2** °wrong, °false, °inaccurate, °incorrect, °erroneous, misleading, °mistaken, distorted: *What you said about Pauline was simply untrue.* **3** °inexact, nonstandard, substandard, °imprecise, °imperfect: *How unfortunate for legend had William Tell's aim been untrue!*

unused *adj.* **1** °(brand) new, untouched, °pristine, °original, °intact, °fresh, first-hand: *He sold me his unused computer for half what it cost.* **2** °disused, °abandoned, °derelict, neglected, given up: *We hid in an old unused factory building.* **3** unconsumed, °left (over), °remaining: *If dissatisfied with this product, return the unused portion for a full refund.* **4** ***unused to.*** °unaccustomed to, °unfamiliar with, inexperienced in *or* at, amateurish at, unpracticed in *or* at, uninitiated in: *I am quite unused to doing that sort of thing.*

unusual *adj.* uncommon, °exceptional, atypical, untypical, °different, unexpected, °singular, °out of the ordinary, °extraordinary, °odd, °peculiar, °curious, °bizarre, °strange, °queer, °remarkable, °unique, freakish, unprecedented, unconventional, °unorthodox, *Slang* off the wall: *It takes an unusual person to be as charitable as Jean. Don't you agree that an egg-laying mammal is a bit unusual?*

unvarnished *adj.* °plain, °simple, °pure, unembellished, straightforward, °straight, °direct, °honest, unelaborated, °naked, °stark, °sincere, °frank, °candid, °outspoken: *My dear, let me tell you the unvarnished truth about Marnie and Jim.*

unveil *v.* °reveal, °expose, uncover, lay bare *or* open, °bare, °bring to light: *Only after Binker died did Connie unveil the truth about his double life.*

unwarranted *adj.* uncalled-for, °unasked, unasked-for, unjustified, indefensible, unjust, unfair, °unconscionable, °unworthy, °improper, °inexcusable, °gratuitous, unmerited, undeserved, unprovoked, °outrageous, °excessive, °unreasonable, unrestrained, intemperate, untempered, °immoderate, undue, °unnecessary: *The police were accused of unwarranted use of force in ejecting the rowdies from the club.*

unwary *adj.* °heedless, °careless, °hasty, incautious, °unguarded, °imprudent, °rash, °foolhardy, °reckless, °thoughtless, °indiscreet, °unthinking, °mindless, unwise: *Many perils await the unwary traveler. It was unwary of you not to lock the doors and windows.*

unwashed *adj.* **1** °dirty, uncleaned, unclean, uncleansed: *Separate your personal things from the unwashed laundry in the hamper.* —*n.* **2** ***the (great) unwashed.*** *Derogatory* °the rabble, °the masses, the mob, the plebs, people (at large *or* in general), the population, the populace, the man *or* woman in the street, Mr. (& Mrs.) Average, the working class(es), most people, the (silent) majority, *US* John Q. Public: *The effort to sell expensive brandy to the great unwashed failed miserably.*

unwelcome *adj.* **1** uninvited, unsought-for, unwished for, undesired, °undesirable, displeasing, unpleasing, °distasteful, unpleasant: *My day was further ruined by the unwelcome arrival of all those bills.* **2** unwanted, rejected, unaccepted, excluded; *persona non grata,* anathema: *Following that episode, Curshaw was made to feel unwelcome at the club.*

unwholesome *adj.* **1** °unhealthy, unhealthful, °detrimental, deleterious, pernicious, insalubrious, unhygienic, insalutary, °harmful, noxious, toxic, °injurious, °destructive: *Fernthwaite finally succumbed to the unwholesome climate and had to be sent home.* **2** °corrupt, °immoral, °bad, °wicked, °evil, °sinful, °perverted; demoralizing, depraved, °degrading, corrupting, perverting: *The social worker said it was wrong for a child to be raised in such an unwholesome atmosphere.* **3** °ill, ailing, °sickly, °sick, °pale, °wan, anemic, pallid, °pasty: *Their unwholesome complexion comes from malnutrition.*

unwieldy *adj.* °awkward, °clumsy, °bulky, oversized, cumbersome, ungainly, unmanageable, unhandy, unmaneuverable: *The huge oil tankers, though economical, often proved unwieldy in the ports they visited.*

unwonted *adj.* infrequent, °unusual, uncustomary, uncommon, °unfamiliar, unprecedented, °rare, °singular, atypical, °abnormal, °peculiar, °odd, °strange, °irregular, unconventional, °unorthodox: *It took some time to accustom myself to the unwonted exercise.*

unworthy *adj.* **1** unequal, meritless, unmerited, substandard, °inferior, second-rate, °menial, °puny, °petty, °paltry, °unprofessional, °mediocre, °despicable, °contemptible, °dishonorable, ignoble, °disreputable, discreditable, °unqualified, °ineligible, unfit, undeserving: *I consider Pendley an unworthy opponent.* **2** ***unworthy of.*** unbecoming to, inappropriate to, unsuitable for, unfit for, out of character for, inconsistent with *or* for, out of place with *or* for, incongruous with *or* for: *That sort of petty haggling is unworthy of your position and of you.*

upbeat *adj.* °positive, °optimistic, °sanguine, °favorable, °cheerful, encouraging, heartening, °buoyant, lighthearted: *Chimberton has a very upbeat attitude toward life despite his age.*

upbraid *v.* °scold, °rebuke, °reprimand, reproach, °berate, °castigate, °chastise, °reprove, chide, censure, °take to task, *Colloq* °tell off, tick off, °dress down, give a dressing-down, give (someone) a piece of (one's) mind, tell (someone) a thing or two, rake (someone) over the coals, jump on *or* all over, °bawl out, *US* °chew out: *The boys were soundly upbraided for going near the railway tracks.*

upbringing *n.* rearing, raising, training, °education, cultivation, nurture, °breeding: *The way one behaves through life is really a matter of upbringing.*

upheaval *n.* °upset, °unrest, commotion, °change, cataclysm, disruption, °disturbance, °disorder, °confusion, °chaos, °furor: *Those were times of drastic political upheaval in Asia.*

uphold *v.* °support, °maintain, °sustain, °preserve, °hold up, °defend, °protect, °advocate, °promote, espouse, °embrace, °endorse, °back, °champion, °stand by: *All demagogues claim to uphold democratic principles.*

upkeep *n.* **1** °maintenance, °repair, °support, °sustenance, °preservation, °conservation, °subsistence, °running, °operation: *The upkeep of the infrastructure came to 10 percent of the annual budget last year.* **2** (operating) costs, °(running) expenses, °outlay, °expenditure, *Brit* °overheads, oncosts, *US* °overhead: *Have you included upkeep in the annual expenses?*

upper *adj.* **1** higher (up), loftier, topmost, more elevated, °uppermost: *Air is less dense in the upper parts of the atmosphere. Our apartment is on an upper floor. Her best notes are sung in the upper register.* **2** higher, upland, more elevated; (more) northerly, northern: *We visited the cataracts on the upper reaches of the Nile. We spent the holiday in upper Canada.* **3** later, more recent: *These strata were laid down in the Upper Cretaceous period.* **4** ***upper case.*** capital letter(s), capital(s), large letters, majuscule (letters *or* characters): *This heading ought to be in upper case.* **5** ***upper crust.*** upper class, elite *or* élite, aristocrats, nobles, blue bloods, °wealthy, *US* Four Hundred: *Gregory thinks of himself as a member of the upper crust and won't associate with us.* **6** ***upper hand.*** °advantage, °control, °authority, °power, °sway, °superiority, °supremacy, °command, dominance, ascendancy, *Colloq* °edge: *Purvis kowtows to whoever has the upper hand.* —*n.* **7** ***on (one's) uppers.*** °poor, °indigent, °destitute, poverty-stricken, *Colloq* °broke: *Stankey has been on his uppers since losing his job.*

upper-class *adj.* **1** elite *or* élite, aristocratic, blue-blooded, well-born, °noble, highborn, patrician, *Colloq* upper-crust: *Eunice comes from an upper-class Irish family.* **2** °high-class, °elegant, °fancy, °luxurious, °first-rate, deluxe, °royal, °regal, °sumptuous, *Colloq* swank(y), ritzy, °posh: *I'll have you know that we stayed only in upper-class hotels on our tour.*

uppermost *adj.* **1** highest, topmost, loftiest, highest, °top: *These curious animals live in the uppermost branches of the trees.* **2** °foremost, °first, most important *or* prominent *or* influential *or* telling, °principal, °paramount, °preeminent, °predominant: *Uppermost in my mind is the safety of the children.*

uppish *adj.* °affected, putting on airs, °snobbish, °conceited, overweening, °self-important, *Colloq* uppity, snooty, high and mighty, °hoity-toity, highfalutin *or* hifalutin *or* highfaluting, stuck-up, on (one's) high horse, *Slang* snotty, *Brit* toffee-nosed: *Even if she is in charge she doesn't have to be so uppish.*

upright *adj.* **1** °erect, °perpendicular, vertical, °on end, straight up and down, °plumb, standup, °standing (up), *Brit* upstanding: *Few upright columns of the Grecian temple remained.* **2** °moral, °principled, high-minded, °ethical, °virtuous, upstanding, °straight, °righteous, straightforward, °honorable, °honest, °just, °trustworthy, unimpeachable, uncorrupt(ed), incorruptible, °decent, °good: *Drumley had long been an upright member of the church council.* —*n.* **3** °post, °pole, column, vertical, °perpendicular: *We need another upright to support the floor over here.* —*adv.* **4** perpendicularly, vertically, upward(s), °straight up (and down): *The javelin was sticking upright out of the ground.* **5** right side up: *Miraculously, the platter with the roast on it landed upright on the floor.*

uprising *n.* rebellion, °revolt, °mutiny, °revolution, insurrection, rising, putsch, coup (d'état): *The government quelled the uprising with water cannons and tear gas.*

uproar *n.* clamor, hubbub, °disturbance, commotion, hullabaloo, brouhaha, °din, °racket, °pandemonium, °tumult, turmoil, pother, °outcry, °outburst, °bedlam, °agitation, °frenzy, °broil, °rumpus, °fuss; affray, °fracas, °brawl; *Colloq* hoo-ha, to-do, *US* hoopla, *Brit* kerfuffle *or Brit* carfuffle *or Brit* kurfuffle: *The an-*

nouncement of the new taxes caused a nationwide uproar.

uproarious *adj.* **1** clamorous, °noisy, deafening, °tumultuous, turbulent, °tempestuous, °excited, frenzied, °rowdy, °riotous, °disorderly, °wild: *We had an uproarious party on New Year's Eve.* **2** °hilarious, °hysterical, °(screamingly) funny, sidesplitting, *Colloq* too funny for words, °killing: *Leslie told his uproarious story about Colonel Cholmondley and the tiger.*

uproot *v.* **1** °transfer, °transplant, °move, °displace; °exile, °banish: *Finney has been uprooted five times in his career with the firm.* **2** deracinate, extirpate, °root out, °dig out, pluck out, tear out; °destroy, °demolish, °ruin, eradicate, annihilate, °kill, °ruin, °devastate, °ravage: *Thousands of trees were uprooted by the hurricane.*

upset *v.* **1** °disturb, °agitate, °distress, unsettle, °put off, °put out, °perturb, disquiet, °fluster, °ruffle, °frighten, °scare, disconcert, °dismay, °trouble, °worry, °bother, discompose, make (someone) nervous: *We once had a nanny who told the children horror stories just to upset them.* **2** °overturn, °capsize, °topple, upend, °tip over, knock over *or* down, invert, turn topsy-turvy *or* upside down, °spill: *That is the third time this week you have upset the cream jug.* **3** °disturb, derange, °disrupt, disarrange, °mess up, disorganize, °snarl up, °jumble, °muddle, *Colloq Scots* kerfuffle *or* carfuffle *or* kurfuffle: *The boys upset the entire house with their horseplay.* **4** °overthrow, °defeat, °beat, worst, thrash, °rout, °conquer, °overcome, win out over, get the better of, get *or* gain the advantage over, °triumph over, be victorious over, vanquish: *The question is whether the challenger will upset the champion in the Wimbledon finals.* **5** °defeat, °ruin, °spoil, °thwart, interfere with, °destroy, °demolish, °mess up, °disturb, *Colloq* throw a *Brit* spanner in *or US* monkey wrench into (the works), *US* discombobulate, *Slang* °screw up, gum up, put the kibosh on, *Taboo slang* fuck up, *Brit* °bugger up: *Your coming early upset my plan to shampoo my hair.* —*adj.* **6** capsized, overturned, upside down, bottom side up, inverted, reversed, toppled, tipped over, °topsy-turvy: *We clung to the bottom of the upset boat till help arrived.* **7** °sick, °queasy: *Eve has an upset stomach and cannot go.* **8** perturbed, °disturbed, disquieted, °disconcerted, °agitated, distressed, °worried, troubled, unnerved, distracted, apprehensive, °nervous, frightened, °scared, °afraid: *We found Vladimir in a terribly upset state at Natasha's disappearance.* **9** disordered, °confused, disorganized, messed up, jumbled, muddled, °disturbed, disarranged: *The house is never so upset as after a party.* **10** °angry, irate, °furious, °beside oneself, °mad, *Colloq* fit to be tied, *Slang* freaked out: *Father was upset because I took the car without his permission.* —*n.* **11** °defeat, upsetting, °conquest, °overthrow, °rout, °thrashing, °triumph, °victory: *We celebrated the upset of the Dolphins by the Rams.* **12** °surprise, unexpected event *or* occurrence: *The upset of the season was when Carridoff won the Derby.* **13** °reversal, °overthrow, °overturn, *bouleversement*: *The change of schedule resulted in a total upset of our plans.*

upshot *n.* °result, °end (result), °outcome, ending, conclusion, °termination, °effect, aftereffect, fallout, °wake, backwash, °repercussion, afterclap, feedback, °resolution, culmination, denouement *or* dénouement, °issue, *Colloq* °payoff, *US* wrap-up: *The upshot of the nasty business was that Bowles was sent to prison for life.*

upstart *n.* parvenu(e), arriviste, nouveau riche, (social) climber, status seeker, °pretender, °nobody: *That upstart husband of Dora's tried to get me to invest in some scheme of his.*

uptake *n.* comprehension, °understanding, apprehension, °grasp, °perception, °insight, perspicaciousness, perspicacity, perceptiveness, °sensitivity: *Billie, very quick on the uptake, immediately knew what the man was after.*

urge *v.* **1** °press, °push, °drive, °force, impel, °speed, accelerate, °hurry, °rush, °hustle, °move, goad, °prod, egg on, °spur: *He urged his steed onward.* **2** °press, goad, °prod, egg on, °spur, °prompt, °induce, °incite, constrain, exhort, °encourage, °demand, °request, °ask, °plead (with), °beseech, °beg, entreat, importune: *We urged the police for months to investigate the goings-on at that house.* **3** °coax, °persuade, °induce, °prevail upon, °campaign (with), °sway, °influence, °talk into, °advise, °suggest, °counsel: *The agents urged the Corbells to sell their house while the market was good.* **4** °argue, °set forth, affirm, °state, °allege, assert, °hold, °advise, °advocate, °demand: *The residents urged that the planning board should deny permission for a supermarket.* —*n.* **5** °pressure, °impetus, °desire, compulsion, impulse, °itch, °longing, yearning, °drive, °fancy, °hunger, °thirst, craving, *Colloq* yen: *We stopped because Corey felt the urge to have some ice cream.*

urgency *n.* imperativeness, °pressure, °stress, °extremity, °importance, seriousness, importunity, °necessity, °need, insistence, exigency, °emergency: *She was asked to phone home on a matter of some urgency.*

urgent *adj.* **1** °immediate, °instant, °imperative, °pressing, compelling, °vital, life-and-death, °important, °serious, °necessary, exigent, °rush, °emergency, °pressing, high-priority: *There was an urgent need to do something at once, so she screamed.* **2** supplicative, begging, °solicitous, °earnest, importunate, °insistent, °loud, clamorous, °active, °energetic, pertinacious, °tenacious, °forceful, °firm: *Both parties were urgent in pressing for tax reform.*

urinate *v.* pass *or* make water, *Technical* micturate; *Babytalk* °(make a) wee, (go) wee-wee, (have a *or US also* take a) pee, (make) pee-pee, do number one, tinkle, piddle; *Euphemistic* go to the men's *or* ladies' (room), go to the lavatory, °excuse (oneself), wash (one's) hands, go to the bathroom, go to the powder room; *Mincing* go to the little boys' *or* girls' room; *Colloq Brit* spend a penny, go to the loo, *Slang* (take *or* have a) piss, *Brit* have a *or* go for a slash: *Benjy had drunk a lot of beer and had to urinate.*

usage *n.* **1** °use, °custom, °habit, °practice, °routine, °convention, °form, °tradition: *British usage reflects decreasing use of the subjunctive mood.* **2** °treatment, °use, °management, handling, °operation, manipulation: *It would appear that the car has been subjected to some rough usage.*

use *v.* **1** °employ, make use of, put into practice *or* operation, practise, utilize, °exercise, bring into play, have recourse to, °resort to, put *or* press into service, put to use, avail (oneself) of; °say, utter, °speak: *If you use a hammer you might mar the surface. Why do you have to use bad language?* **2** capitalize on, turn to account, profit by *or* from, °exploit, utilize, make use of, take advantage of, °manipulate, °maneuver, °handle, °abuse, °misuse, °play, °work: *She used her smile to lure men on. Dorothy never forgave Conrad for the way he used her and her contacts to get his job.* **3** °consume, °eat, °drink, smoke, °take, °partake of, ingest, °inject, *Slang* °shoot (up): *He has never used a habit-forming substance.* **4** °consume, °buy, °purchase, °employ, utilize: *Which kind of detergent do you use?* **5** ***use up.*** °consume, °exhaust, °expend, °run through, °run out of, deplete; °waste, squander, °fritter away, pour down the drain, °throw away: *He used up all the toothpaste. She used up her entire inheritance gambling.* —*n.* **6** °usage,

°application, °employment, utilization; using: *The use of unleaded fuel is recommended.* **7** °function, utility, °application; °advantage, °benefit, °good, °service, °interest, °profit, avail: *Of what use is a radio that doesn't work?* **8** °wear (and tear), utilization, °treatment, °usage, handling: *Long use has worn the workbench smooth.* **9** usability, °usefulness, utility, utilization, °usage, °function, functioning, service(s), serviceability, °power: *He lost the use of his right hand through an accident at work.* **10** °license, °permission, °permit, °privilege: *We paid a high fee for the use of this parking space.* **11** consumption, purchases, buying, °acquisition: *The use of cigarettes has diminished in the last decade.* **12** °advantage, °purpose, °point, °end, °object, °reason, °basis, °ground: *I have tried to help him overcome his habit, but what's the use? Many are the uses of adversity.* **13** °demand, °need, °necessity, °urgency, exigency: *The world has as much use for art as for engineering.* **14** See **usage, 1,** above. **15** ***have no use for.*** execrate, °detest, °abhor, °hate, °despise, °scorn, contemn, °spurn, °reject, °dislike: *We have no use for cowards in this regiment.*

used *adj.* **1** °secondhand, castoff, °old, °worn, *Euphemistic* preowned, *Colloq* hand-me-down, *Brit* reach-me-down: *Would you buy a used car from Arthur Daley?* **2** utilized, employed, occupied; in use: *We slept in a rarely used room.* **3** ***used to.*** accustomed to, habituated to, acclimatized *or* acclimated to, adapted to, hardened *or* toughened *or* inured to *or* against, tempered to, tolerant of; familiar *or* acquainted with: *After all those years, Amanda was used to Bill's shouting.*

useful *adj.* utilitarian, °functional, °serviceable, °practical, usable, of use, °beneficial, salutary, °advantageous, °expedient, °profitable, valuable, °gainful, °helpful, °fruitful, °productive, °effective, °worthwhile: *Does this knob serve any useful purpose? We had some very useful conversations with the bank manager.*

usefulness *n.* utility, applicability, practicability, °purpose, purposefulness, °point, practicality, °benefit, °advantage, expediency, °profit, profitability, value, °gain, °help, fruitfulness, effectiveness, °worth: *I never could see the usefulness of all the forms one has to fill in.*

useless *adj.* **1** °ineffective, °ineffectual, unserviceable, °impractical, °impracticable, unpractical, unavailing, °vain, °pointless, °purposeless, °idle, °futile, unproductive, °unsuccessful, °impotent, effete, °sterile, °barren, abortive, unusable, °bootless, °worthless, *Rare* inutile: *All the pesticides we tried were useless.* **2** °inefficient, °incompetent, unproductive, °ineffectual, °ineffective, °hopeless, °inept: *Galupo has proved totally useless as a watchdog.*

user *n.* **1** consumer, °buyer, purchaser, °owner; °operator: *Users of the equipment report excellent results.* **2** alcohol *or* drug *or* narcotic addict: *The police have been interviewing users to try to identify the dealers.*

user-friendly *adj.* °simple, °practicable, usable, °explicit, °accommodating, understandable: *This user-friendly program comes with a user-friendly manual.*

usual *adj.* °same, °customary, °habitual, °accustomed, °familiar, °well-known, °common, °everyday, established, °traditional, °set, °time-honored, °old, °conventional, workaday, °stock, wonted, °regular, °ordinary, °normal, expected, °routine, °typical, °run-of-the-mill, stereotypic(al), hackneyed, trite, °prosaic, worn-out, shop-worn, °predictable, unexceptional, unoriginal, unremarkable, unimaginative: *The usual answer I get is "Why?" I missed my usual train this morning. They still sing the usual drinking songs.*

usually *adv.* customarily, °as a rule, °generally (speaking), most of the time, °for the most part, most often, mostly, almost always, inveterately, °on the whole, normally, commonly, regularly, predominantly, °chiefly, all things considered, °in the main, °mainly, by and large, as usual, *Colloq* as per usual: *I usually walk home from the office. Johnson usually interviews applicants in the morning.*

utopia *n.* Utopia, °paradise, °heaven, seventh heaven, (Garden of) Eden, °bliss, Cloud-Cuckooland, Nephelococcygia, never-never and, Shangri-La, Cockaigne *or* Cockayne, heaven on earth, °perfection: *He dreams about a Utopia where everyone is rich and there are no taxes.*

utterly *adv.* °completely, °perfectly, °absolutely, °thoroughly, fully, °entirely, °wholly, unreservedly, °totally, unqualifiedly, °out-and-out, °altogether, overwhelmingly, unequivocally, categorically, °definitely, °properly; °extremely, *Brit dialect* °proper; no holds barred, body and soul: *The proposal was utterly defeated in the the council. That was an utterly foolish act. She gave herself to him utterly.*

V

vacancy *n.* **1** °emptiness, °void, °gap, lacuna, hiatus, °blank, deficiency, °opening, °breach, vacuum: *Only after she left did Hubert sense the vacancy in his life.* **2** °(job) opening, °slot, °position, °post, °situation: *You might do well to inquire about the vacancy created by Henfist's departure.* **3** blankness, °emptiness, vacuity, absent-mindedness, inanity, vacuousness, incomprehension, fatuity, unawareness: *I knew by the vacancy of his look that he hadn't any idea of what I was talking about.*

vacant *adj.* **1** °empty, °void, °hollow, unoccupied, untenanted, °uninhabited, °abandoned, °deserted: *The place will remain vacant all summer.* **2** °blank, expressionless, deadpan, °empty, vacuous, °dull, °absent-minded, °inane, uncomprehending, fatuous, °unaware: *After the accident, she would just sit, for hours on end, with a vacant look on her face.* **3** unoccupied, °free, °unused, unutilized, °spare, °extra, °idle, unfilled, unengaged, unspoken-for: *The doctor has a vacant hour tomorrow at ten and could see you then.*

vacate *v.* **1** °leave, °depart (from), withdraw from, °quit, °evacuate, get *or* go out of; °desert, °abandon: *In a fire drill, employees must vacate the building in five minutes.* **2** °give up, °relinquish, °sacrifice, °renounce, let go, °resign, °abdicate, °cede, give up right *or* claim to, °abandon: *We vacated all claims to my aunt's property.* **3** annul, declare null and void, nullify, °void, °repudiate, override, overrule, rescind, °revoke, °recall, °quash, °set aside, invalidate: *The judge vacated the defendant's plea of diminished capacity.*

vade mecum *n.* handbook, °manual, ready reference, °book, °guide: *She always had with her a vade mecum on rare poisons.*

vagabond *n.* **1** gypsy *or* gipsy, °tramp, vagrant, wayfarer, °rover, wanderer, itinerant, °migrant, bird of passage, rolling stone, beachcomber, °derelict, *Chiefly US and Canadian* hobo, *Australian* swagman, *Colloq US* °bum, *Slang US* bindle-stiff: *The police interviewed every vagabond they could find for a clue to the murder.* —*adj.* **2** vagrant, wayfaring, roving, wandering, itinerant, °migrant, °derelict, nomadic, gypsy *or* gipsy, °rambling, roaming, drifting, peripatetic, °transient, peregrinating: *When will he give up his vagabond ways and settle down?*

vague *adj.* **1** °indefinite, °indistinct, °imprecise, °inexact, unclear, °confused, unspecified *or* nonspecified, °general, generalized, unspecific *or* nonspecific, inexplicit *or* unexplicit, ill-defined, °hazy, °fuzzy, °ambiguous, °obscure, amorphous; °shapeless, blurred, blurry, °filmy, °dim, °shadowy, °veiled, bleary, foggy, °misty, cloudy, clouded, hardly *or* barely distinguishable *or* discernible: *I have only a vague recollection of our having met. She must have had a vague idea of your intentions. Leon could almost make out the vague shape of an approaching figure.* **2** undetermined, indeterminate, unfixed, °indefinite, °inexact, unspecified *or* nonspecified, unspecific *or* nonspecific, °ambiguous, °doubtful, °in doubt, °uncertain, °equivocal, °ambiguous: *For security reasons, her time of arrival was left vague.* **3** °veiled, concealed, °hidden, shrouded, °obscure, ill-defined, unspecific *or* nonspecific, inexplicit *or* unexplicit, °ambiguous: *There had been a vague threat of kidnapping, which should not have been ignored.* **4** °subliminal, °subconscious, indefinable, unexplained: *I had a vague sensation of being followed.* **5** °indefinite, °ambiguous, °wishy-washy, undecided, °indecisive, °irresolute, vacillating, wavering, °inconstant, °unsettled, °uncertain, °nebulous, up in the air: *Leonore is vague about her plans to run for reelection.* **6** °vacant, °empty, °blank, expressionless, °empty, vacuous, °dull; puzzled: *By their vague expressions I could see they hadn't understood a word.*

vaguely *adv.* **1** distantly, remotely, indefinitely, dimly, subliminally, subconsciously, inexplicably: *I became vaguely aware of another presence in the room.* **2** ambiguously, imprecisely, inexactly, unclearly, confusedly, confusingly, hazily, fuzzily, nebulously, obscurely: *Kenston claims that he expresses his theories vaguely to allow for broad interpretation.* **3** °idly, vacantly, detachedly, absent-mindedly, dreamily, absently, distractedly: *Manton listened only vaguely to the judge's words.*

vain *adj.* **1** °proud, °conceited, °haughty, °arrogant, °boastful, °egotistical, °cocky, °self-important, vainglorious, narcissistic, *Colloq* bigheaded, swell-headed, stuck-up, swollen-headed: *The Swindells are so vain that they get facelifts every year or so.* **2** °worthless, profitless, °bootless, °pointless, °unsuccessful, °empty, °futile, °useless, unavailing, unproductive, °fruitless, °ineffective, abortive: *Ian sent Carol flowers every day in a vain attempt at winning her back.* **3** ***in vain.*** **(a)** vainly, futilely, unsuccessfully, fruitlessly, bootlessly: *In vain, we sought the help of the police.* **(b)** irreverently, blasphemously, disrespectfully, improperly: *Despite his mother's admonitions, he continues to take the Lord's name in vain.*

valley *n.* glen, dale, dell, vale, dingle, °hollow, coomb *or* coombe *or* combe *or* comb, *Northern Brit and Welsh* cirque *or* corrie *or* cwm, *Scots* strath: *Oh, to stroll once again the cool valleys and the heathered hills of home!*

vanity *n.* **1** °conceit, conceitedness, egotism, narcissism, °arrogance, cockiness, self-importance, vainglory, haughtiness, °pride, self-admiration, self-worship, *Colloq* swell-headedness, bigheadedness: *It is disgraceful what some people do to satisfy their vanity.* **2** vainness, °emptiness, hollowness, worthlessness, futility, unreality, bootlessness, pointlessness, uselessness, °folly, vapidity, silliness, vacuousness, vacuity, foolishness, fatuity, frivolousness: *The vanity of trying to complete the book in a year must be obvious by now.*

vapid *adj.* insipid, flavorless, °tasteless, °bland, °watery, watered-down, °wishy-washy, jejune, °colorless, °unpalatable, °flat, °tame, °lifeless, °boring, °tedious, °tiresome, uninteresting, trite, wearisome, wearying, °humdrum, *Colloq* blah, ho-hum: *That was one of the most vapid performances I have ever seen.*

vapor *n.* **1** °mist, °fog, steam, cloud, smoke, smog, °exhalation: *There arose from the bog a suffocating, miasmic vapor.* **2** ***the vapors.*** morbidity, hypochondria, hysteria, nervousness, °depression, rheuminess, *Archaic* distemper, *Colloq* the pip: *Charlotte suffered an attack of the vapors and retired to her room.*

variable *adj.* °changeable, °protean, changing, °inconstant, varying, wavering, mercurial, °fickle, °capricious, unsteady, unfixed, °unstable, °uncertain, undependable, unpredictable, fluctuating, vacillating, mutable, chameleonic, chameleon-like: *The weather can be quite variable depending on the season.*

variance *n.* **1** °variation, °difference, °disparity, °discrepancy, °disagreement, deviation, inconsistency, divergence, incongruity: *The Augenblick Factor takes account of the minutest variance in blood pressure during time travel.* **2** °disagreement, °misunderstanding, °discord, °difference (of opinion), °dissension, contention, °dispute, dissent, °controversy, °quarrel, °conflict, °argument, °debate, lack of harmony, falling-out, °schism, °rift: *There has never been any variance between us about the disciplining of the children.* **3** ***at variance.*** in dispute, in disagreement, quarreling, in contention, in conflict: *This time, labor is at variance with management in the matter of day care for the children of working parents.*

variant *n.* **1** °alternative, modification, °variation: *"Labour" is a British spelling variant of the word spelled "labor" in the US.* —*adj.* **2** varying, °variable, changing, altering, °unstable, °deviant, deviating, °different, differing; °separate, °distinct: *The law is given variant interpretations depending on the social standing of the suspect.*

variation *n.* **1** °change, changing, °alteration, altering, °variety, varying, modification, modifying, °difference, differing, diversification, °diversity, diversifying, modulation, modulating, conversion, converting, permutation, permuting: *That control allows for variation of the intensity of the light.* **2** °variety, °choice, °novelty, °diversity, departure (from the norm *or* usual), change of pace, divergence, variegation, deviation (from the norm): *This restaurant offers considerable variation in its menu.*

varied *adj.* **1** °diverse, diversified, °mixed, °miscellaneous, assorted, heterogeneous: *I wish the hospital offered a more varied bill of fare.* **2** See **various,** below. **3** See **variegated,** below.

variegated *adj.* multicolor(ed), parti-color(ed), varicolored, many-colored, motley, pied, piebald, brindled, °mottled, polychrome, polychromatic; nacreous, changeant, °opalescent, opaline: *He finds variegated abstracts, like those of Jackson Pollock, most soothing. Plants with variegated foliage add interest in a garden.*

variety *n.* **1** °diversity, diversification, multifariousness, multiplicity, °number, °range, °assortment, °medley, °mixture, °mix, °miscellany, heterogeneity, °choice, °selection, °collection: *Don't you find the variety of foods in supermarkets today to be astonishing? You certainly had a variety of different kinds of people at your party.* **2** °difference, heterogeneity, °discrepancy, °diversity, °disparity, °variation, °contrast: *The animals in Australia display great variety.* **3** °sort, °brand, °make, °mark, °kind, °class, °category, °breed, °type, °order, °genre, species, genus, classification, °strain: *Which variety of laundry soap do you prefer? Rex cultivated only the rarest varieties of orchid.*

various *adj.* **1** °different, a number of, a variety of, diversified, °diverse, °several, °many, numerous, °sundry, heterogeneous, °miscellaneous, *Literary* °divers: *We are having various appraisers in to look at the painting. Catherine held various executive positions in the* *company.* **2** °different, °distinct, °individual: *There are various ways of looking at the problem.*

vary *v.* **1** °change, °alter, °diversify, °transform, reshape, remodel, restyle, °modify, reorganize: *We vary the program each month to make it more interesting.* **2** change off, °switch, °alternate, °fluctuate, vacillate: *The manager's attitude varies from helpful to downright rude.* **3** °depart, °deviate, °differ, °diverge, °shift, veer: *These results do not vary from the average by more than 5 percent.*

vast *adj.* °infinite, °unlimited, °boundless, °limitless, unbounded, interminable, °endless, never-ending, °inexhaustible, indeterminate, °immeasurable, incalculable, measureless; °immense, °enormous, °huge, tremendous, °great, °prodigious, stupendous, °gigantic, °massive, °voluminous, capacious, °colossal, °monumental, mammoth, °jumbo, elephantine, behemoth, cyclopean, Brobdingnagian, titanic, *Literary* vasty, *Colloq US* ginormous, humongous: *Space involves vast distances and the vast time intervals to travel them. Compared with the earth, Jupiter is a vast planet.*

vastly *adv.* immensely, greatly, hugely, enormously, considerably, °substantially, °(almost) entirely, infinitely, °exceedingly, °extremely, very much, *Colloq* worlds: *Your new house looks vastly different from the way I had pictured it. Katinka looks vastly better since her facelift.*

vehicle *n.* **1** conveyance: *The police report that two vehicles were damaged in the crash.* **2** °means, °channel, °mechanism, °carrier, conduit, °agency, °instrument: *Water is the vehicle used in many sprays.*

veil *n.* **1** covering, °cover, °screen, °camouflage, °cloak, curtain, °mask, °shroud: *A veil of silence has been drawn over the meetings. He stole away under the veil of darkness.* —*v.* **2** °cover, °conceal, °hide, °camouflage, °cloak, °mask, °disguise, °shroud, °shield, °obscure: *The writer's identity was veiled from us till recently.*

veiled *adj.* concealed, °hidden, masked, °obscure, unrevealed, covert, disguised, °secret, sub rosa, °subtle: *There seemed to be a veiled threat in the way he said goodbye.*

vein *n.* **1** blood vessel; nervure: *Helmut killed himself by opening a vein in his wrist. The fly's wing reveals a simple pattern of veins.* **2** °streak, °seam, °stripe, striation, stria, °thread, °line: *The reddish vein gives this marble a warm feeling.* **3** °seam, lode, °stratum, °course, °deposit, bed: *The prospector discovered a vein of almost pure gold in the old mine.* **4** °thread, °hint, °suggestion, °touch, °trace, °streak, °line, °strain, °mood, °spirit, °tone, °note, °tenor, °feeling, °attitude, °disposition, °humor, °temper; tendency *or* inclination *or* proclivity toward(s): *Underlying her apparent kindness is a vein of bitter hatred.* **5** °way, °manner, °course, °fashion, °style, °mode, °pattern: *He would have continued in that same vein if the chairman hadn't stopped him.*

velocity *n.* °speed, swiftness, °rapidity, fleetness, quickness, briskness, alacrity, celerity, °pace, rate of speed, miles per hour, mph, kilometers per hour, km/hr: *Our velocity slowed as we neared the outer atmosphere.*

venal *adj.* °corrupt, corruptible, bribable, buyable, purchasable, °mercenary, unprincipled, °dishonorable, °rapacious, °avaricious, °greedy, simoniacal, *Colloq* °crooked, *Slang* °bent: *All we had to do was to find a venal official, cross his palm with silver, and they would release our son.*

vendetta *n.* °(blood) feud, °quarrel, °dispute, °conflict, °rivalry, enmity, °bitterness, hatred, °ill will, bad blood: *Fifty years after leaving Palermo, the vendetta between our families persisted.*

veneer *n.* °gloss, façade, °finish, °pretense, °(false) front, (outward) show *or* display, °appearance, °mask, °guise, °aspect, °superficies: *Beneath that veneer of elegance and refinement lie the heart and soul of a nefarious thief.*

venerable *adj.* °respectable, °honorable, °estimable, respected, honored, esteemed, august, °sedate, °impressive, revered, reverenced, worshiped: *We interviewed a venerable veteran of World War I about Richthofen's Flying Circus.*

venerate *v.* °respect, °honor, °esteem, °revere, °reverence, °worship, °hallow, °adore, °admire, °look up to: *Schweitzer was venerated for his charitable works, especially at Lambaréné.*

veneration *n.* °respect, °honor, °esteem, °reverence, °deference, °homage, °devotion, °worship, °admiration, adoration, idolization, awe: *Their veneration for their hero turned to contempt when his past was revealed.*

venereal *adj.* °sexual; genital; °social, sexually transmitted, gonorrheal *or* gonorrheic, syphilitic: *Reports indicate that venereal disease has been on the decline since the advent of AIDS.*

vengeance *n.* **1** °revenge, retaliation, °retribution, °requital, °reprisal: *Her vengeance for what he had done was to disinherit him.* **2** ***with a vengeance.*** **(a)** violently, °fiercely, ferociously, wildly, vehemently, furiously, forcefully: *The wind blew with a vengeance all night.* **(b)** energetically, to the fullest extent, to the utmost *or* the fullest *or* the limit, (with) no holds barred, enthusiastically, wholeheartedly: *Give Richard a job to do and he goes at it with a vengeance.*

venial *adj.* forgivable, excusable, pardonable, °tolerable, tolerated, °minor, °petty, °insignificant, unimportant, remittable *or* remissible: *That hypocrite considers even his most heinous crimes to be venial.*

venom *n.* **1** °poison, toxin: *The snakes are "milked" of their venom, which is collected for research in antitoxins and pharmaceuticals.* **2** malice, maliciousness, malevolence, °ill will, malignity, °animosity, °hate, hatred, °hostility, °antagonism, °spite, spitefulness, spleen, °rancor, °bitterness, embitteredness, °gall, °poison, poisonousness, °virulence: *Vinnie was totally unprepared for the venom of the verbal attack.*

venomous *adj.* **1** °poisonous, °deadly, toxic, °dangerous, life-threatening, °lethal: *One must be careful of venomous snakes, lizards, and scorpions in the Southwest.* **2** °poisonous, °virulent, malicious, malevolent, malign, °malignant, °savage, baleful, envenomed, °hostile, antagonistic, °spiteful, splenetic, acerbic, °rancorous, °bitter, °embittered, °mean, °vicious: *That venomous little man caused me to lose my job.*

vent *n.* **1** °opening, °slit, °slot, °hole, °aperture, air hole, blow-hole, spiracle, orifice, °outlet, inlet, funnel, flue, duct, °passage; fumarole, fissure, °pipe, mofette *or* moffette: *The tepee has a vent at the peak to let the smoke of the fire escape.* **2** ***give vent to.*** See 3, below. —*v.* **3** °give vent to, °express, verbalize, °air, articulate, °enunciate, °declare, °voice, °announce, °communicate, °pronounce, °proclaim, °reveal, °release, let go, °let loose, allow to become known, °make known, °blurt out, °make public, °broadcast: *Has Margie ever vented her true feelings about your work as a bomb disposal expert? Arnold vents his spleen on any unfortunate who will stand for it.* **4** °discharge, °release, °emit, °eject, °issue, °empty, °dump, °expel, send *or* pour out *or* forth, °throw out: *It is forbidden to vent those fumes untreated into the atmosphere.*

venture *n.* 1 °risk, °chance, hazardous undertaking, °experiment, °speculation, °gamble, °plunge, °fling: *Their new publishing venture turned out to be quite profitable.* —*v.* 2 dare(say), make bold, °hazard, volunteer, °tender, °offer, °broach, °advance, proffer, °put forward: *Ambrose ventured some nasty remark and was asked to leave.* 3 °jeopardize, °risk, °endanger, °hazard, imperil; °gamble, °bet, wager, °plunge, °put down: *Thurman ventured his last chip on the number 4.*

venturesome *adj.* 1 °daring, °bold, °intrepid, °adventurous, °courageous, plucky, adventuresome, °audacious, °fearless, doughty, °brave, °spirited: *Henderson felt that he was being truly venturesome by sailing single-handed around the Isle of Wight.* 2 °risky, °rash, °reckless, sporting, °game, °daredevil: *Taking one's family into the American West was truly a venturesome undertaking—even in the twentieth century.*

verbal *adj.* 1 °spoken, °oral, vocal, said, uttered, expressed, enunciated, articulated, colloquial, conversational, viva voce, word-of-mouth, unwritten: *We had a verbal agreement that you would behave yourself.* 2 word-for-word, °verbatim, °literal: *We have an accurate verbal transcript of the proceedings.* 3 °word, lexical, vocabulary: *She did very well on the verbal parts of the examination but poorly on the mathematics.*

verbatim *adj.* 1 word-for-word, verbatim et literatim (= 'word-for-word and letter-for-letter'), °literal, °exact, °precise, °accurate, °faithful, °strict: *This verbatim translation may be accurate but all the poetic expressiveness is lost.* —*adv.* 2 word-for-word, verbatim et literatim, °literally, °exactly, °precisely, accurately, faithfully, °to the letter, strictly: *To make certain there would be no misunderstanding, the speech was copied verbatim.*

verge[1] *n.* 1 °edge, °border, °boundary, °margin, °brink, °threshold, °brim: *Frances is jealous of John to the verge of insanity.* 2 ***on the verge of.*** about to, ready to, on the (very) point of, preparing to, soon to: *I was on the verge of revealing everything, but suddenly the torture stopped.* —*v.* 3 Often, ***verge on.*** °border (on), °approach, come close *or* near (to), *Technical* be asymptotic to: *Crobeg's idea of comfort verges on what is more commonly thought of as luxury. Her charitable contributions last year verged on zero.*

verge[2] *v.* °incline, °lean, °tend, °extend, °stretch, °turn; °approach, °draw, °move: *The road verges southward after the bridge. I fear that Pengorm's faculties are verging toward decay.*

verify *v.* affirm, °confirm, testify to, °attest (to), bear witness to, vouch for, corroborate, °support, °substantiate, clinch *or* clench, °prove, °demonstrate, °show, °bear out, °authenticate, validate, °certify, °guarantee, °back up, °warrant: *Can you verify the accuracy of Carbunkel's testimony?*

veritable *adj.* °real, °true, °virtual, °genuine, °actual, °legitimate, °authentic: *Furleigh becomes a veritable monster if denied his daily Nesselrode pie.*

vernacular *adj.* 1 °native, °local, regional, °indigenous, autochthonous: *Old Hebrew fell out of use to be replaced by vernacular Aramaic.* 2 °popular, °informal, colloquial, conversational, °ordinary, °familiar, °everyday, °spoken, °vulgar, vulgate; °plain, °simple, straightforward, °easy: *We have both the vernacular and the formal, literary, or clerical levels of language. Can you translate this scientific gobbledygook into vernacular English that we can understand?* —*n.* 3 °jargon, patois, argot, °cant, °idiom, phraseology, °language, °talk, °speech: *Alf occasionally slips into the vernacular of a Billingsgate fishmonger (his former calling).*

versatile *adj.* 1 °adaptable, °resourceful, all-round, all-purpose, many-sided, multipurpose, multifaceted, °flexible, adjustable, °protean, °dexterous, °handy, facile: *Everett is a versatile painter who works in many media and styles.* 2 °variable, °changeable, °protean, changing, °flexible, fluctuating: *Any versatile organization should be able to accommodate staff losses.*

versed *adj.* Usually, **well-versed in:** well-read *or* (well-)informed in *or* (well-)trained *or* (well-)grounded *or* (well-)schooled *or* (well-)educated *or* (well-)tutored *or* learned *or* cultured *or* lettered *or* cultivated *or* literate *or* competent *or* accomplished *or* skilled in, (well-)posted on, knowledgeable in *or* about, proficient *or* experienced *or* practiced *or* expert *or* good in *or* at, conversant *or* familiar *or* (well-)acquainted with: *The applicant appears to be sufficiently well-versed in several subjects to be qualified to teach them.*

version *n.* 1 °form, °variant, °variation, °type, °model, °style, °kind, °variety, °manifestation, portrayal, °adaptation, °rendition, °interpretation, °adaptation, °construct, construction, °conception, °idea: *This is yet another modern designer's version of 1920s' and 1930s' art deco.* 2 °story, °account, °rendering, °rendition, °translation, °interpretation, reading, °understanding, °view, °side: *Skene's version of what happened is completely distorted.*

vertex *n.* °top, °tip, °extremity, °zenith, meridian, apogee, °peak, apex, °acme, °summit, °pinnacle, °crest, °crown, °cap, height(s): *The building is surmounted by a pyramidal dome at the vertex of which is a statue of Osiris.*

vertigo *n.* dizziness, lightheadedness, giddiness, instability, *Colloq* wooziness: *Vertigo suddenly overcame me, and I grabbed a rail to steady myself.*

verve *n.* °spirit, vivacity, vivaciousness, °vitality, °life, liveliness, °animation, °sparkle, °energy, °vigor, °exuberance, briskness, brio, esprit, élan, °dash, °flair, °panache, °flourish, °enthusiasm, zeal, °zest, °gusto, *Colloq* pizazz, zip, vim, get-up-and-go, zing, oomph: *Elena Cardozo's verve adds much to this production of* Carmen.

very *adv.* 1 °extremely, °truly, °really, to a great extent, °exceedingly, greatly, (very) much, °profoundly, °deeply, acutely, unusually, extraordinarily, uncommonly, exceptionally, remarkably, °absolutely, °completely, °entirely, °altogether, °totally, °quite, °rather, hugely, °vastly, *Dialectal* °right, *Brit* °jolly, *Colloq* damn(ed), °terribly, °awfully, darned, *US dialectal* danged, °plumb, *Slang Brit* bleeding, *Chiefly Brit* bloody: *The Bell Inn has a very fine wine list. Thea feels very attached to all her horses. Hubbins is very sure that the murder weapon was hidden in a sandwich. I shall be there very soon.* 2 most, °extremely, certainly, °surely, °definitely, decidedly, unequivocally, unquestionably, °quite, °entirely, °altogether: *If you ask her out, she is very likely to say no.* —*adj.* 3 °exact, °precise, °perfect; °same, selfsame, °identical, °particular: *She wants the very opposite of what you want. I arrived the very day she left. "He is the very model of a modern major-general."* 4 least, °mere, merest, °bare, barest, °sheer, sheerest; utter, °pure, °simple: *The very thought of war makes me ill. The very mention of her name strikes terror into the hearts of the local residents.*

vessel *n.* 1 container, °receptacle, utensil, holder: *Food was stored in vessels neatly arrayed around the sarcophagus.* 2 °craft, °boat, °ship, ark, *Literary* barque *or US also* bark: *The vessel docked gently at the airlock of the space station.*

vestige *n.* °trace, °suggestion, *soupçon*, °hint, glimmer, °inkling, °suspicion, °sign, °evidence, °mark, °token, °scent, whiff, tinge, °taste; °remnant, °scrap, °fragment, °memorial, °residue, °relic, °remains: *Investigators deny that vestiges of an ancient civilization have been found in the Mariana Trench.*

vestigial *adj.* °imperfect, °undeveloped, underdeveloped, °rudimentary, °incomplete: *Anatomists regard the coccyx in man and apes as a vestigial tail.*

vet *v.* °examine, °review, °investigate, °scrutinize, °inspect, °check (out), °look over, °scan; validate, °authenticate; *Colloq* give (something *or* someone) the once-over, °size up: *We engaged an expert to vet the definitions of culinary terms for the new dictionary.*

veteran *n.* **1** old hand, past master, old-timer, trouper, *Colloq* warhorse: *Barlow was a veteran of twenty years' experience in the service.* —*adj.* **2** °experienced, °practiced, °seasoned, °mature, long-serving, battle-scarred: *Did it never occur to this veteran politician that there are degrees of misconduct?*

veto *v.* **1** °stop, °block, °deny, °ban, °turn down, °reject, disallow, °rule out, °quash, °prevent, °prohibit, interdict, °taboo, °outlaw, proscribe, °preclude, *Colloq* put the kibosh on, °kill, nix: *Father vetoed the idea that we go to Brighton for the holiday. Any measure he proposed was summarily vetoed.* —*n.* **2** °denial, °ban, stoppage, °block, °embargo, turndown, °rejection, disallowance, quashing, °prevention, °prohibition, interdiction, °taboo, proscription, preclusion, vetoing, *Colloq* °killing, nixing: *It was the president's veto of the abortion measure that angered the women's groups.*

viable *adj.* sustainable, °supportable, °sensible, °reasonable, °practical, °practicable, °applicable, workable, °feasible, °possible: *We have three days to come up with a viable plan for continuing the research project.*

vibes *n.* pl. vibrations, °feelings, sensations, resonance(s), °rapport, empathy, °sympathy: *I got really good vibes the minute I walked in here, Man.*

vibrate *v.* °quiver, °shiver, °shudder, °fluctuate, °quake, °shake, °tremble, throb, °pulsate, °oscillate, °pulse, reverberate, resonate, *Brit* judder: *The earth began to vibrate, the windows rattled, and the dishes danced off their shelves.*

vicarious *adj.* surrogate, delegated, deputed, commissioned, assigned, °indirect, substituted: *He derived vicarious pleasure from seeing his children do the things that he no longer could.*

vice *n.* **1** immorality, corruption, °evil, badness, depravity, °degradation, degeneracy, iniquity, villainy, venality, evildoing, wickedness, °profligacy, °sin, sinfulness, °transgression: *Would that all vice had been destroyed with Sodom and Gomorrah.* **2** °flaw, °defect, °fault, °imperfection, °blemish, °shortcoming, °failing, °weakness, °frailty, °foible, °infirmity, deficiency: *Rejection without due consideration is a prejudicial vice of editors.*

vice versa *adv.* conversely, contrariwise, to *or* on the contrary, reversed, the other way around: *The older the bird, the tougher its meat, and vice versa.*

vicinity *n.* °area, °neighborhood, °locale, vicinage, environs, locality, °precincts, purlieus, °territory: *All people living in the vicinity of the nuclear plant were tested for radiation poisoning.*

vicious *adj.* **1** °immoral, unprincipled, amoral, barbarous, °corrupt, °evil, °bad, °base, depraved, °vile, °atrocious, execrable, degraded, °degrading, °degenerate, °venal, iniquitous, heinous, odious, °perverted, nefarious, °wicked, flagitious, °devilish, diabolic(al), °fiendish, °monstrous, °profligate, °shameful, °shameless, °abominable, °sinful: *Hitler's vicious, bloody dictatorship lasted only a dozen years.* **2** malicious, °spiteful, °mean, °nasty, °hateful, malevolent, °malignant, °bitter, acrimonious, °rancorous, °venomous, °vindictive, defamatory, °slanderous, °scandalous, *Slang* °rotten, bitchy: *When they argued, they said terribly vicious things to one another.* **3** °savage, °wild, °untamed, °ferocious, °fearful, °brutal, °fierce, °fiendish, bestial, feral, brutish, ravening, *Literary* °fell: *The neighbors complained that Mr. Delanty kept a vicious Rottweiler.*

vicissitude *n.* **1** °change, °mutation, °alteration, changeability, mutability, °variation, variability, °variety, °alternation, °flux, °fluctuation, unpredictability: *Who seeks stability in life often encounters its vicissitude.* **2** ***vicissitudes.*** fluctuations, changes, variations, contrasts, °inconstancy, unpredictability, uncertainties, *Colloq* ups and downs, flukiness: *His success is amazing in light of the vicissitudes he survived.*

victim *n.* **1** sufferer, martyr, °casualty, °scapegoat, sacrificial lamb, injured party: *It is a pity that you had to be the victim of her greed.* **2** °dupe, gull, °fool, °butt, fair game, *Colloq* chump, fall guy, *Slang* °sucker, °sap, *Chiefly US and Canadian* patsy, *US* schnook, schlemiel *or* schlemihl *or* shlemiel: *She was an easy victim for any confidence man.*

victimize *v.* **1** °prey on, °pursue, go after, °pick on, °bully, take advantage of, °persecute, °exploit, °use: *Increasingly, older people are being victimized by unscrupulous relatives.* **2** °cheat, °swindle, bilk, °defraud, °dupe, °hoodwink, °deceive, gull, °fool, °trick, °outwit, °outsmart, outfox, *Colloq* snooker, flimflam, *Slang* suck *or* sucker in, °screw, °shaft, °take (in), rook: *The confidence team used the old Spanish handkerchief trick to victimize unsuspecting tourists.*

victor *n.* °winner, °champion, conqueror, prizewinner: *The victors in the semifinals are pitted against each other to determine the titleholder.*

victorious *adj.* °triumphant, °successful: *Caesar was victorious over the Gauls.*

victory *n.* °triumph, °conquest, °supremacy, °superiority, °success, overcoming, mastery, °winning, quelling, crushing: *Alexander's victory over Darius at Arbela was one of the decisive battles of history.*

vie *v.* °compete, contend, °struggle, °strive: *The brothers constantly vied for their father's affection.*

view *n.* **1** °outlook, °aspect, °prospect, °scene, °perspective, vista, panorama, °spectacle, °picture, °tableau; °landscape, seascape, cityscape: *From the tower we had a view of the devastation below.* **2** °opinion, °point of view, °approach, °position, °judgment, °belief, way of thinking, °conception, °understanding, °impression, °feeling, °sentiment, °notion: *Denham's view is that the country's defenses are inadequate.* **3** °aspect, °angle, °position, °prospect, °perspective, °vision, °representation, °projection: *Here is a termite's-eye view of the timbers in your house.* **4** inspection, °survey, °vision, °sight, °observation, °scrutiny, °examination, contemplation, °study: *The shroud was pulled away, exposing the revolting object to our view.* **5** °aim, °direction, °intent, °intention, °purpose, °objective, °object, °expectation, °prospect, °vision, °hope, °dream: *We cultivated their friendship with a view to being invited aboard their yacht.* **6** ***in view of.*** °in light of, °considering, in consideration of, because of, on account of: *In view of your failure to pass any exam, we must expel you from school.* —*v.* **7** look at *or* upon *or* over, °see, °take in, °watch, °observe, °scrutinize, °examine, °regard, °behold: *From this vantage point one can view the entire valley.* **8** °witness, °see, °watch, °observe, °take in: *We viewed the entire robbery as it took place.* **9** °regard, °consider, °think of, look on *or* upon, °judge, deem, °believe, °hold, °estimate, °rate, °gauge, assess: *Control views Craven as a threat.*

viewpoint *n.* °standpoint, °(point of) view, °attitude, °angle, °slant, °position, °stance, vantage point, °perspective, frame of reference, way of thinking, °context: *From our viewpoint, widening the road only encourages speeding.*

vigilance *n.* watchfulness, alertness, °observance, guardedness, circumspection, attentiveness, °caution: *While the dog remains loose, we must exercise vigilance.*

vigilant *adj.* watchful, °alert, °sharp, °observant, °guarded, circumspect, °attentive, °wakeful, °cautious, °careful, °wary, chary, on one's guard, on the alert, on the lookout, °eagle-eyed, hawk-eyed, Argus-eyed, on the qui vive, on one's toes, with one's eyes open, *Colloq* with one's eyes skinned *or Colloq* peeled: *If you are vigilant, they will never be able to sneak up on you.*

vigorous *adj.* °energetic, °active, °vivacious, °dynamic, °brisk, °lively, °spirited, °robust, °strong, °hardy, °hale, °hearty, °vital, °fit, °lusty, °stalwart, in good *or* fine fettle, spry, °sprightly, resilient, *Colloq* peppy, full of pep, full of get-up-and-go, full of beans: *At sixty-three, he is just as vigorous as he was at forty.*

vigorously *adv.* energetically, actively, vivaciously, dynamically, briskly, spiritedly, robustly, strongly, hardily, heartily, lustily, stalwartly, eagerly, with might and main, °with a vengeance, strenuously, *Colloq* °like mad, like crazy, hammer and tongs: *It may not be good for you to exercise vigorously every day.*

vigor *n.* °vitality, °resilience, °strength, °power, °energy, forcefulness, °force, °stamina, °endurance, mettle, mettlesomeness, °pith, °dynamism, °spirit, liveliness, °animation, °verve, vivacity, °exuberance, brio, briskness, °zest, zealousness, °enthusiasm, °gusto, °eagerness, *Colloq* °spunk, °pep, pizazz, vim, oomph, zing, get-up-and-go: *When attacked, they defended themselves with extraordinary vigor.*

vile *adj.* **1** °base, abject, °contemptible, debased, °degenerate, depraved, °bad, iniquitous, execrable, °atrocious, °sordid, °immoral, amoral, °wicked, °evil, °sinful, hellish, °fiendish, ignoble, °revolting, °despicable, horrid, °horrible, °dreadful, °terrible, °corrupt, °mean, °wretched, °miserable, °degrading, ignominious, °disgraceful, °shameful, °shameless: *In every age, the vilest specimens of human nature are to be found among demagogues.* **2** °disgusting, °nasty, sickening, °nauseous, nauseating, °foul, °loathsome, °offensive, noxious, °repulsive, °repellent, °repugnant: *It is, indeed, a vile disease, but at least it is not contagious. She had prepared some vile cocktail that I refused to drink.*

vilify *v.* °depreciate, devalue, deprecate, °debase, °disparage, denigrate, °diminish, traduce, defame, speak ill of, revile, °slander, °libel, °abuse, defile, °sully, °smear, °tarnish, malign, calumniate, asperse, °run down, decry, *Rare* vilipend, *Colloq US* bad-mouth: *It was shocking the way Rudolph stood up in the council and vilified Dudley.*

villain *n.* °wretch, evildoer, °criminal, °miscreant, blackguard, °rogue, °rascal, cad, scallywag *or* scalawag, malefactor, °scoundrel, dog, cur, viper, reptile, snake in the grass, rat, *Archaic* rapscallion, *Literary* knave, caitiff, *Colloq Brit* bounder, blighter, *Slang* bastard, son of a bitch, *Brit* rotter, *US* s.o.b *or* S.O.B. *or* SOB (= 'son-of-a-bitch'): *They caught the villain they suspect of being the Seven Dials strangler.*

villainous *adj.* **1** treacherous, °perfidious, °dishonest, °unscrupulous, °traitorous, °corrupt, °faithless, °criminal, felonious, °murderous, *Colloq* °crooked, °bent: *Few activities are more villainous than being a double agent.* **2** See **vile, 1,** above.

vindicate *v.* **1** °clear, exonerate, absolve, acquit, exculpate, °excuse: *She was vindicated of any complicity in the affair.* **2** °justify, °support, °uphold, °prove: *Subsequent events vindicated his actions.*

vindictive *adj.* avenging, vengeful, vindicatory, revengeful, retaliatory, °spiteful, unforgiving, splenetic, °resentful, °rancorous, °implacable: *She later turned vindictive, attacking everyone for real or imagined slights.*

vintage *n.* **1** year, crop, °harvest, °origin, °generation: *Which vintage is the better for Bordeaux, 1949 or 1954?* —*adj.* **2** °quality, °choice, °superior, °better, °good, °select, °best, °classic; °aged, °seasoned, mature(d), mellow(ed): *Mark has become an expert in vintage wines.* **3** °antiquated, °old-fashioned, old-fogy(ish), °antique, °bygone, old-time, collector *or* collector's, *Colloq* over the hill: *Clempton drives his vintage tractor to work every day.*

violate *v.* **1** °break, °breach, °disobey, °disregard, contravene, °infringe, °ignore: *He habitually violated the law by carrying a pistol.* **2** °dishonor, °desecrate, °profane, defile, °degrade, °debase, treat irreverently: *Vandals were found to have violated the sacred shrine.* **3** °rape, debauch, °ravish, °ravage, °molest, °attack, °assault, °outrage: *The victim identified the man who had violated her.*

violation *n.* **1** °infringement, °breach, °disregard, disobedience, contravention, °abuse; ignoring, infringing, breaching, disregarding, disobeying, contravening, abusing, violating: *Parking here is a violation of the law. Violation of the rules will be dealt with summarily.* **2** profanation, profaning, °sacrilege, desecration, desecrating, defilement, defiling, °degradation, °degrading, °dishonor, dishonoring, debasement, debasing, violating: *The police regard as serious the violation by graffiti of a place of worship or a cemetery.* **3** °rape, ravishment, molestation, °attack, °outrage, °assault, violating: *The defendant is charged with the violation of three women.*

violence *n.* **1** (brute *or* physical) force, °might, mightiness, °power, °strength, °severity, °intensity, °energy, vehemence, ferocity, ferociousness, fierceness, °fury, °vigor; destructiveness, °virulence: *The violence of the storm continued unabated throughout the night. The violence of the seizures diminished after she was given the medication.* **2** bestiality, brutality, °barbarity, savagery, cruelty, bloodthirstiness, wildness, ferocity, °frenzy, °fury, °passion, fierceness, vehemence, murderousness: *The marauders attacked with a violence that terrified the villagers.* **3 do violence to. (a)** °harm, °damage, °injure: *The scandal did violence to her reputation. Humphry's singing could do violence to any melody.* **(b)** °warp, °twist, °distort: *In reporting that her views are antifeminist, the newspaper did violence to the import of her speech.*

violent *adj.* **1** °wild, °physical, °destructive, vehement, °brutal, brutish, °beastly, °nasty, °cruel, °mean, barbarous, °inhuman, °savage, °fierce, °ferocious, °furious, frenzied, uncontrollable, °untamed, °ungovernable, raging, °raving, irrational, °insane, crazed, *Colloq* fit to be tied: *Tony turned into a violent maniac when he suspected me of seeing another man.* **2** °harmful, °injurious, damaging, °detrimental, °destructive, deleterious, catastrophic, cataclysmic, °ruinous, °devastating: *This was the most violent earthquake to hit Mexico in recorded history.* **3** °acute, °serious, °severe, °extreme, °harsh, °trenchant, °virulent, °intense, °energetic, °forceful, vehement, °passionate, °impetuous, °tempestuous: *One more violent outburst like that and you will be removed from the courtroom.*

virtual *adj.* °effective, °essential; °practical, °understood, accepted: *Traffic on the highway comes to a virtual standstill at certain times of day.*

virtually *adv.* essentially, effectively, °practically, °almost, °to all intents and purposes, for all practical purposes, more or less, °nearly, as good as, °substantially, °in effect, °in essence: *He had virtually won the Brisbane-London sailing race when he passed the Lizard, but then he hit a rock.*

virtue *n.* **1** °morality, high-mindedness, °honor, goodness, justness, righteousness, fairness, °integrity, right-mindedness, °honesty, °probity, uprightness, °rectitude, decency, °worth, worthiness, °nobility, °character, respectability: *You will find the Billingses to be people of unassailable virtue.* **2** virginity, °chastity, chasteness, °honor, innocence, °purity: *As Healey had six daughters, he was kept quite busy protecting their virtue.* **3** °quality, °credit, °strength, good point, °asset: *Opposing the seven deadly sins are the three theological virtues, faith, hope, and charity, and the four cardinal virtues, fortitude, justice, prudence, and temperance.* **4** ***by virtue of.*** by dint of, °owing to, °thanks to, °by reason of, because of, on account of: *Gerry was a part owner by virtue of his investment in the company.*

virtuosity *n.* °(technical) skill, °technique, °ability, °expertise, mastery, °excellence, °brilliance, craftsmanship, °craft, °flair, °dash, élan, éclat, °panache, pyrotechnics, showmanship, °show, staginess, *Colloq* razzle-dazzle: *The concerto was performed with signal virtuosity but little imagination or understanding.*

virtuoso *n.* **1** °master, maestro, °expert, °genius, °talent, °prodigy, old hand, *Colloq* wizard, whizz *or* whiz *or* wiz, whizz-kid *or* whiz-kid *or* wiz-kid, *Chiefly Brit* °dab hand, *US* maven *or US* mavin *or US* mayvin: *Vincent is a veritable virtuoso on the vibraharp.* *—adj.* **2** °masterful, masterly, °expert, °talented, °brilliant, °dazzling, bravura, prodigious, °excellent, °superb, °extraordinary, °exceptional, °superior, °first-rate, °superlative, °matchless, °peerless, °sterling, °marvelous, °remarkable: *His virtuoso performance of "The Flight of the Bumblebee" won him a standing ovation.*

virtuous *adj.* **1** °moral, °honorable, °ethical, °honest, °good, upstanding, high-principled, °upright, °righteous, °pure, uncorrupted, incorruptible, °just, °fair, right-minded, fair-minded, high-minded, °scrupulous, °trustworthy: *Thomas has proved himself a virtuous young man.* **2** °chaste, °innocent, virginal, virgin; °decent, °proper, unsullied, °faithful, °true, uncorrupted: *A virtuous woman makes a virtuous wife.*

virulence *n.* **1** virulency, poisonousness, venomousness, toxicity, noxiousness, deadliness, perniciousness, injuriousness, destructiveness, malignity, malignancy, °violence, balefulness: *The virulence of his affliction was not recognized till it was too late.* **2** virulency, acrimony, acrimoniousness, °bitterness, acerbity, °rancor, spleen, °poison, poisonousness, °venom, venomousness, malignity, malevolence, maliciousness, malice, °spite, °hostility, °resentment, °antagonism, hatred: *We were taken aback by the virulence of the priest's attack on Miss Thompson's morals.*

virulent *adj.* **1** °lethal, life-threatening, °deadly, °fatal, pernicious, septic, °poisonous, toxic, baleful, noxious, °dangerous, °harmful, °injurious, °detrimental, deleterious, °destructive, °unhealthy, °unwholesome: *Chorley died from a virulent form of malaria he caught in the tropics.* **2** °vicious, °venomous, °bitter, °spiteful, °malignant, malign, malicious, malevolent, °poisonous, splenetic, acrimonious, acerbic, acid, mordant, °sarcastic, °nasty, °trenchant, °caustic, antagonistic, °hateful, °hostile: *What did I do to the reviewer of* The Herald *to merit such a virulent review?*

visible *adj.* **1** seeable, perceivable, °perceptible, °discernible, detectable, discoverable, °noticeable, unmistakable *or* unmistakeable, °clear, °obvious, °observable; visual: *Are the people in the car visible through the dark glass? The visible part of the energy spectrum is relatively small.* **2** °obvious, °conspicuous, °evident, °apparent, °prominent, °manifest, °distinct, °patent, well-defined, identifiable: *The ultraconservative influence in the electorate is less visible today.*

vision *n.* **1** eyesight, acutance, °perception, °sight: *The optician said there is nothing wrong with MacKenzie's vision.* **2** farsightedness, °understanding, °imagination, °foresight, foresightedness, °insight: *MacKenzie has brought great vision to this scheme for developing the Australian outback.* **3** °view, °perspective, °perception, envisioning, envisaging, °dream, °idea, °plan, °scheme: *MacKenzie's vision of making the desert bloom seemed impossibly remote.* **4** °phantom, apparition, chimera, °delusion, °hallucination, mirage, °specter, °shade, eidolon, revenant, phantasm, materialization, °illusion, °ghost, wraith: *A vision appeared to MacKenzie as he slept, telling him that her name was Alice Springs.* **5** sight for sore eyes, °(welcome) sight, °dream, °epitome: *To MacKenzie, Alice was a vision of great beauty, and he proposed marriage.*

visionary *adj.* **1** °dreamy, °speculative, unpractical, °impractical, °fanciful, °imaginary, °unrealistic, °unreal, °romantic, °idealistic, unworkable, utopian: *It is a pity that none of MacKenzie's visionary plans ever came to pass.* *—n.* **2** °dreamer, idealist, °romantic, fantast, wishful thinker, Don Quixote: *The world needs more visionaries like MacKenzie.*

visit *v.* **1** (go *or* come to) see, call (in *or* on *or* upon), look in on, stop in *or* by, *Colloq* pop in *or* by, drop in (on), °take in: *I plan to visit Leslie. Did you visit the Smithsonian Institution?* **2** °afflict, °attack, befall, °fall upon, assail, °seize, smite, °scourge, descend upon, °inflict, °affect: *Horrible diseases and tortures visited those who denied God.* *—n.* **3** °stay, °call, °sojourn, °stop, stopover: *The minister will be here for only a brief visit.*

visitation *n.* **1** staying, °calling, visiting, sojourning, stopping (over): *The father has visitation rights to his children on weekends.* **2** °affliction, °ordeal, °trial, °punishment, °disaster, °catastrophe, cataclysm, °calamity, °tragedy, °curse, °scourge, °blight, °plague, °pestilence: *The tyrant succumbed finally to a visitation from the Almighty.*

visitor *n.* caller, °guest, °company; visitant: *May I phone you back after my visitors leave?*

vital *adj.* **1** °imperative, °essential, °necessary, needed, requisite, required, °indispensable, °mandatory, compulsory, °cardinal, °fundamental, °basic, °critical, °crucial, °central, °pivotal: *Air and water are vital for the existence of most known organisms.* **2** °important, °key, °central, °critical, °crucial, life-or-death, °pivotal, °paramount, °main: *The captain must make the vital decision of whether to abandon ship. This is a matter of vital concern for us all.* **3** °lively, full of life, °vivacious, °spirited, °vigorous, °dynamic, °alive, °animated, °brisk, °energetic: *We have lost an extremely vital member of the community, one who took a vital interest in community affairs.* **4** °invigorating, quickening, life-giving, animating, vitalizing, reviving, vivifying, enlivening, rejuvenating: *I could feel the vital energies returning to my limbs.*

vitality *n.* **1** °energy, °life, life force, °vigor, °power, °intensity, °force, liveliness, , vivacity, vivaciousness, °animation, °sparkle, spiritedness, °exuberance, *Colloq* zing, °pep, pizazz, oomph, get-up-and-go, zip, vim: *Godot, there, has the vitality of a forty-year-old.* **2** °stamina, hardiness, °endurance, °energy, °strength, robustness: *Does this society have the vitality needed to take it into the next decade?*

vitalize *v.* °stimulate, °activate, °arouse, vivify, °animate, °awaken, inspirit, invigorate, °enliven, °inspire, °revive, °rejuvenate, innervate, °energize, °fortify, reinvigorate, °renew, °refresh, °charge (up): *The tonic has vitalized Clement to the point where he feels young again.*

vitiate *v.* **1** °spoil, °ruin, °harm, °impair, °mar, °sully, °corrupt, °pervert, °contaminate, °adulterate, °weaken, °degrade, °downgrade, °depreciate, °diminish, °depress, vulgarize, °lower, °reduce, °undermine: *Words appropriated from other languages tend to refine rather than vitiate English.* **2** °debase, deprave, °pervert, °corrupt, °demoralize, defile: *Despite other strengths his entire character was vitiated by overriding avarice.* **3** invalidate, °destroy, °delete, °cancel, nullify, annul, °revoke, °void, abrogate, °abolish, °withdraw, °quash, °suppress: *The invoice was effectively vitiated by the plaintiff's failure to deliver the goods.*

vituperate *v.* °berate, °rate, reproach, revile, °vilify, execrate, °abuse, °denounce, decry, deprecate, °disparage, devalue, °diminish, °put down, °run down, devaluate, °depreciate, °blame, inculpate, censure, find fault with, °attack, assail, °castigate, °scold, °reprimand, °upbraid, °rebuke, chide, °chasten: *From the pulpit he continued to vituperate the vices of the court.*

vituperative *adj.* °abusive, calumniatory, calumnious, °scurrilous, °derogatory, belittling, depreciatory, depreciative, detractory, °contemptuous, damning, denunciatory, denigrating, deprecatory, censorious, aspersive, defamatory, °slanderous, libelous, castigatory, condemnatory, malign, °scornful, °withering, °harsh, °sardonic, °sarcastic, °biting, acid, contumelious, opprobrious, insulting, *Formal* vilipenditory, *Colloq* down-putting: *Why should he have directed this vituperative speech at the very person whom he loves?*

vivacious *adj.* °lively, °spirited, °sprightly, °energetic, °animated, °brisk, °ebullient, °effervescent, °bubbly, °gay, °cheerful, °happy, °blithe, °jaunty, lighthearted, °sunny, °merry, high-spirited, °buoyant, chipper, *Colloq* up, peppy, full of pep, full of beans, zippy: *Had he not gone on the cruise, Michael would never have met the vivacious redhead who became his bride.*

vivid *adj.* **1** °intense, °strong, °brilliant, °fresh, °bright, °dazzling, lucid, °rich, °clear, colorful, °glowing: *Michelle looked splendid in a silk dress of vivid yellow.* **2** °clear, °detailed, °sharp, °realistic, °graphic, true to life, °lifelike, °distinct, °powerful, °strong, °memorable, °dramatic, °striking: *I have a vivid mental picture of our first meeting, Count Cagliostro.* **3** °prolific, °fruitful, °fertile, fecund, inventive, °creative: *That kiss, which never took place, is a figment of Bagley's vivid imagination.*

vocalist *n.* °singer, soloist, choir boy *or* girl, choir member, chorus boy *or* girl, chorus member, chorister, caroler; diva, prima donna, chanteuse; cantor, crooner; *Colloq* songbird, canary, thrush, nightingale: *The vocalist goes on at ten and at midnight.*

vocation *n.* °calling, °trade, métier, °business, °profession, °occupation, °career, °employment, °job, °pursuit, lifework, °line (of work), *Slang* °bag, °thing: *From his diminutive figure, you would never guess his vocation.*

vogue *n.* **1** °fashion, °mode, °style, °look, °taste, °trend, °rage, °craze, last word, *dernier cri,* (latest) thing, *Colloq* °fad, the latest: *Longer skirts were once again the vogue.* **2** °popularity, °favor, °preference, acceptance, currency, °prevalence, fashionableness: *They are capitalizing on the vogue for torn jeans.*

voice *n.* **1** °speech, utterance, articulation, °words, °expression: *Kenting found it difficult to give voice to his innermost thoughts.* **2** °share, °part, °vote, participation, °say, °decision, °option, °turn, °chance: *A junior member of the board, Nerissa had no voice in major decisions.* **3** spokesman, spokeswoman, spokesperson, °representative, °agent, °agency, °instrument; °organ, °medium, °vehicle, forum, °publication: *For all his years in government, Logan had served as the voice of the miners. Though it represents only a few,* The Clarion *views itself as the voice of all the people.* —*v.* **4** °express, utter, articulate, °enunciate, °present, verbalize, put into words, give utterance *or* voice *or* expression *or* vent to, °communicate, convey, °declare, assert, °make known, °reveal, °disclose, °raise, °bring up, °air: *I must voice my misgivings about the step you are planning to take.*

void *adj.* **1** null and void, °invalid, not (legally) binding, inoperative, unenforceable, °ineffectual, °futile, °ineffective, °vain, unavailing, °idle, °useless, °pointless, °bootless: *Owing to irregularities, the election was declared void. If a card is turned face-up, the deal is void.* **2** °empty, °vacant, unoccupied, °unused, unutilized, unfilled, °blank, °clear; °deserted: *It is up to the council to appoint new members as places become void.* **3** ***void of.*** devoid of, without, lacking, °destitute of: *The man was utterly void of imagination. In a minute, the room was void of people.* —*n.* **4** °emptiness, vacantness, vacuum, blankness, nothingness: *Without a word, he leaped off the edge and plunged into the void.* **5** °space, °niche, °slot, °opening, °place, °vacancy, °gap, °emptiness: *Who will fill the void left by Edgar's departure?* —*v.* **6** nullify, annul, °cancel, °delete, disannul, declare *or* render null and void, invalidate, °quash, °vacate, °discharge, °abandon, disestablish, °neutralize, disenact, set *or* put aside, rescind, °reverse, abnegate, abrogate: *They had to void the accusation for lack of evidence.* **7** °evacuate, °discharge, °expel, °emit, °purge, °clear, °empty, °drain, °eject; °pass, excrete, °urinate, °defecate: *The tanker went far out to sea before voiding its tanks. This medication will help in voiding the patient's bowels.*

volatile *adj.* **1** vaporizing, evaporable, evaporative: *A volatile solvent, which evaporates quickly, is the vehicle for thermosetting plastics.* **2** °changeable, °fickle, °flighty, °inconstant, °erratic, °restless, °unstable, °variable, mercurial, °capricious: *Carfield is so volatile that one could never predict what he would do next.* **3** °explosive, hair-trigger, °sensitive, charged, eruptive, °tense, tension-ridden: *Be careful of Christine's volatile temper.*

volition *n.* °(free) will, °choice, °option, choosing, °choice, °discretion, °preference: *She pursued a course of her own volition.*

volley *n.* **1** salvo, bombardment, barrage, cannonade, fusillade, °discharge, °hail, °shower: *A volley of shot struck the tower, which collapsed.* **2** °outpouring, °torrent, °flood, deluge, inundation, °burst, °storm, outbreak: *The cracks of the muleteer's whip were accompanied by a volley of abuse.* **3** °give-and-take, to-and-fro, interaction, reciprocity, °exchange, volleying, crossfire, badinage, bantering: *Samuel enjoyed the volley of dickering that accompanied every sale.*

voluble *adj.* °talkative, °glib, °fluent, loquacious, garrulous, chatty, °profuse, gossipy, °exuberant, long-winded, °bombastic, °windy, °wordy, *Colloq* blessed with the gift of the gab: *The Holdens were quite voluble in their account of every single event—and non-event—that took place on their vacation.*

volume *n.* **1** °amount, °quantity, °supply, °mass, °bulk, °abundance, sum total, aggregate: *The volume of ore from the mine created a pile hundreds of feet high. Her volume of output is greater than yours and mine together.* **2** °capacity, °size, °measure: *The volume of this bottle is not more than a liter.* **3** loudness: *Turn down the volume on that ghetto-blaster.* **4** °book, tome: *I bought a twenty-volume set of Dickens.*

voluminous *adj.* **1** °large, °extensive, °great, °spacious, capacious, °expansive, °roomy, °loose, °ample, °big, °ample, °bulky, cavernous, copious, °massive, °huge, °substantial, tremendous, °enormous, °gigantic, mammoth, °vast: *Rose's voluminous skirt was supported by a farthingale.* **2** oversized, outsized, °ample, billowing: *The voluminous spinnaker filled and we wafted down the estuary at hull speed.*

voluntarily *adv.* °freely, °willingly, spontaneously, of (one's) own free will, on (one's) own (initiative *or* recognizance *or* responsibility), without prompting, without being prompted *or* asked, gratis, gratuitously; by choice, intentionally, purposely, °on purpose, °deliberately: *These students spent their summer vacation voluntarily helping the needy.*

voluntary *adj.* **1** °free, elective, °willing, °spontaneous, °unsolicited, unbidden, °unasked, °gratuitous, contributed: *All work done in behalf of the cause is voluntary.* **2** discretionary *or* discretional, unconstrained, °intentional, °willful, °deliberate, intended, °premeditated, planned, volitional, °optional: *His confession was entirely voluntary, and he was not coerced in any way.*

voluptuous *adj.* **1** °sensual, sensualistic, °sensuous, °luxurious, voluptuary, sybaritic(al), hedonist(ic), pleasure-seeking, pleasure-loving, luxury-loving, (self-)indulgent: *He longed to taste once again the voluptuous delights of the Corinthian court.* **2** °seductive, °attractive, °desirable, °beautiful, °tempting, °inviting, appealing, enticing, alluring, °ravishing, °luscious, °delicious, °gorgeous, °shapely, °buxom, well-proportioned, well-endowed, well-built, *Colloq* curvaceous, °sexy, eye-filling, *Slang* (well-)stacked, busty, *US* built: *As he lolled on silken cushions, voluptuous houris danced round him in tightening circles.*

vomit *v.* spew out *or* up, spit up, belch forth; °regurgitate, °throw up, °gag, retch, °heave, *US* keck, *Colloq* puke, °return (food), *Brit* sick up, *Slang chiefly Australian* chunder, *US* barf, upchuck, toss (one's) cookies, spiff (one's) biscuits: *The ancient steam engine vomited smoke and cinders from its huge stack. The very thought of going on a sea voyage almost made me vomit.*

voracious *adj.* **1** insatiable, °gluttonous, °ravenous, ravening, °rapacious, piggish, °hoggish, cormorant, predacious, edacious, devouring, °greedy, °avaricious, esurient, uncontrollable, °uncontrolled, °unquenchable, °enormous, °prodigious, *US* cormorant: *Is there no satisfying the man's voracious appetite?* **2** °thirsty, °hungry, °desirous, avid, °eager, zealous, °enthusiastic, °fervent, fervid, °ardent, °earnest, °passionate, °devoted: *It was becoming difficult to satisfy the public's voracious demand for gossip.*

vote *n.* **1** ballot, ticket, show of hands; referendum, °plebiscite: *We must have a vote on the issue.* **2** °suffrage, franchise: *Women did not have the vote in those days.* **3** °opinion; voter, elector: *Do you think your platform can win over the liberal vote?* —*v.* **4** opt, °choose, come out (for *or* against), express *or* signify (one's) opinion *or* preference *or* desire: *Many people vote for a party rather than an individual.*

vouch *v.* Usually, **vouch for.** support, °guarantee, °back (up), °endorse, °certify; °uphold, °sponsor, °bear witness, attest to: *My neighbors will vouch for me. Can she vouch for your not having left the house all evening?*

vouchsafe *v.* **1** °offer, °give (up), °yield, °accord, °supply, °grant, °impart, °bestow, deign *or* condescend to give: *The minister vouchsafed no information regarding interest rates.* **2** °permit, °allow, °suffer: *The government vouchsafed the hostages safe passage out of the country.*

vow *v.* **1** °swear, °pledge, °promise, °assure, °state, °declare, give (one's) (solemn) word (of honor): *I vowed to return after the war to care for those who had helped me escape.* —*n.* **2** °oath, °pledge, °promise, °agreement; °(solemn) word (of honor): *She will keep her vow, you may be certain.*

vulgar *adj.* **1** indelicate, °boorish, uncultured, uncultivated, °low, °unrefined, °common, °plebeian, inelegant, unladylike, ungentlemanly, gauche, uncouth, °coarse, °tasteless, °ostentatious, ignoble, low-class, *Colloq* °flash: *Clumbert's wife is known for her vulgar behavior at the dinner table.* **2** °tasteless, indelicate, °indecent, °rude, °crude, °naughty, °dirty, °improper, °off-color, °risqué, ribald, °blue, indecorous, °nasty, °offensive, °gross, °lustful, °obscene, °lewd, °lascivious, licentious, smutty, salacious, scatologic(al), °filthy, °pornographic, *Slang US* raunchy: *He made his money publishing vulgar magazines and videos.* **3** °popular, °vernacular, °ordinary, °everyday, °general, °homespun, commonplace, household, °average: *The dictionary should record the vulgar language as well as the literary.*

vulgarity *n.* **1** coarseness, lack of refinement *or* sophistication, crudeness, rudeness, °indelicacy, tawdriness, baseness, humbleness, unsophistication, gaucherie, gaucheness, ignobility: *When you consider his upbringing, you can scarcely blame him for his vulgarity.* **2** °impropriety, lewdness, grossness, foulness, vileness, filthiness, obscenity, *Slang US* raunchiness: *The magazine was refused an import license because of its vulgarity.*

vulnerable *adj.* exposed, °defenseless, °weak, °sensitive, unprotected, °unguarded, unshielded, °helpless, °powerless: *She felt vulnerable to those who prey on the elderly.*

W

wad *n.* **1** °pad, °mass, °lump, °clod, ball, °plug, chunk, hunk, °block, °pack: *"Why do you keep a wad of cotton in each ear?"—"I beg your pardon?"* **2** °roll, pocketful, °heap, °quantity, °load, *Colloq US* bankroll: *Luke carries a wad of money that could choke a horse.*

waddle *v.* toddle, °shuffle, wobble *or* wabble, °totter, °paddle, °pad, waggle, duckwalk, *Brit dialect* wamble: *He waddled down the road, a shopping bag on each arm.*

wade *v.* **1** ford, °cross, °traverse, °walk, make one's way: *The water is only knee-deep, and you can wade across.* **2** °paddle, °play, °splash: *There is a shallow pool where the little children can wade.* **3** ***wade in*** *or* ***into.*** **(a)** °enter, get in (*or* into), °join (in): *When the fight began, Patrick waded in with arms flailing.* **(b)** °attack, °approach, get *or* set to work, plunge *or* dive into: *I have to wade into an enormous pile of mail that accumulated while I was away.* **4** ***wade through.*** plow *or* plough through, work (one's) way through, hammer *or* pound away at, plod through, peg away at: *You have to wade through that entire book to write your report?*

waffle *v.* **1** Often, ***waffle on.*** °carry on, °jabber (on), °prattle (on), prate, blather (on *or* away), *Colloq* run on, *Brit* witter (on), natter (on), rabbit on, *Slang* run off at the mouth: *He kept waffling on about his new computer and, frankly, I wasn't listening.* **2** °equivocate, hedge, °quibble, °shuffle, tergiversate, hem and haw, prevaricate, beat about *or* around the bush, *Colloq* fudge: *I wish he would stop waffling and say what he means.* —*n.* **3** °talk, °palaver, verbiage, °prattle, twaddle, blather, prolixity, wordiness, °jabber, gibber-jabber, *Colloq* °hot air: *All I got was a lot of waffle when I asked for the best interest rate on a loan.*

waft *v.* **1** °drift, °float, °blow, whiff, be borne *or* carried *or* transported: *The scent of jasmine wafted to me on the warm breeze.* —*n.* **2** °breath, °suggestion, °puff, whiff, °hint: *A waft of cool, fresh air momentarily eased the stench of the dungeon.*

wag[1] *v.* **1** °wave, waggle, °oscillate, °fluctuate, °sway, undulate, °flutter, °flap, °flip, °flicker, °shake, °vibrate, °quiver, °nod, °rock, °dance, wobble, bob, bobble, waver, *Rare* vellicate: *Misty's tail wagged as I approached the house.* —*n.* **2** °wave, waggle, oscillation, °fluctuation, °sway, undulation, °flutter, vellication, °flap, °flip, °flicker, °shake, vibration, °quiver, °nod, wobble, bobble, waver: *The dying animal recognized me and gave a feeble wag of her tail.*

wag[2] *n.* °comedian, °wit, punster, pundit, °joker, jester, °comic, jokester, droll, merry-andrew, °clown, *Colloq* °card: *A wag said that tailors were like storks—known for their big bills.*

wage *n.* **1** Often, ***wages.*** °pay, compensation, emolument, °remuneration, °payment, °fee, °salary, °stipend, recompense, °reward, °earnings; °honorarium: *The strikers wanted a 10 percent increase in their hourly wage.* —*v.* **2** °carry on, °pursue, °conduct, °engage in, °undertake, practise, °prosecute, °proceed with: *History shows that wars are often waged over trifling differences.*

wait *v.* **1** °tarry, °linger, °hold on, °stay, bide (one's) time, mark time, °stand by, *Colloq* cool (one's) heels, °stick around, °sit tight, °hang on, *Brit* °hang about, *US* °hang around: *If you wait till I've finished this, I'll join you for dinner.* **2** be delayed *or* postponed *or* deferred *or* shelved *or* put off, *US* be tabled, *Colloq* be put on ice *or Colloq* on the back burner: *The quarterly sales figures can wait—we haven't yet completed those for this month.* **3** ***wait on*** *or* ***upon.*** °serve, °attend (to), °minister (to): *Rose made such a fuss that the restaurant staff refused to wait on her.* —*n.* **4** °delay, °pause, °stay, °holdup, °interval, °halt, °stop, stoppage, °break, hiatus, lacuna, °gap, °respite, °rest (period), °intermission, discontinuation, °recess: *Because of the long wait between trains, we were late getting home for dinner.*

waiter *n.* waitress, headwaiter, *maître d'hôtel*, °host, hostess, *sommelier* (*des vins*), wine steward, stewardess; cupbearer, Ganymede, Hebe: *Marie asked the waiter for a clean plate.*

waive *v.* **1** °give up, °relinquish, °renounce, °resign, °forsake, °forgo, °cede, °sign away, °surrender, °abandon, °yield, °dispense with: *I waived my right to the inheritance in favor of my daughters.* **2** set *or* put aside, °except, °ignore, °disregard, °overlook, °abandon, °defer, °postpone: *We are happy to waive the rules in your case, madam.*

waiver *n.* renunciation, relinquishment, forgoing, ceding, cession, °resignation, °surrender, abandonment, setting *or* putting aside, deferral, °remission, °postponement: *Her waiver of all rights to the inheritance was duly recorded.*

wake[1] *v.* **1** Often, ***wake up.*** °awaken, °awake, °rouse, waken, °bring around, °stir, bestir (oneself), °get up, °come to, get going: *Wake me when it is over. She didn't wake up till noon.* **2** °awake, waken, °awaken, °animate, °stimulate, °enliven, galvanize, °fire, °quicken, inspirit, °inspire, °activate, liven up, vivify, °kindle, °vitalize, °stir, °arouse, get (someone) going, bring to life: *When I feel this logy it takes a lot to wake me up. Seeing Margo woke a spark I thought had died.* —*n.* **3** vigil, °watch, deathwatch, °funeral: *If you have ever been to an Irish wake, you will know what I mean.*

wake[2] *n.* **1** °track, °trail, aftermath, °path, backwash, °wash, bow wave; °trace, spoor, °scent: *The wake from the passing boats damages the docks. No matter where he went, they followed in his wake.* **2** ***in the wake of.*** following (on *or* upon), after, °subsequent to; as a result *or* consequence of, on account of, because of, °owing to, °by virtue of: *In the wake of the law's enactment, scores of abortions were performed.*

wakeful *adj.* **1** °awake, °sleepless, waking, unsleeping, °restless, °restive, insomniac: *We were wakeful the entire night in anticipation of a visit from St. Nicholas.* **2** watchful, °(on the) alert, on the qui vive, °sharp, °attentive, °vigilant, °wary, °cautious, °observant, heedful, on the lookout: *We remained wakeful, alert to any footstep.*

walk *v.* **1** °advance, °proceed, °move, °go, wend, go *or* make (one's) way by foot, tread, °step, perambulate, °stalk, stride, °tramp, °stroll, amble, °ramble, ambulate, shamble, °pad, °shuffle, °saunter, trudge, trek, °plod, slog, hike, °parade, °promenade, °strut, °swagger, °prance, °march, goose-step, °pace, °trip, °sidle, tiptoe, sashay, °flounce, °stagger, °lurch, °limp, °waddle, °stamp, mince, °slink, °steal, °prowl, skulk, °sneak, °creep, *Colloq* go by *or* ride by shanks' *or* shanks's pony *or* mare, hoof it, foot it, traipse, °pussyfoot, *Slang US* boogie: *Guthrie certainly has an individual way of walking.* **2** °take, convoy, °accompany, °escort, °go with; °conduct, °lead; °empty: *I shall walk you to the station. I have to walk the dog twice a day.* **3** °patrol, trace out, °stalk, °cover, °haunt, °prowl, °wander, °roam, rove *or* range about in *or* on, °frequent: *Her mother could not understand why Maizie walked the streets nightly.* **4** ***walk out.*** **(a)** °leave, °depart, °desert; walk out on: *She just walked out and no one has seen her since.* **(b)** °strike, go (out) on strike, °protest, take industrial action, *Brit* down tools: *Negotiations came to an impasse so the employees walked out.* —*n.* **5** °path, lane, pathway, pavement, footpath, °promenade, esplanade, boardwalk, *Brit* footway, *US* °sidewalk: *We strolled along the pleasant walks that lead through the park.* **6** gait, °step, °carriage, °bearing, stride: *He had great fun imitating Groucho Marx's slinky walk.* **7** constitutional, °stroll, amble; slog, °tramp, hike: *Lettie likes to take a walk in the park after dinner. We had a five-day walk through the jungle before reaching civilization.*

wall *n.* **1** °screen, °partition, divider, °enclosure, separator, bulkhead, °barrier, °obstruction, °obstacle, °impediment, °block, °fence: *Marcie built around herself a wall of silence which no one could penetrate.* **2** barricade, fortification, °protection, °bulwark, breastwork, embankment, °rampart, palisade, stockade: *It was questionable whether the walls would withstand the cannonballs.* **3** ***drive up the wall.*** drive crazy *or* insane *or* mad, °madden, °exasperate, derange, °try, °irritate, °infuriate, °enrage: *The constant caterwauling from the next flat is driving me up the wall.* **4** ***go to the wall.*** °fail, °collapse, be ruined, face ruin, go bankrupt, lose everything, *Colloq* go broke, °go under, °fold (up), *Slang* go bust: *Unicold went to the wall because of increased costs and decreased sales.* —*v.* **5** Often, ***wall up*** *or* ***off.*** °enclose, °partition (off), °close (off), brick up, immure: *The mason was hired to wall up the openings where the windows had been.*

wallet *n.* °purse, pocketbook, *Brit* notecase, *US* billfold: *An anonymous Good Samaritan found my wallet and sent it to me intact, with all the money in it.*

wallow *v.* **1** roll *or* loll about *or* around, °welter, writhe, °tumble, splash *or* plash: *He enjoys watching the pigs wallowing in the mud.* **2** Usually, ***wallow in.*** °luxuriate in, bask in, °revel in, glory in, indulge (oneself) in, give (oneself) up to, succumb to, °take to, °appreciate, °fancy, °enjoy, °like, °love, °savor, *Slang* get a kick *or Slang* a bang *or Slang* a boot from *or Slang* out of: *Now that Gabriel is a big rock star, he simply wallows in all the attention he is getting.* **3** °stumble, °stagger, °lurch, °flounder, °teeter, °totter, falter, °pitch: *Without its en-*

gines, the ship was lifted skyward by a wave one moment only to wallow in a trough the next.

wan *adj.* **1** °white, °sickly, °pale, pallid, livid, °pasty, ashen, bloodless, waxen, wheyfaced, sallow, °colorless, deathly, °ghostly, °ghastly, cadaverous: *I almost wept at seeing the child's wan face against the pillow.* **2** °weary, °weak, °hollow, °feeble, °frail, °ineffectual, °sorry, °pitiful: *She offered no more than a wan smile in response to questions about her family.*

wand *n.* baton, °stick, °staff, °stick: *The magician pointed with his wand and a bottle of gin appeared.*

wander *v.* **1** °walk, °go, °roam, rove, °range, °stray, °ramble, °stroll, °saunter, °meander, °drift, °cruise, °prowl, *Colloq* mosey: *We wandered about the village square, chatting and window-shopping.* **2** °wind, °meander, zigzag, turn this way and that: *A trout stream wanders past our house.* **3** digress, °go off, become absent-minded, go woolgathering, lose concentration *or* focus: *My mind wandered as the speaker droned on.* **4** °deviate, digress, °turn, divagate, °stray, °drift, °depart, go off at a tangent, lose (one's) train of thought, °lapse: *The speaker tended to wander occasionally.*

wane *v.* **1** °decrease, °diminish, grow less, lessen, °decline, °die out, abate, °ebb, °subside, °fade (away), °dim, °taper off, °peter out, °wind down, °weaken: *The waning moon cast its pale light on the dying knight. As his strength waned he could no longer lift his sword.* **2** draw to a close, °end, °terminate: *The day waned into a gloomy evening.* —*n.* **3** °decrease, diminution, lessening, °decline, abatement, °ebb, subsidence, fading, tapering off, petering out, winding down, weakening, deterioration, degeneration: *The 1970s saw the wane in popularity of large, gas-guzzling cars.* ***4 on the wane.*** on the decrease *or* decline *or* ebb, diminishing, decreasing, declining, abating, subsiding, fading, tapering off, petering out, winding down, weakening, deteriorating, degenerating: *At least inflation seems to be on the wane. Since that latest illness, my energy has been on the wane.*

wangle *v.* °scheme, °plot, °work out, contrive, °maneuver, °engineer, °manage, °manipulate, machinate, *Colloq* °fix, °fiddle, °work, °pull off, finagle, °swing: *I hear that you were able to wangle an audience with the pope.*

want *v.* **1** °desire, crave, °wish (for), long for, pine for, °hope (for), °fancy, covet, °hanker after, °lust after, hunger for *or* after, thirst for *or* after, yearn for, *Colloq* have a yen for: *I want you near me. Ignore his crying—he just wants some ice cream. Maybe he's crying because he wants to go.* **2** °need, °lack, °miss, °require, °call for, °demand, be deficient in, be *or* stand in want *or* in need of, necessitate; be *or* fall short of: *This engine wants proper maintenance. The bottle wants only a teaspoonful to be full.* —*n.* **3** °need, °lack, °shortage, deficiency, °dearth, °scarcity, scarceness, insufficiency, scantiness, inadequacy, paucity: *For want of good writers, the literary quarterly diminished in size and finally disappeared.* **4** °appetite, °hunger, °thirst, craving, °desire, °fancy, °wish, °longing, yearning, hankering, °demand, °necessity, °requirement, requisite, °prerequisite, *Colloq* yen: *She gave up trying to satisfy his wants.* **5** °poverty, °need, indigence, homelessness, destitution, °privation, pauperism, penury, neediness, impecuniousness: *The civilized nations are trying to solve the problems of want, which seem to increase daily.*

wanting *adj.* **1** °deficient, °inadequate, not up to par *or* expectations, °insufficient, leaving much to be desired, °unsatisfactory, unsatisfying, °disappointing, second-rate, °inferior, °poor, °shabby, °shoddy, °flawed, °faulty, °imperfect, °incomplete, unfinished, °defective, patchy, impaired, damaged, °broken, °unsound: *These robots were tested at the factory and were found wanting.* **2** °absent, missing, lacking, °short (of), *US and Canadian* °shy (of): *What good is a banjo wanting its strings?*

wanton *adj.* **1** °immoral, °dissolute, °profligate, dissipated, depraved, °loose, °promiscuous, °lustful, licentious, °lecherous, °wild, libidinous, °lewd, °lascivious, °unchaste: *I shouldn't mind your telling me who these wanton women are, Shambleton.* **2** °abandoned, unrestrained, °undisciplined, ungoverned, °ungovernable, unmanageable, °outrageous, °immoderate, intemperate, untempered: *It is not the boy's fault, it is the wanton company he keeps.* **3** °reckless, °rash, uncaring, °lavish, °extravagant, °willful, °heedless, °irresponsible, °careless: *This situation arises from the parents' wanton disregard for their children's moral training.* **4** °wicked, °evil, malevolent, malicious, °merciless, inhumane, °vicious, °cruel, °violent, unjustified, unprovoked, uncalled-for, °purposeless, motiveless, unjustifiable, °arbitrary, °gratuitous: *We were subject to wanton attacks by the hill people.* —*n.* **5** vamp, strumpet, whore, harlot, loose woman, °prostitute, voluptuary, slut, trollop, Jezebel, *Colloq* °tart, *Slang* hooker, working girl, call girl: *Who invited that wanton, with her tight-fitting, shamelessly low-cut dress?*

war *n.* **1** warfare, °combat, °conflict, fighting, °clash, °hostilities, °battle, °struggle, °engagement, °encounter, °strife, contention: *The war, in which scores of millions died, lasted six years.* **2** ***at war.*** fighting, battling, in combat, in conflict; in disagreement, in dispute, in contention, struggling, antagonistic, at daggers drawn: *They are at war because an ambassador refused to retract some silly insult.* —*v.* **3** do battle *or* fight *or* struggle *or* (engage in) combat with *or* against, make *or* wage war with *or* against, take up arms *or* strive *or* campaign *or* tilt against, cross swords *or* contend *or* joust with: *We must continue to war against the forces of evil.*

ward *n.* **1** °district, °division, °precinct, °section, °zone, °quarter: *Perkin was running for reelection as councilor of the Eastgate ward.* **2** °minor, dependent: *The child was made a ward of the court.* —*v.* ***3 ward off.*** °fend off, °repel, avert, °avoid, °block, °thwart, keep away *or* off *or* at bay *or* at arm's length, °check, °repulse, chase away *or* off, °forestall: *This brandy should help ward off the night chill.*

wardrobe *n.* **1** (collection *or* stock of) clothing *or* clothes *or* attire *or* apparel: *At the time, my entire wardrobe consisted of a pair of jeans, a shirt, and a dinner suit.* **2** clothespress, closet, clothes cupboard: *In her wardrobe they found more than three thousand pairs of shoes.*

warehouse *n.* °storehouse, °store, storeroom, depository, stockroom, depot, godown: *The books are kept in a warehouse and shipped to customers as needed.*

wares *n.* pl. merchandise, °goods, commodities, manufactures, °produce, °stock (in trade), supplies, °lines: *Korzeniowski peddled his wares on this corner for twenty years.*

warlike *adj.* combative, °belligerent, bellicose, °aggressive, °pugnacious, °hostile, °bloodthirsty; hawkish, militaristic, jingoistic, warmongering: *Violation of the border was a warlike act that must be punished. There are warlike factions in every government.*

warm *adj.* **1** °heated, °tepid, °lukewarm, °cozy, °comfortable, not uncomfortable, balmy: *We chose a beautiful warm day for our picnic.* **2** passionate, °impassioned, °excited, °animated, °fervent, fervid, °spirited, °lively, °ardent, zealous, °keen, °eager, °emotional, °heated, °intense, irritated, annoyed, vexed, °angry, irate, °furious, °testy, short-tempered, °touchy, °quick-tempered, irascible, °irritable, °stormy, turbulent, °vigorous, °violent,

Colloq worked up, hot under the collar, steamed up: *Charles occasionally became a bit warm on the subject of architecture. The debate grew warm as we joined in.* **3** °amiable, °friendly, °cordial, affable, °pleasant, °genial, °cheerful, °kindly, °hospitable, °hearty; °affectionate, °tender, °mellow, loving, amorous: *After a warm greeting from our host, we went in to meet the other guests. Her generous gifts to charity revealed that she really has a warm heart.* **4** °ardent, °enthusiastic, °earnest, °eager, °sincere: *Our ideas for the new campaign met with warm approval from the client.* **5** uncomfortable, °awkward, unpleasant, °strained, °tense: *The people in this town tend to make things a bit warm for strangers.* **6** Often, ***getting warm.*** close *or* near to making a discovery, about to make a discovery: *From her look as I approached the cabinet, I knew I was getting warm.* —*v.* **7** °heat (up), warm up *or* over: *I warmed myself by the fire.* **8** Often, ***warm to.*** become less antagonistic *or* hostile to *or* toward(s), become enthusiastic *or* supportive of, become excited *or* animated about *or* over, be attracted to *or* toward(s), °like, feel affection for: *Martin never did warm to my idea of using the village notice board for his message.* **9** °stir, °move, please, °delight, make (one *or* someone) feel good: *It warmed me to know that she had at last found someone to love.*

warm-blooded *adj.* **1** *Technical* homoiothermic *or Technical* homoiothermal *or Technical* homoeothermic *or Technical* homoeothermal *or Technical* homeothermic *or Technical* homeothermal: *Birds and mammals are warm-blooded, reptiles and fish are cold-blooded.* **2** °passionate, °ardent, fervid, hot-blooded, °impetuous, *Colloq* °randy: *He tried to excuse his behavior by saying it was what any normal, warm-blooded man would have done.*

warmly *adv.* **1** affectionately, tenderly, °fondly, lovingly: *Linda kissed him warmly to thank him for the gift.* **2** cordially, amiably, amicably, solicitously, warmheartedly: *We were greeted warmly by the manager on our arrival and shown to our rooms.* **3** earnestly, eagerly, fervently, enthusiastically, °well, °kindly: *The directors are warmly disposed to your plan for reorganization.* **4** °vigorously, intensely, °fiercely, intensively, °intently, energetically, doggedly, persistently, zealously, fervently, fervidly, °hotly, ardently, enthusiastically: *We learned that she was being warmly pursued by Interpol.* **5** heatedly, vehemently, vociferously, forcefully, energetically, °vigorously, feverishly, frantically, furiously, angrily, violently: *The sending of a peace-keeping force was being warmly debated in the UN.*

warmth *n.* **1** °heat: *We basked in the welcome warmth of the sun.* **2** cordiality, heartiness, friendliness, geniality, amiableness, kindliness, tenderness, affability, °love: *Mother thrives on the warmth of a family environment and should not be put in a nursing home.* **3** °ardor, effusiveness, °enthusiasm, zeal, excitedness, °fervor, vehemence, °vigor, ebullience, °passion: *I was taken aback by the warmth of the stranger's greeting. The resolution was debated with considerable warmth.* **4** irritation, °annoyance, pique: *The warmth of his reaction to the accusation is understandable.*

warn *v.* **1** °caution, admonish, °advise, °notify, apprise, °inform, give (fair) warning, °alert, give (prior) notice, put (someone) on notice *or* on guard *or* on the alert, make (someone) aware (of), forewarn, °tip off, *Rare* premonish: *The entire population has been warned about the imminent hurricane.* **2** °advise, °counsel, °caution: *You did warn me against investing in llama farms.*

warning *n.* **1** °caution, admonition, °advice, °counsel, caveat, °word (to the wise), °tip, notification, °notice, °threat; °lesson, °example: *Cabister never took seriously his doctor's warning about his cholesterol level. Let that be a warning to you not to drink and drive.* **2** °omen, °sign, °signal, °indication, augury, foretoken, portent, foreshadowing, forewarning, °prophecy: *Sailors take that kind of sky and a falling barometer as warning of a hurricane.*

warp *v.* **1** °twist, contort, °distort, deform, bend out of shape, °wrench, °pervert, misshape: *Using steam, the timber was warped to fit the hull. You certainly have a warped idea of what Yvette does for a living.* —*n.* **2** °twist, contortion, distortion, °bias, deformity, deformation, °bend, °wrench, °perversion, °kink, idiosyncrasy, °quirk, deviation: *There is too much of a warp in this veneer to repair it. Hurley has to overcome a serious warp in his attitude toward food.*

warrant *n.* **1** authorization, °sanction, °reason, justification, °approval, validation, °license, °right, certification, entitlement, °grounds, °cause, °rationale, °basis, °assurance, °carte blanche, °guarantee, °pledge, °security, °charter, °warranty: *We demand to know what warrant may exist for such an action.* **2** writ, °order, affidavit, °paper, °document, credential, °permit, entitlement, °license, °permit, summons, subpoena, mandate, °decree, fiat, edict, ukase: *Do not let them search the premises without a warrant. A warrant has been issued for her arrest.* —*v.* **3** °guarantee, °promise, °assure, ensure *or* insure, °answer for, be answerable for, °certify, vouch for, °underwrite, °back up, °uphold, stand by *or* behind: *The manufacturer warrants these products for one year.* **4** °authorize, °sanction, °justify, °explain, °approve, °verify, validate, °permit, °allow, provide *or* offer grounds *or* justification *or* cause *or* reason for, °call for, necessitate, °entitle, empower, °excuse, °license: *What is it that warrants such accusations of malfeasance?*

warranty *n.* °guarantee, °assurance, °promise, commitment, covenant, °undertaking, °agreement, °pledge, °bond: *This is the maker's warranty that repairs of any defects will be made at no cost to the buyer.*

wary *adj.* °cautious, °careful, on (one's) guard, circumspect, °prudent, apprehensive, chary, watchful, °vigilant, on the qui vive, heedful, °observant, on (one's) toes, *Colloq* cagey, *Slang* °leery (of): *Phoebe is wary of going out with someone she doesn't know.*

wash *v.* **1** wash up, °clean (up), °cleanse, bathe, °shower, douche, douse, °scrub (up), shampoo, soap up, °lather, °launder, °scour, °soak, °rinse, °flush, °wet, °drench, deterge, sponge (off), *Facetious* perform (one's) ablutions, *Archaic* absterge, *Formal* or *literary* lave, *Brit* bath: *Please wash the dishes when you have finished eating. I must wash before I do anything else.* **2** Sometimes, ***wash away*** *or* ***out*** *or* ***off.*** °remove, °move, °transport, °carry, °bear, convey, °deliver, °deposit, °drive, °sweep: *The silt, with the gold dust, is washed downstream. In this process, the impurities are washed away.* **3** °splash, °spatter, splatter, plash, °dash, °beat, °pound, thrash, °break, °toss, °surge, undulate, °rush, °run, °lap, °ripple, °roll, °flow: *We stood watching the sea washing against the breakwater.* **4** Usually, ***wash away*** *or* ***off.*** °erode, wear off *or* away, °remove, °delete, °erase, expunge, °destroy, eradicate, °obliterate, °extinguish, °blot out, °wipe out: *After five thousand years, the action of the waves had washed away all traces of the Vengorian civilization.* **5** Often, ***wash away*** *or* ***out.*** °erode, cut *or* dig *or* wear *or* eat *or* dredge (away *or* out), °excavate, °channel: *The river eventually washed out a new course, fifteen miles to the west.* **6** decontaminate, °purify, °sift, °filter, depurate: *The wastes are thoroughly washed in these tanks before being discharged into the ocean.* **7** overlay, °film, °coat, °paint, °glaze; °plate: *Don't you just adore the way they washed the wall with puce to match the carpet?* **8** °hold up, °stand up, stand the test of time, carry weight, bear scrutiny, prove true, make sense, be believable *or* credible, *Colloq* hold water: *I'm afraid that Helen's explanation of her whereabouts at the time of the murder simply won't wash.* **9** ***wash down.*** °swallow: *Here, wash the pill down with this.* **10** ***wash (one's) hands of.*** stay *or* keep away from, disown, °repudiate, turn (one's) back on, have nothing more *or* further to do with, °get rid of, rid (oneself) of, °desert, °abandon, °leave: *After Neil took the money, Violetwashed her hands of him and his problems.* —*n.* **11** washing, cleaning, cleansing, scrubbing, °scrub, scouring, shampoo, shampooing, bath, bathing, °shower, sponge bath, tub bath; laundering; *Facetious* ablutions; *Colloq Brit* tub, tubbing: *He always likes a good wash and shave before breakfast.* **12** °wave, °wake, °surge, backwash: *The wash from passing ships almost swamped our skiff.* **13** °lotion, °rinse, liniment, °salve, embrocation, emulsion, °preparation; mouthwash, gargle; eyewash, collyrium: *Use this wash twice a day till the condition disappears.* **14** °flow, °wave, °swell, welling, °sweep, °sweeping, ebb and flow, °surge, surging, undulation, rise and fall: *Shellfish in the gap cleanse themselves of impurities in the constant tidal wash running through there.* **15** °coat, coating, °film, overlay, °glaze; plating: *There is a microscopic wash of gold over the tin to lend the bracelet a little cachet.*

washed-out *adj.* **1** °wan, °pale, pallid, °colorless, faded, °lackluster, °flat; blanched, bleached, etiolated: *Gene has that washed-out look because he never gets out in the sun. These jeans look washed-out from lying in the sun.* **2** °exhausted, °spent, °tired, tired-out, °weary, worn-out, °fatigued, drained, *Colloq* dog-tired, bone-tired, done in, all in, fagged out, bushed, *Brit* knocked up, *US* knocked out, *Slang* °beat, *US and Canadian* tuckered out, pooped: *After a hard day in the office, I feel completely washed-out.*

washed-up *adj.* finished, °through, failed, done for, played-out, °over (and done with), *Slang* kaput, *fini*: *After that last fiasco, he's no more than a washed-up has-been.*

washout *n.* °failure, °disaster, debacle, °(total) loss, °fiasco, °disappointment, *Colloq* °flop, °dud, *Brit* damp squib, *US* lead balloon: *His attempts to revive the hula-hoop craze were a washout.*

waspish *adj.* irascible, bad-tempered, foul-tempered, °temperamental, °testy, grouchy, °sensitive, °volatile, °querulous, edgy, °petulant, °spiteful, °peevish, °cantankerous, curmudgeonly, °cross, crabby, crabbed, crotchety, splenetic, grumpy, captious, °cranky, crusty: *Elaine is feeling very waspish today—no one can do anything right.*

waste *v.* **1** squander, °misuse, °throw away, °fritter away, misspend, °splurge, °dissipate, *Slang* °blow: *Why I wasted so much time on him I'll never know. Don't waste your money on such frivolities.* **2** Often, ***waste away.*** °diminish, °deteriorate, °dwindle, °decline, °decay, atrophy, wither, °shrink, °weaken, become debilitated, °fade, become enervated *or* enfeebled, regress, °ebb, °sink: *Victims of this affliction slowly waste away till they disappear completely.* **3** °enervate, enfeeble, emaciate, °gnaw, °destroy, °consume, debilitate, °exhaust, disable: *Arkwright contracted a terrible wasting disease while in the tropics.* **4** assassinate, °murder, °kill, *Slang* °put away, °rub out, *US* ice: *Don't worry, Boss, we'll waste Andy and you won't have no more trouble.* —*n.* **5** °misuse, misapplication, squandering, °dissipation, misemployment, °abuse, °neglect: *What a terrible waste of talent to have such a person doing such lowly work! Do not let that genius go to waste.* **6** wasting, °extravagance, °prodigality, wastefulness, squandering, °indulgence, lavishness, °profligacy, dissoluteness, improvidence, overindulgence: *Government waste is one of our biggest problems.* **7** °refuse, °rubbish, °garbage, °dregs, debris, leavings, °scrap, offscourings, sweepings, °litter, *Archaic* orts, *US and Canadian* °trash: *Techniques are being developed to process the waste in order to avoid using landfills, incineration, or dumping at sea.* **8** wasteland, °desert, wilderness, barrens, °wilds, °emptiness, vastness: *Beyond that system were vast wastes of uninhabited space.* —*adj.* **9** °extra, °leftover, °unused, °superfluous, °worthless, °useless: *The waste food was usually fed to the animals.* **10** °barren, unproductive, unusable, unsalvageable, °useless, unrecyclable, °unprofitable, °worthless: *The waste products of manufacture are a big problem.* **11** ***lay waste.*** °devastate, °destroy, °demolish, despoil, °ruin, °wreck, °ravage, °pillage, °sack, °plunder, °loot, °rob, °strip, °spoil, °gut, °ransack, wreak havoc upon *or* on, °crush, °raze, annihilate, eradicate, extirpate, °wipe out: *These lands and cities, laid waste by the invaders, have never recovered.*

wasteful *adj.* °extravagant, °spendthrift, °profligate, °prodigal, °lavish, °improvident, unthrifty, uneconomical, overindulgent, openhanded, freehanded, penny-wise and pound-foolish: *It was very wasteful of you to buy me that expensive jewelry.*

wastrel *n.* **1** °spendthrift, °profligate, waster, °prodigal, big spender, squanderer: *That wastrel went through his inheritance in a year.* **2** °idler, layabout, malingerer, °loafer, shirker, °good-for-nothing, ne'er-do-well, *Chiefly Brit* drone, *Slang Brit* skiver: *The workhouse is the place for wastrels like you!*

watch *v.* **1** °observe, °regard, °look at, gaze at *or* on, °take in, °contemplate: *Daniel loves to watch the bears at the zoo.* **2** °look after, °tend, °mind, keep an eye on, watch over, °mind, °guard, °care for, °take care of, °safeguard, °protect, °shield, keep safe, °supervise, superintend; chaperon, °accompany, °attend; *Colloq* baby-sit (for), °sit (with): *Could you please watch Suzie while I go shopping?* **3** °observe, °note, °notice, make *or* take note of, °see, pay attention (to), °attend (to), °follow, °(take) heed (of), °examine, °inspect, °scrutinize, °pore over; °eye, peer at; °ogle, make eyes at: *Watch the way I bone this fish. Ted enjoys watching the girls going by.* **4** Often, ***watch (out) for.*** °look for, be on the watch *or* lookout *or* alert *or* qui vive (for), °guard (against), keep an eye open (for), be watchful (for), °note, take note *or* notice of, be vigilant (for *or* of), keep (one's) eyes open (for), keep a (sharp) lookout (for), be prepared *or* ready for, be careful of, °anticipate, await, °wait (for), °look for, *Colloq* keep (one's) eyes peeled *or* skinned (for), keep a weather eye open (for): *We were watching for irregularities in the test results. Watch where you're going! Watch out for that last step!* —*n.* **5** vigil, °surveillance, °observation, °lookout: *Our six-hour watch was rewarded by the appearance of a yellow-bellied sapsucker.* **6** clock, timepiece, pocket watch, wristwatch; chronometer: *I always set my watch by the time signal on the radio.* **7** sentry, °sentinel, °guard, °watchman: *The midnight watch just returned from his rounds.* **8** ***on the watch (for).*** on the alert (for), on the lookout (for), on (one's) guard (for), on the qui vive (for), alert (for *or* to), °awake (to), °observant (of), watchful (of), °cautious (of), °wary (of), °vigilant, circumspect: *We remained on the watch for the slightest movement in the bushes.*

watchman *n.* °(security) guard, °sentinel, sentry, °watch, night watchman, custodian, caretaker; watchdog: *After retiring from the police, he worked as a watchman.*

water *n.* **1** H_2O; distilled water, tap water, drinking water, bottled water, spa water, still water, soda (water), effervescent water, mineral water; sea water, salt water; ditchwater, dishwater, bathwater, *US* branch water, *Fa-*

cetious Adam's ale, *Technical and Latin* aqua; *Technical* heavy water *or Technical* deuterium oxide *or Technical* D_2O; *Brit* fizzy water: *Pure water is odorless, tasteless, and colorless.* **2 *not hold water.*** be illogical *or* unsound *or* invalid, not be sensible, be inconsistent, not make sense, be unbelievable *or* incredible, be indefensible, be unfeasible *or* unworkable, not work, not function, not hold up under *or* bear scrutiny *or* examination, not ring true, ring false, *Colloq* not wash: *Her story that the man appeared in her bedroom when she kissed a frog just doesn't hold water.* **3 *like water.*** lavishly, extravagantly, °freely, wastefully, profligately, openhandedly, liberally, excessively, copiously, unstintingly, unreservedly: *If they won at the races, they would spend the money like water till it ran out.* **4 *make water.*** °urinate, pass water, *Colloq* pee, piss: *When a dog makes water, it marks off a territory bounded by its scent.* **5 *of the first water.*** of superior *or* excellent *or* first *or* top *or* A-one *or* the finest *or* the highest *or* the best quality *or* grade; first-grade, top-grade: *These were not industrial stones but diamonds of the first water.* —*v.* **6** inundate, °flood, °drench, °saturate, °soak, douse, irrigate, hose, °wet, °shower, °splash, °spray, sprinkle, moisten, °damp, °dampen, bedew: *If you don't water the garden, the vegetables will die.* **7** Often, ***water down.*** °dilute, °weaken, °thin out, °adulterate; °cut; mollify, °modify, °soften, °tone down, °qualify: *We no longer go there because they water the drinks. They told the children a watered-down version of what had happened.*

watercolor *n.* aquarelle: *John's watercolors sell as fast as he paints them.*

waterfall *n.* cascade, cataract, fall(s), °chute, Niagara, *Northern Brit* °force, *Scots* linn: *Wasn't it at a waterfall that Holmes disposed of Moriarty?*

watertight *adj.* **1** sealed, waterproof: *The escape hatch has been made watertight.* **2** unassailable, °impregnable, °solid, airtight, °flawless, °faultless, °incontrovertible; without loopholes: *Landry has a watertight alibi for the night of the crime.*

watery *adj.* **1** °weak, dilute(d), watered-down, °tasteless, insipid, flavorless, °bland, °flat, °dull, °thin, runny, pallid, anemic, *Colloq* °wishy-washy: *He asked me in, then gave me some watery tea to drink.* **2** weeping, teary, °tearful, °running, weepy, lachrymose, rheumy: *The child looked up at him with watery eyes and then collapsed, sobbing.* **3** °wet, swampy, boggy, marshy, aqueous, squelchy; soggy, °moist, °damp, °humid; *Colloq* squushy *or Colloq* squooshy: *It took days to make our way across that watery plain to the high ground.*

wave *n.* **1** °swell, undulation, billow, °sea, °heave, °roller, whitecap, white horse; °ripple, wavelet, breaker, comber: *A huge wave tossed me up on the beach.* **2** °surge, °swell, welling up, ground swell, °movement, °flood, upsurge, °uprising, °current, tide: *The period is marked by a wave of materialism and greed that swamped all morality and integrity.* **3** °signal, °sign, gesticulation, °gesture: *Was that wave intended for you or for me?* —*v.* **4** undulate, billow, move to and fro, °flap, °flutter, °quiver, flip-flop, °swing, °sway, °ripple, °oscillate, zigzag, °fluctuate, °shake; °wag, whiffle, wigwag, wiggle, waggle, brandish: *The streamer waved slowly in the quickening breeze. Stop waving that knife at me!* **5** °signal, °sign, °indicate, °signify; °gesture, gesticulate: *Hazel waved goodbye from the train.*

way *n.* **1** °manner, °method, °mode, °fashion, °means, °system, °course (of action), °procedure, °approach, °scheme, °technique, °practice, modus operandi, *Colloq* MO (= 'modus operandi'): *That's no way to talk to your mother! We have ways of making you talk. Is that any way to treat a lady?! It's the wrong way to treat the problem.* **2** °manner, °spirit, °feeling, °sense, °character, °approach, °personality, temperament, °disposition, modus vivendi (= 'lifestyle'), °nature, °technique, °style, °conduct, °habit, behavior pattern, °custom: *He has such a pleasant way about him. Lucinda certainly has a way with children and dogs. You city people may not approve of our down-to-earth country ways.* **3** °path, °road, °street, avenue, °course, °route, °trail, °direction: *Show me the way to go home. It's a long way to Tipperary.* **4** °distance; °route, °trail, °course, °road: *You've come a long way since we last met.* **5** °progress, °passage, °advance, °headway; °speed, °velocity, °motion, °(forward) movement: *We made our way to the front of the crowd. Who has the right of way at the crossing? The tide was too strong for us to get any way on.* **6** °aspect, °respect, °particular, °detail, °point, °sense, °feature: *In certain ways, you remind me of Attila the Hun.* **7** °clearance, pathway, avenue, °scope, °freedom, °opportunity: *When you hear a siren, drive to the side and give way. Make way for the Lord High Executioner!* **8** °condition, °situation: *She prefers the prissy "in a family way" to the straightforward "pregnant" and the vulgar "have a bun in the oven." I saw Luke yesterday and he was really in a bad way.* **9 *by the way.*** °incidentally, °moreover, by the by, parenthetically: *By the way, have I told you how beautiful you look tonight?* **10 *by way of.*** **(a)** via, °through, °by means of: *We drove from London to Oxford by way of Reading.* **(b)** (functioning) as, in (the) way of, in the capacity of, equivalent to, more or less, something like: *Rickles is by way of being an expert on insulting people.* **11 *give way.*** **(a)** °collapse, °break (down), °fail, °cave in, °fall (down), °crumble, °crumple, °disintegrate, °go to pieces: *The cable gave way, causing the bridge to collapse into the river.* **(b)** °yield, °surrender, °retreat, °concede, °withdraw, accede, make concessions, acquiesce, °acknowledge: *Will he give way before the advancing force under General Grouchy?* **12 *under way.*** °proceeding, progressing, °on the move, °moving, advancing, °going, begun, started, °in progress, operating, functioning, at work, *US* °in work, *Colloq* in the works, *US* °in the pipeline: *Once under way, it takes the ship three days to make the crossing. Plans are under way to reclaim the slum area for a park.*

waylay *v.* **1** °ambush, lie in wait for, await, °intercept, pounce upon *or* on, swoop down upon *or* on, accost: *You'll have to waylay the professor before he enters the lecture hall if you want to speak to him.* **2** °attack, °mug, °seize, °assault, accost, °set upon: *He was waylaid by a band of thieves and stripped of his belongings.*

way-out *adj.* **1** °bizarre, °mad, °weird, °crazy, °strange, °odd, °peculiar, freakish, freaky, °eccentric, °queer, °abnormal, °offbeat, °outrageous, °wild, °exotic, esoteric, *Colloq* °kinky, *Slang* kooky, off the wall, far-out, screwy, nutty, *US* flaky, screwball: *The police did not take kindly to the boys' way-out behavior after the party.* **2** °avant-garde, advanced, °original, innovative, °unorthodox, unconventional, °experimental, precedent-setting, °progressive, exploratory, ground-breaking, *Slang* far-out: *Spingle thinks that the way-out art of the 1960s will have great value one day.*

weak *adj.* **1** °feeble, °frail, °fragile, unsubstantial, °flimsy, breakable, frangible, °delicate, °rickety, unsteady, °unsound, °decrepit, °shaky, °infirm: *That chair is too weak for you to stand on.* **2** °frail, °infirm, debilitated, enervated, °delicate, °sickly, anemic, wasted, °decrepit, °puny, effete, worn-out, °tired, °exhausted: *When I last saw Tindell he was so weak he could hardly lift his brandy glass.* **3** unassertive, °retiring, namby-pamby, °spineless, °irresolute, °impotent, °ineffectual, °ineffective, °incompetent, feckless, °inept, °wishy-washy, °timid, °meek, craven, timorous, °cowardly, pusillanimous, lily-livered, chicken-hearted, *Colloq* chicken, yellow: *We know which is the weaker partner in that marriage, don't we?* **4** °feeble, °lame, half-baked, °poor, °miserable, unconvincing, unpersuasive, °empty, °shallow, °flimsy, °hollow, °pathetic, °pitiful, °unbelievable, °untenable: *Everett made some weak excuse for being late. I found Mona's argument against birth control very weak indeed.* **5** weak-minded, dimwitted, dull-witted, slow-witted, °foolish, °feebleminded, °simple, simpleminded, softheaded, °stupid, °dull, moronic, imbecilic, *Colloq* °dumb: *They seek out people of weaker intelligence and make fools of them.* **6** °faint, °dim, °poor, °dull, °pale, faded, °indistinct, °vague, °hazy, °imperceptible, indiscernible, unclear, blurred, blurry, muzzy, wavering, faltering, ill-defined, °feeble, flickering, °subdued: *I could hardly see her face in the weak light of the candle.* **7** °feeble, °subdued, °low, °soft, hushed, muffled, muted, almost inaudible, stifled, °indistinct: *When the victim finally spoke, his voice was very weak.* **8** See **watery, 1,** above. **9** ***weak point.*** See **weakness, 3,** below.

weaken *v.* **1** debilitate, enfeeble, °enervate, emasculate, °mitigate, °moderate, °dilute, deplete, °diminish, lessen, °depress, °lower, °reduce, °sap, °undermine, °exhaust, impoverish: *The continuous torture helped to weaken Errol's resolve.* **2** °fade, °dwindle, °tire, °droop, °sag, °fail, °give way, °crumble, °flag: *My determination weakened when I saw what had happened to the children.* **3** °give in, °relent, acquiesce, °give way, °yield, accede, °consent, °agree, assent, °soften, °bend, ease up, °let up, ease off, °relax: *If you ask very politely, she might weaken and allow you to go out and play.* **4** °water (down), °dilute, °thin (out): *Adding the water so weakened the soup that it tasted like dishwater.*

weakling *n.* °milksop, °baby, mollycoddle, lightweight, namby-pamby, *US and Canadian* milquetoast, *Colloq* sissy *or Brit also* cissy, °loser, cream puff, jellyfish, °pushover, softy *or* softie, *Slang* wimp, twirp, *US* weak sister, schnook, schlemiel *or* schlemihl *or* shlemiel: *If you count on a weakling like Billings, nothing will ever get done.*

weakness *n.* **1** feebleness, °frailty, fragility, °delicacy, delicateness, vulnerability, °infirmity, °decrepitude: *Owing to the weakness of the structure, the building was declared unsafe.* **2** incapacity, irresolution, irresoluteness, °impotence, powerlessness, puniness: *He is ashamed of his weakness in being unable to stop smoking.* **3** °weak point, °foible, °failing, °fault, °shortcoming, °flaw, Achilles' heel, °defect, °imperfection, °liability: *Her one great weakness is her inability to deny her children anything.* **4** soft spot, fondness, °affection, °liking, °preference, °bent, °leaning, °inclination, °fancy, °penchant, predilection, proneness, proclivity, predisposition, °partiality, °appreciation, °appetite, sweet tooth, °taste, °eye: *The two of them share a weakness for good food and wine.*

wealth *n.* **1** affluence, °riches, °money, opulence, °prosperity, °property, holdings, °capital, °assets, wherewithal, °cash: *Most of Castenango's wealth comes from illicit trade in diamonds.* **2** °profusion, °abundance, °bounty, plenteousness, bounteousness, copiousness, °mine, plenitude, fullness, °store, cornucopia, richness: *Janette has a wealth of ideas for situation-comedy plots for TV.*

wealthy *adj.* °rich, affluent, °well-off, °prosperous, well-to-do, °opulent, °comfortable, moneyed, *Colloq* °in the money, on Easy Street, °flush, well-heeled, in clover, *Slang* °loaded, °stinking (rich), filthy rich, *Brit* quids in, rolling in it: *Boxley has a wealthy aunt who sends him a little something now and then.*

wear *v.* **1** be dressed *or* clothed in, dress in, °put on, don, be in, step *or* get into *or* in, °have on, °sport: *I shall wear my new suit tonight.* **2** °display, °show, °exhibit, °have, °adopt, °assume: *I wondered why she wore such a curious expression.* **3** Often, ***wear down*** *or* ***away*** *or* ***off.*** °damage, °impair, °harm, °fray, °erode, abrade, corrode, °rub (off): *The water has worn down the rocks till they are round and shiny. The inscription on the stone is worn away. After years of use, the paint has worn off.* **4** Often, ***wear well.*** °last, °endure, °survive, °hold up, °bear up, °stand up: *I wrote to the makers telling them the shoes have worn well.* **5** °drag, pass slowly, creep by *or* along, go by gradually *or* tediously: *The hours wore on as I waited for the test results.* **6** Often, ***wear out.*** °tire, °fatigue, °exhaust, debilitate, °weary, °enervate, °drain, °burden: *You must be worn out from carrying those heavy books.* **7** °bore, °exasperate, °harass, vex, °annoy, °irritate, °tax, °strain: *I find it wearing to listen to that music all day long.* —*n.* **8** wearing, °use, utilization; attire, garb, clothing, °clothes, °apparel, °dress, °gear: *Did you get much wear out of your new hat? Suzanne is modeling a silver lamé dress for evening wear.* **9** wear and tear, attrition, deterioration, °damage, fraying, chafing, abrasion, °erosion, corrosion, °friction: *As the engine ages, heavier oil is needed to offset the normal wear.*

weary *adj.* **1** °tired, °fatigued, °exhausted, worn-out, drained, °spent, *Colloq* all in, ready to drop, fagged (out), done in, dead (on (one's) feet), frazzled, dead beat, dog-tired, *Brit* knocked up, *US* knocked out, *Slang Brit* whacked, knackered, *US* pooped, zonked (out), °shot: *He is so weary after work that he can scarcely eat dinner.* **2** °boring, °irksome, irritating, °tedious, vexing, annoying, exasperating, °burdensome, wearying, tiring, fatiguing, draining, taxing, wearisome: *We walked many a weary mile before reaching the oasis.* **3** bored, °impatient, °jaded, °blasé, *Colloq* fed up, sick and tired, *Taboo slang* browned off: *I wish he would wipe that weary expression off his smug face.* —*v.* **4** Often, ***weary of.*** °tire (of), be *or* become bored (with *or* by) *or* impatient (with) *or* jaded (with *or* by), *Colloq* be *or* *Colloq* become fed up (with) *or Colloq* sick and tired (with *or Colloq* of): *I soon wearied of her nagging.* **5** °exhaust, °enervate, °fatigue, °tire, debilitate, °drain, °tax, wear *or* tire out: *The long trek across country had wearied us all.*

weather *n.* **1** (meteorological) condition(s), °climate: *The weather changed suddenly at the end of April, and it snowed.* **2** ***under the weather.*** ailing, °ill, °sickly, unwell, °indisposed, °out of sorts, °sick, *Colloq* °poorly, °seedy: *I didn't go sailing because I was feeling a bit under the weather.* —*v.* **3** °stand, °survive, °suffer, bear up against, °endure, °withstand, rise above, ride out, live through, °brave: *Gemma didn't think she could weather much more abuse at the hands of her captors.*

weave *v.* **1** °loom; °braid, plait, °entwine, intertwine, interlace, interweave, °knit (together): *These fabrics were woven by hand. Weave these leather strands together to make a belt.* **2** °blend, °combine, °fuse, °merge, °unite, intermingle, °mesh, °splice, dovetail, °join: *The threads of their lives are inextricably woven together.* **3** °construct, °make, contrive, °build, °create, °fabricate, °compose, °spin, °design: *Coulson was finally caught in the web of his own deceit.* **4** zigzag, crisscross, wend *or* make (one's) way, °dodge, bob and weave, °shift: *Look at that motorcyclist weaving in and out of the traffic!* **5** ***get weaving.*** get started, °get a move on, °hurry (up), °start, *Colloq* °shake a leg, *Brit* get *or* pull (one's) finger out, *Chiefly US* get a wiggle on: *If you're going to get to work on time, you'd best get weaving.*

web *n.* spider's web, cobweb; °net, °network, entanglement, °snare, °trap: *Don't you just hate getting those webs in your hair? Coulson was finally caught in the web of his own deceit.*

wed *v.* **1** °marry, espouse, get married, become husband and wife, say *or* take (one's) (marriage) vows, join *or* unite in holy wedlock *or* matrimony; lead down the aisle, lead to the altar, *Archaic* wive; *Colloq* tie the knot, get hitched, get spliced: *They wed on July 4, 1921. That was when Harry wed Annabel.* **2** °combine, °unite, °ally, °marry, °blend, °merge, °join, °mingle, intermingle, commingle, coalesce, °mix, intermix, °amalgamate, °compound, °alloy, °fuse, homogenize: *As a chef, he is extremely fortunate to be able to wed business and pleasure.*

wedded *adj.* Usually, **wedded to.** intimately *or* obstinately attached *or* connected (to), enamored (of): *He has always been wedded to the notion of living alone on an island.*

wedding *n.* **1** °marriage (ceremony), wedding ceremony, nuptials; confarreation: *Our son's wedding was attended by members of the royal family.* **2** combining, °combination, uniting, °union, joining, °juncture, blending, °blend, allying, °alliance, associating, °association, marrying, °marriage, merging, °merger, mingling, intermingling, commingling, coalescing, coalescence, mixing, °mixture, intermixing, amalgamating, °amalgamation, compounding, °compound, alloying, °alloy, fusing, fusion, homogenizing, homogenization: *Through her wedding of vision to observation she has produced some truly remarkable paintings.*

wedge *n.* **1** °block, chock: *A shoe with a wedge heel is called a "wedgie." We forced wedges into the cracks in the stone to split it.* **2** °separation, separator, °division, °partition, °split, fissure, cleavage: *That woman tried to drive a wedge between me and my husband.* —*v.* **3** °ram, °jam, °stuff, °cram, °crowd, °force, °squeeze, °pack, °thrust: *You couldn't have wedged another person into the back seat of that car using a shoehorn.*

wee *adj.* **1** °tiny, °small, °diminutive, °little, minuscule, midget, °minute, °miniature, Lilliputian, microscopic, *Colloq* itty-bitty, itsy-bitsy, teeny (-weeny), teensy(-weensy): *Johnny caught his wee little finger in the door.* **2** unimportant, °insignificant, trivial, °little, °puny: *It was only a wee thing, of no interest to anyone but me.*

weep *v.* **1** °cry, shed tears, °bawl, blubber, °keen, °sob, °lament, °mourn, °bemoan, °bewail, °moan, °grieve, whine, whimper, mewl, pule, °snivel, *Colloq* blub, boohoo, *Brit and Australian* whinge: *A small child, sitting by herself, was weeping bitterly.* **2** °ooze, seep, exude, °drip: *The wound, which is still weeping, needs bandaging.*

weigh *v.* **1** Sometimes, ***weigh in at*** *or* ***out at.*** *Colloq* tip the scales at: *The sailboard weighed in at 250 pounds. I weigh more than I should.* **2** °consider, °ponder, °contemplate, think on *or* over *or* about, mull over, turn over in the *or* (one's) mind, ruminate over, °chew over, reflect on *or* upon, °brood over, °pore over, °study, °examine: *After weighing the qualifications of the candidates, we voted for you. Weigh your words carefully before responding.* **3** °judge, °estimate, assess, °evaluate, value, °determine: *It is not easy to weigh each person's contribution to the effort.* **4** Usually, ***weigh on*** *or* ***upon.*** lie heavy on, °burden, °depress, °prey on, °oppress, °disturb, °perturb, °upset: *Loneliness weighs heavily on Dave's widow.* **5** °matter, °count, have (an) effect *or* influence, carry weight, be of value *or* account: *Certain factors—like the sizes of the bribes—are likely to weigh heavily with the judges of this particular beauty contest.* **6** ***weigh down.*** °burden, overburden, °load, °overload, °encumber, °tax, overtax, °strain, °trouble, °worry, °depress, °oppress: *He's been weighed down by financial worries since losing his job.*

weight *n.* **1** heaviness, avoirdupois, °mass, tonnage, *Dialect* heft: *They sell peaches by weight but melons are individually priced. The weight of these books must be enormous.* **2** °burden, °load, millstone, onus, °pressure, °strain, albatross, °cross: *He has to support the weight of the entire family's expenses.* **3** °influence, °authority, °power, °substance, °force, °moment, °importance, consequence, °impact, persuasiveness, value, °worth, *Colloq* clout: *Clyde brought down his full weight in favor of the liberal candidate. Clyde's opinion carries little weight with me. He must stop throwing his weight around.* **4** °mass, °majority, preponderance *or* preponderancy, °bulk, °superiority: *The weight of the voters is on the side of reducing the budget. The weight of the evidence is against us.* —*v.* **5** °load, °charge, ballast: *The policemen's truncheons are weighted with lead.* **6** °arrange, °manipulate, °bias, °incline, °slant, *Colloq* °rig: *The statistics are weighted so as to make the advertiser's product look superior.*

weighty *adj.* **1** °heavy, °ponderous, °massive, °huge, °bulky, °substantial, °ample, °large, mammoth, °colossal, °immense, °enormous, °gigantic, °prodigious; corpulent, °fat, °obese, adipose, *Colloq* °hefty: *I found myself face to face with a rather weighty sumo wrestler.* **2** °important, consequential, °significant, °momentous, °grave, °crucial, °portentous, thought-provoking, °provocative: *The lady had weighty motives for wishing me to take on the case of her missing husband.* **3** °influential, convincing, °persuasive, °impressive, °telling, °powerful, °potent, °leading; °forceful: *Influential authorities have raised weighty objections to the proposal.*

weird *adj.* °strange, °odd, °peculiar, °bizarre, °unnatural, °eerie, °queer, °grotesque, freakish, °outlandish, uncanny, °unearthly, otherworldly, °supernatural, preternatural, *Literary* eldritch, *Colloq* spooky, freaky, °kinky, *Slang* far-out, °way-out: *There was something decidedly weird about the creature at the door.*

weirdo *n.* °eccentric, °madman, madwoman, lunatic, °psychotic, *Colloq* °crazy, weirdie, nutcase, oddball, queer fish, °crank, *Slang* °freak, loony, psycho, *Brit* nutter, *US* screwball, nut, kook: *Why belong to a social club whose members are weirdos who refuse to talk to one another?*

welcome *v.* **1** °greet, °hail, °meet, °receive, °accept, offer hospitality (to): *Everywhere she went she was welcomed with open arms.* —*adj.* **2** accepted, °acceptable, well-received, °desirable, °agreeable, gratifying, appreciated: *Norma is always a welcome guest at my house. Ice cream was a welcome relief from the perennial rice pudding.* **3** freely permitted *or* allowed, invited, entitled, suffered: *At that hotel, guests are welcome to dine when they please.* —*n.* **4** °reception, °greeting, salutation: *We gave the ambassador a warm welcome.*

weld *v.* **1** °unite, °combine, °merge, °fuse, °connect, °link, °join; solder, braze, °cement, °bond: *We felt welded together in a brotherhood that would last forever. The brace must be welded along the entire edge.* —*n.* **2** °seam, °joint, °juncture, commissure: *The weld must be filed down smooth.*

welfare *n.* °benefit, °good, °advantage, wellbeing, °prosperity, °(good) fortune, °profit, °interest, °(good) health, °happiness, felicity: *It is up to the parents to look after their children's welfare.*

well[1] *adv.* **1** satisfactorily, sufficiently, adequately, agreeably, nicely, °(well) enough, *Colloq* O.K. *or Colloq* OK *or Colloq* okay: *Alan is doing well these days.* **2** successfully, °famously, marvelously, wonderfully, fabulously, incredibly, splendidly, admirably, spectacularly, excellently, °swimmingly, superbly: *Barbara and David really get along well together.* **3** articulately, understandably, expressively, correctly, accurately, °properly, proficiently, effectively, artistically, poetically; grammatically: *Henry will make a good announcer because he speaks well.* **4** comfortably, luxuriously, prosperously, extravagantly, showily, pretentiously, ostentatiously, sumptuously, grandly, opulently: *How can Rita live so well with no visible income? Living well is the best revenge.* **5** graciously, °kindly, °highly, °favorably, glowingly, approvingly, °warmly, genially, cordially, amiably, kindheartedly, warmheartedly, affectionately, lovingly: *Rudolph has always spoken well of you, Aileen.* **6** skillfully, expertly, adeptly, proficiently, ably: *Paul sings well. Does Anne play well enough to compete professionally?* **7** °far, by a long way, immeasurably, (very) much; far and away, °definitely, °positively, °obviously, °clearly, plainly, °manifestly, °evidently, unquestionably, decidedly, beyond (the shadow of a) doubt, *Colloq* by a long chalk: *Keep well away from the fuel tank before lighting that cigarette. She is well beyond the beginner stage. This painting is well worth what you paid.* **8** good-naturedly, equably, coolly, serenely, calmly, soberly, unexcitedly, sedately: *Burt took the bad news about his car very well.* **9** °likely, °probably, in all probability, doubtlessly, without doubt, not unexpectedly, °indeed: *He might well say he needs no help.* **10** °easily, without difficulty: *Though he gave her the money, he could not well spare it.* **11** °completely, °entirely, °wholly: *Before she could well finish her sentence, he was out of the house.* **12** °thoroughly (cooked), (cooked) through and through, °completely (cooked): *He likes his roast beef well done.* **13** intimately, closely, familiarly, °personally; °thoroughly, °profoundly, soundly, fully: *Do you know Boris well? I learned my lesson well.* **14** °fairly, justly, suitably, °properly, adequately, reasonably, fully, generously, °amply: *James is well paid for his work.* **15** °happily, mercifully, fortunately, luckily: *After the divorce, both said they were well rid of each other.* —*adj.* **16** °healthy, °fit, °hale, °robust, °vigorous, °hearty, in fine *or* good fettle, *Colloq* in good shape: *Wanda feels very well since taking the waters at Montecassini.* **17** °satisfactory, °pleasing, °agreeable, °good, °right, all right, °fine, °proper, O.K. *or* OK *or* okay: *When he phoned the office, they told him all was well.*

well[2] *n.* **1** wellspring, °spring, °fountain, wellhead, fountainhead, fount, °source, reservoir: *The well provided a constant supply of clean water. As his needs demanded, he returned to the well again and again.* —*v.* **2** Often, ***well up*** *or* ***out*** *or* ***forth.*** °flow, °spring, °surge, °rise, °stream, °trickle, brim over, °swell, °start; °gush, °spurt, jet, °spout; °ooze, seep, °leak: *Tears welled up in the child's eyes when she saw the broken doll.*

well-advised *adj.* °prudent, °wise, °sensible, °intelligent, °smart: *You would be well-advised to avoid Chichicastenango at this time of year.*

well-balanced *adj.* **1** °rational, °sane, °sensible, °rational, °reasonable, °levelheaded, °sober, °sound, well-adjusted, cool(headed), *Slang* together: *Ray is as well-balanced as people seem to be these days.* **2** °even, symmetric(al), °harmonious, well-proportioned, °orderly, well-ordered, well-disposed: *Note the well-balanced arrangement of the windows in this Palladian façade.*

well-bred *adj.* well-brought-up, well-mannered, °polite, °decorous, mannerly, °refined, °courteous, °cultivated, °polished, cultured, gentlemanly, °ladylike, °elegant, °suave, urbane, °sophisticated, °gracious, courtly, °genteel, °gallant, °chivalrous: *Jane prefers bikers to the well-bred gentlemen her mother selects for her.*

well-fed *adj.* °plump, chunky, thickset, °chubby, rounded, °rotund, portly, °stout, fleshy, overweight, adipose, °fat, °obese, °gross, *Brit* podgy *or US* pudgy: *On the westward crossing we were seated with a well-fed Austrian couple.*

well-established *adj.* longstanding, °traditional, °set, °venerable, °well-known, accepted, well-founded: *The chairman is having little success convincing employees that* droit de seigneur *is a well-established English custom.*

well-groomed *adj.* °neat, °dapper, °fastidious, °tidy, °trim, °smart, clean-cut, °spruce, natty, well-dressed, *Colloq* °nifty, *Slang US and Canadian* spiffy: *Well-groomed gentleman sought as escort to attractive widow. Hourly rate. Driving license.*

well-informed *adj.* °knowledgeable, °learned, well-read, well-versed, well-educated, literate, °educated, *Colloq* in the know, °wise, *US* vibrating on the right frequency, *Slang* hip *or Slang* hep: *A free press helps to make a well-informed public.*

well-known *adj.* **1** known, °familiar, (well-) established, acknowledged, °customary, °everyday: *Plain salt is a well-known remover of wine stains.* **2** °famous, °noted, °notable, °celebrated, °renowned, °illustrious, famed, °prominent, °eminent, °preeminent: *It was her object to be seen with the well-known society leaders of the moment.*

well-off *adj.* °comfortable, °wealthy, °rich, affluent, °prosperous, well-to-do, *Colloq* well-heeled, *US* well-fixed: *Today, only well-off people can afford to live in what was a miserable slum just a few years ago.*

well-thought-of *adj.* admired, highly regarded, respected, °reputable, venerated, esteemed, revered, looked-up-to, valued: *Timothy is a well-thought-of member of the community.*

well-timed *adj.* °timely, °seasonable, °opportune, auspicious, °favorable, °advantageous, °beneficial: *Tax incentives gave a well-timed boost to the economy.*

welsher *n.* nonpayer, °cheat, cheater, °swindler, *Slang* deadbeat, *US also* welcher: *That welsher never paid the money he lost to me at poker.*

welt *n.* **1** bead, °ridge, °seam, °edge, wale, °stripe: *The cushion has a welt of contrasting color binding it.* **2** °bruise, contusion, °bump, °lump, °scar, weal *or* wale *or* wheal: *The boy's back was covered with red welts where he had been birched.*

welter *n.* **1** °mass, °mess, °jumble, °tangle, °confusion, °mishmash, °muddle, °clutter, *Brit* hotchpotch *or US also* hodgepodge: *Her assignment was to sort out a welter of discarded clothing for the poor. My mind was assailed by a welter of disorganized images.* —*v.* **2** be sunk *or* involved in, °flounder, be bogged down in, be entangled *or* ensnarled in: *Till I found a new secretary, I would have to continue weltering in a sea of correspondence and unfiled papers.*

wet *adj.* **1** °moist, moistened, °damp, dampened, soaked, °soaking, sopping, wringing, dripping, sodden, soppy, saturated, drenched: *During the rainy season, the ground is far too wet for planting.* **2** rainy, raining, teeming, pouring, drizzling, showery: *We have had a very wet spring this year.* **3** °feeble, °weak, °irresolute, effete, namby-pamby, °foolish, °ineffectual, °ineffective, °spineless, timorous, °cowardly: *Abernathy is far too wet to accomplish much if he is voted in.* —*n.* **4** moisture, °water, wetness, dampness, °damp, humidity, °liquid: *The wet stood out on his forehead.* **5** °rain, wetness, °mist, dew, °fog, °damp, humidity: *Come in out of the wet and warm and dry yourself by the fire.* **6** °milksop, softy *or* softie, lightweight, *Colloq* °drip, °loser, *Brit* weed, *Slang* wimp, *US* weak sister: *Why is he such a wet about standing up for his rights?*

wheedle *v.* °coax, °cajole, inveigle, °charm, °beguile, °persuade, °talk; butter up; *Colloq* con, sweet-talk:

Irena always managed to wheedle someone into taking her dancing.

wheel *n.* **1** disc, °ring, annulus, °circle, hoop: *It seems odd that no culture in the western hemisphere ever invented the wheel.* —*v.* **2** °spin, °turn, veer, °swivel, °pivot, °swing, whirl: *Medusa wheeled to face him and he promptly turned to stone.*

whereabouts *n.* pl. or *sg.* **1** °location, °position, °place, °site, °situation, °locale, °neighborhood, °vicinity: *Her present whereabouts are unknown to me. The culprit concealed his whereabouts from the police.* —*adv.* **2** where, in *or* at *or* to what place, whither: *Whereabouts are you going to spend your holiday? Whereabouts are you going?*

whet *v.* **1** °sharpen, hone, °grind, °file, put an edge on, strop: *He whetted the knife on an oiled stone till it was razor-sharp.* **2** pique, °sharpen, °awaken, °arouse, °stimulate, °kindle, °fire, °increase, °excite, °enhance: *The cruise on Bill's yacht whetted my appetite to own a small sailing boat.*

whimsical *adj.* **1** °quaint, fey, °fanciful, °odd, °curious, °unusual, chimeric(al), °queer, °singular, °peculiar, °funny, fantastic(al), pixyish, °playful, puckish, °absurd, °preposterous, *Colloq* °offbeat: *When Andrew's whimsical drawings caught on they were bought by greeting-card makers.* **2** °capricious, °erratic, °eccentric, wavering, °flighty, °unsettled, °fickle, mercurial, wavering, fluctuating, unpredictable, °inconsistent, °volatile, unsteady: *The buses in our village run on what could best be termed a whimsical timetable.*

whip *v.* **1** °beat, thrash, °lash, °flog, horsewhip, °scourge, °switch, cane, birch, flagellate, leather, °spank, strap; °castigate, °chastise, °punish, °discipline; *Slang* tan, *US* wale: *They learned right from wrong because father whipped them when they were wrong.* **2** trounce, °defeat, °beat, °conquer, °overwhelm, °rout, °overcome, °overpower, °thwart, °check, °best, worst, drub, °stop, °outdo, *Colloq* lick, wipe the floor with, °batter, *Slang* °pulverize, clobber, °destroy, °ruin, °murder, °slaughter, °kill, squash, smash, *US* cream: *Karpilova whipped O'Meara 6-0, 6-2 in today's match.* **3** °run, scamper, scoot, °race, °scurry, °scramble, °hurry, °flit, °rush, °dash, dart, *Colloq* zip, zoom, skedaddle: *I whipped round the corner and dived behind a tree to hide.* **4** °beat, °whisk, °fluff up: *Whip the egg whites till they are stiff.* **5** *Nautical* °seize, °bind, °wind, °fasten, °tie: *The end of the line is whipped to prevent its unraveling.* **6** ***whip out.*** yank out, jerk out, °pull (out), whisk out, °present, °exhibit, °flash, °produce: *He suddenly whipped out a gun and started shooting.* **7** ***whip up.*** **(a)** °stir up, °agitate, °arouse, °rouse, °work up, °excite, °incite: *The rabble-rousers whipped the crowd into a frenzy.* **(b)** °improvise, put together *or* assemble *or* prepare quickly *or* hurriedly, *Colloq* knock together, °knock up, *US* slap together: *Tina whipped up a snack before I had to run to catch my plane.* —*n.* **8** °scourge, knout, °lash, cat-o'-nine-tails, rawhide, quirt, horsewhip, bullwhip, cane, birch, °switch, thong, (riding) crop, *Colloq* cat: *We dreaded it when Dr. Hazen eyed the whip hung behind the classroom door.*

whipping *n.* **1** beating, °thrashing, lashing, flogging, horsewhipping, scourging, switching, caning, birching, flagellation, °spanking: *Once he had had a whipping, the boy would not misbehave again.* **2** *Nautical* seizing, binding, tying, winding, °fastening: *The whipping will prevent the end of the line from unraveling.*

whirlpool *n.* maelstrom, vortex, °eddy, whirl, °swirl, *Heraldry* gurges: *As we watched in horror, the canoe was drawn down into the whirlpool and disappeared.*

whirlwind *n.* **1** waterspout, dust devil, cyclone, typhoon, anticyclone, °hurricane, extratropical cyclone, tropical cyclone, *Nontechnical* tornado, *Nautical* white squall, *Australian* willy-willy, *Colloq US* whirly: *The whirlwind picked up the house and put it down across the street, without breaking a dish!* —*adj.* **2** °speedy, °quick, °swift, °sudden, °precipitous, lightning, headlong, °hasty, °rash, °impetuous: *It was one of those whirlwind romances that last till the wind dies down.*

whisk *v.* **1** °rush, dart, °sweep, °brush: *Agatha whisked about the room, aimlessly moving small objects about on tables and shelves.* **2** °speed, °rush, °carry, °whip, °hasten, °hustle, °hurry: *The taxi came and whisked Fran off to the airport.* **3** See **whip, 4,** above. —*n.* **4** °sweep, °wave, °brush, flick: *With a whisk of his hand he dismissed the servants.* **5** °brush, fly-whisk: *King Mbakulé flicked his whisk to signal that the audience was ended.* **6** beater, °whip: *It is quicker to whip cream with a whisk.*

whiskey *n.* whisky, °alcohol, °spirits, John Barleycorn, usquebaugh, Scotch, home brew, mother's ruin (= 'gin'), *Scots* barley-bree, *Brit* Scotch, *US* rye, bourbon, white lightning, white mule, °moonshine, *Colloq* °booze, hooch, rotgut, *US* corn, firewater, *Slang US* sneaky pete, smoke: *Peter is a welcome guest who always remembers to bring a bottle of whiskey.*

whisper *v.* **1** °breathe, °murmur, °mutter, °mumble, °hiss, speak *or* say softly *or* under (one's) breath, °sigh, susurrate: *He whispered sweet nothings in her ear.* **2** °gossip, bruit about, noise abroad, °murmur, °insinuate, °hint, °rumor, °disclose, divulge, °reveal, breathe a word: *We have all heard Corwin whispering of your wife's infidelities.* —*n.* **3** °murmur, undertone, hushed tone(s): *Why do you always speak in whispers when talking about Nesta?* **4** °hint, °suggestion, *soupçon,* °suspicion: *There was never the slightest whisper of gossip about Edwin.*

white *adj.* **1** snow-white, snowy, chalk-white, chalky, ivory, creamy, milky, milk-white, oyster-white, off-white; °silver, hoary: *Many buildings in the tropics are painted white. Do you believe that a person's hair can turn white overnight?* **2** °pale, pallid, °pasty, °wan, wheyfaced, ashen, bloodless, drained, whitish, waxen, °ghastly, °ghostly, anemic, dead-white, deathly white, cadaverous, corpselike: *Her black dress contrasted starkly with her white complexion.* **3** °innocent, °pure, unsullied, stainless, unblemished, °spotless, °immaculate, virginal, °virtuous, undefiled, °chaste: *She came to you with white hands, which you have sought to dirty with your vicious accusations.* **4** Usually, ***White.*** Caucasian, Caucasoid, light-skinned, fair-skinned, pale-complexioned: *The American Indian was—and still is—treated very unfairly by the White man.*

whitewash *v.* °gloss over, °cover up, sugar-coat, °hide, °camouflage, °conceal, °qualify, °minimize, extenuate, °diminish, °play down, downplay, °make light of, °rationalize, °excuse: *We are being accused of whitewashing the dishonest actions of the previous administration.*

whittle *v.* **1** pare (down *or* away), °shave, °trim, °cut, °carve, hew, °shape: *Remember when Grandpapa used to whittle those tiny figures of elves when we were children?* **2** Usually, ***whittle away at*** *or* ***down.*** °pare, °shave, °cut, °trim, °reduce, °diminish, °erode, eat away at: *If he keeps whittling away at his trust fund, he might have to get a job.*

whole *adj.* **1** °entire, °complete, uncut, °full, °intact, unbroken, °total: *I can't believe she ate the whole chicken. Were you here the whole time?* **2** in one piece, °intact, unharmed, undamaged, °unscathed, unimpaired, unhurt, uninjured: *How could anyone emerge whole from that smashup?* **3** °well, °healthy, °sound, °fit, °strong: *The surgeons have made him whole again.*

—*n.* **4** °everything, aggregate, °(sum) total, °totality, °lot, °entirety; °ensemble; *Colloq* whole kit and kaboodle: *I look at the whole and don't bother about petty details.* **5 *on the whole.*** °largely, mostly, °usually, more often than not, °for the most part, in general, °generally, by and large, with few exceptions, all things considered, all in all, °as a rule, °chiefly, °mainly, °in the main, predominantly: *On the whole, our clients collect only originals, not copies or prints.*

wholehearted *adj.* °devoted, dedicated, committed, °earnest, °sincere, °unqualified, °unmitigated, unreserved, °complete, °entire, unstinting, °real, °true, °genuine, °hearty, °heartfelt, °serious, °enthusiastic, zealous, °warm, °fervent, °ardent, °spirited, °eager, °energetic: *Once I agree to help, you can rely on my wholehearted support.*

wholesome *adj.* **1** healthful, °healthy, health-giving, °nutritious, nourishing, °beneficial, °tonic, salutary, salubrious, strengthening, °bracing, °stimulating: *Everyone should regularly engage in wholesome exercise.* **2** °moral, °ethical, °righteous, °upright, °honorable, °decent, °principled, °proper, °fit, °meet: *I am not sure that the child is being brought up in a wholesome atmosphere.*

wholly *adv.* **1** °altogether, °entirely, °absolutely, °quite, °totally, °thoroughly, °completely, *in toto,* fully, in all respects, in every way, all in all, °utterly, unqualifiedly, every inch, 100 percent; °lock, °stock, and barrel; °root and branch; backward(s) and forward(s); from the ground up; *Colloq* bag and baggage; °hook, °line, and sinker; to the nth degree; (the) whole hog, *US* up one side and down the other: *We are wholly on your side in this matter. The company has been wholly reorganized.* **2** °only, exclusively, solely, unexceptionally, categorically, unequivocally, unambiguously, explicitly: *The success or failure of the plan depends wholly on you.*

whoop *n.* **1** °shout, °shriek, °yell, °roar, °bellow, hoot, (battle *or* war) cry, war-whoop, °outcry, °scream, screech, squeal, yelp, yowl, °howl, bark; °cheer, hurrah, huzzah; *Colloq* holler: *At his whoop of joy we knew he had passed the exam.* —*v.* **2** °shout, °shriek, °yell, °roar, °bellow, hoot, °cry (out), °scream, screech, squeal, yelp, yowl, °howl, bark; °cheer, hurrah, huzzah; *Colloq* holler: *We heard the attackers whoop as they descended on the fort.*

whopping *adj.* **1** °huge, °great, °enormous, °colossal, °gigantic, °immense, tremendous, °prodigious, °monstrous, °thumping, mammoth, °massive, Brobdingnagian: *With that whopping Great Dane of yours in the kitchen there's no room to move about.* **2** °flagrant, °outrageous, °extravagant, °terrible, °awful: *Janeson tells such whopping lies I don't know when to believe him.*

wicked *adj.* **1** °evil, °bad, °immoral, unprincipled, °sinful, °impious, piacular, irreligious, °blasphemous, °profane, °sacrilegious, °ungodly, °godless, diabolic(al), °satanic, Mephistophelian, demonic, demoniac(al), hellish, °infernal, accursed, °damnable, °fiendish, °ghoulish: *How wicked must one be to murder one's own child? She has done many wicked things, but this was the worst.* **2** depraved, °dissolute, °villainous, blackhearted, iniquitous, °horrible, horrid, °hideous, heinous, °beastly, °base, °low, °vile, debased, °degenerate, °perverse, °perverted, °corrupt, °foul, °offensive, °abominable, °disgraceful, °shameful, °dreadful, °awful, °gross, °gruesome, °grim, appalling, °grisly, °loathsome, °lawless, °unrepentant, unregenerate, °incorrigible, °criminal, felonious, rascally, knavish, °terrible, egregious, execrable: *The Deluge was regarded as divine vengeance on a wicked world. He was a wicked man, in league with the devil himself.* **3** °dirty, °pornographic, °filthy, °erotic, °obscene, °lewd, °offensive, °indecent, °prurient, smutty, °rude, °taboo, °blue, °coarse, °bawdy, °vulgar, salacious, licentious, °nasty, X-rated, *Colloq US* raunchy: *The townspeople took the wicked books and burned them all.* **4** °vicious, °beastly, °savage, °nasty, °bad, °violent, °mean, °cruel: *Grimbeck has a wicked temper. The boys were subjected to the most wicked treatment.* **5** °naughty, °mischievous, impish, °sly, °devilish, rascally, roguish, scampish, puckish; vexatious, exasperating, annoying, irritating, °irksome, °trying, galling, bothersome: *What trouble have those wicked children got themselves into now? It is wicked of you to make pig noises every time I mention Bertha.* **6** °foul, °offensive, pernicious, baleful, mephitic, °disgusting, °revolting, sickening, °repulsive, °repellent, °nauseous, °repugnant, °rotten, pestilential, noxious: *That liniment the doctor prescribed has a wicked odor.* **7** °expert, °ingenious, °superior, °superb, °superlative, °outstanding, °masterful, masterly, °skillful, deft, °adept: *That chap Gough plays a wicked game of tennis.*

wide *adj.* **1** °spacious, °roomy, °ample, °extensive, °broad: *The road is just wide enough for two cars to pass. If the book is set in wider type it will have more pages.* **2** °broad, °extensive, °comprehensive, encyclopedic *or* encyclopædic, °inclusive, far-reaching, wide-ranging, widespread: *He has wide interests, from music to archaeology to astronomy.* **3** °extreme, °considerable, °substantial, sizable *or* sizeable, °major, °big, °large; widespread: *There have been wide fluctuations in market prices.* **4 *wide of the mark.*** off the mark, astray, °deviant, deviating, off (the) target, not on target, °inappropriate: *Economic forecasts were based on erroneous assumptions and have proved to be quite wide of the mark.* —*adv.* **5** astray, afield, °wide of the mark, off the mark, off (the) target, to one side: *He took careless aim and the shot went wide.* **6** all the way, as much as possible, fully, °completely, to the utmost: *The dentist said, "Open wide." When I heard the drill, I was suddenly wide awake.*

widely *adv.* **1** extensively, °thoroughly, °universally, °everywhere, °generally, by many: *In the days when it was widely believed that the earth was flat, perhaps it was.* **2** to a large *or* a great extent, greatly, °largely, very much, °extremely, considerably, °substantially: *Opinions differ widely regarding the origin of the universe.*

widen *v.* distend, dilate, °spread, °stretch, °enlarge, °increase, °expand; °extend, broaden, °supplement, °add to, augment: *It will be necessary to widen the opening to allow insertion of the device. We are planning to widen the news coverage we provide our readers.*

width *n.* **1** °breadth, wideness, compass, broadness, °span; diameter, °caliber, °bore; °measure; *Nautical* °beam: *What is the standard width of railway track in this country? The width of the cannonball is three inches.* **2** °reach, °scope, °range, °breadth, °extent, extensiveness: *In the years I have known Minnie I have not determined the width of her knowledge.*

wield *v.* **1** °flourish, °swing, brandish, °wave, °handle, °use, °employ: *He wields that machete as if he means business.* **2** °exercise, °have, °employ, °exert, °use, utilize: *Since Juan is the one who wields the power, he is the one to deal with.*

wife *n.* °mate, helpmeet, helpmate, spouse, bride, °partner, *Colloq* better half, the little woman, the missus *or* missis, old lady *or* woman, the ball and chain, *Slang* the trouble and strife: *Doesn't your wife mind your going out every night?*

wild *adj.* **1** undomesticated, °untamed, unbroken, °savage, feral: *The only four-legged animals inhabiting the island were wild pigs.* **2** uncultivated, °uninhabited, °waste, °desert, °desolate, virgin, unpopulated, °empty,

°trackless, °barren, °lifeless; °deserted: *The cottage overlooks an expanse of wild moorland.* **3** °savage, °uncivilized, barbarous, °primitive, °rude, uncultured, uncultivated, brutish, barbaric, °fierce, °ferocious: *Some early explorers studied the wild people they encountered in far-off lands.* **4** °uncontrolled, unrestricted, unrestrained, untrammeled, unbridled, unfettered, unshackled, °free, unchecked, °lively, °impetuous, unconventional, °undisciplined, °disobedient, °insubordinate, °self-willed, wayward, °mutinous, rowdy(ish), °boisterous, °unruly, °tumultuous, turbulent, °tempestuous, °uproarious; uncontrollable, unmanageable, °ungovernable, intractable, unrestrainable: *Pinckney led quite a wild youth. We cannot cope with their wild behavior after they have had too much to drink.* **5** °mad, maniac(al), crazed, °crazy, irrational, distracted, frenzied, °frantic, °distraught, °hysterical, °raving, raging, unhinged, demented, °delirious; °berserk; run amok *or* amuck: *Wormley had a wild look about him. Nellie has been driven wild by the pressures of business.* **6** °exciting, °excited, vehement, °passionate, °romantic, turbulent, °chaotic, °tempestuous, °reckless, madcap: *At the time, Ernest was having a wild love affair with Cardani's widow.* **7** °absurd, irrational, °unreasonable, °extravagant, °fantastic, °imprudent, °foolish, °foolhardy, °impractical, °impracticable, unpractical, unworkable, °ridiculous, °reckless, °silly, °giddy, °flighty, madcap, °outrageous, °preposterous, °bizarre, °strange, °odd, °peculiar, *Colloq* °offbeat: *He lost thousands on some wild scheme for reducing shipping costs by filling bubble packing with helium.* **8** tousled, windblown, °unkempt, disheveled, °untidy, disordered, °disorderly, messed up, *Colloq* mussed up: *Her hair was in wild disarray.* **9** °enthusiastic, avid, °eager, °agog, *Colloq* °crazy, °mad, °daft, dotty, *Brit* potty, *Slang* nutty, nuts: *I'm just wild about Harry, / And Harry's wild about me.* —*n.* **10** Usually, ***wilds.*** wasteland, wilderness, °desert, vastness, °emptiness, *Colloq* sticks, middle of nowhere, back of beyond: *His idea of adventure was to trek through the wilds of Hyde Park.*

wile *n.* Often, **wiles.** trick, °stratagem, °ruse, °artifice, °subterfuge, °dodge, °trap, °snare, °maneuver, ploy, contrivance, °move, gambit, °plot, °scheme, °machination, *Colloq* °(little) game: *You won't get far trying to work your wiles on me, young lady.*

will *n.* **1** °desire, °wish, °longing, °liking, °inclination, °disposition, °drive, purposefulness, °purpose, °intent, °intention, °resolve, commitment, °resolution, °determination; willpower: *Where there's a will there's a way. Some believe that taking medication is against God's will. Marguerite seems to have lost the will to live.* **2** °choice, wishes, °desire, °inclination: *He was forced to submit against his will.* **3** (last will and) testament, last wishes: *In accordance with his will, Josiah was buried at sea.* **4** ***at will.*** as *or* when (one) pleases *or* wishes *or* thinks fit(ting), at (one's) desire *or* whim *or* pleasure *or* discretion: *If he brings his own car, then he can leave at will.* —*v.* **5** °want, °desire, °wish, °choose, see fit, °make, compel, °force, °command, °order, ordain, °require: *When she willed him to appear, there was a flash of lightning and he was there.* **6** °leave, °bequeath, °devise, hand down *or* on, °pass on, °transfer; settle upon *or* on: *My great-uncle Philip willed me his collection of nineteenth-century theater memorabilia.*

willful *adj.* **1** °intentional, °deliberate, °voluntary, °conscious, intended, °purposeful, °premeditated: *The jury decided that such a willful act of vandalism must be punished. She is accused of willful neglect of her children.* **2** °stubborn, headstrong, pigheaded, °obstinate, mulish, °inflexible, adamant(ine), obdurate, intransigent, unyielding, °self-willed, °ungovernable, °recalcitrant, °unruly, °immovable, intractable, dogged, °determined, refractory, uncompromising, wayward, °perverse, °contrary: *Jim is far too willful and has a tantrum if he fails to get his own way.*

willing *adj.* °agreeable, acquiescent, compliant, amenable, consenting, assenting, °passive, complaisant, docile, °ready, well-disposed, °happy, °content, pleased, °delighted, °enthusiastic, avid, °eager, zealous, *Colloq* °game: *Albert might not have thought up the pranks, but he was a willing accomplice.*

willingly *adv.* °readily, °happily, contentedly, °gladly, cheerfully, amenably, agreeably, °freely, passively, docilely, of (one's) own accord *or* free will, on (one's) own, ungrudgingly, by choice, °voluntarily, unhesitatingly, nothing loath, eagerly, enthusiastically, zealously, avidly, *Colloq* at the drop of a hat: *We did not coerce him in any way—he went along quite willingly. You can rely on Hubert to give willingly and generously to support any good cause.*

willowy *adj.* **1** lissom *or* lissome, pliant, lithe, °flexible, °supple, limber, loose-limbed: *Everyone enjoys watching the willowy athletes training for the Olympics.* **2** °slim, °slender, °graceful, sylphlike, svelte, °thin, long-limbed, clean-limbed: *How does Jeanne keep her willowy figure without dieting?*

willy-nilly *adv.* **1** whether one likes it or not, inevitably, °necessarily, of necessity, perforce, whether or no, like it or not, *nolens volens, bon gré mal gré*: *Having borrowed the money, you must, willy-nilly, repay it.* —*adj.* **2** °necessary, °unavoidable, °inevitable, °involuntary: *He joked about a willy-nilly choice, meaning one offering no alternatives.*

wilt *v.* **1** °sag, °droop, wither, °shrink, shrivel (up *or* away), °diminish: *My houseplants wilted because they weren't watered.* **2** °sag, °droop, °bow, °weaken, °sink, °wane, wither, lose courage *or* nerve, °flag, °dwindle, languish: *The lad wilted under the stern glare of the principal.*

wily *adj.* °shrewd, cunning, °crafty, °sly, °artful, guileful, °clever, °foxy, vulpine, °disingenuous, °shifty, °scheming, plotting, °calculating, °designing, °sharp, canny, °deceitful, deceiving, °deceptive, treacherous, °perfidious, °false, double-dealing, °dishonest, underhand(ed), °tricky, °smooth, °slick, °slippery, °oily, unctuous, *Scots and North England* pawky, *Colloq* cagey, °crooked, two-timing, *Slang chiefly Brit* °fly: *He's the wily beggar who sold London Bridge to some Americans, who thought they were buying Tower Bridge.*

win *v.* **1** come (in *or* out) first, carry the day, °conquer, °overcome; °carry off (the palm), finish first (in), achieve first place (in), °triumph (in), be victorious (in), be the victor (in), gain a victory (in), °prevail (in), °succeed (in), take first prize (in), *Colloq* bring home the bacon: *Which team is more likely to win today? We seem to have won the battle but not the war.* **2** °gain, carry off *or* away, attain, °acquire, °get, °obtain, °secure, °procure, °receive, °collect, °net, °bag, °earn, °realize, °pick up, glean: *Look at all those trophies Gillian has won playing golf!* **3** ***win over.*** °influence, °sway, °incline, °persuade, °charm, °prevail upon, °convert, °induce, °bring around, °convince: *Don't think that Gilbert can be won over without a strong argument.* —*n.* **4** °victory, °conquest, °triumph, °success: *So far this season we have recorded six wins and one loss for the Surrey Scouts.*

wind[1] *n.* **1** °breeze, zephyr, °puff, °gust, °breath, °draft, light air, °current (of air): *A gentle wind wafted our little boat across the bay.* **2** puffery, °bombast, rodomontade, °bluster, boasting, braggadocio, vain speech, blather, (idle *or* empty) talk, fustian, °nonsense, twaddle, humbug, °babble, °gibberish, *Colloq* °gab, °hot air, claptrap, hogwash, °rot, hooey, baloney, *Slang Brit* (load of (old)) cobblers: *I went there for advice and all I got was a lot of wind.* **3** gas, flatulence, windiness, flatus, borborygmus, heartburn, *Taboo slang* fart: *I have to take a pill to relieve this wind.* **4** ***before the wind.*** *Nautical* downwind, °off the wind: *We were racing before the wind with every scrap of sail flying.* **5** ***break wind.*** *Taboo slang* fart: *It is rude to break wind.* **6** ***get*** *or* ***have wind of.*** °hear of, °learn of, come to know, °pick up, be made *or* become aware of, °gather, °understand, hear on the grapevine, *Colloq* hear tell of: *We got wind of the company's plans to close this plant.* **7** ***get*** *or* ***have the wind up.*** take fright, become frightened *or* afraid *or* apprehensive: *When I heard a window being raised I got the wind up.* **8** ***in the wind.*** °around, °about, rumored, in the air, detectable, °discernible, discoverable, °imminent, °impending, approaching, °close (at hand), about to happen *or* take place *or* occur, afoot, in the offing, °near, on the way, *Colloq Brit* on the cards, *US* °in the cards: *We knew that a change was in the wind, but never expected the chairman to resign.* **9** ***off the wind.*** See **5,** above. **10** ***on the*** *or* ***a wind.*** *Nautical* upwind, windward, to the wind, into (the teeth *or* the eye of) the wind; near the wind: *Because "Syrena" was a sloop, she could sail closer on the wind than the schooner.* **11** ***put the wind up.*** °scare, °frighten, °alarm: *The leakage at the nuclear plant really put the wind up everyone living in the area.* **12** ***sail close*** *or* ***near to the wind.*** take risks, throw caution to the winds, °play with fire, skate on thin ice, take (one's) life in (one's) hands, *Colloq* stick (one's) neck out, *Slang* go for broke: *One nude scene is highly salacious, and we think the director was sailing very close to the wind.* **13** ***take the wind out of (someone's) sails.*** deflate (someone), disconcert (someone), destroy (someone's) advantage, ruin (someone's) superiority *or* supremacy *or* ascendancy: *It took the wind out of her sails when he announced he was leaving anyway.*

wind² *v.* **1** °turn, °bend, °twist, °snake, worm, °twine, zigzag, slew, °swerve, °loop, °coil, curve, °meander, °ramble, veer: *The river winds along the valley floor.* **2** °reel, °roll, °spiral, °turn, °twist, curl, °coil, °wrap, °twine, wreathe; °crank (up), °wind up: *The streamers are wound around the columns for decoration.* **3** ***wind down.*** **(a)** °taper off, °slow down, °diminish, °reduce, close out, slacken *or* slack off (on), °ease (up on), °decrease, cut back *or* down (on); °wind up: *We were going to wind down the sportswear department, but then business suddenly perked up.* **(b)** °relax, become calm *or* tranquil, °calm down, cool off *or* down, regain (one's) equilibrium, ease up *or* off, *Colloq* unwind, let (one's) hair down, take it easy: *It takes me a couple of days to wind down from the hectic activity in the office.* **4** ***wind up.*** **(a)** °terminate, conclude, come *or* bring to an end *or* a close *or* a conclusion, °end (up), close down *or* up, °finish (up), °wrap up; liquidate, °settle: *After calmly winding up his affairs, Evan sailed off in his boat and was never seen again.* **(b)** end up, °finish (up), become ultimately: *I understand that Sally wound up as a bar girl in Tampico.* **(c)** °excite, innervate, °energize, °stimulate, invigorate, °stir up: *After getting us all wound up about that weekend party in Paris, she called the whole thing off.* **(d)** °agitate, °fluster, disconcert, °ruffle: *Don't let such trifles get you so wound up.* **(e)** See **3 (a),** above.

windfall *n.* bonanza, °godsend, stroke of (good) fortune, serendipitous find, °boon, piece of (good) luck, jackpot, °(lucky) strike: *The sale of the house provided a windfall that saved them from imminent bankruptcy.*

windy *adj.* **1** blustery, blowing, blowy, °breezy, gusting, gusty, °wild, squally, °tempestuous; windswept: *It is quite windy today. I waited at that windy corner for you and you never showed up!* **2** °talkative, long-winded, garrulous, °wordy, verbose, prolix, loquacious, °rambling, °voluble, °fluent, °effusive, °glib, turgid, °bombastic, °pompous, longiloquent: *Potter does tend to be windy at times.*

winner *n.* °victor, °champion, prizewinner, titleholder, conqueror, conquering hero, *Colloq* champ: *The winner will receive a new car.*

winning *adj.* **1** °engaging, °attractive, alluring, captivating, °endearing, °prepossessing, winsome, bewitching, °fetching, °taking, °enchanting, °pleasing, °delightful, charming, °amiable, °friendly, °pleasant, °sweet: *Maggie's winning ways are a tremendous asset to making friends.* **2** °triumphant, conquering, °victorious, °successful: *The winning team is invited to dine at the Red Lion.*

wintry *adj.* **1** hiemal, brumal, hibernal; °icy, snowy, °freezing, frozen, frosty, °cold, °frigid, °bitter (cold), °chilly, chilling, °piercing, °cutting, glacial, Siberian, arctic, hyperborean: *Nanook longed for the wintry winds of home.* **2** °cold, °frigid, °chilly, °cool, chilling, glacial: *She got a very wintry reception from her boyfriend's parents.* **3** °forbidding, °bleak, °dismal, cheerless, °dreary, °harsh, unfriendly, °ugly, °menacing, °ominous, °threatening, °dark: *His comment about her age evoked only a wintry look from Margot.*

wipe *v.* **1** Sometimes, ***wipe off*** *or* ***out*** *or* ***up.*** °rub, clean (off *or* out *or* up), °cleanse; dry (off *or* out *or* up), dust (off), mop (up), swab, sponge (off *or* up): *For God's sake, wipe your nose! Please wipe the crumbs off the table when you've finished eating. Stop crying and wipe up the milk you spilled.* **2** Often, ***wipe off.*** °remove, °erase, take off *or* away, °get rid of: *Wipe that grin off your face, stranger.* **3** ***wipe out.*** °kill (off), annihilate, °massacre, °destroy, °finish (off), °dispose of, eradicate, °obliterate, °exterminate, do away with, °stamp out, °get rid of, °remove, wipe off the face of the earth: *He said he wouldn't be satisfied till he had wiped out childhood diseases.*

wiry *adj.* °muscular, °sinewy, °lean, lank, thin and strong, tough and flexible: *A wiry little man, he gave the appearance of one who had never overeaten.*

wisdom *n.* **1** sagacity, sageness, °judgment, discernment, °reason, °prudence, judiciousness, °(common) sense, °insight, °penetration, sapience, understanding, rationality, clearsightedness, clearheadedness, perspicacity, perspicuity, percipience, °perception, perceptiveness, °intelligence, acuteness, acumen, astuteness, sharpness, shrewdness, longheadedness: *At least she had the wisdom not to shoot the intruder. Allen's wisdom stems from long experience in such matters.* **2** °knowledge, °learning, erudition, °lore, °scholarship, enlightenment: *The wisdom of the ages is locked away in these books.*

wise *adj.* **1** °sage, sagacious, °judicious, °reasonable, common-sensical, °prudent, °sensible, insightful, sapient, °understanding, °rational, °sound, clearsighted, clearheaded, discerning, perspicacious, perspicuous, percipient, °perceptive, °intelligent, °acute, °astute, °sharp, °shrewd, °crafty, °clever, °bright, °quick-witted, °smart, °brilliant, longheaded, *Colloq* brainy: *Among the wisest of men were Solomon, Confucius, Buddha, and, perhaps, Aristotle.* **2** °knowledgeable, °learned, °enlightened, °informed, erudite; (well-)educated, °knowing, well-read, well-versed, °lettered, °scholarly: *We saw to it that they escaped, and the police were never the wiser. Wise heads prevailed at the college on what constituted worthwhile research.* **3** °well-advised, °advisable, °judicious, °sensible, °expedient, °reasonable, °strategic, °tactful, °tactical, °prudent, °politic, °discreet, °diplomatic, well-thought-out, well-considered, °proper, °fitting, °appropriate, °meet: *Do you think it wise to punish the child so severely?* **4** ***put wise (to).*** inform *or* advise *or* warn (of *or* about): *If you must* *know how he found out, it was Myrna who put him wise. Did she also put him wise to where the money is hidden?* **5** ***wise to.*** aware *or* knowledgeable *or* informed of *or* about, sensitive to, on to *or* onto, *Colloq* in the know about: *You are wrong if you think she isn't wise to your little tricks. Get wise to what's going on, man.* —*v.* **6** ***wise up (to).*** become informed *or* aware (of *or* about), °wake up (to), *Colloq* get wise (to) *or Colloq* in the know (about): *If Cooley doesn't wise up, he'll be in deep trouble. Was it you who wised Meg up to the fact that Tom no longer cares?*

wisecrack *n.* **1** °joke, °quip, rejoinder, °witticism, °pun, barb, jest, jibe, *Colloq* °gag; °dig: *Everything is a potential subject for a wisecrack.* —*v.* **2** °joke, °quip, °pun, jibe: *Get serious and stop wisecracking for a change.*

wise guy *n.* wiseacre, smart alec *or* aleck, *Archaic* witling, *Colloq* know-all *or* know-it-all, *Brit* Clever Dick *or* cleverdick, *Slang* smartie-pants, smartie, smart-ass, *US* wisenheimer: *There's always some wise guy waiting to make a clever remark.*

wish *v.* **1** °desire, °want; °yearn, crave, °long, °hope, °hanker, have a mind, °(have a) fancy, °choose, °care: *I wish you to go now. If you had Aladdin's lamp, what would you wish for?* **2** °require, °request, °demand, °order, °specify: *Did you wish lemon or milk in your tea?* **3** foist *or* force *or* thrust *or* impose upon, *Colloq* fob off on *or* upon, palm off on: *The job of treasurer was wished on me when Jock left.* —*n.* **4** °desire, °request, whim, °want, craving, °longing, hankering: *Your every wish is my command.* **5** °desire, °longing, craving, °need, yearning, hankering, °passion, keenness, °thirst, °appetite, °hunger, whim, °urge, °liking, fondness, °fancy, °preference, predisposition, °disposition, °inclination, *Colloq* yen, *Slang* °itch: *She expressed a wish for nightingale tongues, truffles, and champagne.*

wishy-washy *adj.* **1** neither here nor there, undecided, °indecisive, °irresolute, half-hearted, shilly-shallying, tergiversating, vacillating, °uncertain, of *or* having mixed feelings, °of two minds: *Make up your mind and stop being so wishy-washy about everything!* **2** °feeble, °weak, °watery, watered-down, °thin, °vapid, °flat, °bland, runny, diluted, °tasteless, insipid, flavorless, °stale: *Isabel served me some wishy-washy tea and sent me on my way.*

wisp *n.* °shred, °scrap, strand, °thread, snippet, tuft, °lock: *The laboratory has matched the wisp of hair found in the victim's hand with yours, Harbottle.*

wistful *adj.* **1** °melancholy, °mournful, °sad, morose, °sorrowful, disconsolate, heartsick, °forlorn, woeful, °woebegone, °desirous, °longing, yearning: *She gave a last, wistful look at the airplane as it rose into the clouds.* **2** °thoughtful, contemplative, °pensive, °absent-minded, °detached, °absorbed, in a brown study, °preoccupied, meditating, °meditative, °reflective, ruminating, ruminative, °dreamy, dreaming, daydreaming, musing: *Some might have misinterpreted his wistful gaze at her photograph.*

wit *n.* **1** °intelligence, brains, °mind, °(common) sense, °judgment, °understanding, discernment, °wisdom, sagacity, °insight, astuteness, cleverness, *Slang* savvy: *He hasn't the wit to know when he's being insulted.* **2** °humor, drollery, °levity, joking, °repartee, °raillery, facetiousness, waggishness, badinage, °repartee, jocularity, wordplay, paronomasia; °amusement, °entertainment: *Some say that the soul of wit lies in the aptness of incongruity.* **3** °comedian, comedienne, humorist, °comic, °wag, °joker, *farceur, farceuse,* punster, madcap, °zany; parodist, satirist, caricaturist; *Colloq* pundit, °card, °character: *In the face of such a devastating remark, even the club wit was struck dumb.*

witch *n.* **1** sorceress, enchantress, °magician, sibyl, pythoness; warlock: *Dear Margaret was thereafter known as the wicked witch of the west.* **2** °hag, °fury, battle-ax, crone, gorgon, Medusa, ogress, Xanthippe, °shrew, virago, harridan, fishwife, termagant, *Archaic* beldam *or* beldame, *Slang* old bag, °bitch: *How could you stand living with the old witch for so many years?*

withdraw *v.* **1** °draw back, °retract, °pull back, °recoil, shrink back: *He suddenly withdrew into his cocoon whenever she said the name, "Elsie."* **2** °retract, °recall, °take back, °cancel, rescind, °recant, disavow, disclaim, abjure, °void, annul, °go back on, back out (of), back down (on): *If she doesn't withdraw that remark, I shall sue her for slander.* **3** °pull out, °extract, °remove, *Technical* retrude: *Afterward, she withdrew the knife, washed it, and replaced it in the rack. The dentist withdrew the bad tooth.* **4** °retire, °retreat, °go, °repair: *After dinner, we withdrew to the drawing room for brandy and cigars.* **5** °leave, °depart, °go, make (oneself) scarce, °absent (oneself), °retire: *We waited till the servants withdrew before exchanging news about Cora.*

withdrawn *adj.* **1** °reserved, °detached, °distant, °standoffish, °aloof, °shy, diffident, °bashful, °timid, timorous, introverted, °taciturn, °reticent, °silent, °quiet, °retiring, shrinking: *Why is Henry acting so withdrawn tonight?* **2** °remote, °distant, °isolated, °solitary, °hidden, °secluded, °private, °out-of-the-way, °reclusive: *Sarah and Ben have led a withdrawn existence since the children married and moved away.*

withering *adj.* °destructive, °devastating, death-dealing, °murderous, °deadly: *The enemy laid down a withering barrage of machine-gun fire.*

withhold *v.* **1** hold *or* keep back, °retain, °reserve, °restrain, °control, °repress, °check, °hide, °conceal: *He could withhold his anger no longer.* **2** hold *or* keep back, °deduct, °retain, °reserve: *The company is required to withhold a percentage of wages and salaries for taxes.*

withstand *v.* °resist, °oppose, °stand (up to), °face, °defy, confront, °combat, °grapple with, °fight (against), °cope with, hold out against, °weather, °suffer, °survive, °tolerate, °take, °bear, last through, °endure, °brave, *Colloq Brit* °stick: *Can Brian withstand the pressure of his new job? Our small force withstood the attack for days.*

witness *n.* **1** °observer, °onlooker, °spectator, viewer, °eyewitness, °bystander, watcher, *Rare* earwitness: *Were you a witness to what took place here?* **2** deponent, testifier, corroborating witness, corroborator: *The defense will call its witnesses today.* **3** ***bear witness (to*** or ***of).*** °testify (to), °attest (to), be *or* give *or* provide *or* furnish *or* constitute evidence *or* proof *or* testimony (of *or* to), °verify, °confirm, corroborate, °show, °prove: *The bent fireplace poker bears witness to the violence of the attack.* —*v.* **4** °see, °observe, °watch, look on *or* at, °view, °behold, °mark, °note, °notice, °take in, *Colloq* °spot, °catch: *I did witness the way they looked at each other.* **5** countersign, °sign, °certify, °endorse, °substantiate, °document, certificate: *Would you mind witnessing our signatures, Miss Cabot?* **6** See **3,** above.

witticism *n.* °pun, °quip, play on words, bon mot, jest, °joke, °epigram, clever remark, sally, *Archaic* or *literary* conceit, *Colloq* °gag, one-liner: *Clive is just as likely to utter a witticism as say something excruciatingly banal.*

witty *adj.* °ingenious, °subtle, °clever, °humorous, °sarcastic, °sardonic, piquant, °humorous, comic(al), °farcical, °ludicrous, facetious, amusing, jocular, waggish, droll, °funny: *Corbett, not ready with his usual witty riposte, shouted an obscenity at the bus driver.*

wizened *adj.* wrinkled, shrunken, shriveled (up), withered, °gnarled, dried-up, wilted, faded, wasted:

That wizened old crone is his granddaughter, on whom he has cast a spell.

woe *n.* °trouble, °hardship, adversity, °misery, °anguish, tribulation, °calamity, °trial, wretchedness, °grief, unhappiness, °desolation, dolor, °melancholy, °gloom, °depression, °sadness, disconsolateness, °misfortune, °affliction, °sorrow, °distress: *How much more woe could beset one small family?*

woebegone *adj.* troubled, °miserable, anguished, °wretched, grief-stricken, °unhappy, °desolate, °doleful, dolorous, °melancholy, melancholic, °gloomy, °mournful, °sorrowful, depressed, °dejected, °sad, °glum, crestfallen, chapfallen, lugubrious, downcast, disconsolate, °unfortunate, star-crossed, afflicted, distressed, woeful, °forlorn, °downhearted, broken-hearted, °heartbroken, disheartened, *Slang US* bummed out: *Timmy has been so woebegone since his puppy was stolen!*

woman *n.* **1** female, lady; °girl: *Elvira was the first woman to become managing director of a multinational company.* **2** °wife, spouse, bride, ladylove, °sweetheart, lady, °girl, girl friend, °mistress, concubine, °mate, helpmeet, helpmate, °partner, *Colloq* sweetie, better half, little woman, the missus *or* missis, old lady *or* woman, ball and chain, *Slang* trouble and strife: *In this scene, Porgy tells Bess that she is his woman, now and forever.* **3** °lass, °maid, °maiden, °miss, lassie, *Slang* gal, dame, bird, skirt, chick, °bit of fluff, °broad, °piece (of work), °number, baggage, moll, popsy: *The soldiers went into town to try to find some women for an evening's entertainment.* **4** °domestic, housekeeper, °maid, cleaning woman *or* lady, maidservant, chambermaid, handmaiden, abigail, lady-in-waiting, *Brit* charwoman, *Colloq Brit* char, °daily: *They have a woman who comes in every other day.*

wonder *n.* **1** °marvel, °prodigy, °phenomenon, °spectacle, °rarity, °sight, °curiosity, miracle, *Slang* °knockout, stunner, mind-blower, mind-boggler, °trip: *Surely, the wheel must rank high among the wonders of technology. And still he gazed, and still the wonder grew, / That one small head could carry all he knew.* **2** awe, °astonishment, °admiration, °amazement, wonderment, °surprise, stupefaction, °fascination: *Teletransportation, which used to excite so much wonder, might someday be used by all those who commute to work.* —*v.* **3** °ponder, °muse, °meditate, °think, °theorize, conjecture, °puzzle, °query, °question, °inquire, be inquisitive, be curious, ask oneself, °speculate, cudgel (one's) brains: *I wondered if I would be invited to the dance. Have you ever wondered what makes the world go around?* **4** °marvel (at), goggle, °gawk, °gape, °stare, be awed, be thunderstruck, be amazed, be astonished: *We wondered at the circus funambulists.* **5** ***wonder about.*** question *or* doubt the sanity *or* reason *or* reasonableness of: *I wondered about Tammy after she took part in that streaking episode.*

wooded *adj.* sylvan, forested, bosky, tree-covered, woody, timbered: *We rested and had our lunch in the coolness of a wooded glen.*

wooden *adj.* **1** wood, woody, ligneous, xyloid: *The wooden cabinet for the kitchen is almost finished.* **2** °stiff, °rigid, °artificial, °clumsy, °stilted, °unnatural, °awkward, ungainly, spiritless, unanimated, °dead, °lifeless, °dry, passionless, unimpassioned, °impassive, °vacant, °empty, °colorless, expressionless, deadpan: *Ames Caltrop gave a wooden performance as Uncle Vanya.* **3** unintelligent, blockheaded, °stupid, °dull, insensitive, slow-witted, dull-witted, °obtuse, oafish, doltish, tiny-minded, dimwitted, dunderpated, *Colloq* °thick, wooden-headed, knuckleheaded: *Clancy has some pretty wooden ideas about how to run a business.*

woolly *adj.* **1** fleecy, woolen, wool-bearing, laniferous, lanate *or* lanose, lanuginose *or* lanuginous, downy, °fuzzy, °shaggy, flocculent *or* floccose, flocky: *After two days one test tube contained a woolly precipitate.* **2** °hazy, °fuzzy, unclear, obscure(d), foggy, °indistinct, °confused, °vague, cloudy, clouded, °nebulous, ill-defined: *My recollection of the event is a bit woolly after all these years.*

word *n.* **1** °(little) talk, °(brief) conversation, °chat, °discussion, consultation, °dialogue, °huddle, °parley, °tête-à-tête, chitchat, confabulation, °conference, °interview, *Colloq* powwow, confab: *Could I have a word with you before you leave for the day?* **2** °news, °intelligence, °information, °facts, °data, °report, °story, tidings, °account, communiqué, °bulletin, dispatch *or* despatch, °advice, °message, *Colloq* °lowdown, *Slang* info, °dope, *Brit* gen, *US* poop: *Have you had any word about the situation in the Middle East?* **3** °name, °term, designation, locution, appellation, °expression, °phrase: *What is the word for "write" in Greek?* **4** °promise, °pledge, °vow, °oath, (solemn) word of honor, °undertaking, °assurance, °warrant, guarantee *or* guaranty, °warranty: *You have my word that I won't mention this to anyone.* **5** utterance, °expression, °declaration, °statement: *Let me give you a word of warning, my friend.* **6** °suggestion, °hint, scintilla, °bit: *I promise that I won't breathe a word of this to anyone.* **7** °command, °order, °signal, °direction, °instruction, *Colloq US* high sign: *I shall not press the red button till you give the word.* **8** ***in a word.*** succinctly, °briefly, °in brief, in a few words, concisely, °in short, in summary, in sum, not to mince words, to make a long story short, when all is said and done, in the final analysis, not to beat about the bush, *Colloq* in a nutshell: *In a word, I told him why I couldn't go out with him, but that took an hour. His bedside manner was, in a word, menacing..* **9** ***words.*** **(a)** °quarrel, °dispute, °argument, unpleasantness: *Paul and Kitty had words and are not on speaking terms.* **(b)** °lyrics, °book, libretto, °text: *Did Irving Berlin write the words as well as the music to his songs?* —*v.* **10** °put (forth), °say, °couch, °express, °phrase, utter, °state, °term, °style, °set forth: *Had you worded your complaint in politer language, you might have spared yourself a punch in the eye.*

wording *n.* phraseology, °language, phrasing, choice of words, word choice: *From the confused wording of this note I cannot tell whether you are coming to my party or not.*

wordy *adj.* verbose, prolix, °rambling, long-winded; pleonastic, °redundant, garrulous, °windy, °talkative, loquacious: *This contract is too wordy and could be cut to two pages. Try to make your replies less wordy.*

work *n.* **1** °labor, toil, °effort, °drudgery, travail, °exertion, °industry: *Few people make it to the top and stay there without hard work.* **2** °employment, °business, °occupation, °vocation, °calling, °profession, °trade, °line, métier, °career, livelihood, °job, °post, °position, °situation: *Does your work as a mattress tester interfere much with your home life?* **3** °task, °function, °duty, °assignment, °charge, °responsibility, chore, commission, °undertaking, °stint: *When you are finished with that work, clean out the cellar.* **4** °feat, °achievement, °creation, °accomplishment, °opus, handiwork, *oeuvre,* °production, °composition, °piece, masterwork, °masterpiece, *chef d'oeuvre, magnum opus,* °output: *Some of Edwin's works have won prizes.* **5** ***in work.*** in production, °under way, being done, in the works, being planned, in the planning stage(s): *The budget has been approved and the project is already in work.* **6** ***out of work.*** °unemployed, °idle, jobless, °at liberty, between engagements, °available, °free, *Brit* °redundant, *Colloq Brit* on the dole, *US* on *or US* collecting unemployment: *How long have you been out of work?* —*v.* **7** °labor, toil, °exert oneself, °sweat, moil, °slave (away), °peg away, slog (away): *His father worked in the mines from the age of nine till he died of black lung disease at forty.* **8** °till, plough, °farm, °cultivate: *Would he have been healthier and longer-lived had he worked the land?* **9** have a job, hold (down) a post *or* position, earn a living, be employed: *His wife works as a designer in a Parisian fashion house.* **10** °control, °manage, °manipulate, °maneuver, °wield, °handle, °operate, °use, make use of, utilize, °exploit, °deal with, bring into play: *How do you know how to work all those dials and buttons?* **11** °function, °operate, °run, °go, °develop, °turn out, *Colloq* °pan out: *Are you sure that the plan is working to your advantage?* **12** °function, °operate, °run, °go: *The drill works better if you turn on the power.* **13** knead, °mold, °form, °fashion, °shape; °mix, °stir, °incorporate: *Work the clay into long, narrow strips. Try to work the color in with your fingers.* **14** °maneuver, °manipulate, °guide: *See if you can work him over into a corner where you can grab him.* **15** °operate, °use, °employ, put to (good *or* effective) use, °wield, °manipulate, °ply, °apply, °exploit: *Tanya must have worked her magic on Eustace to make him that docile.* **16** °bring about, °effect, °accomplish, carry out *or* off, °make, °produce, °achieve, engender, beget, °create, do, °put through, °execute, °fulfill, °effectuate, °implement, °realize: *I doubt that the new sports center will work many changes in the area.* **17** ***work in.*** find time *or* space for, °include, °insert, °introduce, fit in, squeeze in, °accommodate: *Perhaps the doctor can work you in for an appointment next week.* **18** ***work on.*** °wheedle, °coax, importune, °press, *Brit* pressurize, *US* °pressure; °influence, °persuade, act on, °prevail upon, °induce, °dispose, °urge: *Keep working on him to try to change his mind.* **19** ***work out.*** **(a)** °exercise, do callisthenics, do aerobics, warm up, do setting-up exercises, do (one's) daily dozen, °jog, lift weights, °train, °drill: *I try to work out for an hour every day.* **(b)** Often, ***work out at or to.*** °equal, °total (up to), °result in, °amount to, °come to: *Let's see . . . that works out to one car for every 4.7 people.* **(c)** °clear up, °resolve, °solve, *Slang Brit and New Zealand* suss out: *Thaddeus is old enough to work out his own problems. Can they work out their differences?* **(d)** evolve, °develop, °succeed, °prosper, come out all right, prove satisfactory, go well, be effective, *Colloq* °pan out: *How can you be so sure that everything will work out?* **(e)** °formulate, °work up, contrive, °draw up, °detail, °plan, °develop, °devise, put together, °elaborate, °expand, °enlarge (on): *Colonel Prebble worked out a way for us to escape.* **20** ***work up.*** **(a)** °excite, make excited, °agitate, °inflame, enkindle, °arouse, °rouse, °stir, °move, °animate, °incite, °spur, *Colloq* °fire (up), get (someone) (all) steamed *or* hopped *or* het up: *Those fire-and-brimstone preachers used to get the people all worked up.* **(b)** °prepare, (make *or* get) ready, whip into shape, °develop, come up with, write up, put together, °produce, °turn out: *Can you work up that proposal in time for Monday's meeting?* **(c)** °advance, ascend, °rise, move up *or* ahead *or* on: *In no time at all, Greg worked his way up from assembler to foreman.* **(d)** See **19 (e),** above.

worker *n.* °laborer, working man *or* woman, workman, °hand, °employee, artisan, craftsman, °tradesman, white-collar worker, blue-collar worker, proletarian, breadwinner, wage earner: *It seems unlikely that the workers would strike for longer hours and less pay.*

workmanship *n.* handicraft, °craft, craftsmanship, artistry, °art, °technique, handiwork, °skill, skillfulness, mastery, *US* artisanship: *One cannot help but admire the workmanship in those marquetry tabletops.*

works *n.pl.* **1** °plant, °factory, workshop, °shop, °mill: *We were taken on a tour of the works to see how the marble is carved.* **2** °mechanism, machinery, workings, (moving *or* working) parts; clockwork, *Colloq* innards, °insides, *Slang* °guts: *We once had a clock with all the works exposed.* **3** ***the works.*** **(a)** °everything, °the lot, *Colloq* the whole kit and caboodle, the whole shooting match, everything but *or* including the kitchen sink, *Chiefly US and Canadian* the whole shebang: *We ordered two hamburgers with the works.* **(b)** a thrashing, a beating, a drubbing, a battering, a flogging, a lambasting: *When he refused to reveal the hiding place of the money, they gave him the works.*

world *n.* **1** °earth, planet, °sphere, °globe, terra; °universe, cosmos, °existence, °creation, °life: *Do you believe the world is round? Buckland lives in his own, private world.* **2** °humanity, mankind, °people, the human race, °society, the public, men, humankind, everybody, °everyone, the world at large: *We know what he thinks of the world, but what does the world think of him?* **3** °area, °sphere, °domain, community, °clique, °crowd, °circle, °fraternity, °faction, °set, coterie: *What value does the art world place on Longchamps' paintings?* **4** °period, °time, °age, °era, epoch, time(s): *They have produced an excellent series of programs on the ancient world.* **5** ***bring into the world.*** **(a)** °deliver, °have, give birth to, beget, *Rare or dialectal* °birth: *Grandmother brought sixteen children into the world.* **(b)** °deliver, °midwife: *The town family doctor brought hundreds of children into the world in his long career.* **6** ***for all the world.*** °precisely, °exactly, in all respects, in every respect, in every way, °just: *He behaved for all the world like a man possessed by the devil.* **7** ***on top of the world.*** °ecstatic, °delighted, °elated, °happy, °exultant, °overjoyed, °rapturous, *US* in seventh heaven, *Brit* in the seventh heaven, *Colloq* on cloud nine, *Brit* over the moon: *Grumbleigh is on top of the world since Prudence agreed to marry him.* **8** ***out of this world.*** °marvelous, wonderful, °exceptional, °unbelievable, °incredible, °excellent, °superb, *Colloq* °great, smashing, °fantastic, °fabulous, *Slang* °out of sight, far-out, *Brit* °magic, *US* to the max: *The place we stayed at Ibiza was out of this world!*

worldly *adj.* **1** mundane, °earthly, °terrestrial, °temporal, °physical, °carnal, fleshly, corporeal, °human; °lay, nonspiritual, nonreligious, civic, °secular, °profane: *Theo indulged in worldly pleasures in the fleshpots of the continent. Veronica put aside worldly things and took the veil.* **2** urbane, °suave, °sophisticated, cosmopolitan, worldly-wise, *Slang* with it, °hip, °cool: *He might play the role of a worldly person in the films, but he is really a country bumpkin.*

worn *adj.* **1** °shabby, °threadbare, tatty, °tattered, °ragged, frayed: *When he's around the house, Burt wears an old, worn tweed jacket.* **2** °haggard, °drawn: *He looked anxiously at his mother's thin, worn face.* **3** ***worn out.*** °tired, °fatigued, °exhausted, frazzled, °spent, °jaded, played out, °haggard, °drawn, the worse for wear, *Colloq* dog-tired, all in, done in, *Slang* °dead (on one's feet), °beat, *US* pooped: *I was completely worn out from climbing those stairs.*

worried *adj.* °fearful, apprehensive, °anxious, distressed, °nervous, uneasy, anguished, disquieted, agonized, °agonizing, °distraught, °on edge, on tenterhooks, °ill at ease, troubled, °fretful, °agitated, perturbed, °upset, °suffering: *The two worried mothers waited at the police station for news of their children.*

worry *v.* **1** be anxious, be fearful, be concerned, °fret, agonize, be distressed, be vexed, *Colloq* °stew, bite *or* chew (one's) nails, go *or* get gray, get gray hair, *Slang* sweat blood, *US* sweat bullets: *I worry about you when you stay out so late.* **2** °annoy, °irk, °pester, nettle, harry, °harass, °tease, °bother, °tantalize, °torment, °plague, hector, badger, nettle, °gall, peeve, *Colloq* hassle: *The police have been worrying me the way a dog worries a bone.* —*n.* **3** °concern, °care, °responsibility; °problem, °bother, °trouble, °affliction, irritation, °annoyance, vexation: *I cannot see why that is your worry. I am sorry to hear about your financial worries.* **4** °anguish, °anxiety, uneasiness, unease, nervousness, °distress, apprehension, disquiet, perturbation, °agitation, °upset, °misgiving: *Your associating with that crowd has caused your mother and me a great deal of worry.*

worsen *v.* **1** °increase, exacerbate, °heighten, °intensify, °aggravate: *The recent rains have worsened the drainage problem.* **2** °weaken, °deteriorate, °decline, °degenerate, °decay, °slip, °sink, °slide, °fail, °disintegrate, take a turn for the worse, get worse, go from bad to worse, *Colloq* go downhill: *Her condition continues to worsen.*

worship *v.* **1** °venerate, °revere, °reverence, °extol, °honor, °exalt, °praise, °admire, °adore, adulate, °glorify, deify, °idolize, be devoted to, pay homage to, bow down before, kneel before, °put on a pedestal: *When he arrived, he found the people worshiping huge stone idols.* —*n.* **2** °veneration, °reverence, adoration, °devotion, °homage, °honor, °respect, °esteem, exaltation, °praise, °admiration, adulation, glorification, deification, idolatry: *A healthy respect for security should not be confused with the worship of money.*

worth *n.* °quality, °merit, value, °advantage, °benefit, °good, °importance, °significance, °usefulness: *You have underestimated Roberta's worth to the community.*

worthless *adj.* **1** valueless, unimportant, °insignificant, inessential *or* unessential, °dispensable, °disposable, °paltry: *His contribution to the work of the committee proved utterly worthless.* **2** °pointless, °bootless, °silly, °inane, °vain, unavailing, °useless, °futile, °fruitless, unproductive, °unprofitable: *It is worthless to try to salvage that wreck.* **3** °cheap, valueless, °tawdry, °poor, trashy, rubbishy, °shabby, °wretched, *Colloq* °tinny, crappy, cheesy, *Slang* chintzy: *The deceased's handbag contained only a few items of worthless jewelry.*

worthwhile *adj.* **1** °profitable, justifiable, °productive, °gainful, °rewarding, °fruitful, cost-effective, remunerative: *Would it be worthwhile to buy up that land as a long-term investment?* **2** °useful, valuable, °good, °helpful, °beneficial, °worthy, beneficent, °desirable, °exemplary, °matchless, °honorable, °upright, °sterling, °irreproachable: *Ben's one of the most worthwhile people I met in the navy.*

worthy *adj.* **1** °worthwhile, °deserving, meriting, °meritorious, °praiseworthy, °good, °estimable, °qualified, creditable: *He is a worthy recipient of the honor.* **2** See **worthwhile, 2,** above. —*n.* **3** °dignitary, °personage, °notable, eminence, luminary: *The dinner was attended by all the local worthies.*

wound *n.* **1** °damage, °hurt, °injury, trauma, traumatism; laceration, °puncture, °cut, °gash, °slash, lesion, °bruise, contusion: *We dressed and bandaged the wounds of the victims.* **2** °slight, °damage, °injury, °harm, °blow, °distress, mortification, °torment, torture, °anguish, °pain, °insult: *Hugh Briss is devastated by any wound to his ego.* —*v.* **3** °damage, °harm, °injure, °hurt, traumatize; °cut, °slash, °gash, °lacerate, °slit, °stab, °shoot, *Colloq* wing: *He was wounded in the leg in the war.* **4** °slight, °distress, °damage, °mortify, °insult, °hurt, °pain, °grieve, °offend, aggrieve, °wrong: *I was terribly wounded by the things she said about me.*

wrap *v.* **1** Sometimes, ***wrap up.*** °swathe, swaddle, °bind, °cover, enwrap, °envelop, °wind, enshroud, °shroud, enfold, °fold, °muffle, °enclose, sheathe, encase; °pack, °package, do up, gift-wrap: *Wrap up well—it's cold outside. I've just finished wrapping some Christmas presents.* **2** ***wrapped up in.*** immersed in, submerged in, buried in, absorbed in, engrossed in, bound up in, involved in, occupied with *or* by *or* in, engaged in, dedicated to, devoted to: *While Irma is completely wrapped up in her grandchildren, Henry is wrapped up in his books.* **3** ***wrap up.*** **(a)** °complete, conclude, °finish, °end, bring to a close, °terminate, °wind up, °settle, °tidy up: *We'll be able to wrap things up in an hour and go home.* **(b)** ***Wrap up!*** Be silent!, Be quiet!, Stop talking!, Hold your tongue!, *Slang* Shut up!, Shut your face!, Shut your trap!, Shut your mouth!, *Brit* Put a sock in it!: *He would have kept on wittering away if she hadn't told him to wrap up.* —*n.* **4** °stole, shawl, °mantle, poncho, serape, °cloak, °cape: *Sonia adjusted her evening wrap before going out to dinner.*

wrapper *n.* **1** housecoat, °robe, dressing gown, bathrobe, kimono, negligee, lounging robe, peignoir, *US* housedress: *Peggy threw on a wrapper and went to answer the door.* **2** envelope, °package, packing, wrapping, covering, jacket, °case, casing, container: *Throw the wrapper in the wastebasket.*

wreak *v.* °inflict, °exercise, °exert, °carry out, °bring (to bear), °visit, °effect, °work, unleash, °execute, °impose, °force, °vent, let go: *The tribesmen wreaked vengeance on the hill people for destroying their farms.*

wreck *v.* **1** °destroy, °ruin, °devastate, °demolish, smash, °shatter, °spoil, °dash (to pieces), reduce to nothing, turn into scrap, annihilate: *He drove my motorbike into a tree and wrecked it.* **2** °sink, scuttle, run aground, °founder, °capsize: *The ship was wrecked on a reef near Rarotonga.* —*n.* **3** °hulk, shipwreck, °ruins: *The wreck finally rusted away to nothing.* **4** °mess, °disaster, °ruin; °havoc: *That man made a wreck of her life.* **5** °destruction, °loss, °sinking, devastation, foundering, grounding, °capsize, capsizing, disabling, disablement, wrecking; demolition, demolishing, leveling, tearing down, razing, pulling down, obliteration: *The wreck of the* Titanic *was a major news event in 1912. The wreck of the old tower is scheduled for next month.*

wreckage *n.* debris, °fragments, °remains, rubble, ruin(s): *The wreckage of the building is being searched for survivors.*

wrench *v.* **1** °yank, °twist, °jerk, °force, °pull, °tug, °tear, wring, °rip, wrest: *The gun was wrenched from my grasp, breaking my finger.* **2** °strain, sprain, °twist, overstrain: *After she wrenched her ankle on the ladder, Mary couldn't walk for a week.* **3** °extract, wrest, wring, °force, °pry, °draw: *The only way they could wrench the information from Michael was by torturing him.* —*n.* **4** °yank, °twist, °jerk, °pull, °tug, °rip: *One wrench at the doorknob and it came off in my hand.* **5** °pang, °pain, °agony, torture, °blow, °ache, °throe, °anguish, °sadness, °grief, heartbreak: *The severest wrench at my heartstrings came when we had to part.* **6** *Brit* spanner, shifting spanner, adjustable spanner, *US* monkey wrench: *We need metric wrenches to work on your car.*

wrestle *v.* °battle, °fight, °struggle, tussle, °strive: *After thirty years Garth wrestled his way to the top in the company.*

wretch *n.* **1** °scoundrel, blackguard, worm, °villain, cur, °beast, dog, swine, °rogue, °good-for-nothing, knave, varlet, scalawag *or* scallywag, °rascal, rapscallion, *Archaic* caitiff, whoreson, *Colloq* rat, °stinker, louse, °creep, *Slang* bastard, *Brit* rotter, bounder, blighter, *US* °bum: *That wretch not only stole my wife but asked me for money to support her!* **2** °unfortunate, poor fellow *or* chap, miserable creature, poor devil, pilgarlic, *Slang* poor bastard *or* son-of-a-bitch, *Brit* poor bugger *or* sod, *US* sad sack: *The poor wretch cannot keep a job and lives off others' charity.*

wretched *adj.* **1** °miserable, °awful, °terrible, °atrocious, °deplorable, *Colloq* °lousy, °rotten: *We had the most wretched weather throughout our vacation.* **2** °unhappy, °sad, °miserable, °woebegone, woeful, °dismal, °downhearted, °heartbroken, °brokenhearted, heartsick, °dejected, depressed, melancholic, °melancholy, °mournful, disconsolate, °inconsolable, °doleful, cheerless, crestfallen, °joyless, °desolate: *If you want to see a wretched human being, look at a boy whose dog is lost.* **3** °pitiable, °pathetic, °sorry, °pitiful, hapless, °hopeless, °unfortunate, °miserable: *The poor in these tropical paradises are some of the most wretched people on earth.* **4** °vile, °shameful, °scurvy, underhand(ed), treacherous, °contemptible, °despicable, °base, °low, °mean, °paltry, mean-spirited, detestable: *It was hard to see how the people could support such a wretched little demagogue.*

wriggle *v.* **1** wiggle, wobble, °shake, °tremble, °quiver, °jiggle, waggle, writhe, °twist, °fidget: *I managed to wriggle free of the ropes and run away. The dog was wriggling with excitement at seeing his master.* **2** °twist, °squirm, °snake, worm, writhe, °slither, °crawl: *He wriggled through the narrow opening into the strongroom.* —*n.* **3** wiggle, zigzag, wavy line, *Colloq* squiggle, squiggly line: *The page had no writing on it, only a lot of wriggles.* **4** wriggling, writhing, °squirm, squirming, wiggle, wiggling, shaking, trembling, °quiver, quivering, shimmying, waggle, waggling, twisting, °twist: *Don't confuse that rhythmic wriggle of the hips with dancing.*

wrinkle[1] *n.* **1** crease, °fold, °line, °furrow, crinkle, crow's-foot, corrugation, °pucker, °ridge: *She hasn't a wrinkle on her face—but then she's only nine.* —*v.* **2** crease, °fold, °line, °furrow, crinkle, corrugate, °pucker, °gather, ruck, crimp, °screw up, °rumple, °crumple: *I love the way you wrinkle your nose when you're perplexed.*

wrinkle[2] *n.* °dodge, °gimmick, °device, ploy, °ruse, °scheme, °trick, °idea, °plan, °plot, °stunt, °way, °approach, °technique, °method, *Slang chiefly Brit* wheeze: *Leave it to Edmund to come up with a new wrinkle for fleecing pensioners.*

write *v.* **1** °pen, scribble, °get off, °dash off; indite, inscribe, °make out: *I was writing a note to you when you phoned. She wrote out a check for a thousand dollars.* **2** °correspond (with), send a letter *or* a note *or* a postcard *or US also* a postal card, °communicate (with): *You don't write home often enough. Write if you get work.* **3** °compose, °create, °make up, °author: *No, Addinsell did not write "Stigma Variations."* **4** ***write down.*** **(a)** °register, °list, catalogue, °note, make a note *or* notation, °record, °transcribe, set *or* jot *or* take down, °note, put in writing, put in black and white: *The teacher was known for Irish bulls like "All absent students must write down their names."* **(b)** derogate, decry, °disparage, °put down, °minimize, make little of, °play down, °detract, °belittle: *Cumberland's contributions have been written down by his critics.* **5** ***write off.*** °delete, °cancel, °disregard, °ignore, °forgive, °forget (about), annul, eradicate, °erase: *The bad debts have been written off.*

writer *n.* °author, novelist, littérateur, essayist, man of letters, °scribe, scribbler, wordsmith, freelancer, penny-a-liner, °hack, Grubstreeter, °journalist, newsman, °reporter, °correspondent, member of the fourth estate, (gossip) columnist, stringer, *Brit* paragraphist, *US* paragrapher, *Colloq* pen pusher, pencil pusher, sob sister, *Brit* journo: *These days all those who know how to write call themselves writers.*

writing *n.* **1** handwriting, longhand, °penmanship, °script, calligraphy *or* chirography, scribble: *I sometimes have trouble reading your writing.* **2** Sometimes, ***writings.*** (literary) work(s), °composition, °theme, °book, article, critique, °criticism, °review, °editorial, column, exposé, °essay, °poetry, °poem, °novel, nonfiction, fiction, °document, °letter, correspondence, °publication, *Chiefly Brit journalism* leading article *or* leader, *Chiefly US journalism* op-ed article: *Her writing has been widely published. Poe's writings are collected in this book.* **3** °literature, belles-lettres, °letters: *I don't agree that British writing has declined since 1900.*

wrong *adj.* **1** °improper, unjust, unfair, injudicious, unethical, °terrible, °foul, °awful, °bad, °immoral, °sinful, °evil, iniquitous, °villainous, °wicked, °vile, diabolic(al), °infernal, °fiendish, °corrupt, °dishonest, reprehensible, °abominable, °dreadful, °dishonorable, blameworthy, °naughty, °shameful, °disgraceful, opprobrious, °criminal, felonious, °illegal, °illicit, °unlawful, °illegitimate, *Colloq* °crooked: *They believe that all war is wrong. It was wrong of you to come here while my husband was away.* **2** °mistaken, °in error, °erroneous, °incorrect, °inaccurate, °imprecise, °inexact, fallacious, askew, °false, °wide of the mark; °strange, °odd, °peculiar, °curious; *Colloq* off target *or US also* off the target, off the beam, *Brit* off beam: *The answers to these four questions were all wrong. You certainly have some wrong ideas about me.* **3** °incorrect, °improper, unsuitable: *They must have given us the wrong baby at the hospital.* **4** °inappropriate, inapt, °improper, indecorous, °unseemly, unfitting, °unacceptable, °undesirable, °incongruous, °unbecoming, °out of place, ill-considered, wrongheaded, °imprudent, °misguided, inexpedient, impolitic, infelicitous, *Colloq* out of line: *You saw nothing wrong in inviting girls to stay with you in your rooms? Your mother thinks Anne is the wrong girl for you.* **5** °out of order, not working, °faulty, awry, °amiss, the matter, °defective, °imperfect, °unsound, °flawed, °deficient: *Considering the noise my car makes, there must be something wrong.* **6** °opposite, °reverse, °incorrect, °improper: *That is the wrong direction to the post office.* —*adv.* **7** awry, imperfectly, incorrectly, improperly, inappropriately, °amiss, °badly, wrongly, *Scots and north England and literary* agley, *Colloq* out of sync: *You've wired up the video all wrong.* **8** ***go wrong.*** **(a)** go astray, falter, °fail, °lapse, °err, fall from grace, go to the bad, °deteriorate, go downhill, backslide, regress, retrogress, recidivate: *Her father was convinced that she'd go wrong if she got into the theater.* **(b)** °fail, malfunction, °break down, °miscarry, backfire, °fall through, *Colloq* °flop, °come to grief, go kaput, *Brit* go phut: *The entire plan went wrong when Clara started screaming.* —*v.* **9** °abuse, °mistreat, °injure, °misuse, maltreat, ill-use, °ill-treat, °discredit, asperse, calumniate, malign, °dishonor, °impose upon, take advantage of, °harm, °damage, °oppress: *He believes he was wronged by his commanding officer.*

wry *adj.* **1** distorted, contorted, twisted, °lopsided, °deformed, °crooked, aslant, °one-sided, askew, °bent, tilted, off-center: *He is afflicted by torticollis, which gives his neck a wry appearance.* **2** °dry, droll, °witty, °sardonic, °sarcastic, ironic(al), amusing; °perverse, fey; *Scots* pawky: *Bill has a wry sense of humor that does not appeal to everyone.*

Y

yahoo *n.* °boor, °barbarian, °philistine, °savage, churl, °brute, lout, °bourgeois, lowbrow, oaf, °clod,

°peasant, *Slang* °slob, *Brit* mucker, yob *or* yobbo, *US* rube: *He used every trick to attract the yahoos into the side show.*

yank *v.* **1** °jerk, °jolt, °tug, °wrench, °snatch, °hitch: *He yanked the chair from beneath me just as I sat down.* —*n.* **2** °jerk, °jolt, °tug, °wrench, °snatch, °hitch: *Give two yanks on the rope when you want me to pull you up.*

yap *v.* **1** bark, yelp: *Their dog was yapping away all night.* **2** gabble, °babble, blither *or* blather, °chatter, °jabber, °tattle, °prattle, prate, *Colloq chiefly Brit* witter, natter, *Slang* jaw, run on, *US* run off at the mouth: *I wish she'd stop her yapping and give me some peace.* —*n.* **3** °mouth, *Slang* °trap, °gob, *US* bazoo: *If only Sam would shut his big yap once in a while!*

yardstick *n.* °measure, bench mark, criterion, °standard, °gauge, °basis, °touchstone, °scale, exemplar: *We thought she sang beautifully, but then we had no yardstick against which to judge.*

yarn *n.* **1** °thread, °fiber; strand: *I haven't enough yarn to finish knitting these socks.* **2** °tale, °story, °account, °narrative, anecdote; tall tale, fable, °fabrication, fiction, cock-and-bull story, *Colloq* whopper, *Brit* fishing story, *US* fish story, fish tale: *The old salt had many yarns to spin during the long evenings by the fire.*

yearly *adv.* **1** annually, perennially, every year, once a year, year after year, year in (and) year out, regularly: *We watch the Canada geese migrate north yearly.* **2** per year, per annum, by the year, each year: *What is the interest rate yearly?* —*adj.* **3** annual, °perennial, once-a-year, °regular: *Those yearly migrations told us when to plant the first crop.*

yearn *v.* °long, pine, °ache, °hanker, °itch, °hunger, °thirst, crave, have a craving, °desire, °wish, °want, °fancy, °prefer: *I yearn to see the green-clad hills of home once more/Before my vessel bears me from this mortal shore.*

yell *v.* **1** °shout, °scream, °bellow, °howl, screech, yowl, °roar, °bawl, caterwaul, squall, yelp, *Colloq* holler: *"Stop yelling at your sister!" shouted my father.* —*n.* **2** °shout, °scream, °cry, °bellow, °howl, screech, yowl, °roar, caterwaul, squall, yelp, *Colloq* holler: *I think I heard a yell for help.*

yeomanly *adj.* yeoman, workmanlike, °useful, °staunch, °courageous, °loyal, dedicated, °faithful, °steadfast, unswerving, unwavering, °firm, °sturdy, °reliable, °solid: *Carruthers certainly did a yeomanly job on reorganizing the system.*

yes man *n.* toady, sycophant, toadeater, timeserver, °hanger-on, lickspittle, bootlicker, truckler, °flunky, courtier, jackal, spaniel, lap dog, *Taboo slang Brit* arse-kisser, arse-licker, *US* ass-kisser, ass-licker, *US* brown-noser: *He surrounds himself with yes men, then says he refuses to take no for an answer.*

yet *adv.* **1** as yet, (up) till *or* until now, °so far, thus far, up hitherto, to the present (time): *She hasn't arrived yet.* **2** °still, up to this time, up to now, even now, till *or* until now, to this day: *She was there earlier and I think she's there yet.* **3** °moreover, furthermore, °besides, °further, °still: *He made yet another excuse. They yet have the kitchen to clean.* **4** in the future, in time to come, later, °eventually: *Despite the life I've led, I might yet get to heaven.* **5** °still, °notwithstanding, anyway, anyhow, °nonetheless, °nevertheless, °regardless, in spite of *or* despite everything, just *or* all the same, °even so, after all, *US* still and all: *Despite the life I've led, I might get to heaven yet.* —*conj.* **6** °notwithstanding, in spite of *or* despite it *or* that *or* the fact, °still, °nevertheless, but: *He goes out every night, yet returns every morning.*

yield *v.* **1** °surrender, give up (the fight *or* struggle), °give in, knuckle under, °submit, °cede, cry quits, throw in the towel *or* the sponge, °capitulate, °succumb, raise the white flag: *The captain of the enemy company asked if we would yield.* **2** °give up, °surrender, °give over, hand in *or* over, °abandon, °relinquish, °renounce, °cede: *We agreed to yield our arms to the enemy.* **3** °agree, °consent, °comply, °concede, °relent, assent, °give way, accede, concur: *He did his best to persuade me, and I finally yielded.* **4** °earn, °return, °pay, °bring in, °supply, °generate, °produce, °net: *She has investments that yield more than 15 percent a year.* —*n.* **5** °return, °production, °output, °revenue, takings, °gate, °earnings, °income, °proceeds, °profit, °gain: *The yield from that operation is not even enough to pay the rent.*

yielding *adj.* **1** pliant, flexile, °flexible, °pliable, °soft, °plastic, fictile, °elastic, resilient, °supple, springy, bouncy, spongy, rubbery: *The frame is of a yielding material that gives way under pressure.* **2** °accommodating, docile, °submissive, amenable, °tractable, compliant, °obedient, °flexible, acquiescent, °agreeable, °obliging, °manageable, manipulable: *He wanted a wife with a yielding temperament who would tolerate his foibles.*

young *adj.* **1** youthful, teenage(d), °adolescent, prepubescent, pubescent, °juvenile, °minor, °junior, underage: *Would you believe that she has young children?* **2** °boyish, girlish; °immature, °callow, °green, °inexperienced, °unfledged, uninitiated, °unsophisticated, °childlike, °innocent, naive *or* naïve *or* naïf: *This program may contain material unsuitable for young ears.* **3** °childish, °puerile, °infantile, babyish, *US* sophomoric: *He's too young to invite to the party.* —*n.* **4** °offspring, babies, °issue, little ones, °progeny, °litter, °brood; children: *The cat's young should stay with her for a few weeks.*

youth *n.* **1** °childhood, boyhood, girlhood, young manhood, young womanhood, prepubescence, pubescence, adolescence, salad days; immaturity, minority: *In our youth we do many foolish things we might later regret.* **2** °child, youngster, schoolchild, °teenager, teen, °minor, °juvenile, °adolescent; °boy, schoolboy, °stripling, young boy *or* man, °lad, laddie, whippersnapper; °girl, °lass, lassie, schoolgirl, °maid, °maiden, *Literary* damsel, demoiselle, mademoiselle, *Colloq* kid, (little) shaver, *US and Canadian* tad, *Slang* teeny-bopper, *Brit* sprog: *As youths, they used to spend time at rock concerts.* **3** children, youngsters, juveniles, adolescents, young people, °young, *Colloq* kids: *The youth of today have little respect for their elders.*

yucky *adj.* yukky *or US also* yuchy, °disgusting, °repugnant, °repellent, unappetizing, vomit-provoking, sick-making, ill-making, °nauseous, nauseating, °revolting, °foul, mucky, °revolting, °beastly, °awful, *Slang Brit* grotty: *We get this yucky mess at school that we call "mystery meat."*

yummy *adj.* °delicious, mouthwatering, °luscious, appetizing, °tasty, toothsome, °savory, delectable, ambrosial, *Colloq* scrumptious: *At our school we get the most yummy ice cream.*

Z

zany *adj.* **1** clownish, °mad, °wild, °frolicsome, °sportive, °playful, °gay, °merry, slapstick, °crazy, °funny, comic(al), amusing, °hilarious, °absurd, °nonsensical, °ludicrous, °silly, °foolish, °inane, *Colloq* wacky, loony, madcap, crackpot, nutty, goofy, *US* kooky: *The buskers put on the zaniest show I have seen for some time.* —*n.* **2** °clown, °comic, jester, °fool, °joker, buffoon, °wag, °comedian, merry-andrew, laughingstock, *Slang* nut, *US* screwball: *The two zanies kept the crowd in stitches while pickpockets stole their wallets.*

zap *v.* °destroy, °kill, °slaughter, annihilate, °murder, slay, assassinate, liquidate, °erase; °shoot, electrocute, *Slang* °rub out, °polish off, °knock off, °bump off, snuff (out), °waste, °hit, *US* ice: *The baddies got zapped by the ray gun, one after the other.*

zealot *n.* °fanatic, extremist, °radical, bigot, °maniac, °militant, °terrorist: *We have to deal with suicidal zealots who are willing to die for their cause.*

zealotry *n.* °fanaticism, extremism, radicalism, °bigotry, militantism, terrorism, single-mindedness, monomania, °fervor, °frenzy, hysteria, °obsession, obsessiveness: *Their attacks on civilians show how far their zealotry can take them.*

zenith *n.* meridian, °summit, °acme, apex, °vertex, apogee, high point, °top, °peak, °pinnacle: *The dictator reached the zenith of his power very quickly.*

zero *n.* **1** °none, °nil, null, °nothing, °nought, ought, cipher, *Brit cricket* °duck, *Colloq* nix, *US* goose egg, nada, niente, *Slang US* zilch, *Brit* sweet F.A. *or* Fanny Adams, bugger-all, *Taboo slang* fuck-all: *The visiting team won three games, we won zero.* **2** °(rock) bottom, nadir: *On a scale of one to ten, she is a zero.* **3** °nobody, °nothing, nonentity, *Slang US* nebbish, bupkis: *That fellow she married is a real zero.* —*v.* **4** ***zero in on.*** focus on, pinpoint, °fix on, home in on, concentrate on, bring to bear on: *We finally zeroed in on the problem and had it repaired in no time.*

zest *n.* **1** °spice, °relish, °gusto, °tang, °pepper, ginger, piquancy, pungency, °edge, °bite, °flavor, *Colloq* zing, zip, pizazz: *Put in a little of my homemade sauce to add some zest.* **2** °eagerness, zestfulness, °exuberance, °appetite, °interest, °enthusiasm, °hunger, °thirst: *Where does Sidney get his zest for life?*

zone *n.* °area, °quarter, °district, °region, sector, °section, °sphere, °belt, °territory, °province, °realm, °domain, °precinct, bailiwick, °department, °terrain, °circle, locality, °locale, *Slang US* °turf: *A duty-free zone will allow for quicker transshipment of goods. The northern and southern hemispheres each have a frigid, temperate, and torrid zone.*

zoo *n.* **1** zoological garden, menagerie, *Tiergarten,* °(safari) park: *When I was a child, I enjoyed going to the zoo almost as much as I do now.* **2** madhouse, °mess; °chaos, °pandemonium, °bedlam, *Colloq US* three-ring circus, Chinese fire-drill: *When the three children and my husband get ready in the morning the kitchen is like a zoo.*

INDEX

Guide to Use of the Index

The Index lists in letter-by-letter alphabetical order the words and phrases that appear as synonyms in the Text of the *Thesaurus.*

To find the synonyms for a word, first look it up in the Text; if it is not a main entry there, then look it up in the Index, where it appears followed by an alphabetical listing of all the main entries in the Text where the word or phrase can be found.

The degree sign (°) alongside a main entry in the Index means that the word so marked is a main entry in the Text. For example, the Index entry

> **afford°**, furnish 1; give 1; impart 1; open 25; provide 2; supply 1

shows that *afford* has its own main entry in the Text and that it also appears under the other main entries listed at the sense numbers shown. The absence of the degree mark at the entry

> **abdomen,** gut 2; stomach 1

indicates that *abdomen* appears in the Text under the main entries listed, at the sense number(s) indicated, but has no main entry in the Text.

A phrase consisting of verb + (prepositional or adverbial) particle is listed in alphabetical order at the end of the entry for the verb:

> **work°**, act 9; activity 2; behave; book 1; . . .
> **work as,** function 4
> **work at,** strive 1
> **work (away) (at)** peg 6;
> **work (for),** assist 3; pursue 3; serve 2, 5
> . . .
> **work up,** excite 3; fire 8a; foment; . . .
> . . .

A phrase that begins with a preposition, adverbial particle, any one of certain verbs, or another word that is felt to be semantically "empty" is listed under its most meaningful, or key word, and the expression is inverted; if that key word already has an entry of its own, the phrase appears at the end of the listing for that word, in alphabetical order:

> **heave-ho,** expulsion; sack 3
> **heave-ho, give (someone) the old,** dismiss 1; fire 11; sack 4
> **heave-ho, the (old),** brushoff; dismissal 1; ejection 3; rejection
>
> **going°**, . . .
> **going, get,** activate; excite 1; float 2; get 28c; . . .

incongruous°, . . .
  incongruous, not, natural 9

liking°, affection; affinity 2; . . .
  liking (for), fond 3
  liking for, have a, favor 6; like² 1

ready°, . . .
  ready, at the, ready 7, 9
  . . .
  ready (for), make *or* **be** or **get,** poise 3; provide 5; ready 10

The question of what is semantically "empty" is a touchy matter, better left untreated here. For the purposes of the *Thesaurus,* it includes words like *be, become, do, get, give, make, no, not,* and a few others when they appear in certain contexts. It would be impossible to satisfy every user in the matter of where a given phrase might be found in such a complex listing of phrases, expressions, and sentences, so the best advice is, if a phrase is not found where first sought, look for it under another one of its words.

Set phrases that begin with meaningful words are entered in their full form as main entries:

Blessed Isles *or* **Islands,** . . .

civil affairs, . . .

civil disorder, . . .

civil servant, . . .

handwriting on the wall, . . .

sales agent, . . .

thé dansant, . . .

whited sepulcher, . . .

Phrases in which verbs are followed by their direct objects are listed as main entries; for example,

administer the coup de grâce, . . .

cite (chapter and verse), . . .

cudgel one's brains, . . .

curry favor, . . .

kick the bucket, . . .

If the choice of key word is ambiguous, the phrase is listed under both; for example, *kick the bucket* is listed as above and also as

bucket, kick the, . . .

However, if the object is *one, oneself, someone, something, it,* or a similar pronoun, then the phrase is listed at the end of the entry containing its main verb; for example,

behave°, . . .
  behave like one possessed, . . .
  behave (oneself), . . .

cling°, . . .
  cling together *or* **to one another,** . . .

clothe°, . . .
  clothe (oneself) in, . . .

cut°, . . .
  cut (it) out, . . .
  cut (someone) down to size, . . .

Other phrases are listed under their key words, for example,

daggers drawn, be at, . . .

odds, be at, . . .

loggerheads, be at, . . .

"Jack Robinson," before one can say, . . .

turn around, before one can, . . .

brown study, be in a, . . .

driver's seat, be in the, . . .

straitjacket, be ready for a, . . .

fear, beside oneself with, . . .

snuff, be up to, . . .

false witness, bear . . .

quick, cut to the, . . .

The placement of entries in the Index reflects the editor's view of where they are most likely to be looked up, not any structured grammatical, syntactic, or semantic analysis. The set consisting of **cut to pieces, ribbons, shreds,** can be found sorted under the nouns, accompanied by other verbs like **chop, hack, hew,** etc., and it is hoped that such an arrangement will enable the user seeing such items together—possibly for the first time in any reference work—to manipulate the language in a creative and useful way.

To save space, the variant that appears as "his *or* her" in the Text is shown in the Index as "his . . . ":

set (someone) back on his . . . heels,

Noun phrases like *GI Joe, master of ceremonies, stately home* are listed as main entries in the Index in their proper alphabetical places.

Phrases are listed as subentries under the word that is the key word in the phrase. For example,

cogitate, meditate 1; muse; ponder
  cogitate (about *or* **on** *or* **over),** consider 1; contemplate 2; deliberate 4; mull; reflect 3; speculate 1; think 3

certain°, absolute 4; assertive; bound[3] 4; clear 10; . . .
  certain, a, one 3
  certain, for, doubtless 1
  certain, make *or* **be,** assure 1
  certain, make (someone), assure 3
  certain of, be, know 3; mind 11; rely 1; tell[1] 7
  certain (of), make, ensure 1; guarantee 2; see 5; settle 2

Many verb phrases and other expressions that occur in only one form in the Text are listed as main entries in the Index in the same form, with their key word shown first; for example,

dispelled, be, dissipate 1
  (**dispelled** has no entry of its own)

keyed up, excited 1; tense 2
(**keyed** has no entry of its own)
kibosh on, put the, ruin 9; upset 5; veto 1
(**kibosh** has no entry of its own)
kilter, out of, broken 7; order 13c
(**kilter** has no entry of its own)

Plurals of nouns, past tenses, present and past participles of verbs, etc. are shown in their proper alphabetical places; for example, a sequence of main entries runs (omitting the references),

center°, . . .
centered, . . .
central°, . . .
Central Intelligence Agency, . . .
centralize, . . .
centralized, . . .

installation, . . .
installations, . . .

installment, . . .
installments, in, . . .

It is hoped that the arrangement of the Index as proffered will be appreciated as the user continues to refer to it.

opulent 3; overgrown; plentiful 1; prodigal 2; productive 1; profuse 1, 2, 3, 4; prolific 1; rank[2] 1; rich 9; substantial 1; thick 4
abundant, be, teem[1]
abundant (in), rich 2; fraught 1
abundant in, be, abound 2; fill 6
abundantly, amply 2, 3; freely 4
abuse°, attack 2, 7; batter 3; berate; blaspheme 2; dishonor 1; flak; harm 1, 3; ill 7, 9; ill-treat; indignity; injure 2; injury; insult 1, 2; knock 3d; light[2] 14; manhandle; mishandle 1; mistreat; mistreatment; misuse 2, 4; mock 1; mockery 1; molest 2; oppress 2; oppression; outrage 1, 5; persecute 1; persecution 1; pervert 1; pick 6; pitch[1] 6a; play 15; prostitute 2; prostitution 2; punish 3; punishment 3; sacrilege 1; scandal 3; tongue-lashing; torment 1; use 2; vilify; violation 1; vituperate; waste 5; wrong 9
abused°, downtrodden; put-upon
abusing, violation 1
abusive°, foul 6; harsh 2; injurious 2; nasty 4; outrageous 2; profane 3; prohibitive 2; satirical; scandalous 2; scurrilous; stiff 2; tart[1] 2; vituperative
abusive, be *or* become, rail[2]
abusiveness, severity 1
abut, border 9; butt[2] 1; join 4; meet[1] 4; touch 3
abutting, adjoining; near 5; surrounding
abysm, abyss; depth 5; gulf 2
abysmal°, bottomless; low[1] 6; mean[2] 3
abyss°, crevasse; depth 5; gulf 2; hell 1; pit[1] 3
abyssal, abysmal 2; bottomless; deep 1
academe, instructor
academic°, educational 1; intellectual 2; learned; literary 2; scholar 1; scholarly; studious 2; theoretical 2
academic(al), pure 4
academician, instructor; intellectual 4
academy, institution 2; seminary; society 5
acanthoid, prickly 1; thorny 1
a cappella, unaccompanied
accede, acknowledge 1; comply; consent 1; succumb; way 11b; weaken 3; yield 3
accede (to), accept 2; agree 2, 3; defer[2]; get 1; grant 2; keep 6; submit 1; take 39a; underwrite 2
accelerate, advance 4; dispatch 4; expedite 1; forward 8; hasten 2; hurry 2; pick 8g; precipitate 1; quicken 1, 2; speed 2; step 18b; urge 1
accelerated, fast[1] 1
acceleration, spurt 2
accent°, cadence; dialect; emphasize; expression 4; intonation; point 22; pronunciation; punctuate 2; rhythm; spotlight 3; stress 1, 4; talk 19; tone 2
accent mark, accent 2
accentuate, accent 4; emphasize; play 17a; point 22; punctuate 2; spotlight 3; stress 4
accentuation, accent 1; intonation; pronunciation; rhythm; stress 1
accept°, abide 1; acknowledge 1; admit 1, 3; adopt 1; agree 3; approve 1, 3; assume 1; believe 1; buy 2; concede 1; credit 5; digest 2; embrace 2; face 18a; fall 13b, 15; figure 9; follow 2; have 2; heed 1; honor 8; initiate 2; jump 6; lap[1] 3b; leap 4; obey 1; pass 7; put 30; receive 1; recognize 2, 3; see 3; shoulder 8; stand 3; stomach 3; submit 1; subscribe 1; swallow 2; take 6, 7, 17, 35, 39c; thumb 4; tolerate 2; understand 2, 5; undertake 1; welcome 1
accept as one's own, adopt 1
accept (for), take 8
accept (prematurely *or* without question), leap 3
acceptable°, adequate 2; agreeable 1; bearable; comfortable 4; competent 1; decent 2, 3; good 1; legal 1; legitimate 3; likely 3; OK 2; par 6; passable 1; permissible; plausible 1; presentable 1; presumptive 1; rational 3; reason 6; reasonable 3; regular 8; safe 4; satisfactory; suitable; supportable 1; tolerable 1, 2; welcome 2
acceptable, be, muster 3
acceptable, make, rationalize 1; recommend 3
acceptable to *or* for, be, suit 2
acceptance, acknowledgment 2; admission 2; approval; belief 2; choice 1; consent 4; indulgence 1; passage 9; permission; popularity; prevalence 1; receipt 2; recognition 2; resignation 2; sanction 3; vogue 2
acceptance, find, go 27e
accepted, assumed 3; current 2; foregone; general 1; given 1; going 2; natural 1; official 1; orthodox; popular 1; proper 2; proverbial 1; regular 8; regulation 4; standard 5; understood; undisputed; virtual; welcome 2; well-established
accepted, be, go 27e, 35b
accepted as, be, pass 16a
accepted by, get *or* become, grow 5
access, admission 1; admittance; approach 5; entrance[1] 1, 2; entry 1, 2; gate 1; get 31a; mouth 2; outburst; passage 11; recourse 1; road 1; run 39; splurge 1; spurt 1; touch 9
access to, give, go 4
access to, have, touch 9
accessary. See **accessory**
accessible°, accommodating 2; available; convenient 2; forthcoming 3; free 3; friendly 2; hand 9a; handy 1; home 4b; nearby 2; open 4; outgoing 2; popular 3; public 3; ready 7; reserve 8; sociable; susceptible 2
accessible, be *or* stand, stand 5b
accession, succession 3
accessories, fitting 2; furniture 2; gear 2; paraphernalia; stuff 2; trappings
accessorize, ornament 2
accessory°, accident 3; accomplice; ally 1; associate 3, 6; attachment 4; attendant 1; auxiliary 1, 2, 3; extra 1, 3; fellow 3; fixture 2; further 1; indirect 2; instrumental; nonessential 1; occasional 2; ornament 1; partner; party 4; supplement 1; sympathizer
accessory (to), subsidiary
accident°, casualty 1; catastrophe 2; coincidence 3; disaster; fluke; luck 1; misfortune 2; pileup 1; spill 5
accident, by, unawares 2
accidental°, casual 1; chance 6; circumstantial 2; coincidental; haphazard 1; inadvertent 1; incidental 1; indirect 2; random 1; stray 5
accidentally, chance 5a; incidentally 2; unawares 2
accidie, sloth; sluggishness
acclaim, applause; attention 2; clap 1; credit 4; eulogize; eulogy; extol; fame; glorify 2; hail[1] 2; honor 6; kudos; laud; laurels; ovation; popularity; praise 1, 3; rave 3; renown; tribute 1
acclaimed, celebrated; famous; honorable 4; illustrious; legendary 3; noble 3; notable 1; noted; popular 1; prestigious; prominent 2; renowned
acclamation, applause; eulogy; kudos; laurels; ovation; praise 1; tribute 1
acclimated *or* acclimatized, seasoned 1; used 3
acclimate *or* acclimatize (oneself) (to), accustom; adapt 2; adjust 3; condition 7; orient 3; resign 2
acclimation *or* acclimatization, orientation 2; resignation 2
acclivitous, precipitous 1
acclivity, grade 4; hill 3; incline 4; ramp; rise 13; slope 2
accolade, eulogy; honor 2; kudos; praise 1; prize[1] 1; rave 3
accommodate°, adapt 2; board 7; contain 2; entertain 2; favor 8; fit[1] 7; fix 16a; gear 5; hold 8; house 8, 9; lodge 4; minister 3; oblige 1; orient 3; provide 1; put 28b; quarter 6; reconcile 2; seat 6; serve 1; shape 9; sit 5; square 12; suit 1; supply 1; tailor 2; take 17, 32a; work 17
accommodate (oneself) (to), adjust 3; resign 2
accommodating°, agreeable 2; attentive 2; considerate; decent 4; dutiful; easy 4; elastic 2; favorable 1; gracious; helpful; human 3; kind[1]; obliging; patient 3; respectful; submissive 1; supple 3; sweet 6; user-friendly; yielding 2
accommodation(s), abode; adaptation 2; domicile 1; home 1; housing 1; loan 1; lodging; passage 7; quarter 5; room 3; seating; service 3; settlement 3; shelter 3; space 4
accommodative, adaptable; agreeable 2; favorable 1
accompanied by, fraught 1
accompanied by, be, parallel 4
accompaniment, score 5
accompany°, attend 4, 5; bring 2; call 8b; escort 4; follow 3; go 20a, 39b; join 3; see 6; show 2; squire 1; tag 5; take 18; walk 2; watch 2
accompanying, attendant 1
accomplice°, accessory 2; ally 1; associate 3; cohort 3; partner; party 4; sympathizer
accomplish°, achieve 1, 2; bring 7, 12; carry 10c, 12; complete 5; discharge 5; dispatch 4; effect 7; effectuate; execute 1; finish 2; fulfill 1; implement 2; make 14; negotiate 2; notch 4; perfect 8; perform 1; perpetrate; procure 2; pull 12b; put 27a; reach 4;

accomplish° (*cont.*) realize 1; render 1; work 16
accomplishable, possible 2; practicable
accomplished°, able 2; adept 1; brilliant 3, 4; capable 1; complete 2; experienced 1; expert 2; fine[1] 4; good 12; great 6; learned; masterful 1; old 7; perfect 7; polished 1; practiced 1; proficient; qualified 1; skillful; talented
  accomplished in, versed
accomplishing, consummation 1
accomplishment°, achievement 1, 2, 3; act 1; completion 1; consummation 1; deed 2; discharge 12; effort 3; execution 1; exploit 1; fact 2; feat; fulfillment; perfection 2; performance 1; proceeding 1; realization 2; record 5, 6; skill 2; specialty 1; stroke 7; trick 4; triumph 1; work 4
accomplishments, laurels; stuff 5
accord°, agree 1; agreement 1; award 1; coincide; coincidence 2; comply; correspond 1; extend 4; give 1; go 36a; harmony 1; identity 1; impart 1; match 7; peace 2; provide 2; rapport; reconciliation 1; resemblance; show 10; solidarity; sympathy 2; tally 1; treaty; tune 2; understanding 1; uniformity 1; union 4; unity 1; vouchsafe 1
  accord, in (perfect), unison
  accord, of (one's) own, freely 2; willingly
  accord (with), conform 2; go 39a; square 11
  accord (with), be in, conform 2
  accord (with), in, agreeable 2; consistent 1; harmonious; line[1] 19b; proportional; united 3
accordance, accord 2; coincidence 2; observance 1; tune 2; uniformity 1
  accordance (with), be in, conform 2
  accordance (with), in, agreeable 2; consistent 1; harmonious; line[1] 19b; proportional; united 3
accordant, consistent 1
according (with *or* **to),** step 7
  according (with *or* **to), not,** step 8
accordingly°, consequently; duly 1; hence 1; therefore; thus 2
accost, address 4; buttonhole 1; greet 2; hail[1] 1; molest 2; pick 8j; proposition 3; salute 1; solicit 2; speak 12c; waylay 1, 2
accoucheur, midwife
accoucheuse, midwife
account°, bill[1] 1; bulletin; chronicle 1; chronology; description 2; distinction 2; entry 4; esteem 2; explanation 1; favor 4; history 1, 5; importance 1; mark 8; memoir 1; narrative 1; news 2; portrait; prominence 1; put 16f; recital 2; reckon 2; reckoning 2; record 3; regard 4; rehearsal 2; report 1; representation 3; sake 1; score 1, 8; statement; story[1] 1; tab 2; tale 1; version 2; word 2; yarn 2
  account, be of, weigh 5
  account, call (someone) to, accuse 1; fault 8; scold 1; task 3
  account, of little *or* **no,** immaterial 1; petty 1; puny 1
  account, on no, mean[3] 5; never 1; scarcely 2; shot 11
  account, take into, account 7; allow 6; bargain 4; consider 2; cover 7; figure 11b; reckon 5b; regard 4
  account, taking into, considering; light[1] 11
  account for, account 1; cover 7; explain 2; rationalize 1; reckon 5b; score 11
  account (of), give an, describe 1; narrate; relate 2; report 3
  account of, take little *or* **no,** disregard 1, 2
accountability°, debt 1; fault 3; liability 1; obligation 1; responsibility 1, 4
accountable°, debt 2; fault 5; liable 2; responsible 1, 2; stable 3
  accountable, hold, credit 5; fault 8; pin 5
  accountable for, be, answer 7a
accountableness, accountability
accountant, bookkeeper
accounted for, present[1] 2; speak 13
accounting, account 2; clerical 2; finance 1
account(s), memoir 2, proceeding 2a
accouter, clothe 1; dress 1; equip; outfit 4; rig 1; turn 20d
accouterment(s), attachment 4; equipment; fitting 2; furniture 2; gear 2, 4; kit; outfit 1; paraphernalia; provision 1; regalia; rig 3; robe 2; stuff 2; tack 4; tackle 1; trappings
accredit, assign 4; confirm 1; delegate 2; license 5
accredited, official 1
accrual, increment
acculturate, civilize 1, 2
accumbent, prone 1; prostrate 3; recumbent; supine 1
accumulate°, amass; assemble 2; cluster 3; collect 1; compile; form 11; garner; gather 1; get 50a; heap 3; hoard 2; keep 3; lay[1] 19a; mass 8; mount[2] 6; pile[1] 5, 10; pool 4; salt 8; scrape 6; stack 6; stock 10; store 1; swell 2
accumulation°, collection 1, 2; deposit 4; drift 5; formation 1; group 2; heap 1; hoard 1; mass 1; mountain 2; pack 2; pileup 2; pile[1] 1; profusion; stack 1, 3; store 3; supply 4; tissue
accumulator, collector
accuracy°, precision 1; resolution 5; truth 1
accurate°, authentic; authoritative 2; correct 8; dead 17; deadly 6; definitive 3; exact 1, 2; express 5; factual 2; faithful 2; faultless; graphic 1; meticulous; nice 3; perfect 5; photographic 1; precise 1; proper 2; right 2; square 3; straight 5; sure 3; true 1, 3; trustworthy; truthful; verbatim 1
  accurate, be, nail 5; serve 5
accurately, exactly 1, 2; letter 4; perfectly 3; plumb 4; right 14, 17; straight 14; true 4; truly 2; verbatim 2; well[1] 3
accursed, damnable; diabolic 2; doomed 2; evil 1; lost 7; reprobate 1; wicked 1
accusation°, allegation; charge 6; imputation; rap 6
accuse°, blame 1; charge 11; denounce 1; fault 8; finger 8; impeach 1; incriminate; pin 5; prosecute 1; sue 1
  accuse of being, proclaim 2
  accuse (of *or* **with),** accuse 1, 2; indict
accused, culprit 1; litigant
accustom°, adapt 2; break 18b; condition 7; orient 3; tailor 2
  accustom (oneself) (to), adjust 3
  accustom (to), familiarize
accustomed°, conventional; customary 2; everyday 2; general 2; given 3; habitual 1; ordinary 1; orthodox; regular 1, 7; routine 3; seasoned 1; traditional; usual
  accustomed to, used 3
AC/DC, bisexual 2
ace, dab hand; divine 3; expert 1; fabulous 3; first-rate; gifted; good 2; master 2; proficient; superlative; talented; terrific 2; top 8
acerb, gruff 1; tart[1] 1
acerbate, sour 6
acerb(ic), bitter 1; cutting 2; harsh 3; incisive 2; keen[1] 2; poignant 2; sarcastic; sharp 5; short-tempered; snappish 2; sour 1; tart[1] 1, 2; trenchant; venomous 2; virulent 2
acerbically, sharply 1
acerbity, bitterness 1; gall[1] 1; ill will; sarcasm; virulence 2
acescent, sour 1; tart[1] 1
acetous, tart[1] 1
Achates, friend 1; intimate[1] 5
ache°, anxiety 2; die 5; discomfort 2; distress 1; gripe 3; hurt 2, 6; pain 1; pang 1; smart 7; stab 5; suffer 1; wrench 5; yearn
  ache for, lust 3; pant 2
Acheron, hell 1
achievable, feasible; possible 2; practicable
achieve°, accomplish; bring 7, 12; complete 5; effect 7; effectuate; execute 2; finish 2; fulfill 1; gain 1; hit 6; implement 2; make 14; notch 4; output 3; perfect 8; possess 5; purchase 2; reach 4; realize 1; render 1; work 16
achievement°, accomplishment 1; act 1; consummation 1; deed 2; discharge 12; effort 3; execution 2; exploit 1; feat; fruition; fulfillment; gain 10; output 2; perfection 2; realization 2; stroke 2, 7; success 1; triumph 1; undertaking 2; work 4
achieving, consummation 1
Achilles' heel, weakness 3
aching, anxious 2; pain 1; painful 1; tender[1] 8
  aching (for), desperate 3; starved 2
achromatic, neutral 2
achy, painful 1
acicular, pointed 1
aciculate, prickly 1; thorny 1
aciculiform, pointed 1
acid, caustic 2; cutting 2; devastating 1; embittered; gruff 1; incisive 2, keen[1] 2, poignant 2; pungent 2; sarcastic; scathing; sharp 4, 5; short-tempered; snappish 2; sour 1; tart[1] 2; trenchant; virulent 2; vituperative
  acid head, addict 1
acidic, sarcastic; sour 1; tart[1] 1
acidulated, sour 1; tart[1] 1
acidulent, sour 1
acidulous, sarcastic; sharp 4, 5; short-tempered; sour 1; tart[1] 1; trenchant
aciform, pointed 1

actually° (*cont.*) literally 2; naturally 2; person 2; practice 4; quite 4; really 1, 2, 3; substantially; truly 1, 3; truth 3

actuate, activate; animate 2; drive 1; energize; induce 1; initiate 1; inspire 1; jam 3; motivate; move 6; pep 2; pioneer 2; prod 2; propel

acuity, imagination 2; insight; mentality 1

aculeate, prickly 1

aculeous, pointed 1

acumen, brain 1; capacity 2; depth 3; discrimination 2; insight; judgment 1; mentality 1; profundity 2; tact; wisdom 1

acuminate, pointed 1

acutance, definition 1; resolution 5; vision 1

acute°, agile 2; clear 5, 9; deep 3; desperate 4; excruciating; exquisite 4; extreme 2; farsighted 1; fine[1] 8, 9; grievous 1; heavy 3; high 7; incisive 1; ingenious; intense 1; keen[1] 3, 5, 6; luminous 3; nasty 5; nice 3; nimble 2; penetrating 1; perceptive; piercing 4; poignant 2; pointed 1; profound 3; pungent 3; quick 3; quick-witted; sensitive 3; serious 4; severe 3; sharp 1, 8; shrewd; smart 1; sore 3; tactful; terrible 1; trenchant; violent 3; wise 1

acutely, deeply 2; profoundly; severely 1, 5; sharply 3; very 1

acuteness, brilliance 2; edge 2; gravity 2; ingenuity; insight; penetration 2; sensitivity 3; severity 3; subtlety 1; tact; wisdom 1

ad, advertisement 1; announcement 3; insert 2; pamphlet

adage, byword; epigram 2; maxim; moral 3; motto; phrase 2; proverb; saw

adamant, decided 2; determined 1; dour 2; grim 1; obstinate; perverse 3; recalcitrant; relentless 1; resolute; rigid 2, 4; rocky[1] 2; steely 2; stern 1; stony 2; stubborn; tenacious 1; tough 6; unmoved; unsympathetic; willful 2

adamant, be, persevere

adamant(ine), hard 1; immovable 2; inflexible; rigid 2, 4; rocky[1] 2; steely 2; stern 1; stony 2

Adam's ale, water 1

adapt°, accommodate 1, 2; alter; arrange 3; condition 5, 7; differentiate 2; edit 1; fit[1] 7; gear 5; groom 4; measure 12; modify 1; orient 3; prepare 1; shape 9; square 12; suit 1; tailor 2; tune 3; turn 3

adapt (oneself) (to), adjust 3; conform 1; resign 2

adaptability, elasticity 2; flexibility 2; obedience

adaptable°, elastic 2; flexible 2; obedient; pliable 2; ready-made 2; tractable 2; versatile 1

adaptable (to *or* for), be, lend 3

adaptation°, accommodation 1; alteration; arrangement 4; resignation 2; version 1

adapted, calculated 1; fit[1] 1; proper 1

adapted to, used 3

add°, allow 6; attach 2; enumerate 2; introduce 5; lend 2; mix 3; put 22c

add (on *or* to), add 4; amplify 1; complement 4; compound 3; contribute 2; enhance; enlarge 1; enrich 1; extend 2; heighten 2; maximize 1; suffix 2; supplement 2; tack 7; widen

add (up), calculate; count 1; number 4; reckon 1; run 12; stack 7a; sum 2b; tally 2; total 4

add up to, amount 1a; come 18a; make 8; total 5

added, extra 1; extraneous 1; plus 2; supplementary 1, 2

added attraction, icing 2

added to, plus 1

addendum, addition 3; amendment 2; extension 3; extra 3; insert 2; supplement 1

addict°, demon 2; devotee; enthusiast; fan; fiend 2; freak 4

addicted, given 3

addicted to, fond 3

addiction, abuse 5; habit 2

addictive, hard 11

adding, addition 1

adding up, addition 2

adding machine, calculator

addition°, accessory 1; accident 3; amendment 2; attachment 4; extension 3; extra 3; frill 2; gain 9; increase 4; increment; insert 2; plus 3; reckoning 1; rise 14; shed[1]; suffix 1; supplement 1; tally 3

addition, in, addition 6; besides 1; boot 1; extra 7; further 3; likewise 2; measure 10; moreover; spare 1; still 5

addition of, with the, plus 1

addition to, in, addition 5; besides 2

additional, accessory 3; alternative 1; auxiliary 2; extra 1; extraneous 1; fresh 4; further 1; indirect 2; new 3; nonessential 1; occasional 2; plus 2; spare 1; supplementary 1, 2

additional (to), subsidiary

additionally, addition 6; besides 1; boot 1; extra 7; further 3; measure 10; moreover

additionally to, besides 2

addle, bemuse 1; intoxicate 1; mix 4b; perturb; turn 5

addlebrained, daft 1; mindless 1

addled, groggy; muddy 2; stupid 1

addled, become, spoil 4

addleheaded, stupid 1

addlepated, daft 1; feebleminded; foolish 2; insane 2; mindless 1; senseless 3; stupid 1; thick 6

address°, apply 5; call 11a; dedicate 3; dedication 2; direct 5; greet 2; hail[1] 1; harangue 1; lecture 1; message 2; oration; oratory; salute 1, 3; sermon 2; speak 2, 10, 12c; speech 2; talk 14

address, make *or* deliver *or* give an, lecture 3; talk 6

address oneself to, address 5; set 12a; tackle 3

address to, send 2

addressee, occupant

addresses to, pay (one's), pursue 4

adduce, mention 1; present[2] 7

adept°, able 2; accomplished; capable 1; clever 1; dab hand; experienced 1; expert 2; genius 1; good 12; great 8; handy 3; ingenious; master 2, 4; masterful 1; old 7; perfect 7; polished 1; practiced 1; professional 1; proficient; qualified 1; quick 4; skillful; smart 1; specialist; tactical; talented; wicked 7

adept in, home 5

adeptly, well[1] 6

adeptness, ability 1; capability; ease 2; efficiency 1; finesse 1; ingenuity; knowledge 3; proficiency; prowess 1; skill 1; specialty 1; technique 2

adeptness (with), understanding 3

adequacy, enough 2; fitness 1

adequate°, acceptable 1; ample 4; comfortable 4; competent 1; decent 3; due 3; effectual 1; enough 1; fair[1] 4; fit[1] 2; good 1, 16; OK 3; par 6; passable 1; presentable 1; satisfactory; scratch 5; so-so; sufficient; tolerable 2

adequate, barely, scanty 1

adequate, be, go 21e; serve 2; suffice

adequate for, equal 3

adequate (to), be, measure 15b

adequate to, prove, rise 12a

adequately, enough 3; fairly 1; OK 7; well[1] 1, 14

à deux, face 8; intimate[1] 4; tête-à-tête 2

adhere, attach 5; cement 2; cling 1; stick[1] 5

adhere, cause to, bind 4

adhere to, abide 4; follow 2, 10; keep 6; obey 1; observe 1; pursue 2; stand 5c

adherence, dedication 1; devotion 3; observance 1

adherent, addict 2; admirer 1; devotee; disciple 2; enthusiast; fan; follower 1; friend 4; partisan 1; proponent; sectarian 3; standby 1; supporter 1

adherent (to), observant 2

adherents, party 3; school 2

adhering, observance 1

adhesion, traction

adhesive, cement 1; glue 1; tacky[1]; tenacious 2

adhesiveness, tenacity 2

adieu, farewell 1, 3; goodbye

adieu, bid, exit 3; kiss 3; part 12; see 13

adieu (to), say, leave[1] 1; part 12

adieus *or* adieux, making one's, parting 2

ad infinitum, cease 2

Adios!, farewell 3; goodbye

adipose, fat 2; large 1; oily 1; weighty 1; well-fed

adit, shaft 4

adjacency, presence 1; proximity

adjacent, close 9; flush[2] 1; immediate 2; near 5; nearby 2; present[1] 2; surrounding

adjacent (to), about 9; adjoining; close 20; near 9; neighboring

adjacent (to), be, border 9; join 4

adjectival phrase, phrase 1

adjoin, border 9; join 4; meet[1] 4; touch 3

adjoining°, local 1; near 5; neighboring; surrounding

adjourn, defer[1]; dissolve 3; postpone

adjournment, dismissal 2; dissolution 2; postponement

adjudge, decree 2; determine 1; judge 1; try 1

adjudicate, decide 1; judge 4; officiate; try 4; umpire 2

adjudication, judgment 2

adjudicator, judge 2; peacemaker; umpire 1

adjunct, accessory 1; attachment 4; extension 3; supplement 1; supplementary 1

adjust°, accommodate 1; adapt 2; alter; calibrate; correct 4; differentiate 2; dispose 1; fit[1] 7; fix 3; gear 5; measure 12; modify 1; modulate;

orient 3; overhaul 2; proportion 5; reconcile 2; rectify; reduce 4, 8; regulate 1; repair 1; retouch; scale[3] 4; set 6, 11; settle 3; shape 9; square 12; straighten 1; suit 1; tailor 2; time 23; tune 3
adjust (oneself) (to), resign 2
adjust (to), conform 1
adjustability, elasticity 2; flexibility 2
adjustable, elastic 2; fluid 3; versatile 1
adjustable spanner, wrench 6
adjustableness, flexibility 2
adjusted, calculated 1
adjusting, adaptation 1; adjustment 1
adjustment°, accommodation 1; adaptation 2; alteration; difference 3; overhaul 3; reconciliation 2; regulation 1; repair 2; resignation 2; settlement 3
adjutant, orderly 3
adjuvant (to), subsidiary
ad-lib, extemporaneous; improvise 1; offhand 3, 4; play 12; quip 1; spontaneous 1; unpremeditated
admass, hoi polloi; people 3; public 8
administer°, apply 2; conduct 3; deal 1, 4; deliver 5; direct 1; dispense 2; dose 2; enforce 1; handle 3; head 11; inflict; manage 1; oversee; police 3; portion 4; preside; regulate 2; run 10; supervise; transact
administering, administration 3; direction 1
administer the coup de grâce, finish 4; kill 1
administer things *or* affairs *or* matters, shot 9
administrate, administer 1; preside
administration°, bureaucracy; conduct 2; direction 1; execution 2; executive 2; government 1, 2; leadership; management 1, 2; oversight 2; power 10; regime; reign 1; rein 2; term 3
administration, the, establishment 3
administrative, executive 3; political 1; responsible 3; supervisory
administrator, boss 1; director 1; executive 1; head 2; manager; officer 1; operator 2; politician; superintendent; supervisor
admirable°, beautiful 2; bully 3; desirable 4; divine 3; estimable; excellent; exemplary 2; fine[1] 1, 11; first-rate; glorious 4; good 2; grand 5; laudable; meritorious; noble 5; praiseworthy; prime 2; rare[1] 2; splendid 2; sterling 2; superb
admirably, beautifully 2; perfectly 2; well[1] 2
admiration°, appreciation 3; attachment 3; esteem 3; estimation 2; homage; honor 2; love 1; rave 3; regard 7; respect 1; reverence 1; veneration; wonder 2; worship 2
admire°, adore 1; appreciate 1; esteem 1; follow 9; go 28c; honor 5; idolize; look 12; love 7; respect 6; revere; venerate; worship 1
admired, beloved 1; dear 1; estimable; well-thought-of
admirer°, addict 2; disciple 2; enthusiast; fan; gallant 5; student 2; suitor; supporter 1; sweetheart
admissibility, possibility 1
admissible, acceptable 1; good 1, 3; legal 1; passable 1; permissible; plausible 1; possible 1; presentable 1
admission°, acknowledgment 1; allowance 1; entrance[1] 1; entry 1; initiation 2; profession 2; recognition 2; revelation
admissions, gate 2
admit°, accept 2; acknowledge 1; agree 3; allow 1, 2; clean 8; concede 1; confess; contain 2; face 18a; grant 2; initiate 2; let[1] 5, 7a; permit 1; receive 3; recognize 2; suffer 3; sustain 5; tell[1] 2; tolerate 1
admit of, bear 4
admit defeat, give 13; lose 3
admittance°, admission 1, 3; entrance[1] 1; entry 1; initiation 2; recourse 1
admitted, privileged 2; professed 2; public 4; undisputed
admitting, acknowledgment 1; admission 3
admitting that, though 1
admixture, alloy 1; amalgam; amalgamation
admonish, advise 1; caution 3; correct 2; lambaste 2; lecture 4; notice 7; preach 2; punish 1; rebuke 1; reprimand 2; speak 12a; warn 1
admonishment, caution 1; punishment 1
admonition, advice 1; advisory 2; caution 1; example 3; injunction 2; lesson 3; punishment 1; rebuke 2; reprimand 1; warning 1
admonitory, advisory 1; exemplary 3; ominous 2; reproachful
ado, bother 8; excitement 2; fanfare 2; flap 4; flurry 1; furor 1; fuss 1; hurry 3; racket 1; rumpus; rush 3; song 3a; splash 4; stir 6; tumult; twitter 4
adolescence, childhood; puberty; youth 1
adolescent°, boyish 1; child 2; juvenile 1, 2; minor 3; stripling; teenager; young 1; youth 2
adolescents, youth 3
adopt°, affect[2] 1, 2; assume 1, 2; borrow; embrace 2; favor 6; follow 2; go 29a; resolve 3; strike 12; take 5, 7, 21; wear 2
adoption, passage 9
adorable°, cute 1; darling 4; lovable
adoration, glory 2; love 1; praise 2; reverence 1; veneration; worship 2
adore°, bless 1; delight 2; dote; esteem 1; glorify 2; honor 5; idolize; like[2] 1; love 7; praise 4; revere; venerate; worship 1
adored, beloved 1; darling 3; dear 1; pet[1] 3; precious 2
adoring, fond 1; tender[1] 9
adoringly, fondly
adorn, beautify; become 4; decorate 1; drape 1; dress 2; elaborate 3; embellish 1; enrich 2; grace 6; illuminate 3; illustrate 2; ornament 2; trim 4
adorning, ornamental
adornment, decoration 1; elaboration 1; embellishment 1; flourish 3; ornament 1; trim 5
adornment(s), trappings
adrift, disorientated; end 6; lost 3; sea 4
adroit, adept 1; astute 1; clever 1, 4; cute 2; expert 2; good 12; great 8; handy 3; ingenious; masterful 1; neat 4; nimble 1; perfect 7; quick 4; ready 6; skillful; slick 3; tactful; tactical; talented
adroit in, home 5
adroitly, handily 2
adroitness, art 1, 4; dexterity 1; diplomacy 1; efficiency 1; expertise; facility 1; faculty 1; finesse 1; ingenuity; knack; prowess 1; readiness 2; savoir-faire; skill 1; tact; technique 2; touch 16
adroitness (with), understanding 3
adscititious, accessory 3; indirect 2; nonessential 1; supplementary 2
adulate, adore 3; honor 5; idolize; like[2] 1; love 7; revere; worship 1
adulation, flattery; love 1; praise 2; prostration 2; reverence 1; worship 2
adulatory, flattering 2
adult°, mature 1; racy 2
adulterate°, alloy 2; contaminate; corrupt 4; cut 6; debase 2; degrade 3; dilute 1; doctor 4; foul 12; garble 1; poison 3; pollute 1; vitiate 1; water 7
adulterated, foul 2; impure 2; mixed 1
adulteration, filth 2; impurity 1; pollution
adulterer, libertine 1
adultery, infidelity 2; intrigue 4
adulthood, age 2; majority 2; maturity 1
adulthood, reach *or* attain, grow 6
adumbrate, extinguish 3; obscure 7
advance°, abet 2; appreciation 2; approach 1, 6; benefit 3; better 10; bolster; bring 6, 10b; broach; climb 3; come 1; contribute 2; course 2; cultivate 3; develop 1; development 2; draw 13b; ease 8; elevation 3; encourage 2; evolution; exalt 1; expedite 2; extend 4; facilitate; far 6a; favor 8; foment; forward 1, 6, 8; foster 1; front 10; further 5; gain 3, 7, 9; get 28d, 28e; glorify 1; go 1, 19; growth 3; headway 2; hold 16b; improve 2; improvement 2, 3; increase 2; introduce 2; jump 5, 9; laud; lay[1] 6, 18b; lead 12; leg 4; lend 1; lift 2; loan 1, 2; moot 2; move 1, 7; movement 6; nourish 3; offer 3; oncoming 2; overture; passage 6; place 10; pose 3; posit; precipitate 1; prefer 2; preliminary 1; prepared 1; proceed 1; profit 3; progress 1, 2, 3, 5, 6; progression 1, 2; promote 1, 2, 3; promotion 1, 2; propose 1; push 10, 15; put 5, 17a, 18b, 28; raise 6; recovery 1; revival 3; rise 8; second[1] 9; speed 2; spurt 2; succeed 3; suggest 1; support 1; tender[2] 1; thrive; venture 2; walk 1; way 5; work 20c
advance, in, advance 10; ahead 1; before 2; front 8
advance of, in, before 4
advanced, avant-garde; early 5; far 5a; new 2; precocious; progressive 2; refined 2; sophisticated 2; way-out 2
advanced in years *or* age, elderly 1; old 1
advance(d) position *or* place, lead 11
advance guard, spearhead 2
advancement, advantage 2; development 2; elevation 3; furtherance; growth 3; progress 1, 2, 3; progression 1, 2; promotion 1, 2; recovery 1; revival 3

impressive; moving 1; pathetic 1; prevailing 2
affection°, attachment 3; devotion 3; disorder 3; eye 4; friendship 2; heart 5; illness; inclination 3; liking 1; love 1; passion 3; piety 1; regard 7; thing 4; weakness 4
  affection for, feel, take 38a; warm 8
affectionate°, amiable; brotherly; demonstrative 1; fatherly; fond 1, 3; friendly 2; hearty 1; intimate[1] 1; kind[1]; lovable; maternal; romantic 4; tender[1] 6, 9; warm 3
affectionately, dearly 2; fondly; warmly 1; well[1] 5
affective, emotional 3; pathetic 1
affianced, engaged 1
affiche, promotion 5
affidavit, assertion 1; evidence 2; testimony; warrant 2
affiliate, ally 2; associate 6; attach 2; band[2] 3; branch 2; connect 2; marry 2; syndicate 3; tie 3
affiliate(d), associate 6; fellow 5; related 1; relative 1
  affiliated (to *or* with), akin
  affiliated (with), be, belong 1
affiliation, alliance 1; association 2; fraternity 2; identification 4; liaison 1; link 2; marriage 3; match 3; persuasion 2; relation 1; ring[1] 3; tie 8
affinity°, alliance 1, 2; attachment 3; belonging; fellowship 3; kinship 2; liking 1; love 4; penchant; rapport; relevance; sympathy 2; tendency
  affinity for, feel, take 38a; warm 8
  affinity to *or* toward(s), feeling an, partial 3
affirm, allege; approve 2; attest; certify 1; claim 4; confess; confirm 2; declare 1; establish 3; inspire 2; maintain 3; make 23; plead 3; profess 1; pronounce 2; protest 4; prove 1; ratify; say 1; seal 4; settle 2; stand 5c; state 7; substantiate; support 6; testify; urge 4; verify
affirmation, acknowledgment 1; admission 4; approval; assertion 1, 2; declaration 1; endorsement 1; evidence 1; profession 2; pronouncement 1; sanction 3; seal 2; settlement 5; statement; testimony
affirmative, assertive; emphatic; favorable 2; positive 7
affirmatively, favorably 2
affirmed, given 1; purposeful
affix, anchor 3; apply 1; attach 1, 4; connect 3; fasten 1; fix 1; glue 2; knot 3; pin 3; post[1] 2; secure 8; stick[1] 4, 17b; suffix 1; tack 5
affixing, attachment 2
afflatus, inspiration 1
afflict°, ail 1; anguish 3; beset; blight 3; charge 9; distress 3; hurt 3; inflict; oppress 1; persecute 1; rend 3; scourge 4; seize 3; sicken 2; strike 7, 13; torment 2; trouble 1; visit 2
  afflict with, give 4
afflicted, affected 4, 5; diseased; downtrodden; ill 1; mournful 1; sick 2; smitten 1; unsound 2; woebegone
  afflicted by, -ridden
  afflicted by *or* with, be, catch 5; come 12; get 5
  afflicted (with), stricken
affliction°, agony; ailment; blight 1; bug 2; calamity 2; curse 2; disease 1; disorder 3; distress 1; grief 2; hardship; hell 3; ill 8; illness; infirmity 2; misery 3; ordeal; pain 2; persecution 1; pestilence 2; plague 1; pressure 3; rack 2; scourge 1; smart 8; sorrow 2; suffering; thorn 2; torment 4; trial 4; trouble 5, 8; undoing 2; visitation 2; woe; worry 3
afflictive, painful 2; sorrowful 2; stiff 2
affluence, ease 3; fortune 1; money 2; prosperity; riches; substance 5; wealth 1
affluent, fat 3; favored 2; flush[2] 3; going 1; independent 5; leisured; loaded 4; money 4; opulent 1; prosperous 1; rich 1; strong 11; substantial 4; successful 1; wealthy; well-off
afford°, furnish 1; give 1; impart 1; open 25; provide 2; supply 1
affray, action 6; argument 1; battle 1; conflict 1; disorder 2; disturbance 2; fight 7; fracas 1; fray[1]; riot 1; rumpus; scrap[2] 1; scrimmage; skirmish 1; tumult; uproar
affront, cut 3, 23; dishonor 1, 4; flout; grievance 1; hurt 3; indignity; injure 2; insult 1, 2; offend 1; outrage 2, 4; provoke 3; scandalize; slight 6, 7; slur 1
  affront (at), take, bridle 3
affronted (by), be, bridle 3; mind 15
affronting, foul 6
afghan, spread 14
aficionado, addict 2; admirer 1; devotee; disciple 2; enthusiast; fan; fiend 2; follower 3; freak 4; supporter 1
  aficionado of, be an, follow 9
afield, wide 5
afire, ablaze 1; burning 1; fiery 1, 2; fire 5
  afire, set, kindle
aflame, ablaze 1; burning 1; drunk 2; excited 1; fire 5; incandescent
afloat, buoyant 1
  afloat, set, launch 4
afoot, wind[1] 8
aforementioned, foregoing; preceding; previous 2
aforesaid, foregoing; previous 2
afoul°, foul 18
  afoul of, afoul
afraid°, cowardly; fainthearted 1; fearful 1; insecure 1; panic-stricken; petrified 1; scared; shy 2; tremulous 2; upset 8
  afraid, become, wind[1] 7
  afraid (of), be, dread 1; fear 5
aft, posterior 1; rear[1] 3
after, for 2; last[1] 5; posterior 1, 2; rear[1] 3; subsequent 2; wake[2] 2
  after, be, angle[2]; spoil 5
after all, eventually; however 1; run 48; yet 5
after all is said and done, same 3; ultimately
afterclap, upshot
aftereffect, outcome; repercussion; upshot
aftermath, effect 1; outcome; repercussion; wake[2] 1
  aftermath of, in the, subsequent 2
aftermost, last[1] 1
afters, dessert; sweet 10
afterthought, retrospect
afterward(s), subsequently
Aga, range 4
again, back 15; extra 7; further 3; over 9
again (and again), always 1; frequently 1; often; repeatedly; time 21
against, fight 2; for 4; hostile 1; notwithstanding 2; opposed; resistant 1
  against, be, object 4
agape, goggle-eyed; open 1, 2
age°, date 1; day 2; develop 2; era; generation 2; get 28e; mature 4, 5; mellow 5; page[1] 2; period 2; ripen; season 4; standing 7; time 4; world 4
  age, be one's, mature 4
  age, of, adult 1; mature 1
  age, of advanced, elderly 1
  age, show one's, date 6
  age (group), generation 4
aged°, ancient 3; decrepit 1; doddering; elderly 1; gray 3; high 10; mature 2; mellow 1; moldy; old 1; ripe 1; vintage 2
ageing, elderly 1; timeworn
ageism, intolerance
ageist, intolerant 2
ageless, classic 2; perpetual 1; timeless
agency°, agent 2; bureau 2; cause 2; hand 3; instrument 2; mean[3] 4a; mechanism 3; medium 5; ministry 3; representation 2; tool 2; umbrella 2; vehicle 2; voice 3
agenda, program 1; schedule 1; timetable
  agenda, have on the, have 12b
agent°, ambassador; broker; cause 1, 2; dealer; delegate 1; deputy; envoy; factor 2; go-between; instrument 2; intermediary; liaison 2; merchant 3; minister 2; mouthpiece 2; negotiator; officer 1; operative 4; plant 4; proxy; representative 3, 4; voice 3
agent provocateur, agitator; rabble-rouser; troublemaker
age-old, aged; old 4; time-honored
ages, for, length 4b
ages ago, once 1
agglomerate, stack 6
agglomeration, assortment 2; ensemble 3; group 2; heap 1; hodgepodge; mass 1; medley; pile[1] 1; profusion; stack 3
aggrandizement, appreciation 2; magnification
aggravate°, compound 3; exasperate 2; get 17; inflame 2; irk; magnify 1; tease 1; worsen 1
aggravating, irksome; painful 2
aggravation, annoyance 1; gall[2] 2; plague 2
aggregate, accumulate; alloy 1; amass; amount 1a, 3; cluster 3; combination 3; come 18a; composition 3; ensemble 3; gross 2; heap 3; lump[1] 3; mass 8; muster 2; pile[1] 5; scrape 6; stack 6; store 1; sum 1; total 1; totality; volume 1; whole 4
  aggregate, the, everything
aggregation, accumulation 1, 3; collection 1; combination 3; gathering; group 2; heap 1; knot 2; mass 1; muster 2; pile[1] 1; stack 3
aggression°, assault 1; invasion 2
aggressive°, assertive; belligerent 2; defiant; enterprising; fierce 2; forceful 1; forcible 2;. hostile 3; industrious; make 43; militant 1; offensive 1; officious; pugnacious; pushy; strong 13, 16; strong-arm; warlike
  aggressive, be, liberty 5
aggressively, strong 23

aggressiveness, aggression 1; drive 8; enterprise 2; initiative 2
aggressor°, militant 3
aggrieve, cut 3; hurt 3; sadden; wound 4
aggrieved, hurt 7; sore 5
aggro, aggression 1
aghast, panic-stricken; petrified 2; thunderstruck
aghast at, be *or* **stand,** abhor
agile°, active 3; alert 2; clever 1; dexterous 1; fleet[2]; graceful 1; light[2] 7; lively 1; mobile 4; nifty 2; nimble 1, 2; quick 3; ready 6; sharp 3; sprightly
aging, ancient 3; elderly 1; old 1; timeworn
agitate°, affect[1] 2; disrupt 1; distract 3; disturb 2, 4; excite 1, 2; exercise 3; ferment 2; flurry 2; fluster 1; incite; inflame 1; infuriate; jar[2] 1; jiggle 1; kindle; move 3, 4; perturb; rattle 3; rise 16; ruffle 3; shake 5; shatter 3; stir 1; toss 2, 3; trouble 1; unnerve; upset 1; whip 7a; wind[2] 4d; work 20a
agitate (against), mutiny 2
agitated°, beside 3; disconcerted; distraught; disturbed 1; emotional 4; excited 1; frantic; hectic; impatient 1; jumpy; nervous 1; resentful; restless; rough 2; tumultuous; unsettled 2; upset 8; worried
agitated, become *or* **make less,** calm 5
agitating, incitement 1; shake 9
agitation°, dismay 3; distraction 1; excitement 1; flap 4; flurry 1; fluster 2; frenzy 1; fuss 1; heat 2; hurry 3; incitement 2; motion 1; rampage 1; resentment; ripple 2; sedition; shake 9; storm 3; suspense 2; tempest 2; trouble 7; tumult; twitter 4; unrest; uproar; worry 4
agitator°, rabble-rouser; troublemaker
agitprop, propaganda 1
agley, wrong 7
aglow, ablaze 2; bright 1; fiery 2; glowing 1; luminous 2; lurid 4; radiant 1
agnate, kin 2; kindred 2; related 2
agnomen, nickname
agnostic, faithless 1; godless 2; heathen 1, 2; heretical; infidel; nonbeliever; skeptic; skeptical
agnosticism, skepticism
ago, back 16; past 4
ago, a while, previously
agog°, anxious 2; breathless 3; fanatical; goggle-eyed; impatient 1; keen[1] 1; wild 9
agog (at), be, marvel 1
à-go-go, swinging
agonize, brood 4; fret 1; nail 3; rack 3; sorrow 3; squirm; stew 3; suffer 1; sweat 2; torment 1; worry 1
agonized, worried
agonizing°, excruciating; exquisite 4; hard 5; harrowing; heart-rending; nerve-racking; nightmarish; oppressive 1; painful 1; picnic 3; piercing 4; poignant 1; pungent 3; sore 5; tender[1] 8; traumatic; worried
agonizingly, hard 14; painfully
agony°, anguish 1; distress 1; evil 7; grief 1; hell 2, 3; hurt 6; pain 2; passion 5; picnic 3; rack 2; sorrow 1; suffering; torment 3; unrest; wrench 5
agony aunt *or* **uncle,** scribe 2
agora, square 8
agrarian, pastoral 2; rural 1
agree°, accord 1; check 5; close 4; coincide; commit 4; comply; consent 1; contract 2; correspond 1; cut 9; deign; engage 3; go 36a; match 7; nod 2; play 11; pledge 4; resolve 1; soften 4; stack 7a; string 10a; swear 1; tally 1; term 5; understand 2; undertake 2; weaken 3; yield 3
agree (on *or* **upon),** fix 15; settle 2; strike 8
agree (to), abide 4; accept 2; agree 2; allow 2; approve 1; consent 2; defer[2]; enable 3; fix 2; go 20b; grant 2; hear 3; hold 23; keep 6; make 16; obey 1; OK 5; pass 7; permit 1; stipulate; submit 1; subscribe 1; take 39a; underwrite 2
agree to bear, accept 3
agree (with), agree 4; conform 2; get 28a; hold 23; parallel 4; square 11; subscribe 1; sympathize 2; go 39a; sit 10
agree (with each other), go 6
agreeability, obedience
agreeable°, acceptable 2; amiable; comfortable 3; delicious 2; delightful 1; desirable 2; easy 6; engaging; fair[1] 8; friendly 2; genial; good 1; good-natured; gracious; harmonious; hospitable 1; indulgent; likable; lovely 2; mellow 4; neighborly; nice 1; obedient; obliging; peaceable 2; pleasant 1, 2; ready 2; smooth 6; suave; submissive 1; sweet 4; sympathetic 2; welcome 2; well[1] 17; willing; yielding 2
agreeable, be, get 28a
agreeable, find, like[2] 1
agreeable, make more, sweeten 2
agreeable to, be, sit 10
agreeable (with), be, step 7
agreeable (with), not, step 8
agreeableness, flexibility 2; obedience; readiness 1
agreeably, favorably 1; happily 3; kindly 2; nice 6; readily 1; well[1] 1; willingly
agreed, conventional; fixed 3; OK 1; set 29; straight 10; understood; united 3
agreed(-upon), given 1
agreeing, consistent 1
agreement°, accord 2, 3, 4; approval; arrangement 3; bargain 1; bond 2; charter 1; coincidence 2; consent 4; contract 1; deal 5; deed 3; engagement 3; harmony 1; identity 1; instrument 3; kinship 2; likeness 1; negotiation 2; nod 6; obedience; obligation 2; OK 6; pact; promise 1; proportion 2; protocol 2; reconciliation 2; resemblance; sanction 3; settlement 3, 5; solidarity; stipulation; symmetry; sympathy 2; transaction 1; treaty; truce 2; understanding 1; undertaking 3; uniformity 1; union 4; unity 1; vow 2; warranty
agreement, arrive at *or* **reach** *or* **come to an,** get 50c; term 5
agreement, in, agreeable 2; consistent 1; line[1] 19b; piece 8; united 3
agreement (with), in, according to 1; step 7
agreement (with), out of *or* **not in,** step 8
agribusiness, farming
agricultural, pastoral 2; rural 1
agriculture, farming
agriculturist, farmer
agronomist, farmer
agronomy, farming
ague, chill 2; cold 10
ahead°, before 2, 3; first 4; forward 4; front 8; onward 2; straight 13
ahead (of), advance 10b; before 4
ahead (of), be, lead 4
ahead (of time), advance 10a; early 1; time 10
aid°, abet 1; accommodation 5; advance 2; assist 1; assistance; assistant 1; auxiliary 3; back 2a; backing 1; benefit 1, 3; bolster; boost 2, 5; contribute 2; cooperation 2; ease 8; enable 2; encourage 2; encouragement 2; endowment 1; facilitate; far 6b; favor 8; forward 6; foster 1; further 5; furtherance; hand 2; help 1, 6; kindness 2; largess; militate 2; minister 3; nourish 3; office 5; party 4; patronage 1; patronize 3; philanthropy 2; present[2] 3; profit 3; promote 1; relief 2; relieve 3; sanction 2; second[1] 9; see 11, 14c; service 1; speed 2; start 12; subsidize; subsidy; support 1, 7; supporter 1; tool 1
aid, provide with, see 14c
aid of, be in, boot 4
aid of, with the, mean[3] 3; through 1
aide°, accessory 2; assistant 1; attendant 2; auxiliary 4; help 8; satellite 2; subordinate 2
aide-de-camp, aide; assistant 1; satellite 2; second[1] 8
aiding, auxiliary 1; promotion 1
ailing, diseased; feeble 1; frail 2; ghastly 3; ill 1; indisposed 1; infirm 1; invalid[1] 1; peaked; poorly 2; seedy 2; sick 2; sickly 2; sort 6; tender[1] 1; unhealthy 1; unsound 2; unwholesome 3; weather 2
ailment°, bug 2; disease 1; disorder 3; illness; infirmity 2; trouble 8
aim°, ambition 3; aspiration; aspire; design 7; direct 4; drift 4; end 3; endeavor 1; eye 6; fasten 2; function 1; goal; head 10; idea 4; intend; intent 1; intention; lay[1] 7; level 10; mark 6; mean[1] 1; mission 1; motive 2; object 2; objective 2; plan 4; point 6, 20; prize[1] 3; propose 2; purpose 1, 5; reason 3; seek 2; sight 7; spirit 6; strike 2; target; turn 9; view 5
aim at, aim 2
aim for, go 28e; make 30a; pursue 3
aim that, with the, order 11
aiming, direction 1
aimless°, disjointed 2; end 6; erratic 3; pointless; purposeless; rambling 1; shiftless
aimlessly, helter-skelter 2
aimlessness, emptiness 2
air°, appearance 2; atmosphere 1, 2; aura; bearing 1; broadcast 1; carriage 2; circulate 2; climate 2; display 1; exhalation 2; expose 1; expression 3; feel 11; feeling 6; flavor 2; front 4; guise 1; lay[3]; look 14; manner 2; melody 1; odor 2; open 23; parade 5; piece 4; presence 3; put 18b; radio 2; reveal; semblance 1, 2; song 1; spread 2; strain[1] 8; tone 3; tune 1; vent 3; voice 4
air, be up in the, fire 4

antianxiety drug, tranquilizer
antic, frolic 2; lark 1; prank; trick 2
anticipate°, bargain 4; dread 1; envisage 2; envision; expect 1; fear 7; forecast 1; forestall; hope 3; look 6c, 7a; prevent; provide 5; relish 2; watch 4
anticipated, due 4; eventual 2; forthcoming 2; predictable; prospective
anticipated, as, naturally 1
anticipating, expectant; hopeful 1
anticipation°, dread 2; expectation 1; forecast 2; hope 2; precaution 2; prevention; prospect 2; providence 1; relish 1; suspense 2; thought 3
anticipative, preventive 2
anticipatory, preconceived; preventive 2; sanguine
anticlimax, nonevent
anticlockwise, backward 6
antics, hanky-panky; nonsense 2
anticyclone, high 17; whirlwind 1
antidepressant, drug 2; stimulant 2
antidote°, preventive 4; remedy 1, 2
antihero, protagonist 1
antiheroine, protagonist 1
anti-intellectual, philistine 2
antipasto, hors d'oeuvre
antipathetic, averse; contrary 2; incompatible; negative 1; opposed; opposing
antipathy, animosity; antagonism 1; aversion 1; disgust 3; dislike 2; distaste 1; grudge 1; hate 3; horror 1; hostility 1; ill will; loathing; objection; opposition 1; phobia; rancor; resentment; revulsion
antipodal, polar 2
antipsychotic, tranquilizer
antiquated°, ancient 3; antique 1; date 4; decrepit 2; early 6; extinct 2; musty 2; obsolete; old 2; old-fashioned; out 9; passé; prehistoric 2; quaint 2; stale 2; time 15; timeworn; vintage 3
antique°, ancient 2; antiquated, early 6, fogy, old 2; old-fashioned; passé; quaint 2; ruin 5; timeworn; vintage 3
antiquity, history 6
antireligious, ungodly 1
antiseptic, clean 1; disinfectant; pure 2; sanitary; sterile 2
antiserum, antidote
antisocial, aloof 3; inhospitable 1; misanthropic; private 4; reserved; sullen; unsocial
antithesis, opposite 3; reverse 6
antithetic(al), destructive 2; incompatible; opposing; opposite 2; opposition 3; polar 2
antithetical (to), opposed
antitoxin, antidote
antivenin, antidote
ants in one's pants, fidget 2
antsy, impatient 1; restless; tense 2
A number 1, excellent
anxiety°, alarm 2; anguish 2; care 1; concern 6; desperation 2; dismay 3; distress 1; doubt 3; dread 2; fear 4; foreboding 1; horror 2; load 1; misery 1; misgiving; nerve 4; panic 1; remorse 1; resentment; solicitude; strain[1] 7; stress 3; suspense 2; tension 2; terror 1; trouble 5; unrest; worry 4
anxious°, afraid 1; concerned 2; disturbed 1; eager; edge 5; expectant; explosive 1; fearful 2; fraught 2; hot 4; ill 6; insecure 1; jealous 2; jumpy; keen[1] 1; nervous 1; neurotic; remorseful; shy 2; sinking 1; solicitous 1, 2; tense 2, 3; tremulous 2; worried
anxious, be, fret 1; sweat 2; worry 1
anxious (for), desperate 3
anxiously, fearfully 1; jealously
anxiousness, desperation 2; suspense 2; tension 2; unrest
any, not, little 8; none
anybody, one 4
anyhow, case[1] 5; event 3; however 1, 2; rate[1] 5; regardless 2; same 3; yet 5
any longer, now 2
any minute (now), soon 1
any more, now 2
anyone, one 4
anything, do, finger 6
anything, not, nothing 1
any time, ever 1; sometime 2
anyway, case[1] 5; event 3; however 1; nevertheless; notwithstanding 1; rate[1] 5; regardless 2; same 3; yet 5
any way you look at it, effect 5
A-OK, good 2; neat 5
A-one, good 2; high-class 1; neat 5, splendid 3, top 8
apace, fast[1] 6; flat 17a; quickly 2; readily 3
apache, hoodlum; thug
apart°, aloof 1, disconnected 1; individually; isolated 2; separate 7
apart, driven, estranged
apart from, apart 4; beside 2; besides 2; except 1; exclusive 4; independent 8; irrespective of; short 10
apartheid, prejudice 2; segregation
apartment, cell; chamber 4; flat 15; lodging; pad 3; place 6; room 2, 3
apathetic, aloof 3; blasé 2; callous; casual 3; cold 3; coldblooded 2; coldhearted; cool 4; dead 4; impassive; inattentive; indifferent 1; insensible 2; lackadaisical 2; lethargic 1; listless; lukewarm 2; nonchalant; passive 1; perfunctory 1; phlegmatic 1; rocky[1] 3; slothful; stoical; stolid; supine 2; tepid 2; torpid; unenthusiastic; unmoved; unsympathetic
apathy, indifference 1; inertia; lethargy 1; sloth; torpor
ape, copy 5; crazy 5; echo 4; follow 2; imitate 1; mimic 1; mock 2; monkey 1, 4; parody 3; parrot 2
aperçu, précis; synopsis
aperient, purgative 1, 2
apéritif, hors d'oeuvre
aperture°, chink; gap 1; hole 2; leak 2; mouth 2; opening 1; passage 11; pore[2]; slit 2; vent 1
apex, acme; angle[1] 1; climax 1; extreme 8; head 5; height 2; high 15; maximum 2; peak 2; pinnacle; point 2; spire 2; summit; tip[1] 1; top 1; vertex; zenith
aphonic, silent 5
aphorism, byword; epigram 2; maxim; moral 3; motto; precept 2; proverb; quip 1; saw
aphoristic, compact 3; epigrammatic; proverbial 1; terse 1
aphrodisiac, erotic 3
apiculate, pointed 1
aping, imitation 1
aplenty, galore
aplomb, assurance 4; confidence 2; cool 8; ease 4; philosophy 3; poise 2; presence 5; reserve 4; self-confidence; self-control 2; serenity 2
apocryphal, fictitious 1
apodictic, categorical
apogee, climax 1; extreme 8; height 2; maximum 2; peak 2; top 1; vertex; zenith
Apollyon, devil 1
apologetic°, afraid 2; bad 8; guilty 2; penitent; regretful; remorseful; repentant; sorry 1
apologia, defense 3; reason 1
apologist, advocate 2; proponent
apologize° for, excuse 3
apology, defense 3; excuse 4; reason 1
apophthegm, byword; epigram 2; maxim; moral 3; motto; precept 2; proverb; quip 1; saw
apophthegmatic, epigrammatic; proverbial 1
apoplectic fit, stroke 5
apoplexy, stroke 5
apostasize, recant; secede
apostasy, infidelity 1; relapse 3; secession
apostate, defector; disloyal; dissident 1, 2; heretical; rebel 4; renegade 1, 2; turncoat
apostatical, heretical
apostolic, clerical 1
apostrophe, digression 1
apostrophize, call 11a, 11b; speak 12c
apothecary, druggist; pharmacist; pharmacy 1
apothegm. See **apophthegm**
apothegmatic. See **apophthegmatic**
apotheosize, glorify 2; idealize; idolize; immortalize; pedestal 2
appall°, daunt; disgust 1; dismay 1; frighten; horrify 2; intimidate; petrify 2; scandalize; scare 1; shock 1; sicken 2
appalled, panic-stricken; petrified 2; scared; sick 6
appalling, abysmal 1; atrocious 1, 2; deplorable 2; diabolic 2; disastrous 1; dreadful 2; fearful 3; fearsome; formidable 1; frightening; ghastly 1; grievous 2; grim 3; grisly; hideous 2; horrible 1; lurid 2; monstrous 2; outrageous 2, 3; revolting; shocking 2; terrible 4; tragic; ungodly 2; unmentionable 2; wicked 2
apparatchik, officer 1
apparatus°, contraption; device 1, 2; engine; equipment; fixture 2; furniture 2; gadget; gear 2; implement 1; instrument 1; kit; machine 1; material 4; mechanism 1; outfit 1; paraphernalia; plant 3; regalia; rig 3; set 27; tackle 1; thing 5, 8c; tool 1; trappings; turnout 3
apparel°, clothe 1; clothes; costume; dress 1; ensemble 1; garments; gear 3; get 51e; habit 3; robe 2; turnout 3; wardrobe 1; wear 8
apparent°, clear 8; conspicuous 1; demonstrable; discernible 1; distinct 1; evident; external 3; illusory; manifest 1; marked; observable; obvious; open 13; outward 1; overt; patent 2; perceptible; plain 2; probable; professed 1; prominent 1; public 4; seeming; self-evident; specious; superficial 2; transparent 2; visible 2

arrangement° (*cont.*)
  arrangement, come to *or* **reach an,** term 5
arrangement(s), arrangement 5; passage 7; preparation 1b
arrant, barefaced 2; blatant 1; flagrant; great 11; outright 2; sheer 2; stark 4; thorough 1; unconscionable 1; unmitigated
arras, drapery
array, arrange 1; arrangement 1; assortment 2; combination 1; display 4, 5; dispose 1; drape 1; dress 1, 2; equip; formation 3; line[1] 23d; order 1; parade 2; placement 1; posture 2; range 5; rank[1] 6; review 8; space 5; stack 3; stock 1; string 9
arraying, placement 1
arrear(s), liability 2
  arrears, be in, fall 10
  arrears, in, back 13, 17; debt 2; delinquent 3; due 1
arrest°, capture 1, 2; catch 1; check 1; delay 2; get 8; grab 3; hinder 2; intercept; nab; nip[1] 2; obstruct 1; pick 8h; pinch 4; prevent; prevention; pull 11b; restrain 3; run 27; seize 2; stanch; stay[1] 4, 6; stem[2] 1; stop 2; stunt[2]; suppress 1
  arrest, place *or* **put under,** nab
  arrest, under, arrest 6
arrested, arrest 6
arresting°, magnetic; piercing 2; prevention
arrhythmic, spasmodic 2
arrière pensée, motive 2
arrival°, appearance 1; dawn 2; entrance[1] 3; entry 3; immigrant; landing 2; oncoming 2; receipt 2; return 10
arrive°, appear 1, 3; check 6; come 2, 13c, 17b; dawn 4; draw 11; face 12; get 39b, 39d; land 5; make 32b; place 10; pull 11a; report 5; rest[1] 5; roll 10a; set 19; show 7; succeed 3; turn 20e, 25a
  arrive (at), arrive 3; fix 2; gain 6; get 10, 30c, 49a; hit 6, 9a; make 14; reach 2; strike 16
Arrivederci!, goodbye
arriving, incoming 1; oncoming 1
arrivisme, ambition 1
arriviste, parvenu 1; upstart
arrogance°, air 5; bravado; conceit 1; effrontery; freedom 7; presumption 1; pride 2; side 5; swagger 3; vanity 1
arrogant°, cocky; conceited; disdainful; dogmatic; domineering; haughty; hoity-toity; immodest 2; impudent; lofty 4; macho; majestic 2; masterful 2; overbearing; overconfident 1; pompous 1; positive 5; presumptuous; pretentious 1; proud 2; pushy; self-important; snobbish; supercilious; vain 1
arrogantly, big 9
arrogate, adopt 2; appropriate 2; assume 1; grab 2; help 5; preempt; secure 5; take 3, 7
arrondissement, area 3
arrow, bolt 1; pointer 1
arroyo, canyon; gully; ravine
arse. See **ass**
arsenal, magazine 2; storehouse; thesaurus 1
arsis, rhythm
arsonist, terrorist
arsy-varsy, topsy-turvy 2
art°, craft 1; execution 4; ingenuity; science 2; skill 1; specialty 1; technique 2; trick 3; workmanship
artery, channel 4
artful°, arch 2; astute 1; clever 2; crafty; deceitful; deep 5; designing; dexterous 2; disingenuous; foxy 1; fraudulent 2; insincere; knowing 1; mealy-mouthed; politic 1; scheming; serpentine 1; sharp 6; shifty; shrewd; sly 1; subtle 3; tactical; tricky 1; unscrupulous; wily
artfulness, art 5; artifice 1; craft 2; dexterity 2; finesse 1; subtlety 2; trickery
arthritic, gnarled
article, composition 1; editorial; entity 1; essay 1; exposition 3; feature 2; item 1; paper 4; piece 4; report 1; story[1] 1, 4; theme 2; thing 1; tract[2]; writing 2
article of faith, doctrine; tenet
article of vertu, antique 2
article(s), term 4
articulate, coherent 2; eloquent 1; enunciate 1; express 1; fluent; formulate 1; go 14; mouth 7; phrase 4; proclaim 1; pronounce 1; say 3; set 18c; sound[1] 7; speak 4; state 7; vent 3; voice 4
  articulate in, speak 3
articulated, jointed, oral, verbal 1
articulately, well[1] 3
articulateness, fluency
articulation, accent 3; delivery 4; diction 2; intonation; oratory; pronunciation; speech 1; tongue 2; voice 1
artifact, product 2; relic 1
artifice°, chicanery; deceit 2; deception 2; delusion 1; device 2; evasion 2; feint; finesse 2; fraud 1; game 3; gimmick 1; hocus-pocus 1; intrigue 3; machination; maneuver 1; move 9; pretense 1, 2; racket 2; ruse; stall[2] 2; stratagem; subterfuge; trap 2; trick 1; trickery; wile
artificer, designer 1
artifices, art 6
artificial°, affected 1; camp[2] 1; counterfeit 2; factitious; false 3; fictitious 2; flowery; forced; glossy 2; hollow 4; imitation 5; insincere; labored 2; mannered; mock 3; phony 1; plastic 3; precious 3; quasi 2; rhetorical 2; sham 2; shoddy; spurious; stiff 8; stilted; strained; synthetic; theatrical 2; toy 6; unnatural 2; unreal 3; wooden 2
artificiality, affectation 1
artisan, operative 2; tradesman 2; worker
artisanship, craft 1; workmanship
artist, designer 1; natural 11; specialist
artiste, performer; specialist
artistic, aesthetic 1, 2; creative; elegant 1, 2; great 6; poetic 2
artistically, well[1] 3
artistry, art 1, 2; production 3; skill 1; technique 2; workmanship
artless°, aboveboard 2; camp[2] 1; candid 1; careless 4; childlike; frank 2; honorable 3; ingenuous 1; innocent 4; naive; natural 4; open 15; plain 3; rude 4; rustic 2; simple 3; sincere; transparent 3; unaffected[1]; unsophisticated 1
artlessly, simply 3
artlessness, ease 4; naiveté; simplicity 3
art lover, aesthete; collector
arts°, art 6
artwork, cut 25; picture 1
as, for 13; way 10b
  as for, concerning; for 12
  as for oneself, personally 2
  as if, like[1] 3; quasi 1
  as if, make, make 25, 37e
  as one, shoulder 5
  as soon as, directly 5; immediately 3; once 5
  as though, like[1] 3
  as though, make, make 25, 37e
  as to, concerning
  as well, addition 6; besides 1; boot 1; likewise 2; moreover
  as well as, addition 5; besides 2
  as yet, yet 1
ascend, arise 2; climb 1, 3; get 51c; incline 1; mount[2] 1; rise 3, 6, 10; scale[3] 3; slope 1; top 6; tower 3; work 20c
ascendancy, domination 1; dominion 1; hold 26; influence 1; jurisdiction; might 2; power 2; predominance; preponderance 2; prestige; prevalence 2; rank[1] 2; reign 1; rule 2; sovereignty; success 1; superiority 1; supremacy 1; triumph 1; upper 6
ascendant, dominant 1; mighty 1; predominant; prevalent 1
ascension, progression 1; rise 15
ascent, climb 6; grade 4; incline 4; ramp; rise 13, 15; slope 2
ascertain, calculate; compute; detect 1; determine 2; discover 1; evaluate 2; fathom; find 2; get 31e; hear 2; learn 1; measure 11; perceive 2; receive 5; see 4; tell[1] 8, 9; trace 5
ascertainable, sensible 2
ascertained, determined 2
ascertaining, determination 3; discovery 1
ascertainment, determination 3; measurement 1
ascetic, celibate 2; puritan 2; severe 6; Spartan; temperate 2
ascetically, severely 6; simply 4
asceticism, self-denial 2; severity 6; simplicity 2
ascribe, assign 4; attach 4; attribute 2; credit 6; fix 9; impute; lay[1] 7; put 16c; set 17b
ascription, assignment 4; attribution; credit 2; imputation
aseptic, hygienic; sanitary; sterile 2
asexual, neuter 1; platonic
ashamed°, embarrassed 1; guilty 2; remorseful; repentant; shamefaced 2; terrible 2
  ashamed, be or act, blush; hang 8
  ashamed, feel, small 6
ashen, colorless 1; deadly 4; ghastly 2; gray 1; lurid 3; pale[1] 1; peaked; wan 1; white 2
ashes, embers
ashram, monastery
ashy, pale[1] 1
aside, apart 1; by 8; digression 1
  aside from, apart 4; bar 10; barring; besides 2; independent 8
asinine, absurd 2; crazy 2; daft 1; foolish 2; half-witted; harebrained 2; inane; insane 2; ludicrous; meaningless 1; mindless 1; nonsensical; pointless; preposterous; puerile; senseless 3; silly 1; stupid 2
asininity, folly 1; stupidity 2
as it were, quasi 1; speak 6
ask°, bid 2; call 11a; charge 13; enquire 1; like[2] 2; petition 2; plead 2; poll 3; pose 3; pray 1; press 6; push 4;

question 9; quiz 2; refer 3; urge 2
- **ask (about** *or* **after),** ask 5; query 3
- **ask (for),** ask 6; beg 1; call 8a; claim 3; demand 2, 4; desire 2; hit 8; order 15; plead 1; put 20; request 1; require 1; seek 3; send 6; solicit 1; speak 7c
- **ask for (someone's) cards,** fire 11
- **ask (of),** consult 1
- **ask oneself,** wonder 3

asked, without being, voluntarily
askew, crooked 2; lopsided 1; nonsensical; oblique 1; wrong 2; wry 1
aslant, oblique 1; wry 1
asleep, dormant 1; numb 1
- **asleep, fall,** flake 3a

asleep at the switch *or* **the wheel,** slack 1; unprepared 1
asleep on the job, slack 1
Asmodeus, devil 1
asomatous, unearthly 1
aspect°, angle[1] 2; appearance 2; attitude 1; bearing 1, 3; detail 1; exposure 4; expression 3; face 2; factor 1; feature 1; form 2; format 1; front 4; guise 1; image 6; landscape; look 14; manner 2; outside 2; phase 3; point 10, 14; presence 3; prospect 1; regard 8; respect 4; semblance 1; shape 2; side 3; slant 1; sound[1] 2; thing 2; tone 3; veneer; view 1, 3; way 6
asperity, gall[1] 1; rigor 2; sarcasm
asperse, blacken 2; discredit 1; disgrace 4; foul 14; impeach 2; insult 1; libel 4; tarnish; vilify; wrong 9
aspersion°, discredit 5; disgrace 2; dishonor 5; flak; imputation; indignity; insult 2; jeer 2; libel 2; recrimination; reflection 3; scandal 3; slur 1; smear 4
aspersions on, cast, blacken 2; impeach 2
aspersive, foul 6; sarcastic; scandalous 2; scurrilous; vituperative
asphalt, pave 1; pitch[2]
asphyxiate, choke 1; smother 1; stifle 1
asphyxiated, be, smother 2
aspirant, candidate; pretender; supplicant 2
aspiration°, aim 5; ambition 3; end 3; goal; objective 2
aspiration(s), pretension 1
aspire°, endeavor 1; hope 3; purpose 5; seek 2
- **aspire to,** aim 2; aspire; pursue 3

aspiring, ambitious 1
ass, bottom 1; bum 1; buttocks; clod 2; dolt; dummy 3; fool 1; gawk 1; half-wit; monkey 2; posterior 3; rear[1] 2; sap[1] 2; seat 4; tail 2; twit[2]
- **ass (in), up to the,** replete 1
- **ass (out) of, make an,** laugh 2a

assail, abuse 3; assault 3; attack 1; beset; besiege 3; bombard 2; charge 14; denounce 3; encounter 3; fall 17; go 22, 28d; lace 5a; lay[1] 13; light[2] 14; make 30b; pelt[1] 1; pitch[1] 6a; set 12b; storm 6; turn 19b; visit 2; vituperate
assailant°, aggressor
assassin, cutthroat 1; killer 1; murderer; thug
assassinate, bump 5; cut 13b; dispatch 3; eliminate 4; execute 3; kill 1; murder 3; remove 4; shoot 3; waste 4; zap
assassination, destruction 2; dispatch 7; execution 3; murder 1; removal 2
assault°, aggression 2; attack 1, 6; attempt 3; batter 2; bombard 2; charge 7, 14; come 8; descend 4; fall 17; go 22, 28d; invasion 2; lace 5a; lay[1] 13; light[2] 14; make 30b; mistreatment; mug 5; offensive 5; onset 1; pelt[1] 1; pile[1] 8b; pitch[1] 6b; push 15; raid 2; rough 16; rush 2; set 12b, 21, 24; storm 6; strike 19; thrashing 1; violate 3; violation 3; waylay 2
- **assault sexually,** rape 4

assaulter, assailant
assay, analysis 1; analyze 2; price 5; prove 2; size 3; test 1, 2; try 2
- **assay mark,** hallmark 1

assegai, lance 1
assemblage, accumulation 3; assembly 1; attendance 2; body 6; cluster 2; collection 2; company 2; ensemble 3; fabrication 1; gate 2; gathering; group 1, 2; herd 1; host[2]; knot 2; mass 1; mob 1; muster 2; pack 3; pile[1] 1; rally 1; throng 1; turnout 1
assemble°, accumulate; amass; build 1; call 4; cluster 3; collect 1; compile; construct 1; crowd 3; edit 4; fabricate 1; flock 2; form 7; frame 6; garner; gather 1, 2; get 50, 50a; group 4; herd 3; hoard 2; hold 10; integrate; line[1] 23a; make 1; manufacture 1; mass 8; meet[1] 2; mobilize; muster 1; piece 13; pile[1] 5; prepare 5; produce 1; raise 2, 5, 12; rally 3; round 19; scrape 6; set 23a; sit 2; store 1; summon 1; turn 20a
- **assemble (in** *or* **at),** throng 2
- **assemble (quickly** *or* **hurriedly),** whip 7b

assembled, serried
assembling, organization 1
assembly°, attendance 2; body 6; chamber 1; cluster 2; company 2; composition 2; convention 1; council 1; diet[2]; fabrication 1; flock 1; gathering; group 1; host[2]; house 3; manufacture 3; meeting 2; muster 2; organization 1; pack 3; parliament 2; production 1; rally 1; roundup 1; session 1; throng 1
- **assembly hall** *or* **room,** chamber 2; hall 2

assemblyman, politician; representative 3
assemblywoman, politician; representative 3
assent, belief 2; consent 3; nod 2, 6; permission; sanction 3; soften 4; weaken 3; yield 3
- **assent (to),** accept 2; agree 2, 3; approve 1; enable 3; go 20b; subscribe 1; hear 3

assenting, permissive; willing
assert, allege; announce 3; argue 5; assure 4; attest; charge 11; claim 4; declare 1; insist 2; maintain 3; plead 3; premise 2; profess 1; pronounce 2; protest 4; remark 2; represent 3; say 1, 2; state 7; swear 1; tell[1] 7; testify; urge 4; voice 4
asserted, pretended
assertion°, allegation; argument 2; claim 1; declaration 1; evidence 2; expression 1; position 7; premise 1; profession 2; pronouncement 1; representation 3; statement; story[1] 2; testimony; thesis 1
assertive°, aggressive 2; decided 2; dogmatic; dominant 1; emphatic; insistent; make 43; overbearing; positive 5; self-confident; strong 16, 20
assess, analyze 2; balance 1; calculate; charge 12; criticize 1; estimate 1; evaluate 1; examine 1; gauge 2; judge 5; lay[1] 17b; mark 14; measure 11; orient 3; price 5; rate[1] 6; review 3; size 3; sum 2c; survey 1; tax 3; test 2; think 5a; try 2; view 9; weigh 3
- **assess (as),** take 8
- **assess too highly,** overrate

assessment, charge 3; criticism 1; estimate 3; evaluation 1; examination 1; interview 2; judgment 5; measure 4; measurement 1; review 4; tariff 1; tax 1; thinking 2; toll[2] 1; tribute 2
asset°, advantage 2; beauty 3; blessing 2; plus 3; strength 3; virtue 3
assets°, asset 1; capital 3; estate 2; finance 2; fortune 1; fund 2; good 21b; merit 2; money 2; possession 3; principal 5; property 2; resource 2; stock 5; substance 5; wealth 1
asseverate, allege; announce 3; assure 4; attest; certify 1; declare 1; insist 2; profess 1; pronounce 2; protest 4; say 1; state 7; swear 1; tell[1] 7
asseveration, allegation; assertion 1; declaration 1; expression 1; profession 2; pronouncement 1; statement; testimony
asseverative, assertive; decided 2
asshole, stinker
assiduity, application 3; exertion; industry 2; patience 2; perseverance; persistence; tenacity 1
assiduous, attentive 1; close 14; diligent; earnest 1, 2; enterprising; hard 7; industrious; laborious 2; painful 3; patient 2; strong 5; studious 1; thorough 2
- **assiduous, be,** persevere

assiduously, hard 13; intently; thoroughly 2
assiduousness, exertion
assign°, allot; allow 5; apply 3; appoint 1, 2; appropriate 3; attach 2, 4; attribute 2; award 1; cast 8; charge 9; commit 1; credit 6; dedicate 1, 3; delegate 2, 3; designate 2; destine 2; detail 6; devise 2; devote 1; dispense 1; distribute 1; enable 1; entrust; find 8; fix 9; give 15b, 16; grant 1; impute; job 6; leave[1] 7; make 23, 38b; measure 14; mete; place 18; plant 7; portion 4; post[2] 2; put 2, 8, 16c; refer 2; relegate 3; schedule 2; second[1] 10; send 1; set 17b; sign 11; slot 4; spend 3; station 4; trust 6
- **assign to,** read 4

assignation, appointment 1; date 2; meeting 1
assigned, routine 3; vicarious
assignee, nominee
assignment°, appointment 2, 3; attribution; dedication 1; disposition 3; distribution 1; duty 1; errand 2; function 1; job 2; lesson 1; lot 4; mission 1; office 4; paper 4; portion 2; post[2] 1; project 2; service 6; slot 2; stint 1; task 1; tour 3; work 3

attractive° *(cont.)* endearing; engaging; exquisite 3; fair[1] 7; fetching; fine[1] 10; foxy 2; glamorous 1; good 10; handsome 1; interesting; inviting; likable; lovable; lovely 1; magnetic; nice 1; picturesque 1; pleasant 1, 2; prepossessing; pretty 1; ravishing; seductive; sweet 4; taking; tempting 1; voluptuous 2; winning 1
attractive, make, recommend 3
attractive, make more, sweeten 2
attractive to, be, appeal 2
attractively, beautifully 1
attractiveness, affinity 2; beauty 1; charm 2, 3; fascination; glamour; magnetism
attribute°, accuse 2; aspect 4; assign 4; attach 4; character 2; characteristic 2; credit 6; feature 1; fix 9; impute; lay[1] 7; mark 2; peculiarity 2; point 14; property 4; put 8, 16c; quality 1; respect 4; set 17b; trait
attributes, endowment 3; nature 1; stuff 3
attribution°, credit 2; imputation
attrition, erosion; friction 1; wear 9
attune, tune 3
attuned, sensitive 2
attuned (to), step 7
attuned (to), not, step 8
atwitter, all, overwrought 1
atypical, character 9; eccentric 1; freak 5; odd 1; peculiar 1; queer 1; rare[1] 1; singular 1; unlike 2; unusual; unwonted
au courant **(with),** aware 1; current 4; familiar 4; knowledgeable 1; level 5
audacious°, adventurous; barefaced 2; bold 1, 2; brash 2; brazen; cheeky; cool 6; daredevil 2; daring 2; dauntless; defiant; enterprising; fearless; flagrant; foolhardy; forward 2; heroic 1; impertinent; impudent; indiscreet; intrepid; overconfident 1; pert; presumptuous; rash[1]; shameless; stalwart 3; venturesome 1
audacious, be, liberty 5
audaciously, openly 1
audaciousness, freedom 7
audacity, assurance 3; effrontery; enterprise 2; face 5; freedom 7; gall[1] 2; gumption 2; gut 3a; impertinence; impudence; indiscretion 1; mouth 4; nerve 2; presumption 1; sauce 2
audacity, have the, presume 2
audibly, clearly 3
audience, attendance 2; gate 2; interview 1; turnout 1
audit, attend 1; monitor 3
audition, apply 6
auditorium, hall 2; house 5; theater 1
au fait **(with),** aware 1; current 4; experienced 1; expert 2; familiar 4; informed 2; intelligent; knowledgeable 1; learned; level 5; proficient; smart 2
au fond, essence 3
Auf Wiedersehen!, farewell 3; goodbye
Augean stables, shambles
auger, drill 3; punch[2] 1
aught, naught; nil; zero 1
augment, add 4; amplify 1; compound 3; enhance; enlarge 1; expand 3; extend 2; follow 11b; fortify 3; heighten 2; increase 1; inflame 2; intensify; magnify 1; maximize 1; pad 5; reinforce; supplement 2; swell 3; widen
augmentation, addition 3; expansion 1; extension 1; gain 9; increase 3; increment; reinforcement 2
augur, announce 4; bode; forecast 1; foresee; foreshadow; fortuneteller; mean[1] 3; oracle 1; predict; prognosticate 2; promise 4; prophesy 2; prophet; seer; spell[3] 1; threaten 3
augural, ominous 3
augury, foreboding 2; forecast 2; forerunner 2; harbinger; indication 3; omen; oracle 2; prediction; prophecy 1; sign 6; warning 2
august, ceremonial 2; dignified; grand 2; heroic 6; imperial 2; imposing; lofty 2; magnificent; noble 3; prestigious; proud 3; royal 2; solemn 3; stately; statuesque; venerable
augustness, grandeur 2
au naturel, naked 1
aunt, deviant 3
auntie, deviant 3; homosexual 1
auntie man, deviant 3
Aunt Sally, butt[1]; scapegoat
au pair (girl), servant 1
aura°, air 1, 3, aroma 2, climate 2; halo; mystique; odor 2; presence 3; undercurrent 2
aureate, golden 1, 2
aureola, halo
aureole, glory 4; halo; ring[1] 1
Au revoir!, goodbye
auric, golden 2
auriferous, golden 2
aurora, dawn 1
aurous, golden 2
auspices°, cooperation 2; guidance 1; office 5; oversight 2; patronage 1
auspices of, under the, through 1
auspicious, bright 4; favorable 1; fortunate 2; golden 7; happy 2; hopeful 2; lucky 2; opportune 1; positive 8; promising; propitious; right 3; ripe 3a; rosy 2; seasonable; timely; well-timed
austere, chaste 2; dour 2; frigid 2; grim 4; hard 5; harsh 2; joyless 2; plain 4; puritan 2; rigid 2; severe 1, 2, 6; simple 2; Spartan; stark 3; stern 1; stiff 7; strict 2; temperate 2; tight 3; unadorned; unapproachable 1
austerely, severely 2, 3, 6; simply 4
austereness, simplicity 2
austerity, hardship; rigor 2; severity 1, 2, 6; stoicism
autarchy, despotism; independence 1; supremacy 2
auteur, producer 2
authentic°, actual 1; authoritative 1, 2; bona fide; concrete; factual 1; genuine 1; good 8; hearty 2; historical; honest 2; legitimate 1; lifelike; natural 6; official 1; original 3; perfect 5; precise 1; pure 1; real 1, 2; solid 4; sterling 1; true 1; veritable
authenticate°, check 3, 11; confirm 1; document 2; establish 3; prove 1; ratify; seal 4; show 3; substantiate; support 6; verify; vet
authentication, check 15; hallmark 1; identification 1; proof 1; seal 2; testimony
authenticity, reality 1
author°, compose 2; creator 1; designer 1; father 3, 6; founder[1]; narrator; scribe 2; source 2; write 3; writer
authoritarian°, arbitrary 2; despotic; dictatorial 1, 2; dogmatic; domineering; imperative 2; masterful 2; mighty 1; overbearing; prescriptive; repressive; stern 1; strict 2; totalitarian; tyrannical; tyrant
authoritarianism, monarchy 2; tyranny
authoritative°, absolute 4; authentic; categorical; classical 1; definitive 2; dominant 1; inevitable; influential; orthodox; peremptory 3; persuasive; potent 1; powerful 2; professional 2; responsible 3; solid 7; standard 5
authoritatively, highly 4
authoritativeness, certainty 2
authorities°, authority 4; bureaucracy; power 10
authorities, the, establishment 3
authority°, adept 2; administration 2; auspices; balance 5; carte blanche; chair 2; charter 2; command 7; control 4; crown 2; dab hand; domination 1; dominion 1; expert 1; freedom 4; government 1; grip 2; hand 8; hold 26; intellectual 4; judge 3; jurisdiction; license 1; lock[1] 2; master 2; might 2; nobility 1; oracle 1; pass 24; permit 2; power 2, 3, 6; predominance; preponderance 2; prerogative; privilege; professional 3; protocol 1; pull 21; rank[1] 2; rule 2; sage 2; say 12; say-so; scholar 1; source 3; sovereignty; specialist; supremacy 2; sway 4; upper 6; weight 3
authority of, on the, according to 1
authority over, have *or* **maintain** *or* **wield,** command 2
authorization, approval; authority 1; clearance 2; consent 3; endorsement 1; faculty 4; freedom 4; go-ahead 1; leave[2] 1; liberty 2; license 1; OK 6; pass 24; passage 8; permission; permit 2; power 6; prerogative; privilege; requisition 1; sanction 1; say-so; warrant 1
authorize°, allow 2; appoint 1; approve 1; assign 2; charter 4; confirm 1; consent 2; delegate 2; enable 1; enact 1; endorse 1; entitle 1; legitimate 4; license 4; make 23; OK 5; pass 7; permit 1; requisition 2; sanction 5; sustain 5; warrant 4
authorize (to), let[1] 1
authorized, formal 2; lawful 2; legal 1, 2; legitimate 1; official 1; orthodox; permissible; privileged 2; rightful 1
auto, car 1; machine 2
Autobahn, road 2
autobiography, life 6; memoir 2
autochthon, aboriginal; native 7
autochthonous, domestic 3; indigenous 1; native 2; original 2; vernacular 1
autoclave, sterilize 1
autocracy, despotism; monarchy 2; tyranny
autocrat, despot; dictator; oppressor; tyrant
autocratic, absolute 3; arbitrary 2; authoritarian; despotic; dictatorial 1; domineering; heavy-handed 2; imperative 2; masterful 2;

balls (*cont.*) gibberish; gumption 2; gut 3a; machismo; nerve 1; rot 4; stamina; talk 18; trash 1
 balls up, bugger 3; fluff 4; queer 5
ball-shaped, round 3; spherical
balls-up, hash 2; mess 3; mix-up; muddle 4
 balls-up of, make a, ruin 9
ballyhoo, advertisement 2; fanfare 2; noise 1; promote 4; promotion 4; propaganda 2; puff 7; push 6; racket 1; talk 13
ballyrag, carp
balm, fragrance; help 9; lotion; ointment; salve 1, 2; solace 1
bal masqué, masquerade 1
balmy, calm 3; crazy 1; deranged; fine[1] 2; foolish 2; fragrant; gentle 1; insane 1; mad 1; mental 2; mild 2; pleasant 3; preposterous; soft 3; stupid 2; sweet 2; twist 11; warm 1
baloney, drivel 3; fiddlesticks; flannel 1; nonsense 1; prattle 3; rot 4; rubbish 2; stuff 4; trash 1; wind[1] 2
balsamic, soothing 2
baluster, rail[1] 1
balustrade, rail[1] 1
bamboozle, beguile 1; blind 5; bluff[1] 1; cheat 2; chisel 2; deceive; dupe 3; fool 4; hoax 2; hoodwink; lead 7; mislead; mystify; outsmart; perplex; prey 3b; ride 5a; rip 2b; swindle 1; take 32b; throw 6c; trick 8
ban°, bar 3, 9; boycott 2; embargo 1, 2; exclude 1; exclusion 1; exile 3; expel 2; forbid; freeze 4; outlaw 2; prevent; prevention; prohibit 1; prohibition 1, 2; restraint 1; rule 8; sanction 4; shut 5a; stop 6; taboo 2, 3; veto 1, 2
banal°, common 6; hack[2] 4; humdrum; monotonous; pedestrian 2; prosaic; stale 2; stock 7; tedious; threadbare 2
banality, cliché; tedium
bananas, crazy 1; deranged; mad 1; mental 2
banausic, monotonous
band°, bar 2; belt 2; body 6; braid 2; cabal 2; circle 1; clan 2; cluster 2, 3; cohort 2; combine 1; company 2; complement 2; crew; ensemble 2; flock 1; gang 1; group 1; junta; knot 2; league 1, 3; line[1] 2; pack 3; party 2; ring[1] 1, 3; sling 3; squad; streak 1; stripe 1; strip[1] 1; tape 1, 3; team 1; tie 8, 9
 band (against), gang 3
 band together, ally 2; band[2] 3; club 6; flock 2; pool 4; team 3
bandage, bind 3; dress 3; swathe
bandeau, band[1] 1; ring[1] 1
banderole, banner 1; pennant; streamer
bandido, thief 1
bandit, gangster; outlaw 1; robber; thief 1
bandito, thief 1
banditry, pillage 2
bandleader, leader 2
bandmaster, leader 2
bandstand, stand 18
bandy words, quibble 1; spar[2] 2
bane, curse 2; headache 2; pest; pestilence 2; plague 1; poison 1, 2; ruin 3; scourge 1; thorn 2; torment 4; trial 5; ulcer 2; undoing 2
baneful, black 5; destructive 1; fatal 1; harmful; hurtful 1; mischievous 2; ruinous
bang, boom 1; clap 5; crack 2; crash 3, 4; explosion 1; full 14; hit 1, 10, 12; knock 1; love 6; pelt[1] 4; plumb 4; plump[2] 5; pop 1, 4; report 2; slam 1, 2; slap 6; thrill 1
 bang away (at), peg 6
 bang from *or* **out of, get a,** like[2] 1; love 8; wallow 2
 bang into, hit 3; strike 3
 bang on, exactly 2; faultless; marvelous; precisely 1; right 2
bangtail, nag[2]
banish°, dismiss 2; displace 2; exile 3; expel 2; extinguish 2; isolate; ostracize; punish 2; relegate 1; rid 2; transport 2; uproot 1
banishment, ejection 2; exile 1; punishment 2
banisters, rail[1] 1
bank, bar 4; beach 1; deposit 2; drift 5; heap 3; incline 1; mound 1; pool 2; pot 2; put 12; row[1]; shallow 2; side 1; slope 2; stack 1; storehouse; treasury
 bank on *or* **upon,** count 3; depend 2; go 26b; reckon 4; rely 1; store 5; trust 5
bank card, card 8
banked, oblique 1
banker, factor 3; financier
banking, finance 1; oblique 1
banknote, bill[1] 2; note 4
banknotes, cash 1; money 1
bankroll, back 2a; finance 3; pile[1] 2; roll 18; subsidize; support 4; wad 2
bankrupt, back 7b; break 5; broke; destitute 1; hard 17; impoverished 1; insolvent; pauper; penurious 2; poor 1; rock[1] 3c; ruin 10
bankrupt, go, collapse 2; fail 4; fold 3; go 37b; wall 4
bankruptcy, collapse 6; failure 4; ruin 1
banned, taboo 1; unlawful
banner°, color 2a; flag[1] 1; pennant; sign 4; standard 3; streamer; symbol
banneret(te), flag[1] 1; streamer
bannerol, streamer
banning, prevention; prohibition 1
banquet°, dine; feast 1; function 2; spread 11
banshee, ghost 1; shade 4; shrew
bantam, little 1; miniature; minute[2] 1; slight 3; small 1; tiny; undersized
banter°, chaff 1, 2; fool 5; humor 1; joke 4; raillery; repartee; twit[1]
bantering, volley 3
baptize, call 2; christen 1; entitle 2; name 4; tag 4; title 6
bar°, ban 1; barrier 1, 2; barring; beam 1; block 2, 3; bolt 2; cake 2; counter 2; deterrent; disqualify; dive 3; embargo 1, 2; except 1; exclude 1; exclusion 1; exile 3; expel 2; hamper[1]; handicap 1, 2; hinder 1; hurdle 1; impede; impediment; inhibit; inhibition; isolate; leave[1] 10; lock[1] 1, 3, 8; lounge 3; measure 9; obstacle; obstruct 1; obstruction 1; oppose 2; outlaw 2; preclude; prevent; prevention; prohibit 1; prohibition 1, 2; pub; rail[1] 1; ray 1; restrain 2; rod 1; rule 8; shackle 2; shut 4b, 5a; standard 4; stop 2; streak 1, 4; stripe 1; stumbling block; trammel 1; tribunal
barb, bristle 1; insult 2; prickle 1; quip 1; shaft 3; spine 2; spur 2; squelch 2; thorn 1; wisecrack 1
barbarian°, boor 2; boorish; devil 2; heathen 3; merciless; philistine 1; primitive 2; punk 1; savage 4; uncivilized 1; yahoo
barbaric, atrocious 1; barbarian 3; beastly 1, brutal 1; coldblooded 3; cruel 2; cutthroat 3; deadly 3; ferocious; fierce 1; ghoulish 2; heathen 3; inhuman 1; merciless; monstrous 1; outrageous 2; primitive 2; savage 2, 3; uncivilized 1; wild 3
barbarism, misuse 2; outrage 1; rowdyism; solecism
barbarity°, atrocity 1; enormity; monstrosity 2; outrage 1; violence 2
barbarous, atrocious 1; barbarian 3; beastly 1; brutal 1; coldblooded 3; cruel 2; cutthroat 3; deadly 3; ferocious; fierce 1; inhuman 1; merciless; monstrous 1; murderous 1; outlandish; sanguinary 1; savage 2, 3; truculent; uncivilized 1; ungraceful 2; unmerciful; vicious 1; violent 1; wild 3
barbarousness, barbarity
barbate, hairy 1
barbecue, broil; picnic 1; spread 11
barbed, poignant 2; pointed 1; prickly 1; pungent 2; tart[1] 2; thorny 1
barbel, spine 2; spur 2
barbellate, thorny 1
barber, trim 3
barbiturate, sedative 1; tranquilizer
barbule, spine 2; spur 2
bard, minstrel; poet
bare°, barren 2; blank 1; bleak 3; desert[1] 2; desolate 1; develop 1; disclose 2; dry 1; empty 1, 3; exhausted 2; expose 1; gaunt 2; hard 8; lean[1] 2; lifeless 4; meager 1, 3; mere; naked 1, 4; nude; open 2, 5; plain 4; rude 4; severe 6; show 1; smooth 4; stark 2; strip[2] 1; unadorned; unveil; very 4
 bare, leave, clean 10b
 bare oneself, disrobe
bareass, bare 1
bare-boned, meager 2, 3
bared, bare 4; naked 1
barefaced°, brazen; flagrant; naked 3
barely°, almost; hardly; just 5; little 8; merely 2; narrowly 1; nearly 1; only 3; scarcely 1; severely 6; simply 1
 barely any, little 3
bareness, emptiness 1; severity 6; simplicity 2
barest, very 4
barf, regurgitate; vomit
barfly, alcoholic 2
bargain°, agreement 1; buy 5; contract 1; deal 5; dicker 1, 2; engage 3; engagement 3; find 10; haggle; negotiate 1; negotiation 2; pact; steal 4; transaction 1; understanding 1; undertake 2
 bargain, in *or* **into the,** addition 6; besides 1; boot 1; measure 10; moreover
 bargain for, bargain 4
 bargain price, at a, song 2
bargaining, negotiation 1
bargain-priced, cheap 1; sale 5
barge in, interrupt 1; intrude
baring, exposure 1
bark, hack[1] 2; peel 1; scrape 1; skin 5; thunder 3; vessel 2; whoop 1, 2; yap 1
 bark (at), snap 3b

belief° (*cont.*) faith 1, 2; feeling 2; hope 2; idea 3; impression 1; judgment 4; mind 6; moral 4; opinion 1; persuasion 2; point of view 2; position 7; presumption 3; principle 2; religion; sentiment 2; stand 13; supposition; tenet; thesis 1; thinking 2; trust 1; view 2
  belief, beyond, incredible 1; unthinkable 1
belief(s), ideology; lore 1; philosophy 2
  beliefs, set of, creed; persuasion 2; school 3
believable, good 8; plausible 1; presumptive 1; probable; ready-made 2; reasonable 2; reliable; smooth 8; supportable 2; tenable; thinkable
  believable, be, wash 8
  believable, make, rationalize 1
believably, seemingly
believe°, assume 3; buy 2; consider 3; credit 5; esteem 2; estimate 2; expect 2; fancy 10; feel 4; figure 9; guess 2; hold 7; imagine 2; judge 7; lap[1] 3b; perceive 3; presume 1; say 6; suppose 1; suspect 2; swallow 2; take 8; think 1, 2; view 9
  believe, be led to, gather 4
  believe (in), believe 2; follow 2; lean[2] 4a; store 5; swear 3
  believe (to be), regard 4
believed, reputed
  believed by, as, according to 1
believer, sectarian 4
belittle°, blaspheme 2; debase 1; degrade 2; depreciate 2; deride; diminish 2; disapprove; discredit 1; dismiss 2; disparage 1; downgrade 2; flout; foul 14; impeach 2; laugh 2b; libel 3; lower[1] 4; minimize 2; play 13; put 16g; rag[2]; reduce 7; scoff; take 31b; talk 9a; trivialize; underestimate; write 4b
belittlement, dishonor 4; humiliation; libel 1
belittling, condescending; derogatory; flippant 1; foul 6; haughty; personal 4; vituperative
bell, alarm 1; chime 1
bellbottoms, pants 2
bellboy, page[2] 1
bell (buoy), buoy 1
belle, beauty 2
belles-lettres, letter 3; literature 1; writing 3
bellhop, page[2] 1
bellicose, aggressive 1; belligerent 1, 2; hostile 3; martial 1; militant 1; offensive 1; pugnacious; quarrelsome; truculent; warlike
bellicosity, jingoism
belligerence, aggression 1; fight 9; jingoism
belligerent°, aggressive 1; aggressor; argumentative; defiant; hostile 3; ill 3; martial 1; militant 1, 3; offensive 1; passionate 3; pugnacious; quarrelsome; truculent; warlike
bell-like, clear 11; sweet 3
bellman, page[2] 1
bellow°, bawl 1; blare 1, 2; blast 2; call 1; din 1; exclaim; exclamation; howl 1, 2; low[2]; rant 2; rave 1; roar 1, 3; shout 1, 2; thunder 3; whoop 1, 2; yell 1, 2
bellowing, blatant 2; din 1
bells, set of, chime 1
bells and whistles, accessory 1; elaboration 1; frill 2
belltower, tower 1
belly, bowels; exotic 3; gut 2; paunch; stomach 1; swell 1
  belly (out), protrude
bellyache, gripe 1, 3; squawk 2
bellyaching, gripe 2
bellybutton, navel
belly-up, go, go 37b
belong°, lie[2] 3
  belong to, belong 3; relate 3
  belong (together), go 6
  belong (with each other), accompany 2; go 36a; relate 3
belonging°, fellowship 3
belongings°, effects; estate 2; furniture 1; gear 4; good 21a; luggage; paraphernalia; possession 3; property 1; stuff 2; thing 8c
beloved°, adorable; bosom 4; darling 1, 3; dear 1, 3; favorite 2; gallant 5; love 3; old 8; passion 4; precious 2; sweet 5; sweetheart
below°, beneath 1, 2, 3, 4; downward 2; meantime; subordinate 1; under 1, 2, 7
below-decks, below 2
below-ground, sunken 3; underground 1
below-stairs, below 2
belt°, band[1] 1, bar 2, box[2] 2, 3; drink 1; girth 2; hit 1; jab 2, 4; line[1] 2; pelt[1] 4; punch[1] 1, 2; rap 4; ring[1] 1; scourge 3; sling 3; strike 1; strip[1] 1; swath; tape 1; zone
  belt along, speed 3
  belt out, belt 4
  belt up, hush 1; pipe 7
belting, thrashing 1
bemoan°, bewail; complain; grieve 1; keen[2] 1; lament 1; moan 2; mourn; regret 1; repent; sigh 2; sorrow 3; weep 1
bemuse°, bewilder; confuse 1; daze 2; distract 3; muddle 1
bemused, confused 2; daze 4; dumbfounded; foolish 2
ben, mountain 1
bench, bar 5; chair 1, 2; seat 1; thwart 2; tribunal
  bench mark, example 2; gauge 3; norm 2; oar 2; seminal 1; standard 1; touchstone; yardstick
bencher, lawyer
bend°, angle[1] 1; attach 1; bow 2, 3; buckle 2; dispose 2; distort 1, 2; duck 1; flex 2; fold 1; hump 2; inclination 1, 2; incline 1; lean[2] 2; loop 1, 2; meander 2; relent; sag 1; sink 9; slant 4, 5; snake 4; stretch 3; sway 1; sweep 4, 6; thaw 2; trend 3; turn 11, 27; twist 4, 7; warp 2; weaken 3; wind[2] 1
  bend (down), bow 4; crouch; stoop 1
  bend (to), submit 1; obey 1
bendability, flexibility 1
bendable, elastic 1; flexible 1; pliable 1; supple 1
bender°, carouse 2; drunk 4; orgy 1, 2; spree
  bender, go on a, carouse 1; drink 2; paint 7
bending, inclination 1; inclined 3
bend the elbow, drink 2
bendy, pliable 1
beneath°, below 2, 4, 7, 10, 11; subordinate 1; under 1, 2, 7
Benedict Arnold, snake 2; traitor; turncoat
benediction, blessing 1; glory 2; godsend; grace 5
benefaction, gift 1
benefactor°, backer 2; donor; friend 4; humanitarian 2; patron 1; philanthropist; protector
benefactress, backer 2; donor; friend 4; humanitarian 2; patron 1; philanthropist; protector
beneficence, benevolence 1, 2; bounty 1; charity 1; humanity 3; kindness 1; mercy 1; philanthropy 1; providence 2
beneficent, bountiful 1; charitable 2; generous 1; good 6; gracious; human 3; large 2; merciful; open 14; philanthropic; worthwhile 2
beneficial°, advantageous; benevolent; constructive 1; desirable 4; expedient 2; favorable 1; fruitful 2; gainful; good 13, 14; happy 2; healthy 2; helpful; instrumental; nutritious; opportune 1; positive 4; profitable 1, 2; right 3; therapeutic; useful; well-timed; wholesome 1; worthwhile 2
  beneficial to *or* **for, be,** behoove
beneficiary, heir; recipient
benefit°, account 3; advance 2; advantage 2; aid 2; asset 2; assist 2; assistance; boon; dole 1; enjoyment 2; favor 8; gain 2, 8; gift 1; good 19; icing 2; improve 2; interest 3; pay 2, 4; pension 1; perquisite; plus 3; privilege; profit 2, 3; purpose 3; return 5, 9; sake 1; service 1; use 7; usefulness; welfare; worth
  benefit from, enjoy 2
  benefit of, for the, behalf; for 1, 7
benefit(s), benefit 2; fruit
benevolence°, altruism; charity 2; grace 4; humanity 3; kindness 1; philanthropy 1
benevolent°, benign 1; charitable 2; fatherly; generous 1, 2; good 6; gracious; helpful; human 3; kind[1]; philanthropic; tender[1] 6
benighted°, dark 9; ignorant 2; illiterate
benightedness, ignorance 1; provincialism 2
benign°, amiable; benevolent; fair[1] 5; fatherly; gentle 1; good 6; gracious; harmless; human 3; soft 5
benignant, benevolent; benign 1; human 3; merciful
benignity, humanity 3; kindness 1; mercy 1; philanthropy 1
bennie, stimulant 2
bent°, affinity 2; appetite 1; aptitude 2; attitude 2; bias 1; cast 6; crooked 1, 2; deformed 1, 3; deviant 1; dishonest; disposition 1; drift 3; effeminate; favoritism; fixed 4; flair 1; foul 7; frame 5; fraudulent 2; gift 2; gnarled; habit 2; homosexual 2; inclination 3; instinct; intent 5; knack; leaning; liking 1; love 2; mentality 2; mind 5; odd 1; penchant; position 3; preference 2; rotten 3; shady 2; shapeless 2; shifty; sick 5; slant 2; spirit 2; swish 4; talent 2; tendency; tortuous 1; trend 1; turn 38; venal; villainous 1; weakness 4; wry 1
  bent on *or* **upon, be,** pursue 3; spoil 5
  bent over, double 2
benumb, bemuse 2; daze 1; deaden 1; drug 4; dull 11; numb 2; paralyze 2; petrify 1; stun 1
benumbed, chill 5; dead 2; groggy; insensible 1; numb 1; petrified 1; senseless 2; silly 2
benumbing, stunning 1

bequeath°, devise 2; donate; hand 15a; leave[1] 6; pass 18a; will 6
bequeathed, hereditary 2
bequeathed, be, inherit
bequest°, donation 1; endowment 1; grant 3; inheritance
Be quiet!, hush 1; wrap 3b
berate°, abuse 3; attack 2; chastise; correct 2; damn 1; dress 4; lace 5b; lambaste 2; lash[1] 4; lecture 4; light[2] 14; mind 10; nag[1] 1; rail[2]; rate[2]; rebuke 1; reprimand 2; scold 1; tell[1] 10; twit[1]; upbraid; vituperate
berating, abuse 8; lecture 2; rebuke 2; tongue-lashing
bereavement, loss 1; mourning 2
bereft, forlorn 2
bereft of, destitute 2; starved 3
Bermuda shorts, pants 2; short 15
berserk°, deranged; distraught; frantic; homicidal; hysterical 1; mad 4; maniacal 2; rabid 1; rampage 2; raving 1; wild 5
berserk, go, flip 2; fly 7; rage 4; rampage 3
berserk, going, uncontrolled
berth, appointment 3; cabin 2; dock 1, 2; job 1; land 5; moor[2]; office 4; place 5; position 6; situation 3; space 4
berth, give (something *or* **someone) a wide,** steer 2
beseech°, appeal 1; ask 3; beg 1; hit 8; petition 2; plead 2; pray 1; press 6; request 1; solicit 1; sue 2; urge 2
beseecher, supplicant 2
beseeching, supplicant 1; supplication 2
beset°, assault 3; harass; haunt 2; infest; mob 3; seize 3; smitten 1
beside°, about 9; by 1
beside oneself, beside 3; ecstatic; excited 1; frantic; furious 1; hysterical 1; upset 10
beside oneself (with anger *or* **fury), be,** rage 4
besides°, addition 5, 6; apart 4; barring; boot 1; further 3; independent 8; likewise 2; measure 10; moreover; yet 3
besiege°, beset; bombard 2; siege 2; storm 6
besiegement, siege 1
Be silent!, hush 1; wrap 3b
besmeared, dirty 1
besmirch, blacken 2; blemish 2; dirty 7; discredit 1; disgrace 4; foul 14; libel 4; mire 4; muddy 5; shame 8; smear 1, 2; soil[1] 2; spot 7; stain 5; stigmatize; sully; taint 2
besmirched, dirty 1
besmirchment, filth 2
besom, brush[2] 1
besotted, drunk 1; far 5b; infatuated; stinking 3
besotted by *or* **with,** daft 3
bespatter, dabble 1; muddy 5; spatter; splash 1; spot 7
bespattered, speckled
bespeak, engage 1; indicate 2; promise 4
bespeckle, dapple 2; dot 3; pepper
bespeckled, flecked
bespoke, fitted; speak 13; tailor-made 1
besprinkle, dabble 1; powder 4; spatter; splash 1
besprinkled, flecked
best°, beat 2; better 5; choice 3, 4; defeat 1; elite 2; finish 4; floor 4; flower 2; foremost 1; good 15; great 6; leading 2; native 3; optimum 1, 2; outside 3, 6; outstrip; overcome 1; overpower 1; overthrow 1; peerless; perfect 1, 2, 4; pick 10; pride 3; prime 2; prize[1] 5; rout 2; select 2; shade 8; special 5; superlative; supreme 3; surpass; top 3, 8; triumph 3; vintage 2; whip 2
best, at, ideally 1; just 4; only 3
best, at one's, pink[1] 1
best, do one's, endeavor 1; exert 2; strive 1
best, had, have 10
best, hope for the, finger 3
best, pray for the, finger 3
best, the, end 8b; excellent
best of, get the, outsmart; overcome 1
best bib and tucker, put on one's, dress 5; primp
bested, overcome 2
bested, be, lose 3
bestial, animal 4; beastly 1; brutal 1; ferocious; fierce 1; inhuman 2; libertine 2; savage 2, 3; vicious 3
bestiality, violence 2
Be still!, hush 1
bestir, incite; rouse 2
bestir (oneself), stir 2; wake[1] 1; awake 1; move 12
best-liked, favorite 2
bestow°, award 1; contribute 1; devise 2; dispose 3c; donate; extend 4; give 1; grant 1; heap 4; impart 1; lavish 5; lend 2; pay 3; rain 7; show 10; spare 8; vouchsafe 1
bestow (on), confer 2; present[2] 6
bestowal, award 3; disposition 3; donation 2; endowment 2; presentation 1
best part, feature 2; majority 1; mass 4
bestraddle, mount[2] 3
bestrew, strew
bestride, mount[2] 3
bestseller, novel 2
best-selling, successful 3
best wishes, compliment 2; congratulations; greeting 3; regard 10; respect 5
best years, prime 4
bet°, adventure 4; chance 8; gamble 1, 4; lay[1] 4; play 7; plunge 4; put 6; stake[2] 1, 4; venture 3
bet on, back 1; gamble 2; rely 1
bête noire, fear 3
betide, chance 7; happen 1
betimes, early 3
bêtise, trifle 1; triviality 2
betoken, announce 4; argue 4; bode; denote 2; express 2; foreshadow; imply 2; import 2; indicate 2; involve 2; mean[1] 2; prognosticate 2; promise 4; signify 1; speak 5; stand 7a; symbolize
betokening, symbolic
betray°, blab; deceive; denounce 2; disclose 1; display 1; double-cross; exhibit 1; expose 1; finger 8; give 12b; inform 2; let[1] 7a; manifest 2; register 4; sell 3; shop 3; show 1; stab 2; tell[1] 2; turn 15c
betrayal°, perfidy; stab 6
betrayer, informer; snake 2; traitor; turncoat
betrothal, engagement 2; match 3
betrothed, attached 4; engaged 1; fiancé; girl 2; love 3; sweetheart
better°, advantage 3; amend 1; benefit 3; cap 4; dignify; elaborate 4; enhance; enrich 1; exceed 1; help 2; high-class 1; improve 1; increase 2; lift 2; mend 2; outstrip; perfect 9; player 3; preferential; reform 1; select 2; shade 8; superior 1; surpass; top 3; vintage 2
better, get, come 17a; convalesce; gain 3; improve 3; look 11c; mend 3; pick 8f; progress 6; pull 17; rally 4; recover 2; recuperate
better, had, have 10
better, turn for the, rally 2; recovery 1
better of, get the, best 11; defeat 1; master 8; outsmart; overcome 1; subdue 1; upset 4
better off, better 6
better than average, strong 10
better half, mate 2; wife; woman 2
bettering, recovery 1; refinement 4
betterment, advance 7; advantage 2; amendment 1; elaboration 1; improvement 1; progress 2; recovery 1; refinement 4; reform 3
better part, majority 1; mass 4
betters°, better 9
bettor°, backer 3; player 3
betwixt and between, end 6
bevel, slant 4; slope 2
beverage, brew 4; drink 4; potion
bevies, score 4
bevy, flight[1] 4; flock 1; number 2; pack 3; score 4; throng 1
bewail°, grieve 1; keen[2] 1; lament 1; moan 2; mourn; regret 1; repent; sigh 2; sorrow 3; weep 1
beware°, look 9; mistrust 1
bewhiskered, hairy 1
bewilder°, astound; bemuse 1; bother 3; confuse 1; daze 2; demoralize 3; distract 3; dumbfound; elude 2; entangle 2; floor 5; flummox; flurry 2; fluster 1; fog 4; get 18; knock 6b; mix 4b; muddle 1; mystify; overwhelm 3; perplex; puzzle 1; stagger 2; stick[1] 9; stump 2; stun 2; throw 6c
bewildered, blank 5; confused 2; daze 4; disconcerted; disorientated; dumbfounded; groggy; helpless 2; lost 3; sea 4
bewildering, disconcerting; indefinite 2; inexplicable; mysterious 1; overwhelming 2; paradoxical; perplexing; puzzling
bewilderment, distraction 1; fluster 2; perplexity 1
bewitch°, captivate; catch 11; charm 5; dazzle 1; enchant 1; enrapture; entrance[2]; fascinate; hypnotize; intoxicate 2; jinx 3; possess 3; ravish 1; regale; transfix 2; transport 3
bewitched, charmed 1; doomed 2; infatuated; rapt 1; smitten 2; taken
bewitching, catching 2; darling 4; devastating 2; enchanting; enthralling; glamorous 1; inviting; lovely 1; magic 6; magnetic; prepossessing; provocative 1; ravishing; seductive; spellbinding; taking; winning 1
bewitchment, glamour; magic 3
beyond, above 4; addition 5; advance 10b; besides 2; further 3; independent 8; over 2, 3, 7; ulterior 2
beyond everything, above 6
beyond (someone), be, escape 4
Bhagavad-Gita, Scripture
bias°, bent 5; bigotry; color 4; discrimination 1; drift 3; fanaticism 2; favor 3; favoritism;

blaze the trail, pioneer 2; spearhead 1
blazing, ablaze 1; burning 1, 4; fiery 1; fire 5; full 10; glaring 2; radiant 1; torrid 1
blazon, splash 2
bleach°, fade 1
bleached, washed-out 1
bleak°, cold 4; dark 3; desolate 1; dismal; dreary 1; gaunt 2; harsh 2; heavy 9; joyless 2; lifeless 4; murky; pessimistic; sad 2; somber 2; stark 3; untoward 1; wintry 3
bleakness, rigor 1
blear, film 4
bleary, filmy 2; indistinct 1; vague 1
bleat, gripe 1
bleed, drain 5; extort; fleece; gouge 2; milk; ooze 2; sap[1] 3; squeeze 2, 3; tap[2] 6
bleed for, feel 8; pity 3
bleeding, flaming; very 1
bleep, edit 2; page[2] 2
blemish°, blot 1; bruise 1; deface; defect 2; discredit 5; disgrace 2; dishonor 5; failing 1; fault 1; flaw 1; foible; imperfection; mark 1, 10; nick 1; scandal 3; scar 1, 2; spoil 2; spot 1; stain 2; stigma; stigmatize; sully; taint 1, 2; tarnish; vice 2
blemished, disfigured
blench, bleach 1; blink 3; cringe 1; fade 1; flinch; pale[1] 4; recoil 1; start 5
blench at, dread 1
blend°, alloy 1; amalgam; amalgamate; amalgamation; beat 6; chime 5a; combination 3; combine 2, 3; compound 1, 2, 5; converge; cross 2; fuse; go 6; incorporate; integrate; lump[1] 3; match 7; medley; melt 3; merge; mingle 1; mix 1, 7; mixture 1, 2; neologism; pastiche; solution 3; stir 1; synthesis; unite 1; weave 2; wed 2; wedding 2
blend with, go 39a
blended, general 3
blending, amalgamation; merger; mixture 2; synthesis; wedding 2
bless°, dedicate 2; hallow 1
blessed, favored 2; fortunate 1; godlike 1; golden 5; heavenly 1; holy 1; lucky 1; providential; sacred 1; saintly
blessed with, be, possess 1
Blessed Isles *or* **Islands, the,** heaven 1
blessedness, bliss; paradise 3
blessing°, approval; boon; cooperation 2; glory 2; godsend; grace 5; honor 3; joy 3; recommendation 2
blessing (to), give one's, approve 1
blest, godlike 1; golden 5; sacred 1; saintly
bleu, telegram
bleu(e), rare[2]
blight°, disease 2; mar 2; mold[2]; overshadow 2; pestilence 2; plague 1; poison 2; pollute 1; rot 3; shadow 4; ulcer 2; undoing 2; visitation 2
blighter, rascal; rogue 1; stinker; villain; wretch 1
blighting, pollution
Blimey!, indeed 3
blimpish, reactionary 1
blind°, bluff[1] 3; brute 1; camouflage 1; cover 14; daze 1, 2; fog 3; heedless; pretense 2; screen 3; shade 5; unreasonable 1
blind (to), blind 4; unconscious 2
blind to, be, excuse 1; ignore 1; tune 5
blind alley, impasse
blindfold, blind 5, 6
blinding, dazzling; full 10; glaring 2; strong 19
blind spot, failing 1
blink°, flicker 1; glisten; signal 4; sparkle 1; twinkle 1
blink, go on the, play 17b
blink, on the, blink 6; broken 7; defective 1; faulty; order 13c
blink at, blink 4; disregard 1; overlook 2
blinker, blind 5
blinkered, nearsighted 2
blinking, twinkle 2
blink of an eye, minute[1] 1; moment 1
blious, passionate 3
blip, edit 2
bliss°, delight 3; ecstasy 1; enchantment 2; heaven 1, 3; joy 1; paradise 3; transport 5; utopia
blissful, blithe 1; ecstatic; elated; exalted 3; godlike 2; golden 4; heavenly 2; radiant 2; rapt 2; rhapsodic; seraphic
blissfulness, bliss
blister, bubble 1; lump[1] 2
blistering, hot 1; torrid 1
blistery, torrid 1
blithe°, buoyant 2; carefree; cheerful 1; gay 2; happy 1; jaunty 1; joyful 1; merry 1; radiant 2; sportive; sunny 2; vivacious
blithefulness, merriment
blithely, gaily 2; happily 2
blitheness, bliss; cheer 2; gaiety 1
blither, prattle 1; yap 2
blithesome, breezy 2; cheerful 1; happy 1; joyful 1; merry 1; radiant 2
blithesomeness, happiness; joy 2; merriment
blitz, assault 1; overrun; push 15; raid 1; rush 2; shell 4; storm 6
blitzkrieg, assault 1; invasion 2; push 15; storm 6
blizzard, storm 1
bloat, fill 5; puff 6; pump 4a; swell 1
bloated°, swollen
blob°, gob
bloc, combination 2; faction 1; fellowship 2; front 6; party 3; ring[1] 3; syndicate 1; union 2
block°, area 3; arrest 1; bar 3, 9; barrier 2; besiege 2; brick 1; cake 2; check 1; clog; drag 6; eclipse 1; embargo 1, 2; engraving 1; fill 7; foul 15; frustrate 1; hamper[1]; handicap 1, 2; head 13a; hide[1] 3; hurdle 1; impasse; impede; impediment; intercept; interfere 2; interference 2; jam 2, 4; keep 13; loaf[1] 1, 2; mass 3; obscure 7; obstruct 1, 3; oppose 2; plug 4; preclude; prevent; prevention; preventive 3; prohibit 2; pulley; resist 1; retard 1; setback; snag 1; spike 3; square 7; stamp 6; stay[1] 4; stop 2, 9; stumbling block; stymie; suppress 1; thwart 1; trammel 2; turn 7; veto 1, 2; wad 1; wall 1; ward 3; wedge 1
block, on the, sale 6
block off *or* **up,** besiege 2
block (out), block 4; draft 5; formulate 3; frame 7; outline 3; rough 15; shut 5b; design 2
block(s), trammel 1
block (up), block 5; choke 2; stop 3; stuff 9
blockade, besiege 2; picket 5; siege 1, 2
blockage, eclipse 3; embargo 1; exception 2; impasse; inhibition; jam 4; stay[1] 6; stop 9; stricture 1
block (and tackle), tackle 2
blockbuster, novel 2
blocked, foul 9
blocked (up), congested
blockhead, clod 2; dolt; dummy 3; fool 1; silly 3; twit[2]
blockheaded, foolish 2; thick 6; wooden 3
blocking, eclipse 3; jam 4; prevention; stop 9; suppression
blockish, slow 7
block (of paper), pad 2
bloke, bugger 2; chap; customer 2; devil 3; fellow 1; guy 1; man 1; stick[2] 2
blond(e), fair[1] 3; golden 1
blood, birth 3; bosom 3; dandy 1; extraction 3; family 3; flesh 5; flush[1] 4; gore[1]; house 2; paternity; pedigree; race[2] 2; rank[1] 3; tribe
blood, in the, native 1
blood boil, make (someone's), anger 2; enrage; infuriate; madden 1; outrage 4
blood run cold, make (someone's *or* **the),** horrify 1; terrify
blood up, have *or* **get (someone's),** infuriate
blood-and-thunder, melodramatic
blood bath, carnage; holocaust 2; killing 1; massacre 1; slaughter 2
bloodcurdling, gory; horrible 1; scary
bloodless, gray 1; pale[1] 1; wan 1; white 2
bloodletting, bloodshed; slaughter 2
bloodline, family 3; lineage 1; nationality 2; parentage; pedigree; stock 2; strain[2] 1; succession 4
blood money, payoff 3
blood relation(s), kin 1
blood relative(s), kin 1; relation 3
bloodshed°, gore[1]; hostility 2; killing 1; murder 2; slaughter 2
bloodsoaked, gory
bloodstained, gory
bloodsucker°, parasite
bloodthirstiness, barbarity; violence 2
bloodthirsty°, cruel 2; cutthroat 3; deadly 3; ferocious; fierce 1; ghoulish 2; grim 2; murderous 1; sanguinary 1; savage 2; warlike; homicidal
blood vessel, vein 1
bloody, cutthroat 3; extremely; flaming; gory; murderous 1; perfectly 4; sanguinary 2; savage 2; very 1
bloody-minded, difficult 3; obstinate
bloody-mindedness, obstinacy
bloom, develop 2; flourish 1; flower 1, 3; flush[1] 4; glow 6; grow 1; sprout; thrive
bloomer, error 1; fluff 3; howler; indiscretion 2; misstatement 2; misstep 2; mistake 1; slip[1] 8; solecism
bloomers, pants 1
blooming, flaming; flourishing; fresh 6; sound[2] 2
blooper, fluff 3; misstatement 2; misstep 2; mistake 1; slip[1] 8; solecism
blossom, develop 2; flourish 1; flower 1, 3; grow 1; sprout
blossoming, flourishing
blot°, blemish 3; discredit 5; disgrace 2; dishonor 5; flaw 1, 4; libel 2; mar 2; notoriety; slur 1; smear 3; soil[1] 2; spot 1; stain 1, 4; stigma; taint 1; tarnish
blot out, block 4b; blot 4; delete; eclipse 1; erase 1; extinguish 2; hide[1] 3; obliterate 2; shade 12; strike 5; wash 4
blotch, blot 1; bruise 1; spot 1; stain 1, 4

run 42; runner 3; science 1; section 1; separate 4; shoot 5; split 3; spray[2]; stick[2] 1; stream 1; tributary
branch (off *or* out), diverge 1; diversify; divide 4; sprawl 1; spread 6
branches, brush[1] 1
branchlet, twig[1]
branch water, water 1
brand°, character 1; characterize; denounce 1; description 3; discredit 5; form 3; genre; impression 3; kind[2] 1; label 1, 3, 4; line[1] 16; make 42; mark 2, 13; model 6; mold[1] 2; proclaim 2; scar 1, 2; sign 4; sort 1; stamp 3, 7; stigma; stigmatize; style 7; symbol; variety 3
brandish, dangle 2; exhibit 1; flourish 2; parade 5; shake 4; wave 4; wield 1
brand name, brand 1
brand-new°, fresh 1; hot 5; modern; new 1, 2; recent; spanking 1
brannigan, brawl 1; fracas 1
brash°, adventurous; blunt 2; brassy 1; brazen; brusque; cocky; downright 1; flippant 2; foolhardy; foolish 1; forward 2; immodest 2; impertinent; insolent; outspoken; overconfident 1; pert; presumptuous; pushy; rash[1]; shameless; shortsighted 2
brashly, flat 17b; openly 1; spur 3
brashness, effrontery; face 5; flippancy 2; gall[1] 2; impertinence; mouth 4; nerve 2
brass°, assurance 3; bigwig 2; effrontery; face 5; freedom 7; gall[1] 2; impertinence; management 2; nerve 2; pipe 3; presumption 1; prostitute 1; sauce 2
brassed off, discontented
brasserie, café
brass farthing, damn 4
brass hat(s), bigwig 1, 2
brassiness, impertinence
brass neck, face 5; presumption 1
brassy°, barefaced 2; brash 2; brazen; fresh 8; impertinent; impudent; insolent; pert; pushy
brat, boy 1; guttersnipe; imp
bravado°, bluff[1] 3; bluster 3; bombast
brave°, adventurous; audacious 1; bold 1; challenge 3; courageous; daredevil 2; daring 2; dauntless; defy 1; enterprising; face 14, 18b; fearless; gallant 1; game 8; go 35a; gritty 2; hardy 2; heroic 1; indomitable; intrepid; manly; martial 2; stalwart 3; stand 3, 12a; stout 2; superhuman 1; sustain 3; take 6; venturesome 1; weather 3; withstand
bravery°, chivalry; courage; daring 1; grit; gut 3a; heart 2; manhood 2; nerve 1; pluck 1; prowess 2; spirit 5
Bravo!, bully 4
bravura, showy; virtuoso 2
brawl°, battle 1; conflict 1; disorder 2; dispute 4; disturbance 2; fight 1, 7; fracas 1, 2; fray[1]; quarrel 1, 2; riot 1, 3; roughhouse 2; rumpus; scrap[2] 1, 2; scrimmage; tumult; uproar
brawler, rough 13; rowdy 2
brawling, riotous 1; roughhouse 1; rowdyism
brawn°, power 4; sinew 2; strength 1
brawniness, brawn; strength 1
brawny°, big 2; burly; hefty 2; husky 1; large 1; mighty 2; muscular; robust 1; sinewy; stalwart 1; stout 4; strong 1; sturdy 1; tough 3
bray, blare 1; din 1; grind 1; mill 5; pound[1] 2; powder 3; pulverize 1
braying, din 1
braze, attach 1; cement 2; connect 3; tack 5; weld 1
brazen°, arrogant 1; audacious 2; barefaced 2; blatant 1; bold 2; brash 2; brassy 1; cheeky; cool 6; defiant; flagrant; flippant 2; forward 2; fresh 8; ill-mannered; immodest 2; impertinent; impudent; insolent; open 13; overconfident 1; pert; presumptuous; pushy; shameless; unabashed
brazen (out *or* through), brave 3; defy 1; face 18b
brazenfaced, brazen
brazenly, face 13; flat 17b; openly 1
brazenness, assurance 3; effrontery; flippancy 2; freedom 7; gall[1] 2; impertinence; impudence; nerve 2; presumption 1; sauce 2
breach°, break 25; crack 1; flaw 2; fracture 2, 3; gap 1; gulf 2; hole 2; infringement; jump 12; offense 1; opening 1; rift 1; rupture 1, 3; schism; solecism; split 8; vacancy 1; violate 1; violation 1
breaching, violation 1
breach of faith, betrayal 1
breach of the peace, disorder 2; disturbance 2
bread, cash 1; food; fund 2; money 1
bread-and-butter letter, note 2
breadbasket, stomach 1
breadline, on the, needy
breadroll, roll 17
breadth°, area 4; extension 2; measure 1; measurement 2; scope 1; spread 9; width 1
breadwinner, worker
break°, back 7b, 18; breach 1, 2; check 13; crack 1, 5; cranny; crevice; crush 1; dash 1; dawn 3; degrade 1; disobey; enervate; escape 5; flaw 2; force 8; fracture 1, 2, 3; gap 1; go 25; holdup 2; hole 2; holiday 1; humble 4; infringe 1; interlude; interruption 1, 2; interval 1; jump 12; lapse 2; leak 2; letup; luck 2; lull 1; opening 1, 2; opportunity; overcome 1; pause 1, 2; punctuate 1; quaver 2; recess 2; reduce 6; respite 1; rest[1] 2; rift 1; rupture 1; say 4; schism; secession; separation 1, 3; space 3; splinter 2; split 7, 8; start 12; stop 5, 7; stretch 3; tame 5; tell[1] 2; violate 1; wait 4; wash 3
break, take a, stop 5
break (apart), break 1; burst; come 7; crumble; disintegrate; fall 7; loosen 2; reduce 4; rupture 3; split 1; separate 1
break asunder, break 1; burst; come 7; crumble; fall 7; reduce 4
break away, break 12; get 32a
break away (from), bolt 8; break 15; depart 2; secede; slip[1] 6
break (down), analyze 1; break 1, 16; burst; cave 2a; collapse 1, 2; come 7; crumble; decompose 1; destroy 1; fall 7; founder[2] 2; give 9; miscarry; pack 8b; piece 6; range 7; reduce 4; separate 2; trample 3; way 11a; wrong 8b; collapse 2; decompose 1; destroy 1
break down into, resolve 4
break down (mentally), collapse 4
break forth, break 12, 13
break free, escape 1; get 32a
break (from), disengage
break in on, short 12b
break in *or* into, break 8, 18; chime 5b; chip 4b; discipline 6; disrupt 2; dissolve 2; go 31f; interrupt 1; intervene 1; intrude; shake 6a
break into bits, break 1
break in two, split 1
break loose, break 12; loose 8
break (off) (from), break 10, 19; cease 1; cut 15b; discontinue; dissociate; end 9; interrupt 2; sever 2; snap 1; stop 1
break (off *or* away), disconnect; separate 1
break out, escape 1; get 32a
break out in, go 31f
break out (in *or* into), break 20b; erupt 1, 2; flare 1
break out (of), break 12
break through, breach 4; break 22; peer[2] 2
break up, break 1, 11, 16b, 23; burst; collapse 2; come 7; crumble; decompose 1; destroy 1; disband; disintegrate; disperse 2; dissipate 1; dissolve 3; diversify; divide 1; fall 7; fragment 3; laugh 1; let[1] 8d; part 12; portion 4; pulverize 1; reduce 4; rupture 3; scatter 2; separate 1, 3; sever 2; split 1, 2, 5
break (up) (with), break 24, 24a; disengage; dissociate; jilt; throw 8
break with, break 24; disengage; secede
breakable, brittle 1; crisp 1; delicate 1; flimsy 1; fragile; weak 1
breakage, fracture 1
break an appointment with, stand 10c
break and enter, break 18c
breakaway, schismatic
break bread, dine; eat
breakdown°, analysis 1; collapse 5, 7; destruction 3; dissolution 1; downfall; failure 2; loss 5; miscarriage; prostration 4; ruin 1; shock 2
breakdown lane, shoulder 1
breaker, roller 2; sea 2; surge 2; wave 1
break faith with, betray 1
breakfast, dine; eat; meal 1
break ground°, break 17
breaking, fracture 1; infringement; rupture 1; secession
breaking (apart), parting 1
breaking down, destruction 1; dissolution 1
breaking even, unprofitable 1
breaking (up), destruction 1; dissolution 1; division 1; parting 1; partition 1; rupture 1
breaking and entering, robbery
breakneck°, hurried 1; reckless
break of day, dawn 1
break (one's) promise *or* word, renege 2
breakout, escape 5
breaks, the, fate 1
break the back of°, back 7
break the connection, hang 12
break the ice°, break 17, 21; move 12; set 12a; spearhead 1

break (the law), transgress 2
break the news (about), publish
break the spell, disillusion
breakthrough, inroad 2
break (to smithereens), shatter 1
breakup, breach 2; destruction 3; dissolution 1; divorce 1; parting 1; partition 1; rift 1; ruin 1; rupture 1; separation 1, 3
break wind°, wind[1] 5
breast°, bosom 1; chest 2
breast-feed, nurse 3
breastwork, rampart; wall 2
breath°, air 2; breeze 1; exhalation 1, 2; hint 2; odor 2; puff 1; savor 2; smell 1; sniff 1; spirit 1; suggestion 2; waft 2; wind[1] 1
breath, catch one's, gasp 1
breath, holding one's, limbo
breath, in the same, once 6b
breath, out of, breathless 1
breath, short of *or* **out of,** short-winded
breath, struggle for, gag[1] 2
breath, with bated, expectant
breath away, take one's, breath 3; intoxicate 2; stagger 2; stun 2
breathe°, blow[1] 1; exist 1; heave 3; live 6; puff 4; rumor 2; sigh 1; whisper 1
breathe hard, pant 1
breathe (in), draw 7; inhale
breathe (out), exhale; expire 3
breathe a word, whisper 2
breathe a word, not, tongue 3
breathe (new) life into, animate 1; refresh 1; rejuvenate; resurrect; revive 3
breathe one's last, die 1; expire 2; pass 14a
breather, break 28; pause 2; recess 2; respite 1; rest[1] 2
breathe (some) life into, pep 2
breathing, alive 1; animate 4; live 1; organic 1
breathing one's last, moribund 1
breathing space, interlude; leisure 2; pause 2
breathing spell, recess 2; rest[1] 2
breathless°, agog; short-winded
breath (of air), draft 3
breath of fresh air, like a, refreshing
breathtaking, awesome; dramatic 2; gorgeous 1; marvelous; scenic; sensational 1; superb; terrific 2
breech, bottom 1
breeches, pants 2
breed°, bear 6; bring 15a; description 3; extraction 3; form 3; generate 2; grow 3; hatch 1; kind[2] 1; mate 5; nationality 2; produce 3; proliferate; propagate 1; race[2] 2; rear[2] 1; reproduce 2; sort 3; spawn; tribe; type 1; variety 3
breeding°, background 1; class 3; culture 1; grace 2; propriety 2; refinement 1; reproduction 3; savoir-vivre; stock 2; upbringing
breeding, bad, incivility
breeding ground, hotbed
breeks, pants 2
breeze°, air 2; breath 1; draft 3; gust 1; picnic 2; piece 9; pushover 1; sail 3; snap 12; wind[1] 1
breeze, shoot the, palaver 4; talk 4
breezy°, carefree; rakish; smart 5; windy 1
brevity°, economy 2
brew°, ferment 1; potion
brewing, impending
briar, pipe 2
bribable, accommodating 2; mercenary 2; venal
bribe°, buy 3; corrupt 5; fix 11; get 31d; kickback; pay 10b; payoff 3; rebate 2; venal
bribe, give (someone) a, pay 10b
bribery, fix 18; graft[2]
bribes, accept, line[2] 2
bric-a-brac°, curiosity 3; gewgaw; sundries; trash 2
brick°, block 1; loaf[1] 1
brick up, wall 5
brickbat(s), flak; missile; taunt 2
bridal°, nuptial
bride, mate 2; wife; woman 2
bridegroom, groom 2; husband 1
bride-to-be, fiancé
bridewell, prison
bridge°, span 1, 3; traverse 2
bridle°, bristle 3; contain 3; curb 2; discipline 7; govern 2; inhibit; limit 5; master 8; rein 1, 3; resist 1; restraint 1; shackle 4; subdue 1
brief°, abbreviated; compact 3; concise; curt; fast[1] 1; fleeting; fly-by-night 1; fugitive 3; groom 4; hasty 3; hurried 1; inform 1; instruction 1; little 4; meteoric; momentary; passing 1, 2; post[3] 4; prepare 3; prime 6; short 2, 7; succinct; summary 1, 2; terse 1; thumbnail; transient
brief, in, brief 5; word 8
brief for, hold a, defend 3
brief case, bag 2
briefed, informed 2
briefing, instruction 1; orientation 2; preparation 2; rundown
briefly°, brief 5; short 14; shortly 3; temporarily 2; word 8
briefness, brevity; economy 2
briefs, pants 1
brier, thorn 1
briery, prickly 1
brig, hole 4; jail 1; prison; stir 7
brigade, cohort 1; corps
brigand, gangster; outlaw 1; robber; thief 1
brigandage, pillage 2
bright°, ablaze 2; brilliant 1, 4; broad 2; buoyant 2; cheerful 2; clear 3, 4; dazzling; fair[1] 5; fine[1] 2; fresh 5; full 10; gay 2, 3; glowing 2; golden 3; hopeful 2; hot 8; ingenious; intelligent; keen[1] 6; light[1] 13; liquid 3; lively 4; luminous 1; optimistic; perky; pleasant 3; precocious; propitious; quick 4; radiant 1; ready 6; receptive 2; rosy 2; sharp 3; shiny; silver 3; smart 1; spanking 1; strong 19; sunny 1; vivid 1; wise 1
brighten°, cheer 6; dawn 3; enliven 2; gladden; illuminate 1; lighten[1] 1, 2; light[1] 16, 18; liven; perk up; polish 1; shine 2
brighten (up), tone 6
bright-eyed and bushy-tailed, fresh 5; perky
brightly, clear 17; gaily 1
brightness, blaze 3; brilliance 1; depth 4; gaiety 3; glare 1; gloss[1] 1; glow 2; illumination 1; intelligence 1; light[1] 1; polish 5; radiance 1; sheen; shine 4; sparkle 3, 4; splendor 2
brill, capital 6; good 2; sight 6b; splendid 3
brilliance°, blaze 3; dazzle 3; depth 4; flare 4; gaiety 3; genius 2; glamour; glare 1; glitter 4; glory 3; glow 2; ingenuity; luster 2; polish 5; pomp; radiance 1; sparkle 3; splendor 1, 2; state 3; virtuosity
brilliancy, blaze 3; depth 4
brilliant°, ablaze 2; bright 1, 5, 6; dazzling; fiery 2; fine[1] 4; flamboyant 2; full 10; gay 3; gifted; glaring 2; glorious 4; glowing 2; golden 3; good 2; gorgeous 1; great 6; hot 8; ingenious; jewel 1; light[1] 13; liquid 3; lively 4; luminous 1; magnificent; meteoric 1; nimble 2; quick 4; radiant 1; scintillating 1; sight 6b; smart 1; splendid 1, 2, 3; star 3; strong 19; stunning 2; sunny 1; superb; supreme 4; talented; virtuoso 2; vivid 1; wise 1
brilliantly, gaily 1
brim°, brink 1; edge 1; flow 2; peak 3; periphery 1; rim; side 1; verge[1] 1
brim over, run 33c; spill 1; well[2] 2
brim (with), teem[1]
brimful, full 1; loaded 1; packed
brimful (with), replete 1
brimming, full 1; loaded 1; thick 2
brindle(d), speckled; spotty 1; dapple 1; mottled; variegated
bring°, cart 2; collect 1; deliver 1; lead 2; make 12; reflect 4; run 13; see 6; take 9; transfer 1; transport 1
bring aboard, sign 10b
bring about, achieve 1; arrange 2; brew 2; bring 5, 7, 13b; cause 6; effect 7; effectuate; excite 1; execute 1; found 1; fulfill 1; implement 2; induce 2; lead 6; make 3; motivate; negotiate 2; occasion 5; originate 1; perform 1; precipitate 1; procure 2; produce 2; prompt 5; raise 10; realize 1; rise 17; spark 2; spawn; work 16
bring along (behind), trail 4
bring around, bring 8; convince; revive 1; sway 2; talk 10; wake[1] 1; win 3
bring (back), fetch 1; get 4; recover 3; restore 1, 2, 4; resurrect; retrieve 1; return 3, 4; revive 3
bring (back) together, reconcile 1
bring back to life, refresh 1
bring down, bring 9; defeat 1; depress 2; finish 4; floor 4; humble 4; lower[1] 2; mortify 1; overthrow 1; overturn 2; overwhelm 1; place 13; prostrate 2; raze; reduce 5, 7; rout 2; shame 6; throw 3; topple 1, 2
bring forth, bear 6; breed 2; bring 10; deliver 6; develop 1; draw 4; elicit; fetch 2; hatch 1; produce 2; propagate 1; spawn; throw 9c
bring forward, lay[1] 6; lodge 6; moot 2; produce 4; put 5; raise 8; table 4; throw 9c
bring home, home 11
bring home (to), impress 3; stress 4
bring (in), bring 10b, 11, 13a, 15b; broach; cost 2; earn 2; fetch 3; gain 2; gross 6; import 1; inject 2; introduce 2, 4; mention 1; nab; net[2] 4; pick 2; present[2] 7; produce 4; quote 1;

realize 3; reap 2; touch 8; yield 4; run 27
bring into contact with, expose 3; touch 2
bring low, prostrate 2
bring off, accomplish; achieve 1; bring 12; carry 10c; execute 1; float 3; fulfill 1; manage 3; negotiate 2; perform 1; pull 12b; put 27a
bring on, bring 5, 7, 13; cause 6; induce 2; lead 6; occasion 5; precipitate 1
bring on *or* **upon (oneself),** incur
bring out, bring 10b, 14; circulate 2; develop 1; draw 4; elicit; introduce 4; mean[1] 2; open 23; produce 1, 4; rise 17; throw 9c; trot 2; turn 20a
bring over, turn 3
bring round, convince; get 14; persuade 2; prevail 3; revive 1
bring (someone) down a peg, downgrade 1; shame 6
bring (someone) to book, scold 1
bring to, bring 8a
bring (to bear), wreak
bring to bear (on), enforce 1; exercise 1; exert 1; zero 4
bring to completion, fulfill 1
bring together, assemble 1, 2; close 8b; concentrate 3; gather 1; group 4; integrate; pool 4; raise 5; rally 3; unify
bring up, bring 10b, 15; broach; dig 6; educate; excavate 2; foster 2; introduce 2; lead 10b; mention 1; moot 2; point 21b; present[2] 7; produce 2; quote 1; raise 3, 4, 8, 11; rake[1] 7b; rear[2] 1; refer 1; say 4; suggest 2; table 4; take 39e; throw 9c; touch 8; train 4; voice 4
bring up short, nonplus; puncture 4; stump 2
bring forth young, reproduce 2
bringing about, execution 2
bringing off *or* **about,** performance 1
bringing-up, breeding 1
bring trade to, patronize 2
brink°, border 5; brim 1; edge 1; eve 2; margin 1; point 9; side 1; threshold 2; verge[1] 1
brink of, on the, almost; ready 4
briny, salt 9; sea 1
briny (deep), the, deep 9; drink 7; ocean 1
brio, dash 6; life 7; panache; pep 1; verve; vigor
brisk°, active 1; agile 1; alive 3; breezy 1; energetic; excited 2; expeditious; fast[1] 1; fresh 5, 7; hasty 1; jaunty 1; lively 1; nimble 1; prompt 1; quick 2; rapid; rousing; smart 5; snappy 1; spanking 2, 3; speedy 1; sprightly; stiff 5; sudden; swift; vigorous; vital 3; vivacious
briskly, double 8; fast[1] 6; rapidly 1; swiftly; vigorously
briskness, haste 1; rapidity; speed 1; velocity; verve; vigor
bristle°, bridle 3; prickle 1; spine 2; thorn 1
bristle (with), swarm 3; teem[1]
bristles, pile[2]
bristling, thick 2
bristly, coarse 1; hairy 1; harsh 1; prickly 1
Bristol fashion, shipshape; taut 2
Bristols, bosom 1
brittle°, crisp 1; fragile; stiff 1
broach°, bring 15b; introduce 2; lead 10b; mention 1; moot 2; pose 3; propose 1; raise 8; set 18a; spring 4; tap[2] 6; venture 2
broached, be, come 19a
broad°, ample 1, 2, 6; bawdy; beamy; catholic; comprehensive; deep 1; encyclopedic; expansive 3; extensive 1; full 2, 5, 10; general 3, 4; girl 1; global; heavy 8; inclusive 1; jade 2; large 1, 3; lax 2; liberal 3; loose 5; meager 3; obscene 1; off-color 2; risqué; spacious; squeeze 10; sweeping 1; thick 1; wide 1, 2; woman 3
broadcast°, air 7; announce 1; blab; carry 8; celebrate 4; circulate 2; circulation 2; declare 2; deliver 4; diffuse 3; disperse 1; enunciate 2; give 15c; issue 10; proclaim 1; program 2; pronounce 3; propagate 3; publish; put 9, 23e; radio 2; report 4; scatter 1; send 2; sow; spread 1, 2; tell[1] 2; transmission 2; vent 3
broadcaster, journalist; reporter
broadcasting, circulation 2; issue 6; press 10a; proclamation 1; transmission 2
broad daylight, day 1; daylight 2
broaden, amplify 1; civilize 2; develop 1; enlarge 1; expand 3; extend 2; flare 2; grow 1; increase 1; liberalize 1; maximize 1; progress 6; spread 6; stretch 2; widen
broadening, extension 1; flare 6; growth 1; progress 3; progression 2; spread 8; swell 4
broad in the beam, beamy; fat 1
broadly, abroad 2; amply 1; generally 2
broad-minded, enlightened; liberal 3; tolerant
broad-mindedness, tolerance 1
broadness, breadth 1; width
broadsheet, poster; publication 2
broad-shouldered, muscular
broadside, advertisement 1; fire 3; insert 2; poster; promotion 5; publication 2; sign 5
Broadway, stage 3
Brobdingnagian, big 1; colossal 1; enormous; gigantic; great 1; huge; immense; large 3; massive; prodigious 1; vast; whopping 1
brochure°, advertisement 1; insert 2; leaflet; pamphlet; promotion 5; publication 2; tract[2]
brochures, literature 2
brogue, dialect
broil°, fight 1, 7; uproar
broiling, scorching 1; sweltering; torrid 1
broke°, destitute 1; down-and-out 1; embarrassed 2; hard 17; heel[1] 3; impoverished 1; indigent; insolvent; penurious 2; poor 1; upper 7
broke, dead *or* **flat** *or* **stony,** broke; needy
broke, go, fail 4; fold 3; wall 4
broke, go for, wind[1] 12
broken°, blink 6; defective 1; dud 2; duff; faulty; gentle 2; hurt 8; intermittent; meek 2; order 13c; out 14; piece 7; rough 1; rugged 1; spasmodic 2; split 9; stricken 1; tame 1; wanting 1
broken down, broken 7; order 13c; rock[1] 3b
broken up, mouth 6
broken-down, decrepit 2; dilapidated; leg 7; old 3; ragged 4; ramshackle; rickety; run-down 2; seedy 1; shabby 3; timeworn; tumble-down; unsound 1
brokenhearted°, desolate 3; heartbroken; low[1] 8; miserable 1; sad 1; sorrowful 1; stricken 1; wretched 2; inconsolable; woebegone
broker°, dealer; intermediary; merchant 3; negotiator; trader
brolly, umbrella 1
bromide, cliché; one 5; proverb; tranquilizer; truism
bromidic, tired 3
Bronx cheer, hiss 2; taunt 2
bronze, sculpture 1; statue; sun 2
brooch°, clasp 1; pin 2; plaque 2
brood°, family 2; fret 1; hatch 1; incubate; litter 2; offspring; pout 1; sulk; young 4
brood over *or* **on** *or* **upon,** brood 3; contemplate 2; muse; ponder; see 8; weigh 2
brooding, meditative; moody 1; pensive; preoccupied 1; sullen; thought 1; thoughtful 2
broody, moody 1
brook°, abide 1; accept 4; admit 2; allow 4; bear 4; creek 2; digest 2; feel 5; go 35a; lump[2]; permit 1; put 30; river 1; run 42; stand 3; stomach 3; stream 1; subscribe 1; support 3; take 6; tolerate 1; tributary
brooking, allowance 1
brooklet, stream 1
broom, brush[2] 1
broth°, liquor 2
brothel°, house 6
brother°, clergyman 2; equal 4; monk
brotherhood°, clan 2; club 2; fellowship 1, 2; fraternity 1; friendship 1; love 4; order 9; ring[1] 3; society 1, 5
brotherliness, brotherhood 1; fraternity 2
brotherly°, amicable; fraternal; friendly 1
brought down, prostrate 4
brought low, prostrate 4
brought to (someone's) attention, be, occur 2
brought up, be, arise 3; come 19a
brouhaha, disorder 2; disturbance 2; excitement 2; fanfare 2; fracas 1; furor 1; fuss 1; pell-mell 3; racket 1; row[2] 2; rumpus; scene 3; scrimmage; splash 4; tempest 2; tumult; uproar
browbeat°, bully 2; enforce 2; face 17; hound; intimidate; overawe; pick 6; push 4; railroad 2; threaten 1; tyrannize
brown, dark 8; mousy 1; sun 2
browned off, discontented; disgruntled; weary 3
brownie, goblin; imp
brownish, mousy 1
brownish-gray, mousy 1
brown-nose, cringe 2; cultivate 4; flatter 1; kowtow; play 18; romance 7; truckle
brown-nose(r), flatterer; flunky 2; menial 4; yes man
brown-nosing, flattery; groveling; ingratiating; menial 2; obsequious; servile; servility; submissive 2; timeserving
brown study, fog 2; reverie; thought 1; trance
brown study, be in a, meditate 1; muse

burst° *(cont.)* flare 4; flash 1; flurry 1; gale 2; gush 1, 3; pop 1; round 14; shatter 1; spasm 2; split 5; splurge 1; spurt 1, 3; volley 2
  burst (forth *or* **out),** break 13; erupt 1; flare 3; flash 4; fly 8b; spring 2
  burst in, break 18a
  burst into, fly 8b
  burst out (with), blurt; exclaim
  burst with, swarm 3
bursting, full 1; packed; profuse 2; replete 1; rupture 1; thick 2
  bursting forth, eruption 1
burton, go for a, die 1
burton, gone for a, dead 1
bury°, cache 3; conceal 1; cover 2; dispatch 3; eliminate 4; immerse 2; overwhelm 2; secrete[1]; steep[2] 2; submerge 3
bury the hatchet, make 39d; patch 7
bus, coach 1
bush, plant 1; ring[1] 1
  bush, beat about *or* **around the,** equivocate; fence 4; pussyfoot 2; stall[2] 1; waffle 2
  bush, beating about *or* **around the,** stall[2] 2
  bush, the, stick[2] 3
bushed, dead 9; fatigued; jaded 1; prostrate 5; tired 1; washed-out 2
bushing, ring[1] 1
bush league, amateur 2; minor 2
bush leaguer, amateur 1
bushranger, thief 1
bushwa, flannel 1; nonsense 1
bushwhack, ambush 2
bushy, thick 4
business°, affair 1, 2, 3; art 3; bag 5; calling; career 1; clientele; commerce; company 4; concern 4, 7; custom 3; dealings; employer 2; employment 1; enterprise 3; establishment 2; finance 1; firm 6; function 1; game 4; house 4; industry 1; interest 5; job 3; line[1] 7; matter 2; mercantile; mission 1; occupation 1; office 1; operation 3; outfit 3; palaver 1; part 3, patronage 2; practice 3; proceeding 2a; profession 1; province 3; public 9; pursuit 3; racket 3; subject 1; task 1; thing 8a; trade 1, 2; transaction 1; truck 2; turnover; undertaking 1; vocation; work 2
  business, do, deal 2
  business, give (someone) the, rebuke 1
  business, go out of, close 5; fail 4; fold 3
  business, in, swing 6
  business, send (someone) (on) about his .., pack 7; send 8
  business, transact *or* **do,** trade 5
  business with, do *or* **transact,** patronize 2
business card, card 2
businesslike, hard 6; mechanical 3; methodical; no-nonsense; perfunctory 1; practical 2; professional 2; realistic 1; systematic
businessman, dealer; merchant 3; trader
businesswoman, dealer
buss, kiss 4
bus station, station 3
bust, arrest 2, 4; blow[1] 8b; bosom 1; break 1, 14; breast 1; burst; degrade 1; depression 3; downgrade 1; figure 4; fête 1; insolvent; orgy 1; party 1; pick 8h; pinch 4; pull 11b; raid 1, 2; reduce 6; representation 4; run 27; sculpture 1; seize 2; split 5; statue
  bust, go, fail 4; fold 3; wall 4
bust a gut, exert 2
bust(ed), dud 2; hard 17; order 13c
bustle, activity 1; flurry 1; fuss 1; haste 2; hum 2; hurry 3; industry 3; press 9; ruffle 2; rush 1, 3; stir 6; trot 1; twitter 4
bustling, active 1; alive 4; busy 2; hectic; lively 3
bust-up, party 1
busty, bosomy; buxom 2; full 7; plump[1] 1; voluptuous 2
busy°, active 1; brisk 1; elaborate 2; engage 2; engaged 2; eventful; hectic; live 2; lively 3; move 13b; occupy 3; ornate; overwrought 2; tie 7a
  busy, not, free 3
  busy oneself, mess 4a
busybody°, gossip 3; intruder 2; snoop 2
but, bar 10; barring; except 1, 2; however 1; nevertheless; only 4; same 3; still 8; though 2; time 13a; yet 6
  but for, apart 4; except 1
butch, deviant 2, 3; homosexual 1
butcher°, execute 3; exterminate; hack[1] 1; hash 3; kill 1; killer 1; mangle; massacre 2; mow 2; murder 3, 4; murderer; mutilate 1; slaughter 3
butchering, massacre 1; slaughter 1
butchery, bloodshed; carnage; gore[1]; holocaust 2; killing 1; massacre 1; murder 2; slaughter 1, 2
but (for the fact) that, except 2
butler, servant 1
butt°, bottom 1; buttocks; fag 5; fool 3; join 4; joke 2; jolt 2; keg; monkey 2; object 2; poke 1, 5; posterior 3; ram 2; seat 4; stub 1; stump 1; sucker; target; thrust 1; victim 2
  butt in, butt[2] 2; chime 5b; cut 14; interfere 1; interrupt 1; intervene 1; intrude; meddle; mess 6; poke 3; short 12b; snoop 1
butte, hill 1; mound 1
butt (end), heel[1] 1
butter, spread 12
  butter (up), cajole; cultivate 4; flatter 1; kowtow; make 34b; play 18; romance 7; truckle; wheedle
butterfingered, awkward 1; clumsy 1; thumb 2; ungraceful 1
butterflies (in the stomach), dread 2
butterfly, bug 1
buttering-up, cajolery
buttery, greasy 1; ingratiating; oily 1
buttinsky, busybody; intruder 2; snoop 2
buttocks°, bottom 1; bum 1; posterior 3; rear[1] 2; seat 4; tail 2
button, control 6
  button, on the, dot 2; sharp 10
  button (up), adjust 4
Button your lip!, hush 1
buttress°, backbone 2; bolster; brace 1; confirm 3; follow 11b; fortify 1; inspire 2; mainstay; pier 2; prop 1, 3; reinforce; reinforcement 1; stay[2] 2; strengthen 1; support 1, 2, 8; sustain 2
buttressing, reinforcement 2
butyraceous, greasy 1; oily 1
buxom°, comely; full 7; lusty 1; plump[1] 1; voluptuous 2
buy°, acquire; bribe 2; finger 5b; get 1; lap[1] 3b; obtain 1; order 15; pick 8e; procure 1; purchase 1, 4; shop 4; steal 4; swallow 2; trade 5; use 4
  buy back, redeem 1
  buy from, patronize 2
  buy (off), buy 3; corrupt 5; fix 11; pay 10b; bribe 2
  buy (something) for, treat 5
buyable, venal
buy and sell, deal 2; market 4
buyer°, customer 1; patron 2; trader; user 1
buyers, public 9
buying, purchase 3; sale 2; use 11
buying and selling, trade 1
buy the farm°, farm 2
buzz°, call 3; hum 1, 4; murmur 1; ring[2] 4; telephone 2; thrill 1
  buzz, a, ring[2] 2
  buzz, give (someone) a, telephone 2
  buzz for, buzz 7
  buzz off, depart 1
buzzing, alive 4; buzz 1; hum 4; murmur 1
buzzword, shibboleth
by°, beside 1; part 10; past 3; through 6
  by oneself, alone 1, 3; apart 1; personally 1; single-handed 2
by and by, presently; shortly 1; sometime 2
by and large, chiefly; generally 1; largely; mainly; ordinarily; part 8; usually; whole 5
Bye!, goodbye
Bye-bye!, goodbye
By George!, indeed 3
bygone°, ancient 1; former 2; lost 4; musty 2; old 5; vintage 3
bygones be bygones, let, overlook 2
By Jove!, indeed 3
bylaw, law 1; regulation 2
bypass°, detour 1, 2; diversion 2; get 30b; jump 3; rule 8
by-product, derivative 2; offshoot 3; product 1
bystander°, eyewitness; onlooker; witness 1
by the by, passing 3; way 9
by the bye, incidentally 1; offhand 5
by the way°, incidentally 1; offhand 5; passing 3; way 9
byway, drive 9; road 2; street 1
byword°, maxim; motto; precept 2; saw; shibboleth
Byzantine, complicated; elaborate 2; fancy 1; fussy 2; intricate 1; involved 2; labyrinthine; perplexing; serpentine 2; showy

CA, bookkeeper
cab°, taxi 1
cabal°, combination 2; conspiracy; design 8; faction 1; junta; machine 3; party 3; plot[1] 1, 3; ring[1] 3; sect 2
cabala, occult 3
cabalistic, cryptic 1; mystical; occult 1; profound 1; recondite
cabaret°, club 4
cabbage, money 1
cabbala, occult 3
cabbalism, occult 3
cabbalistic, recondite
cabin°, cottage; hut; lodge 1; shack
cabinet°, board 4; committee; council 2
cabinet officer *or* **member,** minister 2
cable°, flex 1; lead 17; line[1] 11, 13; main 5; rope 1; stay[2] 1; string 1; telegram; transmit 1; twine 1
cablegram, cable 2; telegram

unhurried; uninhibited; unpremeditated
casually, easy 7; incidentally 2; offhand 5; random 2
casualness, ease 4
casualties°, casualty 2a
casualty°, accident 1; fatality 2; victim 1
casuistic, evasive; plausible 2; sophistic; specious; subtle 3
casuistry, falsity 1; subtlety 2
cat, guy 1; lash[1] 1; whip 8
cat on a hot tin roof, like a, edge 5
cat out of the bag, let the, give 12b; leak 5; spill 3; tell[1] 2
cat out of the bag, letting the, telling 2
catachresis, abuse 4; misuse 2
cataclysm, calamity 1; catastrophe 1; disaster; evil 7; fatality 1; ill 8; revolution 2; upheaval; visitation 2
cataclysmic, calamitous; disastrous 1; fatal 2; fateful 2; ill 4; ruinous; violent 2
catacomb, crypt; tomb
cataleptic state, trance
catalog, brochure; detail 5; enroll 2; enumerate 1; group 3; identify 1; include 2; index 1; journal 2; list[1] 1, 2; litany 2; organize 1; range 7; record 1, 3; register 1, 3; roll 13; sort 8; table 3; tabulate; write 4a
cataloging, distribution 3; identification 2; litany 2
catalogue raisonné, list[1] 1
catalyst, agent 2; factor 1; instrument 2
catalytic, instrumental
catamite, prostitute 1
catapult, fire 9; launch 3; put 7; sling 2; toss 1
cataract, fall 25; flood 2; outpouring; river 2; stream 2; waterfall
catastrophe°, accident 1; affliction 2; calamity 1; casualty 1; crisis 2; disaster; distress 2; evil 7; fatality 1; grief 2; ill 8; misery 3; misfortune 2; reverse 8; shame 3; tragedy; visitation 2
catastrophic, calamitous; disastrous 1; evil 3; fatal 2; fateful 2; ill 4; monumental 4; mournful 2; ruinous; tragic; unfortunate 3; violent 2
cat burglar, burglar; robber; thief 1
catcall, hiss 2; jeer 2
catch°, arrest 2; bag 6; bolt 2; brace 3; buckle 1; capture 2; clasp 1; claw 3; come 12; contract 3; detect 2; deterrent; drawback; engage 4; fastening; field 5; find 10; follow 8; foul 13; get 5, 11, 20; grab 1, 3; grapple 1; haul 4; hear 1; hitch 4; hook 1, 6; intercept; interest 7; jewel 2; joker 2; land 6; lock[1] 1; lodge 5; lure 1; mesh 3; nab; net[1] 2; obstacle; obstruction 1; occupy 3; pick 8k; pitfall 2; plum; provision 2; pull 5; rub 11; seize 2, 3; snag 1, 2; snap 4, 5, 10; snare 2; snatch 1; take 1; track 8; trap 4; treasure 2; witness 4
catch (at), pluck 4
catch hold of, lay[1] 11
catch (on), catch 13; grasp 2; learn 3; perceive 2; tumble 3; twig[2]
catch on (to), catch 10; figure 12b; realize 2; understand 1
catch out, trip 6
catch unawares, surprise 2
catch (unprepared), overtake 2
catch (up), catch 14; entangle 1
catch (up) in, involve 3
catch up (to *or* on *or* with), gain 4; overhaul 1; overtake 1
catches, string 7
catching°, capture 1; infectious
catch phrase, byword; phrase 2; saw; shibboleth
catchpole, officer 2
catch (something *or* a bug), sicken 1
catch-22, catch 18; dilemma; fix 17; hole 5; joker 2; perplexity 3; plight
catchword, byword; motto; saw; shibboleth; slogan
catchy, memorable; pretty 2; tuneful
catechism, examination 3
catechize, examine 2; preach 1
catechumen, convert 3; initiate 4; pupil
categorical°, absolute 2; dead 14; definitive 3; direct 9; dogmatic; downright 1; emphatic; explicit 1; express 5; flat 4; peremptory 2; perfect 6; point-blank 1; positive 1; unmitigated; unqualified 2
categorically, absolutely 1; completely 3; dead 18; definitely; expressly 1; flat 16; outright 4; point-blank 3; positively; really 2; utterly; wholly 2
categorization, identification 2; order 1; organization 1
categorize°, bunch 3; class 5; coordinate 1; distinguish 2; distribute 3; divide 5; file 3; grade 6; group 3; identify 1; include 2; label 5; order 16; organize 1; place 16; put 16f; range 7; rank[1] 6; sort 8, 9; stamp 3; tabulate
categorizing, identification 2; organization 1
category°, assortment 1; bracket 3; class 2; description 3; division 3; genre; grade 1; nature 4; order 3; run 44; sort 1; style 1; type 1; variety 3
cater°, provide 1; provision 5
cater (for), cater 2; entertain 2; lay[1] 17a
cater to, cater 2; entertain 2; feed 1; indulge 1; oblige 1; pamper; pander 1; supply 1; tend[2]
catering, provision 1
caterpillar, bug 1
caterwaul, rave 1; scream 1, 3; yell 1, 2
caterwauling, noise 1; puling
cathartic, purgative 1, 2
cathedra, chair 2
cathedral, temple
cathode ray tube screen, monitor 2
catholic°, broad 7; extensive 1; general 3; prevalent 1; public 1; sweeping 1; tolerant
catholicity, breadth 3
cathouse, brothel; house 6
catlike tread, with, silently
catnap, doze 1, 2; nap[1] 1, 3; repose 2; sleep 1
cat on a hot tin roof *or* on hot bricks, fidget 2
cat-o'-nine-tails, lash[1] 1; scourge 2; whip 8
cat's-paw, breeze 1; dupe 2; fool 3; inferior 4; pawn[2]; plaything 2; puppet 2; ripple 1; robot 2; scapegoat; sucker; tool 3
cattle°, stock 4
Caucasian, white 4
Caucasoid, white 4
caucus, cabal 2; council 1, 2; meeting 2; party 3
caught, foul 10
caught, become, lodge 5
caught on the hop, unprepared 1
cauldron, pot 1
caulk, fill 7
cause°, agent 2; bottom 3; breed 2; brew 2; bring 5, 7; call 15; case[1] 2; effect 7; effectuate; entail; excite 1; excuse 4; execute 1; explanation 3; factor 1; generate 3; get 14; give 4, 10; ground 3; have 9; implement 2; induce 2; lead 2, 6; make 2, 3, 5; motivate; motive 1; occasion 2, 5; part 5; procure 2; produce 2; prompt 3; provocation 1; put 2; raise 10; rise 17; root[1] 1; score 8; seed 2; side 3; spawn; subject 3; suit 4; touch 11b; warrant 1
cause, take up a, crusade 2
cause for, provide *or* offer, warrant 4
cause of, be the, bring 5
cause (to), let[1] 2
cause célèbre, issue 4
causerie, tête-à-tête 1
causeuse, lounge 4
caustic°, bitter 1; cutting 2; devastating 1; embittered; gruff 1; incisive 2; keen[1] 2; poignant 2; pungent 2; sarcastic; scorching 2; sharp 4; short-tempered; sour 4; tart[1] 2; trenchant; virulent 2
causticity, edge 3; gall[1] 1
causticness, gall[1] 1
caution°, advise 1; alert 5; calculation 3; care 2; distrust 2; foresight 1; intimate[2]; limit 4b; precaution 2; providence 1; prudence 1; threaten 1; tip[3] 4; vigilance; warn 1, 2; warning 1
caution to the winds, throw, wind[1] 12
cautionary, exemplary 3; ominous 2
cautious°, alert 1; anxious 1; careful 1; conscientious 2; conservative 2; deliberate 3; discreet; distrustful; gingerly 2; guarded; leery; measured 2; mindful; noncommittal; provident 1; prudent 1; reluctant; scrupulous 1; sharp-eyed; shy 2; tentative 2; thoughtful 3; vigilant; wakeful 2; wary
cautious, be, shy 4
cautious (of), watch 8
cautious (of *or* about), be, beware; distrust 1; mind 17; step 12
cautiously, gingerly 1; salt 5; slow 11
cautiousness, calculation 3; precaution 2; prudence 1; suspicion 1
cavalcade, procession 1, 2; review 8; train 3
cavalier, arrogant 1; gallant 4; genteel 2; offhand 1; overbearing
cave°, hole 1; lair 1
cave in, buckle 2; cave 2; collapse 1; fall 14; fizzle 2; sink 2; way 11a
caveat, caution 1; dictate 2; qualification 2; warning 1
cave-in, collapse 5
cavendish, plug 2
cavern, cave 1; hole 1; hollow 7
cavernous, voluminous 1
cavil°, carp; exception 4; fence 4; gripe 1; henpeck; niggle; prod 3; quibble 1, 3; shuffle 3
cavil at, knock 2
cavil (at *or* about), criticize 2; fault 6; pick 4a
caviling, faultfinding 1, 2; gripe 2; hairsplitting; legalistic; overcritical; peevish; sarcastic
cavity°, cave 1; chamber 3; depression 1; excavation; gap 1; hole 1; hollow 7; nook 1; pan 3; pit[1] 1; pocket 2

character° (*cont.*) sort 1; spirit 2; stamp 8; strain[1] 9; stripe 2; structure 1; style 5; temper 1; tenor; texture; timber 3; virtue 1; way 2; wit 3
  character, by, naturally 2
  character, in, character 8; typical 2
  character (for), out of, character 9; unbecoming 1; unnatural 1; unworthy 2
characteristic°, aspect 4; attribute 1; character 2, 8; difference 4; distinctive; exemplary 1; feature 1; flavor 2; hallmark 2; individual 2; mannerism; mark 2; note 5; peculiarity 2; point 14; proper 6; property 4; quality 1; quirk; representative 1; respect 4; seemly 1; sound[1] 2; symptom; touch 18; trait; trick 6; typical 1
  characteristic (of), indicative of; peculiar 2; redolent 2; specific 2; symbolic; symptomatic
characteristics, endowment 3; stamp 5
characterization, character 5; description 1; identification 1; label 2; portrait
characterize°, brand 3; define 2; describe 3; distinguish 2; label 5; mark 13; paint 4; portray 1; proclaim 2; represent 3; sort 9; stamp 3; style 7; typify
  characterize (as), represent 2; think 2
  characterize oneself as, feel 7
characterized by, peculiar 2
characterless, colorless 2; monolithic; nondescript; prosaic
characters, notation 2
charge°, accusation; accuse 2; administration 1; allegation; allege; assault 1, 3; assignment 2; attack 1; attribute 2; bill[1] 3; blame 1; business 1; care 3; come 8; command 1, 6, 7; concern 4; control 4; cost 1; custody 1; decree 2; denounce 1; detail 6; device 3; dictate 2; direct 3; direction 1; duty 1; entrust; errand 2; expenditure; expense 1; fare 2; fee 1; fine[2] 1, 2; forfeit 1; function 1; grievance 2; guidance 1; hire 5; hit 12; impeach 1; imputation; impute; incriminate; instruct 2; interest 6; job 2; lay[1] 7, 17b; lunge 3; make 30b; mission 1; obligation 1; office 4; onset 1; order 14; oversight 2; payment 2; place 4; precept 1; price 1; prosecute 1; protection 2; protégé; province 3; push 15; quotation 2; rap 6; rate[1] 2; responsibility 2; round 15; rush 2, 4; safekeeping; send 1; set 17b; speed 3; stampede 3; stick[1] 8; stint 1; sue 1; tab 2; task 1; tax 1, 3; tell[1] 5; thrill 1; toll[2] 1; tribute 2; trust 3; weight 5; work 3
  charge, take, boss 2; cover 8
  charge, without, free 13
  charge from *or* **out of, get a,** enjoy 1; like[2] 1; love 8
  charge (of), be in, boss 2; govern 1; head 11; manage 1; rule 5; run 10; supervise
  charge (of), have, supervise
  charge of, have *or* **take,** mind 18
  charge (of), take, head 11; care 4; keep 2; see 12a
  charge (up), vitalize
  charge (with), accuse 1; indict
chargeability, responsibility 1
chargeable, responsible 1
charged, electric; explosive 1; hot 9; live 5; loaded 2, 3; momentous; pregnant 2; volatile 3
  charged with, fraught 1
chargé d'affaires, minister 2
charger, mount[2] 8; plate 1
charge(s), check 19; dues
charily, gingerly 1
chariness, mistrust 2; precaution 2
charisma, glamour; magnetism; mystique
charismatic, magnetic
charitable°, benevolent; big 6; bountiful 1; chivalrous; considerate; free 7; generous 1, 2; good 6; good-natured; human 3; kind[1]; large 2; lenient; liberal 1; merciful; open 14; philanthropic; selfless; softhearted; thoughtful 1; tolerant; unselfish
charitableness, altruism; benevolence 1; bounty 1; humanity 3; kindness 1; philanthropy 1
charitably, readily 1
charity°, altruism; benevolence 1; bounty 1; gift 1; grace 3; kindness 1; largess; love 4; mercy 1; philanthropy 1
charivari, jamboree; noise 1
charlady, servant 1
charlatan, adventurer 2; cheat 1; fake 4; fraud 3; hypocrite; impostor; phony 3; quack 1; rogue 1; swindler; thief 2
charlatanism, hypocrisy
charlatanry, hypocrisy
charlie, sap[1] 2
charm°, amulet; appeal 4; attract; beguile 3; bewitch; captivate; catch 11; coax; dazzle 1; disarm 2; enchant 2; enchantment 1, 2; enrapture; entrance[2]; fascinate; fascination; fetish 1; glamour; hypnotize; intrigue 1; lure 1, 2; magic 3; magnetism; mumbo jumbo 2; possess 3; ravish 1; regale; seduce 1; send 4; spell[2] 1, 2; take 20; talisman; transport 3; wheedle; win 3
  charm, like a, charm 4
charmed°, delighted; infatuated; lucky 1; smitten 2; taken
charmer°, roué; siren 2
charming, adorable; attractive; beautiful 1; darling 4; debonair 2; delicious 2; delightful 2; enchanting; engaging; exciting 2; fetching; glamorous 1; idyllic; likable; lovable; magic 6; nice 1; picturesque 1; pleasant 1, 2; prepossessing; pretty 1; provocative 1; ravishing; spellbinding; suave; sweet 4; taking; tasteful; winning 1
charmingly, beautifully 1
charms°, charm 3
chart°, design 5; plan 2; plot[1] 4; profile 3; schematic 2; scheme 2; table 3; trace 6
charted, schematic 1
charter°, enable 1; hire 2, 4; law 2; license 1; patent 1; reserve 2; warrant 1
  charter (out), hire 3; lease 2; let[1] 3; rent[1] 1
chartered, privileged 2
chartered accountant, bookkeeper
charterhouse, monastery
charwoman, servant 1; woman 4
chary, backward 1; careful 1; discreet; distrustful; gingerly 2; leery; reluctant; shy 2; vigilant; wary
Charybdis, eddy 1
chase°, follow 4; heel[1] 5; hunt 1, 3; pursue 1; quest 1; run 21; scribe 3; stalk[1]; take 29b; track 6; trail 7
  chase (after), pursue 4
  chase away *or* **off,** chase 3; flight[2] 2; ward 3
chasing, back 5; pursuit 1
chasm, aperture; breach 3; crevasse; depth 5; gorge 1; gulf 2; pit[1] 3; split 7
  chasm, yawning, abyss
chasmal, deep 1
chasmic, deep 1
chassis, figure 1; frame 1; physique; shell 2
chaste°, celibate 2; clean 4; decent 5; flawless 1; good 5; holy 2; immaculate 2; innocent 2; maiden 2; platonic; pristine 2; pure 3; respectable 3; temperate 2; untarnished; virtuous 2; white 3
chasten°, castigate; chastise; correct 3; humble 4; mortify 1; punish 1; shame 6; vituperate
chastened, shamefaced 2; subdued 2
chasteness, purity 2; virtue 2
chastening, lesson 4; punishment 1; punitive; rod 2; shameful
chastise°, call 7b; castigate; chasten 1; correct 3; discipline 8; flog 1; mind 10; pay 5; punish 1; scourge 4; spank; upbraid; whip 1
chastisement, correction 2; discipline 2; lesson 4; punishment 1; rod 2; thrashing 2; tongue-lashing
chastising, punishment 1
chastity°, celibacy 2; honor 4; purity 2; virtue 2
chat°, chew 2; conversation; converse; dialogue 1; discussion; gossip 1; rag[1] 5a; rap 3, 5; rumor 1; talk 2, 4, 15; tête-à-tête 1; word 1
  chat about, discuss
  chat up, approach 3; chat 3; flirt 1
  chat (with), communicate 2
chateau, castle 2; palace; residence 3
chatelaine, servant 1
chattels, belongings; effects; estate 2; furniture 1; gear 4; good 21a; paraphernalia; possession 3; property 1; stuff 2
chatter°, babble 1; converse; drivel 2; gab 1, 2; gibberish; go 32f; gush 2; jabber 1; jargon 2; palaver 2, 4; patter[2] 2, 3; prattle 1, 2; ramble 3; rattle 6; shiver[1] 2; talk 3, 18; tattle 2; twitter 1, 2; yap 2
chatterbox, gossip 3
chatterer, blabbermouth
chatter(ing), babble 3; chatter 3; prattle 2
chatty, forthcoming 3; talkative; voluble
chauffeur, servant 1
chauvinism, jingoism; prejudice 2
chauvinist, patriot
chauvinist(ic), national 2; nationalistic; patriotic; prejudiced 2
CHE, prison; reform school
cheap°, base[2] 3, 5; common 4; economical 1, 2; empty 4; flashy 1; garish; gaudy; inexpensive; little 6; mean[2] 1; miserly; near 6; penurious 1; petty 2; plastic 3; pulp 4; shoddy; sleazy 1, 2; small 4; sparing 1; tacky[2]; tasteless 1; tawdry; thrifty; tinny 1; worthless 3
  cheap, be, stint 4
  cheap, on the, cheap 7
cheapen, degrade 2; depreciate 1; depress 2; diminish 2; disparage 1; prostitute 2

chide (*cont.*) scold 1; task 3; tell[1] 10; upbraid; vituperate
chiding, lecture 2; lesson 4; piece 10; rebuke 2
chief°, arch 1; bigwig 1; boss 1; capital 5; cardinal; central 2; director 1; dominant 2; employer 1; essential 2; first 1; foremost 1; grand 6; head 2, 9; high 11; lead 18; leader 1; leading 1; main 1; major 1; manager; master 1, 5; monarch 2; overriding; overseer; paramount; predominant; premier 2; prevailing 1; primary 1; prime 2; principal 1, 3; prominent 1; responsible 3; salient; senior; skipper; sovereign 1, 2; staple 1; stellar 2; superintendent; supervisor; supreme 1
chief (executive officer), executive 1; head 2; premier 1; principal 3
chief honcho, boss 1; director 1
chiefly°, above 6; especially 2; large 5b; largely; mainly; part 8; primarily 1; principally; rule 4; usually; whole 5
chief support, mainstay
chieftain, chief 1; leader 1; master 1
chifferobe, bureau 1; cabinet 1
chiffon, thin 5
chiffonier, bureau 1; cabinet 1
child°, baby 1; bugger 2; descendant; human 4; innocent 5; issue 7; minor 3; offspring; person 1; tot; youth 2
child, get with, knock 7c
child, unmarried *or* **unwed and with,** trouble 9b
child, with, expecting; pregnant 1
childbirth, birth 1; delivery 3; labor 4
childhood°, youth 1
childhood, early, infancy 1
childhood, in one's second, senile
childish°, absurd 1; boyish 2; frivolous 2; immature 2; infantile; juvenile 1; mad 2; puerile; silly 1; young 3
childless, barren 1; sterile 1
childlike°, childish; immature 2; ingenuous 1; innocent 4; naive; primitive 3; simple 3; unsophisticated 1; young 2
child minder, minder 1
children, brood 1; family 2; issue 7; posterity; progeny; seed 3; young 4; youth 3
child's play, painless; picnic 2; pushover 1
chill°, brisk 3; chilly 1, 2; cold 1; cool 1, 7, 9; cutting 1; dampen 2; freeze 1; freezing; frigid 1; icy 1, 2; lukewarm 2; nip[1] 4; raw 5; refrigerate
chilled, cold 2; cool 1
chilled (through), chill 5
chilled to the bone, freezing
chilliness, cool 7
chilling, chill 4; cold 3, 4; cool 1; harrowing; icy 1; piercing 3; raw 5; wintry 1, 2
chills, the, shiver[1] 4
chilly°, aloof 3; bleak 2; brisk 3; chill 4, 6; cold 1, 2, 3; cool 1; frigid 1; icy 1, 2; lukewarm 2; raw 5; stiff 7; stony 2; unapproachable 1; unsocial; wintry 1, 2
chime°, jingle 1, 2; peal 1, 2; ring[2] 1, 3; toll[1] 1, 2
chime in, chime 5; chip 4b; interrupt 1
chimera, fantasy 2; hallucination; illusion 2; phantom 2; specter 1; vision 4
chimeric(al), dreamlike; fanciful 2; groundless; ideal 5; imaginary; insubstantial 2; mythical 2; nonexistent; quixotic; romantic 2; shadowy 3; unreal 1; whimsical 1
chiming, chime 2; jingle 2; peal 1; ring[2] 3; toll[1] 2
chimney, stack 4
chimney stack, stack 4
chin, talk 4
china, friend 1; mate 1; pottery
china (plate), intimate[1] 5
Chinese fire-drill, zoo 2
chink°, aperture; crack 1; cranny; crevice; flaw 2; gap 1; jingle 1, 2; leak 2; opening 1; split 7
chinking, jingle 2
chinless wonder, milksop
chintzy, cheap 3; gaudy; miserly; penurious 1; plastic 3; shoddy; sleazy 1; tacky[2]; worthless 3
chinwag, chat 1; conversation; discussion; gossip 1; palaver 3, 4; talk 15
chip°, carve 1; check 18; counter 1; flake 1, 2; flaw 2; fragment 1; mark 10; nick 1, 2; shred 1; sliver; splinter 1
chip in (to *or* **for),** chime 5b; chip 4; pitch[1] 5; stump 4; subscribe 2
chip off the old block, be a, take 29a
chipped, jagged
chipper, gay 2; lively 1; sprightly; vivacious
chippy, prostitute 1; tart[2] 2
chips, cash in one's, check 10d; die 1
chips, in the, flush[2] 3; loaded 4; opulent 1; rich 1
chips and plaice, feature 3
chirography, penmanship; writing 1
chirp°, peep 1, 2; sing 1; twitter 1, 3
chirping, twitter 3
chirpy, lively 1
chirr, chirp 1, 2; peep 2
chirrup, chirp 1, 2; peep 1, 2; twitter 1, 3
chirruping, twitter 3
chisel°, carve 1; chip 3; engrave 1; fleece; gouge 1; sculpture 2; swindle 1
chit, check 17; memorandum; slip[2] 1
chitchat, chat 1, 2; conversation; gab 2; gossip 1; jargon 2; palaver 4; patter[2] 2, 3; rumor 1; word 1
chitter, chirp 1, 2
chivalrous°, fearless; gallant 2; genteel 2; heroic 1; manly; noble 4; well-bred
chivy, harass; nag[1] 1; pester; plague 3; pursue 1; pursuit 1; torment 2
chlorine, bleach 2
chock, wedge 1
chockablock, full 1; loaded 1; packed; replete 1; solid 2; thick 2
chock-full (of), full 1; loaded 1; packed; replete 1; solid 2; thick 2
chocolate, sweet 9
choice°, alternate 5; alternative 2; appointment 2; capital 6; dainty 3; delicious 1, 2; desirable 3; discretion 2; disposition 4; elect 3; election; elegant 1; elite 2; excellent; exclusive 2; exquisite 1, 5; fancy 3; favored 1; favorite 1, 2; fine[1] 1; good 2; option 1, 2; outstanding 1; par 3; pick 9; pleasure 2; precious 2; preference 1; prime 2; prize[1] 5; range 2; rare[1] 1, 2; refusal 2; select 2; selection 1, 3; settlement 5; suffrage; superior 2; supreme 4; variation 2; variety 1; vintage 2; volition; will 2
choice, by, voluntarily; willingly
choice item, tidbit
choice of words, phrase 3; wording
choicest, best 3; flower 2; optimum 2; pick 10; superlative; top 8
choir, ensemble 2
choir boy *or* **girl** *or* **member,** singer; vocalist
choke°, flood 5; foul 15; gag[1] 2; glut 4; satiate 1; smother 1; stifle 1; stuff 9
choke (back) *or* **(down),** choke 4; gulp 2; smother 4; stifle 2; swallow 4
choke off, choke 3
choke (up), clog; stop 3
choked, congested; foul 9; overcrowded; thick 2
choked, be, smother 2
choked, become, silt 2
choke-full, full 1; thick 2
choker, string 4
chokey, jail 1; prison
choking, thick 3
choky, jail 1; prison
choler, anger 1; fury 1; indignation
choleric, bilious; cantankerous; cranky 2; cross 6; embittered; fretful; harsh 3; hasty 4; passionate 3; petulant; pugnacious; quarrelsome; querulous; quick-tempered; snappish 2; sullen; surly; testy; touchy 1
chomp, munch
choo-choo, train 1
choose°, affect[2] 2; appoint 2; assign 3; cast 8; decide 2; designate 2; determine 3; draw 9; elect 1; extract 4; favor 6; fix 15; go 28c; make 23; mind 12; name 5; nominate; pick 1; pitch[1] 7; please 2; plump[2] 3; prefer 1; rather 3; select 1; set 9; settle 2; sift 1; single 4; sort 10b; take 2; vote 4; will 5; wish 1
choosing, choice 1; election; selection 3; volition
choosy°, finicky 1; fussy 1; particular 4; selective
chop°, hack[1] 1, 3
chop, give (someone) the, sack 4
chop, the, sack 3
chop apart, split 1
chop asunder, cleave
chop away *or* **down** *or* **off,** chop 1; cut 4, 13a, 15a
chop in two, cleave; split 1
chop off, lop; sever 1
chop (up), chop 2; cut 17a; mangle
chop-chop, posthaste
chop logic, quibble 1; pedantic 2
choppy, rough 2
chops, mouth 1
chore, assignment 2; drudgery; duty 1; fag 2; function 1; grind 7; job 2, 3; labor 3; office 4; place 4; post[2] 1; stint 1; task 1; thing 6; work 3
choreograph, arrange 2; manipulate 1
chorister, singer; vocalist
chortle, chuckle 1, 2; giggle 1, 2; laugh 1, 3; titter 1, 2
chortling, laughter
chorus, ensemble 2; refrain[2]; sing 1
chorus boy *or* **girl** *or* **member,** singer; vocalist
chosen, elect 2; favored 1; favorite 2; representative 2; select 2; speak 13
chouse, fool 4; swindle 1
chow, food; meat 1; refreshment 1; sustenance 1
chowhound, glutton
chrestomathy, collection 2
Christ, lord 3

civic, municipal; political 1; public 2; worldly 1
civics, politics 1
civil°, amicable; attentive 2; ceremonious 2; courteous; debonair 1; fair[1] 8; genteel 2; interior 2; municipal; national 1; neighborly; obliging; orderly 2; peaceable 2; polite 1; political 1; public 2; respectful; secular; temporal 1
civil affairs, politics 1
civil disorder, disorder 2
civilian, civil 1
civilities, the, propriety 3
civility°, breeding 2; courtesy; deference 1; etiquette; manner 3; refinement 1; respect 2
civilization°, culture 2; society 3
civilize°, educate; refine 2
civilized, civil 3; courteous; cultivated; educated 2; enlightened; orderly 2; polished 2; refined 1; steady 6; suave
civil servant°, politician; servant 2
civil-service employee *or* **worker,** servant 2
clack, prattle 2
clad, face 16
cladding, facing
claim°, allegation; application 4; argue 5; argument 2; call 8a; charge 13; declare 1; demand 2; exact 3; maintain 3; prerogative; pretension 1; profess 1; right 9; say 1; state 7; testimony
claim, lay, profess 2
claim to, give up, vacate 2
claim to, have a, earn 1; merit 3; rate[1] 7
claimant, pretender
claimed, alleged
claim to fame, specialty 1
clairvoyant, fortuneteller; prophet; psychic 3; seer
clambake, picnic 1
clamber (up), climb 1, 4; mount[2] 1; scale[3] 3; scramble 1; shin
clamber up on, mount[2] 3
clamminess, damp 2
clammy°, damp 1; humid; moist 1; muggy; slimy 1; sticky 3; sweltering
clamor, blare 2; din 1; disorder 2; jangle 3; noise 1; outcry; peal 1; racket 1; roar 3; row[2] 2; uproar
clamor over, mob 3
clamoring, outcry
clamorous, blatant 2; boisterous; chaotic 2; demanding 2; loud 1; noisy; obstreperous; tumultuous; uproarious 1; urgent 2
clamp°, brace 3; clasp 3; fasten 1; fix 1; stack 2
clamp down, enforce 1
Clam up!, hush 1
clan°, fellowship 2; folk; fraternity 3; house 2; kin 1; lineage 2; nationality 2; native 4; people 4; race[2] 1; ring[1] 3; tribe
clandestine, furtive 1; illicit 2; inside 4; mystical; oblique 2; private 1; secret 1; stealthy; surreptitious; undercover; underground 2
clandestinely, back 6; incognito 2; private 6; scene 5; secrecy 2; secret 4; secretly; sly 3
clandestineness, privacy 2; secrecy 1; stealth
clang, chime 3; clash 1; jangle 1, 3; peal 1, 2; ring[2] 1, 3
clanger, error 1; flop 4; howler; mistake 1
clanging, chime 2; jangle 3; ring[2] 3
clangor, clash 1; din 1; jangle 3; noise 1; peal 1
clangorous, noisy
clank, clash 1; jangle 1, 3
clanking, jangle 3
clannish, exclusive 2; rarefied 3; sectarian 1
clannishness, fraternity 2
clap°, applaud 1; cheer 7; crack 2; explosion 1; hand 6; noise 1; peal 1
Clapham omnibus, the man *or* **woman on the,** mass 6
clapper°, rattle 8
clappers (of hell), like the, fast[1] 6
clapping, applause; ovation
claptrap, gibberish; hot air; jargon 2; moonshine 2; mouth 3; mumbo jumbo 1; nonsense 1; prattle 3; raving 3; rot 4; stuff 4; talk 18; wind[1] 2
claque, combination 2; crowd 2
clarification, comment 1; definition 2; demonstration 2; explanation 1; exposition 2; illumination 2; interpretation 1; key 4; light[1] 5; refinement 3; solution 1
clarified, refined 4
clarify°, clear 20, 30a, 30c; define 2; explain 1; filter 2; illuminate 2; interpret 1; light[1] 12; purify 1; refine 1; render 8; settle 12; show 1; simplify; solve; sort 10a; spell[3] 2
clarifying, definitive 3; refinement 3
clarion, clear 11; noisy
clarity°, daylight 2; definition 1; purity 1; simplicity 1, 2
clash°, action 6; battle 1; combat 5; compete; conflict 3, 4; disagreement 3; encounter 5; faction 2; fall 18; feud 2; fight 1, 7; jangle 1, 3; jar[2] 2; noise 1; quarrel 2; scramble 4; skirmish 1, 2; struggle 5; war 1
clash with, encounter 3; engage 5
clashing, harsh 1; incompatible; mixed 2; odds 4
clasp°, brace 3; brooch; buckle 1; catch 17; clamp 1, 2; clinch 2; clip[1] 1, 2; embrace 1; fastening; fix 1; fold 2; grapple 1; grasp 1, 3; grip 1, 6; hold 1, 2, 24; hook 1; hug 1, 3; lock[1] 1, 4; press 5; snap 10; snatch 1; squeeze 5, 7; take 1
clasp one another, cling 3
class°, assortment 1; bracket 3; bunch 3; caste; categorize; category; circle 2; course 5; degree 1; denomination 2; description 3; distribute 3; division 3; estate 3; family 4; genre; grade 1, 3, 6; group 3; kind[2] 1; label 5; lesson 2; nature 4; nobility 2; order 3; place 16; position 5; range 7; rank[1] 1, 6; rate[1] 4, 6; reckon 2; run 44; sort 1, 8; sphere 2; stamp 8; station 2; stratum 2; stripe 2; tribe; type 1; variety 3
class as *or* **with, be in the same,** touch 7
class by itself, in a, rare[1] 2
class with, be in a, compare 2
classes, instruction 2
classic°, immortal 3; monumental 1; regular 4; simple 2; standard 6; superb; typical 2; vintage 2
classical°, heroic 5
classics, the, letter 3
classier, superior 1
classification, assortment 1; bracket 3; breakdown 3; category; class 2; denomination 2; distribution 3; division 3; family 4; identification 2; inclusion; label 2; order 1; organization 1; rate[1] 4; sort 1; stamp 8; type 1; variety 3
classified, advertisement 1; confidential
classify, bracket 4; bunch 3; categorize; certify 2; class 5; coordinate 1; distinguish 2; distribute 3; divide 5; file 3; grade 6; group 3; identify 1; include 2; label 5; order 16; organize 1; place 16; range 7; rank[1] 6; rate[1] 6; separate 2; settle 1; sort 8; stamp 3; tabulate
classifying, denomination 3; identification 2; organization 1
classism, intolerance
classist, intolerant 2
classless, democratic
classmate, pal 1
classroom, hall 2
classy, dapper; dressy 2; exclusive 2; flash 6; genteel 2; high-class 2; palatial; plush; posh; sharp 9; sporty; stylish
clatter, chatter 2; din 1; jangle 1, 3; rattle 1, 7
claudication, hobble 3; limp[1] 2
clause, amendment 2; phrase 1; provision 2; stipulation; term 4
claw°, lacerate; nail 2; scratch 1; tear 1
claw (at *or* **away** *or* **out),** scrape 2
clay, dirt 2; earth 2; ground 1; mud; soil[2]
clayey, plastic 1
clay pipe, pipe 2
clean°, brush[2] 4; chaste 1, 2; clarify 2; clean 11a; cleanse 1; clear 19, 20; decent 2; disinfect; empty 6; filter 2; flawless 1; fresh 7; go 34c; gut 4; holy 2; hygienic; immaculate 1; launder 1; neat 1; order 10a; polish 1; preen 1; pure 2; purify 1; refined 4; respectable 3; rinse 1; rub 1; sanitary; scour 1; simple 2; spotless 1; square 4; sterilize 1; sweep 1; tidy 1
clean breast (of), make a, clean 8; confess
clean (off *or* **away),** scrape 2
clean (off *or* **out** *or* **up),** wipe 1
clean (out), clean 10; exhaust 3; flush[1] 2; gut 5; purge 1
clean (up), clean 11; edit 2; neaten; pick 8c; profit 5; refurbish; spruce 2; straighten 3; tidy 4; wash 1
clean-cut, clean 3; trim 2; well-groomed
cleaner, detergent 1; disinfectant; porter[1]; servant 1
cleaners, take to the, fleece
cleaning, detergent 2; refinement 3; rinse 3; wash 11
cleaning man *or* **woman,** servant 1
cleaning woman *or* **lady,** woman 4
clean-limbed, willowy 2
cleanliness, purity 1
cleanly, clean 2; clear 19; freely 5
cleanness, purity 1; simplicity 2
cleanse°, clean 9, 11a; clear 20; disinfect; flush[1] 2; fumigate; launder 1; pound[1] 5; purge 1; purify 1; refine 1; rinse 1; sanctify 2; scour 1; sterilize 1; wash 1; wipe 1
cleansed, clean 2; refined 4
cleanser, detergent 1; disinfectant
cleanshaven, smooth 4

cleansing, detergent 2; refinement 3; rinse 3; wash 11
clear°, apparent 1; bold 3; bright 2; broad 3; clarify 2; clean 11b; coherent 2; conspicuous 1; decided 1; definite 3; defray; direct 9; discernible 1; discharge 6; distinct 1; earn 2; ease 8; easy 1; elementary 1; empty 6, 8; evident; excavate 2; excuse 1; explicit 1; express 5; fair[1] 3, 5, 6; fine[1] 2; focus 2; forgive 2; free 10; fresh 7; gain 2; get 3; glaring 1; graphic 1; hard 12; honor 8; hook 5; innocent 1; intelligible; leap 1; legible; let[1] 6b; liquid 3; luminous 3; make 7; manifest 1; navigable 1; negotiate 3; net[2] 2, 4; noticeable 1; observable; obvious; open 9, 13, 22; outright 1; overt; patent 2; pay 1; perceptible; pictorial 1; plain 1, 2; pleasant 3; positive 1; pound[1] 5; profit 5; pronounced 1, 2; public 4; pure 2; purge 1, 3; purify 2; quit 4; rake[1] 5; realize 3; receive 2; redeem 1; refine 1; self-evident; settle 10, 12; simple 1; smooth 10; stark 4; strong 15; sunny 1; sweep 1; take 22; transparent 1, 2; undisguised; undisputed; unquestionable; vindicate 1; visible 1; vivid 1, 2; void 2, 7
  clear, be, appear 4
  clear, become, penetrate 4
  clear, in the, clear 31
  clear, make, clarify 1; clear 30c; get 26; home 11; indicate 3; put 10; show 1; solve; spell[3] 2
  clear away *or* **out,** clear 25
  clear of, keep *or* **steer,** avoid
  clear off *or* **out,** bugger 5; clean 10b; clear 29; escape 1; evacuate 1; flee 1; fly 2; make 26; purge 2; run 2
  clear out, exhaust 3
  clear (the way), lead 10a
  clear (up), clarify 1; clear 22, 28, 30; explain 1; fix 16c; illuminate 2; interpret 1; resolve 2; settle 3; simplify; solve; sort 10a; square 13; straighten 1; work 19c
clearance°, discharge 13; excuse 5; pass 24; settlement 4; sweep 5; tolerance 2; way 7
clearcut, cut 29a; express 6; clear 7; concrete; conspicuous 1; definite 3; distinct 1; evident; noticeable 1; obvious; overt; positive 1; pronounced 2; public 4; self-evident; specific 1; strong 15
cleared, hook 5
cleareyed, luminous 3; rational 1
clearheaded, luminous 3; rational 1; sober 2; sound[2] 4; wise 1
clearheadedness, judgment 1; wisdom 1
clearing, excuse 5; field 1; settlement 4
  clearing out *or* **away,** purge 4
clearly°, apparently 1; clear 18; completely 3; definitely; easily 2; entirely 2; evidently 1; expressly 1; far 3; manifestly; naturally 1; notably 1; obviously; ostensibly; practically 2; stark 1; undoubtedly; well[1] 7
clearness, clarity 1
clearsighted, wise 1
clearsightedness, wisdom 1
clear the way, make 41
cleavage, fracture 2; rupture 1; wedge 2
cleave°, attach 5; chop 1; cut 15a; divide 1; fracture 3; gash 2; rend 2; rupture 3; section 4; segment 2; sever 1; snap 1; split 1
  cleave (from), disengage
  cleave to, follow 2; press 5
  cleave together, embrace 1; stick[1] 5
cleave to one another, cling 3
cleaving, parting 1
cleft, aperture; chink; crack 1; cranny; crevice; flaw 2; gap 1; gash 1; notch 1; opening 1; ravine; rift 1; rip 3; slit 2; split 7, 9
cleft stick, quandary
clemency, forgiveness 2; grace 3; quarter 4
clement, forgiving; merciful; mild 2
clench, hold 2; nail 7; squeeze 5; verify
Cleopatra, charmer
clergy, ministry 2
  clergy, the, cloth 2
clergyman°, divine 5; father 4; minister 1; pastor; preacher; priest
clergymen, ministry 2
clergywoman, clergyman 1; minister 1; pastor; preacher; priest
clergywomen, ministry 2
cleric, clergyman 1; divine 5; minister 1; preacher; priest
clerical°, pastoral 3; priestly; spiritual 1
clericals, ministry 2
clerk, bookkeeper; salesperson; scribe 1; seller
clever°, able 2; acute 5; arch 2; artful 2; astute 1; bright 6; brilliant 4; capable 1; comic 1; crafty; cute 2; deep 5; dexterous 2; disingenuous; elegant 4; foxy 1; good 12; great 8; handy 2, 3; imaginative 1; ingenious; intelligent; keen[1] 6; knowing 2; neat 4; nifty 3; penetrating 1; politic 1; quick 4; quick-witted; ready 6; resourceful; sharp 3, 6; shrewd; skillful; slick 3; sly 1; smart 1; subtle 3; tactful; tactical; talented; wily; wise 1; witty
Clever Dick, wise guy
cleverly, handily 2
cleverness, ability 2; art 5; craft 1; dexterity 2; faculty 1; finesse 1; gumption 1; ingenuity; intellect 1; intelligence 1; originality; penetration 2; resource 1; skill 1; tact; wit 1
cliché°, idiom 2; maxim; phrase 2; proverb; saw; truism
clichéd, banal; common 6; musty 2; prosaic; routine 4; stale 2; stock 7; threadbare 2; tired 3
cliché-ridden, threadbare 2
click, snap 2, 5, 8
client°, buyer; customer 1; patron 2; regular 12
  client of, be a, patronize 2
clientage, public 9
clientele°, public 9; trade 4
clients, clientele
cliff°, bluff[2] 3; crag; height 3; precipice
climactic, maximum 3
climate°, air 1; atmosphere 2; feel 11; feeling 6; milieu; weather 1
  climate, in the present, now 2
climax°, acme; close 3; consummation 2; crown 5; end 9; head 5; height 2; issue 2; maximum 2; peak 2, 4; pinnacle; showdown; summit
climb°, arise 2; get 46a; rise 3, 6, 7, 10, 15; rocket; scale[3] 3; scramble 1, 5; shin; soar 2; top 6
  climb along, climb 4
  climb down (from), climb 5, 5b; descend 1; get 37a, 41a
  climb down off, get 41a
  climb to, reach 5
  climb (up) (on), climb 1; get 51c; mount[2] 1, 3
climber, upstart
clime, climate 1
clinch°, clip[1] 1; close 4; confirm 2; determine 1; embrace 4; finalize; finish 2; hug 3; nail 7; ratify; seal 4; squeeze 7; tie 5a; verify
cling°, cement 2; hold 19a, 19b; tenacity 2
  cling (to), hang 7a; harbor 3; hug 2; persevere
  cling together, stick[1] 5
  cling together *or* **to one another,** cling 3
clinging, tenacious 2
  clinging (to), tenacious 3
clinic, hospital; infirmary; sanitarium
clink, jail 1; jingle 1, 2; prison; ring[2] 1, 3; stir 7
clinker, dud 1; howler; loser
clinking, jingle 2; ring[2] 3
clip°, brace 3; buckle 1; catch 17; clamp 1; clasp 1, 3; cut 4; diminish 1; embrace 1; fleece; fold 2; hit 1; hook 1; hug 1; jab 2, 4; lop; nip[1] 1; pace 2; pin 2, 3; press 5; prune; punch[1] 1, 2; rate[1] 1; screw 6; shave 1; snip 1; squeeze 5; swipe 3; top 5; trim 3
  clip (together), clamp 2
clipping, cutting 3; extract 6; snip 3
clip (someone's) wings, foil[1]
clique°, brotherhood 2; cabal 2; camp[1] 2; circle 2; clan 2; combination 2; crowd 2; faction 1; fellowship 2; fraternity 1; gang 2; group 1; junta; machine 3; pack 3; party 3; ring[1] 3; school 2; sect 2; set 26; world 3
cliquish, rarefied 3; sectarian 1
clishmaclaver, gossip 1; palaver 2; patter[2] 2
cloaca, drain 1
cloak°, camouflage 1, 2; cape[2]; cover 15; envelop 2; mantle 1, 2, 3; mask 2; muffle 1; obscure 6; pretense 2; pretext 1; robe 1, 3; screen 3; secrete[1]; semblance 2; shroud 1, 3; spread 7; submerge 3; veil 1, 2; wrap 4
clobber, batter 1; beat 1; clothes; effects; equipment; furniture 2; gear 3; hit 1; light[2] 14; manhandle; paraphernalia; pelt[1] 1; pound[1] 1; rout 2; rubbish 3; slaughter 4; stuff 2; tackle 1; thing 8c; whip 2
clock, face 1; mug 2; watch 6
  clock in *or* **on,** come 2; report 5
  clock off *or* **out,** knock 5
clock-watcher, idler
clockwise, forward 4
clockwork, measured 4; works 2
  clockwork, like, swimmingly
clod°, barbarian 2; boor 2; clown 2; clump 1; dolt; fool 1; gawk 1; hulk 2; lump[1] 1; peasant; robot 2; silly 3; wad 1; yahoo
cloddish, clumsy 1; dense 3; dull 1; obtuse 2; opaque 3; provincial 2; rustic 2; slow 7; unrefined 1
cloddishness, simplicity 4
clodhopper, boor 2; clod 2; clown 2; gawk 1; peasant; rustic 3
clodlike, dull 1; stolid
clodpate, fool 1
clodpole, fool 1
clod poll, fool 1

clog°, block 5; bog 2; choke 2; glut 3; hamper[1]; jam 2; obstruct 1; plug 4; stuff 9

clog (up), foul 15; stop 3

clogged, become, silt 2

clogged(-up), foul 9

cloister, isolate; monastery

cloister oneself, retire 1

cloistered, isolated 2; reclusive; secluded 1; separate 7; solitary 1

clone, double 7; duplicate 2, 3; image 2; lookalike; reproduction 2; spit 2; twin 1

clonk, thud

Clootie, devil 1

close°, area 5; bosom 4; by 6; chummy; clammy 2; clasp 3; close 8a, 8b; compact 3; completion 1; cramped; dense 1; enclosure 1; end 2, 9; equivalent 1; expiration; expire 1; faithful 2; familiar 3; fast[1] 5; fill 7; fine[1] 13; finish 5, 9; friendly 1; great 10; greedy 3; halt 1; hand 9a; hard 1; impending; intimate[1] 1; kindred 1; let[1] 8d; lock[1] 4; mean[2] 1; miserly; muggy; narrow 2, 3, 6, 7; near 4, 5, 6, 7, 8; nearby 2; nice 3; old 8; oppressive 2; particular 3; personal 3; petty 2; point-blank 2; present[1] 2; rigid 3; round 18; run 30a; secretive; serried; shut 1; small 4; sparing 1; special 5; sticky 3; stop 6; strict 1; stuffy 1; sultry 1; sweltering; termination 1; thick 8; tight 6; tighten 4; touchy 2; warm 6

close, bring to a, close 3; round 18; stop 1; wind[2] 4a; wrap 3a

close, come to a, stop 1; terminate; wind[2] 4a

close (at hand), convenient 2; nearby 1; ready 7; wind[1] 8

close (by), about 1, 9; available; by 6; close 20; forthcoming 1; hand 9a; handy 1; imminent; local 1; near 1, 5; nearby 1

close (down), close 5, 8a; fail 4; fold 3; knock 5; shut 2; sign 9; wind[2] 4a; halt 2

close in, enclose 1

close in (on), gain 4

close in *or* off, stake[1] 4a

close (off *or* up), block 3; close 1, 6; plug 4; seal 3; shut 4c; stop 3; wall 5

close on *or* onto, close 21

close out, lock[1] 8; wind[2] 3a

close (to), about 2, 6; approximately 1; beside 1; by 1; close 21; fast[1] 8; like[1] 1; near 2, 9; nearly 1; neighborhood 2; ready 4; toward 3

close to being, near 3

close (up), adjust 4; bar 8; block 3; close 1, 5, 6, 8; fail 4; fold 3; plug 4; seal 3; shut 4c; stop 3; wall 5; wind[2] 4a; halt 2

close upon, about 2

close up shop, fail 4

close (with), fight 1; gain 4

closed, close 10; exclusive 2; impermeable; over 6; secure 2; select 3; unopened

closed away, separate 7

closed off, separate 7

closed (up), shut 7

closed-minded, prejudiced 2

closefisted, avaricious; close 18; economical 2; frugal 2; grasping; greedy 3; mean[2] 1; miserly; near 6; penurious 1; petty 2; small 4; sparing 1; thrifty; tight 5

closefistedness, avarice; greed 2; thrift

close-fitting, tight 1

close gaps in, tighten 3

close-knit, close 15

close-lipped, taciturn; tight-lipped

closely, exactly 1; fast[1] 8; hand 10; immediately 2; intently; literally 1; narrowly 2; nearly 2; tight 10, 11; tightly; well[1] 13

close-matched, close 13

close-minded, hidebound; intolerant 2; narrow-minded; nearsighted 2; one-sided 1; parochial; rigid 4; small-minded

close-mindedness, fanaticism 2

closemouthed, close 17; reserved; secretive; silent 2; mum 1; tight-lipped

closeness, affinity 1; familiarity 2; fraternity 2; friendship 1; harmony 2; presence 1; proximity; sympathy 2

close one's eyes to, close 7

close-packed, thick 2

close quarters, clinch 2

close shave, shave 3

closest, immediate 2; present[1] 2; special 5

closet, stealthy; toilet 1; wardrobe 2

closet, out of the, out 3; public 11

closing, expiration; final 1; finish 9; last[1] 4; net[2] 3; parting 3; terminal 1

closing up, stop 9

clot, coagulate; dolt; lump[1] 1; set 3; silly 3; solidify 1; stiffen 1; thicken

cloth°, fabric 1; filter 1; good 21d; material 2; pall[1] 1

cloth, the, cloth 2; ministry 1, 2

clothe°, cover 5; dress 1; equip; get 51e; mantle 3; robe 3

clothe (oneself) in, put 22a

cloth-eared, insensible 2

clothed in, be, have 12a; wear 1

clothes°, apparel; costume; ensemble 1; garments; gear 3; habit 3; outfit 2; rag[1] 3; suit 3; thing 8c; wardrobe 1; wear 8

clothes, best, best 7

clothesbrush, brush[2] 1

clothes cupboard, wardrobe 2

clotheshorse, dandy 1; swell 6; trendy 2

clothes peg, peg 1

clothespress, wardrobe 2

clothier, tailor 1

clothing, apparel; clothes; costume; ensemble 1; garments; gear 3; habit 3; outfit 2; rag[1] 3; robe 2; suit 3; thing 8c; turnout 3; wardrobe 1; wear 8

clothing business, rag[1] 4

clotted, thick 2, 5

cloud, dull 10; film 3; flight[1] 4; fog 1, 3; mist 1; muddy 4; obscure 7; reek 4; shade 12; shadow 4; shroud 3; swarm 1; vapor 1

cloud (over), fade 1; film 4; mist 2

cloud (up), fog 5; mist 2

cloudburst, downpour; rain 1; storm 1

cloud-cuckoo-land, fairyland; utopia

cloud-cuckoo-land, in, forgetful

clouded, black 7; dim 1; dull 5; filmy 2; inauspicious; murky; nebulous; obscure 1; opaque 1; overcast; steamy 2; vague 1; woolly 2

cloudiness, blur 1; gloom 1; obscurity 1

cloudland, fairyland

cloudless, bright 2; clear 1; fair[1] 5; fine[1] 2; pleasant 3; sunny 1

cloudless, become, clear 22

cloud nine, transport 5

cloud nine, on, ecstatic; elated; exalted 3; exuberant 2; exultant; happy 1; joyful 2; overjoyed; pleased; radiant 2; rapturous; world 7

clouds, in the, absent-minded; dreamy 2; forgetful; inattentive

cloudy, dark 5; dim 1; dull 5; filmy 2; gloomy 1; gray 2; hazy 1; heavy 9; imprecise; inauspicious; misty; murky; obscure 1; opaque 1; overcast; steamy 2; vague 1; woolly 2

clough, ravine

clout, batter 1; beat 1; blow[2] 1; box[2] 2, 3; clip[2] 3, 6; effect 2; hit 1; hold 26; influence 1; knock 9; might 2; power 2; pull 21; punch[1] 1, 2; rag[1] 1; rap 4; say 12; slap 1, 4; strike 1; weight 3

clover, in, money 4; opulent 1; prosperous 1; rich 1; wealthy

clown°, boor 2; clod 2; comedian; fool 2; joker 1; panic 2; wag[2]; zany 2

clown around *or* about, clown 3

clowning, fun 2; nonsense 2

clownish, boorish; zany 1

cloy, glut 4; pall[2] 2; satiate 1

cloyed, jaded 2

cloying, excessive 2; gooey 2; gushy; sickly 3; sweet 7

club°, beat 1; brotherhood 2; cabal 2; dive 3; fellowship 2; fraternity 1, 3; group 1; hit 1; league 1; party 3; society 5; staff 1; strike 1; union 2

club, in the, expecting

club, in the (pudding), pregnant 1

club (against), gang 3

club together, club 6

clubbable, friendly 2

clubbiness, fellowship 3; friendship 1

clubby, friendly 1

clubhouse, club 3

club soda, fizz 4

clue°, conception 2; evidence 3; hint 1, 3; idea 2; index 2; indication 1; inkling; key 2; lead 14; line[1] 10; note 6; sign 3; symptom; tip[3] 2; trace 1

clue someone in *or* up, clue 2

clues, indication 4

clump°, bunch 1; huddle 1; lump[1] 1; plump[2] 4

clumsily, roughly 3

clumsiness, embarrassment 1; ineptitude 1

clumsy°, amateur 2; artless 3; awkward 1; heavy 8; heavy-handed 1; hefty 1; hulking; incompetent; inept 1; lame 2; left-handed 1; ponderous 1; rude 4; shambling; stilted; tactless; thumb 2; ungraceful 1; unwieldy; wooden 2

clunk, bump 1; plump[2] 4; thud

cluster°, bunch 1, 2, 4; center 2; clump 2; concentrate 3; crowd 3; gather 2; herd 1; huddle 1, 3; knot 2; mass 8; pocket 3; press 7, 8

clutch°, bunch 2; clasp 5; cluster 1; embrace 4; grab 1, 4; grapple 1; grasp 1; grip 1, 6; hang 7a; hold 1, 2, 19a, 24; lock[1] 4; seize 1; snatch 1, 4; squeeze 5, 7; take 1

clutch (at), pluck 4

clutches°, clutch 2; grasp 3; hand 8; jurisdiction; mesh 2; sway 4

clutch one another, cling 3

clutter°, disorder 1; jumble 2; litter 4; lumber 1; mess 1; welter 1

clutter up, clutter 3; mess 5c

cluttered, disorderly 1; untidy

coach°, brief 6; carriage 1; discipline 6; drill 2; educate; enlighten; groom 4; ground 6; initiate 3; instruct 1; instructor; prime 5; school 4; teach; teacher; train 1, 4; tutor 1, 2
coaching, instruction 2
coachman, servant 1
coadjutor, aide; ally 1
coagulate°, cake 3; jell 1; set 3; solidify 1; stiffen 1; thicken
coagulated, thick 5
coal-black, black 1, 3; pitch-black; swarthy
coalesce, blend 2; combine 3; compound 2; fuse; incorporate; integrate; merge; mix 1; unify; unite 1; wed 2
coalesced, united 1
coalescence, amalgamation; merger; synthesis; wedding 2
coalescing, amalgamation; wedding 2
coalfield, mine 1
coalition, alliance 1; association 1; combination 2; federation; fellowship 2; league 1; merger; organization 3; party 3; ring[1] 3; union 1, 2
coarse°, animal 4; barbarian 3; base[2] 4; bawdy; beastly 1; blue 2; boorish; brassy 1; broad 8; brutal 2; crude 3; dirty 2; earthy; filthy 3; foul 5; gross 3; harsh 1; heavy 8; homespun; immodest 1; impolite; indecent 2; irregular 1; lascivious 2; low[1] 3; nasty 3; obscene 1; plebeian 2; primitive 2; profane 3; prurient 2; racy 2; rough 1, 3; rude 1; rudimentary 2; scurrilous; sexy 2; tactless; tasteless 1; uncivilized 2; ungraceful 2; unrefined 1, 2; unseemly 1; vulgar 1; wicked 3
coarse language, profanity
coarseness, incivility; indelicacy; ribaldry; vulgarity 1
coast°, beach 1; cruise 1; drift 1; glide; roll 3; sail 3; seacoast; slide 1
coastline, coast 1; seacoast
coat°, cloak 1; cover 3; face 16; film 1, 4; glaze 1; paint 2, 5; pelt[2]; plaster; plate 4, 6; powder 4; sheet 5; skin 2; smear 1; spread 7; wash 7, 15
coating, coat 2; exterior 3; facing; film 1; foil[2]; glaze 2; icing 1; paint 2; peel 4; plate 4; scale[2] 2; sheet 5; skin 2; wash 15
coat of arms, device 3; seal 1; stamp 7; symbol
coat rack, rack 1; stand 17
coax°, cajole; drag 2; entice; flatter 3; get 14, 30a; induce 1; lure 1; motivate; prompt 3; tease 2; tempt 1; urge 3; wheedle; work 18
coaxing, cajolery; enticement 1; temptation 1
cobber, comrade; friend 1
cobbler's, gab 2; gobbledegook 1; nonsense 1; drivel 3; gibberish; rot 4; rubbish 2; trash 1; wind[1] 2
cobweb, web
cobwebby, filmy 1
cochlear, spiral 2
cochleate, spiral 2
cock, flannel 1; stack 2; tap[2] 1
  cock up, bugger 3; fluff 4
cock-a-hoop, ecstatic; exultant
Cockaigne, utopia
cockamamie, daft 1; stupid 2
cock-and-bull story, fabrication 3; falsehood; fib 1; lie[1] 2; myth 2; yarn 2
cock a snook at, defy 1; disobey; fly 6; jeer 1; mock 1; scorn 3; snap 6; thumb 8
Cockayne, utopia
cockcrow, dawn 1; morning 1
  cockcrow, at, early 3
cockcrowing, dawn 1
cocker, coddle; friend 1; pamper; pet[1] 5; spoil 3
cockeyed, daft 1; inaccurate; lopsided 1; mistaken 2; nonsensical; one-sided 2; stupid 2
cockiness, vanity 1
cocklebur, thorn 1
cocksure, cocky; confident 2; impudent; opinionated 1; overconfident 1; proud 2
cocktail lounge, bar 6; lounge 3; pub
cockteaser, flirt 3
cock-up, flop 4
cocky°, confident 2; impudent; inflated 1; overconfident 1; proud 2; pushy; vain 1
coconspirator, accessory 2; accomplice; sympathizer
coco(nut), head 1
cocotte, prostitute 1
coddle°, baby 2; cater 2; dote; favor 7; humor 4; indulge 2; make 34b; mother 6; nurse 2; pamper; pet[1] 5; spoil 3
coddled, hothouse 2
code°, law 2; legend 3; notation 2; precept 1; principle 2; regulation 2; symbol
code (of behavior), etiquette
code (of practice), institution 4
code(s) of behavior *or* **conduct,** protocol 1
codeword, symbol
codicil, supplement 1
codification, distribution 3; embodiment 2; order 1; organization 1
codify, coordinate 1; embody 3; form 7; formulate 1; order 16; organize 1; separate 2; standardize; tabulate
codifying, organization 1
codswallop, drivel 3; fiddlesticks; flannel 1; gab 2; gibberish; gobbledegook 1; jargon 2; rot 4; rubbish 2; stuff 4; trash 1
  codswallop, load of, falsehood
coed, miss[2]
coequal, equal 1; even 5; parallel 1, 2; peer[1] 2
coequality, equality 1; parallel 3
coerce, blackmail 2; drive 1; enforce 2; extort; force 7; hustle 4; make 2; oblige 2; pin 4a; pressure 5; push 4, 7; railroad 2; require 1; screw 4; shame 7; twist 6
coercion, duress 1; force 2; pressure 4
coercive, compulsive; forcible 2
coetaneous, contemporary 1
coeval, contemporary 1; current 1
coevality, coincidence 1
coexistent, contemporary 1
coextension, coincidence 1
coextensive (with), even 2
coextensive (with), be, join 4
coffee bar, café
coffee break, break 28
coffee house, café
coffee shop, café
coffer, box[1] 1; casket 1; chest 1; safe 7; trunk 3
coffin°, casket 2
coffin nail, fag 5
cog°, gear 1
cogency, force 4; strength 7
cogent, concise; eloquent 1; forceful 2; good 8; persuasive; potent 2; solid 7; strong 8
cogitate, meditate 1; muse; ponder
  cogitate (about *or* **on** *or* **over),** consider 1; contemplate 2; deliberate 4; mull; reflect 3; speculate 1; think 3
cogitating, preoccupied 1; reflective; thought 1
cogitation, consideration 3; debate 2; reflection 2; speculation 2; thought 1
cogitative, meditative; pensive; philosophical 2; preoccupied 1; reflective; speculative 1; thinking 1
cognate, kin 2; kindred 2; related 1, 2
  cognate (with), like[1] 1
cognition, brain 1; knowledge 1
cognitive, mental 1; psychic 1; psychological
cognizable, sensible 2
cognizance°, device 3; familiarity 1; notice 3; realization 1; recognition 2
  cognizance of, take, admit 3; understand 1
cognizant, aware 1; knowledgeable 1
cognizant (of), privy 2; sensible 4; alive 2; familiar 4
  cognizant of, be, appreciate 3; know 5; mind 9
cogwheel, gear 1
cohere, bond 4; cement 2; hang 11b; stick[1] 5
coherence, connection 2; integrity 2; sense 4; union 4
coherent°, logical 3; organic 3
  coherent, be, hang 11b
cohesive, tenacious 2
cohesiveness, tenacity 2
cohort°, accomplice; aide; attendant 2; corps; friend 2; henchman; mate 1; partner
coif, hairdo
coiffure, hairdo
coil°, coil 1; entwine; garland 2; kink 1; loop 1, 2; roll 9; spiral 1; tangle 1; turn 11, 28; twirl 3; twist 7; wind[2] 1, 2
coiled, spiral 2
coil(s), meander 2
coin°, forge 2; form 8; generate 4; invent 1; make 39c; mint 2; originate 1; token 1; turn 8
  coin it in, rake[1] 4
coinage, neologism
coincide°, accord 1; agree 1; check 5; converge; correspond 1; overlap 2; tally 1; tie 6a
  coincide (with), conform 2; join 4; parallel 4
coincidence°, equality 2; resemblance
coincident, simultaneous
coincidental°, contemporary 1; unpremeditated
coinciding, simultaneous
coin money°, coin 4; rake[1] 4
coin of the realm, cash 1
coin(s), change 4; money 1
coinstantaneity, coincidence 1
coition, intercourse 2; sex 2
coitus, intercourse 2; relation 6a; sex 2
col, pass 23
cola, pop 5
colander, filter 1; riddle[2] 3; screen 4
cold°, bare 2; biting; bitter 6; bleak 2; brisk 3; callous; chill 1, 2, 4, 6; chilly 1, 2; coldhearted; cool 1, 3, 4, 5; cutting 1; dead 4; distant 3; dour 1; freezing; frigid 1, 2, 3; glassy 2; hard 4, 8; heartless; hostile 2; icy 1; impassive; impersonal 2; inaccurate; inanimate; indifferent 1; inhospitable 1; lackadaisical 2; mechanical 3; merciless; nail 4b; negative 2; phlegmatic 1; piercing 3; raw 5; rigor 1; rocky[1] 3; severe 2, 5; stark 2; stiff 7; stony 2; tough 6;

command° (*cont.*)
  command (of), be in, rule 5; shot 9; string 6
commandant, director 2; leader 1
commanded, mandatory
commandeer, appropriate 2; confiscate; grab 2; help 5; requisition 3; seize 5
commandeering, seizure 1
commander, director 2; leader 1; master 1; skipper
commanding, dominant 1; imperative 2; imposing; incumbent 1; magnificent; panoramic; peremptory 1
commandment, law 1; order 4; precept 1
comme ci, comme ça, fair[1] 4; so-so
comme il faut, correct 6, 7; fitting 1; proper 1, 3; seemly 1; tasteful
commemorate°, immortalize; keep 10; memorialize; observe 5; remember 2
commemorated, go 27d
commemoration, celebration 1; feast 2; festival 1; monument 1; occasion 3
commemorative, ceremonial 1; memorial 1; monumental 3; triumphal
commence°, begin 1, 3; break 17; dawn 4; develop 4; embark 2; enter 4; fall 21; inaugurate 1; initiate 1; institute 4; lead 8; move 12; open 20; rise 17; start 1, 2, 4; strike 18
  commence to act, step 11
  commence to take action, step 11
commencement, beginning 1, 2; conception 1; dawn 2; entrance[1] 4; first 6; infancy 2; initiation 1; onset 2; opening 4; origin 2; source 1; start 10; threshold 2
commencing, initial 1
commend, advise 1; applaud 2; approve 2; compliment 3; extol; glorify 2; laud; pat[1] 2; plug 5; praise 3; promote 3; puff 7; recommend 2
commendable, deserving; desirable 4; estimable; exemplary 2; fine[1] 11; good 1; laudable; meritorious; nice 2; praiseworthy
commendation, applause; bouquet 3; compliment 1; credit 4; eulogy; pat[1] 5; praise 1; promotion 3; puff 3; recommendation 2; testimonial; tribute 1
commendatory, complimentary 1; favorable 2
commensurability, comparison 2
commensurate, equal 1, 2; equivalent 1; proportional; relative 2
commensurate with, according to 2; tantamount to
commensuration, measurement 1
comment°, gloss[2] 1; note 3; notice 6; observation 2; pronouncement 1; review 4
  comment on *or* **upon** *or* **about,** comment 4; gloss[2] 2; note 13; observe 4; remark 2; review 3; speak 9
commentary, comment 2; criticism 3; description 2; explanation 1; exposition 3; gloss[2] 1; note 3; notice 6; observation 2; review 4
commentator, journalist; reporter; scribe 2; student 2
commentators, press 10b
commerce°, business 3; dealings; exchange 2; finance 1; industry 1; intercourse 1; passage 10; trade 1; truck 2
commercial, advertisement 1; announcement 3; economic 1; hot 6; mercantile; philistine 2
commercial traveler, merchant 2
commination, menace 3; threat 1
comminatory, threatening
commingle, blend 1; combine 3; fuse; merge; mingle 1; mix 1; scramble 3; stir 1; unite 1; wed 2
commingling, amalgamation; blend 3; merger; wedding 2
comminute, break 23b; grind 1; mill 5; pound[1] 2; powder 3; pulverize 1
comminuted, fine[1] 7; granular
commiserate (with), feel 8; pity 3; respond 2; sympathize 1; understand 5
commiserating (with), sympathetic 1
commiseration, pity 1; sympathy 1
commiserative (with), sympathetic 1
commission, appoint 2; assignment 2; authorize; charge 9; charter 4; committee; cut 20; delegate 2; employ 1; enable 1; engagement 4; errand 2; function 1; job 6; kickback; license 4; make 23; mission 1, 3; office 2, 4; order 7, 15; rake-off; rebate 2; retain 2; royalty 2; sanction 5; send 1; work 3
  commission, out of, broken 7; ill 1; indisposed 1; order 13c; out 14
  commission, put out of, maim
commissioned, vicarious
commissioner, delegate 1; functionary; officer 1; representative 3
commissure, seam 1; weld 2
commit°, apply 5; dedicate 1; deliver 2; devote 1, 2; leave[1] 7; lock[1] 6b; obligate; perpetrate; put 2, 13b; refer 2; relegate 3; tie 7d; trust 6
  commit (oneself) (to), commit 4; decide 2; promise 3; undertake 2
commitment, application 3; assurance 1; contract 1; dedication 1; engagement 1; order 7; project 2; promise 1; resolution 2; subscription 1; undertaking 3; warranty; will 1
committed, devoted 1; earnest 1; heartfelt; intent 5; wholehearted
  committed to, be, have 12b
committee°, board 4; body 6; cabinet 2; council 2; mission 3
commode, cabinet 1
commodious, convenient 1; extensive 2; roomy; spacious
commodities, good 21b; merchandise 1; truck 1; wares
commodity, product 2
common°, accustomed 1; average 3, 4; banal; base[2] 3; conventional; current 2; customary 1; daily 2; everyday 2; familiar 1; field 1; frequent 1; general 1, 2; going 2; green 4; habitual 1; humble 3; humdrum; joint 4; kindred 1; low[1] 3; mean[2] 2; mill 4; mutual 2; natural 1; nondescript; normal 1; ordinary 1, 2; orthodox; plebeian 1; popular 2; prevalent 1; profane 3; prosaic; public 1; reciprocal; regular 1, 11; related 1; simple 5; social 1; tolerable 2; typical 2; undistinguished; united 1, 2; universal 1; usual; vulgar 1
commonality, people 3
commonalty, people 3; populace; public 8; rabble 2
common cold, cold 10
commoner, plebeian 3
commoners, people 3; populace; rabble 2
common(est), prevailing 1
common folk, populace
common herd, herd 2; hoi polloi; people 3
commonly, frequently 2; generally 1; often; ordinarily; popularly; rule 4; usually
common man *or* **woman,** people 3; plebeian 3
commonness, prevalence 1
common-or-garden, common 1; informal 1; ordinary 2
(common-or-)garden variety, mediocre; mill 4; nondescript; ordinary 2; tolerable 2
common people, hoi polloi; people 3
commonplace, average 4; banal; cliché; common 1; conventional; customary 1; daily 2; dead 12; dry 2; everyday 2; familiar 1; hack[2] 4; homely 1; humdrum; indifferent 3; mediocre; monotonous; natural 1; nondescript; ordinary 2; pedestrian 2; philistine 2; phrase 2; plebeian 2; popular 2; prevalent 1; prosaic; proverb; regular 1; saw; small 3; stock 7; threadbare 2; tired 3; trifling; truism; typical 2; undistinguished; vulgar 3
common(s), park 1; food; ration 2
common salt, salt 1
common sense, discretion 1; experience 3; intuition; judgment 1; prudence 1; reason 2; sense 2; enlightened; practical 2; rational 3; realistic 1; sage 1; sensible 1; sound[2] 4
  common sense, use (one's), reason 7
common-sensical, enlightened; judicious; legitimate 3; levelheaded; moderate 1; practical 2; rational 3; sage 1; sensible 1; sound[2] 4; wise 1
commotion, agitation 2; bedlam; bother 8; confusion 2; din 1; disorder 2; disturbance 2; excitement 2; fanfare 2; flap 4; flurry 1; fluster 2; fracas 1; furor 1; fuss 1; hurry 3; incident 2; mayhem; motion 1; noise 1; outcry; pell-mell 3; racket 1; riot 1; row[2] 2; rumpus; rush 3; scene 3; scramble 4; sensation 2; song 3a; splash 4; stir 6; storm 3; tempest 2; tumult; upheaval; uproar
communal, common 5; general 1; joint 4; mutual 2; public 1, 2, 3; social 1; united 2
communicable, catching 1; infectious
communicate°, carry 8; come 16a; correspond 2; deliver 4; express 1; get 26; give 3; impart 2; inform 1; intimate[2]; make 33; mean[1] 2; open 23; put 10, 25a; recount 1; register 5; relate 2; reveal; say 4, 9; send 2; signal 4; signify 1; speak 4, 5; tell[1] 2; transmit 1; vent 3; voice 4

complaint° (*cont.*) grievance 2; gripe 2; groan 4; illness; infirmity 2; moan 1; murmur 2; objection; outcry; peep 3; protest 1; squawk 3; trouble 8
complaint(s), flak
complaisant, accommodating 1; flexible 3; passive 2; supple 3; willing
complement°, enhance; fellow 3; flatter 2; parallel 2; supplement 2
complement each other, go 6
complementarity, parallel 3
complementary, coordinate 3; harmonious; matching 1; mutual 1; reciprocal
complementary (to), subsidiary
complete°, absolute 1, 2; accomplish; achieve 1; ample 4; carry 12; clear 14; clinch 1; close 3; complement 3; comprehensive; cover 9; dead 14, 16; definitive 2; detailed 1; dispatch 4; dramatic 2; elaborate 1; encyclopedic; entire 1, 3; execute 2; exhaustive; file 4; fill 9a; finalize; finish 2; flagrant; flat 4; fulfill 1; full 2, 3, 12; get 46c, 48a; good 11; grand 4; ideal 4; implicit 2; intact; knock 5c; live 7; make 37b, 39a; mature 3; nail 10; negotiate 2; one 2; outright 1; out-and-out; overall; perfect 1, 6, 8; perform 1; positive 9; profound 4; pronounced 2; proper 5; pull 12b; pure 5; put 27a; radical 2; rank² 2; regular 9; right 8; round 5, 18; self-contained 3; serve 6; sheer 2; solid 10; stark 4; strict 1; sweeping 2; thorough 1, 3; thumping 2; top 4; total 2, 3; transact; ultra-; unabridged 1; undivided 1, 2; unmitigated; whole 1; wholehearted; wrap 3a
completed, out 11
completely°, absolutely 2; altogether; blast 4; clean 7; clear 19; cold 11; dead 18, 19; deeply 2; directly 4; downright 2; entirely 1; exactly 2; fairly 3; finally 2; flat 16; full 13, 17, 18; head 8; hook 4; inextricably; length 3, 4c; out 7; outright 4; perfectly 1; quite 1; root¹ 3; simply 2; stark 1; supremely; thoroughly 1, 2; through 7; throughout 3; totally; utterly; very 1; well¹ 11; wholly 1; wide 6
completely (cooked), well¹ 12
completeness, entirety 1; integrity 2; perfection 2
completing, completion 2; consummation 1
completion°, accomplishment 1; achievement 3; close 22; complement 1; consummation 1; end 2; execution 2; finish 9; fruition; fulfillment; maturity 3; perfection 2; performance 1
completion, bring to, round 18
complex, busy 3; complicated; compound 4; detailed 2; difficult 2; elaborate 2; facility 2a; fancy 1; hairy 3; hard 3; heavy 12; intricate 1; involved 2; labyrinthine; maze; mesh 2; network 2; nice 3; obscure 5; perplexing; prickly 4; serpentine 2; sophisticated 2; technical 1; thorny 2; tricky 2
complex, make, complicate 1
complexion, aspect 2; nature 1; strain¹ 9; stripe 2
complexity°, complication 1; depth 2; mesh 2; obscurity 2; perplexity 2; profundity 1; snarl² 3; sophistication 2; tangle 2
compliance, consent 4; deference 2; flexibility 2; fulfillment; obedience; observance 1; resignation 2; sanction 3; submission 1
compliance, in, accordingly 2
compliancy, flexibility 2
compliant, accommodating 1; adaptable; agreeable 2; dutiful; easy 4; flexible 2; manageable; meek 2; obedient; passive 2; patient 1; plastic 2; pliable 2; propitiatory 2; sheepish 1; submissive 1; supple 3; tame 3; tractable 1; willing; yielding 2
compliant (with), observant 2
compliantly, accordingly 2
complicate°, elaborate 3; entangle 2; snarl² 1
complicated°, busy 3; compound 4; detailed 2; elaborate 2; fancy 1; hairy 3; hard 3; intricate 1; involved 2; labyrinthine; nice 3; perplexing; prickly 4; serpentine 2; sophisticated 2; technical 1; thorny 2; tortuous 2; tricky 2
complicated, make, complicate 1
complicatedly, inextricably
complicatedness, complexity 2; perplexity 2; profundity 1
complication°, complexity 1; hurdle 1; kink 3; matter 3; pinch 8; problem 1; ramification 1; snag 1; snarl² 3; tangle 2
complication(s), rigmarole
complications for, cause, bugger 4b
compliment°, congratulate; eulogize; eulogy; extol; flatter 1; pat¹ 2, 5; praise 3; tribute 1
complimentary°, flattering 1; free 4; glowing 3; good 17; gratuitous 1; laudatory; positive 4
complimentary ticket, pass 25
compliment(s), bouquet 3; compliment 2; greeting 3; praise 1; regard 10; respect 5
comply°, consent 1; correspond 1; relent; surrender 2; yield 3
comply (with), abide 4; conform 1; defer²; follow 2; fulfill 2; indulge 1; obey 1; observe 1; satisfy 2; square 12; submit 1
complying, agreeable 2; observance 1
component, accessory 1; detail 1; element 1; factor 1; ingredient; item 1; link 1; organ 1; part 2, 7; section 1; segment 1; unit
component of, be a, compose 1
components, content¹ 2; filling 1; hardware 2; material 3
comport, conduct 6
comport (oneself), behave; act 5
comportment, behavior; carriage 2; conduct 1; guise 1; manner 2; presence 3
compose°, compile; compound 1; construct 2; draft 5; draw 6, 15b; form 8; frame 7; make 39b; mount² 4; originate 1; pen¹ 2; piece 13; produce 1; quell 2; settle 1; tranquilize; weave 3; write 3
compose oneself, compose 4
composed, bland 1; calm 4; collected; cool 2; deliberate 3; dispassionate 1; equable 1; even 4; home 4a; impassive; levelheaded; nonchalant; philosophical 2; phlegmatic 2; poised 1; quiet 4; relaxed; sedate 1; self-contained 1; self-possessed; serene 2; sober 2; staid; straight 8; temperate 1
composed of, be, compose 3; form 9
composing, preparation 3
composite, alloy 1; amalgam; amalgamation; composition 3; compound 4, 5; ensemble 3; general 3; hybrid; mixture 1; pastiche; synthesis
composition°, amalgamation; combination 3; design 6; drawing; essay 1; fiber 3; format 2; formation 2; frame 3; makeup 3; opus; organization 2; paper 4; pattern 6; piece 4; preparation 5; setup 1; substance 1; synthesis; tableau; theme 2; work 4; writing 2
compos mentis, sane
compost, fertilize 2
composure, calm 2; cool 8; patience 1; philosophy 3; poise 2; presence 5; repose 3; sang-froid; serenity 2; temper 2
compound°, alloy 1; amalgam; amalgamate; amalgamation; combination 3; combine 2, 3; complicated; composition 3; elixir 2; enclosure 1; fuse; hybrid; mingle 1; mix 7; mixture 2; pastiche; pound²; preparation 5; run 46; solution 3; synthesis; wed 2; wedding 2
compounding, amalgamation; combination 3; composition 3; mixture 2; synthesis; wedding 2
comprehend°, appreciate 3; catch 10, 13a; conceive 3; cover 7; digest 3; embrace 3; follow 8; get 19; grasp 2; include 1; involve 1; know 1; make 37d; penetrate 5; perceive 2; pierce 3; read 2; realize 2; relate 4; see 2; tumble 3; twig²; understand 1
comprehensibility, clarity 2; simplicity 1
comprehensible, clear 6; coherent 2; evident; intelligible; manifest 1; readable 1; simple 1
comprehension, appreciation 3; conception 2; eye 3; familiarity 1; grasp 5; grip 3; insight; perception 1; realization 1; savoir-faire; uptake
comprehension, beyond *or* **past,** deep 2
comprehension (of), understanding 3
comprehensive°, broad 6; catholic; detailed 1; elaborate 1; encyclopedic; exhaustive; expansive 3; extensive 1; full 2, 12; general 1, 3; grand 4; inclusive 1; intensive; overall; panoramic; radical 2; sweeping 1; thorough 3; total 2; unabridged 2; universal 2; wide 2
comprehensively, depth 6; detail 4; thoroughly 2
compress, abbreviate 1; abridge; contract 4; crowd 5; crush 3; digest 4;

contend in, play 2
contend (with), cope 2; encounter 2, 3; fight 1; grapple 2; oppose 1; play 3; rival 2; tangle 4; tilt 2; grip 5; war 3
contender, antagonist; competitor; contestant; enemy; opponent; player 1; pugilist; rival 1
content°, capacity 1; matter 4; meaning 1; please 1; ready 2; satiate 2; satisfy 1, 3; significance 1; text 1; willing
contented, carefree; comfortable 1, 2; content[2] 3; glad 1; happy 1; pleased
contented (with), proud 1
contentedly, happily 3; willingly
contentedness, content[2] 1, 2
contention, antagonism 2; argument 2; assertion 1; competition 1; conflict 2; contest 2; controversy 1; debate 1; disagreement 3; discord; dissension; encounter 5; faction 2; feud 1; friction 2; position 7; quarrel 1; race[1] 1; rivalry; scramble 4; story[1] 2; strife 1; struggle 5; thesis 1; variance 2; war 1
contention, in, issue 8; variance 3; war 2
contentious, argumentative; belligerent 2; controversial 3; cranky 2; dissident 2; factious; faultfinding 2; hasty 4; legalistic; militant 1; offensive 1; passionate 3; perverse 2; prickly 4; pugnacious; quarrelsome; testy; touchy 1; truculent
contentment, content[2] 1; ease 1; heaven 3; joy 1; pleasure 1; satisfaction 1
content(s), content[1] 2; filling 1; format 2; inside 1; text 1
conterminous (with), be, join 4
contest°, battle 1, 2; bout 3; campaign 2; challenge 1, 3; combat 2, 6; competition 2; debate 3; disagree 2; dispute 1; encounter 5; fight 2, 7; game 2; match 2; meet[1] 7; oppose 1; race[1] 1; rival 2; running 2; scramble 4; skirmish 1; struggle 5; tilt 4; tournament; trial 2
contestable, moot 1
contestant°, belligerent 3; competitor; enemy; entry 5; opponent; player 1; pugilist
contestants, field 3
contested, moot 1
context°, frame 3; matter 4; milieu; setting; viewpoint
contiguity, proximity
contiguous, near 5
contiguous (to), adjoining; neighboring
contiguous (with), be, join 4
contiguousness, proximity
continence, celibacy 2; chastity; temperance 1
continent, celibate 2; chaste 1; temperate 2
contingency, condition 2; eventuality
contingent, cohort 1; provisional 2; qualified 2
contingent on, relative 2; subject 8b
contingent (upon *or* on), be, depend 1; hang 7e; pivot 4; turn 18b
continual°, constant 2; continuous 2; endless 2; eternal 2; everlasting; frequent 1; habitual 2; nagging; nonstop 2; ongoing 1; perennial 1; perpetual 2; persistent 2; recurrent; relentless 2; repetitive; running 3; successive; sustained
continually, always 2; cease 2; daily 3; ever 2; forever 2; frequently 1; night 3; nonstop 3; often; time 11
continuance, existence 2; maintenance 2
continuation, endurance 2; existence 2; maintenance 2; preservation 2; supplement 1
continue°, abide 3; add 3; carry 11a, 12; endure 1; exist 1; extend 1, 2, 3; follow 10; go 15, 19, 32a; hold 20a, 21b; keep 5; last[2] 1; lengthen; live 7; maintain 1; pass 18; perpetuate; persevere; persist 2; preserve 2; proceed 1; progress 5; prosecute 3; pursue 2; push 10; remain 3; resume; soldier 4; spin 5; stand 4; stay[1] 1; stick[1] 7; survive 1; sustain 1; take 39b
continue to be, stay[1] 3
continue with, stick[1] 19
continue without, go 40
continued, ongoing 1; standing 1; sustained
continuing, chronic 1; enduring; frequent 1; nonstop 2; ongoing 1; perennial 1; persistent 2; progressive 1; standing 1
continuity, script 2; unity 1
continuous°, constant 2; continual; endless 2; entire 3; eternal 2; everlasting; flush[2] 1; nagging; nonstop 1, 2; ongoing 1; perennial 1; perpetual 2; persistent 2; progressive 1; regular 5; relentless 2; repetitive; running 3; sequential; solid 4, 10; standing 3; steady 2, 3; successive; sustained
continuously, cease 2; daily 3; end 7b; ever 2; forever 2; intently; night 3; nonstop 3; permanently; time 11
contort, distort 1; screw 7c; twist 2; warp 1
contorted, crooked 2; deformed 1; gnarled; misshapen; wry 1
contortion, twist 9; warp 2
contour, form 2; line[1] 5; outline 1; profile 1; silhouette
contours, shape 1
contour sheet, sheet 1
contraband, booty
contract°, abbreviate 1; abridge; accord 3; agreement 1; apprentice 2; arrangement 3; bargain 1; bond 2; catch 5; charter 1, 3, 5; come 12; deal 5; decrease 1; deed 3; diminish 1; dwindle; engage 3; engagement 3; farm 4; flex 3; form 10; gather 3; get 5; instrument 3; job 6; knit 3; match 3; negotiation 2; obligation 2; pact; pick 8k; pledge 4; project 2; promise 1; pucker 1; purse 4; shrink 1; shrivel; sicken 1; sign 10a; treaty; understanding 1; undertake 2; undertaking 3
contract, put under, sign 10b
contract for, line[1] 23b; order 15; reserve 2
contract (out), let[1] 3
contract with *or* for, engage 1; line[1] 23b
contractile, elastic 1
contraction, abbreviation; abridgment 1; decrease 2; epitome 2
contractions, labor 4
contradict°, deny 1; destroy 4; differ 2; disprove; fight 2; fly 6; rebut; traverse 3
contradiction, confusion 3; denial 1; disagreement 2; paradox; rebuttal
contradictory°, confused 1; contrary 1; destructive 2; discordant 1; incompatible; incongruous; mixed 2; negative 1; opposing; opposite 2; paradoxical; perverse 1; polar 2; puzzling
contradictory, be, differ 2
contradistinction, difference 1
contradistinction to, in, unlike 3
contradistinguish, differentiate 1
contraption°, apparatus; device 1; gadget; gimmick 2; implement 1; instrument 1; invention 2; machine 1; tool 1
contrariety, difference 1; discrepancy; diversity 1
contrarily, contrary 5
contrariness, obstinacy
contrarious, contrary 2
contrariwise, contrary 5; only 4; vice versa
contrary°, cantankerous; destructive 2; different 1; difficult 3; discordant 1; disobedient 2; hostile 1; incompatible; incongruous; insubordinate; negative 1; obstinate; opposing; opposite 2, 3; perverse 1; polar 2; quarrelsome; recalcitrant; reverse 1, 6; self-willed; willful 2
contrary, on the, only 4
contrary, to *or* on the, vice versa
contrary (to), opposed
contrast°, compare 3; comparison 1; differ 1; difference 1; differentiate 1; disparity; distinction 1; diversity 1; oppose 3; opposition 1; pit[1] 6; relief 3; set 13; show 12b; variety 2
contrast with *or* to, in, unlike 3
contrasted with, as, instead 2
contrasting, comparison 1; different 1; dissimilar; distinct 2; opposed; opposing; opposite 2
contrasting (with *or* to), unlike 1
contrastive, different 1
contrastive (with *or* to), different 1; unlike 1
contrasts, vicissitude 2
contravene, break 9; contradict 2; disobey; fight 2; fly 6; infringe 1; traverse 3; violate 1
contravened, broken 5
contravene (the law), transgress 2
contravening, violation 1
contravention, breach 1; infringement; violation 1
contretemps, misfortune 2; run-in
contribute°, afford 2; chip 4a; cooperate 2; donate; far 6b; finger 6; give 1; impart 1; invest 2; lay[1] 18b; lend 2; open 25; pay 11b; pitch[1] 5; provide 1; put 28d, 28e; stump 4; supply 1, 2
contribute (to), advance 2; bring 5; contribute 2; lead 6; make 30c; part 11; participate; patronize 3; subscribe 2
contributed, voluntary 1
contribution, benevolence 2; charity 3; donation 1, 2; endowment 1; gift 1; grant 3; job 2; maintenance 3; offering; payment 2; philanthropy 2; present[2] 2; submission 2; subsidy; tax 1; tribute 2
contributions, largess
contributor, donor; participant 1; party 4; philanthropist
contributory, instrumental; supplementary 2
contrite, apologetic; bad 8; guilty 2; penitent;

contrite (*cont.*) regretful; remorseful; repentant; sorry 1; terrible 2
contriteness, guilt 2; pang 2; remorse 1
contrition, compunction 1; guilt 2; pang 2; penance 1; penitence; regret 2; remorse 1
  contrition, feel, repent
contrivance, arrangement 2; artifice 2; contraption; deceit 2; device 1, 2; dodge 4; expedient 3; fabrication 2; fixture 2; gadget; gimmick 2; implement 1; instrument 1; invention 1, 2; machine 1; mechanism 1; medium 5; subterfuge; thing 5; tool 1; wile
contrive, arrange 2; brew 2; compose 2; conceive 2; create 2; design 2; devise 1; discover 3; draw 6; engineer 4; fix 12; form 8; frame 7; generate 4; get 9; hatch 2; improvise 2; invent 1; make 1; maneuver 3; manufacture 2; muddle 3; originate 1; project 3; scheme 4; stage 5; strike 16; think 6; wangle; weave 3; work 19e
contrived, affected 1; artificial 3; forced; glossy 2; imaginative 2; intentional; labored 2; mannered; premeditated; rhetorical 2; spurious; studied; unnatural 2
contriver, engineer 1; mastermind 2
contriving, calculating; disingenuous; invention 1
control°, administer 1; assurance 4; auspices; authority 1; balance 5; better 8; boss 2; bottle 4a; brake 1; bridle 1, 2; care 3; chain 2; charge 4; charm 5; check 2, 14, 15; clutch 2b; command 2, 3, 5, 7, 8; conduct 2, 3; contain 3; cool 8; curb 1, 2; department 2; direct 1; direction 1; discipline 4, 7; disposition 4; dominate 1; domination 1; dominion 1; drive 2; economy 1; finger 10; fluency; govern 1, 2; government 1; grasp 4; grip 2; guard 2; guidance 1; guide 3; hand 8; handle 3, 4, 6; head 11; helm 2; hold 14a, 15a, 17a, 26; influence 1; inhibit; jurisdiction; keep 1; leadership; lock[1] 2; manacle 3; manage 1, 2, 3; management 1; manipulate 1, 2; master 8; measure 8, 12; mold[1] 6; monopolize; mortify 2; normalize; obsess; occupation 2; operate 2; operation 2; oversee; patent 1; peg 4; philosophy 3; pilot 3; poise 2; police 3; possess 3; possession 1; power 2, 3; predominance; predominate; preponderance 2; preside; prevalence 2; prevent; prevention; providence 2; ration 4; regiment; regulate 1, 2; regulation 1; reign 1, 2; rein 1, 2, 3; remedy 3; repress; repression 2; resist 1; restrain 1; restraint 1, 2, 3; rule 2, 5, 6; run 10; self-control 1, 2; shackle 4; shape 7; steer 1; stem[2] 1; stifle 2; stint 2, 3; string 6; subdue 1, 3; supervise; suppress 3; suppression; supremacy 2; swallow 4; sway 4; tame 6; time 22; upper 6; withhold 1; work 10
  control, go out of *or* **lose,** flare 3; piece 6
  control, have under, dominate 1; finger 10
  control, in, rampant 2; self-contained 1
  control, keep under, contain 3; suppress 3
  control, out of, obstreperous; rampage 2; rampant 1; uncontrolled
  control, under, cool 2
  control, under *or* **in,** line[1] 19c
  control, under (someone's), tame 3; thumb 5
  control (of), understanding 3
  control of, assume *or* **take** *or* **usurp** *or* **gain,** possess 3; take 37
  control (of), be in, control 1; dominate 1; rule 5; string 6
  control of, under (the), subject 8a; under 2
  control of *or* **over oneself, get,** compose 4; simmer 3
  control oneself, compose 4; simmer 3
  control (over), gain, prevail 1; subdue 1
  control over, have *or* **maintain** *or* **wield,** command 2
controllable, gentle 2; manageable; navigable 2
controlled, collected; cool 2; impassive; moderate 1; phlegmatic 2; poised 1; possessed; reserved; sedate 1; self-contained 1, 2; self-possessed; Spartan; steady 4; temperate 1
  controlled by, subject 8a; under 2
controller, master 1; superintendent; supervisor
controlling, dominant 1; influential; master 5; obsessive; possessive 2; predominant; prevention; suppression
control(s), bond 1
control things *or* **affairs** *or* **matters,** shot 9
controversial°, debatable; disputable; doubtful 3; moot 1; problematic; thorny 2
controversy°, argument 1; combat 2; conflict 2; contest 2; debate 1; disagreement 3; dispute 3; faction 2; friction 2; issue 4; misunderstanding 2; quarrel 1; rivalry; variance 2
controvert, contradict 1; deny 1; disprove; rebut; traverse 3
contumacious, contrary 2; defiant; disobedient 2; insubordinate; mutinous 2; obstinate; rebellious 1; recalcitrant; rogue 2; self-willed; unruly
contumaciousness, obstinacy
contumacy, obstinacy
contumelious, contemptuous; derisory; disdainful; lofty 4; regal 1; sarcastic; scornful; vituperative
contumely, contempt; derision; indignity; mockery 1; patronage 3; sarcasm; scorn 1
contuse, bruise 2
contusion, bruise 1; welt 2; wound 1
conundrum, enigma; mystery 1; problem 2; puzzle 4; riddle[1]
conurbation, city
convalesce°, improve 3; mend 3; recover 2; recuperate
convalescence, improvement 2; recovery 1
convalescent, mend 5
convalescent home *or* **facility,** hospital; sanitarium
convalescing, mend 5
convenant, pledge 1; understand 2
convene, assemble 1; call 4; collect 1; gather 2; get 50b; hold 10; mass 8; meet[1] 2; muster 1; raise 5; rally 3; sit 2; summon 1
convenience, accommodate 5; accommodation 3; lavatory; prepared 4; time 8; toilet 1
  convenience, at one's, leisure 3b
convenience(s), facility 2b
convenient°, available; cozy 2; favorable 1; fit[1] 1; hand 9a; handy 1; lucky 2; opportune 2; ready 7; ready-made 2; right 3; seasonable; timely
  convenient (to *or* **for), be,** serve 4; suit 2
convention°, assembly 2; ceremony 2; company 2; conference; council 1; custom 1; etiquette; form 6; formality 1; gathering; get-together; habit 1; meeting 2; motif; muster 2; observance 2; ordinary 3; practice 1; rally 1; ritual 3; rule 3; tradition; usage 1
  convention hall *or* **room,** chamber 2; hall 2
conventional°, bourgeois 1; ceremonious 2; common 1; conservative 2; correct 7; customary 1; everyday 2; formal 1; habitual 1; hidebound; narrow-minded; normal 1; ordinary 2; orthodox; parochial; popular 2; ready-made 3; regular 1, 11; ritual 2; routine 3; sedate 2; set 29; square 6; standard 6; standing 1; staple 2; step 7; stuffy 2; time-honored; traditional; typical 1, 2; usual
conventionalism, convention 2
conventionality, formality 1
conventionalize, fix 10
conventionally, generally 1
convention(s), ceremony 2; code 3; protocol 1
converge°, center 2; collect 1; concentrate 1; focus 4; press 7
convergence, focus 1; meeting 3; press 8
convergent, focal
converging, meeting 3; press 8
conversance, familiarity 1; knowledge 3
conversancy, exposure 3
conversant (with), acquainted 2; familiar 4; home 5; informed 2; versed
conversant with, be, understand 1
conversation°, chat 1; dialogue 1; discussion; gossip 1; interview 1; palaver 3; rap 5; repartee; talk 15; word 1
conversational, verbal 1; vernacular 2
conversation piece, curiosity 3; rarity 1
converse°, chat 2; chew 2; confer 1; opposite 3; palaver 5; rag[1] 5a; rap 3; reverse 1, 6; speak 1
  converse about, discuss
  converse in, speak 3; talk 1
  converse (with), communicate 2; speak 2; talk 12
conversely, vice versa
conversion, adaptation 1; alteration; difference 3; passage 5; transformation; transition 1; translation 1, 2; variation 1
convert°, alter; change 8; differentiate 2; make 9; persuade 2; process 3; redeem 4; reduce 4; render 6; tailor 2; transform; translate 1, 2; turn 3; win 3
  convert into, resolve 4
  convert to, turn 23d
convert(ed) into, be, resolve 4
convertible, changeable 2; liquid 4
converting, variation 1
convey, bear 1; bring 2, 4; carry 1, 2, 3; cart 2; cede; channel 5; communicate 1; conduct 5; deliver 1;

cordial° (*cont.*) respectful; sociable; warm 3
cordial, become (more), thaw 2
cordiality, civility; fellowship 3; hospitality; kindness 1; warmth 2
cordially, kindly 2; warmly 2; well[1] 5
cordon, circle 1; line[1] 6
cordon off, siege 2
cord(s), bond 1
core°, base[1] 3; body 4; bosom 2; bowels; breast 2; center 1; elixir 3; essence 1; fabric 2; focus 1; gist; heart 3; hub; inside 1; interior 6; kernel 2; meat 2; nub 2; nucleus; pith 1; point 5; quintessence; spirit 6; substance 2; theme 1; thick 10
core, to the, home 10; through 7
cork, plug 1, 4; seal 3; stopper; tap[2] 2; top 2
corked-up, pent-up
corkscrew, screw 2; spiral 1, 2
corm, seed 1
cormorant, greedy 1; voracious 1
corn, cure 3; grain 2; lump[1] 2; preserve 3; provender 2; salt 7; sentimentality; whiskey
corncob, pipe 2
corned, salt 10
corner, angle[1] 1; bend 1; buttonhole 1; fix 17; hole 5; monopolize; nook 1; part 6; predicament; turn 27; universe 2
corners, recess 3
cornerstone, essence 1; fundamental 2; groundwork; keystone
corniness, sentimentality
cornucopia, wealth 2
corny, banal; mushy 2; sentimental 2; stock 7
corollary, law 3
corona, glory 4; halo; ring[1] 1
coronach, keen[2] 2; lament 2
coronate, crown 4
coronation, installation 1
coronet, crown 1; ring[1] 1
corporal, mortal 2; physical
corporation, business 4; company 4; employer 2; firm 6; gut 2; office 1; outfit 3; paunch; pot 3; stomach 1; trust 4
corporeal, earthly 2; flesh 3; material 5; mortal 2; physical; real 2; sensible 2; tangible; worldly 1
corporeality, flesh 2; substance 4
corporealization, image 4
corps°, band[2] 1; body 6; cohort 2; crew; group 1; party 2; shift 4
corpse°, body 1; cadaver; remain 4b; stiff 11
corpselike, white 2
corpulence, fat 5; fatness; obesity
corpulent, burly; fat 1; gross 1; heavy 11; large 1; obese; plump[1] 1; rotund 3; stout 1; weighty 1
corpus juris, code 1; law 2
corral, catch 2; crowd 4; enclose 1; enclosure 1; herd 4; pen[2] 1, 2; round 19; stall[1] 3
corralling, roundup 1
correct°, accurate 1; adjust 1; amend 2; appropriate 1; castigate; ceremonious 2; chasten 1; chastise; counteract; cure 2; decorous; discipline 8; due 2; edit 1; ethical; exact 1; expedient 1; factual 2; faithful 2; faultless; fit[1] 1; fix 3; formal 1; good 3; grade 7; honorable 2; impeccable; improve 1; incorrupt 2; ladylike; legitimate 3; make 31a; mark 14; mend 1, 2; OK 2; order 10b; perfect 4, 5, 9; place 11a; polish 2; precise 1; principled; proper 2, 3, 4, 6; punish 1; pure 3; rectify; reform 1; regular 6; remedy 4; retouch; revise 1; right 1, 2, 19; righteous 2; rightful 1; safe 4; square 3; straighten 1; suitable; tasteful; true 1, 3
correct, be, nail 5; serve 5
correcting, adjustment 1
correction°, adjustment 1; amendment 1; discipline 2; punishment 1; reform 3; revision; rod 2
correctional, penal; punitive
correctional institution, prison
corrective, therapeutic
correctly, appropriately; duly 1; exactly 1; precisely 2; properly 1, 2; right 17; true 5; well[1] 3
correctness, accuracy; decorum 2; formality 2; precision 1; propriety 1; purity 1; rectitude; taste 6; truth 1
correlate, compare 3; coordinate 2; relate 1
correlate to *or* with, answer 5; parallel 4
correlated, proportional; related 1
correlation, bearing 3; connection 2; parallel 3; proportion 2; ratio
correlative, coordinate 3; reciprocal
correspond°, accord 1; agree 1; check 5; coincide; compare 2; fit[1] 6; hang 11b; match 7; overlap 2; reciprocate; tally 1; tie 6a
correspond (to), equal 5; parallel 4; resemble; similar 2; square 11
correspond (with), communicate 2; conform 2; contact 4; parallel 4; square 11; write 2
correspondence, accord 4; coincidence 1; equality 2; kinship 2; letter 2; likeness 1; mail 1; note 2; parallel 3; parity 2; proportion 2; ratio; relation 1; resemblance; symmetry; tune 2; uniformity 1; writing 2
correspondent°, coordinate 3; equal 2; journalist; parallel 1; reporter; writer
correspondent (to), like[1] 1; relative 2; synonymous
correspondents, press 10b
corresponding, equal 2; equivalent 1; identical 2; matching 1, 2; parallel 1; proportional; respective; twin 2
corresponding exactly, unison
corresponding (to), like[1] 1; relative 2; synonymous
corridor°, gully; hall 1; lobby 1; passage 4
corrie, valley
corrigendum, correction 1
corroborate, authenticate; bear 9; certify 1; check 3; confirm 3; document 2; establish 3; manifest 3; prove 1; ratify; seal 4; show 3; strengthen 1; substantiate; support 6; verify; witness 3
corroborating witness, witness 2
corroboration, check 15; evidence 1; identification 1; proof 1; reflection 4; seal 2; testimony
corroborator, witness 2
corrode, attack 5; erode; etch 1; gnaw 2; rot 1; wear 3
corroded, pitted; rotten 2
corroding, erosion
corrosion, attack 9; erosion; rot 3; wear 9
corrosive, caustic 1; tart[1] 2; trenchant
corrugate, contract 5; furrow 3; pucker 1; wrinkle[1] 2
corrugation, furrow 1; wrinkle[1] 1
corrupt°, abandoned 2; abusive 3; adulterate; bad 2, 3; base[2] 6; bent 2; bribe 2; buy 3; contaminate; criminal 2; debase 2; decadent 2; degenerate 1; demoralize 2; desecrate; diabolic 2; dirty 3; disgraceful 1; dishonest; dishonorable 2; dissolute; evil 1, 2; filthy 3; fix 11; foul 7; garble 1; get 31d; ill 2; immoral 1; improper 3; jaundiced 1; lawless 2; loose 7; lost 7; mercenary 2; miscreant 2; perfidious; pervert 1, 2; perverted; pollute 2; profane 4; profligate 1; putrid; rancid; rank[2] 2, 3; rotten 1, 3; ruin 11; satanic 1; seduce 2; shameful; sinful; sinister 2; sordid 1; stain 5; taint 2; ugly 2; ungodly 1; unscrupulous; unwholesome 2; venal; vicious 1; vile 1; villainous 1; vitiate 1, 2; wicked 2; wrong 1
corrupted, bent 2; degenerate 1; impure 4
corruptible, accommodating 2; venal
corrupting, ruin 2; unwholesome 2
corruption, abuse 7; degradation 1; dirt 3; evil 6; filth 2, 3; graft[2]; impurity 3; misuse 2; parody 2; perversion 1; pollution; profligacy 1; prostitution 2; rot 3; ruin 2; sin 2; vice 1
corruptive, poisonous 2
corruptness, degradation 1
corsage, buttonhole 2
corsair, pirate 1; robber; thief 3
cortege, escort 1; line[1] 6; march 2; parade 1; procession 1; retinue; suite 3; train 2
coruscate, blink 2; flash 4; radiate 1; shine 1; sparkle 1; twinkle 1
coruscating, brilliant 1; nimble 2; radiant 1; scintillating 1; shiny; twinkle 2
coruscation, flash 1; radiance 1; sparkle 3; twinkle 2
coryza, chill 2; cold 10
Cosa Nostra, syndicate 2; underworld 1
cosh, club 1
cosher, pamper
cosign, enter 8
cosignatory, party 5
cosigner, party 5
cosmetic, flashy 2; superficial 2; token 5
cosmetics, makeup 1; paint 3
cosmic, extensive 1; universal 2
cosmopolitan, cultivated; extensive 1; genteel 3; international; sophisticated 1; suave; worldly 2
cosmopolitanism, sophistication 1; style 4
cosmos, nature 2; universe 1; world 1
cosmos, the, creation 2
cosset, baby 2; cater 2; cherish 2; coddle; humor 4; indulge 2; make 34b; pamper; pet[1] 5
cost°, charge 2; damage 2; expenditure; expense 1, 2; fare 2; fee 1; fetch 3; hire 5; outlay; price 1, 2; quotation 2; subscription 1; toll[2] 2
cost, at any, mean[3] 2b
cost, at great, dear 4; dearly 3
cost, at no, free 13
cost, at *or* below, cheap 7
cost (or obligation), without, free 4
cost (out), price 5

cost-effective, economic 2; economical 1; profitable 1; worthwhile 1
cost-free, free 4
costly, dear 2; expensive; extravagant 3; high 3; hollow 5; invaluable; plush; precious 1; premium 4b; priceless 1; pricey; rich 4; sumptuous
costs, expense 1; overhead 3; upkeep 2
costume°, disguise 3; dress 5a, 6; ensemble 1; garments; getup 1; habit 3; outfit 2; robe 1, 2; suit 3; uniform 3
costume party, masquerade 1
costumier, tailor 1
cot°, cabin 1; cottage; hut
cote, enclosure 1; hut; shed[1]; stall[1] 3
cotemporaneous, parallel 1
cotemporary, parallel 1
coterie, cabal 2; camp[1] 2; circle 2; clan 2; clique; company 2; crowd 2; faction 1; fellowship 2; fraternity 1; gang 2; group 1; junta; pack 3; party 3; ring[1] 3; school 2; set 26; world 3
cotquean, prostitute 1
cottage°, cabin 1; lodge 1; outside 7
cotton (on) (to), get 44c; realize 2; understand 1
couch°, lounge 4; phrase 4; word 10
cough, hack[1] 2
  cough up, pay 11b; stump 4
cough drop, pill 1
coulee, canyon
couloir, pass 23
council°, assembly 2; board 4; body 6; cabinet 2; committee; convention 1; diet[2]; house 3; municipal; parliament 2
councillor, representative 3
counsel°, advice 1; advise 1; adviser; advocate 3; caution 1, 3; counselor; direct 2; enlighten; guidance 2; guide 3, 5; lawyer; preach 2; recommend 1; recommendation 1; suggestion 1; urge 3; warn 2; warning 1
  counsel, keep (one's), tongue 3
  counsel (with *or* **of), take,** confer 1; consult 1
counseling, advisory 1; guidance 2
counselor°, adviser; consultant 2; counsel 3; guide 5; teacher; therapist
counselor(-at-law), advocate 3; counselor; lawyer
count°, calculate; calculation 2; detail 1; enumerate 2; figure 8; include 3; itemize; lord 2; matter 6; number 4; poll 2, 3; put 16f; rank[1] 7; rate[1] 6, 8; reckon 2; score 1; signify 2; tally 3; weigh 5
  count (for), score 12
  count in, involve 1
  count on, anticipate 3; bargain 4; depend 2; gamble 2; go 26b; hope 3; lean[2] 4a; look 6c, 7b; reckon 4; store 5; swear 3
  count on *or* **upon,** count 3; expect 3; fall 9; figure 11a; rely 1; trust 5
  count out, exclude 2; leave[1] 10
  count up *or* **off** *or* **out,** add 2; count 1; tally 2
countable, finite
countenance, abet 2; allow 4; approve 1; authorize; endorse 1; expression 3; face 1; feature 3; front 4; hold 23; look 14; mug 2; permission; permit 1; sanction 2, 6; stand 3; stomach 3; subscribe 1; support 3; tolerate 1
countenancing, allowance 1
counter°, bar 7; check 18; chip 2; contest 3; contradict 2; cover 10; fly 6; foil[1]; frustrate 1; militate 1; oppose 1; respond 1; retaliate; retort 2; stall[1] 2; stand 16; traverse 3
  counter to, go *or* **act,** disobey; traverse 3
  counter to, go *or* **run,** fly 6
counter-accusation, recrimination
counteract°, cancel 3; contradict 2; correct 4; destroy 3; fly 6; frustrate 1; neutralize; offset 1; resist 1
counteractant, antidote; preventive 4; remedy 1
counteraction, backlash; offset 2; opposition 1; reaction 2
counteractive, preventive 2
counteragent, preventive 4
counter-argument, rebuttal
counterattack, oppose 1; recrimination
counterbalance, balance 3; cancel 3; compensate 2; correct 4; counteract; cover 10; neutralize; offset 1, 2; oppose 3; parallel 3; reaction 2
countercharge, recrimination
counterclockwise, backward 6
counterespionage agent, operative 4
counterfeit°, act 8; affected 2; affect[2] 1; artificial 2; assume 4; assumed 2; base[2] 5; bogus; deceitful; deceptive 2; disguise 2; dishonest; dissimulate; duff; erroneous; fabricate 3; factitious; fake 1, 3, 5; false 3; fictitious 2; forge 3; forgery 2; fraudulent 1; glorified 2; glossy 2; hollow 4; imitation 3; mimic 5; mock 3; phony 1, 2; plastic 3; pretended; quack 2; sham 1, 2; spurious; synthetic; unreal 3
counterfeit(ed), false 4
counterfeiter°, phony 3
counterfeiting, forgery 1
counterfoil, check 17; receipt 1; stub 2; tally 5
counterglow, light[1] 1
counterintelligence agent, operative 4
counterirritant, antidote
counter-jumper, salesperson; seller
countermand, cancel 1; reverse 3
countermeasure, preventive 4; remedy 1, 2
countermine, fly 6
counterpane, cover 13; spread 14
counterpart, double 7; equal 4; equivalent 2; fellow 3; image 2; like[1] 7; match 1; mate 3; parallel 2; spit 2; tally 5; twin 1
counterpoise, balance 3; compensate 2; offset 1
counterproductive, prejudicial
countersign, endorse 2; enter 8; execute 2; password; sign 7; underwrite 2; witness 5
countersignature, endorsement 2
counterspy, operative 4
counter-statement, answer 2
countervail, cancel 3; fly 6; militate 1; offset 1
  countervail (against), foil[1]; resist 1
countess, peer[1] 1
counting, calculation 1; inclusion; reckoning 1
  counting, not, apart 4; besides 2; except 1; exclusive 4
  counting up, addition 2
countless, frequent 1; great 2; inestimable 2; limitless; lot 5c; numberless; umpteen; untold 1
countrified, rustic 2
country°, dominion 2; fatherland; land 4; monarchy 1; nation; pastoral 2; patch 3; province 2; provincial 2; rural 1; state 4
  country, the, stick[2] 3
country boy *or* **girl,** rustic 3
country cousin, provincial 3; rustic 3
country-like, green 1
countryman, peasant; rustic 3
countryside, country 3; landscape; nature 3; province 4
  countryside, the, stick[2] 3
countrywide, national 1
countrywoman, peasant; rustic 3
count sheep, rest[1] 6; retire 3
county, genteel 1; high-class 2; local 2
coup, accomplishment 2; hit 11; killing 2; plum; triumph 1
coup de grâce, clincher; kill 9; knockout 1
  coup de grâce, administer *or* **deliver** *or* **give the,** finish 4; kill 1
coup (d'état), revolt 1; revolution 1; uprising
coup de théâtre, sensation 3
coup d'oeil, glance 4; minute[1] 1
couple°, brace 4; connect 3; fix 1; handful 1; hitch 1; join 1; link 3; marry 2; mate 4, 5; pair 1; relate 1; tack 5; tie 1; twin 3
  couple of, a, couple 2
  couple (with), lay[1] 8
coupled, double 1; package 3; related 1
coupler, brace 3; terminal 5
coupling, brace 3; connection 1; link 1; marriage 3; sex 2; terminal 5; union 3
courage°, backbone 3; bottle 2; bravery; character 3; chivalry; confidence 2; daring 1; enterprise 2; fortitude; grit; gumption 2; gut 3a; heart 2; nerve 1; pluck 1; prowess 2; spirit 5; spunk; stamina
courageous°, adventurous; audacious 1; bold 1; brave 1; confident 2; daredevil 2; daring 2; dauntless; enterprising; fearless; gallant 1; game 8; gritty 2; hardy 2; heroic 1; indomitable; intrepid; manly; martial 2; stalwart 3; stout 2; superhuman 1; venturesome 1; yeomanly
courageousness, grit
courier, messenger; runner 2
course°, approach 5; beat 12; chain 1; channel 4; circuit 2; circulate 1; circulation 1; current 6; cycle 1; direction 2; discipline 5; drift 3; flow 1, 5; form 4; hunt 1, 3; journey 2; line[1] 9; mean[3] 4a; medium 5; method 1; mode[1] 1; movement 7; orbit 1; pass 2; passage 3; path 2, 3; plate 2; pour 1; procedure; procession 2; progress 3; progression 3; race[1] 2; road 1; round 12; route 1; run 38; sequence; span 2; spell[1] 1; step 10a; stream 4; succession 2; tack 3; term 3; thread 2; tier; track 2; trail 1; trajectory; trend 1; vein 3, 5; way 3, 4
  course, off, erroneous
  course of, in *or* **over** *or* **during the,** for 10; over 4; through 2
course (of action), measure 6; procedure; process 1; scheme 1; way 1
course (of study), education 3; subject 2
course of treatment, cure 1
court, area 5; bar 5; chase 2; cultivate 4; flatter 1; go 33c, 39b; play 18;

peculiarity 1; quirk; trick 6; twist 10
crotchety, bad 7; cantankerous; capricious; cranky 2; cross 6; gruff 1; irritable; kinky 1; moody 2; nasty 4; peevish; petulant; querulous; sullen; surly; temperamental 1; testy; ugly 4; waspish
crouch°, duck 1; stoop 3
crouch (down), stoop 1
croup, tail 2
crow, bluster 2; boast 2; brag; chuckle 1; glory 5; swagger 2; talk 8
crow (about *or* over), gloat; revel 1
crow flies, as the, directly 1; right 13; straight 12
crowd°, assembly 1; attendance 2; bunch 2, 4; circle 2; clan 2; clique; cluster 2, 3; company 2; concentrate 3; crush 6; faction 1; fill 1; flock 1, 2; flood 6; fraternity 1; gang 1; gate 2; gather 2; group 1; herd 1; hoi polloi; host[2]; huddle 1; hustle 3; jam 1, 5; knot 2; mass 2; mill 6; mob 1; number 2; pack 3, 5; populace; pour 4; press 7, 8; rabble 1; ram 1; set 26; squeeze 4, 9; stuff 6; swarm 2; throng 1; turnout 1; wedge 3; world 3
crowd, the, people 3
crowd (around), mob 3
crowd in *or* into, pile[1] 6; throng 2
crowd on *or* onto, pile[1] 8a
crowd round, besiege 2
crowd together, huddle 3
crowded, alive 4; congested; cramped; dense 2; full 1; loaded 1; mobbed; packed; populous; solid 2; thick 2
crowded with, be, abound 2
crowding, press 8
crowd-pleaser, attraction 2; sensation 3
crowd-puller, attraction 2; sensation 3
crowd(s), many 3; score 4
crowing, chuckle 2; egotistical
crown°, capital 1; complete 6; crest 3; garland 1, 2; glory 4; hit 1; peak 2; pinnacle; ring[1] 1; spire 2; strike 1; summit; tip[1] 1, 2; title 4; top 1, 4; vertex
crowned head, king; monarch 1
crowning, installation 1; maximum 3; supreme 3
crowning blow, clincher
crowning point, pinnacle
crow's-feet, furrow 1
crow's-foot, line[1] 3; wrinkle[1] 1
crow to pluck, grievance 2
crozier, staff 1
CRT, monitor 2; terminal 4
crucial°, acute 2; critical 2; desperate 4; essence 4; fateful 1; fundamental 1; good 2; grave[2] 2; great 4; heavy 3; imperative 1; indispensable 1; key 6; landmark 3; main 3; major 2; momentous; pivotal; pressing; serious 2; strategic; vital 1, 2; weighty 2
crucifix, cross 1
crucify, punish 2; torment 1
crud, filth 1
cruddy, filthy 1
crude°, amateur 2; animal 4; artless 3; barbarian 3; bawdy; beastly 1; bluff[2] 1; boorish; brutal 2; coarse 1, 2; earthy; garish; gaudy; gross 3; harsh 1; homespun; impolite; incomplete; insolent; lewd; low[1] 3; merciless; nasty 3; outspoken; primitive 2; prurient 2; racy 2; raw 2; rough 6, 9; rude 4; rudimentary 2; rugged 4; rustic 2; severe 6; sketchy; tactless; tasteless 1; uncivilized 1; ungraceful 2; unrefined 2; vulgar 2
crudely, poorly 1; roughly 3; severely 6; sketchily
crudeness, indelicacy; severity 6; vulgarity 1
cruel°, abusive 2; atrocious 1; beastly 1; bitter 3, 5; bloodthirsty; brutal 1; coldblooded 3; coldhearted; cutthroat 3; devilish; diabolic 2; ferocious; fiendish; fierce 1; flagrant; ghoulish 2; grim 2; hard 4; harsh 2; heartless; hurtful 2; ill 3; implacable; inhuman 1; mean[2] 4; merciless; monstrous 2; murderous 1; nasty 4; outrageous 2; relentless 1; remorseless 1; rough 5; ruthless; sadistic; sanguinary 1; savage 2; severe 1; stiff 2; unkind; unmerciful; violent 1; wanton 4; wicked 4
cruel and unusual, stiff 2
cruelly, badly 6; roughly 2; severely 2
cruelty, atrocity 1; barbarity; devilry 2; outrage 1; severity 1; violence 2
cruise°, excursion 1; journey 1, 3; navigate 1; passage 7; prowl 2; roam; roll 3; sail 2; tour 4; trip 7; wander 1
crumb°, bit 1; fragment 1; grain 3; little 10; modicum; morsel 2; particle; scrap[1] 1; speck
crumble°, collapse 1; come 7; decay 1b; deteriorate 2; disintegrate; fall 7; grind 1; piece 6; pulverize 1; surrender 2; way 11a; weaken 2
crumbling, decrepit 2; dilapidated; leg 7; old 3; ramshackle; rotten 2; tumble-down; unsound 1
crumbly, crisp 1; rotten 2
crumbs, leftover; remain 4a
crumby, inferior 3
crummy, bad 1; bum 4; inferior 3; pathetic 2; plastic 3; shabby 3; sleazy 2
crumpet, chap; fluff 2; girl 1; head 1
crumple°, buckle 2; collapse 1; crush 2; fall 2; rumple; way 11a; wrinkle[1] 2
crunch°, crush 1; mill 5; munch; payoff 2; pinch 8
crunch, the, scrape 8
crunchy, crisp 1
crusade°, battle 2; campaign 1; drive 10; movement 5; pilgrimage; quest 1
crusader, pilgrim
crush°, back 7b; beat 2; best 11; bow 4; break 4, 7; conquer 1; crumple; crunch 1; defeat 1; demolish 2; demoralize 1; destroy 1; floor 4; grind 1, 5; herd 1; humble 4; jam 5; mangle; mill 5; mob 1; mortify 1; oppress 2; overcome 1; overpower 1; overwhelm 1; pinch 2; pound[1] 2; powder 3; press 1, 3, 8; prostrate 2; pulverize 1, 2; put 16b, 16e; quash 2; quell 1; rout 2; ruin 7; rumple; shatter 2, 3; slaughter 4; squeeze 1, 9; stifle 3; subdue 1; subject 10; subjugate; suppress 2; telescope 3; throng 1; trample 1, 3; waste 11
crush in *or* into, pile[1] 6; throng 2
crush on, have a, adore 3; love 7
crushed, broken 3; brokenhearted; fine[1] 7; heartbroken; prostrate 4; smitten 1; stricken 1; undone[1] 1
crushed, be, break 16c
crusher, mill 1
crushing, overthrow 2; overwhelming 1; pressure 2; suppression; tragic; victory
crust, gall[1] 2; heel[1] 1; nerve 2; sauce 2; scale[2] 2; skin 2
crusty, cantankerous; cross 6; curt; gruff 1; irritable; moody 2; peevish; perverse 2; querulous; short-tempered; snappish 2; sour 4; stern 2; sullen; surly; tart[1] 2; testy; touchy 1; waspish
crux, crunch 2; essence 1; heart 3; keystone; meat 2; nub 2; pass 26; pith 1; pivot 2; point 5; substance 2
cry°, bawl 2; call 13; cheer 4; complain; exclamation; grieve 2; howl 1, 2; interjection; moan 3; plea 1; roar 1, 3; scream 1, 3; shout 2; shriek 1, 2; snivel; sob; squawk 1; weep 1; whoop 1; yell 2
cry (for), plead 1
cry (out), call 1; exclaim; shout 1; sound[1] 8; whoop 2
cry out for, cry 3; demand 3
cry over, bewail; regret 1
crybaby, milksop; sissy
crying, instant 4; lamentation; tearful
crypt°, grave[1]; safe 7; sepulcher; tomb
cryptic°, ambiguous 2; dark 7; hidden; incomprehensible; indefinite 2; mysterious 2; mystical; obscure 2; opaque 2; profound 1; secret 2
cryptogram, code 2; symbol
cryptographic, secret 2
cry quits, surrender 2; yield 1
crystal, glass 1
(crystal-)clear, explicit 1; graphic 1; plain 2
crystal gazer, fortuneteller; psychic 3; seer
crystal gazing, prophecy 1
crystalline, clear 2; transparent 1
crystallization, formation 1
crystallize, jell 2; solidify 1; stiffen 1
crystal set, radio 1
cub, pup
cubbyhole, compartment; sanctum 2
cube, block 1; brick 1; cake 2; chop 2; cut 17a; loaf[1] 1; lump[1] 1; square 7
cubed, square 1
cubic, solid 1; square 1
cubicle, booth 2; cell; compartment; room 2; stall[1] 2
cuckoldry, infidelity 2
cuckoo, crazy 1; deranged; foolish 2; inane; insane 1; mad 1; preposterous; stupid 2; twist 11
cuddle°, caress 1, 2; clinch 2; embrace 1; fondle; hug 1; love 6; pet[1] 4; snuggle
cuddle (up), nestle
cuddly, lovable
cudgel, beat 1; club 1, 5; hit 1; lambaste 1; pound[1] 1; strike 1
cudgel (one's) brains, exert 2; wonder 3
cudgels for, take up the, maintain 4; stick[1] 18; support 1
cue°, hint 3; key 2; lead 14; nod 5; note 6; prod 6; prompt 4, 6; remind; reminder; sign 2; symptom
cuesta, cliff
cuff, clip[2] 3, 6; hit 1; jab 2, 4; knock 9; manacle 2; poke 2, 6; punch[1] 1, 2; rap 4; ring[1] 1; slap 1, 4; strike 1
cuff, off the, extemporaneous; offhand 3, 4; scratch 6; spontaneous 1; unpremeditated

curvet, caper 1, 3; cavort; frisk 1; frolic 3; leap 2; prance 1
curvilinear, round 4; streamlined 1; tortuous 1
curving(s), meander 2
curvy, meandering; tortuous 1
cushion°, deaden 2; insulate 2; moderate 5; pad 1, 4; soften 2; temper 5
cushioning, euphemism
cushiony, soft 1
cushy, fat 4; soft 2
cusp, angle[1] 1
cuspidate, pointed 1
cuss, swear 2
cuss word, expletive 2
custodian, guard 3; guardian; keeper; monitor 1; watchman
custodianship, custody 1; security 4
custody°, captivity; care 3; charge 4; detention; grip 2; hand 8; imprisonment; oversight 2; possession 1; protection 2; safekeeping; security 4; trust 3
custody, in, arrest 6
custody, take into, arrest 2; nab; pick 8h; pinch 4; pull 11b; run 27; seize 2
custom°, clientele; convention 2; form 6; formality 1; habit 1, 2; institution 4; observance 2; patronage 2; policy; practice 1; procedure; public 9; rite; ritual 3; routine 1; specially; trade 4; tradition; usage 1; way 2
customarily, average 2; frequently 2; generally 1; ordinarily; popularly; rule 4; usually
customary°, accustomed 1; average 3; common 1; conventional; correct 7; everyday 2; familiar 2; formal 1; frequent 1; general 2; going 2; habitual 1; natural 1; normal 1; ordinary 1; orthodox; popular 2; prevailing 1; prevalent 1; regular 1; regulation 5; ritual 2; routine 3; set 29; standard 6; standing 1; staple 2; stock 7; time-honored; traditional; typical 2; usual; well-known 1
customary, be, obtain 3
customer°, buyer; chap; client; fellow 1; guest; guy 1; patron 2; regular 12; steady 10
customer of, be a, patronize 2
customer acceptance, demand 6
customer base, market 3
customers, clientele; public 9; trade 4
customization, adaptation 2
customize, accommodate 1
customized, tailor-made 2
custom-made, fitted; tailor-made 1, 2
custom(s), civilization 2; code 3; culture 2; custom 2; duty 3; etiquette; moral 4; morality 2; protocol 1; rule 3; tax 1
cut°, abbreviate 2; abridge; abstract 4; accomplish; adulterate; brief 2; carve 1; chisel 1; chop 1, 3; cleave; curtail; cut 16a, 17a; decrease 2; digest 4; dilute 1; diminish 1; disregard 2; doctor 4; edit 3; engrave 1; engraving 1; etch 1; excavation; execute 1; figure 1; fix 13; flaw 2; form 1; furrow 1, 2; gash 1, 2; groove; gully; hack[1] 1, 3; hairdo; ignore 2; incision; indentation; interest 4; isolate; knife 2; lacerate; lash[1] 2; leak 2; lower[1] 1; lunge 3; make 37c; mangle; mark 10, 15b; mold[1] 2, 4; neuter 2; nick 1, 2; nip[1] 1; notch 1, 3; ostracize; pass 23; percentage; plate 5; print 2; proportion 3; quota; rake-off; rebate 2; rebuff 1, 2; reduce 5; scar 1, 2; scoop 5; score 2, 10; shape 6; share 1; short 2, 12a; shorten 1; skip 2; slash 1, 3, 4, 5; slice 4; slight 5; slit 1, 2; snip 1, 2; sore 7; split 9; stamp 8; sterilize 2; style 1, 8; tailor 2; tear 4; tool 4; trim 3; water 7; whittle 1, 2; wound 1, 3
cut (apart), disconnect; loosen 2; rip 1; sever 1; split 1
cut asunder, cleave; divide 1
cut away, clear 25; undercut 1; wash 5
cut (back *or* down (on)), abridge; abstract 4; bring 9b; clear 25; curtail; cut 5, 8, 11, 13; decrease 1; diminish 1; economize; fell; knock 4b; lower[1] 3; minimize 1; mow 1, 2; pare 2; prune; reduce 1; shorten 1; stem[2] 1; thin 8; wind[2] 3a
cut (dead), cut 12; shoulder 2
cut down, lay[1] 18c
cut in (on), cut 14; interrupt 1; short 12b
cut in two, cleave; sever 1; split 1
cut it, grade 5
cut (it) out, lay[1] 16b
cut loose, disengage; revel 2
cut (off), besiege 2; bottle 4b; break 10; choke 3; clip[2] 1; cut 2, 13b, 15; detach; detached 1; disconnect; dissociate; end 9; hang 12; head 13a; intercept; interrupt 2; isolate; isolated 2; lop; lower[1] 3; mutilate 1; sever 1; shave 1; short 12b; shorten 1; shut 4a, 4b; stop 2; suppress 1; terminate; top 5
cut (on) out, depart 1
cut (out), cut 16; delete; edit 2; eliminate 3; escape 1; excavate 1; exit 3; flee 1; fly 2; give 17a; make 26; omit 1; shut 5b; stop 1; undercut 1
cut (short), clip[2] 2; curtail; interrupt 2; short 12; shorten 1
cut (someone) down to size, foil[1]; place 13
cut (up), carve 2; clown 3; cut 3, 17; divide 1; miserable 1; partition 5; portion 4; section 4
cut up rough, cut 18
cut a hole, drill 1
cut and dried°, cut 29; foregone; predetermined 2
cut and run, depart 1; escape 1; flee 1; fly 2; make 26; run 2; turn 22
cut a rug, dance 1; leg 9b
cut a swath, far 6a; swagger 1
cutback, cut 21; decrease 2; rebate 1; slack 5
cut capers *or* a caper, clown 3; frolic 3
cut corners *or* costs, economize; stint 4
cut didos, clown 3
cut-down, cut 27
cute°, fetching; fine[1] 10; lovable; precious 4; pretty 1; sweet 4
cut edition *or* version, abridgment 3
cutesy, precious 4
cutlass, blade 2
cutlery, silver 1
cutoff, suppression
cut-offs, pants 2
cut-price, cheap 1; sale 5
cutpurse, thief 1
cutter, ship 1
cut the ground from under someone's feet, foil[1]
cutthroat°, butcher 1; killer 1; murderer; thug
cutting°, acute 3; biting; bitter 5, 6; caustic 2; clip[2] 5; dry 3; extract 6; gruff 1; hurtful 2; incisive 2; keen[1] 2; poignant 2; pointed 2; pungent 2; sarcastic; scathing; sharp 5, 8; slip[2] 2; snip 3; tart[1] 2; trenchant; wintry 1
cutting edge, blade 1; spearhead 2
cuttingly, home 10; sharply 1
cutting off, suppression
cutting remark, gibe 2
CV, background 1; history 4; record 6; résumé 2
cwm, pass 23; valley
cycle°, era; orbit 1; pattern 3; procession 2; revolution 3; revolve 1; roll 1; round 11; run 37; sequence; turn 26
cyclic, sequential
cyclical, intermittent; periodic; recurrent; regular 2; spasmodic 2
cyclone, blow[1] 9; eddy 1; hurricane; storm 1; tempest 1; twister; whirlwind 1
Cyclopean, colossal 1; gigantic; great 1; immense; massive; prodigious 1; vast
Cyclops, ogre
cylinder, roll 12; roller 1
cynic, killjoy; misery 4; nonbeliever; skeptic
cynical, disenchanted; distrustful; dry 3; hollow 4; incisive 2; jaundiced 2; pessimistic; sarcastic; sardonic; skeptical; tart[1] 2
cynicism, sarcasm; skepticism
cynosure, focus 1; queen 2
cypher, code 2; cog 2; sign 4
Cyprian, lascivious 1; obscene 1; prurient 1; sensual; unchaste
cyst, lump[1] 2
czar, dictator; monarch 1

dab°, blob; dot 1; drop 2; grain 3; little 10; pat[1] 1, 3; scrap[1] 1; tang 2; tap[1] 1, 2; trifle 2
dabble°, spatter; tinker
dabble at, dabble 2
dabble in, dabble 2; dip 3
dabble (in *or* with), putter
dabble with, dabble 2
dabbler, amateur 1; dilettante
dab hand°, adept 2; expert 1; master 2; professional 3; virtuoso 1
dacoit, thief 1
dad, father 1
dadaistic, camp[2] 1
daddy, father 1
dado, border 4; pedestal 1
daedal, intricate 1; labyrinthine
Daedalean, intricate 1
Daedalian, clever 3; complicated; ingenious; intricate 1; labyrinthine; resourceful; shrewd
Daedalic, intricate 1
daffy, crazy 1; daft 1; inane; senseless 3; unbalanced 2
daft°, absurd 1; crazy 1; deranged; foolish 2; inane; insane 1; mad 1; mental 2; psychotic 1; queer 4; senseless 3; soft 6; stupid 2; twist 11; unbalanced 2; wild 9
daft about, daft 3
daftness, absurdity 1; folly 1
dagger°, blade 2; steel 1
daggers drawn, at, odds 4; opposition 3; war 2
daggers drawn, be at, feud 2
daily°, consistently 1; diurnal; everyday 1; journal 1; maid 2; organ 2; paper 1; publication 2; regular 2; servant 1; sheet 6; woman 4
daily bread, sustenance 1

date (on *or* with), up to, level 5
datebook, calendar 1; diary
dated, antiquated; date 4; extinct 2; obsolete; old 2; old-fashioned; out 9; passé; time 15; timeworn
daub, dab 1, 3; paint 5; plaster; smear 1, 3; spatter; streak 1, 4
daughter, child 1; descendant
daughter(s), issue 7
daunt°, alarm 3; appall; bully 2; demoralize 1; desolate 6; deter; discourage 1; foil[1]; frighten; intimidate; menace 1; overawe; scare 1; threaten 1
daunting, awesome; fearsome; formidable 1; frightening; harrowing; scary
dauntless°, bold 1; brave 1; confident 2; courageous; fearless; gallant 1; gritty 2; heroic 1; indomitable; intrepid; manly; resolute; stout 2
dauntlessness, courage; grit; nerve 1; prowess 2; resolution 1
davenport, couch 1; lounge 4
davit, hoist 2
Davy Jones's locker, bottom 4; deep 9; drink 7; ocean 1; sea 1
Davy Jones's locker, go to, founder[2] 1
dawdle°, bugger 4a; delay 3; drag 4; fool 7b; lag 1; linger 2; roam; shilly-shally; stall[2] 1; tarry 1; time 20; trail 6
dawdler, idler; laggard
dawdling, delay 6; idleness 2; slow 1
dawn°, beginning 1; infancy 2; light[1] 4; morning 1; onset 2; origin 2; threshold 2
dawn, at (the crack *or* break of), early 3
dawn on, dawn 5; hit 5; occur 2; register 7; strike 9
dawn upon, dawn 5; register 7; strike 9
dawning, beginning 1; dawn 1, 2; origin 2
day°, date 1
day before, eve 1
day in (and) dayout, nonstop 3; often
day one, from, originally
day after day, daily 3
day bed, lounge 4; couch 1
daybook, register 1
daybreak, dawn 1; light[1] 4; morning 1
daybreak, at, early 3
day by day, consistently 1
(day)dream, hope 1; dream 1; fancy 6; fantasize; fantasy 2; hallucination; illusion 2; muse; reverie
daydreamer, dreamer
daydreaming, absent-minded; dreamy 2; inattentive; pensive; thoughtful 2; wistful 2
Day-Glo, bright 5
daylight°, day 1; light[1] 1
day one, beginning 1
day(s), era; generation 2; period 2; time 4
days, the (good old), history 6
days gone by, in, formerly; once 1; previously
days of old, in, previously
days of yore, in, once 1; previously; time 12a
days past, in, formerly; previously
days beyond recall, past 6
day school, school 1
days gone by, past 6
days of old *or* yore, history 6; past 6
dayspring, dawn 1; morning 1
daystar, sun 1
days to come, future 1
daytime, day 1; diurnal
day-to-day, diurnal; everyday 1; practice 4
day-tripper, sightseer; tourist; traveler 1
daze°, astonish; distract 3; fluster 1; fog 2; jolt 3; knock 6b; overpower 2; perplex; rock[2] 3; shatter 3; shock 1; silly 2; stun 1, 2; trance
daze, in a, daze 4
dazed, blank 5; confused 2; dizzy 1; dumbfounded; glassy 2; goggle-eyed; groggy; scatterbrained; speechless 2; thunderstruck
dazzle°, amaze; blind 6; breath 3; brilliance 1; captivate; daze 1; flare 1, 4; flash 1, 4; fluster 1; glare 1; life 7; radiance 1; radiate 1; sheen; sparkle 3; splendor 1, 2; twinkle 2
dazzled, daze 4; dumbfounded
dazzling°, amazing; arresting; bright 1; brilliant 1; flamboyant 2; flash 6; full 10; glaring 2; glorious 2; golden 3; gorgeous 1; hot 8; luminous 1; meteoric 1; radiant 1; ravishing; scintillating 1; shiny; splendid 1; strong 19; stunning 2; sumptuous; superb; superlative; twinkle 2; virtuoso 2; vivid 1
deacon, clergyman 1
deactivate, disarm 1; disband; incapacitate; paralyze 1; turn 17a
dead°, boring; casualty 2a; cold 5, 6; defunct 1, 2; dreary 2; extinct 1; fatigued; flat 5, 8; inanimate; jaded 1; late 3; lifeless 1, 2; lost 5; mechanical 3; numb 1; obsolete; old-fashioned; pedestrian 2; plumb 4; prosaic; senseless 2; slow 9; tame 4; tedious; time 15; wooden 2
dead, abode of the, underworld 2
dead, half, moribund 1
dead (and buried), be, push 12
dead (and buried *or* gone), past 1
dead (on one's feet), exhausted 1; ragged 3; sleepy 1; weary 1; worn 3
dead to the world, dead 3; unconscious 1
dead-and-alive, slow 9
deadbeat, welsher
dead duck, failure 3
deaden°, dampen 2; drug 4; dull 11; kill 3; muffle 2; mute 3; numb 2; paralyze 2; soften 3
dead end, impasse; rut 2
deadened, dead 13; dull 8; noiseless; numb 1; senseless 2
deadening, cold 4
deaden (the effect of), silence 4
deadfall, trap 1
dead heat, draw 17; tie 10
dead letter, history 7
deadliness, virulence 1
deadlock°, dilemma; draw 17; impasse; stalemate; tie 10
deadly°, black 5; calamitous; cutthroat 3; destructive 1; evil 3; fatal 1; fateful 2; homicidal; hopeless 1; lethal; macabre; malignant 1; mortal 3, 4; murderous 1; poisonous 1; stodgy; terminal 2; venomous 1; virulent 1; withering
deadly embrace, in, opposition 3
deadpan, empty 5; vacant 2; wooden 2
dead weight, overload 2
dead-white, white 2
deaf°, heedless; insensible 2
deaf (to), unconscious 2
deafening, loud 1; noisy; thunderous; uproarious 1
deal°, administer 3; agreement 1; bargain 1; contract 1; deliver 5; dicker 1, 2; find 10; haggle; negotiate 1; negotiation 2; operation 3; pact; parley 2; strike 2; trade 5; transaction 1; treaty
deal, a good, lot 5a
deal, a great, full 18; lot 5a
deal in, handle 5; market 4; merchandise 2; run 14; sell 2; stock 9
deal (out), allot; dispense 1; dispose 3c; divide 2; dole 3; give 15b; hand 17; measure 14; mete; parcel 4; pass 20b; pay 11a; portion 4; present[2] 6; ration 3; serve 3; split 4; distribute 1; share 3
deal with, attend 2; cope 2; cover 7; deal 4; dispose 3a; face 14, 18b; field 6; grapple 2; grip 5; handle 6, 7; manage 2; meet[1] 5; patronize 2; process 4; reckon 5a; relate 4; speak 10; tackle 3; take 39e; touch 8; treat 1, 2; work 10
deal with carelessly, toy 4
deal a blow to, strike 1
dealer°, broker; merchant 1; seller; trader; tradesman 1
dealing(s), business 3; exchange 2; intercourse 1; passage 10; proceeding 2a; relation 6b; trade 1; transaction 1; truck 2
dealing(s) (with), treatment 1
dealings, have, trade 5
dean, clergyman 1; elder 3; minister 1; principal 4
dear°, adorable; beloved 1; bosom 4; darling 1, 3; dearly 3; expensive; extravagant 3; high 3; intimate[1] 1; invaluable; lovable; old 8; personal 3; precious 1; premium 4b; priceless 1; pricey; prohibitive 2; rich 4; special 5; steep[1] 2; stiff 6; sumptuous; sweet 5; sweetheart; tight 9
dear boy, boy 3
dearest, beloved 1, 2; darling 1; love 3; pet[1] 3; precious 1; special 5
dearly°, dear 4
dear one, love 3
dearth°, absence 2; famine; lack 1; need 5; poverty 2; scarcity; shortage; want 3
deasil, forward 4
death°, doom; end 5; fatality 2; fate 2; finish 10; kill 9; killing 1; loss 7; passing 4
death, do to, overdo 1
death, putting to, destruction 2
death, put to, dispatch 3; end 10; execute 3; kill 1; murder 3; put 16d; slaughter 3
death of, be *or* cause the, carry 10d; kill 1
deathbed, parting 3
deathbed, on one's, dying; moribund 1
deathblow, kill 9
death-dealing, homicidal; withering
death-defying, daredevil 2; superhuman 1
deathless°, classic 2; everlasting; immortal 1; timeless
deathlike, dead 3; deadly 4; macabre
deathly, black 5; dead 3; deadly 4; macabre; murderous 1; wan 1
death notice, obituary
death rattle, rattle 7
death's door, at, dying; moribund 1
deathwatch, wake[1] 3
debacle, disaster; downfall; flood 1; flop 4; rout 1; ruin 1; washout; reverse 8
debar, ban 1; disqualify; exclude 1; expel 2; forbid; freeze 4; isolate; lock[1] 8;

debar (*cont.*) obstruct 3; preclude; prevent; prohibit 1; restrain 2; shut 5a; suspend 3
debark, disembark; land 5; light[2] 13
debarment, exception 2; prevention; prohibition 1, 2
debarring, exclusive 4; prevention; prohibition 1; suspension 1
debase°, adulterate; alloy 2; contaminate; corrupt 3; degrade 2; demoralize 2; desecrate; diminish 2; disgrace 4; dishonor 2; foul 14; humble 4; lower[1] 4; peg 2; poison 3; profane 4; prostitute 2; pull 9c; ruin 11; shame 8; taint 2; take 31b; tarnish; vilify; violate 2; vitiate 2
debased, bad 2; base[2] 5; corrupt 2; decadent 1; degenerate 1; disgraceful 1; dishonorable 1; impure 2, 4; monstrous 2; reprobate 1; rotten 3; sordid 1; ugly 2; unchaste; vile 1; wicked 2
debasement, degradation 2; disgrace 1; filth 2; parody 2; prostitution 2; ruin 2; sacrilege 1; undoing 1; violation 2
debasing, degrading; derogatory; disgraceful 1; ruin 2; violation 2
debatable°, controversial 1; disputable; doubtful 1; implausible; moot 1; open 7; problematic; question 6b; questionable; suspicious 1; theoretical 1; tricky 2; uncertain 3; unresolved
debate°, argue 1, 2, 3; argument 1; contest 2, 3; controversy 1; deliberate 4; disagree 2; disagreement 3; discuss; discussion; dispute 1, 2, 3; fight 4; knock 3c; negotiate 1; quarrel 1; variance 2
  debate with, reason 8
debauch, demoralize 2; dishonor 3; dissipate 4; fling 2; orgy 1; party 1; pervert 2; revel 3; seduce 2; spree; violate 3
debauched, abandoned 2; decadent 2; dissolute; immoral 2; incontinent 2; indecent 2; lascivious 1; lewd; libertine 2; loose 7; obscene 1; outrageous 3; profligate 1; promiscuous 2; prurient 1; sensual
debauchee, degenerate 3; libertine 1; pervert 3; playboy; profligate 3; rake[2]; reprobate 2; roué; seducer; sensualist
debaucher, libertine 1; seducer
debauchery, dissipation 1; excess 2; perversion 2; profligacy 1
debauching, self-indulgent
debilitate, break 6; cripple 2; depress 2; drain 5; enervate; exhaust 2; sap[2]; tire 1; undermine 1; waste 3; weaken 1; wear 6; weary 5
debilitated, broken 3; crippled 1; decrepit 1; delicate 3; exhausted 1; feeble 1; helpless 3; impotent 1; infirm 1; limp[2] 2; low[1] 4; powerless 2; run-down 1; unhealthy 1; weak 2
  debilitated, become, waste 2
debilitating, exhausting 1; killing 3
debilitation, decrepitude 1; exhaustion 2; impairment; impotence 1
debility, decline 7; infirmity 1; prostration 4
debit, charge 3, 12; liability 2; loss 6
debonair°, elegant 1; gay 2; genteel 2, 3; jaunty 2; pleasant 2; polished 2; rakish; suave
debouch, mouth 2
débouché, mouth 2
debouchment, issue 1; mouth 2; outpouring
debris, fragment 2; garbage; junk 1; leftover; litter 1; odds 5; refuse[2]; remain 4a; rubbish 1; ruin 6; scrap[1] 3; trash 3; waste 7; wreckage
debt°, charge 3; liability 2; obligation 3
  debt, in, beholden; debt 2; embarrassed 2
  debt to, be in, owe 1
debt-free, solvent
debug, shake 6a
debunk, explode 2
debut°, initiation 1; onset 2; opening 4; première 1, 2, 3; presentation 3
decadence, decay 3
decadent°, degenerate 1; lecherous; libertine 2
decamp, bolt 8; bundle 3; clear 29; depart 1; drop 13; escape 1; flee 1; flight[2] 3; fly 2; go 2, 23, 31d; leave[1] 1; make 26; move 2; quit 1; retreat 4; run 2; take 34c
  decamp (from), evacuate 2
decampment, escape 5
decanter, bottle 1; jug
decapitate, behead
decarbonated, flat 8
decay°, decline 7; decompose 2; decrepitude 2; degenerate 2; deteriorate 2; disintegrate; disrepair; dissolution 2; ebb 2, 4; fade 2; fail 3; failure 2; fall 24; fester 1; filth 1; mortify 3; putrefy; rot 1, 2, 3; ruin 1; seed 4; spoil 4; stagnate; turn 5; twilight 2; waste 2; worsen 2
decayed, bad 6; decrepit 2; dilapidated; foul 2; moldy; musty 1; old 3; putrid; rancid; rotten 1; seedy 1; unsound 1
decaying, decadent 1; decrepit 2; dilapidated; moldy; putrid; seedy 1
decease, death 1; die 1; expire 2
deceased, dead 1; defunct 1; late 3
deceit°, art 5; craft 2; deception 1; dissimulation; evasion 2; falsity 1; finesse 2; fraud 1; hocus-pocus 1; hypocrisy; lying 1; perfidy; trick 1; trickery
  deceit, practice, game 6
deceitful°, artful 1; crafty; deceptive 2; designing; devious 1; dirty 3; dishonest; disingenuous; disloyal; double 5; evasive; false 2; fraudulent 2; furtive 1; hollow 4; hypocritical; insincere; lying 2; machiavellian; mealy-mouthed; oblique 2; perfidious; phony 1; plausible 2; rotten 3; scheming; shifty; sly 1; sneaky; spurious; traitorous; tricky 1; two-faced; unscrupulous; untrue 1; wily
  deceitful, be, game 6
deceitfully, back 6
deceitfulness, deceit 1; falsity 1; hypocrisy
deceive°, beguile 1; betray 3; blind 5; bluff[1] 1; cheat 2; defraud; disappoint 2; disguise 2; dissimulate; double-cross; dupe 3; equivocate; finesse 4; flummox; fool 4; have 12c; hoax 2; hoodwink; jolly 2; lead 7; leg 8; misinform; mislead; outsmart; ride 5a; rip 2b; seduce 1; string 10c; swindle 1; take 32b; throw 6c; trap 6; trick 8; victimize 2
  deceive (someone), put 26
deceived by, be, fall 13b; labor 7
deceiver, cheat 1; fraud 3; hypocrite; impostor; phony 3
deceiving, deceptive 1; dishonest; hypocritical; shifty; wily
decelerate, brake 2
decency, grace 2; honor 1; integrity 1; kindness 1; morality 1; nobility 1; probity; propriety 2; purity 2; rectitude; shame 4; virtue 1
decent°, chaste 1; clean 4; considerable 1; correct 6; decorous; ethical; fair[1] 4; honest 1; honorable 2; just 2; moral 1; noble 4; presentable 2; proper 3; pure 3; respectable 1, 3; savory 2; seemly 2; solid 5; square 4; straight 4; upright 2; virtuous 2; wholesome 2
decently, honestly 1; properly 1
deception°, act 3; artifice 1; blind 8; bluff[1] 3; camouflage 1; chicanery; deceit 1; delusion 1; disguise 4; dissimulation; dodge 4; evasion 2; feint; finesse 2; foul play; fraud 1, 2; gimmick 1; gobbledegook 2; hanky-panky; hoax 1; hocus-pocus 1; hypocrisy; illusion 1; intrigue 3; masquerade 2; pretense 1, 2; put-on 1; rip-off 2; ruse; subterfuge; swindle 2; trap 2; trick 1; trickery
  deception, practice, game 6
deceptive°, devious 1; dishonest; false 2; fraudulent 2; hollow 4; hypocritical; illusory; insincere; lying 2; oblique 2; plausible 2; sophistic; specious; subtle 3; tortuous 2; tricky 1, 3; wily
deceptiveness, falsity 1; gobbledegook 2; subtlety 2
decern, decree 2
decide°, arrange 2; choose; decree 2; determine 1, 3; dispose 3a; distinguish 1; finalize; fix 2; judge 4; mind 12; persevere; resolve 1, 3; rule 7; see 8; set 9; settle 3; sort 10a
  decide (on *or* upon), decide 2; fix 15; make 16; pitch[1] 7; settle 2; pick 1; take 2
  decide to be, find 9
decided°, bent 4; certain 1; cut 29a; determined 1; emphatic; fated 1; firm 4; fixed 3; grim 1; intent 5; marked; positive 2; pronounced 1, 2; purposeful; resolute; set 29; thick 9
  decided, be, mind 11
decidedly, absolutely 1; definitely; degree 4b; far 2, 3; highly 2; once 3; terribly; truly 3; very 2; well[1] 7
decidedness, decision 3; resolution 1
deciding, settlement 5; ultimate 1
decimal point, dot 1; point 1
decimate, kill 1; massacre 2; mow 2
decimation, killing 1; murder 2
decipher°, figure 12b; interpret 1; make 37d; perceive 2; puzzle 3; read 2; render 6; solve; translate 1
decipherability, simplicity 1
decipherable, intelligible; legible; readable 3
deciphering, solution 1
decipherment, interpretation 1; solution 1; translation 1
decision°, decree 1; determination 2; disposition 4; finding 2; judgment 2; negotiation 2;

INDEX

deflated, flat 6
deflating, derogatory
deflect°, bend 6; distract 1; divert 1, 2; fend 2; intercept; pervert 1; sidetrack; turn 7
deflected, bent 1
deflection, shift 5; slant 3; switch 2
defloration, rape 1; ruin 2
deflower, dishonor 3; outrage 6; rape 4; ravish 2; ruin 11; seduce 2
deflowering, rape 1; ruin 2
defoliate, bare 8; shed[2] 4
defoliated, bare 4
deform, deface; distort 1; mangle; screw 7c; warp 1
deformation, warp 2
deformed°, crooked 2; disfigured; grotesque 1; misshapen; shapeless 2; wry 1
deformity, abnormality 2; freak 1; kink 3; monster 2; warp 2
defraud°, cheat 2; chisel 2; double-cross; dupe 3; fleece; fool 4; gouge 2; hoax 2; hoodwink; misinform; outsmart; prey 3b; ride 5a; rip 2b; screw 6; sting 4; swindle 1; take 27, 32b; trick 8; victimize 2
  defraud (of), beguile 2; rob 3
defrauder, fraud 3
defrauding, rip-off 2; swindle 2
defray°, clear 28; cover 10; meet[1] 5
defrayal, settlement 4
defrost, thaw 1
deft, clever 4; dexterous 1; graceful 1; handy 3; ingenious; masterful 1; neat 4; nimble 1; perfect 7; quick 4; ready 6; talented; wicked 7
deftly, handily 2
deftness, dexterity 1; facility 1; ingenuity; skill 1; touch 16
  deftness (with), understanding 3
defunct°, dead 1; extinct 1; inanimate; past 1
defuse, moderate 5
defy°, beggar 2; brave 3; break 9; challenge 1, 3; combat 6; dare 1; disobey; fight 2; fly 6; mock 1; oppose 1; rebel 2; resist 1; snap 6; stand 12a, 12b; trample 2; turn 12; withstand
defy (the law), transgress 2
dégagé, casual 5; detached 2; perfunctory 1; rocky[1] 3
degeneracy, degradation 1; evil 6; profligacy 1; vice 1
degenerate°, corrupt 2; decadent 2; decay 1b; decline 4; deteriorate 1; deviant 2, 3; die 2; dissolute; flag[2] 2; go 12; immoral 1; impure 4; indecent 2; kinky 2; lecherous; libertine 2; mean[2] 2; obscene 1; outrageous 3; pervert 3; perverted; profligate 1, 3; relapse 2; reprobate 1, 2; rot 2; rotten 3; seamy; seed 4; sink 4; stagnate; ugly 2; vicious 1; vile 1; wicked 2; worsen 2
degenerating, decadent 1; wane 4
degeneration, decay 3; decline 7; degradation 1; ebb 4; impurity 3; relapse 4; wane 3
degenerative, decadent 1
degradation°, discredit 4; disgrace 1; downfall; evil 6; filth 2; humiliation; prostitution 2; ruin 2; scandal 2; shame 2; undoing 1; vice 1; violation 2
degrade°, belittle; corrupt 3; cut 6; debase 1; desecrate; deteriorate 1; diminish 2; discredit 1; disgrace 4; dishonor 2; flout; foul 14; humble 4; lower[1] 4; mortify 1; pervert 2; profane 4; prostitute 2; pull 9c; reduce 6; shame 8; take 31b; tarnish; violate 2; vitiate 1
  degrade oneself, stoop 2
degraded, base[2] 2; corrupt 2; degenerate 1; disgraceful 1; low[1] 12; mean[2] 2; seamy; sordid 1; vicious 1
degrading°, base[2] 2; disgraceful 1; dishonorable 1; menial 1; ruin 2; shameful; unwholesome 2; vicious 1; vile 1; violation 2
degree°, breadth 2; caliber 3; gauge 4; grade 1; incidence; indication 2; measure 5; notch 2; order 3; proportion 4; shade 7; step 5
  degree, to a, degree 4; moderately
  degree, to a limited, partially
  degree, to a slight *or* **minor,** slightly
  degree, to some *or* **a certain,** moderately; part 9; partially; quite 2; rather 1; relatively; slightly; somewhat
  degree, to whatever, however 2
degrees, by, degree 3; piecemeal 1
dehydrate, dry 4; evaporate 1; parch; preserve 3; shrivel
dehydrated, dry 1; thirsty 1
dehydration, evaporation 1
deice, thaw 1
deific, divine 1
deification, worship 2
deiform, divine 1
deify, glorify 2; idealize; idolize; lift 2; pedestal 2; worship 1
deign°, condescend; lower[1] 4; stoop 2
  deign to give, vouchsafe 1
deigning, patronage 3
deipnosophist, conversationalist
deipnosophy, repartee
deity, god
Deity, the, creator 2
deject, chill 8; depress 1; desolate 6; get 37c; sadden; shatter 3
dejected°, bad 9; blue 1; broken 3; brokenhearted; desolate 3; despondent; disappointed; doleful; downhearted; forlorn 1; gloomy 2; heartbroken; heavy 6; hopeless 3; hurt 7; joyless 1; low[1] 8; melancholy 1; miserable 1; moody 1; morbid 3; mouth 6; pessimistic; sad 1; sinking 2; sorrowful 1; subdued 2; unhappy 1; woebegone; wretched 2
dejectedly, sadly 2
dejectedness, melancholy 2; sadness
dejection, depression 2; despair 1; desperation 2; disappointment 2; gloom 2; grief 1; melancholy 2; prostration 3; sadness
de jure, lawful 1; legitimate 2; rightful 1
dekko, glance 4
  dekko (at), take a, glance 1; peek 1
delay°, arrest 1; dawdle; defer[1]; drag 6; fire 4; forestall; gap 1; hamper[1]; hang 6; hesitate 1; hinder 1; hold 18a, 21a, 22b; holdup 2; impede; impediment; interval 1; lag 1; lull 1; moratorium; obstruct 2; pause 1, 2; play 14; postpone; postponement; prevent; prevention; procrastinate 1; put 21a; reprieve 2; reserve 1; respite 2; retard 1; set 16; setback; shilly-shally; sit 8; slack 3b; stall[2] 1, 2; stay[1] 4, 6; string 11b; suspend 1; suspension 2; tarry 1; tie-up 1; time 20; wait 4
  delay, without, demand 8; directly 2; double 8; flat 17b; hastily 1; immediately 1; instantaneously; nail 6; now 3; once 6a; posthaste; promptly; quickly 3; rapidly 2; readily 3; right 13; soon 2; straight 15; summarily 1; time 17
delayed, belated; late 1; slow 5
  delayed, be, fire 4; hang 6, 10; wait 2
delaying, prevention; suspension 2
dele, cancel 2; cut 16a; delete; obliterate 1; omit 1
delectable, dainty 3; delicious 1; luscious; pleasant 1; savory 1; tasty; tempting 2; yummy
delectation°, delight 3; enjoyment 1
delegate°, agent 1; ambassador; appoint 2; assign 3; cast 8; commit 1; deputy; designate 2; detail 6; enable 1; entrust; envoy; farm 4; make 23; minister 2; mouthpiece 2; name 5; proxy; representative 3; send 1; substitute 2; trust 6
delegated, vicarious
delegation, mission 3
delete°, abolish; blot 4b; cancel 2; cross 3; cut 16a; edit 2; eliminate 3; erase 1; forgive 3; neutralize; obliterate 1; omit 1; remove 3; rub 4; scratch 3; strike 5; vitiate 3; void 6; wash 4; write 5
deleterious, bad 2; destructive 1; detrimental; evil 3; harmful; hurtful 1; ill 4; injurious 1; mischievous 2; poisonous 2; prejudicial; ruinous; unwholesome 1; violent 2; virulent 1
deletion, cut 22; omission 1
delftware, pottery
Delhi belly, run 50
deliberate°, balance 1; calculated 2; confer 1; conscious 2; cool 3; debate 4; intentional; measured 2; methodical; moderate 1; muse; parley 2; pitched; premeditated; purposeful; resolute; slow 1; special 3; studied; unhurried; voluntary 2; willful 1
  deliberate (on *or* **over** *or* **about),** consider 1; contemplate 2; discuss; mull; ponder; reflect 3; speculate 1; study 2; think 3
  deliberate (with), consult 1
deliberately°, purpose 4a; voluntarily
deliberateness, purpose 2; resolution 1
deliberating, reflection 2; reflective; thought 1
deliberation, calculation 3; consideration 3; counsel 2; debate 2; discussion; parley 1; purpose 2; reflection 2; resolution 1; speculation 2; thought 1
deliberative, reflective
delicacy°, dainty 4; finesse 3; frailty 1; refinement 1, 2; sensitivity 1; subtlety 1; tact; taste 6; tidbit; weakness 1
delicate°, awkward 5; brittle 2; dainty 1; dodgy; elegant 1; explosive 1; exquisite 1; fastidious; feeble 1; feminine 1; filmy 1; fine[1] 5, 6, 9; finicky 1, 2; flimsy 1, 3; fragile; gingerly 2; hot 10; hothouse 2; light[2] 4, 5; mincing; nice 3, 4; petite; precarious; sensitive 1, 3; sickly 2; silky; slight 3, 4;

delicate° (*cont.*) soft 9, 12; sore 2; squeamish 1; sticky 2; subtle 1; tactful; tender[1] 1, 5; tenuous 1; thin 5; thorny 2; ticklish 2, 3; tiny; tricky 2; unsound 2; weak 1, 2

delicate condition, frailty 1

delicate condition, unmarried *or* unwed and in a, trouble 9b

delicately, gingerly 1

delicateness, weakness 1

delicious°, dainty 3; luscious; mellow 1; pleasant 1; rich 6; savory 1; tasty; tempting 2; voluptuous 2; yummy

delight°, admiration; amuse 2; bliss; carry 9; charm 5; content[2] 1, 4; delectation; distract 2; ecstasy 1; enchant 2; enjoyment 1; enrapture; entertain 1; entrance[2]; exult; feast 3, 6; flush[1] 3, 6; fun 1; gaiety 1; gladden; glee; glory 5; gratification; gratify; gusto; happiness; honor 3; joy 1, 3; love 2; luxury 3; paradise 3; please 1; pleasure 1; radiance 2; rapture; ravish 1; regale; rejoice; relish 1; satisfaction 1; send 4; tickle; transport 3, 5; treasure 2; triumph 2; warm 9

delight (in), admire 1; delight 2; enjoy 1; gloat; like[2] 1; love 8; luxuriate 1; pride 4; relish 2; revel 1; roll 10b; savor 3

delighted°, blithe 1; charmed 3; content[2] 3; ecstatic; elated; enchanted; exuberant 2; exultant; glad 1; gleeful; happy 1; joyful 2; merry 1; overjoyed; pleased; radiant 2; rapt 1, 2; rapturous; rhapsodic; willing; world 7

delighted, be, rejoice

delighted (with *or* about), proud 1

delightedly, happily 2

delightful°, acceptable 2; adorable; agreeable 1; beautiful 2; delicious 2; enchanting; engaging; entertaining; exhilarating 2; glorious 3; golden 4; heavenly 2; lovely 2; nice 1; picturesque 1; pleasant 1; taking; winning 1

delightfully, beautifully 1; nice 6

delimit, bound[1] 2; define 1; limit 6; measure 13; restrict; stake[1] 4b; stunt[2]

delimitation, definition 1; determination 3; restraint 1

delimited, finite

delineate, characterize; define 1, 2; describe 2; design 2; designate 1; detail 5; draft 5; draw 5; explain 1; get 40b; outline 3; paint 4; phrase 4; plan 3; portray 1; project 3; relate 2; represent 3; set 7; specify; spell[3] 2; stake[1] 4b; state 7; tell[1] 4; trace 6

delineated, given 1; graphic 2; specific 1

delineation, definition 1; design 6; explanation 1; relation 4; rendering

delinquency, default 2; misbehavior

delinquent°, derelict 2; disgraceful 2; disobedient 1; guilty 1; perverted; problem 3; punk 1; remiss; slack 1; transgressor; truant 1, 2

deliquesce, dissolve 1; melt 1

delirious°, distraught; drunk 2; ecstatic; mad 1; raving 1; wild 5

delirium tremens, shake 10

deliver°, administer 3; bring 1; catch 6; commit 1; communicate 1; distribute 1; emancipate; enunciate 1; execute 2; extricate; fetch 1; forward 7; give 1; hand 14, 18a; have 8; issue 10; liberate 1; loose 11, 13; make 18; pass 8, 9; pay 3; pipe 5; produce 6; pronounce 3; propagate 1; pump 1; put 24; ransom 3; read 3; recover 3; redeem 2; release 1; relieve 3; render 3, 5, 7; rescue 1; return 6; salvage 1; save 1; say 11; send 2; ship 2; sign 11; snatch 3; spare 6; strike 2; supply 1, 2; take 9; transfer 1; transmit 1; transport 1; turn 15b; wash 2; world 5a, 5b

deliver from, rid 1

deliver (oneself of) a tirade *or* diatribe *or* speech, rant 1

deliver (up), abandon 1; cede

deliverance, delivery 1, 2; freedom 2, 3; liberation; ransom 1; recovery 3; release 3; relief 1; remission 1; rescue 2; resign 1; salvage 2; sell 3; surrender 1; turn 15c

deliver a speech to, address 3

delivered, free 2

deliverer, Messiah; savior 1

delivering, delivery 1; release 3

deliver (the goods), make 31c

deliver the coup de grâce, finish 4; kill 1

delivery°, administration 3; birth 1; diction 2; execution 4; expression 6; freight 1; intonation; issue 6; labor 4; liberation; post[3] 1; presentation 1; pronunciation; receipt 2; recovery 3; supply 5; transfer 2

dell, hollow 7; valley

delphic, ambiguous 2; obscure 2

delude, beguile 1; bluff[1] 1; deceive; defraud; dupe 3; finesse 4; fool 4; hoodwink; misinform; ride 5a

deluded by, be, labor 7

deludedly, blindly

deluge, downpour; drown 1, 2; flood 1, 3, 4; flow 6; flush[1] 5; glut 3; outpouring; overwhelm 2; rain 1; rash[2] 2; satiate 1; shower 2; spate; storm 1; stream 2, 3; submerge 3; surfeit; swamp 2; torrent; volley 2

delusion°, dream 1; fallacy; fancy 6; fantasy 2; hallucination; illusion 1; misconception; phantom 2; vision 4

delusional, blind 3; dreamlike; unrealistic 1

delusionary, dreamlike

delusive, dreamlike; fancy 2; nonexistent; unrealistic 1

delusory, dreamlike; unrealistic 1

deluxe, fancy 3; luxurious 1; palatial; sumptuous; swell 7; upper-class 2

delve, burrow 2; plow 1; plumb 5; root[2]; till[1]

delve into, dig 5; fathom; go 30b; look 8; probe 1; research 2; scan 2

delving, research 1

demagogic, inflammatory

demagogue, agitator; rabble-rouser; speaker; talker 1

demand°, appetite 2; ask 2; bid 3; bidding 2; call 8a, 14; challenge 6; charge 5, 13; claim 1, 3; command 1, 4; condition 2; cry 3; desire 2, 4; dictate 1, 2; drive 1; enforce 2; entail; exact 3; expect 3; expectation 3; have 9; indicate 4; insist 1; lay[1] 10, 17b; look 6a; market 3; need 1, 2; obligation 1; oblige 2; order 4, 14; plead 2; prescribe; press 6; pressure 5; provide 3; provision 2; request 1, 2; require 1; requirement 1, 2; requisition 1, 2; run 43; say 10; screw 4; seek 3; speak 7c; stipulate; stipulation; strain[1] 7; take 16; tax 3; tell[1] 5; urge 2, 4; use 13; want 2, 4; wish 2

demand, in, demand 7; enviable; indispensable 1; popular 1

demand, on, demand 8

demanded, imperative 1; indispensable 1; mandatory; necessary 1; obligatory; order 10b; prerequisite 1

demanding°, choosy; difficult 1, 4, 5; exacting; fussy 1; hard 4; impatient 2; incumbent 1; meticulous; narrow 3; officious; overcritical; painful 3; particular 4; perfectionist 2; precise 2, 3; punishing; rigid 3; rugged 2; selective; severe 1; squeamish 1; stern 1; strait 1; strenuous 1; tough 4; trying

demand note, note 4

demand(s), pressure 3; ultimatum

demarcate, define 1; restrict; stake[1] 4b

demarcation, definition 1; line[1] 4

démarche, maneuver 1

dematerialization, evaporation 2

demean, conduct 6; degrade 2; diminish 2; discredit 1; disparage 1; foul 14; humble 4; lower[1] 4; patronize 1; prostitute 2

demean oneself, condescend; scrape 4

demeaned, be, stoop 2

demeaning, degrading; derogatory; menial 1; shabby 2

demeanor, action 8; appearance 2; attitude 1; bearing 1; behavior; carriage 2; conduct 1; front 4; guise 1; look 14; manner 2; outside 2; performance 4; play 21

demented, crazy 1; delirious 1; deranged; foolish 2; insane 1; mad 1; maniacal 1; mental 2; possessed; psychotic 1; queer 4; senseless 3; unbalanced 2; unsound 3; wild 5

dementedly, madly 1

dementia (praecox), insanity 1; lunacy 1; madness 1; mania 2

demerit, stigma

demesne, estate 1

demigod, god

demilitarize, disarm 1

demi-lune, crescent 1, 2

demimondaine, prostitute 1; tart[2] 2

demise, death 1; leave[1] 6; loss 7; passing 4

demiurge, deity; god

demo, demonstration 2, 3; exhibition

demobilize, disarm 1; disband

democratic°, free 1; representative 2

démodé, extinct 2; obsolete; out 9; passé

demoiselle, girl 1; lass; maid 1; miss[2]; youth 2

demolish°, abolish; blast 5; blot 4b; break 4, 16a; consume 4; desolate 5; destroy 1; devastate 1; devour 2; dispose 3d; disprove; fell; flatten 3; knock 4a; level 9; pillage 1; pull 9a; pulverize 2; ravage 1; raze; rout 2; ruin 7; shatter 1, 2; shred 2; spoil 1; stifle 3;

depone, attest
deponent, witness 2
depopulate, desolate 4
depopulated, uninhabited
deport, banish 1; conduct 6; exile 3; expel 2; isolate; transport 2
  deport (oneself), behave; act 5; play 16
deportation, ejection 2; exile 1
deportee, exile 2
deportment, action 8; bearing 1; behavior; carriage 2; conduct 1; decorum 1; form 6; guise 1; manner 2; performance 4; play 21; presence 3; stance
depose, allege; attest; bring 9a; degrade 1; displace 2; downgrade 1; overthrow 1; overturn 2; put 16b; remove 5; swear 1
deposing, overthrow 2
deposit°, advance 9; dregs 1; dump 1; earnest 3; ground 4; lay[1] 1; lodge 5; mine 2; park 3; pawn[1] 1, 3; pile[1] 1; place 15; pledge 2, 5; plump[2] 2; put 1; security 3; sediment; set 1; silt 1; stack 1; stick[1] 2; stow; tip[2] 3; vein 3; wash 2
depositary, mine 2; storehouse
deposited, become, lodge 5
deposition, allegation; assertion 1; declaration 1; deposit 4; evidence 2; removal 3; representation 3; testimony
depository, mine 2; storehouse; warehouse
depot, installation 3; station 3; stop 8; terminal 3; warehouse
deprave, corrupt 3; demoralize 2; vitiate 2
depraved, abandoned 2; bad 3; corrupt 2; criminal 2; degenerate 1; deviant 2; diabolic 2; dissolute; evil 1; evil-minded 1; filthy 3; ill 2; immoral 1; impure 4; kinky 2; lecherous; libertine 2; low[1] 12; miscreant 2; monstrous 2; outrageous 3; perverted; profligate 1; promiscuous 2; reprobate 1; rotten 3; satanic 1; seamy; sinful; ugly 2; ungodly 1; unwholesome 2; vicious 1; vile 1; wanton 1; wicked 2
depravity, degradation 1; evil 6; impurity 3; perversion 2; profligacy 1; sin 2; vice 1
deprecate, abuse 3; attack 2; belittle; blaspheme 2; debase 1; degrade 2; depreciate 2; diminish 2; disapprove; discredit 1; disparage 1; flout; impeach 2; knock 2; libel 3; minimize 2; play 13; put 16g; regret 1; run 26b; take 31b; talk 9a; trivialize; vilify; vituperate
deprecating, critical 1; personal 4
deprecation, attack 7; libel 1; scorn 1
deprecative, critical 1; foul 6; scornful
deprecatory, abusive 1; critical 1; foul 6; injurious 2; slanderous; vituperative
depreciate°, belittle; blaspheme 2; debase 1; degrade 2; depress 2; deteriorate 1; detract; diminish 2; discredit 1; disparage 1; flout; foul 14; libel 3; minimize 2; put 16g; reduce 7; run 26b; slight 6; stigmatize; take 31b; talk 9a; trivialize; underestimate; vilify; vitiate 1; vituperate
depreciating, critical 1; derogatory
depreciation, dishonor 4; fall 26; humiliation; libel 1; slump 1
depreciative, critical 1; derogatory; foul 6; slanderous; vituperative
depreciatory, abusive 1; critical 1; derogatory; foul 6; vituperative
depredate, gut 5; loot 2; pillage 1; raid 2; rape 5; rob 1
depredation(s), desolation 1; pillage 2; plunder 3; rape 2; ravage 3; robbery
depress°, chill 8; crush 5; debase 1; demoralize 1; depreciate 1; desolate 6; get 37c; hurt 3; pain 5; press 2; prey 3c; push 2; sadden; shatter 3; turn 17b; vitiate 1; weaken 1; weigh 4, 6
depressant, sedative 1
depressed, bad 9; blue 1; brokenhearted; dejected; desolate 3; despondent; disturbed 2; doleful; downhearted; dreary 1; flat 11; forlorn 1; gloomy 2; heartbroken; heavy 6; hollow 2; hopeless 3; hurt 7; joyless 1; low[1] 8; melancholy 1; miserable 1; moody 1; morbid 3; pessimistic; sad 1; sinking 2; soft 7; somber 1; sorrowful 1; sullen; unhappy 1; woebegone; wretched 2
depressing, bleak 1; cold 4; dingy; dismal; doleful; dreary 1; dull 5; funereal; gloomy 3; gray 2; heart-rending; heavy 5; joyless 2; mournful 2; oppressive 1; sad 2; sedative 2; somber 1, 2; sorry 2; stark 3; sullen; sunless; tragic
  depressing air *or* **mood** *or* **atmosphere,** pall[1] 2
depression°, despair 1; desperation 2; dip 5; disappointment 2; distress 1; gloom 2; grief 1; hole 1; hollow 7; hurt 6; impression 3; indentation; melancholy 2; misery 1; pan 3; pit[1] 2; prostration 3; recession; sadness; slot 1; slump 1; vapor 2; woe
deprivation, disadvantage 1; hardship; loss 1; need 4; privation; suspension 1
deprive°, bereave; divest 1; evacuate 1
  deprive (of), beguile 2; refuse[1] 2; rob 3
deprived°, hungry 3; needy
  deprived of, destitute 2; starved 3
  deprived of, be, forfeit 2; go 40; lose 1
deprive the rights of, suspend 3
depth°, breadth 2; expression 4; extreme 8; gulf 2; measurement 2; pith 2; profundity 1; spread 9
  depth, at *or* **to a greater,** further 4
  depth, in, depth 6; detail 4; length 4c
depth(s), bottom 4; bowels; dead 21; depth 5; interior 6; recess 3
  depths, the, ocean 1
depurate, cleanse 2; purge 1; purify 1; sterilize 1; wash 6
depurative, purgative 1, 2
deputation, mission 3
depute, delegate 2, 3; designate 2; enable 1; make 23; send 1; trust 6
deputed, vicarious
deputize, appoint 2; make 23
deputy°, agent 1; alternate 5; ambassador; assistant 2; factor 2; representative 3; substitute 2
deracinate, abolish; cancel 2; pull 18b; uproot 2
deracination, extraction 1
derange, upset 3; wall 3
deranged°, crazy 1; delirious 1; insane 1; mad 1; maniacal 1; mental 2; psychotic 1; queer 4; sick 4; unbalanced 2; unsound 3
derangement, disorder 1; insanity 1; lunacy 1; mania 2
deregulation, freedom 1; laissez faire
derelict°, bum 2; decrepit 2; delinquent 2; dilapidated; disobedient 1; down-and-out 2; homeless 1; hulk 1; ramshackle; stray 4; tramp 4; unused 2; vagabond 1, 2
dereliction, default 1; failure 1; neglect 4; negligence; offense 1; omission 2; oversight 1; sin 1
deride°, depreciate 2; flout; fun 5; gibe 1; hiss 3; jeer 1; laugh 2a; mock 1; parody 3; ridicule 2; satirize; scoff; scorn 4; snap 6; sneer 2; snipe; sport 3; taunt 1; thumb 8; twit[1]
deriding, ridicule 1
de rigueur, correct 6, 7; proper 1, 3; swish 3
  ***de rigueur,* not,** unacceptable
derision°, gibe 2; jeer 2; mockery 1; ridicule 1; sarcasm; scorn 2; sneer 3; taunt 2
derisive, abusive 1; burlesque 4; contemptuous; derisory; disdainful; foul 6; haughty; irreverent 2; left-handed 2; regal 1; sarcastic; sardonic; satirical; scornful
derisiveness, scorn 2
derisory°, abusive 1; foul 6; regal 1; sarcastic; sardonic; scornful
derivation°, derivative 2; extraction 2, 3; family 3; origin 1; pedigree; strain[2] 1
derivative°, offshoot 3; secondary 2
derive°, base[1] 5; deduce; extract 2; get 21; infer; mine 3; originate 2; spin 4; stem[1] 3; take 5
  derive from, proceed 2; spring 3
derived, constructive 2; derivative 1; secondary 2
  derived (from), be, proceed 2
derive pleasure *or* **enjoyment from,** like[2] 1; love 8
derma, skin 1
dermatitis, rash[2] 1
dernier cri, craze; rage 3; rave 4; vogue 1
derogate, belittle; depreciate 2; diminish 2; disparage 1; foul 14; libel 3; look 5; write 4b
derogation, dishonor 4; humiliation; libel 1; shame 2
derogatory°, abusive 1; destructive 2; foul 6; injurious 2; personal 4; scurrilous; vituperative
derrière, bottom 1; bum 1; buttocks; posterior 3; seat 4
derring-do, daring 1
descale, peel 1
descant, chant 1, 2
descend°, climb 5a; decline 3; dip 2; dive 1; drop 6; fall 1; get 37a; incline 1; light[2] 13; plunge 1; rain 6; sag 2; settle 5; shower 3; sink 1, 3, 5, 6; slope 1; slump 2; stem[1] 3; submerge 2; subside 1; swoop 1
  descend (from), get 41a; proceed 2; spring 3
  descend on *or* **upon,** come 8; descend 4; raid 2; visit 2
descendant°, child 1; offshoot 2

despondent° (*cont.*)
gloomy 2; heartbroken; heavy 6; hopeless 3; joyless 1; low[1] 8; melancholy 1; miserable 1; moody 1; morbid 3; mournful 1; mouth 6; pessimistic; sad 1; unhappy 1
despondently, sadly 2
despot°, dictator; disciplinarian; oppressor; tyrant
despotic°, absolute 3; arbitrary 2; authoritarian; bossy; dictatorial 2; domineering; hard 4; heavy-handed 2; imperative 2; masterful 2; overbearing; peremptory 3; positive 5; repressive; severe 1; totalitarian; tyrannical
despotism°, domination 2; monarchy 2; oppression; severity 1; tyranny
desquamate, flake 2; peel 1; shed[2] 4
dessert°, savory 3; sweet 10
destination°, end 3; fate 3; object 2
destine°, appoint 1; cut 16e; design 4
destined°, bound[3] 4, 5; card 11; certain 3; condemn 2; doomed 1; eventual 2; fatal 3; fated 1; inevitable; predetermined 2; prospective; unavoidable
destined for, for 8
destiny°, doom; fate 1; fortune 2; lot 2; luck 1; predestination; providence 2
destitute°, broke; deprived; down-and-out 1; heel[1] 3; impoverished 1; indigent; insolvent; lean[1] 3; low[1] 6; needy; penurious 2; poor 1; rock[1] 3c; straitened; upper 7
destitute of, destitute 2; empty 7; void 3
destitution, misery 2; necessity 3; need 4; poverty 1; privation; want 5
destitution, reduce to, ruin 10
destroy°, abolish; attack 5; back 7b; blast 5, 6; blot 4b; blow[1] 8b; break 4, 16a; bugger 3; bump 5; consume 4, 5; dash 1; deface; defeat 1; demolish 1, 2; desolate 5; devastate 1; devour 2; dispose 3d; disprove; disturb 3, 5; eliminate 4; end 10; erase 2; erode; exterminate; extinguish 2; finish 4; floor 4; hack[1] 1; kill 1, 2; knock 4a; lay[1] 5; level 9; mangle; massacre 2; mess 5b; mishandle 2; mow 2; muck 4; murder 3, 4; mutilate 2; obliterate 2; overrun; overwhelm 1; poison 4; polish 3b; prey 3a; pull 6, 9a; pulp 3; pulverize 2; puncture 4; purge 2; put 16d; queer 5; quench 2; ravage 1; raze; rebut; root[1] 7a; rout 2; rubbish 3; ruin 7, 9; sabotage 3; screw 7b; shatter 2; shred 2; slaughter 3, 4; spoil 1; stamp 4; stifle 3; subvert; sully; sweep 2; tackle 4; taint 2; trample 3; trash 4; uproot 2; upset 5; vitiate 3; wash 4; waste 3, 11; whip 2; wipe 3; wreck 1; zap
destroy one's reputation, blot 3
destroy (someone's) advantage, wind[1] 13
destroyed, broken 1; desolate 2; dilapidated; lost 6; piece 7; rock[1] 3b; spout 4; tatter 2; undone[1] 1
destroyed, be, fall 5, 7; founder[2] 1; perish
destroyer, butcher 2
destruction°, abolition; attack 9; breakdown 1; calamity 1; damage 1; death 3; depredation; desolation 1; dissolution 2; doom; end 5; evil 7; expense 2; fall 24; fate 2; finish 10; havoc 1; holocaust 1; ill 8; loss 7; mayhem; mischief 2; naught; overthrow 2; overturn 3; perdition; pillage 2; ravage 3; ruin 1; sabotage 1; subversion; undoing 1; wreck 5
destructive°, abusive 2; calamitous; caustic 1; detrimental; disastrous 1; evil 3; fatal 2; fateful 2; ferocious; harmful; hurtful 1; ill 4; injurious 1; killing 3; mischievous 2; mortal 3; murderous 1; ruinous; unhealthy 2; unwholesome 1; violent 1, 2; virulent 1; withering
destructiveness, violence 1; virulence 1
detach°, break 19b; disconnect; disengage; dissociate; distance 3; divide 1; divorce 2; free 15; insulate 1; isolate; loosen 2; pull 12a; remove 7; separate 1; sever 1; split 1
detach (from), alienate 1
detachable, separable
detached°, blithe 2; cool 4; cut 26; disinterested; dispassionate 2; distant 3; distinct 2; faraway 2; impassive; impersonal 1, 2; inattentive; indifferent 1; isolated 2; loose 1; neutral 1; nonchalant; objective 1; oblivious; philosophical 2; platonic; remote 7; reserved; rocky[1] 3; secluded 1; sedate 1; self-contained 1; self-possessed; separate 5, 7; standoffish; unattached 1; undone[1] 2; unsocial; wistful 2; withdrawn 1
detached, become, split 1
detachedly, vaguely 3
detaching, detachment 1
detachment°, cohort 1; corps; detail 3; distance 2; indifference 1; objectivity; party 2; reserve 4; section 1
detail°, amplify 1, 3; aspect 4; breadth 2; define 2; describe 2; document 2; element 1; enlarge 2; enumerate 1; expand 4; explain 1; get 40b; item 1; itemize; narrate; particular 5; party 2; point 10; recite 2; recount 2; refinement 2; regard 8; relate 2; report 3; respect 4; specification 2; specify; thing 2; way 6; work 19e
detail, in, depth 6; detail 4; length 4c
detail, unimportant *or* **insignificant** *or* **inconsequential** *or* **trivial** *or* **petty,** triviality 2
detailed°, ample 6; busy 3; circumstantial 3; close 14; elaborate 1; exquisite 2; faithful 3; finicky 2; full 2; fussy 2; graphic 1; intimate[1] 2; keen[1] 4; laborious 2; painful 3; particular 3; photographic 1; scientific; technical 1; thorough 3; vivid 2
detailing, breakdown 3; narration 1; resolution 5
detail(s), data; detail 2; dope 3; fact 3; point 5
detain, arrest 1, 2; buttonhole 1; delay 2; hold 5, 6, 22b; imprison; jail 2; keep 9, 14b; lock[1] 9; restrain 3
detained, belated
detainee, captive 1; prisoner
detect°, determine 2; discover 2; distinguish 3; find 2; hit 9b; make 37a; notice 2; perceive 1; recognize 1; savor 3; scent 3; see 14a; sense 6; spot 6; trace 5
detectable, discernible 1; observable; perceptible; sensible 2; visible 1; wind[1] 8
detection, discovery 2; identification 1; recognition 1; spying
detective°, sleuth
detective story, mystery 3; story[1] 1
détente, reconciliation 1
detention°, arrest 4; captivity; custody 2; imprisonment; justice 2; prison
detention center, prison
deter°, discourage 2; hinder 2; keep 13; repress; shackle 4; stay[1] 4; trammel 2
deterge, cleanse 1; purge 1; wash 1
deteriorate°, decay 1b; decline 4; degenerate 2; die 2; ebb 2; erode; fade 2; fail 3; fall 16; flag[2] 2; go 31e; lapse 4; putrefy; relapse 2; rot 1, 2; seed 4; sink 4; stagnate; suffer 4; waste 2; worsen 2; wrong 8a
deteriorated, decrepit 2; far 5a; ragged 4; seedy 1; sordid 4
deteriorating, decadent 1; rotten 2; wane 4
deterioration, decay 3; decline 7; decrepitude 2; degradation 1; disrepair; ebb 4; failure 2; fall 24; impairment; lapse 3; relapse 4; rot 3; wane 3; wear 9
determinant, factor 1
determinate, determined 2
determination°, backbone 3; bravery; choice 2; decision 1, 3; decree 1; disposition 3, 4; drive 8; energy; enterprise 2; evaluation 2; finding 2; follow-through; fortitude; grit; gut 3a; heart 2; industry 3; judgment 2; manhood 2; measurement 1; negotiation 2; nerve 1; patience 2; perseverance; persistence; pluck 1; purpose 2; push 14; resolution 1, 3; self-control 1; sentence; settlement 5; stability 2; strength 2; tenacity 1; will 1
determination, show, persevere
determine°, appoint 1; arrange 2; assign 2; calculate; check 12b; choose; clinch 1; compute; decide 1; decree 2; define 1; detect 1; diagnose; discover 1; dispose 3a; distinguish 1; divine 4; elect 1; establish 3; estimate 1; evaluate 2; fathom; feel 1; find 2; firm 5; fix 2, 15; gauge 1; get 31e; intend; judge 4; learn 1; limit 7; measure 11; mind 12; pace 4; perceive 2; pin 4b; pitch[1] 7; plot[1] 4; position 9; resolve 1; rule 7; scent 3; see 4; set 7; settle 2; shape 7; specify; tell[1] 7, 8, 9; trace 3; weigh 3
determine to be, find 9
determined°, bent 4; bound[3] 3; certain 1; clear 10; constant 1; decided 2; earnest 1; emphatic; enterprising; firm 4; fixed 2; grim 1; gritty 2; heroic 2; immovable 2; indomitable; inflexible; insistent; intent 4, 5; laborious 3; patient 2; persistent 1; purposeful; relentless 1; resolute; rigid 4; rocky[1] 2; self-willed; set 29;

devotedness (*cont.*) fanaticism 1; loyalty; piety 1, 2; resolution 1
devotee°, addict 2; admirer 1; disciple 2; enthusiast; fan; fiend 2; follower 3; freak 4; partisan 1; student 2; supporter 1
devotees, school 2
devotion°, application 3; attachment 3; dedication 1; enthusiasm 1; faith 3; fanaticism 1; friendship 2; love 1; loyalty; perseverance; piety 1, 2; praise 2; prayer 2; resolution 1; sanctity; veneration; worship 2
devotional, devout 2; solemn 2; spiritual 1
devour°, attack 5; consume 1, 5; dispose 3d; eat; feed 2; finish 3; gnaw 2; gorge 3; gulp 1; prey 3a; swallow 1; tackle 4; take 13
devouring, voracious 1
devout°, devoted 1; godly; hearty 2; holy 2; pious 1; religious 1
devoutness, devotion 1; piety 2
dew, wet 5
dewy, damp 1; moist 1
dewy-eyed, tearful
dexter, right 5
dexterity°, art 1, 4; craft 1; efficiency 1; expertise; facility 1; faculty 1; ingenuity; knack; proficiency; prowess 1; skill 1; tact; touch 16
dexterity (with), understanding 3
dexterous°, adept 1; agile 2; artful 2; clever 1, 4; expert 2; handy 3; ingenious; masterful 1; neat 4; nimble 1; proficient; quick 4; skillful; slick 3; tactful; talented; versatile 1
dexterously, handily 2
dexterousness, ingenuity; prowess 1; tact; technique 2
dextral, right 5
diablerie, devilry 1, 2
diabolic(al), black 6; cruel 2; devilish; dreadful 2; ferocious; fiendish; ghoulish 1; infernal 2; inhuman 2; poisonous 2; satanic 1; serpentine 1; sinister 2; ungodly 1; vicious 1; wicked 1; wrong 1
diabolism, magic 1; sorcery
diacritic, accent 2
diacritical mark, accent 2
diadem, crown 1; ring[1] 1
diagnose°, identify 3
diagnosis, interpretation 2
diagonal, bias 2; line[1] 1; oblique 1
diagram, chart 2; design 5; draft 1, 5; figure 5; pattern 4; plan 2; plot[1] 4; profile 3; projection 3; schematic 2; scheme 2
diagrammatic(al), graphic 2; schematic 1
dial, call 3; control 6; face 1, 7; mug 2
dialect°, cant 2; idiom 1; jargon 1; language 1; lingo; parlance; provincialism 1; speech 3; talk 19; tongue 1
dialectalism, provincialism 1
dialectic, controversial 2
dialectics, logic 1
dialogue°, conversation; discussion; parley 1; rap 5; talk 15; tête-à-tête 1; word 1
diameter, caliber 1; width
diametric, polar 2
diapers, still in, fresh 3
diaphanous, filmy 1; fine[1] 6; flimsy 3; insubstantial 1; see-through; sheer 3; tenuous 1; thin 5; transparent 1
diaphoresis, perspiration; sweat 5
diarrhea, run 50
diary°, calendar 1; chronicle 1; journal 2; memoir 2; record 3; register 1; timetable
diatribe, harangue 1; lecture 1; tirade
dice, chop 2; cut 17a
dice, toss *or* **throw of the,** fate 1
dicey°, dodgy; explosive 1; hazardous; precarious; risky; speculative 2
dichotomize, split 1
dichotomy, split 8
dick, detective; officer 2; operative 3; sleuth
dickens, the, devil 5
dicker°, bargain 3; haggle; negotiate 1
dicky, dodgy; hazardous; ill 1
dictate°, bid 3; bidding 2; charge 5; decree 1, 2; determine 4; direct 3; injunction 2; lay[1] 10; order 4; precept 1; prescribe; regulation 2; tell[1] 5
dictate to, tyrannize
dictator°, despot; disciplinarian; oppressor; tyrant
dictatorial°, absolute 3; arbitrary 2; authoritarian; bossy; despotic; dogmatic; domineering; hard 4; heavy-handed 2; imperative 2; masterful 2; officious; opinionated 1; overbearing; peremptory 3; positive 5; prescriptive; repressive; severe 1; totalitarian; tyrannical
dictatorship, despotism; domination 2; tyranny
diction°, expression 6; language 4; oratory; phrase 3; pronunciation; speech 1; style 5
dictionary°, glossary; literal 2; thesaurus 2
dictum, decree 1; principle 1; pronouncement 2; proverb; saw; say-so; theorem 2
didactic, pedantic 1; prescriptive
diddle, cheat 2; defraud; fleece; swindle 1
diddle (out of), rob 3
dido, caper 2; stunt[1]; trick 2
die°, expire 2; fall 6, 20; farm 2; flag[2] 2; founder[2] 2; go 13; itch 2; miscarry; mold[1] 1; pack 8b; pass 14a; perish; rot 2; sink 4; stall[1] 1; stamp 6; template
die, cause to, carry 10d
die away *or* **down** *or* **off** *or* **out,** die 2, 4; diminish 3; disappear 2; dwindle; end 9; fade 2; fizzle 2; pass 10; peter out; subside 2; taper 2; trail 8; wane 1
die for, desire 1; pant 2
die (out *or* **away),** fail 3; go 33a
die cast, mold[1] 5
died out, extinct 1
diehard, purist; square 9; ultra-
die stamp, stamp 6
diet°, assembly 2; chamber 1; convention 1; fare 3; fast[2] 1, 2; house 3; parliament 2; reduce 3; slim 3
differ°, clash 3; conflict 4; disagree 1; fall 18; quarrel 2; vary 3
differ (from), contrast 2; depart 2
differ (on *or* **about),** dispute 2
difference°, balance 7; change 2; clash 2; combat 3; conflict 3; contrast 3; disagreement 1, 2; discrepancy; disparity; dissimilarity; distinction 1; diversity 1; gap 2; inequality 1; misunderstanding 2; odds 3; rift 1; spread 10; strife 1; tilt 4; variance 1; variation 1; variety 2
difference, make a, matter 6
difference (of opinion), dispute 3; fight 8; quarrel 1; tiff; variance 2
difference(s), difference 2; rivalry
differences between, settle *or* **resolve,** reconcile 1
different°, alternative 1; contrary 1; deviant 1; discordant 1; dissimilar; distinct 2; divergent; divers; diverse; exotic 2; extreme 1, 5; fresh 2; incongruous; kinky 1; new 1, 4; notable 2; noteworthy; novel 1; odd 1; opposite 2; ordinary 4; outlandish; outside 9; pole[2] 3; queer 1; remarkable 3; revolutionary 2; separate 5, 6; several 2; singular 1; special 1; sundry; unfamiliar 1; unrelated; unusual; variant 2; various 1, 2
different, be, differ 1
different (from), unlike 1, 3
differentiate°, contrast 1; discriminate 1; distinguish 1; know 4; set 14a; tell[1] 8
differentiating, judicial 2
differentiation, distinction 1
differently, otherwise 2
differing, discordant 1; divergent; divers; diverse; odds 4; opposite 2; variant 2; variation 1
differing from, unlike 3
difficile, difficult 5
difficult°, arduous 1; awkward 5; choosy; deep 2; demanding 1; deplorable 1; dicey; dodgy; exacting; exhausting 2; fastidious; finicky 1; formidable 3; fussy 1; hairy 3; hard 2, 3, 5; heavy 12; killing 3; labored 1; laborious 1; lean[1] 3; murderous 2; nasty 2; nervous 2; overcritical; parlous; picnic 3; precarious; prickly 4; problem 3; problematic; rebellious 2; rocky[1] 1; rough 7; rugged 2; self-willed; squeamish 1; stiff 9; strained; strait 1; strenuous 1; strong 16; tender[1] 4; thorny 2; ticklish 3; tight 7; toilsome; tough 4, 5; tricky 2; troublesome; trying
difficulties°, difficulty 3; necessity 3; pressure 3; static 2
difficulties for, create, bugger 4b
difficulty°, complication 2; delicacy 3; dilemma; discomfort 1; distress 2; drawback; embarrassment 2; emergency; fix 17; fly 10; grief 2; hardship; headache 2; hitch 4; hole 5; hurdle 1; inconvenience 1; interference 2; issue 4; jam 6; job 4; kink 3; lookout 3; matter 3; mess 3; mischief 2; necessity 3; need 4; nuisance 1; perplexity 1, 2; pinch 8; pitfall 2; plight; predicament; pressure 3; problem 1; profundity 1; quandary; question 2; ramification 1; reverse 8; rub 11; scrape 8; snag 1; snarl[2] 3; sore 6; static 2; stumbling block; trouble 5
difficulty, in (financial), debt 2
difficulty, with no, easily 1
difficulty, without, easily 1; readily 2; swimmingly; well[1] 10
diffidence, fear 1; humility; shame 4
diffident, backward 1; bashful 1; coy;

dipso, alcoholic 2; drunk 3
dipsomania, drunkenness
dipsomaniac, alcoholic 2; drunk 3; soak 3
dire, calamitous; disastrous 1; drastic; dreadful 2; evil 4; fearful 3; fierce 3; frightening; grim 3; infernal 2; macabre; mortal 5; satanic 2; severe 3; sore 3
  dire straits, in, back 10; debt 2; needy; trouble 9a
direct°, aboveboard 2; administer 1; aim 1; artless 1; bare 2; bend 4; blunt 2; boss 2; broad 3, 5; candid 1; categorical; chair 4; channel 5; charge 10; clear 16; coach 3; command 1; concentrate 1; concise; conduct 3, 5; control 1; decree 2; deliver 5; dictate 1; discipline 7; dispense 2; dominate 1; downright 1; emphatic; explicit 2; express 5, 7; fasten 2; fix 4; flat 4; forthright; frank 1; free 9; front 9a; govern 1; guide 1, 2, 3; handle 3; head 11; honest 3; immediate 2; ingenuous 2; instant 4; instruct 2; lay[1] 7; lead 3; level 10, 15; manage 1; manipulate 2; matter-of-fact; moderate 6; mold[1] 6; navigate 2; nonstop 1; obvious; officiate; open 15; operate 2; order 14; outright 1, 2; outspoken; oversee; pilot 3; plain 2, 3; plump[2] 6; point 20; point-blank 1; prescribe; preside; primary 3; raw 6; refer 2; regulate 2; reign 2; round 8; route 2; rule 6; run 10; short 5; show 2; simple 3; sincere; steady 3; steer 1; straight 1, 4, 6; strike 2; supervise; switch 5; tell[1] 5; transmit 2; transparent 3; turn 9; tutor 2; unvarnished
  direct, be, card 13
  direct attention to, get 38; refer 1
  direct oneself to, address 5
  direct to, refer 1
direct affairs, shot 9
directed, bound[3] 5
  directed by, under 2
directing, direction 1; management 1
direction°, administration 1; aim 4; bearing 4; bent 5; bidding 2; care 3; charge 5; command 6, 7; conduct 2; control 4; counsel 1; course 4; dictate 2; discipline 4; drift 3; exposure 4; gist; government 1; guidance 1; injunction 2; instruction 1; lead 13; leadership; line[1] 9; management 1; measure 6; operation 2; order 4; oversight 2; path 3; prescription 1; production 3; providence 2; quarter 3; recommendation 1; reference 1; regime; reign 1; road 1; route 1; rule 1, 2; running 1; step 10a; tack 3; tendency; tenor; thread 2; trend 1; turn 32; view 5; way 3; word 7
  direction, in every, around 2
  direction of, in the, for 6; toward 1
directionless, disjointed 2; erratic 3
directions°, direction 2; formula; manual
  directions (from), in all, about 8; around 10; helter-skelter 2
directive, decree 1; injunction 2; instruction 1; law 1; order 4; precept 1; regulation 2
directly°, dead 20; due 5; expressly 1; face 13; flat 16, 17b; flush[2] 4; full 14; hastily 1; immediately 1, 2, 3; instantaneously; once 6a; openly 2; outright 3, 4; plump[2] 5; point-blank 3, 4; posthaste; promptly; right 13; shoulder 6; slap 6; soon 2; straight 12, 14, 15; summarily 1
direct matters, shot 9
directness, candor 1; honesty 2; simplicity 3
director°, boss 1; employer 1; executive 1; guide 5; head 2; leader 1, 2; manager; master 1; officer 1; operator 2; principal 3, 4; producer 2; showman; superintendent; supervisor
directorate, council 2; direction 1; executive 2; management 2
directorial, responsible 3
directors, board 4; council 2; management 2
directorship, board 4; chair 2; direction 1; executive 2; helm 2; leadership; management 1
directory, council 2; index 1; list[1] 1; register 1; roll 13
direct things, shot 9
diremption, division 1
dire straits°, strait 3
dirge, chant 1; keen[2] 2; lament 2
dirk, dagger; steel 1
dirt°, earth 2; filth 1; grime; ground 1; impurity 2; information; land 2; lowdown; mire 2; mold[3]; muck 2; mud; news 1; pornography; refuse[2]; ribaldry; scandal 3; soil[1] 3; soil[2]; trash 3
dirtiness, impurity 1, 3
dirty°, abusive 1; base[2] 4; bawdy; beastly 2; bedraggled; blue 2; broad 8; coarse 3; dingy; dishonorable 3; disreputable 2; earthy; erotic 3; evil 2; filthy 1, 2, 3; foul 5, 7, 11, 12; gross 3; immodest 1; immoral 2; impure 1, 4; incontinent 2; indecent 2; lascivious 2; lewd; libertine 2; lousy 1; low[1] 3; mangy; mess 5c; mire 4; muddy 1, 5; nasty 1, 3; naughty 3; obscene 1; outrageous 3; pollute 1; pornographic; profane 3; prurient 2; racy 2; rank[2] 3; ratty 2; rough 6; rude 3; sad 3; sensual; sexy 2; shabby 1, 3; sleazy 2; sloppy 1; smear 1; soil[1] 1; sordid 3; spot 7; spotty 1; stagnant; suggestive 2; sully; taboo 1; taint 2; tarnish; untidy; unwashed 1; vulgar 2; wicked 3
  dirty look, give a, frown 1; glare 4
dirty business, conspiracy; foul play
dirtying, pollution
dirty language, profanity
dirty look, frown 3; glare 2; scowl 2
dirty(-minded), evil-minded 1; filthy 3; lecherous
dirty old man, libertine 1; roué
dirty-rotten, rotten 4
dirty trick(s), foul play; machination
dirty word, curse 3; expletive 2; oath 2
dirty work, conspiracy; foul play
Dis, hell 1; underworld 2
disability°, ailment; disease 1; handicap 1; illness; trouble 8
disable, cripple 2; destroy 4; exhaust 2; handicap 2; hurt 4; incapacitate; lay[1] 19b; maim; mutilate 1; paralyze 3; sabotage 3; spike 3; strike 13; undermine 1; waste 3
disabled°, crippled 1; decrepit 1; helpless 1; invalid[1] 1; lame 1; powerless 2
disablement, disability 1; wreck 5
disabling, wreck 5
disabuse, disillusion
  disabuse (of *or* **from),** alienate 1
disabused, disenchanted
disaccord, disagreement 1
disadvantage°, catch 18; detriment; drawback; encumbrance; handicap 1, 2; hurt 5; inconvenience 1; liability 3; loss 3, 5
disadvantaged, deprived; needy
  disadvantaged by, be, labor 7
disadvantageous, detrimental; hurtful 1; inconvenient; prejudicial; ugly 3
disadvantageousness, inconvenience 2
disaffect, divide 3; shake 3; turn 17b
disaffected, estranged
disaffection, dislike 2
disaffiliation, secession
disaffirm, contradict 2; deny 1; reverse 3
disaffirmation, denial 1
disagree°, argue 1; clash 3; conflict 4; differ 2; fall 18; feud 2; fight 4; issue 9; jar[2] 2; protest 3; quarrel 2; row[2] 3; scrap[2] 2
  disagree, cause to, divide 3
  disagree (with), dispute 1; exception 4; tangle 4
disagreeable°, abominable 2; argumentative; bad 4; beastly 2; bitter 2; cantankerous; distasteful; evil 5; foul 8; fretful; frightful 2; harsh 3; horrible 2; nasty 2, 4; picnic 3; pugnacious; quarrelsome; rancid; rank[2] 4; ratty 1; repellent; repulsive; revolting; shabby 2; short-tempered; sour 3; terrible 1, 3; testy; tiresome 2; ugly 2, 4; unacceptable; uninviting; unpalatable; unsavory
disagreeably, painfully
disagreeing, discordant 1; dissident 2; divergent; incongruous; odds 4
disagreement°, argument 1; clash 2; combat 2; conflict 2, 3; controversy 1, 2; difference 1, 2; discord; discrepancy; dispute 3, 4; dissension; division 5; encounter 5; exchange 3; faction 2; feud 1; fight 8; fracas 2; friction 2; gap 2; misunderstanding 2; protest 1; quarrel 1; rift 1; row[2] 1; run-in; scrap[2] 1; strife 1; tiff; variance 1, 2
  disagreement, in, discordant 1; odds 4; variance 3; war 2
disallow, ban 1; contradict 2; deny 2; exclude 1; forbid; keep 13; outlaw 2; prohibit 1; refuse[1] 2; reject 1; repress; shut 5a; veto 1
disallowance, exclusion 1; prohibition 1; veto 2
disallowed, impure 3; inadmissible; pale[2] 3; unlawful
disallowing, prohibition 1
disambiguate, clear 30c, specify
disannul, void 6
disappear°, collapse 2; depart 1; die 2; disperse 2; dissipate 2; dissolve 1; escape 1; evaporate 2; exit 3; fail 3; finger 9; flee 1; get 32a; leave[1] 1; lift 5; melt 3, 4; pass 10, 14b, 17; peter out; powder 2; roll 2; sink 5; slip[1] 6; trail 8
  disappear, make, spirit 10; swallow 3
  disappear (from), drain 6

discomfort° (*cont.*) embarrassment 1; excitement 1; fluster 1, 2; hurt 6; ill 8; inconvenience 1; misery 1; misgiving; pain 1; pang 2; scruple 1; trouble 5

discomforting, sticky 2; ugly 3

discommode, disturb 4; fluster 1; inconvenience 3; put 23a; trouble 1, 3

discompose, discomfit 1; dismay 2; disorder 4; distract 3; embarrass; excite 2; fluster 1; perturb; ruffle 3; shock 1; startle; trouble 1; upset 1

discomposed, agitated; confused 2; disconcerted; embarrassed 1; excited 1

discomposure, embarrassment 1; pain 2

disconcert, agitate 1; bother 3; confuse 1; daunt; demoralize 3; devastate 2; discomfit 1; dismay 1; disrupt 1; distract 3; disturb 4; embarrass; excite 2; floor 5; flurry 2; fluster 1; foil[1]; horrify 2; jar[2] 3; overawe; perplex; perturb; petrify 2; put 21c; rattle 3; rock[2] 3; ruffle 3; shake 5; shame 6; spite 3; surprise 1; throw 4; trip 6; unnerve; upset 1; wind[2] 4d

disconcert (someone), wind[1] 13

disconcerted°, agitated; blank 5; confused 2; dumbfounded; embarrassed 1; excited 1; frantic; insecure 1; upset 8

disconcerted, feel, small 6

disconcerting°, disappointing; disturbing; embarrassing; harrowing; perplexing; unsettling

disconnect°, detach; disengage; dissociate; divide 1; hang 12; remove 7; separate 1; sever 1

disconnected°, broken 6; desultory; detached 1; discrete; disjointed 1; fragmentary; incoherent; independent 2; intermittent; loose 1, 5; odd 2; rambling 1; separate 5

disconnecting, detachment 1

disconnection, detachment 1

disconsolate, bad 9; brokenhearted; dejected; desolate 3; doleful; forlorn 1; heartbroken; heavy 6; hopeless 3; inconsolable; joyless 1; low[1] 8; melancholy 1; miserable 1; moody 1; mournful 1; sorrowful 1; unhappy 1; wistful 1; woebegone; wretched 2

disconsolateness, melancholy 2; woe

disconsonant, incongruous

discontent°, discontented; disgruntled; dissatisfaction 1; dissatisfied

discontented°, disappointed; disgruntled; dissatisfied; sour 4

discontentment, discontent; displeasure 1; dissatisfaction 1; distaste 1

discontinuance, cancellation 2; check 13; expiration; failure 2; pause 2; stay[1] 6

discontinuation, break 27; check 13; dissolution 2; expiration; pause 2; stay[1] 6; suppression; suspension 2; termination 1; wait 4

discontinue°, abandon 4; break 10, 19a; call 10; cease 1; close 5; cut 15b; drop 7; end 9; expire 1; halt 2; interrupt 2; leave[1] 9; lift 3; pause 1; phase 6; quit 3; raise 9; refrain[1] 2; scrub 2; sever 2; shut 2; sign 9; stay[1] 4; stop 1; suppress 1; terminate; turn 17a

discontinue temporarily, suspend 1

discontinued, disused

discontinued, be, lapse 5

discontinuing, suppression; suspension 2

discontinuity, break 27; check 13; gap 1; pause 2

discontinuous, broken 6; discrete; disjointed 2; intermittent; loose 5; odd 2; spasmodic 2

discord°, antagonism 2; conflict 3; disagreement 1; dispute 4; dissension; division 5; faction 2; feud 1; fight 8; fracas 2; friction 2; jar[2] 2; misunderstanding 2; quarrel 1; rivalry; split 8; strife 1; trouble 7; variance 2

discordance, disagreement 1; discord; discrepancy; dissension; noise 1

discordancy, disagreement 1

discordant°, dissident 2; factious; grating 1; harsh 1; incompatible; incongruous; mismatched; noisy; ragged 5; raucous; rough 8; strident

discordant (with), step 8

discount°, allowance 4; dismiss 2; lower[1] 1; mark 15b; rake-off; rebate 1, 3; underestimate

discounted, cut 28

discountenance, disfavor 3; displeasure 1; embarrass; frown 2; horrify 2; rattle 3; shame 6; trouble 1

discountenanced, embarrassed 1

discounting, irrespective of

discourage°, break 7; browbeat; choke 3; dampen 2; daunt; demoralize 1; depress 1; desolate 6; deter; dismay 2; fend 2; frustrate 1, 2; gag[1] 1; get 37c; hinder 2; inhibit; keep 13; militate 1; puncture 4; put 21b; repress; sadden; shackle 4; shake 3; stay[1] 4

discouraged, broken 3; dejected; desolate 3; despondent; disappointed; downhearted; hopeless 3

discouragement, depression 2; despair 1; desperation 2; deterrent; disappointment 2; lesson 3; rebuff 1

discouraging, bad 5; cold 4; disappointing; oppressive 1; prohibitive 1; untoward 1

discourse, address 1; conversation; converse; lecture 1, 3; oration; rap 5; sermon 2; speak 1; speech 2; talk 14

discourse in, speak 3; talk 1

discourse (on *or* upon), hold 16a; speak 10

discourse with, speak 2

discourteous°, abrupt 3; blunt 2; brusque; disagreeable 3; disrespectful; flippant 2; harsh 3; ignorant 4; ill-mannered; impertinent; impolite; irreverent 2; offensive 2; rough 3; rude 1, 2; short 4; short-tempered; tactless; ungracious; unrefined 1; unthinking 2

discourteousness, incivility

discourtesy, disrespect; flippancy 2; impertinence; incivility; indignity; insult 2

discover°, catch 4; check 12b; come 5a; detect 1; determine 2; dig 6; divine 4; expose 1; find 1, 2, 3; finger 5b; get 44c; hear 2; hit 9a; learn 1; light[1] 9; locate 2; make 37a; penetrate 5; perceive 1; pierce 3; root[1] 7b; run 26a; scout 2; see 4; strike 14; stumble 3; surprise 2; trace 5; turn 25c; unearth

discover to be, find 4

discoverable, visible 1; wind[1] 8

discovered, be, light[1] 10

discovery°, find 10; finding 1; location 2; observation 1; protégé; revelation; solution 1

discovery, about to make a, warm 6

discovery procedure, investigation

discredit°, belittle; blacken 2; blast 6; brand 3; degradation 2; degrade 2; depreciate 2; diminish 2; disfavor 2; disgrace 1, 4; dishonor 4; disparage 1; disprove; distrust 1; doubt 1; explode 2; flaw 4; foul 14; humiliation; impeach 2; infamy 1; libel 4; lower[1] 4; notoriety; pull 9c; rebut; scandal 2; shame 2, 8; slur 1; smear 2; taint 1; wrong 9

discreditable, degrading; dishonorable 2; disreputable 1; infamous 1; notorious 1; shabby 4; shameful; unseemly 1; unworthy 1

discrediting, slanderous

discreet°, cautious; conscientious 3; deliberate 3; diplomatic; inconspicuous; judicious; politic 2; provident 1; prudent 1; seemly 2; sensible 1; tactful; wise 3

discreet, be, step 12

discrepancy°, difference 1; disagreement 1; disparity; dissimilarity; diversity 1; gap 2; hole 6; inequality 1; odds 3; variance 1; variety 2

discrepant, contradictory; divergent; incongruous; opposing

discrete°, different 1; distinct 2; diverse; individual 1; particular 1; separate 5; several 2

discrete (to), specific 2

discretion°, age 2; calculation 3; carte blanche; caution 2; diplomacy 1; disposition 4; finesse 3; freedom 4; intelligence 1; judgment 1; pleasure 2; providence 1; prudence 1; savoir-faire; tact; taste 6; volition

discretion, at one's, will 4

discretional, optional; voluntary 2

discretionary, optional; voluntary 2

discriminate°, contrast 1; differentiate 1; distinguish 1; pick 7b; separate 2; tell[1] 8

discriminate against, segregate

discriminating°, acute 5; aesthetic 2; choosy; delicate 5; fine[1] 9; fussy 1; judicial 2; judicious; keen[1] 5, 6; nice 3; particular 4; penetrating 1; politic 2; prudent 1; rational 2; refined 2; selective; tasteful

discrimination°, culture 1; discretion 1; distinction 1; ear 2; eye 3; grace 2; injustice 1; intolerance; judgment 1; penetration 2; prejudice 2; prudence 1; refinement 1; resolution 5; segregation; sense 2; sophistication 1; taste 4

discriminative, judicial 2; judicious; keen[1] 6; nice 3; selective; tasteful

discriminatory, intolerant 2; judicial 2; partial 2; selective

discursive°, diffuse 2; erratic 3; inarticulate 1; purposeless; rambling 1

discuss°, argue 2; bring 15b; comment 4; confer 1; converse; criticize 1; debate 3; dispute 2; get 40b; go 30c; huddle 4;

knock 3c; negotiate 1; palaver 5; parley 2; review 3; speak 10; talk 7, 11; treat 2
discuss (with), consult 1; reason 8
discussed, subject 7
discussion°, conference; conversation; counsel 2; debate 1; dialogue 1; dispute 3; huddle 2; interview 1; negotiation 1; palaver 3; parley 1; rap 5; talk 15; word 1
discussion, under, question 6a; subject 7
disdain, contempt; deride; despise; disregard 2, 3; flout; look 5; mock 1; mockery 1; neglect 1; patronage 3; patronize 1; put 16g; sarcasm; scorn 1, 3; shun; slight 5; snap 6; sneer 2, 3; snobbery; spurn; trample 2
disdain, below *or* beneath *or* beyond, despicable
disdainer, naysayer
disdainful°, arrogant 2; condescending; contemptuous; derisory; haughty; hoity-toity; lofty 4; majestic 2; regal 1; sarcastic; scornful; snobbish; supercilious
disdainfulness, snobbery
disease°, affliction 2; ailment; blight 1; bug 2; disorder 3; illness; infirmity 2; trouble 8; ulcer 2
diseased°, affected 4; ill 1; sick 2; unsound 2
disembark°, get 41a; land 5; light[2] 13
disembarkation, landing 2
disembodied°, immaterial 2
disembogue, discharge 4
disembowel, butcher 3; gore[2]; gut 4
disenable, paralyze 3
disenact, void 6
disenchant, disappoint 2; disillusion; let[1] 4; shake 3; sour 6; turn 17b
disenchanted°, disappointed
disenchantment, disappointment 2; letdown
disencumber, clear 23; deliver 3; divest 1; free 16; lighten[2]; relieve 2
disenfranchise, degrade 1; downgrade 1
disengage°, break 19b; detach; disconnect; dissociate; extricate; free 15; loose 12; separate 1
disengaged, clear 15
disengagement, detachment 1
disengaging, detachment 1
disentangle, decipher 1; detach; disengage; extricate; free 15; separate 1; simplify; solve; straighten 1
disentangled, clear 15; detached 1
disentanglement, resolution 4
disenthrall, disillusion; emancipate; free 14; liberate 1; release 1
disentrance, disillusion
disestablish, void 6
disesteem, disfavor 2; disgrace 1; dishonor 4; dislike 1, 2; disregard 3; shame 2
disfavor°, disapproval; discredit 4; discriminate 2; disgrace 1, 4; dislike 1, 2; displeasure 1; disservice; distaste 1; frown 2; shame 2
disfavor, bring into, discredit 1; foul 14
disfeatured, disfigured
disfigure, batter 3; blemish 1; damage 4; deface; distort 1; flaw 3; mangle; mar 1; mutilate 1; ruin 8; scar 2; spoil 2
disfigured°, crooked 2; deformed 2
disfigurement, blemish 3; blot 1; flaw 1; scar 1
disfranchise, degrade 1; downgrade 1
disgorge, bring 15d; eject 2; heave 4; regurgitate; spew; spout 1; throw 9a
disgorgement, ejection 1
disgorging, ejection 1
disgrace°, brand 3; crush 5; degradation 2; degrade 2; discredit 1, 4; disfavor 2; dishonor 2, 4; embarrass; foul 14; humiliation; infamy 1, 2; libel 1, 3; lower[1] 4; notoriety; pity 2; pull 9c; scandal 1; shame 2, 8; soil[1] 2; stain 5; sully; take 31b; tarnish; undoing 1
disgraced, embarrassed 1
disgraceful°, black 6; criminal 2; deplorable 2; dishonorable 1; disreputable 1; foul 4; glaring 1; infamous 2; mean[2] 3; miserable 4; monstrous 2; notorious 1; outrageous 1; scandalous 1; shameful; unheard-of 3; unmentionable 2; vile 1; wicked 2; wrong 1
disgracefulness, infamy 2
disgruntle, offend 1
disgruntled°, discontented; dissatisfied; indignant; resentful
disgruntled at, be, resent
disguise°, camouflage 1, 2; cloak 3; color 6; conceal 1, 2; cover 15; dissimulate; doctor 4; dress 5a; face 3; front 5; gloss[1] 2, 4; guise 2; hide[1] 2; mantle 3; mask 2, 3; masquerade 2; misrepresent; obscure 6; outside 2; pretense 2; pretext 1; secrete[1]; shade 12; shape 5; shroud 1; veil 2
disguise, in, incognito 1, 2
disguise oneself as, pass 16b
disguised, incognito 1; invisible 2; veiled
disguised as, be, pass 16b; pose 2
disgust°, contempt; dislike 2; distaste 2; nauseate; offend 2; put 21d; repel 2; revolt 3; revulsion; shock 1; sicken 2; turn 17b
disgusted°, nauseated; sick 6
disgusted, easily, squeamish 2
disgusting°, abhorrent; abominable 1; disagreeable 2; distasteful; evil 5; excessive 2; fearful 3; filthy 2; foul 1; ghoulish 2; grisly; gross 5; hideous 1; horrible 1; loathsome; lurid 2; monstrous 1; nasty 1; nauseous; obnoxious; obscene 1; offensive 3; outrageous 3; rank[2] 3, 4; repellent; repugnant; repulsive; revolting; scandalous 1; shocking 2; stinking 2; terrible 5; ugly 2; uninviting; unsavory; vile 2; wicked 6; yucky
dish, bowl[2]; plate 1, 2; platter
dish out, dispense 1; distribute 1; dole 3; expend 1; give 15b; mete; serve 3; shell 5; spend 1
dish up, serve 3
disharmonious, incongruous
disharmonize, clash 4
disharmony, discord; faction 2; friction 2; strife 1
dishearten, chill 8; daunt; demoralize 1; depress 1; desolate 6; discourage 1; dismay 2; get 37c; sadden
disheartened, despondent; disappointed; heartbroken; joyless 1; moody 1; mouth 6; sad 1; sorrowful 1; woebegone
disheartening, cold 4; gloomy 3; joyless 2; mournful 2; off 8; oppressive 1; sad 2
disheartenment, despair 1
dishevel, mess 5a; ruffle 4; rumple; tousle
disheveled, bedraggled; disreputable 2; shaggy; sloppy 1; unkempt; untidy; wild 8
dishonest°, abusive 3; bent 3; bum 4; corrupt 1; criminal 1; crooked 1; deceitful; deceptive 2; devious 1; dirty 3; dishonorable 2; disingenuous; double 5; evil 2; faithless 2; fixed 4; fly-by-night 2; foul 7; fraudulent 2; hypocritical; immoral 1; insincere; jaundiced 1; lawless 2; lying 2; perfidious; phony 1; rotten 3; shady 2; sharp 6; shifty; slippery 2; sneaky; tricky 1; two-faced; unscrupulous; untrue 1; villainous 1; wily; wrong 1
dishonesty, deceit 1; falsity 1; lying 1
dishonor°, default 3; degradation 2; degrade 2; desecrate; discredit 1, 4; disfavor 2; disgrace 1, 4; disparage 1; foul 14; humiliation; indignity; infamy 1; insult 1, 2; libel 1, 3; notoriety; peg 2; pull 9c; ruin 2, 11; scandal 2; seduce 2; shame 2, 8; taint 1; tarnish; violate 2; violation 2; wrong 9
dishonorable°, base[2] 2; corrupt 1; dirty 3; disgraceful 1; dishonest; disreputable 1; evil 2; foul 4, 7; infamous 1; notorious 1; promiscuous 2; scandalous 1; scurvy; shabby 2, 4; shady 2; shameful; sordid 1; unmentionable 2; unscrupulous; untrue 1; unworthy 1; venal; wrong 1
dishonored, shamefaced 2
dishonoring, ruin 2; sacrilege 1; violation 2
dishwater, water 1
disillusion°, let[1] 4; puncture 4; turn 17b
disillusioned, disappointed; disenchanted
disillusionment, letdown
disincentive, deterrent
disinclination, aversion 1; compunction 2; dislike 2; distaste 1; qualm; reluctance
disinclined°, averse; indisposed 2; loath; reluctant; slow 10
disinclined, be, hate 2; shy 4
disinclined to, incapable 2
disinfect°, clean 11b; fumigate; purify 1; sterilize 1
disinfected, clean 1; hygienic; pure 2; sanitary; sterile 2
disinformation, misinformation; propaganda 1
disingenuous°, artful 1; deceitful; foxy 1; insincere; sly 1; sneaky; unscrupulous; wily
disinherit, cut 15d
disintegrate°, break 16c, 23a; collapse 2; come 7; crumble; decay 1b; decompose 1, 2; deteriorate 2; die 2; dissolve 1; fall 7; fragment 3; go 12; piece 6; rot 1; shatter 1; shiver[2]; splinter 2; way 11a; worsen 2
disintegrate into, resolve 4
disintegrated, broken 1; old 3
disintegrating, rotten 2; tumble-down; unsound 1
disintegration, collapse 6; decay 3; dissolution 1; rot 3; ruin 1; separation 3
disinter, dig 6; excavate 2; unearth

dispensary, hospital; infirmary; pharmacy 1
dispensation, administration 3; disposition 3; distribution 2; dole 2; exemption; faculty 4; leave[2] 1; license 1; privilege; share 1; spread 8
dispense°, administer 3; allot; deal 1; diffuse 3; dispose 3c; distribute 1; divide 2; dole 3; dose 2; give 15b; hand 17; measure 14; mete; portion 4; present[2] 6; propagate 3
dispense with, discard 1; dispense 3; rid 2; scrap[1] 4; spare 9; throw 5a; waive 1
dispensing, spread 8
dispersal, dissipation 2; dissolution 2; distribution 2; radiation; rout 1; spread 8
disperse°, break 11, 23a; diffuse 3; disband; dismiss 3; dissipate 1; dissolve 1, 3; distribute 2; evaporate 2; flight[2] 2; measure 14; pay 11a; portion 4; propagate 3; radiate 2; scatter 2; spray[1] 1; spread 1, 4; strew
dispersed, diffuse 1; loose 4; sparse 1
dispersing, spread 8
dispersion, dissipation 2; evaporation 2; spread 8
dispirit, chill 8; daunt; demoralize 1; depress 1; desolate 6; discourage 1; get 37c; sadden
dispirited, blue 1; broken 3; dejected; desolate 3; despondent; forlorn 1; gloomy 2; glum; heartbroken; joyless 1; melancholy 1; mournful 1; mouth 6; sorrowful 1; unhappy 1
dispiritedness, melancholy 2; sadness
dispiriting, bitter 3; cold 4; gloomy 3; joyless 2; oppressive 1; sad 2; trying
displace°, downgrade 1; exile 3; expel 1; lose 1; remove 5; substitute 1; supersede; supplant; transplant; uproot 1
displaced person, emigrant; exile 2; outcast; refugee
displacement, movement 1; removal 3; subversion
display°, air 7; appearance 3; arrangement 1; bear 5; bring 14a; carry 7; demonstrate 2; demonstration 1, 2; develop 3; evidence 4; exemplify 2; exhibit 1; exhibition; expose 1; exposition 1; flash 2; flaunt; float 6; flourish 4; formation 3; hold 22d; indicate 3; indicator; manifest 2; manifestation; model 8; monitor 2; mount[2] 4; open 23; ostentation; pageant; pageantry; parade 2, 5; performance 5; picture 7; pose 5; presentation 2; pretense 1; produce 4, 5; production 3; promenade 4; reflect 2; register 4; reveal; run 19; scene 3; set 20b, 22b; shake 4; show 1, 6, 11, 13, 15; spectacle 1; splendor 1; splurge 1; sport 5; spread 3; stick[1] 17b; strike 12; swagger 3; trot 2; veneer; wear 2
displease°, anger 2; offend 1; outrage 4; turn 17b
displeased, discontented; disgruntled; dissatisfied; resentful
displeased, be, umbrage
displeased at, be, resent
displeasing, distasteful; off 8; offensive 2; repulsive; unsatisfactory; unwelcome 1
displeasure°, anger 1; disapproval; discontent; disfavor 1; dislike 2; dissatisfaction 1, 2; distaste 1; resentment
displeasure, feel, offense 3; umbrage
disport, flaunt
disport (oneself), play 1
disposable°, dispensable; expendable; unnecessary; worthless 1
disposal, dispatch 7; disposition 2, 3, 4; hand 8; order 1; position 1; refusal 2
disposal, at one's, available
dispose°, arrange 1; coordinate 1; give 10; group 4; incline 2; interest 8; lead 2; persuade 1; place 15; position 8; prevail 3; rank[1] 6; set 22b; settle 1; tempt 2; work 18
dispose of, clinch 1; cope 2; demolish 2; devise 2; discard 1; dispatch 3; dispense 3b; dispose 3; dump 2; eliminate 2, 4; finish 3, 4; meet[1] 5; polish 3b, 3c; process 4; remove 4; rid 2; scrap[1] 4; sell 1; settle 10; shred 2; sign 8, 11; throw 5a; wipe 3
disposed°, fit[1] 4; given 3; inclined 1; liable 1; open 18; prepared 3; prompt 2; prone 2; ready 3
disposed, be, incline 3; tend[1]
disposed, be not, shy 4
disposed of, be, go 9
disposed (to), capable 2; likely 4; like[1] 2; subject 6; susceptible 1
disposed toward(s), be, lean[2] 3
disposition°, appetite 1; aptitude 2; arrangement 1; attitude 1, 2; bent 5; bias 1; cheer 1; device 4; discretion 2; distribution 3; fate 3; favor 4; formation 3; frame 5; habit 2; humor 3; inclination 3; love 2; makeup 2; mentality 2; mind 5; mood 1; morale; nature 1; order 1; personality 1; placement 1; position 1; posture 2, 3; scheme 2; spirit 2; taste 3; temper 1; tendency; turn 38; vein 4; way 2; will 1; wish 5
dispossess°, bereave; deprive; divest 1; downgrade 1; evict; expel 1
dispossessed, homeless 1
dispossession, ejection 2; eviction
disproportionate°, excessive 1; exorbitant; extreme 6; inordinate 1; lopsided 2; one-sided 2
disproportionately, fault 7; overly; unduly 1
disprove°, demolish 2; destroy 4; discredit 3; explode 2; rebut
disputable°, controversial 1; debatable; doubtful 1; exceptionable; marginal 2; moot 1; problematic; questionable
disputant, opponent
disputation, controversy 1
disputatious, aggressive 1; argumentative; belligerent 2; controversial 3; factious; legalistic; negative 1; pugnacious; quarrelsome; touchy 1
dispute°, argue 1; argument 1; battle 2; bicker; brawl 1; case[1] 2; cavil 2; challenge 1, 4; clash 2, 3; combat 2; conflict 2; contest 2, 3; contradict 1; controversy 1, 2; debate 1, 3; deny 1; difference 2; disagree 2; disagreement 3; discredit 2; encounter 5; fall 18; feud 1, 2; fight 2, 4, 8; flap 5; fracas 2; fray[1]; friction 2; haggle; issue 4, 9; misunderstanding 2; quarrel 1, 2; query 3; question 5, 10; rivalry; row[2] 1, 3; run-in; scrap[2] 1; spar[2] 2; stand 12a; strife 1; tangle 4; tiff; tilt 4; trouble 7; variance 2; vendetta; word 9a
dispute, in, debatable; issue 8; questionable; variance 3; war 2
disputed, problematic
disputing, negative 1
disquiet, agitate 1; anxiety 1; care 1; concern 6; dissatisfaction 1; doubt 3; dread 2; excitement 1; fluster 1, 2; fuss 1; hurry 3; jar[2] 3; misgiving; move 4; pain 2; perturb; phobia; shake 5; shock 1; solicitude; trouble 1, 5; tumult; unrest; upset 1; worry 4
disquieted, agitated; anxious 1; nervous 1; unsettled 2; upset 8; worried
disquieting, disturbing; harrowing; painful 2; provocative 2; tense 3; trying; ugly 3
disquietude, concern 6; excitement 1; solicitude
disquisition, essay 1; exposition 3; lecture 1; memoir 1; monograph; paper 4; speech 2; talk 14; theme 2; tract[2]
disrate, degrade 1
disregard°, back 9; blink 4; break 9; brush[2] 6; close 7; discount 3; dismiss 2; disobey; excuse 1, 5; forget 3; forgive 1; ignore 1; indifference 1; jump 3; kiss 3; laugh 2b; leave[1] 10; license 3; mind 19; miss[1] 4; neglect 1, 2, 3; negligence; noncompliance; oblivion 2; omit 2; overlook 1, 2; pass 3, 22; put 11; rule 8; scorn 3; skip 2, 4; slight 5, 8; slur 3; snap 6; trample 2; tune 5; violate 1; violation 1; waive 2; write 5
disregarded, broken 5; unsung
disregardful, disobedient 1; oblivious; promiscuous 1
disregarding, even 11; independent 8; noncompliance; violation 1
disrelish, dislike 2; distaste 1; reluctance
disrepair°, in, blink 6; broken 7; order 13c; ramshackle; rickety
disreputable°, base[2] 2; deplorable 2; disgraceful 1; dishonorable 2; doubtful 3; fly-by-night 2; infamous 1; mangy; notorious 1; public 6; scandalous 1; seamy; shabby 4; shady 2; shameful; sleazy 2; sordid 1; unreliable; unseemly 1; unworthy 1
disrepute, degradation 2; discredit 4; disfavor 2; disgrace 1; dishonor 4; infamy 1; notoriety; scandal 2; shame 2
disrepute, bring into, discredit 1; foul 14
disrepute, call into, foul 14
disrespect°, derision; dishonor 4; disregard 3; flippancy 2; freedom 7; impertinence; impudence; indignity; mouth 4; neglect 3; sacrilege 2; sauce 2; slight 5, 7
disrespectable, disreputable 1
disrespectful°, audacious 2; blasphemous; brash 2; cheeky; discourteous; familiar 3; flippant 2; forward 2; fresh 8; ill-mannered; immodest 2; impertinent; impolite; impudent; insolent; irreverent 2; offensive 2;

disrespectful° (*cont.*) pert; profane 1; rough 3; rude 1, 2; sacrilegious; tactless
disrespectfully, vain 3b
disrespectfulness, flippancy 2; sauce 2
disrobe°, divest 2; peel 2; strip[2] 2
disrupt°, break 10; disturb 1; inconvenience 3; interrupt 2; rupture 3; sabotage 3; upset 3
disrupting, inconvenient; tempestuous
disruption, disturbance 1; gap 1; havoc 2; inconvenience 1; interruption 1; mischief 2; rift 1; riot 1; tempest 2; upheaval
disruptive, tempestuous
dissatisfaction°, disappointment 1; disapproval; discontent; displeasure 1; distaste 1; resentment; trouble 7
dissatisfied°, disappointed; discontented; disgruntled; resentful
dissatisfy, disappoint 1; displease; let[1] 4
dissatisfying, disappointing
dissect, analyze 1; decompose 1; scrutinize
dissection, analysis 1; breakdown 3
disseize, evict
disseizin, eviction
dissemble, act 8; conceal 2; cover 2; dissimulate; fake 2; game 6; let[1] 7b; pretend 3
dissembler, pharisee; poseur
dissembling, dissimulation; evasive; hypocritical; insincere; two-faced
disseminate, broadcast 2; carry 8; circulate 2; diffuse 3; disperse 1; dissipate 1; distribute 2; emanate 2; give 15b; hand 17; issue 10; propagate 3; radiate 2; radio 2; release 2; scatter 1; sow; spread 1, 2; throw 7a
disseminating, spread 8
dissemination, circulation 2; distribution 2; issue 6; publication 1; spread 8; transmission 2
dissension°, antagonism 2; difference 2; disagreement 3; discord; faction 2; feud 1; fight 8; friction 2; misunderstanding 2; rivalry; strife 1; variance 2
dissent, disagree 1; disagreement 2; dissension; fight 8; friction 2; protest 1, 3; rebel 2; revolt 2; variance 2
dissent (from), dispute 1
dissenter, dissident 1; naysayer; nonconformist 1; rebel 4
dissentient, dissident 2; nonconformist 2; quarrelsome
dissenting, dissident 2; negative 1
dissentious, negative 1; quarrelsome
dissertation, essay 1; exposition 3; lecture 1; memoir 1; monograph; paper 4; talk 14; theme 2; tract[2]
disservice°, disadvantage 2; grievance 1; turn 34
dissever, sever 1
dissidence, fight 8
dissident°, nonconformist 1, 2; opposing; quarrelsome; schismatic; seditious; subversive 2
dissimilar°, different 1; discordant 1; disproportionate; distinct 2; divergent; diverse; several 2; unrelated
dissimilar, be, differ 1
dissimilar (to), unlike 1, 3
dissimilarity°, contrast 3; difference 1; disagreement 1; discrepancy; disparity; disproportion; diversity 1; inequality 1; odds 3
dissimilitude, difference 1; dissimilarity; diversity 1
dissimulate°, act 8; game 6; let[1] 7b
dissimulating, hypocritical; phony 1
dissimulation°, act 3; deceit 1; deception 1; disguise 4; hypocrisy; masquerade 2
dissipate°, consume 3; diffuse 3; disperse 2; evaporate 2; exhaust 1; expend 2; fritter; lavish 4; lift 5; lose 4; pass 5; run 34b; scatter 2; spend 2; spread 1; waste 1
dissipate money (on), splurge 2
dissipated, abandoned 2; decadent 2; dissolute; epicurean 1; fast[1] 2; misspent; outrageous 3; promiscuous 2; sensual; wanton 1
dissipating, self-indulgent
dissipation°, evaporation 2; excess 2; extravagance 1; prodigality 1; profligacy 1; waste 5
dissipative, profligate 1
dissociate°, distance 3; divorce 2
dissociated, estranged
dissociation, separation 2
dissolute°, abandoned 2; decadent 2; degenerate 1; epicurean 1; fast[1] 2; immoral 1; impure 4; incontinent 2; lecherous; lewd; libertine 2; loose 7; lost 7; obscene 1; outrageous 3; perverted; profligate 1; promiscuous 2; reprobate 1; self-indulgent; sensual; wanton 1; wicked 2
dissolutely, fast[1] 9
dissoluteness, dissipation 1; excess 2; impurity 3; profligacy 1; ribaldry; waste 6
dissolution°, collapse 6; dissipation 1; divorce 1; evaporation 2; ruin 1; solution 4
dissolve°, attack 5; collapse 2; die 2; disband; divorce 2; evaporate 2; melt 1, 3, 4; ruin 7; run 8; sever 2
dissolve into, resolve 4
dissolving, dissolution 2; solution 4
dissonance, jangle 3; noise 1
dissonant, brassy 2; discordant 2; grating 1; harsh 1; incongruous; noisy; raucous
dissuade, argue 6; choke 3; deter; discourage 2; reason 8
dissuasion, deterrent
distance°, alienate 2; background 2; dissociate; gap 1; interval 3; lead 4; range 1; record 5; remove 8; rift 1; way 4
distance, at a, aloof 1; apart 1
distance, at *or* **to a greater,** further 4
distance, go the, endure 1; last[2] 1
distance, quite a, cry 8
distance (oneself) (from), alienate 1; push 8
distant°, aloof 2, 3; back 12; chilly 2; cold 3, 8; cool 4, 5; cry 8; extreme 3; far 9; faraway 1; feeble 3; foreign 1; icy 2; indifferent 1; mechanical 3; nodding; nonchalant; oblivious; off 2, 5; offhand 1; out 10; outlying; perfunctory 1; remote 1, 8; reserved; retiring; self-contained 2; sight 6a; solitary 1; standoffish; unapproachable 1; unsocial; withdrawn 1, 2
distant, become more, recede 2
distant (from), unlike 1
distantly, vaguely 1
distaste°, aversion 1; discontent; disgust 2; dislike 2; displeasure 1; horror 1; phobia
distasteful°, abominable 2; bitter 2; disagreeable 1; disgusting; indecent 2; obnoxious; obscene 2; repellent; repugnant; repulsive, seamy, sour 3, tasteless 1; unacceptable; undesirable 2; uninviting; unlikely 2; unpalatable; unsavory; unwelcome 1
distemper, vapor 2
distend, blow[1] 8e; enlarge 1; expand 1; fill 2, 10a; increase 1; inflate 1; puff 6; stretch 2; swell 1; widen
distended, bloated; protuberant; swollen
distension, expansion 2; swelling
distention, expansion 2
distill, concentrate 2; extract 2
distillate, essence 2; extract 5; extraction 2; liquor 2; summary 1
distillation, breakdown 3; extract 5; extraction 2; refinement 3
distilled, refined 4; terse 1
distilled water, water 1
distinct°, bold 3; clear 5, 7, 11; definite 3; determined 2; different 1; discernible 2; discrete; dissimilar; distinctive; diverse; emphatic; explicit 1; express 5; focus 2; graphic 1; hard 12; independent 2; individual 1; keen[1] 4; legible; manifest 1; noticeable 1, 2; observable; obvious; particular 1; peculiar 1; plain 2; pronounced 1, 2; proper 6; readable 3; self-evident; separate 5; several 2; single 2, 3; singular 3; specific 1; strong 15; unrelated; variant 2; various 2; visible 2; vivid 2
distinct, be, differ 1
distinct (from), unlike 1, 3
distinction°, attention 2; cachet 2; celebrity 1; character 2; class 3; contrast 3; difference 1; dignity 1; elevation 4; excellence; gap 2; glory 1; honor 2, 3; importance 2; luster 2; mark 8; name 2; nobility 1; note 8; odds 3; preeminence 2; prestige; prominence 1; quality 1, 3; refinement 2; renown; superiority 2
distinction, attain, mark 9
distinction, of, considerable 2
distinction(s), laurels
distinctive°, characteristic 1; different 2; diverse; individual 2; notable 2; proper 6; salient; special 1
distinctive of, peculiar 2
distinctively, particularly 1
distinctiveness, distinction 1; diversity 1; eccentricity 1; identity 2; oddity 1; singularity 1
distinctly, apart 2; clear 18; clearly 1, 3; expressly 1; notably 1; sharply 3; simply 5
distinctness, clarity 2; definition 1
distingué, dignified; distinguished 2
distinguish°, accent 4; contrast 1; denote 1; diagnose; differentiate 1; dignify; discriminate 1; find 3; grace 7; know 4; make 37a; mark 13; perceive 1; pick 7b; recognize 4; scent 3; see 1; select 1; separate 2; set 14a; sight 8; single 4; spot 6; tell[1] 8
distinguish oneself, mark 9

diverse° (*cont.*) many 2; miscellaneous; mixed 2; opposite 2; several 2; varied 1; various 1
diverseness, diversity 1
diversification, variation 1; variety 1
diversified, checkered 2; diverse; general 3; manifold; miscellaneous; sundry; varied 1; various 1
diversify°, branch 3; vary 1
diversifying, variation 1
diversion°, amusement 1, 2; delectation; detour 1; dissipation 3; distraction 2; enjoyment 1; entertainment 1, 2; enthusiasm 2; escape 6; excursion 2; fun 1; game 1; hobby; interest 5; misuse 1; pastime; perversion 1; pleasure 1; recreation; relaxation 1; sport 1
diversity°, difference 1; disagreement 1; miscellany; variation 1, 2; variety 1, 2
divert°, amuse 1; beguile 3; busy 4; deflect; delight 1; detour 2; deviate 1; distract 1, 2; entertain 1; fend 2; head 13a; occupy 3; pervert 1; please 1; regale; sidetrack; sublimate; sway 2; switch 4, 5; throw 6c; tickle
  divert from, discourage 2
diverting, delightful 1; entertaining; funny 1; light[2] 11; misuse 1
divertissement, amusement 2; diversion 3; entertainment 2; pastime; sport 1
divest°, bare 8; deprive; evacuate 1
  divest (oneself) of, divest 2; take 34a
divest (oneself) of (one's) clothes *or* **clothing,** strip[2] 2
divide°, allot; branch 3; break 11; cleave; disconnect; disengage; diverge 1; diversify; divorce 2; parcel 4; part 13; partition 6; portion 4; rupture 3; screen 6; segment 2; set 13a; sever 1; slice 4; sort 8, 10b; split 1; tear 1
  divide (up), carve 2; cut 17a; distribute 1; divide 2; partition 5; section 4; separate 1, 3; share 3; split 4
divided, detached 1; disjointed 1; estranged; separate 5; split 9
dividend, cut 20; extra 3; gain 8; icing 2; perquisite; premium 1; share 1
  dividend, as a, measure 10
divider, partition 3; screen 1; wall 1
dividing, division 1; parting 1
  dividing (up), partition 2
dividing line, distinction 1; division 4; separation 2
divination, oracle 2; prophecy 1
divinatory, prophetic
divine°, celestial 1; clergyman 1; deduce; fathom; godlike 1; guess 2; heavenly 1, 2; holy 1; immortal 2; minister 1; pastor; preacher; priest; prognosticate 1; sacred 1; see 3; sense 6; seraphic; solemn 2; spiritual 1; stunning 2; superb; twig[2]
divine intervention, providence 2
(divine) Providence, providence 2
diviner, fortuneteller; oracle 1; prophet
divine spark, psyche
divine utterance, oracle 2
divinity, god; sanctity
divisible, separable
division°, analysis 1; bracket 3; branch 2; bureau 2; category; circle 2; class 2; compartment; corps; department 1; discord; distinction 1; distribution 3; district; family 4; field 4; fleet[1]; fracture 2; gap 1, 2; lot 4; office 2; part 1, 7; parting 1; partition 1, 4; party 3; piece 3; portion 1, 2; proportion 3; province 1; quarter 3; region 1; rift 1; rupture 1; schism; section 1; segment 1; separation 2, 3; share 1; split 7, 8; stage 1; ward 1; wedge 2
divisive, factious; schismatic
divorce°, dissociate; separate 3; separation 1; split 2
divulge, admit 4; air 7; announce 1; babble 2; bare 7; betray 2; blab; blurt; break 2; circulate 2; communicate 1; confess; disclose 1; expose 1; express 2; give 12b; impart 2; inform 1; lay[1] 9; leak 5; let[1] 7a; make 33; mention 2; open 23; publish; register 4; relate 2; report 4; reveal; say 2; show 1; slip[1] 4; tell[1] 2; whisper 2
  divulge suddenly *or* **unexpectedly,** spring 4
divulge all *or* **everything,** spill 3
divulged, be, unfold 2
divulgement, admission 4
divulgence, admission 4; betrayal 2
divulge secrets, tattle 1
divulging, betrayal 2; exposure 1
divvy (up), parcel 4; portion 4
divvying up, partition 2
dizziness, vertigo
dizzy°, daze 4; faint 2; feeble 1; flighty 2; foolish 2; giddy 1; light[2] 8; queer 3; scatterbrained; senseless 3; silly 2; unbalanced 2
  dizzy, be, spin 3
dizzying, dazzling
do, accomplish; act 6; apply 5; bring 12; carry 10c; come 6; commit 3; complete 5; defraud; discharge 5; dispatch 4; effectuate; execute 1; fare 4; fill 8; fulfill 1; fête 1; go 29b; incident 2; make 19, 20; negotiate 2; obey 2; party 1; perform 1, 2, 3; perpetrate; practice 7; prepare 4; prosecute 3; pull 12b; reception 2; render 1; riot 1; run 29b; see 12a; style 8; suffice; take 32b; transact; work 16
  do away with, abolish; bump 5; cancel 2; destroy 2; dispatch 3; dispense 3b; dispose 3d; extinguish 2; kill 1; poison 4; polish 3b; purge 2; put 16d; remove 4; rout 2; scrub 2; wipe 3
  do in, bump 5; consume 5; defeat 1; dispatch 3; kill 1; murder 3; polish 3b; remove 4; ride 5b; rout 2
  do (out of), rob 3; screw 6
  do over, make 38a; modernize; renew 1; renovate
  do too much, overdo 2
  do up, clean 9; fix 16b; refurbish; renovate; revamp; wrap 1
  do well, flourish 1; get 27
  do with, have to, associate 1b; deal 4; regard 5; relate 3; touch 4, 5
  do without, dispense 3a, 3b; forgo 1; spare 9
doable, feasible; possible 2; practicable
dobbin, nag[2]
doc, doctor 1; physician
docent, instructor; teacher
doch-an-doris, drink 6
docile, broken 4; easy 4; feminine 1; flexible 3; gentle 2; manageable; meek 2; obedient; passive 2; pliable 2; sheepish 1; soft 5; submissive 1; tame 1; tractable 1; willing; yielding 2
docilely, easy 3; willingly
docility, flexibility 2; obedience; submission 1
dock°, cut 4; diminish 1; land 5; landing 3; lop; moor[2]; pier 1; prune; sever 1; shorten 1; snip 1; tail 1; trim 3
  dock, put in the, prosecute 1
docket, calendar 1; label 1, 4; paper 2b; roll 13; tag 1
docking, landing 1
doctor°, adulterate; consultant 1; dilute 1; dress 3; fake 1; fix 3; garble 1; instructor; juggle; manipulate 3; neuter 2; patch 6; physician; rig 2; teacher; treat 3
doctor of medicine, physician
doctrinaire, assertive; dogmatic; opinionated 1; pedantic 1; positive 5; sectarian 1; strong 16
doctrinal, orthodox
doctrine°, belief 3; creed; faith 2; idea 3; ideology; institution 4; lore 1; point of view 2; principle 1; religion; tenet
document°, charter 1, 4; chronicle 2; deed 3; dispatch 6; enter 3; file 1; instrument 3; itemize; license 5; minute[1] 4; proof 1; record 1, 3; report 3; slip[2] 1; stamp 2; substantiate; take 31a; warrant 2; witness 5; writing 2
documentation, evidence 1; file 1; good 21c; journal 2; material 4; proof 1; record 4; source 3
  documentation of ownership, title 5
documented, authoritative 1; historical; official 1
document(s), material 4; paper 2a
dodder, hobble 1; limp[1] 1, 2; totter
doddering°, decrepit 1; infirm 2; limp[1] 2; senile
doddle, pushover 1
dodge°, artifice 2; avoid; blind 8; duck 1, 3; elude 1; equivocate; escape 2; evade 1; feint; fence 4; flinch; fraud 2; gimmick 1; loophole; maneuver 1; miss[1] 1; move 9; pretense 2; racket 2; refuge 2; ruse; scheme 3; shirk; shuffle 3, 5; sidestep; steer 2; stratagem; subterfuge; trick 1; weave 4; wile; wrinkle[2]
dodger, truant 1
dodging, evasion 1, 2; reserve 7
dodgy°, crafty; deceptive 2; precarious; risky; tricky 1
dodo, ruin 5
doff, cast 13; divest 2; drop 8; get 41c; peel 3; remove 1; shed[2] 4; take 34a
dog, bag 4; bother 1; dud 1; follow 4; hag; heel[1] 5; hunt 1; pursue 1; run 28a; shadow 7; stalk[1]; tail 4; trace 4; track 6; trail 7; villain; wretch 1
dog (and bone), telephone 1
dog-collar, string 4
dog-eared, timeworn
dogged, determined 1; emphatic; firm 4; grim 1; hard 7; immovable 2; industrious; insistent; laborious 2, 3; obstinate; opinionated 1; patient 2; persistent 1; purposeful; relentless 1; resolute; rigid 4; single-minded;

plain 4; sober 3; tart[2] 2; tedious
drabness, tedium; uniformity 2
Draconian, brutal 1; drastic; extreme 4; harsh 2; severe 4; strong 13
Draconic, brutal 1; severe 4
draft°, air 2; breeze 1; call 12a; design 2, 3, 5; devise 1; draw 6, 15b; drink 4, 6; enlist 1; frame 7; gulp 3; induct 2; mobilize; nip[2]; note 4, 10; outline 2, 3; pattern 4; pen[1] 2; piece 4; potion; project 3; proposal 2; protocol 2; puff 1, 2; recruit 1; rough 14, 15; scheme 1; wind[1] 1
draft, on, tap[2] 4a
draftee, recruit 2
drafting, preparation 3
draftsman, designer 1
drafty, breezy 1
drag°, carry 1; cart 2; connection 3; crawl 2; creep 2; drain 2; draw 1; drip 3; encumbrance; fag 2; force 9; haul 1, 3; hump 3; lug; pain 3; pass 11; pill 2; plague 2; plod 1; power 2; puff 2, 5; pull 1, 22; tow; traction; trail 5; tug 1, 2; wear 5
drag (along), delay 3; trail 4
drag along with *or* **after,** tag 5
drag into, mix 5
drag (on *or* **out),** extend 3; prolong; spin 5
drag (one's feet), shuffle 2
drag one's feet *or* **heels,** delay 3; drag 6; play 14; stall[2] 1
drag out, draw 14a; get 45c; lengthen; prolong; spread 4; string 11b; trot 2
drag through the mud, disgrace 4; smear 2
drag together, rake[1] 8
dragging, shambling
draggle, drag 4
draggletailed, sloppy 1; unkempt; untidy
dragon, monster 1; shrew
dragoon, force 7; push 4; railroad 2
drag queen, homosexual 1
drain°, consume 2; empty 8; enervate; escape 3; evacuate 1; exhaust 3, 4; expend 2; fatigue 2; filter 3; finish 3; milk; ooze 2; percolate; pump 3; sap[1] 3; sap[2]; secrete[2]; secretion; spout 3; strain[1] 5; tap[2] 5, 6; tire 1; undermine 1; void 7; wear 6; weary 5
drain (away), ebb 2
drain (off *or* **out),** draw 3; exhaust 5
drainage, escape 7
drained, beat 13; empty 1; exhausted 1; impoverished 2; pale[1] 1; peaked; prostrate 5; run-down 1; spent 1; tired 1; washed-out 2; weary 1; white 2
draining, escape 7; exhaustion 1; secretion; toilsome; weary 2
dram, draft 4; drink 6; drop 2; modicum; nip[2]; shot 7; sip 2
drama°, piece 4; play 20; production 4; show 14; theater 2, 3
dramatic°, melodramatic; poignant 3; sensational 2; stirring; theatrical 1; thespian 1; vivid 2
dramatic art, drama 2; theater 3
dramatics, drama 3
dramatis persona(e), cast 2; character 5
dramatist°, playwright; scribe 2
dramatize°, magnify 1; play 17a; rub 3; set 20b
dramaturge, dramatist; playwright; scribe 2
dramaturgic(al), dramatic 1
dramaturgist, playwright
dramaturgy, drama 2; theater 2, 3
drape°, drapery; hang 3; string 9
drapery°, drape 2
drastic°, extreme 4; forcible 2; heroic 3; radical 2; stiff 2; strong 13; ultra-
draught, drink 6; nip[2]
draughtsman, piece 5
draw°, attract; attraction 1, 2; base[1] 5; borrow; bring 3; catch 12; charm 3; cut 10; deadlock 1; deduce; derive 1; describe 4; design 3; determine 2; drag 1, 2; earn 2; engage 4; entice; extract 2; fascination; feature 2; gather 4; get 45c; gut 4; haul 1, 3; heave 1; incur; infer; interest 7; invitation 2; magnetism; mine 3; paint 4; picture 7; plot[1] 4; produce 5; profile 4; puff 2, 5; pull 1, 5, 9b, 19, 20, 22; raffle; receive 2; represent 3; sap[1] 3; spell[2] 3; star 2; stimulant 1; take 5; temptation 2; tie 10; tow; trace 6; trail 4; tug 1; verge[2]; wrench 3
draw ahead of, overhaul 1; pull 7
draw away, distract 1; draw 12b; gain 5; pull 7; sidetrack
draw back, draw 10; fall 8; flinch; pull 8a, 14b; retract 1; retreat 4, 5; shrink 2; start 5; withdraw 1
draw close(r), approach 1
draw even *or* **level with,** overtake 1; pull 18c
draw (forth), extract 1, 3; fetch 2
draw (in), draw 11; inhale; involve 3; lure 1; pull 11a; rope 4; tempt 1
draw near(er) *or* **nigh,** approach 1; come 1; draw 13b; near 10
draw (off), drain 4; draw 3, 12; milk; sidetrack; strain[1] 5; tap[2] 5
draw oneself up, bridle 3
draw on *or* **upon,** call 9; command 3; draw 13; screw 7a; seduce 1; summon 2; tap[2] 6
draw out, draw 8, 14; elicit; extend 1; extract 1; lengthen; milk; pluck 2; prolong; pull 3; pump 3; spin 5; spread 3; stretch 2; thin 8
draw to a close, stop 1; wane 2
draw together, assemble 2; close 8b; concentrate 3; contract 4; gather 3; pucker 1; rake[1] 1; round 19; solidify 2
draw (up), collect 2; design 1; draft 5; draw 6, 15; formulate 3; frame 7; make 4, 37c; pen[1] 2; project 3; pull 11a; summon 2; work 19e
draw a bead, level 10; sight 7
draw a blank, forget 1
draw a distinction, discriminate 1
draw and quarter, punish 2
draw attention to, spotlight 3
drawback°, catch 18; complication 2; deterrent; detriment; disadvantage 1; fly 10; hindrance 1; inconvenience 1; joker 2; liability 3; shortcoming
drawers, pants 1
drawing°, attractive; cut 25; design 5; draft 1; figure 5; likeness 2; lot 3; lottery; picture 1; plan 2; scheme 2
drawing back, retreat 1
drawing near(er) *or* **nigh** *or* **close(r),** imminent
drawing out *or* **forth,** exhaustion 1
drawing up, preparation 3
drawing and quartering, punishment 2
drawing card, feature 2; lure 2
drawing lots *or* **straws,** lot 3
drawing pin, pin 1; tack 1
drawing power, pull 20
drawing room, parlor
drawn°, emaciated; even 5; ghastly 2; graphic 2; haggard; peaked; scrawny; sickly 2; sunken 1; worn 2, 3
drawn, be, trail 5
drawn in, be, feature 5
drawn (to), interested 1
draw the fangs *or* **teeth of,** silence 4
draw the line (at), object 4
drayman, carrier 1
dread°, alarm 2; anxiety 1; dismay 3; dreadful 2; fear 1, 5; foreboding 1; fright 1; grim 3; horror 1, 2; macabre; misgiving; panic 1, 3; phobia; terrible 4; terror 1
dreaded, dread 3
dreadful°, abysmal 1; atrocious 1; awesome; calamitous; damnable; deplorable 2; diabolic 2; disastrous 1; dread 3; eerie; evil 1; fatal 2; fearful 3; fearsome; fierce 3; formidable 1; frightening; frightful 2; ghastly 1; grievous 2; grim 3; grisly; horrible 1, 2; infernal 2; macabre; monstrous 1; nightmarish; repulsive; severe 3; shocking 2; terrible 1, 2; tragic; ungodly 2; vile 1; wicked 2; wrong 1
dreadful-looking, ugly 1
dreadfully, awfully; badly 6; severely 1, 5
dreadfulness, monstrosity 2
dream°, ambition 3; aspiration; beauty 2; daydream 1, 2; fancy 6; fantasize; fantasy 2; hallucination; idea 5; ideal 5; make 28; muse; think 1; thought 3; view 5; vision 3, 5
dream, like a, swimmingly
dream of, aspire; long[2]; see 3
dream up, coin 3; conceive 3; create 2; devise 1; envisage 1; fancy 9; form 8; formulate 2; hatch 2; hit 9b; imagine 1; invent 1; make 39c; plot[1] 3; strike 16; think 6
dreamboat, beauty 2
dreamboy, passion 4
dreamed-up, romantic 1
dreamer°, romantic 5; theorist; visionary 2
dreamgirl, passion 4
dreamily, vaguely 3
dreaming, forgetful; wistful 2
dreamland, fairyland; paradise 3
dreamland, in, forgetful
dreamlike°, dreamy 1; quixotic; shadowy 3
dreams, not ine one's wildest, mean[3] 5
dream state, trance
dreamt-up, romantic 1
dreamy°, faraway 2; forgetful; quixotic; visionary 1; wistful 2
drear, dreary 1; stark 3; sunless
dreariness, boredom; desolation 2; tedium
dreary°, bleak 1; boring; colorless 2; dark 3; deadly 5; desolate 3; dingy; dismal; doleful; dour 1; drab; dry 2; dull 5; funereal; gaunt 2; gloomy 2, 3; gray 2; heavy 9; joyless 2; lackluster; leaden 3; lifeless 4; mean[2] 3; monotonous; murky; overcast; pedestrian 2; ponderous 2; sad 2; sober 3; somber 2; stark 3;

dreary° *(cont.)* stodgy; sullen; sunless; tedious; wintry 3
dreck, rubbish 1
dredge, deepen 1; hollow 8; powder 4
  dredge (away *or* **out),** wash 5
  dredge up, dig 6; rake[1] 8; scrape 6; trot 2; unearth
dregs°, deposit 4; ground 4; refuse[2]; residue; rubbish 1; sediment; sludge; soil[1] 3; trash 3; waste 7
dregs (of society), mob 2; rabble 2; riffraff
drench°, drown 1; flush[1] 2; impregnate 2; rinse 1; saturate; soak 1; steep[2] 1; submerge 1; wash 1; water 6
drenched, bedraggled; soaking 2; wet 1
drenching, flush[1] 5; rinse 3; soaking 1
dress°, apparel; clothe 1; clothes; costume; cover 5; equip; face 16; fertilize 2; garments; get 51e; groom 3; gut 4; habit 3; outfit 2; rag[1] 3; robe 1, 3; tool 4; turn 20d; wear 8
  dress down, castigate; dress 4; lambaste 2; lecture 4; mind 10; punish 1; rate[2]; rebuke 1; reprimand 2; scold 1; upbraid
  dress in, get 40a; wear 1
  dress (oneself) in, put 22a
  dress (up), decorate 1; dress 5; embellish 1, 2; get 51e; ornament 2; preen 2; primp; sweeten 2; trim 4
dressed in, be, have 12a; wear 1
dressed in, get, get 40a; put 22a
dressed to kill, dapper
dressed-up, dressy 1
dresser, bureau 1
dressing, salve 1; toilet 2
dressing-down, lecture 2; piece 10; punishment 1; rebuke 2; reprimand 1; sermon 1; tongue-lashing
  dressing-down, give a, reprimand 2; upbraid
dressing gown, robe 1; wrapper 1
dressmaker°, tailor 1
dress rehearsal, rehearsal 1
dressy°, dapper; sharp 9
dribble, drip 1, 2; drivel 1; drop 5; filter 3; run 6; secrete[2]; slaver[1] 1, 2; spit 1; trickle 1, 2
dribbling, secretion
dried-out, stale 1
dried-up, wizened
drift°, current 6; deviate 1; diverge 2; effect 3; float 1; flow 1, 5; gist; hover 1; implication 3; import 3; meaning 2; migrate 2; movement 7; pass 2; point 7; ramble 1; range 8; roam; run 4; sail 3; sense 4; significance 1; straggle; strain[1] 9; stray 1; tendency; tenor; thread 2; trend 1, 3; turn 32; waft 1; wander 1, 4
  drift (of), get the, catch 10; figure 12b; grasp 2; see 2; understand 1
drifter°, bum 2; migrant 1; rover; tramp 4
drifting, end 6; migrant 2; vagabond 2
drill°, bore[1] 2; brace 2; coach 3; discipline 1, 6; educate; exercise 2, 4; groom 4; indoctrinate; initiate 3; instruct 1; instruction 1, 2; lesson 1; maneuver 2; penetrate 1; perforate; pierce 2; practice 2, 6; prime 5; procedure; punch[2] 1, 2; rehearsal 1; routine 1; school 4; sink 7; stick[1] 1; teach; train 4; tutor 2; work 19a
  drill into, pound[1] 4
  drill (out), ream
drill-hole, bore[1] 1
drilling, discipline 1; education 1; exercise 4; instruction 2
drill sergeant, disciplinarian
drink°, booze 1, 2; brew 4; draft 4; have 7; lap[1] 1, 3a; liquor 1; potion; sea 1; shot 7; swallow 1, 5; swill 3; take 13; touch 5; use 3
  drink, the, drink 7; ocean 1
  drink down, drain 4
  drink in, devour 3
  drink (someone's) health, pledge 6
  drink (to), drink 3; pledge 6; toast 3
  drink (up), consume 1; drain 4; finish 3
drinkables, provision 4
drinker, alcoholic 2; drunk 3; soak 3
  drinker, serious *or* **problem,** alcoholic 2
drinking bout, spree
drinking glass, glass 4
drinking water, water 1
drink(s), refreshment 1
drip°, drag 7; drain 6; drop 1, 5; filter 3; leak 4; percolate; sap[1] 2; scrap[1] 1; secrete[2]; secretion; silly 3; trickle 1, 2; weep 2; wet 6
drippiness, sentimentality
dripping, drip 2; secretion; soaking 2; wet 1
drippy, sentimental 2
drive°, ambition 2; banish 2; campaign 1; carry 3; crash 2; crowd 4; dynamism; energy; enterprise 2; excursion 1; exertion; force 8; herd 4; hit 2; hustle 2, 4; impetus; incite; inflame 1; initiative 2; invasion 2; lust 2; maneuver 4; momentum; motivate; movement 5; offensive 5; operate 2; pilot 3; plow 2; power 8; propel; propulsion; provoke 1; purpose 2; push 1, 14; ram 1; ride 4; roll 3; run 13, 17, 36; see 6; shame 7; sink 7; soldier 4; spin 7; spirit 2; spur 4; stimulant 1; street 1; taxi 2; thrust 1, 4; tour 2; trip 3; turn 4, 31; urge 1, 5; wash 2; will 1
  drive at, drive 6; mean[1] 2
  drive away, banish 2; beat 9; chase 3; freeze 4; pack 7; pull 7; rebuff 2; repel 1; shoo 2
  drive back, repel 1; repulse 1; turn 13a
  drive bananas *or* **bats** *or* **batty** *or* **crazy** *or* **insane** *or* **mad** *or* **nuts** *or* **round the bend** *or* **round the twist** *or* **up the wall,** disturb 1; exasperate 1; exercise 3; irritate; madden 1; pester; tease 1; wall 3
  drive home, home 11
  drive (in), inject 1
  drive into, turn 16b
  drive off, beat 9; pack 7; repel 1
  drive (off *or* **away),** flight[2] 2; ward 3
  drive (oneself), exert 2
  drive out, banish 2; cast 14; dispossess; eject 1; exile 3; expel 1; freeze 4; pump 3
  drive up, pull 11a
drivel°, babble 3; froth 2; gab 2; gibberish; gobbledegook 1; jabber 1; jargon 2; moonshine 2; mumbo jumbo 1; nonsense 1; prattle 1, 2; rot 4; slaver[1] 1, 2, 3
driven, possessed
driver, engineer 2; operator 1; servant 1
driverless, runaway 2
driver's seat, helm 2
  driver's seat, be in the, dominate 1; govern 1; shot 9; string 6
driveway, drive 9
driving, motive 3
driving force, propulsion
drizzle, drip 1; mist 1; precipitation; rain 1, 4; shower 1; spray[1] 2; trickle 1
drizzling, wet 2
drizzly, moist 2
droll, comic 1; dry 3; farcical; fool 2; funny 1; humorous; joker 1; ludicrous; panic 2; priceless 2; ridiculous; wag[2]; witty; wry 2
drollery, humor 1; wit 2
drone, buzz 1, 5; hum 1, 4; idler; loafer; murmur 1; wastrel 2
  drone on, go 32f
droning, hum 4; murmur 1
drool, drivel 1; slaver[1] 1, 2; spit 1, 3
droop°, dangle 1; die 2; fade 2; flag[2] 1; sag 1, 2, 3; slouch 1, 2; weaken 2; wilt 1, 2
drooping, flabby 1; limp[2] 1; pendulous; slack 2
droopy, slack 2
drop°, abandon 4; advantage 1; blob; collapse 3; dab 2; dash 7; deposit 1; descend 2; dip 5; discontinue; drapery; drip 1, 2; dump 1; ebb 2, 4; eliminate 1; faint 3; fall 3, 11a, 22, 26, 27; fumble 2; go 27b; jilt; lapse 3, 4; lower[1] 1, 2; lurch[1]; modicum; morsel 1; nip[2]; pass 20a; pendant; pitch[1] 3; plump[2] 1, 4; plunge 1, 3; reduce 5; reject 2; relinquish 1; sag 2, 3; scrap[1] 1; scrub 2; secrete[2]; secretion; set 1; shake 7; shed[2] 3, 4; shower 3; sink 2, 3, 5, 6; sip 2; slash 3; slide 3; slope 2; slump 1, 2; sphere 1; stick[1] 2; tablet 4; taste 1; terminate; throw 8; topple 3; touch 15; trace 2; trickle 1; trifle 2; tumble 1, 2
drop, ready to, tired 1; weary 1
drop away, drop 6
drop back, fall 10; pull 8b
drop behind, trail 6
drop dead, fall 6
drop (down), drop 6; fall 1; flop 1; subside 1; decline 5
drop down on (one's) knees, truckle
drop in (on), drop 12; visit 1; look 11b
drop in *or* **by,** come 16c; pop 2
drop of a hat, at the, immediately 1; offhand 4; willingly
drop (off), decline 3; doze 1; drop 6, 11; flake 3a; nap[1] 1; nod 3; sleep 1; slope 1; taper 2; fall 3
drop (on someone), start 13
drop out, drop 13
drop (out of), leave[1] 4; secede
  drop out of sight, hide[1] 1
  drop to the ground *or* **to (one's) knees,** truckle
drop a brick, err 1
drop a clanger, err 1
drop by drop, gradually
drop-kick, kick 3
droplet, blob; bubble 1; drop 1; sphere 1
drop-off, drop 4; fall 26; slack 5
dropout, hippie
dropped, abandoned 1
dropping, fall 22; omission 1; secretion
droppings, dung; filth 1; muck 1
dropsical, swollen
dross, dirt 1; garbage; refuse[2]; rubbish 1; trash 3
drove, drive 5; flock 1; host[2]; pack 3; swarm 1; throng 1
drove(s), many 3; mass 2; score 4
drown°, drench; flood 4; submerge 3
  drown in red ink, fail 4
  drown one's sorrows, drink 2

drowse, doze 1; nod 3; sleep 1
drowsiness, lethargy 2; torpor
drowsy°, dreamy 3; lethargic 2; sleepy 1
drub, beat 1; floor 4; lambaste 1; lather 4; overcome 1; rout 2; whip 2
  drub into, pound[1] 4
drubbing, loss 5; rout 1; thrashing 1
  drubbing, a, works 3b
drudge, fag 3; hack[2] 1, 2; labor 5; laborer; plod 2; robot 2; slave 2, 3; soldier 4; sweat 3
drudgery°, grind 7; job 4; labor 1; slavery 3; struggle 4; sweat 6; work 1
drug°, antidote; cure 1; doctor 4; dope 2; medicine; narcotic 2; numb 2; pill 1; prescription 2; remedy 1; spike 4
drug addict, user 2
drugged, high 9
druggist°, pharmacist
druggist's, pharmacy 1
drugstore, pharmacy 1
drum, din 2; patter[1] 3; pulsate; roller 1; tap[1] 1
  drum out (of the corps), degrade 1; dismiss 1; eject 3; expel 2; lay[1] 16a; punish 2
  drum up, enlist 2
drummer, merchant 2; peddler; seller
drumming, pulse 1
drunk°, alcoholic 2; bender; carouse 2; far 5b; high 9; loaded 5; orgy 1; soak 3; tight 8; under 6
  drunk, go on a, paint 7
  drunk, make, intoxicate 1
  drunk (as a lord *or* **a skunk),** stinking 3
drunkard, alcoholic 2; drunk 3; soak 3
drunken, drunk 1; stinking 3
dry°, barren 2; boring; cake 3; colorless 2; cure 3; dead 12; dreary 2; fair[1] 5; fine[1] 2; flat 5; heavy 7; humdrum; husky 2; matter-of-fact; monotonous; pedantic 1; pedestrian 2; preserve 3; prosaic; scratchy 2; sober 1; stale 1; thirsty 1; wooden 2; wry 2
  dry (off *or* **out** *or* **up),** dry 5; parch; shrivel; sober 4; wipe 1
dryasdust, boring; colorless 2; heavy 7; matter-of-fact; monotonous; slow 9; stodgy; tedious; tiresome 1
dry-clean, clean 9
dry humor, salt 3
drying (up *or* **out),** evaporation 1
drypoint, engraving 1, 2
dry run, rehearsal 1; trial 1
D.T.'s, shake 10
dual, double 3
dub, call 2; christen 2; designate 4; entitle 2; label 5; name 4; tag 4; term 8; title 6
dubiety, doubt 4; skepticism; suspicion 1
dubiosity, doubt 4
dubious, ambiguous 2; debatable; disputable; disreputable 1; distrustful; doubtful 1, 3; far-fetched; fishy 1; fly-by-night 2; implausible; improbable; incredulous; leery; left-handed 2; marginal 2; obscure 2; precarious; queer 2; questionable; rocky[2]; shady 2; shaky 1; skeptical; speculative 2; suspect 3; suspicious 1; tenuous 2; uncertain 3; unlikely 1
dubiousness, discredit 6; doubt 4; question 2; skepticism; suspicion 1
duchess, peer[1] 1; royal 3
duchy, realm 1
duck°, dip 1; dive 1; dodge 1, 3; elude 1; evade 1; flinch; immerse 1; nil; sidestep; zero 1
  duck (down), stoop 1
  duck out, escape 1
  duck (out of), shirk
duckwalk, waddle
duct, main 5; pipe 1; pipeline 1; shaft 4; spout 3; vent 1
ductile, adaptable; flexible 1; plastic 1; pliable 1; tractable 2
ductility, elasticity 1; flexibility 1
dud°, duff; failure 3; flop 4; loser; misfire 2; nonevent; washout
dude°, dandy 1; guy 1
  dude up, primp
dudgeon, displeasure 2
duds, apparel; clothes; garments; gear 3; rag[1] 3
due, befitting; debt 1; desert[2]; eventual 2; just 3; outstanding 2; payable; prerogative; proper 3; share 1; unpaid 1
  due course, in, eventually; presently; run 48; soon 5
  (due) time, in, run 48; sometime 3
  due to, owe 2; reason 5; thank 4; through 1
  due to the fact that, for 13
duel, battle 1; bout 3; combat 1, 5; encounter 5; fight 7; match 2
duende, magnetism
duenna, companion 3
dues°, custom 2; subscription 1; tax 1; toll[2] 1; tribute 2
duff°, bottom 1; buttocks; dud 2; posterior 3
duffel bag, pack 1
duffer°, mug 3
duke, lord 2; peer[1] 1; royal 3
dukedom, realm 1
dulcet, lyric 3; mellow 2; melodious; musical; pretty 2; silver 4; smooth 7; sweet 3; tuneful
dull°, backward 2; bland 2; blunt 1, 3; boring; brute 1; colorless 2; cut 29b; dampen 2; dark 3; dead 10, 12, 13; deaden 1, 2; deadly 5; dense 3; depress 2; dim 1, 2, 3; dingy; dormant 1; dowdy; drab; dreary 2; drug 4; dry 2; dumb 2; everyday 3; fade 1; faint 1; flat 5, 11, 12; glassy 2; gloomy 1; gray 2; hazy 2; heavy 7; hollow 6; humdrum; inert 3; insane 2; jaded 2; kill 3; lackadaisical 2; lackluster; leaden 2, 3; lengthy; lethargic 1; lifeless 3; literal 3; matter-of-fact; mitigate; moderate 5; monotonous; mousy 1; muddy 2, 4; neutral 2; numb 2; obscure 7; obtuse 2; opaque 3; overcast; pedestrian 2; philistine 2; ponderous 2; prosaic; simple 4; sleepy 2; slow 7, 9; somber 2, 3; stodgy; stolid; stupid 1, 3; tame 4; tasteless 2; tedious; thick 6; thick-skinned; threadbare 2; tiresome 1; tone 5; torpid; vacant 2; vague 6; watery 1; weak 5, 6; wooden 3
dullard, dolt; half-wit
dull as ditchwater *or* **dishwater,** pedestrian 2
dulling, narcotic 1
dullness, boredom; gloom 1; inertia; lethargy 1; opacity 3; simplicity 4; stupidity 1; tedium; torpor; uniformity 2
dull(-witted), blind 2; dim 2; feebleminded; foolish 2; obtuse 2; opaque 3; simple 4; slow 7; stupid 1; thick 6; weak 5; wooden 3
dull-wittedness, simplicity 4; stupidity 1
duly°, properly 2; truly 2
dumb°, backward 2; daft 1; dense 3; dim 2; dull 1; feebleminded; foolish 2; half-witted; inane; inarticulate 3; insane 2; mute 1; nonsensical; simple 4; slow 7; speechless 1; stupid 1; weak 5
dumbbell, dolt; fool 1
dumbfound°, amaze; astonish; astound; daze 1; floor 5; jolt 3; nonplus; overpower 2; overwhelm 3; perplex; petrify 2; rock[2] 3; shock 1; stagger 2; stump 2; surprise 1; throw 4
dumbfounded°, goggle-eyed; petrified 2; speechless 2; thunderstruck; tongue-tied; unprepared 1
dumbfounding, prodigious 2
dumbness, folly 1; silence 2
dumbstricken, speechless 2
dumbstruck, dumbfounded; petrified 2; speechless 2
dummy°, dolt; figurehead; model 1, 5; pawn[2]; silly 3; unreal 3
dump°, discard 1; dispose 3b; dive 3; drop 7; empty 8; hole 3; hovel; jilt; joint 2; shack; throw 5a; tip[2] 3; tumble 2; unload; vent 4
dumps, gloom 2
  dumps, in the, gloomy 2; melancholy 1; moody 1; mouth 6
  dumps, the, depression 2
dumpy°, chubby; short 1; stocky
dun°, dead 10; mousy 1
dunce, clod 2; dolt; dummy 3; fool 1; half-wit; retard 2; sap[1] 2; silly 3
dunderhead, dolt; gawk 1; half-wit
dunderheaded, opaque 3
dunderpate, dolt; gawk 1; half-wit
dunderpated, opaque 3; wooden 3
dune, drift 5; mound 1
dung°, filth 1; muck 1
dungarees, pants 2
dungeon°, hole 4; keep 16; prison; tower 2
dunk, dip 1; duck 2; immerse 1; submerge 1
duo, couple 1; pair 1; team 2
duologue, dialogue 1
dupe°, beguile 1; betray 3; bluff[1] 1; butt[1]; chisel 2; defraud; facsimile; fool 3, 4; hoax 2; hoodwink; misinform; mislead; mug 3; outsmart; pawn[2]; plaything 2; prey 2, 3b; puppet 2; replica; rip 2b; sap[1] 2; scapegoat; score 7; string 10c; sucker; swindle 1; take 32b; tool 3; transcript 2; trap 6; trick 8; victim 2; victimize 2
duped by, be, fall 13b
duplex, flat 15
duplicate°, copy 1, 4; double 6, 7; echo 4; facsimile; identical 1; image 2; imitation 4; likeness 2; mimic 2; monkey 4; parallel 4; pattern 7; picture 2; quote 1; repeat 2, 3; replica; reproduce 1; reproduction 2; run 29b; second[1] 6; spit 2; tally 5; transcribe 1; transcript 2; twin 1, 2
duplicate(d), double 1
duplicating, repetition 2
duplication, copy 1; echo 2; imitation 4; knockoff; repeat 3; repetition 2; replica; reproduction 1; transcript 2
duplication(s), repetition 1
duplicitous, crafty; deceitful; dishonorable 2; disingenuous; equivocal 1; fraudulent 2; insincere; lying 2; scheming; shifty; tricky 1; two-faced; untrue 1

duplicity, art 5; artifice 1; chicanery; craft 2; deceit 1; deception 1; dissimulation; foul play; fraud 1; hanky-panky; hocus-pocus 1; hypocrisy; lying 1; stab 6; trickery
durability, endurance 2; permanence; stability 2; strength 6
durable°, enduring; hardy 1; indestructible; lasting; perennial 1; permanent 1; rugged 3; serviceable 2; solid 6; sound[2] 3; stable 1; stout 3; strong 12; substantial 2; tough 1
durance, captivity; detention; duress 2; imprisonment
duration, age 1; existence 2; length 2; life 4; period 1; space 3; standing 7; term 2; time 2
duration of, for the, for 10
duress°, force 2
during, by 5; for 10; over 4; pending 1; through 2; throughout 1
during the interval, meanwhile 2
dusk°, evening; gloom 1; night 2; obscurity 1; shade 1; shadow 1; twilight 1
duskiness, obscurity 1; shade 1
dusky°, black 1, 4; dark 5; dim 1; dingy; gloomy 1; obscure 1; shadowy 1; somber 2; sunless
dust, clean 9; dirt 1; fluff 1; powder 1, 4; refuse[2]
dust (off), wipe 1
dustball, fluff 1
dust bowl, desert[1] 1
dustbroom, brush[2] 1
dust devil, eddy 1; whirlwind 1
dusted, flecked
dusting, film 1
dust jacket, cover 12
dust storm, storm 1
dust-up, encounter 5; quarrel 1; scrap[2] 1; scrimmage; skirmish 1
dusty, old 3
Dutch courage, bottle 2
duteous, dutiful; obedient
dutiful°, faithful 3; obedient; pious 1; true 2
dutifulness, obedience; piety 1
duty°, assignment 2; business 1; capacity 3; concern 4; custom 2; detail 3; errand 2; faith 3; function 1; job 2; mission 1; obligation 1; office 4; part 3; place 4; post[2] 1; responsibility 2; role 2; stint 1; tariff 1; task 1; tax 1; toll[2] 1; tribute 2; work 3
duty, on, call 16
duty (as *or* for), do, serve 2
duvet, cover 13; spread 14
dwarf°, little 1; miniature; overshadow 1; puny 3; runt; toy 5
dwarfed, stunted; undersized
dwarfish, short 1; undersized
dweeb, jerk 5; square 9
dwell°, abide 2; lie[2] 3; live 8; lodge 3; room 4; settle 4; sit 4; stay[1] 2; stick[1] 7; tarry 2
dwell in, inhabit; occupy 2; populate
dwell on *or* upon, dwell 2; insist 2; labor 6; linger 3; reiterate; rub 3
dweller, citizen 1; denizen; inhabitant; occupant; resident 3
dwelling°, abode; home 1; house 1; housing 1; lodging; place 6; quarter 5; residence 1; resident 1; room 3; shelter 3
dwelling (place), abode; domicile 1; dwelling; home 1; house 1; lodging; quarter 5; residence 1; shelter 3
dwindle°, decay 1a; decline 2; decrease 1; die 2; diminish 3; ebb 2; fade 2; fail 3; fall 3; flag[2] 2; melt 4; peter out; recede 2; remit 2; trail 8; waste 2; weaken 2; wilt 2
dwindling, decrease 2; ebb 4; slack 5
dyad, pair 1
dye, color 1, 3; paint 1, 6; rinse 2, 4; stain 3, 4; tint 2, 3
dyed-in-the-wool, chronic 2; conservative 2; incurable 2; inflexible; out-and-out; positive 9; right 8; ultra-
dying°, death 1; eager; hungry 2; leg 7; loss 7; moribund 1; obsolescent; parting 3; passing 4
dying for, starved 2
dying for, be, pant 2
dying out, moribund 2; passing 4
dyke, deviant 3; homosexual 1
dynamic°, active 1; animated 1; dashing 1; energetic; forceful 1; industrious; live 2; persuasive; powerful 1; productive 1; progressive 2; racy 1; scintillating 2; solid 8; spirited; strenuous 2; strong 5; tireless; vigorous; vital 3
dynamical, dynamic
dynamically, hard 13; vigorously
dynamism°, animation 1; bounce 2; energy; force 1; gut 3b; industry 3; initiative 2; push 14; soul 4; vigor
dynamite, blast 5; blow[1] 8b; explosive 2
dynasty°, clan 1; family 3; house 2; stock 2; succession 4; tribe
dysentery, run 50
dyspepsia, indigestion
dyspeptic, bilious; irritable; quarrelsome; querulous; snappish 2; sullen; surly; touchy 1
dyspnoeal, short-winded
each (and every) night, nightly 1, 3
each (and every) one *or* person, everyone
each (and every) time, always 1
each to each, respectively
each year, yearly 2
eager°, agog; ambitious 2; anxious 2; ardent; breathless 3; crazy 4; dynamic; earnest 2; enterprising; enthusiastic; excited 2; expectant; favorable 2; fervent 2; fiery 3; fire 5; glad 2; great 9; hard 7; hearty 3; hot 4; hungry 2; impassioned; impatient 1; inclined 1; intense 2; intent 5; keen[1] 1; lively 2; mad 6; mood 2; passionate 1; prompt 2; ready 2; ripe 3b; solicitous 2; strenuous 2; strong 5; thirsty 2; voracious 2; warm 2, 4; wholehearted; wild 9; willing
eager (for), desperate 3
eager (for), be, leap 4; spoil 5
eagerly, hard 13; intently; jealously; readily 1; searchingly; vigorously; warmly 3; willingly
eagerness°, anxiety 2; ardor; devotion 3; enterprise 2; enthusiasm 1; excitement 3; fervor; fire 2; flame 2; gusto; heat 2; hurry 3; inclination 4; passion 1; push 14; readiness 1; relish 1; spirit 2; thirst 1; vigor; zest 2
eagle-eyed°, observant 1; sharp-eyed; vigilant
ear°, heed 2
ear (to), lend an, listen 1
eardrop, pendant
earl, lord 2; peer[1] 1; royal 3
earlier, ancient 1; before 1; foregoing; former 1; lower[1] 7; preceding; previous 1; previously; prior 1
earliest, first 2, 4; original 1; prehistoric 1; primary 2; primitive 1; pristine 1
early°, infancy 2; old 5; prehistoric 1; premature 1; remote 8; time 10, 18a
early, too, premature 2; prematurely 1; previous 3
earmark, allot; allow 5; appropriate 3; book 4; characteristic 2; destine 2; feature 1; hallmark 2; label 1, 4; mark 2; peculiarity 2; schedule 2; set 14b; symptom; tag 3
earmarks, stamp 5
earmarks of, have (all) the, resemble; seem
earmarks of, having the, redolent 2
earn°, acquire; bring 11a; come 9b; command 4; deserve; fetch 3; gain 1, 2; get 3; gross 6; harvest 3; make 7, 12, 13; merit 3; mint 2; net[2] 4; obtain 2; profit 5; pull 9b; realize 3; receive 2; return 5; win 2; yield 4
earn a living, work 9
earned, deserved
earnest°, deep 6; devoted 1; devout 3; diligent; eager; emphatic; enterprising; enthusiastic; fervent 2; genuine 2; grave[2] 1; heartfelt; hearty 2; hot 4; impassioned; instant 4; intense 2; intent 4; keen[1] 1; painful 3; passionate 1; poignant 3; real 3; serious 1; sober 2; solemn 1; solicitous 2; strenuous 2; strong 22; truthful; urgent 2; voracious 2; warm 4; wholehearted
earnest, in, earnest 4
earnestly, deep 10; hard 13, 15; intently; seriously 3; sincerely; warmly 3
earnest (money), earnest 3; pledge 2
earnestness, devotion 1, 3; eagerness 1; enthusiasm 1; fervor; heat 2; industry 3; solemnity
earnings°, gain 8; net[2] 1; remuneration 1; return 9; salary; stipend; wage 1; yield 5
earn money, coin 4
ear-piercing, shrill
earshattering, penetrating 2; piercing 1
earshot, sound[1] 3
earshot, within, call 17
ears, prick up one's, listen 1
ears (in), up to the, replete 1
ears open, keep one's, listen 1
earsplitting, loud 1; noisy; penetrating 2; piercing 1; raucous; sharp 7; shrill; thunderous; high 7
earth°, dirt 2; globe 1; ground 1; land 1, 2; mold[3]; soil[2]; world 1
earth, come to, touch 10
earth, on, below 3
earthbound, terrestrial 1
earthenware, pottery
earthiness, ribaldry
earthling, mortal 6; terrestrial 2
earthly°, material 7; mortal 2; physical; temporal 1; terrestrial 1; worldly 1
earthman, terrestrial 2
earthperson, terrestrial 2
earthquake, quake 2
earthshaking, fateful 1; special 2; stunning 2
earthslip, slide 5
earthwoman, terrestrial 2
earthwork, rampart
earthy°, bawdy; gross 3; racy 2; sexual 2; suggestive 2
earwitness, witness 1
ease°, comfort 3; content[2] 2; expedite 2; facilitate; facility 1; familiarity 2; fluency; freedom 4;

edifice, building; house 1; structure 2
edify, civilize 1; discipline 6; educate; enlighten; instruct 1; teach; train 4
edifying, educational 2; informative; instructive
edit°, cut 5; revise 1
  edit, (copy), run 33b
  edit (out), cut 16a; delete; edit 2; eliminate 3; omit 1
edited, cut 27
editing, revision
edition°, issue 5; number 3; revision
editor°, scribe 2
editorial°, writing 2
editorial writer(s), editor; press 10b
editor in chief, editor
educate°, break 18b; bring 15a; civilize 1; condition 6; enlighten; instruct 1; prepare 3; prime 5; rear[2] 1; school 4; teach; train 4; tutor 2
  educate (in), familiarize
educated°, cultivated; enlightened; informed 1; literary 1; professional 1; refined 2; well-informed
education°, culture 1; edification; inspiration 2; instruction 2; knowledge 4; preparation 2; scholarship 1; schooling; tuition; upbringing
educational°, informative; instructive
educationist, teacher
educative, educational 2
educator, instructor; teacher; tutor 1
educe, derive 1
eensy-weensy, minute[2] 1
eerie°, ghostly 1; grim 3; macabre; scary; spectral; unearthly 2; weird
efface, blot 4b; blunt 4; blur 3; cancel 2; delete; erase 1, 2; obliterate 1; remove 3
effect°, accomplish; achieve 1; action 2, 7; bring 7; carry 10c, 12; cause 6; effectuate; end 4; ensure 1; event 2; excite 1; execute 2; exercise 1; float 3; fulfill 1; impact 2; implement 2; impression 2; induce 2; influence 1; issue 2; mark 5; might 2; negotiate 2; obey 2; occasion 5; outcome; perfect 8; perform 1; perpetrate; power 7; procure 2; product 1; punch[1] 3; purpose 3; put 27a; ramification 1; reaction 1; realize 1; reflection 3; repercussion; response; result 1; sound[1] 2; tableau; take 21; tenor; upshot; work 16; wreak
  effect, be in, hold 11; run 15
  effect, be *or* remain in, stand 4
  effect, do for, posture 4
  effect, for, rhetorical 2
  effect, have, run 15
  effect, have (an), get 16; mark 9; move 4; tell[1] 6; weigh 5
  effect, in, effect 5; essence 3; force 6; intent 3; operation 5; practically 1; really 3; virtually
  effect, put into, enforce 1; take 21
  effect, put to, exercise 1; exert 1
  effect, take, act 9; effect 6; take 11
effective°, active 2; beneficial 2; capable 1; dramatic 2; effectual 1, 2; efficacious; efficient; eloquent 1; expedient 2; force 6; forceful 1, 2; fruitful 2; functional 1; going 2; important 3; imposing; influential; mean[2] 6; moving 1; net[2] 3; operation 5; persuasive; potent 2; powerful 2; prevailing 2; profitable 1; smart 4; solid 8; strong 7; sure 3; telling 1; useful; virtual
  effective, be, run 15; tell[1] 6; trick 7; work 19d
  effective, be *or* remain, stick[1] 6
  effective, prove *or* be, take 11
effectively, big 10; effect 5; highly 4; home 10; mainly; out 7; virtually; well[1] 3
effectiveness, action 7; edge 3; effect 2; efficiency 1, 2; force 4; might 2; power 7; punch[1] 3; usefulness
effects°, asset 1; belongings; furniture 1; gear 4; good 21a; paraphernalia; possession 3; property 1; stuff 2; thing 8c
effectual°, active 2; capable 1; effective 1, 2; efficacious; efficient; influential; powerful 2; prevailing 2; strong 7; telling 1
effectuate°, effect 7; execute 2; fulfill 1; perpetrate; realize 1; work 16
effectuation, performance 1; realization 2
effeminate°, camp[2] 1; feminine 2; homosexual 2; mincing; soft 12; swish 4
effervesce, ferment 1; fizz 1; froth 3; gush 2; sparkle 2
effervescence, bubble 2; exuberance 1; fizz 2; foam 1; life 7; pep 1; sparkle 4
effervescent°, bubbly 1, 2; ebullient; exuberant 1; gay 2; irrepressible; perky; scintillating 2; spirited; vivacious
  effervescent, be, gush 2
effete, feeble 1; feminine 2; ineffectual 2; precious 3; soft 12; sterile 3; useless 1; weak 2; wet 3
efficacious°, active 2; beneficial 2; effective 1; effectual 1; efficient; forceful 2; influential; operation 5; persuasive; potent 2; strong 7
  efficacious, prove *or* be, take 11
efficaciousness, efficiency 2
efficacy, effect 2; efficiency 1; force 4
efficiency°, facility 1; output 2
efficient°, capable 1; effective 1; effectual 1; efficacious; expeditious; experienced 1; neat 4; practical 1; professional 1, 2; streamlined 2, 3; strong 6
effigy, figure 4; idol 1; image 1; representation 4; statue
effloresce, flower 3
efflorescence, flower 1; rash[2] 1
effluence, escape 7; outburst; outpouring; stream 2
effluent, exhaust 6; filth 1
effluvium, exhalation 2; fume 3; smell 2; stench
efflux, escape 7; issue 1; outburst; outpouring; stream 2
effluxion, escape 7
effort°, application 3; attempt 2; bother 5; campaign 1; drive 8, 10; endeavor 2; enterprise 1; essay 2; exertion; industry 2; labor 1, 3; main 6; pain 4; pass 27; production 2; push 14; spurt 1; stand 14; strain[1] 7; struggle 4; task 2; trial 3; try 5; undertaking 1; work 1
  effort, greatest *or* maximum, best 8
  effort, make every, shoulder 3; strive 1
  effort (to), make an, endeavor 1; essay 3; exert 2; finger 6; point 17; shoulder 3; strain[1] 4; try 1
effortless°, easy 1; fluent; light[2] 10; painless; runaway 3; smooth 3
effortlessly, easily 1; easy 7; hand 13; handily 1; readily 2; swimmingly
effortlessness, ease 2; facility 1; fluency
effrontery°, arrogance; assurance 3; brass; face 5; front 7; gall[1] 2; impertinence; impudence; nerve 2; presumption 1
  effrontery, have the, presume 2
effulgence, brilliance 1; glare 1; glory 3; glow 2; radiance 1; splendor 2
effulgent, bright 1; brilliant 1; light[1] 13; luminous 1; radiant 1; scintillating 1
effulgently, clear 17
effusion, gush 4; outburst; outpouring; stream 2; torrent
effusive°, demonstrative 1; ebullient; expansive 2; gushy; lavish 2; lyrical 2; rhapsodic; talkative; windy 2
  effusive, be, gush 2
effusiveness, exuberance 2; warmth 3
e.g., example 4; instance 2; like[1] 6; namely; say 15
egalitarian, democratic
egalitarianism, equality 3
egg, hurry 2; seed 1
  egg (on), abet 1; encourage 1; energize; foment; incite; induce 1; motivate; prod 2; prompt 3; provoke 1; push 4; put 29; spur 4; urge 1, 2
egghead(s), highbrow 1; intellectual 3; intelligentsia; scholar 1; scholarly
egg-shaped, oval
egocentric, conceited; egoistic; misanthropic; subjective 2
egoism, self-esteem 1
egoistic°, misanthropic; subjective 2
egoistic(al), egoistic; selfish 1
egotism, arrogance; conceit 1; pride 2; self-esteem 1; snobbery; vanity 1
egotist, showoff
egotistic, egotistical; inflated 1
egotistic(al)°, arrogant 1; boastful; cocky; conceited; haughty; pompous 1; presumptuous; proud 2; selfish 1; self-important; snobbish; vain 1
egregious, bad 1; flagrant; flaming; glaring 1; great 11; grievous 2; infamous 2; monstrous 2; monumental 4; outrageous 2; positive 9; prominent 1; unconscionable 2; wicked 2
egregiousness, infamy 2
egress, exit 1; issue 1; outlet 1
Egyptian underground, underworld 2
eiderdown, cover 13; spread 14
eidolon, phantom 1; vision 4
eighteen, side 4
Einstein, genius 1; prodigy 1
eisegesis, note 3
ejaculate, exclaim
ejaculation, exclamation; interjection
eject°, banish 1; boot 3; cast 14; discharge 2; displace 2; dispossess; emit; empty 8; erupt 1; evict; exclude 3; exhale; exile 3; expel 1; freeze 4; hustle 4; isolate; overturn 2; purge 2; rid 2; spew; spit 1; spout 1; supplant; suspend 3; throw 6a, 7b; turn 20c; vent 4; void 7

emancipate°, deliver 3; free 14; liberate 1; redeem 2; release 1
emancipated, free 2
emancipating, release 3
emancipation, delivery 2; freedom 2; liberation; release 3
emancipator, Messiah; savior 1
emasculate, cripple 2; fix 13; neuter 2; silence 4; sterilize 2; weaken 1
embalm, preserve 3
embankment, wall 2
embargo°, ban 2; bar 3, 9; boycott 1, 2; prohibition 2; restraint 1; veto 2
embark°, get 39a; set 18b; ship 3
embark on *or* upon, board 6; commence 1; embark 2; engage 6; go 29a; launch 1; start 1; undertake 1
embarras de choix, embarrassment 3
embarras de richesse, embarrassment 3
embarrass°, confuse 1; crush 5; devastate 2; discomfit 1; disgrace 3; entangle 2; mortify 1; offend 1; place 13; put 23a; shame 6, 7, 8; show 12c
embarrassed°, ashamed; awkward 3; bashful 1; guilty 2; inhibited; remorseful; repentant; self-conscious; shamefaced 2
embarrassed, be, hang 8
embarrassed, feel, small 6
embarrassing°, awkward 5; disgraceful 1; notorious 1; outrageous 3; shameful; sore 2; sticky 2
embarrassment°, confusion 6; difficulty 3; disgrace 1; humiliation; remorse 1; scandal 1; shame 1
embattled, militant 2
embed, couch 2; engrave 2; implant 2; root[1] 6
embed itself, lodge 5
embedded, entrenched; rooted
embedded, become, lodge 5
embellish°, amplify 3; beautify; decorate 1; elaborate 3; enhance; enrich 2; exaggerate; expand 4; grace 6; illuminate 3; illustrate 2; maximize 2; ornament 2; sweeten 2; tool 4; trim 4
embellished, fancy 1; flamboyant 1; flowery
embellishing, ornamental
embellishment°, decoration 1; elaboration 1; exaggeration; flourish 3; frill 2; ornament 1; trim 5
embellishment(s), trappings
embezzle°, misappropriate 1; pilfer; pocket 4; rip 2b; steal 1; take 3
embezzlement°, rip-off 2; stealing; theft
embezzler, thief 1
embitter, aggravate 2; exasperate 1; sour 6
embittered°, bitter 4; resentful; sour 4; venomous 2
embittered about, feel, resent
embitteredness, venom 2
emblazon, illustrate 2
emblem°, character 1; color 2b; crest 2; device 3; figure 7; mark 2; seal 1; sign 4; stamp 7; standard 3; symbol; token 2
emblematic°, characteristic 1; representative 1; token 4
emblematic (of), symbolic; symptomatic
emblematic of, be, stand 7a
emblematical, emblematic
emblems, regalia
embodiment°, epitome 1; image 4; picture 4; quintessence; soul 3
embodiment of, be the, personify 1
embody°, cover 7; embrace 3; exemplify 1; express 2; flesh 6; include 1; incorporate; involve 1; personify 1; possess 2; represent 4; symbolize; typify
embodying, inclusive 2
embody in words, shape 8
embolden, encourage 1; fortify 2
embolism, stroke 5
embonpoint, fatness; obesity
emboss, impress 2; stamp 2
embouchure, mouth 2; mouthpiece 1
embrace°, adopt 2; caress 1, 2; clasp 2, 4; clinch 2; cling 2, 3; clutch 2a; cuddle 2, 3; fold 2; go 29a; grasp 3; hold 2; hug 1, 3; include 1; incorporate; involve 1; love 6; possess 2; prefer 1; press 5; ring[1] 4; squeeze 5, 7; take 32c, 39c; uphold
embracing, inclusive 2
embrocate, apply 2
embrocation, lotion; ointment; salve 1; wash 13
embroider, amplify 3; blow[1] 8c; decorate 1; embellish 1, 2; exaggerate; lay[1] 14; ornament 2; overstate; sweeten 2; trim 4
embroider (on), maximize 2
embroidered, fancy 1; pompous 2
embroidery, braid 2; embellishment 1; exaggeration; flourish 3; ornament 1; trim 5
embroil, entangle 1; implicate 1
embroiled with, involved 3
embrue, suffuse
embryo, germ 2; seed 1
embryonic, potential 1; primary 2; rudimentary 2; seminal 2; undeveloped
emcee, announcer; host[1] 2; moderator; present[2] 8
emend, amend 2; edit 1; elaborate 4; fix 3; improve 1; mend 2; perfect 9; polish 2; rectify; reform 1; revise 1
emendate, amend 2
emendation, amendment 1; correction 1; reform 3; revision
emerge°, appear 1; break 20a; come 14a, 15a; dawn 4; develop 4; emanate 1; issue 11; light[1] 10; loom 1; materialize 1; originate 2; peer[2] 2; pour 4; result 2; spring 2; start 2; surface 3
emerge (from), proceed 2
emerge suddenly, break 13
emergence°, birth 2; conception 1; dawn 2; infancy 2; issue 1; start 11
emergency°, crisis 2; makeshift 1; necessity 4; need 4; pinch 8; predicament; rush 5; stopgap 2; urgency; urgent 1
emeritus, outgoing 1
emetic, nauseous
emigrant°, exile 2
emigrate°, migrate 1; move 2; stake[1] 2
émigré(e), emigrant; exile 2; refugee
eminence, celebrity 1; dignity 1; distinction 2; elevation 2; excellence; fame; glory 1; grandeur 2; height 3; hill 1; importance 2; mark 8; mountain 1; name 2; note 8; peak 1; preeminence 2; prestige; projection 1; prominence 1; quality 3; rank[1] 2; renown; rise 13; standing 6; status 1; superiority 2; worthy 3
eminent°, big 5; brilliant 3; celebrated; conspicuous 3; distinguished 1; elder 2; elevated 2; exalted 1; famous; glorious 1; grand 2; great 5; heroic 6; honorable 4; illustrious; important 2; legendary 3; lofty 2; noble 3; notable 1; noted; outstanding 1; preeminent 2; prestigious; prominent 2; proud 3; public 6; renowned; salient; singular 2; splendid 2; star 3; sublime 1; successful 3; top 8; well-known 2
eminently°, large 4; preeminently
emissary, agent 1; ambassador; delegate 1; deputy; envoy; messenger; minister 2; representative 3
emission, discharge 11; ejection 1; eruption 1; exhalation 2; exhaust 6; radiation; secretion
emit°, discharge 4; eject 2; emanate 2; erupt 1; exhale; give 8, 14; let[1] 6d, 8c; loose 10; ooze 2; radiate 2; secrete[2]; send 7; shed[2] 3; spew; spout 1; throw 6a, 7a; vent 4; void 7
emitted, be, pass 17
emitting, secretion
emollient, mild 3; ointment; salve 2; soothing 2
emolument, consideration 2; earnings; fee 2; fruit; gain 8; honorarium; perquisite; remuneration 1; salary; stipend; wage 1
emolument(s), benefit 2
emotion°, expression 4; feeling 4; passion 1; sentiment 1, 2; soul 4
emotional°, demonstrative 1; effusive; excitable; expressive 3; fervent 2; full 9; gut 6; impassioned; impulsive; intense 3; lyrical 2; maudlin; moving 1; passionate 1; pathetic 1; piteous; poignant 3; romantic 3; sensitive 2; sentimental 1, 2; sloppy 3; soulful; stirring; strong 22; tender[1] 7; touching; warm 2
emotional attachment, thing 4
emotionalism, sentimentality
emotionalistic, melodramatic
emotionally, badly 9; deeply 2
emotionally upset, problem 3
emotion-charged, stirring
emotionless, blank 4; blasé 2; cold 3; cool 3; dead 4; immovable 2; impassive; insensible 2; rocky[1] 3; stoical; straight 8
emotionlessness, emptiness 3
emotion(s), psychology; feeling 5; sensibility 2
emotive, emotional 3; moving 1; pathetic 1; stirring
empathetic *or* empathic (with *or* to *or* toward(s)), sympathetic 1
empathize, relate 4
empathize (with), feel 8; identify 4; respond 2; sympathize 1; understand 5
empathy, feeling 3; heart 5; identification 4; instinct; rapport; sympathy 1; understanding 2; vibes
emperor, crown 3; king; monarch 1; sovereign 1
emphasis°, accent 1; stress 1, 2; tone 2
emphasis on, put *or* place, point 17
emphasis on *or* upon, lay, enforce 2; stress 4
emphasize°, accent 4; bring 14a; dwell 2; feature 4; home 11; impress 3; insist 2; intensify; play 17a; point 17, 21b, 22;

pump 4c; punctuate 2; rub 3; spotlight 3; stress 4

emphatic°, assertive; broad 3; categorical; dogmatic; dynamic; insistent; peremptory 3; positive 2; strong 16, 20

emphatically, positively

empire, domain 1; kingdom 1; monarchy 1; realm 1; reign 1

empiric(al), experimental 2; practical 1

emplacement, placement 1; position 2

emplacing, placement 1

emplane, get 39a

employ°, apply 3; assume 1; busy 4; draw 13a; embrace 2; engage 1, 2; enlist 2; exercise 1; exert 1; expend 1; fall 9; handle 7; hire 1; job 6; line[1] 23b; manipulate 2; pass 5; retain 2; service 4; sign 10b; take 21, 35a; touch 9; turn 6; use 1, 4; wield 1, 2; work 15

employed, busy 1; engaged 2; used 2

  employed, be, work 9

employee°, hand 4; operative 2; worker

employee(s), help 7; labor 2; rank[1] 5; staff 2

employer°, boss 1; head 2; master 1

employing, employment 2; mean[3] 3

employment°, application 1; calling; career 1; engagement 4; exercise 5; job 1; line[1] 7; occupation 1; office 4; place 5; placement 2; post[2] 1; profession 1; pursuit 3; racket 3; service 4; situation 3; trade 2; use 6; vocation; work 2

empower, authorize; delegate 2; enable 1; entitle 1; license 5; permit 1; trust 6; warrant 4

empowered, privileged 2

empress, crown 3; monarch 1; queen 1; sovereign 1

emptied, empty 1; exhausted 2; spent 2

emptiness°, blank 8; hunger 1; solitude 2; vacancy 1, 3; vanity 2; void 4, 5; waste 8; wild 10

empty°, bare 3; blank 1, 3; clean 10b; clear 24; dead 10; deserted; desert[1] 2; desolate 1; discharge 7; drain 4; dump 1; evacuate 1; exhaust 3, 4; exhausted 2; flat 5; flush[1] 2; fond 2; free 3; futile; glassy 2; gut 5; hollow 1, 3, 5; hungry 1; ill-founded; impoverished 2; lifeless 4; meaningless 1; pale[1] 3; plausible 2; pointless; pound[1] 5; pour 2; pump 3; purposeless; shallow 1; stark 3; superficial 2; tip[2] 3; trackless; uninhabited; unload; vacant 1, 2; vague 6; vain 2; vent 4; void 2, 7; walk 2; weak 4; wild 2; wooden 2

  empty of, empty 7

  empty (out), exhaust 5

empty-headed, dizzy 2; feebleminded; foolish 2; light[2] 8; senseless 3; stupid 1

empty-headedness, stupidity 1

emptying, discharge 11, 14; exhaustion 1

empty talk, exaggeration; mouth 3

empyreal, sublime 1

empyrean, celestial 1; firmament; heaven 2; seraphic; sky 1; sublime 1, 2

emulate, copy 5; echo 4; imitate 1; model 9; pattern 7

emulating, imitation 1

emulation, imitation 1

emulsion, combination 3; solution 3; wash 13

enable°, expedite 2; license 5; permit 1

  enable (to), let[1] 2

enact°, establish 1; make 6; transact

enactment, act 4; decree 1; execution 1; law 1; passage 9

enamel, glaze 1, 2; paint 2

enamor, captivate

enamored, infatuated; smitten 2

  enamored (of), keen[1] 7; sweet 8; wedded

  enamored of, be, care 6b

encamp, camp[1] 3

encampment, camp[1] 1; settlement 1

encapsulate, sum 2a

encapsulation, summary 1

encase, box[1] 2; case[2] 3; package 3; wrap 1

enceinte, expecting; pregnant 1

enchain, enslave

enchant°, bewitch; captivate; catch 11; charm 5; dazzle 1; enrapture; entrance[2]; fascinate; hypnotize; intoxicate 2; possess 3; regale; transfix 2; transport 3

enchanted°, charmed 1, 3; delighted; ecstatic; infatuated; rapturous; smitten 2; taken

enchanted forest, fairyland

enchanté(e), delighted; enchanted

enchanter, charmer; magician 1; sorcerer

enchanting°, catching 2; darling 4; delicious 2; delightful 2; enthralling; fetching; glamorous 1; interesting; lovable; magic 6; seductive; sexy 1; spellbinding; taking; winning 1

enchantment°, charm 2; fascination; glamour; magic 3; rapture; sorcery; spell[2] 1

enchantress, charmer; magician 1; seductress; siren 2; sorcerer; temptress; witch 1

enchase, scribe 3

enchiridion, companion 2; guide 8; manual; schoolbook

encipher, code 4

encircle°, bind 3; circle 3, 4; enclose 1; fence 3; garland 2; mantle 3; orbit 2; ring[1] 4; round 17; siege 2; surround 1; twine 2

encirclement, siege 1

encircling, about 7; around 10

enclose°, box[1] 3; cage 2; circle 4; clasp 4; cover 2; encircle; envelop 1; fence 3; fold 2; frame 8; package 3; pen[2] 2; picket 4; stake[1] 4a; surround 1; wall 5; wrap 1

enclosing, around 10

enclosure°, area 5; cage 1; fence 1; housing 2; pen[2] 1; pound[2]; ring[1] 2; run 46; stall[1] 2, 3; wall 1

encode, code 4

encoded, secret 2

encomiastic, complimentary 1; glowing 3; good 17; laudatory; lyrical 2

encomium, eulogy; praise 1; rave 3; tribute 1

encompass, beset; embrace 3; encircle; enclose 1; include 1; involve 1; ring[1] 4; surround 1; take 32c

encompassing, around 10; comprehensive

encore, repeat 3

encounter°, action 6; battle 1; bout 3; brave 3; brush[2] 3; bump 4; clash 2; combat 1; come 5a; discover 2; engage 5; engagement 5; experience 2, 4; face 14; fight 1, 7; find 1; game 2; happen 3; hit 7; light[2] 15; meeting 1; meet[1] 1, 6; skirmish 1; stand 12a; strike 14; struggle 5; stumble 3; taste 8; tilt 4; war 1

encourage°, abet 1; advise 1; animate 2; arouse 3; ask 6a; assure 2; back 2a; bear 10b; boost 5; cheer 5; confirm 3; cultivate 3; favor 8; flush[1] 3; foment; fortify 2; forward 6; foster 1; fuel 4; incite; induce 1; inspire 2; motivate; nourish 3; nudge 1; patronize 3; pat[1] 2; prod 2; promote 1; prompt 3; provoke 1; pull 10; push 4; put 29; raise 7; reassure; rise 16; root[3]; run 16; sanction 6; second[1] 9; spur 4; stimulate 1, 2; stir 4; strengthen 2; urge 2

encouragement°, boost 2; cheer 3; flush[1] 6; fuel 2; impetus; incentive; incitement 2; inducement; inspiration 2; lift 8; nudge 2; patronage 1; pat[1] 5; progress 3; promotion 1; recommendation 1; sanction 2; shot 12; spur 1; start 12; support 7

encouraging, favorable 1, 2; heartwarming 1; hopeful 2; positive 7, 8; promising; propitious; rosy 2; supportive; sympathetic 2; upbeat

encroach°, intrude

  encroach on, trample 2

  encroach (on *or* upon), infringe 2; presume 3; tie 7b

encroacher, intruder 1

encroachment, aggression 2; inroad 1; invasion 1

encrust, cake 3

encrustation, scale[2] 2

encrypt, code 4

encrypted, secret 2

encumber°, bog 2; burden 2; clog; hamper[1]; handicap 2; hinder 1; interfere 2; load 4; oppress 1; overload 1; stick[1] 8; tax 4; trouble 1; weigh 6

encumbered, debt 2; heavy 13

encumbrance°, debt 1; handicap 1; hindrance 1; impediment; interference 2; liability 3; load 1; overload 2; shackle 2; tax 2

encyclopedic°, comprehensive; exhaustive; general 3; thorough 3; wide 2

encypher, code 4

end°, abolish; abolition; accomplish; aim 5; break 19a; butt[1]; cease 1; close 3, 22; collapse 2; come 15c; complete 5; completion 1; consummation 1; cut 15b; death 1, 2, 3; defeat 2; demolish 2; destination; destroy 2; destruction 3; determination 2; determine 1; die 3; dismissal 2; dissolution 2; doom; drop 7; event 2; expiration; expire 1; extremity 1; fate 2, 3; final 1; finish 1, 5, 9; goal; halt 1, 2; heel[1] 1; idea 4; intent 1; intention; interrupt 2; issue 2; kill 9; lapse 5; last[1] 6; let[1] 8d; lift 3; limit 1, 2; live 7; lot 2; mark 6; measure 8; motive 2; net[2] 3; object 3; overthrow 2; pack 6; pass 10, 14b; passing 4; phase 6; pole[2] 1; polish 3a; purpose 1; raise 9; rear[1] 1, 3; reason 3; remnant 1; result 3; round 18; ruin 1, 3; run 30a; scrub 2; sever 2; stamp 4; stanch; stop 1, 4, 6; stub 1; stump 1; stunt[2];

end° (*cont.*) suppress 1; suppression; take 15; target; terminate; termination 1; tip[1] 1; ultimate 1; use 12; wane 2; wrap 3a
end, at an, exhausted 2; over 6
end, at *or* **in the,** last[1] 5
end, bring *or* **draw** *or* **come to an,** halt 2
end, bring to an, accomplish; close 3; complete 5; demolish 2; destroy 2; end 9; get 46c; knock 5c; round 18; stop 1, 2; terminate; wind[2] 4a
end, come to an, stop 1; terminate; wind[2] 4a
end, in the, eventually; finally 1; really 3; run 48; soon 5; ultimately
end, on, end 7; upright 1
end, the, end 8; limit 4a
end in, reflect 4
end of, make an, demolish 2; destroy 2
end of, making an, destruction 2
end of the day, at the, effect 5; eventually; finally 1; practically 1; really 3; run 48; soon 5; ultimately
end to, bring an, destroy 2
end to, put an, abolish; demolish 2; destroy 2; discontinue; dispatch 3; end 9; exterminate; finish 4; halt 2; kill 1; put 16b; settle 3; sever 2; stamp 4; stay[1] 4; stop 1; stunt[2]; suppress 1; terminate
end to, putting an, destruction 2
end to the other, from one, through 7; throughout 2, 3
end (up), come 13c; end 9; finish 6; pan 6; prove 3; rest[1] 5; turn 20b; wind[2] 4a, 4b
end up at *or* **in,** reach 2
end (up) successfully *or* **satisfactorily,** come 17b
endanger°, adventure 3; chance 8; expose 2; hazard 3; jeopardize; lean[2] 4b; risk 2; threaten 2; venture 3
endangerment, exposure 2; hazard 1
endanger (something), fire 6
endear, attach 3
endearing°, lovable; winning 1
endeavor°, activity 2; attempt 1, 2; effort 2; essay 2, 3; exert 2; pass 27; pretend 2; seek 2; shot 3; strive 1; struggle 1; trial 3; try 1, 5
ended, complete 2; out 11; over 6
endemic, indigenous 1
ending, consummation 1; dissolution 2; end 2; expiration; final 1; finish 9; moribund 2; suffix 1; suppression; terminal 1; termination 1, 2; upshot
end (in view), objective 2
endless°, boundless; constant 2; continual; continuous 2; eternal 1, 2; everlasting; immeasurable; immortal 1; indefinite 4; indestructible; inexhaustible 1; infinite 1; lengthy; nonstop 2; ongoing 1; perennial 2; permanent 1; perpetual 2; protracted; rambling 1; steady 2; tedious; timeless; universal 2; unlimited 2; vast
endlessly, cease 2; ever 2; forever 2; night 3; nonstop 3; permanently
endlessness, eternity
endmost, extreme 3
end of day, night 2
end of the line, terminal 3
end of time, till the, always 2; ever 2; forever 1
endorse°, abet 2; advocate 1; approve 1; authenticate; authorize; back 2a; certify 1; confirm 1; favor 6; initial 2; OK 5; pass 7; praise 3; promote 3; ratify; recognize 3; recommend 2; seal 4; sign 7; subscribe 1; support 6; sustain 5; underwrite 2; uphold; vouch; witness 5
endorsed, official 1
endorsement°, approval; backing 1; clearance 2; indemnity 2; OK 6; passage 9; pat[1] 5; praise 1; recommendation 2; reference 3; seal 2; testimonial
endorsing, passage 9
endow, bless 2; clothe 2; enrich 1; fund 3; supply 1
endow with, award 1
endowed with, be, possess 1, 2
endowment°, award 3; bounty 2; foundation 3; fund 2; grant 3; present[2] 2; scholarship 2; talent 1
endowments°, endowment 3; largess
endpiece, supplement 1
end (result *or* **product),** outcome; resolution 4; result 1; upshot
endue, clothe 2
endurable, bearable; light[2] 10; supportable 1; tolerable 1
endurance°, bearing 2; existence 2; follow-through; fortitude; gut 3b; patience 2; permanence; perseverance; stamina; strength 6; tolerance 3; vigor; vitality 2
endure°, abide 1; accept 4; bear 2, 4; brook[2]; continue 2; digest 2; exist 1; experience 4; feel 5; go 15, 32a, 35a; hang 7c; hold 20a, 22c; labor 7; last[2] 1, 2; linger 1; live 7; lump[2]; meet[1] 6; persevere; persist 2; put 30; receive 4; remain 3; resist 1; see 10; serve 6; soldier 4; stand 3, 10b, 12b; stick[1] 6, 10; stomach 3; suffer 2; support 3; survive 1; sustain 3; take 6; tolerate 2; undergo; wear 4; weather 3; withstand
enduring°, abiding; bearing 2; classic 2; durable; eternal 3; indelible; inextinguishable; invariable 3; irrevocable; lasting; monumental 1; old 4; perennial 1; permanent 1; perpetual 1; stable 2; steadfast; sturdy 2
enduring, be, have 5
enemy°, adversary 1; antagonist; opponent; opposition 2
energetic°, active 1, 3; alive 3; ambitious 2; animated 1; arduous 2; brisk 2, 4; busy 2; dashing 1; dynamic; eager; emphatic; enterprising; enthusiastic; excited 2; exuberant 1; forceful 1; fresh 5; hard 7; hearty 3; industrious; live 2; lively 1, 2; lusty 1; make 43; nifty 2; nimble 1; passionate 1; powerful 1; quick 3; racy 1; rousing; smart 5; spanking 3; spirited; sprightly; stiff 4; strenuous 2; strong 5; tireless; urgent 2; vigorous; violent 3; vital 3; vivacious; wholehearted
energetically, hard 13; hotly; madly 3; vengeance 2b; vigorously; warmly 4, 5
energize°, activate; animate 2; enliven 1; excite 1; fire 8a; fortify 2; inspire 1; kindle; liven 1; pep 2; pump 4b; quicken 3; refresh 1; spark 2; stir 4; strengthen 2; turn 18a; vitalize; wind[2] 4c
energized, eager; electric; excited 2
energizer, stimulant 2
energizing, animation 2; exciting 1; invigorating; rousing
energy°, action 1; activity 1; agency; ambition 2; animation 1; bounce 2; dash 6; drive 8; dynamism; eagerness 1; effort 1; electricity; enterprise 2; exuberance 1; fire 2; force 1; gumption 2; impetus; industry 2, 3; initiative 2; inspiration 1; intensity; life 7; lust 2; main 6; might 1; momentum; panache; pep 1; power 4, 8, 9; push 14; sinew 2; snap 11; soul 4; sparkle 4; spirit 2; stamina; strength 5; thrust 5; verve; vigor; violence 1; vitality 1, 2
enervate°, cripple 2; demoralize 1; depress 2; exhaust 2; fatigue 2; incapacitate; sap[2]; tire 1; waste 3; weaken 1; wear 6; weary 5
enervated, decrepit 1; exhausted 1; fatigued; feeble 1; helpless 3; impotent 1; jaded 1; lethargic 2; limp[2] 2; listless; low[1] 4; run-down 1; weak 2
enervated, become, waste 2
enervating, exhausting 1; killing 3; toilsome
enervation, decrepitude 1; exhaustion 2; fatigue 1; impotence 1; prostration 4
***en face*,** face 8; opposite 1
enfeeble, destroy 4; enervate; incapacitate; waste 3; weaken 1
enfeebled, broken 3; decrepit 1; delicate 3; feeble 1; helpless 3; infirm 1; pale[1] 3
enfeebled, become, waste 2
enfeeblement, impairment
enfold, embrace 1; envelop 1; fold 2; hold 2; lap[2] 3; muffle 1; roll 9; squeeze 5; wrap 1
enforce°, dispense 2; exact 3; force 7; police 3
enforcement, imposition 1
enforcer, bruiser
enforcing, imposition 1
enfranchise, deliver 3; emancipate; entitle 1; free 14; liberate 1; permit 1
enfranchised, privileged 2
enfranchisement, liberation; permission
enfranchising, liberation
engage, beguile 3; book 4; charter 5; commit 4; contract 2; divert 3; employ 1, 3; encounter 3; enlist 1, 2; fight 1; get 1; grip 7; hire 1, 2; hold 4; immerse 2; interest 7; line[1] 23b; lock[1] 4; mesh 4; occupy 3; order 15; play 3; promise 3; reserve 2; retain 2; sign 10b; take 4, 35a; tie 7a; undertake 2
engage (in), deal 4; embark 2; engage 6; enter 8; follow 6; go 29b; hold 10; participate; play 2; prosecute 3; wage 2
engaged°, absorbed; attached 4; busy 1; deep 4; full 6; interested 1; speak 13
engaged in, wrap 2
engage (in battle), attack 1
(engage in) combat with *or* **against,** war 3
engage in fisticuffs, box[2] 1
engage in high jinks, clown 3
engagement°, action 6; appointment 1; battle 1; bond 2; bout 3; brush[2] 3; clash 2; combat 1; conflict 1; date 2; employment 2; encounter 5; fight 7;

game 2; interest 1; meeting 1; performance 2; placement 2; project 2; promise 1; run 38; skirmish 1; tilt 4; war 1
engagement book, diary
engagements, between, work 6
engaging°, absorbing; attractive; darling 4; delicious 2; delightful 2; employment 2; endearing; entertaining; interesting; inviting; likable; lovable; lovely 1, 2; magnetic; pleasant 2; prepossessing; readable 2; scintillating 2; taking; winning 1
engender, bear 6; breed 2; bring 5; cause 6; create 2; father 5; generate 2; induce 2; lead 6; occasion 5; originate 1; raise 10; rise 17; spawn; work 16
engendered, be, rise 11
engine°, machine 2
engineer°, execute 1; maneuver 3; mastermind 1; negotiate 2; wangle
engineered, factitious
engineering, tactic 2
English, in plain *or* **simple,** honestly 2; straight 14
engrain, instill
engrained, ingrained; inherent; rooted
engrave°, carve 1; chisel 1; etch 1, 2; impress 2; scribe 3; stamp 2
engraved, strong 15
engraving°, cut 24, 25; print 2
engraving, wood *or* **steel,** engraving 1
engross, beguile 3; busy 4; distract 2; engage 2; fascinate; grip 7; hold 4; immerse 2; interest 7; occupy 3; transfix 2
engrossed, absorbed; deep 4; full 6; intent 4; interested 1; meditative; preoccupied 1; rapt 1; thoughtful 2
engrossed, become, bury 3
engrossed in, wrap 2
engrossing, absorbing; interesting; riveting
engulf, cover 6; devour 3; drown 1, 2; envelop 1; flood 7; overwhelm 2; plunge 2; submerge 3; swallow 3
engulfed, be, sink 1
enhance°, become 3; complement 4; compound 3; dignify; elaborate 4; embellish 1, 2; enrich 1; flatter 2; fortify 3; glorify 1; grace 6, 7; heighten 2; improve 1; increase 1; lift 2; maximize 2; ornament 2; polish 2; set 20b; touch 12; whet 2
enhancement, amendment 1; appreciation 2; elaboration 1; elevation 3; embellishment 2; exaggeration; gain 9; improvement 1; increase 3; magnification; ornament 1; refinement 4
enhancing, becoming; flattering 1
enigma°, matter 3; mystery 1; paradox; perplexity 3; problem 2; puzzle 4; riddle[1]
enigmatic(al), ambiguous 2; cryptic 2; dark 6; difficult 2; equivocal 2; hard 3; incomprehensible; inexplicable; intricate 2; labyrinthine; mysterious 1; obscure 2; opaque 2; paradoxical; perplexing; profound 1; puzzling; recondite
enjoin, bid 3; charge 10; command 1; decree 2; direct 3; exact 3; instruct 2; order 14; prescribe; prevent
enjoining, prevention
enjoy°, appreciate 1; command 5; delight 2; dig 3; get 40c; have 7; love 8; possess 1; relish 2; savor 3; wallow 2
enjoy oneself, enjoy 3; luxuriate 1, 2; play 1
enjoyable, agreeable 1; comfortable 3; delicious 2; delightful 1; entertaining; fine[1] 3; glorious 3; lovely 2; pleasant 1; readable 2
enjoyably, nice 6
enjoyment°, amusement 1; appreciation 3; bliss; delectation; delight 3; ecstasy 1; entertainment 1; fun 1; glee; gratification; gusto; happiness; joy 1; love 2; luxury 3; merriment; pleasure 1; recreation; relish 1; satisfaction 1; sport 1
enjoyment in, finding, partial 3
enjoyment out of, take the, overshadow 2
enkindle, enliven 1; fire 8a; inflame 1; spark 2; work 20a
enlarge°, add 4; blow[1] 8c, 8d; build 4; double 6; elaborate 4; enhance; exaggerate; expand 1; extend 2; flare 2; gather 5; grow 1; increase 1; inflate 1; liberalize 1; magnify 1, 2; maximize 1; overstate; progress 6; puff 6; scale[3] 5; spread 6; stretch 2; swell 1, 3; widen
enlarge (on *or* **upon),** amplify 1, 3; develop 1; enlarge 2; expand 4; work 19e
enlarged, swollen
enlargement, bulk 1; development 2; elaboration 2; exaggeration; expansion 1; extension 1; flare 6; growth 1; hump 1; increase 3; magnification; progress 3; progression 2; spread 8; swell 4; swelling
enlarging, expansive 1; spread 8
enlighten°, brief 6; civilize 1; discipline 6; disillusion; educate; illuminate 2; inform 1; teach; tutor 2
enlighten (about *or* **as to),** familiarize
enlightened°, aware 1; brilliant 4; educated 1; informed 1; intelligent; judicious; knowledgeable 1; lettered; rational 2; wise 2
enlightening, edification; educational 2; instructive
enlightenment, civilization 1; culture 1; edification; illumination 2; inspiration 1; light[1] 5; schooling; understanding 2; wisdom 2
enlist°, attach 2; book 4; call 12a; employ 1; engage 1; enroll 1; enter 5; hire 1; induct 2; interest 8; mobilize; recruit 1; sign 10a, 10b; take 35a
enlist (in), join 2
enlisted man *or* **woman,** private 5; soldier 1
enlistee, recruit 2
enlisting, employment 2
enlistment, employment 2
enlistment, period of, tour 3
enliven°, animate 1; brighten 1; cheer 6; energize; excite 1; freshen 2; gladden; inspire 1; intoxicate 2; kindle; liven, 1; pep 2; provoke 1; quicken 3; refresh 1; season 3; spice 4; thrill 3; vitalize; wake[1] 2
enlivened, animated 1
enlivening, animation 2; cheerful 2; exhilarating 1; invigorating; refreshing; rousing; thrilling; tonic 2; vital 4
enlivenment, animation 2; refreshment 2
en masse, completely 1
enmesh, entangle 1; foul 13; implicate 1; mesh 3, 4
enmeshed, foul 10
enmeshed, become, mire 3
enmeshed with, involved 3
enmire, mire 3
enmired, be *or* **become,** stick[1] 7
enmity, animosity; antagonism 1; disgust 3; feud 1; grudge 1; hate 3; hostility 1; ill will; rancor; resentment; strife 2; vendetta
ennoble, dignify; exalt 2; glorify 1, 2; honor 6; idealize; immortalize; lift 2; pedestal 2
ennobled, elevated 2; sublime 1
ennoblement, magnification
ennui, boredom; tedium
ennuyant, bland 2; boring; dreary 2; prosaic; slow 9
ennuyé, blasé 1
enormity°, atrocity 1; outrage 1
enormous°, big 1, 2; boundless; colossal 1; exceeding; excessive 1; extensive 2; gigantic; grand 1; great 1, 2, 11; hefty 3; heroic 4; high 4; huge; immense; infinite; jumbo; large 3; limitless; massive; mighty 3; monolithic; monstrous 3; monumental 2, 4; mortal 5; mountainous 2; outrageous 1; princely 1; prodigious 1; spacious; surpassing; thumping 1; towering 2; vast; voluminous 1; voracious 1; weighty 1; whopping 1
enormously, exceedingly; full 18; vastly
enormousness°, extent 1; magnitude 1; mass 5; size 1
enormous number(s), lot 5c
enough°, adequate 1; ample 4; due 3; fill 11; limit 4a; sufficient; well[1] 1
enough, be, go 21e; suffice
enough, not, inadequate 1; insufficient; little 3
enough for, be, cover 10
enounce, enunciate 1; profess 1; spread 2
en passant, offhand 5; passing 3
enplane, embark 1; get 39a
enquire. See **inquire**
enquiring. See **inquiring**
enquiry. See **inquiry**
enrage°, anger 2; exasperate 1; gall[2] 4; inflame 1; infuriate; irritate; madden 1; outrage 4; provoke 3; wall 3
enraged, angry 1; furious 1; indignant; mad 4
enraged, be, offense 3
enraged, become, blow[1] 8a
enraging, provocative 2
en rapport, sympathetic 2
***en rapport* (with), be,** communicate 3; relate 4; sympathize 2
enrapt, fervent 3
enrapture°, bewitch; captivate; carry 9; charm 5; enchant 2; entrance[2]; fascinate; hypnotize; intoxicate 2; ravish 1; regale; send 4; transfix 2; transport 3
enraptured, ecstatic; fervent 3; infatuated; rapt 2; smitten 2
enrapturing, spellbinding
enrich°, elaborate 4; embellish 1; enhance; fertilize 2; fortify 3; grace 7
enrich oneself, coin 4
enriched, rewarding
enriching, rewarding
enrichment, elaboration 1
enrobe, robe 3
enroll°, book 4; check 7; employ 1; engage 1; enlist 1; enter 5; induct 2; interest 8; list[1] 2; mobilize;

enroll° (*cont.*) recruit 1; register 3; sign 10a, 10b; take 35a
  enroll (in), go 29a; join 2
enrollment, initiation 2
ens, entity 2; existence 3
ensconce, establish 2; seat 7
  ensconce oneself in, occupy 2
ensconced in, be, occupy 2
ensemble°, band[2] 2; clothes; company 2; outfit 2; suit 3; whole 4
enshrine, hallow 1; lift 2; revere; sanctify 1
enshroud, envelop 1, 2; muffle 1; roll 9; secrete[1]; smother 3; wrap 1
ensign, banner 1; color 2a; flag[1] 1; pennant; sign 4; standard 3
enslave°, captivate; oppress 2; subject 10; subjugate; tyrannize
enslavement, bondage; captivity; domination 2; oppression; servitude; slavery 1; subjection; tyranny
ensnare, ambush 2; bag 6; catch 2; decoy 2; entangle 2; foul 13; hook 6; implicate 1; mesh 3; net[1] 2; seduce 1; snare 2; trap 4
ensnared, foul 10
ensnarl, confuse 2, 3; entangle 1, 2; mix 4c
ensnarled in, be, welter 2
ensorcell, enchant 1; fascinate; hypnotize; intoxicate 2; transfix 2
ensorcelled, infatuated; smitten 2
ensorcelling, magic 6
ensorcellment, magic 3
ensue, follow 7
ensuing, posterior 2; subsequent 1
ensure°, assure 1, 3; confirm 2; follow 11b; guarantee 2; ratify; seal 4; see 5; warrant 3
ensured, secure 4
entail°, imply 2; involve 2
entailed, inalienable
entangle°, complicate; confuse 2; entwine; foul 13; implicate 1; incriminate; involve 3; jumble 1; lock[1] 4; mesh 3; mix 4c; muddle 2; snarl[2] 1, 2; tangle 3
entangled, foul 10; intricate 1
  entangled, become, catch 7; mire 3
  entangled, be *or* **become,** stick[1] 7
  entangled in, be, welter 2
  entangled with, afoul; involved 3
entanglement, hitch 4; implication 1; liaison 3; mesh 2; morass 2; snarl[2] 3; tangle 1; tie 8; tie-up 1; web
entelechy, reality 1
entente, pact; treaty; understanding 1
enter°, board 6; chronicle 2; come 3, 13d; debut 2; file 3, 4; get 11, 39a, 39b; go 29a, 32d; join 2; list[1] 2; lodge 6; penetrate 1; perforate; permeate; pile[1] 6; post[3] 3; prefer 2; put 16a; register 3; slip[1] 7; submit 2; wade 3a
  enter, allow to, admit 1
  enter (in *or* **into),** engage 6; enter 8; partake 1; participate
  enter (into) the picture, appear 1
  enter on *or* **upon,** begin 1; commence 1; embark 2; enter 4; inaugurate 1; initiate 1; set 12a; undertake 1
  enter one's mind, hit 5
  enter someone's head, occur 2
entering, admittance; incoming 1, 2
enterprise°, action 1; activity 2; adventure 2; ambition 2; business 4; company 4; concern 7; design 5; drive 8; dynamism; endeavor 2; establishment 2; expedition 1; firm 6; gumption 2; house 4; industry 3; initiative 2; operation 3; project 2; push 14; spirit 2; undertaking 1
enterprising°, ambitious 2; bold 1; go-ahead 2; imaginative 1; make 43
entertain°, allow 2; amuse 1; bear 7; date 7; delight 1; distract 2; divert 3; feast 5; flirt 2; fête 2; have 3; hear 3; host[1] 3; occupy 3; please 1; regale; take 23, 36; tickle; treat 4
entertain doubts, doubt 2
entertainer, fool 2; host[1] 2; player 2
entertaining°, delicious 2; delightful 1; fine[1] 3; funny 1; hilarious; light[2] 11; readable 2
entertainment°, amusement 1, 2; attraction 2; cabaret 2; delectation; dissipation 3; distraction 2; diversion 3; enjoyment 1; festival 2; festivity 2; fun 1; fête 1; interest 5; pastime; play 20, 22; pleasure 1; recital 1; recreation; relaxation 1; show 14; sport 1; wit 2
entertain the idea *or* **notion of,** contemplate 3; think 5
enter the lists against, combat 6
enthrall, captivate; catch 14a; charm 5; enchant 2; enrapture; enslave; entrance[2]; fascinate; grip 7; possess 3; ravish 1; subject 10; subjugate; tyrannize
enthralled, rapt 1; smitten 2
enthralling°, devastating 2; enchanting; glamorous 1; magnetic; piercing 2; ravishing; riveting; spellbinding
enthrallment, domination 2; slavery 1; spell[2] 1; subjection; tyranny
enthrone, crown 4; seat 7
enthuse, pump 4b
enthusiasm°, ambition 2; animation 1; appetite 1; ardor; craze; devotion 3; drive 8; eagerness 1; enterprise 2; excitement 3; exuberance 1; fanaticism 1; fervor; fight 9; fire 2; flame 2; furor 2; glow 3; gusto; heart 4; heat 2; inclination 4; inspiration 1; life 7; panache; passion 1, 3; pep 1; push 14; spirit 2; thirst 1; verve; vigor; warmth 3; zest 2
enthusiast°, addict 2; admirer 1; bug 4; demon 2; devotee; fan; fiend 2; follower 3; freak 4; maniac 2; partisan 1; proponent; supporter 1
enthusiastic°, agog; ambitious 1, 2; animated 1; anxious 2; ardent; burning 2; crazy 4; devoted 1; eager; earnest 2; ebullient; effervescent 2; effusive; emotional 1; enterprising; excited 2; exuberant 1; fanatical; favorable 2; fervent 2; fire 5; glowing 3; good 17; great 9; gushy; hearty 3; hot 3, 4; impassioned; intense 2; intent 5; keen[1] 1; lyrical 2; mad 6; passionate 1; positive 7; ready 2; rhapsodic; ripe 3b; rousing; sanguine; strenuous 2; strong 5; voracious 2; warm 4; wholehearted; wild 9; willing
  enthusiastic about, become, get 40c; warm 8
  enthusiastic for, be, pull 10
enthusiastically, favorably 1; happily 2; highly 1, 3; hotly; mad 5; vengeance 2b; warmly 3, 4; willingly
entice°, attract; catch 11; chat 3; decoy 2; induce 1; lead 9b; lure 1; pull 5; rope 4; seduce 1; solicit 2; tempt 1
enticement°, attraction 2, come-on; decoy 1; draw 16; incentive; incitement 2; inducement; invitation 2; motive 1; temptation 2
enticing, attractive; catching 2; exciting 2; flirtatious; inviting; lovely 1; seductive; temptation 1; tempting 1; voluptuous 2
entire°, clear 14; complete 1, 3; dead 14; full 3, 12; gross 2; intact; one 2; overall; perfect 1; radical 2; round 5; self-contained 3; solid 10; total 2; unabridged 1; undivided 1, 2; whole 1; wholehearted
entirely°, absolutely 2; altogether; blast 4; clean 7; clear 19; cold 11; completely 1, 3; dead 18, 19; deeply 2; directly 4; downright 2; full 13, 17; head 8; hook 4; merely 2; out 7; outright 4; perfectly 1; quite 1, 4; root[1] 3; severely 1; simply 1, 2; stark 1; thoroughly 1; through 7; throughout 3; totally; utterly; vastly; very 1, 2; well[1] 11; wholly 1
  entirely, not, partially
entirety°, amount 3; ensemble 3; integrity 2; totality; whole 4
  entirety, in its, full 17
  entirety, the, everything
entitle°, authorize; call 2; designate 4; enable 1; license 5; name 4; tag 4; term 8; title 6; warrant 4
entitled, privileged 1; welcome 3
  entitled to, be, claim 3; deserve; earn 1; merit 3; rate[1] 7
entitlement, license 1; title 5; warrant 1, 2
entitling, denomination 3
entity°, creature 1; existence 3; life 1; object 1; thing 5; unit
entombment, burial; funeral
entourage, company 2; escort 1; parade 1; retinue; suite 3; train 2
entr'acte, interlude; interval 1; rest[1] 2
entrails, gut 1; inside 2
entrain, embark 1; get 39a
entrance°, admittance; bewitch; captivate; dazzle 1; delight 1; enchant 2; enrapture; entry 1, 2, 3; fascinate; gate 1; grip 7; hypnotize; initiation 2; intoxicate 2; mouth 2; penetration 1; ravish 1; regale; threshold 1; transport 3
  entrance, make an, go 32d
entranced, rapt 1; taken
entrance hall, lobby 1
entrancement, fascination
entrancing, catching 2; enchanting; enthralling; glamorous 1; heavenly 2; intoxicating 2; inviting; magic 6; magnetic; piercing 2; provocative 1; ravishing; scintillating 2; seductive; taking
entrant, candidate; contestant; entry 5
entrants, field 3
entrap, ambush 2; bag 6; catch 2; decoy 2; entangle 1; frame 9; hook 6; implicate 1; mesh 3; net[1] 2; seduce 1; snare 2; trap 4
entreat, appeal 1; ask 3; beg 1; beseech; bid 2;

call 7a, 11a; hit 8; petition 2; plead 2; pray 1; press 6; push 4; request 1; solicit 1; sue 2; urge 2
entreating, supplicant 1; supplication 2
entreaty, appeal 3; desire 4; petition 1; plea 1; prayer 1; request 2; suit 5; supplication 1
entrée, admission 1; admittance; entrance[1] 1; entry 1; mouth 2; recourse 1; road 1
  entrée, give, receive 3
entrench, establish 2; root[1] 6
entrenched°, rooted
entrepreneur, producer 2
entrepreneurial, self-made
entropy, chaos
entrust°, charge 9; commit 1; delegate 3; deposit 2; enable 1; give 1, 16; leave[1] 7; trust 6
entry°, admission 1; admittance; approach 5; entrance[1] 1, 2, 3; hall 1; lobby 1; mouth 2; passage 11; penetration 1; submission 2
entryway, entrance[1] 2; entry 2; hall 1; lobby 1; mouth 2
entwine°, lock[1] 4; loop 2; splice 1; twine 2; twist 1; weave 1
enumerate°, count 1; detail 5; figure 8; itemize; list[1] 2; number 4; poll 3; rattle 5; recapitulate; reckon 2; record 1; recount 2; reel 2; specify; tally 2
enumeration, account 2; litany 2; reckoning 1; rehearsal 2; specification 2; tally 3
enunciate°, deliver 4; go 14; mouth 7; proclaim 1; profess 1; pronounce 1; set 18c; sound[1] 7; speak 4; vent 3; voice 4
enunciated, oral; verbal 1
enunciation, delivery 4; diction 2; expression 1; oratory; pronunciation; speech 1
enuretic, incontinent 3
envelop°, clasp 4; cover 2; enclose 1; fold 2; lap[2] 3; mantle 3; muffle 1; ring[1] 4; roll 9; shroud 1; smother 3; surround 1; swathe; wrap 1
envelope, case[2] 2; wrapper 2
enveloping, around 10
envenom, poison 3
envenomed, embittered; venomous 2
envious°, jaundiced 2; jealous 1; resentful
  envious of, feel, resent
enviousness, envy 1
environment°, atmosphere 2; condition 4; context; element 2; habitat; medium 4; milieu; nature 2; setting; surround 2
environmental, green 3
environmentalist°, green 5
environs, environment; milieu; neighborhood 1; outskirts; precinct 1; setting; surround 2; vicinity
envisage°, conceive 3; design 1; envision; expect 1; fancy 9; fantasize; foresee; form 8; imagine 1; picture 6; plan 4; see 3
envisaging, vision 3
envision°, conceive 3; daydream 2; design 1; envisage 1; expect 1; fancy 9; fantasize; foresee; form 8; imagine 1; picture 6; plan 4; see 3
envisioning, vision 3
envoy°, agent 1; ambassador; delegate 1; deputy; forerunner 1; messenger; minister 2; representative 3
envoy extraordinary, envoy; minister 2
envy°, begrudge 1; grudge 2; resentment
enwrap, coil 1; envelop 1; fold 2; roll 9; swathe; wrap 1
eon, age 4
eons ago, once 1
EP, record 7
ephemeral, brief 1; fleeting; fly-by-night 1; fugitive 3; immaterial 2; meteoric; momentary; mortal 1; passing 1; short-lived; temporary; transient
epic, heroic 5; legend 1; legendary 1; myth 1; poetic 1; romance 3; saga; story[1] 1
epicedium, keen[2] 2; lament 2
epicene, neuter 1
epicure°, epicurean 3; gourmet; sensualist; sybarite
epicurean°, epicure; luscious; luxurious 1; self-indulgent; sensualist; sensuous; sybarite
epidemic°, global; pestilence 1; plague 1; rampant 2; rash[2] 2; universal 1
epidermis, skin 1
epigram°, maxim; moral 3; proverb; quip 1; saw; witticism
epigrammatic°, compact 3; concise; proverbial 1; short 3; succinct; terse 1
epilogue, supplement 1
episcopal, clerical 1
episode°, adventure 1; circumstance 2; event 1; experience 2; fact 2; happening; incident 1; matter 2; page[1] 2; patch 4; scene 2, 3
episodic, periodic
epistemology, philosophy 1
epistle, letter 2; missive; note 2
epithet, expletive 2; label 2; nickname; tag 2; title 2
epitome°, abridgment 3; abstract 3; classic 3; embodiment 2; ideal 1; image 2, 4; model 3; paragon; picture 4; prototype 2; queen 2; quintessence; résumé 1; summary 1; synopsis; type 3; vision 5
epitomization, summary 1; synopsis
epitomize, abbreviate 2; abridge; abstract 4; cut 5; digest 4; embody 3; exemplify 1; personify 1; represent 4; stand 7a; sum 2a; symbolize; typify
epoch, age 5; date 1; day 2; era; generation 2; page[1] 2; period 2; time 4; world 4
epoch-making, monumental 1
equable°, dispassionate 1; even 3, 4; mild 1; philosophical 2; steady 4; temperate 1
equably, well[1] 8
equal°, adequate 1, 3; balance 3; coincide; coordinate 3; duplicate 3; equalize; equivalent 1, 2; even 5, 6; identical 2; level 4, 6; like[1] 7; match 1, 6; measure 15a; one 2; par 4; parallel 2; peer[1] 2; reach 5; reciprocate; rival 2; square 2; symmetrical; tie 4; touch 7; uniform 1; work 19b
  equal, without, alone 2; excellent; matchless; peerless
  equal terms, on, square 2
  equal (to), equal 3; like[1] 1; synonymous; tantamount to
  equal (to), be, measure 15b; rise 12a; tie 4
equal-angled, regular 3
equality°, balance 6; parallel 3; parity 1; symmetry; tie 10
equalize°, balance 2; compensate 2; even 13, 14; neutralize; offset 1
equalizer, offset 2
equally, alike 2
equal-sided, regular 3
equanimity, calm 2; patience 1; philosophy 3; poise 2; presence 5; repose 3; restraint 3; sang-froid; self-control 2; temper 2
equanimous, even 4; philosophical 2; phlegmatic 2
equate, equalize; liken; proportion 5
  equate to *or* **with,** parallel 4
equerry, groom 1
equiangular, regular 3
equilateral, regular 3; square 1
equilibrium, balance 6; poise 1
equip°, accommodate 3; appoint 3; condition 5; cut 16d; dress 2; fit[1] 8; furnish 1, 2; outfit 4; prepare 1; provide 1; qualify 1; ready 10; rig 1; school 4; supply 1; turn 20d
equipage, equipment; outfit 1; paraphernalia; regalia; rig 3; robe 2; tack 4; tackle 1; trappings; turnout 3
equipment°, apparatus; facility 2b; fitting 2; fixture 2; furniture 2; gear 2; hardware 2; implement 1; kit; lavatory; outfit 1; plant 3; provision 1; regalia; rig 3; set 27; stuff 2; tack 4; tackle 1; thing 8c; trappings; turnout 3
equipoise, parallel 3; poise 1
equiponderance, balance 6; parallel 3; poise 1
equipped, qualified 1
  equipped, be, qualify 2
equipping, provision 1; supply 5
equitability, objectivity
equitable°, candid 2; deserved; disinterested; dispassionate 2; even 7; fair[1] 1; honest 4; honorable 3; impartial; impersonal 1; indifferent 2; just 1; nonpartisan 2; objective 1; reasonable 3; rightful 2; square 4; straight 4
equitableness, chivalry; equity; honesty 3; indifference 3; justice 1; objectivity
equitably, fairly 2; honestly 1
equity°, bit 4; honesty 3; justice 1; law 2; probity; quota; right 10; share 2
equivalence, equality 2; parallel 3; parity 1; resemblance
equivalent°, coordinate 3; equal 1, 2, 4; even 5; identical 2; level 4; match 1; matching 1; par 4; parallel 1, 2; similar 1
  equivalent (to), like[1] 1; synonymous; tantamount to; way 10b
  equivalent (to), be, match 6
equivocacy, ambiguity 1
equivocal°, ambiguous 1; evasive; indefinite 2; indistinct 2; mealy-mouthed; obscure 2; opaque 2; paradoxical; vague 2
equivocalness, ambiguity 1
equivocate°, evade 2; fence 4; hesitate 2; oscillate; procrastinate 2; pussyfoot 2; quibble 1; shuffle 3; stall[2] 1; waffle 2
equivocating, evasive; mealy-mouthed
equivocation, ambiguity 2; evasion 2; gobbledegook 2; opacity 2; quibble 2
equivoke, quip 1
equivoque, ambiguity 2; epigram 1; pun; quip 1
era°, age 5; date 1; day 2; generation 2; page[1] 2; period 2; time 4; world 4
eradicate, abolish; bury 2; cancel 2; delete; destroy 1;

evolve into, become 2
evolvement, development 2; evolution; growth 1
evolving, evolution; mutation 1; ongoing 2
evzone, guard 3
ewer, jar[1]; jug
ex, prior 1; former 1; late 3; old 6; outgoing 1
exacerbate, aggravate 1; compound 3; fuel 4; inflame 2; magnify 1; worsen 1
exact°, accurate 1; claim 3; command 4; correct 8; dead 17; deadly 6; definite 1; definitive 3; demand 2; detailed 1; determined 2; distinct 1; elaborate 1; even 8; explicit 1; express 5; extort; faithful 2; flat 9; force 7, 9; formal 3; impose 3; literal 1; mathematical; measured 4; meticulous; narrow 3; nice 3; particular 3; pedantic 2; perfect 5; photographic 1; precise 1, 2, 4; prescribe; proper 2; refined 3; religious 2; right 2; rigid 3; round 5; specific 1; square 3; squeeze 2; strict 1; take 12; tax 3; true 3; verbatim 1; very 3
exact (same), same 1
exacting°, choosy; conscientious 2; demanding 1; exact 2; formal 1; fussy 1; hard 4; meticulous; narrow 3; nice 3; overcritical; painful 3; particular 3; perfectionist 2; precise 3; refined 2; rigid 3; scrupulous 1; selective; severe 1; squeamish 1; strait 1; strict 1; tough 4
exactingly, precisely 2
exaction, toll[2] 2; tribute 2
exactitude, precision 1; resolution 5
exactly°, completely 1; dead 20; directly 4; dot 2; due 5; expressly 1; flat 16; full 14; indeed 1; just 5; letter 4; literally 1; nearly 2; outright 4; particular 6; pat[2] 1; perfectly 3; plumb 4; precisely 1, 2; purpose 4b; right 14; sharp 10; slap 6; true 5; verbatim 2; world 6
exactness, accuracy; formality 2; precision 1; resolution 5; rigor 2
exact retribution *or* **payment** *or* **repayment for,** revenge 2
exaggerate°, amplify 2; blow[1] 8c; camp[2] 2; dramatize; embellish 2; inflate 3; lay[1] 14; magnify 1; make 34a; maximize 2; overdo 1; overrate; overreact; overstate; pile[1] 7; talk 8
exaggerated, camp[2] 1; dramatic 3; exalted 2; extravagant 4; heroic 4; immoderate; inflated 1; lavish 3; melodramatic; outrageous 1; pretentious 1; sensational 2; tall 3; theatrical 2
exaggerating, satirical
exaggeration°, burlesque 1; embellishment 2; romance 5; satire 1; tale 2
exalt°, adore 2; celebrate 3; dignify; enhance; exaggerate; extol; glorify 1; hallow 2; honor 6; idealize; idolize; immortalize; laud; lift 2; pedestal 2; praise 4; promote 2; sanctify 1; worship 1
exaltation, ecstasy 2; elevation 4; exaggeration; glory 1, 2; joy 1; praise 2; promotion 2; rapture; transport 5; tribute 1; worship 2
exaltation (of larks), flight[1] 4
exalted°, dignified; divine 2; elevated 2; eminent 1; glorified 1; grand 2; great 7; high 5; honorable 4; imperial 2; lofty 2, 3; magnificent; majestic 1; rarefied 2; regal; sublime 1
exam, examination 2; quiz 1; test 1
examination°, analysis 1; canvass 4; check 15; consideration 3; discussion; experiment 1; exploration; eye 7; inquiry 1; interpretation 2; interrogation; investigation; observance 3; observation 1; perusal; probe 3; profile 3; quiz 1; reconnaissance; research 1; review 5; scan 3; scrutiny; search 4; study 4; survey 3; test 1; trial 1, 2; view 4
examination, not hold up under *or* **bear,** water 2
examine°, analyze 2; canvass 2; check 4, 10b; consider 1; contemplate 2; discuss; experiment 3; explore 2; eye 9; frisk 2; go 30b, 34a; hunt 2; inquire 1; inspect; interview 3; investigate; look 8, 10; monitor 3; mull; observe 2; peer[2] 1; peruse; pore[1]; probe 1, 2; prove 2; pry 1; pump 2; question 9; quiz 2; ransack 1; reconnoiter; reflect 3; report 6; research 2; review 1; rummage 1; scan 2; scout 1; screen 8; scrutinize; search 1; sift 2; sound[3] 1; speak 10; study 3; survey 1; taste 7; test 2; traverse 4; treat 2; try 2; vet; view 7; watch 3; weigh 2
example°, case[1] 1, 4; copy 2; epitome 1; footstep 2; gauge 3; guide 6; ideal 1; illustration 1; image 4; instance 1; lead 13; lesson 3; manifestation; model 2; monument 2; pattern 5; precedent; prototype 2; sample 1; specimen; standard 1; stroke 7; warning 1
example, as an, example 4; instance 2
example, for, example 4; instance 2; like[1] 6; namely; say 15
exasperate°, aggravate 2; anger 2; annoy 1; displease; frustrate 2; gall[2] 4; get 17; harass; inflame 1; infuriate; irk; irritate; madden 2; pester; plague 3; provoke 3; put 23b; rankle; sour 6; tire 2; trouble 2; wall 3; wear 7
exasperated, angry 1; discontented; disgruntled; huff 1; indignant; mad 4
exasperated, be, umbrage
exasperated by, tired 2
exasperating, irksome; nightmarish; painful 2; provocative 2; tiresome 2; trying; weary 2; wicked 5
exasperation, annoyance 1; displeasure 2; gall[2] 2; indignation; rage 1
exasperation, feel, umbrage
excavate°, burrow 2; deepen 1; dig 1; hollow 8; mine 3; scoop 5; sink 7; tunnel 2; undercut 1; unearth; wash 5
excavation°, burrow 1; hole 1; hollow 7; mine 1; pit[1] 1; quarry[2] 1
exceed°, cap 4; excel; far 7; lead 4; outdo; outstrip; overstep; pass 6; run 33c; shade 8; strain[1] 1; surpass; top 3; transcend
exceeding°, above 4; over 2
exceedingly°, awfully; degree 4b; devil 5; eminently; extreme 9; extremely; fault 7; fearfully 2; fiercely; full 15; madly 4; overly; really 4; surpassingly; terribly; vastly; very 1
exceed (the law), transgress 2
excel°, better 11; cap 4; exceed 1; lead 4; outdo; outstrip; overshadow 1; shine 3; surpass; top 3; transcend
excellence°, class 3; dignity 2; distinction 2; glory 3; grace 4; merit 1; nobility 1; perfection 1; preeminence 2; quality 3; superiority 2; supremacy 1; virtuosity
excellent°, admirable; aesthetic 1; beautiful 2; best 1; capital 6; choice 4; classic 2; desirable 3; divine 2, 3; estimable; exceptional 2; exemplary 2; expert 2; exquisite 1, 5; fine[1] 1, 5, 11; first-rate; gifted; glorious 1, 3, 4; golden 5; good 2; gorgeous 2; grand 5; great 6, 12; heavenly 2; ideal 4; imperial 2; laudable; magnificent; marvelous; masterful 1; mean[2] 6; meritorious; neat 5; nifty 3; optimum 2; outstanding 1; par 3; peerless; perfect 2, 5; polished 1; preeminent 2; prime 2; prize[1] 5; professional 2; rare[1] 2; ripping; select 2; splendid 3; sterling 2; superb; superior 2; superlative; supreme 4; talented; terrific 2; top 8; transcendent; unbeatable; virtuoso 2; world 8
excellent, be, shine 3
excellently, beautifully 2; famously; well[1] 2
except°, exclude 2; exempt 1; leave[1] 10; omit 1; waive 2
except (for), apart 4; bar 10; barring; besides 2; exclusive 4; independent 8; short 10
except (for the fact) that, except 2
excepted, exempt 2
excepting, apart 4; bar 10; barring; besides 2; except 1; exclusive 4; short 10
exception°, disapproval; exclusion 2; exemption; liberty 3; nonconformist 1; objection; omission 1; phenomenon 2; protest 1; provision 2; reservation 2
exception, without, always 1; entirely 1; universally
exception of, with the, except 1
exception (to), take, challenge 1; dispute 1; exception 4; fault 6; issue 9; object 4; protest 3; stick[1] 12
exceptionable°, inadmissible
exceptional°, abnormal 1; brilliant 2; choice 4; conspicuous 3; defective 2; different 2; eccentric 1; exceeding; excellent; extraordinary 1; extreme 1; feebleminded; fine[1] 1; first-rate; freak 5; golden 6; great 6; irregular 3; isolated 1; mean[2] 6; neat 5; noteworthy; noticeable 2; odd 1; optimum 2; ordinary 4; outlandish; outstanding 1; out-of-the-way 2; par 3; particular 2; peculiar 1; phenomenal; prime 2; prodigious 2; psychotic 1; queer 1; rare[1] 1; remarkable 1; select 2; signal 3; singular 2; special 1; splendid 1, 3; sterling 2; strange 1; striking; superb; superior 2; superlative; supernatural; terrific 2; unparalleled; unusual; virtuoso 2; world 8
exceptionally, eminently; especially 1; exceedingly; extra 6; extreme 9; extremely; highly 2;

exceptionally (*cont.*) infrequently; particularly 1, 2; really 4; very 1
excerpt°, clip[2] 5; extract 6; passage 2; quotation 1; quote 1; selection 4
excess°, abundance; balance 7; embarrassment 3; exaggeration; excessive 1; extra 2; extravagance 1; exuberance 2; flood 3; flow 6; give 18; glut 1; great 3; leftover 1; needless 1; prodigality 1; profit 1; profligacy 2; remainder 2; residue; rest[2] 1; satiety; superfluity; superfluous; supplementary 2; surfeit; surplus 1, 2
  excess, do to, meal 2
  excess, in, redundant 1
  excess of, in, above 4; galore; over 2
excess(es), dissipation 1; excess 2
excessive°, breakneck; exceeding; excess 3; exorbitant; extravagant 1; extreme 2, 6; fanatical; great 3; gushy; heavy 2; high 2; immoderate; inordinate 1; intense 1; labored 2; lavish 3; needless 1; outrageous 1; pricey; prodigal 1; profligate 2; profuse 3; prohibitive 2; redundant 1; steep[1] 2; stiff 6; superfluous; surpassing; towering 2; ultra-; unconscionable 2; unreasonable 2; unwarranted
  excessive, not, temperate 1
  excessive price, at an, dear 4; dearly 3
excessively, awfully; devil 5; exceedingly; fault 7; madly 4; overly; sky 2; unduly 1; water 3
excessiveness, extravagance 2; prodigality 1
exchange°, alternate 1; alternation; cash 2; change 1, 5; dealings; discussion; give 2; intercourse 1; market 1; mix 4d, 6; passage 10; reciprocate; redeem 3; return 4; reverse 2; rotate 2; sale 1; sell 1; substitute 1; substitution 1; supplant; switch 2, 4; trade 1, 3, 6; transpose; turn 43; turnabout; volley 3
  exchange for, in, for 3
Exchange, the, exchange 4
exchangeable (with), synonymous
exchange blows, spar[2] 1
exchanged, reciprocal
exchanging, substitution 1
exchequer, purse 2; treasury
excise, custom 2; cut 16a, 16b; duty 3; eliminate 3; tariff 1; tax 1; toll[2] 1; tribute 2
excision, cut 22; omission 1
excitable°, emotional 2; fiery 3; frantic; irritable; nervous 1; quick 5; quick-tempered; restless; sensitive 2; temperamental 1
excitation, excitement 1
excite°, agitate 1; animate 1, 2; arouse 2; awake 2; carry 9; delight 1; electrify 2; energize; exalt 3; ferment 2; fire 8a; flurry 2; flush[1] 3; foment; fuel 4; get 16; heat 4; incite; inflame 1; inspire 1; interest 7; intoxicate 2; kindle; madden 1; motivate; move 5; pump 4b; quicken 3; rouse 2; spark 2; spur 4; stimulate 1; stir 4; thrill 3; touch 6; turn 18c; whet 2; whip 7a; wind[2] 4c; work 20a
excited°, agitated; animated 1; bubbly 2; burning 2; crazy 4; delirious 2; distraught; drunk 2; eager; ebullient; ecstatic; effervescent 2; elated; emotional 1; exalted 3; fervent 1; feverish; fiery 3; fire 5; frantic; heated; hectic; high 8; hot 3, 4; tumultuous; uproarious 1; warm 2; wild 6
  excited, make, work 20a
  excited about *or* **over, become,** warm 8
excitedly, mad 5; madly 3
excitedness, warmth 3
excitement°, agitation 2; alarm 2; animation 1; ecstasy 2; electricity; enthusiasm 1; exuberance 1; flurry 1; flush[1] 6; frenzy 1; furor 1; fuss 1; glow 3; heat 2; incitement 2; pell-mell 3; rampage 1; romance 4; rush 3; sensation 2; sparkle 4; spice 2; splash 4; stir 6; suspense 2; thrill 1; tumult; twitter 4
  excitement, feeling of, buzz 4
excite the curiosity (of), intrigue 1
exciting°, delightful 1, 2; electric; eventful; impressive; incitement 1; interesting; intoxicating 2; juicy 2; moving 1; provocative 1; ripping; scintillating 2; sensational 1; sexy 1; steamy 3; stimulating; stirring; thrilling; wild 6
exclamation°, interjection
exclude°, bar 9; block 4b; disqualify; drop 9; eliminate 1; except 3; exile 3; expel 2; filter 2; forbid; freeze 4; isolate; leave[1] 10; lock[1] 8; omit 1; ostracize; outlaw 2; preclude; rule 8; scratch 3; segregate; shoulder 2; shut 5a, 5b; suspend 3
  exclude from, exclude 1
excluded, isolated 2; outside 9; unwelcome 2
excluding, apart 4; bar 10; barring; besides 2; except 1; exclusive 4; independent 8; omission 1; short 10
exclusion°, exception 1; exemption; immunity 1; omission 1; segregation; suspension 1
exclusionary, one-sided 3
exclusive°, elite 2; high-class 2; inside 4; native 3; one-sided 3; only 1; personal 2; private 2; rarefied 3; scoop 2; select 3; sole; special 4; story[1] 4; undivided 2
  exclusive of, barring; besides 2; except 1; exclusive 4; independent 8; short 10
exclusively, alone 4; custom 4; entirely 2; only 2; specially; wholly 2
excogitate, ponder
excogitative, meditative
excommunicate, banish 1; curse 4; ostracize; punish 2
excommunication, punishment 2
excoriate, berate; pan 5; pare 1; skin 3; strip[2] 1
excoriating, scorching 2
excorticate, shell 3
excrement, dirt 1; dung; filth 1; muck 1; soil[1] 3
excrescence, growth 4; hump 1; lump[1] 2; nub 1; pimple; prominence 3; protrusion; swelling; tumor
excrescent, prominent 3; protuberant
excreta, filth 1; secretion
excrete, defecate; discharge 4; pass 12; secrete[2]; void 7
excreting, discharge 11; secretion
excretion, discharge 11; secretion
excruciate, torment 1
excruciating°, acute 3; agonizing; deadly 5; exquisite 4; heart-rending; painful 1; piercing 4; poignant 1; pungent 3; stiff 2
exculpate, excuse 1; forgive 2; justify; pardon 2; purge 3; purify 2, vindicate 1
exculpation, excuse 5; forgiveness 1; pardon 1; remission 1
excursion°, drive 7; expedition 1; flight[1] 1; journey 1; outing; pilgrimage; ride 4; run 36; spin 7; tour 1; trip 3
  excursion, take *or* **make a,** travel 3
excursions, travel 2
excursus, digression 1; excursion 2
excusable, permissible; venial
excusal, pardon 1
excusatory, mitigating
excuse°, alibi 1, 2; call 15; clear 21; defense 3; evasion 2; except 3; explain 2; explanation 2; forgive 1; ground 3; immunity 1; justify; let[1] 6a; out 15; overlook 2; pardon 1, 2; plea 3; pretense 3; pretext 1; purify 2; rationalize 1; reason 4; refuge 2; remission 1; story[1] 3; subject 3; subterfuge; vindicate 1; warrant 4; whitewash
  excuse from, exempt 1
  excuse (oneself), defecate; urinate
excused, exempt 2
excuse-me-for-living, passive 2
excuse(s), allowance 5
  excuses, make *or* **give,** apologize 2
  excuses for, make, rationalize 1
execrable, abhorrent; abominable 1; atrocious 1; awful 2; bad 1; beastly 2; damnable; deplorable 2; diabolic 2; hateful 1; hideous 2; infamous 2; infernal 2; loathsome; obnoxious; obscene 2; outrageous 2; regrettable; repugnant; repulsive; sordid 1; ugly 2; vicious 1; vile 1; wicked 2
execrate, abhor; blaspheme 1; curse 4; damn 3; detest; dislike 1; hate 1; loathe; swear 2; thunder 3; use 15; vituperate
execration, abuse 8; curse 1; dislike 2; horror 1; phobia; revulsion
execute°, accomplish; achieve 1; administer 2; bump 5; carry 12; discharge 5; dispatch 3, 4; dispense 2; effect 7; effectuate; exercise 1; fill 8; fulfill 1; hang 2; implement 2; kill 1; massacre 2; negotiate 2; obey 2; perfect 8; perform 1; perpetrate; punish 2; put 27a; remove 4; render 1; ride 5b; rub 6; shoot 3; slaughter 3; work 16; wreak
executed, complete 2
execution°, course 3; delivery 4; discharge 12; dispatch 7; exercise 5; fulfillment; killing 1; massacre 1; performance 1; punishment 2; removal 2; rendition 1; slaughter 2; touch 18
executioner, butcher 1; hangman; murderer
executive°, director 1; manager; responsible 3; supervisory
executive(s), management 2

exploratory, experimental 1; tentative 1; trial 6; way-out 2
explore°, check 10b, 12a; dig 5; examine 1; hunt 2; inquire 1; investigate; look 8; plumb 5; probe 1, 2; prospect 5; ransack 1; reconnoiter; report 6; research 2; scan 2; scout 1; search 1; treat 2
explorer, pioneer 1
exploring, inquisitive 1
explosion°, blast 3; boom 3; crash 4; discharge 10; eruption 1; fit[2] 3; gale 2; outburst; paroxysm; pop 4; report 2; storm 2; tantrum; thunder 1
explosive°, bomb 1; inflammatory; live 4; temperamental 1; volatile 3
expo, exhibition; exposition 1; show 13
exponent, advocate 2; party 4; proponent; protagonist 2; supporter 1
expose°, air 7; bare 6, 7; betray 2; blab; comment 4; demonstrate 2; develop 1; dig 6; disclose 2; display 1; excavate 2; exhibit 1; express 2; give 12b; lay[1] 9; let[1] 7a; light[1] 9; open 23; rebut; reflect 2; reveal; show 1, 12a; slip[1] 4; subject 9; turn 25c; unearth; unveil
expose (oneself) to, incur
expose to, expose 3
exposé, exposure 1; news 1; revelation; writing 2
exposed, bare 1; bleak 3; defenseless; insecure 2; liable 3; naked 1, 4; open 2, 5, 12, 18, 19; out 4; public 4; raw 4; unguarded 2; vulnerable
exposed, be, light[1] 10
exposed (to), subject 6
exposed to, not, above 5
expose (to danger), endanger
exposition°, comment 1; demonstration 2; display 4; exhibition; explanation 1; extravaganza; fair[2]; representation 3; scheme 1; show 13; spectacle 1; theme 2
expository, explanatory
expostulate, insist 1
exposure°, aspect 3; experience 1; jeopardy; liability 4; manifestation; peril; revelation
expound, enlarge 2; explain 1; lecture 3; rant 1
expound (on), treat 2
express°, air 7; categorical; couch 2; definite 1; deliver 4; direct 9; dispatch 2; embody 1; emphatic; entire 1; enunciate 1; evident; exhibit 1; explicit 1; fast[1] 1; formal 2; forward 7; give 3; imply 2; indicate 3; luminous 3; manifest 2; mean[1] 2; mouth 7; particular 1; pass 9; phrase 4; positive 1; pronounce 1; put 4; quick 1; raise 11; rapid; register 4; say 4; self-evident; set 18a; shape 8; show 1; signify 1; speak 4; special 4; specific 1; state 7; symbolize; take 24; tell[1] 4; turn 8; vent 3; voice 4; word 10
express differently, reword
express (oneself) in, talk 1
express (one's) gratitude *or* **thanks** *or* **appreciation,** thank 1
express (one's) opinion, piece 12; speak 11b
express approval (of), applaud 1, 2
express (one's) opinion *or* **preference** *or* **desire,** vote 4
expressed, verbal 1
expression°, delivery 4; diction 1; embodiment 1; face 2; front 4; idiom 2; look 14; manifestation; phrase 2; speech 1; statement; term 1; tongue 2; voice 1; word 3, 5
expression to, give, voice 4
expressionless, blank 4; empty 5; glassy 2; vacant 2; vague 6; wooden 2
expressionlessness, emptiness 3
expressive°, dramatic 2; eloquent 1; fluent; knowing 1; meaningful 2; mobile 3; pictorial 1; pregnant 2; rhetorical 1; significant 2; soulful
expressively, well[1] 3
expressiveness, diction 1; expression 4; rhetoric 1
expressly°, completely 3; custom 4; exactly 2; particularly 2; purpose 4b; specially
expressway, road 2
expropriate, appropriate 2; confiscate; deprive; grab 2; help 5; misappropriate 1; preempt; requisition 3; strip[2] 4
expropriated, assumed 1
expulsion°, discharge 9; dismissal 1; ejection 1, 2; eruption 1; eviction; exclusion 3; exile 1; purge 4; removal 3; suspension 1
expunge, abolish; cancel 2; delete; eliminate 3; erase 1; obliterate 1; remove 3; rub 4; scratch 3; strike 5; wash 4
expurgate, edit 2
exquisite°, acute 3; beautiful 2; choice 4; dainty 1; delicate 2; elegant 1; excruciating; fine[1] 1, 5; gorgeous 1; heavenly 2; masterful 1; nice 4; noble 5; perfect 2; piercing 4; precious 3; rare[1] 2; rich 4; stunning 2; subtle 1; superb
exquisitely, perfectly 2; richly 1
exquisiteness, delicacy 1; elegance 2; splendor 1; subtlety 1
exsert, protrude
exsiccate, parch
exsiccation, evaporation 1
exsufflate, exhale
extant, remaining 1
extemporaneous°, impulsive; offhand 3; spontaneous 1; unpremeditated
extemporaneously, offhand 4
extemporary, extemporaneous; unpremeditated
extempore, extemporaneous; offhand 3, 4; scratch 6; spontaneous 1; unpremeditated
extemporize, improvise 1; play 12
extemporized, extemporaneous
extend°, amplify 1; continue 3, 5; draw 14a; enlarge 1; expand 1, 3; gather 5; go 8; hold 16b, 20b, 21b; increase 1; jut; lengthen; liberalize 1; offer 3; open 24; pad 5; pass 2; pay 3; perpetuate; progress 6; prolong; protrude; range 6, 8; reach 1; run 12, 33c; spin 5; spread 3, 4, 5, 6; stand 9b; stick[1] 15; stretch 1; string 11a; supplement 2; tender[2] 1; verge[2]; widen
extend (out), overhang 1; project 5
extend over, cover 6, 7; occupy 4; span 3
extend to, join 4
extendable, expansive 1
extend(ed), ample 6; drag 5; general 3; long[1] 1, 2; open 17; protracted
extendible, expansive 1
extending, expansive 1; progression 2; spread 8
extensible, elastic 1; expansive 1; flexible 1
extensile, elastic 1; flexible 1
extension°, addition 3, 4; expansion 1; flex 1; growth 1; increase 3; overhang 3; progress 3; progression 2; projection 1; ramification 2; respite 2; sprawl 3; spread 8; supplement 1; swell 4
extensive°, ample 1, 2, 3, 6; broad 1, 6; comprehensive; deep 1; encyclopedic; exhaustive; expansive 3; full 2; general 1; global; good 16; great 1; immense; inclusive 1; large 3; limitless; long[1] 1, 2; open 9; panoramic; prevalent 1; prodigious 1; spacious; sweeping 1; thorough 2, 3; unabridged 2; unlimited 2; voluminous 1; wide 1, 2
extensively, abroad 2; amply 1; depth 6; far 4; length 4c; thoroughly 2; widely 1
extensiveness, extension 2; prevalence 1; width 1
extent°, amount 3; area 2, 4; bound[1] 1; breadth 2; deal 6; degree 2; depth 1; diversity 2; end 1; expanse; extension 2; fetch 4; gauge 4; incidence; length 1; limit 1, 2; magnitude 1; measure 1, 8; measurement 2; proportion 4; quantity; range 1; room 1; scope 1; size 1; space 3; span 2; spread 9, 10; stretch 5; sweep 7; width 1
extent, to a considerable, degree 4b
extent, to a great, amply 2; deeply 2; largely; very 1; widely 2
extent, to a large, substantially; widely 2
extent, to a limited, partially
extent, to a slight *or* **minor,** slightly
extent, to some *or* **a certain,** far 8b; moderately; part 9; partially; quasi 1; quite 2; rather 1; relatively; slightly; somewhat
extent, to the fullest *or* **greatest,** best 9; full 18; length 3, 4c; vengeance 2b
extent, to whatever, however 2
extenuate, excuse 3; explain 2; whitewash
extenuating°, mitigating; saving 1
extenuation, excuse 4
extenuatory, saving 1
exterior°, external 1, 2; extrinsic; face 6; front 4; outside 1, 5; outward 1; persona; semblance 1; shell 2; superficial 1; surface 1
exterminate°, butcher 3; destroy 1; eliminate 4; end 10; extinguish 2; finish 4; kill 1; massacre 2; mow 2; murder 3; obliterate 2; purge 2; put 16d; remove 4; root[1] 7a; rout 2; slaughter 3; stamp 4; wipe 3
exterminating, massacre 1
extermination, death 3; destruction 2; end 5; finish 10; holocaust 2; killing 1; loss 7; massacre 1; murder 2; purge 4; removal 2; slaughter 2
exterminator, killer 1
external°, exterior 1, 2; extraneous 2; extrinsic; independent 7; outside 5; outward 1; peripheral 2; skin-deep; superficial 1, 2
externalize, embody 1
externally, ostensibly; outwardly
externals, shell 2; superficies

eye open (for), keep an, watch 4
eye to, with an, concerning; for 2
eye to eye (with), see, agree 3; face 8, 13; sympathize 2
eyeball, eye 1; look 1, 10
eye-catching, prominent 1
eye-filling, voluptuous 2
eyeglasses, glass 7; spectacle 3
eyeless, blind 1
eyelet, loop 1; ring[1] 1
eyeopener, surprise 4; bolt 6; bombshell; drink 5; shock 3
eye-opening, educational 2
eyes, sight 1
eyes, sight for sore, vision 5
eyes at, make, ogle 1; watch 3
eyes (in), up to the, replete 1
eyes on, lay, behold; discover 2
eyes on, set, behold
eyes open, with one's, vigilant
eyes open *or* **peeled** *or* **skinned (for), keep (one's),** watch 4
eyes skinned *or* **peeled, with one's,** vigilant
eyes to, shut (one's), overlook 2
eyes (wide) open, with one's, deliberately
eyeshot, sight 2
(eye)sight, eye 2; sight 1; vision 1
eyesore, fright 3; monster 2; sight 4
eyeteeth for, give one's, desire 1; pant 2
eyewash, drivel 3; fiddlesticks; flannel 1; gab 2; gobbledegook 1; moonshine 2; mumbo jumbo 1; nonsense 1; rot 4; rubbish 2; trash 1; wash 13
eyewitness°, bystander; observer; onlooker; spectator; witness 1
eyot, island
eyrie, nest 1

f.a. *or* **FA,** nothing 1
fab, fabulous 3; good 2; splendid 3
fable, fabrication 3; fantasy 3; invention 3; legend 1; myth 1, 2; parable; pretense 2; story[1] 1; yarn 2
fabled, fabulous 1; legendary 1; mythical 1; nonexistent; notorious 2
fabliau, story[1] 1
fabric°, cloth 1; fiber 2; form 3; frame 1; good 21d; material 1, 2; stuff 1; substance 1; texture; tissue
fabricate°, assemble 3; coin 3; conceive 2; construct 2; craft 5; create 2; distort 2; fake 1; forge 1; form 7; generate 4; invent 2; lie[1] 1; make 1; manufacture 1, 2; model 7; prepare 5; produce 1; put 28a; rear[2] 2; spin 2; turn 20a; weave 3
fabricated, bum 3; factitious; false 2; fictitious 1
fabrication°, assembly 3; falsehood; fantasy 3; fib 1; invention 3; lie[1] 2; manufacture 3; misstatement 1; myth 2; pretense 2; production 1; romance 5; tale 2; yarn 2
fabricator, liar; manufacturer; producer 1
fabulous°, amazing; bully 4; extraordinary 2; fanciful 2; fantastic 4; fictional; glorious 2; good 2; grand 5; heroic 5; legendary 2; marvelous; miraculous; nonexistent; portentous 2; prodigious 2; romantic 1; sensational 3; splendid 3; striking; stunning 2; superb; superhuman 1; supernatural; terrific 2; world 8
fabulously, terribly; well[1] 2
façade, affectation 2; camouflage 1; disguise 4; exterior 3; face 3, 6; facing; front 1, 4, 5; gloss[1] 2; guise 2; mask 2; outside 1, 2; pan 2; persona; pose 5; pretense 1, 2; semblance 1, 2; shell 2; superficies; surface 1; veneer
face°, aspect 2; brave 3; color 2b; dare 2; defy 1; encounter 2; endure 2, 3; experience 4; expression 3; exterior 3; feature 3; front 1, 4, 11; grapple 2; grip 5; likeness 3; line[2] 1; look 3, 14; mouth 5; mug 2; outside 1, 2; overlook 3; paint 3; pan 2; persona; plate 6; right 7; semblance 1, 2; side 2; stand 3; superficies; support 3; surface 1; tackle 3; take 35b; trap 3; turn 2; withstand
face, show one's, face 12; turn 25a
face, to one's, face 13; honestly 2
face down, face 17
face down *or* **downward(s),** prone 1
face of, in the, despite; face 9; notwithstanding 2
face of it, on the, face 11; presumably; seemingly
face *or* **faces, make a,** face 10; mug 4
face to face, face 8, 13; tête-à-tête 2
face up to, face 18; tackle 3
face value, at, seriously 3
faced with, be, encounter 2
face-off, showdown
facer, problem 1
face ruin, wall 4
facet, aspect 4; feature 1; phase 4; point 10; side 2
facetious, comic 1; flippant 1; funny 1; humorous; ludicrous; playful 2; witty
facetiously, fun 3; tongue 5
facetiousness, flippancy 1; humor 1; levity; wit 2
facile, eloquent 1; flashy 2; flexible 3; fluent; glib; graceful 1; mobile 3; quick 4; smooth 6; versatile 1
facilitate°, aid 1; assist 2; ease 8; enable 2; expedite 2; favor 8; help 2; make 30c; pave 2; precipitate 1; smooth 10; speed 2
facilitative, favorable 1
facilities°, facility 2; passage 7; toilet 1
facility°, ability 1; accommodation 4; aptitude 2; dexterity 1; ease 2; finesse 1; flair 1; fluency; freedom 4; gift 2; knack; proficiency; prowess 1; readiness 2; skill 1; talent 1; technique 2; toilet 1; touch 16
facing°, exterior 3; face 6, 8; front 1; opposite 1; outside 1
façon de parler, idiom 1; parlance; phrase 3; speech 3; talk 19; tongue 1
facsimile°, copy 1; double 7; duplicate 2; echo 2; image 2; imitation 4; knockoff; likeness 2; match 1; model 1, 10; picture 2; print 2; replica; reproduction 2
fact°, certainty 1; detail 1; existence 1; factor 1; given 4; gospel; incident 1; matter 2; object 1; particular 5; phenomenon 1; reality 1; right 10; rule 3; truth 1
fact, in, actually; fact 4; literally 2; really 3; substantially; truly 1; truth 3
fact, in (point of) (actual), actually; effect 5; fact 4; indeed 1; really 1
fact, the, case[1] 8
fact-finding, research 1
faction°, cabal 2; camp[1] 2; clan 2; cohort 2; crowd 2; front 6; group 1; junta; machine 3; movement 5; part 5; party 3; persuasion 2; ring[1] 3; school 2, 3; sect 2; set 26; side 3; world 3
factional, partisan 3; political 2; sectarian 1
factious°, controversial 2; political 2
factitious°, artificial 3; counterfeit 2; fake 5; false 3; imitation 5; phony 1; spurious
fact of life, fact 2
factor°, agent 1, 2; detail 1; element 1; index 4; ingredient; instrument 2; part 2; proxy; thing 2
factor(s), fact 3
factory°, mill 2; plant 2; works 1
factotum, servant 1
fact(s), data; dope 3; evidence 1; fact 3; information; intelligence 2; knowledge 2; literature 2; material 4; proof 1; story[1] 6; truth 2; word 2; good 21c
facts, the, lowdown
factual°, actual 1; authentic; authoritative 2; correct 8; historical; matter-of-fact; right 2; true 1; truthful
factually, fact 4
facultative, favorable 1; optional
faculties, ability 3
faculty°, ability 1, 3; aptitude 2; art 4; capability; capacity 2; flair 1; genius 3; head 4; ingenuity; instinct; power 1, 5; sense 1; skill 2; talent 1; technique 2
fad°, bug 3; craze; enthusiasm 2; furor 2; mania 1; mode[2]; rage 3; rave 4; style 2; thing 3; trend 2; vogue 1
fad, the, fashion 2
faddist, bug 4
fade°, bleach 1; decline 2; diminish 3; dwindle; flag[2] 2; go 12; melt 3, 4; recede 2; relapse 2; taper 2; waste 2; weaken 2
fade (away *or* **out),** die 2; disappear 1; dissolve 1; ebb 2; evaporate 2; film 4; go 33a; pale[1] 5; pass 10; peter out; sink 5; trail 8; wane 1
faded, dingy; faint 1; light[2] 3; shabby 1; washed-out 1; weak 6; wizened
fading, decay 3; moribund 1; obsolescent; relapse 4; wane 3, 4
fading (away), evaporation 2; passing 1; frail 2
fading (fast), dying
faeces. See **feces**
fag°, deviant 3; drudgery; exhaust 2; hack[2] 2; homosexual 1; menial 3; slave 2
fag (end), heel[1] 1
fag (out), fag 1; fatigue 2; kill 4; tire 1
fag end, rear[1] 1; stub 1
fagged, beat 13; jaded 1
fagged (out), lethargic 2; prostrate 5; ragged 3; spent 1; tired 1; washed-out 2; weary 1; exhausted 1
faggot, deviant 3; homosexual 1
faggoty, effeminate
faggy, effeminate; homosexual 2; swish 4
fail°, back 18; betray 1; collapse 2; decline 4; default 3; die 2;

fail° (*cont.*) disappoint 1; fall 11b, 12, 19, 20; fizzle 2; flag[2] 2; flop 3; fold 3; founder[2] 2; give 9, 15d; go 12, 37b; grief 3; lag 2; let[1] 4; miscarry; misfire 1; omit 2; pack 8b; peter out; relapse 2; run 26c; short 13; sicken 1; sink 4; stall[1] 1; strike 17; wall 4; way 11a; weaken 2; worsen 2; wrong 8a, 8b
fail, not, come 17b; serve 5
fail, without, mean[3] 2b
fail (in), neglect 2
failed, disappointed 1; washed-up
failing°, bug 6; decay 3; decline 7; defect 2; delinquent 2; dying; failure 1; fault 1; flaw 1; foible; imperfection; infirm 1; lapse 1; leg 7; miscarriage; moribund 1; neglect 4; negligence; relapse 4; senile; short 10; twist 10; vice 2; weakness 3
fail to keep an appointment with, stand 10c
failure°, breakdown 1; catastrophe 2; collapse 6; crash 5; default 1; defeat 4; defect 1; disappointment 1; dud 1; fall 24; fault 2; fiasco; flop 4; loser; loss 5; miscarriage; misfire 2; miss[1] 5; naught; neglect 4; negligence; noncompliance; omission 2; oversight 1; relapse 4; ruin 1; shortcoming; slump 1; washout
faineance, indolence; sloth; sluggishness; torpor
fainéant, idle 3; idler; indolent; lackadaisical 1; lazy 1; lethargic 1; shiftless; slack 1; slothful; torpid
faint°, cold 6; collapse 3; delicate 6; dim 1; dizzy 1; dreamy 1; fainthearted 1; feeble 1, 3; fuzzy 2; giddy 1; groggy; hazy 2; helpless 3; inaudible; indistinct 1; light[2] 3, 4; nebulous; obscure 1; outside 8; pale[1] 2; pass 20a; queer 3; remote 5, 6; shadowy 2; soft 4; subtle 2; weak 6
faint, grow, fade 1; trail 8
fainthearted°, afraid 1; cowardly; tame 3
faintheartedness, cowardice
faintness, obscurity 1
fair°, adequate 2; beautiful 1; bright 2; candid 2; clear 1; comely; conscientious 1; considerable 1; decent 3, 4; deserved; disinterested; dispassionate 2; equitable; ethical; exhibition; favorable 1; fine[1] 2, 10; fresh 6; fête 1; good 1, 18; handsome 1; holiday 2; honest 2, 4; honorable 1; impartial; impersonal 1; indifferent 2, 3; ingenuous 1; judicial 3; just 1; legitimate 3; liberal 3; lovely 1; mediocre; mild 2; moderate 3; nonpartisan 2; objective 1; OK 3; open 15; ordinary 2; pleasant 3; pretty 1; reasonable 3; respectable 2; right 1; righteous 1; rightful 2; satisfactory; show 13; straight 4; sunny 1; tidy 3; tolerable 2; tolerant; unprejudiced; virtuous 1
fair, become, clear 22
fair, consider, approve 3
fair, not, foul 11
fair, to be, right 12
fair (enough), passable 1
fair (to middling), adequate 2; decent 3; mediocre; moderate 3; par 6; passable 1; so-so
fair with, be *or* **play,** level 11
fair (and square), equitable; even 7; honorable 3; square 4
fair exchange, give-and-take
fair game, dupe 1; sucker; victim 2
fair-haired, fair[1] 3; golden 6
fair-haired boy *or* **girl,** darling 2; favorite 1; pet[1] 1
fairly°, honestly 1; moderately; pretty 3; quite 2; rather 1; right 12; somewhat; well[1] 14
fairly good, passable 1
fair-minded, equitable; just 1; unprejudiced; virtuous 1
fair-mindedness, equity; justice 1; objectivity
fairness, candor 2; chivalry; conscience; equality 3; equity; honesty 3; honor 1; indifference 3; justice 1; morality 1; objectivity; probity; right 10; sportsmanship; virtue 1
fairness, in, right 12
fair play, equity; justice 1; morality 1; sportsmanship
fair-skinned, white 4
fairy, deviant 3; homosexual 1; imp
fairy-tale, fabulous 1; fancied; fanciful 2; mythical 1; romantic 1
fairy tale *or* **story,** fabrication 3; pretense 2; romance 5; story[1] 1
fait accompli, fact 2
faith°, assurance 2; belief 1; certainty 2; confidence 1; credit 1; hope 2; persuasion 2; reliance; religion; trust 1
faith, bad, betrayal 1
faith, in good, bona fide; honestly 1
faith (in), have, believe 2; credit 5; store 5; trust 5
faith in, place one's, credit 5
faith in(to), put, believe 1; figure 11a; go 26b
faith with, keep, redeem 6
faithful°, authentic; authoritative 2; constant 1; devoted 1; devout 1; dutiful; exact 1; factual 2; fast[1] 5; godly; great 10; holy 2; lifelike; literal 1; loyal; noble 4; photographic 1; pious 1; precise 1; religious 2; special 5; stalwart 4; staunch 1; steadfast; steady 5; strict 1; true 1, 2; trustworthy; truthful; verbatim 1; virtuous 2; yeomanly
faithful to, be, redeem 6; stick[1] 13
faithful to, be *or* **remain,** stand 5a
faithfully, consistently 2; exactly 1; literally 1; verbatim 2
faithfulness, attachment 3; credit 3; dedication 1; duty 2; faith 3; loyalty; precision 1
faithless°, dishonorable 2; disloyal; fickle; insincere; perfidious; promiscuous 2; slippery 2; traitorous; unscrupulous; untrue 1; villainous 1
faithlessness, betrayal 1; inconstancy; infidelity 1; perfidy
fake°, act 3, 8; affected 2; affect[2] 1; artificial 2; assume 4; assumed 2; base[2] 5; bogus; cheat 1; counterfeit 1, 2, 3, 5; deceptive 2; disguise 2; dishonest; dissimulate; duff; fabricate 3; fabrication 3; false 3; falsify; fool 5; forge 3; forgery 2; fraud 2; fraudulent 1; game 6; glorified 2; hypocritical; imitation 3, 5; invention 3; let[1] 7b; mimic 5; mock 3; phony 1, 2; plastic 3; play 10; poseur; pretend 3; pretended; quack 2; quasi 2; rig 2; sham 1, 2; spurious; stage 5; synthetic; theatrical 2; toy 6
fake (down), coil 1
fake it, improvise 1
fake(d), affected 2; artificial 3; counterfeit 2; factitious; false 4; forced; unreal 3
fake(r), cheat 1; fake 4; fraud 3; hypocrite; operator 3; phony 3; poseur; quack 1
fakery, hypocrisy; masquerade 2; pretense 1
faking, hypocritical; pretense 1
falderal, frill 2
fall°, crash 1; descend 2; dip 2; drop 3, 6; err 2; floor 4; flow 2; founder[2] 3; get 6; go 27c; hang 3; lapse 3, 4; overthrow 2; overturn 3; plump[2] 1; plunge 3; rain 5; ruin 1; sag 2, 3; settle 9, 11; shower 3; sink 3; slide 3; slip[1] 2; slope 1, 2; slump 1; spill 5; tackle 2; trip 1; tumble 4; undoing 1
fall, let, shed[2] 2
fall all over, truckle
fall apart, collapse 4; decompose 1; deteriorate 2; disintegrate; fall 7; give 9; grief 3; panic 3; piece 6; separate 1; split 5
fall asleep, nod 3
fall away, fall 4
fall back, fall 8; pull 8b; recede 1; relapse 1; retreat 4
fall back on *or* **upon,** draw 13a; fall 9; resort 3
fall back *or* **away,** ebb 1
fall behind, fall 10; lag 1; trail 6
fall (down), fall 11; flop 1; trip 5; tumble 1; way 11a
fall (down *or* **in** *or* **apart),** collapse 1
fall flat, collapse 2; fail 1; fall 12; flop 3
fall for, fall 13; lap[1] 3b; swallow 2
fall (headlong), pitch[1] 3; plunge 1
fall in, fall 14; line[1] 23c; sink 2
fall in love with, fall 13a
fall in *or* **inward(s),** cave 2a
fall in with, fall 15; fraternize
fall off, decline 3; drop 11
fall (off *or* **away),** decline 2; decrease 1; fall 3, 16; flag[2] 2; lag 2; remit 2; slump 2; suffer 4; taper 2
fall on *or* **upon,** fall 17; lace 5a; light[2] 14; make 30b; pounce 1; set 21, 24; tackle 4; visit 2
fall out (over), differ 2; disagree 2; fall 18; feud 2; fight 4; quarrel 2; row[2] 3
fall (over *or* **down),** topple 3
fall short (of), default 3; fail 1; fall 19; short 13
fall through, collapse 2; fail 1; fall 20; fizzle 2; founder[2] 2; miscarry; misfire 1; wrong 8b
fall to, fall 21
fall together, coincide
fall to *or* **on one's knees,** prostrate 1
fall upon, assault 3; attack 1
fallacious, amiss 1; circular 3; deceptive 2; erroneous; false 2, 5; illusory; inaccurate; incorrect; inexact; misguided; mistaken 2; sophistic; specious; unsound 4; wrong 2
fallacy°, absurdity 2; delusion 2; hole 6; illusion 1
fallen, lost 5, 7
fallen woman, prostitute 1; slattern; tart[2] 2

fall (from grace), sin 3; transgress 1; transgression; wrong 8a
fall guy, dupe 1; fool 3; plaything 2; prey 2; sap[1] 2; scapegoat; score 7; sucker; underdog; victim 2
fall heir to, get 1; inherit
fallibility, frailty 2; imperfection
fallible, human 2
fall ill *or* **sick (with),** get 5; sicken 1
falling, downgrade 4; fall 22
  falling apart, dilapidated; leg 7; shaky 2; tumble-down
  falling back, relapse 3; retreat 1
  falling down, shaky 2
  falling off, decadent 1; decline 6; decrease 2; downgrade 4
falling-off, slump 1
falling-out, argument 1; feud 1; misunderstanding 2; row[2] 1; variance 2
falling star, star 1
falloff, decline 6; drop 4; slack 5; slump 1
fallout, product 1; repercussion; upshot
fall(s), fall 25; waterfall
false°, affected 2; artificial 2; assumed 2; bogus; bum 3; counterfeit 2; deceitful; deceptive 1, 2; disingenuous; disloyal; double 5; duff; erroneous; factitious; faithless 2; fake 5; fictitious 2; forced; fraudulent 1; glossy 2; hollow 4; hypocritical; illusory; improper 1; inaccurate; incorrect; inexact; insincere; insubstantial 2; invalid[2]; lying 2; mistaken 2; mock 3; oblique 2; outward 1; perfidious; phony 1; pretended; seeming; sham 2; slippery 2; sophistic; spurious; synthetic; theatrical 2; unnatural 2; unreal 3; unscrupulous; untrue 1, 2; wily; wrong 2
  false to, be *or* **prove,** betray 1
  false witness, bearing, perjury
false appearance, gloss[1] 2
false colors, mask 2
false display, affectation 2
false face, mask 1
false flag, mask 2
false front, masquerade 2; outside 2
falsehood°, fabrication 3; fib 1; invention 3; libel 2; lie[1] 2; misstatement 1; myth 2; story[1] 3; tale 2
false impression, delusion 2; misunderstanding 1
falsely, seemingly
false move, feint
falseness, falsity 1; hypocrisy; impropriety 1; infidelity 1; perfidy
false notion *or* **idea,** misconception
false scent, misinformation
false step, misstep 1; mistake 2; trip 1, 2
false trail, misinformation
false witness, liar
falsification, fabrication 3; forgery 1; invention 3; lie[1] 2; lying 1; misstatement 1; perversion 1; pretense 2; tale 2; twist 9
falsified, factitious; fraudulent 1; unreal 3
falsifier, liar
falsify°, adulterate; camouflage 2; color 4, 6; counterfeit 4; disguise 2; distort 2; doctor 4; fabricate 3; fake 1; fib 2; fiddle 1; forge 3; garble 1; juggle; manipulate 3; misrepresent; pervert 1; rig 2; twist 2
falsity°, impropriety 1; perfidy
falter, fail 1; flag[2] 2; founder[2] 2; hang 5; hesitate 3; hobble 1; lag 2; limp[1] 1, 2; pause 1; reel 1; sag 2; scruple 2; shilly-shally; stammer 1; stumble 1, 2; totter; trip 5; wallow 3; wrong 8a
faltering, doddering; faint 1; halting; hesitant 2; inarticulate 2; infirm 2; irresolute; limp[1] 2; sag 3; shambling; tremulous 1; weak 6
fame°, celebrity 1; distinction 2; glory 1; honor 2; kudos; laurels; luster 2; name 2; popularity; prestige; prominence 1; renown
famed, celebrated; exalted 1; famous; glorious 1; great 5; honorable 4; illustrious; legendary 3; notable 1; noted; outstanding 1; prestigious; prominent 2; renowned; star 3; successful 3; well-known 2
familial, domestic 1
familiar°, accustomed 1; chum 1; close 15; common 1; everyday 2; frequent 1; friend 1; friendly 1; general 2; homely 1; informal 1; intimate[1] 1, 5; knowledgeable 1; old 8; ordinary 1; outgoing 2; personal 3; regular 1, 7; routine 3; stale 2; standard 6; usual; vernacular 2; well-known 1
  familiar, make, familiarize
  familiar with, acquainted 2; familiar 4; home 5; used 3; versed
  familiar with, be, experience 4; know 1; read 2
  familiar with each other *or* **one another,** acquainted 1
familiar face, steady 10
familiarity°, acquaintance 1; experience 1; exposure 3; fellowship 3; friendship 1; knowledge 3
  familiarity (with), understanding 3
familiarization, orientation 2
familiarize°, accustom; orient 3
  familiarize with, acquaint; familiarize; teach
familiarized, seasoned 1
familiarly, well[1] 13
family°, background 1; birth 3; breed 1; brood 1; clan 1, 2; connection 4; descendant; domestic 1; dynasty; flesh 5; home 8; house 2; kin 1; lineage 2; line[1] 15; parentage; paternity; pedigree; people 2; race[2] 1, 2; root[1] 4; sort 3; stock 2; strain[2] 1; syndicate 2; tribe
  family way, in a *or* **the,** expecting; pregnant 1
  family way, unmarried *or* **unwed and in a,** trouble 9b
family member, relation 3
family tree, family 3; lineage 1; root[1] 4
famine°, hunger 1
famished°, hollow 3; hungry 1; ravenous 1; starved 1
famous°, big 7; brilliant 3; celebrated; distinguished 1; eminent 1; exalted 1; glorious 1; great 5; honorable 4; illustrious; immortal 3; legendary 3; lofty 2; notable 1; noted; notorious 2; outstanding 1; popular 1; prestigious; prominent 2; public 6; renowned; splendid 2; star 3; successful 3; well-known 2
  famous for, be, reputation 2
famously°, well[1] 2
famousness, celebrity 1
fan°, addict 2; admirer 1; awake 2; bug 4; devotee; disciple 2; enthusiast; fiend 2; follower 3; freak 4; inflame 2; maniac 2; partisan 1; supporter 1
  fan of, be a, follow 9
  fan out, spread 3
fanatic°, bug 4; crank 2; demon 2; enthusiast; fanatical; fiend 2; freak 4; maniac 2; partisan 1; purist; puritan 1; radical 4; sectarian 4; supremacist; zealot
fanatic(al), enthusiastic; fervent 1; intense 2; mad 6; passionate 1; rabid 1; radical 3; sectarian 2; ultra-
fanatically, madly 3
fanaticism°, devotion 3; enthusiasm 1; passion 1; zealotry
fancied°, dreamlike; fanciful 2; fictitious 1; ideal 5; illusory; imaginary; insubstantial 2; nonexistent; romantic 1; unreal 1
fanciful°, capricious; dreamlike; extravagant 2; fabulous 1; fancied; fancy 1, 2; fantastic 1, 2; fickle; fictional; fictitious 1; flighty 1; harebrained 1; ideal 5; illusory; imaginary; imaginative 2; improbable; insubstantial 2; legendary 2; mythical 2; nonexistent; quaint 1; quixotic; romantic 1; unreal 1, 2; unrealistic 1; visionary 1; whimsical 1
fancy°, adore 3; assume 3; believe 3; care 6b; conceit 2; daydream 1, 2; desire 1; device 4; dream 1, 2; dressy 1; elaborate 2; elegant 3; enjoy 1; envisage 1; fad; fantasy 1, 2; favor 6; feel 9; flowery; freak 3; fussy 2; go 28c, 29b; guess 2; hanker; high-class 2; hope 1; idea 2; illusion 1; imagination 1; imagine 2; impression 1; inclination 4; intricate 1; kink 4; liking 2; longing; long[2]; love 1; luxuriant 3; luxurious 1; make 28; mind 5; notion 2; ornate; partiality 2; passion 3; phantom 2; picture 6; pleasure 2; prefer 1; preference 2; presume 1; quirk; reckon 3; relish 1, 2; shine 5; showy; suppose 1; supposition; surmise 1, 2; suspect 2; swanky; taste 3; thing 4; thirst 1, 2; upper-class 2; urge 5; wallow 2; want 1, 4; weakness 4; wish 1, 5; yearn
  fancy for, having a, fond 3
  fancy to, take a, prefer 1; shine 5
fancy Dan, swell 6
fancy dresser, dude 1
fancying, fond 3
fancy man *or* **woman,** paramour
Fancy that!, indeed 3
fanfare°, flourish 4; promotion 4; propaganda 2
fanfaron, fanfare 1
fanfaronade, fanfare 1
fan letter, note 2
fanny, bum 1; seat 4
fanny adams, nothing 1
fantabulous, bully 4; fabulous 3; good 2; gorgeous 2; grand 5; great 12; marvelous; splendid 3; superb
fantasies, not in one's wildest, mean[3] 5
fantasize°, daydream 2; make 28; think 1
fantasized, romantic 1
fantasizer, dreamer
fantast, visionary 2
fantastic°, bizarre 1, 2; bully 4; divine 3; dreamlike; excellent;

fantastic° (*cont.*) extraordinary 2; fabulous 2; fanciful 2; fancy 2; far-fetched; formidable 2; good 2; gorgeous 2; grand 5; great 12; grotesque 2; heavenly 2; imaginative 2; insubstantial 2; kinky 1; marvelous; miraculous; odd 1; outlandish; phenomenal; portentous 2; prodigious 2; quaint 1; quixotic; sensational 3; splendid 3; strange 1; superb; superhuman 1; superlative; terrific 2; unreal 1; whimsical 1; wild 7; world 8
fantasy°, daydream 1; dream 1; extravagance 2; fancy 6; hallucination; idea 5; ideal 5; illusion 2; invention 3; reverie; romance 3, 4, 5
fan the flames (of), foment
far°, cold 8; distant 1; well[1] 7
far, by, easily 2; far 3; preeminently
far apart (from), unlike 1
far as one can see, as, evidently 2
far as one can tell, as, evidently 2
far as one is concerned, as, personally 2
far as (someone) is concerned, as, part 10
far as (something *or* someone) is concerned, as, for 12
far away, far 1; sight 6a
far (away) from, not, about 6; close 20; near 1, 9; nearby 1
far down, deep 10
far (from), unlike 1
far off, far 1; off 2
far off, not, near 1; nearby 1; neighborhood 2
far up, high 14
far and away, easily 2; far 3; preeminently; well[1] 7
far and wide°, about 3, 4; abroad 2; around 5; everywhere; far 4; pole[2] 2
faraway°, distant 1; dreamy 2; extreme 3; outlying; preoccupied 2; remote 1; secluded 2; far 9
farce, charade; humor 2; joke 3; mockery 2
farceur *or* **farceuse**, fool 2; panic 2; wit 3
farcical°, absurd 1; funny 1; humorous; hysterical 2; impossible 2; ludicrous; ridiculous; witty
far cry, a, cry 8
fardel, bundle 1
fare°, come 6; diet[1] 1; fee 1; get 28b; hold 22c; live 9; make 37f; passenger; table 1
fare well, feast 4; flourish 1; pan 6; prosper
farewell°, goodbye; parting 2
farewell to, say, kiss 3
far-fetched°, exquisite 2; fancy 2; fishy 1; implausible; improbable; incredible 1; out-of-the-way 2; remote 3; tall 3; unreasonable 1
far-flung, extensive 1; faraway 1; outlying
far gone°, far 5
farm°, cultivate 1; raise 3; spread 13; till[1]; work 8
farm (out), farm 4; job 6; let[1] 3; rent[1] 1
farmer°, peasant; producer 1
farmhouse, farm 1
farming°, pastoral 2
farmland, farm 1
farmstead, farm 1
farm-toun, farm 1
farmyard, enclosure 1
far-off, distant 1; extreme 3; far 9; faraway 1; out 10; outlying; remote 1; secluded 2
far-out, avant-garde; eccentric 1; exciting 1; extraordinary 2; extreme 5; fabulous 3; incredible 2; marvelous; mean[2] 6; miraculous; neat 5; obscure 5; offbeat; outlandish; peculiar 1; ridiculous; sensational 3; sick 5; singular 1; splendid 3; strange 1; superb; way-out 1, 2; weird; world 8
farrago, assortment 2; clutter 1; hash 1; hodgepodge; medley; mess 2; mishmash; mixture 1; pastiche
far-ranging, extensive 1; panoramic; sweeping 1
far-reaching, exhaustive; expansive 3; extensive 1; global; panoramic; unlimited 1; wide 2
far-removed, remote 8
farseeing, farsighted 1
farsighted°, politic 2; provident 1; quick 4; sound[2] 4
farsightedness, foresight 1; forethought; precaution 2; providence 1; prudence 2; vision 2
fart, wind[1] 3, 5
farther, far 9; further 2, 4
farther down, below 1; lower[1] 6
farthest, extreme 3; far 9; outlying; outside 6; ultimate 4
fascinate°, appeal 2; attract; beguile 3; bewitch; captivate; catch 11; charm 5; dazzle 1; delight 1; enchant 2; enrapture; entrance[2]; fix 5; grip 7; hypnotize; interest 7; intoxicate 2; intrigue 1; pull 5; ravish 1; regale; transfix 2; transport 3
fascinated (by), infatuated; interested 1; rapt 1
fascinating, absorbing; catching 2; delicious 2; delightful 2; enchanting; enthralling; glamorous 1; interesting; intoxicating 2; inviting; juicy 2; magic 6; piercing 2; prepossessing; provocative 1; riveting; scintillating 2; seductive; sexy 1; spellbinding; taking
fascination°, appeal 4; charm 2; enchantment 2; glamour; magic 3; mania 1; passion 3; romance 4; spell[2] 1, 3; wonder 2
fascism, tyranny
fascist(ic), repressive; totalitarian; tyrannical
fashion°, adapt 2; beat 5; carve 1; chic 2; construct 2; craft 5; craze; create 2; custom 1; design 2; fabricate 1; fad; fit[1] 7; forge 1; form 4, 7; frame 7; generate 4; genre; make 1; manner 1; model 6, 7; mode[1] 1; mode[2]; mold[1] 4; prepare 5; rage 3; rave 4; school 2; sculpture 2; shape 6; stamp 8; style 1, 2, 8; tailor 2; taste 5; technique 1; thing 3; trend 2; turn 8; vein 5; vogue 1; way 1; work 13
fashion, going *or* passing out of, obsolescent
fashion, in, fashion 4; fashionable; minute[1] 3; modern; popular 1; stylish
fashion, out of, obsolete
fashionable°, becoming; chic 1; contemporary 2; current 3; dapper; dashing 2; date 5; elegant 2; exclusive 2; glamorous 2; latest 2; minute[1] 3; modern; new 2; now 5; popular 1; posh; rakish; sharp 9; smart 3; snappy 2; sporty; stylish; swanky; swell 7; swinging; swish 3; trendy 1
fashionable, become, catch 13b
fashionableness, style 4; vogue 2
fashionably, beautifully 1
fashion industry, rag[1] 4
fashioning, composition 4; fabrication 1
fashion plate, swell 6
fast°, abiding; bound[3] 1; close 10, 15; diet[1] 3; durable; expeditious; express 7; firm 2; firmly 1; fleet[2]; great 10; hasty 1; immovable 1; invariable 1; loose 7; persistent 1; precipitate 3; promiscuous 2; prompt 1; quick 1; quickly 1, 2; rapid; rapidly 1; rooted; secure 2; spanking 3; speedy 1, 3; stable 1; swift; swiftly; tight 1; true 2
fast, be *or* become, stick[1] 7
fast (to), make, bind 1; bolt 11; clamp 2; fix 1; lash[2]; moor[2]; peg 3; secure 8; tie 1
fasten°, anchor 3; apply 1; attach 1; bar 8; bind 1; bolt 10, 11; chain 3; clasp 3; cling 1; clip[1] 1; close 1; connect 3; fix 1, 4, 9; hitch 1; knot 3; lash[2]; nail 7; peg 3; pin 3; rope 3; secure 8; sew; shut 1; stake[1] 3; stick[1] 4; suspend 2; tack 5; tether 2; tie 1; tighten 1; whip 5
fasten on, set 4; single 4
fasten on *or* upon, decide 2; grab 1; take 2
fasten to, suffix 2
fasten (together), clamp 2; join 1; knit 1; link 3; unite 3
fasten (up), adjust 4
fastened, attached 2; bound[3] 1; fast[1] 3; fixed 1; secure 2; shut 7
fastener, brace 3; buckle 1; catch 17; clamp 1; clasp 1; clip[1] 2; fastening; hook 1; nail 1; snap 10
fastening°, attachment 1, 2; brooch; buckle 1; catch 17; clasp 1; nail 1; snap 10; splice 2; tack 2; tie 8; whipping 2
fastidious°, careful 2; choosy; dainty 2; deliberate 3; delicate 5; difficult 5; discriminating; exact 2; faithful 3; finicky 1; fussy 1; gingerly 1, 2; hairsplitting; meticulous; neat 1; nice 3; overcritical; particular 4; pedantic 2; perfectionist 2; precise 3; priggish; prissy; refined 2; religious 2; scrupulous 1; severe 1; squeamish 1; tasteful; well-groomed
fastidiousness, precision 2; refinement 2; severity 1
fasting, fast[2] 2
fast living, dissipation 1
fastness, security 1; stronghold; tower 2
fast one, gag[2] 2
fast-talk, defraud
fast-talking, glib
fastuous, pretentious 1
fat°, beamy; big 2; dumpy; greasy 1; gross 1; heavy 11; large 1; money 4; obese; oily 1; plump[1] 1; rich 6; rotund 3; stout 1; weighty 1; well-fed
fatal°, calamitous; deadly 1; destructive 1; disastrous 1; fated 3; fateful 2; hopeless 1; incurable 1; lethal; malignant 1; mortal 3; murderous 1; poisonous 1; ruinous; severe 3; terminal 2; virulent 1
fatalism, stoicism
fatalities, casualty 2a
fatality°, killing 1
fatally, severely 5
fat cat, dignitary
fate°, chance 1; destine 1; destiny; doom; fortune 2; lot 2; luck 1; predestination; providence 2
fated°, bound[3] 4; card 11; condemn 2; destined 1; doomed 1; fatal 3;

feculence, muck 2
feculent, filthy 1; impure 1; muddy 1; repulsive; rotten 1; sordid 3
fecund, fertile; fruitful 1; pregnant 3; productive 1; prolific 1; rich 9; vivid 3
fecund (in), rich 2
fecundate, fertilize 1; impregnate 1
fed up, discontented; disgruntled; jaded 2; weary 3
fed up (to here) with, tired 2
fed up (with), disgusted; sick 7
fed up (with), be *or* **become,** weary 4
federal, national 1; political 1; state 5
federate, band[2] 3
federated (with), league 2
federation°, alliance 1; association 1; club 2; combination 2; league 1; marriage 3; organization 3; party 3; ring[1] 3; union 1, 2
fee°, admission 5; charge 2; consideration 2; dues; expenditure; expense 1; fine[2] 1; forfeit 1; hire 5; honorarium; interest 6; pay 12; payment 2; price 1; rate[1] 2; rent[1] 2; subscription 1; toll[2] 1; wage 1
feeble°, decrepit 1; delicate 3; doddering; faint 1; fainthearted 2; flabby 2; flimsy 1, 2; fragile; frail 2; helpless 1, 3; impotent 1; ineffectual 2; infirm 1; insubstantial 1; lame 2; limp[2] 2, 3; low[1] 4; pale[1] 3; pathetic 2; peaked; powerless 2; puny 3; shaky 2; sickly 2; slender 2; slight 4; soft 12; spineless 2; tame 3; tender[1] 1; tenuous 2; thin 4; unhealthy 1; unsound 1; wan 2; weak 1, 4, 6, 7; wet 3; wishy-washy 2
feeble imitation, parody 2
feebleminded°, backward 2; crazy 2; daft 1; defective 2; foolish 2; half-witted; insane 2; mindless 1; senile; simple 4; stupid 1; weak 5
feeblemindedness, folly 1; simplicity 4; stupidity 1
feebleness, decrepitude 1; delicacy 2; frailty 1; impotence 1; infirmity 1; prostration 4; weakness 1
feed°, board 7; dine; feast 5; fertilize 2; fuel 4; keep 2; nourish 1; provender 2; spread 11
feed, off one's, ill 1; indisposed 1; moody 1; seedy 2; sort 6
feed on *or* **upon,** feed 3; prey 3a
feedback, reaction 1; response; upshot
feedbag, put *or* **tie on the,** eat
feeder, tributary
feeder (road), turnoff 1
feel°, atmosphere 2; believe 1; climate 2; detect 2; distinguish 3; experience 4; feeling 6; finger 11; flair 1; flavor 2; fumble 1; grip 3; grope; guess 2; handle 2; instinct; perceive 2; perception 2; seem; sense 6; surmise 1; suspect 2; take 8, 23; texture; touch 1, 13, 17
feel for, feel 8; pity 3; respond 2
feel (something) in one's bones, feel 4; sense 6
feel (sorry) (for), apologize 1; pity 3; sympathize 1
feel to be, find 4
feel affinity for, take 38a; warm 8
feel contrition, repent
feel disconcerted, small 6
feeling°, air 1, 3; anticipation 2; atmosphere 2; attitude 2; aura; climate 2; emotion; estimate 4; expression 4; feel 11; fire 2; flavor 2; foreboding 1; grip 3; guess 3; hunch 1; idea 3; impression 1; inspiration 1; instinct; mind 6, 7; mood 1; observation 2; passion 1; perception 2; philosophy 2; position 3; posture 3; premonition; presumption 3; regard 9; sensation 1; sensible 3; sensitivity 2, 3; sentiment 1, 2; sniff 2; soul 4; spirit 7; stand 13; surmise 2; tender[1] 6; thing 4; touch 13, 17; undercurrent 2; vein 4; view 2; way 2
feeling, have a *or* **the,** feel 6; sense 6; suspect 2
feeling, without, numb 1
feeling (that), give one the, sound[1] 5
feeling (that), have a (funny), feel 4; sense 6
feeling(s), feeling 5; heart 4; psychology; sensibility 2; spirit 8, 9a; vibes
feeling(s), bad, rancor; strife 2
feel one's way, orient 3
feel uncomfortable, small 6
feet, extremity 2
feet, get to one's, arise 1; rise 1; stand 10a
feet, on one's, standing 4
feet up, put one's, relax 3; rest[1] 6; retire 3
feign, act 8; affect[2] 1; assume 4; bluff[1] 2; counterfeit 5; dissimulate; fabricate 3; fake 2; fool 5; game 6; let[1] 7b; make 25; play 10; pretend 1, 3; put 22b; strike 12
feigned, affected 2; artificial 3; assumed 2; counterfeit 2; false 4; forced; glossy 2; hollow 4; mimic 5; mock 3; pretended; seeming; spurious; studied; unnatural 2
feigning, act 3; hypocritical; pretense 1
feint°, subterfuge
felicitate, compliment 3; congratulate; hail[1] 2; toast 3
felicitations, compliment 2; congratulations; toast 1
felicitous, appropriate 1; fluent; happy 2; mellow 4; opportune 1; providential
felicity, bliss; content[2] 1; festivity 1; fluency; gaiety 1; glee; happiness; joy 1; merriment; welfare
fell°, bloodthirsty; brutal 1; cut 13a; destructive 1; ferocious; fierce 1; floor 4; grim 2; height 3; hide[2] 1; hill 1; knock 4b; lay[1] 18c; moor[1]; mountain 1; murderous 1; prostrate 2; sanguinary 1; savage 2; skin 1; throw 3; topple 1; vicious 3
fellah, peasant
felled, prostrate 4
fellow°, accomplice; acquaintance 2; associate 2; boy 1; brother; bugger 2; chap; chum 1; cohort 3; companion 1; customer 2; devil 3; dude 2; equal 4; friend 2; guy 1; lad; like[1] 7; man 1; match 1; mate 1, 3; member; partner; steady 9; stick[2] 2; teacher
fellow citizen, brother
fellow clansman, brother
fellow countryman, brother
fellow creature, brother
fellow criminal, accessory 2; accomplice
fellow dancer, partner 2
fellow feeling, love 4; sympathy 2
fellow man, brother
fellowship°, association 3; brotherhood 1; circle 2; clan 2; club 2; companionship; company 1; familiarity 2; fraternity 2; friendship 1; league 1; order 9; scholarship 2; society 1, 5; sympathy 2; union 2
fellow traveler, sympathizer
fellow worker, associate 2; colleague
felon°, criminal 3; miscreant 1; terrorist; transgressor
felonious, black 6; evil 2; illegal; lawless 2, miscreant 2; unlawful; villainous 1; wicked 2; wrong 1
feloniousness, guilt 1
felony, crime; job 5; misdeed; offense 1
female, feminine 1; girl 1; woman 1
femme fatale, seductress; siren 2; temptress
fen, bog 1; flat 14b; marsh; mire 1; morass 1; swamp 1
fence°, barrier 1; compete; enclosure 2; equivocate; evade 2; jump 10; rail[1] 1; shuffle 3; wall 1
fence, be *or* **sit on the,** pussyfoot 2
fence, on the, fence 2; nonpartisan 1
fence (in *or* **off),** keep 14b; partition 6; picket 4; shut 3a; stake[1] 4a
fend for (oneself), fend 1; shift 2
fend off, defend 2; deflect; fence 4; fend 2; forestall; head 13b; hold 18b; prevent; repel 1; repulse 2; turn 7; ward 3
fending off, prevention
fenny, muddy 1
feral, bloodthirsty; cutthroat 3; ferocious; fierce 1; ghoulish 2; grim 2; savage 1; truculent; untamed; vicious 3; wild 1
fermata, pause 2
ferment°, agitation 2; boil[1] 2; brew 1; buzz 2; disorder 2; excitement 2; passion 2; rush 3; sour 5; tempest 2; tumult
ferment, be in, seethe 2
fermented, sour 2
ferocious°, bloodthirsty; brutal 1; cruel 2; cutthroat 3; deadly 3; fierce 1; ghoulish 2; grim 2; homicidal; mad 3; militant 1; ruthless; savage 2; scathing; torrential; truculent; untamed; vicious 3; violent 1; wild 3
ferociously, fiercely; madly 3; vengeance 2a
ferociousness, violence 1
ferocity, fury 2; severity 3, 5; violence 1, 2
ferret, root[2]
ferret about, pry 1
ferret out, dig 6; discover 1; scrounge 1; track 8
ferrule, tip[1] 1
ferry, ship 2; take 9
fertile°, fruitful 1, 2; golden 5; plentiful 2; pregnant 3; productive 1, 2; prolific 1, 2; rank[2] 1; rich 9; vivid 3
fertile (in), rich 2
fertility, fat 6; plenty 2
fertilize°, impregnate 1
fertilizer, dung
ferule, tip[1] 1
fervency, ardor; electricity; feeling 4; fervor; fire 2; glow 3; passion 1
fervent°, animated 1; ardent; burning 2; deep 6; drunk 2; eager; earnest 2; emotional 1; enthusiastic; excited 2; fanatical; feverish; fire 5; heartfelt; heated; hot 3, 4; impassioned; inflammatory; intense 2; keen[1] 1; mad 6; rousing; soulful; strong 5, 22; torrid 2; voracious 2; warm 2; wholehearted

twinkle 1, 2; twinkling 1; vulgar 1; whip 6
flash, in a, hastily 1; rapidly 2; suddenly 1; swiftly
flash, like a, fast[1] 6; shot 10
flash-freeze, freeze 1
flashily, gaily 1
flashiness, glare 3; glitter 4; ostentation
flashing, bright 1; meteoric 1; scintillating 1; shiny; twinkle 2
flashlight, light[1] 2
flashy°, brassy 1; conspicuous 2; extravagant 4; flamboyant 2; garish; gaudy; gay 3; grandiose 1; jaunty 2; loud 2; luxuriant 3; rakish; shiny; showy; sporty; tasteless 1; tawdry
flask, bottle 1; jug
flat°, boring; broke; dead 10, 12; downright 1; even 1; flush[2] 1; hollow 6; horizontal; lackluster; level 1; lifeless 3; matter-of-fact; mousy 1; muddy 3; pad 3; pedestrian 2; peremptory 2; place 6; plain 1; plane 3; point-blank 1, 3; prosaic; regular 5; room 3; small 3; smooth 1; stagnant; stale 1; still 1; stodgy; straight 3; tame 4; tasteless 2; tedious; tiresome 1; uniform 2; vapid; washed-out 1; watery 1; wishy-washy 2
flat (on one's back), recumbent; supine 1
flat out, flat 17
flatfoot, constable; police officer
flatland, plain 6; table 2
flatly, completely 3; flat 16, 17b; point-blank 3
flatmate, friend 3
flatness, uniformity 2
flats°, flat 14
flat (surface), plane 1
flatten°, devastate 1; even 13; fell; knock 6a; lay[1] 18c; press 4; raze; roll 7; rout 2; ruin 7; smooth 9; trample 1
flatten (out), level 8
flatter°, compliment 3; eulogize; flannel 2; get 30a; make 34b; play 18; romance 7
flatterer°, charmer
flattering°, complimentary 1; good 17; ingratiating; menial 2; obsequious; oily 2; servile
flattery°, compliment 1; flannel 1; pat[1] 5; servility
flatties, flat 14a
flatulence, bombast; hot air; rant 3; raving 3; rhetoric 2; wind[1] 3
flatulent, bombastic; pompous 2
flatus, wind[1] 3
flatware, silver 1
flaunt°, camp[2] 2; dangle 2; display 3; exhibit 1; flourish 2; parade 5; promenade 4; show 11; sport 5; trot 2
flaunting, ostentation; ostentatious; pretentious 2
flauntingly, openly 1
flavor°, air 3; aroma 2; life 7; odor 2; salt 6; savor 1; season 3; seasoning; spice 3; suspicion 2; tang 1, 2; taste 1; undercurrent 2; zest 1
flavorful, delicious 1; mellow 1; pungent 1; racy 3; robust 2; savory 1; spicy 1; tasty
flavor(ing), spice 1; flavor 1
flavorless, flat 8; tasteless 2; vapid; watery 1; wishy-washy 2
flavor of the month, favorite 1; modern
flavorous, savory 1; tasty
flavorsome, savory 1; spicy 1; tasty
flaw°, blemish 2, 3; crack 1; defect 2; disadvantage 1; drawback; error 1; failing 1; fault 1; foible; frailty 2; hole 6; impairment; imperfection; kink 3; leak 2; nick 1; shortcoming; taint 1; twist 10; vice 2; weakness 3
flawed°, defective 1; deficient 2; erroneous; false 1, 5; faulty; imperfect; inaccurate; inadequate 1; mistaken 2; poor 4; unsatisfactory; unsound 4; untenable; wanting 1; wrong 5
flawless°, absolute 1; complete 4; exquisite 5; faultless; immaculate 3; impeccable; incorrupt 1; infallible 1; nice 3; perfect 2, 3; polished 1; pure 1; spotless 2; watertight 2
flawlessly, pat[2] 1; perfectly 2, 3
flawlessness, perfection 1; precision 1; purity 1
flaxen, golden 1
flaxen-haired, fair[1] 3
flay, flog 1; lambaste 2; lash[1] 4; pan 5; peel 1; pull 6; skin 3; slam 3; slash 2; strip[2] 1
flay (alive), reprimand 2
flea in his .. ear, send (someone) off with a, lecture 4; reprimand 2; send 8
fleabitten, dapple 1
flèche, spire 1; tower 1
fleck, dot 1, 3; grain 3; pepper; speck; spot 1, 7; streak 1
flecked°, dapple 1; mottled; speckled; spotty 1
fledgling, immature 1; initiate 4; newcomer 2; novice; stripling
flee°, bolt 8; escape 1; flight[2] 3; flit; fly 2; forsake 1; get 32a; heel[1] 4; loose 8; make 26; pull 8b; quit 1; retreat 4; run 2; stampede 3; turn 22
flee from, shun
flee to, take 38b
fleece°, cheat 2; defraud; fool 4; gouge 2; hide[2] 1; pelt[2]; pile[2]; profiteer 2; rip 2b; screw 6; skin 1; sting 4; swindle 1
fleece (of), rob 3
fleecy, fuzzy 1; hairy 1; soft 8; woolly 1
fleeing, flight[2] 1; fugitive 2; run 49b
fleer, flout; parody 3
fleer (at), mock 1
fleet°, expeditious; fast[1] 1; flock 1; navy; quick 1; rapid; speedy 3; swift
fleeting°, brief 1; elusive 2; fly-by-night 1; fugitive 3; hasty 3; intangible; meteoric; momentary; passing 1, 2; perfunctory 1; shadowy 3; short-lived; temporary; transient
fleetingly, briefly 2; suddenly 1; temporarily 2
fleetness, speed 1; velocity
Fleet Street, press 10a
flesh°, meat 1; pulp 1
flesh, in the, flesh 4; person 2; personal 1; personally 1
flesh crawl *or* **creep, make one's,** scare 1; terrify
flesh out, complement 3; elaborate 4; expand 4; flesh 6; make 39a; pad 5
flesh and blood°, flesh 2, 3; kinship 1
flesh-color(ed), pink[1] 2
fleshiness, fat 5
fleshly, animal 4; carnal; earthly 2; mortal 2; outward 1; physical; sensual; sexual 2; temporal 1; worldly 1
flesh-peddler, pander 3; procurer
fleshy, fat 1; obese; plump[1] 1; rotund 3; stout 1; well-fed
flex°, lead 17
flexibility°, elasticity 1, 2; give 18; play 24; resilience; spring 7
flexible°, adaptable; dutiful; easy 3; elastic 1, 2; fluid 3; lax 1; liberal 3; limp[2] 1; mobile 3; open 11; passive 2; plastic 1; pliable 1, 2; receptive 1; soft 1; submissive 1; supple 1; versatile 1, 2; willowy 1; yielding 1, 2
flexibleness, flexibility 1
flexile, flexible 1; soft 1; supple 1; yielding 1
flexuosities, meander 2
flexuous, meandering; pliable 1; soft 1; tortuous 1
flexure, bend 1; sweep 6
flibbertigibbet, gossip 3; scatterbrained
flick, film 2; flash 5; flip 1; flit; movie 1; whisk 4
flick (through), thumb 7; browse
flicker°, blink 1, 2, 5; flare 1; flash 1, 4; flutter 2; glance 2, 5; gleam 1, 2; movement 3; ray 2; shimmer 1, 2; shine 1; spark 1; sparkle 1; twinkle 1, 2; wag[1] 1, 2
flickering, faint 1; flash 1; shiny; twinkle 2; weak 6
flicks, movie 2
flier, insert 2; leaflet; pamphlet; pilot 1; poster; promotion 5; publication 2
flier on, take a, gamble 2
flies°, fly 9; overlap 3
flight°, escape 5; exit 2; flock 1; getaway; hop 4; pass 29; retreat 1; rise 15; stampede 1; takeoff 1
flight, in, aloft; run 49b
flight, put to, chase 3; flight[2] 2; rout 2
flight, take, bolt 8; flee 1; flight[2] 3; fly 1, 2; heel[1] 4; pull 8b; retreat 4; rise 7; run 2; stampede 3
flightiness, extravagance 2
flight of fancy, romance 5
flight path, trajectory
flighty°, capricious; dizzy 2; erratic 1; fickle; frivolous 2; giddy 2; harebrained 2; inconstant; light[2] 8; moody 3; scatterbrained; unstable 1; volatile 2; whimsical 2; wild 7
flimflam, cheat 2; deceit 2; defraud; dupe 3; fiddle 1; fleece; fraud 2; hoax 1; hocus-pocus 1; mislead; prey 3b; swindle 1; victimize 2
flimflam artist *or* **man,** fraud 3; hypocrite; impostor; swindler; thief 2
flimflammer, fraud 3; hypocrite
flimsy°, delicate 1; feeble 2; filmy 1; fine[1] 6; fragile; groundless; immaterial 1; infirm 2; insecure 3; insubstantial 1; lame 2; limp[2] 3; pale[1] 3; ramshackle; rickety; rocky[2]; shaky 1, 2; shallow 1; sleazy 1; slight 4; tenuous 2; thin 4; tinny 1; unreliable; weak 1, 4
flinch°, blink 3; cringe 1; dread 1; jump 2; recoil 1; shrink 2; start 5
flinch (from), pull 8a; hate 2
fling°, affair 4; carouse 2; cast 7; clap 3; dash 2; flounce 2; fly 8a; heave 2; hurl; lark 1; launch 3; lob 1; orgy 2; pitch[1] 1; precipitate 2; project 4; put 7; revel 3; send 3; shoot 2; slap 3; sling 1; spree; throw 1; toss 1, 2; trial 3; try 5; venture 1
fling back, retort 2
fling down, dump 1
fling to the winds, trample 2
flintiness, severity 1

flinty, hard 1; icy 2; rocky[1] 2, 3; severe 1; steely 2; stern 1; unkind; unmerciful
flip°, flippant 2; forward 2; fresh 8; frivolous 2; irreverent 2; peg 5; pert; wag[1] 1, 2
  flip (out), flip 2; fume 1
  flip (through), thumb 7; browse; scan 1; skim 2
flip-flop, wave 4
flip (one's lid *or* **top),** blow[1] 8a; explode 3; flip 2; fly 7; fume 1; rage 4; rave 1
flip one's wig, rage 4
flippancy°, levity; mouth 4
flippant°, forward 2; frivolous 2; irreverent 2; pert; rude 2; satirical
flip side, reverse 7
flirt°, mess 4b; philanderer; play 9b; roué; temptress
  flirt (with), chat 3; flirt 2; toy 4; trifle 3
flirtation, liaison 3
flirtatious°, kittenish; seductive; sexy 1
flirty, flirtatious
flit°, flutter 2; fly 1; leave[1] 1; roll 2; run 1; sail 3; tear 3; whip 3
  flit about, gad
flitter, flutter 2
flivver, rattletrap
float°, buoy 1; drift 1; glide; hover 1; launch 4; poise 3; ride 2; roll 3; sail 3; soar 1; waft 1
floatable, buoyant 1
floating, buoyant 1; migrant 2; movable
floating dock, pier 1
floccose, fuzzy 1; woolly 1
flocculent, fuzzy 1; woolly 1
floccus, fuzzy 1
flock°, assembly 1; cluster 2; company 2; crowd 1, 3; flake 1; flight[1] 4; herd 1, 3; pack 3; press 7; swarm 1, 2; throng 1
  flock (into *or* **to),** throng 2
  flock (together), gather 2; mass 8; huddle 3
flock(s), many 3; score 4
flocky, woolly 1
flog°, beat 1; chastise; hide[2] 2; hit 1; lambaste 1; lash[1] 3; lather 4; paddle 4; peddle; punish 2; scourge 3; sell 2; slash 2; strike 1; switch 3; whip 1
flogging, punishment 2; thrashing 1; whipping 1
  flogging, a, works 3b
flood°, crowd 1, 3; drench; drown 1; flock 2; flow 2, 6; flush[1] 2, 5; glut 3; gush 1, 3; infest; ocean 2; outburst; outpouring; pour 1; rain 3; rash[2] 2; rinse 1, 3; rise 5; river 2; run 6; satiate 1; sea 3; shower 2; spate; spill 4; stream 2, 3, 4, 5; submerge 3; suffuse; surfeit; surge 1, 2; swamp 2; swarm 1, 2; torrent; volley 2; water 6; wave 2
  flood in *or* **into,** pile[1] 6
  flood (over), overwhelm 2
flood(s), many 3
floor°, amaze; astonish; astound; daze 1; devastate 2; dumbfound; fell; flatten 2; knock 4b, 6a; lay[1] 18c; level 13; overpower 2; prostrate 2; stagger 2; story[2]; surprise 1; throw 3
  floor, to the, over 10
floored, daze 4; dumbfounded; prostrate 4; thunderstruck
flooring, floor 1
floor show, cabaret 2
floorwalker, foreman
floozy, bitch 2; jade 2; tart[2] 2
flop°, dud 1; fail 1; failure 3; fall 12, 20; fiasco; flip 1; flutter 1; loser; misfire 1, 2; pad 3; plump[2] 1, 4; strike 17; washout; wrong 8b
flopping, flutter 4
floppy, flabby 1; limp[2] 1; slack 2
floral arrangement, spray[2]
Florence Nightingale, nurse 1
floret, flower 1
florid, brassy 1; flamboyant 1; flowery; garish; gaudy; grandiose 1; luxuriant 3; ornate; overwrought 2; rhetorical 3; rosy 1; showy
floridity, glare 3; glitter 4
floridness, flourish 3; glare 3; glitter 4
floss, fluff 1
flossy, fuzzy 1
flotilla, fleet[1]; navy
flounce°, camp[2] 2; frill 1; fringe 1; ruffle 1; sweep 3; walk 1
  flounce down, flop 1
flounder°, blunder 1; pitch[1] 4; reel 1; squirm; stumble 1; trip 5; wallow 3; welter 2
floundering, incompetent
flourish°, abound 1; boom 2; dangle 2; dash 6; display 3, 5; fanfare 1; flaunt; get 27; grow 1; increase 1; make 31b; ostentation; pan 6; panache; place 10; prosper; root[1] 5; shake 4; shoot 4; stroke 3; succeed 3; swing 1, 3; thrive; verve; wield 1
flourishing°, expansion 1; fresh 6; fruitful 3; going 1; golden 5; hale; healthy 1; lush 1; prosperous 2; rampant 2; rank[2] 1; strong 11; successful 1; swing 3
flout°, break 9; defy 1; deride; disobey; fly 6; gibe 1; jeer 1; mock 1; rebel 2; scorn 3; snap 6; taunt 1; thumb 8
flouted, broken 5
flow°, circulate 1; circulation 1; current 5; discharge 11; drift 3; emanate 1; flood 3, 6; flush[1] 5; follow 7; glide; gush 1, 3; issue 11; movement 1; originate 2; outpouring; pass 2; passage 6; pour 1; progression 3; river 2; roll 2; run 6, 7; sail 3; spate; spout 1; stem[1] 3; stream 2, 3, 4; succession 2; surge 1, 2; swarm 2; torrent; trickle 1; wash 3, 14; well[2] 2
  flow back, ebb 1
  flow from *or* **out of,** drain 6
  flow in, roll 10a
  flow (out *or* **forth),** shed[2] 3
  flow together, fuse
flower°, choice 3; develop 2; flourish 1; gem 2; grow 1; pearl; plant 1; pride 3; put 17b; sprout
flower arrangement, spray[2]
flowerbed, border 6
flower child *or* **person,** hippie
floweret, flower 1
floweriness, flourish 3
flowering, flourishing; growth 1
flowerpot, planter
flowery°, luxuriant 3; ornate; overwrought 2; pompous 2
flowing, circulation 1; fluent; fluid 2; graceful 1; liquid 2; loose 3; smooth 3; streamlined 1
flu, chill 2
  flu, the, cold 10
flub, fumble 2; mistake 1; solecism
fluctuate°, change 7; doubt 2; flicker 2; flutter 1; hesitate 2; oscillate; quaver 1; quiver 1; range 6; seesaw 2; shilly-shally; sway 1; swing 1; vary 2; vibrate; wag[1] 1; wave 4
fluctuating, checkered 2; fitful; flutter 4; hesitant 1; inconstant; indecisive 1; swing 3; unstable 1; variable; versatile 2; whimsical 2
fluctuation°, flutter 4; flux; indecision; quaver 2; swing 3; vicissitude 1; wag[1] 2
fluctuations, vicissitude 2
flue, shaft 4; vent 1
fluency°, facility 1; oratory
fluent°, eloquent 1; glib; liquid 2; smooth 3; voluble; windy 2
fluff°, blow[1] 3; blunder 2; error 1; folly 2; lapse 1
  fluff, bit of, fluff 2, woman 3
  fluff up, fluff 5; whip 4
fluffy°, fuzzy 1; hairy 1; soft 8
fluid°, glib; graceful 1; juice 1; liquid 1, 2; liquor 2; sap[1] 1; solution 3
fluke°, accident 2; coincidence 3; luck 1
flukiness, vicissitude 2
fluky, incidental 1
flume, race[1] 2
flummery, gab 2; nonsense 1
flummox°, confuse 1; mystify; nonplus; puzzle 1; stagger 2; stump 2; stymie; throw 6c
flummoxed, confused 2
flunk, fail 1
flunky°, attendant 2; fag 3; hack[2] 2; inferior 4; menial 3; yes man
fluorescence, illumination 1; light[1] 3
fluorescent, bright 5; light[1] 13; luminous 2
flurry°, daze 3; flap 4; fluster 1, 2; fuss 1; ripple 2; ruffle 2; rush 3; stir 6
flush°, blush; color 5; even 1; glow 3, 6, 7; level 1; loaded 4; money 4; opulent 1; prosperous 1; rich 1; rinse 1; smooth 1; successful 1; suffuse; wash 1; wealthy
  flush out, flush[1] 2
  flush (with), even 2
flushed, drunk 2; feverish; rosy 1
flushing, rinse 3
fluster°, agitate 1; confuse 1; demoralize 3; discomfit 1; distract 3; disturb 4; embarrass; excite 2; flurry 1, 2; fuss 1; irritate; mix 4b; perturb; ruffle 3; shake 5; unnerve; upset 1; wind[2] 4d
flustered, agitated; confused 2; daze 4; disconcerted; embarrassed 1; excited 1; nervous 1; unsettled 2
flute, channel 3; furrow 1, 2; groove
flutter°, bet 1; bother 6; flap 1, 2; flare 1; flicker 2; flit; flurry 1; fluster 1, 2; fly 1; fuss 2; lather 2; quaver 1; ripple 2; shake 4; shiver[1] 2, 3; thrill 2; twitter 4; wag[1] 1, 2; wave 4
  flutter, have a, speculate 2
fluttering, flutter 4
flux°, vicissitude 1
fly°, bolt 8; bug 1; career 2; dash 3; depart 1; escape 1; flap 3; flash 5; flee 1; flit; go 11; hasten 1; hop 2; hurry 1; leave[1] 1; lob 2; make 26; pass 2, 11; pilot 3; race[1] 4; rise 7; roll 3; sail 3; scud; scurry; sharp 6; shoot 1; skim 3; soar 1; speed 3; spring 1; streak 5; tear 3; wily
  fly, let, fling 1; fly 8; heave 2; hurl; loose 13; pitch[1] 1; send 3; shoot 2; sling 1; toss 1
fly apart, explode 1
fly at, come 8; set 21
fly away *or* **off,** fly 2
fly down on, buzz 6
fly in the face of, disobey; fly 6; trample 2
fly into a rage *or* **fury** *or* **temper** *or* **passion,** fly 7
fly off, take 34c
fly off the handle (at), explode 3; flare 3; fly 7; fume 1; rage 4; rave 1; snap 3b; storm 4

harbinger; indication 3; omen; predict; prognosticate 2; warning 2
foretokening, ominous 3
forever°, always 2; cease 2; ever 2; finally 2; forever 1; permanently
forever and a day, ever 2; forever 1
forevermore, forever 1; permanently
forewarn, alert 5; caution 3; forecast 1; predict; prophesy 1; tip[3] 4; warn 1
forewarning, foreboding 2; forecast 2; indication 3; omen; premonition; sign 6; threat 2; warning 2
forewoman, manager; overseer
foreword°, preamble; preface 1; preliminary 3
forfeit°, fine[2] 1; lose 2; penalty; sacrifice 4
forfeiture, fine[2] 1; forfeit 1; loss 1; sacrifice 1, 2
forgather, assemble 1; gather 2; mass 8; meet[1] 2
forge°, beat 5; construct 2; counterfeit 4; create 2; design 2; fabricate 3; fake 1; fashion 5; form 7; formulate 3; frame 7; generate 4; make 1; mold[1] 5; prepare 5
  forge ahead, proceed 1; progress 5
forged, base[2] 5; counterfeit 1; fake 5; false 3; fraudulent 1; mock 3; spurious
forgery°, counterfeit 3; fabrication 3; fake 3; imitation 3; phony 2; sham 1
forget°, bury 2; leave[1] 3; mind 19; neglect 2; omit 2; overlook 1; scrap[1] 4; slide 4; track 5
  forget, do not, mind 9
  forget (about), kiss 3; overlook 2; write 5
forgetful°, oblivious; remiss; senile; thoughtless 2
forgetfulness, negligence; oblivion 2
forging, production 1
forgivable, venial
forgive°, clear 21; excuse 1; let[1] 6a; overlook 2; pardon 2; purge 3; purify 2; understand 5; write 5
forgiven, clear 31
forgiveness°, excuse 5; grace 3; mercy 1; pardon 1; quarter 4; remission 1
forgiving°, charitable 2; human 3; lenient; merciful; mild 1; pardon 1; patient 3
forgo°, abandon 4; decline 1; dispense 3a; forfeit 2; forsake 3; give 17a; miss[1] 1; part 14; pass 22; renounce; repudiate; resist 2; sacrifice 4; shun; sign 8; spare 9; surrender 1; swear 4; throw 5b; waive 1
forgoing, resignation 1; sacrifice 1, 2; waiver
forgone, forfeit 3
forgotten, lost 4; undone[2]
  forgotten by, be, escape 4
fork, diverge 1; separate 4; split 3
  fork out, give 15b; stump 4
  fork out *or* **over** *or* **up,** expend 1; lay[1] 18b; pay 11b; shell 5; spend 1
forkful, morsel 1; mouthful
forlorn°, abandoned 1; brokenhearted; dejected; desolate 1, 3; desperate 6; dismal; doleful; dreary 1; gaunt 2; gloomy 2; hopeless 3; inconsolable; isolated 2; melancholy 1; miserable 1; mournful 1; pessimistic; sinking 2; tragic; unfortunate 2; unhappy 1; wistful 1; woebegone
form°, appearance 2; arrange 1; base[1] 5; beat 5; brew 3; cast 3, 9; ceremony 2; class 4; compose 1; composition 2; conceive 2; condition 3; create 1; custom 1; design 2, 3, 6; devise 1; disguise 3; effect 4; erect 3; establish 1; etiquette; fabricate 1; fashion 5; figure 1; formality 1; format 1; frame 1, 3, 7; generate 4; grade 3; health 1; image 6; jell 2; last[3]; make 1, 39b; makeup 3; manner 3; materialize 1; model 6, 7; mode[1] 1; mold[1] 1, 2, 4, 6; observance 2; order 1; organization 2; organize 2; originate 1; paper 2a; phase 3; physique; procedure; protocol 1; repair 3; rule 3; school 2, 4; sculpture 2; semblance 1; shape 1, 2, 3, 5, 6, 8; silhouette; state 2; structure 1, 3; style 1, 5, 8; taste 5; tradition; trim 6; turn 8, 20a, 37; usage 1; version 1; work 13
  form, in good, nifty 2
  form, in top, pink[1] 1
  form, off, off-color 1; par 5b
  form, take, form 11; loom 1; materialize 1; shape 10a
  form hastily, leap 3
  form (to), give, form 8; frame 7; shape 7
formal°, aloof 2; ceremonial 2; ceremonious 1; civil 3; dignified; distant 3; dressy 1; icy 2; impersonal 2; literary 2; nominal 1; official 2; outward 1; parliamentary; perfunctory 1; polite 1; priggish; proper 2; prudish; reserved; ritual 2; sedate 2; self-contained 2; solemn 3; state 6; stiff 7, 8; stilted
formalist, pharisee; prig; purist
formalistic, pharisaic; priggish
formalities, ceremony 2; etiquette; manner 3; propriety 3; protocol 1; rigmarole
  formalities, the, formality
formality°, ceremonial 3; ceremony 1, 2; convention 2; courtesy; dignity 1; form 6; function 2; gesture 2; observance 2; pageant; propriety 3; protocol 1; reserve 4; rite; ritual 3; sobriety 2
formalized, formal 2
format°, design 6; form 4; getup 2; makeup 3; mold[1] 2; organization 2; posture 2
formation°, cast 3; conception 1; creation 1; establishment 1; evolution; fabrication 1; form 1; generation 3; institution 1; line[1] 14; pattern 6; platoon; production 1; rank[1] 4; structure 1
formative, rudimentary 1; seminal 1
formed of, be, compose 3
former°, ancient 1; bygone; foregoing; late 3; old 5, 6; outgoing 1; past 2; preceding; previous 1, 2; prior 1; sometime 1
formerly°, before 1; late 5; once 1; previously; time 12a
former times, past 6
formidable°, arduous 1; awesome; fearsome; frightening; grim 2; important 3; impressive; mountainous 2; overwhelming 1; potent 1; powerful 2; strong 7, 9; terrible 6
formidableness, difficulty 1
forming, fabrication 1; formation 1; institution 1; production 1
formless, chaotic 1; fluid 3; rough 9; shapeless 1
formlessness, chaos
formula°, composition 3; form 4; index 4; law 3; mumbo jumbo 2; plan 1; prescription 1; recipe 1; routine 1; spell[2] 2; theorem 2
formulaic, formal 1
formulary, formula
formulate°, compose 2; compound 1; conceive 2; construct 2; devise 1; draw 15b; form 8; frame 7; hatch 2; invent 1; line[1] 23a; originate 1; phrase 4; program 4; scheme 4; shape 8; turn 8; work 19e
formulation, cast 3; composition 3, 4; conception 1; fabrication 1; generation 3
fornicator, libertine 1
forsake°, abandon 3; back 9; desert[1] 3; drop 7; emigrate; fail 2; forgo 1; give 17a; go 25; jilt; kiss 3; lurch[1]; maroon; part 14; pass 22; quit 1; recant; relinquish 1; renounce; resign 1; run 32; scrap[1] 4; secede; spare 9; surrender 1; swear 4; throw 8; waive 1
forsaken, abandoned 1; derelict 1; deserted; desolate 1; empty 3; forlorn 2; lonely 3; lonesome 1
forsooth, truly 4
forswear, deny 3; dispense 3a; forgo 1, 2; forsake 3; pass 22; recant; relinquish 1; renounce; repudiate; retract 2; sacrifice 4; swear 4
  forswear oneself, lie[1] 1
forswearing, perjury
forte°, field 4; genius 3; gift 2; line[1] 7; resource 1; skill 2; specialty 1; sphere 3; talent 1
forth, forward 5; onward 2
(forth)coming, imminent; impending; near 4; pending 2; prospective
forthright°, aboveboard 2; broad 5; candid 1; direct 10; explicit 2; frank 1; front 9a; honest 3; ingenuous 2; matter-of-fact; open 15; outspoken; plain 3; plump[2] 6; simple 3; straight 4; transparent 3; truthful
  forthright, be, card 13
forthrightly, honestly 2; openly 2; outright 4; straight 14
forthrightness, candor 1; honesty 2; simplicity 3; sincerity
forthwith, flat 17b; immediately 1; instantaneously; right 13; soon 2; summarily 1; time 17
fortification, bulwark 1; defense 2; rampart; refreshment 2; stronghold; wall 2
fortified, charmed 2
fortify°, confirm 3; defend 2; fence 3; freshen 4; harden 2; inspire 2; lace 4; nourish 3; prepare 6; refresh 1; reinforce; steel 2; strengthen 1, 2; support 1, 2; vitalize
fortifying, bracing; exhilarating 1; refreshing; support 7; tonic 2
fortissimo, loud 1
fortitude°, backbone 3; bravery; determination 1; endurance 1; grit; manhood 2; nerve 1; patience 2; pluck 1; prowess 2; self-control 1; stamina; stoicism; strength 2
fortnightly, journal 1; organ 2; periodical; publication 2; regular 2
fortress, castle 1; stronghold; tower 2
fortuitous, accidental; casual 1; circumstantial 2;

fortuitous (*cont.*) coincidental; fortunate 1; haphazard 1; happy 2; incidental 1; lucky 2; random 1; unforeseen
fortuitously, random 2
fortuitousness, coincidence 3
fortuity, accident 2; coincidence 3; fluke; fortune 2; luck 1
fortunate°, benign 3; happy 2; lucky 1; opportune 1; propitious; providential; seasonable; successful 1
Fortunate Isles *or* **Islands, the,** heaven 1
fortunately, happily 1; well[1] 15
fortune°, accident 2; chance 1; destiny; doom; estate 2; fate 1; hazard 2; lot 2; luck 1; mint 1; money 2; packet 2; pile[1] 2; predestination; property 2; prosperity; providence 2; riches; treasure 1; welfare
  fortune, bad, grief 2
  fortune, ill, grief 2
  fortune, make one's, prosper
  fortune, stroke of (good), godsend; windfall
fortune hunter, adventurer 2
fortune hunting, timeserving
fortune(s), fortune 3; luck 3; story[1] 6
fortuneteller°, oracle 1; prophet; psychic 3; seer
fortunetelling, prophecy 1
forty winks, doze 2; nap[1] 3; repose 2; rest[1] 1; sleep 2
  forty winks, catch, nap[1] 1; sleep 1
  forty winks, get *or* **take,** rest[1] 6; retire 3
  forty winks, have, doze 1
forum, conference; organ 2; voice 3
forward°, advance 2; aggressive 2; ahead 3; audacious 2; barefaced 2; benefit 3; bold 2; brash 2; brassy 1; brazen; champion 4; cheeky; contribute 2; dispatch 2; disrespectful; ease 8; encourage 2; expedite 2; familiar 3; favor 8; foment; foster 1; fresh 8; further 5; immodest 2; impertinent; impudent; intrusive; move 7; nominate; obtrusive; officious; onward 1; pert; presumptuous; promote 1; pushy; recommend 1; remit 1; send 2; shameless; speed 2; support 1; translate 4; transmit 1; transport 1
  forward, be, liberty 5
  forward of, before 4
forwarder, merchant 3
forwarding, translation 4; transmission 1
forward-looking, forward 3; go-ahead 2; sanguine
forward motion, headway 1
forward movement, advance 7
forwardness, freedom 7; impertinence; presumption 1
forwards, forward 4; onward 2
fosse, furrow 1
fossick, dig 6
fossil, ancient 3; fogy; old 2; prehistoric 1; ruin 5; stick-in-the-mud
fossil fuel, fuel 1
fossilize, petrify 3
fossilized, ancient 3; petrified 3
foster°, arouse 3; cherish 2; cultivate 3; encourage 2; entertain 3; foment; forward 6; further 5; harbor 3; nourish 2; nurse 2, 4; patronize 3; promote 1; protect 2; raise 7
fostering, furtherance; promotion 1
foster parent, parent 1
foul°, abominable 1; abusive 1; bad 2; beastly 2; black 6; coarse 3; contaminate; contrary 3; damnable; dark 4; diabolic 2; dirty 1, 4; disgusting; dishonorable 3; distasteful; entangle 1; evil 1, 5; filthy 1; hateful 1; hideous 2; impure 1; indecent 2; infamous 2; lewd; low[1] 12; monstrous 2; nasty 1; nauseous; obnoxious; obscene 1; offensive 3; outrageous 2, 3; perverted; poisonous 2; pollute 1; profane 3; prurient 2; putrid; rank[2] 2, 3, 4; reprobate 1; repugnant; repulsive; revolting; rotten 1; scurrilous; seamy; sexy 2; shocking 2; smelly; soil[1] 2; sordid 3; stagnant; stormy 1; strong 3; taint 2; terrible 1, 4; ugly 2; vile 2; wicked 2, 6; wrong 1; yucky
  foul up, bungle; butcher 4; err 1; fluff 4; foul 16; hash 3; mess 5b
fouled, muddy 1
  fouled up, confused 2
fouling, pollution; sacrilege 1
foul language, profanity
foul matter, filth 1
foulmouthed, coarse 3; evil-minded 1; filthy 3; foul 5; obscene 1; outrageous 3; profane 3; scurrilous
foulness, dirt 3; evil 6; filth 2; impurity 1; ribaldry; vulgarity 2
foul odor, stench
foul play°, conspiracy
foul-smelling, foul 3; rancid; rank[2] 4; smelly; stinking 1
foul temper, temper 3
foul-tempered, waspish
foul-up, error 1; mess 3; mix-up
found°, base[1] 5; begin 2; break 17; build 2; cast 9; erect 3; establish 1; father 6; float 2; ground 5; institute 3; launch 2; organize 2; originate 1; pioneer 2; root[1] 6; set 23b; start 7
  found, be, exist 3; prove 3; rest[1] 7
  found, can be, lie[2] 2
foundation°, base[1] 2; basis 1; bottom 2, 3; derivation; establishment 1; excuse 4; fabric 2; footing 1; ground 3; groundwork; institution 1, 2; keystone; origin 1; pedestal 1; preparation 1a; root[1] 1; start 15; substratum; support 8
  foundation, without, groundless; ill-founded
foundations, element 3b
foundation stone, essence 1
founder°, author; collapse 2; creator 1; fail 1; father 3; flop 3; go 27a; miscarry; sink 1; stock 3; swamp 3; wreck 2
foundering, breakdown 1; wreck 5
founding, establishment 1; formation 2; foundation 3; institution 1; start 15
found object, curiosity 3
foundry, mill 2; plant 2
fount, fountain 1, 2; germ 2; head 6; origin 1; root[1] 1; spring 8; type 2; well[2] 1
fountain°, stream 2; well[2] 1
fountainhead, fountain 1; head 6; origin 1; parent 2; root[1] 1; seat 2; source 1; spring 8; well[2] 1
fountain pen, pen[1] 1
fourflusher, fraud 3; impostor; swindler
Four Hundred, elite 1; nobility 3; upper 5
four-letter word(s), expletive 2; oath 2; profanity
four-sided, square 1
foursquare, ingenuous 2; simple 3
fourth, quarter 1
fourth estate, member of the, journalist; scribe 2; writer
fourth estate, the, press 10a
four times a year, quarterly 2
fox, devil 4; flummox; temptress
foxed, spotty 1
foxiness, craft 2
foxlike, foxy 1
foxy°, artful 1; astute 1; clever 2; crafty; devious 1; disingenuous; furtive 2; insincere; machiavellian; politic 1; scheming; sharp 6; shifty; shrewd; slippery 2; sly 1; smooth 8; subtle 3; tempting 1; tricky 1; wily
foxy lady, temptress
foyer, hall 1; lobby 1; lounge 2; passage 4
fracas°, argument 1; battle 1; brawl 1; conflict 1; disorder 2; dispute 4; disturbance 2; fight 7; fray[1]; incident 2; mayhem; noise 1; quarrel 1; riot 1; row[2] 1; rumpus; scrap[2] 1; scrimmage; skirmish 1; tumult; uproar
fraction, bit 4; hair 2; morsel 2; piece 3; portion 1; section 2; segment 1; shade 6
fractional, part 15
fractionation, breakdown 3
fractious, cross 6; disobedient 1; fretful; insubordinate; naughty 2; perverse 2; prickly 3; pugnacious; quarrelsome; querulous; recalcitrant; rogue 2; testy; unruly
fracture°, break 1, 23b, 25; crack 1, 5; cranny; flaw 2; rupture 1, 3; separation 3; shatter 1; snap 1; split 7
fractured, broken 2; split 9
fracturing, separation 3
fragile°, brittle 1, 2; delicate 1; feeble 1; flimsy 1; hothouse 2; infirm 1; insubstantial 1; slight 4; tender[1] 1; tenuous 1; thin 4; ticklish 2; weak 1
fragileness, delicacy 2
fragility, delicacy 2; frailty 1; weakness 1
fragment°, bit 1; break 1, 23b; chip 1; clip[2] 5; crumb; crumble; dash 1; fall 7; flake 1, 2; gob; grain 3; modicum; morsel 2; part 1; piece 1; portion 1; rag[1] 1; relic 2; remnant 1; scrap[1] 1; segment 1, 2; shatter 1; shiver[2]; shred 1, 2; sliver; snatch 5; snip 3; splinter 1, 2; vestige
fragmentary°, broken 6; incomplete; partial 1; piecemeal 3
fragmentation, separation 3
fragmented, broken 1, 6; incomplete
fragments°, fragment 2; litter 1; odds 5; remain 4a; rubbish 1; ruin 6; wreckage
  fragments, into, piecemeal 2
fragrance°, aroma 1; bouquet 2; fume 3; odor 1; perfume 1; savor 2; scent 1; smell 1
fragrant°, aromatic; redolent 1; rich 8; strong 2; sweet 2
frail°, brittle 2; decrepit 1; delicate 1, 3; doddering; feeble 1; fine[1] 6; flimsy 1; fragile, girl 1, hothouse 2, impotent 1; infirm 1; insecure 3; insubstantial 1; limp[2] 2; low[1] 4; puny 3; rickety; slight 4; soft 12; tender[1] 1; tenuous 2; thin 4; unhealthy 1; unsound 1; wan 2; weak 1, 2
frailness, delicacy 2; infirmity 1
frailty°, defect 1; delicacy 2; fault 1; foible; imperfection; impotence 1; infirmity 1; shortcoming; vice 2; weakness 1

frame°, border 4; coin 3; compose 2; conceive 2; construct 1; couch 2; create 2; design 2; devise 1; draft 5; draw 6, 15b; fabricate 1; fashion 5; father 6; forge 2; form 1, 4; hull 1; make 1, 4, 6; meditate 2; mount[2] 5; phrase 4; physique; rack 1; raise 2; setting; setup 1; shape 7; shell 2; stand 17; support 8
framed, fixed 4
frame of mind°, cheer 1; disposition 1; frame 5; habit 2; humor 3; mentality 2; mood 1; posture 3; spirit 8
frame (of reference), context; viewpoint
framer, author; creator 1; father 3; founder[1]
framework, context; fabric 2; form 4; frame 1, 3; hull 1; organization 2; outline 2; rack 1; scenario; setup 1; shell 2; structure 1
framing, fabrication 1; frame 1
franchise, charter 2, 4; enable 1; liberty 2; patent 1; permission; permit 1, 2; prerogative; privilege; suffrage; vote 2
frangible, brittle 1; crisp 1; delicate 1; flimsy 1; fragile; weak 1
frank°, aboveboard 2; artless 1; bluff[2] 2; blunt 2; broad 5; candid 1; direct 10; dispassionate 2; downright 1; expansive 2; explicit 2; forthright; front 9a; genuine 2; honest 3; ingenuous 2; natural 4; open 15; outspoken; plain 3; raw 6; round 8; simple 3; sincere; straight 4, 6; transparent 3; truthful; uninhibited; unvarnished
  frank with, be, level 11
frankly, aboveboard 1; face 13; freely 1; honestly 2; openly 2; point-blank 3; shoulder 6; sincerely; true 4
frankness, candor 1; freedom 6; honesty 2; naiveté; simplicity 3; sincerity
frantic°, delirious 1; desperate 1, 6; distraught; emotional 4; excited 1; fanatical; fervent 1; feverish; furious 2; heated; hectic; hurried 1; hysterical 1; inflammatory; intense 2; lost 8; overwrought 1; raving 1; stormy 2; tempestuous; tumultuous; uncontrolled; wild 5
frantically, warmly 5
franticness, fanaticism 1
fraternal°, brotherly; friendly 1
fraternity°, body 6; brotherhood 2; circle 2; clan 2; club 2; companionship; fellowship 2; friendship 1; league 1; order 9; ring[1] 3; society 5; union 2; world 3
fraternization, fellowship 1
fraternize°, hobnob; mingle 2; mix 2; socialize; travel 4
  fraternize (with), associate 1b; hang 4b; pal 2; rub 7; shoulder 4
fratricide, murder 1
fraud°, cheat 1; craft 2; deceit 1; deception 1; fake 4; forgery 1; hoax 1; impostor; operator 3; pharisee; phony 2; poseur; quack 1; rip-off 2; sham 1; swindle 2; swindler; trick 1; trickery
fraudulence, deceit 1; falsity 1; forgery 1
fraudulent°, base[2] 5; bogus; bum 3; counterfeit 1; deceitful; deceptive 2; dishonest; fake 5; false 2, 3; foul 7; glossy 2; hollow 4; mock 3; phony 1; pious 2; quack 2; sham 2; spurious; unreal 3
fraught°, explosive 1; momentous; pregnant 2; tense 3
  fraught with, fraught 1
Fräulein, girl 1
fray°, action 6; argument 1; battle 1; brawl 1; conflict 1; disorder 2; disturbance 2; fight 7; fracas 1; quarrel 1; riot 1; scrap[2] 1; scrimmage; skirmish 1; wear 3
frayed, ragged 1; shabby 1; threadbare 1; worn 1
fraying, erosion; wear 9
frazzle, exhaust 2; fray[2]
  frazzle, worn to a, threadbare 1
frazzled, exhausted 1; ragged 1; weary 1; worn 3
freak°, addict 2; demon 2; devotee; eccentric 2; enthusiast; exception 3; fan; fanatic; fiend 2; flip 2; maniac 2; monster 2; odd 1; oddity 2; stray 5; weirdo
  freak out, explode 3; flip 2; rage 4; trip 9
freaked out, upset 10
freakish, abnormal 2; curious 3; freak 5; grotesque 1; irregular 3; kinky 1; monstrous 1; odd 1; outlandish; outré; peculiar 1; phenomenal; queer 1; unnatural 3; unusual; way-out 1; weird
freaky, deviant 1; formidable 2; freak 5; irregular 3; odd 1; offbeat; peculiar 1; way-out 1; weird
freckled, flecked; mottled; speckled; spotty 1
free°, clear 13, 15, 16, 23, 26, 31; complimentary 2; detach; detached 1; discharge 1; disengage; excuse 2; exempt 2; expansive 2; extricate; fair[1] 6; familiar 3; forthcoming 3; frank 1; generous 1; gratuitous 1; hook 5; house 7; independent 1; informal 1; ingenuous 2; large 5a; lavish 2; leisure 3a, 5; let[1] 8b; liberal 1, 3; liberate 1; liberty 4; loose 1, 2, 5, 11; nonpartisan 1; off 6; open 4, 8, 16; optional; out 6; outspoken; profuse 3; public 3; put 24; quit 4; redeem 2; release 1; relieve 2; relinquish 2; rescue 1; save 1; single 1; smooth 3; spare 2, 7; unimpeded; unmarried; unselfish; vacant 3; voluntary 1; wild 4; work 6
  free, cause to go *or* be set, get 41d
  free, for, free 4
  free, set, deliver 3; emancipate; finish 8; free 2, 14; redeem 2
  free from, exempt 1
  free from *or* of, rid 1
  free of, free 11
  free (oneself) of, throw 6b
  free will, of one's own, deliberately; freely 2; voluntarily; willingly
free (and easy), jaunty 1; nonchalant; relaxed; uninhibited; Bohemian; familiar 3; free 8; informal 1
freebie, premium 2; treat 6
freebooter, pirate 1; robber; thief 3
freebooting, pillage 2
freeborn, free 1
free choice, license 1
freed, clear 15; free 2; loose 2
freedom°, exemption; immunity 1; independence 1; laissez faire; leave[2] 1; leeway; leisure 1; liberty 1, 2, 3; license 1; margin 3; partisan 4; pass 24; passage 8; play 24; privilege; release 3; right 9; run 39; scope 2; self-government 1; way 7
freedom fighter, guerrilla; partisan 2; rebel 3
freedom fighters, resistance 3; underground 5
free enterprise, laissez faire
free-flowing, loose 3
free-for-all, brawl 1; disorder 2; fight 7; fracas 1; scramble 4; scrimmage
free hand, freedom 4
freehanded, wasteful
freeholder, landlord 2
freeing, liberation; release 3; rescue 2
freelance, outside 7
freelancer, writer
freeload, scrounge 1
freeloader, bloodsucker; hanger-on; parasite; scrounge 2
freely°, aboveboard 1; free 12; openly 2; readily 1; voluntarily; water 3; willingly
freeman, citizen 1
free of charge, free 4, 13
free pass, pass 25
free rein, carte blanche
free-spending, spendthrift 2
freethinker, individualist; infidel; liberal 4; nonbeliever
freethinking, faithless 1; heretical
free time, freedom 5; leisure 1
free trade, laissez faire
freeway, road 2
freeze°, chill 7; fix 4, 5, 7, 10; harden 1; moratorium; numb 2; paralyze 2; peg 4; preserve 3; refrigerate; set 3; solidify 1; stay[1] 1
  freeze (someone) out, freeze 4; rebuff 2; shoulder 2
  freeze (up), seize 6
freeze-dry, preserve 3
freeze-out, turnoff 2
freeze-up, freeze 5
freezing°, biting; bitter 6; cold 1; cutting 1; frigid 1; icy 1, 2; piercing 3; polar 1; raw 5; wintry 1
freight°, cargo; carriage 3; dispatch 2; ship 2; traffic 1
freightage, carriage 3; freight 1, 2
French leave, take, escape 1; run 2
frenetic, desperate 2; distraught; excited 1; fanatical; feverish; frantic; heated; hectic; hurried 1; overwrought 1; rabid 1; stormy 2; tempestuous
freneticness, excitement 1
frenzied, berserk; chaotic 2; delirious 1; desperate 1; distraught; excited 1; fanatical; fervent 1; feverish; fierce 2; frantic; furious 2; heated; hectic; hysterical 1; inflammatory; intense 2; lost 8; panic-stricken; possessed; rabid 1; rampant 1; raving 1; stormy 2; tempestuous; tumultuous; uncontrolled; uproarious 1; violent 1; wild 5
frenziedly, mad 5; madly 1
frenzy°, ecstasy 2; fanaticism 1; pandemonium; passion 2; rage 2; rampage 1; tumult; uproar; violence 2; zealotry
  frenzy, in a, mad 5; panic-stricken
frequency, incidence; key 3; occurrence 3; prevalence 1
  frequency (with), be *or* vibrate on the same, communicate 3; sympathize 2; tune 4
frequent°, attend 1; common 1; familiar 2; habitual 2; hang 4b; haunt 1; patronize 2; prevalent 1; recurrent; resort 3; walk 3

frequenter, denizen; habitué; patron 2; regular 12; steady 10
frequently°, often; repeatedly; time 21
frequent the company of, go 21c
fresh°, brand-new; brash 2; brassy 1; brazen; breezy 1; brisk 3; clear 4; cool 1; disrespectful; forward 2; fresh 5; further 1; good 7; green 1; hot 5; immature 1; immodest 2; impertinent; impudent; insolent; invigorating; late 2; lush 2; modern; new 1, 3, 6; novel 1; original 4; pert; racy 1; raw 1, 3; recent; refreshing; romantic 4; rude 2; spanking 2; stiff 5; unused 1; vivid 1
  fresh, stay, keep 11
freshen°, air 6; purify 1
  freshen (up), freshen 2; refresh 1; tone 6; top 7
freshening, spanking 2
fresher, initiate 4
freshet, flood 2; stream 1
freshman, initiate 4; newcomer 2; novice
freshness, life 7; mouth 4; novelty 1
fret°, brood 4; care 5; chafe 2, 3; cry 2; fidget 1; fuss 1; gall[2] 3, 4; gnaw 2, 3; grate 3; irritate; nail 3; offend 1; pester; rage 4; stew 3; sweat 2; worry 1
fretful°, anxious 1; discontented; fraught 2; impatient 1; irritable; jumpy; nervous 1; peevish; querulous; restless; sinking 1; testy; worried
fretfulness, nerve 4
fretting, friction 1
fretwork, net[1] 1
Freudian, subconscious 1
Freudian slip, tongue 4; trip 2
friable, brittle 1; crisp 1; rotten 2
friar, clergyman 2; monk
friary, monastery
fribble, fool 7b; putter
  fribble (with), fool 6
friction°, antagonism 2; dissension; strife 2; traction; wear 9
friend°, ally 1; associate 4; boy 3; brick 2; chum 1; cohort 3; comrade; contact 2; date 3; girl 2; intimate[1] 5; mate 1; pal 1; partner; patron 1; playmate; proponent; sweetheart
friend (at court), connection 3; patron 1
friend in need, savior 1
friendless, deserted; desolate 1; forlorn 2; lonely 3; lonesome 1; solitary 1; unpopular
friendliness, affinity 2; attachment 3; familiarity 2; fellowship 3; friendship 2; hospitality; kindness 1; warmth 2
friendly°, accommodating 1; amiable; amicable; bluff[2] 2; brotherly; chummy; considerate; cordial; decent 4; easy 6; engaging; expansive 2; familiar 3; fatherly; forthcoming 3; fraternal; genial; good 6; good-natured; gracious; hearty 1; helpful; homely 1; hospitable 1; intimate[1] 4; kind[1]; likable; mellow 4; neighborly; nice 1; obliging; outgoing 2; peaceable 2; pleasant 2; ready 2; snug; sociable; sunny 2; sweet 4; sympathetic 2; thick 8; warm 3; winning 1
  friendly, become (more), thaw 2
  friendly terms (with), be *or* **get** *or* **become on,** pal 2
  friendly with, be, chum 3; get 28a
  friendly (with), be *or* **become,** pal 2
friends, company 1
  friends (with), be *or* **become,** pal 2
friendship°, association 3; attachment 3; brotherhood 1; companionship; cooperation 2; fellowship 1; fraternity 2; love 1; society 1
frieze, border 4
fright°, alarm 2; dismay 3; dread 2; fear 1; horror 2; nerve 4; panic 1; scare 3; sight 4; terror 1; turn 36
  fright, give (someone) a, scare 1
  fright, take, wind[1] 7
frighten°, alarm 3; appall; browbeat; daunt; dismay 1; horrify 1; intimidate; lean[2] 4b; menace 1; overawe; panic 3, 4; petrify 1; scare 1; shake 5; shock 1; stampede 2; startle; terrify; upset 1; wind[1] 11
  frighten away, shoo 2
  frighten off *or* **from** *or* **away,** deter
frightened, afraid 1; cowardly; fainthearted 1; fearful 1; nervous 1; petrified 1; scared; shy 2; timid; tremulous 2; upset 8
  frightened, be, panic 3
  frightened, become, wind[1] 7
  frightened (of), be, fear 5
  frightened out of one's wits, panic-stricken
frighteners on (someone), put the, frighten
frightening°, awesome; dreadful 2; eerie; fearful 3; fearsome; formidable 1; grim 3; gruesome; hairy 2; harrowing; horrible 1; macabre; menacing; nightmarish; scary; startling; terrible 4, 6; terrifying; touchy 2
frightful°, abominable 2; atrocious 2; awful 2; dreadful 2; eerie; fearful 3; fearsome; formidable 1; frightening; ghastly 1; grim 3; gruesome; horrible 1; lurid 2; macabre; monstrous 1; scary; terrible 4, 6; ungodly 2
frightful-looking, ugly 1; unsightly
frightfully°, fearfully 2; terribly
frightfulness, monstrosity 2
frigid°, chill 4; chilly 1; cold 1, 3; coldhearted; cool 5; cutting 1; dead 4; distant 3; freezing; icy 1, 2; piercing 3; polar 1; raw 5; reserved; severe 2; stoical; stony 2; unapproachable 1; wintry 1, 2
frigidity, chill 3; cold 9; severity 2
frill°, accessory 1; embellishment 2; flounce 1; flourish 3; fringe 1; luxury 4; ornament 1; ruffle 1; trim 5
frilly, luxuriant 3; ornate
fringe°, border 7; edge 1; line[1] 22; trim 5
fringe benefit(s), benefit 2
fringe(s), outskirts; perimeter
fringy, hairy 1
fripperies, trappings; trash 2
frippery, finery; frill 2; ornament 1; sundries; trappings; trash 2
  frippery, bit of, toy 2
frisk°, caper 1, 3; cavort; frolic 3; leap 2; play 1; prance 1; skip 1, 3; sport 4; trip 4
frisky°, frolicsome; jaunty 1; jolly 1; lively 1; perky; playful 1; sportive
frisson, shiver[1] 3; thrill 1
frith, gulf 1
fritter° (away), consume 3; dissipate 3; exhaust 1; fritter; idle 5; loaf[2] 2; lose 4; pass 5; putter; run 34b; spend 2; throw 5b; use 5; waste 1; burn 4
fritter away time, fool 7b
fritz, go on the, play 17b
fritz, on the, blink 6; broken 7; defective 1; faulty; order 13c
frivol (away), fiddle 2
frivolity, festivity 2; flippancy 1; levity; nonsense 2; play 22; raillery
frivolous°, aimless 1; dizzy 2; fickle; flighty 1; flippant 1; fluffy 2; giddy 2; harebrained 2; idle 4; light[2] 6, 8; niggling 2; scatterbrained; shallow 1; stupid 2; trifling
frivolousness, flippancy 1; levity; vanity 2
frizzed, kinky 3
frizzled, crisp 2; kinky 3
frizzy, crisp 2; fuzzy 1; kinky 3
frock, dress 6; robe 1
frolic°, bound[2] 3; caper 1, 3; celebration 3; fool 7a; frisk 1; fun 1; fête 1; jamboree; joke 4; lark 1, 2; leap 2; merry 2; mirth; party 1; play 1; prank; sport 4; spree; trick 2
frolicking, merriment; mirth; sportive
frolicsome°, elfin 1; frisky; jolly 1; mischievous 1; naughty 1; playful 1; sportive; zany 1
frond, blade 3
front°, act 3; affectation 2; blind 8; bravado; camouflage 1; cover 15; disguise 4; effrontery; exterior 3; face 3, 6, 7; facing; figurehead; forward 1; gall[1] 2; gloss[1] 2; guise 2; head 3; lead 11; masquerade 2; movement 5; outside 1, 2; persona; pretense 1, 2; right 7; scapegoat; semblance 1, 2; veneer
  front, at the *or* **in,** ahead 1
  front, in, before 2; first 4; front 8; onward 2
  front for, front 12
  front (of), in, advance 10b; before 4, 5
  front on *or* **onto,** face 15
  front (on) (to), look 3; overlook 3
  front toward, face 15
frontage, exposure 4; front 2
frontier°, barrier 3; border 3; division 4; extremity 1; limit 2; line[1] 4; margin 2
frontier(s), boundary
frontiersman, pioneer 1; settler
frontierswoman, pioneer 1; settler
front (line), frontier; line[1] 14
front (man), puppet 2; figurehead
front matter, preliminary 3
front-office, responsible 3
front-page news, news 4
frore, freezing; frigid 1
frost, freeze 1, 5
frosting, icing 1
frosty, chill 4; chilly 1, 2; cold 1; freezing; frigid 1; piercing 3; standoffish; wintry 1
froth°, bubble 2, 3; ferment 1; fizz 1, 2; foam 1, 2; lather 1, 3
frothing, effervescent 1
frothy, bubbly 1; effervescent 1
frou-frou, frill 2
froward, contrary 2; obstinate
frowardness, effrontery; obstinacy
frown°, glare 2, 4; lower[2] 2; pout 1, 2; scowl 1, 2
  frown on *or* **upon,** disapprove; disfavor 3; frown 2
frowning, black 7; grim 4; stern 2
frowzy, dowdy; stuffy 1; unkempt; untidy
frozen, freezing; frigid 1; icy 1; immovable 1;

fundamentally (*cont.*) merely 1; practically 1; primarily 1; principally; principle 4; substantially; ultimately
fundamental principle, base[1] 2
fundamentals, element 3b; heart 3; rudiments; staple 3; stuff 1
funding, aid 3; backing 2; subsidy; support 9
fund(s), backing 2; capital 3; finance 2; fund 2; mean[3] 4b; money 2; pool 2; purse 2; resource 2; stock 5; treasury
funeral°, burial; wake[1] 3
funeral director, undertaker
funereal°, black 4; dismal; doleful; dreary 1; hopeless 3; mournful 1; murky; sad 2; somber 1; sullen; sunless; tragic
fungus, mold[2]
funk, fright 1
fun-loving, playful 1
funnel, stack 4; vent 1
funniness, humor 1
funny°, comic 1; entertaining; farcical; fishy 1; hilarious; humorous; hysterical 2; ludicrous; odd 1; peculiar 1; priceless 2; queer 1; rich 10; ridiculous; strange 1; uproarious 2; whimsical 1; witty; zany 1
  funny, to be, tongue 5
  funny for words, too, uproarious 2
funny business, deceit 1; fraud 1; hanky-panky; trickery
funny feeling, misgiving; premonition; qualm; sensation 1; suspicion 1
funnyman, card 10; clown 1; comedian; joker 1; panic 2
funster, comedian
fur, hide[2] 1; pelt[2]
furbelow, flounce 1; flourish 3; frill 1; fringe 1; ornament 1
furbish, decorate 2; fix 16b; polish 1
  furbish (up), embellish 1
furfuraceous, scaly 2
furious°, angry 1; black 7; cross 7; desperate 2; dirty 5; fierce 2; heated; indignant; mad 4; rabid 1; resentful; tempestuous; tumultuous; upset 10; violent 1; warm 2
  furious, be, seethe 2
  furious, become, blow[1] 8a; bristle 3; flip 2; stack 5
furiously, fiercely; mad 5; madly 3; vengeance 2a; warmly 5
furiousness, severity 5
furl, roll 9; swirl 2
furlough, free 14; holiday 1; leave[2] 2
furnish°, accommodate 3; administer 3; afford 3; appoint 3; award 1; bless 2; contribute 1; dispense 1; equip; fill 3; fit[1] 8; fix 16a, 16b; give 1; grant 1; lend 2; market 4; offer 2; open 25; outfit 4; present[2] 6; produce 6; provide 1; render 3, 7; replenish; sell 2; stock 9; supply 1, 2
furnishing, administration 3; provision 1; supply 5
furnishings, equipment; fitting 2; furniture 1; kit; regalia; trappings
furniture°, fitting 2; stuff 2; trappings
furor°, bedlam; excitement 2; flurry 1; frenzy 1; fuss 1; hurry 3; mania 1; pandemonium; passion 2; pell-mell 3; rampage 1; scene 3; sensation 2; storm 2, 3; tempest 2; tumult; upheaval
furrow°, channel 3; cranny; crevasse; crevice; gouge 3; groove; hollow 7, 8; knit 3; line[1] 3; plow 1; pucker 1; rut 1; split 7; wrinkle[1] 1, 2
furry, fuzzy 1; soft 8
further°, abet 2; advance 2; assist 2; benefit 3; besides 1; bolster; cultivate 3; ease 8; extra 1; facilitate; foment; forward 6; foster 1; fresh 4; improve 2; increase 2; likewise 2; militate 2; moreover; new 3; precipitate 1; profit 3; promote 1; speed 2; ulterior 2; yet 3
furtherance°, progress 2, 3; promotion 1
furthermore, addition 6; besides 1; further 3; likewise 2; measure 10; moreover; yet 3
furtive, devious 1; illicit 2; mysterious 2; oblique 2; sinister 2; sly 1; sneaky; stealthy; surreptitious
furtively, back 6; private 6; secrecy 2; secret 4; secretly; sly 3
furtiveness, secrecy 1; stealth
furuncle, boil[2]
furuncular, ulcerous
furunculous, ulcerous
fury°, anger 1; bitch 1; frenzy 1; hag; heat 2; indignation; passion 2; rage 1, 2; rampage 1; resentment; scold 2; severity 5; shrew; tantrum; temper 4; violence 1, 2; witch 2
fuse°, blend 2; combine 3; compound 2; connect 3; dissolve 1; fasten 1; fix 1; integrate; marry 2; melt 1; merge; run 8; stick[1] 4; unify; unite 1; weave 2; wed 2; weld 1
  fuse, on a short, cross 6; temperamental 1
  fuse (together), unite 3
fuselage, body 3
fusillade, discharge 10; fire 3; volley 1
fusing, amalgamation; merger; mixture 2; synthesis; union 1; wedding 2
fusion, amalgamation; merger; mixture 2; synthesis; union 1; wedding 2
fuss°, bother 8; disorder 2; excitement 2; fanfare 2; fault 6; fidget 1; flap 4; flurry 1; flutter 2; fracas 1; fret 1; furor 1; hurry 3; lather 2; niggle; racket 1; row[2] 2; rumpus; song 3a; squawk 2; stir 6; sweat 2; uproar
  fuss (about), make a, bother 2; fuss 2
  fuss (at), bother 2
  fuss over, fête 2; gush 2; mother 6
  fuss (with), fiddle 2
fussbudget, fidget 2; perfectionist 1
fussiness, faultfinding 1
fuss-pot, fidget 2; perfectionist 1
fussy°, careful 2; choosy; difficult 5; discriminating; elaborate 2; fastidious; faultfinding 2; finicky 1, 2; hairsplitting; meticulous; narrow 3; niggling 2; ornate; overcritical; overwrought 2; particular 4; pedantic 2; perfectionist 2; precise 3; priggish; prissy; prudish; querulous; scrupulous 1; sedate 2; showy; squeamish 1; strait-laced
fustian, bombastic; grandiose 1; mouth 3; nonsense 1; pompous 2; raving 3; rhetoric 2; rhetorical 3; wind[1] 2
fustigate, beat 1; hit 1
fusty, close 12; musty 1; rancid; stuffy 1
futile°, bootless; fruitless; hollow 5; hopeless 4; idle 4; ineffectual 1; pointless; thankless; unprofitable 2; useless 1; vain 2; void 1; worthless 2
futilely, vain 3a
futility, emptiness 2; madness 2; vanity 2
future°, fate 3; fortune 2; potential 1; predestination; prospect 3; prospective; subsequent 1
  future, in the, before 3; hence 3; yet 4
  future, in the near, soon 1
futurologist, fortuneteller
fuzz, blur 2; constable; fluff 1; officer 2; pile[3]; police officer
  fuzz, the, police 1
fuzzball, fluff 1; police officer
fuzzily, vaguely 2
fuzzy°, dim 1; focus 3; hazy 2; imprecise; indefinite 3; indistinct 1; inexact; misty; muddy 2; nebulous; obscure 2; sketchy; soft 8, 11; vague 1; woolly 1, 2

gab°, babble 1; chat 1, 2; chatter 1; drivel 2; froth 2; jabber 1; jargon 2; patter[2] 2, 3; prattle 1, 2; rap 3; raving 4; talk 3; wind[1] 2
gabbing, patter[2] 2
gabble, babble 1; chatter 1; drivel 2; gab 1; gibberish; go 32f; gossip 4; jabber 1; jargon 2; palaver 4; patter[2] 2, 3; prattle 1, 2; raving 4; yap 2
gabbling, patter[2] 2; prattle 2
gabby, talkative
gabfest, chat 1
gabion, rampart
gad° (about *or* **around),** gad; get 29b; journey 3; knock 3a
gadabout, rover
gadfly, pest; troublemaker
gadget°, apparatus; attachment 4; contraption; device 1; fixture 2; gimmick 2; implement 1; instrument 1; invention 2; machine 1; thing 5; tool 1
gadgetry, apparatus
Gaekwar (of Baroda), sovereign 1
gaff, spar[1]; spur 2
gaffe, blunder 2; error 1; fault 2; folly 2; howler; impropriety 4; indiscretion 2; misstatement 2; misstep 2; mistake 1; peccadillo; solecism; tongue 4
gaffer, boss 1; chief 1; director 1; employer 1; foreman; leader 1; master 1; overseer
gag°, heave 4; joke 1; quip 1; regurgitate; shut 6b; silence 4; story[1] 1; trick 2; vomit; wisecrack 1; witticism
  gag, for a, fun 3
gaga, crazy 1; decrepit 1; smitten 2
gage, gauge 1, 2, 3; hostage; measure 2, 4, 11; pace 4; pawn[1] 1; pledge 2, 5; security 3
gaging, measurement 1
gagman, joker 1
gags, humor 2
gagster, joker 1
gaieties°, gaiety 2
gaiety°, bliss; cheer 2; festivity 1; frolic 1; fun 1; glee; hilarity; joy 2; merriment; mirth; revelry; sparkle 4
gaily°, happily 2
gain°, accomplish; achieve 2; acquire; acquisition 1; advantage 2; appreciation 2; benefit 1, 4; blessing 2; booty; buy 1; carry 6, 10a; conquer 2; derive 1; find 5; get 45d; good 19; hit 6; improvement 2; increase 4; increment; jump 5, 9; killing 2; look 11c; make 7; money 3; net[2] 1; notch 4; obtain 2; pay 12; pick 8f; plus 3; possess 5; prize[1] 2; proceed 3; procure 1; profit 1, 2, 5; prosper; purchase 2; purpose 3;

gather°, accumulate; assemble 1, 2; brush[2] 4; bunch 4; call 4; catch 10; clump 3; cluster 3; compile; concentrate 3; crowd 3; deduce; derive 1; draw 4; enlist 2; flock 2; fold 1, 2, 4; gain 1; garner; get 50a, 50b; group 4; harvest 2; heap 3; hear 2; herd 3; hoard 2; huddle 3; imagine 2; infer; learn 1; make 7; mass 7, 8; meet[1] 2; muster 1; perceive 2; pick 2, 8b; piece 13; pool 4; press 7; presume 1; pucker 1, 2; raise 5; rake[1] 8; rally 3; receive 5; round 19; ruffle 1; scare 2; scrape 6; sit 2; summon 2; surmise 1; swarm 2; take 19; understand 4; wind[1] 6; wrinkle[1] 2
 gather from, understand 3
 gather (in), reap 1
 gather (in *or* at), throng 2
 gather (together), band[2] 3; pack 8a; raise 5
 gather (together *or* up *or* in), amass; bundle 2; collect 1, 2; find 7; herd 4; rake[1] 1, 3; scoop 6
gatherer, collector
gathering°, accumulation 1; assembly 1; attendance 2; bunch 2; cluster 2; collection 1; company 2; convention 1; council 1; flock 1; function 2; get-together; group 1; jamboree; knot 2; meeting 2; muster 2; pack 3; party 1; press 8; rally 1; roundup 1; ruffle 1; throng 1; turnout 1
gathering place, haunt 3
gauche, awkward 1; clumsy 1; incompetent; inept 1; left-handed 1; plebeian 2; rude 4; tactless; tasteless 1; uncivilized 2; ungracious; unrefined 1; vulgar 1
gaucheness, vulgarity 1
gaucherie, fault 2; impropriety 4; ineptitude 1; solecism; vulgarity 1
gaudily, gaily 1
gaudiness, gaiety 3; glare 3; glitter 4
gaudy°, conspicuous 2; extravagant 4; flamboyant 2; flashy 1; garish; gay 3; loud 2; luxuriant 3; overwrought 2; shoddy; showy; tacky[2]; tasteless 1; tawdry
gauge°, calculate; caliber 1; consider 3; estimate 1, 3; estimation 3; evaluate 2; extension 2; fathom; indicator; make 21; measure 2, 4, 11, 12; norm 2; pace 4; pattern 1; plumb 5; rate[1] 6; reckon 2; regard 4; screen 8; size 3; standard 1; view 9; yardstick
gauging, measurement 1
gaunt°, emaciated; haggard; lanky; lean[1] 1; meager 2; peaked; rawboned; scrawny; skinny; spare 3; thin 1
gauze, filter 1
gauzy, filmy 1; fine[1] 6; flimsy 3; see-through; sheer 3; thin 5
gavage, do the, overeat
gawk°, gape 1; ogle 2; rubberneck 1; stare 1; wonder 4
gawking, goggle-eyed
gawky, awkward 2; boorish; clumsy 1; rustic 2
gawp, gape 1; gawk 2; ogle 2; stare 1
gay°, blithe 1; bright 8; cheerful 1, 2; debonair 2; deviant 2, 3; effeminate; frolicsome; gala 2; gleeful; hilarious; homosexual 1, 2; jaunty 1; jolly 1; joyful 1; light[2] 9; lively 1, 4; merry 1; perky; radiant 2; sportive; sprightly; sunny 2; swish 4; vivacious; zany 1
gay blade, swell 6
gay dog, philanderer; playboy; roué
gazabo, chap; fellow 1; guy 1; man 1
gaze°, look 13; ogle 2; sight 2; stare 1, 2
 gaze at, contemplate 1
 gaze at *or* on *or* upon, eye 10; regard 1; watch 1
 gaze on *or* upon, contemplate 1
gazette, journal 1; paper 1; sheet 6
GDP, turnout 2
gear°, apparatus; apparel; clothes; costume; effects; equipment; finery; furniture 1, 2; garments; good 21a; habit 3; kit; luggage; outfit 1; paraphernalia; plant 3; property 1; regalia; rig 3; robe 2; stuff 2; tack 4; tackle 1; thing 8c; trappings; turnout 3; wear 8
 gear (up), fit[1] 8; supply 1
gearbox, gear 1
gearing up, preparation 2
gears, movement 4
geartooth, cog 1
gee-gee, nag[2]
geezer, bugger 2; chap; fellow 1; guy 1; man 1; ruin 5; stick[2] 2
geezers, elderly 2
gegenschein, light[1] 1
Gehenna, hell 1
gel, coagulate; set 3; solidify 1; thicken
gelatinize, jell 1
gelatinous, thick 5
geld, fix 13; neuter 2; sterilize 2
gelid, cold 1
gelignite, explosive 2
gelt, cash 1; money 1
gem°, jewel 1, 2; pearl; pride 3; treasure 2
gems, jewelry
gemstone, gem 1; jewel 1
gen, dope 3; information; intelligence 2; scoop 3; tip[3] 2; word 2
 gen, the, fact 3; rope 2
gendarme, officer 2; police officer
gendarmes, the, police 1
gender, sex 1
genealogical, racial
genealogy, derivation; family 3; lineage 1; line[1] 15; origin 3; pedigree; root[1] 4; stock 2
general°, average 3; broad 4, 6; catholic; common 1, 5; epidemic 1; inclusive 1; indefinite 1; lax 2; loose 5; meager 3; normal 1; ordinary 1; panoramic; pervasive; popular 2; prevalent 1; public 1; rough 10; sweeping 1; universal 1; vague 1; vulgar 3
 general, in, chiefly; generally 1; large 5b; largely; mainly; ordinarily; whole 5
general area, outskirts
general idea, notion 1
generalities°, generality 2
generalization, generality 1
generalization(s), generality 2
generalized, broad 4; general 4; vague 1
generally°, approximately 1; average 2; chiefly; frequently 2; large 5b; largely; mainly; ordinarily; part 8; popularly; primarily 1; rule 4; whole 5; widely 1
generally (speaking), usually
general neighborhood, outskirts
general practitioner, doctor 1; physician
generalship, tactic 2
generate°, bear 6; breed 2; cause 6; create 2; excite 1; father 5; grind 6; induce 2; make 5; mastermind 1; occasion 5; originate 1; output 3; produce 1, 2, 3; propagate 1; rise 17; run 29b; secrete[2]; spawn; stem[1] 3; yield 4
generated, be, rise 11
generation°, era; formation 1; secretion; vintage 1
generative, fertile; productive 2
generosity, altruism; benevolence 1; bounty 1; charity 1; grace 4; heart 5; hospitality; humanity 3; kindness 1, 2; largess; mercy 1; philanthropy 1
generous°, ample 4, 5; benevolent; big 6; bountiful 1; charitable 1; considerate; decent 4; free 7; full 5; good-natured; handsome 2; heroic 2; hospitable 1; human 3; kind[1]; large 2; lavish 2; lenient; liberal 1; merciful; noble 4; open 14; philanthropic; plentiful 1; princely 1; prodigal 2; profuse 4; selfless; softhearted; square 5; substantial 1; sweet 6; tidy 3; tolerant; unselfish
generously, amply 4; freely 4; well[1] 14
genesis, beginning 1; cause 1; conception 1; creation 1; dawn 2; formation 1; fountain 2; generation 3; mother 2; onset 2; origin 2; start 11
genetic, hereditary 1; racial
genial°, amiable; benign 1; cordial; expansive 2; friendly 2; good-natured; hearty 1; hospitable 1; likable; mellow 4; neighborly; nice 1; outgoing 2; peaceable 2; pleasant 2; ready 2; sociable; sunny 2; sweet 4; warm 3
geniality, kindness 1; warmth 2
genially, favorably 1; kindly 2; well[1] 5
genital, sexual 1; venereal
genitalia, genitals; private 7
genitals°, private 7
genius°, ability 2; brain 2; brilliance 2; capacity 2; craft 1; faculty 1; flair 1; forte; gift 2; god; head 4; highbrow 1; immortal 4; ingenuity; intellectual 3; knack; magician 2; master 2; mastermind 2; mind 4; natural 11; power 5; prodigy 1; prowess 1; specialty 1; talent 1; technique 2; touch 16; virtuoso 1
genocide, bloodshed; destruction 2; holocaust 2; killing 1; massacre 1; murder 2; slaughter 2
genre°, class 2; description 3; family 4; form 3; kind[2] 1; nature 4; stamp 8; style 1; type 1; variety 3
gens, tribe
genteel°, dainty 2; debonair 1; decorous; elegant 1; ladylike; pleasant 2; polished 2; proper 3; refined 1; respectable 1; seemly 2; well-bred
Gentile, pagan 1, 2
gentility, breeding 2; decorum 1; elegance 1; propriety 2; refinement 1
gentle°, benign 2; bland 1; dreamy 3; easy 2, 5; feminine 1; good 6; gradual; harmless; human 3; kind[1]; lenient; light[2] 4, 5; low[1] 13; mellow 4; mild 1, 3; passive 2; peaceful 1; refined 1; soft 3, 4, 5; sweet 4; tame 2, 5; tender[1] 5, 6
gentlefolk, gentry
gentleman, fellow 1; man 1; noble 1; squire 2
gentlemanly, chivalrous; courteous; decorous; gallant 2; genteel 2; nice 1;

giant° *(cont.)* monstrous 3; ogre; prodigious 1; tall 1, 2
giantess, giant 1
gibber, babble 1, 3; chatter 1; drivel 2; gab 1; gibberish; jabber 1; patter[2] 2, 3; prattle 1, 2; ramble 3; rattle 6; talk 3
gibbering, prattle 2
gibberish°, babble 3; drivel 3; froth 2; gobbledegook 1; hocus-pocus 2; jargon 2; lingo; mumbo jumbo 1; nonsense 1; patter[2] 2; raving 4; rubbish 2; trash 1; wind[1] 2
gibber-jabber, babble 3; chatter 1; drivel 2; gab 2; prattle 1; talk 3; waffle 3
gibber-jabbering, prattle 2
gibbet, gallows; hang 2
gibbous, protuberant
gibe°, dig 8; fun 5; gag[2] 1; jeer 1, 2; leg 8; quip 1, 2; ridicule 1, 2; shaft 3; taunt 2; twit[1]
  gibe (at), flout; mock 1
gibing, ridicule 1
giddiness, vertigo
giddy°, daft 1; dizzy 1, 2; faint 2; fickle; flighty 1; frivolous 2; harebrained 2; light[2] 8; queer 3; scatterbrained; silly 2; wild 7
  giddy, be, spin 3
gift°, ability 2; accomplishment 3; aptitude 2; benevolence 2; bent 5; blessing 2; boon; bounty 2; capacity 2; dole 1; donation 1; endowment 1, 2; faculty 1; flair 1; forte; genius 3; godsend; grant 3; ingenuity; knack; offering; power 5; present[2] 1; purse 3; skill 2; specialty 1; strength 3; talent 1; technique 2; tip[3] 1; touch 16; treat 6; tribute 2; trick 3
  gift, as a, house 7
gifted°, able 2; accomplished; bright 6; brilliant 4; capable 1; clever 1; divine 2; exceptional 2; golden 6; good 12; great 6; ingenious; intelligent; perfect 7; polished 1; practiced 1; precocious; professional 1; proficient; skillful; talented
  gifted with, be, possess 2
gift of the gab, oratory; rhetoric 1
gift of the gab, blessed with the, voluble
gift(s), ability 3; endowment 3; largess
gift-wrap, paper 3; wrap 1
gift-wrapping, paper 3
gig, engagement 4; launch 6; performance 2; tender[3] 1
gigantic°, big 1, 2; colossal 1; enormous; extensive 2; great 1, 2, 11; heroic 4; huge; immense; jumbo; large 3; massive; mighty 3; monolithic; monstrous 3; monumental 2; prodigious 1; tall 1, 2; thumping 1; towering 1; vast; voluminous 1; weighty 1; whopping 1
giggle°, chuckle 1, 2; laugh 1, 3; snicker 1, 2; titter 1, 2; twitter 2
giggling, laughter
gigolo, paramour
GI (Joe), private 5; soldier 1
gild, embellish 1
gilded, golden 2
gilding, embellishment 2
gild the lily, embellish 1
gill, brook[1]; canyon; gorge 1; gully
gilt, golden 2
gimbal, swivel 2
gimcrack, flimsy 1; gaudy; gewgaw; inferior 3; novelty 2; sleazy 1; toy 2
gimcracks, bric-a-brac
gimlet, drill 3
gimlet-eyed, sharp-eyed
gimmick°, device 1, 2; move 9; novelty 2; tool 1; wrinkle[2]
gimp, limp[1] 2
gimpy, lame 1
gin, ruin 4; snare 1; trap 1
ginger, punch[1] 3; spice 2; zest 1
ginger (beer), deviant 3; homosexual 1
gingerbread, elaborate 2; elaboration 1; fancy 1; fussy 2; ornament 1; ornate
ginger group, faction 1
gingerly°, noncommittal
gink, chap; fellow 1; guy 1; man 1
gin mill, bar 6; pub
ginormous, large 3; massive; prodigious 1; vast
gin palace, pub
gipsy, tramp 4; vagabond 1, 2
gird, bind 3; circle 4; encircle; gibe 1; ring[1] 4; stay[2] 2
  gird oneself, brace 6
girder, beam 1
girdle, belt 1; girth 2; ring[1] 1
girl°, broad 9; child 2; date 3; friend 3; juvenile 2; lass; love 3; maid 1; minor 3; miss[2]; steady 9; teenager; woman 1, 2; youth 2
girl Friday, aide; auxiliary 4; second[1] 8
girl (friend), date 3, flame 3, fluff 2; friend 3; girl 2; love 3; mistress 1; squeeze 10; steady 9; sweetheart; woman 2
girlhood, childhood; youth 1
girlie show, burlesque 2
girlish, young 2
girls' room, lavatory
girt, ring[1] 4
girth°, circuit 1
gismo. See **gizmo**
gist°, core 2; drift 4; import 3; kernel 2; meaning 1; meat 2; nub 2; pith 1; sense 4; significance 1; subject 1; substance 2; tenor; theme 1
git, fool 1
give°, afford 2, 3; allow 5; award 1; bestow; bless 2; carry 8; cede; commit 1; communicate 1; confer 2; contribute 1; delegate 3; deliver 2, 4, 5; devise 2; donate; elasticity 1; extend 4; flex 2; flexibility 1; furnish 1; give 15b; grant 1; hand 14; heap 4; impart 1; lay[1] 18b; leave[1] 8; lend 2; pass 8, 18a; pay 3; place 18; play 24; present[2] 5; provide 2; put 28e; relent; return 6; sign 11; slack 6; spare 8; stretch 4; supply 1, 2; transfer 1; trust 6
give (away), blurt; dispense 1; dispose 3c; give 12; inform 2; let[1] 7a; sell 3; show 12a
give away (the game), leak 5
give away the (whole) show, tell[1] 2
give back, repay; restore 1, 5; return 3; turn 15b
give (be)grudgingly *or* **unwillingly** *or* **reluctantly,** begrudge 2
give dimension to, flesh 6
give forth, afford 3; exhale
give in, fall 5
give in (to), bow 2; capitulate 2; consent 1, 2; defer[2]; give 13; hand 16; obey 1; soften 4; submit 1; surrender 2; turn 15b; weaken 3; yield 1
give in (to defeat), give 17b
give off, emanate 2; emit; give 14; let[1] 6d; radiate 2; secrete[2]; send 7; throw 7a
give (oneself) up to, wallow 2
give (out), administer 3; assign 1; deal 1; dispense 1; dispose 3c; distribute 1; divide 2; dole 3; go 12; hand 17; measure 14; mete; pack 8b; pay 11a; present[2] 6; ration 3; serve 3; shell 5; deliver 1; emanate 2; emit; fail 3; give 14, 15; let[1] 6d; peter out; radiate 2; throw 7a
give (out) on *or* **onto,** face 15
give out (with), give 8; loose 10
give (over), abandon 1; dedicate 1; deliver 2; forfeit 2; give 16; leave[1] 7; yield 2
give (someone) back, restore 2
give substance to, flesh 6
give (to), subscribe 2
give (up), abandon 1, 4; abdicate; afford 2; break 10; capitulate 1; cede; climb 5b; come 5b; concede 2; dedicate 1; deliver 2; devote 2; discontinue; dispense 3a; drop 7; fall 5; forfeit 2; forgo 1, 2; forsake 3; give 5, 13, 16, 17; grant 2; hand 18a; kiss 3; leave[1] 4, 7, 9; lose 3; pack 6; part 14; pass 22; pull 14d; quit 2; refrain[1] 2; relinquish 1, 2; render 3; renounce; resign 1; sacrifice 3, 4; scrap[1] 4; scrub 2; shun; sign 8; spare 9; stop 1; succumb; surrender 1, 2; swear 4; throw 9b; vacate 2; vouchsafe 1; waive 1; yield 2
  give upon, overlook 3
  give up (to), submit 1
  give way (to), bend 5; break 16c; cave 2a, 2b; cede; fold 3; give 9; go 12; obey 1; relent; snap 1; soften 4; succumb; surrender 2; way 11; weaken 2, 3; yield 3
give-and-take°, repartee; volley 3
give a speech to, address 3
giveaway, bargain 2; gift 1; steal 4
give away secrets, tattle 1
given°, disposed; principle 1; prone 2; ready 3; stipulation; understood
  given, a, one 3
  given, be, come 9a; get 2; receive 1
  given, take as, suppose 1
  given up, sacrificial 1; unused 2
  given up, be, go 9
give or take, about 2; approximately 1
giver, donor
give the coup de grâce, finish 4; kill 1
give thumbs down (to *or* **on),** refuse[1] 1; reject 1; thumb 3
give thumbs up (to *or* **on),** thumb 4
give up work(ing), retire 2
give way, yield 3
give way (to), defer[2]
giving, big 6; donation 2; presentation 1; softhearted; unselfish
  giving in, submission 1
  giving off, secretion
  giving (out), assignment 1; distribution 2; partition 2
  giving up, sacrifice 1, 2; self-denial 2
gizmo, contraption; gadget; gimmick 2; instrument 1; invention 2; machine 1; thing 5; tool 1
glabrous, hairless; smooth 4
glacial, chill 4; cold 1; freezing; frigid 1; icy 1; polar 1; wintry 1, 2
glad°, cheerful 1; content[2] 3; ecstatic; elevated 3; exuberant 2; exultant; happy 1; joyful 2; pleased; radiant 2; thankful
  glad, be, rejoice

glad (about), proud 1
glad eye, give (someone) the, ogle 1
glad rags, apparel; clothes; finery
glad rags, put on one's, dress 5; primp
gladden°, cheer 6; comfort 1; content[2] 4; delight 1; enliven 2; feast 6; flush[1] 3; gratify; lighten[1] 2; please 1
gladdening, cheerful 2; exhilarating 2
glad eye, eye 5; leer 2; ogle 3
gladly°, happily 3; readily 1; soon 4; willingly
gladness, bliss; cheer 2; ecstasy 1; glee; happiness; joy 1; radiance 2
gladsome, cheerful 1
Gladstone bag, bag 2
glamorous°, prestigious; romantic 3
glamour°, flair 2; magic 3; prestige; romance 4
glance°, look 13; peek 2; scan 1; sight 5; skim 2
glance, at first, surface 2
glance at, scan 1; sight 5; skim 2
glance through, scan 1; skim 2
glancing, passing 2
gland, ring[1] 1
glare°, flare 4; flash 4; flicker 3; frown 1, 3; gleam 1; light[1] 3; scowl 1, 2; shine 1
glaring°, blatant 1; fiery 2; flagrant; flashy 1; full 10; garish; gaudy; great 11; gross 4; lurid 4; naked 3; obvious; open 13; positive 9; prominent 1; strong 19
glass°, drink 6; mirror 1; pane; telescope 1
glasses°, glass 7; spectacle 3
glasshouse, hothouse 1; prison; stir 7
glassware, glass 1
glassy°, glossy 1; shiny; slick 1; slippery 1; smooth 1, 2
glaze°, gloss[1] 1, 3; icing 1; polish 5; shine 4; spread 7; wash 7, 15
glaze (over), film 4
glazed, glassy 2; glossy 1
gleam°, beam 2; blink 2; dawn 3; flash 1, 4; flicker 3; glance 2, 5; glisten; gloss[1] 1; glow 2, 4; light[1] 3; luster 1; particle; polish 5; radiance 1; radiate 1; ray 1; shaft 2; sheen; shimmer 1, 2; shine 1, 4; sparkle 3; twinkle 1, 2
gleaming, ablaze 2; bright 1; fiery 2; glassy 1; golden 3; luminous 1; radiant 1; scintillating 1; shiny; silver 3; spanking 1; spotless 1; twinkle 2
glean, deduce; derive 1; extract 2, 4; gain 1; gather 1; get 21; harvest 2; heap 3; pick 8b; reap 1; scrape 6; win 2
gleaning, collection 1
glee°, bliss; festivity 1; fun 1; gaiety 1; happiness; hilarity; joy 2; merriment; mirth
gleeful°, cheerful 1; ecstatic; elated; exultant; gala 2; gay 2; happy 1; joyful 1; merry 1; radiant 2; sportive
gleefully, gaily 2; happily 2
glen, hollow 7; valley
glib°, eloquent 1; flashy 2; fluent; greasy 2; offhand 2; oily 2; slick 2, 4; smooth 6; voluble; windy 2
glibness, fluency; oratory
glide°, coast 2; float 1; flow 1; plane 4; roll 2; sail 3; skim 3; slide 1; slip[1] 1; slither; snake 3; stream 4; sweep 3
glide by, elapse
glimmer, blink 2; flare 1; flash 1, 4; flicker 1, 3, 4; ghost 2; glance 2, 5; gleam 1, 2, 4; glisten; glow 4; ray 2; sheen; shimmer 1, 2; spark 1; sparkle 1; suspicion 2; vestige
glimmering, flicker 3; inkling
glimpse, glance 1, 4; peek 1, 2; perceive 1; see 1; sight 5, 8; spot 6; spy 3
glimpse of, catch a, discover 2; see 1; spy 3
glint, flash 1; flicker 1, 3; glance 2, 5; gleam 1, 3, 4; glisten; sheen; shimmer 1, 2; shine 1; spark 1; sparkle 1; twinkle 1, 2
glinting, shiny
glissade, glide; slide 1
glisten°, glance 2; gleam 4; radiate 1; shimmer 1; shine 1; twinkle 1
glistening, glossy 1; luminous 1; radiant 1; scintillating 1; shiny; twinkle 2
glitch, bug 6
glitter°, brilliance 1; dazzle 3; flash 1, 4; glamour; glance 2, 5; gleam 1, 4; glisten; radiate 1; shine 1; sparkle 1, 3; splendor 2; twinkle 1, 2
glittering, bright 1; brilliant 1; golden 3; radiant 1; scintillating 1; shiny; twinkle 2
glitz, glitter 4
glitzy, flashy 1; garish
gloaming, evening; night 2; twilight 1
gloat°, glory 5
glob, blob; clod 1; clump 1
global°, extensive 1; general 1; international; terrestrial 1
globally, everywhere
globate, round 3
globate(d), spherical
globe°, earth 1; orb; round 10; sphere 1; world 1
globelike, round 3; spherical
globe-shaped, round 3; spherical
globe-trot, tour 4; trip 7
globe-trotter, sightseer; traveler 1
globe-trotting, travel 1
globoid, spherical
globose, spherical
globous, spherical
globular, rotund 1; round 3; spherical
globule, blob; bubble 1; drop 1; globe 2; sphere 1
glom, see 1; sight 5
gloom°, dark 11; depression 2; desolation 2; despair 1; desperation 2; melancholy 2; misery 1; night 1; obscurity 1; sadness; shade 1; shadow 1; woe
gloom over, cast a, depress 1
gloomily, sadly 2
gloominess, dark 11; depression 2; despair 1; gloom 1; gravity 3; melancholy 2; sadness; shade 1
gloomy°, black 4; bleak 1; blue 1; brokenhearted; cold 4; dark 3; dejected; desolate 3; dim 1; dingy; dismal; doleful; dour 1; dreary 1; dull 5; dusky 2; forlorn 1; funereal; glum; grave[2] 1; gray 2; heavy 5, 6, 9; hopeless 3; hurt 7; inauspicious; joyless 2; leaden 3; low[1] 8; melancholy 1; miserable 1; moody 1; morbid 3; murky; obscure 1; ominous 1; overcast; pessimistic; portentous 1; sad 1, 2; shadowy 1; sinister 1; solemn 1; somber 1, 2; sorrowful 1; sour 4; stern 2; sullen; sunless; twilight 4; unhappy 1; unpromising; woebegone
gloomy air *or* **mood** *or* **atmosphere,** pall[1] 2
gloomy Gus, killjoy; misery 4
glop, ooze 1
gloppy, slimy 1
glorification, glory 2; magnification; praise 2; reverence 1; tribute 1; worship 2
glorified°, exalted 1; sublime 1
glorify°, bless 1; celebrate 3; dignify; exalt 2; extol; hail[1] 2; hallow 1; honor 6; idealize; idolize; immortalize; laud; pedestal 2; praise 4; revere; sanctify 1; worship 1
glorious°, bright 7; brilliant 2; conspicuous 3; dazzling; divine 3; gallant 3; golden 4; gorgeous 1, 2; heavenly 2; magnificent; majestic 1; marvelous; monumental 1; prestigious; proud 3; splendid 1; stunning 2; sublime 1; superb; triumphal
glory°, dignity 2; distinction 2; fame; honor 2; kudos; laurels; luster 2; nobility 1; pomp; praise 2; rejoice; renown; splendor 1; state 3; triumph 2
glory (in), delight 2; exult; gloat; pride 4; revel 1; wallow 2
gloss°, glaze 1, 2; glossary; luster 1; note 3; polish 1, 5; sheen; shimmer 2; shine 4; splendor 2; translation 1; trot 4; veneer
gloss over, discount 3; disregard 1; gloss[1] 4; jump 3; overlook 2; slide 4; slur 3; smooth 12; whitewash
glossary°, dictionary
glossy°, bright 3; glassy 1; lustrous; radiant 1; shiny; silky; sleek 1; slick 1; smooth 2
glow°, blaze 3; fervor; flame 4; flush[1] 1, 4; gleam 1; glisten; gloss[1] 1; light[1] 3; luster 1; polish 5; radiance 1; radiate 1; sheen; shimmer 1, 2; shine 1, 4; sweat 1
glower, frown 1, 3; glare 2, 4; lower[2] 2; scowl 1
glowering, black 4, 7; dark 5; leaden 3; lurid 4; severe 2
gloweringly, severely 3
glowing°, bright 1; fervent 1; fiery 2; fresh 6; golden 3; impassioned; incandescent; live 3; lively 4; luminous 2; lurid 4; radiant 1; rosy 1; shiny; vivid 1
glowingly, well[1] 5
gloze, deceit 2; flattery
glue°, anchor 3; apply 1; attach 1; cement 1, 2; connect 3; nail 7; stick[1] 4; tack 5
glue (together), join 1; unite 3
gluey, gooey 1; ropy 1; slimy 1; sticky 1; tacky[1]; tenacious 2; thick 5
glueyness, tenacity 2
glum°, blue 1; brokenhearted; dejected; dreary 1; forlorn 1; gloomy 2; gray 2; low[1] 8; melancholy 1; miserable 1; moody 1; morbid 3; pessimistic; sad 1; solemn 1; sullen; unhappy 1; woebegone
glumness, depression 2; melancholy 2
glut°, abundance; excess 1; flood 3, 5; gorge 3; pall[2] 2; profusion; satiate 1; satiety; superfluity; surfeit; surplus 1
glutinous, gooey 1; ropy 1; slimy 1; sticky 1; tenacious 2; thick 5
glutinousness, tenacity 2
glutted, jaded 2; surfeited
glutting, glut 2
gluttonize, devour 1; stuff 8
gluttonous°, epicurean 1; greedy 1; hoggish; ravenous 2; self-indulgent; voracious 1

gluttony°, greed 3
glyph, groove
G-man, officer 2; operative 4
gnarl, distort 1; knot 1; tangle 1, 3
gnarled°, crooked 2; deformed 1; grotesque 1; labyrinthine; misshapen; wizened
gnash, grind 3
  gnash at, snap 3a
gnaw°, bite 1; chew 1; waste 3
  gnaw away (at), erode
gnawing away, erosion
gnome, goblin; motto; proverb; saw
GNP, turnout 2
go°, alternate 1; attempt 2; bounce 2; check 10a; collapse 1; depart 1; drive 3; eagerness 1; essay 2; fling 3; flit; flock 2; flow 1; function 3; get 10; go 36a; head 10; headway 2; hop 2; make 20; migrate 1; mingle 2; move 1; operate 1; perform 2; pull 14c; push 14; range 6; resign 1; roll 2; run 3, 11, 30, 30b; set 18b; shot 3; sink 9; split 6; start 3; stroke 2; travel 3, 4; trial 3; try 5; turn 30; walk 1; wander 1; withdraw 4, 5; work 11, 12
  go, at a, once 6b
  "go", from the word, completely 1; first 8; originally
  go, in one, once 6b
  go, let, abandon 1; discharge 1; dismiss 1; drop 7, 10; emancipate; expel 1; finish 8; fire 11; free 2, 14; lay[1] 16a; let[1] 6a, 8b; liberate 1; loose 11, 12, 13; part 14; relax 1; release 1; relinquish 1, 2; resign 1; sack 4; sacrifice 4; slack 3a; spare 7; vacate 2; vent 3; wreak
  go, let (oneself), thaw 2
  go, on the, active 1; move 13a, 13b, 13c; run 49c
  go aboard, board 6; embark 1; get 39a
  go about, come 4b; essay 3; fall 21; get 29a; go 18; ramble 1; tack 6
  go about *or* **(a)round,** circulate 1; go 21a
  go (abroad *or* **overseas),** journey 3
  go across, cross 5; transit 3; traverse 2
  go after, follow 1; pursue 1
  go against, fly 6; militate 1; traverse 3
  go ahead, headway 2; progress 5
  go (ahead) (of), go 19; pass 2; precede
  go all out, exert 2; strive 1
  go along, progress 5; shape 10b
  go along (with), accompany 1, 2; approve 1; enable 3; fall 15; follow 2, 3; get 28c; go 20; play 8a; second[1] 9; string 10a; sympathize 2; side 10
  go amiss, miscarry
  go ape, rage 4
  go (a)round, turn 4; circle 3; get 29a; orbit 2
  go (a)round *or* **about (together),** fraternize; mix 2
  go (a)round *or* **about (with),** circulate 3; flank 3; fraternize; go 21; pal 2; revolve 1; roll 1; round 16, 17; turn 1
  go as, pass 16b
  go ashore, disembark; land 5
  go astray, err 1; sin 3; wrong 8a
  go at, go 22
  go (at), have a, try 1; attempt 1; endeavor 1; essay 3; tackle 3
  go away, draw 12b; flight[2] 3; get 28c; pull 7, 14b; slip[1] 6
  go (away) (from), abandon 2; bugger 5; depart 1; drop 13; evacuate 2; exit 3; flee 1; fly 2; get 45a; go 23, 31d; leave[1] 1; melt 4; part 12; pass 10, 14, 14b; push 9b; quit 1; ramble 1; retire 1; run 22; shoo 1; take 34c
  go awry, miscarry
  go back, get 33a; recede 1; relapse 1; return 1; revert; turn 13b
  go back and forth *or* **backward(s) and forward(s),** swing 1
  go back on (one's promise *or* **word),** back 18; go 25; renege 2; withdraw 2
  go back (to), go 24
  go backward(s), back 2b; reverse 5
  go bad, decay 2; decompose 2; go 31e; putrefy; rot 1; sour 5; spoil 4; turn 5
  go before, precede
  go (between), intervene 2
  go beyond, exceed 1, 2; excel; far 7; overstep; pass 6; run 33c; surpass; transcend
  go beyond (the law), transgress 2
  go (by), elapse; fly 4; get 11; go 26; lapse 6; pass 1, 11; slip[1] 5
  go (by), let, disregard 1; overlook 2; pass 3, 22
  go by the board, pass 10
  go down, climb 5a; decline 2, 5; descend 1; dip 2; ebb 1; fall 3; founder[2] 1; get 37a; go 27; land 5; light[2] 13; sag 2; set 2; sink 1, 2, 3, 5; submerge 2; subside 1; suffer 4; taper 1
  go downhill, deteriorate 1; seed 4; sink 4; worsen 2; wrong 8a
  go down on, sink 3
  go down to, sink 3
  go down with, contract 3
  go far, far 6
  go first, lead 4; precede
  go for, buy 2; favor 6; fetch 1, 3; get 4; go 28; like[2] 1; make 30b; pick 8i; regard 5
  go (forth), pour 4; issue 11
  go forward, advance 3; go 19; proceed 1
  go (forward(s)), advance 1; progress 5
  go (free), let, rescue 1
  go from, depart 1
  go from *or* **out of,** drain 6
  go in, get 39b
  go in advance (of), precede
  go in for, favor 6; go 29
  go (in *or* **into),** enter 1; turn 16b
  go into, begin 2; dig 5; embark 2; examine 1; get 40b; go 29a, 30; look 8; penetrate 1; probe 1; strike 3; study 3; treat 2
  go in with, side 10
  go make for, come 8
  go (of), let, drop 8; excuse 2; fly 8a; spare 8; surrender 1
  go (off), clear 29; decay 2; decompose 2; depart 1; drop 13; explode 1; flight[2] 3; fly 2; get 41b; go 23, 31, 33, 33a; leave[1] 1, 2; move 2; part 12; pop 1; pull 14b; put 21e; putrefy; ramble 1; retire 1; rot 1; scatter 2; slip[1] 6; spoil 4; sour 5; swear 4; turn 5; wander 3
  go off into, go 31f
  go (on), add 3; advance 3; brew 3; carry 11a; continue 2, 5; go 19, 32; hang 7c; happen 1; place 14; spout 2; talk 3; extend 3; progress 5
  go on (about), brag; hold 16a; lecture 3
  go on (and on), prattle 1; ramble 3
  go on *or* **in,** ride 1
  go (onward(s)), progress 5
  go on (with), continue 1; proceed 1; prosecute 2; take 39b
  go (out), exit 3; get 45a; depart 1; ebb 1; fail 3; get 29b; go 2, 31, 31a, 33; scene 6; slip[1] 6; socialize; step 17b
  go (out) after, apply 6; chase 2; fetch 1; follow 11a; get 4; pick 8i; prey 3b; run 21; seek 1; victimize 1
  go out for, go 29a
  go out of, vacate 1
  go outside *or* **out of doors,** step 17a
  go (out) with, associate 1b; date 7; see 7
  go over, bridge 3; brush[2] 8; come 16a; defect 3; examine 1; frisk 2; go 34; hunt 2; pore[1]; range 8; rehearse 2; run 33b, 33d; scrutinize; span 3; study 3; take 22; transit 3
  go over (again), recapitulate; rehash 1; review 2
  go overboard, far 7; meal 2; overdo 1
  go over the hill, desert[1] 4
  go over the top, far 7
  go over (with a fine-tooth(ed) comb), rake[1] 6; ransack 1
  go past, go 26a; pass 1; take 22
  go pfft, misfire 1
  go phut, misfire 1; wrong 8b
  go places, far 6a; get 27; place 10; rise 8
  go round, bypass 1; circle 3; rotate 1; take 22
  go somewhere, rise 8
  go through, examine 1; experience 4; feel 5; go 34, 34a, 35; meet[1] 6; penetrate 1; puncture 3; rehearse 2; rifle 2; run 33b; scrutinize; search 1; see 10; serve 6; spend 2; stick[1] 6; suffer 2; take 6, 22; transit 3; transmit 2; undergo
  go through (with a fine-tooth(ed) comb), rake[1] 6; ransack 1
  go through with a fine-tooth comb, hunt 2
  go to, attend 1; reach 2, 4; settle 4
  go to and fro, swing 1
  go (together), accord 1; agree 1; coincide; compound 2; converge; fit[1] 6; flock 2; go 33c, 36; marry 2; match 7
  go together with, accompany 2; go 39a
  go too far, far 7; meal 2; overdo 1
  go (to *or* **and) get,** get 4; pick 8i
  go to *or* **at (it),** peg 6
  go to regularly, frequent 2
  go under, collapse 2; dive 1; fail 4; fold 3; founder[2] 1; go 27a, 37; sink 1; wall 4
  go up, arise 2; climb 1, 3; go 38; mount[2] 1; rise 3, 10; scale[3] 3
  go up against, face 14, 18b; tangle 4
  go uphill, rise 6
  go up to, reach 5
  go well, work 19d
  go west, die 1; pass 14a
  go with, chum 3; go 39; join 3; match 7; squire 1; walk 2
  go (with), let, fire 9; fly 8b; loose 10
  go without, forgo 1; go 40
goad, abet 1; drive 1; foment; gall[2] 4; impetus; incentive; incite; induce 1; inflame 1; kindle; motivate; motive 1; nag[1] 1; poke 1; premium 2; prod 3, 5; provoke 1; push 4; put 29; rise 16; rouse 2; signal 2; spur 1, 4; stimulant 1;

good spirits, gaiety 1; glee; merriment
good taste, chic 2; elegance 1
good (to eat), edible
goodwill, affection; benevolence 1; charity 1; favor 1; grace 3; humanity 3; indulgence 1; kindness 1; readiness 1
good wishes, compliment 2; greeting 3; regard 10; respect 5; toast 1
good word, recommendation 2
goody, tidbit
goody-goody°, pharisaic; pious 2; prig; priggish; prude; sanctimonious; self-righteous; strait-laced
gooey°, mushy 2; sentimental 2; slimy 1; sticky 1; tacky[1]
gooeyness, sentimentality
goof, blunder 2; error 1; fault 2; folly 2; lapse 1; misstep 2; mistake 1; peccadillo
  goof around, idle 6
  goof off, idle 6; loaf[2] 1; slack 4
  goof (up), err 1; foul 16b
goof-off, good-for-nothing 2; slacker
goof-up, blunder 2; mistake 1
goofy, crazy 1; foolish 2; inane; mad 1; zany 1
goon, dolt; fool 1; gangster; henchman; hoodlum; punk 1; thug
goop, boor 2; ooze 1
goose, fool 1; silly 3
goose bumps *or* **pimples,** shiver[1] 4
  goose bumps *or* **goose pimples, give one,** scare 1
goose egg, nil; zero 1
goose-step, walk 1
Gordian, labyrinthine; perplexing
Gordian knot, difficulty 2
gore°, stab 1; stick[1] 1; tear 4
gorge°, canyon; crevasse; devour 1; fill 5; glut 3, 4; gully; gulp 1; overeat; pall[2] 2; pass 23; put 13d; ravine; satiate 1; stuff 8
  gorge (oneself), feast 4
  gorge rise, make one's, sicken 2
gorged, full 1; jaded 2; surfeited
  gorged (with), replete 1
gorgeous°, beautiful 1; dazzling; extraordinary 2; glorious 4; heavenly 2; lively 4; lovely 1; ravishing; splendid 1; stunning 2; superb; voluptuous 2
gorgeousness, splendor 1
Gorgon, ogre; bag 4; hag; witch 2
gorgonian, hideous 1
Gorgonize, freeze 3
gorilla, bruiser; tough 8
gormand, glutton
gormandism, gluttony
gormandize, devour 1; feast 4; gorge 3; overeat; put 13d; stuff 8
gormandizer, glutton
gormandizing, gluttonous; gluttony; greed 3; self-indulgent
gormless, backward 2; blind 2; daft 1; feebleminded; insane 2; mindless 1; thick 6
gormless ninnyhammer, gawk 1
gory°, graphic 1; grisly; lurid 2; macabre; sanguinary 1; terrible 4
Goshen, paradise 2
go-slow, strike 20
gospel, given 4
Gospel, spread the, preach 1
gospeler, clergyman 3
Gospel(s), Scripture
gossamer, fine[1] 6; flimsy 3; fluffy 2; insubstantial 1; see-through; sheer 3; tenuous 1; thin 5
gossamer(-like), filmy 1
gossip°, babble 2; blabbermouth; busybody; chat 1, 2; chatter 3; chew 2; conversation; converse; dirt 4; gab 1, 2; grapevine; news 1; palaver 2, 4; patter[2] 2, 3; rag[1] 5a; rap 3; rumor 1; tale 3; talebearer; talk 4, 17, 18; tattle 2; twitter 2; whisper 2
  gossip about, gossip 4
gossip columnist, scribe 2
gossipmonger, gossip 3; talebearer; troublemaker
gossipy, talkative; voluble
gossoon, stripling
got up, factitious
gothic novel *or* **tale,** romance 3
Götterdämmerung, twilight 3
gouge°, dig 1; gore[2]; groove; hollow 8; nick 1, 2; notch 1, 3; pit[1] 5; profiteer 2; score 10; scratch 4; slash 4
  gouge (out), bore[1] 2; excavate 1; scoop 5; scrape 2; scratch 1; undercut 1
goulash, medley; mishmash; mixture 1; stew 1
gourd, head 1
  gourd, out of one's, crazy 1; unbalanced 2
gourmand, epicure; glutton; gourmet; sensualist
gourmet°, epicure; epicurean 2; luxurious 1; sensualist; sybarite
gout, blob
gov, boss 1; chief 1
govern°, chair 4; command 2; control 1; determine 4; direct 1; discipline 7; dominate 1; guide 3; head 11; inhibit; lead 3; manage 1, 2; peg 4; possess 3; preside; regulate 1, 2; reign 2; restrain 1; rule 5, 6; shape 7; supervise
governable, tractable 1
governance, management 1
governess, minder 1; mistress 2; servant 1
governing, direction 1; dominant 1; executive 3; prevalent 1; sovereign 2
government°, administration 2; authority 4; bureaucracy; command 7; conduct 2; crown 2; direction 1; discipline 4; executive 2; guidance 1; management 1; politics 1; power 10; regime; reign 1; state 5
  government, the, establishment 3
governmental, executive 3; national 1; political 1; state 5
government worker, servant 2
governor, boss 1; chief 1; director 1; employer 1; father 1; head 2; jailer; master 1; superintendent; supervisor
governorship, leadership
gown, dress 6
G.P., doctor 1; physician
grab°, abduct; capture 2; catch 1, 9; clasp 5; claw 3; clutch 1; get 8; grapple 1; grasp 1; hang 7a; hook 6; impress 1; intercept; jump 6; lay[1] 11; mesh 3; occupy 3; pilfer; pluck 3; seize 1; snatch 1, 4; take 1
  grab (up), snap 4
grab bag, miscellany
grace°, become 4; bless 2; charm 2; elegance 1; enrich 2; favor 4; finesse 3; fluency; forgiveness 2; piety 2; propriety 2; sanctity; savoir-faire; taste 4
graceful°, dainty 1; delicate 2; elegant 1; exquisite 1; fluent; light[2] 5; lyric 3; neat 3; polished 2; pretty 1; shapely; sleek 2; slender 1; supple 2; tasteful; willowy 2
gracefulness, delicacy 1; grace 1
graceless, heavy-handed 1; stilted; ungraceful 2
graces, the social, propriety 3
gracious°, attentive 2; benevolent; benign 1; chivalrous; considerate; cordial; debonair 1; decent 4; fair[1] 8; gallant 2; genteel 2; gentle 1; good 6; good-natured; hospitable 1; indeed 3; kind[1]; ladylike; merciful, neighborly; nice 1; obliging; pleasant 2; ready 2; refined 1; respectful; suave; sweet 6; well-bred
graciously, favorably 1; kindly 2; readily 1; well[1] 5
graciousness, grace 4; hospitality; kindness 1; mercy 1; readiness 1
gradate, blend 2; graduate 2
gradatim, gradually
gradation, grade 1; notch 2; scale[3] 1; step 5
grade°, blend 2; bracket 3; category; class 1, 4, 5; climb 6; coordinate 1; correct 5; degree 1; denomination 2; distinguish 2; divide 5; downgrade 3; graduate 2; hill 3; incline 4; mark 4, 14; notch 2; place 3, 16; quality 2; ramp; range 7; rank[1] 1, 6; rate[1] 4, 6; roll 7; score 1; screen 8; slope 2; sort 8; stage 1; stamp 8; standing 5; tabulate
  grade, make the, arrive 2; grade 5; make 32a; mark 9; measure 15b; muster 3; qualify 2; rise 8
gradient, downgrade 3; grade 4; hill 3; incline 4; ramp; slant 3; slope 2
grading, mark 4
gradual°, delicate 6; gentle 3; lingering 2; piecemeal 3; progressive 1; slow 1, 2; unhurried
gradually°, degree 3; piecemeal 1; step 9
graduate°, blend 2; calibrate; rank[1] 6; score 10
graduation, scale[3] 1
graft°, blackmail 1; bribe 1; implant 2, 3; line[2] 2; payoff 3; rebate 2; union 3
grain°, bit 1; jot 2; kernel 1; modicum; morsel 2; particle; provender 2; scrap[1] 1; seed 1; shade 3; shred 1; speck; taste 1; texture
  grain, go against the, grate 3; rub 8
grainy, granular; gritty 1; lumpy
gralloch, gut 1
grammar (book), schoolbook
grammatically, well[1] 3
gramophone record, record 7
grand°, big 1; brave 2; dignified; distinguished 2; elegant 3; elevated 2; exalted 1; gallant 3; gorgeous 1; grandiose 2; great 1, 7, 12; heroic 4, 6; high 12; imperial 2; imposing; lofty 2, 4; luxurious 1; magnificent; majestic 1; mighty 3; monumental 1; neat 5; noble 3, 5; palatial; posh; proud 4; regal; royal 2; scenic; solemn 3; splendid 1; stately; statuesque; stunning 2; sublime 1; swanky; swell 7
grandchild, descendant
grandeur°, dignity 1; display 5; elegance 2; elevation 4; glory 3; luxury 1; nobility 1; pageant; pomp; quality 3; solemnity; splendor 1; state 3
grand finale, consummation 2; payoff 2
grandiloquence, bluster 3; bombast; hot air; oratory; raving 3; rhetoric 2
grandiloquent, bombastic; flowery; inflated 2; ornate; pompous 2; ponderous 2;

pretentious 1; rhetorical 3; stilted
grandiose°, bombastic; fancy 2; grand 3; heroic 4; inflated 1; lofty 4; majestic 2; ornate; pompous 1; pretentious 1; rhetorical 3
grandly, well[1] 4
grandstand, show 11; theatrical 2
grandstander, trendy 2
grand total, sum 1
grange, farm 1
granger, farmer
grangerize, illustrate 2
grangousier, glutton
granite(like), hard 1
grant°, accommodation 5; admit 2, 3; afford 3; agree 3; aid 3; allow 1, 5; assign 1; award 1, 3; backing 2; benevolence 2; bestow; bounty 2; cede; concede 1, 2; confer 2; contribute 1; dole 1; donate; donation 1; endowment 1; extend 4; fund 3; gift 1; give 1; impart 1; patent 1; permit 1; philanthropy 2; present[2] 2, 6; recognize 2; scholarship 2; show 10; stipend; subsidy; vouchsafe 1
granted, though 1
  granted, take for, assume 3; presume 1; reckon 4; suppose 1
  granted, take (it) for, imagine 2
  granted, taken for, assumed 3
grant-in-aid, accommodation 5; aid 3; grant 3
granting that, though 1
grants, largess
granular°, gritty 1; lumpy
granulate, grind 1; mill 5; powder 3; pulverize 1
granulated, granular
granule, grain 3; morsel 2
grapevine°, gossip 2; rumor 1
  grapevine, hear on the, wind[1] 6
grapevine telegraph, grapevine
graph, chart 2; profile 3; schematic 2
grapheme, sign 4
graphic°, dramatic 2; eloquent 1; lifelike; lurid 1; photographic 1; pictorial 1; picturesque 2; plain 2; realistic 2; schematic 1; strong 15; vivid 2
grapple°, claw 3; lock[1] 4
  grapple (with), combat 5; deal 4; encounter 3; engage 5; fight 1; grapple 2; grip 5; oppose 1; tackle 3; withstand
grasp°, catch 1, 10, 13a; clasp 2, 5; clutch 1; command 8; comprehend; digest 3; dominion 1; embrace 1; familiarity 1; fathom; figure 12b; follow 8; get 19; grab 1, 4; grapple 1; grip 1, 3, 6; hand 8; hang 7a; hold 1, 19a, 24; know 1; knowledge 1; lock[1] 4; make 37d; master 7; mesh 2; obtain 1; penetrate 5; perceive 2; perception 1; pierce 3; pinch 1; purchase 5; realize 2; relate 4; see 2; seize 1; snatch 1, 4; sway 4; take 1; twig[2]; understand 1; uptake
  grasp at, clutch 2a
  grasp (of), understanding 3
grasping°, avaricious; greedy 2; mercenary 1; possessive 2; rapacious; selfish 1; sordid 2; tenacious 3
grasp one another, cling 3
grass, informer; pasture; sneak 2; talk 5; turf 1
  grass, put out to, retire 2
  grass grow, watch the, loaf[2] 1
  grass (on), inform 2; sell 3; tell[1] 11
grassland, field 1; green 4; pasture; plain 6
grass roots, people 3
grassy, green 1
grate°, gall[2] 3; grating 3; grind 1, 3; jar[2] 3; mesh 1; mill 5; net[1] 1; rake[1] 2; rankle; rasp 3
  grate against, rasp 4; scratch 1
  grate on *or* **upon,** grate 3; rasp 4
grateful°, beholden; obliged 1; thankful
gratefulness, appreciation 1; gratitude; thank 3
grater, mesh 1; rasp 2
graticule, mesh 1
gratification°, content[2] 1; delight 3; ecstasy 1; enjoyment 1; feast 3; joy 1, 3; luxury 3; pleasure 1; satisfaction 1
gratified, content[2] 3; glad 1; happy 1; joyful 2; pleased; proud 1
  gratified by, be, feed 3
gratify°, content[2] 4; delight 1; distract 2; feast 6; humor 4; indulge 1; meet[1] 5; oblige 1; pander 1; please 1; regale; satiate 2; satisfy 1, 2; serve 1; slake; suit 2; tickle
gratifying, flattering 2; heartwarming 2; lovely 2; pleasant 1; rewarding; satisfying; welcome 2
gratifyingly, nice 6
grating°, brassy 2; discordant 2; friction 1; harsh 1; mesh 1; net[1] 1; noisy; ragged 5; rasp 1; raucous; riddle[2] 3; rough 8; scratchy 2; strident; thick 7
gratis, complimentary 2; free 4, 13; gratuitous 1; house 7; voluntarily
gratitude°, appreciation 1; glory 2; thank 3
  gratitude, show, recognize 4
gratuitous°, expletive 1; free 6; groundless; irrelevant; needless 1; nonessential 1; superfluous; unasked; unsolicited; unwarranted; voluntary 1; wanton 4
gratuitously, voluntarily
gratuity, bonus; boon; bounty 3; consideration 2; dole 1; gift 1; perquisite; present[2] 3; tip[3] 1
gravamen, burden 1; nub 2; pith 2; substance 2
grave°, acute 2; bad 11; big 4; carve 1; chisel 1; critical 3; crypt; deplorable 1; desperate 4; dignified; earnest 1; etch 1; funereal; grievous 1; grim 4; heavy 3; high 6; important 1; major 2; momentous; pensive; pressing; scribe 3; sedate 1; self-contained 1; sepulcher; serious 1, 2, 4; severe 2; solemn 1; somber 1; staid; stern 2; subdued 2; terrible 1; tomb; weighty 2
  grave, having one foot in the, elderly 1
  grave, with one foot in the, dying; moribund 1
  grave air *or* **mood** *or* **atmosphere,** pall[1] 2
graveclothes, shroud 2
gravedo, cold 10
gravelly, granular; gritty 1; strident; thick 7
gravely, badly 5; deeply 2; seriously 1; severely 3
graven image, image 1; statue
graveolent, foul 3
gravestone, monument 1; tablet 3; tombstone
gravid, expecting; pregnant 1
gravitas, dignity 1; gravity 3
gravitate, incline 3; settle 11; tend[1]
  gravitate toward(s), lean[2] 3
gravitating, inclined 3
gravitation, attraction 1; gravity 1
gravity°, dignity 1; emphasis; moment 3; pith 2; severity 2, 3; sobriety 2; solemnity
gravy, profit 1; sauce 1
gray°, aged; ancient 3; dead 10; drab; dull 5; elderly 1; heavy 9; leaden 3; mousy 1; murky; old 1; overcast; silver 2; somber 2; stark 3; steely 1; sunless
  gray, go *or* **get,** worry 1
  gray hair, get, worry 1
gray-brown, dingy
grayish, silver 2; steely 1
grayish-brown, mousy 1
grayish-white, silver 2
gray matter, head 4; intelligence 1; mind 1
gray panther, senior citizen
graze, brush[2] 5; feed 2; gall[2] 1; kiss 2; rake[1] 2; scrape 1, 7; scratch 1, 4; skin 5
grease, oil 1, 3; slick 5
greasepaint, makeup 1; paint 3
grease (someone's) palm, bribe 2; fix 11; pay 10b
greasy°, fat 2; oily 1; ratty 2; slippery 1
greasy spoon, café
great°, admirable; ample 1; big 1; bully 4; capital 6; considerable 1; dandy 2; desperate 4; divine 3; eminent 1; exceeding; excellent; extensive 2; fabulous 3; fantastic 4; fine[1] 1, 11; first-rate; glorious 3; good 2, 18; goodly; gorgeous 2; grand 1, 5; gross 1; heavenly 2; high 4, 13, 14; high-class 1; historic; huge; illustrious; immeasurable; immortal 4; intense 1; large 1; legendary 3; long[1] 1; magnificent; marvelous; mean[2] 6; memorable; mighty 3; mortal 5; neat 5; nifty 3; notable 1; powerful 3; profound 3; proud 3; raving 2; sensational 3; spacious; spanking 1; splendid 3; striking; strong 7, 14; sublime 1; substantial 1; superb; superlative; surpassing; sweet 5; swell 8; terrific 2; thumping 1; towering 1; vast; voluminous 1; whopping 1; world 8
  great, not, OK 3
  great guns, go, flourish 1
greatcoat, coat 1
great deal°, deal 6; heap 2; packet 2; pile[1] 3; plenty 1
greater, better 2; major 1
  greater, become, grow 1
  greater part, mass 4
  greater than, above 4; over 2
  greater than, not *or* **no,** only 3
greater part, preponderance 1
greatest, arch 1; best 4; chief 2; full 4; leading 2; main 2; maximum 1, 3; outside 6; sovereign 2; supreme 1, 2, 3; terminal 1; top 8; ultimate 2
  greatest, the, end 8b
greatly, amply 1; awfully; badly 8; dearly 1; deeply 2; exceedingly; full 18; highly 1; materially; profoundly; vastly; very 1; widely 2
greatness, dignity 2; distinction 2; excellence; glory 3; magnitude 1; nobility 1; size 1; superiority 2
great unwashed, herd 2; hoi polloi; mob 2; populace; riffraff
Great White Father, chief 1
greed°, avarice; gluttony; rapacity
greediness, avarice; gluttony; greed 1; rapacity

guidance° (*cont.*) leadership; management 1; oversight 2; providence 2; recommendation 1; schooling; steer 3; tuition
guide°, advise 1; adviser; channel 5; coach 3; companion 2; conduct 3, 4; control 1; counsel 3, 4; direct 2; director 2; ease 7; escort 2, 4; gauge 3; govern 1; handle 3; head 11; index 1, 2; inspiration 2; instruct 1; key 2; landmark 1; lead 1; lesson 3; limit 6; maneuver 4; manual; mark 7; master 3; mold[1] 6; motto; navigate 2; pattern 4; pilot 2, 3; precept 1; principle 1; progenitor 2; recommend 1; regulate 2; rule 1, 6; run 17; see 11; shepherd; show 2; standard 1; steer 1; take 18; teach; teacher; template; train 4; tutor 2; vade mecum; work 14
guidebook, guide 8
guided by, be, follow 2; go 26b
guided missile, missile
guideline, gauge 3; instruction 1; precept 1; rule 1; standard 1
guidepost, landmark 1
guiding, direction 1; influential
guiding light, guide 7
guidon, standard 3
guild, association 1; brotherhood 2; club 2; fellowship 2; fraternity 3; group 1; institute 1; league 1; order 9; ring[1] 3; society 5
guile, art 5; artifice 1; craft 2; deceit 1; dexterity 2; foul play; fraud 1; hypocrisy; intrigue 3; subtlety 2; trickery
guileful, artful 1; astute 1; clever 2; crafty; deceitful; designing; dexterous 2; disingenuous; foxy 1; fraudulent 2; sly 1; tricky 1; wily
guileless, aboveboard 2; artless 1; callow; candid 1; childlike; clear 12; frank 2; honorable 3; ingenuous 1; innocent 4; naive; natural 4; open 15; plain 3; pure 3; rustic 2; simple 3, sincere, transparent 3; unaffected[1]; unguarded 3; unsophisticated 1
guilelessly, simply 3
guilelessness, honesty 2; naiveté; purity 2; simplicity 3
guillotine, behead
guilt°, blame 4; pang 2; regret 2; remorse 1; responsibility 3
guiltiness, guilt 1
guiltless, blameless; innocent 1
guilt-ridden, remorseful; sorry 1
guilty°, fault 5; regretful; remorseful; responsible 4; terrible 2
guilty, find *or* **prove,** condemn 1; convict 1
guilty, not, clear 12, 31; innocent 1
guilty conscience, remorse 1
guinea pig, subject 4
guise°, camouflage 1; disguise 3; front 5; image 6; mask 2; persona; pretense 2; pretext 1; semblance 2; shape 5; veneer
guise of, assume the, pass 16b
guise of, take (on) *or* **assume the,** represent 2
gulch, canyon; gully; ravine
gulf°, breach 2; gap 1; rift 1
gull, butt[1]; chisel 2; dupe 1, 3; fool 3, 4; hoax 2; hoodwink; misinform; mislead; mug 3; outsmart; prey 3b; ride 5a; sap[1] 2; scapegoat; score 7; screw 6; sucker; swindle 1; take 32b; trick 8; victim 2; victimize 2
gullibility, naiveté
gullible°, green 2; innocent 4; naive; susceptible 2; trusting; unsuspecting
gully°, canyon; gorge 1; pass 23; ravine
gulosity, gluttony
gulp°, draft 4; drink 1, 6; gasp 2; gorge 3; nip[2]; swallow 1, 5
gulp (down), bolt 9; consume 1; devour 1; take 13
gulp for air, gasp 1
gum, cement 1; glue 1; stick[1] 4
gum up, upset 5
gum up (the works), queer 5
gumbo, mud
gumboil, ulcer 1
gumminess, tenacity 2
gummy, clammy 1; gooey 1; ropy 1; sticky 1; tacky[1]; tenacious 2
gumption°, assurance 4; bottle 2; enterprise 2; gut 3a; nerve 1; push 14; resource 1; spunk
gumshoe, detective; operative 3; sleuth
gun, pistol; revolver
gun down, dispatch 3
gunfighter, killer 1
gunfire, report 2
gunge, dirt 1; filth 1; muck 2
gung ho, hot 4
gungy, bedraggled; dirty 1; filthy 1, 2; shabby 1
gunk, muck 2; ooze 1
gunky, slimy 1
gunman, cutthroat 1; gangster; terrorist; thief 1; thug
gunpowder, explosive 2
gunsel, cutthroat 1; gangster; henchman; hoodlum; killer 1
gunshot, report 2
gunslinger, gangster
gunyah, hut
gup, gossip 1
gurges, whirlpool
gurgitation, eddy 1
gurgle°, babble 1; flow 1
gurgling, gurgle 2
gurry, gut 1; rubbish 2; trash 1
guru, clergyman 1; guide 5; intellectual 4; master 3; oracle 1; sage 2; teacher; tutor 1
gush°, discharge 4; erupt 1; flow 1, 2, 5; flush[1] 5; pour 1; prattle 1; run 6; spew; spout 1; spurt 3; stream 4; surge 1, 2; well[2] 2
gush over, rave 2
gushiness, sentimentality
gushing, effusive; gushy; outpouring; sloppy 3; sweet 7; torrent
gushy°, sentimental 2; sloppy 3; sweet 7
gussied up, get (all), primp
gussy up, primp
gust°, blast 1; breath 1; breeze 1; puff 1; wind[1] 1
gusting, windy 1
gusto°, eagerness 1; enthusiasm 1; fervor; inspiration 1; panache; passion 1; punch[1] 3; relish 1; sparkle 4; spice 2; verve; vigor; zest 1
gusty, breezy 1; stiff 5; windy 1
gut°, bosom 3; bowels; consume 4; desolate 5; devastate 1; instinctive 1; pot 3; sneaking 1; stomach 1; waste 11
gutless, limp[2] 3; timid
gut(s), assurance 4; bottle 2; bowels; courage; daring 1; determination 1; enterprise 2; face 5; fortitude; gall[1] 2; grit; gumption 2; gut 1, 3; heart 2; inside 2; machismo; manhood 2; nerve 1; perseverance; pluck 1; point 5; resource 1; spirit 5; spunk; stamina; strength 2; works 2
gutsiness, assurance 4; face 5; grit; gut 3a; strength 2
gutsy, brave 1; daring 2; defiant; foolhardy; game 8; gritty 2
gutter, channel 1; drain 1; fail 3; flicker 1; furrow 1
guttersnipe°, ragamuffin
guttural, gruff 2; harsh 1; strident; thick 7
guv, employer 1; head 2
guv'nor, head 2
guy°, boy 1; bugger 2; cable 1; chap; customer 2; devil 3; dude 2; fellow 1, 4; flout; lad; leg 8; love 3; man 1; mock 1; parody 3; ridicule 2; stay[2] 1; steady 9; stick[2] 2; support 8; taunt 1; tease 1
guy, bad, criminal 3
guy Friday, aide; second[1] 8
guywire, support 8
guzzle, drink 1; gulp 1; overeat; swallow 1, 5; swill 3
guzzler, drunk 3
gymnastics, exercise 4
gyp, defraud; fleece; fraud 2; hoax 1, 2; racket 2; rip 2b; swindle 1, 2
gyp (out of), rob 3
gypsy, bum 2; migrant 1, 2; rover; tramp 4; traveler 1; vagabond 1, 2
gyrate°, revolve 1; rotate 1; shake 2; spin 1; swirl 1; turn 1; twirl 1
gyration, revolution 3; shake 8; spin 6
gyve, chain 2, 3
gyve(s), manacle 1; restraint 2; shackle 1

H_2O, water 1
habiliment(s), garments; gear 3; habit 3; trappings
habit°, custom 1; garments; gear 3; institution 4; mannerism; observance 2; practice 1; robe 1, 2; routine 1; rut 2; suit 3; tradition; trick 6; uniform 3; usage 1; way 2
habitat°, element 2; environment; setting
habitation, abode; domicile 1; dwelling; home 1; housing 1; quarter 5; residence 1; shelter 3
habit-forming, hard 11
habit(s), moral 4; morality 2; time 9
habitual°, accustomed 1; chronic 2; conventional; customary 2; everyday 2; familiar 2; fixed 3; frequent 1; general 2; incorrigible; incurable 2; mechanical 2; natural 1; nonstop 2; ordinary 1; popular 2; recurrent; regular 1, 7; relentless 2; ritual 2; routine 3; set 29; standard 6; staple 2; steady 2; time-honored; traditional; usual
habitually, daily 3; frequently 2; generally 1; nonstop 3; often; ordinarily; popularly
habituate, accustom; adapt 2; break 18b; haunt 1; orient 3
habituated, accustomed 1; seasoned 1
habituated to, used 3
habitué°, patron 2; regular 12; steady 10
hack°, accomplish; cab; chop 1; jade 1; journalist; mangle; nag[2]; politician; scribe 2; shape 6; slash 1; slave 2; taxi 1; writer
hack it, grade 5
hack off, cut 15a; lop; mutilate 1; sever 1
hack (out), execute 1
hackles rise, make (someone's), infuriate

handing down *or* on, succession 1
handing out, partition 2
hand in glove°, hand 10, 11; thick 8
  hand in glove, work, play 11
handing over, surrender 3
hand in hand°, hand 10, 11; shoulder 5
handiwork, craft 1; production 2; work 4; workmanship
handle°, attend 2; control 6; cope 2; deal 2, 4; direct 1; drive 2; feel 1; field 6; finger 11; fondle; grip 3, 5; guide 3; knob; manage 1, 2, 3; manipulate 1, 2; massage 3; meet[1] 5; name 1; navigate 2; negotiate 2; nickname; operate 2; oversee; preside; process 4; put 27b; reckon 5a; regulate 2; relate 4; run 10; sell 2; shaft 1; stand 3; stock 9; supervise; tab 1; tag 2; touch 1; transact; treat 1, 2; understanding 3; use 2; wield 1; work 10
handleable, tractable 2
handling, conduct 2; direction 1; guidance 1; management 1; operation 2; service 5; treatment; usage 2; use 8
handmade, homespun
handmaiden, woman 4
hand-me-down, secondhand; stale 2; used 1
hand-off, pass 30
handout(s), bonus; dole 1; gift 1; largess; literature 2; pamphlet; present[2] 3; promotion 5; publication 2
hand over fist°, hand 12
handpick, select 1; pick 1
handpicked, choice 5; select 2
handrail, rail[1] 1
hand(s), help 7; extremity 2; hand 8; oversight 2
  hands, show of, vote 1
  hands, throw up one's, surrender 2
  hands down, easily 1; hand 13
  hands on, get *or* lay *or* put (one's), come 9a; get 31a; grab 1; obtain 1; procure 1
  hands tied (behind one's back), with both, handily 1
hands-down, easy 1
handsel, earnest 3
handset, telephone 1
handsome°, attractive; beautiful 1; brave 2; comely; elegant 1; exquisite 3; fair[1] 7; fine[1] 10; generous 3; liberal 1; lovely 1; prepossessing; statuesque
handsomeness, beauty 1; charm 3
hands-on, practical 3
hand-to-hand encounter, battle 1
hand-to-mouth, frugal 2
handwriting, hand 7; penmanship; script 1; writing 1
(hand)writing on the wall, omen; sign 6
handy°, available; clever 4; convenient 1, 2; hand 9a; ingenious; nearby 2; portable; ready 7; ready-made 2; versatile 1
hang°, drape 1; float 1; hover 1; pivot 4; poise 3; punish 2; soar 1; string 9, 12; suspend 2; swing 2; trick 3
  hang about *or* around *or* out together, fraternize
  hang about *or* around (with), fraternize; go 21c; hang 4; haunt 1; hobnob; hover 2; linger 1; mingle 2; mix 2; pal 2; stick[1] 11; tarry 1; travel 4; wait 1
  hang around at, frequent 2
  hang back, delay 3; drag 6; hesitate 1; lag 1; sit 8; tarry 1; trail 6; withhold 1, 2
  hang back (from), hang 5
  hang (down), dangle 1; droop 1; flag[2] 1
  hang in (there), endure 2; hang 7c
  hang loose, loose 9
  hang of, get the, figure 12b; pick 8d; see 2; understand 1
  hang on, brace 6; depend 1; hang 7; hold 19b, 19c, 20a; linger 1; stick[1] 11; tarry 1; wait 1
  hang on to, hang 7a; keep 1; retain 1; cling 2; hang 7d; hold 1
  hang out, mingle 2
  hang out (at), frequent 2; hang 4a, 9
  hang out in, resort 3
  hang (out) over, overhang 1
  hang out (with), associate 1b; fraternize; go 21c; hang 4b; mix 2; pal 2
  hang over, hang 10; loom 2
  hang together, hang 11
  hang up, hang 12; post[1] 2
  hang upon, hang 7d
hanged, be, swing 2
hanger, hook 1
hanger-on°, flatterer; flunky 2; parasite; satellite 2; yes man
hangers-on, retinue
hang fire°, fire 4; hang 6
hanging, drape 2; drapery; loose 3; pendulous; punishment 2
hanging by a thread, precarious
hanging (fire), abeyance; limbo; pending 2
hang in the air, hover 1
hang one's head, hang 8
hangout, haunt 3; pad 3
hang paper, counterfeit 4; forge 3
hang-up, bug 6; fixation; foible; impediment; obsession; snag 1; thing 4
hanker°, ache 2; die 5; hunger 3; itch 2; long[2]; wish 1; yearn
  hanker after, desire 1; hanker; pant 2; thirst 2; want 1
  hanker for, hanker; thirst 2
  hanker for *or* after, lust 3
hankering, appetite 2; aspiration; desire 3; fancy 7; hope 1; hunger 2; hungry 2; itch 4; longing; stomach 2; thirst 1; thirsty 2; want 4; wish 4, 5
  hankering for, starved 2
  hankering for, have a, hanker
hanky-panky°, deceit 1; foul play; fraud 1; hocus-pocus 1; trickery
hansom (cab), cab
haphazard°, aimless 2; casual 2; chaotic 1; desultory; disorderly 1; erratic 3; fitful; helter-skelter 1; incidental 1; indiscriminate 2; inordinate 2; irregular 2; promiscuous 1; random 1; scratch 6; slipshod; stray 5; uncertain 1
haphazardly, fit[2] 4; random 2
hapless, poor 6; sorrowful 2; tragic; unhappy 2; unsuccessful 2; wretched 3
happen°, appear 3; chance 7; come 4a, 14; develop 4; emerge 1; exist 3; go 31c, 32b; materialize 2; occur 1; pass 13; place 14; result 2; rise 9; transpire 2; true 6; turn 20b; unfold 2
  happen, about to, wind[1] 8
  happen, make, effect 7; effectuate; realize 1
  happen again, recur 1
  happen on *or* upon, come 5a; encounter 1; find 1; happen 3; hit 9a; light[2] 15; meet[1] 1; run 20; strike 14; stumble 3
  happen to, become 5
happening°, adventure 1; affair 3; case[1] 1; circumstance 2; development 1; episode 1; event 1; eventuality; experience 2; fact 2; gala 1; incident 1; occasion 3; occurrence 1; phenomenon 1; progress 4; stroke 9
happenings, thing 8b
happens *or* happened, as it, instance 2
happenstance, coincidence 3; fortune 2; luck 1
happier, better 6a
happily°, gaily 2; gladly; readily 1; soon 4; well[1] 15; willingly
happiness°, bliss; cheer 2; content[2] 1; ecstasy 1; gaiety 1; glee; heaven 3; joy 1; merriment; paradise 3; pleasure 1; radiance 2; satisfaction 1; transport 5; triumph 2; welfare
happy°, appropriate 1; blithe 1; bright 8; carefree; charmed 3; cheerful 1; content[2] 3; delighted; ecstatic; elated; elevated 3; enchanted; exalted 3; exhilarating 2; exuberant 2; gala 2; gay 2; glad 1, 2; gleeful; golden 4; jaunty 1; joyful 1, 2; light[2] 9; mellow 4; merry 1; opportune 1; overjoyed; pleased; propitious; providential; radiant 2; rapt 2; ready 2; rhapsodic; seasonable; sunny 2; vivacious; willing; world 7
  happy, be, rejoice
  happy (with *or* about), proud 1
happy-go-lucky, blithe 2; carefree; casual 5; improvident 1; light[2] 9; nonchalant; relaxed
happy hunting ground(s), heaven 1; paradise 1
  happy hunting ground(s), go to the, die 1; go 13
Happy Isles *or* Islands, the, heaven 1
happy valley, fairyland
harangue°, berate; bluster 1; hold 16a; lecture 1, 3; light[2] 14; preach 2; rant 1; sermon 1; speech 2; talk 14; tirade
haranguer, talker 1
haranguing, bluster 3
harass°, aggravate 2; annoy 2; beset; bother 1; browbeat; bug 7; bully 2; distress 3; disturb 1; exasperate 2; exercise 3; gall[2] 4; gnaw 3; grind 5; heckle; henpeck; hound; irritate; molest 1; nag[1] 1; obsess; oppress 2; persecute 2; pester; pick 4a, 6; plague 3; prod 3; provoke 3; push 4; rack 3; ride 3; tease 1; torment 2; trouble 2; wear 7; worry 2
harassed by, -ridden
harassing, obsessive; painful 2; persecution 2; provocative 2
harassment, torment 4
harbinger°, announce 4; forerunner 1; messenger; omen; precursor 1; prognosticate 2, prophesy 2
harbor°, bear 7; creek 1; entertain 3; house 8; keep 12; lodge 4; nourish 2; nurse 4; oasis 2; port; refuge 1; shelter 4
harbor a grudge, umbrage
hard°, arduous 1; badly 9; bare 2; callous; cruel 1; demanding 1; difficult 1, 4; domineering; dour 2; dull 2; exacting; exhausting 2; firm 1; harsh 2; implacable; labored 1; laborious 1;

hasten° (*cont.*)
move 12a; precipitate 1; quicken 1, 2; race[1] 4; run 1; rush 1; scramble 2; scurry; speed 3; step 16; streak 5; tear 3; trot 1; whisk 2
hastily°, briefly 2; fast[1] 6; pell-mell 1; prematurely 2; quickly 3; run 49a; short 11; shot 10; sketchily; spur 3; suddenly 2; swiftly
hastiness, haste 2; indiscretion 1; speed 1
hasty°, abrupt 1; brash 1; cursory; desperate 2; fast[1] 1; foolhardy; foolish 1; hot-headed; hurried 1; ill-advised 2; impetuous; imprudent; indiscreet; momentary; overconfident 2; passing 2; pell-mell 2; perfunctory 1; precipitate 5; premature 2; quick 2; rapid; rash[1]; rough 10; scratch 6; sketchy; snap 13; speedy 2; sudden; superficial 3; swift; unguarded 3; unthinking 1; unwary; whirlwind 2
hat, cap 1
hatch°, breed 2; brew 2, 3; brood 2; conceive 2; fabricate 2; incubate; make 39c; plot[1] 3; produce 3; scheme 4
hatchet man, cutthroat 1; killer 1
hatching, fabrication 2
hate°, abhor; aversion 2; contempt; despise; detest; dislike 1; ill will; loathe; loathing; rancor; resentment; spite 1; use 15; venom 2
hateful°, abhorrent; beastly 2; bitter 4; black 5; damnable; evil-minded 2; hideous 2; loathsome; lousy 1; malignant 2; obnoxious; rancorous; spiteful; ugly 2; vicious 2; virulent 2
hatless, bare 1
hatrack, rack 1
hatred, animosity; aversion 1, 2; bitterness 2; contempt; disgust 3; dislike 2; feud 1; grudge 1; hate 3; horror 1; ill will; infamy 2; loathing; phobia; rancor; revulsion; spite 1; strife 2; vendetta; venom 2; virulence 2
hat stand, rack 1; stand 17
hat tree, rack 1
haughtiness, air 5; arrogance; distance 2; front 7; pride 2; side 5; snobbery; vanity 1
haughty°, aloof 2; arrogant 2; cocky; condescending; disdainful; distant 3; frigid 2; hoity-toity; lofty 4; majestic 2; overbearing; pompous 1; pretentious 2; proud 2; regal 1; scornful; self-important; snobbish; standoffish; stiff 7; supercilious; vain 1
haul°, cart 2; drag 1; draw 1; freight 2; heave 1; hump 3; loot 1; lug; prize[1] 2; pull 1; ship 2; take 9; tow; trail 4; transfer 1; transport 1; tug 1, 2
haul down, strike 4
haul (someone) over the coals, castigate; dress 4; lambaste 2; mind 10; rebuke 1; reprimand 2; scold 1; tell[1] 10
haul up, raise 1
haulage, transit 2; transport 4
hauler, carrier 1
haunch, flank 1
haunt°, attend 1; frequent 2; gnaw 3; habitat; hang 4b; home 2; nest 2; obsess; resort 3; stalk[1]; study 6; walk 3
haunted, possessed; smitten 1
haunting, obsessive
haunts, territory 2
hauteur, air 5; arrogance; dignity 1; distance 2; pride 2; snobbery
haut monde, elite 1; society 4
have°, bear 5; command 5; conceive 1; enjoy 2; entertain 3; get 5, 22; hold 10, 12; host[1] 3; keep 1, 2; meet[1] 6; possess 1, 2, 4; state 7; stock 9; wear 2; wield 2; world 5a
have, let, spare 8
have (someone) on the carpet, scold 1
haven, harbor 1; hideaway; home 3; oasis 2; port; protection 1; refuge 1; retreat 3; shelter 1
have-not, pauper
haver, hesitate 1; shilly-shally; stall[2] 1
havering, hesitant 1
havers, rubbish 2; trash 1
haversack, pack 1
have the quality of, partake 2b
having said that, same 3
havoc°, desolation 1; destruction 1; mayhem; ruin 1; wreck 4
hawk, belligerent 3; market 4; peddle; sell 2; tout 1
hawker, merchant 2; peddler; seller
hawk-eyed, eagle-eyed; sharp-eyed; vigilant
hawkish, belligerent 1; militant 1; warlike
hawkishness, jingoism
Hawkshaw, detective; sleuth
hawser, cable 1; line[1] 11; rope 1
hay, provender 2
hay, hit the, rest[1] 6; retire 3; sack 2; turn 15a
hayburner, nag[2]
haycock, stack 2
haymaker, punch[1] 2
hayrick, stack 2
hayseed, boor 1; peasant; provincial 3; rustic 3
haystack, stack 2
haywire, order 13c
hazard°, adventure 2, 3; bet 2; chance 8; danger 1; dare 2; endanger; expose 2; exposure 2; gamble 1; jeopardize; jeopardy; lay[1] 4; menace 2; pawn[1] 2; pitfall 2; put 6; risk 1, 2; stake[2] 1, 4; venture 2, 3
hazard, at, stake[2] 3
hazarded, stake[2] 3
hazardous°, adventurous; awkward 4; dangerous 1; desperate 5; explosive 1; hairy 2; menacing; parlous; perilous; precarious; risky; speculative 2; ticklish 2; tight 7; touchy 2; tricky 2; ugly 3
hazardously, dangerously 1
hazardous undertaking, venture 1
haze, blur 2; film 3; fog 1; mist 1; obscurity 1
hazily, vaguely 2
haziness, blur 1; film 3
hazy°, dim 1; dull 7; equivocal 2; faint 1; filmy 2; fuzzy 2; imprecise; indefinite 3; indistinct 1; misty; nebulous; obscure 1, 2; opaque 1; rough 10; shadowy 2; steamy 2; tenuous 2; thick 3; uncertain 3; vague 1; weak 6; woolly 2
head°, addict 1; boss 1, 2; capital 1; category; chief 1, 2; crest 1; devotee; director 1; employer 1; enthusiast; executive 1; facility 2b; first 1; forward 1; front 3, 10; froth 1; grand 6; lavatory; leader 1; loaf[1] 2; lop; manage 1; manager; master 1; mind 3; premier 2; principal 3; representation 4; run 10; sculpture 1; superintendent; supervisor; tip[1] 1; title 3; toilet 1; top 1; trend 3
head, come to a, climax 4
head, get through one's, figure 12b
head, off one's, psychotic 1
head, out of (one's), crazy 1; insane 1; mad 1; raving 1; unbalanced 2; deranged
head, over (someone's), incomprehensible
head, use (one's), reason 7
head above water, keep one's, get 28b; survive 1
head for, head 10; take 38b
head for *or* toward(s), make 30a
head off, head 13; intercept
head spin, make one's, intoxicate 2
head swim, make one's, stagger 2
head to foot *or* toe, from, entirely 1; quite 1; through 7
head (up), command 2; direct 1; govern 1; head 11; lead 3; officiate; preside; rule 6
headache°, drag 7; job 4; lookout 3; nuisance 1; pain 3; plague 2; province 3; trial 5
head cold, cold 10
head covering, cap 1
headed, bound[3] 5
header, spill 5; tumble 4
head for the hills, run 2
head honcho, boss 1; director 1
heading, category; tack 3
headlamp, light[1] 2
headland, bluff[2] 3; cape[1]; height 3; point 8; prominence 2
headlight, light[1] 2
headline, feature 4; title 3
headliner, personage; personality 2; star 2
headlining, stellar 2
headlong, brash 1; breakneck; foolish 1; hot-headed; impetuous; improvident 2; impulsive; precipitate 3; rash[1]; whirlwind 2
head man, chief 1; head 2; director 1
headmaster, head 2; principal 4
headmistress, head 2; mistress 2; principal 4
head of state, premier 1
head or tail, sense 4
head over heels, head 8; topsy-turvy 1
head over heels, go, pitch[1] 4
headquarter, base[1] 6
headquarters, seat 2
head(s) or tail(s) of, make, figure 12b; get 19; grasp 2
head start, advantage 1; edge 4; start 13
headstay, stay[2] 1
headstone, monument 1; tablet 3; tombstone
headstrong, defiant; disobedient 2; grim 1; inflexible; obstinate; recalcitrant; rigid 4; rogue 2; self-willed; unruly; willful 2
head to head, face 8; talk 15
headwaiter, waiter
headway°, advance 7; course 2; progress 1; progression 1, 2; way 5
headway, go *or* make, stem[2] 2
headway, make, go 1; headway 2; look 11c; pick 8f; proceed 1; progress 5
heady, exciting 1; intoxicating 2; stirring; strong 2
heal°, cure 2; doctor 2; knit 2; mend 1, 3; patch 7; recover 2; recuperate; remedy 3

highlights, rundown
high living, dissipation 1
highly°, degree 4b; sky 2; well[1] 5
highly strung, emotional 2; excitable; intense 3; nervous 1; quick-tempered; restless; tense 2; touchy 1
high-minded, conscientious 1; ethical; generous 2; good 5; great 7; honest 1; honorable 1; moral 1; principled; pure 6; scrupulous 2; sublime 1; upright 2; virtuous 1
high-mindedness, equity; good 20; nobility 1; right 10; virtue 1
high-muck-a-muck, dignitary
high noon, midday; noon
high opinion, regard 7
high-pitched, high 7; piercing 1; sharp 7; shrill
high point, climax 1; feature 2; height 2; peak 2; top 1; zenith
high points, rundown
high-powered, dynamic; energetic
high-pressure, drive 1; strong-arm
high-priced, dear 2; expensive; high 3; invaluable; precious 1; premium 4b; priceless 1
high-principled, virtuous 1
high priority, necessary 1; pressing; rush 5; urgent 1
high-quality, choice 4
high-ranking, important 2; noble 2
highroad, road 2; street 1
high seas, ocean 1; sea 1
high sign, gesture 1; word 7
high society, society 4
high-sounding, affected 3; artificial 3; extravagant 4; glorified 1
high-speed, fast[1] 1; rapid
(high-)spirited, effervescent 2; frisky; gallant 1; game 8; gay 2; intense 3; jaunty 1; jolly 1; playful 1; vivacious
high spirits, fun 1; gaiety 1; glee; happiness; hilarity; merriment; mirth
high-strung, emotional 2; excitable; restless; tense 2; touchy 1
high style, style 3
hightail (it), escape 1; flee 1; fly 2; hurry 1; make 26; rush 1; scramble 2; speed 3
highway, road 2
highwayman, outlaw 1; robber; thief 1
highway robbery, rip-off 3
hijack, rob 1; robbery; steal 1
hijacker, thief 1
hijacking, robbery; theft
hike, boost 3; excursion 1; jump 9; march 3; mark 16b; ramble 1, 4; tramp 1, 3; walk 1, 7
  hike (up), hitch 2
hiker, traveler 1
hilarious°, comic 1; funny 1; high 8; humorous; hysterical 2; merry 1; priceless 2; rich 10; ridiculous; uproarious 2; zany 1
hilarity°, festivity 2; gaiety 1; merriment; mirth
hill°, elevation 2; grade 4; height 3; incline 4; knoll; mound 1; peak 1; prominence 2; rise 13; slope 2; stack 1
  hill, over the, elderly 1; old 1; vintage 3
hillbilly, boor 1; peasant; rustic 2, 3
hillock, hill 1; hump 1; knoll; mound 1; prominence 2; rise 13; tell[2]
hills, take to the, fly 2
hilt, handle 1
Himalayan, mountainous 1
himself, boss 1; chief 1; person 2; personally 1
hind, peasant; posterior 1
hinder°, arrest 1; bar 9; block 3; check 1; delay 2; deter; discourage 3; embargo 2; encumber 2; forbid; forestall; frustrate 1; hamper[1]; handicap 2; hobble 2; hold 14a, 22b; impede; inhibit; interfere 2; militate 1; obstruct 2; oppose 2; posterior 1; preclude; prevent; prohibit 2; repress; resist 1; restrain 2; retard 1; set 16; shackle 4; stay[1] 4; stop 2; stunt[2]; stymie; suppress 1; thwart 1; tie 2; trammel 2
hindering, obstruction 2; prevention; preventive 1; repression 2; suppression
hindmost, last[1] 1
hind part, rear[1] 1
hindquarters, bottom 1; bum 1; buttocks; rear[1] 2; seat 4
hindrance°, rub 11
hindrance(s), bar 3; barrier 2; block 2; check 14; deterrent; difficulty 2; drawback; embargo 1; encumbrance; fly 10; handicap 1; hitch 4; hurdle 1; impediment; inconvenience 1; inhibition; interference 2; let[2]; liability 3; obstacle; obstruction 1; overload 2; prevention; preventive 3; repression 2; setback; shackle 2; snag 1; stumbling block; trammel 1
hindsight, retrospect
hinge, base[1] 5; pivot 1, 4; rest[1] 7
  hinge on *or* upon, depend 1; turn 18b
hinged, jointed
hinie, bottom 1; bum 1; buttocks; posterior 3; rear[1] 2; seat 4; tail 2
hint°, appearance 4; aroma 2; bit 2; breath 2; cast 6; clue 1, 2; cue 1; dab 2; dash 7; evidence 3; feeler 2; feeling 2; flash 2; flavor 2; flicker 4; ghost 2; gleam 2; grain 3; idea 2; implication 2; index 2; indicate 3; indication 1; inkling; innuendo; insinuate 1; intimate[2]; lead 14; line[1] 10; make 37e; modicum; note 6; notify 2; odor 2; overtone; particle; point 11; pointer 2; predict; prediction; prompt 6; reference 1; ripple 2; savor 2; say 2; scattering; scrap[1] 1; shade 3; shadow 3; shred 1; sign 3; sniff 2; spark 1; speck; steer 3; strain[2] 2; stroke 8; suggestion 2; suspicion 2; tang 2; taste 1; tint 1; tip[3] 2; touch 15; trace 1; vein 4; vestige; waft 2; whisper 2, 4; word 6
  hint against, discourage 2
  hint (at), announce 2; breathe 4; broach; drive 6; get 31b; imply 1; impute; mean[1] 2; mention 2; partake 2b; promise 4; suggest 2; tell[1] 2
hinterland(s), country 3; interior 7; province 4
  hinterland(s), the, stick[2] 3
hinting (at), indicative of
hip°, aware 1; minute[1] 3; modern; new 2; smart 2; sophisticated 1; swinging; well-informed; worldly 2
  hip, not, square 6
  hip (to), privy 2; sensible 4
  hip to, be, relate 4
hiphuggers, pants 2
Hippocrenian, poetic 2
hippodrome, stadium; theater 1
hippy, plump[1] 1
hipster, hippie
hircine, lascivious 1; lecherous; lewd; libertine 2; sensual
hire°, charter 5; employ 1; employment 2; engage 1; enlist 2; get 1; job 6; line[1] 23b; rent[1] 2; retain 2; sign 10b; take 4, 35a
  hire (out), hire 3; lease 2; let[1] 3; rent[1] 1
hireling, flunky 1; mercenary 3; puppet 2; subordinate 2
hiring, employment 2; placement 2
hirsute, hairy 1; shaggy
His *or* Her Highness *or* Majesty, crown 3
hispid, hairy 1
hiss°, fizz 1, 3; jeer 2; outcry; swish 1, 2; whisper 1
hissing, fizz 3; hiss 1; outcry; swish 2
historic°, ancient 2; early 6; former 2; landmark 3; memorable; monumental 1; old 2; primitive 1; significant 1; traditional
history°, account 4; background 1; chronicle 1; description 2; journal 2; memoir 2; myth 1; narrative 1; past 5; record 6; story[1] 1; tale 1
history-making, monumental 1
histrionic, dramatic 1, 3; grand 3; grandiose 1; melodramatic; theatrical 1; thespian 1
histrionic art, theater 3
histrionic art(s), drama 2
histrionics, drama 3; rant 3; theater 2
hit°, assault 5; batter 1; belt 3; blow[2] 1; box[2] 2, 3; bum 5; bump 1, 3; catch 6; clip[2] 3, 6; dispatch 3; finger 4; get 24; impact 3; jab 2; jolt 2; kill 1; knock 1, 3d; knockout 2; lob 2; lunge 3; nail 9; occur 2; onset 1; overtake 2; pelt[1] 2, 4; penetrate 3; poke 2, 6; punch[1] 1; ram 2; rap 1, 4; run 26d; sensation 3; strike 1, 3, 7, 9; surprise 1; touch 14; zap
  hit, make a, score 13
  hit (back) at, fix 14; pay 5; swipe 1
  hit (by), stricken
  hit (hard), move 4
  hit home, hit 4
  hit it off (with), get 28a
  hit on *or* upon, come 5a; encounter 1; find 1; happen 3; hit 9; invent 1; learn 1; light[2] 15; locate 2; run 20; strike 16; stumble 3; turn 25c
  hit up, bum 5; hit 8
  hit upon, strike 14
hitch°, catch 18; deterrent; drawback; fly 10; hindrance 1; holdup 2; impediment; interference 2; joker 2; kink 3; liability 3; obstacle; obstruction 1; rope 3; rub 11; setback; snag 1; stake[1] 3; stretch 6; thumb 6; yank 1, 2
  hitch, without a, easily 1; swimmingly
  hitch up, hitch 2
hitched, get, marry 1; wed 1
hitch(es), trammel 1
hitchhike, hitch 3; thumb 6
hither and thither, about 3, 4; around 5
hither and yon, about 3, 4; around 5
hitherto, formerly; previously
Hitler, tyrant
hit man, cutthroat 1; gangster; hoodlum; killer 1; murderer; thug
hit-or-miss, haphazard 2; lax 2; random 1; uncertain 1
hit pay dirt, place 10
hit (town), arrive 1
hive, swarm 1
hoard°, accumulate; accumulation 3; amass; cache 2; collection 2; heap 1; husband 2; lay[1] 19a; mass 1; mine 2;

homage°, compliment 1; eulogy; glory 2; honor 2; memory 3; obeisance; praise 2; prostration 2; reverence 1; tribute 1; veneration; worship 2
homage, do, recognize 4
homage to, pay, commemorate; compliment 3; exalt 2; extol; glorify 2; hallow 2; honor 5; memorialize; praise 4; respect 6; salute 2; worship 1
home°, abode; base[1] 4; domestic 1, 3; domicile 1; dwelling; habitat; house 1; institution 3; interior 2; outside 7; pad 3; place 6; quarter 5; residence 1; seat 5; setting; shelter 3
home, at, friendly 1; home 4
home, go, knock 5
home, make one's, settle 4
home, not (at), out 2
home in, make one's, occupy 2
home (in) on, identify 1; pin 4b; set 5; zero 4
home with *or* **in, at,** home 5
home brew, moonshine 3; whiskey
homecoming, return 10
home ground, home 2; turf 2
homegrown, native 2
homeland, country 2; fatherland; land 4
homeless°, stray 4
homeless, the, homeless 2
homelessness, want 5
homelike, homely 1, 2
homely°, modest 2; plain 5; prosaic; simple 2; snug; ugly 1; undistinguished; unsightly
homeowner, resident 3
Homeric, heroic 5
home rule, independence 1; self-government 1
homes, housing 1
home screen, screen 5
homespun°, homely 1; ordinary 2; provincial 2; rude 4; undistinguished; vulgar 3
homestead, dwelling; farm 1; house 1; spread 13
homeward(s), home 9
homework, assignment 2; lesson 1
homey, homely 1, 2
homicidal°, bloodthirsty; cutthroat 3; deadly 3; fierce 1; grim 2; sanguinary 1
homicide, foul play; killing 1; murder 1; murderer; slaughter 2
homiletic, proverbial 1
homily, moral 3; oration; proverb; sermon 2; speech 2; tract[2]
hominid, human 1
hominoid, human 1
homo, deviant 2, 3; homosexual 1
homoerotic, homosexual 2
homogeneity, uniformity 1; unity 2
homogeneous°, solid 4; uniform 1
homogenization, wedding 2
homogenize, standardize; wed 2
homogenizing, wedding 2
homologous (to *or* **with),** like[1] 1; matching 1; parallel 1
homologue, parallel 2
homophile, deviant 3; homosexual 1, 2
Homo sapiens, humanity 1; man 2
homosexual°, deviant 2, 3; effeminate; swish 4
homy, homely 1, 2
honcho, bigwig 1; boss 1; chief 1; executive 1; head 2; overseer; principal 3
honcho, chief *or* **head,** master 1
hone, refine 3; sharpen; whet 1
honest°, aboveboard 2; artless 1; bona fide; candid 1; decent 4; direct 10; ethical; even 7; fair[1] 2; faithful 4; frank 1; free 9; front 9a; genuine 2; good 8; heartfelt; hearty 2; honorable 1, 3; incorrupt 1; ingenuous 1, 2; innocent 1; irreproachable; just 2; level 15; moral 1; natural 4; noble 4; open 15; plain 3; principled; pure 6; raw 6; real 3, 4; reliable; reputable; respectable 1; responsible 2; right 1; righteous 1; round 8; savory 2; serious 3; simple 2, 3; sincere; square 4; straight 4; trustworthy; truthful; unaffected[1]; unvarnished; upright 2; virtuous 1
honest, be, card 13; level 11
honest, to be, right 12
honestly°, fairly 2; naturally 3; quite 4; really 1; right 12; seriously 2; shoulder 6; sincerely; straight 14; true 4; truly 1, 4
honest-to-God, literal 2; solid 4
honest-to-goodness, literal 2
honesty°, candor 1; character 3; credit 3; freedom 6; good 20; honor 1; integrity 1; morality 1; nobility 1; principle 3; probity; purity 2; rectitude; right 10; sincerity; sportsmanship; virtue 1
honesty, in all, right 12; truly 1
honey, dear 3; love 3
honeycomb, perforate; riddle[2] 1
honeyed, flattering 2; sweet 1, 7
honeyed words, pat[1] 5
honeylike, sweet 1
honey-tongued, smooth 6
honky-tonk, dive 3; gaudy; joint 2; pub
honor°, adore 1; celebrate 1, 3; character 3; chivalry; commemorate; compliment 1, 3; conscience; credit 4; dignity 2; discharge 6; distinction 2; esteem 1; eulogize; exalt 2; extol; face 4; fête 2; glorify 2; glory 1, 2; grace 7; hail[1] 2; hallow 1, 2; homage; honesty 1; immortalize; integrity 1; kudos; laud; look 12; luster 2; memorialize; memory 3; name 2; obeisance; pay 1; plaque 3; praise 2, 3; pride 1; principle 3; prize[1] 1; probity; recognition 2; recognize 4; regard 7; renown; respect 6; revere; reverence 1; reward 2; salute 2; self-respect; sportsmanship; toast 1, 3; tribute 1; venerate; veneration; virtue 1, 2; worship 1, 2
honor, on my, indeed 1
honor to, pay, hallow 2
honor upon oneself, bring, mark 9
honorable°, aboveboard 2; clean 4; conscientious 1; decent 4; dignified; distinguished 1; estimable; fair[1] 2; good 5, 9; heroic 2; honest 1; incorrupt 1; ingenuous 2; just 2; lofty 2, 3; meritorious; moral 1; noble 4; principled; prominent 2; proud 3; pure 6; reliable; reputable; right 1; righteous 1; savory 2; scrupulous 2; square 4; straight 4; sublime 1; true 2; trustworthy; upright 2; venerable; virtuous 1; wholesome 2; worthwhile 2
honorableness, dignity 2; good 20; nobility 1; right 10; sportsmanship
honorably, honestly 1
honorarium°, bonus; consideration 2; fee 2; gift 1; pay 12; price 3; wage 1
honorary°, unpaid 2
honored, dear 1; distinguished 1; eminent 1; estimable; exalted 1; famous; glorious 1; honorable 4; illustrious; immortal 3; lofty 2; noble 3; prestigious; privileged 1; prominent 2; proud 1, 3; renowned; time-honored; venerable
honoring, celebration 2
honor(s), laurels; trophy 1
honors, take the, triumph 3
hooch, alcohol; booze 1; drink 5; liquor 1; moonshine 3; spirit 9b; whiskey
hood, bruiser; criminal 3; gangster; hoodlum; miscreant 1; rowdy 2; thug
hoodlum°, bruiser; criminal 3; gangster; henchman; miscreant 1; punk 1; rowdy 2; thug
hoodoo, enchant 1
hoodwink°, beguile 1; betray 3; blind 5; bluff[1] 1; cheat 2; defraud; double-cross; dupe 3; finesse 4; flummox; fool 4; hoax 2; lead 7; mislead; outsmart; prey 3b; pull 15; swindle 1; take 32b; trick 8; victimize 2
hoodwink (someone), put 26
hooey, drivel 3; moonshine 2; mumbo jumbo 1; nonsense 1; prattle 3; rubbish 2; talk 18; trash 1; wind[1] 2
hoof it, dance 1; leg 9b; run 1; walk 1
hoo-ha, fuss 1; tempest 2; uproar
hook°, bend 1; buckle 1; catch 2, 17; clasp 1, 3; completely 1; connect 3; gimmick 1; hook 4; lock[1] 1; make 27; peg 1; pilfer; pocket 4; steal 1; take 3; wholly 1
hook, off the, exempt 2; hook 5
hook on, attach 1
hook or by crook, by, hook 3; somehow
hook (up), fasten 1; hitch 1; put 27c
hook, line, and sinker, completely 1; hook 4; wholly 1
hookah, pipe 2
hook a ride, thumb 6
hooked, mad 6
hooked, become, catch 7
hooked on, fond 3
hooker, bitch 2; catch 18; prostitute 1; provision 2; slattern; tart[2] 2; wanton 5
hookup, marriage 3
hooligan, barbarian 2; bruiser; criminal 3; delinquent 1; gangster; henchman; hoodlum; killer 1; miscreant 1; punk 1; rough 13; rowdy 2; thug; tough 8
hooliganism, rowdyism
hoop, circle 1; loop 1; ring[1] 1; round 10; wheel 1
hoopla, advertisement 2; fuss 1; promotion 4; propaganda 2; uproar
hooplike, circular 1; round 2
hoop-shaped, circular 1; round 2
hooray, cheer 4
hoosegow, jail 1; prison
hoot, damn 4; hiss 2, 3; jeer 2; laugh 1; outcry; roar 2, 4; scream 2; squawk 1; whoop 1, 2
hoot, give a, damn 5
hootch, liquor 1; moonshine 3; spirit 9b
hooting, outcry
hoots (in hell), two, damn 4
hoover, clean 9

imagined, fancied; fanciful 2; fantastic 2; fictitious 1; hypothetical; ideal 5; illusory; imaginary; insubstantial 2; nonexistent; romantic 1; supposed 1
Imagine (that)!, indeed 3
imam, clergyman 1
imbalance, difference 1; disparity; disproportion; inequality 1
imbecile, absurd 1; crazy 2; daft 1; dolt; fool 1; half-wit; jerk 5; natural 12; retard 2; senseless 3; simple 4; stupid 1; twit[2]
imbecilic, absurd 1; crazy 2; daft 1; feebleminded; foolish 2; half-witted; inane; insane 2; mindless 1; nonsensical; preposterous; senseless 3; silly 1; simple 4; stupid 1; thick 6; unreasonable 1; unthinking 1; weak 5
imbecility, folly 1; insanity 2; simplicity 4; stupidity 1
imbed, root[1] 6
imbedded, rooted
imbibe, drink 1; take 13
imbricate, overlap 1
imbricated, scaly 1
imbrication, overlap 3; scale[2] 1
imbroglio, mess 3; predicament; problem 1; riot 1
imbrue, suffuse
imbue, charge 8; impregnate 2; indoctrinate; inject 2; instill; permeate; saturate; steep[2] 2; suffuse
  imbue with, school 4
imbued with, be, possess 4
imitate°, copy 5; counterfeit 4; duplicate 3; echo 4; follow 2; forge 3; knock 5d; masquerade 3; mimic 1; mock 2; model 9; monkey 4; parallel 4; parrot 2; pass 16b; pattern 7; pose 2; represent 2; reproduce 1; satirize; steal 2; take 34b
imitated, secondary 2
imitation°, artificial 1; bogus; burlesque 1; copy 1; counterfeit 1, 3; dummy 2; echo 2; factitious; fake 3, 5; false 3; forgery 2; fraudulent 1; glorified 2; glossy 2; impression 5; knockoff; mimic 5; mock 3; mockery 2; model 10; phony 1, 2; plagiarism; plastic 3; replica; reproduction 2; sham 1, 2; spurious; synthetic; takeoff 2; toy 6
imitative, derivative 1; mimic 5
imitator, mimic 4; parrot 1
immaculate°, celibate 2; clean 2; fair[1] 3; faultless; flawless 1; holy 2; impeccable; incorrupt 1; innocent 2; perfect 3; pristine 2; pure 2; spotless 1; unspoiled; untarnished; white 3
immanent, inherent; intrinsic
immaterial°, disembodied 1; inconsequential; indifferent 4; insubstantial 2; intangible; point 15; shadowy 3
immature°, adolescent 2; boyish 2; callow; childish; crude 2; fresh 3; green 2; gullible; inexperienced; infantile; juvenile 1; mad 2; new 6; premature 1; puerile; raw 3; rudimentary 2; small 1; tender[1] 3; undeveloped; unfledged; young 2
immaturity, inexperience; youth 1
immeasurable°, bottomless; boundless; endless 1; indefinite 4; inestimable 1, 2; infinite; limitless; numberless; prodigious 1; unlimited 2; untold 1; vast
immeasurably, exceedingly; far 3; well[1] 7
immediacy, gravity 2; priority
immediate°, express 7; hasty 1; imminent; impulsive; instant 3; instinctive 2; near 4; present[1] 2; primary 3; prompt 1; quick 2; rapid; ready 5; speedy 1; spontaneous 2; sudden; unhesitating 1; urgent 1
immediately°, cold 11; demand 8; directly 2; double 8; fast[1] 8; flat 17b; hastily 1; instantaneously; nail 6; now 3; once 6a; outright 3; posthaste; promptly; quickly 3; readily 3; right 13; shortly 2; shot 10; soon 2; straight 15; summarily 1; time 17
immediateness, rapidity
immense°, big 1, 2; boundless; colossal 1; enormous; extensive 2; gigantic; grand 1; great 1, 2; huge; immeasurable; inestimable 2; infinite; jumbo; large 3; limitless; massive; monstrous 3; monumental 2; mountainous 2; prodigious 1; spacious; thumping 1; unlimited 2; vast; weighty 1; whopping 1
immensely, highly 1; vastly
immensity, enormousness; extent 1; magnitude 1; mass 5; size 1
immerse°, catch 14a; dip 1; drench; drown 1; duck 2; flood 4; overwhelm 2; plunge 2; sink 8; soak 1; steep[2] 1, 2; submerge 1, 3; swamp 2
immersed, absorbed; deep 4; full 6; preoccupied 2
  immersed in, wrap 2
immersing, soaking 1
immersion, dip 4; plunge 3
immigrant°, alien 2; arrival 2; foreigner; newcomer 1; settler
immigrate, migrate 1
imminent°, forthcoming 1; hand 9a; impending; near 4; oncoming 1; pending 2; pipeline 2; potential 1; prospective; threatening; wind[1] 8
immobile, dormant 1; fixed 1; immovable 1; inactive 1; inanimate; inert 2; sedentary; stagnant; static 1
immobility, inactivity 1; inertia
immobilization, freeze 6
immobilize, fix 5; freeze 3; incapacitate; numb 2; paralyze 1; pin 4c
immobilized, crippled 2; fixed 1; inactive 1
immoderate°, epicurean 1; excessive 1; exorbitant; extravagant 1, 2; extreme 2; fanatical; inordinate 1; lavish 3; outrageous 1; prodigal 1; profligate 2; promiscuous 2; radical 3, 4; self-indulgent; towering 2; ultra-; unconscionable 2; unmitigated; unreasonable 2; unwarranted; wanton 2
immoderately, fault 7; overly; unduly 2
immoderateness, extravagance 2
immoderation, excess 2; extravagance 2; gluttony; prodigality 1
immodest°, barefaced 2; bold 2; brazen; foul 5; immoral 2; improper 3; indecent 1; inflated 1; obscene 2; presumptuous; profane 3; racy 2; rank[2] 3; scandalous 1; sexy 2; shameless; suggestive 2; unchaste; unmentionable 2
immodesty, impropriety 3; indelicacy; presumption 1; ribaldry
immolate, sacrifice 3
immolated, sacrificial 1
immolation, sacrifice 1
immoral°, abandoned 2; bad 3; criminal 2; decadent 2; dissolute; evil 1; fast[1] 2; filthy 3; foul 4; ill 2; improper 3; impure 4; libertine 2; loose 7; naughty 3; obscene 1; outrageous 3; perverted; profligate 1; promiscuous 2; reprobate 1; rotten 3; satanic 1; scandalous 1; sinful; ugly 2; unchaste; unconscionable 1; ungodly 1; unscrupulous; unwholesome 2; vicious 1; vile 1; wanton 1; wicked 1; wrong 1
immorality, evil 6; impropriety 3; impurity 3; profligacy 1; sin 2; vice 1
immorally, badly 4; fast[1] 9
immortal°, celestial 1; classic 2; deathless; enduring; eternal 1; everlasting; god; grand 2; heavenly 1; legendary 3; lofty 2; perennial 2; timeless
immortality, glory 1
immortalize°, commemorate; glorify 1; lift 2; perpetuate
immovable°, constant 1; fast[1] 4; firm 2, 4; fixed 1; inflexible; monolithic; obstinate; parochial; recalcitrant; resistant 2; rigid 4; secure 2; static 1; steadfast; stern 1; tenacious 1; willful 2
  immovable, be *or* become, stick[1] 7
immovably, fast[1] 7; firmly 1
immune°, exempt 2; privileged 2; secure 1
  immune from, free 11
  immune (to), unaffected[2]
immunity°, exemption; freedom 3; privilege; protection 1; tolerance 3
immure, enclose 1; shut 6a; wall 5
immutability, finality; resolution 1
immutable, abiding; changeless 1; constant 3; eternal 3; final 2; fixed 2; hard 8; immovable 2; indestructible; inflexible; invariable 2; irrevocable; permanent 2; perpetual 1; resolute; stable 2; timeless
immutableness, resolution 1
imp°, devil 4; goblin; monkey 3; rascal
impact°, brunt; effect 1, 2, 4; force 1; hit 10; impression 2; influence 1; punch[1] 3; shock 4; strike 3; weight 3
  impact, have an, get 16; score 13
  impact (on), hit 4
  impact with, hit 3
impair°, alloy 2; blemish 1; cripple 2; damage 4; debase 2; deface; drain 5; hurt 1; incapacitate; injure 1; maim; mar 2; overshadow 2; ruin 9; shake 3; strain[1] 3; undermine 1; vitiate 1; wear 3
impaired, defective 1; deficient 2; disabled; disfigured; faulty; hurt 8; invalid[2]; lame 1; under 6; wanting 1
impairment°, blemish 3; damage 1; detriment; disability 1; expense 2; injury; loss 3; sabotage 1
impale°, gore[2]; pierce 1; spike 2; stab 1; stick[1] 1; transfix 1
impalpable, disembodied 1; elusive 2; insubstantial 2; intangible; shadowy 3
imparity, disparity
impart°, betray 2; communicate 1; deliver 5; disclose 1; extend 4; get 26; give 3, 15c; inform 1; instill; lend 2;

make 33; mention 2; recount 1; relate 2; say 4; shed[2] 2; signify 1; tell[1] 2; vouchsafe 1
imparted to, be, rub 5
impartial°, candid 2; detached 2; disinterested; dispassionate 2; equitable; even 7; fair[1] 1; fence 2; honorable 3; indifferent 2; judicial 3; just 1; liberal 3; neutral 1; nonaligned; nonpartisan 2; objective 1; unprejudiced
impartiality, candor 2; chivalry; equality 3; equity; honesty 3; indifference 3; justice 1; objectivity
impartially, fairly 2; honestly 1
impassable, dense 2; impermeable; inaccessible 2; thick 3
  impassable, make, obstruct 1
impasse°, deadlock 1; dilemma; predicament; stalemate
  impasse, bring to an, deadlock 2
  impasse, come to an, deadlock 2
impassion, fire 8a; flush[1] 3; heat 4; inflame 1; rise 16; thrill 3; touch 6; turn 18c
impassionate, tender[1] 7
impassioned°, animated 1; ardent; emotional 1; excited 2; fervent 1; heated; intense 2; keen[1] 1; lyrical 2; passionate 1; romantic 4; tempestuous; tender[1] 7; torrid 2; warm 2
impassive°, blank 4; calm 4; even 4; frigid 3; immovable 2; indifferent 1; insensible 2; listless; passive 1; rocky[1] 3; stoical; stolid; straight 8; unmoved; unsympathetic; wooden 2
impassiveness, indifference 1
impassivity, indifference 1
impatient°, agog; anxious 2; eager; fretful; hasty 4; intolerant 1; irritable; moody 2; petulant; quick 5; quick-tempered; ratty 1; sour 4; temperamental 1; weary 3
  impatient (with), be *or* **become,** weary 4
impeach°, accuse 2; incriminate
impeachment, accusation
impeccable°, flawless 1; immaculate 3; incorrupt 1; irreproachable; polished 1; pure 3
impeccably, perfectly 2
impecunious, destitute 1; hard 17; impoverished 1; indigent; needy; penurious 2; poor 1; short 8
impecuniousness, need 4; poverty 1; want 5
impedance, interference 2; prevention
impede°, bar 9; block 3; bog 2; check 1; clog; delay 2; deter; embargo 2; encumber 2; entangle 1; foil[1]; frustrate 1; hamper[1]; handicap 2; hinder 1; hobble 2; hold 22b; inhibit; intercept; interfere 2; obstruct 2; oppose 2; overload 1; preclude; prevent; prohibit 2; resist 1; restrict; retard 1; set 16; stay[1] 4; stop 2; stunt[2]; stymie; thwart 1; trammel 2
impediment°, rub 11
impedimenta, gear 4; luggage; paraphernalia; stuff 2; thing 8c
impediment(s), bar 3; barrier 2; block 2; charge 1; check 14; deterrent; drawback; embargo 1; encumbrance; fly 10; handicap 1; hindrance 1; hitch 4; hurdle 1; inconvenience 1; inhibition; interference 2; let[2]; liability 3; obstacle; obstruction 1; overload 2; preventive 3; setback; shackle 2; snag 1; stricture 1; stumbling block; trammel 1; wall 1
impeding, obstruction 2; prevention; preventive 1
impel, activate; carry 3; drive 1; force 7; hasten 2; induce 1; make 2; motivate; move 6; occasion 5; press 1; prod 2; prompt 3; propel; provoke 1; push 4; spur 4; thrust 1; urge 1
impelling, motive 3; moving 1
impend, loom 2; overhang 2; threaten 3
impending°, forthcoming 1; imminent; menacing; near 4; pending 2; prospective; threatening; wind[1] 8
impenetrability, opacity 1, 2; perplexity 2
impenetrable, dark 6; deep 2; dense 1, 2; hard 1; heavy 12; impermeable; impregnable; inaccessible 2; incomprehensible; invincible 2; monolithic; occult 2; opaque 1, 2; perplexing; proof 3; recondite; thick 3; tight 1
  impenetrable (to), resistant 3
impenetrable depth(s), abyss
impenetrableness, perplexity 2
impenitent, unrepentant
imperative°, essential 1; exacting; indispensable 1; instant 4; necessary 1; peremptory 1; precondition; prerequisite 1; pronouncement 2; urgent 1; vital 1
imperativeness, urgency
imperceivable, invisible 1
imperceptible°, dim 1; faint 1; feeble 3; inaudible; indistinguishable 2; intangible; invisible 1; light[2] 3; sight 6a; weak 6
  imperceptible, almost, slow 2
imperceptibly, degree 3
imperceptive, blind 2; dull 2; obtuse 2
impercipient, dull 2
imperfect°, defective 1; deficient 2; faulty; flawed; inaccurate; inadequate 1; incomplete; inferior 3; invalid[2]; partial 1; ragged 4; rough 9; rude 4; rudimentary 2; sketchy; unsatisfactory; unsound 1; untrue 3; vestigial; wanting 1; wrong 5
imperfection°, blemish 3; defect 1, 2; failing 1; fault 1; flaw 1; foible; frailty 2; impairment; inferiority 3; kink 3; shortcoming; taint 1; vice 2; weakness 3
imperfectly, amiss 2; sketchily; wrong 7
imperial°, lofty 2; majestic 1; regal; royal 1; state 6; stately
imperil, adventure 3; chance 8; endanger; expose 2; fire 6; hazard 3; jeopardize; lean[2] 4b; overhang 2; risk 2; threaten 2; venture 3
imperiled, endangered; jealous 2
imperilment, exposure 2
imperious, arbitrary 2; arrogant 2; authoritarian; bossy; condescending; despotic; dictatorial 2; dogmatic; domineering; heavy-handed 2; imperative 2; majestic 2; masterful 2; overbearing; peremptory 3; pompous 1; positive 5; prescriptive; tyrannical
imperishable, deathless; immortal 1; indestructible; inextinguishable; perennial 2; permanent 1
impermanent, fly-by-night 1; meteoric; momentary; passing 1; temporary; transient
impermeability, opacity 1
impermeable°, opaque 1; tight 1
impersonal°, detached 2; disinterested; mechanical 3
impersonate, act 7; copy 5; imitate 1; masquerade 3; mimic 3; monkey 4; pass 16b; personify 1; portray 2; pose 2; represent 2
impersonating, imitation 1
impersonation, imitation 1, 2; impression 5; mockery 2; role 1
impersonator, impostor; mimic 4
impertinence°, arrogance; disrespect; effrontery; face 5; familiarity 3; flippancy 2; freedom 7; gall[1] 2; impudence; mouth 4; nerve 2; presumption 1; sauce 2
impertinent°, audacious 2; barefaced 2; bold 2; brash 2; brazen; cheeky; cool 6; discourteous; extraneous 2; flippant 2; forward 2; fresh 8; ill-mannered; immodest 2; impudent; insolent; irrelevant; irreverent 2; offensive 2; pert; presumptuous; pushy; rude 2; tactless
  impertinent, be, liberty 5
imperturbability, patience 1; philosophy 3; poise 2; presence 5; sang-froid; self-control 2; stoicism
imperturbable, coldblooded 2; collected; cool 2; equable 1; even 4; impassive; levelheaded; nonchalant; passive 1; philosophical 2; phlegmatic 2; poised 1; sedate 1; self-contained 1; self-possessed; serene 1; stoical; temperate 1
impervious, dead 4; deaf 2; frigid 2; hard 1; hermetic; impermeable; insensible 2; proof 3; resistant 1; thick-skinned; tight 1
  impervious (to), unaffected[2]; blind 4; incapable 2; resistant 3
imperviousness, tolerance 3
impetration, supplication 1
impetuosity, desperation 1; fury 2; haste 2; heat 2; indiscretion 1
impetuous°, blind 3; brash 1; daredevil 2; dashing 1; desperate 1; foolhardy; foolish 1; hasty 2; hot 3; hot-headed; hurried 1; ill-advised 2; impatient 1; improvident 2; imprudent; impulsive; indiscreet; intense 3; pell-mell 2; precipitate 5; rapid; rash[1]; spontaneous 2; sudden; violent 3; warm-blooded 2; whirlwind 2; wild 4
impetuously, blindly; fiercely; hastily 2; madly 3; pell-mell 1; spur 3
impetuousness, haste 2; indiscretion 1
impetus°, drive 8; incentive; momentum; power 8; propulsion; signal 2; spirit 2; stimulant 1; urge 5
impiety, sacrilege 2; sin 1, 2
impinge on, infringe 2
impious°, blasphemous; diabolic 2; godless 1; heretical; irreverent 1; profane 1; sacrilegious; satanic 1; sinful; ungodly 1; wicked 1
impiousness, sacrilege 2; sin 2
impish, devilish; elfin 1; mischievous 1; naughty 1; playful 1; sly 2; wicked 5
impishness, mischief 1
implacable°, deadly 2; grim 1; hard 4; mortal 4;

suggestible; susceptible 2; tender[1] 3

impressionist, mimic 4

impressions, note 10

impressive°, arresting; conspicuous 2, 3; dramatic 2; effective 2; elevated 2; extraordinary 2; forceful 2; formidable 2; grand 1; grandiose 2; hefty 3; important 2, 3; imposing; magnificent; majestic 1; monumental 1; moving 1; noble 5; persuasive; potent 2; powerful 2; prestigious; remarkable 1; royal 2; salient; scenic; significant 1; solemn 3; splendid 1, 2; stately; statuesque; striking; substantial 2; towering 1; venerable; weighty 3

impressiveness, solemnity

imprimatur, approval; consent 3; endorsement 1; sanction 1

imprimatur, give (something) one's, endorse 1

imprimatur on, put one's, approve 1

imprint, engrave 2; etch 2; implant 1; impress 2; label 1, 4; plant 6; print 1; run 19; seal 1; stamp 2, 7; step 3

imprinted, strong 15

imprison°, cage 2; commit 2; hold 6; jail 2; keep 9; lock[1] 9; punish 2; put 13b; restrain 3; send 5, 9b; shut 6a; trap 5

imprisoned, arrest 6; captive 2; inside 7

imprisonment°, captivity; custody 2; detention; duress 2; justice 2; penalty; punishment 2

improbable°, far-fetched; fishy 1; implausible; incredible 1; off 5; paradoxical; remote 6; tall 3; unlikely 1; unthinkable 2

impromptu, candid 3; extemporaneous; offhand 3, 4; scratch 6; spontaneous 1; spur 3; stopgap 2; unpremeditated

improper°, abusive 2; amiss 1; blue 2; broad 8; bum 3; coarse 3; crude 4; disgraceful 2; dishonorable 3; foul 5; gross 3; illegitimate 3; illicit 2; immodest 1; inadmissible; inappropriate; incongruous; incorrect; indecent 1; ineligible; inept 2; low[1] 3; naughty 3; obscene 1; off-color 2; order 13b; outrageous 3; pale[2] 3; perverse 1; place 12; profane 3; risqué; scandalous 1; sexy 2; shameless; spicy 2; suggestive 2; tasteless 1; unacceptable; unbecoming 2; unprofessional 1; unreasonable 2; unseemly 1, 2; untoward 2; unwarranted; vulgar 2; wrong 1, 3, 4, 6

improper, be, liberty 5

improperly, amiss 3; badly 3, 4, 6; ill 12; poorly 1; turn 42b; unduly 1; vain 3b; wrong 7

improperness, impropriety 1

impropriety°, familiarity 3; slip[1] 8; solecism; vulgarity 2

improve°, advance 2; amend 1; benefit 3; better 10, 11; boost 5; break 3; convalesce; develop 1; dignify; elaborate 4; embellish 1; enhance; enrich 1; gain 3, 7; heal 1; help 2; increase 2; lift 2; look 11c; maximize 1; mend 2, 3; perfect 9; pick 8f; polish 2; profit 3; progress 6; pull 17; rally 4; recover 2; rectify; recuperate; refine 2; reform 1; remedy 4; retouch; revise 1; rise 8; shape 10b; step 18a

improve (with age), mellow 5

improved, better 4, 6a

improvement°, advance 7; advantage 2; amendment 1; correction 1; development 2; edification; elaboration 1; gain 9; growth 3; headway 1; progress 2; promotion 1; rally 2; recovery 1; refinement 4; reform 3; repair 2; revision; revival 3; spurt 2

improvement, show, look 11c; shape 10b

improvidence, extravagance 1; indiscretion 1; prodigality 1; profligacy 2; waste 6

improvident°, extravagant 1; harebrained 1; ill-advised 2; imprudent; indiscreet; lavish 3; prodigal 1; profligate 2; shortsighted 2; spendthrift 2; wasteful

improving, mend 5; promotion 1

improvisation, makeshift 2; stopgap 1

improvise°, formulate 2; invent 1; knock 7a; patch 6; play 12; strike 16; think 6; whip 7b

improvised, extemporaneous; fictitious 2; makeshift 1; stopgap 2

imprudence, desperation 1; indiscretion 1

imprudent°, careless 2; crazy 3; daredevil 2; foolhardy; foolish 1; ill-advised 1; improvident 1; indiscreet; inept 2; mad 2; misguided; rash[1]; reckless; shortsighted 2; tactless; thoughtless 2; unguarded 3; unseemly 2; unthinking 2; untoward 2; unwary; wild 7; wrong 4

imprudently, turn 42b

impudence°, assurance 3; brass; disrespect; effrontery; face 5; familiarity 3; flippancy 2; freedom 7; gall[1] 2; impertinence; mouth 4; nerve 2; presumption 1; sauce 2

impudent°, audacious 2; barefaced 2; bold 2; brash 2; brassy 1; brazen; cheeky; cool 6; disrespectful; familiar 3; flippant 2; forward 2; fresh 8; ill-mannered; immodest 2; impertinent; insolent; irreverent 2; offensive 2; pert; presumptuous; rude 2; shameless

impudent, be, liberty 5

impugn, challenge 1; criticize 2; denounce 3; dispute 1; fault 8; impeach 2

impuissant, feeble 1

impulse, bias 3; fancy 8; impetus; incentive; incitement 2; inspiration 1; momentum; notion 2; occasion 2; propulsion; spirit 7; spur 1; stimulant 1; urge 5

impulsive°, automatic 1; brash 1; capricious; daredevil 2; fanciful 1; giddy 2; hasty 2; impetuous; improvident 2; imprudent; indiscreet; intense 3; involuntary; natural 4; pell-mell 2; precipitate 5; premature 2; quick-tempered; rash[1]; reckless; shortsighted 2; spontaneous 2; sudden; unpremeditated

impulsively, hastily 2; pell-mell 1; spur 3

impulsiveness, indiscretion 1

impunity, exemption; indemnity 2

impure°, filthy 3; foul 2; immoral 1; mixed 1; obscene 1; profane 2; unchaste; unrefined 2

impurities°, impurity 2

imputation°, accusation; aspersion; charge 6; innuendo; reflection 3; slur 1

impute°, accuse 2; attribute 2; credit 6; insinuate 1; lay[1] 7; make 37e; set 17b

impute (to), read 4

in, cage 2; contact 2; contemporary 2; elegant 2; fashionable; home 4b; latest 2; minute[1] 3; modern; now 5; popular 1; stylish; trendy 1

in, be, wear 1

inability°, disability 2; incompetence

inaccessible°, distant 3; frigid 2; out-of-the-way 1; private 2; remote 1; standoffish; unapproachable 2

inaccessible, make, close 2, 6

inaccuracy, error 1; impropriety 1

inaccurate°, careless 3; erroneous; false 1; imprecise; improper 1; incorrect; inexact; loose 5; mistaken 1, 2; off 3; out 12; rude 4; unrealistic 2; untrue 2; wrong 2

inaccurate, be, err 1

inaccurately, badly 3

inaccurateness, impropriety 1

inaction, default 1; idleness 1; inactivity 2; indolence; neglect 4

inactivate, incapacitate; paralyze 1

inactive°, dead 11; dormant 1; dull 3; extinct 3; flat 11; idle 1; inanimate; indolent; inert 1, 3; lackadaisical 1; lazy 1; leaden 4; leisure 3a; lethargic 1; passive 1; phlegmatic 1; quiet 3, 4; silent 4; sleepy 2; slothful; slow 6; supine 2; torpid; unemployed

inactivity°, idleness 1; indolence; inertia; lethargy 1; neglect 4; repose 1; rest[1] 3; slack 5; torpor

inadequacy, dearth; defect 1; imperfection; impotence 2; incompetence; inferiority 2; lack 1; poverty 2; scarcity; want 3

inadequate°, bad 1; deficient 1; disappointing; feeble 2; flimsy 2; hopeless 2; impotent 2; incompetent; ineffective 1; ineffectual 2; insufficient; little 3; low[1] 2; meager 1, 3; miserable 3; pale[1] 3; par 5a; pathetic 2; poor 2, 4; ropy 2; scarce; short 6; slender 3; slight 4; small 4; straitened; thin 2; unsatisfactory; wanting 1

inadequate, be *or* **prove,** fail 1; fall 11b; short 13

inadequate, prove *or* **(turn out to) be,** fall 19

inadequate to, inadequate 2

inadequately, badly 1; poorly 1

inadequateness, scarcity

inadmissible°, pale[2] 3

inadvertence, neglect 3; negligence; oversight 1; slip[1] 8

inadvertent°, absent-minded; accidental; chance 6; thoughtless 2; unguarded 1

inadvertently, chance 5a; unawares 2

inadvisable, ill-advised 1; imprudent; unseemly 2

inadvised, inept 2

inamorata, beloved 2; girl 2; love 3; mistress 1; paramour; sweetheart

indulge° (*cont.*) forgive 1; humor 4; make 34b; mother 6; oblige 1; overdue 1; overeat; pamper; pander 1; pardon 2; pet[1] 5; regale; satiate 1; satisfy 2; spoil 3; stuff 8; suffer 3; tolerate 1
  indulge (oneself) in, indulge 1; luxuriate 1; roll 10b; savor 3; wallow 2
indulged, pet[1] 3; privileged 1
indulgence°, charity 2; excuse 5; fling 2; forgiveness 2; grace 3; kindness 1; luxury 2, 4; mercy 1; office 5; pardon 1; permission; privilege; remission 1; splurge 1; tolerance 1; waste 6
indulgent°, charitable 2; easy 3; fatherly; gentle 1; gracious; kind[1]; lax 1; lenient; merciful; mild 1; obliging; paternal 1; permissive; soft 5; softhearted; tolerant
indulgently, favorably 1
indurate, hard 4; steely 2
industrial, technical 2
industrialist, manufacturer; merchant 3
industrious°, busy 2; diligent; earnest 2; enterprising; hard 7; laborious 3; studious 1; tireless
industriously, hard 13
industriousness, application 3; industry 3
industry°, application 3; drive 8; exertion; labor 1; work 1
indweller, occupant
indwelling, inherent; intrinsic; native 1
inebriant, intoxicating 1
inebriate, intoxicate 1
inebriated, drunk 1; high 9; stinking 3
inebriating, alcoholic 1
inebriety, drunkenness
inedible, revolting; unpalatable
ineffable°, inexpressible; mystical; nameless 3; unmentionable 1
ineffaceable, indelible
ineffective°, bootless; feeble 2; flabby 2; fruitless; futile; hollow 5; impotent 2; impractical 2; incompetent; ineffectual 1; inefficient 1; invalid[2]; lame 2; limp[2] 3; low[1] 5; meaningless 2; nonproductive 2; pale[1] 3; pointless; powerless 1; shiftless; spineless 2; unprofitable 2; unsuccessful 1; useless 1, 2; vain 2; void 1; weak 3; wet 3
  ineffective, make *or* **render,** disturb 5; neutralize
ineffectiveness, impotence 2
ineffectual°, fainthearted 2; feeble 2; flabby 2; fruitless; futile; impotent 2; impractical 2; incompetent; ineffective 1; inefficient 1; irresponsible; limp[2] 3; low[1] 5; meaningless 2; nonproductive 2; pale[1] 3; pointless; powerless 1; spineless 2; tame 3; unsuccessful 1; useless 1, 2; void 1; wan 2; weak 3; wet 3
  ineffectual, make, disturb 5
ineffectualness, impotence 2
inefficacious°, bootless; ineffective 1; ineffectual 1; meaningless 2; unsuccessful 1
inefficacy, impotence 2
inefficiency, incompetence; ineptitude 1
inefficient°, helpless 4; incompetent; ineffective 2; ineffectual 2; inept 1; poor 4; unprofessional 2; unprofitable 2; useless 2
inefficiently, roughly 3
inelastic, firm 1; inflexible; rigid 2; stiff 1
inelegance, indelicacy; simplicity 3
inelegant, awkward 2; broad 8; homespun; hulking; obscene 1; off-color 2; rude 4; tasteless 1; uncivilized 2; ungraceful 2; unrefined 1; vulgar 1
ineligible°, unqualified 1; unworthy 1
  ineligible, declare, disqualify
ineluctable, automatic 2; certain 3; destined 2; eventual 2; fatal 3; fated 3; hard 8; indispensable 2; inevitable; irresistible 2; necessary 2; sure 4; unavoidable
ineluctably, inextricably; necessarily
inept°, artless 3; awkward 1; clumsy 1; heavy-handed 1; helpless 4; hopeless 2; impotent 2; incompetent; ineffective 2; inefficient 1; tactless; useless 2; weak 3
  inept for, inadequate 2
ineptitude°, inability; incompetence
ineptly, badly 1, 3; roughly 3
ineptness, impotence 2; inability; ineptitude 1
inequality°, difference 1; disparity; disproportion; injustice 1; odds 3; prejudice 2
inequitable, one-sided 1; unreasonable 2
inequity, discrimination 1; ill 9; inequality 2; injustice 1
ineradicable, entrenched; indelible; indestructible
inerasable, indelible
inert°, dead 4, 6, 11; dormant 1; inactive 1; inanimate; indolent; leaden 4; lifeless 2; passive 1; slothful; static 1; still 1; supine 2; torpid
inertia°, idleness 1; inactivity 1; indolence; lethargy 1; momentum; power 8; sloth; stupor; torpor
inertness, inactivity 1; inertia; stupor; torpor
inescapable, automatic 2; certain 3; destined 2; fatal 3; fated 3; hard 8; indispensable 2; inevitable; irresistible 2; necessary 2; pervasive; self-evident; sure 4; unavoidable
inescapably, inextricably; necessarily
inessential, dispensable; expendable; extraneous 1; nonessential 2; peripheral 1; petty 1; redundant 1; secondary 1; unnecessary; worthless 1
inestimable°, immeasurable; imponderable; infinite; priceless 1
inevitable°, automatic 2; certain 3; changeless 1; destined 2; eventual 2; fatal 3; fated 3; foregone; necessary 2; secure 4; sure 4; unavoidable; willy-nilly 2
inevitably, necessarily; willy-nilly 1
inexact°, approximate 1; careless 3; erroneous; false 1; general 4; imprecise; improper 1; inaccurate; incorrect; indefinite 1, 2; lax 2; loose 5; rough 10; sketchy; untrue 3; vague 1, 2; wrong 2
inexactitude, impropriety 1
inexactly, vaguely 2
inexactness, impropriety 1
inexcusable°, unconscionable 2; unwarranted
inexhaustible°, abundant 1; bottomless; boundless; infinite 1; unlimited 2; vast
inexorability, necessity 2; severity 1
inexorable, certain 3; grim 1; implacable; inevitable; insistent; irresistible 2; merciless; necessary 3; obstinate; relentless 1; remorseless 2; severe 1; unavoidable
inexorableness, severity 1
inexorably, finally 2; necessarily
inexpedient, ill advised 1; imprudent; inconvenient; inept 2; unhappy 3; unseasonable; unseemly 2; wrong 4
inexpensive°, cheap 1; economical 1; reasonable 3
inexpensively, cheap 5, 7; song 2
inexperience°, ignorance 1; naiveté; provincialism 2
inexperienced°, artless 3; callow; childish; fresh 3; green 2; gullible; ignorant 3; immature 2; innocent 4; naive; new 6; plastic 2; practice 5; raw 3; tender[1] 3; unfledged; unprofessional 2; unsophisticated 1; young 2
  inexperienced in *or* **at** *or* **with,** unaccustomed 2; unfamiliar 2; unused 4
inexpert, amateur 2; artless 3; awkward 1; green 2; incompetent; inefficient 1; inept 1; unprofessional 2
inexpertly, poorly 1; roughly 3
inexpiable, inexcusable
inexplicable°, miraculous; mysterious 2; occult 2; recondite; strange 1; supernatural; unaccountable 1
inexplicably, vaguely 1
inexplicit, imprecise; indefinite 1, 2; vague 1, 3
inexpressible°, ineffable 2; nameless 2; unmentionable 1; untold 3
inexpungible, indelible
inextinguishable°, unquenchable
inextirpable, indelible; indestructible
in extremis, dying; moribund 1
infallible°, certain 2, 3; foolproof; reliable; sure 3
infamous°, atrocious 1; dirty 6; disgraceful 1; evil 1; flagrant; foul 4; monstrous 2; notorious 1; outrageous 2; public 6; scandalous 1; shabby 4; shameful
infamy°, atrocity 1; discredit 4; disgrace 1; notoriety; scandal 2, 3; shame 2
infancy°, childhood
infant, baby 1; child 2; innocent 5; juvenile 2; minor 3; tot
infanticide, murder 1
infantile°, childish; juvenile 1; puerile; young 3
infantryman, private 5; soldier 1
infatuate, captivate; intoxicate 2
infatuated°, crazy 5; mad 6; smitten 2; taken
  infatuated with, daft 3; sweet 8
  infatuated with, be, dote; fall 13a; love 7
infatuation, devotion 3; fanaticism 1; fixation; love 1; passion 3
infeasible, impracticable 1
infect, attack 4; contaminate; corrupt 4; interest 7; poison 3; riddle[2] 2
infected (with), diseased; impure 1; inflamed; -ridden
infected with, become, contract 3; get 5
infection, bug 2; disease 1; inflammation; sore 7
infectious°, catching 1

infecund, impotent 3; infertile; nonproductive 1; sterile 1
infelicitous, evil 4; improper 2; inappropriate; unhappy 3; wrong 4
infelicitousness, impropriety 2
infelicity, impropriety 2; misfortune 1; misuse 2
infer°, deduce; derive 1; determine 2; divine 4; fancy 10; gather 4; imagine 2; judge 7; perceive 2; presume 1; surmise 1; take 19
  infer (from), read 4; understand 3
inferable, implicit 1
inference°, deduction 2; implication 3; interpretation 2; law 3; presumption 3; tenor
inferential, circumstantial 1; constructive 2; implicit 1; logical 1
inferential reasoning, logic 1
inferior°, amateur 2; awful 1; bad 1; base[2] 2, 5; cheap 3; coarse 4; common 4; contemptible; deficient 2; degenerate 1; disappointing; everyday 3; flunky 1; hopeless 2; humble 3; indifferent 3; junior; lousy 2; low[1] 7, 9, 10; mean[2] 2; mediocre; par 5a; petty 1; plebeian 1; poor 4; punk 2; puny 1; ropy 2; secondary 1; shabby 4; shoddy; simple 5; subordinate 2; tinny 1; unprofessional 2; unsatisfactory; unworthy 1; wanting 1
  inferior, treat as (an), patronize 1
  inferior (to), below 9, 10; subordinate 1; under 2
infernal°, atrocious 1; black 6; damnable; devilish; diabolic 1; fiendish; ghoulish 1; satanic 1; ungodly 1; wicked 1; wrong 1
infernal regions, hell 1
inferno, blaze 1; fire 1; hell 1; holocaust 1
inferred, circumstantial 1; constructive 2; presumptive 2
infertile°, barren 1, 2; dead 8; exhausted 3; impotent 3; lean[1] 2; meager 4; nonproductive 1; poor 3; sterile 1
infertility, impotence 3
infest°, blight 3; riddle[2] 2
infestation, blight 1
infested with, -ridden
infested with, be, abound 3
infidel°, heathen 1, 2; heretical; nonbeliever; pagan 1, 2; profane 1
infidelity°, perfidy
infighting, faction 2
infiltrate, enter 2; infest; insinuate 2; permeate
infiltration, invasion 1
infiltrator, intruder 1
infinite°, boundless; endless 1; eternal 1; everlasting; immeasurable; indefinite 4; inestimable 2; inexhaustible 1; numberless; perpetual 1; universal 2; unlimited 2; vast
infinitely, cease 2; vastly
infinitesimal, diminutive; imperceptible 2; little 1; marginal 1; minute[2] 1; slight 2; tiny
infinity, eternity
infirm°, decrepit 1; diseased; doddering; feeble 1; fragile; frail 2; helpless 1; ill 1; impotent 1; insecure 3; invalid[1] 1; low[1] 4; peaked; sick 2; tender[1] 1; unhealthy 1; unsound 1; weak 1, 2
  infirm of purpose, irresolute
infirmary°, hospital
infirmity°, ailment; bug 2; decrepitude 1; delicacy 2; disability 1; disease 1; foible; frailty 1; illness; imperfection; vice 2; weakness 1
inflame°, aggravate 1; enrage; exasperate 1; excite 1, 3; ferment 2; fester 2; fire 8a; flush[1] 3; foment; fuel 4; gall[2] 4; heat 4; incite; infuriate; kindle; madden 1; rise 16; work 20a
inflamed°, angry 2; drunk 2; fervent 1; feverish; heated; hot 3; impassioned; raw 4; sore 1; tender[1] 8; torrid 2
inflaming, incitement 1; inflammatory
inflammable°, explosive 1; flammable
inflammation°, incitement 2; sore 7; ulcer 1
inflammatory°, seditious
inflatable, expansive 1
inflate°, blow[1] 8c, 8e; enlarge 1; exaggerate; expand 1; fill 2; increase 1; magnify 1; maximize 2; overstate; pad 5; puff 6; pump 4a; raise 6; stretch 2; swell 1
inflated°, bloated; bombastic; exalted 2; fancy 4; flowery; grandiose 1; pompous 1, 2; ponderous 2; pretentious 1; rhetorical 3; stilted; swollen
inflatedness, snobbery
inflation, exaggeration; expansion 2; increase 4; swell 4
inflection, accent 1, 3; diction 2; intonation; pronunciation; tone 2
inflexibility, formality 2; obstinacy; perseverance; resolution 1; rigor 2; severity 1; tenacity 1
inflexible°, dour 2; firm 1, 4; fixed 2; formal 1; grim 1; hard 1; hidebound; immovable 2; implacable; incorrigible; merciless; monolithic; obstinate; opinionated 1; persistent 1; perverse 3; precise 2; recalcitrant; relentless 1; resolute; rigid 2, 4; stern 1; stiff 1; stony 2; strict 2; stubborn; tenacious 1; tight 3; tough 2, 6; unkind; willful 2
  inflexible, be, persevere
inflexibleness, resolution 1
inflexibly, precisely 2
inflict°, deliver 5; enforce 1; impose 1; visit 2; wreak
inflicting, imposition 1
infliction, imposition 1
influence°, action 2; affect[1] 3; auspices; bear 8; bias 3, 4; bring 8b; charm 5; clutch 2b; color 4; come 16b; condition 6; connection 3; determine 4; dispose 2; domination 1; effect 2; factor 1; fascination; finger 2, 10; fix 11; get 14, 31d; grip 2; guide 3; hand 3; hold 26; impact 2; importance 2; impress 1; impression 2; incite; incitement 2; incline 2; induce 1; inspiration 2; interest 8; jurisdiction; lead 2; leadership; lobby 3; manipulate 1; mark 5; might 2; mold[1] 6; motivate; motive 1; move 6; nobility 1; part 3; persuade 1; persuasion 1; power 2; predominance; prejudice 3; preponderance 2; pressure 4, 5; prevail 3; prominence 1; prompt 3; pull 21; push 4; rank[1] 2; reach 6; reign 1; say 12; screw 4; shape 7; speak 12b; spell[2] 3; squeeze 11; strike 7; sway 2, 4; tint 4; touch 6, 18; urge 3; weight 3; win 3; work 18
  influence, area *or* **sphere of,** department 2
  influence, have (an), weigh 5
  influence, under the, drunk 1; stinking 3; tight 8; under 6
  influence, use *or* **exert,** pull 16; string 5
  influence on, have (an), concern 2
  influence over, have *or* **maintain** *or* **wield,** command 2
influence, corrupt, poison 2
influence, evil, poison 2
influenceable, susceptible 2
influenced, affected 5; overcome 2; partial 2
  influenced by, be, depend 1; possess 4
influencing, incitement 1; persuasion 1
influential°, active 2; classical 1; considerable 2; dominant 1; effectual 1; great 5; important 3; leading 1; mighty 1; persuasive; potent 1; powerful 2; predominant; prevailing 2; public 6; seminal 1; strong 7; telling 1; weighty 3
  influential, be, tell[1] 6
influentially, highly 4
influenza, chill 2; cold 10
info, dope 3; fact 3; gossip 2; information; intelligence 2; lowdown; news 1; rumor 1; scoop 3; talk 17; word 2
inform°, advise 2; brief 6; clue 2; discipline 6; disclose 1; educate; enlighten; fill 9c; flag[1] 2; ground 6; help 4; instruct 1; notice 7; notify 1; post[3] 4; prime 6; show 4; talk 5; tell[1] 3; warn 1
  inform (about *or* **of),** acquaint; register 5; teach; wise 4
  inform (about *or* **on),** familiarize
  inform against *or* **on,** betray 1; denounce 2; finger 8; give 12b; inform 2; sell 3; shop 3; turn 15c
  inform fully, picture 5
informal°, candid 3; casual 4; easy 6; familiar 3; free 8; homely 1; intimate[1] 4; offhand 1, 3; outgoing 2; scratch 6; sporty; unofficial; vernacular 2
informality, ease 4; familiarity 2
informally, offhand 4, 5
informant, informer; plant 4; source 3; spy 1
information°, advice 2; clue 1; data; direction 2; dope 3; edification; education 2; fact 3; file 1; good 21c; gossip 2; history 2, 4; illumination 2; instruction 1; intelligence 2; knowledge 2; learning; line[1] 10; literature 2; lowdown; material 4; message 1; news 1, 2; notice 5; particular 5; record 4; report 1; revelation; rumor 1; science 1; steer 3; story[1] 4; talk 17; testimony; tip[3] 2; word 2
informational, instructive
informative°, educational 2; forthcoming 3; instructive; significant 2
informatory, instructive
informed°, aware 1; current 4; enlightened; hip; profound 2; rational 2; ripe 2; wise 2
  informed, be, hear 2; understand 4
  informed (about *or* **on),** level 5; privy 2
  informed of, acquainted 2
  informed of, be, know 5; receive 5
  informed of *or* **about,** familiar 4; privy 2; wise 5
  informed (of *or* **about), become,** wise 6

informer°, plant 4; snake 2; sneak 2; spy 1; talebearer

infraction, breach 1; foul 17; infringement; invasion 1; offense 1; peccadillo; sin 1; transgression

infra dig, base[2] 2

infrastructure, base[1] 2; basis 1

infrequent, few 1; occasional 1; rare[1] 1; unwonted

infrequently°, now 4; rarely; seldom

infringe°, break 9; disobey; encroach; intrude; pirate 3; violate 1

infringe on *or* upon, infringe 2; trample 2

infringed, broken 5

infringement°, breach 1; foul 17; invasion 1; offense 1; plagiarism; sin 1; violation 1

infringer, pirate 2

infringe (the law), transgress 2

infringing, plagiarism; violation 1

infuriate°, aggravate 2; anger 2; displease; enrage; exasperate 1; gall[2] 4; inflame 1; madden 1; outrage 4; provoke 3; wall 3

infuriated, angry 1; furious 1; mad 4; rabid 1

infuriated, become, bristle 3; stack 5

infuriating, provocative 2; trying

infuse, brew 1; impregnate 2; insinuate 3; instill; percolate; suffuse; transfuse 2

infuse with, school 4

infusion, liquor 2; solution 3

ingenious°, acute 5; artful 2; astute 1; bright 6; brilliant 4; clever 3; creative; cute 2; dexterous 2; elegant 4; exquisite 2; foxy 1; imaginative 1; master 4; neat 4; original 4; politic 1; productive 2; quick 4; resourceful; shrewd; slick 3; smart 1; subtle 3; wicked 7; witty

ingeniousness, dexterity 2; ingenuity

ingénue, innocent 5

ingenuity°, art 1; craft 1; dexterity 2; genius 2; imagination 1; originality; proficiency; prowess 1; resource 1; skill 1; talent 1

ingenuous°, aboveboard 2; artless 1; candid 1; careless 4; childlike; frank 2; honest 3; honorable 3; innocent 4; naive; natural 4; plain 3; provincial 2; rustic 2; simple 3; transparent 3; unaffected[1]; unsophisticated 1; unsuspecting

ingenuously, honestly 2; simply 3

ingenuousness, candor 1; ease 4; honesty 2; naiveté; provincialism 2; simplicity 3

ingest, swallow 1; take 13; use 3

inglenook, nook 2

inglorious, dishonorable 1; obscure 4; shameful

ingloriousness, obscurity 3

ingraft, implant 3

ingrain, engrave 2; etch 2; instill; plant 6

ingrained°, entrenched; habitual 2; inborn; indelible; inherent; organic 2; rooted

ingratiate oneself with, cultivate 4; play 18

ingratiating°, flattering 2; obsequious; oily 2; propitiatory 2; servile; submissive 2; supple 3

ingredient°, agent 2; element 1; factor 1; item 1; part 2

ingredient of, be an, compose 1

ingredients, content[1] 2; stuff 1

ingress, entrance[1] 2

in-ground, sunken 3

inhabit°, occupy 2; populate; settle 4

inhabitable, habitable

inhabitant°, citizen 1; denizen; inmate; national 3; native 7; occupant; resident 3; tenant

inhabitants, population

inhalation, pull 22

inhale°, draw 7; puff 5; take 13

inhale and exhale, breathe 2

inharmonious, discordant 2; harsh 1; incongruous; mismatched; odds 4; rough 8

inherent°, fundamental 1; implicit 1; inalienable; inborn; indigenous 2; ingrained; intrinsic; native 1; organic 2; radical 1; rooted; sneaking 2

inherently, naturally 2

inherit°, get 1; receive 1; take 5

inherit from, succeed 2

inheritable, hereditary 1

inheritance°, bequest; endowment 1; heritage 1; succession 3

inherited, hereditary 1, 2; inborn; ingrained; inherent; intrinsic; native 1

inheritor, heir

inhibit°, check 2; deter; discourage 3; gag[1] 1; hamper[1]; head 13b; hinder 2; hold 14a; impede; interfere 2; keep 13; manacle 3; oppose 2; preclude; prevent; prohibit 2; repress; resist 1; restrain 2; set 16; shackle 4; silence 4; suppress 1

inhibited°, pent-up

inhibiting, prevention; preventive 1

inhibition°, check 14; impediment; interference 2; prevention; preventive 3; repression 2

inhibitive, preventive 1

inhibitory, preventive 1; prohibitive 1

inhospitable°, hostile 2; intolerant 1; joyless 2; unsocial

inhospitableness, rigor 1

inhuman°, atrocious 1; beastly 1; brutal 1; coldblooded 3; cruel 2; cutthroat 3; deadly 3; diabolic 2; ferocious; fiendish; fierce 1; ghoulish 2; grim 2; hard 4; harsh 2; heartless; merciless; monstrous 1; murderous 1; outrageous 2; ruthless; savage 3; severe 1; unkind; violent 1

inhumane, brutal 1; heartless; inhuman 1; merciless; wanton 4

inhumanity, atrocity 1; barbarity; outrage 1; severity 1

inhumanly, roughly 2; severely 2

inhumation, funeral

inhume, bury 1

inimical, contrary 2; detrimental; exclusive 1; gaunt 2; hostile 2; inhospitable 1; opposing; prejudicial

inimical (to), opposed

inimitability, preeminence 2; superiority 2; supremacy 1

inimitable, incomparable; matchless; model 11; peerless; perfect 2; preeminent 2; special 1; star 3; supreme 4; unique 2; unparalleled

inimitably, perfectly 2

iniquitous, atrocious 1; black 6; blasphemous; criminal 2; dark 4; devilish; diabolic 2; disreputable 1; dreadful 2; evil 1; foul 4; godless 1; grim 3; ill 2; immoral 1; impious; infamous 2; infernal 2; lawless 3; miscreant 2; outrageous 2; perverted; profane 1; profligate 1; reprobate 1; rotten 3; satanic 1; scandalous 1; sinful; ungodly 1; vicious 1; vile 1; wicked 2; wrong 1

iniquitousness, sin 2

iniquity, atrocity 1; devilry 2; error 2; evil 6; harm 2; infamy 2; sin 1, 2; vice 1

initial°, capital 4; early 5; elementary 2; first 2; infancy 2; introductory; maiden 3; original 1; preliminary 1; première 3; primary 2; pristine 1; rudimentary 1; stamp 2

initialed, personalized

initialism, abbreviation; symbol

initially, early 2; first 5, 8; originally; primarily 2

initials, sign 4; stamp 7

initiate°, activate; begin 1; break 17; commence 1, 3; create 2; embark 2; excite 1; father 6; float 2; foment; found 1; generate 3; greenhorn; ground 6; inaugurate 1; install 1; institute 4; introduce 4; invest 3; launch 1; lead 8; learner; newcomer 2; novice; open 20; organize 2; originate 1; pick 3; pioneer 2; produce 2; raise 8; recruit 2; set 23b; spark 2; spearhead 1; start 1, 7; touch 11b

initiate (in), familiarize

initiation°, conception 1; debut 1; generation 3; installation 1; launch 5; onset 2; opening 4; orientation 2; provocation 1; start 11, 14

initiative°, action 1; ambition 2; drive 8; dynamism; enterprise 2; gumption 2; leadership; liberty 3; push 14; resource 1; step 4

initiative, take the, spearhead 1

initiator, aggressor; author; creator 1; father 3; founder[1]

inject°, implant 1; insinuate 3; introduce 5; phase 5; plant 6; transfuse 1; use 3

inject (oneself), insinuate 2

injection, shot 6

injudicious, foolish 1; ill-advised 1; imprudent; indiscreet; inept 2; precipitate 5; rash[1]; reckless; shortsighted 2; tactless; wrong 1

injudiciousness, indiscretion 1

injunction°, caution 1; charge 5; decree 1; dictate 2; law 1; precept 1; prevention; prohibition 2

injure°, abuse 2; blemish 2; bruise 2; damage 4; deface; dishonor 1; get 24; harm 3; hurt 1, 4; ill-treat; impair; insult 1; maim; mishandle 1; mistreat; offense 2; outrage 5; pain 5; punish 3; scar 2; scrape 1; scratch 1; shoot 3; spite 3; spoil 2; sting 2; strain[1] 3; tarnish; violence 3a; wound 3; wrong 9

injured, casualty 2a; disfigured; hurt 7; unsound 2

injured party, victim 1

injurious°, abusive 2; bad 2; destructive 1; detrimental; evil 3; harmful; hurtful 1; ill 4; mischievous 2; prejudicial; ruinous; scandalous 2; traumatic; unhealthy 2; unwholesome 1; violent 2; virulent 1

injuriousness, virulence 1

injury°, bruise 1; damage 1; detriment; disadvantage 2;

insight° (*cont.*) knowledge 1; light[1] 5; mind 1; penetration 2; perception 2; reason 2; understanding 2; uptake; vision 2; wisdom 1; wit 1
  insight (in)to, gain, penetrate 5
insightful, acute 5; astute 2; farsighted 1; intelligent; profound 2; wise 1
insightfulness, profundity 2
insigne, character 1; color 2b; crest 2; device 3; emblem; mark 2; plaque 2; seal 1; stamp 7; standard 3; symbol
insignia, regalia
insignificance, indifference 2; inferiority 1, 2; obscurity 3; triviality 1
insignificant°, empty 4; expendable; feeble 2; frivolous 1; humble 3; idle 4; immaterial 1; incidental 2; inconsequential; inconspicuous; indifferent 4; inferior 2; insubstantial 1; light[2] 4, 6; little 5; marginal 1; meaningless 2; minor 2; minute[2] 2; negligible; niggling 2; nominal 2; nonessential 1; obscure 4; pale[1] 3; paltry; petty 1; pitiful 2; poor 5; puerile; puny 1; remote 6; side 8; slender 3; slight 1; small 2, 4, 5; small-time; sparse 2; superficial 2; tenuous 2; tiny; trifling; unsung; venial; wee 2; worthless 1
  insignificant, not, respectable 2; tidy 3
insincere°, affected 2; artificial 3; counterfeit 2; deceitful; devious 1; disingenuous; double 5; empty 4; factitious; faithless 2; false 4; glossy 2; hollow 4; hypocritical; mannered; pharisaic; phony 1; strained; two-faced; unnatural 2; untrue 1
insincerity, affectation 1; cant 1; falsity 1; hypocrisy
insinuate°, get 31b; hint 3; implant 1; imply 1; impose 2; impute; instill; intimate[2]; libel 4; make 37e; mention 2; phase 5; plant 6; suggest 2; whisper 2
insinuation, dig 8; hint 1; implication 2; imputation; innuendo; libel 2; overtone; reference 1; scandal 3; slur 1; suggestion 2
insipid, bland 2; colorless 2; dead 12; flat 5, 8; humdrum; lackluster; nondescript; pedestrian 2; prosaic; sickly 3; stupid 3; tame 4; tasteless 2; tedious; tiresome 1; vapid; watery 1; wishy-washy 2
insipidity, tedium
insipidness, tedium
insist°, demand 1; require 1; screw 4; swear 1
  insist (on *or* **upon),** claim 3; demand 2; enforce 1, 2; exact 3; lay[1] 10; maintain 3; make 2; persist 1; pressure 5; protest 4; request 1; stipulate
insistence, assertion 2; demand 5; expectation 3; pressure 4; stress 2; urgency
insistent°, assertive; demanding 2; dogmatic; emphatic; officious; peremptory 3; urgent 2
in situ, place 11b
insobriety, drunkenness
insolence, arrogance; assurance 3; brass; disrespect; effrontery; familiarity 3; flippancy 2; gall[1] 2; impertinence; impudence; mouth 4; nerve 2; presumption 1; sauce 2; side 5
insolent°, audacious 2; barefaced 2; brash 2; brassy 1; brazen; cheeky; contemptuous; cool 6; discourteous; disdainful; disrespectful; familiar 3; flippant 2; forward 2; fresh 8; ill-mannered; immodest 2; impertinent; impudent; irreverent 2; offensive 2; pert; presumptuous; pushy; rude 2
insoluble, hard 3; mysterious 1
insolvency, failure 4; poverty 1
insolvent°, broke; destitute 1; embarrassed 2; impoverished 1; needy; pauper; poor 1; straitened
  insolvent, become, fail 4
insomniac, sleepless 1; wakeful 1
insomuch as, considering
insouciance, detachment 2; ease 4; indifference 1; torpor
insouciant, bland 1; blasé 2; blithe 2; carefree; casual 3; comfortable 2; debonair 2; halfhearted; impassive; indifferent 1; lackadaisical 2; listless; lukewarm 2; nonchalant; offhand 1; relaxed; torpid
insouciantly, gaily 2
inspect°, check 4, 10b; contemplate 1, examine 1, explore 2; eye 10; frisk 2; go 34a; inquire 1; look 1, 8; observe 2; peruse; reconnoiter; research 2; review 1; scan 2; scrutinize; search 1; study 3; survey 1; traverse 4; try 2; vet; watch 3
inspection, check 15; examination 1; exploration; inquiry 1; observance 3; observation 1; perusal; reconnaissance; research 1; review 5; scan 3; scrutiny; search 4; survey 3; view 4
inspiration°, encouragement 1; guide 6; imagination 1; impetus; incitement 2; lift 8; promotion 1; spirit 7
inspirational, moving 1; stimulating
inspire°, animate 2; call 9; draw 7; encourage 1; energize; enliven 1; exalt 3; excite 1; fire 8a; foment; generate 3; incite; induce 1; inhale; kindle; motivate; move 6; pep 2; promote 1; prompt 5; pump 4b; raise 10; spark 2; stimulate 1; stir 4; vitalize; wake[1] 2
inspired, fire 5; imaginative 1; impassioned; prophetic
  inspired with, be, possess 4
inspiring, encouragement 1; exciting 1; imaginative 1; moving 1; rousing; stimulating; stirring; sublime 2
inspirit, animate 1, 2; assure 2; cheer 5; encourage 1; energize; enliven 1; exalt 3; excite 1; fire 8a; heat 4; inspire 1; intoxicate 2; kindle; liven 1; perk up; promote 1; pump 4b; reassure; rouse 2; spark 2; spice 4; stimulate 1; vitalize; wake[1] 2
inspirited, drunk 2; eager
inspiriting, exhilarating 2; heartwarming 1; hopeful 2; promotion 1; refreshing; rousing; stirring
inspissate, solidify 1; thicken
inspissated, thick 5
instability, fluctuation; flux; vertigo
install°, establish 2; fit[1] 8; fix 2, 8; inaugurate 2; induct 1; invest 3; let[1] 5; mount[2] 4; seat 7; set 1; site 2; situate; station 4
installation°, admission 2; initiation 2; institution 1
installations, fitting 2
installment, episode 2
installments, in, time 19b
instance°, case[1] 1, 4; example 1; exemplify 1; illustrate 1; illustration 1; manifestation; occurrence 2; pattern 5; prototype 2; quote 1; specimen; typify
  instance, for, example 4; illustration 1; instance 2; like[1] 6; namely; say 15
  instance, in every, universally
instances, in all, universally
instances, in many, often; part 8
instant°, crack 3; flash 3; immediate 1; minute[1] 1; moment 1, 2; point 4; prepared 4; prompt 1; rapid; second[2]; snap 13; spurt 1; twinkling 1; urgent 1
  instant, at this (very), now 1
  instant, in an, swiftly
  instant, in the same, once 6b
instantaneous, hasty 1; immediate 1; instant 3; prompt 1; quick 2; rapid; snap 13; unhesitating 1
instantaneously°, hastily 1; immediately 1; outright 3; promptly; rapidly 2; readily 3; shot 10; suddenly 1
instantaneousness, rapidity
instanter, immediately 1; instantaneously; now 3; rapidly 2; readily 3
instantly, directly 2; immediately 1; instantaneously; now 3; once 6a; outright 3; posthaste; promptly; quickly 3; rapidly 2; readily 3; right 13; short 11; shot 10; straight 15; suddenly 1; summarily 1
instate, inaugurate 2; induct 1; install 1; invest 3; seat 7
instatement, initiation 2; installation 1
instead°, rather 2
  instead of, behalf; for 3; instead 2
instigate, abet 1; excite 1; ferment 2; foment; induce 1; initiate 1; inspire 1; kindle; motivate; precipitate 1; provoke 2; put 29; raise 10; rise 16; touch 11b
instigating, incitement 1
instigation, incitement 2; provocation 1; sedition; spur 1
instigator, aggressor; rabble-rouser; troublemaker
instill°, charge 8; din 2; implant 1; indoctrinate; inject 2; insinuate 3; plant 6; pound[1] 4; school 4; strike 11; teach; transfuse 1
instilled, rooted
instinct°, feeling 2; intuition; perception 2
instinctive°, automatic 1; brute 1; gut 6; immediate 1; impulsive; inborn; involuntary; mechanical 2; spontaneous 2; unconscious 2
instinctively, naturally 2
instinctual, automatic 1; gut 6; inborn; instinctive 1; involuntary; spontaneous 2
institute°, bring 6; erect 3; establish 1; father 6; fix 2; found 1; inaugurate 1; initiate 1; install 1; institution 2; introduce 4; make 6; organize 2; originate 1; pioneer 2; raise 10; school 1; seminary; society 5; start 7
instituting, foundation 3

institution°, admission 2; company 4; establishment 1, 2; formation 2; foundation 3; generation 3; home 3; installation 1; observance 2; organization 3; school 1; seminary; tradition
institutionalize, put 13b
institutor, founder[1]
instruct°, brief 6; charge 10; coach 3; command 1; counsel 4; direct 2, 3; discipline 6; drill 2; educate; enlighten; ground 6; guide 4; indoctrinate; initiate 3; order 14; prescribe; prime 5; require 1; school 4; show 4; tell[1] 5; train 4; tutor 2
instruct (in), familiarize; teach
instructed in, be, learn 2
instructing, direction 1; preparation 2
instruction°, bidding 2; charge 5; command 6; counsel 1; dictate 2; direction 1; discipline 1; dispatch 6; edification; education 1; guidance 2; illumination 2; injunction 2; knowledge 4; lecture 1; lesson 2; order 4, 7; orientation 2; precept 1; preparation 2; prescription 1; schooling; tuition; word 7
instructional, educational 1; instructive
instruction(s), direction 2; formula; instruction 1; manual
instructive°, educational 2; informative
instructor°, coach 2; master 3; schoolteacher; teacher; tutor 1
instructress, mistress 2
instrument°, agent 2; apparatus; deed 3; device 1; document 1; fixture 2; gadget; implement 1; machine 1; mean[3] 4a; mechanism 1; organ 1; paper 2a; pawn[2]; thing 5; tool 1; vehicle 2; voice 3
instrumental°, influential
instrumentalist, player 4
instrumentality, agency; instrument 2; medium 5; tool 2
instrumentation, arrangement 4
instruments, kit
insubordinate°, defiant; disobedient 1; insolent; mutinous 2; naughty 2; rebellious 1; recalcitrant; unruly; wild 4
insubstantial°, disembodied 1; dreamlike; empty 4; feeble 2; filmy 1; flashy 2; flimsy 1, 2; fluffy 2; fragile; insignificant; intangible; meaningless 2; nominal 2; ramshackle; shaky 2; skin-deep; sleazy 1; slight 4; subtle 2; superficial 2; tenuous 2; thin 4
insufferable°, base[2] 6; excruciating; obscene 2; unbearable
insufficiency, absence 2; dearth; defect 1; imperfection; incompetence; lack 1; need 5; poverty 2; scarcity; want 3
insufficient°, deficient 1; disappointing; feeble 2; few 1; inadequate 1; incompetent; ineffective 1; little 3; low[1] 2; meager 1; poor 2; scarce; slender 3; small 4; straitened; thin 2; unsatisfactory; wanting 1
insufficient, prove *or* **(turn out to) be,** fall 19; short 13
insufficiently, badly 1
insular, nearsighted 2; parochial; provincial 2; sectarian 2
insularity, provincialism 2
insulate°, cushion 2; steel 2
insult°, abuse 3; cut 3, 23; dig 8; dishonor 1, 4; disparage 2; flout; indignity; injure 2; knock 10; offend 1; offense 2; outrage 3, 4; provocation 2; provoke 3; put-down; slap 5; slight 6, 7; slur 1; taunt 1, 2; wound 2, 4
insulting, abusive 1; contemptuous; derisory; derogatory; disdainful; foul 6; ill-mannered; injurious 2; insolent; irreverent 2; left-handed 2; offensive 2; outrageous 3; personal 4; rude 2; scurrilous; unflattering 2; vituperative
insuperable, invincible 1
insupportable, heavy 4; insufferable; leg 6; prohibitive 2; question 7; untenable
insuppressible, irrepressible
insurance°, indemnity 2; precaution 1; security 3
insure, ensure 1, 2; underwrite 1; warrant 3
insure against, cover 10
insurgence, mutiny 1; sedition
insurgency, mutiny 1; revolution 1; sedition
insurgent, guerrilla; inflammatory; insubordinate; mutinous 1; rebel 3; rebellious 1; revolutionary 1, 3; seditious; subversive 2; terrorist; traitorous
insurgents, underground 5
insurrection, mutiny 1; revolt 1; revolution 1; sedition; tumult; uprising
insurrectional, insubordinate
insurrectionary, mutinous 1; rebellious 1; revolutionary 1, 3; seditious; subversive 1, 2
insurrectionist, agitator; insubordinate; rabble-rouser; rebel 3; revolutionary 1, 3; seditious; subversive 2; traitorous
insurrectionists, underground 5
insusceptibility, immunity 2
insusceptible, immune
insusceptible to, above 5; incapable 2
intact°, complete 1; entire 2; flawless 1; full 12; maiden 2; pure 3; sound[2] 1; unabridged 1; undivided 1; unused 1; whole 1, 2
intaglio, engraving 1
intake, diet[1] 1; gross 7
intangible°, disembodied 1; dreamy 1; elusive 2; insubstantial 2
integer, number 1
integral°, intact; organic 2
integrate°, amalgamate; combine 1; coordinate 2; embody 3; incorporate; unify
integrated, organic 3
integrating, synthesis
integration, amalgamation; embodiment 2; marriage 3; synthesis
integrity°, character 3; entirety 1; good 20; gut 3b; honesty 1; honor 1; moral 4; morality 1; nobility 1; principle 3; probity; purity 2; rectitude; right 10; self-respect; sportsmanship; unity 2; virtue 1
integument, film 1; skin 1; surface 1
intellect°, aptitude 3; brain 1; brilliance 2; capacity 2; genius 1; head 4; intellectual 3; intelligence 1; mastermind 2; mentality 1; mind 1, 4; prodigy 1; reason 2; soul 1; thought 6; understanding 5
intellectual°, abstract 1; brain 2; fraternal; highbrow 1, 2; learned; mental 1; mind 4; platonic; profound 2; psychic 1; psychological; scholar 1; scholarly; speculative 1; thinking 1
intellectuals, intelligentsia
intelligence°, advice 2; aptitude 3; brain 1; brilliance 2; capacity 2; genius 2; head 4; history 2; imagination 2; information; intellect 1; judgment 1; knowledge 2; logic 3; lowdown; mentality 1; message 1; mind 1; news 1; penetration 2; reason 2; rhyme 2; sense 2; spying; thought 6; understanding 5; wisdom 1; wit 1; word 2
intelligence agent, operative 4
intelligent°, acute 5; advisable; astute 2; bright 6; brilliant 4; clever 1, 3; informed 1; judicious; keen[1] 6; knowing 2; knowledgeable 2; logical 2; nimble 2; penetrating 1; perceptive; politic 2; precocious; quick 4; rational 2; ready 6; reasonable 1; receptive 2; sage 1; sensible 1; sharp 3; shrewd; smart 1; thinking 1; well-advised; wise 1
intelligibility, clarity 2; sense 4; simplicity 1
intelligible°, clear 6; coherent 2; luminous 3; plain 2; readable 1; simple 1
intemperance, abandon 5; dissipation 1; drunkenness; excess 2; gluttony
intemperate, dissolute; epicurean 1; fast[1] 2; immoderate; inclement; inordinate 1; lavish 3; prodigal 1; self-indulgent; towering 2; uninhibited; unwarranted; wanton 2
intemperate, be, overdo 1
intemperately, fast[1] 9
intemperateness, prodigality 1
intend°, aim 3; contemplate 3; design 4; destine 2; drive 6; get 31b; mean[1] 1; plan 4; propose 2; purpose 5; think 5
intended, calculated 1; deliberate 1; destined 1; fiancé; intentional; love 3; premeditated; purposeful; supposed 2; sweetheart; voluntary 2; willful 1
intending, intent 2
intense°, acute 2, 3; ardent; bright 5; burning 2; close 14; deep 3, 4, 6, 8; earnest 1; emphatic; excruciating; exquisite 4; extreme 2; fervent 1, 2; fierce 3; fire 5; forceful 1; full 10; furious 2; heated; heavy 10; hot 3, 4, 8; industrious; intent 4; keen[1] 1, 3; lively 2; mortal 5; passionate 1; piercing 4; poignant 2; powerful 3; profound 3; pungent 3; rich 5; severe 5; sharp 8; soulful; stiff 7; strenuous 2; strong 2, 22; tense 2; torrid 2; towering 2; violent 3; vivid 1; warm 2
intensely, deep 10; deeply 2; fiercely; hard 13, 14; madly 4; profoundly; warmly 4
intensification, progression 2
intensified, heated; intensive; strong 4
intensify°, aggravate 1; build 4; compound 3; concentrate 2; deepen 2; enhance; gather 5; harden 2; heat 4; heighten 2; inflame 2;

short 12b; stay[1] 4; stop 5; suspend 1
interrupted, broken 6; spasmodic 2
interruption°, break 27; check 13; delay 5; disturbance 1; gap 1; holdup 2; interlude; jump 12; lapse 2; letup; lull 1; pause 2; respite 1; space 3; stay[1] 6; suspension 2
  interruption, without, succession 5
intersect, cross 4; meet[1] 4; overlap 2; traverse 2
intersection, angle[1] 1; joint 1; junction; meeting 3
intersperse, punctuate 1; shuffle 1
interstice, interval 3; mesh 1
intertwine, braid 3; entwine; knit 1; lace 3; splice 1; tangle 3; twine 2; twist 1; weave 1
intertwist, tangle 3
interval°, clearance 1; clip[2] 5; delay 5; distance 1; gap 1; interlude; interruption 2; lapse 2; lull 1; meanwhile 1; patch 4; period 1; remove 8; respite 1; rest[1] 2; run 38; snap 9; space 3; span 2; spell[1] 1; spurt 1; term 2; time 1, 2; truce 1; wait 4
intervals, at, once 4; succession 5; time 14
intervene°, break 18a; interfere 1; intrude; meddle; mess 6; step 15
intervener, negotiator
intervening, intermediate 1
intervention, agency; interference 1
interview°, apply 6; canvass 2; poll 3; question 9; screen 8; see 9; tête-à-tête 1; word 1
interweave, entwine; knit 1; lace 3; tangle 3; twine 2; twist 1; weave 1
interweaving, mixture 2; tissue
intestinal, instinctive 1
intestinal fortitude, gut 3a; manhood 2; strength 2
intestines, bowels; gut 1
intimacy, association 3; familiarity 2; fellowship 1, 3; friendship 2; intercourse 2; intrigue 4; sex 2
  intimacy (with), understanding 3
intimate°, announce 2; bosom 4; chummy; close 15; clue 2; confidential; drive 6; express 2; familiar 3; friend 1; great 10; hint 3; impart 2; imply 1; indicate 3; insinuate 1; interior 3; make 37e; mean[1] 2; mention 2; near 7; notify 2; old 8; partake 2b; personal 2, 3, 4; predict; private 3; rumor 2; signify 1; snug; special 5; spicy 3; suggest 2; tell[1] 2; tête-à-tête 3; thick 8
  intimate terms (with), be *or* **get** *or* **become on,** pal 2
  intimate (to), advise 2
intimately, confidence 3; hand 10; immediately 2; tête-à-tête 2; well[1] 13
intimateness, familiarity 2
intimation, clue 1; foreboding 2; hint 1; idea 2; implication 2; import 3; indication 1; inkling; innuendo; lead 14; note 6; notice 5; overtone; prediction; reference 1; shade 3; shadow 3; suggestion 2; threat 2; touch 15; trace 1
intime, intimate[1] 1, 2, 4; tête-à-tête 3
intimidate°, appall; browbeat; bully 2; daunt; deter; discourage 1; dismay 1; enforce 2; face 17; force 7; frighten; get 31d; horrify 1; lean[2] 4b; menace 1; overawe; pick 6; pressure 5; prey 3b; push 7; railroad 2; ride 3; ruffle 3; scare 1; threaten 1; thunder 3; tyrannize; unnerve
intimidated, afraid 1; fearful 2
intimidating, fearsome; formidable 1; frightening; menacing; scary; strong-arm; threatening
intimidation, menace 3; pressure 4; terror 1; threat 1
intimidator, bully 1; oppressor
into, deep 4; for 6; through 4
  into, be, get 40c
  into, be *or* **get,** relate 4
intolerable, beastly 2; grievous 2; heavy 4; inexcusable; insufferable; lamentable; murderous 2; obscene 2; oppressive 1; outrageous 1; repugnant; terrible 1; unbearable
intolerance°, bigotry; discrimination 1; fanaticism 2; prejudice 2
intolerant°, bigoted; hidebound; inconsiderate; narrow-minded; nearsighted 2; one-sided 1; parochial; prejudiced 2; puritan 2; small-minded
  intolerant, be, discriminate 2
intonation°, accent 3; diction 2; expression 4; pronunciation; tone 2
intone, chant 2; hum 3; sing 1
in toto, completely 1; entirely 1; full 17; outright 4; wholly 1
intoxicants, liquor 1
intoxicated, drunk 1; excited 2; high 9; rhapsodic; stinking 3; tight 8
intoxicating°, alcoholic 1; exciting 1; stirring
intoxication, drunkenness; high 16
intractability, obstinacy
intractable, difficult 3; disobedient 1; disorderly 2; fierce 2; grim 1; hard 4; hidebound; implacable; incorrigible; inflexible; monolithic; obstinate; parochial; perverse 3; problem 3; recalcitrant; relentless 1; resistant 2; rogue 2; self-willed; stony 2; stubborn; tough 6; ungovernable; unruly; wild 4; willful 2
  intractable, be, persevere
intransferable, inalienable
intransigence, obstinacy; resistance 1; tenacity 1
intransigent, obstinate; relentless 1; resistant 2; rigid 2; stubborn; tenacious 1; willful 2
  intransigent, be, persevere
intrepid°, adventurous; audacious 1; bold 1; brave 1; courageous; daredevil 2; daring 2; dauntless; fearless; gallant 1; gritty 2; hardy 2; heroic 1; indomitable; manly; stalwart 2, 3; tough 3; venturesome 1
intrepidity, assurance 4; bravery; courage; daring 1; grit; nerve 1; pluck 1; prowess 2; strength 2
intrepidness, bravery; grit
intricacy, complexity 2; complication 1; depth 2; maze; mesh 2; obscurity 2; perplexity 2; profundity 1; sophistication 2; subtlety 1
intricate°, busy 3; complicated; compound 4; detailed 2; difficult 2; elaborate 1; fancy 1; hairy 3; hard 3; involved 2; labyrinthine; nice 3; obscure 5; ornate; perplexing; prickly 4; profound 1; showy; sophisticated 2; technical 1; thorny 2; tortuous 2
  intricate, make, complicate 1
intricately, inextricably
intrigue, liaison 3; love 5a
intrigue, affair 4; cabal 1, 3; conspiracy; dazzle 1; deception 1; design 8; diplomacy 2; engineer 5; faction 2; fascinate; finesse 2; interest 7; machination; maneuver 1, 3; plot[1] 1, 3; romance 1, 4; scheme 3, 4; stratagem; subterfuge; trick 1
intriguer, designer 2; tactician
intriguing, designing; enchanting; enthralling; fetching; glamorous 1; interesting; inviting; juicy 2; machination; picturesque 1; provocative 1; scheming; taking
intrinsic°, essential 2; fundamental 1; ingrained; inherent; instinctive 1; integral 1; native 1; real 5; rooted
intrinsically, substantially
introduce°, bring 10b, 13a, 15b; broach; establish 1; give 6; implant 1; import 1; inaugurate 1; initiate 1, 2; inject 1, 2; insert 1; insinuate 3; install 1; institute 4; lead 10b; mention 1; moot 2; open 25; originate 1; phase 5; pioneer 2; plant 6; precede; preface 2; present[2] 4, 7, 8; produce 4; propose 1, 3; put 18b, 19a; raise 8; set 7; suggest 1; work 17
  introduce oneself to, pick 8j
  introduce suddenly *or* **unexpectedly,** spring 4
  introduce to, expose 3
introduced to, be, meet[1] 3
introducing, imposition 1
introduction, conception 1; debut 1; entrance[1] 1; foreword; imposition 1; initiation 1; installation 1; institution 1; invention 1; orientation 2; preamble; preface 1; preliminary 2, 3; presentation 3
introductory°, elementary 2; initial 1; prefatory; preliminary 1; preparatory 1; rudimentary 1
intromit, inject 1
introspective, thoughtful 2
introverted, shy 1; withdrawn 1
intrude°, break 18a; butt[2] 2; chime 5b; chip 4b; cut 14; encroach; impose 2; insinuate 2; interfere 1; intervene 1; meddle; obtrude; poke 3; pry 2; snoop 1; tamper
  intrude in, interrupt 1; mess 6
  intrude (on *or* **upon),** disturb 1; infringe 2; presume 3
intruder°, outsider; parvenu 1; snoop 2
intruding, interference 1; intrusive; officious
intrusion, disturbance 1; imposition 2; inroad 1; interference 1; interruption 1; invasion 1
intrusive°, curious 2; inquisitive 2; obtrusive; officious; parvenu 2
intrusiveness, curiosity 2
intrust. See **entrust**
intuit, divine 4; feel 4; sense 6
intuition°, anticipation 2; belief 4; depth 3; feeling 2; foreboding 1; hunch 1; instinct; knack; perception 2; premonition; sense 5; sensitivity 3; understanding 2
intuitional, instinctive 1
intuitive, gut 6; instinctive 1; sneaking 1, 2
intumescence, growth 4
intwine, entwine

inundate, besiege 3; bury 5; cover 6; drench; drown 1, 2; flood 4; glut 3; immerse 1; infest; overwhelm 2; shower 4; smother 3; soak 1; steep[2] 2; submerge 1, 3; swamp 2; water 6
inundation, downpour; flood 1; flush[1] 5; spate; torrent; volley 2
inure, condition 7; steel 2
inured, dead 4; dull 2; seasoned 1; thick-skinned; tough 6
  inured to *or* **against,** used 3
inutile, useless 1
invade, breach 4; descend 4; encroach; enter 2; infest; infringe 2; occupy 1; overrun; raid 2
invader, intruder 1; outsider
invalid°, case[1] 3; defunct 2; erroneous; false 1, 5; illegitimate 3; leg 6; patient 4; strike 13; unhealthy 1; unsound 4; void 1
  invalid, be, water 2
invalidate, cancel 1; discredit 3; disprove; neutralize; quash 1; rebut; repeal 1; reverse 3; revoke; undo 3; vacate 3; vitiate 3; void 6
invalidation, cancellation 2; repeal 2
invalided, ill 1
invaluable°, inestimable 1; precious 1; priceless 1; rich 3, 4
invariability, regularity 2; tedium; uniformity 1, 2
invariable°, certain 1; constant 2, 3; equable 2; eternal 3; flat 7; inflexible; level 3; permanent 2; perpetual 1; regular 1; rocky[1] 2; set 30; stable 2; static 1; steady 2; uniform 1
invariably, universally
invariant, invariable 1; rocky[1] 2
invasion°, aggression 2; assault 1; attack 6; inroad 1; push 15; raid 1
invasive, intrusive
invective, abuse 8; foul 6; tirade
inveigh against, attack 2; denounce 3
inveigle, cajole; chat 3; coax; deceive; decoy 2; entice; flatter 3; induce 1; lead 9b; lure 1; trap 6; wheedle
  inveigle (oneself *or* **one's way),** insinuate 2
inveiglement, cajolery
inveigler, fraud 3
inveigling, cajolery
invent°, coin 3; construct 2; create 2; design 2; devise 1; discover 3; engineer 4; fabricate 2; father 6; forge 2; form 8; formulate 2; generate 4; hatch 2; hit 9b; improvise 2; lie[1] 1; make 39c; manufacture 2; originate 1; pioneer 2; project 3; spin 2; strike 16; think 6
invented, fictional; fictitious 1, 2
invention°, discovery 1; fabrication 2; fantasy 3; fib 1; gadget; gimmick 2; innovation 1; lie[1] 2; pretense 2
  invention, certificate of, patent 1
inventive, clever 3; creative; imaginative 1; ingenious; original 4; productive 2; resourceful; shrewd; slick 3; vivid 3
inventiveness, fancy 5; fantasy 1; imagination 1; ingenuity; originality; resource 1
inventor, author; creator 1; designer 1; engineer 1; father 3
inventory, list[1] 1; litany 2; register 1; repertory; reserve 3; roll 13; specification 2; stock 1, 10; store 3; supply 4; table 3
inventorying, litany 2
inverse, reverse 1
invert, capsize; homosexual 1; overturn 1; reverse 2; turn 21b; upset 2
invertebrate, spineless 1
inverted, homosexual 2; reverse 1; topsy-turvy 1; upset 6
invest°, clothe 2; inaugurate 2; induct 1; install 1; put 28e; seat 7; sink 10; spend 3
  invest in, back 1; finance 3; set 23b; underwrite 1
investigate°, analyze 2; canvass 2; check 4, 8, 10b, 12a; examine 1; experiment 3; explore 2; fathom; follow 11a; go 30b, 34a; hunt 2; inquire 1; inspect; look 8; note 11; poke 3; probe 1; pry 1; reconnoiter; report 6; research 2; scan 2; scout 1; scrutinize; search 1; see 4, 12b; sift 2; sound[3] 1; study 3; survey 1; trace 5; traverse 4; treat 2; vet
investigating, research 1
investigation°, analysis 1; canvass 4; examination 1; experiment 1; exploration; inquiry 1; interrogation; probe 3; reconnaissance; research 1; scan 3; scrutiny; study 4, 5; survey 3; test 1
investigative, inquisitive 1
investigator, detective; operative 3; sleuth
investiture, admission 2; initiation 2; installation 1
investment, backing 2; finance 1; interest 4; principal 5; stake[2] 2; stock 5; subscription 1
investor, backer 2; benefactor; financier; friend 4
inveterate, chronic 1, 2; habitual 2; incorrigible; incurable 2; ingrained; steady 5
inveterately, usually
invidious, cutting 2; malignant 2; spiteful
invigilator, monitor 1
invigorant, tonic 1
invigorate, animate 1; energize; enliven 1; excite 1; fortify 2; freshen 2; inspire 1; intoxicate 2; liven 1; pep 2; perk up; quicken 3; raise 13; refresh 1; rouse 2; spice 4; strengthen 2; tone 6; vitalize; wind[2] 4c
invigorated, drunk 2; energetic; fresh 5; passionate 1; perky
invigorating°, bracing; brisk 3, 4; cheerful 2; exciting 1; exhilarating 1; hearty 4; intoxicating 2; refreshing; rousing; scintillating 2; spanking 2; tonic 2; vital 4
invigoration, animation 2; refreshment 2
invincible°, impregnable; indomitable
inviolable, impregnable; inalienable; sacred 2
inviolate, entire 2; intact; sacred 2
invisible°, imperceptible 1; sight 6a
invitation°, bidding 1; call 14; challenge 5; temptation 2
invite, appeal 2; ask 4, 6a; attract; bear 3; bid 2; bring 2; call 4; challenge 2; incur; invitation 1; run 16; seek 3; summon 1; tempt 1
  invite out, take 36
invited, welcome 3
inviting°, attractive; interesting; magnetic; pleasant 1; prepossessing; provocative 1; seductive; sexy 1; tempting 1; voluptuous 2
invocation, litany 1; prayer 2
invoice, bill[1] 1, 3; charge 12; damage 2; reckoning 2
invoke, bring 6; call 7a, 9; evoke; summon 2
involuntarily, protest 2
involuntary°, automatic 1; impulsive; instinctive 2; mechanical 2; spontaneous 2; unconscious 2; unpremeditated; willy-nilly 2
involuted, tortuous 1
involve°, busy 4; catch 14a; come 18c; concern 1, 2; employ 3; engage 2; entail; entangle 1; go 28b; hold 4; immerse 2; implicate 1, 2; imply 2; include 3; incriminate; interest 8; mesh 3; mix 5; occupy 3; regard 5
  involve in *or* **with,** involve 3
involved°, busy 1; complicated; compound 4; concerned 1; deep 4; elaborate 1; engaged 2; hard 3; interested 2; intricate 1; mixed 3; perplexing; profound 1; stake[2] 3; thorny 2; tortuous 2
  involved, become, mire 3; step 15
  involved, make, complicate 1
  involved (in), interested 1; wrap 2
  involved in, be, get 40c, 52; welter 2
  involved in, become, get 40b, 40c, 52; take 39c
  involved in, be *or* **become** *or* **get,** finger 2
  involved (in *or* **with), be,** come 18c; feature 5; part 11; take 10
  involved (in) *or* **with, get** *or* **become,** mess 6; participate
  involved with, involved 3
  involved with, be, concern 1
involvedness, profundity 1
involvement, complexity 2; complication 1; concern 4; experience 1; hand 3; implication 1; inclusion; interest 4; part 3; perplexity 2; profundity 1; stake[2] 2; tie 8
involving, about 11
invulnerability, immunity 1
invulnerable, immune; impregnable; invincible 2; monolithic
inward, interior 1
inward(s), deeply 1; gut 1
iota, bit 2; dot 1; grain 3; jot 2; modicum; particle; scrap[1] 1; shade 3; shred 1; spark 1; speck; stroke 8; suggestion 2; trace 2; trifle 2
I.Q., mentality 1
irascibility, temper 3
irascible, angry 1; bad 7; cantankerous; cranky 2; cross 6; edge 5; fiery 3; fretful; gruff 1; harsh 3; hasty 4; impatient 2; irritable; nasty 4; passionate 3; perverse 2; petulant; pugnacious; quarrelsome; querulous; quick 5; quick-tempered; sensitive 2; short-tempered; snappish 1; temperamental 1; testy; touchy 1; ugly 4; warm 2; waspish
irate, angry 1; cross 7; furious 1; indignant; mad 4; resentful; upset 10; warm 2
ire, anger 1; displeasure 2; fury 1; indignation; outrage 2; rage 1; resentment
ireful, angry 1
iridescent, opalescent

Irish briar, pipe 2
Irish bull, howler
Irish up, get (someone's), enrage; flare 3; infuriate
irk°, annoy 1; bug 7; displease; disturb 1; exasperate 2; gall[2] 4; get 17; gnaw 3; grate 3; inconvenience 3; infuriate; jar[2] 3; madden 2; molest 1; nag[1] 1; pall[2] 1; pester; plague 3; prod 3; provoke 3; put 23b; rankle; rasp 4; rub 8; tire 2; torment 2; trouble 2; worry 2
irked, indignant; resentful; sore 4
irksome°, burdensome; grating 1; inconvenient; nerve-racking; niggling 1; painful 2; provocative 2; tiresome 2; tough 5; troublesome; trying; weary 2; wicked 5
iron, grim 1; press 4; rigid 2; smooth 9; steely 2
iron (out), flatten 1
ironfisted, strict 2
iron hand, oppressor
ironhanded, dictatorial 2; heavy-handed 2; severe 1; tyrannical
iron horse, railroad 1
ironic(al), derisory; dry 3; incisive 2; sarcastic; sardonic; satirical; wry 2
ironically, tongue 5
ironmonger, seller
ironmongery, hardware 1; manacle 1; shackle 1
irons, put *or* **throw** *or* **clap in(to),** imprison; manacle 2
irony, sarcasm; satire 1
irradiate, radiate 2
irrational, absurd 2; arbitrary 1; blind 3; brute 1; circular 3; delirious 1; deranged; disconnected 2; distraught; emotional 4; fantastic 2; foolish 2, 3; gratuitous 2; hysterical 1; inarticulate 1; incoherent; insane 2; instinctive 2; mad 2; neurotic; nonsensical; preposterous; raving 1; senseless 3; silly 1; stupid 2; unreasonable 1; unthinking 1; violent 1; wild 5, 7
irrationality, absurdity 1; extravagance 2; folly 1; insanity 2; lunacy 2; stupidity 2
irrationally, blindly; fault 7; madly 2; unduly 1
irreclaimable, irretrievable 1; lost 6
irreconcilable, incompatible; opposing; opposite 2; pole[2] 3
irrecoverable, irretrievable 1; lost 1
irredeemable, incurable 2; invaluable; irretrievable 2; irreversible; reprobate 1
irrefutability, finality
irrefutable, certain 4; decided 1; final 2; flawless 2; incontrovertible; indisputable; peremptory 1; positive 1; solid 7; strong 8; undisputed; unquestionable
irrefutably, undoubtedly
irregardless, regardless 1, 2
irregular°, abnormal 1, 2; bizarre 2; bumpy; casual 2; changeable 1; chaotic 1; curious 3; desultory; disorderly 1; disproportionate; eccentric 1; erratic 1; exceptional 1; fitful; guerrilla; illegitimate 3; improper 1; inconsistent 2; inordinate 2; intermittent; kinky 1; lopsided 1; occasional 1; odd 2; pale[2] 3; partisan 2, 4; perverse 1; queer 1; ragged 2; rambling 2; reject 4; rough 1; rugged 1; spasmodic 2; sporadic; spotty 3; uncertain 4; unorthodox; unstable 1; unwonted
irregularity, abnormality 1; break 27; cast 5; defect 1, 2; disproportion; eccentricity 1; exception 3; freak 2; impropriety 1; inconstancy; oddity 3; peculiarity 1; perversion 1; turn 27; twist 10
irregularly, fit[2] 4; infrequently; occasionally; random 2
irregulars, resistance 3
irrelevance, indifference 2
irrelevant°, extraneous 2; extrinsic; inapplicable; inappropriate; meaningless 2; peripheral 1; point 15; remote 4; tangential
irreligious, blasphemous; heathen 2; impious; irreverent 1; profane 1; sinful; ungodly 1; wicked 1
irremediable, hopeless 1; incurable 1; irretrievable 2; lost 6
irreparable, hopeless 1; incurable 2; irretrievable 2; irrevocable; lost 6
irrepealable, irrevocable
irreplaceable, invaluable; precious 1; priceless 1
irrepressible°, effervescent 2; enthusiastic; indomitable; inextinguishable; irresistible 1; obstreperous; unquenchable
irreproachable°, blameless; faultless; spotless 2; worthwhile 2
irresistibility, magnetism
irresistible°, forceful 2; inviting; magnetic; overpowering; overwhelming 1; provocative 1; seductive; taking; tempting 1; unavoidable
irresistibly, necessarily
irresolute°, fainthearted 2; fence 2; fickle; hesitant 1; inconstant; indecisive 1; infirm 2; spineless 2; uncertain 2; unresolved; unstable 1; vague 5; weak 3; wet 3; wishy-washy 1
irresoluteness, weakness 2
irresolution, doubt 4; indecision; suspense 1; weakness 2
irrespective of, independent 8
irresponsibility, insanity 2; stupidity 2
irresponsible°, careless 1; derelict 2; fast[1] 2; fickle; flighty 2; fly-by-night 2; foolhardy; frivolous 2; giddy 2; imprudent; insane 2; promiscuous 1; puerile; reckless; shiftless; silly 1; stupid 2; unreliable; wanton 3
irresponsibly, fast[1] 9
irrestrainable, irrepressible
irretrievable°, hopeless 1; irreversible; irrevocable
irretrievably, inextricably
irreverence, dishonor 4; disrespect; flippancy 2; sacrilege 2; sin 2
irreverent°, blasphemous; disrespectful; flippant 2; impious; profane 1; sacrilegious; satirical; sinful
irreverently, vain 3b
irreversibility, finality; resolution 1
irreversible°, final 2; irretrievable 2; irrevocable; resolute
irrevocable°, final 2; flat 10; inevitable; irretrievable 2; irreversible
irrevocableness, finality
irrevocably, deeply 2; finally 2; flat 16
irrigate, rinse 1; water 6
irrigation, rinse 3
irritability, anger 1; sensitivity 1; temper 3
irritable°, angry 1; bad 7; cantankerous; cranky 2; cross 6; edge 5; fiery 3; fretful; gruff 1; hasty 4; impatient 2; moody 2; passionate 3; peevish; perverse 2; petulant; prickly 3; pugnacious; quarrelsome; querulous; quick 5; quick-tempered; ratty 1; sensitive 2; short-tempered; snappish 1; temperamental 1; testy; touchy 1; warm 2
irritant, annoyance 2; bind 6; bother 7; nuisance 1; pest; thorn 2; trial 5
irritate°, aggravate 2; anger 2; annoy 1; bother 1; bug 7; chafe 2, 3; displease; disturb 1; exasperate 2; exercise 3; fret 2; gall[2] 3, 4; get 17; gnaw 3; grate 3; harass; heckle; inconvenience 3; infuriate; irk; jangle 2; jar[2] 3; madden 2; molest 1; nag[1] 1; nerve 3; offend 1; pall[2] 1; persecute 2; pester; pick 4a; plague 3; prod 3; provoke 3; put 23b; rasp 4; ride 3; rub 8; spite 3; tease 1; tire 2; torment 2; trouble 2; wall 3; wear 7
irritated, angry 1, 2; cross 6, 7; discontented; disgruntled; huff 1; indignant; inflamed; mad 4; querulous; ratty 1; resentful; sick 6; sore 1, 4; warm 2
irritating, bitter 2; grating 1; inconvenient; irksome; nerve-racking; niggling 1; painful 2; provocative 2; scratchy 1; tiresome 2; troublesome; trying; untoward 2; weary 2; wicked 5
irritation, anger 1; annoyance 1; bind 6; bother 5; chafe 4; discomfort 2; displeasure 2; dissatisfaction 2; gall[2] 2; indignation; inflammation; itch 3; nuisance 1; pain 3; persecution 2; plague 2; provocation 2; resentment; thorn 2; torment 4; trial 5; trouble 6; warmth 4; worry 3
irritation, create, offense 2
island°, pocket 3
isle, island
Isles *or* **Islands of the Blessed,** heaven 1; paradise 1
islet, island
ism, sect 2
isolate°, dissociate; insulate 1; maroon; ostracize; segregate; shut 4b
isolated°, back 12; deserted; desolate 1; detached 1; out-of-the-way 1; parochial; particular 1; reclusive; remote 2; secluded 1; separate 5, 7; single 2; singular 3; stray 6; withdrawn 2
isolation, privacy 1; retreat 2; seclusion; segregation; solitude 1
isolationist, nationalistic
isometrics, exercise 4
issuance, circulation 2; distribution 2; issue 1, 6; publication 1
issue°, affair 1; bring 14b; business 2; cause 4; child 1; circulate 2; concern 8; descendant; dispense 1; distribute 1; edition; emanate 1; emerge 2; emit; escape 3; event 2; exhaust 5; family 2; flow 2, 3; follow 7; give 8, 15b, 15c; grow 2; impression 4; litter 2; matter 2; measure 14; number 3; offspring; originate 2; pass 9; point 10; posterity; print 1; product 1;

lachrymose (*cont.*) tearful; tragic; unhappy 1; watery 2
lacing, lace 2
lack°, absence 2; dearth; defect 1; famine; go 40; need 1, 5; poverty 2; require 2; scarcity; shortage; want 2, 3
lackadaisical°, careless 1; casual 3; halfhearted; idle 3; impassive; inactive 1; indifferent 1; lukewarm 2; shiftless; slothful; slow 1; supine 2; tardy 2; torpid
lackey, attendant 2; boy 2; fag 3; flunky 1; hack[2] 2; inferior 4; menial 3; servant 1; slave 1; subordinate 2
lacking, absent 2; deficient 1; failing 2; scarce; shy 3; starved 3; unsatisfactory; void 3; wanting 2
  lacking, be, require 2
  lacking, be (found), fail 1; fall 11b
  lacking, prove *or* **(turn out to) be,** fall 19
  lacking (in), destitute 2; empty 7; short 6
lackluster°, colorless 2; deadly 5; dingy; drab; flat 5; lifeless 3; mousy 1; washed-out 1
laconic, compact 3; concise; curt; epigrammatic; short 3; summary 2; terse 1
laconicism, brevity
lacquer, glaze 1, 2
lacuna, break 27; delay 5; discrepancy; gap 1; holdup 2; interlude; interruption 2; interval 3; jump 12; lapse 2; pause 2; skip 4; space 3; split 7; vacancy 1; wait 4
lad°, boy 1; bugger 2; chap; child 2; guy 1; minor 3; stripling; youth 2
ladder, run 18
laddie, child 2; minor 3; youth 2
lade, load 3
laden, heavy 13; loaded 1; momentous
la-di-da, affected 3; genteel 1; haughty; mannered; mincing; supercilious
ladies' (room), go to the, defecate; urinate
ladies and gentlemen, gentry; people 1
ladies' man, blade 4; charmer; dandy 1; rake[2]; roué
ladies of the press, press 10b
ladies' (room), facility 2b; lavatory; toilet 1
ladle, scoop 1, 4
  ladle off, skim 1
la dolce vita, dissipation 1
lady, girl 1; noble 1; peer[1] 1; woman 1, 2
lady-in-waiting, woman 4
lady-killer, charmer; dandy 1; flirt 3; libertine 1; philanderer; playboy; rake[2]; roué; seducer
ladylike°, courteous; decorous; feminine 1; genteel 2; nice 1; proper 3; refined 1; respectful; seemly 2; well-bred
ladylove, love 3; sweetheart; woman 2
lady of the evening *or* **the night,** prostitute 1; slattern; tart[2] 2
lady of the fourth estate, reporter
lady of the house, landlady 1
lady of the press, journalist; reporter; scribe 2
lady's maid, maid 2; servant 1
lag°, convict 2; dawdle; fall 10; flag[2] 2; inmate; insulate 2; linger 2; prisoner
  lag (behind), delay 3; drag 4; trail 6
laggard°, idler; slack 1; slothful; slouch 3; slow 1
laggardness, sluggishness
lagging, slow 1
lagniappe, perquisite; tip[3] 1
lagoon, pool 1
lah-di-dah. See **la-di-dah**
laic(al), civil 1; lay[2] 1; secular; temporal 1
laid, get, score 14
laid-back, easy-going; free 8; lukewarm 2; nonchalant; passive 1
laid low, prostrate 4
laid low (by *or* **with),** flat 3; prostrate 3; stricken
laid off, unemployed
laid up, indisposed 1; sick 2
laid waste, desolate 2
lair°, hideaway; nest 1
lake, pool 1
lakeshore, beach 1
lallapalooza, killer 2
lam, go on the, run 2
lam, on the, run 49b
lamasery, monastery
lambaste°, abuse 3; belabor; box[2] 2; club 5; criticize 2; hit 1; lash[1] 3, 4; lay[1] 13; light[2] 14; pound[1] 1; punch[1] 1; rough 16; slash 2; strike 1
lambasting, thrashing 1
  lambasting, a, works 3b
lambency, glow 1
lambent, bright 1; glowing 1; shiny
lambently, clear 17
Lambeth run(s), run 50
lamblike, passive 2
Lamb of God, savior 2
lambrequin, drapery
lame°, cripple 2; crippled 1; deformed 2; disabled; feeble 2; helpless 1; hurt 4; incapacitate; ineffectual 2; infirm 1; maim; mutilate 1; pale[1] 3; thin 4; weak 4
  lame, go, founder[2] 3
lamebrain, dolt
lamella, scale[2] 1
lamellar, scaly 1
lamellate, scaly 1
lament°, bemoan; bewail; complain; grieve 1; keen[2] 1, 2; moan 1, 2; mourn; mourning 1; murmur 4; regret 1; repent; sorrow 3; weep 1
  lament for, sigh 2
lamentable°, deplorable 1; grievous 2; miserable 3; mournful 2; pathetic 1; piteous; regrettable; sad 3; sorrowful 2; tragic; unfortunate 3
lamentably, painfully; sadly 1
lamentation°, keen[2] 2; lament 2; moan 1; mourning 1
lamenting, lamentation
lamina, flake 1; foil[2]; plate 3; scale[2] 1; sheet 4; skin 2
laminar, scaly 1
laminate, plate 6
lamination, foil[2]; plate 4; sheet 4
lam (on) out (of *or* **from),** beat 8; depart 1; escape 1; exit 3; flee 1; flight[2] 3; fly 2; run 2
lamp, light[1] 2
lamplight, light[1] 1
lampoon°, burlesque 1, 3; caricature 1, 2; derision; fun 5; guy 2; laugh 2a; mimic 3; mock 2; mockery 2; parody 1, 3; ridicule 2; satire 2; satirize; scoff; send 9a; take 34b; takeoff 2; taunt 1
lamppost, standard 4
lampshade, shade 5
lam, take it on the, beat 8; depart 1; escape 1; exit 3; flee 1; flight[2] 3; fly 2; powder 2; run 2; split 6
lanate, hairy 1; woolly 1
lance°, penetrate 1; pierce 1; prick 3; spike 2; stab 1
lancet, lance 1
land°, bag 6; country 2; disembark; dock 2; estate 1; farm 1; field 1; get 39d; ground 1; light[2] 13; nation; property 3; set 17c; settle 5; state 4; territory 1; touch 10
  land at *or* **in,** reach 2
  land on *or* **in** *or* **against,** strike 3
landed, substantial 4
landed estate, estate 1
landed gentry, gentry
landed proprietor, squire 2
landholder, proprietor 1; squire 2
landholding, spread 13
landing place, landing 3; pier 1
landlady°, host[1] 1; landlord 2; proprietor 1
landlocked, interior 4
landlord°, host[1] 1; proprietor 1, 2
landmark°, guide 7; mark 7; seminal 1
Land of Beulah, paradise 2
Land of Nod, be in the, sleep 1
landowner, landlord 2; proprietor 1; squire 2
landowners, gentry
land(s), domain 1
landscape°, prospect 1; scene 4; terrain; view 1
landslide, slide 5
lane, drive 9; road 2; street 1; walk 5
language°, dialect; diction 1; expression 6; idiom 1; lingo; parlance; phrase 3; print 3; prose; speech 1, 3; style 5; talk 19; terminology; tongue 1; vernacular 3; wording
  language, command of the, oratory
  language, pretentious, bombast
languid, feeble 1; inactive 1; indolent; inert 3; lackadaisical 1; lazy 2; leaden 4; lethargic 1; listless; relaxed; slothful; supine 2; tardy 2; torpid
languidness, inactivity 1; indolence; lethargy 1; sloth; sluggishness
languish, droop 2; fade 2; fail 3; flag[2] 2; lounge 1; rest[1] 6; rot 2; sink 4; stagnate; wilt 2
languishing, frail 2
languor, fatigue 1; indolence; inertia; lethargy 1; sloth; sluggishness; stupor; torpor
languorous, fatigued; indolent; inert 3; lackadaisical 1; lazy 2; leaden 4; lethargic 1; relaxed; slothful; supine 2; torpid
lank, lanky; lean[1] 1; skinny; wiry
lank(y), gaunt 1; lean[1] 1; scrawny; skinny; slender 1; spare 3; tall 2; thin 1
lantern, light[1] 2
laodicean, indifferent 1; listless; lukewarm 2
lap°, circuit 2; flap 3; fold 4; gurgle 1; leg 3; overhaul 1; overlap 3; revolution 3; stage 1; wash 3
  lap (over), overlap 1
  lap (up), drink 1; lap[1] 1, 3a; revel 1
lap dog, yes man
lapel, lap[2] 2
lappet, flap 3; lap[2] 2; tab 1
lapse°, default 1, 3; err 2; fall 26; fault 2; indiscretion 2; interval 1; lull 1; misstep 2; offense 1; pause 2; peccadillo; relapse 1, 3; revert; sin 3; solecism; space 3; transgress 1; transgression; trip 2; wander 4; wrong 8a
lapsing, relapse 3
lapsus, lapse 1
lapsus linguae, misstatement 2; trip 2
larboard, left 1, 3
larcenous, lawless 2; predatory 2
larceny, embezzlement; rip-off 1; robbery; stealing; theft

larder, kitchen
lardy, greasy 1
large°, ample 5; big 1, 2, 3, 8; broad 1; bulky; burly; considerable 1; extensive 2; gigantic; good 16; goodly; grand 1; great 1, 2; gross 1; handsome 2; hefty 1; huge; liberal 1; long[1] 1; massive; mighty 3; ponderous 1; roomy; spacious; spanking 1; strong 14; substantial 1, 2; tall 2; tidy 3; voluminous 1; weighty 1; wide 3
  large, at, abroad 2; free 2; large 5; loose 2; out 6
largehearted, large 2
largeheartedness, charity 2
large letter(s), capital 4; upper 4
largely°, amply 2; chiefly; generally 2; mainly; principally; substantially; whole 5; widely 2
largeness, breadth 3; bulk 1; size 1
larger, better 2; major 1
  larger, become, grow 1
larger than life, heroic 4
largess°, bonus; bounty 2; charity 3; dole 1; donation 1; gift 1; philanthropy 2; present[2] 2
largest, best 4; main 2; outside 6
largish, considerable 1
lariat, lasso 1
lark°, amusement 1; caper 2; prank; sport 4; spree
  lark, as a, fun 3
  lark about *or* **around,** fool 7b; lark 2
larrikin, gangster; hoodlum; miscreant 1; rough 13; thug
larva, bug 1
lascivious°, abandoned 2; base[2] 4; bawdy; broad 8; carnal; coarse 3; deviant 2; dirty 2; disgraceful 2; erotic 3; evil-minded 1; filthy 3; immodest 1; immoral 2; improper 3; impure 4; incontinent 2; indecent 2; lecherous; lewd; libertine 2; lustful; obscene 1; profligate 1; prurient 1; scandalous 1; sensual; sexy 2; unchaste; unseemly 1; vulgar 2; wanton 1
lasciviousness, desire 3; impropriety 3; impurity 3; profligacy 1; ribaldry
lash°, beat 1; connect 3; flog 1; hide[2] 2; hit 1; knot 3; lambaste 1; lather 4; punish 2; rod 2; rope 3; scourge 2, 3; slash 2; stake[1] 3; strike 1; switch 1, 3; tie 1, 2; toss 3; whip 1, 8
  lash (at), rack 4
  lash out, fly 8b; lash[1] 4; pitch[1] 6a; snap 3b; swipe 1
  lash up, inflame 1
lashing, punishment 2; thrashing 1; whipping 1
lashings, heap 2
lass°, child 2; girl 1; maid 1; minor 3; miss[2]; woman 3; youth 2
lassie, child 2; lass; maid 1; minor 3; miss[2]; woman 3; youth 2
lassitude, exhaustion 2; fatigue 1; idleness 1; indolence; inertia; prostration 4; sluggishness; stupor
last°, continue 2; endure 1; eventual 1; extend 3; extreme 3; final 1; former 1; go 32a; hold 20a, 22c; keep 11; late 2; live 7; parting 3; past 2; persist 2; prior 1; rear[1] 3; remain 3; see 14b, 14c; serve 6; stand 10b; stick[1] 6; survive 1; ultimate 1, 4; wear 4
  last, at (long), finally 1; length 4a; ultimately
  last (against), resist 1
  last legs, on one's *or* **its,** leg 7; moribund 1; ragged 3; shaky 2
  last (out), go 15
  last resort, as a, always 3
  last through, sit 7; stand 3, 12b; withstand
lasting°, abiding; chronic 1; durable; enduring; eternal 3; fast[1] 5; indelible; monumental 1; old 4; perennial 1; permanent 1; perpetual 1; stable 2
lasting quality, endurance 2; permanence; stability 2
lastly, finally 1
last-minute, unpremeditated
last word, craze; rage 3; rave 4; vogue 1
latch, bolt 2, 10; fastening; lock[1] 1, 3
  latch on to, grab 1; snatch 1
latchkey, key 1
late°, back 13; backward 3; belated; dead 1; delinquent 3; former 1; lost 5; new 2, 7; overdue; past 2; prior 1; recent; slow 5, 12; tardy 1
  late, grow *or* **become,** get 43d
  late, of, late 4, 5
lately°, just 6; late 5, 8
late-model, new 2; recent
latent, abeyance; dormant 2; implicit 1; potential 1; subconscious 1; undeveloped
later, future 2; posterior 2; second[1] 5; subsequent 1; upper 3; yet 4
  later (on), directly 3; subsequently
laterally, sideways
latest°, contemporary 2; current 3; date 5; former 1; fresh 1; hot 5; last[1] 2; minute[1] 3; modern; new 2; prior 1; recent; scoop 3; score 6
  latest, the, fashion 2; vogue 1
(latest) thing, vogue 1
lather°, bother 6; bubble 2; foam 1, 2; froth 1, 4; stew 2; sweat 7; wash 1
  lather, get *or* **work oneself (up) into a,** rage 4
Latin, classical 2
latitude, breadth 3; freedom 4; leeway; license 1; margin 3; range 1; room 1; scope 2; space 1
latitudinarian, broad 7; liberal 2, 4; permissive; tolerant
latrine, lavatory; privy 3; toilet 1
latter, posterior 2
latterly, late 5
lattice, grating 3; mesh 1; net[1] 1
latticework, mesh 1; net[1] 1
laud°, applaud 2; celebrate 3; compliment 3; eulogize; extol; glorify 2; hail[1] 2; honor 6; praise 3; rave 2
laudable°, deserving; estimable; meritorious; praiseworthy; splendid 3
laudation, glory 2; kudos; ovation; tribute 1
laudative, laudatory
laudatory°, complimentary 1; favorable 2; flattering 2; glowing 3; good 17
lauded, immortal 3
laugh°, chuckle 1, 2; giggle 1, 2, 3; joke 1; mockery 3; roar 2; scream 2
  laugh, make, amuse 2
  laugh (at), deride; jeer 1; laugh 2; mock 1; parody 3; ridicule 2; scorn 4; sneer 2; snicker 1
  laugh away, laugh 2b
  laugh off, deride; laugh 2b; trivialize
  laugh up one's sleeve, snicker 1
laughable, absurd 1; crazy 2; farcical; funny 1; humorous; inane; ludicrous; nonsensical; preposterous; rich 10; ridiculous; silly 1; stupid 2; unreasonable 1; unthinkable 2
laughing, laughter
  laughing matter, no, serious 2
laughingstock, joke 2; monkey 2; spectacle 2; zany 2
  laughingstock (of), make a, deride; parody 3; ridicule 2; sport 3
laugh-riot, riot 2
laughter°, chuckle 2; derision; hilarity; mirth; titter 2
launch°, begin 2; commence 1, 3; conception 1; debut 1, 2; deliver 5; embark 2; fire 9; float 2; fly 8a; heave 2; inaugurate 1; initiate 1; institute 3; introduce 4; mount[2] 4; open 20; opening 4; origin 2; originate 1; pioneer 2; pitch[1] 1; precipitate 2; presentation 3; project 4; propel; release 2; set 12a; shoot 2; shot 8; sling 1; spearhead 1; takeoff 1; tender[3] 1; throw 1; toss 1
launching, conception 1; debut 1; origin 2; shot 8
launder°, clean 9; wash 1
laundered, clean 2
laundering, wash 11
laundry list, list[1] 1
laurel(s), decoration 2; trophy 1
lavaliere, pendant
lavatory°, facility 2b; privy 3; toilet 1
  lavatory, go to the, defecate; go 17; urinate
lave, clean 9; rinse 1; wash 1
laving, rinse 3
lavish°, abundant 2; ample 5; blow[1] 4; consume 3; effusive; extravagant 1; free 7; generous 1, 3; grand 1; high 12; improvident 1; liberal 1; lush 1; luxuriant 2; luxurious 1; magnificent; opulent 2; ornate; plentiful 1; plush; posh; princely 1, 2; prodigal 1, 2; profuse 2, 3; rain 7; rich 4; shower 4; spend 2; splendid 1; sumptuous; wanton 3; wasteful
lavishly, amply 4; freely 4; richly 1; unduly 2; water 3
lavishness, extravagance 1; exuberance 2; plenty 2; prodigality 2; profligacy 2; splendor 1; waste 6
law°, act 4; decree 1; fundamental 2; justice 3; measure 7; order 4; precept 1; principle 1; regulation 2; rule 1; standard 1
  law, against the, unlawful
  law, by, legitimate 2
  law, the, justice 2
  law, the (long arm of the), police 1
law-abiding, honest 1; orderly 2; respectable 1; solid 5
law and order, order 8
lawbreaker, criminal 3; delinquent 1; felon; offender; transgressor
law court, bar 5
lawful°, effectual 2; fair[1] 2; formal 2; just 2; legal 1; legitimate 2; official 1; permissible; right 1; rightful 1
lawfulness, justice 3; order 8
lawless°, criminal 1; disorderly 2; riotous 1; rogue 2; unruly; wicked 2
lawlessness, crime; riot 1
lawmaker, politician
lawman, officer 2
lawn°, field 1; green 4; turf 1
law(s), book 3; code 1; generality 2; science 1
lawsuit, case[1] 2; litigation; suit 4; trial 2
lawyer°, advocate 3; counsel 3; counselor; mouthpiece 3
lax°, derelict 2; easy-going; flabby 1; forgetful; indulgent; lazy 1; limp[2] 1;

lax° (*cont.*) loose 6; permissive; promiscuous 2; slack 1; slipshod; slothful; slow 1; soft 5; unprofessional 1
  lax, be, nod 4
laxative, purgative 1, 2
laxity, neglect 4; negligence; oversight 1; permission
laxness, laissez faire; neglect 4; negligence; oversight 1
lay°, amateur 2; civil 1; clap 4; material 7; place 15; popular 3; profane 2; put 1, 8; rest[1] 8; screw 3; secular; set 1, 8, 10; smooth 10; temporal 1; unprofessional 3; worldly 1
  lay about, pelt[1] 1
  lay aside, put 11; salt 8
  lay aside *or* **by,** defer[1]; dismiss 2; garner; heap 3; hoard 2; postpone; put 12; save 3; shelve
  lay at (someone's) door, blame 2; fault 8; pin 5
  lay away, accumulate; deposit 2; hoard 2; put 12; save 3; set 14b; store 1
  lay bare, bare 6, 7; betray 2; excavate 2; lay[1] 9; reflect 2; strip[2] 1; unveil
  lay by, salt 8
  lay (down), deposit 1; define 1; garner; lay[1] 3, 10; plan 3; plot[1] 4; prescribe; provide 3; set 7
  lay down (the law), dictate 1
  lay in, garner; heap 3; hoard 2; lay[1] 12, 19a; stock 10
  lay into, assault 3; lace 5a; lay[1] 13; pitch[1] 6a
  lay it on (thick), boast 2; dramatize; exaggerate; lay[1] 14; overdo 1
  lay low, lay[1] 15, 18c; murder 3; prostrate 2; shelter 5
  lay (off), dismiss 1; eject 3; foist; lay[1] 16; measure 13; sack 4; stop 1
  lay on, fix 16a; lay[1] 17; rough 16
  lay (oneself) open to, incur
  lay open, develop 1; subject 9; unveil
  lay out, block 4a; design 1, 2, 3; expend 1; invest 1; lay[1] 18; order 16; outline 3; pay 11b; plan 3; program 4; run 19; script 4; set 22b; shell 5; space 5; spend 1; spread 3
  lay up, garner; heap 3; hoard 2; lay[1] 19
  lay waste, blast 5; desolate 5; destroy 1; devastate 1; level 9; plunder 1; ravage 1; rout 2; ruin 7; waste 11
layabout, good-for-nothing 2; idler; loafer; truant 1; wastrel 2
layer, coat 2; cover 3; film 1; foil[2]; plate 3, 4; ply; scale[2] 2; seam 2; sheet 4, 5; slice 1; spread 7; stratum 1; streak 1; tier
laying on, imposition 1
laying open, exposure 1
laying waste, depredation; destruction 1; pillage 2
layman, amateur 1
layoff, ejection 3
lay of the land, situation 2
layout, composition 2; design 5, 6; format 1; frame 3; getup 2; order 1; orientation 1; outline 2; pattern 4, 6; plan 1, 2; posture 2; project 1; proposal 2; scenario; schematic 2; scheme 2; setup 1
layover, stop 7
lay person, amateur 1
laze, dawdle; loaf[2] 1; lounge 1
  laze (about), idle 6; rest[1] 6
lazily, idly 1
laziness, idleness 1; inactivity 1; indolence; inertia; lethargy 1; sloth; sluggishness; stupor; torpor
lazing, idleness 2
lazy°, dreamy 3; drowsy; idle 3; inactive 1; indolent; inert 3; lackadaisical 1; lethargic 1; remiss; shiftless; slack 1; slothful; slow 1; soft 3; supine 2
lazybones, good-for-nothing 2; idler; loafer; slouch 3
lea, field 1; meadow; pasture
leach, filter 3; percolate
lead°, bring 2; carry 2; chair 4; channel 5; clue 1; command 2; conduct 4; control 1; direct 2; dispose 2; dominate 1; edge 4; flex 1; give 10; go 4, 5; govern 1; guide 1; head 11, 12; hero 2; heroine; incline 2; induce 1; initiative 1; line[1] 10; mold[1] 6; move 5, 6; odds 2; officiate; pilot 3; plumb 1; precedent; preside; principal 6; protagonist 1; reign 2; rule 6; run 10; see 6; shepherd; show 2; star 2; start 13; string 2; superiority 1; take 18; tempt 2; tether 1; walk 2
  lead, in the, ahead 1; front 8
  lead, male *or* **female,** lead 16
  lead, play *or* **act** *or* **take the,** star 4
  lead, take the, lead 3; pioneer 2; spearhead 1
  lead astray, betray 3; deceive; lead 7; misinform; mislead; pervert 2; ruin 11; seduce 2
  lead balloon, go over like a, fall 12
  lead down the aisle, wed 1
  lead off, lead 8
  lead on, deceive; entice; flirt 1; lead 7, 9; lure 1; tempt 1
  lead out (of one's pants *or* **ass), get the,** run 1
  lead to, cause 6; entail; induce 2
  lead up to, lead 10
lead actor *or* **actress,** lead 16; star 2
lead balloon, dud 1; failure 3; flop 4; loser; nonevent; washout
leaden°, gray 1; heavy 7, 9; inert 3; slow 1; somber 2
leaden-footed, slow 1
leader°, boss 1; brain 2; chair 3; chief 1; director 2; editorial; escort 2; executive 1; figure 6; guide 5; head 2; master 1, 3; moderator; pillar 2; pilot 2; pioneer 1; protagonist 2; queen 2; skipper; string 2; writing 2
leadership°, command 7; conduct 2; control 4; direction 1; executive 2; guidance 1; head 7; helm 2; initiative 2; lead 13; nobility 1; predominance; preponderance 2; regime; reign 1; rein 2; sovereignty; superiority 1; sway 4
leading°, banner 2; capital 5; cardinal; central 2; chief 2; classic 1; dominant 1; essential 2; first 1; foremost 1; forward 1; front 8, 10; grand 6; great 6; head 9; high 11; important 2; influential; lead 18; lofty 2; main 1; predominant; premier 2; primary 1; prime 2; principal 1, 2; prominent 2; responsible 3; sovereign 2; star 3; stellar 2; successful 3; supreme 3; top 8; weighty 3
  leading on, temptation 1
  leading part *or* **role, play** *or* **act** *or* **take the,** star 4
leading article, editorial; writing 2
leading character, protagonist 1
leading light, queen 2; star 2
leading man *or* **actor,** hero 2; lead 16; principal 6; star 2
leading place, lead 11
leading position, head 7; lead 11
leading role *or* **part,** lead 16; principal 6; protagonist 1
leading woman *or* **lady** *or* **actress,** heroine; lead 16; principal 6; star 2
lead-pipe cinch, breeze 2; picnic 2; piece 9; pushover 1
lead (someone) up *or* **down the garden path,** deceive; hoodwink; mislead
lead (the way), pioneer 2; precede; spearhead 1
leaf, blade 3; page[1] 1; plate 3; ply; sheet 2
  leaf (through), thumb 7; scan 1; skim 2
leafless, bare 4
leaflet°, blade 3; brochure; pamphlet; promotion 5; publication 2; tract[2]
leaflets, literature 2
leafstalk, stalk[2]
leafy°, green 1; shadowy 1
league°, alliance 1; ally 2; association 1; body 6; brotherhood 2; cabal 2; class 2; club 2; combination 2; connect 2; federation; fellowship 2; fraternity 3; front 6; group 1; institute 1; marry 2; organization 3; party 3; pool 4; ring[1] 3; society 5; syndicate 3; tie 3; union 2
  league (against), gang 3
  league as *or* **with, be in the same,** touch 7
  league (together), band[2] 3; club 6
  league with, join 2
  league (with), in, hand 10; league 2
leagued (with), league 2
leak°, discharge 4; disclose 1; escape 3, 7; expose 1; exposure 1; give 12b; let[1] 7a; ooze 2; publish; puncture 1; revelation; rumor 2; secrete[2]; secretion; slip[1] 4; spill 4; tell[1] 2; trickle 1; well[2] 2
  leak, take a, go 17
  leak (out), get 29a; leak 6; reveal; come 15a
leakage, escape 7; leak 1; secretion; spill 4
leaking, escape 7; exposure 1; leak 1; secretion
leak-proof, tight 1
lean°, emaciated; incline 1, 3; lanky; list[2] 2; meager 2; prop 2; rarefied 1; rawboned; rest[1] 8; scrawny; slant 3, 4; slender 1; sway 1, 2; tend[1]; thin 1; tilt 1, 3; tip[2] 2; trend 3; verge[2]; wiry
  lean back, recline
  lean (down), stoop 1
  lean on *or* **upon,** enforce 2; lean[2] 4; pressure 5; railroad 2; reckon 4; rely 1; squeeze 3
  lean (over), heel[2]; list[2] 1
  lean to *or* **toward(s),** feel 9
  lean toward(s) *or* **on the side of,** lean[2] 3; near 10; prefer 1
  lean (up) against, touch 3
leaning°, affinity 2; attitude 2; bent 5; bias 1; favoritism; inclination 3; inclined 3; instinct; love 2; partiality 1; penchant; position 3; preference 2; prejudice 1; prone 2; slant 3; taste 3; tendency; trend 1; turn 38; weakness 4

leaning (back), recumbent
leaning toward(s), disposed
lean-to, hut; shack; shed[1]
leap°, bounce 1; bound[2] 1, 3; caper 1, 3; cavort; dance 1; dive 1; frisk 1; hop 1, 3; jump 1, 4, 8; lunge 2; pounce 2; prance 1; shoot 1; skip 1, 3; spring 1, 6
leap, cause to, start 6
leap (at), jump 6; leap 4; pounce 1
leap (on), pounce 1
leap (over), clear 27; hurdle 2
leaps and bounds, by, bound[2] 2; leap 7
learn°, determine 2; discover 1; find 2; gather 4; get 21, 31e, 51f; grasp 2; hear 2; master 7; pick 8d; polish 4; receive 5; run 33d; see 4; soak 2; take 10
learn (about), get 44c; scent 3; study 1
learn of, detect 1; learn 1; wind[1] 6
learned°, academic 1; authoritative 2; deep 3; educated 1; expert 2; highbrow 2; informed 1; knowledgeable 2; lettered; literary 1, 2; philosophical 1; profound 2; scholarly; smart 2; well-informed; wise 2
learned in, versed
learner°, apprentice 1; disciple 1; greenhorn; newcomer 2; novice; pupil; student 1
learning°, culture 1; education 2; illumination 2; knowledge 4; letter 3; lore 2; scholarship 1; schooling; study 5; wisdom 2
lease°, charter 3, 5; farm 4; get 1; hire 2, 4; rent[1] 1, 2; take 4
lease on life, take a new, improve 2
lease (out), hire 3; let[1] 3
leaseholder, occupant; tenant
leash, lead 15; stake[1] 3; string 2; tether 1, 2; tie 9
least, minimal; minimum 1, 2; minute[2] 2; very 4
least, not in the, never 2; scarcely 2
least, to say the, indeed 2
leather, hide[2] 1; whip 1
leathery, hard 1; stringy; tough 2
leave°, abandon 1, 2; approval; beat 8; bequeath; break 15, 24a; bugger 5; check 10a; clear 29; clearance 2; depart 1, 2; deposit 1, 2; desert[1] 3; disembark; drain 6; draw 12b; drop 7, 13; emigrate; evacuate 2; exit 3; fly 2; forsake 1, 2; get 28c, 32a, 41b, 45a; go 2, 23, 31d, 33b; go-ahead 1; liberty 2; license 1; park 3; part 12; permission; pile[1] 9; privilege; pull 14b, 14c; push 9b; put 21e; quit 1, 2; relinquish 1; resign 1; run 22; secede; set 18b; ship 3; slip[1] 6; split 6; stake[1] 2; start 3; step 17a; take 34c; throw 8, 9b; vacate 1; walk 4a; wash 10; will 6; withdraw 5
leave, give, authorize
leave (behind), forget 2; gain 5; overhaul 1; overtake 1; surpass
leave flat, desert[1] 3
leave for, take 38b
leave high and dry, desert[1] 3; run 32
leave of, take, leave[1] 4
leave (of), take one's, depart 1; leave[1] 1
leave off, cease 1; discontinue; give 17a; lay[1] 16b; leave[1] 9; quit 3; refrain[1] 2; shut 2; stop 1
leave out, drop 9; eliminate 1; except 3; exclude 2; forgo 1; jump 3; leave[1] 10; miss[1] 4; omit 1; overlook 1; skip 2
leave (to), give, let[1] 1; permit 1
leave to enter, admittance
leaven, ferment 1
leave (of absence), holiday 1; leave[2] 2
leave-taking, exit 2; farewell 2; parting 2
leaving, exit 2; outgoing 1; parting 2; takeoff 1
leaving aside, exclusive 4
leaving off, omission 1
leaving out, barring; omission 1; short 10
leavings, dirt 1; filth 1; leftover; odds 5; remain 4a; rubbish 1; scrap[1] 2; waste 7
lecher, libertine 1; profligate 3; rake[2]; roué; seducer; sensualist
lecherous°, carnal; evil-minded 1; fast[1] 2; hot 7; immoral 2; impure 4; incontinent 2; lascivious 1; lewd; libertine 2; obscene 1; passionate 2; profligate 1; prurient 1; randy; scandalous 1; sensual; unchaste; wanton 1
lecherously, fast[1] 9
lecherousness, desire 3; impurity 3
lechery, desire 3; profligacy 1
lectern, rostrum
lecture°, address 1, 3; course 5; harangue 2; lesson 1; oration; piece 10; preach 2; rant 1; rebuke 1, 2; reprimand 1, 2; scold 1; sermon 1, 2; speak 12a; speech 2; talk 6, 14; task 3; tell[1] 10
lecture (on), hold 16a
lecturer, instructor; speaker; talker 1; teacher
lecture room *or* **hall,** hall 2
le dernier cri, fashion 2
ledge°, overhang 3; projection 1
ledger, register 1
leech, bloodsucker; flatterer; hanger-on; menial 4; parasite
leer°, eye 5; ogle 1, 3; smirk 1, 2
leeriness, suspicion 1
leery°, distrustful; guarded; reluctant; shy 2; suspicious 2
leery (of), wary
leery of, be, distrust 1
leery of *or* **about, be** *or* **feel,** mistrust 1
lees, deposit 4; dregs 1; ground 4; rubbish 1; sediment
leeway°, clearance 1; give 18; margin 3; play 24; room 1; scope 2; space 1
left°, knock 9; punch[1] 2
left, be, inherit; remain 1, 2
left (out), undone[2]
left (over), remaining 1; unused 3
left-hand, left 1
left-handed°, awkward 1
leftist, left 2; liberal 4; progressive 3; radical 5
left (of center), liberal 2
left-over, extra 2
leftover(s), difference 5; excess 1, 3; leftover; odd 3; odds 5; outstanding 2; remain 4a; remainder 2; remaining 2; remnant 2; residual; residue; rest[2] 1; rubbish 1; scrap[1] 2; spare 1, 2; surplus 1, 2; waste 9
left-wing, left 2
left-winger, liberal 4; progressive 3; radical 5
leg°, post[1] 1; section 3
leg it, leg 5; run 1
leg to stand on, not a, leg 6
leg up, boost 1; hand 2
leg up, a, leg 4
leg up, give a, speed 2
legacy, bequest; heritage 1; inheritance
legal°, effectual 2; formal 2; judicial 1; lawful 1; legitimate 1, 2; official 1; permissible; real 1; regular 6; right 1; rightful 1
legalization, passage 9; sanction 1
legalize, justify; launder 2; legitimate 4; sanctify 3; sanction 5
legal remedy, litigation
legal tender, cash 1; money 1
legate, agent 1; ambassador; delegate 1; deputy; envoy; messenger; minister 2; representative 3
legatee, heir; recipient
legation, mission 3
legend°, device 3; immortal 4; key 4; myth 1; romance 3; saga; story[1] 1; title 3
legendary°, classic 2; fabulous 1; famous; fictional; grand 2; heroic 5; imaginary; mythical 1; notorious 2; superhuman 1
legerdemain, hanky-panky; hocus-pocus 3; magic 2; trick 5
leggy, tall 2
legible°, clear 5; intelligible; readable 3
legion, cohort 1; host[2]; sea 3
legionnaire, mercenary 3
legion(s), mass 2; number 2; score 4
leg iron(s), shackle 1
legislate, enact 1
legislation, measure 7; passage 9
legislative body, house 3
legislator, politician
legislature, chamber 1; diet[2]; house 3; parliament 2
legit, legal 2; permissible; right 1
legitimacy, justice 3
legitimate°, authentic; authoritative 1; bona fide; explain 2; fair[1] 2; genuine 1; good 8; honest 4; just 3; justify; launder 2; lawful 1; legal 2; official 1; permissible; real 1, 3; regular 6; right 1; rightful 1; sanctify 3; straight 4; veritable
legitimately, truly 2
legitimatization, passage 9; sanction 1
legitimatize, explain 2; justify; launder 2; legitimate 4; sanctify 3; sanction 5
legitimization, passage 9; sanction 1
legitimize, justify; launder 2; legitimate 4; sanctify 3; sanction 5
legman, journalist
leg-pull, put-on 1; trick 2
legs, extremity 2
leisure°, ease 1; freedom 5; pastime; relaxation 1; rest[1] 3; spare 2; time 8
leisure, at, idle 2; leisure 3; off 6; unemployed
leisure, at one's, leisure 3b
leisurely, easy 2, 5; measured 1; slow 1, 13; unhurried
leisure(-time) activity, pastime; recreation
leitmotif, motif
leman, love 3
lemon, dud 1; failure 3; flop 4; loser
lemonade, pop 5
lemony, sour 1; tart[1] 1
lend°, advance 6; impart 1; loan 2; provide 2
lend dimension to, flesh 6
lender, factor 3
lend substance to, flesh 6
length°, bolt 4; extension 2; leg 3; measure 1; measurement 2; segment 1; stripe 1
length, at, finally 1; last[1] 7; length 4
length, at full, length 3
length and breadth, expanse

loafing, idle 3; idleness 2; rest[1] 3; truant 2
loam, dirt 2; earth 2; ground 1; land 2; mold[3]; soil[2]
loan°, accommodation 5; advance 9; lend 1
loan of, make a, lend 1
loath°, averse; backward 1; disinclined; guarded; hostile 1; indisposed 2; reluctant; slow 10
loath, be, shy 4
loath (to), be, hate 2; scruple 2
loathe°, abhor; despise; detest; dislike 1; hate 1
loathing°, animosity; aversion 1, 2; contempt; disgust 3; dislike 2; distaste 2; hate 3; ill will; phobia; revulsion; thing 4
loathsome°, abhorrent; abominable 1; base[2] 6; contemptible; damnable; disgusting; distasteful; fearful 3; foul 1; frightful 2; ghastly 1; gruesome; hateful 1; hideous 2; horrible 1; infamous 2; monstrous 2; nasty 1; nauseous; obnoxious; obscene 1; offensive 3; rank[2] 3, 4; repellent; repugnant; repulsive; revolting; satanic 2; shocking 2; terrible 1, 5; ugly 2; vile 2; wicked 2
loathsomeness, infamy 2
lob°, cast 1; fling 1; fly 8a; pitch[1] 1; project 4; put 7; sling 1; toss 1, 6
lobby°, crusade 2; faction 1; hall 1; lounge 2; passage 4
lobbyist, lobby 2
local°, bar 6; indigenous 1; interior 2; native 2, 5; parochial; provincial 1; pub; resident 2, 3; surrounding; topical 2; vernacular 1
local to, peculiar 2
locale°, district; element 2; environment; location 1; neighborhood 1; place 1, 2; point 3; precinct 2; quarter 3; scene 1; setting; site 1; situation 1; spot 2; vicinity; whereabouts 1; zone
localism, provincialism 1
locality, area 3; district; locale; neighborhood 1; quarter 3; region 1; scene 1; site 1; spot 2; vicinity; zone
localize, position 9
locate°, detect 1; discover 1; domicile 2; establish 2; find 2, 6; finger 5b; fix 8; perch 2; place 15; position 9; post[2] 2; put 1; run 19, 26; scout 2; set 1; settle 4; site 2; situate; station 4
locate in, inhabit
located in, be, occupy 2
locating, placement 1
location°, address 2; exposure 4; locale; orientation 1; perch 1; place 1; placement 1; point 3; position 2; quarter 3; scene 1; setting; site 1; situation 1; spot 2; station 1; whereabouts 1
loch, gulf 1; sound[4]
lock°, bar 8; bolt 10; close 1; completely 1; connect 3; fasten 1; fastening; grasp 3; seal 3; shut 1; trap 5; wholly 1; wisp
lock and key, under, captive 2
lock away, lock[1] 5, 6c, 9
lock in, lock[1] 6
lock on, lock[1] 7
lock out, exclude 1; lock[1] 8; shut 5a
lock (up), bar 8; close 8a; imprison; jail 2; knock 5; lock[1] 3, 6c, 9; punish 2; restrain 3; seize 6
lock, stock, and barrel, completely 1; entirely 1; wholly 1
locked (up), captive 2; secure 2; shut 7
locker, compartment; trunk 3
locket, pendant
lock horns (with), tangle 4
lockout, exception 2; exclusion 1
locks, hair 1
lockup, dungeon; jail 1; prison; stir 7
loco, crazy 1; deranged; insane 1; mad 1; mental 2; unbalanced 2
locomotive, engine
locum (tenens), relief 4; substitute 2
locus, place 1
locution, expression 5; idiom 2; phrase 2; term 1; word 3
locutions, terminology
lode, mine 1; seam 2; vein 3
lodestar, guide 7
lodge°, accommodate 4; board 7; cabin 1; camp[1] 4; club 2; cottage; deposit 2; domicile 2; dwell 1; engrave 2; establish 2; fellowship 2; house 8; lay[1] 7; live 8; order 9; plant 6; prefer 2; put 28b; quarter 6; repose 4; room 4; set 1, 11; stay[1] 2; stop 5; take 32a
lodged, be, rest[1] 7
lodged, be *or* **become,** stick[1] 7
lodger, guest; occupant
lodging(s), abode; accommodation 4; domicile 1; dwelling; home 1; hotel; house 1; housing 1; lodging; place 6; quarter 5; room 3
loft, lob 1
loftier, superior 1; upper 1
loftiest, supreme 1; uppermost 1
loftiness, arrogance; dignity 1; elevation 4; nobility 1; snobbery
lofty°, arrogant 2; dignified; elevated 2; exalted 1, 2; generous 2; good 5; grand 1; grandiose 2; great 7; haughty; heroic 6; high 1, 5; hoity-toity; imperial 2; majestic 1; noble 3; pretentious 2; proud 3; rarefied 2; snobbish; stately; stilted; sublime 1; supercilious; tall 1; towering 1
log, book 4; chronology; diary; enter 3; journal 2; minute[1] 2, 4; put 16a; record 1, 3; register 3; stamp 2
log in, check 7; register 6
logbook, journal 2
loggerheads, at, factious; odds 4
loggerheads, be at, quarrel 2
logic°, argument 2; philosophy 1; rationale; reasoning 1; rhyme 2
logic to, apply, rationalize 2
logical°, coherent 1; forceful 2; judicious; legitimate 3; natural 1, 9; philosophical 1; plausible 1; rational 1, 3; reasonable 1; sage 1; sensible 1; sound[2] 4
logical, be, hang 11b; tie 6a
logical basis, rationale
logically, naturally 1
logical positivism, mechanism 4
logjam, tie-up 1
logo, device 3; emblem; label 3; stamp 7; symbol
logorrhoeal, talkative
logorrhoeic, talkative
logo(type), device 3; emblem; sign 4; stamp 7; symbol
loin, flank 1
loiter, dawdle; delay 3; drag 4; fool 7b; hang 4a; hover 2; idle 6; lag 1; linger 1; loaf[2] 1; muck 3; stall[2] 1; stay[1] 5; straggle; tarry 1; time 20; trail 6
loiterer, laggard
loitering, delay 6; idle 3
loll, idle 6; lounge 1; recline; rest[1] 6; slouch 1; sprawl 2
loll about *or* **around,** wallow 1
lollapalooza, killer 2
loller, laggard
lollipop, buoy 1
lollygag, bugger 4a; drag 4; idle 6; loaf[2] 1
lollygagging, idleness 1; slow 1
lone, individual 1; isolated 1; lonely 1; one 1; only 1; reclusive; single 2; single-handed 1; singular 3; sole; solitary 1; stray 6; unique 1
loneliness, seclusion; solitude 2
lonely°, deserted; desert[1] 2; desolate 1; forlorn 2; homesick; isolated 2, 3; out-of-the-way 1; remote 2; secluded 1; solitary 1
loner, independent 9; individualist; misanthrope; nonconformist 1
lonesome°, forlorn 2; homesick; lonely 3; solitary 1
lone wolf, individualist; misanthrope
long°, ache 2; aspire; burn 3; die 5; lengthy; lingering 1; protracted; purpose 5; roundabout 1; wish 1; yearn
long as, as, providing
long chalk, by a, far 3; well[1] 7
long for, desire 1; fancy 11; hanker; miss[1] 2; pant 2; thirst 2; want 1
long run, in the, eventually; finally 1; length 4a; mainly; materially; run 48; sometime 3; ultimately
long shot, by a, far 3; shot 11
long way, by a, well[1] 7
long ago, formerly; once 1
longanimity, stoicism
long-drawn(-out), boring; lengthy; lingering 2; protracted; tedious
longed-for, desirable 1
longest, outside 3, 6
longevity, endurance 2; permanence; standing 7
long face, pout 2
long-faced, solemn 1; unhappy 1
longhair, hippie; scholar 1; square 9
longhair(ed), scholarly
longhand, penmanship; writing 1
longheaded, wise 1; shrewd
longheadedness, wisdom 1
longiloquent, windy 2
longing°, ache 4; anxiety 2; anxious 2; appetite 2; aspiration; desire 3; desirous; eager; eagerness 2; envy 2; fancy 7; homesick; hope 1; hunger 2; hungry 2; inclination 4; itch 4; passion 3; stomach 2; thirst 1; urge 5; want 4; will 1; wish 4, 5; wistful 1
longing for, desperate 3; starved 2
longlasting, chronic 1; durable; enduring; indestructible; permanent 1; serviceable 2; stable 2; staunch 2; tough 1
long-legged, lanky; tall 2
long-limbed, willowy 2
long-lived, chronic 1; old 1; permanent 1; perpetual 1
long-serving, seasoned 1; veteran 2
long-sighted, farsighted 2
long-sightedness, foresight 1; forethought
long standing°, standing 7; chronic 1; stable 2; steady 5; well-established; enduring; old 4; seasoned 1
long-suffering, passive 2; patient 1; stoical
long suit, forte

lucky break, fluke
lucrative, fat 4; gainful; profitable 1; successful 2
lucre, fund 2; money 1; purse 2; riches
lucubrate, cram 2; grind 4; mug 6; overwork 2; slave 3; study 1
Lucullan, epicurean 2
Lucullus, epicure; gourmet
ludicrous°, absurd 1; crazy 2; farcical; foolish 3; funny 1; grotesque 2; humorous; impossible 2; improbable; inane; insane 2; nonsensical; preposterous; queer 1; rich 11; ridiculous; senseless 3; silly 1; stupid 2; unreasonable 1; unthinkable 2; witty; zany 1
ludicrously, madly 2
ludicrousness, absurdity 1; humor 1; stupidity 2
luff, shiver[1] 2
lug°, carry 1; cart 2; drag 1; draw 1; gawk 1; haul 1; hump 3; pull 1; tow
luggage°, bag 2; thing 8c
lugubrious, dismal; doleful; forlorn 1; funereal; gloomy 2; glum; hopeless 3; joyless 2; melancholy 1; moody 1; morbid 3; mournful 1; sad 1, 2; somber 1; stern 2; sullen; tragic; woebegone
lugubriously, sadly 2
lugubriousness, melancholy 2
lukewarm°, cold 3; cool 5; dead 4; halfhearted; indifferent 1; lackadaisical 2; limp[2] 3; listless; tepid 1, 2; unenthusiastic; warm 1
lull°, break 28; calm 5; delay 5; gap 1; interlude; pause 2; plateau 2; quiet 6; rock[2] 1; slack 5; still 9; tranquilize; truce 1
lulling, dreamy 3
lulu, killer 2
lumber°, plod 1; stuff 2; timber 2
lumbering, shambling
luminary, celebrity 2; dignitary; hero 1; idol 2; legend 2; name 3; notable 3; personage; personality 2; somebody 2; star 2; worthy 3
luminescence, light[1] 3
luminescent, luminous 2
luminosity, flare 4; glow 1; illumination 1; light[1] 3; luster 1; radiance 1; splendor 2
luminous°, ablaze 2; bright 1; brilliant 1; fiery 2; glowing 1; light[1] 13; radiant 1
luminously, clear 17
luminousness, glow 2; luster 1; radiance 1; splendor 2
lummox, boor 2; clown 2; duffer; gawk 1
lump°, blob; block 1; bulge 1; bump 2; cake 2; clod 1; clump 1, 3; elevation 2; gob; growth 4; hump 1; loaf[1] 1; mass 3; mouthful; nub 1; pat[1] 4; peasant; piece 1; portion 1; protrusion; swelling; tumor; wad 1; welt 2
  lump it, lump[2]
  lump together, assemble 2; lump[1] 3
lumpen, rustic 2
lumpish, obtuse 2; stolid; stupid 1
lumpishness, stupidity 1
lumpy°, bumpy; gnarled; irregular 1; rough 1
lunacy°, folly 1; insanity 1, 2; madness 1, 2; mania 2
lunatic, crazy 1; delirious 1; deranged; inane; insane 1; mad 1; madman; maniac 1; maniacal 1; mental 2; psychotic 1, 2; weirdo
lunch, dine; eat; meal 1
luncheon, meal 1
lunch-hook, hand 1
lunchroom, café
lune, crescent 1
lunette, crescent 1
lunge°, plow 2; thrust 2, 4
  lunge at, make 30b; snap 3b
lunkhead, dolt
lupine, rapacious
lurch°, careen; founder[2] 3; jerk 2, 4; jolt 4; jump 11; pitch[1] 4; reel 1; rock[2] 2; stagger 1; stumble 1; toss 5; trip 5; walk 1; wallow 3
  lurch, leave in the, desert[1] 3; lurch[1]; run 32
lurcher, mongrel
lurching, shambling
lure°, appeal 4; attract; attraction 2; bring 3; chat 3; come-on; decoy 1, 2; draw 4, 16; entice; enticement 2; incentive; induce 1; inducement; invitation 2; lead 9b; magnetism; motive 1; premium 2; pull 5, 20; rope 4; seduce 1; solicit 2; spell[2] 3; stratagem; take 20; tempt 1; temptation 2
lurid°, conspicuous 2; frightful 2; juicy 2; lascivious 2; prurient 2; pulp 4; rank[2] 3; sensational 2
luring, attractive
lurk°, ambush 2; hide[1] 1; prowl 1; sneak 1; steal 3
lurking, sneaking 1
  lurking about *or* **around,** prowl 3
luscious°, delicious 1; mellow 1; rich 6; savory 1; succulent; tasty; tempting 2; voluptuous 2; yummy
lush°, alcoholic 2; drunk 3; flourishing; juicy 1; luxuriant 1, 2; profuse 2; prolific 1; rank[2] 1; rich 4; soak 3; splendid 1
lust°, desire 3; impurity 3; passion 3; thirst 1
lust after *or* **for,** fancy 11; hanker; lust 3; thirst 2; want 1
luster°, brilliance 1; glare 1; glaze 2; gloss[1] 1; glow 1; polish 5; radiance 1; renown; sheen; shine 4; splendor 2
lusterless, dead 10; drab; lackluster; mousy 1
lustful°, carnal; erotic 1; fast[1] 2; hot 7; immoral 2; impure 4; incontinent 2; lascivious 1; lecherous; lewd; obscene 1; passionate 2; prurient 1; randy; romantic 4; scandalous 1; sensual; sexual 2; sultry 2; torrid 2; vulgar 2; wanton 1
lustfully, fast[1] 9
lustfulness, desire 3; eye 4; impurity 3; lust 1; ribaldry
lustily, vigorously
lustrate, purify 2
lustrous°, bright 3; brilliant 1; clear 3; glossy 1; golden 3; luminous 1; opalescent; pearly; radiant 1; rich 5; scintillating 1; shiny; silky; silver 3; sleek 1
lusty°, bawdy; buxom 1; earthy; racy 2; sexy 2; stalwart 1; stout 4; suggestive 2; sultry 2; vigorous
lusus naturae, monster 2
luxe, plush; posh
luxuriance, prodigality 2
luxuriant°, abundant 2; fertile; flourishing; lush 1; overgrown; plentiful 2; prodigal 2; profuse 2; rampant 2; rank[2] 1; thick 4
luxuriate°, flourish 1; grow 1
  luxuriate (in), luxuriate 1; revel 1; roll 10b; savor 3; wallow 2
luxuries, creature 2
luxurious°, elegant 3; epicurean 1; fancy 3; fruitful 3; grand 1; high 12; lush 3; luxuriant 1; magnificent; noble 5; opulent 2; palatial; plush; posh; princely 2; rich 4; sensuous; soft 13; splendid 1; sumptuous; swanky; swell 7; upper-class 2; voluptuous 1
luxuriously, richly 1; well[1] 4
luxuriousness, elegance 2; grandeur 1; luxury 1; prodigality 2; splendor 1
luxury°, comfort 3; delicacy 4; ease 3; elegance 2; fancy 3; frill 2; indulgence 2, 3; posh; prodigality 2; prosperity; splendor 1; style 3
  luxury, be in the lap of, luxuriate 2
luxury-loving, voluptuous 1
lycée, school 1
Lyceum, school 1
lying°, deceitful; dishonest; evasion 2; false 2; hypocrisy; hypocritical; insincere; perjury; recumbent; two-faced
  lying (down), flat 2; prone 1; prostrate 3; recumbent; supine 1
lymphatic, supine 2
lynch, hang 2; string 12
lynx-eyed, eagle-eyed; sharp-eyed
lyric(al), lay[3]; lyric 1; melodious; musical; poem; poetic 1; pretty 2; sweet 3
lyricist, poet
lyrics°, book 2; lyric 4; word 9b
lyrist, poet

ma, mother 1
macabre°, frightful 2; ghastly 2; ghoulish 2; grim 3; gruesome; horrible 1; lurid 2; morbid 2; sick 5; terrible 4; unearthly 2
macadamize, pave 1
macaroni, dude 1; swell 6
mace, club 1; staff 1
Machiavellian, calculating; designing; insincere; opportunistic; politic 1; scheming; serpentine 1; smooth 8; subtle 3; two-faced
Machiavellianism, diplomacy 2
machinate, cabal 3; engineer 5; maneuver 3; plot[1] 3; scheme 4; wangle
machination°, device 2; diplomacy 2; dodge 4; finesse 2; intrigue 3; maneuver 1; plot[1] 1; politics 2; scheme 3; trick 1; wile
machinator, operator 3
machine°, apparatus; car 1; device 1; engine; gadget; mechanism 1; tool 1
machine copy, duplicate 2
machine-driven, mechanical 1
machinelike, mechanical 2
machine-made, mechanical 1
machine politician, politician
machinery, apparatus; device 1, 2; gear 1; hardware 2; mechanism 2; movement 4; plant 3; works 2
machine screw, bolt 3; screw 1
machine shop, shop 2
machinist, operative 2
machismo°, bravado; bravery; manhood 1
macho°, brave 1; manly
macilent, lean[1] 1
mack, pander 3
macrocosm, universe 1
macula, spot 1
macular, spotty 1
mad°, absurd 1; angry 1; berserk; beside 3; crazy 1, 5; delirious 1; deranged; desperate 2; distraught; disturbed 2; fanatical; flighty 2; foolish 2; frantic; furious 1; hectic; homicidal; hysterical 1; improbable; imprudent;

mad° (*cont.*) inane; indignant; insane 1; maniacal 1; mental 2; mind 14; nonsensical; off 4; pell-mell 2; preposterous; psychotic 1; queer 4; quixotic; rabid 1, 2; rampage 2; raving 1; reckless; romantic 2; senseless 3; sick 4; silly 1; stupid 2; twist 11; unbalanced 2; unreasonable 1; unsound 3; unthinking 1; upset 10; way-out 1; wild 5, 9; zany 1
mad, go, flip 2; fly 7; rage 4
mad about, sweet 8
mad about, be, love 7
mad (about *or* **for),** daft 3; mad 6
madam, procurer
mad as a hatter *or* **a March hare,** crazy 1; deranged; insane 1; mad 1
madcap, daredevil 2; devilish; foolhardy; harebrained 1; hot-headed; impulsive; mad 2; rash[1]; reckless; wild 6, 7; wit 3; zany 1
madden°, anger 2; annoy 1; enrage; exasperate 1; inflame 1; infuriate; outrage 4; provoke 3; wall 3
maddened, crazy 1; furious 1; possessed
maddened, become, bristle 3; flip 2
maddening, provocative 2; trying
made, obliged 2
made (up) of, be, compose 3; form 9
mademoiselle, girl 1; lass; maid 1; miss[2]; youth 2
made-to-measure *or* **-order,** perfect 4; tailor-made 1
made-up, artificial 2; assumed 2; bum 3; false 2; fictional; fictitious 1, 2; imaginary; mythical 2; romantic 1; sham 2; unreal 2
madhouse, bedlam; zoo 2
madly°, head 8; mad 5
madman°, maniac 1; psychotic 2; weirdo
madness°, fanaticism 2; folly 1; insanity 1; lunacy 1; mania 2; stupidity 2
madwoman, madman; maniac 1; psychotic 2; weirdo
Maecenas, friend 4; patron 1; philanthropist; sponsor 1
maelstrom, eddy 1; whirlpool
maenad, hag; shrew
maestro, director 2; genius 1; master 2; professional 3; specialist; virtuoso 1
mafficking, festivity 1; gaiety 2; obstreperous
mafia, syndicate 2
Mafia, the, underworld 1
mafioso, criminal 3; gangster; racketeer; thug
magazine°, journal 1; organ 2; periodical; publication 2; rag[1] 2; review 7; storehouse
maggoty, filthy 1; sordid 3
magic°, admirable; charm 2; divine 3; enchant 1; enchantment 1; excellent; fabulous 3; fascination; glamour; hocus-pocus 3; magnetism; mumbo jumbo 2; mystique; sorcery; spell[2] 1; superb; trick 5; world 8
magical, charmed 1; glamorous 1; magic 4, 6; miraculous; occult 2; psychic 2; supernatural
magician°, charmer; sorcerer; witch 1
magisterial, arbitrary 2; hard 4; heavy-handed 2; imperative 2; judicial 3; lofty 2; majestic 2; masterful 2; overbearing; pompous 1
magistrate, judge 1
magnanimity, benevolence 1; charity 1, 2; heart 5; humanity 3; mercy 1; nobility 1; philanthropy 1; self-denial 1; tolerance 1
magnanimous, benevolent; big 6; bountiful 1; charitable 1; chivalrous; forgiving; generous 1, 2; heroic 2; human 3; large 2; lenient; liberal 1; merciful; noble 4; open 14; philanthropic; profuse 4; selfless; tolerant; unselfish
magnate, dignitary; merchant 3; mogul; personage; tycoon
magnet, lure 2
magnetic°, glamorous 1; magic 6; prepossessing
magnetism°, attraction 1; charm 2, 3; draw 16; fascination; glamour; invitation 2; magic 3; pull 20; spell[2] 3
magnification°, exaggeration
magnificence, brilliance 1; dazzle 3; display 5; glory 3; grandeur 1; luxury 1; pageantry; pomp; preeminence 2; splendor 1; state 3
magnificent°, beautiful 2; bright 7; brilliant 2; dazzling; fine[1] 1; gallant 3; glorious 2; gorgeous 1; grand 1; grandiose 2; imperial 2; imposing; lofty 2; luxurious 1; majestic 1; monumental 1; noble 5; palatial; princely 2; proud 4; regal; royal 2; splendid 1; statuesque; striking; stunning 2; sumptuous; superb; terrific 2; transcendent
magnificently, beautifully 1, 2
magnified, inflated 1
magnify°, aggravate 1; amplify 2; blow[1] 8c, 8d; deepen 2; double 6; enhance; enlarge 1; exaggerate; expand 3; heighten 2; inflate 3; intensify; maximize 1, 2; overstate
magnifying glass, glass 6
magniloquence, bombast; exaggeration; hot air; oratory; raving 3; rhetoric 2
magniloquent, bombastic; heroic 4; inflated 2; pretentious 1; rhetorical 3
magnitude°, amount 2; breadth 2; bulk 1; degree 2; extension 2; extent 1; gauge 4; gravity 2; length 1; mass 5; measure 1; measurement 2; proportion 4; size 1
magnum opus, masterpiece; opus; work 4
magus, magician 1; sorcerer
Mahdi, savior 2
maid°, girl 1; lass; miss[2]; servant 1; woman 3, 4; youth 2
maiden°, first 2; girl 1; lass; maid 1; miss[2]; unmarried; woman 3; youth 2
maidenhead, chastity
maidenhood, chastity
maidenly, pure 3
maidservant, maid 2; servant 1; woman 4
maihem, mayhem
mail°, direct 5; dispatch 2; forward 7; post[3] 1, 2; send 2; transmit 1
Mailgram, cable 2; telegram
maim°, cripple 2; hurt 4; incapacitate; mangle; mutilate 1
main°, broad 4; capital 5; cardinal; central 2; chief 3; dominant 2; essential 2; foremost 1; front 10; fundamental 1; grand 6; head 9; key 6; lead 18; major 1, 2; master 5, 6; ocean 1; overriding; paramount; pipe 1; pipeline 1; predominant; premier 2; prevailing 1, 2; primary 1; principal 1; prominent 1; sea 1; staple 1; stellar 2; vital 2
main, in the, average 2; chiefly; generally 2; large 5b; largely; main 7; mainly; part 8; principally; rule 4; substantially; usually; whole 5
main, the, deep 9; drink 7
main axis, stalk[2]
main film, feature 2
main ingredient *or* **constituent,** basis 2
main issue, nub 2
mainly°, above 6; chiefly; especially 2; generally 1, 2; large 5b; largely; part 8; particularly 2; primarily 1; principally; rule 4; usually; whole 5
main part, body 3, 4, 5; principal 6
mainstay°, anchor 2; backbone 2; pillar 2; prop 3; protagonist 2; stay[2] 1; support 8
main stem, trunk 1
mainstream, current 6
main support, mainstay
maintain°, argue 5; believe 1; carry 4, 5; champion 4; claim 4; conserve 2; continue 3; entertain 3; feed 1; foster 2; harbor 3; have 1; hold 3, 4, 7, 15b, 19b; honor 7; insist 2; keep 1, 2, 4, 8, 12; nourish 1, 2; perpetuate; plead 3; preserve 2; profess 1; pursue 2; regulate 1; run 9; say 1; stand 5c; state 7; subsidize; support 4; sustain 1; uphold
maintainable, tenable
maintained by, as, according to 1
maintaining, maintenance 2; tenacious 3
maintenance°, conservation; keep 15; overhead 3; preservation 1; regulation 1; service 2; subsidy; subsistence 2; support 9; sustenance 2; upkeep 1
maisonette, flat 15
maître d'hôtel, waiter
majestic°, dignified; gallant 3; grand 1; heroic 6; imperial 2; imposing; lofty 2; magnificent; measured 1; palatial; princely 2; proud 4; regal; royal 1, 2; sovereign 3; splendid 1; state 6; stately; statuesque; sublime 2; sumptuous
majesty, dignity 1; grandeur 1, 2; king; splendor 1
major°, arch 1; big 4; capital 5; crucial; dramatic 2; excellent; fateful 1; good 2; great 5, 6; landmark 3; main 1; noticeable 2; pressing; senior; special 6; splendid 3; star 3; substantial 1; wide 3
major-domo, domestic 4; servant 1
majority°, age 2; body 5; bulk 2; mass 4; maturity 1; preponderance 1; rank and file; weight 4
majority, reach one's, grow 6
majority, the (silent), unwashed 2
majuscule, big 8; capital 4; upper 4
make°, assemble 3; beat 3; blackmail 2; brand 1; cause 5; close 4; construct 1; craft 5; create 1; design 2; drive 1; earn 2; effect 7; engineer 4; fabricate 1; fashion 5; fetch 3; force 7; forge 1; form 3, 7; frame 7; gain 2; generate 1; get 3, 11, 14; give 10; gross 6; harvest 3; have 9; incline 2; kind[2] 1; leave[1] 5; line[1] 16; machine 4; manufacture 1; mint 2; model 7; mold[1] 2,

4, 6; move 6; net[2] 4; oblige 2; offer 3; order 14; pin 4a; prepare 4, 5, 7; pressure 5; produce 1; push 4; put 19c; rake[1] 3; reach 4; realize 3; receive 1, 2; render 1, 2; require 1; run 29b; shape 6; sort 1; stamp 8; strike 8, 10; style 1; turn 20a; twist 6; variety 3; weave 3; will 5; work 16
make, on the, make 43; timeserving
make away (with), abduct; appropriate 2; embezzle; make 26, 27; spirit 10; steal 1; embezzle; spirit 10
make back (again), recover 1
make bold, dare 2; venture 2
make (for), make 30; occasion 5; take 38b; come 8
make it, barely, scrape 5
make off, flee 1; go 2; make 26, 35; move 2; run 2
make off *or* **away with,** abduct; appropriate 2; embezzle; make 36; snap 4; spirit 10; steal 1; take 3
make (one's way) through, thread 4
make out (with), cuddle 2; decipher 2; distinguish 3; fare 4; fend 1; fill 9a; get 28b; love 6; make 4, 29, 37; manage 4; notice 2; perceive 1; pet[1] 4; pick 7b; pretend 1; score 14; see 1; spot 6; tell[1] 8; understand 1; write 1
make over, bequeath; cede; deliver 2; dispose 3c; give 1; make 38; pass 18a; sign 11; trust 6; turn 3
make right, right 19
make (up), coin 3; compose 1, 2; compound 1; conceive 2; devise 1; discover 3; fabricate 2; fancy 9; form 9; generate 4; invent 1, 2; make 4, 39; manufacture 2; recoup; retrieve 3; set 9; spin 2; stage 5; think 6; turn 8; write 3
make up (for), balance 3; cancel 3; compensate 1; correct 4; cover 10; make 31a, 40; neutralize; offset 1; outweigh; redeem 5; right 19
make believe°, act 8; believe 3; fake 2; fool 5; make 28; play 10; pretend 1, 3; counterfeit 2; fancied; fanciful 2; fancy 6; fantasy 3; fictitious 2; mimic 5; mock 3; mythical 2; pretended; pretense 1, 2; sham 2; unreal 1, 2, 3
make both ends meet, get 28b
make do°, cope 1; fend 1; get 28b; improvise 2; make 29; manage 4; muddle 3; shift 2
make good (on), compensate 1; flourish 1; follow 10; get 27; make 31, 32a, 40; offset 1; place 10; prosper; prove 1; recoup; redeem 6; replace 3; rise 8; satisfy 4; succeed 3
make it°, arrive 2; grade 5; make 31b, 32; manage 4; muddle 3; pan 6; prosper; reach 2; rise 8; shift 2; squeeze 6; succeed 3
make it big, mark 9
make it snappy, rush 1
make money, coin 4
make one's way, weave 4
make-over, shake-up
maker, author; brand 1; creator 1; manufacturer; producer 1
makeshift°, excuse 6; flimsy 1, 2; rude 4; scratch 6; stopgap 1, 2; temporary
makeup°, composition 2, 5; design 6; disposition 1; fabric 2; fabrication 2; fiber 3; format 2; frame 3; nature 1; organization 2; paint 3; personality 1; psychology; setup 1; structure 1; substance 1; temper 1
make water°, urinate; water 4
making, assembly 3; creation 1; fabrication 1; manufacture 3; preparation 3; production 1
making out, sex 2
making up, toilet 2
makings, ingredient; stuff 1, 3
maladjusted, disturbed 2; neurotic; problem 3
maladroit, awkward 1; clumsy 1; heavy-handed 1; incompetent; inept 1; left-handed 1; tactless; tasteless 1; thumb 2
maladroitly, roughly 3
maladroitness, ineptitude 1
malady, ailment; bug 2; disease 1; disorder 3; illness; infirmity 2
malaise, ailment; concern 6; discontent; dissatisfaction 1; excitement 1; pang 2
malapert, pert
malapropism, howler; misuse 2
malapropos, improper 2; inappropriate; incongruous; inopportune; irrelevant; point 15
malarkey, drivel 3; gobbledegook 1; moonshine 2; mumbo jumbo 1; nonsense 1; rot 4; rubbish 2; stuff 4; talk 18; trash 1
malcontent, disgruntled; killjoy; misery 4; troublemaker
male°, man 1; manly
malediction, abuse 8; curse 1; jinx 1; oath 2
malefaction, offense 1; outrage 1
malefactor, criminal 3; culprit 2; delinquent 1; felon; miscreant 1; offender; sinner; transgressor; villain
malefic, dreadful 2; evil 1; hateful 2; miscreant 2; perverted; sarcastic
maleficent, devilish; diabolic 2; dreadful 2; evil 1; infernal 2
malentendu, misunderstanding 1
malevolence, animosity; devilry 2; grudge 1; hostility 1; ill will; rancor; sarcasm; spite 1; venom 2; virulence 2
malevolent, cutting 2; damnable; devilish; diabolic 2; dirty 3; dreadful 2; evil 1; evil-minded 2; fiendish; hateful 2; ill 3; infernal 2; malignant 2; merciless; miscreant 2; perverted; poisonous 2; rancorous; sarcastic; sinister 2; spiteful; venomous 2; vicious 2; virulent 2; wanton 4
malevolently, ill 12
malformation, abnormality 2
malformed, deformed 1; grotesque 1; misshapen
malfunction, play 17b; wrong 8b
malfunctioning, dud 2; faulty
malice, animosity; devilry 2; grudge 1; hate 3; hostility 1; ill will; rancor; resentment; sarcasm; spite 1; venom 2; virulence 2
malicious, black 6; cutting 2; damnable; diabolic 2; dirty 3; evil-minded 2; fiendish; hateful 2; hurtful 2; ill 3; infernal 2; malignant 2; mean[2] 4; mischievous 2; perverted; poisonous 2; rancorous; sarcastic; sharp 5; spiteful; venomous 2; vicious 2; virulent 2; wanton 4
maliciously, ill 12
maliciousness, sarcasm; spite 1; venom 2; virulence 2
malign, abuse 3; attack 2; blacken 2; blaspheme 2; devilish; diabolic 2; discredit 1; disparage 2; evil 3; ghoulish 1; impeach 2; libel 4; malignant 2; mischievous 2; perverted; rancorous; scurrilous; sinister 2; slander 2; unhealthy 2; venomous 2; vilify; virulent 2; vituperative; wrong 9
malignancy, tumor; virulence 1
malignant°, black 5; destructive 1; devilish; evil 3; fiendish; hateful 2; poisonous 1, 2; rancorous; sharp 5; sinister 2; venomous 2; vicious 2; virulent 2
maligning, abuse 8; foul 6
malignity, outrage 1; rancor; spite 1; venom 2; virulence 1, 2
malingerer, derelict 3; slouch 3; truant 1; wastrel 2
malingering, idleness 2; truant 2
mall, parade 3
malleability, flexibility 2
malleable, adaptable; dutiful; flexible 2; passive 2; plastic 1; pliable 1; tractable 2
malodorous, foul 3; nasty 1; offensive 3; rancid; smelly; stinking 1
maltreat, abuse 2; batter 3; grind 5; harm 3; ill-treat; injure 2; knock 3d; manhandle; mishandle 1; mistreat; molest 2; oppress 2; persecute 1; punish 3; torment 1; wrong 9
maltreated, abused 2; downtrodden
maltreatment, abuse 6; ill 9; injury; mistreatment; oppression; outrage 1; persecution 1; punishment 3; sacrilege 1
mama, mother 1
mama's boy, milksop; sissy
mamma, breast 1; mother 1
mammal, animal 1
mammon, purse 2
mammoth, colossal 1; enormous; giant 1; gigantic; great 1; huge; immense; large 3; massive; monstrous 3; prodigious 1; thumping 1; vast; voluminous 1; weighty 1; whopping 1
mammy, mother 1
man°, bugger 2; chap; chief 1; date 3; dude 2; fellow 1, 4; friend 3; guy 1; hand 4; human 4; humanity 1; love 3; mortal 6; one 4; person 1; physician; piece 5; servant 1; soul 2; steady 9; stick[2] 2
man, the, boss 1; director 1; head 2; leader 1; master 1
man and wife, become, marry 1
man about town, blade 4; dude 1; playboy
manacle(s), bond 1; chain 2; handcuffs; manacle 1; restraint 2; shackle 1, 3; tether 2
manage°, administer 1; afford 1; boss 2; carry 11b; chair 4; conduct 3; control 1, 3; cope 1; direct 1; discipline 7; fare 4; find 8; get 9, 28b; govern 1; guide 3; handle 3, 4; head 11;

manage° (*cont.*) hold 15b; husband 2; lead 3; make 29, 37f; maneuver 3; manipulate 1; mastermind 1; moderate 6; muddle 3; negotiate 2; officiate; operate 2; oversee; preside; process 4; pull 12b; regulate 1, 2; reign 2; rule 6; run 9, 10; scrape 5; see 12a, 14b; shift 2; steer 1; supervise; transact; treat 1, 2; wangle; work 10
manage without, dispense 3b; go 40; spare 9
manageability, flexibility 2
manageable°, bearable; flexible 2; gentle 2; meek 2; plastic 2; pliable 2; portable; submissive 1; tractable 1, 2; yielding 2
management°, administration 1, 2; board 4; command 7; conduct 2; conservation; control 4; direction 1; disposition 4; executive 2; government 1; guidance 1; hand 8; leadership; operation 2; oversight 2; policy; prudence 2; regime; running 1; usage 2
management (of), treatment 1
manager°, boss 1; chief 1; director 1; employer 1; executive 1; foreman; host[1] 1; landlady 1; landlord 1; master 1; officer 1; operator 2; overseer; principal 3; producer 2; proprietor 2; superintendent; supervisor
managerial, executive 3; responsible 3; supervisory
manage things *or* **affairs** *or* **matters,** shot 9
managing, direction 1; management 1; operation 2
managing director, boss 1; executive 1
managing editor, editor
man-at-arms, soldier 1
mandarin, mogul
mandate, act 4; approval; charge 5; command 6; decree 1, 2; delegate 2; dictate 2; injunction 2; law 1; order 4; precept 1; requisition 1, 2; warrant 2
mandatory°, imperative 1; incumbent 1; indispensable 2; obligatory; peremptory 1; regulation 4; vital 1
mandorla, halo
mane, hair 1
man-eater, cannibal; ogre; temptress
manes, ghost 1
maneuver°, artifice 2; deceit 2; deception 2; device 2; drive 2; ease 7; engineer 5; evade 2; exploit 2; feint; finesse 2, 4; finger 10; get 9, 15; gimmick 1; guide 2; handle 4; intrigue 2, 3; machination; manipulate 1; massage 3; move 9; movement 3; navigate 2; negotiate 2, 3; operation 4; pass 29; play 23; proceeding 1; ruse; scheme 3, 4; stage 5; stall[2] 2; stratagem; subterfuge; tactic 1; trick 1; use 2; wangle; wile; work 10, 14
maneuver (oneself *or* **one's way),** insinuate 2
maneuverable, handy 2; navigable 2
maneuverer, operator 3; tactician
maneuvering, diplomacy 2; machination; operation 2; politics 2
maneuver(s), art 6; campaign 1; tactic 2
man Friday, aide; auxiliary 4; second[1] 8
manful, gallant 1; heroic 1; intrepid; male; manly; stalwart 3
manfulness, bottle 2; manhood 1; spirit 5
mangle°, batter 3; crumple; crush 2; garble 1; hack[1] 1; hash 3; hurt 4; lacerate; mishandle 2; misrepresent; murder 4; mutilate 1; rend 1; roller 1; rout 2; smooth 9; tear 1
mangle, put through a, press 4
mangy°, mean[2] 3; seedy 1; shabby 1
manhandle°, abuse 2; beat 1; knock 3d; mishandle 1; mistreat; molest 2; punish 3
manhandling, mistreatment; punishment 3
man-hater°, misanthrope
man-hating, misanthropic
manhood°, majority 2
mania°, bug 3; craze; enthusiasm 1; fad; fanaticism 2; fashion 2; fetish 2; fixation; furor 2; hunger 2; insanity 1; love 5b; lunacy 1; madness 1; obsession; passion 3; rave 4; thing 4
maniac°, fanatic; fiend 2; madman; maniacal 1; psychotic 2; zealot
maniac(al), berserk; fanatical; homicidal; insane 1; mad 1; rabid 1; raving 1; wild 5
manic, insane 1; maniacal 1; raving 1
manifest°, actual 1; apparent 1; bare 2; barefaced 1; clear 8; demonstrable; demonstrate 1; develop 3; display 1; distinct 1; embody 1; evidence 4; evident; exhibit 1; explicit 1; express 2; glaring 1; graphic 1; gross 4; indicate 2; naked 3; noticeable 1; observable; obvious; open 13; out 4; outward 1; overt; patent 2; perceptible; personify 1; physical; plain 2; public 4; register 4, 8; self-evident; sensible 2; show 1; tangible; transparent 2; unquestionable; visible 2
manifest, be, appear 4
manifest, become, appear 1; materialize 1, 2; occur 1
manifest, make, show 1
manifestation°, aspect 4; demonstration 1; display 4; embodiment 1; emergence; evidence 3; expression 2; flash 2; form 1, 3; occurrence 2; presence 2; representation 1; symptom; version 1
manifestly°, apparently 1; clearly 2; evidently 1; notably 1; ostensibly; preeminently; well[1] 7
manifestness, presence 2
manifesto, declaration 2; decree 1; proclamation; pronouncement 1
manifold°, different 3; divers; miscellaneous
manikin, dummy 1
man in the street, hoi polloi; people 3; plebeian 3; public 8
manipulable, manageable; sheepish 1; yielding 2
manipulate°, arrange 2; coax; drive 2; ease 7; engineer 5; exploit 2; feel 1; field 6; finesse 4; finger 10; fix 11; get 9, 15; handle 4, 7; influence 3; juggle; manage 2; maneuver 3, 4; massage 2, 3; operate 2; oversee; play 8b; process 4; rig 2; run 10; stage 5; string 6; tease 2; use 2; wangle; weight 6; work 10, 14, 15
manipulating, machination
manipulation, design 8; direction 1; machination; management 1; massage 1; operation 2; politics 2; usage 2
manipulation (of), treatment 1
manipulative, calculating; loaded 3; shrewd
manipulator, operator 1, 3; tactician
mankind, humanity 1; man 2; society 2; world 2
mankind-hater, misanthrope
manlike, human 1
manliness, bottle 2; machismo, manhood 1, spirit 5
manly°, gallant 1; hardy 2; heroic 1; intrepid; macho; male; stalwart 3; tough 3
man-made, artificial 1; imitation 5; synthetic
mannequin, dummy 1; model 5
manner°, action 8; air 3; appearance 2; approach 7; bearing 1; carriage 2; execution 4; fashion 3; form 3, 4; habit 2; kind[2] 2; look 14; mean[3] 4a; method 1; mode[1] 1; school 2; sort 1, 2; style 1; taste 5; technique 1; tone 3; touch 18; turn 37; vein 5; way 1, 2
manner, in another, otherwise 2
manner, in any, however 2
manner, to whatever, however 2
manner born, to the, elegant 1
manner of, in the, like[1] 5
manner of speaking, in a, speak 6
mannered°, affected 1; camp[2] 1; forced; genteel 1; stiff 7; stilted; theatrical 2; unnatural 2
mannerism°, eccentricity 2; oddity 3; trick 6
mannerliness, decorum 1; grace 2; propriety 2
mannerly, debonair 1; decent 2; decorous; gallant 2; genteel 2; graceful 2; ladylike; orderly 2; polite 1; refined 1; respectful; well-bred
manner of means, by no, mean[3] 5
manner of speaking, parlance; phrase 3; pronunciation; talk 19
manner of talking, parlance
manners°, action 8; behavior; breeding 2; conduct 1; etiquette; form 6; grace 2; manner 3; protocol 1
manners, bad, incivility
mannikin, dummy 1
manor, estate 1
manor (house), castle 2; palace; residence 3
man's, male
manservant, man 3; servant 1
mansion, castle 2; estate 1; palace; residence 3; seat 5
manslaughter, foul play; killing 1; murder 1; slaughter 2
mantel, ledge
mantelpiece, ledge
mantic, ominous 3
mantle°, cape[2]; cloak 1, 2; pall[1] 1; shroud 3; spread 7; suffuse; wrap 4
man to man, shoulder 6
mantra, chant 1
mantrap, siren 2; temptress
manual°, companion 2; guide 8; schoolbook; text 4; vade mecum
manual worker, laborer
manufacture°, assemble 3; assembly 3; create 2; fabricate 1, 2, 3; fabrication 1, 2; fake 1; fashion 5; forge 1; form 7; generate 1; industry 1; machine 4; make 1; output 3; prepare 5; produce 1; production 1; run 29b; turn 20a
manufactured, artificial 1; cut 29c; factitious; false 3, 4; synthetic
manufacturer°, brand 1; producer 1

maturity, come to, ripen
maturity, reach *or* **attain,** grow 6
matutinal, morning 2
maudlin°, drunk 1; gooey 2; mushy 2; romantic 3; sentimental 2; sickly 3; sloppy 3; stinking 3; sweet 7; tender[1] 7
maudlinism, sentimentality
maul, batter 3; knock 3d; lambaste 1; lather 4; manhandle; mishandle 1; mistreat; pound[1] 1; punish 3
mauling, mistreatment; punishment 3; thrashing 1
maunder, prattle 1; ramble 2
maunder (on), spout 2
maundering, prattle 2; rambling 1
mausoleum, crypt; grave[1]; monument 1; sepulcher; tomb
maven, expert 1; master 2; professional 3; virtuoso 1
maverick, independent 9; individualist; misfit; nonconformist 1, 2; oddity 2
maw, mother 1; mouth 1
mawkish, gooey 2; gushy; maudlin; mushy 2; romantic 3; sentimental 2; sickly 3; sloppy 3; tender[1] 7
mawkishness, sentimentality
max, to the, blast 4; world 8
maxim°, belief 3; byword; epigram 2; lesson 3; moral 3; motto; phrase 2; precept 2; principle 1; proverb; saw; tenet; truism
maxim, old, cliché
maximal, exhaustive; maximum 3; outside 6
maximally, blast 4
maximize°, profit 4
maxim(s), code 3
maximum°, climax 1; exhaustive; extreme 2, 7, 8; extremity 3; full 4, 16; height 2; high 15; outside 3, 6; pinnacle; supreme 2; terminal 1; ultimate 2
maximum, at the, blast 4
maxisingle, record 7
maybe°, chance 5b; perhaps; possibly 1
mayhap, maybe; perhaps; possibly 1
mayhem°, destruction 1; havoc 2; injury; rumpus
mayhem, commit, destroy 1
Maytime, spring 9
maze°, network 2; snarl[2] 4; tangle 2
mazelike, labyrinthine; tortuous 1
mazuma, cash 1; money 1
mazy, labyrinthine; meandering; tortuous 1
MC, announcer; host[1] 2; moderator
M.D., doctor 1; physician; head 2
mead, field 1; meadow; plain 6
meadow°, field 1; pasture
meadowland, meadow; pasture; plain 6
meager°, bare 5; diffuse 1; feeble 2; frugal 3; gaunt 1; inadequate 1; insubstantial 1; insufficient; lean[1] 1, 2; limited 2; little 3; measly 1; minute[2] 2; narrow 2; pale[1] 3; paltry; pathetic 2; poor 2; rawboned; remote 6; scanty 1, 2; scarce; short 9; skinny; slender 2; small 4; spare 3, 5; sparse 1, 2; tenuous 2; thin 1
meagerness, poverty 2
meal°, spread 11
meal, sumptuous, banquet 1
meal of, make a, meal 2
meals, board 3; fare 3
mealy-mouthed°, obsequious; pious 2; sanctimonious; self-righteous
mean°, avaricious; average 1; base[2] 1, 2, 3; close 18; coldhearted; common 4; contemptible; denote 2; design 4; designate 3; despicable; destine 2; dirty 3, 6; disgraceful 1; dishonorable 3; drive 6; economical 2; filthy 2; frugal 2; get 31b; grasping; greedy 3; groveling; hateful 2; humble 3; hurtful 2; imply 2; import 2; intend; involve 2; little 6; lousy 1; low[1] 12; mangy; medium 3; menial 1; middle 1; miserable 4; miserly; narrow 7; nasty 4; near 6; norm 1; obscure 4; paltry; penurious 1; petty 2; pitiful 2; plebeian 1; poor 5; propose 2; purpose 5; refer 4; reprobate 1; rotten 4; ruthless; say 9; scurvy; selfish 2; servile; severe 1; shabby 2, 4; shameful; signify 1; simple 5; sleazy 2; small-minded; sordid 1, 3; sparing 1; spell[3] 1; stand 7a; standard 2; stinking 2; symbolize; tight 5; ugly 4; unkind; unmerciful; venomous 2; vicious 2; vile 1; violent 1; wicked 4; wretched 4
meander°, drift 2; mill 6; ramble 2; roam; run 4; saunter; snake 4; sprawl 1; straggle; stray 1; stroll 1, 2; turn 11, 27; twist 4, 7; wander 1, 2; wind[2] 1
meandering°, diffuse 2; digression 2; discursive; erratic 3; indirect 1; serpentine 2
meandering(s), meander 2
meanders°, meander 2
meaning°, definition 2; drift 4; effect 3; force 5; impact 2; implication 3; import 3; matter 4; message 3; point 7; rhyme 2; sense 4; significance 1; spirit 6; substance 3; tenor
meaning of, get the, see 2
meaningful°, eloquent 2; expressive 1; knowing 1; ominous 3; pregnant 2; significant 2
meaningful to, be, speak 12b
meaningfully, notably 2
meaningfulness, matter 4
meaningless°, absurd 2; empty 4; hollow 5; idle 4; nonsensical; pointless; purposeless; senseless 3; superficial 3
meaninglessness, absurdity 1; emptiness 2; triviality 1
mean-minded, narrow-minded
meanness, avarice; greed 2; servility; severity 1
means°, agency; agent 2; asset 1; capability; capital 3; channel 4; expedient 3; form 4; formula; fortune 1; fund 2; instrument 2; manner 1; material 4; mean[3] 4; measure 6; mechanism 3; medium 5; method 1; money 2; path 3; property 2; purse 2; recipe 2; riches; road 1; substance 5; tool 2; vehicle 2; way 1
means, by all, course 6; mean[3] 2
means, by any, possibly 2
means, by no, hardly; ill 13; mean[3] 5; scarcely 2; shot 11
means, by no manner of, mean[3] 5
means, have the, afford 1
means of, by, by 3; mean[3] 3; through 1; way 10a
means or foul, by fair, hook 3; somehow
mean something, matter 6
mean-spirited, mean[2] 1; narrow-minded; servile; shabby 4; wretched 4
mean-spiritedness, servility
meant, destined 1; intentional; supposed 2
meantime°, interval 2; meanwhile 1, 2
meantime, in *or* **for the,** meanwhile 2; temporarily 1
meanwhile°, interval 2; time 16
meanwhile, in *or* **for the,** meanwhile 2; temporarily 1
measly°, meager 1; mean[2] 1; pathetic 2; petty 1; scanty 1; small 4
measurable, slow 2
measure°, act 4; allotment; beat 11; cadence; caliber 3; content[1] 1; degree 2; depth 1; dose 1; draft 4; expedient 3; fathom; gauge 1, 3, 4; index 4; indication 2; judge 5; law 1; length 1; mark 3; melody 1; norm 2; pace 4; pattern 1; plumb 5; portion 2; proceeding 1; proof 2; quantity; rate[1] 1, 6, 8; ration 1; register 8; rhythm; size 2, 3; standard 1; step 4; survey 1, 3; tempo; time 7; touchstone; volume 2; width; yardstick
measure, in some, moderately; part 9; partially
measure, to a certain, rather 1; slightly
measure, to a slight *or* **minor,** slightly
measure, to some, rather 1; somewhat
measure against, compare 3; touch 7
measure for measure, give, score 9
measure of, take the, size 3; sum 2b; survey 1
measure off *or* **out,** divide 2; measure 13, 14; mete; ration 3; section 4; space 5
measure (up) (to), grade 5; match 6; measure 15; muster 3; rival 2; stack 7b; sum 2b; touch 7
measured°, deliberate 2; even 3; rhythmic
measureless, boundless; immeasurable; inestimable 1, 2; inexhaustible 1; infinite; universal 2; unlimited 2; untold 1; vast
measurement°, length 1; measure 4; survey 3
measurement(s), proportion 4; size 1
measures°, measure 6; preparation 1a; provision 3
measuring, measurement 1
meat°, flesh 1; gist; kernel 1; nub 2; pith 1; point 5; spirit 6; substance 2; sustenance 1
meathead, dolt
meaty, forceful 2
mechanic, engineer 3; operative 2
mechanical°, animated 2; automatic 1; impersonal 2; instinctive 2; involuntary; mobile 2; monotonous; perfunctory 1; routine 4; spontaneous 2; technical 2; tedious
mechanicalism, mechanism 4
mechanically, idly 2; rote 2b
mechanical man *or* **monster,** robot 1, 2
mechanism°, agency; contraption; control 6; device 1, 2; engine; gadget; gear 1; implement 1; instrument 1, 2; machine 1; medium 5; movement 4; thing 5; tool 1; vehicle 2; works 2
mechanistic, mechanical 3
medal, decoration 2; plaque 2; trophy 1
medallion, pendant; plaque 1, 2

mess° (*cont.*) foul 16b; hash 3; mishandle 2; muck 4; queer 5; ruin 8; screw 7b
mess (up), blow[1] 3; botch; bugger 3; bungle; butcher 4; clutter 3; complicate 1; confuse 2; err 1; fluff 4; foul 16b; hash 3; mess 5; mishandle 2; muck 4; muddle 2; ruffle 4; ruin 8; rumple; shuffle 1; spoil 1; tangle 3; tousle; upset 3, 5; snarl[2] 1
message°, bulletin; dedication 2; dispatch 6; greeting 2; information; lesson 3; letter 2; meaning 1; memorandum; missive; moral 3; news 2; note 2; oracle 2; sense 4; significance 1; spirit 6; word 2
message, get the, tumble 3
messed up, disorderly 1; disturbed 2; unkempt; upset 9; wild 8
messenger°, ambassador; go-between; orderly 3; runner 2
messenger (boy *or* **girl),** messenger; page[2] 1; runner 2
Messiah°, the, savior 2
messy, bedraggled; confused 1, 3; disorderly 1; disreputable 2; dowdy; loose 4; ragged 4; slipshod; sloppy 1; topsy-turvy 2; unkempt; untidy
metage, measurement 1
metagrobolized, confused 2
metallic, tinny 2
metamorphose, change 6, 8; convert 1; make 9; transform; translate 2
metamorphose into, resolve 4; turn 16a
metamorphosing, mutation 1
metamorphosis, change 3; difference 3; mutation 1; revolution 2; transformation; transition 1; translation 2
metaphor°, image 5; prototype 2; symbol
metaphor, strained *or* **far-fetched,** conceit 3
metaphoric(al), metaphoric; symbolic
metaphorically (speaking), speak 6
metaphrase, translate 1, translation 1
metaphysic(al), abstract 1; psychic 1; supernatural
metaphysical philosophy, philosophy 1
metaphysics, philosophy 1
metastasis, transition 1
metastasize, spread 6
metathesize, transpose
met by, be, meet[1] 6
mete° (out), administer 3; deal 1; dispense 1; distribute 1; divide 2; dole 3; give 15b; hand 17; measure 14; mete; parcel 4; pass 20b; present[2] 6; ration 3
metempsychosis, rebirth
meteoric°, precipitate 3
meter, cadence; indicator; measure 9, 11; rhythm; tempo; time 7
method°, approach 7; form 4; formula; manner 1; mean[3] 4a; measure 3, 6; mechanism 3; medium 5; mode[1] 1; path 3; plan 1; policy; procedure; process 1; recipe 2; road 1; routine 1; scheme 1; science 2; system 2; tack 3; technique 1; touch 18; way 1; wrinkle[2]
methodical°, deliberate 2, 3; even 3; formal 1; hard 6; orderly 1; organic 3; regular 1, 6; scientific; systematic; thorough 2; tidy 2
methodically, exactly 1; thoroughly 2
methodology, method 1; mode[1] 1; procedure; system 2
meticulous°, accurate 2; careful 2; conscientious 2; diligent; elaborate 1; exact 2; faithful 3; fastidious; finicky 1; methodical; narrow 3; nice 3; painful 3; particular 4; pedantic 2; perfectionist 2; precise 2, 3; religious 2; rigid 3; scientific; scrupulous 1; squeamish 1; strict 1; thorough 2
meticulously, narrowly 2; precisely 2; thoroughly 2
meticulousness, care 2; precision 2; rigor 2
métier, calling; career 1; craft 3; employment 1; field 4; job 1; mission 2; occupation 1; profession 1; specialty 1; trade 2; vocation; work 2
meting out, partition 2
metonymic, metaphoric
metonymous, metaphoric
metonymy, metaphor
metrical, even 3; poetic 1
metrics, poetry
metrist, poet
metro, underground 4
metropolis°, city; municipality; town
metropolitan, municipal
metropolitan area, metropolis
mettle, backbone 3; bottle 2; daring 1; enterprise 2; fight 9; fortitude; grit; gumption 2; gut 3a; heart 2; nerve 1; pluck 1; prowess 2; self-control 1; spirit 2; spunk; stamina; vigor
mettled, racy 1
mettlesome, audacious 1; daring 2; enterprising; gallant 1; gritty 2; indomitable; racy 1; spirited; stalwart 3
mettlesomeness, vigor
mewl, cry 2; moan 3; snivel; sob; weep 1
Mexican fox-trot *or* **two-step** *or* **toothache,** run 50
Mexican standoff, deadlock 1; stalemate
mezzo-rilievo, relief 3
MIA(s), casualty 2a
miasma, fume 3; poison 1; reek 3
miasmal, rancid; rank[2] 4; smelly; stinking 1
miasmatic(al), rancid; rank[2] 4; smelly; stinking 1
miasmic, bad 2; poisonous 1; rancid; rank[2] 4; smelly; stinking 1; strong 3
Mickey (Finn), sedative 1
Mickey (Finn), slip (someone) a, drug 4; spike 4
Mickey Mouse, paltry
mickey out of, take the, deride; laugh 2a; mock 1; poke 4; ridicule 2; send 9a; tease 1
micro, diminutive; miniature; minute[2] 1; tiny
microbe°, bug 2; germ 1
microcosm, universe 2
microorganism, germ 1; microbe
microphone, bug 5
microscope, glass 6
microscopic, diminutive; imperceptible 2; little 1; miniature; minute[2] 1; tiny; wee 1
microwave-able, prepared 4
microwave-ready, prepared 4
microzoon, microbe
micturate, urinate
mid, amid; among 1; medium 1; middle 1
midday°, noon
midden, mound 2
middle°, center 1; central 1; core 1; dead 21; heart 3; inside 1; interior 6; intermediate 1; mean[3] 1, 6; medium 1, 3; midst; moderate 3; thick 10
middle of, in the, through 2
middle class, public 8; bourgeois 1
middleman, broker; factor 2; go-between; intermediary; mediator; negotiator
middle-of-the-road, conservative 2; moderate 2
middle-of-the-roader, conservative 3; moderate 4
middle school, school 1
middling, adequate 2; average 4; decent 3; fair[1] 4; indifferent 3; mean[3] 6; mediocre; moderate 3; OK 3; par 5a; passable 1; so-so; tolerable 2; undistinguished
midget, diminutive; little 1; miniature; puny 3; runt; short 1; small 1; tiny; wee 1
midpoint, center 1; medium 3; middle 2; midst
midriff, middle 3
midsection, middle 3
mid-sized, medium 1
midst°, bosom 2; dead 21; middle 2; thick 10
midway, intermediate 1; middle 1
midwife°, world 5b
mien, appearance 2; aspect 2; bearing 1; carriage 2; expression 3; face 2; front 4; guise 1; look 14; manner 2; outside 2; pan 2; presence 3; semblance 1
miff, displease; disturb 1; infuriate; irk; offend 1
miffed, indignant
might°, brawn; energy; force 1; main 6; power 4; sinew 2; stamina; strength 1; violence 1
might and main, with, vigorously
mightily, hard 13
mightiness, might 2; power 4; strength 1; violence 1
mighty°, brawny; forceful 1, 2; formidable 3; important 3; impregnable; massive; mountainous 2; potent 1; powerful 1; sinewy; solid 8; stalwart 1; strong 1; towering 2
mignon(ne), petite; precious 4; pretty 1; small 1; undersized
migraine, headache 1
migrant°, arrival 2; immigrant; vagabond 1, 2
migrate°, emigrate; range 8; stake[1] 2
migration, movement 1
migrator, migrant 1
migratory, migrant 2; rambling 3; traveling
mild°, benign 2, 3; bland 1; calm 3; easy 2, 3; gentle 1; harmless; humble 2; inoffensive; light[2] 4; meek 1; merciful; peaceable 2; quiet 2; smooth 5; soft 3; tame 2, 3; temperate 1; tender[1] 6
mildew, mold[2]; spoil 4
mildewed, moldy; musty 1; rotten 1
mildewy, musty 1; stuffy 1
mildly, quietly 3
mildness, humility
mileage, distance 1; measurement 2
miler, runner 1
milestone, landmark 2
milieu°, climate 2; environment; medium 4; scene 1; setting; site 1; universe 2
militancy, fight 9
militant°, belligerent 1, 3; hostile 3; martial 1; radical 3, 4; soldier 2; zealot
militantism, zealotry
militaristic, warlike
military°, martial 2; service 8
military science, tactic 2
militate° against, militate 1

mirage (*cont.*) illusion 2; phantom 2; vision 4
mire°, dirt 1; muck 2; mud; ooze 1; sludge; soil[1] 3
mirror°, echo 4; follow 2; glass 2; mimic 1; reflect 1; represent 3; reverse 1
mirror image, echo 2; symmetrical
mirrorlike, glassy 1; smooth 2; symmetrical
mirth°, cheer 2; festivity 1; frolic 1; fun 1; gaiety 1; hilarity; merriment; revelry
mirthful, comic 1; gleeful; hilarious; merry 1; ridiculous; sunny 2
mirthfulness, gaiety 1; glee; merriment; mirth
mirthless, solemn 1
miry, muddy 1; mushy 1
misaddress, misdirect
misadventure, accident 1; calamity 1; casualty 1; catastrophe 2; misfortune 2; reverse 8
misadvise, misdirect; misinform
misallied, incongruous; mismatched
misanthrope°, man-hater
misanthropic°, unsocial
misanthropist, man-hater; misanthrope
misapplication, abuse 4; anachronism; embezzlement; misuse 1; waste 5
misapplied, abusive 2
misapply, abuse 1; embezzle; misappropriate 2; misuse 4; pervert 1
misappreciate, miscalculate
misapprehend, misconceive; misinterpret; miss[1] 3; mistake 3; misunderstand
misapprehension, illusion 1; misconception; mistake 1; misunderstanding 1
misappropriate°, embezzle; misuse 4; pilfer; steal 1; take 3
misappropriation, abuse 4; embezzlement; misuse 1
misarticulate, slur 2
misbegotten, illegitimate 2
misbehave°, carry 11c; cut 17b; err 2; play 17b; transgress 1
misbehaved, discourteous
misbehaving, bad 10
misbehavior°, fault 4; incivility; mischief 1; transgression
misbelief, delusion 2
miscalculate°, err 1; misunderstand; slip[1] 3; underestimate
miscalculation, fallacy; misconception; mistake 1; misunderstanding 1
miscarriage°, defeat 4; failure 2; misfire 2; mockery 2
miscarry°, boomerang; fail 1; fall 20; fizzle 2; founder[2] 2; grief 3; misfire 1; wrong 8b
miscellanea, rummage 2; sundries
miscellaneous°, confused 3; divers; diverse; general 3; manifold; odd 2; promiscuous 3; sundry; varied 1; various 1
miscellany°, assortment 2; hodgepodge; mass 1; medley; mess 2; mixture 1; pastiche; potpourri; rummage 2; sundries; variety 1
mischance, accident 1; calamity 1; casualty 1; catastrophe 2; misfortune 2
mischief°, caper 2; devilry 1; evil 7; hanky-panky; harm 1; hocus-pocus 1; ill 8; injury; lark 1; nonsense 2; prank; trick 2
mischief-maker, handful 2; imp; miscreant 1; monkey 3; rascal; troublemaker
mischievous°, arch 3; bad 10; devilish; disobedient 1; elfin 1; evil 3; fatal 2; hurtful 1; miscreant 2; naughty 1; playful 1; sly 2; wicked 5
mischievous, be, misbehave
mischievously, badly 4
mischievousness, devilry 1; mischief 1
misconceive°, misinterpret; misunderstand
misconception°, delusion 2; fallacy; illusion 1; mistake 1; misunderstanding 1
misconduct, error 2; fault 4; guilt 1; misbehavior; misdeed; mishandle 2
misconstrual, misconception; twist 9
misconstruction, misconception; misstatement 1; misunderstanding 1; stick[2] 4; twist 9
misconstrue, garble 1; misconceive; misinterpret; miss[1] 3; mistake 3; misunderstand; pervert 1; twist 2
miscount, miscalculate
miscreant°, criminal 3; delinquent 1; felon; offender; reprobate 2; rogue 1; sinner; transgressor; villain
miscreation, monster 2
misdate, anachronism
misdating, anachronism
misdeed°, crime; fault 4; misdeed; offense 1; peccadillo; sin 1; transgression
misdemeanor, crime; fault 4; misbehavior; misdeed; offense 1; sin 1; transgression
misdesignated, so-called 2
misdirect°, lead 7; misinform; mislead; pervert 1; throw 6c
misdirected, misguided
misdirection, perversion 1
misdoing, misdeed
mise en scène, environment; production 3; scene 2; set 28; setting
misemploy, abuse 1; misappropriate 2; misuse 4
misemployment, abuse 4; misuse 1; waste 5
miser°, stiff 12
miserable°, bad 1; base[2] 2; bitter 3; brokenhearted; dejected; deplorable 1; desolate 3; despicable; despondent; dismal; doleful; downhearted; dreary 1; filthy 2; forlorn 1; gloomy 2; heartbroken; heavy 6; hopeless 3; ill 5; inconsolable; joyless 1; lamentable; lousy 1, 2; low[1] 6, 8; mangy; mean[2] 3; measly 1; melancholy 1; oppressive 1; paltry; par 5a; pathetic 1; piteous; poignant 1; poor 6; rotten 4; sad 1, 3; scurvy; sick 3; sinking 2; sordid 4; sorrowful 1; sorry 2; terrible 3; tragic; unfortunate 2; unhappy 1; vile 1; weak 4; woebegone; wretched 1, 2, 3
miserableness, despair 1; melancholy 2
miserably, sadly 2
miserliness, avarice; greed 2; thrift
miserly°, avaricious; cheap 4; close 18; economical 2; frugal 2; grasping; greedy 3; mean[2] 1; measly 1; narrow 7; near 6; penurious 1; petty 2; selfish 2; small 4; sparing 1; thrifty; tight 5
misery°, affliction 1; agony; anguish 1; calamity 2; care 1; desolation 2; despair 1; desperation 2; distress 1; evil 7; gloom 2; grief 1; hardship; hell 2, 3; hurt 6; ill 8; melancholy 2; mourning 2; ordeal; pain 2; privation; prostration 3; rack 2; sadness; scourge 1; sorrow 1; suffering; torment 3; trial 4; woe
misery, put out of (one's), kill 1; murder 3; put 24
misestimate, miscalculate
misevaluate, miscalculate
misfile, mislay
misfire°, fail 1; failure 3; fizzle 2, 3; miscarry
misfit°, loser; oddity 2
misfortune°, accident 1; affliction 1; blight 2; calamity 2; catastrophe 2; curse 2; disaster; distress 2; evil 7; grief 2; hardship; harm 1; ill 8; mischief 2; misery 3; necessity 4; ordeal; picnic 3; pity 2; reverse 8; scourge 1; sorrow 2; tragedy; trial 4; undoing 2; woe
misgiving(s), compunction 2; distrust 2; doubt 3; dread 2; fear 3, 4; foreboding 1; mistrust 2; pang 2; qualm; scruple 1; suspicion 1; worry 4
misgivings (about), have, doubt 1; scruple 2
misguide, betray 3; lead 7; misdirect; misinform; mislead; throw 6c; trick 8
misguided°, foolish 1; ill-advised 1; inept 2; mistaken 2; off 3; wrong 4
mishandle°, foul 16b; fumble 2; hash 3; ill-treat; screw 7b
mishap°, accident 1; calamity 1; casualty 1; catastrophe 2; disaster; hitch 4; misfortune 2; reverse 8
mishegaas, rigmarole
mishmash°, assortment 2; clutter 1; disorder 1; hash 1; hodgepodge; medley; mess 1; mixture 1; mix-up; muddle 4; pastiche; patchwork; potpourri; stew 1; tangle 2; welter 1
misidentify, mix 6
misidentify as, mistake 4
misinform°, mislead; trick 8
misinformed, mistaken 2
misintelligence, misinformation
misinterpret°, amiss 4; misconceive; miss[1] 3; mistake 3; misunderstand; twist 2
misinterpretation, misstatement 1; misunderstanding 1; stick[2] 4; twist 9
misjudge, miscalculate; misconceive; misinterpret; mistake 3; misunderstand; underestimate
misjudgment, fallacy; misconception; mistake 1; misunderstanding 1
mislaid, lost 1
mislay°, leave[1] 3; lose 1; track 5
mislead°, beguile 1; betray 3; bluff[1] 1; deceive; disappoint 2; double-cross; dupe 3; equivocate; flannel 2; fool 4; hoodwink; lead 7; misinform; seduce 1; throw 6c; trick 8
mislead (someone), put 26
misleading, ambiguous 1; confused 1; deceitful; deceptive 1; devious 1; equivocal 1; erroneous; evasive; false 2; illusory; plausible 2; sophistic; specious; tortuous 2; untrue 2
misled, misguided; off 3
mislocate, mislay
mismanage, blow[1] 3; botch; bungle; foul 16b; hash 3; mishandle 2; muddle 2; screw 7b
mismanagement, miscarriage
mismarriage, misalliance
mismatch, misalliance
mismatched°, incompatible
mismated, mismatched

mock° (*cont.*) lampoon 2; leg 8; mimic 3, 5; parody 3; phony 1; poke 4; put 22e; quasi 2; rag[2]; ridicule 2; satirize; scoff; sham 2; snap 6; sneer 2; snicker 1; spurious; synthetic; take 34b; taunt 1; thumb 8; twit[1]; unreal 3

mock (at), laugh 2a; scorn 4

mock up, rough 15

mock attack, feint

mockery°, burlesque 1; charade; derision; gibe 2; imitation 2; joke 3; parody 1, 2; ridicule 1; satire 1; scorn 2; sneer 3; sport 2; takeoff 2

mockery of, make a, laugh 2a

mock-heroic, burlesque 4

mocking, derisory; disdainful; irreverent 2; ridicule 1; sarcastic; sardonic; satirical; scornful

mock-pathetic, burlesque 4

mock-up, dummy 2; model 1; rough 14

mod, modern; new 2

mode°, execution 4; fashion 1, 3; form 4; habit 1; manner 1; mean[3] 4a; medium 5; method 1; practice 1; procedure; rage 3; style 1, 2; taste 5; technique 1; trend 2; turn 37; vein 5; vogue 1; way 1

model°, carve 1; cast 4; classic 1, 3; classical 1; complete 4; design 5; dummy 1; epitome 1; example 2; exemplary 2; exemplify 2; fashion 5; form 4; gauge 3; guide 6; ideal 1, 4; image 4; last[3]; lead 13; lesson 3; likeness 2; mold[1] 4; nonpareil 1; norm 2; optimum 1; original 5; paragon; pattern 1; perfection 3; picture 4; pose 1; precedent; principle 1; prototype 1, 2; queen 2; quintessence; representation 1, 4; sculpture 2; shape 6; specimen; standard 1; statue; template; type 3; version 1

model after *or* **on,** model 9; pattern 7

modeled after *or* **on, be,** follow 2

Model T, rattletrap

moderate°, alloy 3; chair 1; chasten 2; conservative 2, 3; cool 10; dampen 2; deaden 2; decent 3; easy 5; fresh 7; frugal 1; gentle 1, 3; gradual; judge 6; let[1] 9; liberalize 2; light[2] 10; lower[1] 5; mild 2; mitigate; modest 3; modify 2; modulate; officiate; philosophical 2; phlegmatic 2; qualify 3; quell 2; reasonable 3; reduce 4; relax 1, 2; respectable 2; slack 3b; slow 2; soft 3; soften 2; subdue 3; subside 2; sweeten 2; tame 6; temper 5; temperate 1, 2; tone 5; umpire 2; weaken 1

moderate(d), subdued 1

moderately°, fairly 1; partially; pretty 3; quite 2; rather 1; somewhat

moderating, extenuating

moderation, letup; measure 8; relaxation 2; temperance 1

moderation, in, moderately

moderator°, chair 3; judge 2; mediator; negotiator; umpire 1

modern°, contemporary 2; date 5; fresh 2; latest 2; minute[1] 3; new 2; now 5; recent; streamlined 2; swinging; trendy 1

modernistic, streamlined 2

modernization, innovation 2

modernize°, improve 1; renew 1; renovate

modernized, streamlined 2

modest°, bashful 2; coy; decent 2, 5; delicate 5; homely 1; humble 1; inconspicuous; mean[2] 2; meek 1; mild 1; moderate 1, 3; ordinary 2; poor 5; pure 3; respectable 3; retiring; self-conscious; severe 6; shamefaced 1; shy 1; simple 2; small 4; so-so; spare 5; timid; unobtrusive

modestly, quietly 4; severely 6; simply 4

modesty, humility; propriety 2; purity 2; severity 6; shame 4; simplicity 3

modicum°, shade 3; speck

modifiable, changeable 2; flexible 2

modification, accommodation 1; adaptation 2; alteration; change 3; difference 3; flux; mutation 1; qualification 2; reform 3; regulation 1; revision; transformation; transition 1; variant 1; variation 1

modified, prepared 2; qualified 2

modify°, accommodate 1; adapt 2; adjust 2; affect[1] 3; alloy 3; alter; change 6; condition 5; convert 1; correct 4; differentiate 2; doctor 4; edit 1; fake 1; fit[1] 7; impact 4; influence 3; liberalize 2; make 9; modulate; prepare 7; process 3; proportion 5; qualify 3; reduce 4; regulate 1; relax 2; reverse 4; revise 1; shape 9; square 12; tailor 2; temper 5; tone 5; transform; turn 3; vary 1; water 7

modifying, adaptation 1; mutation 1; variation 1

modish, chic 1; contemporary 2; current 3; dashing 2; elegant 2; fashionable; modern; new 2; nifty 1; smart 3; snappy 2; sporty; stylish; swell 7

modishness, chic 2

modiste, dressmaker; tailor 1

modulate°, change 6; lower[1] 5; modify 2; proportion 5; qualify 3; regulate 1; relax 2; tone 5

modulated, measured 4

modulating, variation 1

modulation, change 3; intonation; pronunciation; regulation 1; tone 2; variation 1

module, terminal 4; unit

modus loquendi, phrase 3

modus operandi, approach 7; formula; method 1; mode[1] 1; practice 1; procedure; recipe 2; system 2; technique 1; way 1

modus scribendi, phrase 3

modus vivendi, way 2

mofette, vent 1

mogul°, merchant 3; personage; tycoon

moil, drudgery; plod 2; slave 3; slavery 3; work 7

moist°, clammy 1, 2; damp 1; humid; juicy 1; lush 2; muggy; steamy 1; sultry 1; watery 3; wet 1

moisten, dampen 1; water 6

moistened, wet 1

moistness, damp 2

moisture, damp 2; wet 4

moisture-laden, moist 2

moistureless, dry 1

mold°, beat 5; cast 3, 4, 9; condition 6; decay 4; design 2; earth 2; fashion 5; fiber 3; figure 4; forge 1; form 1, 7; frame 7; generate 4; image 6; land 2; last[3]; make 1; model 2, 7; nature 4; pattern 4; perfection 3; prepare 5; produce 1; prototype 1; rot 1, 3; school 4; shape 6; sort 1, 9; stamp 6, 8; tailor 2; template; work 13

moldable, plastic 1; suggestible

molder, decay 2; decompose 2; die 2; disintegrate; go 31e; putrefy; rot 1, 2; spoil 4; stagnate; turn 5

moldering, moldy; putrid; rotten 1

molding, border 4; production 1

moldy°, bad 6; foul 2; hack[2] 4; musty 1; off 7; prosaic; putrid; rotten 1; stale 1; stuffy 1

mole, spy 1

molecule, crumb; grain 3; particle; scrap[1] 1; speck

molest°, annoy 2; assault 4; mishandle 1; mistreat; persecute 1; violate 3

molestation, assault 2; mistreatment; persecution 1; violation 3

moll, friend 3; girl 2; prostitute 1; squeeze 10; woman 3

mollification, euphemism

mollify, blunt 4; calm 5; cushion 2; deaden 2; disarm 2; ease 6; humor 4; hush 5; lull 3; melt 2; mitigate; moderate 5; quell 2; salve 3; silence 4; smooth 12; soften 1; still 9; sweeten 2; tame 6; temper 5; water 7

mollifying, mild 3; soothing 2

mollycoddle, baby 2; cater 2; coddle; humor 4; indulge 2; milksop; pamper; pet[1] 5; sissy; spoil 3; weakling

molt, shed[2] 4

molten, liquid 2

mom, mother 1

moment°, bit 3; crack 3; emphasis; flash 3; import 4; importance 1; instant 1, 2; interest 2; juncture 2; matter 4; minute[1] 1; note 8; occasion 1; opportunity; phase 2; pith 2; point 4; second[2]; significance 2; spurt 1; time 3; weight 3

moment, at the, now 1, 2

moment, at the last, finally 1

moment, at this (very), now 1

moment, for the, meanwhile 2; temporarily 2; time 16

moment, in a, once 6a; presently; rapidly 2; soon 1; suddenly 1

moment, in an unguarded, nap[1] 2; unawares 2

moment, of, serious 2

moment, of no, meaningless 2

moment, the, present[1] 4

momentarily, briefly 2; directly 2; instantaneously; promptly; soon 1; suddenly 1

momentary°, brief 1; fleeting; fugitive 3; hasty 3; imminent; meteoric; passing 1; short 7; temporary; transient

moment of truth, crunch 2; showdown

momentous°, banner 2; big 4; critical 2; crucial; eventful; fateful 1; great 4; high 6; historic; important 1; landmark 3; memorable; portentous 1; pressing; serious 2; signal 3; significant 1; solemn 3; special 2; weighty 2

momentousness, gravity 2; solemnity

moments, for a few, briefly 2

momentum°, impetus; power 8; propulsion

mommy, mother 1

monarch°, crown 3; king; lord 1; master 1; queen 1; sovereign 1

monarch, absolute, dictator

monarchism, monarchy 2

mumbo jumbo°, gab 2; gibberish; gobbledegook 1; hocus-pocus 2; lingo; nonsense 1; prattle 3; rigmarole
mummify, preserve 3
mummy, mother 1
mummy's boy, milksop; sissy
Mum's the word!, hush 1; mum 2
munch°, chew 1; crunch 1
Münchhausen, Baron von, liar
mundane, earthly 2; everyday 3; humdrum; material 7; matter-of-fact; outward 1; pedestrian 2; practical 2; secular; small 3; temporal 1; terrestrial 1; worldly 1
municipal°, local 2
municipality°, city; metropolis; town
munificence, bounty 1; charity 1; largess
munificent, bountiful 1; charitable 1; free 7; generous 1; large 2; liberal 1; open 14; philanthropic
munificently, freely 4
munitions, hardware 2
munitions dump, magazine 2
murder°, bloodshed; bump 5; butcher 3; cut 13b; destruction 2; dispatch 3, 7; eliminate 4; execute 3; execution 3; exterminate; foul play; kill 1; killing 1; massacre 1, 2; poison 4; polish 3b; purge 4; removal 2; remove 4; ride 5b; rout 2; rub 6; slaughter 2, 3; waste 4; whip 2; zap
murderer°, butcher 1; cutthroat 1; killer 1; thug
murderess, murderer
murdering, massacre 1; murder 2
murderous°, bloodthirsty; brutal 1; cutthroat 3; deadly 3; fatal 1; ferocious; fierce 1; grim 2; homicidal; sanguinary 1; savage 2; villainous 1; withering
murderousness, violence 2
murder (story), mystery 3
murk, dark 11; gloom 1; obscurity 1; shade 1
murkiness, dark 11; film 3; gloom 1; obscurity 1; opacity 1; shade 1
murky°, dark 2, 5; dim 1; dull 5; dusky 2; filmy 2; gloomy 1; gray 2; indistinct 1; misty; muddy 3; nebulous; obscure 1; opaque 1; overcast; somber 2; thick 3
murmur°, babble 3; breath 2; breathe 4; buzz 1, 5; groan 1, 3; gurgle 1; hum 1, 4; mumble; mutter 1; peep 3; sigh 3; undercurrent 2; whisper 1, 2, 3
murmuration, hum 4; murmur 1
murmured, low[1] 13
murmuring, gurgle 2; hum 4; murmur 1; querulous
murmurous, low[1] 13
muscle(s), brawn; flesh 1; might 1; power 4; pull 21; sinew 1; strength 1
muscular°, brawny; burly; hefty 2; husky 1; large 1; mighty 2; robust 1; sinewy; stalwart 1; strong 1; sturdy 1; tough 3; wiry
muse°, fantasize; meditate 1; wonder 3
muse (about *or* **on** *or* **over)**, brood 3; contemplate 2; mull; ponder; puzzle 2; reflect 3; remember 2; speculate 1; study 2; think 3, 5; turn 21a
mush, face 1; mug 2; pulp 2; tramp 1; trap 3
mushiness, sentimentality
mushroom, proliferate; shoot 4; spread 6; swell 1, 2
mushrooming, explosion 3; spread 8
mushy°, gooey 2; maudlin; romantic 3; sentimental 2; sickly 3; slimy 1; sloppy 3
music, air 4; piece 4; score 5; strain[1] 8
musical°, lyric 1; mellow 2; melodious; poetic 1; pretty 2; show 14; silver 4; sweet 3; tuneful
musicale, recital 1
musicality, melody 2
musician, player 4
musing, daydream 1; dreamy 2; inattentive; pensive; preoccupied 1; reflective; thought 1; thoughtful 2; wistful 2
muss (up), mess 5a; ruffle 4; rumple; tousle
mussed (up), unkempt; untidy; wild 8
mussy, untidy
must°, have 10; precondition; requirement 1
mustard, cut the, measure 15b
muster°, assemble 1; call 4; collect 2; enlist 1; gather 1; list[1] 1; mass 8; meet[1] 7; mobilize; raise 5; rally 1, 3; recruit 1; roll 13; round 19; scrape 6; summon 1; turnout 1
muster, pass, check 10c; go 35b; muster 3; pass 4
muster (up), find 7; summon 2
musts, string 7
musty°, close 12; foul 3; moldy; rank[2] 4; stale 1; stuffy 1
mutability, inconstancy; vicissitude 1
mutable, changeable 1; fickle; fluid 3; inconstant; protean; variable
mutant, freak 1; monster 2; mutation 2
mutate, change 6, 8; convert 1; make 9; transform; translate 2
mutation°, change 3; flux; monster 2; passage 5; transformation; transition 1; vicissitude 1
mutatis mutandis, parallel 1; respectively
mute°, dumb 1; hush 3; inarticulate 3; muffle 2; mum 1; noiseless; shut 6b; silence 3, 4; silent 1, 2, 3; speechless 1; suppress 3; taciturn; tame 6; tight-lipped; tone 5; tongue-tied
muted, delicate 6; dull 8; faint 1; inaudible; low[1] 13; mellow 3; noiseless; soft 4; subdued 1; weak 7
mutely, silently
muteness, silence 2
mutilate°, damage 4; deface; garble 1; hack[1] 1; hurt 4; maim; mangle; mar 1; murder 4; rout 2; tear 1
mutilated, disfigured
mutilation, damage 1
mutineer, rebel 3; revolutionary 3
muting, suppression
mutinous°, defiant; disobedient 2; disorderly 2; factious; inflammatory; insubordinate; rebellious 1; recalcitrant; revolutionary 1; seditious; unruly; wild 4
mutiny°, disobey; rebel 1; revolt 1, 2; revolution 1; rise 4; sedition; strike 15; turn 12; uprising
mutt, mongrel
mutter°, garble 2; mumble; murmur 3, 4; whisper 1
muttered, inarticulate 2
muttering, groan 4; murmur 2
mutual°, common 2; joint 4; reciprocal; related 1; united 1
mutuality, solidarity
mutual understanding, accord 2
muumuu, robe 1; shift 6
muzhik, peasant
muzzle, gag[1] 1, 3; keep 14a; mouth 1; silence 4
muzzy, confused 2; faint 1; groggy; silly 2; weak 6
My goodness!, indeed 3
myopic, nearsighted 1; partisan 3; shortsighted 1; small-minded; unreasonable 1
myriad(s), limitless; lot 5c; many 1, numberless; score 4; unlimited 2; untold 1
myrmidon, follower 2; henchman; robot 2
myself, person 2; personally 1
My stars!, indeed 3
mysterious°, ambiguous 2; cryptic 2; dark 6; deep 2; eerie; funny 2; incomprehensible; mystical; obscure 2; occult 1, 2; opaque 2; paradoxical; profound 1; queer 2; secret 2; supernatural; unaccountable 1; unidentified; unknown 2
mysteriously, secrecy 2; secretly
mystery°, enigma; mystique; obscurity 2; paradox; perplexity 3; puzzle 4; question 2; riddle[1]; romance 3, 4; secrecy 1; secret 3; story[1] 1
mystic, cryptic 1; magic 5; mystical; occult 2; supernatural
mystic(al), cryptic 1; dark 7; deep 2; mysterious 2; occult 1, 2
mystification, opacity 2
mystified, confused 2; daze 4; helpless 2; lost 3; sea 4
mystify°, bewilder; confuse 1; daze 2; distract 3; escape 4; flummox; fog 4; muddle 1; perplex; puzzle 1; stump 2
mystifying, confused 1; funny 2; incomprehensible; indefinite 2; inexplicable; intricate 2; mysterious 1; obscure 2; occult 2; opaque 2; perplexing; profound 1; puzzling; tough 5
myth°, legend 1; story[1] 1
mythic, ideal 5; imaginary; legendary 1; mythical 1, 2
mythic(al), fabulous 1; fancied; fanciful 2; fictional; fictitious 1; ideal 5; imaginary; incredible 1; legendary 2; nonexistent; romantic 1; unreal 2
mythological, heroic 5; imaginary; mythical 1
mythology°, lore 1
mythos, lore 1; myth 1; mythology
myths, lore 1; mythology
My word!, indeed 3

nab°, arrest 2; capture 2; catch 1; get 8; grab 1, 3; grasp 1; hook 6; lay[1] 11; pick 8h; pinch 4; pull 11b; run 27; seize 2; take 1
nabob, bigwig 1; mogul
nacreous, opalescent; pearly; silver 3; variegated
nada, zero 1
nadir, depth 5; extreme 8; minimum 1; zero 2
naff, awful 1; bad 1; indifferent 3; inferior 3
nag°, annoy 2; bitch 1; bother 1, 7; browbeat; carp; dun; gall[2] 4; get 31c; gnaw 3; gripe 1; harass; henpeck; hound; irritate; jade 1, 2; niggle; nuisance 2; pest; pester; plague 3; prod 3; push 4; ride 3; scold 2; shrew; torment 2; trouble 6
nag (at), nag[1] 1; pick 4a
nagging°, demanding 2; gripe 2; sneaking 1
nail°, claw 1; connect 3; convict 1; nab; pin 5;

pull 11b; spike 1; stick[1] 4; tack 1, 5; transfix 1
nail, on the, nail 6; precisely 1
nail down, nail 10; tie 5a
nail on the head, hit the, nail 5
nailbrush, brush[2] 1
naive°, artless 2; benighted; callow; childish; childlike; fond 2; frank 2; fresh 3; green 2; gullible; ignorant 3; immature 2; indiscreet; inexperienced; ingenuous 1; innocent 4; primitive 3; provincial 2; rustic 2; simple 3, 4; square 6; susceptible 2; transparent 3; trusting; unsophisticated 1; unsuspecting; young 2
naively, simply 3
naiveté°, candor 1; indiscretion 1; inexperience; provincialism 2; simplicity 3
naked°, bare 1; nude; raw 7; simple 2; smooth 4; unvarnished
naked, get, strip[2] 2
namby-pamby, colorless 2; cowardly; feeble 2; indecisive 1; limp[2] 3; milksop; sissy; soft 12; weak 3; weakling; wet 3
name°, appoint 2; assign 3; call 2; cast 8; celebrity 2; charge 11; christen 2; define 2; delegate 2; denomination 3; denote 2; designate 1, 2, 4; detail 6; diagnose; entitle 2; enumerate 1; face 4; fame; identify 1, 3; label 2, 5; make 23; mention 1; nominate; notable 3; personage; personality 2; pin 4b; pitch[1] 7; prominence 1; propose 3; put 18a; reckon 2; reel 2; refer 1; reputation 1; specify; stamp 3; star 2; style 7; success 2; tag 2, 4; term 1, 8; title 1, 6; word 3
name for oneself, make a, far 6a
name of, in the, behalf; part 10
name only, in, honorary; nominal 1
name to, lend one's, stand 7b
name brand, brand 1
nameless°, unheard-of 1; unidentified; unknown 1; unsung
namelessness, obscurity 3
namely°, like[1] 6
name names°, inform 2; name 6; sing 3
name tag, tag 1
naming, assignment 3; denomination 3; identification 1; mention 3; reference 1; specification 1
nance, deviant 3; homosexual 1
nancy (boy), deviant 3; homosexual 1; milksop
nanny, minder 1; servant 1
nap°, doze 1, 2; grain 4; nod 3; pile[2]; repose 2; rest[1] 1, 6; retire 3; sleep 1, 2
napery, linen
napping°, nap[1] 2; unprepared 1
napping, catch, surprise 2
narc, officer 2
narcissism, conceit 1; self-esteem 1; snobbery; vanity 1
narcissistic, conceited; egoistic; proud 2; vain 1
narcotic°, dope 2; drug 2; salve 2; sedative 1, 2
narcotic addict, user 2
narcotize, drug 4; dull 11
nark, informer; sneak 2; talebearer
nark (on), inform 2
narrate°, chronicle 2; describe 1; enumerate 1; recite 2; record 2; recount 1; relate 2; report 3; spin 2; tell[1] 1
narration°, account 6; description 2; history 1; recital 2; rehearsal 2; relation 4; tale 1
narrative°, account 4, 6; chronicle 1; description 2; history 1; legend 1; memoir 1; narration 1; novel 2; recital 2; relation 4; report 1; romance 3; saga; story[1] 1; tale 1; yarn 2
narrator°, raconteur
narrow°, acute 1; contract 4; limit 6; limited 2; narrow-minded; near 8; nearsighted 2; parochial; partisan 3; sectarian 2; selfish 2; slender 2; small-minded; strait 1; strict 1; thin 3
narrow (down), narrow 9; taper 1
narrowed, narrow 1
narrow escape, shave 3
narrowing, narrow 1
narrow(-minded), hidebound; intolerant 2; legalistic; little 6; nearsighted 2; one-sided 1; parochial; partisan 3; philistine 2; prejudiced 2; provincial 2; puritan 2; sectarian 2; small-minded; strait-laced
narrow-mindedness, fanaticism 2; intolerance; provincialism 2
narrowness, provincialism 2
narrows°, channel 2; narrow 10; strait 2
narrow-spirited, narrow-minded
narrow squeak, shave 3
nastiness, filth 3
nasty°, awful 2; bad 7; beastly 2; bitter 2, 5; dirty 4, 6; disagreeable 2; disgusting; distasteful; erotic 3; evil 1, 5; evil-minded 1; explosive 1; filthy 1; forbidding 2; foul 1, 2, 8; frightful 2; grisly; harsh 3; horrible 2; hurtful 2; loathsome; low[1] 12; mangy; mean[2] 5; nauseous; obnoxious; obscene 2; outrageous 3; pornographic; rancid; rank[2] 3; repulsive; revolting; rotten 4; ruinous; sarcastic; scathing; scurrilous; seamy; serious 4; shabby 2; sharp 5; sour 3, 4; stinking 2; stormy 1; tart[1] 2; terrible 1; thorny 2; truculent; ugly 2, 4; ungodly 2; uninviting; unpalatable; unsavory; vicious 2; vile 2; violent 1; virulent 2; vulgar 2; wicked 3, 4
nasty look, give a, glare 4
nasty look, glare 2
nasty piece of work, stinker
natal, native 1
nation°, country 1; folk; land 4; monarchy 1; nationality 2; people 4; public 8; race[2] 1; state 4; tribe
national°, extensive 1; home 6; interior 2; native 4, 7; political 1; racial; state 5; subject 5
nationalism, jingoism
nationalist, national 2; nationalistic; patriot
nationalist(ic), national 2; patriotic
national park, sanctuary 3
nationwide, extensive 1; national 1
native°, aboriginal; citizen 1; domestic 3; home 6; inborn; indigenous 1; ingrained; inherent; instinctive 1; intrinsic; local 3; mother 4; national 3; organic 2; original 2; vernacular 1
native to, peculiar 2
native land *or* **country** *or* **soil,** fatherland; land 4; root[1] 4
natives, population
nativity, birth 3
natter, babble 1; chat 1, 2; chatter 1; gab 1; go 32f; gossip 1, 4; gush 2; jabber 1; palaver 2, 4; patter[2] 2, 3; prattle 1; rattle 6; talk 3; tattle 2; yap 2
natter (on), waffle 1
natter on (about), drivel 2; hold 16a
nattering, palaver 2; patter[2] 2
natty, jaunty 2; neat 1; sharp 9; smart 3; snappy 2; spruce 1; trim 1; well-groomed
natural°, artless 1, 2; automatic 1; casual 5; correct 7; crude 1; earthly 2; easy 6; flesh 3; fluent; free 8; habitual 1; homely 1; illegitimate 2; inborn; indigenous 1; informal 1; ingenuous 1; ingrained; instinctive 1; intrinsic; lifelike; mother 4; naive; native 1; normal 1; organic 1, 2; photographic 1; physical; pristine 2; pure 1; radical 1; raw 2; real 1, 2; realistic 2; regular 1; spontaneous 2; typical 2; unaffected[1]; unperfumed; unpremeditated; unrefined 2
natural to, peculiar 2
naturalist, environmentalist
naturalistic, realistic 2
naturally°, absolutely 3; course 6; doubtless 1; indeed 1; necessarily; needless 2; simply 3
naturalness, ease 4; familiarity 2; freedom 6; naiveté
natural philosophy, philosophy 1
nature°, character 2; denomination 2; description 3; difference 4; disposition 1; essence 1; fiber 3; flavor 2; form 3; heart 3; humor 3; kind[2] 2; mold[1] 3; mood 1; personality 1; psyche; psychology; sort 1; stripe 2; structure 1; temper 1; texture; way 2
nature, by, inside 6; naturally 2; nature 5
nature, real, entity 2
nature, in a state of, naked 1
nature lover, environmentalist
naught°, nought; nil; nothing 1; zero 1
naught, come to, collapse 2; fail 1; fall 20; fizzle 2; flop 3; founder[2] 2; grief 3; miscarry; peter out; strike 17; wrong 8b
naught, for, fruitless
naught, reduce to, ruin 7
naught, say, tongue 3
naught, set at, trample 2
naughtily, badly 4
naughtiness, devilry 1; hanky-panky; misbehavior; mischief 1; ribaldry
naughty°, bad 10; devilish; difficult 3; disobedient 1; erotic 3; incorrigible; low[1] 3; mischievous 1; notorious 1; outrageous 3; profane 3; racy 2; risqué; rude 3; sexy 2; sly 2; suggestive 2; undisciplined; unseemly 1; vulgar 2; wicked 5; wrong 1
naughty, be, misbehave
nausea, disgust 2; distaste 2
nauseate°, disgust 1; offend 2; put 21d; repel 2; revolt 3; shock 1d; sicken 2; turn 17b
nauseated°, disgusted; queasy 2; rotten 5; sick 1, 6
nauseated, easily, squeamish 2
nauseating, abominable 1; disagreeable 2; disgusting; distasteful; excessive 2; fearful 3; foul 1; frightful 2; grisly; gross 5; hideous 1; horrible 1; loathsome; monstrous 1; nasty 1; nauseous; obnoxious; obscene 2;

nauseating (*cont.*) offensive 3; outrageous 3; repellent; repugnant; repulsive; revolting; sentimental 2; shocking 2; terrible 5; ugly 2; unsavory; vile 2; yucky
nauseous°, abominable 1; disagreeable 2; disgusted; distasteful; fearful 3; foul 1; frightful 2; hideous 1; horrible 1; monstrous 1; nasty 1; obnoxious; obscene 2; offensive 3; outrageous 3; queasy 2; repellent; repugnant; repulsive; revolting; shocking 2; squeamish 2; terrible 5; ugly 2; vile 2; wicked 6; yucky
nautical°, marine 1; seafaring
naval, marine 1; military 1; nautical; seafaring
naval force(s), navy
naval (task) force, fleet[1]
nave, center 1; focus 1; hub; pivot 2
navigable°, open 9; passable 2
navigate°, guide 2; make 19; maneuver 4; pilot 3; run 17; sail 1
navigational, nautical
navigator°, pilot 2
navvy, laborer
navy°, fleet[1]; military 2; service 8
Nazi, repressive; totalitarian
Nazism, tyranny
near°, about 9; approach 1; approximate 2; beside 1; by 1, 6; close 9, 18, 20; come 1; draw 13b; fast[1] 8; get 49a; greedy 3; hand 9a; mean[2] 1; narrow 7; neighboring; penurious 1; toward 3; wind[1] 8
near as dammit to, as, neighborhood 2
near (by), forthcoming 1; about 1
near and far, abroad 2; everywhere; far 4
nearby°, about 9; around 13; available; beside 1; by 6, 7; close 20; convenient 2; hand 9a; handy 1; immediate 2; local 1; near 1, 5; neighboring; past 3; point-blank 2; present[1] 2; surrounding
nearest, immediate 2; neighboring; present[1] 2
nearing, forthcoming 1; go 32e; impending; nearly 1; oncoming 1, 2; prospective; toward 3
nearly°, about 2, 6; almost; approximately 1; around 1; close 21; go 32e; near 3; neighborhood 2; practically 1; roughly 1; say 14; virtually
nearness, presence 1; proximity
nearsighted°, partisan 3; shortsighted 1; small-minded
near squeak, shave 3
neat°, clean 3; dainty 1; dandy 2; dapper; elegant 4; fabulous 3; fine[1] 1; gorgeous 2; immaculate 1; ingenious; methodical; nice 5; nifty 3; order 10a; orderly 1; scrupulous 1; shapely; shipshape; sight 6b; slick 3; smart 3; splendid 3; spruce 1; straight 7, 9; stylish; swanky; taut 2; tidy 1; tight 4; trim 1; undiluted; well-groomed
neaten°, clean 9; pick 8c; straighten 3
neaten (up), clear 30b; go 34c; groom 3; make 15; neaten; spruce 2; tidy 4
neatness, method 2; order 2
Neat-oh!, bully 4
neb, bill[2]; tip[1] 1
nebbish, loser; nonessential 2; zero 3
nebulous°, cryptic 2; dim 1; hazy 2; indistinct 2; obscure 1; shapeless 1; subtle 2; tenuous 2; vague 5; woolly 2
nebulously, vaguely 2
nebulousness, mystery 2; obscurity 1
necessarily°, willy-nilly 1
necessary°, cardinal; due 3; essential 1; fatal 3; imperative 1; incumbent 1; indispensable 1; key 6; main 3; mandatory; necessity 1; need 3; obligatory; prerequisite 1; staple 1; toilet 1; urgent 1; vital 1; willy-nilly 2
necessary, do *or* **accomplish the,** trick 7
necessitate, demand 3; entail; involve 2; take 16; want 2; warrant 4
necessities, staple 3
necessitous, impoverished 1; indigent; lean[1] 3; needy; penurious 2; poor 1; straitened
necessity°, keystone; must 2; need 2, 3; precondition; prerequisite 2; privation; requirement 1; urgency; use 13; want 4
necessity, of, necessarily; willy-nilly 1
neck, cape[1]; channel 2; cuddle 2; kiss 1; love 6; pet[1] 4
neck and neck, close 13; even 5; level 6; tight 6
neck and neck (with), be, tie 4
neckband, ring[1] 1
necklace, pendant; ring[1] 1; string 4
neck of the woods, part 6
necktie, tie 11
necrologue, obituary
necrology, obituary
necromancer, magician 1; sorcerer
necromancy, magic 1; sorcery
necromantic, magic 5
necropolis, graveyard
necropsy, post-mortem 1
necrose, fester 1; mortify 3
necrosed, ulcerous
necrotic, ulcerous
need°, call 15; cry 3; dearth; demand 3, 6; expect 3; go 40; indicate 4; lack 1, 2; miss[1] 2; necessity 1, 3; poverty 2; privation; require 2; requirement 2; scarcity; stomach 2; take 16; urgency; use 13; want 2, 3, 5; wish 5
need, in, deprived; destitute 2; poor 1
need of, be *or* **stand in,** want 2
need of, in, starved 3
need of *or* **for, have,** need 1
needed, demand 7; due 3; indispensable 1; mandatory; necessary 1; order 10b; vital 1
needful, indispensable 1; necessary 1
needful (of), desperate 3; destitute 2; short 6
needfulness, necessity 2; need 4
neediness, necessity 3; need 4; poverty 1; privation; want 5
needle, aggravate 2; annoy 2; bother 1; exasperate 2; gall[2] 4; index 3; indicator; irk; irritate; molest 1; nag[1] 1; offend 1; pester; pick 6; plague 3; poke 4; prickle 1; prod 3, 5; spine 2; spite 3; tease 1; torment 2
needle, give (someone) the, pester
needle-shaped, pointed 1
needle-sharp, sharp 1
needless°, dispensable; expletive 1; extraneous 1; gratuitous 2; nonessential 1; superfluous; unnecessary
needless to say°, naturally 1; needless 2
needling, incitement 1; torment 4
needy°, broke; deprived; destitute 1; impoverished 1; indigent; lean[1] 3; penurious 2; poor 1; straitened
ne'er-do-well, derelict 3; good-for-nothing 2; idler; loafer; miscreant 1, 2; rascal; reprobate 2; rogue 1; shiftless; wastrel 2
nefarious, black 6; dark 4; evil 1; flagrant; foul 4; immoral 1; lawless 3; machiavellian; miscreant 2; monstrous 2; scheming; sinister 2; vicious 1
nefariousness, evil 6
negate, deny 1; destroy 4; disprove; disturb 5; neutralize; rebut; reverse 3; revoke; rule 8; thwart 1
negating, negative 4
negation, denial 1, 3
negative°, destructive 2; pessimistic; photograph 1
negative, in the, negative 5
negative cash flow, loss 6
negatively, negative 5
neglect°, default 1, 3; disregard 3; fail 2; failure 1; leave[1] 10; negligence; omission 2; omit 2; overlook 1; oversight 1; slack 4; slide 4; slight 8; waste 5
neglect (doing *or* **taking),** forget 2
neglected, abandoned 1; derelict 1; deserted; desolate 1; disused; forlorn 2; ragged 4; ramshackle; shabby 3; undone[2]; unused 2
neglectful, careless 2; delinquent 2; derelict 2; forgetful; heedless; inattentive; promiscuous 1; remiss; slack 1; thoughtless 2; unthinking 2
neglectfulness, neglect 3, 4
negligee, wrapper 1
negligence°, default 1; neglect 4; omission 2
negligent, careless 2; delinquent 2; derelict 2; forgetful; inadvertent 2; inattentive; lax 1; loose 6; perfunctory 2; promiscuous 1; reckless; remiss; slack 1; thoughtless 2; unprofessional 1
negligent, be, nod 4
negligible°, frugal 3; incidental 2; inconsequential; insignificant; little 5; marginal 1; minor 2; petty 1; puny 1; remote 6; slight 1; small 2, 4, 5; tenuous 2; tiny; trifling; unimposing
negotiable, navigable 1
negotiate°, bargain 3; dicker 1; float 3; haggle; leap 1; manipulate 2; palaver 5; parley 2; settle 3; transact
negotiation°, deal 5; dicker 2; diplomacy 2; transaction 1
negotiations, dealings
negotiator°, go-between; mediator
Negro, black 2
negroid, black 2
neighborhood°, area 3; district; local 1, 2; locale; part 6; place 2; precinct 2; proximity; quarter 3; resident 2; site 1; spot 2; surrounding, territory 1, turf 2; vicinity; whereabouts 1
neighborhood, in *or* **into the,** near 1
neighborhood (of), in the, around 8, 13; close 20; neighborhood 2; near 9; nearby 1
neighboring°, adjoining; local 1; near 5; resident 2; surrounding
neighborliness, familiarity 2; friendship 1
neighborly°, brotherly; considerate; friendly 1;

Nostradamus, oracle 1
nostrum, cure 1; elixir 1; medicine; remedy 1, 2
nosy°, curious 2; inquisitive 2; intrusive
  nosy, be, pry 2; snoop 1
Nosy Parker, busybody; gossip 3; intruder 2; snoop 2
Nosy Parkerism, curiosity 2
not, cry 8; never 2
notability, celebrity 1; mark 8; name 2; prominence 1
notable°, banner 2; big 4, 5; celebrity 2; classic 2; considerable 2; conspicuous 3; dignitary; distinguished 1; elevated 2; eminent 1; eventful; exalted 1; excellent; extraordinary 2; extreme 1; famous; figure 6; foremost 1; grand 2; great 5; hero 1; historic; honorable 4; illustrious; important 2; landmark 3; laudable; lofty 2; major 2; memorable; monumental 1; noted; noteworthy; noticeable 2; outstanding 1; particular 2; perceptible; personage; prestigious; prodigious 2; prominent 2; pronounced 1; proud 3; public 6; remarkable 1; renowned; salient; signal 3; significant 1; singular 2; somebody 2; special 1; splendid 2; superior 2; well-known 2; worthy 3
  notable, be, stand 9a
notably°, eminently; especially 1; extra 6; particularly 1, 2; preeminently
notarize, sanction 5
notation°, record 3; reference 1
  notation (of), make a, put 16a; record 1; write 4a
notch°, gorge 1; gully; indentation; nick 1, 2; pass 23; pink[2]; score 2, 10; slot 1, 3
  notch, take (someone) down a, humble 4
  notch up, notch 4
notched°, jagged; ragged 2; serrated
note°, account 5; behold; bill[1] 2; characterize; comment 1; consider 2; cry 6; denote 1; detect 2; dig 4; distinction 2; enroll 2; enter 3; entry 4; expression 4; feel 1, 11; find 3; greeting 2; importance 2; interest 2; item 2; letter 2; line[1] 8; list[1] 2; magnitude 2; mark 8, 12; mean[1] 2; memorandum; mention 1, 3; message 1; mind 16; minute[1] 4; missive; moment 3; notation 1; notice 1, 4; observation 2; observe 4; perceive 1; pipe 6; protocol 2; record 1, 3; refer 1; reference 1; regard 1; register 3, 5; remark 1; renown; represent 3; savor 3; see 1; sight 8; slip[2] 1; spy 3; stress 4; tabulate; take 31a; tone 1, 3; vein 4; watch 3, 4; witness 4; write 4a
  note, of, noteworthy
  note (down), get 37b; jot 1; mark 15a; put 16a
  note of, make (*or* take) (a), account 7; get 37b; heed 1; jot 1; mark 15a; mind 16; minute[1] 4; notice 1; pen[1] 2; put 16a; refer 1; register 3, 5; remark 1; take 31a; watch 3, 4; write 4a
notebook, schoolbook; tablet 1
notecase, wallet
noted°, brilliant 3; celebrated; distinguished 1; eminent 1; exalted 1; famous; foremost 1; given 1; honorable 4; illustrious; important 2; legendary 3; noble 3; notable 1; outstanding 1; prestigious; prominent 2; proud 3; public 6; renowned; well-known 2
  noted for, be, reputation 2
note pad, pad 2
notepaper, paper 3
notes°, cash 1; material 4; minute[1] 2; money 1; note 10; tender[2] 3
noteworthiness, mark 8
noteworthy°, banner 2; big 4, 5; celebrated; classic 2; considerable 2; conspicuous 3; distinguished 1; eminent 1; eventful; excellent; exemplary 2; extraordinary 2; extreme 1; foremost 1; great 5; historic; honorable 4; important 2; landmark 3; laudable; legendary 3; major 2; memorable; monumental 1; notable 1; noticeable 2; outstanding 1; particular 2; prestigious; prime 2; prodigious 2; prominent 1, 2; proud 3; remarkable 1; salient; signal 3; significant 1; singular 2; special 1; splendid 2; superb; superior 2
  noteworthy, be, stand 9a
not guilty, clear 12, 31; innocent 1
nothing°, blank 8; breeze 2; cog 2; naught; nil; nobody 2; nonessential 2; nought; trifle 1; zero 1, 3
  nothing, come to, collapse 2; fail 1; fall 20; fizzle 2; flop 3; founder[2] 2; grief 3; miscarry; peter out; strike 17; wrong 8b
  nothing, do, twiddle 2
  nothing, for, free 4; house 7
  nothing, reduce to, ruin 7; wreck 1
  nothing, Say, mum 2; tongue 3
  nothing but, just 4; mere
nothing loath, willingly
nothingness, oblivion 1; void 4
notice°, account 7; advertisement 1; advice 2; advisory 2; announcement 2; attention 1, 2; behold; bulletin; call 14; cognizance; declaration 2; detect 2; dig 4; discharge 9; discover 2; dismissal 1; distinguish 3; ear 1; find 3; heed 2; interest 1; item 2; look 1; mark 12; note 7, 11; observation 1; observe 3; pamphlet; perceive 1; pipe 6; poster; proclamation; push 16; recognition 2; regard 1, 9; release 4; remark 1; resignation 1; review 4; savor 3; seal 2; see 1; sight 5, 8; sign 1, 5; sit 9; spy 3; warning 1; watch 3; witness 4
  notice, at *or* on short, readily 3
  notice, bring to, throw 9c
  notice, come into, emerge 1
  notice, give (prior), warn 1
  notice, give (someone), fire 11
  notice, put (someone) on, warn 1
  notice (of), give, discharge 2; notice 7; notify 2; resign 1
  notice of, take, mark 12; notice 1; remark 1; watch 4
  notice of, take little *or* no, disregard 1; overlook 2
  notice (to), give, dismiss 1
noticeable°, considerable 1; conspicuous 2; discernible 1; distinct 1; dramatic 2; effective 2; evident; marked; observable; perceptible; prominent 1; pronounced 1; salient; sensible 5; visible 1
  noticeable, be, show 12b; stand 9a
  noticeable, make, bring 14a
noticeably, especially 1; notably 1; ostensibly; quite 2
notification, advice 2; announcement 2; call 14; notice 5; proclamation; reference 1; seal 2; warning 1
notified of, be, receive 5
notify°, advise 2; alert 5; announce 3; fill 9c; inform 1; notice 7; post[3] 4; prime 6; signal 4; tell[1] 3; tip[3] 4; warn 1
notion°, conception 2; eye 6; fancy 8; feeling 2; idea 1, 2, 3, 5; image 3; impression 1; inkling; opinion 1; perception 2; philosophy 2; picture 3; point 12; principle 2; resolution 3; spirit 7; suggestion 1; surmise 2; suspicion 2; theme 1; thesis 1; thought 2; understanding 4; view 2
  notion, vague *or* imprecise, generality 1
notional, abstract 1; ideal 5; imaginary; pure 4; speculative 1
notions, sundries
no to, say, reject 2
notoriety°, attention 2; celebrity 1; fame; infamy 1
notorious°, disgraceful 2; flagrant; infamous 1; noted; public 6; spicy 3
  notorious for, be, reputation 2
notoriousness, notoriety
notwithstanding°, despite; even 11, 12; face 9; for 11; however 1; independent 8; irrespective of; nevertheless; rate[1] 5; regardless 1, 2; same 3; spite 2; still 6, 8; though 2; time 13a; yet 5, 6
  notwithstanding that, though 1
nought. See **naught**
noun phrase, phrase 1
nourish°, cherish 2; feed 1; fertilize 2; forward 6; foster 1; fuel 4; keep 2; mother 5; nurse 3; stimulate 1; strengthen 2
nourished by, be, feed 3
nourisher, mother 3
nourishing, healthy 2; hearty 4; nutritious; wholesome 1
nourishment°, diet[1] 1; food; fuel 3; meal 1; meat 1; provender 1; refreshment 1; subsistence 2; sustenance 1
  nourishment, take (in), eat
nous, gumption 1; intelligence 1; sense 2
nouveau riche, parvenu 1, 2; upstart
nova, star 1
novel°, contemporary 2; different 2; fresh 2; modern; new 1; offbeat; original 4; revolutionary 2; romance 3; unfamiliar 1; writing 2
novelette, novel 2
novelist, author; scribe 2; writer
novella, novel 2
novelty°, change 2; gewgaw; innovation 1; originality; variation 2
novice°, apprentice 1; greenhorn; initiate 4; innocent 5; learner; newcomer 2; pupil; recruit 2
novitiate, initiate 4; novice
now°, instantaneously; just 6; minute[1] 3; presently; present[1] 1, 3, 4; trendy 1

obedient°, broken 4; dutiful; flexible 2; good 4; sheepish 1; submissive 1; tame 1; tractable 1; yielding 2
obedient (to), observant 2; subject 8a
obedient (to), be, serve 2
obeisance°, bow 1; deference 2; homage; prostration 2; reverence 1
obeisance, make, bow 5; scrape 4
obeisant, propitiatory 2
obelisk, tower 1
obese°, beamy; big 2; fat 1; gross 1; heavy 11; large 1; plump[1] 1; rotund 3; stout 1; weighty 1; well-fed
obesity°, fat 5; fatness
obey°, abide 4; comply; conform 1; follow 2; fulfill 2; heed 1; keep 6; listen 2; mark 12; mind 16; observe 1; respect 7; square 11
obeying, observance 1
obfuscate, obscure 7
obfuscated, opaque 1
obfuscation, gobbledegook 2
obit, obituary
obiter dictum, digression 1
object°, acquisition 2; aim 5; ambition 3; bitch 3; butt[1]; cavil 2; design 7; drift 4; effect 3; end 3; entity 1; goal; groan 2; idea 4; intent 1; intention; mark 6; motive 2; objective 2; point 6; protest 3; purpose 1; quarry[1]; reason 3; squawk 2; target; thing 1, 5; use 12; view 5
object of, for the, for 4
object (to), challenge 1; contest 3; disapprove; dispute 1; exception 4; mind 15; oppose 1
objection°, disapproval; grievance 2; gripe 2; opposition 1; protest 1; reservation 2
objection to, have any, mind 15
objection (to *or* against), make *or* raise (an), exception 4
objectionable, disagreeable 1, 2; disgraceful 2; disgusting; dishonorable 3; distasteful; exceptionable; inadmissible; nasty 1, 2; obnoxious; obscene 2; offensive 2; outrageous 3; repugnant; repulsive; revolting, ugly 2, unacceptable; undesirable 2; unlikely 2; unsavory
objective°, aim 5; ambition 3; aspiration; candid 2; design 7; destination; detached 2; disinterested; dispassionate 2; effect 3; end 3; equitable; factual 2; fair[1] 1; goal; idea 4; impartial; impersonal 1; indifferent 2; intent 1; intention; just 1; literal 2; mark 6; mission 1; motive 2; nonpartisan 2; object 3; point 6; prey 1, 2; purpose 1; reason 3; tangible; target; tolerant; unprejudiced; view 5
objectively, fairly 2; honestly 1
objectiveness, justice 1
objective(s), sake 2
objectivity°, candor 2; equity; honesty 3; indifference 3; justice 1
object of vertu, antique 2
objet d'art, antique 2; curiosity 3
objet de vertu, antique 2; curiosity 3
objets d'art, bric-a-brac
objets de vertu, bric-a-brac
objurgate, abuse 3; berate
objurgation, abuse 8; blame 3
objurgative, foul 6
objurgatory, foul 6
oblation, offering; sacrifice 1
obligate°, bind 2; oblige 2; tie 7d
obligated, accountable; beholden; bound[3] 2; indebted; liable 2; obliged 1, 2; thankful
obligated to, be, must 1
obligation°, appreciation 1; assignment 2; business 1; charge 1; contract 1; debt 1; duty 1; engagement 3; faith 3; guarantee 1; liability 2; must 2; office 4; place 4; project 2; resolution 2; responsibility 1; stint 1; stipulation; strain[1] 7; subscription 2
obligation, under, beholden; debt 2; thankful
obligations, string 7
obligatory°, imperative 1; incumbent 1; indispensable 2; mandatory; necessary 1; peremptory 1; prerequisite 1
oblige°, accommodate 5; bind 2; drive 1; force 7; have 9; lock[1] 6b; make 2; obligate; screw 4; serve 1; tie 7d
oblige (with), indulge 1
obliged°, accountable; beholden; bound[3] 2; indebted; supposed 2; thankful
obliged to, be, must 1
obliging°, accommodating 1; amiable; considerate; debonair 1; decent 4, dutiful; gracious; kind[1]; propitiatory 2; respectful; supple 3; yielding 2
obligingly, kindly 2
oblique°, angle[1] 1; equivocal 1; evasive; indirect 1; line[1] 1; roundabout 2; sidelong
obliquely, sideways
obliterate°, abolish; blot 4b; cancel 2; dash 1; delete; desolate 5; devastate 1; devour 2; eliminate 3; erase 1, 2; exterminate; extinguish 2; kill 1; massacre 2; omit 1; remove 3; rout 2; scratch 3; strike 5; wash 4; wipe 3
obliterate from the mind, mind 19
obliteration, death 3; desolation 1; wreck 5
oblivion°, consigned to, limbo
oblivion°, consign to, bury 2; forget 3
oblivious°, absent-minded; deaf 2; heedless; ignorant 2; inattentive; preoccupied 2; unaware
oblivious (to *or* of), unconscious 2
obliviously, idly 2
obliviousness, oblivion 2
obloquy, abuse 8; aspersion; discredit 4; disgrace 1; dishonor 4; humiliation; indignity; infamy 1; jeer 2; libel 1; notoriety; scandal 2; shame 2; slander 1
obnoxious°, abhorrent; bitter 2; disagreeable 1; disgusting; distasteful; foul 1; hateful 1; horrible 2; nasty 1, 4; obscene 2; offensive 3; outrageous 3; pushy; repellent; repugnant; repulsive; revolting; surly; terrible 5; ugly 2; undesirable 2; uninviting; unsavory
obovate, oval
obovoid, oval
obscene°, abusive 1; base[2] 4; bawdy; blue 2; broad 8; coarse 3; dirty 2; earthy; erotic 3; evil-minded 1; filthy 3; foul 5; gross 3; immodest 1; immoral 2; improper 3; impure 4; incontinent 2; indecent 2; lascivious 2; lewd; low[1] 3; nasty 3; naughty 3; obnoxious; outrageous 3; pornographic; profane 3; prurient 2; racy 2; rank[2] 3; rough 6; rude 3; scurrilous; sexy 2; unmentionable 2; unseemly 1; vulgar 2; wicked 3
obscenity, curse 3; dirt 3; expletive 2; filth 3; impurity 3; oath 2; pornography; profanity; ribaldry; vulgarity 2
obscurantism, gobbledegook 2
obscuration, eclipse 3
obscure°, ambiguous 2; blot 4a; blunt 4; blur 3, 4; bury 4; cover 2; cryptic 2; dark 6; deep 2; difficult 2; dim 1, 3; doubtful 1; dull 10; dusky 2; eclipse 1, 2; envelop 2; extinguish 3; feeble 3; film 4; fog 3; fuzzy 2; gloomy 1; hide[1] 3; humble 3; imperceptible 1; incomprehensible; indefinite 2, 3; indistinct 1; intangible; light[2] 3; mantle 3; mask 3; minor 2; misty; muddy 2, 4; nebulous; occult 1; opaque 1, 2; out-of-the-way 1; profound 1; recondite; remote 3; shade 12; shadowy 2; submerge 3; subtle 2; thick 3; twilight 4; unheard-of 1; unknown 1; unsung; vague 1, 3; veil 2; veiled
obscure(d), dim 1; hidden; occult 1; opaque 1; woolly 2
obscurely, vaguely 2
obscuring, eclipse 3; thick 3
obscurity°, background 2; dark 12; depth 2; gloom 1; mystery 2; oblivion 1; opacity 1, 2; perplexity 2; profundity 1; shade 1; shadow 1
obsecrate, beseech; petition 2; pray 1
obsecration, prayer 1; supplication 1
obsequence, servility
obsequies, burial; ceremony 1; funeral
obsequious°, groveling; humble 2; ingratiating; menial 2; oily 2; servile; slimy 2; submissive 2; supple 3; timeserving
obsequious, be, truckle
obsequiousness, servility
observable°, discernible 1; noticeable 1; outward 1; overt; perceptible; sensible 2; visible 1
observance°, celebration 1; ceremonial 3; ceremony 1; devotion 1; discharge 12; feast 2; formality 1; obedience; occasion 3; piety 2; rite; vigilance
observant°, alert 1; astute 2; attentive 1; obedient; perceptive; sharp 3; sharp-eyed; vigilant; wakeful 2; wary
observant (of), observant 2; sensible 4; watch 8
observation°, celebration 1; comment 2; experience 1; exploration; eye 7; note 3; notice 3, 4; observance 1, 3; pronouncement 1; surveillance; thought 2; view 4; watch 5
observations, data
observe°, abide 4; celebrate 1; commemorate; comment 3; conform 1; consider 2; contemplate 1; detect 2; explore 2; eye 10; follow 2; fulfill 1; honor 7; keep 6, 10; look 1; mark 12; meet[1] 5; monitor 3; note 11; notice 1, 2; obey 1; perceive 1; police 3; regard 1; remark 1, 2; savor 3; see 1; sight 8; sit 6a; spy 2; study 3; traverse 4; view 7, 8; watch 1, 3; witness 4
observe on *or* upon, observe 4
observed, empirical
observer°, bystander; eyewitness; onlooker; picket 3; spectator; student 2; witness 1

oddness, eccentricity 1; oddity 1; singularity 2
odds°, chance 3; possibility 1; probability
odds, at, discordant 1; factious; odds 4; pole[2] 3
odds, be at, clash 4; conflict 4; feud 2; quarrel 2
odds, over the, pricey
odds, set at, divide 3; set 13a
odds with, at, afoul
odds and ends°, leftover; litter 1; lumber 1; miscellany; odds 5; remain 4a; rummage 2; sundries; trash 2; truck 1
odds and sods, odds 5
odds-on, likely 3; probable
ode, lay[3]; poem
odious, abhorrent; abominable 1; damnable; diabolic 2; disagreeable 1; forbidding 1; hateful 1; hideous 2; infamous 2; loathsome; mangy; nasty 1; obnoxious; rancid; repulsive; seamy; shabby 4; ugly 2; vicious 1
odium, aversion 1; contempt; discredit 4; disgrace 1; disgust 3; hate 3; horror 1; infamy 2; loathing; revulsion; shame 2
odor°, aroma 1, 2; aura; bouquet 2; flavor 1; fragrance; fume 3; perfume 1; reek 3; savor 2; scent 1; smell 1; sniff 1; tang 1; touch 15
odoriferous, fragrant; smelly; strong 3
odorous, fragrant; redolent 1
odyssey, journey 1
oeillade, ogle 3
oeuvre, opus; production 2; work 4
of course°, absolutely 3; course 6; indeed 1; mean[3] 2a; naturally 1; necessarily; needless 2; obviously
off°, absent 1; absent-minded; amiss 1; bad 6; beside 2; chase 3; distant 2; foul 2; hook 5; mistaken 1; out 12, 14; rotten 1; sick 5; soft 6; sour 2; stale 1; unpalatable
off, be, get 45a; leave[1] 1; shoo 1
off (the mark), erroneous; extraneous 2; inaccurate; inapplicable; misguided; off 3; wide 4, 5
offal, dirt 1; filth 1; garbage; gut 1
off and on, occasionally; sometimes
offbeat°, bizarre 1; curious 3; deviant 1; eccentric 1; extreme 5; grotesque 2; irregular 3; kinky 1; odd 1; ordinary 4; outlandish; peculiar 1; quaint 1; queer 1; singular 1; step 8; strange 1; way-out 1; whimsical 1; wild 7; abnormal 2
off-center, wry 1
off-color°, blue 2; broad 8; dirty 2; foul 5; improper 3; nasty 3; naughty 3; obscene 1; par 5b; profane 3; racy 2; rank[2] 3; risqué; sexy 2; spicy 2; suggestive 2; vulgar 2
offend°, cut 3; disgust 1; dishonor 1; displease; hurt 3; injure 2; insult 1; jar[2] 3; nauseate; offense 2; outrage 4; provoke 3; repel 2; revolt 3; scandalize; sicken 2; sin 3; slight 6; spite 3; transgress 1; turn 17b; wound 4
offended, disgusted; nauseated; sick 6
offended, feel *or* be, umbrage
offended (at), be, exception 4
offended (by), be, bridle 3; mind 15
offender°, criminal 3; culprit 2; delinquent 1; felon; sinner; transgressor
offending, guilty 1
offense°, atrocity 2; crime; cut 23; dishonor 4; error 2; fault 2, 4; indignity; injury; insult 2; misdeed; offensive 4, 5; provocation 2; put-down; sin 1; slap 5; slight 7; transgression
offense, give, offend 1; offense 2
offense (at), take, amiss 4; bridle 3; exception 4; mind 15; offense 3; umbrage
offensive°, abhorrent; abominable 1; abusive 1; assault 1; attack 6; bad 4; base[2] 4; beastly 2; blue 2; campaign 1; coarse 3; crude 4; damnable; derogatory; disagreeable 1, 2; disgusting; distasteful; evil 5; filthy 3; forbidding 1; foul 1, 6; frightful 2; grating 1; gross 3; horrible 2; indecent 2; injurious 2; insolent; invasion 2; lascivious 2; lewd; loathsome; low[1] 3; militant 1; nasty 1; naughty 3; nauseous; obnoxious; obscene 1; outrageous 2, 3; personal 4; pornographic; provocative 2; push 15; pushy; rank[2] 3, 4; repellent; repugnant; repulsive; revolting; rude 2; scurrilous; smelly; sordid 3; spicy 2; suggestive 2; terrible 5; ugly 1, 2; unbecoming 2; undesirable 2; unheard-of 3; uninviting; unpalatable; unsavory; unseemly 1; vile 2; vulgar 2; wicked 2, 3, 6
offensively, strong 23
offensiveness, indelicacy; ribaldry
offer°, afford 3; bid 1; bring 10b; carry 7, 8; communicate 1; declare 1; dedicate 1; donation 2; enter 6; exhibit 1; extend 4; finger 6; give 6, 7; grant 1; hand 16; hold 16b, 20b; introduce 2; lay[1] 6; nominate; open 25; overture; pass 9; pipe 8; plead 3; posit; prefer 2; present[2] 5, 7; produce 4; profess 1; proposal 1; propose 1, 2; propound; provide 2; put 5, 17a, 18, 28h; render 3, 7; return 6; sacrifice 1; sell 2; serve 3; set 18a; stock 9; submit 2; suggest 1; table 4; tender[2] 1, 2; turn 15b; venture 2; vouchsafe 1
offer, on, sale 6
offer (as collateral), put 28d
offer (for), make an, bid 1
offer (up), sacrifice 3
offer a prayer, pray 2
offer hospitality (to), welcome 1
offering°, donation 1; exhibition; gift 1; motion 5; offer 5; presentation 1; present[2] 2; product 2; submission 2; tribute 2
offering (up), sacrifice 1
offerings, line[1] 16
off-guard, catch, surprise 2
offhand°, casual 3, 5; extemporaneous; haphazard 2; hurried 2; impetuous; impulsive; loose 5; nonchalant; oblique 2; perfunctory 1; preoccupied 2; short 4; spontaneous 1, 2; unpremeditated
offhand(ed), flippant 1; offhand 1
offhandedly, idly 2; offhand 5
offhandedness, flippancy 1
office°, appointment 3; branch 2; bureau 2; capacity 3; clerical 2; department 1; duty 1; establishment 2; function 1; ministry 3; mission 1; position 6; practice 3; room 2; study 6
officebearer, officer 1; politician
officeholder, functionary; incumbent 3; officer 1; servant 2
officeholding, incumbent 2
officer°, constable; functionary; police officer
officer of the law, officer 2
offices°, office 3, 5
officeseeker, candidate; nominee
office supplies *or* equipment, stationery
official°, authoritative 1; celebrant; dignitary; director 1; executive 1, 3; formal 1, 2; functionary; incumbent 3; judicial 1; officer 1; orthodox; politician; professional 1; real 1; regular 6; regulation 4; servant 2; standard 5; state 6; umpire 1
officialdom, authority 4; bureaucracy
officialism, bureaucracy
officials, authority 4
officiant, celebrant
officiate°, preside; umpire 2
officiate at, celebrate 1; hold 10
officious°, domineering; intrusive; obtrusive; overbearing; pushy
offing, background 2
offing, have in the, have 12b
offing, in the, card 11; forthcoming 1; impending; near 4; pending 2; pipeline 2; prospect 4; wind[1] 8
offload, dump 1; discharge 7; unload
offloading, discharge 14
off ((one's) rocker *or* trolley), sick 4
off-putting, disconcerting; disgusting; distasteful; disturbing; obscene 2; repellent; repugnant; repulsive; revolting; uninviting
offscourings, trash 3; waste 7
offset°, balance 3; cancel 3; compensate 1, 2; correct 4; cover 10; foil[1]; make 31a; neutralize; oppose 3; parallel 3, 4; redeem 5; run 19
offshoot°, branch 1; derivative 2; descendant; product 1; ramification 2; shoot 5; slip[2] 2; tributary; twig[1]
off somewhere, dreamy 2
offspring°, brood 1; child 1; descendant; family 2; issue 7; lineage 2; litter 2; offshoot 2; posterity; progeny; seed 3; young 4
off-the-beaten-track, isolated 3; secluded 2
off-the-peg, ready-made 1
off-white, white 1
oft, often
often°, always 1; frequently 1, 2; repeatedly; time 21
often, not, seldom
often as not, as, frequently 2
oftentimes, frequently 1, 2, often
ofttimes, frequently 1, 2
ogle°, eye 5; leer 1, 2; watch 3
ogre°, demon 1; devil 2; giant 1; monster 1
ogress, bag 4; devil 2; hag; ogre; witch 2
Ogygian, ancient 2; old 2
oil°, bribe 2; flatter 1; polish 6; slick 5
oilskin (raincoat), slicker 2
oily°, fat 2; greasy 1, 2; ingratiating; mealy-mouthed; pharisaic;

opaqueness, opacity 1

op-ed article, editorial; writing 2

open°, aboveboard 2; accessible; artless 1; bare 2, 6; barefaced 1; begin 2; bluff[2] 2; brazen; broad 2; candid 1; childlike; clear 16; commence 2; cut 1; daylight 2; demonstrative 1; direct 10; dispassionate 2; downright 1; engaging; ethical; expansive 2; explicit 2; fair[1] 6; flagrant; flower 3; forthcoming 3; forthright; frank 1; free 7, 9, 15; friendly 2; front 9a; full 10; genuine 2; hearty 1; honest 3; honorable 3; inconclusive; indecisive 2; ingenuous 1, 2; innocent 4; insecure 2; lance 2; launch 2; level 15; liable 3; liberal 1, 3; obvious; off 6; outspoken; overt; passable 2; plain 3; plastic 2; pleasant 2; preface 2; première 2; public 3, 4; raise 8; raw 4; receptive 1; responsive; right 1; round 8; selfless; simple 3; sincere; smooth 10; square 4; start 1; suggestible; tap[2] 6; transparent 3; undisguised; undo 1, 2; undone[1] 2; unguarded 2; unimpeded; uninhibited; unresolved

open, in the, aboveboard 1; public 11

open, into the, forward 5

open, out in the, aboveboard 1

open (air), in *or* **into the,** out 1, 3

open out, display 2

open (out *or* **up),** expand 1; extend 1; open 24; roll 11; spread 3; unfold 1

open (to), go 4; subject 6; susceptible 1

open to, not, incapable 2

open (up), broach; clear 23; pioneer 2; ream

open wide, gape 2

open (with), be, card 13; level 11

open-air, outdoor

opener, key 1

open fire, blaze 5; shoot 2

openhanded, benevolent; charitable 1; free 7; generous 1; large 2; lavish 2; liberal 1; objective 1; open 14; philanthropic; unselfish; wasteful

openhandedly, freely 4; water 3

openhandedness, charity 1; largess; philanthropy 1

openhearted, candid 1; liberal 1

openheartedness, humanity 3

opening°, aperture; beginning 2; breach 3; break 26, 29; cavity; chink; first 2; gap 1; gate 1; gulf 2; hole 2; initial 1; interval 3; introductory; launch 5; leak 2; mouth 2; nook 1; occasion 1; opportunity; outlet 1; pass 23; passage 11; pore[2]; prefatory; preliminary 1, 2; première 3; preparatory 1; preview; puncture 1; reception 2; slit 2; slot 1, 2; split 7; start 9, 10, 14; vacancy 1, 2; vent 1; void 5

opening (out), expansive 1

opening move, initiative 1

opening (night), première 1

openly°, aboveboard 1; face 13; flat 17b; free 12; freely 1; honestly 2; naturally 3; outright 4; point-blank 3; public 11; seriously 2; shoulder 6; simply 3

open-minded, disinterested; dispassionate 2; enlightened; equitable; hospitable 2; liberal 3; objective 1; open 11; tolerant; unprejudiced

open-mindedness, candor 2; equity; tolerance 1

open-mouthed, goggle-eyed; thunderstruck

openness, candor 1; familiarity 2; freedom 6; honesty 2; naiveté; simplicity 3; sincerity

open sesame, password

open to debate, moot 1; question 6b

open to discussion, moot 1

open to dispute *or* **doubt** *or* **question,** debatable; doubtful 1; moot 1; questionable; suspicious 1; unresolved

open to misconstruction, suspicious 1

openwork, lace 1; net[1] 1

operability, maturity 3

opera glasses, glass 7

operate°, act 9; behave; carry 11b; conduct 3; direct 1; dispense 2; drive 2; function 3; go 3; handle 3; hold 11; maneuver 4; manipulate 2; oversee; perform 2; play 6; preside; regulate 2; run 10, 11; string 6; take 11; trip 8; work 10, 11, 12, 15

operate, begin *or* **start to,** effect 6

operate, cause to, turn 18a

operate against, militate 1

operating, direction 1; functional 2; moving 2; operation 5; serviceable 1; way 12

operating, cease, stall[1] 1

operating, not, blink 6

operating cost(s), overhead 3

operation°, act 1; activity 2; affair 3; agency; application 1; conduct 2; direction 1; employment 3; enterprise 3; exercise 5; job 3; leadership; management 1; maneuver 2; procedure; proceeding 1; process 1; running 1; touch 17; upkeep 1; usage 2

operation, in, effective 3; force 6; progress 4

operation, in (full), swing 6

operation, in *or* **into,** operation 5

operation, put in(to), open 20; use 1

operational, effective 3; functional 2; operable; operation 5

operational, become, effect 6

operational, not, blink 6

operation(s), campaign 1; operation 4

operation(s), cease, close 5; fail 4

operative°, active 2; deputy; effective 1, 3; efficient; force 6; motive 3; operation 5; operator 1; serviceable 1

operative, become, effect 6

operative, be *or* **remain** *or* **prove,** hold 11; take 11

operator°, charmer; cheat 1; devil 4; engineer 2; tactician; tycoon; user 1

operose, arduous 1

ophidian, snake 1

opiate, dope 2; drug 2; narcotic 1, 2; salve 2; sedative 1, 2; tranquilizer

opine, comment 3

opinion°, advice 1; analysis 2; attitude 2; belief 3; comment 2; conviction 2; counsel 1; doctrine; esteem 3; estimate 4; estimation 1; evaluation 2; favor 4; feeling 2; idea 3; judgment 2, 4; mind 6; observation 2; outlook 1; persuasion 2; philosophy 2; point of view 2; position 3; pronouncement 1; respect 1; school 3; sentiment 2; side 3; speculation 1; stand 13; suggestion 1; tenet; thesis 1; thinking 2; understanding 4; view 2; vote 3

opinion, be of the, reckon 3

opinion, in one's own, personally 2

opinion, form an, mind 12

opinion, of like, united 3

opinion of, give one's, review 3

opinionated°, assertive; dogmatic; intolerant 2; jaundiced 1; narrow-minded; nearsighted 2; parochial; partisan 3; positive 5; prejudiced 1; strong 16; ultra-

opinion piece, editorial

opinion poll, poll 2

oppidan, citizen 2

opponent°, adversary 1; antagonist; competitor; contestant; enemy; entry 5; opposition 2; rival 1

opportune°, advantageous; convenient 1; expedient 2; favorable 1; fortunate 2; golden 7; good 13; happy 2; lucky 2; propitious; providential; ripe 3a; seasonable; suitable; timely; well-timed

opportune, be, serve 4

opportunely, happily 1; right 16

opportunist, adventurer 2; opportunistic

opportunistic°, timeserving

opportunities, prospect 3

opportunity°, bout 2; break 29; chance 2; go 41; leisure 1; move 10; occasion 1; opening 2, 3; option 1, 2; possibility 2; prospect 3; say 13; scope 2; season 1; shot 3; start 12; thing 7; time 5; turn 30; way 7

oppose°, combat 6; contest 3; contradict 1; contrast 1; counteract; differentiate 1; discourage 2; disobey; dispute 1; exception 4; fight 2; fly 6; issue 9; jar[2] 2; militate 1; object 4; pit[1] 6; protest 3; resist 1; rival 2; take 35b; thwart 1; traverse 3; withstand

opposed°, adversary 2; averse; contrary 1; destructive 2; discordant 1; disinclined; disobedient 2; hostile 1; incompatible; opposition 3; polar 2; reluctant; resistant 1

opposed to, opposed

opposed to, as, instead 2

opposing°, contrary 1; destructive 2; mixed 2; negative 4; opposed; opposite 2

opposite°, contrary 1, 4; different 1; discordant 1; face 8; incompatible; mixed 2; opposing; polar 2; reverse 1, 6; wrong 6

opposite, be, face 15; front 11

opposite ends of the earth, at, pole[2] 3

opposite (from *or* **to),** opposed; unlike 1, 3

oppositely, contrary 5

opposite number, like[1] 7

opposition°, antagonism 1; antagonist; combat 3; competitor; conflict 3; disagreement 2; hostility 1; interference 2; objection; protest 1; resistance 1; rival 1

opposition, the, enemy; opponent

opposition (to), be in, conflict 4

opposition to, go *or* **act in,** traverse 3

opposition (to), in, contrary 5; face 9; odds 4; opposed; opposition 3

oppress°, afflict; burden 2; depress 1; distress 3;

organize° (*cont.*) put 27b; rally 3; range 7; rank[1] 6; ready 10; regiment; regulate 1; schedule 2; scheme 4; script 4; see 12a; separate 2; set 23b; settle 1; sort 8, 10a; space 5; stage 5; straighten 2; structure 3; tabulate; tidy 4; time 23

organized, coherent 1; methodical; neat 1; organic 3; pitched; sequential; straight 7; systematic; tidy 2

organized crime, underworld 1

organized whole, system 1

organizer, producer 2

organizing, formation 2; foundation 3; organization 1; preparation 3

orgasm, climax 3

orgasmic, ecstatic; rhapsodic

orgiastic, epicurean 1

orgy°, jag; party 1; spree

orient°, guide 1

oriental, orient 2

orientate, guide 1; orient 3

orientation°, aspect 3; attitude 2; bearing 4; exposure 4; point of view 1; posture 3; site 1

orifice, hole 2; mouth 2; opening 1; passage 11; pore[2]; vent 1

origin°, beginning 1; birth 2, 3; bottom 3; cause 1; conception 1; creation 1; dawn 2; derivation; establishment 1; extraction 3; fountain 2; germ 2; head 6; mother 2; onset 2; opening 4; parent 2; parentage; progenitor 2; root[1] 1; seed 2; source 1; spring 8; start 11, 15; vintage 1

original°, avant-garde; clever 3; creative; crude 1; different 2; early 5, 6; eccentric 2; elemental; fanciful 3; first 2; fresh 2; genuine 1; imaginative 1; indigenous 1; ingenious; initial 1; matchless; model 2; native 3; new 1; novel 1; oddity 2; ordinary 4; pattern 1; picturesque 1; première 3; primary 2; prime 3; primitive 1; pristine 1; progenitor 2; prototype 1; revolutionary 2; seminal 1; unused 1; way-out 2

original, not, secondary 2

originality°, fantasy 1; ingenuity; novelty 1

originally°, early 2; primarily 2

originate°, arise 4; begin 2, 3; breed 4; coin 3; compose 2; create 1; dawn 4; design 2; develop 4; devise 1; discover 3; engineer 4; fabricate 2; father 6; flow 3; form 8; formulate 2; found 1; frame 7; generate 1; grow 2; hatch 2; inaugurate 1; initiate 1; institute 4; introduce 4; invent 1; issue 11; launch 2; make 1, 39c; manufacture 1; organize 2; pioneer 2; produce 2; raise 10; rise 9; spring 3; stage 5; start 1, 2; stem[1] 3

originate (in *or* from *or* with), go 24b; proceed 2

origination, beginning 1; birth 2; composition 4; conception 1; discovery 1; establishment 1; fabrication 2; foundation 3; generation 3; initiation 1; institution 1; invention 1; manufacture 3; origin 2; production 1

originative, creative

originator, author; cause 2; creator 1; designer 1; engineer 1; father 3; founder[1]; parent 2; progenitor 2; source 2

origins°, origin 3; root[1] 4

O-ring, ring[1] 1

Orion, hunter

orison, prayer 2

ornament°, attachment 4; bauble; beautify; decorate 1; decoration 1; drape 1; elaborate 3; embellish 1; embellishment 1; enrich 2; flounce 1; frill 1; fringe 1; grace 6; illuminate 3; illustrate 2; jewel 1; motif; novelty 2; pattern 2, 8; pendant; tool 4; trim 4, 5

ornamental°, fancy 1

ornamentation, decoration 1; embellishment 1; flourish 3; frill 2; ornament 1; trim 5

ornamented, elaborate 2; fancy 1; flamboyant 1; flowery

ornaments, finery; jewelry

ornate°, busy 3; complicated; detailed 2; elaborate 2; extravagant 4; fancy 1; flamboyant 1; flowery; fussy 2; intricate 1; labored 2; luxuriant 3; overwrought 2; pompous 2; showy; splendid 1

ornery, testy

orotund, inflated 2; pompous 2; rhetorical 3; round 7

orphan, foundling

orthodox°, conservative 2; conventional; correct 7; normal 1; proper 2; regular 6; standard 6; typical 2

orthodoxy, code 3

orts, dirt 1; leftover; waste 7

oscillate°, flap 1; flicker 2; fluctuate; flutter 1; pulsate; quaver 1; quiver 1; seesaw 2; shake 2; sway 1; swing 1; vibrate; wag[1] 1; wave 4

oscillating, flutter 4; pendulous 1; swing 3

oscillation, flap 2; flutter 4; flux; quaver 2; sway 3; swing 3; wag[1] 2

oscillation(s), fluctuation

oscillatory, pendulous 1

oscitance, negligence

oscitancy, indolence; negligence

oscitant, drowsy; inattentive; sleepy 1

osculate, kiss 1

osculation, kiss 4

Ossianic, flowery; grandiose 1; ornate

ossified, petrified 3

ossify, petrify 3

ostensible, apparent 2; obvious; outward 1; pretended; probable; professed 1; seeming; so-called 2; specious

ostensibly°, apparently 2; evidently 2; outwardly; seemingly; surface 2

ostensive, tangible

ostensively, ostensibly

ostentation°, display 5; frill 2; glare 3; glitter 4; pose 5; pretense 1; pretension 2; show 15; swagger 3

ostentatious°, boastful; camp[2] 1; conspicuous 2; dashing 3; extravagant 4; flamboyant 2; flash 6; flashy 1; gaudy; grand 3; grandiose 1; loud 2; luxuriant 3; overwrought 2; pedantic 1; pompous 1; pretentious 1; showy; theatrical 2; vulgar 1

ostentatiously, well[1] 4

ostentatiousness, splurge 1

ostler, groom 1

ostracism, segregation

ostracize°, banish 1; freeze 4; isolate; segregate; shoulder 2

other, alternate 4; different 1; further 1; second[1] 4, 5; separate 6

other place, hell 1

other side, opposition 2

other than, besides 2; except 1

otherworldly, heavenly 1; mystical; supernatural; unearthly 1; weird

otiose, idle 4; inert 3; lazy 1; shiftless; slack 1

oubliette, dungeon; hole 4; prison

ought, nil; naught

ought (to), have 10; must 1

ounce, grain 3; modicum

Our Lord, savior 2

ourselves, person 2; personally 1

oust, banish 1; bring 9a; cast 14; discharge 2; dismiss 1; displace 2; dispossess; drop 10; eject 1; evict; exclude 3; exile 3; expel 2; fire 11; overthrow 1; overturn 2; purge 2; remove 5; supersede; supplant; topple 2; turn 20c

ouster, discharge 9; ejection 2; eviction; exclusion 3; expulsion; overthrow 2; overturn 3; purge 4; removal 3

ousting, overthrow 2; overturn 3; purge 4; removal 3

out°, absent 1; dead 3, 5; defunct 2; drag 5; extinct 3; forward 5; free 2; insensible 1; obsolete; off 1; old-fashioned; outward 2; passé; sleepy 1

out and about, abroad 3

out-and-out°, absolute 2; bare 2; barefaced 1; dead 14; downright 1; express 5; flagrant; flat 4; great 11; main 4; open 13; outright 2; peremptory 2; perfect 6; positive 9; profound 4; pronounced 2; proper 5; pure 5; radical 2; rank[2] 2; right 8; sheer 2; stark 4; sweeping 1; thorough 1; total 3; ultra-; undisguised; unmitigated; unqualified 2; utterly

out at (the) elbows, heel[1] 3

outback, country 3

outback, the, stick[2] 3

outbalance, outweigh

outbreak, epidemic 2; eruption 1, 2; explosion 2; fit[2] 2; flash 2; outburst; rash[2] 2; spurt 1; tempest 2; trouble 7; volley 2

outburst°, blaze 2; disturbance 2; eruption 1; explosion 1, 2; fit[2] 2; flare 4; flash 2; flurry 1; frenzy 2; furor 1; gale 2; gush 4; outcry; outpouring; paroxysm; passion 2; riot 1; roar 4; round 14; spasm 2; splurge 1; storm 2; tantrum; tirade; torrent; uproar

outcast°, abandoned 1; castaway; derelict 3; down-and-out 2; exile 2; forlorn 2; foundling; homeless 1; lonely 3; lonesome 1; undesirable 1

outcasts, dregs 2; rabble 2

outclass, end 11; outstrip; shade 8; shame 5a; surpass

out (cold), drunk 1; out 8; senseless 1; unconscious 1

outcome°, decision 2; determination 2; effect 1; end 4; event 2; fate 3; fruit; issue 2; judgment 2; payoff 2; product 1; purpose 3; repercussion; resolution 4; result 1; solution 2; upshot

outcrop, prominence 2; rock[1] 2

outcropping, projection 1; prominence 2; rock[1] 2

outcry°, demand 5; exclamation; noise 1; peep 3; racket 1; roar 3; storm 2; uproar; whoop 1

outdated, moldy; old 2; old-fashioned; out 9; prehistoric 2; quaint 2; time 15; obsolete; timeworn

outdistance, exceed 1; gain 5; outstrip; overhaul 1; overtake 1; pass 6; surpass; transcend
outdo°, beat 2; best 11; better 11; cap 4; discomfit 2; end 11; exceed 1; excel; extinguish 3; lead 4; outstrip; pass 6; shade 8; shame 5a; squelch 1; surpass; top 3; transcend; whip 2
outdoor°, outside 5
outdoors, out 1; outside 10
outer, exterior 1; external 1; extrinsic; outlying; outward 1; peripheral 2
outermost, extreme 3; outlying
outer reaches, outskirts
outfall, mouth 2
outfit°, apparatus; clothe 1; clothes; complement 2; condition 5; costume; disguise 3; dress 1, 6; employer 2; ensemble 1; equip; equipment; firm 6; fit[1] 8; furnish 1; garments; gear 2; get 51e; getup 1; house 4; paraphernalia; party 2; platoon; prepare 1; provide 1; rig 1, 3; robe 2; set 27; suit 3; supply 1; tack 4; tackle 1; turnout 3; uniform 3
outfitter, tailor 1
outfitting, provision 1; supply 5
outflank, flank 3; get 30b
outflow, drain 2; escape 7; issue 1; outpouring; spate
outflow(ing), outburst; spate
outfox, outsmart; prey 3b; victimize 2
outgo, drain 2
outgoing°, easy 6; expansive 2; forthcoming 3; issue 1; nice 1; pleasant 2; sociable; sunny 2
outgoings, expenditure; expense 1
outgrowth, fruit; offshoot 3; outcome; prominence 3; ramification 2
out-Herod Herod, overdo 1
outhouse, lavatory; privy 3; toilet 1
outing°, drive 7; excursion 1; journey 1; ride 4; run 36; spin 7; spree; tour 1; trip 3
outlander, alien 2; foreigner; immigrant; newcomer 1; outsider; stranger
outlandish°, absurd 1; alien 1; bizarre 2; curious 3; eccentric 1; erratic 2; exotic 2; extraneous 2; fantastic 1; foreign 2; grotesque 2; impossible 2; kinky 1; odd 1; offbeat; ordinary 4; outré; out-of-the-way 2; peculiar 1; preposterous; prohibitive 2; quaint 1; queer 1; rich 11; ridiculous; singular 1; strange 1; unheard-of 3; unnatural 3; weird
outlandishness, oddity 1; singularity 2
outlast°, resist 1; sit 7; stand 12b; survive 2
outlaw°, ban 1; banish 1; criminal 3; deserter; exile 3; expel 2; felon; forbid; offender; prohibit 1; thief 1; veto 1
outlawed, illegal; taboo 1; unlawful
outlawing, prohibition 1
outlawry, prohibition 1
outlay°, cost 1; expenditure; expense 1; overhead 3; payment 2; price 1; upkeep 2
outlet°, drain 1; exit 1; issue 1; loophole; mouth 2; passage 11; spout 3; store 4; vent 1
outline°, abridgment 3; abstract 3; block 4a; brief 4; circuit 1; conception 3; describe 4; design 2; draft 1, 5; draw 5; drawing; epitome 2; figure 1; lay[1] 18a; line[1] 5; note 10; plan 3; plot[1] 2, 4; profile 1; program 1; project 3; projection 3; proposal 2; prospectus; protocol 2; précis; represent 3; rough 14, 15; roundup 2; rundown; résumé 1; scenario; schedule 1, 2; scheme 1; silhouette; stake[1] 4b; story[1] 5; summary 1; synopsis; tell[1] 4; trace 6
outlining, definition 1; projection 2
outlive, outlast; sit 7; survive 2
outlook°, aspect 3; expectation 4; exposure 4; idea 3; mentality 2; perspective 2; philosophy 2; point of view 1; position 3; posture 3; prospect 1, 2, 3; sentiment 2; thinking 2; view 1
outlying°, back 12; faraway 1; out 10; outward 1; out-of-the-way 1; remote 1
outmaneuver, flank 3; get 30b; outsmart; trick 8
outmanipulate, outsmart
outmoded, antiquated; antique 1; date 4; dead 7; defunct 2; extinct 2; obsolete; old 2; old-fashioned; out 9; passé; time 15
out of, be, run 31
out-of-date, antiquated; antique 1; extinct 2; musty 2; old 2; old-fashioned; passé; slow 8
out-of-door(s), abroad 3; outdoor; outside 5, 10
out of it, confused 2; disorientated; forgetful; hook 5; insane 1; insensible 1; passive 1; preoccupied 2; slow 8; square 6
out-of-pocket (expenses), expense 1
out-of-sight, splendid 3
out of (someplace), be, depart 1
out-of-(the)-body, unearthly 1
out-of-the-ordinary, noteworthy; obscure 5; peculiar 1; remarkable 2; special 1; strange 1; striking; unearthly 3
out-of-the-way°, inaccessible 2; isolated 3; obscure 3; outlying; outré; peculiar 1; remote 2; secluded 2; solitary 1; strange 1; unapproachable 2; withdrawn 2
out-of-towner, provincial 3; rubberneck 2; tourist
out on one's feet, exhausted 1
outpace, exceed 1; excel; pull 7
outperform, outstrip; shade 8; surpass
outplay, outsmart; shade 8
outpost, settlement 1
outpouring°, escape 7; flood 3; outburst; rain 3; spate; spill 4; storm 2; stream 2; torrent; volley 2
output°, harvest 1; product 1, 2; production 1, 2; turnout 2; work 4; yield 5
outrage°, anger 1; atrocity 2; disgust 1, 2; dishonor 1; enormity; grievance 1; horrify 2; indignity; infamy 2; injure 2; injury; insult 1, 2; limit 4b; offend 1; offense 2; persecute 1; persecution 1; provoke 3; sacrilege 2; scandal 1; scandalize; shame 3; shock 1; slight 7; violate 3; violation 3
outraged, disgusted
outrageous°, black 6; camp[2] 1; curious 3; damnable; disgraceful 2; disgusting; dishonorable 3; excessive 1; exorbitant; extravagant 2; extreme 5, 6; fancy 4; flagrant; glaring 1; grievous 2; gross 4; hideous 2; high 2; immoderate; impossible 2; indecent 2; infamous 2; inordinate 1; monstrous 2; obscene 2; preposterous; pricey; prohibitive 2; provocative 2; rank[2] 3; rude 2; scandalous 1; shameful; shameless; shocking 2; sight 6b; spicy 3; tall 3; ultra-; unconscionable 2; unearthly 3; ungodly 2; unheard-of 3; unreasonable 2; unwarranted; wanton 2; way-out 1; whopping 2; wild 7
outrageously, extremely; terribly
outrageousness, enormity; extravagance 2
outrank, exceed 1; excel
outré°, bizarre 2; camp[2] 1; curious 3; extreme 5; offbeat; outlandish; out-of-the-way 2; pale[2] 3; preposterous; queer 1; singular 1
outreach, exceed 1; reach 1
outright°, absolute 2; barefaced 1; categorical; dead 14; direct 9; downright 1; express 5; flat 4; openly 2; out-and-out; peremptory 2; point-blank 1, 3; positive 9; pronounced 2; pure 5; shoulder 6; stark 4; straight 5, 6, 14; total 3; unmitigated; unqualified 2
outrun, exceed 1; lead 4; pass 6; pull 7
outrush, outpouring
outsert, insert 2
outset°, beginning 1; dawn 2; first 6; onset 2; opening 4; origin 2; source 1; start 10; threshold 2
outset, at *or* **from the,** originally
outset, at the, first 5, 8
outshine, eclipse 2; end 11; exceed 1; excel; outdo; outstrip; overshadow 1; shade 8; shame 5a; shine 3; show 12c; surpass; transcend
outshoot, prominence 3
outside°, abroad 3; exterior 1, 2, 3; external 1, 2; extrinsic; face 6; independent 7; out 1; outdoor; outward 1, 2; peripheral 2; remote 4, 6; shell 2; skin 2; superficial 1; superficies; surface 1; ulterior 2
outside (of), bar 10
outsider°, alien 2; exile 2; foreigner; immigrant; newcomer 1; square 9
outsize(d), spacious; swollen; voluminous 2
outskirts°, fringe 2
outsmart°, get 30b; prey 3b; victimize 2
outspoken°, bluff[2] 2; blunt 2; brazen; broad 5; candid 1; direct 10; downright 1; expansive 2; explicit 2; forthright; frank 1; free 9; ingenuous 2; open 15; plain 3; round 8; uninhibited; unvarnished
outspokenly, openly 2
outspokenness, candor 1; honesty 2
outspread, extend 1; flat 2; open 17
outstanding°, big 4; bold 3; bright 7; brilliant 2; capital 6; chief 2; classic 1, 2; conspicuous 2, 3; dominant 2; due 1; effective 2; eminent 1, 2; excellent; exceptional 2; exemplary 2; exquisite 5; extraordinary 1; extreme 1; fine[1] 4, 5; first-rate; gifted; glorious 2; golden 5; good 2; grand 2, 5;

outstanding° (*cont.*) great 5, 6, 12; important 2; laudable; leading 2; magnificent; main 1; major 2; meritorious; monumental 1; notable 1, 2; over 8; par 3; particular 2; payable; phenomenal; polished 1; preeminent 2; prestigious; prime 2; prize[1] 5; prominent 1, 2; rare[1] 2; raving 2; remaining 1; remarkable 2; salient; signal 3; significant 1; singular 2; spanking 1; splendid 2; superb; superior 2; supreme 3, 4; terrific 2; towering 1; unpaid 1; wicked 7
outstanding, be, shine 3
outstanding, not, decent 3
outstandingly, big 10; especially 1; notably 1; particularly 1; preeminently
outstay, sit 7
outstretch, extend 1; reach 1
outstretched, flat 2; open 17
outstrip°, better 11; cap 4; end 11; exceed 1; excel; lead 4; outdo; overhaul 1; overtake 1; pass 6; shade 8; shame 5a; surpass; top 3; transcend
outthink, outsmart
out to lunch, crazy 1; deranged; insane 1; mad 1; psychotic 1
outturn, turnout 2
outvie, transcend
outwait, sit 7
outwardly°, apparently 2; evidently 2; ostensibly; seemingly; surface 2
outward(s), apparent 2; exterior 1; external 1; extrinsic; outside 9; outward 2; seeming; superficial 2
outwear, outlast
outweigh°, outdo; predominate
outwit°, best 11; discomfit 2; dupe 3; finesse 4; foil[1]; get 30b; hoodwink; mislead; outsmart; prey 3b; trick 8; victimize 2
ovariectomize, fix 13; neuter 2; sterilize 2
ovate, oval
ovation°, hand 6; praise 1
oven-ready, prepared 4
over°, above 3, 4; complete 2; give 16; out 11; past 1
over, be, stop 1, 4
over (and over) again, frequently 1; often; repeatedly; time 21
over (the extent of), for 10
over with, over 6
overabundance, abundance; embarrassment 3; excess 1; flood 3; glut 1; pile[1] 3; superfluity; surfeit; surplus 1
overabundant, superfluous; surplus 2
overact, overdo 1
overacted, theatrical 2
overacting, theatrical 2
overactive, hectic
overage, leftover; remainder 2; rest[2] 1; surplus 1
overall°, general 1, 3, 4; gross 2; master 5; panoramic; total 2
overambitious, ambitious 3; grandiose 1
over and above, addition 5; besides 2; further 3
over (and done with), washed-up
overawe°, daunt; discourage 1; face 17; intimidate
overbalance, outweigh
overbalanced, unbalanced 1
overbearing°, arrogant 2; bossy; brusque; condescending; dictatorial 2; dogmatic; domineering; haughty; heavy-handed 2; imperative 2; masterful 2; opinionated 1; overconfident 1; pompous 1; possessive 2; prescriptive; self-important; supercilious; tyrannical; ungracious
overbearing, be, lord 3
overbearingly, strong 23
overblown, exalted 2; inflated 1; tall 3
overburden, encumber 1; oppress 1; overdo 2; overload 1, 2; overwork 1; push 5; strain[1] 1; swamp 2; weigh 6
overcast°, dark 5; dull 5; gloomy 1; gray 2; hazy 1; heavy 9; murky; obscure 1; somber 2
overcharge, clip[2] 4; fleece; overload 1, 2; profiteer 2; sting 4; take 32b
overcharging, rip-off 3
overcoat, cloak 1; coat 1
(over)color, maximize 2
overcome°, back 7a; beat 2; best 11; break 7; bury 5; charm 6; conquer 1, 3; consume 5; crush 4; daze 1, 4; defeat 1; devour 3; discomfit 2; downtrodden; drown 2; finish 4; knock 6b; master 8; mournful 1; outdo; outstrip; outweigh; overpower 1, 2; overwhelm 1, 2, 3; prostrate 2, 4; quell 1; quench 2, repress; shatter 3; slaughter 4; smother 3; squelch 1; stagger 2; stun 2; subdue 2; subjugate; swallow 4; swamp 2; topple 2; triumph 3; undone[1] 1; upset 4; whip 2; win 1
overcome, be, lose 3
overcome (by *or* with), stricken 1
overcoming, victory
overconfidence, front 7; gall[1] 2; presumption 1; pride 2
overconfident°, cocky; cool 6; positive 5; presumptuous; smug
overcook, burn 5
overcoy, prudish
overcrowd, cram 1
(over)crowded, congested
(over)dainty, finicky 1; mincing
(over)decorated, busy 3; fussy 2
overdecoration, elaboration 1
(over)delicate, finicky 1
overdo°, dramatize; embellish 2; exaggerate; exhaust 4; far 7; labor 6; maximize 2; meal 2
overdone, dramatic 3; excessive 2; genteel 1; glorified 1; grand 3; grandiose 1; gushy; hack[2] 4; inordinate 1; labored 2; luxuriant 3; melodramatic; ornate; ponderous 2; precious 3; prosaic; sensational 2; tall 3; theatrical 2
overdose, surfeit; surplus 1
overdramatic, grandiose 1
overdraw, exaggerate; overstate
overdrawn, inflated 1; melodramatic; sensational 2
overdue°, back 13, 17; delinquent 3; late 1; tardy 1
overeat°, stuff 8
overeater, glutton
overeating, gluttony; greed 3
overembellished, labored 2
(over)emotional, maudlin; sentimental 2; sloppy 3
overemotionalism, sentimentality
overemphasis, exaggeration
overemphasize, exaggerate; labor 6; overstate
(over)enthusiastic, effusive; gushy
overenthusiastically, strong 23
overestimate, miscalculate
overexacting, harsh 2
overexcited, overwrought 1
overexert, overwork 1
overexertion, overwork 3
overextend, exceed 2; stretch 3
overfamiliar, familiar 3
overfamiliar, be, liberty 5
overfamiliarity, familiarity 3; freedom 7
overfed, fat 1; surfeited
overfeed, glut 4; overeat
overfill, satiate 1
overflow, abundance; brim 2; excess 1; fill 6; flood 2, 4; flow 2; flush[1] 5; glut 1; gush 2; run 33c; shower 2; spill 1; surfeit; torrent
overflow (with), abound 3; teem[1]
overflow(ing), abundant 1; ebullient; flood 1, 3; flow 6; flush[2] 2; full 9; generous 3; heavy 2; luxuriant 1; packed; profuse 2
overflowing (with), replete 1; rich 2; fraught 1
overfree, familiar 3
overfriendly, familiar 3; romantic 4
(over)fussy, hairsplitting
overgrown°, bloated; lush 1
overhang°, jut; ledge; project 5; projection 1; stand 9b; stick[1] 15
overhanging, protuberant
(over)hastily, prematurely 2
overhasty, premature 2
overhaul°, catch 14b; improve 1; overtake 1; refresh 3; refurbish; renew 1; renovate; revamp; revise 1; revision; service 2; shake-up
overhauling, overhaul 3; revision
overhead°, above 1; aloft; overhead 3; upkeep 2
overhear, eavesdrop
overheat, glow 5
overheated, fiery 1
(over)indulge, feast 4; overdo 1; overeat; pamper; satiate 1; stuff 8
overindulgence, dissipation 1; excess 2; orgy 2; satiety; waste 6
overindulgent, dissolute; epicurean 1; self-indulgent; wasteful
overjoyed°, ecstatic; elated; exalted 3; exuberant 2; exultant; glad 1; gleeful; happy 1; joyful 2; radiant 2; rapt 2; rapturous; rhapsodic; world 7
overjoyed, be, rejoice
overkill, excess 1
overlap°, fold 1, 4; lap[2] 2, 3
overlapped, double 2
overlay, coat 2; face 16; facing; film 1; overlap 1, 3; plaster; plate 6; scale[2] 2; skin 2; spread 7; wash 7, 15
overlie, cover 3; overlap 1
overload°, bow 4; encumber 1; glut 3, 4; oppress 1; overdo 2; overwork 1; swamp 2; tax 4; weigh 6
overloaded, heavy 13; loaded 1; packed
overloaded (with), lousy 4; replete 1
overlong, lengthy; protracted; redundant 2; tedious
overlook°, blink 4; boss 2; close 7; command 5; discount 3; disregard 1; dominate 2; excuse 1; face 15; forget 3; forgive 1; front 11; ignore 1; jump 3; look 3; miss[1] 4; neglect 1; omit 2; pardon 2; pass 3; skip 2; supervise; waive 2
overlook, do not, mind 9
overlooked, unnoticed
overlooking, indulgence 1; pardon 1
overlord, dictator; oppressor; tyrant
overly°, fault 7; sky 2; unduly 1

pacified, become *or* **make,** calm 5
pacifier, dummy 4; peacemaker
pacify, calm 5; compose 4; disarm 2; ease 5; hush 5; lull 3; quell 2; satisfy 1; silence 4; still 9; tame 6; tranquilize
pacifying, propitiatory 1; satisfying; soothing 1
pack°, band[2] 1; batch 2; bundle 1, 2; case[2] 3; cram 1; crowd 1, 4, 5; fill 1; flock 1; gang 1; herd 1; host[2]; huddle 1; jam 1, 5; load 3; mob 1; package 3; packet 1; parcel 3; press 8; ram 1; ring[1] 3; squeeze 4; stack 3; stow; stuff 6, 7; swarm 1; tissue; wad 1; wedge 3; wrap 1
  pack away, eat; overeat; swallow 1
  pack in, stop 1
  pack in *or* **into,** pack 5; pile[1] 6; throng 2
  pack it in, pack 6
  pack off, bundle 3; pack 7
  pack up, pack 8; stuff 9
package°, box[1] 2; bundle 1, 2; case[2] 3; pack 1, 5; packet 1; parcel 1; wrap 1; wrapper 2
package deal, package 2
packed°, alive 4; compact 1; dense 2; full 1; loaded 1; mobbed; overcrowded; populous; serried; solid 2; thick 2, 3
  packed to the gunwales, overcrowded
  packed with, fraught 1
  packed with, be, abound 2
packet°, bundle 1; mint 1; pack 1; package 1; parcel 1; pile[1] 2
  packet, make a, profit 5
packing, wrapper 2
pack of lies, moonshine 2
pack(s), score 4
pact°, accord 3; agreement 1; alliance 1; bargain 1; bond 2; contract 1; instrument 3; negotiation 2; protocol 2; treaty; truce 2; understanding 1
pad°, cushion 1; domicile 1; patch 1; place 6; sneak 1; tablet 1; wad 1; waddle; walk 1
  pad out, pad 5
padding, expletive 3; filling 1; pad 1
paddle°, oar 1; punish 2; spank; waddle; wade 2
paddle one's own canoe, shift 2
paddler, oar 2
paddling, punishment 2
paddock, run 46
paddy, explosion 2; pet[2]; rage 2; tantrum; temper 4
paddywhack, explosion 2; paddle 4; pet[2]; rage 2; spank
padlock, bar 8; close 1; lock[1] 1, 3
padre, father 4; minister 1; priest
paean, chant 1; eulogy; praise 2
pagan°, heathen 1, 2; heretical; infidel; nonbeliever; profane 1
page°, messenger; runner 2; servant 1; sheet 2
pageant°, display 5; entertainment 2; extravaganza; gala 1; sight 3
pageantry°, display 5; glitter 4; glory 3; pomp
pageboy, page[2] 1
paginate, page[1] 3
pagoda, hurt 3; sanctuary 1; temple; tower 1
paid, be, get 3; pull 9b; receive 2
paid to, put, end 9
pail, bucket
pain°, ache 1, 3, 4; affliction 1; agony; ail 1; anguish 1; annoyance 2; bother 7; cut 3; discomfort 2; distress 1; evil 7; fag 2; grief 1; gripe 3; hell 3; hurt 2, 3, 6; ill 8; inconvenience 1; kill 5; offend 1; oppression; pang 1; passion 5; pierce 4; prick 2; rack 2, 3; rankle; rend 3; smart 7, 8; stab 5; sting 2; stress 3; suffering; torment 1, 3; twinge 1, 2; wound 2, 4; wrench 5
pained, hurt 7
painful°, agonizing; arduous 1; bad 11; bitter 2; excruciating; grievous 1; hard 5; harrowing; heart-rending; keen[1] 3; nasty 2, 5; picnic 3; piercing 4; poignant 1; pungent 3; raw 4; severe 4; sore 1, 5; stiff 2; tender[1] 8; thorny 2; traumatic
painfully°, hard 14; severely 4
painfulness, difficulty 1; severity 4; torment 3
pain in the ass (*or* **butt** *or* **rear),** annoyance 2; bother 7; headache 2; job 4; nuisance 1; pain 3; pest; picnic 3; plague 2; thorn 2; trial 5
pain (in the neck), (a), annoyance 2; bind 6; bother 7; drag 7; headache 2; job 4; nuisance 1; pain 3; pest; picnic 3; pill 2; plague 2; thorn 2; trial 5
pain killer, drug 2
painless°, effortless
pains°, care 2; effort 1; endeavor 2; labor 1; pain 4
  pains, take, trouble 4
painstaking°, careful 2; close 14; conscientious 2; deliberate 3; diligent; elaborate 1; exact 2; laborious 2; methodical; meticulous; painful 3; particular 3; rigid 3; scientific; scrupulous 1; severe 1; squeamish 1; studious 1; thorough 2
painstakingly, thoroughly 2
paint°, coat 3; color 3; dapple 1; decorate 2; describe 3; draw 5; exaggerate; picture 7; portray 1; represent 3; spread 7; wash 7
paintbrush, brush[2] 1
painted woman, prostitute 1
painting, likeness 2; picture 1
paint the lily, embellish 1; overdo 1
paint the town red°, carouse 1; celebrate 2; paint 7; revel 2
pair°, bosom 1; brace 4; couple 1; team 2; twin 3
  pair off *or* **up,** match 5; mate 4, 5; pair 2
paired, double 1
pairing, association 2
pal°, boy 3; brick 2; brother; chum 1; comrade; friend 1; intimate[1] 5; mate 1; partner; playmate; shadow 6
  pal (around *or* **up** *or* **about (with)),** associate 1b; chum 2, 3; mingle 2; pal 2; rub 7
palace°, castle 2; residence 3
paladin, gallant 4; guardian; hero 1; protector
palanquin, litter 3
palatable, dainty 3; delicious 1; edible; good 7; luscious; pleasant 1; savory 1; tasty
palate, taste 3
palatial°, grand 1; lush 3; plush; regal; rich 4; sumptuous
palatial home *or* **residence,** palace
palatinate, realm 1
palatine, palatial
palaver°, chat 1; chew 2; conversation; gossip 1; jargon 2; nonsense 1; parley 1, 2; patter[2] 2, 3; prattle 1, 2; talk 4, 15, 18; waffle 3
palavering, palaver 2; prattle 2
palazzo, palace
pale°, circuit 1; colorless 1; deadly 4; dim 1; fade 1; faint 1; frontier; ghastly 2; gray 1; light[1] 14; lurid 3; nebulous; neutral 2; pasty; peaked; pedestrian 2; picket 1; post[1] 1; region 2; sickly 2; soft 9; stake[1] 1; unwholesome 3; wan 1; washed-out 1; weak 6; white 2
  pale, beyond the, extreme 6; pale[2] 3
pale-complexioned, white 4
pale-faced, pasty
palfrey, hack[2] 3; mount[2] 8
paling, pale[2] 1; picket 1; stake[1] 1
palisade, crag; fence 1; pale[2] 1; picket 1; stake[1] 1; wall 2
palisades, bluff[2] 3
pall°, coffin; glut 4; mantle 2; satiate 1; shroud 3
  pall on *or* **upon,** pall[2] 1
  pall over, cast a, depress 1
palliate, ease 6; excuse 3; explain 2; mitigate; relieve 1; salve 3; smooth 12; soften 1; sweeten 2; temper 5
palliating, extenuating; mitigating
palliation, excuse 4
palliative, painkiller; salve 2; soothing 2
pallid, colorless 1; deadly 4; ghastly 2; gray 1; lurid 3; pale[1] 1; pasty; peaked; sickly 2; unwholesome 3; wan 1; washed-out 1; watery 1; white 2
pally, chummy; close 15; friendly 1; thick 8
palm, hook 7; pilfer; pocket 4; take 3; trophy 1
  palm off on, foist; wish 3
palmer, pilgrim
palmist, fortuneteller
palm-reader, fortuneteller
palmy, golden 5
palp, feeler 1
palpable, blatant 1; clear 7; conspicuous 1; distinct 1; evident; live 1; manifest 1; material 5; naked 3; noticeable 1; obvious; open 13; patent 2; perceptible; physical; real 2; self-evident; sensible 2; tangible
palpably, evidently 1; manifestly; materially
palpate, massage 2
palpitate, beat 3; pound[1] 3; pulsate
palpitating, pulse 1; tremulous 1
palpitation, pulse 1
palsied, doddering
palsy-walsy, chummy; close 15; familiar 3; friendly 1; thick 8
palter, fence 4; fib 2; haggle; negotiate 1; quibble 1
paltering, quibble 2
paltriness, triviality 1
paltry°, feeble 2; flimsy 2; frivolous 1; frugal 3; incidental 2; inconsequential; insignificant; insubstantial 1; little 5; meager 1; meaningless 2; measly 1; minor 2; negligible; pale[1] 3; pathetic 2; petty 1; poor 5; puny 1; small 2; tenuous 2; thin 2; tiny; trifling; unworthy 1; worthless 1; wretched 4
pampas, flat 14b; plain 6
pamper°, baby 2; cater 2; coddle; dote; favor 7; humor 4; indulge 2; make 34b; mother 6; nurse 2; pet[1] 5; spoil 3
  pamper (with), indulge 1
pampered, hothouse 2; luxurious 2; soft 12, 13
pamphlet°, brochure; literature 2; publication 2; tract[2]
pan°, bowl[2]; criticize 2; face 1; mug 2; pot 1;

pull 6; rubbish 3; run 26b; slam 3; talk 9a
pan out, pan 6; work 11, 19d
panacea, cure 1; drug 1; elixir 1; medicine; remedy 1, 2
panache°, dash 6; finesse 1; flair 2; spirit 3; style 4; verve; virtuosity
pandect, code 1
pandemic, epidemic 1; global; pestilence 1; plague 1; rampant 2; universal 1
pandemonium°, bedlam; chaos; confusion 2; din 1; disorder 2; fracas 1; hell 1; noise 1; pell-mell 3; racket 1; riot 1; row[2] 2; scramble 4; tumult; uproar; zoo 2
pander°, pimp 1, 2; procurer
pander to, cater 2; indulge 1; kowtow; pander 1; romance 7; solicit 2
panderer, pander 3; pimp 1; procurer; slaver[2] 2
pane°, glass 3; plate 3; sheet 3
panegyric, eulogy; oration; praise 1; tribute 1
panegyric(al), complimentary 1; glowing 3; laudatory; lyrical 2
panegyrize, glorify 2
panel, board 4; committee; council 2; pane; plaque 1; plate 3; sheet 3; tablet 2; terminal 4
pang°, ache 3, 4; gripe 3; hurt 6; kink 2; pain 1; qualm; smart 8; stab 5; throe; twinge 1, 2; wrench 5
Panglossian, idealistic
pangs, agony
panhandle, beg 2
panhandler, beggar 1; bum 2
panic°, alarm 2, 3; dismay 1, 3; dread 2; fear 1; flap 4; fright 1; frighten; horrify 1; horror 2; riot 2; scream 4; stampede 1, 2; terror 1
panicky, fearful 2; jumpy; panic-stricken; pell-mell 2
panic-stricken°, afraid 1; desperate 1; fearful 1; petrified 1
panic-struck, panic-stricken
pannier, hamper[2]
panoply, display 5; gear 2; robe 2; splendor 1; trappings
panorama, pageantry; prospect 1; scene 4; view 1
panoramic°, general 3; scenic
panpipe, pipe 3
pansy, deviant 3; homosexual 1; milksop
pant°, gasp 1; puff 4
pant for *or* after, pant 2
pantalettes, pants 1
pantheistic, heathen 3
panties, pants 1
pantihose, tights
panting, breathless 1; short-winded
pantry, kitchen
pants° down, with (one's), unprepared 1
pantyhose, pants 1; tights
pantywaist, milksop
pap, pulp 2
papa, father 1
papal, clerical 1
paparazzo, photographer; press 10b
paper°, composition 1; disposable 1; document 1; essay 1; exposition 3; instrument 3; journal 1; lecture 1; memoir 1; monograph; organ 2; periodical; publication 2; sheet 6; slip[2] 1; theme 2; tract[2]; warrant 2
paper, commit to, pen[1] 2
paper, put on, pen[1] 2
paperback, book 1
paperhanger, counterfeiter; phony 3
paper money, bill[1] 2; money 1
paper(s), license 2; material 4; paper 2
papers, the, press 10a
Paphian, libertine 2; obscene 1; prurient 1
papule, pimple
par°, parity 1; poise 1; standard 2
par, above, par 3
par, at, par 4
par, below, off 8; par 5; poor 4; poorly 2; run-down 1
par, not up to, low[1] 9; par 5a; poor 4; wanting 1
par, on a, even 5; square 2
par, under, par 5; poor 4
par, up to, par 6; presentable 1; scratch 5
par (with), be on a, compare 2; stack 7b; touch 7
parable°, byword; myth 1; story[1] 1
parabole, metaphor
parabolic(al), metaphoric; mythical 1
parachronism, anachronism
parade°, air 7; area 5; demonstrate 3; demonstration 3; display 3, 5; exhibit 1; file 5; flaunt; flounce 2; line[1] 6; march 1, 2; ostentation; pageant; procession 1; promenade 1, 3, 4; review 8; shake 4; show 11; strut; swagger 1; sweep 3; train 3; walk 1
parade ground, area 5
paradiastole, euphemism
paradigm, classic 3; ideal 1; precedent; standard 1; type 3
paradigmatic, classic 1; exemplary 1
paradisaic(al), celestial 1; heavenly 2; idyllic
paradise°, fairyland; heaven 1, 3; transport 5; utopia
paradisiac(al), celestial 1; heavenly 2; idyllic
parados, rampart
paradox°, absurdity 2; perplexity 3; puzzle 4
paradoxical°, absurd 2; contradictory; incongruous; left-handed 2; perplexing
paraenesis, advice 1; counsel 1
paraenetic(al), advisory 1
paragon°, classic 3; ideal 1; lesson 3; model 2, 3; nonpareil 1; optimum 1; pattern 1; perfection 3; prototype 2; quintessence; standard 1
paragraph, amendment 2; passage 2; text 2
paragrapher, writer
paragraphist, journalist; scribe 2; writer
paralipsis, emphasis
parallel°, compare 2; contemporary 1; coordinate 3; equal 5; kindred 1; level 4; match 1; mate 3
parallel (to *or* with), even 2; like[1] 1
parallelism, harmony 2; kinship 2; likeness 1; parallel 3; parity 2
paralogism, fallacy
paralysis, prostration 4; shock 2
paralytic, cripple 1; far 5b
paralytic attack *or* fit, stroke 5
paralyze°, bemuse 2; daze 1; deaden 1; freeze 3; incapacitate; numb 2; petrify 1; prostrate 2; shatter 3; shock 1; stun 2; terrify; transfix 2
paralyzed, dead 2; far 5b; petrified 1; powerless 2; prostrate 4; speechless 2; thunderstruck
paralyzing, stunning 1; terrifying
paramount°, best 1; capital 5; cardinal; chief 3; dominant 2; foremost 1; head 9; lead 18; leading 1; main 1; major 2; maximum 3; overriding; peerless; predominant; principal 1; sovereign 2; special 6; superlative; supreme 1; towering 1; transcendent; ultimate 2; uppermost 2; vital 2
paramour°, beloved 2; fellow 4; gallant 5; love 3; mistress 1; suitor; sweetheart
paranoiac, psychotic 1, 2
paranoid, psychotic 1, 2
paranormal, supernatural
paraphernalia°, apparatus; effects; equipment; fitting 2; frill 2; furniture 1; gear 2; good 21a; kit; luggage; outfit 1; property 1; regalia; rig 3; robe 2; stuff 2; tackle 1; thing 8c; trappings
paraphrase°, interpret 1; reword; simplify; translate 1; translation 1
paraphrasing, interpretation 1
paraphrasis, paraphrase 1
parasite°, bloodsucker; flatterer; hanger-on; menial 4; satellite 2; scrounge 2
parasol, shade 5; umbrella 1
parasynesis, misunderstanding 1
parasynthesis, combination 3; compound 5
parasynthetic, compound 4
paratactic, coordinate 3
parathesis, combination 3; compound 5
parathetic, compound 4
paravent, screen 1
parcel°, bundle 1; dole 1; pack 1; package 1; packet 1; part 1; patch 2; portion 1; tract[1]
parcel out, allot; carve 2; deal 1; dispense 1; dispose 3c; distribute 1; divide 2; measure 14; mete; parcel 4; portion 4; ration 3; share 3; split 4
parceling out, disposition 3; distribution 1; partition 2
parcel (of land), plot[2]
parch°, dry 4
parched, dry 1; thirsty 1; torrid 1
parching, burning 3; evaporation 1; scorching 1; torrid 1
pardon°, discharge 1; excuse 1, 5; forgive 1; forgiveness 1; free 14; let[1] 6a; overlook 2; purge 3; purify 2; remission 1; spare 7
pardon, beg *or* ask, apologize 1
pardonable, permissible; venial
pare°, chisel 1; lop; peel 1; shave 2; trim 3; whittle 2
pare (down *or* away), abridge; diminish 1; minimize 1; pare 2; prune; whittle 1
parent°, mother 1; raise 4
parentage°, birth 3; extraction 3; family 3; lineage 1; line[1] 15; origin 3; paternity; pedigree; stock 2; strain[2] 1
parental, fatherly
parenthesis, digression 1
parenthetically, incidentally 1; offhand 5; passing 3; way 9
parenthood, maternity 1, 2
paresthesia, hallucination
parfum, perfume 1
pariah, castaway; derelict 3; exile 2; outcast; undesirable 1
pariahs, dregs 2
paring, sliver
pari passu, parallel 1
parish, municipal
parish pump, provincial 2
parity°, equality 1; parallel 3; poise 1
park°, preserve 5; set 1; square 8

path° (*cont.*)
line[1] 9; orbit 1; route 1; run 45; step 10a; swath; tack 3; track 2; trail 1, 2; wake[2] 1; walk 5; way 3
pathetic°, disappointing; forlorn 1; meager 1; measly 1; miserable 3; moving 1; paltry; piteous; poignant 1; poor 6; sad 3; sorry 2; touching; unfortunate 2; weak 4; wretched 3
pathetic(al), tragic
pathfinder, pioneer 1
pathless, trackless
pathogenic, morbid 1
pathological, morbid 1
pathos, expression 4; sentimentality
pathway, path 1; walk 5; way 7
patience°, endurance 1; indulgence 1; kindness 1; perseverance; persistence; philosophy 3; self-control 2; stoicism; tolerance 1
patient°, case[1] 3; client; gentle 1; indulgent; inmate; invalid[1] 2; lenient; meek 1; passive 2; philosophical 2; tolerant
patient (with), be, bear 11; persevere; sit 8
patina, glaze 2; shine 4
patois, cant 2; dialect; idiom 1; jargon 1; language 1; lingo; provincialism 1; talk 19; tongue 1; vernacular 3
pat on the back°, pat[1] 2, 5
patrial, citizen 1
patriarch, elder 3
patriarchal, paternal 1
patrician, genteel 2; noble 1, 2; upper-class 1
patricide, murder 1
patriclinal, paternal 2
patriclinic, paternal 2
patriclinous, paternal 2
patrilateral, paternal 2
patrilineage, paternity
patrilineal, paternal 2
patrilinear, paternal 2
patrimonial, paternal 2
patrimony, heritage 1; inheritance
patriotic°, loyal; national 2; nationalistic
patriotism, loyalty
patroclinal, paternal 2
patroclinic, paternal 2
patroclinous, paternal 2
patrol°, guard 4; picket 3; platoon; police 2; prowl 2; sentinel; walk 3
patrolling, patrol 2
patrolman, constable; patrol 1; police officer
patrolwoman, police officer
patron°, advocate 2; backer 1; benefactor; client; customer 1; employer 1; friend 4; guest; habitué; philanthropist; proponent; protector; regular 12; sponsor 1
patronage°, auspices; backing 1; clientele; cooperation 2; custom 3; favor 3; furtherance; office 5; philanthropy 1; protection 2; trade 4; umbrella 2
patroness, patron 1; philanthropist; protector
patronize°, forward 6; foster 1; frequent 2; further 5; promote 1; resort 3; second[1] 9; sponsor 3; talk 9b
patronizer, patron 2
patronizing, condescending; haughty; lofty 4; patronage 3; snobbish; supercilious
patrons, clientele; public 9; trade 4
patsy, butt[1]; dupe 1; puppet 2; pushover 2; sap[1] 2; score 7; sucker; victim 2
patsy for, be a, fall 13b
patter°, chatter 1, 3; gibberish; jabber 1; jargon 2; prattle 1, 2; repartee; talk 3; tap[1] 2
pattering, tap[1] 2
pattern°, cycle 1; design 1, 5, 6; example 2; figure 7; form 1; formation 3; formula; frame 3; gauge 3; grain 4; habit 1; ideal 1; last[3]; lead 13; method 1, 2; model 2; mold[1] 1, 2; motif; norm 2; order 2; organization 2; original 5; paragon; perfection 3; plan 1; precedent; progenitor 2; prototype 1; routine 1; rut 2; scheme 2; script 4; shape 2; specimen; standard 1; style 1; swing 5; system 1; template; type 3; vein 5
pattern after, model 9
pattern on, model 9; pattern 7
pattern(s), code 3
patty, pat[1] 4; tart[2] 1
paucity, absence 2; dearth; famine; lack 1; need 5; poverty 2; scarcity; shortage; want 3
Paul Pry, busybody; snoop 2
paunch°, pot 3; stomach 1
paunchiness, fat 5; obesity
paunchy, fat 1; heavy 11; obese
pauper°, beggar 1
pauperism, necessity 3; poverty 1; privation; want 5
pauperize, ruin 10
pause°, break 28; check 13; gap 1; hesitate 1; interlude; interruption 2; interval 1; lapse 2; letup; linger 2; lull 1; plateau 2; procrastinate 2; recess 2; respite 2; scruple 2; slack 5; space 3; stammer 1; stop 5; stumble 2; tarry 1; wait 4
pause at, stick[1] 12
pausing, interruption 2
pave°, smooth 10; surface 4
pavement, sidewalk; walk 5
pave the way (for), pave 2; precede
paw, hand 1; manhandle; molest 2
pawky, wily; wry 2
pawn°, cog 2; dupe 2; hostage; plaything 2; pledge 2, 5; puppet 2; tool 3
paws, extremity 2
pay°, afford 1; amends; atone; clear 28; compensate 3; defray; discharge 6; earnings; fee 2; gain 8; give 2, 15b; honor 8; honorarium; lay[1] 18b; meet[1] 5; penance 2; put 28e; remit 1; remuneration 1; reward 1, 4; salary; satisfy 4; settle 10; stipend; wage 1; yield 4
pay back, restore 5
pay back (for), fix 14; pay 5, 8; reimburse; repay; requite 2
pay back (in kind), retaliate
pay (for), finance 3; fund 3; make 31a; pay 9; purchase 1; retrieve 3; spring 5; support 4; treat 4
pay (for), make (someone), pay 5
pay in, deposit 2
pay off, bribe 2; buy 3; pay 1, 4, 5, 7, 8a, 10; redeem 1; score 9
pay out, expend 1; pay 1, 5, 11; share 3; shell 5; spend 1; stump 4
pay (someone) back, get 23; retaliate; revenge 3; score 9
pay (the bill) for, treat 5
pay (up), come 5b; pay 1; stump 4
payable°, due 1; outstanding 2; unpaid 1
(pay) court (to), pursue 4
payload, freight 2
payment°, allowance 2; atonement; consideration 2; cost 1; desert[2]; discharge 13; expenditure; expense 1; fee 2; fruit; gain 8; honorarium; indemnity 1; outlay; pay 12; price 1, 3; ransom 2; remittance; remuneration 1; requital 1; return 11; reward 1; royalty 2; satisfaction 2; settlement 4; stipend; subscription 1; tender[2] 3; term 7a; tribute 2; wage 1
payment for, in, for 3
payoff°, clincher; gain 8; issue 2; kickback; outcome; ransom 2; upshot
payola, bribe 1; graft[2]; kickback; payoff 3; rebate 2
payout, gain 8; payment 2; ransom 2
pay the way for, see 11
PC, police officer; terminal 4
PDQ, double 8; fast[1] 6; hastily 1; posthaste; straight 15; summarily 1
pea-brained, senseless 3
peace°, calm 1, 2; content[2] 2; ease 1; hush 6; inactivity 1; lull 2; order 8; quiet 5; repose 1; serenity 1; silence 1
peaceable°, amicable; dreamy 3; mild 1; orderly 2; pastoral 1; peaceful 1; serene 1
peaceableness, peace 1; serenity 2
peaceably, happily 3; quietly 3
peace and quiet, ease 1
peaceful°, amicable; calm 3; dispassionate 1; dreamy 3; easy 2; equable 1; even 4; gentle 1; home 4a; homely 2; idyllic; inactive 1; mild 1; pastoral 1; quiet 2; relaxed; restful 2; sedate 1; self-contained 1; self-possessed; serene 1; silent 1; smooth 1; soothing 1; sound[2] 6; still 1; subdued 1; tranquil
peacefully, easy 7; quietly 3
peacefulness, calm 1; content[2] 2; ease 1; inactivity 1; lull 2; order 8; peace 1, 2; serenity 1, 2; silence 1; still 3
peace-loving, peaceable 2
peacemaker°, mediator
peacemonger, peacemaker
peace officer, officer 2; police officer
peace of mind, peace 1; quiet 5
peace pipe, pipe 2
peacetime, peace 2
peach, disclose 1; sing 3; tell[1] 2
peach (on), finger 8; inform 2; shop 3
peaches-and-cream, fair[1] 3
peachy, bully 4; fine[1] 1
peacock, braggart; strut
peak°, acme; climax 1, 4; consummation 2; crag; crest 1; extreme 8; head 5; height 2, 3; high 15; maximum 1, 2; mountain 1; pinnacle; point 2; prime 4; prominence 2; spire 2; summit; tip[1] 1; top 1; vertex; zenith
peaked°, pale[1] 1; pointed 1; run-down 1; sickly 2
peakish, pale[1] 1; peaked; sickly 2
peaky, pale[1] 1; peaked; sickly 2
peal°, chime 1, 2, 3; gale 2; ring[2] 1, 3; roll 5, 15; toll[1] 1, 2
pealing, ring[2] 3; thunder 1; toll[1] 2
peanuts, nothing 3; pittance
peanuts, for, cheap 7
pearl°, gem 2; jewel 2
pearllike, pearly
pearly°, filmy 2; gray 1; opalescent; silver 3
pear-shaped, rotund 2; round 7; smooth 7

peasant°, boor 1; clown 2; ordinary 2; rustic 2, 3; yahoo

peasantlike, plebeian 1

peasantry, populace; rabble 2

pea-soup, thick 3

pea-souper, fog 1

pebbly, rocky[1] 1; stony 1

peccadillo°, fault 1; imperfection; indiscretion 2; misdeed; offense 1; sin 1; slip[1] 8

peck, heap 2; kiss 1, 4; pack 2; tap[1] 1, 2

  peck at, pick 4b

  peck away (at), peg 6

pecker, bill[2]

pecking, tap[1] 2

pecking order, order 3

peckish, hungry 1

Pecksniff, pharisee

Pecksniffian, goody-goody; pharisaic

peculate, embezzle; misappropriate 1; steal 1; take 3

peculation, embezzlement; stealing

peculator, thief 1

peculiar°, abnormal 2; bent 2; bizarre 1; character 1; cranky 1; curious 3; deviant 1; different 2; distinct 2; distinctive; eccentric 1; erratic 2; exceptional 1; exotic 2; extraordinary 1; fanciful 3; fantastic 1; fishy 1; foreign 2; funny 2; grotesque 2; improbable; individual 2; irregular 3; kinky 1; local 2; odd 1; offbeat; ordinary 4; outlandish; outré; out-of-the-way 2; pale[2] 3; particular 1; proper 6; quaint 1; queer 1; remarkable 3; sick 5; singular 1; special 1; specific 1; strange 1; unaccountable 1; unaccustomed 1; unearthly 3; unfamiliar 1; unnatural 1; unusual; unwonted; way-out 1; weird; whimsical 1; wild 7; wrong 2

  peculiar (to), peculiar 2; specific 2; symptomatic

peculiar institution, slavery 1

peculiarity°, character 2; characteristic 2; difference 4; eccentricity 2; exception 3; fancy 8; feature 1; foible; freak 3; mannerism; oddity 1, 2, 3; property 4; quirk; singularity 2; trait; trick 6; twist 10

peculiarly, especially 1; particularly 1

pecuniary, economic 1; financial; fiscal; monetary

pedagogic, pedantic 1

pedagogical, educational 1

pedagogue, instructor; scholar 1; schoolteacher

pedal pushers, pants 2

pedant, perfectionist 1; prig; purist

pedantic°, literary 2; pompous 2; priggish; stiff 8; stuffy 2

pedantical, pedantic 1

peddle°, market 4; sell 2; tout 1

peddler°, merchant 2; seller

pedestal°, base[1] 1; standard 4

  pedestal, put *or* **place** *or* **set on a**, idealize; idolize; pedestal 2; worship 1

pedestrian°, banal; mediocre; ordinary 2; ponderous 2; prosaic; ready-made 3; undistinguished

pedicel, stem[1] 1

pedicular, lousy 3

pediculous, lousy 3

pedigree°, class 1; extraction 3; family 3; lineage 1; nationality 2; origin 3; parentage; root[1] 4; stock 2; strain[2] 1; tribe

pedigreed, purebred

pedlar. See **peddler**

peduncle, stem[1] 1

pee, go 17; urinate; water 4

peek°, glance 1, 4; peer[2] 1; pry 1; sight 7

  peek at, sight 5

  peek through, show 5

peekaboo, filmy 1; see-through; sheer 3

peel°, film 1; hull 2, 3; pare 1; rind; shell 3; skin 2, 4; strip[2] 1

  peel (off), peel 1, 3; shed[2] 4; take 34a

peeler, police officer

peeling, peel 4

peel off (one's) clothes *or* **clothing**, strip[2] 2

peep°, chirp 1, 2; glance 1, 4; peek 1, 2; peer[2] 1; pipe 4; sight 7; sing 2; twitter 1, 3

  peep at, sight 5

  peep through *or* **out**, peer[2] 2

pee-pee, urinate

peep(er), busybody; detective; snoop 2

peeping, curiosity 2; nosy; twitter 3

peer°, equal 4; equivalent 2; fellow 2; like[1] 7; lord 2; match 1; noble 1; peek 1; pry 1; royal 3; sight 7

  peer at, watch 3

  peer at *or* **upon**, eye 10

peerage, rank[1] 3; royal 4; royalty 3

peerless°, alone 2; capital 6; excellent; exquisite 5; great 6; incomparable; leading 2; masterful 1; matchless; notable 2; optimum 2; perfect 2; preeminent 2; prime 2; rare[1] 2; sensational 3; splendid 3; star 3; sterling 2; superb; superior 2; superlative; supreme 4; surpassing; top 8; transcendent; unbeatable; unique 2; unparalleled; virtuoso 2

peerlessness, preeminence 2; superiority 2; supremacy 1

peeve, aggravate 2; displease; disturb 1; exasperate 2; gnaw 3; infuriate; irk; irritate; pester; rub 8; sour 6; spite 3; worry 2

peeved, disgruntled; huff 1; indignant; resentful; sore 4

peevish°, bad 7; bilious; cantankerous; cranky 2; cross 6; disagreeable 3; disgruntled; edge 5; fretful; gruff 1; harsh 3; irritable; moody 2; passionate 3; perverse 2; petulant; prickly 3; quarrelsome; querulous; short-tempered; snappish 1; sour 4; sullen; surly; temperamental 1; testy; touchy 1; waspish

peevishness, temper 3

peewee, small 1

peg°, freeze 3; heave 2; hook 1; label 5; leg 1; nip[2]; notch 2; picket 1; pin 1; pitch[1] 1; spike 1; tack 5; tap[2] 2

  peg, take (someone) down a, humble 4

  peg along, peg 6; plod 2

  peg away (at), labor 5; plod 2; wade 4; work 7

pegtops, pants 2

peignoir, robe 1; wrapper 1

pelagic, marine 2; oceanic

pelerine, mantle 1

pelf, fund 2; money 1; purse 2; riches

pelisse, mantle 1

pellet, pill 1; shot 2; tablet 4

pellicle, film 1; skin 2

pell-mell°, disorderly 1; helter-skelter 2; rush 3

pellucid, clear 2; distinct 1; transparent 1

pellucidity, clarity 1

pelmet, drapery

pelt°, batter 1; beat 1; belabor; hail[2] 1; hide[2] 1; patter[1] 2; pound[1] 1; skin 1; strike 1

  pelt along *or* **over**, pelt[1] 3

  pelt down, pelt[1] 2

pen°, cage 1, 2; enclose 1; enclosure 1; jail 1; pound[2]; prison; run 46; script 3; shed[1]; shut 3a; stake[1] 4a; stall[1] 3; stir 7; write 1

  pen in, siege 2

  pen (up), bring 15b; pen[2] 2

penal institution, prison

penalize°, castigate; discipline 8; fine[2] 2; punish 1

penalty°, discipline 2; fine[2] 1; forfeit 1; penance 1; price 2; punishment 2; sanction 4; toll[2] 2

  penalty on *or* **against, impose** *or* **invoke a**, penalize

penance°, atonement; penalty; penitence; punishment 2

  penance, do, penance 2

penchant°, fancy 7; habit 2; inclination 4; leaning; liking 1; partiality 2; talent 2; taste 3; weakness 4

pencil, beam 2; ray 1; shaft 2

pencilmark, line[1] 1

pencil pusher, writer

pencil-thin, thin 3

pend, shell 5

pendent, pendulous

pending°, abeyance; open 6; prospective; unresolved

pendulous°, flabby 1

penetrable, porous

penetralia, recess 3

penetrate°, bore[1] 2; break 22; come 5c; drill 1; enter 2; fathom; get 40b; gore[2]; impregnate 2; infest; percolate; perforate; permeate; pierce 1, 2, 3; plumb 5; puncture 3; riddle[2] 2; see 14a; sink 11; stick[1] 1; suffuse; tunnel 2; understand 1

penetrating°, acute 3, 5; biting; chill 4; cutting 1; deep 3; devastating 1; good 11; hard 6; high 7; incisive 1; intimate[1] 2; keen[1] 5; luminous 3; penetration 1; pervasive; piercing 2, 3; poignant 2; pointed 2; profound 2; pungent 1, 3; raw 5; sharp 3, 7; shrill; strong 2; trenchant

penetratingly, searchingly

penetration°, depth 3; discrimination 2; inroad 1; wisdom 1

peninsula, cape[1]; point 8

penitence°, penance 1; remorse 1

  penitence, show, repent

penitent°, apologetic; guilty 2; regretful; remorseful; repentant; sorry 1

  penitent, feel, repent

penitential, sorry 1

penitentiary, jail 1; prison; stir 7

penknife, blade 2

penman, scribe 2

penmanship°, hand 7; script 1; writing 1

penmark, line[1] 1

pen name, pseudonym

pennant°, banner 1; color 2a; flag[1] 1; standard 3; streamer; symbol

penniless, broke; destitute 1; down-and-out 1; hard 17; indigent; insolvent; needy; penurious 2; poor 1; rock[1] 3c; short 8; straitened

pennon, banner 1; flag[1] 1; pennant; streamer

penny-a-liner, hack[2] 1; scribe 2; writer

penny-ante, frivolous 1; minor 2; paltry

pennypincher, miser

pennypinching, avaricious; miserly; cheap 4; close 18; economical 2; frugal 2; grasping; greed 2; greedy 3; mean[2] 1; near 6; penurious 1; selfish 2; sparing 1; thrifty; tight 5

penny-wise and pound-foolish, improvident 1; wasteful

pen pal, friend 1

pen pusher, writer

pensile, pendulous
pension°, allowance 3; hotel
pension, go on a, retire 2
pension off, pension 2
pensioned off, be, retire 2
pensioner°, senior citizen
pensioners, elderly 2
pensive°, dreamy 2; meditative; preoccupied 1; reflective; serious 1; thinking 1; thoughtful 2; wistful 2
penthouse, flat 15; shed[1]
penurious°, avaricious; broke; cheap 4; close 18; destitute 1; economical 2; frugal 2; grasping; greedy 3; impoverished 1; indigent; lean[1] 3; mean[2] 1; miserly; near 6; selfish 2; sparing 1; thrifty; tight 5
penurious, be, stint 4
penuriousness, avarice; greed 2; thrift
penury, misery 2; necessity 3; need 4; poverty 1; privation; want 5
penury, reduce to, ruin 10
peon, peasant
people°, family 1; folk; humanity 1; inhabit; lineage 2; man 2, 4; one 4; populace; populate; population; public 8; race[2] 1; settle 6; society 2; tribe; unwashed 2; world 2
people, the common, mass 6
peopled, populous
pep°, animation 1; bounce 2; drive 8; dynamism; energy; enterprise 2; fire 2; initiative 2; life 7; salt 2; snap 11; spice 2; vigor; vitality 1
pep up, encourage 1; energize; enliven 1; liven 1; pep 2; perk up; season 3
peplum, flounce 1; ruffle 1
pepper°, pelt[1] 1; punctuate 1; riddle[2] 1; salt 2, 6; spice 2; spirit 3; zest 1
peppery, fiery 3; hot 2; passionate 3; pungent 1; spicy 1
pep pill, stimulant 2
peppy, buoyant 2; dashing 1; energetic; lively 1; perky; racy 1; rousing; vigorous; vivacious
pep talk, encouragement 2
peradventure, maybe; perhaps; possibly 1
perambulate, promenade 3; ramble 1; roam; walk 1
perambulation, tour 2
per annum, yearly 2
perceivable, discernible 1; evident; noticeable 1; observable; perceptible; sensible 2; visible 1
perceive°, appreciate 3; behold; catch 10; comprehend; conceive 3; detect 2; discover 1; distinguish 3; divine 4; feel 1, 3, 4; figure 12b; find 3; get 19; hear 1; hit 9b; make 37a, 37d; note 11; notice 1, 2; observe 3; penetrate 5; realize 2; recognize 2; regard 2; remark 1; savor 3; scent 3; see 1, 2, 14a; sense 6; take 19; tell[1] 8; tumble 3; understand 1
perceive of, perceive 3
percentage°, cut 20; interest 4, 6; kickback; money 3; part 1; piece 3, 11; portion 2; proportion 3; quota; rate[1] 3; ration 1; rebate 2; royalty 2; share 1
perceptible°, conspicuous 1; discernible 1; distinct 1; evident; external 3; noticeable 1; observable; obvious; sensible 2; slight 2; slow 2; tangible; visible 1
perceptibly, clear 18
perception°, appreciation 3; brain 1; capacity 2; cognizance; depth 3; discrimination 2; eye 2, 3; feeling 1; foresight 2; grasp 5; grip 3; head 4; idea 2; identification 1; image 3; insight; intelligence 1; intuition; judgment 1, 4; mind 1, 3; notice 3; opinion 1; penetration 2; realization 1; recognition 2; sensation 1; sense 2; sensitivity 3; sight 2; tact; taste 4; understanding 2, 4; uptake; vision 1, 3; wisdom 1
perceptive°, acute 5; astute 2; clear 9; clever 1; diplomatic; discriminating; eagle-eyed; farsighted 1; hip; incisive 1; intelligent; judicial 2; judicious; keen[1] 6; knowing 2; luminous 3; nice 3; observant 1; penetrating 1; politic 2; quick 4; quick-witted; ready 6; receptive 2; shrewd; smart 2; sound[2] 4; tactful; wise 1
perceptiveness, brain 1; capacity 2; discrimination 2; head 4; insight; intuition; penetration 2; uptake; wisdom 1
perceptual, mental 1
perch°, nest 1; settle 5
perchance, incidentally 2; maybe; perhaps; possibly 1
percipience, brain 1; eye 3; insight; intelligence 1; intuition; judgment 1; mind 1; penetration 2; reason 2; understanding 2; wisdom 1
percipient, acute 5; incisive 1; intelligent; judicial 2; judicious; keen[1] 6; luminous 3; penetrating 1; perceptive; politic 2; shrewd; smart 2; sound[2] 4; wise 1
percolate°, filter 3; strain[1] 5
percolate through, penetrate 2; permeate
peregrinate, journey 3; roam
peregrinating, vagabond 2
peregrination, journey 1; tour 1; trip 3
peregrinations, travel 2
peregrinator, migrant 1
peremptorily, sharply 1; short 11; shortly 3; summarily 2
peremptory°, arbitrary 2; assertive; dogmatic; domineering; emphatic; explicit 1; flat 4; imperative 2; overbearing; positive 2; shabby 2; short-tempered; summary 2
perennial°, constant 2; indestructible; permanent 1; perpetual 1; yearly 3
perennially, yearly 1
perestroika, reform 3
perfect°, absolute 1; accomplished; clear 14; complement 3; complete 4, 6; correct 8; divine 3; exact 1; exquisite 3, 5; faithful 2; faultless; finish 2, 7; flawless 1; heavenly 2; ideal 4; immaculate 3; impeccable; infallible 1; intact; mature 3, 6; model 11; optimum 2; out-and-out; polish 2; polished 1; positive 9; precise 1; profound 4; proper 5; pure 1, 5; refine 2; regular 9; right 2, 8; ripen; stark 4; superb; tailor-made 2; thorough 1; thumping 2; total 3; unmitigated; unqualified 2; very 3
perfect world, in a, ideally 1
perfected, mature 3; perfect 1; practiced 2
perfect example, gem 2; picture 4
perfection°, complement 1; consummation 2; entirety 1; fruition; ideal 2; maturity 3; optimum 1; peak 2; pinnacle; precision 1; purity 1; refinement 4; utopia
perfection, bring to, perfect 8
perfectionist°, meticulous; pedantic 2
perfectly°, altogether; charm 4; completely 1; detail 4; downright 2; exactly 1; full 15; ideally 3; just 5; pat[2] 1; quite 1; supremely; thoroughly 1; totally; utterly
perfervid, fanatical; intense 2
perfidious°, black 6; dirty 3; dishonest; dishonorable 2; disloyal; evil 2; faithless 2; foul 7; hypocritical; insincere; lying 2; machiavellian; renegade 2; slippery 2; traitorous; two-faced; unscrupulous; untrue 1; villainous 1; wily
perfidiously, back 6
perfidiousness, foul play; perfidy
perfidy°, betrayal 1; foul play; infidelity 1
perforate°, bore[1] 2; hole 7; penetrate 1; pierce 2; pink[2]; prick 3; punch[2] 2; puncture 3; riddle[2] 1; stick[1] 1
perforated, pitted
perforating, puncture 2
perforation, flaw 2; hole 2; penetration 1; pore[2]; prick 1; puncture 1, 2
perforator, punch[2] 1
perforce, necessarily; willy-nilly 1
perform°, accomplish; act 6, 9; appear 2; behave; bring 12; carry 10c, 12; celebrate 1; commit 3; discharge 5; enact 2; execute 1; feature 5; follow 10; fulfill 1; function 3; give 7; go 3; implement 2; obey 2; operate 1; perpetrate; play 4; practice 7; prosecute 3; pull 12b; put 22d; redeem 6; render 4; run 11; take 11, 26; transact
perform as, make 10
perform (upon *or* on), play 5
performable, practicable
perform (a function *or* a duty *or* one's duty), serve 2
performance°, act 2, 3; action 1, 7; attraction 2; celebration 1; course 3; deed 1; delivery 4; discharge 12; entertainment 2; execution 1; exercise 5; fulfillment; observance 2; operation 1; palaver 1; presentation 2; production 4; program 2; recital 1; recitation 1; record 5; rendition 1; routine 2; song 3a; spectacle 1; stand 15; technique 2; theater 2; undertaking 2
performance level, touch 17
performance(s), proceeding 2b
performed, complete 2
performer°, player 2, 4; thespian 2
performers, cast 2; company 2
performing, theater 2; thespian 1; undertaking 2
perfume°, aroma 1; bouquet 2; fragrance; odor 1; savor 2; scent 1, 4; smell 1
perfumed, aromatic; fragrant; redolent 1; sweet 2
perfunctorily, sketchily
perfunctory°, careless 1; cursory; hasty 3; hurried 2; mechanical 2; offhand 2; promiscuous 1; quick 2; ritual 2; routine 4; sketchy; summary 2; superficial 3; token 5
perhaps°, chance 5b; maybe; possibly 1; probably

periapt, fetish 1; talisman
pericope, excerpt 1
peril°, adventure 1; danger 1; exposure 2; hazard 1; jeopardy; menace 2; pitfall 2; risk 1; threat 1
perilous°, awkward 4; critical 3; dangerous 1; desperate 5; dodgy; explosive 1; grave[2] 2; hairy 2; hazardous; menacing; nice 4; parlous; precarious; risky; serious 4; tight 7; ugly 3; unhealthy 3
perilously, dangerously 1
perimeter°, belt 2; boundary; circuit 1; edge 1; fringe 2; girth 1; limit 2; margin 1; periphery 1; rim; side 1; silhouette
perimetric, peripheral 2
period°, age 3, 5; bout 1; date 1; day 2; dot 1; era; fit[2] 2; generation 2; interval 1; juncture 2; length 2; measurement 2; page[1] 2; patch 4; phase 1; point 1; round 14; run 38; season 1; session 2; snap 9; space 3; span 2; spell[1] 1; streak 3; stretch 6; style 1; time 1, 2, 4; world 4
period, at any, ever 1
periodic°, fitful; intermittent; occasional 1; recurrent; regular 2; spasmodic 2
periodic(al), journal 1; magazine 1; organ 2; paper 1; periodic; publication 2; rag[1] 2; review 7; sporadic
periodically, occasionally; once 4
period (of time), term 2
peripatetic, migrant 2; pedestrian 1, 3; rambling 3; traveling; vagabond 2
peripheral°, extraneous 1; frivolous 1; nonessential 1; out 10; outlying; outside 7; tangential
periphery°, border 1; circuit 1; edge 1; extremity 1; fringe 2; girth 1; limit 2; margin 1; outline 1; outskirts; perimeter; rim; side 1; silhouette
periphrastic, circular 2; diffuse 2; indirect 1; mealy-mouthed; rambling 1
perish°, die 1, 4; disappear 2; expire 2; fade 2; fall 6; miscarry; pass 14a
perishable, delicate 1; mortal 2
perished, dead 7
perishing, freezing
perjure oneself, lie[1] 1
perjurer, liar
perjury°, lying 1
perjury, commit, lie[1] 1
perk°, bonus; perquisite
perk up, brighten 1; liven 1; pick 8f; rally 4; snap 7
perk(s), benefit 2
perky°, lively 1; smart 5; sprightly
perlaceous, pearly
permanence°, existence 2; perpetuity; stability 2; tenure 2
permanency, permanence; tenure 2
permanent°, abiding; certain 3; changeless 1, 2; deathless; durable; enduring; eternal 3; everlasting; fast[1] 5; immortal 1; indelible; indestructible; invariable 3; irreversible; irrevocable; lasting; monumental 1; perennial 2; perpetual 1; regular 10; stable 2; standing 1; timeless
permanently°, severely 1
permeable, porous
permeate°, charge 8; flood 7; impregnate 2; infest; penetrate 2; percolate; riddle[2] 2; saturate; suffuse; transfuse 2
permeated with, be, possess 4
permeating, pervasive
permeative, pervasive
permissible°, good 3; lawful 2; legal 1; safe 4
permission°, allowance 1; approval; carte blanche; charter 2; clearance 2; consent 3; faculty 4; freedom 4; go-ahead 1; leave[2] 1; liberty 2; license 1; OK 6; pass 24; passage 8; privilege; sanction 1; slip[2] 1; use 10
permission (to), give, let[1] 1; permit 1
permission to enter, admittance
permissive°, easy-going; indulgent; lax 1; lenient; soft 5; tolerant
permissiveness, tolerance 1
permit°, accept 1; admit 2; allow 2, 3, 4; approve 1; authorize; charter 2; consent 2; enable 1, 3; entitle 1; excuse 3; grant 2; license 2, 4; overlook 2; pass 7, 24; patent 1; sanction 5; slip[2] 1; subscribe 1; suffer 3; tolerate 1; use 10; vouchsafe 2; warrant 2, 4
permit, not, refuse[1] 2
permit (to), let[1] 1; permit 1
permitted, free 10; lawful 2; legal 1; permissible; privileged 2
permitting, allowance 1
permutation, transformation; variation 1
permute, transform
permuting, variation 1
pernicious, bad 2; calamitous; damnable; deadly 1; destructive 1; detrimental; evil 3; harmful; hurtful 1; ill 4; injurious 1; malignant 1, 2; mischievous 2; poisonous 1; prejudicial; ruinous; sinister 2; unwholesome 1; virulent 1; wicked 6
perniciousness, virulence 1
pernickety. See **persnickety**
perorate, rant 1
peroration, harangue 1
perpendicular°, erect 1; plumb 2; precipitous 1; sheer 1; straight 2; upright 1, 3
perpendicular to, perpendicular 2
perpendicularly, plumb 3; upright 4
perpetrate°, commit 3
perpetual°, changeless 2; constant 2; continual; continuous 2; endless 2; eternal 2; everlasting; habitual 2; immortal 1; infinite 1; limitless; ongoing 1; perennial 2; permanent 1; persistent 2; relentless 2; running 3; standing 3; steady 2
perpetually, always 2; cease 2; ever 2; forever 2; permanently; time 11
perpetuate°, continue 3; extend 3; maintain 1; preserve 2; spin 5
perpetuating, preservation 2
perpetuation, maintenance 2; preservation 2
perpetuity°, eternity
perpetuity, in, always 2
perplex°, bemuse 1; bewilder; bother 3; confuse 1; daze 2; distract 3; entangle 2; floor 5; flummox; fluster 1; fog 4; get 18; mix 4b; muddle 1; nonplus; puzzle 1; shake 5; stick[1] 9; stump 2
perplexed, blank 5; confused 2; daze 4; disconcerted; dumbfounded; helpless 2; lost 3; sea 4
perplexing°, confused 1; difficult 2; disconcerting; equivocal 2; hard 3; incomprehensible; inexplicable; intricate 2; labyrinthine; mysterious 1; obscure 2; occult 2; opaque 2; paradoxical; puzzling; tough 5
perplexity°, fluster 2
perquisite°, bonus; premium 1
perquisite(s), benefit 2
persecute°, bully 2; grind 5; harass; hound; ill-treat; oppress 2; pester; rack 3; torment 2; victimize 1
persecution°, oppression; torment 4
persecutor, bully 1; oppressor
perseverance°, application 3; determination 1; endurance 1; follow-through; fortitude; industry 2; patience 2; persistence; purpose 2; resolution 1; self-control 1; stability 2; strength 2; tenacity 1
perseverant, insistent; patient 2; resolute; untiring
persevere°, carry 11a; follow 10; go 32a; hang 7c; hold 20a; linger 1; live 7; persist 1; remain 3; see 14b; soldier 4; stick[1] 14, 19
persevere (at), peg 6
persevere (in), continue 1; keep 5; maintain 1; pursue 2
persevering, constant 1; determined 1; diligent; enterprising; insistent; laborious 2; obstinate; persistent 1; relentless 1; resolute; rigid 4; single-minded; stalwart 2; steadfast; tireless; untiring
perseveringly, hard 13
persiflage, banter; raillery; repartee
persist°, carry 11a; continue 2; endure 1; exist 3; follow 10; go 32a; hang 7c; hold 20a; insist 2; last[2] 1; linger 1; live 7; persevere; prosecute 2; remain 3; see 14b; soldier 4; stand 4; stick[1] 14, 19; survive 1
persist (in *or* at *or* with), continue 1, 3; dwell 2; keep 5; maintain 1; peg 6; persist 1; pursue 2; stand 5c; sustain 1
persistence°, application 3; determination 1; drive 8; endurance 1; existence 2; follow-through; force 4; industry 2; maintenance 2; obstinacy; patience 2; permanence; perseverance; purpose 2; resolution 1; self-control 1; stability 2; strength 2; tenacity 1
persistency, obstinacy
persistent°, chronic 1, 2; constant 1, 2, 3; continual; continuous 2; demanding 2; determined 1; diligent; durable; enduring; eternal 2; fixed 2; frequent 1; gritty 2; habitual 2; hard 7; indomitable; industrious; insistent; laborious 2; lingering 1; nagging; nonstop 2; obstinate; officious; ongoing 1; patient 2; perennial 1; permanent 2; perpetual 2; purposeful; recurrent; relentless 2; repetitive; resolute; sneaking 1; stalwart 2; steadfast; steady 2, 5; strenuous 2; strong 5; stubborn; tenacious 1; tireless
persistent, be, persevere; persist 1
persistent (in), tenacious 3
persistently, forever 2; hard 13; hotly; nonstop 3; warmly 4
persisting, enduring; persistent 1; resolute
persisting (in), tenacious 3
persnickety, fastidious; finicky 1; fussy 1; overcritical; particular 4;

pinchbeck, base[2] 5; fake 5; fraudulent 1; plastic 3; shoddy
pinched, drawn; emaciated; gaunt 1; impoverished 1; lean[1] 1; narrow 2; needy; peaked; poor 1; short 8; skinny; thin 1
pinch-hit, sit 6b
pinch-hit (for), stand 8; substitute 1; understudy 2
pinch hitter, alternate 5; understudy 1
pinching, rip-off 1; robbery; theft
pinch (pennies), economize; pinch 5; stint 4
pinchpenny, miser
pine, ache 2; brood 4; die 5; itch 2; yearn
pine for, desire 1; fancy 11; hanker; miss[1] 2; pant 2; sigh 2; want 1
pinguid, fat 2; greasy 1; oily 1
pinhead, dolt; fool 1
pinheaded, thick 6
pinhole, prick 1
pining, homesick
pining (for), desperate 3; starved 2
pinion, shackle 3; tie 2
pinions, restraint 2
pink°, left 2; radical 5; rosy 1
pink, in the, hale; healthy 1; pink[1] 1
pink slip, give (someone) a, dismiss 1; fire 11
pinked, notched
pinkish, pink[1] 2; rosy 1
pinko, radical 5
pink slip, dismissal 1; push 16; sack 3
pin money, allowance 3
pinnacle°, acme; crest 1; extreme 8; height 2; maximum 1, 2; peak 1; perfection 3; prime 4; prominence 2; spire 2; summit; tip[1] 1; top 1; vertex; zenith
pin (one's) faith on *or* upon, trust 5
pin one's hopes on *or* upon, lean[2] 4a
pinpoint, designate 1; diagnose; focus 4; get 22; identify 1; locate 1; name 5; pin 4b; place 15; set 5; zero 4
pinpointing, identification 1
pinprick, prick 1
pins and needles, on, nervous 1; tense 2
pins and needles, be on, sweat 2
pin spotlight, spotlight 1
pint, drink 6
pintle, pivot 1
pinto, dapple 1
pint-sized, minute[2] 1; short 1; slight 3; small 1; tiny
pioneer°, discover 3; early 5; institute 4; introduce 4; originate 1; settle 6; settler; spearhead 1
pioneering, avant-garde
pious°, devout 1; godly; holy 2; pure 6; religious 1
piousness, piety 2
pip, grain 1; kernel 1; pit[2]
pip, the, vapor 2
pipe°, chirp 1; drain 1; main 5; peep 1, 2; pipeline 1; sight 5; sing 1, 2; vent 1
pipe down, pipe 7
pipe up, pipe 8
pipe dream, daydream 1; dream 1; fancy 6; fantasy 2
pipeline°, main 5; pipe 1
pipeline, in the, pipeline 2; way 12
pipe of peace, pipc 2
piping, shrill; trim 5
piping hot, hot 1
Pip! Pip!, farewell 3; goodbye
piquancy, flavor 1; salt 2; savor 1; spice 2; spirit 3; tang 1; zest 1
piquant, epigrammatic; hot 2; pungent 1; racy 3; sharp 4, 5; spicy 1; strong 2; tart[1] 1; witty
pique, anger 1, 2; annoyance 1; displease; disturb 1; exasperate 2; get 17; grudge 1; infuriate; intrigue 1; irritate; madden 2; offend 1; pester; pet[2]; provoke 3; resentment; spite 3; warmth 4; whet 2
pique, create, offense 2
pique, feel, offense 3
pique, in a, querulous
pique, in a (fit of), moody 2
piqued, angry 1; cross 6; discontented; huff 1; indignant; moody 2; querulous; resentful
piqued, be, umbrage
piracy, pillage 2; plagiarism; stealing
pirate°, cutthroat 1; help 5; outlaw 1; plunder 2; robber; steal 2; take 3; thief 3
piratical, predatory 2
pirating, plagiarism; stealing
pirouette°, gyrate; revolution 3; rotate 1; spin 1, 6; swivel 1; turn 26
piscatorial, fishy 2
piscatory, fishy 2
piscina, sink 12
piscine, fishy 2
pish and tush, fiddlesticks; gobbledegook 1; prattle 3
piss, go 17; urinate; water 4
piss off, bugger 5
piss (someone) off, enrage; exasperate 2; infuriate
pissed, drunk 1; far 5b; stinking 3
pissed off, discontented; indignant
pissoir, toilet 1
piste, run 45
pistol°, revolver
pit°, cavity; core 3; depression 1; depth 5; excavation; grain 1; hole 1; hollow 7; indentation; mark 10; mine 1; opening 1; pan 3; pitfall 1; pocket 2; quarry[2] 1; seed 1; shaft 4; sink 13
pit against, oppose 3; pit[1] 6
pit against one another, divide 3
pit (oneself) against, take 35b; play 3
pitapat, patter[1] 2, 3
pitch°, careen; cast 1, 7; climb 6; dash 2; erect 2; field 2; fling 1; fly 8a; head 5; heave 2; hurl; incline 4; intonation; key 3; launch 3; line[1] 18; lob 1, 2; lurch[2] 1, 2; patter[2] 1; plunge 1, 3; put 7; reel 1; slant 3, 4; sling 1; slope 1, 2; speech 2; stagger 1; throw 1; tilt 1, 3; tone 2; toss 1, 5, 6; tumble 1; tune 2; wallow 3
pitch in, pitch[1] 5; shoulder 3; turn 23c
pitch into, pitch[1] 6
pitch on *or* upon, pitch[1] 7
pitch-black°, black 3; swarthy
pitch camp, camp[1] 3
pitch-dark, dark 1; pitch-black
pitcher, jar[1]; jug
pitching, roll 14
pitchpole, pitch[1] 4
pitchpoll, pitch[1] 4
pitchy, sunless
piteous°, heart-rending; pathetic 1; poignant 1; sorrowful 1; tragic
pitfall°, difficulty 2; joker 2; trap 1
pith°, core 2; elixir 3; essence 1; gist; heart 3; implication 3; juice 2; kernel 2; matter 4; meat 2; nub 2; nucleus; point 5; pulp 1; quintessence; significance 1; spirit 6; substance 2; vigor
pithiness, brevity
pithy, compact 3; concise; epigrammatic; expressive 2; forceful 2; meaningful 1; short 3; significant 2; succinct; terse 1; thumbnail
pitiable°, forlorn 1; paltry; pathetic 1; piteous; poignant 1; poor 6; sad 3; sorry 2; tragic; unfortunate 2; wretched 3
pitiful°, forlorn 1; lamentable; miserable 3; paltry; pathetic 1; piteous; poignant 1; poor 6; sad 3; sorry 2; touching; tragic; wan 2; weak 4; wretched 3
pitiless, bloodthirsty; brutal 1; coldblooded 3; coldhearted; cruel 1; cutthroat 2; deadly 3; ferocious; ghoulish 2; grim 2; hard 4; harsh 2; heartless; implacable; inhuman 1; merciless; relentless 1; remorseless 1; ruthless; sanguinary 1; savage 2; severe 1; stony 2; strict 2; unmerciful; unsympathetic
pitilessly, roughly 2
pitilessness, severity 1
pits, the, pit[1] 4
pitted°, bumpy; irregular 1; rugged 1
pitter-patter, patter[1] 2, 3
pity°, feel 8; heart 5; mercy 1; quarter 4; respond 2; shame 3; sympathy 1
pity, show, relent
pity (for), have, sympathize 1
pivot°, center 1; hub; pirouette 2; revolve 1, 2; rotate 1; spin 1; swivel 1, 2; turn 1; wheel 2
pivot on, depend 1
pivotal°, critical 2; crucial; fateful 1; grave[2] 2; key 6; landmark 3; momentous; pressing; vital 1, 2
pivoting, pirouette 1
pix, photograph 1
pixie, goblin; imp
pixilated, curious 3
pixyish, whimsical 1
pizazz, flair 2; glitter 4; snap 11; sparkle 4; spice 2; style 4; verve; vigor; vitality 1; zest 1
placard, advertisement 1; poster; promotion 5; sign 5
placate, calm 5; disarm 2; humor 4; hush 5; mitigate; reconcile 1; satisfy 1
placater, peacemaker
placation, reconciliation 1
placative, propitiatory 1
placatory, propitiatory 1
place°, appointment 3; attach 4; base[1] 6; capacity 3; clap 3; deposit 1, 2; dispose 1; estate 3; fix 8; get 15, 22; group 4; home 1; identify 1; impose 3; install 1; lay[1] 1; locale; locate 1; location 1; make 14; niche 2; office 4; pad 3; perch 1, 2; pitch[1] 2; plant 7; play 7; point 3; pose 1; position 5, 6, 8, 9; post[2] 1, 2; put 1, 8; quarter 3; rank[1] 1; rate[1] 4, 8; recognize 1; residence 1; rest[1] 8; role 2; run 19; scene 1; seat 1, 7; set 1, 11; site 1, 2; situate; situation 1, 3; slot 2, 4; space 1, 4; spot 2; spread 13; square 8; stage 1; standing 5; station 1, 2, 4; stick[1] 2; stow; true 6; void 5; whereabouts 1
place, from this, hence 2
place, have (a), figure 10; obtain 3
place, in, place 11
place, in another, elsewhere
place, out of, awkward 3; extraneous 2; improper 2; inept 2; inopportune; irrelevant; order 13b; place 12; unbecoming 1;

perform 3; performance 2; piece 4; pretend 3; production 4; program 2; put 6; recreation; render 4; script 2; show 9; sit 6a; slack 6; space 1; sport 1, 4; take 26; tolerance 2; use 2; wade 2
play, bring into, exert 1; use 1; work 10
play, putting into, service 10
play along (with), play 8, 11
play around *or* **about,** fool 7a
play (around *or* **about) (with),** tinker
play around (with), fool 6; play 9
play at, dip 3; play 10
play by ear, play 12
play down, belittle; depreciate 2; downgrade 2; minimize 2; play 13; trivialize; whitewash; write 4b
play dumb, mum 2
play false with, double-cross; stab 2
play fast and loose with (someone), string 10b
play for time, play 14; procrastinate 1; stall[2] 1
play (it) by ear, improvise 1
play (itself) out, run 26c
play off (against), oppose 3
play on *or* **upon,** influence 3; manipulate 1; play 15
play (the part *or* **role of),** portray 2; pose 2
play up, carry 11c; emphasize; exaggerate; feature 4; play 17
play up to, flatter 1; kowtow; play 18
play (with), fiddle 2; monkey 5; play 19; toy 3, 4; trifle 3; twiddle 1; finger 11; flirt 2
play a big *or* **an important role** *or* **part,** loom 3
playact, make 28; pretend 3
play-acting, drama 3; masquerade 2
play a joke on, have 12c
play a part (*or* **role) (in),** contribute 2; far 6b; part 11
play a waiting game, sit 8
play ball°, play 11
playboy°, blade 4; flirt 3; philanderer; rake[2]; roué; seducer; sybarite
played, be (being), show 9
played out, exhausted 1; prostrate 5; spent 1; threadbare 2; worn 3
played upon, affected 5
played-out, haggard; washed-up
player°, bettor; contestant; entry 5; performer; thespian 2
players, cast 2; company 2; field 3
playfellow, playmate
playful°, elfin 1; frisky; frolicsome; humorous; jolly 1; kittenish; mischievous 1; naughty 1; sportive; sprightly; whimsical 1; zany 1
playfully, fun 3
playfulness, fun 2; mischief 1
play games°, game 6
playgirl, flirt 3
play host to, treat 4
playhouse, theater 1
playing, performance 2, 3
playing, be (being), show 9
playing card, card 1
playing it cool, noncommittal
playing it safe, noncommittal
playing one's cards close to the chest, noncommittal
playmate°, friend 1; pal 1
play one's cards right°, card 12
play on words, pun; witticism
play the field, play 9b; run 23
play the game°, play 16
play the market, speculate 2
plaything°, bauble; novelty 2; toy 1; trifle 1
playtime, break 28
play with fire°, fire 6; wind[1] 12
playwright°, dramatist; scribe 2
plaza, square 8
plaza de toros, ring[1] 2
plc, company 4; firm 6
plea°, answer 2; appeal 3; argument 2; defense 3; excuse 4; petition 1; prayer 1; request 2; suit 5; supplication 1
plea for, make a, speak 7a
pleach, twist 1
plead°, appeal 1; argue 3; petition 2; sue 2
plead for, advocate 1; cry 3; defend 3; maintain 4; plead 1; request 1; speak 7a
plead in favor of, advocate 1
plead (to), ask 3
plead (with), beg 1; beseech; plead 2; pray 1; reason 8; urge 2
pleader, proponent; supplicant 2
pleading, argument 2; supplication 1, 2
pleasant°, acceptable 2; agreeable 1; amiable; amicable; beautiful 2; comfortable 3; cordial; debonair 1; delicious 2; delightful 1; desirable 2; engaging; entertaining; fair[1] 5; fine[1] 2, 3; friendly 2; genial; good-natured; humorous; likable; lovely 2; luscious; mellow 4; mild 2; nice 1; regular 8; smooth 5; soft 3; sweet 4; sympathetic 2; warm 3; winning 1
pleasant, find, take 38a
pleasantly, kindly 2; nice 6
pleasantry, banter
please°, amuse 1; appeal 2; content[2] 4; delight 1; entertain 1; feast 6; gratify; humor 4; kindly 3; oblige 1; regale; satisfy 1; send 4; suit 2; tickle; warm 9
pleased°, charmed 3; content[2] 3; delighted; enchanted; glad 1; gleeful; happy 1; joyful 2; thankful; willing
pleased, be, rejoice
pleased (as Punch), elated; glad 1; happy 1; pleased
pleased (with), proud 1
pleases, as *or* **when (one),** will 4
pleasing°, acceptable 2; agreeable 1; attractive; beautiful 2; darling 4; delicious 2; delightful 1; desirable 2; engaging; entertaining; heartwarming 2; likable; lovely 2; picturesque 1; pleasant 1; prepossessing; satisfying; shapely; sweet 4; well[1] 17; winning 1
pleasing, find, take 38a
pleasing to, become more, grow 5
pleasingly, nice 6
pleasurable°, agreeable 1; delicious 2; delightful 1; entertaining; glorious 3; heartwarming 2; lovely 2; pleasant 1; readable 2; satisfying
pleasure°, admiration; amusement 1; bliss; content[2] 1; delectation; delight 3; device 4; discretion 2; ecstasy 1; enjoyment 1; entertainment 1; feast 3; fun 1; gaiety 1; glee; gratification; gusto; happiness; honor 3; joy 1, 3; liking 2; love 2; luxury 3; play 22; radiance 2; rapture; relaxation 1; relish 1; satisfaction 1; sport 1
pleasure, at one's, will 4
pleasure, with, gladly; happily 3
pleasure from, take (all) the, overshadow 2
pleasure (in), take, like[2] 1; love 8; relish 2; revel 1; roll 10b
pleasure in *or* **from, take** *or* **derive,** enjoy 1
pleasure-bound, dissolute; self-indulgent
pleasure-loving, voluptuous 1
pleasure-oriented, epicurean 1
pleasure-seeker, sensualist; sybarite
pleasure-seeking, epicurean 1; self-indulgent; voluptuous 1
pleat, fold 1, 4; gather 3; pucker 2
pleating, pucker 2
pleb, plebeian 3
plebeian°, bourgeois 2; common 3; humble 3; low[1] 7; mean[2] 2; unrefined 1; vulgar 1
plebeians, hoi polloi; people 3; public 8
plebeians, the, mass 6
plebiscite°, election; vote 1
plebs, hoi polloi; people 3
plebs, the, unwashed 2
pledge°, assurance 1; commit 1, 4; dedicate 1; dedication 1; devote 2; donate; drink 3; earnest 3; engage 3; engagement 3; guarantee 1, 2; hostage; oath 1; obligate; obligation 2; pawn[1] 1, 3; promise 1, 3; put 28d; resolution 2; security 3; subscribe 2; subscription 2; swear 1; toast 1; undertake 2; undertaking 3; vow 1, 2; warrant 1; warranty; word 4
pledged, engaged 1
plenary, full 2
plenipotentiary, ambassador; delegate 1; minister 2
plenitude, abundance; plenty 2; wealth 2
plenteous, abundant 1; bountiful 2; fertile; fruitful 3; luxuriant 2; opulent 3; plentiful 1; prodigal 2; productive 1; prolific 1
plenteousness, abundance; fat 6; plenty 2; prodigality 2; profusion; wealth 2
plentiful°, abundant 1; ample 4; bountiful 2; fruitful 3; generous 3; lavish 1; liberal 1; opulent 3; prodigal 2; productive 1; profuse 1, 4; prolific 1; rich 9; thick 4
plentiful in, be, fill 6
plentifully, freely 4
plentifulness, abundance; plenty 2; prodigality 2; profusion
plenty°, abundance; comfort 3; ease 3; enough 2; fat 6; fill 11; flow 6; heap 2; many 3; pile[1] 3; prodigality 2; profusion; prosperity; riches; stack 3
plenty of, due 3
pleonasm, expletive 3; tautology
pleonastic, diffuse 2; expletive 1; needless 1; repetitious; wordy
plethora, excess 1; flood 3; flow 6; heap 2; ocean 2; pile[1] 3; profusion; sea 3; superfluity; surfeit
plexure, mesh 1
plexus, mesh 1; network 2; net[1] 1
pliability, flexibility 1
pliable°, accommodating 2; adaptable; elastic 1; flexible 1; limp[2] 1; passive 2; plastic 1; sheepish 1; soft 1; supple 1; tractable 2; yielding 1
pliancy, elasticity 1; flexibility 1
pliant, accommodating 1; adaptable; dutiful; easy 4; flexible 1; fluid 3; obedient; open 11; plastic 1; pliable 1, 2; receptive 1; soft 1; submissive 1; supple 1, 2;

pliant (*cont.*) tame 3; willowy 1; yielding 1
pliant, be, bend 5
pliantness, flexibility 1
pliers, pincers
plight°, difficulty 3; dilemma; lot 2; mess 3; oath 1; pawn[1] 1; quandary; scrape 8; situation 2; strait 3
plighted, engaged 1
plinth, pedestal 1
plod°, tramp 1, 3; walk 1
plod along (*or* away), plod 1, 2
plod through, wade 4
plodder, hack[2] 2
plodding, methodical; rustic 2; slow 1
plonk, stick[1] 2
plonk (down), set 1
plop, plump[2] 2; stick[1] 2
plop down, flop 1
plot°, aspiration; cabal 1, 3; chart 3; conceive 2; conspiracy; design 4, 8; device 2; dodge 4; engineer 5; game 3; intrigue 2, 3; machination; maneuver 1, 3; parcel 2; patch 2; plan 3; reservation 4; scenario; scheme 1, 3, 4; site 1; story[1] 5; stratagem; tactic 1; thread 2; tract[1]; wangle; wile; wrinkle[2]
plot against, gang 3
plotted, put-up
plotter, designer 2; tactician
plotting, crafty; designing; devious 1; disingenuous; forethought; foxy 1; machination; scheming; serpentine 1; sly 1; wily
plough, plow 1, 2; tramp 1; work 8
plouk, pimple
plouky, spotty 2
plow°, cultivate 1; furrow 2; till[1]
plow through, search 1; wade 4
ploy, deceit 2; device 2; dodge 4; feint; game 3; gesture 2; gimmick 1; machination; maneuver 1; move 9; ruse; scheme 3; stratagem; subterfuge; tactic 1; trap 2; wile; wrinkle[2]
pluck°, bottle 2; bravery; courage; daring 1; fight 9; fleece; fool 4; grit; gumption 2; gut 3a; heart 2; jerk 1; manhood 2; nerve 1; perseverance; persistence; pick 2; pull 2; snap 4; snatch 1; spirit 5; spunk; strength 2; swindle 1
pluck, show, persevere
pluck (out), extract 1; pull 3; uproot 2
plucky, bold 1; brave 1; courageous; daring 2; dauntless; fearless; gallant 1; game 7, 8; gritty 2; hardy 2; heroic 1; indomitable; intrepid; manly; stalwart 3; stout 2; venturesome 1
plucky, be, persevere
plug°, advertisement 1; fill 7; get 35; jade 1; jam 2; nag[2]; promote 4; publicize; puff 3, 7; punch[1] 1, 2; push 6; recommend 2; shoot 3; stopper; stuff 9; talk 13; tap[2] 2; tout 1; wad 1
plug (along *or* away) (at), peg 6; plod 2
plug (up), foul 15; plug 4; seal 3; stop 3
plugged, congested
plugged(-up), foul 9
plugging, promotion 4
plugola, bribe 1; kickback; payoff 3; rebate 2
plug-ugly, bruiser; hoodlum
plumb°, erect 1; fathom; flush[2] 4; line[1] 19a; perpendicular 1; probe 2; regular 5; slap 6; sound[3] 1; straight 2; upright 1; very 1
plume, preen 1; primp
plummet°, descend 2; dive 1; drop 6; fall 1; pitch[1] 3; plumb 1; plump[2] 1; plunge 1, 3; slump 2; submerge 2; taper 2
plummy, choice 4; desirable 3
plump°, buxom 1; chubby; dumpy; fat 1; full 7; heavy 11; obese; rotund 3; set 1; stout 1; well-fed
plump down, flop 1
plump for, plump[2] 3
plumpness, fat 5; fatness; obesity
plunder°, booty; depredation; fleece; gut 5; loot 1, 2; overrun; pilfer; pillage 1, 2, 3; prize[1] 4; raid 2; ransack 2; rape 2, 5; ravage 2; rifle 1; rob 1; robbery; spoil 6; strip[2] 5; waste 11
plunderer, thief 1
plundering, depredation; predatory 2; rape 2; robbery
plunge°, bury 3; descend 2; dig 2; dip 1, 4; dive 1, 2; drive 4; drop 6; duck 2; fall 1, 22; hurtle; immerse 1, 2; jab 1; lunge 2, 3; pitch[1] 3, 4; plow 2; plump[2] 1; sink 1, 8; slump 1, 2; sound[3] 2; stab 1; submerge 1, 2; thrust 2; toss 5; trip 5; venture 1, 3
plunge, take a, speculate 2
plunge, take the, break 17
plunge about, flounder
plunge into, wade 3b
plunger, piston
plunk, plump[2] 2, 4, 5; stick[1] 2
plunk down, deposit 1; set 1
plus fours, pants 2
plush°, elegant 3; lush 3; luxurious 1; pile[2]; princely 2; soft 13; splendid 1; sumptuous; swanky; swish 3
plushy, soft 1
plutocrat, financier
ply°, fold 1; operate 2; work 15
PM, premier 1; evening
pneu, telegram
pneuma, psyche; spirit 1
poach, pirate 3
poacher, thief 1
poaching, stealing
pock, mark 1
pocket°, bag 1; diminutive; earn 2; get 3; hole 1; lift 6; make 7; miniature; misappropriate 1; net[2] 4; portable; pouch; purse 1; receive 2; short 2; slight 3; take 3; tiny
pocket, out of, out 5; poor 1
pocketbook, purse 1; wallet
pocketful, wad 2
pocketing, theft
pocket money, allowance 3
pocket(-sized), diminutive; portable; slight 3; small 1; tiny
pocket watch, watch 6
pockmark, mark 1, 10; pit[1] 2, 5
pockmarked, pitted; spotty 2
pockmarks, spot 5
pocky, spotty 2
pococurante, blasé 2; casual 3; indifferent 1; lackadaisical 2; listless; nonchalant; shiftless; slothful; torpid
pococurantism, indifference 1; sloth; sluggishness; torpor
pod, hull 2
podgy. See **pudgy**
podium, platform 1; rostrum; stage 2
poem°, lay[3]; piece 4; rhyme 1; writing 2
poesy, poetry
poet°, scribe 2
poetaster, poet
poetess, poet
poetic°, rhetorical 1
poetic(al), imaginative 2; poetic 1
poetically, well[1] 3
poetry°, rhyme 1; writing 2
po-faced, grave[2] 1; grim 4
pogrom, holocaust 2; massacre 1; slaughter 2
poignancy, salt 2; spice 2; tang 1
poignant°, deep 6; emotional 3; expressive 3; heart-rending; keen[1] 3; moving 1; pathetic 1; piteous; pungent 3; sharp 8; smart 4; tender[1] 7; touching; trenchant
poikilothermic, coldblooded 1
point°, aim 1; angle[1] 1; argument 2; brink 2; business 2; cape[1]; degree 2; design 7; detail 1; direct 4; dot 1; end 3; fasten 2; gist; head 5, 10; idea 4; instant 1; issue 3; item 1; juncture 2; level 10; message 3; moral 3; nub 2; page[1] 2; pith 1; place 1; prominence 2; purpose 1; quarter 3; question 3; reason 3; regard 8; respect 4; significance 1; stage 1; subject 1; substance 2, 3; theme 1; thesis 1; thing 2; thorn 1; time 3; tip[1] 1; topic; turn 9; use 12; usefulness; way 6
point, at any, ever 1
point, beside the, extraneous 2; inapplicable; irrelevant; peripheral 1; point 15
point, off *or* beside the, tangential
point, off the, extraneous 2
point, to a certain, far 8b; partially
point, to a (given), partially
point, to the, brief 2; direct 9; point 18; point-blank 1; relevant; terse 1
point in time, at this, present[1] 3
point of, get the, grasp 2
point of, make a, point 17; stress 4
point of, on the, ready 4
point of, on the (very), verge[1] 2
point out, designate 1; finger 12; identify 1; indicate 1; mention 1; point 21
point to, blame 1; find 2; finger 5c; indicate 1; point 19; refer 1; reflect 2; register 8; spell[3] 1
point to, make (it) a, point 17
point up, emphasize; point 21b, 22; spotlight 3
point-blank°, direct 9; slap 6; straight 6, 14
point by point, detail 4
pointed°, acute 1; epigrammatic; expressive 2; fine[1] 8; keen[1] 2; lively 2; meaningful 2; poignant 2; pregnant 2; sharp 1; smart 4; special 3; trenchant
pointedly, deliberately; expressly 1
pointer°, clue 1; hand 5; hint 1; index 2, 3; indicator; point 11; tip[3] 2
pointing, aim 4; direction 1
point (in time), moment 2
pointless°, aimless 1; bootless; crazy 3; fruitless; hollow 5; hopeless 4; idle 4; meaningless 2; nonproductive 2; point 15; purposeless; senseless 3; silly 1; unprofitable 2; useless 1; vain 2; void 1; worthless 2
pointlessness, emptiness 2; madness 2; vanity 2
point of agreement *or* correspondence, comparison 2
point of departure, base[1] 4; basis 2
point of view°, angle[1] 2; aspect 1; attitude 2; observation 2; outlook 1; phase 4; position 3;

stance; stand 13; standpoint; thinking 2; view 2
point(s), fact 3
point (the finger) (at), accuse 1; blame 1; incriminate; pin 5
poise°, assurance 4; balance 2; confidence 2; cool 8; float 1; grace 1; hover 1; presence 3; proportion 5; repose 3; restraint 3; sang-froid; savoir-faire; self-confidence; self-control 2; serenity 2; sophistication 1
poised°, collected; levelheaded; ready 9a; self-confident; serene 2; steady 6
poised, be, hang 1
poison°, booze 1; contaminate; corrupt 4; debase 2; doctor 4; drug 4; gall[1] 1; pollute 1; prejudice 3; ruin 9; sarcasm; spike 4; ulcer 2; venom 1, 2; virulence 2
poisoning, pollution
poisonous°, bad 2; deadly 1; destructive 1; evil 3; fatal 1; harmful; malignant 1; ruinous; sarcastic; sharp 5; venomous 1, 2; virulent 1, 2
poisonousness, sarcasm; venom 2; virulence 1, 2
poke°, bag 1; dab 3; dig 2, 7; gore[2]; jab 1, 3; nudge 1, 2; pouch; probe 2; prod 1, 4; purse 1; root[2]; sack 1; stab 1; stick[1] 1, 3; thrust 2, 4
poke about *or* around, probe 1; pry 2
poke (along), delay 3; drag 4; lag 1
poke full of holes, explode 2
poke into, pierce 1; probe 1
poke (out), project 5; protrude; stick[1] 15
poke fun (at), deride; fun 5; gibe 1; guy 2; laugh 2a; mock 1; parody 3; poke 4; pull 15; ridicule 2; satirize; scoff; scorn 4; taunt 1
poke one's nose (in(to)), interfere 1; intervene 1; meddle; pry 2; snoop 1
poker-faced, empty 5; reserved; serious 1
pokey, jail 1; prison
polar°, chill 4; cold 1; freezing; frigid 1; icy 1
pole°, bar 1; post[1] 1; rod 1; shaft 1; spar[1]; staff 1; stake[1] 1; standard 4; stick[2] 1; upright 3
pole to pole, from, pole[2] 2
polecat, stinker
polemic, argument 1; debatable; debate 1; dispute 3
polemical, controversial 2; debatable
poles apart, pole[2] 3
police°, authority 4; guard 1; patrol 3
police, the, justice 2
police constable, police officer
policeman, constable; detective; officer 2; police officer
policemen, police 1
police officer(s), officer 2; police 1
policewoman, constable; officer 2; police officer
policewomen, police 1
policing, patrol 2
policy°, experiment 2; habit 1; line[1] 9; platform 2; procedure; rule 3; stand 13; strategy
polish°, brighten 2; charm 2; civilization 1; civilize 1; clean 9; edit 1; elaborate 4; elegance 1; finesse 3; finish 7, 11; fluency; glaze 1, 2; gloss[1] 1, 3; grace 1; grind 2; mature 6; perfect 9; refine 2; refinement 1; refurbish; rub 1; savoir-faire; savoir-vivre; scour 1; sheen; shine 2; smooth 11; sophistication 1; style 4; taste 4
polish off, accomplish; dispatch 3; dispose 3d; eliminate 4; finish 3, 4; kill 1; knock 5c; murder 3; perform 1; polish 3; rout 2; swill 3; zap
polish up, polish 2, 4
polished°, adept 1; bright 3; civil 3; courteous; cultivated; decorous; educated 2; elegant 1; expert 2; fluent; genteel 3; glossy 1; graceful 2; ladylike; lustrous; mature 3; perfect 7; pleasant 2; professional 1; refined 1; shiny; silver 3; smooth 2; sophisticated 1; spotless 1; talented; tasteful; well-bred
polite°, accommodating 1; amicable; attentive 2; ceremonious 2; chivalrous; civil 3; cordial; courteous; debonair 1; decent 2; decorous; diplomatic; dutiful; fair[1] 8; gallant 2; genteel 2; graceful 2; gracious; ladylike; nice 1; obliging; orderly 2; pleasant 2; polished 2; presentable 2; proper 3; refined 1; respectful; suave; tactful; tasteful; well-bred
polite to, be, respect 7
politely, kindly 2; properly 1
politeness, breeding 2; chivalry; civility; courtesy; decorum 1; deference 1; elegance 1; etiquette; grace 2; manner 3; propriety 2; refinement 1; respect 2; tact; taste 6
polite society, society 4
politesse, breeding 2; civility; courtesy; elegance 1; etiquette; formality 3; manner 3; propriety 2; protocol 1; refinement 1; tact; taste 6
politic°, advisable; conscientious 3; diplomatic; expedient 1; judicious; prudent 1; seemly 2; sound[2] 4; tactful; tactical; wise 3
political science, politics 1
politico, politician
polity, nation
poll°, canvass 1, 2, 4; election; plebiscite; register 1; sound[3] 1
pollinate, fertilize 1
pollutant, impurity 2
pollute°, adulterate; alloy 2; contaminate; corrupt 4; debase 2; desecrate; dirty 7; foul 12; poison 3; profane 4; soil[1] 2; sully; taint 2
polluted, bad 2; dirty 1; drunk 1; filthy 1; foul 2; impure 1; sordid 3; stagnant
polluting, pollution
pollution°, filth 1, 2; fume 3; impurity 1, 2
Pollyannaish, optimistic
poltergeist, ghost 1
poltroon, coward; milksop
polychromatic, variegated
polychrome, variegated
polyclinic, hospital
polymath, intellectual 4
polymorphic, protean
polymorphous, protean
polytechnic, technical 2
polytheist, heathen 1; pagan 1
polytheistic, heathen 3; pagan 2
pomace, pulp 2
pomade, lotion; ointment
pomatum, ointment
pommel, batter 1; beat 1; belabor; pelt[1] 1; rough 16; strike 1
pomp°, display 5; glory 3; grandeur 1; pageantry; parade 2; splendor 1; state 3
pomposity, arrogance; hot air; rant 3; raving 3; snobbery
pompous°, affected 3; arrogant 1; bloated; bombastic; condescending; disdainful; flowery; genteel 1; glorified 1; grandiose 1; inflated 2; lofty 4; majestic 2; mannered; official 2; ornate; pedantic 1; ponderous 2; pretentious 1; regal 1; self-important; snobbish; standoffish; stiff 7; stilted; stuffy 2; supercilious; windy 2
pompously, big 9
pompousness, arrogance; snobbery
ponce, pander 3
poncho, cloak 1; wrap 4
poncy, effeminate
pond, pool 1; sea 1
ponder°, balance 1; chew 4; consider 1; deliberate 4; digest 3; imagine 1; meditate 1; mull; muse; revolve 3; study 2; think 5; weigh 2; wonder 3
ponder (about *or* over *or* on), brood 3; contemplate 2; debate 4; meditate 2; puzzle 2; reflect 3; see 8; speculate 1; turn 21a
ponderable, tangible
pondering, preoccupied 1; reflection 2; reflective; speculation 2; thought 1; thoughtful 2
ponderous°, gross 1; heavy 1, 7; hulking; laborious 4; leaden 1; massive; slow 1; stodgy; weighty 1
pong, reek 1, 3; smell 2, 4; stench
pongy, high 10; stinking 1
poniard, blade 2; dagger
pontifical, clerical 1; pompous 1; positive 5
pontificate, harangue 2; lecture 3; preach 2; rant 1; spout 2
pontoon, float 5
pony, nag[2]; trot 4
ponytail, queue
poof, deviant 3; homosexual 1
poofter, deviant 3; homosexual 1
Pooh-Bah, master 1; mogul
pooh-pooh, belittle; deride; dismiss 2; laugh 2b; scorn 3; trivialize
pool°, combine 1; fund 1; kitty; lottery; merge
pooled, united 1
pooling, merger
poop, dope 3; fag 1; gossip 2; information; intelligence 2; rumor 1; scoop 3; score 6; word 2
poop, the, fact 3
poop out, bolt 7
pooped, dead 9; exhausted 1; fatigued; jaded 1; sleepy 1; spent 1; tired 1; washed-out 2; weary 1; worn 3
pooped (out), prostrate 5; ragged 3
poop sheet, promotion 5
poor°, bad 1; barren 2; base[2] 3, 5; broke; bum 4; cheap 3; deprived; destitute 1; disappointing; down-and-out 1; exhausted 3; feeble 2; flimsy 2; frugal 3; hard 17; heel[1] 3; hopeless 2; impoverished 1; indifferent 3; indigent; inferior 3; lame 2; lean[1] 2; lousy 2; low[1] 7, 9; mangy; meager 1, 4; mean[2] 3; mediocre; miserable 3; needy; pale[1] 3; par 5a; pathetic 2; penurious 2; punk 2; ragged 6; remote 6; ropy 2; shabby 2; shoddy; short 8; sleazy 1; slender 2; small 4; sordid 4; straitened; thin 2, 4; unfortunate 2; unhappy 3; unprofessional 2; unsatisfactory; upper 7;

poor° (*cont.*)
wanting 1; weak 4, 6; worthless 3
poor taste, in, tasteless 1; unseemly 1
poorhouse, home 3
poorhouse, in *or* **on the way to the,** needy
poor imitation, parody 2
poorly°, amiss 2; badly 1, 2, 3; frail 2; ill 1, 10, 12; indisposed 1; off-color 1; par 5b; queer 3; seedy 2; sick 2; weather 2
poorly, be *or* **feel,** ail 2
poor-mouth, disparage 2
poor white (trash), peasant
pop°, father 1; pawn[1] 1; protrude; snap 2, 8; spring 4
pop in (on), drop 12
pop in *or* **out** *or* **by,** pop 2; visit 1
pop off, die 1
pop up, surface 3; turn 25b
pop up again, return 2
pope's nose, tail 1
popinjay, dude 1; pup
pop one's clogs, die 1
poppa, father 1
poppet, fluff 2
poppycock, fiddlesticks; gab 2; gibberish; gobbledegook 1; mumbo jumbo 1; nonsense 1; prattle 3; rot 4; rubbish 2; stuff 4; talk 18; trash 1
popsy, fluff 2; girl 1, 2; woman 3
populace°, folk; hoi polloi; mob 2; people 3; population; public 8
populace, the, unwashed 2
popular°, big 7; current 2; demand 7; democratic; famous; favored 1; general 1; hot 6; national 1; noted; orthodox; plebeian 2; public 1; regular 8; social 1; successful 3; vernacular 2; vulgar 3
popular, become, catch 13b
popularity°, celebrity 1; laurels; name 2; prevalence 1; vogue 2
popularized, popular 3
populate°, inhabit; people 5; settle 6
populated, populous
populating, settlement 2
population°, folk; people 4; public 8
population, the, unwashed 2
populous°, overcrowded
porcelain, pottery
porch, portico
porcine, epicurean 1
pore° over, examine 1; inspect; observe 2; pore[1]; read 1; scan 2; watch 3; weigh 2
porn, pornography
porno, pornographic
pornographic°, base[2] 4; blue 2; broad 8; dirty 2; erotic 3; filthy 3; gross 3; immoral 2; indecent 2; lascivious 2; lewd; low[1] 3; nasty 3; naughty 3; obscene 1; outrageous 3; prurient 2; racy 2; rank[2] 3; rough 6; rude 3; sexy 2; suggestive 2; vulgar 2; wicked 3
pornography°, dirt 3; filth 3
port°, harbor 1; left 1, 3
portable°, light[2] 1; mobile 1; movable
portable (radio), radio 1
portend, announce 4; bode; foreshadow; mean[1] 3; predict; prognosticate 2; prophesy 2; spell[3] 1; threaten 3
portent, foreboding 2; forerunner 2; harbinger; indication 3; omen; sign 6; significance 1; threat 2; warning 2
portentous°, fateful 1; important 1; impressive; inauspicious; momentous; ominous 3; pressing; sinister 1; threatening; unpromising; weighty 2
porter°, carrier 1
portfolio, file 1; scrapbook
portière, drapery
portion°, allotment; bit 4; clip[2] 5; cut 20; dole 1; dose 1; draft 4; fragment 1; helping; interest 4; leg 3; lot 1, 4; measure 5; nip[2]; parcel 2; part 1, 2, 7; passage 2; pat[1] 4; percentage; piece 1, 2, 3, 11; plate 2; proportion 3; public 10; quota; ration 1; section 1, 3; segment 1; share 1; slice 2; tract[1]; unit
portion (of), have a, partake 2a
portion out, portion 4
portliness, fat 5; fatness; obesity
portly, dumpy; fat 1; heavy 11; large 1; obese; plump[1] 1; rotund 3; stout 1; well-fed
portmanteau, bag 2; suitcase
portmanteau word, neologism
portrait°, description 1; image 1; likeness 2; picture 1; profile 2; representation 1
portray°, act 7; characterize; describe 3; draw 5; enact 2; express 1; label 5; paint 4; picture 7; play 4; pose 2; render 1; represent 3; show 8; tell[1] 4
portrayal, description 1; history 1; narration 1; narrative 1; performance 3; portrait; relation 4; rendering; representation 1; version 1
portside, left 3
pose°, affectation 2; form 2; masquerade 2; posit; position 1; posture 1, 4; pretense 2; show 11
pose (as), masquerade 3; pass 16b; portray 2; pose 2; represent 2
pose in, model 8
posed, mannered
poser, enigma; model 4; problem 2; puzzle 4; riddle[1]
posh°, elegant 3; fancy 3; genteel 1; grand 1; luxurious 1; palatial; plush; princely 2; refined 1; soft 13; splendid 1; sumptuous; swanky; swell 7; swish 3; upper-class 2
poshness, splendor 1
posit°, conceive 3; moot 2; pose 3; premise 2; presume 1; suppose 2
position°, angle[1] 2; appointment 3; argument 2; arrange 1; aspect 1; assignment 2; attitude 1, 2; base[1] 6; bearing 4; capacity 3; caste; chair 2; character 7; conviction 2; degree 1; dignity 2; draw 15c; engagement 4; estate 3; fix 8; footing 2; fortune 1; game 4; grade 1; idea 3; importance 2; install 1; job 1; lay[1] 1; level 14; locate 1; location 1; mean[3] 4c; mind 6, 7; niche 2; nobility 2; occupation 1; office 4; opening 3; order 3; orientation 1; outlook 1; perch 1, 2; perspective 1, 2; pitch[1] 2; place 1, 3, 5, 8, 9, 15; placement 1; plant 7; point 3; point of view 1; pose 1, 4; posture 1, 2, 3; post[2] 1, 2; precedence; presumption 3; profession 1; prominence 1; purchase 5; put 1; rank[1] 1; rate[1] 4; reputation 1; rest[1] 8; role 2; run 19; scrape 8; seat 3; sentiment 2; set 1, 11; side 3; site 1, 2; situate; situation 1, 3; slot 2, 4; sphere 2; stage 1; stance; stand 2a, 13; standing 5; standpoint; state 1; station 1, 2, 4; tenet; term 7b; terminal 4; vacancy 2; view 2, 3; viewpoint; whereabouts 1; work 2
position, hold (down) a, work 9
position, in, ready 9a
positioning, installation 2; orientation 1; placement 1
position statement, editorial
positive°, absolute 4; assertive; categorical; certain 5; clear 10; confident 1; constructive 1; definite 2; demonstrable; dogmatic; emphatic; explicit 1; express 5; favorable 2; firm 4; flat 4; good 17; incontrovertible; indisputable; optimistic; peremptory 3; photograph 1; print 2; promising; purposeful; self-confident; sure 1; unquestionable; upbeat
positive, be, know 3; mind 11; tell[1] 7
positively°, absolutely 1, 3; certainty 3; clearly 2; course 6; definitely; doubtless 1; entirely 2; exactly 2; expressly 1; fairly 3; favorably 1, 2; flat 16; indeed 1; mean[3] 2a; once 3; perfectly 1; quite 4; really 2; surely 1; thoroughly 1; truly 3; well[1] 7
positiveness, certainty 2; conviction 3; trust 1
posologist, pharmacist
posolutely, doubtless 1
possess°, come 16b; command 5; enjoy 2; haunt 2; have 1, 4; hold 12; obsess; preempt; seize 3
possess oneself of, possess 5; possession 4
possessed°, infatuated
possessed, like one, rapidly 1
possessed of, be, possess 1, 4
possessed with, be, possess 4
possessing, obsessive
possession°, clutch 2b; enjoyment 2; grasp 4; hand 8; occupation 2; seizure 1; tenancy; tenure 1; thing 1; title 5
possession of, assume *or* **take** *or* **usurp** *or* **gain,** take 37
possession of, be in, possess 1
possession of, be the, belong 3
possession of, come into (the), acquire; obtain 1
possession of, get *or* **take,** secure 5
possession of, take, obtain 1; occupy 1; preempt; requisition 3; seize 3, 5; take 1
possession of, take *or* **gain** *or* **get,** come 9a; possess 5; possession 4
possession-oriented, materialistic
possession(s), acquisition 2; asset 1; belongings; effects; estate 2; fortune 1; furniture 1; good 21a; paraphernalia; possession 3; property 1; resource 2; stuff 2; substance 5; thing 8c
possessive°, hoggish; protective; selfish 2
possessor, owner
possess the quality of, partake 2b
possibilities°, field 3; possibility 2; prospect 3
possibilities, realm of, choice 2
possibility°, alternative 2; candidate; chance 3; eventuality; lead 14; opportunity; potential 2; probability; prospect 3; thing 7
possibility, strong *or* **distinct,** likelihood
possible°, card 11; earthly 4; feasible; potential 1; practicable; probable;

prospect 4; specious; tenable; thinkable; viable
possible, if, possibly 1
possible that, it is, perhaps
possibly°, chance 5b; maybe; perhaps; probably; seemingly
possie, position 6
POSSLQ, friend 3; girl 2; love 3; mistress 1; paramour; partner 1
post°, appointment 3; assignment 2; base[1] 4, 6; book 4; capacity 3; direct 5; dispatch 2; engagement 4; forward 7; installation 3; mail 1, 2; occupation 1; officc 4; pale[2] 1; paper 5; picket 1; pier 2; pillar 1; place 5, 15; position 6; profession 1; put 28d; quarter 6; role 2; send 2; service 6; settlement 1; shaft 1; situation 3; stake[1] 1; standard 4; station 1, 4; stick[1] 17b; transmit 1; upright 3; vacancy 2; work 2
post, hold (down) a, work 9
postage, carriage 3
postal card, card 4; line[1] 8; note 2
postal order, draft 2
postal service, post[3] 1
postcard, card 4; line[1] 8; missive; note 2
postdate, anachronism
postdating, anachronism
posted, aware 1; current 4
posted, keep (someone), post[3] 4
poster°, advertisement 1; promotion 5; sign 5
posterior°, bottom 1; bum 1; buttocks; rear[1] 2
posterior(s), seat 4; tail 2
posterity°, descendant; progeny
postgraduate, graduate 1
posthaste°, fast[1] 6; flat 17a; hastily 1; once 6a; quickly 2
postillion, servant 1
post-mortem°, review 6
postpone°, call 10; defer[1]; delay 1; hinder 1; hold 18a, 21a; procrastinate 1; put 21a; reserve 1; shelve; stay[1] 4; string 11b; suspend 1; table 5; waive 2
postponed, abeyance; off 9
postponed, be, fire 4; hang 10; wait 2
postponement°, delay 4; moratorium; reprieve 2; respite 2; stay[1] 6; suspension 2; waiver
postponing, suspension 2
postscript, supplement 1
postulate, conceive 3; doctrine; guess 1, 3; hypothesis; law 3; pose 3; posit; position 7; premise 1, 2; presume 1; principle 1; propound; reasoning 2; speculate 1; suppose 2; supposition; theorem 1; thesis 1
postulated, given 2
postulation, presumption 2; speculation 1
posture°, act 3, 8; attitude 1, 2; bearing 1; camp[2] 2; carriage 2; pose 2, 4; position 1; show 11; stance; stand 13
posturer, poseur
posturing, affectation 1; pretense 1
posy, bouquet 1; bunch 1; spray[2]
pot°, jar[1]; kitty; mug 1; pan 1; pool 2; stomach 1
pot, go to, degenerate 2; deteriorate 1; seed 4; stagnate
potable, drink 4
potables, provision 4
potation, draft 4; drink 4; potion
potbellied, fat 1; obese
potbelly, paunch; pot 3; stomach 1
poteen, moonshine 3
potency, effect 2; force 1; influence 1; might 1; potential 2; power 4; strength 4, 5
potent°, active 2; drastic; dynamic; forceful 1; mighty 1; powerful 1, 2, 3; prevailing 2; rich 9; solid 7, 8; stiff 3; strong 2, 4; telling 1; weighty 3
potentate, crown 3; monarch 1; sovereign 1; tycoon
potential°, capability; capacity 2; dormant 2; faculty 1; lead 14; possibility 2; power 1; promise 2; prospective; seminal 2; timber 3; undeveloped
potentiality, possibility 2; power 1
pothead, addict 1
pother, flurry 1; fuss 1; hurry 3; lather 2; rumpus; rush 3; stew 2; sweat 7; uproar
pothole, pit[1] 2
potholed, bumpy; irregular 1
potion°, elixir 2
potpourri°, assortment 2; confusion 5; hash 1; hodgepodge; medley; mess 2; miscellany; mishmash; mixture 1; pastiche
pot(s), heap 2; mint 1
potted, stinking 3
potter. See **putter**
potter's field, graveyard
potty, crazy 1; deranged; foolish 2; insane 1; mad 1; off 4; queer 4; wild 9
pouch°, pocket 1; purse 1; sack 1
pounce°, jump 1, 8; lunge 2, 3; swoop 1, 2
pounce on *or* **upon,** assault 3; attack 1; come 11; descend 4; jump 6; lace 5a; light[2] 14; make 30b; pounce 1; raid 2; set 21; slam 3; snap 4; waylay 1
pound°, ache 1; batter 1; beat 1, 3; cage 1; crush 1; grind 1; mill 5; pelt[1] 1, 2; pen[2] 1; powder 3; pulsate; pulverize 1; ram 1; run 46; wash 3
pound away at, wade 4
pound into, pound[1] 4
pound out, pound[1] 5
pounding, ache 3; knock 8; pound[1] 7; pulse 1; thrashing 1
pour°, crowd 3; draw 3, 12a; flock 2; flood 6; flow 1; issue 11; lavish 5; pelt[1] 2; rain 4, 5; run 6; shower 3; stream 4, 5; surge 1; teem[2]
pour cats and dogs, teem[2]
pour down the drain, use 5
pour forth, vent 4
pour in, roll 10a
pour into *or* **throughout** *or* **over,** flood 7; suffuse
pour out, vent 4
pour (out *or* **forth** *or* **over),** discharge 4; emit; empty 8; flood 4; run 33c; shed[2] 2, 3; spill 1; spout 1
pourboire, consideration 2; gift 1; present[2] 3; tip[3] 1
pouring, wet 2
pout°, brood 4; lower[2] 2; mouth 5; sulk
pouting, sullen
poverty°, dearth; misery 2; necessity 3; need 4; privation; want 5
poverty, reduce to, ruin 10
poverty level, below the, needy
poverty-stricken, broke; deprived; destitute 1; down-and-out 1; hard 17; impoverished 1; indigent; lean[1] 3; needy; penurious 2; poor 1; rock[1] 3c; sordid 4; straitened; upper 7
powder°, grind 1; mill 5; pound[1] 2; pulverize 1; reduce 4
powder, take a (runout), bolt 8; clear 29; depart 1; escape 1; exit 3; flee 1; flight[2] 3; fly 2; heel[1] 4; make 26; powder 2; run 2; split 6
powdered, fine[1] 7
powder room, facility 2b; lavatory; toilet 1
powder room, go to the, defecate; urinate
powdery, fine[1] 7
power°, ability 1; action 2; agency; agent 2; authority 1; brawn; capability; capacity 2; clutch 2b; command 7; control 4; country 1; dignitary; disposition 4; domination 1; dominion 1; duress 1; dynamism; effect 2; energy; faculty 4; force 1; freedom 4; gift 2; god; grasp 4; grip 2; hand 8; hold 26; importance 2; influence 1; intensity; juice 2; jurisdiction; lock[1] 2; main 6; might 1, 2; momentum; predominance; preponderance 2; prerogative; pressure 1, 4; propulsion; punch[1] 3; rank[1] 2; reign 1; rein 2; right 9; say 12; sinew 2; sovereignty; stamina; strength 1, 5, 6; sway 4; talent 1; tenacity 2; thrust 5; upper 6; use 9; vigor; violence 1; vitality 1; weight 3
power (over), be in, govern 1; rule 5
powered, hot 9
powerful°, active 2; brawny; drastic; dynamic; effective 2; effectual 1; forceful 1, 2; formidable 3; full 11; hefty 2; husky 1; important 3; impressive; influential; intense 1; lusty 2; mighty 1, 2; muscular; overpowering; potent 1; prevailing 2; privileged 3; robust 1; sinewy; solid 5, 7, 8; stalwart 1; stiff 3, 5; strong 1, 2, 7, 9, 17; sturdy 1; telling 1; tough 3; vivid 2; weighty 3
powerfully, deeply 2; highly 4
powerless°, defenseless; feeble 1; impotent 1; ineffectual 2; prostrate 4; spineless 2; unable; vulnerable
powerless (to), incapable 1
powerlessness, disability 2; impotence 1; weakness 2
powers°, power 1, 7
powers that be°, authority 4; power 10
powers that be, the, establishment 3
power (supply), main 5
powwow, discussion; palaver 3, 5; parley 1; talk 15; word 1
PR, plug 3
practicability, feasibility; usefulness
practicable°, constructive 1; feasible; functional 1; operable; possible 2; user-friendly; viable
practical°, advisable; constructive 1; empirical; expedient 1, 2; feasible; functional 1; handy 2; hard 6; helpful; logical 2; no-nonsense; philosophical 2; positive 6; possible 2; rational 3; realistic 1; sensible 1; sober 2; sound[2] 4; useful; viable; virtual
practicality, feasibility; logic 3; purpose 3; usefulness
practical joke, gag[2] 2; lark 1
practical joker, card 10
practically°, almost; close 21; intent 3; nearly 1; neighborhood 2; practice 4; substantially; virtually
practice°, application 1; convention 2; course 3; custom 1; deal 4; discipline 1; drill 4; exercise 5; experience 1;

practice° (*cont.*) follow 6; form 4; formality 1; go 29b; habit 1, 2; institution 4; lesson 2; method 1; observance 2; perpetrate; policy; preparation 4; prepare 2; prevalence 1; procedure; prosecute 3; protocol 1; rehearsal 1; rehearse 1; rite; ritual 3; routine 1; rule 3; run 33d; run-through 1; system 2; tradition; trick 6; usage 1; way 1
  practice, accepted, propriety 3
  practice, in, practice 4
  practice, out of, practice 5
  practice, put into, practice 7; use 1
  practice of, make a, practice 7
practiced°, accomplished; experienced 1; expert 2; neat 4; old 7; professional 1; proficient; qualified 1; skillful; veteran 2
  practiced in *or* **at,** versed
practice(s), code 3; moral 4
practicing, preparation 4
practise, drill 2; exercise 1; practice 6, 7; pursue 2; train 5; use 1; wage 2
practised, learned; perfect 7; practiced 1, 2; seasoned 1
practising, practice 2
practitioner, operator 1
pragmatic, empirical; expedient 1; hard 6; helpful; practical 1; rational 3; realistic 1
pragmatic(al), philosophical 2; positive 6
prairie, plain 6
prairie(s), flat 14b
praise°, applaud 2; applause; bless 1; bouquet 3; celebrate 3; compliment 1, 3; credit 4; eulogize; eulogy; exalt 2; extol; flatter 1; glorify 2; glory 2; hail[1] 2; honor 2, 6; kudos; laud; mention 4; ovation; pat[1] 2, 5; puff 3, 7; rave 2, 3; recommend 2; recommendation 2; tribute 1; worship 1, 2
praised, immortal 3
praiseful, laudatory
praisefully, highly 3
praises (of), sing the, extol; laud; praise 3
praiseworthy°, deserving; estimable; exemplary 2; famous; laudable; meritorious; splendid 3; worthy 1
praising, celebration 2
prance°, camp[2] 2; caper 3; cavort; frisk 1; leap 2; skip 1, 3; strut; swagger 1, 3; walk 1
prang, collision; shell 4
prank°, caper 2; frolic 2; gag[2] 2; giggle 3; lark 1; put-on 1; trick 2
prankish, arch 3; devilish; sportive
pranks, fun 2; nonsense 2
prankster, card 10; joker 1
prat, bottom 1; buttocks
prate, chatter 1, 3; drivel 2; gab 1; jabber 1; palaver 4; patter[2] 2, 3; prattle 1, 2; rattle 6; talk 3; tattle 2; twitter 2; waffle 1; yap 2
prating, palaver 2; prattle 2
prattle°, babble 1, 3; chatter 1, 3; drivel 2; gab 1, 2; gibberish; gossip 1; gush 2; jabber 1; jargon 2; palaver 2, 4; patter[2] 2, 3; rattle 6; talk 3, 18; tattle 2; twitter 2; waffle 3; yap 2
  prattle (on), waffle 1
prattling, palaver 2; talk 18
praxis, habit 1; practice 1; tradition
pray°, appeal 1; ask 3; beg 1; bid 2; solicit 1; sue 2
  pray (for), plead 1; pull 10
prayer°, appeal 3; blessing 1; devotion 1; grace 5; litany 1; suit 5; supplication 1
prayers, say one's, pray 2
praying, prayer 2; supplicant 1
preach°, harangue 2; lecture 3; rant 1
  preach (on *or* **about),** hold 16a
preacher°, clergyman 3; minister 1; missionary; priest
preaching, sermon 1
preachy, pedantic 1
preamble°, foreword; preface 1; preliminary 3
prearrange, arrange 2; fix 12; program 4
(pre)arranged, cut 29a; deliberate 1; fixed 3, 4; given 1; inside 5; predetermined 1; prepared 1; put-up; set 29
prearrangement, fix 18; provision 3; setup 2
prebend, clergyman 1
prebendary, clergyman 1
precarious°, awkward 4; dangerous 1; delicate 4; desperate 5; explosive 1; hairy 2; hazardous; hot 10; insecure 3; nice 4; parlous; rickety; risky; serious 4; shaky 1, 2; slight 4; ticklish 2; tight 7; touchy 2
precarious condition, sore 6
precariously, dangerously 1
precaution°, foresight 1; providence 1; prudence 2
precautional, noncommittal
precautionary, noncommittal; preventive 2
precautions, take, provide 5
precautious, noncommittal
precede°, announce 4; head 12; lead 4, 10a; preface 2
precedence°, predominance; preliminary 2; priority
  precedence (over), take, outweigh
precedency, precedence; priority
precedent°, foregoing; instance 1; lead 13; prototype 1
precedent-setting, avant-garde; landmark 3; seminal 1; way-out 2
preceding°, before 6; foregoing; former 1; late 3; old 6; preliminary 1; preparatory 3; previous 1, 2, 4; prior 2
  preceding the time when, before 8
precept°, charge 5; doctrine; lesson 3; moral 3; motto; principle 1; rule 1; standard 1; tenet; thesis 1
preceptor, instructor
précieux **or** ***précieuse,*** priggish
precinct°, area 3; district; region 1; territory 1; ward 1; zone
precinct(s), limit 3; milieu; neighborhood 1; precinct 1; vicinity
precious°, beloved 1; darling 3; dear 1, 3; expensive; exquisite 5; inestimable 1; invaluable; mincing; old 8; pet[1] 3; priceless 1; priggish; prissy; rich 3, 4; sweet 5, 7
precious stone, gem 1
precious stones, jewelry
precipice°, bluff[2] 3; cliff; crag
precipitancy, haste 2
precipitate°, abrupt 1; brash 1; cause 6; deposit 4; dregs 1; hasty 2; hot-headed; hurried 1; impetuous; impulsive; pell-mell 2; quick 2; rapid; rash[1]; sediment; sink 6; sludge; snap 13; spark 2; speedy 2; sudden
  precipitate (out), settle 11
precipitately, hastily 1; pell-mell 1; sharply 2; summarily 2; swiftly
precipitateness, speed 1
precipitation°, rain 1, 2
precipitous°, abrupt 2; sharp 2; sheer 1; speedy 2; steep[1] 1; whirlwind 2
precipitously, sharply 2
precipitousness, speed 1
précis°, abridgment 3; abstract 3; brief 4; digest 5; epitome 2; outline 2; program 1; rundown; résumé 1; scenario 1; summary 1; synopsis; table 3; telescope 2
precise°, accurate 1; careful 2; circumstantial 3; close 14; correct 8; dead 17; deadly 6; definite 1; definitive 3; delicate 6; detailed 1; determined 2; distinct 1; elaborate 1; even 8; exact 1; explicit 1; express 5; factual 2; faithful 2; fine[1] 9; flat 9; formal 3; graphic 1; literal 1; mathematical; measured 4; meticulous; narrow 3; neat 3; nice 3; particular 1, 3; perfect 5; perfectionist 2; photographic 1; precious 3; proper 2; refined 2, 3; religious 2; right 2; rigid 3; round 5; scientific; scrupulous 1; special 1; specific 1; square 3; strict 1; true 3; verbatim 1; very 3
  precise, be, nail 5
precisely°, completely 1; dead 20; directly 4; dot 2; due 5; exactly 1, 2; flat 16; full 14; just 5; letter 4; literally 1; nearly 2; outright 4; particular 6; pat[2] 1; perfectly 3; plumb 4; purpose 4b; right 14, 17; sharp 10; slap 6; verbatim 2; world 6
preciseness, accuracy; precision 1; rigor 2
precisian, perfectionist 1; prig; purist
precisianism, prudery
precision°, accuracy; definition 1; formality 2; resolution 5; rigor 2
precisionist, perfectionist 1; prig
preclude°, anticipate 1; exclude 2; forbid; forestall; hinder 2; obstruct 3; oppose 2; prevent; prohibit 2; resist 1; rule 8; save 4; suppress 1; veto 1
precluding, short 10; suppression
preclusion, exclusion 2; prevention; suppression; veto 2
preclusive, peremptory 2
precocious°, bright 6; forward 3
precocity, brilliance 2
precognition, anticipation 2
preconceived°, deliberate 1; intentional; jaundiced 1; premeditated; put-up
preconceived notion *or* **idea,** preconception; prejudice 1
preconception°, prejudice 1; presumption 2
preconcerted, put-up
precondition°, prerequisite 2; requirement 1
precooked, instant 5; prepared 4
precursor°, ancestor; forerunner 1; harbinger; pioneer 1; progenitor 2; stock 3
predacious, predatory 1; rapacious; voracious 1
predaciousness, rapacity
predate, precede
predatory°, ferocious; mercenary 1; rapacious
predecessor°, forerunner 1; pioneer 1; predecessor 1; progenitor 2
predecessors, root[1] 4
predestine, destine 1
predestined, bound[3] 4; certain 3; destined 1; doomed 1; eventual 2; fatal 3; fated 1; necessary 3; unavoidable
predetermination, preconception
predetermine, arrange 2; destine 1; fix 12
predetermined°, certain 1; cut 29a; destined 1; determined 2; fatal 3;

fated 1; foregone; limited 1; measured 4; necessary 3; preconceived; set 29; specific 1; unavoidable
predicament°, bind 5; complication 2; difficulty 2, 3; dilemma; embarrassment 2; emergency; fix 17; hole 5; jam 6; mess 3; pass 26; perplexity 3; pinch 8; plight; problem 1; quandary; scrape 8; situation 2; snarl[2] 3; spot 4; strait 3
predicament, in a, trouble 9a
predicate, pose 3; posit; premise 2
predicated, presumptive 2
predication, position 7
predict°, anticipate 2; divine 4; envisage 2; envision; forecast 1; foresee; foreshadow; prognosticate 1; project 6; prophesy 1; say 8; tell[1] 9
predictability, chance 3; regularity 1
predictable°, consistent 2; regular 1; reliable; square 6; usual
predicted, presumptive 2
prediction°, advisory 2; calculation 3; foreboding 2; forecast 2; oracle 2; prognosis; projection 4; prophecy 1, 2; tip[3] 2
predictive, ominous 3; prophetic
predilection, appetite 1; aptitude 2; bias 1; fancy 7; inclination 3, 4; leaning; like[2] 3; liking 1; love 2; partiality 1; passion 3; penchant; preference 2; prejudice 1; talent 2; tendency; thing 4; weakness 4
predilection for, having a, partial 3
predispose, bias 4; incline 2
predisposed, disposed; inclined 1; interested 2; open 18; preconceived; prejudiced 1; prepared 3; prompt 2; prone 2
predisposed, be, incline 3
predisposed to *or* **towards,** capable 2; fond 3; partial 3; susceptible 1
predisposition, bent 5; bias 1; favor 4; favoritism; habit 2; inclination 3; instinct; liking 1; partiality 1; penchant; preconception; preference 2; prejudice 1; talent 2; tendency; weakness 4; wish 5
predominance°, preponderance 2; prestige; prevalence 1, 2; priority; superiority 1
predominancy, predominance
predominant°, dominant 2; influential; main 1; mighty 1; overriding; paramount; popular 2; prevailing 1; prevalent 1; primary 1; prime 2; principal 1; rampant 2; sovereign 2; uppermost 2
predominantly, chiefly; especially 2; generally 1; mainly; primarily 1; principally; usually; whole 5
predominate°, dominate 2; loom 3; prevail 1, 2; reign 3; rule 5
predominate over, predominate
predominately, primarily 1
predominating, main 1; overriding; popular 2; predominant
preeminence°, distinction 2; domination 1; dominion 1; emphasis; excellence; fame; importance 2; lead 12; name 2; perfection 1; precedence; predominance; prestige; priority; prominence 1; sovereignty; status 1; superiority 1; supremacy 1
preeminent°, arch 1; best 1, 3; capital 5; cardinal; central 2; choice 4; distinguished 1; divine 2; dominant 2; elder 2; elevated 2; eminent 1; famous; first 1; foremost 1; grand 2; head 9; leading 2; lofty 2; main 1; major 2; noble 3; notable 2; paramount; peerless; perfect 2; predominant; premier 2; prestigious; primary 1; prime 2; principal 1; prominent 2; singular 2; sovereign 2; splendid 3; star 3; successful 3; supreme 3; top 8; transcendent; uppermost 2; well-known 2
preeminent, be, excel; shine 3
preeminently°, chiefly; large 4; primarily 1
preempt°, beat 2
preempted, assumed 1
preen°, groom 3; primp
preestablished, foregone; predetermined 1
prefabricated, ready-made 1
preface°, foreword; preamble; preliminary 3
prefatory°, introductory; preliminary 1; preparatory 1
prefect, monitor 1
prefer°, choose; fancy 11; favor 6; feel 9; go 28c; lay[1] 7; lean[2] 3; like[2] 2; please 2; rather 3; select 1; side 10; yearn
prefer to, have 11
preferable, right 3; select 2
preferably, better 5; instead 2; rather 2
preference°, appetite 1; choice 1; discretion 2; fancy 7; favor 3; favorite 1; favoritism; inclination 3; leaning; like[2] 3; liking 1, 2; love 2; partiality 1, 2; patronage 4; penchant; pick 9; pleasure 2; precedence; priority; relish 1; selection 1; taste 3; vogue 2; volition; weakness 4; wish 5
preference, show, incline 3; side 7
preference for, have a, approve 3; favor 6; love 8
preference for, show, side 10
preference for, show (a), lean[2] 3; select 1
preference to, in, before 7; instead 2
preferential°, inside 4
preferment, favoritism; promotion 2
preferred, choice 4; favored 1; favorite 1, 2; inside 4; pet[1] 3; right 3; select 2; superior 2
prefix, preface 2
preggers, expecting; pregnant 1
pregnancy, maternity 1
pregnant°, eloquent 2; expecting; expressive 3; meaningful 2; significant 2
pregnant, make, impregnate 1; knock 7c
pregnant, unmarried *or* **unwed and,** trouble 9b
pregnant (with), become, conceive 1
(pre)historic, ancient 2; early 6; former 2; old 2; primitive 1
prejudged, preconceived
prejudgment, preconception; prejudice 1
prejudice°, bent 5; bias 1, 4; bigotry; disadvantage 2; discrimination 1; fanaticism 2; favor 3; favoritism; incline 2; inequality 2; injustice 1; intolerance; leaning; partiality 1; preconception; preference 2; slant 2
prejudice for *or* **against, show,** discriminate 2
prejudiced°, biased; bigoted; interested 2; intolerant 2; jaundiced 1; loaded 3; narrow-minded; nearsighted 2; one-sided 1; opinionated 2; parochial; partial 2; partisan 3; preconceived; preferential; sectarian 2; subjective 1; ultra-
prejudiced, be, incline 3
prejudicial°, detrimental; hurtful 1; interested 2; loaded 3; sectarian 2
prelate, divine 5
prelatic, clerical 1
preliminaries, preliminary 3
preliminary°, foregoing; introductory; prefatory; preparatory 1
prelims°, preliminary 3
prelude, foreword; preliminary 2
premature°, early 4; immature 1; previous 3; undeveloped
prematurely°, early 1; time 10
premeditated°, calculated 2; cool 3; deliberate 1; intentional; measured 2; studied; voluntary 2; willful 1
premeditation, forethought
premier°, chief 2; first 1; head 9; lead 18
premier danseur, lead 16; principal 6
première°, come 15b; debut 1
première danseuse, heroine; lead 16; principal 6
premise°, hypothesis; presumption 2; thesis 1
premised, given 2
premises, accommodation 4; reasoning 2
premiss, hypothesis; premise 1; presumption 2; thesis 1
premium°, bounty 3; extra 3; gift 1; inducement; price 3; prize[1] 1; treat 6
premium, at a, premium 4; scarce
premonish, warn 1
premonition°, feeling 2; foreboding 2; forerunner 2; hunch 1; intuition; misgiving; omen
premonitory, ominous 3; preliminary 1
preoccupation, fixation; foible; life 8; mania 1; obsession; thing 4
preoccupied°, absent-minded; absorbed; deep 4; engaged 2; forgetful; full 6; pensive; wistful 2
preoccupy, consume 5; engage 2; obsess; occupy 3; possess 3
preordain, destine 1
preordained, eventual 2; fatal 3; fated 1; given 1
preowned, used 1
prep, preparation 2
preparation°, instruction 2; orientation 2; practice 2; preliminary 2; prescription 2; production 1; providence 1; provision 3; prudence 2; scholarship 1; schooling; tuition; wash 13
preparation for, in, preparatory 3
preparation(s), arrangement 5; groundwork; preparation 1
preparative, preparatory 1
preparatory°, introductory; prefatory; preliminary 1
preparatory to, preparatory 3
prepare°, arrange 2; break 18b; brew 2; coach 3; condition 5; cut 10, 16e; draw 6, 15b; edit 4; educate; get 51d; groom 4; ground 6; have 6; instruct 1; line[1] 23a; make 15, 17; mobilize; mount[2] 4; poise 3; practice 6; prime 5; process 3; project 3; qualify 1; ready 10; school 4;

problematic°, debatable; doubtful 1; hairy 3; open 7; paradoxical; precarious; thorny 2; tight 7
problematic(al), debatable; difficult 2; moot 1; problematic; questionable; unresolved
problem drinker, drunk 3
problem(s), pressure 3; static 2
proboscis, trunk 4
procedural, parliamentary; ritual 2
procedure°, approach 7; course 3; experiment 2; form 4; formality 1; formula; job 3; line[1] 9; manner 1; mean[3] 4a; measure 6; mechanism 3; method 1; mode[1] 1; operation 3; order 5; palaver 1; path 3; plan 1; policy; practice 1; proceeding 1; process 1; recipe 2; rigmarole; rite; road 1; routine 1; step 4; strategy; system 2; tack 3; technique 1; way 1
  procedure, the, rope 2
proceed°, advance 3; carry 11a; continue 5; drive 3; emanate 1; emerge 2; flow 1; get 28c, 28d; go 1, 19, 32a; head 10; headway 2; hop 2; move 1; originate 2; pass 2, 18; progress 5; push 10; react 1; shape 10a; stem[1] 3; step 11; stream 5; travel 3; walk 1
  proceed (against), sue 1
  proceed before, precede
  proceed first, precede
  proceed from *or* out of, proceed 2; spring 3
  proceed on *or* in, ride 1
  proceed past, pass 1
  proceed toward(s), make 30a
  proceed (with), continue 1; proceed 1; pursue 2; wage 2
proceeding°, action 4; affair 3; incident 1; job 3; matter 2; measure 6; move 13c; operation 3; process 1; progress 4; suit 4; thing 6; transaction 1; way 12
proceeding(s), minute[1] 2; order 5; preparation 1a; proceeding 2; program 3; transaction 2
proceeds°, earnings; gain 8; income; proceed 3; profit 1; receipt 3; return 9; revenue; take 40; yield 5
process°, action 4; course 3; mean[3] 4a; mechanism 3; method 1; organ 1; path 3; prepare 7; proceeding 1; put 27b; screen 8; spur 2; suit 4; system 2
processed, prepared 2
procession°, line[1] 6; march 2; pageant; parade 1; review 8; string 3; succession 2; train 3
processor, producer 1
prochronism, anachronism
proclaim°, announce 1; broadcast 2; circulate 2; declare 1, 2; decree 2; deliver 4; enunciate 2; exclaim; issue 10; maintain 3; notify 2; post[1] 2; pronounce 2, 3; propagate 3; publish; tell[1] 2; testify; vent 3
proclaiming, proclamation 1
proclamation°, announcement 1; assertion 2; circulation 2; declaration 1, 2; decree 1; issue 6; observation 2; order 4; pronouncement 1, 2; publication 1; regulation 3; revelation; statement
proclivity, appetite 1; aptitude 2; bent 5; bias 1; favoritism; flair 1; habit 2; inclination 3; love 2; penchant; preference 2; talent 2; tendency; weakness 4
  proclivity toward(s), vein 4
procrastinate°, dawdle; delay 3; drag 6; linger 4; play 14; stall[2] 1; tarry 1
procrastinating, stall[2] 2
procrastination, stall[2] 2
procreant, sexual 1
procreate, father 5; generate 2; propagate 1; reproduce 2
procreation, generation 1
procreative, sexual 1
procreator, parent 1
proctor, monitor 1
procumbent, prone 1; prostrate 3; supine 1
procure°, acquire; buy 1; come 9a; derive 1; draw 8; earn 2; enlist 2; find 5; gain 1; get 1; harvest 3; have 2; make 7; obtain 1; pander 2; pimp 2; possess 5; purchase 1; reap 2; secure 5; take 1; win 2
procured, derivative 1
procurement, acquisition 1; purchase 3
procurer°, pander 3; pimp 1
procuress, procurer
prod°, abet 1; dig 2; drive 1; excite 1; incentive; incite; induce 1; inflame 1; jab 1, 3; jog 2; motivate; motive 1; move 6; nudge 1, 2; poke 1, 5; press 6; probe 2; prompt 3; push 4; refresh 2; rouse 2; spur 1, 4; stick[1] 3; stimulant 1; stir 4; thrust 1, 4; urge 1, 2
prodding, encouragement 1; incitement 1
prodigal°, extravagant 1; high 12; improvident 1; lavish 3; misspent; profligate 2, 4; rake[2]; spendthrift 1, 2; wasteful; wastrel 1
prodigality°, dissipation 1; excess 2; extravagance 1; exuberance 2; plenty 2; profligacy 2; waste 6
prodigious°, enormous; exceptional 2; formidable 2; great 1, 2; heavy 2; huge; incredible 2; inestimable 2; massive; mighty 3; monstrous 3; monumental 1, 2; mountainous 2; overwhelming 2; phenomenal; portentous 2; vast; virtuoso 2; voracious 1; weighty 1; whopping 1
prodigy°, virtuoso 1; wonder 1
prodromal, preliminary 1
prodrome, preliminary 3
prodromus, preliminary 3
produce°, afford 3; bear 6; breed 2; brew 2; bring 5, 7, 10a, 10b, 11a, 13a, 13b, 14b; cause 6; create 1, 2; cultivate 2; effect 7; fabricate 1; form 7; gain 2; generate 1, 3, 4; good 21b; grind 6; grow 3; harvest 1; induce 2; lead 6; let[1] 8c; make 1, 5; manufacture 1; merchandise 1; mint 2; model 7; occasion 5; originate 1; output 3; perform 3; pound[1] 6; prepare 5, 7; present[2] 5, 7; procure 2; product 2; provide 2; provoke 2; put 17b, 22d; raise 2; realize 1, 3; rear[2] 1; render 1; rise 17; run 29b; spawn; spin 2; stage 4; turn 20a; wares; whip 6; work 16, 20b; yield 4
produced, be, come 15b; result 2; rise 11
producer°, cause 2; manufacturer; showman
produce results, pay 4
produce young, reproduce 2
producing, production 1
product°, calculation 2; preparation 5; production 2
production°, drama 1; entertainment 2; evolution; execution 4; extravaganza; fabrication 1; formation 1; generation 1; industry 1; manufacture 3; opus; output 1; piece 4; play 20; presentation 2; program 2; reproduction 3; show 14; turnout 2; work 4; yield 5
  production, in, work 5
  production (out) of, make a, dramatize
productive°, constructive 1; economic 2; effective 1; effectual 1; efficacious; efficient; fertile; fruitful 1, 2; gainful; golden 5; helpful; persuasive; plentiful 2; positive 4, 6; pregnant 3; profitable 1, 2; profuse 2; prolific 1, 2; rank[2] 1; rewarding; rich 9; streamlined 3; useful; worthwhile 1
  productive (of), rich 2
productivity, efficiency 2; output 2
product(s), fruit; merchandise 1
proem, foreword; preamble; preface 1; preliminary 3
profanation, prostitution 2; sacrilege 1, 2; sin 1; violation 2
profane°, abusive 1; base[2] 4; blaspheme 1; blasphemous; desecrate; earthly 3; foul 5; godless 1; impious; irreverent 1; outrageous 3; pollute 2; prostitute 2; sacrilegious; scurrilous; sinful; temporal 1; ungodly 1; violate 2; wicked 1; worldly 1
profaning, violation 2
profanities, utter, swear 2
profanity°, curse 3; oath 2; sacrilege 2
  profanity, use, swear 2
profess°, declare 1; maintain 3; pretend 1; proclaim 1; protest 4
  profess to be, pose 2
professed°, nominal 1; pretended; seeming; self-styled; so-called 1
professedly, seemingly
profession°, admission 4; art 3; business 1; calling; career 1; craft 3; declaration 1; employment 1; game 4; job 1; line[1] 7; mission 2; occupation 1; practice 3; pursuit 3; racket 3; trade 2; vocation; work 2
professional°, accomplished; clerical 2; experienced 1; expert 1; master 2; perfect 7; proficient; skillful; slick 3; specialist
professional care, service 9
professor, instructor; intellectual 4; scholar 1; schoolteacher; teacher
professorial, pedantic 1
professorship, chair 2
proffer, bid 1; communicate 1; enter 6; extend 4; hand 16; hold 16b, 20b; moot 2; offer 1, 3, 6; prefer 2; present[2] 7; profess 1; proposal 1; propose 1; propound; put 18b; render 3, 7; return 6; submit 2; suggest 1; table 4; tender[2] 1; turn 15b; venture 2
proffering, presentation 1
proficiency°, ability 1; capability; dexterity 1; efficiency 1; facility 1; finesse 1; knack; knowledge 3; prowess 1; qualification 1; readiness 2; science 2; skill 1; technique 2
  proficiency (in), understanding 3
proficient°, able 1; accomplished; adept 1; capable 1; competent 2; efficient; experienced 1; expert 2; good 12; great 8; handy 3; master 4; masterful 1; old 7; perfect 7; polished 1;

proficient° (*cont.*) practiced 1; professional 1, 2; qualified 1; skillful; talented
proficient (in), become, learn 2
proficient in (*or* at), home 5; versed
proficiently, handily 2; well[1] 3, 6
profile°, form 2; line[1] 5; outline 1, 3; portrait; shape 1; silhouette
profit°, account 3; advantage 2; benefit 1, 4; blessing 2; boot 4; gain 2, 8; get 45d; good 19; interest 3; killing 2; money 3; net[2] 1; pay 4, 12; plus 3; prosper; purpose 3; realize 3; reap 2; receive 1; return 5, 9; sake 1; use 7; usefulness; welfare; yield 5
profit by *or* from, enjoy 2; exploit 2; profit 4; use 2
profit to, be of, profit 3
profitability, boom 4; usefulness
profitable°, advantageous; beneficial 1; desirable 4; economic 2; fat 4; fruitful 2; gainful; good 13; helpful; liquid 4; nifty 4; opportune 1; productive 3; rewarding; rich 9; secure 3; solvent; sound[2] 5; streamlined 3; substantial 4; successful 2; useful; worthwhile 1
profitable, be *or* prove, pay 4
profitably, right 16
profitless, bootless; futile; hollow 5; misspent; thankless; unprofitable 1; vain 2
profit-oriented, timeserving
profit(s), fruit; income; proceed 3; profit 1; revenue; take 40
profligacy°, dissipation 1; evil 6; extravagance 1; indulgence 2; prodigality 1; vice 1; waste 6
profligate°, abandoned 2; degenerate 3; dissolute; extravagant 1; fast[1] 2; improvident 1; incontinent 2; lavish 3; lecherous; libertine 1, 2; loose 7; outrageous 3; perverted; prodigal 1, 3; promiscuous 2; rake[2]; reprobate 1, 2; scandalous 1; self-indulgent; sensual; sensualist; sinful; spendthrift 1, 2; vicious 1; wanton 1; wasteful; wastrel 1
profligately, water 3
pro forma, formal 1
proformas, ceremony 2
profound°, abysmal 1, 2; dark 6; dead 15; deep 1, 2, 3, 6; difficult 2; great 11; heartfelt; heavy 12; intense 1; intimate[1] 2; keen[1] 3; poignant 3; pressing; radical 1; recondite; sage 1; scholarly; soulful; strong 7; subtle 2
profoundly°, deep 10; deeply 2; depth 6; downright 2; home 10; very 1; well[1] 13
profoundness, depth 1, 2; profundity 1
profundity°, depth 1, 2, 3, 4
profuse°, abundant 1; effusive; extravagant 1; fertile; heavy 2; improvident 1; lavish 1; luxuriant 1; many 1; opulent 3; plentiful 1; prodigal 2; prolific 1; rampant 2; rank[2] 1; torrential; voluble
profuse (in *or* with), rich 2
profusely, sky 2; unduly 2
profuseness, prodigality 2; profusion
profusion°, abundance; embarrassment 3; exuberance 2; flood 3; many 3; mass 2; ocean 2; plenty 2; prodigality 2; rash[2] 2; sea 3; shower 2; stack 3; superfluity; wealth 2
profusion, in, galore
prog, food; sustenance 1
progenitive, sexual 1
progenitor°, ancestor; father 2; forerunner 1; founder[1]; parent 1; stock 3
progenitors, family 3
progenitrix, mother 1; parent 1; progenitor 1
progeny°, brood 1; child 1; descendant; family 2; issue 7; lineage 2; offspring; posterity; seed 3; young 4
prognosis°, forecast 2; prediction; prophecy 2
prognostic, omen; ominous 3; prophetic
prognosticate°, forecast 1; predict; prophesy 1; say 8
prognostication, foreboding 2; forecast 2; oracle 2; prognosis; projection 4; prophecy 1; sign 6
prognosticator, fortuneteller; oracle 1; prophet
program°, broadcast 4; course 5; enterprise 1; method 1; plan 1; platform 2; policy; project 2; projection 3; proposal 2; prospectus; recipe 2; register 1; routine 1; schedule 1, 2; scheme 1; timetable
programmed, routine 3
progress°, advance 7; boom 2; come 6; course 2; current 6; development 2; evolution; far 6a; flow 1; gain 3, 7, 9; get 27, 28d; go 19; growth 3; headway 1, 2; improve 3; improvement 2; inroad 2; journey 2; look 11c; motion 1; move 1; movement 6; pass 2, 18; passage 1, 6; pick 8f; proceed 1; progression 1, 2; prosper; revival 3; rise 8; shape 10b; step 18a; succeed 3; transit 1; way 5
progress, in, progress 4; way 12
progress, make, look 11c; proceed 1
progress against, go *or* make, stem[2] 2
progressing, move 13c; ongoing 2; onward 1; way 12
progression°, course 2; evolution; passage 5, 6; procession 2; progress 1; run 41; scale[3] 1; sequence; step 5; succession 2; train 3; transit 1
progressive°, avant-garde; forward 3; go-ahead 2; left 2; liberal 2, 4; ongoing 2; onward 1; sequential; slow 2; way-out 2
progressivist, liberal 4
prohibit°, ban 1; bar 9; choke 3; embargo 2; exclude 1; forbid; keep 13; obstruct 3; outlaw 2; preclude; prevent; rule 8; shut 5a; suppress 1; taboo 3; veto 1
prohibited, foul 11; illegal; inadmissible; out 13; pale[2] 3; taboo 1; unlawful
prohibiting, prevention; suppression
prohibition°, ban 2; embargo 1; exclusion 1; injunction 1; prevention; stop 6; suppression; taboo 2; temperance 2; veto 2
prohibitory, prohibitive 1
project°, activity 2; design 5; enterprise 1; job 3; jut; operation 3; plan 1; precipitate 2; proposal 2; protrude; send 3; shoot 2; stand 9b; stick[1] 15; throw 1, 2; undertaking 1
project (out), overhang 1
projected, prospect 4; public 4
projectile°, bolt 1; missile; shell 1; shot 2
projecting, prominent 3; protuberant
projection°, angle[1] 1; bulge 1; hump 1; knob; lap[2] 2; ledge; nub 1; overhang 3; point 8; prognosis; prominence 3; protrusion; relief 3; scheme 1; spur 2; tang 3; view 3
prole, plebeian 3
prolegomenon, foreword; preamble; preface 1; preliminary 3
proles, hoi polloi; people 3
proletarian, bourgeois 2; common 3; mean[2] 2; plebeian 1, 3; worker
proletariat, hoi polloi; low[1] 7; mob 2; people 3; populace; public 8; rabble 2
proletariat, the, mass 6
proliferate°, increase 1; propagate 1; spread 6
proliferate (in *or* with), abound 2; teem[1]
proliferating, increase 5; spread 8
proliferation°, growth 1; increase 3; reproduction 3; spread 8
prolific°, fertile; fruitful 1, 3; lavish 1; luxuriant 2; opulent 3; plentiful 2; productive 1; profuse 2, 3; rank[2] 1; rich 9; vivid 3
prolific (with), be, teem[1]
prolix, boring; diffuse 2; discursive; expletive 1; lengthy; ponderous 2; rambling 1; redundant 2; repetitious; rhetorical 3; talkative; windy 2; wordy
prolixity, rhetoric 2; tautology; waffle 3
prologue, foreword; preamble; preface 1; preliminary 3
prolong°, continue 3; draw 14a; expand 2; extend 3; hold 21b; increase 2; keep 5; lengthen; maintain 1; spin 5; spread 4; sustain 1
prolongation, maintenance 2
prolong(ed), drag 5; lengthy; long[1] 2; protracted; sustained; tedious
prolonged, be, continue 2
prom, dance 2
promenade°, dance 2; parade 1, 3, 4; ramble 4; stroll 1, 2; strut; turn 31; walk 1, 5
prominence°, accent 1; attention 2; cachet 2; celebrity 1; distinction 2; elevation 2; emphasis; excellence; fame; height 3; hill 1; importance 2; lump[1] 2; moment 3; mountain 1; name 2; note 8; prestige; projection 1; protrusion; quality 3; relief 3; renown; rise 13; standing 6; status 1; stress 1; superiority 2; swelling
prominence, bring into, stress 4
prominence to, give, accent 4; spotlight 3
prominent°, big 5; brilliant 3; celebrated; conspicuous 1, 2; distinguished 1; elevated 2; eminent 1; exalted 1; famous; foremost 1; glaring 1; great 5; heroic 6; historic; illustrious; important 2; legendary 3; lofty 2; marked; monumental 1; notable 1; noted; obvious; outstanding 1; patent 2; perceptible; prestigious; principal 2; pronounced 1; protuberant; proud 3; public 6; renowned; salient; singular 2; special 6; splendid 2; star 3; successful 3; visible 2; well-known 2
prominent, be, shine 3; stand 9a
prominently, clear 18; large 4; notably 2; ostensibly; preeminently

promiscuity, impurity 3; profligacy 1
promiscuous°, fast[1] 2; impure 4; indiscriminate 1; loose 7; profligate 1; unchaste; wanton 1
promiscuously, fast[1] 9
promiscuousness, impurity 3
promise°, assurance 1; assure 4; bode; commit 4; contract 2; engage 3; engagement 3; faith 3; guarantee 1, 2; hope 2; mean[1] 3; oath 1; obligation 2; pledge 1, 4; possibility 2; prophesy 2; prospect 2; resolution 2; spell[3] 1; stipulate; stipulation; subscribe 2; subscription 2; swear 1; undertake 2; undertaking 3; vow 1, 2; warrant 3; warranty; word 4
  promise, break (one's), renege 2
  promise of, hold, spell[3] 1
promised, engaged 1
Promised Land, paradise 1
promising°, bright 4; favorable 1, 2; golden 7; hopeful 2; likely 3; positive 8; propitious; right 3; rosy 2
promissory note, note 4
promontory, bluff[2] 3; cape[1]; height 3; hill 1; point 8; prominence 2
promote°, abet 2; advance 2, 5; agitate 2; aid 1; assist 2; back 2a; benefit 3; boost 5; cause 6; champion 4; contribute 2; cultivate 3; develop 1; electioneer; encourage 2; exalt 1; expedite 2; facilitate; favor 8; feature 4; flog 2; foment; forward 6; foster 1; further 5; get 35; glorify 1; improve 2; laud; lift 2; lobby 3; make 30c; merchandise 2; militate 2; nourish 3; patronize 3; plug 5; profit 3; propagate 3; provoke 2; publicize; puff 7; push 6; recommend 2; rip 2a; second[1] 9; sell 4; speed 2; sponsor 3; stand 7b; support 1; talk 13; tout 1; uphold
promoted, be, rise 8
promoter, backer 1; enthusiast; follower 3; party 4; patron 1; proponent; sponsor 1; supporter 1
promoting, encouragement 1; promotion 3; toward 2
promotion°, advertisement 2; elevation 3; encouragement 1; furtherance; patronage 1; plug 3; progress 2, 3; propaganda 2; recommendation 2; succession 3
prompt°, activate; coach 3; cue 1, 2; dispose 2; express 7; foment; give 10; hasty 1; incite; induce 1; inspire 1; jog 2; kindle; lead 2; motivate; move 6; nudge 1; occasion 5; persuade 1; prod 2, 6; produce 2; professional 2; provoke 1; punctual; push 4; put 29; quick 2; raise 10; rapid; ready 5; remind; reminder; rouse 2; speedy 1; spur 4; stimulant 1; stimulate 2; stir 4; tempt 2; timely; unhesitating 1; urge 2
prompted, without being, voluntarily
prompting, incitement 1; occasion 2; recommendation 1; spur 1; suggestion 1
  prompting, without, voluntarily
promptitude, rapidity
promptly°, cold 11; directly 2; hastily 1; immediately 1; instantaneously; nail 6; now 3; once 6a; posthaste; quickly 3; rapidly 2; readily 3; right 13; soon 2, 3; straight 15; summarily 1
promptness, dispatch 5; expedition 2; rapidity; readiness 2; speed 1
promulgate, announce 1; circulate 2; communicate 1; declare 2; deliver 4; enunciate 2; issue 10; make 33; post[1] 2; proclaim 1; pronounce 3; propagate 3; publish; put 17d; report 4; spread 2
promulgating, imposition 1; proclamation 1
promulgation, circulation 2; declaration 2; decree 1; imposition 1; issue 6; proclamation; pronouncement 1; publication 1
prone°, disposed; flat 2; given 3; horizontal; inclined 1, 2; level 2; liable 1; prostrate 3; ready 3
  prone, be, lie[2] 1; tend[1]
  prone (to), subject 6; susceptible 1
proneness, inclination 3; penchant; tendency; weakness 4
prong, point 2; prickle 1; spike 1; spine 2; spur 2; tang 3
pronounce°, declare 2; decree 2; deliver 4; dictate 1; enunciate 1, 2; find 9; go 14; mouth 7; pass 9; proclaim 1, 2; profess 1; rule 7; say 3; sound[1] 7; speak 4; spread 2; vent 3
pronounced°, bold 3; broad 3; decided 1; definite 1; emphatic; great 3; marked; noticeable 2; obvious; oral; prominent 1; salient; strong 15; thick 9
pronouncement°, announcement 1; assertion 1; circulation 2; declaration 1, 2; dictate 2; expression 1; finding 2; observation 2; order 4; publication 1; regulation 3; revelation
pronto, hastily 1; immediately 1; instantaneously; posthaste; quickly 3; readily 3; right 13; soon 2; swiftly
pronunciamento, declaration 2; decree 1; order 4; proclamation; pronouncement 1
pronunciation°, accent 3; delivery 4; dialect; diction 2; intonation
proof°, demonstration 1; evidence 1; experiment 1; good 21c; print 1; reflection 4; test 1; token 2; trial 1
  proof (against), resistant 3
  proof (against), be, resist 1
  proof (of), be *or* **give** *or* **provide** *or* **furnish** *or* **constitute,** witness 3
proof of guilt, conviction 1
proof of purchase, receipt 1
prop°, brace 1; leg 2; pillar 1; post[1] 1; reinforcement 1; rest[1] 4, 8; support 8
  prop (up), bolster; brace 5; buttress; prop 1; reinforce; support 1, 2; sustain 2
propaganda°, advertisement 2; literature 2; promotion 4
propagandize, indoctrinate; propagate 3; push 6
propagate°, breed 3; generate 2; grow 3; post[1] 2; produce 3; raise 3; reproduce 2
propagation, breeding 1; generation 1; reproduction 3
propagative, sexual 1
propel°, boot 3; drive 1; fire 9; fling 1; fly 8a; force 8; hit 2; hurl; project 4; push 1; run 17; send 3; shoot 2; sling 1; throw 1; thrust 1; toss 1
propelling, motive 3
propelling force, propulsion
propensity, aptitude 2; bent 5; bias 1; flair 1; habit 2; inclination 3; instinct; penchant; talent 2; tendency
proper°, adequate 3; advisable; applicable; appropriate 1; becoming; ceremonious 2; character 8; civil 3; correct 6, 8; courteous; decent 1, 4; delicate 5; deserved; due 2; eligible 1; equitable; ethical; expedient 1; fair[1] 2; fitting 1; fit[1] 1; formal 1; genteel 2; genuine 1; good 3, 4, 8; honest 2, 4; honorable 2; impeccable; individual 2; intrinsic; just 3; ladylike; lawful 1; legal 2; legitimate 1, 3; likely 3; logical 2; meet[2]; moral 1; natural 9; official 1, 2; perfect 4; permissible; place 11a; polite 1; presentable 2; principled; prudish; pure 3; puritan 2; real 5; reason 6; reasonable 4; regular 6; relevant; respectable 1, 3; right 1, 2, 8; rightful 1; ripe 3a; safe 4; savory 2; seasonable; sedate 2; seemly 1, 2; square 4; strait-laced; suitable; tasteful; thorough 1; true 3; utterly; virtuous 2; well[1] 17; wholesome 2; wise 3
  proper for, for 9
  proper for, be, become 3
  proper of, be, behoove
  proper (to), befitting; specific 2
  proper (to), be, belong 2
proper behavior, decorum 1
proper form, propriety 2
properly°, appropriately; duly 1; fairly 2; right 12, 17; truly 2; utterly; well[1] 3, 14
properness, propriety 1; purity 2
propertied, bourgeois 1; substantial 4
properties, endowment 3; nature 1
property°, acquisition 2; asset 1; attribute 1; belongings; capital 3; characteristic 2; domain 1; effects; estate 1, 2; feature 1; flavor 2; fortune 1; good 21a; inheritance; interest 5; land 3; mark 2; mean[3] 4c; paraphernalia; peculiarity 2; point 14; possession 3; quality 1; resource 2; respect 4; spread 13; stock 5; stuff 2; substance 5; trait; wealth 1
  property of, be the, belong 3
property owner, proprietor 1
prophecy°, foreboding 2; forecast 2; oracle 2; prediction; prognosis; sign 6; warning 2
prophesier, seer
prophesy°, anticipate 2; envision; forecast 1; foresee; foreshadow; predict; prognosticate 1; tell[1] 9
prophet°, fortuneteller; oracle 1; psychic 3; seer
prophetess, prophet; psychic 3; seer
prophetic°, ominous 3
prophet of doom, killjoy; misery 4; naysayer
prophylactic, preventive 2, 4
propinquitous, present[1] 2
propinquity, presence 1; proximity
propitiate, atone; disarm 2; silence 4
propitiation, atonement; reconciliation 1
propitiative, propitiatory 1
propitiator, peacemaker
propitiatory°, sacrificial 2
propitious°, benign 3; bright 4; expedient 2; favorable 1; fortunate 2;

propitious° (*cont.*) golden 7; good 13; happy 2; hopeful 2; inside 4; lucky 2; opportune 1; positive 8; promising; right 3; ripe 3a; seasonable; timely
propitiously, happily 1
proponent°, advocate 2; follower 3; party 4
proportion°, composition 2; measure 5, 12; parallel 3; parity 2; percentage; piece 3; quota; rate[1] 3; ratio; scale[3] 2, 4; symmetry
  proportion, in, parallel 1; symmetrical
  proportion, out of, disproportionate
proportional°, parallel 1; regular 4; relative 2; symmetrical
proportionate, equal 2; parallel 1; proportional; relative 2; symmetrical
proportioned, equal 2; proportional
proportions°, extension 2; measure 1; proportion 4; size 1
proposal°, conception 3; design 5; measure 7; motion 5; offer 5, 6; overture; premise 1; project 1; projection 2, 3; recommendation 1; resolution 3; suggestion 1; supposition; tender[2] 2
  proposal to, make (a), approach 3
proposals, approach 4
propose°, hold 16b; intend; introduce 2; move 7; nominate; offer 1; plan 4; posit; prefer 2; premise 2; project 3; propound; purpose 5; put 5, 17a, 18a, 18b; recommend 1; set 18a, 18c; submit 2; suggest 1; table 4; tender[2] 1; think 5
  propose to, approach 3
proposed, nominal 1
proposer, proponent
proposition°, approach 3; chat 3; doctrine; hypothesis; law 3; motion 5; offer 6; overture; pass 28; position 7; premise 1; presumption 2; principle 1; proposal 1, 2; question 3; resolution 3; suggestion 1; supposition; tender[2] 2; theorem 1; thesis 1
propositional, nominal 1
propositions, approach 4
propound°, announce 1; enunciate 2; hold 16b; moot 2; move 7; posit; propose 1; put 18b; recommend 1; set 9, 18a
proprieties, ceremony 2; manner 3
  proprieties, the, propriety 3
proprietor°, boss 1; employer 1; host[1] 1; landlady 1; landlord 1; manager; owner; principal 3
proprietorship, business 4; company 4; firm 6; possession 1
proprietress, host[1] 1; landlady 1; owner; proprietor 1
propriety°, civility; decorum 1, 2; elegance 1; etiquette; form 6; formality 3; grace 2; morality 1; purity 2; rectitude; refinement 1; right 10; shame 4; taste 6
propulsion°, thrust 5
propulsive, motive 3
propulsive force, propulsion
prorate, scale[3] 4
prorogue, stay[1] 4
prosaic°, common 1; dead 12; dreary 2; dry 2; everyday 3; flat 5; heavy 7; humdrum; lackluster; literal 3; matter-of-fact; monotonous; ordinary 2; pedestrian 2; philistine 2; stupid 3; tame 4; tedious; threadbare 2; undistinguished; usual
proscribe, ban 1; embargo 2; exclude 1; expel 2; forbid; outlaw 2; prevent; prohibit 1; rule 8; taboo 3; veto 1
proscribed, illegal; narrow 4; taboo 1; unlawful
proscribing, prevention; prohibition 1
proscription, ban 2; embargo 1; exclusion 1; obstruction 2; prevention; prohibition 1, 2; restraint 1; taboo 2; veto 2
prosecute°, administer 2; enforce 1; follow 11b; wage 2
prosecution, execution 1
proselyte, convert 3; disciple 1; newcomer 2; novice
proselytize, convert 2
proselytizer, missionary
Prosit!, bottom 5
prosopography, résumé 2
prospect°, aspect 3; candidate; chance 3; foresight 2; hope 2; horizon; landscape; lead 14; match 4; outlook 1, 2; perspective 1; possibility 1; scene 4; thought 3; timber 3; view 1, 3, 5
  prospect, in, impending; prospect 4
  prospect for, prospect 5
prospective°, forthcoming 2; future 2
prospects°, expectation 4; prospect 3
prospectus°, draft 1; program 1
prosper°, arrive 2; boom 2; flourish 1; get 27; grow 1; make 31b, 32; pan 6; place 10; rise 8; succeed 3; thrive; work 19d
prospering, flourishing; going 1; prosperous 2; successful 1
prosperity°, boom 4; ease 3; fat 6; fortune 1; riches; success 1; wealth 1; welfare
prosperous°, fat 3; favored 2; flourishing; flush[2] 3; going 1; golden 5; leisured; money 4; opulent 1; rich 1; strong 11; substantial 4; successful 1; wealthy; well-off
prosperously, well[1] 4
prosperousness, prosperity
prostitute°, bitch 2; slattern; tart[2] 2; wanton 5
  prostitute oneself, prostitute 2
prostitution°, sacrilege 1
prostrate°, exhaust 2; exhausted 1; fell; flat 2; flatten 2; floor 4; helpless 3; knock 6a; lay[1] 18c; mournful 1; overwhelm 1; prone 1; spent 1; supine 1
  prostrate, be, lie[2] 1
  prostrate oneself (before), bow 5; kowtow; prostrate 1; scrape 4
prostrated, undone[1] 1
  prostrated, be, break 16c
prostration°, bow 1; collapse 7; shock 2
prosy, tedious
protagonist°, champion 2; hero 2; lead 16
protean°, changeable 1; fluid 3; unsettled 1; variable; versatile 1, 2
protect°, bulwark 2; cap 5; care 6a; champion 4; cover 1; defend 1; ensure 2; escort 5; favor 7; fence 3; guard 1; harbor 2; insulate 2; keep 2; look 4; mother 5; muffle 1; patrol 3; police 2; preserve 1; safeguard 2; save 2; screen 7; secure 7; shade 10; shelter 4; shield 2; shroud 1; steel 2; uphold; watch 2
  protect against, cover 10
protected, charmed 2; immune; privileged 2; sacred 2; safe 1, 3, 5; secure 1
protecting, patrol 2; preservation 2
protection°, auspices; care 3; case[2] 2; charge 4; conservation; cover 14; custody 1; defense 1; escort 1; guard 4, 5; housing 1; immunity 1, 2; indemnity 2; insurance; oversight 2; patrol 2; patronage 1; possession 1; precaution 1; preservation 2; preventive 4; providence 2; refuge 1; safeguard 1; safekeeping; safety; sanctuary 2; screen 2; security 1, 4; shade 5; shadow 2; shelter 1; shield 1; trust 3; umbrella 2; wall 2
  protection of, under the, under 4
protection money, protection 3
protective°, fatherly; preventive 2; proof 3
protective coloring *or* **coloration,** camouflage 1
protectively, jealously
protector°, champion 2; escort 1; guard 3; guardian; minder 2; patron 1
protectorate, possession 2
protectress, protector
protégé(e), follower 1
pro tem, provisional 1; temporarily 1; temporary; time 16
pro tempore, time 16
protest°, bitch 3; declare 1; demonstrate 3; demonstration 3; fight 2; gripe 2; groan 2; objection; outcry; peep 3; picket 5; proclaim 1; revolt 2; squawk 2, 3; walk 4b
  protest, under, protest 2
  protest (to *or* **against),** object 4
protestation, outcry; peep 3; protest 1
protester, dissident 1; picket 2
protocol°, ceremony 2; civility; code 3; decorum 2; etiquette; form 6; formality 1; manner 3; policy; program 1; propriety 2; ritual 3
prototype°, classic 3; design 5; epitome 1; example 2; model 2; original 5; paragon; pattern 1; picture 4; precedent; progenitor 2; quintessence; type 3
prototypic(al), classic 1; original 3
protract, draw 14a; lengthen; pad 5; prolong; spin 5; spread 4; string 11b
protract(ed), drag 5; lengthy; lingering 1; long[1] 2
protrude°, bulge 2; jut; overhang 1; pop 3; project 5; stand 9b; start 8; stick[1] 15
protruding, prominent 3; protuberant
protrusion°, hump 1; knob; lump[1] 2; nub 1; overhang 3; projection 1; prominence 3; swelling
protrusive, prominent 3; protuberant
protuberance, bulge 1; bump 2; hump 1; knob; lump[1] 2; nub 1; projection 1; prominence 3; protrusion; swelling; tumor
protuberant°, prominent 3
proud°, cocky; conceited; disdainful; egotistical; haughty; macho; pompous 1; presumptuous; regal 1; swollen; vain 1
  proud of, proud 1
  proud of, be, pride 4

readily°, easily 1; freely 2, 5; gladly; hand 13; handily 1; pat[2] 2; promptly; soon 4; swimmingly; willingly
readiness°, devotion 3; facility 1; maturity 2, 3; preparation 2; preparedness; providence 1; tendency
readiness, be *or* **stand in,** stand 5b
readiness, in, readiness 3; ready 1; reserve 8; ripe 3b
reading, indication 2; interpretation 2; lesson 1; narration 2; perusal; recitation 1; rendition 1; review 5; study 5; understanding 4; version 2
reading room, study 6
reading stand, rostrum
read-through, rehearsal 1
ready°, accessible; available; call 16; cash 1; condition 5; convenient 2; cut 16e; disposed; educate; expectant; expeditious; fit[1] 2, 4; fluent; fund 2; game 7; glad 2; glib; groom 4; handy 1; line[1] 23a; loaded 2; mature 2, 3; mellow 1; mobilize; mood 2; mount[2] 4; order 10a; pat[2] 3; pipeline 2; place 11b; poised 2; prepare 1, 2, 3, 6; prepared 1, 3; prime 5; process 3; prompt 1, 2; purse 2; qualified 1; qualify 1; reserve 8; ripe 1, 2, 3b; school 4; season 2; set 31; smooth 10; tap[2] 4b; unhesitating 1; willing; work 20b
ready, at the, ready 7, 9
ready, be, qualify 2
ready, be *or* **stand,** stand 5b
ready, get *or* **make,** condition 5; set 12a
ready, getting, preparation 4
ready, (the), money 1; purse 2
ready for, line[1] 20
ready for, be, watch 4
ready (for), make *or* **be** *or* **get,** poise 3; provide 5; ready 10
ready to, fit[1] 4; verge[1] 2
ready and willing, prompt 2
readying, preparation 2
ready-made°, instant 5; prepared 4
ready-mixed, instant 5
ready money, fund 2; money 1
ready reference, vade mecum
ready-to-eat, prepared 4
ready-to-serve, instant 5; prepared 4
ready-to-wear, ready-made 1
ready-witted, nimble 2
reaffirm, renew 4; stand 5c
real°, absolute 2; actual 1, 2; authentic; bona fide; concrete; effective 3; factual 1; flesh 3; genuine 1; good 8; hard 8; historical; honest 2; intrinsic; legitimate 1; lifelike; literal 2; live 1; material 5; natural 3, 6; original 3; physical; pure 1; realistic 2; really 4; right 8; solid 4; sterling 1; tangible; unaffected[1]; veritable; wholehearted
real, become, materialize 2
real, make, realize 1
real estate, land 3; property 3
realignment, shake-up
realistic°, actual 1; concrete; factual 1; feasible; graphic 1; hard 6; lifelike; natural 7; philosophical 2; photographic 1; picturesque 2; positive 6; practical 1; raw 6; sensible 1; sober 2; true 1; truthful; unflattering 1; vivid 2
realistic, to be, indeed 1
realistically, practically 2; practice 4
reality°, certainty 1; fact 1; given 4; object 1; substance 4; truth 1
reality, become a, true 6
reality, in, actually; bottom 5; fact 4; indeed 1; reality 2; really 1, 3; substantially; truly 1
reality, make a, realize 1
reality, the, case[1] 8
realizable, feasible; possible 2
realization°, accomplishment 1; achievement 3; appreciation 3; completion 1; consummation 1; embodiment 1; execution 1; fruition; fulfillment; perception 2; perfection 2; undertaking 2
realize°, achieve 1; awake 3; bring 11a; cash 2; comprehend; conceive 3; earn 2; effectuate; embody 1; fetch 3; find 3; follow 10; fulfill 1; gain 2; get 3; grasp 2; implement 2; make 7, 12; net[2] 4; perfect 8; pierce 3; profit 5; purchase 2; recognize 2; see 2; understand 1; win 2; work 16
realized, actual 1; perfect 1
realized, be, materialize 2; penetrate 4; true 6
realizing, consummation 1
really°, absolutely 1; actually; awfully; bottom 5; effect 5; fact 4; fairly 3; flesh 4; indeed 1, 3; literally 2; naturally 2; quite 4; seriously 2; simply 1, 2; sincerely; truly 1, 3, 4; truth 3; very 1
realm°, class 2; country 1; crown 2; domain 1, 2; dominion 2; field 4; habitat; jurisdiction; kingdom 1; nation; province 2; science 1; zone
realty, land 3; property 3
reams, lot 5c
reanimate, refresh 1; rejuvenate; resurrect
reanimation, revival 2
reap°, earn 2; gain 1; harvest 2; heap 3; make 7
reappear, recur 1; return 2
reappearance, return 7
reappearing, recurrent
reappraisal, revision
rear°, back 4, 11; bottom 1; breed 3; bring 15a; bum 1; buttocks; educate; foster 2; posterior 1, 3; raise 4; reverse 7; tower 3; train 4
rear, bring up the, follow 1; trail 6
rear, in *or* **at the,** last[1] 5
rear, to *or* **toward(s) the,** back 14
rear, to the, backward 4
rear (end), bottom 1; bum 1; buttocks; rear[1] 2; seat 4; tail 2
rearing, breeding 1; parenting; upbringing
rearmost, last[1] 1; rear[1] 3
rearrange, adjust 4; make 15; rehash 1; shift 1; shuffle 1; stagger 3; straighten 3; tidy 4
rearrangement, rehash 2; shake-up
rearward(s), back 14; backward 4, 6; posterior 1
reason°, argue 2, 5; call 15; cause 3; defense 3; end 3; excuse 4; explanation 3; ground 3; idea 4; intellect 1; mind 1; motive 1; object 3; occasion 2; plea 3; presumption 4; provocation 1; purpose 1; rationale; right 10; sanity; score 8; seed 2; sense 2; soul 1; subject 3; thought 6; understanding 5; use 12; warrant 1; wisdom 1
reason, for that *or* **this,** hence 1; therefore; thus 2
reason, in, reason 6
reason, within, moderately; reason 6; reasonable 3
reason away, rationalize 1
reason for, give a, account 1
reason for, provide *or* **offer,** warrant 4
reason of, by, for 7; reason 5; thank 4; virtue 4
reason of, question *or* **doubt the,** wonder 5
reason out, rationalize 2
reason with, reason 8
reasonable°, cheap 1; coherent 1; comfortable 4; decent 3; economical 1; enlightened; equitable; fair[1] 4; inexpensive; judicious; just 1, 3; legitimate 3; levelheaded; likely 2; logical 2, 3; moderate 1; modest 3; natural 1, 9; normal 2; plausible 1; possible 1; practical 1, 2; presumptive 1; prudent 1; rational 1, 3; realistic 1; reason 6; respectable 2; sage 1; sane; seemly 1; sensible 1; solid 7; sound[2] 4; stable 3; strong 10; temperate 1; tenable; thinkable; thinking 1; viable; well-balanced 1; wise 1, 3
reasonable, make, rationalize 1
reasonableness, intellect 1; logic 3; sanity
reasonableness of, question *or* **doubt the,** wonder 5
reasonably, cheap 6, 7; enough 3; practically 2; pretty 3; well[1] 14
reasoned, measured 2; reasonable 2; sensible 1
reasoning°, argument 2; deduction 2; logic 1; psychology; rational 1; rationale; reason 2; sound[2] 4; thinking 1, 2; thought 6
reasoning power, understanding 5
reason(s), reasoning 2; sake 2
reassert, renew 4
reassess, review 2
reassessment, review 6; revision
reassurance, assurance 2; encouragement 1; lift 8; pat[1] 5; solace 1
reassure°, assure 2, 3; comfort 1; encourage 1; fortify 2; pat[1] 2; satisfy 3; solace 2
reassuring, encouragement 1; exhilarating 2; favorable 2; helpful; hopeful 2; positive 7; promising; supportive
reata, lasso 1
reawake(n), evoke; refresh 2; resurrect; revive 1, 2
reawakening, rebirth; renaissance; resurgence
rebate°, allowance 4; discount 4; kickback; payoff 3; rake-off
rebate, give (someone) a, pay 10b
rebel°, disobey; dissident 1; mutiny 2; nonconformist 1; revolt 2; revolutionary 1, 3; riot 3; rise 4; strike 15; turn 12
rebel against, rebel 2
rebellion, mutiny 1; revolt 1; revolution 1; sedition; trouble 7; uprising
rebellious°, defiant; disobedient 2; disorderly 2; factious; inflammatory; insubordinate; mutinous 1; nonconformist 2; obstinate; recalcitrant; resistant 2; revolutionary 1; seditious; ungovernable; unruly
rebelliousness, obstinacy; resistance 1
rebirth°, renaissance; resurgence; revival 2
reborn, new 5
rebound°, backlash; boomerang; bounce 1, 3;

recommend° (*cont.*)
  recommend to, counsel 4
recommendable, advisable
recommendation°, advice 1; counsel 1; idea 1; instruction 1; motion 5; plug 3; pointer 2; promotion 3; proposal 1; reference 3; suggestion 1; testimonial
recommended, expedient 2
recompense, allowance 2; amends; atonement; compensate 1, 3; consideration 2; desert[2]; fee 2; fruit; gratification; gratify; honorarium; indemnity 1; offset 1; pay 1, 2, 8a, 12; reciprocate; recoup; reimburse; remuneration 2; repay; reprisal; requital 1; requite 1; restitution 1; retribution; return 11; reward 1, 4; run 40; satisfaction 2; stipend; wage 1
  recompense (for), make 31a
  recompense for, in, for 3
reconcile°, accommodate 2; adjust 1; agree 1; disarm 2; fix 16c; heal 2; patch 7; settle 1, 3; term 5
  reconcile oneself to, stomach 3
  reconcile (oneself) (with *or* **to),** accept 4; adjust 3; approve 3; bear 4; resign 2; square 11
reconciled, be, make 39d
reconciled (to), be *or* **become,** resign 2
reconcilement, reconciliation 1; resignation 2
reconciler, peacemaker
reconciliation°, accord 2; rapprochement; resignation 2; settlement 3; understanding 1
recondite°, cryptic 2; dark 6; deep 2; difficult 2; heavy 12; hidden; incomprehensible; mysterious 2; obscure 5; occult 1; opaque 2; perplexing; profound 1; remote 3; secret 2; subtle 2
reconditeness, depth 2; opacity 2; perplexity 2; profundity 1
recondition, improve 1; overhaul 2, refresh 3, refurbish; renew 1; renovate; retouch; revamp
reconditioning, overhaul 3
reconfirm, renew 4
reconnaissance°, exploration; surveillance
reconnoiter°, explore 1; scout 1; spy 2; traverse 4
reconnoitering, reconnaissance
reconsider, better 7; review 2
reconsideration, retrospect; review 6
reconstruct, modify 1; rehabilitate 2; restore 3
reconstruction, restoration 2
reconversion, restoration 2
record°, book 4; chronicle 1, 2; chronology; description 2; diary; document 1, 2; enroll 2; enter 3, 7; entry 4; file 3; get 37b; high 15; history 1, 4, 5; itemize; jot 1; journal 2; list[1] 1, 2; lodge 6; make 16, 37c; mark 15a; memoir 1, 2; memorandum; minute[1] 2, 4; monitor 3; monument 2; narrative 1; notation 1; note 10, 12; paper 2b; poll 3; portrait; post[3] 3; present[2] 7; proceeding 2a; put 16a; register 1, 3, 4, 5; report 1; roll 13; schedule 1, 2; score 1, 11; set 17a; stamp 2; story[1] 1; table 3; tabulate; take 31a; tale 1; tally 2, 3; tape 4; trace 1, 6; track 4; transaction 2; trophy 2; write 4a
  record, off the, private 1, 6; privileged 4; record 8; unofficial
  record, put on, note 12
record book, journal 2
record-breaking, record 9
recorded, go 27d; historical
recording, narration 1; record 3, 7; tape 2
recount°, chronicle 2; describe 1; detail 5; enumerate 1; narrate; recapitulate; recite 2; record 2; rehearse 2; relate 2; repeat 1; report 3; spin 2; tell[1] 1, 4
recounting, narration 1; recital 2; rehearsal 2; relation 4; story[1] 1
recoup°, find 6; get 33b; pick 8f; recover 1; retrieve 2
recouping, recovery 2
recourse°, expedient 3; option 1; refuge 2
  recourse to, have, draw 13a; fall 9; refer 3; resort 3; take 21; turn 23b; use 1
recover°, convalesce; find 6; get 33b; heal 1; improve 2, 3; mend 3; pick 8f; pull 13, 17; rally 4; reclaim; recoup; recuperate; redeem 1; reform 1; retrieve 2, 3; revive 1; salvage 1; save 1; snap 7, sober 4
  recover, ability to, resilience
  recover (from), come 17a; get 46b
recover consciousness, come 18b
recovered, better 4
recovering, mend 5
recovery°, improvement 2; rally 2; reform 3; restitution 2; revival 3; salvage 2
  recovery, beyond, spout 4
recreant, disloyal; faithless 2
recreate, reproduce 1
recreation°, amusement 1, 2; diversion 3; enjoyment 1; entertainment 1; escape 6; fun 1; game 1; hobby; leisure 2; pastime; play 22; pleasure 1; relaxation 1; sport 1
recreational, leisure 4
recriminate, blame 1; task 3
recrimination°, blame 3
recrudescence, return 7
recruit°, call 12a; employ 1; enlist 1; enroll 1; initiate 4; raise 5; soldier 1
rectangle, square 7
rectangular, square 1
rectification, amendment 1; correction 1; reform 3; revision
rectify°, adjust 1; amend 2; correct 1; cure 2; fix 3; improve 1; make 31a; mend 1, 2; perfect 9; reform 1; remedy 4; revise 1; right 19; straighten 1
rectilinear figure, square 7
rectitude°, character 3; good 20; honesty 1; honor 1; integrity 1; moral 4; morality 1; nobility 1; probity; purity 2; right 10; virtue 1
recto, page[1] 1
rector, clergyman 1; pastor; principal 4
recumbent°, flat 2; prone 1; prostrate 3; supine 1
  recumbent, be, lie[2] 1
recuperate°, convalesce; gain 3; heal 1; improve 3; mend 3; pull 13; rally 4; recover 2
  recuperate (from), come 17a; get 46b
recuperating, mend 5
recuperation, improvement 2; rally 2; recovery 1
recur°, cycle 2; return 2
recurrence, return 7
recurrent°, eternal 2; frequent 1; periodic; perpetual 2; repetitive
recurrently, repeatedly
recurring, frequent 1; nagging; recurrent
recusant, dissident 1; rebel 4
recycle, cycle 2
red, fiery 2; inflamed; left 2; radical 5; rosy 1; tranquilizer
  red, in the, debt 2; embarrassed 2, hard 17, insolvent
  red, see, bristle 3; flare 3; smolder
  red in the face, get, seethe 2
redact, edit 1; revise 1
redaction, revision
redactor, editor
red-blooded, hardy 1; manly; stalwart 3
redcap, porter[1] 1
red carpet for, roll *or* **bring out the,** fête 2
red-carpet treatment, courtesy
redden, blush; color 5; flush[1] 1; glow 7
reddish, rosy 1
(re)decorate, decorate 2; fix 16b; make 38a; modernize; refurbish; rehabilitate 2; renew 1; renovate
redeem°, atone; cancel 1; deliver 3; free 16; honor 8; improve 1; purify 2; ransom 3; reclaim; recoup; recover 1; rehabilitate 1; retrieve 3; salvage 1; save 1; spare 6
redeemer, savior 1
redeeming, saving 1
redemption, ransom 1; recovery 2; salvage 2
redemptional, saving 1
redemptive, saving 1
redemptory, saving 1
redesign, modernize
redesigned, new 5
red-eye, booze 1; drink 5; liquor 1
red-faced, ashamed; embarrassed 1; guilty 2; shamefaced 2
red-handed, catch, surprise 2
red herring, misinformation; pretext 2
red-hot, fiery 1; hot 1; incandescent; live 3
red in the face, embarrassed 1
redirect, divert 1; switch 5
red-letter, historic; special 2
red-letter day, feast 2; festival 2; gala 1; holiday 2
rednecked, narrow-minded
redness, flush[1] 4; glow 3; inflammation; rash[2] 1
redo, modernize; modify 1; reform 1; refresh 3; rehash 1; renew 1; revamp
redoing, rehash 2
redolence, aroma 1; fragrance; odor 1; savor 2; scent 1
redolent°, fragrant; rich 8; sweet 2
  redolent with (*or* **of),** redolent 2; reminiscent
redone, new 5
redouble, intensify
redoubt, bulwark 1
redoubtable, impressive; stalwart 2; strong 9; terrible 6
redound, boomerang
redress, adjust 1; amends; atone; compensate 1; correct 1; correction 1; improve 1; indemnity 1; make 40; offset 1; rectify; redeem 5; relief 1; remedy 2, 4; remuneration 2; reprisal; requital 1; requite 2; restitution 1; retribution; return 11; reward 4; right 19; sanction 4; satisfaction 2; satisfy 4
red tape, bureaucracy; palaver 1; rigmarole
reduce°, abbreviate 1; abridge; belittle; bring 9b; clip[2] 2; concentrate 2; consume 2; contract 4;

retrench, economize
retrench (on), cut 8
retribution°, desert[2]; reckoning 3; reprisal; requital 2; revenge 1; reward 3; sanction 4; vengeance 1
retribution on *or* **against, take,** fix 14
retributive, punitive; spiteful
retributory, spiteful
retrieval, recovery 2, 3; salvage 2
retrieve°, fetch 1; field 5; get 4, 33b; reclaim; recover 1, 3; redeem 1; salvage 1; save 1
retrocede, ebb 1
retrograde, backward 5
retrogress, degenerate 2; relapse 1; wrong 8a
retrogressing, relapse 3
retrogression, relapse 3
retrogressive, backward 5
retrogressively, backward 6
retrude, withdraw 3
return°, bring 11a; cycle 2; earnings; exchange 1; field 5; gain 8; get 33a, 39; gratification; gratitude; income; indemnity 1; make 7, 12; pay 12; plus 3; put 14; realize 3; rebound 2; recall 6; recede 1; receipt 3; reciprocate; recover 3; recovery 3; recur 1; reflect 1; relapse 1, 3; reminisce; renaissance; render 5; repay; replace 3; reply 1; requital 1; respond 1; response; restitution 2; restore 1, 4, 5; resurgence; resurrect; retort 2; retrieve 3; revert; revival 1; reward 1; round 9; run 40; take 40; trade 6; turn 13b, 15b; yield 4, 5
return, in, back 15
return for, in, for 3
return (to), fall 9; go 24a; renew 2
returned, reciprocal
return (food), vomit
returning, recurrent; relapse 3; revival 1
return one's dinner, heave 4; regurgitate
return(s), fruit; net[2] 1; poll 1; proceed 3; profit 1; return 9; revenue
return the favor *or* **compliment,** repay
reunion, reconciliation 1
reunite, reconcile 1
reuse, rehash 1, 2
revamp°, modernize; modify 1; patch 5; refresh 3; refurbish; renew 1; renovate; repair 1; revise 1; shake-up
revamping, repair 2; revision
reveal°, admit 4; air 7; announce 1; babble 2; bare 6, 7; betray 2; blab; blurt; break 2; circulate 2; clean 8; comment 4; communicate 1; confess; denounce 2; develop 1; disclose 1, 2; display 1; evidence 4; exhibit 1; expose 1; express 2; find 2; give 12b, 15c; illuminate 2; impart 2; indicate 2; inform 1; lay[1] 9; let[1] 7a; light[1] 9; make 33; manifest 2; mention 2; narrate; open 23, 25; produce 5; publish; reflect 2; register 4; relate 2; report 4; rumor 2; say 2, 4; show 1, 6, 12a; signify 1; slip[1] 4; solve; speak 4; throw 9c; turn 25c; unveil; vent 3; voice 4; whisper 2
reveal, not, conceal 2
reveal suddenly *or* **unexpectedly,** spring 4
reveal all, card 13
reveal all *or* **everything,** spill 3
revealed, bare 2; open 2; out 4
revealed, be, come 15a; emerge 1; leak 6; light[1] 10; materialize 1; transpire 1; unfold 2
revealed to one, have, learn 1
revealing, abbreviated; exposure 1; expressive 1; forthcoming 3; full 10; immodest 1; informative; instructive; low[1] 11; spicy 3; telling 2
reveal secrets, tattle 1
revel°, bender; carouse 1, 2; celebrate 2; dissipate 4; drunk 4; exult; frolic 2; fête 1; glory 5; lark 1, 2; merry 2; orgy 1; paint 7; rejoice; spree
revel (in), delight 2; devour 3; gloat; luxuriate 1; pride 4; revel 1; roll 10b; savor 3; wallow 2
revelation°, admission 4; betrayal 2; blow[2] 2; bolt 6; bombshell; discovery 2; display 4; exposure 1; illumination 2; inspiration 1; leak 3; narration 1; narrative 1; prophecy 2; publication 1; scoop 3; shock 3; solution 1
revelation, make a, clean 8
revelatory, educational 2; explanatory; spicy 3
reveling, revelry
revelry°, celebration 3; festivity 1; frolic 1; gaiety 2; hilarity; jamboree; merriment; mirth; play 22
revels, gaiety 2; jamboree
revenant, phantom 1; specter 1; vision 4
revenge°, reaction 3; reprisal; requital 2; requite 2; retribution; right 20; vengeance 1
revenge, get, score 9
revenge for, take, revenge 2
revenge (on), take, fix 14; retaliate
revenge on, take *or* **get,** pay 5
revenge oneself (on), even 9; get 23; retaliate; revenge 3
revenged, be, even 9
revengeful, vindictive
revenue°, earnings; gain 8; profit 1; purse 2; return 9; take 4; yield 5
revenue(s), fruit; income
reverberant, resonant; rotund 2; round 7
reverberate, blare 1; echo 3; peal 2; pulsate; resound; ring[2] 1; roll 5; sound[1] 4; thunder 2; vibrate
reverberating, pulse 1; resonant; ring[2] 3; rotund 2; round 7
reverberation, echo 1; jangle 3; peal 1; pulse 1; repercussion; ring[2] 3; roll 15; thunder 1
revere°, admire 2; adore 2, 3; bless 1; commemorate; esteem 1; exalt 2; fear 6; glorify 2; hallow 2; honor 5; idolize; look 12; pedestal 2; praise 4; respect 6; venerate; worship 1
revered, beloved 1; eminent 1; grand 2; precious 2; sacred 1; time-honored; venerable; well-thought-of
reverence°, adore 2; commemorate; devotion 1; esteem 1; exalt 2; fear 2; glory 2; hallow 2; honor 2; idolize; obeisance; piety 2; praise 2, 4; regard 7; respect 2, 6; revere; venerate; veneration; worship 1, 2
reverenced, venerable
reverend, clergyman 1; divine 5; minister 1; pastor; preacher; priest
reverent, devout 1; godly; holy 2; pious 1
reverential, devout 2; holy 2; pious 1; solemn 2
reverie°, daydream 1; dream 1
reverie, be in a, muse
reverie, in a, dreamy 2; pensive
reversal°, about-face; repeal 2; reverse 8; setback; switch 2
reverse°, about-face; back 2b; backward 5; calamity 2; contradict 2; contrary 4; correct 4; destroy 3; disaster; misfortune 2; opposite 3; quash 1; repeal 1; repudiate; retract 2; reversal 1; set 15b; setback; tail 3; turn 3, 13a, 21b; undo 3; void 6; wrong 6
reverse, in, backward 6, 7
reversed, inside 3; reverse 1; topsy-turvy 1; upset 6; vice versa
reversion, relapse 3
revert°, relapse 1; return 1
revert (to), go 24a
reverting, relapse 3
review°, analysis 2; analyze 2; breakdown 3; brush[2] 8; check 11; chew 4; criticism 3; discuss; discussion; exploration; explore 2; gloss[2] 2; go 34a, 34d; investigation; journal 1; judge 5; mull; narrate; notice 6; outline 2; perusal; peruse; polish 4; post-mortem 2; read 1; recapitulate; recount 2; reel 2; rehearse 2; retrospect; roundup 2; run 33d; rundown; résumé 1; stock 6; study 4; sum 2a; summary 1; survey 1; theme 2; traverse 4; treat 2; vet; writing 2
review article, review 4
reviewer, judge 3; scribe 2
reviewing, review 5
revile, abuse 3; attack 2; berate; blacken 2; blaspheme 2; come 11; denounce 3; lace 5b; lambaste 2; rail[2]; rebuke 1; twit[1]; vilify; vituperate
revilement, abuse 8; attack 7; rebuke 2; tongue-lashing
reviling, abuse 8; abusive 1
revise°, alter; amend 2; edit 1; mend 2; modify 1; rectify; reform 1; reword
reviser, editor
revising, revision
revision°, alteration
revisionist, progressive 2, 3
revitalization, rebirth; revival 1
revitalize, animate 1; freshen 2; heal 1; perk up; refresh 1; rejuvenate; renew 1; restore 2; revive 3; tone 6
revitalized, new 5
revitalizing, refreshing
revival°, rally 2; rebirth; recovery 1; refreshment 2; renaissance; restoration 2; resurgence
revivalist, clergyman 3
revive°, arouse 1; awake 2; bring 8a; come 18b; freshen 2; improve 3; perk up; quicken 3; rake[1] 7b; rally 4; recover 2; recuperate; refresh 1; restore 2, 3; resurrect; snap 7; stir 5; vitalize
revivify, rejuvenate
reviving, tonic 2; vital 4
revocation, cancellation 1, 2; recall 5; repeal 2; reversal 3
revoke°, cancel 1; deny 2; quash 1; recall 3; recant; renege 1; repeal 1; retract 2; reverse 3; vacate 3; vitiate 3
revolt°, disgust 1; disobey; mutiny 1, 2; nauseate; offend 2; put 21d; rebel 1; repel 2; revolution 1; riot 3; rise 4; shock 1; sicken 2; strike 15; trouble 7; turn 12; uprising
revolted, nauseated; sick 6
revolted, easily, squeamish 2
revolting°, abhorrent; disagreeable 1, 2;

rigged, factitious; fixed 4
rigging, tack 4; tackle 1
right°, applicable; appropriate 1; authority 1; call 15; charter 2; conservative 1; correct 1, 6, 7, 8; decent 4; desert[2]; deserved; due 2; ethical; expedient 1; faculty 4; faithful 4; fast[1] 8; fit[1] 1, 2; freedom 4; full 14; good 3, 20; honorable 2; justice 3; knock 9; legal 2; legitimate 1; liberty 2; license 1; likely 3; moral 1; morality 1; order 10b; perfect 4, 9; perfectly 4; permissible; place 11a; plumb 4; power 6; prerogative; principled; priority; privilege; proper 1, 2, 3, 4; punch[1] 2; reactionary 1; real 1; reasonable 4; righteous 2; rightful 1, 2; ripe 3a; safe 4; seemly 1; shortly 2; slap 6; square 4; straight 3, 12; suitable; tailor-made 2; title 5; true 1, 3; very 1; warrant 1; well[1] 17
right, be, nail 5
right, consider, approve 3
right, set *or* **put,** adjust 1; correct 1; make 31a; mend 2; patch 7; straighten 1
right, setting *or* **putting,** adjustment 1
right angles (to), at, perpendicular 2
right away, instantaneously; nail 6; outright 3
right (away *or* **off),** directly 2; hastily 1; immediately 1; now 3; once 6a; outright 3; point-blank 3; posthaste; promptly; quickly 3; rapidly 2; readily 3; right 13; soon 2; straight 15
right in the head, sane
right now, immediately 1; now 1
right off the bat, straight 15
right on, perfect 5
right to, give up, vacate 2
right to, have a, earn 1; merit 3
right up one's alley, tailor-made 2
right upstairs, not, insane 1
right and wrong, sense of, conscience; principle 2
right-angled, square 1
right arm, good *or* **strong,** aide
right arm for, give one's, pant 2
righteous°, conscientious 1; ethical; faithful 4; godly; good 5; incorrupt 1; just 2; moral 1; noble 4; perfect 3; principled; pure 6; right 1; saintly; scrupulous 2; simple 3; upright 2; virtuous 1; wholesome 2
righteousness, chivalry; good 20; honor 1; integrity 1; morality 1; nobility 1; probity; rectitude; virtue 1
rightful°, deserved; due 2; honest 4; just 3; lawful 1; legal 2; legitimate 1
rightfully, truly 2
rightfulness, justice 3
right hand, aide; second[1] 8; right 5
right-hand man, aide; henchman; satellite 2
rightist, conservative 1, 3; reactionary 1, 2; right 6
rightly, appropriately; duly 1; properly 2; truly 2
right-minded, conscientious 1; equitable; just 2; pious 1; principled; right 1; sane; sound[2] 4; virtuous 1
rightmindedness, equity; probity; right 10; virtue 1
rightness, morality 1
right(s), claim 2
rights, by, right 12
rights, put to, make 31a
rights, set *or* **put to,** adjust 1; amend 2; settle 1; straighten 1
rights, setting *or* **putting to,** adjustment 1
rights and wrongs, merit 2
right side *or* **hand** *or* **fist,** face 7; right 11
right side up, upright 5
right stuff, capacity 2
right-wing, conservative 1; reactionary 1; right 6
right-winger, conservative 3; reactionary 2
right wrongs, correct 1
rigid°, dead 12; distant 3; dour 2; exact 2; exacting; extreme 4; faithful 3; firm 1, 4; fixed 1, 2; formal 1; frigid 2; hard 1; hidebound; immovable 2; impersonal 2; implacable; inflexible; invariable 1; monolithic; narrow 3; obstinate; parochial; persistent 1; precise 2; prudish; puritan 2; relentless 1; reserved; scrupulous 1; sectarian 2; set 29; severe 1; small-minded; Spartan; staid; stern 1; stiff 1, 7; strict 1; stuffy 2; taut 1; tenacious 1; tense 1; thick 5; tough 6; unkind; wooden 2
rigid, become, fix 7
rigidify, fix 7; stiffen 2
rigidity, distance 2; formality 2; rigor 2; severity 1; tenacity 1
rigidly, bolt 12; firmly 1; precisely 2
rigmarole°, hocus-pocus 2; mumbo jumbo 1; palaver 1; prattle 3
rigor°, precision 1, 2; severity 1
rigorism, rigor 2
rigorous, close 14; conscientious 2; drastic; exact 2; exacting; faithful 3; inclement; inflexible; mathematical; murderous 2; nice 3; painful 3; particular 3; precise 2; religious 2; rigid 2, 3; rugged 2; scrupulous 1; severe 1; Spartan; stern 1; stiff 9; strait 1; strict 1; strong 13; tight 3
rigorous, make more, tighten 3
rigorously, precisely 2; severely 2; tightly
rigorousness, precision 2; severity 1
rigors°, rigor 1
rile, aggravate 2; anger 2; annoy 1; displease; exasperate 1; get 17; inflame 1; infuriate; madden 3; offend 1; outrage 4; provoke 3
riled, indignant; resentful; unsettled 2
rill, brook[1]; creek 2; river 1; run 42; stream 1; tributary
rim°, brim 1; brink 1; edge 1; margin 1; periphery 1; side 1
rime. See **rhyme**
rind°, heel[1] 1; hull 2; peel 4; skin 2
ring°, blare 1, 2; buzz 3; cabal 2; call 14; chime 1, 3; circle 1; clan 2; echo 3; encircle; faction 1; gang 1; garland 2; halo; jangle 1; jingle 1, 2; loop 1, 2; machine 3; peal 1, 2; round 10; set 26; sound[1] 2, 6; surround 1; toll[1] 1, 2; union 2; wheel 1
ring, give (someone) a, ring[2] 2; telephone 2
ring, in a, around 3
ring false, water 2
ring for, buzz 7
ring (out), resound
ring true, conform 2
ring true, not, water 2
ring (up), buzz 7; call 3, 12b; contact 4; look 11b; ring[2] 2; telephone 2
ring-a-ding-ding, ring[2] 1, 3
ringer, double 7; duplicate 2; image 2; lookalike; picture 2; twin 1
ringing, chime 2; jingle 2; peal 1; resonant; ring[2] 3; toll[1] 2
ringleader, chief 1
ringlet, circle 1; lock[2]; ring[1] 1
ringlets, hair 1
ringlike, circular 1
ring-shaped, circular 1; round 2
rink, ring[1] 2
rinse°, flush[1] 2; tint 2, 3; wash 1, 13
rinsing, rinse 3
riot°, disorder 2; fight 7; fracas 1; panic 2; quarrel 1; rampage 1; scramble 4; scream 4; scrimmage; tempest 2; tumult
riot act, read (someone) the, mind 10
riot act (to), read the, castigate; reprimand 2
rioting, riot 1
riotous°, chaotic 2; hectic; inflammatory; noisy; obstreperous; priceless 2; tempestuous; tumultuous; uncontrolled; unruly; uproarious 1
rip°, flaw 2; gap 1; hole 2; lacerate; race[1] 4; rend 2; rent[2]; scurry; slash 4; snag 2; split 7; tear 1, 4; wrench 1, 4
rip, let, fly 8a
rip (apart), pull 4
rip into (someone), scold 1
rip off, cheat 2; defraud; dupe 3; fleece; hook 7; make 27; mutilate 1; pilfer; pocket 4; pull 12a; rip 2; rob 1, 2; sting 4; strip[2] 4; swindle 1; take 3
rip out, pull 3
rip (to pieces *or* **apart** *or* **asunder),** rend 1
rip (up), shred 2; pull 3, 4
rip (current), undercurrent 1
ripe°, high 10; lush 2; mature 2; mellow 1; rancid; ready 1; season 2
ripe for, ripe 3
ripen°, age 6; develop 2; flourish 1; grow 1; mature 5; mellow 5; progress 6; season 4; thrive
ripen into, become 2
ripened, mature 2; mellow 1
ripeness, fruition; maturity 2
ripening, progress 3
rip-off°, fraud 2; robbery; swindle 2; theft
rip-off artist, robber
riposte, answer 1, 4; rebuttal; reply 2; retort 1, 2; squelch 2
ripped, ragged 1
ripper, butcher 1; killer 1
ripping°, capital 6; excellent; fine[1] 1; striking
ripping-off, robbery
ripple°, flow 1; gurgle 1; lap[1] 2; ruffle 2; shimmer 1; wash 3; wave 1, 4
rip-roaring, exciting 1
ripsnorting, capital 6; excellent; admirable; striking
riptide, undercurrent 1
rise°, advance 8; appear 1; appreciation 2; arise 1, 2; boost 1, 3; bristle 2; climb 3; come 19a, 19c; dawn 2; elevation 2; emerge 1; emergence; ferment 1; flow 3; gain 9; gather 5; go 38a; grade 4; growth 3; hill 1; improvement 2; incline 1, 4; increase 4; infancy 2; jump 5, 9; knoll; leap 6; lift 5; mound 1; mount[2] 6; origin 1; originate 2; peak 4; progress 2, 6; progression 1, 2; proliferation; prominence 2; promotion 2; ramp; recovery 1; reply 2; revival 3; slope 1, 2; soar 1, 2; source 1; spring 2; spurt 2; stand 1, 10a; start 11; stir 2;

rolling (*cont.*)

rolling in it *or* **in money** *or* **in dough** *or* **in wealth,** flush[2] 3; loaded 4; money 4; opulent 1; prosperous 1; rich 1; wealthy

rolling pin, roller 1

rolling stock, railroad 1

rolling stone, rover; vagabond 1

roly-poly, fat 1; plump[1] 1; rotund 3

Roman, classical 2

romance°, affair 4; exaggerate; intrigue 4; legend 1; liaison 3; love 5a, 6; novel 2; saga; story[1] 1

roman-fleuve, saga

romantic°, dreamer; ideal 5; idealistic; impractical 1; legendary 2; maudlin; mushy 2; mythical 1; quixotic; sentimental 2; tender[1] 9; unrealistic 1; visionary 1, 2; wild 6

romanticism, sentimentality

romanticist, dreamer; romantic 5

romanticization, exaggeration

romanticize, idealize

romanticized, idealistic

Romeo, charmer; gallant 5; philanderer; playboy; roué; sensualist

romp, bound[2] 3; caper 3; cavort; dance 1; fool 7a; frisk 1; frolic 2, 3; lark 1, 2; leap 2; party 1; prance 1; revel 3; skip 1, 3; sport 4; spree

romping, sportive

rood, cross 1

roof, go through *or* **hit the,** explode 3; fly 7; fume 1; rage 4; rocket

rook, cheat 2; clip[2] 4; defraud; dupe 3; fleece; hoodwink; prey 3b; rip 2b; swindle 1; trick 8; victimize 2

rook (out of), rob 3

rookie, apprentice 1; greenhorn; initiate 4; novice; recruit 2

room°, area 1; board 7; capacity 1; cell; chamber 4; clearance 1; leeway; live 8; lodge 3; margin 3; office 3; partition 4; play 24; scope 2; seating; slack 6; space 1, 4

room, make, make 41

room for, have, seat 6; sit 5

room and board, keep 15

roomer, guest; occupant

roomie, friend 3

roommate, friend 3

room(s), accommodation 4; flat 15; lodging; pad 3; place 6; quarter 5; room 3

room temperature, lukewarm 1

roomy°, open 9; spacious; voluminous 1; wide 1

roost, nest 1; perch 1, 2

roost, come to, settle 5

root°, base[1] 3; basic; cause 1; derivation; germ 2; implant 2; parent 2; plant 6; seed 2; spring 8

root, take, root[1] 5

root (for), applaud 1; pull 10; root[3]; follow 9

root out, discover 1; exterminate; pull 18b; purge 2; root[1] 7; unearth; uproot 2

root to the spot, transfix 2

root up, root[1] 7a

root and branch°, completely 1; entirely 1; root[1] 3; wholly 1

rooted°, entrenched; fixed 1, 2; habitual 1; immovable 1; time-honored

rooter, follower 3; partisan 1

rooting out, purge 4

rootle, root[2]

rootlet, root[1] 2

root(s), nationality 2; parentage; pedigree; root[1] 4; source 1; stock 2; strain[2] 1

roots, put down, settle 4

rootstalk, root[1] 2

rootstock, root[1] 2

rope°, cable 1; cord; lash[2]; lasso 1, 2; line[1] 11; stay[2] 1; string 1; tether 1; tie 1, 9; twine 1

rope, at the end of one's, desperate 6

rope in, defraud; rope 4

rope(s), bond 1

ropes, the, rope 2

rope's end, lash[1] 1

ropy°, rotten 5; stringy; tacky[1]; thick 5; tough 2

roscoe, pistol; revolver

rose, pink[1] 2

roseate, rosy 1

rose-colored, pink[1] 2; rosy 1

rose-red, rosy 1

Rosinante, nag[2]

rosiness, flush[1] 4

roster, list[1] 1; register 1; roll 13

rostrum°, platform 1; stage 2

rosy°, bright 4; fair[1] 3; fresh 6; golden 7; hopeful 2; pink[1] 2; promising; propitious; sanguine; sound[2] 2

rot°, contaminate; decay 2, 4; decompose 2; disintegrate; fester 1; fiddlesticks; gab 2; go 31e; gobbledegook 1; jargon 2; moonshine 2; mortify 3; mumbo jumbo 1; nonsense 1; prattle 3; putrefy; rubbish 2; spoil 4; stagnate; stuff 1; swill 2; trash 1; turn 5; wind[1] 2

rota, list[1] 1; register 1; roll 13

rotate°, alternate 1; cycle 2; go 21b; gyrate; pivot 3; revolve 1; roll 1, 8; screw 5; spin 1; swivel 1; turn 1, 43; twirl 1

rotating, spin 6

rotation, alternation; cycle 1; revolution 3; roll 16; spin 6; turn 26

rotation, be in, alternate 2

rotation, in, alternate 3; turn 40

rote°, routine 3, 4

rote, by, rote 2

rote, learn by, memorize

rotgut, alcohol; booze 1; drink 5; liquor 1; ruin 4; whiskey

rotogravure, print 2

rotted, rotten 1, 2

rotten°, atrocious 2; awful 1; bad 1, 2, 4, 6; beastly 2; bum 4; damnable; dirty 3; dreadful 1; filthy 1; foul 2; impure 1; infamous 2; lousy 1; miserable 2, 3; moldy; musty 1; off 7; offensive 3; off-color 1; poor 4; poorly 2; punk 2; putrid; rancid; rank[2] 4; repulsive; revolting; sad 3; scurvy; seamy; shabby 2; sordid 1; stale 1; stinking 2; strong 3; terrible 1, 2, 3, 5; ugly 2; unsound 1; vicious 2; wicked 6; wretched 1

rottenness, filth 2

rotter, heel[1] 2; rascal; rogue 1; scoundrel; stinker; villain; wretch 1

rotting, decay 4; moldy; putrid

rotund°, fat 1; large 1; obese; plump[1] 1; well-fed

rotundity, fat 5; fatness; obesity

rouche, fringe 1

roué°, degenerate 3; libertine 1; philanderer; playboy; profligate 3; rake[2]; reprobate 2; seducer; sensualist

rough°, approximate 1; back 12; barbarian 3; bluff[2] 1; boisterous; broad 4; brutal 2; bumpy; coarse 1, 2; crude 2, 3; earthy; foul 5, 8; gritty 1; gruff 2; hard 5; harsh 1; homespun; husky 2; incomplete; irregular 1; jagged; loose 5; merciless; miscreant 1; picnic 3; primitive 2, 3; ragged 1, 2, 4, 5; raucous; regrettable; rotten 5; round 6; rude 4; rugged 1, 2; rustic 2; scaly 1; scratch 6; scratchy 2; sexy 2; sketchy; stiff 9; strident; strong 18; surly; tactless; thick 7; thug; thumbnail; tough 3; uncivilized 2; unrefined 1

rough out *or* **in,** block 4a; outline 3; rough 15

rough up, manhandle; mistreat; punish 3; rough 16

rough-and-ready, rough 9; rugged 3

rough-and-tumble, fracas 1; rough 4

roughcast, rough 9

rough handling, mistreatment

rough-hewn, coarse 1; rough 9; rude 4; sketchy

roughhouse°, fracas 1; rumpus

roughhousing, rowdyism

rough idea, rundown

roughing-up, mistreatment

roughly°, about 2; approximately 1; around 1, 11; generally 2; say 14; sketchily

roughneck, bruiser; criminal 3; delinquent 1; hoodlum; miscreant 1; rough 4, 13; tough 8

roughness, indelicacy

rough (sketch), draft 1

rough-spoken, blunt 2; rough 3

roulade, run 47

round, about 1, 8; around 3, 4, 5, 7, 8, 10, 12; beat 12; bout 1; circuit 2; circular 1; cycle 1; even 8; game 2; orbit 1; revolution 3; rotund 2; run 37; spell[1] 1; spherical; tour 2; turn 26, 30

See also **around**

round off *or* **out,** complement 3; complete 6; round 18

round up, assemble 1; catch 2; herd 4; muster 1; pen[2] 2; raise 5; rally 3; round 19; seize 2

roundabout°, circular 2; devious 2; diffuse 2; discursive; equivocal 1; indirect 1; mealy-mouthed; meandering; oblique 2; rambling 1; redundant 2; rotary; serpentine 2; tortuous 2

roundabout, not, downright 1

roundabout way, detour 1

round(ed), obtuse 1; rotund 1; round 4, 7; well-fed

rounded off *or* **out** *or* **up** *or* **down,** even 8; round 6

rounded out, full 7

roundheel(s), prostitute 1; slattern; tart[2] 2

round(ish), full 7

roundlet, ring[1] 1

rounds°, patrol 2; round 12

rounds, make (the), patrol 3

round table, palaver 3

round the bend *or* **twist,** crazy 1; insane 1; mad 1; twist 11

round-the-clock, night 3; nonstop 2, 3; steady 2

roundup°, muster 2

rouse°, activate; agitate 1; animate 2; arouse 1; awake 2; call 6; electrify 2; energize; enliven 1; evoke; exalt 3; excite 1; ferment 2; find 7; fire 8a; foment; freshen 2; heat 4; incite; inflame 1; inspire 1; kindle; motivate; move 5, 6; pierce 4; prod 2; prompt 3; provoke 1; push 4; rally 4; refresh 2; stimulate 1; stir 4, 5; summon 1; wake[1] 1; whip 7a; work 20a

rouse oneself, awake 1

roused, agitated; awake 4; excited 1

rousing°, erotic 3; exciting 1; impassioned;

running (away), fugitive 2; run 49b
running down, pursuit 1
running for, in the, line[1] 20
running off at the mouth, prattle 2; talkative
running on empty, daft 1
running out, expiration
running wild, rampant 2
running cost(s), overhead 3
runny, fluid 2; liquid 2; watery 1; wishy-washy 2
run-of-the-mill°, average 4; common 1, 3; dead 12; dreary 2; everyday 2; general 2; hack[2] 4; mediocre; mill 4; monotonous; normal 1; ordinary 2; pedestrian 2; prosaic; ready-made 3; routine 4; stock 7; tame 4; tolerable 2; typical 2; undistinguished; usual
run one's eye over, peruse
run rings *or* circles around, shade 8
runs, the, run 50
run the gamut, range 6
run the show, govern 1; reign 2; shot 9
run things *or* affairs *or* matters, cover 8; dominate 1; shot 9
run-through°, rehearsal 1; rundown
runtish, undersized
runty, short 1; undersized
runway, chute 2; run 45, 46
rupture°, breach 2, 4; break 1, 25; burst; crack 1, 5; faction 2; flaw 2; fracture 2, 3; parting 1; puncture 3; rend 2; rip 3; schism; separation 3; shiver[2]; split 1, 7, 8; tear 1, 4
ruptured, broken 1; flat 6
rural°, green 1; pastoral 2
ruse°, artifice 2; blind 8; deceit 2; deception 2; delusion 1; device 2; dodge 4; feint; finesse 2; fraud 2; gimmick 1; machination; maneuver 1; move 9; pretense 2; pretext 2; racket 2; refuge 2; scheme 3; stall[2] 2; stratagem; tactic 1; trap 2; trick 1; wile; wrinkle[2]
rush°, assault 3; attack 1; career 2; dash 3, 5; expedite 1; flash 5; flood 3, 6; flow 1, 2, 5; flush[1] 5; fly 3; glow 3; gush 1, 3; haste 2; hasten 1, 2; hit 12; hurry 1, 2, 3; hustle 1; leap 3; leg 9a; lunge 2; move 12a; outburst; pelt[1] 3; pour 1; race[1] 4; run 1; scramble 2, 4; scurry; shoot 1; spate; speed 3; stampede 1, 2, 3; streak 5; stream 2, 3, 4, 5; surge 1, 2; swoop 2; tear 3; torrent; urge 1; urgent 1; wash 3; whip 3; whisk 1, 2
rush, in a, run 49a
rush about *or* around, fuss 2
rush (at), come 8; make 30b
rush (headlong), hurtle
rush (off *or* out *or* away), bolt 8; fly 2
rushed, hectic; hurried 1; perfunctory 1
rushing, torrential
rust, rot 1; stagnate
rusted, rotten 2
rustic°, boor 1; boorish; brutal 2; clown 2; crude 3; homespun; idyllic; pastoral 2; peasant; plebeian 2; provincial 2, 3; rural 1
rusticate, banish 1; drop 13; retire 1
rustication, retreat 2
rusticity, simplicity 3
rustle°, run 14; swish 1, 2
rustling, rustle 2
rusty, practice 5
rut°, furrow 1
rut, in, randy
ruth, grief 1; pity 1; sympathy 1
ruthless°, atrocious 1; bloodthirsty; brutal 1; coldblooded 3; coldhearted; cruel 1; cutthroat 2; deadly 2, 3; ferocious; ghoulish 2; grim 2; hard 4; harsh 2; heartless; implacable; inhuman 1; merciless; monstrous 1; relentless 1; remorseless 1; sadistic; sanguinary 1; savage 2; severe 1; strict 2; unsympathetic
ruthlessly, roughly 2
ruthlessness, barbarity; severity 1
rutted, bumpy
rutting, randy
ruttish, lascivious 1; lewd; libertine 2; obscene 1; prurient 1
rye, whiskey

sabbatical, holiday 1; leave[2] 2
sable, dusky 1; pitch-black; swarthy
sabotage°, interfere 2; jinx 3; sap[2]; subvert; undermine 1
saboteur, guerrilla; subversive 2
saboteurs, underground 5
sabre, blade 2
sac, sack 1
saccharine, gooey 2; ingratiating; mushy 2; romantic 3; sentimental 2; sweet 7
sacerdotal, clerical 1; priestly; spiritual 1
sack°, bag 1; desolation 1; devastate 1; discharge 2; dismiss 1; dismissal 1; displace 2; drop 10; eject 3; expel 1; expulsion; fire 11; gut 5; lay[1] 16a; loot 2; overrun; pillage 1, 2; plunder 1, 3; pocket 1; pouch; raid 2; ransack 2; rape 2, 5; ravage 2; remove 5; rob 1; robbery; strip[2] 5; turn 20c; waste 11
sack, give (someone) the, discharge 2; dismiss 1; drop 10; fire 11; sack 4
sack, give the, eject 3
sack, hit the, rest[1] 6; retire 3; sack 2; turn 15a
sack, the, discharge 9; ejection 3; push 16; sack 3
sack out, rest[1] 6; retire 3; sack 2
sackcloth and ashes, mourning 3
sacking, depredation; dismissal 1; ejection 3; expulsion; pillage 2; plunder 3; rape 2; removal 3; robbery
sacramental, ritual 1; sacred 3; solemn 2
sacred°, divine 1; godlike 1; holy 1; solemn 2; spiritual 1
sacred, make, sanctify 1
sacredness, sanctity
sacred writings, Scripture
sacrifice°, afford 2; expense 2; forgo 1; give 5; loss 1; offering; part 14; price 2; sign 8; slaughter 2; spare 9; undercut 2; vacate 2
sacrificed, sacrificial 1
sacrificial lamb, victim 1
sacrilege°, pity 2; sin 1, 2; violation 2
sacrilegious°, blasphemous; godless 1; impious; irreverent 1; profane 1; sinful; ungodly 1; wicked 1
sacristan, clergyman 1
sacrosanct, inalienable; sacred 2
sad°, bad 8, 9; bleak 1; blue 1; brokenhearted; dark 3; dejected; deplorable 1; desolate 3; despondent; disappointing; dismal; doleful; downhearted; dreary 1; forlorn 1; funereal; gloomy 2, 3; glum; heavy 5, 6; hopeless 3; hurt 7; joyless 1; low[1] 8; melancholy 1; miserable 1; moody 1; morbid 3; mournful 1; mouth 6; pathetic 1; penitent; pessimistic; piteous; poignant 1; regretful; regrettable; somber 1; sorrowful 1; subdued 2; touching; tragic; unhappy 1; wistful 1; woebegone; wretched 2
sadden°, depress 1; desolate 6; get 37c; pain 5
saddened, disappointed; joyless 1; subdued 2
saddening, mournful 2; sad 2
saddle, curse 6; encumber 1; helm 2; impose 4a; tax 4
saddle, be in the, govern 1
saddle with, burden 2; load 4; overload 1; stick[1] 8
saddle horse, hack[2] 3
saddlery, tack 4
sadist, ogre
sadistic°, bloodthirsty; brutal 1; cruel 2; savage 2; sick 5
sadly°, painfully
sadness°, care 1; depression 2; desolation 2; distress 1; gloom 2; grief 1; hurt 6; melancholy 2; misery 1; mourning 2; penitence; regret 2; sorrow 1; woe; wrench 5
sadomasochistic, sick 5
sad sack, loser; wretch 2
safari park, zoo 1
safe°, foolproof; good 8, 13; harmless; immune; impregnable; innocent 3; noncommittal; reliable; secure 1, 3; sound[2] 5; sure 2
safe, make, ensure 2
safe from, free 11
safe and sound, unscathed
safe as houses, safe 3
safe as the Bank of England, safe 3
safe-blower, robber
safe-breaker, robber
safe-conduct, escort 1; pass 24; passage 8
safecracker, robber; thief 1
safe-deposit box, safe 7
safeguard°, bulwark 1; defend 1; defense 1; ensure 2; guard 1, 5; harbor 2; keep 2, 12; precaution 1; preserve 1; preventive 4; protect 1, 2; protection 1; save 2; secure 7; shelter 4; shield 1, 2; watch 2
safeguarding, conservation; patrol 2; preservation 2; protective; security 4
safe house, refuge 1
safekeeping°, care 3; charge 4; conservation; custody 1; preservation 2; protection 1, 2; safety; security 1, 4; trust 3
safely, safe 6
safeness, safety
safer, better 3
safety°, guard 5; indemnity 2; protection 1; sanctuary 2; security 1; shelter 1
safety-deposit box, safe 7
safety measure, precaution 1
safety valve, outlet 1
sag°, dip 2; droop 1, 2; fade 2; flag[2] 1; slouch 1, 2; weaken 2; wilt 1, 2
saga°, legend 1; myth 1; story[1] 1
sagacious, astute 2; clever 3; deep 3; experienced 1; farsighted 1; intelligent; knowing 2; knowledgeable 2; politic 2; profound 2; provident 1; prudent 1; sage 1; shrewd; wise 1
sagacity, brain 1; brilliance 2; depth 3; discretion 1; experience 3; foresight 1; intelligence 1; judgment 1; mind 1; profundity 2; prudence 1;

scandalous° (*cont.*) foul 4; glaring 1; infamous 1; injurious 2; monstrous 2; notorious 1; outrageous 1; shameful; shocking 2; spicy 3; unmentionable 1; vicious 2
scanning, perusal; survey 3
scant, bare 5; insufficient; little 3; low[1] 2; meager 1; measly 1; mere; poor 2; premium 4a; rarefied 1; scanty 1; scarce; sparse 2; thin 2, 7
scantiness, absence 2; dearth; want 3
scantling, beam 1; board 1
scant(y), bare 5; frugal 3; inadequate 1; insufficient; lean[1] 2; little 3; low[1] 2; meager 1; measly 1; poor 2; premium 4a; rarefied 1; rare[1] 1; scarce; short 9; slender 2, 3; small 4; spare 5; sparse 1; thin 7; tight 9
scapegoat°, underdog; victim 1
scapegrace, rascal; rogue 1; scoundrel
scar°, blemish 1, 2, 3; blot 1; cliff; mar 1; pit[1] 5; seam 1; slash 1; spoil 2; stigmatize; taint 1; welt 2
Scaramouch(e), braggart; coward
scarce°, deficient 1; inadequate 1; insufficient; premium 4a; rare[1] 1; scanty 1; short 9; sparse 1; thin 7; tight 9
  scarce, make (oneself), depart 1; escape 1; flee 1; push 9b; take 34c; withdraw 5
scarcely°, almost; barely; hardly; ill 13; just 5; little 7, 8, 9; merely 2; narrowly 1; nearly 1
  scarcely any, few 1
  scarcely (ever), rarely
scarceness, poverty 2; want 3
scarcity°, absence 2; dearth; famine; lack 1; need 5; poverty 2; rarity 2; shortage; want 3
scare°, alarm 3; appall; daunt; dismay 1; fright 2; frighten; horrify 1; intimidate; lean[2] 4b; menace 1, 3; overawe; panic 3, 4; petrify 1; shake 5; shock 1; startle; terrify; turn 36; upset 1; wind[1] 11
  scare from, deter
  scare off, deter; shoo 2
  scare up, scare 2
scarecrow, ragamuffin
scared°, afraid 1; cowardly; fainthearted 1; fearful 1; nervous 1; timid; tremulous 2; upset 8
  scared (of), be, fear 5
  scared out of one's wits, panic-stricken
scarf, muffler; stole
scarfpin, pin 2
scarp, bluff[2] 3; cliff; crag; height 3; rock[1] 2
scarper, bolt 8; depart 1; flee 1; flight[2] 3; fly 2; powder 2; run 2
scarred, disfigured; hurt 8
scary°, dreadful 2; eerie; frightening; ghastly 1; ghostly 1; hairy 2; nightmarish; terrifying
Scat!, shoo 1
scatheless, unscathed
scathing°, caustic 2; cutting 2; pungent 2; sarcastic; scorching 2; sharp 5; tart[1] 2; truculent
scatologic(al), bawdy; foul 5; indecent 2; obscene 1; outrageous 3; rank[2] 3; vulgar 2
scatter°, break 11; broadcast 3; diffuse 3; disband; disperse 1, 2; dissipate 1; distribute 2; litter 4; pepper; seed 5; shed[2] 2; shuffle 1; spray[1] 1; spread 1; stampede 2; strew
scatterbrained°, dizzy 2; foolish 2; frivolous 2; giddy 2; harebrained 2; insane 2; light[2] 8; stupid 2
scattered, diffuse 1; fragmentary; sparse 1
  scattered (about *or* around), loose 4
scattering°, dissipation 2; few 2; stampede 1
scaur, cliff
scavenge, prowl 2
scenario°, drama 1; outline 2; path 3; plan 1; plot[1] 2; scheme 1; script 2; story[1] 5; strategy
scenarist, dramatist; playwright
scene°, episode 2; incident 2; landscape; mount[2] 7; performance 5; place 1; prospect 1; set 28; setting; sight 3; spot 2; tableau; theater 4; view 1
  scene, make *or* do the, scene 6
scenery, nature 3; scene 2, 4; set 28; setting
scenes, behind the, background 3; scene 5
scenic°, panoramic; picturesque 1
scent°, aroma 1; bouquet 2; detect 2; flavor 1; fragrance; odor 1; perfume 1, 2; savor 2; smell 1, 3; sniff 1; touch 15; track 3; trail 2; vestige; wake[2] 1
  scent, throw (someone) off the, misinform; mislead
scented, redolent 1; sweet 2
scepter, staff 1
Schadenfreude, glee
schedule°, calendar 1; list[1] 1, 2; program 1, 4; ration 4; register 1; roll 13; routine 1; shift 4; slot 4; term 7a; time 22, 23; timetable
  schedule, behind, tardy 1
scheduled, bound[3] 5; due 4; regular 1, 2; routine 3; set 29
schedule (of charges), tariff 2
schema, frame 3; scenario; scheme 1, 2
schematic°, scheme 2
scheme°, aspiration; cabal 1; conception 3; conspiracy; design 4, 5; device 2; devise 1; dodge 4; engineer 5; enterprise 1; finesse 2; frame 3; game 3; idea 1; intrigue 2, 3; machination; maneuver 1, 3; meditate 2; method 1; organization 2; path 3; plan 1; plot[1] 1, 3; policy; procedure; program 1; project 1, 3; projection 3; proposal 2; prospectus; purpose 1; racket 2; scenario; schematic 2; stratagem; strategy; system 2; tactic 1; thought 3; vision 3; wangle; way 1; wile; wrinkle[2]
schemer, designer 2; tactician
schemes, art 6
scheming°, artful 1; calculating; crafty; deceitful; deep 5; designing; devious 1; disingenuous; foxy 1; machiavellian; machination; politic 1; serpentine 1; shifty; sly 1; smooth 8; subtle 3; two-faced; wily
schism°, breach 2; break 25; faction 2; rift 1; rupture 1; separation 2, 3; split 8; variance 2
schismatic°, heretical; rebel 4
schizo, insane 1; psychotic 2
schizoid, insane 1
schizophrenia, insanity 1
schizophrenic, insane 1; psychotic 2
schlemiel, loser; sap[1] 2; victim 2; weakling
schlep, carry 1; cart 2
schlimazel, loser
schmaltz, sentimentality
schmaltziness, sentimentality
schmaltzy, melodramatic; mushy 2; sentimental 2
schmooze, talk 4
schmuck, fool 1
schnapps, liquor 1
schnook, loser; sap[1] 2; victim 2; weakling
scholar°, authority 3; disciple 1; expert 1; highbrow 1; intellectual 4; learner; pupil; student 1; thinker
scholarliness, profundity 2
scholarly°, academic 1; authoritative 2; educated 1; highbrow 2; intellectual 2; learned; lettered; literary 1, 2; philosophical 1; profound 2; studious 2; wise 2
scholarship°, aid 3; knowledge 4; learning; letter 3; profundity 2; stipend; wisdom 2
scholastic, academic 1; educational 1; instructor; learned; literary 2; scholarly
scholium, note 3
school°, coach 3; denomination 1; discipline 6; drill 2; educate; faculty 2; flock 1; indoctrinate; institution 2; seminary; side 3; train 4; tutor 2
  school in, school 4; teach
schoolbook°, text 4
schoolboy, boy 1; lad; minor 3; pupil; scholar 2; stripling; student 1; youth 2
schoolchild, pupil; student 1; youth 2
schooled, practiced 2
schoolgirl, lass; minor 3; miss[2]; pupil; scholar 2; student 1; youth 2
schooling°, discipline 1; edification; education 1; instruction 2; knowledge 4; lesson 2; scholarship 1; tuition
schoolmarm, prig; schoolteacher; teacher
schoolmarmish, literary 2; priggish; prissy
schoolmaster, schoolteacher; teacher
schoolmate, boy 3
schoolmistress, mistress 2; schoolteacher; teacher
school (of thought), persuasion 2; sect 2
schoolteacher°, teacher
schooner, drink 6
schtuck, in, trouble 9a
schul, temple
science fiction, romance 3
scientific reasoning, logic 1
scilicet, namely
scintilla, bit 2; crumb; flash 1; flicker 4; ghost 2; gleam 2; grain 3; modicum; particle; ray 2; scrap[1] 1; shade 3; shred 1; spark 1; suspicion 2; trifle 2; word 6
scintillate, blink 2; flash 4; glance 2; radiate 1; shine 1; sparkle 1; twinkle 1
scintillating°, brilliant 1; dazzling; nimble 2; radiant 1; shiny; twinkle 2
scintillation, flash 1; glance 5; light[1] 3; radiance 1; sparkle 3; twinkle 2
scion, cutting 3; descendant; graft[1] 1; implant 3; offshoot 1, 2; shoot 5; slip[2] 2
scion(s), issue 7; progeny
scissile, separable
scission, separation 3
scissors, snip 4
scoff°, flout; food; sustenance 1; taunt 2
  scoff (at), deride; fly 6; fun 5; gibe 1; jeer 1; laugh 2a; mock 1; parody 3; scoff; scorn 4; taunt 1; thumb 8; trivialize
scoffer, skeptic
scoffing, derision; gibe 2; scorn 2; scornful; skeptical

screwed up, confused 2
screwed-up, disturbed 2
screwing, sex 2
screws on *or* **to (someone), put the,** pressure 5; screw 4; squeeze 3
screwup, muddle 4; blunder 2; mess 3; mix-up
screwy, crazy 1; deranged; disturbed 2; flighty 2; foolish 2; inane; insane 1, 2; mad 1; mental 2; nonsensical; odd 1; preposterous; way-out 1
scribble, dash 4; pen[1] 2; scrawl 1, 2; write 1; writing 1
scribbler, hack[2] 1; scribe 2; writer
scribe°, journalist; writer
scrimmage°, fight 7; scramble 4; skirmish 1
scrimp, economize; pinch 5; save 3; stint 4
scrimp (and scrape), scrape 3
scrimping, economical 2; saving 3; thrift; thrifty
scrimpy, meager 1
scrimshank, loaf[2] 1; shirk
scrimshanker, loafer; slacker; truant 1
scrimshaw, scribe 3
script°, hand 7; line[1] 17; paper 4; penmanship; plan 1; scenario 1; writing 1
scriptwriter, dramatist; playwright
scrivener, scribe 1, 2
scroll, roll 12; spiral 1
scrolled, spiral 2
Scrooge, miser
Scroogelike, cheap 4; close 18; penurious 1
scrounge°, beg 2; bloodsucker; bum 5; parasite; rummage 1
scrounge (up), scare 2; scrape 6
scrounger, beggar 1; bloodsucker; hanger-on; parasite; scrounge 2
scrub°, brush[1] 1; brush[2] 4; clean 9; cleanse 1; launder 1; rub 1; scour 1; wash 11
scrub (off *or* **away),** scrape 2
scrub (up), wash 1
scrubbed, clean 2
scrubbing, wash 11
scrubby, shabby 1
scrubwoman, servant 1
scruffy, bedraggled; disreputable 2; mangy; mean[2] 3; scaly 2; seedy 1; shabby 1; sloppy 1; threadbare 1; unkempt
scrum, fight 7; fracas 1; scramble 4; scrimmage; skirmish 1
scrumptious, delicious 1; luscious; tasty; yummy
scrunch, munch; stoop 3
scrunch down, stoop 1
scrunch (up), rumple
scruple°, discredit 6; doubt 2; hesitate 1; misgiving; pang 2; protest 1, 3; qualm; reservation 2
scruple at, stick[1] 12
scruples, conscience; moral 4
scruples (about), have, scruple 2
scrupulous°, accurate 2; careful 2; conscientious 1, 2; diligent; exact 2; faithful 3; finicky 1; laborious 2; meticulous; moral 1; narrow 3; nice 3; painful 3; pedantic 2; perfectionist 2; precise 2, 3; principled; religious 2; rigid 3; squeamish 1; strait-laced; strict 1; thorough 2; virtuous 1
scrupulously, exactly 1; jealously; narrowly 2; precisely 2; thoroughly 2
scrupulousness, precision 2; sportsmanship
scrutinization, research 1
scrutinize°, analyze 2; check 4, 10b; contemplate 1; examine 1; explore 2; eye 9; go 30b, 34a; inquire 1; inspect; investigate; look 1, 8; observe 2; peruse; pore[1]; probe 1; ransack 1; reconnoiter; report 6; research 2; review 1; scan 2; search 1; sift 2; study 3; survey 1; traverse 4; treat 2; try 2; vet; view 7; watch 3
scrutiny°, analysis 1; check 15; discussion; examination 1; exploration; eye 7; inquiry 1; interest 1; investigation; observance 3; observation 1; perusal; probe 3; reconnaissance; research 1; review 5; scan 3; search 4; study 4; surveillance; survey 3; view 4
scrutiny, bear, wash 8
scrutiny, not hold up under *or* **bear,** water 2
scrutiny, pass, check 10c
scud°, run 1; sail 3; scurry
scuff, scrape 1, 7; scratch 1, 4
scuffle, disorder 2; fight 7; fracas 1; fray[1]; quarrel 1; scrap[2] 1, 2; scrimmage
scuffle (with *or* **against),** fight 1
scuffling, shuffle 4
scuff (one's feet), shuffle 2
scull, oar 1; paddle 1, 2
sculler, oar 2
scullery, kitchen
scullery maid, servant 1
scullion, servant 1; slave 1
sculp, sculpture 2
sculpt, carve 1; model 7; mold[1] 4; sculpture 2; shape 6
sculpture°, carve 1; figure 4; image 1; likeness 2; mold[1] 4; shape 6; statue
scum, dregs 2; grime; muck 2; rabble 2; riffraff
scummy, filthy 1; impure 1
scum (of the earth), mob 2
scunner, dislike 2
scunner (to), take a, dislike 1
scuppered, disabled
scurf, flake 1; scale[2] 1
scurfy, scaly 2
scurrility, abuse 8; ribaldry
scurrilous°, abusive 1; base[2] 2; foul 5; injurious 2; lewd; low[1] 3; obscene 1; profane 3; prurient 2; scandalous 2; vituperative
scurrilousness, ribaldry
scurry°, hasten 1; hurry 1; hustle 1; leg 5; patter[1] 1; pelt[1] 3; run 1; rush 1; scramble 2; speed 3; tear 3; whip 3
scurry off *or* **away,** bundle 3
scurvy°, base[2] 6; black 6; contemptible; despicable; dirty 6; filthy 2; hateful 1; infamous 2; lousy 1; mangy; mean[2] 3; miserable 4; obnoxious; seamy; shabby 2; sordid 1; wretched 4
scut, tail 1
scutate, scaly 1
scute, scale[2] 1
scuttle, bucket; flash 5; hasten 1; hurry 1; hustle 1; patter[1] 1; run 1; rush 1; scramble 2; scurry; shoot 1; swamp 3; tear 3; wreck 2
scuttlebutt, dirt 4; gossip 2; news 1; rumor 1; tale 3
scuttling, shambling
scutum, scale[2] 1
scythe, mow 1
sea°, heap 2; marine 1, 2; ocean 1; oceanic; stack 3; wave 1
sea, at, confused 2; disorientated; helpless 2; lost 3; sea 4
sea, put (out) to, sail 2
sea, the, deep 9; drink 7
seaboard, coast 1; seacoast
sea change, revolution 2
sea-chart, chart 1
seacoast°, coast 1
seacock, tap[2] 1
seadog, sailor
seafarer, sailor
seafaring°, marine 1; nautical
seafaring man *or* **woman,** sailor
seagoing, marine 1; nautical
seagoing man *or* **woman,** sailor
seal°, character 1; close 1, 4, 6; confirm 2; connect 3; crest 2; device 3; emblem; execute 2; fill 7; glue 2; lock[1] 3; shut 1; sign 4; stamp 6, 7; tape 3
seal (of approval) on, put one's, endorse 1
seal (of approval) to, set one's, endorse 1
seal (off *or* **up),** plug 4; seal 3
sealed, hermetic; impermeable; tight 1; watertight 1
sealed (up), shut 7
seal (of approval *or* **authenticity),** consent 3; endorsement 1; hallmark 1; sanction 1
seal the doom *or* **fate of,** rout 2
seam°, joint 1; stratum 1; streak 1; union 3; vein 2, 3; weld 2; welt 1
seaman, navigator; sailor
seamanship, navigation
seamstress, dressmaker; tailor 1
seamy°, sordid 3, 4
seaport, port
sear, dry 1; parch; scorch; singe
search°, examination 1; examine 1; exploration; explore 2; frisk 2; hunt 4; inquire 1; inquiry 1; investigation; mine 4; probe 3; pursuit 2; quest 1; rake[1] 6; ransack 1; rifle 2; root[2]; rummage 1; scour 2
search (about *or* **around) (for),** cast 10; hunt 2; look 6b, 11a; prospect 5; pursue 3; quest 2; scout 1; search 2; seek 1; trace 5
search after, quest 2
search (into), probe 1
search of, in, for 2
search of (*or* **for), go in,** hunt 2; pursue 3
search (out), fathom; discover 1
search through, search 1
searching, close 14; hard 6; inquisitive 1; narrow 3; penetrating 1; piercing 2; prowl 3; pursuit 2
searchingly°, intently; narrowly 2
searchlight, spotlight 1
searing, evaporation 1; hot 1; keen[1] 2; scathing; scorching 1
sea robber, pirate 1
sea rover, pirate 1
sea salt, salt 1
seas, the high, deep 9
seascape, prospect 1; scene 4; view 1
seashore, beach 1; coast 1; seacoast
seasick, nauseated; sick 1
seaside, beach 1; coast 1; seacoast
season°, accustom; date 1; flavor 3; mature 5; mellow 5; ripen; salt 6; spell[1] 1; spice 3
season, in, season 2
seasonable°, opportune 2; season 2; well-timed
seasonal, intermittent; regular 2
seasoned°, experienced 2; mature 2; mill 3; practiced 1; professional 1; pungent 1; ripe 1, 2; season 2; veteran 2; vintage 2
seasoning°, flavor 1; salt 2; spice 1, 2
seas over, half, drunk 1
seat°, bottom 1; bum 1; buttocks; chair 1; contain 2; establish 2; invest 3; perch 1; place 9;

posterior 3; root[1] 1; sit 5; space 4; thwart 2
seat, take a, sit 1
seated, sedentary
seated, be, sit 1
seating°, session 1
seating for, have, sit 5
seat (of government), capital 2
seats for, have, sit 5
sea water, water 1
seaworthy, staunch 2
sebaceous, fat 2; greasy 1; oily 1
sec, second[2]
secede°, step 17c
seceding, secession
seclude, maroon; segregate; shut 3a, 4b
seclude oneself, retire 1
secluded°, close 19; isolated 2, 3; lonely 3; obscure 3; out-of-the-way 1; private 2, 4; reclusive; remote 2; separate 7; solitary 1; withdrawn 2
seclusion°, privacy 1; retreat 2; segregation; solitude 1
seclusive, private 4
second°, abet 2; advocate 1, 2; alternate 4; approve 2; assign 3; assist 1; assistant 1; attach 2; back 2a; bit 3; crack 3; flash 3; instant 1, 2; minute[1] 1; moment 1, 2; recommend 2; reject 4; sanction 5; stand 7b; standby 2; stand-in; support 1; twinkling 1; understudy 1, 2
second, at this (very), now 1
second, half a, moment 1
second, in a (split), immediately 1; once 6a; soon 1; suddenly 1
second, to the, dot 2
second thought to, do not give a, mind 19
second (to), subordinate 1; under 2
secondarily, second[1] 11
secondary°, accessory 3; associate 5; auxiliary 2; circumstantial 2; incidental 2; indirect 2; inferior 2; junior; minor 1; nonessential 1; outside 7; peripheral 1; side 8, 9; small 2; supplementary 2
secondary (to), subordinate 1; subsidiary; under 2; below 10
second-best, second[1] 3
second childhood, senility
second-class, inferior 3
seconded, attached 1
seconder, party 4
secondhand°, derivative 1; secondary 2; used 1
second-line, secondary 3
secondly, second[1] 11
secondment, sanction 1; service 6
second nature, habit 2
second-rate, amateur 2; cheap 3; coarse 4; disappointing; inferior 3; lousy 2; low[1] 9; mediocre; par 5a; poor 4; shoddy; unprofessional 2; unworthy 1; wanting 1
seconds, for a few, briefly 2
second-story man, burglar; robber; thief 1
second thought(s), fear 1; qualm; regret 2; scruple 1; suspicion 1
second to none, unique 2
secrecy°, mystery 2; privacy 2; stealth
secrecy, in, confidence 3; secrecy 2
secret°, close 16; confidential; cryptic 1; dark 7; furtive 1; hidden; illicit 2; inside 4; interior 3; intimate[1] 2; isolated 3; knowing 1; mysterious 2; mystical; obscure 3; obscurity 2; occult 1; private 1, 2; privileged 4; profound 1; sneaking 2; stealthy; surreptitious; trick 3; ulterior 1; undercover; underground 2; unofficial; untold 2; veiled
secret, in, private 6; secret 4; secretly
secret, keep, conceal 2; hide[1] 2, 4; mum 2; suppress 3
secretarial, clerical 2
secretary, scribe 1; stenographer
secrete°, cache 3; conceal 1; hide[1] 1; ooze 2; plant 8; stow
secreting, secretion
secretion°, discharge 11
secretive°, close 17; devious 1; furtive 1; secret 1; silent 2; stealthy; surreptitious; tight-lipped
secretively, private 6; secrecy 2
secretiveness, privacy 2; secrecy 1
secret(ly), back 6; incognito 2; private 6; record 8; scene 5; secrecy 2; secret 4; tête-à-tête 2
secret places, recess 3
secrets, reveal *or* **divulge** *or* **give away,** tattle 1
secret service, spying
secret (service) agent, spy 1
sect°, denomination 1; faith 2; order 9; party 3; persuasion 2; school 2; set 26; side 3
sectarian°, partisan 3
sectary, sectarian 3
section°, area 3, 4; branch 2; bureau 2; categorize; category; clip[2] 5; compartment; department 1; district; division 2, 3; leg 3; office 2; parcel 2; part 6, 7; partition 4; passage 2; patch 2; piece 3; place 2; portion 1, 4; precinct 2; province 1; quarter 3; region 1; reservation 4; scene 2; segment 1, 2; spot 2; squad; stall[1] 2; stretch 5; text 2; tract[1]; unit; ward 1; zone
sectionalized, jointed
sectioned, jointed
sectioning, division 1
sector, category; district; division 3; precinct 2; public 10; region 1; section 1; territory 1; tract[1]; zone
secular°, civil 1; earthly 3; lay[2] 1; material 7; outward 1; profane 2; temporal 1; worldly 1
secularization, sacrilege 2
secure°, acquire; anchor 3; assure 1; attach 1; bar 8; base[1] 5; bind 1; bolt 10; buy 1; carry 6, 10a; chain 3; clasp 3; clinch 1; close 1, 10; come 9a; confident 1; connect 3; cozy 1; defend 2; derive 1; draw 8; effect 7; engage 1; enlist 2; ensure 1, 2; establish 2; fasten 1; find 5; firm 2; fix 1; fixed 2; fortify 1; gain 1; get 1; good 8, 13; have 2; impregnable; infallible 2; knot 3; land 7; lash[2]; line[1] 23b; lock[1] 3, 6a; make 13; moor[2]; nail 7, 10; obtain 1; peg 3; pin 3; possess 5; possession 4; procure 1; purchase 1; reliable; reserve 2; rope 3; safe 1, 3; save 2; seal 3; self-confident; shackle 3; shelter 4; shut 1, 3a; snap 4; sound[2] 5; stable 2; stake[1] 3; stay[2] 2; steady 11; sure 2; tack 5; take 1; tether 2; tie 1, 5a; tight 1; tighten 1; win 2
secure(d), bound[3] 1; fast[1] 3; fixed 1; safe 5; tight 1
securely, fast[1] 7; firmly 1; safe 6; surely 2; tight 10
securing, purchase 3
securities exchange, exchange 4
security°, anchor 2; assurance 2; belief 1; comfort 3; earnest 3; guard 5; hope 2; hostage; indemnity 2; insurance; luxury 3; pawn[1] 3; pledge 2; preservation 2; protection 1; rampart; refuge 1; safeguard 1; safety; shelter 1; tenure 2; warrant 1
security guard, guard 3
sedan chair, litter 3
sedate°, calm 4; collected; dignified; drug 4; even 4; grave[2] 1; measured 1; poised 1; reserved; serious 1; sober 2, 3; solemn 1; somber 3; staid; steady 6; straight 8; tranquil; tranquilize; unhurried; venerable
sedately, quietly 4; well[1] 8
sedateness, cool 8; gravity 3; poise 2; propriety 2; sobriety 2; solemnity
sedative°, drug 2; narcotic 1, 2; painkiller; restful 1; tranquilizer
sediment°, deposit 4; dregs 1; ground 4; ooze 1; silt 1
sedition°, faction 2
seditionaries, underground 5
seditionary, subversive 1
seditionists, underground 5
seditious°, factious; inflammatory; insubordinate; mutinous 1; rebellious 1; revolutionary 1; subversive 1; traitorous
seduce°, cajole; captivate; catch 11; charm 5; chat 3; decoy 2; dishonor 3; entice; induce 1; lead 9b; lure 1; make 24; mess 4b; pervert 2; ruin 11; tempt 1
seducer°, charmer; libertine 1
seducing, ruin 2; seductive; temptation 1
seducing, succeed in, score 14
seduction, cajolery; enticement 1; pull 20; ruin 2; temptation 2
seductive°, attractive; desirable 2; enchanting; erotic 1; exciting 2; flirtatious; foxy 2; inviting; kittenish; magnetic; provocative 1; sexy 1; sultry 2; tempting 1; voluptuous 2
seductiveness, magnetism; pull 20
seductress°, charmer; siren 2; temptress
sedulity, exertion; follow-through
sedulous, diligent; hard 7; industrious; laborious 2; painful 3; patient 2; strong 5; studious 1; tireless
sedulous, be, persevere
sedulously, hard 13; thoroughly 2
sedulousness, exertion; follow-through; industry 2; perseverance; tenacity 1
see°, associate 1b; behold; catch 10; city; comprehend; conceive 3; discover 2; distinguish 3; envisage 2; feel 1; figure 12b; find 2, 3; follow 8; get 19; grasp 2; hit 9b; look 1; make 37a, 37d; mark 12; meet[1] 1; note 11; notice 1, 2; observe 3; perceive 1; pierce 3; place 16; realize 2; recognize 2; regard 4; sight 8; spot 6; spy 3; take 19; twig[2]; understand 1, 3; view 7, 8; visit 1; watch 3; witness 4
see, for all to, aboveboard 1; out 3
see, go to, look 11b
see about, see 12
see fit, please 2; will 5
see it through, stick[1] 14
see off, see 13
see red, make (someone), enrage; infuriate; madden 1
see (someone) through, see 14c

sensibility°, delicacy 3; feeling 1, 3; heart 5; sentiment 1
sensible°, advisable; aware 2; coherent 2; enlightened; farsighted 1; feasible; good 13; intelligent; judicious; legitimate 3; levelheaded; logical 2, 3; moderate 1; modest 3; natural 1; perceptive; plausible 1; politic 2; practical 2; presumptive 1; proper 4; prudent 1; rational 3; realistic 1; reason 6; reasonable 1, 4; sage 1; sane; seemly 1; solid 7; sound[2] 4; stable 3; steady 6; temperate 1; thinking 1; viable; well-advised; well-balanced 1; wise 1, 3
sensible of, be, feel 3
sensible of *or* **to,** sensible 4
sensibly, practically 2; right 17
sensitive°, acute 4, 5; aesthetic 2; aware 2; awkward 5; brittle 2; clear 9; considerate; critical 3; dainty 2; delicate 4, 5; difficult 2; diplomatic; dodgy; edge 5; educated 2; emotional 3; explosive 1; feeling 7; hot 10; hothouse 2; human 2, 3; keen[1] 5, 6; mobile 3; nervous 1; nice 4; painful 1; penetrating 1; perceptive; precarious; problematic; raw 4; receptive 2; refined 2, 3; responsive; sensible 3; sore 1, 2; sticky 2; tactful; temperamental 1; tender[1] 1, 4, 5, 8; ticklish 3; touchy 1, 2; tricky 2; volatile 3; vulnerable; waspish
sensitive (to), alive 2; sensible 4; wise 5
sensitiveness, sensitivity 1, 3; understanding 2
sensitivity°, delicacy 3; discrimination 2; ear 2; expression 4; eye 3; feeling 1, 3, 5; grip 3; heart 5; humanity 3; insight; instinct; penetration 2; refinement 1; sensitivity 3; tact; understanding 2; uptake
sensor, feeler 1, 2
sensory, sensuous
sensual°, animal 4; carnal; earthly 2; epicurean 1; erotic 1; hot 7; lascivious 1; lecherous; obscene 1; passionate 2; profligate 1; provocative 1; prurient 1; self-indulgent; sexual 2; sexy 1; sultry 2; voluptuous 1
sensualism, dissipation 1
sensualist°, profligate 3; sybarite
sensualistic, voluptuous 1
sensuality, lust 1; profligacy 1
sensuous°, exciting 2; provocative 1; sexual 2; sexy 1; tempting 1; voluptuous 1
sentence°, commit 2; condemn 1; damn 2; decision 2; passage 2; penalize; penalty; punishment 2; rap 6
sentence, pronounce *or* **pass,** judge 4
sentence structure, style 5
sentencing, punishment 2
sententious, epigrammatic; meaningful 2; pedantic 1; short 3; terse 1
sentience, life 1; penetration 2
sentient, feeling 7; sensible 3
sentiment°, belief 3; emotion; estimate 4; feeling 4; heart 5; idea 3; judgment 4; mind 6; observation 2; opinion 1; perspective 2; philosophy 2; position 3; posture 3; principle 2; regard 9; romance 4; sense 5; soul 4; stand 13; thing 4; view 2
sentimental°, emotional 3; full 9; gooey 2; gushy; lyric 2; maudlin; melodramatic; mushy 2; precious 4; romantic 3; sloppy 3; softhearted; sweet 7; tender[1] 7
sentimentalism, sentiment 1
sentimentalist, romantic 5
sentimentality°, sentiment 1
sentiment(s), heart 4; sensibility 2; sentiment 2; spirit 8, 9a
sentinel°, guard 3; lookout 1; monitor 1; patrol 1; picket 3; watch 7; watchman
sentry, guard 3; lookout 1; patrol 1; sentinel; watch 7; watchman
separate°, aloof 1; analyze 1; apart 1; cleave; come 7; cut 15c; decompose 1; detach; different 1; differentiate 1; disconnect; disconnected 1; discrete; discriminate 1; disengage; dissimilar; dissociate; distance 3; distinct 2; distinguish 2, 4; diverge 1; divergent; diverse; diversify; divide 1, 3, 4; divorce 2; fence 3; filter 2; fracture 3; individual 1; insulate 1; isolate; isolated 1; know 4; loosen 2; outside 9; part 4; part 12, 13; particular 1; partition 5, 6; pick 7b; proper 6; pull 12a; remove 7; rend 2; respective; rupture 3; scatter 2; screen 6; secluded 1, 2; segment 2; segregate; set 14a; sever 1, 2; shut 4b; sift 1; single 3, 4; singular 3; skim 1; snap 1; solitary 1; sort 8, 10b; spin 4; split 1, 2, 3; spread 5; strain[1] 5; tear 1; unattached 1; unrelated; variant 2
separate, be, differ 1
separate (from), break 12, 24a; secede; unlike 1
separate (oneself), break 15
separate (out), screen 8
separate(d), cut 26; detached 1; disconnected 1; disjointed 1; estranged; independent 2; isolated 2; separate 5, 7; split 9; stray 6
separately°, apart 1, 2; individually; respectively; singly
separately from, apart 4
separating, detachment 1; division 1; parting 1
separation°, breach 2; break 25; clearance 1; detachment 1; dissolution 1; distinction 1; division 1, 2; divorce 1; exile 1; extraction 2; fracture 2; gulf 2; opening 1; parting 1; partition 1; remove 8; rift 1; rupture 1; schism; secession; seclusion; segregation; split 7, 8; wedge 2
separatist, schismatic
separator, partition 3; wall 1; wedge 2
separatrix, line[1] 1
sept, tribe
septic, inflamed; poisonous 1; ulcerous; virulent 1
sepulcher°, crypt; grave[1]; monument 1; tomb
sepulchral, funereal; hollow 6
sepulture, burial; funeral
sequel°, outcome; result 1; supplement 1
sequela, outcome
sequence°, chain 1; chronology; cycle 1; order 1; pattern 3; procession 2; progression 3; rank[1] 4; round 11; run 41; string 3; succession 2; train 3
sequence, out of, order 13a; turn 42a
sequentially, turn 41
sequester, confiscate; insulate 1; isolate; segregate; separate 2; shut 4b
sequester oneself, retire 1
sequestered, isolated 2; private 4; reclusive; remote 2; secluded 1; separate 7
sequestrate, confiscate; insulate 1
sequestration, forfeit 1; privacy 1; segregation; seizure 1
seraglio, brothel
serape, wrap 4
seraphic°, divine 1; godlike 1; heavenly 1; pious 1; saintly
serendipitous, accidental; casual 1; haphazard 1; incidental 1; random 1; uncertain 1
serendipitous find, windfall
serendipitously, random 2
serendipity, accident 2; fluke; luck 1
serene°, calm 3, 4; collected; comfortable 1; cool 2; deliberate 3; dispassionate 1; easy 2; equable 1; even 4; home 4a; homely 2; impassive; inactive 1; levelheaded; mild 1; passive 1; pastoral 1; patient 1; peaceful 1; philosophical 2; phlegmatic 2; poised 1; quiet 2; relaxed; restful 2; sedate 1; self-contained 1; self-possessed; silent 1; smooth 1; sober 2; soothing 1; still 1; temperate 1; tranquil
serenely, easy 7; quietly 3; well[1] 8
serenity°, calm 1; content[2] 2; ease 1; inactivity 1; order 8; patience 1; peace 1; philosophy 3; poise 2; quiet 5; repose 3; self-control 2; silence 1
serf, menial 3; peasant; slave 1
serfdom, bondage; servitude; slavery 1
serial, periodical; sequential
serialize, syndicate 4
seriatim, singly
sericeous, silky
series, chain 1; cycle 1; gamut; group 2; procession 2; range 2, 3; rank[1] 4; rash[2] 2; round 11; row[1]; run 41; selection 2; sequence; set 25; stream 3; string 3; succession 2; suite 1; tier; tissue
serious°, acute 2; bad 11; classical 1; conscientious 3; critical 3; deep 2, 6; deplorable 1; desperate 4; dignified; earnest 1, 4; good 2; grave[2] 1, 2; great 4; grievous 1; heartfelt; heavy 3; high 6; major 2; material 6; meaningful 1; momentous; nasty 2, 5; no-nonsense; pensive; pressing; sedate 1; self-contained 1; severe 2; sober 2; solemn 1; somber 1; sore 3; staid; steady 6; stern 2; straight 8; subdued 2; terrible 1; urgent 1; violent 3; wholehearted
seriously°, badly 5, 8; deeply 2; indeed 1; materially; severely 1, 3; sincerely
seriously, not, fun 3; tongue 5
serious-minded, staid
seriousness, gravity 2; moment 3; severity 3; sincerity; sobriety 2; solemnity; urgency
sermon°, address 1; lecture 1; speech 2; talk 14; tract[2]
sermon, deliver a, preach 1
sermonize, harangue 2; lecture 3; preach 2

sheen°, gloss¹ 1; luster 1; polish 5; shine 4
sheep, stock 4
sheepish°, ashamed; bashful 1; coy; embarrassed 1; guilty 2; passive 2; self-conscious; shamefaced 1; shy 1
sheeplike, sheepish 1
sheep's eyes at, make, ogle 1
sheer°, abrupt 2; barefaced 1; blank 6; clear 14; filmy 1; fine¹ 6; flimsy 3; implicit 2; main 4; mere; naked 3; outright 2; positive 9; precipitous 1; pure 5; rank² 2; see-through; sharp 2; stark 4; steep¹ 1; thin 5; thorough 1; transparent 1; ultra-; unmitigated; very 4
  sheer off, swerve
sheerest, very 4
sheet°, film 1; foil²; mantle 2; page¹ 1; pane; paper 1; plate 3
sheet anchor, anchor 2; mainstay
sheets to the wind, three *or* **four,** drunk 1; stinking 3
sheik, flirt 3
Sheik of Araby, sovereign 1
sheila, girl 1
shekels, money 1; purse 2
shelf, bracket 2; ledge; rest¹ 4; shallow 2
  shelf, on the, abeyance; limbo
  shelf, put on the, shelve
shell°, blaze 5; bomb 1, 2; bombard 1; exterior 3; frame 1; hulk 1; hull 2, 3; outside 1; pelt¹ 1; projectile; round 15; skin 2, 4; storm 6
  shell out, expend 1; give 15b; lay¹ 18b; pay 11b; shell 5; spend 1; stump 4
shellac, glaze 1, 2
shellacking, rout 1
shellback, sailor
shell game, swindle 2
shell game artist, thief 2
shelling, fire 3
shelter°, accommodate 4; accommodation 4; bulwark 2; cover 1, 14; defend 1; defense 1; harbor 2; home 3; house 8; housing 1; hut; insulate 1; lodge 1, 4; lodging; mother 5; preserve 1; protect 2; protection 1; quarter 5, 6; refuge 1; retreat 3; safeguard 2; safety; sanctuary 2; save 2; screen 2, 7; secure 7; security 1; shade 5, 10; shed¹; shield 1, 2
  shelter, seek *or* **take,** shelter 5
sheltered, hothouse 2; safe 1; secure 1; snug
sheltering, protective
shelve°, defer¹; delay 1; pension 2; postpone; procrastinate 1; put 21a; slant 4; suspend 1; table 5
shelved, abeyance
shelved, be, fire 4; wait 2
shenanigan, caper 2
shenanigans, hanky-panky; lark 1; mischief 1; nonsense 2; trick 2
Sheol, hell 1
shepherd°, drive 5; escort 4; father 4; guide 1; herd 4; pilot 3; see 11
sherd, chip 1; fragment 1; morsel 2; piece 1; relic 2; scrap¹ 1; splinter 1
sherlock, detective; sleuth
shibboleth°, password; symbol
shield°, cover 1; defend 1; defense 1; envelop 2; guard 1, 5; harbor 2; housing 2; insulate 2; preserve 1; preventive 4; protect 1; protection 1; safeguard 2; save 2; screen 2, 7; secure 7; shade 5, 10; shadow 2; shelter 4; shroud 1, 3; umbrella 2; veil 2; watch 2
shielded, hothouse 2; safe 1; secure 1
shielding, protective
shift°, change 3, 7; deceit 2; displace 1; divert 2; dodge 1; fluctuate; get 28b; motion 1; move 1, 8; movement 1, 6; passage 5; remove 2, 6; second¹ 10; shuffle 3, 5; spell¹ 1; squirm; stint 1; subterfuge; succession 1; swing 4; switch 2, 4; tour 3; transfer 1; transplant; trend 3; turn 2, 29, 30; vary 3; weave 4
  shift (about), shuffle 1
  shift (for oneself), fend 1; manage 4; shift 2
shiftiness, gobbledegook 2; trickery
shifting, changeable 1; desultory; faithless 2; fluid 3; halting; irresolute; motion 1; removal 3
shiftless°, idle 3; indolent; lazy 1; slack 1; slothful
shiftlessness, idleness 1; indolence; sluggishness
shifty°, crafty; deceptive 2; desultory; devious 1; disingenuous; elusive 1; evasive; faithless 2; fly-by-night 2; foxy 1; fraudulent 2; furtive 2; insincere; politic 1; serpentine 1; shady 2; slippery 2; sly 1; smooth 8; sneaky; subtle 3; tricky 1; unscrupulous; wily
  shifty, be, shuffle 3
shillelagh, staff 1
shilling mark, line¹ 1
shilly-shally°, blow¹ 6; dawdle; delay 3; fence 4; hesitate 2; indecision; linger 4; oscillate; procrastinate 2; time 20
shilly-shallying, delay 6; hesitant 1; idleness 2; indecision; indecisive 1; indefinite 5; mind 13; uncertain 2; wishy-washy 1
shimmer°, blink 2; flare 1; flash 1, 4; flicker 1; glance 2, 5; gleam 1, 4; radiance 1; radiate 1; sheen; shine 1, 4; twinkle 1, 2
shimmering, luminous 1; radiant 1; scintillating 1; shimmer 2; shiny; twinkle 2
shimmery, shiny
shimmy, shake 2
shimmying, wriggle 4
shindig, dance 2; fête 1; party 1
shindy, dance 2; party 1
shine°, beam 3; brighten 2; excel; flash 4; glare 1; glaze 1, 2; gleam 1, 4; glisten; gloss¹ 1, 3; glow 4; lighten¹ 2; light¹ 3; polish 1, 5; radiance 1; radiate 1; rub 1; scour 1; shed² 2; sheen; shimmer 1, 2; sparkle 1; splendor 2; twinkle 1, 2
  shine to, take a, like² 1; shine 5
  shine up to, cultivate 4; flatter 1; kowtow
shined, lustrous
shingle, overlap 1; sign 5
shingled, stony 1
shingly, rocky¹ 1; scaly 1; stony 1
shininess, sheen
shining, bright 1; brilliant 1; clear 3; full 10; glassy 1; glossy 1; golden 3; light¹ 13; liquid 3; luminous 1; radiant 1; scintillating 1; silver 3; sleek 1; slick 1; twinkle 2
shinny (up), climb 1; shin
shinplasters, money 1
shin° (up), climb 1; shin
shiny°, bright 3; clear 3; glassy 1; glossy 1; golden 3; luminous 1; lustrous; radiant 1; silky; silver 3; sleek 1; slick 1; smooth 2; spotless 1
ship°, boat; craft 4; dispatch 2; forward 7; send 2; translate 4; transmit 1; transport 1; vessel 2
  ship aboard, board 6
  ship out, ship 3
shipload, cargo; freight 2
shipment, cargo; freight 1; load 2; passage 10; translation 4; transmission 1; transport 4
shipper, carrier 1
shipping, carriage 3; freight 1; passage 10; traffic 1; translation 4; transmission 1; transport 4
shipshape°, neat 1; order 10a; orderly 1; straight 7; taut 2; tidy 1; trim 1
shipwreck, cast 12; hulk 1; wreck 3
shire, local 2
shirk°, duck 3; evade 1; neglect 2; slack 4
shirker, idler; loafer; slacker; truant 1; wastrel 2
shirking, evasion 1; idleness 2; truant 2
shirr, gather 3; pucker 1, 2
shirring, pucker 2
shirt-pocket, portable
shirty, cross 6
shit, defecate; dung; effects; filth 1; flannel 1, 2; furniture 1; gab 2; gear 4; go 17; paraphernalia; stinker; stuff 2; thing 8c
shit creek (without (the vestige of) a paddle), up, trouble 9a
shit-faced, drunk 1
shithouse, privy 3
shitting green, panic-stricken
shitty, stinking 2
shivaree, jamboree; noise 1
shiver°, break 1; crack 5; crumble; crush 1; dash 1; quaver 1; quiver 1, 2; shake 1, 8; shatter 1; shudder 1, 2; totter; tremble 1, 2; vibrate
shivered, broken 1
shivering, chill 5; freezing; shake 8; shiver¹ 4; tremulous 1
shivers, the, shiver¹ 4
shlemiel, loser; sap¹ 2; victim 2; weakling
shlimazel, loser
shlimazl, loser
Shoah, carnage
shoal, bar 4; flat 14b; swarm 1
shoal(s), many 3; score 4; shallow 2
shock°, appall; astonish; astonishment; astound; blow² 2; bolt 6; bombshell; breath 3; brunt; catastrophe 2; daze 1; devastate 2; dismay 2, 3; dumbfound; electrify 1; floor 5; fright 2; frighten; horrify 2; jar² 1, 4; jolt 3, 5; misfortune 2; nonplus; outrage 2; overwhelm 3; petrify 2; revolt 3; rock² 3; scandalize; scare 1, 3; shake 5; sicken 2; stagger 2; startle; stun 2; surprise 1, 3; take 28; terrify; terror 1; turn 36
  shock, give (someone) a, scare 1
shocked, daze 4; dumbfounded; petrified 2; scared; sick 6; speechless 2; thunderstruck
shocker, bolt 6; bombshell; shock 3; surprise 4
shocking°, arresting; awesome; awful 2; disgraceful 2; dramatic 2; dreadful 2; flagrant; frightening; frightful 2; ghastly 1; glaring 1; grievous 2; grisly; gruesome; hideous 2; horrible 1; lurid 1; monstrous 2; obscene 2; outrageous 1, 3; rank² 3; scandalous 1; sensational 1; shameless; sick 5; sight 6b; startling; terrible 4; terrifying; tragic; traumatic;

show° (*cont.*)
turn 20e, 25a; appear 1; come 2
show up again, return 2
show, put on a, posture 4
show biz, stage 3; theater 3
show business, stage 3; theater 3
showdown°, crunch 2; skirmish 1
shower°, clean 9, 11a; hail^2 2; heap 4; lavish 5; pelt^1 1; rain 1, 3, 6, 7; scatter 1; smother 3; splash 1; spray^1 1, 2; storm 1; volley 1; wash 1, 11; water 6
shower (down) on, hail^2 1
showers, precipitation
showers, send to the, dismiss 1; eject 1
showery, wet 2
showily, gaily 1; well^1 4
showiness, flourish 4; frill 2; glare 3; glitter 4; pageantry
show(ing), exhibition; exposition 1; rendering
showing off, ostentation; swagger 3
showman°, daredevil 1
showmanship, virtuosity
shown, be, come 15b; prove 3
showoff°, boast 2; braggart; daredevil 1; poseur; pup; showy; talker 2; trendy 2
showoffish, boastful
show (one's) gratitude *or* **thanks** *or* **appreciation,** thank 1
show (someone) the door, expel 1; fire 11; rebuff 2; reject 2
show-stopper, sensation 3
show the way, pioneer 2
show to be, stamp 3
showy°, brave 2; camp^2 1; conspicuous 2; dashing 3; dramatic 3; elaborate 2; extravagant 4; flamboyant 2; flash 6; flashy 1; flowery; garish; gaudy; gorgeous 1; grandiose 1; jaunty 2; loud 2; luxuriant 3; ostentatious; pompous 1; pretentious 1; pushy; splendid 1; sporty; sumptuous; tawdry; theatrical 2
showy dress, finery
shred°, bit 1; crumb; fragment 1; fray^2; grate 1; modicum; morsel 2; particle; patch 1; piece 1; rag^1 1; remnant 1; rend 1; scrap^1 1; sliver; snip 3; splinter 1; tear 1; wisp
shredded, tattered
shred(s), odds 5; tatter 1
shreds, cut to, rout 2
shreds, in, tatter 2
shreds, into, piecemeal 2
shreds, pull to, pull 6
shrew°, bitch 1; devil 2; fury 3; hag; jade 2; nag^1 2; scold 2; witch 2
shrewd°, acute 5; arch 2; artful 2; astute 1; calculating; clever 2; crafty; cute 2; deep 5; designing; dexterous 2; experienced 1; farsighted 1; foxy 1; incisive 1; ingenious; intelligent; keen^1 6; knowing 1, 2; machiavellian; observant 1; politic 1; provident 1; prudent 1; quick 4; serpentine 1; sharp 3, 6; shifty; sly 1; smart 2; smooth 8; subtle 3; tactical; tricky 1; wily; wise 1
shrewdness, artifice 1; dexterity 2; finesse 1; gumption 1; ingenuity; intelligence 1; judgment 1; mind 1; penetration 2; trickery; wisdom 1
shrewish, short-tempered
shriek°, cry 4; gale 2; scream 1, 3; squawk 1; whoop 1, 2
shrieking, piercing 1
shrift, forgiveness 1
shrill°, brassy 2; grating 2; harsh 1; high 7; noisy; penetrating 2; piercing 1; raucous; sharp 7; strident
shrine, monument 1; sanctuary 1; sanctum 1; temple
shrink°, blink 3; contract 4; cringe 1; decline 2; decrease 1; diminish 1; doctor 1; dry 5; dwindle; melt 3, 4; minimize 1; recede 2; recoil 1; shrivel; start 5; therapist; waste 2; wilt 1
shrink (away *or* **back) (from** *or* **at),** draw 10; dread 1; fear 5; flinch; hang 5; hate 2; hesitate 1; loathe; pull 8a; scruple 2; shirk; shrink 2; shun; stick^1 12
shrink back, withdraw 1
shrink from, abhor
shrinkage, loss 2
shrinking, decrease 2; self-conscious; withdrawn 1
shrive, purify 2
shrivel°, dry 5; fade 2
shrivel (up *or* **away),** decrease 1; dwindle; parch; shrink 1; shrivel; wilt 1
shriveled, emaciated
shriveled (up), thin 1; wizened
shriveling, decrease 2; scorching 1
shroud°, cloak 2, 3; cover 2; dim 1; eclipse 1; envelop 1, 2; fog 5; hide^1 2; keep 14a; mantle 2, 3; mask 3; muffle 1; obscure 6, 7; pall^1 1; roll 9; screen 3; secrete^1; shade 12; submerge 3; swathe; veil 1, 2; wrap 1
shrouded, occult 1; secret 1; vague 3
shrub, plant 1
shrubs, brush^1 1; shrubbery
shrug off, brush^2 6; dismiss 2; laugh 2b; light^2 12; put 11; rid 2
shrunken, emaciated; haggard; lean^1 1; skinny; stunted; thin 1; wizened
shtick, act 2; concern 4; office 4; routine 2
shtuk, in, trouble 9a
shuck, hull 2, 3; pare 1; peel 1; shell 3
shudder°, quake 1; quaver 1; quiver 1, 2; shake 1, 8; shiver^1 1, 3; thrill 2; tremble 1, 2; vibrate
shudder at, abhor; fear 5; loathe
shuddering, shake 8; thrill 2; tremulous 1
shuffle°, drag 3; evade 2; fidget 1; hobble 1, 3; jumble 1; tangle 3; waddle; waffle 2; walk 1
shuffling, shambling; shuffle 5
shufty, glance 4
shufty at, have a, glance 1
shul, temple
shun°, avoid; boycott 1; cut 12; duck 3; flee 2; forgo 1; isolate; ostracize; reject 1; renounce; scorn 3; shirk; shoulder 2; sidestep; steer 2; swear 4
shunned, abandoned 1; forlorn 2; unpopular
shunt, sidetrack
shunted aside, abeyance
shush, hush 1, 2; shut 6b; silence 3
shush (up), pipe 7
shushed, silent 1
shut°, close 1, 2; secure 2; slam 1; unopened
shut away, close 19; separate 7
shut (down), close 5; fold 3; halt 2; shut 2, 4a; stall^1 1
shut in, enclose 1; keep 14b; picket 4; restrain 3; shut 3, 6a; siege 2; stake^1 4a
shut (off), block 3; close 6; reclusive; seal 3; separate 7; shut 4; silence 4; shut 2
shut out, bar 9; exclude 1; isolate; lock^1 8; preclude; shut 4b, 5
shut (up), bar 8; cage 2; close 8a, 10, 19; commit 2; hush 1; imprison; pen^2 2; pipe 7; shut 1, 2, 6; tongue 3; wrap 3b; halt 2; keep 14b; restrain 3
shuteye, doze 2; nap^1 3; repose 2; rest^1 1
shuteye, catch *or* **grab some,** doze 1; rest^1 6; retire 3
shuteye, get some, nap^1 1
shut-in, invalid^1 2
shutout, exception 2; exclusion 1
shutter(s), blind 7
shutters, put up the, close 5
Shut your face!, hush 1; wrap 3b
Shut your gob!, hush 1
Shut your mouth!, wrap 3b
Shut your trap!, hush 1; wrap 3b
shy°, backward 1; bashful 1; cast 1, 7; coy; deflect; fainthearted 1; fearful 2; gingerly 2; hurl; inhibited; lob 1; meek 1; modest 1; mousy 2; peg 5; pull 8a; put 7; reticent; retiring; self-conscious; shamefaced 1; sling 1; start 5; tentative 2; throw 1; timid; toss 1, 6; tremulous 2; withdrawn 1
shy (away) (from), back 3; duck 3; flinch; recoil 1; shrink 2; shun
shy (of), short 6; wanting 2
shyly, fearfully 1; gingerly 1
shy-making, outrageous 3
shyness, humility; shame 4
shyster, lawyer; mouthpiece 3
siamoise, couch 1
Siberia, send (someone) to, banish; isolate
Siberian, cold 1; freezing; frigid 1; icy 1; polar 1; wintry 1
sibilance, fizz 3; hiss 1; rustle 2
sibilate, rustle 1
sibilation, rustle 2
sibling, brother
sibyl, fortuneteller; oracle 1; prophet; psychic 3; seer; witch 1
sibylic, ominous 3
sibylline, ominous 3; prophetic
sic, letter 4; literally 1
sick°, diseased; frail 2; ghastly 3; ghoulish 2; ill 1; indisposed 1; infirm 1; invalid^1 1; low^1 4; morbid 1; off-color 1; poorly 2; psychotic 1; queasy 2; queer 3; rotten 5; unhealthy 1; unsound 2; unwholesome 3; upset 7; weather 2
sick, be *or* **feel,** ail 2; heave 4; throw 9a
sick, make, ail 1; sicken 2
sick, take, sicken 1
sick (and tired) (of), disgusted; jaded 2; tired 2; weary 3
sick and tired (with *or* **of), be** *or* **become,** weary 4
sick list, on the, ill 1; indisposed 1; sick 2
sick (of), disgusted; sick 7
sick (to one's stomach), nauseated; sick 1
sick up, vomit
sick (with), fall, get 5; sicken 1
sick bay, hospital; infirmary
sicken°, disgust 1; glut 4; nauseate; offend 2; pall^2 1; put 21d; relapse 2; repel 2; revolt 3; shock 1; turn 17b
sickened, disgusted; nauseated; sick 6
sickening, disagreeable 2; disgusting; foul 1; grisly; hideous 1; horrible 1; loathsome; nasty 1; nauseous; obnoxious; obscene 2; offensive 3; repellent; repugnant;

skivvies, underclothes; unmentionable 3
skivvy, menial 3; slave 1, 3
skivvying, drudgery
Skoal!, bottom 5
skulduggery, chicanery; foul play; trickery
skulk, creep 3; lurk; prowl 1; slink; sneak 1; steal 3; walk 1
skulking, furtive 1; stealthy
  skulking about *or* **around,** prowl 3
skull, head 1
skunk, rout 2; stinker
sky°, atmosphere 1; firmament; heaven 2
skycap, porter[1] 1
sky-high, towering 1
skylark, frolic 3; lark 2
skylarking, frolic 1; fun 2; play 22
sky pilot, father 4; minister 1
skyrocket, rocket; soar 2
skyscraping, towering 1
skyward, aloft; overhead 1
slab°, block 1; brick 1; cake 2; plank; plaque 1; plate 3; sheet 3; slice 1; tablet 2
slabber, slaver[1] 1, 2
slack°, derelict 2; flabby 1; give 18; inattentive; inert 3; lax 1; lazy 1; leeway; limp[2] 1; loose 3; off 8; remiss; slothful; slow 6; tardy 2
  slack off *or* **up (on),** drop 11; let[1] 10; pay 11c; slack 3; slow 14; wind[2] 3a
slacken, brake 2; decrease 1; diminish 3; drop 11; ebb 2; lag 2; let[1] 9; liberalize 2; loose 12; mitigate; moderate 5; relax 1; remit 2; slack 3a; taper 2; temper 5
  slacken off *or* **up (on),** mitigate; pay 11c; reduce 2; relax 1; slack 3; slow 14; wind[2] 3a
slackening, decrease 2; relaxation 2
  slackening (off), ebb 4
slacker°, derelict 3; idler
slackness, neglect 4; slack 6
slacks, pants 2
slag, trash 3
slain, be, fall 6
slake°, quench 1; satiate 2; satisfy 2
slaked, jaded 2
slaking, refreshing
slam°, jam 3; pull 6; punch[1] 1, 2; ram 2; stroke 1
  slam into, run 26d
slammer, jail 1; prison; stir 7
slander°, abuse 3; aspersion; blacken 2; discredit 1, 5; dishonor 5; disparage 2; impeach 2; insult 1, 2; libel 2, 4; scandal 3; slur 1; smear 2, 4; stigmatize; tale 3; vilify
slanderous°, abusive 1; foul 6; injurious 2; poisonous 2; scandalous 2; vicious 2; vituperative
slang, cant 2; dialect; jargon 1
slanging match, row[2] 1
slant°, angle[1] 1, 2; bias 2; color 4; decline 8; descend 2; distort 2; garble 1; inclination 2; incline 1, 4; lean[2] 2; list[2] 1, 2; outlook 1; phase 4; point of view 1; position 3; prejudice 3; slope 1, 2; sway 2; tilt 1, 3; tip[2] 2; twist 2, 8; viewpoint; weight 6
  slant (downward(s)), decline 3
  slant (upward(s)), rise 6
slanted, oblique 1
slanting, inclined 3; oblique 1
slap°, clap 2, 3, 5; flap 1; full 14; hit 1; pelt[1] 4; plumb 4; precisely 1; slam 2; spank; strike 1
  slap down, squelch 1
  slap on, slap 6
  slap together, throw 9d; whip 7b
slapbang, pell-mell 1
slapdash, careless 2; cursory; haphazard 2; hasty 3; hurried 2; loose 5; makeshift 1; pell-mell 1, 2; slipshod; superficial 3
slaphappy, scatterbrained
slap (in the face), brushoff; cut 23; dig 8; indignity; insult 2; knock 10; shaft 3; slap 5; taunt 2
slap on the wrist, reprimand 1, 2
slap someone's wrist, punish 1; scold 1
slapstick, funny 1; zany 1
slash°, claw 2; cut 1, 8, 19; gash 1, 2; hack[1] 1, 3; incision; knife 2; lacerate; lash[1] 2; line[1] 1; lower[1] 1; mark 15b; reduce 5; rent[2]; rip 3; score 10; slit 1, 2; snip 2; split 7; streak 1, 4; stripe 1; tear 4; wound 1, 3
  slash, have a, urinate
  slash (back), pare 2
slat, flap 1; slap 2
slate, calendar 1; criticize 2; list[1] 1; pan 5; program 1, 4; pull 6; reprimand 2; roll 13; schedule 2; slam 3
slated, card 11
slating, reprimand 1; tongue-lashing
slattern°, jade 2
slatternly, dirty 1; untidy
slaughter°, bloodshed; butcher 3; carnage; destruction 2; exterminate; gore[1]; kill 1; killing 1; massacre 1, 2; mow 2; murder 2, 3; purge 4; removal 2; remove 4; whip 2; zap
slaughterer, butcher 1
slaughtering, massacre 1
slaughterous, sanguinary 1
slave°, attendant 2; boy 2; captive 1; flunky 1; hack[2] 2; labor 5; menial 3; subordinate 2
slave (away), grind 4; overwork 2; plod 2; sweat 3; work 7
slave driver, master 1; oppressor; tyrant
slave girl, slave 1
slaveling, slave 1
slaver°, drivel 1; spit 1
slavery°, bondage; captivity; drudgery; labor 1; servitude
slave ship, slaver[2] 1
slave trade, slavery 2
slave trader, slaver[2] 1, 2
slavey, slave 1
slavish, base[2] 2; groveling; menial 1; obsequious; servile; submissive 2
slavishness, servility
slay, dispatch 3; eliminate 4; execute 3; kill 1; massacre 2; murder 3; polish 3b; remove 4; rub 6; shoot 3; slaughter 3; zap
slayer, killer 1; murderer
slaying, destruction 2; dispatch 7; execution 3; killing 1; massacre 1; murder 1, 2; removal 2; slaughter 2
sleazy°, cheap 3; mangy; seedy 1; sordid 4; tacky[2]
sleek°, glossy 1; silky; slick 1; slippery 1; smooth 2
sleep°, doze 1, 2; repose 2; rest[1] 1, 6; retire 3
  sleep, go to, flake 3a; sack 2; turn 15a
  sleep, put to, kill 1; put 16d
  sleep around, mess 4b; play 9b; run 23
  sleep (with), lay[1] 8
sleeper, tie 12
sleep-inducing, monotonous; narcotic 1; restful 1; sedative 2
sleepiness, lethargy 2; torpor
sleeping, dormant 1; silent 4
sleeping pill, sedative 1
sleepless°, wakeful 1
sleepy°, dormant 1; dreamy 3; drowsy; leaden 4; lethargic 2; slow 9; torpid
sleet, precipitation; storm 5
sleety, foul 8
sleight of hand, dexterity 1; hocus-pocus 3; magic 2; trick 5
slender°, fine[1] 6; lean[1] 1; narrow 1; remote 6; slight 2, 3; spare 3; tenuous 1; thin 1; trim 2; willowy 2
slenderize, reduce 3; slim 3
sleuth°, detective; operative 3
sleuthhound, operative 3
slew, heap 2; number 2; wind[2] 1
slew(s), mass 2; number 2; plenty 1
slice°, cut 2; piece 2; portion 1; rend 2; section 2; segment 1; share 2; slab; slash 4; slit 1
  slice off, sever 1
slicer, slice 3
slick°, devious 1; dexterous 2; disingenuous; flashy 2; fluent; foxy 1; glassy 1; glib; glossy 2; greasy 2; insincere; mealy-mouthed; resourceful; scheming; serpentine 1; shifty; sleek 1; slippery 1, 2; smooth 2, 6, 8; sophisticated 1; subtle 3; wily
  slick down, slick 5
slickness, fluency
slick operator, devil 4; operator 3
slide°, chute 2; coast 2; deteriorate 1; glide; photograph 1; plane 4; roll 2; sag 2, 3; sail 3; skim 3; slip[1] 1; slither; stream 4; worsen 2
  slide, let, neglect 2; omit 2; slide 4
  slide by (*or* **away),** elapse; fly 4
sliding, downward 1
slight°, belittle; cut 3, 12, 23; depreciate 2; dishonor 1, 4; disregard 2; fail 2; faint 1; feeble 1; flimsy 1, 2; fragile; frail 2; frivolous 2; immaterial 1; imperceptible 2; inconsequential; indifferent 4; indignity; injure 2; insubstantial 1; insult 1, 2; light[2] 2, 4, 5, 6; little 1; marginal 1; minor 2; minute[2] 2; neglect 1; negligible; nice 3; nodding; offend 1; offense 2; outrage 3; overlook 1; petite; petty 1; poor 5; put 16e; put-down; rebuff 1, 2; remote 5, 6; shallow 1; short 1; skin-deep; sleazy 1; slender 1, 2, 3; slur 1; small 2, 5; superficial 1, 2; thin 1, 2, 4; tiny; token 5; trifling; trivialize; undersized; wound 2, 4
slightest, jot 2; minimal; minimum 2
slighting, neglect 3; personal 4
slightly°, little 8; moderately; rather 1; somewhat
slim°, diet[1] 3; lean[1] 1; narrow 1; outside 8; remote 6; slender 1, 2, 3; slight 3; spare 3; thin 1; trim 2; willowy 2
  slim (down), reduce 3
slime, dirt 1; filth 1; grime; mire 2; muck 2; mud; ooze 1; sludge
sliminess, servility
slimy°, clammy 1; filthy 1; flattering 2; ingratiating; insincere; muddy 1; obsequious; oily 1; pharisaic; servile; sleek 3; slippery 1; smooth 6; sordid 3; subtle 3
sling°, cast 7; fling 1; fly 8a; heave 2; hurl; launch 3; peg 5; pitch[1] 1; slap 3; string 9; throw 1; toss 1
slingshot, sling 2

slink°, creep 3; lurk; prowl 1; pussyfoot 1; slide 2; slither; sneak 1; walk 1
slinking about *or* **around,** prowl 3
slip°, blunder 2; cutting 3; deteriorate 1; ease 7; error 1; fluff 3; glide; impropriety 4; indiscretion 2; lapse 1, 4; misstep 1; miss[1] 5; mistake 1; offense 1; peccadillo; plane 4; roll 2; runaround; sag 2; sail 3; shoot 5; slide 1, 2; slither; sliver; slump 2; solecism; steal 3; stream 4; stumble 1, 2; tongue 4; trip 1, 2, 5; tumble 4; worsen 2
  slip, give (someone) the, slip[1] 6
  slip, let, disclose 1; give 12b; leak 5; let[1] 7a; lose 4; make 33; reveal; slip[1] 4; tell[1] 2
  slip away, go 11; lapse 6; sink 2
  slip away from, elude 1
  slip away *or* **by,** elapse; fly 4; pass 11; slip[1] 5
  slip away *or* **off** *or* **out,** escape 1; slip[1] 6
  slip back, relapse 1
  slip (by), let, miss[1] 4
  slip in, get 39c; slip[1] 7
  slip into, put 22a
  slip through one's fingers, finger 9
  slip (to), give the, elude 1; lose 5; shake 7
  slip (up), err 1; nod 4; slip[1] 3
  slip up (on), miss[1] 4; overlook 1
slip of the pen, slip[1] 8
slip of the tongue°, misstatement 2; slip[1] 8; tongue 4; trip 2
slip one *or* **something over on (someone),** defraud; misinform; mislead; outsmart
slippery°, deceptive 2; devious 1; elusive 1; foxy 1; glassy 1; greasy 2; insincere; oily 1; precarious; scheming; shady 2; shifty; slick 1; slimy 1, 2; smooth 6; sneaky; tricky 1; unscrupulous; wily
slipping, downgrade 4; downward 1
  slipping away, dying; passing 1
slip road, turnoff 1
slipshod°, derelict 2; haphazard 2; inefficient 2; lax 1; perfunctory 2; poor 4; promiscuous 1; sleazy 1; sloppy 1
slip (someone) something, pay 10b
slip(-up), blunder 2; error 1; fault 2; lapse 1; misstep 2; miss[1] 5; peccadillo; slip[1] 8
slit°, break 26; chink; cleave; crack 1; cut 1; flaw 2; gash 1, 2; hole 2; incision; lance 2; opening 1; rent[2]; slash 1, 4; slot 1, 3; snip 2; split 7; tear 4; vent 1; wound 3
slither°, creep 1; slide 1; slip[1] 1; snake 3; wriggle 2
slithery, greasy 2; oily 1
sliver°, chip 1; crumb; flake 1; fragment 1; modicum; morsel 2; particle; piece 1; portion 1; scrap[1] 1; shred 1; slice 2; slip[2] 1; splinter 1
slob°, boor 2; yahoo
slobber, drivel 1; slaver[1] 1
slobbovian, boor 2; slob
slog, drag 3; drudgery; march 3; plod 1; sweat 3; tramp 1, 3; walk 1, 7
  slog (away), work 7
slogan°, byword; cry 7; device 3; legend 4; maxim; moral 3; motto; phrase 2; precept 2; saw
slogging, drudgery; sweat 6
slop, dirt 1
  slop over, spill 1
slope°, decline 8; descend 2; downgrade 3; drop 4; fall 4, 27; grade 4; hill 3; inclination 2; incline 1, 4; list[2] 2; mound 1; ramp; rise 13; run 45; slant 3, 4; tilt 1, 3
  slope (downward(s)), decline 3
  slope (upward(s)), rise 6
sloping, inclined 3; oblique 1
sloppiness, sentimentality
sloppy°, bedraggled; careless 3; derelict 2; dirty 4; disreputable 2; dowdy; filthy 2; foul 8; gooey 2; gushy; lax 2; loose 5, 6; mushy 2; romantic 3; sentimental 2; slipshod; sweet 7; unkempt; unprofessional 2; untidy
slop(s), garbage; swill 1
slosh over, run 33c
sloshed, drunk 1
sloshy, sloppy 2
slot°, appointment 3; compartment; groove; hole 2; niche 2; opening 1; place 3; split 7; stall[1] 2; track 3; vacancy 2; vent 1; void 5
sloth°, idleness 1; inactivity 1; indolence; inertia; lethargy 1; sluggishness; torpor
slothful°, idle 3; inactive 1; indolent; inert 3; lackadaisical 1; lazy 1; lethargic 1; shiftless; slack 1; supine 2; tardy 2; torpid
slothfulness, idleness 1; inactivity 1; indolence; inertia; sloth; sluggishness; torpor
slouch°, idler; laggard; sprawl 2; stoop 3
  slouch, no, slouch 3
slouching, stoop 3
slough, marsh; mire 1; morass 1; swamp 1
  slough off, mind 19
  slough over, forgive 1
sloven, slattern; slouch 3
slovenly, derelict 2; dirty 1; disreputable 2; dowdy; filthy 2; heel[1] 3; mangy; perfunctory 2; promiscuous 1; slipshod; sloppy 1; threadbare 1; unkempt; untidy
slow°, arrest 3; backward 2, 3; blind 2; bog 2; brake 2; check 1; deliberate 2; dense 3; discourage 3; dull 3; flat 11; gradual; hamper[1]; hinder 1; impede; inert 3; interfere 2; lackluster; lazy 2; lethargic 1; lingering 2; measured 1; obstruct 2; off 8; opaque 3; oppose 2; prevent; remiss; set 16; simple 4; sleepy 2; slothful; soft 7; stay[1] 4; stem[2] 1; stolid; stop 2; stunt[2]; tardy 2; thick 6; torpid
  slow burn, do a, rage 4; simmer 2; smolder
  slow (down *or* **up),** brake 2; delay 2; diminish 3; encumber 2; hold 22b; lag 2; relax 3; retard 1; slack 3b; slow 14; taper 2; wind[2] 3a
  slow (on the uptake), feebleminded; slow 7
slowcoach, laggard
slowdown, strike 20; tie-up 1
slowing, obstruction 2; prevention
slowly, degree 3; gradually; piecemeal 1; slow 11, 13; step 9
slow-moving, lazy 2; slow 1, 3; torpid
slow-paced, slow 1, 3; torpid
slowpoke, laggard
slowup, tie-up 1
slow-witted, backward 2; blind 2; daft 1; dense 3; dim 2; dull 1; feebleminded; foolish 2; obtuse 2; simple 4; slow 7; stupid 1; thick 6; weak 5; wooden 3
slow-wittedness, simplicity 4; stupidity 1
slubberdegullion, slob
sludge°, dirt 1; filth 1; muck 2; mud; ooze 1; silt 1; soil[1] 3
slug, box[2] 2, 3; drink 6; get 24; jab 2, 4; punch[1] 1, 2; rap 4; shot 2, 7; strike 1
slugabed, good-for-nothing 2; idler; loafer
sluggard, good-for-nothing 2; idler; laggard; slouch 3
sluggard(ly), slothful; slow 1
slugger, pugilist
sluggish, dormant 1; drowsy; dull 3; fatigued; flat 11; inactive 1; indolent; inert 3; lackadaisical 1; lazy 2; leaden 4; lethargic 1; listless; phlegmatic 1; slack 1; sleepy 1, 2; slothful; slow 1, 6; stagnant; supine 2; tardy 2; torpid
sluggishness°, fatigue 1; idleness 1; inactivity 1; indolence; inertia; lethargy 1; sloth; torpor
sluice, channel 1; race[1] 2
slum°, hole 3
slumber, doze 1; repose 2; rest[1] 1; sleep 1, 2
slumbering, dormant 1
slumberous, dormant 1; sleepy 1
slumlike, shabby 3
slummy, shabby 3; sleazy 2; sordid 3, 4
slump°, decline 6; depression 3; dip 2, 5; droop 2; eclipse 4; fall 2, 26; flag[2] 2; lapse 4; recession; sag 1, 2, 3; slouch 1, 2; sprawl 2; taper 2; twilight 2
slur°, aspersion; dig 8; discredit 1, 4, 5; disgrace 2, 4; dishonor 5; imputation; insult 2; knock 10; libel 2, 4; scandal 3; slander 1, 2; slight 7; taint 1
  slur on, cast a, denounce 3
  slur over, slur 3
slush, ooze 1
slushiness, sentimentality
slushy, gooey 2; gushy; maudlin; mushy 2; sentimental 2; sloppy 2, 3
slut, jade 2; slattern; tart[2] 2; wanton 5
sly°, arch 2; artful 1; astute 1; calculating; clever 2; crafty; deceitful; designing; devious 1; disingenuous; dry 3; foxy 1; furtive 2; insincere; knowing 1; oblique 2; scheming; serpentine 1; sharp 6; shrewd; slippery 2; smooth 8; sneaky; stealthy; subtle 3; surreptitious; tricky 1; two-faced; unscrupulous; wicked 5; wily
  sly, on the, incognito 2; private 6; secret 4; secretly; sly 3
slyboots, devil 4
sly dog, devil 4
slyly, back 6; secretly; sly 3
slyness, art 5; artifice 1; deceit 1; stealth; subtlety 2; trickery
smack, clip[2] 3, 6; full 14; get 24; hit 1, 10; jab 2, 4; kiss 1, 4; knock 9; pelt[1] 4; poke 2, 6; precisely 1; punch[1] 1, 2; savor 1; slam 2; slap 1, 4, 6; spank; strike 1; stroke 1; tang 2; touch 15
  smack into, collide 2
  smack of, resemble
smacker, kiss 4
smack in the eye, slap 5
smack one's lips over, savor 3
small°, compact 2; diminutive; elfin 2; frugal 3; insubstantial 1; limited 1; little 1, 2, 3, 5; lower[1] 8; low[1] 1, 10; marginal 1; mean[2] 1; miniature; minor 2; minute[2] 1, 2; negligible; nice 3; nominal 2;

small° (*cont.*) overcritical; paltry; petite; pitiful 2; portable; puny 1, 2; remote 6; scanty 2; short 1; slender 2; slight 1, 3; small-minded; small-time; spare 5; stunted; tight 2; tiny; toy 5; undersized; wee 1

small, feel, small 6

small-boned, petite

small change, money 1

small clothes, underclothes; unmentionable 3; pants 1

small craft, boat

smaller, low[1] 10; minor 1

smallest, minimal

small fry, boy 1; cog 2

smallholder, farmer

smallholding, farm 1

small(-minded), little 6; narrow-minded; overcritical; petty 2

smallness, triviality 1

small potatoes, pittance

small print, joker 2

smalls, pants 1; underclothes; unmentionable 3

small-scale, miniature; small 4; small-time

small screen, screen 5; television

small talk, chat 1; gossip 1; patter[2] 2

small-time°, frivolous 1; minor 2; petty 1

small-town, provincial 2

smarminess, servility

smarmy, flattering 2; greasy 2; ingratiating; obsequious; oily 2; pharisaic; pious 2; sanctimonious; servile; sleek 3; slick 2; slimy 2; smooth 6; subtle 3

smart°, ache 1; advisable; brave 2; bright 6; brilliant 4; chic 1; clever 3; dapper; dashing 2; dressy 2; elegant 1; exquisite 3; fashionable; flash 6; glamorous 2; hurt 2; ingenious; intelligent; jaunty 2; keen[1] 6; minute[1] 3; neat 3; nifty 1; nimble 1, 2; penetrating 1; precocious; prick 4; prickle 3; quick 4; quick-witted; rakish; resourceful; sharp 3, 9; shrewd; snappy 1, 2; spanking 1, 2; sporty; spruce 1; stylish; subtle 3; suffer 1; swanky; swell 7; swish 3; tactical; taut 2; tight 4; trim 1; well-advised; well-groomed; wise 1

smart alec, wise guy

smart-aleck(y), fresh 8

smart-ass, wise guy

smart-ass(ed), insolent; pert

smarten (up), dress 1; groom 3; perk up; spruce 2; decorate 1

smartie, wise guy

smartie-pants, wise guy

smarting, ache 3; angry 2; pain 1; painful 1; smart 6, 8; sore 1; tender[1] 8

smartness, ingenuity; panache; style 4

smash, break 4; bump 3; clash 2; collision; crack 5; crash 2, 4; crush 1; dash 1; demolish 1; destroy 1; hack[1] 1; impact 1; knockout 2; pileup 1; pulverize 2; rout 2; ruin 7; shatter 1; slam 2; slaughter 4; whip 2; wreck 1

smash into, collide 2; hit 3; run 26d; strike 3

smashed, broken 1; drunk 1; piece 7; stinking 3

smash (hit), hit 11; knockout 2

smashing, admirable; beautiful 1, 2; capital 6; divine 3; excellent; extraordinary 2; fabulous 3; good 2; gorgeous 2; grand 5; great 12; heavenly 2; impact 1; marvelous; neat 5; outstanding 1; sensational 3; splendid 3; striking; superb; superlative; terrific 2; world 8

smashingly, beautifully 1, 2

smashup, collision; pileup 1

smattering, scattering; tang 2; touch 15

smear°, blacken 2; blemish 3; blot 1; dirty 7; discredit 1, 4; disgrace 2, 4; dishonor 5; foul 14; libel 2, 4; mark 1, 10; mire 4; plaster; rub 2; scandal 3; shame 8; slander 2; slanderous; slur 1; soil[1] 1, 2; splash 3; spread 7; streak 1, 4; taint 2; vilify

smell°, aroma 1, 2; detect 2; distinguish 3; fragrance; fume 3; odor 1; perfume 1; reek 3; savor 2; scent 1, 3; sniff 3; tang 1; touch 15; trail 2

smell of, give 14

smell out, scrounge 1; track 8

smell (to high heaven), reek 1

smell a rat, distrust 1

smelly°, foul 3; rancid; rank[2] 4; stinking 1; strong 3

smidgen, blob; dab 2; dash 7; drop 2; grain 3; jot 2; little 10; modicum; morsel 2; particle; pinch 7; shade 6; speck; spot 3; trifle 2

smile°, lighten[1] 2

smile radiantly, beam 3

smile upon, favor 8

smiling, sunny 2

smirch, blot 1; discredit 1; disgrace 2, 4; dishonor 5; libel 2; muddy 5; scandal 3; stain 1, 5; stigma; sully; taint 2

smirk°, sneer 1

smite, assault 5; batter 1; beat 1; hit 1; lash[1] 3; move 4; poke 2; rend 3; strike 1; visit 2

smithereens, fragment 2

smithereens, in, piece 7

smithereens, smash (in)to, shiver[2]; splinter 2

smitten°, infatuated

smitten (by), stricken 1

smock, bunch 4; shift 6

smocking, ruffle 1

smog, fog 1; fume 3; mist 1; vapor 1

smoggy, hazy 1; thick 3

smoke, cure 3; fag 5; fume 2, 3; preserve 3; puff 5; reek 2, 4; use 3; vapor 1; whiskey

smoke, go up in, collapse 2; fail 1; miscarry

smoke out, discover 1

smoke screen, blind 8; cover 15; film 3

smoke stack, stack 4

smoking gun, evidence 3

smoky, dingy; gray 1; opaque 1; thick 3

smolder°, boil[1] 2; burn 1; fester 2; fume 1; rage 4; seethe 2; simmer 2; stew 3

smoldering, dirty 5; glowing 1

smooch, cuddle 2; kiss 1, 4; pet[1] 4

smooth°, bland 1; calm 3; clean 3; debonair 1; devious 1; disingenuous; ease 8; effortless; eloquent 1; even 1, 13; facilitate; flat 1; fluent; flush[2] 1; foxy 1; gentle 1; glassy 1; glib; glossy 1; graceful 1; greasy 2; grind 2; level 1; offhand 1, 2; oily 1, 2; plain 1; plane 3; plausible 2; polish 1; press 4; quiet 2; regular 5; rub 1, 2; shifty; silky; sleek 1; slick 1, 2, 3, 5; slippery 1; soft 4, 8; spread 7; still 1; straight 3; streamlined 1, 3; suave; tranquil; tuneful; uniform 2; wily

smooth (out *or* away), flatten 1; level 8; roll 7; smooth 9

smooth over, gloss[1] 4; smooth 12

smoothen, smooth 12

smoothie, charmer; devil 4; operator 3

smoothly, easily 1; freely 5; readily 2; swimmingly

smoothness, facility 1; fluency; polish 3; savoir-faire

smooth operator, devil 4; operator 3

smooth-shaven, smooth 4

smooth-spoken, glib; slick 2; smooth 6

smooth talker, charmer

smooth-talking, glib

smooth the way for, pave 2

smooth-tongued, glib

smorgasbord, hors d'oeuvre; mess 2; miscellany; potpourri

smother°, choke 1, 3; gulp 2; keep 14a; kill 3; put 23c; quench 2; rout 2; silence 4; stifle 1; suppress 2; swallow 4

smothered, dead 5

smothering, suppression

smudge, blacken 1; blot 1, 2; mark 1, 10; mire 4; smear 1, 3; splash 3; spot 1, 7; stain 4

smudged, dirty 1

smudgy, dingy

smug°, comfortable 2; conceited; goody-goody; haughty; inflated 1; proud 2; self-important; self-righteous; slick 2; snobbish

smuggle, run 14

smugness, pride 2; self-esteem 1; snobbery

smut, dirt 3; filth 3; impurity 2; mold[2]; pornography; ribaldry

smutch, stain 1, 4

smuttiness, ribaldry

smutty, abusive 1; bawdy; blue 2; coarse 3; dirty 2; evil-minded 1; filthy 3; foul 5; gross 3; immodest 1; immoral 2; indecent 2; lascivious 2; lewd; low[1] 3; nasty 3; naughty 3; obscene 1; outrageous 3; pornographic; profane 3; prurient 2; racy 2; rank[2] 3; rough 6; rude 3; sexy 2; suggestive 2; vulgar 2; wicked 3

smutty language, profanity

snack°, bite 3; eat

snack, have a, dine

snack bar, café

snack(s), refreshment 1

snaffle, pilfer; pocket 4; swipe 2

snafu, bungle; confused 2; fluff 4; foul 16b; hash 2; mess 3; mix-up; muddle 4

snag°, catch 18; clutch 1; complication 2; deterrent; difficulty 2; drawback; entangle 1; fly 10; foul 13; grab 1; hindrance 1; hitch 4; holdup 2; hook 6; hurdle 1; impediment; interference 2; joker 2; kink 3; liability 3; obstacle; obstruction 1; pitfall 2; rub 11; snarl[2] 3; stumbling block; tangle 1, 3

snag(s), trammel 1

snail, laggard

snaillike, slow 1

snake°, coil 1; meander 1; slither; turn 11; twist 4; wind[2] 1; wriggle 2

snake-hipped, slender 1

snake in the grass, snake 2; stinker; traitor; turncoat; villain

snakelike, serpentine 2

snake-oil artist, cheat 1

snaking, serpentine 2

snap°, breeze 2; clap 5; crack 2, 4; flip 1; impulsive; initiative 2; nip[1] 1; photograph 1, 2; picnic 2; piece 9;

social conventions, manner 3; propriety 3
social graces, manner 3
socialist(ic), left 2
socialize°, get 29b, 50b; go 33c; hobnob; mingle 2; mix 2; scene 6; steady 8; step 17b; travel 4
  socialize (with), associate 1b; fraternize; go 21c, 39b; hang 4b; rub 7; see 7; shoulder 4
social security, pension 1
  Social Security, go on, retire 2
societal, public 2; social 1
society°, association 1; body 6; brotherhood 2; circle 2; clan 2; club 2; combination 2; companionship; company 1; federation; fellowship 2; folk; fraternity 1, 3; group 1; humanity 1; institute 1; league 1; order 9; organization 3; people 4; presence 2; public 8; ring[1] 3; sphere 2; union 2; world 2
sock, box[2] 2, 3; clip[2] 3, 6; get 24; hit 1, 10; jab 2, 4; punch[1] 1, 2; rap 4; strike 1
  sock away, salt 8
socking great, hefty 3
socle, pedestal 1
sod, devil 3; earth 2; ground 1; heel[1] 2; land 2; lawn; stinker; turf 1
  sod, piece of, clod 1
  sod off, clear 29
soda, fizz 4
sodality, fraternity 2; order 9; society 5; solidarity
soda (pop), pop 5
soda (water), fizz 4; pop 5; water 1
sodden, sloppy 2; soaking 2; steamy 1; wet 1
so-designated, titular
sodium chloride, salt 1
sodomite, bugger 1; profligate 3
sofa, couch 1; lounge 4; seat 1
soft°, delicate 6; easy 4; faint 1; feminine 1; flabby 1; fluffy 1; gentle 1; inaudible; limp[2] 1; low[1] 13; mellow 1, 3; mushy 1; noiseless; plastic 1; silky; slack 2; smooth 5; soothing 1; tender[1] 2, 5; weak 7; yielding 1
  soft (in the head), feebleminded; foolish 2; soft 6
  soft on *or* **about,** fond 3
  soft spot for, having a, partial 3
softcover, book 1
soft drink, fizz 4; pop 5
soften°, blunt 4; cushion 2; deaden 2; hush 3; liberalize 2; lower[1] 5; mellow 5; melt 1, 2; mitigate; moderate 5; modify 2; modulate; relent; relieve 1; smooth 12; subdue 3; sweeten 2; tame 6; temper 5; thaw 2; tone 5; water 7; weaken 3
  soften the sound of, turn 14b
  soften up, soften 1
softened, mellow 3; soft 4
softening, mild 3
softheaded, weak 5
softhearted°, benign 1; merciful; tender[1] 6; forgiving
softheartedness, mercy 1
softly, quietly 1
softness, expression 4
soft part, pulp 1
soft-pedal, hush 3; mute 3; soften 3; subdue 3; tone 5
soft sell, promotion 4
soft soap, cajolery; enticement 2; flannel 1; flattery; cajole; entice; flannel 2; flatter 1; play 18; romance 7
soft spot, liking 1; partiality 2; weakness 4
softy, sissy; weakling; wet 6
soggy, humid; moist 2; muggy; sloppy 2; watery 3
soi-disant, nominal 1; professed 1; self-styled; so-called 1; titular
soigné(e), cultivated; debonair 1; elegant 1; polished 2; smart 3; smooth 6; sophisticated 1
soil°, blacken 2; contaminate; dirt 1, 2; dirty 7; earth 2; filth 1; foul 12, 14; ground 1; land 2; mire 4; mold[3]; muddy 5; pollute 1; smear 1, 2; spot 7; stain 5; sully; taint 2; tarnish
soiled, bedraggled; dingy; dirty 1; filthy 2; foul 2; impure 1; muddy 1; spotty 1
soiling, pollution
soil stack, stack 4
soirée, party 1; reception 2
sojourn°, abide 2; residence 2; run 36; stay[1] 2, 7; stop 5, 7; tarry 2; visit 3
sojourner, rover
sojourning, visitation 1
Sol, sun 1
solace°, cheer 3, 5; comfort 1, 2; console
solder, attach 1; cement 1, 2; connect 3; stick[1] 4; tack 5; weld 1
  solder (together), join 1; unite 3
soldier°, gangster; militant 3
  soldier on, soldier 4; stick[1] 14
soldierly, martial 2; military 1
soldier of fortune, adventurer 1; daredevil 1; mercenary 3
soldiers, force 3; rank[1] 5; reinforcement 3
soldiery, military 2
sold on, be, sell 5
sole°, exclusive 3; individual 1; lonely 1; one 1; only 1; single 2; solitary 1
solecism°, error 1; fallacy; misstatement 2; misuse 2
solely, alone 4; entirely 2; just 4; merely 2; only 2; particularly 2; proper 7; simply 1; wholly 2
solemn°, ceremonial 2; ceremonious 1; devout 2; dignified; dismal; earnest 1; formal 1, 2; funereal; grave[2] 1; grim 4; measured 1; official 2; sacred 3; sedate 1; self-contained 1; serious 1; sober 2; somber 3; staid; state 6; stately; subdued 2
solemnity°, ceremony 1; dignity 1; gravity 3; rite; severity 2; sobriety 2
solemnization, celebration 1; feast 2; installation 1; rite
solemnize, celebrate 1; commemorate; keep 10; observe 5
solemnness, solemnity
solicit°, appeal 1; apply 7; approach 3; beg 2; canvass 1; desire 2; dun; pander 2; petition 2; pimp 2; plead 2; pray 1; proposition 3; raise 12; request 1; seek 3; sue 2
  solicit (from), ask 3
solicitation, appeal 3; application 4; canvass 3; collection 1; petition 1; plea 1; request 2; suit 5; supplication 1
soliciting, supplication 2
solicitor, advocate 3; lawyer; pander 3
solicitous°, anxious 1; attentive 2; concerned 1; considerate; good 6; paternal 1; supplicant 1; sweet 6; tender[1] 6; thoughtful 1; urgent 2
  solicitous (of *or* **to** *or* **(toward(s)),** sympathetic 1
solicitously, warmly 2
solicitousness, sympathy 1
solicitude°, anxiety 1; care 2; concern 6; consideration 1; fear 4; love 4; thought 4
solid°, actual 1; close 15; compact 1; decided 1; dense 1; fast[1] 4; firm 1, 2; fixed 1; full 1; good 8, 9; hard 1; hearty 4; intact; large 1; material 5; monolithic; physical; pure 1; rocky[1] 2; safe 3; secure 2, 3; solvent; sound[2] 3, 5; splendid 3; square 5; stable 1; stalwart 1; staunch 2; steady 1; stiff 1, 10; stocky; strong 12; sturdy 1; substantial 2, 3; superb; sure 2; tangible; thick 1, 2, 5; watertight 2; yeomanly
solidarity°, fraternity 2; unity 1
Solidarność, solidarity
solidified, hard 1; stiff 1
solidify°, cake 3; firm 5; fix 2, 7, 10; freeze 2; harden 1; jell 1; set 3; stiffen 1; thicken
solidity, body 7; stability 1, 2; strength 6; substance 4
solidly, fast[1] 7; firmly 1; steady 7; surely 2; tight 11
solids, dregs 1
solidus, line[1] 1
solipsistic, egoistic
solitarily, alone 3
solitariness, privacy 1; solitude 1
solitary°, alone 1; deserted; desolate 1; hermit; individual 1; isolated 1; lonely 1; one 1; only 1; private 4; reclusive; secluded 1; separate 5, 7; single 2, 3; single-handed 1; sole; solo 2; unique 1; unsocial; withdrawn 2
solitary confinement, solitary 2
solitude°, privacy 1; retreat 2
solo°, alone 1, 3; lonely 3; single-handed 1, 2; solitary 1; unaccompanied
soloist, singer; vocalist
Solomon, sage 2; thinker
So long!, farewell 3; goodbye
solution°, answer 3; combination 3; determination 2; explanation 3; fluid 1; interpretation 1; liquid 1; remedy 2; resolution 4
solve°, decipher 1; figure 12b; puzzle 3; resolve 2; satisfy 2; sort 10a; work 19c
  solve, try to, tackle 3
solvency, credit 3
solvent°, economic 2; flush[2] 3; liquid 4; strong 11
  solvent, remain, survive 1
solving, resolution 4; solution 1
somatic, physical; tangible
somber°, black 4; bleak 1; dark 3; dim 1; dismal; doleful; drab; dreary 1; dull 5; funereal; grave[2] 1; gray 2; grim 4; heavy 5; leaden 3; morbid 3; mournful 1; murky; obscure 1; overcast; sad 2; serious 1; sober 3; solemn 1; stern 2; sullen; sunless; twilight 4
  somber air *or* **mood** *or* **atmosphere,** pall[1] 2
somberly, sadly 2
somberness, gravity 3
some, divers; few 2; part 1; several 1
somebody°, figure 6; human 4; legend 2; name 3; personage; personality 2; star 2
someday, sometime 2; time 18b
somehow (or other), hook 3; somehow
someone, human 4; somebody 1
somersault, roll 1
somerset, roll 1
something, point 12
  something, do, finger 6; step 11
  something like, way 10b

something of oneself, make, rise 8
some time, soon 5; past 2; previous 1; time 18b
some time ago, once 1; previously
sometimes°, now 4; occasionally; once 4; time 14
someway, hook 3; somehow
somewhat°, degree 4a; fairly 1; moderately; part 9; partially; pretty 3; quite 2; rather 1; relatively; slightly
somewhere else, elsewhere
sommelier (des vins), servant 1; waiter
somnambulant, sleepwalking 2
somnambulation, sleepwalking 1
somnambulism, sleepwalking 1
somnolence, inactivity 1; lethargy 2; torpor
somnolent, dormant 1; drowsy; inactive 1; lethargic 2; narcotic 1; restful 1; sleepy 1; slow 9; torpid
son, child 1; descendant; issue 7
song°, air 4; chant 1; jingle 3; lay[3]; melody 1; one 5; piece 4; poem; refrain[2]; rhyme 1; strain[1] 8; tune 1
  song, for a, cheap 5, 6, 7; song 2
song and dance, line[1] 18; moonshine 2; palaver 1; song 3; speech 2
songbird, singer; vocalist
songlike, lyric 1
song of praise, praise 2
songster, singer
sonneteer, poet
son of a bitch, rogue 1; stinker; villain
son of a gun, rogue 1
Son of God, savior 2
sonority, tone 2
sonorous, deep 7; loud 1; rich 7; rotund 2; round 7
sonorousness, tone 2
sons and daughters, progeny
soon°, directly 3; presently; quickly 3; shortly 1, 2; sometime 2; time 18b
  soon, too, premature 2; prematurely 1; previous 3
  soon to, verge[1] 2
sooner, first 4; rather 2
  sooner, had, have 11
  sooner than, before 7; by 4
sooner or later°, eventually; sometime 3; soon 5; time 18b
soot, dirt 1; grime
soothe, blunt 4; calm 5; charm 6; comfort 1; console; content[2] 4; deaden 2; ease 5; humor 4; hush 5; lull 3; mitigate; moderate 5; quell 2; relieve 1; remedy 3; salve 3; settle 8; silence 3; smooth 12; soften 1; solace 2; still 9; stroke 10; temper 5; tranquilize
soothing°, bland 1; dreamy 3; gentle 1; mild 3; restful 1; sedative 2; smooth 5; soft 4, 9; tender[1] 5
soothsayer, fortuneteller; oracle 1; prophet; psychic 3; seer
soothsaying, prophecy 1
sooty, black 1; dingy; dirty 1; gray 1
SOP, mode[1] 1; procedure; technique 1
sophism, fallacy; quibble 3
sophistic(al), circular 3; deceptive 2; evasive; plausible 2; smooth 8; specious; subtle 2
sophisticated°, blasé 2; chic 1; cultivated; elegant 1; enlightened; experienced 2; genteel 3; highbrow 2; knowledgeable 2; mature 1; mill 3; oily 2; polished 2; refined 1, 2, 3; ripe 2; steady 6; suave; well-bred; worldly 2
sophisticatedness, charm 2; style 4
sophistication°, charm 2; civilization 1; culture 1; experience 3; panache; presence 5; refinement 1; savoir-faire; savoir-vivre; style 4; subtlety 1
sophistry, chicanery; deception 1; evasion 2; quibble 2
sophomoric, childish; puerile; young 3
soporific, boring; drug 2; leaden 4; monotonous; narcotic 1, 2; restful 1; sedative 1, 2; sleepy 2; slow 9; tedious; tiresome 1
soppiness, sentimentality
sopping, sloppy 2; soaking 2; wet 1
soppy, maudlin; romantic 3; sentimental 2; sloppy 2, 3; sweet 7; wet 1
soprano, high 7
sorcerer°, charmer; magician 1
sorceress, charmer; magician 1; siren 2; sorcerer; temptress; witch 1
sorcery°, charm 2; enchantment 1; fascination; glamour; magic 1; occult 3
sordid°, base[2] 2, 4; despicable; dirty 6; filthy 2; foul 4; groveling; low[1] 12; mean[2] 3; nasty 3; scandalous 1; seamy; sleazy 2; sorry 2; ugly 2; vile 1
sordidness, misery 2
sore°, angry 2; chafe 4; heavy 4; indignant; inflamed; inflammation; painful 1; raw 4; scratchy 2; sensitive 1; tender[1] 8; ulcer 1
  sore, make, chafe 2
  sore straits, in, impoverished 1
soreness, ache 3; chafe 4; discomfort 2; pain 1; sensitivity 1
sore (spot), gall[2] 1
sore straits°, sore 6; strait 3
sororicide, murder 1
sorority, club 2; fellowship 2; order 9; ring[1] 3; society 5
sorrow°, care 1; desolation 2; desperation 2; distress 1; evil 7; gloom 2; grief 1; grieve 2; guilt 2; melancholy 2; misery 1; moan 3; mourning 2; pain 5; penitence; pity 1; regret 2; remorse 1; sadden; sadness; woe
sorrow (for *or* over), lament 1; mourn
sorrowful°, brokenhearted; dark 3; dejected; desolate 3; despondent; doleful; forlorn 1; funereal; gloomy 2; guilty 2; heartbroken; heavy 5; hopeless 3; hurt 7; joyless 1; low[1] 8; melancholy 1; miserable 1; mournful 1, 2; mouth 6; pathetic 1; penitent; poignant 1; regretful; remorseful; sad 1; unhappy 1; wistful 1; woebegone
sorrowfulness, melancholy 2; penitence; sadness
sorrowing, mourning 1
sorry°, afraid 2; apologetic; bad 8; base[2] 3; disappointing; guilty 2; low[1] 6; mangy; mean[2] 3; miserable 3; paltry; pathetic 2; penitent; pitiful 2; poor 4; regretful; remorseful; repentant; sad 3; scurvy; sorrowful 1; terrible 2; threadbare 1; wan 2; wretched 3
  sorry (for), be, feel 8; repent
  sorry for, be *or* feel, regret 1
sort°, brand 1; breed 1; bunch 3; categorize; category; character 1, 2; class 2; denomination 2; description 3; distribute 3; divide 5; form 3; genre; grade 6; group 3; kind[2] 1, 2; make 42; model 6; mold[1] 3; nature 4; order 3; peculiar 3; place 16; range 2, 7; rank[1] 6; rate[1] 4; run 44; separate 2; stamp 8; stripe 2; style 1; tabulate; type 1; variety 3
  sort, of a, sort 5
  sort, of the same, piece 8
  sort of, fairly 1; moderately; quasi 2; rather 1; somewhat; sort 7
  sort (out), arrange 1; fix 16c; identify 1; interpret 2; order 16; organize 1; pick 1; puzzle 3; right 19; screen 8; see 12a; separate 1; settle 1; sift 1; sort 8, 10; straighten 1
  sort (someone) out, fix 14
sorted out, straight 7
sortie, action 6; assault 1; charge 7; onset 1; push 15; raid 1
sortilege, magic 1; sorcery
sorting out, interpretation 1; resolution 4
sorts, of, sort 5
  sorts, out of, diseased; ill 1; moody 1; off-color 1; seedy 2; sort 6; sullen; weather 2
so-so°, adequate 2; average 4; decent 3; fair[1] 4; indifferent 3; mediocre; OK 3; ordinary 2; passable 1; tolerable 2; undistinguished
so-styled, titular
sot, alcoholic 2; drunk 3; soak 3
Sotadean, scurrilous
Sotadic, scurrilous
so to speak°, effect 5; quasi 2; speak 6
sotted, drunk 1; stinking 3
sottishness, drunkenness
sough, sigh 1
sought-after, demand 7; desirable 1; enviable; hot 6; popular 1
soul°, bosom 3; breast 2; elixir 3; essence 1; heart 5; human 4; individual 3; life 5, 7; mortal 6; person 1; psyche; spirit 1
soulless, inanimate
soul mate, friend 3
soul-stirring, tender[1] 7; thrilling; touching
sound°, advisable; authoritative 2; blare 2; blast 2; boom 1; chime 3; cry 6; dive 1; durable; effectual 2; entire 2; fathom; flawless 2; go 14; good 8, 13; gulf 1; hale; hardy 1; hearty 5; honest 4; intact; judicious; legal 2; logical 2, 3; mouth 7; noise 1, 2; note 9; OK 4; peep 3; plausible 1; plumb 5; potent 2; practical 1; presumptive 1; rational 1; reasonable 1; reliable; right 2, 4; ring[2] 1; robust 1; roll 5; safe 1, 3; secure 2; seem; sensible 1; sigh 3; solid 3, 6, 7; solvent; stable 1; staunch 2; steady 1; strain[1] 8; strong 11; submerge 2; substantial 2, 3; toll[1] 1, 2; tone 1; tough 1; well-balanced 1; whole 3; wise 1
  sound like, sound[1] 5
  sound like *or* similar to, resemble
  sound mind, of, rational 1; sane
  sound of, decrease the, turn 14b
  sound of, raise, amplify, intensify *or* increase the, turn 25d
  sound off, sound[1] 9
  sound out (*or* off), approach 3; examine 2; interview 3; question 9; sound[1] 7, 8; sound[3] 1

specialized, special 1; technical 1
specially°, custom 4; especially 1; expressly 2; particular 6; particularly 1; purpose 4b
special-occasion, good 15
specialty°, calling; discipline 5; domain 2; field 4; forte; line[1] 7; peculiarity 2; piece 13; profession 1; pursuit 3; sphere 3
specie, cash 1; coin 1; money 1; tender[2] 3
species, breed 1; denomination 2; description 3; family 4; form 3; genre; kind[2] 1; nature 4; sort 1, 3; stamp 8; type 1; variety 3
specific°, antidote; categorical; certain 6; circumstantial 3; concrete; definite 1; detail 1; emphatic; explicit 1; express 5, 6; individual 2; keen[1] 4; local 2; luminous 3; outspoken; particular 1, 5, 6; precise 1; pronounced 2; proper 6; remedy 1; respective; several 2; special 1
  specific, a, one 3
  specific about, be, specify
  specific (to), peculiar 2; specific 2; symptomatic
specifically, detail 4; especially 1; exactly 2; expressly 2; namely; notably 1; particular 6; particularly 2; purpose 4b
specification°, assignment 4; reference 1; stipulation
specifics, detail 2; point 10
specified, given 1; specific 1
specify°, assign 2; define 1; denote 1; describe 2; designate 1; detail 5, 6; enumerate 1; fix 2, 9; formulate 1; identify 1, 3; indicate 1; itemize; mark 11; mean[1] 2; name 5, 6; pin 4b; prescribe; provide 3; recount 2; refer 1; register 8; set 7; signify 1; spell[3] 2; state 7; stipulate; wish 2
specifying, specification 1
specimen°, case[1] 4; copy 2; example 1; illustration 1; pattern 5; sample 1, 3; snatch 5; type 3
specious°, affected 1; deceptive 2; glossy 2; incorrect; plausible 2; seeming; sleek 3; slick 4; sophistic; unsound 4; untenable
speciously, seemingly
speciousness, falsity 1
speck°, bit 2; crumb; dot 1; fragment 1; grain 3; jot 2; little 10; modicum; morsel 2; particle; point 1; scrap[1] 1; shade 3; shred 1; spark 1; spot 1; stain 1; touch 15; trace 2
specked, flecked
speckle, dapple 2; dot 3; pepper; spatter; speck; spot 1, 7; stain 4
speckled°, dapple 1; flecked; mottled; spotty 1
specs, glass 7; spectacle 3
spectacle°, display 5; entertainment 2; extravaganza; glitter 4; pageant; parade 2; phenomenon 2; pomp; presentation 2; prospect 1; sight 3; splendor 1; tableau; view 1; wonder 1
  spectacle of, make a, show 11
spectacles°, glass 7; spectacle 3
spectacular, beautiful 2; brave 2; bully 4; colossal 2; dandy 2; devastating 2; dramatic 3; entertainment 2; extravaganza; fantastic 4; glorious 2; gorgeous 2; great 12; marvelous; meteoric 1; miraculous; monumental 1; prodigious 2; scenic; sensational 1; splendid 1; stunning 2; superlative; swell 8; theatrical 2
spectacularly, beautifully 2; famously; well[1] 2
spectator°, bystander; eyewitness; observer; onlooker; witness 1
specter°, fear 3; fright 3; ghost 1; hallucination; illusion 2; ogre; phantom 1; presence 4; shade 4; vision 4
spectral°, disembodied 1; eerie; ghastly 2; ghostly 1; shadowy 3; supernatural; unearthly 2; unreal 1
spectrum, gamut; scale[3] 1
speculate°, bet 2; divine 4; fantasize; gamble 1; guess 1; presume 1; surmise 1; theorize; wonder 3
  speculate on *or* upon *or* about *or* over, conceive 3; speculate 1
speculated, hypothetical
speculation°, adventure 2; bet 1; chance 4; dream 1; gamble 3; guess 3; hypothesis; surmise 2; venture 1
speculative°, academic 2; experimental 1; groundless; hypothetical; pure 4; tentative 1; theoretical 1; uncertain 3; visionary 1
speculator, bettor; player 3; theorist
speculum, mirror 1
speech°, address 1; diction 2; expression 6; harangue 1; idiom 1; language 1; lecture 1; line[1] 17; lingo; message 2; oration; paper 4; parlance; pronunciation; sermon 2; talk 14, 19; tongue 1; vernacular 3; voice 1
  speech, give *or* deliver a, talk 6
  speech, make *or* deliver *or* give a, lecture 3
speech habit, phrase 3
speechifier, talker 1
speechify, spout 2
  speechify (on *or* about), hold 16a
speechless°, dumb 1; dumbfounded; inarticulate 3; mute 1; overcome 2; petrified 2; thunderstruck; tongue-tied
speechlessly, silently
speechlessness, silence 2
speechmaker, talker 1
speechmaking, oratory
speech (pattern), accent 3; dialect; intonation; pronunciation; speech 1
speed°, advance 4; career 2; clip[2] 7; course 2; dash 3; dispatch 5; expedition 2; facility 1; flash 5; haste 1; hasten 1; hurry 1; hurtle; pace 2; race[1] 4; rapidity; rate[1] 1; rattle 4; rush 1, 3; scud; scurry; shoot 1; stimulant 2; streak 5; tear 3; tempo; urge 1; velocity; way 5; whisk 2
  speed, at, run 49a
  speed, at full *or* top *or* maximum *or* breakneck, double 8; flat 17a; rapidly 1
  speed, put on some, run 1
  speed, with all, fast[1] 6; quickly 2
  speed of light, at the, rapidly 1, 2
  speed (up), dispatch 4; expedite 1; forward 8; hasten 2; hurry 2; pick 8g; precipitate 1; quicken 1, 2; speed 2; step 16, 18b
speedboat, boat
speedily, devil 5; directly 2; fast[1] 6; flat 17a; hand 12; hastily 1; leap 7; posthaste; promptly; quickly 1, 2; rapidly 1; readily 3; right 13; shot 10; soon 2, 3; suddenly 1; summarily 1; swiftly; time 17
speediness, rapidity; speed 1
speed writing, stenography
speedy°, expeditious; express 7; fast[1] 1; fleet[2]; hasty 1, 3; hurried 1; precipitate 3; prompt 1; quick 1; rapid; ready 5; snappy 1; swift; whirlwind 2
spell°, attack 8; bout 1; charm 2; enchantment 1; fit[2] 1, 2; interlude; interval 1; jinx 1; magic 1, 3; mumbo jumbo 2; paroxysm; patch 4; period 1; relieve 4; round 11, 14; run 38; snap 9; space 3; span 2; spasm 2; spurt 1; streak 3; stretch 6; term 2; time 1; tour 3; turn 30
  spell, put *or* have under a, fascinate
  spell on, put a, charm 5
  spell on *or* over, cast a, bewitch; enchant 1; fascinate; hypnotize; intoxicate 2; possess 3
  spell (out), define 2; detail 5; enumerate 1; explain 1; interpret 1; letter 5; put 10; specify; spell[3] 2; translate 3
spellbind, bewitch; dazzle 1; enchant 1; enrapture; entrance[2]; fascinate; fix 5; grip 7; hypnotize; ravish 1; regale; transport 3
spellbinder, talker 1
spellbinding°, absorbing; devastating 2; enchanting; enthralling; interesting; magic 6; magnetic; piercing 2; ravishing; riveting
spellbound, charmed 1; infatuated; rapt 1
  spellbound, hold, fascinate
spelled out, formal 2; specific 1
spend°, blow[1] 4; drain 5; exhaust 1; expend 1; invest 2; kill 7; lavish 4; lay[1] 18b; lead 5; live 7; lose 4; pass 5; pay 11b; put 19b; run 34b; serve 6
spendable, disposable 2
spend a penny, defecate; urinate
spender, spendthrift 1
spending, expenditure; expense 1; outlay
spendthrift°, extravagant 1; improvident 1; prodigal 1, 3; profligate 2, 4; wasteful; wastrel 1
spend time, mingle 2
  spend time at, hang 4b; haunt 1
  spend time together *or* with, fraternize; go 21c; pal 2; run 24
spent°, beat 13; breathless 1; dead 9; empty 1; exhausted 1, 3; haggard; helpless 3; jaded 1; limp[2] 2; listless; lost 2; prostrate 5; ragged 3; run-down 1; tired 1; washed-out 2; weary 1; worn 3
spew°, flow 2; spurt 3
  spew forth *or* out *or* up, eject 2; erupt 1; pour 1; regurgitate; spew; spit 1; spout 1; throw 9a; vomit
sphacelated, ulcerous
sphere°, capacity 3; category; department 2; domain 2; element 2; field 4; globe 2; jurisdiction; kingdom 2; milieu; orb; precinct 2; profession 1; province 3; range 1; reach 7; realm 3; region 2; round 10; scene 1; science 1; scope 1; specialty 1; universe 2; world 1, 3; zone
sphere of action, theater 4
spheric, spherical
spherical°, rotund 1; round 3
spheroid(al), round 3; spherical
spherule, sphere 1
spice°, dash 6; flavor 1, 3; salt 2, 6; season 3;

partition 5; portion 4; separate 1, 2, 3, 4; split 1, 2, 3, 4
split up with, throw 8
split hairs, cavil 2; quibble 1; shuffle 3
split one's sides, laugh 1
split second, minute[1] 1; second[2]; instant 4
splitting, parting 1; rupture 1; separation 3
splitting off *or* **away,** secession
splitting (up), division 1; partition 1
splitting hairs, quibble 2
split-up, divorce 1; partition 1; separation 1
splotch, blot 1; mark 1; smear 3; spatter; splash 1, 3; spot 1; stain 1
splotched, mottled; spotty 1
splotchy, mottled; spotty 1
splurge°, orgy 2; spend 2; waste 1
splurge on, splurge 2
splutter, boil[1] 2; hesitate 3; spit 1
spoil°, baby 2; bitch 4; blemish 2; botch; bungle; cater 2; coddle; contaminate; corrupt 4; damage 4; dash 1; debase 2; decay 2; decompose 2; deface; destroy 4; deteriorate 1; devastate 1; dote; flaw 3; fluff 4; foul 16b; go 31e; hash 3; humor 4; hurt 1; impair; indulge 2; mangle; mar 1; mother 6; murder 4; mutilate 2; overshadow 2; pamper; pet[1] 5; plunder 1; pollute 1; putrefy; queer 5; rot 1; ruin 8, 9; sabotage 3; sour 5; stagnate; stain 5; sully; taint 2; tarnish; turn 5; undermine 1; upset 5; vitiate 1; waste 11; wreck 1
spoil one's reputation, blot 3
spoiled, bad 6; disfigured; foul 2; hothouse 2; moldy; musty 1; putrid; rancid; rotten 1; sour 2; stale 1
spoiling, pollution
spoiling for, be, spoil 5
spoil(s), booty; loot 1; patronage 4; pillage 3; plunder 4; prize[1] 4; spoil 6; trophy 1
spoil (someone's) game, foil[1]
spoilsport°, killjoy; misery 4
spoilt, bad 6; disfigured; moldy; putrid; rotten 1; sour 2
spoken°, oral; verbal 1; vernacular 2
spoken for, attached 4; engaged 1; speak 13; spoken 2
spoken for, not, spare 2
spokesman *or* **spokesperson** *or* **spokeswoman,** agent 1; delegate 1; deputy; mouthpiece 2; proponent; representative 3; voice 3
spoliate, rape 5
spoliation, desolation 1; havoc 1; plunder 3; rape 2
spondulicks, cash 1; money 1
sponge, beg 2; bloodsucker; borrow; bum 5; clean 9; drunk 3; flatterer; hanger-on; parasite; soak 3
sponge (off), wash 1; wipe 1
sponge (up), soak 2; wipe 1
sponge bag, bag 2
sponge bath, wash 11
spongelike, porous
sponger, beggar 1; hanger-on; parasite; scrounge 2
spongy, mushy 1; porous; soft 1; yielding 1
sponsor°, answer 7a; back 2a; backer 1; benefactor; favor 6; patron 1; patronize 3; philanthropist; promote 3; sanction 6; second[1] 9; stand 7b; subsidize; support 4; take 39c; talk 13; underwrite 1; vouch
sponsoring, subsidy
sponsorship, auspices; backing 1, 2; cooperation 2; patronage 1; protection 2; sanction 2; start 12; subsidy
spontaneous°, automatic 1; extemporaneous; free 6; gratuitous 1; immediate 1; impetuous; impulsive; independent 4; instinctive 2; involuntary; natural 2, 4; optional; unasked; unpremeditated; voluntary 1
spontaneously, freely 2; pell-mell 1; voluntarily
spoof, burlesque 1, 3; caricature 1; imitate 2; mock 2; mockery 2; parody 1, 3; put-on 1, 2; satire 2; scoff; send 9a; take 34b; takeoff 2
spoofing, satire 1; satirical
spook, ghost 1; phantom 1; ruffle 3; scare 1; shade 4; specter 1
spooky, eerie; frightening; ghostly 1; scary; spectral; stark 3; unearthly 2; weird
spool, roll 12; tape 2
spoon, kiss 1; scoop 1, 4
spoon out, distribute 1; scoop 5
spoondrift, spray[1] 2
spoon-feed, spoil 3
spoonful, little 10; morsel 1; mouthful; sip 2
spoor, footstep 2; scent 2; step 3; trace 3; track 3; trail 2; wake[2] 1
sporadic°, broken 6; casual 2; fitful; intermittent; irregular 2; occasional 1; odd 2; periodic; piecemeal 3; spasmodic 2; spotty 3; uncertain 4
sporadically, fit[2] 4; infrequently; now 4; occasionally; once 4; piecemeal 1
spore, seed 1
sporran, bag 3; purse 1
sport°, amusement 1, 2; bettor; entertainment 1; flaunt; frolic 1, 3; fun 1, 2; game 1; lark 2; model 8; pastime; play 1, 22; recreation; trick 2; wear 1
sport (of), make, deride; laugh 2a; ridicule 2; satirize; taunt 1; fun 5; mock 1; parody 3; poke 4; sport 3
sport (with), toy 3
sporting, venturesome 2
sporting house, brothel; house 6
sportive°, frolicsome; jolly 1; kittenish; mischievous 1; playful 1; sprightly; zany 1
sportscaster, announcer
sportsman, player 1
sportsmanliness, sportsmanship
sportswoman, player 1
sporty°, jaunty 2
spot°, announcement 3; blot 1, 2; bruise 1; dab 2; dapple 2; detect 2; dilemma; discover 2; dot 1, 3; drop 2; engagement 4; find 2; fix 17; hole 5; jam 6; little 10; locale; location 1; mark 1, 10; modicum; notice 2; particle; perceive 1; perch 1; pimple; pipe 6; place 1, 15; plight; plumb 4; point 3; predicament; quarter 3; scandal 3; scene 1; scrape 8; see 1; shot 7; sight 5, 8; site 1; situate; situation 1; slot 2; slur 1; smear 3; soil[1] 1; speck; splash 1, 3; spotlight 1; spy 3; stage 1; stain 1, 4; station 1, 4; stigma; tarnish; touch 15; trace 2; witness 4
spot, on the, instant 3; nail 6; outright 3; spur 3; trouble 9a
spot, put on the, put 23a
spot on, accurate 3; faultless; marvelous; perfect 4, 5; precisely 1; right 2; true 3
spot (announcement), advertisement 1
spotless°, clean 2; fair[1] 3; flawless 1; holy 2; immaculate 1; impeccable; incorrupt 1; innocent 2; perfect 3; pristine 2; pure 2; shipshape; unspoiled; untarnished; white 3
spotlessness, purity 1
spotlight°, emphasize; feature 4; flaunt; focus 4; play 17a; point 22; stress 4
spotlight, out of the, background 3
spot of bother, brush[2] 3; disturbance 2
spots°, spot 5
spotted, dapple 1; dirty 1; flecked; mottled; speckled; spotty 1
spotty°, mottled; perfunctory 2; sparse 1
spousal, nuptial
spouse, husband 1; mate 2; partner 1; wife; woman 2
spout°, eject 2; erupt 1; flow 2; fountain 1; gush 1, 3; pour 1; prattle 1; run 6; spew; stream 4; tap[2] 1; well[2] 2
spout, up the, drain 3; pregnant 1; spout 4
spouting, eruption 1; outpouring
sprain, strain[1] 6; turn 10; twist 5; wrench 2
sprawl°, recline; spread 6; trip 5
sprawling, rambling 2
spray°, bouquet 1; bunch 1; fountain 1; shower 3; spatter; splash 1, 3; spot 7; water 6
sprayer, spray[1] 3
spread°, breadth 1; circulate 2, 3; circulation 2; coat 3; communicate 1; diffuse 3; diversify; enlarge 1; epidemic 2; expanse; expansion 1; extension 1, 2; feast 1; flare 6; flow 4; gamut; get 29a; grow 1; growth 1; increase 1, 3; lay[1] 1; meal 1; measure 1; noise 3; plaster; progress 3, 6; progression 2; proliferation; propagate 2, 3; put 23e; radiate 2; range 2, 6; riddle[2] 2; root[1] 5; rub 2; scatter 1; scope 2; set 10; shed[2] 2; smear 1; splash 2; sprawl 3; spray[1] 1; straggle; stretch 1, 5; strew; swell 4; widen
spread (about *or* **around),** diffuse 1; distribute 2; measure 14; publish; put 9; spread 2; loose 4
spread (apart), diverge 1
spread (out), broad 1; diffuse 1; disperse 1, 2; display 2; dissipate 1; expand 1; extend 1; flare 2; flat 2; open 17, 24; roll 11; sparse 1; sprawl 1, 2; spread 1, 3, 4, 5; unfold 1
spread over, cover 3; run 33c; suffuse
spread thin, dissipate 2
spread through(out), permeate
spread-eagle(d), flat 2
spreading, circulation 2; expansive 1; extension 1; increase 5; progression 2; rambling 2; spread 8
spread-out, rambling 2
spree°, bender; bout 2; carouse 2; fling 2; frolic 2; fête 1; jag; jamboree; lark 1; orgy 1, 2; party 1; revel 3; revelry; splurge 1
spree, go on a, dissipate 4; paint 7; revel 2
sprig, branch 1; offshoot 1; slip[2] 2; spray[2]; twig[1]
sprightliness, exuberance 1; gaiety 1; life 7; pep 1; snap 11; spring 7
sprightly°, active 3; agile 1; alert 2; alive 3; brisk 2;

stand° (*cont.*)
stand (in) for, relieve 4; represent 1; substitute 1; understudy 2
stand in the way of, obstruct 3
stand on the sidelines, stand 5b
stand out, project 5; protrude; shine 3; show 12b; signify 2; stand 9; stick[1] 15
stand pat, hold 20a
stand (stock-)still *or* **motionless,** freeze 3
stand (up), arise 1; bear 10a; bristle 2; disappoint 2; get 51b; hold 11; keep 11; last[2] 2; rise 1; stand 2, 10; wash 8; wear 4
stand up (against), resist 1
stand up for, champion 4; defend 3; recommend 2; speak 7a; stand 5a, 11; stick[1] 18; support 1
stand (up to), bear 4; defy 1; resist 1; stand 12; withstand; tackle 3
stand up under, stand 3
stand with *or* **beside,** defend 3
stand-alone, self-contained 3
standard°, average 1; banner 1; certain 1; classic 1; classical 1; color 2a; common 1; conservative 2; conventional; correct 7; example 2; flag[1] 1; flat 7; formal 1; gauge 3; guide 6; habitual 1; ideal 1; lead 13; leg 2; mark 3; measure 3; medium 1; model 2; natural 1; nonpareil 1; norm 2; normal 1; ordinary 3; orthodox; par 1, 2, 4; paragon; pattern 1; pennant; perfection 3; pole[1]; popular 2; post[1] 1; precedent; principle 1; proof 2; prototype 2; regular 1; regulation 4; rule 3; set 30; sign 4; staff 1; standing 1; staple 2; stock 7, 8; symbol; systematic; touchstone; traditional; type 3; typical 1; uniform 1; yardstick
standard, up to, presentable 1; scratch 5
standard-bearer, protagonist 2
standardize°, calibrate; equalize; normalize; organize 1; regiment
standardized, systematic
standard operating procedure, mode[1] 1; procedure; technique 1
standard(s), code 3; conscience; ideal 3; moral 4; morality 1
standard(s) of behavior *or* **conduct,** protocol 1
standby°, makeshift 1; stand-in; stopgap 2; substitute 2; temporary
standby, on, call 16
stand-in°, alternate 5; deputy; double 7; relief 4; second[1] 8; substitute 2; understudy 1
standing°, account 5; caste; dead 11; degree 1; dignity 2; end 7a; erect 1; estate 3; face 4; footing 2; grade 1; importance 2; level 14; mark 8; name 2; note 8; par 2; place 3; position 5; prestige; prominence 1; quality 3; rank[1] 1; rate[1] 4; reputation 1; stagnant; station 2; status 1; stratum 2; term 7b
standing, have, rank[1] 7
standing, of a higher, superior 1
standing by, call 16; poised 2
standing (up), upright 1
standoff, deadlock 1; impasse; stalemate
standoffish°, aloof 2, 3; cool 5; distant 3; frigid 2; inhospitable 1; remote 7; reserved; retiring; self-contained 2; stiff 7; unsocial; withdrawn 1; cold 3; unapproachable 1
standoffishness, reserve 4
standpoint°, angle[1] 2; aspect 1; outlook 1; perspective 1; point of view 1; side 3; slant 1; stance; stand 13; viewpoint
standstill°, deadlock 1; halt 1; stalemate; stop 6
standstill, bring to a, deadlock 2; obstruct 1; stop 2; terminate; tie 7c
standstill, come to a, deadlock 2; pull 18a; stall[1] 1
stand the test of time, wash 8
standup, upright 1
stanza, passage 2
staple°, clip[1] 1; connect 3; pin 3; standard 6; stock 7, 8; tack 5
staples°, merchandise 1; provision 4; staple 3
star°, celebrity 2; decoration 2; dignitary; feature 4; hero 1, 2; name 3; personage; personality 2; principal 6; queen 2; somebody 2; stellar 1; success 2; thespian 2
starboard, right 5, 11
starch, bottle 2; enterprise 2; grit; stamina
Star Chamber, tribunal
starched, formal 1
starchy, ceremonious 1
star-crossed, doomed 2; ominous 1; poor 6; portentous 1; sorry 2; tragic; unfortunate 2; unhappy 2; woebegone
stardom, celebrity 1; fame; prestige; renown
stare°, gape 1; gawk 2; gaze 1, 2; glare 2, 4; ogle 2, 3; rubberneck 1; wonder 4
stare at, regard 1
stargaze, fantasize
stargazer, dreamer; fortuneteller; seer
stargazing, absent-minded
staring, glassy 2; goggle-eyed
stark°, blank 6; gaunt 2; hard 12; harsh 2; mere; naked 3; outright 2; plain 4; positive 9; severe 6; simple 2; sorry 2; unadorned; unflattering 1; unvarnished
starkers, bare 1; naked 1; nude; raw 7
starkly, clear 18; clearly 1; outright 4; severely 6; simply 4
stark-naked, bare 1; naked 1; nude; raw 7
starkness, severity 6; simplicity 2
stark raving mad, mad 1
starless, black 3; overcast
starlight, light[1] 1
starring, principal 1; stellar 2
starring role *or* **part,** lead 16
starry-eyed, idealistic; impractical 1; quixotic
start°, activate; attack 3; begin 3; beginning 1, 2; birth 2; blink 3; brew 3; coin 3; commence 1, 2, 3; conception 1; creation 1; dawn 2, 4; embark 2; enter 4; entrance[1] 4; establish 1; excite 1; fall 21; first 6; flinch; foment; found 1; generation 3; germ 2; get 32c; go 18, 29a; inaugurate 1; infancy 2; initiate 1; institute 4; introduce 4; jar[2] 4; jerk 2, 4; jolt 4; jump 2, 11; launch 2, 5; move 12; onset 2; open 20; opening 4; organize 2; origin 2; originate 1, 2; outset; pick 3; pioneer 2; proceed 1; produce 2; propel; provoke 2; pull 8a; put 17c; recoil 1; rise 9, 17; scare 3; set 12a, 23b; source 1; spring 3; strike 18; threshold 2; touch 11b; turn 36; undertake 1; weave 5; well[2] 2
start, at *or* **from** *or* **near the,** early 2; originally
start, at the, first 8
start, from *or* **at the,** first 5; primarily 2
start (from), protrude
start in, begin 1; lead 8; start 2, 3, 4
start (off), begin 1, 2; launch 1; lead 8; start 1, 2, 3
start off into, go 31f
start (on), begin 1
start (out), begin 1; lead 8; set 18b
start to act, step 11
start to finish, from, around 7; completely 1
start to take action, step 11
start (up), arise 4; lead 8; spark 2; start 1; turn 18a
start with, to, first 5
start (with *or* **from),** go 24b; proceed 2
started, way 12
started, get, activate; move 12; weave 5
starter, apprentice 1; hors d'oeuvre; savory 3
starting, original 1
starting point, base[1] 4
startle°, alarm 3; appall; breath 3; daze 1; dismay 2; dumbfound; electrify 1; frighten; horrify 2; jolt 3; scare 1; shock 1; stagger 2; take 28
startled, daze 4; dumbfounded; scared
startling°, dramatic 2; frightening; lurid 1; prodigious 2; shocking 1; sudden; unforeseen
start-off, opening 4
start the ball rolling, break 17; set 12a
start the day, rise 2
startup, start 10; opening 4
starvation, famine; hunger 1
starve (oneself), fast[2] 1
starved°, emaciated; famished; hollow 3; hungry 1; ravenous 1
starved-looking, gaunt 1
starving, famished; hungry 1, 2; meager 2; ravenous 1; starved 1
stash, cache 2; heap 1; stack 1
stash (away), cache 3; deposit 2; heap 3; hoard 2; salt 8; secrete[1]; stack 6; stow
state°, allege; announce 3; assure 4; case[1] 1; claim 4; condition 1; country 1; declare 1; define 2; designate 1; enunciate 2; estate 3; express 1; flap 4; footing 2; form 5; frame 5; grade 1; indicate 3; insist 2; lather 2; lie[2] 6; local 2; maintain 3; mode[1] 2; monarchy 1; nation; national 1; observe 4; phase 3; place 3; plight; political 1; position 4; posture 2; predicament; profess 1; pronounce 2; provide 3; province 1; remark 2; repair 3; represent 3; say 1; secular; set 18a, 18c; shape 4; speak 4; stew 2; tell[1] 2; testify; trim 6; urge 4; vow 1; word 10
state, bad, strait 3
state, in a, frantic
state, poor, strait 3
statecraft, diplomacy 2; politics 1
stated, alleged; explicit 1; given 1
stateliness, dignity 1; splendor 1
stately°, ceremonial 2; dignified; distinguished 2; formal 1; gallant 3; grand 1; imposing; lofty 2; measured 1; noble 5; palatial; proud 4; regal; royal 2; solemn 3; state 6; statuesque
stately home (*or* **residence),** palace; residence 3

stenographic, clerical 2
stenotypist, stenographer
stenotypy, stenography
stentorian, loud 1
step°, act 1; degree 2; footstep 1; ledge; march 1; motion 3; notch 2; pace 1; phase 1; proceeding 1; stage 1; stamp 1; tramp 5; walk 1, 6
  step, in, line[1] 19b
  step, take the first, pioneer 2
  step aside, stand 6
  step at a time, a, step 9
  step down, get 37a; step 14
  step down off *or* **from,** get 41a
  step forward, offer 4
  step in, intervene 1; step 15
  step into *or* **in,** wear 1
  step lively, race[1] 4; run 1; stir 2
  step off *or* **out,** disembark
  step on, trample 1
  step on it, hurry 1; move 12; race[1] 4; run 1; rush 1; speed 3; step 16
  step on the gas, hurry 1; move 12; race[1] 4; run 1; rush 1; speed 3
  step out, paint 7; step 17
  step up, expedite 1; forward 8; intensify; step 18; strengthen 1; swell 3
  step (with), in, step 7
  step (with), out of, incongruous; step 8
step by step°, degree 3; gradually; progressive 1; step 9
stepfather, parent 1
step into the shoes of, follow 5
stepladder, step 10b
stepmother, parent 1
steppe(s), flat 14b; plain 6; table 2
step(s), measure 6; provision 3; step 10
  steps, take, step 11
stereotype, cliché; stamp 6
stereotyped, banal; cold 5; common 6; hack[2] 4; prosaic; ready-made 3; set 30; stale 2; stock 7; threadbare 2; timeworn; tired 3
stereotypic(al), banal; common 6; musty 2; prosaic; ready-made 3; routine 4; timeworn; tired 3; usual
sterile°, barren 1, 2; clean 1; futile; hygienic; impotent 3; ineffectual 1; infertile; lifeless 4; nonproductive 1; pale[1] 3; poor 3; pure 2; sanitary; unsuccessful 1; useless 1
sterility, impotence 3
sterilize°, disinfect; fumigate
sterilized, pure 2; sterile 2
sterilizer, disinfectant
sterling°, excellent; good 2; optimum 2; pure 1; silver 1; superior 2; superlative; virtuoso 2; worthwhile 2
stern°, cutting 2; dour 2; exacting; extreme 4; forbidding 2; gaunt 2; grim 1; hard 4; harsh 2; pungent 2; puritan 2; rear[1] 1, 3; rigid 2; rugged 2; scathing; severe 2; Spartan; strict 2; tight 3; tough 6; unkind
sternly, roughly 2; severely 3; sharply 1
sternness, rigor 2; severity 2
sternutate, sneeze 1
sternutation, sneeze 3
stern-visaged, grim 4
sternway, make, reverse 5
stew°, alcoholic 2; boil[1] 1; bother 6; brew 3; brothel; fret 1; hurry 3; medley; mess 3; mishmash; muddle 4; problem 1; rage 4; scrape 8; seethe 1, 2; simmer 1, 2; smolder; sweat 2; twitter 4; worry 1
  stew, get *or* **work oneself (up) into a,** rage 4
  stew, in a, nervous 1
  stew (over *or* **about),** nail 3
steward, domestic 4; servant 1
stewardess, servant 1; waiter
stewardship, command 7; management 1
stewed (to the gills), drunk 1
stewpot, pot 1
stick°, apply 1; attach 5; bar 1; bind 4; bog 2; bond 4; cement 2; clap 3; cling 1; club 1; connect 3; fasten 1; fix 1; freeze 3; glue 2; gore[2]; hold 13; impale; lodge 5; peg 1; place 15; pointer 1; poke 1; pole[1]; prickle 4; rod 1; run 34a; seize 6; shaft 1; spar[1]; spike 2; stab 1; staff 1; stake[1] 1; sting 1; stomach 3; support 3; tack 5; take 6; tape 3; thrust 2; tolerate 1; transfix 1; twig[1]; wand; withstand
  stick around *or* **about,** linger 1; stick[1] 11; wait 1
  stick at, peg 6; scruple 2; stick[1] 12
  stick by, stick[1] 13
  stick in, insert 1
  stick in (someone's) craw *or* **throat,** scandalize
  stick (into), enter 2
  stick (it *or* **something) (out),** endure 2; persevere; stick[1] 14
  stick on, attach 1
  stick out, bulge 2; jut; overhang 1; pop 3; project 5; protrude; reach 1; see 14b; stand 9b; start 8; stick[1] 15
  stick to, peg 6; persevere; stand 5c
  stick (together), unite 3; hang 11a; stick[1] 5, 16
  stick up, hold 22a; post[1] 2; rob 1, 2; stick[1] 15, 17
  stick up for, defend 3; speak 7a; stand 5a, 11; stick[1] 18; support 1
  stick with, persevere; pursue 2; stick[1] 19
  stick with *or* **at (it),** peg 6
sticker, label 1; tab 1; tag 1
stickiness, sentimentality; tenacity 2
sticking-up, robbery
stick-in-the-mud°, fogy; prig; priggish
stickler, perfectionist 1; purist
stick (one's) neck out, wind[1] 12
stick one's nose (in(to)), meddle; poke 3; pry 2; snoop 1
stickpin, pin 2
sticks, country 3; wild 10
  sticks, the, stick[2] 3
  sticks, up, move 2
stick-to-itiveness, application 3; endurance 1; energy; exertion; follow-through; grit; patience 2; perseverance; resolution 1; stamina; strength 2; tenacity 1
stickup, holdup 1; robbery
stickup man, robber; thief 1
sticky°, awkward 5; clammy 1; delicate 4; gooey 1; humid; muggy; sentimental 2; slimy 1; steamy 1; sultry 1; sweet 7; sweltering; tacky[1]; tenacious 2; thorny 2; tight 7; tricky 2
stiff°, affected 1; body 1; cadaver; ceremonious 1; corpse; dead 12; distant 3; extreme 4; firm 1; forced; formal 1, 3; frigid 2; hard 1; high 2; impersonal 2; inflexible; laborious 1; mannered; obstinate; official 2; parochial; pedantic 1; premium 4b; prudish; recalcitrant; relentless 1; rigid 1; sedate 2; severe 2; smart 6; staid; steep[1] 2; stilted; strained; strong 13; stuffy 2; taut 1; tense 1; thick 5; tough 2; unnatural 2; wooden 2
stiffen°, fix 7; freeze 2; harden 1, 2; jell 1; set 3; square 10; steel 2; strengthen 2; thicken; tighten 4
stiffened, stiff 1
stiffener, brace 1
stiffish, thick 5
stiffly, bolt 12
stiff-necked, narrow-minded; parochial; priggish; puritan 2; relentless 1; self-willed
stiffness, distance 2; formality 2
stifle°, bottle 4a; choke 1, 3, 4; contain 3; dampen 2; fog 3; gag[1] 1; gulp 2; keep 14a; kill 3; muffle 2; mute 3; quench 2; repress; restrain 2; rout 2; shut 6b; silence 4; smother 1, 4; suppress 1; swallow 4
stifled, faint 1; inaudible; low[1] 13; pent-up; weak 7
  stifled, be, smother 2
stifling, close 12; heavy 7; oppressive 2; repression 2; stuffy 1; sultry 1; suppression; sweltering; torrid 1
stigma°, discredit 4; disgrace 2; infamy 1; libel 2; scandal 3; slur 1; spot 1; stain 2; taint 1
stigmatize°, brand 3; denounce 1; discredit 1; disgrace 4; flaw 4; libel 4; proclaim 2; shame 8; smear 2; stain 5; taint 2; tarnish
stigmatized, infamous 1
stiletto, blade 2; dagger; steel 1
still°, calm 3, 5; dead 11; dormant 1; ease 5; even 10, 12; ever 2; gag[1] 1; gentle 1; however 1; hush 2; inactive 1; inanimate; indeed 2; inert 2; kill 3; mitigate; muffle 2; nevertheless; noiseless; passive 1; quiet 1, 2, 4, 6; restful 2; rest[1] 10; serene 1; silence 3; silent 1; soften 3; stagnant; standing 2; static 1; though 2; tranquil; tranquilize; yet 2, 3, 5, 6
still (and all), same 3; yet 5
stillness, calm 1; hush 6; inactivity 1; lull 2; peace 1; quiet 5; serenity 1; silence 1; still 3
still water, water 1
stilly, silent 1; still 2
stilted°, affected 1; forced; formal 1; mannered; pedantic 1; ponderous 2; stiff 7, 8; stuffy 2; unnatural 2; wooden 2
stimulant°, drug 2; spice 2; tonic 1, 2
stimulate°, activate; animate 1, 2; arouse 2; awake 2; electrify 2; encourage 1; energize; enliven 1; exalt 3; excite 1; fire 8a; foment; foster 1; freshen 2; fuel 4; get 16; heat 4; incite; induce 1; inflame 1; inspire 1; intoxicate 2; jog 2; kindle; liven 1; motivate; move 5; nourish 3; pep 2; prod 2; promote 1; prompt 5; provoke 1; pump 4b; push 4; quicken 3; raise 7; refresh 2; restore 2; rise 16; rouse 2; spark 2; spice 4; spur 4; stir 3, 4; thrill 3; touch 6; turn 18c; vitalize; wake[1] 2; whet 2; wind[1] 4c
stimulated, eager; excited 1, 2; fire 5; heated
  stimulated (by), interested 1
stimulating°, bracing; brisk 3, 4; electric; encouragement 1; erotic 1; exciting 1; exhilarating 1; impressive; interesting; intoxicating 2; invigorating; promotion 1; provocative 1; racy 1; readable 2; refreshing; rousing; scintillating 2; sensational 1; stirring;

forced; intentional; measured 2; precious 3; premeditated
studio, study 6
studio (apartment), flat 15
studio couch, lounge 4
studious°, meditative
studiously, hard 13; intently
study°, analysis 1; analyze 2; brush[2] 8; canvass 2, 4; check 4; consider 1; consideration 3; contemplate 2; cram 2; deliberate 4; design 5; dig 5; digest 3; examination 1; examine 1; exploration; explore 2; exposition 3; eye 10; get 51f; go 30b, 34a; grind 4; inquire 1; inquiry 1; inspect; investigate; investigation; look 1, 8, 10; meditate 1; monitor 3; mug 6; mull; muse; note 11; observe 2; paper 4; perusal; peruse; polish 4; pore[1]; portrait; practice 2, 6; preparation 4; prepare 2; probe 1, 3; profile 3; puzzle 2; read 1; refer 3; reflection 2; rehearse 1; report 6; research 2; review 1, 4, 5; run 33b, 33d; sanctum 2; scan 2; schooling; science 1; scout 1; scrutinize; scrutiny; search 4; see 12b; sift 2; survey 1, 3; take 10; test 1; theme 2; traverse 4; treat 2; view 4; weigh 2
studying, preparation 4
stuff°, cloth 1; cram 1; crowd 5; devour 1; effects; fabric 1; fill 1, 5, 7; fluff 2; furniture 1; gear 4; glut 3; good 21a; gorge 3; jam 1; load 3; material 1, 2; matter 1; pack 5; pad 4; paraphernalia; plug 4; preparation 5; ram 1; rig 3; satiate 1; spirit 5; squeeze 4; stow; substance 1; thing 8c; timber 3; truck 1; wedge 3
stuff, bit of, girl 1
stuff, the (right), potential 2; power 1
stuff oneself, overeat
stuff (up), block 5; stop 3; stuff 9
stuff and nonsense, fiddlesticks; gab 2; nonsense 1; stuff 4; talk 18
stuffed, full 1; loaded 1; packed; surfeited
stuffed (up), congested
stuffed (with), replete 1
stuffed shirt, prig; purist; square 9; priggish
stuffiness, prudery
stuffing, filling 1; pad 1
stuffy°, ceremonious 1; close 12; conventional; formal 3; heavy 7; impersonal 2; muggy; narrow-minded; oppressive 2; pedantic 1; pompous 2; priggish; puritan 2; square 6; stiff 7, 8; stodgy; strait-laced; sultry 1; supercilious; sweltering
stultifying, heavy 7
stumble°, blunder 1; bungle; fall 2; flounder; founder[2] 3; fumble 1; hesitate 3; hobble 1; lurch[2] 2; misstep 1; peccadillo; reel 1; slip[1] 2; stammer 1; totter; trip 1, 2, 5; tumble 4; wallow 3
stumble on *or* upon *or* into, come 5a; encounter 1; find 1; happen 3; hit 9a; light[2] 15; meet[1] 1; run 20; strike 14; stumble 3
stumble over, bump 4
stumble upon, discover 3
stumbling, halting
stumbling block°, block 2; deterrent; drawback; hindrance 1; obstacle; obstruction 1
stump°, block 1; campaign 3; canvass 1; elude 2; escape 4; floor 5; flummox; heel[1] 1; leg 1; mystify; nonplus; perplex; puzzle 1; stick[1] 9; stub 1; stymie; thwart 1
stump for, stand 7c
stump up, stump 4
stumpy, chubby; low[1] 1; stocky
stun°, amaze; astonish; astound; daze 1; dumbfound; electrify 1; jolt 3; knock 6b; nonplus; numb 2; overpower 2; overwhelm 3; perplex; petrify 2; rock[2] 3; shatter 3; shock 1; stagger 2; startle; terrify; transfix 2; unnerve
stunned, daze 4; dumbfounded; groggy; panic-stricken; petrified 2; senseless 1; silly 2; thunderstruck; unconscious 1
stunner, beauty 2; knockout 2; phenomenon 2; wonder 1
stunning°, amazing; arresting; awesome; dazzling; devastating 2; noble 5; phenomenal; raving 2; ravishing; sexy 1; striking
stunningly, notably 1
stunt°, caper 2; prank; reduce 1; trick 5; wrinkle[2]
stunted°, low[1] 1; puny 3; short 1; undersized
stuntman *or* stuntwoman, daredevil 1; stand-in
stupefacient, narcotic 1, 2
stupefaction, amazement; astonishment; stupor; surprise 3; torpor; wonder 2
stupefactive, narcotic 1
stupefied, daze 4; goggle-eyed; groggy; panic-stricken; petrified 2; silly 2; torpid
stupefy, amaze; astonish; astound; bemuse 2; daze 1; drug 4; dull 11; intoxicate 1; jolt 3; overpower 2; perplex; petrify 2; rock[2] 3; shatter 3; shock 1; stagger 2; stun 2; unnerve
stupefying, awesome; dazzling; narcotic 1; overwhelming 2; perplexing; shocking 1; stunning 1
stupendous, colossal 2; enormous; gigantic; great 12; heroic 4; huge; immense; massive; monumental 1; prodigious 1; sensational 3; splendid 3; striking; stunning 2; superb; terrific 2; thumping 1; vast
stupid°, absurd 1; backward 2; blind 2; crazy 2; daft 1; dense 3; dim 2; dull 1; dumb 2; empty 5; feebleminded; foolish 2; half-witted; inane; insane 2; mindless 1; nonsensical; opaque 3; pointless; ridiculous; senseless 3; silly 1; simple 4; slow 7; thick 6; thoughtless 2; unreasonable 1; unthinking 1; untoward 2; weak 5; wooden 3
stupidity°, absurdity 1; folly 1; insanity 2; lunacy 2; opacity 3; simplicity 4
stupidly, madly 2
stupor°, fog 2; lethargy 1; shock 2; trance
stupor, in a, groggy
stuporific, narcotic 1, 2
stuporous, lethargic 1
sturdiness, backbone 3; pluck 1; stability 1; strength 1
sturdy°, burly; durable; firm 3; fit[1] 3; hardy 1; healthy 1; husky 1; large 1; mighty 2; muscular; powerful 1; robust 1; rugged 3; secure 2; solid 3, 6; sound[2] 3; stable 1; stalwart 1; staunch 2; steely 2; stocky; stout 4; strong 1, 12; substantial 2; tenacious 1; tough 1, 3; yeomanly
stutter, hesitate 3; slur 2; stammer 1, 2
stuttering, halting; hesitant 2
sty, enclosure 1; hovel; pen[2] 1; shambles; stall[1] 3
Stygian, black 3; dark 1; gloomy 1; pitch-black; sunless
style°, air 3; approach 7; call 2; chic 2; couch 2; dash 6; denomination 3; describe 3; design 6; designate 4; diction 1; edit 1; execution 4; expression 6; fashion 1, 5; finesse 1; flair 2; flavor 2; form 3, 4; format 1; genre; getup 2; key 5; kind[2] 1; language 4; make 42; manner 1; model 6; mode[2]; name 1, 4; panache; phrase 3; practice 1; savoir-faire; school 2; sophistication 1; stamp 3, 8; state 3; stripe 2; tag 4; taste 5; technique 1; term 8; title 6; tone 3; touch 18; trend 2; turn 37; vein 5; version 1; vogue 1; way 2; word 10
style, going *or* passing out of, obsolescent
style, in, minute[1] 3; modern; style 6; stylish
styled, so-called 1
styling, denomination 3
stylish°, becoming; chic 1; contemporary 2; current 3; dapper; dashing 2; dressy 2; elegant 2; exclusive 2; fashionable; glamorous 2; jaunty 2; minute[1] 3; modern; new 2; nifty 1; now 5; popular 1; sharp 9; smart 3; snappy 2; sporty; swanky; swell 7; swish 3; trendy 1
stylishness, chic 2; flair 2; popularity; style 4; taste 4
stylistic, rhetorical 1
stylite, hermit
stymie°, bog 2; dilemma; flummox; frustrate 1; hinder 1; nip[1] 2; perplex; puzzle 1; stump 2; thwart 1
suave°, bland 1; cultivated; debonair 1; elegant 1; genteel 2, 3; glib; graceful 2; mellow 4; oily 2; pleasant 2; sleek 3; slick 2; smooth 6; sophisticated 1; well-bred; worldly 2
suaveness, grace 2; refinement 1; savoir-faire
suavity, charm 2; culture 1; grace 2; refinement 1; savoir-faire; sophistication 1
sub, sit 6b; understudy 1
sub for, relieve 4; substitute 1
subaltern, subordinate 2
subastral, terrestrial 1
subconscious°, instinct; instinctive 1; psyche; psychological; subliminal; unconscious 2; vague 4
subconsciously, vaguely 1
subcontract, farm 4; job 6
subcontract (out), let[1] 3
subcurrent, undercurrent 2
subdivide, branch 3; carve 2; diverge 1; divide 1; partition 5, 6; segment 2
subdivision, branch 2; bureau 2; department 1; family 4; partition 4; portion 1; ramification 2; section 1
subdue°, beat 2; best 11; charm 5, 6; chasten 2; conquer 1; control 3; crush 4; curb 2; dampen 2; defeat 1; demoralize 1; face 17; govern 2; grind 5; humble 4; master 8; mortify 1, 2; muffle 2; mute 3; oppress 2; overcome 1; overpower 1; overwhelm 1; peg 2; put 16b; quash 2; quell 1; quench 2; reduce 7; repress; rout 2; shame 6;

sward, field 1; green 4; lawn; turf 1

swarm°, bristle 4; cluster 2; crawl 4; crowd 1, 3; flight[1] 4; flock 1, 2; flood 6; gather 2; herd 1; host[2]; jam 5; mill 6; mob 1; number 2; pack 3; pour 4; press 7, 8; rabble 1; stack 3; stream 3; throng 1

swarm (into), throng 2

swarm over, infest

swarm (with), teem[1]; abound 3; swarm 3

swarming, alive 4; lively 3; mobbed; overcrowded; populous; prodigal 2; solid 2; thick 2

swarming with, lousy 4

swarming with, be, swarm 3

swarm(s), many 3; mass 2; score 4

swart, black 1; dark 8; dusky 1; swarthy

swarthy°, black 1; dark 8; dusky 1

swash, swagger 1

swashbuckler, adventurer 1

swashbuckling°, flamboyant 2

swat, hit 1, 2, 10

swatch, sample 1

swath°, belt 2; strip[1] 1

swathe°, bind 3; drape 1; envelop 1; muffle 1; roll 9; shroud 1; strip[1] 1; swath; wrap 1

sway°, advantage 1; affect[1] 3; bias 4; careen; command 7; control 4; convince; dangle 1; domination 1; dominion 1; get 14; government 1; guide 3; hold 26; impress 1; induce 1; influence 1, 3; lobby 3; lock[1] 2; lurch[2] 1, 2; might 2; oscillate; persuade 1; power 2; predominance; preponderance 2; pressure 4, 5; prevail 3; prevalence 2; reach 6; reel 1; rock[2] 1, 2; rule 2; say 12; shake 2, 8; sovereignty; stagger 1; supremacy 2; swing 1, 3; talk 10; teeter; totter; upper 6; urge 3; wag[1] 1, 2; wave 4; win 3

sway, holding, rampant 2

sway, in full, rampant 2

sway over, have *or* **maintain** *or* **wield,** command 2

sway (over), hold, command 2; control 1; govern 1; predominate; prevail 1; reign 2, 3; rule 5

swayed, affected 5

swaying, pendulous 1; swing 3

swear°, blaspheme 1; commit 4; declare 1; plead 3; pledge 4; promise 3; tell[1] 7; testify; undertake 2; vow 1

swear (at), abuse 3; curse 5; damn 3; thunder 3

swear by, believe 2; rely 1; swear 3

swear in, induct 1; invest 3; seat 7

swear off, give 17a; renounce; swear 4

swear (to), attest; certify 1; guarantee 2

swearing, profanity

swearing off, self-denial 2

swearing-in, installation 1

swearword, curse 3; expletive 2; oath 2

swear word(s), profanity

sweat°, bother 6; drudgery; flap 4; hurry 3; labor 1, 5; lather 2; perspiration; slave 3; squirm; stew 2; suffer 1; work 7

sweat, get *or* **work oneself (up) into a,** rage 4

sweat, in a, nervous 1

sweat, no, painless

sweat (it *or* **something) out,** endure 2

sweat out, sweat 2

sweat blood, sweat 2; worry 1

sweat bullets, sweat 2; worry 1

sweating, perspiration; sweat 6

sweaty, steamy 1

sweep°, brush[2] 4; carry 6; clean 9; drift 3; expanse; extension 2; extent 1; flood 6; gamut; paddle 1; pass 2; raffle; range 1; sail 3; scan 2; spread 9; stretch 5; sway 1, 3; swing 4; swoop 2; trail 5; wash 2, 14; whisk 1, 4

sweep away, purge 2; sweep 2

sweep down, swoop 1

sweep out, purge 2

sweep up *or* **together,** scoop 6

sweeping°, broad 4; comprehensive; exhaustive; extensive 1; general 3; panoramic; radical 2; wash 14

sweepings, dirt 1; garbage; refuse[2]; rubbish 1; trash 3; waste 7

sweepstake(s), lottery; raffle

sweet°, adorable; dear 3; dessert; gooey 2; lovable; love 3; luscious; lyric 3; mellow 1, 2; precious 4; romantic 3; savory 3; sentimental 2; silver 4; smooth 7, winning 1

sweet on, daft 3; sweet 8

sweeten°, mellow 5

sweetened, sweet 1

sweet F.A., zero 1

sweet Fanny Adams, zero 1

sweetheart°, admirer 2; beloved 2; darling 1; dear 3; fellow 4; flame 3; friend 3; gallant 5; girl 2; love 3; squeeze 10; steady 9; woman 2

sweetie, dear 3; love 3; squeeze 10; sweetheart; woman 2

sweetiepie, dear 3

sweetmeat(s), candy; dainty 4; delicacy 4; sweet 9

sweetness, melody 2; sentimentality

sweet(s), candy; sweet 9

sweet-scented, fragrant; sweet 2

sweet-smelling, fragrant; redolent 1; sweet 2

sweet(-sounding), melodious; sweet 3; tuneful

sweet talk, cajolery; flannel 1; flattery; cajole; entice; flannel 2; flatter 3; seduce 1; wheedle

sweet-talker, flatterer

sweet-talking, ingratiating

sweet tooth, weakness 4

swell°, blow[1] 8e; dandy 1; dapper; dude 1; enhance; enlarge 1; expand 1; fill 2, 10a; fine[1] 1; flare 2; gorgeous 2; grow 1; increase 1; inflate 1; mound 1; mount[2] 6; neat 5; pump 4a; rise 5; roll 14; roller 2; sea 2; sharp 9; sporty; stretch 2; stylish; surge 1, 2; swish 3; wash 14; wave 1, 2; well[2] 2

swell (out), bulge 2; swell 1

swell (up), puff 6; swell 1

swellheaded, egotistical; haughty; inflated 1; self-important; inflated 1; vain 1

swell-headedness, vanity 1

swelling°, bulge 1; bump 2; elevation 2; expansion 2; flare 6; growth 4; hump 1; inflammation; lump[1] 2; nub 1; pimple; prominence 3; protrusion; protuberant; sore 7; swell 4; tumor

sweltering°, hot 1; scorching 1; steamy 1; sticky 3; sultry 1; torrid 1

swept off one's feet, overcome 2; smitten 2

swerve°, careen; deflect; deviate 1; dodge 1; shift 5; wind[2] 1

swerve (off *or* **away),** divert 2

swift°, agile 1; expeditious; express 7; fast[1] 1; fleet[2]; hasty 1; hurried 1; immediate 1; light[2] 7; meteoric; nimble 1; precipitate 3; prompt 1; quick 1, 3; rapid; ready 5; smart 6; spanking 3; speedy 1, 3; sudden; unhesitating 1; whirlwind 2

swiftly°, fast[1] 6; flat 17a; hand 12; hastily 1; leap 7; posthaste; promptly; quickly 1, 2; rapidly 1; readily 3; right 13; shot 10; soon 3; suddenly 1; summarily 1; time 17

swiftness, dispatch 5; expedition 2; facility 1; haste 1; rapidity; speed 1; velocity

swig, draft 4; drink 1, 6; gulp 1, 3; shot 7; sip 2; swallow 1, 5; swill 3

swill°, drink 1; garbage; gulp 1, 3; prattle 3; stuff 4; swallow 1

swill (out), rinse 1

swiller, drunk 3

swim, dip 4; float 1; spin 3

swim, in the, swinging

swim in, luxuriate 1

swimmingly°, well 2

swimming pool, pool 1

swindle°, beguile 1; cheat 2; chisel 2; clip[2] 4; deceit 2; deceive; defraud; double-cross; dupe 3; fiddle 1; fleece; fool 4; fraud 2; gouge 2; hoax 1, 2; hocus-pocus 1; outsmart; prey 3b; racket 2; ride 5a; rip 2b; rip-off 2; screw 6; sting 4; take 27, 32b; trick 8; victimize 2

swindle (out of), beguile 2; rob 3

swindler°, adventurer 2; cheat 1; fraud 3; impostor; rogue 1; slicker 1; thief 2; welsher

swindling, fraud 1; rip-off 2; swindle 2; trickery

swine, heel[1] 2; stinker; wretch 1

swing°, accomplish; cadence; dangle 1; engineer 5; execute 1; extreme 7; float 3; flop 2; flourish 2; fluctuate; flux; hang 1; movement 7; oscillate; rock[2] 1, 2; seesaw 2; shake 2, 8; suspend 2; sway 1, 3; sweep 7; swerve; swipe 3; trend 3; turn 2; wangle; wave 4; wheel 2; wield 1

swing, in full, swing 6

swing at, swipe 1

swinging°, flux; pendulous 1; swing 3

swinging both ways, bisexual 2

swing(s), fluctuation

swinish, epicurean 1; gluttonous; greedy 1; ravenous 2

swipe°, make 27; pilfer; pinch 3; pocket 4; rip 2a; steal 1; stroke 1; take 3

swipe at, swipe 1

swiping, rip-off 1; theft

swirl°, eddy 1, 2; flow 1; gyrate; whirlpool

swish°, dashing 2; dressy 2; flash 6; flourish 2; homosexual 2; rustle 1, 2; swanky

swishing, rustle 2

swishy, homosexual 2; swish 4

switch°, change 1, 2, 5; control 6; convert 2; deflect; divert 1; exchange 1, 2; lash[1] 3; reversal 1; rod 2; rotate 2; seesaw 2; shift 1, 5; substitute 1; substitution 1; swing 4; trade 6; transpose; turn 33; vary 2; whip 1, 8

switch off, shut 2, 4a; turn 17a

talc, powder 1
tale°, account 6; fabrication 3; falsehood; fib 1; invention 3; legend 1; lie[1] 2; myth 1, 2; narration 1; narrative 1; novel 2; romance 3; saga; song 3b; story[1] 1; yarn 2
talebearer°, gossip 3
talemonger, talebearer
talent°, ability 2; accomplishment 3; aptitude 2; art 1; asset 2; bent 5; brilliance 2; caliber 2; capability; capacity 2; craft 1; faculty 1; finesse 1; flair 1; forte; genius 3; gift 2; head 4; ingenuity, instinct; knack; natural 11; power 5; prodigy 1; proficiency; prowess 1; resource 1; skill 1; specialty 1; strength 3; technique 2; timber 3; touch 16; virtuoso 1
talented°, able 2; accomplished; brilliant 4; capable 1; clever 1; exceptional 2; gifted; golden 6; good 12; great 6, 8; ingenious; perfect 7; practiced 1; professional 1; proficient; qualified 1; skillful; strong 6; virtuoso 2
talent(s), ability 3; endowment 3; service 9; stuff 3
taleteller, informer; narrator; talebearer
talisman°, amulet; charm 1; fetish 1
talk°, address 1; argue 3; buzz 2; chat 1, 2; chew 2; conference; conversation; converse; dialogue 1; dirt 4; discussion; gossip 1; hot air; interview 1; lecture 1; lingo; negotiate 1; negotiation 1; news 1; palaver 2, 3, 5; rag[1] 5a; rap 3, 5; sermon 2; speak 1, 3; speech 2; spout 2; tell[1] 2; tongue 1; vernacular 3; waffle 3; wheedle; wind[1] 2; word 1
talk, give *or* **deliver a,** talk 6
talk, have a (little), talk 2
talk, make *or* **deliver** *or* **give a,** lecture 3
talk about, speak 9
talk about *or* **over** *or* **of,** bring 15b; broach; comment 4; discuss; talk 7
talk back (to), answer 6
talk big, bluster 2; boast 2; brag; talk 8
talk down, depreciate 2; disparage 1; minimize 2; talk 9
talk down to, patronize 1; talk 9b
talk freely *or* **unreservedly,** speak 11b
talk in, talk 1
talk (more) loudly *or* **clearly,** speak 11a
talk out of, argue 6; discourage 2
talk (over (with)), confer 1; consult 1; knock 3c; parley 2; reason 8; talk 11
talk (someone) into (something), argue 6; convince; get 14; induce 1; interest 8; persuade 2; prompt 3; sway 2; talk 10; urge 3
talk to *or* **with,** address 3; communicate 2; interview 3; refer 3; see 9; speak 2, 12c; talk 12
talk up, talk 13; tout 1
talkative°, expansive 2; forthcoming 3; lengthy; voluble; windy 2; wordy
talkie, movie 1
talking, speech 1
talking picture, movie 1
talking-to, reprimand 1; sermon 1; tongue-lashing
talking-to, give (someone) a, scold 1
talk(s), parley 1
talky, talkative
tall°, big 3; high 1; lofty 1; towering 1
tallboy, cabinet 1
tallness, height 1
tall story, fib 1; invention 3; romance 5; tale 2
tall tale, fib 1; invention 3; romance 5; story[1] 3; tale 2; yarn 2
tally°, agree 1; bill[1] 1; canvass 4; check 5; coincide; correspond 1; count 1; enumerate 2; figure 8; fit[1] 6; girl 2; love 3; number 4; point 13; poll 1, 3, 4; register 1; score 1, 11
tally (up), reckon 1; tally 2
tally (with), conform 2; equal 5; square 11
talon, claw 1; nail 2
tamable, manageable
tame°, break 8; chasten 2; civilize 2; domestic 2; gentle 2; ineffectual 2; inoffensive; manageable; meek 2; pale[1] 3; passive 2; pet[1] 2; slow 9; soft 5; subdue 1; tractable 1; vapid
tamed, broken 4; tame 1
tamp, dab 1; pack 5; ram 1
tamper°, meddle; poke 3
tamper (with), distort 2; doctor 4; fiddle 1; finger 2; fool 6; juggle; manipulate 3; mess 6; monkey 5; rig 2
tan, sun 2; whip 1
tandem, pair 1; team 2
tang°, flavor 1; nip[1] 4; savor 1; spice 2; taste 1; zest 1
tangent, going off at a, digression 2
tangent, go off at *or* **on a,** stray 2; wander 4
tangential°, peripheral 1; side 8
tangible°, actual 2; concrete; live 1; material 5; object 1; observable; patent 2; physical; real 2; self-evident; sensible 2
tangibles, good 21b
tangle°, catch 7; clutter 2; complicate; disorder 1, 4; entangle 1; entwine; foul 13; hodgepodge; jumble 1, 2; kink 1; knot 1; mesh 2; mess 1; mishmash; mix 4c; mix-up; morass 2; muddle 2, 4; network 2; pastiche; ruffle 4; snarl[2] 1, 2, 3; welter 1
tangle (up), confuse 2; tangle 3; tousle
tangle with, tangle 4
tangled, complicated; foul 10; hairy 3; hard 3; intricate 1; involved 2; kinky 3; labyrinthine; thorny 2
tangled, become, mire 3
tangy, pungent 1; racy 3; sharp 4; spicy 1; tart[1] 1
tankard, mug 1
tanked, drunk 1; stinking 3
tanning, thrashing 1
tan (someone's) hide, spank
tantalize°, flirt 1; mock 1; tease 1; worry 2
tantalizing, inviting; provocative 1; seductive; tempting 1
tantamount, equal 1; equivalent 1
tantamount (to), synonymous
tantivy, rapidly 1
tantrum°, explosion 2; fit[2] 3; outburst; rage 2; scene 3; temper 4
tantrum, have a, fly 7; rage 4
tantrum, throw a, explode 3; flare 3
tap°, bore[1] 2; bug 5, 8; dab 1, 3; drain 4; draw 12a; eavesdrop; jab 1; knock 1, 8; milk; patter[1] 2, 3; pat[1] 1, 3; rap 1, 4; ream; sap[1] 3; screw 7a; stroke 1; touch 14
tap, on, available; ready 9a; reserve 8; tap[2] 4
tape°, stick[1] 4
taper°, light[1] 7
taper off, decline 2; decrease 2; drop 11; dwindle; flag[2] 2; phase 6; taper 2; trail 8; wane 1; wind[2] 3a
tape-record, tape 4
tapering, narrow 1
tapering off, decrease 2; wane 3, 4
tapestry, drape 2; drapery
tapping, tap[1] 2
taproom, bar 6; pub
taproot, root[1] 2
tap-tap, tap[1] 2
tap-tap-tap, patter[1] 3
tap water, water 1
tar, pitch[2]; sailor
tar and feather, punish 2
tardily, late 7; slow 12
tardy°, late 1; overdue; slow 5
tare, allowance 4
target°, aim 5; butt[1]; design 7; destination; focus 1; game 5; goal; intent 1; intention; mark 6; monkey 2; object 2; objective 2; prey 1, 2; purpose 1
target, not on, wide 4
target, off (the), wide 4, 5; wrong 2
target, on, accurate 3
tariff°, admission 5; cost 1; custom 2; duty 3; fee 1; rate[1] 2; tax 1; toll[2] 1; tribute 2
tarmac, pave 1; surface 4
tarn, pool 1
tarnish°, blacken 2; blemish 2; dirty 7; discredit 1, 5; disgrace 4; dull 10; foul 14; mar 2; mire 4; shame 8; smear 2; soil[1] 2; stain 5; sully; taint 2; vilify
tarnished, flawed
tarradiddle, moonshine 2
tarry°, abide 3; delay 3; hang 4a; linger 1; remain 1; sojourn 2; stay[1] 1, 5; stick[1] 11; stop 5; wait 1
tarry over, dwell 2
tarrying, delay 6
tart°, bitch 2; incisive 2; jade 2; prostitute 1; pungent 2; racy 3; sharp 4; short-tempered; slattern; snappish 2; sour 1; terse 2; trenchant; wanton 5
tart up, decorate 1; preen 2; primp
tartan, checkered 1
tartar, scale[2] 2
Tartarus, hell 1
tarted up, get (all), primp
tartlet, tart[2] 1
tartly, shortly 3
Tartuffe, hypocrite; pharisee
Tartuffery, hypocrisy
Tartuffian, pharisaic; sanctimonious; self-righteous
task°, assignment 2; business 2; concern 4; duty 1; errand 2; function 1; grind 7; job 2, 3; labor 3; lesson 1; mission 1; office 4; operation 4; place 4; post[2] 1; responsibility 2; role 2; stint 1; thing 6; undertaking 1; work 3
task, take (someone) to, scold 1
task, take to, punish 1; rate[2]; rebuke 1; reprimand 2; task 3; tell[1] 10; upbraid
task force, fleet[1]
taskmaster, disciplinarian; master 1; oppressor
taskmistress, oppressor
tassel, pendant
taste°, affinity 2; appetite 1; art 2; bite 3; class 3; culture 1; dash 7; discrimination 2; distinguish 3; drink 6; drop 2; ear 2; experience 4; eye 3; fancy 7; fashion 1; favor 4; finesse 3; flair 2; flavor 1, 2; grace 2; grain 3; hint 2;

teetotalism, sobriety 1; temperance 2
telecast, broadcast 1, 4; program 2; send 2
telecasting, transmission 2
telegram°, cable 2
telegraph, cable 3; send 2; transmit 1
telekinetic, psychic 2
Telemessage, telegram
telepathic, psychic 2
telepathist, psychic 3
telephone°, buzz 7; call 3, 12b; contact 4; line[1] 13; look 11b; ring[2] 2; transmit 1
telephone call, call 14
teleplay, script 2
telescope°, glass 6
televise, broadcast 1; send 2; spread 2
television°, press 10a; screen 5
telex, telegram; transmit 1
tell°, advise 2; air 7; announce 1, 3; babble 2; bare 7; betray 2; bid 3; break 2; breathe 4; chronicle 2; communicate 1; direct 3; disclose 1; express 1; fill 9c; hump 1; impart 2; indicate 3; inform 1; instruct 2; let[1] 7a; mound 2; narrate; notify 1; order 14; recite 2; recount 1; rehearse 2; relate 2; rumor 2; say 4; show 4; sing 3; speak 4; spin 2; tattle 1
tell apart, differentiate 1; distinguish 1; pick 7b
tell (of), describe 1; make 33; record 2; report 3; tell[1] 2; touch 8
tell (on), finger 8; inform 2; sell 3; tell[1] 11; turn 15c
tell (someone) a thing or two, reprimand 2; upbraid
tell (someone) off, castigate; dress 4; lecture 4; mind 10; rebuke 1; reprimand 2; scold 1; tell[1] 10; upbraid
tell (someone) where to get lost, rebuff 2
tell (someone) where to go *or* **get off,** rebuff 2; rebuke 1
tell all *or* **everything,** spill 3
tell a soul, Don't, mum 2
teller of tales, liar; narrator
telling°, big 4; devastating 1; effective 2; effectual 1; expressive 3; forceful 2; graphic 1; history 1; influential; meaningful 1; moving 1; narration 1; overpowering; persuasive; pictorial 1; pointed 2; potent 2; powerful 2; predominant; prevailing 2; recital 2; rehearsal 2; relation 4; seminal 1; solid 8; stirring; strong 7; substantial 3; weighty 3
tellingly, home 10
telling-off, lecture 2; reprimand 1; tongue-lashing
Tell no one, mum 2
telltale, blabbermouth; gossip 3; meaningful 2; talebearer
tell the difference, discriminate 1; distinguish 1
tell the world, spread 2
tellurian, terrestrial 1
telluric, earthly 1; terrestrial 1
tell who's who *or* **what's what,** distinguish 1
telly, screen 5; television
temblor, quake 2
temerarious, adventurous; bold 2; foolhardy; indiscreet; presumptuous
temerity, brass; effrontery; gall[1] 2; indiscretion 1; nerve 2; presumption 1
temper°, alloy 3; chasten 2; dampen 2; disposition 1; humor 3; lie[2] 6; mind 5; mitigate; moderate 5; modify 2; modulate; mood 1; pet[2]; qualify 3; rage 2; relax 2; silence 4; smooth 12; soften 2; spirit 2, 8, 9a; subdue 3; sweeten 2; tame 6; tone 3, 5; vein 4
temper, in a, indignant
temperament, disposition 1; humor 3; mentality 2; mind 5; nature 1; personality 1; spirit 2; temper 1; way 2
temperamental°, emotional 2; intense 3; irritable; moody 2; passionate 3; quick-tempered; sensitive 2; snappish 2; sullen; surly; touchy 1; waspish
temperance°, serenity 2; sobriety 1
temperate°, conservative 2; even 4; gentle 1; mild 1, 2; moderate 1; peaceable 2; philosophical 2; phlegmatic 2; quiet 4; serene 2; sober 1; subdued 1
temperately, easy 7; moderately
temperateness, sobriety 2
tempered, proof 3; seasoned 1; subdued 1
tempered to, used 3
tempering, extenuating
tempest°, blow[1] 9; gale 1; passion 2; storm 1
tempestuous°, boisterous; fierce 2; heated; inclement; obstreperous; rough 2; stormy 1, 2; tumultuous; unruly; uproarious 1; violent 3; wild 4, 6; windy 1
tempestuousness, fury 2; severity 5
template°, mold[1] 1; pattern 4
temple°, sanctuary 1
templet, mold[1] 1; pattern 4; template
tempo°, beat 11; cadence; pace 2; rhythm; time 7
temporal°, earthly 3; material 7; mortal 1; outward 1; profane 2; secular; temporary; worldly 1
temporarily°, meanwhile 2; time 16
temporary°, brief 1; fleeting; fly-by-night 1; makeshift 1; meteoric; momentary; passing 1; provisional 1; short 7; short-lived; stopgap 2; transient
temporize, delay 1; hesitate 1; linger 4; play 14; postpone; procrastinate 1; sit 8; stall[2] 1; tarry 1
tempt°, appeal 2; chat 3; decoy 2; dispose 2; entice; interest 8; lead 9b; lure 1; motivate; rope 4; seduce 1; tantalize
temptation°, come-on; enticement 1, 2; incitement 2; invitation 2; lure 2
tempted by, be, rise 12b
tempt fate, endanger; fire 6
tempting°, exciting 2; inviting; provocative 1; seductive; sexy 1; temptation 1; voluptuous 2
temptress°, charmer; seductress; siren 2
tenable°, likely 2; plausible 1; possible 1; presumptive 1; reasonable 2; supportable 2; thinkable
tenacious°, determined 1; gritty 2; industrious; insistent; obstinate; patient 2; persistent 1; purposeful; resolute; rigid 4; stalwart 2; stiff 4; strenuous 2; strong 5; stubborn; tireless; urgent 2
tenacious, be, persevere
tenacious of, tenacious 3
tenaciousness, strength 2; tenacity 1, 2
tenacity°, determination 1; endurance 1; follow-through; fortitude; grit; nerve 1; obstinacy; patience 2; perseverance; persistence; purpose 2; resolution 1; stability 2; strength 2
tenancy°, residence 2; tenure 1
tenant°, inhabitant; occupant; occupy 2; resident 3
tenantless, uninhabited
tenantry, tenure 1
tend°, apply 5; care 4, 6a; cherish 2; cultivate 2; guard 1; incline 1, 3; keep 2; nurse 2; protect 2; sway 2; trend 3; verge[2]; watch 2
tend to, attend 2
tend toward(s), lean[2] 3
tendency°, appetite 1; aptitude 2; attitude 2; bent 5; bias 1; current 6; drift 3; habit 2; inclination 3; instinct; intent 2; leaning; mind 5; movement 7; penchant; strain[1] 9; talent 2; tenor; trend 1; turn 38; undercurrent 2
tendency, have *or* **show** *or* **exhibit** *or* **demonstrate a,** tend[1]
tendency toward(s), vein 4
tendentious, partisan 3
tendentiousness, leaning
tender°, affectionate; benign 1; bid 1; communicate 1; delicate 1; demonstrative 1; devoted 1; emotional 3; enter 6; extend 4; feeling 7; feminine 1; fond 1; gentle 1; hand 16; hold 16b; human 3; launch 6; lenient; lovable; maternal; merciful; offer 1, 2, 3, 5; overture; painful 1; prefer 2; present[2] 7; profess 1; proposal 1; propose 1; put 5, 18a, 18b, 28h; raw 4; render 3, 7; romantic 3; sensitive 1; sentimental 1, 2; softhearted; sore 1, 2; submission 2; submit 2; thoughtful 1; touching; turn 15b; venture 2; warm 3
tenderfoot, greenhorn; initiate 4
tenderhearted, benign 1; feeling 7; gentle 1; good-natured; kind[1]; lenient; merciful; sentimental 1; soft 5; softhearted; tender[1] 6
tenderheartedness, mercy 1; sensitivity 2; sentiment 1; sympathy 1
tenderly, dearly 2; fondly; warmly 1
tenderness, affection; attachment 3; delicacy 2; feeling 3; heart 5; humanity 3; kindness 1; love 1, 4; mercy 1; pity 1; sensitivity 2; sentiment 1; sentimentality; sympathy 1; thought 4; warmth 2
tenderness for, feel *or* **have,** pity 3
tending, inclined 1; prone 2
tending toward(s), disposed
tendon, sinew 1
tendril, fiber 1; offshoot 1; runner 3; twig[1]
tenebrosity, night 1
tenebrous, dark 2; dim 1; dingy; obscure 1; sunless
tenet°, belief 3; creed; doctrine; idea 3; principle 1
tenet(s), ideology; philosophy 2; platform 2
tenor°, current 6; drift 4; effect 3; key 5; note 5; spirit 6; strain[1] 9; style 5; thread 2; tone 3; undercurrent 2; vein 4
tense°, anxious 1; drawn; edge 5; electric; emotional 2; explosive 1; flex 3; fraught 2; intense 3; jumpy; nervous 1, 2; overwrought 1; sinking 1; square 10; stiff 7; strained; taut 1; tight 2; tighten 1,

thanksgiving, glory 2; grace 5; gratitude; thank 3
thank-you note, note 2
that, like, thick 8
that is (to say), like[1] 6; namely
thaumaturge, magician 1
thaumaturgist, sorcerer
thaumaturgy, enchantment 1
thaw°, melt 1, 2
  thaw (out), dissolve 1; thaw 1
theater°, drama 2; hall 2; house 5
  theater, the, stage 3; theater 3
theater art(s), drama 2
theatric(al), camp[2] 1; dramatic 1, 3; melodramatic; ostentatious; thespian 1; unnatural 2
theatricalism, drama 3
theatrics, drama 3; rant 3; theater 2
thé dansant, dance 2
theft°, embezzlement; plagiarism; rip-off 1; robbery; stealing
theme°, base[1] 3; composition 1; content[1] 3; essay 1; exposition 3; gist; matter 4; measure 9; melody 1; motif; note 5; pattern 3; piece 4; plot[1] 2; point 7; question 3; strain[1] 9; subject 1; substance 2; tenor; text 3; thread 2; topic; tune 1; writing 2
themselves, person 2; personally 1
then, previously; thus 2
  then (again), further 3
then and there, outright 3
theorem°, hypothesis; law 3; premise 1; thesis 1
theoretical°, abstract 1; academic 2; assumed 3; experimental 1; hypothetical; mental 1; philosophical 1; pure 4; speculative 1; supposed 1; titular; unreal 2
theoretically, ideally 2; principle 4; supposedly
theoretician, theorist
theorize°, divine 4; premise 2; presume 1; speculate 1; suppose 2; surmise 1; suspect 2; wonder 3
theorized, supposed 1
theorizer, theorist
theory, doctrine; feeling 2; guess 3; hypothesis; idea 5; law 3; opinion 1; presumption 3; rationale; reasoning 2; speculation 1; supposition; surmise 2; thesis 1; thinking 2
  theory, in, ideally 2; principle 4
therapeutic°, medicinal
therapeutist, therapist
therapy°, cure 1; remedy 1; treatment 2
there, be, remain 2
there, up to, packed
there and then, outright 3
therefore°, accordingly 1; consequently; hence 1; thus 2
theretofore, previously
thersitical, abusive 1; foul 6; obscene 1; outrageous 3; profane 3; scurrilous
thesaurus°, dictionary; index 1
thesis°, content[1] 3; doctrine; essay 1; exposition 3; paper 4; position 7; premise 1; rhythm; subject 1; theme 1, 2; theorem 1; thread 2; topic
thespian°, dramatic 1; performer; player 2; theatrical 1
thespian art, theater 3
thespian art(s), drama 2
thespianism, stage 3
thetic(al), arbitrary 2; dogmatic; positive 5
theurgical, magic 5
theurgist, magician 1
theurgy, magic 1
thew, sinew 1
thewless, limp[2] 3
thick°, blind 2; chummy; close 15; compact 1; dense 1, 3; dim 2; dull 1; dumb 2; feebleminded; foolish 2; friendly 1; half-witted; heavy 8; lackluster; lush 1; mindless 1; obtuse 2; opaque 3; profuse 2; simple 4; slow 7; stiff 10; stolid; stupid 1; wooden 3
  thick, be, bristle 4
  thick as thieves, close 15; thick 8
  thick of, in the, amid
  thick with, thick 2
thicken°, cake 3; fix 7; grow 1; jell 1; set 3; solidify 1; stiffen 1
thicket°, brush[1] 2; clump 2; shrubbery; stand 19
thickheaded, dense 3; foolish 2; mindless 1; obtuse 2; opaque 3; simple 4; stupid 1; thick 6
thickheadedness, folly 1; simplicity 4
thickness, body 7; breadth 1; opacity 3; ply; seam 2
thickset, burly; chubby; husky 1; large 1; stocky; stout 1; thick 1; well-fed
thick-skinned°, callous; coldblooded 2; coldhearted; frigid 2; hard 4; insensible 2; obtuse 2; rocky[1] 3; thick 6
thick-skulled, thick 6
thick-witted, dense 3; stupid 1; thick 6
thick-wittedness, stupidity 1
thief°, burglar; intruder 1; robber
thieve, knock 5a; lift 6; pilfer; pinch 3; pocket 4
thievery, embezzlement; plagiarism; robbery; stealing; theft
thieving, dishonest; predatory 2; robbery; stealing
thievish, dishonest
thigh-slapper, scream 4
thimble, ring[1] 1
thimbleful, little 10; sip 2
thimblerigger, swindler; thief 2
thimblerigging, swindle 2
thimble-wit, dolt; dummy 3; sap[1] 2; silly 3
thimble-witted, insane 2; senseless 3; stupid 1; thick 6
thimble-wittedness, stupidity 1
thin°, cut 6; degrade 3; emaciated; feeble 2; fine[1] 6; flashy 2; flimsy 3; fluffy 2; fragile; frail 2; gaunt 1; insubstantial 1; lame 2; lanky; lean[1] 1; meager 2; narrow 1; rarefied 1; rawboned; shallow 1; sheer 3; skinny; slender 1; slight 3; spare 3; subtle 2; taper 1; tenuous 1; watery 1; willowy 2; wishy-washy 2
  thin (down *or* out), dilute 1; taper 2; thin 8, 9; water 7; weaken 4
  thin (on the ground), diffuse 1; premium 4a; rare[1] 1; scanty 2; scarce; short 9; sparse 1; thin 7
thing°, act 2; bag 5; concern 4; craze; entity 1; fixation; item 1; love 5b; matter 2; object 1; obsession; office 4; province 3; rave 4; routine 2; sort 4; specialty 1; sphere 3; style 2; trend 2; vocation
  thing, the (latest *or* newest), rage 3
things°, belongings; effects; furniture 1; gear 4; good 21a; luggage; paraphernalia; possession 3; rig 3; stuff 2, 5; thing 8; time 9
thingumabob, contraption; gadget; gimmick 2; instrument 1; thing 5
thingumajig, contraption; device 1; gadget; gimmick 2; instrument 1; thing 5
thingy, contraption; instrument 1
think°, assume 3; believe 1; concentrate 4; consider 3; esteem 2; estimate 2; expect 2; fancy 10; feel 4; figure 9; find 4; guess 2; hold 7; imagine 2; judge 7; make 21; meditate 1; perceive 3; presume 1; reason 7; reckon 3; say 6, 7; suppose 1; suspect 2; take 8; wonder 3
  think, make one, suggest 2
  think (about *or* over *or* on *or* of), chew 4; consider 1; contemplate 2, 3; debate 4; deliberate 4; digest 3; flirt 2; mull; muse; note 11; play 19b; ponder; puzzle 2; reckon 5a; reflect 3; regard 4; revolve 3; see 8, 12a; speculate 1; study 2; think 3, 5; weigh 2
think back, reminisce
think back on *or* about, remember 2
think back to, recall 1
think better of, better 7
think highly *or* well of, respect 6
think (it) amiss, amiss 4
think it likely *or* probable, suspect 2
think much of, not, frown 2
think no more of, dismiss 2
think nothing of, play 19a
think of, design 1; finger 5a; imagine 1; reckon 2; think 4; view 9
think rationally *or* logically, reason 7
think through, debate 4; puzzle 3; rationalize 2
think (too) little of, underestimate
think twice, better 7; hesitate 1; scruple 2
think twice about, do not, mind 19
think up *or* of, coin 3; compose 2; conceive 2, 3; create 2; design 2; devise 1; envisage 1; fabricate 2; fancy 9; forge 2; formulate 2; hit 9b; imagine 1; invent 1; manufacture 2; mastermind 1; meditate 2; project 3; think 6
thinkable°, plausible 1; possible 1
thinker°, intellectual 3; mind 4
thinking°, estimate 4; estimation 1; imagination 2; judgment 4; mind 8; philosophy 1; point of view 2; psychology; rational 2; reasonable 1; reasoning 1; reflection 2; speculation 2; thought 1; thoughtful 3
think piece, editorial
thinks fit(ting), as *or* when (one), will 4
thin-skinned, sensitive 2; touchy 1
third degree, examination 3; interrogation
  third degree, give (someone) the, pump 2; question 9
third estate, public 8
third party, intermediary
third-rate, mediocre; poor 4
thirst°, ambition 1; anxiety 2; appetite 2; desire 3; hunger 2, 3; itch 2, 4; passion 3; stomach 2; urge 5; want 4; wish 5; yearn; zest 2
  thirst after *or* for, hanker; thirst 2; want 1
  thirst for *or* after, desire 1; lust 3; pant 2
thirsting, hungry 2
  thirsting for *or* after, starved 2

thirst-quenching, refreshing
thirsty°, hungry 3; voracious 2
thirsty (for), desperate 3
thirteen, side 4
thirteen weeks, quarter 2
thistledown, fluff 1
thitherto, previously
thole, peg 1; pin 1
tholepin, peg 1; pin 1
thong, lace 2; lash[1] 1; tie 9; whip 8
thongs, bond 1
thorax, chest 2
thorn°, bristle 1; prickle 1; spine 2
thorn in one's side *or* flesh, nuisance 1; pest; plague 2; thorn 2; trial 5
thorny°, difficult 2; hard 3; prickly 1, 4; sore 2; ticklish 2; tough 5; tricky 2
thorough°, absolute 1; ample 6; careful 2; close 14; complete 3; comprehensive; conscientious 2; dead 14; definitive 2; deliberate 3; detailed 1; diligent; elaborate 1; encyclopedic; entire 1; exact 2; exhaustive; faithful 3; full 2; good 11; intimate[1] 2; laborious 2; meticulous; outright 2; out-and-out; painful 3; particular 3; perfect 6; positive 9; professional 2; profound 4; proper 5; pure 5; radical 2; regular 9; right 8; rigid 3; scientific; sheer 2; solid 8; stark 4; strict 1; studious 1; total 3; ultra-; unabridged 2; unmitigated
thoroughbred, nag[2]; genteel 2; lofty 2; purebred
thoroughfare, passage 3; road 2; street 1
thoroughgoing, exhaustive; intensive; outright 2; out-and-out; painful 3; positive 9; profound 4; proper 5; radical 2; regular 9; right 8; sheer 2; stark 4; sweeping 1; thorough 1; total 3; ultra-; unmitigated
thoroughly°, blast 4; clean 7; cold 11; completely 1, 2; dead 18; deeply 2; depth 6; detail 4; downright 2; entirely 1; full 13, 17, 18; hook 4; length 4c; out 7; outright 1; perfectly 1, quite 1; richly 2; terribly; through 7; totally; utterly; well[1] 13; wholly 1; widely 1
though°, however 1; notwithstanding 3; supposing
thought°, attitude 2; brain 1; concern 5; consideration 3; debate 2; heed 2; idea 1; imagination 2; memory 2; note 7; notion 1; opinion 1; point 12; presumption 3; prospect 2; reflection 2; regard 9; reputed; reverie; sentiment 2; spirit 7; supposition; thinking 2
thought for, without, despite
thought to, give, note 11
thought to, give a, flirt 2
thought to, give some, see 12a
thoughtful°, attentive 2; benevolent; considerate; decent 4; deliberate 3; diplomatic; dreamy 2; earnest 1; gallant 2; gentle 1; intellectual 2; judicious; kind[1]; meditative; merciful; neighborly; pensive; philosophical 2; preoccupied 1; provident 1; reflective; respectful; selfless; serious 1; sweet 6; tactful; tender[1] 6; thinking 1; wistful 2
thoughtfully, kindly 2
thoughtfulness, consideration 1; kindness 1; mercy 1; philanthropy 1; philosophy 3; respect 2; tact; thought 4
thoughtless°, blind 3; blunt 2; brute 1; careless 1, 2; coldhearted; flighty 2; flippant 1; foolhardy; foolish 1; hasty 2; hot-headed; ill-advised 1; imprudent; inconsiderate; indiscreet; insensible 2; merciless; mindless 1, 2; overconfident 2; promiscuous 1; rash[1]; reckless; remiss; selfish 2; shortsighted 2; tactless; unguarded 3; unkind; unreasonable 1; unthinking 1, 2; unwary
thoughtlessly, blindly; hastily 2; idly 2; spur 3
thoughtlessness, indiscretion 1
thought processes, psychology
thought-provoking, expressive 3; intellectual 2; stimulating; weighty 2
thoughts°, mind 8; thought 5
thousand, number 2
thousand(s), many 3; score 4
thousands of, umpteen
thrall, slave 1; slavery 1
thralldom, bondage; captivity; servitude; slavery 1
thrash, batter 1; beat 1; belabor; belt 3; chastise; club 5; crush 4; defeat 1; flap 1; flog 1; floor 4; hide[2] 2; hit 1; lace 5a; lambaste 1; lash[1] 3; lather 4; overthrow 1; paddle 4; pound[1] 1; punish 3; rough 16; rout 2; slash 2; slaughter 4; spank; strike 1; switch 3; toss 3, 4; upset 4; wash 3; whip 1
thrash out, discuss
thrashing°, punishment 3; rout 1; upset 11; whipping 1
thrashing, a, works 3b
thrashing, give (someone) a, beat 1
thread°, fiber 1; lace 3; line[1] 11; plot[1] 2; strain[1] 9; string 1, 8; tenor; theme 1; vein 2, 4; wisp; yarn 1
threadbare°, disreputable 2; prosaic; ragged 1; shabby 1; stale 2; tattered; timeworn; worn 1
threadlike, fine[1] 6; thin 3
threads, apparel; clothes; costume; garments
thready, ropy 1
threat°, danger 1; duress 1; hazard 1; jeopardy; menace 2, 3; peril; warning 1
threaten°, adventure 3; browbeat; daunt; endanger; hazard 3; intimidate; jeopardize; lean[2] 4b; loom 2; lower[2] 1; menace 1; overhang 2; scare 1; shake 6b; thunder 3; undermine 1
threatened, endangered; jealous 2
threatening°, black 4, 7; dangerous 2; dark 5; forbidding 2; heavy 9; imminent; impending; menacing; murky; offensive 1; ominous 1; overcast; portentous 1; sinister 1; strong-arm; wintry 3
three, trio
three-card monte, swindle 2
three-dimensional, solid 1
three-monthly, quarterly 1
three-month period, quarter 2
three or four, couple 2
three-ring circus, zoo 2
threesome, trio
threnody, keen[2] 2; lament 2
thresh, flap 1; flog 1
threshold°, eve 2; line[1] 4; verge[1] 1
thrift°, economy 1; providence 1; prudence 2; saving 3
thriftiness, economy 1; providence 1; thrift
thriftless, improvident 1
thrifty°, economical 1, 2; efficient; frugal 1; penurious 1; provident 2; prudent 2; saving 2; sparing 1
thrifty, be, scrape 3
thrill°, buzz 4; delight 1; ecstasy 2; electrify 2; enrapture; excite 3; flush[1] 3, 6; glow 3; hit 12; intoxicate 2; rapture; rush 4; send 4; sensation 2; shiver[1] 3; tickle; transport 5; turn 18c
thrilled, delighted; delirious 2; ecstatic; elated; enchanted; glad 1; happy 1; overjoyed; pleased; rapturous; rhapsodic
thrilled by, be, love 8
thriller, romance 3; story[1] 1
thrilling°, delightful 1; electric; exciting 1; intoxicating 2; juicy 2; moving 1; ripping; sensational 1; stirring; swell 8
thrive°, abound 1; boom 2; come 19b; flourish 1; get 27; grow 1; make 31b, 37; pan 6; place 10; prosper; rise 8; root[1] 5; succeed 3; triumph 3
thrive on *or* upon, feed 3
thriving, flourishing; going 1; golden 5; healthy 1; luxuriant 2; plentiful 2; prodigal 2; profuse 2; prosperous 2; strong 11; successful 1
throats, at each other's, odds 4
throaty, gruff 2; thick 7
throb, ache 1; beat 3, 11; pound[1] 3; pulsate; pulse 1; rhythm; smart 7; stroke 4; thrill 2; vibrate
throbbing, ache 3; painful 1; pulse 1; rhythm; rhythmic; thrill 2
throe°, fit[2] 1; paroxysm; spasm 1; wrench 5
throes, agony
thrombosis, stroke 5
throne, chair 2; seat 1
throne, sit on the, govern 1
throng°, assembly 1; cluster 2, 3; company 2; crowd 1, 3; flock 1, 2; gather 2; herd 1; host[2]; huddle 1; jam 5; knot 2; mass 2, 8; mill 6; mob 1, 3; pack 3; populace; pour 4; press 7, 8; rabble 1; stack 3; swarm 1, 2; turnout 1
throng together, huddle 3
throng with, abound 3; swarm 3
thronged, mobbed; populous
thronging, alive 4; press 8
throng(s), many 3; score 4
throttle, choke 1; gag[1] 1; mug 5; smother 1; stifle 1
through°, around 7; by 2; direct 6; for 10; mean[3] 3; over 4, 5; owe 2; thank 4; washed-up; way 10a
through with, through 5
through and through, absolute 1; dead 14; definitive 2; hook 4; outright 2; out-and-out; perfect 6; solid 8; thorough 1; through 7; well[1] 12
throughout°, around 2, 12; everywhere; for 10; over 4, 5; pole[2] 2; thoroughly 2; through 2
throughway, road 2
throw°, cast 1, 7; confuse 1; dash 2; deliver 5; fire 9; fling 1; fluster 1; fly 8a; heave 2; hurl; launch 3; lob 1, 2; pass 8, 30; peg 5; pitch[1] 1; project 4; put 7, 21c; reflect 4; rock[2] 3; ruffle 3; send 3; shatter 3; shed[2] 2; shock 1; shoot 2; slap 3; sling 1; sort 9; spread 14; toss 1, 6

tot (up), add 2; figure 8; number 4; sum 2b; total 4
total°, absolute 1; add 2; amount 1a, 3; come 18a; complete 3; consume 4; count 1; dead 14; entire 1; figure 8; flat 4; full 2; grand 4; gross 2, 7; implicit 2; make 8; number 4; outright 1; out-and-out; overall; positive 9; profound 4; pronounced 2; pure 5; quantity; radical 2; score 1; sheer 2; stark 4; sum 1, 2b; sweeping 2; tally 3; thorough 1, 3; totality; unmitigated; whole 1, 4
  total, not, partial 1
  total (up), reckon 1; run 12; tally 2
  total (up to), work 19b
totaling, addition 2
totalitarian°, absolute 3; authoritarian; despotic; dictatorial 1; repressive
totalitarianism, despotism; monarchy 2
totality°, ensemble 3; entirety 1; integrity 2; sum 1; total 1; whole 4
totally°, absolutely 2; altogether; clean 7; clear 19; completely 1, 2, 3; dead 18, 19; downright 2; entirely 1; fairly 3; full 17; hook 4; inextricably; outright 4; perfectly 1; quite 1, 4; root[1] 3; simply 1, 2; stark 1; thoroughly 1; through 7; utterly; very 1; wholly 1
  totally, not, partially
tote, bear 1; carry 1; cart 2; lug
  tote (up), add 2
totem, fetish 1
toting up, addition 2
totter°, hobble 1; limp[1] 1, 2; reel 1; seesaw 2; stagger 1; sway 1; teeter; waddle; wallow 3
tottering, dizzy 1; limp[1] 2; poised 3; ramshackle; rickety; rocky[2]; tumble-down; unsound 1
totting up, addition 2
touch°, affect[1] 2; bit 2; border 9; breath 2; brush[2] 5; bum 5; caress 2; cast 6; concern 2; contact 3; dab 1, 2, 3; dash 7; dexterity 1; execution 4; expression 4; feel 1, 10; feeling 1; finger 4, 11; flash 2; flavor 2; fondle; get 16; handle 2; hint 2; hit 4; impress 1; involve 3; join 4; kiss 2; meet[1] 4; melt 2; modicum; move 3, 4; pat[1] 1, 3; penetrate 3; pierce 4; pinch 7; shade 3; speak 12b; speck; streak 1; stroke 6, 7, 8; suggestion 2; suspicion 2; tang 2; taste 1; technique 2; thought 7; thrill 3; tint 1; trace 1; vein 4
touch, be in, correspond 2
touch, in, current 4; informed 2
touch, soft (*or* easy), mug 3; pushover 2; sucker
touch down, land 5; touch 10
touch off, inflame 1; light[1] 15; set 20a; spark 2; touch 11
touch on, put the, bum 5
touch (on *or* upon), bear 8; bring 15b; broach; concern 1; go 30c; mention 1; note 13; refer 1; speak 10; tell[1] 2; touch 8; treat 2
touch (someone) for, borrow
touch up, edit 1; restore 3; retouch; rinse 2; tint 3; touch 12
touch (up) against, touch 3
touch (with), be, communicate 2
touch with, get in, contact 4; get 12; look 11b; reach 3
touch (with), in, sensible 4
touch with oneself, be in, mind 11
touchable, tangible
touch-and-go, critical 3; explosive 1; precarious; risky; ticklish 1; tight 7; touchy 2; tricky 2; uncertain 3; unhealthy 3
touchdown, landing 1
touch(ed) off, sound[1] 6
touched (by), be, respond 2
touched (in the head), affected 4, 5; deranged; mad 1; off 4; psychotic 1; queer 4; unbalanced 2
touches on, put the final *or* finishing, finish 7; retouch
touchiness, sensitivity 1
touching°, about 11; emotional 3; heartwarming 1; moving 1; neighboring; pathetic 1; poignant 1, 3; tender[1] 7
touchstone°, gauge 3; proof 2; standard 1; yardstick
touchup, tint 2
touch wood, finger 3
touchy°, awkward 5; cantankerous; critical 3; cross 6; delicate 4; dodgy; edge 5; embarrassing; excitable; explosive 1; fiery 3; fretful; hot 10; intense 3; irritable; moody 2; overwrought 1; passionate 3; peevish; perverse 2; prickly 3, 4; problematic; querulous; quick 5; quick-tempered; ratty 1; risky; sensitive 2; short-tempered; snappish 1; sore 2; surly; temperamental 1; tender[1] 4; testy; ticklish 1, 3; tight 7; warm 2
tough°, arduous 1; barbarian 2; brawny; bruiser; burly; callous; criminal 3; demanding 1; difficult 1, 4; domineering; dour 2; durable; exacting; gangster; gritty 2; hard 1, 2, 3, 4, 5, 6; hardy 1; hoodlum; husky 1; laborious 1; merciless; nail 4a; picnic 3; powerful 1; proof 3; punk 1; realistic 1; regrettable; relentless 1; robust 1; rocky[1] 2, 3; rough 4, 5, 7, 11, 13; rowdy 2; rugged 2, 3; ruthless; serviceable 2; severe 4; sinewy; solid 6; sound[2] 3; staunch 2; steely 2; stern 1; stiff 1, 9; stony 2; stout 3; strenuous 1; strict 2; stringy; strong 12, 13; sturdy 1; tenacious 2; thick-skinned; thorny 2; thug; tight 3; toilsome; trying; unkind
  tough it out, stick[1] 14
tough as nails°, nail 4
toughen, harden 2; stiffen 2; strengthen 2; temper 6
toughened, mill 3; seasoned 1; thick-skinned
  toughened to *or* against, used 3
tough going, picnic 3
tough guy, bruiser; tough 8
toughie, bruiser
tough luck, picnic 3; tough 7
tough-minded, realistic 1
toughness, grit; strength 1, 6; tenacity 2
tough nut to crack, problem 2
Tough shit!, tough 7
tough sledding, difficult 1; picnic 3
Tough titty!, tough 7
tour°, beat 12; circle 3; circuit 2; drive 3, 7; excursion 1; expedition 1; explore 1; go 1; journey 1, 3; lap[2] 1; outing; pilgrimage; ramble 4; ride 4; round 12; spin 7; stint 1; travel 3; traverse 1; trip 3, 7
  tour, take *or* make a, travel 3
tour de force, accomplishment 2; feat; masterpiece; stunt[1]
touring, travel 1, 2; traveling
tourism, travel 1
tourist°, arrival 2; rover; rubberneck 2; sightseer; traveler 1
tourista, run 50
tourist house, hotel
tournament°, competition 2; contest 1; game 2; match 2; meet[1] 7; running 2; tilt 4
tourney, contest 1; game 2; match 2; meet[1] 7; tilt 4; tournament
tour (of duty), spell[1] 1; stretch 6; turn 30
tours, travel 2
tousle°, mess 5a; ruffle 4; rumple
tousled, unkempt; wild 8
tout°, recommend 2
***tout de suite*,** immediately 1; instantaneously; soon 2
***tout(e) seule*,** alone 1
tow°, drag 1; draw 1; haul 1; lug; pull 1; trail 4; tug 1, 2
toward°, for 6
towards, toward 1, 2, 3
towel rail, rail[1] 1
towels and washcloths, linen
tower°, keep 16; loom 2; spire 1
  tower over *or* above, dominate 2; overshadow 1; tower 3
towering°, gigantic; heroic 6; high 1; lofty 1; massive; mighty 3; monstrous 3; mountainous 2; tall 1
tow(-haired), fair[1] 3; golden 1
tow-headed, fair[1] 3
town°, city; local 2; municipal; municipality; place 2
  town, go out on the, paint 7
town dweller, citizen 2
township, municipality; town
townsman, citizen 2; local 3
townsperson, local 3
townswoman, citizen 2; local 3
towpath, path 1
toxic, deadly 1; destructive 1; evil 3; fatal 1; harmful; malignant 1; poisonous 1; ruinous; unwholesome 1; venomous 1; virulent 1
toxicity, virulence 1
toxin, poison 1; venom 1
toy°, bauble; flirt 1; gewgaw; little 1; mess 4b; novelty 2; plaything 1; tinker; trifle 1
  toy (with), fiddle 2; finger 11; fool 6; play 19a, 19b, 19c; putter; toy 3, 4; trifle 3; twiddle 1; flirt 2
toy boy, prostitute 1
To your (very good) health!, bottom 5
trace°, bit 2; clue 1; dab 2; dash 7; describe 4; evidence 3; flicker 4; follow 4; footstep 2; ghost 2; gleam 2; grain 3; hint 2; hunt 1; mark 1, 10; monitor 3; outline 3; pursue 1, 2; ray 2; relic 2; remnant 1; rough 15; run 26a; savor 2; scent 1; scrap[1] 1; shade 3; shadow 3, 7; shred 1; sign 3; slight 2; step 3; strain[2] 2; streak 1; suggestion 2; suspicion 2; tang 2; taste 1; thought 7; tint 1; touch 15; track 4, 6, 8; trail 2, 7; vein 4; vestige; wake[2] 1
  trace out, walk 3
  trace to, attribute 2
trace(s), remain 4a; scrap[1] 2; trace 3; track 3
  traces, kick over the, rise 4
tracing, pursuit 1
track°, chase 2; course 1; follow 4, 11a; footstep 2; furrow 1; line[1] 9, 12; lock[1] 7; monitor 3; orbit 1;

track° (*cont.*) path 1, 2; pursue 1; race[1] 2; road 1; run 26a, 45; rut 1; scent 2; shadow 7; step 3; tail 4; trace 5; trail 1, 2, 7; trajectory; wake[2] 1
track, get *or* **go off the,** stray 2
track, off the, cold 8
track, on the wrong, mistaken 1
track, the, race[1] 3
track (down), discover 1; find 2; finger 5b; hunt 1; locate 2; look 11a; quest 2; run 26a; stalk[1]; trace 4; track 8
track of, keep, lock[1] 7; monitor 3; track 4
tracker, hunter
tracking, prowl 3; pursuit 1
tracking (down), hunt 3; location 2
trackless°, uncharted; uninhabited; wild 2
track record, record 5, 6
track(s), trace 3
tracks, make, bugger 5; flee 1; fly 3; make 26; speed 3
tract°, area 3; belt 2; brochure; essay 1; extent 3; field 1; pamphlet; paper 4; parcel 2; patch 2; plot[2]; province 2; region 1; reservation 4; stretch 5; territory 1; theme 2
tractability, flexibility 2; obedience; submission 1
tractable°, adaptable; amiable; easy 4; flexible 2; gentle 2; manageable; meek 2; mild 1; obedient; passive 2; plastic 2; pliable 2; propitiatory 2; receptive 1; sheepish 1; submissive 1; supple 3; tame 3; yielding 2
tractable, be, bend 5
tractableness, flexibility 2
tractile, supple 1
traction°, tension 1
trade°, bargain 3; business 1, 3; calling; career 1; change 5; clientele; commerce; craft 3; custom 3; deal 2; dealings; dicker 1; economic 1; employment 1; exchange 1, 2; game 4; give 2; industry 1; intercourse 1; job 1; line[1] 7; mercantile; merchandise 2; mission 2; occupation 1; passage 10; patronage 2; profession 1; public 9; pursuit 3; racket 3; reciprocate; sale 1, 2; sell 1; specialty 1; switch 2; transpose; truck 2; vocation; work 2
trade (in), market 4; redeem 3; sell 2; stock 9; handle 5
trade on, play 15
trade with, patronize 2
trademark, brand 1, 2; device 3; emblem; hallmark 2; label 3; patent 1; sign 4; symbol; touch 18
trade name, brand 1; label 3; patent 1
trade places, rotate 2
trader°, dealer; merchant 1
tradesman°, dealer; merchant 1; worker
tradeswoman, merchant 1
trading, patronage 2; sale 1, 2
tradition°, convention 2; custom 1; footstep 2; heritage 2; institution 4; legend 1, 2; myth 1; mythology; observance 2; practice 1; procedure; propriety 3; usage 1
traditional°, accustomed 1; certain 1; classical 1; conservative 2; conventional; correct 7; customary 2; familiar 2; habitual 1; hereditary 2; legendary 1; mythical 1; ordinary 1; orthodox; proverbial 1; regular 1; set 30; standard 6; step 7; stock 7; time-honored; usual; well-established
traditionalist, orthodox; reactionary 1, 2; square 9
tradition(s), code 3; lore 1
traduce, abuse 3; blacken 2; disparage 2; libel 4; slander 2; vilify
traffic°, business 3; commerce; deal 2; dealings; exchange 2; intercourse 1; passage 10; patronage 2; sale 1; trade 1, 5; truck 2
traffic in, handle 5; merchandise 2; run 14; sell 2
traffic circle, rotary
traffic jam, tie-up 1
trafficking, commerce; passage 10; sale 1
tragedian, dramatist
tragedy°, calamity 1, 2; catastrophe 2; disaster; distress 2; misfortune 2; ordeal; visitation 2
tragic°, calamitous; deplorable 1; disastrous 1; heart-rending; mournful 2; pathetic 1; poignant 1; unfortunate 3
trail°, climb 2; drag 1; fall 10; follow 4; footstep 2; haul 1; hunt 1; lag 1; path 1; pull 1; pursue 1; run 45; scent 2; shadow 7; spy 2; stalk[1]; tail 4; tow; trace 3, 4; track 2, 3, 6; train 2, wake[2] 1; way 3, 4
trail (along) after, tag 5
trail (behind), drag 4
trail off *or* **away,** trail 8
trailblazer, pioneer 1
trailing, pursuit 1
train°, accustom; aim 1; break 8, 18b; bring 15a; chain 1; coach 3; company 2; condition 6; direct 4; discipline 6; drill 2; educate; exercise 2; groom 4; ground 6; guide 4; indoctrinate; initiate 3; instruct 1; level 10; line[1] 6; parade 1; point 20; practice 6; prepare 2, 3; prime 5; procession 1, 2; progression 3; railroad 1; rear[2] 1; retinue; school 4; sequence; string 3; succession 2; suite 3; tame 5; teach; tutor 2; work 19a
trainable, manageable
trained, broken 4; experienced 1; expert 2; pet[1] 2; practiced 2; professional 1; proficient; qualified 1; seasoned 1; skillful; strong 6; tame 1
trainee, learner; newcomer 2; novice; recruit 2; student 1
trainer, coach 2; instructor; teacher
training, background 1; discipline 1; drill 4; education 1; exercise 4; instruction 2; maneuver 2; orientation 2; practice 2; preparation 2; scholarship 1; schooling; tuition; upbringing
training, out of, soft 12
training ground, seminary
trainload, cargo
train (of thought), thread 2
train station, station 3
traipse, gad; roam; saunter; walk 1
trait°, attribute 1; character 2; characteristic 2; feature 1; mannerism; mark 2; peculiarity 2; point 14; property 4; quality 1; respect 4; strain[2] 2; symptom; trick 6
traitor°, defector; deserter; informer; renegade 1; snake 2; subversive 2; turncoat
traitorous°, dishonorable 2; disloyal; double 5; evil 2; faithless 2; foul 7; inflammatory; perfidious; renegade 2; shifty; subversive 1; villainous 1
traitorously, back 6
traitorousness, betrayal 1; infidelity 1; perfidy
traits, stamp 5
trajectory°, path 2
tramcar, tram
trammel°, chain 2; encumber 2; enslave; hamper[1]; handicap 1, 2; hobble 2; interfere 2
trammel(s), bond 1; shackle 1; trammel 1
tramontane, foreign 1; remote 2
tramp°, beggar 1; bitch 2; bum 2; derelict 3; down-and-out 2; drifter; loafer; pauper; plod 1; ramble 4; rover; slattern; stamp 1; traverse 1; vagabond 1; walk 1, 7
tramp (on *or* **upon),** trample 1
trample°, beat 7; rout 2; stamp 1
trample down, trample 3
trample out, trample 3
trample under foot, oppress 2; trample 3
tramps, dregs 2; homeless 2
trance°, fog 2; stupor
trance, be in a, muse
trance, in a, absent-minded; daze 4; groggy; pensive
trancelike, glassy 2
tranche, slab
tranquil°, calm 3, 4; collected; comfortable 1; dispassionate 1; dreamy 3; easy 2; equable 1; even 4; gentle 1; home 4a; homely 2; inactive 1; levelheaded; mild 1; orderly 2; passive 1; pastoral 1; peaceful 1; philosophical 2; phlegmatic 2; poised 1; quiet 2; relaxed; restful 2; sedate 1; self-contained 1; self-possessed; serene 1; silent 1; smooth 1; sober 2; soft 3; still 1; subdued 1; temperate 1
tranquil, become, wind[2] 3b
tranquil, be *or* **become,** settle 7
tranquilize°, ease 5; hush 5; lull 3; mitigate; quell 2; quiet 6; settle 8; silence 3; tame 6
tranquilizer°, drug 2; narcotic 2; salve 2; sedative 1
tranquilizing, narcotic 1; restful 1; sedative 2
tranquillity, calm 1; content[2] 2; ease 1; hush 6; inactivity 1; leisure 2; lull 2; order 8; peace 1; poise 2; quiet 5; repose 1; self-control 2; serenity 1, 2; silence 1; still 3
tranquilly, easy 7
transact°, float 3; negotiate 1; transact
transaction°, bargain 1; business 3; deal 5; negotiation 2; operation 3; passage 10; proceeding 1; sale 1, 4; truck 2
transactions°, dealings; minute[1] 2; proceeding 2a; record 3; transaction 2
transalpine, foreign 1
transatlantic, foreign 1
transcend°, exceed 1; far 7; outdo; outstrip; overstep; pass 6; run 33c; surpass; top 3
transcendence, perfection 1; predominance
transcendency, predominance, supremacy 1
transcendent°, divine 2; incomparable; predominant; supreme 4
transcendental, occult 2; transcendent
transcendently, supremely
transcribe°, copy 4; minute[1] 4; record 1; render 6; take 31a; translate 1; type 4; write 4a
transcriber, scribe 1
transcript°, copy 1; minute[1] 2

transcription, transcript 1; translation 1
transfer°, carry 2; cede; commit 1; communicate 1; delegate 3; deliver 2; devise 2; displace 1; dispose 3c; disposition 3; exchange 2; give 1, 2; hand 15a, 18a; leave[1] 6; make 38b; move 2, 8; movement 1; pass 8, 18a, 30; relegate 1, 3; removal 3, 4; remove 2, 6; second[1] 10; sell 1; shift 5; sign 11; succession 1; surrender 3; transfuse 1; transit 2; translate 4; translation 4; transmission 1; transmit 1; transplant; transport 1, 4; transpose; uproot 1; will 6
transferable, hereditary 1; movable
transferal, removal 3; succession 1; surrender 3; transit 2; translation 4; transmission 1; transport 4
transference, disposition 3; removal 3; surrender 3; transit 2; translation 4; transmission 1
transferred (to), be, rub 5
transferring, transmission 1
transfiguration, mutation 1; transformation; translation 2
transfigure, convert 1; transform
transfigurement, transformation
transfiguring, mutation 1
transfix°, fascinate; freeze 3; impale; paralyze 1; pierce 1; run 34a; seize 3; stab 1; stick[1] 1
transfixing, riveting
transform°, affect[1] 3; alter; change 6, 8; convert 1; differentiate 2; make 9; modify 1; prepare 7; process 3; rehabilitate 2; renew 1; sublimate; translate 2; turn 3; vary 1
  transform into, become 1; resolve 4
transformable, changeable 2
transformation°, alteration; change 3; difference 3; mutation 1; revolution 2; transition 1; translation 2
transforming, mutation 1
transfuse°, percolate; suffuse
transgress°, blot 3; break 9; disobey; err 2; infringe 1; sin 3
transgressed, broken 5
transgression°, error 2; fault 4; ill 9; infringement; invasion 1; misdeed; offense 1; peccadillo; sin 1; slip[1] 8; vice 1
transgressor°, offender; sinner
transient°, brief 1; fleeting; fly-by-night 1; fugitive 3; meteoric; migrant 1, 2; mortal 1; passing 1; short 7; short-lived; temporary; vagabond 2
transistor, radio 1
transit°, journey 2; motion 1; pass 29; passage 1
  transit, in, move 13a
transition°, journey 2; passage 1; transit 1
transitional, intermediate 1; passing 1; provisional 1
transitory, brief 1; elusive 2; fleeting; fly-by-night 1; fugitive 3; meteoric; momentary; mortal 1; passing 1; shadowy 3; short 7; short-lived; temporary; transient
translate°, decipher 1; figure 12b; gloss[2] 2; interpret 1; render 6; say 3; transcribe 2
translation°, gloss[2] 1; interpretation 1; key 4; transcript 1; trot 4; version 2
transliterate, transcribe 2
transliteration, transcript 1; translation 1
translucent, clear 2; filmy 1; liquid 3; see-through; sheer 3; thin 5
transmissible, catching 1; hereditary 1; infectious
transmission°, broadcast 4; circulation 2; succession 1; transfer 2; translation 4
transmit°, bequeath; broadcast 1; carry 2; communicate 1; conduct 5; dispatch 2; forward 7; give 3; impart 2; pay 3; pipe 5; post[3] 2; project 4; propagate 3; radio 2; register 5; remit 1; send 2; transfer 1; transfuse 1; transport 1
transmittable, catching 1
transmittal, succession 1; transfer 2; transit 2; transmission 1
transmitted (to), be, rub 5
transmitter, bug 5; carrier 2
transmitting, transmission 1
transmogrification, transformation; translation 2
transmogrify, convert 1; transform; translate 2
transmutation, mutation 1; transformation; transition 1; translation 2
transmute, change 8; convert 1; differentiate 2; make 9; sublimate; transform; translate 2
  transmute into, resolve 4
transmuting, mutation 1
transpacific, foreign 1
transparency, clarity 1; photograph 1
transparent°, absolute 2; clear 2; distinct 1; filmy 1; flimsy 3; liquid 3; observable; patent 2; plain 2; see-through; sheer 3; thin 5; undisguised
transpicuous, transparent 1, 2
transpire°, come 4a, 14; elapse; emerge 1; happen 1; leak 6; light[1] 10; occur 1; place 14
transplant°, graft[1] 1; plant 5; uproot 1
transport°, banish 1; bear 1; car 1; carry 1, 2, 9; cart 2; deliver 1; delivery 1; ecstasy 1; enrapture; entrance[2]; exile 3; freight 1; frenzy 1; haul 2; hypnotize; isolate; lug; rapture; ravish 1; run 13; ship 2; spirit 10; take 9; traffic 1; transfer 1; transit 2; translate 4; translation 4; transmission 1; wash 2
transportable, mobile 1, 2; movable; portable
transportation, carriage 3; delivery 1; exile 1; freight 1; traffic 1; transit 2; translation 4; transmission 1; transport 4
transported, ecstatic; elated; exalted 3; fervent 3; overjoyed; rapt 1; rhapsodic
  transported, be, waft 1
transporter, carrier 1
transporting, transmission 1
transports°, transport 5
transposable (with), synonymous
transpose°, reverse 2; shift 1
transubstantiate, translate 2
transubstantiation, translation 2
transudate, secretion
transudation, secretion
transude, percolate; secrete[2]; sweat 4
trap°, ambush 1, 2; bag 6; bottle 4b; box[1] 3; catch 2, 18; decoy 1; enticement 2; frame 9; gimmick 1; hook 2, 6; intercept; joker 2; loaded 3; mesh 2, 3; mouth 1; net[1] 2; pitfall 1; seduce 1; setup 2; snare 1, 2; trip 6; web; wile; yap 3
trapped, become, catch 7
trappings°, equipment; finery; fitting 2; furniture 2; gear 2; kit; outfit 1; paraphernalia; regalia; robe 2; stuff 2; tackle 1; turnout 3
trash°, destroy 1; dirt 1; discard 1; dispose 3b; filth 1; garbage; junk 1, 2; litter 1; lumber 1; nonsense 1; pan 5; prattle 3; rabble 2; refuse[2]; rot 4; rubbish 1, 2, 3; scrap[1] 4; shred 2; stuff 4; throw 5a; truck 1; waste 7
trashy, cheap 3; coarse 4; gaudy; pulp 4; shoddy; sleazy 2; worthless 3
trauma, grief 2; shock 2; wound 1
traumatic°, fraught 2; harrowing
traumatism, wound 1
traumatize, shock 1; wound 3
traumatizing, traumatic
travail, drudgery; effort 1; grief 2; grind 7; labor 1, 4, 5; pain 2; slavery 3; sorrow 2; struggle 4; work 1
travel°, cover 9; cruise 1; drive 3; explore 1; get 10; go 1; hop 2; journey 1, 3; knock 3a; make 19; migrate 1; motion 1; passage 7; progress 5; ramble 1; roam; tour 4; transit 1; trip 7
  travel across, range 8; transit 3
  travel at, make 20
  travel by, get 11
  travel on *or* **in,** ride 1
  travel over, cover 9; transit 3
  travel over *or* **through,** range 8; traverse 1
  travel through, transit 3
traveler°, arrival 2; bum 2; fare 1; migrant 1; passenger; rover; seller; sightseer; tourist
traveling°, migrant 2; mobile 1; motion 1; move 13a; passage 7; rambling 3; step 10a; transit 1; travel 1, 2; traveling
traveling salesman, seller
travels°, step 10a; travel 2
traversable, navigable 1; passable 2
traversal, passage 1; transit 1
traverse°, bridge 3; cover 9; cross 5; describe 4; explore 1; get 46a; make 19; navigate 1; pace 3; passage 1; range 8; transit 1, 3; travel 3; wade 1
traversing, transit 1
travesty, burlesque 1, 3; charade; derision; imitate 2; imitation 2; joke 3; mock 2; mockery 2; parody 2; ridicule 2; satire 2; satirize; take 34b; takeoff 2
trawl, tow; trail 4
tray, platter
treacherous, black 6; crafty; dangerous 2; designing; devious 1; dirty 3; dishonest; dishonorable 2; disloyal; double 5; evil 2; faithless 2; false 2; flagrant; foul 7; hypocritical; immoral 1; inflammatory; insincere; lawless 3; lying 2; machiavellian; perfidious; precarious; renegade 2; scheming; seditious; shifty; sinister 2; slippery 2; sly 1; subversive 1; traitorous; two-faced; unreliable; unscrupulous; untrue 1; villainous 1; wily; wretched 4
treacherously, back 6
treachery, betrayal 1; deceit 1; deception 1; foul play; fraud 1; infidelity 1; perfidy; sabotage 2; sedition; stab 6; subtlety 2
treacly, sweet 7
tread, beat 7; footstep 1; march 1; motion 3; pace 3; plod 1; stamp 1; step 2; tramp 1, 5; walk 1

tread (*cont.*)
tread carefully *or* **cautiously,** step 12
tread (on), trample 1
treading water, limbo
treadmill, rut 2
treason, betrayal 1; perfidy; sabotage 2; sedition
treasonable, disloyal; perfidious; traitorous
treasonous, disloyal; perfidious; renegade 2; subversive 1
treasure°, appreciate 1; cherish 1; dear 3; esteem 1; fortune 1; gem 2; jewel 2; love 7; pearl; plum; pride 3; prize[2]; rarity 1
treasured, beloved 1; darling 3; dear 1; pet[1] 3; sweet 5
treasure trove, mine 2; thesaurus 1
treasury°, mine 2; purse 2; thesaurus 1
treasury note, note 4
treat°, dainty 4; deal 4; doctor 2; dress 3; drug 3; entertain 2; feast 1, 3, 5; get 40b; handle 6, 7; indulgence 3; joy 3; luxury 4; nurse 2; prepare 7; process 3; regard 2; remedy 3; take 39e; tidbit
treat cavalierly *or* **lightly,** play 19a
treat condescendingly, patronize 1
treat in kind, pay 5
treat irreverently, violate 2
treat (of), speak 10
treat seriously, not, play 19b
treat (someone) to (something), indulge 1; spring 5; treat 5
treat thoroughly, exhaust 4
treated, prepared 2
treatise, exposition 3; lecture 1; monograph; paper 4; theme 2; tract[2]
treatment°, drug 1; play 25; reception 1; remedy 1; style 5; therapy 1; twist 8; usage 2; use 8
treatment of, for the, for 4
treaty°, accommodation 2; accord 3; agreement 1; bond 2; negotiation 2; pact; protocol 2; truce 2; understanding 1
treble, high 7
trebuchet, sling 2
trebucket, sling 2
tree, plant 1
tree, out of one's, mad 1
tree-covered, wooded
treeless, open 9
treenail, spike 1
trees, timber 1
tref, impure 3; profane 2
trek, excursion 1; go 1; journey 1, 3; march 3; pilgrimage; ramble 1, 4; run 36; tour 1, 4; tramp 1, 3; travel 3; trip 3, 7; walk 1
trekking, travel 2
treks, travel 2
trellis, grating 3; mesh 1; net[1] 1
trelliswork, mesh 1; net[1] 1
tremble°, cringe 1; flicker 2; flutter 3; quake 1; quaver 1, 2; quiver 1, 2; shake 1, 8; shiver[1] 1, 3; shudder 1, 2; teeter; thrill 2; totter; vibrate; wriggle 1
tremble at, fear 5
trembler, quake 2
trembling, doddering; flutter 4; poised 3; quaver 2; shake 8, 10; shiver[1] 3, 4; thrill 2; tremulous 1; wriggle 4
tremblor, quake 2
trembly, doddering
tremendous, big 1; enormous; fantastic 4; gigantic; great 1, 2; huge; immense; incredible 2; massive; mighty 3; monstrous 3; monumental 2; prodigious 1; vast; voluminous 1; whopping 1
tremendously, fearfully 2; highly 1
tremor, quake 2; quaver 2; quiver 1, 2; shiver[1] 3; thrill 2; tremble 2
tremors, shake 10
trench, channel 1; drain 1; excavation; furrow 1; gouge 3; pit[1] 1
trenchant°, caustic 2; concise; devastating 1; eloquent 1; epigrammatic; incisive 1; keen[1] 2; penetrating 1; pointed 2; pungent 2; sarcastic; sharp 5; smart 4; strong 7; tart[1] 2; violent 3; virulent 2
trencher, plate 1
trencherman, glutton
trend°, craze; current 6; drift 3; fad; fashion 1; mode[2]; movement 7; popularity; rave 4; style 2; swing 4; tendency; thing 3; turn 32; undercurrent 2; vogue 1
trendiness, popularity
trendsetter, pioneer 1
trendy°, chic 1; contemporary 2; current 3; date 5; exclusive 2; fashionable; glamorous 2; minute[1] 3; modern; new 2; now 5; popular 1; snappy 2; sporty; stylish; swinging
trepidation, alarm 2; dismay 3; dread 2; fear 1, fright 1; horror 2
trespass, encroach; err 2; error 2; fault 4; invasion 1; misdeed; offense 1; peccadillo; sin 1, 3; transgress 1; transgression
trespass on, infringe 2
trespasser, intruder 1; outsider; sinner; transgressor
tress, lock[2]
tresses, hair 1
trestle, beam 1; rack 1; rest[1] 4
tret, allowance 4
trews, pants 2
triad, trio
trial°, affliction 1; bind 6; challenge 6; contest 1; distress 2; experience 2; experiment 1; grief 2; hell 2; match 2; misery 3; nuisance 1; ordeal; pain 2; pass 27; pest; predicament; proof 2; run-through 1; sample 3; sorrow 2; suffering; suit 4; tentative 1; test 1; tilt 4; visitation 2; woe
trial, put on *or* **bring to,** prosecute 1
trial run, trial 1
tribade, deviant 3; homosexual 1
tribadic, homosexual 2
tribadistic, deviant 2
tribal, native 4; racial
tribe°, clan 1, 2; folk; nationality 2; people 4; race[2] 1
tribulation, affliction 1, 2; care 1; difficulty 1; grief 2; misery 3; pain 2; sorrow 2; suffering; trial 4; trouble 5; woe
tribulation(s), ordeal
tribunal°, bar 5
tributary°, river 1; stream 1
tribute°, blackmail 1; compliment 1; credit 4; eulogy; homage; memory 3; mention 4; pledge 3; praise 1; rave 3; reward 2; tax 1; toast 1; toll[2] 1
tribute to, pay, commemorate; compliment 3; exalt 2; extol; glorify 2; memorialize; praise 3; salute 2; toast 3
tributes, laurels
trice, flash 3; instant 2; minute[1] 1; moment 1; twinkling 1
trice, in a, once 6a; posthaste; rapidly 2; suddenly 1; swiftly
trichoid, hairy 1
trick°, artifice 2; blind 8; bluff[1] 1; catch 18; cheat 2; chisel 2; deceit 2; deceive; deception 2; decoy 2; defraud; delusion 1; device 2; dodge 4; double-cross; dupe 3; finesse 2, 4; fool 4; fraud 2; gag[2] 2; game 3; gimmick 1; have 12c; hoax 1, 2; hocus-pocus 1; hoodwink; machination; maneuver 1, 3; mislead; move 9; outsmart; prank; prey 3b; put 26; put-on 1; quirk; racket 2; refuge 2; ride 5a; rip 2b; ruse; scheme 3; shuffle 5; stall[2] 2; stratagem; string 10c; stunt[1]; subterfuge; swindle 1; take 32b; trap 2, 6; tricky 3; trip 6; turn 30; twist 10; victimize 2; wile; wrinkle[2]
trick, do the, take 11; trick 7
trick out *or* **up,** embellish 1; primp
trickery°, art 5; artifice 1; camouflage 1; chicanery; craft 2; deceit 1; deception 1; evasion 2; fraud 1; hanky-panky; hocus-pocus 1; intrigue 3; machination; magic 2; racket 2; swindle 2
trickiness, artifice 1
trickle°, drain 6; drip 1, 2; drop 5; filter 3; flow 1; leak 1, 4; ooze 2; percolate; rain 5; run 6; secrete[2]; secretion; well[2] 2
trickling, secretion
tricks, art 6; nonsense 2
tricks (on), not play, serve 5
trickster, adventurer 2; cheat 1; devil 4; fraud 3; impostor; joker 1; phony 3; rogue 1; swindler; thief 2
tricky°, arch 2; artful 1; awkward 5; crafty; deceptive 2; delicate 4; designing; devious 1; dicey; disingenuous; dodgy; elfin 1; elusive 1; evasive; foxy 1; fraudulent 2; hairy 2; hazardous; insincere; loaded 3; machiavellian; politic 1; precarious; problematic; profound 1; scheming; serpentine 1; shady 2; shifty; slippery 2; sly 1; smooth 8; sticky 2; subtle 3; tender[1] 4; tight 7; tortuous 2; wily
tricky Dick, impostor
tried and true, constant 1; safe 3; sure 3
trifle°, bauble; gewgaw; hair 2; mess 4b; modicum; nothing 3; novelty 2; scattering; thought 7; tinker; toy 2; trace 2; triviality 2
trifle away, lose 4
trifle (with), dabble 2; fiddle 2; flirt 2; fool 6; play 19a; putter; toy 3; trifle 3
trifler, dilettante
trifles, trash 2
trifling°, frivolous 1; idle 4; immaterial 1; incidental 2; inconsequential; indifferent 4; insignificant; light[2] 6; little 5; meager 1; meaningless 2; minor 2; minute[2] 2; negligible; niggling 2; nominal 2; paltry; petty 1; pitiful 2; poor 5; puny 1; slender 3; slight 2; small 2, 4, 5; small-time; tenuous 2; tiny; unimposing
trifocals, glass 7
trig, neat 1; neaten; trim 1
trigger, activate; initiate 1; pioneer 2; precipitate 1; set 20a; spark 2; touch 11b; trip 8
triggered, immediate 1
triggerman, killer 1
trill, chirp 1, 2; sing 1; twitter 1, 3
trilling, twitter 3

twisted (*cont.*) misshapen; mistaken 2; perplexing; perverted; shapeless 2; tortuous 1; wry 1
twister°, eddy 1; hurricane
twisting, circular 2; indirect 1; perversion 1; serpentine 2; tortuous 1; wriggle 4
twisting(s), meander 2
twistings and turnings, maze
twist of fate, fluke
twist(s), meander 2
twist (someone's) arm, enforce 2; force 7; pressure 5; screw 4; squeeze 3; twist 6
twit°, chaff 2; deride; dolt; flout; fool 1, 4, 5; gibe 1; half-wit; jeer 1; leg 8; parody 3; poke 4; pull 15; rag[2]; sap[1] 2; scoff; silly 3; taunt 1; tease 1
twitch, fidget 1; flutter 3; jerk 2; jump 11; nip[1] 1; shake 2, 8; shudder 2; spasm 1; switch 1, 3; tweak 1, 2
twitter°, chirp 1, 2; flicker 2; giggle 1, 2; peep 1, 2; prattle 1, 2
  twitter, in a, overwrought 1
twittering, prattle 2; twitter 3
twitting, chaff 1
two, second[1] 11
  two, set of, pair 1
  two and two together, put, reason 7
  two peas in a pod, like, indistinguishable 1
two-a-penny, frivolous 1; minor 2
two-bagger, bag 4; hag
two-bit, cheap 3; frivolous 1; minor 2
two-dimensional, dead 12; flat 5, 13; lackluster; small 3
two-dimensionality, tedium
two-face, hypocrite
two-faced°, crafty; deceitful; dishonest; dishonorable 2; disingenuous; foul 7; hypocritical; insincere; perfidious; shifty; untrue 1
two-facedness, hypocrisy
twofer, pass 25
twofold, double 1, 3
two of a kind, pair 1
twopence, for, cheap 5, 6, 7
twopenny, cheap 3; frivolous 1; paltry
twopenny-halfpenny, paltry; petty 1
two-ply, double 2
twosome, couple 1; pair 1
two-time, deceive; double-cross
two-timer, traitor
two-timing, traitorous; wily
tycoon°, merchant 3; mogul
tying, whipping 2
type°, brand 1; breed 1; cast 4; category; character 1, 2; class 2, 5; denomination 2; description 3; family 4; form 3; genre; kind[2] 1; line[1] 16; make 42; model 6; mold[1] 2, 3; nature 4; norm 2; print 3; race[2] 1; rate[1] 4; run 44; separate 2; sort 1, 2, 3, 9; specimen; stamp 8; standard 1; stripe 2; style 1; variety 3; version 1
  type, of the same, piece 8
typeface, type 2
typescript, paper 4
typewrite, type 4
typhoid Mary, carrier 2
typhoon, blow[1] 9; eddy 1; hurricane; storm 1; tempest 1; twister; whirlwind 1
typical°, average 3; character 8; characteristic 1; distinctive; exemplary 1; natural 1; normal 1; ordinary 1; proverbial 2; regular 1; regulation 5; representative 1; standard 6; thick 9; usual
  typical (of), indicative of; peculiar 2; specific 2; symbolic
typically, average 2; generally 1; ordinarily
typification, soul 3
typify°, embody 2; exemplify 1; personify 1; represent 4; stand 7a; symbolize
typifying, indicative of
typo, misprint
typographical error, error 1; misprint
tyrannical°, absolute 3; arbitrary 2; authoritarian; bossy; despotic; dictatorial 2; domineering; exacting; hard 4; harsh 2; heavy-handed 2; imperative 2; masterful 2; merciless; oppressive 1; overbearing; peremptory 3; repressive; severe 1; strict 2; totalitarian
tyrannically, severely 2
tyrannize°, browbeat; bully 2; grind 5; intimidate; persecute 1; push 7; railroad 2; ride 3; subjugate
  tyrannize (over), oppress 2; tyrannize
tyrannized, downtrodden
tyrannous, tyrannical
tyranny°, despotism; domination 2; monarchy 2; oppression; persecution 1; severity 1
tyrant°, bully 1; despot; dictator; disciplinarian; oppressor
tyro, amateur 1; apprentice 1; greenhorn; initiate 4; learner; newcomer 2; novice; pupil; recruit 2

Übermensch, superhuman 2
ubiquitous, pervasive; prevalent 1; universal 1
ubiquitously, everywhere
ubiquitousness, prevalence 1
ubiquity, prevalence 1
uglify, ruin 8
ugly°, awful 1; explosive 1; forbidding 2; frightful 2; ghastly 1; hideous 1; homely 3; monstrous 1; nasty 4; plain 5; repulsive; seamy; ungraceful 2; unsightly; wintry 3
ukase, declaration 2; decree 1; dictate 2; law 1; order 4; precept 1; pronouncement 2; regulation 3; rule 1; warrant 2
ulcer°, sore 7
ulcerate, fester 1
ulcerated, ulcerous
ulceration, ulcer 1
ulcerative, ulcerous
uliginous, slimy 1
ultimate°, definitive 1, 2; eventual 1; extreme 3; final 1; last[1] 3; net[2] 3; primary 2; supreme 2; terminal 1
ultimately°, eventually; finally 1; last[1] 7; length 4a; run 48; soon 5
  ultimately, become, wind[2] 4b
ultimatum°, challenge 5; dare 3
(ultra)conservative, narrow-minded; prig; priggish; reactionary 1, 2
ultraconventional, hidebound
ultramodern, streamlined 2
ultramontane, remote 2
ultramundane, heavenly 1; supernatural
ululate, howl 1; moan 3
ululating, howl 2
ululation, howl 2
umbilicus, navel
umbrage°, take, offense 3; umbrage
umbrage° (at), take, bridle 3; exception 4
umbral, obscure 1
umbrella°, inclusive 1; shade 5; shelter 2; sweeping 1
ump, umpire 1
umpire°, decide 1; intermediary; judge 2, 6; mediator; moderator; officiate; peacemaker
umpy, umpire 1
unabashed°, audacious 2; barefaced 2; brazen; cool 6; downright 1; forward 2; frank 1; ingenuous 2; shameless
unabashedly, openly 1; shoulder 6
unabated, complete 1; relentless 2; unmitigated
unable°, incompetent; powerless 1
  unable (to), incapable 1
unabridged°, complete 1; full 12
unaccentuated, unaccented
unacceptable°, exceptionable; inadmissible; ineligible; out 13; pale[2] 3; poor 4; taboo 1; unbearable; undesirable 2; unlikely 2; unsatisfactory; unthinkable 2; wrong 4
unacceptably, badly 3
unaccepted, unpopular; unwelcome 2
unaccommodating, contrary 2; difficult 3; inflexible; mean[2] 1, 4
unaccompanied°, alone 1; lonely 1; solo 1, 2
unaccomplished, incomplete; undone[2]
unaccountable°, inexplicable; irresponsible; strange 1; unnatural 3
unaccustomed°, practice 5; strange 2
  unaccustomed to, unaccustomed 2; unfamiliar 2; unused 4
unachievable, impossible 1; impracticable 1
unacknowledged, thankless
unacquainted with, unfamiliar 2
unadaptable, inflexible; rigid 4
unadmitted, sneaking 2
unadorned°, bare 2; blank 2; chaste 2; dry 2; empty 1; matter-of-fact; meager 3; naked 3; natural 3; neat 3; plain 4; severe 6; simple 2; stark 2; understated
unadulterated, absolute 1; clean 1; immaculate 2; literal 2; naked 3; neat 2; primary 5; pure 1; sheer 2; solid 4; straight 9; true 1; undiluted
unadvertised, unheralded
unadvisable, inept 2
unadvised, inept 2
unaffected°, artless 2; homely 1; immune; informal 1, 3; ingenuous 1; insensible 2; naive; natural 4; passive 1; phlegmatic 1; real 3; resistant 1; simple 3; unmoved; unsympathetic
  unaffected by, above 5; blind 4; free 11; unaffected[2]
unaffectedly, naturally 3; simply 3
unaffectedness, ease 4; simplicity 3
unaffiliated, independent 6, 7; neutral 1; nonaligned; nonpartisan 1; unattached 1; unrelated
unafraid, bold 1; brave 1; daring 2; dauntless; gallant 1; indomitable; intrepid; tame 2
unaggressive, meek 2; patient 1, unobtrusive
unagitated, calm 3; quiet 2; silent 1; tranquil
unagreed, open 6
unaided, naked 2; single-handed 1
unaligned, fence 2; independent 4; liberal 2; neutral 1; nonaligned; nonpartisan 1
unalike, different 1; dissimilar; unlike 1
unalleviated, unmitigated
unallied, independent 6; neutral 1; nonaligned; unrelated

unallowable, inadmissible; unacceptable

unallowed, inadmissible

unalloyed, absolute 1; barefaced 1; complete 3; entire 1; implicit 2; naked 3; out-and-out; perfect 6; point-blank 1; pronounced 2; pure 1, 5; rank[2] 2; regular 9; right 8; sheer 2; solid 4; straight 9; thorough 1; total 3; undiluted; undisguised; unmitigated

unalterability, finality; resolution 1

unalterable, changeless 1; constant 3; eternal 3; final 2; firm 4; fixed 2; hard 8; immovable 2; incorrigible; indestructible; invariable 2; irreversible; irrevocable; level 3; permanent 2; resolute; rigid 2; stable 2; unavoidable; uniform 1

unaltered, invariable 3; same 2; straight 9; uniform 1

unambiguous, absolute 4; categorical; clear 6; decided 1; definite 3; definitive 3; direct 9; distinct 1; downright 1; emphatic; explicit 1; express 5; flat 4; forthright; graphic 1; honest 3; intelligible; manifest 1; outspoken; plain 2; plump[2] 6; positive 1; precise 1; pronounced 2; specific 1; straight 5; transparent 2, 3

unambiguously, completely 3; entirely 2; expressly 1; honestly 2; outright 4; perfectly 1; shoulder 6; simply 5; straight 14; wholly 2

unambiguousness, clarity 2; precision 2

unambitious, meek 1; shiftless

unamiable, unsocial

unanimated, mechanical 3; phlegmatic 1; wooden 2

unanimity, accord 2; agreement 2; harmony 1; identity 1; solidarity; union 4; unity 1

unanimous, united 3

unannounced, abrupt 1; precipitate 4; spontaneous 1; sudden; unheralded; unofficial

unanswerable, irresponsible; rhetorical 2

unanswered, unresolved

unanticipated, abrupt 1; accidental; precipitate 4; sudden; unaccustomed 1; unforeseen; unheralded

unapologetic, unrepentant

unappealing, uninviting

unappeasable, implacable; mortal 4

unappetizing, bitter 2; disgusting; uninviting; unpalatable; unsavory; yucky

unapplied, abstract 1

unappreciated, thankless

unappreciative, ungrateful

unappreciativeness, ingratitude

unapproachable°, aloof 3; cold 3; cool 5; distant 3; frigid 2, 3; inaccessible 1; standoffish

unapproved, unauthorized

unapt, extraneous 2; incongruous; unbecoming 1; unreasonable 3; untoward 2

unarguable, certain 4

unarm, disarm 1

unarmed°, clean 6

unarranged, unpremeditated

unartificial, artless 1; innocent 4; natural 3; unaffected[1]

unartistic, matter-of-fact; mechanical 3

unascertainable, uncertain 1

unashamed, blatant 1; brazen; forward 2; shameless; unabashed; unrepentant

unashamedly, openly 1; shoulder 6

unasked°, unwarranted; voluntary 1

unasked-for, free 6; gratuitous 2; unasked; unsolicited; unwarranted

unassailable, flawless 2; invincible 2; spotless 2; watertight 2

unasserted, passive 3

unassertive, passive 2; submissive 1; tame 3; unobtrusive; weak 3

unassertiveness, submission 1

unassimilable, foreign 3

unassisted, alone 1; naked 2; single-handed 1

unassociated, unrelated

unassuming, artless 2; bashful 2; childlike; coy; homely 1; humble 1; inconspicuous; informal 1; meek 1; mild 1; modest 1; retiring; simple 2; sweet 4; unaffected[1]; unobtrusive

unassumingly, quietly 4

unattached°, detached 1; disconnected 1; discrete; free 5; loose 1; single 1; unmarried

unattainable, impossible 1; impracticable 1; inaccessible 1

unattended, alone 1; solitary 1; unaccompanied

unattended to, undone[2]

unattentive, absent-minded

unattested, unfounded

unattractive, homely 3; plain 5; seamy; ugly 1; unenviable; ungraceful 2; uninviting; unlikely 2; unpalatable; unsightly

unauthentic, factitious; spurious; unrealistic 2

unauthorized°, illegal; illegitimate 3; unlawful; unofficial

unavailable, attached 4; inaccessible 1

unavailing, bootless; feeble 2; fruitless; futile; helpless 4; hollow 5; hopeless 4; idle 4; impractical 2; ineffectual 1; meaningless 2; nonproductive 2; unprofitable 2; unsuccessful 1; useless 1; vain 2; void 1; worthless 2

unavoidability, necessity 2

unavoidable°, automatic 2; certain 3; destined 2; eventual 2; fatal 3; fated 3; hard 8; indispensable 2; inevitable; irresistible 2; necessary 2; sure 4; willy-nilly 2

unavoidably, inextricably; necessarily

unavowed, sneaking 2

unaware°, absent-minded; ignorant 2; insensible 2; mindless 2; oblivious; preoccupied 2; uninformed; unsuspecting; vacant 2

unaware (of), blind 4; unconscious 2

unawareness, ignorance 1; oblivion 2; vacancy 3

unawares°, nap[1] 2

unawed, unabashed

unbalanced°, crazy 1; deranged; disproportionate; disturbed 2; insane 1; lopsided 1; mad 1; mental 2; one-sided 2; psychotic 1; queer 4; sick 4; unsound 3

unbar, open 21

unbarred, open 1

unbearable°, excruciating; grievous 2; heavy 4; insufferable; irresistible 1; murderous 2; oppressive 1; overpowering; overwhelming 1; stiff 2; terrible 1

unbeatable°, impregnable; indomitable; invincible 1

unbecoming°, dishonorable 3; dowdy; improper 3; incongruous; indecent 1; undesirable 2; unlikely 2; unprofessional 1; unseemly 1; untoward 2; wrong 4

unbecoming to, unworthy 2

unbefitting, below 11; beneath 4; improper 2; inappropriate; unprofessional 1; unreasonable 3; unseemly 2

unbeknownst, unknown 3

Un bel di, sometime 3

unbelievable°, awesome; colossal 2; dreamlike; extraordinary 2; fabulous 2; fantastic 3; far-fetched; flimsy 2; formidable 2; good 2; implausible; improbable; inconceivable; incredible 1; marvelous; miraculous; monumental 4; phenomenal; preposterous; remarkable 1; sensational 1; splendid 3; superb; superhuman 1; tall 3; terrific 2; thin 4; unheard-of 2; unthinkable 1; weak 4; world 8

unbelievable, be, water 2

unbelievably, pretty 3; terribly

unbeliever, heathen 1; infidel; nonbeliever; pagan 1

unbelieving°, distrustful; faithless 1; godless 2; incredulous; profane 1; suspicious 2

unbelligerent, peaceable 2

unbemoaned, unlamented

unbend, relax 3; thaw 2

unbendable, rigid 1; stiff 1

unbending, firm 4; formal 1; grim 1; immovable 2; inflexible; merciless; monolithic; perverse 3; precise 2; recalcitrant; relentless 1; rigid 1, 2; rocky[1] 2; severe 1; stalwart 2; stiff 1; straight 1; tough 6; unkind

unbesmirched, pure 2

unbetrothed, unmarried

unbewailed, unlamented

unbiased, candid 2; detached 2; disinterested; dispassionate 2; equitable; even 7; factual 2; fair[1] 1; fence 2; honorable 3; impartial; impersonal 1; independent 4; indifferent 2; just 1; literal 2; neutral 1; nonpartisan 2; objective 1; unprejudiced

unbidden, free 6; independent 4; spontaneous 2; unasked; voluntary 1

unbigoted, candid 2; equitable; liberal 3; objective 1; tolerant; unprejudiced

unbind, disengage; free 15; loosen 1; undo 2

unblemished, chaste 1; clear 4; fair[1] 3; flawless 1; immaculate 1, 2; impeccable; innocent 2; intact; pure 2; white 3

unblended, neat 2; undiluted

unblessed, unfortunate 1

unblinking, fixed 2; steady 3

unblock, clear 23; open 22

unblocked, clear 16; navigable 1; open 9; passable 2; unimpeded

unblushing, barefaced 2; shameless; unabashed

unbolt, disengage; open 21; undo 1

unbolted, open 1

unbooked, open 8

unborn, future 2

unbosom, free 16; open 23

unbound, loose 4

unbounded, boundless; endless 1; immeasurable; indefinite 4; inexhaustible 1; limitless; rampant 2; unlimited 2; vast

unbowdlerized, full 12; unabridged 1

unbreakable, indestructible

uncompromising, arbitrary 2; blunt 2; difficult 3; dour 2; emphatic; extreme 4; grim 1, 4; hard 6; hidebound; implacable; inflexible; insistent; point-blank 1; puritan 2; relentless 1; rigid 2, 4; severe 1; stalwart 2; stern 1; strict 2; strong 5; strong-minded; stubborn; sturdy 2; tenacious 1; tight 3; willful 2
uncompromising, be *or* **remain,** persevere
uncompromisingly, downright 2; flat 16; point-blank 3
uncompromisingness, tenacity 1
uncompulsory, optional
unconcealed, bare 2; barefaced 1; broad 3; glaring 1; naked 3; noticeable 1; obvious; open 12, 13; overt; public 5; undisguised
unconcern, detachment 2; ease 4; indifference 1, 2; neglect 3; negligence; torpor
unconcerned, blithe 2; carefree; careless 1; casual 3, 5; cool 4; dead 4; deaf 2; halfhearted; heartless; impassive; inattentive; inconsiderate; indifferent 1; lackadaisical 2; listless; nonchalant; oblivious; offhand 1; perfunctory 1; philosophical 2; phlegmatic 1; supine 2; tepid 2; torpid; unabashed; unkind; unmoved; unsympathetic
unconcerned (by), unaffected[2]
unconcluded, moot 1; open 6
uncondensed, unabridged 1
unconditional, absolute 2; categorical; decided 1; entire 1; explicit 1; flat 4; free 6; full 8; implicit 2; open 4; outright 1; peremptory 2; precise 2; stark 4; total 3; unlimited 1; unqualified 2
unconditionally, absolutely 1; completely 2; dead 18; downright 2; outright 5; totally
unconditioned, absolute 2; free 6
unconfessed, sneaking 2
unconfident, bashful 1; insecure 1; shy 1
unconfined, epicurean 1; free 1; large 5a; limitless; loose 2; out 6
unconfining, loose 3
unconfirmed, pending 2
unconforming, Bohemian; unorthodox
unconformity, abnormality 1
unconfused, clear 7
uncongenial, incompatible; mismatched; standoffish; unlikely 2; unsocial
unconnected, disconnected 1; disjointed 1; extraneous 2; foreign 3; inapplicable; inarticulate 1; incoherent; independent 2, 7; irrelevant; isolated 2; loose 1, 5; outside 9; remote 4; unattached 1; unrelated
unconnected with, beside 2
unconquerable, impregnable; indomitable; invincible 1; irresistible 1
unconscientious, promiscuous 1
unconscionable°, black 6; excessive 1; exorbitant; inordinate 1; outrageous 1; unreasonable 2; unscrupulous; unwarranted
unconscious°, automatic 1; brute 1; dead 3; ignorant 2; inadvertent 1; insensible 1; involuntary; lifeless 2; mechanical 2; oblivious; out 8; psyche; psychological; senseless 1; spontaneous 2; subconscious 1, 2; subliminal; unaware; unpremeditated; unsuspecting
unconscious, knock *or* **render,** knock 6a
unconscious of, blind 4; unconscious 2
unconsciously, idly 2; unawares 2
unconsciousness, faint 4; ignorance 1; oblivion 2; stupor
unconsecrated, profane 2
unconsidered, impulsive; promiscuous 1; rash[1]; unthinking 1
unconsignable, inalienable
unconstrained, absolute 3; casual 5; demonstrative 1; downright 1; forthright; frank 1; free 1, 9; honest 3; incontinent 1; informal 2; liberty 4; natural 4; open 16; smooth 3; unimpeded; uninhibited; unlimited 1; voluntary 2
unconstrainedly, freely 1
unconstraint, freedom 6
unconsumed, remaining 2; unused 3
uncontainable, irrepressible
uncontained, extravagant 2
uncontaminated, clean 1; pure 2; sterile 2
uncontested, positive 1; runaway 3; undisputed
uncontrived, candid 1; simple 3; unpremeditated
uncontrollable, disorderly 2; fierce 2; hysterical 1; incontinent 1; incorrigible; involuntary; irrepressible; irresistible 1; mutinous 2; obstreperous; overwhelming 1; problem 3; rampant 1; recalcitrant; riotous 1; rogue 2; self-willed; tempestuous; ungovernable; unruly; untamed; violent 1; voracious 1; wild 4
uncontrolled°, boundless; disorderly 2; free 1; hysterical 1; incontinent 1; independent 2; inordinate 2; involuntary; obstreperous; promiscuous 2; rampant 1, 2; riotous 1; runaway 2; shameless; tempestuous; undisciplined; uninhibited; unlimited 1; voracious 1; wild 4
unconventional, abnormal 1; bizarre 1, 2; Bohemian; curious 3; different 2; eccentric 1; erratic 2; extreme 5; fresh 2; funny 2; informal 2; irregular 3; kinky 1; novel 1; odd 1; offbeat; ordinary 4; outré; out-of-the-way 2; peculiar 1; quaint 1; queer 1; sick 5; special 1; step 8; unfamiliar 1; unorthodox; unusual; unwonted; way-out 2; wild 4
unconventionality, abnormality 1; eccentricity 1; oddity 1; originality
unconversant with, unfamiliar 2
unconvinced, doubtful 2; unbelieving
unconvincing, far-fetched; feeble 2; flimsy 2; implausible; lame 2; thin 4; weak 4
uncooked, raw 1
uncooperative, disagreeable 3; insubordinate; obstinate; resistant 2; self-willed; unruly
uncooperativeness, noncompliance; obstinacy
uncoordinated, awkward 1; clumsy 1; disconnected 2; incoherent; indiscriminate 2; unrelated
uncordial, cold 3; cool 5
uncork, open 21
uncorrectable, irretrievable 2
uncorroborated, ill-founded
uncorrupt, honest 1; honorable 1
uncorrupt(ed), flawless 1; holy 2; honest 1; honorable 1; incorrupt 2; innocent 2; pristine 2; pure 3; unspoiled; untarnished; upright 2; virtuous 1, 2
uncountable, immeasurable; indefinite 4; infinite; many 1; numberless; untold 1
uncounted, immeasurable; indefinite 4; infinite; numberless; untold 1
uncouple, detach; disconnect; disengage; separate 1
uncoupled, unrelated
uncourageous, shy 2
uncourtliness, incivility; indelicacy
uncourtly, disagreeable 3; ill-mannered; unrefined 1
uncouth, barbarian 3; boorish; brutal 2; coarse 2; crude 3; foul 5; ignorant 4; plebeian 2; primitive 2; profane 3; rough 3; rude 1; rugged 4; shabby 4; tactless; tasteless 1; uncivilized 2; unrefined 1; vulgar 1
uncover, bare 6, 7; detect 1; disclose 2; discover 1; excavate 2; expose 1; find 2; give 12b; hit 9a; lay[1] 9; learn 1; light[1] 9; line[1] 23b; open 21, 23, 25; penetrate 5; reflect 2; root[1] 7b; scout 2; strip[2] 1; turn 25c; undo 2; unearth; unveil
uncovered, bare 1, 2; naked 1; nude; open 2, 5; raw 4; unguarded 2
uncovered, be, light[1] 10
uncovering, discovery 1; exposure 1; revelation
uncoveted, unenviable
uncritical, indiscriminate 1; promiscuous 1
uncrowded, open 9
unctuous, fat 2; flattering 2; glib; gooey 2; greasy 2; ingratiating; insincere; mealy-mouthed; obsequious; oily 1, 2; pharisaic; pious 2; sanctimonious; servile; sleek 3; slick 2; slimy 2; smooth 6; wily
unctuousness, servility
uncultivated, animal 4; barbarian 3; beastly 1; brutal 2; desert[1] 2; gross 3; parochial; philistine 2; primitive 2, 3; provincial 2; rustic 2; savage 3; tasteless 1; uncivilized 1; uneducated; unrefined 1; vulgar 1; wild 2, 3
uncultured, animal 4; barbarian 3; boorish; brutal 2; gross 3; heathen 3; parochial; philistine 2; primitive 2; provincial 2; rough 3; rude 1; rugged 4; rustic 2; tasteless 1; uncivilized 2; uneducated; unrefined 1; vulgar 1; wild 3
uncurbed, incontinent 1; lavish 3; promiscuous 2; uninhibited
uncurl, roll 11; straighten 1
uncurved, straight 1
uncustomary, unorthodox; unwonted
uncut, complete 1; full 12; intact; neat 2; rough 9; shaggy; straight 9; unabridged 1; undiluted; undivided 1; whole 1
undamaged, entire 2; flawless 1; intact; sound[2] 1; unscathed; whole 2
undaunted, bold 1; brave 1; dauntless; gallant 1; heroic 1; indomitable; intrepid; resolute; stalwart 2; stout 2; unabashed
undeceitful, aboveboard 2; candid 1; honorable 3; ingenuous 1
undeceive, disillusion
undeceived, disenchanted

undeceiving, aboveboard 2; candid 1; honorable 3
undeceptive, aboveboard 2; artless 1; ingenuous 1
undecided, debatable; disputable; end 6; fence 2; fickle; hesitant 1; indecisive 1, 2; indefinite 1, 5; independent 6; irresolute; mind 13; moot 1; open 6, 7; pending 2; problematic; uncertain 2; unresolved; unstable 1; vague 5; wishy-washy 1
undecided, be, procrastinate 2
undecipherable, illegible; incomprehensible; recondite
undeclared, implicit 1; sneaking 2; tacit
undecorated, bare 3; blank 2; chaste 2; empty 1; plain 4; severe 6; simple 2; unadorned
undefeatable, unbeatable
undefeated, invincible 1; triumphant
undefended, insecure 2; open 19; unguarded 2
undefiled, celibate 2; chaste 1; clean 1, 4; holy 2; immaculate 2; incorrupt 1; innocent 2; intact; maiden 2; pristine 2; pure 3; untarnished; white 3
undefinable, indefinite 4; ineffable 2
undefined, ambiguous 2; dim 1; dreamy 1; indefinite 1, 3; indistinct 2; lax 2; limitless
undeliberative, candid 1
undemanded, unasked
undemanding, easy 2, 3, 5; easy-going; light[2] 10; soft 2
undemocratic, totalitarian
undemonstrative, aloof 3; cold 3; reserved
undeniable, certain 4; decided 1; demonstrable; flawless 2; hard 8; incontrovertible; indisputable; naked 3; obvious; obviously; positive 1; self-evident; undisputed; unquestionable
undeniably, certainty 3; easily 2; evidently 1; exactly 2; indeed 1; positively; really 2; undoubtedly
undependable, ambiguous 3; capricious; changeable 1; fickle; inconsistent 2; inconstant; irresponsible; moody 3; shaky 1; slippery 2; temperamental 2; unreliable; untrue 1; variable
under°, below 2, 4, 7, 8, 10; beneath 1, 2, 3; subordinate 1
under way, get, fall 21; inaugurate 1; initiate 1; lead 8; move 12; open 20; proceed 1; set 12a, 18b; start 2, 3
under way, get things, start 4
underage, juvenile 1; small 1; young 1
underbrush, brush[1] 1; shrubbery
undercharge, undercut 2
underclothes°, unmentionable 3
underclothing, underclothes; unmentionable 3
undercooked, rare[2]
undercover°, stealthy; underground 2
undercover agent, operative 4; spy 1
undercover work, spying
undercurrent°, buzz 2; current 5; murmur 1
underdeveloped, childish; premature 1; puny 3; undersized; vestigial
underdone, rare[2]
underestimate°, belittle; depreciate 2; disregard 2; minimize 2; miscalculate; trivialize
underfed, emaciated; meager 2; puny 3; rawboned; thin 1
underflow, undercurrent 1
underfunded, short 8
undergarments, underclothes; unmentionable 3
undergo°, accept 4; bear 2; endure 2, 3; experience 4; feel 5; go 35a; meet[1] 6; receive 4; see 10; stand 3; suffer 2; support 3; sustain 3; take 6; taste 8; tolerate 2
undergo, cause to, put 3; subject 9
undergoing, be, have 5
undergraduate, scholar 2; student 1
underground°, beneath 2; partisan 4; resistance 3; sunken 3
underground, go, hide[1] 1
underground fighter, guerrilla; partisan 2
underground (railway), subway 1; underground 4
undergrowth, brush[1] 1; shrubbery
underhand, artful 1; astute 1; hypocritical
underhand, be, game 6
underhand(ed), artful 1; astute 1; corrupt 1; cozy 2; deceitful; designing; devious 1; dirty 3; dishonest; disingenuous; evil 2; foul 7; furtive 1; illicit 2; insincere; oblique 2; scheming; shady 2; shifty; sinister 2; sly 1; sneaky; stealthy; subtle 3; surreptitious; ulterior 1; wily; wretched 4
underhandedly, sly 3
underhandedness, artifice 1; deceit 1; stealth
underived, original 4
underlayer, substratum
underline, accent 4; emphasize; line[1] 1, 21; play 17a; point 22; punctuate 2; spotlight 3; stress 4
underling, assistant 2; attendant 2; cog 2; fag 3; flunky 1; inferior 4; menial 3; subordinate 2
underlining, emphasis
underlying, basic; fundamental 1; intrinsic; primary 1; radical 1; subconscious 1; ulterior 1; ultimate 3
undermanned, short-staffed
undermine°, break 7; destroy 4; get 31d; sabotage 3; sap[2]; shatter 2; subvert; undercut 1; vitiate 1; weaken 1
undermined, shake 3
undermining, impairment; subversion
underneath, below 2, 4, 7; beneath 1, 2, 3; under 1, 2, 7
undernourished, emaciated; meager 2; puny 3; skinny; thin 1
underpants, pants 1
underpass, subway 2; tunnel 1
underpin, sustain 2
underpinning, base[1] 2; basis 1; bottom 2; foundation 1; groundwork; support 8
underplay, trivialize
underprice, undercut 2
underprivileged, deprived; needy
underrate, belittle; depreciate 2; disparage 1; disregard 2; minimize 2; miscalculate; sneer 2; trivialize; underestimate
underscore, accent 4; emphasize; line[1] 1, 21; play 17a; point 22; punctuate 2; spotlight 3; stress 4
underscoring, emphasis
undersea, sunken 2
undersell, undercut 2
undershorts, pants 1
underside, reverse 7
undersize, undersized
undersized°, diminutive; little 1; puny 3; small 1; stunted; thin 1
understaffed, short-staffed
understand°, appreciate 3; awake 3; catch 10, 13a; comprehend; conceive 3; deduce; dig 3; digest 3; divine 4; fancy 10; fathom; figure 12b; find 3; follow 8; gather 4; get 19; grasp 2; hear 1, 2; infer; interpret 2; know 1; learn 1, 3; make 37d; master 7; penetrate 5; perceive 2; pierce 3; presume 1; read 2; realize 2; recognize 2; relate 4; see 2; surmise 1; sympathize 2; take 19; tell[1] 8; tumble 3; tune 4; twig[2]; wind[1] 6
understand, easy to, readable 1
understand, give (someone) to, intimate[2]
understandability, clarity 2; simplicity 1
understandable, clear 6; coherent 2; distinct 1; elementary 1; evident; intelligible; legible; luminous 3; natural 9; plain 2; popular 3; readable 1; simple 1; transparent 2; unprofessional 3; user-friendly
understandable, be, come 5c
understandable, make, communicate 3; rationalize 1
understandable, make (something *or* oneself), put 10
understandably, clear 18; clearly 3; well[1] 3
understanding°, acquaintance 1; agreement 1; appreciation 3; bargain 1; brain 1; capacity 2; charitable 2; charity 2; conception 2; contract 1; deal 5; depth 3; familiarity 1; fatherly; feeling 3; grasp 5; grip 3; heart 5; human 3; humanity 3; idea 2; illumination 2; indulgence 1; indulgent; inference; insight; intellect 1; intelligence 1; intelligent; interpretation 2; judgment 1; kindness 1; kind[1]; knowledge 1; lenient; light[1] 5; maternal; mentality 1; negotiation 2; pact; penetration 2; perception 1; picture 3; rapport; rapprochement; realization 1; reason 2; reconciliation 1; rendition 1; sensitivity 3; supportive; surmise 2; sympathy 1; tact; tactful; twist 8; uptake; version 2; view 2; vision 2; wisdom 1; wise 1; wit 1
understanding, come to *or* reach an, get 50c; term 5
understanding (of), sensible 4; sympathetic 1; understanding 3
understanding (that), with the, providing
understated°, modest 3
understood°, given 2; implicit 1; presumptive 2; silent 3; simple 1; tacit; virtual
understood, be, penetrate 4; sink 11
understood, easily, readable 1
understood, make oneself, communicate 3
understood, make (something *or* oneself), put 10
understructure, foundation 1
understudy°, alternate 5; double 7; relief 4;

unhappiness (*cont.*) resentment; sadness; sorrow 1; woe
unhappy°, afraid 2; bad 9; bleak 1; blue 1; dejected; desolate 3; despondent; disappointed; disgruntled; dismal; dissatisfied; doleful; dreary 1; forlorn 1; funereal; gloomy 2; heartbroken; heavy 6; hopeless 3; hurt 7; joyless 1; low[1] 8; melancholy 1; miserable 1; moody 1; mournful 1; mouth 6; pessimistic; regrettable; resentful; sad 1; somber 1; sorrowful 1; sunless; terrible 3; tragic; unfortunate 2; woebegone; wretched 2
unharmed, intact; safe 1; unscathed; whole 2
unharmonious, grating 1; harsh 1; strident; ungraceful 2
unhealthful, unwholesome 1
unhealthy°, bad 2; delicate 3; diseased; harmful; ill 1; indisposed 1; injurious 1; low[1] 4; morbid 1; par 5b; peaked; run-down 1; sick 2; unsound 2; unwholesome 1; virulent 1
unheard, inaudible; unnoticed
unheard-of°, different 2; extraordinary 1; inconceivable; nameless 1; new 4; obscure 4; outlandish; out-of-the-way 2; phenomenal; strange 1; unaccountable 3; unearthly 3; unknown 1
unhearing, deaf 2
unheated, cold 2; cool 1
unheedful, absent-minded; deaf 2
unheedful (of), unconscious 2
unheeding, absent-minded; inconsiderate; remiss; unconscious 2
unhelpful, shabby 2
unhesitant, resolute; unhesitating 1
unhesitating°, decided 2; deliberate 2; determined 1; immediate 1; nonstop 2; prompt 1, 2; resolute
unhesitatingly, cold 11; firmly 2; flat 17b; immediately 1; outright 4; plump[2] 5; promptly; readily 1; right 13; summarily 1; surely 2; willingly
unhidden, obvious
unhindered, free 10; open 4; unimpeded
unhinged, crazy 1; delirious 1; deranged; insane 1; mad 1; psychotic 1; queer 4; unbalanced 2; wild 5
unhip, square 6
unhitch, disconnect; disengage
unholster, draw 2
unholy, godless 1; impious; irreverent 1; profane 2; satanic 1; sinful
unhonored, unsung
unhook, disconnect; disengage; loosen 1; separate 1; undo 1
unhooked, undone[1] 2
unhoped-for, unforeseen
unhorse, overthrow 1
unhoused, homeless 1
unhurried°, deliberate 2; easy 2, 5; measured 1; slow 1, 3
unhurriedly°, leisure 3b; slow 11
unhurt, safe 1; unscathed; whole 2
unhygienic, unwholesome 1
unidentifiable, nameless 2
unidentified°, incognito 1; nameless 1; unheard-of 1; unknown 1, 2; unsung
unification, amalgamation; embodiment 2; solidarity; synthesis; unity 3
unified, inclusive 1; one 2; united 1
uniform°, changeless 1; consistent 2; constant 3; costume; equable 2; equal 2; even 1, 3; garments; gear 3; habit 3; homogeneous; invariable 1; level 1, 3; measured 3; monolithic; orderly 1; parallel 1; piece 8; plane 3; regular 2, 3, 5; robe 2; same 2; smooth 2; solid 4; steady 2; suit 3; sustained; symmetrical
uniform (with), even 2
uniformity°, equality 1; order 2; parity 1; regularity 1; symmetry; unity 1, 2
uniformly, alike 2; consistently 1, 2; universally
unify°, combine 1; coordinate 2; embody 3; incorporate; join 1; marry 2; solidify 2; stick[1] 16a; unite 1, 2
unify (against), gang 3
unifying, synthesis
unilateral, one-sided 3
unilluminated, dark 1; dusky 2
unimaginable, impossible 1; incomprehensible; inconceivable; incredible 1; unbelievable; unheard-of 2; unlikely 1; unthinkable 1; untold 3
unimaginative, banal; everyday 3; lackluster; literal 3; little 6; matter-of-fact; mediocre; nearsighted 2; pedestrian 2; prosaic; routine 4; shortsighted 2; slow 7; small 3; small-minded; square 6; stodgy; stupid 3; tired 3; usual
unimagined, unheard-of 2
unimpaired, entire 2; flawless 1; intact; sound[2] 1; whole 2
unimpassioned, coldblooded 2; icy 2; philosophical 2; temperate 1; wooden 2
unimpeachability, rectitude
unimpeachable, blameless; correct 8; faultless; flawless 2; impeccable; incorrupt 1; innocent 1; irreproachable; respectable 1; unquestionable; upright 2
unimpeded°, clear 16; free 10; open 4
unimperiled, secure 1
unimportance, indifference 2; inferiority 1, 2; obscurity 3; triviality 1
unimportant, circumstantial 2; expendable; fluffy 2; frivolous 1; humble 3; idle 4; immaterial 1; incidental 2; inconsequential; indifferent 4; inferior 2; insignificant; light[2] 6; little 5; meaningless 2; minor 2; minute[2] 2; negligible; niggling 2; nonessential 1; obscure 4; paltry; peripheral 1; petty 1; pitiful 2; point 15; punk 2; puny 1; secondary 1; shallow 1; side 8; skin-deep; small 2; small-time; superficial 2; thin 2; trifling; unimposing; venial; wee 2; worthless 1
unimportunate, modest 3
unimpressed, blasé 1; lackadaisical 2; unenthusiastic
unimpressed (by), unaffected[2]
unimpressionable, impassive
unimpressive, decent 3; undistinguished; unimposing
unindulgent, intolerant 1
uninfected, clean 1; pure 2; sterile 2
uninfluenced (by), unaffected[2]
uninformed°, benighted; ignorant 1; inexperienced; provincial 2; unaware
uninformed about, unfamiliar 2
uninhabitable, inhospitable 2
uninhabited°, deserted; desert[1] 2; desolate 1; empty 2, 3; free 3; lifeless 4; lonely 2; vacant 1; wild 2
uninhibited°, abandoned 2; bawdy; direct 10; earthy; forthright; frank 1; free 9; honest 3; ingenuous 2; liberty 4; open 16; outspoken; promiscuous 2; rampant 1, 2; raw 6; riotous 2
uninhibited, be, liberty 5
uninitiated, inexperienced; tender[1] 3; young 2
uninitiated in, unaccustomed 2; unfamiliar 2; unused 4
uninjured, intact; safe 1; sound[2] 1; unscathed; whole 2
uninspired, colorless 2; dry 2; dull 4; indifferent 3; lackadaisical 2; mechanical 3; mediocre; ordinary 2; pale[1] 3; pedestrian 2; prosaic; shiftless; small 3; stupid 3; tame 4
uninspiring, colorless 2; dull 4; lifeless 3; pedestrian 2; prosaic; tame 4
uninstructed, uninformed
unintelligent, brute 1; opaque 3; senseless 3; slow 7; stupid 1; wooden 3
unintelligibility, obscurity 2; opacity 2
unintelligible, disconnected 2; illegible; inarticulate 1, 2; incoherent; incomprehensible; indistinct 1; inexplicable; misty; rambling 1; unaccountable 1
unintended, accidental; inadvertent 1
unintentional, accidental; chance 6; inadvertent 1; involuntary; unconscious 2
unintentionally, chance 5a; unawares 2
uninterested, casual 3; cool 4; dead 4; halfhearted; indifferent 1; lackadaisical 2; negative 2; nonchalant; offhand 1; passive 1; perfunctory 1; phlegmatic 1; stolid; supine 2; tepid 2; unenthusiastic
uninteresting, bland 2; boring; colorless 2; dead 12; dreary 2; dry 2; dull 4; flat 5; heavy 7; humdrum; irksome; lackluster; lifeless 3; monotonous; nondescript; pedestrian 2; prosaic; routine 4; slow 9; stodgy; stupid 3; tame 4; tedious; tiresome 1; vapid
uninterrupted, constant 2; continual; continuous 1; direct 7; endless 2; entire 3; eternal 2; even 3; flat 1; nonstop 1, 2; ongoing 1; perennial 1; perpetual 2; plane 3; regular 5; repetitive; running 3; smooth 3; solid 10; sound[2] 6; steady 2; successive
uninterruptedly, end 7b; nonstop 3; succession 5
uninvestigated, unknown 2
uninvited, unasked; unsolicited; unwelcome 1
uninviting°, inhospitable 1, 2
uninvolved, cool 4; detached 2; disinterested; neutral 1; nonpartisan 1; outside 9; passive 1; perfunctory 1; phlegmatic 1; simple 1; unsophisticated 2
union°, alliance 1; amalgamation; association 1; bond 3; brotherhood 2; club 2; combination 1, 2;

imperative 1; important 1; indispensable 1; insistent; instant 4; necessary 1; pivotal; pressing; rush 5; serious 2; sore 3; strong 20

urgently, hard 13

urging, encouragement 1; incitement 1; pressure 4; recommendation 1; spur 1; suggestion 1

urinal, toilet 1

urinate°, go 17; pass 12; void 7; water 4

urn, jar[1]; jug

uropygium, tail 1

usability, use 9

usable, disposable 2; feasible; handy 2; operable; practical 1; ready-made 2; serviceable 1; useful; user-friendly

usage°, convention 2; custom 1; diction 1; expression 6; habit 1; observance 2; phrase 3; practice 1; protocol 1; routine 1; service 5; tradition; use 6, 8, 9

usage (of), treatment 1

use°, account 5; advantage 2; affect[2] 2; application 1; apply 3; assume 1; draw 13a; embrace 2; employ 2; employment 3; enjoy 2; enjoyment 2; exercise 1, 5; exert 1; expend 1; exploit 2; fall 9; function 1; go 32g; good 19; handle 7; manipulate 1, 2; play 15; prey 3b; profit 2, 4; purpose 3; put 23d; refer 3; service 1, 5; speak 3; take 21; talk 1; tap[2] 6; tie 7b; touch 5, 9; treat 1; turn 6; usage 1, 2; victimize 1; wear 8; wield 1, 2; work 10, 15

use, going *or* **passing out of,** obsolescent

use, of, useful

use, put to (good *or* **effective),** apply 3; exercise 1; exert 1; tap[2] 6; touch 9; use 1; work 15

use for, have no, scorn 3; use 15

use of, make, draw 13a; embrace 2; employ 2; enjoy 2; fall 9; take 21; tap[2] 6; touch 9; use 1, 2; work 10

use of, make (good), profit 4; seize 4

use (to *or* **for), be of,** serve 2

use (up), consume 2; dissipate 3; drain 5; exhaust 1; expend 2; finish 3; kill 7; lose 4; occupy 4; pass 5; run 31, 34b; take 39d; use 5

used°, accustomed 2; secondhand

used to, used 3

used (to), get, adjust 3

used (to *or* **for), be,** serve 2

used up, spent 1, 2

used up, be, go 16; run 30b

useful°, advantageous; beneficial 1; constructive 1; convenient 1; effective 1; effectual 1; efficacious; expedient 2; favorable 1; fruitful 2; functional 1; gainful; handy 2; helpful; instrumental; positive 4; practical 1; profitable 2; ready-made 2; serviceable 1; sound[2] 4; worthwhile 2; yeomanly

useful (to *or* **for), be,** behoove; serve 2

usefulness°, advantage 2; good 19; profit 2; purpose 3; service 1; use 9; worth

useless°, bootless; dispensable; dud 2; duff; fruitless; futile; good-for-nothing 1; helpless 4; hopeless 4; idle 4; impracticable 2; impractical 2; incompetent; ineffective 1; needless 1; nonproductive 2; pointless; puny 1; thankless; unprofitable 2; unsuccessful 1; vain 2; void 1; waste 9, 10; worthless 2

uselessness, emptiness 2; incompetence; vanity 2

user°, addict 1

use up one's resources, bolt 7

usher, accompany 1; attend 4; attendant 2; conduct 4; direct 2; escort 1, 4; guide 1; lead 1; see 6; shepherd; show 2

usher in, greet 1; inaugurate 1; introduce 4; phase 5; precede

usherette, attendant 2

using, mean[3] 3; use 6

usquebaugh, whiskey

usual°, accustomed 1; average 1, 3; common 1; conventional; correct 7; customary 1; everyday 2; familiar 2; frequent 1; general 2; going 2; habitual 1; medium 1; natural 1; norm 1; normal 1; ordinary 1; prevailing 1; prevalent 1; proper 2; regular 1, 7, 11; regulation 5; ritual 2; routine 3; set 29; standard 6; standing 1; staple 2; stock 7; traditional; typical 2

usual, as (per), usually

usually°, always 1; average 2; chiefly; frequently 2; generally 1; mainly; often; ordinarily; part 8; popularly; rule 4; whole 5

usurious, predatory 2; rapacious

usurp, appropriate 2; grab 2; help 5; preempt; steal 2

usurpation, plagiarism; seizure 1

usurped, assumed 1

utensil, apparatus; device 1; gadget; implement 1; instrument 1; machine 1; thing 5; tool 1; vessel 1

utensils, gear 2; kit; outfit 1; paraphernalia; thing 8c

utilitarian, expedient 2; functional 1; helpful; instrumental; positive 6; practical 1; profitable 2; serviceable 2; useful

utility, advantage 2; purpose 3; service 1; use 7, 9; usefulness

utilization, application 1; employment 3; enjoyment 2; exercise 5; service 5; use 6, 8, 9; wear 8

utilize, apply 3; employ 2; enjoy 2; exercise 1; exert 1; exploit 2; handle 7; manipulate 1; profit 4; tap[2] 6; use 1, 2, 4; wield 2; work 10

utilized, used 2

utilizing, mean[3] 3

utmost, extreme 3, 7; extremity 3; full 4; maximum 1, 3; outside 3; peak 2; pinnacle; supreme 2; ultimate 2

utmost, do one's, strive 1

utmost, to the, full 18; vengeance 2b; wide 6

utopia°, heaven 3; paradise 2

utopian, dreamer; ideal 5; romantic 1; visionary 1

utter, absolute 2; blank 6; blurt; complete 3; dead 14; deliver 4; enunciate 1; exclaim; express 1; flagrant; give 8; go 14; heave 3; implicit 2; main 4; mouth 7; outright 2; out-and-out; pass 9; perfect 6; phrase 4; positive 9; profess 1; profound 3, 4; pronounce 1; pronounced 2; proper 5; pure 5; put 4; raise 11; rank[2] 2; rattle 5; regular 9; right 8; say 3, 11; sheer 2; sound[1] 7; speak 4; stark 4; tell[1] 2; thorough 1; thumping 2; total 3; unmitigated; unqualified 2; use 1; very 4; voice 4; word 10

utter in, speak 3

utter indistinctly, mumble

utterance, delivery 4; exclamation; expression 1; interjection; observation 2; statement; tongue 2; voice 1; word 5

utterance to, give, voice 4

uttered, oral; verbal 1

utterly°, absolutely 2; altogether; clean 7; clear 19; clearly 2; completely 2; dead 18; downright 2; entirely 1; fairly 3; flat 16; head 8; hook 4; outright 4; perfectly 1; quite 1; root[1] 3; stark 1; thoroughly 1; through 7; totally; wholly 1

uttermost, extreme 3; maximum 1, 3; peak 2; supreme 2; ultimate 4

U-turn, about-face; reversal 1

uxoricide, murder 1

vacancy°, emptiness 1; opening 3; slot 2; void 5

vacant°, bare 3; blank 3; deserted; desert[1] 2; empty 1, 2, 5; feebleminded; free 3; glassy 2; hollow 1; inane; light[2] 8; open 10; stark 3; uninhabited; vague 6; void 2; wooden 2

vacantly, vaguely 3

vacantness, emptiness 1, 3; void 4

vacate°, empty 8; evacuate 2; forsake 1; quash 1; relinquish 1; repeal 1; resign 1; reverse 3; void 6

vacated, deserted; uninhabited

vacation, holiday 1; leave[2] 2; leisure 2, 5; recess 2; respite 1; rest[1] 2; sojourn 1, 2

vacation, on, leisure 3a

vacation, take a, get 32b

vaccinated, immune

vaccination, immunity 2; shot 6

vaccine, preventive 4

vacillate, blow[1] 6; change 7; delay 3; doubt 2; fence 4; fluctuate; flutter 1; hesitate 2; oscillate; procrastinate 2; scruple 2; seesaw 2; shilly-shally; stall[2] 1; vary 2

vacillating, doubtful 2; end 6; fence 2; fickle; flutter 4; hesitant 1; inconstant; indecisive 1; indefinite 5; infirm 2; irresolute; mind 13; rocky[2]; uncertain 2; unresolved; unstable 1; vague 5; variable; wishy-washy 1

vacillation, flutter 4; indecision

vacillation(s), fluctuation

vacuity, emptiness 1, 3; vacancy 3; vanity 2

vacuous, blank 4; colorless 2; empty 5; glassy 2; inane; light[2] 8; meaningless 1; purposeless; stupid 3; vacant 2; vague 6

vacuousness, emptiness 3; vacancy 3; vanity 2

vacuum, clean 9; emptiness 1; vacancy 1; void 4

vacuum-pack, preserve 3

vade mecum°, companion 2; guide 8; manual

vagabond°, bum 2; derelict 3; down-and-out 2; drifter; homeless 1; rover; tramp 4

vagabonds, homeless 2

vagary, fancy 8; freak 3; kink 4; notion 2; quirk

vagrant, aimless 2; beggar 1; bum 2; derelict 3; down-and-out 2; drifter; loafer; migrant 1, 2; pauper; rover; stray 3, 4; tramp 4; vagabond 1, 2

vagrants, homeless 2

vague°, ambiguous 2; broad 6; cryptic 2; dim 1; doubtful 1; dreamy 1; equivocal 2; faint 1; fuzzy 2; general 4; hazy 2;

vastly°, very 1
vastness, size 1; spread 9; waste 8; wild 10
vasty, boundless; vast
vatic, prophetic
vaticinal, ominous 3
vaticinate, forecast 1; predict; prophesy 1, 2
vaticination, forecast 2; prophecy 1
vaticinator, prophet; seer
vault, bound[2] 1, 3; cache 1; cellar; clear 27; crypt; grave[1]; hop 1, 3; jump 1, 8; leap 5; safe 7; spring 1, 6; tomb
  vault (into), leap 1
  vault (over), hurdle 2
vault (of heaven), firmament
vaunt, boast 2; brag; display 3; flourish 2; parade 5; shake 4; swagger 2; talk 8
vaunting, egotistical; ostentatious
Vaya con Dios!, farewell 3
VDT, terminal 4
VDU, monitor 2; terminal 4
vector, carrier 2
vedette (boat), picket 3
veer, careen; deflect; deviate 1; dodge 1; lurch[2] 2; shift 5; sway 2; swerve; trend 3; turn 2, 4; vary 3; wheel 2; wind[2] 1
  veer (from), depart 2
  veer off *or* **away,** tack 6
vegetable, clod 2; plant 1
vegetables, produce 7
vegetate, loaf[2] 1; lounge 1; stagnate
vegetation, growth 2
vehemence, eagerness 1; edge 3; feeling 4; fervor; fury 2; heat 2; intensity; rage 1; violence 1, 2; warmth 3
vehement, burning 2; eager; excited 2; fierce 2; heated; hot 3, 4; impassioned; intense 2; passionate 1; strong 5; tempestuous; torrential; towering 2; violent 1, 3; wild 6
vehemently, fiercely; madly 3; vengeance 2a; warmly 5
vehicle°, agent 2; car 1; machine 2; mean[3] 4a; medium 5; organ 2; tender[3] 2; tool 2; voice 3
veil°, blur 4; camouflage 2; cloak 2, 3; eclipse 1; envelop 2; film 1, 3, 4; gloss[1] 4; hide[1] 2; mantle 2, 3; mask 2, 3; obscure 6; pall[1] 1; pretense 2; pretext 1; screen 7; shade 5, 12; shadow 2; shroud 1, 3; shut 5c; submerge 3
veiled°, hidden; interior 3; invisible 2; mystical; obscure 1; occult 1; subtle 2; surreptitious; vague 1, 3
vein°, mine 1, 2; seam 2; significance 1; strain[1] 9; stratum 1; streak 1; style 5; tone 3
veld, plain 6
veldt, plain 6
velitation, contest 2; disagreement 3; dispute 3, 4
velleity, inclination 3
vellicate, pluck 4; wag[1] 1
vellication, wag[1] 2
velocity°, haste 1; pace 2; rate[1] 1; speed 1; way 5
velvet, in *or* **on,** opulent 1; rich 1
velvety, sleek 1; smooth 2; soft 8
venal°, abusive 3; corrupt 1; foul 7; lawless 2; mercenary 1, 2; rotten 3; timeserving; vicious 1
venality, vice 1
vend, market 4; peddle; sell 1, 2
vendetta°, combat 2; feud 1
vending, sale 1
vendor, dealer; merchant 1; peddler; seller; trader; tradesman 1
veneer°, face 3, 16; gloss[1] 2; semblance 1, 2; sheet 4, 5; skin 2
venerable°, aged; ancient 3; classic 2; elder 2; elderly 1; famous; grand 2; gray 3; illustrious; lofty 2; precious 2; sacred 1; time-honored; well-established
venerate°, admire 2; adore 2; commemorate; esteem 1; exalt 2; fear 6; glorify 2; hallow 2; honor 5; idolize; look 12; praise 4; respect 6; revere; worship 1
venerated, dear 1; holy 1; noble 3; precious 2; sacred 1; time-honored; well-thought-of
veneration°, fear 2; glory 2; honor 2; piety 2; praise 2; prostration 2; regard 7; respect 2; reverence 1; worship 2
venereal°, erotic 2; social 2
Venetian blind, shade 5
vengeance°, reprisal; requital 2; retribution; revenge 1
  vengeance, with a, vengeance 2; vigorously
  vengeance on, take, get 23
vengeful, rancorous; vindictive
vengefulness, rancor
venial°, permissible
venom°, gall[1] 1; grudge 1; ill will; poison 1; rancor; sarcasm; spite 1; virulence 2
venomous°, cutting 2; harmful; malignant 2; poisonous 1; profane 3; rancorous; sarcastic; sharp 5; spiteful; vicious 2; virulent 2
venomousness, rancor; sarcasm; virulence 1, 2
vent°, emit; evacuate 1; exhaust 5; exit 1; mouth 2; outlet 1; pore[2]; wreak
  vent to, give, let[1] 8c; reveal; vent 2, 3; voice 4
ventilate, air 6; expose 1; freshen 3; reveal
vent one's spleen, fly 8b; piece 12
venture°, activity 2; adventure 2, 3; attempt 1; bet 1, 2; dare 2; effort 2; endeavor 2; enterprise 1; essay 2, 3; fling 3; gamble 1, 3; hazard 3; invest 1; jeopardize; operation 3; pawn[1] 2; presume 2; pretend 2; project 2; reckon 3; say 6; sink 10; stake[2] 4; trial 3; try 1, 5; undertaking 1
venturesome°, adventurous; audacious 1; bold 1; daring 2; dauntless; enterprising; fearless; foolhardy; intrepid; manly; rash[1]
venue, locale
veracious, faithful 4; honest 1; precise 1; right 2; sincere; true 1; truthful
veracity, honesty 2; integrity 1
veranda, portico
verbal°, oral; spoken 1
verbalization, expression 1
verbalize, express 1; phrase 4; say 4; vent 3; voice 4
verbatim°, exactly 1; letter 4; literal 1; literally 1; perfectly 3; precise 1; same 2; verbal 2
verbatim et literatim, verbatim 1, 2
verbiage, hot air; talk 18; tautology; waffle 3
verbose, diffuse 2; discursive; expletive 1; lengthy; ponderous 2; rambling 1; redundant 2; talkative; windy 2; wordy
verbosity, rhetoric 2; tautology
verboten, illegal; pale[2] 3; taboo 1; unlawful
verb phrase, phrase 1
verdant, green 1; leafy; lush 1
verdict, decision 2; determination 2; finding 2; judgment 2; resolution 3; sentence
verecund, bashful 2; modest 1; shy 1
Verein, society 5
verge°, border 5; brink 2; edge 1; eve 2; margin 1; perimeter; point 9; shoulder 1; side 1; tend[1]; threshold 2
  verge of, on the, almost; ready 4; verge[1] 2
  verge on *or* **upon,** approximate 2; border 8, 9; join 4; near 10; verge[1] 3
verging, inclined 3
  verging on, almost
verifiable, absolute 4; actual 1; authoritative 2; demonstrable; factual 1; hard 8; historical; real 1; supportable 2; true 1
  verifiable, be, stack 7a
verification, check 15; demonstration 1; evidence 1; hallmark 1; identification 1; proof 1; seal 2; testimony
verified, actual 1; official 1; real 1; true 1
  verified, be, check 10c
verify°, attest; authenticate; certify 1; check 3, 4, 8, 11; confirm 1; determine 2; document 2; establish 3; prove 1; ratify; seal 4; show 3; substantiate; support 6; warrant 4; witness 3
verily, substantially; truly 4
veritable°, absolute 2; authentic; authoritative 2; bona fide; genuine 1; real 1
veritably, fairly 3; substantially
verkrampte, reactionary 1, 2
vermicular, serpentine 2
vermiculate, serpentine 2
vermin, rabble 2
vernacular°, cant 2; dialect; idiom 1; informal 3; jargon 1; language 1, 3; lingo; tongue 1; vulgar 3
vernal, spring 10
vernissage, opening 4; preview; reception 2
versatile°, adaptable; mobile 4
versatility, flexibility 2
verse, jingle 3; passage 2; poem; poetry; rhyme 1; text 2
versed°, adept 1; practiced 2
  versed in, familiar 4
  versed in, be, know 1
versemaker, poet
versification, poetry; rhyme 1
versifier, poet
version°, arrangement 4; edition; issue 5; model 6; paraphrase 1; recital 2; rendering; revision; story[1] 2; twist 8
verso, page[1] 1; reverse 7
vertebrae, spine 1
vertex°, spire 2; tip[1] 1; top 1; zenith
vertical, erect 1; perpendicular 1; picket 1; plumb 2; prop 3; sharp 2; sheer 1; standing 4; straight 2; upright 1, 3
vertically, plumb 3; upright 4
vertiginous, dizzy 1; faint 2; giddy 1; queer 3
vertigo°, **suffer,** spin 3
verve°, animation 1; bounce 2; dash 6; eagerness 1; electricity; energy; fervor; fire 2; flair 2; heart 4; life 7; panache; passion 1; pep 1; push 14; snap 11; vigor
very°, downright 2; eminently; exceedingly; extra 6; extreme 3, 9; extremely; fearfully 2; fiercely; frightfully; full 15; highly 2; level 3; mighty 4; overly; particularly 1; perfectly 1, 4; precise 4; pretty 3; profoundly; quite 1; rather 1; really 4;

improbable; irregular 3; kinky 1; mysterious 1; odd 1; offbeat; ordinary 4; outlandish; outré; out-of-the-way 2; pale[2] 3; peculiar 1; preposterous; queer 1; ridiculous; sick 5; special 1; spectral; strange 1; supernatural; unaccountable 3; unearthly 2; unnatural 3; way-out 1
weirdie, eccentric 2; oddity 2; weirdo
weirdness, eccentricity 1; oddity 1
weirdo°, abnormal 2; character 6; eccentric 2; oddity 2; offbeat; pervert 3
welcher, welsher
welcome°, acceptable 2; embrace 2; greet 1; greeting 1; let[1] 5; receive 3; reception 1; seasonable; see 9; thumb 4
welcoming, cordial; home 4b; hospitable 1; pleasant 2
weld°, attach 1; cement 2; connect 3; fix 1; marry 2; stick[1] 4; tack 5; unify; union 3
  weld (together), join 1
welfare°, charity 3; profit 2; sake 1
  welfare, on, needy
welkin, firmament; heaven 2; sky 1
well°, amply 4; big 10; card 12; easily 3; famously; fine[1] 12; fit[1] 3; fountain 1; fresh 6; healthy 1; highly 1, 3, 4; indeed 3; OK 4; pit[1] 3; properly 1, 2; richly 2; right 16, 17; shaft 4; spring 8; swimmingly; warmly 3; whole 3
  well, all being, hopefully 2
  well, get, come 17a; progress 6; recover 2
  well, not (very), ill 1; par 5b
  well (enough), OK 7
  well (forth *or* up), flow 2; stream 4; surge 1; well[2] 2
  well in, thick 8
  well out, stream 4; well[2] 2
  well up, stream 4; well[2] 2
(well-)acquainted, knowledgeable 1
well-adjusted, normal 2; well-balanced 1
(well-)advanced, forward 3
well-advised°, better 3, 6a; judicious; wise 3
well-aged, ripe 1
well and truly, truly 2
(well-)balanced, levelheaded; rational 1; stable 3; symmetrical; sane
well-behaved, courteous; decorous; good 4; orderly 2; respectful
wellbeing, welfare; ease 1; fitness 2; health 2; sake 1
well-born, elegant 1; ladylike; upper-class 1
well-bred°, courteous; debonair 1; decent 5; elegant 1; gallant 2; genteel 2; ladylike; pleasant 2; polished 2; presentable 2; refined 1
well-brought-up, decent 5; well-bred
well-built, full 7; husky 1; mighty 2; solid 6; sound[2] 3; staunch 2; strong 12; sturdy 1; substantial 2; tough 1; voluptuous 2
well-chosen, choice 5
well-connected, important 3
well-considered, wise 3
well-constructed, solid 6; sound[2] 3; staunch 2
(well-)controlled, scientific
well-crafted, exquisite 1
(well-)defined, clear 5, 11; definite 3; distinct 1; explicit 1; express 5; focus 2; graphic 1; hard 12; precise 1; pronounced 1; specific 1; visible 2
(well-)deserved, just 3
(well-)developed, forward 3
(well-)disciplined, taut 2; tight 4
(well-)disposed, agreeable 2; amiable; benevolent; charitable 2; favorable 2; friendly 1; genial; glad 2; good 6; indulgent; kind[1]; mood 2; neighborly; ready 2; sympathetic 2; well-balanced 2; willing
well done, beautiful 2; congratulations
well-drawn, graphic 1
well-dressed, dapper; well-groomed
(well-)earned, due 2
(well-)educated, knowledgeable 2; learned; lettered; versed; well-informed; wise 2; smart 2
well-endowed, bosomy; buxom 2; voluptuous 2
(well-)established, old 4; strong 9; substantial 3; well-known 1
well-executed, exquisite 1
well-fed°, fat 1
well-fixed, rich 1; successful 1; well-off
well-fortified, impregnable
well-found, flush[2] 3
well-founded, good 8; just 3; solid 7; stable 1; strong 9; substantial 3; well-established
well-groomed°, smart 3; spruce 1; tidy 1; trim 1
(well-)grounded, just 3; learned; reasonable 2; versed
wellhead, fountain 1; well[2] 1
well-heeled, fat 3; flush[2] 3; loaded 4; money 4; opulent 1; prosperous 1; rich 1; successful 1; wealthy; well-off
(well-)illuminated, light[1] 13
(well-)informed, educated 1; enlightened; experienced 1; home 5; judicious; knowing 2; knowledgeable 1; learned; lettered; profound 2; qualified 1; versed
welling, wash 14
  welling (forth), outburst
  welling up, wave 2
welling-up, explosion 3
well-intentioned, innocent 3; kind[1]; sympathetic 1
well-kempt, trim 1
well-kept, tidy 1
well-knit, robust 1
well-known°, big 7; celebrated; common 5; eminent 1; familiar 1; famous; great 5; illustrious; legendary 3; notable 1; noted; notorious 2; old 8; open 12; prestigious; prominent 2; proverbial 1; public 6; regular 9; renowned; stale 2; successful 3; traditional; usual; well-established
(well-)lighted, light[1] 13
(well-)liked, popular 1
(well-)lit, light[1] 13
well-made, exquisite 1; fine[1] 5; solid 6; staunch 2
well-mannered, civil 3; courteous; genteel 2; good 4; graceful 2; gracious; orderly 2; pleasant 2; polished 2; polite 1; presentable 2; refined 1; respectful; well-bred
well-meaning, charitable 2; fatherly; kind[1]; sympathetic 1
well-modulated, smooth 7
well-muscled, muscular; sturdy 1
well-nigh, almost; near 3; nearly 1; practically 1
well-off°, comfortable 2; fat 3; favored 2; flush[2] 3; loaded 4; money 4; opulent 1; prosperous 1; rich 1; wealthy
(well-)oiled, drunk 1
(well-)ordered, coherent 1; neat 1; regular 6; scientific; smooth 3; symmetrical; systematic; tidy 2; trim 1; well-balanced 2
(well-)organized, careful 2; coherent 1; logical 3; neat 1; order 10a; orderly 1; scientific; streamlined 2; systematic; taut 2; tidy 2; tight 4
well-paid, profitable 1
(well-)paying, profitable 1
well-pleased, blithe 1
(well-)posted on, versed
well-proportioned, full 7; good 10; regular 4; shapely; statuesque; symmetrical; voluptuous 2; well-balanced 2
well-provided (with), replete 1
well-provided-for, rich 1
well-read, educated 1; knowledgeable 2; learned; lettered; literary 1; profound 2; smart 2; versed; well-informed; wise 2
well-reasoned, logical 3
well-received, popular 1; welcome 2
(well-)regulated, regular 6; scientific
(well-)ripened, ripe 1
well-rounded, full 7
well-run, streamlined 3
(well-)schooled, versed
(well-)seasoned, spicy 1
well-shaped, full 7
well-skilled, adept 1
well-spent, fruitful 2
(well-)spiced, spicy 1
well-spoken, eloquent 1; fluent
wellspring, beginning 1; cause 1; fountain 1; head 6; mine 2; origin 1; parent 2; root[1] 1; source 1; spring 8; well[2] 1
(well-)stacked, full 7; voluptuous 2
well-stocked (with), replete 1; rich 2
well-substantiated, strong 8
well-suited, seasonable
well-supplied, flush[2] 3; full 1
  well-supplied (with), replete 1; rich 2
well-supported, strong 8
well-thought-of°, good 9; honorable 4; prominent 2; reputable
well-thought-out, deliberate 3; logical 3; measured 2; reasonable 2; sensible 1; studied; wise 3
  well-thought-out, not, snap 13
well-timed°, fortunate 2; happy 2; opportune 2; propitious; seasonable; timely
well-to-do, fat 3; flush[2] 3; leisured; loaded 4; money 4; opulent 1; prosperous 1; rich 1; substantial 4; successful 1; wealthy; well-off
(well-)trained, learned; versed
well-turned-out, dapper; elegant 1; nice 5; shapely; spruce 1; trim 1
(well-)tutored, versed
well-upholstered, plump[1] 1
(well-)versed, experienced 1; home 5; informed 2; learned; lettered; old 7; proficient; seasoned 1; smart 2; strong 6; versed; well-informed; wise 2
well-wishing, benevolent; charitable 2; kind[1]; neighborly
well-worn, timeworn
welsh (on), renege 2
welt°, bruise 1; bump 2
Weltanschauung, philosophy 2
welter°, hodgepodge; wallow 1
wen, elevation 2; growth 4; lump[1] 2
wench, girl 1; maid 1
wend, go 1; walk 1
wend one's way, weave 4
wens, spot 5
West End, the, stage 3
Western, romance 3

word, give the, say 10
word, in a, brief 5; briefly 1; short 14; word 8
word, not say a, tongue 3
Word, spread the, preach 1
word for word, learn, memorize
word (of honor), give (one's) (solemn), commit 4; pledge 4; promise 3; swear 1; vow 1
word with, have a, see 9
wordbook, dictionary; glossary
word choice, diction 1; phrase 3; style 5; wording
word for word, exactly 1; literally 1; perfectly 3; letter 4; literal 1; precise 1; same 2; verbal 2; verbatim 1, 2
word "go", first 6
word group, phrase 2
wordiness, rhetoric 2; tautology; waffle 3
wording°, diction 1; expression 6; language 4; parlance; phrase 3; print 3; style 5; terminology; text 1
wordless, dumb 1; mute 1; speechless 2
wordlessly, silently
word list, glossary
Word of God, Scripture
word (of honor), assurance 1; guarantee 1; oath 1; pledge 1; promise 1; resolution 2; vow 2; word 4
word-of-mouth, oral; verbal 1
wordplay, joke 1; repartee; wit 2
words°, book 2; language 4; line[1] 17; lyric 4; print 3; terminology; text 1; voice 1; word 9
  words, beyond, ineffable 2
  words, have, fight 4; row[2] 3; spar[2] 2
  words, in a few, briefly 1; word 8
  words, in other, like[1] 6
  words, in plain, honestly 2
  words, not to mince, word 8
  words, play on, epigram 1
  words, put into different, reword
wordsmith, scribe 2; writer
word (to the wise), warning 1
wordy°, boring; diffuse 2; discursive; expletive 1; lengthy; ponderous 2; rambling 1; redundant 2; repetitious; rhetorical 3; talkative; voluble; windy 2
work°, act 9; activity 2; behave; book 1; business 1; calling; career 1; cultivate 1; drudgery; duty 1; effort 1; employment 1; engagement 4; exert 2; exertion; exploit 2; fashion 5; ferment 1; function 1, 3; go 3; industry 2; job 1; labor 1, 5; line[1] 7; manipulate 2; mission 1; mold[1] 4; occupation 1; office 4; operate 1; opus; perform 2; plod 2; post[2] 1; practice 3; preparation 4; product 2; production 2; profession 1; project 2; pursuit 3; run 11; service 9; stroke 7; struggle 4; study 5; sweat 6; take 11; task 1; tease 2; till[1]; tool 4; trade 2; trick 7; undertaking 1; use 2; wangle; wreak
work, at, progress 4; way 12
work, begin *or* **start to,** effect 6
work, get *or* **set to,** set 12a; shoulder 3; turn 23c; wade 3b
work, in, pipeline 2; way 12; work 5
work, not, water 2
work, off, off 6
work, out of, idle 2; inactive 2; unemployed; work 6
work, put to, exert 1
work, set *or* **get to,** shoulder 3
work against, militate 1
work (a)round to, lead 10b
work as, function 4
work at, strive 1
work (away) (at), peg 6
work (for), assist 3; pursue 3; serve 2, 5
work hard, shoulder 3
work (hard) (at), persist 1
work in, phase 5; work 17
work like a Trojan *or* **a horse,** slave 3; sweat 3
work on, cultivate 4; influence 3; work 18
work (oneself *or* **one's way),** insinuate 2
work (oneself) (up) into a sweat *or* **lather** *or* **state (over),** stew 3
work (one's way), edge 6
work (one's) way through, wade 4
work out, calculate; close 4; compute; decipher 1; deduce; design 2; devise 1; estimate 1; exercise 2; fathom; figure 8, 12, 12a; formulate 3; get 19; hit 9b; interpret 2; mastermind 1; negotiate 2; pan 6; penetrate 5; practice 6; project 3; puzzle 3; reason 7; reckon 1; resolve 2; set 9; solve; strike 16; train 5; wangle; work 19
work out at *or* **to,** work 19b
work over, pelt[1] 1; pound[1] 1
work together, collaborate; cooperate 1; play 11; stick[1] 16a; team 3
work toward(s), pursue 3
work up, excite 3; fire 8a; foment; incite; inflame 1; infuriate; kindle; mastermind 1; pep 2; pick 3; project 3; prompt 3; provoke 2; reckon 1; rouse 2; set 9; summon 2; turn 18c; whip 7a; work 19e, 20
  work up to, lead 10b
  work with, assist 3
workability, feasibility
workable, feasible; operable; plastic 1; pliable 1; possible 2; practicable; serviceable 1; tenable; tractable 2; viable
workaday, common 1; ordinary 2; prosaic; usual
workbook, text 4
worked up, distraught; excited 1; nervous 1; overwrought 1; panic-stricken; resentful; restless; tense 2; warm 2
worked upon, affected 5
worked up (over), get, flare 3; fly 7; rage 4
worker°, employee; hand 4; laborer; operative 2; operator 1; peasant
worker(s), help 7; labor 2; rank[1] 5; shift 4
work force, shift 4; staff 2
work havoc on, kill 2
work history, résumé 2
workhorse, slave 2
working, active 2; busy 2; functional 1, 2; move 13b; moving 2; operation 1; serviceable 1
  working, not, blink 6; broken 7; order 13c; wrong 5
  working class(es), the, unwashed 2
  working order, put (back) in *or* **into** *or* **restore to,** cure 2; repair 1
  working order *or* **condition, in,** operable
  working out, exercise 4; interpretation 1; settlement 3; solution 1
working class, populace; bourgeois 2; plebeian 1
working girl, prostitute 1; tart[2] 2; wanton 5
working man, laborer; worker
working model, model 1
working order, condition 3; repair 3
working(s), agency; mechanism 2; movement 4; works 2
working woman, worker
workless, idle 2
workman, hand 4; laborer; worker
workmanlike, yeomanly
work of art, masterpiece
work of genius, masterpiece
work one's fingers to the bone, slave 3
workout, exercise 4; practice 2
workplace, office 3
workroom, study 6
work(s), factory; gear 1; mechanism 2; mill 2; movement 4; plant 2; stuff 5; writing 2
  works, give (someone) the, pound[1] 1
  works, in the, pipeline 2; progress 4; way 12; work 5
  works, the, works 3
workshop, mill 2; plant 2; shop 2; works 1
work the land, farm 3
work the runway, strip[2] 3
work-to-rule, strike 20
world°, earth 1; globe 1; nature 2; society 3; universe 2
  world, bring into the, deliver 6; have 8; world 5
  world, for all the, world 6
  world, go up in the, flourish 1; place 10
  world, in the, devil 5
  world, in this, below 3
  world, out of this, fine[1] 1; marvelous; miraculous; superb; world 8
  world, the, creation 2
  world at large, the, outside 4; world 2
  world of one's own, in a, inattentive
  world on fire, set the, far 6a
worldliness, sophistication 1
worldly°, earthly 2; knowledgeable 2; material 7; mortal 2; outward 1; secular; sophisticated 1; suave; temporal 1; terrestrial 1
worldly goods, possession 3
worldly-wise, sophisticated 1; worldly 2
worlds, vastly
worlds apart, pole[2] 3
worldview, philosophy 2
worldwide, extensive 1; general 1; global; international; public 1; universal 1
worm, crawl 1; edge 6; heel[1] 2; slither; snake 3, 4; struggle 3; twist 3, 4; wind[2] 1; wretch 1; wriggle 2
  worm (oneself *or* **one's way),** insinuate 2
  worm (out), extract 3
worn°, blunt 1; haggard; hurt 8; secondhand; shabby 1; timeworn; used 1
  worn out, ragged 3; spent 1
  worn out, become, seed 4
worn-out, beat 13; breathless 1; common 6; dead 9; decrepit 1; drawn; exhausted 3; far 5a; helpless 3; leg 7; limp[2] 2; musty 2; old 3; prostrate 5; run-down 1; seedy 2; shabby 1; stock 7; timeworn; tired 1, 3; usual; washed-out 2; weak 2; weary 1; worn 3
worried°, anxious 1; concerned 2; disturbed 1; insecure 1; nervous 1; queasy 1; shy 2; smitten 1; solicitous 1; tense 2; tremulous 2; upset 8
worrisome, burdensome; harrowing; nerve-racking; niggling 1; nightmarish; painful 2; sneaking 1; tense 3; troublesome; trying

write° (*cont.*)
enter 3; letter 5; make 4; phrase 4; put 4; script 3
write about, mention 1
write (down), designate 1; get 37b; jot 1; make 37c; mark 15a; minute[1] 4; note 12; pen[1] 2; put 16a; record 1; register 3, 5; set 17a; take 31a; write 4; enter 3
write home about, nothing to, mediocre; undistinguished
write of, touch 8
write off, dismiss 2; light[2] 12; obliterate 1; overlook 2; satisfy 4; write 5
write (out), pen[1] 2; make 37c; transcribe 2
write to, contact 4
write (up), pen[1] 2; report 3; work 20b
write "finis" to, finish 2; kill 1
writer°, author; editor; scribe 2
write-up, report 1
writhe, squirm; struggle 3; suffer 1; toss 4; twist 3; wallow 1; wriggle 1, 2
writhing, wriggle 4
writing°, copy 3; penmanship; print 3; prose; script 1
writing, commit to, pen[1] 2
writing, put in, note 12; pen[1] 2; set 17a; take 31a; write 4a
writing room, study 6
writing(s), literature 1; writing 2
written, destined 1; graphic 2; literary 2
written off, to be, spout 4
wrong°, abuse 2; abusive 2; amiss 1, 2; bad 3; careless 3; crime; criminal 2; crooked 1; disgraceful 1; disservice; erroneous; error 3a; evil 1; false 1; grievance 1; guilty 1; harm 2; ill 2, 9; illicit 2; immoral 1; imprecise; improper 1; imprudent; inaccurate; inadmissible; incorrect; inexact; injure 2; injury; injustice 2; invalid[2]; mischief 2; misdeed; misguided; mistaken 1, 2; off 3; offense 1; out 12; perverse 1; place 12; regrettable; sin 1; sinful; tasteless 1; transgression; turn 34; unacceptable; unhappy 3; untrue 2; wound 4
wrong, be, err 1; nod 4
wrong, be in the, err 1
wrong, do, err 2; transgress 1
wrong, get, mistake 3
wrong, get (it (all)), misunderstand
wrong, go, err 1; fail 1; miscarry; misfire 1; sin 3; slip[1] 3; transgress 1; wrong 8
wrong, in the, fault 5; mistaken 1
wrong idea (about), get *or* **have the,** misconceive; misunderstand
wrong way, take the, mistake 3
wrongdoer, criminal 3; culprit 2; delinquent 1; felon; miscreant 1; offender; sinner; transgressor
wrongdoing, error 2; guilt 1; harm 2; misdeed; offense 1; outrage 1; transgression
wronged, hurt 7
wrong end of the stick°, stick[2] 4
wrongful, illegal; sinful
wrongfully, ill 12
wrongfulness, sin 2
wrongheaded, ill-advised 1; imprudent; perverse 1, 3; wrong 4
wrong idea, misunderstanding 1
wrong impression, misunderstanding 1
wrongly, amiss 3; ill 12; wrong 7
wrong move, mistake 2
wrong notion *or* **idea,** misconception
wrong side, reverse 7
wrong side up, topsy-turvy 1
wroth, angry 1; furious 1; indignant; mad 4
wrought up, distraught; excited 1; nervous 1; tense 2
wrought-up, agitated; frantic
wry°, bent 2; dry 3
wryneck, stoop 3
Wunderkind, prodigy 1

X, blue 2; check 16
x out, scratch 3; strike 5
Xanthippe, devil 2; hag; scold 2; shrew; witch 2
xenophobia, intolerance
xenophobic, inhospitable 1; intolerant 2; nationalistic
Xerox, duplicate 3; print 2
Xerox (copy), duplicate 2; facsimile
X-rated, blue 2; filthy 3; nasty 3; naughty 3; pornographic; sexy 2; wicked 3
XXX, blue 2
xyloid, wooden 1

yacht, boat; cruise 1
yachting, nautical
yachting trip, cruise 2
yachtsman, sailor
yachtswoman, sailor
ya(c)k. See **yak**
ya(c)kety-ya(c)k, palaver 4; patter[2] 2, 3; prattle 1, 2
yahoo°, barbarian 2; boor 2; clown 2; philistine 1; punk 1; rough 13; rowdy 2; slob
yak, babble 1; gab 1; palaver 4; patter[2] 2, 3; prattle 1; tattle 2
yammer, bawl 2; groan 2
yammering, groan 4
yank°, hitch 2; jerk 1, 3; pluck 3; pull 2, 19; tug 1, 2; wrench 1, 4
yank out, whip 6
yap°, jabber 1; mouth 1; squawk 2; trap 3
yard, area 5; enclosure 1; pound[2]; run 46; spar[1]
yardage, measurement 2
yardarm, spar[1]
yard goods, good 21d
yard sale, sale 3
yardstick°, gauge 3; mark 3; measure 2; norm 2; pattern 1; precedent; standard 1; touchstone
yare, navigable 2
yarn°, fabrication 3; invention 3; line[1] 11; story[1] 1; thread 1; twine 1
yatter, jabber 1
yaw, toss 5
yawn, gape 2
yawning, deep 1; drowsy; open 2
(yea), verily, truly 4
year, class 4; date 1; grade 3; vintage 1
year, by the, yearly 2
year after year, yearly 1
yearbook, journal 2; periodical
year in (and) year out, yearly 1
yearn°, ache 2; aspire; burn 3; die 5; hunger 3; itch 2; long[2]; purpose 5; wish 1
yearn (for), hanker; miss[1] 2; pant 2; sigh 2; spoil 5; thirst 2; want 1; desire 1
yearning, anxious 2; appetite 2; aspiration; desire 3; desirous; eager; eagerness 2; fancy 7; greed 1; hope 1; hunger 2; hungry 2; itch 4; longing; mania 1; passion 3; stomach 2; thirst 1; thirsty 2; urge 5; want 4; wish 5; wistful 1
yearning (for), desperate 3; starved 2
years, age 4; period 2
years ago, once 1
years gone by, past 6
yell°, bawl 1; bellow 1, 2; call 1, 13; cheer 7; cry 5; exclaim; exclamation; howl 1, 2; rave 1; roar 1, 3; shout 1, 2; squawk 1; thunder 3; whoop 1, 2
yell (out), sound[1] 8
yelling, din 1
yellow, afraid 1; cowardly; fearful 2; golden 1; spineless 3; tame 3; timid; weak 3
yellow(-bellied), cowardly; fainthearted 1; spineless 3; timid
yellow-belly, coward
yellowish, golden 1
yelp, bawl 2; howl 2; shout 2; whoop 1, 2; yap 1; yell 1, 2
yelping, howl 2
yen, desire 3; hunger 2, 3; itch 4; longing; mania 1; passion 3; thirst 1; urge 5; want 4; wish 5
yen for, have a, desire 1; fancy 11; hanker; hunger 3; pant 2; want 1
yeoman, farmer; yeomanly
yes, absolutely 3; OK 1; positive 7
yes, say, nod 2
yes man°, flatterer; flunky 2; hanger-on; inferior 4; menial 4; puppet 2
yesterday, history 6; previously
yesterday's news, history 7
yesteryear, history 6
yet°, even 10, 12; ever 2; further 3; however 1; nevertheless; notwithstanding 1; same 3; still 4, 6; though 2; time 13a
Y-fronts, pants 1
yield°, abandon 1; abdicate; afford 3; bear 6; bend 5; bow 2; bring 10a, 11a; capitulate 1, 2; cave 2b; cede; come 5b; communicate 1; comply; concede 2; consent 1; dedicate 1; deign; deliver 2; earnings; fall 5; fat 6; fetch 3; forgo 2; forsake 3; gain 2, 8; give 9, 11, 13, 17b; hand 18a; harvest 1; haul 4; leave[1] 8; lose 2, 3; open 25; output 1, 3; part 14; proceed 3; produce 2; product 1; profit 1; provide 2; relent; relinquish 1; return 5, 9; revenue; sacrifice 4; secrete[2]; soften 4; spare 8; spawn; succumb; supply 2; surrender 1, 2; take 40; thaw 2; truckle; turn 15b; vouchsafe 1; waive 1; way 11b; weaken 3
yield (to), defer[2]; indulge 1; obey 1; pander 1; submit 1
yield (up), forfeit 2; give 5; render 3; resign 1; sacrifice 3
yield a return, pay 4, 7
yielded, forfeit 3; sacrificial 1
yielding°, accommodating 1; dutiful; flexible 1; meek 2; mild 1; obedience; obedient; passive 2; pliable 2; soft 1; submission 1; submissive 1; supple 3; surrender 3; tractable 1
yielding up, sacrifice 1
yob, barbarian 2; hoodlum; slob; yahoo
yobbo, barbarian 2; hoodlum; slob; yahoo
yodel, sing 1
yoke, couple 1, 3; enslave; join 1; pair 1, 2; slavery 1; team 2; twin 3
yokel, boor 1; clown 2; peasant; provincial 3; rustic 3
yore, of, bygone
young°, adolescent 2; boyish 1; brood 1; childlike; immature 1; issue 7; juvenile 1; litter 2;

little 2; new 7; offspring; progeny; seed 3; small 1; tender[1] 3; unfledged; youth 3
young boy *or* **man,** youth 2
younger, junior; second[1] 5
young fellow *or* **man,** stripling
young gentleman *or* **lady,** child 2
young lady, love 3; miss[2]; teenager
young man, boy 1; fellow 4; lad; love 3; teenager
young manhood, youth 1
young man *or* **woman,** child 2
young people, youth 3
youngster, adolescent 1; boy 1; child 1, 2; lad; minor 3; offspring; stripling; youth 2
youngsters, youth 3
young 'un, stripling
young woman, lass; miss[2]
young womanhood, youth 1
yourself, person 2; personally 1
yourselves, person 2; personally 1
youth°, adolescent 1; boy 1; child 2; childhood; guy 1; juvenile 2; lad; minor 3; prime 4; stripling; teenager
youth custody center, reform school
youthful, adolescent 2; boyish 1; childlike; juvenile 1; little 2; tender[1] 3; young 1
yowl, cry 4; howl 1, 2; rave 1; roar 1, 3; scream 1, 3; sob; squawk 1, 2; whoop 1, 2; yell 1, 2
yowling, howl 2
yoyo, jerk 5; shilly-shally
yuckiness, sentimentality
yucky°, sentimental 2; revolting
yukkiness, sentimentality
yukky, revolting; sentimental 2; yucky
yummy°, delicious 1; luscious; tasty
yuppie, ambitious 3; materialistic

zaftig, full 7
zany°, clown 1; comedian; fool 2; funny 1; joker 1; ludicrous; panic 2; ridiculous; wit 3
zap°, dispatch 3; hit 1; shoot 3
zeal, ambition 2; appetite 1; ardor; devotion 3; drive 8; eagerness 1; enterprise 2; enthusiasm 1; exuberance 1; fanaticism 1; fervor; fight 9; flame 2; gusto; heat 2; inclination 4; inspiration 1; intensity; passion 1; push 14; sparkle 4; spirit 2; verve; warmth 3
zealot°, crank 2; enthusiast; fan; fanatic; maniac 2; partisan 1; puritan 1; radical 4; sectarian 4; supremacist
zealotry°, passion 1
zealous, ambitious 2; ardent; crazy 4; devoted 1; devout 3; dynamic; eager; earnest 2; emotional 1; enterprising; enthusiastic; excited 2; exuberant 1; fanatical; favorable 2; fervent 2; feverish; great 9; hard 7; hearty 3; hot 3, 4; impassioned; intense 2; intent 5; keen[1] 1; mad 6; passionate 1; sanguine; solicitous 2; strenuous 2; strong 5; voracious 2; warm 2; wholehearted; willing
zealously, hard 13; hotly; jealously; warmly 4; willingly
zealousness, gusto; heat 2; passion 1; spirit 2; vigor
zenith°, acme; climax 1; extreme 8; height 2; maximum 2; peak 2; pinnacle; prime 4; summit; top 1; vertex
zephyr, air 2; breath 1; breeze 1; wind[1] 1
zero°, blank 8; cog 2; naught; nil; nobody 2; nonessential 2; nothing 2
  zero in, focus 4
  zero (in) on, finger 5c; zero 4; pin 4b; set 5
zest°, animation 1; appetite 1; bounce 2; dash 6; eagerness 1; enjoyment 1; enthusiasm 1; exuberance 1; fight 9; flavor 1; gusto; life 7; panache; passion 1; pep 1; punch[1] 3; relish 1; salt 2; savor 1; seasoning; spice 2; spirit 2, 3; tang 1; verve; vigor
zestful, ebullient; energetic; passionate 1; racy 1; spicy 1
zestfulness, passion 1; zest 2
zestiness, salt 2; tang 1
zesty, energetic; racy 3; spicy 1
zigzag, devious 2; indirect 1; meander 1; notched; ragged 2; serrated; snake 4; stagger 3; swing 1, 3; tack 6; tortuous 1; turn 11, 27; twist 4, 7; wander 2; wave 4; weave 4; wind[2] 1; wriggle 3
zigzagging, swing 3
zigzag(s), meander 2
zilch, nil; zero 1
zillions, ocean 2
zing, energy; initiative 2; life 7; pep 1; punch[1] 3; push 14; salt 2; snap 11; sparkle 4; verve; vigor; vitality 1; zest 1
zingy, energetic
Zion, heaven 1; paradise 1
zip, bounce 2; drive 8; dynamism; eagerness 1; energy; enterprise 2; fly 9; initiative 2; nil; pep 1; punch[1] 3; push 14; race[1] 4; run 1; salt 2; scurry; shoot 1; snap 11; sparkle 4; speed 3; spice 2; streak 5; tang 1; tear 3; verve; vitality 1; whip 3; zest 1
  zip (up), adjust 4; seal 3
zipper, fly 9
zippy, energetic; vivacious
zits, spot 5
zizz, doze 1, 2; nap[1] 1, 3; repose 2; rest[1] 1; sleep 2
  zizz, take *or* **have a,** rest[1] 6; retire 3; sleep 1
zone°, area 3; belt 1, 2; girth 2; limit 3; precinct 2; province 1; quarter 3; region 1; territory 1; tract[1]; ward 1
zonked (out), weary 1
zooid, animal 3
zoological, animal 3
zoological garden, zoo 1
zoom, career 2; flash 5; rocket; scurry; speed 3; streak 5; sweep 3; tear 3; whip 3
  zoom onto, buzz 6
Z's, catch *or* **log some** *or* **a few,** doze 1; nap[1] 1; rest[1] 6; retire 3; sleep 1